IEEE International Symposium on Circuits and Systems (ISCAS)
May 23 – 26, 2005

International Conference Center
Kobe
Japan

Conference Proceedings

Cosponsored by

The Institute of Electrical and Electronics Engineers, Inc.

Circuits and Systems Society (IEEE CASS)

Science Council of Japan

The Institute of Electronics, Information and Communication Engineers (IEICE)

In cooperation with

**The Institute of Electrical Engineers of Japan (IEEJ)
Information Processing Society of Japan (IPSJ)**

Supported by

**The Telecommunications Advancement Foundation (TAF)
The MEET IN KOBE 21st Century
The TSUTOMU NAKAUCHI FOUNDATION**

PROCEEDINGS OF IEEE INTERNATIONAL SYMPOSIUM
ON CIRCUITS AND SYSTEMS 2005

Copyright and Reprint Permission: Abstracting is permitted with credit to the source. Libraries are permitted to photocopy beyond the limit of the U.S. copyright law for private use of patrons those articles in this volume that carry a code at the bottom of the first page, provided the per-copy fee indicated in the code is paid through Copyright Clearance Center, 222 Rosewood Drive, Danvers, MA 01923. For other copying, reprint or republication permission, write to IEEE Copyrights Manager, IEEE Service Center, 445 Hoes Lane, P.O. Box 1331, Piscataway, NJ 08855-1331. All rights reserved. Copyright © 2005 by the Institute of Electrical and Electronics Engineers, Inc.

IEEE Catalog Number   05CH37618
ISBN: 0-7803-8834-8
Library of Congress: 80-646530

PRINTED IN THE UNITED STATES OF AMERICA

Additional copies can be ordered from

IEEE Service Center
445 Hoes Lane
Piscataway, NJ 08854

1-800-678-IEEE
1-732-981-1393
1-732-981-1721 (FAX)

C3L-J Analog Circuits & Systems for Wireless II (Lecture)
Time: Thursday, May 26, 2005, 14:00 - 16:00
Place: Room 501
Co-Chairs: Mourad El-Gamal, *McGill University*
Wouter Serdijn, *Delft University of Technology, the Netherlands*

14:00
C3L-J.1 A WIDE-BAND LOW-NOISE AMPLIFIER WITH DOUBLE LOOP FEEDBACK..........5353
Miguel A. Martins, *I.S.Técnico/INESC-ID, Portugal*; Koen van Hartingsveldt, Chris Verhoeven, *Delft University of Technology, The Netherlands*; Jorge Fernandes, *IST/INESC-ID, Portugal*

14:20
C3L-J.2 AN INTERFERENCE REJECTION FILTER FOR AN ULTRA-WIDEBAND QUADRATURE DOWNCONVERSION AUTOCORRELATION RECEIVER..........5357
Sumit Bagga, Sandro A. P. Haddad, Koen van Hartingsveldt, Simon Lee, Wouter A. Serdijn, John R. Long, *Delft University of Technology, The Netherlands*

14:40
C3L-J.3 A READ-OUT STRATEGY AND CIRCUIT DESIGN FOR HIGH FREQUENCY MEMS RESONATORS..........5361
Arantxa Uranga, *Universitat Autónoma de Barcelona, Spain*; Humberto Campanella, *Instituto de Microelectrónica de Barcelona, CNM-IMB, Spain*; Nuria Barniol, *Universitat Autónoma de Barcelona, Spain*; Jaume Esteve, Lluis Terés, *Instituto de Microelectrónica de Barcelona, Spain*; Zachary Davis, *EPSON Europe Electronics GmbH*

15:00
C3L-J.4 A LOW-POWER SWITCHED-CURRENT CDMA MATCHED FILTER EMPLOYING MOS-LINEAR MATCHING CELL AND OUTPUT A/D CONVERTER..........5365
Tomoyuki Nakayama, Toshihiko Yamasaki, Tadashi Shibata, *University of Tokyo, Japan*

15:20
C3L-J.5 A RE-CONFIGURABLE HIGH-SPEED CMOS TRACK AND LATCH COMPARATOR WITH RAIL-TO-RAIL INPUT FOR IF DIGITIZATION..........5369
Holly Pekau, Lee Hartley, James Haslett, *University of Calgary, Canada*

15:40
C3L-J.6 FAST-SWITCHING ADAPTIVE BANDWIDTH FREQUENCY SYNTHESIZER USING A LOOP FILTER WITH SWITCHED ZERO-RESISTOR ARRAY..........5373
Sreenath Thoka, Randall Geiger, *Iowa State University, USA*

C3L-K Sensor & Actuator Interface Circuits (Lecture)
Time: Thursday, May 26, 2005, 14:00 - 16:00
Place: Room 502
Co-Chairs: Ralph Etienne-Cummings, *Johns Hopkins University, MD, USA*
          Tadashi Shibata, *The University of Tokyo, Japan*

14:00
C3L-K.1 FULLY INTEGRATED CHARGE SENSITIVE AMPLIFIER FOR READOUT OF MICROMECHANICAL CAPACITIVE SENSORS 5377
Mikko Saukoski, *Helsinki University of Technology, Finland*; Teemu Salo, *VTI Technologies Oy*; Lasse Aaltonen, Kari Halonen, *Helsinki University of Technology, Finland*

14:20
C3L-K.2 FULLY INTEGRATED CHARGE PUMP FOR HIGH VOLTAGE EXCITATION OF A BULK MICROMACHINED GYROSCOPE 5381
Lasse Aaltonen, Mikko Saukoski, Kari Halonen, *Helsinki University of Technology, Finland*

14:40
C3L-K.3 FLEXIBLE HIGH-ACCURACY WIDE-RANGE GAS SENSOR INTERFACE FOR PORTABLE ENVIRONMENTAL NOSING PURPOSE 5385
Marco Grassi, Piero Malcovati, *University of Pavia, Italy*; Andrea Baschirotto, *University of Lecce, Italy*

15:00
C3L-K.4 ADAPTIVE SENSOR RESPONSE CORRECTION USING ANALOG FILTER COMPATIBLE WITH DIGITAL TECHNOLOGY 5389
Mehdi Jafaripanah, Bashir Al-Hashimi, Neil White, *University of Southampton, UK*

15:20
C3L-K.5 A MODULAR RC-ACTIVE NETWORK FOR VIBRATION DAMPING IN PIEZO-ELECTRO-MECHANICAL BEAMS 5393
Massimo Panella, *University of Rome La Sapienza, Italy*; Maurizio Paschero, Fabio Massimo Frattale Mascioli, *University of Rome, La Sapienza, Italy*

15:40
C3L-K.6 COST EFFECTIVE HIGH VOLTAGE DRIVER FOR LARGE CHANNEL COUNT OPTICAL MEMS SWITCH APPLICATIONS 5397
Yuan Ma, Xiqun Zhu, Robert Newcomb, *University of Maryland, USA*

C3L-L Digital Signal Processing Applications I (Lecture)
Time: Thursday, May 26, 2005, 14:00 - 16:00
Place: Room 503
Co-Chairs: Isao Nakanishi, *Tottori University*
Behrouz Nowrouzian, *University of Alberta*

14:00
C3L-L.1 FACE SEGMENTATION BASED ON HUE-CR COMPONENTS AND MORPHOLOGICAL TECHNIQUE 5401
Teerayoot Sawangsri, Vorapoj Patanavijit, Somchai Jitapunkul, *Chulalongkorn Uinversity, Thailand*

14:24
C3L-L.2 A COMBINED INTERPOLATORLESS INTERPOLATION AND HIGH ACCURACY SAMPLING PROCESS FOR DIGITAL CLASS D AMPLIFIERS 5405
Victor Adrian, Bah-Hwee Gwee, Joseph Sylvester Chang, *Nanyang Technolgocial University, Singapore*

14:48
C3L-L.3 TIME-DELAY DIRECTION FINDING BASED ON CANONICAL CORRELATION ANALYSIS 5409
Gaoming Huang, Luxi Yang, Zhenya He, *Southeast University, China*

15:12
C3L-L.4 DWT DOMAIN MULTI-MATCHER ON-LINE SIGNATURE VERIFICATION SYSTEM 5413
Isao Nakanishi, Hiroyuki Sakamoto, Yoshio Itoh, Yutaka Fukui, *Tottori University, Japan*

15:36
C3L-L.5 DIGITAL SYSTEM FOR DETECTION AND CLASSIFICATION OF ELECTRICAL EVENTS 5417
Augusto Cerqueira, Carlos Duque, *UFJF, Brazil*; Moises Ribeiro, *Unicamp*; Rogério M. Trindade, *UFJF, Brazil*

C3L-M Communications Systems I (Lecture)
Time: Thursday, May 26, 2005, 14:00 - 16:00
Place: Room 504
Co-Chairs: Tzi-Dar Chiueh, *National Taiwan University*
           Masaru Kamada, *Ibaraki University*

14:00
C3L-M.1 TAP SELECTION BASED MMSE EQUALIZATION FOR HIGH DATA RATE UWB COMMUNICATION SYSTEMS 5421
Lin Zhiwei, Benjamin Premkumar, A.S. Madhukumar, *Nanyang Technolgocial University, Singapore*

14:20
C3L-M.2 DESIGN OF UWB PULSES BASED ON B-SPLINES 5425
Mitsuhiro Matsuo, Masaru Kamada, Hiromasa Habuchi, *Ibaraki Univeristy, Japan*

14:40
C3L-M.3 OPTIMIZING VERTICAL COMMON SUBEXPRESSION ELIMINATION USING COEFFICIENT PARTITIONING FOR DESIGNING LOW COMPLEXITY SOFTWARE RADIO CHANNELIZERS 5429
A.P. Vinod, E.M-K. Lai, *Nanyang Technolgocial University, Singapore*

15:00
C3L-M.4 A BPSK DEMODULATOR CIRCUIT USING AN ANTI-PARALLEL SYNCHRONIZATION LOOP 5433
You Zheng, Carlos Saavedra, *Queen's University, Canada*

15:20
C3L-M.5 NOVEL SYSTOLIC ARRAY ARCHITECTURE FOR THE DECORRELATOR USING CONJUGATE GRADIENT FOR LEAST SQUARES ALGORITHM 5437
Archana Chidanandan, *Rose-Hulman Institute of Technology, USA*; Magdy Bayoumi, *University of Louisiana at Lafayette, USA*

15:40
C3L-M.6 AN EFFICIENT PRE-TRACEBACK APPROACH FOR VITERBI DECODING IN WIRELESS COMMUNICATION 5441
Yao Gang, Tughrul Arslan, Ahmet T. Erdogan, *University of Edinburgh, UK*

C3L-N Oscillators & PLL (Lecture)
Time: Thursday, May 26, 2005, 14:00 - 16:00
Place: Room 505
Co-Chairs: Francis Lau, *HK Polytechnic University*
           Inoue Yasuaki, *Waseda University*

14:00
C3L-N.1 BIFURCATION ANALYSIS OF ON-CHIP LC VCOS 5445
Marcus Prochaska, Alexander Belski, Wolfgang Mathis, *University of Hannover, Germany*

14:20
C3L-N.2 A SCALABLE DCO DESIGN FOR PORTABLE ADPLL DESIGNS 5449
Chia-Tsun Wu, Wei Wang, I-Chyn Wey, An-Yeu Wu, *National Taiwan University, Taiwan ROC*

14:40
C3L-N.3 A 2.5GHZ PHASE-SWITCHING PLL USING A SUPPLY CONTROLLED 2-DELAY-STAGE 10GHZ RING OSCILLATOR FOR IMPROVED JITTER/MISMATCH 5453
Eva Tatschl-Unterberger, *Infineon Technologies AG, Germany*; Sasan Cyrusian, *Marvell Semiconductor, Inc., USA*; Michael Ruegg, *Miromico AG, Switzerland*

15:00
C3L-N.4 A 360 EXTENDED RANGE PHASE DETECTOR FOR TYPE-I PLLS 5457
Cameron Charles, David Allstot, *University of Washington, USA*

15:20
C3L-N.5 A TWO-CHIP, 4MHZ, MICROELECTROMECHANICAL RESONATOR REFERENCE OSCILLATOR 5461
Krishnakumar Sundaresan, Gavin Ho, Siavash Pourkamali, Farrokh Ayazi, *Georgia Institute of Technology, USA*

15:40
C3L-N.6 A STUDY OF INJECTION LOCKING IN RING OSCILLATORS 5465
Behzad Mesgarzadeh, Atila Alvandpour, *Linköping University, Sweden*

C3L-P   Motion Estimation (Lecture)
Time: Thursday, May 26, 2005, 14:00 - 16:00
Place: Nojiggiku Room, Portopia Hotel
Co-Chairs: Tian-Sheuan Chang, *National Chiao Tung University*
           Gwo Giun Lee, *National Cheng Kung University*

14:00
C3L-P.1   ONE-PASS COMPUTATION-AWARE MOTION ESTIMATION WITH ADAPTIVE SEARCH STRATEGY 5469
Yu-Wen Huang, Chia-Lin Lee, Ching-Yeh Chen, Liang-Gee Chen, *National Taiwan University, Taiwan ROC*

14:20
C3L-P.2   A FRAMEWORK FOR FINE-GRANULAR COMPUTATIONAL-COMPLEXITY SCALABLE MOTION ESTIMATION 5473
Zhi Yang, *Zhejiang University, China*; Hua Cai, *Media Communication Group, Microsoft Research Asia, China*; Jiang Li, *Microsoft Research Asia, China*

14:40
C3L-P.3   FAST SUB-PIXEL INTER-PREDICTION – BASED ON TEXTURE DIRECTION ANALYSIS (FSIP-BTDA) 5477
Hoi-Ming Wong, Oscar C. Au, Andy Chang, *Hong Kong University of Science and Technology, Hong Kong*

15:00
C3L-P.4   EFFICIENT SEARCH AND MODE PREDICTION ALGORITHMS FOR MOTION ESTIMATION IN H.264/AVC 5481
Gwo-Long Li, Mei-Juan Chen, Hung-Ju Li, Ching-Ting Hsu, *National Dong Hwa University, Taiwan ROC*

15:20
C3L-P.5   EFFICIENT VIDEO MOTION ESTIMATION USING DUAL-CROSS SEARCH ALGORITHMS 5485
Xuan-Quang Banh, Yap-Peng Tan, *Nanyang Technolgocial University, Singapore*

15:40
C3L-P.6   RAPID BLOCK-MATCHING MOTION ESTIMATION USING MODIFIED DIAMOND SEARCH ALGORITHM 5489
Xiaoquan Yi, Nam Ling, *Santa Clara University, USA*

C3L-Q   Communication Circuit & System Design for Power (Lecture)
Time: Thursday, May 26, 2005, 14:00 - 16:00
Place: Sumire Room, Portopia Hotel
Co-Chairs: Cesare Alippi, *Politecnico di Milano*
           Koushik Maharatna, *University of Bristol*

14:00
C3L-Q.1   CURRENT MODE MULTI-LEVEL SIMULTANEOUS BIDIRECTIONAL I/O SCHEME FOR CHIP-TO-CHIP COMMUNICATIONS 5493
Ge Yang, *UCSC*; YongSin Kim, *University of California, Santa Cruz, USA*; Sung-Mo Kang, *UCSC*

14:20
C3L-Q.2   ALGORITHM FOR PEAK TO AVERAGE POWER RATIO REDUCTION OPERATING AT SYMBOL RATE 5497
Stefano Marsili, *Infineon Technologies AG, Austria*

14:40
C3L-Q.3   A 3.5-GB/S CMOS BURST-MODE LASER DRIVER WITH AUTOMATIC POWER CONTROL USING SINGLE POWER SUPPLY 5501
Day-Uei Li, Li-Ren Huang, Chia-Ming Tsai, *Industrial Technology Research Institute, Taiwan ROC*

15:00
C3L-Q.4   COMMON-MODE STABILITY IN LOW-POWER LO DRIVERS 5505
Svetoslav Gueorguiev, *Aalborg University, Denmark*; Saska Lindfors, *Helsinki University of Technology, Finland*; Torben Larsen, *Aalborg University, Denmark*

15:20
C3L-Q.5   AN APPLICATION-LEVEL METHODOLOGY TO GUIDE THE DESIGN OF INTELLIGENT-PROCESSING, POWER-AWARE PASSIVE RFIDS 5509
Cesare Alippi, Giovanni Vanini, *Politecnico di Milano, Italy*

15:40
C3L-Q.6   ON THE IMPLEMENTATION OF 128-PT FFT/IFFT FOR HIGH-PERFORMANCE WPAN 5513
Clare Huggett, Koushik Maharatna, *University of Bristol, UK*; Kolin Paul, *Indian Institute of Technology, Delhi, India*

C3L-R   SPECIAL SESSION - Information Assurance & Data Hiding (Lecture)
Time: Thursday, May 26, 2005, 14:00 - 16:00
Place: Tsutsuji Room, Portopia Hotel
Co-Chairs: Alex Chichung Kot, *Nanyang Technological University, Singapore*
            Yun Q. Shi, *New Jersey Institute of Technology*

14:00
C3L-R.1   NEAR-PERFECT COVER IMAGE RECOVERY ANTI-MULTIPLE WATERMARK EMBEDDING APPROACHES 5517
Chao-Yong Hsu, Chun-Shien Lu, *Academia Sinica, Taiwan*

14:20
C3L-R.2   PRIVACY PRESERVING DATA MINING WITH UNIDIRECTIONAL INTERACTION 5521
Chai Wah Wu, *IBM T. J. Watson Research Center, USA*

14:40
C3L-R.3   PROGRESSIVE SCRAMBLING FOR MP3 AUDIO 5525
Wei-Gang Fu, *National University of Singapore*; Wei-Qi Yan, *National University of Singapore, Singapore*; Mohan Kankanhalli, *National University of Singapore*

15:00
C3L-R.4   AVERAGING ATTACK RESILIENT VIDEO FINGERPRINTING 5529
In Koo Kang, Choong-Hoon Lee, Hae-Yeoun Lee, Jong-Tae Kim, Heung-Kyu Lee, *Korea Advanced Institute of Science & Technology, Korea*

15:20
C3L-R.5   MULTIMEDIA DATA ENCRYPTION VIA RANDOM ROTATION IN PARTITIONED BIT STREAMS...5533
Dahua Xie, C.-C. Jay Kuo, *University of Southern California, USA*

15:40
C3L-R.6   A NEW VISUAL CRYPTOGRAPHY USING NATURAL IMAGES 5537
Hyoung Kim, Yongsoo Choi, *Kangwon National University, Korea*

C3P-S Data Conversion Techniques & Applications (Poster)
Time: Thursday, May 26, 2005, 14:00 - 16:00
Place: Reception Hall - Area 1
Chair: Ketaro Sekine, *Tokyo University of Science*

C3P-S.1 DIGITAL SELF-CORRECTION OF TIME-INTERLEAVED ADCS 5541
Pieter Harpe, Athon Zanikopoulos, Arthur van Roermund, *Eindhoven University of Technology, The Netherlands*

C3P-S.2 AUTOMATED DESIGN OF A 10-BIT 80MSPS WLAN DAC FOR LINEARITY AND LOW-AREA 5545
Ankit Seedher, Preetam Tadeparthy, Satheesh Kumar A. S., Anuroop V. T, *Texas Instruments India Pvt. Ltd.*

C3P-S.3 AN ANALOG-TO-DIGITAL CONVERTER WITH GOLOMB-RICE OUTPUT CODES 5549
Walter D. Leon, Sina Balkir, Khalid Sayood, Michael W. Hoffman, *University of Nebraska-Lincoln, USA*

C3P-S.4 10-BIT PROGRAMMABLE VOLTAGE-OUTPUT DIGITAL-ANALOG CONVERTER 5553
Erhan Ozalevli, Christopher Twigg, Paul Hasler, *Georgia Institute of Technology, USA*

C3P-S.5 A BACKGROUND CORRECTION TECHNIQUE FOR TIMING ERRORS IN TIME-INTERLEAVED ANALOG-TO-DIGITAL CONVERTERS 5557
Echere Iroaga, *Stanford University, USA*; Lalitkumar Nathawad, *Atheros Communications Inc., USA*; Boris Murmann, *Stanford University, USA*

C3P-S.6 FEEDFORWARD-TYPE PARASITIC CAPACITANCE CANCELER AND ITS APPLICATON TO 4 GB/S T/H CIRCUIT 5561
Takahide Sato, *Tokyo Institute of Technology, Japan*; Shigetaka Takagi, Nobuo Fujii, *Tokyo Tech., Japan*; Yasuyuki Hashimoto, *ROHM CO.,LTD., Japan*; Kohji Sakata, *Sanyo Electric Co.,Ltd., Japan*; Hiroyuki Okada, *NEC Corporation, Japan*

C3P-S.7 SYNTHESIS OF HYBRID FILTER BANKS BY GLOBAL FREQUENCY DOMAIN LEAST SQUARE SOLVING 5565
Tudor Petrescu, Jacques Oksman, *École Supérieure d'Électricité, France*; Pierre Duhamel, *CNRS/LSS, France*

C3P-S.8 IMPLEMENTATION OF A NOVEL READ-OUT STRATEGY BASED ON A WILKINSON ADC FOR A 16X16 PIXEL X-RAY DERECTOR ARRAY 5569
Vincenzo Ferragina, Piero Malcovati, Fausto Borghetti, Andrea Rossini, *University of Pavia, Italy*; Flavio Ferrari, Nicoletta Ratti, *Alenia Spazio-Laben, Italy*; Giuseppe Bertuccio, *Polytechnic of Milano, Italy*

C3P-T Sigma-Delta Converters IV (Poster)
Time: Thursday, May 26, 2005, 14:00 - 16:00
Place: Reception Hall - Area 2
Chair: Tertulien Ndjountche, *Quebec University at Hull, Canada*

C3P-T.1 A 12-B 10-MSAMPLES/S CMOS SWITCHED-CURRENT DELTA-SIGMA MODULATOR 5573
Guo-Ming Sung, Kuo-Hsuan Chang, Wen-Sheng Lin, *National Taipei University of Technology, Taiwan ROC*

C3P-T.2 A 1.5V MULTIRATE MULTIBIT $\Sigma\Delta$ MODULATOR FOR GSM/WCDMA IN A 90NM DIGITAL CMOS PROCESS 5577
Oguz Altun, *Georgia Institute of Technology, USA*; Jinseok Koh, *Texas Instruments Inc., USA*; Phillip Allen, *Georgia Institute of Technology, USA*

C3P-T.3 TIME-INTERLEAVED MULTIRATE SIGMA-DELTA MODULATORS 5581
Francisco Colodro, Antonio Torralba, M. Laguna, *Universidad de Sevilla, Spain*

C3P-T.4 A DIRECT SYNTHESIS METHOD OF CASCADED CONTINUOUS-TIME SIGMA-DELTA MODULATORS 5585
Ramón Tortosa, José M. de la Rosa, Angel Rodríguez-Vázquez, Francisco V. Fernández, *Instituto de Microelectrónica de Sevilla - IMSE-CNM (CSIC), Spain*

C3P-T.5 A SIXTH-ORDER SUBSAMPLING CONTINUOUS-TIME BANDPASS DELTA-SIGMA MODULATOR 5589
Yuan Chen, Kei Tee Tiew, *Nanyang Technolgocial University, Singapore*

C3P-T.6 HIGH-ORDER SINGLE-LOOP DOUBLE-SAMPLING SIGMA-DELTA MODULATOR TOPOLOGIES FOR BROADBAND APPLICATIONS 5593
Mohammad Yavari, Omid Shoaei, *University of Tehran, Iran*

C3P-T.7 ADAPTIVE PROCESSING APPLIED TO THE DESIGN OF HIGHLY DIGITAL ANALOG INTERFACES 5597
Adão Souza Jr., Luigi Carro, *Instituto de Informática - UFRGS, Brazil*; Jawad Tousaad, *Federal University of Rio Grande do Sul, Brazil*

C3P-T.8 SPURIOUS TONE FREE DIGITAL DELTA-SIGMA MODULATOR DESIGN FOR DC INPUTS 5601
Maciej Borkowski, Juha Kostamovaara, *University of Oulu, Finland*

C3P-U   Analog & Mixed Signal Circuits Synthesis II (Poster)
Time: Thursday, May 26, 2005, 14:00 - 16:00
Place: Reception Hall - Area 3
Chair: Shoji Kawahito, *Shizuoka University*

C3P-U.1 TOWARDS A RIGOROUS FORMULATION OF THE SPACE MAPPING TECHNIQUE FOR ENGINEERING DESIGN 5605
Slawomir Koziel, John Bandler, *McMaster University, Canada*; Kaj Madsen, *Technical University of Denmark*

C3P-U.2 BEHAVIORAL MODELING, SIMULATION AND HIGH-LEVEL SYNTHESIS OF PIPELINE A/D CONVERTERS 5609
Jesús Ruiz-Amaya, José M. de la Rosa, Manuel Delgado-Restituto, Angel Rodríguez-Vázquez, *Instituto de Microelectrónica de Sevilla - IMSE-CNM (CSIC), Spain*

C3P-U.3 DELAY MODELING OF CMOS/CPL CIRCUITS 5613
Yuanzhong Wan, Maitham Shams, *Carleton University, Canada*

C3P-U.4 GROUND BOUNCE CALCULATION DUE TO SIMULTANEOUS SWITCHING IN DEEP SUB-MICRON INTEGRATED CIRCUITS 5617
Mohammad Hekmat, Shahriar Mirabbasi, *University of British Columbia, Canada*; Majid Hashemi, *SiRF Technology Inc.*

C3P-U.5 MODEL-COMPILER BASED EFFICIENT STATISTICAL CIRCUIT ANALYSIS: AN INDUSTRY CASE STUDY OF 4GHZ/6-BIT ADC/DAC/DEMUX AISC 5621
Bo Hu, Zhao Li, Lili Zhou, C-J Richard Shi, *University of Washington, USA*; Kwang-Hyun Baek, Myung-Jun Choe, *Rockwell Scientific, USA*

C3P-U.6 A HIGH-LEVEL DYNAMIC-ERROR MODEL OF A PIPELINED ANALOG-TO-DIGITAL CONVERTER 5625
Kalle Folkesson, Christer Svensson, *Linköping University, Sweden*; Björn Knuthammar, Andreas Dreyfert, *Wavebreaker AB, Sweden*

C3P-U.7 AN EXPLORATIVE TILE-BASED TECHNIQUE FOR AUTOMATED CONSTRAINT TRANSFORMATION, PLACEMENT AND ROUTING OF HIGH FREQUENCY ANALOG FILTERS 5629
Hui Zhang, Preethi Karthik, Hua Tang, Alex Doboli, *Stony Brook University, USA*

C3P-V   Digital Circuits Synthesis & Optimization, Part II (Poster)
Time: Thursday, May 26, 2005, 14:00 - 16:00
Place: Reception Hall - Area 4
Chair: Hsien-Hsin Lee, *Georgia Tech*

C3P-V.1   A FORMAL APPROACH TO THE SLACK DRIVEN SCHEDULING PROBLEM IN HIGH-LEVEL SYNTHESIS 5633
Shih-Hsu Huang, Chun-Hua Cheng, *Chung Yuan Christian University, Taiwan ROC*

C3P-V.2   DESIGN OF MOS CURRENT MODE LOGIC GATES - COMPUTING THE LIMITS OF VOLTAGE SWING AND BIAS CURRENT 5637
Giuseppe Caruso, *Università di Palermo*

C3P-V.3   INTERCONNECT DELAY OPTIMIZATION VIA HIGH LEVEL RE-SYNTHESIS AFTER FLOORPLANNING 5641
Yunfeng Wang, Jinian Bian, Xianlong Hong, *Tsinghua University, China*

C3P-V.4   USING SYMBOLIC COMPUTER ALGEBRA FOR SUBEXPRESSION FACTORIZATION AND SUBEXPRESSION DECOMPOSITION IN HIGH LEVEL SYNTHESIS 5645
Xianwu Xing, *Nanyang Technological University, Singapore*; Ching-Chuen Jong, *Nanyang Technolgocial University, Singapore*

C3P-V.5   BDD DECOMPOSITION FOR MIXED CMOS/PTL LOGIC CIRCUIT SYNTHESIS 5649
Yen-Tai Lai, Yung-Chuan Jiang, Hong-Ming Chu, *National Cheng Kung University, Taiwan ROC*

C3P-V.6   FPGA TECHNOLOGY MAPPING OPTIMIZATION BY REWIRING ALGORITHMS 5653
Wai-Chung Tang, Wing-Hang Lo, Yu-Liang Wu, *The Chinese University of Hong Kong*; Shih-Chieh Chang, *National Chung Cheng University, Taiwan ROC*

C3P-V.7   A PHYSICALLY-DERIVED LARGE-SIGNAL NONQUASI-STATIC MOSFET MODEL FOR COMPUTER AIDED DEVICE AND CIRCUIT SIMULATION PART II - THE CMOS NOR GATE AND THE CMOS NAND GATE 5657
Michael Payton, *The University of Alabama in Huntsville, USA*; Fat Ho, *University of Alabama in Huntsville*

C3P-W   Testing & Verification (Poster)
Time: Thursday, May 26, 2005, 14:00 - 16:00
Place: Reception Hall - Area 5
Chair: Jinian Bian, *Tsinghua University*

C3P-W.1   DOMAIN FAULT MODEL AND COVERAGE METRIC FOR SOC VERIFICATION 5662
Luo Chun, Yang Jun, Gao Gugang, Shi Longxing, *Southeast University, China*

C3P-W.2   VALIDATION ANALYSIS AND TEST FLOW OPTIMIZATION OF VLSI CHIP 5666
Yanzhuo Tan, Yinhi Han, Xiaowei Li, *Chinese Academy of Sciences, China*; Feiyin Lou, *Beijing Micro-Electronics Technology Institute, Beijing*; Yuchuan Chen, *Teradyne Inc., Shanghai*

C3P-W.3   DETERMINISTIC AND LOW POWER BIST BASED ON SCAN SLICE OVERLAPPING 5670
Ji Li, Yinhi Han, Xiaowei Li, *Chinese Academy of Sciences, China*

C3P-W.4   MODELING AND FORMAL VERIFICATION OF DATAFLOW GRAPH IN SYSTEM-LEVEL DESIGN USING PETRI NET 5674
Tsung-Hsi Chiang, Lan-Rong Dung, Ming-Feng Yaung, *National Chiao Tung University, Taiwan ROC*

C3P-W.5   A ROBUST AND CORRECT COMPUTATION FOR THE CURVILINEAR ROUTING PROBLEM 5678
Tan Yan, Hiroshi Murata, *The University of Kitakyushu, Japan*

C3P-W.6   ESTIMATING LIKELIHOOD OF CORRECTNESS FOR ERROR CANDIDATES TO ASSIST DEBUGGING FAULTY HDL DESIGNS 5682
Tai-Ying Jiang, *National Chiao Tung University, Taiwan ROC*; Chien-Nan Liu, *National Central University, Taiwan ROC*; Jing-Yang Jou, *National Chiao Tung University, Taiwan ROC*

C3P-W.7   INSTRUCTION-BASED DELAY FAULT SELF-TESTING OF PIPELINED PROCESSOR CORES 5686
Virendra Singh, Michiko Inoue, *Nara Institute of Science and Technology, Japan*; Kewal Saluja, *University of Wisconsin-Madison, USA*; Hideo Fujiwara, *Nara Institute of Science and Technology, Japan*

C3P-X Blind Signal Processing III (Poster)
Time: Thursday, May 26, 2005, 14:00 - 16:00
Place: Reception Hall - Area 6
Chair: Yujiro Inouye, *Shimane University, Japan*

C3P-X.1 GRADIENT-BASED METHODS FOR SIMULTANEOUS BLIND SEPARATION OF MIXED SOURCE SIGNALS 5690
Sanqing Hu, Derong Liu, *University of Illinois at Chicago, USA*; Huaguang Zhang, *Northeastern University of China*

C3P-X.2 BLIND IDENTIFICATION OF BRAIN MECHANISM IN MEG 5694
Kuniharu Kishida, *Gifu University, Japan*

C3P-X.3 CHAOTIC SIGNAL SEPARATION FROM A LINEAR MIXTURE 5698
Bao-Yun Wang, *Nanjing University of Posts & Telecommunications, China*; Wei Xing Zheng, *University of Western Sydney, Australia*

C3P-X.4 BLIND IDENTIFICATION OF MIMO CHANNELS WITH PERIODIC MODULATION 5702
Ching-An Lin, Yi-Sheng Chen, *National Chiao Tung University, Taiwan ROC*

C3P-X.5 BLIND LOW RATE MULTIUSER DETECTION FOR MULTIRATE MULTICARRIER CDMA SYSTEMS USING ANTENNA ARRAY 5706
Yiwen Zhang, Qinye Yin, Le Ding, *Xi'an Jiaotong University, China*; Ronghai Sun, *Shaanxi Armed Police Force, China*

C3P-X.6 DOA-MATRIX DECODER FOR STBC-MC-CDMA SYSTEMS OVER FREQUENCY-SELECTIVE CHANNEL 5710
Yanxing Zeng, Qinye Yin, Le Ding, Yinkuo Meng, Ying Zhang, *Xi'an Jiaotong University, China*

C3P-X.7 AN ICA BASED APPROACH FOR BLIND DECONVOLUTION OF THREE-DIMENSIONAL SIGNALS 5714
Emanuele Principi, Stefano Squartini, Francesco Piazza, *Università Politecnica delle Marche, Italy*

C3P-X.8 EVALUATING A BLIND CHANNEL ESTIMATION TECHNIQUE THAT USES A HARDWARE EFFICIENT EQUALIZER 5718
Yun Ye, Saman S. Abeysekera, *Nanyang Technolgocial University, Singapore*

C3P-Y   **Blind Signal Processing IV (Poster)**
Time: Thursday, May 26, 2005, 14:00 - 16:00
Place: Reception Hall - Area 7
Chair: Noboru Nakasako, *Kinki University, Japan*

C3P-Y.1   A CONSIDERATION OF BLIND SOURCE SEPARATION USING WAVELET TRANSFORM 5722
Noriyuki Hirai, *Chuo University, Japan*; Hiroki Matsumoto, *Maebashi Institute of Technology, Japan*; Toshihiro Furukawa, *Tokyo University of Science, Japan*; Kiyohi Furuya, *Chuo University, Japan*

C3P-Y.2   AN APPROACH FOR NONLINEAR BLIND SOURCE SEPARATION OF SIGNALS WITH NOISE USING NEURAL NETWORKS AND HIGHER-ORDER CUMULANTS 5726
Nuo Zhang, Xiaowei Zhang, Jianming Lu, Takashi Yahagi, *Chiba University, Japan*

C3P-Y.3   SUBBAND BLIND EQUALIZATION USING WAVELET FILTER BANKS 5730
Amir Minayi Jalil, Hamidreza Amindavar, Farshad Almasganj, *Amirkabir University of Technology, Iran*

C3P-Y.4   AUDIO SOURCE SEPARATION BY SOURCE LOCALIZATION WITH HILBERT SPECTRUM 5734
Md. Khademul Molla, Keikichi Hirose, Nobuaki Minematsu, *University of Tokyo, Japan*

C3P-Y.5   UPLINK CHANNEL ESTIMATION FOR SPACE-TIME BLOCK CODED MULTIPLE-INPUT MULTIPLE-OUTPUT MC-CDMA SYSTEMS 5738
Ke Deng, Qinye Yin, Hongbo Tian, *Xi'an Jiaotong University, China*

C3P-Y.6   AN ALTERNATIVE NATURAL GRADIENT APPROACH FOR MULTICHANNEL BLIND DECONVOLUTION 5742
Massimo Tomassoni, Stefano Squartini, Francesco Piazza, *Università Politecnica delle Marche, Italy*

C3P-Y.7   DECISION FEEDBACK EQUALIZER WITH THE BLIND MATCHED FILTER ESTIMATION 5746
Izzet Özçelik, İzzet Kale, *University of Westminster, UK*; Buyurman Baykal, *Middle East Technical University*

C3P-Y.8   INTEGRATED BLIND ELECTRONIC EQUALIZER FOR FIBER DISPERSION COMPENSATION 5750
Foster Dai, Shengfang Wei, Richard Jaeger, *Auburn University, USA*

C4L-A   **SPECIAL SESSION - Modeling & Analysis of Power/Signal Integrity (Lecture)**
Time: Thursday, May 26, 2005, 16:10 - 18:10
Place: Room 301
Chair: Hideki Asai, *Shizuoka University*

16:10
C4L-A.1   LINEAR AND NONLINEAR MACROMODELS FOR POWER/SIGNAL INTEGRITY 5754
Igor Stievano, Stefano Grivet-Talocia, Ivano Maio, Flavio Canavero, *Politecnico di Torino, Italy*

16:30
C4L-A.2   DELAY EXTRACTION FROM FREQUENCY DOMAIN DATA FOR CAUSAL MACRO-MODELING OF PASSIVE NETWORKS 5758
Rohan Mandrekar, Madhavan Swaminathan, *Georgia Institute of Technology, USA*

16:50
C4L-A.3   PASSIVE APPROXIMATION OF TABULATED FREQUENCY-DATA BY FOURIER EXPANSION METHOD 5762
Yuichi Tanji, *Kagawa University, Japan*; Hidemasa Kubota, *Shizuoka University, Japan*

17:10
C4L-A.4   NOISE GENERATION, COUPLING, ISOLATION, AND EM RADIATION FROM HIGH-SPEED PACKAGE AND PCB 5766
Joungho Kim, Junsp Pak, Jongbae Park, Hyungsoo Kim, *Korea Advanced Institute of Science & Technology, Korea*

17:30
C4L-A.5   ACCURATE AND CLOSED-FORM SPICE COMPATIBLE PASSIVE MACROMODELS FOR DISTRIBUTED INTERCONNECTS WITH FREQUENCY DEPENDENT PARAMETERS 5770
Natalie Nakhla, Ram Achar, Michel Nakhla, *Carleton University, Canada*

17:50
C4L-A.6   MODELING OF POWER DISTRIBUTION NETWORKS WITH SIGNAL LINES FOR SPICE SIMULATORS 5774
Takayuki Watanabe, *University of Shizuoka, Japan*; Hideki Asai, *Shizuoka University, Japan*

C4L-B Communication Coding (Lecture)
Time: Thursday, May 26, 2005, 16:10 - 18:10
Place: Room 401
Co-Chairs: Wael Badawy, *University of Calgary*
          An-Yeu Wu, *National Taiwan University*

16:10
C4L-B.1 SCHEDULING ALGORITHM FOR PARTIALLY PARALLEL ARCHITECTURE OF LDPC DECODER BY MATRIX PERMUTATION 5778
In-Cheol Park, Se-Hyeon Kang, *Korea Advanced Institute of Science & Technology, Korea*

16:34
C4L-B.2 QUANTIZED LDPC DECODER DESIGN FOR BINARY SYMMETRIC CHANNELS 5782
Rohit Singhal, Gwan Choi, Rabi Mahapatra, *Texas A&M University, USA*

16:58
C4L-B.3 LOW COMPLEXITY, HIGH SPEED DECODER ARCHITECTURE FOR QUASI-CYCLIC LDPC CODES 5786
Zhongfeng Wang, *Oregon State University, USA*; Qingwei Jia, *Seagate Technology International*

17:22
C4L-B.4 AN ANALOG/DIGITAL MODE-SWITCHING LDPC CODEC 5790
David Haley, *Cohda Wireless, Australia*; Chris Winstead, *Utah State University, USA*; Alex Grant, *University of South Australia*; Vincent Gaudet, Christian Schlegel, *University of Alberta, Canada*

17:46
C4L-B.5 DIGITAL VLSI OFDM TRANSCEIVER ARCHITECTURE FOR WIRELESS SOC DESIGN 5794
Wei-Hsiang Tseng, Ching-Chi Chang, Chorng-Kuang Wang, *National Taiwan University, Taiwan ROC*

C4L-C SPECIAL SESSION - Design & Applications of Topographic Sensory Microprocessors (Lecture)

Time: Thursday, May 26, 2005, 16:10 - 18:10
Place: Room 402
Chair: Gustavo Linan Cembrano, *CNM*

16:10
C4L-C.1 A ONE-QUADRANT DISCRETE-TIME CELLULAR NEURAL NETWORK CMOS CHIP FOR PIXEL-LEVEL SNAKES 5798
Victor Brea, *University of Santiago de Compostela, Spain*; M. Laiho, *Helsinki University of Technology, Finland*; D.L. Vilariño, *University of Santiago de Compostela, Spain*; A. Paasio, *University of Turku, Finland*; D. Cabello, *University of Santiago de Compostela, Spain*

16:30
C4L-C.2 VARIOUS IMPLEMENTATIONS OF TOPOGRAPHIC, SENSORY, CELLULAR WAVE COMPUTERS 5802
Ákos Zarándy, Péter Földesy, Péter Szolgay, Szabolcs Tökés, Csaba Rekeczky, Tamas Roska, *MTA-SZTAKI, Hungary*

16:50
C4L-C.3 IMPLEMENTATION OF SIMD VISION CHIP WITH 128X128 ARRAY OF ANALOGUE PROCESSING ELEMENTS 5806
Piotr Dudek, *University of Manchester, UK*

17:10
C4L-C.4 DYNAMICALLY COUPLED MULTI-LAYER MIXED-MODE CNN 5810
Mika Laiho, Ari Paasio, *University of Turku, Finland*

17:30
C4L-C.5 SPATIOTEMPORAL PATTERN FORMATION IN THE ACE16K CNN CHIP 5814
Müştak E. Yalçin, *Istanbul Technical University, Turkey*; Johan A. K. Suykens, *Katholieke Universiteit Leuven, Belgium*; Joos Vandewalle, *K.U. Leuven, ESAT-SCD-SISTA, Belgium*

17:50
C4L-C.6 CNN WAVE BASED COMPUTATION FOR ROBOT NAVIGATION ON ACE16K 5818
Paolo Arena, *University of Catania, Italy*; Adriano Basile, *STMicroelectronics, Italy*; Luigi Fortuna, Mattia Frasca, *University of Catania, Italy*; Johan A. K. Suykens, *Katholieke Universiteit Leuven, Belgium*; Guido Vagliasindi, *University of Catania, Italy*; Mustak E. Yalcin, *Istanbul Technical University, Turkey*

C4L-D ASIC Modules (Lecture)
Time: Thursday, May 26, 2005, 16:10 - 18:10
Place: Room 403
Co-Chairs: Bin-Da Liu, *National Cheng Kung University*
Lars Wanhammar, *Linkoping University*

16:10
C4L-D.1 A SINGLE-CHIP FPGA DESIGN FOR REAL-TIME ICA-BASED BLIND SOURCE SEPARATION ALGORITHM 5822
Charayaphan Charoensak, *Nanyang Technolgocial University, Singapore*; Farook Sattar, *Nanyang Technological University, Singapore*

16:30
C4L-D.2 DESIGN AND FPGA IMPLEMENTATION OF FINITE RIDGELET TRANSFORM 5826
Isa Uzun, Abbes Amira, *The Queen's University of Belfast, UK*

16:50
C4L-D.3 A NOVEL METRIC REPRESENTATION FOR LOW-COMPLEXITY LOG-MAP DECODER 5830
Byonghyo Shim, *University of Illinois at Urbana-Champaign, USA*; Hyung G. Myung, *Polytechnic University, USA*

17:10
C4L-D.4 GRADIENT PILE UP FOR EDGE DETECTION ON HARDWARE 5834
André Soares, Leticia Guimarães, Viviane Cordeiro, Altamiro Susin, *UFRGS, Brazil*

17:30
C4L-D.5 AREA-EFFICIENT SYSTOLIC ARCHITECTURES FOR INVERSIONS OVER $GF(2^M)$ 5838
Zhiyuan Yan, *Lehigh University, USA*; Dilip Sarwate, *University of Illinois, USA*; Zhongzhi Liu, *Lehigh University, USA*

17:50
C4L-D.6 A SPEECH RECOGNIZER WITH SELECTABLE MODEL PARAMETERS 5842
Wei Han, Cheong-Fat Chan, Chiu-Sing Choy, *The Chinese University of Hong Kong*; Kong-Pang Pun, *Chinese University of Hong Kong*

C4L-E Test & Analysis (Lecture)
Time: Thursday, May 26, 2005, 16:10 - 18:10
Place: Room 404
Co-Chairs: Sreedhar Natarajan, *MoSys Inc.*
Mohamad Sawan, *Ecole Polytechnique de Montreal*

16:10
C4L-E.1 RECONFIGURABLE MULTIPLE SCAN-CHAINS FOR REDUCING TEST APPLICATION TIME OF SOCS 5846
Jiann-Chyi Rau, Chih-Lung Chien, Jia-Shing Ma, *Tamkang University, Taiwan ROC*

16:30
C4L-E.2 THE IMPROVEMENT FOR TRANSACTION LEVEL VERIFICATION FUNCTIONAL COVERAGE 5850
Wang Zhong-hai, Yi-Zheng Ye, *Harbin Institute of Technology, China*

16:50
C4L-E.3 PARALLELY TESTABLE DESIGN FOR DETECTION OF NEIGHBORHOOD PATTERN SENSITIVE FAULTS IN HIGH DENSITY DRAMS 5854
Ju Yeob Kim, Sung Je Hong, Jong Kim, *Pohang University of Science and Technology, Korea*

17:10
C4L-E.4 EFFICIENT POWER MODEL FOR CROSSBAR INTERCONNECTS 5858
Behrouz Afzal, Ali Afzali-Kusha, *University of Tehran, Iran*; M. El Nokali, *University of Pittsburgh, USA*

17:30
C4L-E.5 COUPLING REDUCTION ANALYSIS OF BUS-INVERT CODING 5862
Rung-Bin Lin, *Yuan Ze University, Taiwan ROC*

17:50
C4L-E.6 ENERGY AND LATENCY EVALUATION OF NOC TOPOLOGIES 5866
Marcio Kreutz, Cesar Marcon, Luigi Carro, *UFRGS, Brazil*; Ney Calazans, *PUCRS, Brazil*; Altamiro Susin, *UFRGS, Brazil*

C4L-F   SPECIAL SESSION - Blind Signal Processing Theory & Applications (Lecture)
Time: Thursday, May 26, 2005, 16:10 - 18:10
Place: Room 405
Co-Chairs: Andrzej Cichocki, *Riken Brain Science Institute, Japan*
          Yujiro Inouye, *Shimane University, Japan*

16:10
C4L-F.1   AN ADAPTIVE SUPER-EXPONENTIAL DEFLATION ALGORITHM FOR BLIND DECONVOLUTION OF MIMO SYSTEMS USING THE MATRIX PSEUDO-INVERSION LEMMA 5870
Kiyotaka Kohno, Yujiro Inouye, Mitsuru Kawamoto, *Shimane University, Japan*

16:30
C4L-F.2   FILTERBANK-BASED BLIND SIGNAL SEPARATION WITH ESTIMATED SOUND DIRECTION 5874
Hyung-Min Park, Chandra Shekhar Dhir, Do-Kwan Oh, Soo-Young Lee, *Korea Advanced Institute of Science & Technology, Korea*

16:50
C4L-F.3   BLIND SIGNAL SEPARATION INTO GROUPS OF DEPENDENT SIGNALS USING JOINT BLOCK DIAGONALIZATION 5878
Fabian Theis, *University of Regensburg, Germany*

17:10
C4L-F.4   BLIND EXTRACTION OF A DOMINANT SOURCE FROM MIXTURES OF MANY SOURCES USING ICA AND TIME-FREQUENCY MASKING 5882
Hiroshi Sawada, Shoko Araki, Ryo Mukai, Shoji Makino, *NTT Corporation, Japan*

17:30
C4L-F.5   INDEPENDENT ARRAYS OR INDEPENDENT TIME COURSES FOR GENE EXPRESSION TIME SERIES 5886
Sookjeong Kim, Seungjin Choi, *Pohang University of Science and Technology, Korea*

17:50
C4L-F.6   BLIND SEPARATION OF A CLASS OF NONLINEAR ICA MODELS 5890
Jan Eriksson, Visa Koivunen, *Helsinki University of Technology, Finland*

C4L-G Networked Sensors & Sensor Circuits (Lecture)
Time: Thursday, May 26, 2005, 16:10 - 18:10
Place: Room 406
Co-Chairs: Mitsuji Muneyasu, *Kansai University*
          Robert Newcomb, *University of Maryland*

16:10
C4L-G.1 A NOVEL APPLICATION SPECIFIC NETWORK PROTOCOL FOR WIRELESS SENSOR NETWORKS 5894
Jichuan Zhao, Ahmet T. Erdogan, Tughrul Arslan, *University of Edinburgh, UK*

16:30
C4L-G.2 HIGH-EFFICIENCY POWER AMPLIFIER FOR WIRELESS SENSOR NETWORKS 5898
Devrim Aksin, *University of Texas at Dallas, USA*; Stefano Gregori, *University of Guelph, Canada*; Franco Maloberti, *University of Pavia, Italy*

16:50
C4L-G.3 A NEW SWITCHED CAPACITOR CIRCUIT FOR PARALLEL-PIXEL IMAGE PROCESSING 5902
Nicola Massari, Nicola Viarani, Massimo Gottardi, *ITC-irst, Italy*

17:10
C4L-G.4 BROADBAND DIELECTRIC SPECTROSCOPY CMOS READOUT CIRCUIT FOR MOLECULAR SENSING 5906
Youngbok Kim, Anuj Agarwal, *Texas A&M University, USA*; Sameer Sonkusale, *Tufts University, USA*

17:30
C4L-G.5 READ-OUT CIRCUIT IN RT-FLUXGATE 5910
Salvatore Baglio, *Università di Catania, Italy*; Adi Bulsara, *SPAWAR, USA*; Pascal Nouet, *Université Montpellier II, France*; Vincenzo Sacco, *Università di Catania, Italy*

17:50
C4L-G.6 INTEGRATED INTERFACE CIRCUITS FOR CHEMIRESISTOR ARRAYS 5914
Carina Leung, Denise Wilson, *University of Washington, USA*

C4L-H State-of-the-Art Media Processing & SoC Design Methodology (Lecture)
Time: Thursday, May 26, 2005, 16:10 - 18:10
Place: Room 407
Co-Chairs: Andreas Demosthenous, *University College London*
Takao Onoye, *Osaka University*

16:10
C4L-H.1 SHD MOVIE DISTRIBUTION SYSTEM USING IMAGE CONTAINER WITH 4096X2160 PIXEL RESOLUTION AND 36 BIT COLOR 5918
Takahiro Yamaguchi, Mitsuru Nomura, Kazuhiro Shirakawa, Tetsuro Fujii, *NTT Network Innovation Laboratories, Japan*

16:30
C4L-H.2 AN IMPLEMENTATION OF JPEG2000 INTERACTIVE IMAGE COMMUNICATION SYSTEM 5922
Jun-Ichi Hara, *Ricoh Company, Ltd., Japan*

16:50
C4L-H.3 DESIGNING AND PACKAGING TECHNOLOGY OF RENESAS SIP 5926
Noriaki Sakamoto, Norihiko Sugita, Takafumi Kikuchi, Hideki Tanaka, Takashi Akazawa, *Renesas Technology Corp., Japan*

17:10
C4L-H.4 SYSTEM LSI DESIGN WITH C-BASED BEHAVIORAL SYNTHESIS AND VERIFICATION 5930
Kazutoshi Wakabayashi, *NEC Corporation, Japan*

17:30
C4L-H.5 APPROACH FOR PHYSICAL DESIGN IN SUB-100NM ERA 5934
Hiroo Masuda, Shinichi Okawa, *Semiconductor Technology Academic Research Center, Japan*; Masakazu Aoki, *Tokyo University of Science, Japan*

17:50
C4L-H.6 AN ADVANCE RTLTOGDS2 DESIGN METHODOLOGY FOR 90NM AND BELOW SYSTEM LSI'S TO SOLVE TIMING CLOSURE, SIGNAL INTEGRITY AND DESIGN MANUFACTURING 5938
Nobuyuki Nishiguchi, *Semiconductor Technology Academic Research Center, Japan*

C4L-J Filters for Communications (Lecture)
Time: Thursday, May 26, 2005, 16:10 - 18:10
Place: Room 501
Co-Chairs: Tony Chan Carusone, *University of Toronto, Canada*
Cosy Muto, *Kuyushu Institute of Technology, Fukuoka, Japan*

16:10
C4L-J.1 CMOS HIGH-LINEAR, WIDE-DYNAMIC RANGE RF ON-CHIP FILTERS USING Q-ENHANCED LC FILTERS 5942
Shengyuan Li, Susanta Sengupta, Huseyin Dinc, Phillip Allen, *Georgia Institute of Technology, USA*

16:30
C4L-J.2 70MHZ CMOS GM-C IF FILTER 5946
Muhammad Qureshi, Phillip E. Allen, *Georgia Institute of Technology, USA*

16:50
C4L-J.3 VOLTAGE-MODE HIGH-ORDER OTA-ONLY-WITHOUT-C LOW-PASS (FROM 215 M TO 705M HZ) AND BAND-PASS (FROM 214 M TO 724M HZ) FILTER STRUCTURE 5950
Chun-Ming Chang, *Chung Yuan Christian University, Taiwan ROC*

17:10
C4L-J.4 APPLICATION OF REVERSE-ACTIVE NPNS FOR COMPACT, WIDE-TUNING F_T-INTEGRATION-BASED FILTERS IN SIGE HBT BICMOS TECHNOLOGY 5954
Phanumas Khumsat, *Prince of Songkla University, Thailand*; Apisak Worapishet, *Mahanakorn University of Technology, Thailand*

17:30
C4L-J.5 A 2V 0.25µM CMOS 250MHZ FULLY-DIFFERENTIAL SEVENTH-ORDER EQUIRIPPLE LINEAR PHASE LF FILTER 5958
Masood Hasan, Yichuang Sun, *University of Hertfordshire, UK*

17:50
C4L-J.6 ANALYSIS OF TRAVELING WAVE AND TRANSVERSAL ANALOG ADAPTIVE EQUALIZERS 5962
Shanthi Pavan, *Indian Institute of Technology, Madras, India*; Shankar Shivappa, *IIT Madras, India*

C4L-K   CAD & Other Tools for Analog Design (Lecture)
Time: Thursday, May 26, 2005, 16:10 - 18:10
Place: Room 502
Co-Chairs: Gordon Roberts, *McGill University, Montreal, Canada*
            Tuna Tarim, *Texas Instruments, Dallas, TX, USA*

16:10
C4L-K.1   RULES FOR SYSTEMATIC SYNTHESIS OF ALL-TRANSISTOR ANALOGUE CIRCUITS BY ADMITTANCE MATRIX EXPANSION 5966
Philip Corbishley, David Haigh, *Imperial College, London, UK*

16:30
C4L-K.2   FAST ITERATIVE METHOD PACKAGE FOR HIGH FREQUENCY CIRCUITS ANALYSIS 5970
Somsak Akatimagool, *King Mongkut's Institute of Technology, North Bangkok, Thailand*

16:50
C4L-K.3   MIXED SIGNAL AND SOC DESIGN FLOW REQUIREMENTS 5974
Tuna Tarim, *Texas Instruments Inc., USA*

17:10
C4L-K.4   ANALOG VLSI CIRCUIT-LEVEL SYNTHESIS USING MULTI-PLACEMENT STRUCTURES 5978
Raoul Badaoui, Ranga Vemuri, *University of Cincinnati, USA*

17:30
C4L-K.5   IMPROVED MODELING OF SIGMA-DELTA MODULATOR NON IDEALITIES IN SIMULINK 5982
Andrea Fornasari, *University of Pavia, Italy*; Piero Malcovati, *University di Pavia*; Franco Maloberti, *University of Pavia, Italy*

17:50
C4L-K.6   ANALYSIS OF SUPPLY AND GROUND NOISE SENSITIVITY IN RING AND LC OSCILLATORS 5986
Volodymyr Kratyuk, Igor Vytyaz, Un-Ku Moon, Kartikeya Mayaram, *Oregon State University, USA*

C4L-L Digital Signal Processing Applications II (Lecture)
Time: Thursday, May 26, 2005, 16:10 - 18:10
Place: Room 503
Co-Chairs: Akira Taguchi, *Musashi Institute of Technology*
Liang Tao, *Anhui University*

16:10
C4L-L.1 MULTICHANNEL SVD-BASED IMAGE DE-NOISING 5990
Yodchanan Wongsawat, K.R. Rao, Soontorn Oraintara, *University of Texas at Arlington, USA*

16:30
C4L-L.2 ANALYSIS OF NONLINEAR RESIDUAL ECHO SUPPRESSORS FOR TELECOMMUNICATIONS 5994
Sen M. Kuo, *Northern Illinois University, USA*; Woon-Seng Gan, *Nanyang Technolgocial University, Singapore*

16:50
C4L-L.3 SMF ROBUST FILTERING IN IMPULSIVE NOISE 5998
Li Guo, Yih-Fang Huang, *University of Notre Dame, USA*

17:10
C4L-L.4 MODELLING OF HIGH-ORDER MECHANICAL PLATE VIBRATION SYSTEMS BY MULTIDIMENSIONAL WAVE DIGITAL FILTERS 6002
Chien-Hsun Tseng, *Southern Taiwan University of Technology, Taiwan ROC*; Stuart Lawson, *University of Warwick, UK*

17:30
C4L-L.5 A COST-EFFECTIVE MEMORY-BASED REAL-VALUES FFT AND HERMITIAN SYMMETRIC IFFT PROCESSOR FOR DMT-BASED WIRE-LINE TRANSMISSION SYSTEMS 6006
Hsiang-Feng Chi, Zhoa-Hong Lai, *National Chiao Tung University, Taiwan ROC*

17:50
C4L-L.6 DEFINING CORRELATION FUNCTIONS AND POWER SPECTRA FOR MULTIRATE RANDOM PROCESSES 6010
Charles Therrien, *Naval Postgraduate School, USA*

C4L-M Communications Systems II (Lecture)
Time: Thursday, May 26, 2005, 16:10 - 18:10
Place: Room 504
Co-Chairs: Hideaki Okazaki, *Shonan Institute of Technology*
An-Yeu Wu, *National Taiwan University*

16:10
C4L-M.1 PILOT TONE DESIGN FOR PEAK-TO-AVERAGE POWER RATIO REDUCTION IN OFDM 6014
Shinji Hosokawa, Shuichi Ohno, Kok Ann Donny Teo, Takao Hinamoto, *Hiroshima University, Japan*

16:30
C4L-M.2 HIGH-SPEED AND LOW-POWER DESIGN OF PARALLEL TURBO DECODER 6018
Zhiyong He, Sébastien Roy, Paul Fortier, *Laval University, Canada*

16:50
C4L-M.3 EFFICIENT VIEW MAINTENANCE IN WIRELESS NETWORKS 6022
Huaizhong Lin, Bo Zhou, Zengwei Zheng, Chun Chen, *Zhejiang University, China*

17:10
C4L-M.4 DIGITAL SIGNAL PROCESSING ENGINE DESIGN FOR POLAR TRANSMITTER IN WIRELESS COMMUNICATION SYSTEMS 6026
Hung-Yang Ko, Yi-Chiuan Wang, An-Yeu Wu, *National Taiwan University, Taiwan ROC*

17:30
C4L-M.5 A NOVEL TECHNIQUE FOR I/Q IMBALANCE AND CFO COMPENSATION IN OFDM SYSTEMS 6030
Jui-Yuan Yu, Ming-Fu Sun, Terng-Yin Hsu, Chen-Yi Lee, *National Chiao Tung University, Taiwan ROC*

17:50
C4L-M.6 ARCHITECTURAL ISSUES IN BASE-STATION FREQUENCY SYNTHESIZERS 6034
Sankaran Aniruddhan, David Allstot, *University of Washington, Seattle, USA*

C4L-N Synchronization & Bifurcation (Lecture)
Time: Thursday, May 26, 2005, 16:10 - 18:10
Place: Room 505
Co-Chairs: Tetsuro Endo, *Meiji University*
Chai Wah *IBM*

16:10
C4L-N.1 INTEGRAL OBSERVER APPROACH FOR CHAOS SYNCHRONIZATIONN WITH TRANSMISSION DISTURBANCES 6038
Guo-Ping Jiang, *Nanjing University of Posts & Telecommunications, China*; Wei Xing Zheng, *University of Western Sydney, Australia*; Wallace K. S. Tang, Guanrong Chen, *City University of Hong Kong*

16:30
C4L-N.2 SYNCHRONIZING CHAOTIC COLPITTS CIRCUITS ADAPTIVELY WITH PARAMETER MISMATCHES AND CHANNEL DISTORTIONS 6042
Cheng Shen, Zhiguo Shi, Lixin Ran, *Zhejiang University, China*

16:50
C4L-N.3 SYNCHRONIZATION IN AN ARRAY OF CHAOTIC SYSTEMS COUPLED VIA A DIRECTED GRAPH 6046
Chai Wah Wu, *IBM T. J. Watson Research Center, USA*

17:10
C4L-N.4 A SUBTLE LINK IN SWITCHED DYNAMICAL SYSTEMS: SADDLE-NODE BIFURCATION MEETS BORDER COLLISION 6050
Yue Ma, *University of Tokushima, Japan*; Chi K. Tse, *Hong Kong Polytechnic University, Hong Kong*; Takuji Kousaka, *The University of Fukuyama, Japan*; Hiroshi Kawamai, *University of Tokushima, Japan*

17:30
C4L-N.5 BIFURCATION AND TRANSITIONAL DYNAMICS IN ASYMMETRICAL TWO-COUPLED OSCILLATORS WITH HARD TYPE NONLINEARITY 6054
Takuya Yoshimura, Kuniyasu Shimizu, Tetsuro Endo, *Meiji University, Japan*

17:50
C4L-N.6 BIFURCATIONS IN MODIFIED BVP NEURONS CONNECTED BY INHIBITORY AND ELECTRICAL COUPLING 6058
Shigeki Tsuji, Tetsushi Ueta, Hiroshi Kawakami, *Tokushima University, Japan*; Kazuyuki Aihara, *University of Tokyo, Japan*

C4L-P Scalable Coding & 3D Video (Lecture)
Time: Thursday, May 26, 2005, 16:10 - 18:10
Place: Nojiggiku Room, Portopia Hotel
Co-Chairs: Lap-Pui Chau, *Nanyang Technological University*
Chun-Jen Tsai, *National Chiao Tung University*

16:10
C4L-P.1 ERROR SENSITIVITY TESTING FOR THE MC-EZBC SCALABLE WAVELET VIDEO CODER 6062
Tamer Shanableh, *American University of Sharjah, UAE*; Tony May, *Motorola Labs, USA*

16:30
C4L-P.2 A MODEL-BASED RATE ALLOCATION MECHANISM FOR WAVELET-BASED EMBEDDED IMAGE AND VIDEO CODING 6066
Ya-Hui Yu, Chun-Jen Tsai, *National Chiao Tung University, Taiwan ROC*

16:50
C4L-P.3 OPTIMAL RESYNCHRONIZATION FOR LAYERED VIDEO OVER WIRELESS CHANNEL 6070
Tao Fang, Lap-Pui Chau, *Nanyang Technolgocial University, Singapore*

17:10
C4L-P.4 SUB-SEQUENCE VIDEO CODING FOR IMPROVED TEMPORAL SCALABILITY 6074
Dong Tian, *Tampere International Center for Signal Processing, Finland*; Miska Hannuksela, *Nokia Research Center, Finland*; Moncef Gabbouj, *Tampere University of Technology, Finland*

17:30
C4L-P.5 SCALABLE MULTIVIEW VIDEO CODING USING WAVELET 6078
Wenxian Yang, *Nanyang Technolgocial University, Singapore*; Feng Wu, Yan Lu, *Microsoft Research Asia, China*; Jianfei Cai, *Nanyang Technolgocial University, Singapore*; King Ngi Ngan, *The Chinese University of Hong Kong*; Shipeng Li, *Microsoft Research Asia, China*

17:50
C4L-P.6 STEREO VIDEO CODING SYSTEM WITH HYBRID CODING BASED ON JOINT PREDICTION SCHEME 6082
Li-Fu Ding, Shao-Yi Chien, Yu-Wen Huang, Yu-Lin Chang, Liang-Gee Chen, *National Taiwan University, Taiwan ROC*

C4L-Q Communication Circuits & Systems II (Lecture)
Time: Thursday, May 26, 2005, 16:10 - 18:10
Place: Sumire Room, Portopia Hotel
Co-Chairs: Shu-Hung Leung, *City University of Hong Kong*
Vinicius Licks, *Pontificia Universidade Catolica - PUCRS, Brazil*

16:10
C4L-Q.1 A GENERALIZED SEMI-BLIND CHANNEL ESTIMATION FOR PILOT-AIDED OFDM SYSTEMS 6086
Ka Yau Ho, Shu-Hung Leung, *City University of Hong Kong*

16:30
C4L-Q.2 JITTER LIMITATIONS ON MULTI-CARRIER MODULATION 6090
Jan H. Rutger Schrader, Eric A.M. Klumperink, *University of Twente, The Netherlands*; Jan L. Visschers, *NIKHEF, The Netherlands*; Bram Nauta, *University of Twente, The Netherlands*

16:50
C4L-Q.3 ESTIMATING THE FADING COEFFICIENT IN MOBILE OFDM SYSTEMS USING STATE-SPACE MODEL 6094
Mihai Enescu, Visa Koivunen, *Helsinki University of Technology, Finland*

17:10
C4L-Q.4 BLOCK-WISE ADAPTIVE MODULATION FOR OFDM WLAN SYSTEMS 6098
Yin-Tsung Hwang, *National Yunlin University of Science & Technology, Taiwan ROC*; Chen-Yu Tsai, Cheng-Chen Lin, *National Yunlin University of Science and Technology, Taiwan ROC*

17:30
C4L-Q.5 A HYBRID SPACE-TIME AND COLLABORATIVE CODING SCHEME FOR WIRELESS COMMUNICATIONS 6102
Ming Ma, Elzinati Masoud, Yichuang Sun, John Senior, *Hertfordshire University, UK*

17:50
C4L-Q.6 AN ANALOG MODULATOR/DEMODULATOR USING A PROGRAMMABLE ARBITRARY WAVEFORM GENERATOR 6106
Ravi Chawla, Christopher Twigg, Paul Hasler, *Georgia Institute of Technology, USA*

C4L-R   VLSI Implementation for multimedia (Lecture)
Time: Thursday, May 26, 2005, 16:10 - 18:10
Place: Tsutsuji Room, Portopia Hotel
Co-Chairs: Ichiro Kuroda, *NEC Electronics*
           Jar-Ferr Yang, *National Cheng Kung University*

16:10
C4L-R.1   A NOVEL LOW-COST HIGH-PERFORMANCE VLSI ARCHITECTURE FOR MPEG-4 AVC/H.264 CAVLC DECODING 6110
Hsiu-Cheng Chang, Chien-Chang Lin, Jiun-In Guo, *National Chung Cheng University, Taiwan ROC*

16:30
C4L-R.2   A HARDWARE-BASED PREDICTIVE MOTION ESTIMATION ALGORITHM 6114
Saku Hämäläinen, Lauri Koskinen, Kari Halonen, *Helsinki University of Technology, Finland*

16:50
C4L-R.3   A MULTIPLICATION-ACCUMULATION COMPUTATION UNIT WITH OPTIMIZED COMPRESSORS AND MINIMIZED SWITCHING ACTIVITIES 6118
Li-Hsun Chen, Oscal T.C. Chen, *National Chung Cheng University, Taiwan ROC*; Teng-Yi Wang, Yung-Cheng Ma, *Industrial Technology Research Institute, Taiwan ROC*

17:10
C4L-R.4   A COST-EFFECTIVE MEDIA PROCESSOR FOR EMBEDDED APPLICATIONS 6122
Wen-Kai Huang, I-Ting Lin, Shi-Wei Chen, Ing-Jer Huang, *National Sun Yat-Sen University, Taiwan ROC*

17:30
C4L-R.5   DESIGN AND IMPLEMENTATION OF A NEW CRYPTOGRAPHIC SYSTEM FOR MULTIMEDIA TRANSMISSION 6126
Jui-Cheng Yen, Hun-Chen Chen, Shu-Meng Wu, *National United University, Taiwan ROC*

17:50
C4L-R.6   LOW-COST IMPLEMENTATION OF A SUPER-RESOLUTION ALGORITHM FOR REAL-TIME VIDEO APPLICATIONS 6130
Gustavo Callicó, Sebastián López, Antonio Núñez, José López, Roberto Sarmiento, *University of Las Palmas de Gran Canaria, Spain*

C4P-S   **Flash ADCs & Comparators (Poster)**
Time: Thursday, May 26, 2005, 16:10 - 18:10
Place: Reception Hall - Area 1
Chair: Sameer Sonkusale, *Texas A&M University, USA*

C4P-S.1   A 0.35µM CMOS COMPARATOR CIRCUIT FOR HIGH-SPEED ADC APPLICATIONS 6134
Samad Sheikhaei, Shahriar Mirabbasi, André Ivanov, *University of British Columbia, Canada*

C4P-S.2   A 4-BIT 5GS/S FLASH A/D CONVERTER IN 0.18µM CMOS 6138
Samad Sheikhaei, Shahriar Mirabbasi, André Ivanov, *University of British Columbia, Canada*

C4P-S.3   A LOW-POWER 4-B 2.5 GSAMPLE/S PIPELINED FLASH ANALOG-TO-DIGITAL CONVERTER USING DIFFERENTIAL COMPARATOR AND DCVSPG ENCODER 6142
Shailesh Radhakrishnan, Mingzhen Wang, Chien-In Chen, *Wright State University, USA*

C4P-S.4   A 1.2GHZ ADAPTIVE FLOATING GATE COMPARATOR WITH 13-BIT RESOLUTION 6146
Yanyi Liu Wong, Marc Cohen, Pamela Abshire, *University of Maryland, USA*

C4P-S.5   AN 8-BIT 160 MS/S FOLDING-INTERPOLATING ADC WITH OPTIMIZIED ACTIVE AVERAGING/INTERPOLATING NETWORK 6150
Meysam Azin, Hamid Movahedian, Mehrdad Sharif Bakhtiar, *Sharif University of Technology, Iran*

C4P-S.6   OFFSET COMPENSATION IN FLASH ADCS USING FLOATING-GATE CIRCUITS 6154
Philomena Brady, Paul Hasler, *Georgia Institute of Technology, USA*

C4P-S.7   A NOVEL APPROACH FOR IMPLEMENTING ULTRA-HIGH SPEED FLASH ADC USING MCML CIRCUITS 6158
Hung Dang, Mohamad Sawan, Yvon Savaria, *École Polytechnique de Montréal, Canada*

C4P-S.8   A 1.8V 3.2µW COMPARATOR FOR USE IN A CMOS IMAGER COLUMN-LEVEL SINGLE-SLOPE ADC 6162
Martijn Snoeij, Albert Theuwissen, Johan Huijsing, *Delft University of Technology, The Netherlands*

C4P-T Pipelined & Algorithmic ADC Design Techniques (Poster)
Time: Thursday, May 26, 2005, 16:10 - 18:10
Place: Reception Hall - Area 2
Chair: Shantanu Chakrabartty, *Michigan State University*

C4P-T.1 DESIGN CONSIDERATIONS OF A FLOATING POINT ADC WITH EMBEDDED S/H 6166
Johan Piper, Jiren Yuan, *Lund University, Sweden*

C4P-T.2 A NEW CCII-BASED PIPELINED ANALOG TO DIGITAL CONVERTER 6170
Yuh-Shyan Hwang, Lu-Po Liao, Chia-Chun Tsai, Wen-Ta Lee, Trong-Yen Lee, Jiann-Jong Chen, *National Taipei University of Technology, Taiwan ROC*

C4P-T.3 A 12BITS/200MHZ RESOLUTION/SAMPLING/POWER-OPTIMIZED ADC IN 0.25μM SIGE BICMOS 6174
Qiong Wu, Albert Wang, *Illinois Institute of Technology, USA*

C4P-T.4 A LOW POWER PIPELINED ANALOG-TO-DIGITAL CONVERTER USING SERIES SAMPLING CAPACITORS 6178
Seong-Hwan Cho, *Korea Advanced Institute of Science & Technology, Korea*; Sungmin Ock, Sang-Hoon Lee, Joon-Suk Lee, *FCI Inc., Korea*

C4P-T.5 A GENERIC MULTILEVEL MULTIPLYING D/A CONVERTER FOR PIPELINED ADCS 6182
Vivek Sharma, *Austria Microsystems, Switzerland*; Un-Ku Moon, Gabor Temes, *Oregon State University, USA*

C4P-T.6 A 10-BIT ALGORITHMIC A/D CONVERTER FOR CYTOSENSOR APPLICATION 6186
Thirumalai Rengachari, *Cirrus Logic Inc., USA*; Vivek Sharma, *Austria Microsystems, Switzerland*; Gabor Temes, Un-Ku Moon, *Oregon State University, USA*

C4P-T.7 AN ADAPTIVE, TRULY BACKGROUND CALIBRATION METHOD FOR HIGH SPEED PIPELINE ADC DESIGN 6190
Zhongjun Yu, Degang Chen, Randall Geiger, *Iowa State University, USA*

C4P-T.8 THE REALIZATION OF A MISMATCH-FREE AND 1.5-BIT OVER-SAMPLING PIPELINED ADC 6194
Shigeto Tanaka, Yuji Ghoda, Yasuhiro Sugimoto, *Chuo University, Japan*

C4P-U   **Emerging Areas in CAD II (Poster)**
Time: Thursday, May 26, 2005, 16:10 - 18:10
Place: Reception Hall - Area 3
Chair: Philippe Coussy, *Univ de Bretagne Sud*

C4P-U.1   **AN 800MBPS SYSTEM INTERCONNECT MODELING AND SIMULATION FOR HIGH SPEED COMPUTING** 6198
Mohammad Sharawi, Daniel Aloi, *Oakland University, USA*

C4P-U.2   **TERNARY WALSH TRANSFORM** 6202
Bogdan Falkowski, Shixing Yan, *Nanyang Technolgocial University, Singapore*

C4P-U.3   **A POST LAYOUT WATERMARKING METHOD FOR IP PROTECTION** 6206
Tingyuan Nie, Tomoo Kisaka, Masahiko Toyonaga, *Kochi University, Japan*

C4P-U.4   **RAPID AND PRECISE INSTRUCTION SET EVALUATION FOR APPLICATION SPECIFIC PROCESSOR DESIGN** 6210
Masayuki Masuda, Kazuhito Ito, *Saitama University, Japan*

C4P-U.5   **MODERN FLOORPLANNING WITH ABUTMENT AND FIXED-OUTLINE CONSTRAINTS** 6214
Chang-Tzu Lin, De-Sheng Chen, Yi-Wen Wang, Hsin-Hsien Ho, *Feng Chia University, Taiwan ROC*

C4P-U.6   **MODELING OF MOS TRANSISTORS BASED ON GENETIC ALGORITHM AND SIMULATED ANNEALING** 6218
Ali Abbasian, Mohammad Taherzadeh-Sani, Behnam Amelifard, Ali Afzali-Kusha, *University of Tehran, Iran*

C4P-V   Placement & Routing II (Poster)
Time: Thursday, May 26, 2005, 16:10 - 18:10
Place: Reception Hall - Area 4
Chair: Xianlong Hong, *Tsinghua University*

C4P-V.1   VLSI BLOCK PLACEMENT WITH ALIGNMENT CONSTRAINTS BASED ON CORNER BLOCK LIST 6222
Song Chen, Xianlong Hong, Sheqin Dong, Yuchun Ma, *Tsinghua University, China*; Chung-Kuan Cheng, *University of California, San Diego, USA*

C4P-V.2   MULTIOBJECTIVE VLSI CELL PLACEMENT USING DISTRIBUTED SIMULATED EVOLUTION ALGORITHM 6226
Sadiq Sait, Mustafa Ali, Ali Zaidi, *King Fahd University of Petroleum & Minerals, Saudi Arabia*

C4P-V.3   A DIVIDE-AND-CONQUER 2.5-D FLOORPLANNING ALGORITHM BASED ON STATISTICAL WIRELENGTH ESTIMATION 6230
Zhuoyuan Li, Xianlong Hong, Qiang Zhou, Yici Cai, Jinian Bian, *Tsinghua University, China*; Hannal Yang, Prashant Saxena, Vijay Pitchumani, *Strategic CAD lab, Intel, USA*

C4P-V.4   A NEW CONGESTION AND CROSSTALK AWARE ROUTER 6234
Chin-Hui Wang, *Chung Yuan Christian University, Taiwan ROC*; Chih-Hung Lee, *Ling Tung College, Taiwan ROC*; Hsin-Hsiung Huang, Yung-Ching Chen, Tsai-Ming Hsieh, *Chung Yuan Christian University, Taiwan ROC*

C4P-V.5   FAST INTEGER LINEAR PROGRAMMING BASED MODELS FOR VLSI GLOBAL ROUTING 6238
Laleh Behjat, Andy Chiang, *University of Calgary, Canada*

C4P-V.6   FLOORPLANNING WITH CLOCK TREE ESTIMATION 6244
Chih-Hung Lee, *Ling Tung College, Taiwan ROC*; Chin-Hung Su, Shih-Hsu Huang, Chih-Yuan Lin, Tsai-Ming Hsieh, *Chung Yuan Christian University, Taiwan ROC*

C4P-V.7   SEGMENTED CHANNEL ROUTING WITH PIN REARRANGEMENTS VIA SATISFIABILITY 6248
Fei He, *Tsinghua University, China*; William N. N. Hung, *Synplicity Inc., California, USA*; Xiaoyu Song, *Portland State University, USA*; Ming Gu, Jiaguang Sun, *Tsinghua University, China*

C4P-W General Image Processing I (Poster)
Time: Thursday, May 26, 2005, 16:10 - 18:10
Place: Reception Hall - Area 5
Chair: Oscar Au, *Hong Kong University of Science and Technology*

C4P-W.1 AN AUTOMATIC FACE RECOGNITION SYSTEM BASED ON WAVELET TRANSFORMS 6252
Abbes Amira, Peter Farrell, *Queen's University Belfast, Canada*

C4P-W.2 HARDWARE REALIZATION OF PANORAMIC CAMERA WITH SPEAKER-ORIENTED FACE EXTRACTION FOR TELECONFERENCING 6256
Kazunori Sugahara, Takao Kawamura, Yukinori Nagase, Takahiko Yamamoto, *Tottori University, Japan*

C4P-W.3 A NOVEL CONTENT-ADAPTIVE INTERPOLATION 6260
Tai-Wai Chan, Oscar C. Au, Tak-Song Chong, Wing-San Chau, *Hong Kong University of Science and Technology, Hong Kong*

C4P-W.4 ARBITRARY SCALE IMAGE ENLARGEMENT WITH THE PREDICTION OF HIGH FREQUENCY COMPONENTS 6264
Shuai Yuan, *Tohoku University, Japan*; Akira Taguchi, *Musashi Institute of Technology, Japan*; Masayuki Kawamata, *Tohoku University, Japan*

C4P-W.5 RESTORATION FROM IMAGE DEGRADED BY WHITE NOISE BASED ON ITERATIVE SPECTRAL SUBTRACTION METHOD 6268
Tetsuya Kobayashi, Tetsuya Shimamura, *Saitama University, Japan*; Tetsuo Hosoya, Yoshitake Takahashi, *Daiichi Radioisotope Laboratories Ltd, Japan*

C4P-W.6 COMPENSATION OF ERRORS GENERATED BY AN ANALOG 2-D DCT 6272
Kati Virtanen, Mikko Pänkäälä, Ari Paasio, *University of Turku, Finland*

C4P-W.7 AN OPTIMAL TONE REPRODUCTION CURVE OPERATOR FOR THE DISPLAY OF HIGH DYNAMIC RANGE IMAGES 6276
Guoping Qiu, Jiang Duan, *University of Nottingham, UK*

C4P-W.8 THE FIR FILTER BANK WITH GIVEN ANALYSIS FILTERS THAT MINIMIZES VARIOUS WORST-CASE MEASURES OF ERROR AT THE SAME TIME 6280
Yuichi Kida, *Ohu University, Japan*; Takuro Kida, *Nihon University, Japan*

C4P-X General Image Processing II (Poster)
Time: Thursday, May 26, 2005, 16:10 - 18:10
Place: Reception Hall - Area 6
Chair: Shipeng Li, *Microsoft Research China*

C4P-X.1 DIRECTIONALLY WEIGHTED COLOR INTERPOLATION FOR DIGITAL CAMERAS 6284
Hung-An Chang, Homer Chen, *National Taiwan University, Taiwan ROC*

C4P-X.2 A NOVEL COLOR INTERPOLATION ALGORITHM BY PRE-ESTIMATING MINIMUM SQUARE ERROR...6288
Jhing-Fa Wang, Chien-Shun Wang, Han-Jen Hsu, *National Cheng Kung University, Taiwan ROC*

C4P-X.3 A RANDOM-VALUED IMPULSE NOISE DETECTOR USING LEVEL DETECTION.................6292
Noritaka Yamashita, Munenori Ogura, Jianming Lu, Hiroo Sekiya, Takashi Yahagi, *Chiba University, Japan*

C4P-X.4 SUPER-RESOLUTION IMAGE RESTORATION FROM BLURRED OBSERVATIONS 6296
Nirmal K. Bose, *Pennsylvania State University, USA*; Michael K. Ng, Andy Yau, *The University of Hong Kong*

C4P-X.5 A NEAREST NEIGHBOR GRAPH BASED WATERSHED ALGORITHM 6300
Wei-Chih Shen, Ruey-Feng Chang, *National Chung Cheng University, Taiwan ROC*

C4P-X.6 JPEG 2000 ENCRYPTION ENABLING FINE GRANULARITY SCALABILITY WITHOUT DECRYPTION..6304
Bin Zhu, *Microsoft Research Asia, China*; Yang Yang, *University of Science & Technology of China*; Shipeng Li, *Microsoft Research Asia, China*

C4P-Y Image Compression II (Poster)
Time: Thursday, May 26, 2005, 16:10 - 18:10
Place: Reception Hall - Area 7
Chair: Gwo Giun Lee, *National Cheng Kung University*

C4P-Y.1 DSVD: A TENSOR-BASED IMAGE COMPRESSION AND RECOGNITION METHOD 6308
Kohei Inoue, Kiichi Urahama, *Kyushu University, Japan*

C4P-Y.2 APPROXIMATE TREATMENT FOR CALCULATION OF THE RATE-DISTORTION SLOPE IN EBCOT 6312
Yue-Xin Zhu, Nan-Ning Zheng, Jing Zhang, Zong-Ze Wu, *Xian Jiaotong University, China*

C4P-Y.3 A SIMPLIFIED ALGORITHM OF JPEG2000 RATE CONTROL FOR VLSI IMPLEMENTATION 6316
Qin Xing, Yan Xiaolang, Ge Haitong, Yang Ye, *Zhejiang University, China*

C4P-Y.4 QUALITY IMPROVEMENT TECHNIQUE FOR JPEG IMAGES WITH FRACTAL IMAGE CODING 6320
Megumi Takezawa, Hirofumi Sanada, Kazuhisa Watanabe, *Hokkaido Institute of Technology, Japan*; Miki Haseyama, *Hokkaido University, Japan*

C4P-Y.5 A SCALABLE ENCRYPTION METHOD ALLOWING BACKWARD COMPATIBILITY WITH JPEG2000 IMAGES 6324
Osamu Watanabe, *Takushoku University, Japan*; Akiko Nakazaki, Hitoshi Kiya, *Tokyo Metropolitan University, Japan*

C4P-Y.6 LOSSLESS IMPLEMENTATION OF MOTION JPEG2000 INTEGRATED WITH INVERTIBLE DEINTERLACING 6328
Takuma Ishida, Shogo Muramatsu, Hisakazu Kikuchi, *Niigata University, Japan*

C4P-Y.7 IMPROVED FAST ENCODING METHOD FOR VECTOR QUANTIZATION BASED ON SUBVECTOR TECHNIQUE 6332
Zhibin Pan, Koji Kotani, Tadahiro Ohmi, *Tohoku University, Japan*

A Wide-Band Low-Noise Amplifier with Double Loop Feedback

Miguel A. Martins [1], Koen van Hartingsveldt [2], Chris J.M. Verhoeven [2] and Jorge R. Fernandes [1]

(1) I.S. Técnico / INESC-ID Lisboa, R. Alves Redol, 9 – 1000 Lisboa, Portugal
(2) Technical University of Delft, Mekelweg 4, 2628 CD Delft, The Neatherlands

Abstract– There is increasing interest in wireless receivers that can work with different standards at different frequencies. The usual way of implementing a low noise amplifier (LNA) block in such a receiver is to use separate narrowband LNAs. In this paper a wideband LNA is presented, with negative feedback by means of a double loop feedback network to achieve simultaneously a low noise figure and input matched impedance for a wide range of frequencies. The designed LNA has a noise figure lower than 3dB, input impedance matched to 50 Ω and a gain above 12dB for a frequency range between 700MHz and 5GHz. The LNA has been designed in 0.18μm CMOS technology from National Semiconductor. It consumes 18.9mW for a supply voltage of 1.8V.

I. INTRODUCTION

One of the demands on present wireless communication systems is the ability to operate with different protocols (GSM, UMTS, WLAN, etc.) using different frequency bands. LNAs with open loop architecture require at the input a resistor (50Ω or other value) to ensure wideband input matching with the antenna, but this degrades the noise performance. This is why most LNAs are narrowband with impedance matching ensured for a single frequency by using inductors. Up to now the most common way to implement a multi-band receiver has been to use different narrowband LNAs which can be switched between them. Due to recent advances in IC CMOS technologies, the transistors' f_T has raised to, at least, one decade above the standard telecommunication frequencies. It is thus possible to explore other topologies, using negative feedback, so far used in lower frequency analog amplifiers. In this paper a wideband LNA is described, with negative feedback, which uses one amplification block and a double loop feedback network to achieve simultaneously input matched impedance and a low noise figure, for a wide range of frequencies. The use of a wideband LNA has advantages over the multi narrowband approach, like smaller area, absence of switches, and maximum flexibility due to the continuous frequency band instead of a limited number of discrete bands.

Negative feedback reduces gain sensitivity, noise and nonlinear distortion. An amplifier with feedback has two basic blocks: the feedback network (with high accuracy) and the amplification block (with high gain). If the amplifying block has enough gain, the overall performance of the LNA depends only on the feedback network.

In Section II the proposed double loop negative feedback amplifier with integrated transformer is presented. In Section III the nonidealities of the feedback network are analyzed, and in Section IV the amplifying block design is presented. In Section V simulation results are presented. Finally, Section VI presents the conclusions.

II. DOUBLE LOOP, NEGATIVE FEEDBACK AMPLIFIER WITH INTEGRATED TRANSFORMER

The feedback network can be implemented with one or more loops [1]. With a single loop it is not possible to have finite impedance matching and low noise at the same time. In [2] a double loop, negative feedback amplifier has been proposed. Here a different topology, shown in Fig. 1, is presented and analyzed in detail.

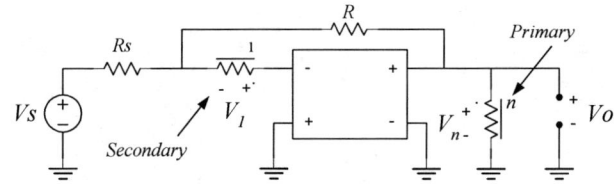

Fig. 1 – Double loop, negative feedback amplifier using transformer

In this circuit, the transformer samples the output voltage, reduces it by a factor n and compares it with the input voltage. Resistor R samples the output voltage, converts it into a current and compares it with the input current. The transformer has:

$$V_1 = \frac{V_n}{n} \quad (1)$$

The difference with respect to the topology proposed in [2], is that, there the current through R is injected at the input of the high-gain amplifier (nullor), while in Fig. 1 there is no current through the transformer secondary. This minimizes the effect of the secondary non-ideal parameters, as shown below. The double loop feedback provides input matched impedance for maximum power transfer. The input impedance is:

$$Z_{in} = \frac{R}{n+1} \quad (2)$$

It is possible to match the amplifier input impedance to the source impedance, by choosing the values of R and n. This topology has an addition of 1 in the denominator relative to the topology presented in [2], which relaxes the specifications of the transformer. From now on, the matching condition ($Z_{in}=Rs$) will always be assumed in all equations.

The noise spectral density due to the feedback resistor, measured at the input is:

$$S_{n,in,R}(f) = 4kTR\left(\frac{Rs}{R}\right)^2 = \frac{4kTRs}{n+1} \quad (3)$$

and the minimum achievable noise factor becomes:

$$F_{min} = 1 + \frac{1}{n+1} \quad (4)$$

This work was supported by POSI and the Portuguese Foundation for Science and Technology (FCT) through project POCTI /38533/ESE/2001 and scholarship BD 12155/2003.

The voltage gain of this amplifier is

$$\frac{V_o}{V_s} = -\frac{n}{2} \quad (5)$$

Equation (4) shows that by increasing n, the noise is reduced and (5) shows that the amplifier voltage gain is proportional to n. Unfortunately, n is limited due to constraints on the realization of the transformer in an integrated circuit.

The next section will be dedicated to the analysis of the influence of the transformer non idealities on the amplifier performance.

III. NONIDEAL FEEDBACK NETWORK ANALYSIS

Fig. 2 displays the amplifier with the transformer model including its main non-idealities: the inductances with their series resistances and parasitic capacitances between the two inductors.

First all capacitances are neglected. The input impedance and the transfer function are, respectively:

$$Z_{in} = R\frac{sL_p}{(n+1)sL_p + nRs_2} \quad (6)$$

$$H(s) = \frac{V_o(s)}{V_i(s)} = \frac{n(n+1)(sL_p + Rs_2)}{2(n+1)sL_p + nRs_2} \quad (7)$$

The secondary inductance L_s and its series resistance Rs_1 do not appear in these equations. This is advantageous with respect to the circuit in [2] and this happens because the input impedance of the high gain amplifier is infinite, i.e. there is no current crossing the secondary. L_p and Rs_2 produce a pole and zero in both equations.

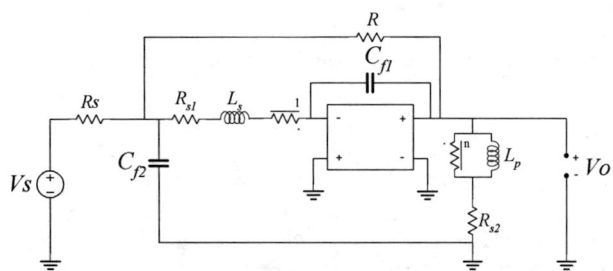

Fig. 2 - Amplifier including inductances L_p and L_s, series resistances Rs_1 and Rs_2 and parasitic capacitances between the two inductors C_{f1} and C_{f2} in the transformer model.

The parasitic series resistances produce extra noise. The total equivalent input spectral noise density is:

$$S_{n,in,eq}(f) \approx 4kT[R + Rs_2 + Rs_1]\left(\frac{1}{n+1}\right)^2 + 4kTRs_1 \quad (8)$$

Considering that $R >> Rs_{1,2}$, these resistances can be neglected in the first term. Rs_2 is not relevant to the overall noise, but the noise contribution of Rs_1 is fully present at the input (there is no attenuation by the topology). The minimum noise factor achievable by this topology is now:

$$F \approx 1 + \frac{1}{n+1} + \frac{Rs_1}{Rs} \quad (9)$$

which shows the importance of keeping Rs_1 with a low value.

The parasitic capacitances C_{f1} and C_{f2} (Fig. 2) between the two inductors will be considered now.

Fig. 3 shows the input impedance $|Z_{in}(\omega)|$ and transfer function $|H(\omega)|$ of the amplifier together with the equations concerning the poles and zeros. The impedance has a zero at the origin and a low frequency pole, both caused by L_p and Rs_2. This pole defines the lower limit of the frequency range in which the input impedance has the desired value used for impedance matching. The upper limit of this range is determined by another pole caused by the source resistance Rs and parasitic capacitances C_{f1} and C_{f2}. Pairs of complex poles and complex zeros are observed in Fig. 3. The influence of the secondary inductor is now present, due to the leakage of current through it and via the parasitic capacitances between the two inductors. Thus, these capacitances are important concerning the transformer implementation.

The transfer function has one pole and one zero at low frequencies. The zero frequency is approximately twice that of the pole frequency. Both are caused by the primary inductance L_p and its series resistance Rs_2. This reduces the voltage gain by the ratio between the pole and zero frequencies. The amplifier bandwidth is determined by a complex pole pair, with frequency determined by the secondary inductor L_s, the parasitic capacitances between the two inductors, and n. This shows again the influence of these capacitances and the importance of reducing their value. An additional pole, caused by the source resistance Rs and the capacitances C_{f1} and C_{f2} is visible, but it is out of the frequency range of interest.

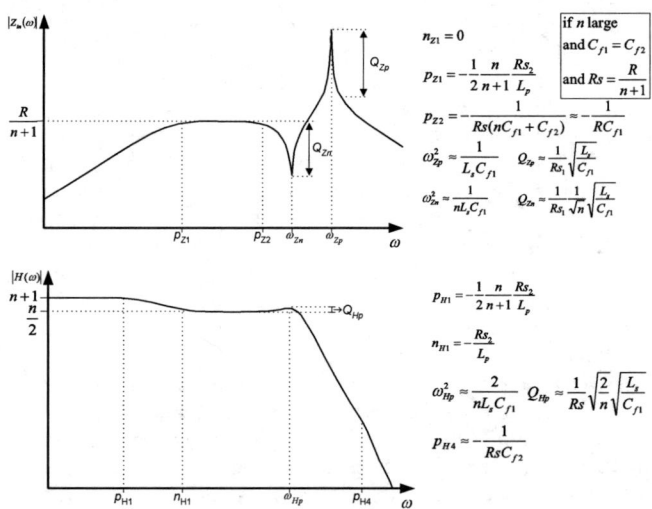

Fig. 3 – Bode plots input impedance $|Z_{in}(\omega)|$ and transfer function $|H(\omega)|$ of the amplifier.

It can be shown that the noise performance is not significantly affected by the presence of the parasitic capacitances.

The inductances between the transformer and substrate resonate with the transformer inductances. It should be ensured that the resonant frequency is outside the frequency range of interest. After this analysis, the amplifying block design will be developed in next section.

IV. AMPLIFYING BLOCK DESIGN

The amplifying block noise figure and gain are the main specifications, the linearity being less important. At the frequencies of interest, only thermal noise is considered. In a

MOS transistor two thermal noise sources are considered: one due to the gate resistance R_g and the other due to the drain current. Their spectral noise density is:

$$\begin{cases} S_{vnRg}(f) = 4kTR_g \\ S_{ind}(f) = 4kT\gamma g_{d0} \end{cases} \quad (10)$$

where R_g is the gate resistance, g_{d0} is the channel conductance, γ is a bias dependent factor, usually equal to 2/3 for long channel devices and greater for short channel devices, k is the Boltzmann constant and T is the temperature in Kelvin [3].

Two equivalent input noise sources, v_n and i_n can be considered at the transistor input, as shown in Fig. 4.

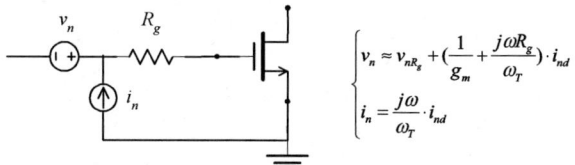

Fig. 4 – Equivalent noise sources at the transistor input

ω_T is the transition frequency, and g_m is the transconductance of the transistor. The spectral noise density, due to the transistor, referred to the input of the amplifier is:

$$S_{n,eq}(f) = 4kTR_g + 4kT\gamma g_{d0}\left[\left(\frac{\omega^2 L_s}{\omega_T} - \frac{1}{g_m}\right) + \left(\frac{\omega}{\omega_T}(Rs_1 + R_g + Rs)\right)\right] \quad (11)$$

Fig. 5 illustrates graphically the equation above.

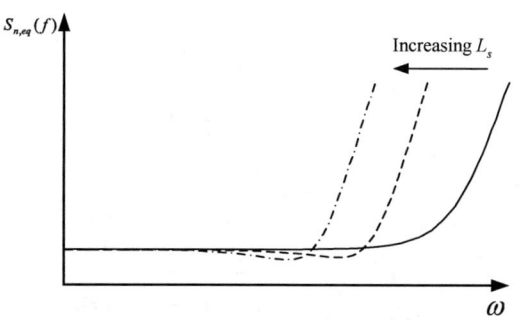

Fig.5 – Spectral noise density caused by the input transistor, reported to the input of the amplifier.

The graphic reveals two different zones. At low frequencies, the noise is constant, while at high frequencies the noise increases at a high rate. In (11) there is a frequency independent term dependent on R_g, which defines the noise at low frequencies. It is important to keep R_g low, by using a wide transistor (dividing the transistor gate into several fingers). All the frequency dependent terms in (11) depend inversely on g_m. ($\omega_T = g_m/(C_{gs} + C_{gd})$). Increasing g_m reduces the noise and is also advantageous because it increases the amplifier gain, thus reducing the noise contribution of the noise sources ahead of the first transistor (Friis Law). A high g_m can be obtained by increasing W/L or by increasing the transistor current. Increasing W (which is also useful to decrease R_g as pointed out above) is limited by the increase of gate capacitance that has a detrimental effect on the bandwidth. The current is limited by the power consumption allowed by the amplifier specifications.

It may happen that a one-stage amplifier is insufficient to satisfy the amplifying block specifications. This will now be discussed.

In a negative feedback amplifier with amplifying block gain A and a feedback factor β the loop gain should be $A\beta \gg 1$ so that the closed loop gain is $\approx \beta^{-1}$, i.e. dependent only on the feedback block. It is thus convenient to exercise the value of the loop gain $A\beta$. Fig. 6 depicts the equivalent circuit used to determine $A\beta$ using a single stage amplifying block. Capacitance C_{Miller} is caused by the Miller effect over the gate-drain capacitance.

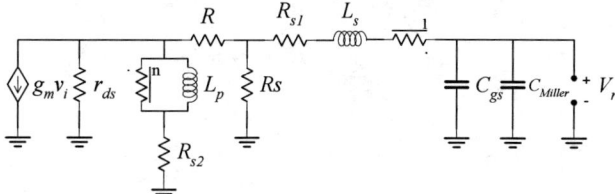

Fig. 6 – Single stage amplifier, equivalent circuit used to determine $A\beta$.

The inductor L_p and the resistor Rs_2 introduce a zero at a frequency below the band of interest; for simplicity, Rs_2 will be neglected, which move the zero to the origin. The circuit has three independent poles, associated mainly to L_p, L_s and C_{gs}. C_{gs} and L_p produce two low frequency poles. L_s produces one pole at high frequencies. Only the low frequency poles are important to determine the loop gain.

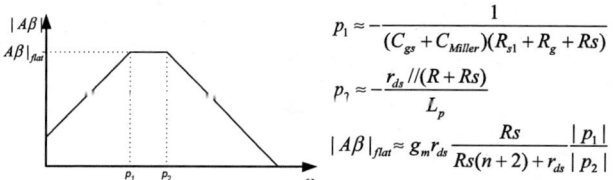

Fig. 7 – Single stage $A\beta$ amplifier gain

It may happen that the loop gain is as represented in Fig. 7, with a flat zone too low, i.e. not greater than 1. In that case, there are several possibilities. One option to increase $A\beta$ is moving pole p_1 to higher frequencies, i.e. p_1 greater than p_2, which is only possible by decreasing C_{gs}, but this is not possible, since the transistor has been sized for low noise and is wide. A cascode topology can also be used to move both poles to higher frequencies, p_1 by reducing the Miller effect and p_2 by increasing r_{ds}. Another solution is the use of a second amplifying stage. This will be analyzed next. The second stage is non-inverting to keep the feedback negative. Its noise is not a concern as explained above (Friis Law). When a 2-stages amplifier is insufficient, a 3-stage amplifier may be considered, but more than two stages may cause stability problems.

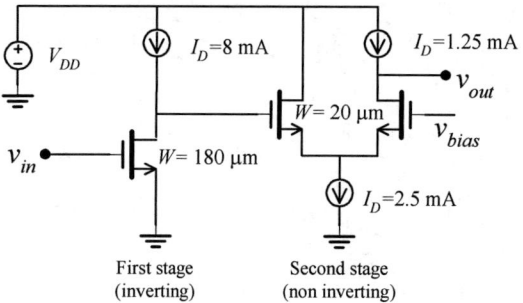

Fig. 8 – Double stage amplifier schematic

The 2-stage amplifying block is represented in Fig. 8. The first stage is inverting and has a high gain (high current and a wide transistor for noise reasons, as explained above). The second stage has less current and the transistors are much smaller (The transistors of the second stage are of the same size). Next section presents the results of the designed amplifier.

V. SIMULATION RESULTS

The amplifier in Fig. 1 has been designed with the two-stage amplifier in Fig. 8 and a custom designed transformer with $L_p = 15.9$ nH, $Rs_2 = 20\,\Omega$, $L_s = 0.3$ nH, $Rs_1 = 2\,\Omega$, $n = 19.1$, $C_{f1} = C_{f2} = 10$ fF and $Rs = 50\,\Omega$. The technology used is 0.18 μm CMOS technology from National Semiconductors and the voltage source is 1.8 V. The results are depicted in the Fig. 9.

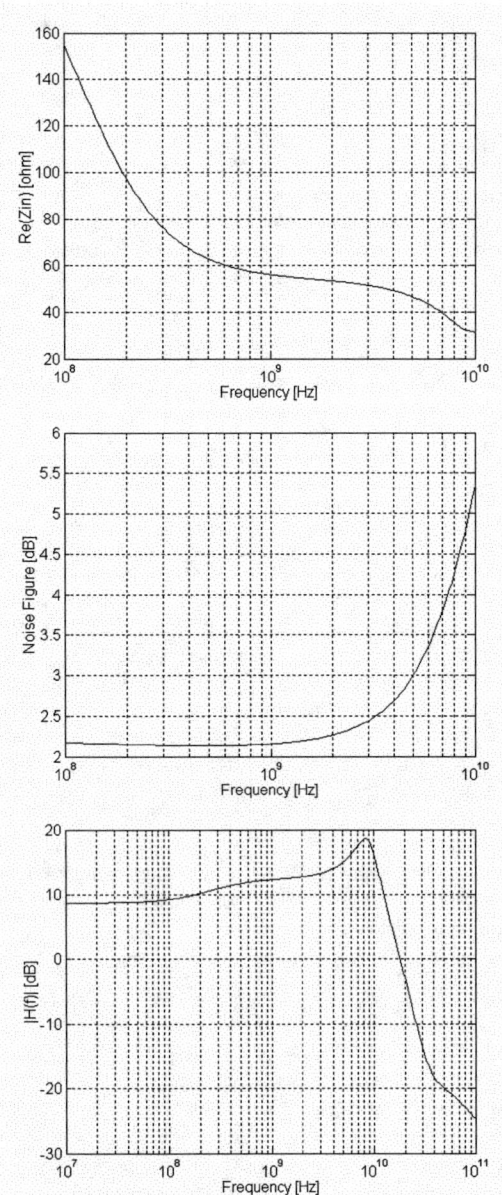

Fig. 9 – Simulated LNA input impedance (first graphic), noise figure (second graphic) and amplifier gain (last graphic).

It is seen that the impedance is close to the value of 50Ω over a large band (from 700MHz to 7GHz the real part of the input impedance is between 40 Ω and 60 Ω). The imaginary part of the impedance is below 20 Ω in that range of frequencies. In the flat zone of the graphic, the amplifier has a noise figure of around 2.2dB, and is below 3 dB till 5 GHz. This frequency value will define the upper boundary of the amplifier bandwidth. The amplifier gain is around 13dB in the band of interest (700MHz to 5GHz). These results are comparable with those of state-of-the-art narrowband LNAs with the advantage of being valid for a wide range of frequencies (almost one decade) which allows for an efficient and elegant solution for multi-standard applications and opens the opportunity to be used for other types of communication protocols.

VI. CONCLUSIONS

A wideband, double loop feedback LNA has been presented, and detailed analysis of several non-idealities has been performed. It was shown that this architecture can simultaneously be matched to a desired input impedance (usually 50 Ω) and have a low noise figure for a wide range of frequencies. The minimum noise factor has also a low value. It has been shown that is also possible to reduce this noise floor by adding an inductor in series with the transformer secondary. The tradeoff is a reduction in bandwidth and an increase in area.

The amplifier has been designed using 0.18μm CMOS technology from National Semiconductors. A wideband, from 700 MHz to 5 GHz has been obtained with values of gain and noise figure comparable to narrowband amplifiers with a similar level of power consumption.

VII. REFERENCES

[1] C. J. M. Verhoeven et al, *Structured Electronic Design – Negative Feedback Amplifiers*, Kluwer Academic Publishers, 2003.

[2] K. van Hartingsveldt et al.; "HF Low Noise Amplifiers with Integrated Transformer Feedback", ISCAS 2002, vol. 2, pp. II-815 – II-818, May 2002.

[3] T. Lee, *The Design of CMOS Radio-Frequency Integrated Circuits*, Cambridge Un. Press, 1998.

An Interference Rejection Filter For An Ultra-Wideband Quadrature Downconversion Autocorrelation Receiver

Sumit Bagga, Sandro A. P. Haddad, Koen van Hartingsveldt, Simon Lee, Wouter A. Serdijn and John R. Long
Electronics Research Laboratory, Faculty of Electrical Engineering, Mathematics and Computer Science,
Delft University of Technology,
Delft, The Netherlands
E-mail: {s.bagga, s.haddad, k.vanhartingsveldt, w.a.serdijn, j.r.long}@ewi.tudelft.nl

Abstract—An analog filter is designed based upon the requirement of an interference rejection filter for the Quadrature Downconversion Autocorrelation Receiver (QDAR). The transfer function of an eight-order elliptic band-pass filter is selected. As a result, a state-space approach (i.e. the orthonormal form [1]) is adopted, which is intrinsically semi-optimized for dynamic range, has low sensitivity, high sparsity and its coefficients can be physically implemented. Each coefficient in the state-space description of the filter is implemented at circuit level using a novel 2-stage gm cell based upon the principle of negative feedback. Simulation results in IBM's Bi-CMOS 0.18 μm technology show that the interference rejection filter requires a total current of 90 mA at a 1.8 V power supply. The 1-dB compression point of the filter is at 565 mV and the SNR is 47.5 dB. On performing a Monte Carlo simulation, it becomes evident that the overall filters transfer response does not suffer from process variations.

Keywords—ultra-wideband, narrowband interference, state-space synthesis, orthonormal filter, quadrature downconversion autocorrelation receiver, analog integrated circuits

I. INTRODUCTION

Although impulse radio ultra-wideband technology promises enhanced data throughput with low-power consumption, it inseparably introduces several challenging design issues. Ultra-wideband systems transmit at very low spectral densities and occupy a large amount of bandwidth, thus it is unequivocal that interference introduced from neighboring narrowband systems is a serious predicament, which could severely hamper or even degrade the overall performance of the system.

Among currently investigated UWB receiver architectures, the transmitted reference scheme proposed by Hoctor and Tomlinson [2] resolves the issue not only of synchronization but also of multipath components. In this scheme, consecutive pulses are transmitted with a predefined delay τ between them. The first pulse acts as a reference, whereas the second pulse is modulated. The autocorrelation receiver correlates the incoming signal with a delayed version of itself. The absolute value of the output after integration is in fact the energy of the pulse while the polarity of the output contains the data. The issue of narrowband interference led us to the design of novel receiver architecture i.e. the quadrature downconversion autocorrelation receiver (QDAR) (see Fig. 1.) [3],

which is capable of operating in the presence of strong narrowband interference. It uses the property of frequency wrapping or in other words, it folds the ultra-wideband frequency spectrum around the origin. At the same time, interferers are positioned outside the band of interest and can be simply removed by the means of a band-pass filter (see Figure 2). Even though the bandwidth reduces significantly and the shape of the transmitter pulse is distorted, the QDAR makes use of the fact that detection with an autocorrelation receiver is feasible as long as the relative polarity and shape of consecutive pulses is preserved.

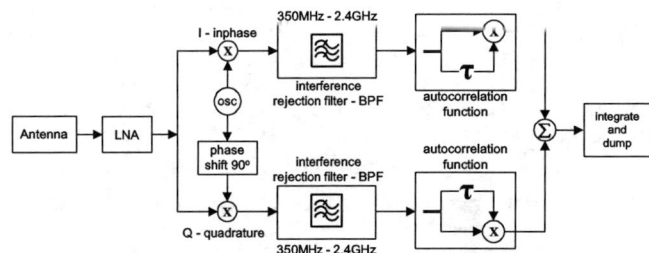

Fig. 1. Quadrature downconversion autocorrelation receiver (QDAR)

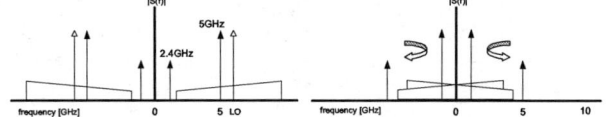

Fig. 2. left) frequency spectrum before downconversion and right) after downconversion

This paper proposes an interference rejection (i.e. an eighth order elliptic band pass) filter to be used in the QDAR (see Fig. 1). Section 2 relates the transformation procedure of the transfer function of the band pass filter into the orthonormal state-space form. Transconductance amplifiers are frequently employed in filters designed for high-frequency applications, and are employed in this particular case of impulse radio ultra-wideband circuit design. Section 3 describes the design and implementation of the 2-stage negative feedback transconductance cell. Simulation data of the transconductance amplifier as well as the overall filter transfer is given in Section 4. Section 5 presents the conclusions.

II. FILTER DESIGN

A. Transfer function and state-space synthesis

The analysis in [3] showed that, after downconversion, the interferer moves adjacent to the band of interest. The 5 GHz interferer appears below 350 MHz and the 2.4 GHz interferer beyond 3.1 GHz, when down-converting with an oscillator frequency of 5.5 GHz.

Trade-offs between slope, attenuation and circuit complexity are taken into consideration prior to choosing the order of the filter. An elliptic filter gives the steepest slope for any given order and is therefore the appropriate choice. The abovementioned extenuates the requirement for an eighth order elliptic band-pass filter. The transfer function (see below) of the interference rejection filter is generated using Matlab. The corner frequencies are set at 350 MHz and 2.6 GHz. The stop-band attenuation is at least 20 dB and the pass-band ripple is 0.5 dB.

$$H(s) = \left[\frac{0.1s^8 - 5.93e - 7s^7 + 1.43e20s^6 - 1.03e15s^5 + 3.82e40s^4 - 3.71e34s^3 + 1.87e59s^2 - 2.8e52s + 1.71e77}{s^8 + 1.62e10s^7 + 5.23e20s^6 + 5.04e30s^5 + 6.53e40s^4 + 1.82e50s^3 + 6.88e59s^2 + 7.64e68s + 1.71e78} \right]$$

Once the desired transfer function is formulated, its state-space description is then determined. A state-space description for a given transfer function is not unique, meaning that many state-space descriptions can implement the same transfer function. Moreover, a state-space description of any filter transfer function should be optimized for dynamic range, sensitivity, sparsity and the coefficient values [1], [4].

B. Orthononormal Ladder Structure

Among known standard state-space descriptions, such as the canonical, the diagonal and the modal, the orthonormal ladder form is notable since it is by definition semi-optimized for dynamic range due to the specific structure of the matrices. Furthermore, since it is derived from a ladder structure, it is intrinsically less sensitive and the matrices are highly sparse. A detailed explanation of the procedure to derive the orthonormal ladder form can be found in [5].

With a state-space approach, the filter can be optimized for dynamic range, sensitivity, sparsity and coefficient values. A low sensitivity suppresses the effect of component variations on the transfer function. It can be proved that a filter that is optimized for dynamic range is also optimized for sensitivity [6]. The sparsity of the matrices directly determines the circuit complexity. State-space descriptions of filters with more zero elements require less hardware and are likely to consume lower power. Thus, it is therefore an important design aspect of state-space filters.

In respect to a fully optimized and fully dense state-space description, the resulting semi-optimal orthonormal filter structure differs only by about 2 dB in dynamic range. The A, B, C, and D matrices of the defined transfer function are as follows:

Matrix A:
$$\begin{bmatrix} 0 & 2.122e9 & 0 & 0 & 0 & 0 & 0 & 0 \\ -2.122e9 & 0 & 6.859e8 & 0 & 0 & 0 & 0 & 0 \\ 0 & -6.859e8 & 0 & 3.933e9 & 0 & 0 & 0 & 0 \\ 0 & 0 & -3.933e9 & 0 & 10.81e9 & 0 & 0 & 0 \\ 0 & 0 & 0 & -10.81e9 & 0 & 10.69e9 & 0 & 0 \\ 0 & 0 & 0 & 0 & -10.69e9 & 0 & 7.77e9 & 0 \\ 0 & 0 & 0 & 0 & 0 & -7.77e9 & 0 & 1.47e10 \\ 0 & 0 & 0 & 0 & 0 & 0 & -1.47e10 & -1.62e10 \end{bmatrix}$$

Matrix B:
$$\begin{bmatrix} 0 & 0 & 0 & 0 & 0 & 0 & 0 & 71732 \end{bmatrix}$$

Matrix C:
$$\begin{bmatrix} 3.048e+4 & -1.64e-10 & -9.022e+4 & 2.476e-11 & 1.026e+5 & 4.307e-11 & 8.596e+4 & -2.253e+4 \end{bmatrix}$$

Matrix D:
$$\begin{bmatrix} 0.09998 \end{bmatrix}$$

C. Scaling –Capacitance and Coefficient Values

Transconductance amplifiers will form the basic building blocks to implement the state-space description of the band-pass filter. The integrators are implemented as capacitors with a normalized value of 1 F. The corresponding matrices A, B, C and D have extremely large coefficients corresponding to large gm values, which are not physically feasible at circuit level. By scaling the capacitors (cap=1 pF), one consequently scales matrices A and C. Coefficients of matrices B and D can too be down scaled by α_1 and α_2 respectively, without affecting the response of the filter.

$$\begin{aligned} A^* &= cap \cdot A \\ B^* &= \alpha_1 B \\ C^* &= \alpha_2 \cdot cap \cdot C \\ D^* &= \alpha_1 \cdot \alpha_2 \end{aligned} \quad (1)$$

The block diagram of the state-space filter is shown in Fig. 3 and has 22 non-zero coefficients.

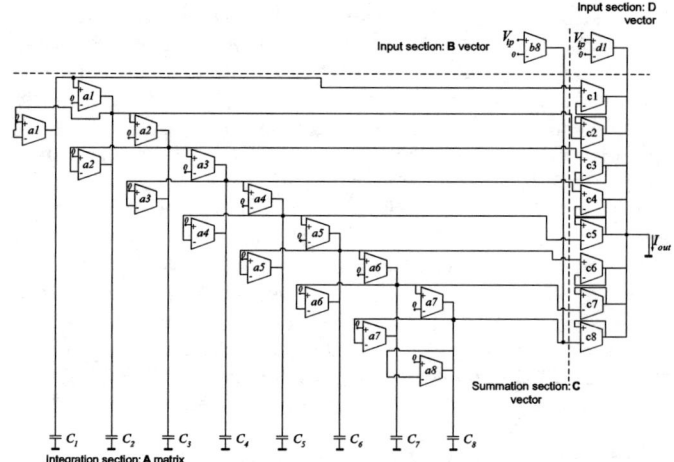

Fig. 3. Complete state-space filter structure

Once the block diagram has been recognized, a transconductance amplifier implements every coefficient.

III. TRANSCONDUCTANCE AMPLIFIER

The transconductance amplifier is implemented using a negative feedback structure consisting of an active circuit, which implements a nullor and a feedback network (see Fig. 4). Theoretically, the nullor is an ideal block that has infinite transfer parameters.

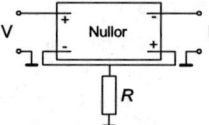

Fig. 4. Negative feedback amplifier with impedance in the feedback network

The orthonormal structure has both positive as well as negative coefficients. Since the structure in Fig. 4 can only implement a negative coefficient, a differential topology is used. Another advantage of using the latter is the cancellation of even order distortion terms that arise from the actual nullor implementation, thus improving linearity.

The nullor (half circuit) is realized using a cascode (CE-CG) stage formed by transistors (Q_1-Q_2) at the input, and a non-inverting cross-coupled differential pair (Q_3-Q_4) at the output. The feedback network is made up of a resistor **R** (see Fig. 5).

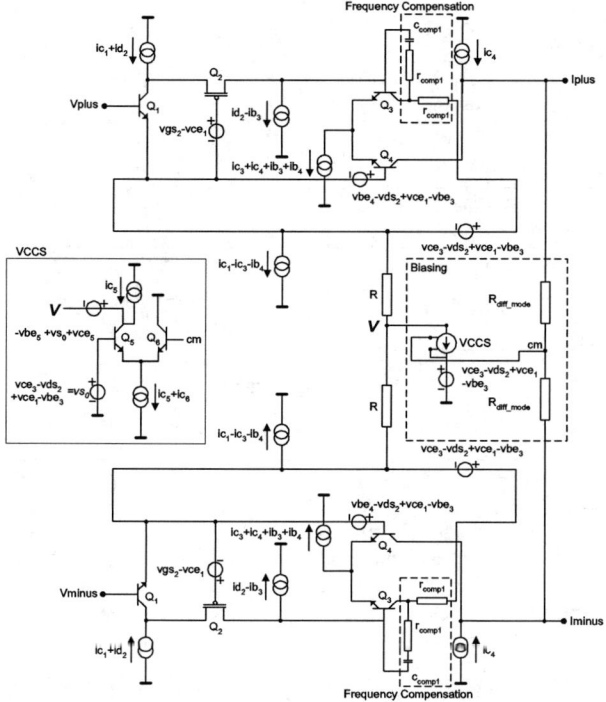

Fig. 5. 2-stage negative feedback gm amplifier

The CB-stage not only presents an output resistance that is larger by $g_{m2}r_{ds2}$ but also reduces Miller's effect of the CE-stage [7]. As compared to a single-stage implementation, a 2-stage nullor improves the loop gain, which yields higher linearity as well as bandwidth [7] at the expense of power consumption. In reference to stability, frequency compensation in the form of pole-zero cancellation is also applied to this transconductance amplifier. For biasing of the differential structure, the common-mode voltage is sensed at the outputs and is compared to the desired reference voltage using a voltage-controlled-current source (VCCS). Its implementation is shown at left in Fig. 5. The output current delivered by the VCCS is then applied to a virtual ground node, **V**.

The small-signal behavior of the transconductance cell will now be analyzed.

A. Small-signal Analysis

The integrating capacitors (Cs) at the input and output have also been taken into account.

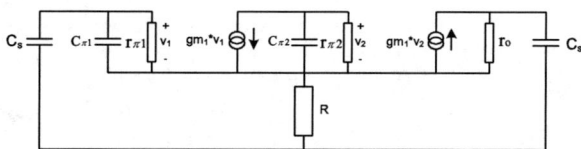

Fig. 6. Small signal model of the gm cell

For negative feedback amplifiers [7], the closed loop transfer (At) can be written in terms of the loop gain $A\beta(s)$ as,

$$At = -\xi \cdot v \cdot At_\infty \cdot \left[\frac{-A\beta(s)}{1 - A\beta(s)}\right] \quad (2)$$

where v and ξ represent the input and output coupling factors and both are assumed to be equal to one.

At_∞ is defined as,

$$At_\infty = -\frac{1}{\beta} \quad (3)$$

where β is the feedback transfer. Now $A\beta(s)$ can be expressed as,

$$A\beta(s) = [-g_{m1}] \cdot \left[\frac{r\pi_1}{1 + s(r\pi_1 \cdot c\pi_1)}\right] \cdot [g_{m2}] \cdot \left[\frac{r\pi_2}{1 + s(r\pi_2 \cdot c\pi_2)}\right] \cdot$$

$$\left[\frac{R}{R + \left(\frac{1}{s \cdot cs} + \frac{r\pi_1}{1 + s(r\pi_1 \cdot c\pi_1)}\right)}\right] \cdot \left[\frac{r_o}{r_o + \frac{1}{s \cdot cs} + \left(R // \left(\frac{1}{s \cdot cs} + \frac{r\pi_1}{1 + s(r\pi_1 \cdot c\pi_1)}\right)\right)}\right] \quad (4)$$

where g_{m1} and g_{m2} are the transistor transconductances of the CE-stage and the differential pair, respectively, $r\pi_1, c\pi_1$ and $r\pi_2, c\pi_2$ are the base-emitter resistances and capacitances of Q_1 and Q_{3-4}, respectively and R is the feedback resistance. Note that for simplicity, terms with Cs (i.e. integrating capacitances of matrix A) will be neglected.

Simplifying by substituting (5) in (4) and assuming that $r_o \gg 1$,

$$\beta f_2 = [g_{m2} \cdot r\pi_2]; \quad Z\pi_1 = \left[\frac{r\pi_1}{1 + s(r\pi_1 \cdot c\pi_1)}\right]; \quad Zs = \left[\frac{1}{s \cdot cs}\right] \quad (5)$$

one obtains,

$$A\beta(s) = [-g_{m1} \cdot \beta f_2] \cdot \left[\frac{R}{R + (Zs + Z\pi_1)}\right] \cdot r\pi_1 \cdot \left[\frac{1}{1 + s(r\pi_2 \cdot c\pi_2)}\right] \quad (6)$$

Substituting (6) in (2) and for large enough loop gains, the transfer is accurately determined by the feedback transfer β.

$$At = -\frac{1}{R} \cdot \left[\frac{-A\beta(s)}{1 - A\beta(s)}\right] \approx -\frac{1}{R} \quad (7)$$

The DC-loop gain poles product (LP_2) predicts the bandwidth of the system and is stated in (7).

$$LP_2 \approx |A\beta(0)| \cdot |p_1| \cdot |p_2| \quad (8)$$

Substituting dc loop gain $A\beta(0)$ (9), closed loop poles p_1 and p_2 ((10) and (11), respectively) in (8),

$$A\beta(0) = [-g_{m1} \cdot \beta f_2] \cdot \left[\frac{R}{R + r\pi_1}\right] \cdot r\pi_1 \quad (9)$$

$$p_1 = \left[\frac{-1}{2\pi(r\pi_2 \cdot c\pi_2)}\right] \quad (10)$$

$$p_2 = \left[\frac{-1}{2\pi(R // r\pi_1) \cdot c\pi_1}\right] \quad (11)$$

the bandwidth, BW, of the proposed 2-stage transconductance amplifier is approximately equal to the geometric mean of the transit frequencies of the respective stages [7].

$$LP_2 \approx f_{T1} \cdot f_{T2} \rightarrow BW \approx \sqrt{LP_2} \approx f_T \quad (12)$$

The influence on the transfer by the non-ideal coupling at the input and output modeled via (ξ) and (v) [7], respectively, needs to be taken into account. This aspect will result in a BW to some extent lower than f_T. The gm-C topology is implemented in IBM's 0.18 μm Bi-CMOS technology. In the same technology, the bias sources in

Fig. 5 are implemented using current mirrors with multiple outputs. All parasitic sources are also accounted for.

IV. SIMULATION RESULTS

Fig. 7 shows the magnitude and phase response of the stand-alone transconductance amplifier, which is used in the band-pass filter. Both the magnitude and the phase demonstrate a relatively flat response up to about 3 GHz.

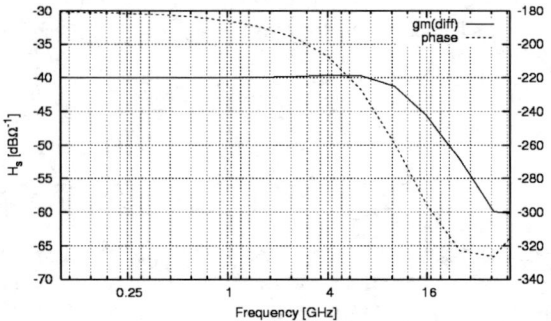

Fig. 7. Magnitude and phase transfer of stand-alone gm cell

From Fig. 8 it is clear that the high-frequency response of the entire filter is preserved at the cost of forfeiting some of the pass-band at lower frequencies. By scaling down the capacitance even lower than 1 pF as well as the coefficients in the matrices A and C, a superior transfer response is attainable because of the trade-off between bandwidth and gain.

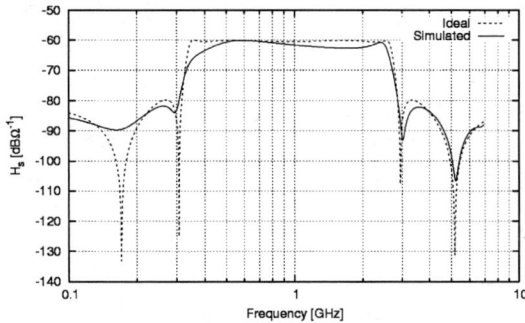

Fig. 8. Transfer of elliptic band-pass filter

Finally, by randomly varying (i.e. 25 iterations) the component tolerances as well as the model parameters between their specified tolerance limits, a Monte Carlo analysis is run in order to estimate the circuit's sensitivity. From Fig. 9 it is inferred that the transfer of the filter is relatively unlikely to show a substantial discrepancy as a result of process variations.

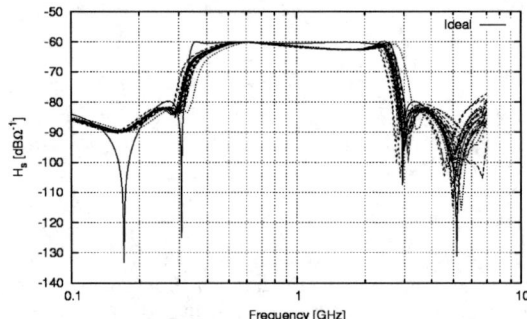

Fig. 9. Sensitivity analyses – Monte Carlo

Table I highlights the simulation parameters of the band-pass filter.

Table 1 Simulation Parameters

Specifications	Simulated (@ 1GHz)
1-dB compression pt. at	565 mV
3-dB compression pt. at	575 mV
Dynamic range at 1-dB compression pt. (SNR)	+47.5 dB
IVIP3 (third-order input referred voltage intercept pt.)	+14 dBV
OIIP3 (third-order output referred current intercept pt.)	-47.6 dBA
Current consumption	90 mA @ 1.8 V
Process	IBM Bi-CMOS 0.18µm

V. CONCLUSIONS

An interference rejection band-pass filter is to be used in the *QDAR* has been presented. An eighth order transfer function for an elliptic filter is selected. Subsequently, an orthonormal state-space approach is adopted, which fulfills the requirements of dynamic range, sensitivity, and sparsity. The coefficients are down scaled in conjunction with capacitance values. Each element of the filter is implemented at the circuit level using a novel negative feedback 2-stage gm amplifier. Simulation results in IBM's Bi-CMOS 0.18 µm technology (see Table 1) show that the interference rejection filter requires a total current 90 mA at a 1.8 V power supply. The 1-dB compression point of the filter is at 565 mV and the SNR is 47.5 dB.

VI. REFERENCES

[1] S.A.P. Haddad, S. Bagga and W.A. Serdijn, "Log-Domain Wavelet Bases," in *Proceedings IEEE International Symposium of Circuits and Systems*, May 2004

[2] R.T. Hoctor and H.W. Tomlinson, "Delay-Hopped Transmitted-Reference RF Communications," *Proceedings of the IEEE Conference on Ultra Wideband Systems and Technologies*, pp. 265-270, May 2002

[3] Simon Lee, S. Bagga and W.A. Serdijn, "A Quadrature Downconversion Autocorrelation Receiver Architecture for UWB," Joint *UWBST and IWUWBS*, May 2004

[4] D. Rocha, "Optimal Design of Analogue Low-Power Systems: A strongly directional hearing-aid adapter," PhD Thesis, Delft University of Technology, 2003

[5] D.A. Johns, W.M. Snelgrove and A.S. Sedra, "Orthonormal Ladder Filters," *IEEE Transactions on Circuits and Systems*, vol. 36, pp. 337-343, March 1989

[6] G. Groenewold,"*Optimal dynamic range integrators*", IEEE Transactions on Circuits and Systems, Volume: 39, Issue: 8, Aug. 1992, Pages: 614 - 627

[7] C.J.M.Verhoeven, A. van Staveren, G.L.E. Monna M.H.L. Kouwenhoven and E. Yildiz, "Structured Electronic Design: Negative-Feedback Amplifiers", *Kluwer Academic Publishers*, 2003

A read-out strategy and circuit design for high frequency MEMS resonators

Arantxa Uranga,
Nuria Barniol
Electronic Engineering Dept.
Universitat Autònoma de Barcelona
Barcelona, Spain
Arantxa.Uranga@uab.es

Humberto Campanella,
Jaume Esteve, Lluis Terés
Instituto de Microelectrónica de
Barcelona, CNM-IMB (CSIC)
Barcelona, Spain
Humberto.Campanella@cnm.es

Zachary Davis
EPSON-CNM BIRD Team
EPSON Europe Electronics GmbH
Zachary.Davis@cnm.es

Abstract—This paper describes the implementation of read-out circuitry to characterize a MEMS resonator in the UHF band. A read-out strategy is proposed and explained, detailing into the requirement issues. Also, the circuit architecture and design are presented. SPECTRE modeling predictions are compared with current characterization results. A transimpedance gain of 70 dB has been verified, with a cut-off frequency of 450 MHz.

I. INTRODUCTION

Integration of MEMS devices with microelectronic circuits is a promise of miniaturization and modular System-On-Chip design for the next few years [1]. Specifically, there is documented evidence of MEMS resonators integrated with circuits to perform specific signal processing functions, e.g. oscillators [2, 3, 4]. These systems provide MEMS-based oscillation in the kHz range, and circuit-aided read out of such functionality. On the other hand, MEMS resonators in the MHz range, particularly in the VHF and UHF bands, have been demonstrated [5, 6]. Also, a VHF MEMS-based oscillator has been documented in [7].

In considering the previous references, a main contribution of this work is to integrate UHF MEMS resonators with CMOS read-out circuitry, this latter with enough bandwidth to amplify the current flowing out of the MEMS. Several circuit architectures and measurement techniques are applicable to perform characterization of a MEMS based resonator. In particular, this work is focused on a linear mode circuit able to develop broadband characterization of the MEMS resonator. The specification and design of the read-out amplifier will include, among other parameters: amplifier architecture, values for passive components, output impedances, dynamic range, bandwidth, and gain range.

In Section II of this paper, a detailed description of a read-out strategy is done. Both read-out requirements and circuit architecture are presented. In Section III a detailed read-out circuit description is done for MEMS and test configurations. Also, characterization results are presented and analyzed in Section IV. Finally, conclusions and future work are commented in Section V.

II. READ-OUT STRATEGY

A. Read-out requirements

The main functionality of read-out circuits is amplification of the MEMS resonator output current. This current is expected to be produced as resonance of the MEMS in a specific mode shape in the hundreds of MHz range and to be very weak, with AC amplitudes in the order of nano-amperes. Regarding these matters, a set of requirements for the read-out amplifier is listed in Table 1.

TABLE I. READ-OUT AMPLIFIER REQUIREMENTS

Issue	Requirement
Dynamic Range	Wide, running from hundreds of nA to units of µA
Bandwidth	Cut-off frequency in the order of 400 MHz
Power	Not restricted
Size	Not restricted
IC Technology	AMS 0,35 µm CMOS
Linearity	Yes
Input mode	Current mode
Feedback configuration	Open/Closed loop
MEMS Integration	Impedance matching not required
Output Matching	Near to 50 Ω

MATLAB and SPECTRE modeling lead to an electrical model of the resonator to have an estimate of the dynamic

---

This work is funded by SEIKO EPSON Corporation (Barcelona R&D Lab.), in a joint R&D initiative among Centro Nacional de Microelectrónica (Barcelona, Spain) and Universitat Autònoma de Barcelona (Spain). Also, this project has been partially supported by CICYT (TIC2003-07237, NANOSYS).

range. In fact, the MEMS device was designed to resonate in the 400 MHz range. The input current mode is justified to couple the current mode signal flowing out the MEMS, which is very likely to have a high mismatching with the input impedance of the amplifier circuit. Additionally, it is highly desired the circuit to have an output impedance near to 50 Ohm, given a network analyzer will be used to characterize the MEMS, then minimizing losses. The linearity aspect can be regarded in two ways. First, if it is talked about necessity of characterize the amplifier, linearity is a must. However, and due to the expected high-Q shape of the MEMS current, distortion would be negligible in a very narrow bandwidth, making of no relevance the linearity issue.

Other aspects concerning maximum size or power consumption are not restricted, only selection of CMOS as the target technology is required. As a conclusion, transimpedance amplifier architecture is chosen as the target configuration to perform read-out of the MEMS resonator. In the following lines, a detailed explanation on transimpedance amplifier architecture and actual implementation is done.

B. Circuit Architecture: transimpedance amplifier

Transimpedance amplifier (TIA) architecture takes an input current signal, and amplifies it through an impedance to generate an output voltage signal. Hence, its main application when needed to couple to output current mode circuits, such as the case of the MEMS resonator.

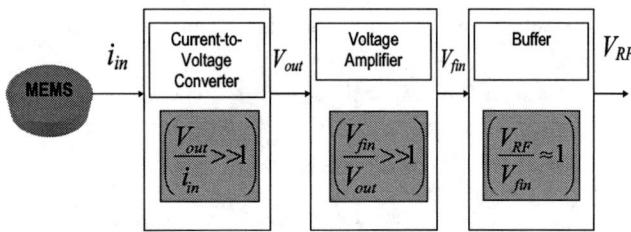

Figure 1. Transimpedance Amplifier: current-to-voltage converter followed by voltage amplifier and buffer

Figure 0

In Fig. 1 it is depicted a block diagram of a transimpedance amplifier including further amplification and buffering. The voltage amplifying stage would allow to achieve a desired Vfin/Vout gain specification. Finally, buffering is required to match the output impedance values of the circuit to those presented by the set-up instrumentation. In the case a 50 Ohm-probe network analyzer is be used to read-out the electrical signal, the output buffer impedance must be designed to be close to this value. The buffer gain VRF/Vfin approximates to unity. Anyway, external matching networks could be integrated to refine impedance matching in the final set-up.

The requirements for a typical TIA are a high bandwidth, high transimpedance gain, adequate power consumption, low noise, low input impedance and small area. Until now, TIAs with a high bandwidth could only be built in GaAS, InP or SiGe technology. As CMOS technology is downscaled to gate lengths of nm, new TIAs designs have appeared offering higher bandwidths. On the other hand, the maximum power supply has also been reduced to prevent the oxide's breakdown but the threshold voltage has not yet been reduced [8-11].

III. CIRCUIT DESIGN

A. Read-out circuit

The configuration proposed is based on a transimpedance amplifier structure (TIA) [12, 13]. In particular, a common source TIA has been selected. As it can be shown on Fig. 2, the structure is formed by an inverting transimpedance amplifier followed by a source- follower, in charge of enhancing the bandwidth of the full amplifier. The feedback resistor has been implemented using a poly resistor provided by the technology (5μm x 35 μm).

Finally, an output matching buffer stage has been designed to drive the 50 ohms input impedance of the network analyzer. It consists on a basic source follower. In order to reduce power consumption at the output stage, the full amplifier is AC coupled to the network analyzer by means of an external capacitive coupling. Input current Iin represents the MEMS current.

B. MEMS-to-Read-out circuit integration

Once designed the read-out configuration, MEMS integration and chip design is the next step. First, the MEMS resonator is integrated into the circuit by properly connection of MEMS signals: AC supply, DC biasing, AC input and grounding to the I/O Pins, and AC output to the read-out circuit. As it has been commented, in the schematic of Fig.2 the input current Iin of the TIA represents the current flowing out the MEMS when electrostatically excited. The die photograph of Fig. 3 points out the location of the embedded MEMS resonator into the TIA layout.

Also, external circuit connections (signal and biasing) are provided by means of 100μm-sided squared pads.

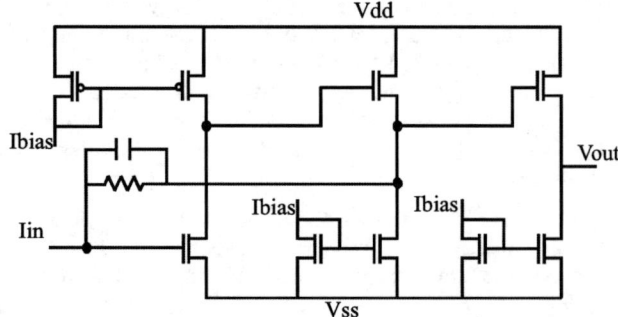

Figure 2. Transimpedance Amplifier: Circuit schematic

C. Test Circuit

A test circuit is necessary in order to demonstrate functionality of read-out circuits, when these are not integrated to the MEMS. In this way, independent characterization of amplification features can be done and, indirectly, an analysis of electrical performance of MEMS when integrated. In order to perform such a testing, a current source is then connected to the input node of the read-out circuit, replacing the MEMS of Fig. 3 by a single-transistor current source. The input voltage node is connected to the gate, and the output current node (drain) goes to the current input node of the transimpedance amplifier.

Figure 3. Die photograph of the integrated MEMS-to-read-out circuit.

In addition to the test circuit integrated with the read-out circuit, a second single-transistor current source has been implemented to measure the AC and DC figures of the current it provides. This information will be useful when characterizing the overall gain of the transimpedance amplifier, due to the fact that only voltage-to-voltage input-to-output interfaces are available.

IV. CHARACTERIZATION AND RESULTS

A. DC Characterization

Initial DC characterization allowed verification of the DC performance as a function of the input DC voltage, along with the transimpedance DC gain. A set-up of DC power supplies, digital voltmeter and a DC probe station was implemented. Measurements were obtained by connecting the DC probes directly to the pad connections provided by the chip.

DC Results are plotted in Fig. 4. The output voltage versus input voltage applied to the current source is presented. By analyzing the slope of the curve, the overall output voltage-to- input voltage gain A_V is calculated.

Three steps were required in order to obtain actual transimpedance DC gain: first, the current-to-voltage gain $A_{I/V}$ for the test circuit was determined, calculating the slope of the DC response of the circuit; second, the overall output voltage-to- input voltage gain A_V was measured. Finally, transimpedance DC gain G_m was calculated as follows:

$$G_m [dB\Omega] = A_V [dB] - A_{I/V} [dBS] \qquad (1)$$

A DC gain of 70 dB was obtained, with dependence on the biasing conditions at the input node of the test circuit.

B. AC Characterization

AC Characterization was performed using a network analyzer (Rhode& Schwartz Model FSQ-26) and a couple of DC power supplies. Two configurations of the set-up were considered, first one with packaged chip and the second one with bonded chip, both directly attached to a PCB.

Measurements were mainly focused on S-parameters, specifically S21. Under the assumption that S21 can be a good approximation of the voltage-to-voltage gain, the transimpedance AC gain can be evaluated in a similar manner as with the DC gain:

$$G_m [dB\Omega] = 33.98 + A_V [dB] - A_{I/V} [dBS], \qquad (1)$$

where 33.98 is the amplitude value in dB of 50Ω.

In Fig. 5 it is plotted the AC response of the transimpedance amplifier. As it can be seen a bandwidth of 450MHz with a gain of 69.5 dBΩ is obtained. Differences between SPECTRE modeling and current characterization are attributed to differences of the load capacitance with respect to its nominal value.

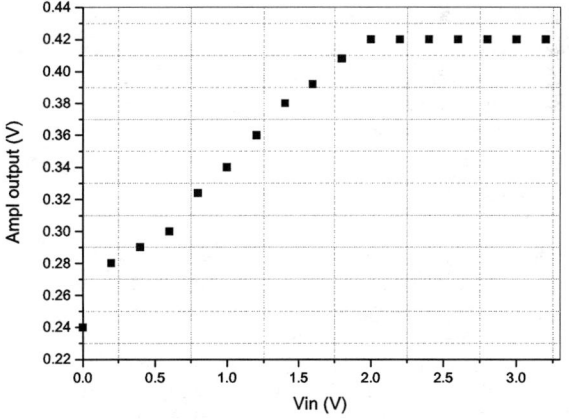

Figure 4. DC gain of the transimpedance amplifier

In order to compute the RMS noise current value, a noise analysis has been performed. A bandwidth of a 30 kHz (corresponding to the measurement instrument) has been used. The CSD present at 400 MHZ is computed into a 30 kHz bandwidth, obtaining an RMS input current noise of 62 pA. Therefore, the maximum peak-to-peak noise signal expected is clearly smaller than the signal wanted to record (range of several hundreds of nA).

$$I_{RMS} = \frac{1}{2\pi} \int_{f_1}^{f_2} (PSD)^2 df = 62 \; pA$$

Table 2 summarizes the main measured results of the implemented amplifier.

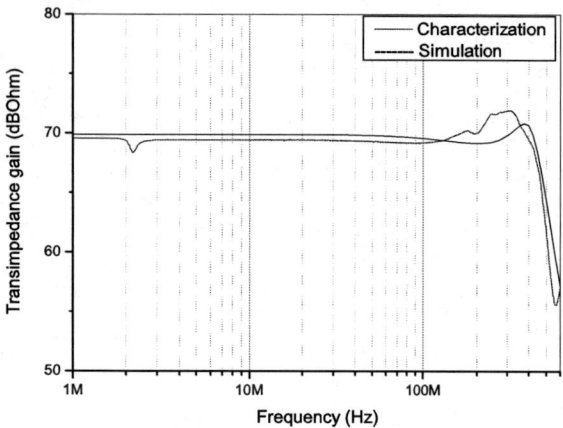

Figure 5. AC simulated and tested transimpedance gain.

V. CONCLUSIONS AND FUTURE WORK

A read-out strategy for MEMS resonators, based on transimpedance amplifier, was proposed. The integrated amplifier has been implemented in a standard 0.35 μm CMOS technology (AMS) and characterization has been performed. Testing results showed an AC gain of 69.5 dBΩ and a bandwidth of 450 MHz. Circuit characterizations along with future MEMS plus the integrated circuit characterization will allow extraction of electrical parameters of the resonator.

REFERENCES

[1] R. Leach, Z. Cui, D. Flack (editors), "Microsystems Technology Standardization Roadmap". Project IST-2001-37682 funded by the EU IST program, 2001.

[2] J. Bienstman., H.A.C. Tilmans, M. Steyaert, R. Puers, "An oscillator circuit for electrostatically driven silicon-based one-port resonators". Proceedings of Transducers'95 and Eurosensors IX, Stockholm, Sweden, June25-29, 1995, pp. 146-149

[3] Y. Satuby et al., "Electrostatically driven microresonator with a CMOS capacitive read out", EEB 95.

[4] C.T.-C. Nguyen, R.T. Howe, "An integrated CMOS micromechanical resonator high-Q oscillator", IEEE Journal of Solid-State Circuits, Volume: 34, Issue: 4, April 1999, pp. 440 – 455

[5] J. R. Clark, W.-T. Hsu, and C. T.-C. Nguyen, "High-Q VHF micromechanical contour-mode disk resonator," Technical Digest, IEEE Int. Electron Devices Meeting, San Francisco, California, Dec. 11-13, 2000, pp. 493-496.

[6] J. Clark, W. Hsu, C. Nguyen, "Measurement techniques for capacitively-transduced VHF-to-UHF micromechanical resonators" Digest of Technical Papers, the 11th International Conference on Solid State Sensors, Actuators and Microsystems (Transducers'01), Munich, Germany, June 10-14, 2001, pp. 1118-1121.

[7] L. Yu-Wei, L. Seungbae, L. Sheng-Shian, X. Yuan, R. Zeying, C.T.-C. Nguyen, "17.7 - 60-MHz Wine-Glass Micromechanical-Disk Reference Oscillator", Digest of Technical Papers IEEE International Solid-State Circuits Conference, 2004. ISSCC. 2004, 15-19 Feb. 2004, pp. 322 – 331.

[8] P. Fay, C. Caneau and I. Adesia, "High-speed MSM/HEMT and p-I-n/HEMPT monolithic photo receivers", IEEE Trans. Microwave Theory Tech., vol. 50, pp.62-67, 2002.

[9] S. S. Mohan, M. Del Mar Hershenson, S. P. Boyd and T. H. Lee, "Bandwidth extension in CMOS with optimised on-chip inductors", IEEE J. of Solid-State Circuits, vol. 35, pp. 346-355, 2000.

[10] M. Kossel, C. Menolfi, T. Morf and T. Toifl, "Wideband CMOS transimpedance amplifier", Electron. Lett., vol. 39, no 7, pp. 587-588, 2003.

[11] C. Kromer, G. Sialm, T. Morf, M. Schmatz, F. Ellinger, D. Erni and H. Jäckel, "A low power 20 GHz 52 dBOhm Transimpedance Amplifier in 80 nm CMOS" IEEE J. of Solid-State Circuits, vol. 39, pp. 885-894, 2004.

[12] C. Kuo, C. Hsiao, S. Yang and Y. Chan, "2 Gbit/s transimpedance amplifier fabricated by 0.35 μm CMOS technologies", Electronics letters, vol. 37, no 19, 2001.

[13] N. Haralabidis, S. Katsafouros and G. Halkias, "A 1 GHz Transimpedance amplifier for chip-to chip optical interconnects", IEEE Int. Symp. Circuits and Systems (ISCAS), 2000, pp. 421-424.

TABLE II. EXPERIMENTAL CHARACTERISTICS OF THE AMPLIFIER

Supply voltage	3.3 V
Total DC gain	69.5 dBΩ
Power consump.	12 mW
Bandwidth	450 MHz
Input Noise CSD	2.28 pA/(Hz)$^{0.5}$
Total integrated noise (30 kHz bandwidth)	62 pA$_{rms}$
Area	600 μm * 450μm

A Low-Power Switched-Current CDMA Matched Filter Employing MOS-Linear Matching Cell and Output A/D Converter

Tomoyuki Nakayama, Toshihiko Yamasaki, and Tadashi Shibata

Department of Frontier Informatics, School of Frontier Sciences
The University of Tokyo
7-3-1 Hongo, Bunkyo-ku, Tokyo, 113-8656, Japan
nakayama@else.k.u-tokyo.ac.jp, yamasaki@hal.k.u-tokyo.ac.jp, shibata@ee.t.u-tokyo.ac.jp

Abstract—A low-power switched-current CDMA matched filter has been developed. The front-end V/I converter has been eliminated by merging the function into each matching cell utilizing the MOS linear I-V characteristics. A low-power A/D converter has also been developed to establish a smooth interfacing to digital back-end processing. The prototype chip was fabricated in a 0.35-μm CMOS technology, and such a low power operation of 1.65mW, including the A/D converter, at 11Mchip/s with 2V power supply has been demonstrated.

I. INTRODUCTION

The direct-sequence code-division multiple-access (DS-CDMA) has been widely used in mobile communication systems. In DS-CDMA systems, a matched filter (MF) is an essential component, in which synchronization is carried out by taking the correlation between the received signal and a pseudo-random noise (PN) code. A number of MF's have been developed using CMOS technologies. In digital implementation [1,2], a large power consumption is a critical issue because an A/D converter is indispensable at the front-end and the total power dissipation increases in proportion to the clock frequency of the system. In order to realize low-power systems, analog CDMA matched filters have been developed using either voltage-mode [3-5] or current-mode techniques [6,7].

In terms of simple circuit implementation, current-mode circuits are preferred because summation/subtraction operations are very easily implemented using the Kirchhoff's current law. In the current-mode implementation of a matched filter, however, a substantial power dissipation occurs in the front-end voltage-to-current converter circuit (V/I converter) which transforms received signals to current signals for following current-mode operations. In order to reduce the power dissipation in the V/I converter, the sub-block architecture was introduced in Ref. [6], where a low-power V/I converter is provided to each sub-block and only one V/I converter is activated at one time. As a result, the power dissipation of 1.95mW at 8Mchips/s has been achieved.

The purpose of this study is to develop a current-mode matched filter in which the front-end V/I converter has been eliminated. For this purpose, the V/I converter function is merged into each matching cell utilizing the MOS linear I-V characteristics in the triode region. In addition, in order to establish a smooth interfacing to the following digital processing, a very low-power A/D converter has been developed. As a result, such a low power operation of 1.65mW at 11Mchips/s has been achieved even including the A/D converter attached at the back end of the analog matched filter. The results were obtained from the fabricated chip which was designed in a 0.35-μm CMOS technology.

II. SYSTEM ORGANIZATION

A. Block Diagram

The block diagram is shown in Fig. 1. The received analog voltage is sequentially stored in an analog memory attached to each matching cell. When all the memories are written with received signals, they are overwritten with new data from the oldest. Therefore, stored data are not shifted for matching. Instead, the PN code sequence is shifted against the stored received signals in order to calculate the correlation. Both "*Activation Flag*", which locates the target cell storing the signal, and the PN code sequence are shifted by shift-registers (*REG*s). The received analog data are converted to current in each cell and the correlation is taken by current summation/subtraction operation. The output current representing the correlation is directly converted to a digital format by the low-power A/D converter also developed in this work.

B. Intra-Cell Voltage-to-Current Conversion and Summation/Subtraction Circuit

In this work, the *V-I* conversion is carried out using the MOS characteristics in the triode region. The schematic of a matching circuit is shown in Fig. 2. Assuming that the NMOS (*M*1) is operating in the triode region, the current flowing in the *M*1 is expressed as

$$I = \beta\left\{(V_{IN} - V_{th})V_D - \frac{1}{2}V_D^2\right\} \quad (1)$$

where V_{IN}, V_{th} and V_D represent the stored voltage, the threshold voltage, and the drain voltage of $M1$, respectively. In the equation, if V_D is constant, the current flowing in $M1$ is proportional to the input voltage. The memorized current is drawn from the positive term node (V_{DP}) or the negative term node (V_{DM}) depending on the PN code. Summation for each term is carried out by taking the wired-sum at the V_{DP} or V_{DM} node. The current flowing in the cell is cut off by $M2$ when it is not used, thus further reducing the power dissipation.

In order to stabilize V_{DP}/V_{DM} for improving the linearity the cascode transistor $M3/M4$ is added above the V_{DP}/V_{DM} as illustrated in Fig. 3. However, just adding the cascode transistor $M3/M4$ is not sufficient for fixing V_{DP}/V_{DM}. Therefore in this circuit, we employed the gain-boosting technique as shown in Fig. 3. The operational amplifier is utilized to make the node of V_{DP}/V_{DM} virtually grounded at V_{amp}. In this structure, by controlling V_{amp}, V_{DP}/V_{DM} can be kept small and stable for good linearity and wide dynamic range of V_{IN}. In addition, the flowing current can also be controlled by V_{amp}. The condition that $M1$ in each cell is in the triode region is given as in the following:

$$V_{IN} - V_{th} > V_D \approx V_{amp} \Leftrightarrow V_{IN} > V_{amp} + V_{th}. \quad (2)$$

The subtraction is carried out by a PMOS current-mirror as shown in Fig. 3. And the offset NMOS current source is added to make the output current ($\Sigma Ip - \Sigma Im$) positive.

C. Low-Power Flash A/D Converter

The structure of the dynamic flash 6-bit A/D converter is illustrated in Fig. 4. It is constructed with two 3-bit A/D converters and current subtraction/multiplication circuits. The input current (I) is firstly converted to a 3-bit digital format (upper 3-bits). And the amount of current, which is decided by the result of the 3-bit A/D converter, is subtracted from I. And the subtraction result is multiplied by 8 and is fed to the 3-bit A/D converter again. In this way, lower 3-bits are also obtained.

In the 3-bit A/D converter, the input current I is copied by the NMOS current-mirror to the seven current comparators (CC's). And the seven copied currents are compared to 1~7 times reference current (I_{REF}) by the CC's in parallel. The reference current is produced by an NMOS current source. Therefore, it can be easily controlled by the external signal. And the digital outputs (upper 3-bits) are obtained by the results of the comparisons. Next, M (which is decided by the upper 3-bits) times the reference current is subtracted from I. And the subtraction result is multiplied by 8 and is fed to the 3-bit-A/D converter as explained before. In this way, lower 3-bits are also obtained using the same circuit, thus achieving the 6-bit A/D conversion of I.

The CC (Current Comparator) is composed of two current-to-delay converters and a delay-detection circuit. Both of them are designed so as to eliminate the direct current which increases the power dissipation. In the CC, the two input currents are firstly converted to delay time by current-

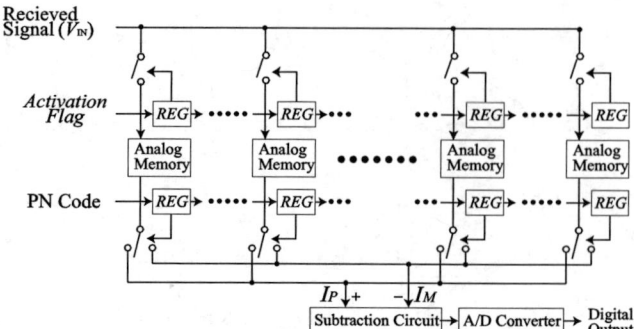

Fig. 1. Block diagram of the MF.

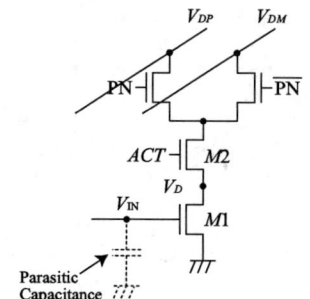

Fig. 2. Schematic of the matching cell.

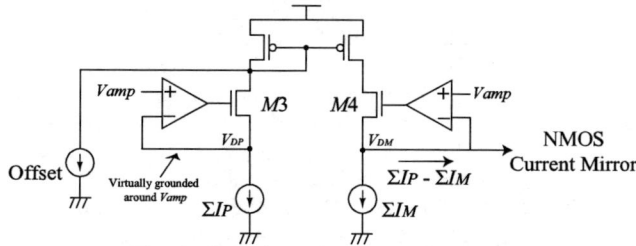

Fig. 3. Gain-boosting cascode circuit and subtraction circuit.

to-delay converters and the two delay times are detected and compared by the delay detection circuit. The schematic of the current-to-delay converter is illustrated in Fig. 5. The node of $Vcap$ is firstly charged up to V_{DD}. And then, the ACT is turned on. In the operation, the $M5$ must be activated in the saturation region in order to copy and evaluate the current accurately. In this circuit, the transistor $M5$ operates in the saturation region until the inverter turns off because the node $Vcap$ is $V_{DD}/2 \sim V_{DD}$. Therefore the current is converted to the turn-on delay time of the inverter, thus eliminating the direct current completely.

The schematic of the two-input delay-detection circuit is illustrated in Fig. 6. The outputs of the two current-to-delay converters are fed to the delay-detection circuit and the first coming signal is sensed. Dynamic logic gates are employed in order to realize a quick response for timing detection and eliminate the direct current.

III. CHIP DESCRIPTION

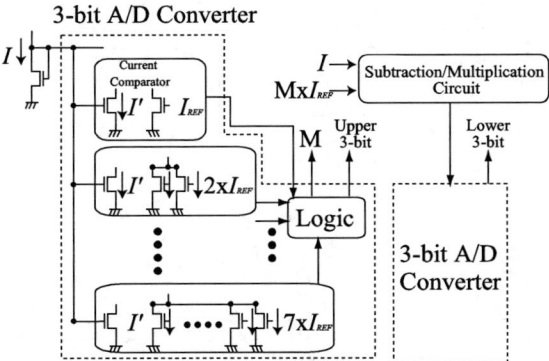

Fig. 4. Block diagram of 6-bit A/D converter.

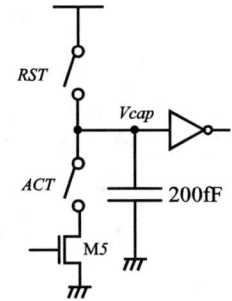

Fig. 5. Schematic of the current-to-delay converter.

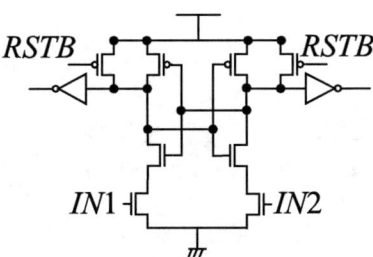

Fig. 6. Schematic of the delay-detection circuit.

Two types of the current-mode MF for 256-chip length correlation were designed and fabricated in a 0.35-μm double-polysilicon triple-metal CMOS technology. One is a current output type (TYPE-I), without an output A/D converter and the other is a digital output type (TYPE-II) having an output A/D converter. The chip photomicrographs are shown in Figs. 7 and 8. The chip area is 0.35mm² for TYPE-I and 0.49mm² for TYPE-II.

IV. MEASUREMENT RESULTS

Measurement results obtained from fabricated chips of both TYPE-I and TYPE-II are presented in Figs. 9-12.

A. Current Output Type

Fig. 9 shows the characteristics of the current output peak amplitude of the MF, not including an A/D converter. The supply voltage was 2V. Here, the linearity between the input voltage amplitude (V_{IN}) and the detected current peak

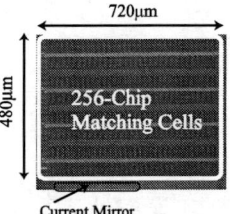

Fig. 7. Chip photomicrograph of TYPE-I.

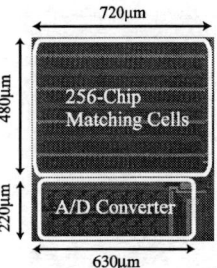

Fig. 8. Chip photomicrograph of TYPE-II.

amplitude is shown. V_{IN} is the voltage input to $M1$ in Fig. 2. The correlation was taken at 256-chip length for varing values of V_{amp} (defined in Fig. 3). The input voltage (V_{IN}) must be higher than V_{amp} (less than 0.2V) + V_{th} (around 0.7V) in order to drive $M1$ (in Fig. 2) in the triode region. Therefore, the minimum input voltage was set to 0.9V. As shown in the figure, a good linearity and the programmability of the output peak current by changing V_{amp} are observed. In the following experiments, V_{amp} was fixed at 0.2V in order to achieve a wide output current range and good linearity.

B. Digital Output Type

The digital output peak amplitude in the correlation value of 256-chip-length MF including the A/D converter is shown as a function of the input voltage in Fig. 10. As shown in Fig. 10, the A/D converter has at least 4-bit accuracy, which is sufficient to detect the peak amplitude.

Fig. 11 demonstrates the digital output peak amplitude gain as a function of the chip rate. Peaks were detected properly up to 11Mchip/s chip rate. Fig. 12 shows the power dissipation characteristics. The power in the analog part contains that of the A/D converter. The total power dissipation at 11Mchip/s operation is 1.65mW.

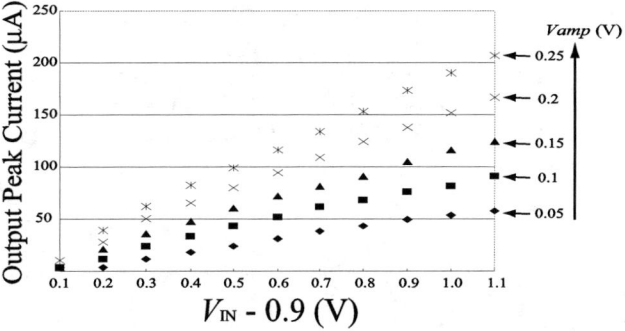

Fig. 9. Measured results showing linearity between the input voltage and output current peak amplitude (TYPE-I).

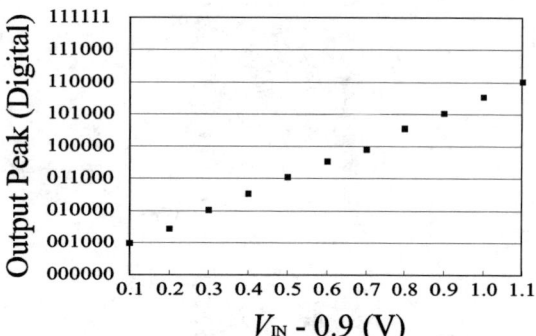

Fig. 10. Measured results showing linearity between the input voltage and digital-output-peak amplitude (TYPE-II).

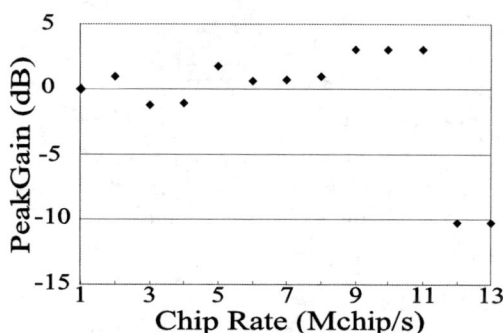

Fig. 11. Measured digital-output-peak amplitude as a function of chip rate.

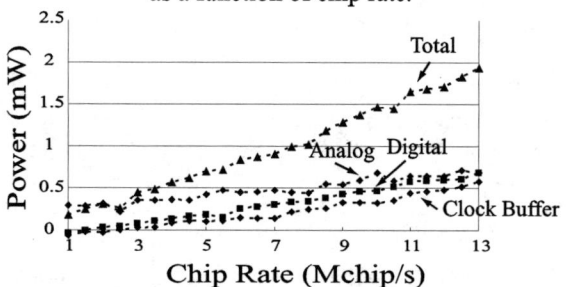

Fig. 12. Measured power dissipation as a function of chip rate.

V. Conclusions

An intra-cell V-I conversion architecture and a low-power dynamic flash A/D converter have been developed to build a low-power switched-current CDMA matched filter. The chip designed in a 0.35-μm CMOS technology occupies a very small area of 0.49mm^2. The measurement of the fabricated chip demonstrated that a 256-chip MF having an output A/D converter dissipates only 1.65mW at 11Mchip/s operation under 2V power supply.

Acknowledgment

The authors would like to thank Prof. Atsushi Fukasawa and Prof. Yumi Takizawa for their valuable discussion about this work. The VLSI chips in this work were fabricated through VLSI Design and Education Center (VDEC), the University of Tokyo with the collaboration by Rohm Corp. and Toppan Printing Corp.

References

[1] T. Yamada, S. Goto, N. Takayakma, Y. Matsushita, Y. Harada, and H. Yasuura, "Low-Power Architecture of a Digital Matched Filter for Direct-Sequence Spread-Spectrum Systems," IEICE Trans. Elec., vol. E86-C, no. 1, pp. 79-88, Jan. 2003.

[2] J. Wu, M. Liou, H. Ma and T. Chiueh, "A 2.6V, 44-MHz all digital QPSK direct-sequence spread spectrum transceiver IC," IEEE JSSC, vol. 32, pp. 1499-1510, Oct. 1997.

[3] D. Senderowicz, S. Azuma, H, Matsui, K. Hara, S. Kawama, Y. Ohta, M. Miyamoto, and K. Iizuka, "A 23-mW 256-Tap 8Msample/s QPSK Matched Filter for DS-CDMA Cellular Telephony Using Recycling Integrator Correlators," IEEE Int. Solid-State Circuit Conference, pp. 354-355, Feb. 2000.

[4] M. Sakai, T. Sakai, and T. Matsumoto, "A Low Power Matched Filter for DS-CDMA Based on Analog Signal Proc-essing," IEICE Trans. Fund., vol. E86-A, no. 4, pp. 752-757, Apr. 2003.

[5] T. Yamasaki, T. Fukuda, and T. Shibata, "A Floating-Gate-MOS-Based Low-Power CDMA Matched Filter Employing Capacitance Disconnection Technique," VLSI Circuits Dig. Tech. Papers, pp. 267-270, 2003.

[6] T. Yamasaki, T. Nakayama, and T. Shibata, "A Low-Power Switched-Current CDMA Matched Filter with On-Chip V-I and I-V Converters," VLSI Circuits Dig. Tech. Papers, pp. 214-217, Jun. 2004.

[7] K. Togura, H. Nakase, K. Kubota, K. Masu, and K. Tsubouchi, "Low Power Current-Cut Switched-Current Matched Filter for CDMA," IEICE Trans. Elec., vol. E84-C, no. 2, pp. 212-219, Feb. 2001.

A Re-configurable High-Speed CMOS Track and Latch Comparator with Rail-to-Rail Input for IF Digitization

Holly Pekau, *Student Member, IEEE*, Lee Hartley, *Member, IEEE,* and James W. Haslett *Fellow, IEEE*

Abstract— A re-configurable high speed track and latch comparator with rail-to-rail input range is designed in a $0.18\mu m$ CMOS process. The comparator architecture consists of parallel PMOS and NMOS differential pairs followed by a regenerative latch. The bias current of the differential pairs and the duty cycle of the clock can be adjusted so that the comparator consumes the minimum amount of power for a given input voltage sensitivity and a given clock frequency. The comparator is suitable as a building block for a re-configurable analog-to-digital converter in an IF digitizing software radio receiver.

I. INTRODUCTION

High-speed low-power analog-to-digital converters (ADCs) are key building blocks for modern wireless communications systems. In classical receiver architectures, ADCs are used to digitize signals that have already been downconverted to baseband frequencies. Recent receiver designs have used ADCs to digitize signals at an intermediate frequency (IF) in order to reduce the receiver complexity and to allow the receiver to be used for multiple standards. To facilitate the implementation of such a multi-standard IF sampling receiver while maintaining the cost benefit of small chip area and low power consumption, the ADC should be made reconfigurable for optimized operation at a variety of sampling rates and resolutions. Adaptive optimization of ADCs in terms of power consumption and resolution has been suggested for radio receivers, where the received signal strength depends strongly on geographical location [1] and several recent works have proposed reconfigurable ADCs using various architectures [1] [2], [3].

In this work we propose a comparator with improved reconfigurability that is suitable for a wide range of ADC's used in IF sampling applications. The proposed comparator is based on a standard architecture that consists of a differential pre-amplifier followed by a regenerative latch. The comparator can be configured in two different ways. The input bias voltage of the clock buffer sets the duty cycle of the clock and therefore allows the power consumption of the comparator to be reduced since DC power is only consumed in the track phase. Two external resistors set the differential pre-amplifier currents and allow the resolution and the maximum operating frequency of the comparator to be set.

II. CIRCUIT DESIGN

The proposed comparator circuit has been designed in a $0.18\mu m$ CMOS process. The schematic and die micrograph are shown in figures 1 and 2. The comparator consists of parallel NMOS and CMOS differential pairs followed by a regenerative latch. Previous authors [4] [5] have achieved GHz sampling frequencies and low power consumption with resolutions in the range of 6 bits using similar architectures.

Parallel NMOS and PMOS differential pairs are used to obtain a rail-to-rail input voltage range. A simple clock buffer consisting of three cascaded inverters with a variable externally supplied input DC voltage and a sinusoidal clock input signal is designed to generate an on-chip clock signal with fast transition times and adjustable duty cycle.

III. SIMULATED AND MEASURED RESULTS

A. Transient Waveforms and Comparator Speed

The simulated and measured transient output of the comparator for a clock frequency of 100MHz, an input frequency of 10MHz, an input common mode voltage of 900mV, and an input differential (overdrive) voltage of 20mV are shown in Figures 3 and 4. The simulation and measurement is done with a 50Ω oscilloscope load so voltage division with the output buffer PMOS on-resistance causes the equivalent logic high voltage to be 350mV rather than the supply voltage of 1.8V.

Manuscript received October 4, 2004. H. Pekau, L. Hartley, and Dr. J.W. Haslett are with the University of Calgary, Department of Electrical and Computer Engineering, Calgary, Alberta, Canada, T2N 1N4 (e-mail: haslett@enel.ucalgary.ca)

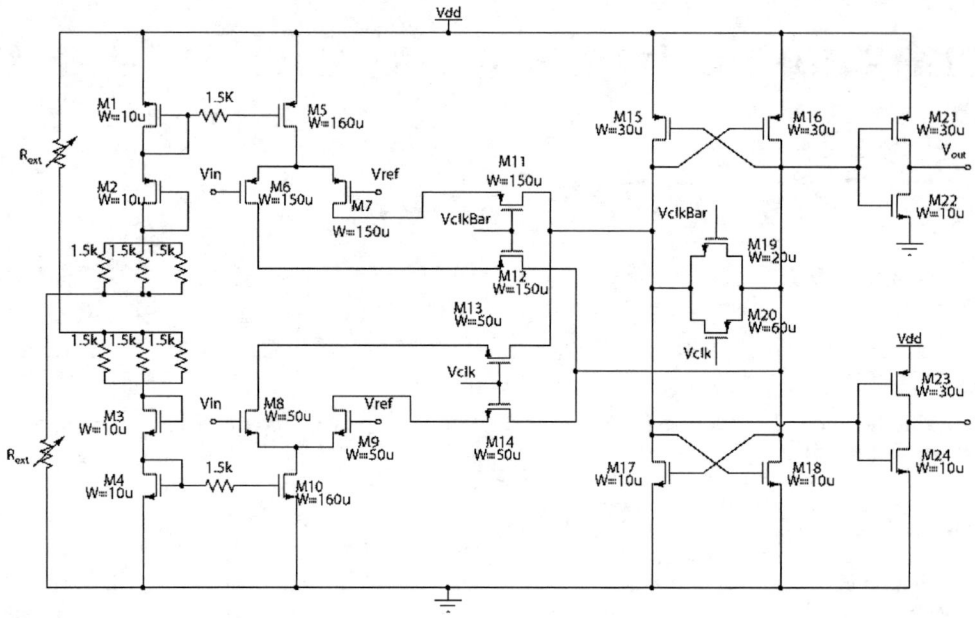

Fig. 1. Schematic of the proposed comparator circuit (all device lengths are $0.18\mu m$)

The maximum operating frequency of a track-and-latch style comparator depends on the latch mode time constant, the track mode 3dB bandwidth, and the input overdrive voltage. An approximation of the latch-mode time constant, τ_{LTCH}, based on an ideal model is [6]

$$\tau_{LTCH} = \frac{C_L}{G_m}, \quad (1)$$

where C_L is the load capacitance at the regenerative nodes G_m is the transconductance of each inverter. Transient simulation predicts a latch mode time constant of 77ps, which agrees with calculations based on the preceding equation.

The track mode 3dB bandwidth of the comparator is approximately given by

$$f_{3dB} = \frac{\omega_{3dB}}{2\pi} = \frac{1}{2\pi RC} = 1.8 GHz. \quad (2)$$

AC simulations show a track mode 3dB bandwidth of 1.3GHz to 1.6GHz at differential pair bias currents of 1mA to 3.2mA. These results agree reasonably well with the simple predicted result from the preceding equation. Since the regenerative nodes of the circuit are not externally accessible, the track mode bandwidth cannot be measured.

B. Input Sensitivity

The equivalent number of bits resolvable by a flash type ADC using a comparator with a minimum resolvable input voltage of V_{min} and a maximum input common mode voltage range of V_{CM-FS} is

$$N_{COMP} = log_2\left(\frac{V_{CM-FS}}{V_{min}}\right). \quad (3)$$

For IF sampling applications, a typical range of A/D converter resolutions is 5 bits to 8 bits[1].

The minimum resolvable input voltage (sensitivity) of a comparator is determined by the track and latch mode time constants and settling times, the clock frequency, and the input offset voltage due to device mismatch, V_{OS}. Since the track mode time constant and the input offset voltage vary with the bias currents of the differential pairs and the input common mode voltage, the minimum resolvable input voltage will also vary with these parameters.

To verify these predictions in simulation, a typical mismatch condition of 0.2% absolute width mismatch applied to pairs of devices in such a way as to generate maximum input offset voltage is used. This typical mismatch condition is based on the work of Maxim et al[7]. The simulated input offset voltages at bias currents of 1mA and 4mA with an input common mode voltage of 900mV are 1mV and 0.25mV respectively. At common mode voltages closer to the supply rails, the maximum simulated input offset voltage is 5mV. The measured offset voltage is less than 2mV for a bias current of 1.3mA and an input common mode voltage of 900mV.

Figures 5 and 6 contain plots of the simulated and

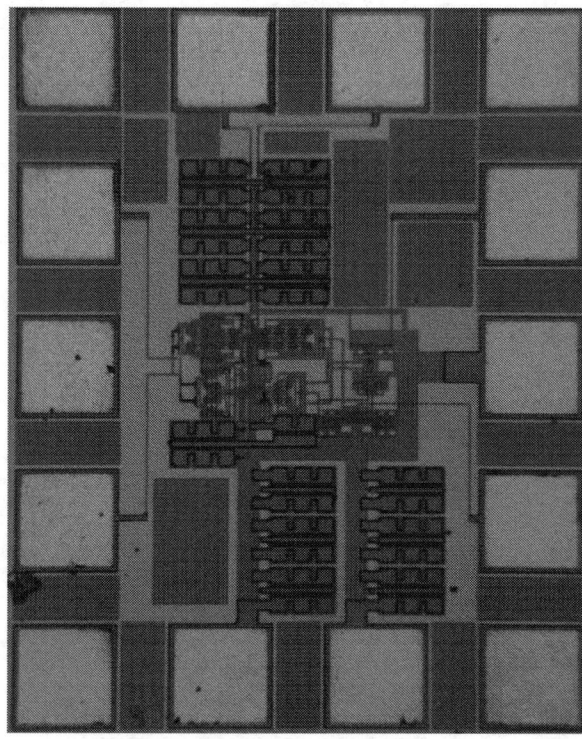

Fig. 2. Die micrograph of the comparator circuit (width = $500\mu m$)

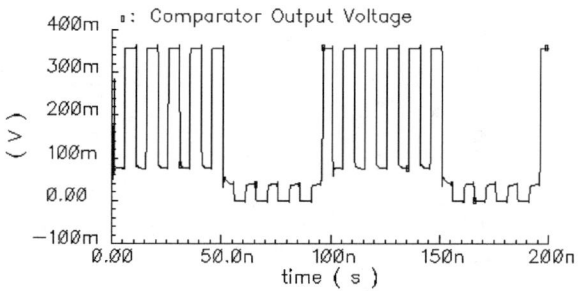

Fig. 3. Comparator simulated transient output waveform

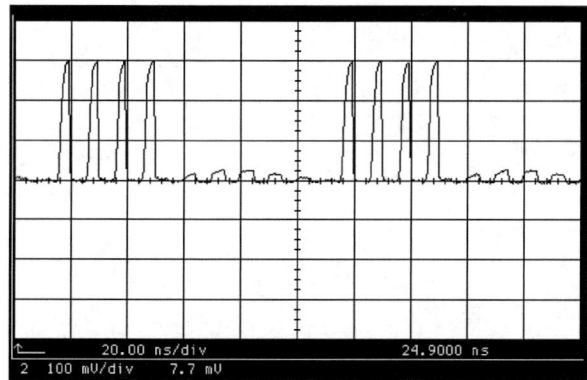

Fig. 4. Comparator measured transient output waveform

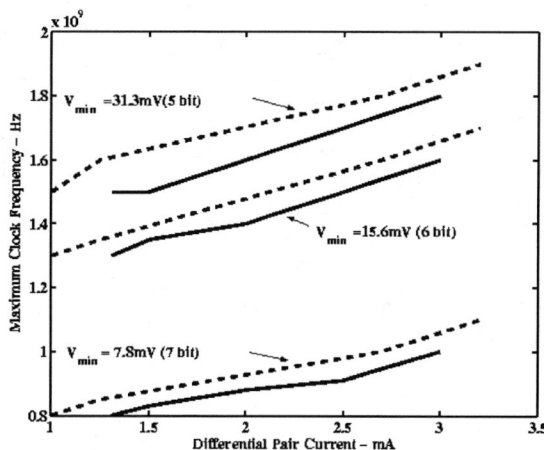

Fig. 5. Simulated and measured maximum clock frequency versus differential pair bias current for various resolutions with a 50% clock duty cycle

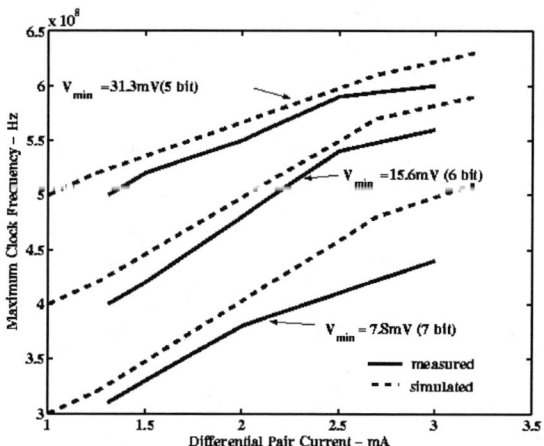

Fig. 6. Simulated and measured maximum clock frequency versus differential pair bias current for various resolutions with a 25% clock duty cycle

measured maximum frequency of operation for different input overdrive voltages corresponding to 5, 6, and 7 bit resolutions for a full scale voltage of 1V as the differential pair bias currents are varied. It can be seen that the maximum operating frequency can be increased by increasing the input differential pair bias current. It can also be seen that the maximum resolution is reduced for the lower clock duty cycle of 25% because the resolution becomes increasingly limited by the track mode time constant and the track mode settling time. At the lowest clock frequency of 100MHz, measurements show that 9 bit comparator resolution can be achieved, but this drops off rapidly to 7 as the clock frequency increases.

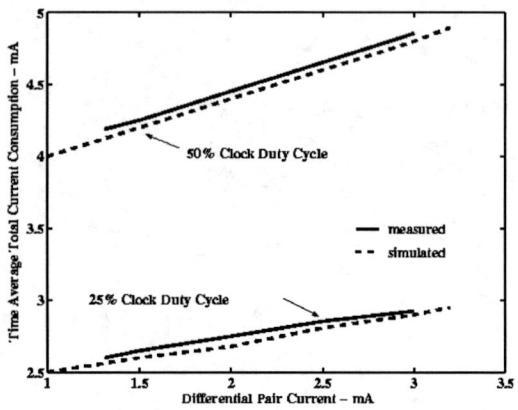

Fig. 7. Simulated and measured time average comparator current consumption vs input differential pair current

C. Power Consumption

With the exception of the small amount of DC current drawn by the reference current mirrors, the proposed comparator architecture only draws current in the track phase. Ideally, the current drawn will be reduced by 50% with a 25% clock duty cycle. Figure 7 shows the simulated and measured time average total current consumption of the comparator including the unloaded output buffers as the bias current is increased for duty cycles of 50% and 25%. It can be seen that reducing the duty cycle reduces the time average current consumption, but the reduction is less than 50% because significant switching current is drawn by the inverters of the regenerative latch and the clock buffers.

IV. Conclusion

A novel reconfigurable track and latch comparator with rail-to-rail inputs with 1.8GHz performance at 5 bit operation, 1.2GHz performance at 6 bit resolution, and 600MHz performance at 7 bit resolution is proposed. The power consumption of the comparator can be minimized for a given minimum resolvable input voltage and clock frequency by adjusting the current of the input stage and reducing the clock duty cycle for low clock frequencies and/or high overdrive voltages to reduce the percentage of time the comparator spends in the power consuming track mode. The time average power consumption of the comparator is configurable in the range of 1.5mW to 9mW, including the power consumed by the output buffers. The comparator is suitable as a building block in flash and pipelined ADCs for IF sampling applications.

V. Acknowledgements

This work was supported by the Natural Sciences and Engineering Research Council of Canada, the Alberta Informatics Circle of Research Excellence, TRLabs, the University of Calgary, and the Canadian Microelectronics Corporation.

References

[1] J. Yoo, D. Lee, K. Choi, and J. Kim, "A power and resolution adaptive flash analog-to-digital converter," in *2002 International Symposium on Low Power Electronics and Design*, May 2002, pp. 233–236.

[2] P. Setty, J. Barner, H. Burger, and J. Sonntag, "A 5.75b 350MSample/s or 6.75b 150MSample/s reconfigurable flash ADC for a PRML read channel," in *ISSCC*, Feb. 1998, pp. 148–150.

[3] J. Terada, Y. Matsuya, F. Morisawa, and Y. Kado, "8-mW, 1-V, 100-MSPS, 6-BIT A/D converter using a transconductance latched comparartor," in *The second IEEE Aisa Pacific Conference on ASICs*, 2000, pp. 53–56.

[4] K. Uyttenhove and M. Steyaert, "A CMOS 6-bit 1 GHz ADC for IF sampling applications," in *IEEE Conference on Microwave Theory and Techniques*, 2001, pp. 2131–2134.

[5] C. J. Fayomi, G. W. Roberts, and M. Sawan, "Low power/low voltage high speed CMOS differential track and latch comparator with rail-to-rail input," in *ISCAS*, 2000, pp. V653–V656.

[6] D. A. Johns and K. Martin, *Analog Integrated Circuit Design*. New York, NY: Wiley, 1997.

[7] A. Maxim and M. Gheorghe, "A novel physical based model of deep-submicron CMOS transistor mismatch for monte carlo simulation," in *2001 IEEE International Symposium on Circuits and Systems*, vol. 5, 1997, pp. 511–514.

Fast-Switching Adaptive Bandwidth Frequency Synthesizer using a Loop Filter with Switched Zero-Resistor Array

Sreenath Thoka
Department of Electrical and Computer Engineering
Iowa State University
Ames, IA 50011 USA
sree@iastate.edu

Randall L. Geiger
Department of Electrical and Computer Engineering
Iowa State University
Ames, IA 50011 USA
rlgeiger@iastate.edu

Abstract— **Secondary glitches caused by the switching of bandwidth in adaptive-bandwidth frequency synthesizers, are studied and a simple solution based on stepped-bandwidth switching is proposed to reduce their effects on the switching time of the synthesizer. Behavioral simulations using GSM specifications indicate nearly 35% improvement in switching time using the proposed solution.**

I. INTRODUCTION

Frequency synthesizers are ubiquitous in today's wireless communication systems. In the design of PLL frequency synthesizers, there exists a trade off between the settling time and spur level at the output [1] [2]. Though theory allows a maximum of 1/10 of the reference frequency for the loop bandwidth, practically the bandwidth is several orders of magnitude smaller than this limit in order to meet the noise specification. Further in Integer-N synthesizers, the reference frequency cannot exceed the channel spacing. These limitations result in elongated switching times for conventional synthesizers.

A common solution to offset the limitation on the loop bandwidth set by the noise requirement is to use an adaptive bandwidth synthesizer [3]. In this approach, high loop bandwidth is used during frequency jumps and the loop bandwidth is restored to its nominal value after the frequency has settled to its new value. The loop bandwidth in high bandwidth mode is still limited by the reference frequency, the theoretical maximum being 1/10 of the reference frequency. The switching of bandwidth is accomplished by switching an element in the loop filter usually a resistor. If the synthesizer has to settle to a very small frequency error in a very short time, it becomes important to make the transition from high bandwidth to low bandwidth as smooth as possible; otherwise the glitches arising from switching of bandwidth contribute significantly to the overall switching time [4]. An applicable case is the synthesizer used in a GSM base station transmitter which has a switching time specification of less than 10μs for a frequency jump of 75MHz with 0.1ppm tolerance. This work focuses on studying the non-ideal effects in the elements of the loop filter that can cause the transition from high bandwidth to low bandwidth to be not as smooth as desired. A solution is then proposed to counter these effects.

This paper is organized as follows. In section II, the basic principle of an adaptive bandwidth synthesizer is explained in terms of loop parameters and also the secondary glitch effects are studied. In section III, the proposed solution for controlling the secondary glitches is described. In section IV, verification of the proposed scheme through behavioral level simulations is presented followed by conclusions.

II. BASIC PRINCIPLE OF AN ADAPTIVE PLL SYSTEM

The switching time of an under damped type-II second-order PLL is given by

$$t_s \cong \frac{\ln\left(\frac{\Delta f_o}{\varepsilon\sqrt{1-\varsigma^2}}\right)}{\omega_n \varsigma} \quad (1)$$

where Δf_o is the output frequency step, ε is the frequency error tolerance, ω_n is the natural frequency and ζ is the damping factor of the loop. Thus, the settling time for a given frequency step and error tolerance depends on the natural frequency and the damping factor. The switching time can also be expressed in terms of the change in VCO tune voltage corresponding to the frequency step Δf_o in which case the frequency error tolerance ε will be replaced with the equivalent tune voltage error tolerance. The VCO gain sensitivity relates the equivalent parameters in the two expressions. Since a closed-form expression for switching time of a type-II third-order loop shown in fig. 1 is difficult to obtain, we will neglect the effect of loop filter pole in this discussion. This is a valid assumption since the pole can be added only after locking is complete if so chosen. The natural frequency and the damping factor (ignoring C_p) are given by:

$$\omega_n = \sqrt{\frac{I_{cp}K_{vco}}{2\pi NC_z}} \quad (2)$$

$$\varsigma = \frac{1}{2}\tau_z\sqrt{\frac{I_{cp}K_{vco}}{2\pi NC_z}} \quad (3)$$

where $\tau_z = R_zC_z$.

Equations (1), (2), (3) indicate that to reduce the settling time by a factor of β, the natural frequency has to be increased by a factor of β while keeping the damping factor constant. In terms of the loop component parameters this is equivalent to increasing the charge pump current I_{cp} by a factor of $β^2$ and reducing the zero resistor R_z by a factor of β. The open loop bandwidth increases by a factor of β with the reduction in R_z. Even though noise requirements limit the loop bandwidth to be narrower, it can be made wider during the locking process and then restored after lock is established. This is the basic principle of operation of the adaptive PLL of fig. 1. The lock signal generated by a lock-detect circuit not shown, or a predetermined timing signal is utilized for the adaptive bandwidth control.

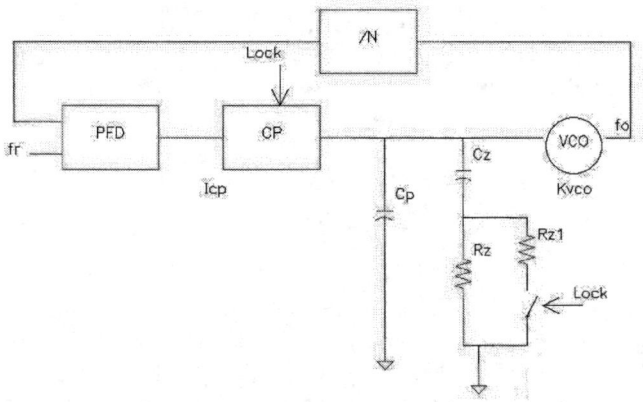

Figure 1 Adaptive Bandwidth Synthesizer

The adaptive system described above can be employed to achieve significant improvement in the lock time as long as the theoretical limit on loop bandwidth, which is 1/10 of reference frequency, is not exceeded during the high-bandwidth mode. However, as discussed in section I it is important to limit the level of secondary glitches when switching from high bandwidth to low bandwidth mode. Switching a memory less element such as a resistor inside a filter that's completely settled should not give rise to transient behavior. Thus the sources of the secondary glitches are the parasitic capacitances associated with the switch element (typically an NMOS). At the time of frequency jump, the switch in fig. 1 is closed to reduce the effective resistance in the loop filter by a factor β. When the settling is complete the switch has to be opened to restore the bandwidth. The transitions on the voltage signal controlling the on/off states of the MOS switch cause glitches on the tune voltage line. The size of the parasitic gate-drain capacitance of the MOS switch determines among other things, the level of these glitches. If the tune voltage has already settled to within the tolerance window at the time of turning off the MOS switch, then the glitches can cause the tune voltage to swing outside of the window, thus increasing the over all lock time of the synthesizer.

III. PROPOSED SCHEME

As described in section II, minimizing the gate-drain capacitance of the MOS switch helps to restrict the level of glitches to within the tolerance window. A second factor that determines how fast the VCO tune voltage settles after a glitch occurs, is the effective bandwidth after switching. Thus if a relatively large MOS switch is used to switch from 8x bandwidth (high bandwidth) to 1x bandwidth (nominal bandwidth), then the size of the MOS switch and the 1x bandwidth after switching together contribute to significantly large switching time. Instead in this paper, a scheme is proposed wherein bandwidth is reduced from 8x to 1x in several steps using multiple MOS-switched resistors in parallel with the nominal zero-resistor R_z to form a switched-resistor-array. The largest sized MOS switch in the array would still be smaller than the single MOS switch used in the one-step approach. To keep the damping factor constant and only increase the natural frequency the following relationship has be valid at each of the bandwidth steps.

$$\sqrt{I_{cp}} \times R_z = constant \quad (4)$$

One of the factors determining the effectiveness of the proposed scheme is the pattern used for the steps. There are several possibilities such as binary stepping of I_{cp}, linear stepping of I_{cp} or the Fibonacci pattern for I_{cp}. Binary stepping and the Fibonacci pattern result in successive bandwidths being close to each other near the nominal bandwidth and far apart near the high bandwidth. Linear stepping of bandwidth results in equal-sized MOS switches and is also an optimum choice. Though the different patterns were simulated, the results are presented in section IV only for the binary pattern. Note that once the high bandwidth and the type of pattern are chosen, the no. of steps will be fixed. For, example if the high bandwidth is 8 times the nominal bandwidth and if binary stepping of I_{cp} is chosen, then the no. of steps will be six corresponding to I_{cp} multipliers of 64, 32, 16, 8, 4, 2, 1. The second factor determining effectiveness is the time at which each step occurs. In the two-step case presented in section IV, the time for the next step is chosen to be the time of the first zero-crossing of the tune voltage in the current step. +/- 15% variation of this time has also been considered to demonstrate the robustness of the proposed scheme.

IV. BEHAVIORAL SIMULATION

The proposed scheme is verified through behavioral simulation of a synthesizer using GSM specifications. In order to focus primarily on the non-idealities of the switches used to adjust the loop filter resistor, the simulation setup uses ideal components except for the NMOS transistors, which. In all the branches of the switched-resistor-array, containing a fixed resistor and a MOS switch, the on-resistance of the switch is always chosen to be 1/10 of the fixed resistance. Firstly, comparison is done between the switching times obtained in the following two cases.

Case1: High bandwidth switched to nominal bandwidth in one step

$$8x \rightarrow 1x$$

Case2: High bandwidth switched to nominal bandwidth in two steps

$$8x \rightarrow sqrt(8)x; \quad sqrt(8)x \rightarrow 1x$$

The simulation setup used for the two cases is shown in fig. 2. The list of common specifications is given below:

Frequency jump: 1710 MHz - 1785 MHz

Reference Frequency: 26 MHz

Nominal charge pump current (1x): 0.213 mA

Nominal loop bandwidth (1x): 40 kHz

Intermediate loop bandwidth (sqrt(8)x): 113 kHz

High loop bandwidth (8x): 320 kHz

Phase margin: 46.4°

VCO gain: 37.5 MHz/V

Loop filter zero: 16 kHz

Loop filter pole: 100 kHz

Nominal R_z: 2.5 K

C_z: 3.98nF

C_p: 758pF

In the one-step case, the bandwidth is switched at 10.5μs, whereas in the two-step case, the first step occurs at 10.5μs and the second at 11.43μs. Fig. 3 shows the locking profile where the VCO tune voltage settling to the tolerance window is shown against time. The total switching time is 18.21μs for the one-step case and 11.75μs for the two-step case. This is nearly 35% improvement in the switching time.

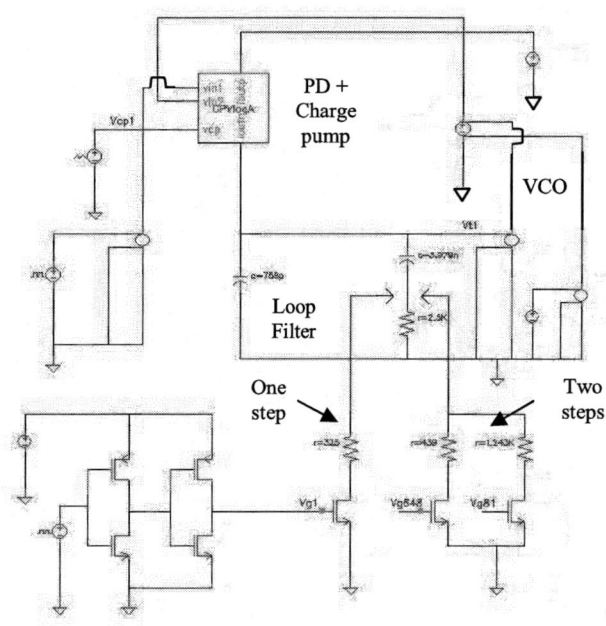

Figure 2 Simulation setup to compare one step vs two-step bandwidth switching

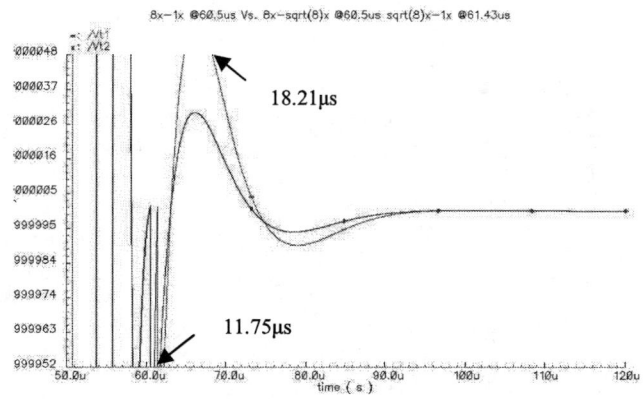

Figure 3 Locking profile for one- and two-step bandwidth switching

It was mentioned in section III that the time at which switching of bandwidths has to occur is determined based on the zero-crossings of the tune voltage. To demonstrate robustness of the proposed scheme, simulations are performed with +/- 15% variations in these times. Table 1 shows the results of these simulations. It can be observed that with the variations included, there is still 35% improvement in the total switching time.

Table 1 Switching time Variations

One step at t_{sw}	Total switching time	Two steps; first step at t_{sw1}; second step at t_{sw2}		Total switching time
$t_{sw}(\mu s)$	$t_s(\mu s)$	$t_{sw1}(\mu s)$	$t_{sw2}(\mu s)$	$t_s(\mu s)$
10.5	18.21	10.5	11.43	11.75
			11.29	11.84
			11.57	11.67
10.14	18.46	10.14	11.43	11.43
			11.29	11.29
			11.57	11.57
10.36	19.11	10.36	11.43	11.52
			11.29	11.61
			11.57	11.57

Finally, a synthesizer with six bandwidth steps is simulated. The first step occurs at 8.1μs and time between subsequent steps is 0.5μs. Simulation setup is shown in fig. 4 and the results are shown in fig. 5. I_{cp} and R_z at each step are multiplied by factors as shown below.

I_{cp} multiplier: $64 \to 32 \to 16 \to 8 \to 4 \to 2 \to 1$

R_z fraction: $1/8 \to 1/\sqrt{32} \to 1/4 \to 1/\sqrt{8} \to 1/2 \to 1/\sqrt{2} \to 1$

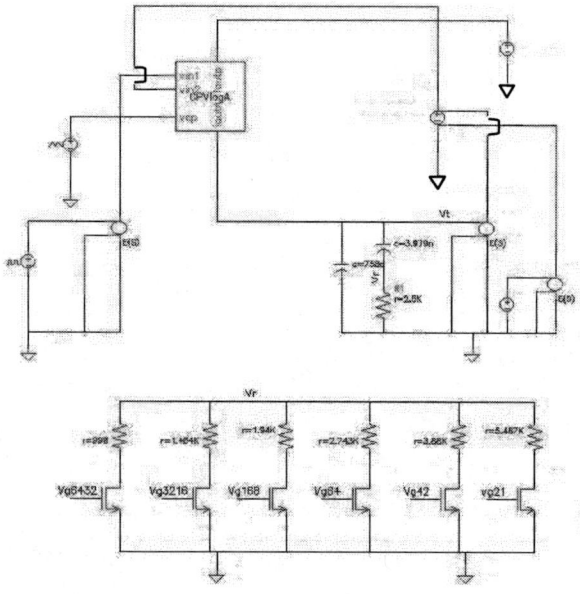

Figure 4 Simulation setup for six step bandwidth switching

The total switching time in this case is 10.62μs.

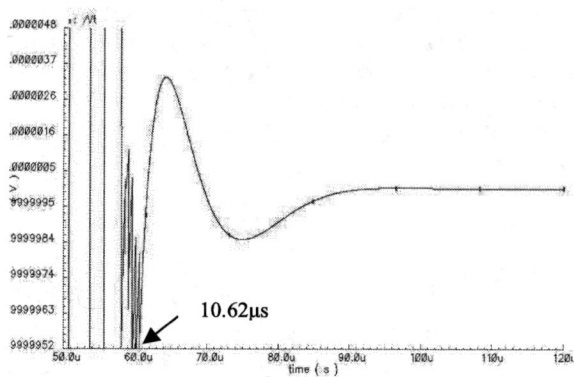

Figure 5 Locking profile for six step bandwidth switching

CONCLUSIONS

In adaptive bandwidth synthesizers, the effect of secondary glitches sometimes dominates the overall switching time. The stepped bandwidth approach presented in this paper achieves significant improvement in overall switching time for the cases studied. It is a simple and practical on-chip solution. A rigorous optimization in the areas mentioned earlier, would make this solution viable for many applications.

REFERENCES

[1] Yiwu Tang et al., "A low noise fast settling PLL with extended loop bandwidth enhancement by new adaptation technique," Proceedings of ASIC/SOC Conference, 2001

[2] Tai-Cheng Lee, et al., "A stabilization technique for phase locked loop synthesizers," JSSC Vol. 38, Issue 6, June 2003

[3] Keaveney, M et al., "A 10μs fast switching PLL synthesizer for GSM/EDGE base-station," ISSCC 2004 Digest of Technical Papers

[4] Vaucher, C.S., "An adaptive PLL tuning system architecture combining high spectral purity and fast settling time," JSSC Vol. 35, Issue 4, April 2000

Fully Integrated Charge Sensitive Amplifier for Readout of Micromechanical Capacitive Sensors

Mikko Saukoski, Lasse Aaltonen, Kari Halonen
Electronic Circuit Design Laboratory
Helsinki University of Technology
Espoo, Finland

Teemu Salo

VTI Technologies Oy
Vantaa, Finland

Abstract—This paper presents a charge sensitive amplifier (CSA) for readout of micromechanical capacitive sensors. Transfer function and noise performance together with noise optimisation are studied from a theoretical point of view. The feedback resistors of the CSA are implemented as long-channel MOS transistors to set the -3 dB corner frequency well below 1 kHz. The common-mode feedback is designed to keep the input common-mode level constant to achieve an accurate gain from the change in capacitance to output signal. A CSA was designed and implemented for the readout of a bulk micromachined gyroscope with a resonance frequency of 8-10 kHz and signal bandwidth of 100 Hz. A measured resolution of 2.47–3.00 aF (rms) in capacitance change integrated over the signal band is achieved. The CSA tolerates leakage currents of over 5 nA at its input.

I. INTRODUCTION

Sensors are needed in all systems in which movements or other physical or chemical phenomena have to be detected. At the moment the market area is rapidly growing as electronic controlling systems are getting more common in very different applications, including automotive, industrial and consumer products. Gyroscope, which is an angular velocity sensor, may be the most critical sensor in automotive controlling systems like ESP (Electronic Stability Program), but it also has many other potential adaptations.

With capacitive sensors, the sensor works as a capacitor whose capacitance varies relative to the physical quantity being sensed. For example, if the sensor is realised as a simple plate capacitor, the measurand can give excitation on the plate area, the distance between the plates or the permittivity of the insulator between the plates. Micromechanical capacitive sensors are capacitive sensors that are manufactured using micromachining techniques.

The readout of a capacitive sensor involves the conversion of the sensor capacitance or its changes to an electrical signal such as voltage, current or frequency. This paper presents a charge sensitive amplifier (CSA) that is used to convert the change in sensor capacitance to voltage.

The CSA is implemented as a lossy charge integrator that integrates charge flowing into and out of a constant-voltage biased capacitive sensor when its capacitance changes.

The paper consists of three main topics. First, the theory consisting of the CSA transfer function and the noise performance together with possibilities for noise optimisation is presented. Then, the gyroscope and the designed and implemented CSA are described. Finally, the measurement results of the implemented CSA are presented.

II. THEORY

A. Transfer Function

A CSA with a differential capacitive sensor represented by capacitors C_{Dp} and C_{Dn} connected to its input is shown in Fig. 1. In the following analysis, it will be assumed that $R_{fp}=R_{fn}=R_f$ and $C_{fp}=C_{fn}=C_f$.

A voltage $V_B=V_{CM}-V_{MID}$ is set over the sensor by connecting the common center plate to voltage V_{MID} and letting the common-mode feedback of the operational amplifier set the input voltage to V_{CM}. If the sensor capacitances are approximated to be linear, they can be

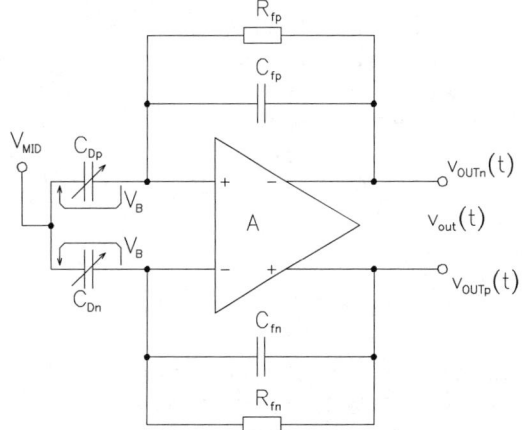

Figure 1. Charge sensitive amplifier with a differential capacitive sensor connected to its input.

The project under which this work was done is funded by National Technology Agency of Finland (TEKES) and VTI Technologies Oy. Design tools were provided by Cadence Design Systems and Mentor Graphics.

written as $C_{Dp}=C_D+\Delta C_D/2$ and $C_{Dn}=C_D-\Delta C_D/2$, and the s-plane transfer function from the change in capacitance $\Delta C_D(t)$ to the output voltage $v_{OUT}(t)=v_{OUTp}(t)-v_{OUTn}(t)$ is given by

$$H(s) = \frac{V_{out}(s)}{\Delta C_d(s)} = \frac{s \cdot V_B/C_f}{s+1/(R_f C_f)}. \quad (1)$$

From this equation, which is of single-pole highpass type, two design issues can be recognised. First the -3 dB corner frequency is given by $1/(2\pi R_f C_f)$, which means that if the signal frequency is low, passive feedback components become large in their value. Secondly, if an accurate gain is required, the voltage V_B over the sensor must also be accurate enough, as gain is directly proportional to it.

B. Noise

The resolution of capacitive sensor readout is limited by noise that comes mainly from two different sources. The sensor itself always creates noise associated with the dissipations that are caused for example by mechanical and viscous losses [1]. The CSA (like any other type of read-out electronics) also creates additional noise. In this discussion, attention is paid only to the noise contribution of the CSA.

The noise created by the CSA comes from two different sources: the operational amplifier and the feedback resistors. In addition, the leakage currents drawn from the CSA input limit the best achievable noise performance.

In order to be able to determine the resolution of the CSA, the noise voltage at the output is reduced to capacitance change by dividing with the transfer function given in (1). The result will be called noise capacitance, and its spectral density will be given in units of F/Hz$^{1/2}$ (or F^2/Hz). When integrated over the signal bandwidth, the minimum detectable rms change in sensor capacitance, or the resolution of the CSA, is achieved.

The noise contribution of the operational amplifier comes mainly from the differential input pair. The channel thermal noise of these MOS transistors causes a noise capacitance, the density of which is given by

$$C_{n,th}^2 = 2 \cdot \frac{1+\left[\omega R_f\left(C_f+C_D+C_{inW}\cdot W\right)\right]^2}{\left(V_B R_f \omega\right)^2} \cdot \frac{8}{3}kT\left(2\mu C_{OX}\frac{W}{L}I_D\right)^{-1/2}. \quad (2)$$

Similarly, the channel flicker (1/f) noise causes a noise capacitance, the density of which is given by

$$C_{n,f}^2 = 2 \cdot \frac{1+\left[\omega R_f\left(C_f+C_D+C_{inW}\cdot W\right)\right]^2}{\left(V_B R_f \omega\right)^2} \cdot \frac{KF'}{C_{OX}^2 WL f}. \quad (3)$$

Here, f is the frequency, $\omega=2\pi f$ the corresponding angular frequency, k is the Boltzmann constant, T the absolute temperature, W and L the dimensions of the input transistor, μ the carrier mobility in the channel, C_{OX} the capacitance of the gate oxide per unit area, I_D the drain current and KF' the flicker noise coefficient. C_{inW} is the capacitance per unit width seen at the gate of the input transistor, defined as

$$C_{inW} = CGDO + CGSO + 0.67 \cdot C_{OX} \cdot L, \quad (4)$$

where CGDO and CGSO are gate-drain and gate-source overlap capacitances per unit width, respectively. Both (2) and (3) ignore the leakage currents present at the CSA input.

By differentiating (2) relative to W, equating the derivative to zero and solving for W, a transistor width

$$W \approx (C_f + C_D)/(3C_{inW}) \quad (5)$$

is achieved. Similarly, by differentiating (3) relative to W, equating the derivative to zero and solving for W, a transistor width

$$W \approx (C_f + C_D)/C_{inW} \quad (6)$$

is achieved. Equation (5) implies that to minimise the thermal noise, the external capacitance (C_f+C_D) connected to the operational amplifier input should be three times the capacitance seen at the input ($C_{inW}\cdot W$), while (6) implies that to minimise the 1/f noise, the external capacitance connected to the input should equal the input capacitance [2]. The final optimal input capacitance is determined by the ratio of the magnitudes of these noise contributions. Further, by substituting (4) and (5) to (2) and (4) and (6) to (3), it can be seen that to minimise the contribution of the channel thermal noise, a minimum L is optimal, whereas to minimise the contribution of the channel 1/f noise, a maximum L is optimal. Again, the ratio between these noise contributions determines the final optimum L.

The noise capacitance contribution of the feedback resistors is given by

$$C_{n,Rf}^2 = \frac{8kT}{V_B^2 R_f \omega^2}. \quad (7)$$

If the leakage currents are modelled by connecting resistors R_{Dp} and R_{Dn} with resistances $R_{Dp}=R_{Dn}=R_D$ in parallel with C_{Dp} and C_{Dn}, correspondingly, their noise capacitance contribution is given by

$$C_{n,RD}^2 = \frac{8kT}{V_B^2 R_D \omega^2}. \quad (8)$$

As can be seen from (7) and (8), the noise capacitance is reduced as the voltage V_B over the sensor is increased. However, this also increases the leakage current that has to be supplied from the CSA output. If the voltage drop over the feedback resistors R_{fp} and R_{fn} is limited to be $V_{OFF,max}$ at its maximum, and if the leakage resistance is $R_{D,min}$ at its minimum (limited by the structure and the manufacturing process of the sensor together with ESD protection devices), then the product of V_B and R_f is limited to be

$$V_B R_f = R_{D,min} V_{OFF,max} \quad (9)$$

at its maximum. Hence, when V_B is increased, R_f has to be decreased. By summing (7) and (8) and substituting V_B solved from (9) to the sum, an absolute minimum for the noise caused by R_{Dp}, R_{Dn}, R_{fp} and R_{fn} given by

$$C_{n,R,min}^2 = \frac{8kTR_f}{R_{D,min}^2 V_{OFF,max}^2 \omega^2}\left(1+\frac{R_f}{R_{D,min}}\right) \quad (10)$$

is achieved. By substituting R_f solved from (9) to the sum of (7) and (8), an equivalent formula for the minimum given by

$$C_{n,R,min}^2 = \frac{8kT}{R_{D,min} V_B^2 \omega^2}\left(1+\frac{V_B}{V_{OFF,max}}\right) \quad (11)$$

is achieved. From (10) and (11), it can be seen that minimum noise is achieved with maximum possible V_B. This voltage is ultimately limited by the operating voltages and properties of the electronics and the sensor element, such as the pull-in voltage or other electromechanical parameters.

III. DESIGN AND IMPLEMENTATION

A charge sensitive amplifier was designed and implemented for the readout of the bulk micromachined gyroscope shown in Fig. 2, with a resonance frequency of 8-10 kHz and signal band of 100 Hz. The -3 dB corner frequency had to be set below 1 kHz in order to get a negligible gain error and phase shift. To achieve this, the size of the capacitors C_{fp} and C_{fn} was chosen to be 10 pF. This gives a minimum size of about 16 MΩ for the feedback resistors, which is not realisable with passive components with reasonable silicon area. High-ohmic resistors could also be implemented with active implementations based on multiple transistors, and with a resistor T network. The problem with both of these is that their noise contribution is higher than with simple resistors.

In order to get a high resistance with low noise and reasonable area, the feedback resistors were implemented with single long-channel FETs biased in the linear region with constant V_{DG}, as shown in Fig. 3. The differential

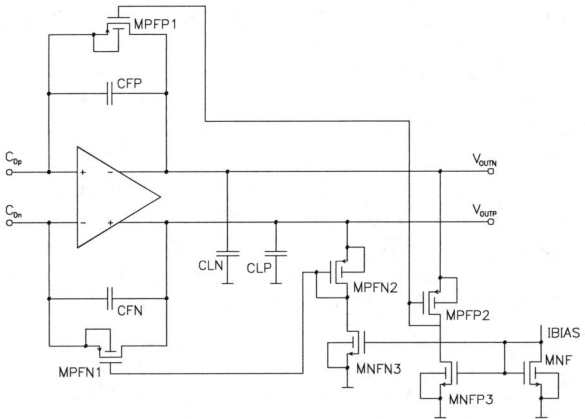

Figure 3. Implementation of the high-ohmic feedback resistors with long-channel FETs.

conductance between the drain and source of MPFP1 (MPFN1) can be shown, based on the quadratic model for MOS transistor [3], to be a constant independent of the voltage V_{SD} over it. When simulated with a more accurate MOSFET model, it can be seen that the differential conductance increases as V_{SD} becomes more positive. When V_{SD} becomes more negative, the differential conductance first decreases, until the pn-junction between the drain and bulk of MPFP1 (MPFN1) starts to conduct. When this happens, the differential conductance increases rapidly.

The aspect ratios (W/L) for MPFP1 and MPFN1 were chosen to be 1 μm/250 μm, and for MPFP2 and MPFN2, 10 μm/250 μm. With a bias current of 20 nA flowing through MNFP3 and MNFN3, a simulated resistance of 53 MΩ is achieved with V_{SD} over MPFP1 (MPFN1) equal to 0 V.

Another design issue is caused by external leakage currents that have to be supplied from the CSA input. With the sensor element and the CSA on separate chips, the level of these currents can be quite high. Current flows from the operational amplifier output through the feedback resistors, causing a voltage drop. If the output common-mode (CM) voltage level is kept constant, the voltage V_B over the sensor varies, thus changing signal gain, as can be seen from (1).

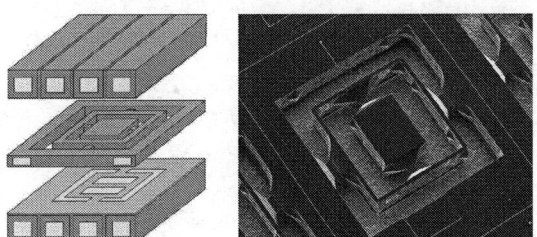

Figure 2. a) Schematic drawing of the gyroscope, b) SEM image of the structural wafer.

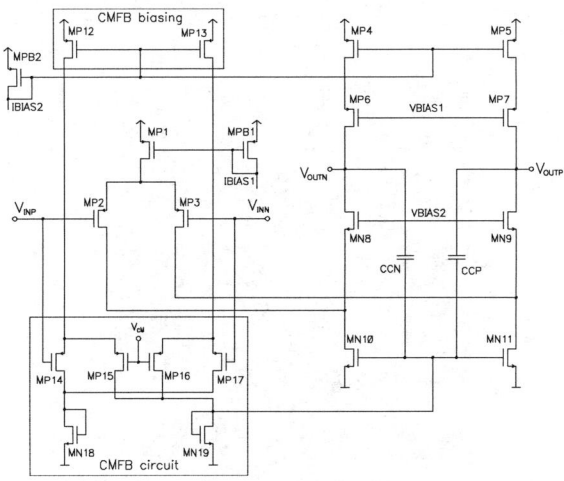

Figure 4. Implementation of the operational amplifier.

In this design, the problem was addressed by connecting the input of the continuous-time common-mode feedback (CMFB) circuit to the input of the operational amplifier, as shown in Fig. 4, instead of its output. Now, when the CM voltage at the input tends to change due to leakage currents, the change is sensed and the CM level at the output is controlled. This affects the CM level at the input through the feedback resistors, thus keeping the input CM voltage constant. The stability of the CMFB loop is ensured with capacitors CCP and CCN. These together with capacitors CLP and CLN in Fig. 3 also ensure the stability of the whole CSA.

IV. Measurements

The CSA was realised with a 0.7 μm double-metal, double-poly CMOS process. A micrograph of the chip is shown in Fig. 5, with the CSA at the top right corner. The silicon area consumed by the CSA is about 0.42 mm^2.

Transfer function was measured by connecting 1 pF capacitors to CSA inputs and feeding input signals through them. The measured transfer function with feedback resistor biasing current of 20 nA is shown in Fig. 6. The -3 dB point is at 300 Hz, indicating that the real value of the feedback resistor is about 53.1 MΩ which agrees well with the simulated value.

The output noise spectrum is shown in Fig. 7. With CSA gain being 250 GV/F in the signal band, the noise density reduced to capacitance is $2.47 \cdot 10^{-19}$ F/Hz$^{1/2}$ at 10 kHz. When integrated over the 100 Hz signal band centered at 10 kHz, a resolution of 2.47 aF (rms) is achieved. When the signal band is centered at 8 kHz, the noise density is $3.00 \cdot 10^{-19}$ F/Hz$^{1/2}$ and the resolution is 3.00 aF (rms).

Leakage current tolerance was verified by connecting the CSA input to the positive (+5 V) power supply through 0.5 GΩ resistors and verifying the CSA performance. Next, the current consumption of the CSA was measured to be 850 μA. Finally, the CSA was connected in a system with the gyroscope and the functionality was verified.

V. Conclusions

Charge sensitive amplifier design for the readout of

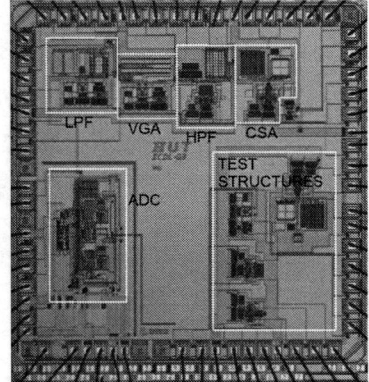

Figure 5. A micrograph of the implemented chip.

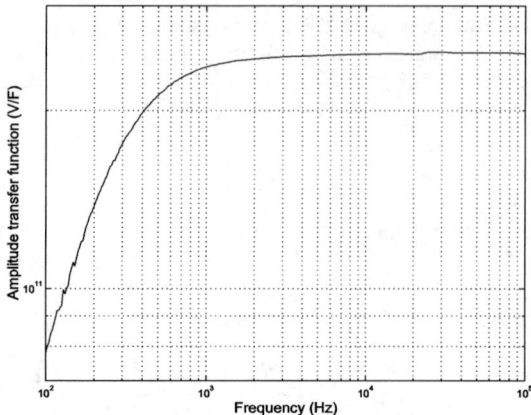

Figure 6. Measured amplitude transfer function of the CSA.

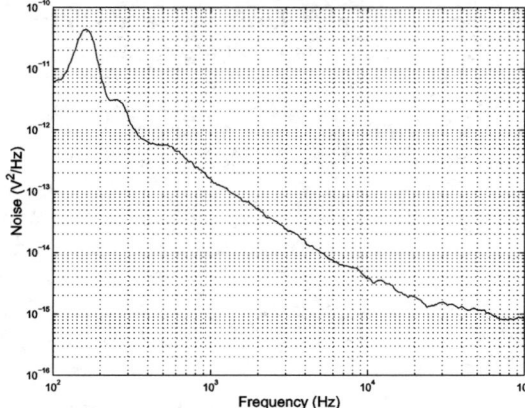

Figure 7. Measured noise at the CSA output.

micromechanical capacitive sensors was introduced. The transfer function from the change in capacitance to the output voltage was presented. The noise contributions from different components of the CSA were analysed, and ways to optimise noise performance were studied. After that, a CSA designed for the readout of a micromechanical gyroscope with a resonance frequency of 8-10 kHz and signal bandwidth of 100 Hz was presented. The CSA achieves a measured resolution of 2.47–3.00 aF (rms) integrated over the signal band and tolerates leakage currents of over 5 nA.

Acknowledgment

The authors wish to thank VTI Technologies Oy for providing the sensor elements for measurements.

References

[1] Thomas B. Gabrielson, "Mechanical-thermal noise in micromachined acoustic and vibration sensors," IEEE Transactions on Electron Devices, vol. 40, no. 5, pp. 903-909, May 1993.

[2] Zhong Yuan Chang, and Willy M.C. Sansen, "Low-noise wide-band amplifiers in bipolar and CMOS technologies," Kluwer Academic Publishers, 1991.

[3] Harold Shichman, and David A. Hodges, "Modelling and simulation of insulated-gate field-effect transistor switching circuits," IEEE Journal of Solid-State Circuits, vol. SC-3, no. 3, pp. 285-289, September 1968.

Fully Integrated Charge Pump for High Voltage Excitation of a Bulk Micromachined Gyroscope

Lasse Aaltonen, Mikko Saukoski, Kari Halonen
Electronic Circuit Design Laboratory
Helsinki University of Technology
Espoo, Finland

Abstract—This paper describes a fully integrated excitation electronics, which enables quick start-up or powerful continuous stage drive of a resonating sensor. The system of a micromechanical resonator, readout circuitry and high voltage drive electronics is connected to form a positive feedback from the capacitive readout of the sensor to the drive electrodes. The charge pump circuit converts the low voltage signal at the resonant frequency to a 20 V differential square wave. The functionality is verified by forming a positive feedback loop and measuring the start-up time of a bulk micromechanical gyroscope. Measured charge pump is implemented within chip area of 0.46 mm² and draws 0.9 mA from a 5 V supply.

I. INTRODUCTION

Micromechanical sensors with fully integrated interface electronics have enabled the vast use of low-cost micro-electro-mechanical systems (MEMS). One of the major targets for low cost sensors is the automotive industry, which requires a variety of sensors for different applications. From the automotive solution point of view, especially when gyroscopes are considered, one of the important features is the sensor start-up time, the time it takes the system to be operational after the power is switched on. Gyroscopes can be used for example for vehicle chassis control.

In the operational mode, relatively much energy is stored in bulk micromachined gyroscope's resonator. Like bulk micromechanical resonating sensors in general the tested gyroscope requires powerful drive signal to rapidly transfer the required energy to the resonator. A micromechanical resonator can be excited by using capacitive coupling. The force applied to the mechanical element is calculated as negative gradient of the interconnect capacitance energy,

$$\overline{F} = -\frac{\partial E}{\partial x}\overline{u}_x = -\frac{\partial C_{int} U^2}{2\partial x}\overline{u}_x, \qquad (1)$$

where U is the applied voltage and x is the displacement in the direction of x-axis. The size of the capacitor C_{int} can be calculated from the geometry of the sensor. The voltage U has both dc and ac component

$$U = U_{DC} + U_{AC}\sin(\omega_0 t), \qquad (2)$$

where ac component is at the resonance frequency ω_0. By combining (1) and (2) the force at the resonance frequency is given approximately as

$$\overline{F_{x\omega_0}} \approx \psi U_{DC} U_{AC} \sin(\omega_0 t)\overline{u}_x, \qquad (3)$$

where ψ is a constant dependent on sensor geometry. According to (3) to excite the resonator at resonance frequency both dc and ac voltage components are required.

II. POSITIVE FEEDBACK

The positive feedback is formed by reading the change in sensor capacitance and connecting the voltage mode signal amplified and 90 degrees phase shifted to the sensor excitation electrodes. The block diagram is presented in Fig. 1. The first block in the readout chain is a charge sensitive amplifier followed by a high pass and a low pass filter. Phase shift is implemented with a derivator. After phase shifting, the signal is converted to square wave with a comparator, which controls the charge pump. Charge pump generates differential 20 V square wave to the excitation electrodes of the sensor.

The 90 degree phase shift between the read signal and excitation voltage is desired but due to numerous non-idealities there is always some phase error present. In this case the phase shift in the chain will be $90° + \varphi_{err}$. The effect of the phase error to the oscillation amplitude and power can

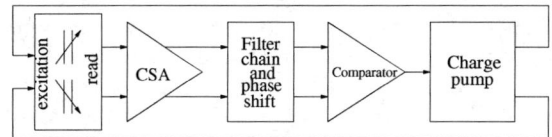

Figure 1. Block diagram of the sensor excitation system.

The project under which this work was done is funded by National Technology Agency of Finland (TEKES) and VTI Technologies Oy. The Design tools were provided by Cadence Design Systems and Mentor Graphics.

be evaluated by dividing the excitation voltage into in-phase and out-of-phase quadrature components. The quadrature component has the same phase as the resonator mass movement and will thus deliver no energy to the sensor. Due to the high Q-value of the sensor, the square wave drive signal can be considered as sinusoidal with the same frequency as the fundamental frequency of the square wave. Thus by assuming that the power is transferred at the fundamental frequency and in correct phase we can write for the drive signal

$$D\sin(\omega_0 t + \varphi_{err}) = B\sin(\omega_0 t) + C\cos(\omega_0 t), \quad (4)$$

where loop will have exactly 90 degrees phase shift if phase error φ_{err} is zero. Otherwise phase error will cause attenuation (B/D) of $\cos(\varphi_{err})$. This result indicates that small phase errors have negligible effect on resonator beam vibration amplitude. The amplitude C of the quadrature component can be neglected.

III. CHARGE PUMP

Considering the chip area one of the most effective ways of creating voltages higher than the supply are voltage doublers [1]. However, as the drive voltages must be above 20 V, the transistors, which are used as switches, often cannot tolerate full scale gate-source voltages. This would lead to very complex clock generating circuits and start-up problems if doublers were used.

To avoid aforementioned problems traditional Dickson-type charge pumps with diodes can be used. When designing a fully integrated structure the traditional architecture developed by Dickson [2] with output voltage given as

$$V_{cp} = (N+1)(V_{Supply} - V_d), \quad (5)$$

can result in many stages N if the desired output voltage V_{cp} is much higher than the supply voltage V_{Supply} reduced by a diode drop V_d. The equation excludes the effect of stray capacitances and assumes zero resistive load. In order to optimize the output voltage versus chip area, the voltage can be increased by two different pumps by using the first stage voltage for the generation of second stage clock signals. The block diagram of this kind of voltage boosting structure is shown in Fig. 2. The blocks CP1 and CP2 are basic Dickson-type pumps, where N denotes the number of diode stages in one pump. The blocks required for a second output are drawn in the picture with a dotted line. To be able to fully utilize the voltage Vmid at the output of the first stage one must boost the swing of the clock that is used to pump the voltage of the second stage. This can be easily done by using a cross coupled differential pair presented in Fig. 2. The circuit can be controlled by low voltage clock signals clk2 and xclk2 as long as their voltage level is high enough to turn the NMOS-transistors M3 and M4 on. In addition transistors M1 and M2 must be able to tolerate gate-source voltages as high as –Vmid. The medium voltage clock signals are then used as clocking signals for the second stage. If the supply voltage used is large enough the diode voltage drop can be

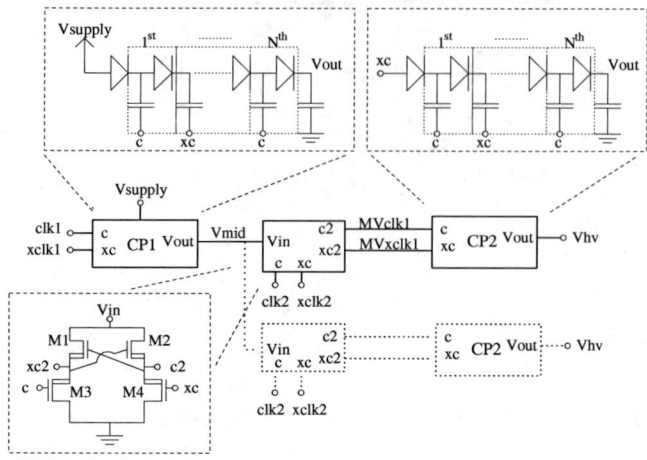

Figure 2. Block diagram of the presented two stage charge pump structure.

tolerated and Dickson charge pumps can be used in both stages of the pump. Voltage gain of the two stage pump of Fig. 2 is given as

$$V_{cp} = (N_2 + 1)(V_{mid} - V_d). \quad (6)$$

V_{mid} can be solved from equation

$$V_{mid} = N_1(V_{Supply} - V_d - xN_2 V_{mid}) + (V_{Supply} - V_d), \quad (7)$$

where N_1 and N_2 denote the number of diode stages and

$$x = I_L /(CN_2 f_1 V_{mid}) = C_p f_2 /(Cf_1), \quad (8)$$

where I_L is the load current drawn by the second stage, C the capacitance of one of the first pump stage capacitors, f_1 the first stage pumping frequency, C_p is the parasitic capacitance per capacitor at the second stage and f_2 the pumping frequency of the second stage. By substituting V_{mid} solved from (7) to (6) and assuming that $V_d \ll V_{mid}$ in (6) we get

$$V_{cp} \approx (N_1 + 1)(N_2 + 1)\left(\frac{V_{Supply} - V_d}{1 + N_1 N_2 x}\right),$$

which assumes zero load current of the second stage and that only the parasitic capacitances load the first stage.

A. Optimization

Let us define the unity area constant to be the relative chip area per one diode and one capacitor. For low voltages higher capacitor densities can be used and therefore first and second stage will have different constants y denoting the relative area. There is also the level shifter, an extra diode and a capacitor at the output of the first stage, which will have a small constant contribution to final area. The approximation of the relative chip area is thus given as

$$A(N_1, N_2) = N_1 y_1 + N_2 y_2 + y_3, \quad (9)$$

where N_1 and N_2 are the number of diode stages in the first and second pump stage and y_1 and y_2 are the constants which define the relative size of a single diode stage of a pump. If we now solve N_2 from (6), substitute the result into (9) and differentiate the resulting equation and set the derivative to be zero we get the optimal number of diode stages in the first pump. Result is given as

$$N_1 = \sqrt{\frac{y_2}{y_1} \frac{V_{cp}}{V_{Supply} - V_d - xV_{cp}}} \frac{V_{Supply} - V_d}{V_{Supply} - V_d - xV_{cp}}. \quad (10)$$

The accurate evaluation of relative chip areas consumed by the charge pumps is very difficult. Rough evaluation of the required N_1 can be obtained by setting constants y_1 and y_2 equal. This enables bigger capacitance in the first stage to compensate for second stage parasitics. For the designed structure V_{cp} was required to be at minimum 20 V, supply is 5 V, diode drop is 0.66V and constant x is 0.015. Using (10) we get the results N_1 equals 1 and N_2 2. Number of diode stages in a single stage pump implementation is 5. The difference between the number of diode stages will become more noteworthy with low supply voltages. Instead of area optimization more usually the output voltage of the first stage is limited by some other constraint such as the gate-source voltage of the level shifter PMOS-transistor or voltage limitation of the first stage capacitors. When two simultaneously used outputs are required, the second stage needs to be doubled and the constants y_2 and A must be multiplied by two for correct optimization.

The second stage of the implemented charge pump is presented in Fig. 3. This stage with clock signal level shifter is doubled in order to generate two independent high voltage square waves. Fig. 3 also shows the switches that control the high voltage output Vhv. The usage of floating switches would have resulted in difficult control signal generation because transistors cannot tolerate full scale gate-source voltages. This is avoided by the use of non-floating switches and by resetting the whole second stage when zero voltage output is required. This is done by turning clock signals clk2 and xclk2 of the level shifter in Fig. 2 off and both transistors M1 and M2 of Fig. 3 on. As the medium voltage clock signal node MVxclk1 stays low and MVclk1 high, no current will flow through M1 or M2 in continuous state. The pump structure presented in Fig. 3 also enables a high impedance output mode, when M2 is off and M1 on. This way, after the quick start-up of the resonating sensor is performed, the pump can be turned completely off and continuous mode excitation of the sensor can be performed with accurately controlled low voltage signal.

B. The Effect of Charge Pump Output Impedance

The effective output impedance of the pump will define the maximum switching speed of Vhv. Due to the large impedance of the charge pump the rise time of the output square wave will be quite long. The fall time will be short, because the voltage is reset with a switch. Fourier series will show the effect of slow rise time to the amplitude of the component at the fundamental frequency. For the periodic waveform of Fig. 4 the Fourier coefficients are calculated as

$$a_n = \frac{2}{T} V_0 \int_0^{T/2} (1 - e^{-t/\tau}) \sin(n\omega t) dt, \quad (11)$$

where V_0 is the maximum output voltage of the charge pump, T is the period and ω is the angular frequency of the square wave. At the fundamental frequency, we get

$$a_1 = \frac{2V_0}{\pi} \left[1 - \frac{2\pi^2}{(T/\tau)^2 + 4\pi^2} (1 + e^{-T/(2\tau)}) \right], \quad (12)$$

where the time constant τ is formed by the load capacitor and the effective output impedance of the charge pump. If the sine function is replaced with cosine function in (11), the resulting quadrature component has no effect on sensor energy. The attenuating effect of rise time given by equation (12) is 9.1 % when comparing fundamental component amplitudes with zero rise time and when T/τ is 10.

C. Clocking of the Charge Pump

To be able to use the charge pump structure described above the clock frequency must be about two decades higher than the frequency of the square wave. The medium voltage Vmid clock can have lower frequency than the first stage clock signals. This will considerably lower the component sizes at the first Dickson-type charge pump stage. The implemented clock generator has the second stage clock frequency divided by four. Size of the charge pump load capacitor will have effect on the required clock frequency. Accurate clocking frequency is not required as the level of high voltage signals would still have temperature and process dependency. In addition, phase noise properties are of no importance. Clock frequency must however be adequate to obtain high enough worst case voltage.

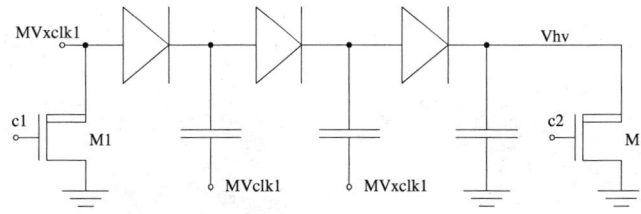

Figure 3. The second part of the charge pump.

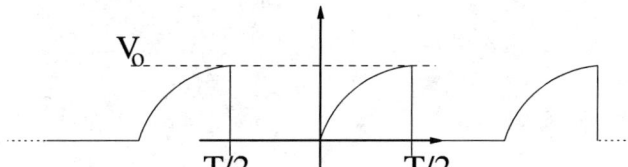

Figure 4. High voltage signal waveform.

IV. MEASUREMENTS

The excitation loop was implemented with a 0.7 μm double-metal, double-poly high voltage BiCMOS process that offers high-ohmic polysilicon resistors, good-quality analog capacitors and floating diodes. The micrograph of the circuit is shown in Fig. 5. The chip area consumed by the pump is 0.46 mm^2. The used supply voltage is 5 V and the current consumption of the charge pump is 0.9 mA in the operational mode. The resonator which was used for testing the excitation electronics was a bulk micromachined gyroscope. The SEM picture of the sensor is in Fig. 5.

The functionality of the charge pump was tested by applying input signal directly to the comparator. The pump response to a 5 kilohertz signal is shown in Fig. 6. Figure shows both outputs, which form a differential high voltage square wave. Measurement shows that the system can be used at frequencies above 10 kilohertz before the signal component at the fundamental frequency, given by (12), is considerably attenuated. Unloaded DC output was measured to be 21.7 V. The result given by (6) is 23.3 V. The difference in voltages results from process variations and various parasitic resistors and capacitors.

The output impedance of the charge pump was measured by loading the amplifier resistively and measuring the steady state output voltage. The impedance was calculated to be 500 kΩ, low enough to tolerate leakage currents and to drive the micromechanical resonator at 10 kilohertz frequency. The sensor excitation was detected by measuring the amplified CSA output during the start-up. Reset signal of the charge pump is removed after the supply voltage is connected. After this the excitation loop is closed and mechanical element starts to vibrate with increasing amplitude. Measured signal at start-up is presented in Fig. 7. Loop is closed at time instant 0 s, after which signal amplitude increases. After 230 ms, the mechanical vibration amplitude has reached the desired level, the pump is shut down and amplitude is kept constant. The start-up times with charge-pump and supply voltage limited signals can be compared by examining the force applied to the sensor. The reference voltage of the sensor is 5 V. The DC-component of the pump signal is 10 V and the fundamental frequency component of the differential square wave is $\pi/2*20$ V. Respectively, the supply limited

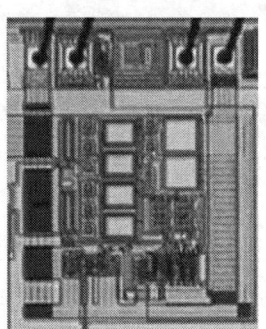
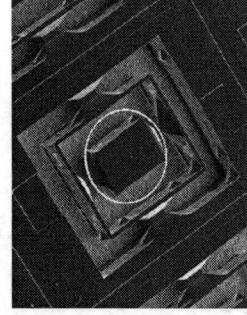

Figure 5. Left: Microphotograph of the implemented charge pump. Right: SEM picture of the gyroscope. Encapsulating wafers with electrodes are anodically bonded to both sides of the structural wafer. The excited resonator is encircled in the picture.

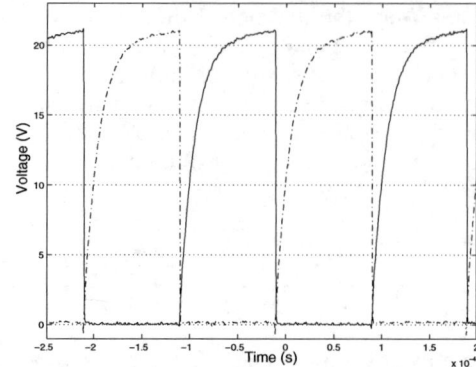

Figure 6. Measured charge-pump output waveforms at 5 kHz frequency.

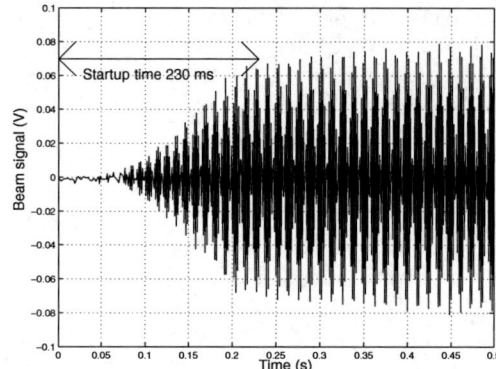

Figure 7. Measured gyroscope resonator signal during start-up. Signal is folded due to low sampling rate of the oscilloscope.

signal has DC level of 2.5 V and resonance frequency component of $\pi/2*5$ V. According to (3) the force applied by the charge-pump is eight times higher compared to the force excited by the supply limited start-up signals.

V. CONCLUSIONS

Fully integrated charge pump for the excitation of a resonating sensor was presented. The system was realized within chip area of 0.46 mm^2 and start-up of a bulk micromachined gyroscope was measured. The excitation force attained was eight times higher than that achieved by supply limited excitation. The measured results indicate that with small chip area contribution the drive of a micromechanical resonator can be boosted considerably.

ACKNOWLEDGEMENT

The authors wish to thank VTI Technologies Oy for providing the sensor elements.

REFERENCES

[1] Chi-Chang Wang and Jiin-chuan Wu. "Efficiency Improvement in Charge Pump Circuits" IEEE Journal of Solid-State Circuits, Vol. 32, No. 6, pp.852-860 June 1997.

[2] John F. Dickson. "On-Chip High-Voltage Generation in NMOS Integrated Circuits Using an Improved Voltage Multiplier Technique." IEEE Journal of Solid-State Circuits, Vol. SC-11, No. 3, pp. 374-378, June 1976.

Flexible High-Accuracy Wide-Range Gas Sensor Interface for Portable Environmental Nosing Purpose

M. Grassi, P. Malcovati
University of Pavia
Dept. of Electrical Engineering
Pavia, Italy
E-Mail: {marco.grassi, piero.malcovati}@unipv.it

A. Baschirotto
University of Lecce
Dept. of Innovation Engineering
Lecce, Italy
E-Mail: andrea.baschirotto@unile.it

Abstract — **In this paper we present a flexible high-precision wide-range front-end for resistive sensors. The output of a programmable continuous-time transresistance stage is processed by a differential switched-capacitor incremental A/D converter. A digital signal processor reconfigures the interface leading to a resolution of 0.1% over a range of 5 decades, as required by pattern recognition algorithms for sensors arrays in environmental nosing applications. Particular care has been taken in reducing the power consumption.**

I. INTRODUCTION

Today's gas sensors for environmental monitoring purpose are able to detect the concentration of many different types of gases with a precision near to 1 ppm. This result can be achieved thanks to the fact that a complete gas sensor device is actually an array of several different unitary elements, whose output signals are properly processed with suitable digital pattern recognition algorithms. In the case of this research work [1], an array of four sensors is considered. Thus, even if each of the four devices is optimized to detect a particular kind of gas, i.e. CO_2, CH_4, H_2 and SO_2, they provide important information also on the concentrations of other gas types. The sensors behave electrically as resistors whose resistance value (R_{sens}) is the combination of three components:
- the baseline R_{bl} (which mainly depends on the fabrication technique);
- the deviation from the baseline ΔR_{bl}, due to technological and aging spread and temperature;
- the resistance variation ΔR_{gas}, which depends on gas concentration, negative for most gas types. Its value can be as large as a couple of decades from the actual baseline value.

Depending on sensor type, technology spread and sensor age, the baseline usually varies from a very low value, i.e. 10kΩ, to a very high one, i.e. 10MΩ and the sensor resistance has to be measured with a 0.1% precision in order to be sure to detect every gas type presence with appropriate accuracy.

The large baseline variation together with the large measurement accuracy would require, without range compression [2], to use a front-end with a resolution so high to be unpractical. For this reason alternative read-out structures have to be preferred. For instance, an oscillator can be used [3], whose main time constant is function of the sensor impedance: this technique allows the measurement of higher resistance values but it is slower and of course affected by the parasitic capacitance of the sensor layer. Fig. 1 shows an example of logarithmic compression of the value of R_{sens}. The current $I_{sens}=V_{REF}/R_{sens}$ flowing trough the sensor is mirrored into the diode D1, while a reference current I_{REF} flows through a matched diode D2. Due to the exponential behavior of these devices the difference between the two voltage drops across the diodes is:

$$V_0 = -V_T \cdot \ln\left(R_{sens} \cdot \frac{I_{REF}}{V_{REF}}\right) \quad (1)$$

leading to a logarithmic compression of the R_{sens} value. This circuit is able to measure resistance values with a precision up to 1% over 5 decades using a single scale.

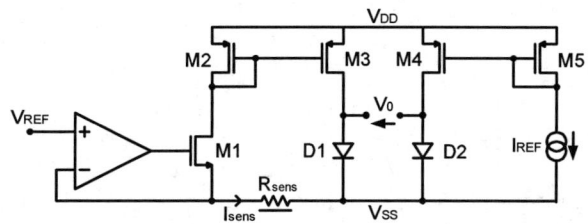

Fig. 1 – Example of logarithmic compression of R_{sens} and I_{sens} values

In this paper the design of the 4-sensor array interface is

described. The read-out structure of Fig. 2 includes a single channel for each sensor, thus allowing the four sensor resistance values to be measured separately. Each channel consists of a trans-resistance stage that transforms the current Im_i into a voltage. This voltage is then digitalized by a high-linearity incremental A/D converter.

The critical requirement of the large resolution has been satisfied by using a dedicated digital signal processing (DSP) section. This block reconfigures the trans-resistance front-end in order to match its features with the electrical parameters of the sensors. The complete four-channel interface circuit has been designed in a 0.35μm CMOS technology and consumes about 20 mW from a single 3.3V supply voltage.

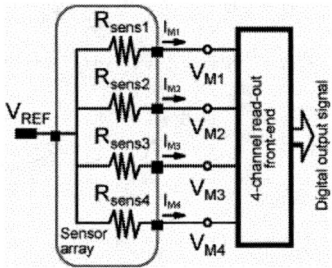

Fig. 2 – The sensor array input device of the electronic nose

II. THE HALF-LOGARITHMIC APPROACH

The input stage of each channel is the trans-resistance amplifier shown in Fig. 3. In this circuit, the sensor resistance (R_{sens}) is connected between V_{REF} and the operational amplifier (opamp) virtual ground. The output voltage V_0 is given by:

$$V_o = a \cdot V_{REF} + b \cdot V_{offset}$$
$$a = -R_f /(R_{bl} + \Delta R_{bl} + \Delta R_{gas})$$
$$b = 1 + [R_f /(R_{bl} + \Delta R_{bl} + \Delta R_{gas})] \quad (2)$$

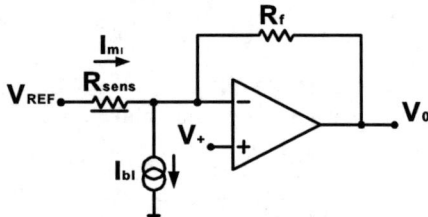

Fig. 3 – Transresistance input stage with baseline cancellation solution

The output voltage V_0 deviates from its nominal value ($V_0 = -V_{REF} \cdot R_f / R_{sens}$) due to several non-idealities: among them the most important ones are the opamp offset (V_{offset}) and sensor resistance baseline deviation (ΔR_{bl}). All these effects can be reduced by means of the controlled current I_{bl}, as shown in the schematic, which forces V_0 to deviate from the ideal behavior given by (2). The actual behavior of V_0 is then described as follows, where also the correction terms of the proposed scheme are included:

$$V_o = a \cdot V_{REF} + b \cdot V_{offset} - I_{bl} \cdot R_f \quad (3)$$

A proper value of I_{bl}, which ideally should be $I_{bl}=(V_{REF}-V_{offset})/(R_{bl}+\Delta R_{bl})$, allows us to maximize the output voltage accuracy. The critical point of this structure is related to the large variation of the nominal value of the sensor resistance, that also requires a large dynamic range for the I_{bl} control. In order to prevent the above mentioned problems, we developed the circuit shown in Fig. 4. The entire front-end includes for each channel a programmable trans-resistance amplifier, whose output is connected to a switched capacitor single-ended to differential adapter, which is the input block of a 13 bits incremental ADC, operating at a sampling frequency of 1MHz (providing a full conversion every 8 ms). All the front-end features will be controlled by a DSP.

The large spread of the sensor resistance baseline is divided in multiple half-decade sub-ranges by changing the value of the feedback resistor R_f. A further fine offset correction is possible by varying the current I_{cal} sunk by the generator from the virtual ground, i.e. in order to have exactly zero output in the absence of gas. As shown in Fig. 4, the regulated current source is realized with an 8 bits buffered resistive DAC (DAC1) and a programmable resistor R_{DAC}, that also has to be swapped. In this design resistors have been designed such that $R_{DAC}=R_f$, in order to operate with opamp gain and feedback factors of the same order of magnitude over the whole dynamic range of the sensor resistance, improving opamp design. Furthermore a good component matching ($R_{DAC}=R_f$) is guaranteed.

Besides, if a fine regulation of the voltage reference V_{REF} is also performed, it is possible to reduce independently the gain-error of each scale available in the front-end. This is possible by applying an additional buffered DAC (DAC2), also controlled by the DSP block.

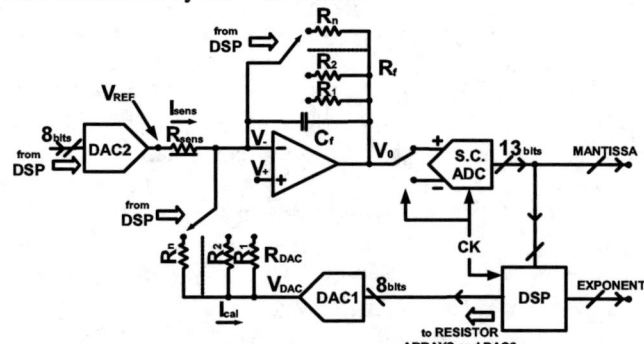

Fig. 4 - Autorange and error-correction programmable front-end

The absolute resistance measurement is performed in two steps:

- Structure calibration with respect to the actual sensor features: a series of coarse sub-measurements is performed during which the ADC is used at reduced resolution (i.e. 6 bits) with a full-scale that is increased for each

measurement by changing first $R_{DAC}=R_f$ and second V_{DAC}. The first couple {R_{DAC}, V_{DAC}} for which the ADC is not saturated is used for the fine measurement.
- Complete (13bit) measurement of the resistor value, with the {R_{DAC}, V_{DAC}} settings obtained previously.

The measurement result in digital form will be composed by the mantissa (supplied by the ADC) and by the modulus (supplied by the DSP).

III. DESIGN DETAILS

The front-end is designed using a 0.35μm CMOS technology and operates with a 3.3V supply. This supply voltage limits the possible amplitude of the voltages adopted in the circuit. As a consequence a larger accuracy in the measurement is required. The sensor is biased with (V_{REF} − V^-) = 500mV, with V^- = 2V. Both resistors R_{DAC} and R_f are stepped using half-decade ratios (i.e. 300Ω, 1kΩ, ..., 3MΩ, 10MΩ). In this way for any R_{sens} value there is always at least a choice of {R_{DAC}, V_{DAC}} that leads to an absolute voltage gain between V_0 and V_{REF} included in the interval [1/3; 3]. Notice that due to the not strictly monotonic composite action of the logarithmic step of the resistances and of the linear 8-bits DAC1, there are at least two couples {R_f, V_{DAC}} that give an acceptable voltage gain, i.e. the nearest $R_f = R_{DAC}$ to the sensor composite baseline $R_{b1}+\Delta R_{b1}$ and V_{DAC} that leads to zero ADC output, compensating also the opamp offset. The advantage of having overlapped scales is then shown in Fig. 5.

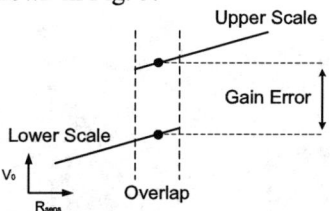

Fig. 5 – Overlapped scales help correcting system non idealities

Assuming a dynamic swing of ±400 mV and $R_{DAC}=R_f$, the whole scale of interest for the sensor resistance can be covered if DAC1 voltage references are set to 2.40V and 1.60V. For every chosen R_{DAC}, the value of the sensor resistance can vary half-decade over and half-decade below it. The entire output dynamic range of the trans-resistance stage is read with a 100mV safety overrange by the switched-capacitor input integrator of the ADC. Moreover the operational amplifier used for the sensing must settle with a precision of 150μV if a tolerance of 0.1% is needed for the resistance value. This means that the amplifier requires a gain of more than 80 dB and the incremental ADC must have a resolution of at least 13 bits, which is a reasonable performance. In the above analysis the technology spread for integrated resistances (R_f and R_{DAC}) has not been considered. This can be compensated with a dedicated circuitry, which performs a preliminary calibration phase taking into account the mismatch between the internal components and a precise (0.1%) external resistor. A reasonable trade-off between internal resistors matching parameter and area occupancy has been instead to make the built-in resistors arrays with about 1% precision.

Finally, regarding noise performance, since the flicker component is dominant, the white noise of the integrated resistors and of the sensor itself should be kept much below the equivalent desired resolution by filtering with the capacitor $C_f \gg$ 300 fF, i.e. C_f = 3pF, in order to widely respect the kT/C rule. The total power consumption of the single read-out channel is about 5mW (from a single 3.3V supply), and doubles when sensing minimum resistance values.

Finally, the switched capacitor incremental analog-to-digital converter (Fig. 6) consists of a fully-differential resettable integrator followed by a comparator whose output is connected to a single-bit digital accumulator. The entire architecture is driven by a control-logic unit that also provides useful information like the end of conversion signal (EOC), parallel and serial outputs, overflow. The sample circuit of the integrator has also the function of single-ended to differential adapter block.

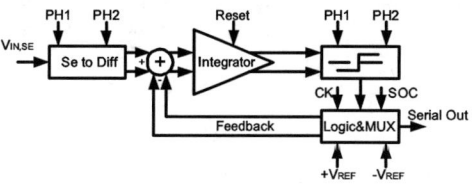

Fig. 6 – Architecture of the incremental A/D converter

IV. SIMULATION RESULTS

The opamp designed for the transimpedance block is a low-noise two-stage PMOS input pair compensated amplifier; the total bias current needed is about 250μA and the DC gain reaches 100dB with a gain-bandwidth product of 100kHz. The maximum current that the amplifier is able to sink, for low resistance values, is 3.5mA and the total average (RMS) output noise is lower than 200μV for every scale.

Using half-decade scales if a local resolution of 0.1% is needed, the output noise must be kept below 250μV with also some security room for the additional ADC possible inaccuracy if the dynamic range is V_{dyn}=800mV.

Concerning linearity, in frequency domain, the programmable transimpedance amplifier features a spurious-free dynamic range (SFDR) above 90 dB and a signal-to-noise and distortion ratio (SNDR) of 80 dB, which means 13 bits in terms of equivalent digital resolution, as appears in Fig. 7.

In Fig. 8, on the other hand, the full response of the stage for a given scale is shown. Considering the output on the vertical axis and the sensor conductance on the horizontal axis the ideal relation turns linear. The deviation from ideal behavior is smaller than 100µV for every point of each available scale.

Using Matlab® linear fitting over transistor parametric simulation results it is possible to calculate the effective non-linear component of the output of the transimpedance stage for a given scale [4] canceling offset and gain-error from the estimation, also shown in Fig. 8.

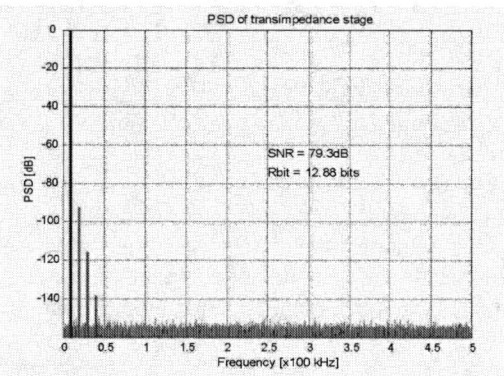

Fig. 7 – Power Spectral Density of the amplifier output applying a full-scale single tone at its input (f_{in}=10 kHz)

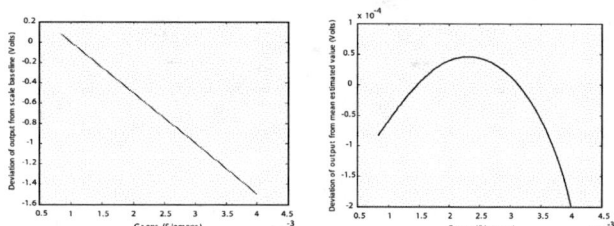

Fig. 8 – Output of the transimpedance stage as a function of the sensor conductance (on the left) and deviation from ideal behavior of the stage (on the right), both with (R_{FS}=1kΩ)

V. CONCLUSIONS

A high-precision wide-range front-end for portable environmental monitoring has been presented, whose simulated specifications are summarized in Tab. 1.

The first silicon prototype will be available for testing in the next few months. Fig. 9 shows the layout of the entire chip, which occupies an area of 6.5 mm². The most critical block in terms of matching and, as consequence, of area is the couple of programmable resistor arrays (1.8 mm² each), reported in Fig. 10.

The front-end will be reconfigured by a DSP section in order to achieve the necessary resolution of 0.1% for the needed resistance ranging in [100Ω-10MΩ].

Furthermore, exploiting the wide dynamic range of the front-end during calibration phase, it is possible to make a fast query of any sensor of an array setting up the read-out circuitry for the element of interest at the moment. Indeed, if for every instant measurement the read value or the configuration data for each sensor is written in a register, the successive measurement will be very straight-forward, avoiding multiple iterations by the DSP, which also leads to save further time and energy.

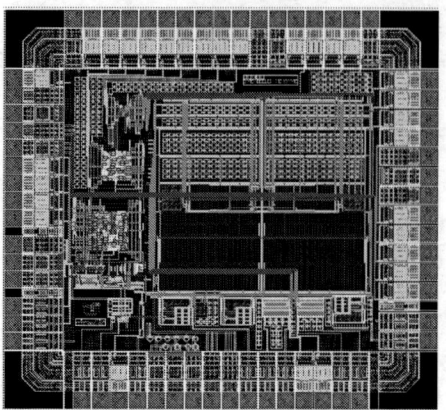

Fig. 9 – Complete layout of the first silicon prototype

Input range	100 Ω – 10 MΩ
Gas sensing accuracy	1 p.p.m. (after pattern recognition)
Front-end accuracy	0.1%
A/D conversion time	8 ms
Max. read-out rate	100 Hz
Average power consumption	< 5 mW/channel
Max. noise contribution	< 220 µV
Max. channel distortion	< -78 dB
Technology & Power Supply	AMS 0.35 µm; 3.3 V

Tab. 1 – Performance summary

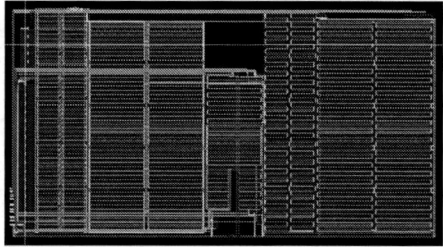

Fig. 10 – Layout of each programmable resistors array (no dummy resistors plotted)

ACKNOWLEDGMENT

The activity reported in this paper has been carried out within the PRIN project no. 2003091427 funded by the Italian Government.

REFERENCES

[1] "Portable system for ambient gas monitoring with smart A/D front-end improving sensors resolution and accuracy", PRIN project no. 2003091427, (homepage: http://ims.unipv.it/prin03).

[2] D. Barrettino et al., IEEE Sensors Journal, Volume 4, Issue 1, Feb 2004, Pages 9-16.

[3] A. Flammini et al., IEEE IMTC '03 Proceedings, Volume 1, May 2003, Pages 726-731.

[4] M. Grassi et al., (translated) "High precision and wide range sensing circuit optimized for resistive gas sensors devices", GMEE-Italy Proceedings, Sep. 2004.

ADAPTIVE SENSOR RESPONSE CORRECTION USING ANALOG FILTER COMPATIBLE WITH DIGITAL TECHNOLOGY

M. Jafaripanah, B. M. Al-Hashimi and N. M. White

ESD Group, School of Electronics and Computer Science
University of Southampton, SO17 1BJ, UK
{mj01r,bmah,nmw}@ecs.soton.ac.uk

ABSTRACT

An analog adaptive filter for response correction of a load cell sensor is presented. The filter employs only transistors and therefore it can be integrated using digital CMOS technology, which is suitable for System-on-Chip applications. To achieve adaptive compensation over a wide range of measurand, a novel $CMOS$ multiplier was developed. The analog adaptive filter has been simulated using $0.35\mu m$ 3.3V BSim3v3 CMOS foundry models and found to perform effective compensation.

1. INTRODUCTION

Some sensors, such as load cells, have an oscillatory output, which needs time to settle down. It is therefore necessary to determine the value of the measurand while the output is still in oscillation. Load cells are used in a variety of industrial weighing applications such as vending machines and checkweighing systems. Since the measurand contributes to the load cell response characteristics, a compensation filter is required to track variation in measurand whereas a simple, fixed filter is only valid at one specific load value. A number of methods, including digital adaptive techniques [1] and artificial neural network [2] have been reported for sensor compensation, which basically employ digital signal processing chips to implement the required filtering algorithms. Recently, analog adaptive techniques [3, 4] have been used to perform effective sensor response correction, with the main benefits of smaller size, lower complexity and lower power consumption compared to digital techniques.

In recent years, the quest for smaller and cheaper electronic systems has led manufactures to integrate systems onto a single chip (Systems on Chip, SoC). In the sensor research community, efforts have focused on making silicon-based sensors and circuit designers investigate techniques to develop CMOS compatible analog electronic circuits [5, 6, 7] because this is dominant processing technology for integrated circuits and systems. Despite the effectiveness of the op-amp based compensation filter reported in [4], the filter is not compatible with digital CMOS technology since it contains resistors and capacitors. This limits its applications in SoCs, and therefore, the motivation of this research is to develop and implement an analog filter for senor compensation, which is compatible with CMOS technology. It should be noted that traditionally the switched-capacitor technique has been employed extensively to integrate the analog portion of mixed-signal chips. However, switched-capacitors are not fully compatible with the digital CMOS process and as technology advances further, the drawbacks of switched-capacitor are becoming more significant [8]. The switched-capacitor technique requires high quality capacitors usually implemented using two polysilicon layers. The second layer is not required in wholly digital circuits and often it is not available, particularly in deep submicron technology used to fabricate SoCs.

This paper shows that it is possible to design and implement an analog adaptive filter capable of effectively correcting the sensor response without the use of floating capacitors. The proposed filter consists entirely of transistors and therefore it is suitable for integration using standard digital CMOS process (single polysilicon). The filter is designed using switched-current (SI) techniques, which exploit the ability of a MOS transistor to maintain its drain current, when its gate is open-circuited, through the charge stored on the parasitic gate oxide capacitance, and without the explicit need for designed capacitors [9]. SI techniques are increasingly being applied to sensor applications [5, 7]. The application of SI to dynamic sensor compensation has not been addressed in the literature, and is therefore the main aim of this paper.

2. SENSOR RESPONSE CORRECTION

The general principle of sensor response correction, in order to eliminate oscillatory sensor output, involves cascading a filter, having the reciprocal characteristic of the sensor, with it(Fig.1). Therefore, the transfer function of the whole system is *"unity"*, which means that any changes in the input transfer to the output without distortion. It has been shown that the load cell can be modelled as a 2nd order system [1]:

$$G(s) = \frac{X(s)}{F(s)} = \frac{\frac{1}{m+m_0}}{s^2 + \frac{c}{m+m_0}s + \frac{k}{m+m_0}} = \frac{A}{s^2 + \frac{\omega_0}{Q}s + \omega_0^2} \quad (1)$$

Where m is the mass being weighed, m_0 is the effective mass of the sensor, c is the damping factor, k is the spring constant, and $F(t)$ is the force function. Equation (1) shows that m affects all characteristics of the sensor such as gain factor, A, quality factor, Q, and natural frequency, ω_0. Equation (1) yields a pair of complex conjugate poles $a \pm jb$ where

$$a = -\frac{c}{2(m+m_0)} \quad \text{and} \quad b = \sqrt{\frac{k}{(m+m_0)} - \frac{c^2}{4(m+m_0)^2}} \quad (2)$$

Thus the zeros of the adaptive filter, which are the poles of the sensor, can be found. The parameter m is unknown in the

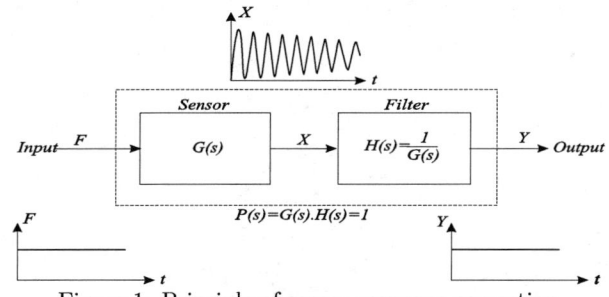

Figure 1: Principle of sensor response correction

first instance when a new measurement begins. Therefore the parameters of the adaptive filter can not be set to appropriate values in order that the filter behaves as an inverse system. Hence, an adaptive rule is required to modify the parameters of the adaptive filter according to the value of measurand, m. Usually, in classic adaptive techniques, an adaptive algorithm, such as least mean squares (LMS) method, updates the parameters of the adaptive filter to minimize a cost function. However, (1) shows that, for a load cell, the suitable filter has a pair of conjugate zeros, $z_{1,2} = a \pm jb$, where, a and b can be considered as the parameters of adaptive filter and the relationship between them and the load is expressed in (2). The adaptive compensation operation is shown in Fig.2.

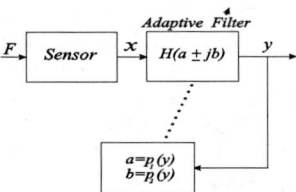

Figure 2: Adaptive sensor response correction diagram

So far the zeros of the 2nd order compensation filter have been examined. In order that the analog filter can be realized, it is necessary to add at least two poles to the filter. The values of these poles can be determined practically. These poles are selected so that the output of the filter quickly reaches its steady-state value with minimum oscillation. The transfer function of the compensation filter is:

$$H(s) = \frac{(m+m_0) \cdot s^2 + c \cdot s + k}{s^2 + 87.5s + 0.0004} \qquad (3)$$

The transfer functions of the compensation filter, (3), is a biquadratic function. The problem is how to make it adaptive and from design simplicity point of view, it is necessary to have only one filter component to track changes in m without any influence on the other characteristics of the filter. How to achieve this with $CMOS$ transistor alone circuits is discussed in the next section.

3. ADAPTIVE COMPENSATION FILTER

Amongst the various biquadratic structures, the integrator based biquad [9], shown in Fig.3, has been chosen. It consists of current mirrors, memory cells and switches, which are driven by two non-overlapping clock pulses, ϕ_1 and ϕ_2. With this biquad, it is possible to track variation in the measurand by controlling a single filter parameter. The s-domain transfer function of the biquad circuit is:

$$H(s) = \frac{[\frac{4\alpha_6+2\alpha_5-\alpha_1\alpha_3}{D}]s^2 + [\frac{4\alpha_5}{T.D}]s + [\frac{4\alpha_1\alpha_3}{T^2.D}]}{s^2 + [\frac{4\alpha_4}{T.D}]s + [\frac{4\alpha_2\alpha_3}{T^2.D}]} \qquad (4)$$

where $D = 2\alpha_4 - \alpha_2\alpha_3 + 4$, T is the clock period and each α_i is the ratio of two currents in the filter circuit. Comparing (4), with the compensation filter transfer function, (3), gives:

$$\frac{4\alpha_6 + 2\alpha_5 - \alpha_1\alpha_3}{D} = (m+m_0) \qquad (5)$$

$$\frac{4\alpha_5}{T.D} = c \qquad (6)$$

$$\frac{4\alpha_1\alpha_3}{T^2.D} = k \qquad (7)$$

$$\frac{4\alpha_4}{T.D} = 87.5 \qquad (8)$$

$$\frac{4\alpha_2\alpha_3}{T^2.D} = 0.0004 \qquad (9)$$

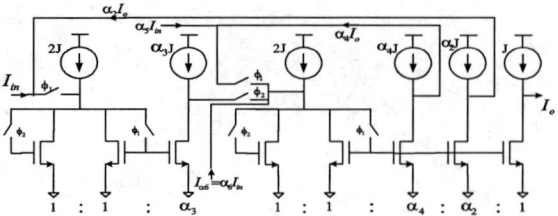

Figure 3: Integrator-based biquadratic filter [9]

Equation (5) confirms that only one filter parameter, α_6, is proportional to m. Examining Fig.3 shows that α_6 is a coefficient for the filter input current ($I_{\alpha_6} = \alpha_6 I_{in}$) and in the adaptive case it should be $I_{\alpha_6} = \alpha_6(m)I_{in}$. In our current-mode filter, the output current, I_o, displays the load cell measurand, m, therefore having α_6 proportional to m is equivalent to control the filter input current by a variable gain proportional to the filter output current. This means that a current multiplier is needed to make an adaptive compensation filter. This is clarified schematically in Fig.4.

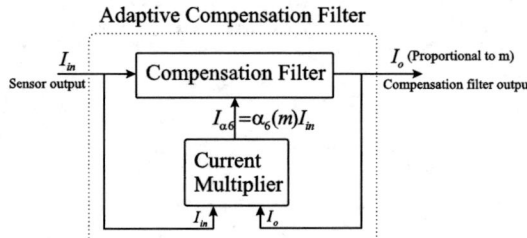

Figure 4: Adaptive compensation filter block diagram

The design procedure involves determining the parameters of a fixed filter, α_i, and then using a current multiplier block (Fig.4) to make it adaptive. From experimental data for a particular load cell [2] the damping factor c, spring constant k, and the effective mass of the load cell m_0, are 3.5, 2700 Pa, and 0.1 kg, respectively. In addition, m is considered to be 1 kg, which is an arbitrary choice. The filter parameters, α_i, are implemented by $[\frac{W}{L}]$ of the transistors in the current mirrors. The clock period, T, affects the spread of the transistor sizes for the filter. With $T = 10^{-2}s$ and using (5) to (9), the parameters of the compensation filter can be calculated as shown in table 1. Choosing $T = 10^{-2}s$ provides parameter spread of 25:1, while for $T = 10^{-5}s$ it is 3000000:1, which is clearly impractical. It is worth noting that the load cell output is oscillatory with low frequency (less than $30Hz$), therefore $T = 10^{-2}s$ is a reasonable choice.

Table 1: Compensation filter parameters, $T = 10^{-2}s$

α_1	α_2	α_3	α_4	α_5	α_6
1	1	0.4	1.4	0.0519	1.556

To find how α_6 is related to m, it is calculated for different values of m from $0.1kg$ to $1kg$ by using (5). Table 2 shows α_6 for different values of m. It is possible to express the values in table 2 as a linear relationship between α_6 and m:

$$\alpha_6 = 1.4816m + 0.074 \qquad (10)$$

To ensure the correct operation of the transistors in the filter circuit, it is assumed that a load cell measurand of $m = 1kg$ corresponds to $10\mu A$ output current. Therefore, with reference to Figs.4 and using (10), to have an adaptive compensation filter, a current multiplier with the following relationship between its input and output is required.

$$I_{\alpha_6} = \alpha_6 \cdot I_{in} = (0.14816 I_o + 0.074) \cdot I_{in} \qquad (11)$$

Table 2: Filter parameter α_6 for different values of m

$m[kg]$	0.1	0.3	0.5	0.7	0.9	1
α_6	0.222	0.519	0.815	1.111	1.407	1.556

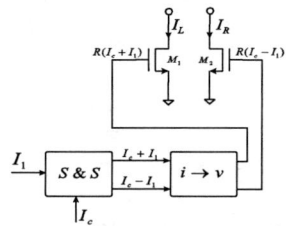

Figure 5: Current gain cell [10]

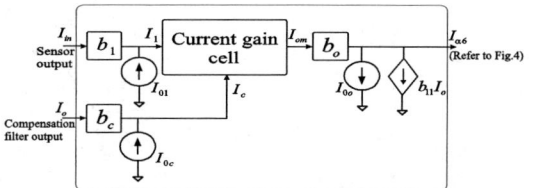

Figure 6: Proposed multiplier for adaptive compensation

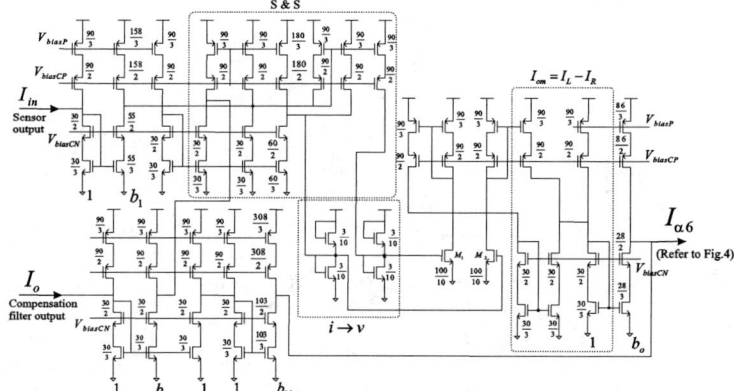

Figure 7: Transistor-level realisation of Fig.6

where I_{in} is input to the filter (output of the sensor), I_o is output of the filter and $I_{\alpha 6}$ is the current required for the second integrator in the filter (Fig.3). The implementation of (11) with $CMOS$ transistors is discussed next.

3.1. Proposed Current Multiplier for Adaptive Filter

One approach to achieve multiplication of two signals x and y is accomplished by evaluating their quadratic terms: $(x+y)^2 - (x-y)^2 = 4xy$. With this base, in [10] a $CMOS$ current gain cell is proposed (Fig.5). It consists of a summer and subtractor (S & S), a linear current to voltage convertor, and two matched MOS transistors M_1 and M_2, which are assumed to perform voltage-to-current conversion with a squaring characteristic. I_1 is the input current to be amplified and I_c is a current for the gain control. These two currents are applied to the input nodes of S & S. After $i \longrightarrow v$ conversion, both voltages, $R(I_c + I_1)$ and $R(I_c - I_1)$ are applied to M_1 and M_2, respectively. Applying square-law characteristic to M_1 and M_2 yields:

$$I_{om} = \frac{\beta}{2}[R(I_c+I_1)-V_t]^2 - \frac{\beta}{2}[R(I_c-I_1)-V_t]^2 = 2\beta R(I_c R - V_T)I_1 \quad (12)$$

Where I_{om} is the output of the current gain cell, V_t is threshold voltage and β is a process dependent constant.

To evaluate this current gain cell, transistor level simulation was performed using MOS models including all high-order effects and realistic parameters of $0.35\mu m$ CMOS process. The following linear relationship of the cell input-output was obtained for $8 < I_1 < 55\mu A$ and $0.2 < I_c < 8\mu A$:

$$I_{om} = 0.087(I_1 + 33)(I_c + 0.263) \quad (13)$$

Out of the above ranges, there is a nonlinear input-output relationship because some of the transistors in the S & S are leaving their saturation region. Whereas, for the adaptive compensation filter, a current multiplier is needed with the input-output relationship of (11) and the input current ranges of $0 < I_{in} < 20\mu A$ and $0 < I_o < 10\mu A$, which are correspond to $0 < m < 1kg$. Since the load cell output (I_{in}) is oscillatory, with the steady-state value of $10\mu A$, its peak could be as much as $20\mu A$. To achieve these features, the block diagram depicted in Fig.6 is proposed, which contains current mirrors (b_1, b_c, b_o and b_{11}) and constant current sources (I_{01}, I_{0c} and I_{0o}). The aim of b_1, I_{01} and I_{0c} are to bring the range of the multiplier input currents to the operating range of the current gain cell. The following equations can be obtained from Fig.6:

$$I_1 = b_1 I_{in} + I_{01} \quad (14)$$

$$I_c = b_c I_o + I_{0c} \quad (15)$$

$$I_{\alpha 6} = b_o I_{om} - I_{0o} - b_{11} I_o \quad (16)$$

In order to find the appropriate values for b_c, b_0, b_{11} and I_{0o}, the combination of (14), (15), (16) and (13) gives a relationship for $I_{\alpha 6}$, which should be made equivalent to (11).

The $CMOS$ realisation of the block diagram of Fig.6 is shown in Fig.7. The size of squaring characteristic transistors, M_1 and M_2 are equal and it is $\frac{W}{L} = \frac{100\mu m}{10\mu m}$ and $\frac{W}{L}$ of transistors in $i \longrightarrow v$ convertors are $\frac{3\mu m}{10\mu m}$. The other part of the circuit, including S & S, are composed of current mirrors. The structure of the current mirrors and the size of their transistors are designed such that to be compatible with the the compensation filter, which is explained in the next section.

3.2. CMOS Adaptive Compensation Filter

Current mirrors and memory cells in the filter affect the performance of switched-current circuits and numerous designs have been proposed to have improved cells [9]. In this work, to improve transmission errors, due to the finite input conductance and nonzero output conductance of transistors, high compliance cascode memory cell and current mirror have been used. The bias currents for the memory cells and NMOS mirrors are chosen to be $J = 100\mu A$, which are provided by PMOS mirrors of the same type and bias voltages (V_{biasP}, V_{biasCP} and V_{biasCN}) can be produced by a separate bias generation circuit. Different bias currents in the all current mirrors can be altered easily by scaling all transistor widths. In addition, the circuit needs to be preceded by a sample-and-hold with multiple scaled output currents. All the filter switches are implemented by $NMOS$ transistors. The complete adaptive compensation filter is shown in Fig.8, in which the box marked " X " denotes the proposed multiplier circuit of Fig.7 needed to achieve adaptive operation. As it can been seen, the adaptive compensation filter consists of entirely of transistors without using capacitors and resistors, which retains the important advantage of being compatible with digital $CMOS$ process. The combination of relatively large size memory transistor and minimum geometry switch transistor is chosen primarily to keep the effect of charge injection errors at a reasonable level. However, the use of a reasonably large memory transistor also has other performance benefits, such as lower output conductance, better matching and improved current mirror resolution. Hence, the aspect ratio of the memory transistors is considered to be $\frac{W}{L} = \frac{30\mu m}{3\mu m}$ and for the $NMOS$ switch $\frac{W}{L} = \frac{2\mu m}{0.35\mu m}$.

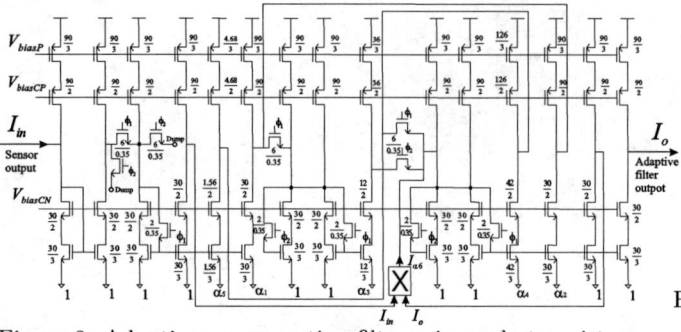

Figure 8: Adaptive compensation filter using only transistors

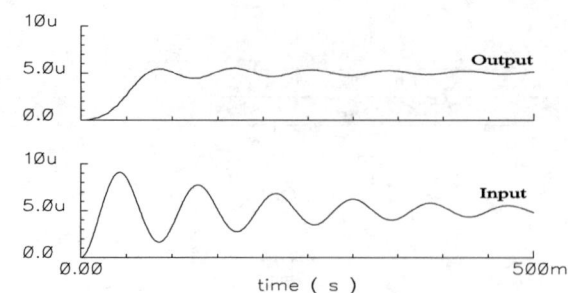

Figure 11: Input and output of non-adaptive filter (m=0.5kg)

4. TEST AND RESULTS

To test the compensation filter, the output signal of the load cell is needed. For CMOS transistor level simulation with Cadence, a VerilogA behavioral modelling facility has been used, which can produce load cell oscillatory current. A step excitation current is applied to the load cell model whose output has been applied to the compensation filter and according to the amplitude of excitation current, the value of the $(m+m_0)$ has been changed in the VerilogA model.

The circuit in Fig.8, including the multiplier circuit of Fig.7, was simulated using Cadence with $0.35 \mu m$ 3.3V BSim3v3 CMOS foundry models. Fig.9 shows the load cell output and the compensation filter output for $m=0.1 kg$, which corresponds to $1 \mu A$ excitation current. To illustrate the capability of the filter in tracking changes in m, Fig.10 shows the results for $m=1kg$ corresponding to $5\mu A$ excitation current. Clearly these results show that fully $CMOS$ adaptive compensation filter can be used to correct the oscillatory output of the load cell sensor. To indicate the effectiveness of using an adaptive filter, a fixed filter was also used for compensation. When the excitation current is $5\mu A$, a fixed filter suitable for $m=0.1 kg$ have been used and the input and output of the filter are depicted in Fig.11, which shows that the fixed filter is unable to perform the sensor response correction.

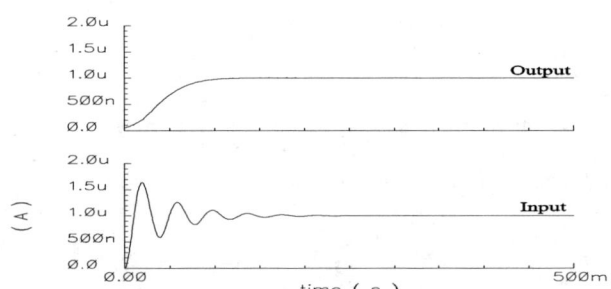

Figure 9: Input and output of the adaptive filter for m=0.1kg

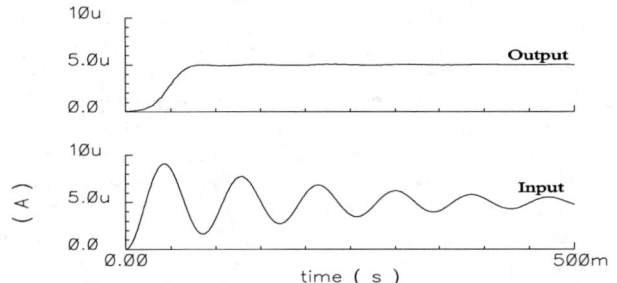

Figure 10: Input and output of the adaptive filter for m=0.5kg

5. CONCLUDING REMARKS

This paper has shown that it is possible to perform effective response compensation of dynamic sensors using switched-current techniques. The proposed analog adaptive filter employs only transistors without using passive elements and therefore compatible with digital $CMOS$ process, which makes it suitable for SoC applications. Adaptive response correction is achieved by developing a new $CMOS$ current multiplier circuit. Transistor-level simulations of the adaptive compensation filter, with realistic Spice transistor models, confirm the effectiveness of the proposed technique.

6. REFERENCES

[1] W J Shi N M White and J E Brignell. 'Adaptive filters in load cell response correction'. *Sensors and Actuators A*, A 37-38:280–5, 93.

[2] A A Yasin and N M White. 'The application of artificial neural network to intelligent weighing systems'. *IEE proc.- Science, Measur. and Tech.*, 146:265–9, Nov. 99.

[3] M. Jafaripanah B. M. Al-Hashimi and N. M. White. 'Load cell response correction using analog adaptive techniques'. pages IV752–5, ISCAS, 2003.

[4] M. Jafaripanah B. M. Al-Hashimi and N. M. White. 'Application of analog adaptive filters for dynamic sensor compensation'. *To be appear in IEEE Trans. on Instru. and Measur.*, 54(1), Feb. 2005.

[5] A. Graupner J. Schreiter S. Getzlaff and R. Schuffny. 'CMOS image sensor with mixed signal processor array'. *IEEE J. Solid-State Circuits.*, 38(6):948–57, June 2003.

[6] M. Sergio N. Manaresi F. Campi R. Canegallo M. Tartagni R. Guerrieri. 'A dynamically reconfigurable monolithic CMOS pressure sensor for smart fabric'. *IEEE J. Solid-State Circuits*, 38(6):966–75, June 2003.

[7] C. Rubio S. Bota J. G. Macias J. Samitier. 'Modelling, design and test of a monolothic integrated magnetic sensor in a digital CMOS technology using a switched current interface system'. *Analog Integrated Circuits and Signal Processing*, 29:115–26, 2001.

[8] J. B. Hughes A. Worapishet and C. Toumazou. 'Switched-capacitors versus switched-currents'. pages 409–412. ISCAS, 2000.

[9] C. Toumazou J. B. Hughes and N. C. Battersby, editors. *'Switched-Current: An Analogue Technique for Digital Technology'*. IEE Peter Peregrinus Ltd., London,UK, 93.

[10] Z. Wang. 'Two CMOS large current-gain cells with linearly variable gain and constant bandwidth'. *IEEE Trans. on circ. and syst. I: Fundamental theories and Applications*, 39(12):1021–4, Dec. 92.

A Modular RC-Active Network for Vibration Damping in Piezo-Electro-Mechanical Beams

Massimo Panella
INFO-COM Department
University of Rome "La Sapienza"
Via Eudossiana, 18
00184 Rome, Italy
Email: panella@infocom.uniroma1.it

Maurizio Paschero
INFO-COM Department
University of Rome "La Sapienza"
Via Eudossiana, 18
00184 Rome, Italy
Email: paschero@infocom.uniroma1.it

Fabio Massimo Frattale Mascioli
INFO-COM Department
University of Rome "La Sapienza"
Via Eudossiana, 18
00184 Rome, Italy
Email: mascioli@infocom.uniroma1.it

Abstract—A modular RC-active circuit is proposed in this paper to emulate the behavior of a fourth order transmission line, which is able to perform multimodal control of vibrations in piezo-electro-mechanical systems. The basic idea of this work is to consider the line as the cascade of basic sections and to design a RC-active circuit whose voltage signals emulate actual voltages and currents of the section. The principal advantages of this approach, with respect to the straightforward synthesis of each inductor, are the saving of one amplifier for each section, the opportunity to connect the piezo-electric capacitance to high impedance nodes and the possibility to obtain a scheme very suitable for integration purposes.

I. INTRODUCTION

Vibrations often represent a serious drawback in several day life activities (i.e. acoustic pollution, reduced safety and durability of materials and buildings, etc.). In order to solve these problems, damping of vibrations on mechanical structures become an essential topic to study. This problem can be efficiently solved using several competencies pertaining to different engineering fields. A classic example is the synergy between structural engineering and circuit theory in the design of electro-mechanical structures, which represent an up-to-date and effective solution to reduce mechanical vibrations by using piezoelectric transducers. Such elements are bonded to, or embedded in a vibrating structure to produce a voltage when strained or conversely to strain when exposed to a voltage.

An emerging and promising approach is based on the use of piezo-electro-mechanical (PEM) systems, where a sufficiently large number of uniformly distributed piezoelectric transducers is bonded to the mechanical structure. In this case, an electric network is used to interconnect transducers among them instead of using an independent resonant shunt, i.e. grounded inductors for each element, as previously proposed in the literature [1]. In this way, it is possible to design a network having a larger number of resonance frequencies. In order to maximize the energy exchange between the mechanical structure and the electrical network, those frequencies should match with the natural vibration modes of the hosting structure [2]. Once the energy is catched in the network, it can be thermically dissipated by means of properly designed resistors.

Historically, several circuit topologies were proposed in order to interconnect the array of piezo-electric actuators. In [2] the authors proposed to connect the actuators by means of floating inductors realizing the classical scheme of a second order transmission line. This strategy requires lower inductance values with respect to the ones needed in the classical approaches. However, those inductance values are still too big to be realized using passive technology. Moreover, the second order circuit can be optimally tuned on only one vibration mode of the mechanical structure.

A more complex topology, which is able to overcome some of the previous drawbacks, was proposed in [3]. We will refer to this circuit, shown in Fig. 1, as fourth order line. This circuit is composed by positive and negative floating inductors, while capacitors represent the loads related to the piezo-electric actuators. It is able to perform an optimal multimodal tuning of every vibration modes of the structure. Anyway, such an approach still requires the RC-active synthesis of floating inductors. This is the major problem in designing the electric circuit of a PEM structure, since the RC-active implementation of floating inductors is much more critical with respect to grounded inductors [4], [5].

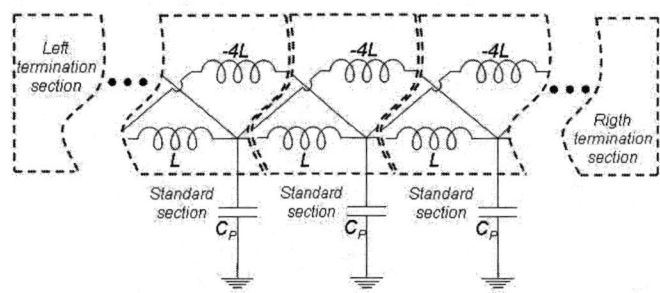

Fig. 1. Fourth order line as the cascade of sections.

In this paper, we propose a new strategy to synthesize the fourth order transmission line, alternatively to the 'straightforward' RC-active synthesis of each inductor, which produces a network denoted in the following as SF network. The proposed strategy is based on the idea of considering the whole network as the cascade of basic sections, as shown in Fig. 1. In Section II, we propose the synthesis of a modular RC-active circuit emulating the behavior of the fourth order line. The synthesis is based on the normalization of the currents flowing through the section with respect to a properly chosen admittance. Using this approach, we do not need to synthesize each inductor. Moreover, the complexity of the network is reduced compared to the SF synthesis, since we obtain a 20% saving in the number operational amplifiers (OPA) needed for the implementation of each basic section. The validity of the proposed approach will be proved in Section III, where we will test the behavior of the designed circuit when it is connected on a simply supported PEM beam. We will show that the proposed approach allows a more accurate emulation of the fourth order line.

II. THE TENSION NORMALIZED FOURTH ORDER LINE

In the following, we will derive a RC-active network able to perform the control of voltages on an array of piezo-electric capacitances, thus emulating the fourth order inductors's network shown in Fig. 1. First of all, we consider the inductors's network as a cascade of basic cells, each of them driving one piezo-electric capacitance. Referring to Fig. 2, the i-th section is externally connected through pins characterized by the voltages of the driven capacitors (i.e. V_i, V_{i-1} and V_{i+1}), by the currents flowing through the branches of the section (i.e. I_{i-1}^C, I_i^B, I_{i+1}^C, I_{i+1}^B) and by the current driving the i-th capacitor (i.e. I_i).

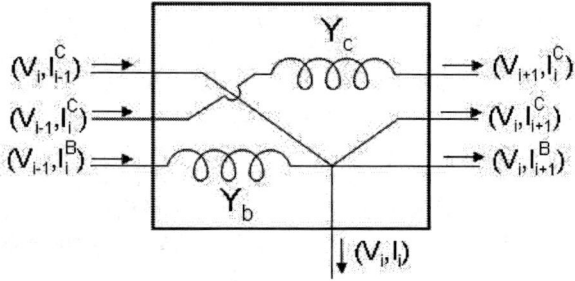

Fig. 2. Basic section of the fourth order line.

Using this notation, the behavior of the basic section can be characterized by the following equations:

$$I_i = \left(I_{i-1}^C + I_i^B - I_{i+1}^C - I_{i+1}^B\right), \quad (1)$$
$$I_i^B = (V_{i-1} - V_i) Y_B, \quad (2)$$
$$I_i^C = (V_{i-1} - V_{i+1}) Y_C, \quad (3)$$

where

$$Y_B = \frac{1}{sL}, \quad Y_C = -\frac{1}{4sL}.$$

If we perform a tension normalization, dividing by the admittance Y_B the branch currents I_j^K, $K = \{B, C\}$, $j = \{i-1, i, i+1\}$, we can introduce the virtual normalized voltages related to the actual currents; they can be defined as $\overline{V}_j^K = I_j^K / Y_B$. Furthermore, scaling the actual voltages V_j by the β factor, we can rewrite equations (1), (2) and (3) in the following form:

$$I_i = \left(\overline{V}_{i-1}^C + \overline{V}_i^B - \overline{V}_{i+1}^C - \overline{V}_{i+1}^B\right) T_A, \quad (4)$$
$$\overline{V}_i^B = (\beta V_{i-1} - \beta V_i) T_B, \quad (5)$$
$$\overline{V}_i^C = (\beta V_{i-1} - \beta V_{i+1}) T_C, \quad (6)$$

where

$$T_A = Y_B = \frac{1}{sL},$$
$$T_B = \frac{Y_B}{\beta Y_B} = \frac{1}{\beta},$$
$$T_C = \frac{Y_C}{\beta Y_B} = -\frac{1}{4\beta}.$$

We will prove in the following that a value of the β factor different by the unity will allow us to design a more robust and simpler circuit.

Looking at equations (4), (5) and (6), we note that the transfer function T_A can be associated with a block having four virtual input voltages and one output real current, i.e. I_i, driving the piezo-electric capacitance. Anyway, we will show how this block can also be used to obtain the scaled real voltage βV_i. Conversely, transfer functions T_B and T_C can be associated with blocks having two real input voltages and one virtual output voltage.

The previous blocks must be interconnected among them, in such a way the resultant scheme will reproduce the behavior of equations (4), (5) and (6). It can be easily realized that the scheme shown in Fig. 3 is able to produce the desiderate connections. We remark that some connections in this scheme are duplicated, in order to simplify the connection between adjacent sections of the line. Moreover, the twofold function of block T_A is evidenced by embedding the piezo-electric capacitance into the scheme of Fig. 3.

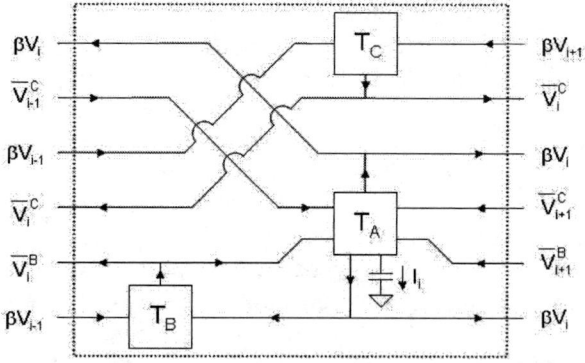

Fig. 3. Blocks diagram implementing the basic section of the line.

The design of the three blocks related to equations (4), (5) and (6) is quite simple, even if the values of the components used in the corresponding circuits should be carefully chosen to obtain a suitable behavior (i.e. dynamic, bandwidth, etc.) of the whole system emulating the ideal fourth order line. The implementation of blocks T_B and T_C is trivial and can be obtained by means of a standard differential stage using one OPA per each block. The implementation of block T_A is more complex and it can be obtained using the scheme shown in Fig. 4. In this scheme, the first stage realizes a differential integrator, which is controlled by the linear combination of the four virtual input voltages. The R_c resistor is used to reduce the DC gain of the stage, with respect to the infinite DC gain of T_A due to pole located at $s = 0$ in (4). The second stage realizes a VCCS and hence the output current I_i will be

$$I_i = \frac{1}{sR_1R_2C}\left(\overline{V}_{i-1}^C + \overline{V}_i^B - \overline{V}_{i+1}^C - \overline{V}_{i+1}^B\right). \quad (7)$$

Thus, comparing (4) to (7), the simulated inductance is equal to $L = R_1R_2C$.

The current I_i flows through the capacitance C_P and produces the actual voltage V_i of the piezo-electric actuator. The key point is that the VCCS stage can also be used as buffer of V_i since, if R_2 is suitably chosen taking account of the characteristic of the load C_P, then V_i is doubled at the output of the second OPA in Fig. 4, i.e. $\beta = 2$. A great advantage of this design is in the opportunity to connect the piezo-electric capacitance on high impedance nodes, in fact those nodes are the only ones through which the network is externally connected to the PEM beam. This approach to the synthesis of the fourth order line will be denoted in the following as 'tension normalized' (TN) network. It is interesting to

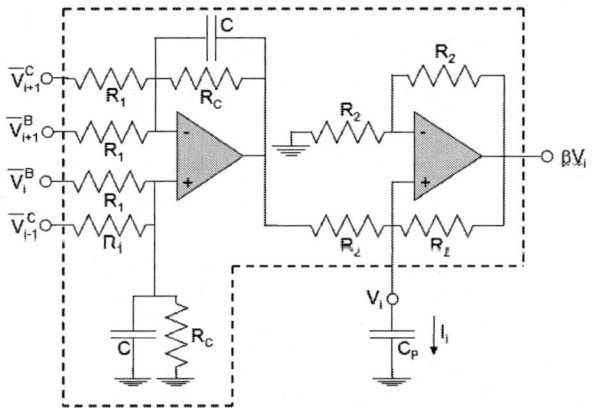

Fig. 4. Implementation of the transfer function T_A.

estimate how much OPA's can be saved in the design of a basic section of the line, using the proposed TN strategy instead of the SF synthesis of each inductor. The SF synthesis of a positive floating inductor require three OPA's [6], while a negative floating inductor can be implemented by using two OPA's [4]. We do not consider the design of the termination sections, which depends on the constraint conditions at the edges of the hosting beam. It is reasonable to ignore the contribution of those sections to the network complexity; in fact, it can be easily realized that the contribution of the termination sections to the complexity of the circuit decreases as the number of piezo-electric actuators increases. Performing some simple computations, we can note that the TN strategy requires four OPA's per each section shown in Fig. 2, while the SF synthesis requires five OPA's per section. Thus, the proposed approach allow us to save one OPA for each piezo-electric actuator and thus a 20% relative saving.

III. ILLUSTRATIVE TESTS

The results illustrated in this Section pertain to a PEM aluminum beam with a rectangular cross-section of 1.95×0.19 cm^2. Five commercial piezo-electric actuators, having rectangular cross-section of 1.78×0.027 cm^2, are bonded uniformly distanced to the top of the beam. Each actuator, whose length is l_p, is centered around a module of the beam of length Δ; a further margin of $\Delta/2$ is left before the first and after the last module, i.e. the overall length of the beam is $l = 6\Delta$. The parameters of the electro-mechanical model associated with the PEM beam are summarized in Table I, where [7]: λ is the mass density per unit length; K_M is the bending stiffness of the beam; C_P, K_{mm}, and K_{me} are the parameters defining the piezo-electric actuator.

TABLE I
PARAMETERS OF THE ELECTRO-MECHANICAL MODEL OF THE BEAM

l	$2.736 \cdot 10^{-1}$ m	l_p	$3.560 \cdot 10^{-2}$ m
Δ	$4.560 \cdot 10^{-2}$ m	C_P	$7.431 \cdot 10^{-8}$ F
λ	$1.641 \cdot 10^{-1}$ Kg/m	K_{mm}	$7.554 \cdot 10^{1}$ mN
K_M	$7.802 \cdot 10^{-1}$ m^{2}N	K_{me}	$6.710 \cdot 10^{-4}$ mN/V

The PEM beam is simply supported at both the edges. This constraint, from a mechanical point of view, yields the conditions of zero vertical displacement and zero bending moment at both the edges. From an electrical point of view, suitable termination sections have to be designed in order to reproduce equivalent conditions in the fourth order line. The simply supported condition considered in the paper can be obtained using only one grounded negative inductor, which means only one OPA using either the SF or the TN approach. Anyway, it is possible to design suitable electrical terminations emulating any mechanical constraint.

In order to realize the optimal dissipation of the vibration's energy, we need to shunt a resistor R to each negative floating inductor of the ideal fourth order line. Following the optimization criterion used in [7], we can estimate the optimal inductance value to be $L = 1.9$ H and the optimal resistance to be $R = 6.381$ KΩ. In order to emphasize the multimodal damped behavior of the whole electromechanical structure, an external concentrated force $F(t)$ is applied between the third and the fourth actuator and the velocity $v(t)$ at the same point is measured. The frequency response is obtained considering the magnitude of the mechanical mobility $A(\omega) = \left|\frac{v(\omega)}{F(\omega)}\right|$, which has been evaluated using a sinusoidal $F(t)$ with a frequency sweep of 60 seconds from 10 Hz to 10 KHz. We also consider the time evolution of the PEM beam, by considering the deflection velocity $v(t)$ at the same point when an impulsive $F(t)$ (about 1 ms wide) is applied.

In Fig. 5 we show the frequency and time behavior of the freely vibrating beam (gray line), when every piezo-electric actuator is connected to ground. We also plot in this figure the behaviors of the PEM beam when it is connected to the ideal fourth order line (black line). It can be noted the presence of five resonant frequencies; the first one, at 57.8 Hz, is associated with the dominant mechanical mode of the beam. As expected, the fourth order line is able to realize an optimal multimodal damping, since every resonance peak of the mobility is attenuated and the vibrations are damped in about 500 ms.

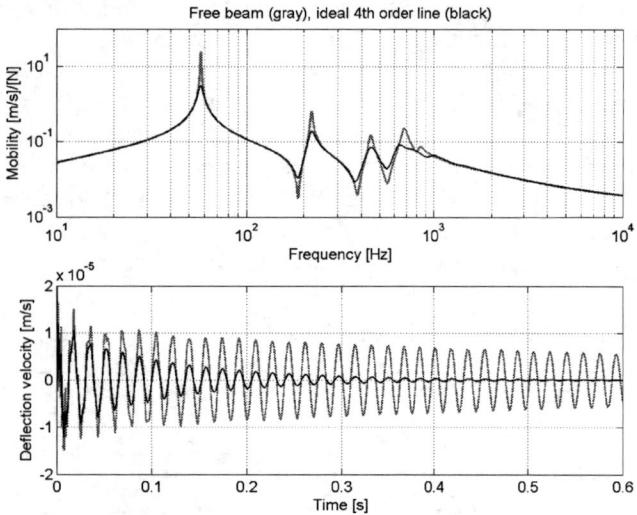

Fig. 5. Frequency and time behavior of the PEM beam: freely vibrating (gray); controlled by the ideal fourth order line (black).

In Fig. 6(a) we make a comparison in the frequency domain between the ideal fourth order line and the RC-active network obtained by the SF synthesis of each inductor. In this figure we have plotted the relative error of the magnitude of the mechanical mobility, defined as $\left| \frac{A_{ID}(\omega) - A_{SF}(\omega)}{A_{ID}(\omega)} \right|$, where subscript 'ID' refers to the ideal line and subscript 'SF' to the SF network. Similarly, the comparison between the ideal fourth order line and the proposed TN network is shown in Fig. 6(b). We would like to note that the proposed TN network is more accurate, especially in correspondence of the first mode, where the relative error of the SF network is two orders of magnitude larger than the TN one.

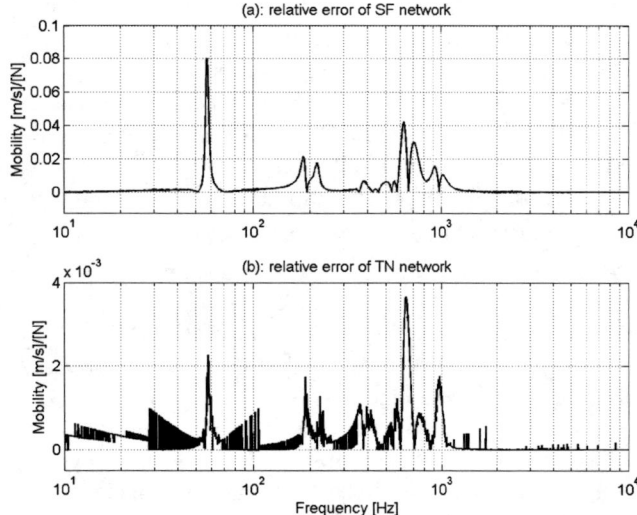

Fig. 6. Relative error of mechanical mobility: (a) SF network; (b) TN network.

The better performance of the TN network is also proved by the error time evolution defined as the absolute error between the velocity $v(t)$ of the ideal fourth order line and the one obtained using the SF or TN network; such errors are plotted in Fig. 7(a) and Fig. 7(b), respectively. Even if both SF and TN networks are quite accurate, achieving the damping of vibrations still in 500 ms, the error of the TN network is clearly smaller compared to the SF network.

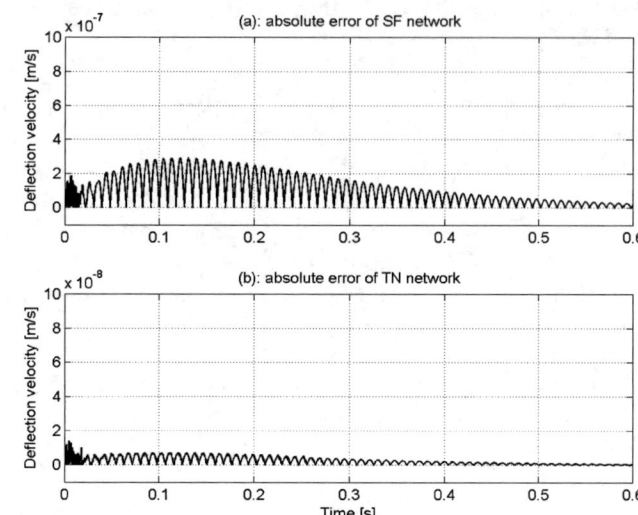

Fig. 7. Absolute error of the time response: (a) SF network; (b) TN network.

IV. CONCLUSION

A modular RC-active circuit has been proposed for emulating the behavior of a multimodal RL network for vibration damping of PEM beams. The circuit is designed on the basis of a tension normalization procedure. With respect to previous approaches, the proposed circuit performs better, in particular considering that it accurately approximates an ideal RL network in the whole frequencies range of interest, in particular at the dominant resonant frequency where the largest amount of energy should be dissipated. Moreover, the proposed circuit avoids the use of floating inductors, and saves 20% of operational amplifiers needed for the implementation. This design technique allows the connection of the piezoelectric actuators at high impedance nodes, in such a way the polarization of the circuit has a greater robustness with respect to the electrical loads induced by the mechanical deformation of the structure.

REFERENCES

[1] S. O. R. Moheimani, "A survey of recent innovations in vibration damping and control using shunted piezoelectric transducers," *IEEE Trans. on Control Systems Technology*, vol. 11, pp. 482–494, 2003.
[2] F. dell'Isola and S. Vidoli, "Damping of bending waves in truss beams by electrical transmission lines with PZT actuators," *Archive of Applied Mechanics*, vol. 68, pp. 626–636, 1998.
[3] S. Alessandroni, F. dell'Isola, and M. Porfiri, "A revival of electric analogs for vibrating mechanical systems aimed to their efficient control by PZT actuators," *Int. J. of Solids and Structures*, vol. 39, pp. 5295–5324, 2002.
[4] L. T. Bruton, *RC-active Circuits: Theory and Design*. Englewood Cliffs, NJ: Prentice-Hall, 1980.
[5] W. Kiranon and P. Pawarangkoon, "Floating inductance simulation based on current conveyors," *Electronics Letters*, vol. 33, pp. 1748–1749, 1997.
[6] G. J. Deboo, "Application of a gyrator-type circuit to realize ungrounded inductors," *IEEE Trans. on Circuit Theory*, pp. 101–102, 1967.
[7] C. Maurini, F. dell'Isola, and D. D. Vescovo, "Comparison of piezoelectric networks acting as distributed vibration absorbers," *Mechanical System and Signal processing*, vol. 18, pp. 1243–1271, 2004.

Cost Effective High Voltage Driver for Large Channel Count Optical MEMS Switch Applications

Yuan Ma[*], Xiqun Zhu[&] and Robert W. Newcomb[%]

Electrical and Computer Engineering Department
University of Maryland
College Park, MD 20742

[*]yuanma@eng.umd.edu, [&]xiqun@eng.umd.edu, [%]newcomb@eng.umd.edu

Abstract— In this paper, a simple cost effective high voltage driving architecture suitable for system on chip integration in the optical switch application is introduced. A simple principle circuit implementation of this architecture is presented and simulated in PSPICE. The simulation results are explained.

I. INTRODUCTION

With the increasing demand for communication network capacity, major long haul telecommunication transmission has mostly migrated to optical transmission and the number of channels being transmitted is growing rapidly to thousands. Therefore, large channel count switches are needed [1]. Due to the great advantages of all-optical switches over optical-electrical-optical switches, large channel count all-optical switches, mostly MEMS based systems, are under development [2]. To meet the special requirements of MEMS drivers and the trend of system-on-chip integration, we introduce a very simple circuit architecture, which can easily be integrated at a very large scale.

II. MIRROR DRIVER REQUIREMENTS

An optical switch can consist of a large number of mirrors. For example, a 1024 by 1024 (which stands for 1024-channel input and 1024-channel output) optical switch has 2048 mirrors with 1024 of them input mirrors and 1024 of them output mirrors.

In order to control the independent x and y axis movement, each mirror has two independent groups of actuators correspondingly. One group controls x-axis direction movement and the other one controls y-axis direction movement. Two separately located bottom actuators in each group control the up and down movement [1]. So totally four actuators are needed for each mirror. Correspondingly, as each actuator needs a separate analog driving voltage, each mirror needs four driving analog voltage circuits belonging to two independent groups. In this way, there will be four outputs from a DAC (digital to analog converter) for each mirror to provide the driving analog voltage.

One conventional way to reduce the number of DACs needed is to introduce digital control bits. If we make use of two digital bits, one of them is used to select the x-and y-axis and another bit to select the up or down movement, then we need only two outputs instead of four from a DAC for each mirror. Thus, we reduce the number of DACs by half. As a DAC is very expensive and being used a lot in the system, we reduce the cost for the whole system.

In this way, a 1024 by 1024 optical switch needs 4096 analog voltages as well as 4096 bits of digital control signals. This has put the demand that the driving circuits need to be simple so that they can be small in size and low in power consumption. Therefore the total size and power consumption should be within a reasonable range. Also the circuit should be easy to be integrated with other parts of the system. Ideally, the whole system, mirror and control circuits could be integrated in a chip.

Unfortunately, the driving voltages involved here are quite high, up to two or three hundred volts. In response to this demand, some companies have already developed high voltage driver devices and made them commercially available. For example, there is a 16-channel high voltage driver from Agere System [4]. That device has 16 independent channels. Each channel has two independent voltage outputs up to 295 volts. It also has 16-bit digital control signals to select two outputs of each channel. A 1024 by 1024 optical switch needs 128 (128=2048/16) such devices, just for the high voltage driving. This makes the whole system very complex and contradicts with the requirement of compact size for optical switches.

The high voltage driver proposed in section V of this paper will have a very simple analog circuit architecture and will eliminate the need of a digital control signal. Therefore it can significantly reduce the complexity of the high voltage driver needed for large channel count optical switches. This driver also makes it possible to integrate all circuits in one single IC device to implement the whole function, which needs 128 (128=2048/16) devices if using Agere 16 channel components. It also answers the call of the trend of integrating electronic circuits with MEMS devices. So it will be possible to have the full switch with MEMS mirrors and their drivers all in one chip.

III. CONTROL SCHEME OF OPTICAL SWITCH

Before we get into the actual driver circuits, we look at the control scheme of optical switches. As shown in Figure 3.1, there is a closed-loop control to optimize the system performance.

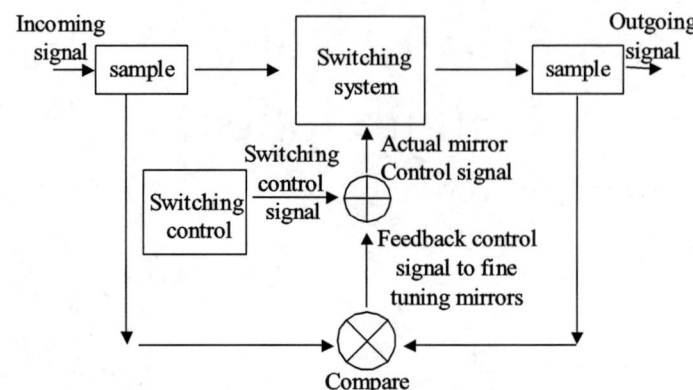

Figure 3.2 closed loop control scheme for optimal mirror control

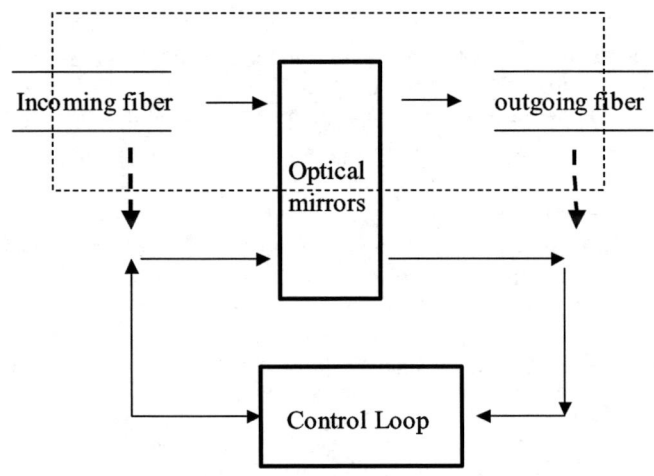

Figure 3.1 the basic system configuration of MEMS based all-optical switch

The incoming signals are sampled and compared with the sample output. The results of the comparison are used to fine tune the mirrors to the optimal tilt angle to minimize the overall insertion loss of the signal. The control loop is outlined in Figure 3.2.

As shown in Figure 3.2, mirrors in the optical switch system are controlled by a switch control module, which sets the tilt angle for each mirror to implement the channel switch requirements of the system. Say channel i is switched to channel j, and channel j switched to channel k, etc.. After the switch settling and before the next switching, (it can be as short as milli-seconds or as long as years), the closed loop control system is activated to fine tune each mirror based on the feed back from the samples of input and output signals to compensate for any noise as well as system parameters drifting over time.

The actual mirror control signal, a combination of switch control signal and fine tuning signal (normally very small compared with the former), will be amplified to two to three hundred volts to drive the mirrors. The circuits that make this amplification can be the high voltage driver circuit introduced below.

IV. HIGH VOLTAGE DRIVER FOR MEMS

Our design of the high voltage driver circuit has the block diagram shown in Figure 4.1.

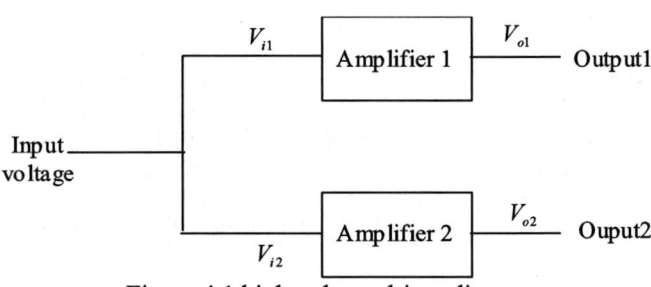

Figure 4.1 high voltage driver diagram

It consists of two amplifiers with each input port being an output port from a DAC. The two amplifiers have the same output voltage range, which is the desired voltage to drive micro mirrors in the optical switch system. These two amplifiers are used to drive two actuators, which belong to the two independent groups.

Here is the working principle of the circuit shown in Figure 5.1. Assume the gains of the first (including Q1) and the second amplifiers (including Q2) are g_1 and g_2, respectively, and their desired output driving voltage range is (V_{o_l}, V_{o_r}) for both outputs, where Vo_l and Vo_r are the left and right boundary of the desired mirror driving voltage range. That is

$$V_{o_1} \in (V_{o_l}, V_{o_r}), \text{ and } V_{o_2} \in (V_{o_l}, V_{o_r}) \quad (1)$$

where V_{o_1}, V_{o_2} are output voltages of the first and the second amplifiers, respectively. Then the input of both amplifiers will range from

$$V_{s_n} + \frac{V_{o_l}}{g_n} \text{ to } V_{s_n} + \frac{V_{o_r}}{g_n}, n \in \{1, 2\} \quad (2)$$

where V_{s_n} is the input DC shift of amplifier n. Therefore

$$V_{i_1} \in (V_{s_1} + \frac{V_{O_l}}{g_1}, V_{s_1} + \frac{V_{O_r}}{g_1}), \text{ and}$$

$$V_{i_2} \in (V_{s_2} + \frac{V_{O_l}}{g_2}, V_{s_2} + \frac{V_{O_r}}{g_2}) \quad (3)$$

where V_{i_1}, V_{i_2} are the input voltages of the first and the second amplifiers.

The first and the second amplifier are designed to activate to mutually exclusive input signal ranges. Therefore, while one amplifier is actuated, the other will be cut off, and vice versa. That is

$$V_{i_1} \notin (V_{s_2} + \frac{V_{O_l}}{g_2}, V_{s_2} + \frac{V_{O_r}}{g_2}) \quad (4)$$

$$V_{i_2} \notin (V_{s_1} + \frac{V_{O_l}}{g_1}, V_{s_1} + \frac{V_{O_r}}{g_1}) \quad (5)$$

Or

$$(\frac{V_{O_l}}{g_1}, \frac{V_{O_r}}{g_1}) \cap (\frac{V_{O_l}}{g_2}, \frac{V_{O_r}}{g_2}) = \phi = empey\ set \quad (6)$$

With this feature, we can assemble the control signal in the following way:

We shift the desired input signal for the first amplifier into the range $(V_{s_1} + \frac{V_{O_l}}{g_1}, V_{s_1} + \frac{V_{O_r}}{g_1})$, and shift the desired input signal for the second amplifier into the range

$(V_{s_2} + \frac{V_{O_l}}{g_2}, V_{s_2} + \frac{V_{O_r}}{g_2})$, and add the two ranges together.

The resulting combined input signal range is

$$(V_{s_1} + \frac{V_{O_l}}{g_1}, V_{s_2} + \frac{V_{O_r}}{g_1}) \cup (V_{s_2} + \frac{V_{O_l}}{g_2}, V_{s_2} + \frac{V_{O_r}}{g_2}) \quad (7)$$

Therefore, with one input signal, the circuit can be used to drive two different actuators. An automatic distinction between two outputs is implemented and the need of a one bit digital selecting signal is eliminated.

V. IMPLEMENTATION OF THE HIGH VOLTAGE DRIVER AND ITS SIMULATION RESULTS

In this section, a design example is presented. Assume we have the following parameter requirements:
$V_{o_l} = 0V$, $V_{o_r} = 300V$

Please note here that when $V_{o_r} = 300V$, the voltage on the mirror body itself is 300V also, which means the mirror is not activated when the amplifier's output is 300V.

Let $g_1 = 60$, $g_2 = 60$, $V_{s_1} = 1V, V_{s_2} = -4V$ Then

$$V_{i_1} \in (V_{s_1} + \frac{V_{o_l}}{g_1}, V_{s_1} + \frac{V_{o_r}}{g_1})$$
$$\in (1 + \frac{300}{60}, 6 + \frac{0}{60}) \in (1, 6) \quad (8)$$

$$V_{i_2} \in (V_{s_2} + \frac{V_{o_l}}{g_2}, V_{s_1} + \frac{V_{o_r}}{g_2})$$
$$\in (-4 + \frac{300}{60}, -4 + \frac{0}{60}) \in (1, -4) \quad (9)$$

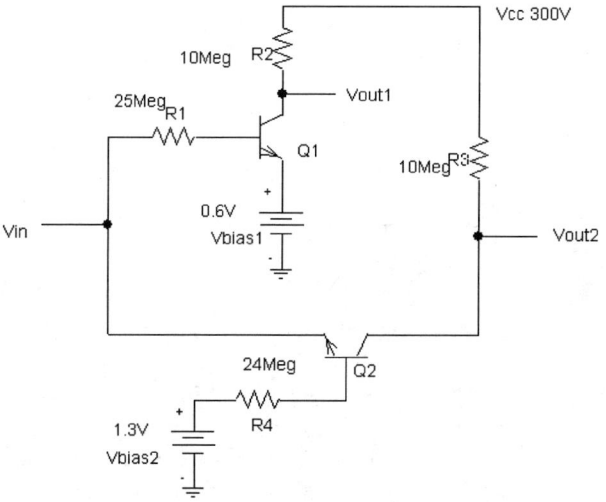

Figure 5.1 Schematics of the multi-channel high voltage driver circuit

One of the simplest implementations of the diagram in Figure 4.1 is shown in Figure 5.1, where only two transistors Q1 and Q2, along with four resistors are used to drive the two independent actuators. Both Q1 and Q2 share the same input Vin, but are actuated and cutoff at different input ranges. Properly biased, this circuit can generate the high voltages needed by the actuators, which are labeled Vout1 and Vout2 here.

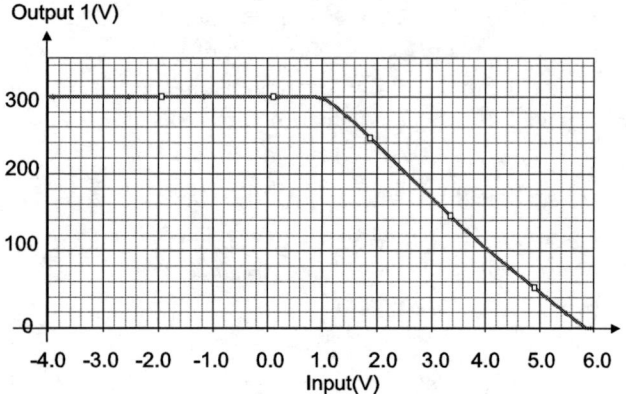

Figure 5.2 DC sweep (Vin vs. Vout1) of the driver circuit

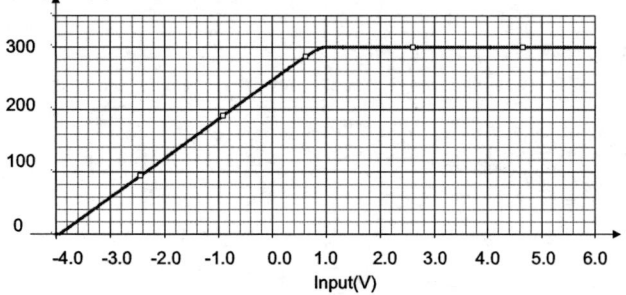

Figure 5.3 DC sweep (Vin vs. Vout2) of the driver circuit

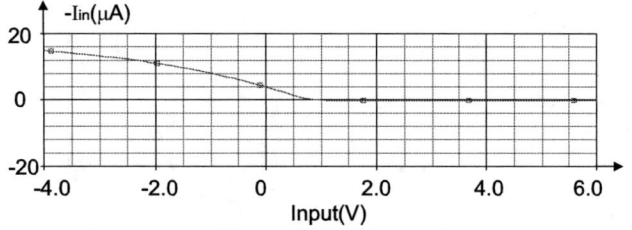

Figure 5.4 input current $-I_{in}$ vs. Vin

The PSPICE simulation results for the circuit shown in Figure 5.1 are displayed in Figures 5.2 and 5.3, assuming that Vcc in Figure 5.1 is 300V. The desired output range for both actuators is (0V, 300V). When the input changes from 1V to -4V, which is the desired input signal range for the second amplifier, Vout2, the output from Q2, changes from 300V to 0V, almost linearly with the input change. At the same time, the output of the first amplifier, Vout1, the output of Q1, keeps constant at 300V, meaning no driving signal is applied to the first actuator. When the input changes from 1V to 6V, which is the desired input range for the first amplifier, the output of the first amplifier, Vout1, the output from Q1, drops linearly from 300V to 0V, driving the first actuator, while the output of the second amplifier output from Q2 keeps the constant voltage level at 300V, which means no driving signal is applied to the second actuator.

The input to this driver circuit in an actual system comes from a DAC. The output impedance of this DAC is normally in the range of 50-5K Ω. When Vin is low, the current flowing into the input terminal of the driver is $I_{in} = I_{E2} = -(I_{b2} + I_{c2})$. Both I_{b2} and I_{c2} are limited due to the large values of R_4 and R_3. The simulation result shows that the maximum $|I_{in}|$ is less than 20µA, as shown in Figure 5.4.

VI. CONCLUSIONS

In this paper, a high voltage driver circuit, which meets the special requirements of large channel count MEMS based optical switches, is presented. The circuit is very simple in architecture, so can be small in size even being duplicated several thousand times, which is normally the case of real optical switches. Therefore, it can be easily integrated with other parts of the system and meet the trend of integrating the whole system on a chip.

VII. ACKNOWLEDGEMENT

Thanks to ISCAS reviewers for their constructive comments.

REFERENCES

[1] P. B. Chu, S. Lee and S. Park., "MEMS: The path to large optical crossconnects", *IEEECommunication Magazine*, Vol. 40, Issue 3, Mar. 2002, pp. 80-87
[2] J. Kim, et.al.,"1100x1100 port MEMS-Based Optical Crossconnect with 4-dB Maximum loss", *IEEE Photonics Technology Letters*, Vol. 15, No.11, Nov. 2003, pp. 1537-1539
[3] L. Y. Lin, E.L. Goldstein, "Opportunities and chanllenges for MEMS in lightwave communications", *IEEE Journal on Selected Topics in Quantum Electronics*, Vol.8, No.1, Jan/Feb 2002,pp. 163--172
[4] Y. Ma, K. Timko, "Multichannel, high-voltage IC amplifiers for optical MEMS", Vol.18, Issue 8, *Lightwave*, July, 2001,pp. 85-88

Face Segmentation Based on Hue-Cr Components And Morphological Technique

Teerayoot Sawangsri, Vorapoj Patanavijit, and Somchai Jitapunkul
Department of Electrical Engineering
Chulalongkorn University
Bangkok, Thailand
teerayoot.s@student.chula.ac.th, vorapoj.p@student.chula.ac.th, and somchai.j@chula.ac.th

Abstract—The paper proposes a novel algorithm to dynamically define the Region of Interest (ROI) videophone application. The algorithm uses the color information Hue and Cr to find the skin-color pixels and also use range of threshold obtained from red and blue components in normalized RGB color space to remove nonskin-color pixels because the human skin tends to have a predominance of red and nonpredominance of blue. Post-processing is used to remove such noises by a morphological operator. Moreover, the algorithm performs temporal filtering to remove skin-color pixels that immediately appear from frame to frame by using object tracking process to perform as memory for collecting skin-color objects obtained from previous frame to guide the next frame. The experimental results confirm the effectiveness of the proposed algorithm.

I. INTRODUCTION

The demand for applications of the digital video communications, such as videophone and videoconferencing, has increased considerably then coding for very low rate channels requires novel compression schemes. The content-based coding offers a more flexible and intelligent approach than traditional methods embodied by H.263/H.263+ and MPEG-4 standard.

The developed video coder improves the quality of perceptually important image regions, relative to traditional coding standard, at the expense of perceptually insignificant portions of the image.

Faces are an important component of image sequences (i.e. the head and shoulders videophone and videoconferencing) and faces have some distinct features from background therefore a skin segmentation plays an important role in recent color-based approaches to human face segmentation [5] and [6].

In these approaches, regions of the input image that have skin colors are segmented to give an initial estimate of the face locations. These segmented skin regions will be further processed in the later stages of a face segmentation that utilizes several color spaces to label pixels as skin including normalized RGB [1], HSV [5] and YCbCr [2] and [6].

This paper proposes a skin segmentation algorithm for color image sequences. The proposed algorithm uses a combined human skin color model to classify skin-colors and nonskin-colors. The model is based on the nonparametric histogram color model [2], [3], [5] and [6]. Compared to many existing skin segmentation approaches, the proposed algorithm combine two color spaces for skin segmentation and use property of skin color to set threshold for removing some noises that similar to skin-color.

This paper is organized as follows: Section II. explains the human skin color model. Section III. presents the skin segmentation algorithm. Section IV. discussed experimental results and Section V. concludes this paper.

II. HUMAN SKIN COLOR MODEL

A human skin color model is used to decide either a pixel is skin color or nonskin-color. This model is characterized by a classification algorithm and a color space used to represent pixel color. Color spaces used in skin color segmentation include YCbCr, HSV and normalized RGB.

The skin color model used in this paper is based on the nonparametric histogram color model [2], [5] and [6]. Jones and Rehg [8] compare the performance of histogram and parametric method, mixture models, for skin detection, they find histogram models to be superior in accuracy and computational cost.

There are many color models used for the modeling of human skin-color. Liu [4] has contended that there does not exist a single color models that work best for all kinds of image sequences therefore it is very important to choose the suitable color space to be combined for the modeling of human skin-color. Sobottka and Pitas [5] have operated for the HSV color space that is compatible with human color perception. They have considered hue and saturation as discriminating color information that describes the human skin color. This information can be defined as a priori and used subsequently as reference for any skin color. They denote R_S and R_H as the respective ranges of Hue and Saturation values that correspond to skin-color and they have chosen the parameters as follows : $R_S = [0.23\ 0.68]$ and $R_H = [0°\ 50°]$. Consider an input image sequence of $M \times N$ pixel. Hence, we can obtain a reference map by following criterion.

$$\text{Map}_{HS(x,y)} = \begin{cases} 255, & Hue \in R_H \cap S \in R_S \\ 0, & otherwise \end{cases} \quad (1)$$

where x = 1,2, .. , M and y = 1,2, .. , N.

The paper shows the reference maps in Fig. 1b-1d that match all skin-color regions besides some noise pixels that are similar to skin-color however Hue (H) is not reliable for the discrimination task when the saturation is low [9]. Because saturation defines the relative purity or the amount of white light mixed with a hue that is a color attribute as a pure color, skin-color pixels of Saturation component are wider distribution than hue component. Hence, we just adopt the color information of the hue component for the modeling of human skin-color for decrement effect from Saturation component.

Chai and Ngan [6] have found that a skin-color region can be identified by the presence of a certain set of chrominance (i.e. Cr and Cb) values narrowly and consistently distributed in the YCbCr color space. They denote R_{Cr} and R_{Cb} as the respective ranges of Cr and Cb values that correspond to skin-color and they set R_{Cr} = [133 173] and R_{Cb} = [77 127].

$$\text{Map}_{Cb\&Cr(x,y)} = \begin{cases} 255, & Cb \in R_{Cb} \cap Cr \in R_{cr} \\ 0, & otherwise \end{cases} \quad (2)$$

where x = 1,2, .. , M/2 and y = 1,2, .. , N/2.

The paper shows the reference maps in Fig. 2b-d. Notice that Fig. 2b match all skin-color regions but it is very over segmentation because many background can not be eliminated by Cb value. Fig. 2c and Fig. 2d match all skin-color regions besides some noise pixels that are similar to skin-color.

Moreover, Mei-Juan Chen and Ming-Chieh Chi [7] found an interesting problem, though Chai and Ngans' method is well but if the skin color region of Map_{Cb} is poor even though Map_{Cr} is still strong to present the region of skin-color, the skin-color region of $\text{Map}_{Cb\&Cr}$ will be poor. Moreover, the distribution of skin areas is consistent across different races in the Cb and Cr color spaces [4]. Then, we can only adopt the color information of Cr component for the modeling of human skin-color.

Because there does not exist a single color models that work best for all kinds of image sequences, this paper proposes a combined color spaces that are suitable for skin pixel classification under different conditions, i.e. different races and varying illuminations. We use color information of Hue and Cr to define skin-color and estimate a reference map in YCbCr and HSV color space. Moreover, R_{Cr} and R_H is denoted as the respective ranges of Cr and Hue values that correspond to skin-color and R_{Cr} = [133 173] [6] and R_H = [0° 50°] [5].

$$\text{Map}_{Cr.\&.Hue(x,y)} = \begin{cases} 255, & Cr \in R_{Cr} \cap Hue \in R_H \\ 0, & otherwise \end{cases} \quad (3)$$

For Hue color component, where x = 1,2, .. , M and y = 1,2, .. , N. For Cr color component, x = 1,2, .. , M/2 and y = 1,2, .. , N/2.

The reference maps result is shown in Fig. 3b. Notice that Fig. 3b matches all skin-color regions and have a little some noise pixels that are similar to skin-color.

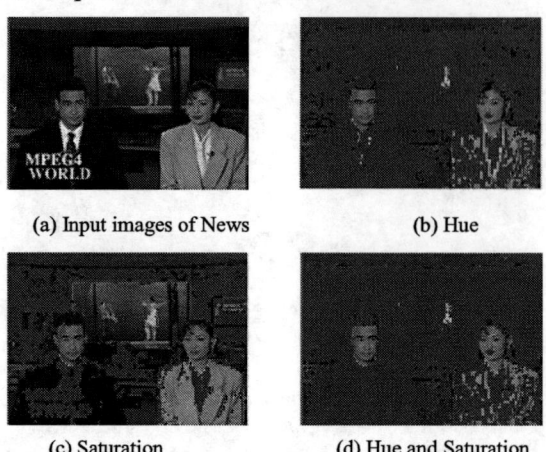

(a) Input images of News (b) Hue

(c) Saturation (d) Hue and Saturation

Figure 1. Result of different color components reference maps in YCbCr color space.

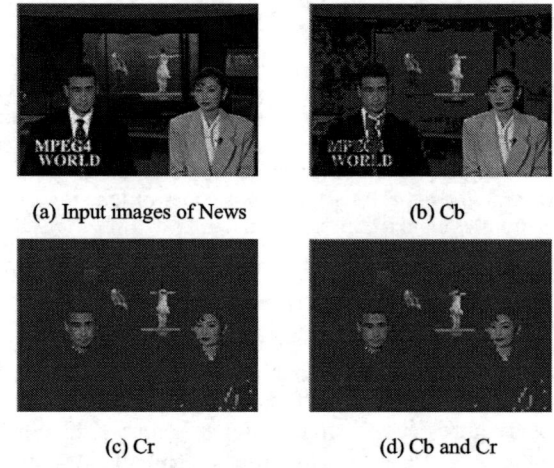

(a) Input images of News (b) Cb

(c) Cr (d) Cb and Cr

Figure 2. Result of different color components reference maps in YCbCr color space.

(a) Input images of News (b) Proposed

Figure 3. Result of proposed color components reference maps in YCbCr and HSV color space.

III. SKIN-COLOR SEGMENTATION

As shown in Fig. 4, the proposed method contains three stages

A. Color Segmentation

The first stage of the color segmentation algorithm involves the use of color information in a low-level region segmentation process to define region of interest. The aim is to classify pixels of the input image sequence into skin color and nonskin-color. To do so, the paper have devised a skin-color reference map thus we can easily discriminate the region of interest and noninterest via the proposed reference map that this paper use color information of Hue and Cr component to define skin-color.

B. Blackground Elimination

This section presents an algorithm to remove some noise pixels that are similar to skin-color, pseudo-skin-color, from first state. It is a fact that human skin tend to have a predominance of red and nonpredominance of blue. Furthermore, Mei-Juan Chen and Ming-Chieh Chi [7] found that skin-colors and pseudo-skin-colors locate at the same range, $R_{Cr} = [133\ 173]$, in Cr component but these two colors are different in RGB color space. The blue and red histogram of skin-color pixels of Fig. 2a is shown in Fig. 5a and Fig. 5b, respectively. The statistical results show that the histograms of red and blue for skin-color pixels in a normalized RGB color space are narrowly distributed because the normalization was employed to minimize the dependence on the luminance values then the paper can utilize the characteristic to remove pseudo-skin-color. And we can set R_{blue} and R_{red} as a threshold selected empirically from the histogram of samples and check that if each pixel from first state has the value of Blue in range of R_{blue} and has the value of Red in range of R_{red} that is real skin-color pixel. On the other hand, if the each pixel has the value of Blue in range of R_{blue} but has not the value of Red in range of R_{red} then this pixel is classified to be pseudo-skin-color pixel and we set $R_{blue} = [0.15\ 0.3]$ and $R_{red} = [0.4\ 0.5]$.

C. Post-processing and Object Tracking

Next, the process extracts the large skin-color objects from the image by using morphologic closing operations that are commonly used to smooth, fill in, and remove objects in an image sequence. Morphologic closing is equivalent to a dilation followed by an erosion. Nevertheless, these operations can only reduce some noise areas but cannot completely remove noise from the mask. In order to remove the remaining noise, the image is classified into 4x4 pixels block and these blocks are then categorized into skin-color and nonskin-color blocks. If there are more than 8 pixels in the block, it will be assigned to full pixel block, otherwise it will be assigned as empty block then each block that has the same value will be considered as a pixel. Finally, the algorithm performs temporal filtering to remove skin-color pixels that immediately appear from frame to frame by using the object tracking process that performs as memory for collecting skin-color object obtained from previous frame to guide the next frame. If the skin-color in next frame cover the skin-color in previous frame more than 50% then skin-color regions will be not removed, otherwise some regions that immediately appear and cover less than 50% of previous frame, will be removed. As a result, the output mask of this step should contain the face region with minimal or no noise and the next section will show the experimental results.

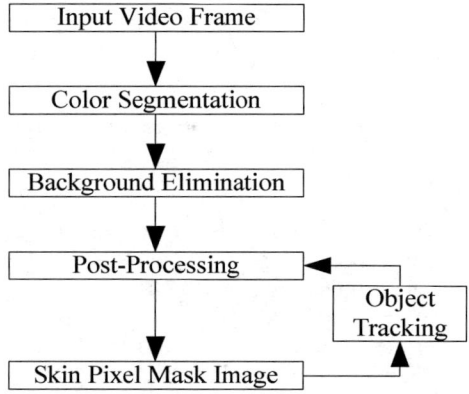

Figure 4. Block diagram of videophone skin-color segmentation algorithm

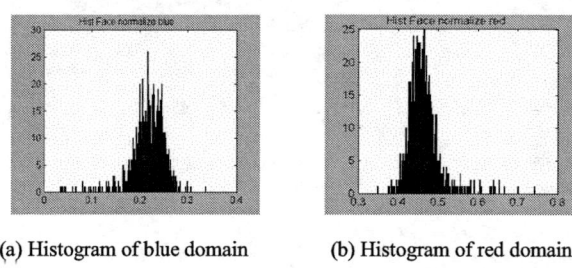

(a) Histogram of blue domain (b) Histogram of red domain

Figure 5. The histograms of Blue and Red domain of Fig. 2a in normalized RGB color space.

IV. EXPERIMENTAL RESULTS

The experimental results are performed using three videophone sequences: News, Grandma and Carphone. These sequences are video standard sequences in QCIF format, which can be categorized by moving characteristic. First, News sequence has less movement because only their heads and shoulders move. Second, Grandma sequence has very slowly movement along the sequence. Finally, Carphone sequence has movement very much. At the first stage of the algorithm color segmentation, the reference maps result produced by the proposed criterion is shown in Fig. 6d, Fig. 7d and Fig. 8d that are compared with the Cb and Cr method [6] in Fig. 6b, Fig. 7b and Fig. 8b and Hue and Saturation method [5] in Fig. 6c, Fig. 7c and Fig. 8c.

The performance of the skin-color segmentation methodology is evaluated quantitatively. To quantitatively measure the accuracy of the proposed methodology, each frame was manually segmented into skin and nonskin pixels. The manually segmented images serve as a reference to which the automatically segmented images are compared. The detection rate and false alarm rate are expressed as a percentage, which represent in equation (4) and (5) respectively.

$$DR = \frac{TP}{TP + FN} \quad (4)$$

$$FAR = \frac{FP}{TP + FP} \quad (5)$$

where :

TP, True Positive, is defined as a group of pixels that are foreground and the system distinguish that a group of pixels are foreground.

FP, False Positive, is defined as a group of pixels that are background but the system distinguish that a group of pixels are foreground.

FN, False Negative, is defined as a group of pixels that are foreground but the system distinguish that a group of pixels are background. The detection rates and false alarm rates for Fig. 6 – Fig. 8 video sequences are shown in Table I and Table II respectively. These results show that the proposed segmentation algorithm could provide the face segmentation more efficient and it also has less affect from the fast or slow moving object in video sequences.

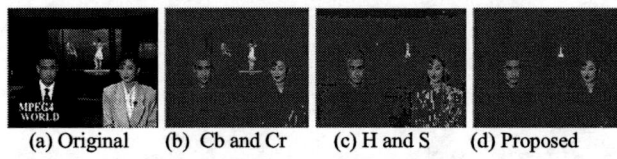

(a) Original   (b) Cb and Cr   (c) H and S   (d) Proposed

Figure 6. Comparison of different skin-color segmentation method of News sequence

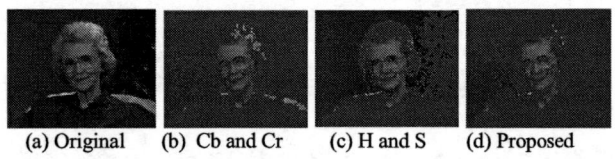

(a) Original   (b) Cb and Cr   (c) H and S   (d) Proposed

Figure 7. Comparison of different skin-color segmentation method of Grandma sequence

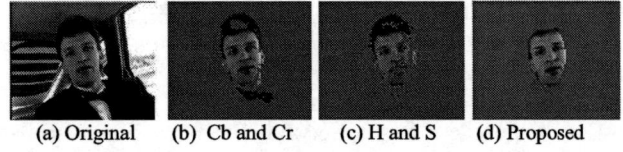

(a) Original   (b) Cb and Cr   (c) H and S   (d) Proposed

Figure 8. Comparison of different skin-color segmentation method of Carphone sequence

TABLE I
DETECTION RATES.

Sequence	Detection Rate		
	CbCr	HS	Proposed CrHue
Car phone	87.56 %	70.48 %	72.09%
Grandma	70.11 %	76.79 %	71.63 %
News	90.17%	90.97 %	93.79 %

TABLE II
FALSE ALARM RATES.

Sequence	False Alarm Rate		
	CbCr	HS	Proposed CrHue
Car phone	27.90 %	29.51 %	12.43%
Grandma	29.88 %	43.20 %	20.36%
News	13.20 %	19.02%	6.82%

V. CONCLUSION

This paper proposed a novel algorithm for skin segmentation of color image sequence. The proposed algorithm uses a combined human skin color model based on the nonparametric color model to classify skin-colors and nonskin-colors. We use property of skin color to set threshold for removing some noises that similar to skin-color because the human skin tend to have a predominance of red and nonpredominance of blue and we use morphologic closing operations to smooth, fill in, and remove objects in an image sequence.

Finally, the object tracking process performs as memory for collecting skin-color objects obtained from previous frame to guide the next frame in order to remove skin-color pixels that immediately appear from frame to frame. The experimental results show the satisfying subjective test results.

ACKNOWLEDGMENT

The authors would like to express the grateful thanks to grant from government research and development in cooperative project between Department of Electrical Engineering and Private Sector for Research for supporting this work and development under Chulalongkorn University.

REFERENCES

[1] J.L. Crowley and J.M. Berard, "Multi-Modal Tracking of Faces for Video Communications," Proc. IEEE Conf. on Computer Vision and Pattern Recognition, pp. 640 – 645, 1997.

[2] D. Tancharoen, S. Jitapankul, P. Kittipanya-ngam, S. Chompon, and H. Kortrakulkij, "Semantic Object Segmentation for Content Based Video Coding," Proceedings Vol. XIII, Image, Acoustic, Speech and Signal Processing: Part II, World Multiconference on Systemics, Cybernetics, and Informics, pp. 579-583, July 2001.

[3] L. Sigal., S. Sclaroff and V. Athitsos, "Skin Color-Based Video Segmentation under Time-Varying Illumination," IEEE Trans. on Pattern Analysis and Machine Intelligence, Vol. 26, pp. 862-877, July. 2004.

[4] J, Liu and Y.-H. Yang, "Multiresolution Color Image Segmentation," IEEE Transaction on Pattern Analysis and Machine Intelligence, Vol.16, No.7, pp. 689-700, 1994.

[5] K, Sobottka and I. Pitass, "Face localization and facial feature extraction based on shape and color information," IEEE Int. Conf. on Image Processing, Vol. III, pp. 483-486, 1996.

[6] D. Chai and K. N. Ngan, "Face segmentagion using skin-color map in videophone applications," IEEE Trans. on Circuits and System for video Technology, Vol.9, No.4, pp. 551-564, 1999.

[7] Mei-Juan, Ming-Chieh Chi, Ching-Ting Hsu and Jeng-Wei Chen, "ROI Video Coding Based on H.263+ with Robust Skin-Color Detection Technique," IEEE Trans. on Consumer electronics, Vol.49, No.3, pp. 724-730, 2003.

[8] M.J. Jones and J.M. Rehg, "Statistical Color Models with Application to Skin Detection," Proc. IEEE Conf. on Computer Vision and Pattern Recognition, Vol. 1, pp. 274-280, 1999.

[9] Jongmoo Choi, Sanghoon Lee, Chilgee Lee and Juneho Yi, "A Real-Time Face Recognition Systems Using Multiple Mean Faces And Dual Mode Fisherfaces," IEEE Symposium on Industrial Electronics, Vol.3, pp. 1686-1689, 2001.

A Combined Interpolatorless Interpolation and High Accuracy Sampling Process for Digital Class D Amplifiers

Victor Adrian, Bah-Hwee Gwee and Joseph S. Chang
Centre of Integrated Circuits and Systems
School of Electrical and Electronic Engineering
Nanyang Technological University, Singapore 639798
E-mails: victorad@pmail.ntu.edu.sg and ebhgwee@ntu.edu.sg

Abstract—We propose a novel combination of an 'interpolatorless' interpolation technique with a high accuracy sampling process for low-power low-distortion digital Class D amplifiers. This novel technique has an advantage of reducing the computational complexity of the usual two-step method (input signal interpolation cum sampling process). The complexity of this proposed technique is approximately equivalent to the complexity of the sampling process computation. Our proposed technique employs a 2nd-order Lagrange Interpolation polynomial and a Newton-Raphson method. We propose an initial guess for the Newton-Raphson method that produces an accurate estimation of the sampling process output in only one iteration. The simulated complete Class D amplifier features low-power dissipation (~ 56 μW, estimated @ 1.1 V, 0.35 μm CMOS) and relatively high Signal-to-Noise+Distortion Ratio (85.6 dB FS @ 4 kHz bandwidth), thereby rendering it appropriate for a hearing aid device.

I. INTRODUCTION

An important parameter for the commercial viability of portable audio medical and consumer devices such as hearing instruments (hearing aids), mobile phones, etc, is the life-span of the batteries in these devices. The life-span of the batteries in these devices is, in part, dependent on the power-efficiency of the power amplifiers (amps) embodied therein, rendering linear conventional analog Class A, B, and AB amps inappropriate. It is well established that although these linear amps have a maximum power-efficiency between 50-78% (bridge output), their practical power efficiencies are substantially lower, of the order of 20%, at nominal voltage swings, or where the crest factor is high. Furthermore, when the analog amps are connected to a digital processor, a digital-to-analog converter (DAC) is required. To circumvent the poor power efficiency of the linear amps and the need for a DAC, digital Class D amps are now routinely employed; Class D amps typically feature >90% power efficiency (η) over a large modulation index (M) range (signal swing). Specifically, in power-critical devices such as hearing instruments, the η is particularly pertinent and for high η, the quiescent current needs to be small. In a Class D amp, the high η is largely due to the output transistors operating in the ohmic and cut-off regions (as opposed to the active region in linear analog amps).

A digital Class D amp typically comprises of two parts. The first part modulates the input data according to a modulation technique and generates output pulses while the latter is an output stage that drives a load (a low-pass filter and an output transducer). Of the 3 modulations (Pulse Width Modulation [1], Pulse Density Modulation, and Click Modulation), the Pulse Width Modulation (PWM) is prevalent due to the simplicity of the hardware and relatively low switching frequency. These attributes are advantageous in power-critical and small form-factor-critical applications such as hearing instruments where aesthetics is paramount for their acceptance. For these reasons, we will only consider the PWM approach for realizing a digital Class D amp in this paper.

In a Class D amplifier based on the PWM approach, the ideal PWM employs the Natural Sampling (NS) process [1] that compares the input signal to a triangular carrier signal. It is well established [1] that this sampling process outputs a PWM signal with zero harmonic distortion, that is the Total Harmonic Distortion (THD) = 0. However, it is nearly impossible to use the ideal NS process practically in a digital realization because a continuous sampling is required, that is the sampling rate is ideally infinite.

In a practical realization, the sampling process in digital PWM involves a mathematical algorithm to digitally emulate the NS process. This is done by computing the intersection point of the modulating signal sampled by the Uniform Sampling process and the carrier signal. This signal sampling process is sometimes termed the cross-point deriver [2] and as its name implies, the process simply involves estimating the cross-point or the intersection point.

The reported sampling processes for the algorithmic PWM method include the Linear Interpolation [2, 3], Pseudo-Natural PWM [2], Static-Filter PWM, Weighted PWM and its variants, Derivative PWM, Parabolic Correction PWM, Prediction Correction PWM and more recently, our earlier proposed Delta-Compensation PWM [4], optimized Linear Interpolation [5], and combined Lagrange Interpolations [6] sampling processes. However, to achieve a high linearity, most of these algorithms require either intensive computations or a high oversampling rate (OSR). In view of the power-critical hearing instrument application, the Delta-Compensation and optimized Linear Interpolation may be appropriate. However, their non-linearity may be unacceptable if the sampling frequency, f_s, is low. For example, the optimized Linear Interpolation has a THD of ~ −75 dB FS @ M=0.9, f_s=48 kHz, input frequency f_{in}=997 Hz, output pulse frequency f_o=48 kHz. Our earlier proposed combined Lagrange Interpolations somewhat improves the THD performance (~ −85 dB FS @ M=0.9, f_s=48 kHz, f_{in}=997 Hz, f_o=96 kHz) with small computation and output frequency overheads. Put simply, it is highly desirable to obtain a sampling process that features low THD with low computation, low f_s, and low f_o.

One technique to reduce the harmonic non-linearity is by interpolating the input data samples, effectively oversampling the input at a frequency higher than the original sampling frequency. Interpolation is commonly employed in Delta-Sigma data converters to spread the quantization noise energy over a wider frequency range, thereby reducing the noise floor in the bandwidth of interest. In the context of the PWM sampling process, interpolation has a further effect of providing finer samples of the input signal. This translates to a better estimation of the intersection point or higher linearity because of the smaller distance between samples available to the sampling process. Interpolation also has an effect to shift the carrier frequency to higher frequency, thus lessening the effect of the fold-back distortions from the intermodulation of input and carrier signal frequencies. Interpolation, nonetheless, suffers from the drawback for the need of an interpolation filter and this can be costly in terms of computation.

In this paper, we propose a novel technique that combines the Lagrange Interpolation and the Newton-Raphson sampling process [2], but without the need to implement any actual interpolation. For brevity, we will abbreviate our proposed technique the 'LAG-NR'.

The proposed technique features very low harmonic distortion and this distortion is below the noise floor (quantified later). We take advantage of this by tolerating an increase in the quantization noise, and this allows us to reduce the clock frequency – reducing the clock frequency in the pulse generator is paramount for low power operation. We do this by reducing the number of bits (in the pulse generator) from the previous 8 bits [6] to 6 bits, equivalent to reducing the clock frequency from ~12.3 MHz to ~3 MHz. In power dissipation terms, the quiescent power dissipation of the pulse generator is reduced from 5.6 µW to 1.3 µW.

II. REVIEW OF BASIC PWM SAMPLING PROCESSES

In this section, we will briefly review two fundamental PWM sampling processes: Natural Sampling (NS) and Uniform Sampling (US), to delineate the basics of these PWM sampling processes. Fig. 1 depicts these sampling processes, where S_0 and S_1 are the previous and current magnitudes of the sampled inputs, and T is the sampling period. As depicted in Fig. 1, NS has an exact intersection point, S_{NS}, between the modulating input signal and the carrier signal. This point will determine the width of the PWM pulse (t_{NS}), and being the exact intersection that translates to an exact pulse width, NS has theoretically zero THD in the baseband of the PWM pulses. The NS, however, as mentioned earlier, has a drawback that it can only be realized using an analog sampling process. This is because at the intersection point with the carrier waveform, the amplitude of the input modulating signal must be precisely known. However, in digital domain, the amplitude of the input modulating signal at this intersection is not usually known because the input modulating signal is time and magnitude quantized. Put simply, an infinite sampling rate and resolution is required to obtain a complete and smooth contour of the input modulating signal.

The US process on the other hand, is a simple sample-and-hold process. From Fig. 1, as the pulse width of US (t_{US}) differs significantly from NS (t_{NS}), the US process has an intolerably high THD (~ −31 dB FS @ M=0.9, f_s=48 kHz, f_{in}=997 Hz, f_o=48 kHz).

From the preceding review, the basic idea of algorithmic PWM sampling processes is to overcome the linearity problem of US process, thereby emulating the NS process, without excessive computation and hardware costs.

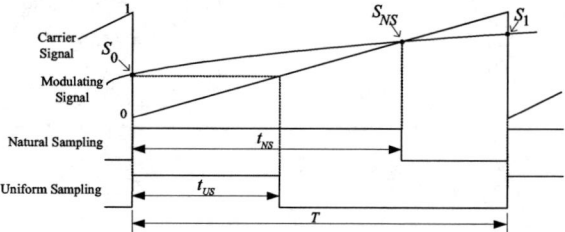

Figure 1. Natural Sampling (NS) and Uniform Sampling (US) processes

III. PROPOSED LAG-NR TECHNIQUE

In the digital filter design for the interpolator [7], a polynomial function can be used as an interpolator to interpolate the data as many times as the OSR. However, in the usual two-step methods (interpolation + sampling process), the common interpolator design is zeros-insertion followed by a Finite Impulse Response (FIR) filter. In our proposed technique, we will employ the polynomial technique to merge the interpolation and the sampling process together. We note that the two processes by themselves are rather lengthy and computationally complex. By means of our proposal to combine them into one process, we reduce the overall complexity to approximately equivalent to the complexity of the sampling process computation.

We will now describe in detail our proposed LAG-NR technique. The LAG-NR technique starts by first deriving a polynomial function which contour passes through the input samples. We have previously reported [6] a sampling process design that uses a 2nd-order Lagrange Interpolation [8]. Here, we will generalize the design to Lagrange Interpolation of order Q–1; the design for interpolator filter using Lagrange Interpolation was described in [7], albeit not in the context of PWM sampling process. Fig. 2 shows the construction of the polynomial. The polynomial $p(t)$ is constructed by Q input samples and the corresponding order of this polynomial is Q–1. The S_1 in Fig. 2 is the current sample while S_0 to $S_{-(Q-2)}$ are the previous samples. The amplitude, f, axis is normalized to 1. The time axis, t, is normalized to 1 at sample S_1 and further decremented by 1 at each previous sample. Mathematically, the generalized Lagrange Interpolation polynomial [8] of order Q–1 is:

$$p(t) = \sum_{i=-(Q-2)}^{1} G_i(t) f_i \quad (1)$$

where

$$G_j(t) = \prod_{\substack{i=-(Q-2) \\ i \ne j}}^{1} \frac{(t-t_i)}{(t_j-t_i)} \quad (2)$$

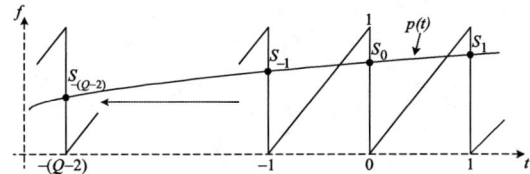

Figure 2. Polynomial construction of the signal arc from the samples

The region of interest of the interpolation lies between the 0 to 1 time period or correspondingly between S_0 and S_1 samples. The samples S_{-1} to $S_{-(Q-2)}$ are only necessary to facilitate in creating the polynomial. We now consider an interpolation of L-times as shown in Fig. 3 where there are L new interpolated data between S_0 and S_1. The new interpolated data samples, $S_{1/L}$ to $S_{(L-1)/L}$ can be obtained by replacing t in (1) and (2) with the locations of the data samples in the normalized time axis, i.e. $1/L, 2/L, \ldots, (L-1)/L$. Note that the interpolation step using a polynomial effectively increases the number (frequency rate) of data samples as in the interpolation using zeros-insertion + FIR filter of the two-step method. The difference, however, is that in the two-step method, there is no polynomial $p(t)$ as in Fig. 3, that can be used to directly estimate the intersection points. To estimate an intersection point between two data samples, the two-step method first uses two or more data samples to create a new polynomial and subsequently uses an algorithmic sampling process, such as Newton-Raphson method, to estimate the intersection point. In total, there will be L polynomials for an L intersection points.

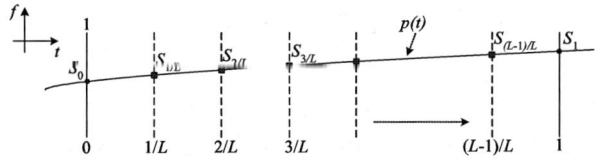

Figure 3. L-times interpolation

We will now explain how we can directly estimate the intersection points to make an efficient one-step 'interpolatorless' interpolation cum sampling process technique. We observe that to generate PWM pulses, we only require the intersection points. In other words, we do not need to generate the $S_{1/L}$ to $S_{(L-1)/L}$ data samples as they will be discarded after we have estimated the intersection points. Hence, instead of using the polynomial $p(t)$ to first interpolate and generate $S_{1/L}$ to $S_{(L-1)/L}$ data, we instead use $p(t)$ to directly find all intersection points inside the 0 to 1 period as depicted in Fig. 4. The intersection points R_1 to R_L can be estimated by intersecting $p(t)$ with 'virtual' carrier waveforms $c_1(t)$ to $c_L(t)$.

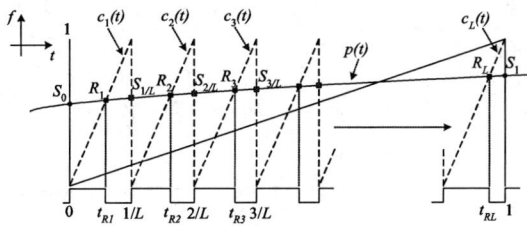

Figure 4. Estimating the L intersection points directly using the polynomial $p(t)$ and the carrier waveforms $c_1(t)$ to $c_L(t)$

From Fig. 4, we note that the carrier waveforms are parallel straight lines with the same gradient but shifted from one another by a constant value. With both t and f normalized to 1, we can easily show that the carrier waveforms can be expressed as:

$$c_i(t) = L \cdot t + k, \; i = 1, 2, \ldots, L \quad (3)$$

where
$$k = 1 - i \quad (4)$$

The intersection points are hence where the magnitudes of two waveforms are equal, i.e.:

$$e_i(t) = p(t) - c_i(t) = 0, \; i = 1, 2, \ldots, L \quad (5)$$

To estimate the time, t_{R1} to t_{RL}, of the intersection points, we use the Newton-Raphson method, a simple method that has a fast convergence rate. As a matter of interest, note that a reported sampling process [2] using the Newton-Raphson method employed the more complex two-step methods where it interpolated the data samples to L-times and subsequently, created L polynomials. It further used different normalization values (–0.5 to 0.5) that may lead to additional computations.

The Newton-Raphson method converges on a condition that an initial guess of the time is close enough to the actual time of the intersection point. We propose to use an initial guess of the time where the previously estimated time is added by 1 interpolation period ($1/L$). An exception is for the first initial value (i=1), where we need to prevent this value from overflowing by subtracting it with the normalized time value (=1). Hence, the initial guess can be written as:

$$t_{init(i)} = t_{R(i-1)} + 1/L - M \quad (6)$$

where
$$M = 1$$
$$t_{R0} = t_{RL} \text{ of previous period}$$
$$\Big\} \; i = 1$$
$$M = 0, \qquad\qquad i = 2, \ldots, L$$

Using (6) as the initial guess, we can now express the Newton-Raphson method [8] for the proposed LAG-NR as:

$$t_{R(i)} \approx t_{init(i)} - \frac{e_i(t_{init(i)})}{e_i'(t_{init(i)})}, \; i = 1, 2, \ldots, L \quad (7)$$

The close estimation of the time and the fast convergence of the Newton-Raphson method enable us to obtain an accurate estimation of the time of the intersection point in one iteration, thereby reducing the number of computation steps and the resultant complexity. The final (estimated) intersection point value, that is the PWM data, can hence be obtained by substituting $t_{R(i)}$ in (7) into t in (3).

IV. COMPLETE CLASS D AMPLIFIER

We depict the complete Class D amplifier embodying the LAG-NR technique in Fig. 5. We use a 2nd-order Lagrange Interpolation polynomial and an interpolation of 2 times (L=2) for the LAG-NR technique to keep the computation simple. The remaining sections of the complete Class D amplifier are similar to our previous design [6]. The input data is a 16-bit digital signal sampled at f_s=48 kHz. It is subsequently processed by the LAG-NR which outputs a 16-bit PWM data at twice the original frequency. As a consequence of the oversampling, the quantization noise is now spread over twice the original Nyquist frequency. The quantization noise in the bandwidth is further shaped by a 3rd-order noise shaper [9]. To reduce the power dissipation of the pulse generator, we quantize the 16-bit data to 6 bits instead of 8 bits [6]; note that as previously delineated, this reduces the quiescent power dissipation of the complete amplifier. The pulse generator converts data from the noise shaper to a PWM pulse width equivalent to the 6 bits data. We use our proposed frequency doubler [5] to reduce the fast-clock frequency of the pulse generator by half to $2^{6-1} \times 96$ kHz (~3 MHz). The PWM output is buffered by a full-bridge output stage to drive a transducer load modeled as a 300 Ω resistor. The pulses are finally filtered by a 2nd-order low-pass filter with a cutoff frequency of 4 kHz.

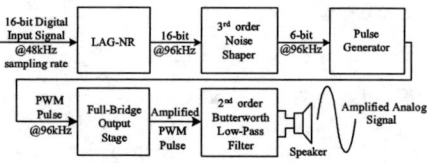

Figure 5. Complete Class D amplifier embodying the LAG-NR technique

V. SIMULATION RESULTS

We verify the performance of our Class D amplifier embodying our proposed LAG-NR technique using MATLAB simulations. We use a 16-bit f_{in}=997 Hz sinusoidal signal sampled at f_s=48 kHz in the simulation. The modulation index of the sine wave signal is set to near full swing at M=0.9. The PWM output signal is windowed with a Blackman-Harris window for its low side-lobes characteristics in the frequency spectrum. Fig. 6 shows the frequency spectrum of the PWM output signal (before the low-pass filter). Note that the harmonic distortion components are not visible in the bandwidth of interest as they are either equal or below the noise floor. Also, note that if the spectrum is observed after the low-pass filter, the noise floor improves beyond the 4 kHz bandwidth.

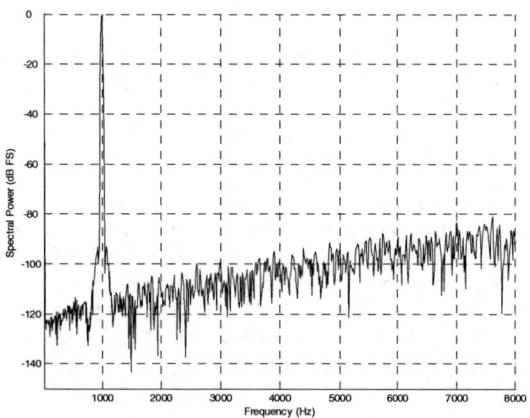

Figure 6. PWM spectrum for Class D amplifier embodying the LAG-NR technique

We summarize in Table 1, the performance of the Class D amplifier using the LAG-NR technique. Although the hardware for the proposed process is more complex than our simplistic Delta-Compensation [4], optimized Lagrange Interpolation [5], and combined Lagrange Interpolations [6], however, it features much better performance (if 8-bits is used in our proposed method here, the SNDR is better). When compared to the usual two-step method, our technique requires less computations and features simpler hardware. Our technique also results in relatively high SNDR which is more than sufficient for a hearing instrument. The complete Class D amplifier features low power-dissipation for a power-efficient portable hearing instrument.

TABLE I. PERFORMANCE SUMMARY OF A CLASS D AMPLIFIER DESIGN USING LAG-NR TECHNIQUE OF 2ND-ORDER LAGRANGE INTERPOLATION POLYNOMIAL AND L=2

Resolution of PWM output	6 bits
Input sampling frequency	48 kHz
Output frequency	96 kHz
Fast-clock frequency	3.072 MHz
Signal-to-Noise+Distortion Ratio (SNDR, integrated noise floor) @ M=0.9, BW = 20 Hz – 4 kHz	85.6 dB FS
Approximate complexity of hardware components and their operating rate for calculating an intersection point	5 adders @ 48 kHz, 7 adders @ 96 kHz, 3 multipliers @ 96 kHz, 1 divider @ 96 kHz
Estimated overall power dissipation @ 1.1 V, 0.35 µm CMOS	~ 56 µW

VI. CONCLUSIONS

We have described our proposed LAG-NR technique which is a novel combination of an 'interpolatorless' interpolation and a sampling process. We have also described our proposed initial guess in the estimation of the intersection time using. We have verified our design using computer simulations. We have shown that our Class D amplifier has low non-linearities and low power-dissipation. It is hence appropriate for power-critical hearing aid devices.

REFERENCES

[1] H.S. Black, Modulation Theory, New Jersey: Van Nostrand, 1953, pp. 263-281.

[2] J.M. Goldberg, and M.B. Sandler, "New High Accuracy Pulse Width Modulation based Digital-to-Analogue Convertor/Power Amplifier," IEE Proceedings - Circuits, Devices and Systems, vol. 141, no. 4, pp. 315-324, 1994.

[3] P.H. Mellor, S.P. Leigh, and B.M.G. Cheetham, "Reduction of Spectral Distortion in Class D Amplifiers by an Enhanced Pulse Width Modulation Sampling Process," IEE Proceedings-G, vol. 138, no. 4, pp. 441-448, 1991.

[4] B.H. Gwee, J.S. Chang, and H. Li, "A Micropower Low-Distortion Digital Pulse Width Modulator for a Digital Class D Amplifier," IEEE Transactions on Circuits & Systems II, vol. 49, no. 5, pp. 1-13, 2002.

[5] B.H. Gwee, J.S. Chang, V. Adrian, and H. Amir, "A Novel Sampling Process and Pulse Generator for A Low Distortion Digital Pulse-Width Modulator for Digital Class D Amplifiers," Proceedings of the 2003 International Symposium on Circuits and Systems, vol. 4, pp. 504-507, 2003.

[6] V. Adrian, B.H. Gwee, and J.S. Chang, "A Novel Combined First and Second Order Lagrange Interpolation Sampling Process for A Digital Class D Amplifier," Proceedings of the 2004 International Symposium on Circuits and Systems, vol.3, pp.233-236, 2004.

[7] R.E. Crochiere, and L.R. Rabiner, Multirate Digital Signal Processing, New Jersey: Prentice-Hall, 1983.

[8] E. Kreyszig, Advanced Engineering Mathematics, 8th edition, John Wiley & Sons, 1999.

[9] S.K. Tewksbury, and R.W. Hallock, "Oversampled, Linear Predictive and Noise-Shaping Coders of Order N>1," IEEE Transactions on Circuits & Systems, vol. CAS-25, no. 7, pp. 436-447, 1978.

Time-Delay Direction Finding Based on Canonical Correlation Analysis

Gaoming Huang [1,2],
1. Institute of Electronic Engineering,
Naval Engineering University
Wuhan, China
redforce@sohu.com

Luxi Yang [2], Zhenya He [2]
2. Department of Radio Engineering,
Southeast University
Nanjing, China
{lxyang, zyhe}@seu.edu.cn

Abstract—Direction finding is an important component of Electronic Warfare (EW) systems. Direction finding precision may directly affect signal sorting, recognition, location, and jamming decisions, etc. Thus, it is an urgent task to improve direction finding precision. To solve the problems of time-delay direction finding's sensitivity to interference, this paper proposes a novel technique based on Canonical Correlation Analysis (CCA) to improve time-delay direction finding performance and completely analyzes the algorithm. The implementation of this time-delay direction finding method is simple and the computational complexity is low. The precision of direction finding by the new method is much better than that by conventional way. Simulations show that applying CCA to time-delay direction finding is efficient and feasible.

I. INTRODUCTION

In the modern EW signal environments, the signals are dense and the types of which are also complicated. This makes the task of signal processing in electronic reconnaissance very hard. As the position of emitters is relatively fixed or the variety of speed relative to signals is very low, direction finding become a key technology in current electronic reconnaissance. At the same time, there are more demands for direction finding such as high precision, high resolving ability, multi signal simultaneous processing and real-time direction finding, etc. An effective novel method should been acquired to solve these problems in direction finding, which may be very helpful to signal sorting and recognition in electronic reconnaissance, especially to passive location [1].

Time-delay direction finding is an important method featured the high sensitivity, high precision and excellent real-time. Time-delay direction finding method can obtain the arrival of azimuth (AOA) of electric wave by the measurement of the difference of arrival time to each direction finding antenna, which is the time delay. As the bad anti-jamming property and the carrier wave must have determinate modulate, time-delay direction finding has not been broadly applied at present.

Canonical Correlation Analysis (CCA) is an important method of multivariate statistical since it was proposed by H.Hotelling in 1936 [2], which can find the basic vectors from two sets of variables. There are detail descriptions of CCA in [3][4]. CCA has been applied in some preliminary work [5][6][7] in recent years. Uncorrelated components can be obtained by CCA, in addition, have maximum spatial or temporal correlation within each component. Then we can apply CCA to seek the correlate components of the data from double receiving antennas. After general correlation of the two canonical components, we can obtain the time delay and complete the computation of azimuth. The structure of this paper is as follows: In the Section 2, we will analyze the issue of time-delay direction finding and pose the problem. In the Section 3, the direction finding algorithm will be analyzed. Then some experiments of the algorithm applied in this paper are conducted in the Section 4. Finally a conclusion is given.

II. PROBLEM FORMULATION

The model of time delay between the signals received by two groups of separate antennas is shown as Fig.1.

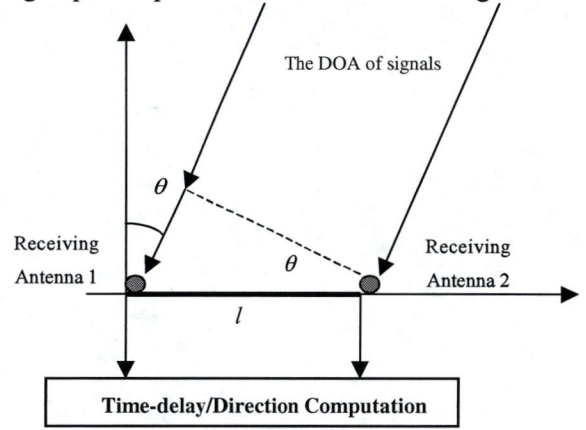

Figure. 1. Signal receiving model

Supposing that receiving station has m receiving channels, the mutual distance is l. If have $n(n \leq m)$ narrow band signals, $r(\theta_i)$ is the antenna respond to narrow band signal on the direction θ_i, then $r(\theta_i) = \left[1, e^{j\phi_i}, \ldots, e^{j(m-1)\phi_i}\right]^T$, where $\phi_i = 2\pi l \sin\theta_i / \lambda$, λ is the wave length. Receiving signal model could be described as:

$$x(t) = H \cdot \sum_{i=1}^{n} r(\theta_i) s_i(t) + n(t) = A \cdot s(t) + n(t) \quad (1)$$

where $x(t)$ is the receiving signals of antenna. A is a $m \times n$ mixing matrix, which is the product of antenna responding function $R = [r(\theta_1), \ldots, r(\theta_n)]$ and the mixing matrix H during the signal transmitting process. $s(t)$ are the source signals including radar signals, interference signals etc. $n(t)$ are $m \times 1$ dimension noise signals, which include exterior noise, inner noise and electricity noise etc. The signals are assumed as mutually independent and independent with noise in the following analysis. The key problem of direction finding based time delay is how to eliminate or reduce the affect of noise, interference and mixing signals to time delay computation by correlation, which is the main problem that this paper want to solve.

III. Direction Finding Algorithm

Direction finding algorithm includes three steps: canonical correlation analysis, time delay estimation, and azimuth computation. Before all these steps, the receiving signals should be normalized to avoid the influence to following processing.

A. Canonical Correlation Analysis

The main difference between CCA and the other three methods is that CCA is closely related to mutual information [5]. Hence CCA can be easily motivated in information based on tasks and our natural selection.

Consider two sets of input data $x_1, x_2, \ldots, x_p$ and $y_1, y_2, \ldots, y_q$, $p \leq q$, we attempt to find the coefficient $a = (a_{i1}, a_{i2}, \ldots, a_{ip})$ and $b = (b_{i1}, b_{i2}, \ldots, b_{iq})$ by the idea of principle components. The two sets of data can be written as combination of some pairs of variables ξ_i and η_i, which can be described as:

$$\begin{cases} \xi_1 = a_{11}x_1' + \cdots + a_{1p}x_p' & \eta_1 = b_{11}y_1' + \cdots + b_{1q}y_q' \\ \cdots \\ \xi_p = a_{p1}x_1' + \cdots + a_{pp}x_p' & \eta_p = b_{p1}y_1' + \cdots + b_{pq}y_q' \end{cases} \quad (2)$$

where x', y' are the standardization value of x, y respectively. ξ_1, η_1 are the first pair of canonical variable, the correlation coefficient can be describe as $r_{\xi_1\eta_1}$, brief written as r_1. ξ_2, η_2 are the second pair of canonical variable, the correlation coefficient can be describe as $r_{\xi_2\eta_2}$, brief written as r_2. Then p pairs of canonical variables and p canonical correlation coefficients can be obtained.

Mutual independent variables can be obtained by the method of canonical correlation. Here we introduce a theorem [4] (which will be proven in appendix A).

Theorem Let $x = (x_1, x_2, \ldots, x_p)^T$, $y = (y_1, y_2, \ldots, y_q)^T$ are two sets of random variables, $\text{cov}(x) = \Sigma_{xx}$, $\text{cov}(y) = \Sigma_{yy}$, $\text{cov}(x,y) = \Sigma_{xy}$, $\text{cov}(y,x) = \Sigma_{yx}$, Σ_{xx}, Σ_{yy} are positive definite, then $\Sigma_{xx}^{-1}\Sigma_{xy}\Sigma_{yy}^{-1}\Sigma_{yx}$ and $\Sigma_{yy}^{-1}\Sigma_{yx}\Sigma_{xx}^{-1}\Sigma_{xy}$ have the same non-zero latent roots $\lambda_1^2 \geq \lambda_2^2 \geq \cdots \geq \lambda_r^2 > 0$. If their mutual orthogonal identity eigenvectors are $\alpha_1, \alpha_2, \ldots, \alpha_r$ and $\beta_1, \beta_2, \ldots, \beta_r$ respectively, then $a_i^* = \Sigma_{xx}^{-1/2}\alpha_i$, $b_i^* = \Sigma_{yy}^{-1/2}\beta_i$ $(i=1,2,\ldots,r)$ are the ith pair of canonical correlation variables, λ_i is the ith canonical correlation coefficient.

By the theorem, canonical correlation variable and coefficients can be obtained by the follow steps:

Step1: computing the correlations of the two sets of variables as:

$$\Sigma = \begin{bmatrix} r_{x_1 x_1} & \cdots & r_{x_1 x_p} & r_{x_1 y_1} & \cdots & r_{x_1 y_q} \\ \vdots & \ddots & \vdots & \vdots & \ddots & \vdots \\ r_{x_p x_1} & \cdots & r_{x_1 x_p} & r_{x_p y_1} & \cdots & r_{x_p y_q} \\ r_{y_1 x_1} & \cdots & r_{y_1 x_p} & r_{y_1 y_1} & \cdots & r_{y_1 y_q} \\ \vdots & \ddots & \vdots & \vdots & \ddots & \vdots \\ r_{y_q x_1} & \cdots & r_{y_q x_p} & r_{y_q y_1} & \cdots & r_{y_p y_q} \end{bmatrix} = \begin{bmatrix} \Sigma_{xx} & \Sigma_{xy} \\ \Sigma_{yx} & \Sigma_{yy} \end{bmatrix} \quad (3)$$

The right matrix in (3) is a partitioned matrix.

Setp2: Computing the canonical correlation coefficients r_i. Firstly, we compute two matrices L and M, where $L = \Sigma_{xx}^{-1}\Sigma_{xy}\Sigma_{yy}^{-1}\Sigma_{yx}$, $M = \Sigma_{yy}^{-1}\Sigma_{yx}\Sigma_{xx}^{-1}\Sigma_{xy}$, secondly, we compute the eigenvalue λ_i of matrix L and M, we can obtain $\lambda_1 \geq \lambda_2 \geq \cdots \geq \lambda_i \geq \cdots \geq \lambda_p$, then $r_i = \sqrt{\lambda_i}$ can be obtained.

Step3: Computing the canonical variables ξ_i and η_i. First computing the eigenvectors of matrix L about λ_i, we can obtain the coefficient matrix A. Second computing the eigenvectors of matrix M about λ_i, we can obtain the coefficient matrix B. Thirdly we introduce standardization variables x_j' and y_j', then the v canonical variables ξ_i and η_i can be obtained.

Then canonical variables ξ_i and η_i can be obtained in turns, which is the basis of time delay estimation.

B. Time delay Estimation

After the canonical variables ξ_i and η_i having been obtained, the following step is to estimate the time delay $\Delta\tau$ of the two independent receiving signals, viz. the canonical variables ξ_i and η_i. To a same emitter, the receiving time is different as the as the different receiving position. In this case, we must consider the resemble property of the two signals must be considered during the time varying. Delaying $\Delta\tau$ to $s_1(n)$ and change it to $s_1(n-\tau)$, then the coefficient $R_{s_1 s_2}(\tau)$ can be written as:

$$R_{s_1 s_2}(\tau) = \sum_{n=-\infty}^{\infty} s_1(n) s_2(n-\tau) \quad (4)$$

When τ varying from $-\infty$ to $+\infty$, $R_{s_1 s_2}(\tau)$ is a function of τ, call $R_{s_1 s_2}(\tau)$ as the coefficient of $s_1(n)$ and $s_2(n-\tau)$, τ is the time delay of $s_2(n-\tau)$. When $|R_{s_1 s_2}(\tau)|$ reach the maximal value at τ_0, then τ_0 is the time difference $\Delta\tau$ of the two signals.

C. Azimuth Computation

After $\Delta\tau$ has been obtained, the computation of azimuth is relatively simple. From Fig.1, the following equations can be obtained: $\Delta l = l\sin\theta$, $\Delta\tau c = \Delta l = l\sin\theta$, where $c = 3\times 10^8 m/s$ is the electric wave transmitting speed, the DOA can be obtained as:

$$\theta = \arccos\left((c\cdot\Delta\tau)/l\right) \quad (5)$$

From (5) we can see that the precision of DOA decided by l and the measure precision $\Delta\tau$, where $\Delta\tau = (l\sin\theta)/c$. Partial derivative of $\Delta\tau$ can be written as:

$$d(\Delta\tau) = \frac{\partial(\Delta\tau)}{\partial\theta}d\theta + \frac{\partial(\Delta\tau)}{\partial l}dl + \frac{\partial(\Delta\tau)}{\partial c}dc,$$

where l, c are all constant, so the result of partial derivative is $d(\Delta\tau) = \frac{l}{c}\cos\theta d\theta$, then:

$$d\theta = (c\cdot d(\Delta\tau))/(l\cos\theta) \quad (6)$$

From (6) we can know that the precision of direction finding is limited by the base line length l and the measure precision $\Delta\tau$, it also relate to the measure angle. In order to improve the performance of direction finding, usually 4 groups of antenna are applied to cover the whole azimuth.

IV. SIMULATIONS

In order to verify the validity of this direction finding algorithm proposed in this paper, here a series of experiments have been conducted of direct correlation, GCC and CCA method. Background of the experiments assumed as: there are two groups of receiving antennas and each includes two antenna units. Distance between the antennas is $30m$, the relative direction of DOA to really north way is $30°$. From the receiving model established in Fig.1 that the time delay is $50ns$, if the sampling interval is $1ns$, then the delay is 50 sampling intervals. Source signals include one FM noisy jamming and which can be generated using the corresponding steps from [8]. The other signal is a FM signal, described as: $s_{FM}(t) = \cos(2\pi f_0 t + \phi(t))$, where f_0 is the carrier frequency, $\phi(t)$ is a modulating component. Simulation steps as: first taking a direct correlation to the filtered receiving signals in the interference background (shown as in Fig.2). From the result as Fig.3, we can see that there are easy to have two extremums, so as to GCC method.

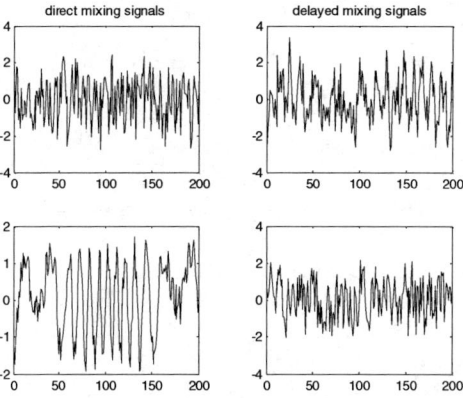

Figure 2. Mixing signals

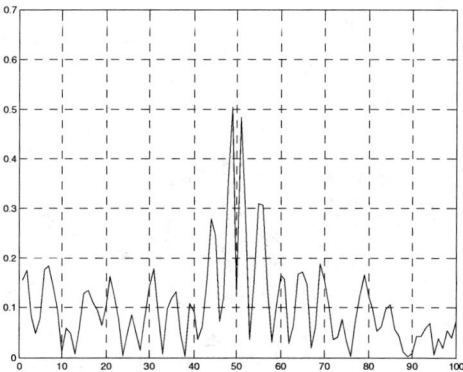

Figure 3. Direct correlation results

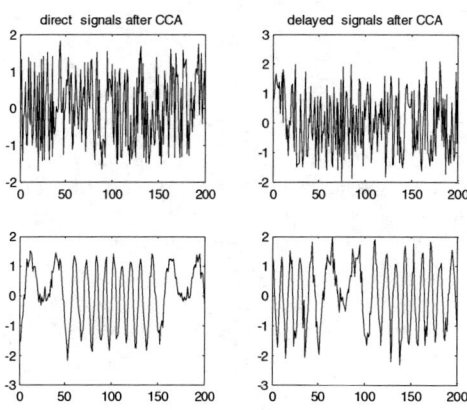

Figure 4. CCA results

Then conduct correlation processing to the canonical variable results (shown as Fig.4) by CCA. The correlation result as Fig.5, the time delay is just 50 sampling intervals, which is accuracy and stable.

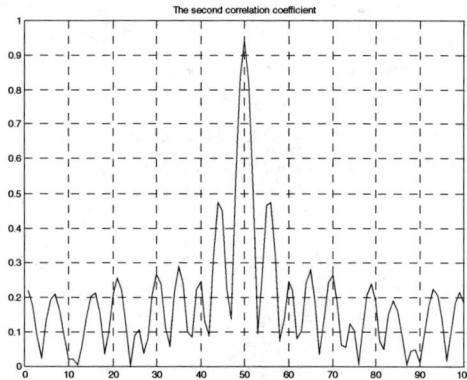

Figure 5. Canonical variables correlation results

A lot of experiments have been conducted by the changing the impact of additive noise. Results as in Fig.6 show that the estimation of time delay after CCA processing is more accuracy and stabilization than direct correlation and GCC. The anti-jamming property is also excellent.

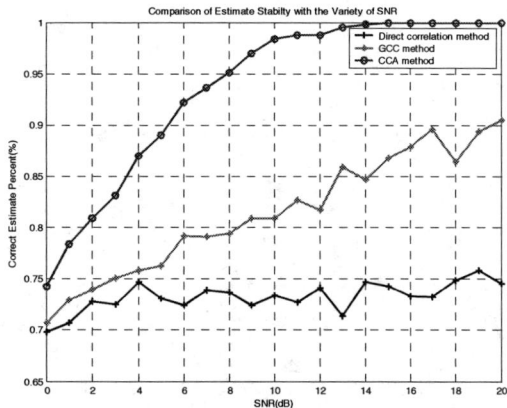

Figure 6. Performance comparison

V. CONCLUSION

This paper proposes the application of CCA method to time-delay direction finding, which can not only effectively overcome the sensitivity to interference of this direction finding method, but also avoid the localization of time-delay direction finding and improve the precision of the direction finding method, so as to enhance the location precision. The important contribution of this direction finding method proposed by this paper can improve the applicability of time-delay direction finding, which will play an important role in military and civilian affairs.

ACKNOWLEDGMENT

This work was supported by NSFC (60496310, 60272046), National High Technology Project (2002AA123031) of China, NSFJS (BK2002051) and the Grant of PhD Programmers of Chinese MOE (20020286014).

REFERENCES

[1] L.Z. Hu. Passive Locating, National Defence Industry Press, Beijing, 2004.
[2] H. Hotelling. "Relations between two sets of variates". Biometrika, 28, pp. 321-377, 1936.
[3] T. W. Anderson. An Introduction to Multivariate Statistical Analysis. John Wiley & Sons, second edition, 1984.
[4] R.T. Zhang, K.T. Fang. An Introduction to Multivariate Statistical Analysis. Science House, 1982.
[5] M. Borga. Learning Multidimensional Signal Processing. PhD thesis, Linköping University, Sweden, SE-581 83 Linköping, Sweden, 1998. Dissertation No 531, ISBN 91-7219-202-X. 1998.
[6] F. R. Bach. M. I. Jordan. "Kernel independent component analysis", Journal of Machine Learning Research, 3, pp. 1-48, 2002.
[7] C. Fyfe and P. L. Lai. "Ica using kernel canonical correlation analysis". ICA2000, pp. 279-284, 2000.
[8] D. Curtis Schleher. Electronic Warfare in the Information Age, Artech House, 1999.

Appendix A: Proof of theorem

Computing the variable is to choose a, b in the condition of $a^T \Sigma_{xx} a = 1, b^T \Sigma_{yy} b = 1$, then we will obtain the maximum value of $a^T \Sigma_{xy} b$. Taking Lagrangean function as the assistant function as:

$$\Phi(a,b,s,t) = a^T \Sigma_{xy} b + s\left(a^T \Sigma_{xx} a - 1\right) + t\left(b^T \Sigma_{yy} b - 1\right) \quad (A1)$$

In order to maximize Φ, Φ must content with following condition:

$$\begin{cases} \partial \Phi / \partial a = \Sigma_{xy} b + 2s\Sigma_{xx} a = 0 \\ \partial \Phi / \partial b = \Sigma_{yx} a + 2t\Sigma_{yy} b = 0 \end{cases} \quad (A2)$$

Then we can obtain:

$$2sa = -\Sigma_{xx}^{-1} \Sigma_{xy} b \quad (A3)$$

$$2tb = -\Sigma_{yy}^{-1} \Sigma_{yx} a \quad (A4)$$

Equation (A3) left multiplies $a^T \Sigma_{xx}$, then $2s = -a^T \Sigma_{xy} b$. Equation (A4) left multiplies $b^T \Sigma_{yy}$, then $2t = -b^T \Sigma_{yx} a$, so $s = t$. a, b can be written as:

$$\begin{cases} a = \Sigma_{xx}^{-1} \Sigma_{xy} \Sigma_{yy}^{-1} \Sigma_{yx} a / 4s^2 \\ b = \Sigma_{yy}^{-1} \Sigma_{yx} \Sigma_{xx}^{-1} \Sigma_{xy} b / 4s^2 \end{cases} \quad (A5)$$

By (A5), we can know that a, b are the eigenvectors of $\Sigma_{xx}^{-1} \Sigma_{xy} \Sigma_{yy}^{-1} \Sigma_{yx}$, $\Sigma_{yy}^{-1} \Sigma_{yx} \Sigma_{xx}^{-1} \Sigma_{xy}$ and they have the same eigenvalue.

DWT Domain Multi-matcher On-line Signature Verification System

Isao Nakanishi
Faculty of Regional Sciences
Tottori University
4-101 Koyama-minami, Tottori, 680-8551 Japan
Email: isao@rstu.jp

Hiroyuki Sakamoto, Yoshio Itoh and Yutaka Fukui
Faculty of Engineering
Tottori University
4-101 Koyama-minami, Tottori, 680-8552 Japan
Email: {itoh, fukui}@ele.tottori-u.ac.jp

Abstract—This paper presents a multi-matcher on-line signature verification system which fuses the verification scores in pen-position parameter and pen-movement angle parameter at decision level. Features of pen-position and pen-movement angle are extracted by the sub-band decomposition using the Discrete Wavelet Transform (DWT). In the pen-position, high frequency sub-band signals are considered as individual features to enhance the difference between a genuine signature and its forgery. On the other hand, low frequency sub-band signals are utilized as the features for suppressing the intra-class variation in the pen-movement angle. Verification is achieved by the adaptive signal processing using the extracted features. Verification scores in the pen-position and the pen-movement angle are integrated by using a weighted sum rule to make total decision. Experimental results show that fusion of pen-position and pen-movement angle can improve verification performance.

I. INTRODUCTION

Single biometric systems may not be always applicable because of unacceptable performance and inability to operate on a large user population. Multiple biometric systems can overcome these limitations. Five scenarios of the multiple biometric system have been proposed in [1], that is, multi-sensor system, multi-modal system, multi-unit system, multi-impression system, and multi-matcher system. Among of them, the multi-matcher system which uses multiple representation and matching algorithm for the same input biometric signal is the most cost-effective way to improve the performance of the biometric system [1].

We have proposed the on-line signature verification system in the Discrete Wavelet Transform (DWT) domain [5]. This system utilized only pen-position parameter since it was detectable even in portable devices such as the Personal Digital Assistants (PDA). A time-varying signal of pen-position parameter was decomposed into sub-band signals by using the DWT. Verification was achieved by using the adaptive signal processing in each sub-band. Total decision for verification was done by combining such verification results. Verification rate of about 95% was obtained, which was improved by about 10% comparing with a time-domain verification system.

In this paper, we introduce multi-matcher scheme into our on-line signature verification system. In addition to pen-position parameter, pen-movement angle parameter is processed by the DWT and the adaptive signal processing to obtain verification results in sub-bands. The pen-movement angle parameter is derived from the pen-position parameter; therefore, the proposed system requires no additional sensor. While high frequency sub-band signals are treated as individual features in the pen-position to enhance the difference between a genuine signature and its forgery, low frequency sub-band signals are utilized as the features in the pen-movement angle for suppressing the intra-class variations in signatures of one individual. Verification results of both the pen-position and the pen-movement angle are integrated at total decision level.

II. ON-LINE SIGNATURE VERIFICATION IN DWT DOMAIN

The on-line signature is digitized with the electronic pen-tablet. Especially, we utilize only pen-position parameter since it is provided even in such as the PDA for handwriting or pointing. Actually, the pen-position parameter consists of discrete time-varying signals of x and y coordinates, which are $x^*(n')$ and $y^*(n')$, respectively. $n'(= 0, 1, \cdots, N_{max}-1)$ is a sampled time index. N_{max} is the total number of sampled data. As the on-line signature is a dynamic biometrics, each writing time is different from the others. This results in the different number of sampled data even in genuine signatures. Moreover, different writing place and different size of signature cause variations in pen-position parameter. To reduce such variations, pen-position data are normalized in general. The normalized pen-position parameter is defined as

$$x(n) = \frac{x^*(n) - x_{min}}{x_{max} - x_{min}} \cdot \alpha_x \quad (1)$$

$$y(n) = \frac{y^*(n) - y_{min}}{y_{max} - y_{min}} \cdot \alpha_y \quad (2)$$

where $n(= 0 \sim 1)$ is a normalized sampled time index given by $n = n'/(N_{max}-1)$. x_{max} and y_{max} are maximum and minimum values of $x^*(n)$ and $y^*(n)$, respectively. α_x and α_y are scaling factors for avoiding underflow calculation in sub-band decomposition described later.

However, such normalization makes the difference between a genuine signature and its forgery unclear. In addition, the on-line signature is relatively easy to forge if the written signature is known. Easiness of imitating pen-position data decreases the difference between the genuine signature and the forgery

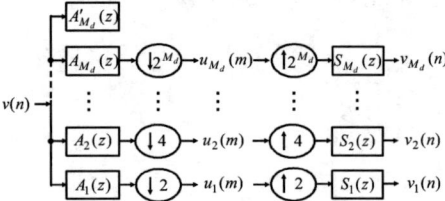

Fig. 1. Parallel structure of sub-band decomposition by DWT.

further. Therefore, it is not easy to distinguish between the genuine signature and the forgery by using the time-varying signal of a pen-position parameter.

In order to enhance the difference between a genuine signature and its forgery, we have proposed to verify the on-line signature in DWT domain [5]. In the following, $x(n)$ and $y(n)$ are represented as $v(n)$ for convenience. The DWT of the normalized pen-position $v(n)$ is defined as

$$u_k(m) = \sum_n v(n)\overline{\Psi_{k,m}(n)} \quad (3)$$

where $\Psi_{k,m}(n)$ is the wavelet function and $\overline{}$ denotes the conjugate. k is a frequency (level) index.

Moreover, it is well known that the DWT corresponds to the octave-band filter bank. Fig.1 shows a parallel structure of the sub-band decomposition. M_d is decomposition level. The synthesized signal $v_k(n)$ in each sub-band is called *Detail*. The *Detail* is the signal in high frequency band and so it contains differences between signals. Therefore, we consider the *Detail* as an enhanced individual feature in pen-position.

Results by sub-band decomposition of pen-position parameter are omitted for duplication. The difference between a genuine signature and its forgery can be enhanced in the DWT domain. Please refer to [5] in detail.

III. Fusion of Pen-position and Pen-movement Angle

In this paper, we propose to use not only pen-position parameter but also pen-movement angle parameter to improve verification performance.

A. Pen-movement Angle

We define pen-movement angle parameter $\theta(n)$ in Fig.2 and Eq.(4).

$$\theta(n) = \begin{cases} \tan^{-1}\frac{\Delta y(n)}{\Delta x(n)}, & \Delta x(n) > 0 \\ \tan^{-1} sgn(\Delta y(n)) \cdot \frac{\pi}{2}, & \Delta x(n) = 0 \\ \tan^{-1}\frac{\Delta y(n)}{\Delta x(n)} + \pi, & \Delta x(n) < 0, \Delta y(n) \geq 0 \\ \tan^{-1}\frac{\Delta y(n)}{\Delta x(n)} - \pi, & \Delta x(n) < 0, \Delta y(n) < 0 \end{cases} \quad (4)$$

$$\Delta x(n) = x(n) - x(n-s), \quad \Delta y(n) = y(n) - y(n-s) \quad (5)$$

where s presents amount of time shift.

The pen-movement angle parameter is derived from both x and y coordinates; therefore, it essentially has two-dimensional characteristics. As a result, the pen-movement angle parameter brings more obvious individual feature of the on-line

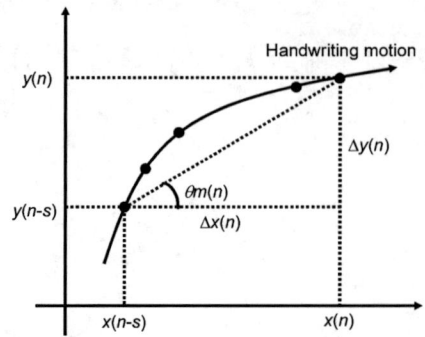

Fig. 2. Definition of pen-movement angle $\theta(n)$.

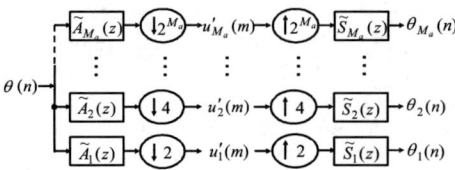

Fig. 3. Sub-band filter bank for extracting *Approximation*s.

signature than the pen-position parameter which is actually in one-dimensional. However, it is confirmed that the pen-movement angle parameter has large intra-class variation in signatures of one individual. For utilizing the pen-movement angle parameter in verification, some reduction method of the intra-class variation is required.

B. Suppressing Intra-class Variations

While a *Detail* is the high frequency band signal in sub-band decomposition by the DWT, the signal in low frequency band is called *Approximation* and so it contains similarity between signals. For suppressing the intra-class variations, we extract the *Approximation* as an enhanced similarity of the pen-movement angle parameter.

Fig.3 shows the sub-band filter bank for extracting the *Approximation*s. $\tilde{A}_k(z)$ and $\tilde{S}_k(z)$ where $k = 1, \cdots, M_a$ are the synthesis filter and the analysis filter, respectively. M_a is decomposition level in the pen-movement angle.

Examples of the *Approximation* are shown in Fig.4. Fig.4(a) indicates that the similarity of two genuine signatures is extracted in *Approximation*s. On the other hand, Fig.4(b) shows that the difference between a genuine signature and its forgery is kept even in *Approximation*s. These comparisons suggest that the verification using the *Approximation* can suppress the intra-class variation keeping the difference of the genuine signature and its forgery.

C. Multi-Matcher Verification System

By using pen-position and pen-movement angle parameters, we propose a new multi-matcher on-line signature verification system. Fig.5 shows a system overview. Pen-position, actually x and y coordinates and pen-movement angle are separately processed in verification block. The verification block is common to processing of pen-position and pen-movement angle.

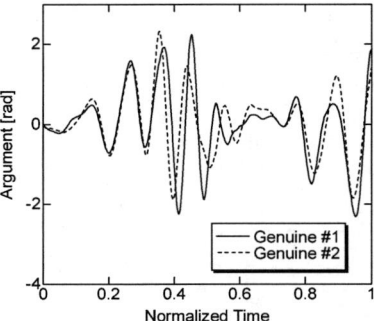

(a) Comparison between genuine signatures

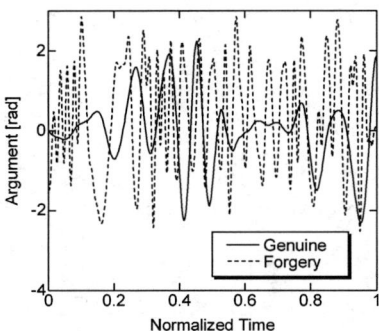

(b) Comparison between a genuine signature and its forgery

Fig. 4. Examples of *Approximation* in pen-movement angle.

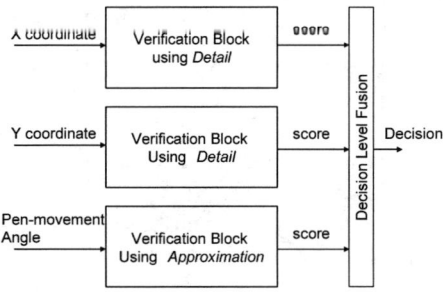

Fig. 5. Proposed multi-matcher verification system.

Fig. 6 describes the verification block, where pen-information, that is, pen-position and pen-movement angle parameters are decomposed into *Detail*s or *Approximation*s and then they are verified with templates using the adaptive signal processing at each level.

Before verification, templates must be enrolled in order to be compared with input signatures. As the template, T genuine signatures which have equal number of strokes are prepared and then their pen-position and pen-movement angle parameters are decomposed into sub-band signals by the DWT each other. Decomposition level is decided after examinations of those genuine signatures. Extracted T *Detail*s for the pen-position and T *Approximation*s for the pen-movement angle are averaged at the same level each other.

By the way, if the number of strokes in an input signature is different from that in a template, it is natural to consider

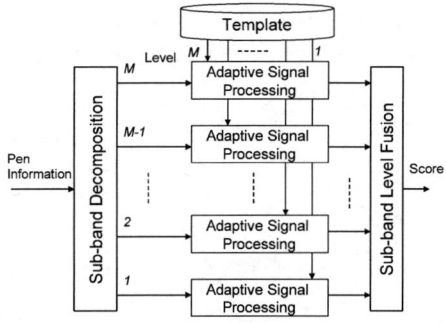

Fig. 6. Verification block

the input signature as a forgery. However, not all genuine signatures have the same number of strokes. We adopt the dynamic programming (DP) matching method to identify the number of strokes in an input signature with that in a template. The procedure of the stroke matching is omitted for lack of space. It is described in detail in [5].

D. Verification Using Adaptive Signal Processing

After enrollment phase, verification phase is executed. The verification is achieved by using the adaptive signal processing. The purpose of the adaptive signal processing is to reduce the error between the input signal and the desired signal sample by sample. When an input signal is of a genuine signature, the error between the input and its template becomes small; therefore, adaptive weights are expected to converge close on 1. Inversely, if the input signature is a forgery, adaptive weights converge far from 1. In this way, the verification can be achieved by examining whether converged value is nearly 1 or not [5].

As the adaptive algorithm, we use a new kind of steepest descent algorithm defined as follows.

$$w_k(n+1) = w_k(n) + \mu E\left[e_k(n)\rho_k(n)\right] \quad (6)$$

$$e_k(n) = t_k(n) - w_k(n)\rho_k(n) \quad (7)$$

$$E\left[e_k(n)\rho_k(n)\right] = \frac{1}{N_{tmp}} \sum_{l=0}^{N_{tmp}-1} e_k(r-l)\,\rho_k(r-l) \quad (8)$$

$$\mu = \mu_0 / \left\{ E\left[|\rho_k(n)|\right] \right\}^2 \quad (9)$$

$$E\left[|\rho_k(n)|\right] = \frac{1}{I} \sum_{l=0}^{I-1} \rho_k(n-l) \quad (10)$$

where I is the number of sampled data in an input *Detail* or *Approximation*. N_{tmp} is the number of sampled data in a template. μ is a step size parameter which controls the convergence in the adaptive algorithm. The step size parameter is normalized by input power as shown in Eqs.(9) and (10), so that convergence is always guaranteed. μ_0 is a positive constant.

The verification is done in all sub-bands in parallel. After enough iterations for convergence, $w_k(n)$ is averaged in past N_{tmp} samples and then we obtain the converged value w_k. A verification score is obtained by fusing several sub-band level results. In this paper, such sub-band level fusion is achieved by averaging the converged values.

$$\text{Score}^x = \frac{1}{L_d} \sum_{k=M_d-L_d+1}^{M_d} w_k^x$$
$$\text{Score}^y = \frac{1}{L_d} \sum_{k=M_d-L_d+1}^{M_d} w_k^y$$
$$\text{Score}^\theta = \frac{1}{L_a} \sum_{k=M_a-L_a+1}^{M_a} w_k^\theta \quad (11)$$

where w_k^x, w_k^y and w_k^θ respectively denote the converged values of x, y and pen-movement angle at level k. L_d is the used level number in decision fusion of pen-position and L_a is that of pen-movement angle.

E. Decision Fusion

As shown in Fig.5, verification results are fused and then we obtain a total decision. There have been proposed many fusion methods such as the sum rule, the minimum score, the maximum score and so on [6]. In this paper, we employ the sum rule in which scores by pen-position and pen-movement angle are weighted and then summed. The total decision for verification is defined as

$$\text{TC} = c_x \cdot \text{Score}^x + c_y \cdot \text{Score}^y + c_\theta \cdot \text{Score}^\theta$$
$$c_x + c_y + c_\theta = 1, \quad c_x > 0, c_y > 0, c_\theta > 0 \quad (12)$$

c_x and c_y denote the weighting factors for x and y coordinates, respectively and c_θ is that for pen-movement angle. These are determined in preliminary examinations. When the TC is greater than a threshold, an input signature is decided to be genuine.

IV. EXPERIMENTAL RESULTS

Four subjects were requested to sign their own signatures and then we obtained 118 genuine signatures. Five genuine signatures for each subject were used to make a template and the remaining 98 genuine signatures were used for verification. Five subjects were required to counterfeit the genuine signature 10 times each, so that 200 forgeries were prepared in total. Other conditions of simulation are summarized as follows.

- Scaling parameter: $\alpha_x = \alpha_y = 100$
- Wavelet function: Daubechies8
- Number of signatures for making a template: $T = 5$
- Upper limit decomposition level: $M_d^{max} = 8$
- Decomposition level in pen-movement angle: $M_a = 4$
- Number of processed level: $L_d = 4, L_a = 2$
- Step size constant: $\mu_0 = 0.0001$
- Number of iterations: 10^5
- Time shit in pen-movement angle: $s = 8$
- Weighting factor: $c_x = 0.4, c_y = 0.2, c_\theta = 0.4$

Fig. 7 shows the variation of False Rejection Rate (FRR) and False Acceptance Rate (FAR) with total decision threshold. In general, verification performance is estimated by Equal

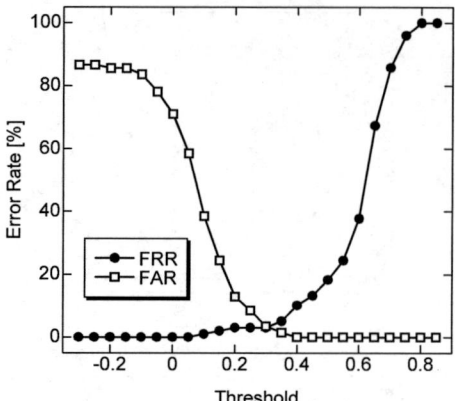

Fig. 7. Variation of FAR and FRR with decision threshold.

Error Rate (EER) where the FRR is equal to the FAR. The EER was about 3.5% when the threshold value was about 0.30. Verification rate was about 96.5%. Comparing with our conventional on-line signature verification method in which only pen-position is processed [5], the verifiaction performance was improved by about 1.5%.

V. CONCLUSION

We have presented a multi-matcher on-line signature verification system. Both pen-position and pen-movement angle parameters were decomposed into sub-band signals by the DWT. Moreover, high frequency sub-band signals called *Detail* were extracted as individual features in pen-position to enhance the difference between a genuine signature and its forgery while low frequency sub-band signals called *Approximation* were used for pen-movement angle to suppress the intra-class variation of signatures in one individual. Verification results of pen-position and pen-movement angle were combined at decision level. We demonstrated that the proposed multi-matcher scheme improved the performance of the on-line signature verification system in the DWT domain by about 1.5%.

REFERENCES

[1] S. Prabhakar, A. K. Jain, "Decision-level Fusion in Fingerprint Verification," *Pattern Recognition*, vol.35, pp.861-874, 2002.
[2] Y. Sato and K. Kogure, "Online Signature Verification Based on Shape, Motion, and Writing Pressure," *Proc. of 6th International Conference on Pattern Recognition*, pp.823-826, 1982.
[3] M. Yoshimura, Y. Kato, S. Matsuda, and I. Yoshimura, "On-line Signature Verification Incorporating the Direction of Pen Movement," *IEICE Trans.*, vol.E74, no.7, pp.2083-2092, July 1991.
[4] Y. Komiya, T. Ohishi and T. Matsumoto, "A Pen Input On-line Signature Verifier Integrating Position, Pressure and Inclination Trajectories," *IEICE Trans. Inf. & Syst.*, vol.E84-D, no.7, pp.833-838, July 2001.
[5] I. Nakanishi, N. Nishiguchi, Y. Itoh, and Y. Fukui, "On-line Signature Verification Method Based on Discrete Wavelet Transform and Adaptive Signal Processing," *Proc. of Workshop on Multimodal User Authentication*, Santa Barbara, USA, pp.207-214, Dec. 2003.
[6] M. Indovina, U. Uludag, R. Snelick, A. Mink, and A. Jain, "Multimodal Biometric Authentication Methods: A COTS Approach," *Proc. of Workshop on Multimodal User Authentication*, Santa Barbara, USA, pp.99-106, Dec. 2003.

Digital System for Detection and Classification of Electrical Events

Augusto S. Cerqueira, Carlos A. Duque and Rogério M. Trindade
Electrical Engineering Department
UFJF
Juiz de Fora, Brazil

Moises V. Ribeiro
DECOM/FEEC
UNICAMP
Campinas, Brazil

Abstract—This paper describes an algorithm to detect and classify electrical events related to power quality. The events detection is based on monitoring the statistical characteristics of the energy of the error signal, which is defined as the difference between the monitored waveform and a sinusoidal wave generated with the same magnitude, frequency, and phase as the fundamental sinusoidal component. The system novel feature is the event recognition based on a neural network that uses the error signal as input. Multi-rate techniques are also employed to improve the system operation for on-line applications. Software tests were performed showing the good performance of the sytem.

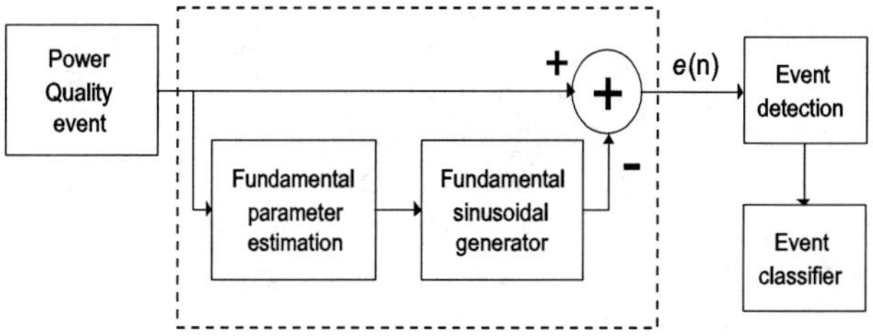

Fig. 1. The system block diagram

I. INTRODUCTION

POWER QUALITY (PQ) issues have emerged as a research field of exponentially demanding attention to electrical utilities and end-users after the worldwide energy market deregulation due to many reasons [1]. Among them, deserve attention the increased use of electronic devices, non-linear loads and microprocessor-based solutions in the commercial centers and industrial plants. The increased number of independent electrical power producers (IPP's) with poorly controlled synchronization together with the market competition also contributes for power quality problems.

In this context, several tools have been employed to measure and diagnose PQ problems. The main PQ events are transients, long and short-duration voltage variations, flickers, interruptions, waveform distortions, unbalances, etc [2], [3].

As far as the PQ event monitoring solutions are being concerned, the use of digital signal processing and computational intelligence techniques [4] have been widely applied to detect disturbance occurrences, estimate the harmonic contents, evaluate PQ parameters, and classify the PQ problems. Several solutions based on wavelet transform, FFT, and computational intelligence have been directly applied to the monitored signal to detect PQ disturbances [3]–[5]. However, previous works [6], [7] presented a different perspective for PQ event detection and compression by using an engineering principle, namely *divide and conquer*. Basically, this principle states that a good approach to solve a complex problem is to break it into several simple problems.

We pointed out in [6], [7] that when the monitored PQ event is split into stationary and non-stationary components, the detection and compression problems can be efficiently solved.

In this work, we propose a digital system for detection and classification of electrical events on power quality with two new contributions to the previous one. The first contribution is the use of multi-rate technique in the detector structure of the system and the second contribution is the inclusion of one classifier stage in the system. The new approach on the classifier is the use of the error signal instead of the original monitored signal.

Next section presents a overview of the digital system and its multi-rate implementation. On Section III, the classification method is described. The results are presented on Section IV and conclusions are derived on Section V.

II. DIGITAL SYSTEM FOR DETECTION AND CLASSIFICATION OF POWER QUALITY EVENTS

The complete block diagram of the proposed system is presented in Fig. 1. The samples of the monitored waveform are represented by the Power Quality Event block. From these samples, the Fundamental Parameter Estimation block estimates the frequency, the phase and the amplitude of the fundamental component. After the parameters estimation, the system generates the fundamental signal to be compared with the original waveform, on the Sinusoidal Generation block, generating the error signal $e(n)$. The Event Detection block implements the error innovation concept to detect the presence of PQ disturbances. Finally, using the error signal and the some information from the event detection block, the event classification is performed on the Event Classifier block.

The blocks surrounded by a dashed box in Fig. 1 work as a notch filter centered in the fundamental frequency (60 or 50 Hz). The goal is extract the fundamental component of the input signal and estimate its parameters (amplitude, phase

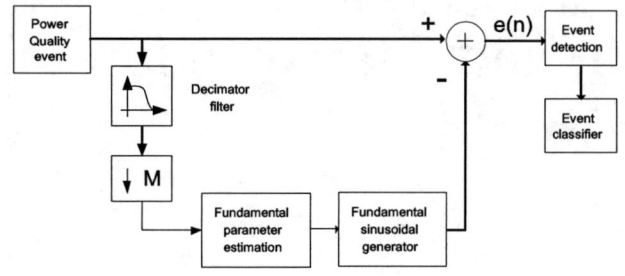

Fig. 2. The multi-rate version of the detection system.

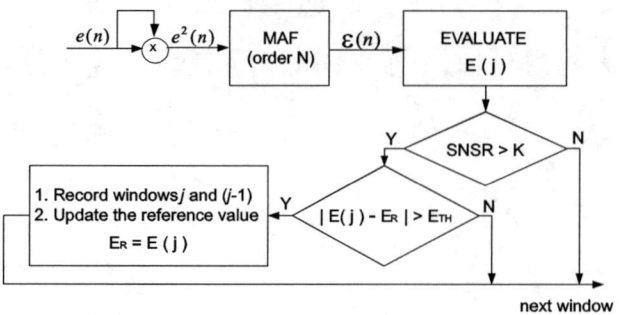

Fig. 3. The innovation concept applied to detect disturbances.

and small frequency deviations). These parameters are used to update the sinusoidal generator. The projected Sinusoidal Generator does not respond to transients in the input signal but only when new fundamental stead state is reached. More details for these blocks are presented in [6], [7].

Some PQ events have high frequency components, e.g capacitor switching and notch, so the digital system needs to work at high samplng rate to detect these events. On the other hand, the fundamental parameter are related with 60 Hz (or 50 Hz) frequency and a low pass filtering is a good choice to improve the fundamental parameters estimation. This is a good structure to apply the multi-rate approach.

The implementation of the multi-rate technique to the system reduces the computer complexity of the algorithm and save processing time. The new system can be seen in Fig. 2. The error signal is generated at high frequency. Note that the decimator filter and the down sampling device reduce the sampling rate by a M factor, so the parameters of the fundamental component are estimated in lower frequency. One important feature of this system is that the up-sampler and the interpolating filter are not necessary once the increasing of the fundamental sinusoidal generator sampling rate can be implemented directly inside the generator. The implementation was done with 256 samples per cycle in a high sample rate and the down-sampling factor was M=8.

The event detection algorithm is based on the Error Energy Innovation Concept described in [6]. Fig. 3 illustrates the main block in the event detector. First of all, the RMS of the error signal is generated ($\epsilon(n)$) using a Moving Average Filter (MAF). The N samples of $\epsilon(n)$ are collected in a frame (or window) and the mean of the j-th frame is computed. Then, the Stationary to Non-Stationary Ratio (SNSR) value is computed using Eq. (1).

$$SNSR = 10 \cdot \log_{10}\left(\frac{\left\{\frac{1}{N}\sum_{n=0}^{N-1}\hat{A}_0(n)\right\}^2}{E(j)}\right) \quad (1)$$

where $\hat{A}_0(n)$ is the estimated amplitude for the fundamental component.

The SNSR informs the relationship between the energy of the fundamental component and the non-stationary component. If the SNSR is higher than a threshold value K, the level of non-stationarity is high and a possible event can exist in the j-th window. The K threshold is related with the environment signal to noise rate (SNR). The next step consists in verify if this error window has new information about the error signal or not. The practices have shown that the first order statistic has being sufficient to detect a new event when the error signal is used.

When a new event is detect, its respective window is recorded and the reference energy E_R is update as shown in Fig. 3.

The next step consists in classify the events detected. This issue will be discussed on the following section.

III. EVENT CLASSIFICATION

To perform the event classification task, a feed-forward multilayer neural network was used, based on the good performance achieved by the neural networks on classification problems [8].

The new concept applied on PQ event classification is the use of the error signal to perform this task. The motivation for this approach comes from the fact that for some PQ events (transients, notchs, etc), the major part of the signal energy comes from the fundamental sinusoidal signal, which does not contributes to the classification problem. Using the error signal we avoid this problem.

Another issue that was taken into account, when we applied the error signal on the classifier, is the computational simplicity of the proposed solution. This solution avoids the use of the wavelet transform, as proposed in several works in this area [9], [10].

The classification process starts with the error window given by the event detection structure, each window has 1024 samples. The events are aligned from the beginning by a simple Moving Average Filter (10 coefficients) based algorithm that detects the sample where each event starts. Once the event start is detected, the signal is truncated with 512 samples, which means two periods of the fundamental sinusoidal signal.

To further reduce the computational complexity and processing time, the 512 samples signal is down-sampled by a factor of 8 resulting into a 64 samples signal. This signal is finally presented to the neural network previously trained for event classification. Then, the outputs of the neural network can be used for PQ event classification. The classifier block diagram can be seen on Fig. 4.

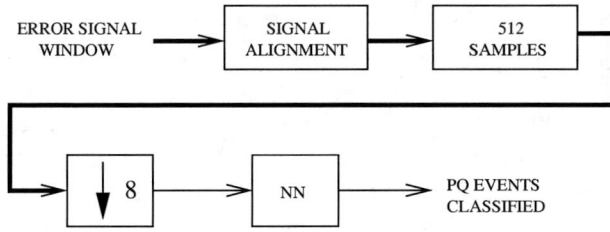

Fig. 4. The classifier block diagram

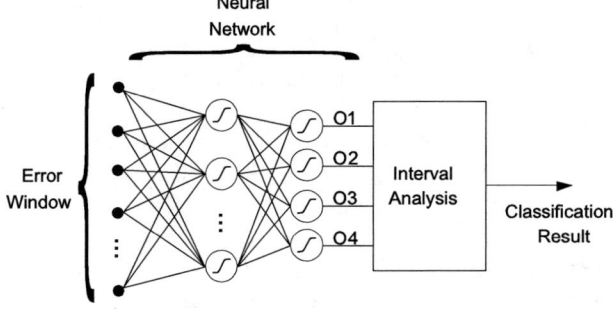

Fig. 5. The neuron network architecture for PQ event classification.

It is important to emphasize that the down-sampling operation introduces aliasing on the original error signal because we avoided the use of the non-aliasing filter to reduce the processing time. On Section IV, it could be seen that the aliasing doesn't harm the classifier efficiency.

The neural network architecture (see Fig. 5) is based in three layers, the input layer, one hidden layer and the output layer. The input layer is composed by the 64 samples of the error signal normalized by the peak value. The activate function used in each neuron was the hyperbolic tangent.

IV. RESULTS

A. Events Generation and Detection

To check the system performance it was necessary to generate by software PQ events according to their features described in [2]. The events generated were: oscillatory events, sags, swells, notches and spikes. The features of each event are randomly defined during the event generation. This results into a population of statistically independent events that could be grouped in classes to check the system performance on detection and classification. To all signals generated was added a white noise with signal to noise ratio (SNR) of 25 dB.

Once the PQ events are presented to the system, the detection stage generates two outputs: the signal window, which contains 1024 samples of the original signal containing the events detected; and the error window, which contains the corresponding 1024 samples of the error signal generated by the system.

Typical signal windows recorded are show in Fig. 6, each one has 1024 samples. The events are: an oscillatory event in (a), sag in (b), swell in (c) and the presence of notches and spikes in (d) and (f) respectively.

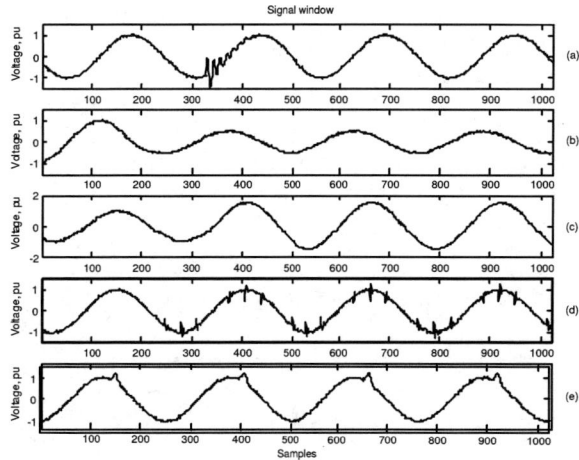

Fig. 6. Typical signal window recorded by each event kind: (a) oscillatory, (b) sag, (c) swell, (d) notch and (e) spike.

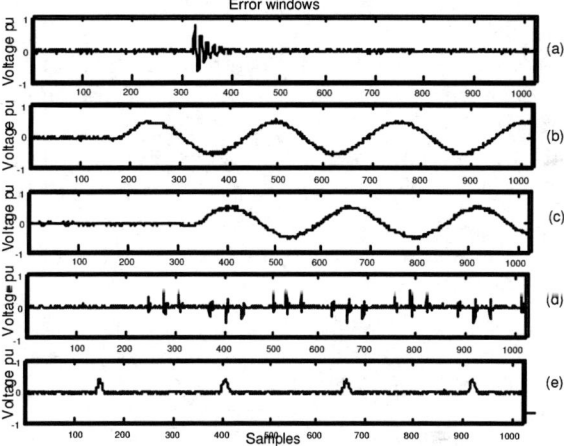

Fig. 7. Error window recorded by each event shown in Fig. 6: (a) oscillatory, (b) sag, (c) swell, (d) notch and (e) spike.

The error windows related to the signal windows in Fig. 6 are shown in Fig. 7.

B. Classification

To the classification problem, four event classes were defined: the oscillatory events (Class 1), sags and swells (Class 2), notchs (Class 3) and the spikes (Class 4). The sags and swells were grouped into the same class because their error signals have similar patterns (see Fig. 7). This restriction could be easily solved using the amplitude information of the original waveform after the classification result given by the neural network (NN). The number of generated events per class were 120.

A example of the signals presented to the NN can be seen in Fig. 8. As explained on last section, the error signals are first aligned and then down-sampled by a factor of 8 resulting in a 64 samples signal (see Fig. 4). Again, if we compare the sag and the swell events we see the same behavior.

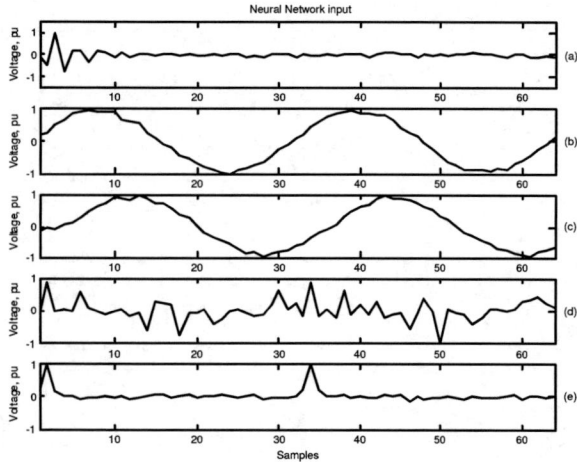

Fig. 8. A example of the neural network input signals corresponding to the events shown in Fig. 7: (a) oscillatory, (b) sag, (c) swell, (d) notch and (e) spike.

The NN was designed with 4 neurons on the output layer, one for each class, and the number of neurons on the hidden layer was defined after several tests. The goal of these tests was to minimize the number of neurons on the hidden layer keeping a good discrimination performance. The tests had shown that 8 neurons on the hidden layer is a good compromise between performance and time processing. Refer to Fig. 5 for details on the NN architecture

The classification problem requires a supervised training method, therefore, for the events of the Class 1 the target output for the NN is $[1, -1, -1, -1]$, the Class 2 target output is $[-1, 1, -1, -1]$, the Class 3 target output is $[-1, -1, 1, -1]$ and finally for the Class 4 the target output is $[-1, -1, -1, 1]$.

To the training procedure, 120 events of each class were generated and divided randomly in two sets: the training set with 80 events and the test set with 40 events. The weights (synapses values) of the NN were adjusted using the back-propagation algorithm and the training stop criterion was based on the global efficiency achieved on the test set to preserve the generalization feature of the neural network.

One event were considered to be corrected classified if the output neuron corresponding to the target 1 shows a greater value than the other ones (target -1). Therefore, the efficiency for a particular class is the number of correct events classified over the number of events on the set. The global efficiency were defined as the product of the individual efficiencies. The global efficiency were 100 % for both training and test sets.

After the training procedure, 100 events for each class were independently generated to check the system performance and generalization of the classifier. The weights of NN are fixed after the training procudure. The efficiency for class 1 was 99 %, for class 2 100 %, class 3 100 % and for the class 4 the efficiency was 99 %. The global efficiency was 98 %. The results had confirmed the generalization capacity of the proposed classifier.

V. CONCLUSIONS

This paper presented an algorithm to detect and classify electrical events related to power quality. The main contributions in this work are the classification method adopted and the event detector multi-rate approach for on-line implementations.

The multi-rate implementation of the detector system reduces the computational burden and improves the fundamental parameter estimation of the monitored signal, once the low-pass filter decimator reduces the high frequency components that could deteriorate the estimator performance.

A neural network applied directly to the error signal was used for the classification of the power quality events. This solution shown good results with reduced computational complexity. For on-line implementations, the neural network could also be trained in place to adapt to the typical events presented on that particular environment, showing the flexibility of the proposed solution.

The next steps are related to the on-line implementation of the proposed system and the increase of the number of classes to the classification problem,

ACKNOWLEDGMENT

The authors would like to thank CAPES and CPNq (from Brazil) for the support, the colleagues from the electrical engineering department of UFJF for the fruitful discussions and DECOM/FEEC/UNICAMP for the collaboration with this work.

REFERENCES

[1] J. Arrillaga, N. R. Watson, and M. H. J. Bollen, "Power quality following deregulation," in *Proc. of the IEEE*, vol. 88, no. 2, Feb. 2000, pp. 246–261.
[2] R. C. Dugan, M. F. McGranaghan, and H. W. Beaty, *Electrical Power Systems Quality*. McGraw-Hill, 1996.
[3] O. Poisson, P. Rioual, and M. Meunier, "Detection and measurement of power quality disturbances using wavelet transform," *IEEE Trans. on Power Delivery*, vol. 15, no. 3, pp. 1039–1044, 2000.
[4] W. R. A. Ibrahim and M. M. Morcos, "Artificial intelligence and advanced mathematical tools for power quality applications: a survey," *IEEE Trans. on Power Delivery*, vol. 17, no. 2, pp. 668–673, 2002.
[5] S. Santoso, W. M. Grady, E. J. Powers, J. Lamoore, and S. C. Bhatt, "Characterization of distribution power quality events with fourier and wavelet transforms," *IEEE Trans. on Power Delivery*, vol. 15, no. 1, pp. 247–254, 2000.
[6] M. V. Ribeiro, C. A. Duque, and J. M. T. Romano, "An improved method for signal processing and compression in power quality evaluation," *IEEE Trans. on Power Delivery*, vol. 19, no. 2, pp. 464–471, 2004.
[7] F. R. Ramos, M. V. Ribeiro, J. M. T. Romano, and C. A. Duque, "On signal processing approach for event detection and compression applied to power quality evaluation," in *Proc. of the IEEE 10th International Conference on Harmonic and Quality of Power*, 2002, pp. 133–138.
[8] R. P. Lippmann, "A critical overview of neural network pattern classifiers," *IEEE workshop on Neural Network for Signal Processing*, pp. 266–275, 1991.
[9] S. Santoso, E. Powers, W. M. Grady, and A. C.Parsons, "Power quality disturbance waveform recognition using wavelet-based neural classifier-part 1: Theoretical foundation," *IEEE Trans. on Power Delivery*, vol. 15, no. 1, pp. 222–228, 2000.
[10] ——, "Power quality disturbance waveform recognition using wavelet-based neural classifier-part 2: Application," *IEEE Trans. on Power Delivery*, vol. 15, no. 1, pp. 229–235, 2000.

Tap Selection Based MMSE Equalization for High Data Rate UWB Communication Systems

Lin Zhiwei, Benjamin Premkumar and A.S.Madhukumar
School of Computer Engineering, Nanyang Technological University
Block N4, Nanyang Avenue, Singapore 639798
Email: zhiweilin@pmail.ntu.edu.sg, {asannamalai, asmadhukumar}@ntu.edu.sg

Abstract—In this paper, the performance of a tap selection based Minimum Mean Square Error (MMSE) equalization technique for high data rate Ultra Wideband (UWB) systems is analyzed. This technique is shown to significantly outperform the conventional uniformly spaced equalizer with the same number of taps. In addition, the impact of channel estimation error on the performance of tap selection based Decision Feedback Equalization (DFE) is analyzed. This channel estimation based DFE is shown to perform well with only small amount of training symbols. It can then be adapted effectively using Least Mean Square (LMS) algorithm for improved performance. The numerical analysis and simulation results show that tap selection based equalization technique is a promising approach for practical high performance UWB system implementation with reduced complexity.

I. INTRODUCTION

Impulse Radio (IR) technique [1], which uses very short duration baseband pulses, is one of the most promising approaches for UWB systems. Compared to multicarrier Orthogonal Frequency Division Multiplexing (OFDM) based UWB method [2], IR has the advantages such as very low power consumption, simple transceiver design and robustness due to its resolvable multipath. For high data rate IR based UWB systems, channel response may span over multiple symbol frames. Conventional Rake receiver suffers from performance degradation due to severe Intersymbol Interference (ISI). An MMSE equalization based receiver is superior for ISI mitigation. But this involves large computational complexity and requires large amount of training symbols due to the large delay spread for UWB channel. A key challenge is to develop a high performance equalizer with manageable complexity.

Non-uniformly spaced equalizer has been discussed for sparse multipath channels such as high-definition television test channel [3], "typical urban" and "hilly terrain" profiles [4]. In these works, the performance of the equalization is discussed using simulations and is based on various intuitive tap selection techniques. These prior works motivate us to consider tap selection based equalization for UWB channels. The UWB indoor channel is not as sparse as those channels discussed in the prior works. Therefore, it is important to have instructive performance analysis in order to evaluate the effectiveness for using tap selection based equalization for UWB systems.

In this paper, the performance of a tap selection based MMSE equalization technique for high data rate UWB systems is analyzed for the first time using greedy method. In addition, the impact of channel estimation error on tap selection based DFE technique is discussed. The performance is evaluated based on channel estimation techniques using Least Squares (LS) and Matching Pursuit (MP) algorithms for UWB indoor channels.

II. SYSTEM MODELS

A UWB system with bipolar data modulation is considered. Let $d(n) \in \{\pm 1\}$ be the data bit stream and $w(t)$ be the pulse waveform.

The transmitted signal is represented as follows.

$$s(t) = \sum_{n=-\infty}^{\infty} d(n) w(t - nT_f) = x(t) * w(t) \quad (1)$$

where $x(t) = \sum_{n=-\infty}^{\infty} d(n) \delta(t - nT_f)$ and T_f is the symbol duration.

Applying a tapped-delay-line based multipath channel model $c(t) = \sum_{p=1}^{n_L} \alpha_p \delta(t - (p-1)\Delta\tau - \tau)$, the received signal is given by,

$$r(t) = c(t) * s(t) + n(t) = h(t) * x(t) + n(t) \quad (2)$$

where $h(t) = c(t) * w(t)$ denotes the generalized Channel Impulse Response (CIR), $\Delta\tau$ is the sampling duration, τ denotes the channel delay and $n(t)$ is the AWGN signal.

For simplicity, assuming that the sampling rate $\frac{1}{\Delta\tau}$ is an integer multiple of the symbol repetition rate $\frac{1}{T_f}$. Let $n_\tau = T_f/\Delta\tau$ and the number of symbols affected by ISI be $n_{ISI} = \lceil n_L/n_\tau \rceil$, where n_L denotes the maximum length of the multipath channel counted in samples. Then we have,

$$\mathbf{y}(n) = \sum_{k=n-n_{ISI}+1}^{n} \mathbf{h}_k^n x(k) + \mathbf{m}(n) \quad (3)$$

where $x(k) = d(k)$ is the transmitted symbol at the *k-th* frame. The vector $\mathbf{y}(n) = [y^{(1)}(n), \cdots, y^{(n_\tau)}(n)]^T$ denotes the multi-tap received signal samples and $\mathbf{m}(n)$ is the corresponding AWGN signal vector. The vector $\mathbf{h}_k^n$ denotes the channel transmission matrix from the *k-th* transmitting frame to the *n-th* receiving frame and is expressed as,

$$\mathbf{h}_k^n = \mathbf{h}(n-k) = \begin{bmatrix} h^{(1)}(n,k) & \cdots & h^{(n_\tau)}(n,k) \end{bmatrix}_{1 \times n_\tau}^T \quad (4)$$

Let $h^{(p)}(n,k) = h_q$, where $q = p + (n-k)n_\tau$ and h_q is the *q-th* path of the sampled generalized CIR $h(t)$.

To acquire more multipath energy and diversity, $x(n)$ is detected after a delay of n_{sym} frames. A UWB system model extended from (3) can be expressed as,

$$\mathbf{y} = \mathbf{H}\mathbf{x} + \mathbf{m} \quad (5)$$

where $\mathbf{y} = [\mathbf{y}^T(n), \cdots, \mathbf{y}^T(n+n_{sym}-1)]^T$, $\mathbf{x} = [x(n-n_{ISI}+1), \cdots, x(n), \cdots, x(n+n_{sym}-1)]^T$ and the channel transmission matrix in block Toeplitz form is represented as,

$$\mathbf{H} = \begin{pmatrix} \mathbf{h}(n_{ISI}-1) & \cdots & \mathbf{h}(0) & \cdots & 0 \\ \vdots & \ddots & \ddots & \ddots & \vdots \\ 0 & \cdots & \mathbf{h}(n_{ISI}-1) & \cdots & \mathbf{h}(0) \end{pmatrix}_{N_R \times N_T} \quad (6)$$

where $N_R = n_\tau n_{sym}$, $N_T = n_{ISI} + n_{sym} - 1$.

Consider Successive Interference Cancellation (SIC) for the previously detected symbols, (5) can be re-written as

$$\mathbf{y}' = \begin{pmatrix} \mathbf{H}' & \mathbf{H} \end{pmatrix} \begin{pmatrix} \mathbf{x}' \\ \mathbf{x} \end{pmatrix} + \mathbf{m} \quad (7)$$

$$\mathbf{y} = \mathbf{y}' - \mathbf{H}'\mathbf{x}' = \mathbf{H}\mathbf{x} + \mathbf{m} \quad (8)$$

where $\mathbf{x} = [x(n), \cdots, x(n+n_{sym}-1)]^T$ and $\mathbf{x}' = [x(n-n_{ISI}+1), \cdots, x(n-1)]^T$.

III. TAP SELECTION BASED MMSE EQUALIZATION PERFORMANCE ANALYSIS

From UWB system model (8), assume that the desired symbol for detection is the *first* element in vector $\mathbf{x}$, the Mean Square Error (MSE) under MMSE detection criterion can be derived as follows,

$$MSE^{(1)} = \left\{ E\left\{ (\mathbf{x} - \hat{\mathbf{x}})(\mathbf{x} - \hat{\mathbf{x}})^T \right\} \right\}_{1,1} = \left\{ (\mathbf{H}^T \mathbf{R}_m^{-1} \mathbf{H} + \mathbf{I}_{N_T})^{-1} \right\}_{1,1} \quad (9)$$

where the subscript (i,j) in $\{\mathbf{A}\}_{i,j}$ denotes the element from the *i-th* row and *j-th* column of the matrix $\mathbf{A}$.

In the presence of AWGN signal vector $\mathbf{m}$ with covariance of $\mathbf{R}_m = \sigma_m^2 \mathbf{I}$, the SIR is given by [5],

$$SIR^{(1)} = \frac{1}{MSE^{(1)}} - 1 = \frac{1}{\sigma_m^2 \left\{ (\mathbf{H}^T \mathbf{H} + \sigma_m^2 \mathbf{I}_{N_T})^{-1} \right\}_{1,1}} - 1 \quad (10)$$

Let n_L be the total number of taps within the observation window that are over sampled and n_S be the number of selected taps for non-uniformly spaced equalization. For the optimal way of selecting n_S taps out of total n_L taps, an exhaustive search should be made which requires the MMSE detection performance to be evaluated for $\binom{n_L}{n_S}$ possible combination of the tap subsets. This is not feasible for large n_L. So a greedy algorithm based method is developed for the performance analysis of tap selection based equalization.

Tap selection can be considered as forming a transmission matrix $\mathbf{H}_S$ by choosing a subset of rows from $\mathbf{H}$ to maximize the $SIR^{(1)}_{\mathbf{H}_S}$ for MMSE detection. From (10), this is equivalent to minimizing an objective function defined as $\{\mathbf{A}_S\}_{1,1} = \left\{ (\mathbf{H}_S^T \mathbf{H}_S + \sigma_m^2 \mathbf{I}_{N_T})^{-1} \right\}_{1,1}$.

Suppose that at the *(n-1)-th* step, a selected transmission matrix with $(n-1)$ rows is denoted as $\mathbf{H}_{n-1}$. Define,

$$\mathbf{A}_{n-1} = (\mathbf{H}_{n-1}^T \mathbf{H}_{n-1} + \sigma_m^2 \mathbf{I}_{N_T})^{-1} \quad (11)$$

To obtain $\mathbf{H}_n$ at the *n-th* step, one more row denoted as $\mathbf{h}_s$ will be selected and appended to the matrix $\mathbf{H}_{n-1}$. Also notice that the rows in the transmission matrix may be re-ordered. Then we have,

$$\mathbf{H}_n^T \mathbf{H}_n = \begin{pmatrix} \mathbf{H}_{n-1}^T & \mathbf{h}_s^T \end{pmatrix} \begin{pmatrix} \mathbf{H}_{n-1} \\ \mathbf{h}_s \end{pmatrix} = \mathbf{H}_{n-1}^T \mathbf{H}_{n-1} + \mathbf{h}_s^T \mathbf{h}_s \quad (12)$$

$$\mathbf{A}_n = (\mathbf{H}_n^T \mathbf{H}_n + \sigma_m^2 \mathbf{I}_{N_T})^{-1} = (\mathbf{A}_{n-1}^{-1} + \mathbf{h}_s^T \mathbf{h}_s)^{-1} \quad (13)$$

Applying matrix inversion lemma [6], i.e., $(\mathbf{A} + \mathbf{U}\mathbf{V}^T)^{-1} = \mathbf{A}^{-1} - \mathbf{A}^{-1}\mathbf{U}(\mathbf{I} + \mathbf{V}^T \mathbf{A}^{-1} \mathbf{U})^{-1} \mathbf{V}^T \mathbf{A}^{-1}$, an iterative formula is obtained with reduced computation cost for matrix inversion.

$$\mathbf{A}_n = \mathbf{A}_{n-1} - \mathbf{A}_{n-1}\mathbf{h}_s^T (1 + \mathbf{h}_s \mathbf{A}_{n-1} \mathbf{h}_s^T)^{-1} \mathbf{h}_s \mathbf{A}_{n-1} \quad (14)$$

Thus, an incremental iterative algorithm is developed for tap selection as follows.

1) Start from $n = 0$, set $\mathbf{A}_0 = \frac{1}{\sigma_m^2} \mathbf{I}_{N_T}$.
2) At step n, using (14), one more row $\mathbf{h}_s$ is selected by $s = \arg\min_k \left\{ \{\mathbf{A}_n\}_{1,1} \right\} = \arg\min_k \left\{ \{ (\mathbf{A}_{n-1}^{-1} + \mathbf{h}_k^T \mathbf{h}_k)^{-1} \}_{1,1} \right\}$, where k denotes the unselected row index for $\mathbf{H}$.
3) Set $n = n+1$, continue looping to select the taps until $n = n_S$.

Based on the IEEE UWB channel model recommendation [7], the performance analysis for tap selection based MMSE equalization is illustrated in Figure 1, where the channel model used is CM2 with a sampling duration of $0.167ns$ and total number of samples $n_L = 512$. It is shown that the greedy method is able to achieve near optimal tap selection. Since the Bit Error Rate (BER) performance curve by tap selection (with merely 64 taps) is already quite close to the optimal MMSE detection performance when applying all taps (totally 512 taps for CM2).

On the other hand, the tap selection based MMSE equalization significantly outperforms the conventional uniformly spaced MMSE equalization (with either 64 equally spaced taps over the entire channel length or 64 earliest taps selected), as shown in Figure 1.

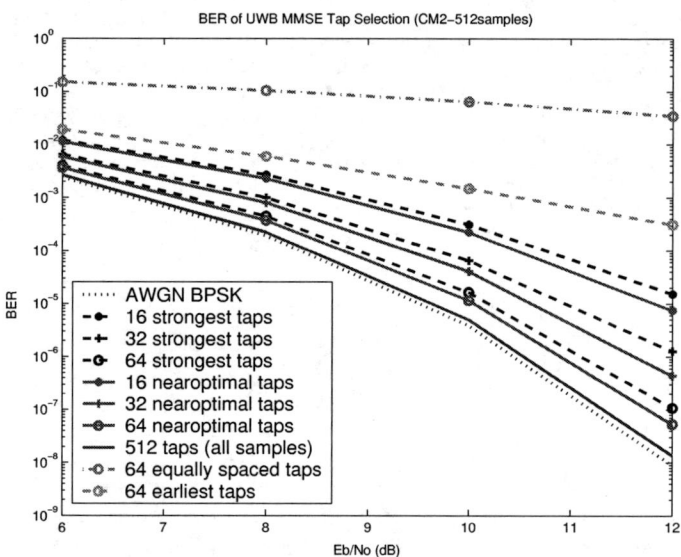

Fig. 1. Tap selection based MMSE equalization performance analysis

An intuitive way to make the tap selection is to simply choose n strongest paths from CIR. It is observed in Figure 1 that the performance gap between greedy method based tap selection and the strongest path based tap selection is limited. This indicates that the strongest path based tap selection is an effective way for real time implementation of tap selection with limited performance tradeoff.

IV. UWB CHANNEL ESTIMATION

Perfect channel knowledge has been assumed for tap selection based UWB receiver in the above analysis. In practice, CIR needs to be estimated for system implementation. Imperfect channel estimation inevitably results in performance degradation. It is therefore important to discuss the receiver performance based on the channel estimation techniques.

From UWB system model (5), an alterative representation of the signal model can be written as,

$$\mathbf{X}\mathbf{h} + \mathbf{m} = \mathbf{y} \quad (15)$$

where $\mathbf{h}$ denotes the generalized CIR and $\mathbf{m}$ is AWGN signal. Assume the length of transmitted training symbols as $(n_x n_{ISI})$, where n_x is a scale multiplier, the observation window will span over total $(n_x n_L)$ samples. A Toeplitz matrix that consists of training symbols is then given by,

$$\mathbf{X} = \begin{pmatrix} x(n) & 0 & \cdots & 0 & x(n-1) & 0 & \cdots \\ 0 & x(n) & 0 & \cdots & 0 & x(n-1) & \cdots \\ \vdots & & \ddots & & \vdots & & \\ x(n+1) & 0 & \cdots & 0 & x(n) & 0 & \cdots \\ \vdots & & & & \vdots & & \\ x(n+n_{ISI}-1) & 0 & \cdots & \cdots & x(n+n_{ISI}-2) & 0 & \cdots \\ \vdots & & & & \vdots & & \end{pmatrix}_{n_x n_L \times n_L} \quad (16)$$

A. Least Squares (LS) Based Channel Estimation

LS based estimation [8] for CIR can be obtained as,

$$\hat{\mathbf{h}} = \mathbf{X}^\dagger \mathbf{y} = (\mathbf{X}^T \mathbf{X})^{-1} \mathbf{X}^T \mathbf{y} \quad (17)$$

Let $\mathbf{h} = \hat{\mathbf{h}} + \Delta \mathbf{h}$, where $\mathbf{h}$ denotes the true CIR and $\Delta \mathbf{h}$ is the channel estimation error. Then

$$\Delta \mathbf{h} = \mathbf{h} - \mathbf{X}^\dagger (\mathbf{X}\mathbf{h} + \mathbf{m}) = \mathbf{X}^\dagger \mathbf{m} \quad (18)$$

Assume $E[\mathbf{m}] = 0$ and $E[\mathbf{m}\mathbf{m}^T] = \sigma_m^2 \mathbf{I}$. Then, the mean and covariance of the channel estimation $\hat{\mathbf{h}}$ can be derived as

$$E[\hat{\mathbf{h}}] = E[\mathbf{X}^\dagger \mathbf{y}] = E[\mathbf{X}^\dagger (\mathbf{X}\mathbf{h} + \mathbf{m})] = \mathbf{h} \quad (19)$$

$$cov(\hat{\mathbf{h}}) = E[\Delta \mathbf{h} \Delta \mathbf{h}^T] = E[(\mathbf{X}^\dagger \mathbf{m})(\mathbf{X}^\dagger \mathbf{m})^T] = \sigma_m^2 (\mathbf{X}^T \mathbf{X})^{-1} \quad (20)$$

Suppose that the input signal is uncorrelated, i.e., $E[x(i)x(j)] = \delta_{ij}$. In practice, an m-sequence can be applied to approximate the uncorrelated training symbol sequence. From (16) we obtain

$$\mathbf{X}^T \mathbf{X} = \left[\sum_k E[x_{k,i} x_{k,j}]\right] = n_x n_{ISI} \mathbf{I}_{n_L} \quad (21)$$

From (20), we then have the covariance matrix for LS channel estimation given by,

$$cov(\hat{\mathbf{h}}) = \frac{\sigma_m^2}{n_x n_{ISI}} \mathbf{I}_{n_L} \quad (22)$$

Let $\hat{\mathbf{y}} = \mathbf{X}\hat{\mathbf{h}}$. The MSE per sample for LS estimation is given by,

$$MSE_{LS} = \frac{1}{n_x n_L} E[\|\mathbf{y} - \hat{\mathbf{y}}\|^2] = (1 + \frac{1}{n_x}) \sigma_m^2 \quad (23)$$

where using (22), $E[\|\Delta \mathbf{h}\|^2] = trace\{cov(\hat{\mathbf{h}})\} = \frac{n_L \sigma_m^2}{n_x n_{ISI}}$.

From (23), it is noticed that increasing the number of training symbols ($n_x n_{ISI}$) will reduce the LS channel estimation error. Tradeoff should be made since the computational cost involved in LS estimation will increase and data transmission throughput will decrease as well.

B. Thresholded LS (THLS) Based Channel Estimation

LS based channel estimation assumes that all the taps have nonzero values. This in fact affects channel estimation accuracy and hence channel equalization performance, in the presence of sparse CIR. Based on IEEE channel model recommendation, UWB indoor channel with long delay spread can be considered as sparse due to its clustering effect at very high sampling rate ($\Delta \tau = 0.167ns$). By thresholding the LS channel estimation, the MSE of channel estimation can be effectively reduced and this improves the performance for channel estimation based equalization. This is described as follows.

From (22), it does not make sense to estimate a path h_k when the energy of the path is very low ($\|h_k\|^2 < \frac{\sigma_m^2}{n_x n_{ISI}}$). By simply thresholding those paths with very low energy to zero, the variance for path estimation will be limited by $\sigma_{\hat{h}}^2 \leq \frac{\sigma_m^2}{n_x n_{ISI}}$. Then from (23), MSE for channel estimation will be bounded by $MSE_{THLS} \leq (1 + \frac{1}{n_x}) \sigma_m^2$. Thus, thresholding does help to improve LS channel estimation performance.

As LS channel estimation involves large computational complexity ($\mathcal{O}(n_L^3)$) due to matrix inversion as given by (17). A Matching Pursuit algorithm which is computationally more efficient is discussed in Section IV-C.

C. Matching Pursuit (MP) Based Channel Estimation

Matching Pursuit [9] is a greedy algorithm for subset selection, which progressively refines the signal approximation with an iterative procedure. It has been effectively applied to sparsely distributed multipath channel estimation based on "hilly terrain" delay profile [10]. In this paper, MP algorithm is applied to channel estimation for UWB indoor channel. Our numerical simulation shows that MP achieves slightly better channel estimation and equalization performance as compared to THLS, especially when the number of training symbols is limited. This is due to the reason that MP algorithm estimates the channel path orderly by starting estimation from the strongest path, which is supposed to be the most reliable path estimation, then the residual is computed to continue the estimation of the next strongest path. In addition, MP algorithm involves less computational complexity ($\mathcal{O}(n_L^2 n_S)$, $n_S \ll n_L$). Assume that nonzero paths are detected by MP algorithm first, then LS algorithm can be applied for re-estimation of these nonzero paths. The details of using MP algorithm for channel estimation is described in [10].

V. Channel Estimation Based DFE

From (8), to formulate the channel estimation error, let $\mathbf{H} = \hat{\mathbf{H}} + \Delta \mathbf{H}$ and $\mathbf{H}' = \hat{\mathbf{H}}' + \Delta \mathbf{H}'$, where for simplicity, the subscript S using in $\mathbf{H}_S$ to denote tap selection is dropped. Then, we have

$$\mathbf{y} = \mathbf{y}' - \hat{\mathbf{H}}' \mathbf{x}' = \hat{\mathbf{H}} \mathbf{x} + \mathbf{w} \quad (24)$$

where $\mathbf{w} = \mathbf{m}_{\Delta H x} + \mathbf{m}_{\Delta H' x'} + \mathbf{m}$, $\mathbf{m}_{\Delta H x} = \Delta \mathbf{H} \mathbf{x}$ and $\mathbf{m}_{\Delta H' x'} = \Delta \mathbf{H}' \mathbf{x}'$. Assume Gaussianity for interference signal $\mathbf{m}_{\Delta H x}$ and $\mathbf{m}_{\Delta H' x'}$ which are due to channel estimation error. Then we have

$$\mathbf{R}_w = cov(\mathbf{w}) = \mathbf{R}_{\Delta H} + \mathbf{R}_{\Delta H'} + \sigma_m^2 \mathbf{I} \quad (25)$$

where we have assumed $E[\mathbf{x}\mathbf{x}^T] = E[\mathbf{x}'\mathbf{x}'^T] = \mathbf{I}$ and $E[\mathbf{x}'\mathbf{x}^T] = \mathbf{0}$.

In addition, from (6), $\mathbf{R}_{\Delta H}$ and $\mathbf{R}_{\Delta H'}$ can be evaluated as follows.

$$(\mathbf{R}_{\Delta H})_{i,j} = E\left[\sum_k \Delta \mathbf{H}(i,k) \Delta \mathbf{H}(j,k)\right] = \sum_k E[\Delta h(i') \Delta h(j')] \quad (26)$$

where $E[\Delta h(i') \Delta h(j')]$ can be obtained from the corresponding element of the channel estimation covariance matrix given in (22). Based on the evaluation from (26), it can be observed that the interference contributions from $\mathbf{R}_{\Delta H}$ and $\mathbf{R}_{\Delta H'}$ can be many times less than the AWGN term $\sigma_m^2 \mathbf{I}$, as long as reasonable enough number of training symbols is provided for channel estimation.

Following from the orthogonality principle [8], let $\hat{x}(n) = \mathbf{f}_{est}^T \mathbf{y}$, an MMSE equalizer based on channel estimation is obtained as,

$$\mathbf{f}_{est}^T = \{\hat{\mathbf{H}}^T (\hat{\mathbf{H}} \hat{\mathbf{H}}^T + \mathbf{R}_w)^{-1}\}_1 \approx \{\hat{\mathbf{H}}^T (\hat{\mathbf{H}} \hat{\mathbf{H}}^T + \sigma_m^2 \mathbf{I})^{-1}\}_1 \quad (27)$$

From (24) we obtain

$$\hat{x}(n) = \mathbf{f}_{est}^T (\mathbf{H}\mathbf{x} + \Delta \mathbf{H}' \mathbf{x}' + \mathbf{m}) = B_1 x(n) + I_{ISI} + I'_{ISI} + \tilde{m} \quad (28)$$

where $B_1 = (\mathbf{f}_{est}^T \mathbf{H})_1$, $\sigma_{\tilde{m}}^2 = \|\mathbf{f}_{est}^T\|^2 \sigma_m^2$.

The residual ISI term after equalization is given as $I_{ISI} = \sum_{k=2}^{n_{sym}} B_k x(n+k-1)$, $B_k = (\mathbf{f}_{est}^T \mathbf{H})_k$ for $k = 1, \cdots, n_{sym}$.

The residual ISI term due to channel estimation error $\Delta\mathbf{H}'$ which affects SIC is given as $I'_{ISI}=\sum_{k=1}^{n_{ISI}-1} B'_k x(n-n_{ISI}+k)$, $B'_k = (\mathbf{f}_{est}^T \Delta\mathbf{H}')_k$ for $k=1,\cdots,n_{ISI}-1$.

Applying Gaussian approximation to the interference terms, the SIR and BER performance for MMSE based DFE is given by,

$$SIR_{est}^{(1)} = \frac{B_i^2}{\sum_{k=2}^{N_T} B_k^2 + \sum_{k=1}^{N'_T} B'^2_k + \sigma_{\tilde{m}}^2} \quad (29)$$

$$P_{b,est}^{(1)} = Q(\sqrt{SIR_{est}^{(1)}}) \quad (30)$$

VI. NUMERICAL ANALYSIS AND SIMULATION RESULTS

In this section, the BER performance of tap selection based DFE is evaluated. Simulation result is given to show the effectiveness of tap selection based equalizer for UWB channel.

For simplicity, CM2 (0-4m, Non-Line-of-Sight (NLOS)) [7] is chosen with a sampling duration of $0.167ns$. The symbol rate for UWB BPSK transmission is set at $375MHz$, in the presence of severe ISI ($n_{ISI} = 32$). UWB channel estimation is implemented by MP algorithm with $(2 \times n_{ISI} = 64)$ training symbols. Only 32 strongest paths are selected out of total $n_L = 512$ samples of the channel for tap selection based equalization for all the curves depicted in Figure 2. A DFE (32 feedforward $taps$ + 31 feedbackward $taps$) is obtained based on the channel estimation (denoted as Ch-Est-DFE), where the feedforward filter is estimated by (27) and the feedbackward filter is then obtained by $\mathbf{f}_{back}^T = -\mathbf{f}_{est}^T \hat{\mathbf{H}}'$. LMS algorithm is applied to adapt the DFE with additional 128 training symbols (denoted as Ch-Est+LMS-DFE). In practice, decision directed adaptation by LMS can be implemented without the need for additional training symbols. The result is also compared with a DFE $(32 + 31\; taps)$ directly trained from the same amount of $(64 + 128 = 192)$ training symbols by Recursive Least Squares (RLS) algorithm (denoted as RLS-DFE).

From Figure 2, it is observed that

1) The performance analysis using (29) is consistently agreed with the simulation result for Ch-Est-DFE using (27).
2) Tap selection based Ch-Est-DFE is much more efficient than a directly RLS or LMS based DFE without tap selection. It requires much less training symbols to estimate the DFE with reasonable quick start performance. This is especially important for UWB channel with large delay spread which implies long training sequence is required for RLS or LMS equalizer implementation using all taps.
3) The simple LMS algorithm can be effectively applied to adapt the DFE after a quick start by channel estimation. Performance of Ch-Est+LMS-DFE can be improved by decision directed adaption.
4) With limited amount of training symbols, a Ch-Est+LMS-DFE performs better than a RLS-DFE.
5) The performance gap between the simulation curve using Ch-Est+LMS-DFE and the analysis curve using DFE by assuming perfect CIR is within $1.5dB$ in terms of SNR for BER at 10^{-5}.

VII. CONCLUSIONS

This paper analyzes the tap selection based MMSE equalization performance for high data rate UWB system in the presence of severe ISI using greedy method. Based on this analysis, a non-uniformly spaced MMSE equalizer that significantly outperforms the conventional uniformly spaced equalizer is developed for UWB

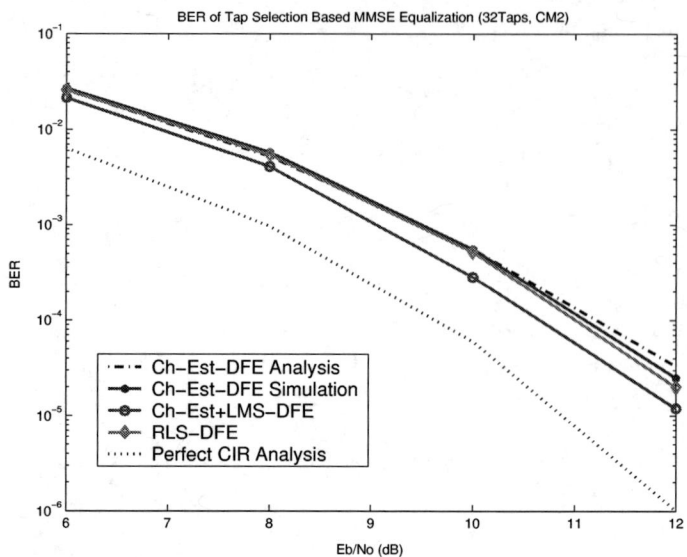

Fig. 2. Channel estimation based DFE performance (CM2, 32 taps selected)

channel. With tap selection for less than one tenth of the total paths of CIR, the complexity of the equalizer is greatly reduced with limited performance degradation when compared to that obtained with optimal MMSE detection using all taps. In addition, the impact of channel estimation error on the performance of tap selection based DFE is evaluated. The analysis and simulation results show that a channel estimation based DFE with tap selection is able to achieve a quick start using only small amount of training symbols. LMS can then be effectively applied to adapt the DFE for improved performance. This tap selection based Ch-Est+LMS-DFE is a promising method for high performance UWB system implementation with a low complexity receiver structure.

REFERENCES

[1] Scholtz, R., "Multiple Access with Time-Hopping Impulse Modulation". *IEEE Military Communications Conference: Communications on the Move*, 1993, vol. 2, pp. 447-450.
[2] Anuj Batra et al., 03268r1P802-15/TG3a-Multi-band-CFP-Document.doc, http://grouper.ieee.org/groups/802/15/pub/.
[3] Frederick K. H. Lee and Peter J. McLane, "Design of Nonuniformly Spaced Tapped-Delay-Line Equalizers for Sparse Multipath Channels". *IEEE Transactions On Communications*, Vol. 52, No. 4, April 2004, pp. 530-535.
[4] Sirikiat Ariyavisitakul, Nelson R. Sollenberger and Larry J. Greenstein, "Tap-Selectable Decision-Feedback Equalization". *IEEE Transactions On Communications*, Vol. 45, No. 12, December 1997, pp. 1497-1500.
[5] N. Al-Dhahir and J. M. Cioffi, "Fast Computation of Channel-Estimate Based Equalizers in Packet Data Transmission". *IEEE Transactions On Signal Processing*, Vol. 43, No. 11, November 1995, pp. 2462-2473.
[6] Gene H. Golub and Charles F. Van Loan *Matrix Computations*, The Johns Hopkins University Press, Third Edition, 1996.
[7] J. Foerster, P802.15-02/490r1-SG3a-Channel-Modeling-Subcommittee-Report-Final.doc, Feb. 2003, http://grouper.ieee.org/groups/802/15/pub/.
[8] Simon Haykin, *Adaptive Filter Theory*, Prentice Hall, Fourth Edition, 2002.
[9] Stephane G. Mallat and Zhifeng Zhang, "Matching Pursuits With Time-Frequency Dictinaries". *IEEE Transactions On Singal Processing*, Vol. 41, No. 12, December 1993, pp. 3397-3414.
[10] Shane F. Cotter and Bhaskar D. Rao, "Sparse Channel Estimation via Matching Pursuit With Application to Equalization". *IEEE Transactions On Communications*, Vol. 50, No. 3, March 2002, pp. 374-377.

Design of UWB pulses based on B-splines

Mitsuhiro Matsuo
Graduate School of
Science and Engineering,
Ibaraki University,
Hitachi, Ibaraki 316-8511, JAPAN.
Email: nm03017f@cis.ibaraki.ac.jp

Masaru Kamada
Department of Computer
and Information Sciences,
Ibaraki University,
Hitachi, Ibaraki 316-8511, JAPAN.
Email: kamada@cis.ibaraki.ac.jp

Hiromasa Habuchi
Department of Computer
and Information Sciences,
Ibaraki University,
Hitachi, Ibaraki 316-8511, JAPAN.
Email: habuchi@cis.ibaraki.ac.jp

Abstract— The present paper discusses construction of UWB pulses on the basis of B-splines having the following properties: (i) The B-splines are time-limited piecewise polynomials. (ii) They are rectangular pulses when their order is one and they converge to band-limited functions at the limit that their order tends to infinity. (iii) There are an analog circuit and a fast digital filter for the generation of B-splines. A constrained minimization technique is proposed for designing pulses so as to comfort the FCC spectral mask and satisfy basic requirements for UWB pulses.

I. INTRODUCTION

A principal approach to the construction of UWB pulses employs a window function to modulate a sinusoidal signal. The root-raised-cosine window is classical. The Gaussian window yields the UWB pulses related to Gabor wavelet. With the Gaussian-modulated pulses, effective techniques for multipath compensation [1] and interference mitigation [2] by means of sub-carrier pulses have been proposed.

Another approach lets a digital stimulus directly drive the circuit system including the transmission antenna [3]. Its simple circuit structure is a remarkable property.

At the cost of circuit complexity, a different kind of transmission pulses are shaped by feeding digitally designed signals to a spectrum-masking filter and an integrator [4] so that the received signals be orthogonal multi-codeword pulses. An advanced design of the transmission pulses employs Hermite orthogonal polynomials [5] and it is approximately implemented in terms of sinusoidal series [6].

The present paper discusses a little different construction of UWB pulses. The basis functions employed for designing pulses are B-splines [7] having the following properties: (i) The B-splines are time-limited piecewise polynomials. (ii) They are rectangular pulses when their order is one and they converge to band-limited functions at the limit that their order tends to infinity [8]–[10]. (iii) There are an analog circuit [11] and a fast digital filter [12] for the generation of B-splines.

A constrained minimization technique is proposed for design of the pulses. The pulses are designed so that their energy in the GPS and wireless LAN bands is minimized under the condition of having no direct current components and of being orthonormal to each other. We need a rather large number of shifted versions of the B-splines to have the resulting pulses conform the FCC spectral mask [13]. An example set of four proper pulses is obtained by using eleven shifted B-splines.

Contribution of this paper is to show the possibility of constructing UWB pulses in terms of such a new ingredient as B-splines. Performance analysis including estimation of error rates under the influence of jitter, white noise, and other interfering signals are yet to be investigated by further studies.

II. PRELIMINARIES FOR B-SPLINES

The B-spline [7] of order m having the knot interval $T > 0$ is defined by

$$\varphi_m(t) := T \int_{-\infty}^{\infty} \left(\frac{\sin \pi f T}{\pi f T}\right)^m e^{i 2\pi f(t - \frac{m}{2}T)} df, \quad m = 1, 2, 3, \cdots \quad (1)$$

and is a piecewise polynomial of degree $m - 1$ [7]. Its Fourier transform is

$$\hat{\varphi}_m(f) := T \left(\frac{\sin \pi f T}{\pi f T}\right)^m e^{-i 2\pi f \frac{m}{2} T}, \quad m = 1, 2, 3, \cdots. \quad (2)$$

The B-spline is a time-limited function [7] such that

$$\varphi_m(t) = 0, \quad t \notin (0, mT). \quad (3)$$

Its integration is a constant [8] such as

$$\int_{-\infty}^{\infty} \varphi_m(t) dt = T \quad (4)$$

and its derivative is given [8] by

$$\frac{d}{dt}\varphi_{m+1}(t) = \varphi_m(t) - \varphi_m(t - T). \quad (5)$$

For a natural number L, the B-spline and its shifted versions $\varphi_m(t - kT)$, $k = 0, 1, 2, \cdots, L$ can be generated by an analog circuit [11] illustrated in Fig.1. The circuit in Fig.1 is based on the recurrence formulae [8]

$$\varphi_1(t) = \begin{cases} 1, & 0 \le t < T \\ 0, & \text{elsewhere,} \end{cases} \quad (6)$$

$$\varphi_m(t) = \int_0^{mT} [\varphi_{m-1}(t) - \varphi_{m-1}(t - T)] dt, \quad m = 2, 3, \cdots. \quad (7)$$

We start from the first order B-splines $\varphi_1(t), \varphi_1(t-T), \cdots, \varphi_1(t-(L+m-1)T)$, which are rectangular pulses as represented by (6). Integration of the difference of lower-order B-splines yields higher-order B-splines as prescribed by (7). Then $(m-1)$-fold differential integration results in the B-splines of order m out of which we take $\varphi_m(t), \varphi_m(t-T), \cdots, \varphi_m(t-LT)$. This circuit operates periodically by the period of $(L+m-1)T$. There is also an oversampling digital filter [12] that simulates the analog circuit efficiently [14].

III. PERSPECTIVE OF INTENDED UWB SYSTEM

A perspective of the UWB system intended in this paper is illustrated in Fig.2. Denote message bits by $u_l \in \{0, 1\}$, $l = 1, 2, \cdots, L$. For their transmission, let us construct UWB pulses as linear combinations of the B-splines of order m, i.e.

$$\psi_l(t) = \sum_{k=0}^{L} c_{l,k} \varphi_m(t - kT), \quad l = 1, 2, \cdots, L, \quad (8)$$

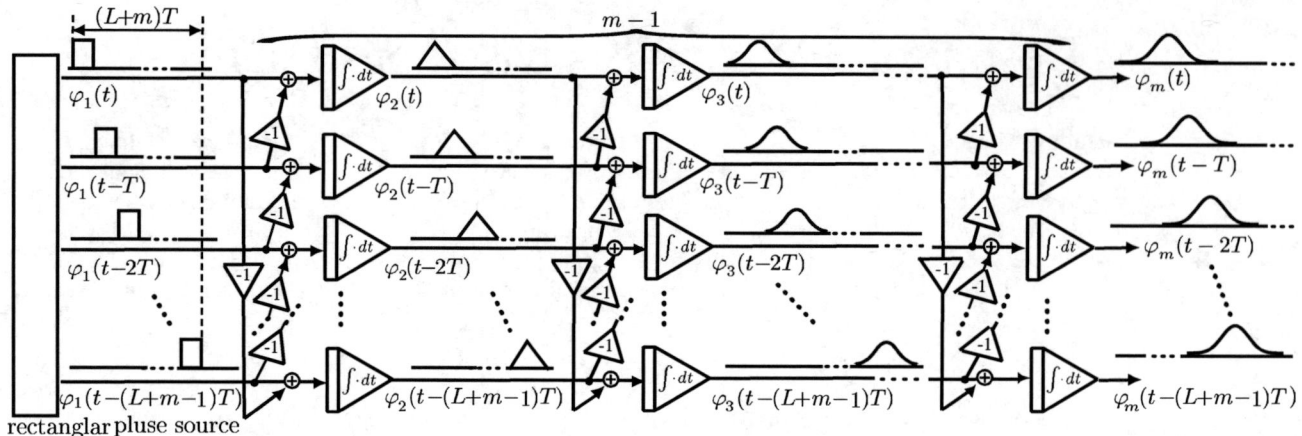

Fig. 1. Analog circuit for the generation of B-splines.

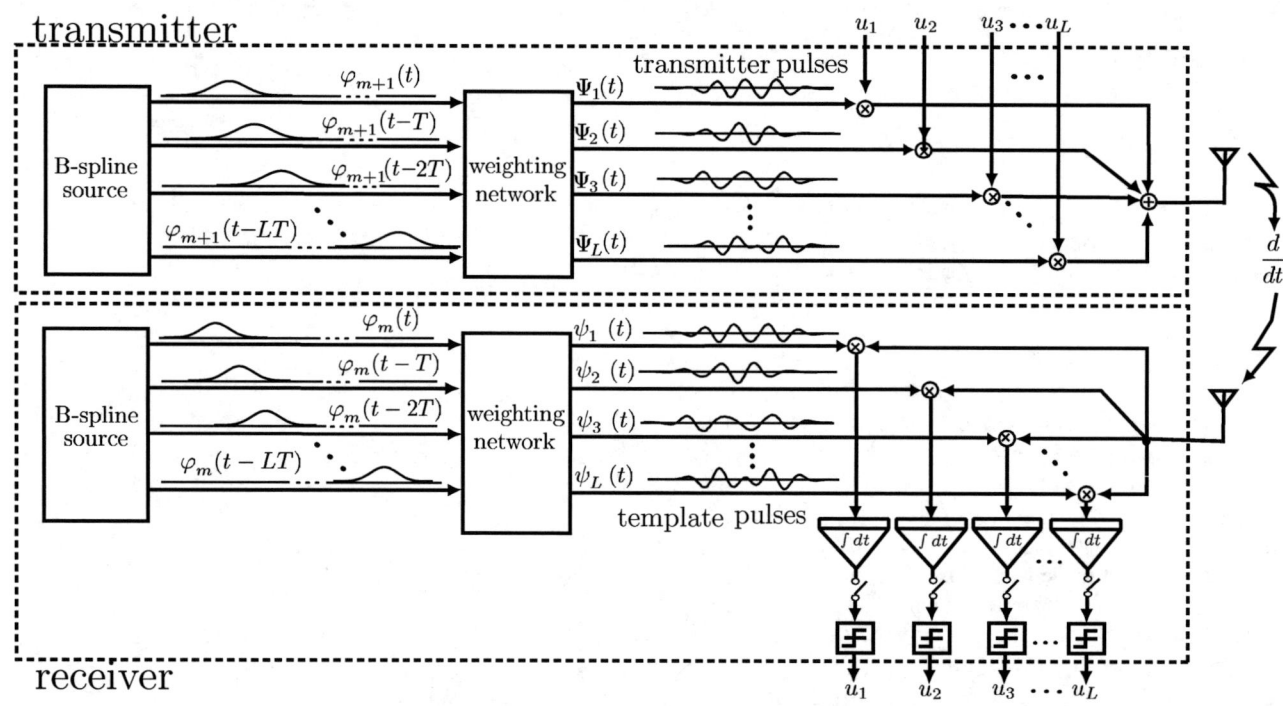

Fig. 2. A perspective of the UWB communications system based on B-splines.

where coefficients $c_{l,k}$ are real numbers.

Since the channel between transmitting and receiving antennas can be regarded as differentiation with respect to time [3], [15], the transmitter pulses $\Psi_l(t)$ should be the integral of the receiver pulses $\psi_l(t)$, i.e.

$$\Psi_l(t) = \int_0^t \psi_l(\tau)d\tau, \quad l = 1, 2, \cdots, L. \quad (9)$$

Because radio waveforms must not have any direct current component, the receiver pulses must satisfy the condition

$$\int_{-\infty}^{\infty} \psi_l(t)dt = 0, \quad l = 1, 2, \cdots, L. \quad (10)$$

This condition is equivalent to

$$c_{l,0} + c_{l,1} + \cdots + c_{l,L} = 0 \quad (11)$$

by (4) and (8). Under this condition, it follows from (5) and (8) that

$$\Psi_l(t) = \sum_{k=0}^{L} d_{l,k}\varphi_{m+1}(t - kT), \quad l = 1, 2, \cdots, L, \quad (12)$$

where

$$d_{l,k} = \sum_{p=0}^{k} c_{l,p}, \quad k = 0, 1, 2, \cdots, L; l = 1, 2, \cdots, L. \quad (13)$$

In the transmitter part of Fig.2, the B-splines $\varphi_{m+1}(t - kT)$, $k = 0, 1, 2, \cdots, L$, of order $m + 1$ is generated, and their weighted sums $\Psi_l(t)$, $(l = 1, 2, \cdots, L)$ in the form of (12) are generated as the transmitter pulses. Each pulse is modulated by a message bit u_l to form the transmission signal

$$Y(t) = \sum_{l=1}^{L} u_l \Psi_l(t). \quad (14)$$

to be fed to the antenna.

At the receiving antenna, we obtain its derivative

$$y(t) := \frac{d}{dt}Y(t) = \sum_{l=1}^{L} u_l \psi_l(t). \quad (15)$$

In the receiver part of Fig.2, the B-splines $\varphi_m(t - kT)$, $k = 0, 1, 2, \cdots, L$, of order m are generated, and their weighted sums $\psi_l(t)$, $(l = 1, 2, \cdots, L)$ are generated as template pulses. Since the inner product of the received signal and the template is

$$\int_{-\infty}^{\infty} y(t)\psi_p(t)dt = \sum_{l=1}^{L} u_l \int_{0}^{(m+L)T} \psi_l(t)\psi_p(t)dt, \ p = 1, 2, \cdots, L, \quad (16)$$

the message bits are detected by the correlation

$$\int_{0}^{(m+L)T} y(t)\psi_p(t)dt = u_p, \ p = 1, 2, \cdots, L \quad (17)$$

if the template pulses are orthonormal such that

$$\int_{-\infty}^{\infty} \psi_l(t)\psi_p(t)dt = \begin{cases} 1, & l = p \\ 0, & l \neq p. \end{cases} \quad (18)$$

The above can be summarized as follows: Design of transmitter as well as template pulses in terms of the B-splines is to determine $c_{l,k}$ under the basic conditions that the template pulses $\psi_l(t)$ must not have direct current component, i.e. satisfying (11), and that they are orthonormal to each other, i.e. satisfying (18). Then the template pulses $\psi_l(t)$ are given by (8) and the transmitter pulses $\Psi_l(t)$ are determined by (12) and (13).

IV. DESIGN OF UWB PULSES IN ACCORDANCE WITH FCC REGULATION

Pulses satisfying the FCC spectral mask in addition to the basic conditions (11) and (18) shall be designed in this section. We might wish to optimize the coefficients so that the energy of pulses be maximized under the conditions (11), (18), and the spectral mask. However, it is quite difficult to identify the feasible set for the coefficients satisfying the spectral mask. Alternatively, let us determine the coefficients so that the frequency characteristics

$$\hat{\psi}_l(f) := \int_{-\infty}^{\infty} \psi_l(t) e^{-i2\pi ft} dt, \ l = 1, 2, \cdots, L \quad (19)$$

of the pulses be minimized in the GPS and wireless LAN bands but have non-zero energy under the conditions (11) and (18). In addition, we may wish the pulses have a notch at some frequency f_n.

Substituting (8) for $\psi_l(t)$ in (19), we have

$$\hat{\psi}_l(f) = \sum_{p=0}^{L} c_{l,p} \hat{\varphi}(f) e^{-i2\pi fpT}, \ l = 1, 2, \cdots, L. \quad (20)$$

Under the conditions (11) and (18), the coefficients $c_{l,k}$ shall be determined successively so that the energy

$$Q_l = \int_{0.96 \times 10^9}^{3.1 \times 10^9} |\hat{\psi}_l(f)|^2 df + \int_{-0.96 \times 10^9}^{-3.1 \times 10^9} |\hat{\psi}_l(f)|^2 df$$

$$= 2 \int_{0.96 \times 10^9}^{3.1 \times 10^9} |\hat{\psi}_l(f)|^2 df, \ l = 1, 2, \ldots, L \quad (21)$$

of pulses in the domains $[0.96 \times 10^9, 3.1 \times 10^9]$ and $[-0.96 \times 10^9, -3.1 \times 10^9]$ is minimized and that we have a notch at $f = \pm f_n$, i.e.

$$\hat{\psi}(f_n) = \hat{\psi}(-f_n) = 0. \quad (22)$$

The trivial solution is $c_{l,0} = c_{l,1} = \cdots = c_{l,L} = 0$, which implies the pulse has no energy. To avoid this trivial solution, let us fix one of the coefficients be non-zero in advance of optimization. The solution depends on which coefficient is fixed. Let us fix $c_{l,0}$ as

$$c_{l,0} = 1 \quad (23)$$

for example.

For designing the first pulse $\psi_1(t)$, we have only to incorporate the condition (11). The minimization of Q_l for $l = 1$ under the constraints of (22), (23) and the condition (11) is equivalent to

$$2 \int_{0.96 \times 10^9}^{3.1 \times 10^9} |\hat{\psi}_1(f)|^2 df + \mu_0 \hat{\psi}_1(f_n) + \mu_1 \hat{\psi}_1(-f_n) + \lambda_0 \sum_{p=0}^{L} c_{1,p} \to \min., \quad (24)$$

where μ_0, μ_1 and λ_0 are Lagrange indefinite multipliers. The minimizer must satisfy the linear equations

$$\frac{\partial}{\partial c_{1,k}} \left(2 \int_{0.96 \times 10^9}^{3.1 \times 10^9} \left| \sum_{p=0}^{L} c_{1,p} \hat{\varphi}_m(f) e^{-i2\pi fpT} \right|^2 df + \lambda_0 \sum_{p=0}^{L} c_{1,p} \right.$$

$$\left. + \mu_0 \sum_{p=0}^{L} c_{1,p} \hat{\varphi}_m(f) e^{-i2\pi fpT} + \mu_1 \sum_{p=0}^{L} c_{1,p} \hat{\varphi}_m(-f) e^{-i2\pi fpT} \right) = 0,$$

$$k = 1, 2, \ldots, L \quad (25)$$

in addition to (11), (22) and (23). Solving (11), (22), (23) and (25), we have $c_{1,0}, c_{1,1}, \cdots, c_{1,L}$.

For successively designing the pulses $\psi_l(t)$, $l = 2, 3, \cdots, L$, we have to incorporate the orthonormality condition (18) in addition to (11). The minimization problem (21) for $l \geq 2$ under the conditions (11), (18), (22) and (23) is equivalent to

$$2 \int_{0.96 \times 10^9}^{3.1 \times 10^9} |\hat{\psi}_l(f)|^2 df + \mu_0 \hat{\psi}_l(f_n) + \mu_1 \hat{\psi}_l(-f_n) + \lambda_0 \sum_{p=0}^{L} c_{l,p}$$

$$+ \sum_{p=1}^{l-1} \lambda_p \sum_{q=0}^{L} c_{l,q} \sum_{r=0}^{L} c_{p,r} 2 \int_{0}^{(m+L)T} \varphi_m(t-qT)\varphi_m(t-rT) dt \to \min., \quad (26)$$

where $\mu_0, \mu_1, \lambda_0, \lambda_1, \cdots, \lambda_{l-1}$ are indefinite multipliers. Solving the linear equations

$$\frac{\partial}{\partial c_{l,k}} \left(2 \int_{0.96 \times 10^9}^{3.1 \times 10^9} \left| \sum_{p=0}^{L} c_{l,p} \hat{\varphi}_m(f) e^{-i2\pi fpT} \right|^2 df + \lambda_0 \sum_{p=0}^{L} c_{l,p} \right.$$

$$+ \mu_0 \sum_{p=0}^{L} c_{l,p} \hat{\varphi}_m(f) e^{-i2\pi fpT} + \mu_1 \sum_{p=0}^{L} c_{l,p} \hat{\varphi}_m(-f) e^{-i2\pi fpT}$$

$$\left. + \sum_{p=1}^{l-1} \lambda_p \sum_{q=0}^{L} c_{l,q} \sum_{r=0}^{L} c_{p,r} \int_{0}^{(m+L)T} \varphi_m(t-qT)\varphi_m(t-rT) dt \right) = 0,$$

$$k = 1, 2, \ldots, L, \quad (27)$$

for $c_{l,0}, c_{l,1}, \cdots, c_{l,L}$ under the constraints (11), (18), (22) and (23), we have the coefficients for the pulses.

Example pulses for $m = 4$, $T = 0.61 \times 10^{-10}$ are plotted in Fig.3 where the spectrum density is scaled for the convenience of comparison with the FCC spectral mask. In the case we do not require any notches, we have obtained four orthonormal pulses complying with the FCC spectral mask at the cost of $L + 1 = 11$ shifted

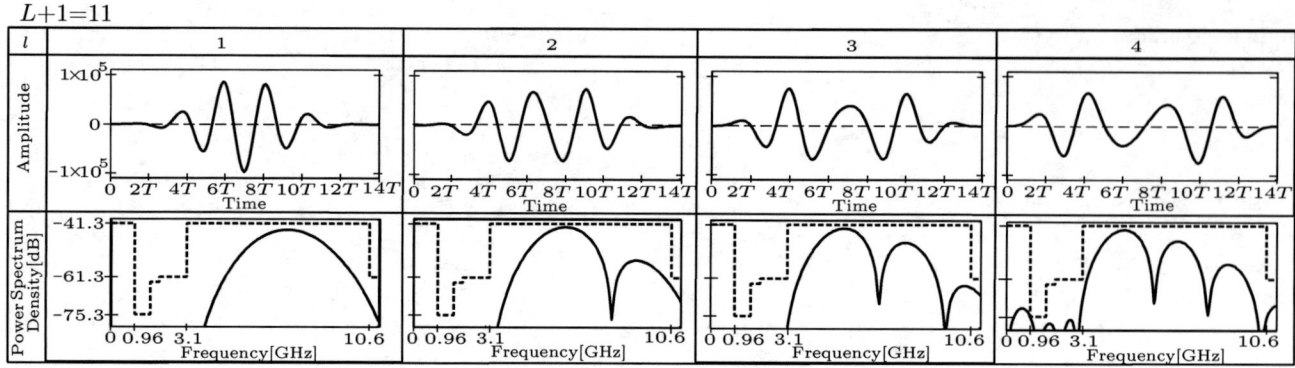

Fig. 3. Pulses which comply with the FCC spectral mask

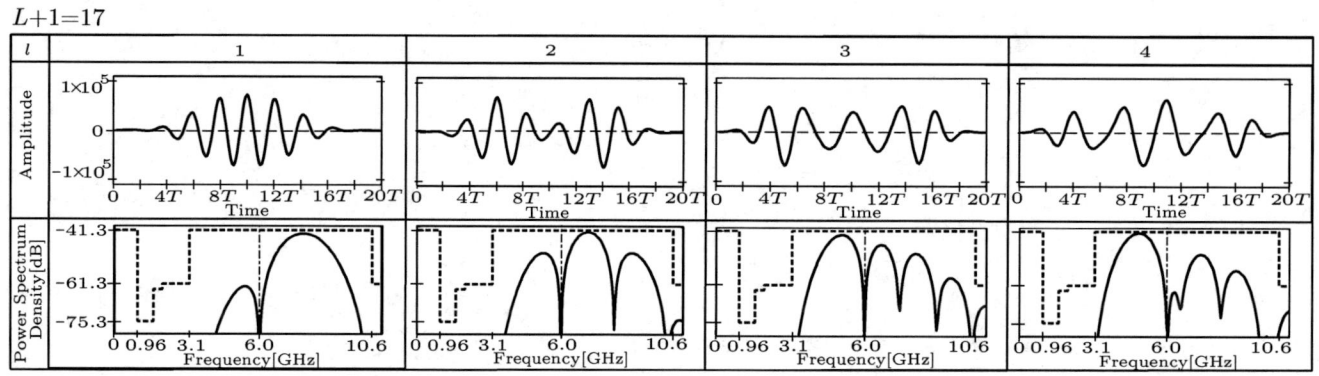

Fig. 4. Pulses complying with the FCC spectral mask and having a common notch.

versions of the B-splines. Pulses having a notch may be robust against multipath environment with specific lags. Those in Fig.4 have a common notch at $f_n = 6$ GHz. We need $L + 1 = 17$ shifted B-splines to obtain those pulses.

V. CONCLUSION

Employing the B-splines as basis functions instead of the sinusoidal functions, we attempted to construct UWB pulses. A constrained minimization technique was proposed for designing pulses in accordance with the FCC regulation. An example set of four proper pulses was obtained by using eleven shifted B-splines. By this example, the possibility of constructing UWB pulses in terms of B-splines was shown. Performance analysis including estimation of error rates under the influence of jitter, white noise, and other interfering signals are yet to be investigated by further studies.

REFERENCES

[1] K. Ohno, S. Sonobe, R. Nishibori, T. Ikebe and T. Ikegami, "Multipath compensation by using template processing in UWB radio," *CD-ROM Proc. of the IEICE National Convention*, A-5-38, Mar. 2004.

[2] K. Ohno, T. Ikebe and T. Ikegami, "A proposal for an interference mitigation technique facilitating the coexistence of bi-phase UWB and other wideband systems," *CD-ROM Proc. for Internal workshop on ultra wideband systems joint with conference on ultra wideband systems and technologies(Joint UWBST & IWUWBS 2004)*, WA2-5.

[3] S. Yoshizumi, T. Terada, J. Furukawa, Y. Sanada and T. Kuroda, "All digital transmitter scheme and transceiver design for pulse-based ultra-wideband radio," *Proc. IEEE Ultra Wideband Systems and Technologies(UWBST'03)*, pp.438-442, 2003.

[4] B. Parr, B. Cho, K. Wallace and Zhi Ding, "A novel ultra-wideband pulse design algorithm," *IEEE Communications Letters*, vol.7, no.5, pp.219-221, 2003.

[5] G. Abreu, C. Mitchell and R. Kohno, "On the orthogonality of Hermite pulses for ultra wideband communications," *CD-ROM Proc. of the Wireless Personal Multimedia Communications (WPMC 2003)*, TA6-4.

[6] K. Taniguchi and R. Kohno, "Design and analysis of template waveform for receiving UWB signals," *CD-ROM Proc. of the Internal workshop on ultra wideband systems joint with conference on ultra wideband systems and technologies(Joint UWBST & IWUWBS 2004)*, WA5-3.

[7] I. J. Schoenberg, "Contributions to the problem of approximation of equidistant data by analytic functions," *Quart. Appl. Math.*, vol.4, Part A, pp.45-99; Part B, pp.112-141, 1946.

[8] I. J. Schoenberg, *Cardinal Spline Interpolation*, CBMS Regional Conference Monograph, No.12, SIAM, Philadelphia, 1973.

[9] M. Unser, A. Aldroubi and M. Eden, "Polynomial spline signal approximations: Filter design and asymptotic equivalence with Shannon's sampling theorem," *IEEE Trans. Info. Th.*, vol.38, no.1, pp.95-103, 1992.

[10] M. Kamada, K. Toraichi and R. Mori, "Spline function approach to digital signal processing," *Intl. J. Systems Sci.*, vol.19, no.12, pp.2473-2490, 1988.

[11] M. Kamada, K. Toraichi and R. E. Kalman, "A Smooth signal generator based on quadratic B-spline functions," *IEEE Trans. Signal Process.*, vol.43, no.5, pp.1252-1255, 1995.

[12] K. Ichige, M. Kamada, R. Ishii, "A simple scheme of decomposing and reconstructing continuous-Time signals by B-Splines," *IEICE Trans. Fundamentals*, vol.E81-A, no.11, pp.2391-2399, 1998.

[13] "Revision of Part 15 the Commission's rule regarding ultra-wideband transmission systems," FCC, ET Docket 98-153, 2002.

[14] K.Ichige and M.Kamada, "An approximation for discrete B-splines in time domain," *IEEE Signal Process. letters*, no.4, pp.82-84, 1997.

[15] Moe Z. Win, Robert A. Scholtz, "Impulse radio: How it works," *IEEE Commun. Letters*, vol.2, no.2, pp.36-38, 1998.

OPTIMIZING VERTICAL COMMON SUBEXPRESSION ELIMINATION USING COEFFICIENT PARTITIONING FOR DESIGNING LOW COMPLEXITY SOFTWARE RADIO CHANNELIZERS

A.P.Vinod and E.M-K.Lai

School of Computer Engineering, Nanyang Technological University
Nanyang Avenue, Singapore 639798
Email: {asvinod, asmklai}@ntu.edu.sg

ABSTRACT

The complexity of finite impulse response (FIR) filters used in the channelizer of a software defined radio (SDR) receiver is dominated by the complexity of the coefficient multipliers. A method for designing low complexity channel filters by optimizing vertical common subexpression elimination (VCSE) using coefficient partitioning is presented in this paper. Our algorithm exploits the fact that when multiplication is implemented using shifts and adds, the adder width can be minimized by limiting the shifts of the operands to shorter lengths. Design examples of the channel filters employed in the Digital Advanced Mobile Phone System (D-AMPS) receiver show that the proposed method offers considerable full adder reduction over the VCSE methods.

1. INTRODUCTION

The most computationally intensive part of an SDR receiver is the channelizer since it operates at the highest sampling rate [1]. It extracts multiple narrowband channels from a wideband signal using a bank of FIR filters, called channel filters. Low power and high-speed FIR filters implemented with the minimum number of adders are required in the channelizer. Among the approaches for reducing the number of adders in the multipliers of FIR filters, the CSE techniques in [2]-[4] produced the best hardware reduction since it deals with multiplication of one variable (input signal) with several constants (coefficients). However, the methods in [2]-[4] have not addressed the issue of minimizing the complexity of each adder of the multiplier, which is significant in low power and high-speed implementations. In our recent work [5], we have analyzed the complexity of implementation of FIR filters in terms of the number of full adders (FA's) required for each multiplier. A vertical super-subexpression elimination (VSSE) method for optimizing the VCSE method in [4] to implement low-complexity channel filters using minimum number of FA's has been proposed in [5]. This technique is based on the extension of conventional two-nonzero bit (2-bit) vertical common subexpressions (VCS) in [4] to form three-nonzero bit and four-nonzero bit vertical super-subexpressions (called 3-bit and 4-bit VSS, respectively). The main limitation of the method in [5] is its dependence on the statistical distribution of shifts between the 2-bit VCS in the canonic signed digit (CSD) representations of FIR filter coefficients. Moreover, the routing complexity of the filters designed using the method in [5] is higher than that of the 2-bit VCSE techniques in [2]-[4] as the former method has more number of subexpressions.

In this paper, we show that low complexity coefficient multipliers can be realized by combining three techniques: an efficient coefficient partitioning algorithm, the pseudo floating-point (PFP) representation and the VCSE, which reduces the number of FA's. The FA reduction techniques proposed in this paper do not employ VSS used in [5] and hence they do not have the dependence on statistical distribution of shifts between the 2-bit CS. The problem that we address here is how to minimize the number of FA's required in each adder of a given minimum-adder multiplier filter structure. Though we use the VCSE [4] and VSSE [5] techniques for comparison, our algorithm can also optimize the coefficient multipliers designed using other methods to further minimize the number of FA's.

The paper is organized as follows. In section 2, we provide a brief review of the complexity analysis of coefficient multipliers. Our coefficient partitioning method is presented in section 3. In section 4, we illustrate the implementation of channel filters for the D-AMPS standard using our method and provide comparisons. Section 5 provides our conclusions.

2. MULTIPLIER COMPLEXITY

Definition 1 (Range): The range is analogous to the wordlength, which is equal to the number of bits of an operand (input signal shifted corresponding to the positional weights of the nonzero terms of the coefficient form the *operands* of the adders). For example, if x_1 is an 8-bit quantized signal (as assumed throughout the paper), the range of the operand, $x_1 \gg 6$, is fourteen. (Note that *range* is same as *span* in [5]. In this paper, we use the term *span* in the PFP representation. Correspondingly, r_n is the range of the nth operand, which is same as s_n in [5]).

Case I: Odd number of operands: The number of FA's, (N_o), required to compute the output corresponding to a coefficient with n (for n odd) operands can be determined using the expression [5]:

$$N_o = r_2 + a_1 r_3 + 2r_4 + a_3 r_5 + r_6 + a_5 2r_7 + 3r_8 + a_7 r_9 + r_{10} + a_9 2r_{11} + 2r_{12} \quad (1)$$

where a_i's are equal to zero except $a_{n-2} = 1$.

Case II: Even number of operands: The number of FA's, (N_e), required to compute the output corresponding to a coefficient with n operands is given by [5]:

$$N_e = r_2 + 2r_4 + c_0 r_6 + 3r_8 + c_1 r_{10} + 3r_{12} \qquad (2)$$

where $c_0 \equiv \begin{cases} 2, \text{ for } n=6 \\ 1, \text{ elsewhere} \end{cases}$ and $c_1 \equiv \begin{cases} 2, \text{ for } n=10 \\ 1, \text{ elsewhere} \end{cases}$.

Note that (1) and (2) are same as in [5], except that we use the notation r_n here instead of s_n and *range* is same as *span* in [5].

The coefficients $h(0)$ and $h(1)$ of an FIR filter expressed in 12-bit CSD form shown in Fig. 1 is used as an example to illustrate the VCSE method and our optimization. The numbers in the first row of Fig. 1 represent the number of bitwise right shifts.

	1	2	3	4	5	6	7	8	9	10	11	12
$h(0)$	0	0	0	1	0	-1	0	0	-1	0	1	0
$h(1)$	0	0	0	1	0	-1	0	0	-1	0	0	1

Fig. 1. VCS in filter coefficients.

In direct implementation, (i.e., the implementation using shifts and adds and without VCSE) the outputs of the filter taps are given by (3) and (4).

$$y(0) = 2^{-4} x_1 - 2^{-6} x_1 - 2^{-9} x_1 + 2^{-11} x_1 \qquad (3)$$

$$y(1) = 2^{-4} x_1[-1] - 2^{-6} x_1[-1] - 2^{-9} x_1[-1] + 2^{-12} x_1[-1] \qquad (4)$$

where $[-k]$ represents a delay of k units. For both (3) and (4), n is 4 (even). The ranges r_2 and r_4 in (3) are 14 and 19 respectively. Using (2) the number of FA's required to compute (3) in direct method is $r_2 + 2r_4$, i.e., 52 FA's. Similarly, the ranges r_2 and r_4 in (4) are 14 and 20 respectively and the number of FA's required to compute (4) in direct method is 54. Thus, a total of 106 FA's are needed to compute (3) and (4) using direct implementation.

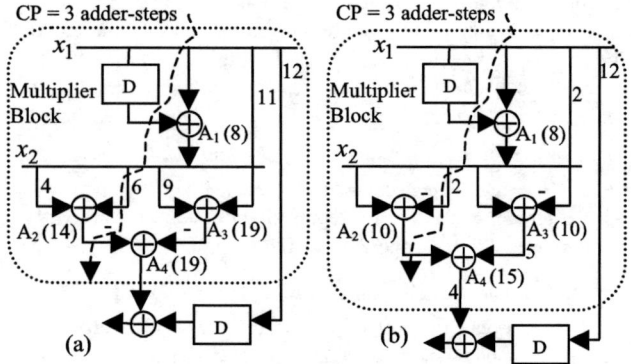

Fig. 2. Coefficient multiplier structures using VCSE (a) and our CPM (b).

The objective of VCSE algorithm is to identify multiple identical bit patterns that exist across the coefficient set and eliminate redundant computations by forming VCS from the bit patterns. The 2-bit VCS, [1 1] shown encircled in Fig. 1 is given by $x_2 = x_1 - x_1[-1]$. Using VCS, the output of the filter can be expressed as

$$y = 2^{-4} x_2 - 2^{-6} x_2 - 2^{-9} x_2 + 2^{-11} x_1 + 2^{-12} x_1[-1] \qquad (5)$$

Fig. 2(a) shows the multiplication structure using VCSE. The numerals adjacent to the data path in Fig. 2 represents the number of bitwise right shifts. The numerals in brackets alongside the adders indicate the number of FA's used in the adder. Thus, the number of FA's required for the multiplier block using VCSE method [4] is 60 in this case, which is a reduction of 43.4% over the direct method.

3. THE COEFFICIENT PARTITIONING METHOD

The key idea in our approach is to reduce the *ranges* of the operands so that the adder width can be reduced which in turn minimizes the number of FA's. To achieve this, firstly the coefficients are encoded using the PFP representation and then partitioned for further reduction of range.

Definition 2 (Pseudo floating-point (PFP) representation): The CSD representation for the ith filter coefficient of wordlength B is $h_i = \sum_{j=0}^{B-1} 2^{a_{ij}}$. The PFP representation of h_i is [6]

$$h_i = 2^{a_{i0}} \cdot \sum_{j=0}^{B-1} 2^{a_{ij} - a_{i0}} = 2^{a_{i0}} \cdot \left[\sum_{j=0}^{B-1} 2^{c_{ij}} \right] \qquad (6)$$

where $c_{ij} = a_{ij} - a_{i0}$. The term a_{i0} is known as the *shift* and the upper limit value, $(a_{i(B-1)} - a_{i0})$, is known as the *span*. Instead of expressing the coefficients using B-bit CSD, it can be expressed as a (*shift, span*) pair using fewer bits. For example, the PFP form of the coefficient $h(0)$ in the example in Fig. 1 is $2^{-4}(2^0 - 2^2 - 2^5 + 2^7)$. The term 2^{-4} is the *shift* part (implying 'right shift by 4'), and the bracketed term is the *span* part. Note that the shift operation can be performed after the addition of all the terms of the span part. The cost of shifts is negligible as they can be hardwired. This reduces the effective wordlength of the coefficient to that of the span (7 bits), which in turn reduces the ranges of the operands. Using (2), the number of FA's required to implement the coefficient multipliers in Fig. 1 when the coefficients are coded using PFP is 82. Though this FA requirement is less than that of direct implementation, it must be noted that the PFP implementation needs more FA's when compared with the VCSE method in Fig. 2(a). We shall now show that by combining the PFP coding scheme with the VCSE and then partitioning the resulting expression, considerable reduction of FA's can be achieved.

3.1 FA Reduction Using Coefficient Partitioning

The basic idea is to reduce the range of the span part of PFP by partitioning it into two sub-components, called *sub-filters*.

Definition 3 (Order): The most significant bit of a filter coefficient represented in CSD form is defined as the order of the coefficient. For instance, the order of a coefficient $h(n) = 2^{-6} + 2^{-8} + 2^{-11} + 2^{-14} + 2^{-16}$ is 2^{-6}.

Firstly, the CSD coefficient is expressed using VCS and the resulting expression is then coded using PFP representation. Let M represents the span of the PFP representation. The span part is partitioned into two sub-components (sub-filters) of length $M/2$ (or two sub-components of lengths $\lfloor M/2 \rfloor$ and $\lceil M/2 \rceil$ if M is odd). The latter sub-component is then scaled by its *order* to reduce its span. The 'partitioned and scaled' versions of the PFP coefficients thus obtained can be added using fewer numbers of FA's since their ranges are reduced. Consider the same example shown in Fig. 1. Using PFP, the filter output corresponding to the nonzero bits of $h(0)$ and the VCS formed by $h(0)$ with $h(1)$ obtained in VCSE method can be expressed as $2^{-4}(x_2 - 2^{-2}x_2 - 2^{-5}x_2 + 2^{-7}x_1)$. In this case, the span (M) is 7 and the shift is 4. Partitioning the span part into two sub-filters, $h_1(n)$ and $h_2(n)$, we have

$$h_1(n) = x_2 - 2^{-2}x_2 \text{ and } h_2(n) = -2^{-5}x_2 + 2^{-7}x_1 \qquad (7)$$

where $h(n)$ is the sum of $h_1(n)$ (MSB half) and $h_2(n)$ (LSB half). The LSB sub-filter is further scaled by its order, 2^{-5}, and expressed as $h_2(n) = -2^{-5}(x_2 - 2^{-2}x_1)$. Fig. 2(b) shows the implementation of the filter taps using our coefficient partitioning method (CPM). When compared with the VCSE method in Fig. 2(a), the adders A_2 and A_3, have shorter widths since the ranges of their operands are shorter. The shift 2^{-5} of $h_2(n)$ and that of the final expression $2^{-4}(x_2 - 2^{-2}x_2 - 2^{-5}x_2 + 2^{-7}x_1)$ are performed after the addition stages as shown alongside the data paths at the outputs of adders A_3 and A_4 respectively. Thus, our method requires only 43 FA's to implement the filter tap, which is a reduction of 28.3% over the VCSE method [4]. Note that both methods have identical critical path (CP) lengths (3 adder-steps) and hence their multiplier delays are same.

In order to meet the stringent adjacent channel interference specifications of wireless communications standards, higher order FIR filters are needed in SDR channelizers. It has been observed that the reduction rates offered by our method increases with the filter order. This can be explained by considering the numerical property of the *end-coefficients* of higher-order FIR filters.

Definition 4 (End-Coefficients): We designate the first $N/4$ coefficients, $h(0)$ to $h(\lceil N/4 \rceil - 1)$, of an FIR filter with N taps as the *end-coefficients*. For example if N is 40, the coefficients, $h(0)$ to $h(9)$, of one half of the symmetric set, $h(0)$ to $h(19)$, are called its end-coefficients. As the filter order increases, the side-lobes of the impulse response decrease and hence the magnitudes of the end-coefficients of $h(n)$ will also decrease. Due to their lower magnitudes, most of the nonzero bits of the CSD representations of end-coefficients occur in the LSB part. In conventional implementation, this will lead to the use of longer *shifts* which will in turn increase the *ranges* of the operands and correspondingly the number of FA's. On the other hand, the use of shorter *shifts* in our method results in considerable reduction of FA's. Therefore, our method offers considerable FA reduction in the channelization application where higher order FIR filters are needed.

We also examined the adder complexity reduction achieved by partitioning the coefficient into more than two sub-components. If x_2 and x_3 are the VCS obtained from the input x_1, and x_{k_j} represents the data from the set $\{x_1, x_2, x_3\}$ that has to be shifted corresponding to the position of the j-th CSD bit, the general expression for filter output corresponding to a coefficient $h(n)$ of wordlength B is

$$y(n) = \sum_{j=1}^{z}(s_j 2^{-p_j})(x_{k_j}) \qquad (8)$$

where $s_j \in \{-1, 0, 1\}$, $p_j \in \{0, 1, \ldots B\}$, and z is the number of nonzero digits. If p_{s_1} is the shift, (8) can be expressed in PFP form as

$$y(n) = 2^{-p_{s_1}} \sum_{j=1}^{z}(s_j 2^{-(p_j - p_{s_1})})(x_{k_j}) \qquad (9)$$

Partitioning $h(n)$ into n sub-components at equal intervals (i.e., $n_1, n_2, \ldots n$), (9) can be written as

$$y(n) = 2^{-p_{s_1}}\left[\sum_{j_1=1}^{n_1} s_{j_1} 2^{-(p_{j_1} - p_{s_1})} x_{k_{j_1}} + \right.$$
$$2^{-p_{s_2}}(\sum_{j_2=1}^{n_2} s_{j_2} 2^{-(p_{j_2} - p_{s_1} - p_{s_2})} x_{k_{j_2}}) + \ldots$$
$$\left. \ldots + 2^{-p_{s_n}}(\sum_{j_n=1}^{n} s_{j_n} 2^{-(p_{j_n} - p_{s_1} - p_{s_n})} x_{k_{j_n}})\right] \qquad (10)$$

In this case, the widths of the adders in the intermediate-stages of the multiplier are larger since the multiple inner shifts, $(2^{-p_{s_2}}, 2^{-p_{s_3}}, \ldots, 2^{-p_{s_n}})$, in (10) need to be performed prior to the intermediate additions. Hence, each of these intermediate-stage adders would require more FA's. On the other hand, when the coefficient is partitioned into two sub-components, only one inner shift operation exists (i.e., $2^{-p_{s_2}}$) and this is done just before the final-stage adder of the multiplier. Therefore, the widths of the adders in the preceding stages that compute the sum of the bracketed term of $2^{-p_{s_2}}$ are less and only the final-stage adder requires the highest width. Hence, partitioning a coefficient into two halves offers the best reduction of FA's than partitioning into multiple parts.

The steps of our CPM are as follows.

Step 1: Set $k = 0$. Identify the VCS [1 1], [1 –1], [1 0 1] and [1 0 –1] and their negated versions in the CSD representation of $h(k)$. Express the output corresponding to $h(k)$ using VCSE.

Step 2: Express the VCSE output corresponding to $h(k)$ in PFP. Set $M = span$.

Step 3: Partition the span part into two sub-filters of length $M/2$. Scale the latter sub-filter by its *order*.

Step 4: Increment k. If $k \neq N$, go to Step 1. Otherwise, terminate the program.

4. DESIGN EXAMPLES

The FIR filters employed in the D-AMPS Channelizer [7] are considered. The sampling rate of the wideband signal chosen is 34.02 MHz as in [7]. The channel filters extract 30 kHz D-AMPS channels from the wideband signal after downsampling by a factor of 350. The pass-band and stop-band edges are 30 kHz and 30.5 kHz respectively. The peak pass-band ripple specification is 0.1 dB. The peak stop-band ripple (PSR) specifications at different frequencies and respective filter lengths (N) are chosen to be as in the D-AMPS standard. These parameters are shown in Table I.

Table I Specifications of the D-AMPS channel filters

PSR (dB)	-48	-65	-85	-96
N	260	610	940	1180

The reduction of FA's over the direct implementation in designing the channel filters whose coefficients are coded using 16-bit CSD, for different filter lengths are shown in Fig. 3. For the filter with 1180 taps (corresponding to the most stringent blocking specification), our method (CPM) offers a reduction of 71%, whereas the reductions offered by the VSSE [5] and the VCSE [4] methods are 56.7% and 33.8% respectively. The average reduction of FA's for different filter lengths achieved using the VCSE [4] is 30.5% and the VSSE [5] is 50.2%. On the other hand, our method offers an average reduction of 64%.

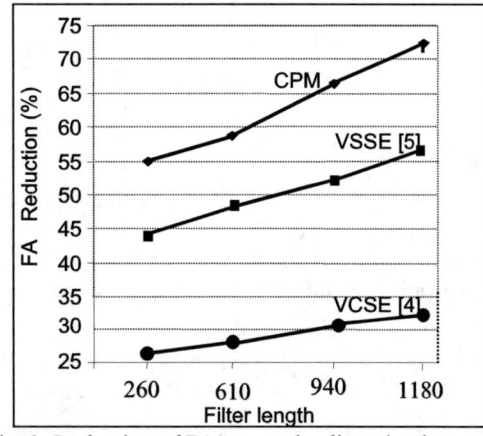

Fig. 3. Reduction of FA's over the direct implementation in designing the D-AMPS channel filters for different filter lengths.

Further, we examine the number of FA's needed to employ the filter bank channelizer, where extraction of each channel requires a separate narrowband filter. The wideband signal considered for channelization consists of 1134 D-AMPS channels, each occupying 30 kHz. We analyzed the requirement of adders to implement the filters for extracting 70, 141, 283, 567, and 1134 channels. The number of filter taps chosen is 1180 and the coefficient wordlength considered is 16 bits. Fig. 4 depicts the FA reduction achieved using different optimization methods over the direct implementation as a function of the number of extracted channels. The average reduction of FA's offered by our CPM is 54.8% whereas the reductions achieved using the VCSE [4] and the VSSE [5] methods are 32.7% and 41.8% respectively. Note that the reduction rate offered by our method increases when the number of channels extracted increases.

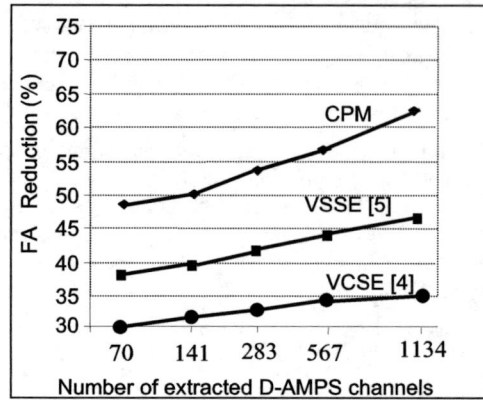

Fig. 4. Reduction of FA's to implement the D-AMPS channel filters for different number of channels extracted.

5. CONCLUSIONS

We have proposed a coefficient partitioning technique to efficiently implement low-complexity channel filters for SDR receivers. The design examples show that our method offers average FA reductions of 22% over the VCSE method [4] and 13% over the VSSE method [5]. Though we used the common subexpression techniques to compare our method, it must be noted that our algorithm can also be applied to reduce the FA requirement of minimum-adder FIR filter coefficient multipliers designed using other methods. Therefore, our approach offers a more general solution to multiplier complexity reduction.

6. REFERENCES

[1] J. Mitola, *Software Radio Architecture.* New York: Wiley, 2000.

[2] R. I. Hartley, "Subexpression sharing in filters using canonic signed digit multipliers," *IEEE Trans. Ckts. Syst. II*, vol. 43, pp. 677-688, Oct. 1996.

[3] M. M. Peiro, E. I. Boemo, and L. Wanhammar, "Design of high-speed multiplierless filters using a nonrecursive signed common subexpression algorithm," *IEEE Trans. Ckts. Syst. II*, vol. 49, no. 3, pp. 196-203, March 2002.

[4] Y. Jang and S.Yang, "Low-power CSD linear phase FIR filter structure using vertical common subexpression," *Electronics Letters,* vol. 38, no. 15, pp. 777-779, July 2002.

[5] A. P. Vinod, E. M-K. Lai, A. B. Premkumar and C. T. Lau, "Optimization method for designing filter bank channelizer of a software defined radio using vertical common subexpression elimination," *Proceedings of the IEEE International symposium on Ckts. Syst.,* vol. 4, pp. 437-440, Vancouver, Canada, May 2004.

[6] A. P. Vinod, A. B. Premkumar and E. M-K. Lai, "An optimal entropy coding scheme for efficient implementation of pulse shaping FIR filters in digital receivers," *Proceedings of the IEEE International Symposium on Ckts. Syst.,* vol. 4, pp. 229-232, Bangkok, Thailand, May 2003.

[7] K. C. Zangi, R. D. Koilpillai, "Software radio issues in cellular base stations," *IEEE Journal on Selected Areas in Communication,* vol. 17, no. 4, pp. 561-573, April 1999.

A BPSK Demodulator Circuit using an Anti-Parallel Synchronization Loop

You Zheng and Carlos E. Saavedra
Department of Electrical and Computer Engineering
Queen's University
Kingston, Canada
2yz2@qlink.queensu.ca

Abstract—**A novel anti-parallel loop carrier synchronization method for BPSK demodulation is proposed and demonstrated in this work. The method contains an anti-parallel dual loop, which locks the carrier by its upper loop and lower loop alternately, according to the data bits contained in the received BPSK signal. Simulation and experimental results are shown along with BER performance.**

I. INTRODUCTION

Satellite communications have unique advantages over other types of communications, such as large coverage over geographical areas and capability of broadcasts directly to the public and end-to-end connections directly to users. Currently, several satellite systems use Binary Phase Shift Keying (BPSK) modulation. One the most well-known systems using BPSK is the INMARSAT satellite network, which is used primarily by ocean-going vessels. Another important satellite network using BPSK modulation is the Global Positioning System (GPS). To reduce the cost of a satellite receiver it is essential that new methods for the demodulation of BPSK signals be found which can reduce the size and the complexity of the overall system.

To demodulate BPSK, circuits such as the squaring loop [1][2], the remodulator loop [3], and the Costas loop [4] are widely used. The PLL-based demodulators are all coherent detectors. There also exist 'non-coherent' BPSK demodulators that use pulse detection by differentiating the incoming modulated signal [5], and others that use encoding at the source such as in the Differential BPSK (DBPSK) case, and that only require a delay element and a mixer at the receiver [1].

An anti-parallel carrier synchronization method for BPSK demodulation is proposed in this work, resulting in a new BPSK demodulator. This method has a simple structure and contains easily-integrated elements, and thus is competitive with the Costas loop. In the section II, the proposed method is described with its operation and mathematical analysis. The demonstration of the proposed method with simulation and experimental results is presented in Section III, and Section IV concludes this work.

II. CIRCUIT OPERATION

The proposed carrier synchronization method is shown in Figure 1. It contains a dual loop with a 180° phase shifter in the lower loop, so-called anti-parallel loop here. The phase detectors used for this anti-parallel loop utilize two multiplier-type detectors, plus a DC offset V_{dc} at each detector. The DC offsets are introduced using two voltage summers and they play important role for the operation of this anti-parallel loop. Moreover, there are two switches at the VCO input and a control circuit: a comparator and an inverter.

The concept here is that with proper control of the switches, an anti-parallel dual loop with 180° phase difference can offer the locking to a received BPSK signal, which switches its phase between 0° and 180° in accordance with the data. For example, when the received BPSK signal is at 0° phase, the upper switch closes and the lower switch opens, and therefore the output of the upper loop is fed to the VCO and it operates like a single PLL (the upper loop works as demodulation loop in this case); When the received BPSK signal switches its phase to 180°, the upper switch opens and the lower switch closes, and thus the lower loop will operate as the locking loop.

Proper control of the switches is required for the above operation. The control circuit of the switches requires a voltage difference between its two inputs in order to produce the control signal. If the two DC offsets were removed from Fig. 1 and the upper loop locked first, the upper detector output would be zero (assuming the input carrier frequency was equal to the centre frequency of the VCO). The phase difference between its two inputs is -90° at this time,

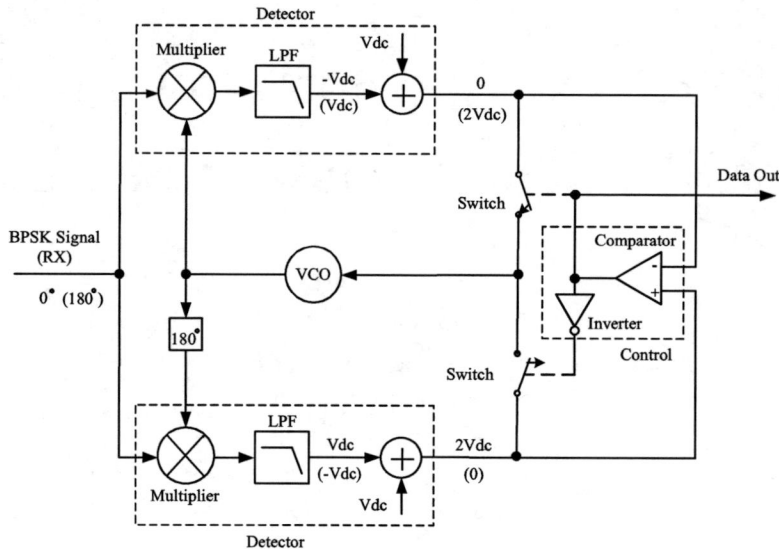

Figure 1. Circuit diagram of the proposed BPSK demodulator

according to the characteristic of the multiplier-type detector. At the same time, due to the 180° phase shifter at the lower loop, the input phase difference for the lower detector was 90°, which results in a zero output at the lower detector, too. Because the inputs of the control circuit come from these two detector outputs, the control circuit cannot properly distinguish these two zero inputs, and would fail to give the proper control signal for the switches.

With the DC offsets introduced, the multiplier-type detectors can produce the different outputs in this anti-parallel loop and meet the above requirement. In Fig. 1, if the upper loop locks first, the VCO's phase is driven to let the upper detector output a zero voltage. A $-V_{dc}$ voltage is produced from the upper multiplier-type detector in order to cancel the DC offset V_{dc} at the output of the upper voltage summer. At the same time, the voltage from the lower multiplier-type detector is $+V_{dc}$, which is the inverse of that of the upper multiplier-type detector due to the 180° phase difference between the two loops. As a result, the output of the lower detector will be $2V_{dc}$. Thus, this $2V_{dc}$ voltage output at the lower loop and the zero output at the upper loop are fed to the comparator in the control circuit together to produce a high-voltage signal to close the upper switch (the inverter produces a lower-voltage signal to open the lower switch). These switch states are exactly what we need. When the phase of the received BPSK signal switches 180°, the lower detector outputs zero instead and the upper detector outputs $2V_{dc}$. The control signal after the comparator is inverted and turn on the lower switch only, then the lower detector's output (zero) is fed to the VCO to lock its phase.

The input of the VCO above keeps at zero whenever the received BPSK signal switches its phase in accordance with the data. Therefore, the data information contained in the BPSK signal is removed by switching the locking loop between the two loops, and then the carrier is recovered. It can be noted that the control signal from the comparator switches between high-voltage and low-voltage in accordance with the phase of the received BPSK signal. It is exactly the demodulated data signal. Thus, a coherent BPSK demodulator is realized based on this anti-parallel method illustrated in Fig. 1.

The following mathematical analysis of the proposed demodulator gives more operation details for the proposed demodulator. Assume that the received BPSK signal has the form

$$S(t) = A_1 \cos(\omega_c t + \theta_1 + \varphi) \quad (1)$$

where θ_1 represents the received carrier phase, and φ bears the data information and switches between 0° and 180°. This received signal is multiplied by $A_2\cos(\omega_c t+\theta_2)$ and $-A_2\cos(\omega_c t+\theta_2)$, which are the outputs from the VCO and the 180° phase shifter, respectively. The two multiplier-type detector outputs after the low-pass filters (the high frequency products of the multipliers are removed by the low-pass filters) are

$$U(t) = k_d \cos(\theta_e + \varphi) \quad (2)$$

$$L(t) = -k_d \cos(\theta_e + \varphi) \quad (3)$$

where $k_d = (A_1 A_2/2)$ is the phase detector gain and $\theta_e = \theta_1 - \theta_2$ is the initial phase difference between the carrier and the VCO. The multipliers are assumed to have a unit gain to simplify the analysis. Note that these two outputs are inversed with each other, which explains the inversion of the outputs between the upper and the lower multiplier-type detectors in the previous description. After summed with the DC offsets V_{dc} at the voltage summers, the final detector outputs are

$$U'(t) = k_d \cos(\theta_e + \varphi) + V_{dc} = \pm k_d \cos(\theta_e) + V_{dc} \quad (4)$$

$$L'(t) = -k_d \cos(\theta_e + \varphi) + V_{dc} = \mp k_d \cos(\theta_e) + V_{dc} \quad (5)$$

where the data information φ alternates between 0° and 180°, resulting in a "±" sign for the cosine functions. As we can see, the two outputs in the equations (4) and (5) alternates oppositely between two voltage values

$$V_1 = k_d \cos(\theta_e) + V_{dc} \text{ and } V_2 = -k_d \cos(\theta_e) + V_{dc} \quad (6)$$

Since the configuration of the control circuit in the anti-parallel loop only allows the detector output with smaller value to pass the switches and enter the VCO, if the initial value of $\cos(\theta_e)$ before locking is negative, the first value V_1 in the equation (6) is the smaller one and it is selected as the error voltage to drive the VCO, regardless of which loop it is from. Thus, the data information φ or the "±" sign in the equations (4) and (5) is eliminated by the control from the control circuit. When the VCO is locked, its input error voltage V_1=0 volt, which results in

$$k_d \cos(\theta_e) = -V_{dc} \quad (7)$$

Substituting the above value to the expression in the equation (6) yields the other detector output

$$V_2 = -k_d \cos(\theta_e) + V_{dc} = V_{dc} + V_{dc} = 2V_{dc} \quad (8)$$

The V_1 (i.e. zero) and V_2 (i.e. $2V_{dc}$) are then fed to comparator to produce the data output and the proper control signal as well.

In the above analysis, the VCO is locked to the upper loop first as the result of $\cos(\theta_e)$<0. However, if the initial value of $\cos(\theta_e)$ before locking is positive, the locking state in the above case will be reversed, i.e. V_2 is the smaller one and is chosen for the error voltage to drive the VCO. When the VCO is locked, V_2=0 volt and V_1=$2V_{dc}$. In this case, the comparator outputs a low voltage in order to turn on the lower switch only, which results in an inversion on the data output compared to the last case. This data inversion phenomenon exists at other coherent demodulators too and can be further solved by differential coding/decoding technique [6].

The DC offsets determine the detector outputs in the proposed demodulator, and they cannot exceed the maximal output of the multiplier-type detector, k_d, according the equations (4), (5) and (6). The proposed demodulator still works if there is difference between these two DC offsets. In this case, the detector of the demodulation loop will output (V_{dc1}+V_{dc2}) instead of $2V_{dc}$, where V_{dc1} and V_{dc2} represents the two different DC offsets. The detector output of the locking loop still keeps zero in this case. For the case that there was a small deviation on the carrier frequency of the input BPSK signal, the two detectors will output a small voltage of δ and ($2V_{dc}$-δ) instead of zero and $2V_{dc}$ respectively, when the loop is locked. Note that the two detectors still have different outputs to ensure the demodulator operation till δ exceeds V_{dc}. Other mismatches between the two loops, such as the detector-gain mismatch and the loop-filter mismatch, could come from circuit implementation. The detector-gain mismatch has similar effect as the DC-offset mismatch described above. The loop-filter mismatch causes different locking speeds between the two loops, and the highest data rate is mainly determined by the lower-speed loop.

The proposed demodulator is similar to the Costas loop, but gets rid of the third multiplier used in the Costas loop, which, usually realized on Gilbert cell in IC implementation, would double the number of the loop filters, because Gilbert cell requires differential inputs. Moreover, the used 180° phase shifter is more easily-integrated than the 90° phase shifter. The switches can be realized on two N-MOSFETs, which was demonstrated in the experiment as described below. The DC offsets can utilize DC bias in the multipliers.

III. SIMULATION AND EXPERIMENTAL RESULTS

The proposed BPSK demodulator was simulated using the simulator Advanced Design System (ADS) from Agilent Technologies. For demonstration purposes, the center frequency of the VCO was set to 133 kHz. A pseudo random bit sequence (PRBS) data with a rate of 10 kbps was used in this simulation, which is slightly higher than the data rate 9.6kbps used in INMARSAT systems [7]. Two RC LPFs with cutoff frequency of 14.4kHz were used for the loop filters in the phase detectors. The detector gain and the gain constant of VCO were optimized to achieve a damping factor of 0.7 for the PLL [8]. The detector gain was 1volt/rad and the DC offsets were set to 0.3 volt.

Fig. 2 shows the data contained in the received BPSK signal, the detector outputs and the successfully-demodulated data from the simulation, respectively. The initial phase difference between the carrier and the VCO was 220° for this simulation, resulting in a negative value for $\cos(\theta_e)$, thus there was no inversion on the demodulated data according to

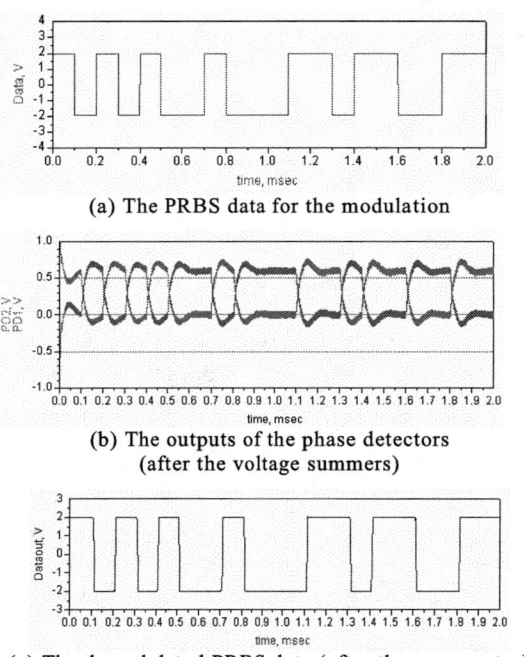

(a) The PRBS data for the modulation

(b) The outputs of the phase detectors (after the voltage summers)

(c) The demodulated PRBS data (after the comparator)

Figure 2. The simulation results for θ_e = 220°

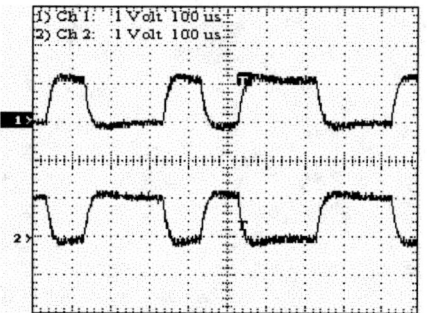

(a) The outputs of the detectors for V_{dc}=0.5V (after the two amplifiers)

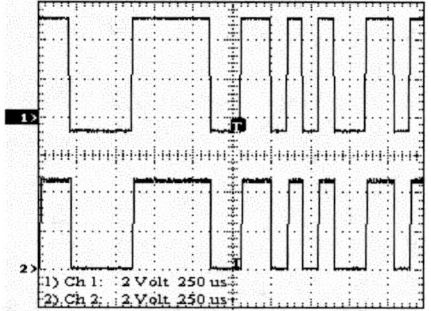

(b) The PRBS data for modulation (Channel 1) and the demodulated data (Channel 2)

Figure 3. The captured waveforms in the experiment

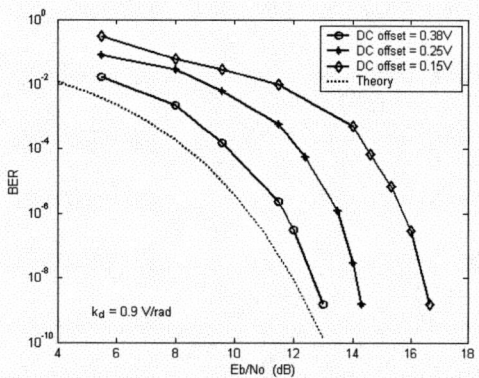

Figure 4. BER versus E_b/N_0 on different DC offset values.

the analysis in the previous section. The detector outputs alternated between 0 volt and 0.6 volt, i.e. $2V_{dc}$, as predicted in the previous description. Other simulations with different values of $\cos(\theta_e)$ were also carried out and the results were consistent to the analysis of this demodulator.

Based on the above simulations, a test circuit was built using packaged integrated circuit components. Two four-quadrant analog multipliers were chosen for the phase detectors, which had an additional summing input and was utilized for the DC offsets needed in this demodulator. The VCO and LPFs had same parameters as in the simulation. An amplifier with a gain of -1 was used for the 180° phase shifter. The switches were implemented using two N-MOSFETs with symmetric configuration. The detector gain was 0.9 volt/rad and the damping factor of the loops was also optimized to 0.7 in the experiment.

The BPSK signal generator used for the test was implemented using a multiplier circuit to multiply a carrier signal with a PRBS NRZ data coming from an HP 3764A Digital Transmission Analyzer (DTA). The NRZ data from the DTA was transformed to be symmetric data (±2 volt) before it went into the multiplier. Fig. 3 presents the experimental waveforms captured from a test with a DC offset of 0.5 volt. Fig. 3(a) gives the outputs of the two detectors. Note that the two outputs offset with each other and their voltage levels were 0 volt and $2V_{dc}$=1 volt, which were exactly expected according the analysis. Fig. 3(b) gives the PRBS data (Channel 1) used for test and the successfully-demodulated data (Channel 2) from the BPSK demodulator. It needs to be pointed out that the two figures shown in Fig. 3 were captured at different times and thus contained different data information (the PRBS data varies with time). The case that the demodulated data would contain an inversion of the modulating PRBS data was also observed sometimes when the circuit was reset in the experiments. Fig. 4 shows the measured bit error rate (BER) versus bit energy to noise density (E_b/N_0) in several DC offset values, as well as an ideal curve for the BPSK demodulator from theory. The measured BER curve got close to the ideal curve when the DC offset went up. It indicated that BER performance can be improved by use of as large DC offset as allowed in the proposed demodulator.

IV. CONCLUSION

An anti-parallel synchronization method for BPSK demodulation has been demonstrated both in simulation and experiment. Due to its simple structure and easily-implemented elements, this demodulator is competitive with other BPSK demodulation techniques such as the Costas Loop or the remodulator loop.

REFERENCES

[1] Lawrence E. Larson, *RF and Microwave Circuit Design for Wireless Communications*, Artech House, Boston, London, 1997.

[2] L. E. Franks, "Carrier and bit synchronization in data Communication- a tutorial review," *IEEE Trans. on Comm.*, vol.COM-28, pp.1107-1121, August 1980

[3] T. Shimamura, "On False-Lock Phenomena in Carrier Tracking Loops," *IEEE Trans. on Comm.*, vol.28, pp.1326-1334, Aug. 1980.

[4] J. Costas, "Synchronous Communications," *IEEE Trans. on Comm.*, vol.5, pp. 99-105, Mar. 1957.

[5] A. Tan and C. E. Saavedra, "A Binary Phase Shift Keying Demodulator using Pulse Detection," *IEEE Int. Conference on Electrical and Electronics Engineering*, pp. 49-52, 2004.

[6] R. E. Ziemer and R.L. Peterson, *Digital Communication and Spread Spectrum Systems*, MacMillan Publishing Company, New York, 1985.

[7] INMARSAT LTD., Maritime Communications and Safety, http://maritime.inmarsat.com/

[8] Roland E. Best, *Phase-Locked Loops, Design, Simulation, and Applications*, McGraw-Hill, Toronto, 2003.

Novel Systolic Array Architecture for the Decorrelator using Conjugate Gradient for Least Squares Algorithm

Archana Chidanandan
Computer Science and Software Engineering
Rose-Hulman Institute of Technology
Terre Haute, IN 47803, USA
Email: chidanan@rose-hulman.edu

Magdy A. Bayoumi
Center for Advanced Computer Studies
University of Louisiana at Lafayette
Lafayette, LA, USA
Email: mab@cacs.louisiana.edu

Abstract—The Decorrelator is a linear transform that can be used in multi-user detection receivers in cellular networks to increase the capacity of the base-station receiver. It has been shown that the Decorrelation operation can be performed using iterative methods that do not require computation-intensive inverse matrix computation. In this paper, we propose a high throughput systolic array architecture for the Decorrelator receiver, which meets the throughput demands of 3G mobile communications systems.

I. INTRODUCTION

Multi-user detection algorithms increase the performance of the base station receivers in cellular networks by taking advantage of the correlative structure of the incoming received signal. The optimal multi-user detection algorithm was proposed by Verdu [1] in 1986. The optimal algorithm suffers from the problem of being too complex to implement as the number of users grows. A number of sub-optimal algorithms have since been proposed.

In this paper, we propose an architecture to implement the Decorrelator which is a linear sub-optimal multi-user detection algorithm and show that the throughput of the architecture can meet the requirements of 3G mobile communications systems. In Section II, we describe the Decorrelator. In Section III, we show why it is difficult to implement it for time-varying systems. In Section IV, we describe the Conjugate Gradient method for least squares(CGL) and the proposed architecture for the implementation of the Decorrelator using the CGL method. In Section V, we study the performance of the architecture in terms of throughput. Finally, we conclude that the novel implementation of the architecture makes it suitable for the Decorrelator to be used for 3G cellular base-station receivers.

II. THE DECORRELATOR RECEIVER

Sub-optimal multi-user algorithms can be classified as linear and non-linear. In non-linear algorithms, interference estimates are obtained and then removed from the received signal. The enhanced received signal is then used to estimate the symbols.

In linear algorithms, a linear transform is applied to the soft outputs of the conventional detector to produce a set of new decision variables with multi-access interference (MAI) greatly decoupled [2].

Consider,

$$Y = Rb + n \qquad (1)$$

where Y is the soft output of the matched filters, and R is the matrix of cross-correlations of the spreading waveforms. Then,

$$b' = TY \qquad (2)$$

where T is the linear transform. There are two linear algorithms proposed for Multi-user detection: Minimum Mean Squared Error(MMSE) Filter [3] and the Decorrelator [2]. The Decorrelator detector is chosen such as to remove all multi-access interference, i.e. interference generated as a result of a number of users using the medium simultaneously. The Decorrelation operation is as follows:

$$r = Sh + w \qquad (3)$$

where h = Ab, r is the received signal, S is the spreading codes matrix, A is amplitudes of the received signals, w is AWGN noise, and b is the bits transmitted.

The output of the bank of matched filters is given by:

$$y = Rh + \overline{w} \qquad (4)$$

where $R = S^T S$ and is the cross-correlation matrix. The output of the Decorrelator is given by:

$$Z = R^{-1}(Rh + \overline{w}) \qquad (5)$$

As we can see, the inverse of the cross-correlation matrix needs to be determined for the Decorrelation operation.

III. THE S MATRIX

Let us examine the spreading codes matrix S. It is of the size $(W+1)N_s \times WK$, where W=2P+1 is the detector memory length and P is an integer, N_s is the number of samples per symbol and K is the number of users [4].

$$S = \begin{bmatrix} S_u & 0 & \cdots & 0 \\ S_l & S_u & \cdots & 0 \\ 0 & S_l & \cdots & 0 \\ \vdots & \vdots & & \vdots \\ 0 & \cdots & \cdots & S_u \\ 0 & \cdots & \cdots & S_l \end{bmatrix} \epsilon \mathcal{R}^{(W+1)N_s \times WK} \quad (6)$$

In the above matrix, S_u and S_l are matrices with the first N_s samples and the last N_s samples of the received signal, respectively. Also,

$$\frac{S_u}{S_l} = \begin{bmatrix} | & | & & 0 \\ | & | & | & \\ | & | & | & \\ & | & | & \\ 0 & & | & \end{bmatrix} \epsilon \mathcal{R}^{2N_s \times K} \quad (7)$$

The above matrix is of size $2N_s \times K$, with the columns bars being the samples of the signature waveforms and the zeros representing their delays. The correlation matrix R will thus be of size $WK \times WK$.

It is evident that the matrix S is a sparse and large matrix. For time varying systems, the matrix S changes for every bit symbol. Also, whenever a new user enters or a user leaves the system, and when a mobile user changes its position with respect to the base-station, the matrix S changes. The matrix R and its inverse also have to be computed every time S changes. Since all 3G and current mobile communication systems use the time-varying model, it would seem that the Decorrelator would be unsuitable.

IV. SYSTOLIC ARRAY ARCHITECTURE FOR THE CGL METHOD

The modified version of the Conjugate Gradient method [5] called CG for Least Squares (CGL) [6] is an iterative technique suitable for implementing the Decorrelator [4]. Notably, it does not require the inverse to be determined at all. The CGL algorithm states, assuming that $y = Rh + \overline{w}$, if $R = S^T S$ and $y = S^T r$ then the solution to $\hat{h} = ||Sh - r||_2$ is a solution to h in the above equation. Note that, with the CGL, the input to the iterative method is 'y' the received signal and not the output of the bank of matched filters. Also, the correlation matrix $R = S^T S$ need not be computed at all.

Figure 1 gives the steps in the computation of the CGL method. Recall that K is the number of users, while in Figure 1, k is the iteration number. The algorithm makes an initial guess for the solution h in equation 3. It has been shown in [4], that using the matched filter estimates as the initial estimate greatly reduces the number of iterations required for the algorithm to converge to the correct solution. r and S, in the algorithm, correspond to the received signal and the spreading code matrix, respectively, in equation 3.

For the analysis of the architecture, we use the following typical values [4]: K = Number of users = 10, N = Processing gain = 31, N_s = Number of samples per symbol = 62, W = Detector memory length = 7, I = Number of iterations = 6.

Initializations:
h_0 = initial guess
$g_0 = r - Sh_0$
$p_0 = l_0 = S^T h_0$
for k = 0, 1, 2 ...

$if(p_k = 0)$
  stop; h_k is the solution
else
  $q_k = Sp_k$
  $a_k = \frac{(l_k^T l_k)}{(q_k^T q_k)}$
  $h_{k+1} = h_k + a_k p_k$ and $g_{k+1} = g_k - a_k p_k$
  $l_{k+1} = S^T g_{k+1}$
  $b_k = \frac{(l_k^T l_k)}{(l_k^T l_k)}$
  $p_{k+1} = l_{k+1} + b_k p_k$

Fig. 1. The CGL method

A. Why Systolic Arrays?

Systolic arrays are suited for applications that have the following features [7]:

- High throughput and large processing bandwidth.
- Algorithm can be implemented on arrays consisting of a few types of simple processing elements.
- Algorithms are characterized by repeated computations of a few types of relatively simple operations that are common to many input data items.
- The pattern of generation and usage of data by different operations displays some regularity and uniformity, such that the resulting communication requirements can be met by the localized interconnections.

For the CGL algorithm and the S matrix, it is evident that the matrix is large. S is made up of smaller identical matrices and therefore we can develop a architecture for the smaller matrix and re-use that for the other matrices. The CGL algorithm exhibits characteristics that are suited for systolic array architectures as listed above. Therefore, we propose a systolic array architecture to implement the matrix-vector operations in the CGL algorithm for the Decorrelator when it is applied to the base-station receiver.

B. Architecture for the CGL algorithm

The CGL algorithm can be split into the following four stages:

1) determining the vector q_k
2) determining the scalar a_k and partially determining g_{k+1} and h_{k+1}
3) determining the vector l_{k+1} and the rest of vector g_{k+1} and h_{k+1}
4) determining the scalar b_k and the vector p_{k+1}

Since the matrix-vector is the most computational intensive stage, it will result in the largest delay. Stages 2 and 4 are therefore designed to fully utilize the time available, and in the process the hardware requirements can be minimized. It is not our aim to keep the size of the architecture at a minimum as in base-station receiver design, area is not a major restriction.

In the next section, we discuss stages 1 and 3 of the architecture, and estimate the delay and thence the throughput of the entire architecture.

1) Stage 1: Generation of q_k: The matrix-vector product Sp_k determines the vector q_k where q_k is of the size WK × 1, p_k is of the size WK × 1, and S is of the size (W+1)N_s × WK. By splitting S into W=7 matrices, we also split q_k into W vectors of size K × 1 each. We have to now multiply, a $2N_s$ × K matrix with a K × 1 vector. For the values chosen, this translates to vectors q_k and p_k of size 70 × 1 each. The vectors are split into vectors of size 10 × 1 respectively.

We make the following observations about the matrix, which can be exploited to obtain a high throughput systolic array.

- Rows 1 to (K - 1), i.e., 1 to 9 and N_s + 1 to (N_s + K - 1), i.e., 63 to 71 have fewer than K values in each row and form lower and upper triangular matrices respectively.
- Rows K to (N_s + K), i.e., 10 to 62 have K=10 values each.
- The last N_s - K + 1 rows, i.e, 72 to 124 are all zeros and need not be used in the computations.
- The calculation of the product for rows 1 to K is similar to the calculation of the product for rows N_s + 1 to (N_s + K - 1). The input p_k however must be sent in the reverse order for the upper triangular matrix.
- The time required to calculate the product for the rows 1 to K and N_s + 1 to (N_s + K - 1) respectively will be the least delay that can be expected. This delay is K - 1 as that is the number of multiplications that will be required to determine the product of the (K - 1)th row and the (N_s + K - 1)th row.
- In all, N_s - K + 1 rows have K values each.

A systolic array architecture is proposed for the matrix-vector computation in the CGL algorithm for the Decorrelator for the base-station receiver. The minimum delay that can be obtained has been determined to be the delay in determining the product of the K-1 rows, which is (K - 1) × delay of the processing element of the systolic array.

However, the delay has to be balanced against the load on the inputs p_k to the processing elements of the systolic arrays. Therefore, the delay will be K + r where r ≪ K. For our example, the value of K = 10 and the value of "r" is chosen as 2 so that we have a maximum delay of K + r = 12. Therefore, the delay of each stage will be K + r or the delay of 12 PEs in our example.

Figure 2 shows the processing element to be used. In Figures 3 and 4, we see the systolic arrays for rows 13 to 15 and rows 1 to 12 respectively. For rows 60 to 71, a systolic array similar to Figure 4 can be used. However, the inputs 'p_k' will be fed in the descending order of indices, and similarly the input values of S. Rows 16 to 59 will be implemented in arrays identical to that shown in Figure 3.

2) Stage 3: Generation of l_{k+1}: To implement the vector product, $l_{k+1} = S^T g_{k+1}$ where for the chosen values l_{k+1} is of the size 70 × 1, g_{k+1} is of the size 496 × 1, and S^T is

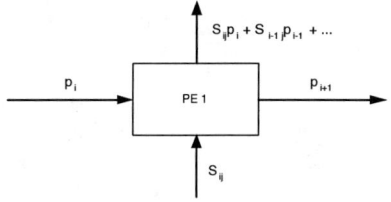

Fig. 2. Processing element (PE1) for the systolic array

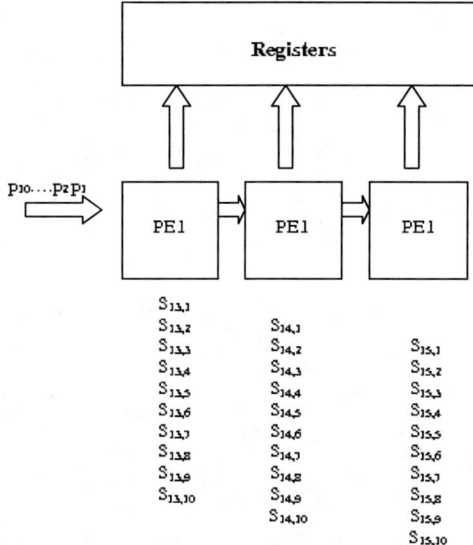

Fig. 3. Processing element (PE1) for the systolic array

of the size 70 × 496. The matrix S^T is split into W smaller modules, and the resultant smaller matrix will be of the form $(\frac{S_u}{S_l})^T$. The following observations are made about the matrix:

- the last N_s - K + 1 columns are zero.
- When multiplying the above matrix with vector 'g', only the first N_s + K - 1 values of 'g' need to be used, as the values in the remaining columns are zero.
- For the second window, values starting from the N_s + 1 to $2N_s$ + K - 1 values will be used and so on.

For the above multiplication, the above matrix is split along columns similar to stage 1. We must balance the delay in this stage with the delay of stage 1, and hence, the matrix is split along the same columns numbers as the row numbers in S. The overall delay will be that of 12 PE1 units.

V. PERFORMANCE OF THE SYSTOLIC ARRAY ARCHITECTURE

The systolic array implementation of the matrix-vector product results in a delay of K + r where r ≪ K. In our example K + r = 12. We summarize the delays of each stage here:

- Stage 1 has a delay of K + r. Stage 2 must wait for one window to be processed before it can start processing the data. For W windows, the total time required to process the data in stages 1 and 2 is (W + 1) × (K + r).

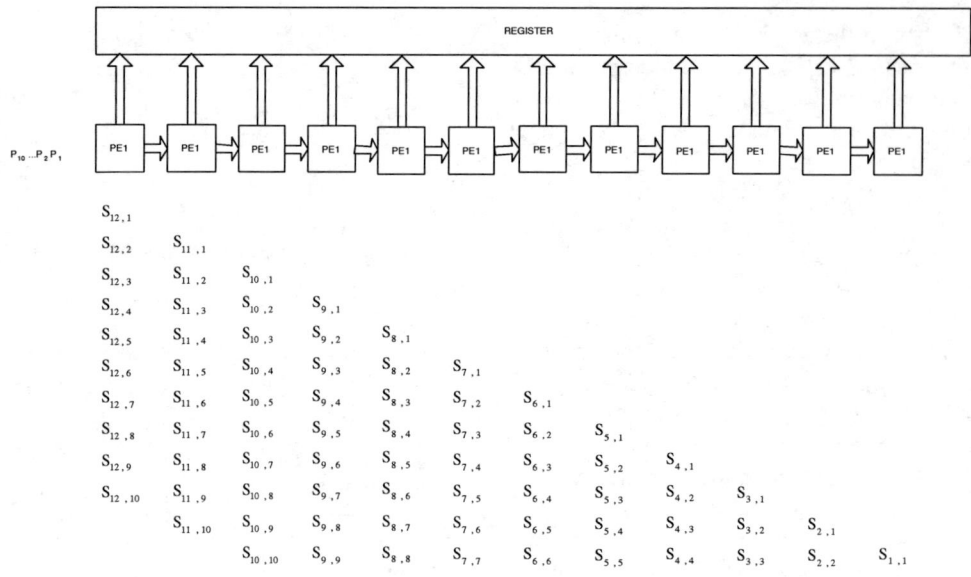

Fig. 4. Systolic array for rows 12 to 1 of the matrix $\frac{S_u}{S_l}$

- Stage 3 has a delay of K + r. Stage 4 must wait for one window to be processed before it can start processing the data. For W windows, the total time required to process the data in stages 3 and 4 is (W + 1) × (K + r).
- Since stages 3 and 4 have to await the completion of all W windows by stages 1 and 2, the total time for one iteration of the CGL algorithm is 2(W + 1) (K + 2).
- Since stages 1 and 2 have to await the completion of all W windows by stages 3 and 4 before starting the computations of the next iteration, the total time for I iterations of the CGL algorithm is 2(W + 1) (K + 2)I × delay of PE1.

Therefore, the total delay of the architecture for K users with W window samples and a processing gain of N_s is 2(W + 1) (K + 2)I. The number of iterations required for the CGL algorithm to converge is typically 6 [8]. Therefore, the total delay is of the order of O(WK).

The architecture was written in Verilog HDL. ModelSim XE II 5.7c was used to verify the functionality of the architecture. For the FPGA implementation, the Spartan xc3s5000 device was chosen with a speed grade of four for simulation. The design was synthesized and mapped to the target device using the tools that are part of the Xilinx 6.1i suite.

From our FPGA simulations, we get the delay of PE1 in the systolic array to be 6.496 ns. With this value of delay, the total delay for the proposed architecture is estimated as follows: $2 \times (7+1) \times (10+2) \times 6 \times 6.496 \times 10^{-9} = 7483.39 ns$

This results in a maximum frequency of 133.6 KHz for the receiver. The throughput is therefore estimated to be: frequency × number of users(K) × window length(W) = $133.6 KHz \times 10 \times 7 = 9.4 Mbps$

3G systems require a bit rate varying between 1.4Kbps to 2Mbps [9]. As is evident, the performance of our proposed architecture for the Decorrelator goes well beyond the stated requirements.

VI. CONCLUSION

The Decorrelator can be implemented using the CGL algorithm, which removes the need to determine the inverse of the cross-correlation matrix. We have proposed a systolic array architecture that can be used to implement the CGL method for the Decorrelator. The throughput of the systolic array architecture has been shown to be well above the requirements of 3G cellular systems.

ACKNOWLEDGMENT

The authors acknowledge the support of the U.S. Department of Energy (DoE), EETAPP program DE97ER12220, the Governor's Information Technology Initiative.

REFERENCES

[1] S. Verdu, "Minimum probability of Error for asynchronous Gaussian multiple-access channels," *IEEE Trans. Inform. Theory*, vol. IT-32, no. 1, pp. 85–96, Jan. 1986.
[2] R. Lupas and S. Verdu, "Near-far resistance of multiuser detectors in asynchronous channels," *IEEE Trans. Commun.*, vol. COM-38, pp. 496–508, Apr. 1990.
[3] Z. Xie, R. Short, and C. Rushforth, "A family of sub-optimum detectors for coherent multiuser communications," *IEEE J. Select. Areas Commun.*, vol. 8, pp. 683–690, May 1990.
[4] M. Juntti, B. Aazhang, and J. Lilleberg, "Iterative implementation of linear multiuser detection for dynamic asynchronous CDMA systems," *IEEE Trans. Commun.*, vol. 46, no. 4, pp. 503–508, Apr. 1998.
[5] G. Golub and C. V. Loan, *Matrix Computations 2nd ed.* Baltimore: The Johns Hopkins University Press, 1989.
[6] J. Stoer and R. Bulirsch, *Introduction to Numerical Analysis*. New York: Springer-Verlag, 1980.
[7] J. Fortes and B. Wah, "Systolic arrays - from concept to implementation," *IEEE Computer*, vol. 20, no. 7, pp. 12–17, July 1987.
[8] M. Juntti, M. Latvao-aho, and M. Heikkila, "Performance comparison of PIC and decorrelating multiuser receivers in fading channels," in *Proceedings of the 1997 Global Telecommunications Conference (GLOBECOM97)*, Phoenix, U.S.A., Nov. 1997, pp. 123–127.
[9] "3G Information," Internet draft, Nov. 2002. [Online]. Available: http://www.fcc.gov/3G

An Efficient Pre-Traceback Approach for Viterbi Decoding in Wireless Communication

Yao Gang[1], Tughrul Arslan[1,2], Ahmet T Erdogan[1,2]
[1]School of Electrical Engineering, Edinburgh University, Edinburgh EH9 3JL, UK
[2]Institute for System Level Integration, Livingston, EH54 7EG, UK
Email: y.gang@ed.ac.uk

Abstract— An efficient *pre-traceback* architecture for the survivor path memory unit (SMU) of Viterbi Decoder(VD) targeting wireless communication applications is proposed. Compared to the conventional traceback approach which is based on three kinds of memory access operations: *decision bits write, traceback read* and *decode read*, the proposed architecture exploits the inherent parallelism between the *decision bit write* and the *decode traceback* operation by introducing a *pre-traceback* operation. The proposed *pre-traceback* approach reduces the survivor memory read operations by 50%. As a result of the reduction of the memory access operations, the size of the survivor memory as well as the decoding latency is reduced by as much as 25%. Implementation results show that the *pre-traceback* approach achieves up to 11% energy efficiency and 21% area saving compared to the conventional traceback architecture for typical wireless applications.

I. INTRODUCTION

Viterbi Algorithm (VA) with high constraint length K is a widely used error detection / correction scheme such as GSM, the voice channels of 3G and *IEEE 802.11a* wireless LAN. The well-known VA has been described in literature extensively [1] [2] [3]. The data path of the VD is composed of three major components: Branch Metric Calculation Unit (BMU), Add Compare Select (ACS) and Survivor Memory Unit (SMU).

ACS attracts most of the research efforts in the field of VD since the existence of feedback loop of the butterfly-like processor makes ACS unit the bottleneck of speed of the whole VD system [2] [3] [4]. However, in [5], it has been shown that SMU contributes over half of the power consumption and energy due to the extensive large memory access operations. Two basic approaches used to record survivor paths are register exchange and traceback in SMU of VD [3]. Up to now, only VDs with $K \leq 5$ implemented using register exchange have been reported in literature. In this paper, we focus on the traceback approach implementing SMU of the VD in high constraint length wireless applications. The techniques in previous work [5] [6] come at an expense of the decoding degradation to some extent.

The remainder of the paper is organized as follows: Section II briefly describes the conventional traceback. Section III deals with the proposed *pre-traceback* in which the algorithm as well as the implementation architecture is presented. Moreover, an example is also illustrated. Finally, in section IV the synthesis implementation results are examined to verify the analysis models and proposed architecture.

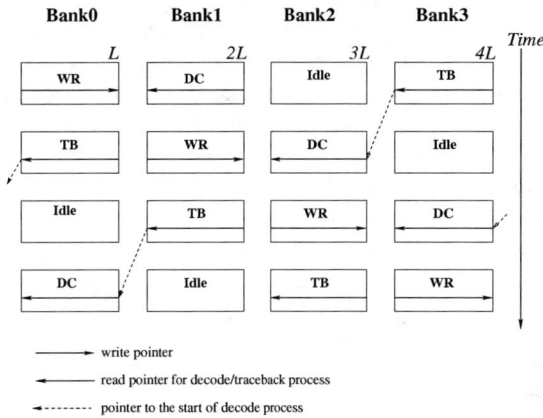

Fig. 1. Architecture Of Conventional Traceback Approach

II. CONVENTIONAL TRACEBACK

Generally, all the VDs with conventional traceback approach employ some variations of the k−pointer traceback architecture [7]. Here k refers to the number of the read pointers to access the survivor memory. The classical implementation of the trace back scheme is based on the property of unification [8]. That is, if we follow all survivor sequences back M stages, they all merge to the same state. Similarly, if we choose an arbitrary state at stage X and trace back to $X - M \gg L$ stages, we will reach the same maximum likelihood correct state.

The survivor path memory can be regarded as a circular 4 blocks of memory with L columns each as shown in Fig 1. Three operations *decision-vector write, traceback-read* and *decode-read* work in parallel to manipulate the decision bits vectors in different memory banks:

- *decision bit-write (WR)* writes the updated decision bit vectors from ACS unit into the survivor memory in an increased memory address order.
- *traceback-read (TB)* recursively estimates the previous state S_{n-1} according to the current state S_n and the associated decision bit d_n^s. For radix-2 applications, the estimation rule can be described as follows:

$$S_{n-1} = \{d_n^s, S_n \gg 1\} \quad (1)$$

The initial state is arbitrary according to the unification

property and the recursion will be repeated for consecutive L iterations.
- *decode-read (DC)* performs the similar read operation as the *traceback* operation defined by equation 1 except that the initial state is the output state estimated by *TB* in the last L time slot.

Every L time interval, the *TB* begins the trace back read from an arbitrary state, and the *DC* starts with the state determined by the *TB* process in the last L time interval. Since the decoded bits generated by the *DC* is in the reverse order, a simple two-stack Last In First Out (LIFO) scheme is used to perform bits order reversal. Each stack with a size of L depth, *DC* process pushes the reversal decoded bits stream into one stack while the decoded bits stored in the other stack are popped. Hence, the overall latency of traceback method, which is the time delay between the writing of the decision bit of *WR* process and the time when the corresponding bit is popped out of the LIFO, is $4L$. A key observation is that every decision bit in the survivor memory written by the *WR* will be read by *TB* and *DC* respectively. In other words, every decision bit in the survivor memory will be read two times. The redundant memory read operations are the main optimization target our proposed architecture try to address.

III. Proposed Pre Trace Back

A. Pre-Traceback Algorithm Description

The main task of the *TR* in conventional traceback approach is to get the initial start state for the *DC* through a recursive M shift and survivor memory read operations. For simplicity, we let the traceback length $M = L$. At time instant $L + M$, for $\forall i \in N = \{0, 1, 2, \cdots m\}$, where N is the set of trellis state and $m = 2^{K-1}$, S^i_{kL+M} will converge to the maximum likehood state $S^{max-like}_{kL}$. Furthermore, during the time interval $[kL, (k+1)L]$, for $\forall i \in N$, we refer the pre state of i at the time instant kL as the *target trace back state*. Naturally, the goal of *TR* which starts at time instant $(k+1)L$ is to locate the *target traceback state* at time instant kL, which is the initial start state of the *DC* operation.

We propose an approach involving a *pre-traceback* operation through which the start of the *DC* can be obtained directly through a pointer register indicating the *target trace back state* instead of estimating the *DC* start state through a recursive *TR* operation. In our proposed approach, $\forall i \in N$, a pointer register which is denoted as S^i_n indicating the *target trace back state* of i at time instant n. At time instant n, in parallel with *WR* writes the decision bits into the survivor memory, the pointer register S^i_n is updated concurrently as follows:

$$S^i_n = S^{\{d^i_n, i \gg 1\}}_{n-1} \quad (2)$$

Here, d^i_n represents the decision bits of state i at time instant n. The exponential number, $\{d^i_n, i \gg 1\}$, represents the index number of the temporal adjacent pre state. The operation defined in equation 2 implies a multiplex select operation. It is apparent that the *pre-traceback* operation defined in equation 2 is performed in the forward direction opposed to the conventional traceback operation which is done in backward direction. According to the unification property of the trellis graph, if L is set large enough, the pointer register S should point to the same state because of the emergence of the state. This fact is confirmed by our simulation. Hence, at time instant $(k+1)L$, for trellis state $\forall i \in N$, $S^i_{(k+1)L}$ should point to the maximum likelihood state:

$$S^i_{(k+1)L} = S^{max-like}_{kL} \quad (3)$$

which is the *target track back state* and the start state of the *DC*. As soon as *WR* complete $L + 1$ columns of decision bit update, the *target traceback state* of the last column is available at the same time. Hence, the corresponding *DC* can be performed from the *target traceback state* which is estimated by the *pre-traceback* operation directly. In our proposed *pre-traceback* approach, *pre-traceback* and *DC* are done in forward direction and backward direction respectively whereas in the conventional approach *TB* and *DC* are both done in backward direction. In summary, compared to the conventional traceback scheme which is based on three types of memory access operations(*WR*, *TB* and *DC*), only two types of operations are necessary in the *pre-traceback* approach, *WR* and *DC*:

- *decision bit-write (WR)* : In addition to update decision bit vectors in the survivor memory, the pointer register are also updated as described in equation 2. The update of the survivor memory and the pre-traceback of the pointer register are done in parallel.
- *decode-read (DC)* process performs two steps of operation:
  - First, select an arbitrary state and look for the *target traceback state* in the pointer register.
  - Second, from the start state, performs the iterative read operation for decoding.

B. Example

Further insight into the operations of the *pre-traceback* approach can be obtained through an example shown in Fig 2, where a four state trellis with the constraint length $K = 3$ is described. Each node corresponds to a state at a given time instant and each branch corresponds to a survivor path transition decided by ACS unit. The trellis state set N is indexed as $\{00, 01, 10, 11\}$. The corresponding decision bit associated with the survivor path transition is labelled under the node. For simplicity, a small time interval from kL to $kL+5$ is described, where $k = \{0, 1, 2, \cdots\}$. The *WR* writes 5 columns of the decision bit vectors into the survivor memory in an increasing address order. In conventional traceback scheme, at time instant $kL+5$, in order to get the pre state of the 2(10) state at time instant kL, a series of shift operations specified as equation 1, are performed iteratively. At time instant $kL+5$, the decision bit of state 10 is 0, hence the pre state is 01. Similarly, the pre state of 01 is 00. After 5 consecutive memory read and shift operations, 10 is obtained, which is the pre state at time instant kL of the state 10 at time instant $kL+5$.

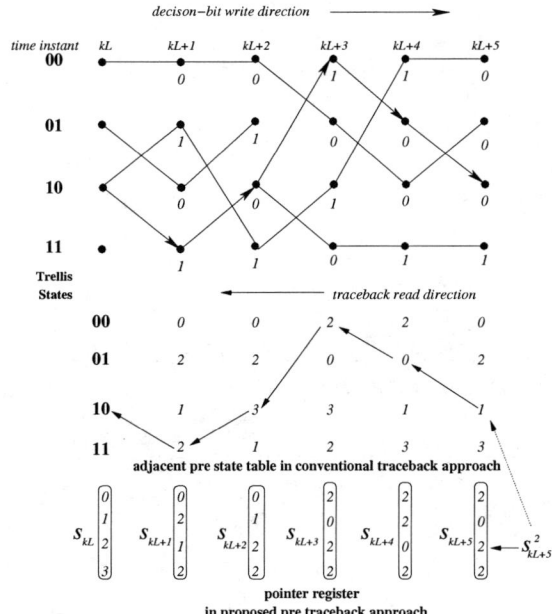

Fig. 2. Example: Comparisons Of Operations Of Conventional Traceback and Proposed Pre-Traceback Approach

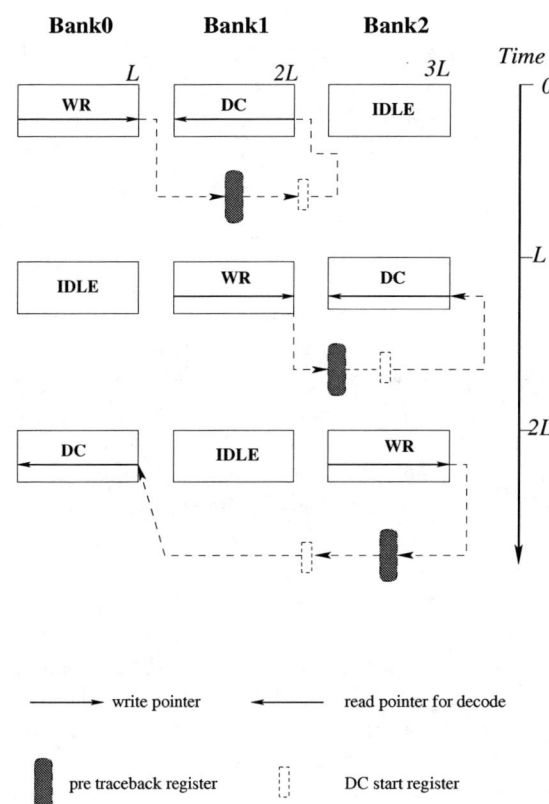

Fig. 3. Proposed Pre-Traceback Architecture

The process of the traceback is in fact the process of building an adjacent pre state table. It can be seen that the traceback can only be done iteratively in that the conventional approach only maintains the information adjacently between time instant $kL+n$ and time instant $kL+n+1$.

In our proposed *pre-traceback* approach, *TB* and *DC* are done in forward and backward direction respectively. During the *WR*, a pointer register which always points to the pre state of time instant kL of all the possible trellis state is employed. At time instant $kL+1$, the pointer register is updated as $\{00, 10, 01, 10\}$, all of which point to the pre state at time instant kL. It should be noted it is the same as the pre state table as shown in the conventional approach in the sense that the two time instant are adjacent. Whereas at time instant $kL+2$, the pointer register of state 10 denoted as S_{kL+2}^2 points to state 10 which is the target pre state at time instant kL compared to the adjacent pre state 11. Similarly, the whole column of register are updated as $\{00, 01, 10, 10\}$, which points to the pre state at time instant kL. Consequently, at time instant $kL+5$, the pre state of the state 10, which is also state 10, can be obtained directly. Through the similar *pre-traceback* operation, at time instant $(k+1)L$, the *target trace back* state of the state i, which is at time instant kL can be directly located instead of the recursive memory access operations in the conventional approach.

C. Architecture

The survivor memory is divided into three banks, with L columns for each bank, as shown in Fig 3. Only two types of operations: *WR* and *DC* work in parallel to access the survivor memory concurrently. In addition to the pointer register mentioned in the above section, another register is used to store the intermediate *DC* start state, which is simply referred to as *DC start register*. *WR* updates the decision vector columns in an increasing order. Correspondingly, the pointer register is updated based on the general description in equation 2. However, as we only need to know the *target traceback state* L stages before, the operation of *pre-traceback* is based on the block with the size of L. It should be noted that at time instant kL, where $k = \{1, 2, \cdots\}$, *pre-traceback* for the time interval $[(k-1)L, kL]$ and $[kL, (k+1)L]$ is overlapped. A simple trick is employed with the aim to share the pointer register as well as overcoming the problem of overlap. It is apparent to observe that at time instant kL, for the *pre-traceback* operation at time interval $[kL, (k+1)L]$, the ith pointer register should always be initialized as i. So by just plugging the initial state number into equation 2 at time instant $kL+1$, the pointer register between the two blocks can be shared. The *target traceback state* in the pointer register is shifted into *DC* start register at every kL time instant. Specifically, the operation of the pointer registers can be summarized as follows:

$$S_n^i = \begin{cases} i, & \text{if } n = 1 \\ i\{d_n^i, i \gg 1\}, & \text{if } n = kL+1, \text{ where } k = 1, 2, \cdots \\ S_{n-1}^{\{d_n^i, i \gg 1\}}, & \text{otherwise} \end{cases}$$

(4)

At each kL time instant, with an initial start state determined by pointer registers, *DC* begins a consecutive memory read

TABLE I

SIZE AND LATENCY OF TRACEBACK AND PRE-TRACEBACK ARCHITECTURE

	Survivor Memory Depth	Decoding Latency
trace back	$4L$	$4L$
pre-traceback	$3L$	$3L$

operation with a decreasing address. A LIFO is also required to perform the bit reverse order. The overall latency including LIFO is only $3L$ opposed to $4L$ in the conventional approach. A simple architectural comparison is summarized in TABLE I. Compared to the conventional traceback approach, the proposed *pre-traceback* approach has 25% improvement in both memory size efficiency and latency reduction. Although a similar size and latency can also be achieved with the conventional traceback scheme according to [8], it must involve some form of dedicated clocking design strategies. Specifically, in [3], read pointer must work 3 times faster than write pointer and there is a stringent constraint on the phase relationship between the write pointer and the read pointer. This obviously improves the design efforts compared to the proposed *pre-traceback* approach which adopts a simple clocking strategy that the read and write pointers work at the same frequency.

IV. IMPLEMENTATION RESULTS

Two tyical VDs in wireless communication field are implemented in UMC $0.18u$ standard cell environment:

- 3 bits, 8 level soft decision inputs
- code rates $R = 1/2$
- constraint length: $K = 7$ and $K = 9$, which corresponds to 64 and 256 states in the trellis graph

The BMU and ACS are exactly the same. For the convenience of the implementation, the truncation length L, is empirically set to 64. The proposed *pre-traceback* architecture takes advantages of the conventional architecure under the assumption of the overhead introduced by the pointer registers is negliable compared to the cost reduction by the survivor memory. The results presented in TABLE II summarizes the area synthesis results of the architecture specification of $K = 7$ and $K = 9$. For the typical wireless communication applications, i.e. wireless LAN where $K = 7$, the area savings in SMU block and the whole VD are 37.9% and 21.9% respectively.

10^4 random patterns were simulated at the clock frequency of 10 MHz for each VD, and the switching activities of each node were captured and then back annotated into the circuit. Power dissipation was then estimated for the synthesized gate-level circuits using Synopsys tools. and TABLE III, where $\hat{\eta}_{smu}$ represents the energy efficiency of the SMU unit and $\hat{\eta}_{vd}$ represents the energy efficiency of the whole VD. For typical applications in wireless LAN and 3G where $K = 7$ and $K = 9$ respectively, the energy efficiency of SMU is 22.8% and 21.6% respectively and the energy efficiency of the whole VD is 11.9% and 10.8% respectively.

TABLE II

SYNTHESIZED AREA RESULT: CONVENTIONAL TRACEBACK AND PROPOSED PRETRACEBACK ARCHITECTURE(mm^2)

K	7	9
SMU(*Conventional Traceback*)	2.24	8.98
SMU(*Propose PreTraceback*)	1.41	5.71
SMU Block Savings	37.9%	36.4%
VD(*Conventional Traceback*)	3.76	15.4
VD(*Proposed PreTraceback*)	2.94	12.1
VD Savings	21.9%	21.3%

TABLE III

ENERGY EFFICIENCY OF PROPOSED PRE-TRACEBACK APPROACH

K	7	9
$\hat{\eta}_{smu}$	22.8%	21.6%
$\hat{\eta}_{vd}$	11.9%	10.8%

V. CONCLUSION

An efficient *pre-traceback* architecture for the survivor memory unit for Viterbi Decoder (VD) targeting wireless applications is presented. The memory read operations are reduced by 50% by introducing the *pre-traceback* operation. The survivor memory size and the decoding latency is reduced by 25% compared to the conventional 2−pointer traceback algorithm. The combined reduction of survivor memory access operations as well as the survivor memory size leads to significant energy efficiency and area reduction.

REFERENCES

[1] P. J.Black and T. H.Meng, "A 140-mb/s 32-state,radix-4 viterbi decoder," *IEEE Journal Of Solid-State Circuits*, vol. 27, p. 1877 to 1885, December 1992.

[2] I.Kang and A.Willson, "Low power viterbi decoder for cdma mobile terminals," *IEEE Journal Of Solid-State Circuits*, vol. 33, p. 473 to 482, March 1998.

[3] Y.-N. Chang, H. Suzuki, and K. K.Parhi, "A 2-mb/s 256-state 10-mw rate 1/3 viterbi decoder," *IEEE Journal Of Solid-State Circuits*, vol. 35, p. 826 to 834, June 2000.

[4] K.Parhi, "An improved pipelined msb-first add-compare select unit structure for viterbi decoders," *IEEE Transactions on Circuits and Systems I:Regular Papers*, vol. 51, p. 505 to 511, March 2004.

[5] R. Henning and C. Chakrabarti, "An approach for adaptively approximating the viterbi algorithm to reduce power consumption while decoding convolutional codes," *IEEE Trans. On Signal Processing*, vol. 52, p. 1443 to 1451, May 2004.

[6] Stanley.J.Simmons, "Breadth-first trellis decoding with adaptive effort," *IEEE Transaction on Communications*, vol. 38, p. 3 to 12, January 1990.

[7] G. Feygin and P.G.Gulak, "Architecture tradeoffs for survivor sequence memory management in viterbi decoders," *IEEE Trans. On Communications*, vol. 41, p. 425 to 429, March 1993.

[8] R. Cypher and C. Shung, "Generalized trace back techniques for survivor memory management in the viterbi algorithm," in *IEEE Global Telecommunications Conference,GLOBECOM*, p. 1318 to 1322, December 1990.

Bifurcation Analysis of On-Chip LC VCOs

Marcus Prochaska, Alexander Belski, Wolfgang Mathis
University of Hanover
Institute of Electromagnetic Theory and Microwave Technology
Hannover, Germany
prochaska@tet.uni-hannover.de

Abstract—Since the beginning of the last century the problem of oscillator circuit design is a main subject for analog circuit designers. In recent years the study of differential VCOs has received attention because of their importance for application. In this paper we present an analytic analysis of the steady state oscillation of integrated differential VCOs which is based on a nonlinear model of the oscillator. Starting from the Andronov Hopf Theorem we analyze the stability as well as we apply methods to predict the amplitude.

I. Introduction

Recent developments in the field of semiconductor technology lead to the realizability of integrated LC-tank VCOs (Fig. 1). Due to the easy implementation of the differential operation and relatively good phase noise, LC VCOs play a decisive role for application [1]. This holds especially for high frequency systems like UMTS receivers or flash ADCs. In this paper a bifurcation analysis of integrated LC oscillators is presented. By using symbolic algorithms, which are based on geometric methods, we analyze the stability as well as we calculate an approximate amplitude. By using these techniques we can provide a functional dependence of the results on circuit parameters. Thus, our analysis can directly be used for the design of LC-Tank oscillators. The initial point of our analysis is a nonlinear model since a nonlinear gain element is necessary to compensate the damping as well as for the stabilization of the amplitude. In contrast, mostly linear models are used for the design of oscillators which leads to the restriction that the amplitude is not computable. Of course, designers are able to estimate the amplitude of the oscillation under several assumptions. For example, if we imply that the current in the differential stage switches quickly from one side to another, the amplitude is proportional to I_0 until it is limited by V_{CC} [1]. Contrary to estimations our approach needs less simplification and so we can achieve a higher accuracy. Since the nonlinearity of an oscillator is an integral part of its functionality [2], it is suitable to model electrical oscillators by using a system $\dot{x} = f(x)$, where $f: \mathbb{R}^n \to \mathbb{R}^n$, $x \in \mathbb{R}^n$. The vector x corresponds to time depending currents or voltages of the circuit while f is a nonlinear vector field containing the influence of the gain element. We exclude

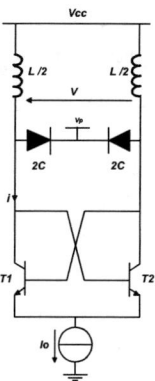

Figure 1. Cross-coupled LC-Tank oscillator

descriptions with differential-algebraic equations but most of the results can be extended to the so-called index 1 case [3]. It is well-known that a nonlinear dynamic system is stable, if its linear part has only eigenvalues with negative real parts. If at least one eigenvalue has a positive real part, a circuit is unstable. In those cases a system is called hyperbolic. It is an also well-establish condition for a steady state oscillation that a system has a pair of conjugate complex eigenvalues with vanishing real parts. Related criteria were presented by Barkhausen and Nyquist [4]. In this case a circuit model is called a non-hyperbolic system. The Hartman Grobman Theorem tells us that, if a system is non hyperbolic, we cannot neglect the nonlinearity [5]. So the necessity of a nonlinear analysis of oscillators is originated in the Hartman Grobman Theorem. On this account we present a nonlinear design methodology for integrated LC-Tank oscillators.

II. Andronov Hopf Theorem

In order to apply the Andronov Hopf Theorem it is useful to get an overview of the possible solutions. A planar nonlinear dynamical system has primarily two different types of solutions: equilibrium points and periodic solutions. Equilibrium points are the roots of the nonlinear field $f(x)$ while periodic solutions are close trajectories in the state space. If a periodic solution is isolated, i.e. isolated in the sense that there are no other close trajectories in its neighborhood, this curve will be called a limit cycle. The relationship between an equilibrium point and a limit cycle is given by the Andronov Hopf Theorem. Originally this theorem was proved by Andronov in 1935 for the analysis of tube oscillators [6]. In this context we consider a dynamic system which additionally depends on the parameter μ:

$$\dot{x} = f(x, \mu) \qquad (1)$$

Let $f(x,\mu)$ be a C^3 vector field such that $f(0,0)=0$ and $D_x f(0,0)$ has a pair of imaginary eigenvalues $\lambda_\pm(\mu=0)= \pm j\omega_0$. A so-called Hopf Bifurcation occurs when a pair of eigenvalues crosses the imaginary axis. Comprising, the Hopf Theorem describes the birth of a limit cycle depending on μ in the neighborhood of zero and under the following conditions:

- $D_x f(0,0)$ has a pair of conjugate complex eigenvalues
- all other eigenvalues possess negative real parts
- $\dfrac{d}{d\mu}\Re\lambda(\mu)\Big|_{\mu=0} > 0$
- the equilibrium point is asymptotic stable

If these conditions are satisfied, there is a stable equilibrium point for $(-\mu,0)$ and a stable limit cycle for $(0, \mu)$. The main conclusion of the Andronov Hopf Theorem is the condition that an asymptotic stable equilibrium is necessary for a stable limit cycle, i.e. there exists a variation of the parameter μ, which leads to a stable limit cycle, if the equilibrium point is asymptotic stable. Thus, the Andronov Hopf theorem is the basis for the operating mode of oscillators [7].

III. Circuit Model

For a nonlinear analysis of LC-Tank oscillators modeling of the tuning diodes, the bipolar transistors and the monolitic inductors is necessary. In order to get an adequate approximation of our model we shape the transistors by the Ebers-Moll model. We assume that both transistors are the same and that their parameters are sufficiently known. Furthermore, we neglect the nonlinearity of tuning diodes which we approximate as ideal capacitors. The on-chip inductors perform a critical role in integrated RF circuits, since their Quality factors are much lower than that of off-chip components. Since the model of integrated inductors depends primarily on the IC process and the implementation, it is not easy to specify a generally applicable model. Moreover, a suitable inductor model leads to heavy calculations of high-dimensional dynamical systems. For example, an adequate equivalent inductor model of an Octo-Coil, which can be used for the implementation of a fully integrated LC tuned VCO using a 0.12μm CMOS process, consist of two inductors, four capacitors and five resistors in a double-π-circuit [8]. That is why we represent the on-chip inductors by ideal electrical devices. Strategies for the modelization of integrated inductors are given in [9]-[11]. In [12] we have presented some ideas for the reduced order modeling of high dimensional oscillators. In Fig. 2 a useful

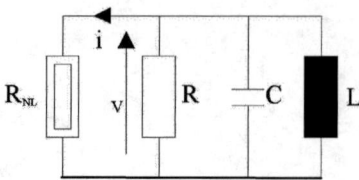

Figure 2. Equivalent circuit of LC-Tank oscillators

nonlinear model for differential VCOs is shown [13]. In order to gain such equivalent circuit of cross-coupled LC-Tank oscillators, first we calculate the base-emitter voltage of the transistors. After that, we are able to find an equation for the current

$$i(v) = \frac{\alpha_F I_0}{1+e^{v/V_T}}\left(1+\frac{1}{\beta_F}e^{v/V_T}\right) + \frac{I_S}{\alpha_R}(e^{v/V_T}-e^{-v/V_T}) + \frac{I_S}{1+e^{v/V_T}}\left[\alpha_F\left(e^{-v/V_T}-e^{2v/V_T}\right)+\left(4-3\alpha_F-\frac{1}{\alpha_F}\right)\left(1-e^{v/V_T}\right)\right], \qquad (2)$$

which enables us to define a simple equivalent circuit. R_{NL} represents the nonlinearity of the bipolar transistors. So we can calculate the model of the oscillator:

$$\begin{pmatrix} \dfrac{dv}{dt} \\ \dfrac{di_L}{dt} \end{pmatrix} = \begin{pmatrix} -\dfrac{1}{RC} & -\dfrac{1}{C} \\ \dfrac{1}{L} & 0 \end{pmatrix}\begin{pmatrix} v \\ i_L \end{pmatrix} + \begin{pmatrix} \dfrac{1}{C} \\ 0 \end{pmatrix} i(v) \qquad (3)$$

IV. Stability of the Equilibrium

The Hopf Theorem tells us, that the linear part of an oscillatory system, which is given by the Jacobian $D_x f(0)$ evaluated at the equilibrium point, has to have a pair of conjugate complex eigenvalues with vanishing real parts. The equilibrium point of (3) is given by $i_L^* = i(I_0)$ and $v^*=0$. In order to guarantee the existence of a pair of pure imaginary eigenvalues, we have to calculate a multidimensional power series of the nonlinear vector field of (3). We find

$$\begin{pmatrix} \dfrac{dv}{dt} \\ \dfrac{di_L}{dt} \end{pmatrix} = \begin{pmatrix} -\dfrac{1}{RC}+\dfrac{i'(I_0)}{C} & -\dfrac{1}{C} \\ \dfrac{1}{L} & 0 \end{pmatrix}\begin{pmatrix} v \\ i_L \end{pmatrix} + \begin{pmatrix} \dfrac{1}{C}i(I_0) \\ 0 \end{pmatrix}, \qquad (4)$$

where only the linear part of the expansion is treated. If the condition $-1/RC + i'(I_0)/C = 0$ holds, (4) has eigenvalues $\lambda_{1,2} = \pm j\omega_0 = \pm j/\sqrt{(LC)}$. So the first condition given by the Hopf Theorem is fulfilled. In this case we find

$$I^* = I_0 = -\frac{2\big((2\alpha_R V_T - 4I_S R + 4I_S R\alpha_R)\alpha_F - I_S R\alpha_R\big)}{\alpha_R \alpha_F (2\alpha_F - 1) R}. \qquad (5)$$

If we choose $I_0 = I^*$ the equilibrium point of (4) is in the origin and so a necessary condition for the applicability of the Hopf is also satisfied. The next point, we have to prove that the equilibrium point is asymptotic stable. For the further calculation it is useful to simplify the linear and nonlinear part of our given system. The initial point for the simplification is (4). The idea is to choose a coordinate transformation so as to simplify the terms of the vector field. In order to simplify the linear part of the vector field, we use the Jordan form theorem, which transfers $\dot{x} = f(x)$ to $\dot{y} = Jy + T^{-1}\tilde{f}(Ty)$ with $J = T^{-1}AT$, $x = Ty$ and $\tilde{f}$ the nonlinear part of f. The invertible transformation T diagonalizes the linear part or at least puts it into Jordan form. The initial point of the analysis of stability of our oscillator is (4), which can be rewritten in the form

$$\begin{pmatrix} \dfrac{dv}{dt} \\ \dfrac{di_L}{dt} \end{pmatrix} = \begin{pmatrix} \mu(I_0) & -\dfrac{1}{C} \\ \dfrac{1}{L} & 0 \end{pmatrix} \begin{pmatrix} v \\ i_L \end{pmatrix} + \begin{pmatrix} \dfrac{1}{C}\gamma v^3 \\ 0 \end{pmatrix}, \quad (6)$$

where the bifurcation parameter μ is given by

$$\mu(I_0) = \dfrac{i'(I_0)}{C} - \dfrac{1}{RC} \quad (7)$$

and the constant

$$\gamma(I_0) = -\dfrac{1}{48}\dfrac{I_0}{V_T^3} + \dfrac{1}{24}\dfrac{I_0}{V_T^3}\alpha_F + \dfrac{1}{3}\dfrac{I_S}{\alpha_R V_T^3} \\ -\dfrac{1}{2}\dfrac{I_S}{V_T^3}\alpha_F + \dfrac{1}{6}\dfrac{I_S}{V_T^3} - \dfrac{1}{24}\dfrac{I_S}{\alpha_F V_T^3}. \quad (8)$$

So we can calculate the Jordan normal form

$$\dot{x} = \begin{pmatrix} \mu & \omega_0 \\ -\omega_0 & \mu \end{pmatrix} x + \begin{pmatrix} 0 \\ -\dfrac{1}{C}\gamma v^3 \end{pmatrix}. \quad (9)$$

After that, in order to simplify the nonlinear terms we try to find a sequence of coordinate transformations $x = y + h_k(y)$ which remove terms of increasing degree from the Taylor series [14]. This equation is a so-called near identity transformation. The terms which cannot be eliminated are called resonant terms. A dynamic system which has eigenvalues $\alpha(\mu) \pm j\omega(\mu)$ can be expressed in the so called Poincaré normal form

$$\dot{y} = \begin{bmatrix} \alpha(\mu) & \omega(\mu) \\ -\omega(\mu) & \alpha(\mu) \end{bmatrix} y + \sum_{i=1}^{\infty} (y_1^2 + y_2^2)^i \begin{bmatrix} a_i & b_i \\ -b_i & a_i \end{bmatrix} \begin{bmatrix} y_1 \\ y_2 \end{bmatrix}, \quad (10)$$

where the coefficients a_i and b_i are 0 for even i. The reader should note that the equilibrium is asymptotical stable, if the so-called Poincare coefficient a_1 is negative ($a_1 < 0$). So the Poincare normal form is a primary key for fulfilling the Andronov Hopf Theorem. The calculation of the Poincare normal form of (9) gives

$$\dot{y} = \begin{bmatrix} 0 & \omega \\ -\omega & 0 \end{bmatrix} y + (y_1^2 + y_2^2) \begin{bmatrix} -\dfrac{3}{8}\gamma & 0 \\ 0 & -\dfrac{3}{8}\gamma \end{bmatrix} \begin{bmatrix} y_1 \\ y_2 \end{bmatrix} + \ldots \quad (11)$$

We get the Poincare coefficient $a_1 = -3/8\gamma$. For $\gamma > 0$ the equilibrium is asymptotic stable. So we can guarantee the asymptotic stability of the equilibrium point. In order to demonstrate the birth of a limit cycle vividly, we analyze the product of the voltage and current $iv = f(v, I_0)$ of the nonlinear resistor R_{NL} which is shown in Fig. 3. If we choose $I_0 > I^*$, iv has locally a negative slope. When iv is interpreted as a power the negative slope indicates a negative differential resistor in the neighborhood of the zero solution. This leads to an oscillation since the loss of the circuit will be compensated. It must be pointed out that (5) also satisfied the condition $d(\Re\lambda(\mu)|_{\mu=0})/d\mu > 0$ given by the Hopf Theorem which can be tested by an easy computation.

V. PREDICTING THE AMPLITUDE

In order to calculate the amplitude of a sinusoidal oscillator it is suitable to transform the reduced system to polar coordinates. Starting from the Jordan form we obtain the following system:

$$\begin{bmatrix} \dot{\Theta} \\ \dot{r} \end{bmatrix} = \begin{bmatrix} \omega \\ 0 \end{bmatrix} + f(\Theta, r) \quad (12)$$

Since the equation $\dot{r} = 0$ is mostly a function of Θ, i.e. both equations of (12) are coupled; we cannot calculate the amplitude directly. To produce a relief we use an average technique – a perturbation method. Our approach of Lie series averaging is based upon the results of [15]. Like the normal form transformation, averaging uses a near identity transformation to simplify the given system. The transformation is to be chosen to transform the original system into the so called averaged system. By means of this method it is our goal to eliminate the action of Θ in the second equation of (12). The method hinges on the identification of a small parameter ε which marks the perturbation. We assume that a linear oscillator is perturbed by a small nonlinearity. So we get in Cartesian coordinates for a planar system with $\mu = 0$:

$$\dot{y} = \begin{bmatrix} 0 & \omega \\ -\omega & 0 \end{bmatrix} y + \varepsilon f_S(y) = J y + \varepsilon f_S(y) \quad (13)$$

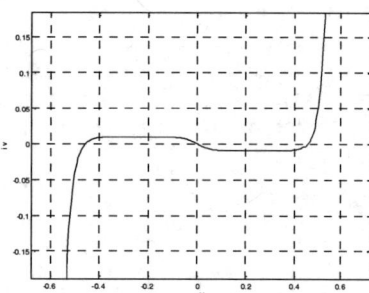

Figure 3. Analysis of the power $i\,v$

In polar coordinates we can write the following system, where mostly both equations are coupled

$$\begin{bmatrix}\dot{\Theta}\\ \dot{r}\end{bmatrix}=\begin{bmatrix}\omega\\ 0\end{bmatrix}+\varepsilon\begin{bmatrix}R(\Theta,r)\\ T(\Theta,r)\end{bmatrix}. \quad (14)$$

By using a suitable transformation we find the averaged system

$$\begin{bmatrix}\dot{\overline{\Theta}}\\ \dot{\overline{r}}\end{bmatrix}=\begin{bmatrix}\omega\\ 0\end{bmatrix}+\varepsilon\begin{bmatrix}\overline{R}(\overline{r})\\ \overline{T}(\overline{r})\end{bmatrix}. \quad (15)$$

Here we have assumed that already one averaging gives a usable result. But sometimes a sequence of coordinate transformations is necessary. Especially in this case our approach, which is based upon Lie series, is advantageous. So we can transform (9) to polar coordinates:

$$\begin{bmatrix}\dot{\Theta}\\ \dot{r}\end{bmatrix}=\begin{bmatrix}\omega_0\\ 0\end{bmatrix}+\varepsilon\begin{bmatrix}r^2\sin\Theta\cos\Theta\\ r\mu+\mu\cos^2\Theta+\ldots\end{bmatrix} \quad (16)$$

Since both equations of (16) are coupled averaging is necessary. We calculate

$$\begin{bmatrix}\dot{\overline{\Theta}}\\ \dot{\overline{r}}\end{bmatrix}=\begin{bmatrix}1\\ \varepsilon\overline{r}\mu-\varepsilon\dfrac{3}{8}\gamma\overline{r}^3\end{bmatrix}, \quad (17)$$

where already one averaging removes terms depending on Θ. The roots of $\dot{\overline{r}}=0$ are

$$\overline{r}_1=0,\ \overline{r}_2=\pm\sqrt{\dfrac{8}{3}\dfrac{\mu}{\gamma}} \quad (18)$$

The value $\overline{r}_1$ defines the averaged amplitude for an unstable equilibrium. If the Andronov Hopf Theorem is fulfilled, the oscillator has the amplitude $|\overline{r}_2|$, which depends on μ and the parameters of the transistors.

VI. Conclusion

In this work we have presented a complete bifurcation analysis of integrated differential LC oscillators. We have shown the analysis of the stability of the circuit as well as the calculation of an approximate amplitude. Moreover, we have presented the results in an analytic form - designers are able to implement differential VCOs by means of the given results even if other models of the components of the circuit are required. In this case possibly the calculations are more complicated. However, our methodology can be used in exact the same manner. So it represents a guide for the analysis and the design of electrical oscillators whereas the nonlinearities of circuit components are an integral part of the design process.

References

[1] A. Hajimiri and T. H. Lee, "Design Issues in CMOS Differential LC Oscillators," IEEE J. of Solid State Circuits, Vol. 34, No 5, May 1999

[2] W. Mathis, "Nonlinear electronic circuits – An overview," Proc. 7th. MIXDES 2000, Gdynia, Poland

[3] Q. Zeng, "Hopf Bifurcation in Differential Algebraic Equations and Applications to Circuit Simulation," Int. Series of Num. Math., Vol. 93, 1990, pp. 45-58

[4] B. Parzen, "Design of Crystal and Other Harmonic Oscillators," Wiley & Sons, 1983

[5] D. W. Jordan and P. Smith, "Nonlinear Ordinary Differential Equations," Oxford University Press, 1999

[6] W. Mathis, "Historical Remarks to the History of Electrical Oscillators," In: Proc. MTNS-98 Symposium, July 1998, IL POLIGRAFO, Padova 1998, pp. 309-312

[7] A.I. Mees and L.O. Chua, "The Hopf Bifurcation and its Application to nonlinear Oscillations in Circuits and Systems," IEEE Trans. on Circuits and Systems, Vol. 26, 1979, pp. 235-254

[8] G. Konstanznig, T. Pappenreiter, L. Maurer, A. Springer and R. Weigel, "Design of a 1.5V, 1.1mA Fully Integrated LC-tuned Voltage Controlled Oscillator in a 4 GHz-Band using a 0.12μm CMOS-Process," Asian-Pacific Microwave Conference, Nov. 2002, pp. 1471-1474

[9] P. Arcioni, R. Castello, L. Perregrine, E. Sacchi and F. Svelto, "An innovative modelization of loss mechanism in silicon integrated inductors," IEEE Trans. on Circuits and Systems, Vol. 46, No. 12, pp. 1453-1460, Dec. 1999

[10] R.L. Brunch, D.I. Sanderson and S. Ramon, "Quality factor and inductance in differential IC implementations," IEEE Mircrowave Mag., Vol. 3, No. 2, pp. 82-92, June 2002

[11] M. Danesh and JR. Long, "Differential driven symmetric microstrip inductors," IEEE Trans. on Microwave Theo. and Tech., Vol. 50, No. 1, pp. 332-341, Jan. 2002

[12] M. Prochaska and W. Mathis, "On limit cycles in singularly perturbed electrical circuits," Proc. of the 16th Int. Sym. on Math. Theory of Networks and Systems (MTNS), Leuven 2004

[13] A. Buonomo and A. Lo. Schiavo, "Determining the oscillation of differential VCOs," Proc. IEEE Intern. Sym. on Circuits and Systems (ISCAS 2003), Vol. 3, pp. 25-28, May 2003

[14] W. Mathis, "Transformation and Equivalence," in: The Circuits and Filters Handbook, Ed. W.K. Chen, CRC Press, Boca Raton, 1995

[15] W. Mathis and I. Voigt, "Applications of Lie Series Averaging in Nonlinear Oscillations," Proc. IEEE Intern. Symp. of Circuits and Systems (ISCAS 1987), Philadelphia, May 1987

A Scalable DCO Design for Portable ADPLL Designs

Chia-Tsun Wu, Wei Wang, I-Chyn Wey, and An-Yeu (Andy) Wu

Graduate Institute of Electronics Engineering, and Department of Electrical Engineering,
National Taiwan University, Taipei 106, Taiwan, R.O.C.

Abstract—A novel Digital Controlled Oscillator (DCO) design methodology is presented in this paper. The new design methodology includes a scalable DCO architecture and the developed design flow. With precise analysis in early stage, the design effort of DCO can be reduced significantly. The proposed DCO architecture has the characteristics of, high resolution, flexible operating range, and easy design. The design is suitable for high performance clock generator in System on a Chip (SoC) application.

I. INTRODUCTION

The Phase-Locked Loop (PLL) is a widely used circuit for clock generator as system clock or Clock Data Recovery (CDR). There are many digital building blocks or digital Intellectual Properties (IP) integrated in one chip for SoC application. Traditionally, a PLL is made as an analog or mixed-signal building block [8]. However, to integrate an analog PLL in a digital noisy SoC environment is difficult. Therefore, All-Digital Phase-Locked Loop (ADPLL) is designed using digital design techniques and very suitable for being integrated into SoC chips.

A typical diagram of an ADPLL is shown in Figure 1. An ADPLL consists of major blocks: Phase Frequency Detector (PFD), Loop Filter (LF), DCO, and Frequency Divider (FD). The DCO is the most critical component in ADPLL design [1]. As an oscillator, building circuits in DPLL such as voltage controlled oscillator (VCO) and current controlled oscillator (CCO) have been widely studied. However, ring-based inverter chains are mostly adopted in DCO architectures.

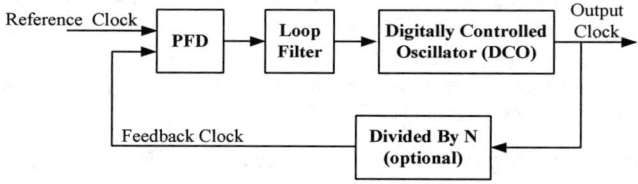

Figure 1. A typical ADPLL Block diagram.

Traditional DCOs are suggested with full custom approach [4][7]. To incorporate the DCO in the HDL simulation, design DCO with full custom design technique has to simulate DCO as an analog component to get reliable estimates of its behavior with SPICE level simulation [1][7].

In addition, variable number of inverters may be used for implementing a variable delay. In [2][3], this technique is used for coarse acquisition. Only the delay of inverters for coarse-acquisition results in hundred picoseconds and gives an inaccurate and unstable phase lock for high-frequency applications. Interesting structures composed of coarse-acquisition and fine-tune are developed to enhance resolution. The improved delay cell is used with selective logic gates and shunted driving cells [1][2] to enhance the resolution. Embedded lookup tables specify loop period relation with respect to control code. But the dependency possibly changes because of different process, voltage, and temperature (PVT) variation cases. Furthermore, lookup tables imply more hardware cost.

In this paper, new DCO architecture and developed design flow are both presented to achieve scalability, high performance and wide operating range. The proposed methodology can be analyzed in early stage and easy design. The prototype chip is designed in pure standard-cell library and implemented in UMC's 0.18 1P6M CMOS process. The chip operates in the range of 140 MHz to 1030 MHz. The measured P-P jitter and RMS jitter are 143 ps and 30 ps, respectively.

II. METHODOLOGY OF STANDARD-CELL BASED DCO

The DCO is the most challenging block in ADPLL design, especially with standard-cell library. In this section, a novel architecture of DCO is presented in Figure 2. Our goal is to design DCO capable of handling various standard-cell libraries and implement VLSI architecture in pure standard-cell design flow. The novel DCO design methodology presents process-independent topologies.

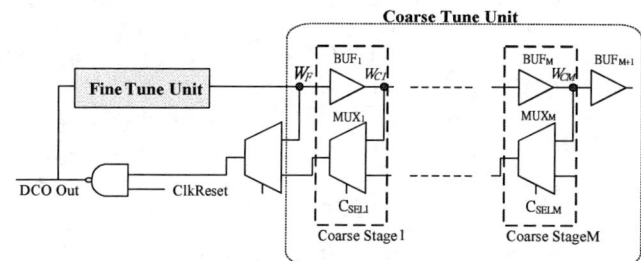

Figure 2. Proposed DCO architecture

It is desirable to decompose DCO into coarse and fine inputs. The oscillating period mainly consists of delays of Fine-Tune-Unit (FTU) and Coarse-Tune-Unit (CTU). Once the first stage of CTU is constructed, the following stages are duplicated. A BUF_{M+1} is added to balance wire load for the last coarse-tune stage. Total loop delay (τ_{DCO}) is shown as eq. (1), where τ_{FTU} is the total gate delay of fine-tune stage and also the minimum DCO period when switch off all coarse-tune stages. The τ_{CTU} is the total gate delay of CTU. When N coarse-tune stages are switched on, the τ_{CTU} can be described as eq. (2), where τ_c is the timing delay of one coarse-tune stage.

$$\tau_{DCO} = \tau_{FTU} + \tau_{CTU} \quad (1)$$

$$\tau_{CTU} = \tau_c \cdot N \quad (2)$$

If there are M coarse-tune stages constructed for CTU, the DCO operates at the minimum frequency when N equals to M. Base on the proposed architecture, we develop a new design flow for DCO design methodology in this section.

A. Coarse-Tune-Unit Architecture

As depicted in Figure 2, direct inspection of the fan-in and fan-out loading in the equivalent circuit of coarse-tune stage gives in eq. (3). The wire load (W_F, W_{C1}, W_{C2}, W_{CM-1}, W_{CM}) for each coarse-tune stage is identical to assure equal coarse-acquisition step. The impedance load of one buffer and multiplexer are $Z_{BUFinput}$ and $Z_{MUXinput}$, respectively. The consistence allows the CTU architecture very suitable for standard-cell based design flow.

$$W_F = W_{C1} = W_{C2} = W_{CM-1} = W_{CM} = Z_{BUFinput} + Z_{MUXinput} \quad (3)$$

Compared to traditional designs [2][3], the main advantage of the coarse-tune architecture is to reserve the highest frequency output. When oscillating at the highest frequency, the fan-out of FTU is only the first coarse-tune stage but not all coarse-tune stages. In particular, when increase coarse-tune stages for CTU to extend the DCO oscillating range, the architecture proves no overloaded wire that limits maximum frequency. The robust and linear characteristic of CTU architecture allows simple formulation of loop period. As discussion in [1], CTU stage should be designed carefully to minimize rising and falling time to assure step accuracy to reduce the next stage's propagation delay.

B. Fine-Tune-Unit Architecture

A new systematic method of improving resolution is proposed for DCO fine-tune-unit. The main idea is using the delay difference of paths. For example, the capacitances and output strengths of different pins are close for a NAND gate in standard-cell library. The timing delay difference from different input pins to the same output pin approximates to the intrinsic delay difference.

TABLE I. CELL LIBRARY DATA OF 2-INPUT NAND GATE

Paths	Intrinsic Delay							
	XL		X1		X2		X4	
A → Y↑	54	85	46	73	41	65	41	65
A → Y↓	31		27		24		24	
B → Y↑	71	107	62	94	58	88	58	87
B → Y↓	36		32		30		29	

Complete intrinsic-delay data of 2-input NAND gate is listed in TABLE I which is abstracted from a 0.25 um standard-cell library datasheet. Use those timing differences for basis of fine acquisition. With investigating two paths of different/same NAND gates for possible timing difference, we obtain following differences from TABLE I for NAND gates: 1, 2, 3, 6, 7, 8, 9, 12, 13, 14, 15, 19, 20, 21, 22, 23, 29, 34, and 42ps.

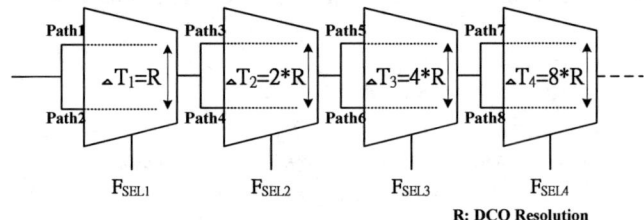

Figure 3. Proposed FTU architecture

$$\tau_{FTU} = \tau_{FC} + \sum_{i=1}^{K} F_{SELi} \cdot R \cdot 2^{i-1} \quad (4)$$

Figure 3 shows the proposed FTU architecture based on our idea of path delay difference. The total loop delay of FTU can be described in eq. (4), where K is the total stages of FTU, F_{SELi} is the i-th control bit, τ_{FC} is the FTU minimum period, and R is the DCO resolution. There are versatile combinations of logic gate to construct a 2-to-1 multiplexer, for example three NAND or NOR gates. We investigate two paths of gates for possible timing difference. Arithmetically, the proposed work can enhance resolution to 1ps. With the proposed design flow in section II-D, the decision of FTU hardware implementation can be decided in early stage for different resolution or specification without wasted trial time in backend simulation.

From our proposed DCO architecture in Figure 2, FTU will limit the highest oscillating frequency. A high speed four-path FTU in Figure 4 can reduce the minimum closed-loop delay time. Compare to Figure 3, the modified architecture can achieve higher frequency output without hardware cost penalty. An implement and measurement results can be shown in section III for this modified architecture.

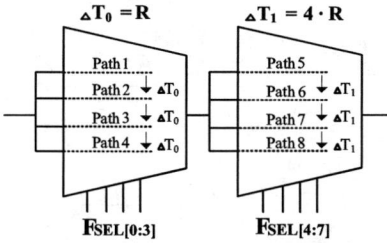

Figure 4. A modify FTU to achieve higher frequency output

C. Hardware Complexity of DCO Controller

The control complexity of system influences overall hardware cost. More control wires in DCO also may induce higher possibility of glitch due to different latency of each control line. TABLE II shows the hardware overhead for DCO controller. We deduce some parameters to clarify the tradeoff between resolution and control complexity. Suppose M is the total number of coarse-tune stages in CTU, S is the one coarse-tune stage delay step and R is the DCO resolution. As the DCO architecture in [3], M is also the number of CTU control lines and S is the number of FTU control lines. S is typically more than one hundred picoseconds in 0.18um process and larger in elder process.

TABLE II. CONTROLLER HARDWARE COMPLEXITY

Design	JSSC99 [3]	ISCAS01 [1]	JSSC04 [7]	ours
DCO wordlength	$M+S/R$	$2*M+(S/R/4)*M$	$M*S/R$	$M+(\log_2 S/R)$
Lookup Tables	N	Y	N	N

The proposed work shows many hardware cost can be reduced substantially. Since the controller issue control code in binary order for FTU, this hardware is significantly less complex, in terms of monotonic increasing/decreasing than a lookup table and it is more efficient for future applications for low cost and high speed.

D. The Proposed Design Flow

As illustrated in Figure 5, the developed design flow that enables the designer to construct hardware based on the proposed architecture. After parsing standard-cell data, the delay differences are pre-calculated by Matlab program. Iterations start with detailed design of crucial FTU components which result in the maximum frequency and resolution. And it works up towards the hardware decision of CTU components that extends the minimum frequency and frequency range. Refinements of jitter reduction due to voltage variation are both applied to individual FTU and CTU components.

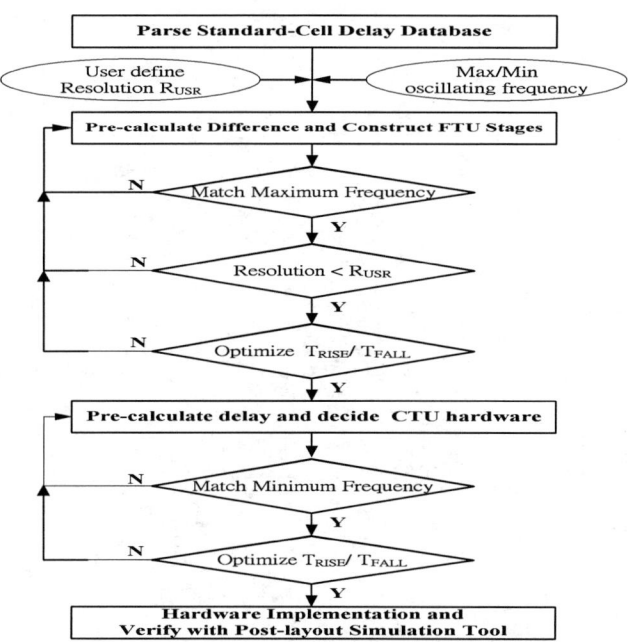

Figure 5. Proposed DCO designe flow.

III. IMPLEMENTATION AND CHIP MEASUREMENTS

Based on the proposed methodology, a prototype ADPLL design using UMC 0.18 um 1P6M CMOS technology is implemented as Figure 6. The DCO resolution is about 22 ps. The implementation consists of 16 CTU stages, 4 FTU stages, and $(1+16)*2^4$ acquisition steps; the oscillating frequency operates from 140 MHz to 1030 MHz. The DCO core area is 345 um × 56 um.

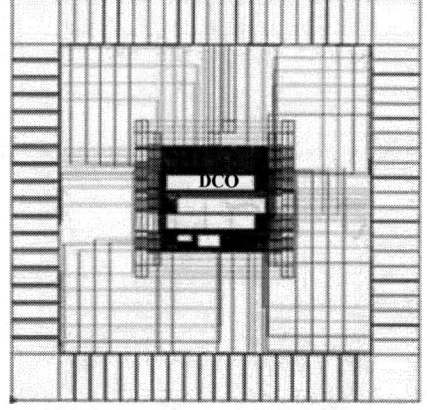

Figure 6. Chip layout of the prototype ADPLL and DCO module

Post-layout simulations of period and bandwidth according to different control code in three corner cases (SS, TT, FF) are depicted in Figure 7 and Figure 8, respectively.

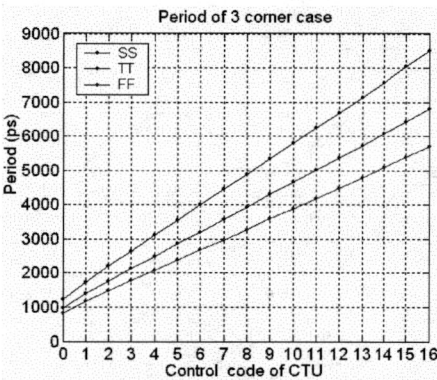

Figure 7. CTU delay linerity

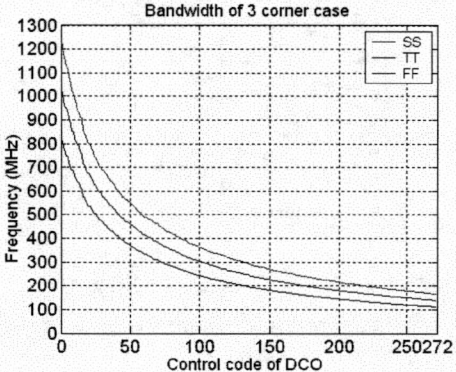

Figure 8. DCO oscillating bandwidth versus control code

The jitter measurement of chip output is performed on Agilent 86100B Infiniium DCA Wide-Bandwidth Oscilloscope as Figure 9. Due to pad limitations, output clock is divided by 4 and the measured frequency is about 960MHz. The DCO operates at high frequency and both in wide range. TABLE III shows the performance comparisons to recent works.

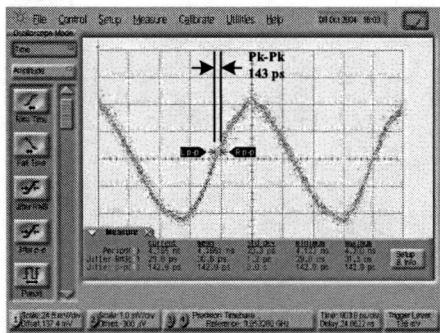

Figure 9. Measured waveform and jitter of the prototype ADPLL oscillating at 960 MHz

TABLE III. COMPARISONS OF STANDARD-CELL DESIGN

Design	JSSC04 [7]	ISCAS02 [2]	ISCAS01 [1]	Ours
Process	0.35	0.35	0.25	0.18
F_{min}(MHz)	152	40	170	140
F_{max}(MHz)	366	545	650	1030
$Jitter_{P-P}$ (ps)	1200	340	324	143
$Jitter_{RMS}$ (ps)	--	39	25	30

IV. CONCLUSION

We propose a frequency-scalable DCO design methodology for standard-cell library based design flow. After establishing the database of gate delay, paths delay differences are used for basis of FTU acquisition step. The proposed CTU reserves both linear acquisition step and high oscillating frequency even when operating frequency range extends. With arithmetic and systematic method to construct CTU and FTU hardware in early stage, designer can save wasted trial iterations in the backend flow. Following issues on voltage variation, the optimization of frequency distortion is addressed. The measured P-P jitter and RMS jitter are 143 ps and 30 ps, respectively. The ADPLL design can be easily incorporated into modern SoC design flow due to its digital feature.

ACKNOWLEDGMENT

The authors thank Chip Implementation Center (CIC) for post-layout simulation and IC fabrication.

REFERENCES

[1] J. Jong and C. Lee, "A novel structure for portable digitally controlled oscillator," IEEE International Symposium on Circuits and Systems, vol.1, pp.272 – 275, May 2001.

[2] C.Chung and C. Lee, "An all-digital phase-locked loop for high-speed clock generation," IEEE International Symposium on Circuits and Systems, vol. 3, pp.26-29, May 2002.

[3] T. Hsu, B. Shieh and C. Lee, "An all-digital phase-locked loop (ADPLL)-based clock recovery circuit," IEEE Journal of Solid-State Circuits, vol. 34, pp.1063-1073, Aug 1999.

[4] J. Chiang and K. Chen, "A 3.3 V all digital phase-locked loop with small DCO hardware and fast phase lock," IEEE International Symposium on Circuits and Systems, vol. 3, pp.554-557, May 1998.

[5] Pialis and K. Phang, "Analysis of Timing Jitter in Ring Oscillators Due to Power Supply Noise," IEEE International Symposium on Circuits and Systems, vol. 1, pp.I-685 – I-688, May 2003.

[6] A. Abidi and R. G. Meyer, "Noise in Relaxation Oscillators," IEEE Journal of Solid-State Circuits, Vol. 18, pp.794-802, December 1983.

[7] T. Olsson and P. Nilsson, "A digitally controlled PLL for SoC applications," IEEE Journal of Solid-State Circuits, vol. 39, pp.751-760, May 2004.

[8] Zhinian Shu; Ka Lok Lee; Leung, B.H.," 2.4-GHz ring-oscillator-based CMOS frequency synthesizer with a fractional divider dual-PLL architecture," IEEE Journal of Solid-State Circuits, vol:39, pp.452-462, March 2004.

A 2.5GHz Phase-Switching PLL using a supply controlled 2-delay-stage 10GHz Ring Oscillator for improved Jitter/Mismatch

Eva Tatschl-Unterberger[1], Sasan Cyrusian[2], Michael Ruegg[3]
Infineon Technologies North America Corporation
Santa Cruz, CA, USA
[1] now with Infineon Technologies Austria, Villach, Austria
[2] now with Marvell Semiconductors, Scotts Valley, CA, USA
[3] now with Miromico AG, Zuerich, Switzerland

Abstract—**A fully integrated ring oscillator PLL for hard disk channel applications is presented. A number of 16 equidistant phases of the output clock, programmable in 0.4% steps between 200 MHz and 2.5 GHZ are achieved by the use of a phase-switching fractional N architecture. Phase mismatch is optimized by the use of a novel two delay stage ring oscillator running at 4x output frequency (800 MHz-10 GHz) and a subsequent divider chain. Jitter and Area consumption are improved by solely controlling the VCO via its power supply. The proposed VCO's jitter/power/number of stages relationship behaves analogue to a single ended structure although the ring delivers 4 differential clock phases. The PLL was built in standard 0.12 μm CMOS technology. It achieves a phase noise performance of -96 dBc/Hz @ 1 MHz offset on a 1.6 GHz signal. The integrated jitter in the measured band (10 kHz-10 MHz) is 3.8 ps. The PLL consumes 0.06 mm² only.**

I. INTRODUCTION

Today's data storage and data link applications demand for fully integrated synthesizer PLLs with very low accumulated jitter, fast lock time and a high degree of output frequency programmability. Additionally multiple output clocks with exactly equidistant phase differences and 50% duty cycle are required for the use in subsequent circuits. Chip cost and therefore the requirement for minimum area consumption dominates the design. Robustness against noise becomes more and more of an issue because of the integration of multiple blocks into complex SOCs with read channels, high speed digital controllers and interface circuits such as 3 Gbit/s SATA on one chip.

The following sections outline the measures taken to fulfill the above criteria on a fully integrated PLL for a read write channel application on system and circuit level as well as the measurements of the silicon.

II. PLL SYSTEM OVERVIEW

Fig. 1 outlines the architecture of the presented PLL. It runs from a low cost 20-40 MHz crystal oscillator (fref) and supplies 16 equidistant output clocks with 22.5° phase difference at full speed (200 MHz-2.5 GHz). To achieve 0.4% output frequency granularity in combination with fast lock time a phase switching fractional N counter architecture (:N-P/16) in the feedback path [1] is used. As described in section III this principle avoids ΔΣ modulation.

The heart of the PLL is formed by a novel 2-stage ring oscillator that runs at 4x the output frequency (4xVCO). Two divide-by-2 units (:2) form an entity with the VCO acting as decoupling and division circuits at the same time. The idea was to optimize power consumption, mismatch and jitter by trading a 2 delay stage 4xVCO and two 2x dividers against the commonly used 4 stage 2xVCO with identical phase noise performance.

In order to minimize jitter the VCO is solely controlled by its power supply. To achieve the necessary robustness against power supply noise created on the SOC a special three stage buffer amplifier/regulator (Reg) from 3.3 V is employed.

The charge pump features 8 selectable currents and a very wide output range between 0.15 and 1.35 V. The loop contains a fully integrated single ended passive RC-C filter (LPF) and a dead zone free phase frequency detector (PFD).

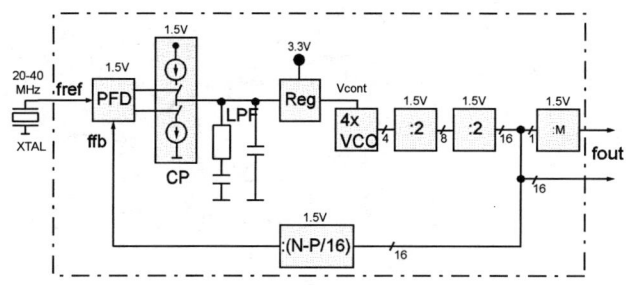

Figure 1. PLL Architecture

III. FEEDBACK DIVIDER

The fractional division factor is achieved by the use of the divider structure pictured in Fig. 2.

16 equidistant clock phases with 22.5° shift are shipped to a phase multiplexer. On every switch_en impulse the multiplexer rotates glitch free for exactly one phase. As shown in Fig. 3 a slightly shorter period on the composite clock (fcom) is produced.

The P-counter circuit controls the number of rotations during one fref cycle, i.e. during one count to N. The resulting composite clock which consists of P shortened and N-P regular periods is counted by a standard asynchronous :N counter.

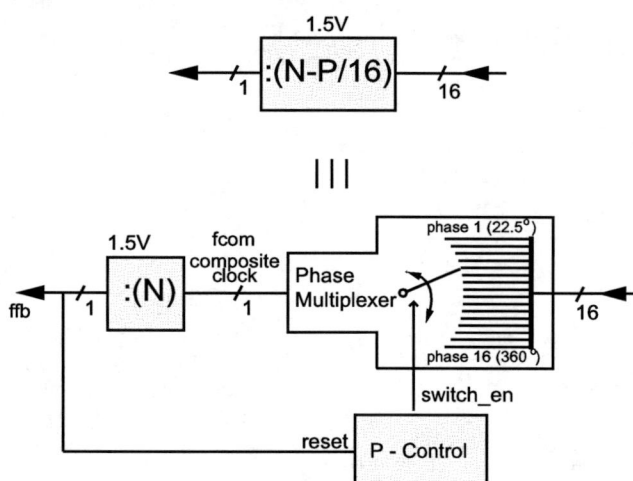

Figure 2. Multi phase switching feedback divider

P rotations cause the feedback period to be

$$T_{fb} = N \cdot T_{fout} - P \cdot \frac{T_{fout}}{16} \qquad (1)$$

and because the PLL forces $T_{fb} = T_{fref}$ it follows that:

$$f_{out} = f_{ref} \cdot (N - \frac{P}{16}) \qquad (2)$$

When the phases are equally spaced and P is kept constant, a constant fractional feedback period T_{fb} is generated and compared to the reference clock fref. $\Delta\Sigma$ modulation is therefore not necessary. Phase mismatch becomes an important factor in the spurious performance of the PLL.

IV. 4xVCO

Hard disk applications are well served with ring oscillators since they combine the advantages of minimal area and therefore minimal cost with a large frequency tuning range.

A ring oscillator usually has two inputs that control the output frequency, a very high gain control voltage input (K_{vco}), and the supply connection which embodies a parasitic high gain input. A power supply regulator is commonly used to provide the necessary decoupling. A standard differential ring oscillator contains a number of biasing transistors to facilitate the frequency control via the control voltage. These transistors are usually on during the complete cycle of oscillation and introduce a significant amount of noise into the ring. By controlling the VCO solely by its power supply as proposed in this article there is no need for these biasing transistors and therefore the noise performance can be improved.

The commonly used power supply regulator is replaced by a regulator type that follows the loop filter voltage as introduced in section VI. A welcomed side effect of the regulation/buffering of the control voltage is the low impedance point that is created at the combined VCO control input. This makes it possible to view the control voltage off chip by hooking it to a test bus, with greatly reduced noise coupling risk.

A. VCO Circuit

The delay cell consists of an NMOS input pair and a PMOS input pair as well as a PMOS positive feedback structure. The positive feedback allows to maintain the necessary 90° phase shift per delay cell by shifting the pole onto the imaginary axes. Viewing the large signal behaviour, one can see the positive feedback as hysteresis that provides the extra phase shift for a total of 90° per delay cell. The work of [3] and [4] presents similar delay cells but the use of additional transistors for the basic cell as well as the control input in these implementations cause extra noise that can be avoided using the structure presented in this paper.

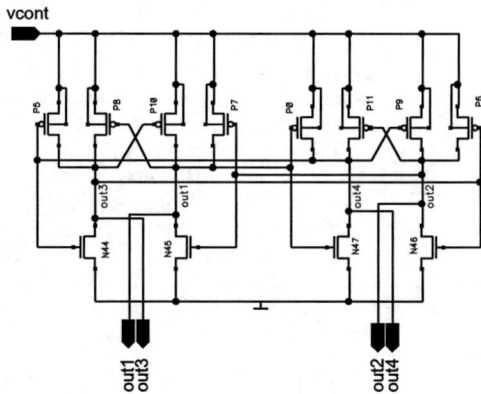

Figure 4. Schematic 4x VCO

The NMOS input pair in the proposed circuit offers high gain during high input voltages, the PMOS input pair during low input voltages. Fig. 4 shows the schematic of the proposed VCO.

The VCO has a wide operating frequency range from 3.2 GHz @ vcont=0.2 V on fast silicon to 10GHz for vcont=1.55 V on slow

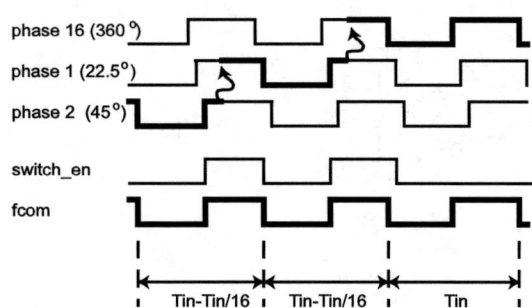

Figure 3. Clock shortening by phase switching

silicon. K_{vco} is highest at low frequencies (3.2 GHz) where it ranges between 7.5 GHz/V on slow and 15 GHz/V on fast silicon and lowest at high frequencies (10 GHz) where it ranges between 3 GHz/V on slow and 10 GHz/V on fast silicon. Please note that the effective K_{vco} for the application $K_{vcoeff} = K_{vco}/4$ because of the division by 4 described in the next section.

The VCO consumes 10 mA @ 6.4 GHz on slow silicon and shows a phase noise of -102 dBc/Hz @ 1 MHz offset measured on the output signal of 1.6 GHz, therefore resulting into -90 dBc/Hz @ 1 MHz offset on the internal 6.4 GHz signal.

B. System Considerations

The ring oscillator proposed consists of two delay cells. In order to achieve the number of output phases necessary for the application it is run at 4x frequency. Usually the overall jitter/power relationship of differential rings can be optimized by using a smaller number of stages like it is shown in [2]. It is quite remarkable that this specific ring shows opposite behavior. A larger number of stages is beneficial for the jitter/power relationship.

This can be explained since the Power consumption of this ring is only a function of the frequency of oscillation and not of the number of stages used. We can see that because current only flows through the stage that is switching in one moment of time. The ring must therefore show the behavior of a single ended ring (more stages better jitter/power ratio) although it offers an even number of phases.

Fig. 5 confirms this theory by measurement. Three different size rings of the kind shown in Fig. 4 were put on a test chip and measured. Power can be reduced by half when the number of stages is doubled. A larger number of stages is therefore beneficial for the jitter assuming a fixed power budget.

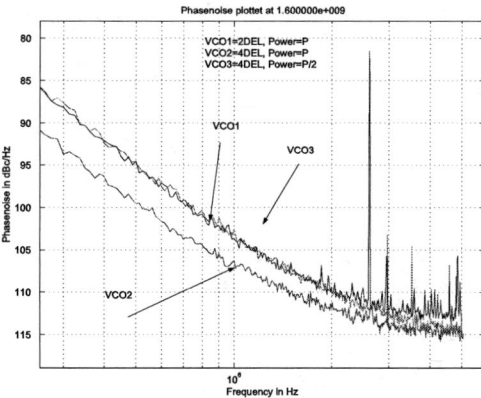

Figure 5. Measured performance of three test VCOs with different number of stages and different power on 0.12μm silicon.

Nevertheless mismatch, the second important parameter for the overall jitter performance in a phase switching PLL improves by the use of a smaller number of bigger components and therefore with a smaller number of delay stages. Since the ring itself is usually only a small contributor to mismatch this effect is negligible. The decoupling circuit on the other hand is the main source of phase mismatch. Since the use of differential dividers as decoupling like we describe in section V offers good mismatch performance we still opted for the smaller number of delay stages. Also the VCO is not dominating the total phase noise performance for this PLL as shown in Fig. 10.

V. DIVIDE-BY-TWO CIRCUITS

The 4xVCO and the divide-by-two (:2) circuits proposed in this paper form an entity.

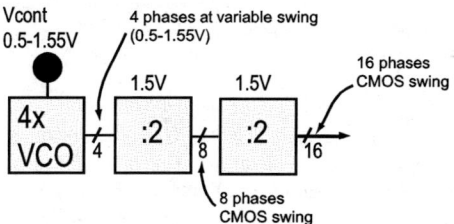

Figure 6. 4xVCO plus divide-by-two circuits

Since the VCO is controlled by its power supply (vcont) as pictured in Fig. 1 and 4 its output swing is greatly variable. It is necessary to use decoupling circuit to reach CMOS levels and to suppress amplitude modulation noise. For the proposed circuit it was possible to collapse the divide-by-two and the amplitude normalization inside the first divide-by-two (:2) stage.

The clock signals from the VCO supply the enable inputs and switch the latches between store and transparent mode. These inputs are designed in NMOS logic so that the possibly very small swing on the VCO clocks is enough to control the operating mode of the latches. Since the output signal swings are mainly dependent on the power supply (1.5V) the desired level shift occurs.

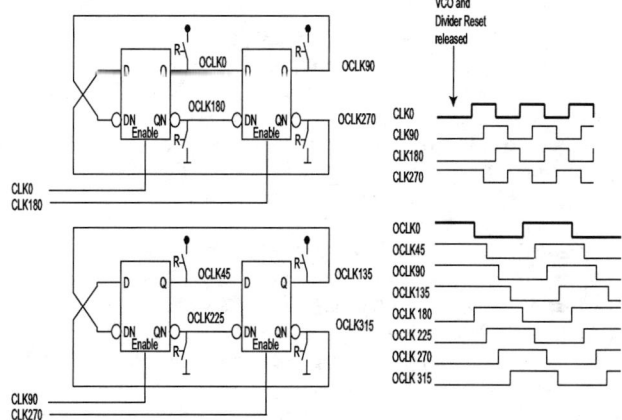

Figure 7. First divide-by-two stage – Reset scheme

To ensure the correct order of the phases produced by the divider a reset circuit had to be employed. The VCO reset circuit always starts the oscillation with the same phase (clock0). Three of the four latches are therefore in store mode, only latch 4 is transparent. Once clk0 starts up latch 1 gets transparent and the division starts correctly. The second divide-by-two stage is built out of 8 latches in the same way.

VI. 3-STAGE VCO BUFFER AMPLIFIER/REGULATOR

A high speed ring oscillator usually possesses a very high gain from the control voltage to the output of the VCO, often in the order of Kvco=1..4 GHz/V. In a similar way the gain from the supply to the output is very high making the circuit very sensitive to noise on the power supply. The outlined 4xVCO architecture collapses both sensitivities since the power node coincides with the control voltage. Extra care has been taken to filter power supply noise.

Noise on the power supply can be viewed as a correlated noise input to the VCO. As described in [2] correlated noise is up-converted to VCO noise slightly different than random device noise originating inside the VCO. It is suppressed around the frequency of oscillation and uneven multiples of it. Low frequencies and noise at even multiples of the oscillation frequency is transferred to the output and experiences up-conversion resulting in a -20 dB/decade slope. For a PLL we experience some relive from the filtering of noise within the loop bandwidth. The maximum transfer of noise to the output occurs around the loop bandwidth and we require excellent suppression in this region. The regulator presented shows attenuation of more than 90 dB around 1 MHz as shown in Fig. 8.

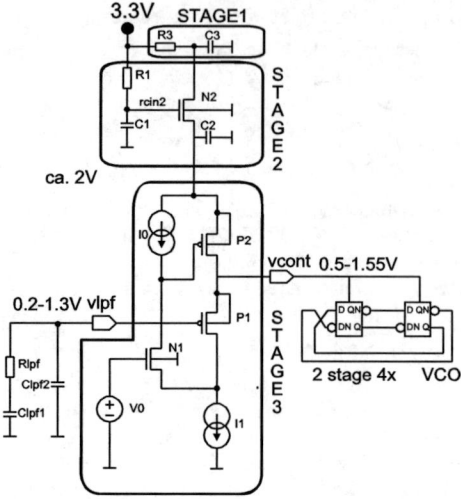

Figure 9. Schematic of the 3-stage regulator

The VCO shows a performance of -102 dBc/Hz, the PLL -96 dBc/Hz @ 1 MHz offset. As we can see, the total phase noise is not dominated by the VCO. The integrated jitter in the measured band (10KHz-10MHz) equals to 3.8 ps. (1σ accumulated jitter)

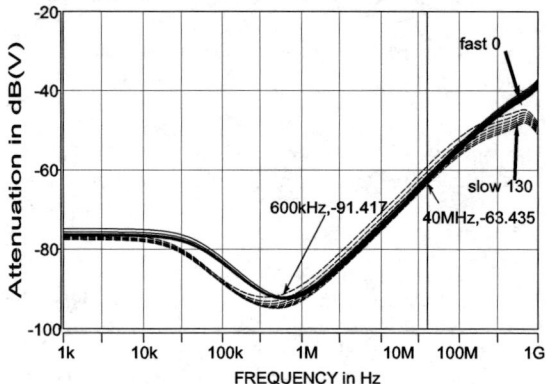

Figure 8. Voltage regulator: PSRR

The level of correlated noise on the supply caused by other circuits on the chip can be as high as 10 mV at certain frequencies. Prominent correlated noise frequencies are 20/40 Mhz originating from the reference clock, fractions of the reference frequency resulting from the fractional division factor, low frequency spurs originating from regulators on the PCB (kHz Region), supply noise from other PLLs on chip, etc.

The dotted line in Fig. 10 indicates the spurious level at the PLL output for 10 mV disturbance on the VDD and the noise attenuation provided by the regulator as shown in Fig. 8. One can see that the +20 dBc/Hz slope of the regulator is compensated by the -20 dBc/Hz property of the noise up-conversion in the VCO.

As pictured in Fig. 9, the proposed regulator comprises three stages. The second stage, the filtered input NMOS, takes about -20 dB out of the higher frequencies. It starts to kick in at around 30 kHz and operates until the parasitic capacitance of N2 starts to dominate which happens well above 1GHz. The third stage accounts for the extra suppression at lower and medium frequencies. Very high frequencies are damped by the first stage, the combinations of R3/C3 and C2. Extra care was also taken to filter the bias currents for the regulator to inhibit noise coupling through this path.

VII. Experimental Results

The proposed circuit was fabricated using Infineon's 0.12 μm CMOS Process. The PLL uses an area of 0.06 mm². Fig. 10 shows the simulated and the measured phase noise performance on an 1.6 GHz output clock. The reference frequency is 40 MHz, the division factor 40. The loop bandwidth was turned down to approx. 750 kHz for this measurement. The simulated phase noise using a Matlab model provides a very good match to the measurement.

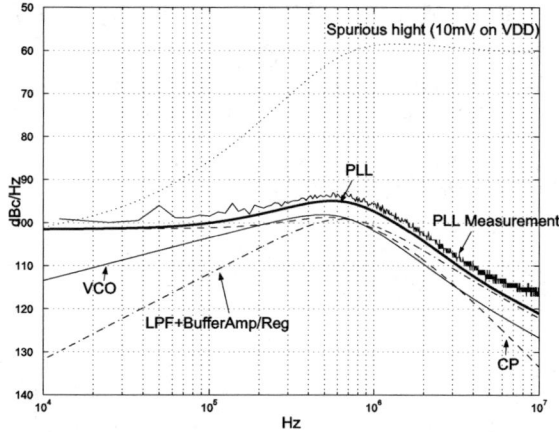

Figure 10. Measured and simulated Phase noise of the PLL @ 1.6GHz

Acknowledgment

The authors would like to thank Minoo Nasseri for excellent mask design and Harvey Newman for his help measuring the device.

References

[1] S. Cyrusian, S Mehrgardt, "Frequenzteiler-schaltung und Phasenregelkreis mit einer solchen Frequenzteilerschaltung", Patent DE19844953 A1, 1998

[2] A. Hajimiri, T. Lee, "Low Noise Oscillators", Kluwer Academic Press, 1999

[3] W. Yan, H Loung, "A 2-V 900-MHz Monolithic CMOS Dual-Loop Frequency Synthesizer for GSM Receivers" IEEE Journal of Solid State Circuits, Vol 36, No2, February 2001, pp. 204-215

[4] L Dai, R Harjani, "Design of High Performance CMOS Voltage Controlled Oscillators" Kluwer Academic Publishers, Boston, 2003

A 360° Extended Range Phase Detector for Type-I PLLs[1]

Cameron T. Charles and David J. Allstot
Department of Electrical Engineering
University of Washington
Seattle, Washington 98195
Email: ccharles@ee.washington.edu

Abstract— This paper presents control circuitry for extending the range of a tri-state phase-frequency detector in a type-I charge-pump phase-locked loop. The extended range phase detector allows the entire range of the voltage controlled oscillator to be used, independant of loop filter gain and other design variables. A phase-locked loop containing the proposed control circuitry has been simulated by SPECTRE with BSIM3v3.2 models, and the expected operation has been confirmed. The phase-locked loop has been fabricated in the IBM 7RF 0.18 μm CMOS process.

I. INTRODUCTION

The wireless communications industry has been experiencing rapid growth for the past several years, much of which has been driven by the increased availability of low-cost, low-power wireless systems. The reduced cost of wireless communications systems allows them to be incorporated in a wider range of products and applications, and this growth in turn encourages research into new architectures and design techniques which allow further reductions in cost and power.

One of the most obvious ways to achieve these objectives is to make use of architectures which have reduced complexity and fewer components, which in turn allows for greater integration and high volume manufacturability. One architecture which is attractive from this standpoint, and has grown in popularity in recent years [1]-[3] is that of the direct-conversion transmitter. The direct-conversion architecture eliminates the need for the additional voltage controlled oscillator (VCO) and the off-chip surface acoustic wave (SAW) filters which are required in a superheterodyne architecture.

One technique for performing the modulation in a direct-conversion transmitter is through indirect modulation of the VCO by applying the baseband data to the control of a frequency synthesizer [4]. This approach is pictured in figure 1. This scheme eliminates additional components such as D/A converters and mixers which are required in more conventional designs, leading to reduced power consumption and higher integration. In this approach, the bandwidth of the modulation data is limited by the bandwidth of the phase-locked loop (PLL) used as a frequency synthesizer. One way to reduce this limitation is to apply a pre-filter to the modulation data, with a frequency response that is the inverse of the PLL transfer function [5]. This compensation filter emphasizes the

[1]Research supported by a grant from Intel Corp.

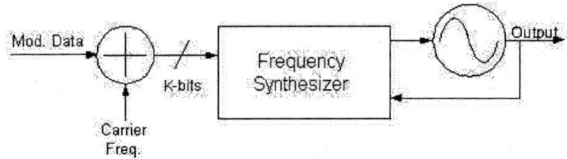

Fig. 1. Frequency modulation upconversion through indirect modulation of a VCO.

high frequency components of the modulation data signal, and allows the transmit bandwidth to exceed the PLL bandwidth, as shown in figure 2.

Most modern PLLs employ type-II loops, which have an integrator in the loop filter that provides infinite gain at DC. This leads to zero steady state phase error and reduces the negative effects of reference feed-through on the output spectrum. There are a number of advantages to using a type-I PLL for the transmit scheme described previously. Firstly, the lack of an additional integration in the loop filter reduces the complexity of the loop dynamics, and simplifies the implementation of the compensation filter. Second, type-I PLLs do not exhibit jitter peaking, which corrupts the output spectrum of the transmitted signal in type-II PLLs [6]. Type-I PLLs also have the advantage of faster settling times.

A drawback of type-I PLLs is that they typically have insufficient gain in their loop filters to span the entire range of the VCO [7]. For charge-pump PLLs, this necessitates adding coarse tuning and control circuitry to cover the full VCO range. One way to implement this is to extend the range of the phase detector beyond the ±360° which is normal for a tri-state phase-frequency detector (PFD), and use this to control the coarse tuning circuitry [8]. This work uses the concept of an extended range phase detector (ERPD) introduced in [8], and presents simplified control circuitry with PFD operation in multiples of the ±360° range, and fewer coarse tuning steps. Section II describes the operation of the control circuitry, Section III provides details on the subcircuit implementations, Section IV presents simulation results, and finally conclusions are drawn.

II. CONTROL CIRCUITRY OPERATION

Extending the range of the phase detector is acheived by adding digital control circuitry to allow the phase error to

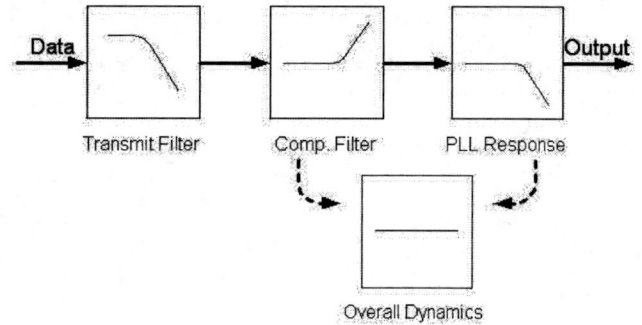

Fig. 2. Using a compensation filter to extend the bandwidth of a PLL used for direct upconversion.

extend beyond ±360° which is normal for a tri-state PFD. A system level schematic of the PLL with the added control circuitry is shown in figure 3. The PLL is assumed to be type-I, so there is no integration performed in the loop filter. The blocks contained by the dashed line are the components which have been added to extend the range of the phase detector, the remaining blocks are the standard PLL components.

The operation of the control circuitry is as follows. The 360° detector blocks monitor the UP and DN outputs of the tri-state PFD, and they trigger their outputs if the phase error indicated by the UP (DN) signal approaches 360° (-360°). The operation of the 360° detector is shown in figure 4. The 360° detector monitors the gaps between pulses on the UP (DN) lines. When this gap shrinks below some threshold, the detector triggers its output, signalling the control register that the phase error has reached 360°. This causes the control register to increment a counter which keeps track of the total phase deviation.

The output of the control register controls the current digital-to-analog converter (DAC). When the control register is incremented, the output of the current DAC is increased by a unit amount slightly greater than double the current output of the charge-pump for a 360° phase error. Choosing the DAC current output to be greater than the full scale charge-pump current output adds hysteresis to the system, which prevents the control circuitry from entering a state where the DAC current is continually switching on and off. The output of the current DAC is summed with the charge-pump output and fed into the loop filter.

The control register output also controls the pre-PFD block. This is simply a multiplexor which selects between delayed versions of the F_{ref} signal to feed into the tri-state PFD. The delay is chosen to be slightly greater than the threshold at which the 360° detector is triggered. After the control register is incremented (decremented), the pre-PDF switches to a more (less) delayed version of F_{ref}, which forces the tri-state PFD to move past 360° of phase error and reset. If this additional circuitry was not present, the 360° detector would be prone to false triggering when the steady-state phase error was close to the triggering threshold.

The overall effect of the control circuitry on the operation of the PLL is pictured in figure 5. The double sided arrow on the vertical axis denotes the control voltage range that would

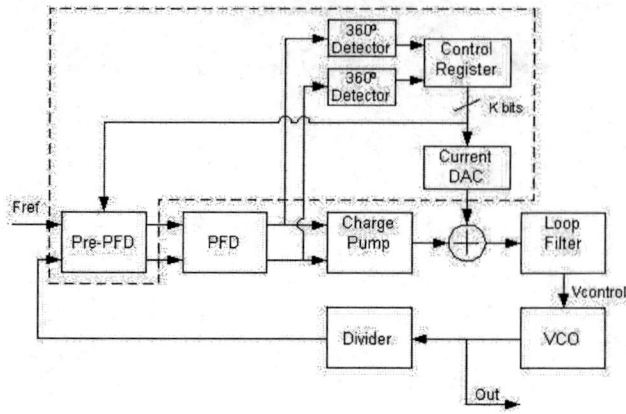

Fig. 3. System level schematic of PLL with extended range phase detector.

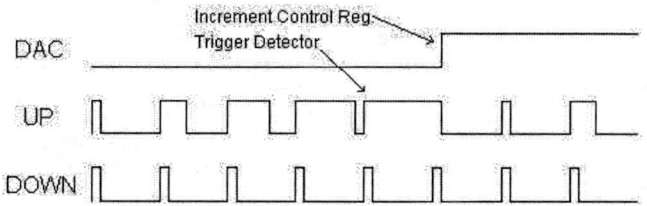

Fig. 4. Waveforms demonstrating operation of 360° detector.

be achievable without the extended range phase detector. As the phase error increases past 360°, the CP output resets, but the DAC current is incremented, so the cumulative effect is that the control voltage can ramp up continually, limited only by the number of bits in the control register.

III. CIRCUIT IMPLEMENTATIONS

A. 360° Detector

The schematic for the 360° detector is shown in figure 6. The circuit is very simple, consisting of a delay cell (implemented using a number of inverters) and a D-type flip-flop. The circuit samples the duration of the gaps between pulses on the UP (or DN) outputs of the tri-state PFD. If

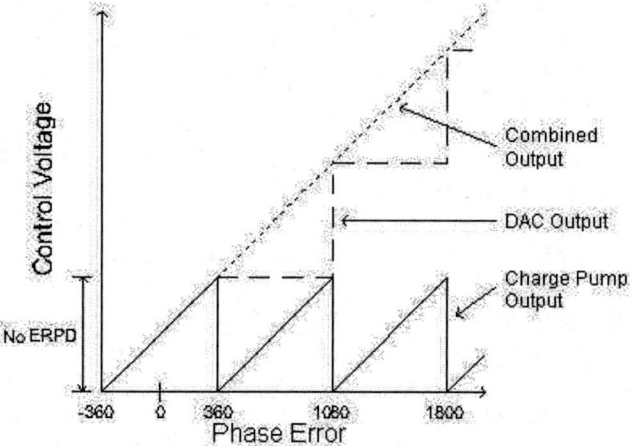

Fig. 5. Effect of extended range phase detector on control voltage.

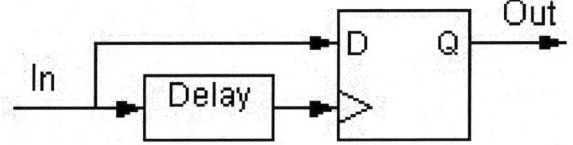

Fig. 6. Schematic of 360° phase detector.

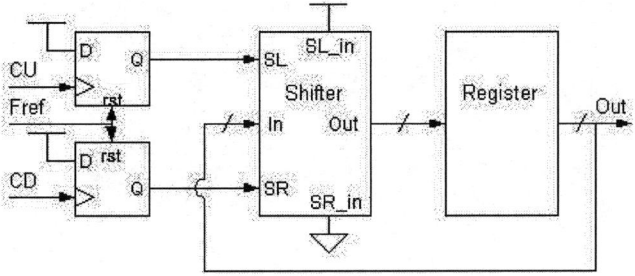

Fig. 7. Schematic of control register.

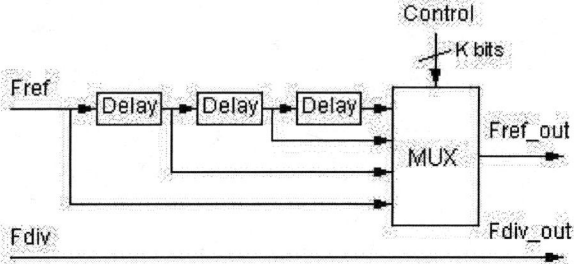

Fig. 8. Schematic of Pre-PFD block.

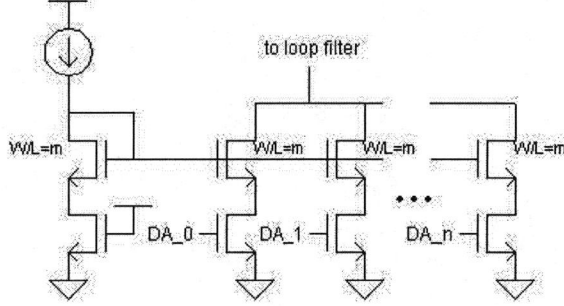

Fig. 9. Schematic of current DAC.

the gap becomes smaller than the delay of the delay cell, the output goes high, indicating to the control register that the phase error is approaching 360 degrees. Choosing the delay for the delay cell is a trade-off between having the charge-pump range as close as possible to the ideal ±360° (small delay) and not failing to detect any phase wrap-arounds for rapidly changing frequency outputs (large delay). For this work, the compromise chosen was:

$$t_{delay} = \frac{1}{10 \cdot F_{ref}} \quad (1)$$

B. Control Register

The schematic for the control register is shown in figure 7. The CU (count up) and CD (count down) inputs come from the outputs of the 360° detectors. The register holds the current state of the phase error, using a unary counting scheme. The unary scheme was chosen because the number of phase states will not be very large for any practical PLL, and it simplifies the design of the current DAC. The D flip-flops control the shifter, which shifts in 1's from one side and 0's from the other side. On each rising edge of F_{ref}, the output of the shifter is loaded into the register. If the CU flip-flop goes high, indicating an increase in phase shift, an additional 1 is shifted into the output of the shifter. This new value is loaded into the register, incrementing the output, which is then used to control the current DAC and the pre-PFD.

C. Pre-PFD

The schematic for the pre-PFD is shown in figure 8. This block is simply a multiplexor which chooses between delayed versions of the F_{ref} signal to provide to the tri-state PFD. The delay blocks are implemented using inverters, and the delay for each block is slightly greater than the delay used in the 360° detector, so that once the detector increments the control register, the pre-PFD will delay the F_{ref} signal, causing the tri-state PFD to reset. This prevents multiple triggering of the 360° detector for one phase wrap-around.

D. Current DAC

The schematic for the current DAC is shown in figure 9. The current DAC is a simple circuit consisting of a number of equally sized current mirrors which are switched on or off according to the output of the control register. The unit current for each mirror is chosen as slightly more than double the current output of the charge pump for a 360° phase error. This provides hysteresis in the switching illustrated in figure 5, and prevents the current DAC from switching on and off for steady state phase errors near 360°.

IV. SIMULATION RESULTS

A PLL using the proposed extended range phase detector circuitry was simulated by SPECTRE using the BSIM3v3.2 models from the IBM 7RF 0.18 μm CMOS process. The expected behaviour was confirmed, and the operation of the PLL for an excursion over the full range of the control voltage is shown by the signals in figure 10. The control to the divider was swept from low to high to cover the full range of VCO output frequencies. The PLL design required two DAC current steps in order to cover the full range of the VCO control voltage. The transitions in the DAC output current (corresponding to the changes in the DAC control bits shown in the upper plot) can be seen in the disturbances in the control voltage (shown in the lower plot). The simulated PLL has been fabricated in the IBM 7RF 0.18 μm CMOS process, although test results were not available at the time of this publication. The layout of the fabricated PLL is shown in figure 11.

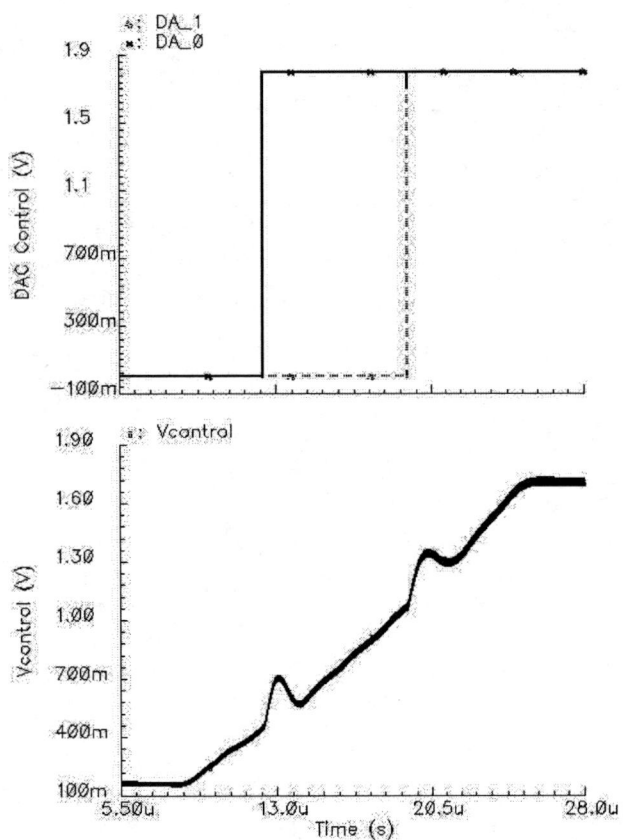

Fig. 10. Simulation showing control voltage and DAC control over control voltage range.

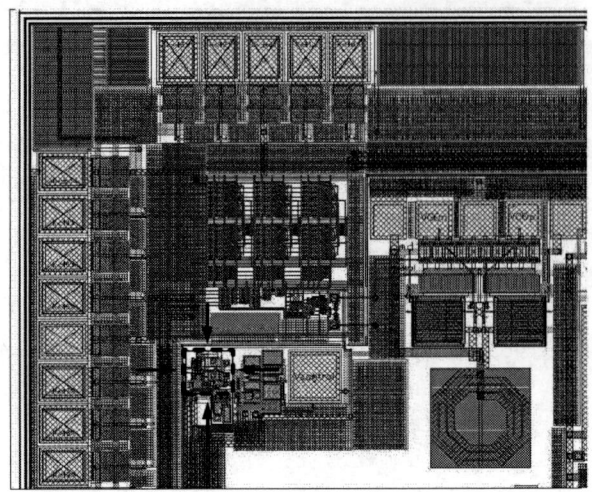

Fig. 11. Layout of the fabricated PLL. The phase detector and control circuitry is denoted by arrows and enclosed by the dashed box.

The noise contribution of the additional circuitry was investigated by performing SPECTRE Pnoise analysis on the PFD with and without the addition of the Pre-PFD block. There was concern that the Pre-PFD block could introduce significant phase noise which could cause problems in PLLs with a high divide ratio, since the PFD phase noise appears at the PLL output multiplied by N (the PLL divide ratio). The Pnoise simulations revealed that the phase noise with the Pre-PFD block is within 1 dB of the phase noise without it. In a PLL used for indirect modulation, the PFD noise at critical frequencies is typically negligible compared to other noise sources [9] (such as $\Sigma - \Delta$ quantization noise), so the noise added by the Pre-PFD block should not pose any problems.

A possible limitation of the proposed scheme is revealed by the disturbances on the control voltage at $13\mu s$ and $20\mu s$ in figure 10. These disturbances will cause some non-ideality in the output spectrum of the PLL. The negative edge of the disturbance could also be a source of instability in the PLL, but if the DAC currents are chosen properly the built in hysteresis should prevent this from being a problem. Due to time constraints, simulations were not performed to investigate the impact of these disturbances. A test chip has been fabricated and measurements will determine the extent of the non-idealities in the PLL output spectrum.

V. CONCLUSION

Circuitry to extend the control voltage range of a type-I charge-pump PLL has been presented. The control circuitry makes use of an extended range phase detector, and is based on ideas presented in [8], but makes several improvements. The DAC current switches after 360° instead of 180°, which results in fewer DAC transitions, and fewer control voltage disturbances as seen in figure 10. The reduced number of states to be stored in the coarse tuning register also results in a simplified current DAC design. Simulated results indicate that the circuits perform as expected, and do not add significant phase noise to the PLL.

ACKNOWLEDGMENT

The authors would like to thank Dr. Hossein Zarei for guidance on the project which inspired this work.

REFERENCES

[1] P. Zhang et al., "A 5-GHz Direct-Conversion CMOS Transceiver," *IEEE J. Solid-State Circuits*, vol. 38, pp. 2232-2238, Dec. 2003.
[2] Y. Jung et al., "A 2.4-GHz 0.25-μm CMOS Dual-Mode Direct-Conversion Transceiver for Bluetooth and 802.11b," *IEEE J. Solid-State Circuits*, vol. 39, pp. 1185-1190, Jul. 2004.
[3] G. Brenna et al., "A 2-GHz Carrier Leakage Calibrated Direct-Conversion WCDMA Transmitter in 0.13 μm CMOS," *IEEE J. Solid-State Circuits*, vol. 39, pp. 1253-1262, Aug. 2004.
[4] T. Riley, M. Copeland, and T. Kwasniewski, "Delta-Sigma Modulation in Fractional-N Frequency Synthesis," *IEEE J. Solid-State Circuits*, vol. 28, pp. 553-559, May 1993.
[5] M. Perrott, T. Tewksbury, and C. Sodini, "A 27-nW CMOS Fractional-N Synthesiser Using Digital Compensation for 2.5-Mb/s GFSK Modulation," *IEEE J. Solid-State Circuits*, vol. 32, pp. 2048-2060, Dec. 1997.
[6] S.T. Lee et al., "A 1.5V 28mA Fully-Integrated Fast-Locking Quad-Band GSM-GPRS Transmitter with Digital Auto-Calibration in 130nm CMOS," in *IEEE ISSCC Dig. Tech. Papers*, Feb. 2004, pp. 188-189.
[7] M. Perrott, *PLL Design Using the PLL Design Assistant Program*. http://www-mtl.mit.edu/perrott, 2002.
[8] S. Willingham et al., "An Integrated 2.5GHz $\Sigma\Delta$ Frequency Synthesizer with 5 μs Settling and 2MB/s Closed Loop Modulation," in *IEEE ISSCC Dig. Tech. Papers*, Feb. 2000, pp. 200-201.
[9] M. Perrott, *Techniques for High Data Rate Modulation and Low Power Operation of Fractional-N Frequency Synthesizers*, PhD Thesis, Massachusetts Institute of Technology, Sept. 2002.

A Two-Chip, 4-MHz, Microelectromechanical Reference Oscillator

Krishnakumar Sundaresan, Gavin K. Ho, Siavash Pourkamali and Farrokh Ayazi
School of Electrical and Computer Engineering
Georgia Institute of Technology, Atlanta, GA 30332
krishnak@ece.gatech.edu, ayazi@ece.gatech.edu

Abstract — The paper describes a 4-MHz temperature compensated reference oscillator based on a capacitive silicon micro-mechanical resonator. The design of the resonator has been optimized to offer large quality factors (22000), while maintaining tunability in excess of 3000ppm for fine tuning and temperature compensation. Oscillations are sustained with a CMOS amplifier and temperature compensation is performed with a novel resonator bias generator. When interfaced with the bias circuit, the oscillator exhibits a temperature drift of 380ppm over a 90°C range, a 6X improvement in stability over an uncompensated oscillator. The sustaining amplifier and compensation circuitry were fabricated in a 2P3M 0.5μm CMOS process. The oscillator is designed to prototype highly stable, low phase-noise reference oscillators integrated at the chip or package level.

I. INTRODUCTION

The reference oscillator is usually one of the hardest blocks in a system to integrate on Silicon. While quartz crystals offer excellent temperature stability and phase-noise performance, the inability to integrate them with silicon increases the size of systems that rely on clocking schemes. Silicon micromechanical resonators have been proven to offer excellent stability and quality factors in excess of 10000 over a wide range of frequencies [1,2] making them suitable for reference oscillators. The potential for electrostatic tuning of these resonators without using noisy varactor diodes is an attractive option for electronic fine tuning and temperature compensation of the resonant frequency. Further, these devices also offer the possibility of integration at the chip or package level with standard electronics.

While micromechanical resonators have been proven to offer excellent quality factors, the temperature stability of these resonators (typically -25ppm/°C) [2] is largely inferior to that of quartz. Conventional resonator temperature compensation schemes focus on localized resonator heating or stress compensation techniques [3,4]. The idea of electrostatic compensation is largely unexploited due the need for large voltages, conventionally unavailable in CMOS.

In this work, we demonstrate a 4-MHz electrostatically temperature compensated resonator-oscillator operating with a 5V supply. A voltage multiplier [5] circuit is used to generate a voltage of approximately 24V; this voltage is applied to a tuning slope generator based on a diode chain to realize a temperature variant bias for the resonator. The oscillation frequency, operating under this bias, drifted by only 380ppm over 90°C, an improvement of 6X over an uncompensated oscillator.

II. SYSTEM BLOCK DIAGRAM

Fig. 1 shows the complete system block diagram of the implemented resonator oscillator system. Oscillations are started up and sustained with the amplifier and the inverting buffer that maintain unity loop gain and zero phase shift. The gain of the amplifier is varied with a voltage controlled MOS resistor to minimize resonator saturation due to large drive amplitudes. The resonator is biased with a voltage in the range of 20V (with a negative temperature coefficient) generated from a charge pump and passing it through a diode chain circuit. The temperature coefficient of the circuit is trimmed for process variations by varying the gate voltage of a MOS transistor in the diode chain. The clock ripple on the bias voltage is minimized by using a MOSFET-C filter at the output of the charge pump.

III. MICROMECHANICAL RESONATOR DESIGN

The characteristics of the resonator are dictated by the system-level requirements: to achieve low phase noise, low power consumption, and reduced frequency-temperature drift, the micromechanical resonator must have high quality factor, low motional impedance, and high tunability for electrostatic frequency tuning.

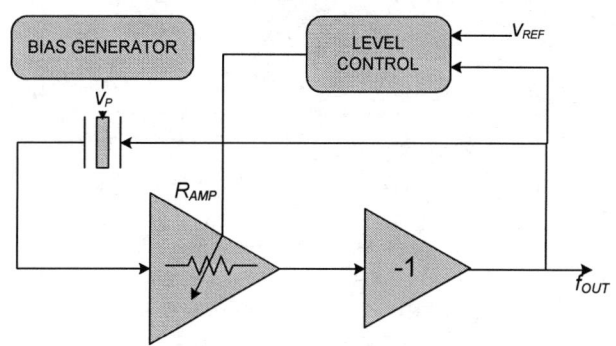

Figure 1. System Block Diagram

Support for resonator fabrication was provided by the DARPA NMASP program. IC Fabrication support for was provided by MOSIS.

These three requirements are typically difficult to achieve simultaneously with a micromechanical resonator. Flexural mode capacitive beam resonators in the frequency of interest are highly tunable, but have high impedances and low Q. Conventional extensional mode capacitive resonators can have high Q and lower impedances, but suffer from low tunability. Piezoelectric resonators are another alternative; they have lower impedances, but no effective tuning technique has been demonstrated for these resonators.

The new I^2 resonator was designed to meet all the above requirements simultaneously [6]. The structure resonates predominantly in an extensional mode. Large flanges are placed at the ends of the extensional beams for increasing the capacitive transduction area (Fig. 2). The choice 4MHz resonator with 225nm gaps fabricated using the HARPSS-on-SOI process [2] is shown in Fig. 3.

The resonator in Fig. 3 had a resonant frequency of 4.08MHz with a Q of 22000. The impedance was measured to be approximately 10kΩ with a polarization voltage of 20V, and a tuning coefficient of -10ppm/V^2 was extracted for the device (Fig. 4).

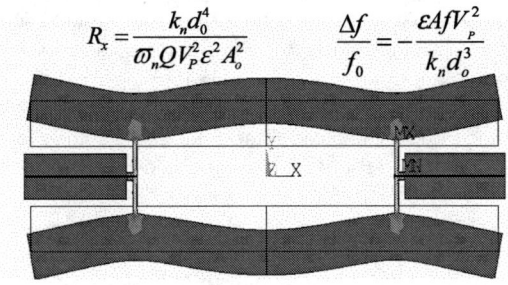

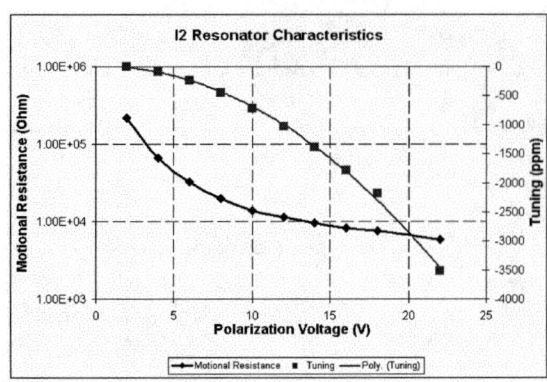

Figure 4. Impedance and tuning characteristic of the 4-MHz I^2 resonator

IV. INTERFACE CIRCUIT ARCHITECTURES

A. Sustaining Amplifier

The sustaining amplifier for the oscillator is based on a CMOS trans-impedance amplifier with a folded-cascode OTA as the core. The schematic of the sustaining amplifier is shown in Fig. 5. An inverting buffer is used at the output of the amplifier to drive an off-chip load and also to maintain zero loop phase shift. The amplifier is implemented in a shunt-shunt feedback configuration to minimize Q-loading [7]. To maintain optimal phase-noise performance, it is essential to control the drive amplitude of the resonator to minimize non-linearity in the resonator [8]. This is achieved by controlling the trans-impedance gain by varying the control voltage of the feedback MOS resistor M15.

B. Temperature Compensating Bias Circuitry

The block diagram of the temperature compensating bias circuitry is shown in Fig. 6. A large voltage is first generated by the charge pump, which is based on the Dickson charge pump architecture shown in Fig. 7 [5].

Figure 2. Design of a 4MHz extensional mode resonator for low impedance and high tunability [6]

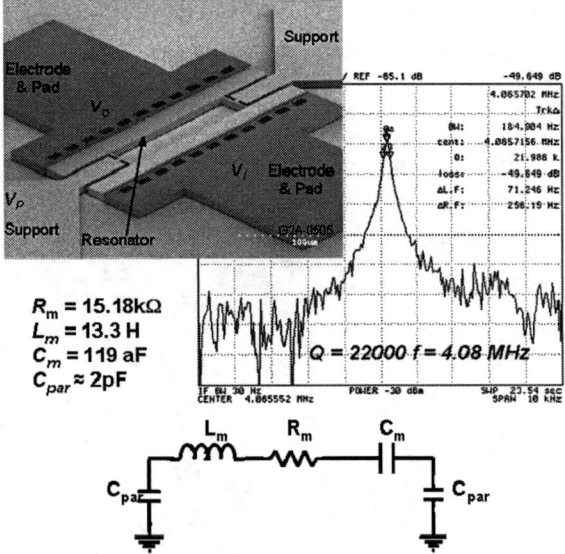

Figure 3. SEM, equivalent circuit parameters and open-loop frequency response of the 4MHz I^2-resonator

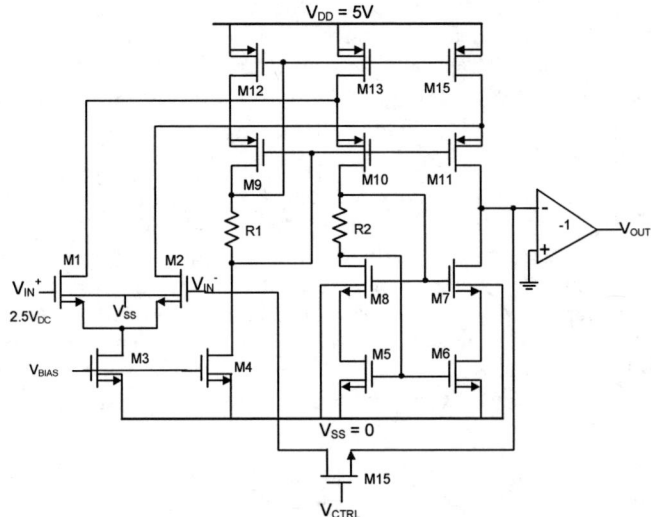

Figure 5. Schematic diagram of sustaining amplifier

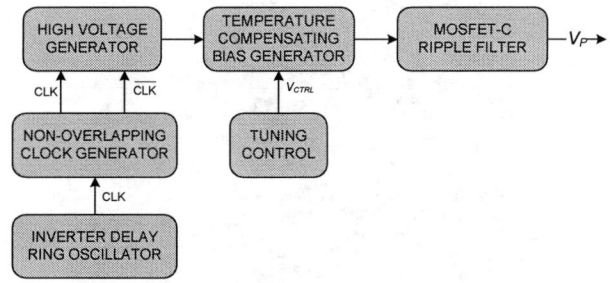

Figure 6. Block Diagram of the Temperature Compensating Bias Generator for the I2- resonator

The voltage generated by the charge pump is used to drive the diode chain circuit shown in Fig. 8, which generates the required negative temperature coefficient. The diodes used in the charge pump use the p$^+$-n-well junction of a PMOS transistor. The clocks for the charge pump are realized on the chip using an inverter delay ring oscillator and a logic circuit to generate non-overlapping clocks. The circuit diagram for the clock generator is given in Fig. 9.

The diode chain circuit of Fig. 8 creates a negative temperature coefficient of 2.5mV/°C for each diode due to the temperature variation of V_{BE}. Stacking a set of 25 diodes, we can get a slope of 62.5mV/°C. The resistors (R1 and R2) set the voltage range and the MOS transistor (M1) is used to trim the output voltage and the temperature coefficient for process variations. Although the generated voltage is large (24V), the voltage across each of the diodes and the transistor is much smaller. Hence, there is no risk of break down in any of the diodes. The circuit after voltage trimming, gives a variation of 16-22V for a 100°C change in temperature. The variation of the charge pump voltage with temperature is shown in Fig. 10.

An important consideration in the biasing of the resonator is the elimination of clock ripple on the bias voltage. The charge pump output voltage can contain significant clock ripple which can modulate the signal and add spurs to the spectrum. The ripple is minimized by using a low pass filter with a small cut-off frequency (typically 1kHz as compared to a clock frequency of 1MHz).

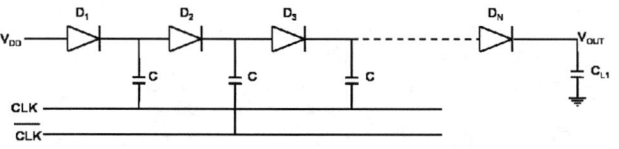

Figure 7. Charge pump used in the bias cell

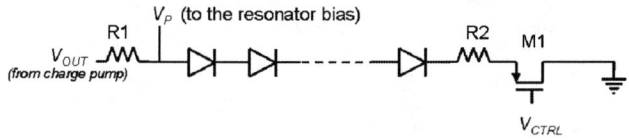

Figure 8. Tuning circuit for the Bias generator

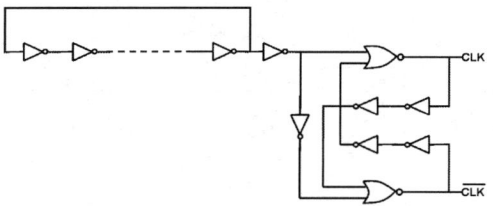

Figure 9. Clock generator for the charge pump

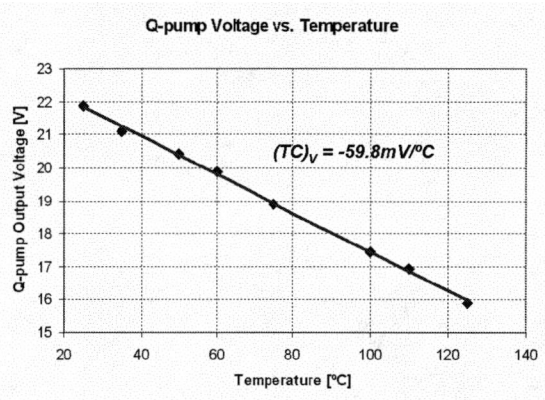

Figure 10. Variation of Resonator Bias Circuit Voltage with Temperature

V. MEASURED RESULTS

The interface IC was fabricated in a 2P3M 0.5μm CMOS process. The op-amp and the compensation circuit were tested for functionality. The gain-bandwidth of the op-amp was measured to be 175 MHz, large enough to satisfy the gain requirements to sustain oscillations.

After open-loop characterization, the resonator and the IC were interfaced with bond wires and tested under a *Desert Cryogenics* vacuum probe station with temperature control. The functionality of the oscillator was verified. A typical oscillator waveform (after external amplification) and the frequency spectrum are shown in Fig. 11. The temperature coefficient of the compensated oscillator was measured to be 4.2ppm/°C, a 6X improvement over the uncompensated case. Fig. 12 shows the temperature variation of frequency for both compensated and uncompensated cases. A summary of the resonator and IC specifications is shown in Table 1.

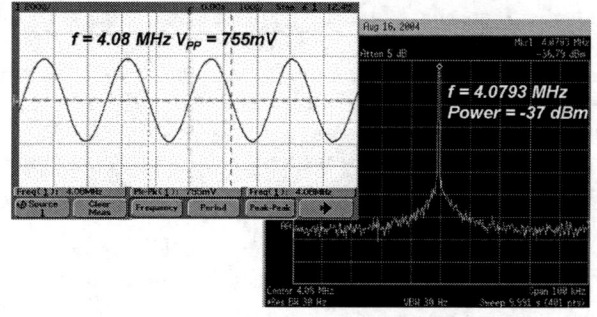

Figure 11. Oscillation waveform from the interfaced resonator and IC

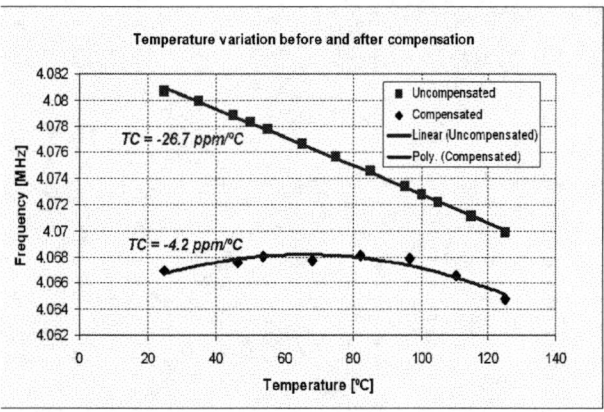

Figure 12. Temperature variation of compensated and uncompensated resonator-oscillator

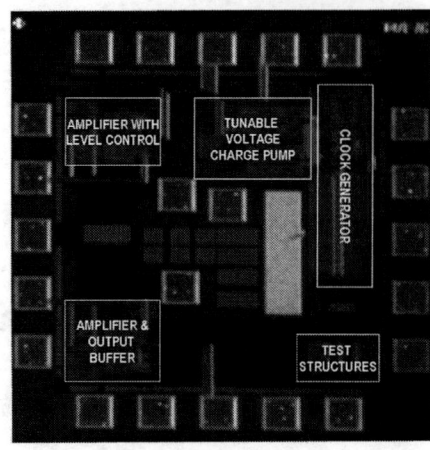

Figure 13. Die picture of fabricated interface IC

TABLE I. SUMMARY OF MEASURED RESULTS

Specification	Performance
Resonant frequency	4.08 MHz
Unloaded Quality factor	22000
Temperature Coefficient	-26.7 ppm/°C
Circuit Specifications	
Amplifier Gain Bandwidth product	175 MHz
Tuning voltage range (V_{clk} = 2.5V)	16-22 V
Charge pump clock	1 MHz
Ripple filter -3dB frequency	1 kHz
Total power consumption	1.8 mW
Oscillator specifications	
Frequency variation w/ temperature (25°C-115°C)	380ppm
Tuning range	2400 ppm

As we can observe from Fig. 12, the temperature variation of the resonant frequency follows a linear relation. The tuning characteristic with V_P is proportional to V_P^2 and is given by (1)

$$\frac{\Delta f}{f_0} = -\frac{\varepsilon A f V_P^2}{k_n d_o^3} \quad (1)$$

where A is the area of the electrode, d_0 the electrode gap, f_0 is the natural frequency (with zero V_P), Δf the difference between f_0 and the operating frequency and k_n the stiffness of the resonator. Thus a linear variation of V_P with temperature cannot achieve accurate compensation over a wide temperature range. We may notice in fig. 12 that the frequency at the extreme temperatures is almost equal indicating that the compensation technique is optimized over the entire temperature range and further performance gains are difficult. However, with a parabolic compensation voltage described by (2), we can achieve improvements of the order of 50X over the uncompensated oscillator.

$$V_P^2 = A + BT \quad (2)$$

A circuit to generate the appropriate temperature profile has been designed and submitted for fabrication. Simulation results indicate an overall variation of 53ppm over -25°C to 125°C, a 75X improvement over the uncompensated case.

VI. CONCLUSIONS

We have a demonstrated an electronically temperature compensated reference oscillator based on a high Q micromechanical resonator. The temperature coefficient was measured to be 4.2ppm/°C, a 6X improvement over an uncompensated oscillator. Techniques to improve the temperature coefficient further have been identified. The oscillator is designed to prototype fully integrated reference oscillators for communication systems.

REFERENCES

[1] S. Pourkamali, A. Hashimura, R. Abdolvand, G. K. Ho, A. Erbil, and F. Ayazi, "High-Q single crystal silicon HARPSS capacitive beam resonators with self-aligned sub-100nm transduction gaps", IEEE/ASME J. MEMS, vol. 12, no. 4, pp. 487-496, August 2003.

[2] S. Pourkamali, Z. Hao and F. Ayazi, "VHF Single Crystal Silicon Capacitive Elliptic Bulk-Mode Disk Resonators—Part II: Implementation and Characterization," IEEE/ASME J. MEMS, vol. 13, no. 6, pp. 1054-1062, Dec. 2004.

[3] W. T. Hsu and C. T.-C. Nguyen, "Geometric stress compensation for enhanced thermal stability in micromechanical resonators," Proc. of IEEE Intl. Ultrasonics Symposium, pp. 945-948, 1998.

[4] M. Hopcroft et al, "Active temperature compensation for micromachined resonators," Technical Digest, Solid-state Sensor, Actuator and Microsystems Workshop, June 2004, pp. 364-367.

[5] J. Dickson, "On-chip high-voltage generation in NMOS integrated circuits using an improved voltage multiplier technique," IEEE J. Solid-State Circuits, vol. 11, no. 6, pp. 374-378, June 1976.

[6] G.K. Ho K. Sundaresan, S. Pourkamali and F. Ayazi "Low-motional-impedance highly-tunable I^2 resonators for temperature-compensated reference oscillators," Proc. of IEEE MEMS 2005, pp. 116-120, Jan 2005.

[7] C.T-C. Nguyen and R.T. Howe, "An integrated high-Q CMOS micromechanical resonator-oscillator," IEEE J. Solid-State Circuits, vol. 34, no. 4, pp. 440-455, April 1999.

[8] V. Kaajakari, et al, "Nonlinearities in single-crystal silicon micromechanical resonators", Intl. Conference on Solid-state Sensors, Actuators and Microsystems, pp. 1574-1577, June 2003.

A Study of Injection Locking in Ring Oscillators

Behzad Mesgarzadeh and Atila Alvandpour
Division of Electronic Devices
Department of Electrical Engineering
Linköping University
Linköping, Sweden

Abstract—This paper presents an analysis about injection locking phenomenon in CMOS ring oscillators. In this analysis Adler's equation in injection locking has been proved for a three-stage ring oscillator and behavior of this kind of oscillators in locked condition with respect to phase noise and jitter reduction has been analyzed.

I. INTRODUCTION

Oscillatory systems with environmental coupling and close frequencies can have interaction resulting in changes in their phase and frequency. This phenomenon has been called "injection locking". For the first time it was observed in the 17th century, by the Dutch scientist Christian Huygens who is known for inventing the pendulum clock. He noticed that the pendulums of two clocks, which were hung close two each other on the wall, moved in unison [1]. Injection locking can be useful in electrical application reducing jitter and phase noise of the oscillators [2]-[4].

From integrated circuit design point of view, in previous studies, injection locking has been analyzed for LC tank-based oscillators [5]. This paper describes injection locking in CMOS ring oscillators and presents an analytical formulation showing the phase and frequency relationships. In section II we introduce a linear model for ring oscillators. Section III presents a mathematical analysis of injection locking in ring oscillators. In section IV, phase noise and jitter reduction behavior of injection-locked ring oscillators is discussed and in Section V, this behavior is verified by circuit level simulation.

II. RING OSCILLATORS

A three-stage ring oscillator has been shown in Fig. 1(a). Fig. 1(b) shows a simplified linear model of the ring oscillator, which we have utilized for our mathematical analysis. The model comprises equivalent output resistance and load capacitance for each inverter stage. The linear model yields a first-order transfer function for each stage expressed as $-A/(1+s/\omega_0)$, where $\omega_0 = \sqrt{3}/(RC)$ and $A=2$ (since each stage should contribute 60° phase shift). Consequently, open-loop transfer function for a three-stage ring oscillator is:

$$H(j\omega) = -\frac{A^3}{(1+j\frac{\sqrt{3}\omega}{\omega_0})^3}. \quad (1)$$

We utilize the equation (1) as a frequency domain model for a three-stage ring oscillator.

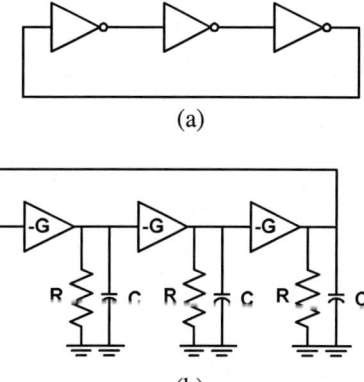

Fig.1. Ring oscillator: (a) a block diagram and (b) a linear model

III. INJECTION LOCKING

To perform a first-order analysis, the injection in a ring oscillator can be modeled as shown in Fig. 2 where except fundamental other harmonics are neglected. In this model we assume that ω_i is located at an offset $\Delta\omega$ with respect to the carrier frequency (free-running frequency) of the ring oscillator (ω_0). Then:

$$\omega_i = \omega_0 + \Delta\omega. \quad (2)$$

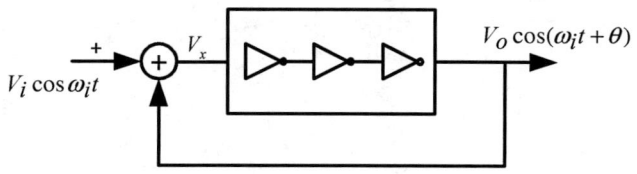

Fig.2. A first-order model of injection into a ring oscillator.

In steady state the total phase shift in the output can be calculated by:

$$\varphi + \xi = \theta. \quad (3)$$

Where ξ is the phase of V_x and φ is the total phase shift, which is applied to V_x in the inverter stages. From [5] the relationship between injection ratio (S) and output phase can be written as:

$$\tan(\theta - \xi) = S \sin\theta. \quad (4)$$

$$S = \frac{V_i}{V_o} = \sqrt{\frac{P_i}{P_o}}. \quad (5)$$

Combining (3) and (4) yields:

$$\tan\varphi = S \sin\theta. \quad (6)$$

In order to have a simple mathematical formulation we assume that the ring oscillator has three stages of inverters but the principle is the same for more number of stages. If each of the inverters contributes in one third of the total phase shift in the ring oscillator then:

$$\varphi = 3\alpha. \quad (7)$$

Where α is the phase shift in one of the stages in the ring oscillator. Then from (1):

$$\tan\alpha = \frac{\sqrt{3}\omega_i}{\omega_0}. \quad (8)$$

From (2) and (8):

$$\tan\alpha = \sqrt{3}(1+\lambda). \quad (9)$$

$$\lambda = \frac{\Delta\omega}{\omega_0}. \quad (10)$$

λ can be called as relative lock range. From (7) and (9):

$$\tan\varphi = \tan(3\alpha) = \frac{3\tan\alpha - \tan^3\alpha}{1 - 3\tan^2\alpha}$$

$$= \frac{3\sqrt{3}(1+\lambda) - 3\sqrt{3}(1+\lambda)^3}{1 - 9(1+\lambda)^2} = \frac{3\sqrt{3}(\lambda^3 + 3\lambda^2 + 2\lambda)}{9\lambda^2 + 18\lambda + 8}$$

Since $\lambda \ll 1$ then:

$$\tan\varphi \approx \frac{3\sqrt{3}}{4}\lambda. \quad (11)$$

In a ring oscillator for an operating point in a frequency close to carrier, the phase shift in each stage will be close to $\pi/3$ and then with a first-order approximation for $y=f(x)$ and by replacing $f(x)=\tan(x)$ in $x_0 = \pi/3$:

$$f(x_0 + \Delta x) = f(x_0) + \Delta x \frac{df}{dx}\bigg|_{x=x_0} \quad (12)$$

$$\alpha = \frac{\pi}{3} + \Delta x. \quad (13)$$

$$\tan\alpha = \tan(\frac{\pi}{3} + \Delta x)$$

$$= \sqrt{3} + \Delta x.(1 + \tan^2\frac{\pi}{3}) = \sqrt{3} + 4\Delta x. \quad (14)$$

Combining (9) and (14) yields:

$$\Delta x = \frac{\sqrt{3}\lambda}{4}. \quad (15)$$

From (13) and (15):

$$\alpha = \frac{\pi}{3} + \frac{\sqrt{3}}{4}\lambda. \quad (16)$$

On the other hand the phase shift of the ring oscillator (φ) can be written as a function of the operating point frequency ($y = \varphi(\omega)$). For frequencies close to the carrier frequency, from (2) and (12) we can write:

$$\varphi(\omega_i) = \varphi(\omega_0 + \Delta\omega) = \varphi(\omega_0) + \Delta\omega\frac{d\varphi}{d\omega}. \quad (17)$$

For a free-running oscillator in ω_0 the total phase shift is π then:

$$\varphi(\omega_i) = \pi + \Delta\omega\frac{d\varphi}{d\omega}. \quad (18)$$

Combining (7), (16) and (18) results in:

$$\pi + \Delta\omega\frac{d\phi}{d\omega} = 3(\frac{\pi}{3} + \frac{\sqrt{3}}{4}\lambda).$$

Hence:

$$\Delta\omega\frac{d\varphi}{d\omega} = \frac{3\sqrt{3}}{4}\lambda. \quad (19)$$

From (11) and (19):

$$\tan\varphi = \Delta\omega\frac{d\varphi}{d\omega}. \quad (20)$$

Combining (6) and (20) yields:

$$S \sin\theta = \Delta\omega\frac{d\varphi}{d\omega}. \quad (21)$$

We define Q for an oscillator as:

$$Q = \frac{\omega_0}{2}\frac{d\varphi}{d\omega}. \quad (22)$$

Then by combining (21) and (22) we can write:

$$\Delta\omega = \frac{\omega_0}{2Q}S \sin\theta. \quad (23)$$

Replacing corresponding equations from (2) and (5) yields:

$$\omega_i - \omega_0 - \frac{\omega_0}{2Q}\cdot\sqrt{\frac{P_i}{P_o}}\sin\theta = 0. \quad (24)$$

This equation formulates the injection locking in the ring oscillators. It is interesting to note that (24) is the same equation which was derived by Adler [6] when an oscillator locks to the injected signal, and θ remains constant with respect to time ($d\theta/dt = 0$).

IV. PHASE NOISE AND JITTER REDUCTION

A. Phase noise

In any of practical oscillators, there are fluctuations in both amplitude and phase of output signal because of different sources of noise. Phase variation due to existence of noise is called *phase noise*. There are many studies on formulation of the phase noise in the oscillatory systems. In some of these studies, phase noise in time domain has been modeled [7], while there are studies in which a frequency domain model was presented [8], [9]. In order to have a better understanding about phase noise, first we will discuss a simple formulation in frequency domain. Suppose that there is a free-running oscillator, which its carrier frequency is ω_0. Furthermore we assume that the noise is injected as an input to the oscillatory system as shown in Fig.3. Output spectrum for this system is shown in Fig. 4. The spectrum exhibits "skirts" around the carrier frequency.

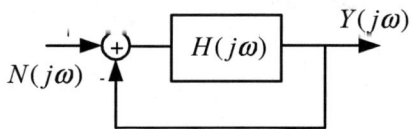

Fig.3. Noise is injected to the oscillatory system

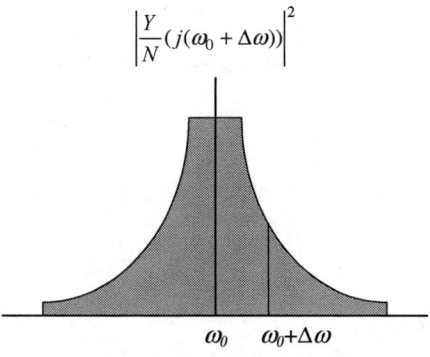

Fig.4. Output spectrum of an oscillatory system.

According to Fig.4 the noise components close to carrier frequency has significant power. According to Leeson's equation [10] the noise shaping function for the oscillatory system shown in Fig. 3, can be expressed as:

$$\left| \frac{Y}{N}(j(\omega_0 + \Delta\omega)) \right|^2 = \frac{1}{4Q^2}\left(\frac{\omega_0}{\Delta\omega}\right)^2. \quad (25)$$

This equation can be used for estimating the phase noise in a free-running ring oscillator. Using (23) and denoting $(\omega_0/2Q)S$ by ω_L yields:

$$|\Delta\omega| \leq \frac{\omega_0}{2Q}S = \omega_L. \quad (26)$$

$$\omega_0 - \omega_L \leq \omega_i \leq \omega_0 + \omega_L \quad (27)$$

In (27) a frequency range, in which output of oscillator can be locked to the injected signal, is determined. In this condition, assuming the ring oscillator as a black box, as shown in Fig.5, we can calculate the closed-loop transfer function.

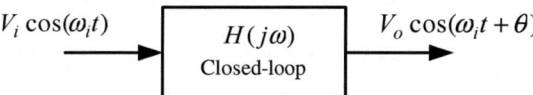

Fig.5. Ring oscillator after injection locking.

In Fig.5, we assume that injected signal exhibits very low phase noise comparing with the ring oscillator. Using Fourier transform, $|H(j\omega)|$ in lock range (determined by (27)) for Fig.5 can be calculated:

$$|H(j\omega)| = \left|\frac{F(V_o \cos(\omega_i t + \theta))}{F(\cos(\omega_i t))}\right| = \left|\frac{V_o}{V_i}e^{-j\omega\frac{\theta}{\omega_i}}\right| = \frac{1}{S}. \quad (28)$$

According to Fig.3, noise is an input to the closed-loop transfer function. Then using (28) yields the noise shaping function in the locked condition:

$$\left|\frac{Y}{N}(j\omega_i)\right|^2 = |H(j\omega)|^2 = \frac{1}{S^2}. \quad (29)$$

From equation (29) it is obvious that if the injection ratio is increased, total phase noise for an injection-locked ring oscillator can be decreased. To gain more insight, we can compare the phase noise for a free-running ring oscillator with an injection-locked one. We assume a ring oscillator under injection where the injected signal satisfies (27) and it is located at an offset $\Delta\omega$ with respect to carrier. We denote the phase noise of this oscillator by PN_1. At the same time for a free-running ring oscillator, the phase noise component located at an offset $\Delta\omega$ with respect to carrier can be calculated using (25) and we denote it by PN_2. The ratio of phase noises for these two oscillators can be reached by dividing (29) by (25):

$$\frac{PN_1}{PN_2} = \frac{4Q^2}{S^2}\left(\frac{\Delta\omega}{\omega_0}\right)^2 = \left(\frac{\Delta\omega}{\omega_L}\right)^2. \quad (30)$$

Using (30), we can make some conclusions. First of all, when the frequency of the injected signal is located at the lock range ($\Delta\omega < \omega_L$) the phase noise will be decreased comparing with the phase noise of a free-running oscillator.

Secondly, for the frequencies close to carrier, the phase noise reduction reaches its maximum but for the frequencies close to the lock range edges ($\omega_0 \pm \omega_L$), the reduction is minimum. For an offset of ω_L with respect to the carrier frequency ($\Delta\omega=\omega_L$), the phase noise of the injection-locked oscillator is the same as that of a free-running oscillator.

B. Jitter

Fig. 6 shows a technique to inject a periodic signal into a ring oscillator. Injection locking phenomenon can be useful to reduce jitter at the output of ring oscillators. According to a z-domain model for an injection-locked ring oscillator, expressed in [3], the structure shown in Fig.6 can operate as a first-order low-pass filter for the output phase, with a single pole located at:

$$p = \frac{\ln(\frac{1}{1+S})}{T_i}. \quad (31)$$

Where T_i is period of injected signal and S is injection ratio which for structure shown in Fig. 6 can be expressed as:

$$S = \frac{W_1}{W_2}. \quad (32)$$

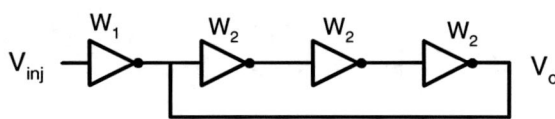

Fig.6. Injection into a ring oscillator.

According to (31) decreasing S can locate the pole (p) low enough to filter out high-frequency and large cycle-to-cycle jitter. On the other hand S cannot be arbitrarily low, because the injection strength is determined by size of injecting buffer (W_1). A small size for this buffer compared to W_2, will reduce injection capability. In this case, the capability for correcting of zero crossings in each period will be reduced.

V. SIMULATION RESULTS

Fig.7 shows jitter reduction behavior utilizing the injection locking. In this simulation jitter is generated by injecting noise in the power supply for a five-stage ring oscillator. The simulation has been done using 0.18 μm, 1.8V CMOS process. The free-running frequency of a five-stage ring oscillator is 3.9 GHz. In Fig.7 four different waveforms are shown. Waveform number 1 is related to free-running oscillator output which its peak-to-peak jitter is 6ps. In the other waveforms, output of injection-locked ring oscillator (Fig.6) has been presented. In the waveforms number 2-4 injection frequencies are 3.8 GHz, 3.7 GHz and 3.6 GHz, respectively. Fig.7 shows that the peak-to-peak jitter is decreased by approaching to the carrier frequency.

VI. CONCLUSIONS

An analytical method to formulate the injection locking phenomenon for CMOS ring oscillators has been presented. According to the obtained results, the phase noise in an injection-locked ring oscillator becomes independent of frequency variation. Also low-pass filter behavior of ring oscillator under injection reduces cycle-to-cycle jitter. Thus this technique can be used in order to reduce phase noise and jitter for clock generators based on the ring oscillators in practical applications.

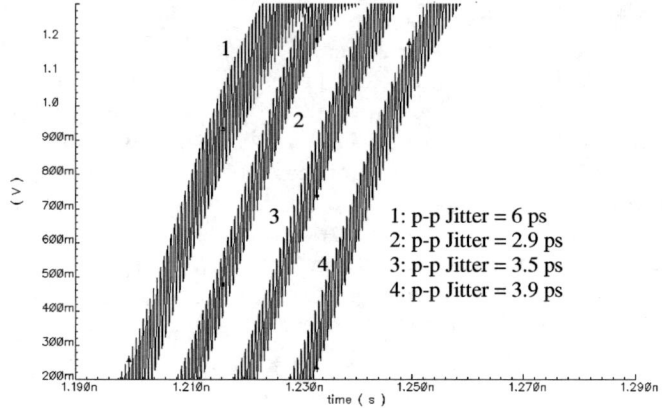

Fig.7. Output of the ring oscillator for different injection frequencies. The waveform number 1 is output of a free-running oscillator with f_O=3.9 GHz and 2-4 are output of injection-locked oscillator with 3.8 GHz, 3.7 GHz and 3.6 GHz injection frequency, respectively.

REFERENCES

[1] A. E. Siegman, Lasers, Mill Valley, CA: University Science Books, 1986.

[2] M. -J. Edward Lee et al., "Jitter transfer characteristics of delay-locked loops - theories and design techniques," *IEEE J. Solid-State Circuits*, vol. 38, pp. 614-621, April 2003.

[3] H-T Ng et al., "A second-order semidigital clock recovery circuits based on injection locking," *IEEE J. Solid-State Circuits*, vol. 38, pp. 2101-2110, Dec. 2003.

[4] S. Verma, H. R. Rategh, T. H. Lee, "A unified model for injection-locking frequency dividers," *IEEE J. Solid-State Circuits*, vol. 38, pp. 1015-1027, June 2003.

[5] B. Razavi, "A study of injection locking and pulling in oscillators," *IEEE J. Solid-State Circuits*, vol. 39, pp. 1415-1424, Sept. 2004.

[6] R. Adler, "A study of locking phenomena in oscillators," *Proc. IEEE*, vol. 61, pp. 1380-1385, Oct. 1973.

[7] A. A. Abidi and R. G. Meyer, "Noise in relaxation oscillators," *IEEE J. Solid-State Circuits*, vol. SC-18, pp. 794-802, Dec. 1983.

[8] B. Razavi, "A study of phase noise in CMOS oscillators," *IEEE J. Solid-State Circuits*, vol. 31, pp. 331-343, Mar. 1996.

[9] A. Hajimiri, S. Limotyrakis and T. H. Lee, "Jitter and phase noise in ring oscillators," *IEEE J. Solid-State Circuits*, vol. 34, pp. 790-804, June 1999.

[10] D. B. Leeson, "A simple model of feedback oscillator noise spectrum," *Proc. IEEE*, vol. 54, pp. 329-330, Feb. 1966.

One-Pass Computation-Aware Motion Estimation with Adaptive Search Strategy

Yu-Wen Huang, Chia-Lin Lee, Ching-Yeh Chen, and Liang-Gee Chen

DSP/IC Design Lab
Graduate Institute of Electronics Engineering and Department of Electrical Engineering
National Taiwan University, Taipei, Taiwan
Email: yuwen@video.ee.ntu.edu.tw

Abstract— A computation-aware motion estimation algorithm is proposed in this paper. Its goal is to find the best block matching results in a computation-limited and computation-variant environment. Our new features are one-pass flow and adaptive search strategies. The prior scheme allocates more computation to the macroblock with the highest distortion in the entire frame step by step. This implies that random access of macroblocks is inevitable, and the search pattern must be determined in advance. The random access flow requires a huge size of memory for all macroblocks to store the up-to-date minimum distortions, best motion vectors, and searching steps. On the contrary, the one-pass flow can not only significantly reduce the memory size but also effectively use the context information of neighboring macroblocks to achieve faster convergence and better quality. Moreover, to improve video quality when computation resource is still sufficient, the search strategy is allowed to adaptively change from diamond search to three step search, and then to full search. Last but not least, traditional block matching speed-up methods are combined to provide much better computation-distortion curves.

I. INTRODUCTION

Motion estimation (ME) is the heart of video encoders to remove temporal redundancy within video sequences. The block matching algorithm (BMA) is adopted by all of the existing video coding standards. Full search block matching algorithm (FSBMA) produces the best video quality but demands the most computation. Many fast BMAs, such as three step search (TSS) [1] and diamond search (DS) [2], have been proposed to speed up the FSBMA with acceptable loss of video quality or with sacrifice of simplicity and regularity.

Usually, ME is implemented with a hardware accelerator. The rapid improvements in processors and fast BMAs make the software encoder a feasible solution, too. However, when the encoder has to support a wide range of applications (e.g. QCIF (176×144) and CIF (352×288), 15 frames/s (fps) and 30fps), traditional BMAs will face two problems. First, a traditional BMA stops only when subsequent search points are all examined, and the searching process of a frame cannot be interrupted when the allowed time interval is passed, so real-time constraints may be violated. Second, once the BMA procedure is finished, it cannot be extended when extra computation is still available, so better video quality cannot be achieved.

Recently, the computation-aware (CA) concept is more and more important. In software implementations, processors may have to support video coding of different frame rates, frame sizes, and search ranges. In hardware implementations, even if the frame rate, frame size, and search range have been clearly determined, the computation resource (e.g. operating frequency) may still be adjusted according to the battery power for portable devices. The goal of CA BMAs is to find the best block matching results in a computation-limited and computation-variant environment.

The authors of [3] are pioneers of CA BMAs. They contributed a novel scheme, which allocates more computation to the macroblocks (MBs) with the highest distortion in the entire frame step by step, as shown in the Fig. 2(d) of [3]. The main concept is that the larger the initial distortion, the more likely the distortion can be significantly reduced, and thus the more computation should be allocated. It is very simple and effective. Nevertheless, there are three problems in their scheme. First, random access of MBs is inevitable, requiring a huge size of memory for all MBs to store the up-to-date minimum distortions, best motion vectors (MVs), and searching steps. The advantage of MV predictors cannot be applied. For example, the predictive diamond search (PDS) [4] outperforms DS in both speed and quality. Second, the search pattern must be determined in advance. The advantage of adaptive search strategy cannot be applied, either. For instance, PDS is better in small motion cases, but TSS is better in large motion cases. The third problem is the poor hardware feasibility since it was intended for software. The distortion sorting operations can be easily implemented as hash tables or lists in software, but they are too expensive in hardware. The random access flow and enormous memory size are also harmful for hardware.

In this paper, a one-pass CA BMA with adaptive search strategy is presented. The ME is done MB by MB to solve the mentioned problems. The rest of this paper is organized as follows. In Section II, motion analysis is reported. In Section III, proposed algorithm is described. Simulation results are shown in Section IV. Finally, Section V gives a conclusion.

II. MOTION ANALYSIS

In this section, motion analysis is done in four aspects, as described in the following subsections. Four QCIF 30fps standard video sequences, Foreman, Silent, Stefan, and Weather, will be used in the statistics with search range as [-16,+15].

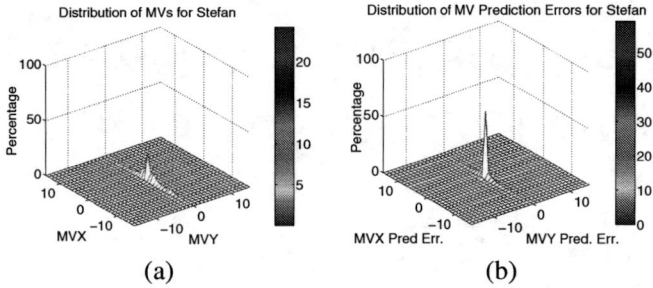

Fig. 1. Statistics of motion for Stefan; (a) MVs; (b) MV prediction errors.

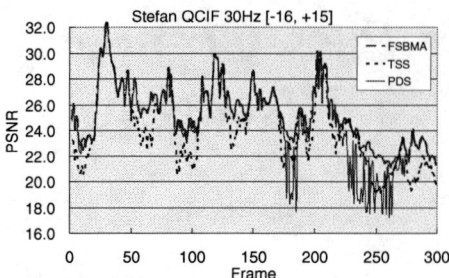

Fig. 2. Comparison of different search patterns.

Foreman and Stefan are videos with large motion, while Silent and Weather are videos with small motion.

A. Motion Vector Predictor

MV predictors exploit the spatial correlation of neighboring MBs. Figure 1(a) and 1(b) show the distribution of MVs and that of MV prediction errors, respectively. FSBMA and the medium prediction from the left, top, and top right MBs are considered in the statistics. The distribution of MV prediction errors is much more concentrated around the origin than that of MVs, and the peak value at the origin increases from 24% to 59%. Starting from MV predictors makes PDS significantly better than DS in convergence speed and video quality.

Supplementary advantage of MV predictors is to support the rate-distortion optimized mode decision [5], known as Lagrangian method. Not only the distortion but also the MV costs are jointly considered in the mode decision. It is reported that 1dB PSNR gain can be achieved. However, in our experiments, we only use sum of absolute differences (SAD) as the matching criterion for generality because MV costs are dependent on entropy coding and quantization parameters.

B. Different Search Patterns

Different search patterns have different merits and thus should be combined into one CA BMA. Figure 2 compares FSBMA, TSS, and PDS. Among all frames, FSBMA gives the best quality (motion compensated PSNR). On average, PDS is better than TSS. However, when the camera pans very fast, TSS is better than PDS. The results are quite reasonable. When the motion field is small and regular, MV predictor works well, and the diamond pattern can quickly find a good match. As for TSS, the first step search points are dispersed, making final results tend to be trapped in local minima. On the contrary, when the motion field is large and complex, MV predictors do not work well, and the diamond pattern moves slowly toward the best MVs with a high probability of being trapped in local minima. In this case, TSS first glances the entire search area and has better chances to focus on the vicinity of global minimum.

C. PDS versus FSBMA

When the allocated computation for an MB has not been used up, a CA BMA will continue. However, if the global minimum distortion has been reached, searching more candidates is a waste. Therefore, there should be some detection to check if the optimal MV is reached for early termination of an MB. Thus, the saved computation can be utilized for later MBs. Table I lists the conditional probabilities of identical MVs between PDS and FSBMA. The smaller the distance from MV predictor to the final MV, the more likely the global distortion minimum is reached. Therefore, the MV differences (MVDs) defined in Table I can be used to skip BMA operations after PDS.

D. TSS versus FSBMA

Table II lists the conditional probabilities of identical MVs between TSS and FSBMA. After the first step search, if the best MV is the origin, it is very possible that the optimal MV will be found. Hence, the best MV right after the first step search can be used to stop the BMA operations after TSS.

E. Summary

The motion analysis is summarized as follows.
- MV predictors can achieve faster speed and better quality.
- PDS is suitable for small and regular motion fields.
- TSS is suitable for large and complex motion fields.
- PDS tends to reach the global minimum distortion when the MV predictor is close to the final MV.
- TSS tends to reach the global minimum distortion when the best MV of the first step is the origin.

TABLE I
PERCENTAGES OF IDENTICAL MVs BETWEEN PDS AND FSBMA.

Sequence	MVD$\leq$0	MVD$\leq$1	MVD$\leq$2	MVD$\leq$3
Foreman	97.41	94.17	79.84	80.27
Silent	99.51	97.66	92.24	91.30
Stefan	97.11	91.49	80.55	79.87
Weather	99.96	99.31	90.44	96.56

MV difference=MVD=$|MVx-MVPx|+|MVy-MVPy|$
MV=(MVx,MVy), MV predictor=MVP=(MVPx,MVPy)

TABLE II
PERCENTAGES OF IDENTICAL MVs BETWEEN TSS AND FSBMA.

Sequence	MV1st==0	MV1st!=0
Foreman	92.76	7.24
Silent	99.17	0.83
Stefan	93.97	6.03
Weather	99.72	0.28

MV1st: best MV after 1st step search

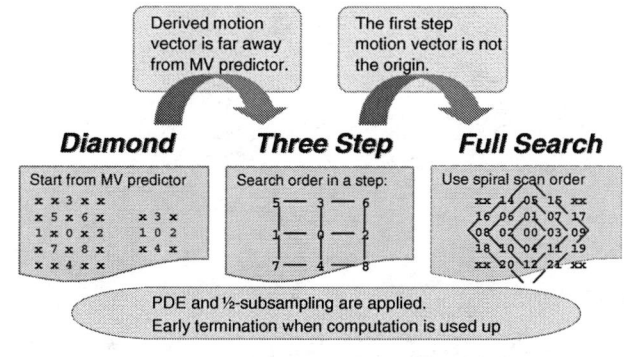

Fig. 3. Proposed adaptive search strategy.

III. PROPOSED ALGORITHM

In this section, our one-pass CA BMA will be proposed in four viewpoints as the following subsections.

A. Adaptive Search Strategy

Figure 3 illustrates our adaptive search strategy. First, PDS is selected as the initial search pattern for an MB. Second, when the PDS ends with available computation left for current MB, the search pattern is switched to TSS. Finally, FSBMA is adopted if TSS is finished with extra computation resource left.

In general, PDS is better than TSS in speed and quality, except for scenes with large and complex motion. In addition, CA DS and CA TSS performs better than CA FSBMA in the computation-distortion (C-D) plots, as stated in [3]. When the BMA is relatively abundant in computation resource, FSBMA still can improve the results. Based on the above reasons, we combine the three search strategies in this way.

As the analysis of Section II summarizes, detection of global minimum is employed. If the final MV of PDS is close to the MV predictor, TSS will not continue. If the best MV of the first step in TSS is the origin, FSBMA will not proceed.

B. Computation Allocation

In [3], the computation pool for the entire frame is determined with the constraints of video smoothness and the computation economy. However, for real-time bidirectional communication applications in which low latency is required, ME must be finished in time for every frame, and the frame computation pool must not exceed the reciprocal of frame rate (e.g. 1/15 sec for 15fps videos). In this paper, we focus on the MB-level computation allocation. The frame computation pool is taken as a given parameter.

Figure 4 is our computation allocation program. The new concept is dividing the computation resource into a base layer (BL) and an enhancement layer (EL). The BL guarantees the least computation for each MB. The EL allows each MB to receive additional computation according to the MB-level adjustment and early stop criteria. As shown in Fig. 4, the target search points per MB (MB_Tar_SPts) and that in BL (MB_Tar_SPts_BL) are user-defined. Afterwards, the frame target search points (FM_Tar_SPts) and that in BL (FM_Tar_SPts_BL) can be obtained from multiplying MB_Tar_SPts and MB_Tar_SPts_BL, respectively, with total number of MBs in one frame (TotalMB). The frame target search points in EL (FM_Tar_SPts_EL) is the result of subtracting FM_Tar_SPts_BL from FM_Tar_SPts.

At the MB-level in Fig. 4, the concept of allocating more resource to MBs with larger distortions is adopted. The average minimum SAD of previous MBs (AvgMinSAD) is obtained as the accumulated minimum SAD (AccMinSAD) divided by the number of processed MBs (DoneMB). The allocated search points for an MB (MB_Alloc_SPts) is MB_Tar_SPts_BL plus the EL part, which is a product of two items. The first item denotes the future average search points per MB in EL, and is the left available computation pool of EL (Left_FM_Tar_SPts_EL) divided by the number of MBs that have not been processed (LeftMB). The second item denotes the ratio of initial distortion of current MB (InitSAD) to AvgMinSAD.

- User definition
  - MB_Tar_SPts
  - MB_Tar_SPts_BL
- Frame level computation allocation
  - FM_Tar_SPts   = MB_Tar_SPts* TotalMB
  - FM_Tar_SPts_BL= MB_Tar_SPts_BL* TotalMB
  - FM_Tar_SPts_EL= FM_Tar_SPts - FM_Tar_SPts_BL
- Macroblock level computation allocation
  - AvgMinSAD = AccMinSAD/ DoneMB
  - MB_Alloc_SPts = MB_Tar_SPts_BL+ (Left_FM_Tar_SPts_EL/ LeftMB) * (InitSAD/ AvgMinSAD)

Fig. 4. Proposed computation allocation.

- Frame layer computation allocation
- Initialize AccMinSAD, DoneMB, LeftMB, and Left_FM_Tar_SPts_EL
- Loop MBs
  - Initial block matching (MV=0) and find InitSAD
  - MB layer computation allocation
  - Block matching motion estimation
    - Adaptive search strategy (PDS, TSS, FSBMA)
    - Terminate when MB_Actual_SPts >= MB_Alloc_SPts
    - Terminate when quasi-optimal MV is found
  - Update AccMinSAD, DoneMB, LeftMB, and Left_FM_Tar_SPts_EL

Fig. 5. Macroblock procedure.

C. Macroblock Procedure

Figure 5 shows the macroblock procedure. The one-pass flow denotes that BMA is processed for MBs one at a time. Before entering the loop of MBs, frame level computation allocation and variable initialization are required. Inside the loop, the first step is to compute the SAD at the origin to find InitSAD for MB layer computation allocation. Then, adaptive search strategy determines the next search points. As long as the number of actual searched points reaches MB_Alloc_SPts, or the quasi-optimal MV (detection of global minimum distortion) is found, the BMA is terminated, and some variables are updated for the next MB.

D. Combination with Traditional Speed-up Methods

For each search point, partial distortion elimination (PDE) is applied to eliminate redundant SAD computation. Besides, 1/2-subsampling is also adopted.

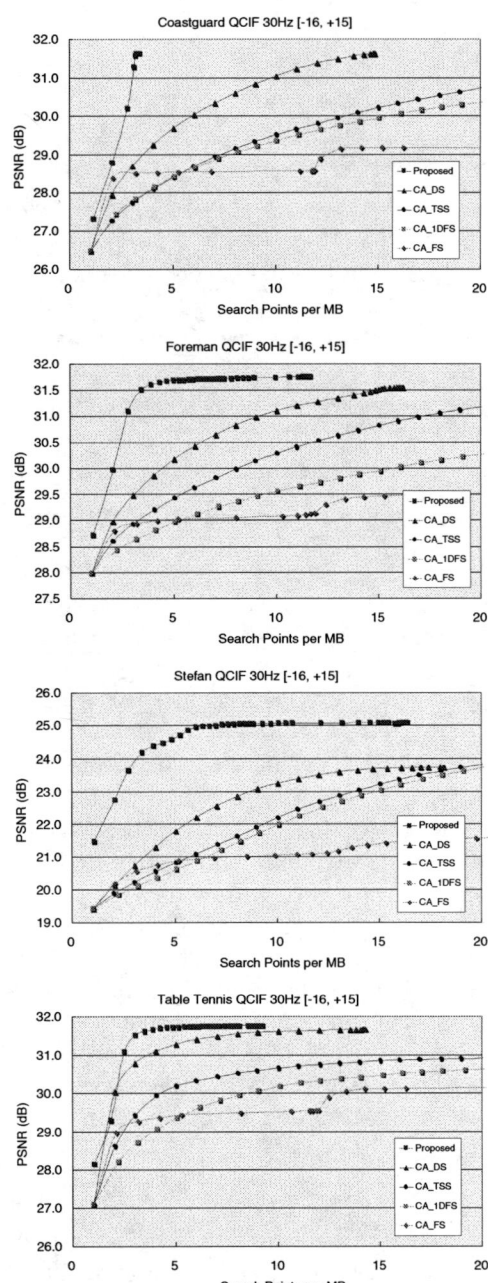

Fig. 6. Computation-distortion curves.

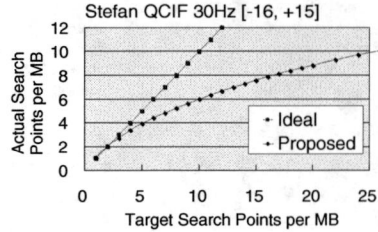

Fig. 7. Capability of proposed computation control.

Therefore, further increasing MB_Tar_SPts will not increase the actual search points. Furthermore, the best video quality of our CA BMA is only 0.1–0.2dB lower than that of CA_FS, and is better than those of remaining CA BMAs. However, this cannot be completely represented by Figure 6 because CA_FS reaches the best quality with many more search points.

Figure 7 shows the capability of the proposed computation control. The number of actual search points is never larger than that of target search points, which meets the real-time constraints. When the computation resource is little, the available computation will be exhausted. When the computation resource is abundant, the resource may not run out due to the detection of global minimum distortion.

In fact, if PDE and 1/2-subsampling are applied to [3], our algorithm cannot win so much, and even a small part of the CA_DS C-D curve may move to the upper left side of the proposed curve. The information of entire frame is indeed good for computation allocation. However, only our one-pass method can be benefited from Lagrangian mode decision, which enhances a lot of quality. Our strength also includes high hardware feasibility and much less memory requirement.

IV. SIMULATION RESULTS

Figure 6 shows the C-D curves of the proposed algorithm and others in [3]. CA_DS, CA_TSS, CA_1DFS, and CA_FS are abbreviated from CA DS, CA TSS, CA one dimensional full search, and CA FSBMA, respectively. Many sequences were tested, but only Coastguard, Foreman, Stefan, and Table Tennis are shown due to the limited space and similar trends of C-D curves. The C-D performance of the proposed algorithm is significantly better than those of others. The average actually used computation of our algorithm cannot exceed a certain value for each sequence because our CA BMA early terminates when detecting that all MBs have reached the optimal MVs.

V. CONCLUSION

We presented a computation-aware motion estimation. The main idea is to convert the processing flow from random access to one-pass for hardware feasibility. Moreover, motion vector predictors and adaptive search strategy can thus be utilized for faster speed and better quality. Detection of global minimum distortion is also proposed to early stop the unnecessary computation. Simulation results show that the provided computation-distortion performance is relatively better.

REFERENCES

[1] T. Koga, K. Iinuma, A. Hirano, Y. Iijima, and T. Ishiguro, "Motion compensated interframe coding for video conferencing," in *Proc. Nat. Telecommun. Conf.*, 1981, pp. C9.6.1–C9.6.5.
[2] S. Zhu and K. K. Ma, "A new diamond search algorithm for fast block matching motion estimation," in *Proc. of IEEE Int. Conf. Image Processing (ICIP'97)*, 1997, pp. 292–296.
[3] P. L. Tsai, S. Y. Huang, C. T. Liu, and J. S. Wang, "Computation-aware scheme for software-based block motion estimation," *IEEE Trans. Circuits and Syst. Video Technol.*, vol. 13, no. 9, pp. 901–913, Sept. 2003.
[4] A. M. Tourapis, O. C. Au, and M. L. Liu, "Highly efficient predictive zonal algorithms for fast block-matching motion estimation," *IEEE Trans. Circuits Syst. Video Technol.*, vol. 12, no. 10, pp. 934–947, Oct. 2002.
[5] T. Wiegand, H. Schwarz, A. Joch, F. Kossentini, and G. J. Sullivan, "Rate-constrained coder control and comparison of video coding standards," *IEEE Transactions on Circuits and Systems for Video Technology*, vol. 13, no. 7, pp. 688–703, July 2003.

A FRAMEWORK FOR FINE-GRANULAR COMPUTATIONAL-COMPLEXITY SCALABLE MOTION ESTIMATION

Zhi Yang [*], *Hua Cai* [†], *and Jiang Li* [†]

[*] College of Computer Science and Technology, Zhejiang University, Hang Zhou, China
[†] Media Communication Group, Microsoft Research Asia, Beijing, China

ABSTRACT

This paper presents a novel motion estimation (ME) framework that offers fine-granular computational-complexity scalability. In the proposed framework, the ME process is first partitioned into multiple search passes. A priority function is used to represent the distortion reduction efficiency of each pass. According to the predicted priority of each macroblock (MB), computational resources are then allocated effectively in a progressive way to achieve fine-granular computational-complexity scalability. Experiments show that our proposed scheme achieves progressively improved performance over a wide range of computational capabilities.

1. INTRODUCTION

With the rapid development of wired and wireless networks, more and more users are seeking real-time video communication services. However, real-time video coding faces a big challenge from computational complexity, especially for mobile devices such as Pocket PCs and handheld PCs, which are of weak computational capability and short battery lifetime. Because of the complexity constraint, many highly efficient but complex algorithms cannot be used directly for real-time video coding. Although one can simplify the algorithms to meet a specific scenario (e.g., a given video resolution and bit rate for a certain device), it is not a cost-effective way since there are so many different scenarios. Also, conventional encoders cannot adapt well to the varying computational requirements of video contents. Therefore, it is highly desirable to have a computational-complexity scalable video encoder that can offer a trade-off between coding efficiency and the embedded available computational performance.

In video coding systems motion estimation (ME) plays a key role in removing temporal correlation between video pictures. All the video standards that have so far emerged, such as H.263 and MPEG-1/2/4, are based on the ME in the encoding loop [1]. Meanwhile, ME is a very critical module of the encoder since it consumes most of the computing time. There are significant advances in fast ME techniques in recent years for alleviating the heavy computation load, such as the new three-step search (NTSS) [2], the diamond search (DS) [3], the circular zonal search (CZS) [4], and the predictive algorithm (PA) [5]. However, despite the significant speedups, ME still consumes the largest amount of computational resources, especially in real-time video encoding.

Computational-complexity scalable ME has been studied to further reduce the complexity of fast ME [6][7]. It also provides a proper trade-off between motion accuracy and time consumption such that it can adapt to the available computational resources dynamically. In Lengwehasatit's method [6], a partial-distance metric is used within the motion search process to eliminate unlikely candidates through a thresholding process that enables computation scalability. And in Mietens' method [7], complexity scalability is obtained by scaling the number of the processed motion-vector (MV) fields and the number of vector evaluations.

Different from previous works, in this paper, we present a novel ME framework that offers fine-granular computational-complexity scalability [1]. In the proposed framework, the ME process is first partitioned into multiple search passes. A priority function is used to represent the distortion reduction efficiency of each pass. According to the predicted priority of each MB, computational resources are then allocated effectively in a progressive way. As a result, the ME process can be stopped at any time with a progressively improved performance, and thus scalability is achievable. Furthermore, the proposed scheme can be easily integrated with many existing fast ME algorithms, such as NTSS [2], DS [3], CZS [4], and PA [5].

The rest of the paper is organized as follows. The new ME framework is presented in Section 2. In Section 3, the prediction of the priority for each search pass is discussed. Experimental results are shown in Section 4 and conclusions are drawn in Section 5.

2. PROPOSED ME FRAMEWORK

In the ME process, a recursive temporal prediction loop is employed to find the best MV that optimizes the rate-distortion performance. Usually, the prediction loop works as follows: for a given starting point, first check a number of candidate MVs and then select one MV from the candidates as the new starting point. The loop will repeat until either all of the candidate MVs have been checked or the stop condition is satisfied. Therefore, the ME process of a certain MB (say, the ith MB) can be naturally partitioned into multiple search passes: $Pass_i(1)$, $Pass_i(2)$, ..., and so on. The jth pass of the ith MB, i.e. $Pass_i(j)$, would simply check $N_i(j)$ candidate MVs and determine the new starting point for $Pass_i(j+1)$. Fig. 1 illustrates an example of pass partitioning for the DS algorithm [3], where the ME process is partitioned into five passes, each having 1, 8, 5, 3, and 4 candidate MVs respectively.

After pass partitioning, the MV prediction of a certain MB is deployed in a progressive way. If all passes are searched, the encoder will get the best MV that is equivalent to that of the conven-

---

[*]The work presented in this paper was carried out in Microsoft Research Asia.

[1]To simplify terminology, the word *scalability* refers to *fine-granular computational-complexity scalability* hereinafter.

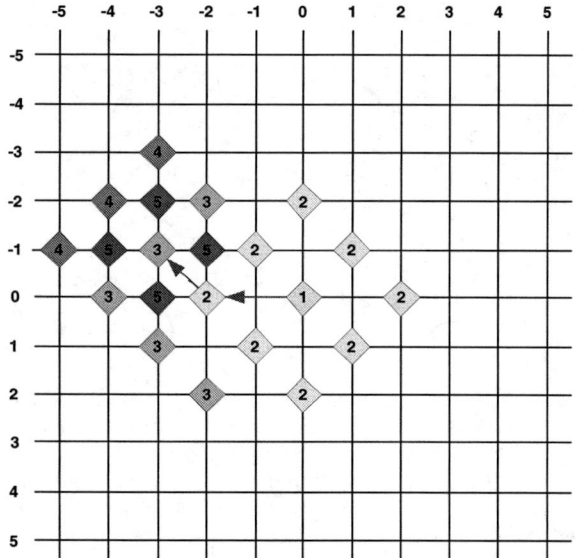

Figure 1: An example of pass partitioning for the DS algorithm. Numbered points are those candidate MVs that will be checked in the corresponding pass.

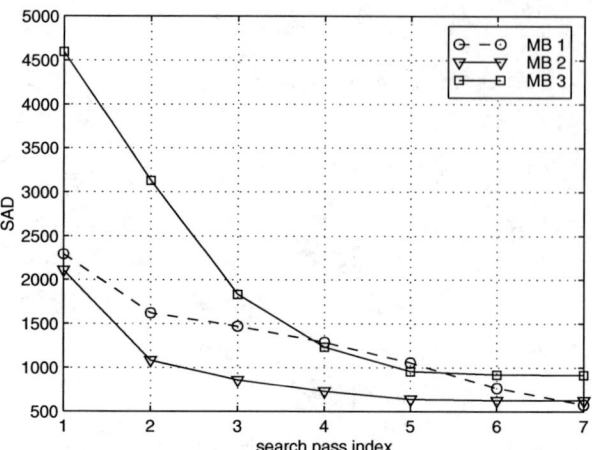

Figure 2: Distortion vs. number of passes (obtained with DS algorithm for the *Foreman* (CIF) sequence).

tional ME schemes. Meanwhile, the encoder also has the freedom to stop the MV prediction at any of these passes. Moreover, by selecting passes among a frame of MBs, scalability is achievable.

The simplest way to attain scalability is to select passes according to their indexes, that is, from the first pass of all MBs all the way down to the last pass of all MBs. As a result, all MBs have the same chance to refine their own MVs. However, its performance is not good since the search efficiencies of different passes, which are of the same pass index but from different MBs, might be quite different. Fig. 2 clearly demonstrates that different MBs might have quite different search efficiencies.

Instead of uniformly allocating computational resources to each MB, in our proposed framework, a more sophisticated approach is considered based on the priority of each pass. We measure the search priority of $Pass_i(j)$ by the reduced distortion per computing time:

$$P_i(j) = \frac{\Delta D_i(j)}{N_i(j)} = \frac{\Delta \text{SAD}_i(j)}{N_i(j)} \quad (1)$$

where $\Delta D_i(j)$ and $N_i(j)$ denote the distortion reduction and the number of checked MVs of $Pass_i(j)$ respectively. Particularly, we use the sum of absolute differences (SAD) in this paper to measure the distortion, and $\Delta \text{SAD}_i(j)$ is the reduced SAD of the ith MB after performing $Pass_i(j)$. Furthermore, we use $N_i(j)$ to represent the consumed time since in most ME algorithms the computing time for each candidate MV is invariant (i.e., the distortion calculating is of constant complexity). Also note that only the predicted value of $P_i(j)$, denoted as $\hat{P}_i(j)$, can be obtained in practical systems since $\Delta D_i(j)$ cannot be calculated without finishing $Pass_i(j)$. Details of the priority prediction will be discussed in Section 3.

Now the computational resources can be allocated as follows: a priority table which contains N priority elements is first created for a video frame consisting of N MBs. Each priority element represents the predicted priority of the current unperformed pass for a certain MB. Next, the MB which has the largest priority is selected for one pass of ME. After finishing that pass, the new priority of its next pass is then predicted and the corresponding priority element is updated as well. This resource allocation process continues recursively until either the available computing time is consumed or there is no new pass left for any of the N MBs.

Ideally, the highest resource utilizing efficiency can be attained by the above method if the priority can be obtained accurately and the priority function for any MB is convex. However, the priority function is not always convex in practice. To make our algorithm more robust, we also take into account the non-convex case when predicting the priority.

On the other hand, it is critical to limit the implementation complexity for both priority prediction and table maintenance. In particular, we have observed that the most time consuming work for the table maintenance is to find the largest priority element from the table, especially for a large table. In one of our implementations, we separate the priority table into several sub-tables with different priority ranges, and all the MBs belonging to the largest prioritized sub-table are searched at every resource allocation step. Doing this can significantly reduce complexity, only with a slight performance loss.

3. PRIORITY PREDICTION

We have discussed a new ME framework in Section 2 that offers scalability. By smartly determining the priority of each ME search pass, an encoder can adapt to the limited/varying computational resources with efficient resource utilization. In this section, we present an effective and robust priority prediction method which does not require complex computation.

Usually, in fast ME algorithms, prior search passes are more likely to catch the optimal MV than the subsequent passes. This is the main reason that fast ME algorithms can dramatically reduce complexity. It also implies that the distortion reduction efficiency of a prior pass is usually greater than that of the subsequent ones. In other words, from the statistical point of view, the distortion reduction function for a ME process has a decreasing slope. Based

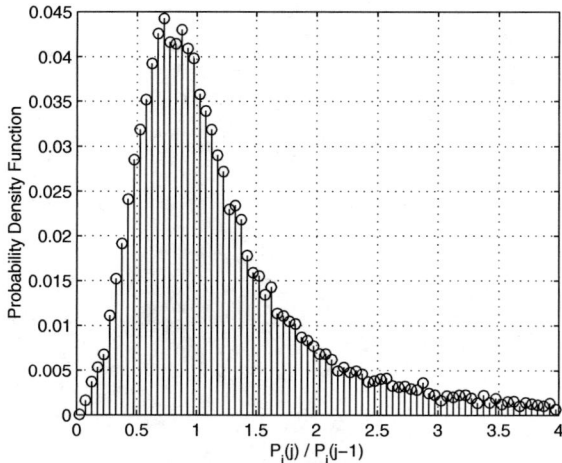

Figure 3: Probability density function of $P_i(j)/P_i(j-1)$.

on this assumption, the priority of $Pass_i(j)$ can be predicted as:

$$\hat{P}_i(j) = \begin{cases} \infty, & \text{if } j = 1 \\ \alpha \text{SAD}_i(j-1)/N_i(j), & \text{if } j = 2 \\ min\left(\beta P_i(j-1), \alpha \text{SAD}_i(j-1)/N_i(j)\right), & \text{else.} \end{cases} \quad (2)$$

In the above, only $\Delta D_i(j)$ needs to be predicted whereas $N_i(j)$ is explicit before performing $Pass_i(j)$. The priority is first set to infinity because we have no knowledge about each MB at the first pass of the ME process. Then, the predicted distortion reduction of the second pass is simply obtained from the current SAD scaled down by a pre-determined factor α. As for the remaining passes, priority $\hat{P}_i(j)$ is related to the actual priority of the previous pass, $P_i(j-1)$, scaled down by a pre-determined factor β. Meanwhile, a minimum value is chosen to ensure that $\hat{P}_i(j)$ is always within a reasonable range.

Note that in Eqn. (2) we assume the distortion reduction efficiency is convex for any ME process. However, it is not always convex in practice. Although the above method might still be robust for many non-convex cases where there is not any inefficient pass(es) (which has very small priority) before efficient pass(es), it cannot handle an inefficient pass which will block the search process. To make our algorithm more robust, we check the priority with another item:

$$\hat{P}_i^*(j) = max\left(\hat{P}_i(j), \frac{SAD_i(j-1) \cdot \gamma^K}{N_i(j)}\right) \quad (3)$$

where K denotes the number of consecutive inefficient passes prior to the current pass; and γ is a pre-determined scaling factor. From above equation, if consecutive inefficient passes are detected for a certain MB, it is believed that a global or near-global minimum is found, and thus low priority should be set for that MB.

From our testing, we observed that the performance is not very sensitive to the choice of parameters α, β, and γ. Hence we just set α and γ to 1/4 and 1/8 respectively throughout our experiments. As for β, its value can be calculated by averaging over the probability density function (PDF) of $H\left(\frac{P_i(j)}{P_i(j-1)}\right)$:

$$\beta = \int_0^{+\infty} x \times H(x) dx. \quad (4)$$

To improve robustness, only large $P_i(j)$ is used for collecting the PDF. Moreover, β can be either calculated beforehand or updated frame by frame. Fig. 3 shows one example of the PDF obtained from the *Foreman* (CIF) sequence. The corresponding β equals 0.95, which is used throughout our experiments. It can also be seen from Fig. 3 that the non-convex case (i.e. $P_i(j)/P_i(j-1) > 1$) happens very often.

4. EXPERIMENTAL RESULTS

Many experiments have been performed to evaluate extensively the performance of our proposed framework. The standard test sequences *Foreman*, *Carphone*, and *News* of CIF resolution at 30 fps are used as our test set. Only the first frame is encoded as an **I** frame and all others as **P** frames.

We implemented our framework upon the MPEG-4 encoder [8]. The DS algorithm [3] is used for ME with integer motion accuracy. The parameters α, β, and γ are invariant throughout the experiments. The priority table is split into 16 sub-tables as discussed in Section 2 in order to reduce the complexity of table maintenance. To better evaluate the proposed framework, two other schemes are also implemented as benchmarks for comparison. In the first benchmark (named '*Benchmark 1*') the computational resources are uniformly allocated to different MBs. In the second benchmark (named '*Benchmark 2*'), we assume that the priority is known beforehand. It serves as a performance upper bound of our framework since the priority has to be predicted in practice (although it can only achieve sub-optimal performance due to the non-convex cases, we still treat it as the upper bound).

We first evaluate the resulting SAD for a randomly selected frame after checking a limited number of MVs. It is clear from Fig. 4 that, without considering the priority, the performance (achieved by 'Benchmark 1') is bad. On the contrary, our proposed framework achieves very good performance which is also close to the upper bound (i.e. 'Benchmark 2'). Similar performance is also observed from other frames.

We then evaluate the PSNR values of different sequences under a rate of 1024 kbps. As shown in Fig. 5, the simple 'Benchmark 1' scheme may suffer more than one dB loss compared with the performance upper bound. But our proposed framework significantly reduces the big gap, especially for the *News* sequence where our performance is very close to the upper bound. This demonstrates the effectiveness of our priority prediction approach used in the framework. It can also be seen that, by stopping MV prediction at any point on these curves, the encoder can easily adapt to the limited/varying computational resources only with a slight performance loss. This indicates that good embedded available computational performance is achievable.

5. CONCLUSIONS AND DISCUSSIONS

Computational-complexity scalability is an important yet practical topic in real-time video encoding. This paper presents a novel ME framework that offers fine-granular computational-complexity scalability. By partitioning the ME process into multiple search passes and prioritizing each pass according to its search efficiency, computational resources can be efficiently allocated in a progressive manner. Thus good embedded available computational performance is achievable. Good results have been observed in our experiments.

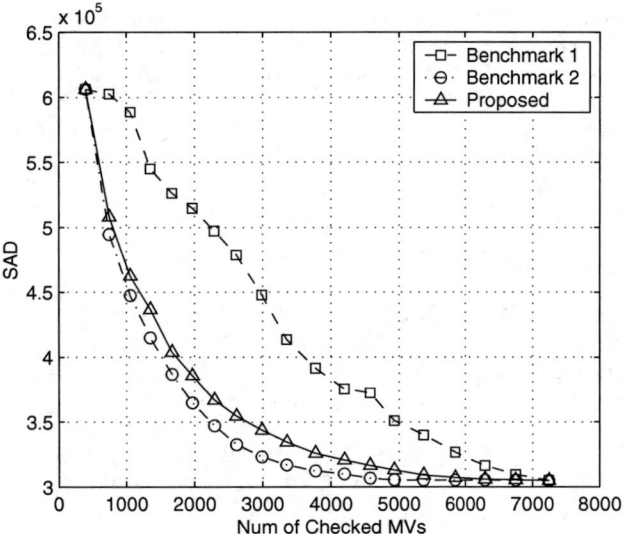

Figure 4: SAD vs. number of checked MVs (obtained from the 71th frame of the *Foreman* sequence).

Many of the popular fast ME algorithms can be integrated with the proposed framework, as long as they can be partitioned into multiple search passes. Also, we believe that the performance of the new framework could still be improved by using a better priority prediction approach.

6. REFERENCES

[1] V. Bhaskaran and K. Konstantinides, *Image and video compression standards – algorithms and architectures*, Kluwer Academic Publishers, second edition, 1997.

[2] R. Li, B. Zeng, and M.L. Liou, "A new three-step search algorithm for block motion estimation," *IEEE Transactions on Circuits and Systems for Video Technology*, vol. 4, no. 4, pp. 438–442, Aug. 1994.

[3] S. Zhu and K.K. Ma, "A new diamond search algorithm for fast block-matching motion estimation," *IEEE Transactions on Image Processing*, vol. 9, no. 2, pp. 287–290, Feb. 2000.

[4] A.M. Tourapis, O.C. Au, and M.L. Liou, "Highly efficient predictive zonal algorithms for fast block-matching motion estimation," *IEEE Transactions on Circuits and Systems for Video Technology*, vol. 12, no. 10, pp. 934–947, Oct. 2002.

[5] A. Chimienti, C. Ferraris, and D. Pau, "A complexity-bounded motion estimation algorithm," *IEEE Transactions on Image Processing*, vol. 11, no. 4, pp. 387–392, Apr. 2002.

[6] K. Lengwehasatit and A. Ortega, "Computationally scalable partial distance based fast search motion estimation," *Proc. ICIP'2000*, vol. 1, pp. 824–827, Sept. 2000.

[7] S. Mietens, P.H.N. de With, and C. Hentschel, "Computational-complexity scalable motion estimation for mobile MPEG encoding," *IEEE Transactions on Consumer Electronics*, vol. 50, no. 1, pp. 281–291, Feb. 2004.

[8] Microsoft Corporation, "ISO/IEC 14496 (MPEG-4) Video Reference Software, Version 2.2.0," July 2000.

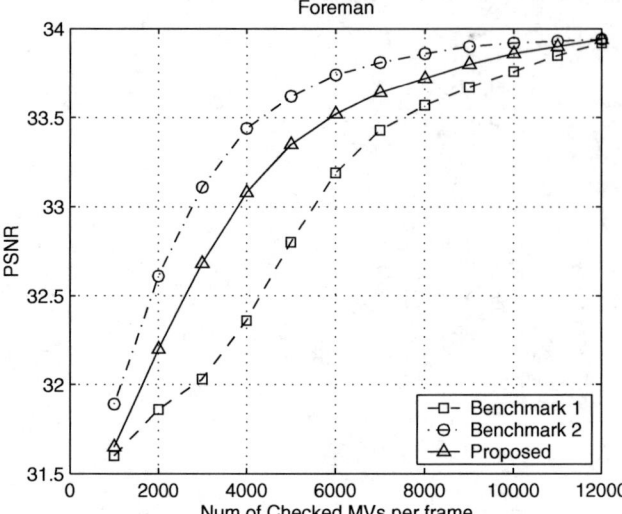

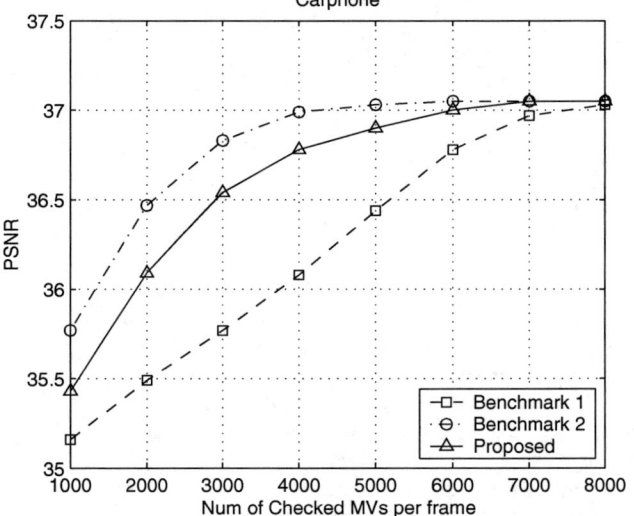

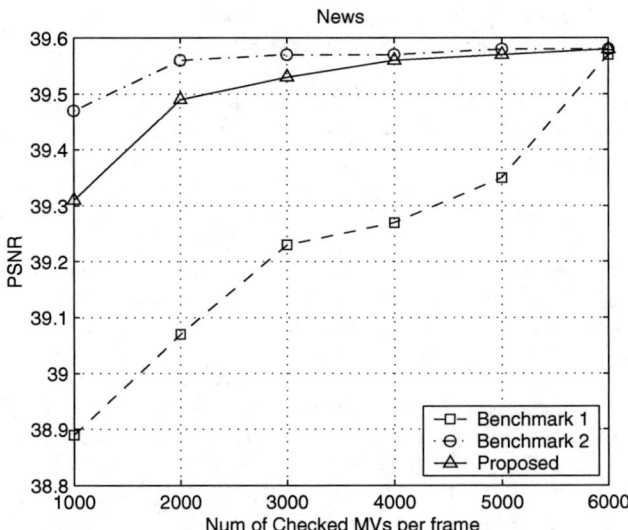

Figure 5: Comparative evaluation of the proposed framework.

Fast Sub-Pixel Inter-Prediction – Based on Texture Direction Analysis (FSIP-BTDA)

Hoi-Ming Wong, Oscar C. Au, Andy Chang
The Hong Kong University of Science and Technology, Hong Kong
Email: hoimingw@ust.hk, eeau@ust.hk, eecax@ust.hk

Abstract—*Motion Estimation (ME) is a core part of modern video coding systems to exploit temporary redundancy in a video. Motion Estimation is typically performed firstly with integer pixel accuracy and then at sub-pixel accuracy, which includes half-pixel and quarter-pixel accuracy. When sophisticated fast integer-pixel accuracy motion-search algorithms are used to decrease the number of search points for integer-pixel motion search, thus sub-pixel motion search became another important processing bottleneck in the encoding process. The conventional method is to search 8 half-pixel positions around the Motion Vector (MV) obtained from Integer pixel Motion Search, then do motion search in the same way on 8 quarter-pixel positions around the MV obtained from half-pixel motion search, therefore totally 16 search points are needed. The proposed algorithm, Fast Sub-Pixel Inter-Prediction – Based on Texture Direction Analysis (FSIP-BTDA), successfully optimizes the sub-pixel motion search on both half and quarter-pixel accuracy, and improves the processing speed with low PSNR penalty.*

1. INTRODUCTION

In modern video coding standard such as H.261/3/4[2] and MPEG-1/2/4, block matching motion estimation (BMME) and compensation is done to exploit the temporal correlation and to reduce the redundancy between adjacent frames in video sequences so as to achieve high compression efficiency. In the latest standard H.264[2], BMME is performed in quarter-pixel accuracy rather than in integer-pixel as in older standards. Quarter-pixel accuracy is especially useful for low bit-rate video coding. The most common distortion measure used in BMME is the sum of absolute difference (SAD), which requires no multiplication and can achieve similar performance as the mean square error (MSE).

Sub-pixel motion estimation and compensation is useful if an object in a video does not move in integer-pixel displacement. If it moves to some half pixel locations which is quite probable, half-pixel motion estimation and compensation would be needed. The latest video coding standard, H.264, performs ME on quarter-pixel accuracy. In the conventional brute force quarter-pixel search, 8 quarter-pixel positions are searched around a selected half-pixel position with lowest SAD for each search block. For a frame of size P x Q and a frame rate of T fps, the amount of computation (addition) for full quarter-pixel search (FQPS) is:

$$8*T*[(P \times Q)/(N*N)]*(2N^2-1) \text{ op/sec}$$

For T = 30, P = 352, and Q = 288 which is CIF format, this amounts to 4.8×10^7 operation/s. Note that this has not included the computation needed to compute the sub-pixel values. FQPS can consume up to 30% of the total computational power of an encoder when sophisticated fast integer-pixel motion search such as PMVFAST is used.

Most of existing fast sub-pixel motion search algorithms such as PPHS[1], 2SS[4] and SQIA[7] are focused only on half-pixel or quarter-pixel motion search, but not combine them together. In this paper, we propose a novel fast sub-pixel search algorithm called Fast Sub-Pixel Inter-Prediction – Based on Texture Direction Analysis (FSIP-BTDA), which reduces the total number of search points for both half and quarter-pixel motion search with little PSNR penalty by analyzing the characteristic of the current block's texture.

In section 2 we will first introduce the FQPS algorithm. In section 3 we will discuss the algorithm we chose for integer-pixel motion search, which is performed prior to the proposed FSIP-BTDA. The typical texture direction characteristic of a video will be discussed in section 4. We will introduce each step of FSIP-BTDA in section 5, section 6 will shows some mathematic analysis and simulation result. Section 7 is a brief conclusion.

2. FULL QUARTER-PIXEL SEARCH (FQPS)

Let I be the integer-pixel position obtained from integer-pixel motion search. Let h1, h2, h3, ..., h8 be the eight half-pixel positions around I. FQPS computes the SAD for all these 8 positions as shown in Fig. 1 and find the best one, say h2, with lowest SAD. Then it computes the SAD for the eight quarter-pixel positions q1, q2, q3, ..., q8 around the best half-pixel location. Finally, the position, say q8, with lowest SAD is chosen.

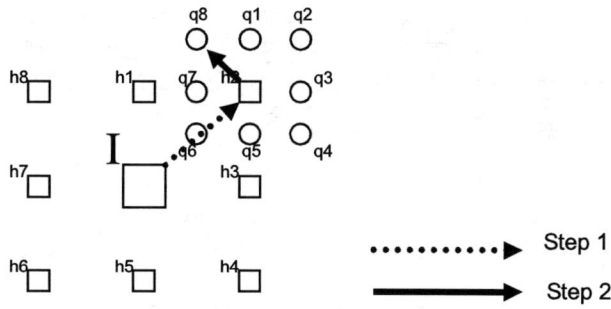

Fig. 1 An illustration of FQPS

3. REQUIREMENT OF INTEGER-PIXEL MOTION SEARCH ALGORITHMS

The proposed FSIP-BTDA algorithm needs information from integer-pixel motion estimation (IME). The performance of IME can affect the performance of FSIP-BTDA. In this paper, we use PMVFAST for IME because the resulting motion field tends to be smoother and less noisy than brute force integer-pixel full search with essentially similar perceptual quality. Also, unlike other sub-pixel motion search algorithms, FSIP-BTDA requires only the final integer motion vector (MV) for each search block. Thus full search is not necessary for IME. Fig. 2(a) shows the motion field of full search which is chaotic, and Fig. 2(b) shows the motion field of PMVFAST [3] which is smoother and more reasonable.[5]

Fig. 2(a) and (b) Motion Field of Full Search and PMVFAST

4. MODELING SUB-PIXEL MOTIONS SEARCH USING TEXTURE DIRECTION CHARACTERISTIC

Let the MV_{refine} be the refined MV of sub-pixel motion search after the IME, and MV_{final} be the final MV after performing sub-pixel motion search, i.e.

$$MV_{refine} = MV_{final} - MV_{IME}$$

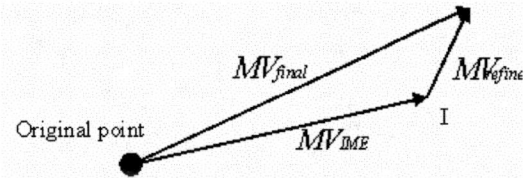

Fig. 3 An illustration for $MV_{refine} = MV_{final} - MV_{IME}$

We cannot predict MV_{refine} with MV_{IME} only, so we need to look for other information for fast sub-pixel motion search. Typical video sequence contains a lot of edges, within a single macro-block these edges can be assumed to be parallel as shown in Fig.4. We can look at the relationship between the direction of the edges and MV_{refine}. Since the variance of pixels along the edge is much smaller than the pixels perpendicular to the edge as shown in Fig. 4, so the SAD difference for the former direction will be much larger than the latter one. The Delta-SAD surface can be modeled as a rotated parabolic function[1][4]:

$$\Delta_{SAD}(x' - x'_p, y' - y'_p) = A(x' - x'_p)^2 + B(y' - y'_p)^2 \quad (1)$$

Where

1. (x_p, y_p) is the matched point by FQPS
2. $(x', y') = RightRotation(x, y, \theta)$
3. **B >> A**

Where θ is the incline angel of the direction of the edges. As mentioned before, since the SAD increase much slower in the direction parallel to the edge than that perpendicular to the edges we can assume **B>>A**. The RightRotation function transforms x and y so that the first term of equation (1) represents the SAD change parallel to the edges, and the second term represents the SAD change perpendicular to the edges. Fig. 5 shows a typical SAD error surface. From the above observation, the sub-pixel motion is biased towards the direction perpendicular to the edges in order to reduce the total number of search points.

In some cases, no matter how to search to sub-pixels, it is impossible to find a good match. E.g. if the object in the referenced frame is at full-pixel position, but it is at half pixel position in current frame, the sub-pixel motion search can easily find the good match from referenced frame by interpolating half-pixel points. In contrast, if the object in the referenced frame is at half-pixel position but at Integer-pixel position in current frame, then the good match will never be found. By this observation and experiments, about half of the sub-pixel motion search can get good prediction.[8]

Fig. 4 A block in foreman sequence (CIF)

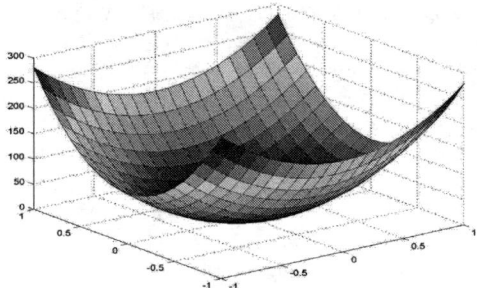

Fig. 5 A typical error surface

5. FAST SUB-PIXEL INTER-PREDICTION – BASED ON TEXTURE DIRECTION ANALYSIS (FSIP-BTDA)

Based on directional characteristic of a video, FSIP-BTDA predicts the direction of the texture and performs sub-pixel motion search on the direction predicted. In reality, direction is hard to be predicted. Instead of predicting the direction precisely prior to searching, FSIP-BTDA performs prediction and searching at the same time. For convenience, all half or quarter-pixel positions will be called as quarter-pixel positions in this section.

Step 1:

Fig. 6(a) shows a part of the sub-pixel interpolation plane. The circles are Integer-pixel positions while the one which at the middle with "I" is the Integer-pixel positions that return from IME. All crosses are half or quarter-pixel positions. The first step of FSIP-BTDA is to search 4 quarter-pixel positions around the Integer-pixel position "I". Now FSIP-BTDA is going to decide which direction it will search. Suppose the square with "L" is the quarter-pixel position with the lowest SAD amount all searched points, and the quarter-pixel point with "1" has the lower SAD than that with "2". After step 1, the gray arrow indicates the direction for next step to search. Anyway, if all SADs of these 4 quarter-pixel points are larger than the "I", then the search will terminate and return the "I", i.e. $MV_{final} = MV_{IME}$

Step 2:

According to the direction predicted in step 1, FSIP-BTDA searches point "A" and "B". If "A" or "B" has the smallest SAD amount all searched points, then go on to search along the upper arrow or lower arrow, otherwise, FSIP-BTDA will terminate and return the positions that with lowest SAD. Fig. 6(b) shows this step.

Step 3:

According to Fig. 6(b) and Step 2, if the point "A" has the lowest SAD, then search "b", "c" and "d", otherwise search "a" and "b". And then the FSIP-BTDA terminates. For each step, A SAD threshold T can be applied to early terminating, i.e. if Current lowest SAD < T, then terminate. The threshold for FSIP-BTDA is:

$$T = AverageNeighborSAD + 128 \quad (2)$$

Where the AverageNeighborSAD is the average SAD value for the block that on the left, top and top-right position of the current block.

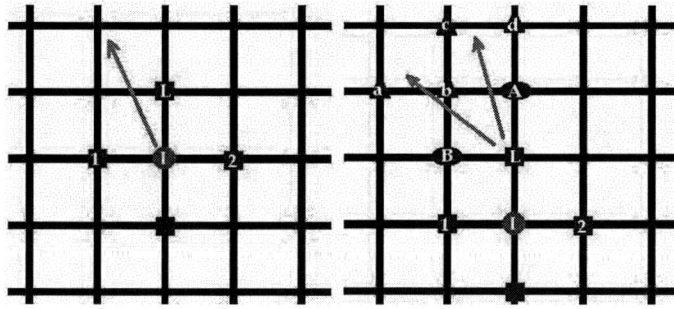

Fig. 6(a) The step 1   Fig. 6(b) The step 2 and 3

6. MATHEMATICAL ANALYSIS AND SIMULATION RESULT FOR FSIP-BTDA

6.1 Mathematical Analysis

Fig. 7 shows part of the interpolation frame, point "I" is the Integer-pixel points that return from IME. The FSIP-BTDA will search perpendicularly to the edges, so FSIP-BTDA and FQPS will reach the same perpendicular location to the edges. As a result, $y = y_p$, and equation (1) will become:

$$\Delta_{SAD}(\Delta_x) \cong A(\Delta_x)^2 = (\sqrt{A}\Delta_x)^2 \quad (3)$$

Where $\Delta_x = x - x_p$. Line P-F represents equation (3), "P" is the point with lowest SAD, which will be return by the FQPS. The FSIP-BTDA algorithm will reach the point "F" instead of "P". Most likely, the SAD returned from FSIP-BTDA will be bigger than the FQPS, but the number of bits to encode FSIP-BTDA MV is smaller due to the smooth motion field of FSIP-BTDA. Therefore the total cost will be similar to the FQPS while the number of search points is reduced by a large factor.

The incline angel of the edges' direction is equally distributed from 0 to 2π, now we only consider the horizontal case that is $\theta = 0$. Let ϕ be the angel of line P-I and the direction of the edge P-F. The first movement of FSIP-BTDA is always towards the points "P" because FSIP-BTDA always searches 4 quarter-pixel points before search along the direction predicted, let τ be the direction predicted by FSIP-BTDA. When $\phi = \pi/2$, the MV of FQPS is exactly same as the MV of FSIP-BTDA so as their SAD. The expected value of Δ_{SAD} per pixel can be expressed in terms of r and ϕ, where $r = |MV_{refine}|$, i.e.

$$E[\Delta_{SAD}(r,\phi)] \cong \frac{1}{R}\frac{4}{2\pi}\int_0^{\pi/2}\int_0^R \Delta_{SAD}(r,\phi)drd\phi \quad (4)$$

Where $\quad \Delta_{SAD}(r,\phi) = \{\sqrt{A}[r-1]\cos(\phi)\}^2 \quad (5)$

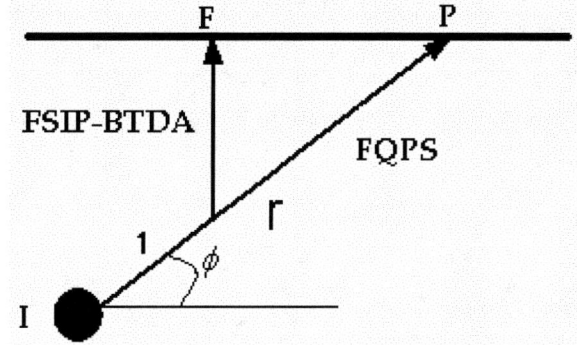

Fig. 7 MV_{ref} of FSIP-BTDA and FQPS

After solving the equation, we got:

$$E[\Delta_{SAD}(r,\phi)] \cong \frac{A}{2}\left(\frac{R^2}{3} - R + 1\right) \quad (6)$$

From experiment simulation, less than 2% of FQPS MVs go beyond 3, so we can set R = 3 to be the search range of FSIP-BTDA. When R = 3, $E[\Delta_{SAD}(r,\phi)] \cong 0.5A$, from section 4, on average sub-pixel motion search can get half benefit on SAD gain[8], and by experiment, $E[A] \approx 1$, so the approximated SAD increase comparing to FQPS per pixel is 0.5A/2 = 0.25.

6.2 Simulation Result

The proposed FSIP-BTDA algorithm was embedded into the H.264 JM8.2 encoder [6] and was tested using Intel Pentium 4 PC with Windows XP. Table 2 shows the comparisons between the proposed FSIP-BTDA, FQPS and 2SS in speed up factor and their corresponding PSNR for various video testing sequences. PPHS and SQIA are not chosen because they are focused on only half-pixel or quarter-pixel motion search, but not both. Simulation results show the proposed FSIP-BTDA algorithm performs similarly well in each sequences with different bitrate, because for low bitrate testing, the MV cost saving in FSIP-BTDA in obvious, and for high bitrate testing, the direction of the texture became more clear so that FSIP-BTDA achieves high accuracy in direction prediction. Compared to FQPS, FSIP-BTDA achieves a speed up factor ranging from 2.48 to 3.79, which is significantly higher than 2SS. The average PSNR loss of FSIP-BTDA compared with FQPS is 0.05969DB which is slightly better than 0.07569DB in 2SS. The Full simulation result is shown in table 1, 2 and 3.

7. CONCLUSION

In this paper we proposed a novel fast sub-pixel motion search algorithm called Fast Sub-Pixel Inter-Prediction – Based on

Texture Direction Analysis (FSIP-BTDA). The algorithm performs half and quarter-pixel motion search by predicting the direction of the texture in a video. it is easy for implementation and has very small overhead because it just compares the SADs to predict the direction rather than computing complex formula, and we can easily extend it to 1/8 or higher accuracy of motion estimation by slightly modifying the algorithm.

8. ACKNOWLEDGEMENT

This work has been supported in part by the RGC grant HKUST6203/02E, and the ITF grant ITS/122/03, and the grant DAG03/04.EG10 of the Hong Kong Special Administrative Region, China.

9. REFERENCES

[1] Cheng Du, Yun He, and Junli Zheng, "PPHS: A Parabolic Prediction-Based, Fast Half-Pixel Search Algorithm for Very Low Bit-Rate Moving-Picture Coding", *IEEE Transactions on Circuits and Systems for Video Technology,* Vol. 13, No.6, June 2003.

[2] (2003) Study of Final Committee Draft of Joint Video Specification (JTU-T Rec. 11.264, ISO/IF C 14 496-10 AVG). *ITU-T Video Coding Experts Group and ISO/IEC Moving Picture Experts Group.*

[3] A.M. Tourapis, O.C. Au, M.L. Liou, "Predictive Motion Vector Field Adaptive Search Technique (PMVFAST) Enhancing Block Based Motion Estimation", *Proc. Of SPIE Conf. On Visual Communication and Image Processing,* Vol.4310, pp.883-892, Jan 2001.

[4] Bo Zhou; Jian Chen; A fast two-step search algorithm for half-pixel motion estimation, Electronics, *Circuits and Systems, 2003. ICECS 2003. Proceedings of the 2003 10th, IEEE International Conference on,* Volume: 2, 14-17 Dec. 2003, Pages:611 - 614 Vol.2.

[5] http://www.ee.ust.hk/~eeau/demo_fastMotion.htm.

[6] Joint Video Team (JVT), International Telecommunication Union.

[7] Hoi-Ming Wong, Oscar C. Au, Jinxin Huang, Shiju Zhang, Winnie N. Yan, Sub-Optimal Quarter-Pixel Inter-Prediction Algorithm (SQIA), Hong Kong University of Science & technology, Hong Kong, submitted to *ICASSP2005.*

[8] A. Chang, Oscar C. Au, Y. M. Yeung, "A novel approach to fast multi-frame selection for H.264 video coding", *Circuits and Systems, 2003. ISCAS '03. Proceedings of the 2003 International Symposium on,* Volume: 2, 25-28 May 2003 Pages:II-704 - II-707 vol.2.

Sequences		64kbps			128kbps			192kbps			256kbps			320kbps		
		FQPS	2SS	FSIP-BTDA	FQPS	2SS	FSIP-BTDA	FQPS	2SS	FSIP-BTDA	FQPS	2SS	FSIP-BTDA	FQPS	2SS	FSIP-BTDA
News	PSNR	30.7	30.66	30.63	34.07	34.04	34.04	36.12	36.06	36.02	37.45	37.39	37.36	38.45	38.4	38.37
	Speed up	1	2	3.56	1	2	3.4	1	2	3.29	1	2	3.23	1	2	3.19
Weather	PSNR	27.89	27.81	27.8	32.69	32.57	32.52	34.95	34.87	34.83	36.43	36.34	36.31	37.58	37.49	37.43
	Speed up	1	2	3.74	1	2	3.51	1	2	3.47	1	2	3.43	1	2	3.39
Sean	PSNR	34.37	34.23	34.25	36.92	36.84	36.86	38.39	38.33	38.34	39.55	39.49	39.51	40.44	40.39	40.4
	Speed up	1	2	3.32	1	2	3.2	1	2	3.14	1	2	3.11	1	2	3.09
Hall	PSNR	33.2	33.16	33.17	35.28	35.26	35.24	36.32	36.3	36.28	36.96	36.93	36.94	37.44	37.39	37.41
	Speed up	1	2	3.54	1	2	3.42	1	2	3.33	1	2	3.31	1	2	3.28

Table 1 Simulation results and comparisons between FQPS, 2SS, and the proposed FSIP-BTDA for low bitrate CIF sequences

Sequences		256kbps			320kbps			384kbps			448kbps			512kbps		
		FQPS	2SS	FSIP-BTDA	FQPS	2SS	FSIP-BTDA	FQPS	2SS	FSIP-BTDA	FQPS	2SS	FSIP-BTDA	FQPS	2SS	FSIP-BTDA
Children	PSNR	28.47	28.33	28.44	29.44	29.29	29.39	30.52	30.41	30.47	31.23	31.16	31.19	32.2	32.05	32.01
	Speed up	1	2	3.79	1	2	3.72	1	2	3.69	1	2	3.67	1	2	3.65
foreman	PSNR	32.21	32.11	32.11	33.12	33.01	33.02	33.82	33.75	33.74	34.46	34.35	34.36	35	34.87	34.86
	Speed up	1	2	2.85	1	2	2.79	1	2	2.75	1	2	2.71	1	2	2.7
Paris	PSNR	31.55	31.48	31.44	32.5	32.48	32.44	33.34	33.33	33.26	34.07	34.04	34.01	34.7	34.69	34.61
	Speed up	1	2	3.59	1	2	3.52	1	2	3.46	1	2	3.43	1	2	3.41
Coast_guard	PSNR	28.02	27.99	27.98	28.72	28.66	29.67	29.35	29.31	29.28	29.9	29.86	29.83	30.43	30.38	30.38
	Speed up	1	2	3.01	1	2	2.95	1	2	2.89	1	2	2.85	1	2	2.82
Table	PSNR	31.85	31.75	31.76	32.91	32.79	32.78	33.69	33.61	33.56	34.37	34.29	34.3	35.01	34.95	34.92
	Speed up	1	2	3.37	1	2	3.28	1	2	3.24	1	2	3.22	1	2	3.2

Table 2 Simulation results and comparisons between FQPS, 2SS, and the proposed FSIP-BTDA for median and high bitrate CIF sequences

Sequences		32kbps			64kbps			96kbps			128kbps			160kbps		
		FQPS	2SS	FSIP-BTDA	FQPS	2SS	FSIP-BTDA	FQPS	2SS	FSIP-BTDA	FQPS	2SS	FSIP-BTDA	FQPS	2SS	FSIP-BTDA
foreman	PSNR	26.23	26.21	26.18	30.23	30.06	30.04	32.23	32.13	32.21	33.68	33.51	33.51	34.72	34.61	34.62
	Speed up	1	2	2.9	1	2	2.62	1	2	2.55	1	2	2.51	1	2	2.48
Coast_guard	PSNR	25.72	25.62	25.56	28.12	28.07	28.05	29.5	29.45	29.45	30.63	30.61	30.61	31.48	31.43	31.49
	Speed up	1	2	3.16	1	2	3	1	2	2.89	1	2	2.82	1	2	2.78
Children	PSNR	19.68	19.67	19.73	24.17	24.04	23.91	26.56	26.47	26.54	28.27	27.98	28.36	29.3	29.23	29.21
	Speed up	1	2	3.4	1	2	3.16	1	2	3.16	1	2	3.16	1	2	3.16
Hall	PSNR	34.51	34.5	34.56	37.37	37.33	37.32	38.97	38.87	38.89	39.4	39.31	39.26	40.27	40.21	40.22
	Speed up	1	2	3.19	1	2	3.07	1	2	3.02	1	2	3.03	1	2	3

Table 3 Simulation results and comparisons between FQPS, 2SS, and the proposed FSIP-BTDA for QCIF sequences

Efficient Search and Mode Prediction Algorithms for Motion Estimation in H.264/AVC

Gwo-Long Li, Mei-Juan Chen, Hung-Ju Li and Ching-Ting Hsu
Department of Electrical Engineering
National Dong Hwa University
Hualien, Taiwan, R.O.C.

Abstract—H.264 video coding standard adopts different coding schemes like variable block sizes and multiple reference frames for motion estimation. Hence, H.264/AVC provides gains in compression efficiency of up to 50% over a wide range of bit rates and video resolutions compared to previous standards. However, these features incur a considerable increase in encoder complexity, mainly regards to motion estimation and mode decision. In this paper, the efficient methods aiming at speed up of motion estimation and mode decision are proposed. Experimental results show that our proposed algorithms can save the coding time about 88% compared with the reference software JM80 while keeping the video quality.

I. INTRODUCTION

Motion estimation is the most important component in the video encoder. It directly affects the encoding speed and image quality. In H.264 [1] standard, motion estimation invokes multiple prediction modes, multiple reference frames and variable block sizes to achieve more accurate prediction and higher compression efficiency. However, the computation load of motion estimation increases greatly in H.264 because of the new features. According to statistics, it consumes about 70% of the whole encoding time.

There are many fast algorithms proposed in terms of speeding up the mode decision [2], multi-frame [3] and search pattern of motion estimation. A fast reference frame selection method for motion estimation algorithm named FFS is proposed by Hsu et.al [4]. In this scheme, the best reference frames of neighboring blocks are employed to determine the best reference frame of the current block. In this paper, we propose multi-directional rood pattern search (MDRPS) and fast mode selection (FMS) algorithms to improve the speed of H.264.

II. MULTI-DIRECTIONAL ROOD PATTERN SEARCH (MDRPS)

The proposed multi-directional rood pattern search (MDRPS) for integer motion estimation consists of three stages: 1)search center prediction, 2)multi-directional rood pattern search and 3)refined local search. For the search center prediction stage, a new search center prediction method is used in order to find the predicted motion vector for more accuracy. Then, the multi-directional rood pattern search is exploited in term of the search center to find the best match position. Finally, in the refined local search, a small diamond pattern and a small cross pattern are used repeatedly and unrestrictedly, until the final motion vector in found or the search boundary is reached. In the following, the detail of the proposed MDRPS is described.

1. Search Center Prediction

In order to obtain an accurate motion vector prediction of the current block, two factors are necessary to be considered. One is the choice of the neighboring blocks whose motion vectors will be used to calculate the predicted motion vector, and another is the method used for computing the predicted motion vector. The motion vectors of neighboring blocks include the information of integer pixel motion vector and quarter pixel motion vector. The predicted motion vector is defined as the integer pixel precision by truncation operation in H.264. This operation is rough and not accurate enough. However, we could extract the predicted integer pixel motion vector by using following equations:

if $pred\_mv>0$, $integer\_pred\_mv = (pred\_mv+1.6)/4$ (1)
if $pred\_mv<0$, $integer\_pred\_mv = (pred\_mv-1.6)/4$ (2)

where the *integer_pred_mv* is the proposed integer pel motion vector of current block and *pred_mv* is the predicted motion vector of current block in fractional pixel accuracy from neighboring blocks in H.264. For example shown in Fig.1, assume that the point P indicates the predicted motion vector with fractional accuracy. In the H.264 standard, the integer pixel motion vector is extracted by truncation operation. Therefore, the point A will be selected to be the predicted integer pixel motion vector. However, after the calculation of proposed method as shown (1) and (2), the point D will be selected to become the predicted integer pixel motion vector. Through the extraction of proposed method, the predicted integer motion vector will be more precise and improve the coding efficiency.

2. Multi-directional rood pattern search

The search patterns for MDRPS algorithm are illustrated in Fig. 2. These patterns consist of one initial checking pattern and four directional rood patterns which are diamond search pattern (DSP), horizontal rood search pattern (HRSP), vertical rood search pattern (VRSP), diagonal rood search pattern (DRSP) and anti-diagonal rood search pattern (ADRSP). These directional search patterns are designed for covering more widely search directions and the real-world motion vector distribution.

The initial search pattern consists of five categories: center, horizontal, vertical, diagonal and anti-diagonal. In Fig. 2(a), the center is marked as "0", the horizontal points are marked as "1" and "5"; the vertical points are marked as "3" and "7"; the diagonal points are marked as "4" and "8"; the anti-diagonal points are marked as "2" and "6".

The main idea of the proposed algorithm is first choosing one of the four directional search patterns by the position with minimum Rate-Distortion-Cost (RDC) occurs in the initial search pattern. Then the directional search pattern for next search step is chosen by the position with minimum RDC in current search pattern. Fig. 3 shows the switching strategies of the MDRPS algorithm. The MDRPS algorithm is summarized as follows.

Step 1: The initial diamond search pattern (DSP) is centered on the origin of the search window and set as the current search pattern. The RDC on all the nine positions in the current search pattern are checked. If the minimum RDC occurs at the central position then stop the search; the motion vector points to the central position. Otherwise, the decision is as follows:

$$\begin{cases} \text{if min RDC is at '0', Stop} \\ \text{else if min RDC is at '1' or '5', goto Step2} \\ \text{else if min RDC is at '3' or '7', goto Step3} \\ \text{else if min RDC is at '4' or '8', goto Step4} \\ \text{else if min RDC is at '2' or '6', goto Step5} \end{cases} \quad (3)$$

Step 2: The horizontal rood search pattern (HRSP) is centered on the position with minimum RDC in previous search stage, checking all five positions shown in Fig.3 (a). If the position with minimum RDC occurs at the center "1", then terminate the search process. Else if the position with minimum RDC occurs at "3", then go to Step 2; otherwise, go to Step 3.

Step 3: The vertical rood search pattern (VRSP) is centered on the position with minimum RDC, checking all five points shown in Fig.3 (b). If the position with minimum RDC occurs at the center "1", then terminate the search process. Else if the position with minimum RDC occurs at "2", then go to Step 3; otherwise, go to Step 2.

Step 4: The diagonal rood search pattern (DRSP) is centered on the position with minimum RDC, checking all five points shown in Fig.3(c). If the position with minimum RDC occurs at the center "1", then terminate the search process. Else if the position with minimum RDC occurs at "2", then go to Step 4; otherwise, go to Step 5.

Step 5: The anti-diagonal rood search pattern (ADRSP) is centered on the position with minimum RDC, checking all five points shown in Fig.3 (d). If the position with minimum RDC occurs at the center "1", then terminate the search process. Else if the position with minimum RDC occurs at "2", then go to Step 5; otherwise, go to Step 4.

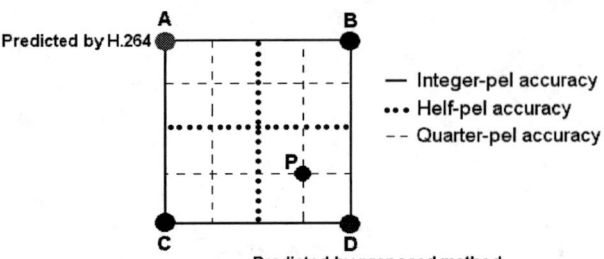

Fig. 1 Illustration of motion vector predictor

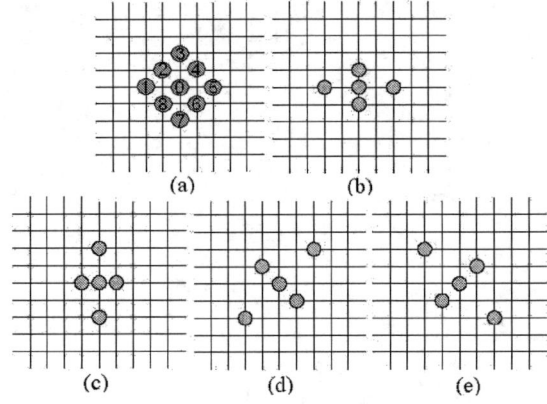

Fig. 2 The search patterns for MDRPS algorithm; (a) initial search pattern, (b) horizontal rood search pattern (HRSP), (c) vertical rood search pattern (VRSP), (d) diagonal rood search pattern (DRSP), (e) anti-diagonal rood search pattern (ADRSP).

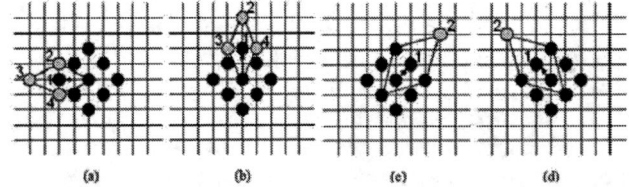

Fig. 3 The switching strategies of MDRPS; (a) horizontal search strategy, (b) vertical search strategy, (c) diagonal search strategy, (d) anti-diagonal search strategy.

3. Refined local search

In multi-directional rood pattern search procedure, the size between central position and the other positions is either two or one. It means there are some positions are around central position with size one not been checked. It is possible that the un-checked positions are better than current position. For this reason, the motion vector found in MDRPS must be refined to improve the performance. Two types of compact search patterns are used in our algorithm to refine the motion vector. One is the small diamond search pattern (SDSP) and the other is the small cross search pattern (SCSP) shown in Fig. 4(a) (b), respectively. The two types of refined search pattern are used in different cases. If the switching strategy of MDRPS is horizontal or vertical, the small cross pattern is adopted in the refined local search. If the switching strategy of MDRPS is diagonal or anti-diagonal, the small diamond pattern is adopted in the refined local search. The search steps will stop only if the best point locates at the center point.

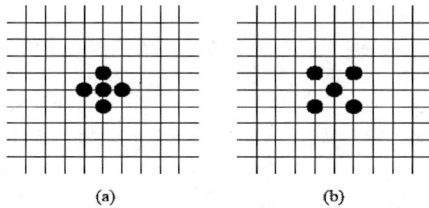

Fig. 4 The refined search pattern; (a) small diamond pattern; (b) small cross pattern

III. FAST MODE SELECTION ALGORITHM (FMS)

As mentioned before, H.264 standard allows variable block size to be used in motion estimation. It includes seven different block sizes namely 16×16, 16×8, 8×16, 8×8, 8×4, 4×8 and 4×4, respectively. For each macroblock, the motion estimation and rate-distortion optimization are performed to find the best block sizes resulting in heavy computational load. But in the real word video sequences, they usually have high spatial correlation. The block size of current block has high probability similar with neighboring coded blocks. Since, the block size can be predicted from neighboring coded blocks. Therefore, we propose a fast mode selection algorithm to predict the block size accurately. It can predict one best block size for motion search rather than search for all block size.

In this proposed algorithm, the best predicted block size for current block is determined by using $J(m, \lambda_{motion})$ values of neighboring blocks. The best inter prediction mode with the minimum $J(m, \lambda_{motion})$ value is selected for encoding. And the criterion is shown as following expression:

$$J_i = \min(J_{BLOCK\_A}, J_{BLOCK\_B}, J_{BLOCK\_C}) \quad (4)$$
$$i \in \{BLOCK\_A, BLOCK\_B, BLOCK\_C\}$$

The predicted block mode = MODE(i)

Fig. 5 illustrates the two situations in our proposed algorithm. Fig. 5(a) depicts the relative location between current macroblock and neighboring macroblocks when performing the 16×16 modes. Let E be the current macroblock, let A be the macroblock immediately to the left of E, let B be the macroblock immediately above E, let C be the macroblock above and to the right of E and let D be the macroblock above and to the left of E. Fig. 5(b) depicts the relative location between current 8×8 block and neighboring 8×8 blocks when performing the 8×8 modes. Fig. 5(b) illustrates in the case of 8x8 modes similarly as Fig. 5(a). If the block C is unavailable (the current block E located in the right boundary of the frame), the block C will be replaced by block D.

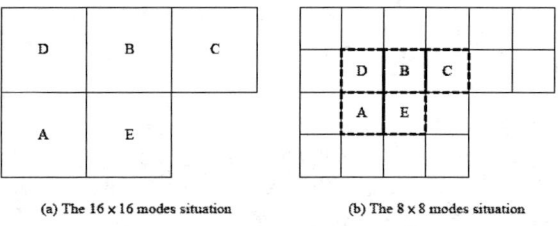

Fig. 5 The relative location between current and neighboring blocks for 16×16 and 8×8 modes

The proposed mode selection method is summarized as follows.

A. If the macroblock/block locates in the first row or the first column, the block is calculated by exhaustively testing all the seven variable block sizes.
B. If the macroblock/block doesn't locate on the boundary of the picture, the block modes of neighboring blocks which are left, up, and up-right to the current block in the current frame will be considered as the predicted mode candidates. The best block mode of the current block can be selected using these three J values of the neighboring blocks and the block mode with the minimum value is selected for encoding as shown in (4).
C. If the BLOCK_C is unavailable, the block modes of the other three blocks locate at left, up and up-left to the current block in the current frame will be considered as the block mode candidates. The three values are checked in this case. And the best predicted mode with the minimum value is selected. The method is similar to (4), but the neighboring blocks are BLOCK_A, BLOCK_B and BLOCK_D.
D. After performing 16×16 modes and 8×8 modes, the rate distortion cost is used to select the best inter prediction mode of this current macroblock.

IV. EXPERIMENTAL RESULTS

The performance of the proposed algorithms is compared. We compare the total bits, PSNR and coding speed in our simulation for various sequences. Analysis has been performed in JVT reference software JM80 [5] on eight video sequences in QCIF (176×144) format, each representing a different class of motion. These sequences include 80 frames of Carphone, Claire, Container, Foreman, Mother and Daughter, News, Silent and Suzie.

In Table 1, 2 and 3, the average bits, PSNR and average time of individual approach and the combined algorithms are compared, respectively. As shown in Table

1, the average bits of proposed algorithm don't increase significantly. In Table 2, the proposed MDRPS and FMS algorithms have objective quality loss between 0.04 and 0.10 dB. The insignificant loss is not obvious in human view. However, the average processing speed is improved significantly. As shown in Table 3, each of our proposed algorithms provides the time saving about 60%-70% compared with JM80. Table 1-3 also shows the results of combined algorithms (MDRPS+FMS and MDRPS+FMS+FFS). The FFS indicates a fast frame selection algorithm described in [4]. The tables show that the combined three algorithms cause the PSNR loss (between 0.52 dB and 0.6dB), but the time consumption can be saved 97% in average compared with JM80. This improvement is useful for real-time application.

V. CONCLUSION

In this paper, two fast motion estimation algorithms MDRPS and FMS are proposed for H.264/AVC. Experimental results show that the two proposed algorithms save the coding time about 88% while keeping the quality. The proposed approaches can be applied to the real-time implementation of H.264 encoder.

REFERENCES

[1] Joint Video Team (JVT) of ISO/IEC MPEG and ITU-T VCEG, "Draft ITU-T Recommendation and Final Draft International Standard of Joint Video Specification (ITU-T Rec. H.264 | ISO/IEC 14496-10 AVC)," document JVT-G050d35.doc, March 2003.

[2] A. Ahmad, N. Khan, S. Masud and M.A. Maud, "Efficient block size selection in H.264 video coding standard," *Electronics Letters*, Vol.40, pp. 19-21, January 2003.

[3] Yi-Yen Chiang, Mei-Juan Chen and Hung-Ju Li, "Efficient Multi-Frame Motion Estimation Algorithm for MPEG-4 AVC/JVT/H.264 Standard," *IEEE International Symposium on Circuits and Systems*, Vol.3, pp. 737-740, May 2004.

[4] Ching-Ting Hsu, Hung-Ju Li and Mei-Juan Chen, "Fast Reference Frame Selection Method for Motion Estimation in JVT/H.264," *IEICE Transactions on Communications*, Vol.E87-B, No.12, pp.3827-3830, December 2004.

[5] Joint Video Team software JM80, April 2004 http://bs.hhi.de/~suehring/tml/download/

Table 1 Comparison of average bits of the proposed algorithms

	Average bits (bits/frame)								
	JM80	MDRPS		FMS		MDRPS+FMS		MDRPS+FMS+FFS	
	Avg. bits	Avg. bits	Variation	Avg. bits	Variation	Avg. bits	Variation	Avg. bits	Variation
Carphone	6590.29	6590.09	-0.003%	6519.18	0.013%	6530.60	-0.042%	6535.78	-0.83%
Claire	6630.50	6638.06	0.114%	6650.13	0.296%	6642.70	0.184%	6645.36	0.22%
Container	6374.40	6375.10	0.011%	6379.95	0.087%	6378.09	0.058%	6399.49	0.39%
Foreman	6558.93	6559.13	0.003%	6559.59	0.010%	6563.32	0.067%	6610.47	0.79%
Mthr_dot	6512.96	6503.58	-0.144%	6523.58	0.163%	6524.94	0.184%	6537.77	0.38%
News	6536.53	6512.80	-0.363%	6542.74	0.095%	6524.30	-0.187%	6507.15	-0.49%
Silent	6436.59	6441.09	0.070%	6436.01	-0.009%	6449.72	0.204%	6477.99	0.64%
Suzie	6626.83	6626.87	0.061%	6634.02	0.169%	6627.79	0.075%	6635.28	0.13%

Table 2 Comparison of PSNR of the proposed algorithms

	PSNR (dB)								
	JM80	MDRPS		FMS		MDRPS+FMS		MDRPS+FMS+FFS	
	Avg. PSNR	Avg. PSNR	Diff.	Avg. PSNR	Diff.	Avg. PSNR	Diff.	Avg. PSNR	Diff.
Carphone	39.90	39.80	-0.10	39.86	-0.04	39.71	-0.19	39.31	-0.60
Claire	47.29	47.25	-0.04	47.24	-0.05	47.17	-0.12	46.76	-0.53
Container	41.14	41.07	-0.07	41.06	-0.08	41.05	-0.09	40.62	-0.52
Foreman	37.27	37.18	-0.09	37.23	-0.04	37.14	-0.13	36.69	-0.58
Mthr_dot	39.38	39.34	-0.04	39.29	-0.09	39.24	-0.14	38.83	-0.55
News	41.14	41.06	-0.08	41.07	-0.07	41.01	-0.13	40.62	-0.52
Silent	39.20	39.15	-0.05	39.14	-0.06	39.10	-0.10	38.66	-0.54
Suzie	40.24	40.14	-0.09	40.13	-0.11	40.12	-0.12	39.64	-0.60

Table 3 Comparison of average time of the proposed algorithms

	Average time (seconds/frame)								
	JM80	MDRPS		FMS		MDRPS+FMS		MDRPS+FMS+FFS	
	Avg. time	Avg. time	Time-saved	Avg. time	Time-saved	Avg. time	Time-saved	Avg. time	Time-saved
Carphone	3.18	1.08	66%	1.12	65%	0.39	88%	0.06	98%
Claire	2.50	1.00	60%	0.87	65%	0.35	86%	0.07	97%
Container	3.16	1.00	68%	1.23	61%	0.35	89%	0.10	96%
Foreman	3.47	1.10	68%	1.26	64%	0.40	88%	0.10	97%
Mthr_dot	3.11	1.07	66%	1.06	66%	0.39	88%	0.06	98%
News	2.60	0.98	62%	0.90	65%	0.35	87%	0.10	96%
Silent	2.82	1.03	63%	0.94	67%	0.36	87%	0.08	97%
Suzie	3.79	1.17	69%	1.39	63%	0.43	89%	0.07	98%

EFFICIENT VIDEO MOTION ESTIMATION USING DUAL-CROSS SEARCH ALGORITHMS

Xuan-Quang Banh and Yap-Peng Tan
School of Electrical & Electronic Engineering
Nanyang Technological University, Singapore

Abstract—We present in this paper two new and efficient block-matching algorithms (BMAs) for fast motion estimation that is commonly used in motion-compensated video compression. The proposed algorithms are improvement of existing BMAs and consist of three effective steps: 1) Initial search center prediction, 2) Early search termination, and 3) Dual-cross search. Extensive simulation results and comparative analysis have shown that the proposed algorithms outperform conventional algorithms, like the three-step search, orthogonal search and diamond search, as well as the newly proposed algorithms, like the hexagonal search and adaptive rood pattern search, in terms of both peak signal-to-noise ratio and the number of search points evaluated.

I. INTRODUCTION

Block-matching algorithm (BMA) is a commonly used technique in video compression for exploiting temporal redundancy between neighboring frames. It has been adopted by leading video compression standards such as MPEG 1/2/4 [1] and H.261/263/264 [2].

The most widely referenced block-matching algorithm is probably the full search (FS) algorithm, an algorithm that finds the optimal reference block—the one with the minimum sum-of-absolute-difference (SAD)—by exhaustively evaluating all possible candidate blocks within a search window in the reference frame. Being exhaustive, the FS is computationally intensive, making it unsuitable for many real-time video applications, particularly those using software for video compression.

To reduce the computational complexity, many fast BMAs, including the three-step search (TSS) [3], orthogonal search (OS) [4] and diamond search (DS) [5], have been proposed to obtain sub-optimal solution by limiting the number of search points to be checked for the best-matched blocks in the reference frame. Among these algorithms, the DS has been shown outperforming many others, in terms of search speed and video quality [5].

However, the DS does not perform as well for sequences containing large or complex motion. A number of algorithms have been proposed to enhance the performance of DS algorithm. Some of them have achieved very promising results, in either reducing the computational complexity like the hexagonal search (HEXBS) [6] or reducing the computational complexity and enhancing the video quality like the adaptive rood pattern search with zero-motion prejudgment (ARPS-ZMP) [7]. Due to space constraint, interested readers are directed to the relevant references for the details of these algorithms.

Based on the insights gained from our examination and comparison of the above-mentioned BMAs, we propose in this paper two improved algorithms, referred to as *Dual-Cross Search* (DCS) and *Adaptive Dual-Cross Search* (ADCS), to address the limitations of the existing fast BMAs. The difference between the two proposed methods mainly lies in how the threshold value for early terminating the search of optimal motion vector is determined. As the names suggest, DCS uses a fixed-value threshold, while ADCS uses one that is adaptive to local video content. As we shall see later, these two algorithms not only can reduce markedly the computational complexity, but also achieve superior video quality, as compared with the existing state-of-the-art fast BMAs.

II. THE PROPOSED ALGORITHMS

Our two proposed algorithms make use of three schemes to achieve superior performance: 1) Initial search center prediction, 2) Early search termination, and 3) Dual-cross search.

A. Initial Search Center Prediction

As most of the motion vectors (MVs) are well correlated with that of their neighboring blocks, our proposed algorithms make use of the neighboring MVs to predict the *initial search center* (ISC) of the current block. To minimize the computational cost and storage requirement in the prediction, for each current block we shall consider only the zero motion vector (0, 0) and the motion vectors of its left and above adjacent blocks, denoted as (mv_x^L, mv_y^L) and (mv_x^A, mv_y^A), to predict the ISC.

B. Early Search Termination

It was reported in [7] that early termination of the search for the optimal motion vector could reduce substantially the cost of block-matching without degrading much the video quality. In our proposed algorithms, the early search termination (EST) is achieved by comparing the SAD between the current block and the block corresponding to the predicted ISC in the reference frame against a threshold T. If the SAD is smaller than threshold T, the search stops and outputs the ISC as the optimal solution.

We have considered two approaches for deciding the value of threshold T used in the proposed EST scheme: (i) In the proposed DCS algorithm, we set threshold T as a constant equal to 512 (i.e., $T = 512$). (ii) In the proposed ADCS algorithm, we set threshold T adaptively based on the correlation of local video motion and content. Specifically, the threshold T for each block is calculated as follows: If the predicted ISC is not one of the two adjacent motion vectors, (mv_x^L, mv_y^L) and (mv_x^A, mv_y^A), set threshold $T = 512$. Otherwise, the threshold is set as $T = \max\{\rho \times \text{SAD}_{adj}, 512\}$, where ρ is a proper scale parameter and SAD_{adj} is the minimum SAD of the adjacent block whose motion vector was selected as the ISC, and value 512 is included to sustain the minimum video quality. Note that, the matching distortion SAD_{adj} should not be confused with the SAD of the current block obtained at the predicted ISC. For example, if (mv_x^L, mv_y^L) is chosen as the ISC of the current block, then SAD_{adj} is obtained from its left adjacent block.

The main idea behind the adaptive thresholding is that if the two adjacent blocks have the same or similar motion vectors, they will have similar block-matching distortion as well, under the assumption that the local video motion and video content are generally highly correlated. Hence, setting the threshold T based on the adjacent SAD can terminate the unnecessary search more effectively for video with complex motion or scene. The scale parameter ρ is included to allow some difference between two adjacent SADs. Conceivably, the larger the scale parameter ρ, the higher the threshold value T, and the more likely the search can be terminated early, probably at the cost of lower video quality.

To select a proper scale parameter ρ so that the computational cost can be reduced and the video quality can be maintained as much as possible, we conducted empirical studies using different values: 0.9, 1.0, 1.1, 1.2, 1.3, and 1.4. Four test sequences (Football, Foreman, Stefan, and Tennis) that show significant reduction in PSNR as compared with the FS were used in the experiments. Two additional sequences (Mobile & Calendar and Mother & Daughter) which contain mainly small motions or stationary scenes were also included in the experiments for comparison.

Table 1 and Table 2 record the average PSNR differences and the number of search points required. The tables show that the number of search points reduces steadily when the scale parameter ρ increases, while the PSNR results, as compared with that of the FS, only decrease slightly (less than 0.1 dB) when ρ is between 0.9 and 1.2.

In practice, the scale parameter can be set based on the application's requirements in trading off the computational complexity and video quality. For sequences with mainly slow motion and/or stationary scene, a large scale parameter can be used to reduce the computational cost without degrading much the video quality. For the experimental results reported in the following section, we have used $\rho = 1.2$ in the proposed ADCS algorithm.

C. Dual-Cross Search

Even when the predicted ISC coincides with the optimal solution (the point with the minimum SAD), the DS still needs to evaluate as many as 13 search points—9 points in the outer large diamond and 4 points in the inner small diamond—to confirm the optimality. In our proposed algorithms, we only need to evaluate 4 search points to verify it. Furthermore, as the proposed ISC prediction scheme is rather accurate, we found that it is more efficient to first evaluate the endpoints of a small (2×2) cross pattern before those of a large (4×4) cross pattern.

In the first stage of the proposed dual-cross search, the SADs at the four endpoints of a 2×2 cross pattern centered at the predicted ISC are examined and compared with that at the cross center (see Fig. 1). If the minimum SAD is found at the cross center, the search stops and the cross center is selected as the optimal solution. Otherwise, the four endpoints of a 4×4 cross pattern will be examined. This cross search is repeated until the minimum SAD is found at the cross center. It should be noted that the 2×2 cross search is applied only once, while the 4×4 cross search may be performed several times. Furthermore, the 4×4 cross search is necessary in some cases as performing only the 2×2 cross search has two limitations: First, the search could be trapped at local minima. Second, in case the ISC prediction fails (i.e., the optimal solution locates far away from the predicted ISC), more points need to be checked by the 2×2 search before it can arrive at the global minimum. The 4×4 cross search is then followed by a final refinement step.

Table 1. Average PSNR difference (in dB, as compared with that of the FS algorithm) achieved by the proposed ADCS algorithm

Sequence	ADCS using different scale parameter ρ					
	0.9	1.0	1.1	1.2	1.3	1.4
Football	0.07	0.08	0.09	0.11	0.13	0.14
Foreman	0.04	0.04	0.05	0.05	0.07	0.08
Stefan	0.17	0.18	0.22	0.27	0.42	0.59
Tennis	0.11	0.11	0.13	0.16	0.22	0.27
M&C	0.00	0.00	0.01	0.02	0.06	0.09
M&D	0.00	0.00	0.00	0.00	0.01	0.01
Average	0.06	0.07	0.08	0.10	0.15	0.20

Table 2. Number of search points checked by the proposed ADCS algorithm

Sequence	ADCS using different scale parameter ρ					
	0.9	1.0	1.1	1.2	1.3	1.4
Football	233716	216654	199567	186142	175110	166435
Foreman	498100	472785	445712	419901	396177	376094
Stefan	581103	537885	492587	451772	414585	380953
Tennis	145113	132743	119526	107652	98442	92246
M&C	156439	144918	134555	125918	118982	113325
M&D	216414	209669	202073	195384	188817	182529

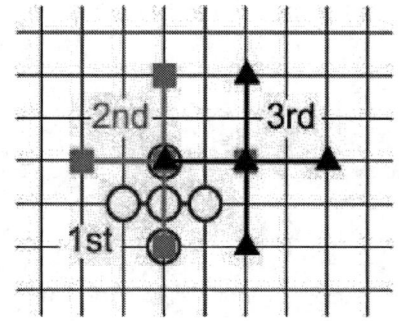

Fig. 1. Proposed Dual-Cross Search.

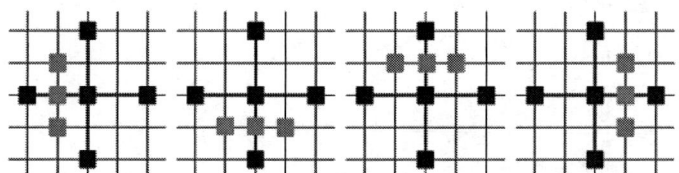

Fig. 2. Three intermediate points checked in the final refinement step.

For the final refinement, unlike the conventional cross or diamond search, our proposed algorithms do not examine the four adjacent neighbors around the cross center of minimum SAD. With the assumption that the SAD increases monotonically when the search point moves away from the optimal solution, out of the four cross endpoints, the one nearest to the optimal solution will have the lowest SAD. Therefore, we identify the endpoint that has the lowest SAD and examine the SADs of three intermediate search points residing between this endpoint and the 4×4 cross center (see Fig. 2). Out of these three additional search points and the current cross center, the one that has the minimum SAD is selected as the final solution.

D. Summary of the Proposed Algorithms

The proposed ADCS algorithm can be summarized as follows:

Step 1. Initial search center prediction
Compute the SADs at three points: (0, 0), (mv_x^L, mv_y^L) and (mv_x^A, mv_y^A). The one with the lowest SAD is selected as the ISC for the remaining search.

Step 2. Early Search Termination:

2.1 Set threshold value T: If the predicted ISC is not one of the two adjacent motion vectors, (mv_x^L, mv_y^L) and (mv_x^A, mv_y^A), set $T = 512$. Otherwise, set $T = \max\{\rho \times \text{SAD}_{adj}, 512\}$, where $\rho = 1.2$ and SAD_{adj} is the minimum SAD of the adjacent block whose motion vector has been selected as the ISC.

2.2 Compare the SAD at the ISC against threshold T:
If $\text{SAD}(ISC) < T$
    Solution = ISC;
    Stop;
Else
    Go to Step 3.

Step 3. 2×2 Cross Search:
If the minimum SAD is found at the cross center,
    Solution = the cross center;
    Stop;
Else
    Go to Step 4.

Step 4. 4×4 Cross Search:
Repeat 4×4 cross search until the minimum SAD is found at the cross center.

Step 5. Final Refinement:
Determine the cross's endpoint with the lowest SAD. Examine the intermediate three points residing between the cross center and the endpoint. The cross center or one of the three intermediate points having the minimum SAD is selected as the final solution.

Different from the proposed ADCS algorithm, the proposed DCS algorithm simply sets threshold $T = 512$; the other steps remain the same. In summary, our proposed algorithms can exit from the search for optimal motion vector at three possible stages: after the Early Search Termination (Step 2), after the 2×2 Cross Search (Step 3), and after the Final Refinement (Step 5).

III. EXPERIMENTAL RESULTS

We have conducted a series of experiments to evaluate the performance of the proposed BMAs. In the experiments, we used the Test Model 5 (TM5) encoder provided by the MPEG Software Simulation Group at MPEG.org [8]. Our test sequences include ten popular sequences of size 352×288 pixels (Coastguard, Flower, Foreman, Mother & Daughter (M&D), News, Silent, and Stefan) or 352×240 pixels (Football, Mobile & Calendar (M&C), and Tennis). Comprising different amounts of motion and spatial details, these test sequences have been widely used in video compression research and standardization activities. Sample frames of these test sequences are shown in Fig. 3.

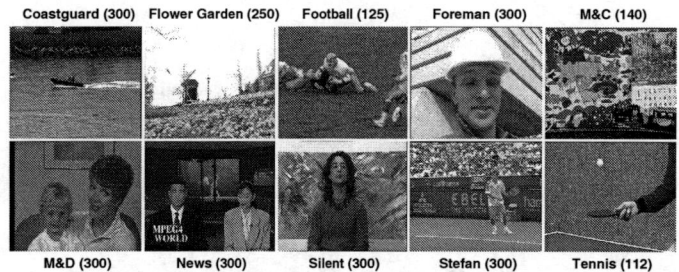

Fig. 3. Test sequences (number of frames shown in parentheses).

In the experiments, the block size was chosen as 16×16 pixels, the search window was set to 31×31 pixels, the group-of-picture (GOP) structure was IPPPPPPPPP, the frame rate was fixed at 30 frames/sec, and the coding bit-rate was set to 1.5 Mbps. To evaluate the performance of our proposed algorithms as compared with other existing algorithms, including the FS, TSS, OS, DS, HEXBS and ARPS-ZMP, we used the PSNR and number of search points evaluated as the measures of video quality and computational complexity, respectively.

The experimental results are listed in Tables 3, 4, and 5. The results show that the proposed algorithms outperform all the existing fast BMAs under comparison. In particular, both the proposed algorithms achieve better PSNR than the conventional TSS, OS, DS as well as the newly proposed HEXBS, ARPS-ZMP algorithms, while checking only a fractional number of points. More specifically, compared with the

FS algorithm, the two proposed algorithms markedly improve the search speed, on average, by 228.4 and 309.0 times, respectively. This computational gain ranges from 100-200 times for fast-motion sequences (e.g. Football, Foreman) to more than 300 times for sequences containing low motion and stationary background scene (e.g. M&D, Silent). The gain is achieved at the cost of only on average 0.04-0.07 dB lower PSNR as compared with the FS algorithm. Comparing with the DS and ARPS-ZMP algorithms, our proposed algorithms achieve higher PSNRs for most sequences at a much lower computational cost. On average, the DCS and ADCS algorithms are 1.34 and 1.81 times faster than the ARPS-ZMP, respectively. To see that the proposed algorithms perform consistently better than the existing algorithms over the test sequence, Fig. 4 shows the frame-by-frame PSNR results of the first-half of the Flower Garden sequence obtained by the FS, DS, ARPS-ZMP, and the proposed ADCS algorithms.

Table 3. Average PSNR results (in dB) obtained by different BMAs

Sequence	FS	TSS	OS	DS	HEXBS	ARPS-ZMP	DCS	ADCS
Football	34.52	34.34	34.34	34.50	34.49	34.52	34.51	34.51
Foreman	29.38	28.81	28.86	29.30	29.16	29.23	29.36	29.34
Stefan	29.49	29.20	29.13	29.41	29.37	29.27	29.44	29.38
Tennis	37.06	36.57	36.47	36.73	36.57	36.92	37.03	37.01
M&C	24.85	24.73	24.74	24.83	24.78	24.85	24.86	24.83
M&D	43.84	43.79	43.78	43.83	43.83	43.84	43.84	43.84
Stefan	41.53	41.50	41.48	41.52	41.52	41.52	41.51	41.51
Tennis	39.09	39.03	39.01	39.07	39.06	39.03	39.07	39.06
M&C	31.80	30.88	30.96	30.71	30.70	31.45	31.66	31.53
M&D	33.48	32.98	32.96	33.41	33.29	33.13	33.39	33.32
Average	34.50	34.18	34.17	34.33	34.28	34.38	34.47	34.43

Table 4. Gain ratio of the number of search points required by different BMAs as compared with that of the FS algorithm

Sequence	TSS	OS	DS	HEXBS	ARPS-ZMP	DCS	ADCS
Football	28.2	53.5	52.5	67.1	98.6	160.1	297.1
Foreman	28.3	53.6	54.4	68.7	128.5	192.2	278.1
Stefan	28.2	53.3	52.0	67.2	83.0	115.4	170.7
Tennis	28.2	53.5	46.3	62.2	102.4	161.6	221.4
M&C	28.3	53.4	66.8	80.9	114.3	177.7	283.8
M&D	28.3	53.6	64.0	78.4	331.5	400.5	475.7
Stefan	28.4	53.7	68.0	81.5	391.5	456.0	515.7
Tennis	28.4	53.7	64.5	78.7	238.8	312.9	378.2
M&C	28.2	53.5	45.5	59.7	94.2	136.9	205.7
M&D	28.2	53.3	56.8	70.9	125.5	171.1	263.5
Average	28.3	53.5	57.1	71.5	170.8	228.4	309.0

Table 5. Average number of search points per motion vector

Algorithm	Method	Min.	Max.	Average
Existing	FS	961	961	961
	TSS	33	33	33
	OS	17	17	17
	DS	13	47	16.5
	HEXBS	11	35	13
	ARPS-ZMP	1	45	7
Proposed	DCS	1	33	5
	ADCS	1	33	3.5

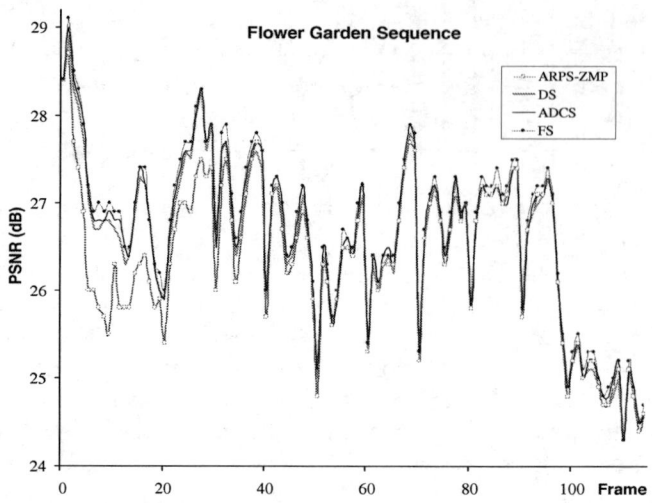

Fig. 4. Frame-by-frame PSNR performance of the first 115 frames of the Flower Garden sequence.

IV. CONCLUSIONS

We have presented in this paper two new fast block-matching motion estimation algorithms, called Dual-Cross Search (DCS) and Adaptive Dual-Cross Search (ADCS), for video compression. The proposed algorithms make use of three effective schemes to achieve superior performance: 1) exploiting the correlation among adjacent motion vectors; 2) making use of the adaptive early search termination scheme to reduce the number of search points without degrading much the video quality; and 3) locating the optimal motion vector quickly by the dual-cross search scheme.

As for the coding performance, the proposed ADCS algorithm speeds up the search by a factor of 3.3-7.6 as compared with the DS algorithm and 1.3-3.0 with the ARPS-ZMP algorithm, while attaining better PSNR performance. For real-time applications, ADCS is preferred because of its low computational cost (1.35 times faster than the DCS) and good PSNR results (only 0.07 dB lower than that of FS) on average.

REFERENCES

[1] ISO/IEC JTC1/SC29/WG11,"ISO IEC CD 13818: Information Technology," MPEG-2 Committee Draft, December 1993.
[2] International Telecommunication Union,"Video codec for audiovisual services at px64 kbits," ITU-T Report/Rec. H.261, March 1993.
[3] T. Koga, K. Iinuma, and A. Iijima,"Motion-compensated interframe coding for video conferencing," *Proc. NTC81*, pp. C9.6.1-9.6.5, New Orleans, LA, 1981.
[4] A. Puri, H. M. Hang, and D. L. Schilling, "An efficient block matching algorithm for motion compeansated coding," *ICASSP'87*, pp 1063-1066, Dalas, TX, 1987.
[5] S. Zhu and K. K. Ma, "A new diamond search algorithm for fast block matching motion estimation," *Proc. of Int. Conf. Information, Communications and Signal Processing*, vol. 1, pp. 292-6, 1997.
[6] C. Zhu, X. Lin, and L. P. Chau, "Hexagon-based Search Pattern for Fast Block Motion Estimation," *IEEE Trans. on Circuits and Systems for Video Technology*, vol. 12, no. 5, May 2002.
[7] Y. Nie and K. K. Ma, "Adaptive Rood Pattern Search for Fast Block-Matching Motion Estimation," *IEEE Trans. on Image Processing*, vol. 11, no. 12, 2002.
[8] MPEG Software Simulation Group (MMSG), "Test Model 5 (TM5)." [Online] Available: http://www.mpeg.org/MPEG/MMSG/.

Rapid Block-Matching Motion Estimation Using Modified Diamond Search Algorithm

Xiaoquan Yi and *Nam Ling*
Department of Computer Engineering, Santa Clara University
Santa Clara, California 95053, USA
Email: {xyi, nling}@scu.edu

Abstract - Due to the considerable computational complexity of full-search (FS) in motion estimation, many suboptimal but fast block-matching algorithms (BMAs) have been developed. Among them, diamond search (DS) series is the most promising method. To further reduce complexity and improve performance, we propose a *modified diamond-search* (MODS) algorithm for rapid block matching based on the well-known DS algorithm. A novel *fine granularity halfway-stop* (FGHS) method based on a dynamic block distortion threshold is also proposed. To avoid being trapped to local optima, unlike some small DS methods, MODS adaptively starts with a relatively large search pattern for high motion blocks which are automatically determined via the first block matching distortion. The threshold is obtained via a linear model utilizing already computed distortion statistics. Experiments show that the proposed algorithm achieves less search points with no significant PSNR degradation when compared to that of FS and other fast BMAs.

I. INTRODUCTION

In many video coding applications, motion estimation (ME) and motion compensation (MC) are commonly applied with significant improvements in bit rate reduction. Block-matching algorithm (BMA) is a popular method and has been adopted by many international video coding standards. BMA is to seek for the best-matched block from the reference frame within a search window. Based on a block distortion measure (BDM), the displacement of the best-matched block is described as motion vector (MV) to the block in the current frame. The best match is usually evaluated by a cost function based on a BDM such as sum of absolute differences (SAD). The most obvious candidate for a search technique is the exhaustive or full search (FS). The number of search points (SPs) is defined in terms of search window (Ω) as: $(2\Omega +1)^2$. Being exhaustive, the FS demands very high computational complexity, making real-time software-based video encoding impractical. This drawback has motivated many suboptimal but faster search strategies such as new three-step search (N3SS) [1], four-step search (4SS) [2], diamond search (DS) [3], [4], cross-diamond search (CDS) [5], hexagon based search (HEXBS) [6], new CDS (NCDS) [7], and kite CDS (KCDS) [8], etc. Among them, many recently developed fast BMAs utilize the characteristics of center-biased MV distribution. Statistically, more than 80% of the blocks can be regarded as stationary or quasi-stationary blocks [1]-[8]. This center-based characteristic can even be found in the fast-motion sequences. DS introduced a diamond shape searching pattern and unrestricted searching steps. NCDS and KCDS are further fitted cross-center-biased MV distribution property, employing a small cross-shaped search pattern in the first and second steps. Assuming $\Omega = 7$, for the best case (i.e. near-stationary video), aforementioned BMAs have the minimum search points shown in Table I. Those BMAs always examine a fixed number of SPs in the initial step regardless whether the current block has a zero MV (ZMV) or not. Technically speaking, those SPs are redundant and can be avoided to further reduce computational complexity without much impact on video fidelity.

Another observation is that the halfway-stop technique employed in many fast BMAs [1]-[8] only stops if the minimum is found in the center after finishing a *complete* search step which usually consists of several SPs. There are two problems associated with this approach: 1) easily missing true MV in the case of local optimum occurring in the center and 2) searching redundant points if the MV is not coincident with the last point being searched. To circumvent this halfway-stop problem, we propose a *fine granularity halfway-stop* (FGHS) method in which the search can be stopped at any point whenever the true minimum or near-true minimum has been located, based on a dynamic block distortion threshold. We introduce a simple linear model to estimate the dynamic threshold by utilizing already computed distortion statistics on per block-base. Tourapis *et al.* [9] proposed adaptive thresholding parameters which are calculated by a function of a few minimum distortions from spatial and temporal neighboring blocks. Our dynamic threshold calculation is different with [9] in the sense that ours utilizes the information of the current block whereas [9] does not.

TABLE I – MINIMUM SPS FOR BMA'S IN THE BEST CASE ($\Omega = 7$)

BMA	FS	N3SS	4SS	DS
SP	225	17	17	13
BMA	HEXBS	CDS	NCDS	KCDS
SP	11	9	5	5

DS consistently uses a large diamond search pattern (LDSP) through the whole searching process except in the last step which is a small diamond search pattern (SDSP). On the contrary, NCDS and KCDS always use an SDSP during the first two steps, followed by one or more LDSPs, and finally SDSP. Although NCDS and KCDS are particularly fast for slow-motion videos, they degrade video quality greatly for high-motion videos due to their fixed small starting search pattern. Nie and Ma [10] introduced an adaptive rood pattern (ARP) in which the ARP's size is dynamically determined for each block, based on the available MV of its left block. However, our experiments show that the blocks often do not have similar MV with their left-neighboring blocks especially for complex motion videos. Besides, the rood pattern in [10] is less efficient than DS due to its sparse shape when the rood arm stretches too long.

In this paper, we aim to develop a very simple yet efficient fast BMA to further reduce ME searching complexity and improve video quality. The next section shows center-biased MV distribution characteristics of natural videos. Section III is devoted to our detailed proposed algorithm. Extensive experiments are offered in Section IV. This paper concludes with Section V.

II. CENTER-BIASED MV DISTRIBUTION CHARACTERISTICS

It is widely known that the block motion field of natural video is usually gentle and smooth. Nie and Ma [10] suggested a zero-motion prejudgment (ZMP) to speed up the search. Without exception, we will further utilize both zero- and center-biased-MV characteristics, however in an improved manner. The ZMP [10] uses a fixed threshold of SAD 512 for ZMV decision. We use an adaptive threshold to decide ZMV because we observed many ZMVs have SAD values larger than 512. To demonstrate the property of the global minimum MV distribution, by applying FS with spiral block-matching scheme, we used twelve MPEG sequences which are listed in Table II. Among them, "Akiyo", "Grandma", "Claire", "Missa", and "Salesman" are regarded as low-motion videos. "Foreman", "Coastguard", and "Carphone" are medium-motion videos. "Stefan" and "Football" are high-motion videos. Zooming, fast moving object, and panning can be found in "Stefan" sequence. Some irregular motion appears in "Flowergarden" and "Mobile" such as rotation. After an in-depth analysis on the motion vector probability (MVP) distributions, we found that the zero MVP (ZMVP) is considerably high for gentle videos as depicted in Fig. 1 (a). Even in the case of high motion videos, ZMVP is more than 20% as shown in Fig. 1 (b). In the next section, we shall discuss our proposed algorithm that takes the advantage of zero- and center-biased-MV characteristics. However our algorithm is not just limited to this type of applications.

TABLE II – VIDEO SEQUENCES USED FOR ANALYSIS

Format	Frames	Sequences
QCIF	200	Akiyo, Claire, Salesman, Grandma, Missa, Foreman, Carphone, Coastguard, Stefan
CIF	100	Mobile, Flowergarden, Football

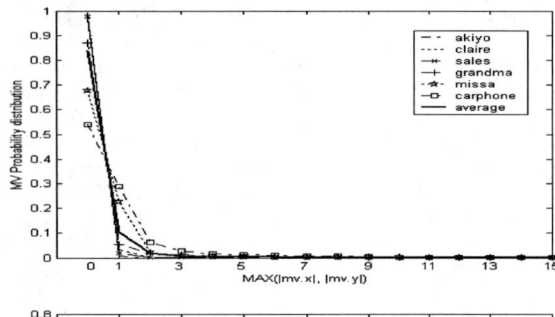

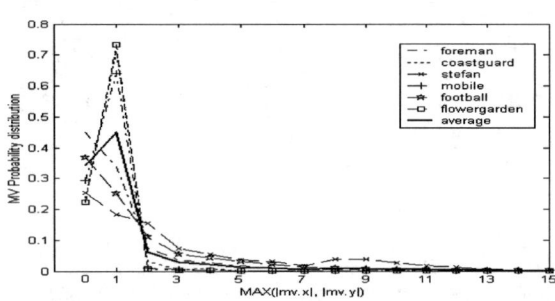

Fig. 1. Motion vector probability distributions in percentage (top) Slow-motion videos and (bottom) Medium-high-motion videos using FS with ±61 search area (only those of MVs with absolute distance of 0-15 are shown).

III. PROPOSED MODIFIED DS (MODS) ALGORITHM

To illustrate our algorithm, let the search window be Ω and the displacement with respect to current block located at (x, y) be (u, v). The SAD, for the current block q between current frame n and reference frame $n-\tau$, is used for the BDM which is obtained through

$$SAD_{n,q;u,v} = \sum_{i=0}^{N-1}\sum_{j=0}^{N-1} | f_n(i,j) - f_{n-\tau}(i+u, j+v) | \quad (1)$$

where $-\Omega \leq u, v \leq \Omega$ and N is the block size, $f(\cdot)$ is the pixel intensity.

A. FGHS via a Linear-Model Based Dynamic Threshold

A key improvement of MODS is a novel *fine granularity halfway-stop* (FGHS) strategy as plotted in Fig. 2 (fine solid line). During the search process, many conventional BMAs only stop if the minimum BDM found is in the center, after finishing a *complete* search step which usually consists of several SPs as shown in Fig. 2 (dotted lines). Other BMAs keep advancing search points by three or more new neighboring points before stopping even though the best MV is not coincident with the last point. In our algorithm, those redundant search points are avoided via a dynamic block difference threshold. Ideally, the dynamic threshold for the current qth block ($T_{FGHS,q}$) should be very close to the final minimum SAD as using FS. With this goal in mind, $T_{FGHS,q}$ is chosen adaptively to the content of block motion via a linear model. Let $SAD_{n,q;u,v}$ be the SAD of the current qth block in the nth frame. We use two terms to obtain $T_{FGHS,q}$,

namely, SAD ratio ($SADR_{n,q;0,0}$) and average minimum SAD ($ASAD_{n,q;\min}$). The $SADR_{n,q;0,0}$ is the ratio of $SAD_{n,q;0,0}$ to the average of SAD of the same location between the current frame and the reference frame up to the qth block in a frame, or mathematically

$$SADR_{n,q;0,0} = SAD_{n,q;0,0} \Big/ (\frac{1}{q}\sum_{m=1}^{q}SAD_{n,m;0,0}). \quad (2)$$

Since the minimum SAD of the qth block is not known yet, $ASAD_{n,q;\min}$ is calculated using all the blocks prior to the qth block as

$$ASAD_{n,q;\min} = \frac{1}{q-1}\sum_{z=1}^{q-1}SAD_{n,z;\min}. \quad (3)$$

The dynamic threshold $T_{FGHS,q}$ is computed using a linear model with $SADR_{n,q;0,0}$ and $ASAD_{n,q;\min}$. Mathematically,

$$T_{FGHS,q} = SADR_{n,q;0,0} * ASAD_{n,q;\min} * \lambda + \varepsilon \quad (4)$$

where λ is a constant scale factor and is set as 0.5, ε is a constant factor and is set as $1.8N^2$ empirically. Those parameters are carefully selected so that no or minimum premature endings happen. Plugging (2) and (3) into (4), the threshold is expressed as

$$T_{FGHS,q} = SAD_{n,q;0,0}\frac{q}{q-1}\frac{\sum_{z=1}^{q-1}SAD_{n,z;\min}}{\sum_{m=1}^{q}SAD_{n,m;0,0}}\lambda + \varepsilon. \quad (5)$$

Approximately, $q/(q-1)$ is close to 1. Eq. (5) is then simplified as

$$T_{FGHS,q} = SAD_{n,q;0,0}\frac{\sum_{z=1}^{q-1}SAD_{n,z;\min}}{\sum_{m=1}^{q}SAD_{n,m;0,0}}\lambda + \varepsilon. \quad (6)$$

If the $SAD_{n,q,u,v}$ of the current block is less than $T_{FGHS,q}$, we immediately stop the search process for this block and return the MV that has the minimum SAD so far. It is noted that the threshold is computed only once per block. The summations are accumulative so only two additions are required for each block. Since $SAD_{n,q;\min}$ and $SAD_{n,q;0,0}$ have already been estimated in the original DS algorithm, the extra operations are one multiplication, three additions, one division, and one right-shift ($\lambda = 0.5$ can be translated to $\gg 1$) in (6). Thus, the overhead of obtaining the threshold is negligible compared to the computation required by each block distortion calculation.

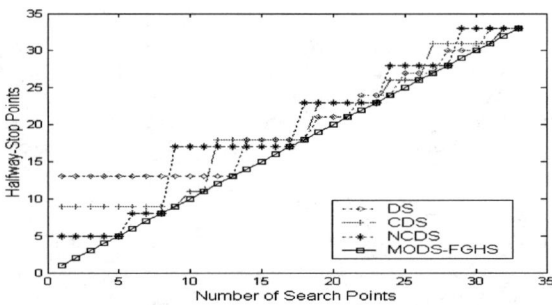

Fig. 2. Halfway-stop search points for conventional fast BMAs (large stair-step lines) and for the fine-granularity halfway-stop search points (straight line).

During the block-matching process, if the $SAD_{n,q,u,v}$ of the current matching is less than $T_{FGHS,q}$, we immediately stop the searching process for this block and return the motion vector that has the minimum SAD so far. Our FGHS is briefly summarized in pseudo-code as the following:

```
Compute T_FGHS,q using (6);
FOR each block in a frame DO
  Compute SAD_n,q;u,v using (1);
  IF (SAD_n,q;u,v < T_FGHS,q) THEN
    Immediately stop and return the MV which has the minimum SAD;
  ELSE
    Continue;
  ENDIF
DONE
```

Extensive simulations for videos with a wide range of motion content confirm that it is possible to achieve both good performance and speedup through FGHS. This is due to the accurate $T_{FGHS,q}$ by using the linear model.

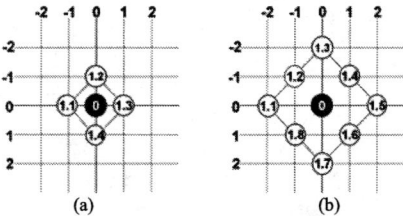

Fig. 3. Motion adaptive diamond search patterns. (a) SDSP. (b) MDSP.

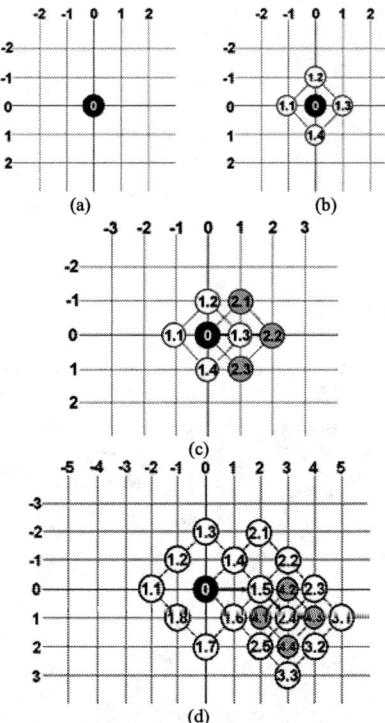

Fig. 4. Examples of the MODS: (a) initial-step-ZMV-stop with MV(0,0); (b) first sub-step stop with 1.1, 1.2, 1.3 or 1.4; (c) SDSP, second sub-step stop 2.1, 2.2, or 2.3 etc; (d) MDSP, a search path for MV(4,1), possible third, fourth, or final sub-step stop 3.1, 3.2, 4.1, or 4.2 etc.

B. Motion Adaptive Small/Medium Diamond Search Pattern

In favor of the low motion videos, NCDS and KCDS always start with a small diamond search pattern (SDSP) as depicted in Fig. 3(a). This starting may degrade video quality greatly for high- or irregular-motion sequence because it may be easily trapped to a local optimum. To remedy this inadequacy of NCDS and KCDS, our MODS begins the search with a medium diamond search pattern (MDSP), as shown in Fig. 3(b), if the $SAD_{n,q;0,0}$ of current block is greater than an MDSP threshold T_{MDSP} (e.g. $34N^2$ in our experiment). In other words, a large $SAD_{n,q;0,0}$ indicates the best match is relatively far from the center. By adaptively choosing SDSP/MDSP after the initial step, MODS becomes very effective and robust for any motion content video while retaining the merits of NCDS and KCDS. Even a larger diamond search pattern (LDSP) can be a starting choice if it is necessary. From our experiments, LDSP provides little speedup gain. For simplicity, we do not elaborate LDSP here.

C. MODS Algorithm Development

Our complete MODS consists of the following steps:
Initial Step (Starting – ZMV/SDSP/MDSP decision): Compute $SAD_{n,q;0,0}$ using (1). If the $SAD_{n,q;0,0}$ is less than the threshold $T_{FGHS,q}$, then the search stops (Initial Step ZMV Stop) as shown in Fig. 4(a); else if the $SAD_{n,k,u,v}$ is greater than T_{MDSP}, go to Step (iii) with an MDSP; else, go to Step (i) with an SDSP.

Step (i) (Small Diamond Shape Pattern -- SDSP): Examine four search points of the SDSP one-by-one located at the center of search window as shown in Fig. 4(b). If the $SAD_{n,q;u,v} < T_{FGHS,q}$, the search stops immediately (First Sub-Step Stop); otherwise go to Step (ii). To avoid being trapped into local optima, unlike other fast BMAs, our search does not stop here even if the minimum SAD occurs at the center of the SDSP.

Step (ii) (SDSP): With the vertex (minimum SAD point) in the previous SDSP as the center, a new SDSP is formed as in Fig. 4(c). Again, the SDSP search points are checked one-by-one. If the $SAD_{n,q;u,v} < T_{FGHS,q}$, the search stops immediately (Second Sub-Step Stop). If the minimum SAD search point occurs at the center of this SDSP, the search safely stops (Second Full-Step Stop); otherwise go to Step (iii).

Step (iii) (Medium Diamond Searching Pattern -- MDSP): An MDSP is formed by repositioning the minimum SAD found in previous step as the center of MDSP. The MDSP search points are checked one-by-one. If the $SAD_{n,q;u,v} < T_{FGHS,q}$, the search stops immediately (Third Sub-Step Stop); If the new minimum SAD search point is at the center of the newly formed MDSP, then go to Step (iv) for converging to the final solution; otherwise, this step is repeated as depicted in Fig. 4(d).

Step (iv) (Ending – Converging step): With the minimum SAD search point in the previous step as the center, an SDSP is formed. The SDSP search points are checked one-by-one. If the $SAD_{n,q;u,v} < T_{FGHS,q}$, the search stops immediately (Final Sub-Step Stop); Otherwise, identify the new minimum SAD search point for the SDSP, which is the final solution for the motion vector.

D. Analysis of MODS Algorithm

From previous sub-sections, the best case occurs at the initial step in which only one SP is performed. In addition, with our FGHS method, the search can be stopped at any point (e.g., 0, 1.1, 2.2, 3.1 or 4.1, etc) once the threshold is satisfied. Those methods make MODS the fastest BMA. The adaptive SDSP/MDSP selection provides more accurate MV for medium- and high-motion videos. Besides, from our simulations, MODS provides smoother motion fields hence demanding less MV difference (MVD) bits than many fast BMAs. The additional computational complexity of MODS itself is very low, while much more computations are reduced due to reducing the number of search points (as in Table III). Specifically, in addition to the DS, the extra computational complexity requires a few operations to obtain the threshold per block. For each SP, a few comparisons are added depending on the stop points. In summary, the overhead of MODS computation is small when compared to the many computations required by additional SPs using other methods.

IV. SIMULATION RESULTS

Numerous experiments have been conducted to justify the performance of our MODS algorithm. The peak-signal-to-noise-ratio (PSNR) was used as a performance measure. The MV search is based on the luminance component with the search block of 16×16, and the search window (Ω) was ±7 pixels. The simulation was performed with nine sequences in QCIF (each with 200 frames) and three sequences in CIF (each with 100 frames). Due to limitation in paper length, only some of them are shown here. We compare our MODS against FS, N3SS [1], 4SS [2], DS [3] [4], CDS [5], HEXBS [6], NCDS [7], and KCDS [8] in terms of PSNR and average number of search points per block (SPs). The PSNR difference (Δ PSNR) of reconstructed video was used, which is the PSNR difference between that of FS and that of a fast BMA. Due to boundary blocks, the average SPs per block is less than $(2\Omega + 1)^2$, i.e. 184.6 vs. 225 for FS with QCIF. The speedups of MODS are quite significant for all slow-, medium-, and high-motion videos as tabulated in Table III. Particularly it achieves near *one* SP for gentle video, which is 12 times faster than that of DS, 7.7 times faster than that of CDS, and 4.3 times faster than those of NCDS and KCDS, with comparable resulting PSNR. One slow-motion example of "Salesman" is given in Fig. 5. The performance improvement for high motion video is evident when compared to those of NCDS and KCDS. Subjective frame quality is reflected using a high-motion example of "Stefan" in Fig. 6. MODS is also very efficient in fractional-pel motion estimation search. Typically in that of a fractional-pel search, integer-pel search is performed first, followed by refined fractional-pel search, where the integer-pel search contributes to most of the computational complexity. MODS under other search windows ±16, ±32, and ±64 consistently shows its superiority over any other suboptimal BMAs.

V. CONCLUSION

In this paper, we proposed several effective techniques to reduce motion estimation computational complexity with no significant visual quality degradation. Our new algorithm MODS is simple, robust, and easy to implement. MODS shows distinct improvements in terms of speedup. Further experiments show that our algorithm also results in smoother motion field, hence demanding less motion vector difference bits when compared to that of many well-known fast BMAs. Our algorithm can be applied to video applications, such as video conferencing and video phoning due to its high speedup and superior performance.

REFERENCES

[1] R. Li, B. Zeng, and M. L. Liou, "A new three-step search algorithm for block motion estimation," *IEEE Trans. Circuits Syst. Video Technol.*, vol. 4, pp. 438-442, Aug. 1994.

[2] L. M. Po and W. C. Ma, "A novel four-step search algorithm for fast block motion estimation," *IEEE Trans. Circuits Syst. Video Technol.*, vol. 6, pp. 313-317, June 1996.

[3] J. Y. Tham, S. Ranganath, M. Ranganth, and A. A. Kassim, "A novel unrestricted center-biased diamond search algorithm for block motion estimation," *IEEE Trans. Circuits Syst. Video Technol.*, vol. 8, pp. 369-377, Aug. 1998.

[4] S. Zhu and K.-K. Ma, "A new diamond search algorithm for fast block-matching motion estimation," *IEEE Trans. Image Processing*, vol. 9, pp. 287-290, Feb. 2000.

[5] C. H. Cheung and L. M. Po, "A novel cross-diamond search algorithm for fast block motion estimation," *IEEE Trans. Circuits Syst. Video Technol.*, vol. 12, pp. 1168-1177, Dec. 2002.

[6] C. Zhu, X. Lin, and L.-P. Chau, "Hexagon-based search pattern for fast block motion estimation," *IEEE Trans. Circuits Syst. Video Technol.*, vol. 12, pp. 349-355, May 2002.

[7] C. W. Lam, L. M. Po, and C. H. Cheung, "A new cross-diamond search algorithm for fast block matching motion estimation," *IEEE International Conference on Neural Networks and Signal Processing*, pp. 1262-1265, Nanjing, China, Dec. 2003.

[8] C. W. Lam, L. M. Po, and C. H. Cheung, "A novel kite-cross-diamond search algorithm for fast block matching motion estimation," *IEEE ISCAS*, vol. III, pp. 729-732, May 2004.

[9] A. Tourapis, O. C. Au, and M. L. Liou, "Highly efficient predictive zonal algorithm for fast block-matching motion estimation," *IEEE Trans. Circuits Syst. Video Technol.*, vol. 12, pp. 934-947, Oct. 2002.

[10] Y. Nie and K.-K. Ma, "Adaptive rood pattern search for fast block-matching motion estimation," *IEEE Trans. Image Processing*, vol. 11, pp. 1442-1449, Dec. 2002.

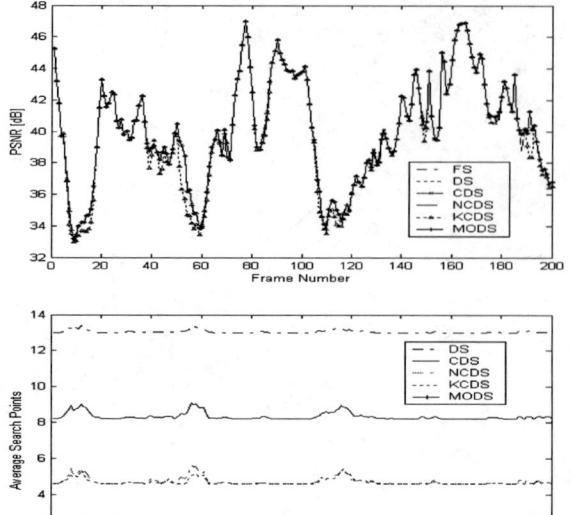

Fig. 5. Frame by frame comparisons for "Salesman" sequence. Top: Average PSNR. Bottom: Average search points per block.

Fig. 6. Resulting images (left) and absolute value of the error frame (right) of "Stefan" frame 128. The error is amplified by a factor of 5 and truncated to 255. (a) KCDS is used, PSNR = 18.27 dB. (b) MODS is used, PSNR = 26.55 dB.

TABLE III - SIMULATION RESULTS OF BMAs IN TERMS OF Δ PSNR (dB) AND NUMBER OF SPs PER BLOCK (FS SHOWS ACTUAL PSNR)

Sequence	Format	Δ PSNR	Block Matching Algorithms								
	Frames	SP	FS	N3SS	4SS	DS	CDS	HEXBS	NCDS	KCDS	**MODS**
Akiyo	QCIF	Δ PSNR	44.94	0	0	0	0	-0.02	0	-0.15	**0**
	200	SP	184.6	14.68	14.66	13.01	8.222	9.65	4.620	4.620	**1.140**
Carphone	QCIF	Δ PSNR	33.70	-0.07	-0.18	-0.1	-0.15	-0.39	-0.17	-1.11	**-0.37**
	200	SP	184.6	16.75	15.61	14.63	11.00	10.36	7.428	6.441	**3.116**
Foreman	QCIF	Δ PSNR	32.40	-0.11	-0.23	-0.18	-0.26	-0.65	-0.29	-1.76	**-0.44**
	200	SP	184.6	16.88	15.89	14.81	11.09	10.51	7.544	6.731	**3.904**
Stefan	QCIF	Δ PSNR	25.21	-0.19	-0.46	-0.49	-0.54	-0.53	-0.71	-2.91	**-0.64**
	200	SP	184.6	19.28	16.72	16.74	14.62	11.70	11.39	7.638	**7.469**
Flower garden	CIF	Δ PSNR	25.63	-0.01	-0.12	-0.02	-0.01	-0.11	-0.02	-7.28	**-0.04**
	100	SP	204.3	18.47	17.13	15.28	12.98	11.53	9.079	6.088	**5.756**
Mobile	CIF	Δ PSNR	23.89	-0.03	-0.17	-0.08	-0.05	-0.25	-0.06	-1.15	**-0.10**
	100	SP	204.3	17.99	16.20	14.07	11.40	10.70	7.207	7.276	**5.318**
Football	CIF	Δ PSNR	23.36	-0.37	-0.57	-0.60	-0.69	-0.79	-0.76	-2.46	**-0.76**
	100	SP	204.3	20.93	18.40	17.81	15.89	12.61	12.68	10.20	**10.20**

CURRENT MODE MULTI-LEVEL SIMULTANEOUS BIDIRECTIONAL I/O SCHEME FOR CHIP-TO-CHIP COMMUNICATIONS

Ge Yang, YongSin Kim, and Sung-Mo Kang

Univ. of California, Santa Cruz
Department of Electrical Engineering
1156 High Street, Santa Cruz, CA 95064

ABSTRACT

We have developed a novel current mode multi-level simultaneous bidirectional I/O scheme for high throughput chip-to-chip communications. Data are represented by multi-level currents on the transmission line instead of multi-level voltages. The current mode scheme consumes less power and is faster by reducing the voltage swing on the transmission line. Simulations based on 0.18um CMOS SPICE parameters show that the data rate is as high as 4Gb/S/Pin, and the power consumption is only 34mW.

1. INTRODUCTION

The chip-to-chip communication is limited by the interconnection bandwidth between chips. Most CMOS chips drive unterminated lines with full-swing CMOS drivers and use CMOS gates as receivers. Such full-swing CMOS interconnect must ring-up the line, and hence has a bandwidth that is limited by the length of the line rather than the performance of the semiconductor technology [1]. Thus high speed I/O circuits are important to alleviate the communication bottleneck between IC chips. Power consumption is also becoming a critical design issue since parallel-processing computer servers, network switching fabrics, and CPU-memory interfaces require hundreds of high speed links be integrated on a single chip [2, 3].

To improve the I/O data rate, simultaneous bidirectional signaling with a high frequency clock generator was employed [4]. However, high frequency clock generator requires small clock jitter and low power consumption. Multi-level signaling with a current DAC was proposed [5], which requires more symbol-margin between voltages levels and improved impedance matching. Multi-level simultaneous bidirectional signaling was implemented [6], however full Vdd (1.8V for 0.11um process) was used for higher voltage margin, which leads to large power consumption.

2. BASIC IDEA

Although the multi-level simultaneous bidirectional signaling implemented in work [6] can improve the I/O data rate, full Vdd swing on the transmission line increases the switching power at the pins.

We propose to use the current mode for multi-level simultaneous bidirectional signaling. Data are represented by multi-level currents on the transmission line instead of multi-level voltages. The current mode scheme consumes less power and is faster by reducing the voltage swing on the transmission line. The current mode 1Gb/S bidirectional I/O has been proposed [7]. However only one bit was sent and received at the same time.

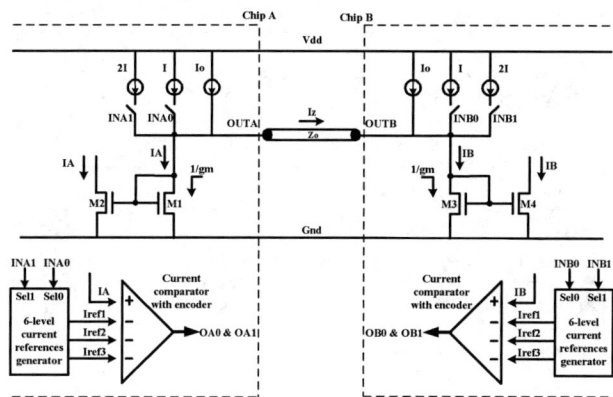

Figure 1. Proposed Multi-level Current Mode Simultaneous Bidirectional I/O Scheme

The proposed Multi-level Current Mode Bidirectional I/O Scheme is shown in Figure 1. Chip A and chip B are connected through a transmission line. INA1 and INA0 are the 2-bit outgoing data on chip A, and OA0 and OA1 are the received 2-bit data (from chip B). INB1 and INB0 are the 2-bit outgoing data on chip B, and OB0 and OB1 are the received 2-bit data (from chip A). The current source I and the current source 2I are switched by the 2-bit outgoing data on each chip. If we apply KCL to node Gnd, node OUTA, and node OUTB, we can get the following equations:

$$Iz = \frac{INA1*2I + INA0*I - INB1*2I - INB0*I}{2} \quad (1)$$

$$IA = IB = \frac{INA1*2I + INA0*I + INB1*2I + INB0*I + Io}{2} \quad (2)$$

Iz is the current flowing in the transmission line. IA and IB are the currents flowing through M1 and M3, respectively. M1 and M3 are diode-connected, and act as termination resistors for the transmission line. Impedance control is needed for M1 and M3, which is not shown in Figure 1. Table 1 shows the current flowing through the transmission line for all 16 possible input combinations. We can see from Table 1 that there are seven current levels on the transmission line. Table 2 shows the current flowing through the termination transistor. For outgoing data, say (0, 0) from chip A, there are four possible current levels, corresponding to the four possible outgoing data from chip B. In order to recover the data received at the chip, three reference currents Io+0.25I, Io+0.75I, and Io+1.25I are needed. Table 3 shows the reference currents required by different outgoing data. Six current references are used. We can see from Figure 1, IA is duplicated using the current mirror. The reference current generator is capable of generating the six current references, and 3 of the references are selected by the outgoing data (INA1, INA0). Then IA and the three selected current references are fed to the current comparator with encoder. The output data (OA1, OA0) are the recovered data coming from chip B.

Table 1. Transmission line current

Iz	(INB1,INB0)=(0, 0)	(0,1)	(1,0)	(1,1)
(INA1,INA0)=(0,0)	0	-0.5I	-I	-1.5I
(INA1,INA0)=(0,1)	0.5I	0	-0.5I	-I
(INA1,INA0)=(1,0)	I	0.5I	0	-0.5I
(INA1,INA0)=(1,1)	1.5I	I	0.5I	0

Table 2. IA & IB

IA=IB	INB1,INB0=(0, 0)	(0,1)	(1,0)	(1,1)
INA1,INA0=(0,0)	Io	Io+0.5I	Io+I	Io+1.5I
(0,1)	Io+0.5I	Io+I	Io+1.5I	Io+2I
(1,0)	Io+I	Io+1.5I	Io+2I	Io+2.5I
(1,1)	Io+1.5I	Io+2I	Io+2.5I	Io+3I

Table 3. Reference currents

Reference I	Ref1	Ref2	Ref3
(INA1,INA0)=(0,0)	Io+0.25I	Io+0.75I	Io+1.25I
(0,1)	Io+0.75I	Io+1.25I	Io+1.75I
(1,0)	Io+1.25I	Io+1.75I	Io+2.25I
(1,1)	Io+1.75I	Io+2.25I	Io+2.75I

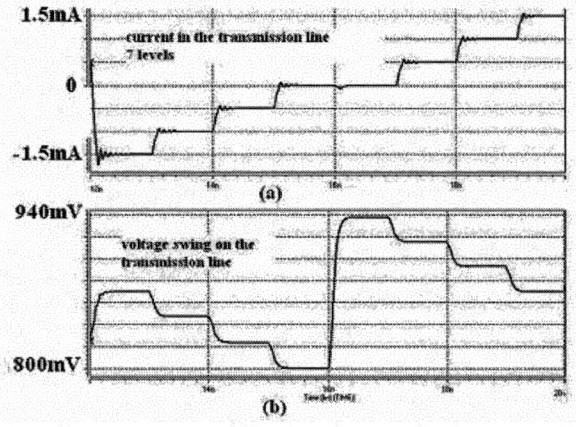

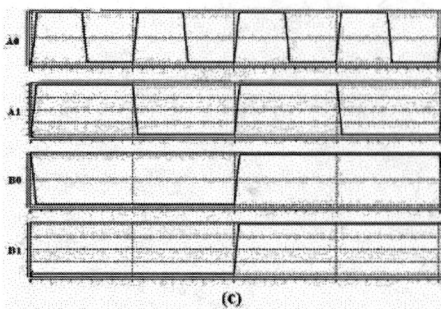

Figure 2. (a) Seven current levels on the transmission line (b) low voltage swing at one end of the transmission line (c) inputs

Figure 2(a) shows the seven current levels on the transmission line. Figure 2(b) shows the voltage swing on one end of the transmission line. The voltage swing is between 800mV and 940mV. Figure 2(c) shows the two-bit input waveforms on the two chips. In the simulation, ideal current sources were used.

3. IMPROVED I/0 STRUCTURE TO REDUCE CURRENT GLITCHES

3.1 Current glitches on the transmission line

Figure 3 shows the circuit used to switch the current source on and off depending on the data IN. PMOS transistors M1 and M2 are biased so that they are both in saturation and they provide current I. Transistors M3 and M4 are used to switch this current source. When the data IN is high, M4 is off, M3 is on, the bias voltage is applied to the gate of M1, and the current source is on. When the data IN is low, M3 is off, and M4 is on. M4 turns off M1, and the current source is disabled.

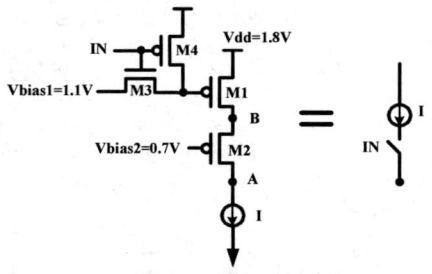

Figure 3. Current source controlled by data (IN)

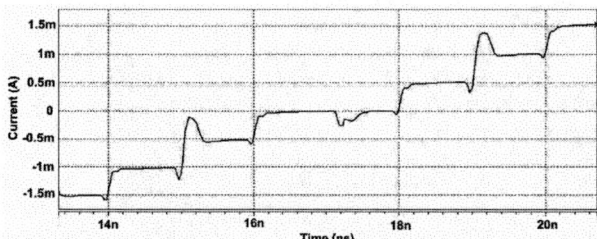

Figure 4. 7 current levels on the transmission line with glitches

We applied the current source circuit in Figure 3 to the proposed I/O scheme in Figure 1, and Figure 4 shows the seven current levels on the transmission line. However, there are some glitches on the transmission when the current level is about to change. We can analyze how the glitch happens around time 15ns. Figure 5 shows the input patterns. INB0 and INB1 are low from 14ns to 16ns. INA0 switches from low to high at 15ns, while INA1 switches from high to low at 15ns. INA0 controls a 1mA current source and INA1 controls a 2mA current source. This means at 15ns, a 1mA current source is switching on, while a 2mA current source is switching off. The glitch happens due to the race between the two current sources switching simultaneously in opposite directions.

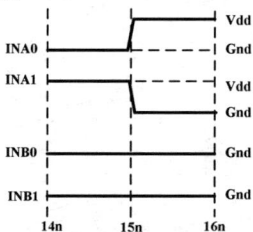

Figure 5. Input patterns

3.2 Improved I/O structure to reduce current glitches

We need to find a way so that when the outgoing data (INA0, INA1) change, the current sources switch in the same direction, either both off or both on. We propose to add one more current source, and the three current sources are control by X2, X1, and X0, which are derived from INA0 and INA1.

Table 4. Original current source control scheme

INA1 (controls 2mA)	INA0 (control 1mA)	Total current
0	0	0
0	1	1mA
1	0	2mA
1	1	3mA

Table 4 shows the original current source control scheme. When the pattern of (INA1, INA0) change between (1,0) and (0,1), the glitch problem happens. Table 5 shows the proposed current source control scheme using 1 more current source. We can see from Table 5 that when the pattern of (X2, X1, X0) changes, no current sources switch opposite directions. This helps to solve the glitch problem.

Table 5. Proposed current source control scheme without glitches

X2 (1mA)	X1 (1mA)	X0 (1mA)	Total current
0	0	0	0
0	0	1	1mA
0	1	1	2mA
1	1	1	3mA

We can also easily derive the following relations:

$$X2 = INA1 \bullet INA2 \quad (3)$$
$$X1 = INA1 \quad (4)$$
$$X0 = INA0 + INA1 \quad (5)$$

The modified Multi-level Current Mode Bidirectional I/O Scheme is shown in Figure 6. Simulation for the proposed current sources control scheme shows that the glitch problem is gone, the seven levels on the transmission line is very clean.

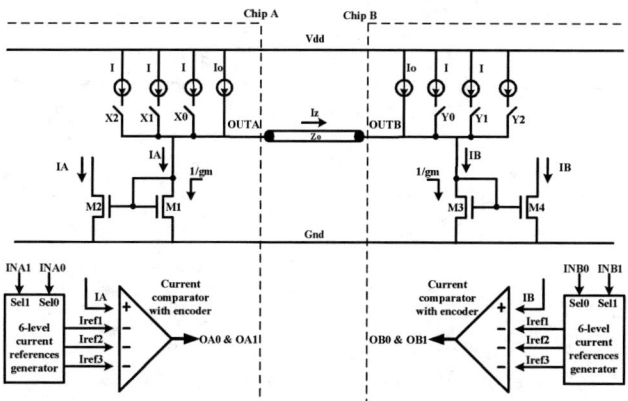

Figure 6. Modified Multi-level Current Mode Bidirectional I/O Scheme

Figure 7 shows the current waveforms on the transmission line of the improved I/O structure.

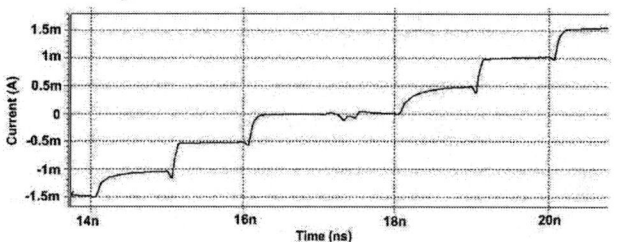

Figure 7. Seven current levels without glitches

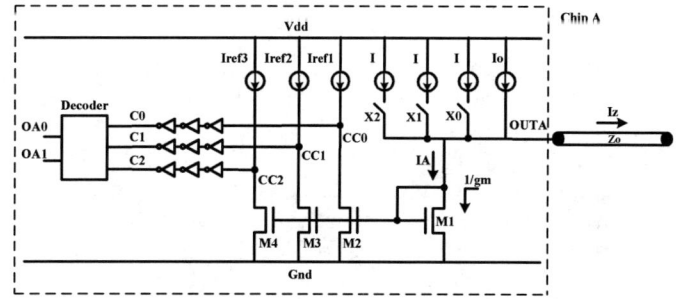

Figure 8. I/O circuits on Chip A

4. OTHER CIRCUIT BLOCKS

We next examine how the current comparison is done. Figure 8 shows the I/O circuits on Chip A. IA is mirrored to M2, and M2 is connected to current reference Iref1. Assume when IA=Iref1, the voltage at node CC0 is V0. If IA>Iref1, Vcc0=V0-ΔV. If IA<Iref1, Vcc0=V0+ΔV. The inverters connected to CC0, CC1, and CC2 are designed to flip at V0, so the comparison results can be recovered with full swing at nodes C0, C1, and C2. The decoder is used to recover the data coming from Chip B.

Table 6. Encoder Inputs & Outputs

(C2, C1, C0)	(OA1, OA0)
(0, 0, 0)	(0, 0)
(0, 0, 1)	(0, 1)
(0, 1, 1)	(1, 0)
(1, 1, 1)	(1, 1)

Table 6 shows the encoder inputs and outputs. We can derive the logic relations from the truth table:

$$OA0 = C0 \bullet C1 \quad (6)$$
$$OA1 = C0 \bullet \overline{C1} \bullet \overline{C2} + C0 \bullet C1 \bullet C2 \quad (7)$$

Table 3 shows the reference currents we need for different outgoing data. In order to avoid current glitches in the current references, the same method we used for the switched current sources was applied. Figure 9 shows how we can obtain those reference currents without glitches.

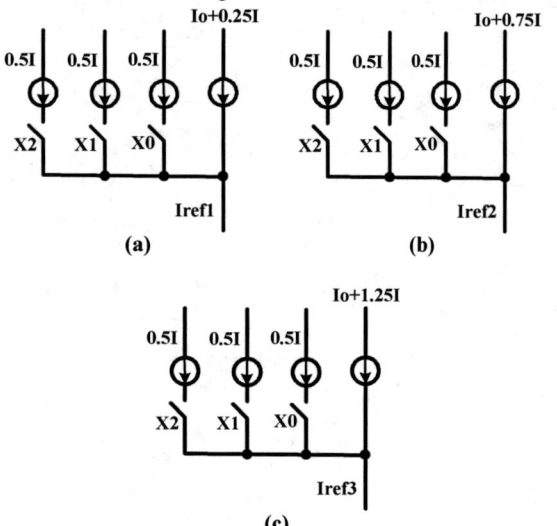

Figure 9. Current references (a) Iref1 (b) Iref2 (c) Iref3

5. SIMULATION RESULTS

We used 0.18um CMOS SPICE parameters for the simulation. 1m cable model was used for the transmission line. Vdd was 1.8V, and the temperature was 120°C. Figure 10 shows the outgoing data at Chip A and Chip B. Figure 11 shows the recovered data at Chip B and Chip A. Simulations have confirmed the correct operation of the proposed multi-level bi-directional current mode I/O scheme. The performance is 4Gbit/S/Pin, and the power consumption is 50mW.

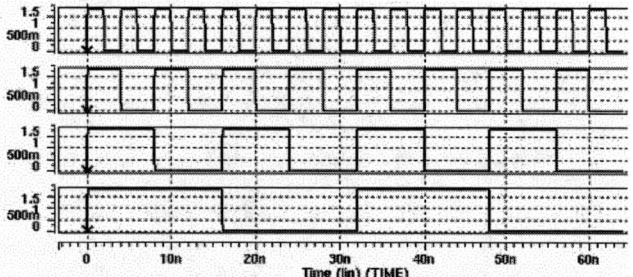

Figure 10. Outgoing data from Chip A and Chip B

Figure 8 shows the I/O circuits on chip A. The power consumed by the current comparison circuits (Irefs, M2, M3, and M4) is the static short circuit power. If we want to reduce this power, we need to reduce the reference currents and the currents flowing through M2, M3, and M4. I and Io can not be changed, since we need to keep the termination transistor M1 in saturation and have a Z=50Ω. Thus we scaled the transistor width of M2, M3, and M4 to 50% of M1, so that the mirrored currents are reduced to one half of IA. At the same time, the current references are also scaled by 50%. The current was not scaled too much, since the difference between Iref and IA was also scaled.

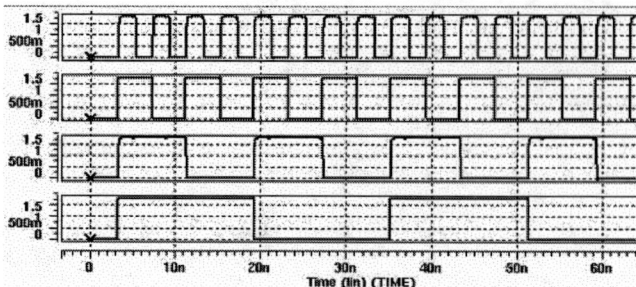

Figure 11. Received data at Chip B and Chip A

The power consumption was reduced from 50 mW to 34mW, which is around 32% power reduction. The performance was not changed, it was still 4Gb/S/Pin. The total transistor width of the I/O circuits is 2100um.

6. SUMMARY

We have developed a novel current mode multi-level simultaneous bidirectional I/O scheme for chip-to-chip communications. We have designed I/O circuits for high speed links with higher data rate, lower power consumption. Simulations based on 0.18um CMOS SPICE parameters show that the data rate is as high as 4Gb/S/Pin, and the power consumption is only 34mW.

REFERENCES

[1] M.-J.E. Lee et al., "Low-power area-efficient high-speed I/O circuit techniques", IEEE Journal of Solid-State Circuits, Volume: 35, Issue:11, Nov. 2000, Page(s):1591–1599

[2] E. Reese et al., "A phase-tolerant 3.8 GB/s data-communication router for multi-processor super computer backplane," IEEE ISSCC Dig. Tech. Papers, Feb. 1994, Page(s): 296–297

[3] M. Galles et al., "Spider: A high-speed network interconnect," IEEE Micro, Volume:17, Jan./Feb. 1997, Page(s): 34–39

[4] H. Wilson et al., "A six-port 30-GB/s nonblocking router component using point-to-point simultaneous bidirectional signaling for high-bandwidth interconnects", IEEE Journal of Solid-State Circuits, Volume: 36, Issue:12, Dec. 2001, Page(s): 1954 – 1963

[5] D.J. Foley et al., "A low-power 8-PAM serial transceiver in 0.5-μm digital CMOS", IEEE Journal of Solid-State Circuits, Volume:37, Issue:3, March 2002, Page(s):310 – 316

[6] Jin-Hyun Kim et al., "A 4Gb/s/pin 4-Level Simultaneous Bi Directional IO Using a 500MHz Clock for High-Speed Memory", 2004 IEEE International Solid-State Circuits Conference, Tech. Papers, Feb 2004, Page(s): 412 – 413

[7] J.-Y. Sim et al., "1 Gb/s current mode bidirectional I/O buffer", 1997 Sym. On VLSI Circuits, 1997, Page(s): 121–122

Algorithm for Peak to Average Power Ratio Reduction operating at Symbol Rate

Stefano Marsili
Infineon Technologies Austria AG
Siemensstrasse 2, A-9500 Villach, Austria
Email: stefano.marsili@infineon.com

Abstract—Multi-tone or multi-code modulation schemes like OFDM or WCDMA have the disadvantage of a large *Peak to Average Power Ratio (PAPR)*. Efficient transmission of such high data rate signals is limited by the non-linearity of the *Power Amplifier (PA)*. This causes interference within and outside the signal band. Without any measures this interference can be reduced only with a sufficient input back off of the PA. In order to achieve a better efficiency in the PA, in addition to pre-distortion methods, the signal PAPR should be reduced. This work shows a new algorithm which limits the PAPR without introducing any additional out-of band emission. Simulations for the WCDMA downlink signal are presented and compared to a state of the art solution.

I. INTRODUCTION

In modern high data rate communication systems, the transmitted signal is obtained as superposition of orthogonal sequences randomly modulated in both amplitude and phase. Examples are the *Orthogonal Frequency Division Multiplex (OFDM)* system like WLAN or DVB-T and the *Wideband Code Division Multiple Access (WCDMA)* system for the third generation mobile communication.

The sum of random and independent signals, results in a Gaussian distributed process that is characterized by large *Peak to Average Power Ratio (PAPR)*. If k was the number of orthogonal sequences, the maximum theoretical PAPR respect to a single sequence could increase up to factor k. This large dynamic range makes the design of a power efficient analog transmit path very difficult, especially for the PA. To ensure linearity, this must operate backed-off from its saturation region at least by the PAPR and therefore very inefficiently. This increases cost, power consumption and size of the transmitter.

For efficient implementations of the analog transmitter part, the PAPR has to be reduced. In this work a new method operating at symbol rate is proposed. In sec. II a system

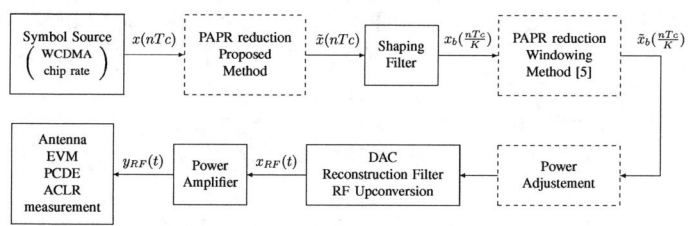

Fig. 1. WCDMA downlink system overview. The dashed boxes can be inserted for PAPR reduction.

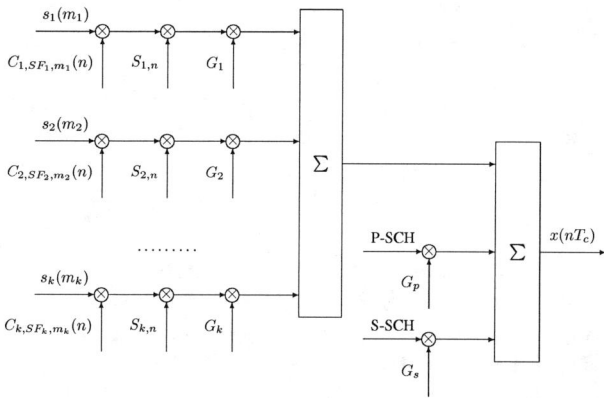

Fig. 2. Block diagram of the WCDMA modulation scheme according to [1].

overview of the WCDMA system is given and the minimum power back off allowed in the system is derived by simulation. In sec. III the algorithm is described in detail and in sec. IV simulation results are presented and compared to methods in the literature.

II. SYSTEM OVERVIEW

The communication system that was considered in this work is sketched in Fig. 1, where the solid line blocks represent the system before any PAPR compensation. For simulation purpose it was selected the WCDMA modulation scheme [1] but the model is suited also for different systems like OFDM.

The first block in the chain is the symbol source. It provides complex symbols at the rate F_c. In WCDMA F_c is the *chip rate* of 3.84 MHz. The complex symbols are then interpolated by the shaping filter, that in [1] is a root raised cosine filter with 22% roll off. Depending on the RF architecture chosen (super-heterodyne or direct conversion), the signal is digital to analog converted by one or two DAC and then up-converted to the final RF carrier frequency f_0.

Before the antenna the signal enters the PA. Its efficiency is higher when operating close to saturation, but in order to maintain spectral requirements and good signal quality at the receivers, it is necessary to back off the operation point and therefore an over sized PA is required.

Fig. 2 shows in more detail the structure of the complex symbol source for the WCDMA downlink signal [1]. User and control complex channels $s_1(m_1)\ldots s_k(m_k)$ with different rates are first spreaded with different spreading factors $SF_1 \ldots SF_k$ to the same chip rate. The k signals so obtained are then scrambled with complex sequences $S_{i,n}$, their power levels are adjusted by factors G_i and they are added together with two further synchronization signals to the output $x(nT_c)$.

The standard [2] specifies several test cases for the conformance requirements of a base station. For simulations it was selected the *Test Model 1 (TM1)* with 64 user channels operating at spreading factor 128. When sizing the PA, referring only to the maximum theoretical PAPR of TM1 would be too pessimistic, therefore it was preferred to refer to the PAPR value where its *complementary cumulative distribution function (ccdf)* assumes a probability of 10^{-4}. For TM1 signal this PAPR(ccdf=10^{-4}) value is 9.5 dB at the shaping filter output (see Fig. 6). For distortion-less transmission, the PA must be backed off at least by this value from saturation. The standard [2] allows anyhow some distortion and specifies signal quality requirements in terms of

1) *Error Vector Magnitude (EVM)* which should be below 17.5%,
2) *Peak Code Domain Error (PCDE)* that should be smaller than -33 dB for a SF=256 and
3) *Adjacent Channel Leakage power Ratio (ACLR)* which should be larger than 45 dB for the adjacent channel at 5 MHz and than 50 dB for the channel at 10 MHz.

For a more precise evaluation of the minimal back off required, it is necessary to introduce in simulation a PA model. A classical third order non linear model was chosen, in which the output of the PA, $y_{RF}(t)$, can be expressed according to the equation

$$y_{RF}(t) = G \cdot x_{RF}(t) - a_3 \cdot x_{RF}^3(t), \quad (1)$$

where G is the small signal gain of the PA, a_3 is the coefficient of the third order non linearity and $x_{RF}(t)$ is the input signal to the PA. This latter can be expressed in terms of the complex baseband signal $x_b(t)$ according to the known equation

$$x_{RF}(t) = \Re\{x_b(t) \cdot e^{j2\pi f_0 t}\} = \Re\{[x_{re}(t) + jx_{im}(t)] \cdot e^{j2\pi f_0 t}\}, \quad (2)$$

where f_0 is the RF carrier frequency. With the aid of (2) it is possible to rearrange (1) and to derive a complex baseband representation of the PA output signal $y_{RF}(t)$. As the terms

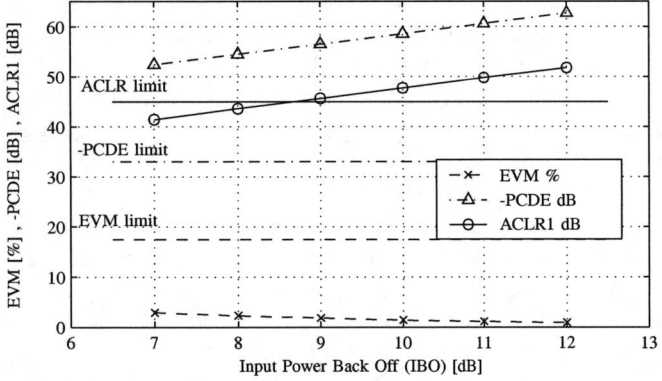

Fig. 3. Influence of the input power back off on the ACLR, EVM and PCDE for the uncompensated TM1 signal.

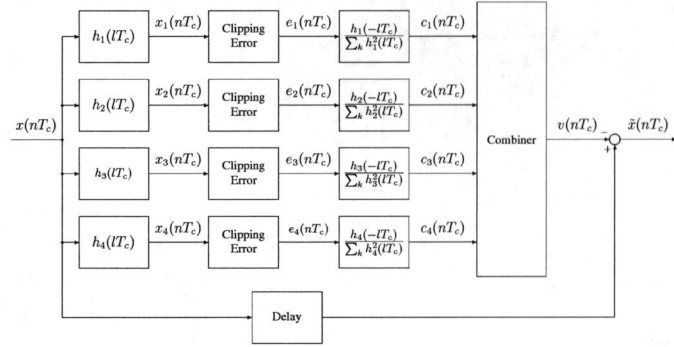

Fig. 4. Block diagram of the PAPR reduction algorithm operating at the input of the shaping filter (symbol rate).

modulated around three time the carrier frequency $3f_0$ are not relevant for the current analysis, this can be finally reduced to

$$y_b(t) = y_{re}(t) + j \cdot y_{im}(t)$$
$$y_{re}(t) = [G \cdot x_{re}(t) - a_3 \cdot \frac{3}{4} \cdot (x_{re}^3(t) + x_{re}(t) \cdot x_{im}^2(t))] \quad (3)$$
$$y_{im}(t) = [G \cdot x_{im}(t) - a_3 \cdot \frac{3}{4} \cdot (x_{im}^3(t) + x_{im}(t) \cdot x_{re}^2(t))].$$

A typical design parameter for PA is the *1 dB compression point*. This input power level defines the beginning of the saturation region, where the average output power for a pure sinusoidal signal is 1 dB smaller than the output power of the extrapolated linear PA. In this work the *Input power Back Off (IBO)* expresses the ratio between the input power at the 1 dB compression point and the average input power of $x_{RF}(t)$.

With the aid of the PA baseband model (3) it was measured the minimal IBO required in order to be conform to [2]. The results are presented in Fig. 3, where the solid line represents the ACLR, the dashed line the EVM and the dashed dotted line the absolute value of the PCDE. The corresponding limits are also reported on the diagram. The system results to be limited by the ACLR and the IBO should not be smaller than 8.7 dB in order not to violate the emission limits on the adjacent channels.

For this IBO value the margins on the EVM and PCDE remain very large an it would be even possible to work with IBO below 5 dB without infringing the specifications.

In the next sections it will be described a new algorithm which trades some of this enormous margin available for EVM and PCDE against performance improvement in the ACLR in order to reduce the limitation on the IBO and consequently to relax the specifications on the PA.

III. PROPOSED METHOD

Several algorithms for PAPR reduction can be found in the literature. Those methods can be grouped in

- *additive* - where signal peaks are compensated by subtracting pulses with desired spectral properties [3] [4],
- *multiplicative* - where the peaks are attenuated by windowing the signal around them [5] [6].

The solution considered for this work belongs to the additive methods but the signal is corrected before the shaping filter in Fig. 1. Applying the correction in that point has the advantage that the compensated signal exhibits exactly the same spectrum as the uncompensated one and therefore no additional out of band interference appears in the PA input $x_{RF}(t)$.

A method where peaks detected at the shaping filter input were simply attenuated was already presented in [7]. However, the time location of the power peaks after many interpolation stages does not necessarily correspond to the symbol rate position. New peaks can arise and escape the correction procedure. In order to improve this, it is necessary to predict the peaks at the output of the interpolation stages and correct the signal accordingly.

Fig. 4 shows a block diagram of the new method proposed. The signal from the symbol source $x(nT_c)$ enters the first filter bank which has the task of predicting the output of the interpolation stages for different time phases during one symbol interval T_c. Simulations show that 4 phases are sufficient to predict most of the peaks. A similar result was obtained in [8] [9] where an oversampling factor four proofed to be sufficiently accurate.

The four phases were selected uniformly distributed over one symbol interval. For simulations the effects of the shaping filter only were considered and Fig. 5 shows its impulse response (solid line) together with the coefficients selected for the four predictor filters $h_1(\cdot)\ldots h_4(\cdot)$. Eight coefficients per filter gives already performance close to the optimal.

It is straightforward to include further effects, for instance the reconstruction filter that follows the DAC, once it is known an equivalent impulse response from the symbol source to the particular point of interest where a PAPR reduction is desired.

For each of the predicted signal phases $x_i(nT_c)$, it is then checked whether the magnitude is exceeding a desired threshold δ. In this case an error signal $e_i(nT_c)$ is computed according to the equation

$$e_i(nT_c) = \begin{cases} 0 & \text{if } |x_i(nT_c)| \leq \delta \\ x_i(nT_c) - \frac{x_i(nT_c)}{|x_i(nT_c)|} \cdot \delta & \text{if } |x_i(nT_c)| > \delta. \end{cases} \quad (4)$$

These error signals need to be corrected at output of the shaping filter. The target of the algorithm is anyhow to inject a correction signal just after the symbol source. Therefore the error signals need to be computed back at the input of the interpolation stages. As this correction is seen by the receiver as an additional noise source and it degrades EVM and PCDE

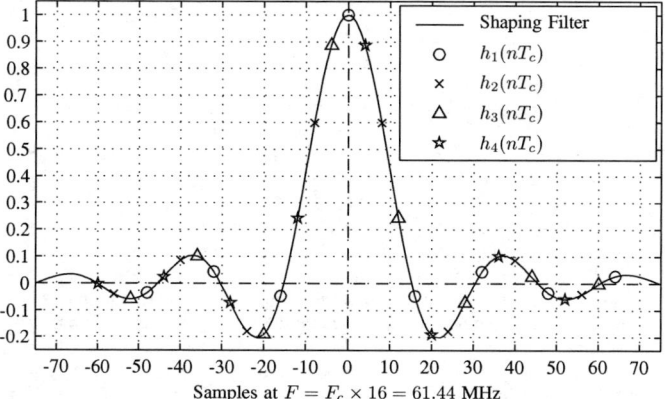

Fig. 5. Impulse response of the RRC shaping filter and of the four predictors.

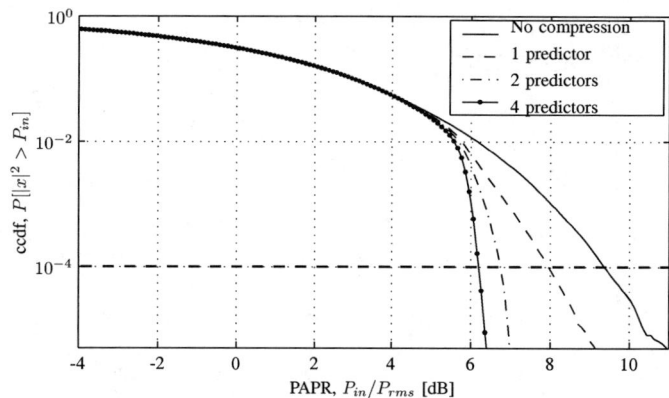

Fig. 6. PAPR Complementary Cumulative Distribution Function (*ccdf*) for TM1 signal with 64 user channels.

values the correction signal $c_i(nT_c)$ with the minimum power is desired. This can be computed by filtering the errors $e_i(nT_c)$ with the filters

$$\hat{h}_i(lT_c) = \frac{h_i(-lT_c)}{\sum_l h_i^2(lT_c)} . \quad (5)$$

These filters are the time reversed replica of the prediction filters normalized by their energy (matched filter).

Once the four equivalent corrections at symbol rate are available, it is necessary to combine them together to a single correction signal $v(nT_c)$. Normally more than one phase needs to be corrected at the same time and more than one correction signal is different from zero. The best results were obtained by extracting among the four signals the real and the imaginary part with the largest magnitude, namely

$$v_{re}(nT_c) = \Re[c_1(nT_c)]$$
if $|\Re[c_2(nT_c)]| > |v_{re}(nT_c)|$ then $v_{re}(nT_c) = \Re[c_2(nT_c)]$
$\ldots$
if $|\Re[c_4(nT_c)]| > |v_{re}(nT_c)|$ then $v_{re}(nT_c) = \Re[c_4(nT_c)]$

$$v_{im}(nT_c) = \Im[c_1(nT_c)] \quad (6)$$
if $|\Im[c_2(nT_c)]| > |v_{im}(nT_c)|$ then $v_{im}(nT_c) = \Im[c_2(nT_c)]$
$\ldots$
if $|\Im[c_4(nT_c)]| > |v_{im}(nT_c)|$ then $v_{im}(nT_c) = \Im[c_4(nT_c)]$

$$v(nT_c) = v_{re}(nT_c) + j \cdot v_{im}(nT_c) .$$

Finally the correction signal $v(nT_c)$ is subtracted from a delayed version of the symbols sequence $x(nT_c)$. The delay should compensate the latency introduced by the filters and the processing time of the blocks.

IV. SIMULATION RESULTS

The correction capability of the new algorithm was first proofed at the shaping filter output. It was first investigated the number of predictor paths needed. Fig. 6 illustrates the improvement in the ccdf of TM1 for the same threshold value δ for 1, 2 or 4 predictors. In order to obtain the same compensation level with one predictor only, it would be necessary to reduce the threshold by 2 dB or more, introducing therefore larger noise power in the signal. The single predictor case represents approximately the method in [7] where only a reduced number of peaks can be corrected. More than four predictors do not add any relevant performance improvements.

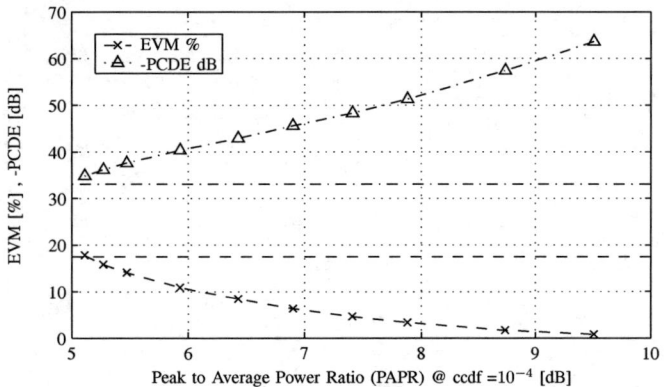

Fig. 7. ACLR and PCDE performance of the new algorithm as a function of the PAPR(ccdf=10^{-4}) at the shaping filter output

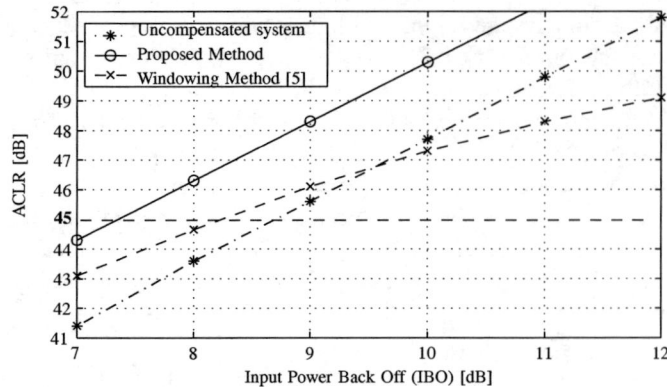

Fig. 9. Influence of the input power back off on the ACLR before and after compensation. The IBO can be reduced up to 1.35 dB with the new method

For the four predictors case, it was then progressively decreased the threshold δ and measured the PAPR(ccdf=10^{-4}), the EVM and PCDE. The ACLR is constant and depends only on the shaping filter design. Fig. 7 shows that it would be possible to reduce the PAPR(ccdf=10^{-4}) below 5.3 dB before violating the requirement on the EVM and PCDE.

The results presented in the rest of this section refer to the complete system of Fig. 1. In simulations it was possible to use alternatively the new algorithm presented in sec. III or the windowing scheme that was described in [5]. Beside one of the two PAPR reduction blocks it was necessary to introduce in the system a power adjustment block. In fact due to the peak compensation a reduction of the in-band power occurs and therefore the PA output would not be at the desired level.

The test with the PA model (3) was repeated using alternatively the two methods. For a fear comparison, the two PAPR reduction blocks were configured to the same level of in-band distortion (EVM =16 %, PCDE =-36 dB), which referring again to Fig. 7 corresponds for the new algorithm to a PAPR(ccdf=10^{-4}) of 5.3 dB at the shaping filter output. The IBO was progressively reduced until for both algorithms the ACLR limit was reached. Fig. 8 shows the spectrum at the PA output for the two methods at the same IBO level (8 dB). As the correction for the new algorithm is applied at the shaping filter input, the spectrum results less distorted.

Fig. 9 illustrates the ACLR results for both algorithms and for the uncompensated system. For the system [5] the limit of 45 dB is reached for IBO of 8.25 dB with an improvement of 0.45 dB to the uncompensated signal. The algorithm is not convenient for ACLR larger than 47 dB. For the new algorithm the limit is reached for IBO of 7.35 dB with an improvement of 1.35 dB compared to the uncompensated signal. This improvement remains constant for all ACLR values. The two diagrams are in fact parallel due the choice of injecting the correction directly at symbol rate.

V. CONCLUSIONS

In this work a new algorithm for the reduction of the PAPR has been presented. Its performance in terms of back off of a PA model were simulated for a WCDMA test signal.

The new approach takes advantage of the correction performed at symbol rate, before the interpolation stages. Therefore, when compared to different methods in the literature, no additional out of band radiation is present up to the PA input, which allows a larger back off reduction with the same amount of in band distortion.

REFERENCES

[1] *Spreading and modulation (FDD)*, 3rd Generation Partnership Project, TS Group Radio Access Network 25.213, Rev. 5.13, Mar. 2003.
[2] *Base Station (BS) conformance testing (FDD)*, 3rd Generation Partnership Project, TS Group Radio Access Network 25.141, Rev. 6.20, June 2003.
[3] M. Lampe and H. Rohling, "Reducing out-of-band emission due to nonlinearities in ofdm systems," in *IEEE 49th Vehicular Technology Conference*, vol. 3, Huston, May 1999, pp. 2255–9.
[4] N. Hentati and M. Schrader, "Additiv algorithm for reduction of crestfactor (aarc)," in *5th International OFDM-Workshop*, Hamburg, Sept. 2000, pp. 27.1–.5.
[5] O. Väänänen, J. Vankka, and K. Halonen, "Effect of clipping in wideband cdma system and simple algorithm for peak windowing," in *Proc. World Wireless Congress*, San Francisco, May 2002, pp. 614–9.
[6] M. Pauli and H. Kuchenbecker, "Minimization of the intermodulation distortion of a nonlinearly amplified ofdm signal," *Wireless Personal Communications*, vol. 4, no. 1, pp. 93–101, Apr. 1997.
[7] B. Hedberg, B. Hermansson, and C. Nyström, "Amplitude limitation in cdma system," International Patent WO 99/53 625, Apr. 8, 1998.
[8] M. Sharif, M. Gharavi-Alkhansari, and B. Khalaj, "New results on the peak to average of ofdm signals based on oversampling," in *Proc. IEEE International Conference on Communication*, vol. 2, New York, Apr. 2002, pp. 866–71.
[9] A. Aggarwal and T. Meng, "Minimizing the peak to average power ratio of ofdm signals via convex optimization," in *Proc. IEEE Global Telecommunications Conference*, vol. 4, San Francisco, Dec. 2003, pp. 2385–9.

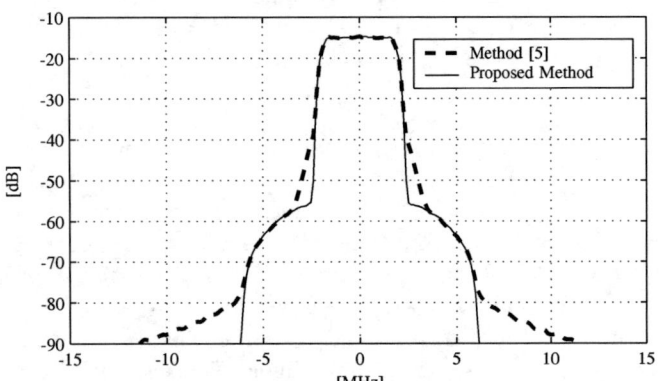

Fig. 8. Power spectral density at the PA output. *Dashed* - windowing method [5]. *Solid* - new method at symbol rate

A 3.5-Gb/s CMOS Burst-Mode Laser Driver with Automatic Power Control Using Single Power Supply

Day-Uei Li, Li-Ren Huang, and Chia-Ming Tsai
SoC Technology Center
Industrial Technology Research Institute
Hsinchu, Taiwan, R.O.C.
davidli@itri.org.tw

Abstract—A 3.5-Gb/s burst-mode laser driver with automatic power control (APC) fabricated in 0.25um 1P5M CMOS process is presented. With the proposed waveform shaping circuit, the driver could operate under a single 3.3 V power supply without sacrificing the output headroom, and it could output 6mA to 60mA modulation current with negligible overshooting. Measurements on mounted chips show clear electrical eye diagrams operating over 3.5-Gb/s data rate with typical rise/fall times (20% to 80%) of 84/97ps. Moreover, optical eye diagram is also demonstrated by connecting the driver with a 1.25-Gb/s 1310nm laser diode. The laser turn on/off time is about 30/20ns, much less than the criteria (512ns) that IEEE 802.3ah standard has proposed.

I. INTRODUCTION

The concept of passive optical network (PON) was first proposed into the full service access networks (FSAN) infrastructure. The result of this first effort was the 155 Mb/s PON system specified in the ITU-T G.983 series of standards. This system has become known as the B-PON system by using ATM as its bearer protocol. Over the years, several PON standards have been proposed for different PON systems with higher bit rates, such as the ATM-based PON (APON) enhanced to support 622-Mb/s bit rates, the Gigabit PON (GPON) initiated by the FSAN group for data rates over 1-Gb/s, and the Ethernet PON (EPON) that concentrates on standardizing an 1.25-Gb/s symmetrical system. All these PON systems require special burst-mode transmitters for the upstream direction. As the bit rates increase from 155-Mb/s to over 1-Gb/s nowadays, the timing specs on the time division multiple access (TDMA) timings, such as laser turn on/off times, become strict. Moreover, the design concept, the power consumption, the cost, and, the stability over wide temperature range are essential issues for designing a competitive burst-mode transmitter [1]-[3].

In this paper, we will present a burst-mode laser driver operating over 3.5-Gb/s with complete functions of data retiming control, pulse-width control, waveform shaping control, and automatic power control (APC). The IC is developed for direct modulation of laser diodes in over 1-Gb/s passive optical network. It delivers a 0mA to 80mA laser bias current as well as 6mA to 60mA modulation current. With an innovative waveform shaping circuit, only one power supply is required to drive high speed and large laser modulation current. The driver IC presented here consumes only 145mW at 3.3V power supply. The optical link between this driver and a commercial packaged laser will be demonstrated in this work.

II. DESCRIPTION OF CIRCUIT DESIGN AND LAYOUT

Fig. 1 shows the circuit diagram for the burst-mode laser driver. The driver contains a pair of 50-Ω input single-to-differential level translator circuits for both the clock and the data path; a master-slave D flip-flop for retiming; a 2 to 1 MUX with a select signal RET to select the latched or direct data; a pulse-width control stage to compensate the duty cycle distortion caused by laser non-linearity; an adaptive waveform shaping stage for compensating the duty cycle distortion when the output current is low; an APC control for delivering constant laser power; one pair of source followers for DC level shifting; a laser module for emitting light-wave signal and returning the monitored current to the APC loop; an open-drain switched differential pair and a laser bias control circuit.

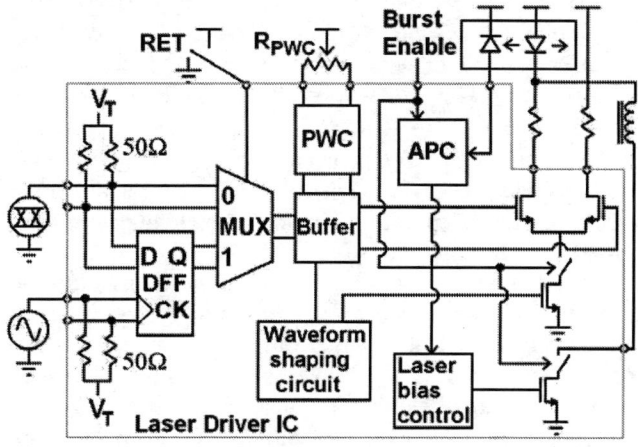

Figure. 1. Laser driver functional diagram.

The input common mode voltage V_T can be set on-chip or can be programmable by voltage division method using off-chip passive components. Thus input buffers are eliminated for low-power consideration. The modulation current can be programmable from 6mA to over 60mA. When working in PON system, all subscribers on the burst-off state turn off their transmitters to avoid accumulated dark currents from interfering with the signal from the one on the burst-on state. Therefore, the modulation and bias current of the driver should be switched off during the burst-off state. Fig. 2(a) shows a conventional current source used for generating modulation current. For burst mode application, a switch controlled by the Burst Enable (B_EN) is inserted at the point A, as shown in Fig. 2(b). The RC constant seen at the point A is quite small and therefore fast switching can be achieved.

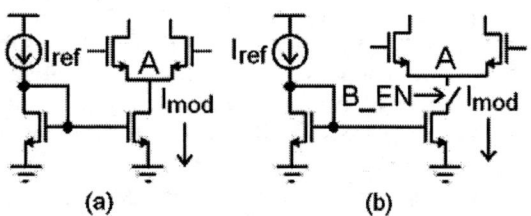

Figure 2. Current sources used for (a) the continuous-mode operation and (b) the burst-mode operation.

The proposed driver also contains a waveform shaping circuit. Its idea is to adaptively control the output voltage swing of the buffer stage to avoid duty cycle distortion when large voltage-swing driving small modulation output. Fig. 3(a) shows a conventional design proposed by Lucent [4]. Although this design could effectively reduce the overshooting and duty cycle distortion when driving small output current, it has problems of limited output headroom and low speed. For example, if the transistor M4 in Fig. 3(a) is conducted corresponding to the signal high, its gate voltage will be $V_G = V_{DD}$, while its drain voltage will be $V_D = V_{DD} - V_{fL} - R_d \cdot I_m - R_L \cdot (I_m + I_b) - \Delta V_l$, where V_{fL} is the laser forward voltage, R_d the damping resistor, R_L is the laser internal resistance, I_m is the modulation current, I_b is the bias current, and ΔV_l (= $l \cdot \Delta I_m / \Delta t$) is the voltage drop due to the laser package parasitic inductance l. If we take I_m = 30mA, I_b = 10mA, R_d = 15Ω, R_L = 5Ω, l = 1nH, and signal rise time (20% ~ 80%) = 100ps, the drain voltage is V_D = 3.3 − 1.2 − 15 · 0.03 − 5 · 0.04 − (1nH) · (60% · 0.03)/(100ps) = 1.27V which is much less than the gate overdrive voltage. Under such condition, the transistor M4 operates in the deep triode region. Its switching speed is then greatly reduced, and is impossible to drive large output current. To solve this problem, we might need large driving current to enhance speed or to add another power supply higher than 3.3V to offer larger output headroom. A better way improved from [5] is to add a source follower between the gain stage and the output stage as shown in Fig. 3(b). In this design, the gate voltage of the transistor M4 is lower by an M8 V_{GS} voltage drop. The output headroom is therefore enhanced. The controlled signal V_C is generated by a current comparator and designed in a range that the equivalent resistance of the shunt transistor is linear such that it will adaptively be adjusted and changes the overall gain following the variation of the modulation current. For lower-power consumption, the current source in the source follower could be designed as $I_{sf} = I_o + k \cdot I_m$, where I_o is a fixed current source and k < 1/5.

The die microphotograph of the realized 3.5-Gb/s burst-mode driver IC is shown as Fig. 4. The chip was fabricated by using standard TSMC 0.25μm CMOS process. Its dimensions, including the bonding pads were 1.2 mm × 1.0mm.

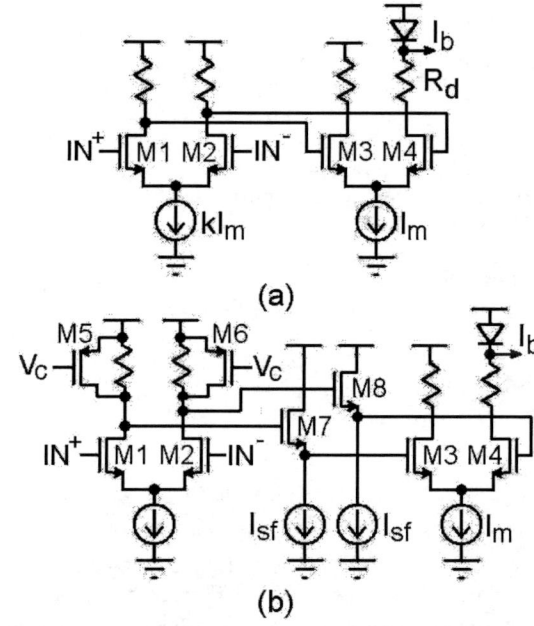

Figure 3. Waveforms shaping circuits used for (a) the conventional design and (b) the proposed design.

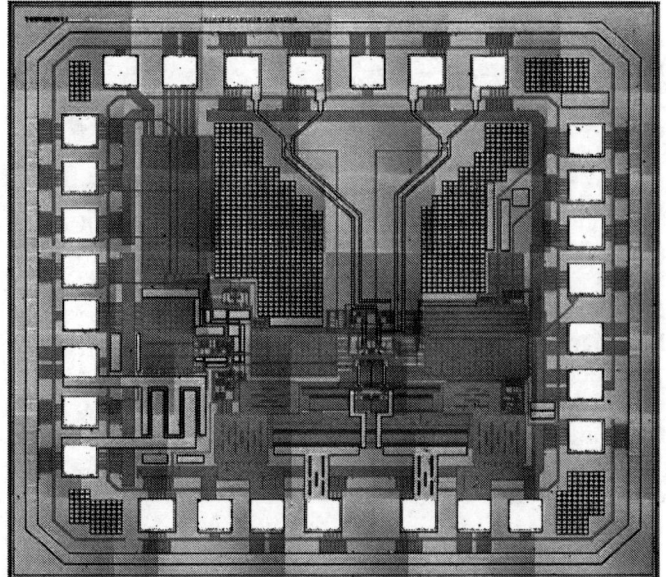

Figure 4. Driver IC Photomicrograph (1.2 mm × 1.0 mm).

III. ELECTRICAL TEST RESULTS

The driver consumes 44mA at 3.3V power supply excluding the bias and modulation currents. The chip is tested on-board with a 500mV, differential pseudo-random (2^{31}-1 NRZ) bit sequence. Fig. 5(a) shows the output eye diagram for modulation current of 60mA operating at 3.5-Gb/s. The diagram shows clear eye with rise/fall time of about 84/97ps. Fig. 5(b) and (c) show the output eyes for modulation current of 6mA operating at 3.5-Gb/s with and without the adaptive waveform control, respectively. Fig. 6(a) and (b) show the output eye diagrams for the direct and latched data operating at 2.5-Gb/s data rate. The measured RMS jitters are 10.2ps and 2.8ps for the direct and latched data, respectively. This result illustrates how the function of retiming control works to reject the input pattern-dependent jitter.

Fig. 7(a) and (b) show the eye diagrams at 2.5-Gb/s data rate with maximum and minimum width pulse control, respectively. With proper choice of the off-chip resistor R_{PWC} (Fig. 1), the adjustable pulse width can be up to about 140ps, which is large enough to compensate the duty cycle distortion caused by the laser non-linearity. And within this tuning range, the mark and space levels are well controlled without detectable increase in edge jitter.

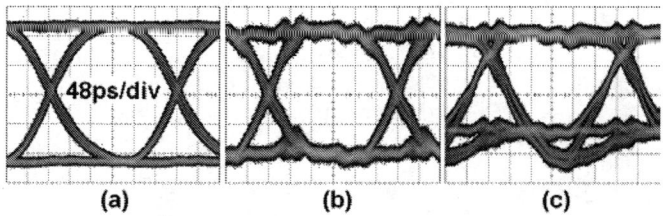

Figure 5. Output eye diagram operating at 3.5-Gb/s with modulation current (a) I_m = 60mA, (b) I_m = 6mA with adaptive waveform control, and (c) I_m = 6mA without adaptive waveform control.

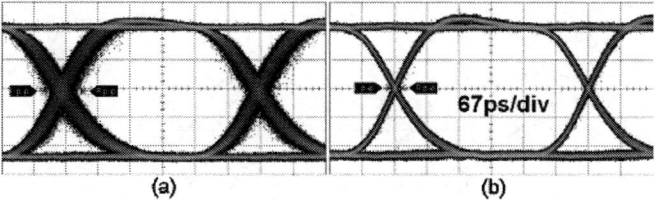

Figure 6. Output eye diagrams for the (a) direct and (b) latched data operating at 2.5-Gb/s.

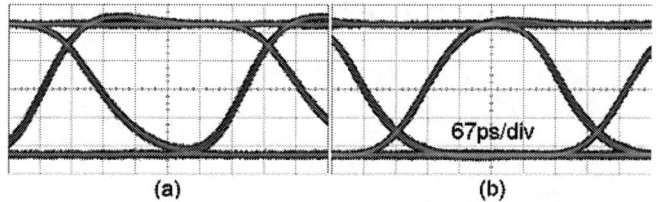

Figure 7. Eye diagrams showing pulse width control. (a) Maximum width. (b) Minimum width.

Fig. 8 shows the block diagram for measuring the modulation turn on delay. Two pairs of differential data and clock are generated from the pattern generator (PG) and connected to the driver. The trigger output of the PG is connected to a function generator (FG) to produce a burst enable signal. The FG also produces a trigger signal connecting to the trigger input of the oscilloscope. The output signal from the laser driver therefore could be compared with the burst enable signal for obtaining the turn on/off delay. Since the turn on/off delay is independent to the data rate, we use 1.25-Gb/s here for obtaining clear timing diagram. Fig. 9(a) and (b) show the timing diagrams for the signal Enable and the modulation current during the transitions of burst on and off, respectively. It is shown the driver turn on/off delay is less than 6/5ns.

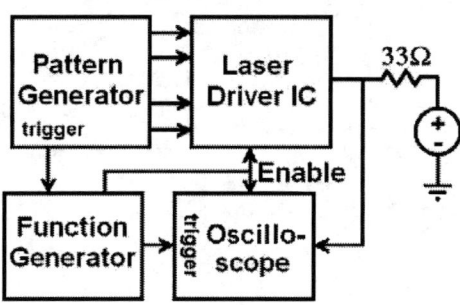

Figure 8. Modulation current turn on/off delay measurement setup.

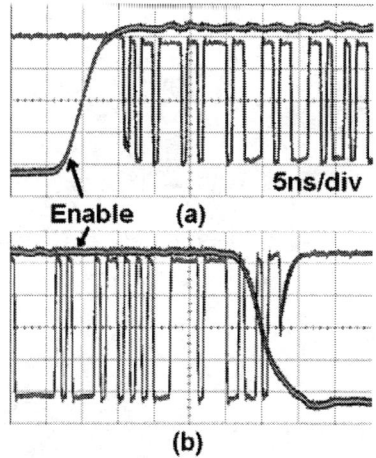

Figure 9. Timing diagrams for Enable and I_m during the transitions of (a) burst on and (b) burst off.

IV. OPTICAL TEST RESULTS

Fig. 10 shows the laser turn on/off time measurement setup using the presented laser driver IC. This driver IC directly modulates a 1.25-Gb/s 1310nm-wavelength Fabry-Perot laser diode (FP-LD). A golden 1310nm O/E converter is applied to receive optical data streams emitted from the laser. The converted electrical signal is measured in the oscilloscope. When the signal Enable is set to high, and the trigger output from pattern generator is directly connected to the oscilloscope, laser optical eye diagram can be seen in the oscilloscope. And the output off-chip damping resistor we use here is 15Ω. Fig. 11(a) and (b) show the optical eye diagrams for FP-LD at 1.25-Gb/s before filtering and after filtering with a 1.25-Gb/s fourth-order Bessel-Thomson filter, respectively. Fig. 12(a) and (b) show the timing diagrams for the signal Enable and the laser output during the transitions of burst on and off, respectively. The laser turn on/off time is less than 30ns, which is quite smaller than the criteria for laser turn on/off time proposed by IEEE 802.3ah standard. Fig. 13(a) and (b) show the timing diagrams for different modulation current. The automatic power control fixed the laser output power no matter how we change the modulation current.

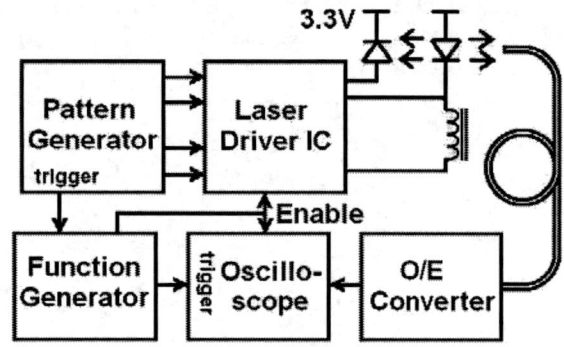

Figure 10. Laser turn-on/off time measurement setup.

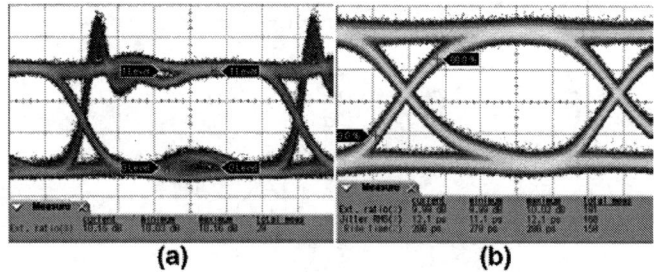

Figure 11. Optical eye diagrams (a) without filtered (b) with filtered.

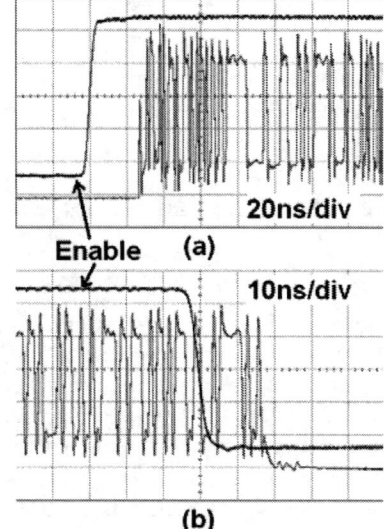

Figure 12. Timing diagrams for Enable and laser output during the transitions of (a) burst on and (b) burst off.

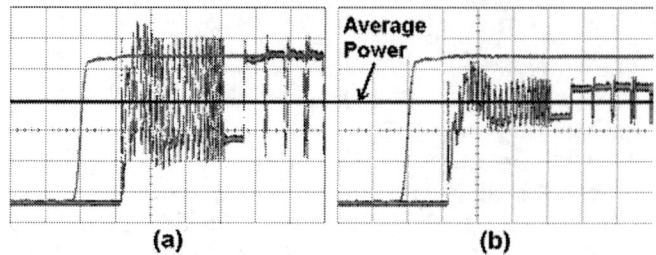

Figure 13. Timing diagrams for Enable and laser output with (a) large and (b) small modulation current.

V. OPTICAL TEST RESULTS

A burst-mode laser driver IC with automatic power control capable of waveform shaping, fast laser turning on/off, retiming control, and adjustable pulse-width control for over 1-Gb/s EPON has been demonstrated. The circuit can be operated over 3.5-Gb/s, and provides a very high-quality drive signal with over 60mA modulation current. The rise/fall time (20% to 80%) is about 84/97ps. The adjustable pulse width can be up to about 140ps. The modulation current turn on/off time is less than 6ns. With our proposed waveform shaping circuit, the driver IC could utilize only one power supply to drive a commercial laser diode without losing its operating speed. Without considering the laser bias and modulation current, the total supply current is typically 44mA. When driving an un-cooled laser, a clear optical eye diagram is shown. And the laser turn on/off time is about 30/20ns, which is quite smaller than the laser turn on/off time criteria (512ns) that the IEEE 802.3ah standard has defined.

REFERENCES

[1] E. Sackinger, Y. Ota, T. J. Gabara, and W. C. Fischer, "Low power CMOS burst-mode laser driver for full service access network application," Proceedings of The Pacific Rim Conference on Lasers and Electro-Optics CLEO/Pacific Rim '99, Vol. 2, pp. 519 – 520, 30 Aug.-3 Sept. 1999.

[2] J. Bauwelinck, Y. Martens, X. Z. Qiu, P.Ossieur, K. Noldus, J. Vanderwege, E. Gilon, P. De Meulenaere, and A. Ingrassia, "Design of a generic and high performance CMOS burst mode laser driver," 2002. Proceedings of 9th International Conference on Electronics, Circuits and Systems, Vol. 3, pp. 1115 – 1118, Sept. 2002.

[3] P. Ossieur, K. Noldus, X. Z. Qiu, J. Bauwelinck, Y. Martens, J. Vanderwege, E. Gilon, and B. Stubbe, "Fast and accurate level monitoring circuitry for a burst-mode CMOS laser driver," Proceedings of IEEE Circuits and Systems for Communications Conference ICCSC '02, pp. 94 – 97, 26 – 28 June 2002.

[4] US Patent US6021143, "Dynamic control for laser diode drivers," 2000.

[5] C.-M. Tsai, L.-R. Huang, D.-U. Li, and C.-Fu. Chang, "10Gb/s single-ended laser driver in 0.35μm SiGe BiCMOS technology," in Proc. 29th European Solid-State Circuits Conference ESSCIRC '03, pp. 289 – 292, 16 – 28 Sept. 2003.

Common-mode Stability in Low-power LO Drivers

Svetoslav Gueorguiev*, Saska Lindfors†, Torben Larsen*
*Radio Frequency Integrated Systems and Circuits Division
Aalborg University, 9220 Aalborg, Denmark
Email: sg@kom.auc.dk
†Electronics Circuits Design Laboratory
Helsinki University of Technology, Finland

Abstract—A resonator loaded LO buffer is shown to be potentially common mode unstable, when it is connected to drive a double balanced current steering mixer. The stability margin is found to be inversely proportional to the quality factor of the resonator load and, therefore, to the power consumed by the LO buffer. A known solution to the problem is compared with two proposed ones. The approaches differ in terms of current consumption, occupied voltage headroom and area.

I. INTRODUCTION

In wireless communication systems mixers are widely used to perform frequency translation in both reception and transmission. The inputs of a mixer are the signal to be translated and a Local Oscillator (LO) signal, which determines how much in frequency the input signal is translated. In general, the oscillator output cannot be connected directly to the mixer but an intermediate amplifier stage is used to provide isolation and to drive the mixer. Double balanced current steering mixers are a popular topology, because they do not require a very large LO signal and can be designed to provide power gain. In addition, they can be insensitive to amplitude noise in the LO signal and also have good port-to-port isolation (Fig. 1).

Since minimizing power consumption is very important in wireless devices, which are often mobile and, hence, battery operated, a great deal of effort has been invested in finding circuit structures with a low power consumption. One technique for reducing the power consumption of a current steering mixer is to use a resonator load. This is most practical in up-conversion, where the output is at RF and a large output power is highly desirable.

The LO driver of an up-conversion mixer can also consume a significant amount of power. In particular, when the up-conversion is to RF, either from baseband or intermediate frequency (IF), the driver must amplify at radio frequencies. A resonator load can be used in the driver as well to save power (Fig. 2) [1], [2]. However, in the case of current steering mixer, the real part of the LO port common-mode input impedance is in general negative at some high frequencies. When connected in parallel with the resonator load of the driver, it can easily give rise to a common-mode instability, unless some counter measures are taken.

In section II the LO port common-mode impedance of a double balanced current steering mixer and the mechanism causing the potential instability is discussed in more detail. A known circuit technique for stabilization is presented in section III along with two proposed solutions. A comparison is made in section IV and the paper is concluded in section V.

II. COMMON-MODE IMPEDANCE OF A CURRENT STEERING MIXER'S LO PORT

The circuit equivalent of the current steering mixer and its driving stage for the common-mode component of the LO signal are shown in Fig. 3. In order to estimate the common-mode load of the driving stage the input impedance Z_{in} at the LO port of the mixer is calculated. For that purpose, the loss of the resonator load is modeled with a parallel resistor. The combined input transistor output impedance is modeled with

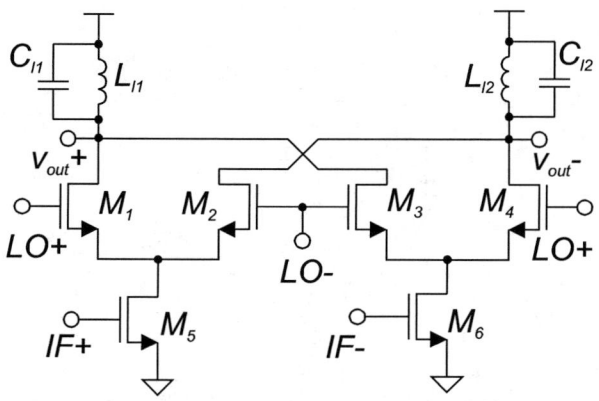

Fig. 1. A CMOS double balanced current steering up-conversion mixer

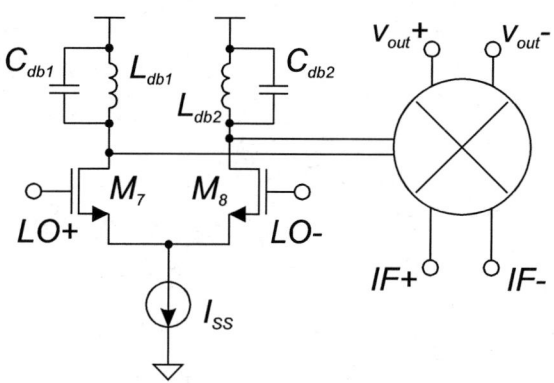

Fig. 2. The double balanced CMOS current steering upconversion mixer from Fig. 1 with its driving circuitry

a resistor and a capacitor, which includes also the parasitics at the sources of $M_1 + M_2 + M_3 + M_4$. Observing the resulting schematic in fig. 4 we note that the transistors M_1, M_2, M_3 and M_4 form a capacitively loaded (C_{out}) source follower with respect to the common-mode signal. It is well known that the input impedance of such a circuit has a negative real part at high frequencies [3]. The impedance Z_1 can easily be derived and its real part is given by:

$$\Re(Z_1) \approx -\frac{g_m}{w^2 C_{gs} C_{out}} \quad (1)$$

Since the drain of the circuit is loaded with a resonator rather than connected to supply, as in a source follower, the impedance Z_2 also affects the real part of the input. Unfortunately, the resulting expression is rather complex and due to the narrow band resonator model, it is not even very accurate over a wider frequency range. For simplicity, the input admittance is simulated with Spice using a more accurate model for the inductor. The simulated real part of Y_{in} is plotted in fig. 5 for different values of R_l. We note that, with the resonator included, the real part of input admittance is still negative over a wide frequency range. Furthermore, the real part of Y_{in} becomes more negative when the loss of the mixer load resonator is reduced. This is very undesirable, because it is in contradiction with the requirement of high mixer gain or low power consumption.

A negative real part in an input admittance is not necessarily a problem. However, when it is connected in parallel with the resonator load of the driver circuit (Fig. 6), it may cancel the loss in the resonator resulting in oscillation much the same way as in LC-oscillators. For proper functioning the condition

$$R_{cm} < \frac{1}{\Re(Y_{in})} \quad (2)$$

must be met. This can be ensured by making the Q-value of the resonator load sufficiently low. However, this will reduce the LO amplitude at the driver output, because the load impedance at resonance frequency is lowered. In order to restore the LO output amplitude more power has to be consumed in the driver circuit. In order to resolve this trade-off the load for the common-mode component of the LO signal has to be decoupled from the load for the differential component. In

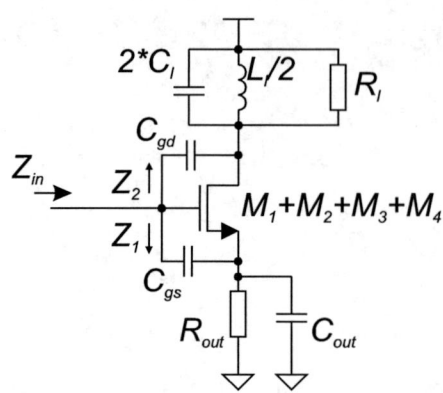

Fig. 4. The model of the mixer for calculating its common mode impedance from the local oscillator port

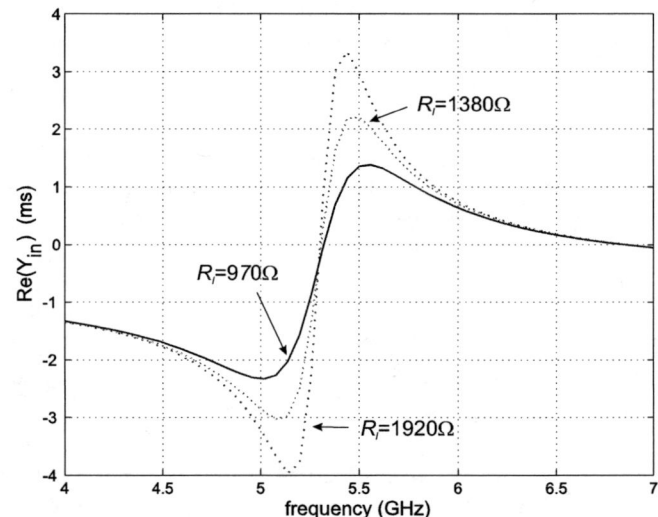

Fig. 5. The real part of Y_{in} versus frequency for different values of R_l; $C_{gd} = 30fF, C_{gs} = 200fF, C_{out} = 600fF, 2C_l = 270fF, L_l/2 = 3nH, g_m = 0.005$

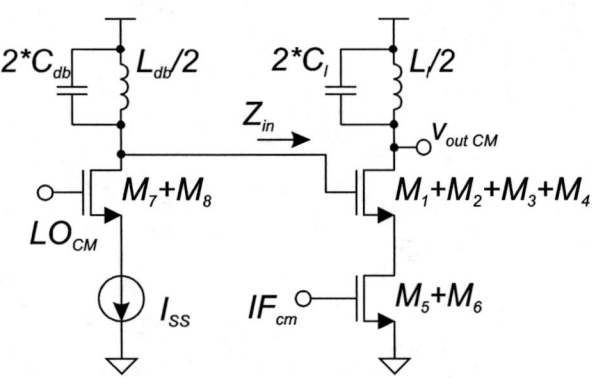

Fig. 3. The structure of Fig. 2 for the common mode signal

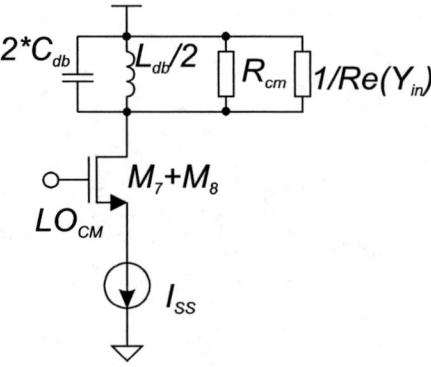

Fig. 6. The buffer with the *oscillating* load

the following section three techniques for achieving this are discussed.

III. METHODS FOR COMMON MODE STABILIZATION

A. Common biasing resistor

The first method to decouple the load for the common-mode component from the differential one is well known and is shown in fig. 7 [2]. Point A is a virtual ground and the differential signal does not see the resistor R_b. However, for the common-mode signal the Q of the resonator is reduced. The differential load at the frequency of resonance is:

$$R_{diff} = 2R_{db} \quad (3)$$

and for the common mode is:

$$R_{cm} \approx \frac{R_{db}w^2 L_{db}^2}{2(w^2 L_{db}^2 + 2R_{db}R_b)} \quad (4)$$

The advantage of this solution is that the resistor R_b can be used to bias the switches of the mixer lower than the positive supply. If the LO frequency is high compared to the f_T of the transistors, the bias current in the driver stage needs to be high causing the resistor to occupy substantial voltage headroom. Depending on the particular mixer topology this can be a limitation, when the supply voltage is low.

B. Transformer load

This elegant solution is shown in fig. 8. It exploits the fact that by utilizing a transformer the inductance of the tank is different for the common-mode and for the differential signal. The resistance for differential signal is:

$$R_{diff} = 2R_{db} \quad (5)$$

and for the common mode signal it is:

$$R_{cm} \approx \frac{R_{db}\omega_2^2}{2\omega_1^2} \frac{(L_{db} - M)^2}{(L_{db} + M)^2}, \quad (6)$$

where

$$\omega_1 = \frac{1}{\sqrt{C_{db}(L_{db} + M)}} \quad (7)$$

is the resonance frequency for the differential signal and

$$\omega_2 = \frac{1}{\sqrt{C_{db}(L_{db} - M)}} \quad (8)$$

is the resonance frequency for the common mode signal. Furthermore, the common mode resonance of the driver load is shifted away from the frequency range, where the negative notch of $\Re(Y_{in})$ lies (Fig. 5). This effect of frequency shift can also be achieved by introducing a differential capacitor or inductor in the tank. However in practice the capacitances C_{db} are dominated by the input capacitances of the mixer and introducing a differential capacitor for a given operating frequency would mean a lower L_{db} and subsequently lower R_{db}. Also, a higher resonance frequency for the common mode leads to a higher R_{db} at that frequency, which is something undesired. Introducing a third differential inductor is obviously not a practical solution. The presented solution has all the advantages of using transformer loads and in fact it does not cost any extra hardware compared to a two inductor solution. Nevertheless, the design of transformers is an iterative and time consuming process.

C. Differential negative resistance

The most obvious solution is to connect parallel resistors (R_{c1}, R_{c2}) to the resonators to lower the load's common mode resistance to fulfill the condition of equation 2. Unfortunately, the differential resistance becomes lower as well, which must be compensated by a higher bias current in the driver transconductor. The solution proposed here is to restore the differential resistance by connecting a floating negative resistance to the load (Fig. 9).

The common mode load resistance at resonance is given by:

$$R_{cm} = \frac{R_c R_{db}}{2(R_c + R_{db})} \quad (9)$$

and the differential one:

$$R_{diff} = \frac{-2R_c R_{db} R}{2R_c R_{db} - R(R_c + R_{db})} \quad (10)$$

In order to ensure proper functioning the denominator has to be always negative. The second term of the polynomial expression has to be bigger than the first one taking into account the

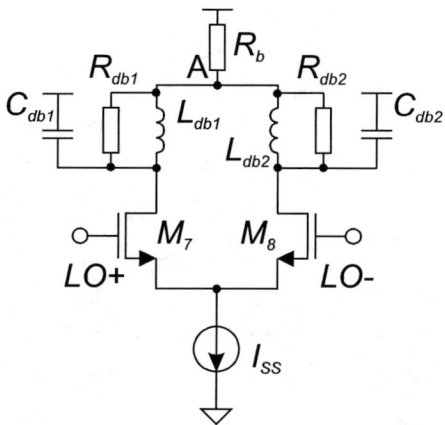

Fig. 7. Decoupling of the differential load from the common mode one by means of biasing resistor R_b

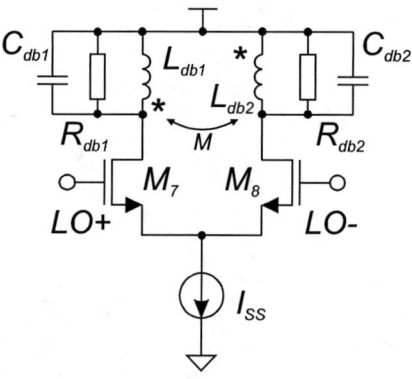

Fig. 8. Decoupling of the differential load from the common mode one by means of electromagnetically coupled inductors

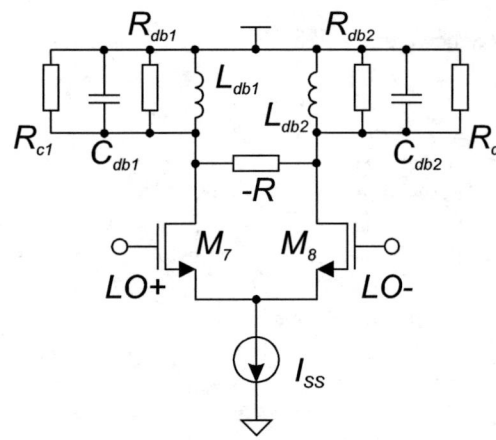

Fig. 9. Decoupling of the differential load from the common mode one by means of differential negative resistance

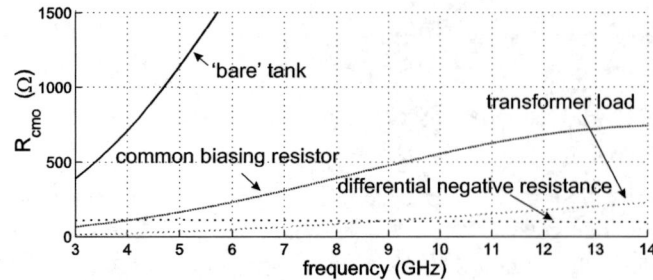

Fig. 10. The real part of the output common mode impedance R_{cmo} versus frequency for the presented solutions

process spread and the temperature. The differential negative resistance can be implemented by using well known techniques such as a differential cross coupled pair [4].

For completeness it should be noted that combinations between the presented solutions are also possible.

IV. COMPARISON BETWEEN THE PRESENTED METHODS

In order to facilitate the design choice between the presented solutions a first order comparison between them was made in terms of power consumption. For that purpose three test buffers were designed in 0.25u CMOS process. They were meant for driving an earlier designed 5.2 GHz Wireless LAN (WLAN) I/Q modulator. The comparison is based on the following assumptions:

- A LO frequency equal to the center frequency of the resonator load for the differential signal is fixed ($f = 5.2\ GHz$).
- The inductive part of the load is fixed, meaning that the inductance of the first and the third solution is equal to the sum of the inductance and the mutual inductance of the second one.
- The losses associated with the inductive part are the same.
- The load resistance for the differential signal R_{diff} is equal for all the solutions at the resonance frequency.
- Equal voltage gain of all buffers.

According to the simulations the common biasing resistor and the transformer load solutions consume 4.86 mW, providing common mode resistance of $R_{cmo} = 185\Omega$ (including the output impedances of the transistors) at the corresponding resonance frequencies of 5.2GHz and 12.3GHz. The biasing resistor R_b is found to be 11Ω. The negative differential resistance from Fig. 9 was implemented as a cross coupled differential pair. The total power consumption of the corresponding solution is 9.18mW, but it provides $R_{cmo} = 113\Omega$. If the ratio between the consumed power and the corresponding R_{cmo} provided is calculated it appears that the negative differential resistance solution is slightly less power efficient. The results for R_{cmo} are shown in fig. 10.

In practice apart from the power consumption other issues play an important role as well. The RF transformers are area efficient and eliminate the problem of the uncertain electromagnetic coupling between two single inductances on a chip. On the other hand their design requires specialized expertise and it is long and labor intensive process. Most design kits available don't provide models for them at all. The common biasing resistor solution provides quite some freedom in adjusting R_{cm}, if there is no voltage headroom limitation. Its use is limited in the advanced CMOS processes, because of the low voltage supply. Also in some applications it is possible to have a large DC current through the resistor R_b making the voltage drop over it excessively high. The differential negative resistor solution has a load that does not have almost any voltage drop over it, but the negative resistor itself consumes power.

V. CONCLUSION

The power consumption of a mixer LO driver can be reduced substantially by utilizing a resonator load. However, due to the negative real-part in the LO port input impedance of a CMOS double balanced current steering mixer, it has a potentially unstable common-mode level. Three solutions to the problem were presented. A common-mode series resistor is mainly limited by the additional voltage drop, the transformer approach is time-consuming to design and the differential negative resistor circuit consumes additional current, while providing additional stability margin.

ACKNOWLEDGMENT

The authors would like to thank all members of the RISC group for the fruitful discussions, especially Ole Kiel Jensen.

REFERENCES

[1] Araya R. Behzad et al.,"A 5-GHz direct-conversion CMOS transceiver utilizing automatic frequency control for IEEE 802.11a wireless LAN standard," *IEEE J. Solid-State Circuits*, vol. 38, no. 12, pp. 2209-2220, Dec. 2000
[2] Ting-Ping Liu and Eric Westerwick, "5-GHz CMOS radio transceiver front-end Chipset," *IEEE J. Solid-State Circuits*, vol. 35, no. 12, pp. 1927-1933, Dec. 2003
[3] Behzad Razavi, "Design of Analog CMOS Integrated Circuits", McGraw-Hill, 2000.
[4] Behzad Razavi, "RF microelectronics", McGraw-Hill, 1998.

An Application-Level Methodology to Guide the Design of Intelligent-Processing, Power-Aware Passive RFIDs

Cesare Alippi
Dipartimento di Elettronica e Informazione
Politecnico di Milano
P.za L. da Vinci 32, 20133 Milano, Italy
Email: alippi@elet.polimi.it

Giovanni Vanini
Dipartimento di Elettronica e Informazione
Politecnico di Milano
P.za L. da Vinci 32, 20133 Milano, Italy
Email: vanini@elet.polimi.it

Abstract— We provide a methodology for reducing the power consumption in wireless sensor networks based on wireless active nodes (also implementing RFID reading ability) and sensor/processing-integrated passive RFIDs. Information acquired by the low cost, redundant, passive units is suitably compressed with an integrated sensor fusion algorithm based on neural networks, hence reducing data to be transmitted to (and stored in) the reading unit. The methodology, by exploiting the application needs, also identifies the minimum number of bits to be transmitted to the reader (communication bit rate) as well as those required at each neuron output (information which immediately impacts on its complexity and electronics design) composing the sensor fusion algorithm. Simulation results applied to a comfort monitoring application in buildings show the effectiveness of the method with an energy saving factor of about 55% in the considered application.

I. INTRODUCTION

RFID tags are simple and cheap electronic devices able to send their identification code to a central reader once enabled [1]; their market shows an increasing trend which will make them pervasive provided a continuously decreasing fabrication costs. Attempts have been recently made to augment traditional passive RFIDs with sensorial potentialities; pioneer academic works in this direction integrate a temperature sensor on the RFID chip [2]. On the industrial front, Crosslink and Bridgestone have announced an RFID tire-monitoring system based on temperature and pressure readings [3] while Elektrobit proposes RFID temperature and concrete stress sensors in its catalogue. Again, Phase IV Engineering Inc.provides bolus and injectable lifetime temperature RFIDs for cattle and pets to automatically monitor the animals health. Surely, by enhancing current functioning with sensor acquisition potentiality we offer a new dimension to RFID applications. An increased reading distance (engineers of Trolley Scan [4] claim that a credit card sized tag can be read at an effective 11 meter distance with a 1W reader, or 8 meter with $300\mu W$ reader, even when the transponders are attached to metal objects) will further push the interest of using Sensor RFID (SRFID) in complex applications, e.g., envisaging the monitoring of good quality (temperature and luminosity for wine, temperature and humidity for cheese, etc), homeland and local environments.

Passive sensor RFIDs have further pushed the research towards the design of very low power consuming electronics to be deployed in the tag hence granting low power data acquisition abilities. Some main achievements have been reached in ADC design to keep as low as possible power consumption with 31pJ/8 bit sample at 1V supply and 100Ksample/s in ultra low power ADC [5]. Moreover, recent achievements in low sensitivity transponders [6] (that means less energy consumption of the readers, given the same reading range) allow us to foreseeing a full integration of the readers in handheld devices, i.e., PDA, notebook or active nodes of a more sophisticated Wireless Sensor Network in the very close future. Thanks to electronics advances we can hence envision sophisticated RFID-like devices integrating sensorial and processing ability in addition to more traditional operational functionalities. The target environment we envisage in this paper is composed of an active wireless sensor network (WSN) augmented with RFID reading abilities. As such, each active unit reads data coming from passive RFID sensors units, locally processes data according to the specific needs and transmits information along the network distributed application. A general passive tag is considered which enables onboard a multi-sensor ability and, once inquired, sends the processed data to the reader along with its identification code.

In the paper we also envision for the first time a local processing activity to be carried out directly on the tag. The processing can be surely intended as a filtering activity but, provided that the bit transmission to the reader accounts for most of energy consumption, it should be better intended as a sensor fusion activity: acquired information is compressed so that the content to be delivered to the reader represents the minimum information needed to solve the application (hence reducing the power requirements). Based on these reasonable assumptions, here we suggest a novel high level design methodology for reducing the total consumed energy both on SRFIDs and reader sides, yet granting the transfer of relevant sensorial information from SRFIDs to the reader.

The structure of the paper is as follows. Section II de-

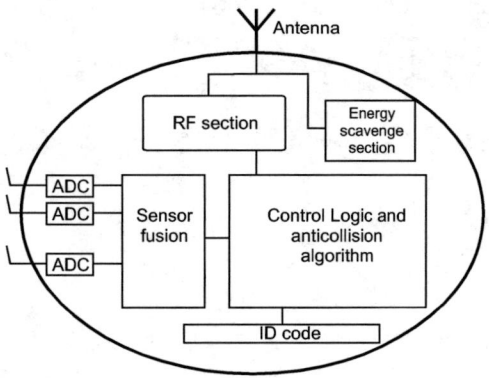

Fig. 1. Logical schematization of a typical RFID tag with sensor acquisition and data fusion capability

scribes the functional model for the system (tags and reader) while section III formalizes the energy reduction optimization problem based on a model for the whole wireless distributed application. Simulated results are finally applied in section IV to a real data acquisition problem requiring a robust estimate of the thermal comfort indexes for rooms in a given building.

II. Local Passive RFID-like Sensor Networks: The Envisaged Framework

We consider a sensor network in which active devices, wirelessly connected so as to constitute a proper primary network, are augmented with an additional RF section to behave also as a RFID reader.

On the passive tag side we can think of an integrated circuit with an external antenna designed to maximize the transponder sensitivity. Each tag, whose logic structure is given in figure 1, possesses mainly an RF section, an energy scavenge section, sensors – possibly integrated in the chip – and chosen according to the application needs, internal ADCs to convert the acquired sensorial data into discrete quantities stored in registers (say 10 bits). Since transmission is the main energy-consuming phase, a processing section is envisaged to extract and compress information to be sent to the reader, i.e., to implement a sensor fusion algorithm. The tag also contains an internal register holding its unique ID code: differently from standard RFID applications the code is here rather short, since with 8 bits we identify 256 different SRFIDs. The word to be transmitted is hence the juxtaposition of ID code, compressed value coming from the sensor fusion processing section and some checksum bits. Such a word is then modulated by the modulation section according to the envisaged backscattering mechanism and sent back to the reader. Conversely, a demodulating section receives command/configuration words to be inserted in the control register to handle the transmitting protocol and the anti-collision algorithm.

III. An Application-level Methodology to Reduce Energy Consumption

The energy consumption reduction method is based on the following assumption: we assume, as happens for every sensor network node, that the most power consuming operation of the readers is data communication; furthermore the communication interface towards the SRFIDs (the reading signal) needs more power than the one with the primary network, due to the back-scattering modulation on the passive tag. We consider also that for enlarging the communication distance (with the back-scattering mechanism), at a given incoming power, the needed energy for internal computation must be as low as possible so as to maximize the reflected power.

Therefore, the method we suggest aims at reducing the overall energy consumption at a high abstraction level by

1) implementing a computational intelligence-based sensor fusion algorithm allowing the SRFID to transmit only the relevant information to the reader (e.g., the comfort index instead of temperature, luminosity and humidity data). We reduce energy consumption by reducing the number of data instances to be transmitted;
2) reducing the number of bits to be transmitted to the reader (which is a main cause for energy consumption);
3) reducing the complexity of sensor fusion algorithm at the word-length level. We reduce the complexity and hence the power consumption of the circuit implementing the sensor fusion algorithm.

The designer, after having optimized the application with the suggested methodology directly at a high abstract level (application-level) can move the design activity to lower, more traditional, design-optimization levels involving low-power electronics.

The sensor fusion algorithm to be selected as required by point 1 must be simple and easily implementable in the RFID chip. Neural network-based sensor fusion algorithms are particularly appealing methods which associate a straightforward implementation to a proven effectiveness [7]; as such, we surely suggest them for solving the sensor fusion task and will be considered in the following.

Despite the fact that points 2 and 3 are associated with different energy optimization aspects, they can be transformed in a word-length optimization problem for the sensor fusion algorithm. In fact, reduction of the number of bits to be transmitted coincides with the reduction of the word-length of the sensor fusion output. In addition, word-length dimensioning can be seen as a perturbation problem where errors introduced e.g., by quantization (of the interim variable of the neural network) can be abstracted by suitable perturbations. The energy consumption reduction problem can be finally cast in a performance/perturbation problem which requires minimization of the word-length for variables involved in the sensor fusion algorithm yet granting enough information at the computation output to solve the application. In particular, our problem can be modeled as follows.

Denote by x_i the generic input signal acquired by the SRFIDs and represented over the maximum resolution n_{max} provided by the ADCs and by $y = f(x_i, x_2, \ldots, x_N)$ the sensor fusion function (neural network) receiving N sensorial inputs and providing a real value y. We consider an additive perturbation model for the generic neuron unit; as such, all sources of errors internal to the neuron and associated with its subsequent implementation are abstracted by the truncation error limiting the resolution of the output value for the neuron [8]. Let n_i be the number of bits associated with the representation of the output value of the i-th neuron according to the additive perturbation model and group such resolutions in the $\theta = \{n_1, n_2, \ldots, n_p\}$ vector (the cardinality p is equal to the number of hidden units augmented by the output neuron). Resolutions different from the maximal one will induce a perturbed computation $y_p = f(x_1, x_2, \ldots, x_N, n_1, n_2, \ldots, n_p)$ which, in turn, will affect the application accuracy. Such discrepancy in accuracy can be quantified by means of a generic loss function $u(x, \theta)$ whose values depend both on the application and the considered quantization vector θ. In several applications $u(x, \theta)$ is a Mean Squared Error (MSE) loss function but any other function can be considered, provided it is Lebesgue Measurable with respect to the input space. We further adopt an auxiliary scalar function $g(\theta)$ defined over θ, whose minimum will be intended as the optimal θ_0 configuration. For instance, we can consider a linear function $g(\theta) = \sum_{i=1}^{p} n_i$ or, when we want to give more relevance to some bits, even the weighed sum $g(\theta) = \sum_{i=1}^{p} w_i n_i$.

The optimization problem can be formalized by requiring the optimal bit vector θ_0 granting that the accuracy loss $u(x, \theta_0)$ induced at the processing section output (before modulation) is below a given accuracy loss γ tolerated by the application specifications, i.e.,

$$\begin{cases} \theta_0 = \mathrm{argmin}_{\theta \in \Theta}(g(\theta)) \\ u(x, \theta_0) < \gamma \qquad \forall x \in X \end{cases} \quad (1)$$

The optimization phase can be tackled by means of genetic algorithms, evolutionary computation, tabu search, simulated annealing or the most favored designer technique. For generic non linear functions the adoption of evolutionary methods is in general preferred (in the experimental section we will consider a straightforward genetic algorithm minimization procedure) whenever the complexity of the search space does not allow for an exhaustive search.

IV. EXPERIMENTAL RESULTS

The methodology proposed in the previous section has been applied to the practical problem of finding the thermal comfort index of a given environment. The theoretical background and the experimental data for this application are taken from a research project of the American Society of Heating, Refrigerating and Air-Conditioning Engineers (ASHRAE), whose aim was to study the comfort in different ventilated offices [9][10]. The project consists in measuring different physical information of the environment under monitoring (air temperature, mean radiant temperature, plane radiant asymmetry, dew point temperature, wet bulb temperature, relative humidity, air velocity and others) and associate them with comfort indexes as provided by office occupants or computed from readings according to some physiologically-based issues [10]. We select the Predicted Mean Vote (PMV) index, that predicts the mean value of votes of a large group of people on a 7-points thermal sensation scale: +3 (hot), +2(warm), +1 (slightly warm), 0 (neutral), -1 (slightly cool), -2 (cool), -3 (cold). This comfort index takes into account the main factors related to the steady-state thermal balance of the body, based on environmental parameters (air temperature, mean radiant temperature, air velocity, relative humidity), activity level and clothing level. We focus on a data-set of 589 records collected during the summer in the Kalgoorlie-Boulders office buildings -Western Australia. We consider only temperature and humidity sensor information to evaluate an approximated PMV which is, nevertheless, very close to the measured one (this is probably due either to a relevance issue or/and to a reduced dynamic for the not considered acquisitions).

In our framework we consider passive SRFID devices integrating temperature and humidity sensors (e.g., such sensors can be integrated by using capacitive sensing die set in thermoset polymers that interacts with platinum electrodes, as in [11]) and compressing the information by locally estimating the PMV by means of an artificial neural network which, in addition to the positive data compression effect also provides an estimate of the comfort index function. Whenever activated, the tag transmits its comfort index to the active node of the wireless sensor network according to the model presented in previous sections. In this way, a single reading unit covers several offices while several SRFIDs per room allow us for obtaining a robust measure of the comfort index (we can average acquired data within a room) and get rid of faulty devices. The computationally intelligent data compression algorithm is based on a feed-forward single hidden unit layer with linear output which must be designed to properly estimate the PMV based on the available sample data and be, at the same, as simple as possible for area and power consumption reasons (it must be implemented on the RFID). Once the best neural network is identified off-line the next step requires its implementation in the RFID.

We identified in a two hidden units layer the optimal/minimal neural networks solving the problem with acceptable accuracy; training was perfected by using a quasi-Newton Levenberg-Marquardt learning algorithm applied to the available training dataset of 400 input-output pairs while cross-validation was applied to 189 verification pairs. A Mean Squared Error (MSE) function was considered and led to the appreciably low MSE value of 0.227 on the validation set. Accuracy results made viable the estimation of the comfort index on RFID and, at the same time, halves the number of information to be sent to the reader with an immediate power consumption reduction of 50%.

Further optimization is achieved by reducing the number of bits to be transmitted (quantization of the networks output) and the complexity of the neural network by acting on the number of bits of the outputs of the hidden units (refer to section III). Quantization affects:
- input signals, which are represented with a 10 bits resolution within a [0;50] range for the temperature and a [0;100] range for humidity to deal with the office environmental conditions;

TABLE I
MEAN VALUE AND VARIANCE OF THE BITS OF THE QUANTIZED VARIABLE
IN THE NETWORK

	n_1	n_2	n_3
Mean	5	4.68	6.32
Variance	0	0.22	0.22

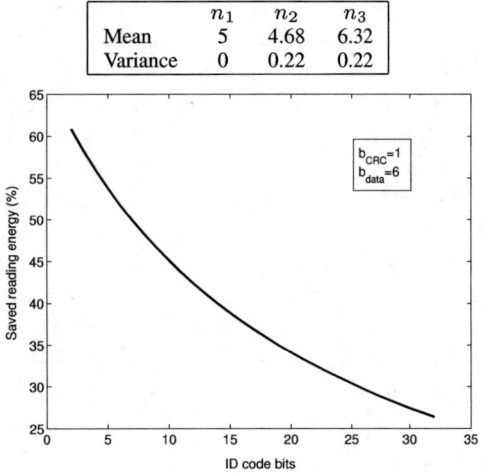

Fig. 2. Energy saving on the reader side w.r.t. the number of ID bits.

- the two hidden units outputs, which are represented with n_1 and n_2 bits (2 optimization parameters);
- the networks output, ranging in [-3;+3] interval, which is represented with n_3 bits (1 optimization parameter).

As such, the number of parameters to be optimized (vector θ) is 3. We set the tolerated error in accuracy on the comfort index up to 2%, which gives $\gamma = 0.12$. Parameters optimization was carried out by considering a traditional genetic algorithm applied to the $\sum_{i=1}^{p} n_i$ fitness function (crossover probability=0.6; mutation probability=0.05; population size=50; reproduction epochs=200) even if the reduced searched space could have been explored in this case with an exhaustive search ($p = 3$). Genetic optimization was applied 50 times to characterise the averaged value and variance of the optimal solution found within a genetic evolution trial; results are given in table I. Finally, the optimal configuration is $\theta = [556]$.

With the obtained results we wish to quantify the reduction in energy consumption after the optimization at the reader side. Here we assume that the total energy involved in the communication can be expressed as

$$E_t = P_M \cdot (n_{tag} t_{tag} + t_{oh}(n_{tag})) = E_{ec} + E_{oh},$$

where P_M is the imposed transmission power (500mW), n_{tag} is the number of tags to be read in the current monitoring area, t_{tag} is the time required for a single reading of each tag (we are implicitly assuming a sequential reading for tags) and t_{oh} is the time required for synchronization issues or anti-collision algorithm (tag arbitration). This time is strongly dependent on the number of tags in the reading area. We split energy consumption in the two terms: $E_{ec} = P_M \cdot (n_{tag} t_{tag})$ (effective data reading), and $E_{oh} = P_M \cdot (t_{oh}(n_{tag}))$(different contributes). The time needed for a single reading equalises the ratio between the bits of the transmitted packet and the bit-rate, that is

$$t_{tag} = \frac{b_t}{br} = \frac{b_{IDcode} + b_{data} + b_{CRC}}{br},$$

where b_{IDcode} is the number of bits of the ID code, b_{data} is the number of bits coming from the sensor measurements (or the sensor fusion) and b_{CRC} is the number of bits associated with the error detection/correction code.

The saved energy, defined as the difference between the consumed energy for data transmission E_{ec} before and after the optimisation procedure is given in figure 2. In the first case the number of bits to be sent is the sum of the sampled and converted sensorial data (temperature and humidity over the full resolution range), while in the second is the quantized output of the neural network. We consider a single-bit parity check code ($b_{CRC} = 1$) for different ID code length (x-axis) with a resolution depth at the ADC output set to 10 bits. By considering that for this kind of applications a maximum of 32 local tags can be sufficient, the resulting saved energy is about 55%.

V. CONCLUSION

The paper suggests a methodology for reducing the power consumption in a wireless sensor network based on wireless active nodes and passive sensor/processing-augmented RFIDs. The proposed optimization allows the designer for decreasing the needed energy both on the reader side, by diminishing the received data instances (sensor fusion) and data resolution depth, and on the tag side, by reducing its internal complexity. Simulation results applied to a comfort monitoring in buildings (evaluated by processing temperature and humidity data acquired and processed on the SRFID) show an energy saving factor of about 55% at the reader side when considering a passive sub-network of 32 passive RFID-based sensors.

REFERENCES

[1] V. Stanford, "Pervasive computing goes the last 100 feet with RFID systems," *IEEE Perv.Comp.*, vol. 2, no. 2, pp. 9–14, Apr.–June 2003.

[2] A. Nambi, S. Nyalamadugu, S. Wentworth, and B. Chin, "Radio frequency identification sensors," Auburn University, 200 Broun Hall Alabama 36849-5201, USA, Tech. Rep., 2003.

[3] (2003, Sept.) "New RFID Tire Sensor for Trucks". RFID Journal. [Online]. Available: http://www.rfidjournal.com/article/articleview/552/1/1/

[4] (2004, Feb.) "trolleyscan deliver passive transponder system can be read at 11 meters range even when attached to metal", press release. Trolley Scan(Pty) Ltd. [Online]. Available: http://trolleyscan.co.za/pressrel.html

[5] M. D. Scott, B. E. Boser, and K. S. J. Pister, "An ultralow-energy ADC for smart dust," *IEEE Journal of Solid-State Circuits*, vol. 38, no. 7, pp. 1123–1129, July 2003.

[6] M. Marsh, "New design for energy transfer to passive transponders (ecotag(®))," Patent 6,621,467(US),ZA 2001/9659,(EU)(TW)(CA).

[7] J. M. E. Hoff, "Learning phenomena in networks with adaptive switching circuits," Ph.D. dissertation, Dept. Electrical Eng., Stanford Univ., CA, June 1962. [Online]. Available: http://www.ece.udel.edu/~qli

[8] C. Alippi, "Selecting accurate, robust and minimal feedforward neural networks," *IEEE Transactions on Circuits and Systems*, vol. 49, no. 12, pp. 1799–1810, Dec. 2002.

[9] R. de Dear, "A global database of thermal comfort field experiments," *ASHRAE Transactions*, vol. 104, no. 1, 1998. [Online]. Available: http://atmos.es.mq.edu.au/~rdedear/ashrae_rp884_home.html

[10] K. Cena and R. de Dear, "Field study of occupant comfort and office thermal environments in a hot, arid climate," *ASHRAE Transactions*, vol. 105, no. 2, pp. 204–217.

[11] Humidity and moisture sensors. Honeywell International Inc. [Online]. Available: http://content.honeywell.com/sensing/prodinfo/humiditymoisture/

On the Implementation of 128-Pt FFT/IFFT for High-Performance WPAN

Clare Huggett*, Koushik Maharatna*, Kolin Paul**
*University of Bristol, Bristol, UK
**Indian Institute of Technology, Delhi, India
Email: {ch9346, Koushik.Maharatna}@bristol.ac.uk; kolin@cse.iitd.ac.in

Abstract - This paper deals with the efficient realization of a 128-pt FFT/IFFT processor for application in IEEE 802.15.3a standard. The 128-pt FFT/IFFT architecture has been designed by devolving it into one 8-pt and one 16-pt FFT. The 16-pt FFT was decomposed again and two separate 128-pt FFT algorithms have been developed, *viz.*, 8x4x4 and 8x2x8. Their relative merits and demerits have been analyzed from the algorithm as well as implementation point of view. The architectures have been prototyped on a Virtex II FPGA. The results indicate that the 8x2x8 architecture is better suited for the above mentioned purpose.

I. INTRODUCTION

High performance Wireless Personal Area Networks (WPAN) is currently the focus of research and development. IEEE 802.15.3a standard is intended for the development of such a WPAN system with a target data rate of up to 480 Mbps. Although the standardization committee has not yet finalized the standard, the most popular proposal under consideration uses a Multi-band Orthogonal Frequency Division Multiplexing (OFDM) based physical layer (PHY) [1] implementation. FFT/IFFT is one of the most computationally intensive components in such an OFDM-based PHY implementation. In this paper we concentrate on the efficient implementation of the FFT/IFFT processor for the above mentioned standard.

Proposal [1] states that a 128-pt FFT/IFFT has to be performed within 312.5 ns. This massive computation rate can only be supported either at a very high frequency (achievable with the 90 nm technology) or using massive parallelism. However, since the target application is of a mobile and portable nature, the second approach may lead to significant power consumption which is in direct contradiction with the application goal. Thus, algorithm level reformulations as well as innovative design techniques have to be developed to simultaneously satisfy the power and timing constraints.

This paper reports a high-throughput low-power 16-bit fixed point 128-pt FFT/IFFT implemented on a FPGA. It may be noted that there are implementations [2] which report a higher throughput but they do so by having parallelism and duplicating functional units like multipliers. This of course gives rise to "big" circuits which consume unnecessary large amounts of power. Our contribution in this paper represents a "sequential" design which operates at the required frequency and is extremely low power. We achieve this by totally avoiding "complex multipliers" and replacing them by simple pipelined shift-and-add units which helps us to achieve high operating frequency at low power. We propose two different algorithmic reformulation and subsequent architectures for computing 128-pt FFT. The architectures are targeted towards 90 nm process. Since the process has just been announced, we choose to illustrate and prototype our systems on the Virtex II family [3] of FPGA which are built with a 130 nm process. We then scale the FPGA performance suitably to estimate the ASIC performance following certain previous comparisons [4] established in literature. The remainder of the paper is structured as follows: Section II deals with the algorithmic reformulation and comparison of the proposed methods with the existing approaches. In Section III the respective architectures have been described while Section IV provides the implementation details of the architectures on FPGA and compares their relative performances. We summarize our contribution and draw some conclusions in Section V.

II. THEORETICAL GROUNDWORK

The N-point Discrete Fourier Transform (DFT) of a complex data sequence B_k is defined by [5]

$$A(r) = \sum_{k=0}^{N-1} B(k) W_N^{rk} \quad (1)$$

where $r, k \in \{0, 1, ..., N-1\}$ and $W_N = e^{-j2\pi/N}$, known as the twiddle factor.

Now considering $N = MT$, $r = s+Tt$ and $k = l+Mm$, equation (1) can be rewritten as

$$A(s+Tt) = \sum_{l=0}^{M-1} W_M^{lt} \left(W_{MT}^{sl} \sum_{m=0}^{T-1} B(l+Mm) W_T^{sm} \right) \quad (2)$$

For the present problem considering $M = 8$ and $T = 16$ one may write

$$A(s+16t) = \sum_{l=0}^{7} W_8^{lt} \left(W_{128}^{sl} \sum_{m=0}^{15} B(l+8m) W_{16}^{sm} \right) \quad (3)$$

Equation (3) suggests that the 128-pt. DFT can be computed by first computing an 8-pt DFT on the appropriate data slot (described by equation (3)), then multiplying them by 105 non-trivial complex twiddle factors and computing the 16-pt DFT on the resultant data with appropriate data reordering.

The 16-pt DFT can be decomposed further using the same method. In the present work, we have explored two different types of decomposition of it viz., 4x4 and 2x8. The important point to be noted here is that for realization of the 8-pt DFT using a Decimation-In-Time (DIT) FFT algorithm, explicit multiplications are not required. The constants to be multiplied for the first two columns of the FFT flowgraph are either 1 or j. In the third column, the multiplication of $1/\sqrt{2}$ can be realized using hard-wired shift-and-add operations. Similarly, using a radix-4 FFT for realization of 4-pt DFT, one does not need explicit multiplication either. As shown in Table 1, using these decompositions, a significant reduction in the number of required complex multiplications can be achieved compared to the existing techniques.

Algorithm	Non-trivial complex multiplication
Radix-2	258
Radix-4	204
Split-Radix	186
8x4x4	168
8x8x2	152

Table1. Comparison of the multiplicative complexity of the proposed schemes with other existing schemes

Apart from the reduction of complex multiplications, another advantage of these decompositions is that of the 105 non-trivial twiddle factors required in the proposed schemes, only 16 are unique. This can be exploited to reduce the overall size of twiddle RAM which is one additional advantage compared to the other existing schemes.

The IFFT can be performed using the same structure. First, the real and the imaginary parts of the input have to be swapped and the computation proceeds as a forward FFT. At the output, the real and imaginary parts are swapped back and the output is scaled by 128.

III. ARCHITECTURE

The generalized architecture of the 128-pt FFT can be derived from equation (3) and is shown in Fig. 1. The processor consists of an input unit, an 8-pt FFT, a multiplier unit, a 16-pt FFT, a resorting unit, an output unit and a master control counter.

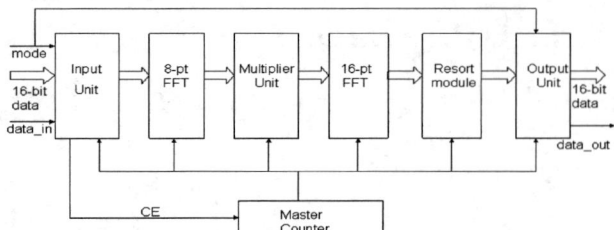

Fig. 1. Block diagram of the proposed FFT processor

A. Input Unit

The input unit consists of an input register bank having 16-bit wordlength that can store 113 complex samples. It is equipped with two single bit signals "*mode*" and "*data_in*". While the first one enables the processor to understand which mode of operation is needed (whether FFT or IFFT), the second one indicates the presence of valid data at the input. After the assertion of the *data_in* signal, serial data is inputted at every clock cycle at the 112th position of the input register bank and at every clock cycle the complex data having index i is shifted to the $(i-1)^{th}$ position where $i \in \{1, .. 112\}$. The register bank is provided with eight pairs of 16-bit fixed wired outputs corresponding to the registers having index $8j$, where $j \in \{0, …, 7\}$. When the input buffer is filled, the appropriate data (see equation (3)) gets self aligned with these hard-wired outputs and is delivered in parallel to the first 8-pt FFT unit. This operation is continued in every cycle. A 7-bit counter controls the serial input of the data in the register bank, which is enabled with the assertion of the *data_in* signal. When the 112th position of the input register bank is full a signal *CE* is asserted to enable the master control counter. The subsequent operations are controlled by the master counter. This method of downward shifting of data in conjunction with its self-alignment with the hard-wired outputs reduces the number of multiplexed signals by a factor of 112 and 14 at the input and output of the input unit respectively. This massive reduction of wiring allows a more efficient layout of the processor.

B. 8-pt FFT Unit

It has been discussed in Section 2, that to implement the 8-pt FFT one does not need to use conventional digital multipliers. This fact opens up the scope to implement a fully parallel 8-pt FFT unit. In our implementation we realized a pipelined 8-pt FFT architecture following the 8-pt DIT FFT flowgraph where each complete column of the 8-pt FFT is computed in a single clock cycle. Thus, a complete 8-pt FFT can be computed in three clock cycles. The multiplications by the factor $1/\sqrt{2}$ in the third column are realized using hard-wired shift-and-add operations and thus require minimal hardware.

C. Multiplier Unit

It has already been mentioned in Section 2 that to carry out 105 non-trivial complex multiplications our architecture needs only 16 unique sets of twiddle factors. The complete operation can be carried out by changing the signs or swapping the real and imaginary coefficients of them. However, in the steady state, the data from the 8-pt FFT unit arrives at every clock cycle, consequently a single complex multiplier unit cannot be used to take the full speed advantage offered by the parallel 8-pt FFT. On the other hand, employing 8 complex multipliers will increase the area and power consumption of the entire system. However, since the twiddle factors used to multiply each of the incoming samples are known *a priori*, one can realize each of them using a simple shift-and-add technique. As an example, a constant 0.99895 can be decomposed to $1-2^{-10}-2^{-12}$ with 16-bit accuracy. Thus the multiplication of any input by this constant becomes a series of additions/subtractions of right shifted values of the input quantity. This approach has previously been used in [6] which resulted in a significant performance gain in terms of power consumption and area. Since the target architecture in the present case must run at a high frequency to satisfy the required data rate, we realized each of these 16 constants in a pipelined way and arranged them in parallel. The multiplier unit also has a temporary storage register bank that can hold 128 complex data. Again, hard-wired connections are used like the input unit and the data after multiplication enters the registers at locations 127 – 120. At every cycle, the data is down shifted as a block of eight until the register bank is full. This approach saves signal wire multiplexing by a significant amount. To deliver the data to the 16-pt FFT unit once again the hard-wiring and data self-alignment strategy used in the input unit has been employed and thereby substantially reducing the number of signal multiplexing again.

D. 16-pt FFT

Using the decomposition schemes proposed in Section 2, the 16-pt FFT module is realized using two different kinds of further decomposition which resulted in two different architectures *viz.*, *Arch*1 (4x4 decomposition) and *Arch*2 (2x8 decomposition) for 128-pt FFT. *Arch*1 contains two 4-pt FFTs where each of the 4-pt FFT is realized using radix-4 butterfly. On the other hand, *Arch*2 contains a 2-pt and 8-pt FFT. Once again, the 8-pt FFT module is realized following the similar strategy described for the first 8-pt FFT module. Both 16-pt FFT architectures have a multiplier unit, which operates in a similar fashion to the main multiplier unit. Here only three pairs of unique constants are required. Once the data has been multiplied, it is put directly into a register bank able to hold 16 complex samples. The inputs to this block are hard-wired and the data is entered at either every 4th / 8th register depending on the architecture. At every cycle the data is down shifted until the buffer is full. At this point another data sample is ready to be input, so rather than hold-up this data a second 16 register buffer is used. The data is transferred to this buffer and is down shifted at every clock cycle in blocks of 4 (*Arch*1) or 8 (*Arch*2) until all data has been passed onto the next module.

E. Resort Unit

To make the reshuffling at the output to form the full 128-pt FFT simpler it was spilt into two stages. The resort module takes the data directly from the 2nd FFT in the 16-pt FFT module and reorders it to form a set of 16 data. This is done by hardwiring the output of the second FFT to every $4^{th}/2^{nd}$ register and down shifting it at every clock cycle.

F. Output Unit

The output has a dual structure of the input unit. It consists of a complex bank of 128 registers and a 7-bit counter. The data is hard-wired into this register bank at every 8th register position and is down shifted at every clock cycle until the buffer is full. Once full, the buffer is emptied from the 0th register while continuing down shifting of the data until it is empty. When the first sample is outputted the signal *data_out* is asserted high and is held high until the buffer is exhausted. This indicates valid data at the output. If the IFFT has to be performed, the real and imaginary parts of the data are swapped and are shifted to the right by 7 bits (scaling by 128). As with the input module this is also indicated by the signal *mode*.

G. Master Control Counter

The 7-bit master counter is responsible for synchronizing the entire system. It is turned on with the assertion of the signal *CE* and it stops counting when the last set of data has reached to the output buffer. The master counter is used to generate an enable signal for the modules at different time instants. This easily translates to implement clock gating in the entire circuitry with no additional design effort.

IV. IMPLEMENTATION AND PERFORMANCE EVALUATION

The architectures described in the previous section have been coded in VHDL. The parallel-to-parallel 128-pt FFT is performed in 73 and 104 clock cycles for *Arch*1 and *Arch*2 respectively. The architectures were synthesized using the Xilinx tool chain. The target board was Alphadata ADM-XRC II which has a Virtex II 6000 series FPGA. The Virtex II FPGA represents a product from 130 nm technology. The results of post place and route are shown in Table 2. It is clear that the second architecture is more area and power efficient while runs almost at the same frequency as the first one. The number of clock cycles required for *Arch*2 can be brought down from 104 to 56 by simply duplicating the 2-pt FFT butterfly structure four times in the 16-pt FFT module. Since the basic 2-pt FFT is very simple, this added parallelism will result in a minimal hardware overhead. The maximum operating frequency required to satisfy the goal can be tailored further by adding another degree of parallelism with minimum area overhead.

Parameters	*Arch*1	*Arch*2
# of slices	20751	20580
# of slice Flip Flops	22437	23171
# of 4-input LUTs	32604	32245
# of IO	67	67
Clock load	23203	23905
Maximum frequency (MHz)	63	62
Power Estimate (mW)	712	337

Table2. Comparison of *Arch*1 and *Arch*2 implementation

The maximum operating frequency is limited by the routing resources available in the FPGA. The critical paths have up to 25 – 37 % of routing involved and hence limit the highest frequency achievable.

When implemented on a same process technology, it has been reported [4] that the ASIC performance scales up to 2 - 2.5 times compared to an FPGA implementation. If we scale the performance to a higher process technology (90 nm), the throughput is expected to be about at least 4 times. Hence we expect that our proposed architectures can easily meet the performance requirement of computing the 128-pt FFT in 312.5 ns in 90 nm process (ASIC).

During implementation, we optimized our architectures by removing the set-reset logic for many of the internal Flip-Flops of the computation blocks. This resulted in area savings to a great amount (1:7). The 16 unique twiddle factors are synthesized in a LUT and thus, avoiding the requirement of RAM. The floorplan of *Arch*2 is shown in Fig. 2.

The power estimates, using XPower, are also presented in Table 2. While it is impossible to accurately scale exact power dissipation figures for ASIC, from design experience these estimates show that if implemented in 90 nm technology we can expect the architectures to dissipate at least about 1/4 of the power. Furthermore, each of the sub blocks operates at a precise value of the control counter. Thus, significant power reduction is further possible by employing clock gating while implementing the proposed architectures in ASIC.

V. CONCLUSION

Two potential architectures for realizing 128-pt FFT/IFFT processor for IEEE 802.15.3a standard have been proposed and implemented on a Virtex II FPGA. Though it is natural that the FPGA implementation gives much lower maximum frequency of operation compared to the ASIC implementation, it sets a baseline for choosing an appropriate architecture. In our implementation it is found that the architecture based on 8x2x8 decomposition of the 128-pt FFT performs better in terms of area compared to the architecture based on 8x4x4 decomposition while both of them give significant reduction of algorithmic complexity compared to the existing schemes.

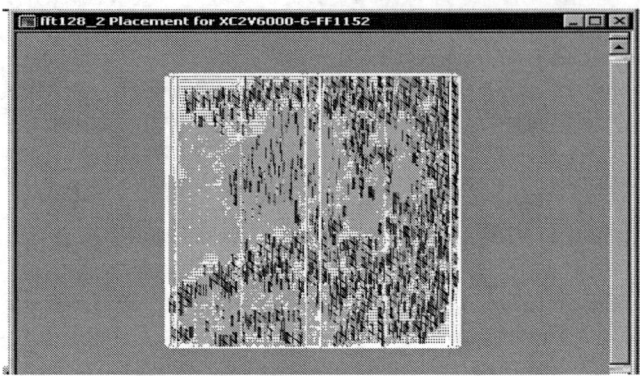

Fig. 2: Floorplan of *Arch*2

In terms of power, *Arch*2 shows approximately 52% less power consumption than that of *Arch*1. The number of required clock cycles of *Arch*2 can be brought down further by adding another degree of parallelism in its 16-pt FFT module with very nominal area overhead. We expect that when implemented in the form of ASIC using 90 nm technology, both of the architectures will satisfy the timing constraint and *Arch*2 will consume much less power compared to the *Arch*1 and thus is a better choice.

REFERENCES

[1] http://grouper.ieee.org/ groups/802/15/pub/2003/ May03/ 03142r1P802-15_TG3a-TI-CFP-Document.doc
[2] I. S. Uzun and A. A. Bouridane, "FPGA Implementations of Fast Fourier Transforms for Real-Time Signals and Image Processing", *Proceedings of IEEE International Conference on Field-Programmable Technology (FPT)*, 2003.
[3] http://www.xilinx.com/virtex2
[4] R. J. Peterson and B. L. Hutchings, "An assessment of the suitability of FPGA-Based system design for use in DSP", *5th Intl Workshop on FPL and Application*, Oxford, England August 1995.
[5] A. M. Despain, "Very fast Fourier transform algorithms hardware for implementation", *IEEE Trans. Comput.*, vol. C-28, no. 5, pp. 333 – 341, May 1979.
[6] K. Maharatna, E. Grass and U. Jagdhold, "A 64-point Fourier transform chip for high-speed Wireless LAN application using OFDM", *IEEE J. Solid-State Circuits*, vol. 39, no. 3, pp. 484 – 493, March 2004.

Near-Perfect Cover Image Recovery Anti-Multiple Watermark Embedding Approaches

Chao-Yong Hsu
Institute of Information Science
Academia Sinica
Taipei, Taiwan 115, ROC
Email: cyhsu@iis.sinica.edu.tw

Chun-Shien Lu
Institute of Information Science
Academia Sinica
Taipei, Taiwan 115, ROC
Email: lcs@iis.sinica.edu.tw

Abstract— Robustness is a critical requirement for a watermarking scheme to be practical. Especially, in order to resist geometric distortions a common way is to locally insert multiple-redundant watermarks in the hope that partial watermarks could still be detected. However, there exist watermark-estimation attack (WEA), such as the collusion attack, that can remove watermarks while making the attacked data further transparent to its original. Another kind of attack is the copy attack, which can cause protocol ambiguity within a watermarking system. The aim of this paper is to propose an efficient cover data recovery attack, which is more powerful than the conventional collusion attack. To this end, we begin by gaining insight into the WEA, leading to formal definitions of optimal watermark estimation and near-perfect cover data recovery. Subject to these definitions, an exquisite collusion attack is derived. Experimental results verify the effectiveness of the proposed watermark estimation and recovery algorithm.

I. INTRODUCTION

Robustness is known to be a critical issue affecting the practicability of a watermarking system. In the literature, robustness is usually examined with respect to removal attacks or geometrical attacks or both. However, there indeed exist attacks that can defeat a watermarking system without sacrificing media quality. In particular, the collusion attack [10], [11], which is a removal attack, and the copy attack [4], which is a protocol attack, are typical examples of attacks that can achieve the aforementioned goal. The common step used to realize a collusion or copy attack is watermark estimation, which is easily accomplished by means of denoising. Consequently, we call both the collusion attack and copy attack watermark-estimation attacks (WEAs) [5]. In this study, we particularly focus on the collusion attack.

The aim of the collusion attack is to collect and analyze a set of watermarked media data so that an unwatermarked copy can be constructed to create the false negative problem. In digital watermarking, a collusion attack naturally occurs in video watermarking because a video is composed of many frames, and one way of watermarking a video is to embed the same watermark into all the frames. This scenario was first addressed in [11]. However, we argue that [5] the collusion attack is not exclusively applied to video watermarking. In the past few years, image watermarking with resistance to geometrical attacks has received much attention because even a slight geometrical distortion may disorder the hidden watermark bits and disable watermark detection. In view of this fact, some researches [2], [12], [9], [14] inserted multiple redundant watermarks into an image in the hope that robustness could be maintained as long as partial watermarks existed. Commonly, various kinds of image units, such as blocks [14], meshes [2], or disks [12], are extracted as carriers for embedding. Taking advantage of this unique characteristic, we propose to treat each image unit in an image like a frame in a video; in this way, collusion attacks can be equally applied to those image watermarking methods that employ a multiple redundant watermark embedding strategy. Therefore, once the hidden watermarks are successfully removed by means of a collusion attack, the false negative problem occurs even though no geometrical attack is imposed on stego images. Of particular interest are possible fidelity improvements of attacked images as a result of a collusion attack.

When the hidden watermark is estimated and removed by means of collusion, it is necessary to check the presence or absence of a watermark. A simple way is to calculate a correlation (e.g., cross-correlation) and compare it against a threshold to make the final decision about the existence of a watermark. However, one may argue that this does not imply that the hidden watermark has been "optimally estimated and removed" by means of such a simple cross-correlation. This is because an "optimal" watermark detector [1], [8], which is usually based on exploiting the statistic characteristic of a host content, may be able to discover the hidden watermark. In order to address this issue, we don't evaluate the optimal estimation/removal of a watermark from the viewpoint of a watermark detector. On the contrary, we investigate how an embedded watermark could be "sufficiently" estimated/removed. In this paper, we propose a new watermark estimation and cover data recovery method. The comparison of our method with perceptual remodulation [13] is also evaluated.

II. HOW WATERMARK COULD BE COMPLETELY REMOVED?

Let **W** represent the original watermark with its energy extended by means of either a constant factor or a human visual system to enhance robustness. From an attacker's perspective, the energy of each watermark value must be accurately predicted so that the previously added watermark energy can be completely subtracted to create an ideally unwatermarked image. If this goal could be achieved, it is said that watermark removal is effective without leaving sufficient residual watermark. Consequently, an estimated watermark's energy is closely related to the accuracy of the watermark removal attack. To better explain our point, several motivating scenarios are shown in Fig. 1, which illustrates the energy variations of (a) an original watermark; (b)/(d) an estimated watermark (illustrated in gray-scale); and (c)/(e) a residual watermark generated by subtracting the estimated watermark from the original watermark. We can observe from Fig. 1(a)~(c) that even though the watermark's sign bits are fully obtained (Fig. 1(b)), the residual watermark signal (Fig. 1(c)) still suffices to reveal the encoded message due to the original watermark's energies cannot be completely discarded. Furthermore, if the sign of an estimated watermark value is different from its original one, then any additional energy subtraction will not be helpful in improving removal efficiency. On the contrary, watermark removal in terms of energy subtraction operated in the opposite (wrong) polarity will undesirably damage the media data's fidelity. Actually, this corresponds to adding a watermark with higher energy into cover data without satisfying the masking constraint, as shown in Fig. 1(d). After Fig. 1(d) is subtracted from Fig. 1(a), the resultant residual watermark is illustrated in Fig. 1(e). By comparing Figs. 1(a) and (e), it is highly possible to reveal the existence of a watermark.

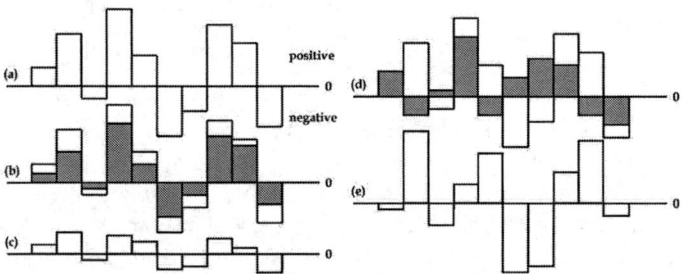

Fig. 1. Watermark estimation/removal illustrated with energy variation: (a) original embedded watermark with each white bar indicating the energy (determined using perceptual masking) of each watermark value; (b) gray bars show the energies of an estimated watermark with all the signs being the same as the originals (a); (c) the residual watermark obtained after removing the estimated watermark (b); (d) the energies of an estimated watermark with most the signs being opposite to those in (a); (e) the residual watermark derived from (d). In the above examples, sufficiently large correlations between (a) and (c), and between (a) and (e) exist, indicating the presence of a watermark.

The observations from Fig. 1 motivate us to formulate the definitions of "optimal watermark estimation" and "near-perfect cover data recovery." This implies that we try to recover a cover data from its stego version. If this goal can be achieved, even optimal watermark detector will fail to detect the hidden watermark; otherwise false positive will appear.

Definition 1 (Optimal Watermark Estimation): Given an original embedded watermark signal $\mathbf{W}$ and its approximate version $\mathbf{W}^e$ estimated from the stego image $\mathbf{X}^s$, the necessary condition for the optimal estimation of $\mathbf{W}$ as $\mathbf{W}^e$ is defined as

$$BER(sgn(\mathbf{W}), sgn(\mathbf{W}^e)) \leq \tau, \quad (1)$$

where τ is watermarking algorithm- and application-dependent, and the sign function, $sgn(\cdot)$, is defined as

$$sgn(t) = \begin{cases} +1, & \text{if } t \geq 0, \\ -1, & \text{if } t < 0. \end{cases}$$

Basically, Definition 1 is naturally derived from Fig. 1 in that the "polarity of each watermark value is particularly crucial. To assist our later analysis, we use Θ to denote the set of indices satisfying $sgn(W^e(i)) = sgn(W(i))$ in Eq. (1). This is the first step, where the existence of a watermark may be efficiently eliminated if most sign bits of the watermark can be obtained by an attacker. Beyond this step, however, to avoid leaving a residual watermark (as illustrated in Fig. 1(c)) that can reveal the hidden watermark, accurate estimation of the energy of $\mathbf{W}^e$ is absolutely indispensable. In addition to Eq. (1), watermark removal can be completely achieved if the watermark energy to be subtracted is also larger than or equal to the added energy, i.e., $mag(W^e(i)) \geq mag(W(i))$, where $mag(t)$ denotes the magnitude $|t|$ of t. Therefore, it is said that $\mathbf{W}^e$ is an optimal estimation of $\mathbf{W}$ if and only if

$$BER(sgn(\mathbf{W}), sgn(\mathbf{W}^e)) \leq \tau \text{ and}$$
$$mag(W^e(i)) \geq mag(W(i)) \ \forall i \in \Theta. \quad (2)$$

Definition 2 (Near-Perfect Cover Data Recovery): Under the prerequisite that Definition 1 (Eq. (2)) is satisfied, it can be said that $\mathbf{X}^r$ is a near-perfect recovery of the cover image $\mathbf{X}$ if

$$PSNR(\mathbf{X}, \mathbf{X}^r) \approx \infty, \quad (3)$$

where $\mathbf{X}^r = \mathbf{X}^s - sgn(\mathbf{W}^e)mag(\mathbf{W}^e)$, $\mathbf{X}^s = \mathbf{X} + sgn(\mathbf{W})mag(\mathbf{W})$, and $sgn(\mathbf{v})$ and $mag(\mathbf{v})$ are two vectors representing the sign and magnitude of the elements in a vector $\mathbf{v}$, respectively.

It is noted that ideally Eq. (3) is satisfied only if $mag(\mathbf{W}^e) \approx mag(\mathbf{W})$; otherwise, even if the watermark values have been completely removed based on $mag(\mathbf{W}^e) >> mag(\mathbf{W})$, the quality of the attacked/recovered image would be undesirably degraded. Typically, evaluation of $mag(\mathbf{W}^e)$ can be achieved by means of either averaging or remodulation. It should be noted that if the residual watermark (Fig. 1(c)) becomes empty or negatively correlated with the hidden watermark (Fig. 1(a)), then even optimal watermark detector is unable to detect the hidden watermark. Definition 2 has specified how a cover data could be recovered in a near-perfect manner. In Sec. III, a near-perfect cover data recovery algorithm will be described.

III. A NEAR-PERFECT COVER DATA RECOVERY ALGORITHM

"Near-perfect" here means that the hidden watermarks can be mostly removed with a high probability (say 90%) so that the recovered data is more similar to the cover data than the stego data. Under this circumstance, it is not necessary to worry about the detection ability of optimal watermark detector; otherwise, they will run the risk of raising the false positive problem. Here, we shall take the block-based multiple self-reference watermarking method [14] as an example to explain the performance of our algorithm in removing the hidden watermarks. However, it should be noted that our algorithm can be extended to other methods [2], [9], [12] that adopt the similar concept of multiple redundant watermark embedding.

In the following, the method [14] where the watermark embedded in each image block is a bipolar sequence is briefly described. This watermark $\mathbf{W}$ is flipped and copied in each direction to produce a symmetric signal, which is repeated over the entire image. In the embedding process, both the expanded watermark signal and the cover image are first decomposed using wavelet transform. Then, the watermark signals are embedded into the cover image in the wavelet domain through linear additive modulation together with a perceptual masking model called "noise visibility function (NVF)" [14]. The NVF is basically a wavelet-based content-adaptive visual model so that the degree for each wavelet coefficient that can be modified without raising perceptual difference can be defined. Let $NVF_{k,l}(m,n)$ denote the masking threshold for the wavelet coefficient at the position (m,n) of subband k,l (where k denotes scale and l denotes orientation), and let $x_{k,l}(m,n)$ and $y_{k,l}(m,n)$ denote the cover and stego image wavelet coefficients, respectively. They are related as

$$\begin{aligned} y_{k,l}(m,n) &= x_{k,l}(m,n) \\ &+ (1 - NVF_{k,l}(m,n)) \cdot w_{k,l}(m,n) \cdot S^e_{k,l} \\ &+ NVF_{k,l}(m,n) \cdot w_{k,l}(m,n) \cdot S^f_{k,l}, \end{aligned} \quad (4)$$

where $w_{k,l}(m,n)$'s denote the watermark wavelet coefficient, $S^e_{k,l}$ and $S^f_{k,l}$ denote the embedding strength for non-flat and flat regions, respectively.

Now, the proposed near-perfect cover data recovery algorithm based on the collusion estimation of watermark's signs and NVF-based estimation of watermark's magnitudes is described as follows. Let $\mathbf{W^e}$ be the watermark estimated by means of collusion. As pointed out in Fig. 1 and Definition 2, accurate estimation of watermark's magnitudes is crucial to completely remove the hidden watermarks. In fact, we would rather remove more watermark energy than it should be so that the watermark energy can be more guaranteed to be eliminated. Let $NVF^s_{k,l}(m,n)$ denote the masking threshold for a stego image. The wavelet coefficient for the recovered image $\mathbf{X^r}$ based on Definition 2 can be derived as

$$\begin{aligned} z_{k,l}(m,n) &= y_{k,l}(m,n) \\ &- [(1 - NVF^s_{k,l}(m,n)) \cdot w^e_{k,l}(m,n) \cdot S^e_{k,l} \\ &+ NVF^s_{k,l}(m,n) \cdot w^e_{k,l}(m,n) \cdot S^f_{k,l}](1 + \epsilon_{k,l}(m,n)), \end{aligned} \quad (5)$$

where $w^e_{k,l}(m,n)$'s denote the estimated watermark wavelet coefficient and $\epsilon_{k,l}(m,n)$'s are used to more guarantee that the hidden watermark can be completely removed. By substituting Eq. (4) into Eq. (6) and assuming that the recovered image is equal to the cover image; i.e., $z_{k,l}(m,n) = x_{k,l}(m,n)$ for all k, l, i, and j, the desired parameters, $\epsilon_{k,l}(m,n)$'s, can be derived. To simplify analysis, we further assume that $\mathbf{W} = \mathbf{W^e}$; i.e., their watermark wavelet coefficients satisfy $w_{k,l}(m,n) = w^e_{k,l}(m,n)$ for all k, l, i,, and j. In this case, $\epsilon_{k,j}(m,n)$ can be ideally derived as

$$\epsilon_{k,l}(m,n) = \frac{NVF_{k,l}(m,n) - NVF^s_{k,l}(m,n)}{NVF^s_{k,l}(m,n) - \frac{S^e_{k,l}}{S^e_{k,l} - S^f_{k,l}}}. \quad (6)$$

In Eq. (6), $NVF_{k,l}(m,n)$'s are unknown since no cover data is available in a blind detection scenario to obtain its NVF. However, they can be approximately estimated if $\mathbf{X^s} - sgn(\mathbf{W^e})mag(\mathbf{W^e})$, as described in Definition 2, is used to obtain an approximate cover image. It should be noted that when $\epsilon_{k,l}(m,n)$'s are equal to zero, this algorithm degenerates to watermark remodulation [13]. If watermark's energy is estimated by means of averaging, this algorithm degenerates to conventional collusion attack.

In this section, we have derived how the watermark magnitude can be estimated to achieve complete watermark removal. In order to evaluate the performance of "near-perfect cover data recovery," it is best to compare the recovered image with the cover image to check how many watermark bits still survive in the recovered image.

IV. EXPERIMENTAL RESULTS

In our experiments, ten varieties of standard cover images of size 512×512 were used for watermarking. In this study, Voloshynovskiy et al.'s block-based image watermarking approach [14] was chosen as the benchmark due to its strong robustness and computational simplicity. However, we would like to particularly emphasize that the proposed scheme is readily applied to other watermarking algorithms that implement the similar principle of embedding multiple redundant watermarks [2], [9], [12]. Wiener filter was used to perform denoising-based blind watermark extraction.

In order to verify how the hidden (content-independent) watermark could be removed by means of the proposed optimal watermark estimation algorithm (Sec. III), the survived watermark of the obtained recovered image was extracted using the cover image so that we can accurately check how many correct watermark bits still remain. Table I shows the BER values, which were obtained from comparing the original watermark and the extracted watermarks, and the PSNR values, which were calculated between the cover image and the recovered/stego image. As we can see from Table I that if $\epsilon_{k,l}(m,n) = 0$ is used, this corresponds to perceptual remodulation [13]. It is observed that PSNRs have been increased and most BERs fall into the interval between 50% ~ 60%, which means that a significant part of watermark values is not completely removed.

However, if $\epsilon_{k,l}(m,n) \neq 0$ is adopted, BERs can be increased averagely as high as 0.9 except for some very smoothing images, which implies that our estimation and recovery algorithms are able to remove almost all the watermark

TABLE I

Validation of our estimation and recovery scheme. BER is computed between the original and the extracted watermarks. PSNR is computed between the cover and the recovered/stego images.

	X_1^s	X_2^s	X_3^s	X_4^s	X_5^s	X_6^s	X_7^s	X_8^s	X_9^s	X_{10}^s
Stego image PSNR (dB)	38.15	37.92	37.96	37.51	37.61	37.94	38.13	38.98	37.74	37.94
	X_1^r	X_2^r	X_3^r	X_4^r	X_5^r	X_6^r	X_7^r	X_8^r	X_9^r	X_{10}^r
Recovered image using $\epsilon_{k,l}(m,n) = 0$ [13] PSNR (dB)	45.46	53.25	53.16	43.10	56.20	48.18	52.79	45.05	54.83	53.25
BER (%)	68.8	48.4	48.4	43.8	57.8	78.1	78.1	89.1	60.9	57.8
	X_1^r	X_2^r	X_3^r	X_4^r	X_5^r	X_6^r	X_7^r	X_8^r	X_9^r	X_{10}^r
Recovered image using $\epsilon_{k,l}(m,n) \neq 0$ PSNR (dB)	53.15	55.30	54.38	59.65	58.28	53.44	53.31	49.06	56.08	54.23
BER (%)	82.1	85.7	85.1	89.6	89.4	90.5	86.3	80.5	87.8	84.0
	X_1^r	X_2^r	X_3^r	X_4^r	X_5^r	X_6^r	X_7^r	X_8^r	X_9^r	X_{10}^r
Recovered image using $2 \times \epsilon_{k,l}(m,n)$ PSNR (dB)	51.69	53.53	52.90	57.12	56.06	51.52	52.16	47.65	54.45	52.34
BER (%)	88.5	92.8	93.0	93.8	94.1	93.7	92.8	81.8	92.8	90.9

bits in a stego image. In addition, the obtained PSNRs are further improved than those obtained using [13] such that the recovered image can be more similar to its cover version. Since $\epsilon_{k,l}(m,n)$'s are approximated derived in Eq. (6), if $2 \times \epsilon_{k,l}(m,n)$ is heuristically adopted, we show that (by comparing those results obtained using $\epsilon_{k,l}(m,n) = 0$ and $\epsilon_{k,l}(m,n) \neq 0$) the BERs can be further increased and the PSNRs are moderate.

Under the circumstance that the proposed near-perfect cover data recovery algorithm is used, we are confident based on Table I that even the so-called "optimal watermark detector" [1], [8] is difficult to detect the survived (but few) watermark bits to sufficiently claim the existence of a watermark; otherwise, false positive probability is easy to occur. This validates our claim that efficient elimination of previously added watermark energy is indispensable to really remove the hidden watermark. In addition to the effective watermark removal, the recovered images were found to be more transparent to their cover ones than the stego images in terms of PSNR values. These experimental results demonstrate the performance of the proposed cover data recovery algorithm in defeating the multiple redundant watermark embedding methods that were originally addressed to tolerate geometric distortions.

V. Conclusion

Although multiple watermarks can be embedded into an image to withstand geometrical distortions, they are unfortunately vulnerable to collusion and copy attacks, and the desired geometric invariance is lost. In this study, we have proposed an efficient watermark estimation and recovery algorithm (which is regarded as an exquisite collusion attack) that can eliminate almost all watermark values. To cope with the watermark estimation attack (WEA), an anti-disclosure content-dependent watermark (CDW) with resistance to WEA has been investigated in [5]. In our recent paper [7], the proposed CDW has been combined with geometric-invariant image hash [3], [6] to obtain a mesh-based content-dependent watermarking scheme that can resist both the geometric and estimation attacks.

References

[1] M. Barni, F. Bartolini, A. E. Rosa, and A. Piva, "A New Decoder for the Optimum Recovery of Nonadditive Watermarks," *IEEE Trans. on Image Processing*, Vol. 10, No. 5, pp. 755-766, 2001.

[2] P. Bas, J. M. Chassery, and B. Macq, "Geometrically invariant watermarking using feature points," *IEEE Trans. Image Processing*, Vol. 11, No. 9, pp.1014-1028, 2002.

[3] C. Y. Hsu and C. S. Lu, "A Geometric-Resilient Image Hashing System and Its Application Scalability," *Proc. ACM Multimedia and Security Workshop*, pp. 81-92, Germany, 2004.

[4] M. Kutter, S. Voloshynovskiy, and A. Herrigel, "The Watermark Copy Attack," *Proc. SPIE: Security and Watermarking of Multimedia Contents II*, Vol. 3971, 2000.

[5] C. S. Lu and C. Y. Hsu, "Content-Dependent Anti-Disclosure Image Watermark," *Proc. 2nd Int. Workshop on Digital Watermarking (IWDW)*, LNCS 2939, pp. 61-776, Seoul, Korea, 2003.

[6] C. S. Lu, C. Y. Hsu, S. W. Sun, and P. C. Chang, "Robust Mesh-based Hashing for Copy Detection and Tracing of Images," *Proc. IEEE Int. Conf. on Multimedia and Expo: special session on Media Identification*, Taipei, Taiwan, 2004.

[7] C. S. Lu, S. W. Sun, and P. C. Chang, "Robust Mesh-based Content-dependent Image Watermarking with Resistance to Both Geometric Attack and Watermark-Estimation Attack," to appear in *Proc. SPIE: Security, Steganography, and Watermarking of Multimedia Contents VII (EI120)*, 2005.

[8] A. Nikolaidis and I. Pitas, "Asymptotically Optimal Detection for Additive Watermarking in the DCT and DWT Domains," *IEEE Trans. on Image Processing*, Vol. 12, No. 5, pp. 563-571, 2003.

[9] J. S. Seo and C. D. Yoo, "Localized image watermarking based on feature points of scale-space representation," *Pattern Recognition*, Vol. 37, pp. 1365-1375, 2004.

[10] K. Su, D. Kundur, D. Hatzinakos, "Statistical Invisibility for Collusion-resistant Digital Video Watermarking," *IEEE Trans. on Multimedia*, 2004.

[11] M. D. Swanson, B. Zhu, and A. H. Tewfik, "Multiresolution Scene-Based Video Watermarking Using Perceptual Models," *IEEE Journal on Selected Areas in Communications*, Vol. 16, No. 4, pp. 540-550, 1998.

[12] C. W. Tang and H. M. Hang, "A Feature-Based Robust Digital Image Watermarking Scheme," *IEEE Trans. Signal Processing*, Vol. 51, No. 4, pp.950-958, April 2003.

[13] S. Voloshynovskiy, S. Pereira, A. Herrigel, N. Baumgartner, and T. Pun, "Generalized Watermarking Attack Based on Watermark Estimation and Perceptual Remodulation," *Proc. SPIE: Security and Watermarking of Multimedia Contents II*, Vol. 3971, San Jose, CA, USA, 2000.

[14] S. Voloshynovskiy, F. Deguillaume, and T. Pun, "Multibit Digital Watermarking Robust against Local Nonlinear Geometrical Distortions," *Proc. IEEE Int. Conf. on Image Processing*, pp. 999-1002, 2001.

Privacy preserving data mining with unidirectional interaction

Chai Wah Wu

IBM T. J. Watson Research Center, P. O. Box 218, Yorktown Heights, NY 10598, U. S. A.
e-mail: chaiwahwu@ieee.org

Abstract— Privacy concerns over the ever-increasing gathering of personal information by various institutions led to the development of privacy preserving data mining. Two main approaches to privacy preserving data mining have emerged in recent years. The first approach protects the privacy of the data by perturbing the data through a random process. The second approach uses cryptographic techniques to perform secure multi-party computation. While the second approach is generally viewed as superior due to its strong assurance of privacy, there are reasons why it might not be appropriate in certain applications. For instance, in some cases, the data collection phase usually does not allow for complicated processing, such as ballots cast via short message service. In other cases, such as paper surveys, the data flow is unidirectional from the survey taker to the survey collector. The requested data is sent once with no further interaction. In both these cases, the complicated computations and iterative interaction between data originators and data collectors that are required for secure multi-party computation cannot be used.

We show how in these cases random perturbation of data can be a useful approach to privacy preserving data mining. In particular, we study a data perturbation scheme which was shown to have asymptotically small privacy loss and information loss. We illustrate the ideas using an example of a privacy preserving paper-based survey where no computation is done by the users. Finally, we apply this method to privacy preserving association rules mining.

I. INTRODUCTION

The proliferation of data mining algorithms to extract statistics and trends from large sets of data results in the need for data mining algorithms which preserve the privacy of the owners of this data. In privacy preserving data mining (PPDM), the goal is to perform data mining operations on sets of data without disclosing the content of the sensitive data. In [1] a review of the state of the art in this rapidly evolving field is presented. Since the results of the mining tell us something about the data (otherwise, there would have been no interest in performing the mining) some information about the original data is leaked to the mining results. Thus the first questions to be asked are: how much privacy is leaked by knowing the mining results, and what type of mining operations result in a small amount of such privacy breach. In this paper, we will not answer these questions, as it can involve social and policy issues, but assume that some form of mining is predetermined. Thus there is a nontrivial baseline privacy breach (under some suitable definition) inherent in the data mining results. Our goal is to determine what the algorithms should be to limit the actual privacy breach to be close to the baseline.

In this paper, we consider the basic framework consisting of n users, each with a piece of data x_i. The goal of the *data mining agent* is to compute the aggregate data $h = f(x_1, \ldots, x_n)$ without having the users reveal what the exact values of x_i's are. The baseline privacy breach is the amount of information about x_i that is inherent in the aggregate data h. We are interested in scenarios where there is no interaction among users and minimal interaction between the users and the data mining agent, as is the case in surveys conducted over the internet for example.

Secure multi-party computation [2]–[4] has been shown to be an important component in implementing PPDM. In this approach, cryptographic techniques are used to allow a party to perform computation on a set of numbers without knowing the actual values of these numbers. The computation usually involves several iterations of a protocol between the data mining agent and the users. There are several scenarios why this approach is not appropriate. First of all, since some information about the values x_i are leaked to the computed data h anyway, it may be overkill to use cryptographic techniques to ensure zero privacy breach in the computation of h. This is especially true since cryptographic techniques are usually computation-intensive. Secondly, in some applications, such as taking a survey using short message service (SMS) or paper questionnaires, there is no interaction between the users and there is little interaction between the users and the data mining agent, making it infeasible or impractical to use an iterative protocol.

The purpose of this paper is to study the usefulness of random perturbation to PPDM in these cases. We first restrict ourselves to the scenario where the goal of data mining is to generate an estimate of the distribution of the collected data points. In this case the problem can be formulated as follows.

The data mining agent would like to construct an estimate of the distribution $f_X(x)$ of n data points x_i from n users, without actually knowing the values x_i. The n points x_i are assumed to be i.i.d. samples from a random variable X. In the random perturbation approach, this mining is accomplished by having the users send modified values z_i to the data mining agent and estimating $f_X(x)$ using only the values z_i. Two concepts quantify the performance of a PPDM scheme. Privacy loss describes the amount of information about x_i that is leaked to the data mining agent due to the transmission of z_i. Information loss describes the additional loss of precision in estimating the aggregate data h due to the use of modified data z_i instead of the actual data x_i.

Using a signal processing approach, it was shown in [5] that the Expectation-Maximization reconstruction algorithm in [6] is equivalent to applying the Richardson-Lucy deblurring algorithm to the histogram estimator. Furthermore, it was shown that simple addition of random noise (i.e. $z_i = x_i + n_i$, where n_i are i.i.d. samples from a random variable Y) as proposed in [6], [7] can prevent the distribution to be estimated accurately, depending on the spectral characteristics of f_X and f_Y. The tradeoff between privacy loss and information loss is exhibited in maximizing both the support of f_Y and the support of the Fourier transform of f_Y.

A novel PPDM data perturbation scheme was also presented in [5] which solves this problem. The purpose of this paper is to study and extend this scheme and illustrate how it can be applied in a survey with paper questionnaires where the interaction is unidirectional. In Section VI we apply this method to privacy preserving association rules mining.

II. DATA PERTURBATION SCHEME

We summarize here the data perturbation scheme in [5]. We assume that $x_i \in \mathbf{R}^m$. The range of X is partitioned into k regions $R_1, \ldots R_k$. The measure of R_i is denoted as m_i. Let I_{R_i} be the indicator function of R_i, i.e. $I_{R_i}(x) = 1$ if $x \in R_i$ and 0 otherwise. For each data point x_i, construct the k-vector $\chi(x_i) = (\frac{1}{m_1} I_{R_1}(x_i), \ldots, \frac{1}{m_k} I_{R_k}(x_i))$. We also generate a k-vector y_i consisting of k i.i.d. samples from a random variable W with mean μ. The perturbed data is then defined as the k-vector $z_i = \chi(x_i) + y_i$.

To estimate $f_X(x)$, we simply compute the average of z_i and subtract μ. In other words, let y_{ij} and z_{ij} be the j-th component of the k-vectors y_i and z_i respectively. Then $\tilde{\theta}_j \stackrel{\text{def}}{=} \frac{1}{n}\sum_{i=1}^n z_{ij} - \mu = \frac{1}{m_j n}\sum_{i=1}^n I_{R_j}(x_i) + \frac{1}{n}\sum_{i=1}^n y_{ij} - \mu$ converges to $\theta_j \stackrel{\text{def}}{=} \frac{1}{m_j n}\sum_{i=1}^n I_{R_j}(x_i)$ as $n \to \infty$ by the law of large numbers[1]. Since a histogram estimate of f_X is precisely $\sum_j \theta_j I_{R_j}$, this allows us to construct an estimate of f_X.

As the samples y_i are generated independently, there is no interaction between users. Furthermore, the datum z_i is sent from the user to the data mining agent unidirectionally with no further interaction. This data flow is schematically shown in Fig. 1.

The perturbation random variable W is a discrete random variable on the natural numbers with distribution $P_W(j) = \frac{a-1}{a^{j+1}}$ for $j = 0, 1, \ldots$ where $a > 1$.

III. PROPERTIES OF DATA PERTURBATION SCHEME

A. Privacy loss

Privacy loss describes how much information about x_i is leaked through z_i. We will compute privacy loss similar to [5]. Let $x_i \in R_{j(i)}$. Given z_i, we construct an estimate $\tilde{j}(i)$ of

[1]If n is known and the y_{ij}'s are known by the trusted system designer, but not the data mining agent, then we can simply set μ to be $\frac{1}{n}y_{ij}$, and a very large n is not necessary.

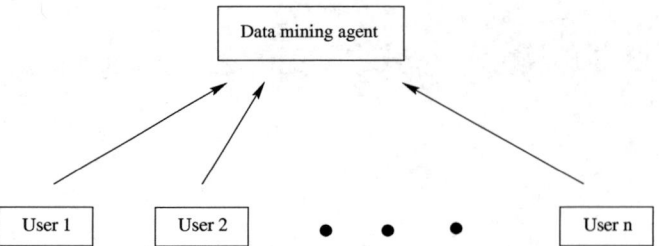

Fig. 1. Data flow between users and data mining agent in data perturbation scheme. There is no interaction between users and data are sent once from the users to the data mining agent.

the value of $j(i)$, the index of the region where x_i belong. Let $P_i = P(\tilde{j}(i) = j(i)|z_i)$, i.e. the probability that the estimate $\tilde{j}(i)$ is correct. The privacy loss is then defined as $L = \sum_i P(z_i) P_i$.

Let e_j be the j-th unit vector. In particular, if $x \in R_j$, then $\chi(x) = e_j$. Let us assume that $\tilde{j}(i)$ is the maximum a posteriori estimate (MAP) of $j(i)$, i.e. $\tilde{j}(i) = \arg\max_j P(\chi(x_i) = e_j|z_i)$. Now $P(\chi(x_i) = e_j|z_i) = \frac{P(\chi(x_i) = e_j, y_i = z_i - e_j)}{\sum_j P(\chi(x_i) = e_j, y_i = z_i - e_j)}$. Recall that $P_W(j) = \frac{a-1}{a^{j+1}}$. If $z_{ij} = 0$, then $P(y_i = z_i - e_j) = 0$. Otherwise

$$P(y_i = z_i - e_j) = \frac{(a-1)^k}{a^{z_{ij}} \prod_{k' \neq j} a^{z_{ik'}+1}}$$
$$= \frac{(a-1)^k}{a^{(k-1+\sum_{k'} z_{ik'})}} \stackrel{\text{def}}{=} \kappa_i$$

which is independent of j. Let J_i be the set of indices j such that $z_{ij} = 0$. Since x_i and y_i are independent, this means that $\sum_j P(\chi(x_i) = e_j, y_i = z_i - e_j) = \kappa_i c_i$ where $c_i = P(\chi(x_i) = e_j, j \notin J_i)$. Therefore if $j \notin J_i$, then $P(\chi(x_i) = e_j|z_i) = P(\chi(x_i) = e_j)/c_i$. Thus $\tilde{j}(i) = \arg\max_{j \notin J_i} P(\chi(x_i) = e_j)$. Since $P_W(0) = \frac{a-1}{a}$, by choosing a close to 1, we can ensure that J_i is small. This in turn implies that P_i and consequently the privacy loss is small as the number of regions increases and the measures of the regions R_k decrease.

B. Baseline privacy loss

Let j_b be the MAP estimate of $j(i)$ knowing only f_X, i.e. $j_b = \arg\max_j P(\chi(x_i) = e_j)$. We define Q_i as the probability that x_i is in R_{j_b} given z_i, i.e. $Q_i = P(j_b = j(i)|z_i)$ and we define the baseline privacy loss as $\sum_i P(z_i) Q_i$.

Q_i is always smaller or equal to P_i. In particular, $Q_i = 0$ if $j_b \in J_i$ and $Q_i = P_i$ if $j_b \notin J_i$. This means that the baseline privacy loss is always smaller than the privacy loss. If J_i is empty then the privacy loss is equal to the baseline privacy loss. Thus for $a \approx 1$, the privacy loss is close to the baseline privacy loss.

C. Information loss

Estimating f_X using the samples x_i already incurs a error in determining f_X. Information loss describes the *additional* loss in precision in estimating f_X from z_i due to the introduction of y_i. In particular, when $y_i = 0$, the information loss should also vanish, which is not the case for the definition of information loss in [6]. Let us assume that we estimate f_X using the histogram method. As in [5] we define the information loss as the statistical difference between the reconstructed histogram of f_X constructed using z_i and constructed using x_i using the same partition of the range of X. More precisely, the information loss is defined as:

$$\frac{1}{2} E \int |hist(x_i) - hist(z_i)|$$

where $hist(x_i) = \sum_j \theta_j I_{R_j}$ is the histogram reconstructed using x_i and $hist(z_i) = \sum_j \tilde{\theta}_j I_{R_j}$ is the histogram reconstructed using z_i. Note that in this definition, the information loss does vanish when $y_i = 0$. The information loss can be written as:

$$\frac{1}{2} E \sum_j |\theta_j - \tilde{\theta}_j| m_j = \frac{1}{2} E \sum_j \left| \frac{1}{n} \sum_{i=1}^n y_{ij} - \mu \right| m_j$$

which converges to 0 for large n.

Thus we have shown that this PPDM scheme has small privacy loss close to the baseline privacy loss if we make the perturbation large ($a \approx 1$) and small information loss if the number of samples n is large.

IV. AN EXAMPLE OF PRIVACY PRESERVING DATA MINING WITH UNIDIRECTIONAL INTERACTION

Consider an example where paper-based surveys are distributed to a group of users to determine the distribution of answers to a multiple-choice question. To preserve the privacy of the survey takers, the above data perturbation technique can be used as follows. The survey is designed as shown in Fig. 2. Each survey card consists of k columns of dots, one column for each possible answer. Within each column some dots will be punched out. The person taking the survey is instructed to pick the column corresponding to his or her answer and punch out a dot that has not yet been punched out. If all dots in the chosen column has already been punched out, the survey-taker does nothing. Of course, care must be taken that the mechanism the survey-taker uses to punch out a dot is the same as the mechanism used to generate the prepunched dots.

Suppose that each column has v dots. It is clear that this is very similar to the data perturbation technique presented earlier. The only difference is that z_{ij} is bounded to be less than or equal to v. In other words, $z_{ij} = q(\frac{1}{m_j} I_{R_j}(x_i) + y_{ij})$, where $q(x) = x$ if $x \leq v$ and $q(x) = v$ otherwise (Fig. 3). We will also use the notation $z_i = q(\chi(x_i) + y_i)$. To estimate the distribution of the answers, μ_q is subtracted from the average number of punched dots in each column where μ_q is defined as the mean of $q(W)$. Let us study this scheme in more detail.

A. Privacy loss

The analysis is similar to Section III-A. Given z_i and e_j, Let $U_i = \{y_i : q(y_i + e_j) = z_i\}$. Then $P(\chi(x_i) = e_j|z_i) = \frac{P(\chi(x_i)=e_j, y_i \in U_i)}{\sum_j P(\chi(x_i)=e_j, y_i \in U_i)}$. If $z_{ij} = 0$, then $P(y_i \in U_i) = 0$. If $1 \leq z_{ij} < v$, then $P(y_i \in U_i) = \kappa_i$. If $z_{ij} = v$, then

$$P(y_i \in U_i) = \frac{(a-1)^k \beta_i}{\prod_{k' \neq j} a^{z_{ik'}+1}} = \frac{\kappa_i}{a-1}$$

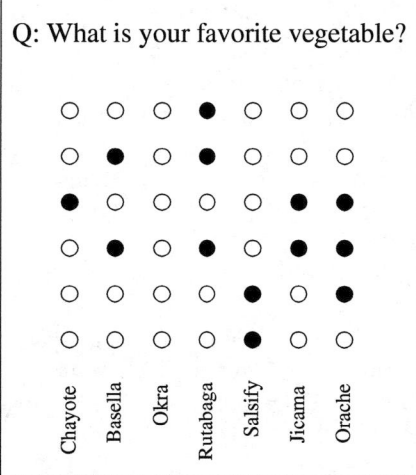

Fig. 2. Example of a questionnaire. Filled circles represent punched out holes. In this case $k = 7$ and $v = 6$.

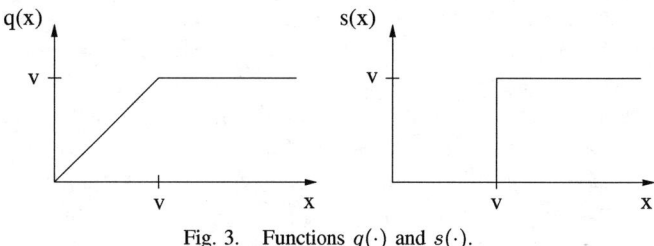

Fig. 3. Functions $q(\cdot)$ and $s(\cdot)$.

where $\beta_i = \sum_{i=v}^\infty \frac{1}{a^i} = \frac{1}{a^v(a-1)}$. Let J_i^v be the set of indices j such that $z_{ij} = v$, and let $c_i = P(\chi(x_i) = e_j, j \notin J_i \cup J_i^v)$ and $g_i = P(\chi(x_i) = e_j, j \in J_i^v)$. This means that $\sum P(\chi(x_i) = e_j, y_i \in U_i) = c_i + \frac{g_i}{a-1}$. Therefore if $j \notin J_i \cup J_i^v$, then $P(\chi(x_i) = e_j|z_i) = \frac{(a-1)P(\chi(x_i)=e_j)}{(a-1)c_i+g_i}$. If $j \in J_i^v$, then $P(\chi(x_i) = e_j|z_i) = \frac{P(\chi(x_i)=e_j)}{(a-1)c_i+g_i}$. Similar to Section III-A, for a close to 1 and a large v, J_i and J_i^v can be ensured to be small which results in the privacy loss to be small when the measures of the regions are small.

B. Information loss

$$\tilde{\theta}_j - \theta_j = \sum_{i=1}^n q(y_{ij} + I_{R_j}(x_i)) - I_{R_j}(x_i) - \mu_q = \alpha_1 + \alpha_2$$

where $\alpha_1 = \frac{1}{n} \sum_{i=1}^n q(y_{ij}) - \mu_q$ which vanishes for large n and $\alpha_2 = \sum_{i=1}^n q(y_{ij} + I_{R_j}(x_i)) - I_{R_j}(x_i) - q(y_{ij})$. The quantity α_2 can be written as $\frac{1}{n} \sum_{i=1}^n I_{R_j}(x_i) s(y_{ij}) \leq \frac{1}{n} \sum_{i=1}^n s(y_{ij})$ where s is the step function defined by: $s(x) = 1$ if $x \geq v$ and $s(x) = 0$ otherwise (Fig. 3). For large n, $\frac{1}{n} \sum_{i=1}^n s(y_{ij})$ converges to the mean of $s(Y)$ which is small for large v. Thus the information loss is small for large n and v.

This analysis shows that v needs to be large to obtain small information and privacy loss.

V. Tradeoff between information loss and privacy loss

When the number of data samples n is small, there is a tradeoff between privacy loss and information loss. To get small privacy loss, a should be chosen close to 1. On the other hand, when a is close to 1, y_{ij} have a larger variance and thus more samples are needed to limit information loss. Furthermore, for the scheme in Section IV-A, v needs to be large when a is close to 1 in order for the privacy loss and information loss to be small.

Since the regions R_i are independent, the parameters a and v can be chosen to be different for each region. This is useful in cases where the requirement of information loss is different for different regions.

VI. Privacy preserving association rules mining

The above method can also be used for privacy preserving association rules mining if the number of rules of interest is not excessive. We use the definitions in [8]. Let $\mathcal{I}$ be a set of items. Each user's data is a transaction x_i which is a subset of $\mathcal{I}$. An association rule is an implication of the form $A \Rightarrow B$, where $A \subset \mathcal{I}$, $B \subset \mathcal{I}$ and $A \cap B = \emptyset$. The rule $A \Rightarrow B$ holds with confidence c if c percent of the transactions that contain A also contain B. The rule $A \Rightarrow B$ has support s if s percent of the transactions contain $A \cup B$. The goal is to find rules which have high support and confidence.

To each rule $A \Rightarrow B$ corresponds a 2-vector which maps a transaction x_i as follows: $\chi_R(x_i) = (v_1, v_2)$ where

$$v_1 = \begin{cases} 1 & \text{if } A \subset x_i \\ 0 & \text{otherwise} \end{cases}, \quad v_2 = \begin{cases} 1 & \text{if } A \cup B \subset x_i \\ 0 & \text{otherwise} \end{cases}$$

It is clear that this scheme allows us to compute the support and the confidence of the rule easily. Summing the values of v_2 of all the transactions and dividing it by the number of transactions gives the support of the rule, whereas the sum of v_2 divided by the sum of v_1 gives the confidence of the rule. To achieve PPDM, these vectors are perturbed as before. The same analysis as before shows that information loss can be made small if the number of transactions is large.

For r rules, these 2-vectors are concatenated and each user sends a $2r$-vector to the data mining agent. Depending on the rules, a shorter vector might suffice. For instance, for the 2 rules $A \Rightarrow B$ and $A \Rightarrow C$, only a 3-vector describing membership of the sets A, $A \cup B$ and $A \cup C$ in x_i is needed. When $\mathcal{I}$ has p items and the rules of interest are all the rules, a $(2^p - 1)$-vector is generated for each transaction x_i. Each component of this vector is either 0 or 1 depending on whether each of the $2^p - 1$ nonempty subsets of $\mathcal{I}$ is contained in x_i.

VII. Conclusions

We study the usefulness of random perturbation in privacy preserving data mining scenarios where the data flow is only from the users to the data mining agent. For these scenarios, secure multi-party computation which involves complicated processing and iterative protocols is not appropriate. We study a random perturbation scheme suitable for use in such scenarios and where no computation is done by the users, such as when a survey is conducted using paper questionnaires. We show that the privacy loss and information loss can be made small in this case. Finally we apply this scheme to privacy preserving association rules mining when the number of rules is not too large.

References

[1] V. S. Verykios, E. Bertino, I. N. Fovino, L. P. Provenza, Y. Saygin, and Y. Theodoridis, "State-of-the-art in privacy preserving data mining," *SIGMOD Record*, vol. 33, no. 1, pp. 50–57, 2004.

[2] Y. Lindell and B. Pinkas, "Privacy preserving data mining," in *Advances in Cryptology-crypto 2000*, 2000.

[3] W. Du and M. J. Atallah, "Privacy-preserving cooperative scientific computation," in *14th IEEE Computer Security Foundations Workshop*, 2001.

[4] M. Kantarcioglu and C. Clifton, "Privacy-preserving distributed mining of association rules on horizontally partitioned data," in *ACM SIGMOD Workshop on Research Issues in Data Mining and Knowledge Discovery*, 2002.

[5] C. W. Wu, "Privacy preserving data mining: a signal processing perspective and a simple data perturbation protocol," in *IEEE Workshop on Privacy Preserving Data Mining, International Conference on Data Mining*, 2003. Available electronically at http://www.cis.syr.edu/~wedu/ppdm2003/papers/2.pdf.

[6] D. Agrawal and C. C. Aggarwal, "On the design and quantification of privacy preserving data mining algorithms," in *Symposium on Principles of Database Systems*, 2001.

[7] R. Agrawal and R. Srikant, "Privacy-preserving data mining," in *Proc. of the ACM SIGMOD Conference on Management of Data*, pp. 439–450, ACM Press, May 2000.

[8] R. Agrawal and R. Srikant, "Fast algorithms for mining association rules," in *Proceedings of the 20th VLDB conference*, 1994.

Progressive Scrambling for MP3 Audio

Wei-Gang Fu, Wei-Qi Yan, Mohan S. Kankanhalli
School of Computing
National University of Singapore
Singapore 117543

Abstract – Audio scrambling can be employed in audio distribution for the purpose of guaranteeing the confidentiality. Electronic commerce in audio products would be facilitated by the development of solutions which can ensure security/privacy, efficiency in compressed domain to reduce the bandwidth requirements, flexibility of implementing progressive audio quality control and computationally inexpensive to be used in real-time systems. This paper proposes an effective progressive scrambling algorithm that works with MP3 audio. We perform multiple-level scrambling based on keys to produce a set of audio outputs of differing qualities. During decoding, the keys provided and rounds of descrambling performed decide the quality of the audio output. Our experiments show that the proposed algorithm is effective, fast, simple to be implemented, and at the same time it does provide flexibility to control the quality of audio output progressively. We also present some experimental results to show its utility in real-time systems.

I. INTRODUCTION

Audio scrambling aims to minimize the residual intelligibility of the original audio and control the access to only authorized users. It is similar to but not a direct application of normal cryptographic techniques. The main advantage is that scrambled audio are still validly formatted which can be played by the corresponding players.

With the wide-spread use of the MP3 audio format, the need for developing a new scrambling algorithm that can work in the compressed domain has become important. The past research is based on permutation in either the frequency or temporal domain [1] for normal audio data. However, not much work has been done in the compressed domain. MP3 is mainly used for online music distribution. There is a need for allowing the music owners to have flexible control over their music. An algorithm that allows them to provide music of varying quality to different users and which at same time, is able to protect the copyright for the MP3 files that can be very useful.

In this paper, we propose a progressive algorithm that does multiple rounds of scrambling of MP3 audio. We reconstruct the MP3 outputs of different qualities based on number of keys provided and rounds of scrambling performed. Each de-scrambling is based on the results of the previous step. This work is one part of our research program on media security. We had earlier developed a compressed-domain video scrambling technique [10]. In this paper, we focus on the audio domain.

The rest of paper will be structured as follows: section II will introduce the background, section III will present our proposal, section IV provides the results and the security analysis. The final section draws the conclusion and outlines the future work.

II. BACKGROUND

A. Overview of MP3

MPEG audio is a family of open standards for portable audio with the high commercial value which includes MP2, MP3 and AAC (Advanced Audio Coding). MP3 stands for MPEG (Moving Picture Experts Group Technology) Layer-III. Its file structure, as shown in Fig. 1, can be expressed in this scheme (TAGs are optional):

[TAG v2] Frame1 Frame2 Frame3,…,[TAG v1]

An MP3 file is made up of frames having a constant duration of 0.026 second. The size of one frame (in Bytes) varies according to the *BitRate*. The first four bytes of each frame are the frame header and the rest is audio data. MP3 frame header consists of information about frame, such as *BitRate*, *SampleRate* etc, each of them has its own characteristics. The MP3 specification (ISO –11172-3) does not specify on how the MP3 encoding shall be done. It only specifies the resulting output format. Developers are supposed to develop their own algorithms to meet the requirements [2]. There are many different MP3 encoders available, amount which the most popular standard is LAME. These MP3 encoders perform lossy compression by discarding the information that does not significantly contribute to the signal perception.

On top of that, these MP3 encoders minimize coding redundancy using "Huffman Coding". Huffman coding is a byte-substitution scheme that replaces the more frequently used values with shorter bytes and less frequently used values with longer bytes, by doing this it can achieve an overall reduction in the data volume.

B. Related Work

Many algorithms have been proposed for audio scrambling.

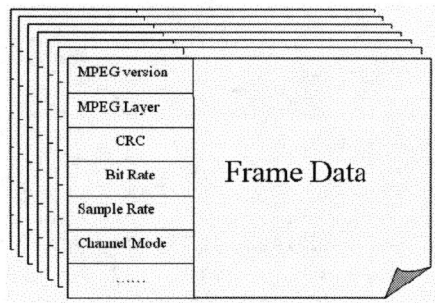

Fig. 1. Frame structure of MP3

Borujeni [1] presents a fast Fourier transform (FFT) related algorithms to deal with the permutation of FFT (Fast Fourier Transform) coefficients in a speech encryption system, which is able to provide a high level of security. Matsunaga et al [3] present another FFT based algorithm which is accompanied by adaptive dummy spectrum insertion and commanding operation, and prove that the dummy spectrum insertion will further enhance the security.

Farkash S. et al. [4] present an algorithm that modifies the Gabor representation of the input speech signal. The proposed scheme can perform any invertible modifications, and it does not need the commonly used permutation operations. Hadamard approaches are widely employed in audio scrambling [5][6]. Milosevic et al. [7] present a new algorithm that makes use of Hadamard matrices to perform scrambling. It not only changes the sequence of the data elements, but also the values. Thus it enhances security, but at the cost of increased complexity.

Wavelet transform tends to create low-value coefficients at the finer scale. Ma et al. [8] use wavelet to scramble analog signals in 2D space, it is highly secure in temporal and spatial domain. Kadambe [9] uses adaptive wavelet for speech scrambling. However, there has been no work on compressed domain audio scrambling.

Kankanhalli et al [10] have proposed a joint encryption and compression framework that works on compressed-domain video (MPEG video). The algorithm shuffles the Huffman code words and provides reasonable security level in video scrambling. In this paper, our work is different from other audio scrambling algorithms in two aspects – it works directly in the compressed domain and it supports progressive scrambling.

III. PROPOSED ALGORITHM

The idea behind our proposal is to scramble the Huffman code words in the MP3 file. However, instead of directly shuffling the Huffman code word table, we operate on the actual MP3 data, this is to eliminate the repeats of values, i.e. one value in the original audio will be shuffled to different values, which will break the statistical correlation between the source audio and the output. This is very important for thwarting statistical cryptanalysis.

Progressive quality levels are attained by conducting multiple scrambling. A few rounds of scrambling are required and the quality degrades gradually with each scrambling. By performing the inverse descrambling processes, audio with different qualities will be generated.

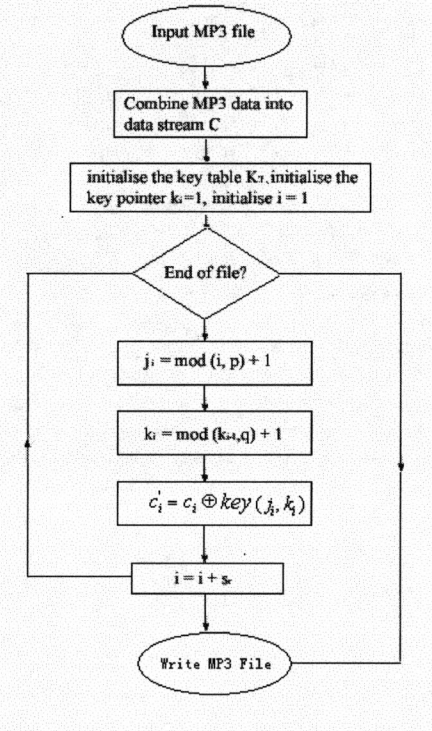

Fig. 2. Flowchart of proposed algorithm for MP3 audio scrambling

The procedure of MP3 audio scrambling/descrambling is explained in Fig. 2. Given an MP3 audio, we ignore the frame headers and only operate on the audio data $C = \{c_i, i = 1,2, ... , n\}$, n is the length of the audio stream. A sample rate s_r needs to be predefined before each round of scrambling. To make the quality degradation in a linear way, the sampling rate $s_r(l)$ needs to be set at an exponential increment rate for different rounds.

$$s_r(l) = a^l, a \in Z \quad (1)$$

where l is the scrambling level number, $l = 1,2, ...,L$. L is the total number of scrambling rounds to be conducted. Also a k_T will be initialized; the format of k_T is shown as follows:

$$K_T = (key(i,j))_{p \times q} \quad (2)$$

The key table generation makes use of special techniques to ensure the uniqueness and randomness of the key tables generated. For the keys used in this scheme, we employ the Arnold matrix [11] to generate random $p \times q$ matrix using an input random generated seed, here we assign $p=q$. The Arnold matrix is shown as follows:

$$A = \begin{bmatrix} 1 & 1 & \cdots & 1 \\ 1 & 2 & \cdots & 2 \\ \cdots & \cdots & \cdots & \cdots \\ 1 & 2 & \cdots & p \end{bmatrix} \quad (3)$$

We first generate a random matrix $R=(r_{i,j})_{p \times q}$

$$r_{i,j} = rand(range), i = 1,2,...,p; j = 1,2,...,q \quad (4)$$

where $rand(\cdot)$ is a function to generate a random number sequence based on the seed $range$. Then we generate our K'_T by doing a matrix product of the Arnold matrix [11] by:

$$K_T = \begin{bmatrix} 1 & 1 & \cdots & 1 \\ 1 & 2 & \cdots & 2 \\ \cdots & \cdots & \cdots & \cdots \\ 1 & 2 & \cdots & p \end{bmatrix} \bullet R \pmod{p} \quad (5)$$

We use the Arnold matrix to transform the keys in the key table to the different kinds of choices for different users. The transformation based on this matrix can ensure the validity of the keys since they are bounded.

For the i-th data item c_i in the original audio, we do a modulo division of the current position i with the row length p of the key table and the remainder decides the row number j of the key to be used.

$$j_i = \mod(i, p) + 1 \quad (6)$$

The column number k_i will cyclically range from the first column to the maximum column number q.

$$k_i = \mod(k_{i-1}, q) + 1 \quad (7)$$

After we select the key $k_i = key(j_i, k_i)$, we perform an XOR operation with the MP3 audio datum c_i. XOR is chosen due to the speed and simplicity of its implementation. The changed datum c'_i is written into the corresponding position of the output stream:

$$c'_i = c_i \oplus key(j_i, k_i) \quad (8)$$

where $key(j_i, k_i)$ is the element in matrix K_T, $\oplus$ is the bit-wise XOR. This procedure is repeated till the end of the data stream and the output is also an MP3 stream. The keys used will be stored and they will be distributed to the authorized consumers for descrambling.

The descrambling process is exactly similar to the scrambling procedure. For descrambling level l, a MP3 file generated from descrambling level $l-1$ will be used as an input. We apply the same process and use the same K_T that has been used in the scramble round l.

With the implementation of the scrambling process as described above, the quality of the output MP3 file depends on how many descrambling levels involved. For example, during scrambling if an MP3 audio has been scrambled 5 times using 5 different keys, a user needs to have all the 5 keys and perform 5 rounds of descrambling before they can perfectly restore the original audio. If the user has only 4 keys or performs only 4 rounds of descrambling, (s)he will get an audio with a small degradation of the original quality.

IV. RESULTS AND ANALYSIS

In this section, we conduct experiments to evaluate the effectiveness of the proposed algorithm and prove the feasibility of our scheme.

Figure 3 depicts the waveform for an audio file at different scrambled levels. Figure 3(a) is the waveform of source MP3 audio, we scramble the audio 5 times and obtain Fig. 3(f). We descramble Fig. 3 (f) to get Fig.3 (e), which is of slightly better quality, we further generate Fig. 3 (d) to Fig.3 (b) by performing additional rounds of descrambling. The waveforms show that the output is gradually reconstructed and eventually we will get an audio that is exactly same as the source.

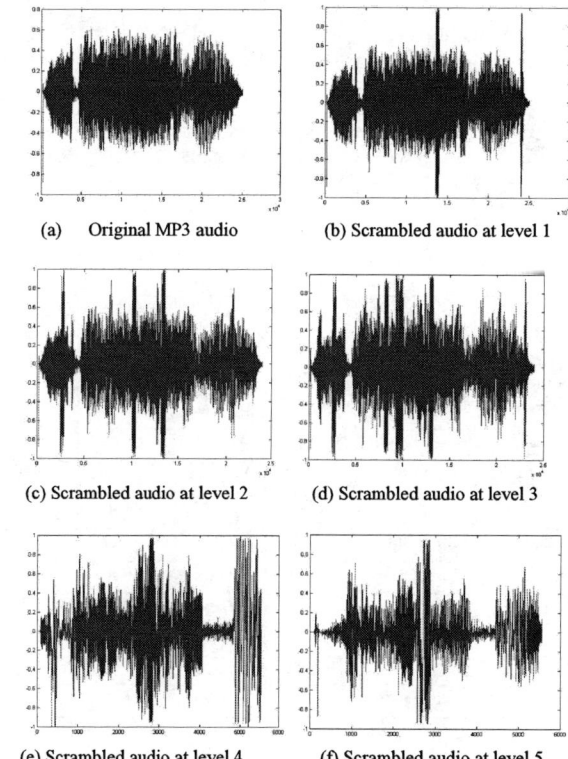

(a) Original MP3 audio (b) Scrambled audio at level 1

(c) Scrambled audio at level 2 (d) Scrambled audio at level 3

(e) Scrambled audio at level 4 (f) Scrambled audio at level 5

Fig. 3. Original and scrambled MP3 audio waveforms at different scrambled level

Subjective evaluation is a reasonable approach to measure the scrambling levels. In our work of MP3 audio scrambling, we conduct an evaluation of four audios of different categories, namely classic, pop, jazz and rock music.

Each music clip has been scrambled to five levels of degraded qualities. Seven investigators participated in this poll, the feedbacks with the average scores are listed in Table I.

Table I: Audiences' score for different quality levels
(6 for the best quality – 1 for the worst quality)

Quality	POP	JAZZ	ROCK	CLASSIC
Level 6	6.00	6.00	6.00	6.00
Level 5	4.14	4.57	5.00	5.00
Level 4	3.86	3.71	3.29	3.86
Level 3	3.86	3.00	3.00	2.86
Level 2	1.71	2.29	2.14	1.71
Level 1	1.29	1.57	1.57	1.57

Table I shows that most of the listeners are able to perceive the gradually decrease in quality as the amount of scrambling increases. They quite accurately identify the correct sequence of the scrambled audio, especially in the better quality levels. We also calculate the computation time required for scrambling/descrambling at each level, which is shown in Table II.

As our scrambling has been done at different sampling rates for different levels and the amount of computation is closely related to the number of samples, the time taken increases with the sampling rate. (e.g. for classic music time consumption for scrambling increases from 0.94 second at level 1 to 4.59 seconds at level 5).

Table II. Time consumption for MP3 scrambling / descrambling

Audio Size /length	Operations	POP 3kb / 2sec	JAZZ 23 kb / 22sec	ROCK 10kb / 9 sec	CLASSIC 32kb / 32 sec
Level 1	Scrambling	0.82 sec	0.91sec	0.84 sec	0.94 sec
	Descrambling	0.81sec	0.90sec	0.85 sec	0.94 sec
Level 2	Scrambling	1.02sec	1.01 sec	0.92 sec	1.12 sec
	Descrambling	0.79sec	0.97 sec	0.84 sec	1.04 sec
Level 3	Scrambling	1.27sec	1.52 sec	1.34 sec	1.58 sec
	Descrambling	0.84sec	1.04 sec	0.91 sec	1.15 sec
Level 4	Scrambling	1.3sec	1.67 sec	1.43 sec	1.86 sec
	Descrambling	0.87sec	1.23 sec	0.99 sec	1.41 sec
Level 5	Scrambling	1.55sec	3.57 sec	2.25 sec	4.59 sec
	Descrambling	1.07sec	3.07 sec	1.77 sec	4.1 sec

The result shows that our method is fast. For an MP3 audio around 30 seconds, it needs only 8.7 seconds for the scrambling and descrambling processes, which is much shorter than the music duration. Thus the proposed algorithm can potentially be applied in real-time systems.

In our scheme, the security of our scheme relies on the key table. Range of values allowed in the key table determines the key space. The amount of calculation for attacker to break the whole MP3 file of length L by brute force is calculated as following: given an MP3 audio of duration 30 seconds and data length of 3.2×10^4 bytes, assume that the key table allows random values within the range from 1 to 16 and the key table is of size 16×16. An adversary has to complete 32000×16^{256} tries for bit permutation before they can restore the perfect audio. This will take more than 10^{250} years at a rate of 10^8 instructions per second. However, this can potentially be reduced by studying audio coherence for domain-specific cryptanalysis. Since the purpose of this proposal is to protect music clips, the security level appears to be sufficient. Incidentally, this scheme has no size blow-up. The original and scrambled audios are of exactly the same size.

V. CONCLUSION

In this paper, we have proposed a progressive audio scrambling and descrambling scheme. Our experiments show that the effectiveness of the proposal. It is a fast and simple solution, yet it can provide sufficient security for musical products. Our future work will be on enabling traitor tracing in the scrambling algorithm and on utilizing watermarks as keys to manage audio copyright.

Acknowledgement: Wei-Qi Yan's work in this paper is partially supported by a fellowship from Singapore Millennium Foundation (SMF).

REFERENCES

[1] S.E Borujeni., "Speech encryption based on fast Fourier transform permutation," in Proc. of The 7th IEEE International Conference on Electronics, Circuits and Systems, Dec. 2000, vol. 1, pp. 290–293.

[2] Y. Wang, W.D. Huang and K. Jari, "A framework for robust and scalable audio streaming," in Proc. of ACM Multimedia 2004, Oct. 10-16, 2004.

[3] A. Matsunaga, K. Koga, and M. Ohkawa, "An analog speech scrambling system using the FFT technique with high level security," IEEE Transactions on Selected Areas in Communications, 7(4), May 1989, pp. 540–547.

[4] S. Farkash, S. Raz, and D. Malah, "Analog speech scrambling via the Gabor representation," in Proc. of the 17th Convention of Electrical and Electronics Engineers in Israel, 1991, pp. 365 – 368.

[5] V. Senk, V.D. Delic, and V.S. Milosevic, "A new speech scrambling concept based on hadamard matrices," IEEE Signal Processing Letters, Jul. 1997, vol. 4, pp. 161–163.

[6] Y. Wu and B. P. Ng, "Speech scrambling with hardamard transform in frequency domain," in Proc. of IEEE ICSP, 2002, pp. 1560-1563.

[7] V. Milosevic, V. Delic, and V. Senk, "Hadamard transform application in speech scrambling," in Proc. of 13th International Conference on Digital Signal Processing, 1997, pp. 361 – 364.

[8] F.L. Ma, J. Cheng and Y.M. Wang, "Wavelet transform based analogue speech scrambling scheme," Electronics Letters, 32(8), Apr. 1996, pp.719 – 721.

[9] S. Kadambe and P. Srinivasan, "Application of adaptive wavelets for speech coding," in Proc. of the IEEE-SP International Symposium on Time-Frequency and Time-Scale Analysis, Oct.1994, pp. 632–635.

[10] M.S. Kankanhalli and T.G. Teo, "Compressed domain scrambler and descrambler for digital videos," IEEE Transactions on Consumer Electronics, 48(2), May 2002, pp. 356–365.

[11] D.X. Qi, J.C. Zou, X.Y. Han, "A new class of scrambling transformation and its application in the image information covering," Science in China (series E), 43(3), Jun. 2000, pp.305-311.

Averaging Attack Resilient Video Fingerprinting

In Koo Kang, Choong-Hoon Lee, Hae-Yeoun Lee, Jong-Tae Kim and Heung-Kyu Lee

Dept. of computer science,
Korea Advanced Institute of
Science and Technology,
Daejon, Korea
{ikkang|chlee|hytoiy|jtkim|hklee}@mmc.kaist.ac.kr

Abstract — This paper addresses an effective technique for digital fingerprinting for video. While orthogonal modulation technique is a straightforward and widely used method for digital fingerprinting, it shows limitations in computational costs, signal effectiveness, and robustness against averaging attacks. Code modulation technique that uses averaging collusion resilient fingerprint code derived from BIBD is a robust method against averaging attacks. The strategy described in [2] requires determining threshold values to detect embedded fingerprint code, but threshold changes frequently according to contents and the number of colluders, so that it is difficult to decide threshold. In this paper, we propose a new fingerprint code based on GD-PBIBD theory that is more efficient in designing fingerprint code and describe an effective method to embed and detect fingerprint code, in which the merits of orthogonal modulation and coded modulation techniques are combined. To analyze the performance, we apply our method to video applications using spread-spectrum watermarking technique on spatial domain and the results show the proposed method is effective to fingerprint code embedding and detection.

I. INTRODUCTION

Digital fingerprinting is a kind of copyright protection technique that traces illegal distribution of copyrighted contents. A unique fingerprint code that identifies a recipient is embedded into host content for selling or distributing. When copies of the content are found from illegal route, the content seller can identify the traitor who had distributed the content illegally by detecting the embedded fingerprint code.

The averaging attack is a serious problem in the digital fingerprinting. Since digital fingerprinting technique embeds slightly different codes according to the customers, several fingerprinted contents, but based on identical content are available to attackers. The averaging attack is an attempt to remove the embedded fingerprints by using the several copies of the content. For example, adversaries can estimate the original content without fingerprints by averaging several fingerprinted copies. When the number of the copies is small, a normal watermarking method that embeds orthogonal signal as a watermark can overcome the averaging attacks. However, as the number of copies increasing, the limitation is clear.

One approach to resists the averaging attack is to use averaging-resistant fingerprint code [1,2]. In these methods, the fingerprint code that is not affected by averaging is designed. Boneh [1] presented the code system named "Frameproof code", that should be satisfied for preventing false detection of a recipient who did not join in averaging attack. Stinson [3] proposed a fingerprint code system that satisfies "Frameproof code" by using combinatorial design theory first time. Trappe [2] presented a fingerprinting system for images using BIBD (Balanced Incomplete Block Design) code derived from combinatorial design system proposed by Stinson.

In [2], v fingerprint patterns for all bit sequence are generated and added into one pattern prior to embedding into host content. While patterns generated using this method are robust to averaging attacks when applied to images, they can be vulnerable in case of videos due to compression effect of video codec which causes severe drop of correlation coefficient values when code extraction step. Furthermore, when even one bit is extracted wrongly, innocent users who were actually not involved in coalition can be reported as pirates.

The code efficiency, as well as the averaging resistance, is an important factor for the fingerprinting code design. The code efficiency refers to number of recipients that can be handled by code length. The higher code efficiency, the better content fidelity can be achieved since less bits of information are embedded as well as better robustness.

In this paper, we propose a digital fingerprint code comes from GD-PBIBD (Group Divisible Partially Balanced Incomplete Block Design), also one of combinatorial design theories. Using GD-PBIBD, we can control elements replication numbers in blocks more flexibly, thus we can make more blocks that will be transformed to fingerprint code than BIBD, based on same number of elements. The proposed code has higher efficiency than that of the code derived from BIBD. We also introduce a video fingerprinting system using the proposed fingerprint code. By experiments, we show the efficiency of the code and performance of the proposed video fingerprinting system in aspect of averaging attacks.

```
              S1    S2    S3    ...                                                                                              S16   S17   S18
    User1:  0000  0000  1111  1111  1111  1111  1111  1111  1111  1111  1111  1111  1111  1111  1111  1111  1111  1111
    User2:  1111  1111  1111  1101  1111  0111  1110  1111  1111  1111  0111  1011  1011  1111  1101  1111  1111  1110
    User3:  1111  1111  1111  1110  1111  1011  1111  0111  1011  1111  1111  1101  0101  1111  1110  1111  1111  1111
    User4:  1111  1111  1111  1111  1111  1101  1111  1011  1101  1111  1111  1110  1110  1111  0111  0111  1011  1111
    User5:  1111  1111  1011  1111  1111  1111  0111  1110  1111  1111  1111  1111  1111  1011  1111  1101  1010  1111
    User6:  1111  1111  0101  1111  1111  1111  1111  1110  1111  0111  1011  1111  1111  1011  1111  1101  1110  1111
    User7:  1111  1111  1110  1111  0011  1111  1111  1111  1111  1011  1101  1111  1111  1101  1111  1110  1111  0111
    User8:  1111  1111  1111  0111  1101  1111  0011  1111  1111  1101  1110  1111  1111  1110  1111  1111  1111  1111
                                                        ...
```

Figure 2. Part of 72×89 fingerprint code based on GD-PBIBD design (S*no* represent segment #).

II. AVERAGING RESILIENT FINGERPRINT CODE

We tried to find out design theory focusing on code efficiency that is also important factor as well as averaging resiliency. Code efficiency refers to the number of recipients that can be handled by designed code length. In [2], the fingerprinting code BIBD admits 20 persons by using 16bit code. Therefore, the code efficiency is only 1.25. The higher code efficiency, the better content fidelity and robustness can be achieved since less bits of information are embedded for the same performance as lower code.

We found the fingerprint code which is resilient against averaging attack and presents high code efficiency ratio using GD-PBIBD (Group-Divisible Partially Balanced Incomplete Block Design). ($v,b,r,k,\lambda_1,\lambda_2$) GD-PBIBD is a also combinatorial design which divides v elements into b blocks [4,5]. To create blocks, first GD-PBIBD classifies v elements into two groups according to group divisible association scheme as an intermediate stage. Any two elements which are in the same group are placed λ_1 times in blocks and which are in different groups are found λ_2 times in blocks. In BIBD, there is only one group and any two elements are presented only λ_1 times in blocks. Thus, using GD-PBIBD we can control the number of elements presence in blocks more flexibly and consequently we can make more blocks than BIBD, based on same number of elements. Moreover, GD-PBIBD has averaging resiliency up to k-1 blocks which is the same number as that of BIBD. Fig.1 shows a code efficiency comparison, each of which shows a needed number of bits for codes distributing to 100 customers according to colluder numbers from 1 to 7 in BIBD and GD-PBIBD case.

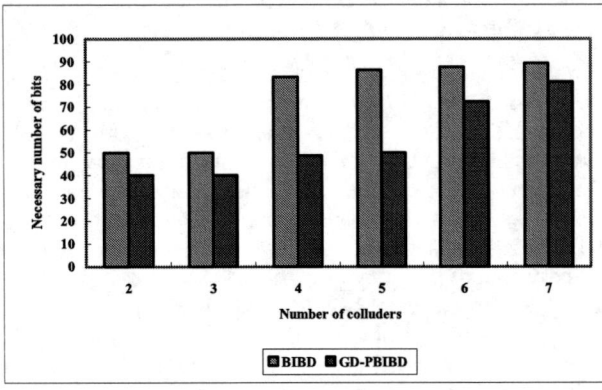

Figure 1 : Code efficiency ratio comparison

More specific information can be found in [5].

Once b blocks from v elements are completed, we transform the element blocks to incidence matrix that actually presents fingerprint code set. An incidence matrix $M_{ij} = m_{ij}$ is $v \times b$ matrix which is created as follows:

$$m_{ij} = \begin{cases} 0 : if\ i^{th}\ symbol\ belongs\ to\ j^{th}\ blocks \\ 1 : otherwise \end{cases} \quad (1)$$

Each column C_j of M_{ij} is a unique v-bit sequence that will be distributed to each customer j as a fingerprint code and, we can distinguish b customers uniquely with averaging resiliency up to k-1 customers.

We assume, when a sequence of fingerprint is averaged, that the detected binary sequence is the logical AND of the fingerprint codes C_j. Once the detector extracts bit sequence even after averaging, detector can find out which user is in coalition, by comparing bit position of a result sequence whose value is 1 with all columns C_j in incidence matrix M_{ij}. The only columns of colluders in M_{ij} have value 1 at the same position as result sequence. Consequently, detector can trace pirates involved in coalition out of innocent users.

III. HYBRID MODULATION TECHNIQUE

As explained before, coded modulation technique described by [2] shows severe problems during the detection of embedded fingerprinting code, in which threshold values are used to determine the embedded code, but it is impossible or difficult to select efficiently. For example, when seven fingerprint codes are averaged, six same codes and one different code are possibly averaged. In this case of averaging, it is difficult to detect the one different code by thresholding because the one code signal will be attenuated by the six code signals. In this section, we will describe an efficient technique to embed and detect fingerprinting code where we can select threshold values based on false positive analysis in watermarking.

For the convenience of explanation, we will describe our proposed hybrid modulation technique using (72,89,9,8,0,1) GD-PBIBD design. In this design, 89 users can be identified by a vector whose length is 72bits. In other words, one user can be represented by a 72bits length vector. If coded modulation technique proposed by [2] is used for embedding a vector, 72 watermark patterns must be embedded for each code. Fig. 2 shows part of (72,89,9,8,0,1) GD-PBIBD.

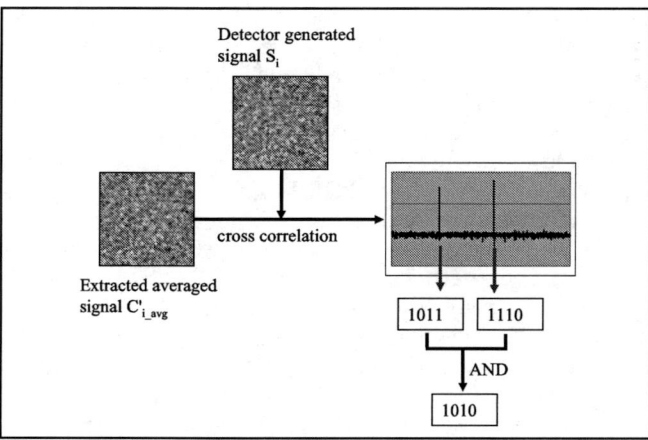

Figure 3 : Detection scheme for each segment

A. Embedding technique

We will describe the way to efficiently embed M bits length fingerprinting code.

    A. Split M bits length fingerprinting code into segments by grouping N bits.
    B. Embed N bits grouped code in each segment using orthogonal modulation method.

In case of coded modulation method, M patterns must be generated and embedded for M bits length fingerprinting code. However, for the efficiency, we divided M bits length fingerprinting code into several segments by grouping N bits. For each segment, a pattern based on N bits code is generated and embedded by using orthogonal modulation method. In this case, there are M/N segments so that only M/N patterns are used for orthogonal modulation method. For example, in (72,89,9,8,0,1) GD-PBIBD design, we split 72 bits length fingerprinting code into 18 segments by grouping 4 bits so that we only have to embed 18 patterns not 72 patterns.

When the number of grouping bits N is high, we can increase the efficiency of embedding. However, maximum 2^N patterns can be used to collude and that control maximum correlation values. If correlation is used as detection statistics, theoretical maximum correlation value is inverse-proportional to N. If normalized correlation is used, then theoretical value is inverse-proportional to sqrt(N).

In our experiment using GD-PBIBD design, we group 4 bits unit and use normalized correlation as detection statistics so that there are 16 cases. If all cases are used for averaging attacks, theoretical correlation value is 0.25. Fortunately, in each segment, GD-PBIBD design has maximum 10 cases so that theoretical correlation becomes 0.32 and it will be helpful to increase robustness in detection steps.

B. Detection technique

The embedded code is detected as follows.

    A. For each segment,
        A.1 Detect all patterns embedded by using orthogonal modulation.
        A.2 Determine N bits code of retrieved patterns and apply logical AND operation.
    B. Construct M bits length code by concatenating the result codes from all segments.
    C. Detect colluders by comparing the M bits code with fingerprinting code.

In the detection phase, we can determine the embedded N bits code of each segment by extracting the embedded orthogonal modulation pattern by using correlation based pattern detector. If collusion attack has been applied, several orthogonal patterns may be extracted from one segment. In this case, we firstly construct the bits code for each extracted orthogonal pattern, and make colluded bits code by applying bit "AND" operation with the constructed bits codes. Then, the "AND"-ed bit codes for all segments are concatenated. Using the concatenated codes, we can find out the colluders by inspecting the reference fingerprinting code by using the colluder finding method described in Sec. II. Fig. 3 depicts this procedure more specifically. For example, when we insert an orthogonal signal representing (1111), a 4 bits grouped code, into a segment during embedding step, it can be easily retrieved by orthogonal modulation method in detection stage. If three segments where orthogonal signals representing (1111), (1010), and (0011) respectively are colluded by averaging attacks, we can detect three signals representing (1111), (1010), and (0011) and hence acquire the resulting code (0010) by applying logical bits AND operation. The concatenated code from each segment is used for tracing colluders.

Similar to the method by [2], our method also requires threshold values which used to retrieve the embedded pattern. However, we can easily select threshold values by analyzing false positive error probability of orthogonal modulation method.

C. Experiments and performance discussion

We carried out experiments to analyze the performance of our hybrid modulation method by using the fingerprint code constructed by (72,89,9,8,0,1) GD-PBIBD theory. This code can accommodate 89 users, the length of a code vector for one user is 72 bits, and maximum 7 collusions can be traceable.

6 CIF format sized (352×288) videos extracted from DVD are used as host signals and blind watermarking strategy on spatial domain was applied to embed / detect orthogonal signals into / from host signals. For imperceptibility, Noise Visibility Function (NVF) was adapted to every raw frame of video sequences and normalized correlation was used as detection statistics. The error probability of normalized correlation-based detector follows a Gaussian distribution model [6]. By using 100 random orthogonal signals and 100 random raw images, we analyzed detection

statistics and selected the reliability of detector as 10^{-4} by using threshold value as 0.124.

Figure 4 : 18 segmented blocks of CIF size image

As mentioned previously, we split a 72 bits length fingerprint code into 18 segments by grouping 4 bits unit. In each segment, there are maximum 10 cases of 4-bits. In theoretical analysis, the output of normalized correlation-based detector would be 0.32 and this value is higher than threshold values to determine the resulting code from each segment such that our modulation method will be efficiently trace colluders.

To insert 18 orthogonal signals into host signals, we divided images into 18 blocks as shown in Fig. 4 whose size is 64×64 pixels and a 1024 length orthogonal signal representing a 4 bits code from one segment was inserted into each block. PSNR of images after fingerprinting was over 39db and imperceptible to native eyes.

For each image, we had generated 100 illegal videos by colluding 1 to 7 users respectively and tried to trace colluders. For each colluding case, table I summarizes colluder detection results from 100 illegal copies of 6 videos. In most of experiments, we could trace colluders successfully and colluded orthogonal signals could be retrieved accurately except 7 colluding cases. However, when 7 users colluded, the performance was decreased because the applied watermarking scheme had attenuated the embedding strength of the orthogonal signal for imperceptibility and compression effect of video scheme so that the embedded signals were severely removed by colluding in some parts of images.

IV. CONCLUSION

Generally, considering fingerprint code can increase code efficiency and resiliency to averaging attacks. However, to cover many users, large code size must be considered so that effective modulation technique will be essential. In this paper, we described a fingerprint code based on GD-PBIBD design that would have code efficiency comparing to the conventional anti-collusion code and also proposed a hybrid modulation technique for embedding and detection of fingerprinting code in which large fingerprint code can be easily handled. We conducted practical experiments on video applications using our GD-PBIBD fingerprint code and hybrid modulation technique. We got good performance on extracting fingerprint code precisely even under averaging and compression attacks and hence colluders were detected correctly. We convinced that the proposed fingerprint code from GD-PBIBD design and the proposed hybrid modulation technique can provide a solution of the problem caused by averaging collusion, as well as compression attacks in video application. Our future research is increasing robustness in other attacks and finding more efficient fingerprinting code optimized for the proposed hybrid fingerprinting technique.

	Number of extracted colluders						
	1	2	3	4	5	6	7
1 colluders	600	-	-	-	-	-	-
2 colluders	-	600	-	-	-	-	-
3 colluders	-	-	600	-	-	-	-
4 colluders	-	-	-	600	-	-	-
5 colluders	-	-	-	-	600	-	-
6 colluders	-	-	-	-	-	600	-
7 colluders	-	-	-	-	1	75	524

Table I. Performance of video fingerprinting under averaging attacks.

REFERENCES

[1] D. Boneh and J. Shaw, "Collusion-secure fingerprinting for digital data," IEEE Trans. On Information Theory, Vol.44, pp.1897-1905, Sept. 1998.

[2] W. Trappe, M. Wu, Zhen Wang, K.J.R. Liu, "Anti-collusion Fingerprinting for Multimedia," IEEE Trans. On Signal Processing, April, 2003.

[3] D. R. Stinson and R. Wei, "Combinatorial properties and constructions of traceability schemes and frameproof codes," J. of Discrete mathematics, Jan. 1997.

[4] C. J. Colbourn and J. H. Dinitz, The CRC Handbook of Combinatorial Designs, CRC, 1996.

[5] Willard H. Clatworthy, Tables of two-associate-class partially balanced designs, National Bureau of Standards Washington, D.C., U.S., 1973.

[6] I. J. Cox, M.L. Miller, and J.A. Broom, Digital Watermarking, Morgan Kaufmann Publishers: San Francisco, CA, 2002, Chapter 5.

Multimedia Data Encryption via Random Rotation in Partitioned Bit Streams

Dahua Xie and C.-C. Jay Kuo
Integrated Media Systems Center and Department of Electrical Engineering
University of Southern California, Los Angeles, CA 90089-2564
E-mails: dahuaxie@usc.edu, cckuo@sipi.usc.edu

Abstract— A new method to encrypt a compressed multimedia bit stream, which is the output of any compression system using an entropy coder in the last stage, is proposed in this research. The basic idea is to partition a coded bit stream into random-sized blocks and perform a random rotation in each block. The proposed encryption scheme demands low computational overhead and does not increase the size of compressed bit stream. Security analysis to both the ciphertext-only attack and the known/chosen-plaintext attack shows that the proposed algorithm achieves high security.

I. INTRODUCTION

Advanced multimedia compression techniques and the wide availability of high-bandwidth networks have contributed to the exploding growth of multimedia applications recently. As a consequence, multimedia data security has emerged as an important research topic, receiving more and more attention in both academia and industry.

One fundamental problem in multimedia data security is multimedia data confidentiality, which protects multimedia content from being accessed by unauthorized parties. Due to the astronomical size of multimedia data and processing speed requirements in real-time applications, traditional cryptographic ciphers tend to add a significant amount of overhead in achieving the encryption goal. An efficient yet secure multimedia data encryption algorithm that meets the stringent performance requirements imposed by real-time multimedia applications is desirable and will be the focus of our current work.

Most previous research was based on the idea of selective encryption. The work can be classified into two categories. The first one works in the compressed domain, which encrypts a chosen subset of coefficients in this domain, in the hope that these encrypted parts would be critical to the semantic meaning of the content. The resulting schemes encrypted the magnitude and the sign of selected DCT coefficients and the motion vectors [1],[2],[3],[4]. The second category performs selective encryption in the bit stream domain [5]. A promising new direction in this field is to embed encryption into entropy coding using random change of multiple coding parameters [6] [7]. This approach has the salient feature of very low computation overhead while achieving a high security.

In this paper, we propose another method to encrypt the compressed bit stream by adding enough randomness to the bit stream creation process. Security analysis to both the ciphertext-only attack and the known/chosen-plaintext attack is performed. Simply speaking, the redundancy-free nature of the bit stream ensures the security of this scheme to the ciphertext-only attack. The resistance to known/chosen-plaintext attack is enhanced due to large number of alias keys that encipher the same plaintext to the same ciphertext. The resulting algorithm is very efficient and maintains high security strength.

The rest of paper is organized as follows. In Section II, we describe the main idea and present a novel bit stream encryption scheme. Security analysis of our scheme under common attacks is provided in Section III. Finally, concluding remarks are given in Section IV.

II. PROPOSED ENCRYPTION ALGORITHM

A. Properties of Coded Bit Streams

Before addressing the issue of designing an efficient encryption algorithm, let us first take a close look at the generation process of a bit stream so as to understand its structure better. The general workflow of most image/video compression algorithms is as follows. Input video frames first go through several processing steps, mainly motion search, transform coding (DCT), and quantization. The results of these processing steps, such as quantized DCT coefficients and motion vectors, are often called quantities (or parameters) in the compressed domain. To further reduce the data size, these parameters enter an entropy coder, where they are converted to a consecutive 0-1 binary string called the bit stream. A bit stream has the following two important characteristics.

First, it contains little redundancy. The operations in the beginning stages such as motion search, DCT transform and quantization have already removed a significant amount of redundancies in each individual frame and between successive frames. The resultant parameters are compressed furthermore by an entropy coder, which helps to squeeze out remaining redundancy. Thus, we can treat a bit stream as a statistically random 0-1 binary string (though not in the strictest sense).

Second, a bit stream is always uniquely decodable and there should be no strong statistical correlations between the compressed domain parameters. This is again due to the fact that these parameters are almost redundancy-free. For instance, given DCT coefficients of a particular 8x8 block, it does not help much to guess the values of neighboring blocks.

In summary, a bit stream can be viewed statistically random with its compressed domain parameters nearly correlation-free. This property has been confirmed by previous study on the statistical behavior of MPEG bit streams [5]. This bit stream structure suggests a mechanism to design an efficient bit stream encryption scheme. That is, in order to protect the underlying video content, it is not necessary to perform heavyweight secret key encryption algorithms to encrypt the bit stream as done in [5]. Instead, we may use simple yet efficient techniques to add enough randomness to the target bit stream to achieve the encryption effect.

B. Random Rotation in Partitioned Blocks

Our idea is to exploit the order of bits in the entire bit stream. By default, the 0-1 bits output from the entropy coder are sequentially concatenated to form the bit stream. Here, we propose two techniques to encrypt the bit stream. First, the 0-1 output bits are grouped into blocks of a random size. Second, we alter the order of individual bits in each of these blocks before they are written into the bit stream. Both the length of bit blocks and the change of order will be selected according to a random sequence, which becomes the encryption key.

Many operations can be used to alter the bit order in a block. A permutation on all bits shuffles the bit order most thoroughly but it requires a lot of computation. To reduce the complexity and facilitate the bit stream formation, we restrict the bit manipulation to a simple *left rotation*. For a block of n bits $A = (a_1 a_2 \ldots a_n)$, an r-bit left rotation transforms A into $(a_{r+1} a_{r+2} \ldots a_n a_1 a_2 \ldots a_r)$ by cutting the first r bits and put them at the end of A. The main reason to use this simple operation is that it can be easily merged into the algorithm that prepare the bit stream for the final output, thus adding minimum computation overhead. Second, although left rotation is a trivial operation, our analysis in Sec. III shows that combined with random partition, it leads to an attack complexity exponential in the size of encrypted bit stream.

We can express the above concept mathematically as follows.

Definition 1: Let $A = (a_1 a_2 \ldots a_N)$ be a bit stream of N bits. The (p, r) *rotation in partition* of A, denoted $RPB(A, p, r)$ with $p = (p_1 p_2 \ldots p_m)$ and $r = (r_1 r_2 \ldots r_m)$, is obtained by the following 2 operations.

1) partition A into m blocks A_i with length p_i, $i = 1, 2, \ldots m$, $\sum_{i=0}^{m} p_i = n$
2) perform an r_i-bit left rotation on each component block A_i, $i = 1, 2, \ldots m$

Our encryption scheme enciphers a bit stream A into $RPB(A, p, r)$ with the partition key p and rotation key r. The component p_i's and r_i's are obtained from a pseudo-random bit sequence, which can be easily generated by a pseudo-random bit generator (or a stream cipher). For ease of processing, we impose an upper bound on p_i, requiring that $p_i < 2^b$ for some positive integer b. The detailed encryption algorithm is described below.

"Random Rotation in Partitioned Block" Encryption

1) Select a secure pseudo-random bit generation algorithm. Generate a random number s as the seed. The output sequence z is grouped into b-bit blocks to produce a random number in the range $0 \sim 2^b - 1$.
2) Obtain two random numbers p and r' from z. Scale r' into range $0 \sim p$ by computing $r = \lceil \frac{p \times r'}{2^b - 1} \rceil$
3) Save the first r bits output from entropy coding.
4) Accumulate next $p - r$ bits output and append the r bits in Step (3) to the end.
5) Write the block of p bits obtained in Step (4) to the final bit stream.
6) Go to Step (2) until no more bits are output from entropy coding. Update the random sequence z when it is used up.

The secret seed s is the encryption key and $A' = RPB(A, p, r)$ is the ciphertext bit stream. On the receiving side, sequence z with component partition key p and rotation key r can be generated using the same encryption key. It is readily checked that operation $RPB(A', p, p-r)$ recovers the plaintext bit stream A.

III. SECURITY AND PERFORMANCE ANALYSIS

To evaluate the security strength of the encryption scheme proposed in Sec. II, we consider two types of attacks; namely, ciphertext-only attack and known/chosen-plaintext attack. For the ciphertext-only attack, the amount of computation required to break the proposed algorithm is estimated. For the known/chosen-plaintext attack, we describe the effect of alias keys and show how they can help resist the attack. Finally we discuss several performance issues such as computation efficiency and impact to compression ratio.

A. Security under ciphertext-only attack

In a ciphertext-only attack, the cryptanalyst has only certain ciphertexts available for analysis. In the previous section, we mentioned that the bit stream is a random binary sequence without much statistical irregularity and the compression domain parameters are almost correlation-free. Thus, an attack would have to resort to a brute-force exhaustive search. In this case, the computational complexity is closely related to the total number of possible random rotations in possible partitions $RPB(A, p, r)$ for a given bit stream A of length N. We have the following conclusion.

Lemma 1: For a ciphertext bit stream of N bits, the complexity of an exhaustive search to break the random rotation in partition encryption is larger than 2^N

Proof

Let $A = (a_1 a_2 \ldots a_N)$ be a given N-bit bit stream. We use $R(N)$ to denote the total number of all possible rotations in partitions $RPB(A, p, r)$ of A, each one of which corresponds to a different way to partition and rotate the bit stream A. Apparently $R(1) = 1$ and we define $R(0) = 1$ for the ease of notation. Clearly, an N-bit plaintext can be encrypted in $R(N)$ possible ways. On the other hand, since partition and rotation are 1-to-1 reversible operations, an N-bit ciphertext can also be decrypted in $R(N)$ possible ways. The complexity of an exhaustive search thus equals $R(N)$.

While an exact expression of $R(N)$ may be difficult to obtain, we derive a recursive equation of $R(N)$ and establish that $R(n) > 2^N$ for $N \geq 6$ in appendix I. Thus, we conclude that the complexity of exhaustive search exceeds 2^N, where N is the length of the ciphertext bit stream. ∎

As the above lemma shows, random rotation in partition has a very nice property that the total number of possible encryptions grows exponentially with the size of bit stream. This would thwart any brute-force attack if the bit stream is long enough. For instance in the state-of-the-art video compression standard such as H.264, it would cost no less than $1 \sim 2$ kilo-bits to encode a CIF-size (352x288) video frame. It is thus practically impossible to perform an exhaustive search attack.

B. Security under known/chosen-plaintext attack

Under the known-plaintext attack, the cryptanalyst has several plaintext/ciphertext pairs to study. The goal is to discover partition p and rotation r used to encrypt the data. In this case, the cryptanalyst has certain advantages over the ciphertext-only attack since the comparison of the plaintext with the corresponding ciphertext reveals some important key information or even allows the guess of the correct key.

However, the random rotation in partition encryption bears another nice property. For a given plaintext/ciphertext pair, there exists more than one keys that encipher the same plaintext to the same ciphertext. We call these keys *alias keys*. But only one of them is the correct key. Let us consider an illustrative example:

plaintext: $A = (01011101)$
ciphertext: $A' = (00111011)$
key 1: $p_1 = (1,7), r_1 = (0,1)$
key 2: $p_2 = (3,4,1), r_2 = (2,1,0)$

It can be easily verified that both keys 1 and 2 converts the plaintext A into the same ciphertext A'. They are thus an example of alias keys. The cryptanalyst cannot distinguish which key is correct by simply observing the values of p and r. We stress that the concept of alias keys is associated with a particular ciphertext (assuming a fixed plaintext). Two alias keys for one ciphertext may not be alias keys for another ciphertext. Discussion on alias keys is not meaningful without the context of one particular plaintext/ciphertext pair.

Under the known/chosen-plaintext attack, the security of the RPB encryption relies on the total number of alias keys for a plaintext/ciphertext pair. We have the following lemma about the number of alias keys for a general plaintext A.

Lemma 2: Let $A = (a_1 a_2 \ldots a_N)$ be a stream of N bits containing Z 0's and $Alias(A,C)$ denote the number of alias keys for the plaintext/ciphertext pair (A,C). Then, there exists a ciphertext A' such that

$$Alias(A, A') > \left\lceil 2^N / \binom{N}{Z} \right\rceil \qquad \text{(III.1)}$$

In a statistically average sense, a random plaintext A contains half 0's ($Z = N/2$). When the plaintext length N is large enough, we have

$$Alias(A, A') > \sqrt{\pi N / 2} \qquad \text{(III.2)}$$

Lemma 2 establishes the existence of alias keys for any plaintext A. The quantity $\sqrt{\pi N/2}$ is however a conservative estimate of the number of alias keys. We consider the statistical average, and use $A(N)$ to denote the average number of alias keys for a general N-bit plaintext. Our analysis shows that the size of $A(N)$ grows exponentially with the ciphertext length N as described in the following lemma.

Lemma 3: $A(N) \sim c^N$ for sufficiently large N, where $c > 1$ is a constant.

Thus, we arrive at the conclusion that the computational complexity of a known/chosen-plaintext attack grows exponentially with the ciphertext length N since a cryptanalyst has to examine $A(N)$ possible alias keys to find out the correct key. Due to space limitation proofs of Lemma 2 and 3 are omitted.

C. Performance evaluation

The performance of the proposed encryption scheme is discussed below.

1) *encryption cost*

The proposed encryption scheme does not perform actual cryptographic operations on the bit stream. Instead, we just select bit blocks of a random size and delay their writing in the bit stream by a random amount. It can be easily implemented with the following trivial modification to the process of bit stream writing. First, save the chosen r-bit block. Next, continue to write $p-r$ bits to the bit stream. Finally, write the saved r-bit block to the bit stream. In many practical implementations, an output buffer is used to collect a large chunk of bits before they are written to the bit stream in one stroke. In this case, we just fill the buffer according to the above-mentioned modification. Thus, our algorithm incurs negligible computational overhead.

2) *comparison with stream cipher*

The proposed RPB encryption is similar to a stream cipher in that a stream cipher or a pseudo-random bit generator is used to produce the random partition key and rotation key. Since the XOR operation in a stream cipher costs only one CPU instruction, one may argue that it might not worth the effort to apply the new scheme due to a small computational gain. However, a stream cipher is known to be vulnerable under the known-plaintext attack. That is, one plaintext/ciphertext pair suffices to reveal the pseudo-random sequence. While the proposed encryption algorithm can resist known/chosen-plaintext attack due to the huge number of alias keys.

3) *impact to compression ratio*

Our algorithm has no influence on the compression ratio of the associated multimedia compression system. The encryption only alters the order of certain bit blocks.

The size of encrypted bit stream is exactly the same as that of original bit stream without encryption.

IV. CONCLUSION

A novel method to encrypt the compressed multimedia bit stream called the random rotation in partition encryption was proposed. The new scheme adds low computational overhead and can be effortlessly embedded into the operation of writing the bit stream. The redundancy-free nature of the bit stream ensures the security of this scheme to the ciphertext-only attack. The resistance to known/chosen-plaintext attack is enhanced due to large number of alias keys (exponential with ciphertext length) that encipher the same plaintext to the same ciphertext.

APPENDIX I
PROOF OF LEMMA 1

Let $A = (a_1 a_2 \ldots a_N)$ be an N-bit bit stream. We denote by $R(N)$ the total number of possible rotations in partitions $RPB(A, p, r)$ of A. Assume B is the maximum block size allowed in partitioning A. Notice that this restriction implies any rotation in partition cannot start with bit beyond a_B. Thus, all $R(N)$ possible rotation in partition $RPB(A, p, r)$ can be classified into the following B categories.

1) those starting with a_1;
2) those starting with a_2;
$\vdots$
B) finally, those starting with a_B.

We denote the total number of each category by $R_1(N)$, $R_2(N)$, ..., $R_B(N)$. Note this classification is mutually-exclusive and all-inclusive, meaning that any possible resultant bit stream $A' = RPB(A, p, r)$ must belong to one and only one of the above categories. Thus, we have

$$R(N) = \sum_{i=1}^{B} R_i(N) \quad (\text{I}.1)$$

Now let us look at each of the above categories. In category 1), a_1 is fixed and we are left with $N-1$ bits after a_1 which we can freely partition and rotate. Thus $R_1(N) = R(N-1)$. In category 2), it must be true that the first $2 \leq k \leq B$ bits are chosen as a block and a 1-bit left rotation is performed. This is the only way A' can start with a_2. If $k = 2$ (A' starts with $a_2 a_1$), we have $N-2$ bits left over and total number of possible rotation in partition is clearly $R(N-2)$. $k = 3$ (A' starts with $a_2 a_3 a_1$) corresponds to $R(N-3)$. Finally for $k = B$ we have the number $R(N-B)$. Notice that this classification with different values of k is again mutually-exclusive and all-inclusive. Hence we end up with:

$$R_2(N) = \sum_{k=N-B}^{N-2} R(k)$$

Continuing the same line of reasoning we have the following equation:

$$R_i(N) = \sum_{k=N-B}^{N-i} R(k)$$

Finally, summing up $R_i(N)$, we arrive at

$$R(N) = \sum_{i=1}^{B} R_i(N)$$
$$= R(N-1) + \sum_{i=2}^{B} \sum_{k=N-B}^{N-i} R(k) \quad (\text{I}.2)$$
$$= R(N-1) + \sum_{k=N-B}^{N-2} (N-1-k) R(k)$$

Rearranging the terms of the above equation, we obtain another recursive relationship of $R(N)$:

$$R(N) = 2R(N-1) + \sum_{k=N-B}^{N-3} R(k) \quad (\text{I}.3)$$

If we define the following recursive sequence

$$S(N) = \begin{cases} 2S(N-1) & N > 0 \\ 1 & N = 0 \end{cases}$$

the solution is apparently $S(N) = 2^N$.

From the above definitions of $R(N)$ and $S(N)$, it is clear that if $R(N_0) > S(N_0)$ for some N_0, then $R(N) > S(N)$ for all $N > N_0$. Now, it is readily checked that $R(6) = 65 > S(6) = 64$. Thus, we come to the conclusion that $R(N) > 2^N$ for $N \geq 6$. This completes the proof. ∎

ACKNOWLEDGMENT

This research has been funded in part by the Integrated Media Systems Center, a National Science Foundation Engineering Research Center, Cooperative Agreement No. EEC-9529152. Any Opinions, findings and conclusions or recommendations expressed in this material are those of the author(s) and do not necessarily reflect those of the National Science Foundation.

REFERENCES

[1] L. Tang, "Methods for encryption and decrypting MPEG video data efficiently," *Proceedings of the 4th ACM International Conference on Multimedia*, Boston, MA, Nov 18-22, pp. 219-230, 1996.
[2] C. Shi and B. Bhargava, "A fast MPEG video encryption algorithm," *Proceedings of the 6th ACM International Conference on Multimedia*, Bristol, UK, Sep 1998.
[3] C. Shi, S.-Y. Wang and B. Bhargava, "MPEG video encryption in real-time using secret key cryptography," *1999 International Conference on Distributed Processing Techniques and Applications (PDPTA'99)*, Las Vegas, NV, Jun 28 - Jul 1, 1999.
[4] W. Zeng, and S. Lei, "Efficient frequency domain digital scrambling for content access control," *Proc. ACM Multimedia 99*, pp. 285-294, Orlando, FL, Oct 1999.
[5] L. Qiao and K. Nahrstedt, "A new algorithm for MPEG video encryption," *Proc. of the 1st international conference on imaging science, systems, and technology*, Las Vegas, NV, Jul 1997.
[6] C. Wu and C.-C. J. Kuo, "Efficient Multimedia Encryption via Entropy Codec Design," *SPIE international symposium on electronic imaging*, San Jose, CA, Jan 2001.
[7] D. Xie and C.-C. J. Kuo, "Efficient multimedia data encryption based on flexible QM coder," *Security, Steganography, and Watermarking of Multimedia Contents VI, Proc. of the SPIE*, vol 5306, pp. 696-704, San Jose, CA, Jan 2004.

A New Visual Cryptography Using Natural Images

Hyoung Joong Kim, Yongsoo Choi
Dept. Instrumentation/Control Engineering
Kangwon National University
Chunchon, Korea
khj@kangwon.ac.kr

Abstract—In this paper an (n, n) visual cryptography scheme without dithering is proposed. This scheme takes n grayscale input images to cover a target image across n grayscale images and produces n grayscale output images which are very close to the input images, respectively. Since the output images are visibly innocuous and natural, it may be easy to pass visual inspection, which is a very desirable property in terms of the steganography aspect. This method is different from the existing schemes from the fact that it keeps the input images almost intact.

I. INTRODUCTION

Visual cryptography is a kind of cryptographic technique that shares a visual secret among n participants by breaking up an image into n shares so that only he or she who has all of n shares can decrypt the image by overlaying each of the shares over each other. In 1994 Naor and Shamir [8] propose the visual cryptographic technique first. A generalized version of the visual cryptography is the (k, n)-threshold visual cryptography that encodes a target image into n shares such that any k or more shares enable the visual recovery of the hidden image. However, by inspecting less than k shares one cannot gain any information on the secret image. The 2-out-of-2 visual cryptography scheme can be thought of as a private key system. One of the two shares will be a private share and the other serves as a public share. A visual cryptography that reveals the target image by stacking meaningful images is called the Extended Visual Cryptography [1]. Nakajima and Yamaguchi [7] introduce two definitions: the sheets are the output images and the target image is the resulting image reconstructed by stacking the sheets all together. An access structure is a rule, which defines how to share a secret. Tzeng and Hu [9] define the general access structure by three components: P is the set of participants, F is a collection of forbidden sets, and Q is a collection of the qualified sets. An element of a forbidden set or a qualified set represents a sheet held by the corresponding participant. Stacking all the forbidden sheets of a forbidden set cannot reveal any information about the target image while stacking all the sheets of a qualified set can reproduce the target image.

The basic model of the conventional visual cryptography assumes that on each transparency a ciphertext indistinguishable from white noise is printed. However, noise-like sheets seem to be suspicious and thus are susceptible to attacks by wardens in the middle. Naor and Shamir [8] have mentioned an extension of the visual cryptography scheme that conceals the very existence of the secret message (that is, target image), which is important from the point of secret communications like steganography. Therefore, producing meaningful sheets like pictures rather than random dots is important.

In the black-and-white scheme (that is, with binary images) the pixel is *black* if the number of black subpixels is more than a constant threshold t, and *white* if the number of black subpixels is less than the threshold when the transparencies are stacked together. The threshold visual cryptography is a visual cryptography based on the threshold value used as a criterion of determining black dots or white dots. However, white or black color is a logical concept. Note that, in case of grayscale images, the pixel value 0 represents the darkest black and the value 255 the brightest white color.

Most of the visual cryptographic schemes need to expand pixels; that is, the pixel of the target images is reproduced by m subpixels on the sheets, where $m \geq 2$. Consequently, the sheet is m times the size of the target image, and that leads to not only distortion of images but also inconvenience of carrying large size of sheets and waste of the storage space. This situation is more serious for grayscale or color images. The parameter m is called the *pixel expansion*, and the case of "$m=1$" refers to situation that the size of the sheets is same to the target. A few of studies have been done on this situation including Hou et al. [2], Hou and Tu [3], and Ito et al. [4]. The existing schemes are mostly based on the half-toning or dithering methods to expand the binary or grayscale images. However, half-tone images are still unnatural and low in visual quality.

This work was in part supported by the MSRC-ITRC under the auspices of IITA, Korea, and BK21 project.

Previous works on the extended visual cryptography has dealt with binary images such as text. Recently, Hou et al. [2], Ito et al. [4], and Nakajima and Yamaguchi [7] have handled natural images for visual cryptography. Generally it has been believed that visual cryptography suffers from severe deterioration in image quality. Thus, the existing schemes are not free from image degradation. Lin and Tsai [6] convert the grayscale image into an approximate binary image by using dithering technique first, and then the existing visual cryptography schemes for binary images are applied to accomplish the work of creating sheets. Thus, even though this scheme takes grayscale images, it produces sheets of random dots. Hou et al. [2], Ito et al. [4], and Nakajima and Yamguchi [7] have taken grayscale images and produced relatively low quality sheets due to dithering.

One of the main contributions of this paper is to present (n, n) visual cryptography scheme with $m=1$, where sheets are almost visibly innocuous and natural and their PSNR is sufficiently high. The proposed scheme is absolutely free from dithering operations. Moreover, the target image is robust against malicious attacks such as JPEG compression.

II. Proposed Method

The proposed algorithm in this paper consists of two major phases: the encoding phase and the decoding phase. In the encoding phase, n grayscale images and one target image are processed to produce n innocuous grayscale sheets which are almost close to the input images. Since the sheets are so innocuous and natural that warden may pay little attention as Alice and Bob have want. In the decoding phase, the n sheets are stacked to reproduce target image. The proposed scheme keeps the target image intact as long as the sheets are kept to be intact, that is, free from malicious attacks.

A. Encoding Phase

Let there be n input images P_k, $k = 1, ..., n$. Let $P_k(i, j)$ denote the pixel of P_k in pixel position (i, j). Now, we introduce a new value $P(i, j)$ such that

$$P(i, j) = \sum_{k=1}^{n} P_k(i, j) \bmod v \quad (1)$$

where v is the grayscale value. When the image is an 8-bit gray-level, the value v is 256. The main idea of the encoding phase is to make the value $P(i, j)$ equals zero before generating sheets. It is obvious that $P(i, j)$ rarely be zero. Therefore, we have to modify the values so that

$$\sum_{k=1}^{n} \{P_k(i, j) \pm x(i, j)\} \equiv 0 \bmod v \quad (2)$$

Equation (2) is a necessary step to hide target image across the input images. The Equation (2) is the preliminary step for encoding.

Let $T(i, j)$ be the target image. The encoding step is very simple. Encoding step is just modifying the pixel values so that

$$\sum_{k=1}^{n} \{P_k(i, j) \pm x(i, j) \pm y(i, j)\} \equiv T(i, j) \bmod v \quad (3)$$

The values $x(i, j)$ and $y(i, j)$ should be chosen carefully so that sheets do not have noticeable artifacts. Needless to say, human visual system characteristics for each input image are taken into consideration to choose $x(i, j)$ and $y(i, j)$. Another important fact is the basic requirement of the secret sharing schemes: *stacking all n sheets reveals the target, but any combination of less than n sheets reveals nothing close to the target image.* Thus, choosing $x(i, j)$ and $y(i, j)$ well is a key of the encoding scheme. Let S_k be the kth sheet image. Then, the sheet image is obtained as follows:

$$S_k(i, j) = P_k(i, j) \pm x(i, j) \pm y(i, j) \quad (4)$$

The following example may illustrate how the proposed encoding scheme works. Let the pixel in $(1, 1)$ position of 5 input images be 100, 110, 120, 60, and 150, respectively. $P(1, 1) \equiv \{102 + 110 + 120 + 60 + 150\} \bmod 256 \equiv 30$ as given in Equation (1). Then, we can subtract 6 from each pixel value. So the resulting pixel values are given as 96, 104, 114, 54, and 144, respectively. Now, let the target image value in position $(1, 1)$ be 33. Thus, the pixel values of sheets are 101, 111, 121, 61, and 151, respectively.

B. Decoding Phase

The decoding phase is also very simple. Simply computing the following Equation (5) produces the target image.

$$T(i, j) = \sum_{k=1}^{n} S_k(i, j) \bmod v \quad (5)$$

The sheets have been designed to produce the target. In addition, even though the sheets are attacked, it is obvious from the Equations (3), (4) and (5) that the target image can be revealed with slight degradation in image quality.

III. Experiments

In this experiment the target image is "Tank" (grayscale image with 128×128 pixels) as shown in Fig. 1 and 30 grayscale images with 128×128 pixels are used as input images, where $n = 30$. Fig. 2 shows the input images.

Figure 1. Target image "Tank" of TIFF format.

Encoding phase follows the modification scheme according to the Equation (3). After the encoding phase, the average PSNR is 34.02 dB. The average PSNR (that is, the average image quality of sheets) depends on the types of input images and target image. For example, the darker the target image is, the higher the PSNR is. In addition, pixel value modification algorithm also affects the average PSNR. Needless to say, the extracted image at the decoding phase based on the Equation (5) is exactly the same as the target image. It is obvious from the statement given in Section II.

Figure 2. Input images (30 TIFF format images)

Additional experiment shows what happens if the sheets are attacked. (Of course, robustness is not a main part of this paper. However, some day robustness against attacks will be a hot issue in the visual cryptography.) The target images extracted from the sheets after compression by JPEG coding scheme is shown in Fig. 3. Note that all image formats are TIFF. It shows that JPEG compression can degrade the image quality of the extracted target image. The degree of the degradation of the image quality is proportional to the degree of compression. Table I shows the average PSNR of the sheet images after JPEG compression. Our compression tool is MATLAB version 6. The PSNR is computed based on the errors (that is, the differences between input image and sheet image). Table II lists the target image's PSNR computed based on the differences between the target image and the extracted target images from the compressed sheets. Fig. 3 shows that the simple encoding scheme proposed in this paper is quite robust against compression attack. Extracted target images show the shape of tank clearly up to 90% of compression even though the encoding mechanism has nothing to do with counterattack measures.

TABLE I. Average PSNR of sheet images after JPEG compression

JPEG Compression Quality Factor [%]	Average PSNR [dB]
100	33.9957
95	33.4934
90	32.6124
85	31.6737
80	30.8445

(a) 100% quality factor (b) 95% quality factor

(c) 90% quality factor (d) 85% quality factor

Figure 3. Extracted targets from the compressed sheets

Figure 4. Extracted image from 29 sheets

TABLE II. PSNR of targets extracted from the compressed sheets

JPEG Compression Quality Factor [%]	Average PSNR [dB]
100	43.5830
95	26.4876
90	19.9451
85	16.8890
80	14.8821

Another important requirement is the property of the secret sharing: *stacking all n sheets reveals the target, but any combination of less than n sheets reveals nothing close to the target image*. Thus, it is important to see if any combination of sheets can reveal target-like image. Fig. 4 shows an image extracted from 29 sheets, which is not similar to the target at all. Note that the extracted image looks like a synthetic image or other natural images corrupted severely. From the experiment we find that any image extracted from forbidden sets is not similar to the target. However, we cannot conclude that there is no possibility that an image from the forbidden sets is similar to the target. However, some combination can show negative photo effect [5], which violates the requirement of secret sharing. Such a negative effect can be avoidable by modifying the embedding and decoding schemes slightly.

IV. CONCLUSIONS

In this paper a new (n, n) visual cryptography scheme without dithering is proposed. This scheme takes n grayscale input images to cover a target image across n grayscale images and produces n grayscale output images which are very close to the input images, respectively. This technique does not expand pixels to hide target images across the sheets. One contribution is its simplicity of the encoding scheme and decoding scheme. The most important contribution is that the sheets are so close to the input images and keep high visual quality that it makes the sheets visibly almost innocuous and natural. This fact is very important from the perspective of steganography to fool the wardens in the middle. In addition, note that in spite of the simplicity of the proposed scheme it is shown that the scheme is relatively robust against JPEG compression attack.

REFERENCES

[1] G. Ateniese, C. Blundo, A. de Santis, and D. Stinson, "Visual cryptography for general access structures," Information and Computation, vol. 129, no. 2, pp. 86-106, 1996.

[2] Y. C. Hou, C. F. Lin, and C. Y. Chang, "Visual cryptography for color images without pixel expansion," *Journal of Technology*, vol. 16, no. 4, pp. 595-603, 2001.

[3] Y. C. Hou, and S. F. Tu, "A visual cryptographic technique for chromatic images using multi-pixel encoding method," private communication, 2004.

[4] R. Ito, H. Kuwakado, and H. Tanaka, "Image size invariant visual cryptography," *IEICE Transactions on Fundamentals*, vol. E82-A, no. 10, pp. 2172-2177, 2002.

[5] H. J. Kim, and Y. S. Choi, "Secret sharing by natural image visual cryptography," WIAMIS 2005 (to appear).

[6] C.-C. Lin, and W.-H. Tsai, "Visual cryptography for gray-level images by dithering techniques," Pattern Recognition Letters, vol. 24, issue 1-3, pp. 349-358, 2003.

[7] M. Nakajima, and Y. Yamaguchi, "Extended visual cryptography for natural images," *Proceedings of the.10th International Conference in Central Europe on Computer Graphics, Visualization and Computer Vision*, University of West Bohemia, Czech Republic, pp. 303-340, 2002.

[8] M. Naor, and A. Shamir, "Visual cryptography," *Lecture Notes in Computer Science*, vol. 950, Springer-Verlag, pp. 1-12, 1995.

[9] W. G. Tzeng, and C. M. Hu, "A new approach for visual cryptography," *Designes, Codes and Crryptography*, vol. 27, pp. 207-227, 2002

Digital Self-Correction of Time-Interleaved ADCs

Pieter Harpe, Athon Zanikopoulos, Arthur van Roermund

Mixed-signal Microelectronics Group, Eindhoven University of Technology,
Eindhoven, The Netherlands, email: p.j.a.harpe@tue.nl

Abstract—A well known problem of time-interleaved analog-to-digital converters is the matching between the individual channels of the converter. Any mismatch between the channels affects the accuracy of the converter adversely. The random mismatch between the channels originates mainly from the mismatch of components like transistors and capacitors. To achieve a certain degree of matching between the channels, the sizes of the individual components have to be chosen accordingly. Especially for high-resolution converters, this means that physically large transistors are required, resulting in a large chip area, increased power consumption and reduced conversion speed.

Instead of increasing sizes to achieve a certain accuracy, one can also start with an analog circuit that is relatively inaccurate from itself (allowing physically small devices), and use a digital post-correction technique afterwards to correct for the actual deviations of each component. With this method, a high accuracy can be obtained while the requirements for the components are relaxed significantly. Although these techniques have been available for single-channel converters for many years, techniques correcting the mismatch between several channels are scarce. In this paper, an existing algorithm for single-channel pipelined converters is extended to include inter-channel correction as well, requiring almost no additional hardware.

I. INTRODUCTION

Mismatch of components is one of the main bottlenecks for the overall accuracy of analog-to-digital converters (ADCs). As the mismatch of components is inversely proportional to their physical size, this problem is traditionally solved by increasing the size of all critical components until the required accuracy is achieved. However, with this method, the required chip area grows exponentially with the required effective number of bits (ENOB). This problem can be alleviated by using digital post-correction, e.g. as in [1]. This technique starts with an ADC having a low initial accuracy, thus physically small devices can be used. Then, post-correction is used to correct for mismatch of components, thereby improving the accuracy significantly. The algorithm is performed in the digital domain after the actual AD conversion. Therefore, it is simple to implement and it requires only minor changes in the analog architecture. Moreover, no accurate calibration references are required. A disadvantage of the post-correction techniques currently available is that they work only for single-channel AD converters, like pipelined or multi-step approaches. For time-interleaved structures, using several converters in parallel to obtain a higher effective sample frequency, correction techniques are lagging behind. The existing techniques for time-interleaved converters either intervene in the analog domain

[2] or they require expensive digital operations (like multiplications) operating at the sample frequency of the converter [3], [4]. In this paper, an available technique for single-channel pipelined AD converters is modified as to support time-interleaved structures as well. As opposed to existing techniques, no additional analog or digital hardware is required during normal operation.

In section II a single-channel pipelined converter is introduced as a test-vehicle. The influence of an available correction algorithm on the accuracy is shown. In section III, it is shown that this algorithm is not capable of improving the performance of time-interleaved converters. In section IV, a novel approach for correction of the inter-channel mismatch in time-interleaved converters is introduced and verified with simulations. Lastly, conclusions are drawn in section V.

II. PIPELINED CONVERTER

A. Architecture

In this section, a pipelined converter including a digital post-correction algorithm according to [1] is introduced. The influence of mismatch of components is modelled and simulated on behavioral level. Without loss of generality, this example considers a pipeline based on 1.5-bit-per-stage blocks, while the overall target accuracy is set to 12 bits. The general structure of the pipeline is given in fig. 1: it is composed of a sample-and-hold (S&H) stage, a concatenation of k basic blocks (BBs) and a post-correction algorithm. This algorithm is used to combine the digital outputs of each cell, correct for several circuit impairments, and to provide a valid output code. In this example, the output code is 12-bit large. The number of stages k was fixed to 18, as this turned out to be sufficient for the given target.

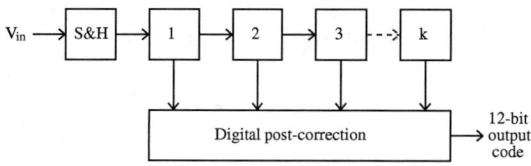

Fig. 1. Example of a pipelined ADC with digital post-correction.

Each BB (fig. 2) is composed of a *sub-ADC*, a *sub-DAC* and a *sample-and-hold amplifier* (*SHA*). The SHA itself is composed of a summing node, an amplifier and a S&H-stage.

The work is sponsored by Stichting Technische Wetenschappen.

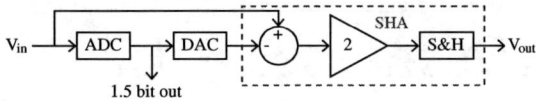

Fig. 2. Basic block of the pipelined ADC.

B. Digital Post-Correction Algorithm

The principle of the selected post-correction algorithm is explained briefly in this section. A detailed explanation can be found in [1], while [5] gives a thorough analysis of the influence of the algorithm on several important error sources.

The basic idea of a post-correction algorithm is that most errors can be corrected *afterwards* (in the digital domain), as long as the analog output of each block remains in the valid input range of the next block. The algorithm performs a simple mapping-function from the uncorrected digital code from the pipeline to the corrected output code of the converter.

In the situation where no post-correction is applied, the code $c_j \in \{0, 1, 2\}$ produced by each block j ($1 \leq j \leq k$) contributes with a value $(c_j - 1)2^{-j}$ to the output-code of the ADC. However, when the stage is not ideal (e.g. due to mismatch of components), these fixed contributions do not match anymore with the actual transfer curves.

The solution provided is to make the contribution of each possible code of each block *variable*, and to match all variables as good as possible to the actual transfer curves of the basic blocks. For each block, three variables or *weights* are required (one for each output code of each sub-ADC), so for the complete pipeline, $3k$ weights are needed. When all weights are chosen optimally, all constant and linear deviations of the transfer curves are corrected for.

The correction algorithm requires a measurement phase at start-up. During this phase, one by one, the weights of each block are determined based on measurements of the corresponding transfer curve. The measurements are transferred to the digital domain using the ADC itself. After this phase, the converter is available for normal operation. All measurements are performed *relative* to each other, instead of measuring values in an *absolute* sense. This has the important advantage that no additional accurate reference-sources are required.

An important modification in our implementation is that all stages are calibrated, instead of only the first (most significant part) of the pipeline [1]. Because of that, the digital corrected output code is based completely on *variable* weights, instead of a combination of variable and fixed weights. This modification is the key to the solution provided for time-interleaved structures (section III).

The simulation results in the next section are intended to illustrate the effect of the algorithm described here on the accuracy of the converter.

C. Simulation Results

Simulations of a pipelined ADC were performed on system level. The error sources taken into account are: random deviations of the comparator levels in the sub-ADCs, random deviations of the sub-DAC levels, random offset, gain-error and third-order distortion of the SHA. Based on the design strategy from [5], a maximum value for each of these error sources was derived in order to achieve 12-bit accuracy after application of the digital post-correction algorithm.

Simulations were performed to derive the INL curve (fig. 3). As a reference, the results for the same converter, but with post-correction *disabled* are given as well. The significant influence of the correction algorithm is clear: without correction, the maximum INL is around 32 LSB, yielding an effective accuracy around 6-bit only. With correction, the INL (max. 0.7 LSB) corresponds to 12-bit accuracy.

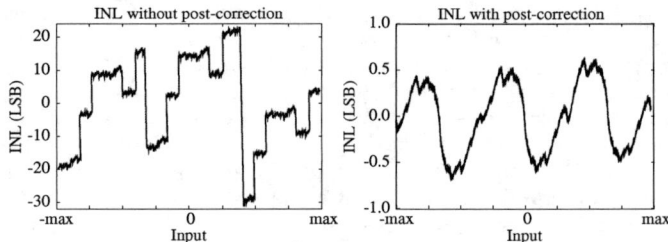

Fig. 3. INL for a pipelined ADC without (left) and with (right) post-corr.

As a second benchmark, a time-domain analysis was performed with a full-scale sine at $f_{in} = 0.051 f_s$. Results (after FFT) are given in fig. 4. Spurious tones are reduced considerably when post-correction is applied. The remaining distortion-components are due to the non-linearity of the SHA, which is not corrected by the algorithm. From these figures the SINAD and ENOB were derived, yielding 41.3 dB / 6.6 bit without correction and 71.6 dB / 11.6 bit with correction.

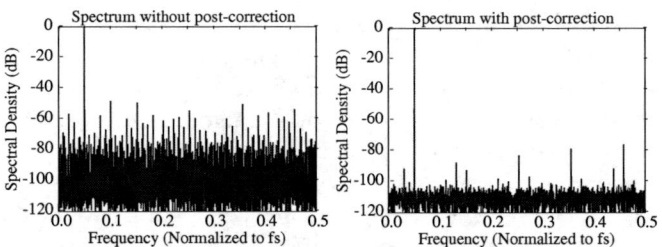

Fig. 4. Output spectral density for a pipelined converter without (left) and with (right) post-correction. The input signal is a single tone at $0.051 f_s$.

III. TIME-INTERLEAVED CONVERTER

A. Architecture

The idea of time-interleaving is to use p converters in parallel running at a sample frequency f_s/p each, and to combine their in- and outputs in a time-interleaved fashion, to achieve an effective sample frequency f_s. With this technique, the effective sample frequency increases with a factor p, while the accuracy of each converter can be maintained. However, to obtain a certain target accuracy (12-bit in this example) not only the individual converters should be 12-bit accurate, the matching between all parallel converters has to be 12-bit accurate as well. Otherwise, the inter-channel matching will limit the overall accuracy.

Now, it seems a logical step to use post-corrected pipelined converters in a time-interleaved structure. However, the simulation results in the next paragraph show that the performance of such a converter is actually pretty poor due to the inter-channel mismatch of the converters.

B. Simulation Results

For these simulations, a time-interleaved converter with two channels is considered. The values of the error sources are chosen as in section II-C. The post-correction method is applied separately to the two channels. This is necessary as the deviations of each channel are chosen randomly, hence require different weights.

The output frequency spectrum for an input tone at $f_{in} = 0.051 f_s$ is given in fig. 5. The left picture is comparable to the single-channel result (fig. 4, left). A SINAD of 36.1 dB (ENOB 5.7 bit) is achieved. After post-correction (right picture), it can be observed that many distortion components decreased due to the correction of the individual pipelines. However, the tone at $\frac{1}{2}f_s - f_{in}$ remains. This component arises from the mismatch between the two channels of the time-interleaved converter, and limits the performance to 5.3-bit effective resolution (SINAD is 34.0 dB).

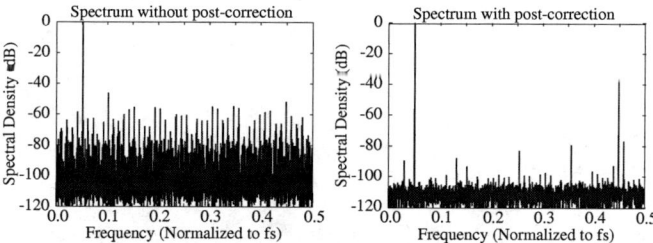

Fig. 5. Output spectral density for a 2-channel time-interleaved ADC without (left) and with (right) post-correction. The input signal is a tone at $0.051 f_s$.

IV. INTER-CHANNEL MISMATCH CORRECTION

A. Problem description

The reason why post-correction does not improve the performance of the time-interleaved converter is due to mismatch between the channels. Remember that in this example, an individual pipeline without post-correction has a intrinsic resolution of around 6-bit only (section II-C). After correction, the accuracy becomes 12-bit, which means that when the converter is described by (1) (with V_{in} the analog input and C_{out} the digital output), the maximum deviation of the actual ADC curve from this model coincides with $\frac{1}{2}$LSB of a 12-bit converter. However, as the post-correction algorithm performs a *relative* instead of an *absolute* calibration, the spread in offset O and gain G in (1) still corresponds to 6 instead of 12-bit operation. Therefore, the inter-channel mismatch in the time-interleaved converter is also of the order of 6-bit.

$$C_{out} = G \cdot V_{in} + O \qquad (1)$$

B. Algorithm

The new post-correction algorithm for time-interleaved converters presented here, is composed of two parts. The first step is to measure all individual pipelines i ($1 \leq i \leq p$), using the algorithm described before, and to determine their initial weights ω. Then, each pipeline can be described by model (2), showing that each individual pipeline is linear but also that each pipeline has a different G and a different O.

$$C_i = G_i \cdot V_{in} + O_i \qquad (2)$$

The second step of the algorithm is to correct for the inter-channel mismatch, which is a function of G_i and O_i. To prevent the requirement of an accurate reference source, the channels are not matched in an *absolute* sense, but *relative* to one of the channels (e.g. channel 1). As the mismatch of each channel has only two degrees of freedom (being G_i and O_i), two well-selected measurements for each channel provide enough information to perform the correction.

First of all, a measurement phase is initiated. During this phase, two reference voltages V_a and V_b are required, serving as input signal to the time-interleaved converter. For each channel i the output-codes for voltage V_a and V_b are determined:

$$\begin{cases} C_{a,i} &= G_i \cdot V_a + O_i \\ C_{b,i} &= G_i \cdot V_b + O_i \end{cases} \qquad (3)$$

Then, for each channel a gain and offset correction is calculated based on the measurements:

$$\begin{cases} G_{cor,i} &= (C_{b,1} - C_{a,1}) \big/ (C_{b,i} - C_{a,i}) \\ O_{cor,i} &= \frac{1}{2}\{(C_{a,1} + C_{b,1}) - G_{cor,i}(C_{a,i} + C_{b,i})\} \end{cases} \qquad (4)$$

Combining (3) with (4) yields:

$$\begin{cases} G_{cor,i} &= \frac{G_1}{G_i} \\ O_{cor,i} &= O_1 - \frac{G_1}{G_i} O_i \end{cases} \qquad (5)$$

Most notably is that this result is independent of the actual values of the reference voltages $V_{a,b}$. Therefore, there is no constraint on their value or their accuracy. Even slow drifts of the references (e.g. due to thermal effects or ageing) are unimportant as long as each channel experiences the same *average* voltage. For example, a linear drift $V_a(t) = V_a(0) + \xi \cdot t$ is completely compensated when all channels (1 up to p) are measured in the order: $1, 2, \cdots, p-1, p$ first, and then in the order $p, p-1, \cdots, 2, 1$, as then, the average reference value becomes equal for all channels.

Now, the corrected code C'_i follows from the uncorrected code C_i using the gain and offset correction:

$$C'_i = G_{cor,i} \cdot C_i + O_{cor,i} = G_1 \cdot V_{in} + O_1 \qquad (6)$$

From this result, it can be seen that all channels now have a global transfer function, matched to the first channel. The simulation results in section IV-E confirm this result.

C. Implementation

The intention of this work is to integrate the additional correction technique in the system with as little additional hardware as possible. A possible implementation is given here.

First of all, two references (V_a and V_b) are required. As stated before, their actual value is not really important. A convenient solution is to use the two references from one of the sub-ADCs. In that case, just a few switches are required to connect these references to the input node of the converter.

Secondly, the actual correction $C'_i = G_{cor,i} \cdot C_i + O_{cor,i}$ (6) has to be implemented. One way is to implement this function directly in hardware after the first post-correction algorithm. However, this requires a multiplication and a summation for each produced sample. A better approach is to take the effect of (6) into account in the variable weights. Notice that the uncorrected output code C_i of channel i is given by the summation of the weights of the k stages in the pipeline ($C_i = \sum_{j=1}^{k} \omega_{i,j}$), hence the corrected code C'_i equals:

$$\begin{aligned} C'_i &= G_{cor,i} \cdot C_i + O_{cor,i} = G_{cor,i} \cdot \sum_{j=1}^{k} \omega_{i,j} + O_{cor,i} \\ &= \sum_{j=1}^{k} \left(G_{cor,i} \cdot \omega_{i,j} \right) + O_{cor,i} = \sum_{j=1}^{k} \omega'_{i,j} \text{ , with:} \end{aligned} \quad (7)$$

$$\begin{cases} \omega'_{i,j} = G_{cor,i} \cdot \omega_{i,j} & \text{when } j \neq 1 \\ \omega'_{i,j} = G_{cor,i} \cdot \omega_{i,j} + O_{cor,i} & \text{when } j = 1 \end{cases} \quad (8)$$

Now, the weights $\omega_{i,j}$ have to be replaced just a single time by the weights $\omega'_{i,j}$, while no additional hardware is required anymore during normal operation. After replacement of the weights, the output code of the converter becomes C'_i directly.

Summarizing, the additional correction algorithm requires a few analog switches and a onetime-only algorithm. During normal operation, no additional circuitry is needed.

D. Matching Accuracy

The inter-channel matching accuracy that can be obtained after post-correction is limited by the accuracy of the measurements and the processing. The measurement accuracy is determined by the accuracy of the measurement device: the individual pipelined ADCs composing the time-interleaved ADC. However, this is not an additional limitation for the overall accuracy, as the accuracy of the time-interleaved ADC is limited to the accuracy of the individual channels anyway.

The accuracy of the processing is determined by the representation resolution of the digital variables. The representation resolution of $G_{cor,i}$ and $O_{cor,i}$ can be chosen abundantly, as these variables are used only temporarily and hence require resources only during the measurement phase. On the other hand, the resolution of the weights ω is critical for the achievable accuracy and for the required hardware resources during operation. However, the required representation resolution remains the same as in case of a single-channel converter, and hence does not require additional hardware resources. In the worked-out example, all weights have a resolution of 19-bit (due to the 18 1.5-bit stages).

E. Simulation Results

A two-channel time-interleaved converter as in section III-B is considered here. This time, both the original correction algorithm for the individual channels and the new extension for inter-channel matching are applied. The INL curve is given in fig. 6 (left). It can be seen that the converter is composed of two channels, however the matching is accurate enough to achieve 12-bit overall accuracy. The frequency plot of the time-domain analysis is also given in fig. 6, showing the effectiveness of the novel correction technique. A SINAD of 71.5 dB is achieved, yielding an ENOB of 11.6 bit.

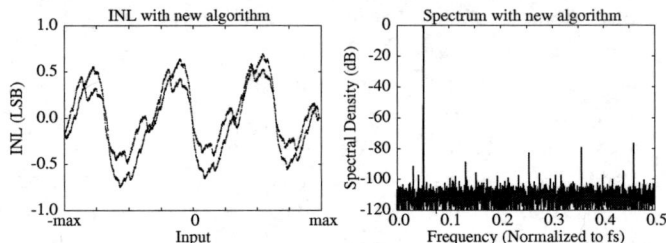

Fig. 6. INL (left) and output spectral density (right) for a 2-channel time-interleaved converter with the new post-correction technique applied.

The simulations confirm that the overall accuracy becomes limited again by the accuracy of the individual channels, as the achieved result (11.6 bit) equals the result of a single-channel ADC (section II-C) and the intermodulation product at $f_{in} = 0.051 f_s$ is not visible anymore in fig. 6.

V. Conclusion

A digital post-correction algorithm was presented, correcting for the inter-channel mismatch of time-interleaved analog-to-digital converters. With this technique, the channel-matching becomes independent of the matching of components like transistors, etc. When combined with an existing post-correction algorithm for single-channel pipelined converters, this new technique hardly costs any additional hardware. Dependent on the actual implementation of the converter, just a few additional switches are required for each channel and a minor modification of the digital calibration algorithm has to be applied. Simulations on a behavioral model showed that with this new technique, the overall accuracy of the converter is not limited anymore on the inter-channel matching, but on the accuracy of each individual channel.

References

[1] A. N. Karanicolas, H.-S. Lee, and K. L. Bacrania, "A 15-b 1-MSample/s digitally self-calibrated pipeline ADC," *IEEE J. Solid-State Circuits*, vol. 28, pp. 1207–1214, Dec. 1993.

[2] K. Dyer, D. Fu, S. Lewis, and P. Hurst, "Analog background calibration of a 10b 40MSample/s parallel pipelined ADC," in *Proc. IEEE ISSCC 1998*, San Francisco, CA USA, Feb. 5–7 1998, pp. 142–143, 427.

[3] D. Fu, K. Dyer, S. Lewis, and P. Hurst, "Digital background calibration of a 10b 40MSample/s parallel pipelined ADC," in *Proc. IEEE ISSCC 1998*, San Francisco, CA USA, Feb. 5–7 1998, pp. 140–141, 426.

[4] V. Hakkarainen, L. Sumanen, M. Aho, M. Waltari, and K. Halonen, "A self-calibration technique for time-interleaved pipeline ADCs," in *Proc. IEEE ISCAS 2003*, vol. 1, May 25–28 2003, pp. 825–828.

[5] P. J. A. Harpe, "Design of a high-speed, high-resolution pipelined AD converter," Master's thesis, Technische Universiteit Eindhoven, Eindhoven, The Netherlands, Jan. 2004.

Automated Design of a 10-bit, 80Msps WLAN DAC for linearity and low-area

Ankit Seedher, Preetam Tadeparthy, Satheesh Kumar A.S., Anuroop V.T.
Broadband Silicon Technology Center
Texas Instruments (India) Pvt. Limited
Bangalore, India
Email: seedher@ti.com, preetam@ti.com, satheesh@india.ti.com, anuroop@ti.com

Abstract—The paper presents the design of a 10-bit, 80MSPS current steering D/A converter(DAC) using only digital thin-oxide CMOS transistors. A large part of the design is automated reducing the design-cycle-time. To combat systematic gradients on the wafer, we propose using global combinatorial optimization techniques such as simulated annealing and genetic algorithms to obtain a nearly optimal randomized array without any additional area penalty. Additionally, a simple yet elegant technique is used in the current-to-voltage conversion amplifier following the DAC to improve its phase margin in the presence of process variations without expending extra power. The DAC fabricated in a 0.13 μ digital CMOS technology shows an INL and DNL of 0.4 and 0.5 LSBs respectively, an SFDR of $\geq$ 70 dB, occupying 0.26 mm^2 and expending 1.25mA static power.

I. INTRODUCTION

We present the design of a D/A converter where the design of the DAC core has been automated and some novel techniques are used to obtain good linearity with minimal area and power expenditure. We illustrate these techniques with the design of a 10-bit 80MSPS WLAN DAC. Although the design of higher performance DACs might require additional techniques(calibration [1] or DEM [2] for static linearity or quad-switching for dynamic linearity [1]), the techniques illustrated in this paper can still be applied there for spatial linearity and reduced design effort.

II. AUTOMATION OF THE DAC CORE DESIGN

The current-steering DAC with an 8-2 segmentation is followed by a current-to-voltage conversion amplifier (I-V amplifier) loads connected as feedback resistors.

A. Static Linearity

1) Random Mismatch: For a typical broadband signal, the linearity specification is in terms of a distortion floor for a discrete-multi-tone(DMT) signal. As such, the matching requirement is derived by running statistical simulations with a sample DMT signal to see the affect of mismatch on the distortion floor. The sizes from the current source transistors to give the required matching can then be derived from the mismatch constants for the process using the well-known Pelgrom's equations [3].

2) Spatial Mismatch: Oxide thickness and doping variations across the wafer create deterministic spatial mismatches between current sources. These effects can be modelled as a combination of linear and quadratic gradients(equation 1). [1] Knowing the exact values of α and β is not required and the placement strategy can be devised with any reasonable values.

$$I = 1 - \alpha(x+y) - \beta(x^2 + y^2) \quad (1)$$

Fig. 1(a) shows the INL and DNL if no placement strategy is adopted. To reduce the INL in presence of linear gradients each source can be divided into four parts and the four parts placed symmetrically about the x and y axes. To reduce DNL inside each of these four quadrants, succesive current sources can be placed symmetrically inside each individual quadrant. To completely remove the quadratic error, each current source would need to be further sub-divided four times and the sixteen sub-parts comprising a current source placed in a very area inefficient fashion. This can be seen from the fact that if there is a one-dimensional gradient of $y = 1 - x^2$ spread over say $x \in [-1, 1]$, the average value of 2/3 occurs for $x = \pm\frac{1}{\sqrt{3}}$. This would require each source divided into four parts placed symmetrically over $[-1, -1/\sqrt{3}], [-1/\sqrt{3}, 0], [0, 1/\sqrt{3}]$ and $[1/\sqrt{3}, 1]$. Since these are not equal length segments, placing equal number of current sources in each of them would be wasteful of space. Extending this to two dimensions, sixteen sub-parts of each current source would have to be spread over a large area. Instead these sixteen parts can be placed symmetrically about the origin by dividing each quadrant obtained in the quadrant switching scheme further into four quadrants reducing the residual quadratic INL. The quadratic shape in the INL can be gotten rid of by randomizing the ordering of current sources in each quadrant(still placing successive current sources symmetrically inside each quadrant/sub-quadrant) [4]. Using 16 sub-quadrants however, increases the routing complexity considerably and therefore has an area penalty. We propose finding a nearly optimal randomization of current sources so that splitting of current sources into 16 sub-quadrants is not required.

---
[1]The quadratic gradient can be assumed to have a peak at the center of the current source array using a shift of origin that would only result in a change in the linear coefficients.

Algorithm 1 Simulated Annealing Algorithm
1: Temperature = InitialTemperature = $10.\sigma_{INL}$
2: currentPlacement = randomInitialPlacement
3: currentScore = maxINL(currentPlacement)
4: **while** Equilibrium Not Reached **do**
5:      selectedCurrentSrces = select(atRandom)
6:      trialPlacement = move(selectedCurrentSrces)
7:      trialScore = maxINL(trialPlacement)
8:      **if** trialScore $\leq$ currentScore **then**
9:          currentScore = trialScore
10:         currentPlacement = trialPlacement
11:     **else if** rand $\leq e^{-\frac{trialScore - currentScore}{Temperature}}$ **then**
12:         currentScore = trialScore
13:         currentPlacement = trialPlacement
14:     **end if**
15:     Temperature = Alpha.Temperature
16: **end while**

Algorithm 2 Genetic Algorithm Based Algorithm
1: **for** i = 1 : populationSize **do**
2:      Generate a Randomized Array, A(i)
3:      Evaluate and assign fitness to A(i) {Fitness(i) = (meanINL + 5σINL) - INL(i)}
4: **end for**{Initial Population now ready}
5: **for** i = 1 : numberofGenerations **do**
6:      Perform Inversion on population with very low prob Pinv
7:      **for** i = 1 : numberofOffspringArrays **do**
8:          Select parents with probability $\propto$ fitness
9:          Perform Crossover between selected parents to get offspring current source array
10:         Mutate offspring array with very low prob Pmut
11:         Evaluate Fitness of offspring array
12:         Include offspring array in population {Population bloats temporarily}
13:     **end for**
14:     Choose new population {Survival of the fittest!}
15: **end for**
16: Choose best current source array form the population

There are a total of (n/2)!¹ possible random permutations for an n-element array. Finding an optimal randomized array by an exhaustive search is therefore an NP-complete problem. Simulated Annealing(SA) and Genetic Algorithms(GA) lend themselves very well to finding a near-optimal solution to such problems in finite time.

The generic SA algorithm [5][2](Algorithm 1) is adapted to obtain an optimal current source array. Several distinct random arrays are evaluated to get the σ of the INL. The variable InitialTemperature is then chosen as $10.\sigma_{INL}$ [6]. The temperature decrement *Alpha* is chosen such that the INL decreases uniformly with the natural log of temperature [7]. This is done to replicate the actual physics of annealing and therefore approach better convergence. When a large number of iterations lead to no change in the INL³, equilibrium is assumed to have reached and the simulation is stopped to get the optimal array.

The solution using GA (Algorithm 2) [8] starts with a population of several⁴ randomized arrays. Fitness for each array in the population is calculated as $(mean_{INL} + 5\sigma_{INL})$ - INL, where $mean_{INL}$ and σ_{INL} are the mean and standard deviation of the INL for a large number of such randomized arrays, thereby assigning higher fitness to an individual with lower INL. The algorithm was run for 100 generations with a crossover probability of 0.4. The two parents for each offspring current source array in the next generation are chosen with a probability proportional to the fitness of the arrays in the present generation and Partially Mapped Crossover(PMX) [9] is used to obtain the offspring.

---
[1]n! possible random permutatations if we dont place successive sources symmetric to the origin of each array
[2]known as the Metropolis algorithm.
[3]Which means that the temperature in the equivalent physical annealing model has cooled sufficiently and the probability of choosing an inferior solution is approaching zero.
[4]The optimal size of the population varies with the array size, accuracy desired and run-time. We use a population size of 20-40 for the 256 element array.

Using several iterations and accepting the array that gives lowest INL is a less effective solution as such an algorithm tends to get stuck in local minima only accepting new states that lower the cost(INL). On the other hand, SA occasionally accepts an inferior solution (step 11 in Algorithm 1) while GA relies on a few inferior individuals(current source arrays with higher INL) to avoid false optima. As can be seen from Figs. 1(b), 1(c) and 1(d), using SA gives an INL that is comparable to the Q^2 switching scheme with only a Quad placement strategy. Using GA gives similar results. Therefore, a randomized array is obtained using SA for the DAC being designed. Since the linear gradients are typically the largest, it was decided to still mirror the array in four quadrants to compeletely cancel them and tackle the quadratic and higher errors using SA/GA. These methods can however, be effectively used to get a current source array that is not divided into quads/sub-quads. Since SA/GA dont make any assumption of the gradient shape, a gradient other than equation 1 can also be handled.

3) Layout Automation: A script is developed that takes the array map file generated using the methods outlined above and places and routes the entire current source array(Algorithm 3). This script is written in a technology independent way and lends itself to easy reuse for different CMOS technology nodes and different types of array map files.

B. Dynamic Linearity

The switches in the current source unit cell are not placed inside the randomized array to reduce timing skew between switches and routing complexity. To obtain reduced routing capacitance at the common source node of the switches(for lower glitch and higher high frequency current source impedance), the cascode transistor of the current source is also placed

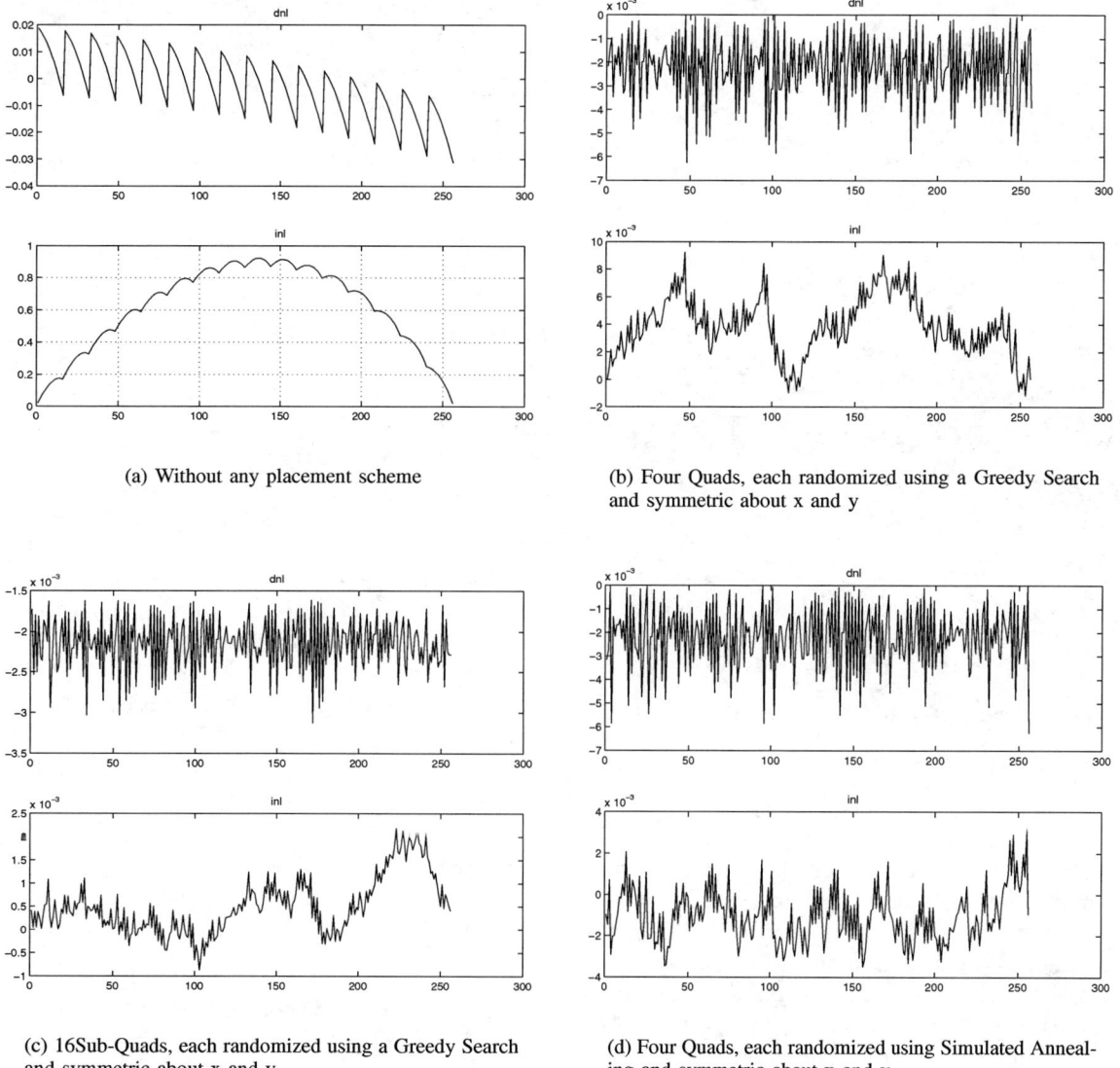

Fig. 1. Various placement schemes. The Greedy Search uses the same number of iterations as the SA algorithm to get a fair comparison. A generic linear+quad gradient is used for illustration.

Algorithm 3 Automated Layout of Current Source Array
Require: Technology File, DRC Rules
Require: Properties and Sizes of Current Source Unit Cell
1: Read the Placement Map File()
2: Pick and place the Unit Cells()
3: Routing of Cells()
   Calculate Vertical Line Requirements()
   Calculate Horizontal Line Requirements()
   Calculate Horizontal and Vertical Line Locations()
   Place Horizontal and Vertical Lines over the unit cells()
   Calculate Optimal Via Locations()
   Calculate Optimal Via Types()
   Place the vias()
   Bring out drains of current sources as per map file
4: Save Layout and CellView Properties()

---

outside the array. Latches before each switch reduce timing skew between switches. The cascode, switch pair and latch are laid-out as single unit cells placed symmetrically below the current source array. The clock to the latches is fed using a balanced clock tree. Latches from the standard digital libraries are tweaked to obtain a make-before-break switching. Only core digital CMOS transistors are used for the design. Inspite of lower r_{ds} of the digital core transistors, it does not degrade the dynamic performance due to increased high frequency impedance(smaller sizes are required for the same matching, hence lower device capacitances).

III. CURRENT-VOLTAGE CONVERSION AMPLIFIER

The use of an I-V amplifier at the output of the DAC reduces the swing at the output of the DAC current sources

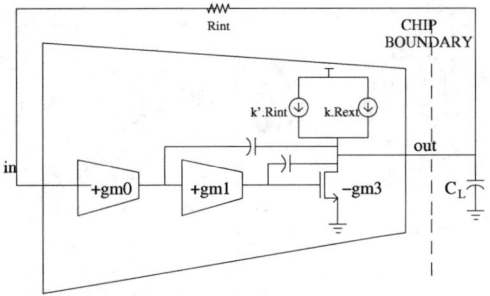

Fig. 2. Simplified single-ended representation of I-V amp

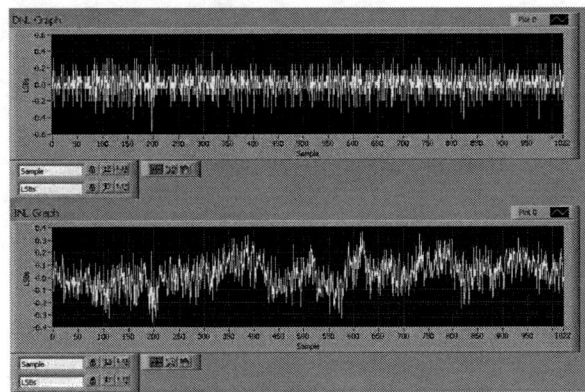

Fig. 3. Obtained DNL and INL from fabricated chip

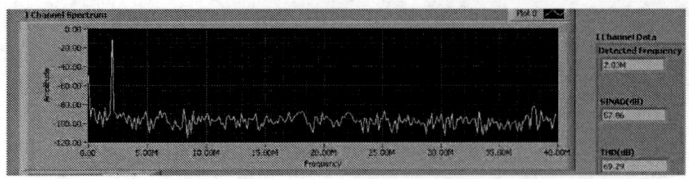

Fig. 4. Obtained single-tone spectrum from fabricated chip

due to the virtual ground seen at the amplifier input. A three stage nested-miller fully-differential amplifier is used to give good linearity. The first two stages are fully differential pairs while the last stage is a pair of common-source stages[1] (Fig 2). For biasing the DAC current sources, the input common-mode of the amplifier is 0.4V while the output common-mode is 1V for interfacing with an on-board radio. This means large common-mode and swing currents inversely proportional to the feedback resistor have to be supplied by the third stage. Since the feedback resistor is an on-chip silicide block resistor[2], it can vary ± 20% and can lead to starvation of the input transistor in the third stage reducing its transconductance. If a constant bias current(k.Rext) is used for the third stage, it leads to a degradation in phase margin since the non-dominant pole(g_{m3}/C_L) is pulled in. This can be avoided by using current proportional to the feedback resistor(k'.Rint) in the output stage in addition to the constant bias current generated from an external relatively constant resistor. This current tracks the feeback resistor and the constant bias current maintains a constant gm, therefore the non-dominant pole location does not move maintaining a good phase margin with process variations. This reduces power consumption as we only bias the Class-A output stage for the nominal corner and not for the corner where the feedback resistor is minimum.

---

[1] Assuming sufficient common-mode rejection from the first two stages.

[2] To maintain constant voltage swing since the DAC current is also invresely proportional to a Silicide Block Resistor

IV. SILICON RESULTS AND CONCLUSION

The sizing of current sources for mismatch, the placement algorithms for spatial linearity, and the layout of the DAC core are all completely automated in the form of a technology-independent toolbox than can be used to generate a complete laid-out DAC core starting from specifications considerably reducing the total design time of the DAC. The toolbox permits the user to try most of the array placement techniques reported in literature in addition to the ones suggested in this paper and to mix-and-match these techniques.

A 10-bit(8-2 segmentation) chip designed using this toolbox (with SA used to optimize the array for spatial gradients) and fabricated in a 0.13μ digital CMOS process shows a DNL and INL of 0.5 and 0.4 LSBs respectively(Fig. 3). The INL profile is completely randomized and the peak value of 0.5 LSB appears largely from random mismatches only thus verifying the efficacy of the proposed placement schemes. For signal frequencies up to 10MHz, an SFDR of ≥ 70 dB is seen(Fig. 4). The current source array alone occupies 0.165 mm^2 and the complete DAC occupies 0.26 mm^2. A chip microphotograph has been omitted for space constraints and shall be presented at the conference.

ACKNOWLEDGMENT

The authors thank S. Mathur, G. Chandra, S. Ramadoss, J. Joy and K. Krishnan for constructive suggestions.

REFERENCES

[1] W. Schofield, D. Mercer and L. St. Onge, "A 16b 400MS/s DAC with $\leq$ -80dB IMD to 300MHz and $\leq$-160dBm/Hz Noise Power Spectral Density," *ISSCC Dig. Tech. Papers*, 2003, pp. 90-91.
[2] I. Galton, "Spectral Shaping of Circuit Errors in Digital-to-Analog Converters," *IEEE Transactions on Circuits and Systems-II*, vol. 44, No. 10, pp. 808-817, October 1997.
[3] M. Pelgrom, A. Duinmaijer and A. Welbers, "Matching properties of MOS transistors," *IEEE J. Solid-State Circuits*, vol. SC-24, pp. 1433-1439, Oct. 1989.
[4] G. Ven der Plas, J. Vandenbussche, W. Sansen, M. Steyaert and G. Gielen, "A 14-bit intrinsic accuracy Q^2 random walk CMOS DAC," *IEEE J. Solid-State Circuits*, vol. 34, pp. 1708-1717, Dec. 1999.
[5] S. Kirkpatrick, C. D. Gelatt and M. P. Vecchi, "Optimization by Simulated Annealing," *Science*, vol. 220, pp. 671-678, May 1983.
[6] S. White, "Concepts of Scale in Simulated Annealing," *Proc. Intl. Conf. Computer Design*, pp. 646-651, 1984.
[7] M. D. Huang, F. Romeo, and A. L. Sangiovanni-Vincentelli, "An Efficient General Cooling Schedule for Simulated Annealing" *Proc. Intl. Conf. Computer-Aided Design*, pp. 381-384, Nov., 1986.
[8] P. Mazumdar and E. Rudnick, "Genetic Algorithms for VLSI Design, Layout and Test Automation," Prentice Hall, December 1998.
[9] D. E. Goldberg and Jr. R. Lingle, "Alleles, loci and the traveling salesman problem," *Proc. International Conference on Genetic Algorithms and Their Applications*, pp. 154-159, 1985.

# An Analog-to-Digital Converter with Golomb-Rice Output Codes

Walter D. Leon, Sina Balkır, Khalid Sayood, Michael W. Hoffman
Department of Electrical Engineering,
University of Nebraska-Lincoln, 209N Walter-Scott Engineering Center,
Lincoln, NE 68588-0511, USA
daniel@torpedo.unl.edu

Abstract— An analog-to-digital converter with data compression capabilities is described. By sharing circuits between an integrating converter and a Golomb-Rice encoder it is possible to jointly perform the tasks of quantization and coding. The Golomb-Rice codes are generated during the conversion cycle by employing a shift register and a digital multiplexer. The final codeword is read out serially from the shift register. The converter can also work in a non-compressing mode. This design provides a compact circuit suitable for on-sensor compression. Simulations at the system and transistor level corroborate the validity of the design.

I. INTRODUCTION

Advances in VLSI technology have allowed the integration of silicon sensors and signal processing circuits on the same chip. Integrating an analog-to-digital converter (A/D) with a sensor enables the digital processing, transmission and storage of the sensor output. This is the scenario in remote sensing and telemetry applications. In such applications, it is desirable to incorporate a data compression mechanism to ease the requirements on transmission bandwidth or storage space.

Data compression schemes assume as an input the output of a digital source with known statistical properties. The output of the source is a set of symbols with a given probability of occurrence. Hardware implementations of popular compression algorithms such as the Huffman coding [1], Lempel-Ziv coding [2], binary arithmetic coding [3], and the Rice algorithm [4] have been reported in the literature. These complex hardware implementations typically preclude the direct integration of an A/D and a compression algorithm at the sensor level.

Research efforts addressing this integration issue have been also reported. A 12-bit A/D with a simplified Huffman encoder is presented in [5]. The design relies on a charge-redistribution digital-to-analog (D/A) converter to perform a successive approximation algorithm. To save on the number of storage elements, the input is coarsely divided into three regions, with a codeword assigned to each region. Even though it provides a sub-optimal Huffman encoding, the design reduces power consumption by not generating unnecessary bits. The complete A/D occupies an area of 3 mm^2 based on a 0.6 μm CMOS technology. The concept of vector quantization applied to the design of an A/D is outlined in [6]. However, the complexity of the design grows with the dimensionality of the input vector as it requires one scalar A/D per dimension and the respective storage space for the binary tree employed.

A novel integration of an A/D with a Golomb-Rice encoder is presented in this paper. The integration has been possible by sharing the common circuitry between an integrating A/D and the encoder. The result is a highly compact and linear A/D with Golomb-Rice output codes that is capable of providing cost-efficient compression for analog sources. Adaptive compression is also achievable as the design allows dynamically changing the encoder parameters. The validity of the design has been verified through transistor level simulations.

II. GOLOMB-RICE CODES

Golomb-Rice codes have inherent low-complexity. This complexity advantage has led the JPEG committee to adopt them as part of the JPEG-LS standard. The Consultative Committee for Space Data Systems (CCSDS) recommends the use of Golomb-Rice codes to compress the data of on-board scientific instruments [7].

A. Preliminaries

The Golomb codes were initially proposed in [8] as a means to encode the run lengths of favorable events of a binary source. Golomb's results are most applicable when $p_o^m = 1/2$, where p_o is the probability of the favorable events and m an integer. Golomb also noticed that the special case when the parameter m is a power of two accepts simple encoding and decoding algorithms. Later, Gallager [9] showed that Golomb's codes are Huffman codes for a source of nonnegative integers, n, with geometric distribution, i.e., $p(n) = (1-p_o)p_o^n$. It follows that Golomb codes are optimal for encoding geometric sources. The Golomb codes for the special case $m = 2^k$, with k being a positive integer, were independently proposed by Rice [10] and are easily obtained by the following procedure; to encode a non-negative integer n simply concatenate the unary code for $\lfloor n/m \rfloor$ with the natural binary code for $(n \bmod m)$, where $\lfloor x \rfloor$ is floor function. The length of each codeword is $k + 1 + \lfloor n/m \rfloor$ bits. This set of codes are commonly referred to as the Golomb-Rice codes.

In practice, most sources of interest do not follow a geometric distribution. However, it has been shown in [11] that the adaptive Golomb-Rice codes are optimal for *digital* sources with a Laplacian distribution. Moreover, the decorrelated output of most sources of interest closely follow a Laplacian distribution, making the proposed techniques applicable for high compression rates.

As the Golomb-Rice encoding accepts only nonnegative integers, a suitable mapping is required to encode signed numbers. We define such a mapping for signed analog values below.

B. Joint Quantization/Coding of a Laplacian Source

The proposed A/D jointly performs the quantization and encoding of its input. Consider a zero mean random variable x with a Laplacian probability distribution function (pdf) $f_X(x)$, i.e.:

$$f_X(x) = \frac{1}{\sqrt{2\sigma^2}} e^{-\frac{\sqrt{2}}{\sigma}|x|} \quad (1)$$

where σ^2 is the variance of x. A uniform quantization of this random variable is shown in Fig. 1, where Δ is the quantization step size. The probability of x being in a quan-

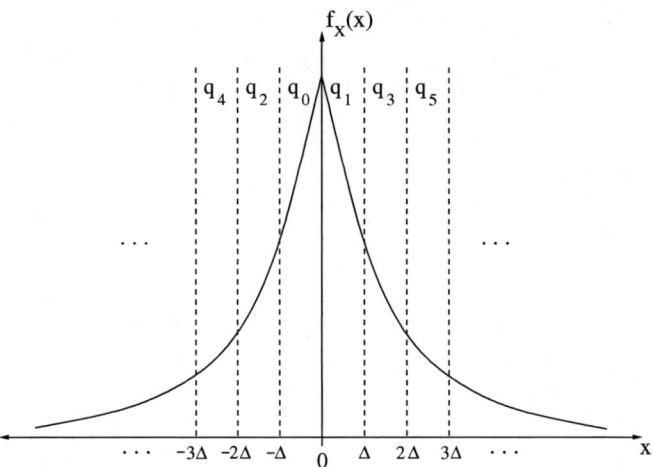

Fig. 1. Uniform quantization of a Laplacian distributed signal.

tization interval q_{2i+1}, $(i = 0, 1, 2, \cdots)$ is readily obtained by integrating the pdf, which yields

$$p(q_{2i+1}) = \tfrac{1}{2} e^{-\frac{\sqrt{2}}{\sigma} i \Delta} (1 - e^{-\frac{\sqrt{2}}{\sigma} \Delta}), \quad i = 0, 1, 2 \cdots \quad (2)$$

Notice that $p(q_{2i}) = p(q_{2i+1})$. Letting $p_o = e^{-\frac{\sqrt{2}}{\sigma}\Delta}$, one can write $p(q_{2i}) = p(q_{2i+1}) = \tfrac{1}{2}(1 - p_o) p_o^i$. If the quantization intervals q_{2i} and q_{2i+1} are jointly represented by the codeword $\tilde{c}_i$, the probability of occurrence of this codeword is $(1 - p_o) p_o^i$. Since this is now a geometric distribution, the optimal assignment for $\tilde{c}_i$ is the one provided by the Golomb encoding algorithm [9]. A sign bit b_s is necessary to uniquely determine the quantization interval in which the input x lies. The sign bit will be set to "1" if x is in the interval q_{2i+1} and set to "0" if x is in the interval q_{2i}. The final codeword is simply the concatenation of $\tilde{c}_i$ and b_s. This assignment essentially uses the LSB of the codeword as a sign bit.

To obtain a simple estimate of the optimal parameter m needed in the algorithm, recall the Golomb condition $p_o^m = \tfrac{1}{2}$. Taking the logarithm of both sides and restricting to the special case $m = 2^k$ yields

$$k = \left\lceil \log_2 \left(\frac{\log 2}{\Delta} \frac{\sigma}{\sqrt{2}} \right) \right\rceil \quad (3)$$

where $\lceil x \rceil$ is the smallest integer greater than or equal to x. Notice also that the expected value of $|x|$ is given by

$$E\{|x|\} = \int_{-\infty}^{\infty} \frac{1}{\sqrt{2\sigma^2}} |x| e^{-\frac{\sqrt{2}}{\sigma}|x|} dx = \frac{\sigma}{\sqrt{2}} \quad (4)$$

Combining this result with (3) gives

$$k = \lceil \log_2 (B \cdot E\{|x|\}) \rceil, \quad B = \frac{\log 2}{\Delta} \quad (5)$$

Equation (5) shows a simple way to estimate the value of k.

III. A/D ARCHITECTURE

The Golomb-Rice encoding can be performed by a binary counter. The encoding steps are summarized as follows:

```
1) Reset the counter;
2) If counter value equals input n then
   go to 6);
3) Increment counter by one;
4) If the k^th bit of the counter, b_{k-1}, goes
   from "1" to "0", output a "1";
5) Go to 2);
6) Output a "0" followed by the counter bits
   b_0 ... b_{k-1};
```

Step **4)** generates the unary code for $\lfloor n/2^k \rfloor$. The counter used to generate the Golomb-Rice codes can actually be shared with an integrating A/D. This key concept is illustrated in Fig. 2. The main building blocks of the dual-slope integrating A/D with compressed output are, a voltage integrator, a comparator, an $(N-1)$-bit digital counter, a negative-edge N_{sr}-bit shift register, and a $(N+1)$-to-1 multiplexer. It is assumed in this design that the input signal V_e lies in the range from 0 to V_{ref} volts and it is centered around $\frac{V_{ref}}{2}$. Hence, it can be expressed as $V_e = \frac{V_{ref}}{2} + v_e$, where $-\frac{V_{ref}}{2} \le v_e \le \frac{V_{ref}}{2}$. The A/D has a resolution of N bits with a quantization step size Δ given by $\frac{V_{ref}}{2^N}$ when uniform quantization is performed. The final codeword is read out serially at the output of the shift register. The conversion cycle starts when the **reset** signal goes low. The conversion has two phases. In phase I the switch **S1** is closed and the switch **S2** is open allowing the integrator to integrate V_e. V_o, the output of the integrator during this phase is given by

$$V_o(t_1) = \frac{V_{ref}}{2} - \frac{1}{RC} \int_0^{t_1} v_e \, dt = \frac{V_{ref}}{2} - \frac{v_e t_1}{RC} \quad (6)$$

where t_1 is the time since the beginning of the conversion. Simultaneously, the digital counter counts up. After 2^{N-1} clock cycles, the carry output of the counter is pulsed, toggling the T flip-flop FF2, which in turn opens the switch **S1** and closes the switch **S2**. This event marks the end of phase I

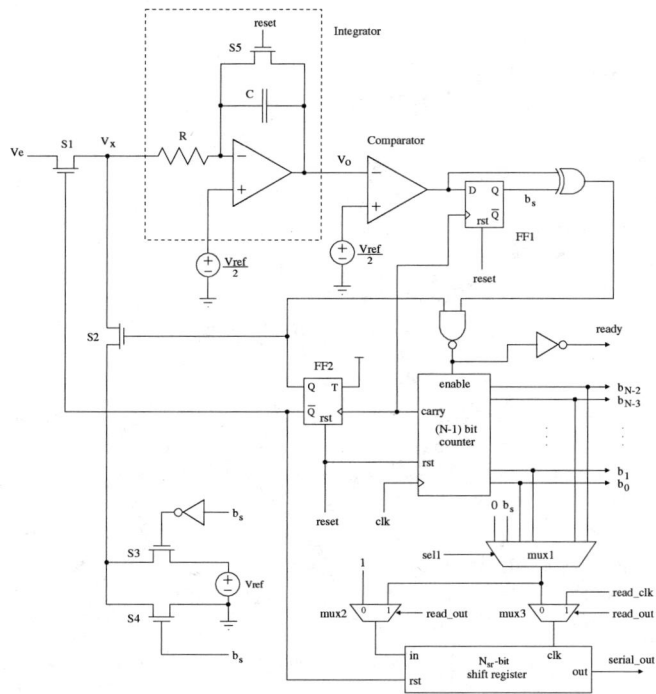

Fig. 2. Block diagram of the A/D.

and the beginning of phase II. At this point the voltage at the output of the integrator is given by

$$V_{o,I} = \frac{V_{ref}}{2} - \frac{2^{N-1}T_{clk}\, v_e}{RC} \quad (7)$$

where T_{clk} is the period of the counter clock. The rising edge of the carry signal also stores the output of the comparator on the D flip-flop FF1. The Q output of FF1 is the sign bit and will constitute the least significant bit of the conversion b_s. If $v_e > \frac{V_{ref}}{2}$, b_s is set to "1", otherwise it is set to "0". Notice that this bit reflects the polarity of v_e and also assigns even values to inputs with negative v_e and odd values to inputs with positive v_e (this is consistent with the notation used in Fig. 1). The voltage V_x is set to ground or V_{ref} through the switches **SW3** and **SW4** depending whether b_s is set to "1" or "0". The output of the integrator in phase II is given by

$$V_o(t) = \begin{cases} V_{o,I} + \frac{1}{RC}\frac{V_{ref}}{2}t_2 & \text{if } v_e > 0; \\ V_{o,I} - \frac{1}{RC}\frac{V_{ref}}{2}t_2 & \text{if } v_e \le 0. \end{cases} \quad (8)$$

where t_2 is the time since the beginning of phase II. After the generation of the carry signal, the counter value is set to zero and the counting process starts over again. To generate the unary code, the selection input **sel1** of the multiplexer **mux1** is set to $k+1$, with k being the Golomb-Rice code parameter described previously. During conversion, the **read_out** input is set to "0". This setting causes the shift register to shift in a "1" every time counter bit b_{k-1} goes from "1" to "0". The conversion cycle ends when V_o reaches $\frac{V_{ref}}{2}$ causing the comparator output to flip. The logic formed by the XOR and NAND gates detects this change and stops the counter. Solving (8) for t_2 with the condition at the end of conversion, $V_o(t) = \frac{V_{ref}}{2}$, yields

$$T_{conv} = \frac{|v_e|}{\frac{V_{ref}}{2}} 2^{N-1}T_{clk} + T_1 \quad (9)$$

where, $T_1 = 2^{N-1}T_{clk}$. The value of the counter at the end of the conversion is also a measure of the time $T_{conv} - T_1$:

$$2^{N-2}b_{N-2} + \cdots + 2^1 b_1 + 2^0 b_0 = \left\lfloor \frac{T_{conv} - T_1}{T_{clk}} \right\rfloor \quad (10)$$

Thus, the binary word $b_{N-2}b_{N-3}\cdots b_1 b_0$ is the digital representation of $|v_e|/(V_{ref}/2)$. The unary code for $\lfloor n/2^k \rfloor$, with $n = 2^{N-2}b_{N-2}+\cdots+b_0$, is already in the shift register and the binary code for ($n \bmod 2^k$) is simply $b_{k-1}\cdots b_0$. The output signal **ready** is set to "1" signaling the end of conversion.

The next step is the read out process. In this step, the input **read_out** is set to "1" causing the output of the multiplexer **mux1** to be redirected to the shift register input through **mux2**. By successively incrementing the selection input of **mux1** and toggling the input **read_clk** the remaining part of the Golomb-Rice code is shifted into the register. The codeword is available in a serial fashion at the output of the shift register **serial_out**. The converter output has the form "$11\cdots 10$", $b_s b_0 b_1 \cdots b_{k-1}$. The choice of the length N_{sr} of the shift register is a trade-off between hardware complexity, input range, and compression ratio. If the shift register is too short, for some values of the input v_e, the register will get full before the end of the conversion, limiting the input range in this manner. The range could be increased by increasing the value of the parameter k, reducing the length of the unary code, but an increment of k beyond its optimal value will impact the compression ratio negatively. To compute the probability of overloading the shift register, let I_j be the union of the intervals $(-2^{k-1}(j+1)\Delta, -2^{k-1}j\Delta]$ and $[2^{k-1}j\Delta, 2^{k-1}(j+1)\Delta)$ with $j = 0, 1, 2, \cdots, 2^{N-k}-1$. If v_e lies in I_j, the converter will produce a unary code of length j. The probability of overloading the shift register is then given by

$$\Pr\{j > N_{sr}\} = \sum_{l=N_{sr}+1}^{2^{N-k}-1} p(I_l), \text{ where} \quad (11)$$

$$p(I_l) = 2\int_{2^{k-1}\Delta l}^{2^{k-1}\Delta(l+1)} f_X(x)\, dx = e^{-\frac{\sqrt{2}}{\sigma}\Delta 2^{k-1}l}\left(1 - e^{-\frac{\sqrt{2}}{\sigma}\Delta 2^{k-1}}\right) \quad (12)$$

If the shift register is long enough such that $N_{sr}+1 > 2^{N-k}-1$, the probability of overloading the shift register is zero, otherwise substituting (12) into (11) yields

$$\Pr\{j > N_{sr}\} = p_o^{2^{k-1}(N_{sr}+1)} - p_o^{2^{N-1}} \quad (13)$$

With the optimal value for k given by (5), a converter with a reference voltage $V_{ref} = 5V$ and a shift register length $N_{sr} = 8$, has a maximum probability of overloading of 0.0109 when $0.01 < \sigma < 1$.

Fig. 3 shows the average codeword length of the Golomb-Rice codes for different values of σ. Each point in this plot was

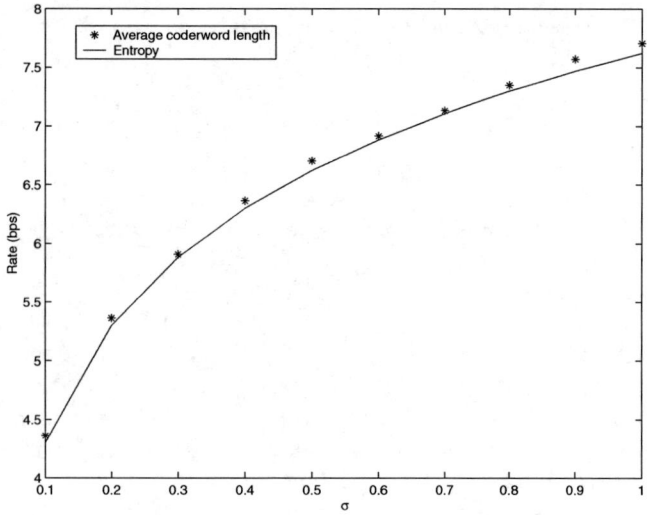

Fig. 3. Average codeword length and source entropy vs. σ.

estimated by quantizing and encoding 10^5 random variables with Laplacian distribution. The quantization has 8 bits of resolution and a uniform step size. The solid line in the graph is the first-order entropy of the quantized source. The code parameter k was estimated according to (5). It can be seen from this plot that good compression performance over a wide input variance range can be achieved.

IV. HARDWARE IMPLEMENTATION

A continuous-time integrator has been utilized in the realization of dual-slope converter due to its simplicity. The digital section has a gate count of less than 100. The overall converter architecture including the passive components occupies a layout area of 0.09 mm^2 based on a 0.5 μm CMOS target technology.

The circuit implementation of Fig. 2 was simulated at the transistor level in Hspice using the target technology for a typical speech/audio application with a conversion resolution of $N = 8$ bits. The counter clock frequency has been set at 20 MHz which in the worst case gave a conversion speed of 73,000 conversions/sec including the read out overhead, validating the proposed approach.

V. CONCLUSIONS

An analog-to-digital converter with data compression capabilities has been presented. The design allows joint quantization of an analog signal and entropy coding of its digital representation. This joint operation is achieved by sharing circuits between an integrating A/D and a Golomb-Rice encoder, yielding a highly compact converter able to compress data. Simulations at the system and transistor level validate the design. Thus, the proposed architecture is suitable for fully integrated compact implementations that encapsulate A/D conversion, compression, and sensor interfacing for a variety of on-sensor data compression applications.

ACKNOWLEDGMENT

This work was supported by Catalyst Foundation grant "Focal Plane Video Compression".

REFERENCES

[1] H. Park and V.K. Prasanna, "Area efficient VLSI architectures for Huffman coding," *IEEE Trans. Circuits and Systems-II*, vol. 40, no. 9, pp. 568-575, Sept. 1993.

[2] I.A. Shah, O. Akiwumi-Assani, and B. Johnson, "A chip set for lossless image compression," *IEEE J. Solid State Circuits*, vol. 26, no. 3, pp. 237-244, March 1991.

[3] Shiann Rong Kuang, Jer Min Jou, Ren Der Chen, and Yen Horng Shiau, "Dynamic pipeline design of an adaptive binary arithmetic coder," *IEEE Trans. Circuits and Systems-II*, vol. 48, no. 9, pp. 813-825, Sept. 2001.

[4] J. Venbrux, Pen-Shu Yeh, and M.N. Liu, "A VLSI chip set for high-speed lossless data compression," *IEEE Trans. Circuits and Systems for Video Technology*, vol. 2, no. 4, pp. 381-391, Dec. 1992.

[5] R. Peck and D. Schroeder, "A low-power entropy-coding analog/digital converter with integrated data compression," *European Solid-State Conf.*, pp. 173-176, Sept. 2003.

[6] D. Martinez, "Time-adaptive vector A/D conversion," *IEEE Trans. Circuits and Systems-II*, vol. 45, no. 10, pp. 1420-1424, Oct. 1998.

[7] R. F. Rice, "Lossless Coding standards for Space Data Systems," *Asilomar Conf. Signals, Syst. and Computers*, vol. 1, pp. 577-585, Nov. 1996.

[8] S. W. Golomb, "Run-length encodings," *IEEE Trans. Inform. Theory*, vol. 12, no. 3, pp. 399-401, July 1966.

[9] R. G. Gallager and D. C. van Voorhis, "Optimal codes for geometrically distributed integer alphabets," *IEEE Trans. Inform. Theory*, vol. IT-21, pp. 228-230, Mar. 1975.

[10] R. F. Rice, "Some practical universal noiseless coding techniques," *JPL Publication 79-22*, Jet Propulsion Laboratory, Pasadena, California, Mar. 1979.

[11] P.-S. Yeh, R.F Rice, and W.H. Miller, "On the optimality of a universal noisless coder," *Proc. of the AIAA Computing in Aerospace Conference*, San Diego, Oct. 1993.

10-Bit Programmable Voltage-Output Digital-Analog Converter

Erhan Ozalevli
School of Electrical and
Computer Engineering
Georgia Institute of Technology
Atlanta, Georgia 30332–0250
Email: ozalevli@ece.gatech.edu

Christopher M. Twigg
School of Electrical and
Computer Engineering
Georgia Institute of Technology
Atlanta, Georgia 30332–0250
Email: ctwigg@ece.gatech.edu

Paul Hasler
School of Electrical and
Computer Engineering
Georgia Institute of Technology
Atlanta, Georgia 30332–0250
Email: phasler@ece.gatech.edu

Abstract—This paper describes an implementation of a compact and low-power 10-bit Floating-Gate Digital-to-Analog Converter (FGDAC). Nonvolatile Floating-Gate voltage references are utilized to build a charge amplifier DAC architecture. This novel implementation eliminates the large element spread and resolution trade-off in the traditional design of charge amplifier voltage-output DACs. The FGDAC was fabricated in a 0.5μm CMOS process and its total area is $0.0522mm^2$. The presented experimental data shows that INL and DNL values less than $\pm 0.5LSB$ ($0.68mV$) are easily achievable. This structure will enable digital to analog conversion with programmable linearly or nonlinearly spaced levels.

I. INTRODUCTION

Traditional DAC designs are driven by their applications and are generally subject to constraints imposed by the trade-off between speed, accuracy, and area. This is especially the case for embedded on chip systems where die area tends to be a major concern. Depending upon the application, speed and/or resolution is often sacrificed for reduced area. However, smaller DAC designs are prone to process variations and random mismatches. It has been shown that random mismatches can be compensated by increasing the area of components [1], which exemplifies the trade-off between the area and effective DAC resolution. Also, it has been previously reported that introducing programmable analog elements into the DAC implementation allows for improved resolution while maintaining small area requirements [2]. The epot [3] is an ideal device for obtaining a dynamically reprogrammable, nonvolatile, on-chip voltage reference which can be easily integrated with traditional DAC architectures in standard CMOS processes. The core of the epot is the floating-gate transistor, which has been successfully integrated into various earlier DAC implementations [4].

In the case of charge amplifiers, traditional designs require highly matched, scaled capacitor values, which pose an area concern [5]. The key element in these structures is the binary-weighted capacitor array, which was first presented by McCreary and Gray in their DAC implementation [6]. In order to eliminate the large element spread in these architectures, DACs based on two stage capacitor arrays [7], and C-2C ladders [8] were proposed. In this paper, we propose the use of epots to compensate for capacitor mismatch by adjusting the

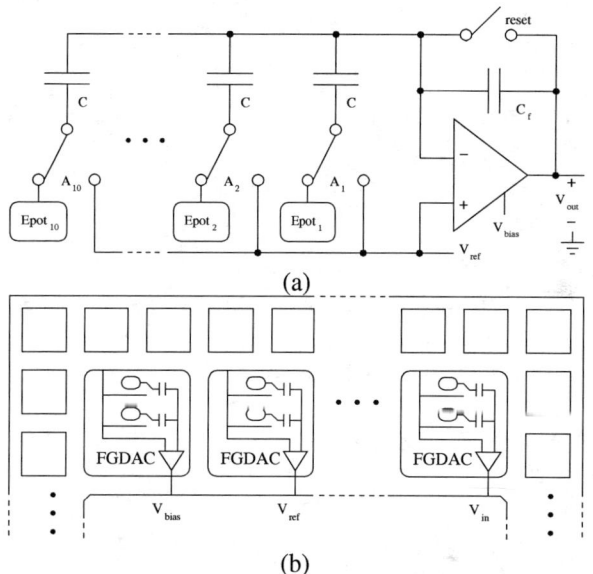

Fig. 1. Circuit architecture for a 10-bit charge amplifier FGDAC and its potential application for the parallel/serial processing of digital inputs. (a) The FGDAC circuit is implemented by employing epots in a charge amplifier structure. Stable bias voltages can be accurately obtained by using epots. Reference voltages for each bit can be programmed both to scale the input voltages and to minimize the effect of the mismatch between capacitors. By using the circuit topology illustrated above, an FGDAC with equal input capacitors, $C_1 = C_2 = C_3 = ... = C_{10} = C_f/2 = 140fF$ is built. (b) An application of FGDACs. The small die size of the FGDAC allows for multiple FGDACs for generating many reference voltages and analog input signals on a single chip. In this way, the effects of noise and distortion resulting from additional external hardware can be minimized in the analog signals.

corresponding voltage for each bit, and to scale the voltages instead of scaling the capacitor values. This implementation results in a compact, low-power voltage-output DAC. Earlier results from a 4-bit DAC [9] demonstrated the feasibility of the epot integration into a charge amplifier architecture. In this paper, experimental static and dynamic results from a 10-bit version of the epot based Floating-Gate DAC (FGDAC) structure are presented.

II. CIRCUIT ARCHITECTURE OF FGDAC

The FGDAC is composed of several sub-systems including an operational amplifier, digital switches, and epots as illus-

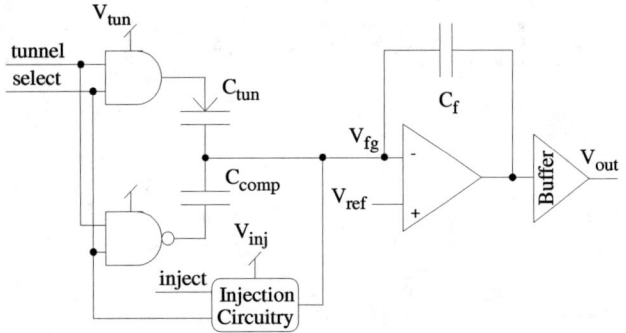

Fig. 2. Circuit schematic of the epot. Charge at the floating node is used to program the voltage output of the epot. A charge amplifier structure is employed to amplify the effects of the charge at the floating node. Charge is decreased on the floating gate through hot-electron injection and increased on the floating gate by utilizing tunnelling quantum mechanical phenomena. Digital control circuitry makes epot programming very user-friendly. The tunnel, select, and inject lines are the digital signals used for digital control of the epots. C_{tun} is the tunnelling MOSCAP used for tunnelling and C_{comp} is the compensation capacitor employed to minimize the switching effect of the tunnelling voltage. The supply voltage of the amplifier is 5V, while the tunnelling voltage is kept at 16.5V and injection voltage is set to 5.8V. The buffer is realized by employing a source follower with a current-sink load to be able to drive large capacitive loads at the output.

trated in Figure 1a. It is designed to realize low-power and compact DACs that can be integrated with larger systems as shown in Figure 1b. While the design of the FGDAC is slightly different, it is functionally same as a traditional charge DAC. The digital input word controls the desired voltage output by switching the individual capacitors between the reference voltage and the corresponding epot output. This results in a charge on the input side, which is amplified by the charge amplifier structure to produce a voltage output that can be expressed as

$$V_{ref} - V_{out} = \frac{1}{C_f} \sum_{i=1}^{n} a_i C_i (V_i - V_{ref}) \qquad (1)$$

where V_{ref} is the reference voltage, C_f is the feedback capacitor, V_i is the epot output voltage, and a_i is the digital input bit. In this implementation, equal size input capacitors were used together with a feedback capacitance that was double the input capacitor value. The area used for an individual block is summarized in Table I. The major contribution of area comes

TABLE I
AREA USED FOR FGDAC AND ITS COMPONENTS.

Epots	Cap./switches	DAC w/o Epots	DAC with Epots
32,247 μm^2	8,867 μm^2	0.0199 mm^2	0.0522 mm^2

TABLE II
PROGRAMMED EPOT VOLTAGES (BUFFER OUTPUTS).

Epot1	Epot2	Epot3	Epot4	Epot5
2.5948V	2.5966V	2.5997V	2.6162V	2.6296
Epot6	Epot7	Epot8	Epot9	Epot10
2.6718	2.7860V	2.9340V	3.2698V	3.9393V

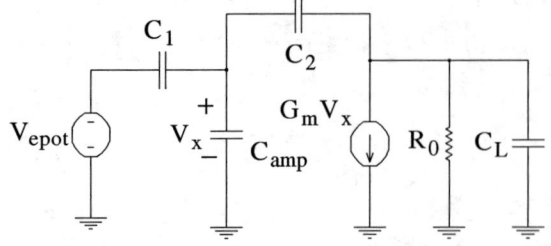

Fig. 3. Equivalent circuit of the DAC structure in Figure 1 when only one bit is enabled. C_L, C_{amp}, R_0 and G_m are the load capacitance, input capacitance, open-loop output impedance and transconductance of the op amp, respectively.

from the epots, which can be further reduced by eliminating the additional digital selection and compensation circuitry. The compensation circuitry is utilized to minimize the effect of the digital selection circuit on the epot output voltage during programming. This part can be removed to reduce the area, but this complicates the controllability of the epot output.

The input voltages are adjusted to their value, as shown in Table II, by using epots in order to obtain the scaled charge normally accomplished through capacitor sizing. Since capacitor size is not a restraining issue in this structure, for a given power consumption, DAC's speed can be maximized by reducing the capacitor area. In this implementation, the unit capacitor value is set to $140 fF$.

Epots provide a user-friendly method of analog programming and can easily be configured into a large array structure. Hot-electron injection and Fowler-Nordheim tunnelling are used to adjust the output voltage by changing the charge on the floating node. The schematic of the epot is illustrated in Figure 2. Programming is controlled via digital signals, which allows the epot to be adjusted to within 100 μV of the target. Also, it has been previously reported that the analog values can be successfully retained by epots for long-term operations [10].

Voltage references are generally required to have low noise, and epots have been shown to be suitable for low noise applications [3]. For static measurements, flicker noise becomes the main concern. These epots exhibit 3.6 $\frac{\mu V}{\sqrt{Hz}}$ flicker noise. In addition, the thermal noise should also be suppressed to operate at high speeds with high SNR, and epots exhibit 2.2 $\frac{\mu V}{\sqrt{Hz}}$ thermal noise. Moreover, epots are required to drive large capacitive loads when integrated into the FGDAC. Therefore, the epot voltages are buffered by utilizing a two-transistor source follower with a current sink load. While other types of output amplifiers could be chosen, this follower is employed due to its simplicity and compactness. Depending on the power consumption requirement, the output resistance of the buffer can be tuned to allow operation at different speeds.

The output amplifier of the FGDAC is a 9-transistor OTA operating in subthreshold. In this region, this micro-power amplifier yields very high transconductance, bandwidth (BW) and gain for a given current level [11]. The time constant of the FGDAC can be found by using the equivalent circuit

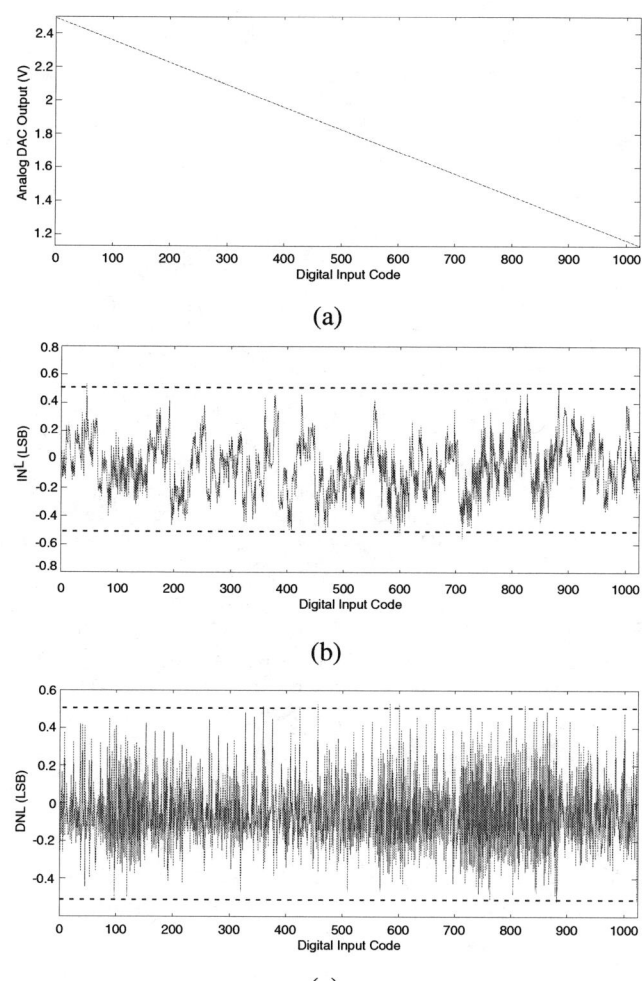

Fig. 4. Experimental results obtained to characterize the static behavior of the 10-bit FGDAC. The DAC element is cleared before measuring an entire transfer characteristic. (a) Output response of the FGDAC to 10-bit digital input code. The voltage output is a linear function of the digital input word. (b) INL characterization results for 10-bit digital input code. INL is in the range of ± 0.5 LSB. (c) DNL measurements of the FGDAC. Similarly, DNL is obtained to be between ± 0.5 LSB.

illustrated in Figure 3 and can be shown [12] as

$$\tau_{FGDAC} = \frac{(C_L + C_2)(C_1 + C_{amp}) + C_L C_2}{G_m C_2} \quad (2)$$

In this design, C_L is around 400fF (very small since FGDAC is buffered before driving the signal off-chip), and the bias current is set to 6uA (biased in the moderate inversion region), giving very high GB while consuming very low power, in the μW range.

III. INL / DNL LINEARITY

The transfer characteristic of the FGDAC is illustrated in Figure 4a and used to compute the static linearity. Epots are programmed to obtain a full linear range of 1.36 V that can be easily increased by programming the epots to a different set of voltages. From the results of this transfer function, INL and DNL are computed and shown in Figure 4b, c, respectively.

Within this limited full-scale range, the DAC yields 10-bit linearity with less than 0.68mV quantization error.

The limits of the FGDAC performance in these experiments are mainly caused by the practical issues with epots and the experimental set-up. The epot voltages are programmed with a resolution of $100\mu V$ within a 1.5V range; higher DAC linearity would require tighter programming resolution as well as lower epot noise levels. High resolution and controllability of the epots makes this implementation realizable for higher resolutions. Our primary limitation for higher linearity from this IC comes from the second-harmonic distortion of our output buffer used for driving signals off-chip. For on-chip implementations, the linearity of the buffers will not be an issue. Also, flicker noise in the signal path is another limiting factor for static measurements. Therefore, amplifiers as well as epots have to be designed to exhibit low flicker noise. If the epots are designed to be smaller, the programming accuracy becomes degraded. The reason is that the effect of the charge at the floating node of the epot becomes higher as the feedback capacitance becomes smaller. Therefore, this limitation can be improved by increasing C_f in the epot. In addition, electron traps are created through the injection and tunnelling processes and cause the amount of charge at the floating node to be changed. This sets another limit for the operation, but again can be easily minimized by increasing the capacitance at the floating node.

IV. DYNAMIC BEHAVIOR AND LINEARITY

Dynamic measurements of the FGDAC are obtained by testing the performance of the DAC for sinusoidal and MSB step inputs. The results prove that the structure of the FG-DAC is also very suitable for operating at high frequencies. Theoretically, the conversion speed of a traditional charge amplifier DAC design is limited by the output amplifier, and this is also true for the FGDAC. For better performance and smaller area, the input and feedback capacitors of the FGDAC are optimized, and therefore their effect on the speed are effectively minimized. This architecture can be assumed to have two poles that are determined by the epots and the output amplifier. Since the epot output is buffered before supplied as an input, the primary pole comes from the amplifier. As is the case for static measurements, the primary limitations in these experiments are the test equipment and buffers of the FGDAC chip.

First, a step input at 1.5 MHz is applied to the MSB of the FGDAC, since the large swing of this bit is the main limiting factor for the speed of the FGDAC. The step response is illustrated in Figure 5a. The rise time of the step response is around 160ns which is a known limitation of the output buffer. It is also realized that other on-chip buffers and the test equipment caused additional distortion, but the FGDAC still exhibited promising performance over this range. Simulation results (Figure 5b) show expected settling times around 60ns, which confirms that the output buffer is limiting the experimental performance. Second, we applied a 9-bit sinusoidal input and obtained a 7-bit linear sinusoidal output

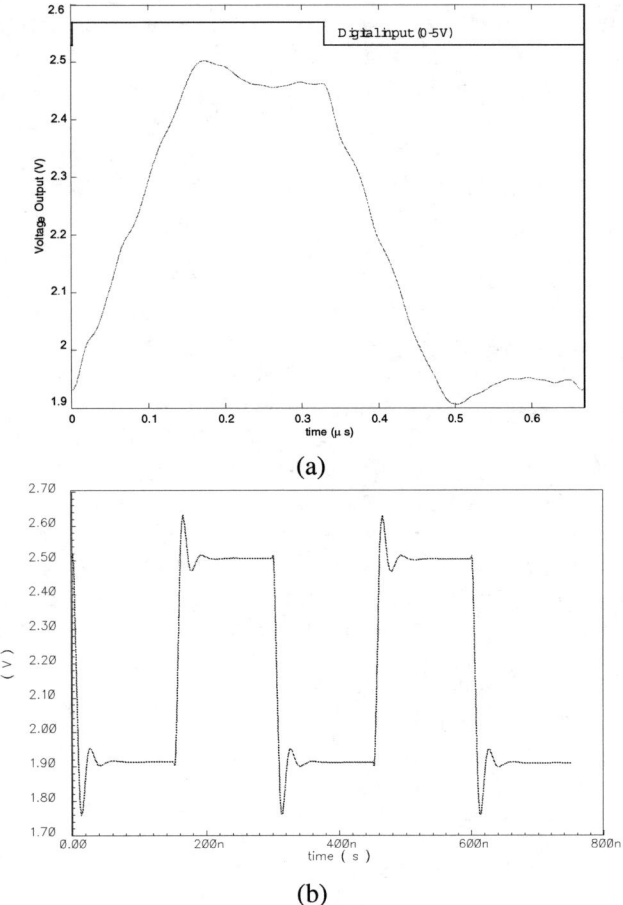

(a)

(b)

Fig. 5. Step response of the most significant bit of the FGDAC. (a) Experimental IC measurements with the MSB digital input is switched at 1.5MHz. We see a rise-time of roughly 160ns, which is consistent with the limitations of our on-chip buffering circuitry for this application. (b) Simulation results of the FGDAC for an MSB digital input. The rise time is in the range of 10ns and settling time is around 60ns confirms that the output buffer is limiting the experimental measurements.

as shown in Figure 6. Again the main factor in the degradation of the linearity is the buffer and can be minimized for on-chip system level implementations.

V. CONCLUSION

In this article, we described an implementation of an FG-DAC and showed that it can yield a compact and low-power DAC or DAC array. This structure can be used for a wide range of embedded system applications where power and area become one of the main concerns. The results illustrate the flexibility and programmability of this architecture, which can be leveraged to create linear or non-linear output voltage spacing. Dynamic re-calibration can also be achieved using this programmability feature to accommodate varying operating conditions.

REFERENCES

[1] M.J.M. Pelgrom, A.C.J. Duinmaijer, and A.P.G. Welbers, "Matching properties of MOS transistors," vol. SC-24, pp. 1433–1439, 1989.

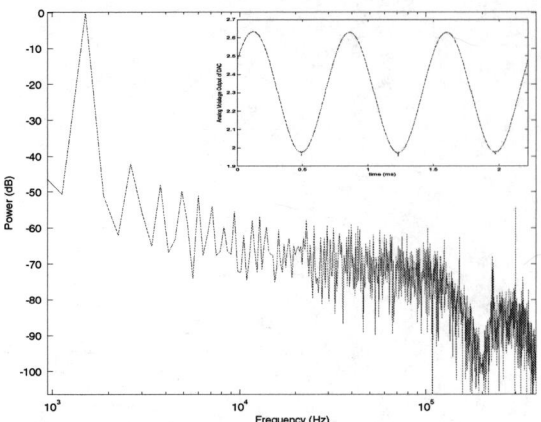

Fig. 6. Dynamic measurements from the FGDAC: Sinusoidal output response of the FGDAC (1.5kHz sinusoidal waveform), measured frequency spectrum of the FGDAC were created by switching the digital inputs of the FGDAC at 50kHz. Linearity is limited to 7 bits due to nonlinearity caused by the output buffer amplifier driving the signals off chip.

TABLE III
PARAMETERS OF THE FGDAC.

Process	0.5μm, 2 poly
Power supply	5V
Power consumption	150μW
Linearity (INL/DNL)	10-bit
SFDR	7-bit
Epot Programming Resolution	100μV
Programming Mechanisms	Hot-Electron Injection and Electron Tunneling
DAC area	0.0522 mm^2
Input capacitor	140 fF each

[2] Guillermo Serrano and Paul Hasler, "A Floating-Gate current DAC array," in *IEEE Symposium on Circuits and Systems*, Vancouver, OK, May 2004.

[3] R.R. Harrison, J.A. Bragg, P. Hasler, B.A. Minch, and S. Deweerth, "A CMOS Programmable Analog Memory Cell Array using Floating-Gate Circuits," *IEEE Trans. on Circuit and Systems*, 2001.

[4] A.J. Ramirez-Angulo, J.; Lopez, "MITE circuits: the continuous-time counterpart to switched-capacitor circuits," *IEEE Trans. on Circuit and Systems II*, 2001.

[5] P.E. Allen and D.R. Holberg, *CMOS Analog Circuit Design*, Oxford University Press, Oxford, 2002.

[6] J.L. McCreary and P.R. Gray, "All-MOS charge redistribtion Analog-to-Digital conversion tecniques-part 1," *IEEE Journal of Solid-State Circuits*, vol. SC-10, December 1975.

[7] Y.S. Yee, L.M. Terman, and L.G. Heller, "A two-stage weighted capcitor network for D/A-A/D conversion," *IEEE Journal of Solid-State Circuits*, vol. SC-14, August 1979.

[8] J.L. McCreary, "Matching properties, and voltage and temperature dependence of MOS capacitors," vol. SC-16, pp. 608–616, December 1981.

[9] Erhan Ozalevli, Paul Hasler, and Farhan Adil, "Programmable Voltage-Output, Floating-Gate Digital-to-Analog Converter," in *IEEE Symposium on Circuits and Systems*, Vancouver, OK, May 2004.

[10] Ethan Farquhar, Chris Duffy, and Paul Hasler, "Practical issues using epot circuits," in *International Symposium on Circuits and Systems*, Phoenix, AZ, May 2002.

[11] J. Fellrath and E. Vittoz, "Small signal model of MOS transistors in weak inversion," pp. 315–324, 1977.

[12] B. Razavi, *Design of Analog CMOS Integrated Circuits*, McGraw-Hill Companies, Inc., 2000.

A Background Correction Technique for Timing Errors in Time-Interleaved Analog-to-Digital Converters.

Echere Iroaga and Boris Murmann
Dept. Of Electrical Engineering
Stanford University
Stanford, CA, USA
eiroaga@stanford.edu, murmann@stanford.edu

Lalitkumar Nathawad
Atheros Communications, Inc.
Irvine, CA, USA
nathawad@atheros.com

Abstract—A background correction scheme for timing mismatch in time-interleaved Analog-to-Digital Converters (ADCs) is presented in this paper. The architecture is based on the use of an extra ADC channel and an input ramp signal to estimate the timing errors, and digital interpolation to correct the output digital codes. Simulated results demonstrate a 35dB improvement in SFDR and a 20dB improvement in SNDR for a 10-bit converter with an over-sampling ratio greater than 2X.

I. INTRODUCTION

Time-interleaved ADCs have become the architecture of choice for achieving conversion speeds in excess of the raw technology limits. In time interleaved converters, several sub-converter channels are operated in parallel. Each sample of the input signal is processed by one of the sub-converters in turn, achieving an N fold increase in speed, where N is the number of channels that are time-interleaved. Thus, this architecture exploits parallelism to improve speed.

The performance of time-interleaved ADCs is typically limited by the mismatches between the interleaved channels. In particular, the mismatches in gain, offset and timing introduce either fixed pattern noise or distortion in the output spectrum [1].

II. PROPERTIES OF PATH MISMATCH

Gain mismatch amongst the different paths of interleaved converters result in an amplitude modulation of the input signal. This manifests itself as aliases of the input spectrum at integer multiples of the individual channel sampling frequency f_s/N. Depending on the value of N, these aliases could exist in the frequency band of interest making it impossible to filter out. Offset mismatches can be thought of as an additive signal with a period of f_s/N. This signal then appears in the output spectrum as tones at integer multiples of f_s/N. Similarly, timing errors can be thought of as phase modulation of the input signal and produce a similar spectrum as gain mismatches. However, the aliases from timing mismatches are also dependent on the input signal frequency. A detailed derivation of the properties for each of these mismatches is shown in [7]. Of all these error sources, the error power due to timing mismatch becomes more significant at higher signal frequencies.

Timing errors can be broken down into either clock skew or clock jitter [5]. The tones generated by clock skew reduce SFDR. The random nature of clock jitter, on the other hand, causes errors, that are spread over the Nyquist bandwidth of the converter. This raises the noise floor and reduces SNR. It should be noted that unlike deterministic clock skew, errors due to clock jitter are more fundamental and do not arise from the time-interleaving architecture, but are also present in single path converters.

Figure 1 shows the simulated SFDR degradation for a 3-channel 10-bit ADC with a 100MHz input signal and timing skew in one of the channels. The SFDR in the absence of timing skew is 85dB, as is predicted in [11]. From Figure 1, it can be seen that a timing skew on the order of 2ps will cause a SFDR degradation of up to 15dB. This paper proposes a scheme to correct for these errors caused by static timing mismatches.

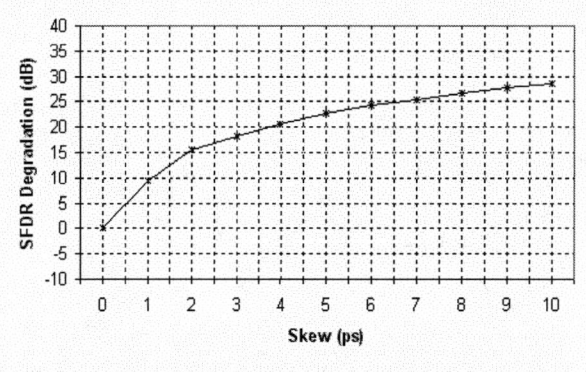

Figure 1. SFDR degradation vs. Timing skew

III. CORRECTION METHODS

There has been a lot of attention paid to improving the performance of interleaved analog-to-digital converters by correcting for gain and offset mismatches [2],[3],[4]. Not as much has been done, however, in improving errors due to static timing mismatch. Many proposed correction methods use a single front-rank sample and hold, running at the full sampling rate, to eliminate the need for timing correction [2]-[4]. In these architectures, the sample and hold limits the overall sampling rate. As a result, increasing the number of interleaved channels beyond a certain point is of no benefit. The maximum speed of the converter becomes limited by technology, and to achieve higher operating speeds, the sample and hold must be built in a faster, more expensive technology. The benefits of time-interleaving are fully realized when the speed is limited by the settling of the amplifiers in each channel. In this case, the interleaving of more channels results in an increase in overall sampling rate and is less dependent on technology. To remove the stringent requirements on the sample and hold, each channel can be designed to have its own dedicated sample and hold running at 1/N the overall sampling rate. With the use of per-channel sample and hold circuitry, the errors due to timing mismatch between the channels must be considered.

The techniques used to compensate for mismatch errors in time-interleaved converters can be grouped into two main approaches:

1. Measure the mismatch between the channels and adjust each channel, in the analog domain, until better matching is achieved.

2. Measure the mismatch between channels and use this information to correct the output digital codes without adjusting the path.

With the increased digital computing power available due to aggressive technology scaling, the second approach has gained practicality. The use of digital signal processing in the backend to correct for mismatches allows for portability to new technology nodes and can result in better performance at comparable power levels. Timing mismatches, in particular, are suitable for this approach.

In order to have the distortion power from timing skew less than the quantization noise, the skew Δ_t, must be bounded by (1) [8]:

$$|\Delta_t| \leq \sqrt{\left(\frac{N}{N-1}\right) \cdot \left(\frac{2}{3\omega_0^2}\right) \cdot \left(\frac{1}{2^{2m}}\right)} \quad (1)$$

where N is the number of interleaved channels, m is the bit-resolution of the ADC and ω_0 is the frequency of the input signal. With an input signal of 100MHz and with 3 interleaved channels, the timing skew must be reduced to below 2ps to achieve 10-bit accuracy. At these levels, it becomes much more difficult to correct the timing edges, requiring complex variable delay circuitry that may add jitter to each clock phase. Thus, it is more convenient to measure the clock skew and correct the output codes.

There are several drawbacks to existing schemes for timing mismatch correction. One method of timing mismatch compensation involved adding a ramp with a known slope to the input signal and obtaining the skew values from the channel output [5]. When used in the background, the input ramp takes up some of the input dynamic range. In addition, any frequency component of the input at integer multiples of the channel sampling rate will manifest itself as a fixed timing offset. Furthermore, this method assumes certain statistical properties of the input signal. Other approaches have been to digitally estimate the timing skew from correlation between samples [6], [8]. This scheme also relies on certain spectral characteristics. Several schemes for gain and offset correction have also been investigated [2],[3],[4]. These schemes however, are not easily extended to include timing mismatch correction.

IV. PROPOSED ARCHITECTURE

A continuous calibration scheme, which corrects for timing mismatch, is now presented in this paper. The timing correction requires the addition of an extra calibration ADC as well as a digital interpolation filter. The scheme lends itself to being incorporated with various gain and offset correction algorithms previously presented (e.g. [3]). The use of an extra calibration ADC removes any statistical restrictions on the input signal and its spectrum, except that it is band limited to $f_s/2$ or less, where f_s is the overall sampling rate of the ADC.

In the proposed architecture, a periodic calibration signal is applied to each channel to extract timing information, which is then used to correct the output digital codes. A block diagram of the ADC architecture is shown in Figure 2.

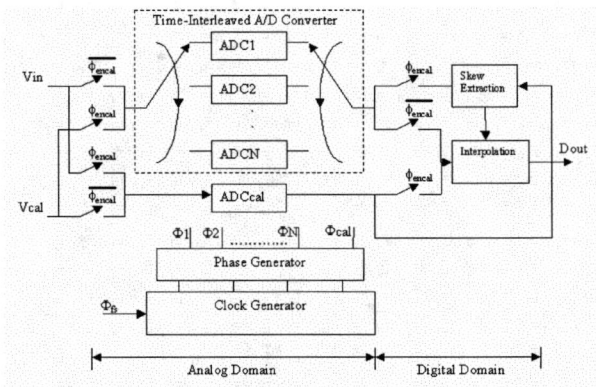

Figure 2. Self-Calibrating Time-Interleaved ADC

The sub-converters *ADC1..ADCN* are driven by time-interleaved clocks, $\Phi_1..\Phi_N$ each running at f_s/N. *ADCcal* is driven by Φcal, which runs at $f_s/(N+1)$. Ideally, $\Phi_1..\Phi_N$ are phase aligned with the global clock, Φ_{fs}, such that the input signal is sampled at uniform time intervals. *ADCcal* will then sample at exactly the same time as one of the interleaved sub converters but at $N/(N+1)$ the rate. The sampling rate of *ADCcal* and the other sub-converters are relatively prime, causing the *ADCcal* sampling times to cycle through the other sub-converters clock phases. This timing relationship for a system with *N*=3 is illustrated in Figure 3.

The algorithm for extracting the phase information is as follows. Consider the case where the sampling time for *ADCcal* occurs at the nominal sampling time of *ADC1*. *ADCcal* is placed in the data path to process the input and a known calibration signal, *Vcal*, is applied to *ADC1*. The output of *ADC1* is fed into the skew extraction circuitry and is used to extract the timing skew, Δt_1, in Φ_1. The timing skew in each of the other clock phases can be extracted in a similar manner.

Figure 4 shows an illustration of the ramp calibration signal with slope *S* applied to *ADCi*. The timing error Δt_i can then be extracted from the output digital code of *ADCi* (V_{adci}) as given by (2).

$$\Delta t_i = \frac{(V_{adci} - V_{out\Delta t=0})}{S} \quad (2)$$

As long as the slope of the ramp is known, an accurate estimation of Δt_i will be achieved. Since Δt_i is small, the ramp is only required to be linear over a small range.

The timing skew of all the sub-converters is referenced to the *ADCcal* sampling instant. Therefore, the $\Phi encal$ must be periodically kept low to extract the digital output code that corresponds to $\Delta t = 0$ ($Vout_{\Delta t=0}$) from *ADCcal*. Once this value is obtained, the interleaved sub-converters can be calibrated in turn.

It should be noted, however, that the timing skew extraction algorithm must filter the clock jitter in order to accurately obtain clock skew values. Taking a running average of Δt_i accomplishes this. The extracted skew values are then fed into the digital interpolator (Figure 2).

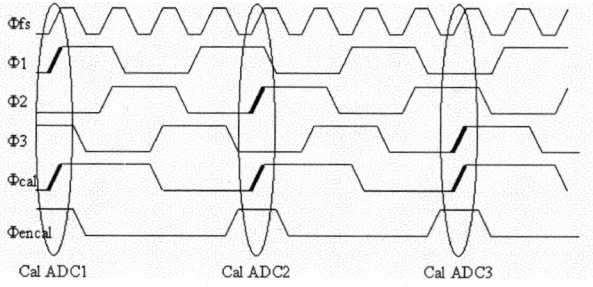

Figure 3. Timing relationship between sub-converter clocks for N=3

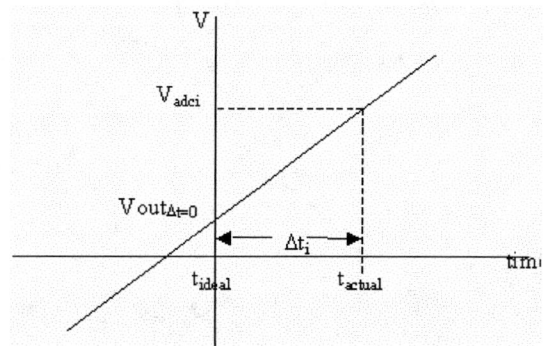

Figure 4. Timing extraction from a ramp

In the digital interpolator, Δt_1 is mapped to the digital codes from *ADC1*, Δt_2 to *ADC2* and Δt_3 to *ADC3* such that the data pairs in the interpolator are of the form: $(D_1, t_s+\Delta t_1)$, $(D_2, 2t_s+\Delta t_2)$ etc., where D_1 obtained from the output from *ADC1* and D_2 is obtained from *ADC2* and t_s is the overall sampling period. The interpolator keeps track of the last *K* output samples, $D_1..D_K$, and their timing information. *K* is chosen to be odd so that every cycle, the middle sample, $D_{(K-1)/2}$, is corrected by interpolation using the surrounding *K-1* samples, just as in [5]. An algorithm based on Neville's method [9] is used for the interpolation. The interpolation algorithm requires $K(K-1)$ multiplications and $K(K-1)/2$ subtractions per output [10], with some of these operations being performed at a rate less than Φ_{fs}.

There are several advantages of this scheme. Firstly, it allows the calibration to run in the background without interrupting the operation of the ADC and eliminating the need for a dedicated calibration sequence. Also, any drift in the timing edges over time is captured and corrected. Secondly, the fact that the calibration ramp signal is applied to the ADC when it is swapped out of the data path avoids the ramp taking up part of the input dynamic range as occurs in [5]. It also eliminates any restrictions on the input signal. The extra ADC increases the area of the converter by less than $1/N$, where *N* is the number of interleaved sub-converters, making this scheme very practical for architectures where over 4 converters are interleaved.

V. SIMULATION RESULTS

A 3-channel time-interleaved ADC using the above scheme was modeled in MATLAB. Each of the sub-converters is an ideal 10-bit ADC. For this simulation, thermal noise was ignored. The interpolation was accomplished by using Neville's method with 11 points (*K*=11) and the interpolated values were re-quantized to 10 bits. Figures 5 and 6 show the frequency spectrum with and without timing correction at 5X over-sampling and 5% timing skew between each successive channel. The figures show that the errors due to timing mismatch are significantly reduced by the proposed scheme.

As the input frequency gets closer to the Nyquist rate, the accuracy of the interpolation algorithm is degraded. This results in incomplete correction of the timing mismatch error and a reduction in the attainable SFDR and SNDR as is shown in Figures 7 and 8.

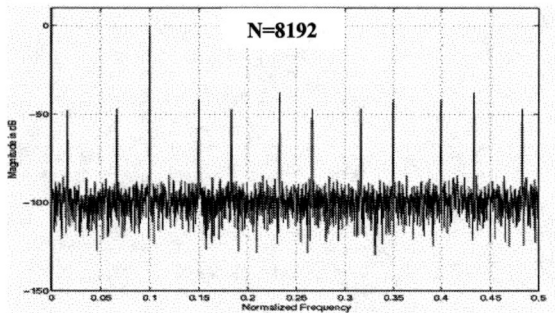

Figure 5. ADC Output spectrum without correction

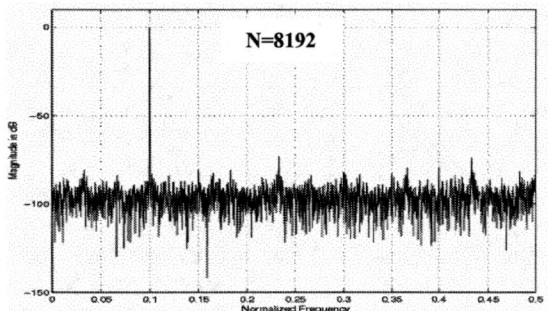

Figure 6. ADC Output spectrum with correction

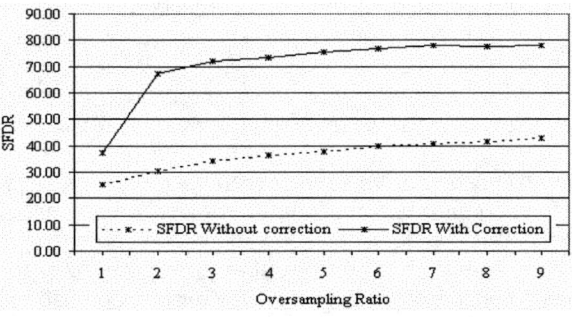

Figure 7. SFDR versus Oversampling ratio

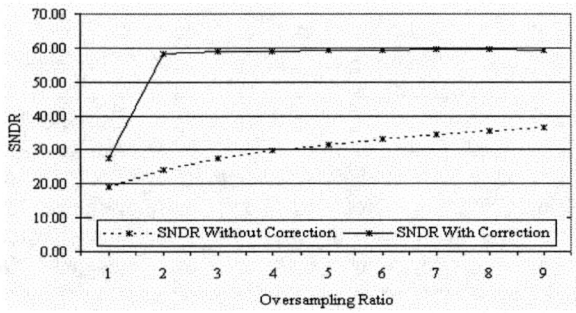

Figure 9. SNDR Versus Over sampling ratio.

VI. CONCLUSION

A time-interleaved ADC architecture with background timing error compensation is proposed. The architecture is based on the use of an extra ADC channel with an input ramp signal to estimate the timing errors, and digital interpolation to correct the output digital codes. The compensation scheme does not assume any statistical restrictions on the input signal and can be combined with existing gain and offset calibration techniques. Simulation results show an improvement in SFDR of over 35B and an SNDR improvement of over 20dB for a 10-bit converter with at least 2X over-sampling.

REFERENCES

[1] W.C Black, Jr, and D.A. Hodges, "Time interleaved Converter arrays," IEEE J. Solid-State Circuits, Vol. 15, no. 6 , pp. 1022-1029, Dec 1980.

[2] Huawen Jin, Edward Lee and Marwan Hassoun, "Time-interleaved A/D Converter with Channel Randomization", IEEE Int. Symposium on Circuits and Systems, June 1997.

[3] D. Fu, K.C. Dyer, S.H. Lewis and P.J. Hurst, "A Digital Background Calibration Technique for Time-Interleaved Analog-to-Digital Converters", IEEE J. Solid-State Circuits, Vol. 33, no. 12, pp. 1904-1911, Dec. 1998.

[4] K.C. Dyer, D. Fu, S.H. Lewis and P.J. Hurst, "An analog Background Calibration Technique for Time-Interleaved Analog-to-Digital Converters", IEEE J. Solid-State Circuits, Vol. 33, no. 12, pp. 1912-1919, Dec. 1998.

[5] Huawen Jin and Edward K.F. Lee, "A Digital-Background Calibration Technique for Minimizing Timing-Error Effects in Time-Interleaved ADCs," IEEE Trans. Circuits and Systems II, Vol.47, no. 7, pp. 603-613, July 2000.

[6] S. M. Jamal, D. Fu, M. P. Singh, P. J. Hurst, S. H. Lewis", "Calibration of Sample-Time Error in a Two-Channel Time-Interleaved Analog-to-Digital Converter", IEEE Trans. Circuits and Systems I, Vol. 51, no. 1, pp 130-139, January 2004.

[7] C.S.G. Conroy, "A high speed parallel pipeline A/D converter technique in CMOS", Ph.D. dissertation, Univ. of California at Berkeley, Berkeley, 1994.

[8] S.M Jamal, "Digital Background Calibration of Time-Interleaved Analog-to-Digital Converters", Ph.D. dissertation, Univ. of California, Davis, 2001.

[9] R. L. Burden et. al., Numerical Analysis: Prindle, Weber & Schmit, 1981, pp. 88-92.

[10] Huawen Jin and Edward K.F. Lee, "A Digital Technique For Reducing Clock Jitter Effects In Time-Interleaved A/D Converter," IEEE Int. Symposium on Circuits and Systems, Vol. 2, pp. 330-333, June 1999.

[11] Hui Pan, Asad A. Abidi, "Spectral Spurs due to Quantization in Nyquist ADCs", IEEE Trans. Circuits and Systems I, Vol. 51, no. 8, pp 1422-1439, August 2004.

[12] K.C. Dyer, D. Fu, P.J. Hurst and S.H. Lewis "A Comparison of Monolithic Background Calibration in Two Time-Interleaved Analog-to-Digital Converters", Proc. IEEE Int. Symposium on Circuit and Systems, Vol . 1, no. 12, pp. 13-16, May. 1998.

FEEDFORWARD-TYPE PARASITIC CAPACITANCE CANCELER AND ITS APPLICATON TO 4 GB/S T/H CIRCUIT

Takahide Sato[†] Shigetaka Takagi[†] Nobuo Fujii[†]
Yasuyuki Hashimoto[††] Kohji Sakata[†††] Hiroyuki Okada[††††]

Tokyo Institute of Technology[†] ROHM CO.,LTD.[††]
SANYO Electric Co., Ltd.[†††] NEC Corporation[††††]

ABSTRACT

This paper proposes a novel parasitic capacitance cancellation method. Since parasitics are canceled by feedforwading signals, a circuit using the proposed cancellation method is always stable unlike a conventional method using a negative impedance converter. This cancellation method is applicable to a balanced-type circuit driving capacitors with source followers. As an example it is applied to implementation of a high-speed track-and-hold circuit (T/H circuit). Thanks to this method implementation of a 4-Gb/S T/H circuit with 6-bit accuracy is confirmed through HSPICE simulations with 90-nm CMOS process under 0.9-V power supply voltage.

1. INTRODUCTION

Implementation of high-speed circuits always meets a problem of parasitic capacitors which limit their operation speed. A simple method to overcome this problem is use of a negative impedance converter. Negative capacitors realized by negative impedance converters will cancel out the parasitic capacitors. Unfortunately implementation of a negative impedance converter has another problem, that is, stability problem. Some deviation of element values will cause oscillation of a circuit if a negative impedance converter is used in the circuit. The stability problem is due to feedback loop in a negative impedance converter.

This paper proposes a novel cancellation method of parasitic capacitors in a balanced-type circuit. The method is based on feedforwarding signals unlike a negative impedance converter. Therefore a balanced-type circuit using the method is always stable regardless of element-value deviation. The method is applied to a very high speed track-and-hold circuit to verify its effectiveness.

2. CONVENTIONAL PARASITIC CANCELLATION FOR BALANCED-TYPE CIRCUITS

A schematic for balanced-type circuits is shown in Fig. 1. In a balanced-type circuit voltage signals at corresponding

This work was supported by Semiconductor Technology Academic Research Center.

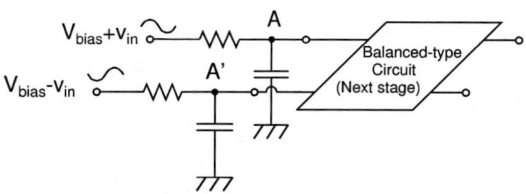

Fig. 1. Balanced-type circuit

nodes like the nodes A and A' in Fig. 1 have the same amplitude and the opposite polarity, though the two nodes are biased with the same potential. Balanced-type circuits as well as imbalance-type circuits are mainly affected by parasitic capacitors which make large time constants. In other words operation speed is limited by such parasitic capacitors.

A simple method to enhance the operation speed is cancellation of such parasitic capacitors. A negative impedance converter (NIC) is usually used for the parasitic cancellation [1][2].

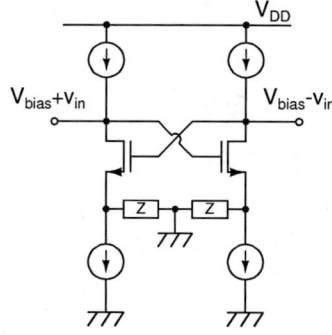

Fig. 2. balanced-type negative impedance converter (NIC)

In particular a balanced-type NIC shown in Fig.2 is very useful because it works well even at high frequencies. As shown in Fig.3 the balanced-type NIC terminated with capacitors, C_c, whose values are the same with parasitic ones will cancel out the parasitic capacitors almost perfectly. Unfortunately a balanced-type circuit will oscillate if terminal stability conditions of the NIC are not satisfied. This is a fatal problem when a NIC is used for the cancellation.

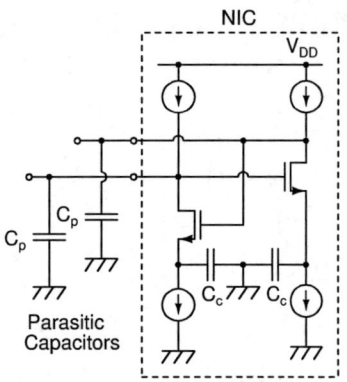

Fig. 3. Parasitic capacitance cancellation using NIC

3. FEEDFORWARD-TYPE PARASITIC CANCELLATION

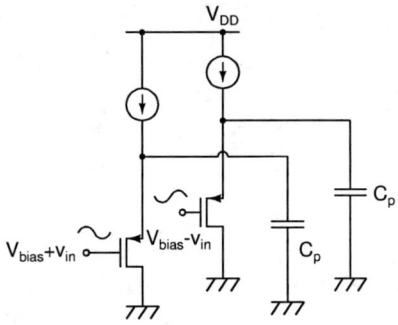

Fig. 4. Two source followers driving capacitors

Consider two source followers driving capacitors as shown in Fig.4. The two source followers are assumed to be used in a balanced-type circuit. The output voltage of the source followers have the same amplitude and the opposite polarity. The NIC terminated with capacitors, as shown in Fig.3, can be connected directly to the capacitors in order to cancel out the capacitors. In this case oscillation may happen because of a deviation of capacitances. To overcome this problem another pair of source followers shown in Fig.5 (A) and (B) are added and used to drive the gates of the two MOSFETs in the NIC. Insert of the source followers removes the feedback loop from the NIC since voltages between C_p is not used as input voltages of the source followers. Therefore the proposed circuit is always stable. It is obvious that the proposed parasitic capacitance canceler acts as a negative capacitors, $-C_c$, like the conventional parasitic capacitance canceler using NIC. The negative capacitors are connected in parallel with C_p. Therefor the total capacitance becomes $C_p - C_c$.

Table 1. Specifications of the T/H circuit (Fig.7)

Power supply voltage	0.9 V
Signal amplitude (differential)	500 mV$_{P-P}$
Resolution	6 bits
Maximum signal frequency	2 GHz
Maximum clock frequency	4 GHz

4. APPLICATION TO HIGH-SPEED T/H CIRCUIT

As an example the cancellation method is applied to an implementation of a high-speed T/H circuit. One of the most simple T/H circuits is shown in Fig.6. A T/H circuit is often used in front of an analog-to-digital converter (ADC) for its proper operation [3]. Therefore an input capacitor of an ADC is a parasitic load of a T/H circuit and mainly limits its operation speed. Cancellation of the input capacitor will drastically improve the operation speed of a T/H circuit.

Figure 7 shows a T/H circuit using the proposed cancellation method. Specifications of the T/H circuit is summarized in Table 1. This circuit is simulated by HSPICE with 90-nm CMOS process parameters. 2-GHz input signal and 4-GHz clock are used for the simulations. Simulation results are shown in Figs.8 to 11 and Table 2. Simulation results of the simple T/H circuit illustrated in Fig.6 is also illustrated in the same figures.

Figure 8 shows input and output voltages of the T/H circuits when the simple T/H circuit and the proposed T/H circuit dissipate the same power, that is, about 34mW. Not only the power consumption at the T/H circuit but also that at the clock generator are considered. The clock generator is realized by many CMOS inverters connected in cascade to enhance its fanout. The both T/H circuits hold the input voltages when they are crossing zero. Figure 9 is an enlarged figure of Fig.8 when the T/H circuits are in the hold phase. The simple T/H circuit shown in Fig.6 holds its output voltage within a 6-bits accuracy for 61 ps. On the other hand the T/H circuit using the proposed method holds it for 74 ps. It corresponds to 17.4-% extension of the conventional value.

Figure 10 and 11 show the simulation results when two T/H circuits occupy equal chip area (320 μm^2). The other conditions are the same with the previous simulations. It is obvious from Fig.10 and 11 that the proposed circuit shows faster response even when their chip area is equal. Figure 11 shows transitional response of the T/H circuits in the hold phase. As shown in Fig.11 holding time (6-bits accuracy) of the proposed circuit is extended to 11 % longer than that of the simple T/H circuit.

Table 2 summarizes the chip area and the total power dissipation of Fig.6 and Fig.7 when their output voltages are kept 6-bits accuracy for 74ps. The chip area of the proposed circuit is about 320 μm^2. 26 % of chip area of the conventional one is saved thanks to the proposed para-

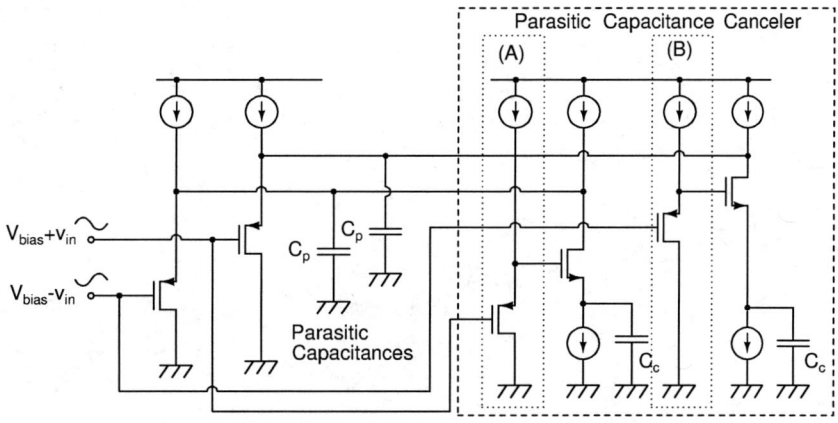

Fig. 5. Proposed parasitic capacitance cancellation circuit

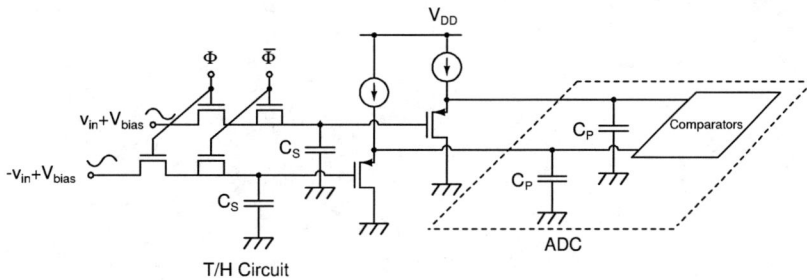

Fig. 6. Simple T/H circuit

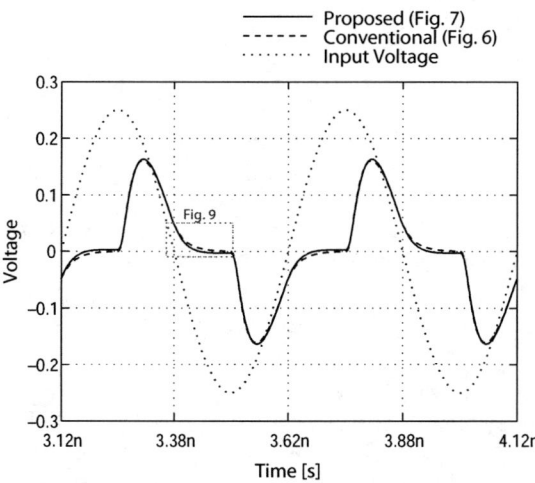

Fig. 8. Input and output characteristics of the T/H circuits when the power dissipation is 34 mW

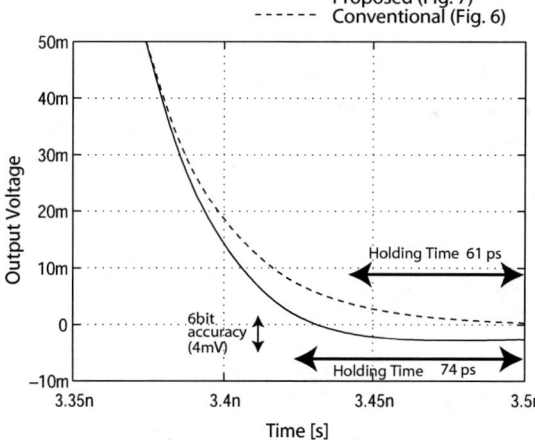

Fig. 9. Output voltage of T/H circuits in hold phase (Enlarged figure of Fig.8)

sitic capacitance canceller. The power consumption of Fig.6 is 34.1 mW. It implies 60 % of that of the conventional one. Because of the parasitic capacitance canceller the parasitic capacitor is drastically reduced. Smaller source followers, switches and clock generators are required to drive the smaller parasitic capacitors and realize smaller chip area and smaller power dissipation. The reduction of the occupied chip area and the power dissipation of the these circuits are quite larger those of the parasitic capacitance canceller. Therefore the proposed circuit can save the chip area and power dissipation drastically.

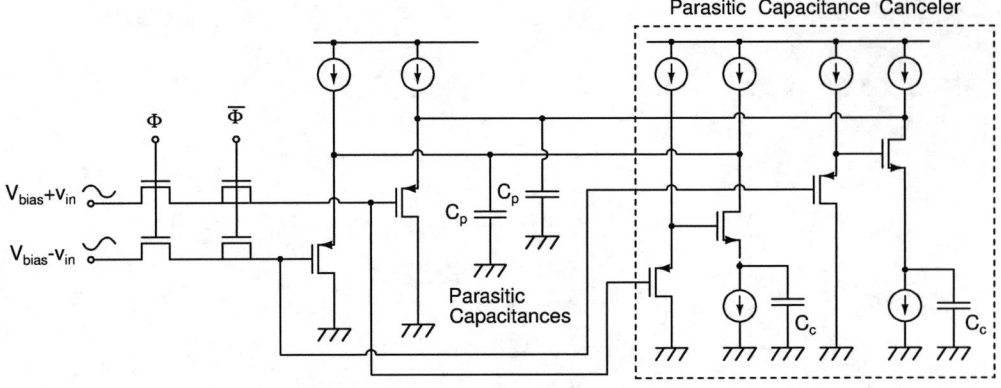

Fig. 7. T/H circuit using proposed cancellation method

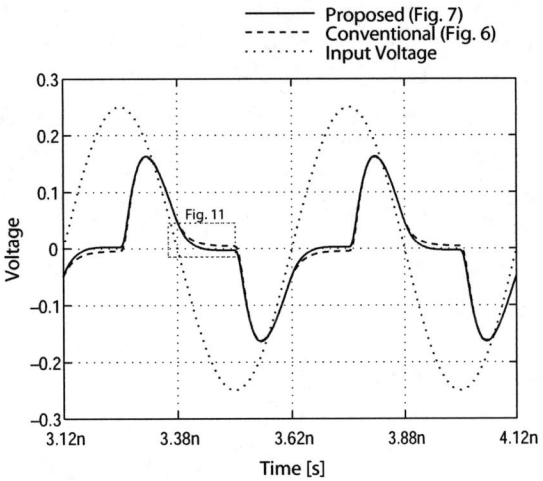

Fig. 10. Input and output characteristics of the T/H circuits when the chip area is 320 μm^2

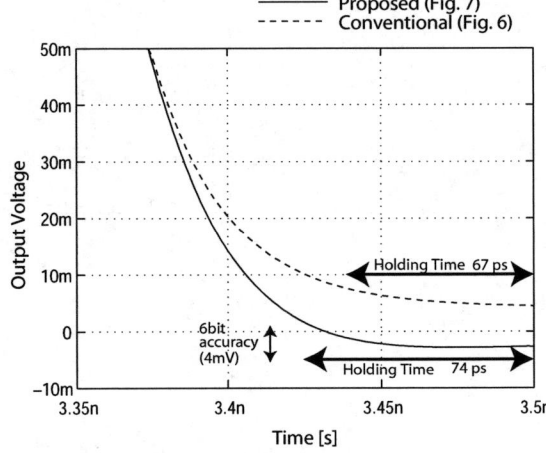

Fig. 11. Output voltage of T/H circuits in hold phase (Enlarged figure of Fig.10)

Table 2. Total chip area and power consumption of Fig.6 and Fig.7

	chip area	power dissipation
Fig.6	431 μm^2	57 mW
Fig.7	320 μm^2 (74 %)	34.1 mW (60 %)

5. CONCLUSION

This paper has proposed a novel parasitic cancellation method for a balanced-type circuit based on feedforwarding signals. A balanced-type circuit using this method is always stable because a feedback loop, which is cause of oscillation, has been removed. This method has been applied to an implementation of a T/H circuit. It has been confirmed by HSPICE simulations that 4-Gb/S and 6-bit accuracy performance is obtained by the proposed method. Thanks to the parasitic capacitance canceller the chip area and the total power consumption are reduced to 74 % and 60 % of those for the conventional T/H circuit respectively.

Chip implementation to verify the effectiveness of the proposed method has been left as a future work.

6. REFERENCES

[1] H. Hagiwara, et al. "A Monolithic Video Frequency Filter Using NIC-Based Gyrators," IEEE Journal of Solid-state Circuits, Vol. 23, No. 1, February 1988

[2] T. Sato, et al. "4GB/s Track and hold circuit using parasitic capacitance canceller," Proc. of European Solid-State Circuits Conference, pp.347–350, 2004

[3] L.Y.Nathawad, et al. "A 20 GHz Bandwidth, 4b Photoconductive-Sampling Time interleaved CMOS ADC," International Solid State Circuit Conference, Digest of Technical papers, pp. 320-321, 2003

Synthesis of Hybrid Filter Banks by Global Frequency Domain Least Square Solving

Tudor Petrescu, Jacques Oksman
École Supérieure d'Électricité
Department of Signal Processing
and Electronic Systems
91192, Gif sur Yvette, France
Email:firstname.lastname@supelec.fr

Pierre Duhamel
CNRS/LSS
École Supérieure d'Électricité
91192, Gif sur Yvette, France
Email: pierre.duhamel@lss.supelec.fr

Abstract—This paper introduces a simple method for designing Hybrid Filter Banks (HFB) by solving a linear equation system directly derived from the perfect reconstruction conditions. The design starts with the knowledge of the analysis filters. FIR synthesis filters result after simple linear algebra computations so that the output of the HFB is as close as possible to a sampled and delayed version of the HFB input. In the case that only a smaller part of the input signal potential bandwidth is of interest, it is shown that the use of weighting functions stressing a certain frequency interval considerably improves the performance of the HFB.

I. INTRODUCTION

Hybrid Filter Banks are a good option for implementing analog to digital (A/D) converters [1], [2]. Time-interleaved converters can be seen as special cases of HFB converters with delays as analysis filters. It is shown that gain and phase mismatch errors and timing errors introduced by time-interleaved structures are attenuated in the case of HFB A/D converters [2]. Different methods of designing HFB have been proposed. Starting from a Digital Filter Bank (DFB) prototype (which has digital analysis and digital synthesis filter banks) with perfect reconstruction and using a discrete to continuous time transform, Velazquez [2] transformed each digital filter of the analysis bank to a continuous-time correspondent. Oliaei [3] also started from a digital prototype and found the continuous-time analysis filters by simply putting $TH_m(j\Omega) = H_m(e^{j\omega})$ for $|\Omega T| < \pi$. In [4], [5], [6] or [7], starting from the knowledge of the analog filters, the synthesis filters are found after minimizing different error criteria. Pinheiro et al. present [8] a design algorithm based on the minimization of the noise energy, derived from a time domain analysis of the HFB. The method proposed in this paper starts with the knowledge of analog filters (for the sake of analog electronics feasibility) and directly finds the coefficients of the discrete synthesis filters by the least square resolution of a linear equations system resulting from the perfect reconstruction conditions.

A block diagram of an HFB is shown in Figure 1. $x(t)$ is the input analog signal. It is supposed to be band limited to π/T by some external filter. $H_m(s)$ are the continuous-time analysis filters and $F_m(z)$ are the discrete-time synthesis filters, with $m \in \{1, 2, ..., M\}$. $y(n)$ is the digital output of

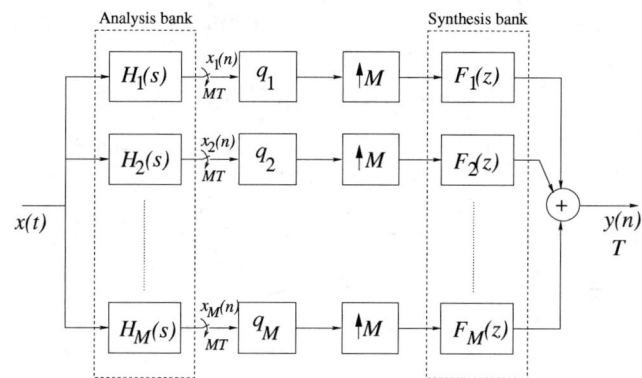

Fig. 1. Hybrid filter bank

the filter bank. The blocks $q_1, ..., q_M$ are the branch quantizers. After analog filtering, the branch signals $x_m(t), m \in \{1, 2, ..., M\}$ are sampled at a rate $2\pi/MT$. This is M times lower than the necessary Nyquist rate $2\pi/T$ and, since the filters' frequency characteristics are not perfectly band-limited, this undersampling generates aliasing. The Fourier transform of the output is [4], [9]:

$$Y(e^{j\omega}) = \sum_{p=-\infty}^{\infty} X(j\Omega - j\frac{2\pi p}{MT}) T_p(e^{j\omega}) \quad (1)$$

where

$$T_p(e^{j\omega}) = \frac{1}{MT} \sum_{m=1}^{M} F_m(e^{j\omega}) H_m(j\Omega - j\frac{2\pi p}{MT}) \quad (2)$$

$$\omega = \Omega T.$$

Since $x(t)$ is bandlimited to π/T, (1) can be rewritten as follows:

$$Y(e^{j\omega}) = \sum_{p=-(M-1)}^{M-1} X(j\Omega - j\frac{2\pi p}{MT}) T_p(e^{j\omega}) \quad (3)$$

because, for $-\pi < \omega \le \pi$, only $2M - 1$ terms have non zero contributions [9]. Moreover, it can be shown [10],[11] that for every particular Ω_0 with $|\Omega_0| < \pi/T$ the sum in (3) contains only M terms with non zero contributions. Perfect reconstruction means the output is only a scaled, delayed and sampled version of the input. Therefore, perfect reconstruction conditions will be [4], [9]:

$$T_p(e^{j\omega}) = \begin{cases} ce^{-j\omega d}, & p = 0 \\ 0, & p \in \mathcal{P}. \end{cases} \quad (4)$$

$d \in \mathbb{R}, d > 0$ is the filter bank's delay, $c \in \mathbb{R}$ is a scale factor and $\mathcal{P} = \{-(M-1), \ldots -1, 1, \ldots, M-1\}$ (the set of consecutive numbers between $-(M-1)$ and $M-1$ except 0). $T_p(e^{j\omega})$ for $p \in \mathcal{P}$ represent the aliasing functions whereas $T_0(e^{j\omega})$ is the distortion function. Section II-A introduces a new, direct method for HFB design. In Section II-B it is shown how the use of weighting functions (to stress a smaller frequency spectrum of interest) considerably improves HFB performance. Section III presents simulation results showing the validity of the presented method and section IV summarizes the results.

II. Synthesis method

Two cases will be studied. In the first case (Section II-A) the design will be done on the entire input signal frequency band. In the second case (Section II-B), the design will be done only on a part of the frequency band. In each case, the same synthesis method described below will be used.

A. Design of synthesis filters - case I

The proposed approach starts with the knowledge of the analog transfer functions.

Let the synthesis filters be N-length FIR filters. Therefore, their Fourier transforms will be:

$$F_m(e^{j\omega}) = \sum_{n=0}^{N-1} f_n^{(m)} e^{-j\omega n}, \quad m \in \{1, 2, \ldots, M\}. \quad (5)$$

FIR filters coefficients $f_n^{(m)}$, $m \in \{1, 2, \ldots, M\}$, $n \in \{0, 1, \ldots, N-1\}$ are to be computed. Perfect reconstruction conditions (4) are written for each of the K frequency values equally distributed in $-\pi < \omega \leq \pi$ interval, using (2) for $T_p(e^{j\omega})$ and (5) for $F_m(e^{j\omega})$:

$$T_p(e^{j\omega_k}) = \begin{cases} ce^{-j\omega_k d}, & p = 0 \\ 0, & p \in \mathcal{P} \end{cases} \quad (6)$$

$k \in \{1, 2, \ldots, K\}$, where:

$$T_p(e^{j\omega_k}) = \frac{1}{MT} \sum_{m=1}^{M} F_m(e^{j\omega_k}) H_m(j\Omega_k - j\frac{2\pi p}{MT}). \quad (7)$$

Equations (6) for a given ω_k and all $p \in \mathcal{P}$ are constituted of $2M-1$ complex expressions or, equivalently, $2 \times (2M-1)$ real equations. Hence, a linear system of $2 \times (2M-1) \times K$ real equations and $M \times N$ unknown variables results. For a matrix form of the above equation system, the definitions of different vectors follow:

$$\mathbf{e}_k = \begin{bmatrix} 1 & e^{-j\omega_k} & \cdots & e^{-j(N-1)\omega_k} \end{bmatrix}^T \quad (8)$$

$$\mathbf{E}_k = \mathbf{e}_k^T \mathbf{I}_N, \quad k \in \{1, 2, \ldots, K\} \quad (9)$$

$$\mathbf{E} = \begin{bmatrix} \mathbf{E}_1^T & \mathbf{E}_2^T & \cdots & \mathbf{E}_K^T \end{bmatrix}^T \quad (10)$$

where $\mathbf{A}^T$ denotes the transpose of a matrix $\mathbf{A}$ and $\mathbf{I}_N$ represents the identity matrix of order N. The FIR filter coefficients vector is:

$$\mathbf{f} = \begin{bmatrix} f_0^{(1)} & f_1^{(1)} & \cdots & f_{N-1}^{(1)} & \cdots & f_0^{(M)} & f_1^{(M)} & \cdots & f_{N-1}^{(M)} \end{bmatrix}^T. \quad (11)$$

The matrix of the shifted versions of the filter $H_m(j\Omega)$ is:

$$\mathbf{H}_m = \begin{bmatrix} \mathbf{H}_m^{(0)} \\ \mathbf{H}_m^{(1)} \\ \vdots \\ \mathbf{H}_m^{(M-1)} \\ \mathbf{H}_m^{(-1)} \\ \vdots \\ \mathbf{H}_m^{-(M-1)} \end{bmatrix}. \quad (12)$$

The matrix $\mathbf{H}_m^{(p)}$ may be seen in (25) for typographic constraints, and $H_m^{(p)}(j\Omega)$ is the $2\pi p/MT$ shifted version of $H_m(j\Omega)$:

$$H_m^{(p)}(j\Omega) = H_m(j\Omega - j\frac{2\pi p}{MT}). \quad (13)$$

Eventually, with the following five notations:

$$\mathbf{H}_{em} = \mathbf{H}_m \mathbf{E} \quad (14)$$

$$\mathbf{H}_c = \begin{bmatrix} \mathbf{H}_{e1} & \mathbf{H}_{e2} & \cdots & \mathbf{H}_{eM} \end{bmatrix} \quad (15)$$

$$\mathbf{t}_c = c \begin{bmatrix} e^{-j\omega_1 d} & e^{-j\omega_2 d} & \cdots & e^{-j\omega_K d} & \mathbf{0}_{(2M-2)K} \end{bmatrix}^T \quad (16)$$

$$\mathbf{H} = \begin{bmatrix} \operatorname{Re}\{\mathbf{H}_c\} \\ \operatorname{Im}\{\mathbf{H}_c\} \end{bmatrix} \quad (17)$$

and

$$\mathbf{t} = \begin{bmatrix} \operatorname{Re}\{\mathbf{t}_c\} \\ \operatorname{Im}\{\mathbf{t}_c\} \end{bmatrix} \quad (18)$$

the matrix form of the perfect reconstruction conditions written for the given set of frequency values is:

$$\mathbf{H}\mathbf{f} = \mathbf{t}. \quad (19)$$

In (16), $\mathbf{0}_{(2M-2)K}$ is a 1 by $(2M-2)K$ vector filled with zeros. $\operatorname{Re}\{\mathbf{A}\}$ and $\operatorname{Im}\{\mathbf{A}\}$ denote the real and the imaginary part of a matrix $\mathbf{A}$ respectively. In the general case, for a reasonable FIR filter length, $2 \times (2M-1) \times K > M \times N$ so the system in (19) is overdetermined and inconsistent. However, a least square solution can be found [12]:

$$\mathbf{f} = (\mathbf{H}^T \mathbf{H})^{-1} \mathbf{H}^T \mathbf{t}. \quad (20)$$

This solution minimizes the square sum of the error vector (the error vector's Euclidean norm) $\mathbf{\Delta} = \|\mathbf{H}\mathbf{f} - \mathbf{t}\|^2$.

B. Design of synthesis filters - case II

Analog filters cannot be band limited. Supposing that the input signal is band limited to π/T the HFB design procedure may lead to discontinuities of the synthesis filters frequency responses which are not reasonably implementable as finite length FIR filters. That is the reason why we chose to slightly reduce the input band of interest (i.e. slightly oversample) in order to ease the synthesis filter design by means of a weighting function relaxing the constraints near the discontinuity frequencies:

$$W(j\Omega) = \begin{cases} r_1, -\eta\frac{\pi}{T} < \Omega < \eta\frac{\pi}{T} \\ r_2, \text{otherwise} \end{cases} \quad (21)$$

and $r_1, r_2 \in \mathbb{R}, r_1, r_2 > 0$ and $r_1 \gg r_2$. Perfect reconstruction conditions are then rewritten for $-\pi < \omega < \pi$:

$$T_p^W(e^{j\omega}) = \begin{cases} ce^{-j\omega d}W(j\Omega), p = 0 \\ 0, p \in \mathcal{P} \end{cases} \quad (22)$$

where

$$T_p^W(e^{j\omega}) = \frac{1}{MT}\sum_{m=1}^{M} F_m(e^{j\omega})H_m(j\Omega - j\frac{2\pi p}{MT}) \cdot W(j\Omega - j\frac{2\pi p}{MT}). \quad (23)$$

If (13) is rewritten as follows:

$$H_m^{(p)}(j\Omega) = H_m(j\Omega - j\frac{2\pi p}{MT})W(j\Omega - j\frac{2\pi p}{MT}) \quad (24)$$

all the matrices and the synthesis method exposed in Section II-A remain unchanged. In the next section it will be shown that much better results are obtained when using this method.

III. SIMULATION RESULTS

An eight channel HFB was designed. High frequency implementation is only feasible for very simple structures built upon integrated LC circuits or SAW devices [6], [7]. That is why, basic resonators were used as analysis filters. Their transfer functions are:

$$H_m(s) = \frac{\frac{\Omega_m}{Q_m}s}{s^2 + \frac{\Omega_m}{Q_m}s + \Omega_m^2}, m \in \{2, ..., M\}. \quad (25)$$

This corresponds to a parallel LC structure with

$$\Omega_m = \frac{1}{\sqrt{L_m C_m}}, \quad Q_m = \frac{R_{pm}}{\Omega_m L_m}.$$

R_{pm} includes the parasitic resistance of the inductance L_m. For $H_1(s)$, low pass filters are usually considered. Simple RC circuits are used:

$$H_1(s) = \frac{\Omega_1}{s + \Omega_1}, \quad \Omega_1 = \frac{1}{R_1 C_1}. \quad (26)$$

The resonators were chosen with equal bandwidths. Eight $N = 128$-length FIR filters are then computed from equation (20). In Case I, the average aliasing is below -46 dB and peak aliasing is below -21 dB, but within 96% of the input signal bandwidth, the peak aliasing is below -40 dB. Peak distortion is of 2.54 dB and average distortion is 0.01 dB. In 96% of the input signal bandwidth, peak distortion is 0.12 dB and average distortion is 0.001 dB. Figure 2 shows the magnitude of the distortion function in Case I and *Phase difference* stands for the phase of the ratio

$$R = \frac{T_0(e^{j\omega})}{ce^{-j\omega d}} \quad (26)$$

with $d = 64$. Ideally, for an HFB with no phase distortion, this phase difference must be either 0 (if d is the system delay) or constant (if d is an arbitrary constant). The average deviation from linear phase is 0.0078 radians in Case I. For Case II, FIR filters coefficients were computed again for the same analysis bank as in Case I. Equation (20) was solved for a band reduction factor $\eta = 0.9375$ (for 93.75% of the initial input signal bandwidth). The performances are tremendously improved: the peak aliasing being below -112 dB, and the mean aliasing below -120 dB. The mean distortion is $5 \cdot 10^{-8}$ dB and peak distortion is $1.3 \cdot 10^{-5}$ dB. The mean deviation from linear phase in Case II is only $6.9 \cdot 10^{-7}$ radians. Table I summarizes the results for the magnitude of the distortion function and for the phase difference (the phase of the ratio in (26)). In Table I, the mean distortion and the peak distortion for Case I are given for 96 percent of the input signal frequency bandwidth. Since the mean deviation from linear phase (Mean Phase Difference) in Case II is of $6.9 \cdot 10^{-7}$ radians, the filter bank has virtually constant group delay of 64 samples, because $d = 64$. Figure 3 shows, as an example, the magnitude of the first aliasing function in both Cases I and II. All the other aliasing functions are similar. The aliasing energy must be smaller than the quantization noise energy so that aliasing does not affect the overall ADC resolution. Considering a 6 dB/quantization bit scheme the HFB obtained in Case II

$$\mathbf{H}_m^{(p)} = \begin{bmatrix} \underbrace{H_m^{(p)}(j\Omega_1) \cdots H_m^{(p)}(j\Omega_1)}_{N} & \underbrace{0 \cdots 0}_{N} & \underbrace{0 \cdots 0}_{N} \cdots & \underbrace{0 \cdots 0}_{N} \\ \underbrace{0 \cdots 0}_{N} & \underbrace{H_m^{(p)}(j\Omega_2) \cdots H_m^{(p)}(j\Omega_2)}_{N} & \underbrace{0 \cdots 0}_{N} \cdots & \underbrace{0 \cdots 0}_{N} \\ \vdots & \vdots & \vdots & \ddots & \vdots \\ \underbrace{0 \cdots 0}_{N} & \underbrace{0 \cdots 0}_{N} & \underbrace{0 \cdots 0}_{N} \cdots & \underbrace{H_m^{(p)}(j\Omega_K) \cdots H_m^{(p)}(j\Omega_K)}_{N} \end{bmatrix} \quad (25)$$

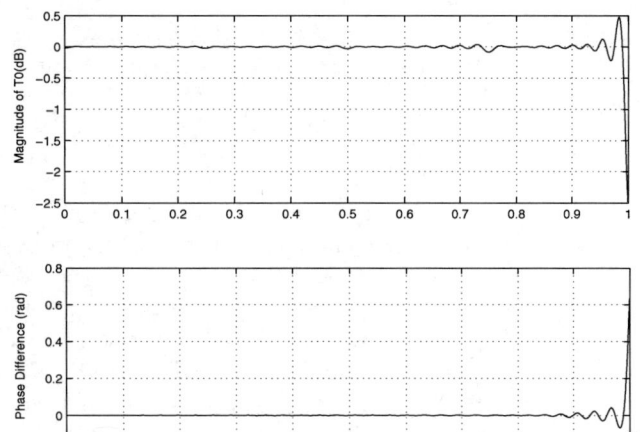

Fig. 2. Magnitude and the phase of the ratio $R = \frac{T_0(e^{j\omega})}{ce^{-j\omega d}}$ (Phase Difference) for the distortion function in Case I - Synthesis on the entire input signal potential bandwidth

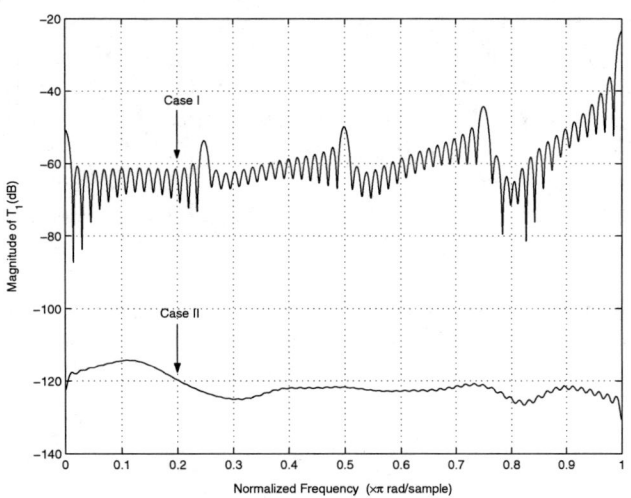

Fig. 3. Magnitude of the first aliasing function for the two cases: Case I - Synthesis on the entire input signal potential bandwidth, Case II - Synthesis on 93.75 percent of the input signal potential bandwidth

TABLE I
COMPARISON OF METHODS

	Mean Distortion (Magnitude) (dB)	Mean Phase Difference (rad)	Peak Distortion (Magnitude) (dB)	Peak Phase Difference (rad)
Case I	0.001	0.0078	0.12	0.634
Case II	$5 \cdot 10^{-8}$	$6.9 \cdot 10^{-7}$	$1.3 \cdot 10^{-5}$	$3.6 \cdot 10^{-6}$

could be used for an 18 bits ADC converter. Using really simple analog filters, as is done in this work, is a requirement for real-life implementation. An open question remains about the interest of more complex structures. Higher order analysis filters lead to reducing ADC mismatch-induced distortion [2] but, it may be supposed that the higher the order of analog filters, the higher the errors caused by the limited precision of the analog implementation.

IV. CONCLUSIONS

A new method for designing HFB was proposed. It consists in searching the least squares solution of an overdetermined linear equation system resulting directly from the perfect reconstruction conditions. Tests were performed in order to prove the validity of the method. If the design method is applied only for a smaller part of the input signal bandwidth by use of weighting functions (in case the input signal's energy lies in a smaller frequency band) the HFB performance is improved. From about -40 dB of peak aliasing in the initial case, -112 dB of peak aliasing were obtained when synthesizing on 93.75 percent of the initial signal bandwidth. The peak distortion was improved from 2.54 dB to $1.3 \cdot 10^{-5}$ dB. The average deviation from a linear phase is 0.0078 radians in the Case I and $6.9 \cdot 10^{-7}$ radians in the Case II. These figures show that wide band, high frequency, high resolution HFB ADCs are possible.

REFERENCES

[1] A. Petraglia and S.K. Mitra, "High speed A/D conversion incorporating a QMF bank," *IEEE Transactions on Instrumentation and Measurement*, vol. 41, pp. 427–431, June 1992.

[2] S. R. Velazquez, T. Q. Nguyen, and S. R. Broadstone, "Design of hybrid filter banks for analog/digital conversion," *IEEE Transactions on Signal Processing*, vol. 46, no. 4, pp. 956–967, April 1998.

[3] O. Oliaei, "Asymptotically perfect reconstruction in hybrid filter banks," in *Proceedings of IEEE International Conference on Acoustics, Speech, and Signal Processing*, May 1998, vol. 3, pp. 1829–1832.

[4] S. R. Velazquez, T. Q. Nguyen, S. R. Broadstone, and J. K.Roberge, "A hybrid filter bank approach to analog to digital conversion," in *Proceedings of IEEE International Symposium on Time-Frequency and Time-Scale Analysis*, October 1994, pp. 116–119.

[5] P. Löwenborg, H. Johansson, and L. Wanhammar, "A class of two-channel hybrid analog/digital filter banks," in *Proceedings of IEEE Midwest Symposium on Circuits and Systems*, August 1999, vol. 1, pp. 14–17.

[6] T. Petrescu and J. Oksman, "Synthesis of hybrid filter banks for A/D conversion with implementation constraints -optimized frequency response approach-," in *Proceedings of IEEE Midwest Symposium for Circuits and Systems*, December 2003.

[7] T. Petrescu, C. Lelandais-Perrault, and J. Oksman, "Synthesis of hybrid filter banks for A/D conversion with implementation constraints - mixed distortion/aliasing optimization -," in *Proceedings of IEEE International Conference on Acoustics, Speech, and Signal Processing*, May 2004, vol. 2, pp. 997–1000.

[8] Pinheiro M.A.A., Batalheiro P.B., Petraglia A., and Petraglia M.R., "Improving the near-perfect hybrid filter bank performance in the presence of realization errors," in *Proceedings of IEEE-SP International Conference on Acoustics, Speech, and Signal Processing*, May 2001, vol. 2, pp. 1069–1072.

[9] P. Löwenborg, H. Johansson, and L. Wanhammar, "On the frequency response of M-channel mixed analog and digital maximally decimated filter banks," in *Proceedings of European Conference on Circuit Theory and Design*, September 1999, vol. 1, pp. 321–324.

[10] P. P. Vaidyanathan and V.C. Liu, "Classical sampling theorems in the context of multirate and polyphase digital filter bank structures," *IEEE Transactions on Acoustics, Speech and Signal Processing*, vol. 36, no. 9, pp. 1480–1495, September 1998.

[11] J.L. Brown Jr., "Multi-channel sampling of low-pass signals," *IEEE Transactions on Circuits and Systems*, vol. CAS-28, no. 2, pp. 101–106, February 1981.

[12] G. Strang, *Linear algebra and its applications*, Academic Press, Orlando, second edition, 1980.

Implementation of a Novel Read-Out Strategy Based on a Wilkinson ADC for a 16x16 Pixel X-Ray Detector Array

V. Ferragina, P. Malcovati, F. Borghetti, A. Rossini
Dept. of Electrical Engineering, University of Pavia, Italy
{vincenzo.ferragina, piero.malcovati, fausto.borghetti, andrea.rossini}@unipv.it

F. Ferrari, N. Ratti
Alenia Spazio – Laben
Vimodrone (MI), Italy
{ferrari.f, ratti.n}@laben.it

G. Bertuccio
Dept. of Electronics
Polytechnic of Milano, Italy
giuseppe.bertuccio@polimi.it

Abstract— In this paper we propose a Wilkinson type A/D converter as well as all the digital logic required for reading-out a 16x16 array of X-ray detectors. The proposed A/D converter architecture and read-out strategy allows us to handle an event rate as large as 10^6 event/s over the whole array and 10^4 event/s over the single row of the array with a resolution of 10 bits, consuming only 77 mW from a 3.3 V power supply. The A/D converter and the logic are embedded in an ASIC to be bump-bonded on top of the detector, which includes also the front-end electronics required for processing the sensor output signals. This work was done in the framework of an ESA research activity.

I. INTRODUCTION

A large variety of scientific and industrial applications nowadays require X and gamma ray imaging and spectroscopic detectors. This has led to a significant effort aimed to the development of the related readout electronics [1, 2, 3]. One of the most challenging goals in this kind of applications is the implementation of spectroscopy-grade front-end analog-digital circuits for large format pixel arrays, due to constraints in the area and power consumption available for each channel. A solution for reading-out such pixel arrays is to bump-bond an application specific integrated circuit (ASIC) onto the detector, matching the pixel size with the front-end channel. This, however, is not sufficient since the area available for each channel is barely enough for implementing the conventional signal processing functions required to transform the charge packets produced by the detectors into voltage pulses. Therefore, in all the ASICs published so far the A/D conversion of the detector output has been performed with a separate integrated circuit or directly in the remote acquisition system. However, for the next generation of detectors, due to a larger number of pixels and more stringent constraints on the power consumption, the A/D conversion of the detector output must be performed on-chip, thus requiring specific circuit techniques to achieve a conversion rate sufficient to read-out all the excited channels of the array for a given event rate with low power consumption. Basically there are two possible solutions for this problem. The first is the use of a small number (possibly one) of A/D converters operated at very high speed, which can serve the whole pixel array. The second is the use of large number of slow and simple A/D converters, each serving a small subset of the pixel array. The most promising A/D converter architecture for implementing the latter solution is the Wilkinson type A/D converter [4], since it guarantees linearity performances suitable for spectroscopic applications (±0.03% of the full scale).

In this paper we present such an A/D converter as well as all the digital logic required for reading out the output of a 16x16 array of X-ray detectors. The A/D converter achieves 10 bits of resolution and allows the detector array to operate with an event rate as large as 10^6 event/s (10^4 event/s over a single row of the array), consuming only 77 mW from a 3.3 V power supply.

II. READ-OUT ASIC ARCHITECTURE

The block diagram of the whole ASIC where the proposed converter is embedded is shown in Fig. 1. The ASIC can be divided in three main parts: the front-end electronics (containing one read-out cell per pixel), the back-end electronics and the auxiliary services.

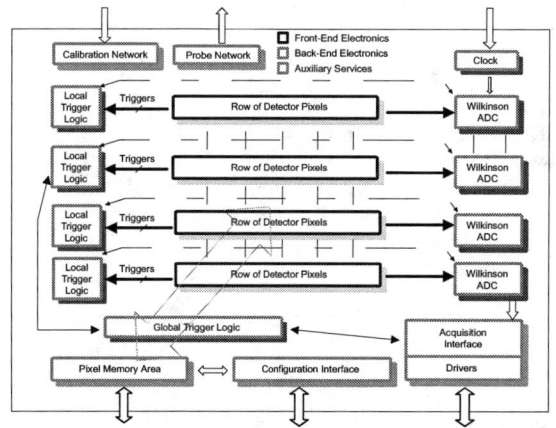

Figure 1. Block diagram of the complete ASIC.

The back-end section of the ASIC performs three main functions: processing the trigger signals produced by the front-end section in order to providing to the external acquisition system the information that some events have been detected; performing the A/D conversion of the analog data produced by the front-end section; delivering outside the chip the digital converted energy value and the location of the occurred events. When a front-end cell asserts its trigger output, the back-end section schedules the analog output of the pixel for conversion on the A/D converter. At the same time, it starts the handshake with the external controller to deliver the converted data. Until the acquisition system acknowledge, the ASIC continues to acquire events from the detector and stores them on then analog output memory of each pixel. Two different modes of operation are foreseen, a slow event rate mode and a fast event rate mode. In slow event rate mode the acquisition of new events is inhibited by disabling the front-end electronics during A/D conversion, while in fast event rate mode the acquisition continues during A/D conversion. Upon acknowledgement by the acquisition system, the ASIC starts to convert and transfer the data (in slow event rate mode the acquisition is stopped at this point). The functions performed by the back-end section of the ASIC are reported in detail in the following paragraphs.

A. Trigger acquisition and processing

The Trigger signals produced by the matrix of front-end circuits (RPC) have to be acquired by the back-end section of the ASIC and processed in order to produce a global trigger signal to be delivered to the external acquisition interface. Moreover, the back-end section of the ASIC has to produce the address of each triggered pixel (row and column) and schedule the A/D conversion of the analog output signals of the triggered RPCs in the correct sequence through a suitable priority logic. FIFO registers have to be included in order to de-randomize the events arriving with a Poisson distribution. The address and energy value associated with each triggered pixel have finally to be delivered to the external acquisition interface.

B. A/D conversion

The A/D conversion of the analog output of each RPC is performed by means of a Wilkinson A/D converter with 10-bits of resolution. In order to reach the best compromise between complexity (and hence power consumption) and event rate, we decided to use one A/D converter channel for each pixel row. In the considered case, therefore, 16 channels are implemented on the ASIC.

C. Interaction with the external acquisition system

The acquired science data, once processed, can be delivered to the external world through a dedicated interface (the acquisition interface) which is based on a double handshake mechanism that allows us to comply with the different peculiarities of the two different mode of operation of the ASIC.

local trigger logic (LTL, Fig. 3), that has in charge the prioritization of the channel processing inside the group itself and the A/D converter dedicated to the group. These local logics are managed and supervised by the global trigger logic (GTL, Fig. 4) that provides a prioritization scheme for "scanning" all the LTLs and schedules through a FIFO register the A/D conversions and the transmission of the results to the acquisition interface.

A clock generation block is needed because the ADC precision is directly related to the input clock precision in terms of frequency stability and jitter. An external clock source will feed the ASIC with a stable differential low voltage swing and sinusoidal clock that will be amplified on-chip into a sharp rail-to-rail clock. This solution should reduce the contribution of the clock to the substrate noise. The clock will be converted in proximity of the region where it is really needed.

The operation of the system is the following:
1) When an event occurs in a pixel, the front-end electronics rises the trigger (*Triggers*) signal.
2) The local trigger logic (LTL) detects the trigger signal and rises the *LocalTrigger* line. In the meantime, the LTL decodes the address of the triggered pixel (if more pixels belonging to the same LTL are triggered, the address of the highest priority pixel is decoded).
3) The global trigger logic (GTL) detects the *LocalTrigger* signal (or signals if triggers arrive from different LTLs) and produces a *GlobalTrigger* signal.
4) The acquisition interface receives the *GlobalTrigger* signal, starts the handshaking with the external world and eventually rises the *TriggerAck* line.
5) The GTL detects the *TriggerAck* and rises the *Convert* line for the LTLs having the *LocalTrigger* signal active. In slow acquisition mode the front-end electronics of the whole pixel matrix is disabled.
6) The LTLs that receive the *Convert* signal address the triggered pixel (*Address*) and rise the start of conversion (*SOC*) line for the corresponding A/D converter.
7) The A/D converters that receive the *SOC* signal start the conversion. When the conversion finishes each A/D converter rises the end-of-conversion (*EOC*) line and delivers the data (*Data*).

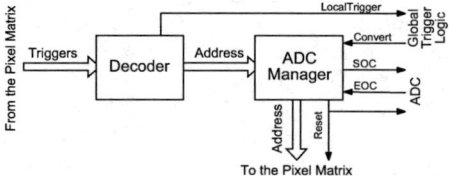

Figure 3. Block diagram of the local trigger logic.

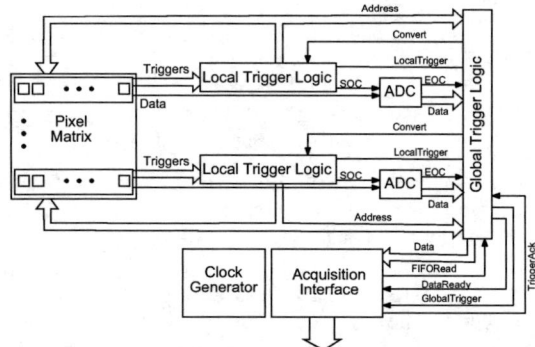

Figure 2. Block diagram of the back-end section of the ASIC.

III. BACK-END ASIC ARCHITECTURE

The simplified block diagram of the ASIC back-end section is shown in Fig. 2. Each row of RPC in the ASIC is connected to a local processing structure, which include the

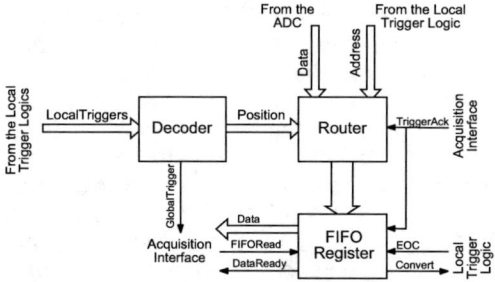

Figure 4. Block diagram of the global trigger logic.

8) The GTL detects the *EOC* signals, stores the converted data and the corresponding pixel address of each A/D converter in the FIFO register and rises the *DataReady* line. If the FIFO is full, the GTL produces an error signal. The acquisition interface accesses the FIFO register through the *FIFORead* signal delivering the data to the output pins. When an element of the FIFO has been read by the acquisition interface, it becomes free again for new data. When the FIFO is empty the *DataReady* signal is lowered.
9) The LTL detects the *EOC* signal and resets the addressed pixel (*Reset*).
10) The operations from 2 to 9 are repeated until no more trigger signals are present.

IV. LOCAL TRIGGER LOGIC

The local trigger logic, whose block diagram is shown in Fig. 3, consists of decoder block and an ADC manager block. The decoder block receives the trigger signals from its associated group of pixels, identifies the first signal which is active (the signal with highest priority), determines its address and produces the *LocalTrigger* signal. The ADC manager block receives the *Convert* signal form the global trigger logic and produces the *SOC* signal for the corresponding A/D converter as well as the address for connecting the analog output of the triggered pixel to the A/D converter. When the ADC manager receives the *EOC* signal from the A/D converter, rises the *Reset* signal for resetting the addressed pixel of the front-end electronics as well as the A/D converter itself.

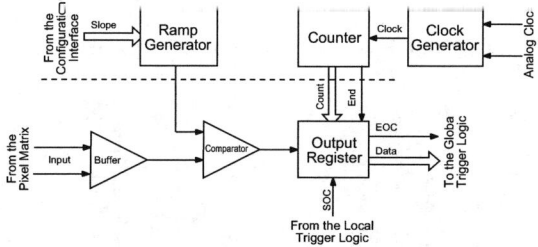

Figure 5. Block diagram of the Wilkinson A/D converter.

V. GLOBAL TRIGGER LOGIC

The block diagram of the global trigger logic is shown in Fig. 4. The circuit consists of a FIFO register, a decoder block and a router block. The decoder block receives the *LocalTrigger* signals from the different local trigger logics, identifies the first signal which is active (the signal with highest priority), determines its *Position* (which represents also the row address of the considered pixel if the pixels are grouped by row) and produces the *GlobalTrigger* signal. The router block, when a *TriggerAck* signal is received, assigns the first free element of the FIFO register to the considered group of pixels (identified by their *Position*). When the *EOC* signal arrives the data and address of the considered pixel are hence written in the FIFO register. The FIFO register, besides storing the converted energy value (*Data*) and the *Address* of the triggered pixels, receives the *TriggerAck* signal and, if there is an empty element, produces the *Convert* signal for the addressed group of cells. When the FIFO contains data the *DataReady* signal is active. The *FIFORead* signal is used to read-out the FIFO register.

VI. WILKINSON ADC

The block diagram of the Wilkinson A/D converter is shown in Fig. 5. The circuit consists of an input buffer, a comparator, an output register, a ramp generator a counter and a clock generator. The ramp generator, the counter and the clock generator are common to all of the A/D converters, while there is an input buffer, an output register and a comparator associated with each local trigger logic.

The comparator compares the pixel output signal, which is a constant voltage proportional to the incident photon energy, with the ramp produced by the ramp generator. When the output register is enabled by the *SOC* signal received from the local trigger logic and the comparator changes state because the ramp crossed the input signal, the content of the counter is copied in the output register. When the counter reaches full scale, the output register produces the *EOC* signal. With a clock frequency of 50 MHz, the Wilkinson A/D converter requires 20 µs of conversion time to achieve 10 bits of resolution (42 µs in the worst case, considering also 2 µs of dead-time between two consecutive ramps). The slope of the ramp generator is controlled with 7 bits (from 0.92 V/ms to 116 V/ms, nominal value 58 V/ms), in order to guarantee the required gain in spite of any process variations, as shown in Fig. 6.

The schematic of the comparator is shown in Fig. 7. The circuit consists of a differential pair with mirrored load followed by an inverter. A regenerative loop (MP2 and MP3) has been introduced to speed up the response of the comparator and to introduce a few millivolts of hysteresis. The operational amplifier used both in the ramp generator and in the buffer is based on a folded-cascode structure with nMOS input pair followed by a source follower to guarantee low output impedance.

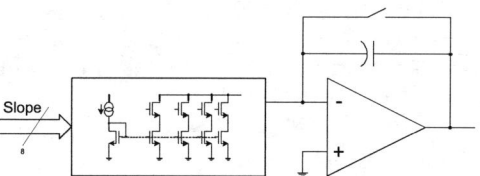

Figure 6. Schematic of the ramp generator

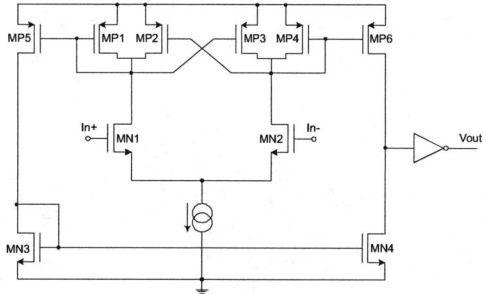

Figure 7. Schematic of the comparator

VII. SIMULATION RESULTS

The proposed Wilkinson A/D converter, as well as all the digital logic has been implemented using a 0.35 µm CMOS technology and simulated at transistor level. The achieved conversion time (Fig. 8) ranges from 20 µs to 42 µs, depending on the amplitude of the input signal and on the timing of the *SOC* signal with respect to the ramp. In fact, to avoid conversion errors due to the asynchronicity of the ramp and the *SOC* signal, we always wait the beginning of the first ramp after the rising edge of the *SOC* signal for starting the actual conversion cycle (two complete ramps are required in the worst case to accomplish the conversion). The non-linearity error of the ramp is reported as a function of time in Fig. 9. The circuit is extremely linear (below 0.1 LSB at 10 bits of resolution or 0.01% of the full-scale), thus allowing to achieve an excellent linearity in the whole A/D converter. The achieved differential (DNL) and integral (INL) non-linearity of the A/D converter, obtained with 2048 simulation points are shown in Fig. 10 and Fig. 11, respectively. Both parameters are always bound between − 0.5 LSB and +0.5 LSB as expected from a 10-bit A/D converter with the considered number of simulation points (the quantization error of the simulation with 2048 points at 10 bits of resolution is ±0.5 LSB). The total power consumption of the 16 A/D converters, required by the actual topology, operated from a 3.3V power supply, including the digital logic, is 77 mW, of which 13 mW are independent of the number of converters, while 64 mW are proportional to the number of converters. Therefore, the additional power consumption for any additional converter (e.g. for larger pixel arrays) is only 4 mW.

VIII. CONCLUSIONS

In this paper we presented a Wilkinson type A/D converter embedded in an ASIC used for reading-out a 16x16 array of X-ray detectors. The results of the simulations on the whole ASIC confirmed the validity of the proposed A/D converter architecture for the considered application. Indeed, the A/D converter allows us to handle an event rate as large as 10^6 event/s over the whole array and 10^4 event/s over the single row of the array with a resolution of 10 bits, consuming only 77 mW from a 3.3 V power supply. The ASIC has been integrated in a 0.35 µm CMOS technology. The measurements on the chip are in progress.

ACKNOWLEDGMENTS

The authors would like to acknowledge the European Space Agency for supporting this activity and in particular Dr. D. Martin for his precious suggestions and help.

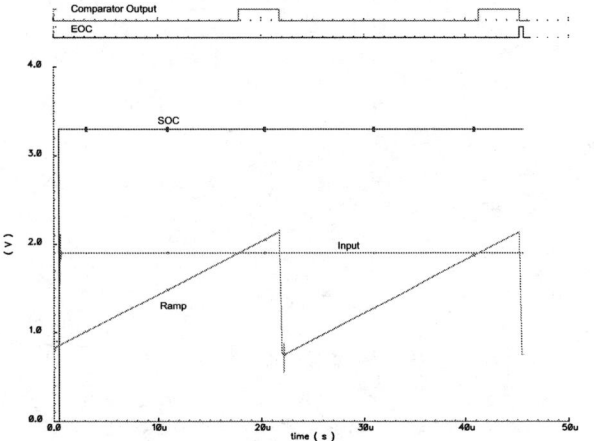

Figure 8. Simulated conversion cycle of the Wilkinson A/D converter.

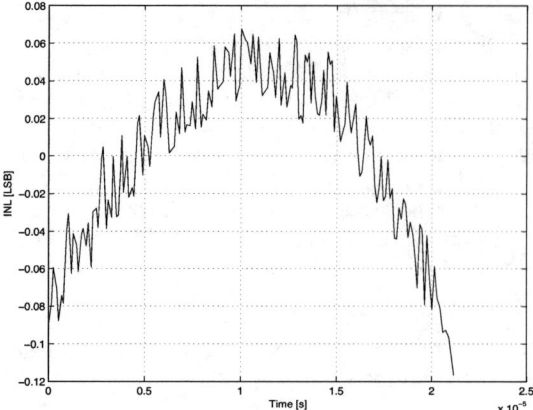

Figure 9. Non-linearity error of the ramp.

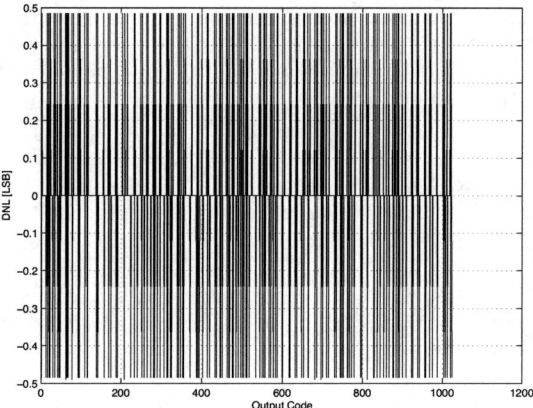

Figure 10. DNL of the Wilkinson A/D convereter.

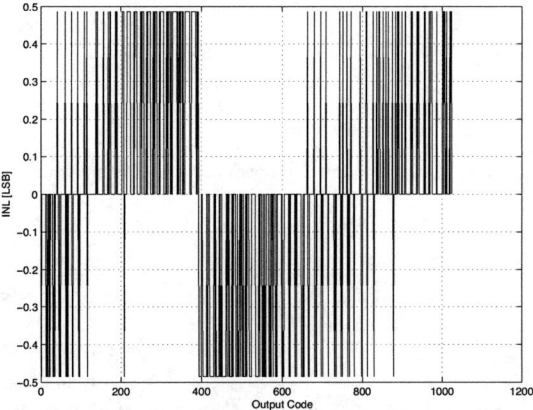

Figure 11. INL of the Wilkinson A/D converter.

REFERENCES

[1] Proceeding of the 11th International Workshop on Room Temperature Semiconductor X and Gamma Ray Detectors and Associated Electronics, in R. James, P. Siffert ed., Nuclear Instruments and Methods A, vol. 458, Elsevier Science, 2001.
[2] M. Prydderch, P. Seller, "A 16 channel analogue sparse readout IC for Integral", IEEE Nucl. Sc. Symp. and Med. Imag. Conf., 1, pp. 65-68, 1994.
[3] K. Tsukada, H. Ikeda, T. Kamae, T. Takahashi, H. Murakami, "Peak hold monolithic integrated circuit with built-in shaping amplifier for hard X-ray detector", IEEE Trans. on Nucl. Sc., 40, pp. 724-728, 1993.
[4] M. Emery, S. Frank, C. Britton Jr., A. Wintenberg, M. Simpson, M. Ericson, G. Young, L. Clonts, M. Allen, "A multi-channel ADC for use in the Phenix detector", IEEE Trans. on Nucl. Sc., 44, pp. 374-378, 1997.

A 12-B 10-MSAMPLES/S CMOS SWITCHED-CURRENT DELTA-SIGMA MODULATOR

Guo-Ming Sung
Department of Electrical Engineering,
National Taipei University of Technology,
1, Sec. 3, Chung-Hsiao E. Rd., Taipei 106, Taiwan
Email: gmsung@ntut.edu.tw

Kuo-Hsuan Chang
Institute of Mechatronic Engineering,
National Taipei University of Technology,
1, Sec. 3, Chung-Hsiao E. Rd., Taipei 106, Taiwan

Wen-Sheng Lin
Department of Electrical Engineering,
National Taipei University of Technology,
1, Sec. 3, Chung-Hsiao E. Rd., Taipei 106, Taiwan

Abstract—This paper presents a design of fully differential second-order delta-sigma modulator. A current feedback technique is used in the proposed switched-current feedback memory cell (FMC) to decrease the input impedance and to improve the transmission error in memory cell. Furthermore, the entire memory cell is designed in a coupled differential replicate (CDR) form to eliminate the clock feedthrough (CFT) error. In this paper, the SDM is simulated with TSMC 0.35μm CMOS process technology. And that, the simulation results reveal that the peak signal to noise plus distortion ratio (SNDR) is 75 dB at 10.24MHz sampling rate with 40kHz bandwidth, and the power dissipation is 16mW.

Key word: delta-sigma modulator, switched-current, memory cell, clock feedthrough error.

I. INTRODUCTION

In recent years, the trend on circuit design shows that the digital CMOS process technique is the first choice for commercial application, but there is an analog signal in nature. It is necessary to develop a transformed circuit for the digital application. The sampling data technique is a good transformed circuit. This technique includes either switched-capacitor (SC) or switched-current (SI). Among those, SI technique is a good choice to build a sampling data system than that of SC technique especially in digital CMOS process. The reason is that the needful capacitor of SC circuit will expend more area and more high-end process to improve the linearity of SC technique [1].

Even though the SI technique is superior to the SC technique on cost down, there are several non-ideal effects must be pointed out in SI technique. The first effect comes from the mismatch at fabrication. This effect can be improved by process technology or by enlarging the area of transistors. The next one is the charge feedthrough error induced by charge injection. And the third effect is transmission error. Those effects will cause a lot of sampling errors and seriously constrict the system performance. That is, the induced errors will move the poles of the noise transfer function out of the unit circle at the z-domain and then make the system unstable [2].

Up to now, some novel techniques have been developed to overcome those errors in SI circuit. Firstly, they use the dummy cell, including dummy switch and dummy circuit, to reduce the CFT error [3]. Secondly, a modified cell, which includes the regulator cascode circuit or S²I, is proposed to improve the transmission error [1]. Unfortunately, those modified techniques made the circuit design more complex. In this paper, we propose a simple feedback switched-current memory cell (FMC) to improve the transmission error without additional cascode transistors or switches. Besides, we propose a coupled differential replicate (CDR) circuit to minimize the non-ideal effect from CFT and to increase the resolution of the second-order Δ-Σ modulator.

II. MODULATOR STRUCTURE

The building block of the second-order switched-current delta-sigma modulator is shown in Fig. 1. It consists of two discrete time integrators, a 1-bit current quantizer, two 1-bit current DACs, and a RS flip-flop. Each integrator is implemented with two memory cells, named as SH1 and SH2, and with a positive feedback loop. Note that SH1 and SH2 must operate with non-overlapped clocks, $\phi 1$ and $\phi 2$.

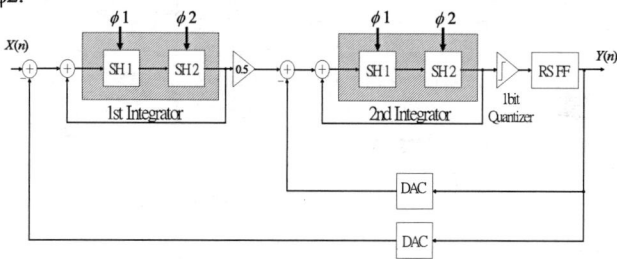

Fig. 1 Building block of the second-order SI Δ-Σ modulator.

III. PROPOSED SI MEMORY CELL

In SI circuit, both the transmission error and CFT error are the main non-ideal effects. Fig. 2 shows the simple dynamic current transmission model of the SI memory cell and its small signal equivalent circuit [2].

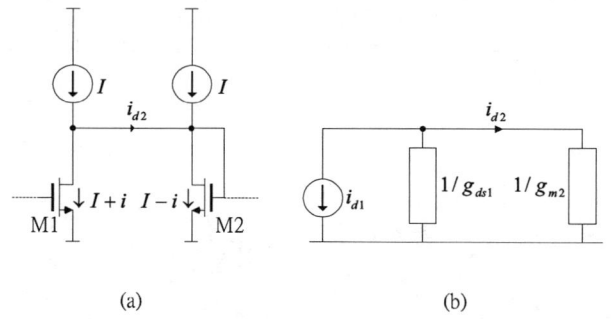

Fig. 2 (a) the simple dynamic current transmission model of the SI memory cell, and (b) its small signal equivalent circuit.

From the small signal equivalent circuit, as shown in Fig. 2(b), the drain current ratio between M_1 and M_2 is given by

$$\frac{i_{d2}}{i_{d1}} = -\frac{g_{m2}}{g_{m2}+g_{ds1}} = -\frac{1}{1+g_{ds1}/g_{m2}} = -\frac{1}{1+\varepsilon} \quad (1)$$

where ε is the transmission error which is defined as the ratio of input conductance g_{ds1} to output conductance g_{m2}. In order to reduce the transmission error, we have to increase the output impedance ($1/g_{ds1}$) or to decrease the input impedance ($1/g_{m2}$).

A traditional first generation (FG) SI memory cell is shown in Fig. 3 [4], where α is the width-to-length ratio (W/L) and ϕ is the switch. In this circuit, the input impedance is

$$r_{in} = \frac{1}{g_{m1}} \quad (2)$$

As equation (2) mentioned, the input impedance is too high for SI memory cell.

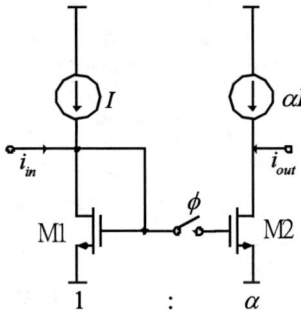

Fig. 3 The traditional FG SI memory cell.

One technique can be used to decrease the input impedance of SI memory cell and is shown in Fig. 4 [1]. When the sampling phase ϕ_1 is on, the input impedance is given by

$$r_{in} = \frac{1}{\frac{1}{r_{o1}}+A\cdot g_{m1}} \approx \frac{1}{A\cdot g_{m1}} \quad (3)$$

where r_{o1} and g_{m1} are the output resistance and transconductance of M1, respectively. A is defined as the gain of the operational amplifier (OP). However, not only the operating frequency will be restricted by the cut-off frequency of the operational amplifier, but also the circuit is more complex with an OP.

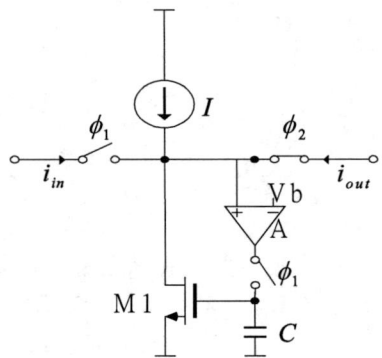

Fig. 4 One technique to decrease the input impedance for SI memory cell [1].

Based on the current feedback technique as shown in Fig.5, we proposed a feedback switched-current memory cell to reduce the input impedance.

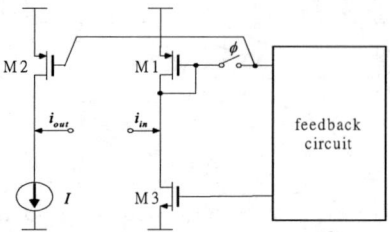

Fig. 5 The technique to decrease the input impedance for FG SI memory cell.

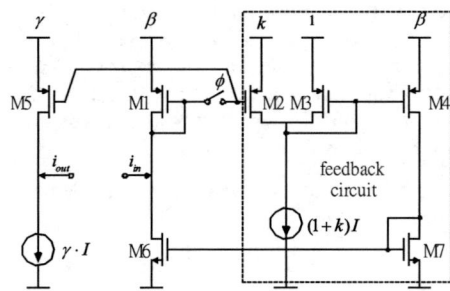

Fig. 6. The proposed feedback switched-current memory cell.

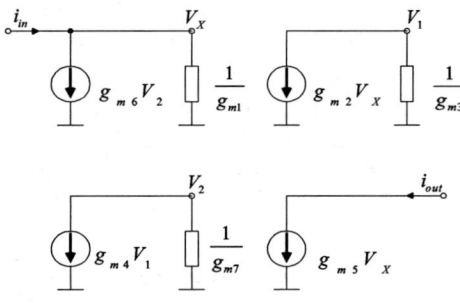

Fig. 7. The simplified small signal equivalent circuit of the proposed memory cell.

Fig. 6 shows the whole circuit of the proposed feedback switched-current memory cell. Note that the constants β, γ, κ are the width-to-length ratios (W/L). Fig. 7 shows the simplified small signal equivalent circuit of the proposed memory cell. As the sampling phase ϕ is on, the current transfer function is given by

$$\frac{i_{out}}{i_{in}} = \frac{g_{m5}}{g_{m1}+\frac{g_{m6}g_{m4}g_{m2}}{g_{m7}g_{m3}}} = \frac{\gamma}{\beta}\cdot\frac{1}{1+k} \quad (4)$$

Thus, the input impedance can be determined as

$$r_{in} = \frac{1}{(1+k)g_{m1}}. \quad (5)$$

In order to hold the condition of $|I_{IN}|=|I_{OUT}|$, we can properly adjust the width-length ratio of β, γ and k as

$$\gamma:\beta:k = 1:0.5:1 \quad (6)$$

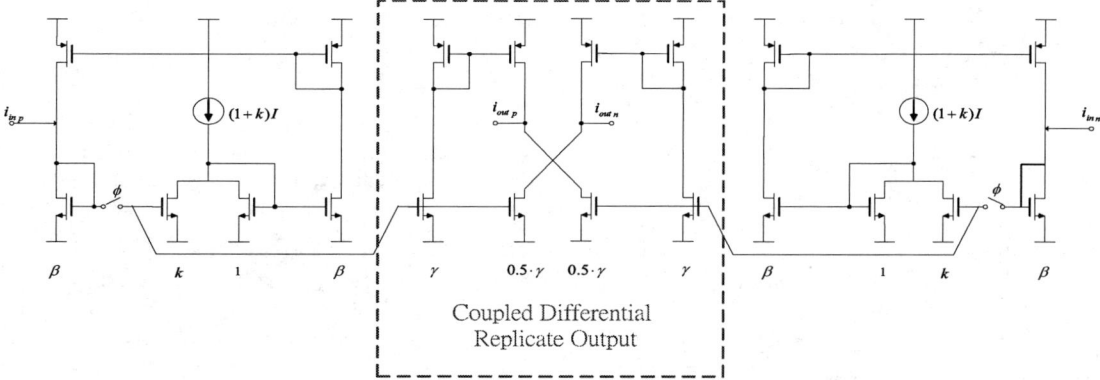

Fig. 8 The proposed whole circuit of CDR FMC.

At this simplified ratio, leading to the input impedance is given by

$$r_{in} = \frac{1}{2 \cdot g_{m1}} \qquad (7)$$

Note that the input impedance of the proposed memory cell is half of the traditional FG SI memory cell. Thus, the transmission error is reduced and the accuracy is improved. Furthermore, the operation speed is also higher due to the less time constant.

Finally, a coupled differential replicate (CDR) circuit is taken to eliminate the clock feedthrough (CFT) error. It is imbedded into the FMC circuit to establish a symmetrical architecture. Fig. 8 shows the proposed whole circuit of the CDR FMC.

IV. OTHER MODULATOR SUBCIRCUITS

A. 1-bit current quantizer

The popular means of decreasing the input impedance and speeding up the transition time is to use a 1-bit current-mode quantizer concluding a source follower input stage, as shown in Fig. 9 [5]. The circuit not only operates with low input resistance $1/g_m$, but also has the sufficient gain to amplify the small input voltage variations by applying the feedback to the gates of the input stage. During operation, the output voltage V_{out} is low when the input current i_{in} is feed into the quantizer, whereas it is high as the input current flows outside. Normally, a CMOS inverter is used to inversely change the output state.

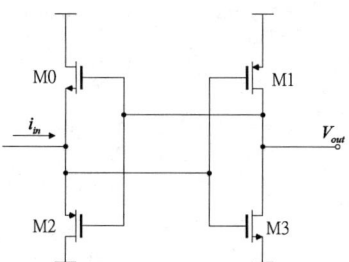

Fig. 9 1-bit current quantizer.

B. 1-bit current DAC

The proposed 1-bit DAC is a simple cascode current mirror with two switches, D and $\overline{D}$, as shown in Fig. 10. Both switches are controlled by the digital input signal. When the switch D is on, the output current I_{out} is equal to I_{DA}, whereas the I_{out} is equal to $-I_{DA}$ as the switch D is off.

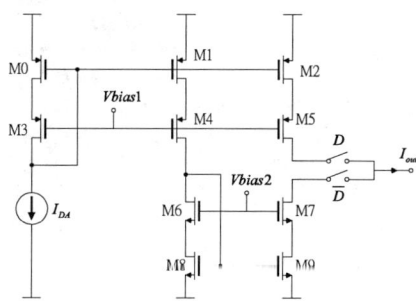

Fig. 10 A 1-bit cascode DAC.

C. Accurate reference current generator

To prevent the current mismatch, it is necessary to add an accurate reference current to the DACs and the CDR FMCs. The proposed circuit is shown in Fig. 11 [6], where the op is used to make sure that the reference voltage V_{ref} is across the reference resistor R_{ref}. Then, the reference current is given by

$$I_{ref} = \frac{V_{ref}}{R_{ref}}. \qquad (8)$$

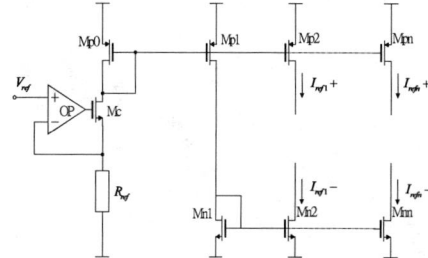

Fig. 11 An accurate reference current generator.

V. SIMULATION RESULTS

To find the step response of the proposed switched-current feedback memory cell (FMC), a step input signal whose values range from -10μA to +10μA is used. The simulated result, as shown in Fig. 12, presents that the proposed FMC is more accurate than the traditional FG memory cell. That is, the rising time of the proposed FMC is shorter than that of the traditional FG memory cell from -9μA to +9μA. The proposed FMC not only expands its bandwidth, but also speeds up its response time.

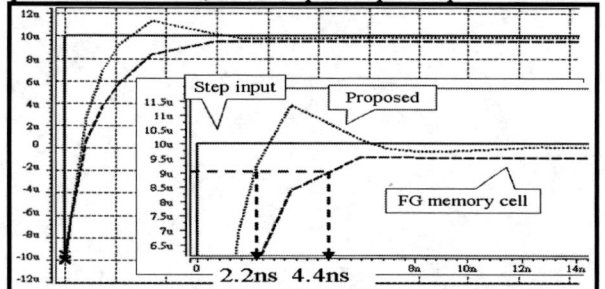

Fig. 12 The step response of the proposed FMC and the traditional FG memory cell.

Finally, all the proposed circuit is used to build up a second-order SI Δ-Σ modulator, as shown in Fig. 1. Within the input conditions that the input sine waveform operates with 10kHz and 30μA, the sampling frequency is 10.24MHz, and the supplied voltage is 3V, Fig. 13 depicts the input signal and the output modulator code of the proposed modulator. Also, Fig. 14 shows the output power spectrum of the proposed second-order Δ-Σ modulator. As indicated, the proposed modulator has a SNDR of 75dB within the bandwidth of 40 KHz and an oversampling ratio (OSR) of 128. The effective number of bits (ENOB) is calculated as 12 bits [7].

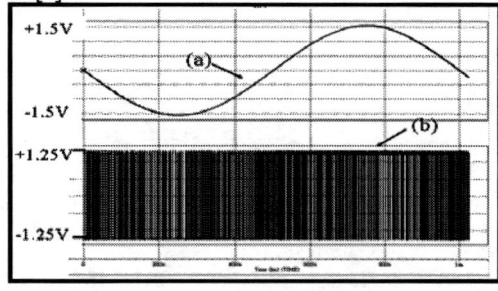

Fig. 13 The simulated results of the proposed second-order SI Δ-Σ modulator. (a) Input signal waveform (V_{p-p}=3V), and (b) Output modulator code.

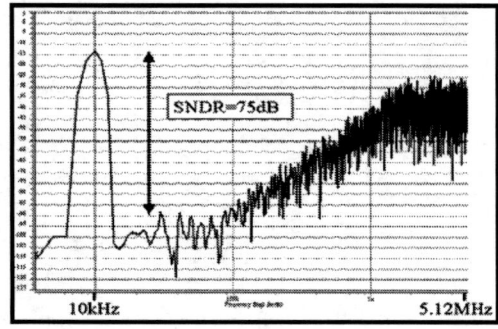

Fig. 14 The output power spectrum of the proposed modulator.

VI. CONCLUSSION

Table I presents the simulated results of the proposed second-order SI Δ-Σ modulator. By comparing those data with several lately articles [8-9], the proposed one operates with about 2 times larger bandwidth, however, with smaller area. Note that the SNDR is about 75dB and the dynamic range (DR) is about 70 dB at 10.24MHz sampling rate with 40kHz bandwidth, and the power consumption is 16 mW. That is, high resolution can be obtained in the proposed circuit even though the modulator spends more power.

In this paper, we present a fully differential second-order delta-sigma modulator, which uses a current feedback to decrease the input impedance and to improve the transmission error in the switched-current feedback memory cell. Also, a coupled differential replicate (CDR) circuit is done to eliminate the clock feedthrough (CFT) error. Furthermore, the entire memory cell is designed in a coupled differential replicate (CDR) form to eliminate the clock feedthrough (CFT) error.

Table I: Simulated results of the proposed second-order SI Δ-Σ modulator.

Year	2004
CMOS Process(μm)	0.35
SI type	FG
Supply Voltage (V)	2.5
f_s (M Hz)	10.24
OSR	128
f_B (k Hz)	40
Power Consumption (mW)	16
SNDR	75 dB
DR	70 dB

VII. ACKNOWLEDGMENT

The author would like to thank the ERSO and National Science Council of the Republic of China, Taiwan, for financially supporting this research (NSC 92-2213-E-027-041). CIC, Taiwan, is appreciated for fabricating the test chips.

REFERENCES

[1] M. Gustavsson, J. J. Wikner and N. N. Tan, CMOS Data Converters for Communications, Kluwer Academic Publishers, Boston, 2000.

[2] B. E. Jonsson, Ericsson Radio Systems AB, Switched-Current Signal Processing and A/D Conversion Circuits Design and Implementation, Kluwer Academic Publishers, Boston, 2000.

[3] N. Tan, Microelectronics Research Center Ericsson Components AB Sweden, Switched-Current Design and Implementation of Oversampling A/D Converters, Kluwer Academic Publishers, 1997.

[4] J. B. Hughes, N. C. Bird, and L. C. Macbeth, "Switched current – A new technique for analog sample-data signal processing", IEEE International Symposium on Circuits and Systems, pp. 1154-1187, 1989.

[5] H. Traff, "Novel approach to high-speed CMOS current comparators", Electron. Lett., 28, pp.310-2. Jan. 1992.

[6] P. R. Gray and R. G. Meyer, Analysis and Design of Analog Integrated Circuits, 4th ed., John Wiley and Sons Inc., 2001.

[7] Y. Geerts, M. Steyaert and W. Sansen, Design of Multi-Bit Delta-Sigma A/D Converters, Kluwer Academic Publishers, Boston, 2002.

[8] R. Rodriguez-Calderon, J. Santana-Corte and F. Sandoval-Ibarra, "Reducing non-idealities on switched-current sigma-delta modulators", in Proceedings of the Fourth IEEE International Caracas Conference on Device, Circuits and Systems, pp. C019-1–C010-5, 17-19 April 2002.

[9] M. Loulou, D. Dallet and P. Marchgay, "A 3.3V switched-current second order sigma-delta modulator for audio applications", in Proceedings of the 2000 IEEE International Symposium on Circuits and Systems, Geneva, Vol. 4, pp. 409-412, 28-31 May 2000.

A 1.5V MULTIRATE MULTIBIT ΣΔ MODULATOR FOR GSM/WCDMA IN A 90nm DIGITAL CMOS PROCESS

Oguz Altun[1], Jinseok Koh[2], Phillip. E. Allen[1]

[1]Georgia Institute of Technology, Atlanta, Georgia 30332, USA
[2]Wireless Analog Technology Center, Texas Instruments Inc., Dallas, TX 75243, USA

ABSTRACT

Multirate multibit Sigma Delta (MM-ΣΔ) modulator structure is used for low power A/D converter design for wireless applications. New integrator structures and a dedicated timing scheme are proposed for a very low power implementation. A dual-mode second order MM-ΣΔ modulator targeting GSM/WCDMA applications is designed. In Spice simulations, the prototype design achieves 71.4dB peak Signal-to-Noise+Distortion (SNDR) in the 200kHz GSM band and dissipates 1.1mA of total current from a 1.5V supply. This dual-mode design also achieves 45.7dB SNDR in the 1.94MHz WCDMA band with 1.9mA of total current consumption.

1. INTRODUCTION

In high performance wide band ΣΔ modulators, the trade-off between high modulator order (L) and high oversampling ratio (OSR) [1] becomes especially difficult. For a fixed OSR of 32, the theoretical increase of SNR by going from a second order to a third order modulator is about 21dB. But higher order single loop structures have stability problems in addition to increased chip area and power consumption. On the other hand, with every doubling of OSR, SNDR increases by 3(2L+1) dB. However, for a fixed Nyquist rate (f_N), doubling the OSR generally translates into opamps that are roughly twice as fast. Consequently, both of these approaches suffer from increased total current consumption. This is highly undesirable in mobile applications where the battery life is a primary concern.

Multibit quantization has been a successful method to alleviate these problems. In multibit ΣΔ modulators, Signal-to-Noise ratio (SNR) improves about 6c dB where c represents quantizer resolution in bits. The problem is flash A/D converters are normally used as quantizers, thus requiring comparator banks. Every additional bit requires doubling the number of comparators and hence the area of the quantizer. The area and current consumption of these quantizers may become prohibitive even with c=4.

MM-ΣΔ modulators are proposed to overcome the area and power penalties of multibit quantization [3]. Proper selection of OSR in each integrator of a ΣΔ modulator became another architectural decision to be considered in the design of high-resolution, low-power, high-speed ΣΔ modulators [2].

In particular, consider an Lth-order MM-ΣΔ modulator with first integrator OSR of M_1 and upsampling of N in the remaining stages. Also, consider a conventional c-bit L-th order ΣΔ modulator with an OSR of $M=M_1$. These two modulators have the same inband quantization noise power (QNP) provided [2]:

$$c = \log_2\left(N^{(L-0.5)} + 1\right) \quad (1)$$

Equation (1) shows the relation between the quantizer resolution in a conventional modulator and the upsampling ratio in a MM-ΣΔ modulator for an L-th order system.

In Section (2), the modulator architecture will be presented and system level issues will be discussed. Section (3) describes the optimization of circuit parameters through behavioral simulations. Then, in Section (4), Spice simulation results obtained from the prototype design will be presented.

2. MODULATOR ARCHITECTURE

To achieve our design specifications, we determine that a 2nd order loop with a quantizer resolution of 2 bits is needed using the theory of conventional ΣΔ modulators. The required OSR is 48 and 10 for GSM and WCDMA, respectively. From equation (1), with L=2 and c=2, we find the upsampling ratio N of the corresponding MM-ΣΔ modulator as 2. Thus, the modulator in Fig 1 is designed for optimum power and area as well as good performance.

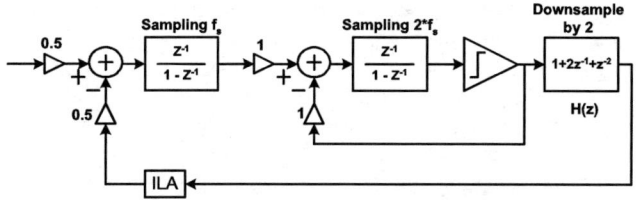

Figure 1. Block diagram of the multirate multibit ΣΔ modulator.

In this structure, the quantizer is only a single-bit comparator. Since the second integrator and comparator is running twice as fast as the first integrator, the output needs to be downsampled before it is fed back to the first integrator. A second-order digital FIR filter is used as a low pass filter to avoid aliasing. After this filter, the data is downsampled. Single-bit comparator and the FIR filter along with an upsampling in the second integrator replaces a bulky flash A/D converter if we were to design a second order modulator with conventional multibit quantization. In order to remedy the DAC non-linearity problem, a dynamic element-matching (DEM) block using the individual level averaging (ILA) algorithm is included in the feedback.

2.1 Proposed integrator structures and timing

In an ideal model of the modulator, it is assumed that the first integrator output settles to its final value almost instantaneously [2, 3]. However, in reality, if amplifiers settle much faster than needed, then they consume more current than is really needed. It is a good practice in terms of saving power to design the amplifier such that it utilizes the whole half period with some margin for slewing and settling of the amplifier output.

Therefore, in the modulator depicted in Fig 1, there is an important problem that needs to be addressed for the circuit implementation of this architecture. The second integrator samples the output of the first integrator twice in one clock period $T=1/f_s$. If the first integrator uses the whole half clock cycle $T/2$ to slew and settle to its final value, then one of the samples that the second integrator is receiving from the first integrator output will not be a valid sample. So, for correct operation, the settling behavior should be such that the first integrator output settles to its final value in about $T/4$ time. This case is depicted in Fig 2. However, then we will have to design an amplifier for the first integrator that is able to run with a $2*f_s$ clock, which means extra current consumption.

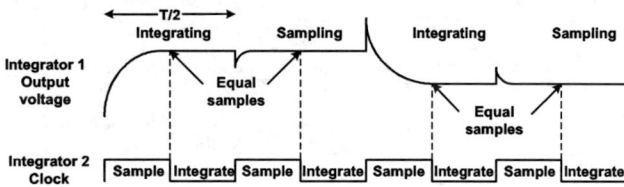

Figure 2. Sampling and integration of both integrators if the first integrator were to settle within $T/4$ time.

A power efficient method is proposed in Fig 3 to implement the sampling and integration operations. First note that in the case depicted by Fig 2, the two samples received by the second integrator from the first integrator output in the same clock period are exactly the same provided the first integrator settles fast enough. In Fig 3, we let the first integrator use the full half clock period for integration (during N2 is high in Fig 4). At the same time, the sampling switches of the second integrator are ON and the sampling capacitors of the second integrator are charging. The second integrator has two sampling capacitors instead of only one. After the first integrator output settles to its final value, we take a sample of the first stage output on to these two sampling capacitors (falling edge of N3 in Fig 4).

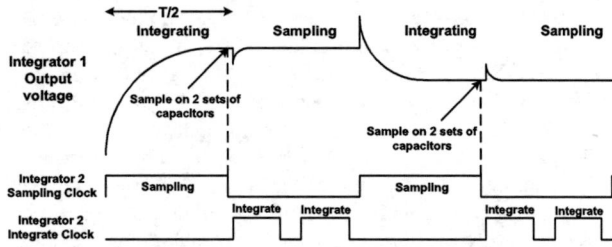

Figure 3. Proposed sampling and integration scheme for a power efficient implementation.

After the data is stored on the sampling capacitors, we integrate the charge stored on one of these capacitors immediately (during N4 is high in Fig 4). After integration, the comparator following the second integrator makes a comparison and determines the new feedback bit. This new feedback bit is used to determine the direction of charge flow when we integrate the charge stored on the second sampling capacitor (during N5 is high in Fig 4).

This new arrangement for sampling and integration operations is critical for a power efficient design of the system. Fig 4 shows the timing of all clock signals used in the system. Fig 5 and 6 illustrate the details of the first and second integrators.

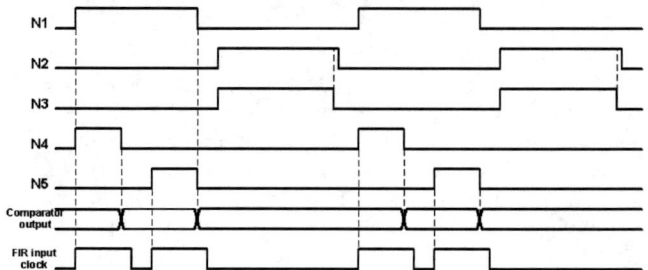

Figure 4. Timing arrangement of the clock signals in the system.

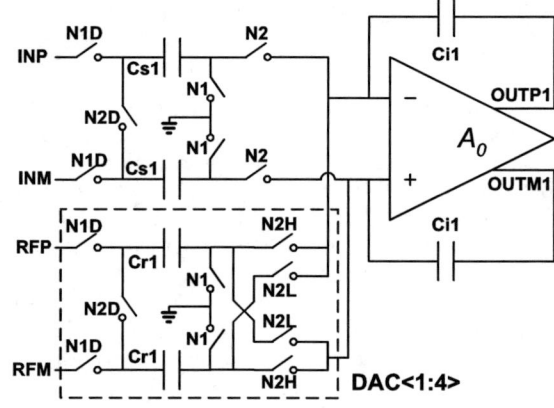

Figure 5. Structure of the first integrator in the proposed design.

3. CIRCUIT OPTIMIZATION

3.1 Integrator Design

The first integrator of a $\Sigma\Delta$ modulator is generally the most critical block in the system. Because all the subsequent blocks benefit from the noise shaping and their nonidealities can be tolerated to a greater extent. A significant portion of the total power consumption also takes place in the first integrator. Therefore, the optimization of the design of the integrators is critical for an efficient design. For this purpose, a very detailed non-ideal behavioral model of the system is constructed in Simulink using models presented in [4].

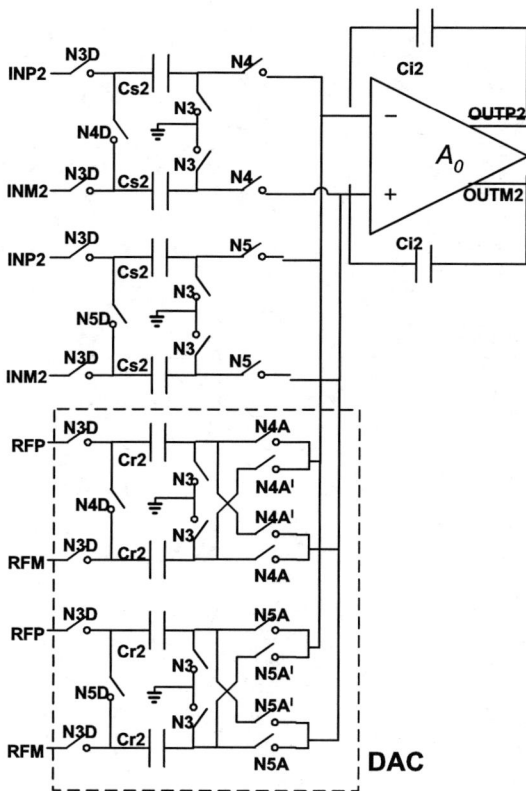

Figure 6. Structure of the second integrator in the proposed design.

In Fig 7, the DC gain of the first integrator amplifier is swept across a range of values and the sensitivity of modulator SNDR with respect to DC gain is presented. From this plot, the minimum required DC gain for the first amplifier is found to be around 45dB. Below 45dB, the rate of SNDR degradation quickly increases whereas the rate of SNDR improvement above 45dB is minimal.

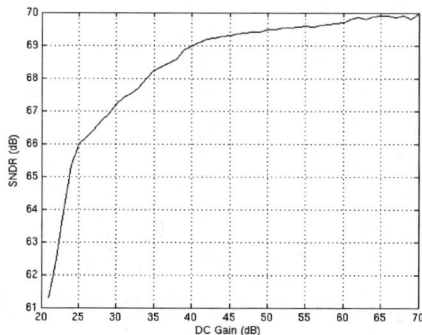

Figure 7. Optimization of amplifier DC gain

Fig 8 shows contour plots to determine the gain bandwidth (GBW) requirement of both amplifiers. Note that first integrator GBW is on the x-axis while the second integrator GBW is on the y-axis. Slew rate of both amplifiers are set to 60V/usec in this experiment. A 30kHz input sinusoid is used as an input and the SNDR is calculated within 270kHz to include all high order harmonics. As seen from this plot, first integrator amplifier must be designed such that its worst case GBW is more than 40MHz. This minimum GBW for the second integrator amplifier is more than 45MHz to get the best performance out of this system.

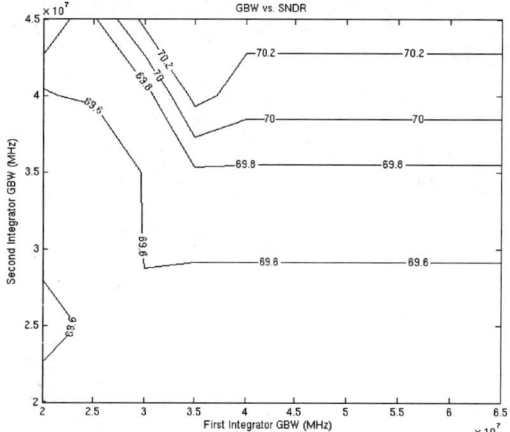

Figure 8. Optimization of the GBW of both amplifiers

Fig 9 shows the histogram of the outputs of both integrators when a –6dB sine wave with respect to the reference voltage is applied to the input of the modulator. X-axis shows the output swing as a ratio of the reference voltage (Vref=0.75V). According to this, first integrator output is within $\mp$ 0.6V and the second integrator output is within $\mp$ 0.1V in normal operation. The voltage supply is 1.5V in this design. It is therefore possible to have cascode configuration as the output stage and still meet the output swing requirements. Since we have moderate DC gain and GBW requirements, a single-stage amplifier like the folded cascode amplifier used in [5] is chosen for implementation. Unlike two-stage amplifiers, folded cascode will have less number of legs to bias. Hence it will be power efficient as well.

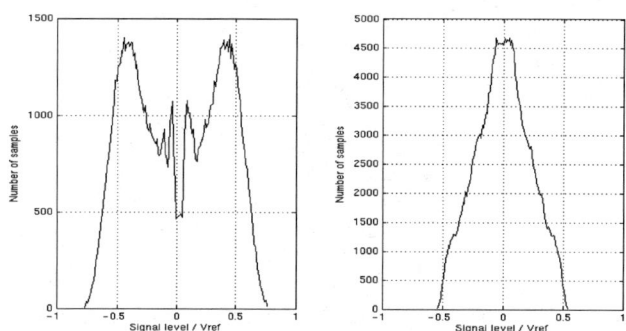

Figure 9. Histogram of integrator outputs

4. RESULTS

The modulator is designed using a 90nm digital CMOS process. The chip layout occupies 0.165mm² excluding the pads. The peak SNDR is 71.4dB and 45.7dB in GSM and WCDMA modes, respectively. In GSM mode, total current consumption is 1.1mA. Out of this 1.1mA, 1mA is dissipated on analog side and the remaining 100uA is dissipated on digital side. In WCDMA mode, the bias currents of the amplifiers are increased by 50% and the sampling clock is doubled with respect to the GSM mode. Total current consumption in WCDMA mode of operation is 1.9mA. 1.7mA of this total current is dissipated on analog side and the remaining 200uA is dissipated on digital side. The oversampling ratio in the first integrator is 48 and 10 in GSM and WCDMA modes, respectively.

Fig 10 shows the 8192-point FFT of the modulator output spectrum obtained from Spice simulations. The input signal frequency is 30kHz with amplitude of –6dB with respect to the reference voltage (Vref=0.75V).

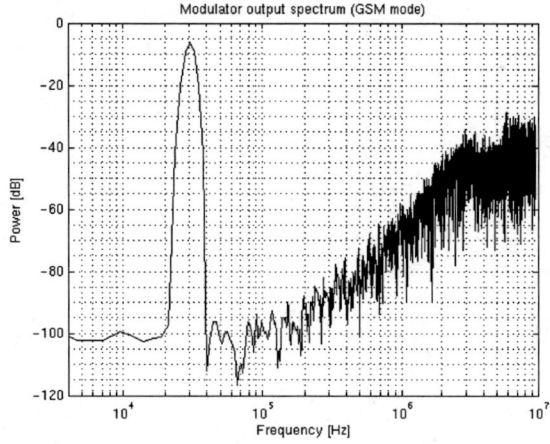

Figure 10. Modulator output spectrum in GSM mode

Fig 11 shows the SNDR versus the input signal power obtained from Spice simulations. The input is a sinusoidal signal with frequencies of 30kHz and 400kHz in GSM and WCDMA, respectively. In this plot, the input power is increased with a step of 10dB from –30dB to –10dB. Then, the step size is reduced to 1dB. Peak SNDR occurs at an input power of –4dB for GSM and –3dB for WCDMA.

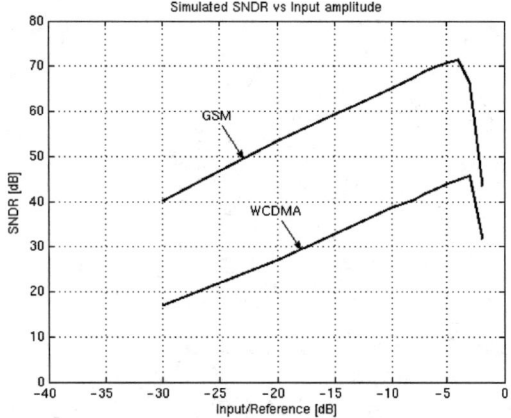

Figure 11. SNDR vs. input signal power in GSM mode

The complete list of specifications and results is presented in Table 1.

	GSM	WCDMA
Signal bandwidth	200kHz	1.94MHz
Oversampling ratio	48	10
Peak SNDR	71.4dB	45.7dB
Current consumption	1.1mA	1.9mA
Voltage supply	1.5V	
Vin (p-p) differential	0.375V	
Technology	90nm CMOS	
Chip area	0.165mm^2	

Table 1. Performance summary

5. CONCLUSION

A multirate multibit dual-mode switched-capacitor $\Sigma\Delta$ modulator is implemented for mobile communication applications. In GSM mode, it achieves 71.4dB of peak SNDR across a 200kHz bandwidth. In WCDMA mode, across a 1.94MHz signal band, peak SNDR is 45.7dB. The modulator is designed in a 90nm all digital CMOS technology without analog add-ons and the performance is verified through extensive Spice simulations. The prototype is laid out for fabrication occupying 0.165mm^2 total area. With a total current consumption of 1.1mA and 1.9mA in GSM and WCDMA modes, respectively, this design shows that MM-$\Sigma\Delta$ modulators are good candidates for low power implementation of $\Sigma\Delta$ A/D converters for wireless applications.

6. REFERENCES

[1] J. Candy and G. Temes, "Oversampling methods for A/D and D/A conversion," *Oversampling Delta-Sigma Data Converters*, pp 1-29, IEEE Press, 1992.

[2] F. Colodro and A. Torralba, "Multirate $\Sigma\Delta$ Modulators," *IEEE Trans. on Circuits and Systems-II: Analog and Digital Signal Proc.*, vol. 49, no. 3, Mar 2002.

[3] F. Colodro, A. Torralba, F. Munoz, and L. G. Franquelo, "New class of multibit sigma-delta modulators using multirate architecture," *Electronic Letters*, vol. 36, no.9, pp 783-785, April 2000.

[4] P. Malcovati, S. Brigati, F. Francesconi, F. Maloberti, P. Cusinato, and A. Baschirotto, "Behavioral Modeling of Switched-Capacitor Sigma-Delta Modulators," *IEEE Trans. on Circuits and Systems-I: Fundamental Theory and Appl.*, vol. 50, no. 3, Mar 2003.

[5] G. Gomez and B. Haroun "A 1.5V 2.4/2.9mW 79/50dB DR Sigma Delta Modulator for GSM/WCDMA in a 0.13um Digital Process," *Solid State Circuits Conference, Digest of Technical Papers*, vol. 1, pp. 306-307, Feb 2002.

Time-Interleaved Multirate Sigma-Delta Modulators

Colodro F., Torralba A, and M. Laguna.
Dpto. de Ingeniería Electrónica, Univ. Sevilla,
Camino de los Descubrimientos s/n
41092, Sevilla (SPAIN)
e-mail: pcolr@gte.esi.us.es

Abstract— A new strategy for the implementation of multirate Sigma-Delta Modulators (SDM's) is proposed in this paper. In multirate SDM's, the first integrator is clocked at a rate lower than the rest of the integrators in the forward path. In the new architecture, each integrator clocked at the high rate is replaced by two parallel integrators operating in interleaving mode and clocked at the same low rate as the first one. The new architecture has several nice features. Firstly, all integrators operate at the same low rate, which simplifies the clock generation circuit. Secondly, the delayed cross-paths in time-interleaved (TI) SDM's, which are difficult to implement, are not present in the proposed architecture. Thirdly, the high-rate Sample&Hold (S&H) circuit at the input of TI-SDM's is replaced by a low-rate one in the proposed modulator. Finally, the first integrator is clocked at low rate and is implemented as a single-path module. Accordingly, the modulator is simplified in silicon area and complexity. Furthermore, the architecture is very robust to mismatch between paths.

I. INTRODUCTION

In the past, Sigma-Delta (SD) modulation was extensively applied to the design of Analog-to-Digital Converters (ADC) for the instrumentation and audio industries [1]. Recently, the high operating speed of the modern CMOS technologies has released the bandwidth constraints of the Sigma-Delta modulators (SDM's), and SD-based ADC's are usually embedded in receiver architectures for narrow and wide band applications [2]. A high Signal-to-(Noise+Distortion) Ratio (SNDR) can be achieved using modulators with a high Oversampling Ratio (OSR), or with a high-order Noise Transfer Function (NTF). In the first case, the high frequency requirements of the amplifiers preclude its wideband application, while the design of high-order SDM's is concerned with extreme problems of stability. To achieve high SNDR at low OSR using low-order modulators, several architectural solutions have been proposed. Among them: multibit, time-interleaved, and multirate modulators.

The main advantage of multibit SDM's is that the quantization noise power at the modulator's output dramatically decreases when compared to a single-bit modulator, typically by 6 dB per additional bit. Furthermore, in single-loop multibit modulators, due to their more robust stability characteristic, an additional improvement in the Signal-to-Noise ratio (SNR) can be achieved by increasing the attenuation of the quantization noise in the baseband by proper selection of the error transfer function [3]. Unfortunately there are some drawbacks, the most important concerning the accuracy of the multibit Digital-to-Analog converter (DAC) in the feedback path. Several attempts to relax this problem have been proposed, based on dual-quantization [4]-[7], or dynamic element matching [8]-[14].

Another architectural solution to achieve high SNDR at low OSR is time-interleaving. In [15]-[16], parallelism is applied to all integrators of the modulator, including the first one. However, the performance of the overall modulator strongly depends on the first integrator. Accordingly, replicating this one is more critical in terms of silicon area and power consumption than replicating the rest of integrators. Furthermore, the parallel paths are mutually connected by gain and delay blocks, and the existence of delayed cross-paths severely affects the complexity of their implementation. In [17]-[18], the same type of time interleaving originally proposed to Nyquist converters is applied to SD ADC's. Now each parallel branch has a complete SD ADC including the modulator and the decimation filter. High resolution is obtained by extensive parallelism, which implies an increase in the complexity.

II. MULTIRATE SIGMA-DELTA MODULATORS

For the sake of simplicity the original idea proposed in this paper is presented for the second-order SDM. Nevertheless, the same strategy can be applied to high-order modulators. The multirate modulators [6],[19] are reviewed briefly in this section. The 2nd-order one is depicted in Figure 1. The first and second integrators are clocked at the rate of $f_S/2$ and f_S, respectively. The output of the single-bit quantizer is fed back directly to the input of the second integrator. The anti-aliasing filter $H(z)$ shown in the figure is necessary before the quantizer output is down-sampled and fed back to the modulator input. A simple comb filter can be used to avoid that aliasing deteriorates the performances of the modulator.

$$H(z) = (1+z^{-1})^2 / 4 \qquad (1)$$

As the input of $H(z)$ is a binary word, the output can be represented using 5 levels.

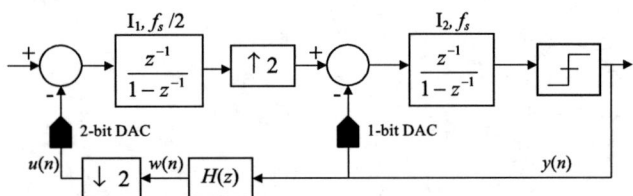

Figure 1. Second-order multirate Sigma-Delta modulator

Figure 2 shows the performances obtained from simulation of multirate and conventional modulators. Note that it is necessary to specify two values of OSR for the multirate modulator, one for the first integrator and the other one for the second integrator. All simulations in this paper have been done in the time domain using SIMULINK$^@$. As it could be expected, the SNDR curve of the multirate modulator is placed in an intermediate position between the SNDR curves of the conventional modulators clocked at the low and at the high OSR's. The behavior of the modulator at low amplitudes cannot be explained by the linear model of the quantizer [19].

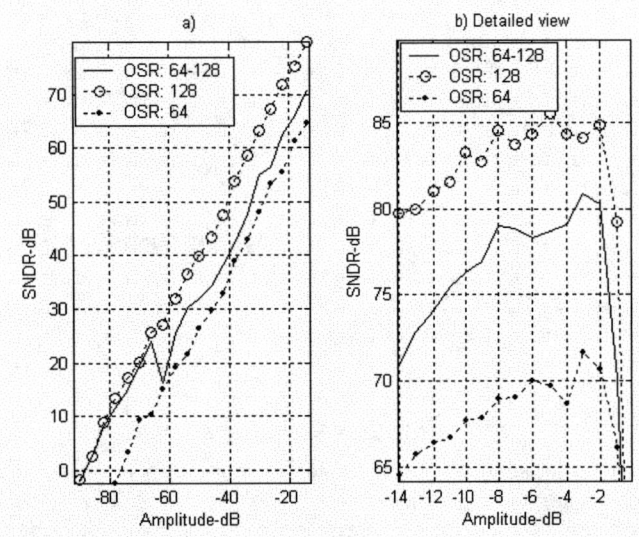

Figure 2. SNDR curves for the multirate (OSR=64 for the first integrator and 128 for the second one) and conventional modulators (OSR = 64 and OSR =128 in both integrators).

Multirate modulators achieve better SNDR when the sampling frequency of the integrators, excluding the first one, is increased. The speed constraints of their corresponding amplifiers are relaxed because any non-ideal effect due to these amplifiers is shaped by the transfer function of the first integrator. In conclusion, the integrator that operates at the lowest frequency, i.e., the first one, imposes the speed constraints of the overall architecture, while increasing the sampling frequency of the other integrators causes an increase of the SNDR at a low cost [19].

III. THE TIME-INTERLEAVED MULTIRATE SIGMA-DELTA MODULATOR

The second integrator of the architecture in Figure 1 has been replaced by the equivalent time-interleaved module, as can be seen in Figure 3.a). Note that the high-frequency quantizer of the modulator in Figure 1 has been replaced by two low-frequency quantizers located in the two new paths. These transformations do not affect the performances of the new architecture as compared to the original multirate SDM. In an electronic implementation using switched-capacitor techniques [15], the top and bottom integrators are clocked in phase and counter-phase at the frequency $f_S/2$, respectively. With this arrangement, it is not necessary to dedicate additional circuitry in the implementation of both delays. Nevertheless, the low-rate delay and the unity gain terms in the cross paths have to be implemented using additional circuitry.

The inner up-samplers and down-samplers of the architecture depicted in Figure 3(a) can be removed taking into account the signals which flow thorough different paths from the source points (A, C and E) to the sink points (B and D). Every path is compounded of a up-sampler, a delay of n samplers at the rate f_S, and a down sampler (that is: $\uparrow 2$ - z^{-n} - $\downarrow 2$). If n is an odd number, the path is an open circuit (no signal flows through the path). In the case that n is a even number, the path can be replaced by a delay of $n/2$ samplers at the rate $f_S/2$ (that is, the path $\uparrow 2$ - z^{-n} - $\downarrow 2$ is replaced by $z^{-n/2}$). If all these transformations are carried out, the architecture of the Figure 3.(b) is obtained. Finally, the elements shown in Figure 3(b) can be arranged in such a way that the modulator proposed in this paper is obtained (). The performances of the proposed modulator are identical to those of the original multirate modulator in Figure 1. Accordingly, the solid line in Figure 2 corresponds to the SNDR of the proposed time-interleaved modulator.

The proposed architecture is characterized by the following features. Firstly, the delayed cross-paths that appear in the forward paths of the time-interleaved SDM's [15]-[16] have been moved to the feedback path reducing the complexity of the analog circuitry. Note that there are not unit delays in the analog circuitry, either. Secondly, in the proposed architecture, the first integrator is not time-interleaved. As the modulator performances are strongly affected by this integrator, which usually determines power consumption and silicon area, the proposed architecture is simpler than the conventional time-interleaved SDM's. Note that in the last case, it is necessary duplicating all integers, including the first one.

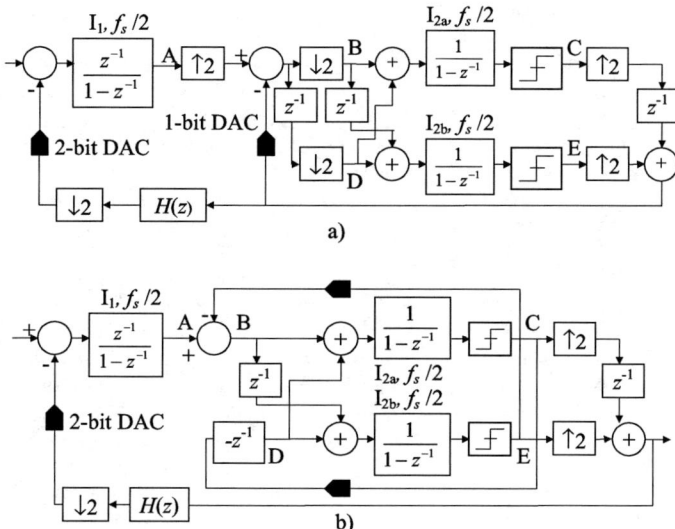

Figure 3 a) Multirate SDM where the second integrator has been replaced by a time-interleaved structure. b) The top figure without the inner up- and down-samplers, obtained by block diagram manipulation.

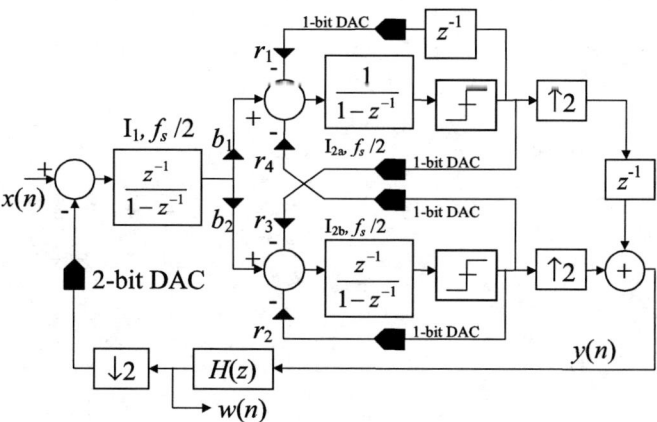

Figure 4. The proposed architecture (b1 = b2 = r1 = r2 = r3 = r4 = 1).

Thirdly, the errors produced by mismatch between both interleaved paths are shaped by the inverse transfer function of the first integrator so that, as it will be shown later, the proposed architecture is more robust than other parallel and time-interleaved modulators against circuit non-idealities. Fourthly, not only the proposed architecture inherits the advantages of the multirate SDM of Figure 1 concerning the low frequency operation of the first integrator and its high SNDR, in the proposed architecture the three integrators are clocked in phase at the low rate which simplifies the design of the integrators I_{2a} and I_{2b} and the clock generation circuit, as compared to the original multirate modulator. Furthermore, in other TI-SDM's the input signal has to be sampled using a S&H circuit clocked at the high rate. In the proposed modulator the input signal is sampled at the low rate.

IV. INFLUENCES OF NON-IDEALITIES IN THE MODULATOR

A. Finite gain of the integrators

A low DC gain in the first integrator can produce a degradation of the modulator performances [1]. As shown by simulation, the degradation of the SNDR for the proposed architecture is less than 3 dB for DC gains greater than 40 dB. The SNDR degradation is similar to that of a conventional modulator whose integrators are clocked at the same rate than the integrators of the proposed modulator and less than the conventional one with a double rate.

B. Coefficient mismatch

Time-interleaved and parallel modulators are very sensitive to mismatch between parallel signal paths. Mismatch produces a degradation of the SNDR because of the quantization and the distortion noise around the frequency $f_s/2$ are folded back into the band of interest [15], where f_s is the effective sampling rate. The effect of mismatch can be reduced by changing the coefficient values in the cross paths [15]. Following the same procedure that in [19], the Noise Transfer Function (NTF) can be calculated for $r_3=r_4=1$

$$NTF_1(z) = \frac{Y(z)}{E(z)} = \frac{(1-z^{-1})^2}{1-z^{-1}+z^{-3}/8+z^{-4}/8} \quad (2)$$

where $E(z) = z^{-1} \cdot E_a(z^2) + E_b(z^2)$. $E_a(z)$ and $E_b(z)$ are the low-rate quantization noises in the top and bottom quantizers in Figure 4, respectively. In the general case, for $r_3=r_4=k$,

$$NTF_k(z) = \frac{(1-z^{-1})^2 \cdot (1+z^{-1})}{1+(k-1)\cdot z^{-1} - k\cdot z^{-2} + z^{-3}/8 + z^{-4}/4 + z^{-5}/8} \quad (3)$$

The new function $NTF_k(z)$ has a pole and a zero more than $NTF_1(z)$. For values of k slightly less than the unity, the four poles of $NTF_1(z)$ are slightly shifted from their original position and the new pole is close to the zero $z = -1$. For instance, if k=0.95, the poles are shifted to $0.7867 \pm j \cdot 0.3170$ and $-0.2866 \pm j \cdot 0.3174$, and the new pole is located at -0.9502. Now, the noise floor around the frequency $fs/2$ is attenuated by the new zero before the mismatch produces aliasing.

The degradation of the SNDR for the proposed modulator and for the time-interleaved SDM of [15] are shown in Figure 5. The coefficients that affect the performance of the proposed modulator are r_1, r_2, r_3, r_4, b_1, and b_2 (Figure 4). Their values were taken following a normal distribution with mean equal to the nominal value and standard deviation of 2% and 1%. The nominal values for coefficients r_3 and r_4 are $r_3=r_4=k=0.95$. For the time interleaved SDM in [15], the coefficients chosen for the Montecarlo analysis has been those that produce mismatch between both paths. A normal distribution has been used with a standard deviation of 0.1%. A similar strategy to reduce the mismatch effect has been also adopted (as explained in [15], the nominal value of the coefficients in the crossed path has been changed from 1 to 0.95). Each point in Figure 5 was calculated as the average of 10 different simulations. As it can be seen in Figure 5, the proposed modulator is much more robust against mismatch than the time-interleaved SDM of [15]. Note that the SNDR degradation of the proposed modulator for a coefficient deviation of 1% is less than 2 dB.

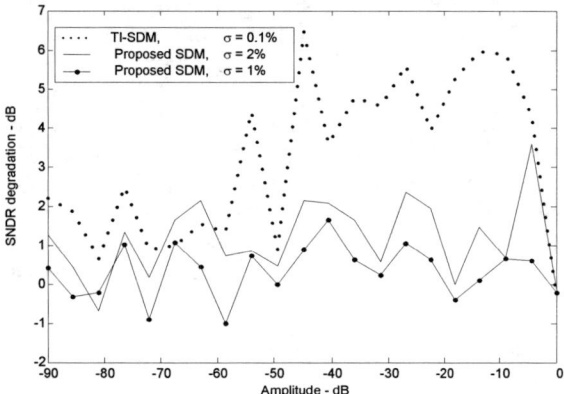

Figure 5 SNDR degradation due to coefficient mismatch versus input signal amplitude.

V. CONCLUSIONS

In this paper, time-interleaving has been applied to multirate SDM's, showing that, in a second-order multirate SDM with the second integrator sampled at twice the frequency of the first integrator, the second integrator can be implemented using a time-interleaved structure without delayed cross-paths. This structure favourably compares to other time-interleaving approaches in terms of a higher SNDR, a simpler implementation and less sensitivity against mismatches and other circuits non-idealities. Besides, unlike the original multirate SDM, it operates with a single frequency clock. Finally, using this approach, higher order SDM's can be built with similar good properties. Analytical expressions have been obtained and simulations in the time-domain have been carried-out to support them.

ACKNOWLEDGMENT

Authors would like to acknowledge financial support from the Spanish CICYT under project TIC2002-04323-C03-01.

REFERENCES

[1] S.R. Norsworthy, R.Schreier and G.C. Temes, eds. *Delta-Sigma Data Converters*, IEEE Press: New York, 1997.

[2] I. Galton, "Delta-Sigma Data Conversion in Wireless Transceivers," *IEEE Trans. Microwave Theory Tech.*, Vol. 50, no 1, January, 2002.

[3] P. Ju, and D.G. Vallancourt. "Quantization noise reduction in multibit oversampling S-D AD convertors," *Electron. Lett.*, vol. 28, no. 12, pp. 1162-1164, 4th Jun. 1992.

[4] T.C. Leslie and B. Singh, "An improved sigma-delta modulator architecture," in *Proc. of the IEEE Int. Symp. Circuits Syst.*, vol. 1, pp. 372-375, 1990.

[5] A. Hairapetian, G.C. Temes, and Z.X. Zhang, "Multibit Sigma-Delta modulator with reduced sensitivity to DAC nonlinearity," in *Electron. Lett.*, vol. 27, no. 11, pp 990-991, 23rd May 1991.

[6] F. Colodro, and A. Torralba, "Multirate single-bit modulators," *IEEE Trans. Circuits Syst. II*, vol. 49, no. 9, pp. 629-634, Sept. 2002.

[7] F. Colodro, A. Torralba, and J.L. Mora "Robust Dual-Quantisation Multibit Sigma-Delta Modulator," *Electron. Lett.*, Vol. 39, May 2003.

[8] R.J. Van de Plassche and D.Goedhart, "A monolithic 14-bit D/A converter," *IEEE J. Solid-State Circuits*, vol. SC-14, no. 3, pp. 552-556, Jun. 1979.

[9] L.R. Carley, "A noise-shaping coder topology for 15+ bit converters," *IEEE J. Solid-State Circuits*, vol. 28, no. 2, pp. 267-273, Apr. 1989.

[10] B.H. Leung, S. Sutarja, "Multibit A/D converter incorporating a novel class of dynamic element matching techniques," *IEEE Trans. on Circuits Syst. II*, vol. 39, no. 1, pp. 35-51, Jan. 1993.

[11] R.T. Baird, and T.S. Fiez, "Improved ΔΣ DAC linearity using data weighted averaging," in *Proc. of the IEEE Int. Symp. Circuits Syst.*, vol. 1, pp. 13-16, 1995.

[12] R. Schreier and B. Zhang, "noise-shaped multibit D/A convertor employing unit elements," *Electron. Lett..*, vol. 31, no. 20, pp. 1712-1713, 28th Sept. 1995.

[13] R.W.Adams, T.W.Kwan, "Data-directed scrambler for multi-bit noise shaping D/A converters," U.S. Patent no. 5,404,142, Apr. 4, 1995.

[14] I. Galton, "Spectral shaping of circuit errors in digital-to-analog converters," *IEEE Trans. on Circuits Syst. II*, vol. 44, pp. 808-817, Oct. 1997.

[15] R. Kohini-Poorfard, L.B. Lim, and D.A. Johns, "Time-Interleaved oversampling A/D Converters: Theory and Practice," *IEEE Trans. Circuits Syst. II*, vol 44, pp. 634-645, Aug. 1997.

[16] M. Kozak, M. Karaman, and I. Kale, "Efficient Architectures for Time-Interleaved Oversampling Delta-Sigma Converters," *IEEE Trans. Circuits Syst. II*, vol. 47, pp. 802-810, Aug. 2000.

[17] I. Galton, and H.T. Jensen, "Oversampling Parallel Delta-Sigma Modulator A/D Conversion", *IEEE Trans. Circuits Syst. II*, vol. 43, pp. 801-810, Dec. 1996.

[18] Eshraghi A. and Fiez T.S., "A Time-Interleaved Parallel SD A/D Converter", *IEEE Trans. Circuits Syst. II*, vol. 50, pp. 118-129, Mar. 2003.

[19] F.Colodro y A. Torralba, "Multirate Sigma-Delta Modulators", *IEEE Trans. Circuits Syst. II*, Vol. 49, No 3, pp 170 –176, Mar. 2002.

A DIRECT SYNTHESIS METHOD OF CASCADED CONTINUOUS-TIME SIGMA-DELTA MODULATORS

Ramón Tortosa, José M. de la Rosa, Angel Rodríguez-Vázquez and Francisco V. Fernández

Instituto de Microelectrónica de Sevilla – IMSE-CNM (CSIC)
Edificio CICA-CNM, Avda. Reina Mercedes s/n, 41012- Sevilla, SPAIN
Phone: +34 95 5056666, Fax: +34 95 5056686, E-mail: {tortosa|jrosa|angel|pacov}@imse.cnm.es

ABSTRACT

This paper presents an efficient method to synthesize cascaded sigma-delta modulators implemented with continuous-time circuits. It is based on the direct synthesis of the whole cascaded architecture in the continuous-time domain instead of using a discrete-to-continuous time transformation as has been done in previous approaches. In addition to place the zeroes of the loop filter in an optimum way, the proposed methodology leads to more efficient architectures in terms of circuitry complexity, power consumption and robustness with respect to circuit non-idealities.[†1]

1. INTRODUCTION

$\underline{C}$ontinuous-$\underline{T}$ime (CT) $\underline{S}$igma-$\underline{D}$elta $\underline{M}$odulators ($\Sigma\Delta$Ms) have demonstrated to be an attractive solution for the implementation of $\underline{A}$nalog-to-$\underline{D}$igital (A/D) interfaces in systems-on-chip integrated in deep-submicron standard CMOS technologies [1]. Although most reported $\Sigma\Delta$Ms have been implemented using $\underline{D}$iscrete-$\underline{T}$ime (DT) circuits, the increasing demand for broadband data communication systems has motivated the use of CT techniques. In addition to show an intrinsic antialiasing filtering, CT $\Sigma\Delta$Ms provide potentially faster operation with lower power consumption than their DT counterparts [2][3].

In spite of their mentioned advantages, CT $\Sigma\Delta$Ms are more sensitive than DT $\Sigma\Delta$Ms to some circuit errors, namely: clock jitter, excess loop delay and technology parameter variations [2][3]. The latter are specially critical for the realization of cascaded architectures. This has forced the use of single-loop topologies in most reported silicon prototypes even thought low oversampling ratios (< 12) are needed [4][5], whereas very few cascaded CT $\Sigma\Delta$M $\underline{I}$ntegrated $\underline{C}$ircuits (ICs) have been reported [6].

However, the need to achieve medium-high resolutions ($> 12 \text{bits}$) within high signal bandwidths ($> 20\text{MHz}$) while guaranteeing stability, has prompted the interest in proper methods for the synthesis of high-order cascaded CT $\Sigma\Delta$Ms [7]-[9]. These methods are based on applying a DT-to-CT transformation to an equivalent DT topology that fulfils the required specifications. In most cases, the use of such a transformation is normally translated into an increase of the analog circuit complexity with the subsequent penalty in silicon area, power consumption and sensitivity to parameter tolerances.

This paper presents a direct synthesis method of cascaded CT $\Sigma\Delta$Ms which, dispensing with the DT-to-CT equivalence, allows to reduce the number of analog components and to efficiently place the zeroes/poles of the noise transfer function, thus yielding to more robust architectures than using a DT-to-CT transformation.

2. CASCADED CT $\Sigma\Delta$ MODULATORS

Fig.1 shows the conceptual block diagram of a m-stage cascaded CT $\Sigma\Delta$M. Each stage, consisting of a single-quantizer CT $\Sigma\Delta$M, re-modulates a signal containing the quantization error generated in the previous stage. Once in the digital domain, the outputs, y_i, of the stages are properly processed and combined (by the cancellation logic) in order to cancel out the quantization errors of all the stages, but the last one in the cascade. This latter error appears at the overall modulator output shaped by a function of order equal to the summation of the order of all the stages.

Cascaded CT $\Sigma\Delta$Ms are normally synthesized from equivalent (well-known) DT systems and use the same digital cancellation logic [8]. This DT/CT equivalence can be guaranteed because the overall open loop transfer function of each stage in Fig.1 is in fact a DT system [2]. Thus, in the case of a rectangular impulsive response of the $\underline{D}$igital-to-$\underline{A}$nalog $\underline{C}$onverter (DAC), it can be shown that the equivalent DT loop filter

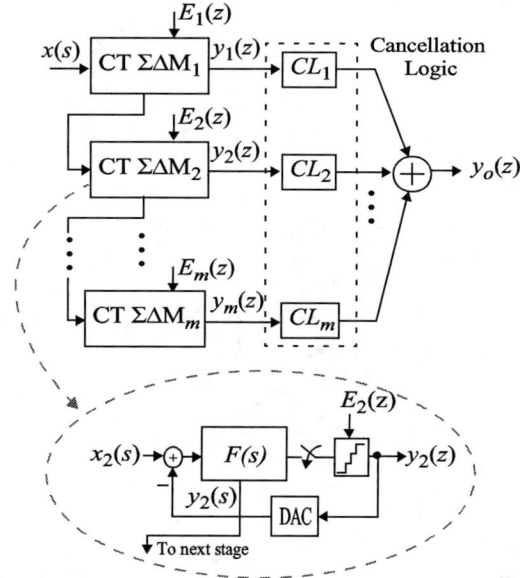

Figure 1. Conceptual block diagram of a cascaded CT $\Sigma\Delta$M.

[†1]. This work has been supported by the Spanish Ministry of Science and Education (with support from the European Regional Development Fund) under contract TEC2004-01752/MIC.

transfer function is given by [10][11]:

$$F(z) = \sum_{p_i} \text{Re}\left(\frac{F(s)}{s} \cdot \frac{e^{m_1 T_s \cdot s}}{z - e^{T_s \cdot s}}\right) - \sum_{p_i} \text{Re}\left(\frac{F(s)}{s} \cdot \frac{e^{m_2 T_s \cdot s}}{z - e^{T_s \cdot s}}\right) \quad (1)$$

where $f_s = 1/T_s$ is the sampling frequency; $m_1 = 1 - t_d/T_s$; $m_2 = 1 - (t_d + \tau)/T_s$; t_d and τ are respectively the time delay and pulse width of the DAC waveform; p_i are the poles of $F(s)/s$ and $\text{Re}(x)$ stands for the residue of x.

However, in order to get a functional CT ΣΔM while keeping the cancellation logic of the original DT ΣΔM, every state variable and DAC output must be connected to the integrator input of later stages [8], thus increasing the number of analog components, i.e transconductors, amplifiers and DACs. As an illustration, Fig.2(a) shows a cascaded 2-1-1 CT ΣΔM obtained from an existing DT ΣΔM [12]. Note that at least eight scaling coefficients (k_{g2-9}) and their corresponding signal paths are needed to connect the different stages of the modulator. The number of integrating paths can be reduced – as illustrated in Fig.2(b) – if the whole cascaded ΣΔM is directly synthesized in the CT domain as proposed in the next section.

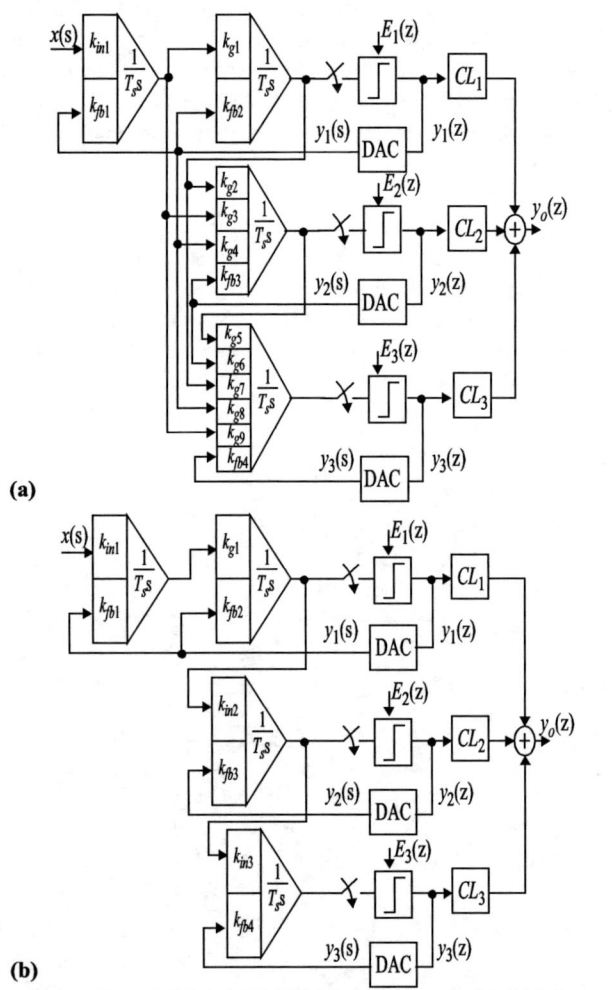

(a)

(b)

Figure 2. Cascaded 2-1-1 CT ΣΔM architecture obtained (a) from an equivalent DT ΣΔM (b) using the proposed method.

3. PROPOSED METHODOLOGY

The idea of dispensing with the DT-to-CT transformation was previously reported in [3] for single-loop architectures. However, in the case of cascaded architectures, the cancellation logic functions (not present in single-loop ΣΔMs) must be included in the synthesis procedure in order to get an optimum architecture.

Let's consider the more general case of the m-stage cascaded CT ΣΔM shown in Fig.1. The overall output, y_o, is given by:

$$y_o(z) = \sum_{k=1}^{m} y_k(z) CL_k(z) \quad (2)$$

where $y_k(z)$ and $CL_k(z)$ represent respectively the output and partial cancellation logic transfer function of the k-th stage.

If the modulator input, $x(t)$, is set to zero, it can be shown that the output of each stage can be written as:

$$y_k(z) = \frac{E_k(z) + \sum_{i=1}^{k-1} Z\left\{L^{-1}[H_D F_{ik}]\big|_{nT_s}\right\} y_i(z)}{1 - Z\left\{L^{-1}[H_D F_{kk}]\big|_{nT_s}\right\}} \quad (3)$$

where Z stands for the Z-transform, L^{-1} is the inverse Laplace transform, $H_D \equiv H_{DAC}(s)$ is the transfer function of the DAC, and

$$F_{ij} \equiv F_{ij}(s) = \frac{\text{Input Quantizer } j}{y_i(s)} \quad (4)$$

represents the transfer function from $y_i(s)$ to the input of j-th quantizer.

Using the notation $Z\left(L^{-1}(H_D F_{km})\big|_{nT_s}\right) \equiv Z_{km}$, the output of each stage is given by:

$$y_k(z) = \frac{E_k(z) + \sum_{i=1}^{k-1} Z_{ik} y_i(z)}{1 - Z_{kk}} \quad (5)$$

and the output of the modulator can be written as:

$$y_o = \sum_{k=1}^{m} y_k CL_k = \sum_{k=1}^{m} \left(\frac{E_k}{1 - Z_{kk}} + \frac{1}{1 - Z_{kk}} \sum_{i=1}^{k-1} Z_{ik} y_i\right) CL_k \quad (6)$$

The partial cancellation logic transfer functions can be calculated by imposing the cancellation of the transfer function of the first $m - 1$ quantization errors $E_k(z)$ in (6). This gives:

$$CL_k(z) = \frac{-Z_{km} CL_m}{1 - Z_{mm}} = \frac{-Z\left\{L^{-1}[H_D F_{km}]\big|_{nT_s}\right\} CL_m(z)}{1 - Z\left\{L^{-1}[H_D F_{mm}]\big|_{nT_s}\right\}} \quad (7)$$

where the partial cancellation logic transfer function of the last stage, $CL_m(z)$, can be chosen to be the simplest form that preserves the required noise shaping.

It is important to mention that the design equations (2)-(7) do not only take into account the single-stage loop filter transfer functions (F_{ii}), but also the inter-stage loop filter transfer functions ($F_{ij}, i \neq j$). The latter are continuous-time integrating paths appearing only when the modulator stages are connected to form the cascaded $\Sigma\Delta M$ and must be included in the synthesis methodology to obtain a functional modulator with minimum number of inter-stage paths.

Therefore, the following procedure can be used in a systematic methodology for the synthesis of cascaded CT $\Sigma\Delta Ms$[†2]:

- First, the poles of different transfer functions ($F_{ij}(s)$) are optimally placed in the signal bandwidth for given specifications. Scaling is needed in order to optimize the dynamic range of each integrator. This process is carried out entirely in the CT domain and no equivalence to an existing DT modulator needs to be imposed.
- Second, once the individual stages are designed and optimized, cancellation logics are calculated using (7).

4. SYNTHESIS EXAMPLE

For illustrative purposes, the 2-1-1 CT $\Sigma\Delta M$ of Fig.2(b) is synthesized using (2)-(7) to achieve 16-bit resolution in a 750 kHz bandwidth, with a sampling frequency of 48MHz (oversampling ratio, $M = 32$) [12]. Although in the proposed methodology the modulator in Fig.2(b) would be entirely designed in the CT domain, a slightly different approach will be used in this particular case in order to facilitate the comparison of the performance of both modulators in Fig.2. The coefficients of the first stage ($k_{in1}, k_{g1}, k_{fb1}, k_{fb2}$) are taken to be equal in both systems and are obtained from a DT-to-CT transformation of the first stage of a DT $\Sigma\Delta M$ in [12]. The rest of coefficients in Fig.2(b) are taken such that the time constant of the integrators is the inverse of the sampling frequency:

$$k_{in1} = -k_{fb1} = 1/4; \quad k_{fb2} = -3/8$$
$$k_{g1} = k_{in2} = -k_{fb3} = k_{in3} = -k_{fb4} = 1 \quad (8)$$

Hence, the single-loop and inter-stage transfer functions are given by:

$$F_{13} = \frac{(sT_s k_{fb2} + k_{fb1})k_{in2}k_{in3}}{(sT_s)^4}$$

$$F_{23} = \frac{k_{fb3}k_{in3}}{(sT_s)^2} \quad (9)$$

$$F_{33} = \frac{k_{fb4}}{sT_s}$$

and the partial cancellation logic transfer functions can be calculated using (7). This gives:

[†2]. In this procedure, the modulator order, oversampling ratio and number of bits of internal quantizers are assumed to be determined for given specifications from well-known expressions [1].

$$CL_1 = \frac{-Z\{L^{-1}[H_D F_{13}]|_{nT_s}\}}{\left(1 - Z\{L^{-1}[H_D F_{33}]|_{nT_s}\}\right)} CL_3$$

$$CL_2 = \frac{-Z\{L^{-1}[H_D F_{23}]|_{nT_s}\}}{\left(1 - Z\{L^{-1}[H_D F_{33}]|_{nT_s}\}\right)} CL_3 \quad (10)$$

From (8)-(10) and using a Non-Return-to-Zero (NRZ) DAC, the following cancellation logics are derived:

$$CL_1 = \frac{z^{-1}}{48}(7 + 29z^{-1} - 7z^{-2} - 5z^{-3})$$
$$CL_2 = z^{-1}(1 + z^{-1})(1 - z^{-1})^2 \quad (11)$$
$$CL_3 = 2(1 - z^{-1})^3$$

where CL_3 is chosen to have three zeroes at DC, corresponding to the zeroes contributed by the first three integrators.

In order to verify the proposed methodology, both modulators in Fig.2 were simulated using SIMSIDES, a SIMULINK-based time-domain behavioral simulator for $\Sigma\Delta Ms$ [13]. Fig.3 shows two ideal output spectra of the modulators

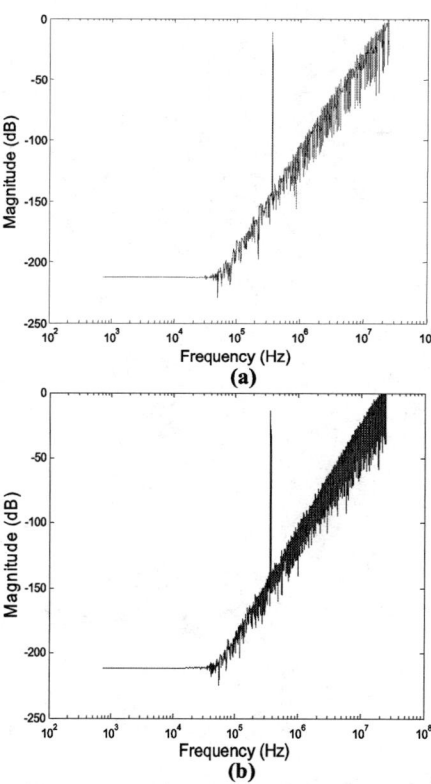

Figure 3. Output spectrum of a cascaded 2-1-1 CT $\Sigma\Delta M$ obtained from: (a) an equivalent DT $\Sigma\Delta M$; (b) the proposed synthesis method.

showing a similar performance. The effect of mismatch on the Signal-to-Noise Ratio (*SNR*) was also simulated. For this purpose, maximum values of mismatch were estimated for a 0.18 μm CMOS technology and both modulators in Fig.2 were simulated considering a Gm-C implementation. The results are shown in Fig.4, where the *SNR* loss is represented as a function of the standard deviation of the transconductances (σ_{gm}) and capacitances (σ_C). For each point of these surfaces, 150 simulations were carried out using random variations with the standard deviation given in the diagrams. The value of *SNR* loss represented in Fig.4 stands for the difference between the ideal *SNR*, i.e with no parameter variation, and the *SNR* with 90% of the 150 simulations above it. It is shown that the lower analog component count in Fig.2(b) is reflected in a lower variance of the modulator coefficients, leading to a better behaviour in terms of sensitivity to mismatch.

The same reasoning can be applied to the requirements in terms of DC gain of the individual transconductors. Since the number of transconductors connected to the same node is higher in the previous method, the equivalent impedance associated with these nodes tends to be lower and the requirements in terms of the individual transconductor output impedance is higher. Therefore, a higher DC gain is needed. Fig.5 illustrates this point. Note that the *SNR*-peak starts dropping at a DC gain $\cong 8$ dB higher in the case of the system designed following the previous method.

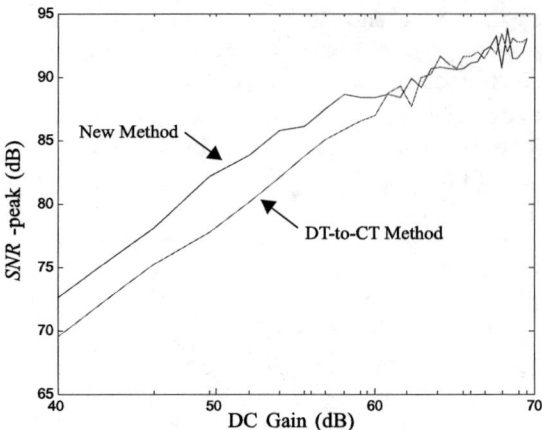

Figure 5. Effect of DC gain on the *SNR*-peak.

CONCLUSIONS

In this paper a new methodology of synthesizing cascaded continuous-time ΣΔ modulators has been presented. It is demonstrated that more efficient topologies in terms of circuit complexity can be generated if the design is directly done in the continuous-time domain and the cancellation logic transfer function is taken into account in the synthesis procedure. Behavioral simulations considering critical error mechanisms validate the presented approach.

REFERENCES

[1] A. Rodríguez-Vázquez, F. Medeiro and E. Janssens (Editors): *CMOS Telecom Data Converters*. Kluwer, 2003.
[2] J.A. Cherry and W.M. Snelgrove: *Continuous-Time Delta-Sigma Modulators for High-Speed A/D Conversion*. Kluwer, 2000.
[3] L. Breems and J.H. Huijsing: *Continuous-Time Sigma-Delta Modulation for A/D Conversion in Radio Receivers*. Kluwer, 2001.
[4] M. Moyal, M. Groepl, H. Werker, G. Mitteregger, J. Schambacher: "A 700/900mW/Channel CMOS Dual Analog Front-End IC for VDSL with Integrated 11.5/14.5dBm Line Drivers". *Proc. of the 2003 IEEE Int. Solid-State Circuits Conf.*, pp. 416-417.
[5] S. Patón, A. Di Giandoménico, L. Hernández, A. Wiesbauer, T. Pötscher and M. Clara: "A 70-mW 300-MHz CMOS Continuous-Time ΣΔ ADC With 15-MHz Bandwidth and 11 Bits of Resolution". *IEEE Journal of Solid-State Circuits*, Vol. 39, pp. 1056-1063, July 2004.
[6] L. J. Breems: "A Cascaded Continuous-Time ΣΔ Modulator with 67dB Dynamic Range in 10MHz Bandwidth". *Proc. of the 2004 IEEE Int. Solid-State Circuits Conf.*, pp. 72-73.
[7] C.-H. Lin and M. Ismail: "Synthesis and analysis of high-order cascaded continuous-time Sigma-Delta modulators". *Proc. of the 1999 IEEE Int. Conf. on Electronics, Circuits and Systems*, pp. 1693-1696.
[8] M. Ortmanns, F. Gerfers, and Y. Manoli: "On the Synthesis of Cascaded Continuous-Time Sigma-Delta Modulators". *Proc. of the 2001 IEEE Int. Symposium on Circuits and Systems*, pp. 419-422.
[9] O. Oliaei: "Design of Continuous-Time Sigma-Delta Modulators with Arbitrary Feedback Waveform". *IEEE Transactions on Circuits and Systems-II*, Vol. 50, pp. 437-444, August 2003.
[10] O. Shoaei: *Continuous-Time Delta-Sigma A/D Converters for High Speed Applications*. PhD Thesis, Carleton University, 1995.
[11] H. Aboushady, M. Louerat: "Systematic Approach for Discrete-Time to Continuous-Time Transformation of ΣΔ Modulators". *Proc. of the 2002 IEEE Int. Symposium on Circuits and Systems*, Vol. 4, pp. 229-232.
[12] G. Yin; W. Sansen: "A High-Frequency and High-Resolution Fourth-Order ΣΔ A/D Converter in BiCMOS Technology". *IEEE Journal of Solid-State Circuits*, Vol. 29, pp. 857-865, August 1994.
[13] J. Ruiz-Amaya, J.M. de la Rosa, F. Medeiro, F.V. Fernández, R. del Río, B. Pérez-Verdú and A. Rodríguez-Vázquez: "An Optimization-based Tool for the High-Level Synthesis of Discrete-time and Continuous-Time ΣΔ Modulators in the MATLAB/SIMULINK Environment". *Proc. IEEE Int. Symp. Circuits and Systems*, Vol V., pp. 97-100, 2004.

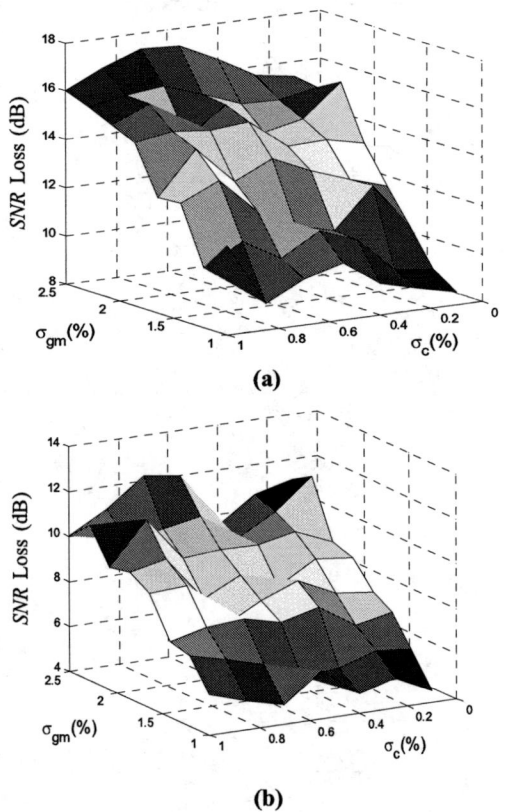

Figure 4. Effect of mismatch on the *SNR* of a cascaded 2-1-1 CT ΣΔM obtained from: (a) an equivalent DT ΣΔM; (b) proposed method.

A Sixth-order Subsampling Continuous-time Bandpass Delta-Sigma Modulator

Yuan Chen, Kei-Tee Tiew
Centre for Integrated Circuits and Systems
School of Electrical and Electronic Engineering
Nanyang Technological University
Singapore 639798
Ch0005an@ntu.edu.sg , EKTTiew@ntu.edu.sg

Abstract—A sixth-order bandpass (BP) delta-sigma modulator (ΔΣM) that uses subsampling relative to the radio frequency (RF) or intermediate frequency (IF) of a receiver, while oversampling the signal bandwidth is described in this paper. The continuous-time (CT) filter that works at RF/IF frequency together with the discrete-time (DT) filter operating at lower frequency provide both noise shaping to quantization noise and subsampling noise. The transfer function derivation with finite Q-factor bandpass filter is analyzed. Pulse-shaped DAC is used to reduce the DAC clock jitter and realize the frequency translation. Because the sampling frequency is much lower than the RF/IF frequency, the complexity and power consumption of the subsequent digital signal processing stage is greatly reduced

I. Introduction

Recent years have seen an increasing interest in using bandpass delta-sigma modulators placed very close to the front-end of a receiver to provide relatively narrowband, high-resolution analog-to-digital conversion in the front-end analog signal processing, so as to push more signal processing functions into the digital domain for its programmability and low-voltage low-power operation.

Normally for bandpass delta-sigma modulators, the sampling frequency is four times the center frequency, resulting in a very high sampling frequency. Sampling at such high frequencies requires expensive processes and increases the complexity and power consumption of the digital signal processing stage. This problem can be solved by using a continuous-time architecture employing subsampling with respect to the IF signal [1] as depicted in Fig.1. This enables sampling at a frequency much lower than the center frequency.

The operation of the subsampling CT ΔΣM in [1] can be explained by constructing the resulting spectra of the signals inside the ΔΣM. The ΔΣM is sampled at a rate of f_s. The center frequency of the continuous-time filter is tuned to $f_0=(N-0.25)f_s$, which is much higher than f_s. When an input signal with a frequency f_{in} around f_0 passes through the continuous-time filter and is subsampled by the sample-and hold (S/H), the input signal will be aliased to frequency f_a according to:

$$f_a = \pm(f_{in} - N \cdot f_s). \qquad (1)$$

f_a is around $0.25f_s$, which realizes the signal frequency down-conversion. N is called the subsampling factor. The output signal at f_a is up-converted to f_{in} by the mixer in the feedback path, thereby increasing both the efficiency of the DAC at the frequency f_{in} and the maximum possible input amplitude.

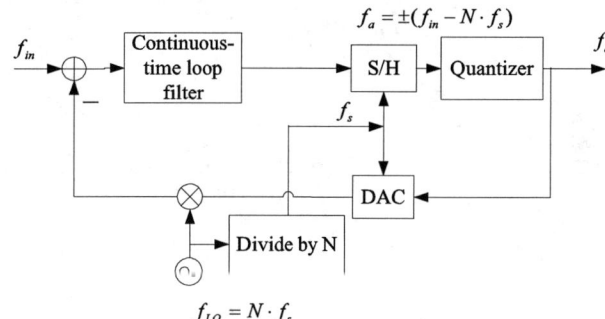

Figure 1. Block diagram of the subsampling CT ΔΣM in [1]

Subsampling operation will introduce more noise than conventional sample-and-hold [2]. In the subsampling CT ΔΣM, the subsampling operation is inside the loop and hence the sampling errors are attenuated by the loop.

One of the problems of this architecture is that the requirement for the CT filter is very high. The subsampling operation will reduce the effective Q-factor of the CT filter and change the noise transfer function (NTF) of the quantization noise and subsampling noise, thereby degrading the modulator performance.

In this paper, the transfer function with finite Q-factor bandpass filter is derived and a new architecture is proposed to reduce the requirement for CT filter and increase the modulator performance.

II. Transfer Function Derivation with Finite Q-Factor Bandpass Filter

For the subsampling ΔΣM depicted in Fig.1, from the DAC in the feedback to the S/H is internally a discrete-time system as shown in Fig.2.

A continuous-time bandpass filter is given as :

$$\hat{H}(s) = \frac{A(s-B)}{s^2+s\omega_0/Q+\omega_0^2}, \qquad (2)$$

where A is a constant gain factor and AB is the coefficient for s^0 in the numerator. Assuming the local oscillator signal is $\sin \omega_l t$ and the mixer output is the multiplication of the NRZ DAC output and the local oscillator signal.

$$M(t) = \sin \omega_l t [u(t) - u(t - T_s)] \quad (3)$$

This can be transformed to the s-domain as

$$M(s) = \frac{\omega_l}{s^2 + \omega_l^2} \cdot (1 - e^{-sT_s}) \quad (4)$$

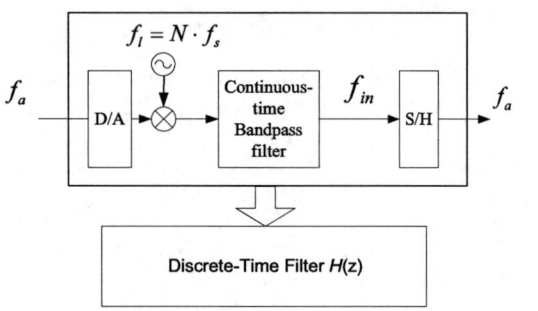

Figure 2. Equivalent DT transfer function from the DAC to S/H

Using impulse-invariant transformation [3], the equivalent DT transfer function from the DAC to S/H is then

$$H(z) = ZL^{-1}\left[\frac{A(s-B)}{s^2 + s\omega_0/Q + \omega_0^2} \cdot \frac{\omega_l}{s^2 + \omega_l^2} \cdot (1 - e^{-sT_s})\right] \quad (5)$$

Solving eq.(5), when $Q \gg 1$

$$H(z) \approx g_t \cdot \frac{-k_1 z^{-1} - k_2 z^{-2}}{1 + a^2 z^{-2}} \quad (6)$$

Where

$$a = e^{-\frac{\omega_0}{2Q} \cdot T_s}, \quad (7)$$

From eq.(7) it can be seen that a determines the poles of $H(z)$ and how much noise shaping can be provided by this loop filter.

For the conventional CT bandpass $\Delta\Sigma M$ without subsampling, $f_0 = 0.25 f_s$, then

$$a = a_1 = e^{-\frac{2\pi f_s}{8Q} \cdot \frac{1}{f_s}} = e^{-\frac{\pi}{4Q}}. \quad (8)$$

For the subsampling bandpass $\Delta\Sigma M$, $f_0 = (N-0.25)f_s$,

$$a = a_2 = e^{-\frac{2\pi f_s (N-0.25)}{2Q} \cdot \frac{1}{f_s}} = e^{-\frac{\pi(N-0.25)}{Q}} \quad (9)$$

Rewriting eq.(9)

$$a_2 = e^{-\frac{\pi(N-0.25)}{Q}} = e^{-\frac{\pi}{4Q^*}} \quad (10)$$

Where

$$Q^* = Q/[4(N - 0.25)]. \quad (11)$$

It can be seen from eq.(11) that after subsampling, the effective Q-factor of the CT filter is reduced by a factor of $1/4(N-0.25)$. It is this filter with a Q-factor of Q^* that provides the transfer function zeros for the quantization noise and noise introduced by the subsampling operation.

III. THE PROPOSED SUBSAMPLING BANDPASS $\Delta\Sigma M$

From the preceding analysis, the effective Q-factor of the CT filter after the subsampling operation is reduced and the filter may not suppress the quantization noise adequately.

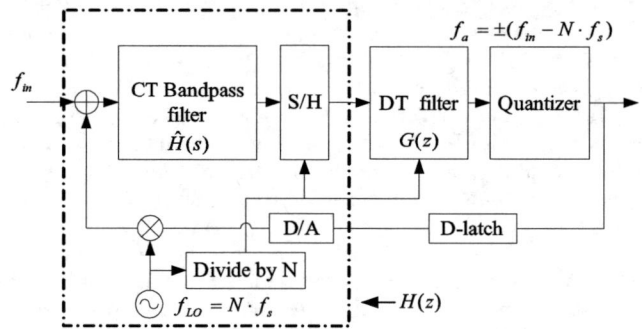

Figure 3. Block diagram of the proposed subsampling BP $\Delta\Sigma M$

To overcome this problem, a novel subsampling $\Delta\Sigma M$ is proposed. The modulator is operating at 70MHz IF under 40MHz sampling frequency (subsampling factor $N=2$) with 200kHz bandwidth. The 70MHz IF was down-converted to 10MHz by subsampling. The block diagram is depicted in Fig.3. A fourth-order DT filter follows a relatively low-Q (e.g. Q=30) second-order CT bandpass filter to give additional noise shaping for the quantization noise. Different from the $\Delta\Sigma M$ in [4], after subsampling, the IF signal is down-converted to a lower IF instead of to baseband by a pair of I-Q mixers. So it requires less complex circuit and does not suffer from I-Q mismatch problem. The subsampling operation is performed after the CT filter and the subsampling noise will be attenuated by the loop. A latch is placed before the DAC in the feedback loop to eliminate the nonzero excess loop delay and signal-dependent delay of the quantizer.

A. Local Oscillator

As the mixer is placed within the feedback loop after the DAC, any additional distortion introduced by the mixer will add directly to the distortion in the output of the $\Delta\Sigma M$. However, the modulator is less sensitive to the DAC jitter due to mixing with selected local oscillator (LO).

The jitter effect with different local-oscillator signal is shown in Fig.4. The left one is rectangle wave oscillator, if the NRZ DAC pulse is mixed with the rectangle wave oscillator; it forms a RZ pulse. For a timing error Δt, the charging error Δq is a big fraction of the total DAC power. If the NRZ DAC pulse is mixed with the sinusoidal wave oscillator, the DAC pulse is transformed to a sinusoidal pulse. With the same timing error Δt, the charging error Δq is smaller than the conventional NRZ/RZ pulse. However, sinusoidal signal has the highest slope at the zero crossing. Hence any loss of synchronization between the sinusoidal signal and clock signal will greatly increase the amplitude of the pulse at the sampling instants and hence increase the charging error Δq. In the proposed ΔΣM, a raised cosine oscillator as shown on the right in Fig.4 is used. First and second-order insensitivity to clock jitter are expected for the raised cosine oscillator one since the pulse has both zero value and zero slope at the sampling instants [5].

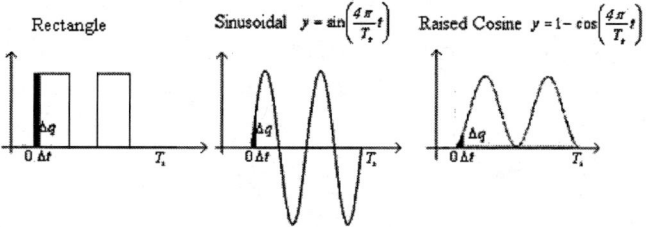

Figure 4. DAC jitter effect with different local-oscillator signals

So a raised cosine oscillator is used in the proposed ΔΣM to reduce the clock jitter sensitivity. To realize the frequency up-conversion, the frequency of the LO $f_l=2f_0=80$MHz for a subsampling factor $N=2$. There are two raised cosine pulses in one sampling period. The phase noise is expected to be improved due to the averaging effect [5]. So with a well-designed mixer, a better SNDR can be achieved compared to the conventional CT ΔΣM without mixer in the feedback path.

B. BP ΔΣM Structures & NTF

The bandpass ΔΣM uses the cascade-of-resonate-in-feedforward (CRFF) structure. Compared to a ΔΣM using cascade-of-resonate-in-feedback (CRFB) structure, the CRFF structure provides more noise shaping for the subsampling noise when their NTFs for quantization noise are the same.

Referring to Fig.3, the NTFs for quantization noise and subsampling noises are:

$$NTF_q = \frac{1}{1+H(z)G(z)z^{-1}} \quad (12)$$

$$NTF_s = \frac{G(z)}{1+H(z)G(z)z^{-1}} \quad (13)$$

The second-order CT filter and fourth-order DT filter form a sixth-order noise shaping for the quantization noise and second-order noise shaping for the subsampling noise. These NTFs can be designed by designing the equivalent DT filter, from the DAC to S/H, $H(z)$ and the following DT filter $G(z)$. In the proposed ΔΣM, the Q-factor of the CT filter is 30 to reduce the CT filter requirement. The inband rms gain of the NTF_q and NTF_s is −63dB and −8dB, respectively.

The noise shaping for the subsampling noise is much lower than the noise shaping for the quantization noise, because it is mainly decided by the relatively low Q CT filter. However, as the subsampling noise is at a much lower level compared to the quantization noise when the bandwidth is narrow [6], the lower level of noise shaping is acceptable.

IV. SIMULATION RESULTS

In MATLAB simulation, the peak SNDR is 85dB with a signal bandwidth of 200kHz, 40MHz sampling frequency when the input signal amplitude is −7dBFS and all the models are ideal. The 70MHz IF signal is down converted to 10MHz and the band-reject noise shaping is centered at 10MHz.

The inband output spectrum of the proposed ΔΣM with subsampling noise is depicted in Fig.5. Fig. 6 shows the SNDR versus input level plot of the ideal proposed ΔΣM and the proposed ΔΣM with the added subsampling noise.

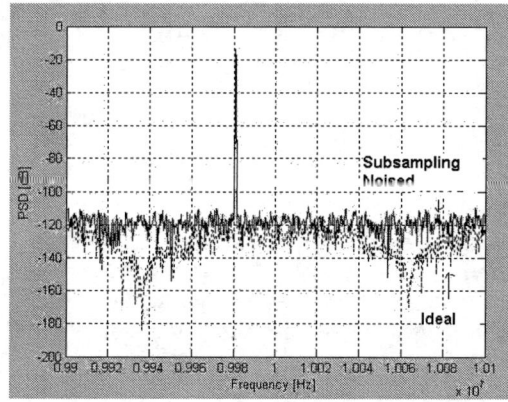

Figure 5. Inband output spectrum of the proposed ΔΣM with subsampling noise

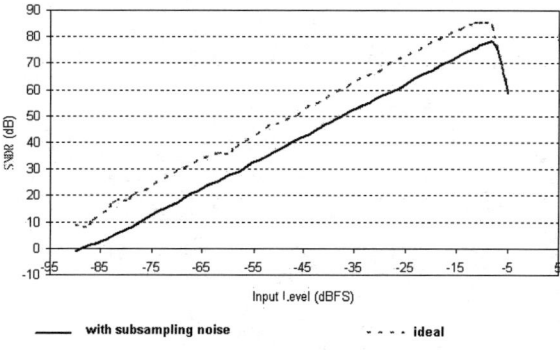

Figure 6. SNDR versus input level plot of the ideal/ subsampling noised proposed ΔΣM

The continuous-time filter provides -8dB noise shaping to the subsampling noise. To completely suppress all the subsampling noise under the quantization noise floor, a Q=100 CT filter is necessary based on the simulation. However, when higher Q-factor CT filter is used, the matching requirement is also more stringent. The performance degradation for the center frequency mismatch is shown in Table 1. Although when Q is 30, the noise shaping is not enough to suppress all the noise in S/H, the requirement for the continuous-time filter is relaxed.

The coefficient spread of the DT filter is very low at around 3.7 and also this DT filter architecture is quite insensitive to variation in coefficients. The SNDR drop is less than 1dB when the coefficient mismatch is 2%.

TABLE I. PEAK SNDR VS. CENTER FREQUENCY MISMATCH OF THE CT BANDPASS FILTER WITH DIFFERENT Q (WITH SUBSAMPLING NOISE)

Q factor	100	30
0 Mismatch	87dB	78dB
0.5% Mismatch	84dB	77.5dB
1% Mismatch	77dB	75.5dB

Thanks to the raised cosine oscillator, the proposed $\Delta\Sigma M$ is very insensitive to the DAC clock jitter. The SNDR performance is almost the same with 1% DAC clock jitter. With the same clock jitter in DAC, the SNDR of the $\Delta\Sigma M$ with sinusoidal oscillator drops 12dB, compared to that of 23dB drop in the SNDR of the $\Delta\Sigma M$ with rectangle oscillator! The simulation results are shown in Fig.7. The subsampling noise is not introduced in these DAC jitter simulations.

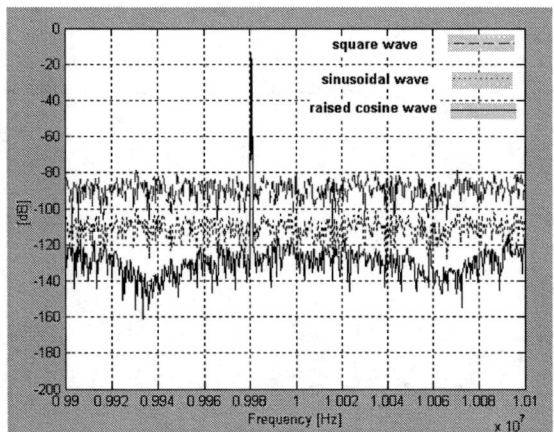

Figure 7. Inband output spectrum of the 1%DAC jittered proposed $\Delta\Sigma M$ with different local-oscillator

Higher SNDR can be achieved by increasing the Q-factor of the CT filter; it will increase both the noise shaping for quantization noise and subsampling noise. The plot of peak SNDR of the $\Delta\Sigma M$ with/without the DT filter versus Q-factor of the CT filter is depicted in Fig.8. The subsampling noise is included in this simulation. It can be seen from Fig.8 that when the Q-factor is not very high, the noise shaping provided by the CT filter cannot suppress the quantization noise adequately. So the additional noise shaping provided by the DT filter is very necessary. The relatively low Q CT filter can suppress the noise introduced by the S/H which is much lower than the quantization noise, due to the fact that the S/H is placed after the CT filter in the proposed $\Delta\Sigma M$.

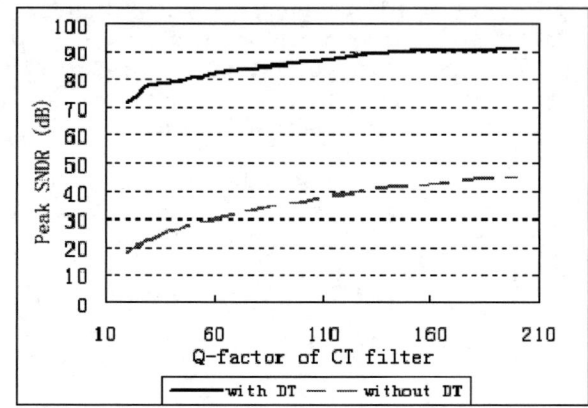

Figure 8. Peak SNDR vs. Q-factor of CT bandpass filter

V. CONCLUSIONS

A new subsampling continuous-time bandpass delta-sigma modulator is presented based on the derived transfer function. The second-order continuous-time bandpass filter operates at high frequency and after subsampling the fourth-order discrete-time loop filter operates at low frequency. These two filters working together provide noise shaping both to the quantization noise and the noise introduced by the S/H. Pulse-shaped DAC is used to reduce the clock jitter sensitivity and realize the frequency translation. The subsampling technique reduces the complexity and power consumption needed in the discrete-time loop filter and subsequent digital signal processing stages.

REFERENCES

[1] A.I.Hussein, W.B. Kuhn, "Bandpass $\Delta\Sigma M$ employing undersampling of RF signals for wireless communication," IEEE Trans. on Circuit and System II, vol. 47, July 2000, pp 614-620.

[2] D.H.Shen, "A 900-MHz RF frond end with integrated discrete-time filters," IEEE J.Solid-State Circuit, vol.31, Dec 1999, pp 1945-1954.

[3] F. M. Gardner, "A transformation for digital simulation of analog filters," IEEE Trans on Commun., vol. 44, July 1986, pp. 676–680.

[4] Hai Tao, J.M. Khoury, "A 400-Ms/s frequency translating bandpass sigma-delta modulator " IEEE J.Solid-State Circuit, vol.34, Dec 1999, pp. 1741-1752.

[5] S.Luschas; H.S. Lee, "High-speed sigma delta modulators with reduced timing jitter sensitivity," IEEE Trans on Circuits and Systems II, vol. 49, Nov. 2002, pp 712-720.

HIGH-ORDER SINGLE-LOOP DOUBLE-SAMPLING SIGMA-DELTA MODULATOR TOPOLOGIES FOR BROADBAND APPLICATIONS

Mohammad Yavari and Omid Shoaei

IC Design Lab, ECE Department, University of Tehran, Tehran 14395-515, Iran
E-mail: myavari@ut.ac.ir

ABSTRACT

This paper presents novel low-voltage high order single loop sigma-delta modulator structures for wideband applications. The proposed architectures employ the technique of double-sampling to double the effective oversampling ratio (OSR) without increasing the sampling frequency. To alleviate the quantization noise folding into the inband frequency region which is a result of the mismatch between the sampling capacitors of the feedback's digital-to-analog converter (DAC) paths, a zero is placed at the half of the sampling frequency of the modulator's noise transfer function (NTF). The problem of this additional zero in the NTF is solved through the proposed modulator architectures with a very simple design procedure.

1. INTRODUCTION

Sigma-delta ADCs are the main candidates for high resolution due to their inherent immunity to the circuit non idealities. In order to employ them in broadband applications, a low oversampling ratio (OSR) must be considered. But, the reduction of OSR decreases the accuracy of the modulator, drastically. So, novel modulator structures are needed to alleviate the reduction of resolution in low OSR applications.

A useful approach in switched-capacitor realizations of the modulators is to employ the double-sampling technique. In this method, the circuit operates during both phases of the clock. Hence, the effective sampling frequency of the system is twice that of the clock frequency. This results in doubling the OSR and/or the available time for settling of the integrators if OSR is fixed. Fig. 1 shows a single-ended double-sampling integrator. In this circuit, two distinct capacitors are used to sample the input signal. The capacitor C_{S1} is used to sample the input signal at the phase ϕ_1 while at this interval the stored charge on C_{S2} is transferred into the integrating capacitor, C_I. At the next phase, C_{S2} samples the input signal and C_{S1} transfers its stored charge into C_I. Hence, in both phases of the clock, the sampling and integrating is performed resulting in doubling the effective sampling rate of the system. Unfortunately, double-sampling sigma-delta modulators are sensitive to the mismatch between the two sampling capacitors, C_{S1} and C_{S2}. This results in folding the quantization noise into the signal band through the paths of the feedback DAC. Several techniques have been proposed to alleviate the quantization noise folding in doubled-sampled modulators such as employing fully floating method in the feedback DAC paths and/or placing a zero at the half of the sampling frequency of the NTF, etc [1-3].

The effect of placing a zero at the half of the sampling frequency ($f_s/2$) of the NTF should be considered in the design of the modulator's NTF. Its effect is not considerable in the first and second order modulators. But, the NTF of high order structures is affected drastically using both above-mentioned techniques resulting in ineffective and even unstable modulators. So, the design of a stable high order double-sampling modulator with a zero at $f_s/2$ would be complicated. In [4] a systematic procedure has been proposed to design the stable double-sampling modulators. But, this technique is complex and results in the modulator structures that are not efficient for wideband applications mainly due to the fact that they employ several multibit DACs in the modulator's loop which results in more area, power, and circuit complexity.

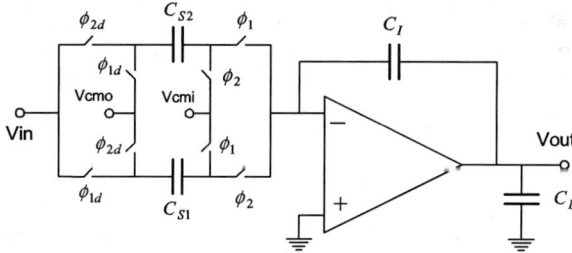

Figure 1: Single-ended double-sampling integrator.

In this paper, novel sigma-delta modulator topologies employing the double-sampling technique are proposed. The proposed modulator structures employ finite impulse response (FIR) NTF with a zero at $f_s/2$. The paper is organized as follows. Section (2) describes the derivations of the proposed modulator structures. In section (2) the circuit requirements to realize the proposed modulators are also described. Section (3) provides simulation results. The conclusions are given in section (4).

2. PROPOSED MODULATOR TOPOLOGIES

The general structure of a sigma-delta modulator is shown in Fig. 2 where $H(z)$ is the loop transfer function and its NTF is given by

$$NTF(z) = \frac{1}{1+H(z)} \qquad (1)$$

In this structure the signal transfer function (STF) is unity. Unity gain STF in a sigma-delta ADC has many advantages such as the followings. First, the effects of the circuit non-idealities such as the limited opamp DC gain and nonlinearities are reduced since only the quantization noise is processed by the integrators [5]. Second, the dynamic range is increased because the quantizer tracks the input signal through a direct path to the its input. Third, the integrators need small output swings since only the quantization noise is processed by them.

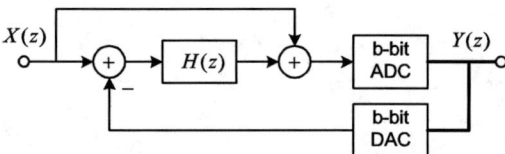

Figure 2: General structure of a sigma-delta modulator.

The loop transfer function, $H(z)$, can be obtained in term of the NTF as follows:

$$H(z) = \frac{1 - NTF(z)}{NTF(z)} \quad (2)$$

The NTF of the proposed double-sampling sigma delta modulators is considered as FIR filter as follows:

$$NTF(z) = \begin{cases} (1-z^{-1})(1+z^{-1}) \prod_{i=1}^{M} (1 - \alpha_i z^{-1} + z^{-2}) & \text{if } L \text{ is odd} \\ (1-z^{-1})^2 (1+z^{-1}) \prod_{i=1}^{M-1} (1 - \alpha_i z^{-1} + z^{-2}) & \text{if } L \text{ is even} \end{cases} \quad (3)$$

where $M = [L/2]$ and L is the order of the modulator and is considered greater than two. For the second order structure only the first two terms of the NTF is assumed. A zero is placed at $f_s/2$ of the NTF through the term of $(1+z^{-1})$ to reduce the folding effect of the quantization noise into the signal band which is a result of the paths gain mismatch between the feedback DAC capacitors as discussed above. The term $(1-z^{-1})$ is needed to realize the input stage of the modulator as an integrator without any local feedback DAC. For odd order modulators only one zero of the NTF is placed at DC and the other zeros are located in the inband frequencies to shape out the quantization noise aggressively. For even order structures two zeros of the NTF are located at DC and the others at the inband frequencies. Another zero at DC is needed in even order modulators compared to the odd order structures in order to remove the requirement of the local feedback DAC around first pair of integrators. It is worth mentioning that the local feedback DAC around the first pair of integrators enhances the kT/C noise of the modulator, directly.

The loop transfer function, $H(z)$, of the proposed modulators is obtained with relations (2) and (3). For example, $H(z)$ of the second and third order modulators is as follows, respectively:

$$H_2(z) = \frac{z^{-1}}{1-z^{-1}} \frac{1+z^{-1}-z^{-2}}{1-z^{-2}} = \frac{z^{-1}}{1-z^{-1}} \left(1 + \frac{z^{-1}}{1-z^{-2}}\right) \quad (4)$$

$$H_3(z) = \frac{z^{-1}}{1-z^{-1}} \left(\alpha + \frac{\alpha(\alpha-1)z^{-1} + \alpha(\alpha-2)z^{-2} + (1-\alpha)z^{-3}}{1 + (1-\alpha)z^{-1} + (1-\alpha)z^{-2} + z^{-3}} \right) \quad (5)$$

As it is seen, the loop transfer function, $H(z)$, of the proposed modulators can be realized as an integrator and an infinite impulse response (IIR) filter. For example, the realization of the proposed third order modulator is shown in Fig. 3 where only one integrator and a third order IIR filter is used for its implementation. Although such realization of the proposed FIR NTF double-sampling modulators can be obtained using switched-capacitor circuits in CMOS technologies, but the number of the modulator coefficients becomes large for high order structures. So this realization will be more sensitive to the mismatch between the coefficients. The other drawback of this realization is the existing of three paths terminating to the input of the quantizer which demands a delay free gain stage. So, an efficient implementation which needs lower number of coefficients and minimum terminating paths to the quantizer input is proposed in the following. This realization uses only one first order IIR filter and more integrators to reduce the number of modulator coefficients.

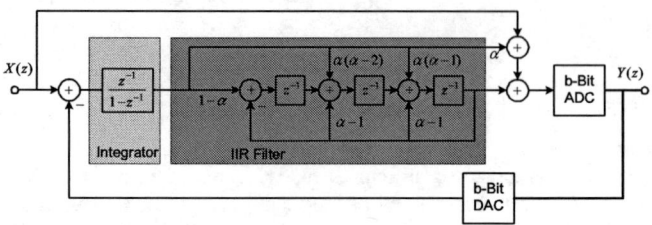

Figure 3: Third order double-sampling modulator.

The proposed NTF for double-sampling modulators can also be implemented efficiently as the combination of the integrators and only one first order IIR filter as shown for the third, fourth and fifth order modulators in Figs. 4, 5 and 6, respectively. In this realization, the summer at the input of the quantizer can be implemented with a passive switched-capacitor circuit as proposed in [5]. But, in the passive realization of this summer a loss of two is obtained where it should be compensated by the quantizer. This means that the voltage reference of the quantizer should be reduced by two making the design of the quantizer especially in low voltage circuits complicated. However, the realization of a 4-5 bit flash ADC with one bit higher accuracy which is used in the loop of sigma-delta modulators is not very complex. The realization of high order modulators can be obtained with a straight forward extension of the proposed structures. As it is seen to realize the even order modulators (see Fig. 5), the structure is somewhat different and needs other delays in the modulator feedforward paths coefficients. However, the term $(1+z^{-1})$ can be realized using fully floating technique as proposed in [1] without any effective extra circuit requirement.

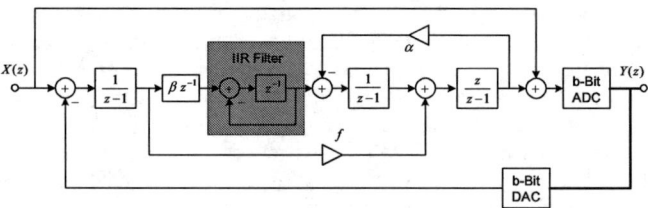

Figure 4: The proposed third order double-sampling modulator.

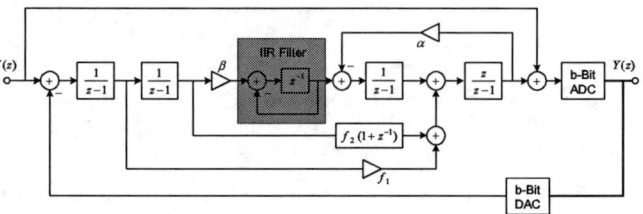

Figure 5: The proposed fourth order double-sampling modulator.

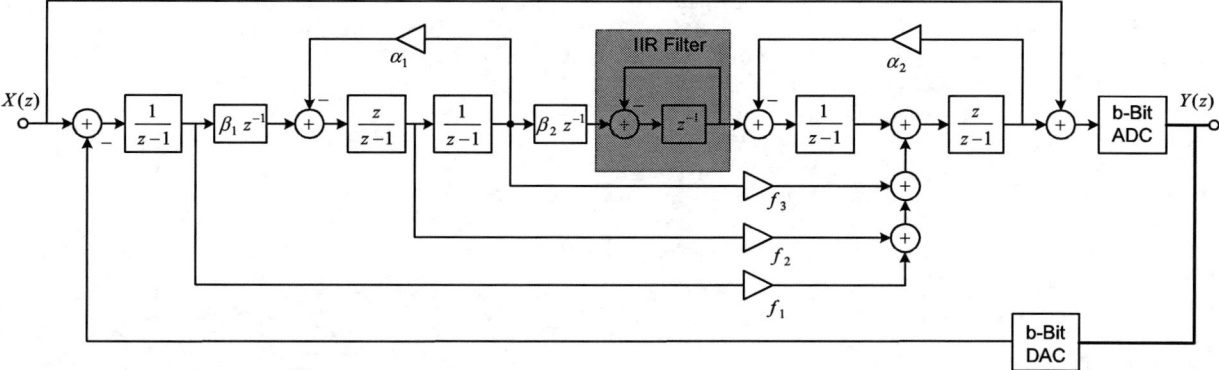

Figure 6: The proposed fifth order double-sampling modulator.

The coefficients of the proposed modulators to obtain the FIR NTF for the third, fourth and fifth order structures are as follows:

$$\beta = 1, \quad f = 2 - \alpha \qquad \text{for} \quad L = 3 \qquad (6)$$

$$\beta = 1, \quad f_1 = 3 - \alpha, \quad f_2 = 1 \qquad \text{for} \quad L = 4 \qquad (7)$$

$$\begin{aligned} &\beta_1 = \beta_2 = 1, \quad f_1 = 4 - \alpha_1 - \alpha_2 \\ &f_2 = f_3 = 3 - \alpha_1(4 - \alpha_1) \end{aligned} \qquad \text{for} \quad L = 5 \qquad (8)$$

The coefficients α_i are needed to place some of NTF's zeros at the inband frequencies. The design method proposed by Schreier [6] can be used to obtain the values of α_i in order to place the inband zeros at the optimal points. The other coefficients are obtained such that the NTF become an FIR filter as given in the relations (6)-(8) for the proposed third, fourth and fifth order modulators.

To realize the IIR filters, the generic low-Q biquad structure can be used [7]. The unit delays of the feedforward paths can be implemented with appropriate designing of the clock phases in switched-capacitor circuits without any extra circuit [8].

3. SIMULATION RESULTS

To show the usefulness of the proposed modulator topologies, their system level architectures taken into account the circuit non-idealities were simulated. Simulation results of the proposed third, fourth and fifth order structures shown in Figs. 4, 5 and 6 are only presented here.

To get the optimal location of the inband zeros, MATLAB with Schreier's Delta-Sigma Toolbox [9] was used. The proposed modulators were designed for OSR = 8 and b = 5. The values of α = 0.092, α = 0.109 and α_1 = 0.125, α_2 = 0.045 were obtained for the third, fourth and fifth order modulators, respectively. The other coefficients of the modulators were determined from the relations (6), (7) and (8). Then, signal scaling was performed to limit the output swing of the integrators and the other blocks for a real implementation. A gain is placed in the input signal path of the first integrator to achieve the overload level factor of one for all three structures with a safety margin. The value of this gain is 0.875, 0.8 and 0.6 for the third, fourth, and fifth order architectures, respectively. The resultant ideal maximum SNDR for a 0-dBFS sinusoidal input is about 73 dB, 84.4 dB and 98 dB for the third, fourth and fifth order modulators, respectively. Figure 7 shows the output spectrum of the simulated modulators.

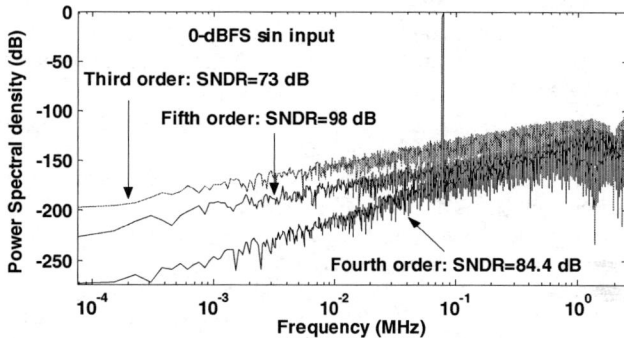

Figure 7: Power spectral density.

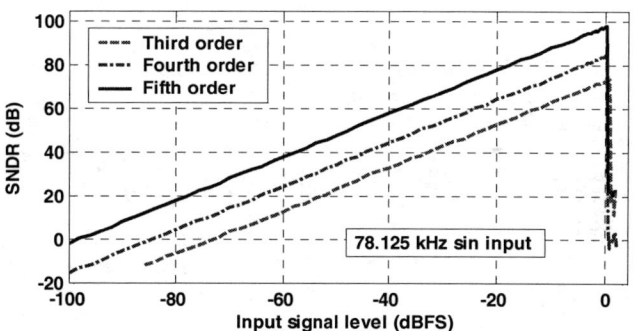

Figure 8: Dynamic range: SNDR vs. input signal level.

In Fig. 8 SNDR is shown versus the input signal level of the proposed modulators. As it is seen the proposed modulators are stable for an input signal with amplitude greater than of the feedback reference level. This is mainly due to a direct path from the input signal to the quantizer input which results in a wide input signal range.

Simulation results also show that the SNDR degradation versus the mismatch error of the modulator coefficients is negligible even with a coefficients mismatch of 5% if the gain of paths is completely matched.

The main concern of double-sampled modulators is the quantization noise folding due to the paths gain mismatches of the feedback DAC capacitors. Figure 9 shows SNDR versus the mismatch of the sampling paths of the first integrator. As it is seen the SNDR degradation of the proposed third order modulator is negligible even with a 1% mismatch between the sampling paths. With a mismatch of 0.3%, SNDR of the proposed fourth order modulator degrades about 1 dB. However, the proposed fifth order is more sensitive to the mismatch between the sampling paths. With a mismatch of 0.05% between the sampling paths of the first integrator, SNDR degrades about 4 dB. The paths mismatch of the remaining blocks does not affect the SNDR of the fifth order modulator considerably even with a 0.2% mismatch. However it should be noted that since the sampling capacitors of the first integrator are large due to the circuit noise considerations, a paths mismatch of about 0.05% is realizable as well [2].

The other circuit requirements of the proposed architectures such as the amplifier finite DC gain are more relaxed and simulation results show that about 40 dB DC gain for the first integrator is sufficient to prevent any SNDR degradation with an enough margin (about 10 dB). The proposed architectures have small output swings compared to the conventional distributed feedback and weighted feedforward architectures [10] due to the unity-gain STF and also feedforward paths.

In these simulations ideal DAC elements have been assumed. However, in the real implementations, dynamic element matching (DEM) such as data weighted averaging (DWA) and calibration or correction techniques can be used to correct the DAC errors [10, 11]. It is worth mentioning that only one DAC is used in the proposed modulators and it is located at the input of the first integrator. The value of the sampling and feedback capacitors of this integrator are determined due to the kT/C noise considerations and their sizes are large in the high-resolution and low-voltage applications.

The main features of the proposed modulator structures are summarized as follows.

1. Due to the single loop structure they are suitable for very low-voltage applications because the proposed topologies demand the relaxed analog circuit requirements.

2. High speed applications can be obtained using low speed clock frequencies. This is achieved using the double-sampling technique.

3. Only one multibit DAC is used in the loop of the modulator. This decreases the circuit complexity, greatly.

4. FIR NTF is used to achieve the aggressive shaping of the quantization noise and a very simple design of the modulator transfer function. Most of the NTF zeros are placed at the inband frequency region to shape out the quantization aggressively.

5. Only two paths are terminated to the input of the quantizer. This results in the realization of the summer preceding the quantizer as a passive switched-capacitor circuit.

4. CONCLUSION

In this paper, novel single loop sigma delta modulator topologies for broadband applications in the low-voltage environments using only a low OSR were proposed. To alleviate the quantization noise folding effect into the signal band, an FIR NTF with an additional zero at $f_s/2$ was used. Unity-gain STF was employed to decrease the modulator's sensitivity to the circuit non-idealities. Only one multibit DAC is needed in the feedback loop which greatly decreases the circuit implementation complexity. To compensate the errors resulted from the DAC unit elements mismatch, DAC linearization techniques such as DWA can also be used.

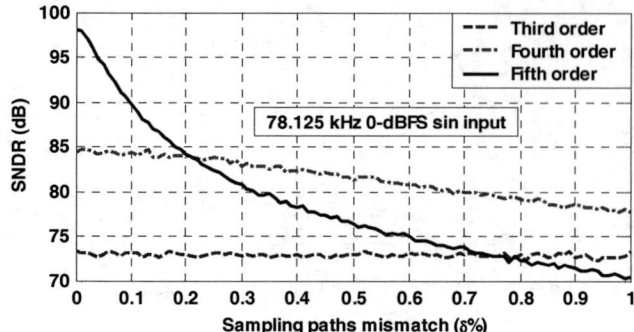

Figure 9: SNDR versus the mismatch of the sampling paths. $\delta = (C_{S1} - C_{S2})/(C_{S1} + C_{S2})$

5. REFERENCES

[1] D. Senderowicz et al., "Low-voltage double-sampled $\Sigma\Delta$ converters," *IEEE J. Solid-State Circuits*, vol. 32, no. 12, pp. 1907-1919, Dec. 1997.

[2] P. Rombouts et al., "An approach to tackle quantization noise folding in double-sampling $\Sigma\Delta$ modulation A/D converters," *IEEE Transactions on Circuits and Systems—II*, vol. 50, no. 4, pp. 157-163, April 2003.

[3] T. Burmas et al., "A second-order double-sampled delta-sigma modulator using additive-error switching," *IEEE J. Solid-State Circuits*, vol. 31, no. 3, pp. 284-293, March 1996.

[4] P. Rombouts et al., "Systematic design of double-sampling $\Sigma\Delta$ ADC's with modified NTF," *Proc. ISCAS*, vol. 1, pp. 401-404, 2004.

[5] J. Silva, U. Moon, J. Steensgaard, and G. Temes, "Wideband low-distortion delta-sigma ADC topology," *IEE Electronics Letters*, vol. 37, no. 12, pp. 737-738, July 2001.

[6] R. Schreier, "An empirical study of high-order single-bit delta-sigma modulators," *IEEE Transactions on Circuits and Systems—II*, vol. 40, no. 8, pp. 461-466, Aug. 1993.

[7] R. Jiang and T. Fiez, "A 14-bit $\Delta\Sigma$ ADC with 8× OSR and 4-MHz conversion bandwidth in a 0.18μm CMOS process," *IEEE J. Solid-State Circuits*, vol. 39, no. 1, pp. 63-74, Jan. 2004.

[8] A. Hamoui and K. Martin, "A 1.8-V 3MS/s 13-bit $\Delta\Sigma$ converter with pseudo data-weighted-averaging in 0.18-μm digital CMOS," in *Proc. CICC*, pp. 7.3.1-7.3.4, Sept. 2003.

[9] R. Schreier, *Delta-Sigma Toolbox, Version 6*, available: http://www.mathworks.com/matlabcentral/fileexchange

[10] S. R. Northworthy, R. Schreier, and G. C. Temes, *Delta-Sigma Data Converters*, Piscataway, NJ: IEEE Press, 1997.

[11] R. T. Baird and T. Fiez, "Linearity enhancement of multibit delta-sigma A/D and D/A converters using data weighted averaging," *IEEE Transactions on Circuits and Systems—II*, vol. 42, no.12, pp. 753-762, Dec. 1995.

Adaptive Processing Applied to the Design of Highly Digital Analog Interfaces

Adão A. de Souza Jr., Luigi Carro.
Instituto de Informática / PPGC
Federal University of Rio Grande do Sul - UFRGS
Porto Alegre, Brazil.
{adaojr, carro} @ inf.ufrgs.br

Jawad Tousaad.
Departamento de Engenharia Elétrica
Federal University of Rio Grande do Sul - UFRGS
Porto Alegre, Brazil.
jawad@eletro.ufrgs.br

Abstract— This work proposes the use of complex adaptive processing as means to reduce the analog to digital design gap on mixed-signal systems. Linear and non-linear adaptive filters are used to compensate for the non-idealities on the analog acquisition path allowing the use of low cost analog blocks. Several different digital architectures are investigated to increase resolution and reduce non-linearities for a target two-tone acquisition application. A prototype board is designed and the theoretical and measured results are analyzed.

I. INTRODUCTION

Many times, in order to fulfill current mixed-signal design application requirements, one must face the need to incorporate analog acquisition capabilities in an otherwise totally digital system. For this kind of mixed-signal applications, several highly digital analog acquisition architectures have been proposed in the literature ([1-3]). Their common features include the ability to explore the higher switching frequencies of digital devices and the use of minimal analog blocks that could be easily implemented on standard digital processes [4].

Another prominent trend on mixed signal application is the migration of analog tasks to the digital block, in order to reduce production costs [5]. The availability in the digital domain of very high processing power at low integration costs can be singled out as the main driving force in this change of design methodology. Using more complex digital models it is now possible to deal around several design issues that, on the past, created serious constraints for the analog designer, easing design, validation, prototyping and test phases [6].

In this work analog signal acquisition is approached from a system modeling perspective. One-bit comparators and low-cost random reference blocks are used to acquire the analog signal statistics. Self-configuration and adaptive processing are proposed to avoid the acquisition non-idealities these blocks present. The main idea is that in any mixed-signal design it should be possible to use a low cost sampler, followed by a application specific DSP conditioning block, with all the benefits previously mentioned. In this paper the target application is the acquisition of a two-tones periodic signal.

The article is organized as follows: section two presents the acquisition architecture and the mathematical framework for the modeling of a statistical sampler. Then, two different adaptive strategies are used to increase the overall acquisition performance. The system acquisition block is assembled and its results are used to verify the theoretical models. Section five makes some final remarks and points out new directions on the research.

II. ANALOG ACQUISITION AS SYSTEM MODELING

The target low-cost sampling architecture is composed by a reference generator, with high-amplitude noise, and at least one comparator block, followed in the digital domain by a pulse averaging block (figure 1). In fact, this structure can be repeated β times with the combined outputs providing a tradeoff between resolution and input bandwidth. It is possible to exchange a lower number of averages M with using additional parallel acquisition blocks in a way that keeps the product $\beta \cdot M$ as a constant [6].

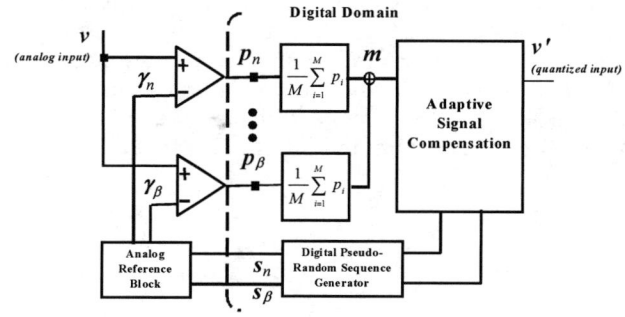

Figure 1. Analog acquisition as a digital modeling problem: digital and analog blocks division.

Also one can see that the reference generator is digitally controlled. More specifically, it is a pseudo-random multilevel signal generator proposed elsewhere [7], that transforms a digital uncorrelated sequence into a multilevel analog signal using just a few analog components and switches. The important thing to keep in mind is that the resulting noise has a non-uniform distribution that depends upon the real values of the analog components in the generator. This variation is one of the non-ideal features of the implementation in a digital technology that our compensation block should deal with.

Adaptive digital blocks have access only to the pulse-stream output of the comparators (p_n), with or without a pre-averaging block. Also the compensation blocks have the original digital random sequences (s_n) used in the generators and the ideal component values of the analog blocks available. Everything in between should be seen as the unknown system to model. In the following sections the architecture behavior is seen in detail.

A. Hard-limiter and Reference Model

In order to define he kind of adaptive model to be used one must understand the expected behavior of our system to model. It is possible to estimate this through quantization theory [8]. For each sampler the input signal is compared with a random reference with amplitude higher than the dynamic range and a bandwidth several times the maximum input signal. Assuming the input v is constant in the averaging interval T_a, it is possible to show that the output average m will be approximate by (1).

$$m(v) \approx \frac{-q \cdot \int_{-\infty}^{-v} P_\gamma d\Omega + q \cdot \int_{-v}^{\infty} P_\gamma d\Omega}{\int_{-\infty}^{+\varpi} P_\gamma d\Omega}. \quad (1)$$

Where $P\gamma$ is the PDF (Probability Distribution Function) for the reference generator. It is also easy to see from (1) that for a non-uniform $P\gamma$ the mapping between the analog input v and the pulse averages m will be non-linear and determinate by the shape of $P\gamma$. As the average estimator will have itself a probabilistic distribution one can estimate the scattering of the acquired data using the Central Limit Theorem [9]. In fact for any reference distribution shape one should expect the relationship given in (2) to hold.

$$\|v - m\| \propto \frac{1}{\sqrt{\beta \cdot M}}. \quad (2)$$

Now figure 2, depicts the reference pseudo-random generator employed. A simple charge-discharge periodic signal sweeps the dynamic range at each sampling interval

---

One author of this paper is sponsored by a grant from CAPES

Ts (in order that the above mentioned relations hold, one must design $Ts << Ta$). As the digital clock is several times higher than the sampling frequency, a digital pseudo-random sequence s determinates the exact time where the reference should be sampled.

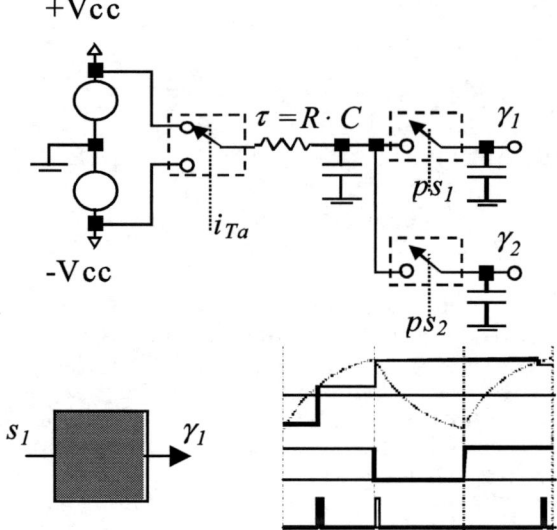

Figure 2. Reference multilevel random signal generator architecture.

This architecture has several advantages to be implemented in a CMOS technology. First, the analog reference signal to be used may be of any kind, as long as it sweeps the whole input dynamic range. This allows the use of small passive components with a high variability or even of non-linear components. Also, it is possible to generate several uncorrelated references just using different sets of sample and hold switches with different digital sampling sequences s. This eases the bandwidth constrains of the acquisition, since it allows a larger number of parallel comparators to be used.

If one knows the PDF of s, and the charge and discharge constants (τ_{LH} and τ_{HL}) of the circuit, it is also possible to determinate the ideal output distribution of this generator. In fact, this is a similar problem to a well-know relationship found in a kind of analog noise generator [10]. One should take the low-to-high and the high-to-low cycles separately. Assuming s is split into an even index sequence (s_0) and an odd index sequence (s_1), the relationship (3) holds between the input sequence and the output PDF of the generator.

$$P_\gamma \propto \left\| \frac{\partial F_{LH}^{-1}}{\partial t}(s_0) \right\| + \left\| \frac{\partial F_{HL}^{-1}}{\partial t}(s_1) \right\|. \quad (3)$$

This assumes F_{LH} is the function describing analog reference during charge cycle and F_{HL} in the discharge cycle. Also, sampling sequence (s_0) begins in the low-to-high transition. This result will be important when studying the hybrid approaches in section 3.3.

B. System Behavior

Using the already mentioned relationships it is possible to better illustrate the behavior of our analog acquisition block. Figure 3 shows the $v \times m$ plot for different parameters in the analog reference one can clearly see its non-linear behavior.

The input/output relationship is a well-defined monotonous non-linear function. In fact, since this $v \times m$ is defined only by the circuit electric features it would be possible to determinate an algebraic inverse expression in order to linearize the results in m [11]. What makes this difficult in general is the variation from the specified values in the analog components of the generator.

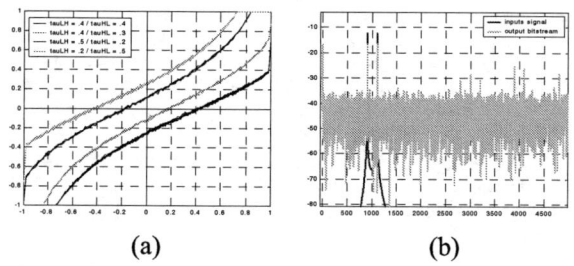

(a) (b)

Figure 3. Non linear behavior on the statistical sampler : (a) Input/ Output Mapping. M = 1000, β = 10. (b) Harmonic distortion for a two-tones acquisition (F1@.99kHz, F2@1.1kHz). M= 1, β=1, τ_{LH} = .4, τ_{HL} = .2.

III. ADAPTIVE PROCESSING MODELS

Two architectures where used in order to test the capabilities of an adaptive filtering block to improve the acquisition performance for a two-tones acquisition application:

- a self-correlation based architecture to reduce quantization noise and increase resolution, without the need for a configuration reference;
- a hybrid approach where a pre-linearization is performed by changing the sampling probability in the digital block of the generator using information from the ideal system behavior before the application of the adaptive filter.

A. Self-Correlation Adaptive Filter Topology

As it was shown the result in m have a high discretization noise level. One first approach to reduce this noise, and thus increase output resolution on a periodic two-tone signal is to use a self-correlation delay filter architecture. In this setup the input to the filter and the reference signal are the same bit-stream taken with delay from the input line (Figure 4).

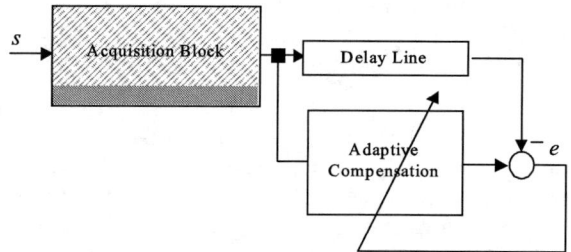

Figure 4. Delay line filter architecture.

The idea behind this architecture is that the signal will be preserved, since it keeps a strong correlation between samples. At the same time the noise is uncorrelated and after the filter convergence it will be strongly reduced. Table 1 shows the main features of a single tone acquisition with this architecture. The data was pre-processed by averaging with M taps. Results are around three bist higher in resolution, while non-linearities also show a strong improvement. Table 1 show the output characteristics for a single-tone acquisition with post-processing, one accan see that it is possible to achieve resolutions over eigth bits.

Without compensation statistical sampler resolution is about four bits (ENOB ≈ 4) due to averaging of the bistream with M = 40. In table 2 only the increase in peformance due to adaptive filtering for a two-tones acquisition is shown. For instance. on table 2 the upper leftmost improvement result is +3.3187, our total ENOB is actually 7.362 .

TABLE I. SINGLE TONE ACQUISITION WITH STATISTICAL SAMPLING AND SELF-CORRELATION DELAY ADAPTIVE FILTER.

Time constants	Acquisition performance improvement	
	ENOB	SINAD
τ_{LH} = .3, τ_{HL} = .4	7.2503	45.4071
τ_{LH} = .4, τ_{HL} = .4	8.4898	52.8686
τ_{LH} = .4, τ_{HL} = .5	7.1555	44.8362

TABLE II. TWO-TONES ACQUISITION PERFORMANCE AS A FUNCTION OF THE SIGNAL BANDWIDTH.

τ_{LH} = .32, τ_{HL} = .32.	Parameter variation (Δ)		
	F1@ 0.2kHz F2 @0.7kHz	F1 @ 0.7kHz F2 @1.2kHz	F1 @1.2kHz F2 @1.7kHz
ENOB	+3.3187	+2.6585	+2.0404
SINAD	+19.9789	+16.0040	+12.2833
THD	-24.2950	-31.858	-33.793

B. Hybrid Approach

In this approach one uses the knowledge the designer possesses about the ideal behavior of our generator to produce a more linear acquisition before subjecting the output bit-stream to the adaptive filter. The idea is to shape the distribution of s so our output reference γ is more uniform since a uniform distribution warrants a linear mapping between v and m. Expression 3 points to the way this can be accomplished. Two mapping tables are defined in the generation of *s*. These tables are loaded with the values of the expected derivatives of F_{LH} and F_{HL} in relation to *s* [12].

However, since this shaping should be made considering the ideal values of τ_{LH} and τ_{HL} it is clear that the resulting output will still be non-linear, although with much lower values of total harmonic distortion. The same adaptive filter topology of section 3.1 is then used to take care of the remaining non-idealities. Table 3 summarizes the results.

TABLE III. HYBRID MODEL RESULTS, PARAMETER MISMATCH IS ASSUMED AS A 20% VARIATION.

$\tau_{LH}=.32$, $\tau_{HL}=.32$.	Parameter variation (Δ)		
	F1@ 0.2kHz F2 @ 0.7kHz	F1 @ 0.7kHz F2 @ 1.2kHz	F1 @1.2kHz F2 @1.7kHz
ENOB	+3.7984	+3.1155	+2.9207
SINAD	+22.9166	+18.996	+15.3943
THD	-29.4030	-36.233	-37.1787

IV. PROTOTYPING AND MEASUREMENTS

A prototype of the analog acquisition system was designed. Analog blocks were assembled with standard discrete components. Four LM311 comparators performed the acquisition, and data was gathered using the digital channels of a HP54645 oscilloscope. The digital control of the generator (including uniform sampling mapping for the hybrid approach) was described using VHDL and synthesized using EPLD prototyping boards. Reference sampling frequency was set to 200kHz. Each generator semi-cycle was divided into 256 time-slices selected trough the lower bits of a 14-taps LFSR (Linear Feedback Shift Register) pseudo-random generator. Two references γ where generated using LFSRs with different seeds. These two signals where transformed into four with analog inversion in order to feed four parallel comparators in the input.

Four sets of signals where acquired: one (1kHz) and two-tones (.9Khz and 1.1KHz) signals with a simple non-uniform generator and with a digitally uniformized generator (for the hybrid approach). Also, to configure the adaptive block a grounded input acquisition was performed using both kinds of generator. Table 4 shows the results for the synthesis of the generators. The results of the adaptive filter strategies using the acquired data are summarized in table 4.

TABLE IV. ACQUIRED DATA RESULTS.

Measurement	Parameter variation (Δ)	
	ENOB gain	*SINAD gain*
1 - tone, simple	+3.0365	+18.2798
2 - tones, simple	+2.0594	+12.3978
1 - tone, hybrid	+3.7811	+22.7625
2 - tones, hybrid	+2.8118	+16.9272

V. DISCUSSION OF RESULTS AND FUTURE WORK

Three different adaptive modeling strategies where suggested to provide application specific analog signal acquisition with a rather low analog area overhead. From the examined models the better performance achieved for both overall noise reduction and non-linearity reduction is the hybrid approach employing non-uniform stochastic reference sampling and adaptive processing of the output.

Many problems with the adaptive model were consequence of the chosen topology, which demands a careful choice in its parameters. In future works other non-linear modeling architectures should be tested. Also, given the good results obtained with the hybrid approach with relatively low cost in digital area, an algorithm that adaptively updates the uniformization table should be studied.

VI. REFERENCES

[1] R. F. Wolffenbuttel, W. Kurniawan, "Stochastic analog-to-digital converter based on the asynchronous sampling of a reference triangle". Instrumentation and Measurement, IEEE Transactions on , v.38, n.1 ,. pp. 10-16. Feb, 1989.

[2] G. Cawemberghs, "A nonlinear Noise-Shaping Delta-sigma Modulator with On-Chip Reinforcement Learning". Analog Integrated Circuits and Signal Processing, 18. Kluwer Academic Publishers, 1999.

[3] T. Watanabe, T. Mizuno, and Y. Makino, "An All-Digital Analog-to-Digital Converter With 12-(V/LSB Using Moving-Average Filtering", IEEE Journal of Solid-States Circuit, vol. 38, n.1, pp. 120-125, January 2003

[4] J. O. Mainardi, A. A. de Souza Jr, L. Carro, A. A. Suzin, "A Comparison of Totally Digital ADCs for SoCs". In: ISCAS 2004.

[5] G. W. Roberts, "DFT Techniques for Mixed-Signal Integrated Circuits" in Circuits And Systems In The Information Age, IEEE Press, pp. 251-271, June 1997.

[6] L. Carro, A. A. de Souza Jr., M. Negreiros, G. Jahn, D. Franco, Circuit Level Considerations for Mixed-Signal Programmable Components. In: IEEE Design and Test of Computers. January-February 2003, p. 76-84.

[7] A. A. de Souza Jr., L. Carro, "A Highly Digital, Low-Cost Design of Statistic Signal Acquisition in SoCs". Design Automation and Test in Europe, DATE 2004. Paris.

[8] M. G. C. Flores, M. Negreiros, L. Carro, A. A. Susin, "A Noise Generator for Analog-to-Digital Converter Testing". Proceedings of the 15 th Symposium on Integrated Circuits and Systems Design,

[9] M. J .M. Pelgrom, A. C. J. Duinmaijer, A. P. G. Welbers, "Matching Properties of MOS transistors". IEEE solid-State Circuits, v.24, pp. 1433-1440, October, 1989.

[10] U. G. Gujar, R. J. Kavanagh, "Generation of Random Signals with Specified Probability Density Functions and Power Density Spectra". pp. 716-719. In: IEEE Transactions on Automatic Control. December, 1968

[11] B. Widrow, I. Kollár, and M.-C. Liu, "Statistical theory of quantization," IEEE Trans. Instrum. Meas., v. 45, pp. 389-396. June, 1996.

[12] A. Papoulis. "Probability, Random Variables, and Stochastic Processes", 2nd ed. New York: McGraw-Hill, 1984

Spurious Tone Free Digital Delta-Sigma Modulator Design for DC Inputs

Maciej Jan Borkowski and Juha Kostamovaara
Department of Electrical and Information Engineering
Electronics Laboratory, P.O.Box 4500
90014 University of Oulu, Finland
maciej.borkowski@ee.oulu.fi

Abstract—A method for spurious tone free delta sigma modulator design is presented in this paper. The method is based on an analysis of modulator state space orbits and it is applicable to digital modulators working with DC inputs. Modulator is designed to maintain a very long controllable sequence lengths. This results in smooth distribution of the quantization noise power and prevents it from concentrating into spurious tones. The sequence length is controlled by modulator scaling and applying initial conditions each time DC input changes. A list of suitable initial conditions and corresponding sequence lengths is given for MASH and error feedback modulators. This method is extended further to a more flexible case in which reloading initial conditions is no longer necessary. It is shown that when modulator resolution is increased by one bit and initial conditions applied at startup, the modulator spontaneously enters very long orbits each time DC input changes. In return for the small hardware increment the spectrum remains smooth and is further decreased by $8\,\mathrm{dB}$.

I. INTRODUCTION

Digital delta sigma modulators (digital DSMs) are commonly used in data conversion systems and frequency synthesis applications. DSMs are nonlinear chaotic systems and despite of they common use still pose serious analytical challenges. DSMs can be divided into two groups: analog DSMs (discrete time, continuous amplitude) and digital DSMs (discrete time, discrete amplitude). This work is focused on digital DSMs.

A common problem found in both types of modulators is they tonal behavior. One of the most common methods for randomizing modulator behavior and breaking the sequence length is dither [1], [2]. The drawbacks to using dither are the need for extra hardware, additional noise introduced to the system and possible problems with modulator stability. An interesting solution to the tonal behavior appearing in the analog domain is the use of chaos [3], [4]. Another effective technique used in the digital domain is based on loading predefined initial conditions or setting the least significant bits of modulator registers [5], [6]. In recent work [7], [8] this approach has been given a theoretical background. The analysis shows that the quantization noise spectrum is smooth and tone free if irrational initial condition is applied to the first modulator stage. This analysis is a continuation of the general analytical approach [9]. Irrational initial condition cannot be realized in fixed point digital DSM, therefore authors recommend using odd initial conditions, which approximately act as irrational numbers. Simulation confirms the theoretical findings. This analysis draws from more general approach and adopts it, using an approximation, to the needs of digital DSMs.

However, what distinguishes the digital DSMs is their fully deterministic nature. Digital DSM do not suffer from implementation imperfections and realize their algorithms with 100 % accuracy. Due to the finite bus widths all arithmetic operations are performed on integer numbers. Modulator input and internal states are often interpreted as rational numbers when normalized with respect to the quantization step. It has been long known that when modulators operate with rational input and initial conditions they behavior is very different. In such case modulators always produce finite length sequences [10], [11]. Indeed, digital DSMs are finite state machines and must return to a state once visited.

This work is based on exploration of digital modulator deterministic behavior. It recognizes that periodical behavior is inherently present in digital DSMs. Therefore, instead of avoiding it, this work shows how that the sequence length can be controlled. Each time modulator is started with a DC input it enters a state space orbit. It can be guaranteed that modulator always enters very long orbits by proper selection of the initial condition. The orbit length can be controlled regardless of the modulator input. The orbit length, and consequently the output sequence length, is controlled by applying an initial condition and modulator scaling. The initial conditions that force modulator toward long orbits provide good internal randomization. Therefore when maximum, or almost maximum, length sequences are generated, quantization noise spectrum remains smooth and tone free. When sequence length is controlled, modulator spectrum can be fully characterized for all input DC levels.

This approach has been further extended to avoid reloading initial conditions each time DC input is changed. This work shows that modulator can spontaneously enter a very long orbit each time new input DC segment arrives. As a result spurious tones never appear when input signal varies slowly, or remains at DC. Such desired modulator behavior is achieved by increasing modulator resolution by 1 bit and applying selected initial conditions at power-up. Additionally, this novel design methodology further improves the quantization noise spectrum by $8\,\mathrm{dB}$.

II. ARCHITECTURES

In this work two well known architectures have been studied: MASH depicted in Fig. 1 and error feedback modulator (EFM) depicted in Fig. 2. The architectures are shortly described here in order to avoid ambiguities and allow repetition of the results. Both modulators share the same noise transfer function:

$$NTF = \frac{(z-1)^m}{z^m} \qquad (1)$$

The set of coefficients required for implementing an m-th order EFM can be calculated using the following formula:

$$\sum_{i=1}^{m} A_i z^{-i} = 1 - NTF \qquad (2)$$

The first order EFM named as EFM1 is a basic building block of MASH, see Fig. 1 and Fig. 3. EFM is constructed using a uniform quantizer conveniently described by the floor function:

$$y = \lfloor e/2^n \rfloor \qquad (3)$$

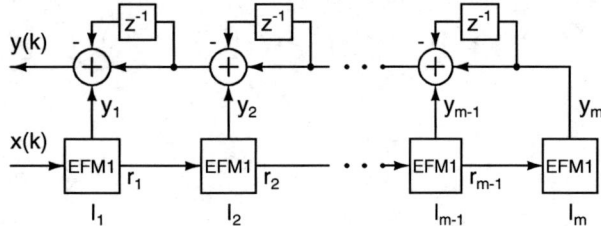

Fig. 1. MASH architecture

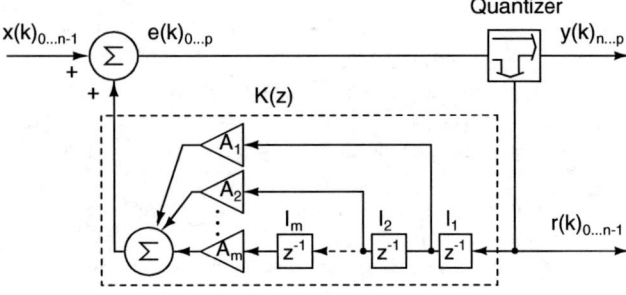

Fig. 2. EFM architecture

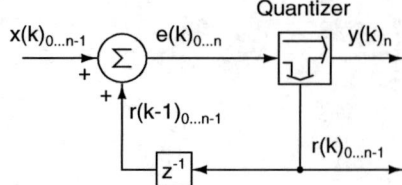

Fig. 3. EFM1 - first order error-feedback modulator

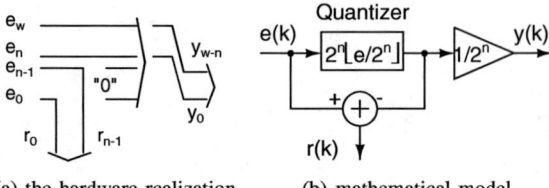

(a) the hardware realization (b) mathematical model

Fig. 4. Binary weighted quantizer - bus split

In digital DSMs such quantizer is most easily realized by a bus split. The principle is depicted in Fig. 4. The signal entering the quantizer, $e(k)$ is split into two parts: $r(k)$ and $y(k)$. The quantization error signal $r(k)$ is formed directly by n least significant bits of $e(k)$. The modulator output $y(k)$ is formed by shifting down the remaining most significant bits of $e(k)$. The shift operation in Fig. 4(a) corresponds to the division by 2^n in the mathematical model Fig. 4(b). The modulator output DC value is a fractional number in the range $(0,\ldots,1)$. Therefore when $r(k)$ is n-bit wide, the input signal $x(k)$ is n-bit wide as well.

In this work the bus width n of the quantization error signal $r(k)$ acts as a modulator scaling parameter. All remaining bus widths can be easily derived when the bus width n is given.

III. CASE 1: RELOADING INITIAL CONDITIONS

It has been known that proper selection of initial conditions can guarantee spurious free spectrum [6]–[8]. To guarantee proper modulator behavior initial conditions are reloaded each time DC input is changed. This section develops the previous findings further by introducing the concept of controllable sequence lengths. Digital DSM is a finite state machine (FSM). It is very well known that when a FSM works with a constant input it must enter a state space orbit after a possible transient state [12]. In general, the output sequence length depends on the modulator input and initial conditions. There are, however initial conditions, for which the sequence length does not depend on the input DC value.

A. Sequence length based analysis

Output sequence length can be easily found by tracking a state space orbit. When the end of the orbit is found it is possible to extract a repetitive output sequence. Autocorrelation function has been used to confirm that the extracted sequence is equal to exactly one period.

1) Simulation results: The modulator simulation has shown that it is possible to control modulator sequence length. The sequence length can be controlled by loading initial conditions and modulator scaling. The sets of *preferred* initial conditions are defined in Table I. They can be applied to corresponding modulator registers, see Fig. 1 and Fig. 2. When a modulator is started from such initial condition, the *guaranteed* output sequence length depends on the modulator bus width n and does not depend on the modulator DC input. The relationship between the sequence length L_s and the bus width n is presented in Table II.

Table II collects so-called *guaranteed* sequence lengths L_s and *maximum* sequence lengths. The maximum sequence length refers to the longest sequence that can be produced by a modulator with corresponding order and bus width. If a modulator is started from the initial conditions collected in Table I the sequence length will never be shorter than L_s. For modulator order 3 and 5 the guaranteed sequence length is equal to the maximum sequence length. For each input DC level the sequence length is exactly known. For modulator order 2 and 4 it is impossible to find one initial condition that gives always the maximum sequence length for all inputs. Therefore depending on the modulator input the sequence length will be between the guaranteed and the maximum value.

TABLE I
INITIAL CONDITIONS

Modulator order m	MASH	EFM sum odd
2	I_1 odd	$I_1 + I_2$
3	I_1 odd	$I_1 + I_3$
4	I_1 odd	$I_1 + I_2 + I_3 + I_4$
5	I_1 odd	$I_1 + I_5$

TABLE II
SEQUENCE LENGTH

Modulator order m	Guaranteed sequence length L_s	Maximum sequence length
2	2^{n-1}	2^{n+1}
3	2^{n+1}	2^{n+1}
4	2^{n+1}	2^{n+2}
5	2^{n+2}	2^{n+2}

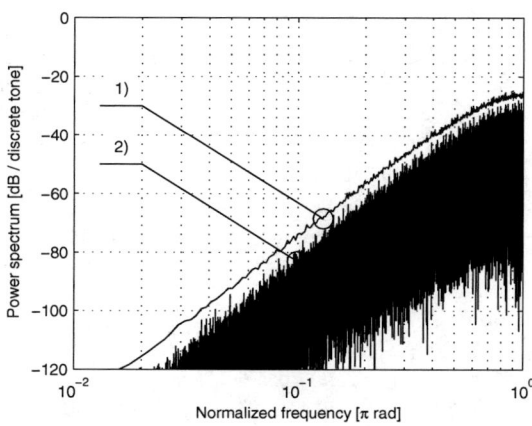

Fig. 5. Discrete power spectrum for EFM. Sequence length $L_s = 2^{17}$. 1) Envelope, 2) Arbitrary DC input

TABLE III
CONTENT OF MODULATOR REGISTERS

k	$r(k-1)$	$r(k-2)$	$r(k-3)$	$x(k)$
1	O	E	E	E
2	O	O	E	E
3	E	O	O	E
4	E	E	O	E
5	O	E	E	E

O - odd number, E - even number

B. Simulation scope

The simulation has been performed in two stages; stage 1 - analysis, and stage 2 - verification. In **stage 1** DSMs were simulated for all combinations of initial condition and input DC value. This exhaustive simulation has been repeated for modulator bus width $n = 2, 3, 4$. At this stage the initial conditions presented in Table I have been found and corresponding equations, collected in Table II, formulated. In **stage 2** the collected data have been verified for both modulators in a broader scope. It has been fully verified for the the odd initial condition $I_1 = 1, I_2, \ldots, I_n = 0$ within the bus width range $2, \ldots, 17$. It has been further verified for an arbitrarily selected group of initial conditions from Table I within the same bus width range and higher. The equations collected in Table II as well as the spectrum presented in Fig. 5 have been always found valid.

C. Quantization noise

Controlling modulator sequence length allows complete modulator characterization. One period of a periodical signal carries full information about that signal. Therefore to fully characterize modulator one full sequence should be collected for each DC input number. The spectrum for all inputs can be compared and the worst case level, so called an *envelope* can be obtained. The envelope and an arbitrary DC input spectrum for EFM running with $L_s = 2^{17}$ are depicted in Fig. 5. The envelope for the same order and sequence length MASH overlaps with EFM, therefore only one is presented.

The envelope shown in Fig. 5 shows that spectrum is smooth and free from spurious tones for all DC inputs. The initial conditions that force modulator toward long orbits provide good internal randomization. When maximum, or almost maximum length sequences are generated, the quantization noise spectrum remains smooth and tone free. Therefore long sequence lengths can be associated with good modulator randomization.

IV. CASE 2: INITIAL CONDITIONS APPLIED AT POWER-UP

It is possible to extent method presented in III so that the initial conditions are loaded only once, when modulator operation begins. When *preferred* initial condition is applied at start-up modulator enters a very long orbit. In the new control scheme DC level can change at any given time. Therefore the second orbit is entered from an unknown, arbitrary initial state. If it happens to be a very short sequence spurious tones appear. It has been numerously reported that spurious tones are especially likely to occur with slowly varying or DC inputs [2]. This section suggests a solution to this problem.

A. Input signal restriction

The described problem can be effectively eliminated if modulator always spontaneously enters a very long orbit. It can be shown that if modulator is once loaded with *preferred* initial conditions, and the input signal is even, the registers remain all the time within preferred initial conditions. Restricting the input signal to even numbers requires setting the LSB input bit to "0". This can be done without any loss of accuracy. If, for example, the input signal is 10 bit wide the modulator input should be extended to 11 bits and the LSB set to "0". This procedure slightly increases hardware consumption, but in return brings a number of benefits.

1) Proof for MASH: The MASH architecture, shown in Fig. 1 is composed of a chain of EFM1 blocks depicted in Fig. 3. EFM1 is in fact a simple accumulator, where the 1-bit output $y(n)$ is a carry signal. If an initial condition for an accumulator is odd $r(-1) = 2a + 1$, and the input is always even $x(n) = 2b(n)$, then the accumulator content remains odd. This can be shown by studying the recurrence $r(n) = r(n-1) + x(n)$. The solution at the discrete time instant $n \geq 0$ can be expressed as:

$$\begin{array}{rcl} r(n) & = & r(-1) + \sum_{i=0}^{n} x(i) \\ & = & 2\left(a + \sum_{i=0}^{n} b(i)\right) + 1 \end{array} \quad (4)$$

The real accumulator truncates the MSBs of $r(n)$, but this does not change the fact that $r(n)$ always remains odd. This proof shows that if the first register in MASH is loaded with an odd initial condition and the modulator input LSB is set to "0" then the register's content remains odd. This satisfies the conditions stated in Table I and therefore, MASH always spontaneously enters very long orbits.

2) Proof for Error Feedback Modulator: This section presents the proof for a third order EFM. The authors have similarly repeated the proof until the fifth order. More general proofs would require a general approach toward the preferred initial conditions, sequence length and spectrum purity, therefore are beyond the scope of this work.

A general EFM architecture is depicted in Fig. 2. For the third order EFM the error signal $e(k)$ can be calculated as a function of the quantization error $r(k)$:

$$e(k) = 3r(k-1) - 3r(k-2) + r(k-3) \quad (5)$$

The content of the EFM registers $I_1 = r(k-1)$, $I_2 = r(k-2)$, $I_3 = r(k-3)$ as a function of the time step k is presented in Table III. The registers content is described either as an odd number ("O") or even number ("E"). The modulator is started from a *preferred* initial condition and the input signal is always even $x(k) = E$. It can be seen that after five clock cycles the situation returns to the starting

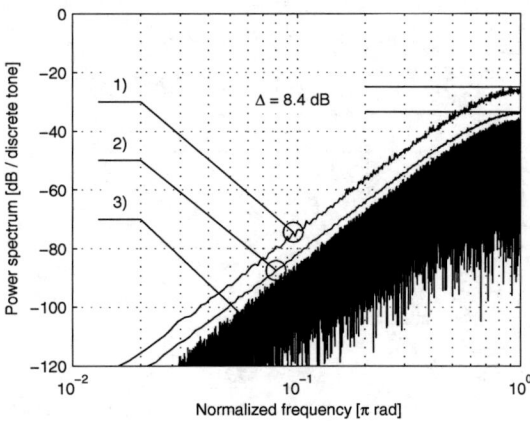

Fig. 6. Discrete power spectrum for EFM. Sequence length $L_s = 2^{17}$. 1) Envelope for all inputs, 2) Envelope for even inputs, 3) Arbitrary even DC input.

point. Furthermore, at all times the content of the register remains within the set of preferred initial conditions. The sum of I_1 and I_3 is always odd, see Table I. As previously, this proof shows EFM always spontaneously enters very long orbits when started from *preferred* initial conditions and the input LSB is set to "0". It is assumed that the digital numbers within the architecture are represented in two's complement code.

B. Quantization noise

It is interesting to note that modulators with even input outperform modulators with full scale input with regard to the spectrum quality. Fig. 6 compares spectrum envelopes for both cases. When input is restricted to even numbers the envelope appears 8 dB lower despite that in both cases the quantization noise variance is approximately the same. As modulator bus width increases, the approximation is more accurate and the quantization noise variance approaches the uniformly distributed white noise variance. Unassigned LSB and always present *preferred* initial condition enhance modulator internal randomization resulting in more compact spectrum.

V. DISCUSSION

The control method has been presented on the example of MASH and EFM. It can be applied to other architectures if sets of *preferred* initial conditions can be found by repeating the procedure described in III-B. This procedure requires that a DSM model is scalable with respect to the bus width. When a DSM is modelled with only a few bits of resolution a complete simulation, for all combinations of input and initial conditions, is a matter of seconds or minutes on an ordinary desktop PC. In addition, to simplify the sequence length search, the model should return the orbit length. In practice it is simply achieved by storing an internal state and waiting until the state is revisited.

Similarly, the extension to the CASE 2 has been based on the set of preferred initial conditions. When modulator input is even the set possesses the property of positive invariance; that is all subsequent sets lie within the original set. Exhaustive simulation has been used to find the positive invariant sets and as a result the input signal restriction. The search procedure has been inspired by a method previously used to prove modulator stability [13].

VI. CONCLUSIONS

The significance of this work lies in exploring the deterministic character of digital DSMs. Instead of attempting to eliminate modulator periodical behavior this work aims at controlling it. This work supports and extends the results presented in [7], [8], where it has been shown that odd initial condition applied to the first modulator register guarantees smooth and spurious tone fee spectrum. This work shows that odd initial conditions allow also sequence length control.

In this work a complete list of *preferred* initial conditions for second to fifth order MASH and EFM is given. When modulator is started from *preferred* initial conditions it generates a sequence not shorter than a guaranteed length L_s. The sequence length L_s depends on the modulator bus width, and does not depend on the modulator DC input. This work provides the relationships between L_s and the bus width for MASH and EFM architectures. The relationships were derived experimentally for modulator order from two to five and tested in a a broad scope of bus widths. When modulator is forced to generate maximum, or almost maximum, sequence lengths the spectrum remains smooth and free from spurious tones. Furthermore, knowing the sequence length for each input, it is possible to fully characterize the modulator for all DC inputs using simulation or an automated measurement.

This approach have been further extended. It has been analytically shown that when modulator is started from *preferred* initial conditions at power-up, and input remains even, modulator state always remains within the set of *preferred* initial conditions. This guarantees that modulator always spontaneously enters orbits not shorter than L_s. Consequently, spurious tones never appear when the input signal is varying slowly or remains at DC. Unassigned LSB and always present *preferred* initial condition enhance modulator internal randomization and improve spectrum for all DC inputs by 8 dB.

REFERENCES

[1] W. Chou and R. M. Gray, "Dithering and its effects on sigma-delta and multistage sigma-delta modulation," *IEEE Trans. Inform. Theory*, vol. 37, pp. 500–513, May 1991.

[2] S. R. Norsworthy, R. Schreier, and G. C. Tames, *Delta-Sigma Data Converters*. New York: IEEE Press, 1997.

[3] R. Schreier, "On the use of chaos to reduce idle-channel tones in delta-sigma modulators," *IEEE Trans. Circuits Syst. I*, vol. 41, pp. 539–547, Aug. 1994.

[4] M. Motamed, S. Sanders, and A. Zakhor, "The double loop sigma delta modulator with unstable filter dynamics: stability analysis and tone behavior," *IEEE Trans. Signal Processing*, vol. 43, pp. 549–559, Aug. 1996.

[5] B. Miller and R. J. Clonley, "A multiple modulator fractional divider," *IEEE Trans. Instrum. Meas.*, vol. 40, pp. 578–583, Jun 1991.

[6] M. A. Kozak and I. Kale, "A pipelined noise shaping coder for fractional-N frequency synthesis," *IEEE Trans. Instrum. Meas.*, vol. 50, pp. 1154–1161, May 2001.

[7] ——, *Oversampled Delta-Sigma Modulators Analysis, Applications and Novel Topologies*. Kluwer Academic Publishers, 2003.

[8] ——, "Rigorous analysis of Delta-Sigma modulators for Fractional-N PLL frequency synthesis," *IEEE Trans. Circuits Syst. I*, vol. 51, pp. 1148–1162, June 2004.

[9] W. Chou, P. W. Wong, and R. M. Gray, "Multistage sigma-delta modulation," *IEEE Trans. Inform. Theory*, vol. 35, pp. 784–796, Jul 1989.

[10] V. Friedman, "The structure of the limit cycles in sigma delta modulation," *IEEE Trans. Commun.*, vol. 36, pp. 972–979, Aug 1988.

[11] R. M. Gray, "Oversampled sigma-delta modulation," *IEEE Trans. Commun.*, vol. 35, pp. 481–489, May 1987.

[12] S. W. Golomb, *Shift register sequences*. Laguna Hills, California: Aegean Park Press, 1982.

[13] R. Schreier, M. V. Goodson, and B. Zhang, "An algorithm for computing convex positively invariant sets for delta-sigma modulators," *IEEE Trans. Circuits Syst. I*, vol. 44, pp. 38–44, Jan 1997.

Towards a Rigorous Formulation of the Space Mapping Technique for Engineering Design

Slawomir Koziel, John W. Bandler
Simulation Optimization Systems Research Laboratory
ECE Department, McMaster University
Hamilton, L8S 4K1, Canada
koziels@mcmaster.ca, bandler@mcmaster.ca

Kaj Madsen
Informatics and Mathematical Modelling
Technical University of Denmark
Lyngby, DK-2800, Denmark
km@imm.dtu.dk

Abstract—This paper deals with the Space Mapping (SM) approach to engineering design optimization. We attempt here a theoretical justification of methods that have already proven efficient in solving practical problems, especially in the RF and microwave area. A formal definition of optimization algorithms using surrogate models based on SM technology is given. Convergence conditions for the chosen subclass of algorithms are discussed and explained using a synthetic example, the so-called generalized cheese-cutting problem. An illustrative, circuit-theory based example is also considered.

I. INTRODUCTION

Mapping (SM) technology involves well-established and efficient optimization methods [1-4]. The main idea behind SM is that direct optimization of a so-called (accurate but computationally expensive) fine model of interest is replaced by the iterative optimization and updating of a corresponding so-called coarse model (less accurate but cheap to evaluate). Provided their misalignment is not significant, SM based algorithms are able to converge after only a few evaluations of the fine model. SM was originally demonstrated on microwave circuit optimization [1], where fine models may be based on electromagnetic simulators, while coarse models are physics-based circuit models.

SM techniques have recently solved modeling and optimization problems in a growing number of areas, not only RF and microwave design [1-4] but also structural design [5], vehicle crashworthiness design [6], magnetic systems [7], and others. For a review see [4].

Although SM algorithms have been developed that solve practical optimization problems, the unified formulation and theory of SM, including convergence proofs (with the exception of a subclass of algorithms based on trust-region methods [8]) is lacking. Besides the theoretical importance of a unified formulation and classification of SM methods, there are two important questions an SM theory should be able to answer: (i) how good the coarse model should be (compared with the fine model) in order to make an SM algorithm converge, (ii) how to design better (more efficient) algorithms. We try to answer these questions for algorithms based on the so-called Output Space Mapping (OSM).

This work was supported in part by the Natural Sciences and Engineering Research Council of Canada under Grant OGP0007239 and Grant STGP269760, and by Bandler Corporation.

II. OPTIMIZATION USING SURROGATE MODELS

We are concerned with a class of optimization algorithms that use surrogate models. Let $\boldsymbol{R}_f: X_f \to R^m$ denote the response vector of the so-called fine model of a given object, where $X_f \subseteq R^n$. Our goal is to solve

$$\boldsymbol{x}_f^* = \arg\min_{\boldsymbol{x} \in X_f} U(\boldsymbol{R}_f(\boldsymbol{x})) \qquad (1)$$

where U is a suitable objective function and $\boldsymbol{R}_{spec} \in R^m$ is a given specification vector. In many engineering problems, we are concerned with so-called one-sided specifications: if $\boldsymbol{R}_f = (R_{f.1},..,R_{f.m})$, $\boldsymbol{R}_{spec} = (R_{sp.1},...,R_{sp.m})$, and $I_l, I_u \subset \{1,2,...,m\}$ are such that $I_l \cap I_u = \varnothing$, then we require that $R_{f.i} \leq R_{sp.i}$ for $i \in I_u$ and $R_{f.i} \geq R_{sp.i}$ for $i \in I_l$. Typically, U is defined as

$$U(\boldsymbol{R}_f) = \max\left\{\max_{i \in I_u}(R_{f.i} - R_{sp.i}), \max_{i \in I_l}(R_{sp.i} - R_{f.i})\right\} \qquad (2)$$

In some problems, U can be defined by a norm, i.e.,

$$U(\boldsymbol{R}_f) = \| \boldsymbol{R}_f - \boldsymbol{R}_{spec} \| \qquad (3)$$

We shall denote by X_f^* the set of all $\boldsymbol{x} \in X_f$ satisfying (1) and call it the set of fine model minimizers.

We consider the fine model to be expensive to compute and solving (1) by direct optimization to be impractical. Instead, we use surrogate models, i.e., models that are not as accurate as the fine model but are computationally cheap, hence suitable for iterative optimization. We consider a general optimization algorithm that generates a sequence of points $\boldsymbol{x}^{(i)} \in X_f$, $i=1,2,...$, and a family of surrogate models $\boldsymbol{R}_s^{(i)}: X_f \to R^m$, $i=0,1,...$, so that

$$\boldsymbol{x}^{(i+1)} = \arg\min_{\boldsymbol{x} \in X_f} U\left(\boldsymbol{R}_s^{(i)}(\boldsymbol{x})\right) \qquad (4)$$

and $\boldsymbol{R}_s^{(i+1)}$ is constructed using suitable matching conditions with the fine model at $\boldsymbol{x}^{(i+1)}$ (and, perhaps, some of $\boldsymbol{x}^{(k)}$, $k=1,...,i$). If the solution to (4) is non-unique we may impose regularization. We may match responses, i.e.,

$$\boldsymbol{R}_s^{(i)}(\boldsymbol{x}^{(i)}) = \boldsymbol{R}_f(\boldsymbol{x}^{(i)}) \qquad (5)$$

and/or match first order derivatives

$$\boldsymbol{J}_{\boldsymbol{R}_s^{(i)}}(\boldsymbol{x}^{(i)}) = \boldsymbol{J}_{\boldsymbol{R}_f}(\boldsymbol{x}^{(i)}) \qquad (6)$$

where $\boldsymbol{J}_{\boldsymbol{R}_s^{(i)}}$ and $\boldsymbol{J}_{\boldsymbol{R}_f}$ denote jacobians of the surrogate and fine models, respectively. More precisely, we try to define models so that conditions such as (5) and (6) are satisfied.

III. SM Based Surrogate Models

The family of surrogate models $\{R_s^{(i)}\}$ can be implemented in various ways. SM assumes the existence of a so-called coarse model that describes the same object as the fine model: less accurate but much faster to evaluate. Let $R_c : X_c \to R^m$ denote the response vectors of the coarse model, where $X_c \subseteq R^n$. In the sequel, we assume for simplicity that $X_c = X_f$. By x_c^* we denote the optimal solution of the coarse model, i.e.,

$$x_c^* = \arg\min_{x \in X_c} U(R_c(x)) \qquad (7)$$

We denote by X_c^* the set of all $x \in X_c$ satisfying (7) and call it the set of coarse model minimizers. In the SM framework, the family of surrogate models is constructed from the coarse model in such a way that each $R_s^{(i)}$ is a suitable distortion of R_c, such that given matching conditions are satisfied.

A. Original SM Based Surrogate Model

The original SM approach assumes the existence of the mapping $P : X_f \to X_c$ such that $R_c(P(x_f)) \approx R_f(x_f)$ on X_f at least on some subset of X_f which is of interest. The proximity of R_c and R_f is measured using a suitable metric; in the ideal case we have $R_c(P(x_f)) = R_f(x_f)$. For any given $x_f \in X_f$, $P(x_f)$ is defined using parameter extraction:

$$P(x_f) = \arg\min_x \| R_c(x) - R_f(x_f) \| \qquad (8)$$

The surrogate model $R_s^{(i)}$ can be defined as

$$R_s^{(i)}(x) = R_c(P(x^{(i)}) + B^{(i)} \cdot (x - x^{(i)})) \qquad (9)$$

for $i = 0, 1, \ldots$, where P is defined by (8) and $B^{(i)}$ is an approximation of $J_P(x^{(i)})$, the jacobian of P at $x^{(i)}$, obtained, e.g., by solving a Parameter Extraction (PE) problem of the form $B^{(i)} = \arg\min_B \| J_{R_f}(x^{(i)}) - J_{R_c}(P(x^{(i)}) + B \cdot (x - x^{(i)})) \cdot B \|$.

B. Output SM Based Surrogate Modeling

The Output Space Mapping (OSM) aims at reducing misalignment between the coarse and fine models by adding a difference (residual) between those two to R_c. Let us define function $\Delta R : X_f \cap X_c \to R^m$ as

$$\Delta R(x) = R_f(x) - R_c(x) \qquad (10)$$

We construct surrogates that use (local) models of ΔR, denoted as ΔR_m. A generic surrogate model defined by OSM is

$$R_s^{(i)}(x) = R_c(x) + \Delta R_m(x, x^{(i)}) \qquad (11)$$

We consider the zero-order model $\Delta R_m(x, x^{(i)}) = \Delta R(x^{(i)})$ which leads to the surrogate

$$R_s^{(i)}(x) = R_c(x) + \Delta R(x^{(i)}) \qquad (12)$$

Model (12) is based on the matching condition (5).

The second model is a first-order approximation of ΔR of the form $\Delta R_m(x, x^{(i)}) = \Delta R(x^{(i)}) + J_{\Delta R}(x^{(i)}) \cdot (x - x^{(i)})$, where $J_{\Delta R}(x^{(i)})$ denotes the jacobian of ΔR at $x^{(i)}$. This leads to

$$R_s^{(i)}(x) = R_c(x) + \Delta R(x^{(i)}) + J_{\Delta R}(x^{(i)}) \cdot (x - x^{(i)}) \qquad (13)$$

Instead of the exact jacobian (usually unavailable) we can use its approximation produced by the Broyden update. Model (13) is based on matching conditions (5) and (6).

C. Implicit Space Mapping Based Surrogate Modeling

Implicit Space Mapping (ISM) assumes that the coarse model depends on additional parameters, i.e., we have $R_c : X_c \times X_p \to R^m$ where $X_p \subseteq R^p$ is the domain of such preassigned parameters.

An ISM optimization algorithm aims at predistortion of the coarse model by adjustment of its preassigned parameters x_p so that, at the current point $x^{(i)}$, the fine and coarse model response vectors coincide. The predistorted model becomes a surrogate which, in turn, is optimized in order to obtain the next point $x^{(i+1)}$. Thus, the surrogate model defined by ISM is

$$R_s^{(i)}(x) = R_c(x, x_p^{(i)}) \qquad (14)$$

where $x_p^{(i)}$ is determined by solving a PE problem of the form

$$x_p^{(i)} = \arg\min_x \| R_f(x^{(i)}) - R_c(x^{(i)}, x) \| \qquad (15)$$

The model (14), (15) is based on the matching condition (5).

IV. Convergence Properties of OSM Algorithm

In this section we examine the convergence properties of the algorithm (4) using the OSM based surrogate model (12). We give sufficient conditions for convergence. We assume that $X_f = X_c = X \subseteq R^n$. We also assume that X is a closed subset of R^n. We denote by Y_R the range of $\Delta R = R_f - R_c$, i.e., $Y_R = \Delta R(X) \subseteq R^m$.

Definition 1. Let $R_c^* : Y_R \to R^n$ be the function defined as

$$R_c^*(R) = \arg\min_{x \in X} U(R_c(x) + R) \qquad (16)$$

Theorem 1. Suppose that X is a closed subset of R^n and
(i) the function R_c^* is Lipschitz continuous on Y_R, i.e.,

$$\| R_c^*(R_2) - R_c^*(R_1) \| \leq L_C \| R_2 - R_1 \| \qquad (17)$$

for any $R_1, R_2 \in Y_R$, where $R_c^*(R_2)$ is the solution to (16) which is closest in norm to $R_c^*(R_1)$, whereas $L_C \in R_+$,

(ii) the function ΔR is Lipschitz continuous, i.e.,

$$\| \Delta R(y) - \Delta R(x) \| \leq L_R \| y - x \| \qquad (18)$$

for any $x, y \in X$, where $L_R \in R_+$,
(iii) L_R and L_C are such that $L_R L_C < 1$.
Then, for any $x^{(0)} \in X$ the sequence $\{x^{(i)}\}$ defined by (4), (12) is convergent, i.e., there is $x^* \in X$ such that $x^{(i)} \to x^*$ for $i \to \infty$. Moreover, for any $\bar{x} \in X_f^*$ we have the estimate

$$U(R_f(\bar{x})) \leq U(R_f(x^*)) = U(R_c(x^*) + \Delta R(x^*)) \qquad (19)$$

A proof of Theorem 1 is given in the Appendix.

This result shows that there are basically two fundamental (and natural) requirements for convergence of OSM based algorithms: (i) regularity of the perturbed coarse model optimal solution with respect to the perturbation vector, and (ii) similarity between the fine and coarse model in terms of the difference between their first-order derivatives.

Let $U_{\min}$ be defined as $U_{\min} = \min_{x \in X_f} U(R_f(x))$. We have

Corollary 1. Suppose that for any $R \in Y_R$ there is $x_R \in X$ such that $U(R_c(x_R) + R) \leq U_{\min}$ and assumptions of Theorem 1 are satisfied. Then $x^* \in X_f^*$, where $x^* = \lim_{i \to \infty} x^{(i)}$ is the limit of the sequence $\{x^{(i)}\}$ defined by algorithm (4), (12).

Proof. Convergence of the sequence $\{x^{(i)}\}$ follows from Theorem 1. Assumption of the corollary implies, that $U(R_c(x^{(i+1)}) + \Delta R(x^{(i)})) \leq U_{\min}$ for any $i = 1, 2, \ldots$. In the limit ($i \to \infty$), we have $R_c(x^{(i+1)}) + \Delta R(x^{(i)}) \to R_c(x^*) + \Delta R(x^*) = R_f(x^*)$ so $U(R_f(x^*)) = U_{\min}$ and $x^* \in X_f^*$.

Under similar assumptions (and the requirement that the jacobian of function ΔR is bounded and Lipschitz continuous on X) one can show convergence of the algorithm (4) using the OSM based surrogate model (13). It can also be shown that the convergence rate of algorithm (4), (13) is much better (we have $\|x^{(i+2)} - x^{(i+1)}\| \leq C_2 \cdot \|x^{(i+1)} - x^{(i)}\|^2$) than algorithm (4), (12) (we have $\|x^{(i+2)} - x^{(i+1)}\| \leq C_1 \cdot \|x^{(i+1)} - x^{(i)}\|$ in this case).

We consider an illustration of the convergence conditions. We analyze a generalized "cheese-cutting" problem [4]. The fine model is the irregular two-dimensional object of Fig. 1a. Its upper edge is defined by a positive function f. The corresponding coarse model is a rectangle of height H (Fig. 1b). Both models use the design parameter x (length).

We aim at finding x^* so that the area of the irregular object in Fig. 1a equals A_{opt}. The fine model $R_f : X \to R$ is given by

$$R_f(x) = \int_0^x f(t) dt \qquad (20)$$

The coarse model $R_c : X \to R$ is given by

$$R_c(x) = Hx \qquad (21)$$

In the sequel, we check the assumptions of Theorem 1 for R_f and R_c defined above assuming $f(x) = H + \sin(x)\exp(-x/5)$, $H = 2$, $X = [0, 10]$. Objective function U is defined by a norm (cf. (3)). In particular

(i) we have $Y_R = \{y \in R : y = \int_0^x (f(t) - H) dt, x \in X\}$ and $R_c^*(R) = \arg\min_x \|R_c(x) + R - A_{opt}\| = (A_{opt} - R)/H$. Thus, $\|R_c^*(R_2) - R_c^*(R_1)\| \leq |(A_{opt} - R_2)/H - (A_{opt} - R_1)/H| \leq H^{-1} \|R_2 - R_1\|$, i.e., assumption (i) is satisfied with $L_C = H^{-1}$.

(ii) since $\Delta R(x) = \int_0^x (f(t) - H) dt$, we have

$$\|\Delta R(y) - \Delta R(x)\| \leq \left|\int_y^x (f(t) - H) dt\right| \leq \sup_{t \in X} |f(t) - H| \cdot \|y - x\|$$

i.e., assumption (ii) is satisfied with $L_R(x) = \sup\{|f(t) - H| : t \in X\}$. This estimate is pessimistic, because the local Lipschitz constant (i.e., the constant valid within some neighborhood of x) is usually much lower.

(iii) we have $L_R L_C = \sup\{|f(t)/H - 1| : t \in X\}$. In the worst case we have $L_R L_C < 1$ if $f(x)/H < 2$ on X. For the assumed data we have $L_R L_C < 0.5$, enough to ensure convergence.

Note that in our example, the assumptions of Corollary 1 are satisfied. Indeed, let ΔR be arbitrary except it is not larger than A_{opt}. Then $\|R_c(x_{\Delta R}) + \Delta R - A_{opt}\| = 0 = U_{\min}$ is satisfied for $x_{\Delta R} = (A_{opt} - \Delta R)/H$. Thus, we have ensured convergence to the optimal solution of the fine model.

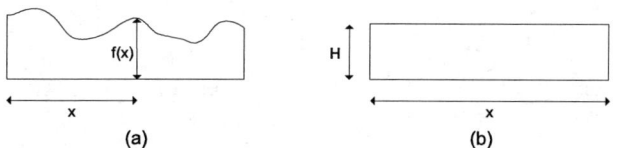

Fig. 1. The fine (a) and coarse (b) model in a two-dimensional generalized cheese-cutting problem.

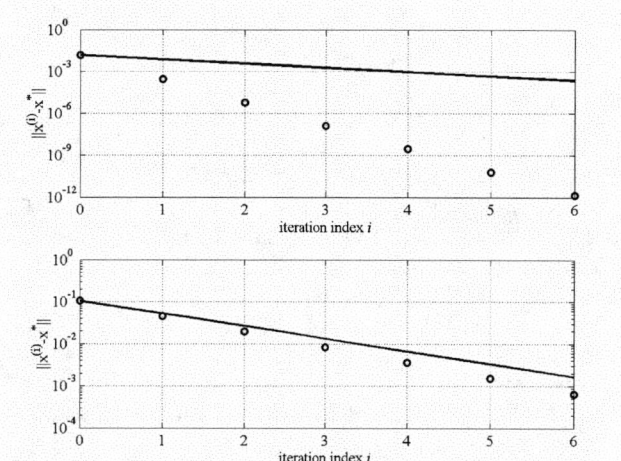

Fig. 2. Generalized cheese-cutting example: lower limit for the convergence rate (solid line), and actual convergence (circles) for $A_{opt} = 10$ (upper graph), and $A_{opt} = 2$ (lower graph).

Fig. 2 shows convergence of algorithm (4), (12) with the starting point being the coarse model optimal solution for $A_{opt} = 10$ (upper graph) and $A_{opt} = 2$ (lower graph). Solid lines denote a lower limit for the convergence rate (i.e., assuming $L_R L_C = 0.5$). Circles denote the actual convergence rate, which is faster for $A_{opt} = 10$ because in this case, all iterations except $x^{(0)}$ are located in the interval $[4.65, 4.70]$ in which we have $\sup\{|f(t) - H| : t \in [4.65, 4.7]\} < 0.05$, which gives $L_R L_C < 0.025$ (the starting point is $x^{(0)} = 5$, and the final solution $x^* = 4.64663401$). For $A_{opt} = 2$ convergence is slower, due to the fact that the local Lipschitz constant around the optimal solution $x^* = 0.77715486$ is about $L_R L_C \approx 0.5$.

V. EXAMPLE

We discuss the application of OSM based optimization algorithms to active filter design. This is a synthetic problem that aims at illustrating the performance of algorithm (4) with OSM based surrogate models (12) and (13). We consider a second-order OTA-C [9] low-pass filter (Fig. 3) implementing a Butterworth transfer function of the form

$$H_0(s) = (s^2 + \sqrt{2} s + 1)^{-1} \qquad (22)$$

A coarse model of the filter is the transfer function formula assuming ideal transconductors and no parasitic elements:

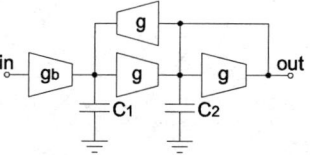

Fig. 3. Diagram of the second-order OTA-C low-pass filter.

$$H_c(s) = \frac{g_b g}{s^2 C_1 C_2 + s C_1 g + g^2} \quad (23)$$

A fine model is the transfer function formula that takes into consideration parasitic conductances g_o and capacitors C_p:

$$H_f(s) = \frac{g_b g}{s^2 C_1^* C_2^* + s\left(C_1^*(2g_o + g) + 2C_2^* g_o\right) + g^2 + 2g_o(g + 2g_o)} \quad (24)$$

where $C_i^* = C_i + C_p$, $i=1,2$. We use normalized elements with fixed $g=1$, $g_o=0.1$, $C_p=0.1$. Optimization variables are C_1, C_2 and g_b. The optimization problem is to find C_1, C_2 and g_b so that the difference between the fine model response vector $\mathbf{R}_f$ and the target vector $\mathbf{R}_{spec}$ is minimal w.r.t. the l_2 norm, where $\mathbf{R}_f = (|H_f(j\omega_1)|, \ldots, |H_f(j\omega_m)|)$, and $\mathbf{R}_{spec} = (|H_0(j\omega_1)|, \ldots, |H_0(j\omega_m)|)$. We set $m=21$ and $\omega_1=0.0$, $\omega_2=0.1$, …, $\omega_{21}=2.0$. Optimization variable values that correspond to the optimal solution of the coarse model are $C_1=1.4142$, $C_2=0.7071$, $g_b=1.0$.

We have applied algorithms (4), (12) and (4), (13) to find the fine model solution assuming $\mathbf{x}^{(0)} = (1.4142, 0.7071, 1.0)$. We consider (i) the surrogate is constructed directly using $\mathbf{R}_c$ as in (12) or (13); (ii) the coarse model is improved using the so-called input SM [4]. In particular, at the i-th iteration we use $\bar{\mathbf{R}}_c^{(i)}(\mathbf{x}) = \mathbf{R}_c(\mathbf{B}^{(i)} \cdot \mathbf{x} + \mathbf{c}^{(i)})$, where $\mathbf{B}^{(i)}$ and $\mathbf{c}^{(i)}$ are $n \times n$ and $n \times 1$ matrices, respectively, obtained using multipoint PE of the form $(\mathbf{B}^{(i)}, \mathbf{c}^{(i)}) = \arg\min_{\mathbf{B},\mathbf{c}} \sum_{k=1}^{i} \| \mathbf{R}_f(\mathbf{x}^{(k)}) - \mathbf{R}_c(\mathbf{B} \cdot \mathbf{x}^{(k)} + \mathbf{c}) \|$.

This improvement reduces misalignment between the coarse and fine models so that the Lipschitz constants in convergence conditions are smaller. As a result we can expect a better convergence rate of the algorithm.

Fig. 4 shows the convergence properties of our algorithms (w.r.t. the l_2 norm of the difference between $\mathbf{R}_{spec}$ and $\mathbf{R}_f$). The optimal fine model solution is $\mathbf{x}^* = (1.157, 1.006, 1.390)$. The results show that preconditioning of the coarse model by input SM and using a higher-order approximation of $\Delta \mathbf{R}$ (i.e., model (13) versus (14)) both improve the convergence rate of the algorithm. This complies with the theory of Section IV.

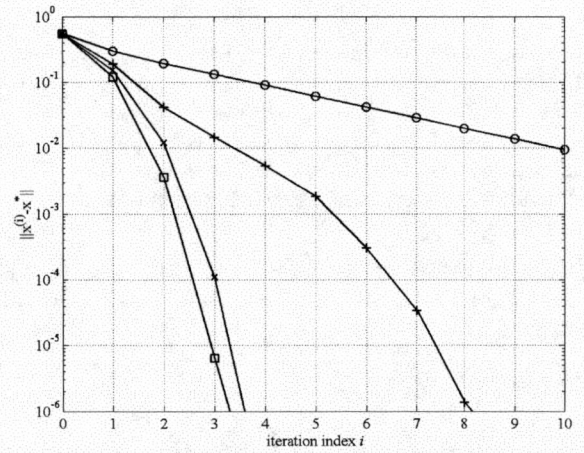

Fig. 4. Convergence of our OSM algorithms for the OTA-C design problem: algorithm (4), (12) without (o) and with (+) coarse model improvement, and algorithm (4), (13) without (×) and with (□) improvement.

VI. CONCLUSIONS

A rigorous formulation of optimization algorithms using SM-based surrogate models is presented. Properties of a subclass, utilizing Output Space Mapping, are investigated. Theoretical results are followed by examples that explain the assumptions imposed on the coarse and fine model to ensure algorithm convergence. The paper is a first step in an SM theory that aims not only at a theoretical justification of SM methods but also at guidelines for designing more efficient algorithms.

APPENDIX

Proof of Theorem 1. Take any $\mathbf{x}^{(0)} \in X$. Define $\{\mathbf{x}^{(i)}\}$ according to (4), (12), i.e., $\mathbf{x}^{(i+1)} = \arg\min_{\mathbf{x} \in X} U(\mathbf{R}_c(\mathbf{x}) + \Delta \mathbf{R}(\mathbf{x}^{(i)}))$ for $i=1,2,\ldots$. From Definition 1 we have that

$$\mathbf{x}^{(i+1)} = \mathbf{R}_c^*(\Delta \mathbf{R}(\mathbf{x}^{(i)})) \quad (A1)$$

We would like to obtain an estimate for $\| \mathbf{x}^{(i+2)} - \mathbf{x}^{(i+1)} \|$. Using (A1), and assumption (i)-(iii) we get

$$\| \mathbf{x}^{(i+2)} - \mathbf{x}^{(i+1)} \| = \| \mathbf{R}_c^*(\Delta \mathbf{R}(\mathbf{x}^{(i+1)})) - \mathbf{R}_c^*(\Delta \mathbf{R}(\mathbf{x}^{(i)})) \| \leq$$
$$\leq L_C \| \Delta \mathbf{R}(\mathbf{x}^{(i+1)}) - \Delta \mathbf{R}(\mathbf{x}^{(i)}) \| \leq q \| \mathbf{x}^{(i+1)} - \mathbf{x}^{(i)} \| \quad (A2)$$

where $q = L_C L_R < 1$. Now, for any $j > i$ we have

$$\| \mathbf{x}^{(j)} - \mathbf{x}^{(i)} \| \leq \| \mathbf{x}^{(i+1)} - \mathbf{x}^{(i)} \| + \ldots + \| \mathbf{x}^{(j)} - \mathbf{x}^{(j-1)} \| \leq$$
$$\leq (1 + q + q^2 + \ldots + q^{j-i-1}) \cdot \| \mathbf{x}^{(i+1)} - \mathbf{x}^{(i)} \|$$
$$\leq \frac{1 - q^{j-i}}{1-q} \| \mathbf{x}^{(i+1)} - \mathbf{x}^{(i)} \| \leq \frac{q^{i+1}}{1-q} \| \mathbf{x}^{(1)} - \mathbf{x}^{(0)} \| \quad (A3)$$

which is arbitrarily small for sufficiently large i, i.e., $\{\mathbf{x}^{(i)}\}$ is a Cauchy sequence. Thus, there is $\mathbf{x}^* \in X$, $\mathbf{x}^* = \lim_{i \to \infty} \mathbf{x}^{(i)}$. Estimate (19) is obvious. This ends the proof of the theorem.

REFERENCES

[1] J.W. Bandler, R.M. Biernacki, S.H. Chen, P.A. Grobelny and R.H. Hemmers, "Space mapping technique for electromagnetic optimization," *IEEE Trans. Microwave Theory Tech.*, vol. 42, pp. 536–544, Dec. 1994.

[2] J.W. Bandler, R.M. Biernacki, S.H. Chen, R.H. Hemmers and K. Madsen, "Electromagnetic optimization exploiting aggressive space mapping," *IEEE Trans. Microwave Theory Tech.*, vol. 43, pp. 2874–2882, Dec. 1995.

[3] J.W. Bandler, Q.S. Cheng, N.K. Nikolova and M.A. Ismail, "Implicit space mapping optimization exploiting preassigned parameters," *IEEE Trans. Microwave Theory Tech.*, vol. 52, pp. 378–385, Jan. 2004.

[4] J.W. Bandler, Q.S. Cheng, S.A. Dakroury, A.S. Mohamed, M.H. Bakr, K. Madsen and J. Sondergaard, "Space mapping: the state of the art," *IEEE Trans. Microwave Theory Tech.*, vol. 52, pp. 337–361, Jan. 2004.

[5] S.J. Leary, A. Bhaskar and A. J. Keane, "A constraint mapping approach to the structural optimization of an expensive model using surrogates," *Optimization Eng.*, vol. 2, pp. 385–398, 2001.

[6] M. Redhe and L. Nilsson, "Using space mapping and surrogate models to optimize vehicle crashworthiness design," 9th AIAA/ISSMO Multidisciplinary Analysis and Optimization Symp., Atlanta, GA, 2002, Paper AIAA-2002-5536.

[7] H.-S. Choi, D. H. Kim, I. H. Park and S. Y. Hahn, "A new design technique of magnetic systems using space mapping algorithm," *IEEE Trans. Magn.*, vol. 37, pp. 3627–3630, Sept. 2001.

[8] K. Madsen and J. Søndergaard, Convergence of hybrid space mapping algorithms, *Optimization and Engineering*, vol. 5, pp. 145–156, Kluwer Academic Publishers, 2004.

[9] Y. Sun (Editor), *Design of High Frequency Integrated Analogue Filters*, The Institution of Electrical Engineers, London, 2002.

BEHAVIORAL MODELING, SIMULATION AND HIGH-LEVEL SYNTHESIS OF PIPELINE A/D CONVERTERS

Jesús Ruiz-Amaya, José M. de la Rosa, Manuel Delgado-Restituto and Angel Rodríguez-Vázquez

Instituto de Microelectrónica de Sevilla – IMSE-CNM (CSIC)
Edificio CICA-CNM, Avda. Reina Mercedes s/n, 41012- Sevilla, SPAIN
Phone: +34 95 5056666, Fax: +34 95 5056686, E-mail: {ruiz|jrosa|mandel|angel}@imse.cnm.es

ABSTRACT

This paper presents a MATLAB® toolbox for the time-domain simulation and high-level sizing of pipeline analog-to-digital converters. SIMULINK® C-coded S-functions are used to describe the behavioral models of all building blocks, including their main circuit errors. This approach significantly speeds up system-level simulations while keeping high accuracy – verified with HSPICE – and interoperability of different subcircuit models. Moreover, their combined use with an efficient optimizer makes the proposed toolbox a valuable CAD tool for the high-level design of broadband communication analog front-ends. As a case study, an embedded 0.13μm CMOS 12bit@80MS/s A/D interface for a PLC chipset is designed to show the capabilities of the presented tool.[†1]

1. INTRODUCTION

The exponential increase of the capabilities of digital CMOS circuits – fuelled by the evolution of process technologies towards deep submicron – is prompting the integration of complete electronic systems onto a single chip. In such Systems on Chip (SoC), most of the signal processing is carried out by digital circuitry, whereas the role of analog circuits basically reduces to implement the necessary signal conditioning and data conversion interfaces [1][2]. In spite of this apparently minor role, the design of high-performance analog circuitry (usually, in adverse digital-oriented technologies) most often represents an important bottleneck for a short time-to-market deployment [3].

This problem is aggravated in modern telecommunication applications, like Very high-rate Digital Subscriber Line (VDSL) and Power Line Communication (PLC), where data converters targeting 12-14bit resolution at conversion rates of 40-80 MSamples/second (MS/s) are needed [1]. Although such data rates are easily achievable with flash or folding/interpolation Analog-to-Digital Converters (ADCs), their area and power consumptions become so significant at resolutions beyond 10 bit, that makes their deployment not competitive at least for SoC applications [4][5]. On the other hand, the use of ΣΔ modulator topologies is neither a viable solution for high signal bandwidths (beyond 20MHz) because of the prohibitive sampling frequencies which are required to achieve medium-high resolution [6][7].

In this scenario, pipeline ADCs have demonstrated to be a good alternative for interfaces requiring medium-high resolution at video-range conversion rates and beyond [8][9]. This has motivated the interest for CAD tools which can optimize and shorten the synthesis procedure of such ADCs [10]-[14]. Most of them are based on an iterative optimization procedure in which the design problem is translated into a cost function minimization problem that can be evaluated through numerical methods. Evaluation of the cost function is normally performed by means of equations [11][12][14], so that very short computation times are obtained. As a drawback, this approach results in closed tools because equations must be changed every time the topology is changed.

This paper aims at palliating this problem by using simulation instead of equations for cost function evaluation. To this end, a complete toolbox for the high-level synthesis of arbitrary pipeline ADCs has been developed in the MATLAB® environment [15]. The embedded simulator uses SIMULINK® S-functions [16] to model all required subcircuits including their main non-idealities. This approach considerably reduces computational costs as compared to using standard library blocks as in [13]. For all subcircuits, the accuracy of the behavioral models has been verified by HSPICE. Additionally, the toolbox includes an efficient hybrid optimizer which uses statistical techniques for design space exploration and deterministic techniques for fine tuning [17]. Other important features of the platform are a friendly Graphical User Interface (GUI), high flexibility for tool expansion[†2] and wide signal processing capabilities [15].

As a case study on the use of the proposed synthesis tool, a 0.13μm CMOS 12-bit@80MS/s pipeline ADC for PLC is synthesized and designed. Different experiments show the effectiveness of the proposed methodology.

2. BEHAVIORAL MODELING OF PIPELINE ADCs USING SIMULINK® C-CODED S-FUNCTIONS

Fig.1(a) shows the conceptual block diagram of a generic pipeline ADC, consisting of an arbitrary cascade of k stages and a Sampled-and-Hold (S/H) circuit at the front [1][2][11]. Each stage re-

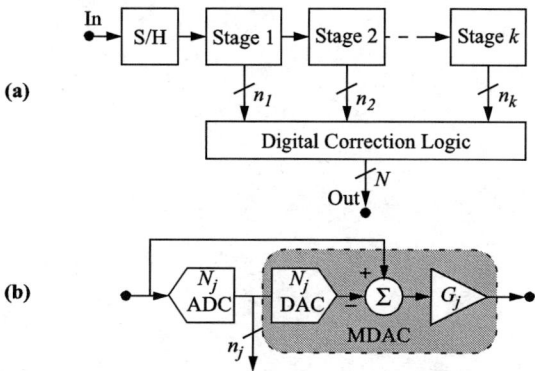

Figure 1. Generic pipeline ADC architecture. (a) Conceptual block diagram; (b) structure of a single stage.

[†1]. This work has been supported by the MEDEA+ (A110 MIDAS) Project.

[†2]. Indeed, the toolbox has been already extended to cover other converter topologies, such as full flash ADCs and current-steering Digital-to-Analog Converters (DACs).

solves partial code words of length n_j, $j = 1, ..., k$, which are all re-ordered and combined at the digital correction block to obtain the N-bit output of the converter. The inner structure of a pipeline stage comprises four blocks, as illustrated in Fig.1(b): a flash sub-ADC with $N_j \leq 2^{n_j}$ output codes, a sub-DAC with N_j output levels, a substractor, and a S/H residue amplifier with gain G_j. The latter three blocks are implemented in practice by a single subcircuit which is often referred to as Multiplying DAC (MDAC).

All the critical blocks in the pipeline architecture, namely, S/H circuit, sub-ADCs and MDACs, have been modelled and coded in the proposed toolbox including their most important error mechanisms. As an illustration, Fig.2(a)[†3] shows the conceptual schematic of a MDAC block which operates with two non-overlapped clock phases. Its model in the proposed toolbox includes the most critical error mechanisms which are computed according to the flow diagram in Fig.3. The flow graph has two branches corresponding to the two clock phases. During the sampling phase, the input-equivalent thermal noise, (Δv_n), is calculated and added to the voltage stored at the sampling capacitors C_i. This is computed taking into account the finite switch on-resistance effects (r_{on}). Next, an iterative procedure is started to calculate the output voltage v_o by solving the equivalent circuit of Fig.2(b), which models the effects of finite and non-linear opamp DC-gain ($A_v(v_x) = G_m(v_x)/g_{ox}$), opamp offset (not shown in Fig.2(b) for simplicity), non-linear capacitors, and the opamp dynamics (comprising both linear incomplete settling and slew-rate limitation), parasitic capacitances (C_p, C_b, C_{load}), output range limitations and charge injection error. Note that a two-pole model (in-

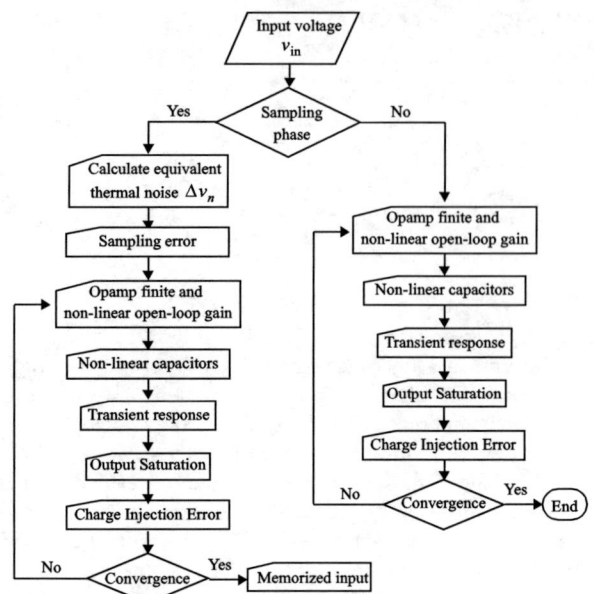

Figure 3. Flow diagram of the MDAC model.

cluding parameters g_m, C_{load}, g_o, g_{mh}, C_h, g_{oh}) using Miller compensation (C_c) has been developed for the opamp. During the residue-amplification phase, a similar procedure is applied to solve the equivalent circuit shown in Fig.2(c).

As the value of state signals are important only at the end of each clock phase, a set of finite difference equations have been generated to describe the operation of real pipeline sub-circuits. These equations have been codified in C and incorporated as S-functions into the SIMULINK® environment [16]. This approach allows to drastically speed up the simulation CPU-time[†4] (up to 2 orders of magnitude) as compared to previous approaches – based on the use of SIMULINK® elementary blocks [13]. Moreover, S-functions are more suitable for implementing a more detailed description of the circuit. As an example of the accuracy of the behavioral model, Fig.4 compares the transient response of a 2-bit MDAC for a constant input voltage by using HSPICE and our model showing a good agreement.

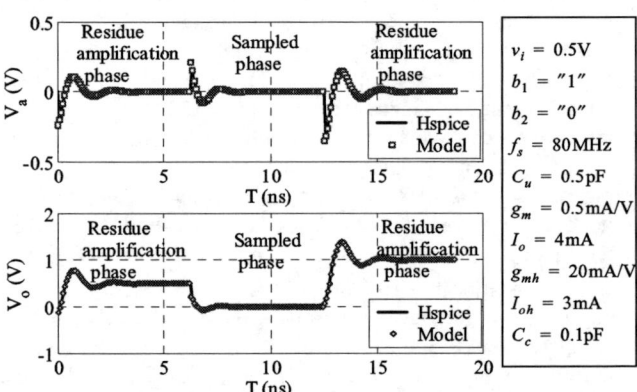

Figure 4. Transient response of a 2-bit MDAC considering a constant input voltage: comparison between HSPICE and two-pole behavioral model.

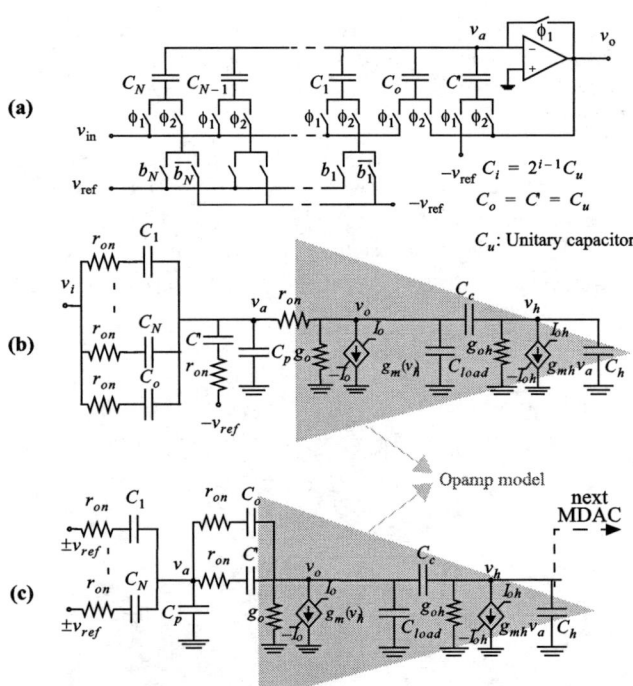

Figure 2. MDAC (a) schematic, (b) equivalent circuit in sampling phase and (c) equivalent circuit in residue amplification phase.

---

†3. For illustration purposes, schematics are shown in its single-ended version, although actually the fully-differential structures have been modelled.

†4. A 32738-samples simulation takes 2-3 seconds. All simulations shown in this paper were done using a PC with an AMD XP2400 CPU@2GHz @512MB-RAM.

3. DESCRIPTION OF THE SYNTHESIS TOOLBOX

The models described above have been included in a SIMULINK®-based simulator. This simulator, used for performance evaluation, is combined with a statistical optimizer for design parameter selection as described below.

3.1 Optimization procedure

Deterministic optimization methods, like those available in the MATLAB® standard distribution [15], are not suitable for synthesis purposes because they are strongly dependent on the initial conditions. However, initially designers may have little or no idea of an appropriate design point and hence, the optimization procedure is quickly trapped in a local minimum. For this reason, we developed an optimizer which combines an adaptive statistical optimization algorithm inspired in simulated annealing (local minima of the cost function can then be avoided) with a design-oriented formulation of the cost function (which accounts for the modulator performances). Moreover, an integrated approach is addressed: statistical techniques are applied for wide design space exploration whereas deterministic techniques are used for fine-tuning of best solutions found by the previous techniques. Unlike conventional simulated annealing procedures, in which the control parameter – commonly named temperature – follows a predefined temporal evolution pattern, the implemented global optimization algorithm dynamically adapts this temperature to approximate a predefined evolution pattern of the acceptance ratio (accepted movements / total number of iterations). This idea prevents excessively high temperatures which will make convergence difficult and inappropriately low temperatures which can make the algorithm to stuck on a local minimum. The amplitude of parameter movements through the design space is also synchronized with the temperature for improved convergence.

The optimizer has been integrated in the MATLAB®/SIMULINK® platform by using the MATLAB® engine library [15], so that the optimization core runs in background while MATLAB® acts as a computation engine. The optimization core is very flexible, in so far as the cost function formulation is very versatile: multiple targets with several weights, constraints, dependent variables, and logarithmic grids are permitted. This optimization procedure has been extensively tested with design problems of $\Sigma\Delta$ modulators involving behavioral simulators [18] as well as electrical simulators [1].

3.2 Implementation in the MATLAB® environment

The proposed tool has been conceived as a MATLAB® toolbox for the simulation and synthesis of Nyquist-rate data converters, including flash, pipeline ADCs and current-steering DACs. Fig.5 shows some parts of the toolbox comprising a GUI to allow the designer to browse through all steps of the simulation, synthesis and post-processing of results.

4. CASE STUDY: A 12-bit@80MS/s ADC FOR PLC

In order to illustrate the capabilities of the proposed toolbox, the high-level design of a 0.13μm CMOS 12-bit@80MS/s pipeline ADC for PLC will be described. The specifications are shown in Table 1 and the objective consists on achieving those specifications with the minimum area and power consumption. In addition, this design is planned to be implemented without using calibration. For that reason, capacitor mismatch is a critical issue. In fact, this limitation forces us to optimize the capacitor sizes not only in terms of thermal noise and dynamics considerations but also in terms of minimum capacitance area needed to achieve the required mismatch. Another critical parameter considered in the optimization is the resolution per stage.

Taking into account these factors, a wide exploration of several architectures has been carried out with the proposed synthesis toolbox. The optimum architecture was a 7-stage pipeline with the next resolution-per-stage: 3-2-3-2-3-3-2. Redundant sign digit coding was used in order to relax the requirements for the comparators in the flash quantizers. The results of the high-level synthesis for the first stages as well as the requirements for the opamps are summarized in Table 2. The optimization procedure for a given architecture required about 4000 iterations of 16384-clock cycles taking about 2 hours of CPU-time.

Finally, Table 3 shows a summary of the converter performance from where it can be deduced that the specifications are fully satisfied. MonteCarlo analysis has been carried out taking into account both capacitor and resistor mismatch in order to characterize typical and worst cases of different figures. As an illustration, Fig.6(a) shows an histogram of a MonteCarlo analysis of the MTPR ((case (a) in Table 1)) comprising 1000 simulations and Fig.6(b) shows the output spectrum of one of those simulations.

CONCLUSIONS

A complete MATLAB® toolbox for the high-level synthesis and verification of pipeline ADCs has been described. The combination of an efficient SIMULINK®-based time-domain behavioral simulator and an advanced statistical optimizer allows to efficiently map system-level specifications into building-block specifications in reasonable computation times. Critical design issues such as the resolution-per-stage are optimized in terms of power consumption and silicon area. As a case study, a 0.13μm CMOS 12bit@80MS/s ADC for PLC front-end has been designed and analysed using the proposed toolbox.

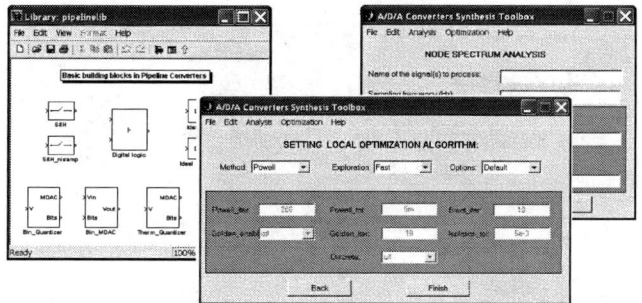

Figure 5. Some parts of the data converter MATLAB® synthesis toolbox.

Table 1: Specifications for pipeline ADC

Specifications: 12bit@80MS/s	
Multi-Tone Power Ratio (MTPR) (a) 15 tones on-1 tone off (b) 120 tones on - 8 tones off (c) 240 tones on - 16 tones off	≥ 56dB
Effective Number Of Bits (ENOB)	≥ 9.2 bits
Differential Input Range	2 Vp-p
Power Supply	3.3V

REFERENCES

[1] A. Rodríguez-Vázquez, F. Medeiro and E. Janssens (Editors): *CMOS Telecom Data Converters*. Kluwer Academic Publishers, 2003.
[2] M. Gustavsson, J. J. Wikner and N. N. Tan: *CMOS Data Converters for Communications*. Kluwer Academic Publishers, 2000.
[3] G. G. E. Gielen and R. A. Rutenbar: "Computer-Aided Design of Analog and Mixed-Signal Integrated Circuits". *Proceedings of the IEEE*, Vol. 88, pp. 1825-1852, December 2000.
[4] K. Uyttenhove, J. Vandenbussche, E. Lauwers, G. G. E. Gielen and M. S. J. Steyaert: "Design Techniques and Implementation of an 8-bit 200-MS/s Interpolating/Averaging CMOS A/D Converter". *IEEE Journal of Solid-State Circuits*, Vol.38, pp. 483-494, March 2003.
[5] K. Uyttenhove and M. S. J. Steyaert: "A 1.8-V 6-bit 1.3-GHz Flash ADC in 0.25-μm CMOS". *IEEE Journal of Solid-State Circuits*, Vol. 38, pp. 1115-1122, July 2003.
[6] Y. Geerts, M. Steyaert, and W. Sansen, "A High Performance Multibit ΔΣ CMOS ADC". *IEEE Journal of Solid-State Circuits*, Vol. 35, pp. 1829-1840, December 2000.
[7] A. Di Giandomenico, S. Paton, A. Wiesbauer, L. Hernández, T. Pöster and L. Dörrer: "A 15-MHz Bandwidth Sigma-Delta ADC with 11 bits of Resolution in 0.13μm CMOS". *Proc. of the 2003 European Solid-State Circuits Conference*, pp.233-236.
[8] A. Shabra and H.-S. Lee: "Oversampled Pipeline A/D Converters With Mismatch Shaping". *IEEE Journal of Solid-State Circuits*, Vol. 37, pp. 566-578, May 2002.
[9] W. Yang, D. Kelly, I. Mehr, M. Sayuk and L. Singer: "A 3-V 340-mW 14-b 75-Msamples/s CMOS ADC with 85-dB SFDR at Nyquist Input". *IEEE Journal of Solid-State Circuits*, Vol. 36, pp. 1931-1936, Dec. 2001.
[10] E. Liu and A. Sangiovanni-Vincentelli: "Verification of Nyquist Data Converters Using Behavioral Simulation". *IEEE Trans. on Computer-Aided Design of Integrated Circuits and Systems*, Vol. 14, pp. 493-502, April 1995.
[11] J. Goes, J. C. Vital and J. E. Franca: *Systematic Design for Optimization of Pipelined ADCs*. Kluwer Academic Publishers, 2001.
[12] P. T. F. Kwok and H. C. Leung: "Power Optimization for Pipeline Analog-to-Digital Converters". *IEEE Trans. on Circuits and Systems - Part II*, pp. 549-53, May 1999.
[13] E. Bilhan, P. C. Estrada-Gutierrez, A. Y. Valero-Lopez and F. Maloberti: "Behavioral Model of Pipeline ADC by Using Simulink". *Proc. of 2001 Southwest Symposium on Mixed-Signal Design*, pp. 147-151.
[14] R. Lotfi, M. Taherzadeh-Sani, M. Yaser-Azizi and O. Shoaei: "Systematic Design for Power Minimization of Pipelined Analog-to-Digital Converters". *Proc. of 2003 IEEE Int. Conf. Computer-Aided Design*, pp. 371-374.
[15] The MathWorks Inc.: "MATLAB®. The Language of Technical Computing. V.6.5 R.13", July 2002.
[16] The MathWorks Inc.: "Simulink®. Model-based and System-based Design. V.5 R.13", July 2002.
[17] F. Medeiro, B. Pérez-Verdú, and A. Rodríguez-Vázquez: *Top-Down Design of High-Performance Modulators*, Kluwer Academic Publishers, 1999.
[18] J. Ruiz-Amaya, J.M. de la Rosa, F. Medeiro, F.V. Fernández, R. del Río, B. Pérez-Verdú and A. Rodríguez-Vázquez: "An Optimization-based Tool for the High-Level Synthesis of Discrete-time and Continuous-Time ΣΔ Modulators in the MATLAB/SIMULINK Environment". *Proc. IEEE Int. Symp. Circuits and Systems*, Vol V., pp. 97-100, 2004.

Table 2: High-level synthesis for the pipeline ADC

Block		Parameter	Requirement
S/H		Sampling capacitor (pF)	<5.5
		Switch on-resistance (Ω)	<60
	Opamp	Eq. load (pF)	11.8
		Slew-Rate (V/μs)	>932
		GB (MHz)	>135
		DC-gain (dB)	>60
		Noise PSD (nV/$\sqrt{Hz}$)	<6
1st. stage (3 bits)	Flash quant.	Comparators Offset (mV)	<100
		Comparators Hysteresis (mV)	<100
	MDAC	Unitary capacitor (pF)	1.1
		Switch on-resistance (Ω)	<40
	Opamp	Eq. load (pF)	32.9
		Slew-Rate (V/μs)	>107
		GB (MHz)	>218
		DC-gain (dB)	>81
		Noise PSD (nV/$\sqrt{Hz}$)	<3.3
2nd. stage (2 bits)	Flash quant.	Comparators Offset (mV)	<100
		Comparators Hysteresis (mV)	<100
	MDAC	Unitary capacitor (pF)	1
		Switch on-resistance (Ω)	<40
	Opamp	Eq. load (pF)	22.2
		Slew-Rate (V/μs)	>180
		GB (MHz)	>201
		DC-gain (dB)	>72
		Noise PSD (nV/$\sqrt{Hz}$)	<8
3rd. stage (3 bits)	Flash quant.	Comparators Offset (mV)	<100
		Comparators Hysteresis (mV)	<100
	MDAC	Unitary capacitor (pF)	0.75
		Switch on-resistance (Ω)	<60
	Opamp	Eq. load (pF)	15
		Slew-Rate (V/μs)	>107
		GB (MHz)	>146
		DC-gain (dB)	>70
		Noise PSD (nV/$\sqrt{Hz}$)	<14
4th. stage (2 bits)	Flash quant.	Comparators Offset (mV)	<100
		Comparators Hysteresis (mV)	<100
	MDAC	Unitary capacitor (pF)	0.5
		Switch on-resistance (Ω)	<200
	Opamp	Eq. load (pF)	13.4
		Slew-Rate (V/μs)	>112
		GB (MHz)	>143
		DC-gain (dB)	>66
		Noise PSD (nV/$\sqrt{Hz}$)	<63

Table 3: Simulation results.

Results	Typical case	Worst case
ENOB (bits) (f_i@34MHz)	10.13	9.7
INL(12bits) (LSB)	-2.60/2.83	-
DNL(12bits) (LSB)	-0.59/0.66	-
MTPR (case a)	59.62	55.96
MTPR (case b)	61.48	57.95
MTPR (case c)	62.15	60.16
Power Consumption (mW)	230	

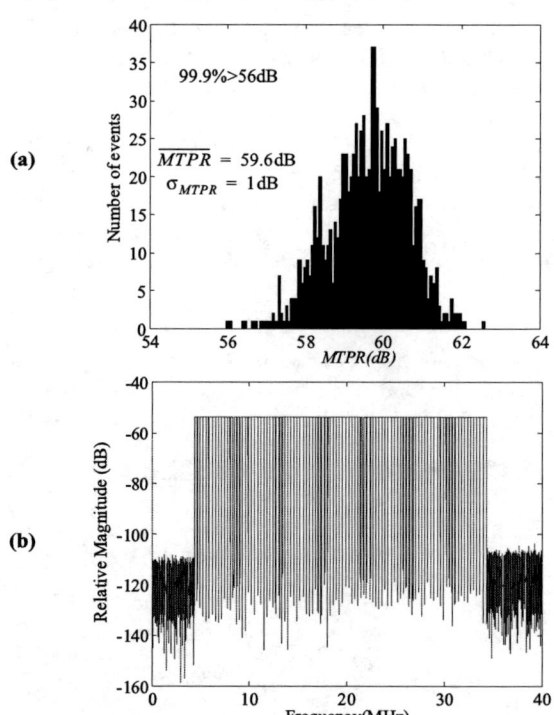

Figure 6. MTPR (case (a) in Table 1): (a) histogram of a MonteCarlo analysis; (b) output spectrum.

Delay Modeling of CMOS/CPL Logic Circuits

Yuanzhong Wan
Department of Electronics
Carleton University
Ottawa, ON K1S 5B6
Canada
ywan@doe.carleton.ca

Maitham Shams
Department of Electronics
Carleton University
Ottawa, ON K1S 5B6
Canada
shams@doe.carleton.ca

Abstract — Effective optimization methods aimed at achieving maximal speeds in single-technology logic circuits are widely available but systematic ways suitable for circuits involving mixtures of logic families are not. In this paper, the combination of standard CMOS with CPL is examined with an eye to finding the best structure and the best insertion points for CMOS buffers intended to improve a CPL chain's propagation time and drive capability.

I. INTRODUCTION

One method of enhancing the performance of digital CMOS circuits is to use combinations of logic styles, one such being Conventional CMOS (C-CMOS) with Complementary Pass-transistor Logic (CPL). The idea behind combining these two logic forms is to realize the best properties of each in the resulting application - the superior current-drive capabilities and noise margins of C-CMOS with the speed and low power consumption of CPL.

This paper begins with a brief description of some methods for increasing the current-drive capability of a CPL chain while reducing its propagation lag. The more important characteristics of each method are reviewed. Then the transistor-size calculations, an esential part of the optimization process, are summarized, and finally the numerical experiments which allowed the optimal structure/placement to be determined are reviewed.

II. MODELING OF DELAY IN MIXED-LOGIC CIRCUITS

A. C-CMOS/CPL logic circuits

1) The C-CMOS Circuit

C-CMOS is still the mainstream IC technology, and the optimization of digital CMOS circuits remains a major focus of research.

A C-CMOS gates (Figure 1) is in general a series connection of an N block and a P block, only one of which is conducting in the steady state. These gates have good current drive capability. (But because of lower carrier mobility in PMOS transistors, the P block must be physically larger if this capability is to be symmetrical, with '1' source current being the same as '0' sink current.)

2) CPL Circuit

CPL is a widely used alternative to C-CMOS. A CPL device consists of an NMOS pass-transistor network connected to cross-coupled PMOS transistors to achieve level restoration, to compensate for voltage drops along the chains, as shown in Figure 2. A CPL circuit needs fewer transistors than does its C-CMOS equivalent to perform a given function, so that in general that there will be associated with it a lower overall capacitance with a resulting lower power consumption at high speeds. In a CPL circuit, the rising delay is the worst-case one since the rise-time at the drain of an NMOS transistor is longer than the fall time (We say that "NMOS transistors are slower at transferring a logic '1' than in transferring a logic '0' ".)

The CPL chain's input signals are supplied by transistor gate and source drives (Figure 2). The characteristics of these will in general differ, and so they will be given the separate names terms 'G-Drive' and 'S-Drive'.

Shams *et. al* [3] have presented three comprehensive models to be used for the estimation and optimization of propagation times in three worst-case situations: a critical-path involving the G-Drive, another one associated with the S-Drive, and the third where both drives are involved. The S-Drive critical-path

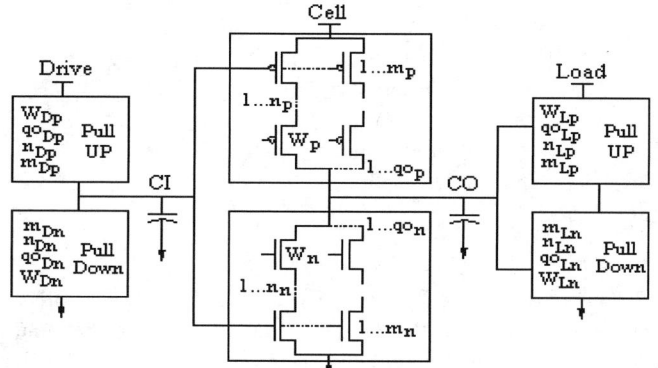

Figure 1. A C-CMOS Cell Gate Connected with Drive and Load Gate

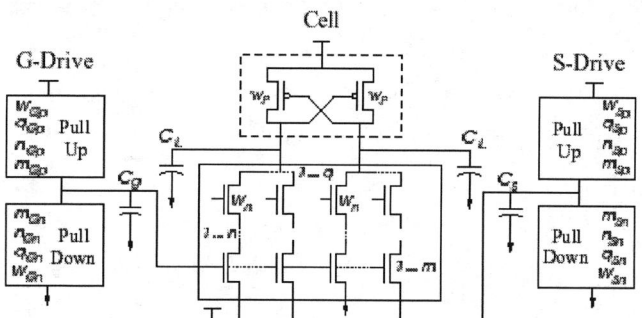

Figure 2. A CPL Cell Gate Connected with S-Drive and G-Drive

tends to be the most important of the three, and so it will get most of the attention in the next sections.

A CPL gate consists of an NMOS pass-transistor network connected to cross-coupled PMOS transistors to achieve level restoration and to compensate for ohmic voltage drops along the chains. The higher speed and lower power consumtion of a CPL circuit is due its use of fewer transistors than CMOS to perform a given function [4].

B. Delay modeling of CPL chain

The CPL chain's input signals are supplied by transistor gate and source drives. The characteristics of these will in general differ, and so they will be given the separate names terms 'G-Drive' and 'S-Drive'. Shams *et. al* [3] have presented three comprehensive models to be used for the estimation and optimization of propagation times in three worst-case situations: one involving the critical-path containing the G-Drive, another associated with the S-Drive, and a third path containing both drives. The S-Drive critical-path tends to be the most important of the three, and so it will get most of the attention in what follows.

The delay associated with a chain of pass-transistors is directly related to the performance of CPL devices; this attribute is related to the poor current drive capacity and the poor noise margins of CPL chains. Some of the ways [2] that have been proposed to deal with these problems are outlined below. We studied four of these, namely the use of a level restorer, insertion into the CPL chain of a latch alone, insertion of a buffer alone, and finally the use of a latch/buffer combination.

1) Level restorer (br)

A commonly-used way of solving the voltage drop and drive problems is to place the construct of Figure 3(A) in each branch. This 'level-restorer' uses feedback to compensate for accumulated voltage-level degradation in a differential CPL chain. The worst-case delay here is the rising delay on the path of the NMOS transistors from the S-Drive buffer input to the CPL output. Usually, the current I_p which feeds back from the inverter output through to the PMOS device charges the node capacitance after one inverter delay as shown in Figure 3(B), after which it causes the output to be pulled up from $V_{DD} - V_T$ to V_{DD}, so it's use doesn't lead to a rising-delay reduction.

Furthermore, the presence of the PMOS transistor adds a diffusion capacitance to the total load capacitance, reducing the CPL speed. To lessen this effect, the size of the PMOS device

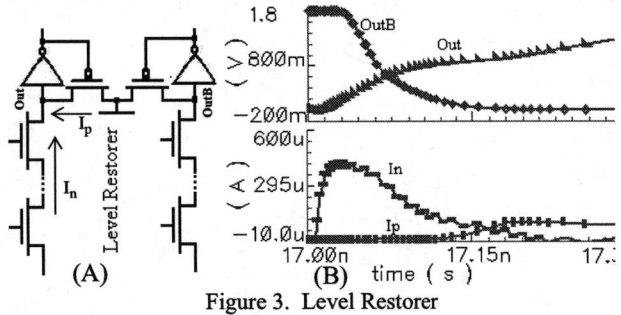

Figure 3. Level Restorer

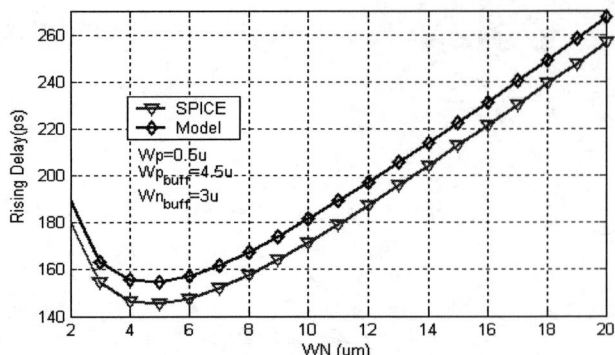

Figure 4. Rising Delay with Level Restore

should be kept as small as possible. The equations used to predict the rising propagation time effects of the level-restorer mentioned above, on the path between the S-Drive buffer's input through to cell gate's output, are:

$$Delay_{Rise} = (1+S_{nf})DS_{CellRise} + DS_{LoadFall} + DS_{DriveRise} \quad (1)$$

$$DS_{CellRise} = \left(\frac{v_{pr}}{W_{Dp}} + \frac{v_{nr}Y_{nr}}{W_n}\right)(qo_n W_n Cd_{nr} + W_p Cd_{pr} + CL_{Rise}) \quad (2)$$

$$DS_{DriveRise} = \frac{v_{pr}}{W_{Dp}}(qi_n W_n Cd_{nr} + CS_{Rise}) \quad (3)$$

$$DS_{LoadFall} = \frac{v_{nf}}{W_{Ln}}CO_{Fall} \quad (4)$$

The quantities appearing in them are:

$DS_{CellRise}/DS_{DriveRise}$:rising step delay of the cell/drive gate.
$DS_{LoadFall}$:falling step delay of load buffer.
S_{nf} :NMOS falling input slope effect factor.
v :unit resistance per unit $1/W$, which combines all process parameters and minimum length.
v_{pr}/v_{nr} :PMOS/NMOS rising unit resistance.
v_{nf} :NMOS falling unit resistance.
Y_{nr} :NMOS rising degradation factor of cell-gate which quantifies the effect of a number of series-connected transistors.
qi_n/qo_n :Number of NMOS branches connected to an input/output node.
Cd_{nf}/Cd_{pf} :NMOS/PMOS falling unit diffusion capacitance.
Cd_{nr}/Cd_{pr} :NMOS/PMOS rising unit diffusion capacitance.
Cg_{nf}/Cg_{pf} :NMOS/PMOS falling unit gate capacitance.
Cg_{nr}/Cg_{pr} :NMOS/PMOS rising unit gate capacitance.
CL_{Rise} :Total node rising capacitance at the output of the cell gate except for the cell-gate-related capacitances.
CS_{Rise} :Total node rising capacitance at the drive gateoutput, or at the input of the cell-gate, excluding cell-gate-related capacitances.
CO_{Fall} :Total falling capacitance at the load buffer output.

In Figure 4 are shown model-predicted and SPICE-simulated rising delays.

2) Latch insertion (l)

Cross-coupled PMOS transistors may used to obtain both a level restoration effect and a rising delay improvement (Figure 5(a)). As in the previous case, the worst-case delay is the rising delay. Figure 5(b) shows the overlap currents: I_n

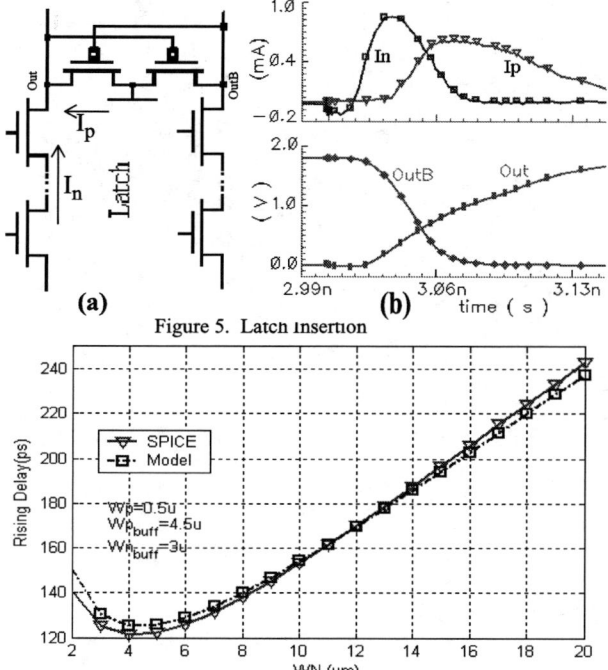
Figure 5. Latch insertion

output buffer output are:

$$Delay_{Rise} = (1+S_{nf})DS_{CellRise} + DS_{LoadFall} + DS_{DriveRise} \quad (8)$$

$$DS_{CellRise} = \left(\frac{v_{pr}}{W_{Dp}} + \frac{v_{nr}Y_{nr}}{W_n}\right)(qo_n W_n Cd_{nr} + CL_{Rise}) \quad (9)$$

$$DS_{DriveRise} = \frac{v_{pr}}{W_{Dp}}(qi_n W_n Cd_{nr} + CS_{Rise}) \quad (10)$$

$$DS_{LoadFall} = \frac{v_{nf}}{W_{Ln}} CO_{Fall} \quad (11)$$

In Figure 8 are compared the delays determined by SPICE simulation to those predicted by our model.

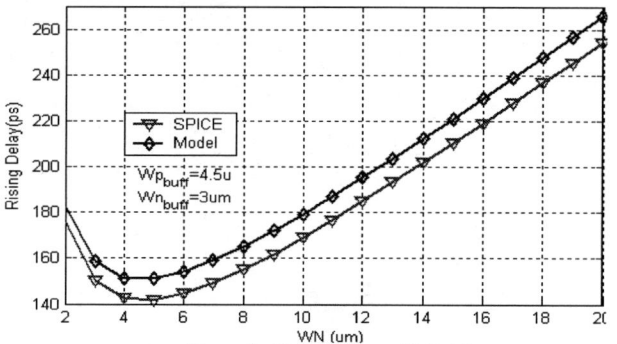
Figure 6. Rising Delay with Latch Insertion

Figure 8. Rising Delay with Buffer

from the S-Drive through the CPL chain and I_p from the PMOS latch, which is determined by the falling delay. The current I_p begins to charge the output capacitance soon after the current I_n drops. This helps to increase the output voltage but does little to improve the rise time.

To simplify the delay model then, the relatively unimportant I_p is ignored and the PMOS transistor size is set to a minimum. The resulting equations are

$$Delay_{Rise} = DS_{CellRise} + DS_{DriveRise} \quad (5)$$

$$DS_{CellRise} = \left(\frac{v_{pr}}{W_{Dp}} + \frac{v_{nr}Y_{nr}}{W_n}\right)$$
$$\times (qo_n W_n Cd_{nr} + W_p(Cd_{pr} + Cg_{pr}) + CL_{Rise}) \quad (6)$$

$$DS_{DriveRise} = \frac{v_{pr}}{W_{Dp}}(qi_n W_n Cd_{nr} + CS_{Rise}) \quad (7)$$

Figure 6 shows the rise time delay predictions of SPICE and of our model.

3) Buffer insertion (b)

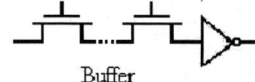
Figure 7. Buffer Insertion t

This method is the most common way of dealing with combined long delay and poor current drive (Figure 7). An inverting buffer is connected directly to each of the differential branches in hopes of lessening output voltage drop while increasing the output current drive. However, this approach suffers from low speed and large power dissipation because of the voltage ($V_{DD} - V_T$) at the buffer's input is sufficient to partially turn on both sides of the buffer simultaneously.

The rising delay will again be the worst-case delay in this case. The prediction equations for the drive buffer input to the

4) Combined latch and buffer insertion (lb)

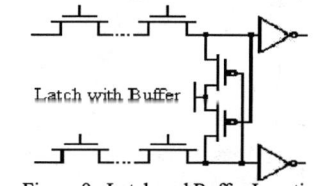
Figure 9. Latch and Buffer Insertion t

The last of the four methods studied for attacking the voltage-drop and current-drive problems uses a combination of the previous two ideas to produce higher speed and higher-voltage signals at CPL output nodes (Figure 9). Yet again, it is the rising delay which must command most of our attention..

The equations listed here can be used to model the rising delay along the path from the S-Drive input to the load buffer output:

$$Delay_{Rise} = (1+S_{nf})DS_{CellRise} + DS_{LoadFall} + DS_{DriveRise} \quad (12)$$

$$DS_{CellRise} = \left(\frac{v_{pr}}{W_{Dp}} + \frac{v_{nr}Y_{nr}}{W_n}\right)$$
$$\times (qo_n W_n Cd_{nr} + W_p(Cd_{pr} + Cg_{pr}) + CL_{Rise}) \quad (13)$$

$$DS_{DriveRise} = \frac{v_{pr}}{W_{Dp}}(qi_n W_n Cd_{nr} + CS_{Rise}) \quad (14)$$

$$DS_{LoadFall} = \frac{v_{nf}}{W_{Ln}} CO_{Fall} \quad (15)$$

Figure 10 depicts the modelled and SPICE-simulated rising delays:.

III. DELAY OPTIMIZATION OF C-CMOS/CPL COMBINATION

The set of models of Shams *et al* [3] may be used to predict the relative delays between drive gates, cell gates and load gates, they were used by us to develop a more-accurate for-

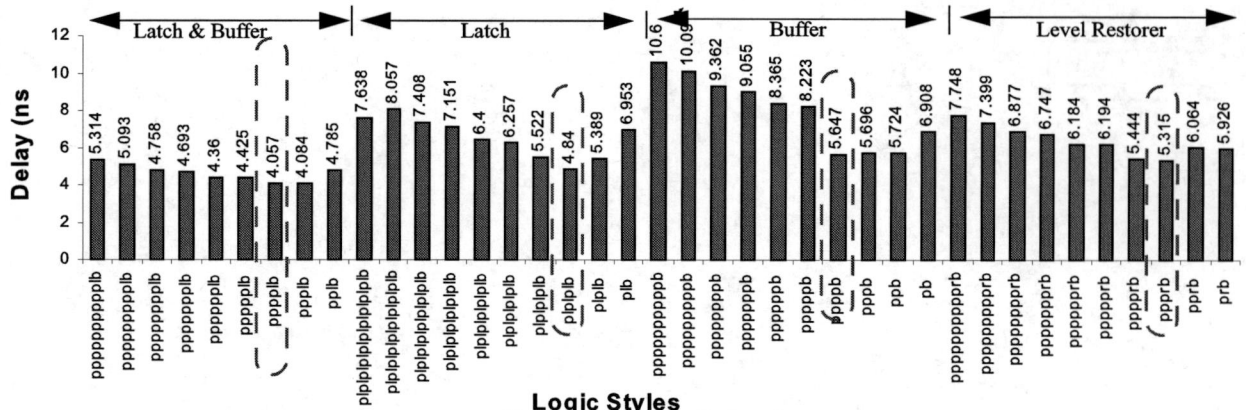

Figure 11. Optimal Buffer Insertion

TABLE I. Optimal Transistor Sizing of CPL Chain

Style	Model(μ m) W_p/W_n	SPICE(μ m) W_p/W_n
rb	0.5/4.2	0.5/5
l	2.2/4.4	2.2/4.4
b	0/4.1	0/5
lb	0.5/4.4	0.5/4.4

Figure 10. Rising Delay with Latch and Buffer

mula for computing optimal sizes of the transistors in a CPL chain involving an S-Drive. On simplifying the models of the previous section and then considering the derivatives $(\partial Delay_{Rise})/(\partial W_n) = 0$. we found the optimal size formula:

$$W_n = \sqrt{\left(\frac{v_{nr} Y_{nr}}{v_{pr} Y_{Dpr}}\right)\left(\frac{W_p(Cd_{pr} + Cg_{pr}) + CL_{Rise}}{(2(n_n-1) + qi_n + qo_n)Cd_{nr}}\right)} W_{Dp} \quad (16)$$

The quantities appearing above are:

Y_{nr}/Y_{Dpr} : NMOS/PMS rising degradation factor of cell-gate/drive-gate which quantifies the effect of a number of series-connected transistors.

W_p/W_{Dp} : PMOS width of the cell/drive gate.

n_n : The number of series-connected NMOS transistor in the cell gate

The expression (16) can be used not only for a CPL-chain-with-latch which involves an S-Drive, but also for one with a level-restorer and buffer that involves an S-Drive. One simply notes that the expression $w_p \cdot (C_{pdr} + C_{gdr})$ is equal to $w_p \cdot C_{dpr}$ in the case of the level restorer and zero in the case of a buffer.

One can use the expression (16) plus the formulas from [3] to compute optimal transistor sizes for a mixed logic block. We used these formulae to derive the optimal transistor sizes appropriate to the four voltage-drop/current-drive cases of the previous section. Our findings are listed in Table 1. The calculated values lie in the vicinity of the minima appearing in the previous figures.

IV. OPTIMAL BUFFER INSERTION

Long chain CPL circuits are often found in circuits such as adders or deep multiplexers. The most common approach for dealing with long delays in these devices is to restore voltage level and drive current by inserting into the CPL chain every n positions buffers made of transistors having the widths determined above.

We studied the effect of inserting a latch, buffer, level-restorer or latch-and-buffer combinations into a 100 transistor CPL chain at every n^{th} point for a series of n values ($n = 1, 2,..., 10$). TSMC 0.18um technology was assumed. Each situation was optimized using the formulas (16) and [1] in a simulation. The results obtained are summarized in Figure 11, where 'p' indicates a pass-transistor, 'l' a latch, 'b' a buffer and 'rb' indicates a level-restorer. The simulation showed that the best value of n to be about 4 for all cases - refer to the dashed cycles in Figure 11. The best of the four best cases uses the latch with buffer approach.

V. CONCLUSION

In this paper, a enhanced design methods-mixed logic circuits design was proposed. The most promising mixed logic candidate turned out to be a combination of C-CMOS and CPL. We used delay models [3] to obtain design-influencing results which were essentially equivalent to those accomplished by simulation to find that a 4-bit CPL structure using latch with buffer is the best structure and the best insertion points.

REFERENCES

[1] M. Shams, "Modeling and optimization of CMOS logic circuits with application to asynchronous design," PhD Thesis, University of Waterloo, 1999.

[2] J.M. Rabaey, A. Chandrakasan, B. Nikolic, "Digital integrated circuits: a design perspective," 2nd.Ed., Prentice-Hall, 2002.

[3] M. Shams, and M. Elmasry, "Delay optimization of CMOS logic circuits using closed-form expressions," IEEE International Conference on Computer Design, Oct. 1999.

[4] K. Yano, Y. Sasaki, K. Rihino, and K. Seki, "Top-down pass-transistor logic design," IEEE J. Solid-State Circuit, June 1996.

Ground Bounce Calculation due to Simultaneous Switching in Deep Sub-micron Integrated Circuits

Mohammad Hekmat[1], Shahriar Mirabbasi[1], Majid Hashemi[2]

[1]Department of Electrical and Computer Engineering, University of British Columbia, Vancouver, BC, Canada
Email: {mohammad, shahriar}@ece.ubc.ca

[2]SiRF Technology Inc., San Jose, California, USA
Email: mhashemi@sirf.com

Abstract—Supply and ground variation due to switching noise is an important issue in digital and mixed-mode integrated circuits. In this paper, an approach for calculating the supply and ground bounce is presented in which the effects of parasitic elements of package and bond wires are considered. The proposed method leads to a system of linear equations whose analytical solution can be used to predict the behavior of supply and ground variations. SPICE simulations are used to verify the accuracy of the approach. The importance of modeling of the package parasitics and the dependence of the switching noise on parasitic elements are also discussed.

I. INTRODUCTION

Advancements in Silicon-based technologies, in particular CMOS, have made the implementation of complete mixed signal systems on a single chip feasible. One of the major design challenges associated with this high level of integration is to cope with the excessive amount of supply and ground noise. An obvious negative implication of the increased noise level is the decrease in the noise margin of the system. Therefore, calculating the amount of noise due to switching of the gates is of crucial importance in estimating the noise margins and designing noise-aware digital and mixed-signal circuits.

Package parasitics and transient currents due to switching activities in digital circuits are the main cause of supply and ground noise. With the advent of new generations of CMOS technology, higher levels of perturbation are expected due to the increased clock speeds and faster switching of digital gates that lead to larger transient currents. Therefore, voltage drops across parasitic elements of the package and bond wires introduce a substantial amount of noise on the digital power lines. Furthermore, each new CMOS technology entails scaling down of the supply voltage that in turn, increases the relative ratio of the magnitude of the noise to that of available power supply, thus reducing the noise margin of the system.

For the past two decades, many researchers have focused on modeling the supply/ground bounce phenomenon and several closed-form formulas have been developed to predict the maximum value of switching noise on suplly lines [1]-[3]; however, owing to nonlinear behavior of MOS devices, simplifying assumptions are required to find an analytical solution. In [1] the power supply variation is approximated with a straight line that as shown in our results may not be a realistic assumption. In [2] Taylor series expansions have been used to linearize the governing equations; however, a first order system is assumed, neglecting the effects of higher order terms in the amplitude of noise. In [3] an application specific model of MOS device, similar to the model presented in this paper, has been used, but the resistance in power and ground lines is ignored. The amount of voltage drop across the resistance of the bond wires, as will be shown in this paper, can be significant with larger switching currents and neglecting these effects in advanced technologies would underestimate the amount of noise.

This paper is organized as follows: The model of transistor that is used in the proposed approach is described in Section II. Switching noise generated by a single gate is discussed in Section III. Extensions of the results of Section III to the case of simultaneously switching gates are presented in Section IV. Comparisons of the results of this work to those of SPICE simulations and previous works as well as discussion on the behavior of the ground bounce are given in Section V. Finally, the concluding remarks are provided in Section VI.

II. MOS DEVICE MODELING

Early works in the field of switching noise used MOS quadratic I-V characteristic [1]. However, the model that is now widely used for short-channel devices is the α-power law model that accounts for velocity saturation effects [4]. According to this model the drain current of a MOSFET device in saturation region can be calculated using the following equation:

$$I_D = K(V_{GS} - V_T)^\alpha \qquad (1)$$

where V_T is the threshold voltage of the MOS device and α is a parameter that depends on the technology and is close to one for deep sub-micron devices. Using this formula for the purpose of switching noise analysis will result in a system of nonlinear differential equations that is difficult to

---
This work is partially supported by SiRF Technology, Inc. and National Science and Engineering Research Council (NSERC) of Canada.

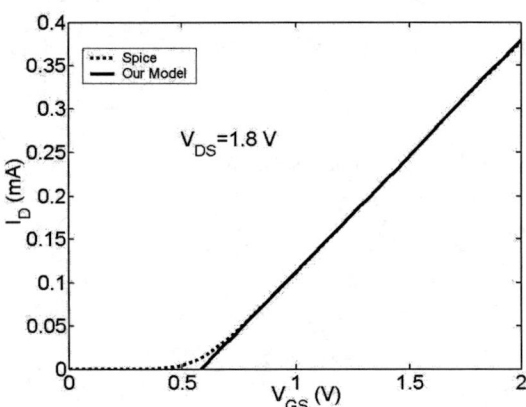

Fig. 1. I-V characteristic of NMOS device

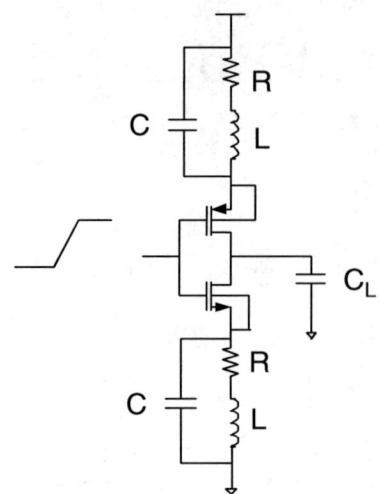

Fig. 2. A typical CMOS inverter and package model

solve analytically, if not impossible. Researchers have used different approximation techniques to simplify and/or linearize the resulting equations. In this paper a linear approximation of the MOS device characteristic is used that leads to a system of linear differential equations. The I-V characteristic of an NMOS as well as its linear approximation is shown in Figure 1. In this figure, SPICE simulations for a $0.18\mu m$ CMOS technology are used to obtain the NMOS I-V characteristic. As can be seen from the figure, the linear approximation shows a good agreement with actual I-V characteristic especially in the active region. Since during the switching of a gate from high to low, the switching NMOS transistor will most likely remain in active region, it is expected that the approximation gives accurate results. This is true particularly for output drivers where due to the large capacitive load the output voltage remains constant during the switching of the transistor. Using data obtained from Figure 1, the linear approximation of the MOS device can be written as:

$$I_D = K_1(V_{GS} - V_{GS0}) \qquad \text{for } V_{GS} \geq V_{GS0} \qquad (2)$$

where K_1 is a constant and V_{GS0} is the point that the approximation line crosses the V_{GS} axis. Note that V_{GS0} is not necessarily equal to V_T of the MOS device; in fact, it is always larger than device threshold voltage. Reasonable changes of V_{DS} have a minor effect on I-V characteristic and are ignored in our calculations.

III. SWITCHING NOISE IN A SINGLE GATE

Transient currents in logic gates introduce noise on both V_{DD} and ground lines. Due to the similarity of switching noise calculations on supply and ground lines, only the corresponding derivations for ground lines are presented in this section. The results can be easily modified to obtain formulas for power supply noise. A typical CMOS inverter is shown in Figure 2. In this figure, package and bond wire parasitics of the power supply and ground lines are represented using a lumped RLC model. C_L is the capacitive load of the inverter gate and the input signal is modeled with a ramp signal changing from ground to supply with a finite risetime:

$$V_{in} = \frac{t}{t_r}V_{DD} \qquad \text{for } 0 \leq t \leq t_r \qquad (3)$$

To derive an analytical formula for ground variations due to switching, the noise on the ground line is assumed to be zero at the beginning of the transition. After the input signal reaches V_{GS0} the NMOS transistor starts conducting. The following equations capture the behavior of the circuit:

$$I_N = K_1(V_{in} - V_n - V_{GS0}) \qquad (4)$$

$$I_L = I_N - C\frac{dV_n}{dt} \qquad (5)$$

$$V_n = RI_L + L\frac{dI_L}{dt} \qquad (6)$$

where I_N is the current through the NMOS transistor, I_L is the current through the parasitic inductance and V_n is the noise voltage at the substrate node. Combining equations (4)-(6) results in:

$$LC\frac{d^2V_n}{dt^2} + (LK_1 + RC)\frac{dV_n}{dt} + (1 + RK_1)V =$$
$$LK_1\frac{dV_{in}}{d_t} + RK_1V_{in} - RK_1V_{GS0} \qquad (7)$$

This equation is a second-order linear differential equation that can be solved analytically. It should be noted that this linear equation is a result of our linear approximation to the I-V characteristic of the MOS device. The analytical solution of the differential equation has the following form:

$$V_n = V_h + V_p \qquad (8)$$

where, V_h is the general solution of the homogenous equation and V_p is the particular solution.

$$V_h = d_1 e^{\lambda_1} + d_2 e^{\lambda_2} \qquad (9)$$

where λ_1 and λ_2 are the roots of the characteristic equation that may be complex and d_1 and d_2 are constants that are calculated based on the initial conditions of the circuit. Assuming a linear ramp approximation as in Equation (3), V_p has the following form:

$$V_p = at + b \quad (10)$$

where:

$$a = \frac{RK_1 V_{DD}}{(1+RK_1)t_r}$$

$$b = \frac{\left(K_1(\frac{LV_{DD}}{t_r} - RV_{GS0}) - (LK_1 + RC)a\right)}{(1+RK_1)} \quad (11)$$

IV. SIMULTANEOUS SWITCHING OF N GATES

It is a common practice in large circuits to replace individual logic blocks with a large inverter or a chain of inverters that generate the same amount of switching current. This practice will decrease the computational complexity of simulations and is used in many switching-noise macromodeling approaches [6]. The derivations of the previous section can be extended to the case of N simultaneously switching gates or equivalently a large inverter with N times the width of the single inverter. The equations can be derived in the same manner. The final results are:

$$LC\frac{d^2V_n}{dt^2} + (NLK_1 + RC)\frac{dV_n}{dt} + (1+NRK_1)V =$$
$$NLK_1\frac{dV_{in}}{dt} + NRK_1V_{in} - NRK_1V_{GS0} \quad (12)$$

The solution of this differential equation can be used to find the maximum peak-to-peak value of the ground bounce. The characteristic equation of this differential equation is:

$$LC\lambda^2 + (NLK_1 + RC)\lambda + (1+RNK_1) = 0$$
$$\Delta = (NLK_1 - RC)^2 - 4LC \quad (13)$$

Depending on the value of parameters and N, the discriminator of the quadratic equation, Δ, can be either positive, negative, or zero. We define N_{crit} as the value of N for which the equation has two equal roots:

$$N_{crit} = \frac{2\sqrt{LC} + RC}{LK_1} \quad (14)$$

For $N > N_{crit}$ the equation have two real roots. This happens for large values of N. It should be noted that as K_1 increases (for example when using larger transistors), N_{crit} decreases, which is the case for large output drivers. In situations where $N < N_{crit}$, the behavior of noise is different and ringing due to the on-chip parasitic capacitance and inductance is observed that should be accounted for. Since output drivers, because of their large transient currents, are the major contributors to switching noise, here we solve the equation assuming $N > N_{crit}$. In case of real roots, both λ_1 and λ_2 are negative numbers, resulting in a decaying exponential term. It can be shown that under the circumstances mentioned above the maximum value of the switching noise will always occur at the end of the input transition time and its value can be calculated from the following formula:

$$V_{n_{max}} = (-a_n t_0 - b_n + \frac{a_n}{\lambda_1})e^{\lambda_2(t_r - t_0)} + a_n t_r + b_n \quad (15)$$

where λ_1 and λ_2 are real roots of the characteristic equation and $|\lambda_1| > |\lambda_2|$. t_0 is the time when the transistor starts conducting and can be approximated by the following formula:

$$t_0 = \frac{V_{GS0}}{V_{DD}} t_r \quad (16)$$

In Equation (15), a_n and b_n are modified versions of a and b calculated for the case of N switching gates using the following formula:

$$a_n = \frac{RNK_1 V_{DD}}{(1+NRK_1)t_r}$$

$$b_n = \frac{\left(NK_1(\frac{LV_{DD}}{t_r} - RV_{GS0}) - (NLK_1 + RC)a_n\right)}{(1+NRK_1)} \quad (17)$$

V. SIMULATION RESULTS AND DISCUSSION

SPICE simulations are performed in a CMOS $0.18\mu m$ technology to evaluate the accuracy of the proposed model. The results are shown in Figures 3 to 5. As can be seen from Figure 3, the proposed approach closely follows the simulation results especially when the transistor is fully conducting. The maximum value of simultaneous switching noise (SSN) as a function of the number of switching gates is plotted in Figures 4 and 5. For the purpose of comparison, the results of two other works are also included in these figures. As can be seen, there is a large deviation between the results predicted by the approach presented in [1] and the SPICE simulation results. This deviation can be attributed to the fact that in [1] the effect of velocity saturation is not taken into account. In the other work (i.e., [5]) the effects of package capacitance and resistance are ignored. To make a fair comparison the values of package resistance and capacitance are set to zero in Figure 4. As can be seen from this figure, the results of the proposed model are within 2% of those of SPICE simulations. However, the accuracy decreases as the number of gates increases; this is due to the increase in the error in the approximation used in determining the initial conditions for solving the differential equation.

Equation (15) can be used to analyze the effect of several parameters such as power supply voltage, parasitic inductance of package and bond wires, and the number of simultaneously switching gates, on the maximum amount of noise in the circuit. While parasitic resistance has been neglected in simultaneous switching noise calculations in [1], [3], and [5], simulation results show that as the CMOS technology scales

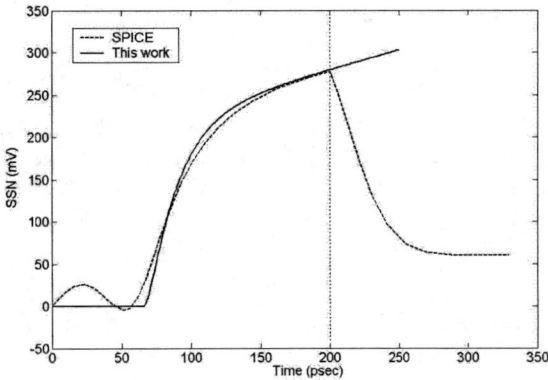

Fig. 3. SSN voltage with $L = 1nH$, $R = 2\Omega$, $t_r = 200ps$

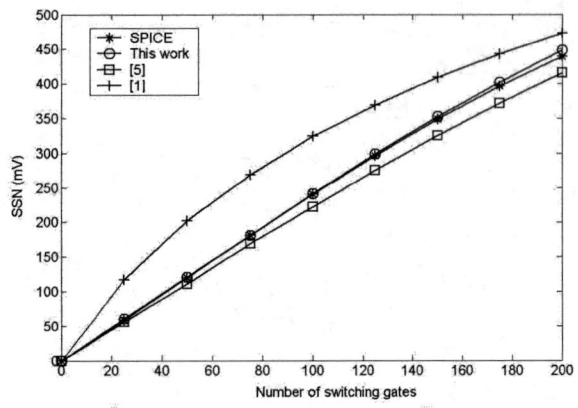

Fig. 4. Comparison of the results with other works and SPICE ($R = 0$, $t_r = 200ps$, $L = 1nH$, $C = 0$)

down and integration level and transient currents increase the voltage drop across this resistance is not negligible and should be included in the calculations. As an example ignoring the resistance in Figure 5 for the case of 100 gates switching simultaneously leads to 12% error in predicting the maximum magnitude of switching noise. As can be inferred from equation (15), this resistance results in a term that increases linearly with time, hence this term is more significant in slower parts of the circuit such as output drivers that have the largest switching time and contribute more to noise.

Equation (15) confirms the previous result observed in [1] that SSN peak value is not a linear function of the inductance. This can be considered as a consequence of the built-in negative feedback of MOS device, which does not allow the noise to increase unlimitedly. The built-in negative feedback is due to the I-V characteristic of the device that decreases the current as the voltage of the source of the MOS transistor increases, thus decreasing the amount of transient current through the parasitics.

VI. CONCLUSION

An approach for calculating simultaneous switching noise including all parasitics of bond wire and package pins is presented. The model gives an intuition of how the parasitics in power supply lines can affect the supply variations. Ground bounce calculated based on the analytical formula derived in this paper matches that of SPICE simulations with a good accuracy. The results show that the parasitic resistance and capacitances are of significant importance especially in advanced CMOS technologies with higher levels of integration and transient currents and lower power supplies. Furthermore, as discussed in the paper, the effect of parasitic resistances is more significant in slower parts of the circuit with larger transient current, e.g., output drivers so its effect should be taken into account in those cases.

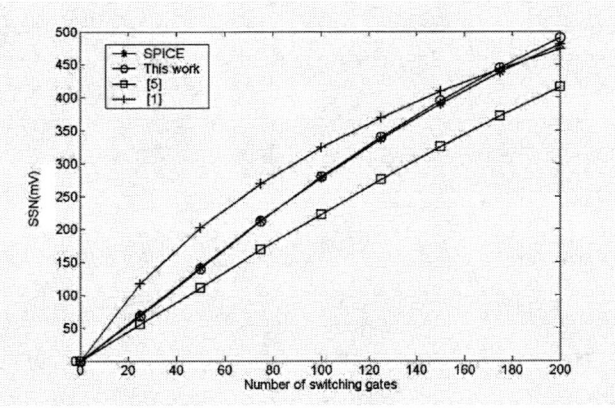

Fig. 5. Comparison of the results with other works and SPICE ($R = 2\Omega$, $t_r = 200ps$, $L = 1nH$, $C = 100fF$)

REFERENCES

[1] A. Vaidyanath, B. Thoroddsen, and J.L. Prince, "Effect of CMOS Driver Loading Conditions on Simultaneous Switching Noise," IEEE Transactions on Components, Packaging, and Manufacturing Technology, Vol. 17, Nov. 1994.

[2] K.T. Tang, and E.G. Friedman, "Simultaneous Switching Noise in On-chip CMOS Power Distribution Networks," IEEE Transactions on VLSI systems, Vol. 10, pp. 487-493, Aug. 2002

[3] L. Ding, P. Mazumder, "Accurate Estimating Simultaneous Switching Noise by Using Application Specific Device Modeling," Proceedings of Design, Automation and Test Conference in Europe, pp. 1038-1043, Mar. 2002.

[4] T. Sakurai, and A.R. Newton, "Alpha-Power Law MOSFET Model and its Application to CMOS Inverter Delay and other Formulas," IEEE Journal of Solid-State Circuits, Vol. 25, pp. 584-594, Apr. 1990.

[5] S.R. Vemuru, "Accurate Simultaneous Switching Noise Estimation Including Velocity-saturation Effects," IEEE Transactions on Components, Packaging, and Manufacturing Technology, Vol. 19, pp. 344-349 May 1996.

[6] N.K. Verghese, T.J. Schmerbeck, and D.J. Allstot, Simulation Techniques and Solutions for Mixed-Signal Coupling in Integrated Circuits, Norwell, MA: Kluwer Academic Publishers, 1995

Model-Compiler based Efficient Statistical Circuit Analysis: An Industry Case Study of a 4GHz/6-bit ADC/DAC/DEMUX ASIC*

Bo Hu, Zhao Li, lili Zhou, C-J Richard Shi
Dept. of Electrical Engineering
University of Washington, Seattle, WA 98195, USA
{hubo,lz2000,llzhou,shi}@ee.washington.edu

Kwang-Hyun Baek, Myung-Jun Choe
Rockwell Scientific Company
1049 Camino Dos Rios, Thousand Oaks, CA 91360, USA
kbaek@ieee.org, mchoe@rwsc.com

Abstract---A new, efficient, and reliable approach to speeding up large-scale mixed signal circuit simulation is proposed for full-chip electrical statistical analysis. This approach has been verified with an industry 4GHz/6-bit ADC/DAC/DEMUX ASIC design. The key idea is to retain those devices that need to model statistically and their surrounding circuitry at the transistor level and to replace the other parts of the circuit by analog behavioral models. Instead of using unproven VHDL-AMS or Verilog-A simulators not optimized for statistical electrical analysis, a state-of-the-art model compiler MCAST is used to compile analog behavioral models automatically into designers' preferred circuit simulator. With the model compiler, digital and analog circuits at the behavioral level can be simulated efficiently and seamlessly with analog circuits at the transistor level by a SPICE-like circuit simulator. In this way, full-chip mixed-signal circuit simulation is orders of magnitude faster than that at the transistor level. Further, the simulation accuracy is kept as close to that of transistor-level simulation as possible. This approach has been applied successfully to dynamic nonlinearity analysis of the ADC/DAC/DEMUX ASIC caused by device mismatching due to process variations, which was previously an extremely time consuming task.

I. INTRODUCTION[*]

For high-speed deep-submicron mixed-signal circuits, statistical analysis is a very important tool to study the impact of process variations on the chip performance and manufacturing yield [1][7][9]. To be accountable, Monte Carlo simulation is typically needed and it requires many runs of electrical circuit simulation at the transistor level, an extremely time consuming process. Due to the pressure of fast time to market, it is virtually impossible to carry out chip-level statistical analysis in the design flow. In practice, designers often would use high level programming languages such as MATLAB [14] to model a circuit and conduct some preliminary analysis on process variations. This approach is fast, but quite inadequate, since it often oversimplifies the circuit. For example, in the case of an industry ADC/DAC/DEMUX ASIC design, only static nonlinearity was studied before.

With process variations are modeled as statistical electrical parameters, electrical simulation at the transistor level is required for statistical analysis. A natural approach to speeding up full-chip mixed-signal electrical simulation for statistical analysis is to retain those devices needed to be modeled statistically, as well as their surrounding circuitry, at the transistor level while replacing the other parts of the circuit by high level analog behavioral models. Then the recently emerging VHDL-AMS [13] or Verilog-A simulators can be explored.

However, there are several difficulties of this approach. First, analog designers often rely heavily (only) on their golden SPICE circuit simulators with golden device models that have been quantified for years. Second, existing VHDL-AMS models are often not optimized for statistical electrical analysis at the transistor level as existing SPICE simulators do. Finally, the essential question is whether such mixed-level modeling can indeed reduce the simulation time so to enable statistically meaningful Monte Carlo simulation while achieving sufficient accuracy so that dynamic nonlinearity of such circuits like 4GHz ADC/DAC/DEMUX due to device mismatching can be accurately modeled.

In this paper, we describe an efficient, reliable, and practical approach to speeding up large-scale mixed signal circuit simulation for full-chip electrical statistical analysis. Using this approach, the non-critical blocks, most of which are digital blocks in a mixed signal design, are replaced by simpler analog behavioral models, which have the same functionalities but much easier to evaluate during SPICE simulation; in the same time, the other important analog blocks are still at the transistor level to preserve the influence of process variations. Analog behavioral models are developed in high-level behavioral languages such as VHDL-AMS and Verilog-A, and are compiled directly to a SPICE-like circuit simulator, using a state-of-art model compiler MCAST [10]. Like ADMS [5], MCAST is a powerful tool that allows the fast implementation of device models, and has been used to successfully compile industry-grade device models such as BSIM3 [11] and BSIMSOI [12] into the SPICE3 simulator [6][8]. Through an application programming interface, MCAST can support any SPICE-like simulator such as SPECTRE [15]. In addition, with any device model can be implemented to a circuit simulator with MCAST, new statistical device models can also been incorporated.

The paper is organized as follows. Section II describes the architecture of an ADC/DAC/DEMUX ASIC chip. The details of behavioral modeling for statistical analysis are discussed in Section III. Section IV presents the simulation results for mismatch analysis. Section V concludes the paper.

---

*This research was supported by DARPA NeoCAD Program under Grant No.N66001-01-8920, an NSF CAREER award, and Rockwell Scientific.

II. ARCHITECTURE OF AN ADC/DAC/DEMUX DESIGN

The 4GHz/6-bit ADC/DAC/DEMUX ASIC, which is the core part of an advanced Σ-Δ ADC, is fabricated with the BJT technology. It has seven major blocks: input reference ladder, preamplifier, comparator, clock-driver circuit, encoder, DAC, and DEMUX. Figure 1 is the top-level block diagram of the whole chip.

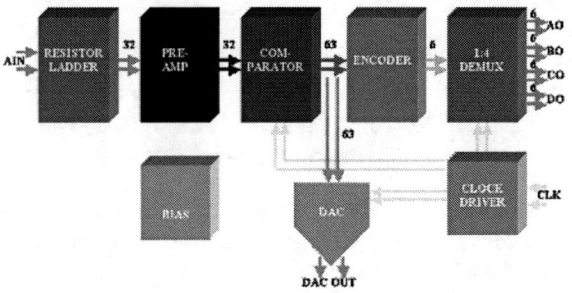

Figure 1. The block diagram of the ADC/DAC/DEMUX ASIC.

The comparator and encoder blocks have more transistors than other blocks. However, except the first level of the comparators, these two blocks are largely digital circuitry. These digital blocks are generally less sensitive to process variations, but take a large amount of simulation time. On the other hand, the input ladder and the DAC blocks are relatively small, but more sensitive to process variations and have a greater impact on the overall ADC and DAC nonlinearity. In this work, we focus on how to efficiently simulate the impact of mismatching in the input ladder and DAC blocks due to process variations on the ADC and DAC linearity performance.

III. MODELING AND SIMULATION METHODOLOGY

We model the behavior of digital gates with VHDL-AMS. The VHDL-AMS description will capture the electrical behavior of the gates. The level of abstraction is up to the designer. It could be as detailed as the transistor level or as simplified as a handful of equations to describe the input and output relationship and loading conditions.

After modeling the behavior of one specific digital gate in VHDL-AMS, we use MCAST to compile the VHDL-AMS description into a new device of a SPICE-like simulator; and it could then be instantiated in the SPICE netlist. An example flow of model compiler based mixed-level circuit simulation is illustrated in Figure 2; in this case, the target circuit simulator is open-source SPICE3. We note that the compiled code can be also linked to any commercial circuit simulator through a model application programming interface.

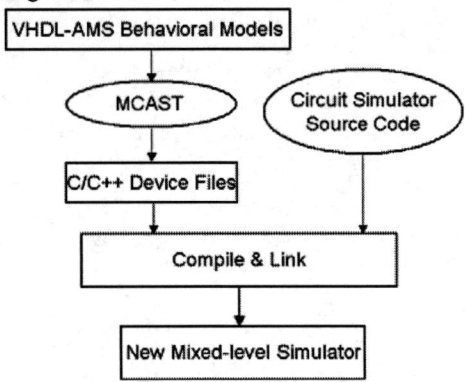

Figure 2. The flow graph of model compiler based simulation.

The "new" behavioral device is similar to other built-in devices such as resistors and capacitors, and can be as sophisticated as BSIM3 and BSIMSOI models. For example, suppose we model the inverter gate in VHDL-AMS. After compilation, it would be used as follows:

N1 in out inverter (instance parameters …)

Once we have built up a set of digital gates and analog behavioral models into devices for the SPICE simulator, we can use those as primitives to build up the comparator and encoder. It needs to be pointed out that this is one of many ways to do high level behavioral modeling for sub-blocks in the circuit. With model compiler MCAST, the whole block of encoder and comparator could be modeled and compiled as one device.

Since we aim at analog level of accuracy, and use the SPICE simulator, the requirement to achieve the robust simulation convergence is as important as the accuracy itself. Ideal digital signals only have two values: 0 or 1. Any intermediate values are not defined. However, simulating circuits with purely digital devices using SPICE often cause the convergence failure. The reason is that purely digital devices tend to have dis-continuous behavior, and will lead to infinite derivates and cause the Newton-Raphson (N-R) [6][8] simulation algorithm used by SPICE to fail.

To overcome the convergence problem, digital circuit blocks in this design are modeled using analog behavioral models. Smoothing functions are used to soften the transition of signals so that the signals will have finite derivatives with respect to terminal voltages to help SPICE to converge.

The following is an example of an inverter model. It models the input and output relationship using arctan as the smoothing function.

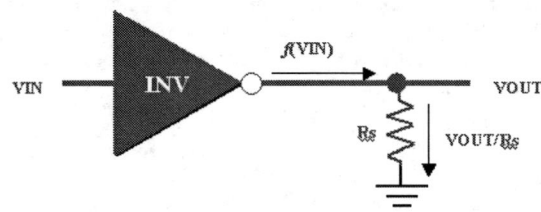

Figure 3. Inverter diagram.

```
    -- define the model parameter for inverter
-- device: input voltage threshold, output
-- voltage low and high values.
generic(
    ...
    in_mid  : real := -1.540; -- input
threshold
    out_hi  : real := -1.315; -- output high
    out_lo  : real := -1.705; -- output low
    ...
);
...
-- define the I(v) relationship
    out_mid := (out_hi + out_lo)/2;
-- following is just one inverting function
-- fvin is the current source equals to the
-- invert of vin
    fvin := out_mid - ((out_hi - out_lo)/m_pi)
    *arctan(k*(vin-in_mid+a*(vin-in_mid)**3));
-- one implicit output resistor of 1 Ohm
-- is added at output. Total current flowing
```

```
-- out of the out node equals:
-- vout/1Ohm - fvin
    iout := -fvin+vout;
...
```

In Fig. 3 a voltage controlled current source (VCCS) with a small resistor Rs is used to model the inverting behavior. This generally leads to sparse circuit matrices and thus faster simulation speed than other modeling techniques.

In our implementation, all digital circuit blocks are replaced by their analog behavioral models. An example of analog behavioral model for NOR2 is shown below.

```
...
out_mid := (out_hi + out_lo)/2;
-- get fvin1 = INV(vin1)
fvin1 := out_mid + ((out_hi - out_lo)/m_pi)*
arctan(k*(vin1-in_mid1+a*(vin1-in_mid1)**3));
-- get fvin2 = INV(inv2)
fvin2 := out_mid + ((out_hi - out_lo)/m_pi)*
arctan(k*(vin2-in_mid2+a*(vin2-in_mid2)**3));
-- get vo3 = finv1 AND finv2
vo3 := out_hi - (fvin1 - out_lo)*(fvin2 -
out_lo)/(out_hi - out_lo);
-- one implicit output resistor of 1 Ohm added
ioutq := -vo3+vout;
...
```

Once we have the basic digital gates properly modeled and compiled into a circuit simulator as new devices, the encoder, which turns the thermometer code into the binary code, can be implemented using a netlist of such devices.

The comparator has a gain stage and 3 stages of latches inside with bubble correction functions. The gain stage is modeled as a differential amplifier.

```
...
generic(
...
    mpgain : real := 3.0;    -- gain
    mpvhi: real := -2.2;     -- output roof
    mpvlo : real := -2.6;    -- output bottom
    mpalpha : real := 100.0
);
x1 := vinp - vinn;   -- differential input
ca := (mpvhi + mpvlo)/2;
cb := (mpvhi - mpvlo)/m_pi;
cc := mpgain/cb;
x2 := cb*arctan(cc*(x1+mpalpha*x1*x1*x1));
ioutpq := -ca-x2+voutp;
ioutnq := -ca+x2+voutn;
...
```

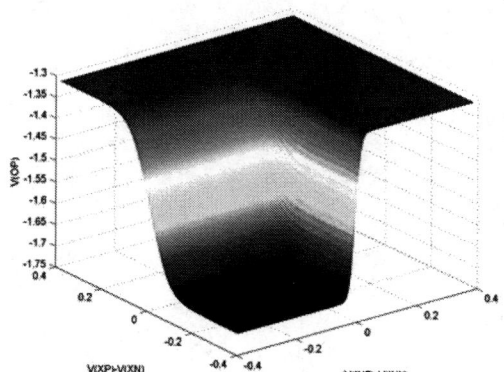

Figure 4. The comparator transfer curve using the behavioral model.

We simulated the input and output transfer curve of the comparator using the transistor-level circuit netlist and using the behavioral model above. Figure 4 show the result using the behavioral model. Figure 5 shows the relative error between the results of using the transistor netlist and using the behavioral model. It can be seen that the maximum error is less then 0.7%.

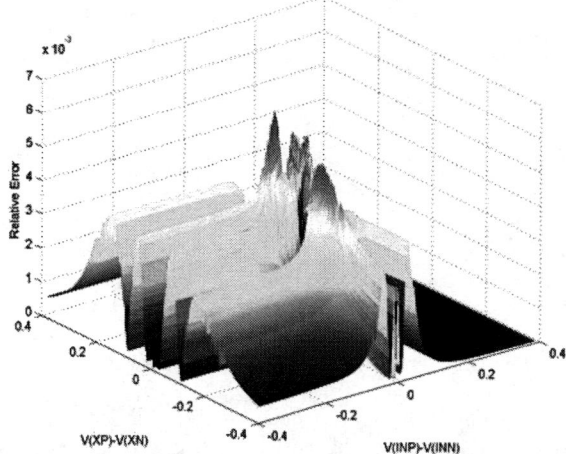

Figure 5. Relative error: behavioral model vs transistor netlist.

The latch can be modeled using AND2 and OR2 gates shown in Fig. 6. It is a 2-stable circuit, which is utilized here to store the previously latched value, and could have **Hold** and **Pass** operations, based on the clock signals. The capacitor in the middle of AND2 and OR2, combined with the output resistance of the AND2 gate, provides the necessary delay to help improve the convergence.

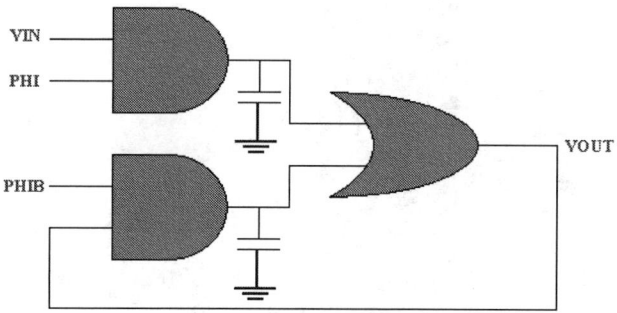

Figure 6. The schematic of a latch.

IV. SIMULATION RESULTS

For ADC design, the nonlinearity can be measured by using static specifications, such as differential nonlinearity (DNL) and integral nonlinearity (INL), and/or using dynamic specifications, including signal-to-noise and distortion ratio (SNDR), spurious free dynamic range (SFDR), and total harmonic distortion (THD). The state-of-art design practice is to use MATLAB modeling and simulation to calculate the static nonlinearity.

The input to the ADC is a sine waveform at frequency of 85.6 MHz with sampling frequency (a.k.a clock frequency) at 1GHz. The ratio of input frequency and sampling frequency is 1/11 to improve the sampling quality. This would yield 11 samples each period of input. To accumulate 128 samples would require the simulation of 12 input cycles, which is equivalent to simulating the circuit's transient behavioral for about 140ns.

The simulation was carried out on a SPARC workstation with

900 MHz CPU. The CPU time statistics was collected in Table 1.

TABLE I.

Running statistics	Original netlist all devices in BJT	Behavioral level partially BJT
Matrix Size	24063	8737
LU time	4441 sec.	223 sec.
Solve time	1565 sec.	77 sec.
Load time	4362 sec.	331 sec.
Simulation time	10693 sec.	688 sec.
Speed up	15 x	

It can be seen that the overall simulation speed up is 15 times. Previously, a single run of the simulation takes about 3 hours, and after behavioral modeling, the time is dramatically cut down to about 11 minutes, which makes the statistical analysis become feasible.

After inserting behavioral models, we performed statistical analysis by perturbing several most important parameters. Those parameters are: input ladder's resistor, 2% variation of Gaussian distribution. Is (saturation current), 3% variation; beta (current gain): 5%; rb (base resistance), 1%, re (emitter resistance), 0.2%, respectively. We also included 5% current gradient in statistical analysis, which represents the current difference due to layout positions of 63 current mirrors for the DAC.

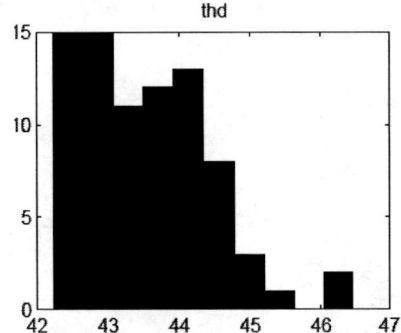

Figure 7. THD histogram of mismatch analysis of ADC using behavioral models.

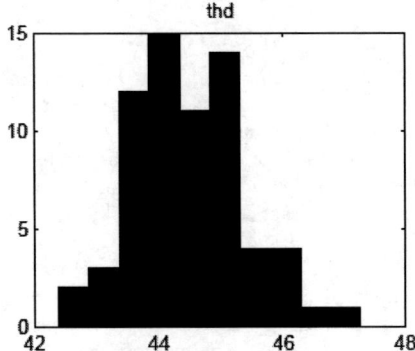

Figure 8. THD histogram of mismatch analysis of ADC without using behavioral models.

The SNDR, THD, and SFDR distribution obtained from SPICE simulation with behavioral models (Fig. 7) and without (Fig. 8) are comparable in terms of accuracy. For THD, the mean and standard derivation values are 43.52 and 0.91, respectively, for using behavioral models, and 44.50 and 0.94, respectively, for using the original netlist. Overall, they match with each other quite well, with only small discrepancy due to the use of behavioral models.

V. CONCLUSION

We described an efficient approach of statistical analysis using behavioral modeling for mixed signal circuit design and model-compiler based circuit simulation. We have shown that the technique can reduce the simulation time by orders of magnitude while preserving reasonable accuracy, and thus enable statistical analysis. We have applied this approach to a 4GHz/6-bit ADC/DAC/DEMUX ASIC design.

VI. REFERENCE

[1] S. Borkar, T. Karnik, S. Narendra, J. Tshanz, A. Keshavarzi, and V. De, "Parameter variations and impact on circuits and microarchitecture", *Proc. IEEE/ACM Design Automation Conf.*, pp. 338-342, June 2003.

[2] H. Carter, "Modeling and simulating semiconductor devices using VHDL-AMS", *Proc. IEEE/ACM Intl. Workshop on Behavioral Modeling and Simulation*, Oct. 2000.

[3] D. A. Johns and K. Martin, *Analog Integrated Circuit Design*, John Wiley & Sons, 1997.

[4] K. Kundert, "Automatic model compilation – An idea whose time has come", *The Designer's Guide*, May 2002.

[5] L. Lemaitre, C. C. McAndrew, and S. Hamm, "ADMS - Automatic device model synthesizer", *Proc. IEEE Custom Integrated Circuits Conf.*, pp. 27–30, May 2002.

[6] L. W. Nagel, *SPICE2 – A computer Program to Simulate Semiconductor Circuits*, Univ. of California, Berkeley, ERL Memo ERL-M520, May 1975.

[7] S. R. Nassif, "Delay variability: Source, impacts and trends", *Dig. Tech. Papers ISSCC*, pp. 368-369, Feb. 2000.

[8] T. L. Quarles, *The SPICE3 Implementation Guide*, Univ. of California, Berkeley, ERL Memo 89/44, April 1989.

[9] C. Visweswariah, "Death, taxes and failing chips", *Proc. IEEE/ACM Design Automation Conf.*, pp. 343-347, June 2003.

[10] B. Wan, B. P. Hu, L. Zhou, and C.-J. R. Shi, "MCAST: An abstract-syntax-tree based model compiler for circuit simulation", *Proc. IEEE Custom Integrated Circuits Conference*, pp. 249-252, Sept. 2003.

[11] *BSIM3v3 MOSFET MODEL User's Manual*, U. C. Berkeley, Dec. 2001.

[12] *BSIMSOI3.1 MOSFET MODEL User's Manual*, U. C. Berkeley, Feb. 2003.

[13] *IEEE 1076.1, VHDL-AMS Language Reference Manual*, IEEE standard association.

[14] *MATLAB 6.5 Manual*, The MathWorks Inc.

[15] *SPECTRE Circuit Simulator User Guide*, v4.4.6, Jun. 2000, Cadence Design Systems Inc.

A High-Level Dynamic-Error Model of a Pipelined Analog-to-Digital Converter

Kalle Folkesson and Christer Svensson
Electronic Devices
Dept. of Electrical Engineering
Linköping University
SE-581 83 Linköping, Sweden

Björn Knuthammar and Andreas Dreyfert
Wavebreaker AB
S:t Persgatan 19
SE-601 86 Norrköping, Sweden

Abstract—This work presents a fast and accurate high-level model of a pipelined analog-to-digital converter implemented in MATLAB. Mechanisms causing dynamic errors, such as settling time of slew-rate-limited amplifier, are analyzed and parameters to model these identified. All parameters are associated with actual physical properties and all simulations are validated by comparison to measured data in both time and frequency domains.

I. INTRODUCTION

In communication applications, the requirements for analog-to-digital converters (ADCs) are high and increasing. High resolution at high frequencies is needed and to achieve that the pipelined architecture is dominating because of its superior balance between speed and resolution. To be able to design high-performance ADCs it is important to have accurate models of the performance-limiting effects. At high clock frequencies the performance in terms of effective number of bits (ENOB) tends to decrease due to dynamic errors in the sampling and quantization processes [1]. Examples of such effects are settling times in switched circuits and metastability in comparators. Circuit-level simulations give accurate results, but are much too time consuming. To be able to perform complete system simulations on complex communication systems, high-level modeling is necessary and a detailed model is needed [2]. When designing these high-level models it is important to verify that an introduced error effect gives the correct behavior in both time and frequency domains. Previous work has shown that errors can show the correct behavior in one domain, but fail to accurately describe another [3].

This work presents a MATLAB model targeting a pipelined ADC. The effects of various model parameters on dynamic errors are investigated to gain better understanding of the dominating mechanisms causing the speed limitation. The simulations are validated by measurements on an experimental ADC, presented in section II. The model implementation is described in section III, measurements and simulations are presented in section IV, and some conclusions are drawn in section V.

II. THE ADC

The ADC used as a target for the model is a differential 40 MHz 10-bit pipelined ADC manufactured in a 0.18 μm CMOS process and designed to meet the requirements of an IEEE 802.11a/b/g WLAN and WCDMA transceiver. Figure 1 shows a block diagram of a pipelined converter. It consists of a number of pipelined stages each including a track-and-hold circuit (T/H), low-resolution sub-ADC and digital-to-analog converter (DAC), summing circuit, and an amplifier to provide inter-stage gain. The incoming signal is simultaneously sampled by the T/H circuit and converted by the sub-ADC. The converted signal is the stage output, but it is also converted back to an analog voltage by the DAC and subtracted from the sampled voltage. The residue signal is sent to the next stage after being amplified to compensate for the change in significance level. The stages are clocked so that when one stage is in amplification mode the next is in sampling mode and vice versa. Adding the stage outputs gives the total ADC output starting with the most significant bit(s) from the first stage. Usually the stage outputs are added with an overlap giving a redundancy that is used for digital error correction. The ADC used in this work has 9 1.5-bit stages yielding a 10-bit output after error correction. Normally, as in this case, the T/H, DAC, sum, and amplify functions are preformed by the same circuit, a multiplying DAC (MDAC). A switched-capacitor MDAC architecture is shown in figure 2. During the first clock phase, ϕ, the circuit is in sampling mode and the input is sampled to capacitors

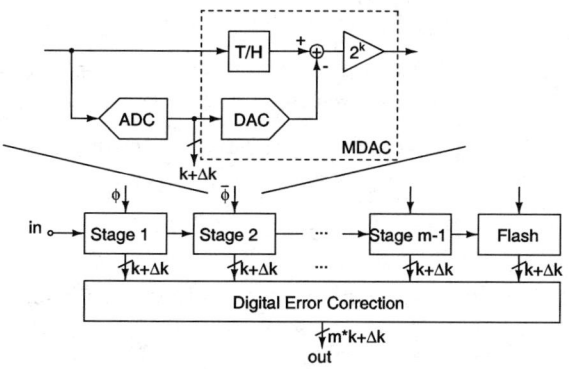

Figure 1. Pipelined converter block diagram.

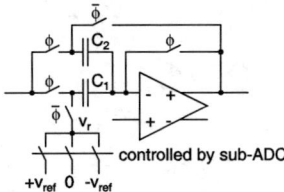

Figure 2. Switched-capacitor MDAC architecture.

C_1 and C_2 while the amplifier outputs are connected to the inputs with negative feedback to provide offset cancellation in the amplification phase. In amplification mode, during the $\bar{\phi}$ phase, C_2 is connected in a feedback loop and C_1 to a reference voltage, v_r, determined by the current sub-ADC output. For equally sized C_1 and C_2 this gives an ideal output of $2v_{in} - v_r$. The performance-limiting effects in this circuit are well known: capacitance mismatch and, more importantly, amplifier settling affected by finite gain and bandwidth and limited slew rate [4].

III. THE MODEL

The model is implemented in MATLAB with the objective to model all performance-limiting errors as realistically as possible to reduce the risk of overlooking effects. For this reason, a time-domain mathematical behavioral approach based on explicit equations that describe the physical behavior of the circuits is used. This makes the model both accurate and fast. Also, it is completely transparent as opposed to many other high-level models, which use predefined blocks in some high-level description language. This model gives total control of the model parameters and makes it easy to choose which of them to include in a specific simulation. The drawback is that if there is a need to include e.g. high-order settling effects, the model equations become very complex and difficult to handle. However, as will be shown in section IV, the simple equations presented in this section are enough to accurately describe the principal behavior of the ADC.

To derive the model equations for amplification and settling, consider figure 3, which shows two consecutive pipeline stages: the first one in amplification mode and the second in sampling mode. If $v_{s,p}$ and $v_{s,n}$ are the voltages sampled to the positive and negative amplifier inputs during sampling mode, the positive output in amplification mode, $v_{a,p0}$, is calculated as:

$$v_{a,p0} = \frac{1}{2}\left[\left(1+\frac{C_1}{C_2}\right)v_{s,p} - \left(1+\frac{C_1}{C_2}\right)v_{s,n} - \frac{C_1}{C_2}\cdot subref \cdot v_r\right] \quad (1)$$

where *subref* is [-1 0 1] depending on sub-ADC output. If consideration is taken to finite amplifier open-loop gain, A, the expression is modified to:

$$v_{a,pA} = \frac{v_{a,p0}}{1+\dfrac{C_1+C_2+C_{p,n}}{A\cdot C_2}} \quad (2)$$

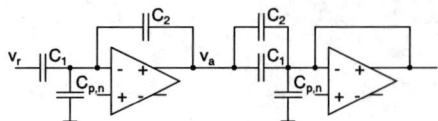

Figure 3. Two consecutive pipeline stages.

To include the effect of settling in an amplifier which initially is limited by slew rate, the following expression is used:

$$v_{a,p} = v_{a,pA}\left(1-e^{-(t-t_{SR})/\tau}\right) + \frac{I_{max}}{C_{tot}}t_{SR}\,e^{-(t-t_{SR})/\tau} \quad (3)$$

where t_{SR} is the time when the amplifier stops being limited by slew rate, I_{max} the maximum current it can deliver, and C_{tot} the total capacitance:

$$C_{tot} = C_1+C_2 + \frac{C_2(C_1+C_{p,n})}{C_1+C_2+C_{p,n}} \quad (4)$$

It is simply the standard equation for RC settling starting at the time t_{SR} with the initial condition $(I_{max}/C_{tot})\cdot t_{SR}$, the voltage after charging time t_{SR} at the maximum rate I_{max}/C_{tot}. Hence, the amplifier settles with its maximum rate for as long as it is slew-rate limited and then continues with RC settling. The time constant, τ, see e.g. [5], is:

$$\tau = \frac{(C_1+C_{p,n})(C_1+C_2)+(C_1+C_2)C_2+C_2(C_1+C_{p,n})}{g_m C_2} \quad (5)$$

and t_{SR} is the time when the RC settling rate equals the maximum rate. For the positive path it is:

$$t_{SR} = -\tau \ln\left(\frac{I_{max}}{C_{tot}}\cdot\frac{\tau}{v_{s,p}}\right) \quad (6)$$

Model parameter values are obtained from specified design parameters, estimated parasitics, amplifier circuit simulations, and typical process parameters.

Errors mechanisms implemented in the model include:

- quantization and clipping
- thermal noise
- capacitor mismatch
- process variations of parameters such as threshold voltage, transconductance, and bias current
- settling incorporating nonideal amplifier gain and bandwidth as well as limited slew rate
- comparator metastability

Thermal noise is simply modeled as a random addition to the input signal. Comparator metastability is modeled as a random addition to the comparator outputs, which, for small output values, causes the stage output and DAC input bits to be independently random.

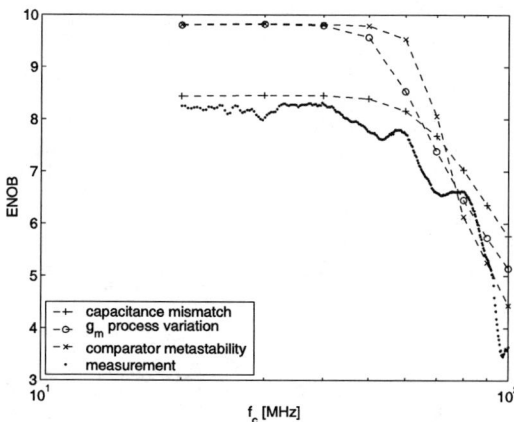

Figure 4. Frequency-domain comparison.

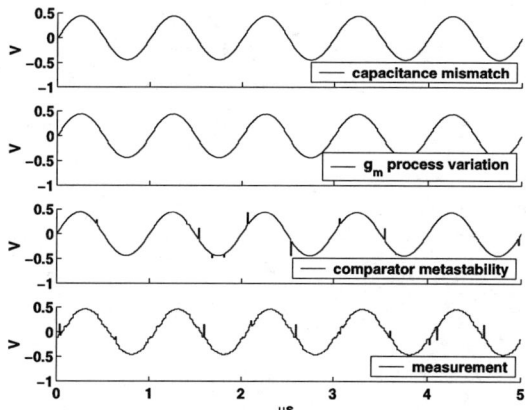

Figure 5. Time-domain comparison for 100 MHz clock.

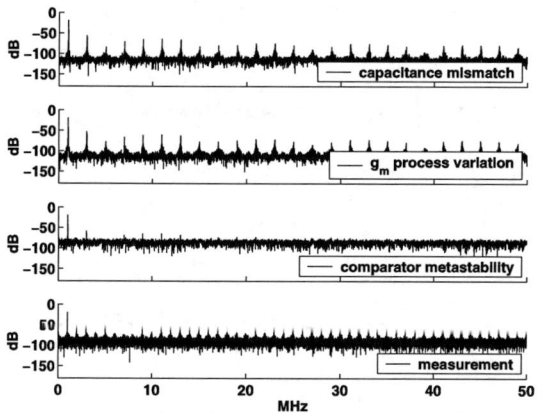

Figure 6. Spectra comparison for 100 MHz clock.

IV. Measurements and Simulations

Figures 4 through 6 show different views of measured and simulated data for a few simulation cases to be discussed below. In the simulations only one of the investigated model parameters is included at a time to give a clear view of how that particular parameter affects the performance. Figure 4 shows ENOB, calculated from signal-to-noise-and-distortion ratio found by an analysis of FFT transformed data, as a function of clock frequency, f_c. The input frequency is 1 MHz for all simulation cases and the same evaluation software is used for simulations and measurements. To include all relevant dynamic effects, the clock frequency is increased until the effective resolution is down to only a few bits. Figure 5 shows the data in time domain for a clock frequency of 100 MHz, the highest frequency used in the sweep. Since the performance at that frequency is so bad, a distortion of the waveform is visible. Figure 6, finally, shows FFT data and thus gives a view of nonlinearities such as harmonics. As for the time domain plot, the clock frequency here is 100 MHz. By observing measured data in these plots, several distinct features that model simulations must match can be identified. The most obvious is the ENOB decrease as a function of f_c. In addition to this, the time domain waveform distortion needs to be matched, as do the harmonics in the FFT plot. The FFT plot also shows a noise floor higher than the thermal level. The effect of several model parameters was investigated and a set of three that together give the same behavior as measured data was found. These were then tuned, within reasonable process spread, to give as good a match as possible. The first parameter is mismatch between C_1 and C_2. The offset cancellation scheme of this design includes a switch, which in amplification mode is connected in parallel with C_2. The parasitic capacitance introduced by that switch causes an amplifier mismatch, which leads to a gain error. The second parameter is the amplifier transconductance, g_m, which directly affects the time constant according to equation 5 and the third is comparator metastability.

The plots in figure 4 show that adding the capacitance mismatch gives the correct ENOB for low frequencies. To get the correct behavior for intermediate frequencies, however, the effect of a g_m decreased by 20% needs to be added. This is a large, but not unthinkable error. Most likely, however, effects not included in this simulation case, such as decreased I_{max} and increased capacitance values, also contribute and the actual g_m decrease is not as large. Finally, to model the very steep performance decrease for high frequencies, the comparator metastability effect is needed. The cause of the variations seen in the measured data has not been determined, but it is likely to be high-order effects in the amplifier settling or possibly ringing on the test board. These effects are not included in the model. Figure 5 shows that although all simulation cases show some hints of the waveform distortion seen in the measurement, it is only the metastability effect that can explain the spikes. Figure 6 shows that both the mismatch and decreased g_m cases yield the odd harmonics seen in the measurement. The metastability, on the other hand, gives the correct noise level while the other simulations are off by 20 dB. The measured spectrum shows not only the odd, but also the even harmonics. This suggests that there is some imbalance between positive and negative paths in the differential ADC or on the test board. This has not been included in the model at this time.

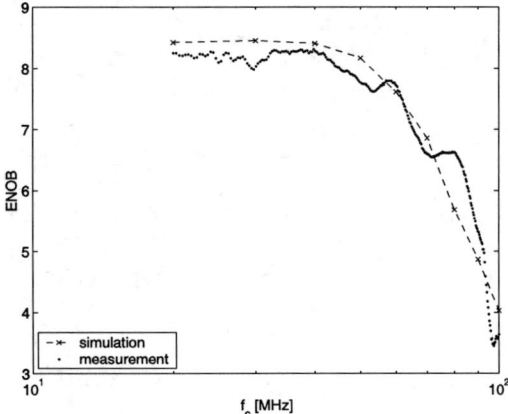

Figure 7. Frequency-domain comparison, all errors.

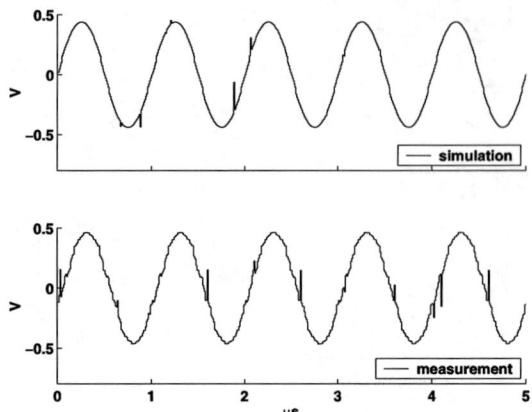

Figure 8. Time-domain comparison for 100 MHz clock, all errors.

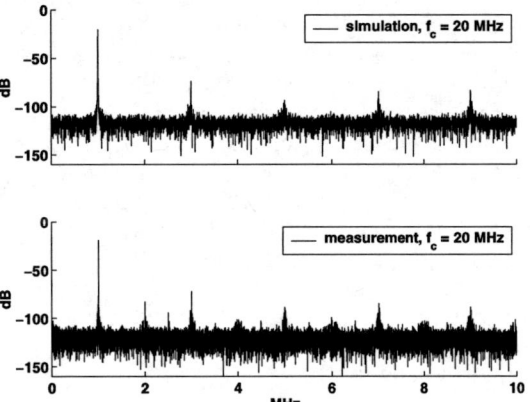

Figure 9. Spectra comparison for 20 MHz clock, all errors.

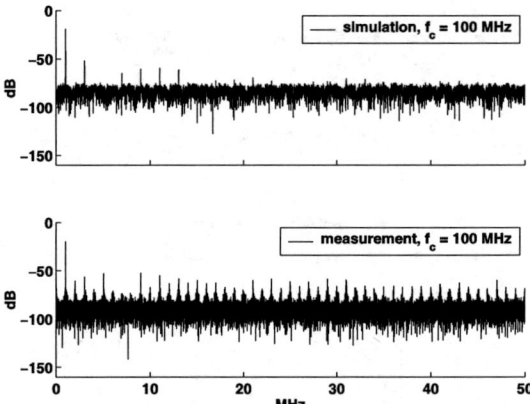

Figure 10. Spectra comparison for 100 MHz clock, all errors.

In figures 7 through 10 measured data are compared to simulations where all three parameters are included. Figure 7 shows that the principal frequency domain behavior matches well and figure 8 does the same thing for the time domain behavior. Figure 9 shows the spectra for a low clock frequency. The harmonics match well even though there is a hint of even-order peaks visible in the measurement. In figure 10 the spectra for a high clock frequency are shown and the match of the increased noise level is found to be good. The harmonics are somewhat weaker in the simulation and are therefore partly drowned out by the noise floor. In conclusion, an accurate fit to the behavior of measured data is achieved using parameters well within a reasonable spread from typical values. Without detailed knowledge of the process outcome, this is as good as can be expected due to large process variations and uncertainty in parameter values. Simulation time for one sweep such as those presented in figure 4 is about 50 s on a 1-GHz Pentium III. The minimal SPICE simulations to produce a similar plot would take over a month.

V. CONCLUSIONS

A model has been presented, which by tuning of only a few physically relevant parameters gives an accurate match to measured data from a pipelined ADC, in time as well as frequency domain, over a wide frequency range. The mathematical behavioral approach based on explicit equations makes it fast and thus well suited to be included as a part in full system simulation. The transparency of the model and the physical relevance of the parameters also make it ideal to test error hypotheses and to investigate the effects of specific design choices.

REFERENCES

[1] R. Walden "Analog-to-digital converter survey and analysis" in *IEEE Journal on Selected Areas in Communications*, vol. 17, no. 4, April 1999

[2] K. Folkesson, J.-E. Eklund, and C. Svensson, "Relevance of using single-tone tests to characterize ADCs for ADSL modems", in *Proc. NORCHIP*, 2002

[3] K. Folkesson, J.-E. Eklund, and C. Svensson, "Modeling of dynamic errors in algorithmic A/D converters", in *Proc. IEEE International Symposium on Circuits and Systems*, pp. 455-458, 2001

[4] E. Bilhan, P. C. Estrada-Gutierrez, A. Y. Valero-Lopez, and F. Maloberti "Behavioral model of pipeline ADC by using SIMULINK", in *Proc. Southwest Symposium on Mixed-Signal Design*, pp. 147-151, 2001

[5] B. Razavi, "Design of analog CMOS integrated circuits", McGraw-Hill, 2001

An Explorative Tile-based Technique for Automated Constraint Transformation, Placement and Routing of High Frequency Analog Filters

Hui Zhang, Preethi Karthik, Hua Tang, Alex Doboli
Department of Electrical and Computer Engineering
State University of New York at Stony Brook
Stony Brook, NY 11794-2350, U. S. A.
Email: {huizhang,preethik,htang,adoboli}@ece.sunysb.edu

Abstract— This paper presents an original methodology and algorithms for designing high-frequency analog filters. Filters are realized to meet AC performance specifications as closely as possible. Addressed design steps include parameter sizing, placement and routing. The explorative design method, based on tabu search, takes into account layout parasitics of the circuit. The synthesis strategy alternates between (1) a uniform sampling of the solution space and (2) a focused search of attractive regions. We also discuss the usage of a layout representation based on symbolic tiles for analog system synthesis. Experiments on filters demonstrate the quality of our solution.

I. INTRODUCTION

Currently, there is a shortage of efficient CAD tools for sizing and layout design of high performance analog circuits and systems [7]. With analog layout tools, the goal is not merely to design the circuit within a given chip area but also to keep *minimal the performance degradation due to layout parasitics* [3] [6] [9]. Most of the analog layout techniques follow an optimization-based place-and-route approach, where layout is generated by an optimization process e.g. simulated annealing, genetic algorithms etc driven by a cost functions expressing criteria such as performance degradation, wire length, area etc [1] [3] [6] [9]. However, it is possible that the fixed circuit parameters do not leave enough performance margins to accommodate the layout-induced performance degradations. In this case, the final design would be incorrect. Typical examples include high-frequency filters that incorporate capacitors of the same order of magnitude (tens/hundreds of fF) as interconnect parasitics. Parasitic capacitors are an integral part of the signal processing performed by the passive elements. Costly re-iterations through circuit sizing and layout generation are needed to produce a constraint satisfying design. *The solution is to combine the parameter sizing process with the layout design step to improve design quality and convergence of the CAD algorithms.*

A second specifics of the current layout design methods is that placement and routing are distinct and subsequent steps [3] [9] [10] [11]. This works well for medium range frequencies. *At high frequencies, however, placement must be tightly integrated with routing to allow accurate evaluation of performance degradation due to layout parasitics.* In spite of the merits of existing analog layout tools, new approaches are compulsory for sizing, placement and routing of high frequency analog circuits.

This paper proposes an original synthesis algorithm for high-frequency filter design. The method combines (1) sizing of individual components (constraint transformation), (2) placement and (3) routing based on an (4) improved circuit representation. The addressed problem is difficult as it involves three combined design steps, each having a large complexity. For example, a 6th order filter (one of the examples in Section 4) requires about 190 design variables for combined sizing, placement and routing. Similar to other work [5], our experiments with traditional optimization algorithms such as simulated annealing showed poor convergence and long execution times for a significant number of applications. Another difficulty originates in the nature of the optimization problem. For example, the best design point for a 100-400 MHz bandpass filter is located in a narrow solution region, for which the neighboring points present poor values of the cost function. Thus, adjacent regions might not predict the existence of high quality solution points. This results in gradient-based search not being effective. We observed for analog-to-digital converters(ADC) a similar characteristic of the solution space. This suggests that the exploration strategy for synthesis must perform (1) a uniform sampling of the solution space with (2) a step of focused search of attractive regions. This strategy is difficult to implement in typical random search algorithms e.g. simulated annealing, genetic algorithms, which significantly rely on random moves.

This paper discusses a tabu search algorithm [8] for analog filter synthesis through combined constraint transformation, placement and routing. The tabu search algorithm is used to methodically explore attractive regions of the solution space to find their local optima. This is possible because of the deterministic nature of the technique [8]. Solution space sampling is achieved by a component that executes a number of random moves until the solution point is significantly perturbed. This is different from *aspiration* and *diversification* steps [8] of tabu search. For example, diversification attempts to explore parameters which rarely change. For analog synthesis, however, this is not an efficient strategy because these parameters might have secondary effects on the system performance. Experiments show that superior quality filters are designed with the proposed synthesis methodology than with traditional approaches, which separate sizing, placement and routing. In fact for filters operating at $n \times 100$ MHz or above, the combined methodology is compulsory as separate sizing, placement and routing result in very poor solutions.

The synthesis algorithm uses *symbolic tiles* [12] as a compact representation of the layout. Though this approach has been used before [3] [9], our representation is much simpler and more compact, thus increases the efficiency of the design algorithms. A smaller number of tiles offers the advantage of faster methods for tile swapping and tile moving, which have a linear complexity with the number of tiles [12]. In Section 3, we offer examples to support this conclusion. Constraints on the relative position of tiles are used to express symmetries and matching. Another difference as compared to previous work [3] [9] is that our tiles are *soft* (their sizes and aspect ratio can change). This is a consequence of constraint transformation being part of the synthesis loop. For keeping the complexity of tile managing methods low, complete knowledge about left, right, top

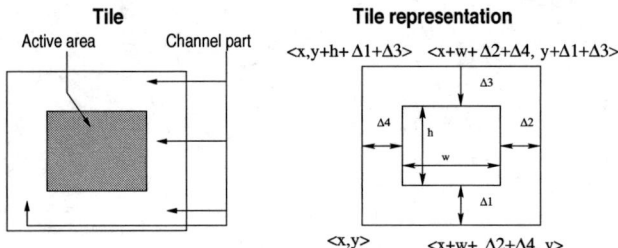

Fig. 1. **Tile definition**

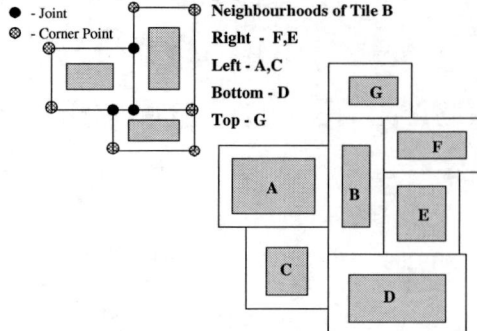

Fig. 2. **Tile relations**

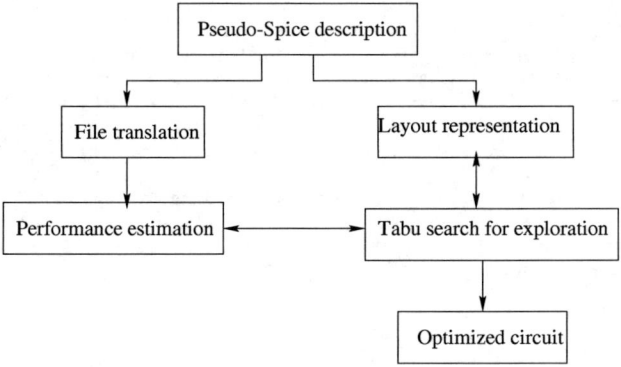

Fig. 3. **Flowchart of the filter design method**

and bottom neighbors of each tile must be *explicitly* available. We decided to store the neighborhood information as distinct O trees [13] for each of the four directions. Other representations such as sequence-pairs [1] or B* trees [2] are not efficient for our problem as they implicitly offer the neighborhood information.

This paper is organized as follows. Section 2 details the basic representation adopted in our approach. Section 3 discusses the design methodology. Section 4 presents our experiments. Finally, Section 5 provides conclusions and plans for future work.

II. TILE BASED LAYOUT REPRESENTATION

A tile [12] based representation is adopted for our layout. The tile is the basic building block for the layout. Figure 1 depicts the tile representation. It represents both active blocks and channels. The active part of the tile is the actual component, and the channel part is the portion of the channels surrounding the active region. The widths of the channel part is denoted by Δ_i, i=1,4. A layout is a collection of tiles. The used definition reduces the number of tiles for a layout as it decreases the number of tiles needed to express empty spaces.

Tiles can be of three types: (1) active tiles, (2) empty tiles, (3) margin tiles The active tile is a tile which represents an electrical component such as a resistor, a capacitor, an op-amp etc. A tile which does not represent an electrical component is called an empty tile. The tiles at the four borders of the layout are called margin tiles. The layout size is determined from the number of tiles present.

A tile is defined by its four *corner-points*. Figure 2 shows corner-points as gray bubbles. A corner-point is the pair $(x,y) \in R^2, 0 < x < w_{max}, 0 < y < h_{max}, (x,y) \in T$. T is the set of all tiles. A corner-point belongs to at least one tile. The set of all corner-points is denoted by CP. A *joint* of a tile is a corner-point which meets an adjoining tile at a point other than its corner-point. Or it can be alternately defined as a point which is a corner-point of two or more tiles and is not the corner-point of at least one tile. A joint is defined as the pair $(x,y) \in CP, (x,y) \subseteq T \& \exists t \in T$ for which (x,y) is not a member but one of its corner-points has its x- or y coordinate equalt to x or y. The left part of Figure 2 illustrates joints as black bubbles. A *sequence* is an ordered list of corner-points. Sequences are used during the routing stage. The relative positions of the neighboring tiles is defined by neighborhood relationships. Four neighborhood relations exist for each tile, as shown in Figure 2. The width of the Δ-s is determined during the routing phase by the number of nets which are to be routed through the channels. These relationships are stored as distinct O trees [13] for each of the four directions. This helps in immediate retrieval of the neighbors of a tile.

Advantages of the proposed tile representation

Since the layout dimensions are flexible, the resizing of blocks can be easily handled by a move in the optimization algorithm known as domino move, which will be introduced in the next section. This representation is better than earlier tile based representations as the same circuit representation can be used for both the placement and routing stages.

Wiring the circuits is also simplified as depending on the path of the wire, the Δ-s of the tiles through which the wire is routed is correspondingly increased. In the case that there are tiles with no wires running through their channel the tile space is decreased, thus allowing optimization in area. Because of the Δ-s, we need not at every stage ensure that there is sufficient space in the layout for routing.

The resizing of the components is a major advantage of our solution. Resizing helps us manipulate the specification according to our requirements. Device merging cannot be handled by our software as the devices are represented by black boxes. A modeling of the thermal and substrate effects can be easily added to our current solution to further improvise it.

III. COMBINED SIZING, PLACEMENT AND ROUTING

A. Exploration using Tabu Search

Figure 3 presents the flowchart of the method for combined parameter sizing, placement and routing for design of high-frequency filters. The input to the placement tool is a SPICE-like format description of the circuit. This input file is parsed and a data structure is created which holds the information of the blocks and their connections. At the same time, tiles are also created corresponding to the various blocks.

The design flow is done by the tabu search algorithm [8] where in the tiles are (1) moved, (2) routed or (3) their dimensions changed depending on their functionality and the requirement to match the system specifications. Figure 4 presents the pseudocode for the tabu-search exploration method. A single iteration of the tabu search

```
PROCEDURE system_synthesis IS
    while the final iteration number is not reached do
        if initial routing is not done then
            do initial routing for layout L;
        for (D = each direction of LEFT, DOWN, RIGHT, UP) do
            save current design L;
            resize_tile T by amount AMT in direction D;
            fix_routing ();
            Cost = calculate cost function ();
            save current design L;
            resize_tile T by amount -AMT in direction D;
            fix_routing ();
            Cost = calculate cost function ();
        for any possible pair of tiles T1 <> T2 do
            save current design L;
            swap_tiles T1 and T2;
            fix_routing ();
            Cost = calculate cost function ();
        find the min. cost, and change the design L to get this min. cost;
        fix_routing ();
        if cost is within bounded range for a fixed # of moves then
            execute random moves on the important parameters;
END PROCEDURE
```

Fig. 4. **Tabu-search based exploration method**

```
PROCEDURE domino_move IS INPUT: T - tile id in the layout which will
be resized, moved or swapped
D - the direction of change
CHG_AMT - the quantity by which the tile is to be changed
BEGIN
    make L_COPY as the copy of the layout L;
    if CHG_AMT ¿ 0
        add all neighbors of tile T on direction D to set N,
        add all neighbors of every tile in set N to N;
        set all members of N to UNCHANGED;
        add the tiles which are not the members of N but are the
            neighbors of some members of N on reverse direction of D;
        add T to set M;
        change T on direction D by amount CHG_AMT;
        while N <> ∅ do
            NT = a tile removed from N;
            if NOT all neighbors of NT on reverse direction
                of D are CHANGED then
                add NT back to N;
            else
                move NT on direction D by the minimum distance
                    to make sure NT does not overlap with any tiles of M;
            if NT <> ∅ then add NT to M;
END PROCEDURE
```

Fig. 5. **Domino_move algorithm**

involves increasing or decreasing a tile's dimensions in all the four directions and swapping the tiles to obtain a solution which has the least cost. The tabu search algorithm is used as it avoids being trapped in a local minima [8]. Traditionally simulated annealing (SA) has been used in analog layout tools [3]. Simulated annealing is a weaker search heuristic. Repeatedly annealing is very slow, especially if the cost function is expensive to compute. The subjective nature of choosing SA parameters (temperatures & step size) and the complicated tuning restricts the value of the algorithm. Tabu search on the other hand is much quicker, and does not have the limitations of SA.

The evaluation of the effectiveness of each move and how it affects the layout is determined by the cost function:

$$\text{Cost} = \alpha \times area + \beta \times freq\_resp\_error,$$

where α and β are weighting factors. The cost function considers layout area as well as the error of the obtained transfer function with respect to the desired AC response at different frequencies (term $freq\_resp\_error$).

Each tile is assigned a set width and height depending on the component it is assigned to. At the initial stage each block is empty i.e. the models for the resistors, capacitors, op-amps etc. in the block have not been included at this stage. These empty tiles will be replaced by macro-models at a later stage.

All the moves are tried on a copy of the layout, called the shadow layout. The moves are replicated on the original layout only if they are successful.

Given that the number of nets is small, routing is done using the Dijkstra's shortest path algorithm [4].

The tabu search algorithm was expanded with a solution space sampling step (last if statement in Figure 4). If the cost function remains within a fixed bounded range for a number of iterations then the algorithm attempts to move to a different solution region. This is done by performing a number of random moves, predominantly resizings of the important parameters of the design. Important param-

eters have a dominant impact on the filter performance. For example, the first gm and C of the third order elliptic filter in Section 5 are more important than the rest of the parameters. Knowledge on the importance of parameters is provided as input. This strategy prevents from modifying less important parameters, which keeps exploration basically in the same region.

We also experimented with traditional strategies for escaping from local optima such as aspiration and diversification [8]. Aspiration attempts to move parameters in a tabu status, if they result in the best solution ever reached until that step. Diversification defines a penalty that encourages the changing of parameters, which are rarely modified. None of the two strategies improved the convergence of tabu search. Diversification actually offered worse results as a significant number of moves was wasted to modify unimportant parameters of the design.

B. Tile Based Placement

The tile based placement functions are move_tile, swap_tiles and resize_tile. They rely on a basic function called the domino_move function.

Domino Move: In the domino effect, to accommodate an expansion in a tile in the layout, its neighbors will have to be moved to avoid overlapping. This movement of tiles moves towards the edges of the layout stopping either when the move is completely absorbed by the empty tiles or at the border is called domino effect. The domino move function which implements this phenomenon is an important function in placement as all other routines will rely on this function. Figure 5 introduces the pseudocode for this step.

IV. EXPERIMENTS

Experiments were developed to observe the practicality of our tool in designing high-frequency filters. The results were analyzed based on the quality of AC response, the layout and its compactness. The accuracy of the parasitics generation by the program is also

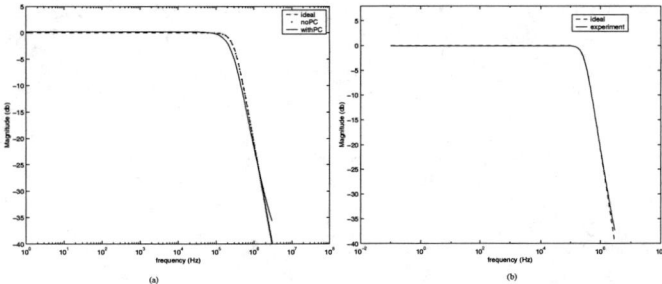

Fig. 6. 2nd order Filter Results (a) without parasitics (b) with parasitics

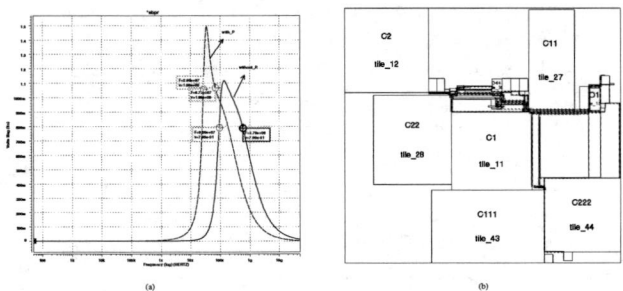

Fig. 7. (a)Response of the 6th order elliptic filter with and without considering parasitic (b)Layout

considered. Experiments were carried out on filters of various orders. Second order, third order elliptic and sixth order low-pass and bandpass filters were used for different bandwidth requirements. The filters are presented in [14]. The filters were optimized for the 3-dB points and frequency response by the resizing the components of the layout. The experiments were conducted on a SUN 80 workstation. The optimization for the 2-nd and 3-rd order filters involved about 30,000 iterations and ran for about 36 hours. The 6-th order filter required about 30,000 iterations and 70 hours.

Figure 6(a) presents the frequency responses for a 300 KHz second order filter when parasitics and coupling capacitances are not considered during the optimization. We see from these results that the ideal filter response and the response of the filter synthesized without considering the capacitances are almost identical. For the layout obtained by this method, when capacitances are considered, we obtain the graph of with_P&C which we can see is clearly different from the required response. The 3dB point is shifted by about 100kHz. This indicates the impact of layout parasitics on the performance of the design and also the need for the inclusion of parasitics and coupling capacitances during the synthesis stage. Figure 6(b) shows the filter characteristics when the layout parasitics were considered during synthesis. Results are much better in this case. Similar results were obtained for the 6-th order filter, also. Figure 7(a) presents the frequency response of the sixth order bandpass filter with the bandwidth of 100-400 MHz. Note that layout parasitics has a significant impact on the filter performance. These results motivate that constraint transformation, placement and routing must be integrated for synthesis of high frequency filters. Traditional approaches (which separate the three steps) will result in very poor designs.

Figure 7(b) shows the layout obtained for the 6-th order filter, the larger example. The largest blocks are the capacitors and the smaller blocks represent resistors and op-amps. The placement shows that the most communicating devices are placed close together thus also reducing wire length and improving routing. The results can be further improved by varying the macromodels currently being used. Reduction of wire length also improves performance by reducing the parasitics in the circuit.

V. CONCLUSIONS AND FUTURE WORK

In this paper we have presented a novel algorithm for combined parameter sizing, placement and routing for the design of analog filters. The optimization based on tabu search takes into account parasitics of the circuit. The synthesis strategy alternates between (1) a uniform sampling of the solution space and (2) a focused search of attractive regions. We use a layout representation based on symbolic tiles. As compared to other tile representations, the described layouts have a smaller complexity, which improves the efficiency of the CAD algorithms. The representation is used for uniform expression of sizing, placement and routing. The effectiveness of our tool is presented by experimenting with three high-frequency filters.

Future work includes extension of the tool to other applications. To adapt this tool for layout of converters and other circuits, some more input parameters are required. These changes can be incorporated without major variations in the program. Low-power, low-voltage circuits may also designed by varying the models and adopting low power op-amps and OTAs. Faster algorithms have to be used to decrease the time for the design.

ACKNOLEDGEMENT

This work was funded in part by Defense Advanced Research Projects Agency (DARPA) and managed by the Sensors Directorate of the Air Force Research Laboratory, USAF, Wright-Patterson AFB, OH 45433-6543.

REFERENCES

[1] F. Balasa et al, "Module Placement for Analog Layout Using the Sequence-Pair Representation", *Proc. of 36th DAC*, pp. 274-279, 1999.
[2] Y. Chang et al, "B* Trees: - A New Representation for Non-Slicing Floorplans", *Proc. of DAC*, 2000.
[3] J. Cohn et al, "Analog Device-Level Layout Automation", *Kluwer*, 1994.
[4] T. H. Cormen et al, "Introduction to algorithms", *McGraw Hill*, 1990, pp.527-531.
[5] N. Dhanwada, R. Vemuri, "Hierarchical Constraint Transformation using Directed Interval Search for Analog System Synthesis", *Proceedings of DATE*, 1999, pp. 328-335.
[6] Sree Ganesan et al, "FAAR: A Router for Field-Programmable Analog Arrays", *Proc. of 12th Intl. Conf. on VLSI Design*, pp.556-563, January 1999.
[7] G. Gielen, R. Rutenbar, "Computer Aided Design of Analog and Mixed-signal Integrated Circuits", *Proc. of IEEE*, Vol. 88, No 12, Dec 2000, pp. 1825-1852.
[8] F. Glover, "Tabu search - part II", *ORSA Journal of Computing*", pp.4-32
[9] K. Lampaert, G. Gielen, W. Sansen, "Analog Layout Generation for Performance and Manufacturability", *Kluwer*, 1999.
[10] E. Malavasi et al, "Area Routing for Analog Layout", *IEEE Trans on CAD*, Vol.12, No.8, August 1993, pp.1186-1197.
[11] E. Malavasi et al, "Automation of IC Layout with Analog Constraints", *IEEE Transactions CAD*, Vol. CAD-15, no. 8, pp. 923-942, August 1996.
[12] J. Osterhout, "Corner Stitching: A Data-Structuring Technique for VLSI Layout Tools", *IEEE Trans. on CAD*, Vol. CAD-3, No. 1, pp. 87-100, January 1984.
[13] Y.Pang et al, "Block Placement with Symmetry Constraints based on the O-tree Non Slicing Representation", *Proc. of 37th DAC*, pp. 464-467, 2000.
[14] B. N. Ray et al, "Efficient Synthesis of OTA Network for Linear Analog Functions", *IEEE Trans. on CAD*, Vol. 21, No. 5, May 2002, pp. 517-533.

A FORMAL APPROACH TO THE SLACK DRIVEN SCHEDULING PROBLEM IN HIGH-LEVEL SYNTHESIS

Shih-Hsu Huang and Chun-Hua Cheng

Department of Electronic Engineering,
Chung Yuan Christian University
Chung Li, Taiwan, R.O.C.

ABSTRACT

With the advent of deep sub-micron era, there is a growing need to consider the design closure problem in high-level synthesis. Previous work has shown that the slack (in terms of the number of clock cycles) is an effective means of tolerating variations in operation delays. In this paper, we propose an integer linear programming (ILP) approach for the slack driven scheduling problem. Our objective is to maximize the total useable slack under the design constraints (timing and resource). Compared with previous work, our approach has the following two advantages: first, our approach guarantees the optimality; secondly, our approach is more suitable for the design space exploration.

1. INTRODUCTION

The design flow of modern VLSI designs is a highly complex procedure. At the higher-level design abstraction, the more amount of improvement in the design objective is possible. However, there are a lot of uncertainties at the higher-level design abstraction. These uncertainties can make a design not meet the design constraints after the design flow is completed. Especially, with the advent of deep sub-micron era, the design closure is becoming harder and harder to achieve.

Operation scheduling is a very important task in high-level synthesis. Although typical scheduling algorithms [1-3] can find a feasible schedule that meets the design constraints (timing and resource), they [1-3] do not consider the tolerance to uncertainties. Thus, Memik, Srivastava, Kursun and Sarrafzadeh defined the slack driven scheduling problem [4]. The slack of an operation is defined as the maximum amount of additional delay (in terms of the number of clock cycles) that an operation can tolerate without violating any design constraint (timing and resource). Given an initial scheduled data flow graph, which meets the design constraints, their approach [4] is to heuristically transforming the initial schedule to another schedule with more useable slack while meeting the design constraints. The main drawback of their work is that they do not discuss the interaction between operation scheduling and slack optimization. However, in fact, their result is related to the initial schedule.

Different from a post-processing approach [4], this paper investigates the simultaneous application of operation scheduling and slack optimization. An integer linear programming (ILP) approach is proposed to model the slack driven scheduling problem. Compared with [4], our approach has the following two advantages: first, our approach guarantees maximizing the total useable slack under the design constraints; secondly, our approach is more suitable for the design space exploration.

2. PROBLEM DESCRIPTION

A data flow graph is a directed acyclic graph consisting of a set of vertices and edges. A vertex represents a particular operation. The delay of each operation o_i is D_i clock cycles, where $D_i \geq 1$. A directed edge $o_i \rightarrow o_j$ represents a data dependency relation, which means that operation o_j can be executed if and only if operation o_i has completed its execution. Given the design constraints, the objective of typical scheduling algorithms is to assign each operation to a proper control step to start the execution. Note that a control step corresponds to a clock cycle.

The slack of an operation [4] is defined as the maximum amount of additional clock cycles that an operation can tolerate without violating any design constraint (timing and resource). For a valid schedule, the slack of any operation should be non-negative. The total useable slack is defined as the summation of the slack of each operation. The objective of the slack driven scheduling problem is not only to assign each operation to a proper control step under the design constraints but also to maximize the total useable slack.

In this paper, we use an ILP approach to model the slack driven scheduling problem. Note that, although Hwang, Lee and Hsu [2] also used ILP formulations to solve the typical scheduling problem, their approach cannot be directly applied to the slack driven scheduling problem.

3. ILP FORMULATIONS

In our ILP formulations, we use the notation $x_{i,j,s}$ to denote a binary variable (i.e., an 0-1 integer variable). Binary variable $x_{i,j,s} = 1$, if and only if operation o_i is scheduled into control step j and the slack of operation o_i is exactly s clock cycles; otherwise, binary variable $x_{i,j,s} = 0$. Clearly, we have $1 \leq i \leq n$, $1 \leq j \leq t$ and $0 \leq s \leq t-1$, where n is the number of operations in the data flow graph and t is the total number of control steps. Thus, intuitively, the total number of binary variables is $n \cdot t^2$. However, in fact, from the ASAP (as soon as possible) and ALAP (as late as possible) schedules, we can find that a lot of binary variables are redundant since their values are definitely 0. Therefore, we can prune these redundant binary variables without sacrificing the accuracy of the solution.

The constants used in our ILP formulations are as below.

- The value n denotes the number of operations in the data flow graph.
- The delay of each operation o_i corresponds to D_i clock cycles.
- The value E_i denotes the earliest possible control step of operation o_i. Note that we can use the ASAP calculation to determine the value E_i for each operation o_i.

- The value L_i denotes the latest possible control step of operation o_i. Note that, given the total number of control steps, we can use the ALAP calculation to determine the value L_i for each operation o_i.
- We use FU_k to denote functional unit of type k, and we say $o_i \in FU_k$ if and only if operation o_i can be executed by FU_k.
- The value M_k is the number of functional units of type k.

The slack driven scheduling problem is formulated as follows.

Maximize $\sum_{i=1}^{n} \sum_{j=E_i}^{L_i} \sum_{s=0}^{L_i-j} s \cdot x_{i,j,s}$ (Formula 1)

Subject to
For each operation o_i and $1 \leq i \leq n$

$\sum_{j=E_i}^{L_i} \sum_{s=0}^{L_i-j} x_{i,j,s} = 1$ (Formula 2)

For each data dependency relation $o_i \to o_l$ and $1 \leq i \leq n$, $1 \leq l \leq n$

$\sum_{j=E_i}^{L_i} \sum_{s=0}^{L_i-j} (j+D_i+s-1) \cdot x_{i,j,s} < \sum_{j=E_l}^{L_l} \sum_{s=0}^{L_l-j} j \cdot x_{l,j,s}$ (Formula 3)

For each control step c and each type of function unit FU_k

$\sum_{o_i \in FU_k} \sum_{j=E_i, s=c-(j+D_i-1)}^{c} \sum_{}^{L_i-j} x_{i,j,s} \leq M_k$ (Formula 4)

Formula 1 defines the objective function. Formula 2 states the constraint that every operation must be scheduled to a control step. Formula 3 ensures that the data dependency relationships are preserved. Formula 4 states that the number of resources, type k, used in any control step should be less than or equal to the allocated resources M_k.

We use the HAL example as shown in Figure 1 to illustrate the ILP formulations. The delay of each operation is 1 control step, i.e., $D_i = 1$ for $i = 1, 2, ..., 11$. The timing constraint is 5 control steps; in other words, the total number of control steps is 5. Figure 1 (a) and (b) show the ASAP and ALAP schedules of this example. According to the ASAP and ALAP schedules, we can prune all the redundant binary variables. Table 1 tabulates all the necessary (i.e., irredundant) binary variables associated with each operation.

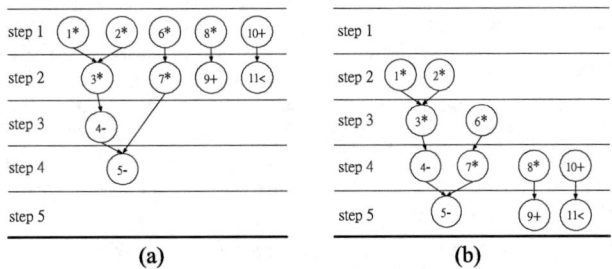

Figure 1: HAL example. (a) ASAP schedule. (b) ALAP schedule.

Assume that there are two types of functional units: the multiplier (FU_1), which can execute the multiplication operation, and the ALU (FU_2), which can execute other operations. Now we can construct the ILP formulations as below.

Formula 1. Using operation o_6 as an example, there are six binary variables associated with operation o_6. If binary variable $x_{6,1,2}$ is true, the slack of operation o_6 is two clock cycles; if binary variables $x_{6,1,1}$ or $x_{6,2,1}$ are true, the slack of operation o_6 is one clock cycle; if binary variables $x_{6,1,0}$ or $x_{6,2,0}$ or $x_{6,3,0}$ are true, the slack of operation o_6 is zero. Thus, Formula 1 is shown in the following.

Table 1: The binary variables associated with each operation.

Operation	Associated Binary Variables
o_1	$x_{1,1,0}, x_{1,1,1}, x_{1,2,0}$
o_2	$x_{2,1,0}, x_{2,1,1}, x_{2,2,0}$
o_3	$x_{3,2,0}, x_{3,2,1}, x_{3,3,0}$
o_4	$x_{4,3,0}, x_{4,3,1}, x_{4,4,0}$
o_5	$x_{5,4,0}, x_{5,4,1}, x_{5,5,0}$
o_6	$x_{6,1,0}, x_{6,1,1}, x_{6,1,2}, x_{6,2,0}, x_{6,2,1}, x_{6,3,0}$
o_7	$x_{7,2,0}, x_{7,2,1}, x_{7,2,2}, x_{7,3,0}, x_{7,3,1}, x_{7,4,0}$
o_8	$x_{8,1,0}, x_{8,1,1}, x_{8,1,2}, x_{8,1,3}, x_{8,2,0}, x_{8,2,1}, x_{8,2,2}, x_{8,3,0}, x_{8,3,1}, x_{8,4,0}$
o_9	$x_{9,2,0}, x_{9,2,1}, x_{9,2,2}, x_{9,2,3}, x_{9,3,0}, x_{9,3,1}, x_{9,3,2}, x_{9,4,0}, x_{9,4,1}, x_{9,5,0}$
o_{10}	$x_{10,1,0}, x_{10,1,1}, x_{10,1,2}, x_{10,1,3}, x_{10,2,0}, x_{10,2,1}, x_{10,2,2}, x_{10,3,0}, x_{10,3,1}, x_{10,4,0}$
o_{11}	$x_{11,2,0}, x_{11,2,1}, x_{11,2,2}, x_{11,2,3}, x_{11,3,0}, x_{11,3,1}, x_{11,3,2}, x_{11,4,0}, x_{11,4,1}, x_{11,5,0}$

Maximize { $x_{1,1,1} + x_{2,1,1} + x_{3,2,1} + x_{4,3,1} + x_{5,4,1} + x_{6,1,1} + 2x_{6,1,2} + x_{6,2,1} + x_{7,2,1} + 2x_{7,2,2} + x_{7,3,1} + x_{8,1,1} + 2x_{8,1,2} + 3x_{8,1,3} + x_{8,2,1} + 2x_{8,2,2} + x_{8,3,1} + x_{9,2,1} + 2x_{9,2,2} + 3x_{9,2,3} + x_{9,3,1} + 2x_{9,3,2} + x_{9,4,1} + x_{10,1,1} + 2x_{10,1,2} + 3x_{10,1,3} + x_{10,2,1} + 2x_{10,2,2} + x_{10,3,1} + x_{11,2,1} + 2x_{11,2,2} + 3x_{11,2,3} + x_{11,3,1} + 2x_{11,3,2} + x_{11,4,1}$ }

Formula 2. Using operation o_6 as an example, there is exactly one binary variable is true among all the six binary variables associated with operation o_6. Thus, we have $x_{6,1,0} + x_{6,1,1} + x_{6,1,2} + x_{6,2,0} + x_{6,2,1} + x_{6,3,0} = 1$. All the constraints due to Formula 2 are listed in the following.

$x_{1,1,0} + x_{1,1,1} + x_{1,2,0} = 1$;
$x_{2,1,0} + x_{2,1,1} + x_{2,2,0} = 1$;
$x_{3,2,0} + x_{3,2,1} + x_{3,3,0} = 1$;
$x_{4,3,0} + x_{4,3,1} + x_{4,4,0} = 1$;
$x_{5,4,0} + x_{5,4,1} + x_{5,5,0} = 1$;
$x_{6,1,0} + x_{6,1,1} + x_{6,1,2} + x_{6,2,0} + x_{6,2,1} + x_{6,3,0} = 1$;
$x_{7,2,0} + x_{7,2,1} + x_{7,2,2} + x_{7,3,0} + x_{7,3,1} + x_{7,4,0} = 1$;
$x_{8,1,0} + x_{8,1,1} + x_{8,1,2} + x_{8,1,3} + x_{8,2,0} + x_{8,2,1} + x_{8,2,2} + x_{8,3,0} + x_{8,3,1} + x_{8,4,0} = 1$;
$x_{9,2,0} + x_{9,2,1} + x_{9,2,2} + x_{9,2,3} + x_{9,3,0} + x_{9,3,1} + x_{9,3,2} + x_{9,4,0} + x_{9,4,1} + x_{9,5,0} = 1$;
$x_{10,1,0} + x_{10,1,1} + x_{10,1,2} + x_{10,1,3} + x_{10,2,0} + x_{10,2,1} + x_{10,2,2} + x_{10,3,0} + x_{10,3,1} + x_{10,4,0} = 1$;
$x_{11,2,0} + x_{11,2,1} + x_{11,2,2} + x_{11,2,3} + x_{11,3,0} + x_{11,3,1} + x_{11,3,2} + x_{11,4,0} + x_{11,4,1} + x_{11,5,0} = 1$;

Formula 3. Using the data dependency relation of $o_1 \to o_3$ as an example, operation o_3 can be executed if and only if operation o_1 has completed its execution. If operation o_1 is schedule into control step 1 with zero slack, operation o_3 can be scheduled into control step 2 with the slack of at most one clock cycle. If operation o_1 is scheduled into control step 1 with the slack of one clock cycle or operation o_1 is scheduled into control step 2 with zero slack, operation o_3 can only be scheduled into control step 3 with zero slack. Thus, we have $x_{1,1,0} + 2x_{1,1,1} + 2x_{1,2,0} < 2x_{3,2,0} + 2x_{3,2,1} + 3x_{3,3,0}$. All the constraints due to Formula 3 are listed in the following.

$x_{1,1,0} + 2x_{1,1,1} + 2x_{1,2,0} < 2x_{3,2,0} + 2x_{3,2,1} + 3x_{3,3,0};$

$x_{2,1,0} + 2x_{2,1,1} + 2x_{2,2,0} < 2x_{3,2,0} + 2x_{3,2,1} + 3x_{3,3,0};$

$2x_{3,2,0} + 3x_{3,2,1} + 3x_{3,3,0} < 3x_{4,3,0} + 3x_{4,3,1} + 4x_{4,4,0};$

$3x_{4,3,0} + 4x_{4,3,1} + 4x_{4,4,0} < 4x_{5,4,0} + 4x_{5,4,1} + 5x_{5,5,0};$

$x_{6,1,0} + 2x_{6,1,1} + 3x_{6,1,2} + 2x_{6,2,0} + 3x_{6,2,1} + 3x_{6,3,0} < 2x_{7,2,0} + 2x_{7,2,1} + 2x_{7,2,2} + 3x_{7,3,0} + 3x_{7,3,1} + 4x_{7,4,0};$

$2x_{7,2,0} + 3x_{7,2,1} + 4x_{7,2,2} + 3x_{7,3,0} + 4x_{7,3,1} + 4x_{7,4,0} < 4x_{5,4,0} + 4x_{5,4,1} + 5x_{5,5,0};$

$x_{8,1,0} + 2x_{8,1,1} + 3x_{8,1,2} + 4x_{8,1,3} + 2x_{8,2,0} + 3x_{8,2,1} + 4x_{8,2,2} + 3x_{8,3,0} + 4x_{8,3,1} + 4x_{8,4,0} < 2x_{9,2,0} + 2x_{9,2,1} + 2x_{9,2,2} + 2x_{9,2,3} + 3x_{9,3,0} + 3x_{9,3,1} + 3x_{9,3,2} + 4x_{9,4,0} + 4x_{9,4,1} + 5x_{9,5,0};$

$x_{10,1,0} + 2x_{10,1,1} + 3x_{10,1,2} + 4x_{10,1,3} + 2x_{10,2,0} + 3x_{10,2,1} + 4x_{10,2,2} + 3x_{10,3,0} + 4x_{10,3,1} + 4x_{10,4,0} < 2x_{11,2,0} + 2x_{11,2,1} + 2x_{11,2,2} + 2x_{11,2,3} + 3x_{11,3,0} + 3x_{11,3,1} + 3x_{11,3,2} + 4x_{11,4,0} + 4x_{11,4,1} + 5x_{11,5,0};$

Formula 4. Consider that there are four multiplication operations o_3, o_6, o_7 and o_8 can be scheduled into control step 3. However, the maximum number of multiplication operations that can be scheduled into control step 3 is constrained by the number of multipliers (i.e., M_1). Thus, we have $x_{3,2,1} + x_{3,3,0} + x_{6,1,2} + x_{6,2,1} + x_{6,3,0} + x_{7,2,1} + x_{7,2,2} + x_{7,3,0} + x_{7,3,1} + x_{8,1,2} + x_{8,1,3} + x_{8,2,1} + x_{8,2,2} + x_{8,3,0} + x_{8,3,1} \leq M_1$. Suppose that we are given two multipliers and two ALUs; in other words, $M_1 = 2$ and $M_2 = 2$. All the constraints due to Formula 4 are listed in the following.

$x_{1,1,0} + x_{1,1,1} + x_{2,1,0} + x_{2,1,1} + x_{6,1,0} + x_{6,1,1} + x_{6,1,2} + x_{8,1,0} + x_{8,1,1} + x_{8,1,2} + x_{8,1,3} \leq 2;$

$x_{1,1,1} + x_{1,2,0} + x_{2,1,1} + x_{2,2,0} + x_{3,2,0} + x_{3,2,1} + x_{6,1,1} + x_{6,1,2} + x_{6,2,0} + x_{6,2,1} + x_{7,2,0} + x_{7,2,1} + x_{7,2,2} + x_{8,1,1} + x_{8,1,2} + x_{8,1,3} + x_{8,2,0} + x_{8,2,1} + x_{8,2,2} \leq 2;$

$x_{3,2,1} + x_{3,3,0} + x_{6,1,2} + x_{6,2,1} + x_{6,3,0} + x_{7,2,1} + x_{7,2,2} + x_{7,3,0} + x_{7,3,1} + x_{8,1,2} + x_{8,1,3} + x_{8,2,1} + x_{8,2,2} + x_{8,3,0} + x_{8,3,1} \leq 2;$

$x_{7,2,2} + x_{7,3,1} + x_{7,4,0} + x_{8,1,3} + x_{8,2,2} + x_{8,3,1} + x_{8,4,0} \leq 2;$

$x_{10,1,0} + x_{10,1,1} + x_{10,1,2} + x_{10,1,3} \leq 2;$

$x_{9,2,0} + x_{9,2,1} + x_{9,2,2} + x_{9,2,3} + x_{10,1,1} + x_{10,1,2} + x_{10,1,3} + x_{10,2,0} + x_{10,2,1} + x_{10,2,2} + x_{11,2,0} + x_{11,2,1} + x_{11,2,2} + x_{11,2,3} \leq 2;$

$x_{4,3,0} + x_{4,3,1} + x_{9,2,1} + x_{9,2,2} + x_{9,2,3} + x_{9,3,0} + x_{9,3,1} + x_{9,3,2} + x_{10,1,2} + x_{10,1,3} + x_{10,2,1} + x_{10,2,2} + x_{10,3,0} + x_{10,3,1} + x_{11,2,1} + x_{11,2,2} + x_{11,2,3} + x_{11,3,0} + x_{11,3,1} + x_{11,3,2} \leq 2;$

$x_{4,3,1} + x_{4,4,0} + x_{5,4,0} + x_{5,4,1} + x_{9,2,2} + x_{9,2,3} + x_{9,3,1} + x_{9,3,2} + x_{9,4,0} + x_{9,4,1} + x_{10,1,3} + x_{10,2,2} + x_{10,3,1} + x_{10,4,0} + x_{11,2,2} + x_{11,2,3} + x_{11,3,1} + x_{11,3,2} + x_{11,4,0} + x_{11,4,1} \leq 2;$

$x_{5,4,1} + x_{5,5,0} + x_{9,2,3} + x_{9,3,2} + x_{9,4,1} + x_{9,5,0} + x_{11,2,3} + x_{11,3,2} + x_{11,4,1} + x_{11,5,0} \leq 2;$

After solving these ILP formulations, we have that $x_{1,2,0} = x_{2,1,0} = x_{3,3,0} = x_{4,4,0} = x_{5,5,0} = x_{6,2,0} = x_{7,3,1} = x_{8,1,0} = x_{9,2,3} = x_{10,1,0} = x_{11,2,1} = 1$ and the values of other variables are 0. Figure 2 (a) gives the corresponding schedule. The slacks of operations o_7, o_9 and o_{11} are one, three and one clock cycles, respectively. The total useable slack corresponds to five clock cycles.

Figure 2 (b) shows the tolerance of our schedule to the variations in operation delays. Even though operations o_7, o_9 and o_{11} become two, four and two clock cycles, respectively, our schedule can still work well.

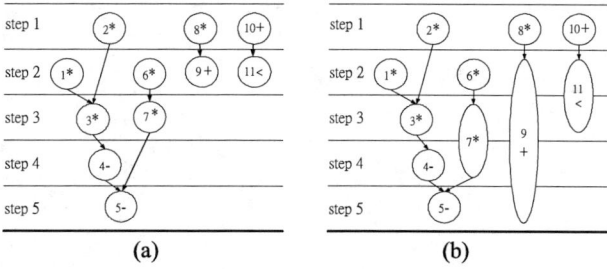

Figure 2: (a) Our schedule. (b) The tolerance to variations in operation delays.

4. EXPERIMENTAL RESULTS

We use the Extended LINGO Release 8.0 to solve the ILP formulations on a personal computer with P4-2.4GHz CPU and 512M Bytes RAM. Five benchmark circuits, including AR filter [5], HAL [2], bandpass filter [6], FIR filter [7] and elliptic filter [8], are used to test the effectiveness of our approach.

In our experiments, we assume that the delay of each operation is 1 control step. Given the number of multipliers (*#muls*), the number of ALUs (*#alus*) and the number of control steps (*#steps*), we can derive the ILP formulations for each benchmark circuit. Table 2 describes the problem complexities of each benchmark circuit with respect to different given conditions. The column *#vars* denotes the number of variables used in the ILP formulations. The column *#cons* denotes the number of constraints used in the ILP formulations. The column *CPU* denotes the CPU time (in seconds) required to solve the ILP formulations. With further analysis, we find that, for every given condition, the required CPU time is only few seconds. Therefore, our approach works well in practice.

In fact, for each given condition, our approach guarantees maximizing the total useable slack, i.e., our approach guarantees achieving the optimal result. To further validate the usefulness of our approach, we also implement the design methodology of [4] for comparisons. Note that [4] does not discuss how to find an initial schedule; in our experiments, we use the solutions of [2] as the initial schedules of [4]. Table 3 tabulates the comparisons of the total useable slacks (in terms of the number of clock cycles) between our approach and the design methodology of [4]. We find that, for most conditions, the design methodology of [4] does not achieve the optimal results; in other words, for most conditions, our approach achieves larger total useable slacks than the design methodology of [4].

It is also noteworthy to mention that our approach is more suitable for the design space exploration than the design methodology of [4]. Our approach guarantees that the total useable slack monotonically increases as the design constraints are relaxed. However, on the other hand, due to the lack of the interaction between operation scheduling and slack optimization, the design methodology of [4] does not have this property.

In the following, we use the AR filter as an example to point out a drawback inherited in the design methodology of [4]. As shown in Table 3, under the condition that *#muls* = 7, *#alus* = 5 and *#steps* = 9, the design methodology of [4] achieves the slack of 20 clock cycles; on the other hand, under the condition that *#muls* = 7, *#alus* = 5 and *#steps* = 10, the design methodology of [4] only achieves the slack of 17 clock cycles. In other words, even though the design constraints are further relaxed, the design methodology of [4] may lead to a worse result. Therefore, the design methodology of [4] is not suitable for the design space exploration.

Table 2: Problem complexities under different conditions.

Circuit	Given Condition			ILP Complexity		CPU Time
	#muls	#alus	#steps	#vars	#cons	
AR filter	6	4	8	126	49	< 1
	6	4	9	234	73	< 1
	6	4	10	348	75	2
	7	5	8	126	49	< 1
	7	5	9	234	73	2
	7	5	10	348	75	3
HAL	2	2	5	67	28	< 1
	2	2	6	110	30	< 1
	2	2	7	164	32	< 1
	3	3	5	67	28	< 1
	3	3	6	110	30	< 1
	3	3	7	164	32	1
bandpass filter	3	3	9	151	76	< 1
	3	3	10	256	78	< 1
	3	3	11	390	80	1
	4	4	9	151	76	< 1
	4	4	10	256	78	< 1
	4	4	11	390	80	< 1
FIR filter	1	1	9	158	41	< 1
	1	1	10	225	43	< 1
	1	1	11	305	45	< 1
	2	1	9	158	41	< 1
	2	1	10	225	43	< 1
	2	1	11	305	45	< 1
elliptic filter	3	4	14	92	75	< 1
	3	4	15	208	104	1
	3	4	16	334	108	1
	4	5	14	92	75	< 1
	4	5	15	208	104	1
	4	5	16	334	108	1

5. CONCLUSIONS

This paper investigates the slack driven scheduling problem in high-level synthesis. Different from a post-processing approach [4], we investigate the simultaneous application of operation scheduling and slack optimization. An ILP approach is proposed to maximize the total useable slack under the design constraints. Experiments with benchmark circuits consistently show that our approach achieves the optimal results within only few seconds. Compared with [4], the main advantages of our approach include the following two aspects: first, our approach guarantees the optimality; secondly, our approach is more suitable for the design space exploration.

6. ACKNOWLEDGMENTS

This work was supported in part by the National Science Council of R.O.C. under contract number NSC 92-2220-E-033-001.

7. REFERENCES

[1] P. G. Paulin and J. P. Knight, "Force Directed Scheduling for the Behavioral Synthesis of ASICs", IEEE Trans. on Computer-Aided Design of Integrated Circuits and Systems, vol. 8, no. 6, pp. 661—679, 1989.

[2] C.T. Hwang, J.H. Lee and Y.C. Hsu, "A Formal Approach to the Scheduling Problem in High Level Synthesis", IEEE Trans. on Computer-Aided Design of Integrated Circuits and Systems, vol. 10, no. 4, pp. 464—475, 1991.

[3] P. Faraboschi, J.A. Fisher and C. Young, "Instruction Scheduling for Instruction Level Parallel Processors", Proc. of the IEEE, vol. 89, no. 11, pp. 1638—1659, 2001.

[4] S.O. Memik, A. Srivastava, E. Kursun and M. Sarrafzadeh, "Algorithmic Aspects of Uncertainty Driven Scheduling", Proc. of IEEE International Symposium on Circuits and Systems, vol. 3, pp. 763—766, 2002.

[5] J. Ramanujam, S. Deshpande, J. Hong and M. Kandemir, "A Heuristic for Clock Selection in High-Level Synthesis", Proc. of IEEE Asia and South Pacific Design Automation Conference, pp. 414—419, 2002.

[6] C.A. Papachristou and H. Konuk, "A Linear Program Driven Scheduling and Allocation Method Followed by an Interconnect Optimization Algorithm", Proc. of IEEE/ACM Design Automation Conference, pp. 77—83, 1990.

[7] D. Shin and K. Choi, "Low Power High Level Synthesis by Increasing Data Correlation", Proc. of IEEE International Symposium on Low Power Electronic Design, pp. 62—67, 1997.

[8] M. Balakrishnan and P. Marwedel, "Integrated Scheduling and Binding: A Synthesis Approach for Design Space Exploration", Proc. of IEEE/ACM Design Automation Conference, pp. 68—74, 1989.

Table 3: Comparisons between our approach and [4].

Circuit	Given Condition			Total Useable Slack	
	#muls	#alus	#steps	Ours	[4]
AR filter	6	4	8	15	9
	6	4	9	23	16
	6	4	10	31	16
	7	5	8	16	10
	7	5	9	25	20
	7	5	10	33	17
HAL	2	2	5	5	2
	2	2	6	9	9
	2	2	7	13	12
	3	3	5	9	9
	3	3	6	14	12
	3	3	7	19	16
bandpass filter	3	3	9	8	7
	3	3	10	14	9
	3	3	11	20	13
	4	4	9	10	8
	4	4	10	17	13
	4	4	11	24	16
FIR filter	1	1	9	2	1
	1	1	10	4	1
	1	1	11	6	2
	2	1	9	10	9
	2	1	10	13	3
	2	1	11	16	8
elliptic filter	3	4	14	9	5
	3	4	15	14	9
	3	4	16	19	13
	4	5	14	11	10
	4	5	15	16	15
	4	5	16	21	20

DESIGN OF MOS CURRENT MODE LOGIC GATES – COMPUTING THE LIMITS OF VOLTAGE SWING AND BIAS CURRENT

Giuseppe Caruso

Dipartimento di Ingegneria Elettrica,
Università di Palermo,
viale delle Scienze 90128 Palermo, Italy

ABSTRACT

Minimizing a quality metric for an MCML gate, such as power-delay product or energy-delay product, requires solving a system of nonlinear equations subject to constraints on both bias current and voltage swing. In this paper, we will show that the limits of the swing and the bias current are affected by the constraints on maximum area and maximum delay. Moreover, methods for computing such limits are presented.

I. INTRODUCTION

MOS Current Mode Logic (MCML) has been proved to be preferable to complementary CMOS in mixed signal environment [1]. The main reason for that is the large amount of switching noise introduced by complementary CMOS. This adversely affects the accuracy of the analog section in mixed signal IC's. In contrast, MCML produces a small switching noise because the current supplied by V_{DD} during switching is almost constant. Moreover, owing to their differential structure, MCML gates exhibit a better noise immunity with respect to complementary CMOS logic.

The optimum design of MCML gates involves minimizing a chosen quality metric such as energy-delay product or power-delay product. Since the delay of an MCML gate depends on the voltage swing ΔV and the bias current I_B, minimizing each of those metrics requires solving a system of nonlinear equations subject to constraints on both I_B and ΔV. The present paper investigates the dependence of I_B and ΔV on the constraints on area and delay. More specifically, we will show that the constraint on the area may affects the lower limit of ΔV and the upper limit of I_B whereas the constraint on the delay sets a lower limit on I_B. Moreover, methods will be presented for computing such limits. The method for finding the minimum bias current is based on the concept of *crossing point current*. It is worth noting that computing the minimum current is the same as minimizing the power dissipation for a given delay.

II. PRELIMINARY CONCEPTS

A. MCML inverter

Figure 1 shows the schematic of an MCML inverter. The circuit is designed so that the bias current I_B provided by the constant current source is steered into one of the branches, depending on the sign of the input signal v_{id}.

Using the analytical expression of the static characteristic, it is possible to derive the expression for the noise margin [4]:

$$NM = \Delta V \left(1 - \frac{\sqrt{2}}{A_V}\sqrt{1-\frac{1}{\sqrt{2}A_V}}\right), \qquad (1)$$

where $\Delta V = I_B R_D$ is the output single-ended voltage swing and A_V is the small signal gain.

Noise margin is set on the basis of considerations about the internal noise due to process variations, and the external noise such as crosstalk [2]. Given NM, ΔV and I_B, it is possible to dimension the PMOS and NMOS devices [4].

The NMOS devices are dimensioned so that they operate in the saturation region. That requires $\Delta V < V_{THN}$. The PMOS loads must operate in the linear region [5], which requires $\Delta V < V_{DD} - |V_{THP}|$. Hence, $\Delta V_{max} = \min(\Delta V_{THN}, V_{DD} - |V_{THP}|)$.

Eq. (1) shows that NM is a lower limit for ΔV. By plotting ΔV versus A_V for a given noise margin, we can see that the voltage swing begins to saturate for $A_V>5$. Thus, a reasonable upper limit on the gain can be set in the range 5÷7.

Finally, an upper limit I_{Bmax} on the bias current is set by the constraint on maximum power dissipation ($P_{max} = V_{DD} I_{Bmax}$).

B. Delay model

The propagation delay of an MCML gate can be computed by using the method of the open-circuit time constants that gives

$$\tau_{PD} = 0.69 R_D (C_{int} + C_L) = 0.69 \frac{\Delta V}{I_B}(C_{int} + C_L), \qquad (2)$$

where R_D is the resistance of the PMOS load, C_L is the external capacitive load, consisting of the input capacitance of the driven gates and the wire capacitance, and C_{int} is the intrinsic or self-loading capacitance, formed by the parasitic capacitances of the PMOS and NMOS devices.

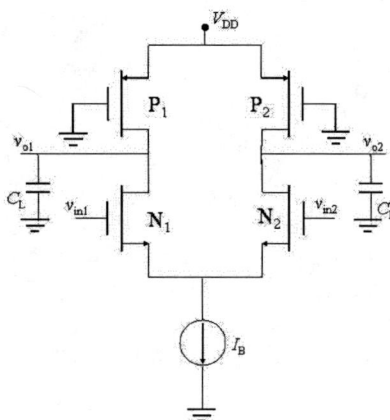

Fig. 1. MCML inverter

Eq. (2) has been derived under the assumption that the input signal has null rise and fall times. In [4] it has been shown that the intrinsic capacitance has the general form

$$C_{int} = aI_B + \frac{b}{I_B} + c. \quad (3)$$

Coefficients a, b and c depend on some parameters of the device model, the supply voltage V_{DD}, ΔV and A_V. Moreover, their values depend on the range where the bias current is located. Three ranges can be identified, namely, low current (L), medium current (M) and high current (H). Usually, the current is located in range M or H. [4]

For reasons that will be apparent in the following, we prefer to make the dependence on A_V and ΔV explicit. Thus, Eq. 3 can be rewritten as follows for ranges M and H

$$C_{int}^M = a_t \left(\frac{A_V}{\Delta V}\right)^2 I_B + \frac{b_t \Delta V}{I_B} + c^M. \quad (4)$$

and

$$C_{int}^H = \left(a_t \left(\frac{A_V}{\Delta V}\right)^2 + \frac{a'_t}{\Delta V}\right) I_B + c^H \quad (5)$$

In Eqs. (4) and (5), a_t, a'_t, b_t, c^M and c^H depend only on technological parameters and V_{DD}. Combining Eqs. (4), (5) and (2) yields

$$\tau_{PD}^M = 0.69 \left(a_t \frac{A_V^2}{\Delta V} + b_t \frac{\Delta V^2}{I_B^2} + \frac{c^M + C_L}{I_B} \Delta V \right). \quad (6)$$

and

$$\tau_{PD}^H = 0.69 \left(a_t \frac{A_V^2}{\Delta V} + a'_t + \frac{c^H + C_L}{I_B} \Delta V \right). \quad (7)$$

C. Crossing point current

Crossing point current $I_{B,cp}$ is defined as the current for which the delays corresponding to two different values of voltage swing, ΔV_1 and ΔV_2, are equal each other, that is $\tau_{PD}(I_{B,cp},\Delta V_1) = \tau_{PD}(I_{B,cp},\Delta V_2)$.

As an example, Figure 2 shows the curves of the delay versus I_B for ΔV_1=0.22V, ΔV_2=0.43V, NM=0.15V, C_L=50fF and a 0.25μm CMOS technology [3]. By solving equation

$$\tau_{PD}(I_B,\Delta V_1) = \tau_{PD}(I_B,\Delta V_2), \quad (8)$$

and by performing a few approximations valid for practical capacitive loads, we can derive the approximate expression of $I_{B,cp}$

$$I_{B,cp} \cong \frac{C_L(\Delta V_2 - \Delta V_1)}{a_t \left(\frac{A_{V1}^2}{\Delta V_1} - \frac{A_{V2}^2}{\Delta V_2}\right)}. \quad (9)$$

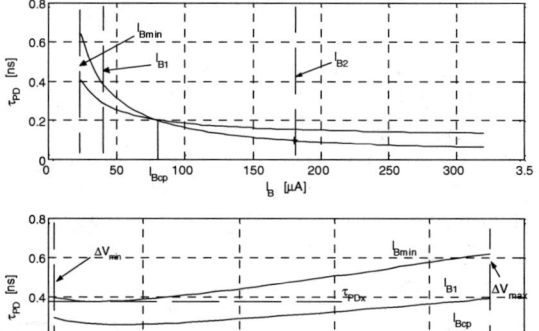

Fig. 2. Delay vs bias current for two voltage swing and delay versus swing for $I_{B,cp}$, I_{B1} and I_{B2}

III. DEPENDENCE OF ΔV AND I_B ON THE CONSTRAINTS ON AREA AND DELAY

A. Area constraint

Both the upper limit of I_B and the lower limit of ΔV may be affected by the constraint on the maximum area.

It is difficult to predict the area of an MCML gate before realizing its layout [5], therefore we will take the sum of the gate area of all PMOS and NMOS devices as a measure of it, that is

$$AREA = 2(W_n L_n + W_p L_p). \quad (10)$$

Using the expressions of the width and length of the PMOS and NMOS devices for regions M and H derived in [4], we obtain:

$$AREA^M = K_1\left(\frac{A_V}{\Delta V}\right)^2 I_B + K_2 \frac{\Delta V}{I_B} + K_3, \quad (11)$$

and

$$AREA^H = K_1\left(\frac{A_V}{\Delta V}\right)^2 I_B + K_4 \frac{I_B}{\Delta V} \quad (12)$$

where,

$$K_1 = 2\frac{L_{n,m}^2}{\mu_n C_{ox}}, \qquad K_2 = 2W_{p,\min}^2 \mu_p C_{ox}(V_{DD} - |V_{Tp}|),$$

$$K_3 = -W_{p,\min} \mu_p C_{ox}(V_{DD} - |V_{Tp}|)R_{DSW}10^{-6},$$

$$K_4 = \frac{2L_{p,\min}^2}{\mu_p C_{ox}(V_{DD} - |V_{TP}|)K_5} \quad \text{and}$$

$$K_5 = \left[1 - \frac{R_{DSW}10^{-6}}{L_{P,\min}} \mu_p C_{ox}(V_{DD} - |V_{TP}|)\right].$$

Figure 3 shows the area versus the bias current for a few values of the voltage swing and NM=0.15V. We see that in order to reduce the area, we should use the smallest current and the largest voltage swing allowed by all constraints.

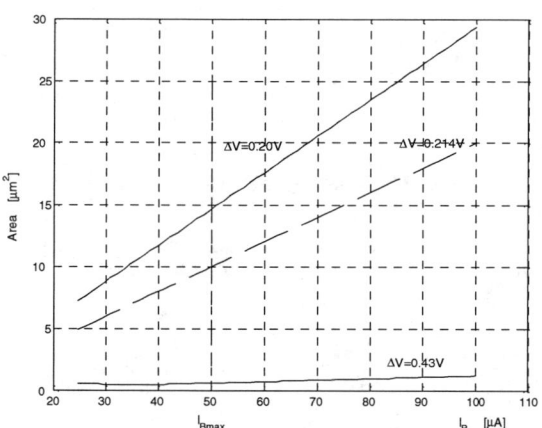

Fig. 3. Area vs bias current for three values of the voltage swing

The constraint on the maximum area may set a lower limit on the voltage swing and an upper limit on the bias current. Suppose, for instance, $AREA_{\max}$=20μm² and $I_{B\max}$=50μA. In such a case, as we can see in Figure 3, we can use any voltage swing between 0.20V and 0.43V. Now suppose $AREA_{\max}$=10μm². In this case, we have two options. If we want to leave unchanged the maximum current, we have to increase the minimum voltage swing. We can see in Figure 3 that this is about 0.214V. The exact value can be computed by Eq. (11) if $I_{B\max}$ is in range M or Eq. (12) if it is in range H. If, instead, we want to leave unchanged ΔV_{MIN}, we have to reduce the maximum current to about 35μA. The exact value can be computed by setting $AREA$=10μm² and ΔV=0.20V in Eq. (11). If the resulting $I_{B\max}$ is located in range H, we have to use Eq. (12) to achieve the correct value. Since the delay decreases quickly as the current increases, the first option is preferable if high performance is a primary design goal.

B. Delay constraint

Figure 2 shows that the curves of the delay versus the swing move upper ward with I_B decreasing and exhibit a minimum. Obviously, $I_{B\min}$ corresponds to the curve that is tangent to the line τ_{PDx}. The analytical evaluation of $I_{B\min}$ is a two-folded operation. First, we have to solve the nonlinear equation system

$$\begin{cases} \left.\dfrac{\partial \tau_{PD}}{\partial \Delta V}\right|_{I_B} = 0 \\ NM = \Delta V\left(1 - \dfrac{\sqrt{2}}{A_V}\sqrt{1 - \dfrac{1}{\sqrt{2}A_V}}\right) \end{cases} \quad (13)$$

subject to constraints

$$\begin{cases} \Delta V_{\min}(I_B) \leq \Delta V \leq \Delta V_{\max}(I_B) \\ I_B \leq I_{B\max} \end{cases} \quad (14)$$

The outcome is an expression, $\Delta V_{opt}(I_B)$, that allows us to compute the swing for which τ_{PD} for a given I_B achieves the minimum. Next, $I_{B\min}$ is computed by solving the equation obtained by substituting ΔV for $\Delta V_{opt}(I_B)$ in the expression of the delay and by putting $\tau_{PD} = \tau_{PDx}$. To perform the derivative of τ_{PD} with respect to ΔV, we have to derive an expression of A_V as a function of ΔV and to combine it with the expression of the delay. This implies inverting the second equation of the system (13). Unfortunately, this involves the solution of a third degree equation, therefore it is not possible to obtain an easily manageable expression for the gain. Moreover, the limits of the swing depend on I_B (see the proof of Proposition 1). As a result, we are compelled to compute $I_{B\min}$ by a numerical method. The method presented in this paper for evaluating $I_{B\min}$ is based on the following two propositions. Let ΔV_1 and ΔV_2 be the swings corresponding to the intersection points of the curve for a fixed I_B with a line parallel to the ΔV axis. Moreover, let us define $\Delta V_{diff} = \Delta V_2 - \Delta V_1$ and $\Delta V_{av} = (\Delta V_2 + \Delta V_1)/2$.

Proposition 1: The swing, ΔV_{opt}, for which the curve of τ_{PD} versus ΔV for a fixed I_B achieves the minimum is the limit

$$\Delta V_{opt} = \lim_{\substack{\Delta V_{diff} \to 0 \\ I_B = I_{B,cp}(\Delta V_1, \Delta V_2)}} \Delta V_{av}$$

Proof: Suppose the line parallel to the ΔV axis is moved down ward. Obviously, ΔV_{diff} decreases. By definition of crossing point current, it is clear that $I_{B,cp}(\Delta V_1, \Delta V_2)$ always equals I_B. As soon as the line is tangent to the curve, $\Delta V_{diff} = 0$ and $\Delta V_{av} = \Delta V_1 = \Delta V_2$. Consequently, $\Delta V_{opt} = \Delta V_{av}$.

Proposition 2: $I_{B,cp}(\Delta V_1, \Delta V_2)$ increases if ΔV_{av} is increased, while keeping ΔV_{diff} constant.
This proposition is given without proof for latch of space.

Minimum_I_B($\Delta V_{min}, \Delta V_{max}, \tau_{PDx}, NM$)
1. **if** $\tau_{PDx} < \tau_{PD}(I_B \to \infty, \Delta V_{max})$
2.   Constraint on maximum delay cannot be met;
3.   **return**
4. **end**
5. $\Delta V_{diff} = 0.001(\Delta V_{max} - \Delta V_{min})$;
6. $\Delta V_1 = \Delta V_{min}$, $\Delta V_2 = \Delta V_1 + \Delta V_{diff}$;
7. $I_{B,cp} = I_{b,cp}(\Delta V_1, \Delta V_2)$;
8. **if** $\tau_{PDx} > \tau(I_{B,cp})$
9. $\Delta V_{opt} = \Delta V_{min}$,
10. $I_{Bmin} = I_B(\tau_{PDx}, \Delta V_{min})$;
11.   **return**
12. **end**
13. **while** $\tau_{PDx} < \tau(I_{B,cp})$
14. $\Delta V_1 = \Delta V_2$, $\Delta V_2 = \Delta V_1 + \Delta V_{diff}$;
15. $I_{B,cp} = I_{b,cp}(\Delta V_1, \Delta V_2)$;
16. $\Delta V_{opt} = \Delta V_{av}$, $I_{Bmin} = I_{B,cp}$;
17. **end**

Fig. 4. Procedure minimum_I_B

An outline of the procedure, called **minimum_I_B**, that computes I_{Bmin} is shown in Fig. 4. It has been implemented within the MATLAB environment.

In each iteration (steps from 13 to 17), ΔV_{av} is increased, while keeping ΔV_{diff} constant and $I_{B,cp}(\Delta V_1, \Delta V_2)$ is computed. By Proposition 2, $I_{B,cp}$ increases. The procedure stops as soon as $\tau(I_{B,cp}) < \tau_{PDx}$. Since ΔV_{diff} is very small, $\Delta V_{av} \cong \Delta V_{opt}$ by Proposition 1. Notice that ΔV_{opt} is lower than ΔV_{min} if the test in step 8 is successful.

IV EXPERIMENTAL RESULTS

Let us assume $NM = 0.15V$, $A_{Vmax} = 5$, $\tau_{PDx} = 1.9ns$, $AREA_{max} = 8.1 \mu m^2$ and $P_{max} = 0.125mW$ ($I_{Bmax} = 50\mu A$). By setting $\Delta V_{max} = 0.43V$, the NMOS devices operate in saturation. Moreover, using Eq. 1 we obtain $\Delta V_{min} = 0.2V$.

Owing to the constraint on the area, ΔV_{min} or I_{Bmax} must be changed. The first row of Table 1 shows the minimum bias current, the corresponding swing and the simulated delay for the original I_{Bmax}. The second row shows the same data for the original ΔV_{min}. Finally, we can see that the simulated delays are about 20% higher than τ_{PDmax}.

Table 1. Experimental results

I_{Bmax} [μA]	ΔV_{min} [V]	I_{Bmin} [μA]	ΔV_{IBmin} [V]	$\tau_{PD,spice}$ [ns]
50	0.22	4.5	0.22	2.29
28	0.20	4.23	0.20	2.29

V. CONCLUSIONS

In this paper we have showed that the constraints on maximum area and delay affect the limits of both ΔV and I_B. More specifically, we have shown that the constraint on the delay sets a lower limit on I_B and, consequently, on the power dissipation for the given maximum delay, whereas the constraint on the area may affect the upper limit of I_B and the lower limit of ΔV. Moreover, methods for computing such limits have been presented.

VI. REFERENCES

[1] D. Allstot, S. Chee, S. Kiaei, and M. Shrivastawa, "Folded source-coupled logic vs. CMOS static logic for low-noise mixed-signal ICs," *IEEE Trans. Circuits Syst. I*, vol. 40, pp. 553-563, Sept. 1993.

[2] S. Bruma, "Impact of on-chip process variations on MCML performance," *IEEE Proceedings of SOC Conference*, Sept. 2003.

[3] J.M. Rabaey, A. Chandrakasan, and B. Nikolic, *Digital Integrated Circuits – A Design Perspective*, Prentice Hall Electronics and VLSI series, Charles S. Sodini,, series editor, 2003D.

[4] M. Alioto, and G. Palumbo, "Design strategies for source coupled logic gates," *IEEE Trans. Circuits Syst. I*, vol. 50, pp. 640-654, May 2003.

[5] J. Musicer, "An analysis of MOS current mode logic for low power and high performance digital logic," http://bwrc.eecs.berkeley.edu/Publications/2000/Theses/Anlys_MOS_curnt_mode_logic/

Interconnect Delay Optimization via High Level Re-synthesis After Floorplanning[1]

Yunfeng Wang
C.S. Dept.
Tsinghua University
Beijing, China
Wangyf00@mails.tsinghua.edu.cn

Jinian Bian
C.S. Dept.
Tsinghua University
Beijing China
bianjn@tsinghua.edu.cn

Xianlong Hong
C.S. Dept.
Tsinghua University
Beijing China
hxl-dcs@tsinghua.edu.cn

Abstract — With the progress of manufacturing technologies, more transistors can be integrated into one chip, which validates the System-On-Chip (SoC) technology. Besides, as the feature size of integrated circuits scales down into super deep sub-micron level, interconnect delay has played a dominant role in total delay of the circuit. A new technology which improves the performance of the circuit in high-level synthesis phase by utilizing high level re-synthesis after floorplan is presented in this paper. The techniques presented in this paper are demonstrated by experiments.

I. INTRODUCTION

As the feature size of integrated circuit scales down into deep sub-micron level, more and more transistors can be integrated into one chip. The complexity of the circuit has been enlarged rapidly, which drives designers to move to higher levels of abstraction to enable larger systems to be described and more powerful computer aided design tools to be applied[6]. High level synthesis (HLS), as a higher level abstraction design phase of integrated circuits, can translate the functional description of a circuit into the register transfer level (RTL) description of the circuit, which will effectively improve the design speed of the circuit.

Current deep submicron regime of integrated circuits brings profound effect on design techniques. In particular, as the feature size of integrated circuits decreases continuously, interconnect delay is playing a dominant role in total delay of a circuit [5]. In traditional design flow of integrated circuits, high-level synthesis does scheduling and allocation, and then floorplanning determines the actual positions of modules in a physical design. Since floorplanning is separated from high level synthesis, no interconnect information can be supplied to synthesis process. This communication-less flow will cause serious problems. Because these two phases of design flow are based on different delay estimation model, the result of high-level synthesis may be totally wrong for floor-planning, especially in timing aspect. To solve this problem, a co-operation between the two phases is necessary.

In this paper, we propose an algorithm that performs re-synthesis after floorplanning, which is used to optimize the interconnect delay of the circuit. As we mentioned before, a co-operation between high level synthesis and floorplanning is needed to consider the interconnect delay in high level synthesis phase. Some previous researchers have addressed the problem of incorporating physical design information in high level synthesis. The algorithm 3D scheduling[1] considers floor planning in high level synthesis, it decides the shape and position of each functional unit on the floor-plan concurrently as operations are scheduled and functional units are allocated. GB[2] does not consider scheduling problem, it combines binding with one-dimensional floor planning, and translates the problem to a two-dimension grid placement to minimize the interconnect wire lengths, however the usage of one-dimensional floor-planning is very limited. BINET[7] is a binding algorithm that performs incremental binding and floor planning on a previously scheduled result based on a network flow model. Midas[3] incorporated floor planning and formulates high level synthesis using a data-transfer model. Prabhakaran et al.[8] provided a simulated annealing based algorithm that combines scheduling, binding and floor planning. For each step of the simulated annealing of scheduling and binding, a constructive timing driven floor-planning algorithm is performed to evaluate floor plan area and time delay. William E. Dougherty and Donald E. Thomas presented an algorithm to unify behavior synthesis and physical design together[6], which allows behavioral and physical decisions to be made simultaneously. The algorithm is a constructive technique, and a quadratic placement if used to determine placement locations for efficiency considerations. Because the algorithm in [6] is a constructive algorithm, the search of solution space may be limited.

However, the constructive algorithm will limit the search of solution space, and the estimation of interconnect delay will be unprecise when simultaneously doing high level synthesis and physical design.

In this paper, a virtual-force balance based reallocation and rescheduling algorithm is presented. The algorithm optimize the interconnect delay of the circuit after floorplanning by changing the topology of the circuit based on the position information of functional units. The example of a fir filter and a real world IDCT decoder is used to test the algorithm.

The paper is organized as following: In the next section, the representation and the estimation of a solution of high-level synthesis in presented. The re-synthesis algorithm used to optimize the timing of the circuit after floorplanning is presented in Section III. The experimental results and conclusions are given in the last section.

---
[1] This work is supported by NSFC 90407005 and partially supported by NSFC 60236020 and 90207017

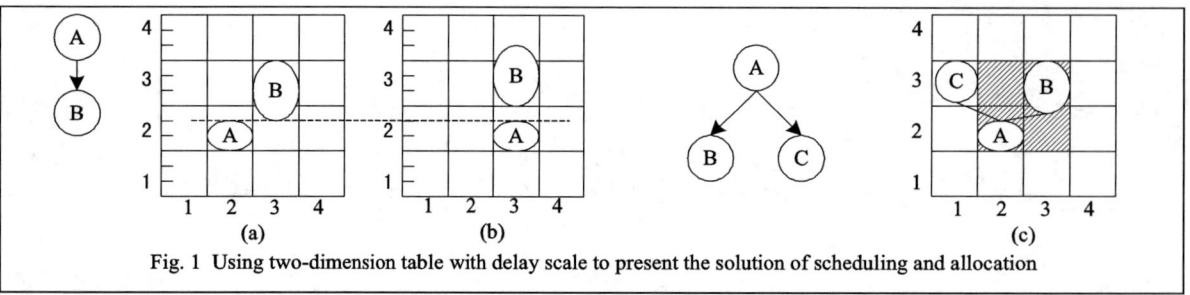

Fig. 1 Using two-dimension table with delay scale to present the solution of scheduling and allocation

II. REPRESENTATION AND ESTIMATION OF HLS SOLUTIONS

A. Representation of the result of scheduling and allocation

A two-dimension grid with delay scales is used to represent the result of scheduling and allocation, as shown in Fig. 1, which was first presented in [2]. The columns of the grid stand for control-steps of scheduling, while the rows stand for the functional units to be used in the circuit. The result of scheduling and allocation can be considered as the placement of the two-dimension grid.

According to the grid, a new placement can be considered as a new solution of scheduling and allocation. In Fig. 1 (a), operation A is placed in row 2, column 2, which means operation A is scheduled in step 2 and allocated to functional unit 2. When operation A is moved from column 2 to column 3, as shown in Fig. 1 (b), it means that the operation is allocated to functional unit 3 instead of functional unit 2. For each cell in two-dimension grid, it can be occupied by only one operation.

Chained operations can also be represented by the two-dimension table. As shown in Fig. 1 (c), operation B and C are chained operations, then, functional unit 2 and 3 are both occupied in control step 2 and 3.

B. Estimation of a HLS Solution

Here we consider delay and area for solution estimation. That is, the optimization result of high-level synthesis is a tradeoff between chip delay and chip area.

For a synchronous circuit, the total delay and area can be calculated as below:

$$A = \frac{1}{k} \times \sum_i A_i \quad (0 < k < 1) \quad (1)$$
$$D = d_s \times s$$

where A is the total area of the circuit, A_i is the area of functional unit i and k is the usage ratio of the chip; D is the total delay of the circuit, d_s is the length of a control step and s is the number of total control steps used by the circuit. Because the usage ratio of the chip can not be got until floorplan is finished, in HLS before floorplanning, k is set to be an expected ratio. Then, the total cost of a solution can be calculated as:

$$C = w_A \cdot A + w_d \cdot D \quad (2)$$

where w_A and w_d are the weight of area and total delay respectively.

The high-level synthesis algorithm tries to minimize the cost of the solution under the total area and total delay constraint, formulized as following:

$$\min \quad C = w_A \cdot A + w_d \cdot D$$
$$A \leq total\_area \quad (3)$$
$$D \leq total\_delay$$

III. RESYNTHESIS AFTER FLOORPLANNING

In traditional high-level synthesis, no position information of any functional unit is available. However, since the interconnect delay is not negligible in sub-micron regime, timing violation occurs constantly in the circuit designed through traditional high-level synthesis. Therefore, incorporating floorplanning information is beneficial for high-level synthesis. Here we propose a virtual-force balance based reallocation and rescheduling algorithm.

A. Local path of an Operation/Chained operations

In traditional high level synthesis phase, the delay of one operation is only considered as the delay of the functional unit which assigned to the operation, and the delay of chained operations are considered as the sum of functional unit delay. In our approach, the delay of interconnect wires between functional units and registers are also calculated into the delay of operations. In order to formulate our algorithm, a new principle of local path is defined as following:

For each operation O, assume that its inputs are from register set $R_{in} = \{r \mid r \text{ is the input register of } O\}$, and its output is from register set $R_{out} = \{r \mid r \text{ is the output register of } O\}$, Then, the interconnects in $R_{in} \times \{O\} \cup R_{out} \times \{O\}$ and the functional unit which is assigned to O are so called the *local path* of O.

The delay of each local path can be calculated as following:

$$d_o = d_f + d_{wi} + d_{wo}$$
$$d_{wi} = \max\{d_{wire} \mid wire \in R_{in} \times \{O\}\} \quad (4)$$
$$d_{wo} = \max\{d_{wire} \mid wire \in R_{out} \times \{O\}\}$$

where d_f is the functional unit delay of operation O.

Equation. (4) can be easily applied on chained operations if we consider chained operations as a single operation. For example, if operation A and B are chained together, then the d_f of this

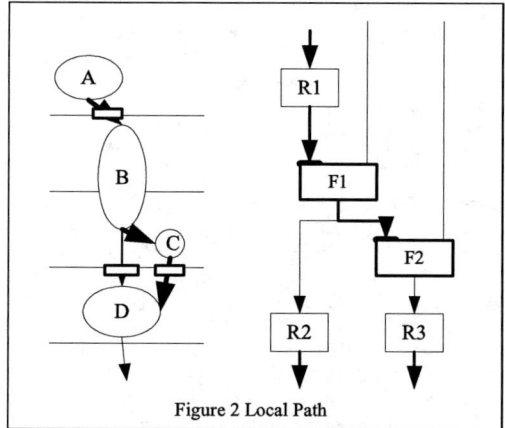

Figure 2 Local Path

chained operation can be calculated as following:

$$d_f = d_{f_A} + d_{f_B} + d_{w_{AB}} \quad (5)$$

where d_{f_A} is the functional unit delay of A, and d_{f_B} is the functional unit of B, $d_{w_{AB}}$ is the delay of interconnect wire between A and B.

As shown in Fig. 3, the path $R1 \to F1 \to R2$ is the local path of operation B, and the path $R1 \to F1 \to F2 \to R3$ is the local path of B and C. Obviously, the delay of a local path will violate the timing constraint if it meets the following condition:

$$d_o > [ES(O) - BS(O) + 1] \times c \quad (6)$$

where d_o has the same meaning as in (4), and $ES(O)$ denotes the end step of O, $BS(O)$ denotes the begin step of O, c is the length of each control step.

B. Reallocation and Rescheduling after Floorplanning

A force-balance based algorithm, FIDER (**F**orce-balance based **I**nterconnect **DE**lay driven **R**eallocation and rescheduling algorithm), is used to do reallocation and rescheduling after floorplan. This algorithm optimized the interconnect delay of each local path by replace the operations in the two-dimension grid.

The main idea of FIDER is illustrated in Fig. 3. Assume that operation A is allocated to function unit 2, while B is allocated to function unit 4. We found that $f2$ is too far from $f4$ for satisfying the timing constraint, and $f3$ is just the very position for A, FIDER will reallocate A to $f3$. Thus, a new solution which has less-delay critical local path is got without doing floorplan again. If another operation C has been allocated to $f3$, FIDER will lock A to $f3$ and try to find another valid place for C, and iteratively. If the iteration does not finish within finite cycles, a rescheduling process will be triggered.

If we consider registers to be a kind of function units, and the process of storing data into a register to be a special operation *store (store operation is always assigned to a register)*, the virtual force acting on an operation O can be defined as follows.

Suppose operation i is allocated to function unit f, and its output (or input) is from (or to) a store operation s (or other kinds of operation), while operation s is allocated to function unit r, then a virtual force which acts on the operation i is defined as follows:

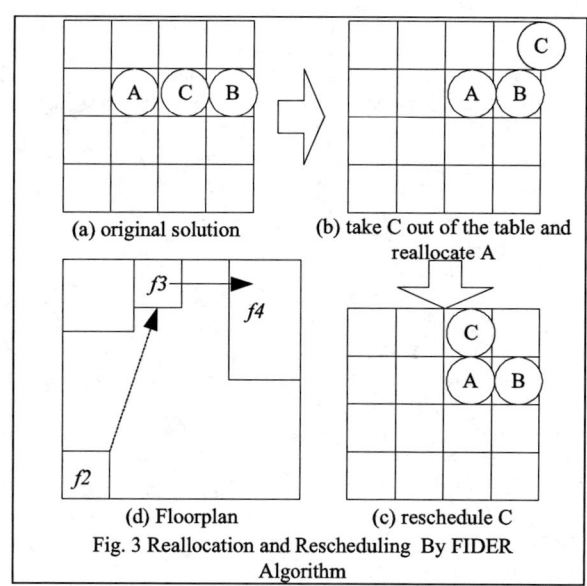

Fig. 3 Reallocation and Rescheduling By FIDER Algorithm

$$\vec{F}_{i,s} = \vec{P}_r - \vec{P}_f \quad (7)$$

where $\vec{P}_r$ and $\vec{P}_f$ are the position vectors of function unit f and r respectively, $\vec{F}_{i,s}$ is the force acts on operation i. Note that all positions and forces are presented in vector.

The total virtual force that acts on the operation i is calculated by

$$\vec{F}_i = \sum_j w_{i,j} \times \vec{F}_{i,j} \quad (8)$$

where $\vec{F}_i$ is the total force the operation received, $\vec{f}_{i,j}$ is the force between operations i and j, $w_{i,j}$ is the weight of each force.

For each operation i in a critical local path, FIDER tries to search for a cell in the two-dimensional grid such that the total force acting on the operation will be minimized. If the best cell has been occupied by another operation j, FIDER will allocate the operation i to the cell, then try to find another function unit for j.

IV. EXPERIMENTAL RESULTS AND CONCLUSIONS

The VHDL descriptions of fir11 benchmark and IDCT (a real world JEPG decoder) are used as the test case to check the algorithms we proposed. The benchmark is firstly compiled into a HCDFG[4] representation, and then a traditional scheduling and allocation approach launched. A CBL based floorplanning algorithm[9] is used to get the result of floorplan.

The basic information of fir11 benchmark is given in TABLE 1. We set the length of a control step to be $2ns$. Two kinds of multipliers are used in synthesis, and the delay of operation MULTIPLY is assumed to be $3.2ns$ on multiplier1, and $3.5ns$ on multiplier2. The experimental results using different adders are shown in TABLE 3. As shown in TABLE 3, when delay of ADD is about 70% of the length of one step, total delay of the circuit is reduced by 7.69%. When delay of ADD is less than half the length

of one step, the reduction of total delay is great. And the total area is increased slightly. Sever reasons may cause this result:
1. The area is increased because more function units are needed. The use of chaining operations makes some function units un-reusable during certain control steps.
2. Less variables need to be stored in registers, which causes a reduction of register number and saves some area.
3. The total delay is reduced because the number of total control steps is reduced.

From the experimental result, we can see that when the delay of add is 70% of the length of control step, the reduction of total delay is not so obviously. This is caused by the delay of interconnect wires. The delay of interconnect wires violates the timing of the critical local path, thus more control steps has to be added to keep the correctness of the circuit.

When the delay of add is about 40% of the length (it is true because the delay of interconnect wires has count the 70% of the total delay in deep sub-micron level IC chip), the reduction of the circuit is reduced obviously, because more operations can be combined into chaining operations. In this case, the interconnect delay reduction caused by our re-synthesis algorithm will be very efficient.

A real world IDCT is also used to check the effect of our algorithm. The purpose clock period is 1ns. In order to reduce the effect of interconnect delay in synthesis phase, only 0.9ns in each control step can be scheduled for functional units in synthesis phase, the other 0.1ns is reserved for interconnect delay. The experimental result is shown in TABLE 2, and the result of floorplanning is shown in Fig. 4. As shown in TABLE 2, the target frequency is not achieved because of the existence of interconnect delay. The re-synthesis algorithm effectively erase this timing violation of interconnect delay. The reduction of clock period is 7.93%.

Experimental results show that the algorithm is efficient. The total delay of the circuit can be reduced obviously by operations chaining, while the area of the circuit is increased tinily. The condition assumed in this paper is similar to the real world constraint, which indicates that the algorithm is useful in real world design.

REFERENCE

[1] J.P.Weng and A.C.Parker, "3D Scheduling: High Level Synthesis with Floorplanning" 28th ACM/IEEE Design Automation Conference, pp. 668-673, 1991.
[2] H. Jang and Barry M. Pangrle, "A Grid-Based Approach for Connectivity Binding with Geometric Costs" ICCAD-93, pp. 94-99, 1993.
[3] S.Tarafdar, M.Leeser et al., "A Data-Centric Approach to High-Level Synthesis", IEEE Transactions on Computer-Aided Design of Integrated Circuits and Systems, Vol.19, No.11, November 2000
[4] Qiang Wu, Yunfeng Wang, Jinian Bian, Weimin Wu and Hongxi Xue, "A Hierarchical CDFG as Intermediate Representation for Hardware/Software Codesign", Proceeding of ICCCAS'02, Chengdu, China, 2002.
[5] Jason Cong "An Interconnect-Centric Design Flow for Nanometer Technologies" Proceedings of The IEEE, vol 89, No.4, April 2001, p505-527
[6] William E. Dougherty and Donald E. Thomas, "Unifying Behavior Synthesis and Physical Design" Design Automation Conference, 2000. Proceedings 2000. 37th , June 5-9, 2000 pp. 756 – 761
[7] M.Rim, A.Majumdar, et al., "Optimal and Heuristic Algorithms for Solving the Binding Problem" IEEE Transactions on VLSI Systems, vol. 2, pp. 211-225, June 1994.
[8] P.Prabhakaran and P.Banerjee, "Simultaneous Scheduling, Binding and Floorplanning in High-level synthesis", Proceedings of the IEEE International Conference on VLSI Design, pp. 428-434, 1998.
[9] Yunchun Ma, Xianlong Hong, Sheqin Dong, Yici Cai, Chung-Kuan Cheng, Jun Gu, "Floorplanning with abutment constraints based on corner block list", INTEGRATION, the VLSI journal, pp. 65-77, vol. 31, 2001

TABLE 1 Basic Information of Example

	# of add/minus	#of multiply	# of negative
Fir11	10	11	6

TABLE 2 Synthesis Result of IDCT

# of Functional Units	112
# of Control Steps	125
Objective Clock Period	1ns
Actual Clock Period before Re-synthesis	1.08617ns
Actual Clock Period after Re-synthesis	1ns

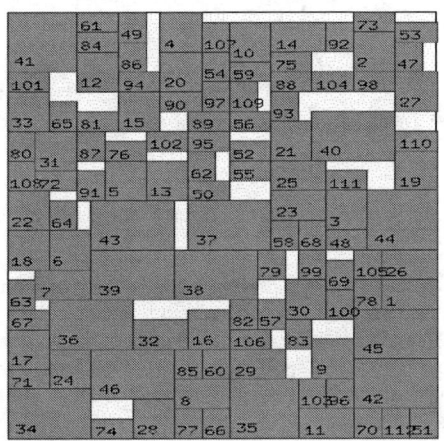

Fig. 4 The floorplanning result of IDCT

TABLE 3 Experimental Results

	Delay of adder is 1.4*ns*		Delay of adder is 0.9*ns*	
	Area(μm^2)	Delay(ns)	Area(μm^2)	Delay(ns)
ACLA&FIDER	2185	24	2185	16
Traditional	2175	26	2175	26
Ratio	100.46%	92.31%	100.46%	61.54%

Using Symbolic Computer Algebra for Subexpression Factorization and Subexpression Decomposition in High Level Synthesis

Xianwu Xing and Ching-Chuen Jong
Center of Integrated Circuits and Systems, Nanyang Technological University
School of EEE, Nanyang Avenue, Singapore 639798

Abstract—In High Level Synthesis, signals can usually be regarded as symbols. With modern Symbolic Computer Algebra, behavioral-level circuit descriptions consisting of symbols can be transformed into more efficient codes for hardware implementation. In this paper, two high level transformation techniques, Subexpression Factorization and Subexpression Decomposition, which are based on modern Symbolic Computer Algebra, are proposed to optimize both area and Critical Path Delay (CPD). The Subexpression Factorization tries to find an optimal combination of subexpressions of the input polynomial for hardware implementation. The Subexpression Decomposition tries to achieve the best decomposition of a polynomial for the mapping of complex components.

I. INTRODUCTION

Datapath intensive applications, such as DSP and multimedia applications, have a large share of the market of digital circuits. As the problem size and optimization dimensions increase, more techniques should be used at a higher level of the design abstraction. Usually performed before high level synthesis tasks, high level transformation techniques can partially or totally change the implementation of circuits while maintaining the system functions, thus they give a significant impact on the resultant circuit quality.

The application of Symbolic Computer Algebra in high level synthesis is previously proposed in [4]-[6], in which a complex component mapping technique is proposed. In this work, we further address the problems in using symbolic computer algebra for hardware implementation. Here, we propose to use two high level transformation techniques, subexpression factorization and subexpression decomposition, in high level synthesis. The subexpression factorization technique factorizes subexpressions of a polynomial so that the hardware implementation is optimized. The subexpression decomposition technique decomposes a polynomial for achieving an optimal mapping of complex components, such as square, MAC (multiplier/accumulator), sine, cosine.

There are two foundations that the proposed techniques are based on. The first one is Symbolic Computer Algebra, which has witnessed its many applications in fields like control system, but has not been fully explored in high level synthesis. The Maple system is used here as the Symbolic Computer Algebra tool to perform the necessary symbolic manipulations. The other is the polynomial formulation of the input data flow segment and the availability of library components, which has been elaborated in [1]-[3].

Although polynomial factorization has been used in [4]-[6], no effort was made to find the best combinations of subexpressions for factorizations when the whole polynomial cannot be factorized. In this paper, a subexpression factorization technique is proposed to find an optimal combination of subexpressions according to predefined cost functions, while preventing the computational complexity from exploding, which is achieved by the introduction of a new graph, called the Term Dependency Graph (TDG).

This paper is organized as follows. Section II formulates the problem of subexpression factorization and presents the algorithm. In Section III, subexpression decomposition is proposed to avoid the problems raised by decomposing the whole polynomial during the mapping of complex components. Section IV presents the experimental results. Conclusion and future work are discussed in section V.

II. SUBEXPRESSION FACTORIZATION

A. Problem formulation

Our work addresses the problem when the whole input polynomial may not be factorized, but some subexpressions can be. Consider the following polynomial P_e:

$$\begin{aligned} P_e &= a^2 + 2\,a\,b + b^2 + a^2\,b + x^2\,y + x^2\,z + w \\ &= (a+b)^2 + a^2\,b + x^2\,(y+z) + w \\ &= a^2 + b\,(2\,a + b + a^2) + x^2\,(y+z) + w \\ &= a\,(a + 2\,b) + b\,(b + a^2) + x^2\,(y+z) + w \end{aligned}$$

The main objective of the proposed subexpression factorization technique is to find an optimal combination of subexpressions for hardware implementation while preventing the computational complexity from being exponential. To reduce the exploration space, the number of subexpression candidates for factorization needs to be reduced. There are many algorithms to test the irreducibility of polynomials [7]-[9]. Unfortunately, as far as we know, all of them have restrictions or high complexity. So it is neither efficient nor effective to implement them. Therefore, a heuristic algorithm is proposed here to reduce the number of candidates without missing many good solutions.

B. Algorithm

Polynomial factorization is a complex problem. However, when hardware implementation is concerned, it is usually sufficient to consider only the relationship between the terms. A new graph, called Term Dependency Graph (TDG), is proposed for this purpose.

Definition 1: A Term Dependency Graph G (V, E) is an undirected graph, in which each vertex $v_i \in V$ corresponds to a term of the input polynomial. There is an edge $(v_i, v_j) \in E$ if and only if there is at least one common variable between vertex v_i and v_j. The set of the common variables, called CV, is the weight of (v_i, v_j).

For example, the TDG of the polynomial $P_e = a^2 + 2ab + b^2 + a^2 b + x^2 y + x^2 z + w$ is shown in Fig.1.

Definition 2:

(a) AV of a TDG is a set of all the variables appear in the graph. For P_e, AV = {a, b, x, y, z, w}.

(b) ACV of a TDG is a set of all the variables in the CVs of all edges. For P_e, ACV = {a, b, x}.

(c) CVG of a TDG is a set of common variables, each of which appears in the CV of every edge. For P_e, CVG = Φ.

Several criteria are applied in the heuristic algorithm:

Criteria 1: If a subgraph is not connected, it will not be selected as a candidate for factorization.

Criteria 2: If a subgraph has its CVG ≠ Φ, it will be selected as a candidate for factorization.

Criteria 3: If a subgraph has its ACV ≠ AV, it will not be selected as a candidate for factorization.

Criteria 4: Subexpressions will be selected in descending size. That is, longer subexpressions will be selected and evaluated first.

We note that these criteria do not always stand in the mathematical point of view. However, from the experimental results, it can be perceived that satisfactory results can be obtained when hardware implementation is concerned.

The algorithm is presented as Algorithm 1. It is based on the subgraph selection and update of TDG. Subexpression selection is performed based on the above criteria, and factorization is performed with function calls to Maple. The evaluation of factorization results is based on a predefined cost function, which uses information from either the whole polynomial or only the subexpressions.

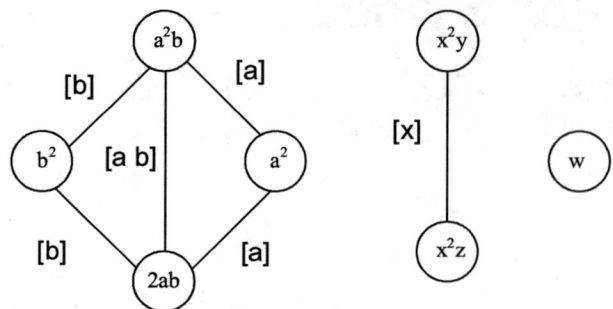

Figure 1.   TDG of P_e

Algorithm 1. Subexpression Factorization

//Input: the TDG G (V, E) of the input polynomial.
//Output: factorization result.
while (E ≠ Φ)
{
    find a connected subgraph $G_s (V_s, E_s) \subseteq G (V, E)$;
    for (i = |V_s|; i >1; i − −)
    {
        C =candidate_selection ($G_s (V_s, E_s)$, i);
        S = factorization (C);
        if (S ≠ Φ)
        {
            evaluate and store the best result;
            update G (V, E);
            break;
        }
    }
}
report stored result;
// End of Algorithm 1

III. SUBEXPRESSION DECOMPOSITION

Currently, mapping of complex components depends on the skills of designers. To achieve automation, polynomial decomposition is introduced in [4]-[6].

A. Polynomial Decomposition

Polynomial decomposition is to simplify a polynomial with a set of equations (called side relations in the Maple system). Consider the following code fragment in Maple:

> sine:=x-1/6*x^3+1/120*x^5-1/5040*x^7:

> f:=x^8+x-1/6*x^3+121/120*x^5-1/5040*x^7:

> simplify(f,{sine=z});

$$x^5 + 5040x^2 - 840x^4 + 42x^6 - 5040x\,z + z$$

The Maple function "simplify" performs the polynomial decomposition, which simplifies the function "f" with an equation "sine = z". Note that the variable "sine" represents the polynomial formulation of a sine function, and the simplification process is actually a mapping process of the sine function. Consequently, polynomial decomposition provides a way to automate the mapping of complex components.

B. Subexpression Decomposition

The example in the previous section also exposes the defect of decomposing the whole polynomial, because the best mapping should be $f = x^8 + \text{sine} + x^5$. The undesirable Maple output is caused by the decomposition of the term x^8.

To solve this problem, subexpression decomposition is proposed, in which certain subexpressions are selected and decomposition is performed. An example is shown in Maple codes:

> sine:=x-1/6*x^3+1/120*x^5-1/5040*x^7:

> f:=x^8+x-1/6*x^3+121/120*x^5-1/5040*x^7:

> selected_subexpr:=-1/5040*x^7:

> remaining_subexpr:=f-selected_subexpr:

> result:=simplify(selected_subexpr,{sine=z}):

> f:=result+remaining_subexpr;

$$f := x^8 + x^5 + z$$

C. Algorithm

A modified branch-and-bound algorithm is used. Polynomial decomposition will only decompose the term whose order is not lower than the highest order of the side relations. As a result, only the terms with orders higher than or equal to the highest order of the side relation are the candidates for subexpression selection.

IV. EXPERIMENTAL RESULTS

A. Subexpression Factorization

In our experiment, the input polynomial is synthesized with only adders and multipliers. All the signals are set to be 16 bits. Synopsys Design Compiler is used to estimate the area and delay of a 16-bit adder and a 16-bit multiplier and the result is shown in Table 1.

The polynomials used in the experiment are shown in Table 2. The first five polynomials are from [6]. The others are typical for subexpression factorization.

Synopsys Design Compiler is also used to synthesize the benchmarks. The synthesis results are reported in Table 3.

Table 1. Delay and Area of Library Elements

Library Element	Delay	Area
Addition	177	10.24
Multiplication	3140	18.56

Table 2. Benchmark Polynomials

P1	$a^2 - b^2$
P2	$b^3 + b\,a^2\,c$
P3	$1 - 1/2\,x^2 + 1/24\,x^4 + x + y\,z$
P4	Gabor-Transform* (used in neural system)
P5	PSK* (used in digital communication)
P6	$a\,c + a\,d + b\,c + b\,d + b^2\,d^2$
P7	$a^2 + 2\,a\,b + b^2 + x^2 + 2\,x\,y + y^2 + 1$
P8	$a\,x + a\,y + a + x + y + 1$
P9	$a^2 - b^2 + a - b$

* Gabor-Transform = $1 - a^2 - b^2 + a^2 b^2 + 1/2\,a^4 + 1/2\,b^4 - 1/6\,a^6 - 1/2\,a^4 b^2 - 1/2\,a^2 b^4 - 1/6\,b^6 + 1/24\,a^8 + 1/6\,a^6 b^2 + 1/4\,a^4 b^4 + 1/6\,a^2 b^6 + 1/24\,b^8$.

*PSK = $1 - 0.5\,x^2 - x\,y - 0.5\,y^2 + .041667\,x^4 + .166668\,x^3 y + .250002\,x^2 y^2 + .166668\,x\,y^3 + .041667\,y^4$.

The subexpression factorization achieves an average area reduction by a factor of 43.8% among the 9 benchmarks, and the best is up to 80%. On the other hand, an average increase of 4.7% in Critical Path Delay (CPD) is reported. We contribute this increase to the fact that factorization will change a long polynomial into several shorter polynomials, and thus diminish the opportunity to perform tree height reduction. Note that the increase in CPD can be prevented by adjusting the cost function.

B. Subexpression Decomposition

In this experiment, estimations of the area and delay of library elements reported in [6] are used. The area and delay of all the elements are normalized by those of an adder. As shown in Table 4, the library elements used here are adder, square, multiplier, MAC, Sine and Cosine.

In Table 5, an average improvement of 24.0% in area and 2.4% in CPD are observed. This is because complex components are usually pre-optimized. For example, a square module is a pre-optimized multiplier in which the two inputs are the same.

Table 3. Synthesis Results for Subexpression Factorization

Benchmark polynomial	Without Factorization		With Factorization		Improvement (%)	
	Area	CPD	Area	CPD	Area	CPD
P1	4798	27.31	3401	28.80	29.1	−9.6
P2	13940	41.30	10893	29.93	21.9	−35.1
P3	7269	85.88	6800	55.80	6.5	1.0
P4	47833	106.22	14613	85.05	69.4	7.4
P5	33097	65.53	6839	98.36	79.3	2.2
P6	21831	41.40	12492	64.07	42.8	0.2
P7	15957	31.80	6425	41.32	59.7	−7.2
P8	6830	29.74	3267	34.10	52.2	4.7
P9	5156	28.68	3426	28.34	33.6	−5.6
Average					*43.8*	*−4.7*

Table 4 Normalized delay and area of library components

Library Element	Delay	Normalized Delay	Area	Normalized Area
Adder	7.54	1.00	15090	1.00
Square	7.89	1.05	89814	5.95
Multiplier	10.17	1.35	133401	8.84
MAC	17.28	2.29	142554	9.45
Sine	45.21	6.00	625218	41.43
Cosine	45.37	6.02	622849	41.28

Table 5. Results for Subexpression Decomposition

Benchmark polynomial	Without Complex Components		With Complex Components		Improvement (%)	
	Area	CPD	Area	CPD	Area	CPD
$cos(x) + x^5 + x^3 + x$	95.40	7.40	73.08	7.02	23.4	5.1
$sin(x) + x^3 + x$	91.40	6.05	57.83	7.00	36.7	−15.7
$2 cos(x) + 3 sin(x)$	140.60	7.40	101.39	8.37	27.9	−18.0
$sin(x) cos(x)$	201.48	8.40	91.55	7.37	54.6	12.3
$x^{10} + cos(x)$	84.56	7.40	68.97	7.02	18.4	5.1
Average					24.0	2.4

C. Comparison

The proposed algorithms are also compared with existing work. The results are reported in Table 6. The column "existing work" refers to [6], and the column "proposed technique" refers to an integration of the two transformation techniques proposed in this paper.

In Table 6, an average area improvement of 20.1% over the existing work is observed, with only 0.3% of CPD increase. This achievement is mainly due to subexpression factorization, because the proposed technique tries to find the best subexpression combination for factorization.

V. CONCLUSION AND FUTURE WORK

Two high level transformation techniques, subexpression factorization and subexpression decomposition, are proposed. There are two foundations for them. The first is polynomial formulation for the input data flow segment and the components from the available library. The second is symbolic computer algebra.

Table 6. Comparison

Benchmark Polynomials	Existing Work		Proposed Technique		Improvement (%)	
	Area	CPD	Area	CPD	Area	CPD
B1	10.84	2.35	10.84	2.35	0.0	0.0
B2	30.19	4.69	30.19	4.69	0.0	0.0
B3	41.24	5.58	35.30	5.63	14.4	−0.9
B4	96.61	9.41	51.09	9.97	47.1	−6.0
B5	42.28	7.02	25.85	6.63	38.9	5.6
Average					20.1	−0.3

Subexpression factorization tries to find the best subexpression combination for factorization to optimize predefined cost functions. A new graph, Term Dependency Graph is designed and heuristic algorithms are used to reduce the computational complexity. An average area improvement of 43.8% is reported on several experimental polynomials, at the expense of 4.7% increase in CPD.

The proposed subexpression decomposition technique uses subexpression decomposition instead of decomposing the whole polynomial to compromise the inherent defect of polynomial decomposition for hardware implementation. Experimental results show an average improvement of 24% in area and 2.4% in CPD.

Our future work includes the integration of word length and presumed scheduling and binding information into the estimation process to obtain more precise estimation results. The proposed algorithms will also be extended for multi-expression problems. Finally, as factorization can reduce the number of operations substantially [10], and there are many power-optimized complex components, the two proposed techniques will be modified and evaluated as transformation techniques for power optimization.

REFERENCES

[1] J. Smith and G. De Micheli, "Polynomial methods for component matching and verification", in Proc. Int. Conf. Computer-Aided Design, 1998.

[2] J. Smith and G. De Micheli, "Polynomial methods for allocating complex components", in Proc. Design Automation, Test Eur. Conf., 1999.

[3] J. Smith and G. De Micheli, "Polynomial Circuit Models for Component Matching in High-Level Synthesis", IEEE Trans. VLSI, vol. 9, Dec 2001.

[4] A. Peymandoust and G. De Micheli, "Using Symbolic Algebra in Algorithmic Level DSP Synthesis". Proceedings of the Design Automation Conference, pp. 277-282, 2001.

[5] A. Peymandoust and G. De Micheli, "Symbolic Algebra and Timing Driven Data-Flow Synthesis". ICCAD 2001, November, 2001.

[6] A. Peymandoust and G. De Micheli, "Application of Symbolic Computer Algebra in High-Level Data-Flow Synthesis". IEEE Trans. CAD, Vol. 22, Sept 2003.

[7] J. Mott, "Eisenstein-type irreducibility criteria". In Zero-dimensional commutative rings (Knoxville, TN, 1994), 307-329, Lecture Notes in Pure and Appl. Math., 171, Dekker, New York, 1995.

[8] ShuHong Gao, "Absolute Irreducibility of Polynomials via Newton Polytopes". J. of Algebra 237 (2001), 501-520.

[9] E. Kaltofen, "Polynomial Factorization: a Success Story". ISSAC'03, August 3-6, 2003.

[10] E. Macii, M. Pedram, and F. Somenzi, "High-Level power modeling, estimation, and optimization". IEEE Trans.CAD, Nov. 1998.

BDD Decomposition for Mixed CMOS/PTL Logic Circuit Synthesis

Yen-Tai Lai, Yung-Chuan Jiang, and Hong-Ming Chu
Department of Electrical Engineering,
National Cheng Kung University,
Tainan 70101, Taiwan
ytlai@casdc.ee.ncku.edu.tw

Abstract—Logic synthesis plays a major role in design automation. A logic function can be represented by a binary decision diagram (BDD). In this paper, we propose a technique to construct a BDD whose nodes can be implemented by CMOS logics and pass-transistor logics (PTL) in a cell library. The conventional synthesis flow needs three cell libraries: CMOS cell library, PTL cell library, and CMOS remapping pattern. To simplify the synthesis flow, we decompose the logic function to two kinds of functions and map them to PTL and CMOS cells, respectively. The cell library contains high speed cells and low power cells. The experimental results show that our approach has better performance and use less area than conventional CMOS technology mappings.

I. INTRODUCTION

A logic function can be represented by a binary decision diagrams (BDD). Gates in the library are then used to implement the nodes in the BDD. In the past, synthesis for BDD used PTL [1], [7]. This is because PTL is suitable for a multiplexer. The variable *s* associated with the node in the Figure 1(a) is the control signal in multiplexer in the Figure 1(b). Thus, a BDD to represent a logic function can be mapped to PTL directly.

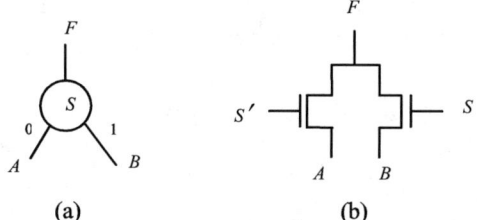

Figure 1. Implementing a BDD node with a PTL gate.

Pass transistor logic (PTL) has the capability to implement a logic function with a smaller number of transistors, smaller delay, and less power dissipation. Several researches on realizing circuits with PTL have been published during the last few years [1], [2]. It is shown that this technology is viable. Unfortunately, a function may require BDD representation of exponential size if it is mapped to PTL [7]. Even more, a monolithic BDD circuit may have long chains of transistors and we must put level-restoring logics at the PTL output gates. These level-restoring logics will slow down the PTL gates and increase power consumption as well.

Static CMOS logic style has long been used to realize a very large scale integration (VLSI) system because of ease of use and well-developed synthesis methods. We may use PTL and CMOS gates simultaneously. The conventional synthesis flow needs three cell libraries: CMOS cell library, PTL cell library, and CMOS remapping pattern. There are approaches to transform a Boolean equation into a BDD which can be implemented with AND/OR and XOR (XNOR, MUX) gates [3], [4]. They use static CMOS gates to realize the AND/OR gates and uses PTL gates to realize the XOR gates. This strategy is proved feasible by a fabricated chip shown in the reference [5].

The BDD transformation strategy must consider the types of cells being used because these cells will implement the nodes in the BDD. In other words, the transformation strategy of BDD and the types of cells in the library are interrelated to each other.

To simplify the flow, we decompose the logic function to two kinds of functions and map them to PTL and CMOS cells, respectively. Section II presents a new synthesis flow and describes BDD decomposition technique. With power being increasingly an important factor in high-performance VLSI designs, a great deal of effort has been made to explore low-power design options without sacrificing performance. Section III presents a cell library and new level restoring logic. Section IV presents experimental results. Finally, the conclusions are discussed in Section V.

II. BDD DECOMPOSITION

The transformation of BDD is a sequence of decomposition with respect to a variable. We must discuss the characteristic of a function related to a variable and then the strategy of BDD transformation. Let $f(x_1, x_2, ..., x_i, ..., x_n)$ be an Boolean function. It has Shannon expression:

$$f = x_i f_{xi} + x_i' f_{xi'} \text{ where } f_{xi} = f|_{xi=1} \text{ and } f_{xi'} = f|_{xi=0}.$$

We define:

- A function $f(x_1, x_2, ..., x_i, ..., x_n)$ is *positive*(*negative*) *unate* in x_i if $f_{xi} \geq f_{xi'} (f_{xi} \leq f_{xi'})$. Otherwise, it is *binate* (or *mixed*) in that variable.

- A function is *positive*(*negative*) *unate* if it is *positive*(*negative*) *unate* in all support variables. Otherwise, it is *binate* (or *mixed*).

If f is a binate function in x_i, x_i is called an *binate variable of function* f. Consider an example $f = x_1x_2 + x_2'x_3'$. Obviously, f is a

---

This work was supported by the National Science Council of Taiwan under Grant No. NSC 93-2215-E-006-018.

binate function in x_2. With respect to x_2, $f_{x2}=x_1$ and $f_{x2}'=x_3'$. Equivalently, $f = f_{x2}$ if $x_2 = 1$ and $f = f_{x2}'$ if $x_2 = 0$. We can consider x_2 as a selection signal; x_2 will select either the value of f_{x2} or the value of f_{x2}' as the value of f. In the BDD such a selector can be implemented with an inverter and a PTL gate. Hence, a PTL gate is good for the implementation of a binate function in a variable. Consider another example $g = x_1x_2$. It is seen that g is an unate function and CMOS gate is good for such an unate function.

Based on this observation, we first inspect a function to see if it is unate in all variables. If it is not, there is at least one binate variable and we decompose the given function as Shannon expression and replace the vertex corresponding to the given function with a selection vertex and two new vertices. The procedure for BDD decomposition is as follows:

Procedure BBD decomposition (f_0)
Push f_0 onto stack T ;
While (T is no empty) begin
  POP a function f from T
  Simplify function f.
  If (f is binate in some variables) begin
    select a binate variable x_i ;
    Use x_i as a cofactor and express function f as:
      $f = x_i f_{xi} + x_i' f_{xi'}$;
    Decompose the vertex corresponding to f and map it to a PTL gate;
    Push f_{xi} and $f_{xi'}$ onto stack T ;
  End of if
  Else // function f is unate
    map it to static CMOS;
End of while

In the above procedure, we need an algorithm to find binate variables. There can be more than one binate variable, we then need a strategy to select a binate variable such that the resultant BDD has the optimal depth. The selection procedure is shown as follows:

1) Calculate the number of appearance of every variable x_i in the function f.
2) Label the variables which contain both polarities (xi and xi') in product terms.
3) Select the variables that are labeled and have the largest number of appearances.
4) If more than one variable is selected, then for every selected variable x_i count the number of appearance of x_i in $f_{xi'}$ and f_{xi}, say l_i and r_i, respectively.
5) Select the variable which has the smallest $|l_i - r_i|$.

Consider an example:

$$F = a'bce' + a'cde' + bc'd'e + abc'de' + ac'd'e + a'cd'e.$$

It is seen that F is unate in b and binate in the other variables. We use the selection procedure to pick up a variable as shown in Figure 2. It is seen that both c and e have the largest number of appearance which is 6. Therefore, we have to calculate $|l_c - r_c|$ and $|l_e - r_e|$. Again, $|l_c - r_c|$ and $|l_e - r_e|$ are equal. We select variable c as decompose element. In Shannon expression, the function F is re-written as

$$F = c[a'be' + a'de' + a'd'e] + c'[bd'e + abde' + ad'e]$$
$$= cF_c + c'F_{c'}.$$

Recursively, we select binate variables to decompose F_c and $F_{c'}$. The final Boolean equation is written as:

$$F = c[e(a'd') + e'(a'b + a'd)] + c'[e(bd' + ad') + e'(abd)].$$

Let $f_1 = a'd'$, $f_2 = a'b + a'd$, $f_3 = bd' + ad'$, and $f_4 = abd$. Since f_1, f_2, f_3, and f_4 are unate functions, they are not decomposed further. They are mapped to static CMOS directly. The function F can be re-written as:

$$F = c[e(f_1) + e'(f_2)] + c'[e(f_3) + e'(f_4)].$$

It can be expressed with a BDD as shown in Figure 3.

```
*  a  5
   b  3
*  c  6   <3, 3>
*  d  5
*  e  6   <3, 3>
```

Figure 2. Example of selecting a binate variable.

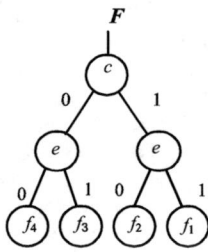

Figure 3. Example of BDD decomposition.

III. MIXED COMS/PTL CELL LIBRARY

Power consumption and circuit performance of PTL-based circuits vary. The well known LEAP library is composed of PTL cells [1]. The cells in the LEAP library are not suitable for implementing high speed combination logic circuit because they are designed for low power consumption system. Hence, we explore a new cell library which consists of CMOS and PTL cells. Level restoring circuits are required in PTL logics and discussed in this first. Performance oriented PTL are proposed in this section.

A. Level Restoring Ciruit

Both NMOS only and PMOS only PTL has a common problem, that is, the voltage decays seriously when passing through a long chain of transistors. NMOS can not pass good "1" and PMOS can not pass good "0". In order to solve this problem, level restorers are used to recovery the output voltage level [8]. Unfortunately, a level restorer consumes power and reduces speed of the PTL gate. Because NMOS only PTL is good for high speed circuit. Thus, we develop a new level restoring circuit which has high speed and uses less area. The new level restoring circuit is depicted in Figure 4.

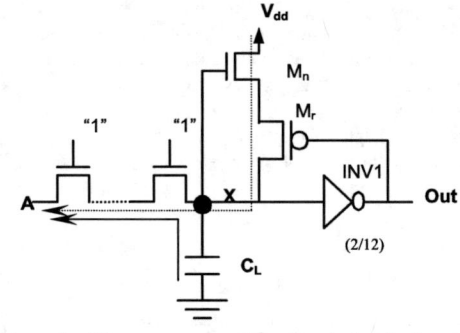

Figure 4. The new proposed level restoring logic gate.

In Figure 4, the level restorer is composed of M_n and M_r, where M_n is a minimum size transistor. When input signal A goes from V_{dd} to G_{ND}, a leakage current occurs. This leakage current is from drain of M_n to input A. When 0 arrives at node X (C_L discharge to ground), the transistor M_n is in cut-off region. Thus, the leakage current is cut off. Compare to the level restoring in [1], $(W/L)_p/(W/L)_n$ of $INV1$ in this level restoring logic reduce to 2/12 and has no additional inverter. When $A = 1$, the voltage of node X is in a level of power high. This is because we cascade a transistor M_n. Thus, we must enlarge NM_H of $INV1$ to make sure the voltage level in output is correct. This approach is not only maintains the function correctly but also reduce the area of $INV1$. Besides, this approach can reduce propagating delay because the input signal A need not pass $INV2$ in [1] to cut off the leakage current.

B. Proposed Cell library

In order to compare the effect of various level restoring circuits in PTL. We use two symbols to represent LEAP cells and this proposed cells respectively as shown in Figure 5(a) and 5(b).

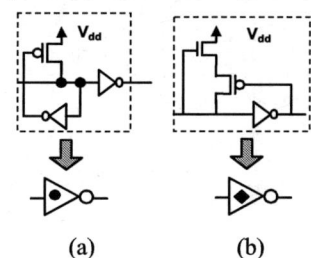

Figure 5. a) A LEAP level restoring circuit, and b) a new circuit.

A standard cell library has NAND, NOR, AOI, OAI, MUXI, and XOR cells. For each type of gate we compare various implementations with each other. The implementations are CMOS, LEAP cell, new cell, and transmission gate (TG). The LEAP cells are proposed in [1]. The new LEAP cells are similar to the LEAP cells except that the level restoring circuits are replaced new proposed circuits. All implementations except for CMOS are of PTL family. For every basic function, 15 cells are connected in a series. The propagating delay, power consumption, and transistor count are measured and listed in the Table I~VI.

TABLE I. COMPARISONS OF NAND GATES

	CMOS	LEAP	New cell	TG
Delay (ns)	3.6	3.7	3.5	4.7
Power (μW)	6.852	26.469	69.25	8.514
Transistor count	2N+2P	4N+3P	5N+3P	4N+4P

TABLE II. COMPARISONS OF NOR GATES

	CMOS	LEAP	New cell	TG
Delay (ns)	4.6	3.95	3.4	4.7
Power (μW)	7.202	28.34	68.82	8.339
Transistor count	2N+2P	4N+3P	5N+3P	4N+4P

TABLE III. COMPARISONS OF AOI (AND-OR-INVERTER) GATES

	CMOS	LEAP	New cell	TG
Delay (ns)	6	8.9	4.75	6.9
Power (μW)	10.34	53.93	90.83	10.86
Transistor count	5N+5P	8N+5P	8N+4P	7N+7P

TABLE IV. COMPARISON OF OAI (OR-AND-INVERTER) GATES

	CMOS	LEAP	New cell	TG
Delay (ns)	6	10.1	3.95	4.65
Power (μW)	10.34	57.36	87.48	7.419
Transistor count	5N+5P	8N+5P	8N+4P	7N+7P

TABLE V. COMPARISON OF MUXI GATES

	CMOS	LEAP	New cell	TG
Delay (ns)	6	4.15	3.65	4.8
Power (μW)	10.34	28.91	66.38	8.514
Transistor count	8N+8P	4N+3P	5N+3P	4N+4P

TABLE VI. COMPARISON OF XOR GATES

	CMOS	LEAP	New cell	TG
Delay (ns)	4.6	4.15	3.5	4.9
Power (μW)	9.947	31.029	71.02	11.02
Transistor count	8N+8P	5N+4P	6N+4P	5N+5P

From Table I to VI, it is seen that CMOS is suitable for NAND, NOR, AOI, and AOI gates. On the other hand, PTL is good for MUXI and XOR. In the PTL family, the speed of new cells is the fastest and the transmission gate has the lowest power consumption. Minimal area is our major objective. Thus, our synthesis strategy is as follows:

- NAND, NOR, AOI, OAI:

  Using static CMOS cells

- MUXI, XOR:

  Using new LEAP cells (for high speed)
  Using transmission gates (for low power)

Based on this strategy, the cell library for selection functions is shown in Table VII. Those cells replace the selection cells (MXI, MX, XOR,...etc.) in the standard CMOS cell library.

TABLE VII. PROPOSED CELL LIBRARY OF NEW LEAP CELLS

High speed (nY cells)		Low power (cY cells)	
nY1		cY1	
nY2		cY2	
nY3		cY3	

Consider the example in Figure 3. The leaf nodes are mapped to CMOS cells and the nodes in the other two levels are mapped to PTL gates. An inverter must be inserted at the top of the tree as shown in Figure 6. It is a level restoring inverter. The final synthesis result is shown in Figure 7(a). Compare it with the conventional CMOS mapping as shown in Figure 7(b). The mixed PTL/CMOS mapping has 48 transistors ($21P + 27N$) and the conventional CMOS mapping has 86 transistors ($43P + 43N$). This result proves that this approach can reduce overall area (cost). In this example, the mixed PTL/CMOS mapping uses about half area of the conventional CMOS mapping.

IV. EXPERIMENTAL RESULTS

A BDD decomposition program is written in C langrage. Several random logic functions from MCNC91 are used to verify our approach. First, decompose these functions with decomposition program and generate a binate function and some unate functions. Second, based on binate function to build BDD and the others (unate functions) are written by Verilog. Based on a 0.25 1P5M technology file, the leaf nodes of the BDD are synthesized by Synopsys design compiler and the other nodes are divided into nY cells and cY cells. Finally, the integrated circuits are simulated with HSPICE. We compare these results with conventional CMOS technology mapping which synthesis by Synopsys design compiler without decomposition. The results are shown in Table VIII. It is seen that our approach can improve area 38% and 34% with nY cells and cY cells respectively. Obviously, the total area depends on functional characteristic. The more multiplex functions, the more areas are saved. In the Table IX, the total speed improvements are almost the same between nY cell and cY cell mapping, about 14% speed gain with both nY cell and cY cell mapping.

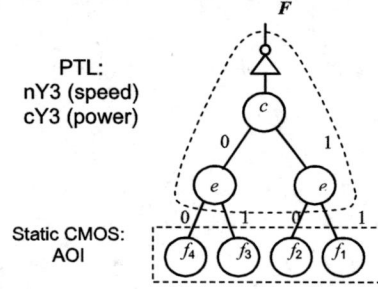

Figure 6. Mapping of a BDD in Figure 3.

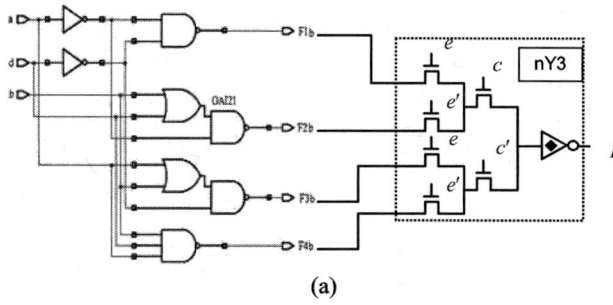

(a)

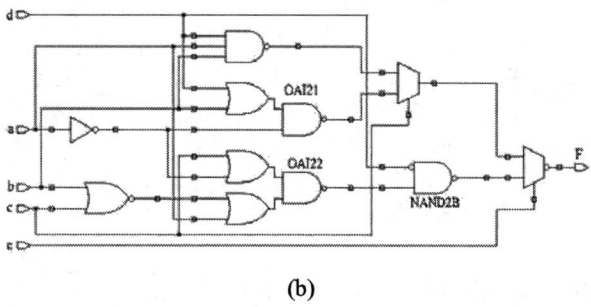

(b)

Figure 7. a) Mixed PTL/CMOS mapping. b) Conventional CMOS mapping.

V. CONCLUSIONS

In this paper, we presented a mixed CMOS/PTL synthesis method and compared the results of cell library using the proposed synthesis method with their conventional static CMOS cell library. A new cell library is proposed for multiplex function which can optimize speed or power consumption. Compared to the conventional CMOS cell library, the number of cells does not increase. In addition, a decomposition technique was proposed to obtain a BDD with minimal length of transistor chain. Also, this technique prevents the size of BDD grow exponentially. Experimental results show the average improvements in area are 38% and 34% in nY cell and cY cell mapping. The average improvements in speed are 13.9% and 13.8% in nY cell and cY cell mapping.

TABLE VIII. EXPERIMENT RESULTS OF CMOS AND OUR MIXED CMOS/PTL METHODS WITH FOCUS ON THE TOTAL AREAS

MCNC91 benchmarks	Transistor count				
	CMOS	(Mix.) nY	Improv	(Mix.) cY	Improv
cm152	122	32	73%	40	67%
parity	180	78	56%	86	52%
cm151	128	46	64%	54	57%
cm82	84	90	-7%	90	-7%
b1	28	30	-7%	30	-7%
mux	220	86	60%	106	51%
tcon	176	96	45%	96	45%
cm150	174	64	63%	84	51%
pcle	288	212	26%	212	26%
cmb	206	186	9%	186	9%
Average			38%		34%

TABLE IX. EXPERIMENT RESULTS OF CMOS AND OUR MIXED CMOS/PTL METHODS WITH FOCUS ON THE TOTAL DELAYS

MCNC91 benchmarks	Propagating delay (ns)				
	CMOS	(Mix.) nY	Improv	(Mix.) cY	Improv
cm152	1.02	0.64	37%	0.67	34%
parity	1.5	1.11	26%	1.15	23%
cm151	1.06	0.74	30%	0.77	27%
cm82	0.69	0.8	-17%	0.71	-3%
b1	0.23	0.27	-19%	0.24	-6%
mux	1.8	1.42	21%	1.53	15%
tcon	1.44	1.25	13%	1.29	10%
cm150	1.42	0.92	35%	1	29%
pcle	2.35	2.16	8%	2.2	6%
cmb	1.68	1.59	5%	1.62	3%
Average			13.9%		13.8%

REFERENCES

[1] K. Yano, Y. Sasaki, K. Rikino, and K. Seki, "Top-down pass-transistor logic design," *IEEE J. Solid-State Circuits*, vol. 31, June 1996, pp. 792–803.

[2] M. Tachibana, "Synthesize pass transistor logic gate by using free binary decision diagram," in *Proc. Int. ASIC Conf.*, Sept. 1997.

[3] C. Yang, V. Singhal, and M. Ciesielski, "BDD decomposition for efficient logic synthesis," in *Proc. Int. Conf. Computer Design*, Austin, TX, Oct. 1999.

[4] C. Yang and M. Ciesielski, "BDS: a BDD-based logic optimization system," *IEEE Trans, Computer-Aided Design*, vol. 21, July 2002, pp. 866-876.

[5] S. Yamashita, K. Yano, Y. Sasaki, Y. Akita, H. Chikata, K. Rikino and K. Seki "Pass-Transistor/CMOS Collaborated Logic: The Best of Both Worlds" in *Proc. Sym. VLSI Circuits*, june 1997, pp.31-32.

[6] G. D. Micheli, Synthesis and Optimization of Digital Circuits. New York: McGraw-Hill, 1994.

[7] P. Buch, A. Narayan, and A. Richard Newton, A. Sangiovanni-Vincentelli, "Logic Synthesis for Large Pass-transistor Circuits" in *Proc. IEEE/ACM Int. Conf. Computer-Aided Design*, San Jose, CA, Nov. 1997, pp. 663–670.

[8] Jan M. Rabaey, Anantha Chandrakasan, and Borivoje Nikolic, Digital Integrated Circuits. Prentice-Hall, 2002.

[9] R. Zimmermann and W. Fichtner, "Low-Power Logic Styles: CMOS versus Pass-Transistor Logic," *IEEE J. of Solid-State Circuits*. vol. 32, July 1997, pp. 1079-1090.

FPGA Technology Mapping Optimization by Rewiring Algorithms

Wai-Chung Tang, Wing-Hang Lo, Yu-Liang Wu
Dept. of Computer Science and Engineering
the Chinese University of Hong Kong
Shatin, Hong Kong
Email: {wctang, whlo, ylw}@cse.cuhk.edu.hk

Shih-Chieh Chang
Dept. of Computer Science and Information Engineering
National Chung Cheng University
Chia-Yi, Taiwan R.O.C.
Email: scchang@cs.ccu.edu.tw

Abstract— Rewiring algorithms provide a new style of logic transformations by replacing a target wire with its alternative wire while maintaining the functionality of the circuit. In this paper, these algorithms will be used to minimize the number of LUTs used to map a given circuit with Flowmap. The proposed approach is to evaluate each alternative wire with Flowmap and choose the first one which can reduce the number of LUTs by 1 or more. Despite its simplicity, it can efficiently transform the circuit to one suitable to be mapped with Flowmap and used in FPGA. Experimental result shows that the proposed approach can reduce up to 17% of the LUTs in a circuit without any depth increment.

Keywords: rewiring, FPGA, technology mapping.

I. Introduction

Field Programmable Gate Array (FPGA) has become an important approach in ASIC design and rapid prototyping in recent years due to its fast turn around time and lower cost. One popular architecture for the programmable logic blocks inside the FPGA is to use lookup-tables (LUTs). A K-input LUT (K-LUT) is a programmable logic device that can implement any Boolean function with K or less input variables. In order to implement a Boolean network on an FPGA, the gates in the Boolean network have to be packed into the K-LUTs of the FPGA. Technology mapping for LUT-based FPGAs is the problem of transforming a given Boolean network into a functionally equivalent LUT network.

A lot of efforts was spent on the study of the technology mapping problem for LUT-based FPGA over the last decade. Earlier research mainly used combinatorial optimization techniques like bin packing [1] and flow computation [2] [3] to cover the Boolean network with K-LUTs. One of the major breakthroughs in this area was the Flowmap algorithm proposed by Cong *et al.* [3]. The Flowmap algorithm gives a depth-optimal solution to the LUT mapping problem in polynomial time by using max-flow min-cut technique to partition the Boolean network into K-LUTs. However, these approaches only perform limited logic transformations on the Boolean network and the solution space is usually limited by the general structure of the network.

More recent techniques like SLDMap [4] try to enlarge the solution space by combining logic decomposition with technology mapping. This approach performs technology mapping over a set of different decompositions of the given network and chooses the best solution. The resulting area and delay are improved over the solution of Flowmap. It is obvious that there is still much room for improvement if different logic transformations is considered on the given network.

Rewiring [5] [6] [7] [8] [9] [10] [11] is a technique to replace a wire, called the target wire, with a different alternative wire in a circuit without changing the circuit functionality at the primary outputs. The idea is to add a redundant wire to the circuit so that the target wire becomes redundant and thus removable. It can be viewed as a form of functionally equivalent logic transformation on the circuit. Rewiring techniques have found applications in areas like logic synthesis, circuit partitioning [12] [13] and routing [14], in which wires with undesirable properties are replaced.

In this work, possibility of incrementally transforming the Boolean network by rewiring is investigated so that a better mapping solution can be arrived. This paper is organized as follows: Section II gives a review on the Flowmap algorithm and different rewiring algorithms, section III defines the problem focused by this paper, section IV describes clearly the approach applied while the experiment results are shown in section V.

II. Review of Flowmap and Rewiring Algorithms

A. The Flowmap Algorithm

In Flowmap, a Boolean network is represented as a direct acyclic graph. A cut $(X, \overline{X})$ is a partition of nodes such that source node s is in X and sink node t is in $\overline{X}$. Node cut size of a cut $(X, \overline{X})$ is the number of nodes in X that are adjacent to some nodes in $\overline{X}$. The cut is k-feasible if and only if its node cut size is less than k. Each node v has a label $l(v)$. The method of assigning labels will be discussed later. Height of a cut $h(X, \overline{X})$ is the maximum label in X.

The Flowmap algorithm consists of two phases: the labeling phase and the mapping phase.

The labeling phase computes the label for each node in topological order starting from the primary inputs of the network. Here we denote the sub-network formed by a node v and its predecessors as N_v. The label of a node v, $l(v)$ is defined as the height of the minimum height k-feasible cut in

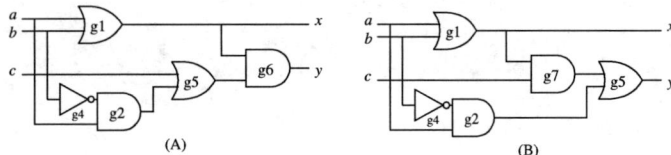

Fig. 1. (A)Original Circuit (B) Modified Circuit

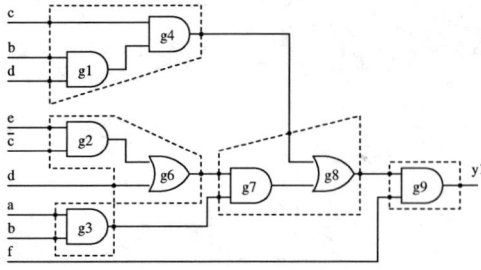

Fig. 2. Initial Mapping Solution using Flowmap

the sub-network N_v plus 1. That is

$$l(t) = \min_{(X,\overline{X})\ is\ k-feasible} h(X,\overline{X}) + 1 \quad (1)$$

Therefore, the computation of label $l(v)$ is equivalent to the search of a minimum height k-feasible cut in N_v. This can be solved in polynomial time using network flow method. [3] gives the details on how to compute the minimum height k-feasible cut in N_v. The minimum height k-feasible cut in N_v for each node v is saved for the computation in the next phase.

The mapping phase generates the optimal mapping solution from the information obtained in the labeling phase. Let $(X_v, \overline{X_v})$ be the cut computed for node v, since the cut is k-feasible, the number of inputs into nodes $\overline{X_v}$ is less than or equal to k, thus $\overline{X_v}$ can be implemented with one k-input LUT. The algorithm starts from primary outputs, implementing $\overline{X_v}$ of each node, and then moves to the inputs of the $\overline{X_v}$ nodes and implements the corresponding $\overline{X_v}$, until only primary inputs are left. [3] proves that the LUT network thus formed is depth-optimal with respect to the input network.

B. Rewiring Algorithms

Rewiring refers to the process of replacing a certain wire with another wire in a circuit without changing the functionality of the circuit.

Consider the example circuit shown in figure 1, we target at wire $g1 \rightarrow g6$. It is possible to maintain the circuit functionality by

1) Replacing $c \rightarrow g5$ with the output of an AND gate whose inputs are $g1$ and c respectively.
2) Removing $g1 \rightarrow g6$.

The modified circuit is shown in figure 1. It is easy to verify that the output function of the circuit does not change after the operations.

There are two major rewiring techniques, namely ATPG (Automatic Test Pattern Generation)-based and graph-based. RAMBO [5], REWIRE [6] [7] are ATPG-based while GBAW [8] [9] is graph-based. In this work, we mainly used REWIRE to locate target wire/alternative wire pairs which were then used to perform incremental transformation on the network.

REWIRE is based on redundancy identification by testing of stuck-at faults using logic implication [5]. REWIRE tries to add wires that force the implication on the stuck-at fault test of the target wire to become inconsistent, that is, after the addition of the wire, the target wire's stuck-at fault will become untestable and hence redundant. It then ascertains the added wire itself is also redundant so that its addition will not change the circuit functionality. Thus the added wire is able to replace the target wire in the circuit without altering the function of the circuit. REWIRE also screens out wires that cannot possibly become redundant after addition so that less redundancy identification operations are needed.

III. PROBLEM DEFINITION AND EXAMPLES

A Boolean network is assumed to be decomposed or synthesized into a 2-bounded equivalent network before applying the proposed optimization procedure since both Flowmap and rewiring algorithms work better with 2-bounded network.

Most of the notations used are the same as those in [3]. Additionally, given a Boolean network N, denote the number of total LUTs and the maximum depth of the mapping solution of N in Flowmap by $lut(N)$ and $depth(N)$ respectively.

This paper focus on reduction of $lut(N)$ without any trade-off in $depth(N)$ through transformation of the network using alternative wires found in rewiring algorithms. In other words, it is a problem to find a "good" set of alternative wires in which after the addition of these alternative wires, together with the removal of the corresponding target wires, the original network N will be transformed into N' with $lut(N') < lut(N)$ and $depth(N') \leq depth(N)$.

An example is shown to illustrate the problem. Consider the circuit C shown in figure 2, the mapping solution generated by Flowmap is shown in dashed box (here $k = 3$), in total $lut(C) = 5$ and $depth(C) = 3$. To reduce number of LUTs used, the circuit is transformed by

1) Removing $g6 \rightarrow g7$.
2) Adding $g6 \rightarrow g3$, thus $g7$ is removed and $g3$ becomes a 3-input AND gate.

The resultant circuit, denoted by C', is shown in figure 3. When Flowmap is used to map C', it gives a solution $lut(C') = 4$ and $depth(C') = 3$.

IV. REWIRING-BASED LUT MINIMIZATION

A LUT L is defined to be *fully-utilized* if $input(L) = k$. And a LUT is said to be *m-utilized* if $input(L) = m$, here the term utilization will be used in the meaning of $input(L)$.

It is note that when $k = 5$, the number of fully-utilized LUTs in the mapping solutions of Flowmap is only 50% of all LUTs, and on average around 20% of them are 2-utilized.

One advantage of rewiring is to add in some new wires to LUTs of utilization from 2 to $k - 1$. Obviously adding a wire to a LUT may increase the depth of circuit or the same

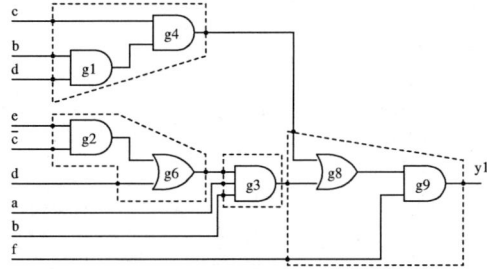

Fig. 3. Final Mapping Solution after Modification

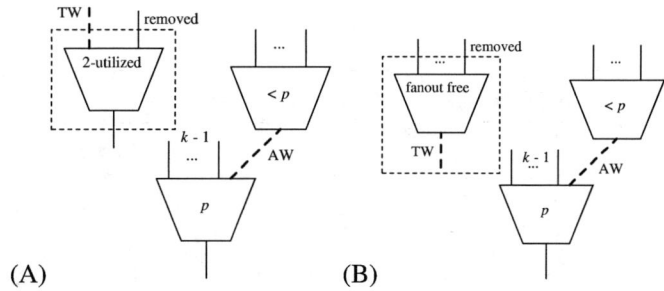

Fig. 4. Reduction of LUT by rewiring

number of LUTs will be used after the addition. The following lemma tries to identify alternative wires whose addition to the network will not alter the depth but increase the utilization of the LUT.

Lemma 1: Given a target wire w_t, an alternative wire $w_a = (s_a, d_a)$ and a LUT L covering d_a, if $l(s_a) < l(L)$, then with addition of w_a, $input(L)$ will
1) increase by 1, if w_t is not in L, or
2) remain the same, if w_t is in L,
and $l(L)$ remain unchanged.

Proof: Let $(X_L, \overline{X_L})$ be the minimum height height $k-feasible$ cut of L, and any node $u \in X_L$ has $l(u) < l(L)$.

Consider the case that w_t is not in L. Clearly, $d_a \in \overline{X_L}$, then after addition of w_a, there is an augmenting path from s_a to d_a and the maximum flow will be increased by 1. Flowmap algorithm will search for the maximum flow and finally includes s_a in X_L, increase $input(L)$ by 1.

When w_t is in L, when it is removed, exactly one path will be removed as well. Yet another augmenting path will be created with addition of w_a, as a whole the maximum flow will keep unchanged, which means $input(L)$ will be the same after the transformation.

In both case, since $l(s_a) < l(L)$, $p \neq l(s_a)$ and thus the label for L will remain unchanged. ∎

In order to reduce the number of LUT, the target wire has to be chosen carefully. Two theorems will be given to show that alternative wire is useful in reducing LUTs.

The first idea is to take away one input wire to a 2-utilized LUT so that immediately such LUT can be removed, as shown in figure 4(A). Another idea is to take the output of a LUT to be the target wire, so that when its alternative wire can be inserted into a existing LUT, one single LUT is saved. This idea is illustrated in 4(B).

A. First-Choice LUT Optimization Approach

In order to investigate the usefulness of alternative wires, a simple but productive approach is proposed. It is to accept an alternative wire when it can reduce $lut(N)$ by at least 1. Using this approach, not all alternative wires for the network have to be found. Instead, the network will be transformed once an alternative wire is found to be useful. And this can reduce the total runtime of the optimization process.

Given a Boolean network, every wire in the network is considered as target wire and its corresponding alternative

Algorithm *FIRST-CHOICE-LUT-MIN*
Input: Boolean network N
Output: LUT minimization network N'
1. **for** each wire $w \in N$
2.   **do** finding a set of alternate wires A of N
3.   **for** each AW a in A
4.   **do** transform N according to a
5. $N_f \leftarrow flowmap(N)$
6.   **if** $lut(N_f) < lut(N)$
7.   **then** $N \leftarrow N_f$
8.   goto step 1
9. (∗ no more possible reduction ∗)
10. **return** N

Fig. 5. First-choice LUT Minimization Algorithm

wires will be computed by rewiring algorithms. And every alternative wire is then added to the network with the removal of the target one. The modified network will be mapped with Flowmap once again to check if less number of LUTs can be used without any increase in depth. If so, this target and alternative wire pair will be accepted immediately and the procedure repeats from its beginning. The algorithm is summarized in figure 5.

V. EXPERIMENTAL RESULT

The first-choice approach algorithm described in IV-A is implemented using C language on Linux platform with routines from UC Berkeley SIS library. In all experiments, k is set to 5. The implementation is tested with several MCNC benchmark circuits. All the following experimental results are obtained from a PC running Redhat Linux 9.0 with 2.4 GHz CPU and 1024 MB RAM.

REWIRE is used to locate alternative wires. The reduction over initial mapping solution generated by Flowmap will be shown and the inputs to different approaches are identical.

Table I shows the number of LUTs of different benchmark circuits when no LUT optimization applied (solely Flowmap) and first-choice LUT optimization applied. On average the total number of LUTs can be reduced by 17.6%, this shows alternative wires are effective in reducing number of LUTs and this proves the motivation to apply rewiring technique in technology mapping for FPGA.

The result is compared with the result generated by Cutmap[15] and FlowSyn[16]. Table I also shows the performances of different methods. As seen, the performance

TABLE I
COMPARISON BETWEEN DIFFERENT TECHNOLOGY MAPPING OPTIMIZATION ALGORITHMS

	Initial		FCLM with REWIRE[1]				FlowSYN			CutMap		
Circuit	Depth	# LUT	Depth	# LUT	% Red	Runtime	Depth	# LUT	% Red	Depth	# LUT	% Red
5xp1	3	30	3	25	16.6%	159.95 s	2	19	36.7%	3	26	13.3%
pcler8	3	47	3	35	25.5%	9.23 s	3	36	23.4%	3	33	29.8%
ttt2	3	61	3	49	19.7%	378.75 s	3	47	23.0%	3	44	6.4%
f51m	3	28	3	22	21.4%	172.10 s	3	17	39.3%	3	27	3.6%
term1	4	77	4	58	24.7%	1145.09 s	4	64	16.9%	4	61	20.8%
C499	4	78	4	74	5.1%	763.24 s	4	66	15.4%	4	66	15.4%
C880	8	148	8	133	10.1%	2917.22 s	8	127	14.2%	8	126	14.9%
alu2	9	173	9	143	17.3%	8940.36 s	9	152	12.1%	9	145	16.2%
Average					17.6%				22.6%			15.1%
Total		642		539	16.0%			528	17.8%		528	17.8%

of first-choice LUT minimization (FCLM) is comparable to both FlowSYN and CutMap, and in some circuits like 5xp1, ttt2, it performs better than CutMap. Its reduction is the best among the others in alu2. Since the focus of this paper is on LUT minimization, current FCLM is not targeted on depth reduction. In the future more work will be done to apply rewiring for depth optimizaton as well and it should be a good area for investigation.

Besides, the runtime of FCLM can be further improved by using faster rewiring algorithms instead of REWIRE applied in this paper. For example, RAMFIRE can be used as rewiring engines and speed up the implementation. Theoretically, the speed up can be up from 20 to 30 times of the current result. However, due to the technical problems, no such tool can be put to experiments in this paper.

VI. CONCLUSION

A new alternative method to reduce the number of LUTs used to map a circuit is proposed in this paper, as a new application of rewiring algorithms. With the idea to increase the utilization of the LUTs, rewiring algorithms find their places to suggest useful alternative wires which can reduce the number of LUTs by modifying the circuit structure. The idea is based on a trial-and-evaluate method with Flowmap to find out alternative wires which can reduce the number of LUT without having penalty in depth. With incremental transformations on the network, experimentally, it is possible to reduce up to 17% of LUTs of a circuit on average.

Experimental result also shows that the power of the proposed approach is comparable with FlowSYN and CutMap. It is hoped that, as part of the future work, one can explore the possibilities of using rewiring algorithms to further improve the results from the state-of-the-art technology mapping algorithms like FlowSYN or CutMap and allow rewiring to be used complementarily with the tools available to optimize the mapping solution.

ACKNOWLEDGEMENT

All the experiments were carried out successfully with the support of the RASP technology mapping package, thanks to the provision of the tools by UCLA CAD group.

[1]Experiments carried out with locally-developed REWIRE implementation

REFERENCES

[1] R. J. Francis, J. Rose, and K. Chung, "Chortle: A technology mapping program for lookup table-based field programmable gate arrays," in *27th ACM/IEEE Design Automation Conference*, 1990, pp. 613–619.

[2] K. C. Chen, J. Cong, Y. Ding, A. B. Kahng, and P. Trajmar, "Dag-map: Graph-based fpga technology mapping for delay optimization," in *IEEE Design and Test of Computers*, Sept. 1992, pp. 7–20.

[3] J. Cong and Y. Ding, "Flowmap: An optimal technology mapping algorithm for delay optimization in lookup-table based fpga designs," *IEEE Trans. Computer-Aided Design*, vol. 13, pp. 1–12, June 1994.

[4] G. Chen and J. Cong, "Simultaneous logic decomposition with technology mapping in fpga designs," in *FPGA*, 2001, pp. 48–55.

[5] L. A. Entrena and K. T. Cheng, "Combinational and sequential logic optimization by redundancy addition and removal," *IEEE Trans. Computer-Aided Design*, vol. 14, no. 7, pp. 909–916, July 1995.

[6] S. C. Chang, L. P. P. P. van Ginneken, and M. Marek-Sadowska, "Fast boolean optimization by rewiring," in *Proc. Int'l Conf. Computer-Aided Design*, Nov. 1996, pp. 262–269.

[7] ——, "Circuit optimization by rewiring," *IEEE Trans. Comput.*, vol. 48, no. 9, pp. 962–969, Sept. 1999.

[8] Y. L. Wu, W. Long, and H. Fan, "A fast graph-based alternative wiring scheme for boolean networks," in *International VLSI Design Conference*, 2000, pp. 268–273.

[9] Y. L. Wu, C. N. Sze, C. C. Cheung, and H. B. Fan, "On improved graph-based alternative wiring scheme for multi-level logic optimization," in *Proc. IEEE International Conference on Electronics, Circuits and Systems*, Lebanon, Dec. 2000, pp. 654–657.

[10] C. W. Chang and M. Marek-Sadowska, "Single-pass redundancy addition and removal," in *ICCAD*, 2001, pp. 606–609.

[11] ——, "Who are the alternative wires in your neighborhood (alternative wires identification without search)," in *Great Lakes Symposium on VLSI*, 2001, pp. 103–108.

[12] Y. L. Wu, C. C. Cheung, D. I. Cheng, and H. B. Fan, "Further improve circuit partitioning using gbaw logic perturbation techniques," *IEEE Trans. VLSI Syst.*, vol. 11, no. 3, pp. 451–460, June 2003.

[13] D. I. Cheng, C. C. Lin, and M. Marek-Sadowska, "Circuit partitioning with logic perturbation," in *Proc. Int. Conference on Computer Aided Design*, Nov. 1995, pp. 650–655.

[14] S. C. Chang, K. T. Cheng, N. S. Woo, and M. Marek-Sadowska, "Post-layout logic restructuring using alternative wires," *IEEE Trans. Computer-Aided Design*, vol. 6, pp. 587–596, June 1997.

[15] J. Cong and Y. Hwang, "Simultaneous depth and area minimization in lut-based fpga mapping," in *Proc. ACM/SIGDA International Symposium on FPGAs*, 1995.

[16] J. Cong and Y. Ding, "Beyond the combinatorial limit in depth minimization," in *Proc. IEEE Int'l Conf. on Computer-Aided Design*, 1993, pp. 110–114.

A PHYSICALLY-DERIVED LARGE-SIGNAL NONQUASI-STATIC MOSFET MODEL FOR COMPUTER AIDED DEVICE AND CIRCUIT SIMULATION PART-II THE CMOS NOR GATE AND THE CMOS NAND GATE

Michael Walter Payton and Fat Duen Ho*

Department of Electrical and Computer Engineering
The University of Alabama in Huntsville
Huntsville, Alabama, USA 35899

ABSTRACT.

The primary goal of this work is to develop a low-level physics-based nonquasi-static MOSFET model that can be extended to the simulation of high-level CMOS logic circuits. In this part of our papers (Part II), the results of using our model described in the companion paper [1] to simulate the CMOS NOR gate and NAND gate are presented. The numerical methods discussed in [1] are applied in the simulations for the NOR gate and the NAND gate. In addition, a bisection root finding algorithm is used to calculate any junction voltage that appears between two devices connected in series. The results compared well with those obtained from the SPICE Level 3 and SPICE Level 7 (BSIM 3.1) for a wide range of device geometries and circuit loading conditions. The results show that our model is capable of accurately simulating the transient response of devices with channel lengths as small as 0.33 μm and for switching frequencies approaching 1GHz.

1. Introduction.

The primary goal of this work is to develop a physics-based nonquasi-static MOSFET model that can be extended to the simulation of CMOS logic circuits with a CAD software application. Many sources were found that presented some form of nonquasi-static MOSFET model [2],[3],[4]. However, other than the theses of Phillips [5] and Darty [6], none applied the low-level device model to the high-level simulation of circuits containing multiple transistors.

Two separate computer aided design programs were developed to execute device and CMOS logic gate simulations using the nonquasi-state MOSFET model presented in Payton's thesis [7]. The first program is designed to allow for transient or dc simulation of an individual MOSFET device. This program provides detailed terminal current, surface potential, and channel potential calculations for each simulation time step. In addition, this program returns information detailing how the device is affected by charge sharing, velocity saturation, and channel length modulation during each time step of a simulation. The second computer aided design program was developed to execute simulations of CMOS inverters, CMOS NOR gates, and CMOS NAND gates. This program calculates the terminal currents, terminal voltages, and output capacitance of a logic gate at each simulation time step. The program allows the user to input the physical parameters for both the NMOS and PMOS processes that define a CMOS logic gate.

Like the CMOS inverter in Part 1 of our papers [1] the NOR gate and the NAND gate are simulated using the CAD programs developed in our work. The results predicted by our model compare favorably with those obtained from the industry standard SPICE Level 3 and SPICE Level 7 (BSIM 3.1) models for various load capacitances and device scaling factors.

2. The Simulations Of The CMOS NOR Gate

The schematic diagram of a CMOS NOR gate is shown below in Fig. 1. From Fig. 1 it is obvious that the load current of the CMOS NOR gate is defined by the following relationship:

$$i_L(t) = -i_{Dn1}(t) - i_{Dn2}(t) - i_{Dp2}(t) \qquad (1)$$

The output voltage is related to the load current by the following differential equation:

$$i_L(t) = C_L(t)\frac{dV_{OUT}(t)}{dt} \qquad (2)$$

Equation 2 can be used to relate the NOR gate's output voltage to the load current defined above in equation 1. As was the case for the CMOS inverter, the device capacitances of the NOR gate's NMOS and PMOS transistors are implicitly lumped into the load capacitance

$$\begin{aligned}C_L = &\, C_{gdn1} + C_{gdn2} + C_{gdp2} + C_{bdn1} + C_{bdn2} + C_{bdp2} + \\ &\, C_{gdOn1} + C_{gdOn2} + C_{gdOp2} + C_{jdn1} + C_{jdn2} + C_{jdp2} + \\ &\, C_{jswdn1} + C_{jswdn2} + C_{jswdp2} + FO(C_{gn} + C_{gp}).\end{aligned}$$
$$(3)$$

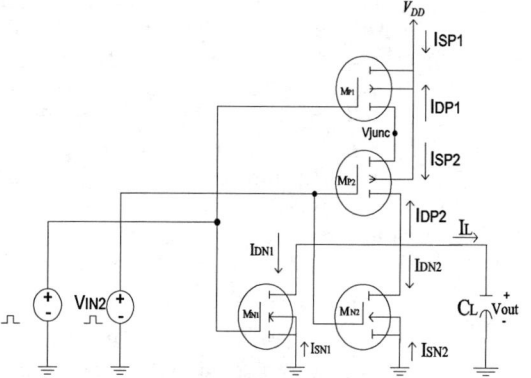

Figure 1: CMOS NOR Gate Circuit

*Author to whom correspondence should be addressed.

The same methodology used to simulate the transient behavior of the CMOS inverter is applied to the transient analysis of the CMOS NOR gate. At each time step of a NOR gate transient simulation, the nonquasi-static modeling process depicted in [7] is used to calculate the terminal currents and device capacitances of each NMOS and PMOS transistor within the NOR gate circuit. Once the MOSFET terminal currents and device capacitances have been calculated for a simulation time step, equation 1 can be used to calculate the NOR gate's load current and equation 3 can be used to calculate the gate's load capacitance for the time step. Equation 2 can then be solved using Euler's method (just as it was for the inverter) to predict the output voltage for the NOR gate for the next simulation time step.

The relative difficulty in modeling the NOR gate as compared to the inverter is the presence of the two PMOS transistors in series. The junction voltage V_{junc} between the drain terminal of M_{p1} and the source terminal of M_{p2} must be determined at each time step in order to calculate the terminal currents of the two PMOS transistors. From Figure 1 it is obvious that $i_{Dp1}(t)$ and $i_{Sp2}(t)$ are related by the following function:

$$f = i_{Dp1}(t) + i_{Sp2}(t) = 0 \qquad (4)$$

Furthermore, it is also obvious from Figure 1 that the value of the junction voltage should fall within the following interval:

$$(V_{OUT} - 0.2V_{DD}) \leq V_{junc} \leq (V_{DD} + 0.2V_{DD}) \qquad (5)$$

where the $0.2V_{DD}$ terms allow the junction voltage to exceed the conventional bounds of V_{DD} and V_{OUT} due to capacitive coupling. Thus, the junction voltage is determined by applying a bisection root finding technique [8] to the function defined in equation 4. The bisection method involves the evaluation of the function f (defined in equation 4) at the interval endpoints (defined in equation 5) and at the interval midpoint. Once it is determined on which half of the interval the function f evaluates to zero (indicated by a sign change in f), that half of the interval then becomes the new interval. This process continues until f is evaluated to be equal to zero within a specified tolerance. For each evaluation of f, the nonquasi-static modeling process depicted in [7] is executed to calculate the terminal currents of both M_{p1} and M_{p2}. The process of simulating the transient response of a CMOS NOR gate is presented in [7].

As was the case for the CMOS inverter, the device parameters which were used in [1] are used here for both the CMOS NOR gate and the NAND gate. In addition, as was the case for the CMOS inverter, the switching frequency of the baseline CMOS NOR gate (scale factor 1) is evaluated as the fan-out is increased from 1 to 40. For each simulation, a 5 V pulse is applied to V_{IN1} while $V_{IN2}=0$ V. Table 1 provides a comparison of the fall time, rise time, and maximum switching frequency calculated by the NQS model to the corresponding values calculated by the SPICE Level 3 model for each fan-out value. Table 2 provides a comparison between the NQS results and the SPICE Level 7 results for each fan-out value.

Table 1: CMOS NOR Gate Switching Speed for Various Fan-out Values NQS vs. SPICE Level 3

Fan-out	NQS f_{MAX} (MHz)	Level 3 f_{MAX} (MHz)	% Error NQS vs. Level 3
1	133.690	141.804	5.72
2	86.655	90.310	4.05
4	50.994	52.140	2.20
6	36.127	36.678	1.50
8	27.972	28.376	1.42
10	22.826	23.095	1.16
20	11.895	11.936	0.34
40	6.074	6.079	0.08

Table 2: CMOS NOR Gate Switching Speed forVarious Fan-out ValuesNQS vs. SPICE Level 7 (BSIM3.1)

Fan-out	NQS f_{MAX} (MHz)	Level 7 f_{MAX} (MHz)	% Error NQS vs. Level 7
1	133.690	138.236	3.29
2	86.655	85.353	-1.53
4	50.994	48.326	-5.52
6	36.127	34.083	-6.00
8	27.972	26.108	-7.14
10	22.826	21.129	-8.03
20	11.895	10.933	-8.80
40	6.074	5.540	-9.64

The fan-out results produced for the CMOS NOR gate are very similar to those produced for the CMOS inverter. The NOR gate results calculated by the NQS model compare very well with the SPICE Level 3 results. As was the case for the CMOS inverter, the NOR gate switching speeds predicted by the NQS model grow closer in value to those predicted by SPICE Level 3 as the fan-out increases. While the overall agreement between the NQS model and SPICE Level 7 is reasonable for the NOR gate, the switching frequencies calculated by the two models begin to diverge slightly at higher fan-out values. The divergence between the NQS model and SPICE Level 7 at higher fan-out values is consistent with the behavior documented previously for the CMOS inverter [1].

Just like the CMOS inverter, the device dimensions of the CMOS NOR gate are scaled by factors ranging from 1 to 15 while the fan-out is held constant at 20. A 5 V pulse is applied to V_{IN1} while $V_{IN2} = 0$ V for each simulation. The fall time, rise time, and maximum switching frequency of the NQS NOR gate are calculated for each scale factor and compared to the corresponding values calculated by SPICE Level 3 and Level 7. Table 3 below compares the results obtained from the NQS model to those obtained from SPICE Level 3. Table 4 provides a comparison of the NQS results to the SPICE Level 7 results.

Table 3: CMOS NOR Gate Switching Speed for Various Scaling Factors NQS vs. SPICE Level 3

Scale Factor	NQS f_{MAX} (MHz)	Level 3 f_{MAX} (MHz)	% Error NQS vs. Level 3
1	11.895	11.936	0.34
2	41.667	41.010	-1.60
3	76.805	75.775	-1.36
4	116.550	113.804	-2.41
5	158.730	152.975	-3.76
6	221.877	205.846	-7.79
7	273.373	246.488	-10.91
10	453.926	403.388	-12.53
15	794.281	668.896	-18.75

The scaling results produced for the CMOS NOR gate are very consistent with those obtained for the CMOS inverter. The results of Table 3 show that the NQS model again compares well with SPICE Level 3 for channel lengths as small as 0.71 um (scale factor 7). Table 4 shows that the NQS model compares favorably with the SPICE Level 7 model for channel lengths as small as 0.33 um (scale factor 15).

Table 4: CMOS NOR Gate Switching Speed for Various Scaling Factors NQS vs. SPICE Level 7 (BSIM3.1)

Scale Factor	NQS f_{MAX} (MHz)	Level 7 f_{MAX} (MHz)	% Error NQS vs. Level 7
1	11.895	10.933	-8.80
2	41.667	37.043	-12.48
3	76.805	72.606	-5.78
4	116.550	115.875	-0.58
5	158.730	160.333	1.00
6	221.877	217.155	-2.17
7	273.373	280.348	2.49
10	453.926	458.295	0.95
15	794.281	884.956	10.25

The Simulation Of The CMOS NAND Gate

The schematic diagram of a CMOS NAND gate circuit is shown in Figure 2. From Figure 2, it is obvious that the load current of the CMOS NAND gate is defined by the following relationship:

$$i_L(t) = -i_{Dn2}(t) - i_{Dp1}(t) - i_{Dp2}(t) \quad (6)$$

Equation 2 can again be used as it was for the NOR gate to relate the NAND gate's output voltage to the load current defined above in equation 6. The device capacitances of the NAND gate's NMOS and PMOS transistors are lumped into the load capacitance as follows:

$$\begin{aligned}C_L = &\ C_{gdn2} + C_{gdp1} + C_{gdp2} + C_{bdn2} + C_{bdp1} + C_{bdp2} + \\ &\ C_{gdOn2} + C_{gdOp1} + C_{gdOp2} + C_{jdn2} + C_{jdp1} + C_{jdp2} + \\ &\ C_{jswdn2} + C_{jswdp1} + C_{jswdp2} + FO(C_{gn} + C_{gp}).\end{aligned} \quad (7)$$

The same process used to calculate the transient behavior of the CMOS inverter and NOR gate is once again employed to predict the transient behavior of the CMOS NAND gate. At each time step of a NAND gate

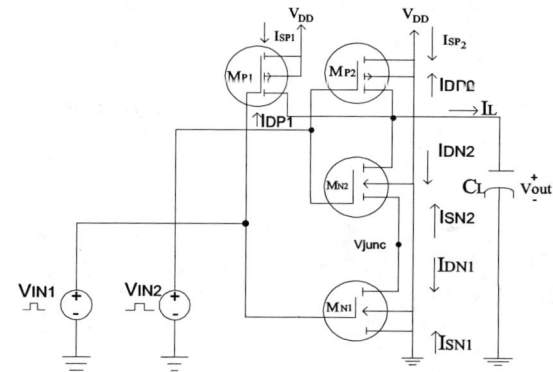

Figure 2: CMOS NAND Gate Circuit

transient simulation, the nonquasi-static modeling process depicted in Payton's thesis [7] is used to calculate the terminal currents and device capacitances of each NMOS and PMOS transistor within the NAND gate circuit. Once the MOSFET terminal currents and device capacitances have been calculated for a simulation time step, equation 6 can be used to calculate the NAND gate's load current and equation 7 can be used to calculate the gate's load capacitance for the time step. Equation 2 is then solved using Euler's method to calculate the NAND gate's output voltage for the next simulation time step.

As was the case for the CMOS NOR gate, a junction voltage V_{junc} located between two series connected transistors must be calculated at each time step of the NAND gate simulation. For the NAND gate, the junction voltage appears between the drain terminal of M_{n1} and the source terminal of M_{n2}. The junction voltage must be determined at each simulation time step before the terminal currents of the two NMOS transistors can be calculated. From Figure 2, it is obvious that $i_{Dn1}(t)$ and $i_{Sn2}(t)$ are related by the following function:

$$f = i_{Dn1}(t) + i_{Sn2}(t) = 0 \quad (8)$$

Furthermore, it is also obvious from Figure 2 that the value of the junction voltage should fall within the following interval:

$$(0 - 0.2V_{DD}) \leq V_{junc} \leq (V_{OUT} + 0.2V_{DD}) \quad (9)$$

where the $0.2V_{DD}$ terms allow the junction voltage to exceed the conventional bounds of 0 V and V_{OUT} due to capacitive coupling. The NAND gate's junction voltage is determined by using the same bisection root finding technique that was used to calculate the NOR gate's junction voltage. For the NAND gate, the bisection root finding technique is applied to the function defined in equation 8. The bisection method evaluates the function f at the interval endpoints defined in equation 9 and at the interval midpoint. Once it is determined on which half of the interval the function f evaluates to zero, that half of the interval then becomes the new interval. This process continues until f is evaluated to be equal to zero within a specified tolerance. The simulation flow chart which illustrates the process of simulating the transient response of the CMOS NAND gate is shown in Payton's thesis [7].

As was done previously for the CMOS inverter and NOR gate, the switching frequency of the baseline CMOS NAND gate (scale factor 1) is evaluated as the fan-out is increased from 1 to 40. For each simulation, a 5 V pulse is applied to V_{IN1} while V_{IN2}=5 V. Table 5 provides a comparison of the fall time, rise time, and maximum switching frequency calculated by the NQS model to the corresponding values calculated by the SPICE Level 3 model for each fan-out value. Table 6 provides a comparison between the NQS results and the SPICE Level 7 results for each fan-out value.

Table 5: CMOS NAND Switching Speed for Various Fan-out Values NQS vs. SPICE Level 3

Fan-out	NQS f_{MAX} (MHz)	Level 3 f_{MAX} (MHz)	% Error NQS vs. Level 3
1	156.986	157.480	0.31
2	95.969	95.557	-0.43
4	53.850	52.549	-2.48
6	37.439	36.918	-1.41
8	28.686	28.457	-0.80
10	23.256	23.051	-0.89
20	11.945	11.841	-0.88
40	6.054	6.021	-0.55

Table 6: CMOS NAND Switching Speed for Various Fan-out Values NQS vs. SPICE Level 7 (BSIM3.1)

Fan-out	NQS f_{MAX} (MHz)	Level 7 f_{MAX} (MHz)	% Error NQS vs. Level 7
1	156.986	149.209	-5.21
2	95.969	89.839	-6.82
4	53.850	49.324	-9.18
6	37.439	34.382	-8.89
8	28.686	26.348	-8.87
10	23.256	21.345	-8.95
20	11.945	10.878	-9.81
40	6.054	5.500	-10.07

The fan-out results produced for the CMOS NAND gate serve to reconfirm the fan-out results obtained previously for the CMOS inverter and NOR gate. The NAND gate results calculated by the NQS model compare very well with the SPICE Level 3 results. As was the case for the inverter and NOR gate, the NAND gate switching speeds predicted by the NQS model grow closer in value to those predicted by SPICE Level 3 as the fan-out increases. The overall agreement between the NQS model and SPICE Level 7 model is also reasonable for the NAND gate. Like the inverter and NOR gate before, the NAND gate switching speeds calculated by the NQS model diverge slightly from the switching speeds calculated by SPICE Level 7 at higher fan-out values.

Just like the CMOS inverter and NOR gate, the device dimensions of the CMOS NAND gate are scaled by factors ranging from 1 to 15 while the fan-out is held constant at 20. For each simulation, V_{IN1}=5 V while a 5 V pulse is applied to V_{IN2}. The fall time, rise time, and maximum switching frequency of the NQS NAND gate are calculated for each scale factor and compared to the corresponding values calculated by SPICE Level 3 and Level 7. Table 7 below compares the results obtained from the NQS model to those obtained from SPICE Level 3. Table 8 provides a comparison of the NQS results to the SPICE Level 7 results.

Table 7: CMOS NAND Switching Speed for Various Scaling Factors NQS vs. SPICE Level 3

Scale Factor	NQS f_{MAX} (MHz)	Level 3 f_{MAX} (MHz)	% Error NQS vs. Level 3
1	11.999	11.971	-0.23
2	41.820	40.627	-2.95
3	77.101	73.632	-4.71
4	116.686	113.366	-2.93
5	158.479	152.022	-4.25
6	222.124	206.526	-7.55
7	266.241	248.324	-7.22
10	453.104	383.142	-18.26
15	766.284	657.895	-16.48

Table 8: CMOS NAND Switching Speed for Various Scaling Factors NQS vs. SPICE Level 7 (BSIM 3.1)

Scale Factor	NQS f_{MAX} (MHz)	Level 7 f_{MAX} (MHz)	% Error NQS vs. Level 7
1	11.999	10.875	-10.34
2	41.820	36.858	-13.47
3	77.101	69.779	-10.49
4	116.686	113.366	-2.93
5	158.479	156.470	-1.28
6	222.124	208.812	-6.38
7	266.241	267.594	0.51
10	453.104	448.029	-1.13
15	766.284	829.187	7.59

The scaling results produced for the CMOS NAND gate lead to the same conclusions as the results obtained from the CMOS inverter and NOR gate. The results of Table 7 show that the NQS model again compares well with SPICE Level 3 for channel lengths as small as 0.71 µm (scale factor 7). Table 8 shows that the NQS model compares favorably with the SPICE Level 7 model for channel lengths as small as 0.33 µm (scale factor 15).

4. Conclusion

The results presented in this paper demonstrate that the nonquasi-state MOSFET model developed in this work can be effectively used to extend detailed physics-based transistor modeling to the simulation and design of CMOS logic gate circuits, such as the NAND gate and the NOR gate. Gate switching speeds predicted by the nonquasi-static model compare well with those obtained from the industry standard SPICE Level 3 and SPICE Level 7 (BSIM 3.1) models for various load capacitances and device scaling factors. The results of our simulations show that the model developed in this work is capable of accurately simulating the transient response of devices with channel lengths as small as 0.33 µm for switching frequencies approaching 1 GHz.

5. References

[1] Payton, M.W., and Ho, F.D., "*A Physically-Derived Large-Signal Nonquasi-Static MOSFET Model for Computer-Aided Device and Circuit Simulation Part-I,*" submitted to 2005 IEEE International Symposium on Circuits and Systems.

[2] Bagheri, M., and Tsividis, Y.,"A Small Signal dc-to-High-Frequency Nonquasistatic Model for the Four-Terminal MOSFET Valid in All Regions of Operation," *IEEE Transactions on Electron Devices*, Vol. ED-32, No. 11, 1985, pp. 2383-2391.

[3] Chai, K., and Paulos, J., "Comparison of Quasi-Static and Non-Quasi-Static Capacitance Models for the Four-Terminal MOSFET," *IEEE Electron Device Letters*, Vol. EDL-8, No. 9, 1987, pp.377-379.

[4] Chai, K., and Paulos, J., "Unified Nonquasi-Static Modeling of the Long-Channel Four-Terminal MOSFET for Large- and Small-Signal Analyses in All Operating Regimes," *IEEE Transactions on Electron Devices*, Vol. 36, No. 11, 1989, pp. 2513-2520.

[5] Phillips, A. B., "*A Non quasi-static Method for the Computer Aided Design of CMOS Logic Circuits,*" Master's Thesis, The University of Alabama in Huntsville, School of Graduate Studies, Huntsville, Alabama, 2000.

[6] Darty, M. A., "*A Quasi-static Method for the Computer Aided Design of CMOS Logic Circuits,*" Master's Thesis, The University of Alabama in Huntsville, School of Graduate Studies, Huntsville, Alabama, 1998.

[7] Payton, M.W., "*A Physically-Derived Large-Signal Nonquasi-Static MOSFET Model for Computer Aided Device and Circuit Simulation,*" Master's Thesis, The University of Alabama in Huntsville, 2004.

[8] Press, William H., Teukolsky, Saul A., Vetterling, William T., and Flannery, Brian P., *Numerical Recipes In C. The Art of Scientific Computing,* Cambridge University Press, New York, New York, 1992.

Domain Fault Model and Coverage Metric for SoC Verification

Luo Chun, Yang Jun, Gao Gugang, Shi Longxing
National ASIC System Engineering Research Center
Southeast University
Nan Jing 210096 P.R.China
Email: {luochun, dragon}@seu.edu.cn, gaogugang@jpi.gov.cn, lxshi@seu.edu.cn

Abstract— An innovative domain fault and coverage metric for SoC verification is proposed. Domain fault model is based on a geometrical analysis of the domain boundary and takes advantage of the fact that points on or near the boundary are most sensitive to domain errors. The purpose of this paper is to present an efficient fault model and coverage metric for measuring the completeness and quality of verification approach. The domain coverage metric has been implemented using VPI (Verilog Procedural interface) and has been applied to verification of SoC(Syetem on Chip) design. Our domain coverage tool works smoothly with simulator and vector generator. The results showed that the domain fault model is accurate and efficient, the domain coverage metric is powerful to find the potential boundary faults of control path.

I. Introduction

With the increase in size and the complexity of integrated circuit designs, it has become imperative to address critical verification issue at early stages of the design cycle. The process of verifying correctness of integrated circuit designs consumes between 60% and 80% of design effort[1]. Furthermore, the verification engineers outnumber designers, with this ratio reaching two or three to one for the most complex design, in our projects. The time-to-market is dominated by the verification process in which full automation has not been achieved. According to the ITRS2003[2], one of the most challenges of IC designs in the following 6 years(2004-2009) is to increase verification capacity.

A number of approaches exists for verifying a design before production, popular approaches include physical prototyping, emulation, hardware acceleration, simulation, and formal verification. Physical prototyping is an expensive approach and not an economically effective verification approach for IC design. Formal verification promises the ability to prove that a given design meets its specification. However, as the design size grows, most techniques are affected by the state explosion problem. Emulation system use reconfigurable hardware (typically FPGA) to implement essentially equivalent functionality in hardware[3]. However limitation on the visibility into an emulated system makes the debug of the hardware description very difficult, especially in the early development stage. Hardware acceleration for verification comprises custom hardware that is dedicated to specific simulation application. As a result, this approach is extremely expensive and the reuse ability is very poor.

Simulation is still the dominating verification approach[3]. In this paper we focus on the simulation based verification approach. A key problem in the simulation verification approach is how to measure the completeness of the verification approach. Some coverage metrics inherited from software testing, such as statement coverage, branch coverage, expression coverage[1], toggle coverage[4], path coverage and domain coverage[5] etc. Other kind of coverage metrics are taken from manufacturing testing, such as state coverage, transition coverage[4] and observation-based coverage[6] etc. The statement coverage and the branch coverage are too simple to reveal the faults of the RTL HDL description. Expression coverage is a little more effective than the statement and the branch coverage. Toggle coverage is just used to measure some special variables coverage.

Domain coverage was proposed by White and Cohen[7], which focuses on the detection of control flow errors of software. This method appeals to our intuition in that it provides a formal approach for satisfying the often suggested guideline that boundary conditions be tested. As the increase in size and the complexity of integrated circuit, we mostly design IC at the register-transfer-level. An important issue in RTL hardware verification is to check the correct implementation of specified functions and to determine the presence error. We inherit the domain methodology from the software testing. To some extent, the domain methodology is more suitable for verification of Verilog RTL hardware description, because the analysis of paths in process of RTL description is simpler than the software. Paper[5][8] proposed a weak domain coverage for the validation of behavioral VHDL description. The weak domain coverage only tests separately predicate's boundary but does not test path's boundary, so the weak domain coverage is not an effective metric to measure the completeness and quality of validation.

This paper presents a domain fault model and domain coverage metric for the verification of Verilog RTL Description. The metric has been applied to measure the completeness and quality of the verification approach of several benchmark circuits and the SoC under design. The remainder of this paper is organized as follow: section 2 presents the domain fault model, section 3 proposes the domain coverage metric and the experiments result, section 4 concludes this paper..

II. DOMAIN FAULT MODEL

A. Terminology

In Verilog, a program can be described by modules executed concurrently. A module consists of a number of concurrently executing processes. Process can run continuously like always or continuous assignment, or run only once like initial process.

The control flow graph (CFG) of process (always or continuous assignment) is a directed graph $G=<B, V, E, D>$, where V is the set of nodes of G, and E is the set of the edges. Nodes represent statements, and the edges represent possible flows of the control between statements. If a node of G has more than one edge leaving, then the edge represents the outcomes of decision statement, such as if, if-else or case statement. They are thought of as being labeled with appropriate predicates. If and if-else statements have two outgoing edges, case statement is rewritten in terms of if-else statements. B and D are specific nodes, denoting respectively begin node and end node of the process.

The CFG of a process is shown in figure 1. *TriggerLevel* and *FullLevel* are constants, *FifoCounter* and *Time* are variables. There are four pathes from always begin node to the always end node, which path to be executed depends on the value of variables *FifoCounter* and *Time*. If *Time<TimeOut* and *FifoCounter>=TriggerLevel&&FifoCounter<FullLevel* both are true, as the shadowed domain the in figure 2. The path, from begin node, through nodes 5,7 and 9, to end node, will be executed. Every path has a executing condition, which is termed *path condition*. Predicate node (if, if-else) has two out-going edges, the Boolean value of predicate's expression determines which edge will be executed. If the expression's Boolean value is true, left edge will be executed, else right edge. In another word, while left edge is to be executed, the expression's value is true; while right edge is to be executed, the expression's negative value(!expression) is true. So, a path corresponds to a path condition, if and only if a condition is satisfied, the path can be executed. A path condition is the conjunction of all individual predicate encountered along the path. The expression of a predicate may be a complex expression, which includes logical operator (&& ,and ||). Papers[7][9][10] assumed that all the expressions of predicates are simple expressions. In RTL description, this assuming is not suitable, because complex expression is widely used to describe the behavior of hardware. In this paper, we do not limit that predicate's expressions are only simple expressions. For complex expressions, we apply a method to translate them into simple expression.

A path condition defines a path domain, which is the set of variables making path condition satisfied and causing the path to be executed. A path domain is surrounded by a *boundary*. Section of the boundary is called *border*, which corresponds to a simple expression of the path condition. A border of a domain may either be *closed* or *open* with respect to that domain. A closed border belongs to the path domain and comes from a simple expression that contains $>=$, $<=$, or

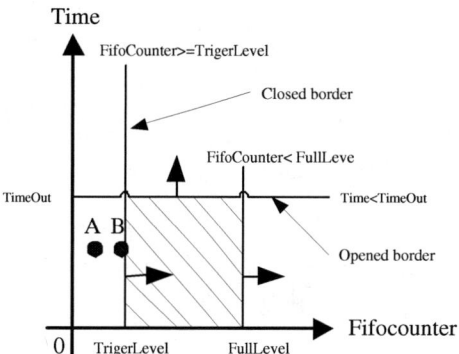

```
3   always @ ( FifoCounter or Time )
4   begin
5     if( Time < TimeOut )
6     begin
7       if( FifoCounter == 0 )
8         FifoIndicator = `EMPTY;
9       else if( FifoCounter >= TriggerLevel
10              && FifoCounter < FullLevel )
11        FifoIndicator = `ALMOSTFULL;
12      else if( FifoCounter >= FullLevel )
13        FifoIndicator = `FULL;
14      else
15        FifoIndicator = 2`b00;
16    end
17    else
18      FifoIndicator = `TIMEOUT;
19  end
```

Fig. 1. Process and CFG

Fig. 2. Path Domain

$==$. An open border is not part of the path domain and comes from a simple expression that contain $!=$, $>$, or $<$. The given border, which corresponds to a simple predicate expression, is the border to be tested and may be correct or not. When a domain error occurs along a path, it may be thought of as being caused by a border that is different from the correct, then we say that a border shift has occurred. In order to find whether the given border is correct, we need a strategy to select a number of test points. Papers[7][9][10] have proposed different strategies for selecting test points. In this paper, we propose a new strategy. The first type of point required by domain testing is known as *ON point*, which lies on the given border. The second type of point required is knows as *OFF point*, which lies slightly off the given border on the open side of it. The distance between the border and the OFF point is as small as possible. The conception is illustrated in Figure 2, point A and B are respectively OFF point and ON point. Small arrows are used to indicate that which side of the border is closed.

B. Domain Fault Model

RTL description mostly use vector to present variables, so we should first map the vector type variable reg[msb:lsb] and wire[msb:lsb] to scalar type variable. The following formulas are used to map vector to scalar.

For reg[msb:lsb]:

$$\|reg[msb:lsb]\| =$$
$$reg[0] \times 2^0 + reg[1] \times 2^1 + ... + reg[l] \times 2^l,$$
$$where: l = msb - lsb$$

For wire[msb:lsb]:

$$\|wire[msb:lsb]\| =$$
$$wire[0] \times 2^0 + wire[1] \times 2^1 + ... + wire[l] \times 2^l,$$
$$where: l = msb - lsb$$

After map the vector to scalar, the way to handle the RTL description is the same to the software testing. All the variables in the RTL description are mapped to plus integers, so we only need to consider the plus integer space and ignore the minus integer space.

Domain errors are caused by an incorrectly specified predicate. Domain errors are divided into two types: *Domain Fault I* and *Domain Fault II*.

Definition 1

Domain Fault I: A fault in predicate's simple expression, which causes a border shifted from its correct position, is termed Domain Fault I.

Domain Fault II: A operator fault in predicate's simple expression is termed Domain Fault II.

Domain Fault I is introduced by a designer, if she/he writes an incorrect predicate, such as substituting correct predicate if(A[3:0] > 5) with wrong predicate if(A[3:0] > 3). Domain Fault II is introduced by a designer, if she/he writes an incorrect operator, such as substituting correct operator if(A[3:0] > 5) with incorrect operator if(A[3:0] >= 5).

III. DOMAIN COVERAGE METRIC

Coverage Metric is used to measure the quality and the completeness of the verification approach. In order to obtain the domain coverage, we should generate the test points before simulation, monitor value of variable for the purpose of checking whether the test point being activated during the simulation, calculate the domain coverage after simulation.

For revealing the potential design errors, a set of test points are needed. White and Cohen has proposed a strategy for selecting test points, known as N×1 strategy. The strategy requires N ON-points and one OFF-point. Paper[9] proposed two alternative test points selecting strategy, known as N×N strategy and V×V strategy. The N×N strategy needs N on-points and N off-points. The V×V strategy needs V ON-points and V OFF-points. A simplified domain testing strategy was proposed in paper[10], which requires a constant number of test points(i.e. the number of required points is independent of the dimension or the type of the border or the number of vertices on the given border). Here we proposed more efficient test points selecting strategy.

For testing a border, which corresponds to a simple expression containing equality or nonequality (==, !=) operator, one ON point and two OFF points are required. Such as, for detecting if(reg[2:0] == 3'b111), requires one ON point and two OFF points.

For testing a border, which corresponds to a simple expression containing inequality ($<, >, <=, >=$) operator and N variables, one ON point and one OFF points are required. such as, for detecting $if(reg\_a[2:0] + reg\_b[2:0] < 3'b111)$, requires one ON point and one OFF point.

For testing a border, which corresponds to a simple expression no containing any relation operator ($==, !=, <, >, <=, >=$), such as predicate if(A[1]&B[1]), one ON point (A[1]&B[1] true) and one OFF point (A[1]&B[1] false) are required.

Before simulation, first we analyses the structure of the RTL source code, get all the processes of the DUV (Devive Under Verification) and obtain all paths of every process. Second, we generate the path condition for every path, and translate the path condition into simple expression. Third, we generate the testing points for every simple expression. Last, we get the total number of points needed.

During simulation, we monitor the value of all variables, for checking whether test points are activated. If the point has been activated, we mark the point.

After simulation, we sum up the number of test points that have been activated during simulation, and calculate the DUV's coverage by the following formula:

$$domain\ coverage = \frac{\sum activated\_test\_points}{\sum test\_points}$$

The purpose of coverage metric is to help designers to find out design errors as early as possible. Therefore the metric are always used together with the HDL simulator, and results are profiled and then analyzed to derive the coverage. Programming Language Interface(PLI)[11] provides a means for Verilog users to access and modify data in an instantiated Verilog data structure dynamically.

We implemented the domain coverage tool, using about 8600 lines C code through the PLI2.0(VPI). Compared with PLI1.0, VPI has a substantial advantages[11]. Figure 3 illustrates the relation of callbacks and Verilog simulator. User only need put a system task or function in the Verilog source description. Our domain coverage tool has been built on the Candence NcVerilog5.0/Verilog-xl5.0 simulator and Synopsys VCS 7.1.1 simulator. Our tool can work smoothly with Candence's simulator, but there is a problem when works with the VCS simulator, because the VCS7.1.1 simulator does not support statement callback and can not obtain expression's value through function vpi_get(). At present, our tool can only work with Candence simulator.

Our experiment was carried out under the environment: SUN Fire-v880 server, SUN OS5.8 NC-Verilog5.0 simulator and Specman Elite4.2. Test vector is generated by constrained random, using Specman. DUVs used in the experiment include

TABLE I
COVERAGE COMPARISION

DUV	# of Line	# of Path	Domain Cov.		Stmt. Cov.(%)	Path Cov.(%)
			# test point	Cov.(%)		
MPEG	1150	383	6588	56.33	100.00	83.26
DMAC	8316	622	2565	57.59	100.00	86.66
EMI	5414	1976	69803	0.14	100.00	11.59
LCDC	6292	1594	3689	15.61	100.00	82.12
PWM	1003	142	274	27.74	100.00	90.85
AC97	3438	423	2039	42.82	100.00	80.61
RTC	1731	142	274	27.74	100.00	90.85
ARBITER	255	38	435	53.33	100.00	87.53

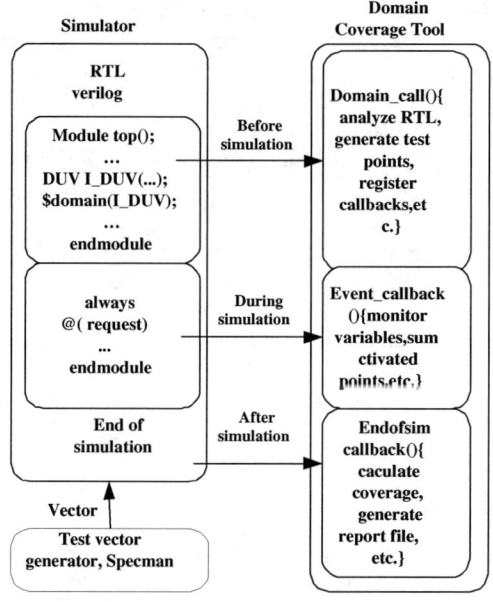

Fig. 3. Framework of Domain Coverage Tool

standard benchmark circuit and SoC under design. The standard benchmark circuit is texas97-benchmarks-MPEG System Decoder[12]. SoC circuits include a DMA controller EMI (external memory interface), LCD controller, PWM, AC97 controller, RTC and a arbiter circuit. We have compared the domain coverage with statement coverage and path coverage. and the results are showed in table I. The table illustrates that the statement coverage and path coverage both reach sufficient level, however the domain coverage is still low. So we generated vectors aiming at increasing the domain coverage. In the process of enhancing the domain coverage, we found some potential faults.

IV. CONCLUSION

The importance of verification continues to grow with the size of design. In this paper we proposed domain fault model and domain coverage for SoC verification. For gaining enough confidence for taping out bug free design, we adopted the coverage metrics: first applying the simplest statement coverage metric, then using the path coverage metric and functional coverage metric, last employing sophisticated domain coverage metric. Results show that the domain coverage can detect some design faults that other coverage can not.

ACKNOWLEDGMENT

The authors thank the colleagues who help to run simulation and analyses the data. This work was supported by National Science Foundation of China under grant 60176018 and Hi-Tech Program of Science and Technology of China (863) under grant 2003AA1Z1340.

REFERENCES

[1] J. Bergeron, Ed., *Writing Testbenchs: Functional Verification of HDL Models*. London: Kluwer Academic Publishers, 2000.
[2] ITRS. (2003) International technology roadmap for semiconductors,2003 edition. [Online]. Available: http://www.itrs.net
[3] G. Peterson, "Predicting the performance of soc verification technologies," in VHDL International Users Forum Fall Workshop, Orlando, FL, USA, 10 2000, pp. 17–24.
[4] *VCS / VCS MX Coverage Metrics User Guide. Version 7.1.1*, Synopsys, 2004.
[5] Q. Zhang, "Validation of behavioral hardware descriptions," Ph.D. dissertation, University of Massachusetts Amherst, 2003.
[6] B. E. Min, "Register-transfer-level design verification: Coverage and acceleration," Ph.D. dissertation, Texas A&M University, 2001.
[7] E. I. White, L. J. Cohen, "A domain strategy for computer program testing," *IEEE Transaction on Software Engineering*, vol. SE-6, no. 3, pp. 247–257, May 1980.
[8] I. Qiushuang Zhang Harris, "A domain coverage metric for the validation of behavioral vhdl descriptions," in *Test Conference, 2000 Proceedings International*, 2000, pp. 302–308.
[9] J. L.A.Clarke and R. D, J, "A close look at domain testing," *IEEE Transaction on Software Engineering*, vol. SE-8, no. 4, pp. 380–390, July 1982.
[10] E. J. B.Jeng, "A simplified domain-testing strategy," *ACM Transactions on Software Engineering and Methodology*, vol. 3, no. 3, pp. 254–170, July 1994.
[11] *Standard Verilog Hardware Description Language*, IEEE Std. 1364-2001, 2001.
[12] texas97-benchmarks. [Online]. Available: http://www-cad.eecs.berkeley.edu/index.html

Validation Analysis and Test Flow Optimization of VLSI Chip

Yanzhuo Tan[1,2], Yinhe Han[1,2], Xiaowei Li[1,2], Feiyin Luo[3], Yuchuan Chen[4]

(1. Institute of Computing Technology, Chinese Academy of Sciences, Beijing, 100080)
(2. Graduate School of Chinese Academy of Sciences, Beijing, 100039)
(3. Beijing Micro-Electronics Technology Institute, Beijing, 100076)
(4. Teradyne Inc., Shanghai, 201206)

Abstract: This paper gives a validation analysis on a high-performance general-purpose processor of 0.18um process and based on this analysis a test flow optimization algorithm is presented. The fault detection capacity of different test items is first analyzed. Then the validation information can be reused to generate a test item efficiency table. Based on this table, a tradeoff between test item efficiency and test time is achieved as the heuristic object for optimizing our test flow, which can save much test time of faulty chips. Compared to the dynamic programming algorithm, the complexity of our heuristic ordering algorithm has been decreased from $O(dn2^n)$ to $O(dn^3)$. Several experimental results have shown that this algorithm is efficient.

Key words: validation analysis, at-speed test, dynamic programming algorithm, heuristic ordering algorithm

1. Introduction

The rapid development of the semiconductor technology drives higher integration and frequency of today's ICs, which not only add design difficulties but also test costs. The SIA roadmap [1] predicts that if present trends continue, costs of testing a single transistor would be equivalent to its manufacture costs by the year of 2014. The capacity of at-speed test is required for test equipments to verify ICs' performance. However, such test equipments are costly and involve higher test costs.

The at-speed test generally uses functional patterns, which will pose such problems as follows [2]. Firstly functional patterns mainly cover functional paths of ICs but not physical defects, so the coverage for physical failures is limited. Secondly such patterns are generated by functionally simulating all the netlist, which is time-consuming, e.g. according to our experimental results, one 80MB pattern for an 800,000 gate level chip needs about 15 hrs. Also the transformation of the simulated patterns needs much time, so the whole test process slows down. Thirdly functional patterns are usually huge. For the pattern configuration or execution on test equipments, it is hard and time-consuming.

To reduce the test time and add the fault coverage, the design for test ability (DFT) technology is widely used in VLSI chip designs, including scan design, logic/memory built-in self-test (LBIST/MBIST), boundary scan, analog self-test, etc. However these structural tests are generally low-frequency and mainly aim at stuck-at faults. Therefore test engineers can not assure whether those chips only passing low-frequency structural tests are good-working. So for the final production test both structural and functional patterns are generally involved, where structural ones are the main part for stuck-at faults and functional ones help find out more timing faults [2,3], e.g. path delay faults, transition delay faults, cross-talks due to the deep sub-micron process, etc. The at-speed test can also help find many un-defined faults. This paper will give some experimental data and a general analysis for the efficiency of each test pattern.

The test cost correlates with the test time [4]. For faulty chips in the production test, as long as any fault is found test equipments will stop at once and no longer continue. So the test time is determined by the moment when patterns first find faults, i.e. the time is determined by the position in the whole test flow where faults are first found. Thus based on their test efficiency we can order the test patterns or items in the flow to reduce the test time for faulty chips and then the total test time. Based on dynamic programming techniques [5,6] have proposed some test ordering optimization algorithms for fault-driven analog ICs. However they are based on the theoretic fault models, so they are infeasible. [7] formalized the optimization problem as the Lagrange relaxation which was complex. [8] used the fault signature to optimize the test flow and this method was not feasible.

This paper will give a greedy heuristic ordering algorithm to optimize test flows. Compared to the dynamic programming algorithm, the heuristic ordering algorithm is easier because it is based on a new data structure and new heuristic information from the validation analysis in our validation tests. The experiments also show it is very effective.

2. Details of the validation analysis

2.1 Background of ICs

In this paper we conduct some experiments on a commercial 32-bit general-purpose CPU chip with the following characteristics:
- 850,000 gates (expect memories)
- TSMC 0.18um process
- 13,000 register cells
- 5 independent RAM modules, totally about 40% chip area
- 200MHz work frequency

The DFT techniques applied to this chip are shown as follows [9]:
- Full scan design with 16 scan chains
- For 5 RAM modules 5 BIST logic circuits with 1.3% area of the whole chip
- JTAG: 3 general commands including SAMPLE, TEST and BYPASS, which are fully compatible with the IEEE1149.1 standard

Test patterns of this chip include both functional and

structural ones as Table1 has shown. Note that the parametric test includes both DC and AC testing.

Table 1 Breakdown of each test pattern set

Category	Faults detected	Fault coverage	Test depth (cycles)	Test frequency (Hz)	Test time(s)
continuity test	continuity faults of interfaces	100%	--	--	0.5
functional test	Stuck-at	60%	9191832	200M 133M 20M	0.6
scan test	stuck-at	97.3%	3492990	20M	0.3
parametric test	-		--	200M	1.5
MBIST	stuck-at, transition	100%	7184	20M	0.0035

The test equipment used in our validation analysis is Catalyst C400e of Teradyne Inc., totally 256 test channels, with 14MB memory each and the highest frequency is 400MHz.

2.2 Efficiency analysis of each test

Totally 569 chips were tested and Table2 shows the number of faulty chips found by each test item, e.g. 10 chips passed all the four tests except the memory BIST. By comparison structural patterns were more effective which could find 63.9% of the total. Functional patterns were also effective but with twice test time of structural patterns and after functional tests, 36 still were not found out. Moreover we found MBIST was a very effective test item, which could found 57 faulty chips in very short time (about 0.0035s).

Table 2 Breakdown of coverage analysis

		Scan test				
		pass	pass	fail	fail	
A C T E S T	pass	--	10	22	3	pass
	pass	14	22	301	1	fail
	fail	142	0	0	1	pass
	fail	15	2	18	18	fail
		pass	fail	pass	fail	
		Memory BIST				

for different test items

2.3 At-speed test

Chips perform differently under functional tests with different frequency. We examined 108 samples which failed in functional tests. As shown in Fig.1, 88 faulty chips were "found out" by the functional test of 133MHz, i.e. 81.5% coverage of faulty chips. And the functional test of 200MHz could find 108 faulty chips, i.e. 100% coverage of faulty chips. So the latter (200MHz) could find more faulty chips (20 chips) than the former (133MHz). It is shown that functional tests with higher frequencies (e.g. the at-speed test) can help to enhance the test coverage.

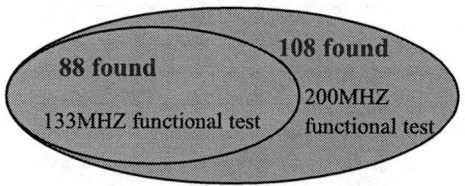

Fig.1 Validation analysis under different frequencies

3. Optimization techniques of the test flow

3.1 Preliminaries

As the above analysis, test time is determined by the size of each test items. The elements of test time can be formally described as follows:
- Set of all the test items $IS=\{I_1, I_2, ..., I_n\}$
- Set of test time corresponding to each test item $TS=\{T_1, T_2, ..., T_n\}$
- Set of all the chips to be tested $CS=\{C_1, C_2, ..., C_q\}$
- Set of faulty chips $FCS=\{FC_1, FC_2, ..., FC_p\}, FC \in CS$

In fact a test flow is an ordered set of test items, e.g. a test flow is $TF\{I_1, I_2, ..., I_n\}$. The test mode used is "Stop as soon as faults are found". Thus the corresponding set of the first failing test items for p faulty chips is $FS=\{f_1, f_2, ..., f_p\}, f_i \in IS$. So the test time in the sample space of q chips is:

$$T_{total} = (q-p) \times \sum_{i=1}^{n} T_i + \sum_{i=1}^{q} \sum_{j=1}^{f_i} T_j$$

The first part $(q-p) \times \sum_{i=1}^{n} T_i$ is the test time of normal-working chips and the second part $\sum_{i=1}^{p} \sum_{j=1}^{f_i} T_j$ is the test time of all the faulty chips. Once test patterns and frequencies are determined, the first part is fixed. So to reduce the total test time, it is necessary to minimize the value of the second part, which is close related with the location of each test item in the whole test flow. Therefore the test flow optimization problem is equivalent to seek a proper order to achieve the minimal T_{total}.

First we need to partition the whole test flow into several smaller test items. Each item is a set of patterns with similar features. Here are some related definitions:

Def.1 Efficiency factor of **test item I**:
  $Y(I)$ = No. of faulty chips that I can find out

Def.2 Eigen-efficiency factor of **test item I**:
  $E(I)$ = No. of faulty chips that only I can find out

The data from the validation analysis needs to be transformed into the format of optimal algorithms. This paper gives a new data structure—a table of test item efficiency to record the related information. Assuming a test flow is a set with n test items, i.e. Flow=<$I_1, I_2, ..., I_i, ..., I_{n-1}, I_n$>, the test item efficiency table shown as Table3. In the table each row/column represents a test item. $Y(I_i, I_j)$ is the number of faulty chips found by I_i and I_j. Obviously $Y(I_i, I_i)$ is the eigen-efficiency factor of the test item I $E(I)$. Also test time is very important for test flow optimization, which can be recorded by a simple list as shown as Table4.

Table 3 Test item efficiency table

Test item	I_1	I_2	...	I_i	...	I_{n-1}	I_n
I_1	$E(I_1)$	$Y(I_1, I_2)$		$Y(I_1, I_i)$		$Y(I_1, I_{n-1})$	$Y(I_1, I_n)$
I_2	$Y(I_2, I_1)$	$E(I_2)$		$Y(I_2, I_i)$		$Y(I_2, I_{n-1})$	$Y(I_2, I_n)$
⋮							
I_i	$Y(I_i, I_1)$	$Y(I_i, I_2)$		$E(I_i)$		$Y(I_i, I_{n-1})$	$Y(I_i, I_n)$
⋮							
I_{n-1}	$Y(I_{n-1}, I_1)$	$Y(I_{n-1}, I_2)$				$E(I_{n-1})$	$Y(I_{n-1}, I_n)$
I_n	$Y(I_n, I_1)$	$Y(I_n, I_2)$		$Y(I_n, I_i)$		$Y(I_{n-1}, I_{n-1})$	$E(I_n)$
Efficiency	$Sum(I_1)$	$Sum(I_2)$		$Sum(I_i)$		$Sum(I_{n-1})$	$Sum(I_n)$

Table 4. Test time assumption of each test items

Test item	I_1	I_2	...	I_i	...	I_{n-1}	I_n
Test time	$T(I_1)$	$T(I_2)$	...	$T(I_i)$	...	$T(I_{n-1})$	$T(I_n)$

3.2 Heuristic test flow optimization

To optimize the test flow, two factors considered are test item efficiency and test time. How to get a proper tradeoff between them is a research-worthy problem, which can be solved by the dynamic programming algorithm using a multi-step decision process. However it is an exhausted-searching method with an exponential complexity--$O(dn2^n)$. So a polynomial-time algorithm is required. We present an ordering algorithm based on the heuristic information.

Def.3 Efficiency rate of test item I:

$$F(I) = Y(I)/T(I)$$

Where $Y(I)$ is the **efficiency factor** of test item I and $T(I)$ is the time cost of test item I.

According to this definition, $F(I)$ is actually a parameter involved both the test efficiency and the test time. So for a test item, the larger $F(I)$ is, the higher its efficiency to find faulty chips. The following lemma will show that $F(I)$ is a proper parameter to locally optimize test flows.

Lemma 1 The efficiency rate $F(I)$ can lead to a locally optimized test item order.

Proof Assuming I_i and I_j are two test items whose efficiency factors are $Y(I_i)$ and $Y(I_j)$ respectively and test time cost are $T(I_i)$ and $T(I_j)$ respectively. Let N be the number of faulty chips.

If the test order is $I_i I_j$, the total test time of all the faulty chips is:

$T_{ij} = T(I_i)*N + T(I_j)*(N - Y(I_i))$
$\quad = N*(T(I_i) + T(I_j)) - T(I_j)*Y(I_i)$;

If the test order is $I_j I_i$, the total test time of all the faulty chips is:

$T_{ji} = T(I_j)*N + T(I_i)*(N - Y(I_j))$
$\quad = N*(T(I_i) + T(I_j)) - T(I_i)*Y(I_j)$;

Assuming $F(I_i) > F(I_j)$, then

$$Y(I_i)/T(I_i) > Y(I_j)/T(I_j)$$

So $T_{ij} < T_{ji}$. (End)

From the above proof, it is shown that for two test items I_i and I_j in the test flow, if $F(I_i) > F(I_j)$, the order "$I_i I_j$" is proper to decrease the total test time, i.e. by selecting the efficiency rate $F(i)$ as the heuristic information, a locally optimal solution can be achieved easily. Although for the whole test flow optimization we can not get the overall-optimal solution by such heuristic information, an almost-overall optimal solution can be reached. So we can take advantage of the heuristic searching algorithm based on test efficiency rate $F(I)$ to optimize our test flow, i.e. $F(I)$ is the heuristic factor. The optimization algorithm is described as follows:

Input: test item efficiency table
Output: OPTIMIZED_FLOW

1. Select a test item I_i with the biggest test efficiency rate from the test item efficiency table ;
2. Put the selected item I_i in step 1 into the OPTIMIZED_FLOW ;
3. Remove the responding row and column of I_i in step 1 from the test item efficiency table ;
4. Update the test item efficiency table and re-compute the test efficiency of the rested test items ;
5. Jump to step 1 until all the test items have been put into the OPTIMIZED_FLOW.

♦ **Algorithm complexity analysis:** the time complexity mainly consists in two points: searching the test item with the biggest test efficiency rate and updating the test item efficiency table. The complexity of the former is $\sum_{k=0}^{n-1}(n-k) = (n(n-1)/2)$ and the complexity of the latter is $\sum_{k=0}^{n-1}(n-k)^2 = \frac{1}{6}[n*(n+1)*(2n+1)]$. Thus the time complexity of the whole algorithm is $O(dn^3)$.

4. Results

We have implemented the dynamic programming ordering algorithm and the heuristic ordering algorithm respectively in C language. For the former we used a branch algorithm with higher efficiency than the original one in this paper. To prove the algorithm efficiency, we designed our experiments by using the data from our validation analysis of 54 actual chips. Firstly we select 54 faulty chips to perform the validation analysis. It is noted that the test items selected in this experiment are different from the ones in Table 1. To further reduce the test time, they are much smaller test items in the experiment, e.g. the functional patterns in Table 1 will be partitioned into 7 different test items according to different functions, the scan test pattern set of stuck-at faults will be partitioned into 6 different test items and the parameter test pattern set will be partitioned into 5 different items. Then a new test flow with 19 test items is achieved. Secondly we verify the 54 chips on the test equipment by using the test items in the new flow respectively. Then according to the test data report, we can get the test item efficiency table. Lastly the dynamic programming algorithm and the heuristic ordering algorithm in this paper are carried out. The results are shown as Table5.

Table5 Comparison of the two algorithms

Category	Dynamic programming	Heuristic	Original
Complexity	$O(dn2^n)$	$O(dn^3)$	–
Running time	1 s	1 s	–
Test time	60.06 s	60.06 s	70.02 s

It can be seen that the heuristic ordering algorithm is the best because it not only gets the optimal solution, but also has the low complexity. Running the dynamic programming algorithm and the heuristic ordering algorithm are both fast (about 1s)since the experimental test flow has fewer test items. However SOC tests will make a test flow involved more test items and partition the flow into much smaller test items The correlation between the partition and test time is shown as Fig.3, where the analysis is based on the above 54 faulty chips, the partition is based on different functions and the heuristic ordering algorithm is used in the test flow.

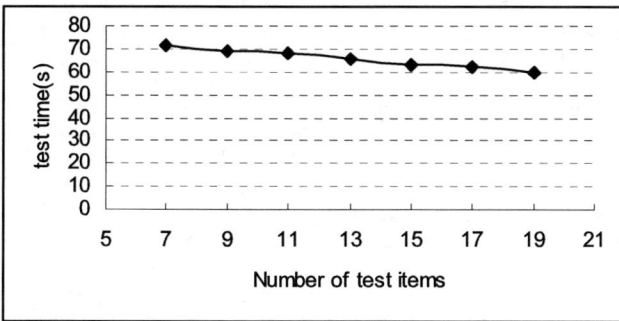

Fig.3 Correlation between the partition and test time

In Fig.3 the number of test items changes from 7 to 19 and Y axis shows the variation of test time with the different partition. From the curve, it is shown that the smaller test items are, the least test time is cost. So in practice test engineers tend to partition a test flow into more test items. In the future due to the complex multi-core structure, SOC tests will lead to a greater number of test items. So the heuristic ordering algorithm is more advantageous than the dynamic programming algorithm as Table6 shows. In this experiment we used the dynamic programming algorithm to get the optimal solution and the heuristic algorithm to get locally-optimized solution. Then we can use the difference of the above two solutions to describe the difference between the two algorithms, that is: $D_h = (T_h - T_d) / T_d$, where D_h, T_h and T_d respectively represent the difference, the test time of the heuristic solution and dynamic programming solution.

As shown in Table6, the smaller D_h is, the heuristic ordering algorithm works better. When the number of test items is small, the time cost of the dynamic programming algorithm is acceptable. However, when the number of test items exceeds 100 or more, it is very slow to use the dynamic programming algorithm. By comparison, thrice time cost can be saved by using the heuristic ordering algorithm, except that the test time of its local-optimization solution is about 5% larger than the optimal one.

Table6 Comparison of algorithms with more test items

Test flow optimization cases	No. Of Test items	Dynamic programming		Heuristic		D_h
		Running Time (s)	Test time (s)	Running Time (s)	Test time (s)	
case1	50	356	42.80	1.2	43.32	1.2%
case2	50	378	31.67	1.5	36.75	16.0%
case3	100	28K	93.34	10.2	93.88	0.6%
case4	100	31K	112.79	10.2	116.90	3.6%
case5	200	745K	267.45	40.1	267.45	0%
case6	200	813K	334.09	64.3	356.98	6.59%

5. Conclusion

This paper gives the validation analysis for a type of CPU chip in the deep sub-micron process, which provides the reliable criteria for selecting proper test patterns. Based on the analysis, we generate a new data structure—the test item efficiency table and present a heuristic ordering algorithm to optimize our test flow. By experimental analysis, it is proved that the heurist ordering algorithm is more advantageous, especially for current SOC tests.

References

[1] A. Allan et al. 2001 Technology Roadmap for Semiconductors [J], Computer, 35(1), 2002, 42～53
[2] B. G. West, Accuracy Requirements in At-Speed Functional Test [A], In: Proceeding of International Test Conference, Atlantic City, NJ, USA, 1999, 780～787.
[3] A.Krstic, J.Liou, K.T.Cheng, etc. On Structural vs. Functional Testing for Delay Faults. In: Proceeding of International Symposium on Quality Electronic Design, 2003.
[4] J. Rurino. Test economics in the 21 century [J], IEEE Design & Test of Computer, 1997, 13(5), 41～44
[5] S. D. Huss, R.S. Gyurcsik. Optimal ordering of analog integrated circuit test to minimize test time[A], In: Proceeding of Design Automation Conference, 1991, 494~499
[6] L. Milor, A. L. S. Vincentelli. Minimizing production test time to detect faults in analog circuits[J]. Transaction on Computer Aided Design, 13(6), 1994, 796～813
[7] T. R. Chen and T. S. Chang et. al , Scheduling for IC sort and test with preemptiveness via Lagrangian relaxation[J], IEEE Trans. On System, Man, Cybernetics, 25(7), 1995, pp. 1249～1256
[8] Y. H. Lee, C. M. Krishna, Optimal scheduling of signature analysis for VLSI testing[J], IEEE Trans. On Computers, 40(3), 1991, pp. 336～341
[9] H. Li, X. L, et.al. DFT Techniques for a general-purpose CPU[A], In: Proceeding of Conference of CAD/CG, 2002, 365～370

Deterministic and Low Power BIST Based on Scan Slice Overlapping

Ji Li Yinhe Han Xiaowei Li

Institute of Computing Technology, CAS, Beijing, 100080, China
Graduate School of the CAS, Beijing, 100039, China
Leeji1212@ict.ac.cn

Abstract

This paper presents a new deterministic pattern generation structure that can be used in conjunction with any LFSR reseeding scheme. The proposed scheme utilizes Scan slices overlapping to reduce the number of specified bits and the number of transitions at the same time. Thus, it can significantly reduce test power and even further reduce test storage. A decoder is used to generate control signals. Experimental results indicate that the proposed method significantly reduces the switching activity by 80% and only needs relatively small test data storage.

1. Introduction

When the process of very large-scale integrated circuits (VLSI) scales down into deep sub-micron, the complexity of circuit designs has greatly increased. Testing of a chip has posed some new challenges, especially with the emergence of System-on-chip (SOC). Traditional test methods based on the automatic test equipment (ATEs) are becoming unacceptable. Built-in self-test(BIST) offers an attractive solution to these challenges.

In recent years, research on deterministic BIST has become a hot topic, which can provide very high test coverage and relatively short test time. The architecture of deterministic BIST mainly consists of three parts: test data storage, stimulus decoder, and response compactor. Many kinds of decoders have been proposed to reduce test data storage, such as LFSR[1-3], decoder of encoding compression[4-8], XOR network[9] and the other customized counters[10]. Among these decoder schemes, LFSR reseeding[1-3] generate deterministic test cubes by expanding seeds, providing significant reduction of test data storage and bandwidth. Therefore LFSR reseeding technique has been studied extensively, and supported by many commercial tools including TestKompress by Mentor Graphics[11] and SmartBIST by Cadence[12].

While LFSR reseeding is powerful for test data compression, it is not so good for power consumption. BIST implementation randomly fills don't care bits in the test cubes, which can result in excessive switching activities when the filled test patterns are shifted into the scan chains, which may give rise to severe hazards to the circuit reliability. One method has been proposed in [13] for reducing test power for LFSR reseeding, which considers together the problems of test data compression and low power test. In this scheme, two LFSRs are used. The main LFSR generates the test cube through conventional reseeding. An extra "masking" LFSR is used to generate a set of mask bits. Test power is reduced because the output of the two LFSRs are ANDed or ORed, thus reducing the transition probability. However, the test data storage for this scheme is greatly increased compared with conventional LFSR reseeding because it requires storing an extra set of seeds for the extra "masking" LFSR.

A novel parallel CWD (pCWD) approach is presented in [14] for lowering test power by shortening wrapper scan chains and adjusting test patterns. In [14], a two-phase process on test pattern: "partition" and "fill", is presented. Our method proposed in this paper expands Han's work, and applies relative idea to BIST. A key feature of the proposed approach is that it reduces the number of specified bits and the number of transitions at the same time. Thus, the total test storage and power dissipation can be reduced, since the amount of compression for LFSR reseeding depends on the number of specified bits and excessive transitions likely result in excessive power dissipation.

2. Scan Slices Overlapping

There are many don't care bits in test cubes. Conventionally, in BIST environment, the don't care bits are filled with random values, which result in excessive switching activity when they are shifted into a scan chain. Observing the X-bit distribution in test cubes can help us to find a method to handle this problem. Mintest patterns are examined. We get an observations from the result: The density of X-bit in circuit under test is high. To S13207, the X-bit occupies up to 93.2% of total test bits. "High density" of X-bit can make two consecutive scan slices in test patterns have a high probability to be overlapping. <u>Scan slice</u> is defined to be the set of inputs applied to the scan chain inputs at a scan cycle. <u>Scan slices overlapping</u> means that scan slices are same. See the example in Figure 1, the pattern contains 10 scan slices. Each slice consists of 4 bits. The first and second slice in the original pattern are:{{1 X X 1},{X X 1 1}}. If the second and third bits in the first slice and the first and second bits in the second slice are assigned to 1, they will be {{1 1 1 1}, {1 1 1 1}} and overlapping.

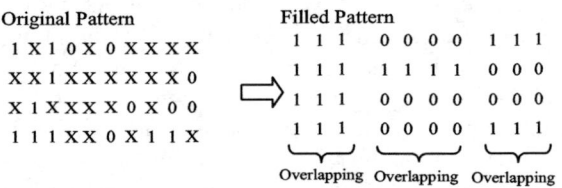

Figure 1. Slices overlapping in test patterns

3. Test Pattern Generation

Rather than using LFSR reseeding to directly encode the specified bits as in conventional LFSR reseeding, we can divide the test cube into blocks and slices of a block overlap each other. We call this kind of block as <u>overlapping block</u>. Then, for each test cube, when calculating the seed of this cube by solving equations, we can only take the first slice in every block into account.

---

*This paper is supported by the National Natural Science Foundation of China(NSFC) under grant No.90207002 and 60242001, key technique project of Beijing (No. H020120120130) and Basic Research Foundation of Institute of Computing Technology, Chinese Academy of Sciences(No. 20036160).

That is, when the LFSR expands a seed into a test pattern to randomly fill the scan chains, only the value of the first slice in every block is generated from the LFSR, and the value of the rest slices can be got by directly shifting the value of the first slices of blocks. After the scan chains have been filled, the test pattern is applied to the circuit under test and the response is shifted out to MISR. The process is then repeated to generate the next test pattern.

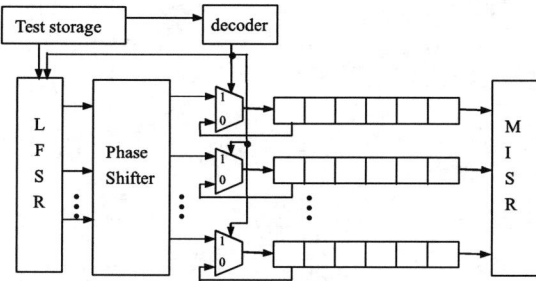

Figure 2. Hardware implementation

The hardware implementation for the proposed scheme is shown in Figure 2. Compared to the STUMPS structure, additional hardware overhead consists of one 2-to-1 MUX per scan chain and one decoder. The decoder is used to generate the 0 or 1 control signals to control the behavior of MUX. When the first slice of overlapping block is shifted into scan chains, the value of control signal is 1, so that the scan chains are loaded from the LFSR. And when the rest slices of overlapping block are shifted into the scan chains, the value of control signal is 0, so that the last value shifted into the scan chains is repeatedly shifted into the scan chains and the data from the LFSR is ignored. We use the control signals to gate the clock of the LFSR so that the power of the LFSR and the phase shifter is reduced at the same time, which leads to the reduction of the total power during the test.

4. Pattern partition

According to the proposed architecture, the pattern is needed to be partition into some overlapping slice sets. Obviously, in order to utilize the overlapping of consecutive scan slices to reduce the specified bits and transitions furthest, we should minimize the number of blocks. The partitions which can lead to the minimal number of blocks are not unique, and different partitions schemes may result in different results. A simple example is shown. See Figure 3, the original pattern contains 7 scan slices. Because of incompatibility between the third scan slice and the fifth scan slice, the original pattern should be divided into two overlapping blocks and the incompatible two scan slices should be contained in these two blocks respectively. Note that the forth scan slice is compatible with the two incompatible scan slice at the same time, which results in two different partition schemes as shown in Figure 3. We can see, after combination of compatible scan slices, total number of specified bits in the first partition scheme is 7, more than that in the second partition scheme which is reduced to 6.

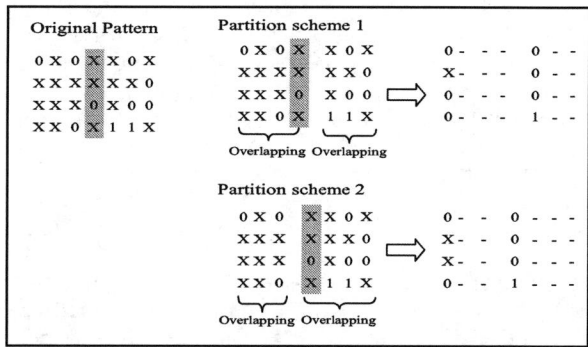

Figure 3. An illustration of different partition schemes

It's possible to find the best partition schemes by simulating all possible partition situations. However, this is time consuming work. In this paper, we use a simple heuristic method to get partition of overlapping blocks. The algorithm is described as Algorithm 1.

```
while( partition of scan slices is not finished )
    Judge if the current scan slice is compatible with frontal
    compatible sets.
    If (compatible)
        Jump to 1
    Else
        An overlapping block has been found, and all don't
        care bits in this block are specified.
1: continue the next scan slice
endwhile
```

Algorithm 1. pattern partition algorithm

An example of the pattern partition procedure is shown in Figure 4. Figure 4-(1) shows the original pattern. In Figure 4-(2), the original pattern is divided into overlapping blocks by the algorithm 1. In Figure 4-(3), in every overlapping block, the specified bits are combined into the first slice and the assignments of all slices are equal. Thus, the LFSR only need to generate the specified bits in the first slice of every block after partition and combination, and don't care (what the symbol '-'stand for) the rest slices.

This approach reduces the number of transitions in the scan chains and also reduces the total number of specified bits that must be generated by the LFSR as compared with conventional LFSR reseeding. It can be seen that the total number of specified bits in the original pattern is 16, and the total number is reduced to 10 after partition in this example.

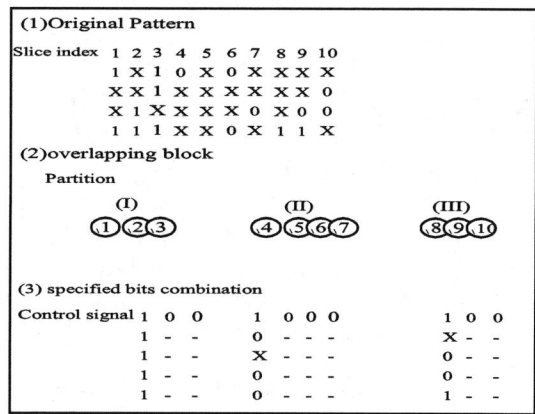

Figure 4. Pattern partition.

Besides the reduction of number of specified bits, the total number of transitions is also greatly reduced in this way. Now we make a simple analysis about reduction of transitions. Because the don't care bits in each test cube get filled with random values by the LFSR, the transitions may happen to every neighbor scan cell in original pattern. When the pattern is divided into overlapping blocks, the transitions only happen between blocks. Obviously, the total number of transitions can be reduced after pattern partition.

It must be noticed that the control signals become the one part of total test storage. Therefore in order to reduce test storage, the control signals should be compressed. This can be done using fixed run-length encoding.

5. Experimental results

To validate the efficiency of the proposed method, experiments were performed on the largest ISCAS 89 benchmark circuits. Test sets used in this experiment were obtained using Mintest dynamic compaction [15].

Table 1 presents the results. Each test cube is divided into overlapping blocks, and the number of blocks is shown in each case. The number of specified bits and the number of transitions required for the proposed scheme is compared with the original test cubes. The number of transitions is calculated as described in [16]. Column NP, NC and NB respectively show number of patterns, the number of scan chains and the corresponding number of overlapping blocks. Column NCB shows the total number of control signals before compression. Column T_D shows the size of the original test sets. The total number of specified bits after our method is applied to the test cube is shown in column partNSB. The last column of storage gives the percentage of compression. Finally, Column testing Power lists the original number of transitions(origNT), the number of transitions after dividing and filling(partNT), and the reduction of transitions(Pred%). As can be seen, our scheme yields great reduction in the total number of transitions, which will result in a significant reduction of power dissipation. And preferable test data compression is also obtained. It should be noted there is a tradeoff between the storage compression and the power reduction. When the number of scan chains is added, the test data compression increases, but the power reduction decreases. Figure 5 shows the change of test data compression and the power reduction with the number of scan chains.

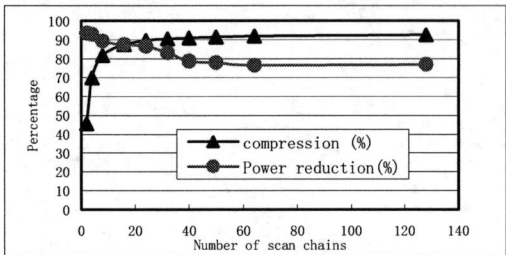

Figure5. Circuit s13207: Comp% vs. Pred%

Table 2 provides a test data compression comparison between the proposed scheme and the best previously published results[5-8], whose compression ratios are all relative to the size of the dynamic compacted test set from MinTest. Column T_D gives the size of precomputed test set. With the exception of one case when proposed scheme has lower compression rates than VIHC (s35932), the table clearly shows that the proposed method leads to better compression ratios than previous approachs.

Finally a comparison of the previous approaches [4,13] and the proposed scheme is given in Table3. To the each scheme, the first column shows the required storage and the second column shows the power reduction. Note that the power consumption resulting from the test pattern generation is not included in the power computing list in the table(the same to [4] and [13]), since this part power consumption is relatively very small to the power

Table1.Experimental Result for Proposed Scheme

Circuit				Data Storage				Test Power		
Name	NP	NC	NB	NCB	T_D	partNSB	Comp%	origNT	partNT	Pred%
s9234	160	4	2785	9920	39273	7970	54.45	587162	188445	67.91
		8	1881	4960		9102	64.20	292826	117485	59.88
		16	1250	2560		9713	68.75	150211	68102	54.66
		32	773	1280		10097	71.04	70027	40687	41.90
s13207	236	4	3053	41300	165200	8549	69.83	7174481	513681	92.84
		8	2266	20768		9518	81.67	3614877	393556	89.11
		16	1557	10384		10289	87.49	1786081	224177	87.45
		32	1064	5192		10645	90.42	869403	147262	83.06
s15850	126	4	3272	19278	76986	9199	63.02	2896184	505570	82.54
		8	2409	9702		10606	73.63	1447180	336952	76.72
		16	1597	4914		11603	78.55	736383	212328	71.17
		32	1007	2520		12092	81.02	378865	124083	67.25
s38417	100	4	260	41600	164736	25256	59.42	79814	20621	74.16
		8	9145	20800		31847	68.05	14347971	3580890	75.04
		16	7009	10400		38127	70.55	7249499	2270429	68.68
		32	5032	5200		44014	70.13	3678937	1406398	61.77
s38584	137	4	3319	50142	199104	26474	61.52	1804967	856030	52.57
		8	8798	25071		29882	72.39	17907510	3453106	80.72
		16	6046	12604		32292	77.45	8959395	2105740	76.50
		32	4013	6302		33688	79.92	4499316	1314616	70.78

Table 2. Test Data Compression Comparison for Same Test Set

Circuit	T_D	Test Data Compression (%)				
		Golomb [5]	FDR [6]	VIHC [7]	Selective Huffman[8]	Proposed
S5378	23754	40.70	48.19	51.52	55.1	**70.38**
S9234	39273	43.34	44.88	54.84	54.2	**71.04**
S13207	165200	74.78	78.67	83.21	77.0	**90.42**
S15850	76986	47.11	52.87	60.68	66.0	**81.02**
S35932	28208	N/A	10.19	**66.47**	N/A	53.23
S38417	164736	44.12	54.53	54.51	59.0	**70.13**
S38584	199104	47.71	52.85	56.97	64.1	**79.92**

Table 3. Result comparing proposed scheme with previous schemes

Circuit	Dual-LFSR reseeding[13]		Alternating run-length code[4]		Proposed	
	Storage	Pred%	Storage	Pred%	Storage	Pred%
S9234	19440	24.35	21612	76.30	12273	54.66
S13207	11803	25.26	32648	93.68	15837	83.06
S15850	14518	25.14	26306	85.27	16517	71.17
S38417	66234	24.90	64976	81.35	48527	61.77
S38584	23835	24.70	77372	83.52	44896	70.78

resulting from the transitions. When compared to dual-LFSR reseeding proposed in [13], the proposed scheme is much more effective and obtains up to 60% better power reduction (s13207). Because dual-LFSR reseeding scheme uses different test sets, the optimality of the ATPG and compaction procedures used to obtain the test sets strongly affects the results, and 1000 pseudo-random patterns are applied first. Even though all of this, our method needs less storage in s9234 and s38417. When compared to alternating run-length code in [4], the reduction of test power for the proposed scheme is not as much as the method proposed in [4] which is based on run-length encoding and there is still 20% gap, but the storage is much less than the latter.

6. Conclusions

LFSR reseeding is an attractive approach for compressing test data. We have shown that the proposed deterministic BIST scheme provides a way to reduce test power for LFSR reseeding. By utilizing *Scan slices overlapping* combined with LFSR reseeding, it can reduce transitions during test(up to 90%) and specified bits in test cubes. Experimental results indicate that the proposed method significantly reduces test power and even further reduces test storage. The proposed technique can be combined with other techniques, such as partial reseeding and seed encoding, and thus the better results can be obtained.

Reference

[1] B. Koenemann, "LFSR-coded test patterns for scan designs",Proc.of ETC, pp. 237-242, 1991
[2] S.Hellebrand, J.Rajski, S.Tarnick, S.Venkataraman, and B.Courtois, "Built-in test for circuits with scan based on reseeding of multiple-polynomial linear feedback shift registers", IEEE Trans. On Comput., vol.44, pp. 223-233, 1995.
[3] C.V.Krishna, Nur A. Touba, "Reducing Test Data Volume Using LFSR Reseeding with Seed Compression,"Proc. of ITC, pp. 321-330, 2002.
[4] Chandra, A., and K. Chakrabarty, "Reduction of SOC Test Data Volume, Scan Power and Testing Time Using Alternating Run-length Codes," Proc. of DAC, pp. 673-678, 2002.
[5] A. Chandra and K. Chakrabarty, "System-on-a-chip test data compression and decompression architectures based on Golomb codes," IEEE Trans. CAD, vol. 20, pp. 355–368, March 2001.
[6] A. Chandra and K. Chakrabarty, "Frequency-directed run-length (FDR) codes with application to system-on-a-chip test data compression," Proc. of VTS,pp.114-121, Apr.2001.
[7] P. T. Gonciari, et.al, "Improving compression ratio, area overhead, and test application time for system-on-a-chip test data compression /decompression," Proc.of DATE.,pp. 604–611, 2002.
[8] A. Jas., et.al, "An efficient test vector compression scheme using selective Huffman coding," IEEE Trans. CAD, Volume: 22,pp. 797- 806, June 2003.
[9] I. Bayraktaroglu and A. Orailoglu, "Test Volume and Application Time Reduction Through Scan Chain Concealment", Proc. of DAC, pp. 151-155, 2001.
[10] H.-G.Liang, et.al, "Two dimensional test data compression for scan-based deterministic BIST," Proc. ITC., pp. 894–902, 2001.
[11] J. Rajski., et.al, "Embedded deterministic test for low cost manufacturing test" , Proc. of ITC, pp. 301-310, 2002.
[12] B.Koenemann, et.al, "A SmartBIST variant with guaranteed encoding," Proc. Of VTS, pp. 325-330, 2001.
[13] P.M.Rosinger, et.al, "Low Power Mixed-Mode BIST Based on Mask Pattern Generation Using Dual LFSR Re-seeding," Proc. of ICCD, pp. 474-479, 2002.
[14] Y. Han, et.al, "Wrapper scan chains design for rapid and low power testing of embedded cores", Proceedings of the 13th Asian Test Symposium, Pingdong, pp.9-14,Nov.,2004
[15] I. Hamzaoglu and J.H. Patel,"Test set compaction algorithms for combinational circuits,"Proc. of CAD, pp.283-289, Nov. 1998.
[16] R. Sankaralingam, et.al, "Static compaction techniques to control scan vector power dissipation," Proc. of ATS., pp. 35-40, 2000.

Modeling and Formal Verification of Dataflow Graph in System-Level Design Using Petri net

Tsung-Hsi Chiang
Dept. of Electrical and Control Engineering
National Chiao Tung University
Hsinchu, Taiwan R.O.C.
aries.ece89g@nctu.edu.tw

Lan-Rong Dung
Dept. of Electrical and Control Engineering
National Chiao Tung University
Hsinchu, Taiwan R.O.C.
lennon@cn.nctu.edu.tw

Ming-Feng Yaung
Dept. of Electrical and Control Engineering
National Chiao Tung University
Hsinchu, Taiwan R.O.C.
mifon.ece92g@nctu.edu.tw

Abstract—Formal verification in system-level, which also means architecture verification, is different from functional verification in RTL level. DSP algorithms need high-level transformation to achieve optimal goals before mapping on a silicon. However, suitable CAD tool is absent to support the simulation and verification in high-level. This paper presents a novel modeling and high-level verification methodology based on Petri net (PN) model. By proposed method, a system of DSP algorithm in the form of FSFG is transformed into PN model. Moreover, verification methods which include static and dynamical phases are applied in PN domain. At last, we introduce our software implementation, called HiVED, to show the experimental results.

I. INTRODUCTION

With increasing enormous size and design complexity of DSP system, verifying the correctness of hardware synthesis is becoming more and more important, especial in VLSI design. In various verification techniques, formal verification which ensures 100% coverage of function and system model correctness has gained large attention. Formal verification means using a mathematical proof to check whether the system satisfies the required specification. In [1], [2], the authors give excellent survey of major trends of formal verification techniques which include equivalence checking [2] and model checking [3]. Equivalence checking is used to proof the functional equivalence of two design representations modeled at the same or different levels of abstraction. Model checking is an automatic technique for checking a model of the design that is satisfied for a given specification covers all the properties that the system should hold. However, most verification techniques focus on the RTL or gate level. In the behavioral or high-level domain, well-established and practical verification techniques do not exist.

In our study, we put verification beyond the RTL or gate level and focus at high-level. The proposed high-level verification techniques, which include static and dynamical phases, are based on Petri net (PN) model. Two-phase verification with respect to the system design flow is showing in Figure 1. In static phase, many iterative or recursive applications can be represented by fully specified flow graph (FSFG). In order to achieve the iteration period bound (IPB) and obtain perfect FSFG, graph-based high-level transformations [4], such as retiming and unfolding techniques [5], [6], are employed for shortening the iteration period (IP) as much as possible. Once the intermediary result produced, it is transformed and verified in PN domain.

In dynamical phase, PN model is used to verify the schedule schemes of given FSFG. In order to map DSP algorithm on a hardware circuit, scheduling algorithms are used to find suitable schedules of FSFG. For high performance issues, such as smaller silicon area, higher throughput rate, and lower power consumption, these algorithms start from as-late-as-possible (ALAP) to as-soon-as-possible (ASAP) schedule time and assign the starting time of each task of FSFG [7]. The execution of all tasks is referred to as an iteration. Within an iteration, the execution sequences are different from various scheduling algorithms, but the original behavior is not changed. Once a schedule scheme of FSFG is produced from scheduling algorithms, the permanent behavior of system and schedule scheme are transformed and verified in PN domain as well in static phase.

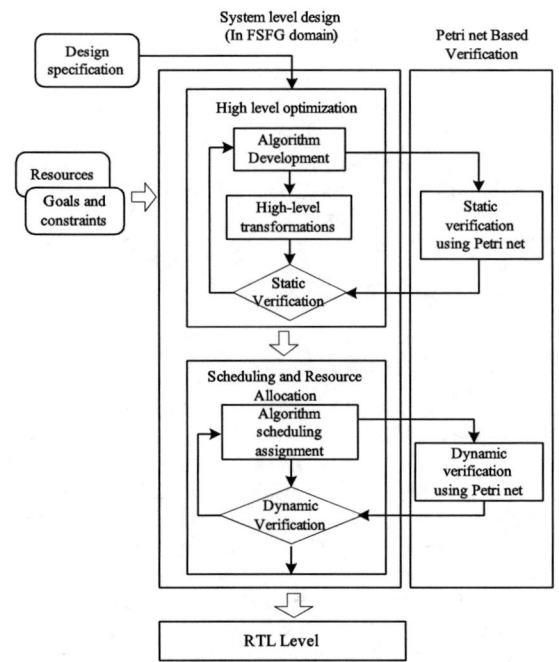

Fig. 1. System-level design flow and the PN-based verification

In system-level design flow, static and dynamical phases attempt to achieve perfect FSFG from the original one and find an optimal static schedule of design. The design procedure requires complex recursive processes and evaluations. However, one may suffer from several design faults at high-level in that. In order to unveil high-level faults, designers have to apply verification techniques in their design flow.

PN-based verification can be considered as a model checker with capability of checking a number of properties that should be hold under system specification. In PN domain, the inputs to the model checker include a characteristic matrix describing the system and the system properties described in matrix equations. By applying mathematics theory and matrix manipulation, the model checker reports whether the design is valid. When comparing with FSFG, designing DSP algorithm at FSFG domain and verifying it at PN domain provides double-checking verification solution. Additionally, PN theory also provides primary property analysis techniques of system, such as reachability, liveness, safeness, boundedness, and reversibility [8] that FSFG can not be offered.

This paper is organized as follows. First, we describe the background and the methodology of transformation from FSFG to PN model. Then, PN based high-level verification flow is presented in Section 3. In Section 4, we introduce our verification software application tool called HiVED (High-Level Verification Engine for DSP) and show some experimental results. Finally, we make some concluding remarks in the last section.

II. DEFINITION AND TRANSFORMATION

In the following section, we will discuss some useful properties and proposed transformation technique to transform a FSFG into PN model.

A. Fully specified signal-flow graph (FSFG)

Fully specified signal-flow graph (FSFG) or DFG is a natural paradigm for describing DSP algorithms. A FSFG $G_{FSFG}(V, E, D)$, where $V = \{v_1, \ldots, v_m\}$ and $E = \{e_1, \ldots, e_n\}$, is a three-tuple directed and edge-weighted graph which contains a vertex set V, a directed edge set E, and a ideal delay set D. Vertex set V represents atomic operation of functional units. Directed edge set E describes the direction of flow of data between functional units. Inter data dependencies between functional units are represented by weighted edges. Figure 2, for instance, shows a DSP algorithm of second-order IIR filter.

B. Petri Net

A Petri net $G_{PN}(P, T, W, M_0)$ is a four-tuple [9], where $P = \{p_1, \ldots, p_n\}$ and $T = \{t_1, \ldots, t_m\}$ are finite sets of places and transitions, W is the weighted flow relation, and M_0 is the initial marking. A marking is a function $M : P \rightarrow \mathbf{N}$. If $M(p_i) = k$ for place p_i, we will say that p_i is marked with k tokens. If $W(u, v) > 0$, then there is an arc from u to v with weight $W(u, v)$. For a node u in $P \cup T$, $^\bullet u$ (the pre-set of u) is specified by:

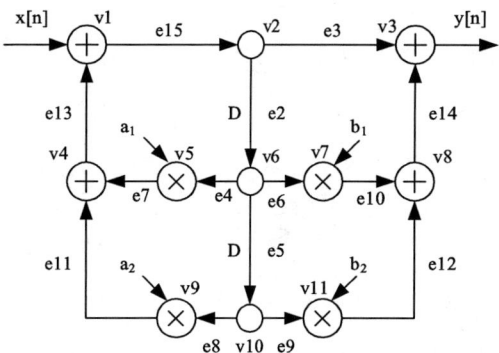

Fig. 2. A DSP algorithm of second-order IIR filter

$^\bullet u = \{v \in P \cup T | W(v, u) > 0\}$ and $u^\bullet$ (the post-set of u) is specified by: $u^\bullet = \{v \in P \cup T | W(u, v) > 0\}$. A PN can *execute* by firing *enabled* transitions. A transition t is *enabled* at marking M (denoted by $M[t\rangle$) if $\forall p \in {}^\bullet t : M(p) \geqslant W(p, t)$. Once a transition t is enabled at a marking M, it may fire and then reach a new marking M' (denoted by $M[t\rangle M'$). The occurrence of t lead to a new marking M', defined for each place p by

$$M'(p) = M(p) - W(p, t) + W(t, p). \quad (1)$$

A sequence of transitions $\sigma = t_1 \cdots t_{k-1} \in T^*$ is a *firing sequence* from a marking M_1 to a marking M_k if and only if there exist markings $M_2, \ldots, M_{k-1}$ such that

$$M_i[t_i\rangle M_{i+1}, \text{ for } 1 \leqslant i \leqslant k-1. \quad (2)$$

Marking M_k is said to be *reachable* from M_0 if and only if there exists a firing sequence $\sigma : M_0[\sigma\rangle M_k$. $[M\rangle$ is the set of markings reachable from M by firing any sequence of transitions, i.e., $M' \in [M\rangle \Leftrightarrow \exists \sigma \in T^* : M[\sigma\rangle M'$. $[M_0\rangle$ is the set of all markings reachable from M_0.

Matrix representation of PN is defined by *incidence matrix* A (also called the characteristic matrix), which is a $|P| \times |T|$-matrix with entries

$$A_{ij} = W(t_j, p_i) - W(p_i, t_j). \quad (3)$$

The matrix representation usually gives a complete characterization of PN. Let $x_j = \{t_j\} = (\ldots, 0, 1, 0, \ldots)$ be the unit $|T| \times 1$ column vector which is zero everywhere except in the j-th element. Also, let μ is the $|P| \times 1$ column vector respected to a marking M_0 with entries $\mu_i = M_0(p_i)$. The transition t_j is represented by the column vector x_j. A transition t_j is enabled at a marking M_0 (denoted by $M_0[t_j\rangle$) if $\mu \geqslant A \cdot x_j$. And the result of firing enabled transition t_j in a marking μ_0 is represented by

$$\delta(\mu_0, t_j) = \mu_0 + A \cdot x_j. \quad (4)$$

For a sequence of transition firing $\sigma : M_0[\sigma\rangle M_k$ and $M_i[t_i\rangle M_{i+1}, 1 \leqslant i \leqslant k-1$, we have

$$\begin{aligned}\delta(\mu_0;\sigma) &= \delta(\mu_0;t_1 t_2 t_3 \ldots t_{k-1}) \\ &= \mu_0 + \sum_1^{k-1} A \cdot x_j \\ &= \mu_0 + A \cdot f(\sigma).\end{aligned} \quad (5)$$

The vector $f(\sigma)$ is firing vector of the sequence $\sigma = t_1 \ldots t_{k-1}$. The i-th element of $f(\sigma)$, $f(\sigma)_i \in \mathbf{N}$, is the number of times that transition t_i fires in the σ.

C. Transformation from FSFG to PN model

The FSFG is attractive to algorithm developers because it directly models the equations of DSP algorithm. Yet, it does not sufficiently unveil the dynamical behavior and the implementation limits in terms of the degree of parallelism and the memory requirement. Thus, we use Petri net to model DSP algorithms. It also allows us to discover the characteristic of the target architecture and to observe the dynamical behavior of the algorithm.

The FSFG $G_{FSFG}(V, E, D)$ of a DSP algorithm can be modeled as PN $G_{PN}(P, T, W, M_0)$ by applying the following rules:

1) Functional element set V is transformed into the transition set T whose elements have computational power.
2) The edge set E is transformed into the place set P denoting the system states.
3) Since each place in PN has only one output, the pseudo transition of each fork edge will be added as source duplicators.
4) In static phase, the delay element set D of edge in FSFG is corresponded to the number of tokens of place in PN. In dynamical phase, moving tokens between places represents the executing order of functional element set V in FSFG.

PN tokens in static and dynamical phases are quite different. In static phase, tokens can be represented as delay elements of FSFG, thus moving tokens between places can be seen as retiming delay elements. In dynamical phase, moving tokens represents the execution orders of function element of V. For instance, Figure 3 shows the PN model of the second-order IIR filter in Figure 2.

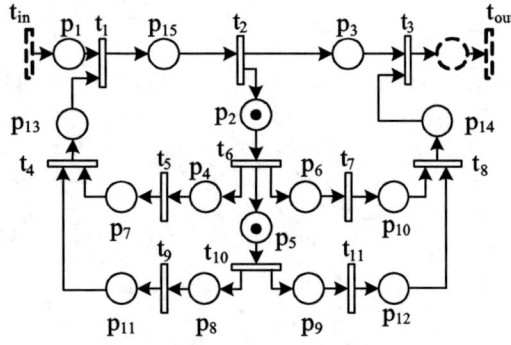

Fig. 3. A PN model of second-order IIR filter

III. HIGH-LEVEL VERIFICATION

The proposed PN-based two-phase high-level verification is described in this section.

A. Static verification

In static verification, Figure 4, the DUV (design-under-verification) to be verified are system design optimization algorithms, such as retiming, unfolding, and loop shrinking methods [4]–[6]. System design optimization techniques are used to reduce the critical path delay, silicon area, number of latches, and power reduction. However, these algorithms have a lot cost that limits their use. In applications, if we could tell whether the optimization algorithm results are correct quickly we could reduce the computation of these algorithms. To simply show our methodology, we address retiming verification in static verification phase.

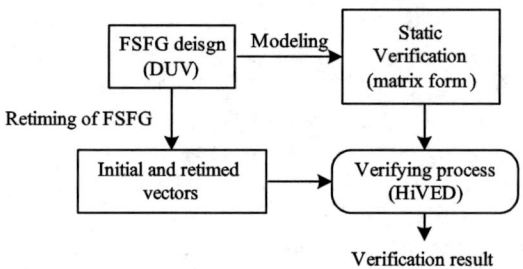

Fig. 4. Static phase verification

As mentioned in section 2, FSFG can be transferred into PN model using proposed method. Since, retiming problem can be transferred to the token state reachability problem, thus it is possible to analyze retiming problem using PN theory. For a given initial token state M_0 and static characteristic matrix D, after firing transition vector x the finial token state becomes M' and we have and we have $M' \in [M\rangle \Leftrightarrow \exists \sigma \in T^*: M[\sigma\rangle M'$. Thus, the matrix equation can also be expressed as:

$$\mu' = \mu_0 + A \cdot x. \quad (6)$$

The delays of FSFG can also be scaled by a positive integer number n without affecting the functional behavior of the FSFG, thus it becomes:

$$\begin{aligned}\mu' &= n \cdot \mu_0 + A \cdot x, \text{ (where } n \in \mathbf{N}) \text{ or} \\ A \cdot x &= (\mu' - n \cdot \mu_0).\end{aligned} \quad (7)$$

Using linear system theory, equation $A \cdot x = y$ has a unique solution x if and only if A has *full rank*. In our case, $y = (\mu - n \cdot \mu_0)$ is known and A is the characteristic matrix of PN, we can check the rank of A and determine whether exists a unique solution x.

B. Dynamical verification

Dynamical verification process at system level is shown in Figure 5. In our approach, the design under verification (DUV) is modeled as a dynamic PN model. The scheduling

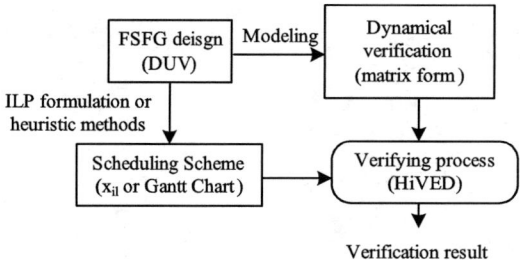

Fig. 5. Dynamical phase verification

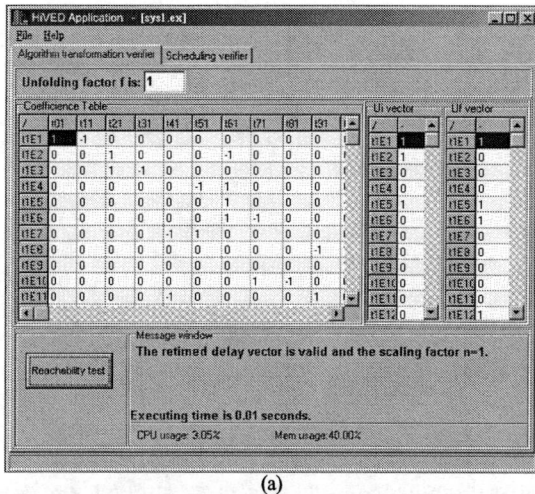

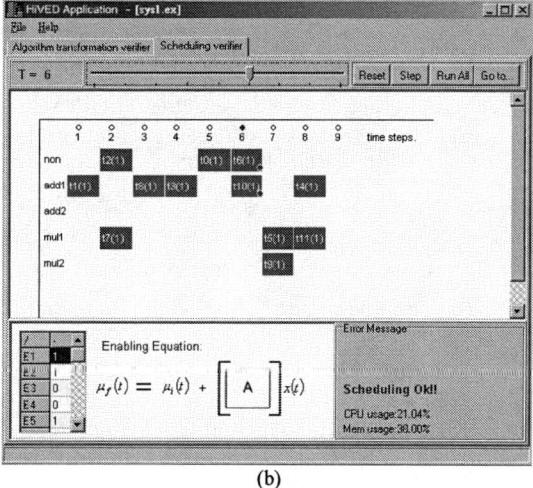

Fig. 6. (a) Static phase verification, and (b) dynamical phase verification of second-order IIR filter by using HiVED verification software.

scheme that needed to be verified can be either solved by Integer Linear Programming (ILP) [7], [10] or found by heuristic method. The solution of scheduling scheme is in binary decision variable matrix X or Gantt chart form. Once the schedule solution is found, we can obtain firing vector from schedule solution to apply reachability checking in dynamic verification phase as showing in previous section.

IV. IMPLEMENTATION

In order to prove our methodology, we develop verification software HiVED (High-Level Verification Engine for DSP) in this study. The software application is implemented by using Borland C++ Builder on Microsoft Windows XP. In static phase, as shown in Figure 6(a), the inputs to HiVED are PN characteristic matrix of second-order IIR filter, Figure 2, the initial and retimed vectors. In a few time, HiVED will report the retimed vector is correct or not. In dynamical verification phase, Figure 6(b), the inputs to HiVED are PN characteristic matrix and schedule schemes from various schedule algorithms. In the center of HiVED workplace, it shows the schedule from input file. After push the aided buttons, HiVED will show the result step by step animatedly.

V. CONCLUSION

In this study, we explored the modeling and formal verification of dataflow graph in system level design using PN model. The verification flow includes two phases, static and dynamical. In static phase, proposed methodology checks system architecture design using PN characteristic matrix equations. In dynamical phase, a FSFG schedule also can be verified by checking reachability of PN firing sequence.

The main contribution of our work is to put formal verification beyond RTL level by using PN model. PN theory also provides analysis methods that FSFG can not offer. The implementation software HiVED also proves our verification methodology. We believe to have contributed fostering the evidence that high-level verification to be widely used in DSP system design beyond RTL level.

REFERENCES

[1] A. Gupta, "Formal hardware verification methods: a survey," *Formal Methods in System Design*, vol. 1, pp. 151–238, 1992.
[2] C. Kern and M. Greenstreet, "Formal verification in hardware design: a survey," *ACM Transactions on Design Automation of E. Systems*, vol. 4, pp. 123–193, Apr. 1999.
[3] E. M. Clarke, O. Grumberg, and D. A. Peled, *Model checking*. The MIT Press, 1999.
[4] K. Parhi, "Algorithm transformations for concurrent processors," *In Proceedings of the IEEE*, vol. 77, no. 12, pp. 1879–1895, Dec. 1989.
[5] K. Parhi and D. Messerschmitt, "Static rate-optimal scheduling of iterative data-flow programs via optimum unfolding," *IEEE Transactions on Computers*, vol. 40, no. 2, pp. 178–195, Feb. 1991.
[6] L.-F. Chao and E. H.-M. Sha, "Scheduling data-flow graphs via retiming and unfolding," *IEEE Transactions on parallel and distributed systems*, vol. 8, no. 12, pp. 1259–1267, Dec. 1997.
[7] L.-R. Dung and H.-C. Yang, "On multiple-voltage high-level synthesis using algorithmic transformations," *IEICE Transactions on Fundamentals*, 2004.
[8] C. Girault and R. Valk, *Petri nets for systems engineering*. Springer-Verlag, 2003.
[9] W. Reising and G. Rozenberg, *Lectures on Petri nets I: Basic Models*. Springer-Verlag, 1998.
[10] K. Ito, L. E. Lucke, and K. K. Parhi, "Ilp-based cost-optimal dsp synthesis with module selection and data format conversion," *IEEE Transactions on Very Large Integration Systems*, vol. 6, no. 4, pp. 582–594, Dec. 1998.

A Robust and Correct Computation for the Curvilinear Routing Problem

Tan Yan and Hiroshi Murata
Graduate School of Environmental Enginneering
The University of Kitakyushu
1-1, Hibikino, Wakamatsu-ku, Kitakyushu, 808-0135, Japan
Email: {yantan, hmurata}@env.kitakyu-u.ac.jp

Abstract— Curvilinear routing has been a long standing problem in layout design. Several algorithms have been proposed based on real numbers but none of them consider the numerical error caused by digital computation which might tamper the robustness and correctness. This paper shows that the robustness and the correctness can be guaranteed if the input is given in integers, by presenting a novel implementation for the algorithm proposed by Duncan et al. [3].

I. INTRODUCTION

Curvilinear routing has been a critical problem for MCM routing[5], Hybrid-IC routing[8] and IC package routing[9]. More recently, many modern circuit packaging methods such as 2.5D system[1] and sytem-in-package(SiP)[2] integrate chips onto substrates that support curvilinear planar routing. The *Curvilinear Routing Problem* (CRP) is to transform given topological paths into physical detailed routing with minimum Euclidian length while preserving the topology of the paths as well as separating the wires by given distances. An example is shown in Fig. 1.

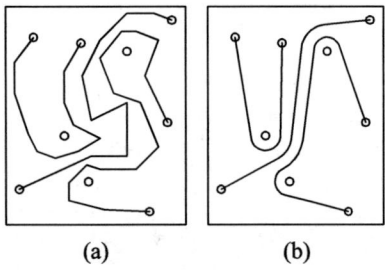

Fig. 1. An Example of the Curvilinear Routing Problem: (a) Input feature points and topological paths (b) Output detailed routing.

The work presented in this paper was motivated by our experience in the curvilinear routing of industrial designs. Algorithms commonly used to rely heavily on division; with floating point numbers, there are unavoidable roundoff errors. For large and complex designs, these errors can accumulate, possibly resulting in incorrect results and poor algorithmic stability.

Rather than trying to bound the degree of error, we instead have developed an approach to avoid error entirely. A key observation is that if the input to our algorithm contains only integers, every required computation can be performed using the ratio of two integers. The range of values that can be represented by a pair of integers covers the entire set of values needed by all possible computations; this is not true for floating point values, even if one uses double precision.

The rest of this paper is organized as follows: Section II presents the background and prior work. Section III gives a brief description of the wire growing process and the FED algorithm. Our robust and correct implementation is then presented in Section IV and its limitation is given in section V. Section VI extends our implementation to a more general model. A conclusion is given in Section VII.

II. BACKGROUND & PREVIOUS WORK

After the concept of the *sketch* was introduced by Leiserson et al. [6] to describe the topology of a planar routing, CRP was first formulated by Liu et al. [7][10] using a *wire growing process* as the definition of topological changelessness.

The input of a CRP is a planar layout which consists of a set of feature points O, a set of 2 terminal nets represented by polygonal paths P and an uniform separation rule w indicating the minimum distance between two paths. The output of the problem is expected to be a set of curves C with minimum Euclidian length which preserve the topology of the input and satisfy the separation rule, or an error message if such curves do not exist. The output curves consist of arcs centered at feature points and straight segments connecting those arcs.

Liu et al. also propose one solution using the funnel method that routes wires sequentially. Gao et al. [4] give a detailed description of the funnel method and later Hama et al. [5] improve the average performance of the algorithm. By directly simulating the wire growing process, Duncan et al. [3] propose the *Fat Edge Drawing* (FED) algorithm which simultaneously routes all the wires. Their algorithm reduces the worst case complexity by the use of bundling technique.

Their algorithms are all based on real numbers but in practice, the routing problem is not necessarily formulated in real number domain. This paper focuses on a subproblem of CRP. The Integer CRP (ICRP) is the same as CRP except that the input $\{O, w, P\}$ is given in integer domain. Note that the computation of ICRP is still in real domain, discussion on robustness and correctness of such algorithms is non-trivial.

An implementation of a CRP algorithm is said *robust* if it does not produce any exception during the process, such as dividing a number by zero or square-rooting a negative number. The implementation is also said *correct* if the sequences of the arcs implied in the output curves C are correct. Note that here we mean topological correctness. Inexactness in the coordinates of the output curves is inevitable due to the limitation of digital computers.

This paper presents a robust and correct implementation of the FED algorithm for ICRP. The main idea is to show that the order of the events during the wire growing process can be determined using integer arithmetics only.

III. WIRE GROWING PROCESS

In [10], the maintenance of the topology is defined by a wire growing process which regards wires as strips with certain width and let these strips grow from zero width to the required separation following some rules. Later in [3], Duncan et al. consider every wire as a virtual *Fat Edge* with expected width same as the separation distance w. The final routing of each wire goes through the the central line of the corresponding fat edge. The FED algorithm simulates the growing process of such edges and a bundle structure is introduced to reduce the time complexity.

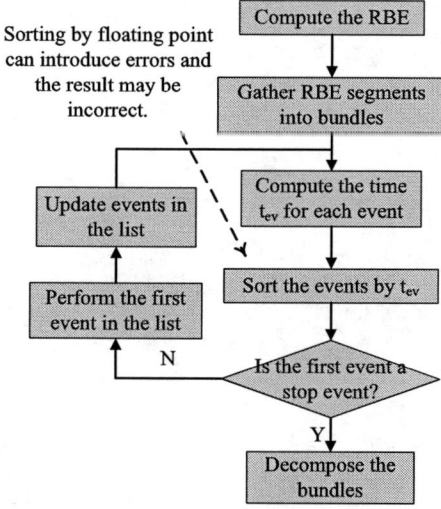

Fig. 2. Overview of the FED algorithm. Correctness of the results depends on the order of the events.

A sketch of the FED algorithm is shown in Fig. 2: (Step 1) Compute the set of Rubber-Band Equivalents(RBE) P' of the input topological paths P. This could be done easily by the funnel method[6]. (Step 2) Break the rubber-band equivalents into segments and classify them into a set of bundles B, each of which collects all the segments sharing same starting and ending feature points. (Step 3) Simultaneously increase the thickness of each bundle in B as follows: A *time* parameter $t \in [0, 1]$ is introduced to control the growth. The widths of the bundles are proportional to t and reach the required separation rule while t reaches 1. During the increase of t from 0 to 1, the growing process may cause one bundle to be split or two bundles to be merged, which are called *events*.

This step is further divided into: (Step 3a) Compute the time t_{ev} of the first event for each bundle and store it to a event list L. (Step 3b) Sort L. (Step 3c) Perform the first event in L. If this event is a Stop Event, i.e. two arc-like bundle hit each other and thus neither of them can further grow, then report an error message and terminate the algorithm. (Step 3d) Update L. (Step 3e) Terminate Step 3 when t hits 1 or else goto (3c). (Step 4) Decompose bundles into segments and arcs and reorganize them along each wire. An illustrative example of such a process is shown in Fig. 3.

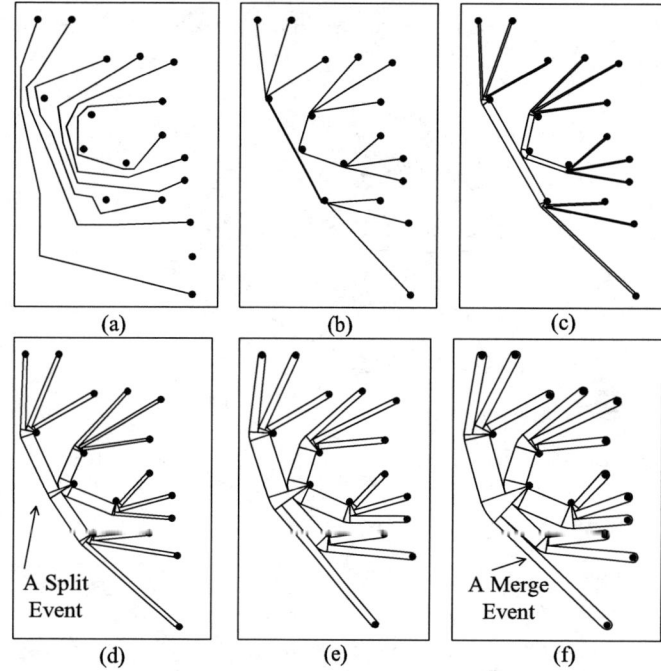

Fig. 3. The Wire Growing Process: (a) Input Topological Path. (b) Rubber-band Equivalents. (c) Bundles and Wire Growing Process. (d) Splitting Event. (e) Keep Growing. (f) Merge Event.

IV. INTEGER COMPUTABILITY OF THE ALGORITHM

As shown in Fig. 2, using floating point number in the computation might result in an incorrect output. To avoid it, we use only integer arithmetics in which the robustness and the correctness are automatically secured.

Definition 1: A problem is called *integer computable* if the correct output can be achieved using only integer arithmetic. It's trivial that a rational number can be represented by the division of two integers. Then we have:

Lemma 1: Arithmetic and sorting on rational numbers are integer computable. □

The most critical part in the FED algorithm is in Step 3b, sorting the events by their times. Unfortunately, the time of an event can be irrational. However, it satisfies an important property as follows.

Lemma 2: The square of time t_{ev}^2 for arbitrary event is rational.

Proof : There are three kinds of events: Split, Merge and Stop event. Let us consider the split event and merge event first.

A split or merge event happens at time t_{ev} when a thick bundle touches three circles. Let L be the thick bundle with width w and A, B and C be the three circles with radii r_A, r_B and r_C respectively. The radii of the circles and the width of the line are proportional to each other:

$$\frac{r_A}{c_a} = \frac{r_B}{c_b} = \frac{r_C}{c_c} = \frac{w}{c_w} = t_{ev} \quad (1)$$

where c_a, c_b, c_c and c_w are constants indicating the expected separation rules of A, B, C and L. We also use A, B and C to denote the center of circles A, B and C when not ambiguous. Each of the circles is tangent to either side of the thick line. We examine t_{ev} in the following by classifying the situation into two cases:

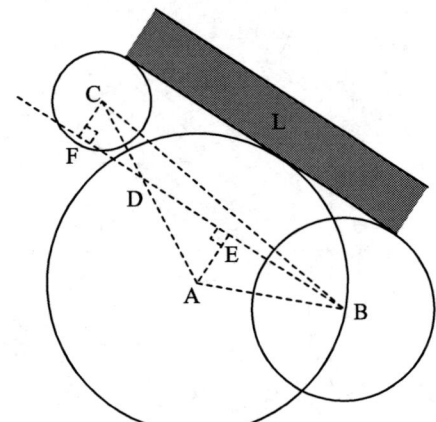

Fig. 4. Case 1 in the Proof of Lemma 2

Case 1: Three circles lie at the same side of the line. Without loss of generality, we assume $c_a \geq c_b \geq c_c$. Let us first discuss the subcase $c_a > c_b > c_c$ which also means $r_A > r_B > r_C$. Through B we draw a line BD parallel to L. It will intersect AC at D. From C and A, we draw two perpendicular lines to BD and denote their feet as F and E respectively. Then we have:

$$|CF| = r_B - r_C = (c_b - c_c) \cdot t_{ev} \quad (2)$$
$$|AE| = r_A - r_B = (c_a - c_b) \cdot t_{ev} \quad (3)$$

From the similarity of $\triangle CDF$ and $\triangle DEA$ we have:

$$\frac{|CD|}{|DA|} = \frac{|CF|}{|AE|} \quad (4)$$

By (2), (3) and (4) we can calculate D's coordinates as:

$$x_D = \frac{(c_b - c_c) \cdot x_A + (c_a - c_b) \cdot x_C}{c_a - c_c} \quad (5)$$

$$y_D = \frac{(c_b - c_c) \cdot y_A + (c_a - c_b) \cdot y_C}{c_a - c_c} \quad (6)$$

Then we get the linear equation of line BD:

$$(y_D - y_B) \cdot x + (x_B - x_D) \cdot y + (x_D y_B - x_B y_D) = 0 \quad (7)$$

The distance between C and line BD is $|CF|$ which means:

$$|CF| = \frac{|(y_D - y_B)x_C + (x_B - x_D)y_C + (x_D y_B - x_B y_D)|}{\sqrt{(y_D - y_B)^2 + (x_B - x_D)^2}} \quad (8)$$

By (2), (5), (6) and (8) we have (12). It can be seen that t_{ev}^2 is a rational number.

Now we discuss the subcase when two constants share the same value. One such subcase is $c_b = c_c$. If this happens, we will have $|CF| = 0$ and D coincides with C. Then we know that $|AE|$ is the distance from point A to line BC:

$$|AE| = \frac{|(y_C - y_B)x_A + (x_B - x_C)y_A + (x_C y_B - x_B y_C)|}{\sqrt{(y_C - y_B)^2 + (x_B - x_C)^2}} \quad (9)$$

By (3) and (9) we have:

$$t_{ev}^2 = \frac{|(y_C - y_B)x_A + (x_B - x_C)y_A + (x_C y_B - x_B y_C)|^2}{(c_a - c_b)^2 ((y_C - y_B)^2 + (x_B - x_C)^2)} \quad (10)$$

The claim also holds. Other subcases could be discussed in a similar way.

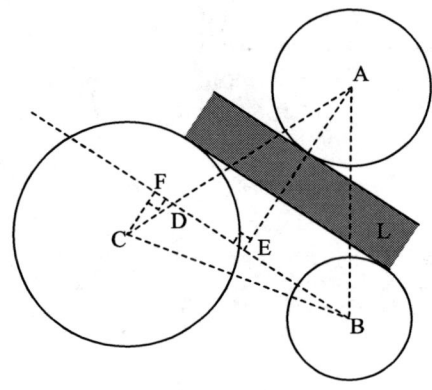

Fig. 5. Case 2 in the Proof of Lemma 2

Case 2: Two circles lie at the same side of the line while the other lies at the other side. Without loss of generality we assume B and C lie at the same side and $c_c \geq c_b$ which also means $r_C \geq r_B$. Similar to Case 1 we have the result (13). Then the claim also holds for this case.

A Stop event happens at time t_{ev} when two circles touch each other. Suppose the two circles are centered at A and B with radii r_A and r_B respectively. Then we have:

$$\frac{r_A}{c_a} = \frac{r_B}{c_b} = t_{ev} \quad (11)$$

It's obvious that the distance between A and B, $|AB|$, is the square root of a rational number given that A and B's coordinates are integers. Thus, $t_{ev}^2 = |AB|^2/(c_a + c_b)^2$ is rational. Thus the claim holds for all events. □

Now we have the conclusion:

Theorem 1: The FED algorithm can be implemented for Integer CRP with robustness and correctness.

Proof (sketch) : In the FED algorithm, Step 1 is naturally integer computable since the funnel method involves only integral operations. For Step 2, we merge all the RBE segments sharing the same starting and ending feature points into one bundle. Such operations are obviously integer computable.[1]

---

[1] They sorts the orientations of the segments around one feature point in order to reduce the time complexity in [3]. If we use the slopes to represent such orientations, sorting the slopes is also integer computable.

$$t_{ev}^2 = \frac{(((c_b - c_c)y_A + (c_a - c_b)y_C)(x_C - x_B) + ((c_b - c_c)x_A + (c_a - c_b)x_C)(y_B - y_C) + (c_a - c_c)(x_B y_C - y_B x_C))^2}{(c_b - c_c)^2(((c_b - c_c)y_A + (c_a - c_b)y_C - (c_a - c_c)y_B)^2 + ((c_a - c_c)x_B - (c_b - c_c)x_A - (c_a - c_b)x_C)^2)} \quad (12)$$

$$t_{ev}^2 = \frac{\begin{array}{c}(((c_c - c_b)y_A + (c_a + c_w + c_b)y_C)(x_A - x_B) + ((c_c - c_b)x_A \\ + (c_a + c_w + c_b)x_C)(y_B - y_A) + (c_a + c_w + c_c)(x_B y_A - y_B x_A))^2\end{array}}{\begin{array}{c}(c_a + c_w + c_b)^2(((c_c - c_b)y_A + (c_a + c_w + c_b)y_C - (c_a + c_w + c_c)y_B)^2 \\ + ((c_a + c_w + c_c)x_B - (c_c - c_b)x_A - (c_a + c_w + c_b)x_C)^2)\end{array}} \quad (13)$$

The output of Step 2 is a Compact Routing Structure[3] which maintains the topology and structure of all the bundles. It represents a bundle by the labels of its comprising wires, the labels of its depending feature points, the tag how it winds around such points and its expected width. All these items are either labels or integers and is thus easily managed by integral operations. Step 3 is based on such a structure and maintaining it in Step 3c(performing the events) is naturally integer computable. Lemma 2 shows that the square of event time t_{ev}^2 can be sorted using only integers and thus Step 3a & 3b are also integer computable. Therefore, Step 1 to 3 are integer computable and the sequence of arcs can be obtained correctly with robustness. It's obvious that Step 4 can be performed robustly without changing the sequence of arc of Step 3. Thus the whole process of the FED algorithm can be implemented with robustness and correctness. □

V. Limitation of Our Implementation

During integral computations in the two critical equations (12) and (13), multiplying an integer might cause overflowing. To prevent its occurrence, a 10-bit input range might require 120-bit integers during computation in the worst case. But practical cases always allow us to have larger input range. Moreover, integer overflowing is easy to detect, unlike the roundoff error in floating point number, with which one can hardly tell how many bits are enough for robustness.

VI. A More General Model

In this section, we extend the CRP to a more general routing model and implement the FED algorithm on it. Instead of a uniform separation rule w, we assume each wire p_i has its own width w_i and separation distance s_i. The distance $Dist(p_i, p_j)$ between the central lines of two wires p_i and p_j must satisfy:

$$Dist(p_i, p_j) \geq w_i/2 + w_j/2 + \max(s_i, s_j) \quad (14)$$

In Step 2 of the FED algorithm, when we bind wire segments $\{p_1, p_2, \ldots, p_m\}$ together to make a bundle, the expected width of the bundle is decided as follows: (1) The expected distance $Dist(p_i, p_{i+1})$ between central lines of two adjacent segments p_i, p_{i+1} in the bundle is calculated as in (14) if the segments belong to different wires or is 0 otherwise. (2) The expected width of the whole bundle is the sum of all the expected distance between adjacent segments: $Width = \sum_{1 \leq i \leq m-1} Dist(p_i, p_{i+1})$. In Step 3, we will handle the case when two bundles touch each other.(Please refer to the proof of Lemma 2.) In this case, an additional spacing rule SW between such two bundles is considered when we sort the events. Here, $SW = Dist(p_i, p_j)$ where p_i and p_j are the two outmost segments of the two bundles respectively which touch each other.

It's obvious that the robustness and correctness of our implementation is maintained in this general model given that the wire width w_i and the separation distance s_i are integers.

VII. Conclusion

This paper describes a robust and correct implementation of the FED algorithm for Integer Curvilinear Routing Problem. In order to avoid the roundoff error of floating point numbers, the implementation involves only integer arithmetics. The robustness and correctness is naturally guaranteed. The author also suggests that the technique used in this paper might also be applicable to many other computations in Euclidian geometry if the given input and required output are in integer domain. Instead of bounding the error of floating point number computation, one might use only integer arithmetics instead.

Acknowledgment

The author would like to thank Prof. Patrick H. Madden from State University of New York at Binghamton for the helpful discussion.

References

[1] Y. Deng and W. Maly, *2.5D System Integration: A Design Driven System Implementation Schema*, Proc., ASPDAC, pp. 450-455, 2004.
[2] S. K. Pienimaa, J. Miettinen and E. Ristolainen, *Stacked Modular Package*, IEEE Trans on Advanced Packaging, Vol. 27, no. 3, pp. 461-466, 2004.
[3] C. Duncan, A. Efrat, S. Kobourov and C. Wenk, *Drawing Graphs with Fat Edges*, Proc., 9th International Symposium on Graph Drawing, pp. 162-177, 2001.
[4] S. Gao, M. Jerrum, M. Kaufmann, K. Mehlhorn, W. Rülling and C. Storb, *On Continuous Homotopic One Layer Routing*, Proc., 4th Annual Symposium on Computational Geometry, pp. 392 - 402, 1988.
[5] T. Hama and H. Etoh, *Curvilinear Detailed Routing with Simultaneous Wire-Spreading and Wire-Fattening*, IEEE Trans on CAD, Vol. 18, no. 11, pp. 1646-1653, 1999.
[6] C. E. Leiserson and F. M. Maley, *Algorithms for Routing and Testing Routability of Planar VLSI Layouts*, Proc., 17th Annual ACM Symposium on Theory of Computing, pp. 69 - 78, 1985.
[7] E. Liu, *Two dimensional IC layout compaction*, Ph.D dissertation, Univ. of Calgary, Nov. 1986.
[8] H. Murata and Y. Kajitani, *Interactive terminal sliding algorithm for Hybrid IC planar layout*, ISPJ Journal, Vol.35, No.12, pp.2806-2815, 1994. (in Japanese)
[9] H. Murata, *A New Routing Design Methodology For Multi-Chip IC Packages*, Proc. MWSCAS, pp. 473-476, 2004.
[10] J. Valainis, S. Kaptanoglu, E. Liu and R. Suaya, *Two-Dimensional IC Layout Compaction Based on Topological Design Rule Checking*, IEEE Trans on CAD, Vol. 9, no. 3, pp. 260-275, 1990.

Estimating Likelihood of Correctness for Error Candidates to Assist Debugging Faulty HDL Designs

Tai-Ying Jiang
Dept. of Electronics Engineering
National Chiao Tung University
Hsinchu, Taiwan, ROC
giani@eda.ee.nctu.edu.tw

Chien-Nan Jimmy Liu
Dept. of Electrical Engineering
National Central University
Chung-Li, Taiwan, R.O.C.
jimmy@ee.ncu.edu.tw

Jing-Yang Jou
Dept. of Electronics Engineering
National Chiao Tung University
Hsinchu, Taiwan, ROC
jyjou@ee.nctu.edu.tw

Abstract—*Debugging priority* is a helpful technique to assist debugging faulty HDL designs [9]. However, *debugging priority* obtained by sorting *confidence score* is not good enough due to the inaccuracy in estimating likelihood of correctness for error candidates. Therefore, we developed Refined Confidence Score for deriving better *debugging priority*.

I. INTRODUCTION

Once functional mismatches occur during verification, Design Error Diagnosis and Correction (DEDC) is required. Most of previous approaches on DEDC are mainly for the implementation described at gate-level or even lower level. They can be roughly divided into simulation-based approaches [1,2,3] and symbolic approaches [4,5,6]. In comparison, symbolic approaches are more accurate but suffer memory explosion while handling large circuits.

If design errors occur in the design stage of modeling circuit behavior using HDLs, DEDC is traditionally conducted by manually tracing faulty HDL codes and is time-consuming. Thus, some researches try to reduce error candidates to save the debugging time [7,8]. However, since not every error candidate in the reported list is a design error, searching bugs in the report blindly may still waste trials in testing correct statements before getting any bug.

Jiang and et al not only reduces the number of error candidates but also report error candidates in an order, from the most suspected to the most innocent one [9] (referred to as *debugging priority*). Searching design errors guided by *debugging priority* can derive errors in fewer trials as compared to searching blindly. Unfortunately, *Confidence Score* (CS) developed in [9] is not so accurate (a detailed discussion is given in section II) that *debugging priority* can not be suggested well and the saved time is limited.

Therefore, we try to develop Refined Confidence Score (RCS) to more accurately estimate likelihood of correctness for each error candidate to derive a better *debugging priority*. Experimental results show that the *debugging priority* sorted by our RCS is often better than the one sorted by CS and the computation time is acceptable.

II. MOTIVATION

In this section, we use an example to demonstrate the proposed techniques in [9] and the flaw of its Confidence Score (CS). Suppose the HDL code the designer wants to write is the code fragment in Figure 1(a) but statement *S11* is carelessly written into "assign a = PI1;". Applying the test stimulus in Figure 1(c) to simulate the HDL design gets the waveform in Figure 2. A mismatch between the expected value of PO1 (Expected_PO1) and the simulated value (PO1) occurs at the time instance *t=55*. We apply the techniques proposed in [9] to first obtain error space and then calculate CS for reporting error candidates in *debugging priority*.

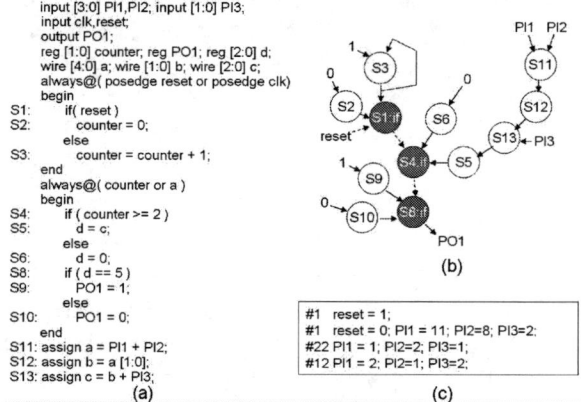

Figure 1. An Example

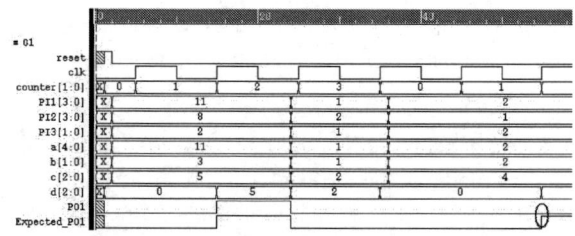

Figure 2. Waveform

By taking the intersection of the *relation space* and executed statements, we get the *error space* {S1,S3,S4,S5, S8,S10,S11,S12,S13}. Then, we use *sensitized statements* (SSs) in the time period from t=1 to t=45 to calculate CS. SSs are found by conducting Depth First Search (DFS) backtracing starting from Primary Outputs (POs). For example, SSs at the time instance t=1 are S8, S10, S4, S6, S1, and S2. S8, S10, S4, and S1 are in the *error space* and thus get one point of CS at t=1. After accumulating CS, we sort the error candidates according to their CS and derive the report list with *debugging priority* as shown in Figure 3. Numbers in the round brackets ( ) are CSs.

```
S12 (2) : assign b = a [1:0];
S13 (2) : assign c = b +PI3;
S11 (2) : assign a = PI1;  ← the design error
S5  (2) : d = c;
S3  (5) : counter = counter +1;
S10 (5) : PO1 = 0;
S1  (6) : if ( reset )
S4  (6) : if ( counter >= 2 )
S8  (6) : if ( d == 5 )
```

Figure 3. Report with debugging priority

The design error S11 is not placed at the rank of first, but the third. Searching design errors according to this *debugging priority* wastes two trials in testing the two correct statements S12 and S13. Inaccuracy of CS is because operations in the HDL code may mask erroneous values. At the time instance when S11 gets its first point (t=15), "a[1:0]" masks the erroneous value (11=5'b01011) on a and generates the output value 3 (2'b11), no matter the value of *a* at t=15 is correct (19=5'b10011) or wrong (11=5'b01011). At the time S11 gets its second point (t=25), similar masking error situation happens as well. Masking error situations can become worse if design errors hide behind more of these operations that mask errors, such as ">", "[:]", and e.t.c..

III. REFINED CONFIDENCE SCORE CALCULATION

If no operations could mask erroneous values, expected values appearing at POs ensure that the SSs are very possibly correct. Consequently, for an error candidate Si in error space, we define our Refined Confidence Score of Si, RCS(Si), as the likelihood that an erroneous value on Si's left-hand variable u is propagated to POs (Likelihood Of Propagating Error of u, LOPE(u)). The higher LOPE(u) is; the less probably an erroneous value on u is masked and the more probably Si is correct.

A. Work Overview

Given the faulty HDL implementation, the dumpfile obtained from simulation, *error space* (a set of *error candidates*), and the clock cycle in which functional mismatches occur at primary outputs or registers (Error-Occurring Clock cycle (EOC)), we calculate our RCS for each error candidate in the *error space* and obtain *debugging priority*. In this paper, the given *error space* is found by the technique proposed in [9]. However, one can apply some other techniques to derive an *error space*.

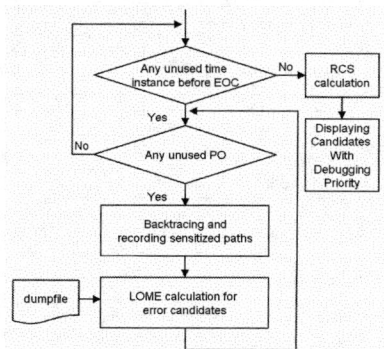

Figure 4. The flow of RCS calculation

The flow of RCS calculation is shown in Figure 4. After constructing CDFG for each PO, for each time instance $t=ti$ before EOC, we conduct DFS backtracing from each PO and record the *sensitized path* (SP) when reaching an error candidate. After completing traversing one PO, we calculate the likelihood that an erroneous value on each error candidate's left-hand variable u is masked at $t=ti$ (Likelihood of Masking Error of u at $t=ti$, LOME(u)@$t=ti$) and traverse another. Continue the above process until each time instance before EOC is applied. Finally, we calculate RCS for each error candidate using the formulas derived in the next section and output the report of error candidates displayed with *debugging priority*.

B. Formulas Derivation

We say that an erroneous value w on a variable u is masked by some statements in some SPs from u to a primary output POi at a time instance $t=ti$ if the function of these SPs generates the expected value of POi at $t=ti$ (**EV**) with w.

$$f(w, v_1, v_2, ..., v_n) = \mathbf{EV} \qquad (1)$$

, where variables $v1$ to vn are other involved variables.

The set of all these w is defined as Masked Value Set (MVS) from u to POi as described by formula (2). For a HDL with m primary outputs $\{PO_1, PO_2, ..., PO_m\}$, since an erroneous values on u is masked on its way to each PO at $t=ti$. Masked Value Set of u at $t=ti$ (MVS(u)@$t=ti$) should be the values in the intersection set of each $MVS_{u \to POi}$.

$$MVS_{u \to POi} = \{w \mid f(w, v_1, v_2, ..., v_n) = \mathbf{EV}\} \qquad (2)$$

Example:

We take the example in Section II to illustrate what MVS is. At time instance $t=15$, by testing each value of a one by one exhaustively, we have MVS($S11$)@$t=15$ is {3,7,11, 15,19,23,27,31}. Each value in this MVS can result in the expected value of PO1 at $t=15$ (1).

The bigger size MVS(u)@$t=ti$ is; the more likely an erroneous value is in MVS(u)@$t=ti$ and masked at $t=ti$. Therefore, the LOME(u) at $t=ti$ (Likelihood of Masking Error of u at $t=ti$ (LOME(u)@$t=ti$)) can be estimated by formula (3).

$$LOME(u)@t = t_i = \frac{(|MVS(u)@t = t_i| - 1)}{2^n} \qquad (3)$$

, where n is the bitwidth of variable u.

Suppose that time instances before EOC are $\{t_1, t_2, ..., t_n\}$. LOME of u is estimated as described in formula (4) since erroneous values on u should be masked at each time instance in $\{t_1, t_2, ..., t_n\}$.

$$LOME(u) = \prod_{i=1}^{n} \frac{(|MVS(u)@t = t_i| - 1)}{2^n} \qquad (4)$$

LOPE of u should be the complement of LOME of u as described in formula (5). Our RCS calculation problem turns into the problem of deriving MVS for each error candidate's Left-Hand Variable (LHV).

$$LOPE(u) = 1 - LOME(u) = 1 - \prod_{i=1}^{n} \frac{(|MVS(u)@t = t_i| - 1)}{2^n} \qquad (5)$$

C. A Quick Approximated Masked Value Set Computation

With the recorded SPs and the dumpfile, we compute MVS for each error candidate's LHV. First, we introduce the key theorem to compute exact MVS in the case of single SP. Then, we explain approximated MVS derivation for the case of multiple SPs.

Assume there is a single SP from vertex v_1's LHV u to POi $\{v_1, v_2, ..., v_i, v_{i+1}, ..., v_n, PO_i\}$ at time instance $t=t1$. The function of vertices v_2 to v_i is function f, the function of vertex v_{i+1} to v_n is function g, and the entire function from v_1 to POi is function h as illustrated in Figure 5. In other words, function h is the cascade of function f and g.

$$h(u, a_1, ..., a_m, b_1, ..., b_n) = g(f(u, a_1, ..., a_m), b_1, ..., b_m) \quad (6)$$

, where $\{a1, a2, ..., am\}$ and $\{b1, b2, ..., bn\}$ are other operands that are not on the recorded SP.

Theorem 1 *For a value w on u, w is in $MVS_{v_1 \to POi}$ if and only if function f can generate an output value that is in $MVS_{vi \to POi}$ with the input value w.*

$$MVS_{v_1 \to PO_i} = \{w \mid f(w, a_1, a_2, ..., a_m) \in MVS_{vi \to PO_i}\} \quad (7)$$

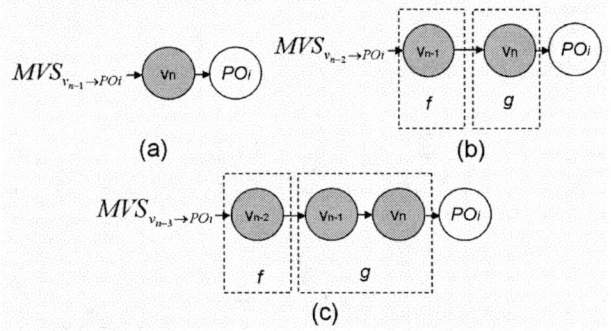

Figure 5. An inductive scheme to compute MVS

To compute $MVS_{v_1 \to POi}$, Theorem 1 suggests a quick inductive scheme. We first check the type of vertex v_n and find a suitable formula in TABLE I according to v_n's type to compute $MVS_{v_{n-1} \to POi}$ as shown in Figure 6(a). Next, by taking v_n's function as function g and v_{n-1}'s function as function f, we again apply a suitable formula in TABLE I to compute $MVS_{v_{n-2} \to POi}$ by inheriting the previously calculated $MVS_{v_{n-1} \to POi}$ as shown in Figure 6(b). We can obtain $MVS_{v_{n-3} \to POi}$ similarly as shown in Figure.6(c). Continue the above processes in an inductive scheme until $MVS_{v_1 \to POi}$ is computed. The notation "$\{..p..\}$" in TABLE I means p is one element in $MVS_{v_i \to PO_i}$. "n" is the bitwidth of v_{i-1}. "Bj" is the value on the jth branch of a conditional vertex. "b_1" is the value of the other operand that is not on our SP if vertex v_i is a binary operation.

We use the example in Section II to illustrate the MVS computation. At $t=15$, SSs from S11 to PO1 are S8, S4, S5, S13, S12, and S11. $MVS_{PO1 \to PO1}$ is the value of PO1 at $t=15$ and is $\{1\}$. The *conditional vertex if:S8* has two branches, the FALSE branch $b0$ and the TRUE branch $b1$ (*CASE* may have more than two branches). Since only the value on branch $b1$ is in $MVS_{PO1 \to PO1}$, we find that the evaluation result of "d==5" is TURE using the formula in the second row in TABLE I. By applying the formula in the 19th row, we obtain $MVS_{S4 \to PO1} = \{5\}$. Applying the formula in the first row can easily get $MVS_{S5 \to PO1} = \{5\}$ since S5 is on the taken branch of S4. $MVS_{S13 \to PO1} = \{5\}$ is obtained using the formula in the third row. $MVS_{S12 \to PO1} = \{3\}$ is obtained using the formula in the 6th row with the parameter $b1$ equals to 2 (PI3=2 at $t=15$). Finally, we derive $MVS_{S11 \to PO1} = \{3,7,11,15,19,23,27,31\}$ by applying the formula in the 5th row with $n=5$, $i=1$, and $j=0$. The result of $MVS_{S11 \to PO1}$ is exactly the same as the one in section II obtained by exhaustively testing each value of a. According to formula (3), LOME$(a)@t=15$ is $7/32 = 0.21875$.

TABLE I. MVS FORMULAS

Operation	$MVS_{v_{i-1} \to PO_i}$ if $MVS_{v_i \to PO_i} = \{...p...\}$
conditional vertex	$\{p\}$ is v_{i-1} is on the taken branch
conditional vertex	$\{j \mid B_j \in MVS_{v_i \to PO_i}\}$ if v_{i-1} is a control signal
$v_i = v_{i-1}$	$\{p\}$
$v_i = \sim v_{i-1}$	$\{2^n - p\}$
$v_i = v_{i-1}[i:j]$	$\{\bigcup_{k=0}^{2^{n-i-1}} \{p \cdot 2^j + k \cdot 2^{i+1}\}\}$
$v_i = v_{i-1} + b_1$	$\{p - b_1\}$
$v_i = v_{i-1} * 0$	$\{0 \sim 2^n - 1\}$
$v_i = v_{i-1} * b_1, b_1 > 0$	$\{\lfloor p/b_1 \rfloor\}$
$v_i = v_{i-1} - b_1$	$\{p + b_1\}$
$v_i = b_1 - v_{i-1}$	$\{b_1 - p\}$
$v_i = v_{i-1} >> b_1$	$\{[p \cdot 2^{b_1} \sim q \cdot 2^{b_1} + 2^{b_1} - 1]\}$
$v_i = b_1 >> v_{i-1}$	$\{[0 \sim 2^n - 1]\}$ if $p = 0$. Otherwise, $\{b_1\}$
$v_i = v_{i-1} << b_1$	$\{\bigcup_{k=0}^{2^{n-b_1}-1} \{[k \cdot 2^{b_1} + \lfloor p/2^{b_1} \rfloor]\}\}$
$v_i = b_1 << v_{i-1}$	$\{[0 \sim 2^n - 1]\}$ if $p = 0$. Otherwise, $\{b_1\}$
$v_i = v_{i-1} > b_1$	$\{[b_1 + 1 \sim 2^n - 1]\}$
$v_i = v_{i-1} >= b_1$	$\{[b_1 \sim 2^n - 1]\}$
$v_i = v_{i-1} < b_1$	$\{[0 \sim (b_1 - 1)]\}$
$v_i = v_{i-1} <= b_1$	$\{[0 \sim b_1]\}$
$v_i = v_{i-1} == b_1$	$\{b_1\}$
$v_i = v_{i-1} != b_1$	$\{0 \sim 2^n - 1\} - \{b_1\}$

If the number of recorded SPs is multiple, we simply leave each error candidate's MVS to be the entire universe and use other single SPs recorded in backtracing other POs to obtain MVS by taking intersection of all the MVSs obtained in the case of single SPs. This approximated MVS computation obtains an upper bound approximated MVS that offer a lower bound RCS estimation. In other words, we adopt a conservative estimation for this complex RCS calculation to save computation time.

After calculating each error candidate LOME at each time instance before EOC, we calculate RCS for each error candidate according to formula (5). The report of error candidate displayed with *debugging priority* is given in Figure 7. Numbers in the round ( ) are RCSs. We may notice that the design error S11 is exactly placed at the rank of first and no other statements get the same RCS points as S11. Searching design error according to this *debugging priority* can intermediately find the design error S11 and save the two trials in testing correct statements, as compared to the *debugging priority* in Figure 3.

```
S11 (0.78125) : assign a = PI1;   ← the design error
S3  (0.99902) : counter = counter +1;
S1  (0.99976) : if ( reset )
S10 (1.00000) : PO1 = 0;
S5  (1.00000) : d = c;
S4  (1.00000) : if ( counter >= 2 )
S8  (1.00000) : if ( d == 5 )
S12 (1.00000) : assign b = a [1:0];
S13 (1.00000) : assign c = b +PI3;
```

Figure 6   Debugging priority using RCS

IV. EXPERIMENTAL RESULTS

In this section, we show our experimental results on five designs written in Verilog HDL. The design Mx2X2 is a design for the 2x2 matrix multiplication. The design FSM is a simple mealy finite state machine used to control traffic lights. Div16 is a 16-bit integer divider. Rankf is a rank filter. FFT performs Fast Fourier Transform. The number of lines in each HDL code is given in the column "Line".

For each HDL design, we arbitrarily change some statements in the design to create multiple designs errors. Total 50 experimental cases are conducted to compare the *debugging priority* sorted according to our RCS and the *debugging priority* sorted according to CS. We define "The rank of design errors in the report" as "1-(its rank in the reported error list) / (total number of statements in the reported error list)" for demonstrating the advantage of our RCS. The column "100~80 (%)" records the number of times that our design errors are placed in the first 20% in the reported error list. The column "80~50 (%)" records the number of times that the created design errors are put in the first 20%~50% of the reported error list. The column "50~ (%)" records the number of times that our created design errors are put in the last 50%. From the recorded experiment results in the above three columns, we can see that with the *debugging priority* sorted according to our RCS, design errors are indeed placed in the front of the report list more often. Thus, finding design errors with the guidance of our *debugging priority* sorted according to RCS can save more time. Moreover, in the column "AVG Computation Time (s)", we may see that the computation time to derive this more accurate RCS is acceptable.

TABLE II.   **EXPERIMENTAL RESULTS**

Design	Lines	The rank of design errors in the report (CS / RCS)			AVG. Computation Time (s) (CS/ RCS)
		100~80 (%)	80~50 (%)	50~ (%)	
Mx2X2	80	38 / 45	12 / 5	0 / 0	0.8 / 1.0
FSM	113	42 / 39	5 / 11	3 / 0	0.7 / 0.9
Div16	235	19 / 37	22 / 8	9 / 5	1.3 / 1.7
Rankf	656	20 / 43	28 / 7	2 / 0	3.1 / 4.0
FFT	1530	13 / 30	22 / 20	15 / 0	4.3 / 5.3

V. CONCLUSION

We develop a quick estimation technique to accurately estimate likelihood of correctness for each error candidate. Better *debugging priority* can be obtained by sorting error candidates according to our *refined confidence score* in a decreasing order. This *debugging priority* can indeed further save the time spent on locating design errors in the report list of error candidates and the time to derive this more accurate estimation is just little more than the time in computing *confidence score*.

REFERENCES

[1] P.Y Chung, Y.M. Wang, and I. N. Hajj, "Diagnosis and correction of logic design errors in digital circuits", in Proceeding of Design Automation Coference, 1998.

[2] S.Y. Huang, and K. T. Cheng, "Error tracer: design error diagnosis based on fault simulation techniques", in IEEE Transaction of CAD, 1999.

[3] D. W. Hoffmann, and T. Kropf, "Efficient Design error correction of digital circuits", in Proceeding of Int'l Conference on Computer Design, 2000.

[4] D. Brand, "Incremental synthesis", In Proceeding of Int'l Conference on Computer Aided Design, 1992.

[5] M. Tomita, and H. H. Jiang. "An algorithm for locating logic design errors", in Proceeding Design Automation Conference, 1994.

[6] H. T. Liaw, J. H. Taih, and C.S. Lin, "Efficient automatic diagnosis of digital circuits", in Proceeding of Int'l Conference on Computer Aided Design, 1990.

[7] V. Boppana, I. Ghosh, R. Mukherjee, J. Jain, and M. Fujita, "Hierarchical error diagnosis targeting RTL circuit", in Proceeding of Int'l Conference on VLSI Design, 2000.

[8] C. H. Shi, and J. Y. Jou, "An efficient approach for error diagnosis in HDL design", in Proceeding of ISCAS, 2003.

[9] T.Y. Jiang, C. N. Liu, and J. Y. Jou, "Effective error diagnosis for RTL designs in HDLs", in Proceeding of Asia Test Symposium, 2002.

Instruction-Based Delay Fault Self-Testing of Pipelined Processor Cores

Virendra Singh[1,3], Michiko Inoue[1], Kewal K Saluja[2], Hideo Fujiwara[1]

[1]Nara Institute of Science & Technology, Takayama, Ikoma, Nara – 630-0192 Japan
[2]University of Wisconsin-Madison, USA
[3]Central Electronics Engineering Research Institute, Pilani, India
{virend-s, kounoe, fujiwara}@is.naist.jp, saluja@engr.wisc.edu

Abstract—Although nearly all modern processors use pipelined architecture, yet no method has been proposed in literature to model these for the purpose of test generation. This paper proposes a graph theoretic model of pipelined processors and develops a systematic approach to delay fault testing of such processor cores using the processor instruction set. Our methodology consists of using a graph model of the pipelined processor, extraction of architectural constraints, classification of paths, and generation of tests using a constrained ATPG. These tests are then converted to a test program, a sequence of instructions, for testing the processor. Thus, the tests generated by our method can be applied in functional mode of operation and can also be used for self-test. We applied our method to two example processors, namely a 16 bit five stage VPRO pipelined processor and a 32 bit pipelined DLX processor, to demonstrate the effectiveness of our methodology.

I. INTRODUCTION

In modern high performance processors, it is no longer sufficient to target conventional stuck-at faults, instead delay faults and cross talk faults are becoming increasing important. At-speed testing using external tester is almost infeasible because of its inherent accuracy limitation and cost. Due to the need of design change, possibility of excessive power consumption, and high area and performance overhead, hardware-based self-test (BIST) is also not a feasible solution. Software-Based Self-Test (SBST) is an alternate to BIST which uses processor instructions and functionality in order to test processor core in functional mode.

A number of approaches have been proposed for testing non-pipelined processors targeting stuck-at faults but a very few [2,3] approaches have been proposed to test pipelined processors. Chen [2] proposed a template based methodology, whereas Kranitis[3] proposed an approach which targets functional blocks. Though pipelined processors are studied in [2,3], the pipelined behavior is not considered, instead the focus is on functional blocks; and also faults in the controller are not explicitly addressed. Researchers in [4,5,6] proposed SBST approaches targeting delay faults for non-pipelined processors. However, Lai's [4] approach does not provide details about testing the controller, whereas [5, 6] provide an efficient graph theoretic model based approach, it is also limited to non-pipelined processors, though it can handle architectural registers and an FSM based controller.

This work was supported in part by Semiconductor Technology Academic Research Center (STARC) under the Research Project and in part by Japan Society for the Promotion of Science (JSPS) under Grants-in-Aid for Scientific Research B (2) (No. 15300018).

The paper is organized as follows. Section II lists the contributions of this paper and describes an overview of our methodology. Section III and IV describe the test generation methods for datapath and controller. Section V describes the instruction sequence generation process. Section VI presents experimental results to demonstrate the effectiveness our methodology, and finally we conclude with section VII.

II. CONTRIBUIONS AND OVERVIEW OF THE PROPOSED APPROACH

To the best of our knowledge, no approach has been proposed in literature for testing pipelined processors and targeting delay faults. We believe this is the first work towards modeling of pipeline behavior for testing of a microprocessor in functional mode.

The main contributions of this work are:

1. Develop a graph theoretic model for pipeline behavior using the RT level description of the processor,
2. Provide a systematic approach to test the processor based on the model.

This paper presents a unified approach to test normal and bypassing/forwarding paths in the datapath by using a graph model of the behavior of the datapath and the controller. A hierarchical approach is presented for the test generation, which classifies paths at RT level and extract constraints for potentially testable paths to generate test vectors at gate level using constrained ATPG. Path delay fault model [7] is used in this work.

Unlike a non-pipelined processor, in which one instruction must finish execution before the execution of the next instruction, in a pipelined processor multiple instructions can be in various stages of execution. This makes its behavior more complex. These stages can be viewed as independent hardware units and all the stages execute instructions concurrently. In order to support concurrent execution of instructions, necessary data and control signals are carried along as an instruction progresses in the pipeline stages. Simultaneous execution of multiple instructions can lead to data, control and structural hazards. Data bypassing is a commonly used technique to resolve data hazards and stalling is used for the unresolved hazards. Data flows from the first pipeline stage to the last pipeline stage during the normal execution (without any hazard). It is very difficult to separate datapath and controller part in a pipelined processor as every pipeline stage carries all the data and control signals required by the pipeline stages ahead of it. Nonetheless, our model defines them clearly, yet for testing the paths in the datapath part and the control part are treated separately. The data transfer activities between the architectural registers and data and address (memory address and register address) part of the pipeline registers are assumed be in the datapath and the remaining paths are considered in the control part.

A graph theoretic model called pipeline instruction execution graph (PIEG), has been developed that is constructed by using instruction set architecture and RT level description. It is based on instruction execution graph (IE-Graph) introduced by us in [5,6], and similar work in [1] for non-pipelined processors. Our present model classifies paths as functionally testable (FTP), functionally untestable (FUTP), potentially functionally testable (PFTP), and parity check functionally untestable (PCFTP). After the classification, it extracts constraints for the PFTP and PCFTP paths. First, constraints on the control signals in one or more relevant pipeline stages are extracted and then the constraints on justifiable data in the data registers or pipeline registers under the control constraints are extracted. PCFTPs are further classified FUTPs or PFTPs. A combinational constrained ATPG is used for the test vector generation for the PFTPs. We can get test sequences without using ATPG for FTPs, and test sequence is not needed for FUTPs. For testing the controller, the constraints on the legitimate values for a group of control signals are extracted by using RT level description. PIEG is used along with these constraints for further extraction of control and data constraints for target control paths, and their classification. Constraint ATPG is used to generate the test vectors. Finally, an instruction sequence to apply the generated test vectors, is generated by using the knowledge of the control signals of various pipeline stages and the PIEG.

III. DATAPATH

In this section we consider the paths that are responsible to transfer data between architectural registers or data and address between pipeline registers, which are significant in number. Other paths are considered in the control part. Datapath of a pipelined processor can be modeled by PIEG which captures the pipeline behavior, and is used for constraint extraction, path classification, and instruction sequence generation. Nodes of the PIEG are: (i) registers, (ii) part of registers which can be independently readable and writeable, (iii) equivalent registers (set of registers which behave identically for all instructions, like register file), (iv) two special nodes, IN and OUT, which model the external world such as memory and IO devices, and (v) data and address (memory address and register address) part of pipeline registers. A directed edge is drawn between two nodes iff there exists at least one instruction responsible to transfer data (with or without manipulation) between the corresponding two registers. Each edge is marked with a 4-tuple [<*instruction set*>, <*stage from, stage to*>, <*distance*>, <*logic type*>]. This 4-tuple signifies that the instructions from the <*instruction set*> are responsible for the transfer of data from <*stage from*> to <*stage to*> through the logic specified by <*logic type*>, and the pair of instructions for delay testing must be separated by the number of cycles specified by the <*distance*>. Informally, the number of pipeline stages bypassed by a path is referred to as *distance*.

We use our 16 bit, 5 stage pipelined processor VPRO design [9], which has 24 instructions as an example processor to explain the concepts. A partial PIEG of VPRO is shown in Figure 1, and complete PIEG and other details are given in [9].

Logic classification for <*logic type*> was based on our observation that many paths directly transfer data to the next stage using simple interconnects or through multiplexers. Keeping this in mind we classified logic into three types, namely interconnect (I), multiplexers (M), and processing logic (L). This classification simplifies the test generation process.

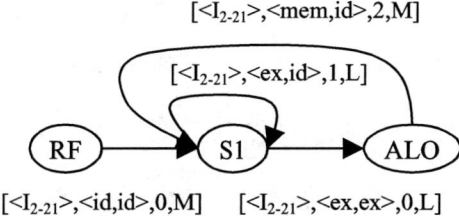

Figure 1 Partial PIEG of VPRO processor

Instructions, which have identical behavior in a given stage, are defined as *equivalent instructions* for that stage. For example, ADD and INC are the equivalent instructions in the EX stage of the VPRO processor. We can use these equivalent instructions to reduce the cardinality of the instruction set marked on each edge, which in turn reduce the constraint extraction and test generation effort.

We assume that any instruction can be followed by any other instruction in a pipeline stage with the exception of those instructions, which always need stall after the execution, such as unconditional jumps.

To test a path from a register R_i to a register R_o, we must create a transition at R_i and capture the transferred data at R_o. We need two data transfers to R_i to make a transition at R_i, and one data transfer from R_i to R_o along the target path. Though there may be a bypass/forwarding path (with one or more distance) to R_i, there must be another normal path (zero distance path) which brings the same values as the bypass path, and hence we only consider the normal path for data transfer to R_i. We also allow the propagation of data to R_o through normal paths except from R_i.

In order to generate the test vectors to be applied in functional mode, we need to extract architectural constraints. There are two types of constraints, i) control constraints, ii) data constraints. Control constraints are the constraints on control signals, which are responsible to transfer data between two nodes. These are obtained from PIEG. Data constraints are the constraints on justifiable data under the control constraints.

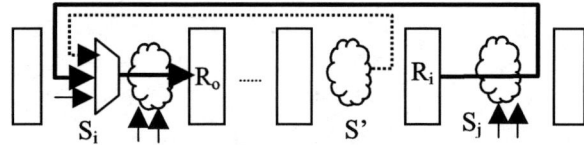

Figure 2 Target path and pipeline stages

A path from register R_i to register R_o, marked with [<I_{set}>, <S_j, S_i>, d, LT], where LT∈{I, M, L}, can be tested by a test instruction sequence IP_1, IP_2, ID_1, ID_2, ...I_D_{d-2}, IS_1, and IS_2. To test a path we need constraints for both stages S_j and S_i in two consecutive cycles (see Fig. 2). Instruction pair IP_1, IP_2 creates a transition at register R_i and allows it to propagate in S_j stage, hence provide the control constraints for the stage S_j. Instruction pair IS_1, IS_2 propagates it in S_i stage and finally instruction IS_2 latches the result in register R_o; hence provide the control constraints for the stage S_i. Other instructions ID_1, ...ID_{d-2} are dummy instructions between IP_2 and IS_1 to excite the path. Though there may be another data from another stage (ex. data from S' in Fig. 2), we assume that such data is transferred through a MUX and does not affect data transfer along the target path. Therefore, ATPG does not care these values and we do not need to extract their constraints.

Control constraints are extracted as instruction pairs (IP_1, IP_2) and (IS_1, IS_2). Instruction pair (IP_1, IP_2) must be marked on any zero distance (d = 0) in-edge of R_i and instruction pair (IS_1, IS_2) must be marked on the target path (edge between R_i and R_o). Note that IP_2 = IS_1 if d=1, and IP_1 = IS_1 and IP_2 = IS_2 if d=0.

Data Constraints for the three different logic types are extracted as follows:
(1) when logic type is interconnect 'I':
These paths are generally used to carry data to the next stage and always have zero distance (d = 0). R_o has only one edge, and that is from R_i; hence, it will not observe any data constraint. These paths can always be tested as interconnects test, therefore they are classified as FTP.
(2) when logic type is multiplexer 'M':
These paths pass through a set of MUXs and behave as interconnects if control signals are properly assigned. Therefore, under the control constraints (proper assignment of MUX select signals), data constraints are not applicable, as other paths to R_o will automatically be disintegrated with the proper assignment of MUXs control signals.
These paths are classified as *FUTP* if these are marked with d=1 and have a self-loop because a transition cannot be launched. Otherwise, paths are classified as FTP and these paths can be tested as interconnect test.
(3) when logic type is processing logic 'L':
This includes the paths which pass through the combinational logic. Let an edge between two registers R_i and R_o be marked with [<I_{set1}>, <S_j, S_i>, d, L]. Following edges and registers must be considered: i) All the in-edges to R_o with *distance* d and logic type 'L' (having some instructions common with I_{set1}), ii) all the in-edges to R_o with zero distance, logic type 'L', and have some instruction common with I_{set1}, iii) all the zero distance (d=0) in-edges to R_i.
All those registers which have out-edge to R_o (with distance d (same as target path distance), logic type 'L', and some instruction common with the target path) provide the data constraints for the propagation of created transition in S_j stage. All those registers which have out-edge to R_o (with zero distance, logic type 'L', and have some instruction common with the target path) provide data constraints for the propagation of the created transition in S_i stage. Figure 3 shows the edges and nodes which are needed to be considered. Note that $I_{set1} \cap I_{set2} \neq \phi$, $I_{set1} \cap I_{set3} \neq \phi$.

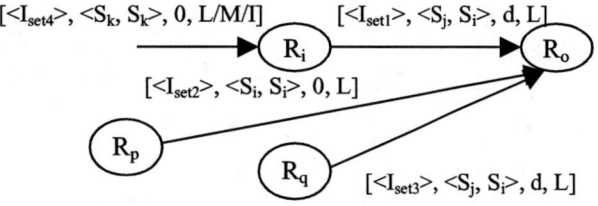

Figure 3 Edge consideration for constraint extraction

We consider different distance cases separately:
when d = 0 (Normal paths inside a pipeline stage):
We have to find out the data constraints for all those registers which have zero distance (d=0) in-edge to R_o with logic type 'L' using PIEG and RTL description. Let R_o has an in-edge from register R_p which is marked with common instructions with the target edge. If selected instrcution IS_2 is not marked on any of zero distance in-edge of R_p, then R_p must have constant value across two time frames (under IS_1 and IS_2). Otherwise, the register R_p do not observe any data constraints.

when d = 1 (paths across the pipeline stages, i.e. forwarding paths)
(a) Paths from bit *i* to bit *i* of register R_i in case of self-loop
These paths can be functionally testable only when there is odd inversion parity exists in the path; otherwise, these paths are functionally untestable. These paths are declared as *PFTUP*. Many paths of such kind exist in the circuit, such as paths in the pass logic of ALU, paths in shifter, paths in logic operation block of ALU etc.
Constraints must be extracted for the registers which have unity delay edge to R_o under IP_1 and IS_1 instructions, and for register which has zero distance edge to R_o under IS_1 and IS_2 instructions, as explained in d = 0 case.
(b) For other cases, paths are PFTP and data constraints can be obtained as stated above.

Similarly, constraints can be obtained for d >1 case. Details are given in [9].
Inversion parity test program is used to further classify PCFTP paths into FUTP or PFTP. The above stated classification can also be used to simplify the circuit for ATPG. Constraint ATPG is used for test vector generation for all the PFTP paths by using extracted constraints.

IV. CONTROLLER

In order to execute an instruction, the instruction is decoded by the decode unit (in decode stage) which dispatches control signals along with the required data for the pipeline stages ahead. Therefore, each pipeline stage does have control signals that are not structured in nature but often form a small group. In our approach, small grouping is used to find constraints. Therefore, we need to extract two types of constraints: i) constraints on the legitimate value of the group of control signals, ii) constraints on inter group signals in a pipeline stage.
(i) Constraints on the legitimacy of signals: Every possible value of a small group of signals is not valid. Therefore, we need to extract all the legitimate values. For example, comparator control (*comp_ctrl*) signals in VPRO are grouped in a group of 3 bits, and legitimate values are <0XX, 10X, and 110>.
(ii) Constraints on inter group signals: We extract these constraints in terms of instructions, i.e., map to the instruction which can generate the particular combination and all possible combinations are extracted. For example, in VPRO when ALU ctrl (*alu_ctrl*) signal is 0000 the comparator control (*comp_ctrl*) signal must be 000. Here onwards we will discuss how we can use these constraints for the test generation.

The part of a pipeline register, which carries the control signals is called control register. There are paths between control register (CR) to control register, control register to data register (DR), or data register (such as IR) to control register. The paths between CR-to-CR are used to carry the control signals for the pipeline stages ahead. These paths are connected directly and can be tested as interconnects. Paths from CR to DR are the paths which pass through the combinational logic. We construct a table which shows the transition at some bit in CR with instructions after exclusion of equivalent instructions.
Let there be a path between a bit *i* of control register C_i, and data register R_o. Constraints can be extracted in the following manner:
(i) when C_i and R_o are in the same stage:
It needs an instruction sequence of two instructions (IS_1, IS_2). The instruction pairs that can produce a transition at bit *i* and also marked on the in-edge of the register R_o can be the test instructions (IS_1, IS_2). All the data registers that have zero distance out edge to R_o (have some common instructions with the selected potential instruction pairs) are needed to check for data constraints. Data constraints can be obtained in the same way as we obtain for datapath.

(ii) when C_i and R_o are in different stages:

Instruction sequence (IP_1, IP_2, ID_1, ……ID_{d-2}, IS_1, IS_2) is needed to apply test vector. The instructions which can produce the transition at bit i of C_i can act as IP_1, IP_2. Constraints on the registers which have out-edge to R_o (with distance = d) must be considered under IP_1, and IP_2. The instructions which are marked on the in-edge of R_o (with distance = d) can act as IS_1, IS_2, and data constraints on those registers which have zero distance out-edge to R_o must be considered under the control constraints of IS_1, IS_2 instructions.

V. Test Instruction Sequence Generation

The generated test vector pairs as explained above are assigned to control signals and registers. A sequence of instructions is needed to apply these test vectors. A sequence of instructions which is responsible to launch the transition, propagate the launched transition, and latch the result, provided that desired data are available in the appropriate registers, is called test instruction sequence. These data are made available by the justification instruction sequence. Finally, the result must be transferred to memory by a sequence of instructions called observation sequence.

It is clear from the earlier discussion that if an edge between registers R_i and R_o is marked [<I_{set}>, <S_j, S_i>, d, LT], then we need a test instruction sequence (IP_1, IP_2, ID_1, ……ID_{d-2}, IS_1, IS_2) to apply the test vectors provided that test vectors are available in desired registers. Instructions IP_1 (when d > 0) and IP_2 (when d > 1) are decided by the control signals of the stage S_j, and instructions IS_1 and IS_2 are decided by the control signals of S_i stage. If there are more than one potential candidates for these instructions then we must select easy to observe instruction (such as STORE) for IS_2, and easy to justify instruction for the rest. Once IP_1, IP_2, IS_1, IS_2 instructions are decided, we fill the rest of the instructions by NOP instructions, but these can be later on replaced by the justification instruction for IS_1 and IS_2 to reduce the number of instructions.

VI. Experimental Results

VPRO processor was synthesized using 2345 gates and 268 sequential elements, and pipelined DLX processor [8] was synthesized with 34,347 gates and 1898 sequential elements. Complete PIEGs for both the processors are constructed by using instruction set architecture and RT level description. Note that the PIEG is extracted manually in this work but this can be automated. PIEG is used for the constraint extraction and the path classification. A constrained ATPG is developed for delay faults as commercially available ATPG doesn't handle the required constraints. Results for VPRO and DLX processors for the Non Robust (NR) and Functional Sensitizable (FS) [7] tests are shown in the Tables 1 and 2 respectively. Less than 1% paths are classified as PCFTP which are further classified as FUTP. The results show that only a small number (about 24%) of paths are functionally testable. However, we achieve 100% fault efficiency.

VII. Conclusion

A systematic approach for the delay fault testing of a pipelined processor cores using their instruction set has been presented. A graph theoretical model has been developed to model the complex pipeline behavior. This model can efficiently extract the constraints under which a processor can be tested. This model also assists the test instruction sequence generation process. Some paths can be declared as functionally untestable paths at the early stage. We would like to extend this model for the more complex processors such as super-scalar architecture in future.

Table 1 Results for VPRO processor

	Datapath		Controller	
	NR	FS	NR	FS
No. of paths	112,752	112,752	98,786	98,786
No. of faults	225,504	225,504	197,572	197,572
No. of functionally testable paths	32,134	52,092	27,512	42,282
No. of functionally untestable paths	193,370	173,412	170,060	155,290
Fault coverage (%)	14.2	23.1	13.9	21.4
Fault efficiency (%)	100	100	100	100

Table 2 Results for pipelined DLX processor

	Datapath		Controller	
	NR	FS	NR	FS
No. of paths	372,459	372,459	190,542	190,542
No. of faults	744,918	744,918	381,084	381,084
No. of functionally testable paths	148,718	185,247	57,502	89,974
No. of functionally untestable paths	596,200	559,671	323,582	291,110
Fault coverage (%)	19.9	24.8	15.0	23.6
Fault efficiency (%)	100	100	100	100

References

[1] S.M. Thatte and J. Abraham, "Test generation for Microprocessors", IEEE Trans. on Computers, Vol. C-29, No.6, June 1980, pp. 429-441.

[2] L. Chen, S. Ravi, A. Raghunath, and S. Dey, "A Scalable Software-Based Self-Test Methodology for Programmable Processors", Proc. of the Design Automation Conference 2003, pp. 548-553.

[3] N .Kranitis, G. Xenoulis, A. Paschalis, D. Gizopolous, Y. Zorian, "Application and Analysis of RT-Level Software-Based Self-Testing for Embedded Processor Cores", Proc. of International Test Conference, 2003, pp 431-440.

[4]. W.-C. Lai, A. Krstic, and K.-T. Cheng, "Test Program Synthesis for Path Delay Faults in Microprocessor Cores", Proc. of International Test Conference 2000, pp 1080-1089.

[5] V. Singh, M. Inoue, K.K. Saluja, and H. Fujiwara, "Instruction-Based Delay Fault Testing of Processor Cores", Proc. of the International Conference on VLSI Design 2004, pp 933-938.

[6] V. Singh, M. Inoue, K.K. Saluja, and H. Fujiwara, "Delay Fault Testing of Processor Cores in Functional Mode", IEICE Trans. on Information & Systems, Vol. E-88D, No. 3, pp 1-9.

[7]. A. Krstic and K.-T. Cheng, Delay fault testing for VLSI circuits, Kluwer Academic Publishers, 1998.

[8] J.L. Hennesy, and D.A. Patterson, Computer Architecture: A Quantitative Approach, Morgan Kaufmann Publishers, 1996.

[9] V. Singh, M. Inoue, K.K. Saluja, and H. Fujiwara, "Instruction-Based Delay Fault Self-Testing of Pipelined Processor Cores", NAIST Technical report. http://isw3.aist.nara.ac.jp/IS/TechReport/2004006.

Gradient-Based Methods for Simultaneous Blind Separation of Mixed Source Signals

Sanqing Hu, Derong Liu
Department of Electrical and Computer Engineering
University of Illinois at Chicago
Chicago, IL 60607, USA
Tel: (312) 355-4475, Fax: (312) 996-6465
E-mail: dliu@ece.uic.edu

Huaguang Zhang
School of Information Science and Engineering
Northeastern University
Shenyang, Liaoning 110004, P. R. China
Tel: (86) 24-83687762, Fax: (86) 24-83671498
Emails: hg_zhang@21cn.com

Abstract—This paper presents gradient-based methods for simultaneous blind separation of arbitrarily mixed source signals. We consider the regular case where the mixing matrix has full column rank as well as ill-conditioned cases. Two cost functions based on fourth-order cumulants are introduced to simultaneously separate all separable single sources and all inseparable mixtures. By minimizing the cost functions, two gradient-based methods are developed. Our algorithms derived from gradient-based methods are guaranteed to converge. Finally, simulation results show the effectiveness of our methods.

Index Terms–Blind source separation, cumulants, gradient-based methods, ill-conditioned cases, independence

I. INTRODUCTION

In the present paper, we consider linear instantaneous mixtures of sources. In this case, the general model of n sources and m observable mixtures is described by

$$\mathbf{y}(t) = \mathbf{A}\mathbf{x}(t) \quad (1)$$

where $\mathbf{x}(t) = [x_1(t), \cdots, x_n(t)]^T$ is a vector of mutually independent unknown sources with zero means, $\mathbf{y}(t) = [y_1(t), \cdots, y_m(t)]^T$ is a vector of measured mixed signals, and $\mathbf{A} = [a_{ij}]$ is an $m \times n$ unknown constant mixing matrix. As usual, it is assumed that at most one source signal has Gaussian distribution. The mixing matrix $\mathbf{A}$ either has full column rank or belongs to one of the following four ill-conditioned cases [12]:

Case 1) The number of sensors equals that of sources, i.e., $m = n$, but the mixing matrix $\mathbf{A}$ is singular;
Case 2) The number of sensors is less than that of sources;
Case 3) The number of sensors is larger than that of sources, but the column rank of $\mathbf{A}$ is deficient;
Case 4) The number of sources is unknown, and the column rank of $\mathbf{A}$ is deficient.

Every case above implies that $\mathbf{A}$ is not full column rank. In many references dealing with blind separation problem, the mixing matrix $\mathbf{A}$ is assumed to be nonsingular ($m = n$) or have full column rank ($m > n$). However, in practice, there is no guarantee that the mixing matrix $\mathbf{A}$ is always square and nonsingular. In particular, when the number of sources is not known *a priori*, there is no guarantee that we can always choose $m = n$ and that the mixing matrix has full column rank. Hence, the ill-conditioned cases are more often encountered.

We use the following general blind separation model:

$$\mathbf{z} = \mathbf{By} = \mathbf{Cx} \quad (2)$$

where $\mathbf{z}$ is an L-dimensional (or m-dimensional) output vector, $\mathbf{B} = [b_{ij}]$ is an $L \times m$ (or $m \times m$) blind separation matrix, $L \leq m$, and $\mathbf{C} = \mathbf{BA}$. Our task is to determine $\mathbf{B}$ such that all output components of $\mathbf{z}$ are mutually independent where some components correspond to theoretically separable sources up to certain scales while other components correspond to inseparable mixtures.

In the literature there are two classes of approaches to recover original sources from instantaneous mixtures: the sequential extraction approach (see, e.g., [5], [12], [13], [15]) and the simultaneous separation approach (see, e.g., [2]–[4], [6]–[11], [14]). The main difference between the two approaches is that the former recovers the sources one by one from available mixtures, while the later separates all separable sources simultaneously. The main disadvantage of the sequential approach is that the extraction errors will inevitably accumulate from component to component. Hence, simultaneous source separation is more desirable.

II. IDENTIFIABILITY AND OPTIMAL PARTITION

For n sources $x_1, \cdots, x_n$, let $X = \{x_1, \cdots, x_n\}$. Now partition X into L disjoint subsets $X_1, \cdots, X_L$, i.e., $X_i \cap X_j = \emptyset$ for $i \neq j$ and $\bigcup_{i=1}^{L} X_i = \{x_1, \cdots, x_n\}$. Let p_i be the number of all sources in X_i. Then, $p_1 +$

$\cdots + p_L = n$. Let $z_i = \sum_{x_j \in X_i} c_{ij} x_j$. Obviously, z_i is a mixture of all sources in X_i, $i = 1, \cdots, L$.

Definition 1: Mixtures $z_1, \cdots, z_L$ are said to be separated if there exists a matrix $\mathbf{B}$ such that $\mathbf{C} = \mathbf{BA}$ satisfies $c_{ij} \neq 0 \ \forall x_j \in X_i, i = 1, \cdots, L$, and all the other entries of $\mathbf{C}$ are equal to zero.

Remark 1: If there exists an $L \times m$ matrix $\mathbf{B}$ such that $\mathbf{C} = \mathbf{BA}$ has the following form:

$$\mathbf{C} = \begin{bmatrix} C_1 & \cdots & O \\ \vdots & \ddots & \vdots \\ O & \cdots & C_L \end{bmatrix} \quad (3)$$

or, if there exists a nonsingular $m \times m$ matrix $\tilde{\mathbf{B}}$ such that $\mathbf{C} = \tilde{\mathbf{B}}\mathbf{A}$ has the following form:

$$\mathbf{C} = \begin{bmatrix} C_1 & \cdots & O \\ \vdots & \ddots & \vdots \\ O & \cdots & C_L \\ O & \cdots & D_{(L+1)L} \\ \vdots & \cdots & \vdots \\ O & \cdots & D_{mL} \end{bmatrix} \quad (4)$$

where $C_1 = [c_{11}, \cdots, c_{1p_1}]$, $C_2 = [c_{2(p_1+1)}, \cdots, c_{2(p_1+p_2)}], \cdots, C_L = [c_{L(n-p_L+1)}, \cdots, c_{Ln}]$ with all entries nonzero and each O a matrix or vector of zeros with proper dimensions, $D_{iL} = k_i C_L$ with k_i a constant for $i = L+1, \cdots, m$, then from $\mathbf{z} = \mathbf{Cx}$ we see that mixtures $z_1 = c_{11}x_1 + \cdots + c_{1p_1}x_{p_1}, \cdots, z_L = c_{L(n-p_L+1)}x_{n-p_L+1} + \cdots + c_{Ln}x_n$ are separated.

We denote A_i to be an $m \times p_i$ submatrix composed of p_i columns of $\mathbf{A}$ and denote $\bar{A}_i$ to be the $m \times (n - p_i)$ submatrix composed of the remaining $n - p_i$ columns of $\mathbf{A}$. We now state our identifiability analysis results, Theorems 1 and 2.

Theorem 1: There exists an $L \times m$ matrix $\mathbf{B}$, where $L \leq m$ and $b_{ii} = 1, i = 1, \cdots, L$, such that mixture z_1 of p_1 sources, $\cdots$, mixture z_L of p_L sources can be separated simultaneously if and only if there exist L submatrix pairs of $\mathbf{A}$ denoted by $(A_1, \bar{A}_1), \cdots, (A_L, \bar{A}_L)$ such that $\text{rank}(\bar{A}_i) < \text{rank}(\mathbf{A})$ and $\text{rank}([\mathbf{a} \vdots \bar{A}_i]) = \text{rank}(\bar{A}_i) + 1$ for any column vector $\mathbf{a}$ of A_i, $i = 1, \cdots, L$, where A_i and A_j have no common column for $i \neq j$, and $A_1, \cdots, A_L$ form $\mathbf{A}$ by combination.

The conditions in Theorem 1 guarantee that there exists an $L \times m$ matrix $\mathbf{B}$ such that the matrix $\mathbf{C}$ has the form of (3) where

$$\mathbf{B} = \begin{bmatrix} 1 & -f_{12} & \cdots & -f_{1L} & -f_{1(L+1)} & \cdots & -f_{1m} \\ -f_{21} & 1 & \cdots & -f_{2L} & -f_{2(L+1)} & \cdots & -f_{2m} \\ \vdots & \vdots & \ddots & \vdots & \vdots & \cdots & \vdots \\ -f_{L1} & -f_{L2} & \cdots & 1 & -f_{L(L+1)} & \cdots & -f_{Lm} \end{bmatrix} \quad (5)$$

The set of matrices $(A_1, \cdots, A_L)$ satisfying the conditions in Theorem 1 is said to be a partition of the mixing matrix $\mathbf{A}$.

Theorem 2: There exists a nonsingular $m \times m$ matrix $\mathbf{B}$ such that mixture z_1 of p_1 sources, $\cdots$, mixture z_L of p_L sources can be separated simultaneously if and only if there exists L submatrix pairs of $\mathbf{A}$ denoted by $(A_1, \bar{A}_1), \cdots, (A_L, \bar{A}_L)$ such that $\text{rank}(\bar{A}_i) < \text{rank}(\mathbf{A})$ and $\text{rank}(A_i) = 1, i = 1, \cdots, L$, where A_i and A_j have no common column for $i \neq j$, and $A_1, \cdots, A_L$ form $\mathbf{A}$ by combination.

We state that the nonsingular matrix $\mathbf{B}$ has the following form:

$$\mathbf{B} = \begin{bmatrix} I_L & O \\ O & B_L \end{bmatrix} \tilde{\mathbf{B}},$$

$$\tilde{\mathbf{B}} = \begin{bmatrix} I_{L-1} & O \\ O & B_{L-1} \end{bmatrix} \begin{bmatrix} I_{L-2} & O \\ O & B_{L-2} \end{bmatrix} \cdots \begin{bmatrix} I_2 & O \\ O & B_2 \end{bmatrix} B_1 \quad (6)$$

$$B_i = \begin{bmatrix} 1 & -f_{i(i+1)} & \cdots & -f_{im} \\ -f_{(i+1)i} & 1 + f_{(i+1)i}f_{i(i+1)} & \cdots & f_{(i+1)i}f_{im} \\ \vdots & \vdots & \ddots & \vdots \\ -f_{mi} & f_{mi}f_{i(i+1)} & \cdots & 1 + f_{mi}f_{im} \end{bmatrix}$$

$$i = 1, \cdots, L,$$

where I_i is an $(i-1) \times (i-1)$ identity matrix, $i = 2, \cdots, L$. Moreover, $\tilde{\mathbf{B}}$ is such that $\mathbf{C} = \tilde{\mathbf{B}}\mathbf{A}$ has the forms of (4).

III. COST FUNCTIONS AND THE GRADIENT-BASED METHODS

Theorem 5 of [12] states that if one of the outputs z_i in (2) is pairwise independent to all the rest, then $c_{kj} = 0$ if $c_{ij} \neq 0, k = 1, \cdots, i-1, i+1, \cdots, L$ (or m), $j = 1, \cdots, n$.

Under the conditions of stated above, either z_i corresponds to a single source and other components in $\mathbf{z}$ do not contain the source, or z_i corresponds to a mixture of several sources and other components in $\mathbf{z}$ do not contain these sources. The principle of simultaneous blind source separation is based on the pairwise independence of all outputs of a separation model. We introduce two cost functions similar to those in [12] and [14] as criteria for simultaneous blind separation.

When $\mathbf{B}$ is an $L \times m$ matrix, we define a cost function using fourth-order cumulants as

$$J_1 = \sum_{j=2}^{L} \text{Cum}_{2,2}^2(z_1, z_j) + \sum_{j=3}^{L} \text{Cum}_{2,2}^2(z_2, z_j) + \cdots$$

$$+ \sum_{j=L-1}^{L} \text{Cum}_{2,2}^2(z_{L-2}, z_j) + \text{Cum}_{2,2}^2(z_{L-1}, z_L). \quad (7)$$

When $\mathbf{B}$ is an $m \times m$ matrix, we define a cost function using fourth-order cumulants as

$$J_2 = \sum_{j=2}^{m} \text{Cum}_{2,2}^2(z_1, z_j) + \sum_{j=3}^{m} \text{Cum}_{2,2}^2(z_2, z_j) + \cdots$$
$$+ \sum_{j=L}^{m} \text{Cum}_{2,2}^2(z_{L-1}, z_j), \quad (8)$$

where the last term of (8) will disappear if $L = m$. It is easy to see that if z_k and z_j are pairwise independent (for $k = 1, \cdots, L-1, j = 2, \cdots, L$ in (7) or for $k = 1, \cdots, L, j = 2, \cdots, m$ in (8), and $k < j$), then $J_1 = 0$ or $J_2 = 0$.

Our blind separation using (2) can be converted to solving the following minimization problem:

$$\min_{\mathbf{B} \in R^{L \times m}} J_1 \quad (9)$$

or

$$\min_{\det(\tilde{\mathbf{B}}) \neq 0} J_2. \quad (10)$$

A. The Gradient-Based Method 1

In this part we will focus on the minimization problem (9) with an $L \times m$ matrix $\mathbf{B}$. The minimization problem (9) becomes

$$\min J_1 \text{ with } \mathbf{B} \text{ having the form of (5)}.$$

From (5), it is easy to see that the minimization problem (9) is converted to the following optimization problem:

$$\min J_1(f_{12}, \cdots, f_{L(L-1)}, \cdots, f_{Lm}) \triangleq \min J_1(f)$$

where $f \triangleq [f_{12}, \cdots, f_{L(L-1)}, \cdots, f_{Lm}]^T$. Letting the time derivative of the variable be directly proportional to the negative gradient of $J_1(f)$ with respect to the vector variable f, we have the following updating equation

$$\frac{df}{d\tau} = -\mu \frac{\partial J_1(f)}{\partial f}, \quad f(0) = f_0 \quad (11)$$

where μ is a positive scaling constant and f_0 is the initial value of f.

Equation (11) is hereby called "Gradient-Based Method 1" for simultaneous blind source separation.

B. The Gradient-Based Method 2

In this part we will focus on the minimization problem (10) with the constraint of nonsingular matrix $\tilde{\mathbf{B}}$. Due to the use of J_2 in stead of J_2, we use $\tilde{\mathbf{B}}$ in (6) instead of $\mathbf{B}$. The constrained minimization problem (10) becomes

$$\min J_2 \text{ with } \tilde{\mathbf{B}} \text{ having the form of (6)}.$$

From $\tilde{\mathbf{B}}$ in (6), it is easy to see that the constrained minimization problem (10) is converted to the following unconstrained minimization problem:

$$\min J_2(f_{12}, \cdots, f_{(L-1)m}, \cdots, f_{m(L-1)}) \triangleq \min J_2(f)$$

where $f \triangleq [f_{12}, \cdots, f_{(L-1)m}, \cdots, f_{m(L-1)}]^T$. Letting the time derivative of the variable be directly proportional to the negative gradient of $J_2(f)$ with respect to the vector variable f, we have the following updating equation

$$\frac{df}{d\tau} = -\mu \frac{\partial J_2(f)}{\partial f}, \quad f(0) = f_0 \quad (12)$$

where μ is a positive scaling constant and f_0 is the initial value of f.

Equation (12) is called "Gradient-Based Method 2" for simultaneous blind source separation.

Our algorithms have the following two advantages over the sequential blind extraction algorithm developed in [12]. (i) The number of variables involved in equation (11) or (12) is less than that involved in the algorithm in [12]. For example, when we consider to separate all single sources of nonsingular mixing matrix $\mathbf{A}$, the number of variables involved in equation (11) or (12) is $n^2 - n$; while the number of variables involved in the algorithm in [12] is $2^2 + 3^2 + \cdots + n^2$. (ii) In the algorithm in [12], one has to compute a matrix inversion in each step which demands for high computational complexity. In equation (11) or (12) we do not need to compute any matrix inversion.

IV. SIMULATION RESULTS

In this section simulation results are presented using an example. To verify the separation results and show the convergence behavior of the present methods, we use performance index similar to that used in [13].

Example 1: Consider the ill-conditioned Case 2 with five sources and four observable mixtures where $x_1(t)$ and $x_2(t)$ are two benchmark male speech signals (8000 samples) from the second and third signals of the data file Speech10.mat downloaded from [1], $x_3(t) = \sin(3(n_1(t) - 0.5))$, $x_4(t) = n_2(t)$, $x_5(t) = n_3(t)$, n_1 is a uniform white noise with values in $[0, 1]$, and n_2 and n_3 are two uniform white noises with values in $[-0.5, 0.5]$ independent of n_1. These five sources are all sub-Gaussian. The mixing matrix is assumed to be

$$\mathbf{A} = \begin{bmatrix} 1 & 0.3 & 0.1 & 0.3 & 0.2 \\ 0.2 & 2 & 0.2 & 0.6 & 0.4 \\ 0.1 & 0.8 & 0.3 & 0.9 & 0.6 \\ 1 & 2 & 2 & 6 & 4 \end{bmatrix}.$$

Existing simultaneous blind separation methods, such as the EASI algorithm [3], the JADE algorithm [4], the

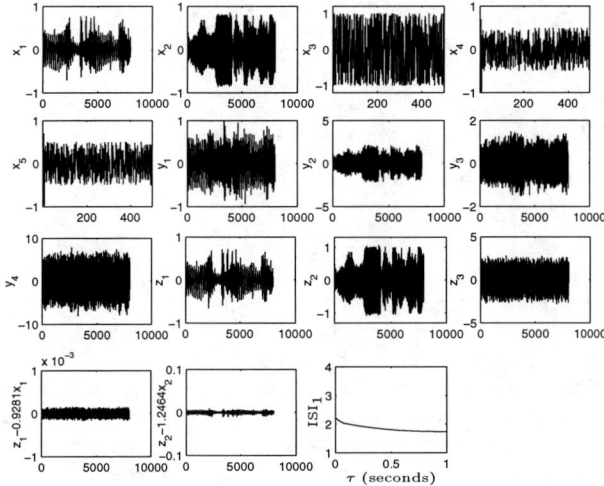

Fig. 1. Simultaneous blind separation in Example 1 using the Gradient-Based Method 1

the horizontal axis is time t. The separated signals (z_1, z_2, z_3), obtained using the Gradient-Based Method 1, are shown in the third row. The first and second subplots in the fourth row show the calibrated deviations $z_1 - 0.9281x_1$ and $z_2 - 1.2464x_2$, which imply that the sources x_1 and x_2 are separated. The performance index ISI_1 is shown in the last subplot of Figure 1, where the horizontal axis is time τ in seconds. One can see that equation (11) converges to an equilibrium in less than 1 second. Figure 2 shows the results by using the Gradient-Based Method 2. The separated signals z_1, z_2, z_3 are shown in the first row. The first and second subplots in the second row show the calibrated deviations $z_1 - 0.8816x_1$ and $z_2 - 2.2009x_2$, which imply that the sources x_1 and x_2 are separated. The performance index ISI_2 is depicted in the last subplot of Figure 2. One can see that equation (12) converges to an equilibrium in less than 3 seconds.

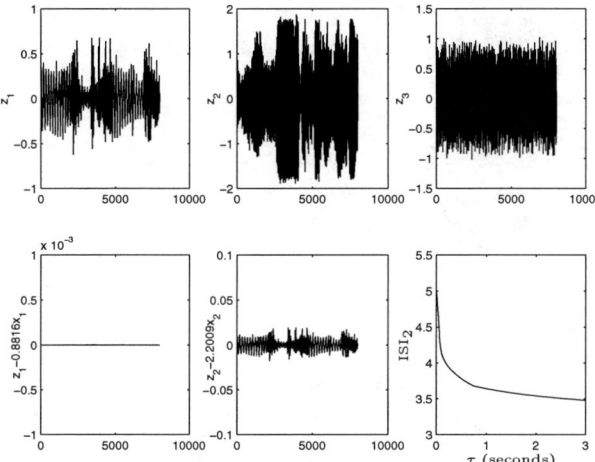

Fig. 2. Simultaneous blind separation in Example 1 using the Gradient-Based Method 2

deflation algorithm [7], and the ICA algorithms [9], [10], cannot be applied to deal with this case.

To apply the Gradient-Based Method 1, we let the separation matrix $\mathbf{B}$ have the form of (5). The initial value of the vector $[f_{12}, f_{13}, f_{14}, f_{21}, f_{23}, f_{24}, f_{31}, f_{32}, f_{34}]^T$ is chosen randomly. Letting $\mu = 10^9$ and using the Gradient-Based Method 1, we obtain $J_1(f^*) = 7.0837 \times 10^{-17}$.

To apply the Gradient-Based Method 2, we let the separation matrix $\tilde{\mathbf{B}}$ have the form of (6). The initial value of the vector $[f_{12}, f_{13}, f_{14}, f_{21}, f_{31}, f_{41}, f_{23}, f_{24}, f_{32}, f_{42}]^T$ is chosen randomly. Letting $\mu = 10^9$ and using the Gradient-Based Method 2, we obtain $J_2(f^*) = 7.3606 \times 10^{-16}$.

In Figure 1, five sources $(x_1, x_2, x_3, x_4, x_5)$ and four observable mixtures (y_1, y_2, y_3, y_4) are plotted where

REFERENCES

[1] http://www.bsp.brain.riken.jp/ICAbookPAGE/benchmarks.php.
[2] X.-R. Cao and R.-W. Liu, "General approach to blind source separation," *IEEE Transactions on Signal Processing*, vol. 44, pp. 562–571, Mar. 1996.
[3] J.-F. Cardoso and B. H. Laheld, "Equivariant adaptive source separation," *IEEE Transactions on Signal Processing*, vol. 44, pp. 3017–3030, Dec. 1996.
[4] J.-F. Cardoso and A. Souloumiac, "Blind beamforming for non Gaussian signals," *IEE Proceeding-F: Radar and Signal Processing*, vol. 140, pp. 362–370, 1993.
[5] A. Cichocki, R. Thawonmas, and S.-I. Amari, "Sequential blind signal extraction in order specified by stochastic properties," *Electronics Letters*, vol. 33, pp. 64–65, Jan. 1997.
[6] P. Comon, C. Jutten, and J. Herault, "Blind separation of sources, part II: Problems statement," *Signal Processing*, vol. 24, no. 1, pp. 11–20, 1991.
[7] N. Delfosse and P. Loubaton, "Adaptive blind separation of independent sources: A deflation approach," *Signal Processing*, vol. 45, no. 1, pp. 59–83, 1995.
[8] S. Fiori, "Extended Hebbian learning for blind separation of complex-valued sources," *IEEE Transactions on Circuits and Systems–II: Analog and Digital Signal Processing*, vol. 50, pp. 195-201, 2003.
[9] A. Hyvarinen, "Fast and robust fixed-point algorithms for independent component analysis," *IEEE Transactions on Neural Networks*, vol. 10, pp. 626-634, 1999.
[10] A. Hyvarinen and E. Oja, "Independent component analysis: algorithms and applications," *Neural Networks*, vol. 13, pp. 411-430, 2000.
[11] C. Jutten and J. Herault, "Blind separation of sources, part I: An adaptive algorithm based on neuromimetic architecture," *Signal Processing*, vol. 24, no. 1, pp. 1–10, 1991.
[12] Y. Li and J. Wang, "Sequential blind extraction of instantaneously mixed sources," *IEEE Transactions on Signal Processing*, vol. 50, pp. 997-1006, May 2002.
[13] Y. Li, J. Wang, and J. M. Zurada, "Blind extraction of singularly mixed source signals," *IEEE Transactions on Neural Networks*, vol. 11, pp. 1413–1422, Nov. 2000.
[14] A. Mansour and C. Jutten, "Fourth-order criteria for blind sources separation," *IEEE Transactions on Signal Processing*, vol. 43, pp. 2022-2025, Aug. 1995.
[15] E. Moreau and O. Macchi, "Self-adaptive source separation–Part II: Comparison of the direct, feedback, and mixed linear network," *IEEE Transactions on Signal Processing*, vol. 46, pp. 39–50, Jan. 1998.

BLIND IDENTIFICATION OF BRAIN MECHANISM IN MEG

Kuniharu Kishida

Department of Information Science, Faculty of Engineering, Gifu University
1-1 Yanagido Gifu 501-1193 Japan
kishida@cc.gifu-u.ac.jp

ABSTRACT

By the decorrelation method of BSS two components related to a SEF evoked magnetic field are selected from MEG data. The remixing matrix is applied to the two selected components to retrieve SEF MEG signals. Brain mechanism of SEF MEG data is identified equivalently by an innovation model form the viewpoint of statistical inverse problem. If the innovation model has a feedback structure corresponding to brain regions, transfer functions between two regions are evaluated via the innovation model. Feedback paths of transfer functions are checked by taking advantage of scaling transformation of SEF MEG data. Since transfer functions of real paths are invariant for scaling transformation, paths or routes corresponding to brain mechanism are diagnosed by examination of identified transfer functions.

I. INTRODUCTION

The somatosensory evoked field (SEF) in the present paper is considered to be a typical example of dynamical mechanism, since SEF is one of cerebral evoked fields. Electric stimulation to any part of the body evokes a cortical SEF. In clinical studies, the median nerve at the wrist is usually stimulated. Recent studies of SEFs detected by averaging MEG as in Fig. 1 have identified dipole sources activated in sensory cortices after median nerve stimulation.

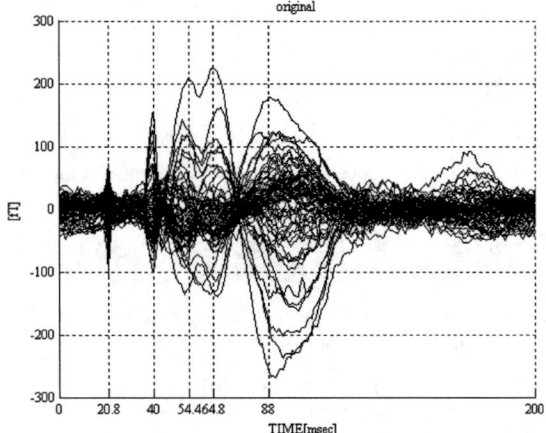

Fig.1 Averaged waveform MEG (Stimulus time is 20.8.)

Subject of Fig.1 was a 23-year-old healthy male volunteer. The right median nerve was stimulated electrically with a constant voltage, square-wave pulse of 0.2 msec duration delivered at the wrist. Stimulus frequency, f_p, was 5Hz, and stimulus intensity was adjusted to the lowest level that would produce a twitch of the thumb. SEFs were recorded with a 64-channel whole-head MEG system equipped with third-order SQUID gradiometers (NeuroSQUID Model 100; CTF Systems Inc.). MEG data were digitized with sampling frequency f_s=1250 Hz. Data of 200 msec duration were recorded for each of 500 trials and analyzed off-line as a single sweep, N=125000.

In the conventional MEG approach [1] averaged waveforms have been examined along the left side of Fig.2. The main topics in brain science are to know a mechanism between brain regions; however, we have no effective tool to attack them in the MEG analysis. Instead of averaged waveforms of the first moment statistically, correlation functions of the second moment are used in the present paper for extraction of brain dynamics. We will obtain a mechanism of SEF corresponding to two brain regions in terms of transfer functions by using blind source separation and identification method based on feedback system theory along the right hand side of Fig. 2.

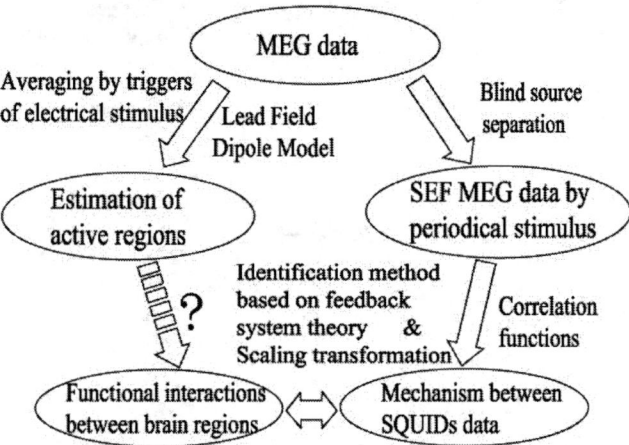

Fig. 2 Relationship between conventional analysis and new approach in MEG.

II. DECORRELATION METHOD OF BSS AND EQUIVALENT CLASS OF SEF MEG

To retrieve SEF MEG signals, we have used the second-order correlations for periodic types of source separation, so called decorrelation method. The decorrelation method under the stationary assumption was developed by Molgedey and Schuster [2], Murata, et al. [3], and briefly summarized in [4]. For steady MEG data we have tried the BSS method based on temporal structure to select 5Hz periodical components. In the BSS method we assume that MEG vector data $\underline{y}(n)=(y_1(n),\ldots,y_{64}(n))^T$ at discrete time n consist of underlying 64 sources $\underline{s}(n)=(s_1(n),\ldots,s_{64}(n))^T$ which are mutually statistically independent, and that the sources are mixed by a linear mixing matrix as

$$\underline{y}(n)=A\underline{s}(n), \qquad (1)$$

where each BSS component $s_i(n)$ as zero mean. In the BSS method k_N correlation matrices, $M(\tau_k)=1/N\Sigma_{n=1}^{N}\underline{y}(n)\underline{y}(n+\tau_k)^T$,

($k=1, 2, ..., k_N$), are diagonalized simultaneously, where $\underline{y}(n)$ is the orthonormalized vector of $y(n)$ and τ_k is defined by $\tau_k = [f_s k/f_p]$, $k=1, 2,..., k_N$. Here [...] rounds the value to the nearest integer. A Jacobi-like algorithm proposed by Cardoso and Souloumiac [5] was used to solve approximately the simultaneous diagonalization problem on k_N correlation matrices.

After we have found the mixing matrix A, we can obtain BSS source vector, $s(n)=A^{-1}y(n)$. From their peaks of power spectral density functions and periodical peaks of autocorrelation functions we selected two BSS components related to the 5Hz periodical time structure, $s_6(n)$ and $s_{13}(n)$. Evoked MEG data have been obtained by $y^e(n)= A^e s^e(n)$, where $s^e(n):=(s_6(n)\ _{13}(n))^T$, and $A^e:=(A_{*,6}\ A_{*,13})$ consists of two column vectors of the mixing matrix A. Here $A_{*,i}$ is the i-th column vector of A. Their isofield maps are shown in Figs. 3 and 4. As mentioned in [6] [7], we can have a dipole pattern of the primary somatosensory cortex (SI) in the left hemisphere of Fig. 4, which corresponds to electrical stimulus on the right hand. In Fig. 3 there is a weak dipole pattern can be found near right SI of the right hemisphere.

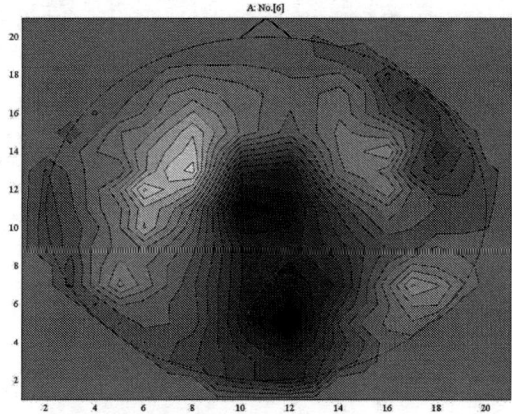

Fig. 3   Isofield map of $A_{*,6}$

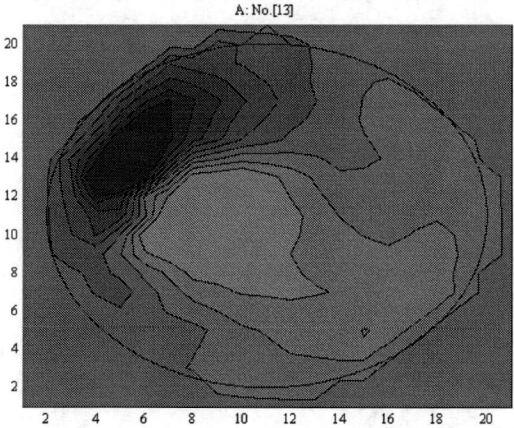

Fig. 4   Isofield map of $A_{*,13}$

Since averaged waveforms of $y^e(n)$ are almost equal to those of $y(n)$ as in Fig. 5, our blind source separation of SEF MEG data has succeeded. If we put $A_{*,a}=cA_{*,6}+dA_{*,13}$, $A_{*,b}=cA_{*,6}-dA_{*,13}$, $2s_a(n)=s_6(n)/c+s_{13}(n)/d$, and $2s_b(n)=s_6(n)/c-s_{13}(n)/d$, evoked MEG

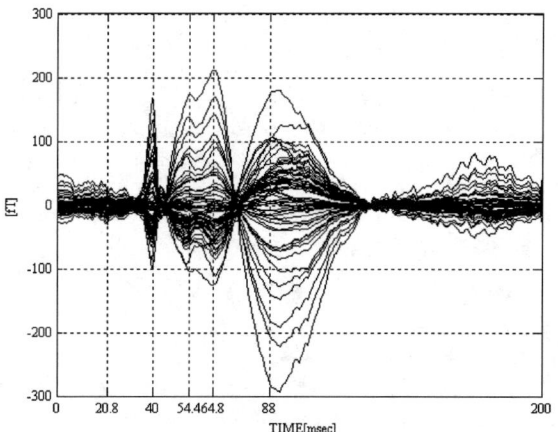

Fig. 5   Averaged waveform of SEF MEG (y^e)

data $y^e(n)=A_{*,6}s_6(n)+A_{*,13}s_{13}(n)$ can be rewritten as $y^e(n)=A_{*,a}s_a(n)+A_{*,b}s_b(n)$, though $s_a(n)$ and $s_b(n)$ are not independent. If we can rewrite $f^e(n):=(f_i(n)\ f_j(n))^T=(s'_a(n)\ s'_b(n))^T$ as in a similar manner, there exists a 2x2 nonsingular matrix U such that

$$y^e(n)= A^e s^e(n) = A^e UU^{-1} s^e(n) = Lf^e(n), \qquad (2)$$

where L is the lead field matrix containing both SIs and $f(n)$ is a correlated noise vector of dipole currents at two positions i and j in Fig. 6. From Eq. (2) it should be noted that one of equivalent classes of SEF MEG data could be identified in SEF MEG.

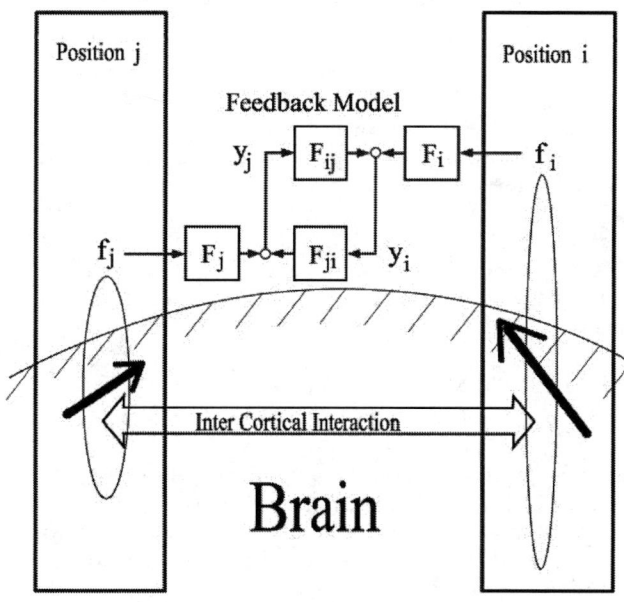

Fig. 6 Mechanism of active regions with two dipoles and its feedback model representation used for identification

III. IDENTIFICATION OF TRANSFER FUNCTIONS

To investigate an inter-cortical mechanism of SEF, transfer functions between two positions will be calculated from the viewpoint of statistical inverse problem. Let us write SEF MEG signals of SQUID sensors at two positions i and j as $y_i(n)$ and

$y_j(n)$ included in $y^e(n)$. We will examine dynamical properties between SQUID channels instead of brain neural currents, since dynamical properties between SQUID channels are expected to be similar to those between brain currents, if their positions are not near. Hereafter we analyzed two channels of SEF MEG data at 25(L42) and 57(R42) of SQUID sensors, since their sensors were located as in Fig. 1 of [4]. Sensor L42 is near left SI in Fig. 4, and sensor R42 is near right SI in Fig 3. Let us $y_i(n)$ and $y_j(n)$ at L42 and R42 sensors be described by a feedback model:

$$y_i(n) = F_{ij}(z^{-1})y_j(n) + F_i(z^{-1})f_i(n) \quad (3)$$
$$y_j(n) = F_{ji}(z^{-1})y_i(n) + F_j(z^{-1})f_j(n),$$

where z^{-1} is the backward time shift operator.

A minimum phase innovation model of $y_i(n)$ and $y_j(n)$ was identified from AEF MEG time series data by the next steps [8]:
1) calculatation of correlation functions of $y_i(n)$ and $y_j(n)$,
2) sigular value decomposition of Hankel matrx, of whcih elements were arranged by correlation function matrices,
3) selection of a stable solution of matrix Riccati equation,
4) determination of three coefficient matrices of innovation model, A, B, and C as mentioned in Appendix A of [4].

Then, a closed loop transfer function matrix,
$$G(z^{-1})=C(I-Az^{-1})^{-1}B = \{G_{ij}(z^{-1})\}, \quad (i,j=1,2) \quad (4)$$
was obtained from the minimum phase innovation model of two variables $y_i(n)$ and $y_j(n)$, which can be rewritten into the representation of feedback model structure:

$$y_i(n) = F_{ij}(z^{-1})y_j(n) + F_i(z^{-1})\gamma_i(n) \quad (5)$$
$$y_j(n) = F_{ji}(z^{-1})y_i(n) + F_j(z^{-1})\gamma_j(n),$$

where $\gamma(n)$ is the innovation. Finally, transfer functions $F_{ij}(z^{-1})$ and $F_{ji}(z^{-1})$ were calculated from the closed loop transfer functions by our blind identification based on feedback system theory under the identifiable condition of feedback structure [8]. After a model reduction, transfer functions between $y_i(n)$ and $y_j(n)$ were determined by
$$F_{ij}(z^{-1}) := G_{12}(z^{-1})/G_{22}(z^{-1}) \text{ and } F_{ji}(z^{-1}) := G_{21}(z^{-1})/G_{11}(z^{-1}). \quad (6)$$

From 12x12 Hankel matrix the bode diagrams of $F_{ij}(z^{-1})$ and $F_{ji}(z^{-1})$ of order 9 are shown by solid lines in Figs. 7 and 8. As in the case of auditory evoked field [9] the present approach was also applicable for the other planar type of SQUID.

IV. SCALE TRANSFORMATION AND DIAGOSIS OF FEEDBACK PATH

There remains a freedom of scaling of SEF MEG data in identification. Scaling transformations have been used in treatment of our blind identification method to obtain transfer functions correctly [10]. Scaling transformations of $y_i(n)$ and $y_j(n)$ are described by
$$y_i(n)=\alpha y_{si}(n) \text{ and } y_j(n)=\beta y_{sj}(n), \quad (7)$$
where α and β are parameters.

$$y_{si}(n) = F_{ij}^s(z^{-1})y_{sj}(n) + F_i^s(z^{-1})f_i(n) \quad (8)$$
$$y_{sj}(n) = F_{ji}^s(z^{-1})y_{si}(n) + F_j^s(z^{-1})f_j(n).$$

Here, scaled transfer functions between channels are defined by
$$F_{ij}^s(z^{-1}) = \frac{\beta}{\alpha}F_{ij}(z^{-1}), \quad F_{ji}^s(z^{-1}) = \frac{\alpha}{\beta}F_{ji}(z^{-1}),$$
$$F_i^s(z^{-1}) = \frac{1}{\alpha}F_i(z^{-1}), \text{ and } F_j^s(z^{-1}) = \frac{1}{\beta}F_j(z^{-1}).$$

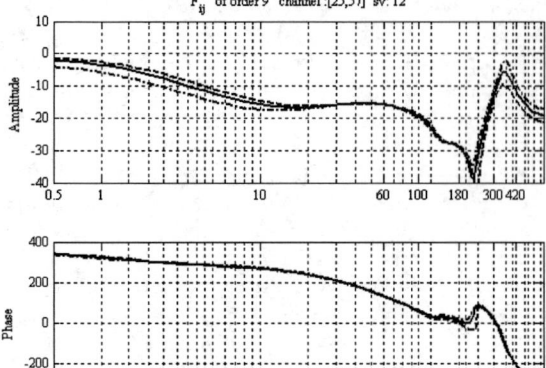

Fig.7 Bode diagram of $F_{ij}(z^{-1})$

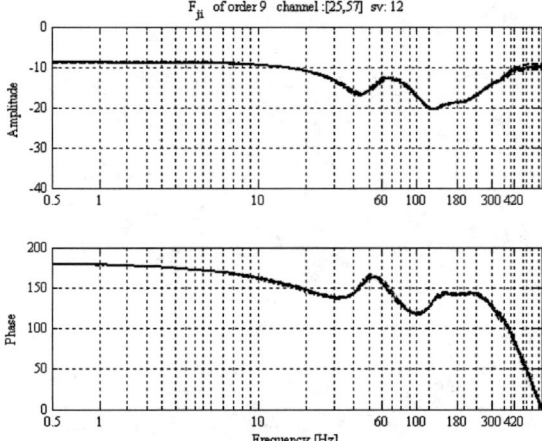

Fig. 8 Bode diagram of $F_{ji}(z^{-1})$

Original transfer functions between channels can be estimated from inversely scaled transfer functions:

$$F_{ij}(z^{-1}) = \frac{\alpha}{\beta}F_{ij}^s(z^{-1}), \quad \text{and} \quad F_{ji}(z^{-1}) = \frac{\beta}{\alpha}F_{ji}^s(z^{-1}). \quad (9)$$

Scaling parameter β was slightly changed as 0.9 or 1.1 with $\alpha=1$ in Eq. (7). Scaled transfer functions from scaled SEF MEG data were evaluated by Eq.(6) with 4 steps mentioned in Sec. III, and transfer functions estimated by the inverse scaling of Eq. (9) are shown in Figs. 7 and 8 by a dotted-broken line for β=0.9 and a broken line for β=1.1. In the scaling transformation $F_{ij}(z^{-1})$ is not robust for small changes of β as in Fig. 7. Since transfer functions of real path are invariant for small changes in the scaling transformation, the feedback path may not exist. On the other hand, $F_{ji}(z^{-1})$ has robustness for small changes as in Fig. 8.

The comparison of the averaged waveform and a predicted waveform via identified transfer functions can check goodness of identification. The solid line in Fig. 9 is an averaged waveform $av(y_i)$ of $y_i(n)$. The dotted-broken line is a predicted waveform given by $F_{ij}(z^{-1})av(y_j)$. There is little consistency in Fig. 9. On the other hand, the comparison of $av(y_j)$ with predicted waveform $F_{ji}(z^{-1})av(y_i)$ is shown in Fig. 10. It is concluded that the feedback path $F_{ji}(z^{-1})$ is reliable from Figs. 8 and 10. We can find the self-consistence between Figs. 9 & 10 and Figs. 7 & 8. Hence, the feedback path from L42 to R42 exists in brain regions of SIs.

Impulse response function of identified $F_{ji}(z^{-1})$ from L42 to R42 is shown in Fig. 11. It can be understood that the transfer response from L42 in left hemisphere to R42 in right hemisphere is within 20 msec. This result is the same as that of the different subject of 32-year-old healthy man in [7].

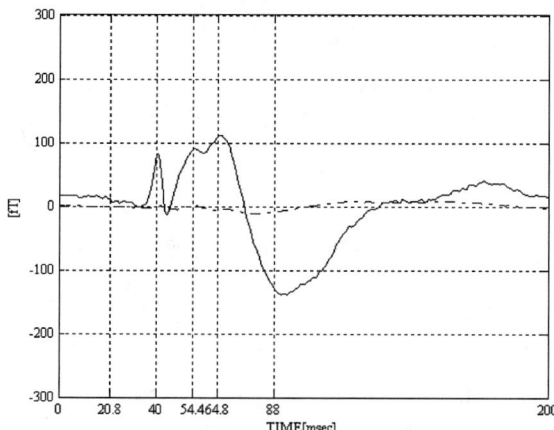

Fig. 9 Comparison of av(y_i) with waveform predicted via F_{ij}

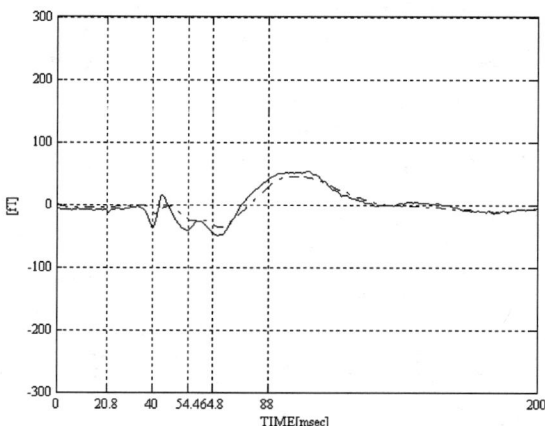

Fig. 10 Comparison of av(y_i) with waveform predicted via F_{ji}

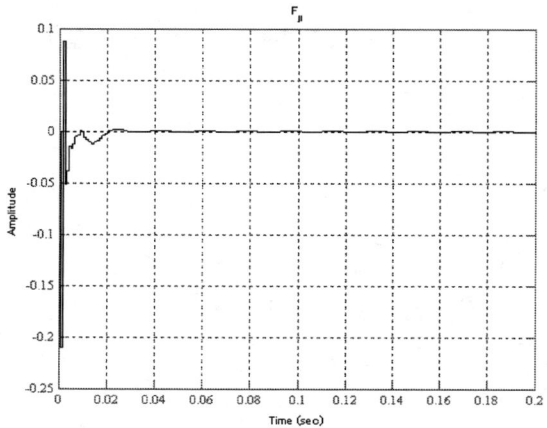

Fig. 11 Impulse response of F_{ji} from L42 to R42 in SEF

V. CONCLUSION

The dynamical mechanism of SEF has been obtained in Figs. 8 and 11 from identification method based on feedback system theory and BSS. The path or route from L42 to R42 is reasonable from human brain point of view. Hence, a new diagnosis of paths between active regions in brain could be introduced in MEG analysis by using scaling transformation of evoked MEG time series data.

VI. ACNOWLEDGEMENT

The author would like to thank Prof. K. Shinosaki for providing MEG data. Thanks also for Mr. K. Kato for his help in programming. This work was supported in part by Grant Aid for Scientific Research (No.14580346) of Japan society of the Promotion of Science.

VII. REFERENCES

[1] R. Hämäläinen, R. Hari, R. J. Ilmoniemi, J. Knuutila, and O. V. Lounasmaa, "Magnetoencephalography - theory, instrumentation, and applications to noninvasive studies of working human brain", *Rev. Mod. Phys.*, vol. 65, pp. 413-497, April 1993.

[2] L. Molgedey and H. G. Schuster, "Separation of a mixture of independent signals using time delayed correlations", *Phys. Rev. Lett.*, vol. 72, pp.3634-3637, 1994.

[3] N. Murata, S. Ikeda and A. Ziehe, "An approach to blind source separation based on temporal structure of speech signals", *Neurocomputing*, vol. 41, pp. 1-24, 2001.

[4] K. Kishida, H. Fukai, T. Hara, and K. Shinosaki, "A new approach to blind system identification in MEG data", *IEICE Transactions on Fundamentals of Electronics, Communications and Computer Sciences*, vol. E86-A, pp. 611-619, 2003.

[5] J. F. Cardoso and A. Souloumiac, "Jacobi angles for simultaneous diagonalization", *SIAM J. Math. Anal. Appl.*, vol. 17, no. 1, pp. 161-164, 1996.

[6] S. Makeig, T. P. Jung, A. J. Bell, D. Ghahremani and T. J. Sejnowski, "Blind separation of auditory event-related brain responses into independent components", *Proc. Nat. Acad. Sci. USA*, vol. 94 pp. 10979-10984, 1997.

[7] K. Kishida, K. Kato, K. Shinosaki, and S. Ukai, "Blind identification of SEF dynamics from MEG data by using decorrelation method of ICA." *Proceedings of the fourth International Symposium on Independent Component Analysis and Blind Signal Separation (ICA2003)*, Nara Japan, pp. 185-190, 2003.

[8] K. Kishida, "Contraction of information and its inverse problem in reactor system identification and stochastic diagnosis", In: Lewins J, Becker editors. *Advances in Nuclear Science and Technology*, Plenum Press, vol. 23, pp. 1-68, 1996.

[9] K. Kishida, Y. Ohi, M. Tonoike, and S. Iwaki, "A new dynamical approach to auditory evoked magnetic field by blind identification", *Proceedings of the 14th International Conferenceon Biomagnetism (Biomag2004)*, pp. 513-514, 2004.

[10] K. Kishida, "Identification of transfer functions and statistical inverse problems," *Prog. Nucl. Energy*, vol.43, pp. 297-303, 2003.

CHAOTIC SIGNAL SEPARATION FROM A LINEAR MIXTURE

Bao-Yun Wang
Department of Information Engineering
Nanjing University of Posts and Telecommunications
Nanjing 210003, China
e-mail: bywang@njupt.edu.cn

Wei Xing Zheng
School of QMMS
University of Western Sydney
Penrith South DC NSW 1797, Australia
e-mail: w.zheng@uws.edu.au

Abstract—In this paper the problem of blind separation of chaotic signals from an instantaneous linear mixture is considered. This problem is formulated as the extraction of each chaotic source. A new algorithm is provided to complete the separation task. A noisy chaotic signal is extracted by solving an optimisation problem. Then, a noise reduction technique is exploited to refine the extracted signal. The numerical simulation results demonstrate that the proposed approach can efficiently separate chaotic signals from a linear mixture.

I. INTRODUCTION

The blind source separation/extraction problem is to recover sources from their mixtures without the prior knowledge of the mixing channels and the sources [1], [2]. In general, there are two classes of approaches for conducting the signal separation: the simultaneous separation approach and the extraction approach. Compared with simultaneous blind source separation, blind source extraction has some advantages: it requires weaker solvability conditions on the mixing matrix; and it is more flexible in extracting one or some interested sources [2].

Over the past few years, many approaches have been developed to solve the blind source separation or extraction problem. Most of these approaches involve either higher-order statistics or second-order statistics. However, for chaotic signals, the statistics based approaches may not be applicable to completing the separation task since the statistics of the chaotic signal does not contain the inherent deterministic nonlinear dynamics.

Blind separation of mixed chaotic signals may find its application in multiuser communication system, in which several users transmit their chaotic signal through the same channel [3]-[6]. The chaotic signals in the mixture may be produced by different dynamic systems. This assumption corresponds to the independence of random signals assumed in [1], [2]. Therefore, instead of the independence, the inherent property of chaotic signals may be exploited to perform the separation.

This paper is concerned with blind separation of chaotic signals from a linear mixture. This problem is firstly formulated as the separation of each chaotic source. Then a new algorithm is developed to conduct the separation task. Specifically, a noisy chaotic signal is firstly extracted by solving an optimisation problem. Secondly, a noise reduction technique is employed to refine the noisy chaotic signal. Simulation results are given which show that the proposed approach is able to extract a specific chaotic signal from a linear mixture in an efficient way.

II. PROBLEM DESCRIPTION

Assume that p components $\mathbf{s}(t) = [s_1(t), \cdots, s_p(t)]^T$ are jointly observed through the m sensors $\mathbf{x}(t) = [x_1(t), \cdots, x_m(t)]^T$, and

$$\mathbf{x}(t) = \mathbf{A}\mathbf{s}(t) \quad (1)$$

where $\mathbf{A}$ is an $m \times p$ mixing matrix. Here, we assume that all p sources are chaotic signals. The k-th source $s_k(t)$ is produced by a known chaotic system, i.e.,

$$s_k(t) = f_k(s_k(t-1)) \quad (2)$$

where $f_k(\cdot)$ is a nonlinear function and its parameters are all in chaotic regions. In (1), the measurement is assumed to be noiseless. If it is noisy, we can consider the noise as one source signal.

This work was supported in part by a Research Grant from the Australian Research Council and in part by a Research Grant from the University of Western Sydney, Australia.

Our objective is to separate the mixture $\mathbf{x}(t)$ into the chaotic sources $\{s_k(t)\}$. Specifically, we seek a matrix $\mathbf{W} \in R^{p \times m}$ such that

$$\mathbf{Y} = \mathbf{WX} \qquad (3)$$

where $\mathbf{X}$ is a matrix formed by collecting of N samples, i.e., $\mathbf{X} = [\mathbf{x}(1), \cdots, \mathbf{x}(N)]$ and the k-th row of $\mathbf{Y}$ is very close to $\mathbf{s}_k = [s_k(1), \cdots, s_k(N)]$ in the sense that $\|\mathbf{Y}(k,:) - \mathbf{s}_k\|^2 \leq \varepsilon$, where ε is a very small positive number.

In the remaining part, we make the following assumptions:

Assumption 1: The number of sources p is no larger than the number of measurements m.

Assumption 2: The nonlinearity of the chaotic system $f_k(\cdot)$ is known.

Assumption 3: The mixing matrix $\mathbf{A}$ may not be square matrix but must be of full column rank.

At the end of this section, it should be noted that there exist some differences between our problem and the previous description of chaotic signal separation. In our problem, there are many sensors. In contrast, References [3]-[5] all considered the problem of separating of a sum of several chaotic signals into the component signals.

III. BLIND SEPARATION OF CHAOTIC SIGNAL

In this section, we first discuss the extraction condition. Under the assumptions given in Section II, we can always achieve a successful separation. Since the mixing matrix is of full column rank, its pseudo-inverse $\mathbf{A}^{\#}$ exists. Thus, our task in this section is to seek the estimate of $\mathbf{A}^{\#}$.

Denote the separating matrix by $\mathbf{W}$. By acting on the observation matrix $\mathbf{X}$, we obtain the estimated chaotic signals

$$\mathbf{Y} = \mathbf{WX} \qquad (4)$$

The signal at the k-th row should satisfy the known nonlinear map

$$s(t) = f_k(s(t-1))$$

Hence, we define the following criterion function

$$\varepsilon(\mathbf{W}) = \sum_{k=1}^{p}\sum_{t=1}^{N}(\mathbf{Y}_k(t) - f_k(\mathbf{Y}_k(t-1)))^2 \qquad (5)$$

where $\mathbf{Y}_k$ denotes the k-th row of $\mathbf{Y}$.

Now our problem becomes an unconstrained optimisation problem:

$$\mathbf{W} = \arg\min_{\mathbf{W} \in R^{p \times m}} \varepsilon(\mathbf{W}) \qquad (6)$$

Evidently, $\varepsilon(\mathbf{W})$ is a nonlinear non-negative function of unknown parameters $\mathbf{W}$. There may exist many local minima for this problem.

A closer look reveals that the criterion defined in (5) can be decomposed into p optimisation problems

$$\mathbf{w}_k = \arg\min_{\mathbf{w}} \varepsilon_k(\mathbf{w}) \qquad (7)$$

where

$$\varepsilon_k(\mathbf{w}) = \sum_{t=1}^{N}(\mathbf{Y}_k(t) - f_k(\mathbf{Y}_k(t-1))) \qquad (8)$$

This decomposition indicates that the separation of chaotic sources can be fulfilled by extracting each component.

IV. BLIND EXTRACTION OF CHAOTIC SIGNAL FROM A LINEAR MIXTURE

We now discuss the problem (7) in this section. Without loss of generality, we will consider to extract the first chaotic source, where the subscript will be omitted.

For this problem, apparently, there are at most two global minima: a trivial solution $\mathbf{w} = \mathbf{0}$ if $f(0) = 0$, and the desired solution, $\mathbf{w} = \mathbf{A}^{\#}(1,:)$, i.e., the first row of the pseudo-inverse of the mixing matrix $\mathbf{A}$. As is well known, there exists the ambiguity in the separation problem since $\mathbf{x} = \mathbf{As} = (\mathbf{AP})(\mathbf{P}^{-1}\mathbf{s})$, where $\mathbf{P}$ is a non-singular matrix. However, in our problem, the desired extracting vector is unique. The multiple of the optimal vector $c\mathbf{w}_{opt}$ will not minimize the objective function, i.e., $\varepsilon(c\mathbf{w}_{opt}) \neq 0$ due to the nonlinearity of chaotic dynamics. Based on the above discussion, we have the following proposition.

Proposition: The objective function defined in (5) has at most two global minima. If $f(0) = 0$, there exists a trivial solution $\mathbf{w} = \mathbf{0}$. Otherwise, a unique global optimum corresponds to the desired extracting vector.

Since the objective function is multimodal, the gradient type algorithm often fails to reach the global optimal solution. In [7], the optimization problems involved are very similar as that given in (7). Here we will make use of the

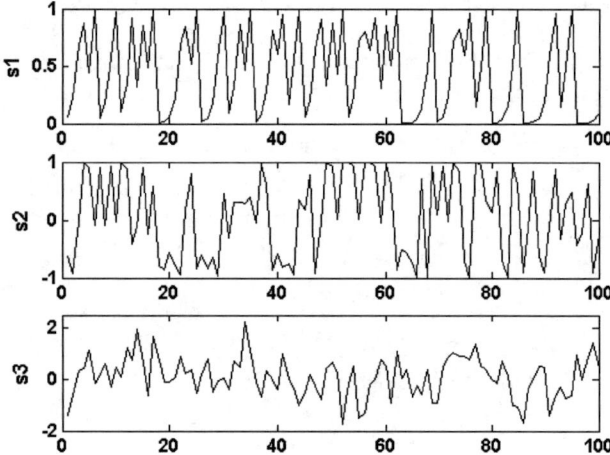

Fig. 1 Three source signals, s1--the logistic chaotic signal; s2--Chebysev chaotic signal; s3--white Gaussian signal

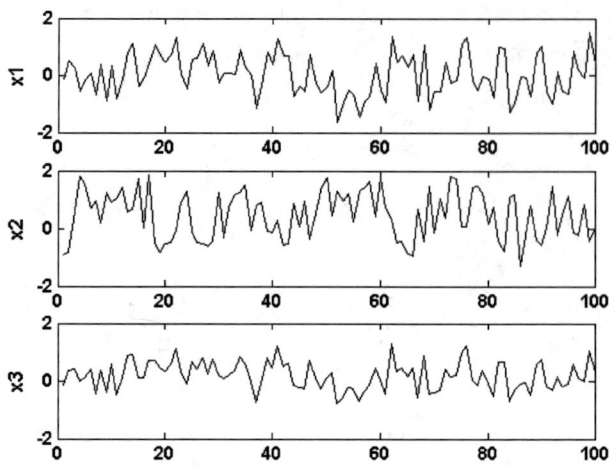

Fig. 2 Three measurements by mixing three sources in Fig. 1

adaptive simulated annealing algorithm (ASA) [8], which is a fast version of the famous simulated annealing technique. The main drawback of the ASA is still its convergence speed. The computational cost in the ASA algorithm is mainly in evaluating the objective function. Fortunately, our objective function is very simple since the number of parameters in our objective function is often not too large. Thus, the convergence speed should be very fast. The simulation results given in the next section will also support this analysis.

After extraction, the recovered chaotic signal may still contain some weak noise in it. We can exploit the noise reduction technique to refine the noisy chaotic signal. Several approaches have been proposed to reduce the noise in a chaotic signal. Here, we will adopt the smoothing algorithm developed in [9].

In the following, we take the logistic map as our illustrative example. Suppose that the noisy chaotic signal sequence is $\{y(n), n = 1, \cdots, N\}$. The corresponding symbolic sequence is obtained according to

$$u(n) = \begin{cases} 0, & y(n) < 0.5 \\ 1, & y(n) \geq 0.5 \end{cases} \quad (9)$$

The two inverse mappings are thus given by

$$f_u^{-1}(y) = \frac{1 + (2u-1)\sqrt{1 - 4y/\lambda}}{2}, \quad u = 0, 1 \quad (10)$$

where λ is the parameter of the logistic map. For L-lag smoothing, the estimate of the true chaotic signal at the n-th instant is

$$\hat{s}(n) = f_{u(n)}^{-1} \circ f_{u(n+1)}^{-1} \circ \cdots \circ f_{u(n+L-1)}^{-1}(\omega) \quad (11)$$

where ω is a point in the domain of $f_{u(n+L-1)}^{-1}$.

Based on the above discussion, the proposed approach can be summarized as follows:

(0) Initialization: $\mathbf{w}(0)$ is randomly generated.

(1) Solve the optimisation problem (7) by using the adaptive simulated annealing algorithm.

(2) Perform the noise reduction on the extracted signal by using (11).

V. NUMERICAL SIMULATIONS

In this section, we will present a numerical example to illustrate our approach. The mixing matrix used in the example is randomly generated in interval $[-1, +1]$ and given by

$$\mathbf{A} = \begin{bmatrix} 0.0905 & -0.8657 & 0.5060 \\ 0.8490 & 0.9396 & 0.2637 \\ 0.4108 & -0.5536 & 0.3659 \\ 0.1486 & 0.5461 & 0.0449 \end{bmatrix}$$

The chaotic sequence generated by the logistic map

$$s(t) = 4s(t-1)(1 - s(t-1))$$

is mixed with other signals, including a Chebysev chaotic signal and a Gaussian distributed signal with zero mean and unit variance. Fig. 1 shows the three sources and the mixed measurements are depicted in Fig. 2.

In running the adaptive simulated annealing (ASA) algorithm, we terminate the ASA algorithm when the objective function is below a small constant. In Fig. 3, the errors in the extracted chaotic signal are depicted. The error in the smoothed chaotic signal is shown in Fig. 4. It can be seen that the error induced by the extracting vector is still not satisfactory. It is due to our terminal condition. We can expect a much better extracting vector, but it will cost more time. Fortunately, although there is an error in the estimate of the chaotic signal, the noise reduction technique will reduce the error significantly.

At the same time, we also present the following commonly used performance index

$$I = \frac{\|\mathbf{c}\|^2 - \max(\mathbf{c}_i^2)}{\max(\mathbf{c}_i^2)}$$

where $\mathbf{c} = [c_1 \cdots c_p]$ and $\mathbf{c} = \mathbf{w}^T \mathbf{A}$. The performance index is presented in the caption of Fig. 3.

VI. CONCLUSION

In this paper blind separation of chaotic signals from a linear mixture has been addressed. The problem has been formulated as the extraction of each chaotic source. A new algorithm has been provided to complete the extraction task. In order to determine the extracting vector, the formulated optimisation problem is solved by the adaptive simulated annealing algorithm. Then, the noisy extracted chaotic signal is refined by exploiting the noise reduction technique. Our approach is applicable to both the mixture of several chaotic signals and the mixture of chaotic signal and other type of sources. The simulation results have illustrated that the proposed approach can extract the specific chaotic signal very accurately. As a final remark we note that the computational complexity of the ASA may be a limitation of our approach.

REFERENCES

[1] P. Comon, "Independent component analysis, A new concept?" *Signal Processing*, vol. 36, no. 3, pp. 287-314, 1994.

[2] Z. Malouche and O. Macchi, "Adaptive unsupervised extraction of one component of a linear mixture with a single neuron," *IEEE Transactions on Neural Networks*, vol. 9, no. 1, pp. 123-138, 1998.

[3] Y. V. Andreyev, A. S. Dmitriev, E. V. Efremova and A. N. Anagnostopoulos, "Separation of chaotic signal sum into components in the presence of noise," *IEEE Transactions on Circuits and Systems Part I*, vol. 50, no. 5, pp. 613-618, 2003.

[4] D. F. Drake and D. B. Williams, "Tracking multiple chaotic systems from a single observed sequence," in *Proc. 6th IEEE Digital Signal Processing Workshop*, pp. 75-78, 1994.

[5] M. Cifici and D. B. Williams, "An optimal estimation algorithm for multiuser chaotic communications systems," in *Proc. IEEE International Symposium on Circuits and Systems*, vol. 1, pp. 397-400, 26-29 May 2002.

[6] T. Lo, H. Leung and J. Litva, "Separation of a mixture of chaotic signals," in *Proc. IEEE International Conference on Acoustics, Speech, and Signal Processing*, vol. 3, pp. 1798-1801, 7-10 May 1996.

[7] Z. Zhu and H. Leung, "Identification of linear system driven by chaotic signals using nonlinear prediction," *IEEE Trans. Circuits and Systems, Part I.*, vol. 49, no. 2, pp. 170-180, 2002.

[8] S. Chen and B L. Luk, "Adaptive simulated annealing for optimisation in signal processing applications," *Signal Processing*, vol. 79, no. 1, pp. 117-128, 1999.

[9] C. Ling, X. F. Wu and S. G. Sun, "A general efficient method for chaotic signal estimation," *IEEE Transactions on Signal Processing*, vol. 47, no. 5, pp. 1424-1428, 1999.

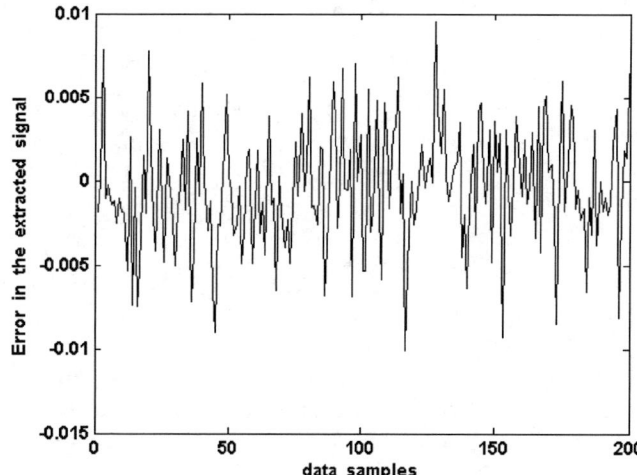

Fig. 3 Error in the extracted signal, $I = 2.625 \times 10^{-6}$

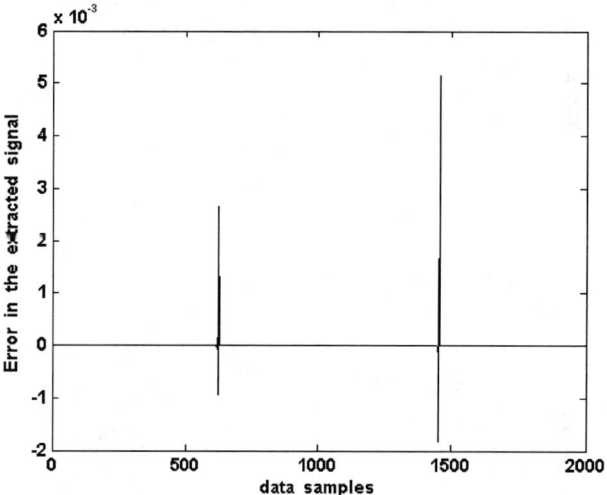

Fig. 4 The error in the smoothed signal

Blind Identification of MIMO Channels With Periodic Modulation[†]

Ching-An Lin and Yi-Sheng Chen
Department of Electrical and Control Engineering
National Chiao-Tung University
Hsinchu, Taiwan, R.O.C.
calin@cc.nctu.edu.tw   yschen.ece90g@nctu.edu.tw

Abstract

We propose a method for blind identification of MIMO FIR channels with periodic modulation. The method exploits linear relation between the covariance matrix of the received data and the "product matrices" of the channel. The proposed algorithm requires solving linear equations and computing eigenvalues and eigenvectors of a Hermitian matrix. We show that the identifiability condition depends on the modulating sequences alone and the method is robust with respect to channel order overestimation. A method for optimal selection of modulating sequences is proposed, which takes into account the effect of noise and the numerical error in estimating the covariance matrix of the received data. Simulations are used to demonstrate the performance of the proposed algorithm.

1 Introduction

We propose a method for blind identification of MIMO FIR channels with periodic modulation as a means to induce cyclostationarity [1]. The method exploits the linear relation between the covariance matrix of the received data and the "product matrices" of the channel, and compute the "product matrices" by solving a set of linear equations. The channel coefficients are then obtained (to within a unitary matrix ambiguity) by computing the positive eigenvalues and the associated eigenvectors of a Hermitian matrix. We show that the set of linear equations relating the "product matrices" of the channel and the covariance matrix of the received data can be further arranged into decoupled groups. The arrangement reduces computations and improves accuracy of the solution; it also leads to very simple identifiability conditions, which depend on the modulating sequences alone. The method is robust with respect to channel order overestimation. The proposed optimal selection minimizes the effects of channel noise and error in autocorrelation matrix estimation. The paper generalizes the results for the SISO case [2].

The paper is organized as follows. Section 2 is

[†]Research sponsored by the National Science Council under grant NSC-93-2213-E009-041

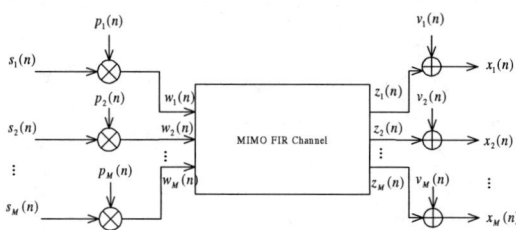

Figure 2.1: MIMO Channel Model

the problem statement and formulation. Section 3 establishes the identifiability conditions and proposes the MIMO blind identification algorithm. Section 4 discusses the optimal selection of the modulating sequences. Simulation results are given in Section 5. Section 6 concludes this paper.

2 Problem Statement and Formulation

Consider the linear MIMO FIR model with M transmitters and M receivers shown in Figure 2.1, where the source symbol sequences $s_1(n), s_2(n), \cdots, s_M(n)$ are modulated respectively by the M periodic sequences $p_1(n), p_2(n), \cdots, p_M(n)$ with period N. For $i,j = 1, 2, ..., M$, the discrete time model can be described by

$$x_i(n) = \sum_{j=1}^{M} \sum_{l=0}^{L} h_{ij}(l) w_j(n-l) + v_i(n) \quad (2.1)$$

where $w_j(n) = p_j(n) s_j(n)$, $h_{ij}(l)$ is the impulse response between the jth transmitter and the ith receiver, and $v_i(n)$ is the noise associated with the ith receiver. (2.1) can be written more compactly as

$$x(n) = \sum_{l=0}^{L} H(l) w(n-l) + v(n), \quad (2.2)$$

where $w(n) = P(n) s(n)$, $H(l) \in \mathbf{C}^{M \times M}$ is the impulse response matrix with ijth element $h_{ij}(l)$, $H(L) \neq 0$, $P(n) = \text{diag}[p_1(n), p_2(n), \cdots, p_M(n)] \in \mathbf{R}^{M \times M}$ is a diagonal matrix, and $x(n)$, $w(n)$, $s(n)$, and $v(n) \in \mathbf{C}^M$ are vector signals formed by stacking the respective scalar signal together, e.g. $x(n) = [x_1(n) \; x_2(n) \; \cdots \; x_M(n)]^T \in \mathbf{C}^M$. In addition, some basic assumptions are made as follows.

(**A1**): $s(n)$ and $v(n)$ are zero mean vector sequences with covariance matrices $E[s(k)s(j)^*] = \delta_{kj} \cdot \text{diag}[\sigma_{s_1}^2, \sigma_{s_2}^2, \cdots, \sigma_{s_M}^2] = \delta_{kj} \cdot \Sigma_s^2 \in \mathbf{R}^{M \times M}$, $E[v(k)v(j)^*] = \delta_{kj} \cdot \text{diag}[\sigma_{v_1}^2, \sigma_{v_2}^2, \cdots, \sigma_{v_M}^2] = \delta_{kj} \cdot \Sigma_v^2 \in \mathbf{R}^{M \times M}$, $E[s(k)v(j)^*] = 0 \in \mathbf{C}^{M \times M}$, $\forall k, j$, where δ_{kj} is the Kronecker delta.

(**A2**): An upper bound $\hat{L}$ of the channel order L is known and the period $N > \hat{L} + 1$.

(A3): $\text{rank}[H(0)^T\ H(1)^T\ \cdots\ H(L)^T]^T = M$.

Due to periodic modulation, (2.2) is a periodically time-varying system. In order to obtain a time-invariant representation, we define block signal $\bar{x}(n) = [x(Nn)^T, x(Nn+1)^T, \cdots, x(Nn+N-1)^T]^T \in \mathbf{C}^{MN}$, and let $\bar{w}(n), \bar{s}(n), \bar{v}(n)$ be similarly defined. Then $\bar{w}(n) = G\bar{s}(n)$, where $G = \text{diag}[P(0), P(1), \cdots, P(N-1)] \in \mathbf{R}^{MN \times MN}$ is a block diagonal matrix. Thus, $\bar{x}(n)$ can be written as

$$\bar{x}(n) = H_0 G \bar{s}(n) + H_1 G \bar{s}(n-1) + \bar{v}(n) \quad (2.3)$$

where H_0 is a $MN \times MN$ block lower-triangular Toeplitz matrix with $[H(0)^T\ H(1)^T\ \cdots\ H(L)^T\ 0_M\ \cdots\ 0_M]^T \in \mathbf{C}^{MN \times M}$ as its first block column(i.e., the first M columns), and H_1 is a $MN \times MN$ block upper-triangular Toeplitz matrix with $[0_M\ \cdots\ 0_M\ H(L)\ H(L-1)\ \cdots\ H(1)] \in \mathbf{C}^{M \times MN}$ as its first block row(i.e., the first M rows). Here 0_M is a zero matrix of dimension $M \times M$. The following matrix operations will be used in the derivation of the main result. For any $mn \times mn$ matrix $B = [B_{k,l}]_{0 \leq k,l \leq n-1}$, where $B_{k,l}$ is a block matrix of dimension $m \times m$, define $\Gamma_j[B] = [B_{0,j}^T\ B_{1,j+1}^T\ \cdots\ B_{n-1-j,n-1}^T]^T$ for $0 \leq j \leq n-1$, i.e., $\Gamma_j[B]$ is the matrix formed from the jth block super-diagonal of B. Let $A \in \mathbf{C}^{m \times m}$ and $B = [B_{ij}]_{1 \leq i \leq p,\ 1 \leq j \leq q}$ be a $pm \times qm$ complex matrix with $B_{ij} \in \mathbf{C}^{m \times m}$. We define $A \circ B = [AB_{ij}]_{1 \leq i \leq p,\ 1 \leq j \leq q}$. Note that $A \circ B \in \mathbf{C}^{pm \times qm}$.

3 Channel Identification

In this section, we show that any channel satisfying (A.3) is identifiable by an appropriate choice of modulating sequences and the proposed method is robust with respect to channel order overestimation.

3.1 Identifiability Condition

If we consider the noise free case, then (2.3) becomes

$$\bar{x}(n) = H_0 G \bar{s}(n) + H_1 G \bar{s}(n-1) \quad (3.1)$$

With assumption (**A1**), the covariance matrix of $\bar{x}(n)$ can be written as

$$R_{\bar{x}}(0) = E[\bar{x}(n)\bar{x}(n)^*] = H_0 G_w^2 H_0^* + H_1 G_w^2 H_1^* \quad (3.2)$$

where $G_w = \text{diag}[P(0)\Sigma_s, P(1)\Sigma_s, \cdots, P(N-1)\Sigma_s] \in \mathbf{R}^{MN \times MN}$ is a block diagonal matrix. Let $J \in \mathbf{R}^{MN \times MN}$ be the matrix whose first block sub-diagonal are all $M \times M$ identity matrix I_M, i.e., $\Gamma_1[J^T] = [I_M\ I_M\ \cdots\ I_M]^T \in \mathbf{R}^{M(N-1) \times M}$, and all remaining blocks are zero. Write H_0 and H_1 as $H_0 = \sum_{k=0}^{L} H(k) \circ J^k$ and $H_1 = \sum_{m=1}^{L} H(m) \circ (J^T)^{N-m}$. Then $H_0 G_w^2 H_0^*$ and $H_1 G_w^2 H_1^*$ can be written as

$$H_0 G_w^2 H_0^* = \sum_{k=0}^{L}\sum_{l=0}^{L}(H(k)H(l)^*) \circ (J^k G_w^2 (J^T)^l) \quad (3.3)$$

$$H_1 G_w^2 H_1^* = \sum_{m=1}^{L}\sum_{n=1}^{L}(H(m)H(n)^*) \circ ((J^T)^{N-m} G_w^2 J^{N-n}) \quad (3.4)$$

From (3.2), (3.3) and (3.4), we see that $R_{\bar{x}}(0)$ can be written as a weighted sum of the known matrices $J^k G_w^2 (J^T)^l$ and $(J^T)^{N-k} G_w^2 J^{N-l}$ with the unknown matrices $H(k)H(l)^*$ as weighting factors. Proposition 3.1 shows that $J^k G_w^2 (J^T)^l$ and $(J^T)^{N-k} G_w^2 J^{N-l}$ have special structures that allows decomposition of (3.2) into a group of decoupled equations.

Proposition 3.1: Let $0 \leq k, l \leq L$ be two non-negative integers. Then

(a) For $l = k+j$, where $0 \leq j \leq L-k$, both $J^k G_w^2 (J^T)^l$ and $(J^T)^{N-k} G_w^2 J^{N-l}$ are block upper triangular with only the respective jth block upper diagonals nonzero, and $\Gamma_j[J^k G_w^2 (J^T)^l]$ and $\Gamma_j[(J^T)^{N-k} G_w^2 J^{N-l}]$ are $(N-j)M \times M$ matrices shown as (3.5) and (3.6), respectively.

$$[\underbrace{0_M \cdots 0_M}_{k\ \text{blocks}}\ \underbrace{P(0)^2 \cdots P(N-1-k-j)^2}_{N-k-j\ \text{blocks}}]^T \Sigma_s^2 \quad (3.5)$$

$$[\underbrace{P(N-k)^2 \cdots P(N-1)^2}_{k\ \text{blocks}}\ \underbrace{0_M \cdots 0_M}_{N-k-j\ \text{blocks}}]^T \Sigma_s^2 \quad (3.6)$$

(b) For $l < k$, both $\Gamma_j[J^k G_w^2 (J^T)^l]$ and $\Gamma_j[(J^T)^{N-k} G_w^2 J^{N-l}]$ are block lower triangular with zero block diagonal. □

From Proposition 3.1, the jth block super-diagonal part of (3.2), $0 \leq j \leq L$ can be written as

$$\Gamma_j[R_{\bar{x}}(0)] = M_j F_j \quad \forall\ 0 \leq j \leq L \quad (3.7)$$

where $M_j \in \mathbf{R}^{M(N-j) \times M(L-j+1)}$ is defined as

$$M_j = \Sigma_s^2 \circ \begin{bmatrix} P(0)^2 & P(N-1)^2 & \cdots & P(N-L+j)^2 \\ P(1)^2 & P(0)^2 & \cdots & P(N-L+j+1)^2 \\ P(2)^2 & P(1)^2 & \cdots & P(N-L+j+2)^2 \\ \vdots & \vdots & \vdots & \vdots \\ P(N-3-j)^2 & P(N-4-j)^2 & \cdots & P(N-L-3)^2 \\ P(N-2-j)^2 & P(N-3-j)^2 & \cdots & P(N-L-2)^2 \\ P(N-1-j)^2 & P(N-2-j)^2 & \cdots & P(N-L-1)^2 \end{bmatrix} \quad (3.8)$$

and

$$F_j = [(H(0)H(j)^*)^T\ \cdots\ (H(L-j)H(L)^*)^T]^T \quad (3.9)$$

We note that M_0 is a block circulant matrix and M_j is obtained from M_0 by deleting its last jM rows and last jM columns.

Since $N > L+1$, the $(L+1)$ equations in (3.7) are overdetermined and they are also consistent. If M_j is full column rank, then the solution can be obtained as

$$F_j = (M_j^T M_j)^{-1} M_j^T \Gamma_j[R_{\bar{x}}(0)] \quad (3.10)$$

Let Q be the Hermitian matrix defined by $\Gamma_j[Q] = F_j$ for $j = 0, 1, \cdots, L$, and let the channel matrix $H = [H(0)^T\ H(1)^T\ \cdots\ H(L)^T]^T$. Clearly we have

$$Q = HH^* \quad (3.11)$$

Since H is full rank, Q has rank M. Since Q is Hermitian and positive semidefinite, Q has M positive eigenvalues, say, $\lambda_1, \cdots, \lambda_M$. We can expand Q as

$$Q = \sum_{j=1}^{M}(\sqrt{\lambda_j}v_j)(\sqrt{\lambda_j}v_j)^* \quad (3.12)$$

where v_j is a unit norm eigenvector of Q associated with $\lambda_j > 0$. We can choose the channel matrix to be

$$\hat{H} = [\sqrt{\lambda_1}v_1\ \sqrt{\lambda_2}v_2\ \cdots\ \sqrt{\lambda_M}v_M] \in \mathbf{C}^{M(L+1) \times M} \quad (3.13)$$

up to a unitary matrix ambiguity.

Since M_j is determined by the modulating sequences and the source symbol variances, by appropriately selecting the modulating sequences, we can make M_j full rank (assuming that $\Sigma_s^2 > 0$), and any channel satisfying (A.3) can be identified. We thus have established the following identifiability condition.

Theorem 3.2: The MIMO channel described in (2.2) satisfying the assumptions (**A1**), (**A2**), and (**A3**) is identifiable, up to a unitary matrix ambiguity, if and only if each matrix M_j, $j = 0, 1, \cdots, L$, defined in (3.8), is full rank. □

3.2 Channel Order Overestimation

If only an upper bound $\hat{L} \geq L$ is available with $N > \hat{L} + 1$, then following the same process given in Section 3.1, the corresponding $M(\hat{L}+1) \times M(\hat{L}+1)$ matrix Q can be similarly constructed as in (3.11). The last $(\hat{L} - L)$ block columns (i.e., $(\hat{L} - L)M$ columns) of Q are zero, so are the last $(\hat{L} - L)$ block rows. Hence again, Q is of rank M and has M positive eigenvalues and the eigenvectors associated with the positive eigenvalues all have the form $v = [q^T \; 0 \; \cdots \; 0]^T \in \mathbf{C}^{M(\hat{L}+1)}$ where $q \in \mathbf{C}^{M(L+1)}$. Thus, we can determine the actual channel order and impulse response matrix, up to a unitary matrix ambiguity, from the M eigenvectors associated with the M positive eigenvalues of Q.

3.3 Identification Algorithm

We summarize the proposed method as the following algorithm.

1) Select the modulating sequences $p_1(n)$, $p_2(n)$, $\cdots$, $p_M(n)$ such that each matrix M_j defined in (3.8) is full column rank.

2) Estimate the autocovariance matrix $R_{\bar{x}}(0)$ via the time average
$$\hat{R}_{\bar{x}}(0) = \frac{1}{K} \sum_{i=1}^{K} \bar{x}(i)\bar{x}(i)^* \quad (3.14)$$
where K is the number of data block.

3) Compute the product channel coefficients F_j, $0 \leq j \leq L$, using (3.10).

4) Form the matrix Q as in (3.11), and obtain the channel impulse response (3.13) by computing the M largest eigenvalues and the associated eigenvectors of Q.

4 Optimal Selection of the Modulating Sequences

In this section, we discuss the selection of modulating sequences so as to minimize the channel estimation error due to noise and error in the estimation of $R_{\bar{x}}(0)$.

4.1 Optimality Criterion

Now we consider the general case that the noise is present and discuss the selection of the modulating sequences $p_1(n), p_2(n), \cdots, p_M(n)$. From (2.3) and assumption (**A1**), the covariance matrix of the received signal is
$$R_{\bar{x}}(0) = H_0 G_w^2 H_0^* + H_1 G_w^2 H_1^* + \Sigma_v^2 \circ I_{MN} \quad (4.1)$$

From (4.1) and (3.2), we see that noise has only contribution to the main (block) diagonal entries of $R_{\bar{x}}(0)$. Therefore the $(L+1)$ decoupled groups of equations in (3.7) remain unchanged, except for the $j = 0$ group, which becomes
$$\Gamma_0[R_{\bar{x}}(0)] = M_0 F_0 + B\Sigma_v^2 \quad (4.2)$$
where $B = [I_M \; I_M \; \cdots \; I_M]^T \in \mathbf{R}^{MN \times M}$.

Since Σ_v^2 is unknown, F_0 can not be solved from (4.2). Instead, we compute the least squares solution $\hat{F}_0$ of equation (4.3) to get (4.4)
$$\Gamma_0[R_{\bar{x}}(0)] = M_0 F \quad (4.3)$$
$$\hat{F}_0 = F_0 + (M_0^T M_0)^{-1} M_0^T B\Sigma_v^2 = F_0 + Z \quad (4.4)$$

(4.4) is the actual solution F_0 plus a perturbation term due to noise. The perturbation term Z is the least squares solution of the equation $M_0 Z = B\Sigma_v^2$.

We note that if every column of $B\Sigma_v^2$ is orthogonal to every column of M_0, then $Z = 0$, which implies $\hat{F}_0 = F_0$. But that is impossible since the entries of M_0 are positive and $B\Sigma_v^2$ are nonnegative. Therefore, we seek to appropriately choose the modulating sequences $p_1(n), p_2(n), \cdots, p_M(n)$ such that every column of $B\Sigma_v^2$ is as close to being orthogonal to that of M_0 as possible. To this end, we first define q_{ki} and b_i shown below as the columns of M_0 and $B\Sigma_v^2$, respectively:

$$M_0 = \left[\underbrace{q_{01} \; q_{02} \; \cdots \; q_{0M}}_{M_0(:,1:M)} \; \cdots \; \underbrace{q_{L1} \; q_{L2} \; \cdots \; q_{LM}}_{M_0(:,LM+1:(L+1)M)} \right] \quad (4.5)$$

$$B\Sigma_v^2 = [\Sigma_v^2 \; \Sigma_v^2 \; \cdots \; \Sigma_v^2]^T = [b_1 \; b_2 \; \cdots \; b_M] \quad (4.6)$$

Then, due to the special structure of the block circulant matrix M_0 and $B\Sigma_v^2$, it is easy to check that q_{ki} is orthogonal to b_j, $q_{ki}^T b_j = 0$, for $j \neq i$ and each $q_{ki}^T b_i$ assumes the same value. Thus we only need to consider the relation between columns of q_{0i} and b_i, for $i = 1, 2, \cdots, M$. Define
$$\gamma_i = \frac{q_{0i}^T b_i}{\|q_{0i}\|_2 \|b_i\|_2} \quad i = 1, 2, \cdots, M \quad (4.7)$$

Since γ_i is nonnegative and by Cauchy-Schwarz inequality, $0 \leq \gamma_i \leq 1$, $\gamma_i = \cos\theta_i$ for some $|\theta_i| \leq \pi/2$. We can think of θ_i as the angle between q_{0i} and b_i. To minimize γ_i is to maximize the angle θ_i. Based on this point of view, we formulate the optimal selection problem as minimizing γ_i subject to
$$\frac{1}{N} \sum_{n=0}^{N-1} |p_i(n)|^2 = 1 \quad (4.8)$$
$$|p_i(n)|^2 \geq \delta > 0, \quad \forall 0 \leq n \leq N-1 \quad (4.9)$$

Roughly, constraint (4.8) normalizes the power gain of the modulating sequence of each transmitter to 1; constraint (4.9) requires that at each instant, the power gain has a lower bound. Note that the problem of selecting each of the M modulating sequence is identical to the SISO case considered in [2], and each can be solved separately. Thus, from [2], the optimal modulating sequences $p_i(n)$ is a two-level sequence with a single peak in one period. More specifically,

$$p_i(n) = \begin{cases} \sqrt{N(1-\delta)+\delta}, & n = m_i \\ \sqrt{\delta}, & n \neq m_i, \; 0 \leq n \leq N-1 \end{cases} \quad (4.10)$$

From (4.10), we have N optimal modulating sequences for the ith transmitter, each defined by the peak location m_i. These N sequences are equivalent in the sense that they yield the same γ_i.

4.2 On Selection of m_i

We now consider the selection of m_i. We know different choices of m_i result in different matrix M_j, and different M_j will affect the numerical computation of F_i, $i = 0, 1, \cdots, L$, in (3.10), since each $M_j^T M_j$ may have different condition numbers. If the condition number is large, then the matrix $M_j^T M_j$ is ill-conditioned and the computation in (3.10) and (4.4) is sensitive to data error. Let
$$\bar{\mu} = \max_{0 \leq j \leq L} \kappa(M_j^T M_j) \quad (4.11)$$
where $\kappa(A)$ is the condition number of A. Our goal is to choose $\{m_i\}_{i=1}^M$ so as to minimize the largest condition number of the corresponding matrices $M_j^T M_j$, $j = 0, 1, \cdots, L$. Since each transmitter has N possible optimal modulating sequences, the M transmitters may result in N^M different $\bar{\mu}$. The following result

shows that some choices of $\{m_i\}_{i=1}^M$ are to be avoided since they result in some M_j being rank deficient and thus $\bar{\mu} = \infty$.

Proposition 4.1: At least one M_j, $0 \leq j \leq L$, is not full rank if and only if $N - L + 1 \leq m_i \leq N - 2$ for some $1 \leq i \leq M$. □

Hence if we choose, for each i, either $0 \leq m_i \leq N - L$ or $m_i = N - 1$, then each M_j is full rank and the channel is identifiable. The following result shows that if $\sigma_{s_i}^2$ are all the same for $i = 1, 2, \cdots, M$, we can classify the $(N - L + 2)^M$ choices into 3 groups that are relevant to the optimal choice of $\{m_i\}_{i=1}^M$.

Proposition 4.2: Assume that, for $i = 1, 2, \cdots, M$, $\sigma_{s_i}^2 = 1$. Then
(a) If $0 \leq m_i \leq N - L - 1$, $i = 1, 2, \cdots, M$, then each of the $(N - L)^M$ choices results in the same $\bar{\mu}$, defined in (4.11), which is denoted by $\bar{\mu}_1$.
(b) If either $m_i = N - L$ or $m_i = N - 1$, $i = 1, 2, \cdots, M$, then each of the 2^M choices results in the same $\bar{\mu}$ which is denoted by $\bar{\mu}_2$.
(c) $\bar{\mu} > \bar{\mu}_1$ for all the remaining choices of $\{m_i\}_{i=1}^M$. □

From Proposition 4.2, we only need to consider case (a) and (b). Moreover, for $j = 0, 1, \cdots, L$, since the eigenvalues of $M_j^T M_j$ for $0 \leq m_i \leq N - L - 1$, $i = 1, 2, \cdots, M$, are the same as those of $m_i = 0$ for all i, we will use $m_i = 0$ for all i to represent case (a). Similarly, $m_i = N - 1$ for all i can be used to represent case (b). Hence the optimal selection of m_i can be simply selected from one of two cases: $m_i = 0$ (the first position) or $m_i = N - 1$ (the last position) for all $i = 1, 2, \cdots, M$.

5 Simulation Results

In this section, we use the 2×2 FIR channel given in [3](p.105) to illustrate the performance of the proposed method. The two input source symbols are i.i.d. QPSK signals with the same variances. The channel noises are white Gaussian distribution. The channel normalized root-mean-square error (NRMSE) is given by

$$NRMSE = \frac{1}{\|H\|_F}\sqrt{\frac{1}{I}\sum_{i=1}^{I}\|\widehat{H}^{(i)}U^* - H\|_F^2} \quad (5.1)$$

where I is the number of Monte Carlo runs, and $\widehat{H}^{(i)} = [\widehat{H}^{(i)}(0)^T \ \widehat{H}^{(i)}(1)^T \ \cdots \ \widehat{H}^{(i)}(L)^T]^T$ is the estimate of channel impulse response matrix H. The ambiguity matrix U is found by least square method. The signal-to-noise ratio (SNR) is defined as

$$SNR = \frac{\frac{1}{N}\sum_{n=0}^{N-1} E[\|z(n)\|_2^2]}{E\|v(n)\|_2^2} \quad (5.2)$$

For the results below, the number of Monte Carlo runs is fixed at $I = 100$, and SNR is fixed at 10 dB. The effects of selecting modulating sequences on the performance of our identification method are demonstrated for $N = 4$ and $\delta = 0.5878$. The modulating sequences are chosen based on (4.10). The two possible peak positions are $m_i = 0$ and $m_i = 3$ for $i = 1, 2$, by computation, the corresponding $\bar{\mu}$ are 4.6641 and 22.0646, respectively. Thus $m_i = 0$, $i = 1, 2$ is a better selection. Figure. 5.1 shows that there is a 6 dB difference in NRMSE between the two choices.

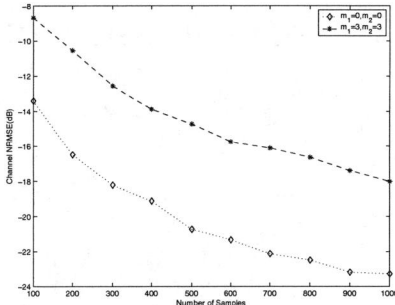

Figure 5.1: Channel NRMSE: Different choices of m

For each upper bound $\widehat{L}$, $0 \leq (\widehat{L} - L) \leq 6$, we choose $N = L + 2$, and 1000 samples for simulation. The modulating sequences are chosen as (4.10) with $m_1 = m_2 = 0$. Figure 5.2 shows that the proposed method is quite robust to channel order overestimation since the NRMSE still maintains a low value (about -16.5dB) when $(\widehat{L} - L) = 6$.

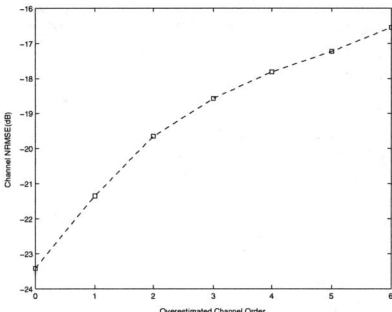

Figure 5.2: Channel NRMSE v.s. $(\widehat{L} - L)$

6 Conclusions

We propose a method for blind identification of FIR MIMO channels. The identifiability condition depends only on the modulating sequences which can be optimally selected to reduce the effect of noise and error in estimating the covariance matrix of the received data. Simulation results show that the method yields good performance. Although only channels with equal number of inputs and outputs are considered, the method can be extended to general MIMO channels [4].

References

[1] Z. Wang and G. B. Giannakis, "Wireless Multi-carrier Communications", *IEEE Signal Processing Magazine*, vol. 17, no. 3 pp. 29-48, May 2000.

[2] C. A. Lin and J. W. Wu, "Blind identification with periodic modulation: A time-domain approach", *IEEE Trans. Signal Processing*, vol. 50, no. 11 pp. 2875-2888, Nov. 2002.

[3] H. Bölcskei, R. W. Heath, Jr., and A. J. Paulraj, "Blind channel identification and equalization in OFDM-based multiantenna systems", *IEEE Trans. Signal Processing*, vol. 50, no. 1, pp. 96-109, Jan. 2002.

[4] C. A. Lin and Y. S. Chen, "Blind Identification of MIMO Channels With Periodic Modulation", submitted to *IEEE Trans. Signal Processing*.

Blind Low Rate Multiuser Detection for Multirate Multicarrier CDMA Systems Using Antenna Array

Yiwen Zhang, Qinye Yin, Le Ding
School of Electronics and Information Engineering
Xi'an Jiaotong University, Xi'an 710049, China
Email: zeven@mail.edu.cn

Ronghai Sun
Shaanxi Armed Police Force
Xi'an 710054, China
Email: srhai@sina.com

Abstract—In the paper, a space-time low rate blind multiuser detector for synchronous uplink multirate multicarrier CDMA systems with antenna arrays at the base station is proposed to mitigate multiple access interference and introduce spatial diversity. After high rate physical users are modelled as several corresponding low rate virtual users, the space-time signature vectors of the low rate virtual users are derived. And then these space-time signature vectors are estimated blindly based on subspace projection algorithm to realize the multirate multiuser detection without computing the ambiguous complex factor between the true and the estimated signature vectors. The proposed scheme can increase system performance and capacity due to introducing spatial domain processing. Simulations results demonstrate that the performance of the proposed scheme is close to that of minimum mean-square error multiuser detector with perfect channel information and that it will increase the system capacity about two times to employ two-element antenna array.

I. INTRODUCTION

Multirate communication has recently become an active research topic as an efficient solution to wireless multimedia communications. There are two schemes, variable spreading length (VSL) and multicode access, which can be used to support multirate services in CDMA systems [1], [2]. A multirate CDMA system can be viewed as a traditional single-rate CDMA system after the various-rate physical users are modelled as the corresponding single-rate virtual users. So the proposed multiuser detection algorithms for single-rate systems can be used for multirate systems, like decorrelating detector, MMSE detector and the corresponding blind versions [3], [4].

VSL CDMA systems have the similar performance to multicode CDMA systems but are more simple in physical-layer implementation. So it get more extensively studied than multicode CDMA systems [3]. There are two classes of typical multiuser detection schemes for VSL CDMA systems, low rate (LR) detector and high rate (HR) detector. LR detector treats HR users as some equivalent LR virtual users, whereas HR detector treats each LR users as a corresponding HR virtual user and then the traditional multiuser detectors are applied to the equivalent single rate system. LR detector will result in serious multiple access interference (MAI) for the bad orthogonality between the spreading codes of virtual LR users created by it.

Antennae arrays processing has been considered as a powerful method to reduce MAI and to introduce spatial diversity. It is often adopted as a flexible way to increase the system performance and capacity. [2] illuminates that the nonorthogonality between the spreading codes of HR and LR users will affect the system performance by analyzing multirate near far resistance (NFR) performance of maximum-likelihood detector. To mitigate cochannel interference, [4] introduces antenna arrays at the base station, and then proposes space-time blind linear multiuser detectors for dual-rate synchronous direct sequence CDMA systems. However, the system discussed in [4] is assumed in a flat-fading channel environment.

In this paper, the synchronous VSL multicarrier (MC) CDMA systems over frequency-selective Rayleigh fading channel is investigated. We employ multiple independent receiving antenna elements at the base station to reduce the impact of MAI on the multirate system performance and improve multirate system capacity. After HR physical users are modelled as corresponding LR virtual users, the space-time signature vectors of the virtual LR users including the temporal and spatial domain information are constructed. Based on subspace projection algorithm, the space-time signature vectors are estimated blindly. On condition that the differential modulation scheme is employed at the transmitter, a space-time LR multiuser detector is proposed based on the estimated signature vectors without computing the ambiguous complex factor between the true signature vector and the estimated version. Finally, the simulation results validate the performance of the proposed scheme in term of bit error rate (BER) and the system capacity.

II. SYSTEM MODEL

In this section, a dual-rate uplink VSL MC-CDMA system over frequency-selective Rayleigh fading channel is considered. There are M_L LR users and M_H HR users in the system. The bit rate of high-rate users is Q times that of low-rate users, Q is an integer. In the paper, we only consider the dual-rate system, so Q equals 2 for the discussion later. The energy normalized time-domain spreading sequences are $G_L \times 1$ vector $\mathbf{c}_{k,L}$ for the kth LR user, $k \in [1, M_L]$, and $G_H \times 1$ vector $\mathbf{c}_{k,H}$ for the kth HR user, $k \in [1, M_H]$. G_L and G_H denote the spreading gains of the LR and HR users respectively, and $G_L/G_H = Q$. The number of subcarriers is $G = G_L$ and cyclic prefix (CP) are inserted to eliminate inter-symbol interference. The transmitter structures for LR and HR users are shown in Fig. 1. An antenna arrays with P independent elements is employed at the detector. We assume that the symbol synchronization among all HR and LR users is made using timing synchronization techniques.

The complex baseband channel impulse response vector for the pth receiving antenna element of LR user k and HR user

Partially supported by the National Natural Sciences Foundation (No. 60272071) and the Research Fund for Doctoral Program of Higher Education (No. 20020698024 & No. 20030698027) of China.

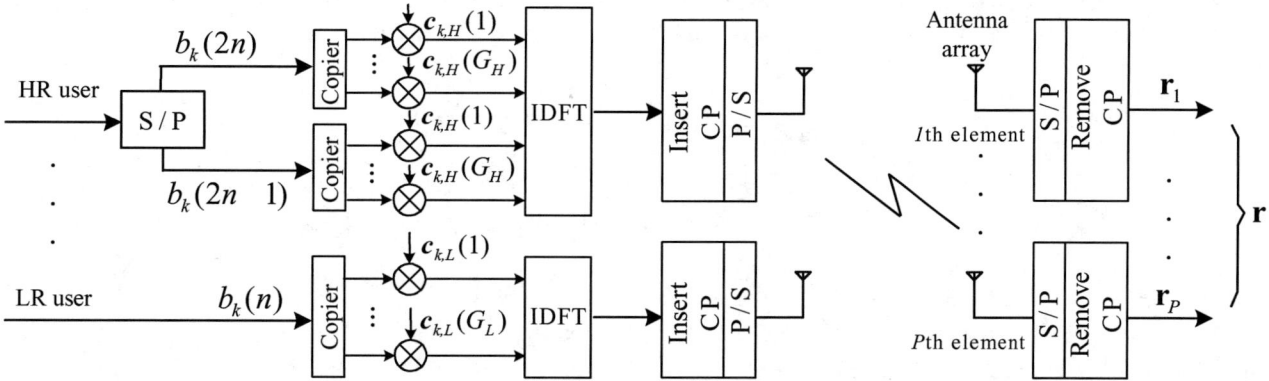

Fig. 1. uplink system model of dual-rate MC-CDMA with multiple antennae at the receive ends

k are $\mathbf{h}_{k,L}^{(p)} = [h_{k,L}^{(p)}(0), h_{k,L}^{(p)}(1), ..., h_{k,L}^{(p)}(L_{ch})]^T$ and $\mathbf{h}_{k,H}^{(p)} = [h_{k,H}^{(p)}(0), h_{k,H}^{(p)}(1), ..., h_{k,H}^{(p)}(L_{ch})]^T$, respectively. $(L_{ch}+1)$ is the length of finite impulse response channel, and we let $(L_{ch}+1) < G_H$ in the paper. Different channel impulse response vectors are assumed to be independent each other.

Consider the block diagram of the mobile transmitter in Fig. 1. At the output ends of IDFT, the kth LR user produces a sequence of the baseband MC-CDMA signal in the nth LR bit interval

$$\mathbf{x}_{k,L} = \mathbf{F}_{IDFT}\mathbf{c}_{k,L}b_{k,L}(n) \quad (1)$$

where $\mathbf{F}_{IDFT}$ is the $G \times G$ IDFT transformation matrix with elements defined by $(\mathbf{F}_{IDFT})_{m,n} = G^{-1/2}\exp(-j\frac{2\pi}{G}(m-1)(n-1))$. $b_{k,L}(n)$ is the nth channel symbol produced according to the M-ary modulation scheme of LR user k.

In a single LR bit interval, each HR user can be modelled as 2 virtual LR users whose processing gain is G_L. The jth virtual user for HR user k transmits bit $b_{k,H}(2n+j)$ using signature vectors $\mathbf{c}_{k,j,H}$, $j=0,1$. $\mathbf{c}_{k,1,H} = [\mathbf{c}_{k,H}^T, \mathbf{0}^T]^T$, $\mathbf{c}_{k,2,H} = [\mathbf{0}^T, \mathbf{c}_{k,H}^T]^T$, $\mathbf{0}$ is $G_H \times 1$ zero vector. $b_{k,H}(n)$ is the nth symbol of HR user k. So in the nth LR bit interval, the baseband signal vector produced by the kth HR user is

$$\mathbf{x}_{k,H} = \mathbf{F}_{IDFT}\mathbf{c}_{k,0,H}b_{k,H}(2n) + \mathbf{F}_{IDFT}\mathbf{c}_{k,1,H}b_{k,H}(2n+1) \quad (2)$$

For the simplicity of denotation, let $\mathbf{c}_{k,L}^{(F)} = \mathbf{F}_{IDFT}\mathbf{c}_{k,L}$, $\mathbf{c}_{k,j,H}^{(F)} = \mathbf{F}_{IDFT}\mathbf{c}_{k,j,H}$. $\mathbf{c}_{k,L}^{(F)}(i)$ and $\mathbf{c}_{k,j,H}^{(F)}(i)$ denote the ith element of $\mathbf{c}_{k,L}^{(F)}$ and $\mathbf{c}_{k,j,H}^{(F)}$, respectively.

At the pth receiving antenna element, the received noiseless baseband data vector of user k in the nth LR bit interval after series-parallel converting and removing CP can be modelled as

$$\mathbf{r}_{k,L}^{(p)}(n) = \mathcal{C}_{k,L}\mathbf{h}_{k,L}^{(p)}b_{k,L}k(n) \quad k \in [1, M_L] \quad (3)$$

where $\mathcal{C}_{k,L}$ is $G \times (L_{ch}+1)$ Toeplitz matrix defined as

$$\mathcal{C}_{k,L} = \begin{bmatrix} c_{k,L}^{(F)}(1) & c_{k,L}^{(F)}(G) & \cdots & c_{k,L}^{(F)}(G-L_{ch}+1) \\ c_{k,L}^{(F)}(2) & c_{k,L}^{(F)}(1) & \cdots & c_{k,L}^{(F)}(G-L_{ch}+2) \\ \vdots & \vdots & \ddots & \vdots \\ c_{k,L}^{(F)}(G) & c_{k,L}^{(F)}(G-1) & \cdots & c_{k,L}^{(F)}(G-L_{ch}) \end{bmatrix} \quad (4)$$

For HR users k, the received noiseless baseband data vector at the pth receiving antenna element in the nth LR bit interval can be modelled as

$$\mathbf{r}_{k,H}^{(p)}(n) = \mathcal{C}_{k,0,H}\mathbf{h}_{k,H}^{(p)}b_{k,H}(2n) + \mathcal{C}_{k,1,H}\mathbf{h}_{k,H}^{(p)}b_{k,H}(2n+1)$$
$$k \in [1, M_H] \quad (5)$$

where $\mathcal{C}_{k,0,H}$ and $\mathcal{C}_{k,1,H}$ are Toeplitz matrix defined like (4), constructed from $\mathbf{c}_{k,0,H}^{(F)}$ and $\mathbf{c}_{k,1,H}^{(F)}$ respectively.

Concatenating the received data vectors $\mathbf{r}_{k,L}^{(p)}(n)$ and $\mathbf{r}_{k,H}^{(p)}(n)$ on P antenna elements, an extended data vector can be obtained

$$\begin{cases} \mathbf{r}_{k,L}(n) = \mathbf{C}_{k,L}\mathbf{h}_{k,L}b_{k,L}(n) & k \in [1, M_L] \\ \mathbf{r}_{k,H}(n) = \mathbf{C}_{k,0,H}\mathbf{h}_{k,H}b_{k,H}(2n) + \\ \qquad\qquad \mathbf{C}_{k,1,H}\mathbf{h}_{k,H}b_{k,H}(2n+1) & k \in [1, M_H] \end{cases} \quad (6)$$

where $\mathbf{C}_{k,L} = \mathbf{I}(P) \otimes \mathcal{C}_{k,L}$, $\mathbf{C}_{k,j,H} = \mathbf{I}(P) \otimes \mathcal{C}_{k,j,H}$, $\otimes$ denotes the Kronecker product; $\mathbf{I}(n)$ is an $n \times n$ identity matrix; $\mathbf{h}_{k,L} = [\mathbf{h}_{k,L}^{(1)T}, \mathbf{h}_{k,L}^{(2)T}, \cdots, \mathbf{h}_{k,L}^{(P)T}]^T$, $\mathbf{h}_{k,H} = [\mathbf{h}_{k,H}^{(1)T}, \mathbf{h}_{k,H}^{(2)T}, \cdots, \mathbf{h}_{k,H}^{(P)T}]^T$.

When there are M_L active LR and M_H active HR users in the system, we can write received data vector perturbed by additive white Gaussian noise $\mathbf{n}$ as

$$\mathbf{r}(n) = \sum_{k=1}^{M_L} \mathbf{C}_{k,L}\mathbf{h}_{k,L}b_{k,L}(n) + \sum_{k=1}^{M_H}\sum_{j=0}^{1} \mathbf{C}_{k,j,H}\mathbf{h}_{k,H}b_{k,H}(2n+j) + \mathbf{n} \quad (7)$$

where $\mathbf{n}$ is a vector of independent identically distributed (i.i.d) complex zero-mean Gaussian noises.

III. LR BLIND MULTIUSER DETECTION

A. Blind estimation of space-time signature vectors

We can see (7) as a single rate system with K virtual LR users

$$\mathbf{r}(n) = \sum_{k=1}^{K} \mathbf{s}_k b_k(n) + \mathbf{n} = \sum_{k=1}^{K} \mathbf{C}_k \mathbf{h}_k b_k(n) + \mathbf{n} \quad (8)$$

where $K = M_L + 2M_H$ denotes the number of virtual users in the system. $s_k = C_k h_k$ denotes the space-time signature vector of the kth virtual user. $C_k h_k b_k(n)$ is defined in detail as

$$C_k h_k b_k(n) = \begin{cases} C_{k,L} h_{k,L} b_{k,L}(n) & k \in [1, M_L] \\ C_{k-M_L,0,H} h_{k-M_L,H} b_{k-M_L,H}(2n) \\ \quad\quad k \in [M_L+1, M_L+M_H] \\ C_{k-M_L-M_H,1,H} h_{k-M_L-M_H,H} b_{k-M_L-M_H,H}(2n+1) \\ \quad\quad k \in [M_L+M_H+1, M_L+2M_H] \end{cases} \quad (9)$$

Performing eigen decomposition on the covariance matrix of $r(n)$ can obtain

$$R = rr^H = E_s Y_s E_s^H + E_n Y_n E_n^H \quad (10)$$

where Y_s is the diagonal matrix containing K largest eigenvalues. E_s is the signal subspace in which column vectors are the eigenvectors corresponding to the diagonal elements in matrix Y_s. The column vectors in E_s form an orthonormal basis of the signal subspace.

Because the range space of E_s contains all of the 1-D subspaces spanned by the space-time signature vectors of all users, we have the subspace constraint: $s_k \in R(E_s)$. $R(A)$ denotes the range space spanned by the column vectors of A. From the definition of s_k, we have another subspace constraint: $s_k = C_k h_k \in R(C_k)$. Therefore

$$s_k \in R(E_s) \cap R(C_k) \quad (11)$$

Because C_ks are different and FIR channels are independent each other, when the dimension of the complement space of $R(E_s)$ equals or exceeds the dimension of $R(C_k)$ minus one, the two constraint subspaces generally intersect at a unique line in the overall observation space [5]. It means that when the inequality

$$K \leq (G - L_{ch})P - (L_{ch}+1)P + 1 \quad (12)$$

is true, the following equation does generally hold

$$R(s_k) = R(E_s) \cap R(C_k) \quad (13)$$

The orthogonal projection matrix into the subspace $R(C_k)$ is given by $P_k = C_k(C_k^H C_k)^{-1} C_k^H$, $(\cdot)^{-1}$ denotes Pseudo-inverse. The projection matrix into E_s is $P = E_s E_s^H$ since the columns of E_s are orthonormal.

[7] gives us that: In a Hilbert space H, if G_i is a closed subspace for $i = 1, 2, \cdots, m$, and if their intersection G_0 is non-empty, let P_i denotes the projection operator onto the set G_i, then for every $x \in H$, $P_m P_{m-1} \cdots P_1 x$ converges strongly to the projection of x onto G_0 [7].

By executing alternating projection $\hat{s}_k^{[l+1]} = P_k P \hat{s}_k^{[l]}$ iteratively, the arbitrary initial vector but nonzero $\hat{s}_k^{[0]}$ will finally converge to $\hat{s}_k$, a complex scaled version of s_k. In the case of convergence, we have

$$\hat{s}_k = \hat{s}_k^{[\infty]} = P_k P \hat{s}_k^{[\infty]}$$

that is

$$(P_k P - I)\hat{s}_k = 0 \quad (14)$$

So the final estimation of $\hat{s}_k$ is the eigenvector of matrix $(P_k P - I)$ corresponding to zero eigenvalue.

B. LR blind linear multiuser detector

Based on the blind linear minimum mean-square error (MMSE) multiuser detector in [6], when the differential binary phase shift keying (DBPSK) modulation scheme is employed at the transmitter, the blind multiuser detector for the kth virtual user is

$$\hat{b}_k(n) = sgn[Re(w_k^H r(n)/w_k^H r(n-1))] \quad (15)$$

where w_k denotes the weight vector of detector. In [6], $w_k = E_s(Y_s)^{-1} E_s^H s_k / (s_k^H E_s(Y_s)^{-1} E_s^H s_k)$, where $s_k^H E_s(Y_s)^{-1} E_s^H s_k$ is an integer and can be neglected. From (15), we can see that the complex scaled w_k has no effect on symbol decision. So $\hat{s}_k$ can be used to substitute s_k directly needlessly to worry about the ambiguous complex factor between them. Finally, the weight vector w_k in (15) becomes

$$w_k = E_s(Y_s)^{-1} E_s^H \hat{s}_k \quad (16)$$

The symbol of physical LR user k is $\hat{b}_k(n), k \in [1, M_L]$. Since the symbols of HR user are divided into two sub symbol stream, the symbols of a physical HR user is composed of the decision symbols of the two corresponding virtual LR users.

IV. SIMULATIONS

The numeric simulations were executed to evaluate the performances of the proposed blind detector and the advantage of using antenna arrays. MC-CDMA system with the DBPSK modulation mode is employed in our simulations. The basic system parameters are $L_{ch} = 4$, $G_L = 32$, $G_H = 16$, $Q = 2$ and the length of CP is 5. The random sequences is used as the temporal spreading codes and the spreading codes are energy normalized. The antenna array with one or two elements which are independent each other is considered. 2000 Monte Carlo trials are performed for each simulation. In each trial, 128 symbol samples are used to estimate the correlation matrices of the received data vector, and 1000 LR symbols and 2000 HR symbols are detected.

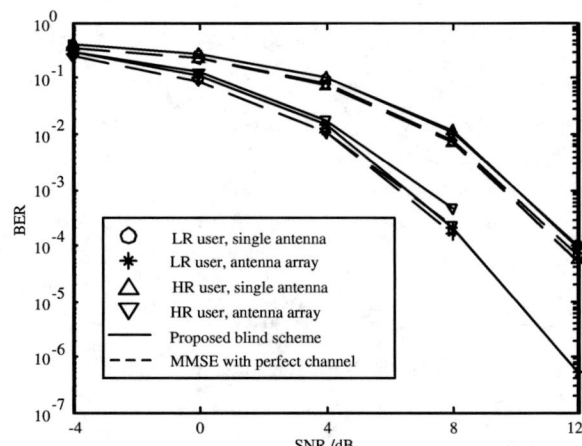

Fig. 2. The Multirate BER performance for the desired LR and HR users versus SNR.

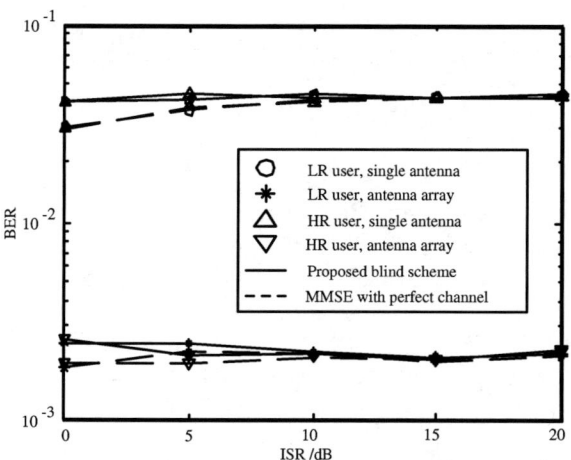

Fig. 3. The NFR performance, BER of desired users versus the power ratio of the interference users to the desired user.

of all users is fixed to 6dB. The number of interference users in the system increases continuously. The BER of the desired user indicates that there are different influence to the system performance between adding a LR user and adding a HR user. The system can contain more physical LR users than physical HR users. It is also shown that the system capacity are increased about two times after using two-element antenna array.

V. CONCLUSIONS

We have proposed a blind LR space-time detection scheme for multirate MC-CDMA system with antenna arrays at the receiver over frequency selective-fading channel. The space-time signature vectors of virtual users are estimated blindly employing subspace projection algorithm which will get a reliable estimation even in situations when training signal based estimator is overwhelmed by MAI. Without computing the complex ambiguous factors between the true space-time signature vectors and the corresponding estimated versions, we execute the symbol detection based on the estimated signature vector directly, which save the computation. The simulation results demonstrate that the proposed scheme has the close performance to MMSE multiuser detector with perfect channel information. And the space-time detector with multiple receiving antennae can improve the system performance and increase the system capacity distinctly. At last, although we only consider a dual-rate system in the paper, our derivations and conclusions can be easily extended to the system with more classes of rates.

A. Multi-user detection performance

For Fig. 2, the numbers of active HR and LR users are all fixed to 3, and the mean power of all users is identical. The SNR (Signal-Noise Ratio) of all users varies from -4dB to 12dB. The BER in Fig. 2 shows us that there are close performance between the proposed detector and the MMSE detector with perfect channel information. And we can see that the proposed detector with two-elements antenna array outperforms the purely temporal counterpart with single antenna.

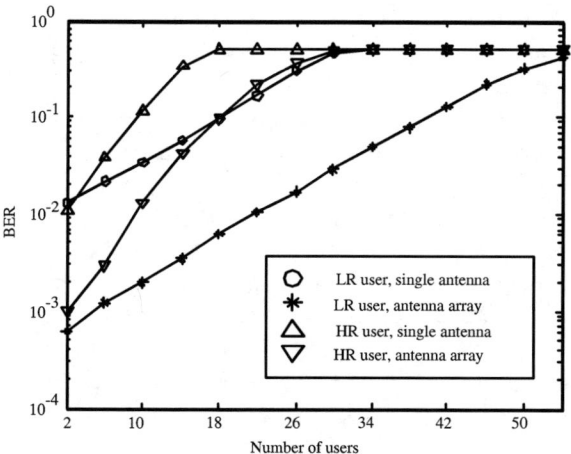

Fig. 4. The system capacity, BER of desired users versus the number of all users in the system.

For Fig. 3, the SNR of the desired LR and HR users is fixed to 6dB and the power ratio of the interference user to the desired user changes form 0dB to 20dB. It is shown that the power of the interference users has almost no effect on the BER of the desired users in Fig 3.

B. System capacity

Fig. 4 gives the BER of the desired user versus the number of the active users in the system. In this simulation, the interference users in the system are all LR users or HR users, and the SNR

REFERENCES

[1] M. J. Juntti, "System concept comparisons for multirate CDMA with multiuser detection," in *Proc. VTC'98*, vol. 1, 1998, pp. 36-40.
[2] Z. Guo and K. B. Letaief, "Performance of Multiuser Detection in Multirate DS-CDMA Systems," *IEEE Trans. Commun.*, vol. 51, pp. 1979-1983, Dec. 2003.
[3] M. Saquib, R. Yates and N. Mandayam, "Decorrelating detectors for a dual-rate synchronous DS/CDMA channel," in *Proc. VTC'96*, vol. 1, 1996, pp. 377-381.
[4] L. Huang, F. C. Zheng and M. Faulkner, "Space-Time Multirate Blind Multiuser Detection for Synchronous DS/CDMA Systems," in *Proc. GLOBECOM'01*, vol. 1, 2001, pp. 146-150.
[5] D. J. Sadler and A. Manikas, "Blind Reception of Multicarrier DS-CDMA Using antenna Arrays," *IEEE Trans. Wireless Commun.*, vol. 2, pp. 1231-1239, Nov. 2003.
[6] X. Wang, H. V. Poor, "Blind Multiuser Detection: A Subspace Approach," *IEEE Trans. Inform. Theory*, vol. 44, pp. 677-690, Mar. 1998.
[7] H. Stark and Y. Yang, *Vector Space Projections*, 1st ed. New York: Wiley, 1998.

DOA-Matrix Decoder for STBC-MC-CDMA Systems Over Frequency-Selective Channel

Yanxing Zeng, Qinye Yin, Le Ding, Yinkuo Meng and Ying Zhang
School of Electronics and Information Engineering, Xi'an Jiaotong University, P.R.China, 710049
Email:zyanxing@hotmail.com, qyyin@mail.xjtu.edu.cn

Abstract— A space-time block coded multicarrier (MC)-CDMA system equipped with a uniform linear array (ULA) at the base station over frequency-selective channel is studied. A blind decoder providing closed-form solutions of transmitted symbol sequences for all active users in one macrocell is developed. The decoder employs DOA-Matrix method to obtain a set of signal subspaces. From them, the information-bearing symbol sequences of multiple users are estimated by exploiting the structure of space-time block coding (STBC). The performance and computation complexity of the proposed decoder are evaluated by computer simulations.

I. INTRODUCTION

Space-time block coded MC-CDMA (STBC-MC-CDMA) system is one of the most promising schemes for next-generation wireless communications. To coherently decode the STBC signals, [1,2] proposed a subspace-based channel estimation algorithm and developed a MMSE receiver with the estimated channels. However, it may be difficult or costly to estimate the channel accurately in mobile fading environment. So [3] gave an alternative subspace-based receiver that directly estimates the transmitted symbol sequence without channel information by exploiting the structure of STBC. All of them assumed the channels between all transmit antennas and receive antennas are independent. But in a macrocell environment, the base station is deployed above the surrounding scatters. The received signals at the base station result from the scattering process in the vicinity of the mobile station and the multipath components at the base station are thus restricted to a small angular range. Therefore, it is usually assumed the receive antennas at the base station are correlated, meanwhile the transmit antennas at the mobile station are uncorrelated [4]. Under the circumstances, we claim that the aforementioned algorithms are not optimal because they didn't explore the correlations between receive antennas.

In this paper, we apply a full correlated ULA to the base station of STBC-MC-CDMA systems in macrocell and propose a blind decoder to detect the STBC symbols without uplink channel estimation by use of direction of arrival Matrix (DOA-Matrix) method [5]. The algorithm separates multiple users with different impinging DOAs in spatial field and obtains a set of signal subspaces, each of them spanned by the transmitted symbol sequences of an individual user. From

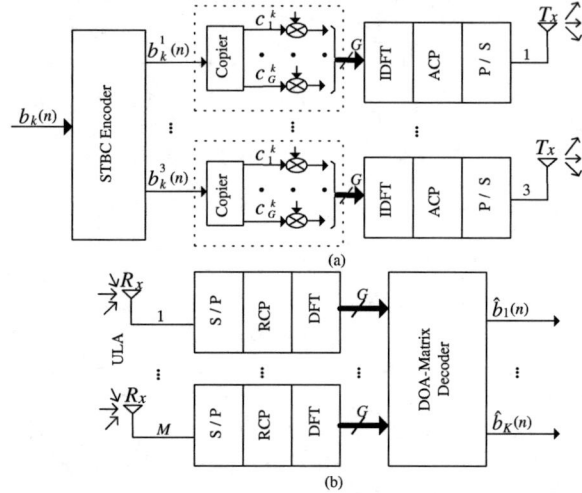

Fig. 1. The baseband model of the STBC-MC-CDMA system equipped with a ULA at the base station (a) user transmitter (b) base station receiver

these signal subspaces, the original symbol sequences of multiple users are estimated by exploiting the structure of STBC. Computer simulations show the proposed decoder outperforms the conventional MMSE receiver based on channel estimation [1,2].

The paper is organized as follows. In Section II, we describe the system model. In Section III, we introduce the DOA-Matrix method to separate multiple co-channel users. Then Section IV presents the STBC decoding algorithm. Finally, computer simulation results are given in Section V.

Notation: $(\cdot)^*$, $(\cdot)^T$, and $(\cdot)^H$ denote the complex conjugate, transpose and the conjugate transpose of matrix.

II. SYSTEM MODEL

Fig. 1 displays the baseband model of the STBC-MC-CDMA system equipped with a ULA at the base station, where the subscript k represents the number of user. The antenna separation of the ULA with M antenna elements is half a carrier wavelength. It is assumed that K users uniformly distributed around a macrocell site. All K active users share the same set of subcarriers. As shown in Fig. 1, the input data symbols of each user are coded by Tarokh's rate 1 space-time block encoder [6] from real orthogonal design, with three transmit antennas (specified in Section IV). When MC-CDMA systems over wireless finite impulse response (FIR)

Partially supported by the National Natural Sciences Foundation (No. 60272071) and the Research Fund for Doctoral Program of Higher Education (No. 20020698024 & No. 20030698027) of China.

channels, a usual approach for combating the resultant inter-block interference (IBI) is via adding cyclic prefix (ACP) to each transmitted data block. Meanwhile, by removing CP (RCP) at the beginning of each received data block, the IBI can be eliminated. For the sake of simplicity, we assume that symbols from different users are synchronized for the uplink.

A frequency domain spreading code is assigned to the k-th user's three transmit antennas, which is defined as a vector $\mathbf{c}^k = [c_1^k, \cdots, c_G^k]^T$. At the n-th symbol interval, we define the frequency domain IBI-free uplink received data vector on the reference element 1 of the base station antenna array from the m-th ($m = 1, 2, 3$) transmit antenna of the k-th user as (reference (5) in [2])

$$\mathbf{x}_{1,k}^{m,n} = \mathrm{diag}(\mathbf{c}^k)\boldsymbol{\eta}_{1,k}^{m,n} b_k^m(n) = \mathbf{f}_{1,k}^{m,n} b_k^m(n) \quad (1)$$

where $\boldsymbol{\eta}_{1,k}^{m,n}$ is the frequency domain response [2] of the FIR channel between the m-th transmit antenna of the k-th user and the reference element 1, with dimension of $G \times 1$. $\mathbf{f}_{1,k}^{m,n} = \mathrm{diag}(\mathbf{c}^k)\boldsymbol{\eta}_{1,k}^{m,n}$, $b_k^m(n)$ is the transmitted symbol from the m-th antenna of the k-th user.

On the reference element $r(r = 1, \cdots, M)$ of the base station antenna array, the received data vector consisting of received signals from all K active users over each subcarrier is

$$\mathbf{x}_r^n = \sum_{k=1}^{K} \sum_{m=1}^{3} \mathbf{x}_{r,k}^{m,n} + \mathbf{n}_r^n = \mathbf{F}_r^n \mathbf{b}^n + \mathbf{n}_r^n \quad (2)$$

where $\mathbf{b}^n$ consists of transmitted symbols from all transmit antennas of K users, namely, $\mathbf{b}^n = [b_1^1(n), b_1^2(n), b_1^3(n), \cdots, b_K^3(n)]^T$; $\mathbf{n}_r^n$ is a vector of independent identically distributed complex zero-mean Gaussian noises with variance σ_n^2; $\mathbf{F}_r^n$ is defined as $\mathbf{F}_r^n = [\mathbf{f}_{1,1}^{1,n} e^{j\chi(r-1)\sin\theta_1^n}, \mathbf{f}_{1,1}^{2,n} e^{j\chi(r-1)\sin\theta_1^n}, \mathbf{f}_{1,1}^{3,n} e^{j\chi(r-1)\sin\theta_1^n}, \cdots, \mathbf{f}_{1,K}^{3,n} e^{j\chi(r-1)\sin\theta_K^n}]$, and $\chi = 2\pi d/\lambda$. Notations λ, d, and θ_k^n represent the wavelength of Radio Frequency (RF) carrier, the inter-element spacing and the incident angle with respect to the array normal, respectively. We assume θ_k^n denotes the DOA of the main cluster from user k's three transmit antennas and there is only one main DOA for each user in the paper. However, this simple system model can be readily extended to a more practical situation where the number of DOA of each user is larger than one [7].

III. DOA-Matrix Method for Co-channel User Separation

Concatenating uplink received data vectors from element 1 to $M - 1$, an extended data vector can be obtained, that is

$$\mathbf{u}^n = \begin{bmatrix} \mathbf{x}_1^{n^T} & \cdots & \mathbf{x}_{M-1}^{n^T} \end{bmatrix}^T = \mathbf{A}^n \mathbf{b}^n + \mathbf{n}_{head}^n \quad (3)$$

where $\mathbf{u}^n$ is a $(M-1)G \times 1$ vector, $\mathbf{n}_{head}^n$ is given by $\mathbf{n}_{head}^n = [\mathbf{n}_1^{n^T}, \cdots, \mathbf{n}_{M-1}^{n^T}]^T$; $\mathbf{A}^n$ is a $(M-1)G \times 3K$ matrix, and is given by

$$\begin{aligned}\mathbf{A}^n &= \begin{bmatrix} \mathbf{F}_1^{n^T} & \cdots & \mathbf{F}_{M-1}^{n^T} \end{bmatrix}^T \\ &= [\mathbf{a}_1^n \mathbf{f}_{1,1}^{1,n} \quad \mathbf{a}_1^n \mathbf{f}_{1,1}^{2,n} \quad \mathbf{a}_1^n \mathbf{f}_{1,1}^{3,n} \quad \cdots \quad \mathbf{a}_K^n \mathbf{f}_{1,K}^{3,n}] \end{aligned} \quad (4)$$

where $\mathbf{a}_k^n$ is an $(M-1) \times 1$ steering vector for the k-th user, and is defined as $\mathbf{a}_k^n = [1, e^{j\chi\sin\theta_k^n}, \cdots, e^{j\chi(M-2)\sin\theta_k^n}]^T$; $\otimes$ denotes the Kronecker product.

When concatenating the received data vectors from element 2 to M, another extended data vector can be obtained,

$$\mathbf{y}^n = [\mathbf{x}_2^{n^T} \quad \cdots \quad \mathbf{x}_M^{n^T}]^T = \mathbf{A}^n \boldsymbol{\Phi}^n \mathbf{b}^n + \mathbf{n}_{tail}^n \quad (5)$$

where $\mathbf{n}_{tail}^n$ is given by $\mathbf{n}_{tail}^n = [\mathbf{n}_2^{n^T}, \cdots, \mathbf{n}_M^{n^T}]^T$; $\boldsymbol{\Phi}^n$ is a $3K \times 3K$ diagonal matrix, and is given by

$$\boldsymbol{\Phi}^n = \mathrm{diag}(e^{j\chi\sin\theta_1^n}, e^{j\chi\sin\theta_1^n}, e^{j\chi\sin\theta_1^n}, \cdots, e^{j\chi\sin\theta_K^n}) \quad (6)$$

Terms on the main diagonal of the above matrix are associated with users' DOAs. Hence, we call them DOA items.

Under the assumption that the channel is constant during several tens of MC-CDMA symbols, two sets of extended data vectors $\mathbf{u}^n$ and $\mathbf{y}^n$, which are corresponding to successive L MC-CDMA symbols, can be aggregated into two matrices as

$$\mathbf{U}^n = [\mathbf{u}^n \quad \cdots \quad \mathbf{u}^{n+L-1}] = \mathbf{A}^n \mathbf{B}^n + \mathbf{N}_{head}^n \quad (7)$$

$$\mathbf{Y}^n = [\mathbf{y}^n \quad \cdots \quad \mathbf{y}^{n+L-1}] = \mathbf{A}^n \boldsymbol{\Phi}^n \mathbf{B}^n + \mathbf{N}_{tail}^n \quad (8)$$

where $\mathbf{U}^n$ and $\mathbf{Y}^n$ are $(M-1)G \times L$ matrices; $\mathbf{B}^n$ is a $3K \times L$ matrix, and is defined as $\mathbf{B}^n = [\mathbf{b}^n, \cdots, \mathbf{b}^{n+L-1}]$; $\mathbf{N}_{head}^n$ and $\mathbf{N}_{tail}^n$ are written as $\mathbf{N}_{head}^n = [\mathbf{n}_{head}^n, \cdots, \mathbf{n}_{head}^{n+L-1}]$ and $\mathbf{N}_{tail}^n = [\mathbf{n}_{tail}^n, \cdots, \mathbf{n}_{tail}^{n+L-1}]$ respectively.

For brevity, we omit the superscript n in the following discussion.

By performing matrix transpose operation, we can obtain $\mathbf{X} = \mathbf{U}^T = \mathbf{B}^T \mathbf{A}^T + \mathbf{N}_{head}^T$ and $\mathbf{Z} = \mathbf{Y}^T = \mathbf{B}^T \boldsymbol{\Phi} \mathbf{A}^T + \mathbf{N}_{tail}^T$.

Now, the auto-correlation matrix of $\mathbf{X}$ and the cross-correlation matrix between $\mathbf{Z}$ and $\mathbf{X}$ are defined as

$$\begin{aligned}\mathbf{R}_{XX} &= \mathrm{E}[\mathbf{X}\mathbf{X}^H] = \mathbf{B}^T \mathrm{E}[\mathbf{A}^T \mathbf{A}^*] \mathbf{B} + \sigma_n^2 \mathbf{I} \\ &= \mathbf{B}^T \mathbf{R}_{AA} \mathbf{B} + \sigma_n^2 \mathbf{I} = \mathbf{R}_{XXO} + \sigma_n^2 \mathbf{I} \end{aligned} \quad (9)$$

$$\begin{aligned}\mathbf{R}_{ZX} &= \mathrm{E}[\mathbf{Z}\mathbf{X}^H] = \mathbf{B}^T \boldsymbol{\Phi} \mathrm{E}[\mathbf{A}^T \mathbf{A}^*] \mathbf{B} \\ &= \mathbf{B}^T \boldsymbol{\Phi} \mathbf{R}_{AA} \mathbf{B} \end{aligned} \quad (10)$$

where $\mathrm{E}[\cdot]$ denotes ensemble average; $\mathbf{R}_{AA}$ is a $3K \times 3K$ matrix, which denotes the auto-correlation matrix of space-time channels; $\mathbf{R}_{XXO}$ is an $L \times L$ matrix; $\mathbf{I}$ is an $L \times L$ identity matrix.

When space-time channels of different users' different transmit antennas are uncorrelated, $\mathbf{R}_{AA}$ is nonsingular, and the rank of $\mathbf{R}_{XXO}$ equals to $3K$. Performing eigen decomposition on $\mathbf{R}_{XXO}$ can obtain $\mathbf{R}_{XXO} = \sum_{l=1}^{3K} \mu_l \mathbf{v}_l \mathbf{v}_l^H$, where μ_l and $\mathbf{v}_l$ are eigenvalues and corresponding eigenvectors of $\mathbf{R}_{XXO}$, respectively.

We define an auxiliary matrix by $\mathbf{R}_{ZX}$ and $\mathbf{R}_{XXO}$ as in [5]

$$\mathbf{R} = \mathbf{R}_{ZX} \mathbf{R}_{XXO}^+ \quad (11)$$

where $\mathbf{R}_{XXO}^+$ is the Penrose-Moore pseudo-inverse of $\mathbf{R}_{XXO}$, and is defined by $\mathbf{R}_{XXO}^+ = \sum_{l=1}^{3K} \frac{1}{\mu_l} \mathbf{v}_l \mathbf{v}_l^H$.

Theorem 1: Given $\mathbf{B}$ is row full-rank, $\mathbf{R}_{AA}$ is nonsingular, then $\mathbf{R}\mathbf{B}^T = \mathbf{B}^T \boldsymbol{\Phi}$.

The detailed proof can be found in [7]. Based on Theorem 1, $3K$ eigenvalues can be obtained via eigen decomposition on matrix $\mathbf{R}$. These eigenvalues are DOA items associated with K different users. According to the definition of $\mathbf{\Phi}$ in equation (6), $3K$ eigenvalues in deed include K different values, which are $e^{j\chi\sin\theta_1^n},\ldots,e^{j\chi\sin\theta_K^n}$, and each of them with multiplicity of three. Moreover, each of the three eigenvectors corresponding to the same eigenvalue $e^{j\chi\sin\theta_k^n}$ is exactly the linear combining of the transmitted symbol sequences associated with three transmit antennas of an individual user. From this interesting observation, we define a matrix consisting of the three eigenvectors corresponding to the same eigenvalue $e^{j\chi\sin\theta_k^n}$ as $\widetilde{\mathbf{T}}_k$, which can be represented as

$$\widetilde{\mathbf{T}}_k = [\mathbf{t}_k^{1,n} \quad \mathbf{t}_k^{2,n} \quad \mathbf{t}_k^{3,n}]\mathbf{F} \tag{12}$$

where $\widetilde{\mathbf{T}}_k$ is a column full-rank $L\times 3$ matrix; $\mathbf{F}$ is an unknown full-rank 3×3 matrix; $\mathbf{t}_k^{m,n}$ is the transmitted symbol sequence from the m-th transmit antenna of the k-th user, and is defined as $\mathbf{t}_k^{m,n} = [b_k^m(n),\cdots,b_k^m(n+L-1)]^T$; n denotes the symbol interval.

Remarks: Via eigen decomposition on matrix $\mathbf{R}$, we have obtained the following two results. Firstly, we got the closed-form solutions of the DOAs for K users, which are the K different eigenvalues of $\mathbf{R}$. Secondly, we obtained a set of signal subspaces $\widetilde{\mathbf{T}}_k(k=1,\cdots,K)$ and observed that each of them is spanned by the transmitted symbol sequences associated with three transmit antennas of an individual user. Since the eigenvalue of $\mathbf{R}$ is associated with its eigenvector, the DOA of an individual user is associated with its signal subspace spanned by transmitted symbol sequences correspondingly. In this sense, we say it can be automatically paired between the signal subspace of a user and corresponding DOA.

Then, by performing singular value decomposition (SVD) on $\widetilde{\mathbf{T}}_k$, we can obtain

$$\widetilde{\mathbf{T}}_k = [\mathbf{U}_{k,s} \quad \mathbf{U}_{k,o}]\begin{bmatrix}\mathbf{\Sigma}_k \\ \mathbf{0}\end{bmatrix}\mathbf{V}_{k,s}^H \tag{13}$$

where $\mathbf{U}_{k,s}$ is an $L\times 3$ matrix; $\mathbf{U}_{k,o}$ is an $L\times(L-3)$ matrix; $\mathbf{\Sigma}_k$ is a 3×3 matrix; $\mathbf{0}$ is an $(L-3)\times 3$ zero matrix; $\mathbf{V}_{k,s}^H$ is a 3×3 matrix.

Because $\mathbf{U}_{k,o}^H \perp \text{Range}\{\widetilde{\mathbf{T}}_k\}$, we have

$$\mathbf{U}_{k,o}^H[\mathbf{t}_k^{1,n} \quad \mathbf{t}_k^{2,n} \quad \mathbf{t}_k^{3,n}] = \mathbf{U}_{k,o}^H\mathbf{T}_k = \mathbf{0} \tag{14}$$

where $\mathbf{0}$ is an $(L-3)\times 3$ zero matrix; $\mathbf{T}_k$ is an $L\times 3$ matrix, and is defined as $\mathbf{T}_k = [\mathbf{t}_k^{1,n},\mathbf{t}_k^{2,n},\mathbf{t}_k^{3,n}]$.

By now, we have obtained a set of signal space matrices $\widetilde{\mathbf{T}}_k$. From these matrices, we will obtain closed-form solutions of the original information-bearing symbol sequences for multiple users by exploiting the structure of STBC in next Section.

IV. STBC Decoding

For each user, the input of its encoder is a group of four successive data symbols $\{x_1,x_2,x_3,x_4\}$, and the output of its encoder is a 4×3 coded symbol matrix [6]

$$\mathbf{C} = \begin{bmatrix} x_1 & x_2 & x_3 \\ -x_2 & x_1 & -x_4 \\ -x_3 & x_4 & x_1 \\ -x_4 & -x_3 & x_2 \end{bmatrix} \tag{15}$$

where the m-th ($m=1,2,3$) column of $\mathbf{C}$ is the transmitted symbol sequence of the m-th transmit antenna in four successive symbol intervals.

Remarks: It is not difficult to extend our algorithm to many non-rate 1 STBC encoder and many redundant linear precoders with arbitrary transmit antennas. See [7] for more examples.

Based on the encoder defined in (15), the coded symbol matrix $\mathbf{C}$ and the input symbol vector $\mathbf{c} = [x_1,x_2,x_3,x_4]^T$ have the following relationship:

Theorem 2: Denote an $l\times 4$ matrix $\mathbf{U} = [\mathbf{u}_1,\mathbf{u}_2,\mathbf{u}_3,\mathbf{u}_4]$, where $\mathbf{u}_j$ ($j=1,\cdots,4$) is the j-th column of $\mathbf{U}$. With the columns of $\mathbf{U}$, construct a $3l\times 4$ matrix $\bar{\mathbf{U}}$

$$\bar{\mathbf{U}} = \begin{bmatrix} \mathbf{u}_1 & -\mathbf{u}_2 & -\mathbf{u}_3 & -\mathbf{u}_4 \\ \mathbf{u}_2 & \mathbf{u}_1 & -\mathbf{u}_4 & \mathbf{u}_3 \\ \mathbf{u}_3 & \mathbf{u}_4 & \mathbf{u}_1 & -\mathbf{u}_2 \end{bmatrix} \tag{16}$$

Then, $\bar{\mathbf{U}}\mathbf{c} = \text{vec}(\mathbf{U}\mathbf{C})$, where $\text{vec}(\mathbf{A})$ represents the vectorization of matrix $\mathbf{A}$, that is, stacking the columns of $\mathbf{A}$ one by one as a long vector.

We can prove Theorem 2 by simple substitution. So, the product of a matrix $\mathbf{U}$ and the coded symbol matrix $\mathbf{C}$ can be transformed to the product of a matrix $\bar{\mathbf{U}}$ and the input symbol vector $\mathbf{c}$.

From the previous Section, we know $\mathbf{T}_k$ is an $L\times 3$ coded symbol matrix of the k-th user. Let $L=4B$, and split $\mathbf{T}_k$ into B sub-matrices, we get

$$\mathbf{T}_k = [\mathbf{T}_{k,1}^T \quad \mathbf{T}_{k,2}^T \quad \cdots \quad \mathbf{T}_{k,B}^T]^T \tag{17}$$

where $\mathbf{T}_{k,i}(i=1,\cdots,B)$ is a 4×3 coded symbol matrix associated with the i-th group of four successive input symbols $b_k(n+4i-4),\cdots,b_k(n+4i-1)$. Here we define $\mathbf{x}_{k,i}$ as $\mathbf{x}_{k,i} = [b_k(n+4i-4),\cdots,b_k(n+4i-1)]$.

$\mathbf{U}_{k,o}^H$ is an $(L-3)\times L$ matrix, and it is also split into B sub-matrices, that is,

$$\mathbf{U}_{k,o}^H = [\mathbf{U}_{k,1} \quad \mathbf{U}_{k,2} \quad \cdots \quad \mathbf{U}_{k,B}] \tag{18}$$

where $\mathbf{U}_{k,i}$ is an $(L-3)\times 4$ matrix. Thus, equation (14) can be represented as

$$\mathbf{U}_{k,o}^H\mathbf{T}_k = \mathbf{U}_{k,1}\mathbf{T}_{k,1} + \cdots + \mathbf{U}_{k,B}\mathbf{T}_{k,B} = \mathbf{0} \tag{19}$$

Based on Theorem 2, we have

$$\bar{\mathbf{U}}_{k,i}\mathbf{x}_{k,i}^T = \text{vec}(\mathbf{U}_{k,i}\mathbf{T}_{k,i}) \tag{20}$$

where $\bar{\mathbf{U}}_{k,i}$ is a $3(L-3)\times 4$ matrix, and is given by

$$\bar{\mathbf{U}}_{k,i} = \begin{bmatrix} \mathbf{u}_{k,i}^1 & -\mathbf{u}_{k,i}^2 & -\mathbf{u}_{k,i}^3 & -\mathbf{u}_{k,i}^4 \\ \mathbf{u}_{k,i}^2 & \mathbf{u}_{k,i}^1 & -\mathbf{u}_{k,i}^4 & \mathbf{u}_{k,i}^3 \\ \mathbf{u}_{k,i}^3 & \mathbf{u}_{k,i}^4 & \mathbf{u}_{k,i}^1 & -\mathbf{u}_{k,i}^2 \end{bmatrix} \tag{21}$$

where $\mathbf{u}_{k,i}^j(j=1,\cdots,4)$ is the j-th column of $\mathbf{U}_{k,i}$.

Successively, equation (14) can be transformed to

$$\text{vec}(\mathbf{U}_{k,o}^H \mathbf{T}_k) = \bar{\mathbf{U}}_k \mathbf{x}_k = \mathbf{0} \tag{22}$$

where $\bar{\mathbf{U}}_k$ is a $3(L-3) \times L$ matrix, and is defined as $\bar{\mathbf{U}}_k = [\bar{\mathbf{U}}_{k,1}, \cdots, \bar{\mathbf{U}}_{k,B}]$; $\mathbf{x}_k$ is an $L \times 1$ vector, and is defined as $\mathbf{x}_k = [\mathbf{x}_{k,1}, \cdots, \mathbf{x}_{k,B}]^T$.

Now, by performing SVD on $\bar{\mathbf{U}}_k$, we can easily obtain the estimation of the original input symbol sequence $\hat{\mathbf{x}}$ for the k-th user, which is just the right singular vector associated with the smallest singular value. Obviously, by constructing $\bar{\mathbf{U}}_k$ based on (22) for different user, we can get the input symbol sequences for multiple users.

V. SIMULATION RESULTS

In this Section, computer simulation results are presented to evaluate the performance of the proposed algorithm. DBPSK modulation mode is used in our simulations. Hadamard codes with length $G = 32$ are assigned to different users. We assume a rich scattering environment near the mobile station and generate the FIR channel coefficients (reference (3) in [2]) as independent identically distributed (i.i.d) complex Gaussian random variables with zero-mean and variance $1/(L_{ch}+1)$. The signal to noise power ratio (SNR) per receive antenna is defined as SNR=$10 \log_{10} 1/\sigma_n^2$ in dB. We use samples within 80 MC-CDMA symbols to estimate the auto- and the cross-correlation matrices of the uplink-received data sequence, which are the approximate estimation of auto- and cross-correlation matrices in ensemble-average sense.

The length of FIR channels is fixed to 7 and the number of active users is fixed to 5. The performance of the proposed decoder is evaluated by BER. The performance of the MMSE receiver with estimated channel information [1,2] is also presented for comparison. Fig.2 shows the BER versus the SNR with antenna array of 2 and 4 elements, respectively. It can be seen that our algorithm outperforms the conventional one. We can also see that the BER decreases as the number of receive antenna array element increases. Reason behind this is that the antenna array gain of ULA enhances the SNR of received signals when the number of antenna array element increases.

Fig.3 shows the computation load versus the number of users, when antenna array element is fixed to 2, 4 and 5, respectively. From this figure, it is clear that the computation load linearly increases when a new user is added. Furthermore, the computation load does not increase too much with the increase of M. This is desirable since the system performance can be greatly improved by increasing M at the cost of slight increases of the total computation load. Finally, the computation complexity of the proposed algorithm is $O(L^3)$ flops and that of [1,2] is $O((M-1)^3 G^3)$ flops. Since $L < (M-1)G$ in general, the proposed algorithm has obvious superiority of computation.

REFERENCES

[1] W. Sun, H. Li and M. Amin, "MMSE detection for space-time coded MC-CDMA," Proc. IEEE ICC 2003, Anchorage, USA, May 2003, vol. 5, pp.3452-3456.

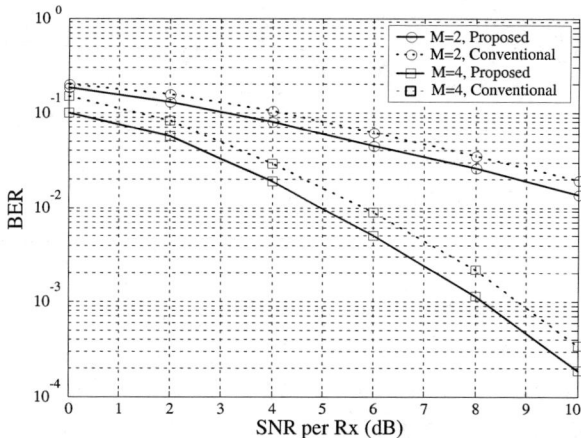

Fig. 2. BER versus SNR

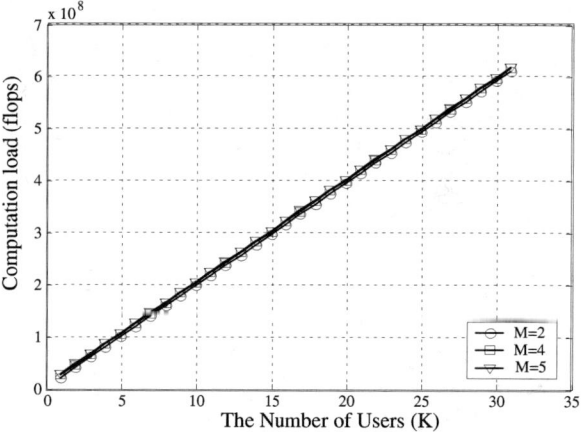

Fig. 3. Computation complexity

[2] X. Wu, Q. Yin and Y. Zeng, "Downlink channels identification for space-time coded multiple-input multiple-output MC-CDMA systems," Proc. IEEE ICASSP 2003, Hong Kong, April 2003, pp.417-420.

[3] A. L. Swindlehurst and G. Leus, "Blind and semi-blind equalization for generalized space-time block codes," IEEE Trans. Signal Processing, vol. 50, Oct. 2002, pp. 2489-2498.

[4] L. Dai, S. Sfar and K. Letaief, "Receive antenna selection for MIMO systems in correlated channels," Proc. IEEE ICC 2004, Paris, France, July 2004, vol. 5, pp.2944-2948.

[5] Q. Yin, R. Newcomb and L. Zou, "Estimating 2-D angles of arrival via two parallel linear array," Proc. IEEE ICASSP 1989, Glasgow, Scotland, May 1989, pp. 2803-2806.

[6] V. Tarokh, H. Jafarkhani and A. R. Calderbank, "Space-time block codes from orthogonal designs," IEEE Trans. Inform. Theory, vol. 45, July 1999, pp. 1456-1467.

[7] Y. Zeng and Q. Yin, "Direct Decoder of Uplink Space-Time Block Coded MC-CDMA Systems", IEICE Trans. Com., vol.E88-B, no.2, Feb. 2005.

An ICA based approach for Blind Deconvolution of Three-dimensional Signals

E. Principi, S. Squartini and F. Piazza

Dipartimento di Elettronica,
Intelligenza Artificiale e Telecomunicazioni
Università Politecnica delle Marche
60131 Ancona, Italy
Email: e.principi@deit.univpm.it, sts@deit.univpm.it, upf@deit.univpm.it

Abstract—This work presents a new algorithm based on Independent Component Analysis (Infomax approach) for blind deconvolution of three-dimensional signals. In this sense it represents an extension of what done so far for one and two dimensional signals. Such a method has been also successfully implemented in the frequency domain to alleviate the computational burden, due to the efficient calculation of convolutive terms. Experimental tests carried out on synthetic source signals, with sub-gaussian and super-gaussian distributions, confirm the validity of the idea.

I. INTRODUCTION

Several application in digital signal processing area share the purpose of improving the quality of available signals. The answer to such a need is represented by algorithms proposed to enhance and/or restore audio [1], video [2] and image signals [3]. The present work can find a place in this field as its scope is developing an algorithm for blind deconvolution of 3D signals, which could be video, 3D images or even audio arrays. Up to author's knowledge, the existing approaches in the literature deal with image sequences and are based on the 3D extension either of the NAS-RIF algorithm [4] or of the homomorphic deconvolution [5].
We shall propose an algorithm based on *Independent Component Analysis* (ICA): precisely, it will be an exstension of the Infomax approach [6] to the 3D blind deconvolution problem. Looking at the system depicted in Fig. 1, we define the following quantities: $s(i,j,k)$ is the original 3D signal, $h(i,j,k)$ is the impulse response of blurring filter, $x(i,j,k)$ is the blurred signal, whose dimensions are $N_{ix} \times N_{jx} \times N_{kx}$, $w_t(i,j,k)$ is the impulse response of deconvolution filter at time instant t, whose dimensions are $N_{iw} \times N_{jw} \times N_{kw}$, $u_t(i,j,k)$ is the deconvolution output signal at time instant t, $g(\cdot)$ is a non-linear function and $y(i,j,k)$ is the signal available at the output of the nonlinear block. The relationships between the involved variables are described by the following:

$$\begin{aligned} u_t(i,j,k) &= w_t(i,j,k) * x(i,j,k) \\ &= w_t(i,j,k) * h(i,j,k) * s(i,j,k) \\ &= g_t(i,j,k) * s(i,j,k) \end{aligned} \quad (1)$$

where $g_t(i,j,k)$ is the global system at learning time step t and $*$ is the 3D convolution operator, which operates as follows:

$$\begin{aligned} u_t(i,j,k) &= w_t(i,j,k) * x(i,j,k) \\ &= \sum_{m=0}^{N_{iw}-1} \sum_{n=0}^{N_{jw}-1} \sum_{l=0}^{N_{kw}-1} w_t(m,n,l) x(i-m,j-n,k-l) \end{aligned}$$

The goal is to find a proper deconvolution filter such that the global system is equal to a 3D Kronecker delta, up to allowed delay indeterminancy. This ensures that the original signal is restored. As in the original INFOMAX algorithm, the basic assumptions used to make the problem tractable are: the signal $s(i,j,k)$ must be i.i.d. and non-gaussian.

II. THE 3D BLIND DECONVOLUTION ALGORITHM

A. Objective function and learning rule

As stated in [6], [7] the blind deconvolution problem can be treated as a blind signal separation problem, as follows:

$$\mathbf{u} = \mathbf{W}\mathbf{x} \quad (2)$$

where $\mathbf{u}$ and $\mathbf{x}$ are vectors of dimensions equal to $M = N_{ix} \cdot N_{jx} \cdot N_{kx}$. u_p is the $p-th$ element of $\mathbf{u}$, corresponding to the $u_t(i,j,k)$ element where $p = (i+1)+j \cdot N_{ix}+k \cdot N_{ix} \cdot N_{jx}$. The matrix $\mathbf{W}$ is an $M \times M$ upper triangular one, whose diagonal entries are all equal to $w_t(0,0,0)$. Being J the Jacobean of the transformation $\mathbf{u} = \mathbf{f}(\mathbf{W}\mathbf{x})$, we can derive the following objective function to maximize, by just adopting the same 1D Infomax procedure [6]:

$$\begin{aligned} \phi\big(w_t(i,j,k)\big) &= \ln|\det \mathbf{J}(\mathbf{x})| \\ &= M \ln|w_t(0,0,0)| + \sum_{p=1}^{M} \ln|f'\big(u_p(t)\big)| \end{aligned} \quad (3)$$

Here we employed the natural gradient approach [7], [8] for deriving our algorithm learning rule in order to get an improved convergence speed rather than that one attainable through the classic stochastic gradient. According to this and to the gradient descent technique, we have that the increment of $w_t(m,n,l)$ is proportional to the natural gradient of the objective function (3). To get this we differentiate (3) respect with the new differential variables in the Riemannian space

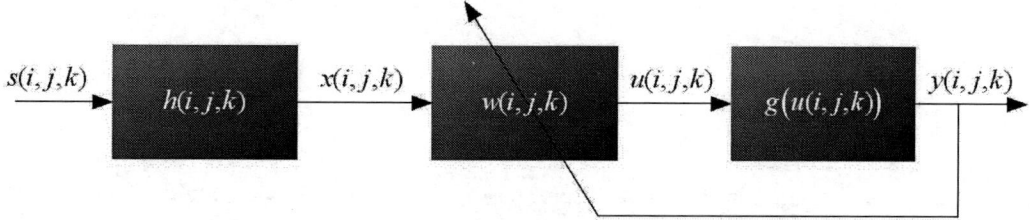

Fig. 1. Blurring and deconvolution scheme.

where we are searching our optimal solution. These variables are given by (in the Z-domain):

$$dY_t(z_1, z_2, z_3) = \sum_{m=0}^{N_{iw}-1} \sum_{n=0}^{N_{jw}-1} \sum_{l=0}^{N_{kw}-1} dy_t(m,n,l) z_1^{-m} z_2^{-n} z_3^{-l} \quad (4)$$

and they satisfy the following relation:

$$dY_t(z_1, z_2, z_3) = dW_t(z_1, z_2, z_3) W_t^{-1}(z_1, z_2, z_3) \quad (5)$$

the Z-transform of $w_t(m,n,l)$ being:

$$dW_t(z_1, z_2, z_3) = \sum_{m=0}^{N_{iw}-1} \sum_{n=0}^{N_{jw}-1} \sum_{l=0}^{N_{kw}-1} dw_t(m,n,l) z_1^{-m} z_2^{-n} z_3^{-l} \quad (6)$$

Looking at the two terms appearing in (3), we can obtain the following through simple algebraic calculations:

$$\frac{d \ln |w_t(0,0,0)|}{dy_t(m,n,l)} = \delta_0(m,n,l)$$

$$\frac{d\left(\sum_{p=1}^{M} \ln |f'(u_p(t))|\right)}{dy_t(m,n,l)} = \quad (7)$$

$$= \sum_{i,j,k} \frac{f''(u_t(i,j,k))}{f'(u_t(i,j,k))} u_t(i-m, j-n, k-l)$$

being $\delta_0(m,n,l)$ the Kronecker delta. Equation (5) allows us to right multiply both terms by $w_t(z_1, z_2, z_3)$. In this way we can derive the final formula for the natural gradient based learning rule to update the deconvolution filter coefficients:

$$w_{t+1}(m,n,l) = w_t(m,n,l) + \mu(t) \Big[M w_t(m,n,l) - \\ - \sum_{i,j,k} \varphi(u_t(i,j,k)) v_t(i-m, j-n, k-l) \Big] \quad (8)$$

where $\mu(t)$ is the time varying stepsize, $\varphi(u_t(i,j,k)) = -f''(u_t(i,j,k))/f'(u_t(i,j,k))$ and

$$v_t(i,j,k) = \sum_{q_1, q_2, q_3} u_t(i+q_1, j+q_2, k+q_3) w_t(q_1, q_2, q_3) \quad (9)$$

and taking into account that:

$$q_1 = 0 \cdots N_{iw}-1, q_2 = 0 \cdots N_{jw}-1, q_3 = 0 \cdots N_{kw}-1$$

B. Algorithm implementation in the frequency domain

The learning rule described by (8) is computationally heavy, that is why it seemed convenient to develop a suitable algorithm implementation in the frequency domain, according to the approach followed in [9]. In particular, the *overlap and save* technique in 3D case has been used for the FFT based calculation of convolutions. On purpose, the original signal has been divided into blocks as shown in Fig. 2: their dimensions are the same of the deconvolution filter. The number sequence in Fig. 2 informs us about the block elaboration direction. Then, the updating equation becomes:

$$w_{t+1}(m,n,l) = w_t(m,n,l) + \mu_M(t) \Big[w_t(m,n,l) - \\ - \frac{1}{M} \sum_{i,j,k} \varphi(u_t(i,j,k)) v_t(i-m, j-n, k-l) \Big] \quad (10)$$

where $\mu_M(t) = \mu(t) M$ and the index variability is given by:

$$i = i_{min} \cdots i_{max}, j = j_{min} \cdots j_{max}, k = k_{min} \cdots k_{max}$$

Such limits define the block dimensions and they are related to the learning time step t through:

$$\begin{aligned}
i_{min} &= ((t N_{iw}))_{N_{ix}}; \\
i_{max} &= (((t+1)N_{iw} - 1))_{N_{ix}}; \\
j_{min} &= (([t/N_{br}] N_{jw}))_{N_{jx}}; \\
j_{max} &= ((([t/N_{br}] + 1)N_{jw} - 1))_{N_{jx}}; \\
k_{min} &= [t/N_{bs}] N_{kw}; \\
k_{max} &= ([t/N_{bs}] + 1) N_{kw}; \\
N_{br} &= \frac{N_{jx}}{N_{iw}}; \quad N_{bs} = \frac{N_{ix} N_{jx}}{N_{iw} N_{jw}}
\end{aligned} \quad (11)$$

where $((a))_N$ stands for the modulus after division operation, while $[a]$ for the rounding to the lowest integer operation. It must be observed that the block number coincides with the time instant t.

Equation (10) is clearly non-causal and it can be transformed into a causal one by introducing delays in non-causal terms. The non-causal version of the algorithm is here dealt with, suitable for off-line applications as those considered later on.

The convolution terms in the learning rule equation are the following:

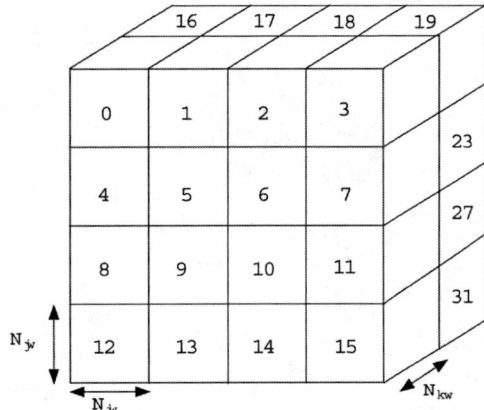

Fig. 2. Block decomposition of a 3D cubic signal. The sequence number stands for the order which has to be followed for block elaboration

- $u_t(i,j,k) = w_t(i,j,k) * x(i,j,k)$
- $c(r_1, r_2, r_3) = \dfrac{1}{M} \sum_{i,j,k} \varphi\big(u_t(i,j,k)\big) \cdot$
  $\cdot u_t(i-r_1, j-r_2, k-r_3)$ (12)
- $\nabla w_t(m,n,l) = \sum_{q_1,q_2,q_3} \gamma(m-q_1, n-q_2, l-q_3) \cdot$
  $\cdot w_t(q_1, q_2, q_3)$

where $\gamma(m,n,l) = \delta(m,n,l) - c(m,n,l)$. The dimensions of the 3D FFT operations, to be used for convolution calculation, depend on the number of blocks involved in (8). Such dimensions are $4N_{iw} \times 4N_{jw} \times 4N_{kw}$. This can be easily understood if we look at (9) and (8), where the independent variables of $u_t(i',j',k')$ are bounded by the following limits:

$$i_{min} - 2(N_{iw}-1) \leq i' \leq i_{max} + N_{iw} - 1;$$
$$j_{min} - 2(N_{jw}-1) \leq j' \leq j_{max} + N_{jw} - 1; \quad (13)$$
$$k_{min} - 2(N_{kw}-1) \leq k' \leq k_{max} + N_{kw} - 1;$$

This means that there 4 blocks involved for each independent variable (the current block t, the two preceding and the one following), resulting in the total amount of $4^3 = 64$ blocks, namely a superblock.

Let us introduce some further insights on the used notation:
- $\mathbf{A} = \text{FFT}(\mathbf{a}, [M, N, P])$ is the FFT of the original signal $\mathbf{a}$ calculated over $M \times N \times P$ points;
- t stands for the learning time step, while B_t is the variable to note the t-th block of a signal;
- $\otimes$ is the matrix element-wise product.

The steps of the algorithm version in the frequency domain are:

1) FFT of the deconvolution filter impulse response: $\mathbf{W}_t = \text{FFT}(\mathbf{w}_t, [4N_{iw}, 4N_{jw}, 4N_{kw}])$
2) FFT of the blurred filter impulse response: $\mathbf{X}_{B_t} = \text{FFT}(\mathbf{x}_{B_t}, [4N_{iw}, 4N_{jw}, 4N_{kw}])$
3) calculation of the convolution between $x(i,j,k)$ and the filter $w_t(i,j,k)$: $\mathbf{U}_t = \mathbf{X}_{B_t} \otimes \mathbf{W}_t$

$$u_t(i,j,k) = \begin{cases} 0 & \text{if } 0 \leq \alpha \leq N_{\alpha w} - 2 \\ \text{IFFT}(\mathbf{U}_t) & \text{otherwise} \end{cases}$$

where $\alpha = i,j,k$, $\mathbf{U}_t = \text{FFT}(\mathbf{u}_t, [4N_{iw}, 4N_{jw}, 4N_{kw}])$.

4) calculation of the non-linearities for block t: $\mathbf{\Phi}_{B_t} = \text{FFT}(\varphi_{PAD}(\mathbf{u}_t), [4N_{iw}, 4N_{jw}, 4N_{kw}])$ where $\varphi_{PAD}(\mathbf{u}_t)$ has dimensions $4N_{iw} \times 4N_{jw} \times 4N_{kw}$ and

$$\varphi_{PAD}(\mathbf{u}_t) = \begin{cases} \varphi(\mathbf{u}_t) & \text{if } \alpha_{min} \leq \alpha \leq \alpha_{max}, \\ 0 & \text{otherwise} \end{cases}$$

where $\alpha = i,j,k$.

5) calculation of the correlation term: $\mathbf{C} = \mathbf{\Phi}_{B_t} \otimes \mathbf{U}_t^* / M$
6) weight update ($\mathbf{1}$ is the matrix of ones):

$$\nabla \mathbf{W}_t = (\mathbf{1} - \mathbf{C}) \otimes \mathbf{W}_t$$
$$\nabla \mathbf{w}_t = \text{the first block } N_{iw} \times N_{jw} \times N_{kw}$$
$$\text{of IFFT}(\nabla \mathbf{W}_t)$$
$$\mathbf{w}_{t+1} = \mathbf{w}_t + \mu_M \nabla \mathbf{w}_t$$

C. Computational cost

The computational cost of the algorithm can be evaluated by using the number of real multiplications performed in one single learning step. The cost of the frequency based algorithm is compared to that one of the time based algorithm (block sequence version). Such a version is analogous to the frequency based counterpart described in the previous subsection, with the only difference that the convolution terms are calculated directly by (12).

It can be shown that the ratio R between the number of real multiplications in the time based algorithm and those in the frequency based one is:

$$R = \frac{35a - 22b + 14c - 9}{768 + 192 \log_2(64a)}$$

where $a = N_{iw} N_{jw} N_{kw}$, $b = N_{iw} N_{jw} + N_{iw} N_{kw} + N_{jw} N_{kw}$ and $c = N_{iw} + N_{jw} + N_{kw}$. Assuming that $N_{iw} = N_{jw} = N_{kw}$, we have that when $N_{iw} = 6$ the ratio becomes $R \cong 1.6$, confirming the improved speed achieved through the implementation in the frequency domain.

III. EXPERIMENTAL RESULTS

In this section we show the experimental results obtained by applying our algorithm to synthetic signals. We are referring to generalized gaussian distribution, modeled by:

$$p(s) \propto \exp\left(-\frac{|s|^S}{S \cdot \Xi(|s|^S)}\right)$$

being Ξ the expectation operator, while the non-linear function at the end of the deconvolution system in Fig. 1 is:

$$f(u) = \frac{|u|^{S-2} u}{\Xi(|u|^S)}$$

The sources involved for experiments have sub-gaussian (parameter $S = 6$) and super-gaussian ($S = 1$) distribution respectively. The blur is a $5 \times 5 \times 5$ gaussian filter whose standard deviation is $\sigma = 0.75$, and it is given by:

$$h(i,j,k) = \frac{1}{\sigma^3 \sqrt{8\pi^3}} \exp\left(-\frac{i^2 + j^2 + k^2}{2\sigma^2}\right) \quad (14)$$

The source dimensions are $508 \times 508 \times 508$, the stepsize μ is $5 \cdot 10^{-3}$ (not time-varying), in both tests performed. The deconvolution filter ones are $32 \times 32 \times 32$ and its starting values are:

$$w_0(i,j,k) = \begin{cases} 1 & \text{if } i = N_{iw}/2, j = N_{jw}/2, k = N_{kw}/2, \\ 0 & \text{otherwise} \end{cases}$$

The results are depicted in Fig. 3 and show the inter-symbolic interference (ISI) curves respect with the iteration steps. The ISI formula is:

$$\text{ISI} = \frac{\sum_{i,j,k} |g(i,j,k)|^2 - \max_{i,j,k}|g(i,j,k)|^2}{\max_{i,j,k}|g(i,j,k)|^2}$$

where $g(i,j,k) = h(i,j,k) * w(i,j,k)$ is the global system impulse response at the end of training.

Looking at curves in Fig. 3 and Fig. 4 our expectations find a confirmation. Indeed it can be observed that the algorithm converges after 700 iterations in the first case and after 500 in the second one, yielding low ISI values, -25dB and -29dB respectively, corresponding to good deconvolution performances.

IV. CONCLUSIONS

A new ICA based algorithm for blind deconvolution of 3D signals has been developed. Two are the versions implemented, one in the time and the other in the frequency domain. The former let us achieve a consistent reduction of the computational cost, by using 3D FFT operation for calculation of convolutive terms. The overall approach has been applied to recover synthetic signals (with sub-gaussian and super-gaussian distribution) blurred by 3D synthetic filters. Experimental results confirm the effectiveness of the proposed idea.

Further works concern the algorithm applicability to real world signals (video, audio source array, 3D images). This is far to be straightforward, since such signals typically do not satisfy the assumptions made to let our algorithm work. Some pre-processing based solutions at blurred signal level (as prewhitening) are actually under study.

Another challenging idea is to involve long video sequences and then using non-cubic blocks extended along the time axis. In this way we can think to recover the iid assumption, generally not valid for real 3D signals, as it happens in the monodimensional case when one wants to deal with speech signals. Of course, such an approach strictly requires to reduce the algorithm complexity and the relative computational burden somehow.

REFERENCES

[1] S. J. Godsill, P. J. W. Rayner, S. H. Godsill. *Digital Audio Restoration: A Statistical Model Based Approach*. Springer Verlag, 1998.
[2] A. Kokaram. *Motion Picture Restoration*. Springer-Verlag, 1998.
[3] H. C. Andrews, B. R. Hunt. *Digital Image Restoration*, New Jersey: Prentice-Hall, Inc., 1977.
[4] M. Mignotte, J. Meunier. Three-Dimensional Blind Deconvolution of SPECT Images. In *IEEE Transactions on Biomedical Engineering*, vol. 47, no. 2, January 2000.
[5] T. Taxt. Three-Dimensional Blind Deconvolution of Ultrasound Images. In *IEEE Transactions on Ultrasonics, Ferroelectrics, and Frequency Control*, vol. 48, no. 4, July 2001.
[6] A. J. Bell, T. J. Sejnowski, An information-theoretic approach to blind separation and blind deconvolution, *Neural Computation*, 7:1129-1159, 1995.
[7] A.Cichocki, and S.Amari, *Adaptive Blind Signal and Image Processing*. Wiley&Sons, West Sussex, England, 2002.
[8] S. Amari, Natural Gradient Works Efficiently in Learning, *Neural Computation*, vol. 10, pag 251-276, 1998.
[9] M. Joho, and P. Schniter, Frequency-domain realization of a multichannel blind deconvolution algorithm based on the natural gradient, *ICa 2003*, vol. 1, pp. 543-548, Nara, Japan, 2003.

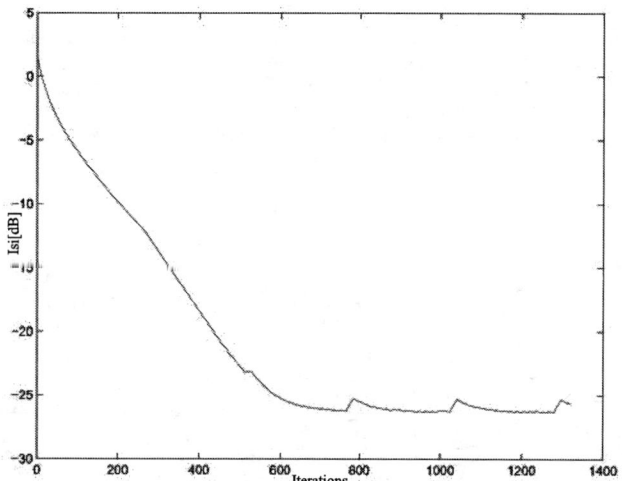

Fig. 3. ISI curve in sub-gaussian source case study.

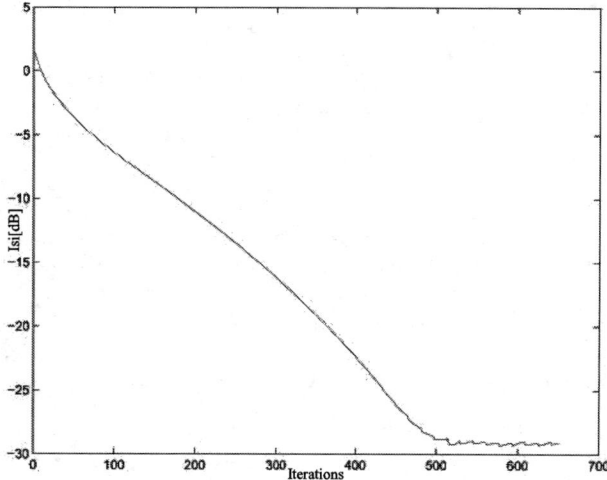

Fig. 4. ISI curve in super-gaussian source case study. sources.

> # Evaluating a Blind Channel Estimation Technique that uses a Hardware Efficient Equalizer

Yun Ye
School of Electrical and Electronic Engineering
Nanyang Technological University, Nanyang Avenue
SINGAPORE, 639798
yeyun@pmail.ntu.edu.sg

Saman S. Abeysekera
School of Electrical and Electronic Engineering
Nanyang Technological University, Nanyang Avenue
SINGAPORE, 639798
esabeysekera@ntu.edu.sg

Abstract—The blind parameter estimation of a non-minimum phase (NMP) channel having a finite impulse response (FIR) is studied in this paper. An efficient and reliable blind estimation algorithm is proposed which is based on the combination of second order statistics (SOS) and the kurtosis of the signals. SOS based methods provide efficient estimation of channel zeros from a very small number of samples. As the SOS based methods are phase blind, the kurtosis is used to resolve the ambiguity in system zero locations. It is also shown that the equalizer output could be exploited recursively to improve the estimation accuracy using finite alphabet (FA) properties. It is noted that as all the available information for blind channel parameter estimation are used, the proposed method can achieve a very high accuracy. Performance of the estimation method is also discussed.

I. INTRODUCTION

Blind equalization and estimation of communication channels have been challenging problems attracting intensive research over the past two decades. A comprehensive review of blind channel estimation can be found in [1]. Parameter estimation algorithms using second-order statistics (SOS) usually require smaller data lengths and achieve smaller estimation variance than the other methods [2]. However, SOS based method presumes that the zeros of the channel are inside the unit circle. That is, the SOS based methods are phase blind. In avoiding this, it is shown in [3] that the channel phase information can be extracted from the second-order cyclostationary statistics (SOCS). But oversampling and multichannel configuration (diversity) are required for SOCS based methods, and when a single sensor is used these conditions cannot always be satisfied. Thus, SOCS based methods are not suitable for baud rate spaced systems [4]. On the other hand, higher order statistics (HOS) can be used even in NMP channel estimation if input signal is not Gaussian, as shown in [4] and references therein. But HOS methods have high complexity and difficulty in implementation. Lii and Rosenblatt [5] noted that combing SOS with HOS can provide the solution to blind channel estimation problem. Reference [6] extends this idea by using the full fourth order cumulant to locate zero positions. However, a single point of the fourth order cumulant is not robust enough to identify the correct zeros. Use of other signal information can also correctly locate the zeros that are initially estimated by a SOS based method.

The well-known Shalvi-Weinstein criterion [7] maximizes the kurtosis subjected to the condition that the power of the equalizer input and output are equal. In this paper, we use a simplified version of this criterion such that the normalized kurtosis at the transmitter and the receiver being made to equal. We propose a low complexity method that uses a simple zero-forcing equalizer to find the correct channel parameters and recover the transmitted signal. The zero locations are efficiently located using the normalized kurtosis (one point of the fourth order culumant at zero lags) of the equalizer output and thus avoids the use of complicated full (or slices of) HOS cumulants. Furthermore, using FA characteristics, the equalizer output signal is recursively used to improve the initial estimation. The FA characteristic of the transmitted signal has shown to be extremely useful in [8] and reference therein. However, these reported algorithms are complicated and realizing low complexity techniques for blind channel estimation and equalization is still a challenging research [9]. The recursive use of the output has similarity to the iterative least squares with projection (ILSP) in [8] which exploits FA property to provide channel parameter and recover original signal. However, the proposed algorithm here is different to these in that it does not require oversampling (as in the use of an antenna array) and does not have initial value problems as of ILSE and ILSP. Only a single iteration of a least square computation is utilized in this algorithm. The comparisons show that the proposed method achieves better performance than reported techniques due to the recursive use of the equalizer output.

The main contribution of the work is the proposal and evaluation of a simple, low complexity blind channel estimation technique. The method also incorporates a simple equalizer thus a separate equalizer implementation is unnecessary.

II. CHANNEL MODEL

A slowly time-varying channel can be considered as a time-invariant system if it is varying slower than the baud rate [1]. For a single-user system, the discrete-time baseband model is a time-invariant single-input single-output (SISO) system described as,

$$y(n) = \sum_{k=0}^{L} x(n-k)h(k) + w(n) \qquad (1)$$

where $\{y(n)\}$ is the equalizer input signal; $\{x(n)\}$ is the unknown input, assumed to be a zero-mean, non-Gaussian, independent and identically distributed (i.i.d.) signal. The constellation of $\{x(n)\}$ is supposed to be known $\{w(n)\}$ is an additive white Gaussian noise having variance σ^2, and L is the channel order that is assumed to be known. (Reference [10] gives a comprehensive summary of channel order estimation.) Suppose $\{h(k)\}$ denote the channel impulse response, then the Z-transform of the channel transfer

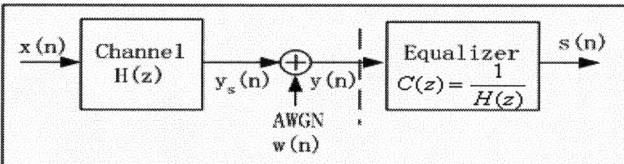

Figure 1. Zero-Forcing Equalizer

function is given by,

$$H(z) = 1 + h_1 z^{-1} + \cdots + h_L z^{-L} = \prod_{i=1}^{L}(1 - b_i z^{-1}) \quad (2)$$

where $\{b_i\}$ are the zeros of the channel. It is assumed here that no zeros exist on the unit circle. Since $\{x(n)\}$ is non-Gaussian, the channel is identifiable.

III. Proposed Estimation Algorithm

A. Initial Zero Estimation Via Second Order Statistics

It is noted that a minimum phase zero inside the unit circle ($|b_i|<1$) and its reciprocal on the unit circle non-minimum phase zero ($|b_i|>1$) yield the same autocorrelation sequence. All the minimum phase zeros can be directly obtained from the autocorrelation function (ACF) as,

$$(\hat{b}_1, \hat{b}_1, \cdots \hat{b}_L) = \min\_ph\_roots\left(\sum_i \hat{r}_i z^{-i}\right) \quad (3)$$

where $\hat{r}_i$ is the ith lag of the autocorrelation sequence given by,

$$\hat{r}_i = \frac{1}{N-i} \sum_{n=1}^{N-i} y(n) y(n+i) \quad (4)$$

and N is the number of samples. Alternatively, the minimum phase zeros can be obtained as,

$$(\hat{b}_1, \hat{b}_1, \cdots \hat{b}_L) = roots\left(\sum_i \hat{h}_i z^{-i}\right) \quad (5)$$

where $\{\hat{h}_k\}$ is estimated by any other sophisticated SOS based method. Among the many such algorithms, the best reported in the literature is the overparametrized signal (OS) approach [11]. Other simpler SOS methods, such as Durbin in [2], can also be used in zero estimation. For each zero estimated via the SOS based method, we note that there would be an alternative maximum-phase zero [5], and thus using $(\hat{b}_1, \hat{b}_2, \cdots, \hat{b}_L)$, a NMP system can be obtained by reciprocating any number of zeros on the unit circle. For an Lth order system, there are 2^L such configurations. For example, the zeros can be $\varphi = (\hat{b}_1, 1/\hat{b}_2, \cdots, \hat{b}_L)$ or $\varphi = (1/\hat{b}_1, \hat{b}_2, \cdots, \hat{b}_L)$. Here we select the correct φ by observing the output of an appropriate equalizer.

B. Zero-Based Zero-Forcing Equalizer

Consider a simple zero-forcing equalizer based on the zeros estimated in previous section A. For the zero-forcing equalizer in Fig.1, $H(z)C(z)=1$, and the equalizer can be implemented as a combination of parallel FIR filters given by,

$$C(z) = \frac{1}{H(z)} = \sum_{k=0}^{L} \frac{a_k}{1 - b_k z^{-1}} \quad (6)$$

where $\{b_k\}, \{a_k\}$ are the poles (zeros of the channel) and residues respectively. The minimum phase branches with ($|b_i|<1$) can be easily implemented approximately by an M length FIR filter. The maximum phase branches with ($|b_i|>1$) can be realized in FIR form with a delay D. The parallel filter structure has efficient hardware implementation and low computation load because only $2 \times L$ branches are needed to realize 2^L different configurations required to realize the equalizer as noted in III.A. (When the zero b_i is close to the unit circle, the recursive Laguerre equalizer proposed in [12] would be useful for better efficiency.)

C. Zero Location using The Kurtosis

As shown in Fig.1, the equalizer output signal $s(n)$ can be expressed as,

$$s(n) = x(n) * h(n) * c(n) + w(n) * c(n) \quad (7)$$

(7) can be rewritten as $s(n) = x(n) * g(n) + c(n) * w(n)$ with the vector $g(n) = h(n) * c(n) = (g_1, g_2, \cdots)$ denoted as **g**. The sufficient condition to achieve equalization is that **g** be a unit vector having only one nonzero component equal to the phase shift. This condition means that the inverse filter output has the same probability distribution information as that of the input signal [7]. As a further simplification of this condition, here we propose to use the kurtosis of the distributions for comparison. (Kurtosis is the fourth order cumulant of the signal at zero lags.) Note for brevity, we use x_n for $x(n)$, s_n for $s(n)$, w_n for $w(n)$ in the following deduction. The kurtosis of the equalizer output s_n is

$$K(s_n) = Cum_{4s} = E\{|s_n|^4\} - 2E^2\{|s_n|^2\} - \left|E\{s_n^2\}\right|^2 \quad (8)$$

A similar form exists for the input signal x_n. Using the i.i.d assumption on the input signal x_n, we get

$$K(s_n) = K(x_n) \sum_l |g_l|^4 + K(w_n) \sum_l |c_l|^4 \quad (9)$$

$$E(|s_n|^2) = E(|x_n|^2) \sum_l |g_l|^2 + E(|w_n|^2) \sum_l |c_l|^2 \quad (10)$$

Note that $K(w_n) = 0$ as the kurtosis of a Gaussian signal is zero. The relationship between the normalized kurtosis of s_n and x_n can be expressed in (11) after simplification,

$$\frac{K(x_n)}{E^2\{|x_n|^2\}} - \frac{K(s_n)}{E^2\{|s_n|^2\}} - \frac{K(s_n)}{E^2\{|s_n|^2\}} \frac{\phi^2 + 2\phi\theta}{(\phi+\theta)^2}$$
$$= \frac{K(x_n)}{E^2\{|x_n|^2\}} \left|1 - \frac{\sum_l |g_l|^4}{(\sum_l |g_l|^2)^2}\right| \left|\frac{\theta}{(\phi+\theta)^2}\right| \quad (11)$$

where $\phi = E\{|w_n|^2\} \sum_l |c_l|^2$ a constant, depends on the noise power and $\theta = E\{|x_n|^2\} \sum_l |g_l|^2$ is a constant because the equalizer impulse response $\{c_l\}$ are chosen to have the same autocorrelation function for different configurations. Therefore, the minimization of left side of (11) is the minimization of

$\left|1-\sum_l |g_l|^4 / (\sum_l |g_l|^2)^2\right|$. Note that for the vector **g**, (11) is minimized if and only if **g** has at most one nonzero component, $\left|\sum_l |g_l|^4 = (\sum_l |g_l|^2)^2\right|$, i.e., **g** is a proper equalizer transfer function. As the constellation of transmitted (e.g. communication) signal is known, based on (11), a cost function can be used as the criterion for equalization given by

$$J(\varphi) = \left| \frac{K(x_n)}{E^2\{|x_n|^2\}} - \frac{K(s_n|\varphi)}{E^2\{|s_n|\varphi|^2\}} - \xi \right| \quad (12)$$

This makes the equalization easier and simpler when compared with, for example (16) in [12]. In equation (12) the offset ξ can be evaluated once the signal to noise ratio is known. We note that $J(\varphi)$ has following properties: nonnegative; $J(\varphi) \geq 0$, and $J(\varphi)$ is minimized if and only if the input normalized kurtosis is closest to normalized output kurtosis. That is when the constellation of the equalizer output is similar to the constellation of the input signal, i.e. correct equalization. Thus, using the kurtosis, the minimum-phase channel zeros, which were initially estimated by an SOS method can be correctly located.

D. Recursive Estimation via Least Square (LS)

The equalizer output $\{s(n)\}$ can be used recursively to provide a better estimation. We project every data $\{s(n)\}$ to the closest symbol in FA set, to obtain $\{v(n)\}$. Combination of $\{v(n)\}$ and the observed noisy signal $\{y(n)\}$ would then provide the coefficients $\{h(k)\}$ of the unknown system. To do this, replace $\{x(n)\}$ by the equalizer output $\{v(n)\}$ in (1), to obtain,

$$y = v * \hat{h} + \varepsilon \quad (13)$$

In order to estimate the parameters $\{h(k)\}$, the estimation error can be minimized in a least square sense to obtain,

$$\hat{H} = (V^T V)^{-1} V^T Y \quad (14)$$

where $\hat{H} = [\hat{h}_0, \hat{h}_1, \cdots \hat{h}_L]^T$, $Y = [y(n_1), y(n_1+1), \cdots, y(n_2)]^T$, $V = [v_{n_1}, v_{n_1+1}, \cdots v_{n_2}]^T$, and $v_i = [v(i+L), v(i+L-1), \cdots v(i)]$. With known $\{v(n)\}$ and $\{y(n)\}$, the coefficients $\{\hat{h}_k\}$ can be obtained, and usually, $n_2 - n_1 = 10*L$, is sufficient to provide an high accurate estimation. By not using all the samples, the computation complexity of the LS method can be reduced.

IV. Estimator Performance

A. Cramer–Rao Lower Bound (CRLB)

The CRLB cannot be determined for blind SISO NMP channel estimation as the channel phase is unknown. However, we note that the CRLB for blind MP channel estimation can be readily evaluated. Thus, we can use this CRLB for a comparison of estimator performance. When the MP zeros are real, given that C_{zz} is the $L \times L$ autocorrelation matrix of an AR process $\{z(n)\}$ with parameters equal to the channel parameters $\{h(n)\}$, the CRLB of parameter estimation is given by [2] as,

$$\text{var}(\hat{h}_n) \geq \frac{\sigma^2}{N} \left[C_{zz}^{-1}\right]_{ii} \quad i=1,2,\ldots L \quad (15)$$

B. Equalizer Output Noise Variance

Using another approach, we use the equalizer output noise variance as a criterion to evaluate the estimation performance as,

$$\text{var}(s_n - x_n) \quad (16)$$

This method of comparison is justified as the variance in equation (16) finally determines the error rate of a communication system.

V. Simulation and Comparison Results

To illustrate the performance of the proposed method, we first define the SNR (Signal to Noise Ratio) as $SNR = 10\log 10(P_{y_s}/P_w)$ (dB) and evaluate the estimation accuracy via the normalized mean square error (MSE) given by,

$$MSE = \sum_{i=0}^{L}(h(i) - \hat{h}(i))^2 / \sum_{i=0}^{L} h(i)^2 \quad (17)$$

A QPSK sequence with zero mean is used as the input signal $x(n)$.

The first example (Channel I) is a second order MP system with zeros at 0.5 and 0.3. The system transfer function is $H(z) = 1 - 0.8z^{-1} + 0.15z^{-2}$. In Fig. 2, channel I is used to compare the performance of the proposed method with the CRLB for blind MP system estimation. Note that the LS method using ACF or OS is better than the CRLB, as the LS technique uses extra information of the FA property.

Second Example (Channel II): It is an example channel used in [4]. The system transfer function of this third order NMP model is $H(z) = 1 - 0.9z^{-1} + 0.385z^{-2} + 0.771z^{-3}$ with system zeros 0.6 and $0.75 \pm j0.85$. The proposed algorithm is compared with the generalized weighted slice (GEWS) algorithm via using HOS slices for SISO FIR channel estimation [4]. GEWS is used for comparison as it is arguably the best HOS technique reported in the literature [4]. From Fig. 3, the proposed initial estimation using OS method has better performance than GEWS algorithm. And the LS method using ACF or OS outperforms the GEWS algorithm and converges to a very low value rapidly and shows the best performance.

Third Example (Channel III): This MP model with zeros $0.7 \pm j0.7$, is also used in [4]. The system transfer function is $H(z) = 1 - 1.4z^{-1} + 0.98z^{-2}$. The magnitude of the channel zeros is 0.9899 and thus are extremely close to the unit circle. In Fig. 4, the data length for estimation varies from 100 to 6400 samples, and the proposed method with LS and the OS algorithm outperforms the GEWS algorithm and has better performance when smaller samples are used. However, using ACF for initial zero estimation does not provide good performance as zeros are too close to the unit circle.

Finally from Fig. 5, it can be seen that the output noise variance resulting from the estimation of proposed LS method is close to that resulting from an ideal equalizer, and thus the proposed method provides the best means of equalization for the channel.

VI. Conclusion

In this paper, a novel blind FIR channel parameter estimation technique is proposed. It takes advantage of efficient SOS based methods in conjunction with the known kurtosis of the transmitted signal. A simple linear zero-forcing equalizer with low hardware complexity is used in the estimator. The equalizer output is used recursively to achieve a very low estimation error. Simulation results are presented to demonstrate the superiority of the proposed algorithm. At low values of data samples, the proposed method is far superior to HOS techniques. This is not surprising as HOS techniques need larger data samples (or low noise) to achieve better performances. Furthermore, if the zeros are not very close to the unit circle (if fading is not severe in the channel), any simple SOS based method, such as ACF, can be used in the algorithm to obtain an extremely low complexity equalizer (for example results as in Fig.2). Noting that if channel zeros are close to the unit circle, linear equalizers would not result in proper equalization, other efficient equalizer architectures are under investigation.

REFERENCES

[1] J. K. Tugnait, L. Tong, and Z. Ding, "Single-user Channel Estimation and Equalization," *IEEE Signal Processing Magazine*, vol. 17, no. 1, pp. 17-28, May. 2000.

[2] S. M. Kay, "Modern Spectral Estimation: Theory and Application." Englewood Cliffs, NJ: PTR Prentice Hall, 1988.

[3] L. Tong, G. Xu, and T. Kailath, "Blind Identification and Equalization Based on Second-Order Statistics: A Time Domain Approach," *IEEE Trans. Inform. Theory*, vol. 40, no. 2, pp. 340-349, Mar. 1994.

[4] J. Liang and Z. Ding, "FIR Channel Estimation Through Generalized Cumulant Slice Weighting," *IEEE Trans. Signal Processing*, vol. 52, no. 3, pp. 657- 667, Mar. 2004.

[5] K.-S. Lii and M. Rosenblatt, "Deconvolution and Estimation of Transfer Function Phase and Coefficients for Non-Gaussian Linear Process," *The Annals of Statistics*, vol. 10, pp. 1195-1208, 1982.

[6] J. K. Tugnait, "Identification of Non-Minimum Phase Linear Stochastic Systems," *Automatica*, vol. 22, no. 4, pp. 457-464, 1986.

[7] O.Shalvi and E.Weinstein, "New Criteria for Blind Deconvolution Of Nonminium Phase Systems (Channels)" *IEEE Trans. On Infor. Theory*, vol.36, No.2 Mar. 1990.

[8] S. Talwar, M. Viberg, and A. Paulraj, "Blind Estimation of Multiple Co-Channel Digital Signals Using an Antenna Array," *IEEE Signal Processing Letters*, vol. 1, no. 2, pp. 29-31, Feb. 1994.

[9] C. Y. Chi, C. Y. Chen, C. H. Chen, and C. C. Feng, "Batch Processing Algorithms for Blind Equalization using Higher Order Statistics," *IEEE Signal Processing Magazine*, pp. 25-49, Jan. 2003.

[10] P. Stoica, T. McKelvey, and J. Mari, "MA Estimation in Polynomial Time," *IEEE Trans. Signal Processing*, vol. 48, no. 7, pp. 1999-2012, Jul.2000.

[11] P. Stoica and Y.Selen, "Model-Order Selection," *IEEE Signal Processing Magazine*, vol. 21, pp. 36-47, Jul. 2004.

[12] S. S. Abeysekera, "Implementation of a Zero-Forcing Residue Equalizer Using a Laguerre_Filter Architecture," Proceeding of International Symposium on Circuits and Systems, 2004, pp.III 385-388 Vancouver, Canada, May. 2004.

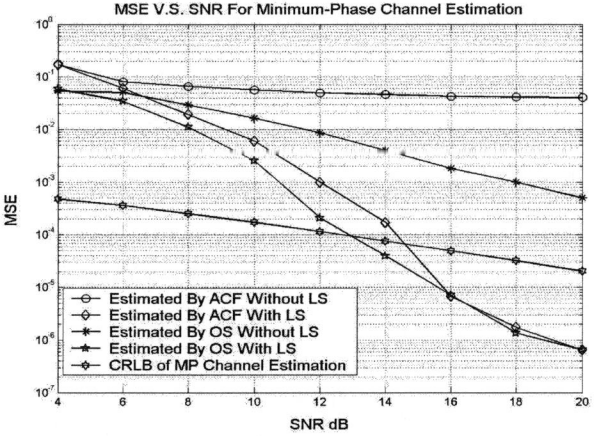

Figure 2. Channel I. Performance comparison of CRB with the proposed method for blind MP channel estimation. N=800 for each run. 200 runs

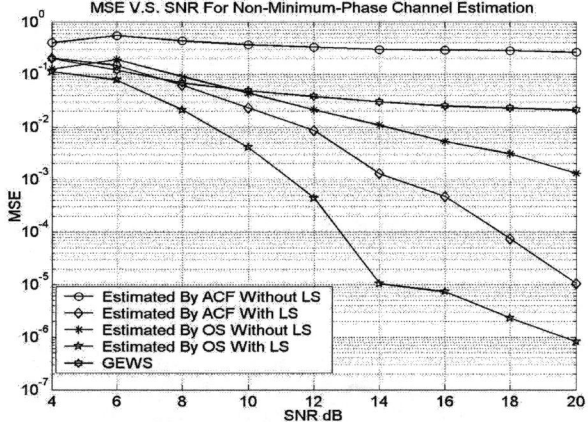

Figure 3. Channel II. Performance comparison of GEWS with the proposed method. 200runs, N=800 for each run.

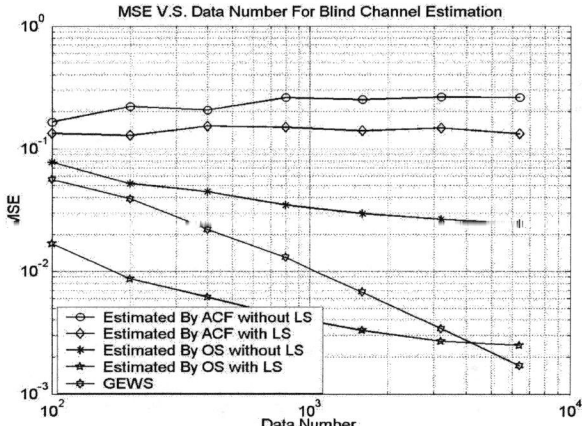

Figure 4. Channel III. Perfromance comparison of proposed method with GEWS. Sample number varies from 100 to 6400. 200 runs. SNR=20dB.

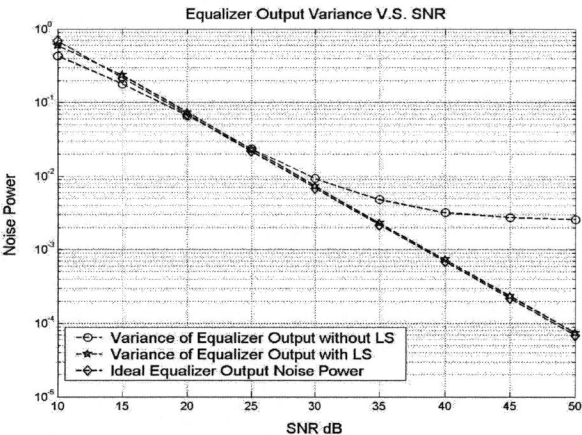

Figure 5. Channel II. Comparison of the equalizer output noise variance based on estimated parameters with the ideal equalizer noise variance. N=800 for each run. 200 runs.

A CONSIDERATION OF BLIND SOURCE SEPARATION USING WAVELET TRANSFORM

Noriyuki Hirai[1] Hiroki Matsumoto[2] Toshihiro Furukawa[3] Kiyoshi Furuya[4]

Dept. of information engineering[1,4]
Chuo Univ.1-13-27,kasuga
bunkyo-ku,Tokyo,112-8551,Japan
E-mail: nori550218@hotmail.com[1]

Dept. of information engineering[2]
Maebashi institute of technology

Dept. of management science[3]
Tokyo university of science

Abstract:

In this paper, we propose a new method of blind source separation for convolutive mixtures. We use wavelet transform which can get informations from the two side of time and frequency when we change observed signals into time-frequency domain. Our idea is using a new estimating equation of separating matrix for time delay. And we aim at improvement of separation performance. We show the result of desktop simulation artificially controlled data.

1. INTRODUCTION

We have the capability that can talk even in the environment where various sound exists simultaneously. The capability which is excellent in such human being's hearing is known as "cocktail party effect". And blind source separation(BSS) is known as method for realizing this capability in engineering. The BSS within the framework of independent component analysis(ICA) has attracted a great deal of attention in engineering field. The BSS has a wide range of applications including speech recognition, hands-free telecommunication systems and high-quality hearing aids.

BSS is the problem to separate independent sources from given mixed signals where the mixing process is unknown. We want to extract each source from the mixed signals using some technique. Even if the mixing process is unknown, we can separate the sources. On BSS, we have instantaneous mixtures which does not take time delay into consideration and convolutive mixtures which take time delay into consideration. And Frequency domain ICA are known as a method for convolutive mixtures[1]. In this paper, we take notice of frequency domain ICA for convolutive mixtures. And we propose the method of raising a separation performance using wavelet transform which can get informations from the two side of time and frequency, and using a new estimating equation of separating matrix for time delay.

The time signals are transformed to time-frequency signals, and we apply ICA algorithm to the signals of each frequency component. However, since we have to consider time delay, we need to extend the updating equation of ICA. This algorithm cannot solve the ambiguity of permutation and scaling for separated signals, when we reconstruct the time-frequency signals into separated time signals. In particular, we use the envelope of each frequency signals to group the sources.

This paper is organized as follows: in section 2 and 3, we describe basic approach to BSS of instantaneous mixtures. In section 4, we describe an algorithm of BSS for convolutive mixtures, in section 5, we describe wavelet transform which can get informations from the two side of time and frequency. In section 6, we propose an algorithm for BSS of convolutive mixtures using wavelet transform. In section 7 the result of our algorithm will be shown. Finally, we give a brief summary and concluding remarks in section 8.

2. BASIC BLIND SOURCE SEPARATION PROBLEM

In this section, we formulate the basic problem of BSS. The model of separation algorithm is shown by figure 1.

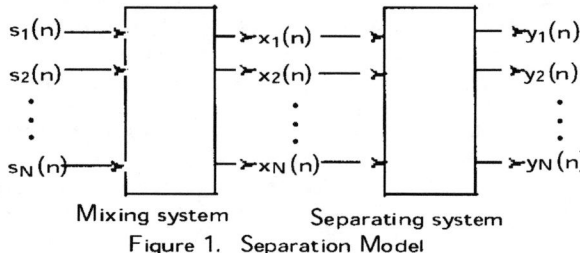

Figure 1. Separation Model

source signals are denoted by a vector

$$s(n) = (s_1(n), \cdots, s_N(n))^T \quad n = 0, 1, 2, \cdots \quad (1)$$

We assume the source signal $s(n)$ to be zero mean. And each component of $s(n)$ is assumed to be independent of each other, i.e. the joint distribution of the signals is factorized by their marginal distribution.

$$p(s_1(n), \cdots, s_N(n)) = p(s_1(n)) \cdots p(s_N(n)) \quad (2)$$

Observed signals are represented by

$$x(n) = (x_1(n), \cdots, x_N(n))^T \quad n = 0, 1, 2, \cdots \quad (3)$$

and they correspond to the recorded signals at sensors or microphones. In instantaneous mixtures, we assume that observed signals are linear mixtures of source signals:

$$x(n) = As(n) \quad A = \begin{pmatrix} a_{11} & \cdots & a_{1N} \\ \vdots & \ddots & \vdots \\ a_{N1} & \cdots & a_{NN} \end{pmatrix} \quad (4)$$

where A is called mixing matrix and $n \times n$ real valued matrix. The goal of blind source separation is to find a separating matrix W such that the components of the separated signals

$$y(n) = Wx(n) \quad (5)$$

$$y(n) = (y_1(n), \cdots, y_N(n))^T \quad n = 0, 1, 2, \cdots \quad (6)$$

are mutually independent, without knowing the mixing matrix A and the probability distribution of source signals $s(n)$. Ideally we expect the separating matrix W to be the inverse matrix of the mixing matrix A. But since we lack information about the source signals, there remains indefiniteness of scaling factors and permutation factors:

$$WA = PD \quad (7)$$

where D is a diagonal matrix which represents scaling factors, and P is a permutation matrix.

3. THE STANDARD OF INDEPENDENCY

Since BSS poses a problem which estimates the separating matrix W from Section 2, it can regard as one of the optimization problems. When we deal with an optimization problem, it is necessary for us to set up the standard for optimization. In this section, we describe Kullback-Leibler divergence used as the standard of optimization. Joint destribution of separated signals $y(n)$ are represented by

$$p(y) = p(y_1, y_2, \cdots, y_N) \quad (8)$$

If observed signals is correctly separated by the separating matrix W, each component y_i of separated signals y are independent. In this case, joint destribution of separated signals $y(n)$ and marginal distribution are represented by

$$p(y) = \prod_{i=1}^{N} p(y_i) \quad (9)$$

And Kullback-Leibler divergence is defined as follows.

$$KL(W) = \int p(y) \log \frac{p(y)}{p(y_1)p(y_2)\cdots p(y_N)} dy \quad (10)$$

If Kullback-Leibler divergence is used, the BSS problem which solves for separation matrix W which fills Equation (10) will change a problem which solves for separation matrix W which makes Kullback-Leibler divergence the minimum.

The optimal separation matrix W is obtained by the following updating equation for separation matrix W.

$$W_{n+1} = W_n + \eta(I - \phi(y)y^T)W_n \quad (11)$$

where n is an index for the updating, I is an identity matrix, η is a step size parameter, and $\phi(t)$ is a nonlinear function.

4. BLIND SOURCE SEPARATION FOR CONVOLUTIVE MIXTURES

In this section, we foumulate the BSS of convolutive mixtures. We introduce frequency domain ICA algorithm[1].

In convolutive mixtures, the observed signals can be represented as

$$x_i(n) = \sum_{j=1}^{N} \sum_{m} a_{ij}(m) s_j(n-m) \quad (12)$$

Let $a_{ij}(m)$ be a unit impulse response from source j to sensor i with time delay m. We write this relation in matrix form as

$$x(n) = A * s(n) \quad (13)$$

where A is called a filter matrix and $*$ denotes the convolution. First, we apply the short-time Fourier transform to observed signals

$$\hat{x}(k; n') = \hat{A}(k)\hat{s}(k; n') \quad n' = 0, WL, 2WL, \cdots \quad (14)$$

$$k = 0, \frac{1}{WL} 2\pi, \cdots, \frac{WL-1}{WL} 2\pi \quad (15)$$

where k denotes the frequency and WL denotes the number of points in the discrete Fourier transform, n' denotes the window position. $\hat{x}$, $\hat{A}$ and $\hat{s}$ denote each the signals by Fourier transforme of x, A and s.

If we fix the frequency as k for spectrograms,

$$\hat{x}_k(n') = \hat{x}(k; n') \quad (16)$$

is a time series of n'. Hence convolutive mixtures problem for time domain is changed into instantaneous mixtures at each frequency k and we can apply basic BSS algorithm.

But we have to note that $\hat{x}_k(n')$ is complex values, and therefore the method in Section 3 should be extended to complex domain and use following updating equation[3].

$$W_{n'+1}(k) = W_{n'}(k) + \eta[\text{diag}[\Phi(\hat{u}_k(n'))\hat{u}_k(n')^H]$$
$$- \Phi(\hat{u}_k)(n')\hat{u}_k(n')^H] W_{n'}(k) \quad (17)$$

$$\Phi[\hat{u}_k(n')] = \tanh[re(\hat{u}_k(n'))] + j \tanh[im(\hat{u}_k(n'))] \quad (18)$$

where n' is an index for the iteration, H is conjugate transpose. After this algorithm is applied, we have an estimated signals $\hat{u}_k(n')$ whose components are mutually independent for each frequency k.

$$\hat{u}_k(n') = W(k)\hat{x}_k(n') \quad (19)$$

Since BSS algorithm cannot solve the scaling and permutation problem, in each frequency channel, the estimated signals have the ambiguity of scaling and permutation. So we need to solve these two problems.

To solve the ambiguity of scaling, we disassemble the spectrograms exploiting the independent components at each frequency channel. Let us define split spectrograms by

$$\hat{v}_k(n'; i) = W(k)^{-1} \begin{pmatrix} 0 \\ \vdots \\ 0 \\ \hat{u}_{i;k}(n') \\ 0 \\ \vdots \\ 0 \end{pmatrix} \quad (20)$$

where index i denotes the dependence of the spectrograms at k on the i-th independent component of $\hat{u}_k(n')$. In order to obtain $\hat{v}_k(n'; i)$, we apply $W(k)$ and $W(k)^{-1}$, and therefore $\hat{v}_k(n'; i)$ does not have an ambiguity of scaling.

To solve the ambiguity of permutation, we define an moving average operator for estimating the envelope of time series by

$$^2\hat{v}_k(n'; i) = \frac{1}{2M+1} \sum_{n''=n'-M}^{n'+M} \sum_{j=1}^{N} \hat{v}_{j;k}(n''; i) \quad (21)$$

where M is a positive constant. Also we define its inner product and norm as

$$^2\hat{v}_k(i) \cdot {}^2\hat{v}_k(j) = \sum_{n'} {}^2\hat{v}_k(n'; i) {}^2\hat{v}_k(n'; j) \quad (22)$$

$$||{}^2\hat{v}_k(i)|| = \sqrt{{}^2\hat{v}_k(i) \cdot {}^2\hat{v}_k(i)} \quad (23)$$

We solve the permutation by sorting based on the correlation of envelops as follows

- Sort k in order of the weakness of correlation between independent components in k. This is done by sorting in increasing order of

$$sim(k) = \sum_{i \neq j} \frac{{}^2\hat{v}_k(i) \cdot {}^2\hat{v}_k(j)}{||{}^2\hat{v}_k(i)|| \cdot ||{}^2\hat{v}_k(j)||} \quad (24)$$

$$sim(k_1) < sim(k_2) < \cdots < sim(k_N) \quad (25)$$

- For k_1, assign $\hat{v}_{k_1}(n'; i)$ to $\hat{y}_{k_1}(n'; i)$ as it is:

$$\hat{y}_{k_1}(n'; i) = \hat{v}_{k_1}(n'; i) \qquad i = 1, 2, \cdots, N \quad (26)$$

- For k_h, find the permutation $\pi(i)$ which maximizes the correlataion between the envelope of k_h and the aggregated envelope from k_1 through k_{h-1}. This is achieved by maximizing

$$\sum_{i=1}^{N} \hat{v}_{k_h}(\pi(i)) \cdot \left(\sum_{j=1}^{h-1} \hat{y}_{k_j}(i) \right) \quad (27)$$

within all the possible permutation π of $i = 1, \ldots, N$.

- Assign the appropriate permutation to $\hat{y}_{k_h}(n'; i)$

$$\hat{y}_{k_h}(n'; i) = \hat{v}_{k_h}(n'; \pi(i)) \quad (28)$$

As a result, we can solve the ambiguity of permutation and obtain separated spectrograms

$$\hat{y}(k; n'; i) = \hat{y}_k(n'; i) \quad (29)$$

Finally, applying the inverse Fourier transform, we obtain a set of separated signals.

5. APPLICATION OF WAVELET TRANSFORM

5.1 Wavelet transform

The fourier transform analyzes signals using the time waveform which had an infinite spread in time. On the other hand, the wavelet transform analyzes signals by shifting or scaling an small time waveform called mother wavelet. Therefore, the wavelet transform can detect a sudden change of signals. i.e. wavelet transform is suitable for the analysis of non-stationary signals. Also, wavelet transform has ambiguity about selection of mother wavelet. Our blind source separation algorithm use Haar function calculable only with the real number.

Haar wavelet is the function made by Haar and is making the group of scaling function and mother wavelet. Each scaling function and mother wavelte are defined as follows.

$$\phi(n) = \begin{cases} 1 & (0 \leq n < 1) \\ 0 & (\text{otherwise}) \end{cases} \quad (30)$$

$$\bar{\phi}(n) = \begin{cases} 1 & (0 \leq n < \frac{1}{2}) \\ -1 & (\frac{1}{2} \leq n < 1) \\ 0 & (\text{otherwise}) \end{cases} \quad (31)$$

We use this function when we change observed signals of time domain into the signal of time-frequence domain.

5.2 Apply to blind source separation

There are the following advantages by applying wavelet transform to blind source separation of convolutive mixtures.

- By using scaling function of Haar, the expression of wavelet transform is expressed by simple calculation of addition and multiplication. For this reason, unlike frequency domain ICA which used the both sides of the real number and an imaginary number, it can treat only with the real number.

- By using wavelet transform, it becomes possible about signals to get informations from the two side of time and frequency. Furthermore, by taking a decomposition level into consideration, we can cope with a signal with large time change or frequency change.

From the above-mentioned thing, when performing blind source separation problem, it is possible that it is significant to use wavelet transform.

6. PROPOSED METHOD USING WAVELET TRANSFORM

In this section, we propose a new method of blind source separation algorithm using wavelet transform.

First, we apply the wavelet transform of Haar function to observed signals and change observed signals of time domain into the signals of time-frequence domain. When observed signals are decomposed to the arbitrary levels j by the wavelet transform, observed signals can be represented as follow.

$$\hat{x}^{(j)}(L_k; n') = \frac{1}{2}(\hat{x}^{(j-1)}(2n') + \hat{x}^{(j-1)}(2n' + 1)) \quad (32)$$

$$\hat{x}^{(j)}(H_k; n') = \frac{1}{2}(\hat{x}^{(j-1)}(2n') - \hat{x}^{(j-1)}(2n' + 1)) \quad (33)$$

where L_k and H_k denote each frequency of low and high level, j is level of wavelet transform, n' is time component on wavelet. $\hat{x}$ denotes the observed signals changed by wavelet transform.

Next, We fix a level j and perform separation processing in Section 2 at each level j. However, since the components on wavelet contain delay components, the BSS algorithm to the characteristic containing the delay components is needed. Therefore, updating equation which made KL divergence the evaluation function is extended as follows at each k.

$$W^{(j)}_{(k; n'+1; p)} = W^{(j)}_{(k; n'; p)} + \eta(p)(I - A(u^{(j)}(k; n' - p)))$$
$$\cdot u^{(j)}(k; n' - p)^T) W^{(j)}_{(k; n'; p)} \quad (34)$$

$$A(u^{(j)}) = \tanh(u^{(j)})$$

where p denotes the time. Decomposed frequency k assume $0 \sim 2^j - 1$ from low to high.

Also $u^{(j)}(k; n')$ are estimated signals by separation process.

$$u^{(j)}(k; n') = \sum_p W^{(j)}_{(k; n'; p)} x^{(j)}(k; n' - p) \quad (35)$$

By this process, we can obtain the separated signals independently at each frequency k.

But since this algorithm also cannot solve the scaling and permutation problem, in each frequency k, the estimated signals $u^{(j)}(k; n')$ have the ambiguity of scaling and permutation. So we make use of method in Section 4. Therefore we can obtain the separated signals without the ambihuity of scaling and permutation. And we apply inverse wavelet transform to the separated signals.

$$\hat{y}^{(j)}(k; 2n') = \hat{y}^{(j+1)}(L_k; n') + \hat{y}^{(j+1)}(H_k; n')) \quad (36)$$

$$\hat{y}^{(j)}(k; 2n') = \hat{y}^{(j+1)}(L_k; n') - \hat{y}^{(j+1)}(H_k; n')) \quad (37)$$

And we finally obtain a set of separated signals on time domain.

7. SIMULATION RESULTS

In this section, we show the result of the proposed blind source separation algorithm using wavelet transform. First, the sources are mixed on the computer and our algorithm was applied to those mixed data. Since the true sources were available, we can evaluate the performance of the algorithm.

Figure 2 is source signals separately recorded by computer.

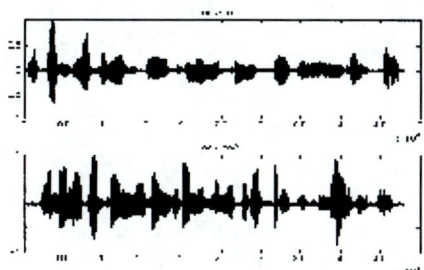

Figure 2. Source Signals

And convolutive mixtures signals can be defined below like Equation (12).

$$x_i(n) = \sum_{j=1}^{X} \sum_{m} a_{ij}(m) s_j(n-m) \quad (38)$$

We mixed these source signals where the matrix $A_p(n)$ is shown below

$$A_p(n) = \begin{pmatrix} a_{11}(n-p) & a_{12}(n-p) \\ a_{21}(n-p) & a_{22}(n-p) \end{pmatrix} \quad (39)$$

Figure 3 is convolutive mixtures signals on a computer.

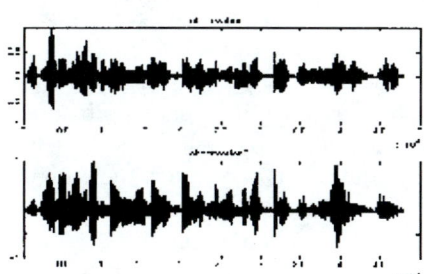

Figure 3. Observed Signals of Convolutive Mixtures

Also since we know the true sources and the mixing rates, we can evaluate the performance using the SNR (Signal to Noise Ratio) which is defined as

$$error_i(n) = y_i(n) - s_i(n) \quad i = 1, 2, \cdots, N \quad (40)$$

$$SNR_i = 10 \log_{10} \frac{\sum_n s_i(n)^2}{\sum_n error_i(n)^2} \quad (41)$$

The blind source separation algorithm in Section 6 was applied to convolutive mixture signals. The result is shown in Figure 4. Furthermore, SNR is shown in Table 1.

This result showed that separation performance of the algorithm using wavelet transform improved from conventional algorithm in Section 4.

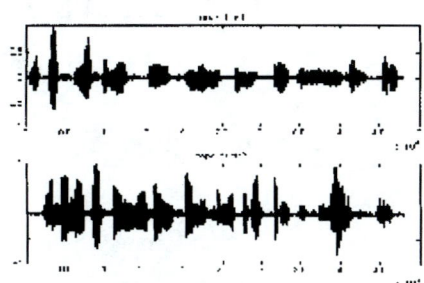

Figure 4. Separated Signals

Table 1. the result of simuration

SNR	conventional algorithm	proposed algorithm
SNR_1	6.735	10.861
SNR_2	6.114	8.733

8. CONCLUTION

We proposed a blind source separation algorithm using wavelet trasform for convolutive mixtures. Our new idea is using a new estimating equation of separating matrix for time delay. Also our algorithm has advantage that it can calculate only real value for separate process. The algorithm worked very well for the data mixed on the computer compared with conventional algorithm on the computer simulations. Some future works are as follows:

First, It is how to decide mother wavelet in case we use wavelet transform. Since there are some bases, it is necessary to decide the base for which it was most suitable in it by simulation.

Second, it is the problem of the separation accuracy at each frequency when performing wavelet transform. Generally, a part of frequency domain does not have sufficient compotnet. So, in this domain, since blind source sepataion and the ambiguity of permutation cannot perform well, it is thought that aggravation of separation accuracy will be influenced. It is necessary to devise the highly efficient blind source separation algorithm to frequency with few such component.

References

[1] Noboru Murata, Shiro Ikeda, and Andreas Ziehe : An Approach to Blind Source Separation Based on Temporal Structure of Speech Signaols , Neurocomputing, Vol.41, Issue 1-4, pp.1-24, (2001)

[2] Noboru Murata and Shiro Ikeda, \An on-line algorithm for blind source separation on speech signals," In Proceedings of 1998 International Symposium on Nonlinear Theory and its Applications (NOLTA'98) , pp.923-926, Crans-Montana, Switzerland, September 1998

[3] H. Sawada, R. Mukai, S. Araki, S. Makino, "A Polar-Coordinate based Activation Function for Frequency Domain Blind Source Separation," in Proc. of ICA2001 (International Conferenece on Independent Component Analysis and Blind Signal Separation), Dec.2001. R. Mukai, S. Araki, S. Makino, "Separation and Dereverberation Performance of Frequency Domain Blind Source Separation in a Reverberant Environment," in Proc. of IWAENC 2001, pp.127-130.

An Approach for Nonlinear Blind Source Separation of Signals with Noise Using Neural Networks and Higher-Order Cumulants

Nuo Zhang, Xiaowei Zhang, Jianming Lu and Takashi Yahagi
Graduate School of Science and Technology
Chiba University, Chiba-shi, 263-8522 Japan
Email: zhang@graduate.chiba-u.jp

Abstract—In this paper, we propose a robust approach for blind source separation when observations are contaminated with Gaussian noise and nonlinear distortion. A radial basis function networks (RBFN) is employed to estimate the inverse of nonlinear mixing matrx. We utilize an novel cost function which consists of mutual information and higher-order cumulants of signals. Compared with moments, higher-order cumulants can provide a clearer form and more information of signals. Thus the proposed method has not only the capacity of recovering the nonlinearly mixed signals, but also removing high-level Gaussian noise from transmitted signals. Through the simulation and analysis of artificially synthesized signals, we illustrate the efficacy of this proposed approach.

I. INTRODUCTION

Blind source separation is a class of methods that recover unaccessible independent original signals from nonlinear mixtures. It has received a great deal of attention recently. So far there are a number of algorithms have been proposed for blind source separation problem and have been applied to the area of speech signal processing, communication and medical signal processing (MEG, ECG and EEG).

The main study of blind source separation considers about the instantaneous linear mixture by now [1]. Nonlinear blind source separation is a much more recent research topic. Generally, a nonlinear mixing model is more realistic and accurate than linear model and suitable for many practical situations. Moreover, there are also many Gaussian noise are present in the real-world and it has been studied within traditional independent component analysis method by researchers [2]. It is of important significance to address the topic of nonlinear blind separation with noise in depth.

However blind separation of signals in nonlinear mixtures has been very sparsely studied because of non-uniqueness of separation. Neural networks are valuable tools to deal with a variety of nonlinear problems occurred in many practical domains and have been studied extensively. Hence several blind source separation methods have been derived from neural networks naturally. This can be found in following literatures: Kohonen's self-organizing map (SOM) is utilized to extract sources from nonlinear mixture (but the computational complexity of this kind of approaches grow exponentially) [3], an information BP algorithm for training of separating system [4]. One drawback is slow convergence rate yield from the highly nonlinear relationship between the output and learning weights of the network. In contrast, an RBF network has several advantages over a multilayer perceptron in terms of natural learning manner. An approach using RBF network proposed by Tan *et al.* [5] can recover source signals better, and the convergence rate is very fast. However, this method degrades greatly when high-level noise is present with nonlinear distortion. The disadvantage of this approach is that it does not consider the influence of noise in the cost function.

The purpose of this paper is to utilize higher-order statistics to obtain a robust method with RBF network when both nonlinear distortion and Gaussian noise are present. Note that, cumulants usually present in a clear form, where more information can be provided by higher-order cumulants. In addition, we used some useful properties of cumulants, which can not be provided by moments. (They will be described in the following sections). We introduced three- and four-order cumulants together with mutual information into a novel cost function. The minimization of the proposed cost function can recover source signals as good as original method. Furthermore, it outperforms the original method when the high-level Gaussian noise is present in nonlinear mixtures. Thus, the proposed approach can also be used to reduce noise from data. According to the theoretical analysis and computer simulation results, it is shown that the proposed algorithm has better performance than original approach when facing with Gaussian noise in nonlinear mixtures.

II. NONLINEAR MIXING SYSTEM MODEL WITH NOISE

The noise introduced in this paper is Gaussian noise. Denoted $x(t) = [x_1(t), x_2(t), \cdots x_n(t)]^T$ the vector of observed random variables, and $s(t) = [s_1(t), s_2(t), \cdots s_n(t)]^T$ the independent source vector, the noisy nonlinear mixing model can be expressed as

$$x(t) = f[s(t) + G] \qquad (1)$$

where G is Gaussian noise, f is an unknown multiple-input and multiple-output (MIMO) mapping function.

We further assume the noise is independent with the source components. For simplicity and convenience, we also assume the dimension between original signals and mixtures is equal to each other, without the loss of generality.

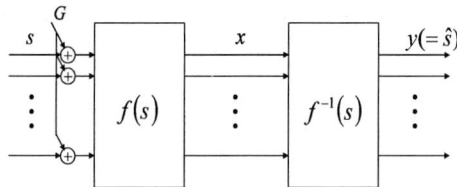

Fig. 1. System model.

Figure. 1. shows the mixing model and blind source separation system expressed in Eq. (1). The purpose of the separation system is to recover the original signals only from their observations $x(t)$ and remove the Gaussian noise. Obviously, this problem is difficult to solve. Hence we assume $f(\cdot)$ is componentwise invertible and its inverse $f^{-1}(\cdot) = \left(f_1^{-1}(\cdot), f_2^{-1}(\cdot), \cdots, f_n^{-1}(\cdot)\right)$ exists. This paper will employ an RBF network to solve the problem due to its capacity of approximating an arbitrary function [6].

It is also difficult that how to recover the mixing matrix when the mixtures are degraded by Gaussian noise. A class of noise reduction method based on higher-order cumulants has been proposed [7]. Considered the better properties of higher-order cumulants, we applied it into the cost function in this paper, what will be introduced in the following sections.

III. RADIAL BASIS FUNCTION NETWORKS

The RBF network is very different from multilayer perceptrons with sigmoidal activation function in which it utilizes neurons with radial basis function that are locally responsive to input stimulus in the hidden layer.

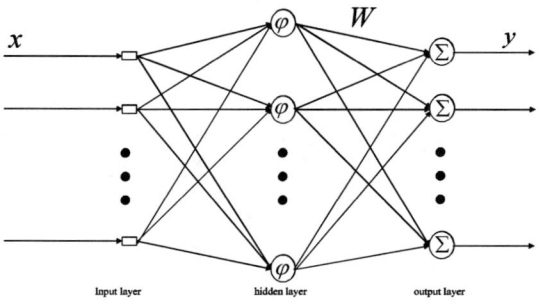

Fig. 2. The RBF network.

As shown in Fig. 2., an RBF neural network consists of two layers: hidden layer and output layer with linear neurons. A hidden RBF neuron is usually implemented using a Gaussian kernel function, which has two parameters, what are center, and width. The whole relationship between input and output of an RBF neural network can be given by:

$$y(x) = \sum_{j=0}^{M} w_j \phi_j(x), \quad (2)$$

and

$$\phi_j(x) = \exp\left(-\frac{1}{\sigma_j^2}(x - \mu_j)^2\right) \quad (3)$$

where w_j is the connecting weight between the j-th hidden RBF neuron and linear output neuron, $\phi_j(x)$ is the response function of the j-th hidden RBF neuron, μ_j and ϕ_j are the center and width of the j-th hidden RBF neuron, respectively, M indicates the number of hidden RBF neurons in the network. There are a number of learning strategies for RBF networks such as randomly selecting the radial basis function centers and employing unsupervised or supervised procedures for selecting the radial basis function centers. In this paper we will use the RBF network to construct the inverse mapping of the nonlinear mixing function in Fig. 1. In the following, we will focus on the learning rules of the parameters in our proposed algorithm.

IV. THE PROPOSED COST FUNCTION AND HIGHER-ORDER CUMULANT

Recently, a few unsupervised RBF based methods have been proposed [5]. They make use of the characteristics of neural networks to solve the nonlinear mixing problem.

This paper will study the problem by adopting the higher-order cumulants in the cost function when high-level noise is present. In the original method, the higher-order moments of sources and estimations are required to be the same, which yields the drawback that the moment information of sources have to be known, which is too rigorous in real world. The proposed higher-order cumulants based method is not constrained to the severe condition. Accordingly, the proposed algorithm can separate more general kinds of signals with high-level noise and without prior information of source signals. Thus it is more flexible and applicable.

A. The Definition of Cumulant and its Properties

Since the datum we used in this paper are assumed to be zero mean random vector, the definition of cumulants up to order four, can be given as follows

$$\begin{aligned} cum_i(y) &= 0, \\ cum_{ij}(y) &= E\{y_i y_j\}, \\ cum_{ijk}(y) &= E\{y_i y_j y_k\}, \\ cum_{ijkl}(y) &= E\{y_i y_j y_k y_l\} - E\{y_i y_j\} E\{y_k y_l\} \\ &\quad - E\{y_i y_k\} E\{y_j y_l\} - E\{y_i y_l\} E\{y_j y_k\}. \end{aligned} \quad (4)$$

In addition, cumulants have two properties. 1. If the sets of random variables $(x_1, x_2, ..., x_n)$ and $(y_1, y_2, ..., y_n)$ are independent, then $cum\{x_1 + y_1, x_2 + y_2, \cdots x_n + y_n\} = cum\{x_1, x_2, \cdots x_n\} + cum\{y_1, y_2, \cdots y_n\}$. 2. If the sets of random variables $(x_1, x_2, ..., x_n)$ are Gaussian distributed, then $cum\{x_1, x_2, \cdots x_n\} = 0$.

Obviously, utilized the conclusion the two properties above to the output of the separation system, we can draw the following relation in theory:

$$cum(y + G) = cum(y). \quad (5)$$

Therefore, by applying such property of higher-order cumulants into our cost function to estimate RBF networks' parameters, we can conclude that there is no effect on estimations from Gaussian noise, such that better separation results can be obtained.

Higher-order cumulants also have another property that it is helpful to measure the dependence among random vector. For example, four-order cumulant is a four-dimensional array whose entries are given by the fourth-order cross-cumulants of the data. This can be considered as a four-dimensional matrix, since it has four different indices. The four-order cumulants contain all the fourth-order information of the data. The diagonal elements characterize the distribution of single components. The off-diagonal elements of cross-cumulants (all cumulants with $ijkl \neq iiii$) characterize the statistical dependencies among components. If and only if, all components are statistically independent, the off-diagonal elements vanish, and the cumulants (of all orders) are diagonal. In the next section, we will derive the cost function with higher-order cumulants.

B. Cost Function with Higher-Order Cumulants

The general measure for the degree of the dependence among random variables is mutual information. Several mutual information based algorithms have been presented for linear mixtures separation in literature. However it is well known that only mutual information itself is not sufficient to recover source signals from nonlinear mixtures. In order to solve this problem we introduce higher-order cumulants into the cost function, by minimizing which we can make the output of separation system as independent as possible. Moreover, utilizing the properties described above, this approach is robust to high-level Gaussian noise.

Three- and four-order cumulants are introduced into our cost function, in order to recover the original signals from their nonlinear mixtures when Gaussian noise is present, given as follows:

$$J(W) = I(y) + \sum_{ijk \neq iii} cum_{ijk}(y)^2 + \sum_{ijkl \neq iiii} cum_{ijkl}(y)^2 \quad (6)$$

where $I(y)$ is mutual information of the separation system outputs, $\sum_{ijk \neq iii} cum_{ijk}(y)^2$ and $\sum_{ijkl \neq iiii} cum_{ijkl}(y)^2$ are square sum of non-diagonal three-order and four-order cumulant of outputs, respectively.

Mutual information $I(y)$ in Eq. (6) is defined as

$$I(y) = \sum_{i=1}^{n} H(y_i) - H(y) \quad (7)$$

where $H(y)$ is joint entropy of random vector y; and $H(y_i)$ is entropy of random variable y_i, the i-th component of vector y.

This cost function consists of two parts: mutual information and square of cumulants of system outputs. Note that, these two items are always nonnegative, and vanish when both of them are minimized. Therefore, the proposed algorithm can recover original signals through minimizing the cost function.

The proposed algorithm contains two steps: first, it uses traditional k-means algorithm for the selection of the centers and widths of RBF neurons; then updates the weights in output layer of neural network by utilizing gradient decent of cost function.

After estimating the centers and widths of hidden layer using k-means algorithm, the weights of output layer can be updated by minimizing the cost function. In this paper, we utilize gradient descent manner to obtain the unsupervised learning rule of output layer weights of RBF network. The gradient of the cost function of Eq. (6) with respect to output layer weights is shown as follows:

$$\frac{\partial J(w)}{\partial w} = \frac{\partial I(y)}{\partial w} + \sum_{ijk \neq iii} 2cum_{ijk}(y) \frac{\partial cum_{ijk}(y)}{\partial y} \frac{\partial y}{\partial w} + \sum_{ijkl \neq iiii} 2cum_{ijkl}(y) \frac{\partial cum_{ijkl}(y)}{\partial y} \frac{\partial y}{\partial w}. \quad (8)$$

By using the Gram-Charlier expansion method to approximate the pdf of y_i, we obtain the gradient decent of mutual information as follows

$$\frac{\partial I(y)}{\partial w} = \sum_{i=1}^{n} \frac{\partial H(y_i)}{\partial y_i} \frac{\partial y_i}{\partial w} - \left| \frac{\partial g(x,w)}{\partial w} \right|^{-1} \frac{\partial}{\partial w} \left| \frac{\partial g(x,w)}{\partial w} \right|. \quad (9)$$

From definitions in (4) and (8), we can simply derive gradient decent of the two latter items in Eq. (6).

From Eq. (4), (6)-(9) we finally derive the updating rule of output layer weights:

$$w_{k+1} = w_k - \lambda \frac{\partial J(w_k)}{\partial w_k} \quad (10)$$

where $0 < \lambda < 1$ is learning rate.

C. Description of Learning Rule

In this paper, the simulation results are analyzed based on the root-mean-squares (RMS) error, which also is employed to stop iteration of the unsupervised learning:

$$RMSE = \sqrt{\frac{1}{N} \sum_{t=1}^{N} (s(t) - \hat{s}(t))^2} \quad (11)$$

where $\hat{s}$ is recovered source signals, N denotes the number of signals.

The proposed algorithm can be summarized as follows:

a. Initialize the centers, widths and weights of RBF network by using small random number. The termination condition, a small positive number ε, of learning is also determined.
b. Update the centers and widths of RBF network's hidden layer by using k-means algorithm.
c. Use Eq. (8), Eq. (9) and Eq. (10) to update output layer weights of RBF network.
d. Repeat step b until the RMSE between s and $\hat{s}$ satisfies the condition ($RMSE < \varepsilon$).

V. SIMULATION AND RESULT ANALYSIS

In this section we present and discuss simulation results of the proposed algorithm. The proposed algorithm is performed with artificial signals: sinusoid and modulated sinusoid signal corrupted by additive Gaussian noise. In order to compare the performance of proposed algorithm with the traditional method, we use the same RBFN with 2 inputs, 6 hidden neurons and 2 outputs, and 5000 samples of signals. The mixing process is described as follows:

$$\begin{pmatrix} x_1 \\ x_2 \end{pmatrix} = A_2 \begin{bmatrix} (\cdot)^3 \\ (\cdot)^3 \end{bmatrix} A_1 \begin{pmatrix} s_1 \\ s_2 \end{pmatrix} \quad (12)$$

where $A_1 = \begin{pmatrix} 0.25 & 0.86 \\ -0.86 & 0.25 \end{pmatrix}, A_2 = \begin{pmatrix} 0.5 & 0.9 \\ -0.9 & 0.5 \end{pmatrix}$.

We first performed a simulation without influence of noise to verify the separation ability of proposed algorithm. Since the proposed algorithm has the same good result as that of the original method, we do not show the results here for brief. Then we used mixtures obtained from Eq. (12) in the following simulation. For a more general problem, we considered sources corrupted by Gaussian noise with 10dB SNR (Eq. (1)). Fig. 3. shows two source signals. They are nonlinearly mixed with noise into mixtures, illustrated in Fig. 4. From Fig. 5. and Fig. 6., we can see that the proposed algorithm can obtain better separation results compared with original method.

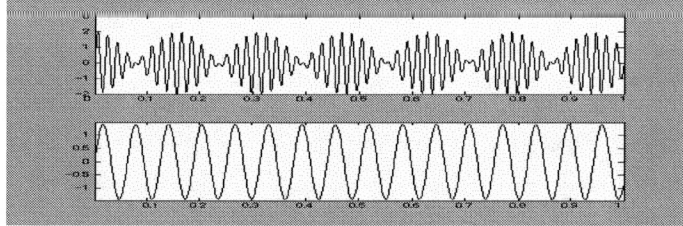

Fig. 3. Source signals.

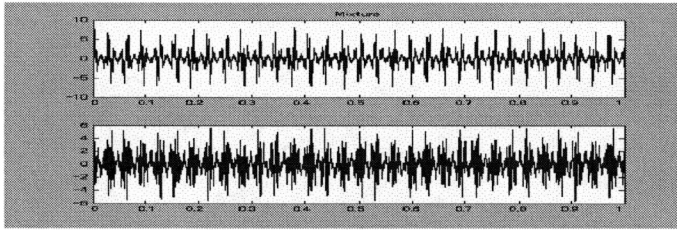

Fig. 4. Mixtures.

Furthermore, we used RMSE to analyze the separation results of the proposed and original algorithms. The RMSE curves is shown in Figure. 7. The proposed method has the same convergence rate with original method. Although both of them are converged at 1000 iterations, we obtained better estimations from the proposed algorithm due to the robustness to Gaussian noise of higher-order cumulants.

VI. CONCLUSION

A nonlinear separation method using higher-order cumulants, which is robust to noise, has been proposed in this paper. The comparisons of performance of the proposed and original algorithm has been studied in the presence of strongly nonlinear distortions and Gaussian noise. Simulation results showed that the proposed algorithm can obtain clearer estimations of sources corrupted by noise from nonlinear mixtures.

By using higher-order cumulants, the proposed algorithm can separate mixtures without the requirement of prior information of sources. Therefore, it is more applicable to high-level noise in real problems.

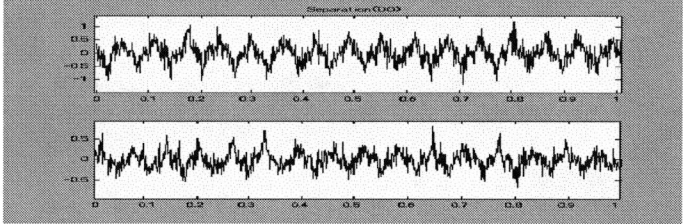

Fig. 5. Separation results by original method.

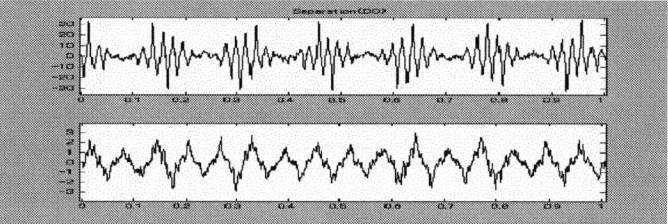

Fig. 6. Separation results by proposed method.

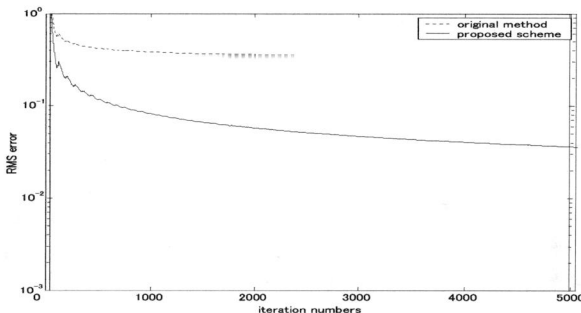

Fig. 7. Convergence curves.

REFERENCES

[1] C. Jutten and J. Herault, "Blind separation of sources, part I: An adaptive algorithm based on neuromimetic architecture," *Signal Processing*, vol. 24, pp. 1–20, 1991.
[2] A. Hyvärinen, "Independent component analysis in the presence of gaussian noise by maximizing joint likelihood," *Neurocomputing*, vol. 22, pp. 49–67, 1998.
[3] M. Herrmann and H. H. Yang, "Perspectives and limitations of self-organizing maps in blind separation of source signals," *Progress in Neural Information Processing: Proceedings of ICONIP'96*, pp. 1211–1216, 1996.
[4] S. A. H. H. Yang and A. Cichocki, "Information backpropagation for blind separation of sources from nonlinear mixture," *Proc. IEEE ICNN, TX*, pp. 2141–2146, 1997.
[5] J. Ying Tan; Jun Wang; Zurada, "Nonlinear blind source separation using a radial basis function network," *Neural Networks, IEEE Transactions on*, vol. 12, no. 1, pp. 124–134, Jan 2001.
[6] W. A. Light and E. W. Cheney, "Interpolation by piecewise-linear radial basis functions," *Journal of Approximation Theory*, vol. 64, no. 1, pp. 38–54, Jan 1991.
[7] N. C. R. S. Bhattacharya and S. Sinha, "2-D signal modelling and reconstruction using third-order cumulants," *Signal Processing*, vol. 62, no. 1, pp. 61–72, Oct 1997.

Subband Blind Equalization using Wavelet Filter Banks

Amir Minayi Jalil
Electrical Engineering Department
Amirkabir University of Technology
Tehran, Iran
minayijalil@aut.ac.ir

Hamidreza Amindavar
Electrical Engineering Department
Amirkabir University of Technology
Tehran, Iran
hamidami@aut.ac.ir

Farshad Almasganj
Biomedical Engineering Department
Amirkabir University of Technology
Tehran, Iran
almas@aut.ac.ir

Abstract— In this paper we propose an approach to decrease the computational cost and improve the convergence rate of blind equalizers using wavelet filter banks. Subband adaptive filtering is known for its improved convergence rate over the conventional least mean square LMS algorithm, even with lowering the computational cost; on the other hand, blind equalizers in many applications such as mobile communication channels and digital subscriber lines suffer from the poor convergence rate. We propose the subband equalization method using the wavelet filter banks to improve the convergence rate and decreasing the computational complexity of such equalizers and discuss its advantage over other subband adaptive filtering (SAF) schemes. This discussion is done over an important category of blind equalizers; cyclostationarity based approach that is used in the case of blind fractionally spaced equalization (FSE) of channels.

I. INTRODUCTION

Adaptive equalizers almost require an initial training sequence, however, there are some applications where the adaption process should take place during regular data transmission without any extra training sequences imbedded in the data stream. This is known as blind equalization. Generally speaking, there are four important approaches to the problem of blind equalization, Bussgang approaches that minimize a chosen cost function over all possible choices of the equalizer coefficients in an iterative fashion, the higher order statistics (HOS) method that is using the higher order cumulants or their discrete Fourier transforms known as polyspectra [1], [4], the approaches that exploit statistical cyclostationarity information [8], and the algorithms that are based on the maximum likelihood criterion. Though, Zhou and Viberg [6] have recently proposed a first order statistical method for blind channel estimation. Each one of these blind methods has an LMS type adaptive version which is the most popular and widely used adaptive algorithm, appearing in numerous communication applications such as speech enhancement, radar signal processing, image processing and etc. This is not only because of the ease of implementation, the computational simplicity, and the efficient use of memory, but its robustness in the presence of numerical errors caused by finite-precision arithmetic, analytically characterizable behavior and versatility to perform also in the transform domain allowing constraints on the adaptive filter [3]. The LMS algorithms using transformations and related filter banks such as discrete cosine, discrete Fourier, and wavelet techniques have been proposed to improve the convergence rate by reducing the eigenvalue spread of the autocorrelation matrix of the input data in each subchannel, we also follow this path, though, so far, the equalizations have been non-blind, we use subband LMS equalization to equalize a channel blindly, but the problem arises about the selection of the best transformation. Here we show the advantage of wavelet transform and its related filter banks over other transforms and discuss FSE category of blind equalization using wavelet filter banks.

It is well known that second-order statistics of the received signal sequence provide information on the magnitude of the channel characteristics, but not on the phase. However this statement is not correct if the autocorrelation function of the received signal is periodic, as is the case for a digitally modulated signal. This cyclostationarity property of the received signal forms the basis for fractionally spaced equalizers. In this paper, we address the equalization problem by a subband approach to reduce computational complexity and to improve convergence speed. We discuss, why amongst other possibilities of subband processing the oversampled approach is particularly appealing to significantly reduce computational complexity and improve convergence speed.

The paper is organized as follows: In section 2, providing the problem formulation, the convergence behavior of conventional and subband adaptive LMS algorithms are discussed. In section 3 we focus on the selection of the best transformation and then in section 4, the new scheme for blind FSE is introduced and its performance is evaluated.

II. MATHEMATICAL FRAMEWORK

It is well known that the convergence behavior of conventional LMS-type adaptive filters depends on the eigenvalue spread of the underlying process, and hence to the flatness of the power spectral density of the underlying process, i.e. the rate of convergence of filter coefficients toward their optimum value, at a given frequency depends on the value of the power spectral density of the underlying process at that frequency relative to all other frequencies, but the convergence rate is generally slow due to a number of issues. First, the channel to be equalized may have considerable length, which in turn leads to a long equalizer impulse response. For example, some mobile communication channels are modelled as impulse responses of up to 20 μsec length, while the sampling period is on the orders of nanosecond on the time scale. Similarly, equalizers for digital subscriber lines (ADSL) may require filters of more than one hundred taps. Second, the potential presence of fractional delays and non-minimum phase parts makes the equalization process to be particularly hard. Besides, this can result in a large computational complexity, this will in general cause slow convergence for LMS-type adaptive algorithms. Finally, if the channel exhibits large spectral dynamics, a typical characteristics for non-stationary channels, the input signal to the filter will have a large eigenvalue spread and the convergence speed of LMS-type adaptive algorithms is further slowed down. Therefore we are motivated with the use of SAF. In an SAF system, both input and desired signals, are decomposed into decimated frequency bands. By operating adaptive filters independently in these subbands, both the update rate and the length of the adaptive filters can be greatly reduced leading to a lower computational complexity. Furthermore, the subband decomposition performs a whitening of the input signal, resulting in an improved convergence behavior.

Let's consider the subband equalization scheme as follows [2]: The sequence $x(n)$ denotes the input symbols to the equalizer that are passed through a pair of identical analysis filter banks to be partitioned into subbands and decimated to lower rate. This lower rate is $1/L_i$ of the full band rate. Thus the subband adaptive filters are running at this lower rate. To generate the adaptive filter output at the full band, the outputs from the subband filters are combined through a synthesis filter bank.

III. WAVELET FILTER BANKS

The selection of the proper filter bank is the subject we pursue here, such that we have a flatter power spectral density PSD in each subband after the transformation is fulfilled. Wavelet transform is closely relate to the tree structured digital filter banks [9]. This gives rise to nonuniform filter bandwidths and nonuniform decimation ratios in subbands. On the other hand, it is known that wavelet analysis filters are "constant-Q" filters; i.e., the ratio of the bandwidth to the center frequency of the band is constant, therefore, using wavelet transformation, the input signal is partitioned into finer bands in low frequencies where the energy of most communication signals is concentrated on. Hence by using such a transform it is more likely to obtain a flat spectrum in each subband.

Figure 1 shows the power spectral density of an AR(2) process, before splitting into subbands. We pass this process through a Daubechies-type wavelet filter banks and split it into three subbands. Figures 2 and 3 show the power spectral density of the result from two subbands. The autocorrelation function of the input signal is assumed to have an eigenvalue spread of 10. The capability of the wavelet transform to gain flatter PSD in each subband is apparent from these figures.

IV. BLIND FSE USING SUBBAND FILTER BANKS

Now, we focus on the blind FSE [8] using subband filter banks. As indicated in [5], blind LMS-type FSE suffers from a relatively high mean square error and low convergence rate. But we will show that by using our method a significant improvement in both convergence rate and MSE will be achieved. Here, we assume that the sequence $W(n)$ denotes the information symbols; T denotes the symbol duration, $h_c(t)$ is the "composite" channel, $n_c(t)$ is the additive noise and P is an integer denoting the amount of oversampling. Passing the received signal through a wavelet filter bank is equivalent to taking wavelet transform of the channel input and then passing the result from a set of subchannels, which are filtered and decimated versions of the composite channel impulse response (figure 4). But it's obvious that these subchannels have significantly shorter impulse responses. Therefore we can use equalizers of shorter length in each subchannel. We show the received signal of each subchannel with $x_j(t)$.

If $x_j(t)$ is sampled at the rate of $t = nT_s/P$, the received data is expressed as:

$$y_j(n) = \sum_\ell W_j(\ell) h_{cj}\left(\frac{nT_s}{P} - \ell T_s\right) + n_{cj}\left(\frac{nT_s}{P}\right) \quad (1)$$

It is convenient to write the input-output relationship as an equivalent discrete-time system

$$y_j(m) = \sum_{\ell=-\infty}^{\infty} W_j(\ell) h_j(m - \ell P) + n_j(m) \quad (2)$$

It is assumed that the $h_j(n)$ is an FIR filter of order L_{hj}. In many scenarios, the periodically correlated signals are conveniently represented by vector stationary processes; i.e., $y_{ji}(m) = y_j(mp - i)$, hence, (2) is expressed as

$$y_{ji}(m) = \sum_{\ell=-\infty}^{\infty} W_j(\ell) h_{ji}(m - \ell) + n_{ji}(m), \quad (3)$$

where we use the notations, $y_{ji}(m) = y(mP - i)$, $n_{ji}(m) = n(mP - i)$, and $h_{ji}(m) = h(mP - i)$ for $i = 0, \cdots, P - 1$. We can rewrite the input-output equation in matrix notation

$$Y_{jN}(n) = H_j W_{j,N+L_{hj}}(n) + N_{jN}(n), \quad (4)$$

with the following notations

$$\begin{aligned} Y_{jN}(n) &= [y_j(nP), \cdots, y_j(nP - P + 1), \\ &\quad y_j((n-1)P), \cdots, y_j((n-N)P + 1)]^T \\ W_{j,N+Ln}(n) &= [W_j(n), \cdots, W_j(n - N - L_h + 1)]^T \end{aligned}$$

And H_j is the $NP \times (N + L_{hj})$ block Toeplitz matrix

$$H_j = \begin{bmatrix} h_j(0) & h_j(1) & \cdots & h_j(L_h) & \cdots & 0 \\ 0 & h_j(0) & \cdots & \ddots & \ddots & \vdots \\ 0 & 0 & \ddots & \vdots & \ddots & \vdots \\ \vdots & \ddots & \ddots & \vdots & \vdots & \vdots \\ 0 & 0 & \cdots & \cdots & h_j(L_h - 1) & h_j(L_h) \end{bmatrix}$$

Our goal is to find each equalizer coefficients (figure 5) such that the cost function defined as below is minimized.

$$J_j = E\left\{ \left| \hat{W}_j - W_j(n) \right|^2 \right\}, \quad (5)$$

and assuming that the equalizer is FIR filter of order L_{gj} then the equalized signal is computed by

$$\hat{W}_j(n) = \sum_{i=0}^{P-1} \sum_{\ell=0}^{L_{gj}} g_{ji}(\ell) y_{ji}(n-\ell). \quad (6)$$

The cyclic LMS blind equalizer updates the equalizer's estimate at each symbol through the applications of the gradient descent algorithm, i.e.,

$$\hat{g}_j(T) = \hat{g}_j(T-1) - 0.5\,\mu\,\nabla \hat{J}_j(T) \quad (7)$$

where $g(T)$ is the equalizer tap vector at time, and $\nabla \hat{J}(T)$ is the instantaneous approximation at time T to the gradient of the cost function. By taking the derivative with respect to the unknown equalizer coefficients and setting them to zero and performing some simplifications, we have [5]:

$$\nabla \hat{J}_j(T) = Y_{j,N}(T) Y'_{j,N}(T) g_{j,0}(T) - \sigma^2_{j,w} H_j(:,1) \quad (8)$$

Figures 6 and 7 show an example of the convergence rate in subbands. Here the equalization is performed in three filtered and decimated versions of the input signal, i.e. the input signal is decomposed into three subbands. so the number of samples is different in figures 6 and 7. These figures are achieved for the case of an ill-conditioned input with an eigenvalue spread of 10. We have chosen the Daubechies-2 wavelet filter banks. Then, we get the resultant MSE of figure 8. It is apparent that the new method shows an improvement in the convergence rate besides the decreasing of computational cost. It is noticeable to remind that the amount of decrease in the computational cost is proportional to the length of required equalizer and the amount of improvement in the convergence rate is proportional to the number of subbands which are selected in the decomposition of the input signal.

V. CONCLUSIONS

In this paper we proposed the wavelet filter banks to decrease the computational cost of blind adaptive equalization algorithms and improve the convergence rate of such equalizers and discussed its advantage over other filter banks. The discussion was performed on an important category of blind equalization; the cyclostationarity based approach that is used in blind FSE. The new method shows its capability in many cases; when the channel has considerable length, at the presence of fractional delays and non-minimum phase parts in the channel and when the received data is highly correlated.

REFERENCES

[1] B. Baykal, "Blind channel estimation via combining autocorrelation and phase estimation," *IEEE Trans. on Circuits and Syst.*, Vol. 51, NO. 6, pp. 1125-1131, June 2004.
[2] B. Farhang-Boroujeny, "Adaptive Filters Theory and Applications," New York: Wiley, 1998.

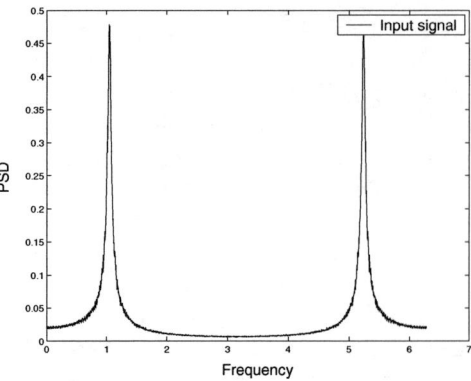

Fig. 1. power spectral density of an AR(2) process, before splitting into subbands.

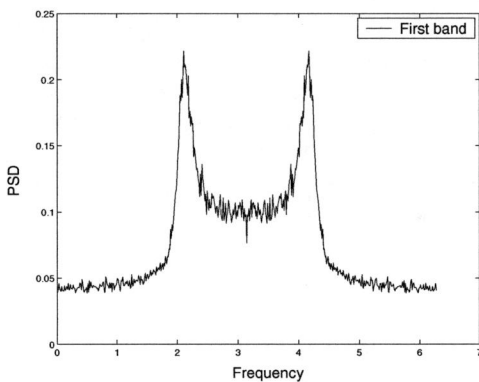

Fig. 2. PSD of the filtered and decimated input signal in the first subband.

[3] B. Rafaely, S. J. Elliott, "A computationally efficient frequency-domain LMS algorithm with constraints on the adaptive filter," *IEEE Trans. on Signal Processing*, Vol. 48, NO. 6, pp. 1649-1655, June 2000.
[4] D. Hatzinakos, C. L. Nikias, "Blind equalization using tricepstrum based algorithm," *IEEE Trans. on Comm.*, Vol. 39, pp. 669-682, May 1991.
[5] G. B. Giannakis, and S. Halford, "Blind fractionally-spaced equalization of noisy FIR channels: direct and adaptive solutions," IEEE Transactions on Signal Processing, vol. 45, pp. 2277-2292, September 1997.
[6] G. T. Zhou, M. Viberg, T. McKelvey. "A First-Order Statistical Method for Blind Channel Estimation" IEEE Signal Processing Letters, Dec 2002.
[7] H. Nguyen, B. C. Levy, "Blind and semi-blind equalization of CPM signals with the EMV algorithm," *IEEE Trans. on Signal Processing*, Vol. 51, NO. 10, pp. 2650-2664, Oct. 2003.
[8] J. K. Tugnait, "On the blind identifiability of multipath channels using fractional sampling and 2nd-order cyclostationary statistics," *IEEE Trans. on Info. Theory*, Vol. 41, pp. 308-311, Jan. 1995.
[9] P.P. Vaidyanathan, "Multirate Sustems and Filter Banks," Prentice-Hall, Inc., Englwood Cliffs, New Jersey.
[10] S. Weiss and R.W. Stewart, "On Adaptive Filtering in Oversampled Subbands." Shaker Verlag, Aachen, Germany, 1998.

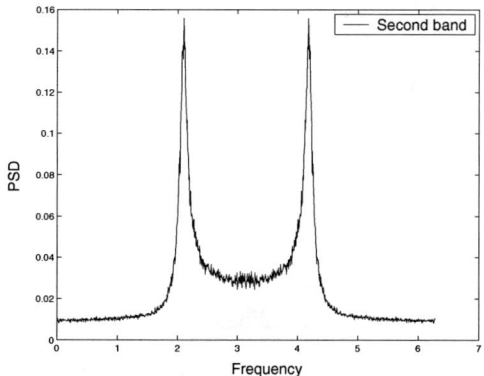

Fig. 3. PSD of the filtered and decimated input signal in the second subband.

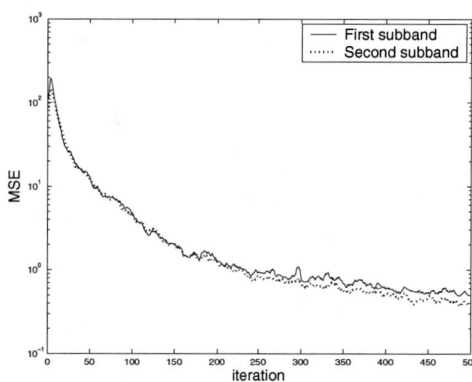

Fig. 6. MSE for subband blind FSE in the first and second subband.

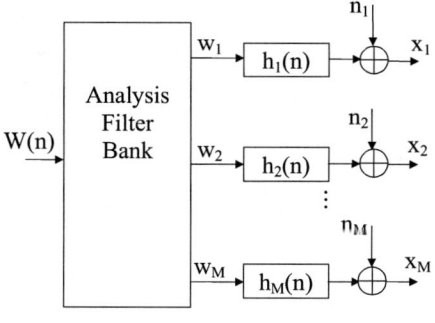

Fig. 4. Representation of the channel, when using subband equalization scheme.

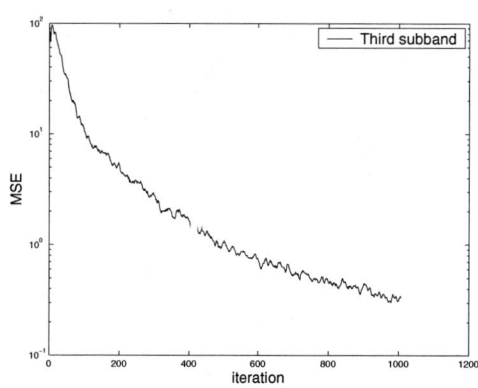

Fig. 7. MSE for subband blind FSE in the third subband.

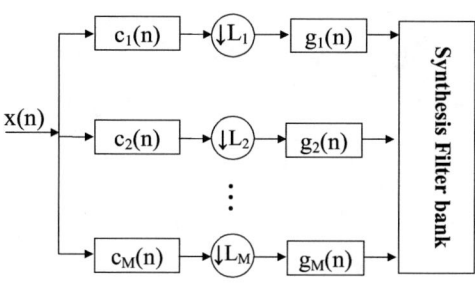

Fig. 5. Representation of the equalization scheme in each subband.

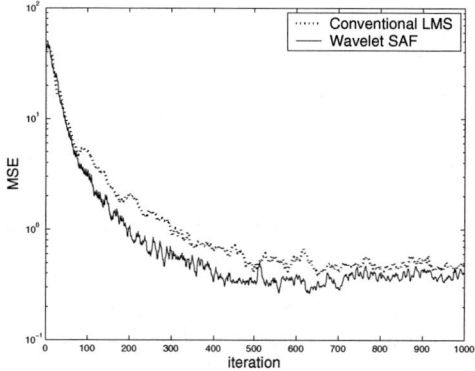

Fig. 8. Comparison of MSE for subband blind FSE and Conventional blind FSE.

Audio Source Separation by Source Localization with Hilbert Spectrum

Md. Khademul Islam Molla[1]
[1]Graduate School of Frontier Sciences
The University of Tokyo
7-3-1 Hongo, Bunkyo-ku, Tokyo 113-0033, Japan
Email: molla@gavo.t.u-tokyo.ac.jp

Keikichi Hirose[2], Nobuaki Minematsu[1]
[2]Graduate School of Information Science and Technology
The University of Tokyo
7-3-1 Hongo, Bunkyo-ku, Tokyo 113-0033, Japan
Email: {hirose, mine}@gavo.t.u-tokyo.ac.jp

Abstract—This paper presents a technique to separate the audio signals from their binaural mixtures based on localizing the sources in the space of interaural differences. Two interaural differences *ITD* (interaural time difference) and *ILD* (interaural level difference) are used as the principal cues to localize and segregate the sources. Hilbert spectrum is employed to decompose the mixture signals into time-frequency (*T-F*) space. The sources of the mixtures are considered as disjoint orthogonal in the *T-F* space. Hilbert spectrum has a better *T-F* resolution than Fourier based method and hence it produces a better disjoint orthogonality of the sources. The separation efficiency as presented in experimental results using our proposed algorithm is noticeable in this research area.

I. INTRODUCTION

The separation of mixed audio signals has many potential applications including robust speech recognition, music trascription, speaker separation from recorded meeting and video conferencing. The present research trend is to reduce the number of mixture signals. The researches on single mixture source separation [1,2] produce some application specific results. They have the limitations in robustness and it is very difficult to separate more than two sources using single mixture. The cocktail-party effect is a crucial situation for humanoid robotics to segregate and recognize a particular sound. In such situation human has the ability to keep the attention to a single audio source in an adverse acoustical condition. The location of the source helps human auditory to be separated from the interfering sounds. In multi-source listening situation, human exploits spatial characteristics of source signals by the mechanism of binaural hearing.

The human localization ability is simulated for speech segregation in [3,4]. They have only focused on the separation of one target source (speech). The location based separation is also applied in [5, 10] using *FFT* based time-frequency (*T-F*) representation cosidering the sources as disjoint orthogonal. Whereas *FFT* based method (*STFT* short-time Fourier transform) is only acceptable for disjoint orthogonality consideration of speech signal but not well suited for all types audio signals.

This paper presents a technique to detect, discriminate and separate individual audio source from their binaural mixtures using some spatial localization cues. In human audition, *ITD* and *ILD* are introduced between two ears' binaural signals. The *ITD* is the main localization cue at low frequency (<1.5kHz) and *ILD* dominates the high frequency range [3]. Measured head related transfer functions (*HRTFs*) introduce natural combination of *ITD* and *ILD* in binaural mixture. The empirical mode decompostion (*EMD*) together with Hilbert transform [2, 6] is used for T-F representation of the binaural mixture signals. The *T-F* is clustered in *ITD/ILD* space to localize the source signals. The sources properly localized in *ITD/ILD* space are segregated and recostructed using some reverse transformations. *EMD* based *T-F* represenation has better *T-F* resolution and hence more suitable for source disjoint orthogonality consideration.

The proposed method can blindly separate the sources (at fixed position) from the mixture of more than two sources. The separation model is described in scetion 2, section 3 presents the derivation of source disjoint orthogonality, some experimental results are presented in section 4 and finally section 5 contains some discussion and conclusions

II. PROPOSED MODEL DESCRIPTION

The schematic diagram of proposed binaural separation model is shown in Figure 1. It consists of two basic stages: (1) source localization in *ITD/ILD* space and (2) source separation and reconstruction. Each stage is described in detail in the following sub-sections. The inputs to the system are the monaural signals presented at different but fixed locations. The binaural signals are obtained by convoluting the monaural signals with measured *HRTFs* (from a KEMAR dummy head of MIT media laboratory under anechoic condition) corresponding to the direction of incidence [6]. The mixtures can be defined as the convolutive sum with *HRTFs*:

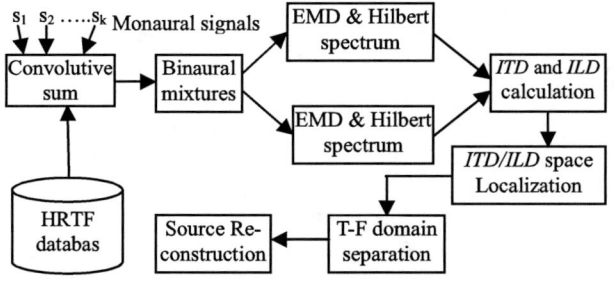

Figure 1. Schematic diagram of the proposed method

$$x_j(t) = \sum_k h_{jk} * s_k \qquad (1)$$

where j =l, r (denoting left and right respectively), s_k is the kth source and h_{jk} denotes the measured *HRTF* at the kth source position. We have proposed a localization based source separation method considering source disjoint orthogonality with Hilbert spectrum which is a fine resolution T-F representation.

A. Hilbert Spectrum

The principle of the *EMD* method is to decompose a time domain signal into a sum of oscillatory functions called intrinsic mode functions (*IMFs*) [?]. Each *IMF* satisfies two conditions: (i) in the whole data set the number of extrema and the number of zero crossing must be same or differ at most by one, (ii) at any point, the mean value of the envelope defined by the local maxima and the envelope defined by the local minima is zero. The first condition is similar to the narrow-band requirement for a stationary Gaussian process and the second condition adapts a global requirement to a local one, and is necessary to ensure that the instantaneous frequency will not have redundant fluctuations as induced by asymmetric waveforms [6]. Another way to explain how *EMD* works is that it extracts out the highest frequency oscillation that remains in the signal. Thus, locally, each *IMF* contains lower frequency oscillations than the one extracted just before. There exist many algorithmic approaches of *EMD* [7]. At the end of *EMD*, any signal $x_r(t)$ can be represented as:

$$x_r(t) = \sum_{i=1}^{n} imf_i(t) + r_n(t) \qquad (2)$$

where n is the number of *IMF*s and $r_n(t)$ is the final residue signal. Every *IMF* is a real valued signal. Analytic signal method is used to calculate the instantaneous frequency (*IF*) of the *IMF*s. The analytic signal corresponding to i^{th} *IMF* is defined as: $imf_i(t) + jH[imf_i(t)] = a_i(t)e^{j\theta_i(t)}$, where $H[]$ is the Hilbert transform operator, $a_i(t)$ and $\theta_i(t)$ are instantaneous amplitude and phase respectively. The instantaneous frequency $\omega_i(t)$ can easily be computed as the change of $\theta_i(t)$ with respect to time t as: $\omega_i(t) = d\theta_i(t)/dt$.

Hilbert spectrum $H(\omega,t)$ describes the joint distribution of signal amplitude as a function of frequency and time. To build $H(\omega,t)$, the *IF* of each *IMF* is first scaled according to the given frequency bins. Then for every $imf_i(t)$, if $\omega_i(t)$ is the corresponding *IF*, we represent the time-frequency plane as the triplet $\{t, \omega_i(t), a_i(t)\}$ where $a_i(t)$ is the amplitude of the analytic signal associated to $imf_i(t)$. $H(\omega,t)$ can produce the instantaneous (even at every sampling time) spectra of nonlinear and non-stationary signals.

B. Source Localization in ITD/ILD Space

Hilbert spectrum (*HS*) is a fine resolution time-frequency representation. If $H_L(\omega,t)$ and $H_R(\omega,t)$ are the Hilbert spectrum of binaural mixtures $x_l(t)$ and $x_r(t)$ respectively, the *ITD* and *ILD* can easily be computed with a simple division [5, 8] as:

$$\{ITD(\omega,t_f), ILD(\omega,t_f)\} = \left[\frac{1}{\omega}\angle \frac{H_L(\omega,t_f)}{H_R(\omega,t_f)}, 20\log\left|\frac{H_R(\omega,t_f)}{H_L(\omega,t_f)}\right|\right] \qquad (3)$$

The average relative energy and phase are calculated within the time frame t_f (1ms length with 0.5ms overlapping) yielding the *ITD* and *ILD*.

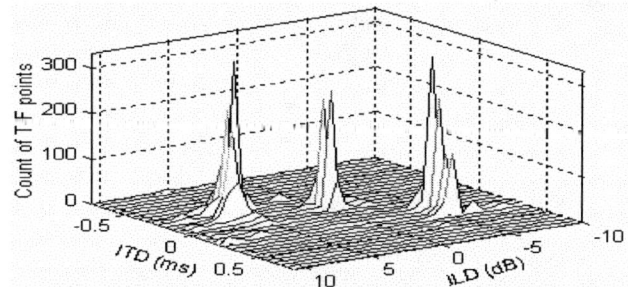

Figure 2. *ITD/ILD* Space Localization of three sources

The values of computed *ITD* and *ILD* are quantized into the discrete levels (50 levels) and the histogram h(*ITD*, *ILD*) is constructed by mapping T-F points into quantized *ITD/ILD* space. In h(*ITD*, *ILD*), we observe that each source is properly localized at specific region in *ITD/ILD* space. Figure 2 shows the *ITD/ILD* space localization of three sources (two speech signals and flute sound located at −40°, 30° and 0° azimuths respectively with all at 0° elevation). It is clearly noticeable that the strong peaks correspond to distinct active sources.

C. Source Separation and Re-Synthesis

Some further processing is necessary to smooth the histogram such that every source produces only one peak with some surrounding points. The individual source signal is separated by deriving corresponding masking function for T-F domain. A binary mask (by nullifying T-F points of interfering sources) is computed to collect the T-F points (from H_L or H_R) corresponding to each peak region (for every source) in joint *ITD-ILD* space. If $M_k(\omega,t)$ be the binary mask of the kth source, the *HS* of kth source $H_k(\omega,t) = M_k(\omega,t)H_L(\omega,t)$ or $H_k(\omega,t) = M_k(\omega,t)H_R(\omega,t)$.

During the Hilbert transform the real part of the signal remains unchanged. The time domain signal of kth source is reconstructed by filtering out the imaginary part from the *HS* and summing over frequency bins as [2]:

$$s_k(t) = \sum_{\omega} H_k(\omega,t) \cdot \cos[\phi(\omega,t)] \quad (4)$$

where $\phi(\omega,t)$ is the phase matrix of H_L (or H_R). The phase matrix is saved during the construction of Hilbert spectrum to be used in re-synthesis.

III. DISJOINT ORTHOGONALITY OF THE SOURCES

In order to better measure a signal at a particular time and frequency (ω,t), it is natural to desire that Δ_t and Δ_ω be as narrow as possible. In Fourier based *T-F* representation Δ_t and Δ_ω has to satisfy an uncertainty inequality $\Delta_t \Delta_\omega \geq \frac{1}{2}$ which is the trade-off of the selection of time-frequency resolution in *STFT*. The simple definition of disjoint orthogonality of audio sources says that not more than one source signal is active at the same time and with same frequency. This is a very hard definition to comply with the audio signals. Some assumption relaxes this definition as in [5, 10]. They have called two functions $f_1(t)$ and $f_2(t)$ as w-disjoint orthogonal if for a given window function $w(t)$, the supports of the windowed Fourier transforms with $w(t)$ of $f_1(t)$ and $f_2(t)$ are disjoint. If $F_1(\omega,t)$ and $F_2(\omega,t)$ are the windowed Fourier transform of the signals $f_1(t)$ and $f_2(t)$, the W-disjoint Orthogonality assumption can be stated as: $F_1(\omega,t)F_2(\omega,t) = 0; \forall \omega,t$.

We are not considering the window function to measure the disjoint orthogonality (*DO*) here as no window function is required in computing the Hilbert spectrum. Hence we are simply calling it disjoint orthogonality by dropping the *w* term. The signal to interference ratio (*SIR*) is used as basis to measure the *DO*. The SIR for the jth source signal is,

$$SIR_j = \sum_{\omega} \sum_{t} \frac{X_j(\omega,t)}{Y_j(\omega,t)}; Y_j(\omega,t) \neq 0$$

$$Y_j(\omega,t) = \sum_{\substack{i=1 \\ i \neq j}}^{N} X_i(\omega,t) \quad (5)$$

where *N* is the number of audio signal considered to be disjoint orthogonal, $X_j(\omega,t)$ is the *T-F* representation (using *STFT* or Hilbert spectrum) of the jth signal. The dimension of *T-F* representation using *STFT* and *HS* may be different, hence the *DO* is defined as the percentage over the whole *T-F* region. It is achieved by dividing the SIR_j with the total number of *T-F* points used to calculate SIR_j. Finally the average disjoint orthogonality (*ADO*) is the average of all *SIR*s of individual signal as: $ADO = \frac{1}{N}\sum_{j=1}^{N} SIR_j$.

The same process is applied to measure *ADO* (between 0 to 1) for *STFT* and *HS* based *T-F* representation of the experimental audio signals. We have presented some experimental results to compare *STFT* and Hilbert spectrum as the *T-F* representation tools of audio signals in terms of source disjoint orthogonality.

IV. EXPERIMENTAL RESULTS

We have used the binaural mixtures of three audio streams of two male speech (sp1 and sp2) and flute sound (ft) to test the efficiency of our algorithm. Each monaural recording is upsampled to 44.1 kHz to match with the *HRTF*. The binaural mixtures are obtained using equation (1) and then down-sampled to 16kHz. Placing the sound sources at various locations (azimuth, elevation) produces different binaural mixtures for the test purpose. Such three mixtures are produced as: m1{sp1(-40°, 0°), sp2(30°, 0°), ft(0°, 0°)}, m2{sp1(20°, 10°), sp2(0°, 10°), ft(-10°,10°)}, m3{sp1(40°, 20°), sp2(30°, 20°), ft(-20°, 20°)} and the separation result is presented bellow. The origin (0° azimuth, 0° elevation) is considered at the front of the listener.

The average value of short time energy ratio between original and separated signal is used here as the criterion to measure the separation efficiency. It is termed as *OSSR* (original to separated signal ratio) [2] and defined by

$$OSSR = \frac{1}{T}\sum_{t=1}^{T} \log 10 \left(\frac{\sum_{i=1}^{w} s_{original}^2(t+i)}{\sum_{i=1}^{w} s_{separated}^2(t+i)} \right) \quad (6)$$

where $s_{original}$ and $s_{separated}$ are the original and separated signal respectively, *w* is frame length (10 ms) and *T* is the number of frames. In the case for zero energy in a particular window, no *OSSR* measurement is performed. If the two signals are similar, *OSSR=0* and any other value (positive or negative) is a measure of their dissimilarity. Table 1 shows the average *OSSR* of each signal for every mixture. Smaller deviation of *OSSR* from 0 indicates the higher degree of separation. It also presents comparative experimental results of separation using *HS* and *STFT* though *STFT* depends on many factors.

Table 1: The experimental results of audio source separation using *HS* (with 257 frequency bins) and *STFT* (30ms Hamming window, 20ms overlapping and 512 point *FFT*).

Mixtures	T-F	OSSR of sp1	OSSR of sp2	OSSR of ft
m1	HS	-0.0271	0.0213	0.0264
m1	STFT	0.0621	-0.0721	-0.0531
m2	HS	0.0211	-0.0851	-0.0872
m2	STFT	0.0824	0.1202	0.1182
m3	HS	0.0941	-0.0832	0.0225
m3	STFT	-0.1261	0.1092	-0.0821

The separation efficiency depends only on the apart angle between the sources locations but not on the signal contents.

The separation accuracy is better for larger apart angle between the sources. It is suggested to keep apart angle among the locations of the sources not less than 10° (both azimuth and elevation).

Each one of the three audio signals is converted to *T-F* space using *HS* and *STFT* separately to produce some experimental results of *DO*. Figure 3 shows *ADO* of *HS* and *STFT* (using Hamming and Hanning window with 60% overlapping) as a function of the number of frequency bins. Figure 4 presents the comparison between *HS* and *STFT* as a function of window overlapping.

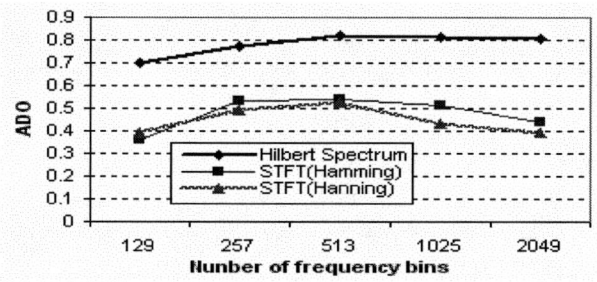

Figure 3. ADO of HS and STFT as a function of frequency bins

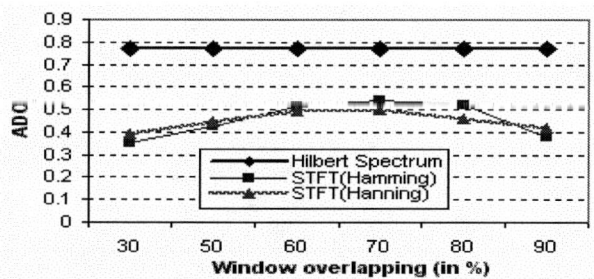

Figure 4. ADO of HS and STFT as a function of window overlapping

The variation of *ADO* of *HS* is very small (only for the number of frequency bins), whereas many factors affect the *ADO* of *STFT*. It is also a crucial decision to determine the suitable parameters for better source separation with *ADO* consideration. Hence the *ADO* of the audio signals of *HS* is better than that of *STFT* based *T-F* representation. It is obvious to produce better source separation efficiency by the proposed method with *HS* as the *T-F* representation.

V. DISCUSSION AND CONCLUSIONS

We have proposed an audio source separation technique by localization of the sources in the domain of binaural cues. *HRTF* is employed to introduce the binaural cues of human auditory system. It is considered that the sources are disjoint orthogonal [5, 10] in their *T-F* domain. The better the *DO* of the sources produces better separation on the basis of *DO* consideration. Hilbert spectrum is used for *T-F* representation of the binaural mixture signals. In [5] it is argued that the source disjointness depends on many factors including window size, window type, number of overlapping samples and also for the number of *FFT* points to produce the spectrogram. Hilbert spectrum is not affected by any of the mentioned factors as presented in experimental results. Another potential issue of the improvement of *DO* of *HS* is that some crossed terms are introduced in *STFT* with window overlapping and *HS* is free of such scenario. In [3] the authors have used the binaural cues to estimate the ideal binary mask to separate speech signal from interfering noise. The source with higher contribution at any *T-F* point is considered as the target signal (speech). They proposed a training based energy ratio function measuring the relative strength between the target source and the acoustic interference at each T-F point. The individual filter output of the filterbank (with 128 of gammatone filters) is divided into 20ms time frame with 10ms overlap that correspond the T-F unit. This consideration is hard to be used for separating more signals individually. In our system *ITD/ILD* space localization is used to separate more than two sources. Being independent of signal content the localization cues can be used to segregate sequence of voiced and unvoiced components originating from the same location. The separation performance does not depend on the spectral nature of the target sources.

The specialty of the Hilbert spectrum is that the time resolution can be as precise as the sampling period and the frequency resolution depends on the choice (it should not be the power of 2 as in Fourier method) up to Nyquist frequency. Hence it serves as the potential *T-F* representation for the consideration of source disjoint orthogonality. The robust analysis of disjoint orthogonality of various audio sources and the separation of moving sources are the main concern as the future works of this research.

REFERENCES

[1] Sam T. Roweis, "One Microphone Source Separation", Neural Information Processing Systems, pp. 793-799, 2000.

[2] Md. Khademul Islam Molla, Keikichi Hirose, Nobuaki Minematsu, "Audio Source Separation from the Mixture using Empirical Mode Decomposition with Independent Subspace Analysis", To appear in the Proc. of ICSLP2004.

[3] Nicoleta Roman, Deliang Wang, Guy J. Brown, " Speech segregation based on sound localization", Acos. Soc. of America, 114(4): 2236-2252, 2003

[4] Johannes Nix, Volker Hohmann, "Enhancing sound sources by use of binaural spatial cues", Workshop on Consistent & Reliable Acoustic Cues (CRAC'01), 2001.

[5] Matthias Baeck, Udo Zolzer, "Real-Time Implementation of Source Separation Algorithm", DAFx-03, London, UK, 2003.

[6] N. E. Huang etl., "The empirical mode decomposition and Hilbert spectrum for nonlinear and non-stationary time series analysis", Proc. Roy. Soc. London A, Vol.454: 903-995, 1998.

[7] P. Flandrin, G. Rilling, P. Goncalves, "Emperical Mode Decomposition as a filter bank", IEEE Sig. Proc. Letter, (in press), 2003.

[8] S. Srinivasan, N. Roman, D. Wang, "On binary and ratio time-frequency masks for robust speech recognition", To appear in the Proc. of ICSLP2004.

[9] http://sound.media.mit.edu/KEMAR.html

[10] Scott Rickard, Ozgur Yilmaz, "On the approximate W-disjoint orthogonality of speech", ICASSP2002, Orlando, Florida, USA, 2002

Uplink Channel Estimation for Space-Time Block Coded Multiple-Input Multiple-Output MC-CDMA Systems

Ke Deng, Qinye Yin, Hongbo Tian

Institute of Information Engineering, School of Electronics and Information Engineering
Xi'an Jiaotong University, Xi'an 710049, China
E-mail: denke@mail.china.com, qyyin@xjtu.edu.cn, jhk@mailst.xjtu.edu.cn

Abstract—We propose a space-time block coded (STBC) multicarrier code division multiple access (MC-CDMA) scheme, where multiple antennas are distributed arbitrarily at the base station. Unlike other blind STBC-MC-CDMA schemes, ours does not decrease the maximum number of available users after linearly precoding at the transmitter. By exploiting the eigen value decomposition (EVD) and the redundant structure of the STBC, we estimate the uplink vector channel impulse response (VCIR) blindly. Computer simulation results demonstrate the validity and effectiveness of the proposed VCIR estimation algorithm.

I. Introduction

The space-time coding (STC) is an effective coding technique that uses transmit diversity to combat the detrimental effects in wireless fading channels by combining the signal processing at the receiver with coding techniques appropriate to multiple transmit antennas to achieve higher data rates [1]. Among various space-time codec schemes, the space-time block code (STBC) [2], especially the Alamouti's scheme [3], has the remarkably low decode complexity and capability of achieving the channel capacity.

Till now, there are some researches on the multiuser scenario. [4] applied the STBC into the multicarrier code division multiple access (MC-CDMA) system, where 1 user occupies M spreading code (M is the number of the transmit antennas of 1 user). Consequently, the maximum number of available users decreases M times after using multiple-transmit antennas.

A STBC-MC-CDMA scheme, with multiple antennas distributed arbitrarily at the base station, is proposed. Each user only occupies 1 spreading code, therefore, the maximum number of available users, unlike [4], does not decrease. By exploiting the eigen value decomposition (EVD) and the redundant structure of the STBC, we obtain a closed-form solution for the uplink vector channel impluse response (VCIR) estimation. Computer simulations illustrate both the validity and the performance of the proposed algorithm.

II. System model

A. Time-Domain Model of MC-CDMA System with Single Antenna

It is assumed that K users are randomly distributed around a cell site. The number of subcarriers equals to the length of spreading code G. All active users share the same set of subcarriers. The MC-CDMA scheme does the spreading spectrum (SS) operation in the frequency domain [5], and the spreading code vector for the k-th user is $\tilde{\mathbf{c}}_k (k = 1, \ldots, K)$ with length G.

The MC-CDMA signal can also be interpreted as a DS-CDMA one, where the time-domain spreading code vector $\mathbf{c}_k$ is the inverse discrete Fourier transform (IDFT) of $\tilde{\mathbf{c}}_k$. Since the IDFT is an isometric transform, the IDFT-transformed time-domain spreading code $\mathbf{c}_k$ is also orthogonal.

By exploiting the cyclic prefix (CP), we can convert the time-domain linear convolution of the finite impulse response (FIR) to a cyclic convolution. Consequently, the so-called time-domain signature waveform describing the symbol level transmission property is the cyclic convolution of the channel FIR and the IDFT-transformed spreading code.

For clarity, the FIR vector for user k is described as $\mathbf{h}_k = [h_k(0) \ \ldots \ h_k(L)]^T$. The corresponding matrix-vector representation of the cyclic convolution is

$$\mathbf{w}_k = \mathbf{C}_k \mathbf{h}_k \quad (1)$$

where $\mathbf{w}_k$ is the time-domain signature waveform vector for user k with length G, $G \times (L+1)$ $\mathbf{C}_k$ is given by

$$\begin{bmatrix} c_k(0) & c_k(G-1) & \ldots & c_k(G-L) \\ c_k(1) & c_k(0) & \ldots & \vdots \\ \vdots & \ddots & \ddots & c_k(G-1) \\ \vdots & & \ddots & c_k(0) \\ c_k(G-2) & \vdots & \ddots & \vdots \\ c_k(G-1) & c_k(G-2) & \ldots & c_k(G-L-1) \end{bmatrix} \quad (2)$$

Accordingly, at the base station of the MC-CDMA system, the received data vector during the n-th symbol from the k-th user is

$$\mathbf{x}_k(n) = \mathbf{w}_k b_k(n) \quad (3)$$

where $b_k(n)$ represents the n-th transmit symbol from the user k; the received data vector $\mathbf{x}_k(n)$ with length G excludes the samples corresponding to CP.

B. Equivalent Spatial-Temporal Model for STBC-MC-CDMA System

We use an Alamouti-like scheme at the mobile station because of its simple structure and remarkable performance.

Partially supported by the National Natural Science Foundation (No. 60272071) and the Research Fund for Doctoral Program of Higher Education (No. 20020698024) of China.

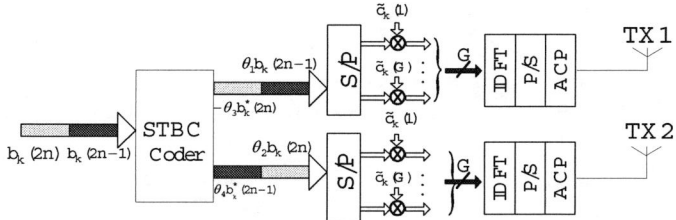

Fig. 1. The discrete baseband model of user k

The discrete baseband model of proposed scheme is illustrated in Fig. 1, where each user has two transmit antennas. $\theta_i, (i = 1, \ldots, 4)$ are the coefficients in the precoding, which are constant and can ensure the effectiveness of the blind uplink channel estimation. The number of the antennas at the base station is M, the FIR channel vector from antenna m of the base station to the antenna i of user k is denoted by $\mathbf{h}_{k,m}^i = [h_{k,m}^i(0), \ldots, h_{k,m}^i(L)]^T$. And all $\mathbf{h}_{k,m}^i$ are linearly independent because of the rich scattering environment.

Then, the signature waveform between the antenna i ($i=1,2$) of user k and the antenna m of the base station can be written as

$$\mathbf{w}_{k,m}^i = \mathbf{C}_k \mathbf{h}_{k,m}^i \qquad (4)$$

where $\mathbf{w}_{k,m}^i$ is with length G.

In the two successive symbol periods $2n-1, 2n$, the $2G \times 1$ recieved signal vector can be constructed without the noise

$$\begin{aligned}\mathbf{x}_{k,m}(n) &= \begin{bmatrix} \theta_1 \mathbf{w}_{k,m}^1 & \theta_2 \mathbf{w}_{k,m}^2 \\ \theta_4^* \mathbf{w}_{k,m}^{2*} & -\theta_3^* \mathbf{w}_{k,m}^{1*} \end{bmatrix} \begin{bmatrix} b_k(2n-1) \\ b_k(2n) \end{bmatrix} \\ &= \mathbf{W}_{k,m} \begin{bmatrix} b_k(2n-1) \\ b_k(2n) \end{bmatrix}\end{aligned} \qquad (5)$$

where $\mathbf{W}_{k,m}$ is with dimensions $2G \times 2$.

Concatenating received data vectors from antenna 1 to antenna M for user k, we can obtain an extended received data vector with length $2MG$

$$\mathbf{x}_k(n) = [\mathbf{x}_{k,1}^T(n) \quad \ldots \quad \mathbf{x}_{k,M}^T(n)]^T = \mathbf{W}_k \mathbf{b}_k(n) \qquad (6)$$

where $\mathbf{W}_k = [\mathbf{W}_{k,m}^T(1) \quad \ldots \quad \mathbf{W}_{k,m}^T(M)]^T$, with dimensions $2MG \times 2$, representing the uplink spatial-temporal signature waveform matrix of user k; $\mathbf{b}_k(n) = [b_k(2n-1) \quad b_k(2n)]^T$.

Considering all K users, the extend received data vector in the multiuser scenario can be expressed as

$$\mathbf{x}(n) = \sum_{k=1}^{K} \mathbf{x}_k(n) = \sum_{k=1}^{K} \mathbf{W}_k \mathbf{b}_k(n) = \mathbf{W}\mathbf{b}(n) \qquad (7)$$

where W with dimensions $2MG \times 2K$ is the whole uplink spatial-temporal signature waveform matrix, defined as $\mathbf{W} = [\mathbf{W}_1 \quad \ldots \quad \mathbf{W}_K]$. $2K \times 1$ matrix $\mathbf{b}(n) = [\mathbf{b}_1^T(n) \quad \ldots \mathbf{b}_K^T(n)]^T$ represents the transmitted symbols from all users.

Considering the thermal noises, (7) can be rewritten as

$$\mathbf{y}(n) = \mathbf{W}\mathbf{b}(n) + \mathbf{v}(n) \qquad (8)$$

where every element of the noise vector v is the independent identically distributed (i.i.d.) complex zero-mean Gaussian noise with variance σ_v^2.

III. VCIR estimation

Based on the aforementioned equivalent space-time channel model, we present a VCIR estimation approach for the uplink STBC-MC-CDMA system with multiple antennas distributed arbitrarily at the base station.

A. Estimation of the VCIR

Let $\mathbf{R}_y$ be the covariance matrix of $\mathbf{y}(n)$. We have

$$\mathbf{R}_y \triangleq E\{\mathbf{y}(n)\mathbf{y}^H(n)\} = \mathbf{W}\mathbf{W}^H + \sigma_v^2 \mathbf{I} \qquad (9)$$

where we assumed that the user symbols $\mathbf{b}(n)$ are i.i.d. and drawn from a unit-energy constellation.

It is obviously that W has full column rank unless at least one channel is trivial, i.e., $\mathbf{h}_{k,m}^i = \mathbf{0}$. Therefore, the eigenvalue decomposition (EVD) of $\mathbf{R}_y$ is given by

$$\mathbf{R}_y = \mathbf{U}_s \mathbf{\Sigma}_s \mathbf{U}_s^H + \sigma_v^2 \mathbf{U}_o \mathbf{U}_o^H \qquad (10)$$

where column vectors of $\mathbf{U}_s$ are associated with $2K$ nonzero largest eigen values and span the signal subspace defined by columns of W, while column vectors of $\mathbf{U}_o$ are associated with smallest eigen values and span the orthogonal complement subspace of the aforementioned signal subspace [6]. Therefore, the following equation holds, i.e.

$$\mathbf{U}_o \perp \mathbf{W} = 0 \Rightarrow \mathbf{U}_o^H \mathbf{W}_k = 0 \qquad k = 1, 2, \ldots, K \qquad (11)$$

$\mathbf{U}_o^H$ can be divided into $2M$ matrices by column, where every "slice" is with dimensions $(2MG - 2K) \times G$. With the definition $\mathbf{U}_o^H = [\mathbf{U}_1 \quad \mathbf{U}_2 \quad \ldots \quad \mathbf{U}_{2M}]$, (11) can be rewritten as

$$[\mathbf{U}_1 \quad \mathbf{U}_2 \quad \ldots \quad \mathbf{U}_{2M}] \begin{bmatrix} \theta_1 \mathbf{w}_{k,1}^1 & \theta_2 \mathbf{w}_{k,1}^2 \\ \theta_4^* \mathbf{w}_{k,1}^{2*} & -\theta_3^* \mathbf{w}_{k,1}^{1*} \\ \vdots & \vdots \\ \theta_1 \mathbf{w}_{k,M}^1 & \theta_2 \mathbf{w}_{k,M}^2 \\ \theta_4^* \mathbf{w}_{k,M}^{2*} & -\theta_3^* \mathbf{w}_{k,M}^{1*} \end{bmatrix} = 0 \qquad (12)$$

(12) is equivalent to the following (13)

$$\begin{bmatrix} \theta_1 \mathbf{U}_1, \theta_4^* \mathbf{U}_2, \ldots, \theta_1 \mathbf{U}_{2M-1}, \theta_4^* \mathbf{U}_{2M} \\ -\theta_3^* \mathbf{U}_2^*, \theta_2^* \mathbf{U}_1^*, \ldots, -\theta_3^* \mathbf{U}_{2M}^*, \theta_2^* \mathbf{U}_{2M-1}^* \end{bmatrix} \begin{bmatrix} \mathbf{w}_{k,1}^1 \\ \mathbf{w}_{k,1}^{2*} \\ \vdots \\ \mathbf{w}_{k,M}^1 \\ \mathbf{w}_{k,M}^{2*} \end{bmatrix} = 0 \qquad (13)$$

From 2.1 we have $[(\mathbf{w}_{k,1}^1)^T, (\mathbf{w}_{k,1}^{2*})^T, \ldots, (\mathbf{w}_{k,M}^1)^T, (\mathbf{w}_{k,M}^{2*})^T]^T = \mathbf{A}_k[(\mathbf{h}_{k,1}^1)^T, (\mathbf{h}_{k,1}^{2*})^T, \ldots, (\mathbf{h}_{k,M}^1)^T, (\mathbf{h}_{k,M}^{2*})^T]^T$, where $\mathbf{A}_k$ with dimensions $2MG \times 2MG$ is defined as $\mathbf{A}_k = \mathrm{diag}\{\mathbf{C}_k, \mathbf{C}_k^*, \ldots, \mathbf{C}_k, \mathbf{C}_k^*\}$. If $\widetilde{\mathbf{U}}$ is defined as the left matrix in (13) with dimension $4(MG - K) \times 2MG$, we have

$$\widetilde{\mathbf{U}} \mathbf{A}_k \left[(\mathbf{h}_{k,1}^1)^T \quad (\mathbf{h}_{k,1}^{2*})^T \quad \ldots \quad (\mathbf{h}_{k,M}^1)^T \quad (\mathbf{h}_{k,M}^{2*})^T \right]^T = 0 \qquad (14)$$

where $k = 1, 2, \ldots, K$.

When $4(MG - K) > 2M(L+1)$ holds, that is, $K < M[G - (L+1)/2]$, the above linear equation set is overdetermined. For different matrix $\mathbf{A}_k$, by solving with the least-squares (LS) fitting (14), we can estimate the corresponding VCIR $\hat{\mathbf{h}}_{k,m}^i$ for $k = 1, \ldots, K$, respectively.

When $K \geq M[G - (L+1)/2]$, the linear equation set (14) becomes underdetermined. Therefore, the VCIR $\mathbf{h}_{k,m}^i$ ($k = 1, \ldots, K$) is unable to be estimated uniquely.

B. Ambiguous Complex Coefficient Identification and Multiuser Detection

Because of the usage of the subspace method, an uncertain complex coefficient exists between the estimated VCIR and the original one. We further use the finite alphabet property of transmitted symbols to estimate the ambiguous complex coefficient so that the original VCIR is obtained. The matrix form of (8) is

$$\mathbf{Y} = \hat{\mathbf{W}}\mathbf{\Gamma}\mathbf{B} + \mathbf{N} \quad (15)$$

where $\hat{\mathbf{W}}$ denotes the estimated signature waveform matrix; the ambiguous complex coefficient matrix $\mathbf{\Gamma}$ with dimensions $K \times K$ is described as $\mathbf{\Gamma} = |\mathbf{\Gamma}|e^{j\mathbf{\Phi}}$; $\mathbf{Y} = [\mathbf{Y}(1), \ldots, \mathbf{Y}(N)]$ and $\mathbf{B} = [\mathbf{b}(1), \ldots, \mathbf{b}(N)]$.

We identify the modulus matrix $|\mathbf{\Gamma}|$ and the phase angle matrix $e^{j\mathbf{\Phi}}$ as in [5], that is

$$|\mathbf{\Gamma}| = \sqrt{\hat{\mathbf{W}}^+ \cdot (\mathbf{R}_{yy} - \sigma_n^2 \mathbf{I}) \cdot (\hat{\mathbf{W}}^+)^H} \quad (16)$$

$$e^{j\mathbf{\Phi}} = \sqrt{\mathrm{diag}(\mathbf{z})} \quad (17)$$

where $(\cdot)^+$ represents the Penrose-Moore pseudo-inverse; the vector $\mathbf{z}$ is the column-wise average of matrix $[(\hat{\mathbf{W}}|\mathbf{\Gamma}|)^+\mathbf{Y} \circ (\hat{\mathbf{W}}|\mathbf{\Gamma}|)^+\mathbf{Y}]$ (($\circ$) denotes the Hadamard product).

Using the estimated ambiguity coefficient matrix $\mathbf{\Gamma}$, we obtain the spatial-temporal signature waveform matrix as $\hat{\mathbf{W}}\mathbf{\Gamma}$.

Based on the estimated equivalent spatial-temporal signature waveform, the equivalent spatial-temporal minimum mean square error (MMSE) multiuser detector is defined as

$$\hat{\mathbf{B}} = ((\hat{\mathbf{W}}\mathbf{\Gamma})(\hat{\mathbf{W}}\mathbf{\Gamma})^H + \sigma^2\mathbf{I})^{-1}(\hat{\mathbf{W}}\mathbf{\Gamma})\mathbf{Y} \quad (18)$$

where $\hat{\mathbf{B}}$ is the estimated matrix for the transmitted symbol matrix B.

C. Choice of θ_i

Although the closed-form solution of all $\mathbf{h}_{k,m}^i$ is given, they are not uniquely determined when θ_i are arbitrarily chosen. For instance, we choose $\theta_i = 1$, which is the original Alamouti's scheme. The following equation holds

$$\begin{bmatrix} \mathbf{U}_1, \mathbf{U}_2, \ldots, \mathbf{U}_{2M-1}, \mathbf{U}_{2M} \\ -\mathbf{U}_2^*, \mathbf{U}_1^*, \ldots, -\mathbf{U}_{2M}^*, \mathbf{U}_{2M-1}^* \end{bmatrix} \begin{bmatrix} \mathbf{w}_{k,1}^1 & \mathbf{w}_{k,1}^2 \\ \mathbf{w}_{k,1}^{2*} & -\mathbf{w}_{k,1}^{1*} \\ \vdots & \vdots \\ \mathbf{w}_{k,M}^1 & \mathbf{w}_{k,M}^2 \\ \mathbf{w}_{k,M}^{2*} & -\mathbf{w}_{k,M}^{1*} \end{bmatrix} = 0 \quad (19)$$

(19) shows that the dimension of the solution space of (14) is 2 for each user, therefore the uplink VCIR can not be determined under this condition. In order to guarantee the estimation of the VCIR, we have the following theorem.

Theorem 1: the dimension of the solution space of (14) is 1 if $\frac{\theta_2 \theta_3}{\theta_1 \theta_4}$ is not a real number.

The proof is in the Appendix.

The joint spatial-temporal multiuser detector can be summarized as follows

1) Find the proper θ_i according to theorem 1.
2) Perform EVD operation on $\mathbf{R}_y$, and then obtain $\widetilde{\mathbf{U}}$ from (13).
3) Construct $\mathbf{A}_k$ for each active user, and estimate the VCIR by (13) and (14).
4) Estimate $|\mathbf{\Gamma}|$ and $e^{j\mathbf{\Phi}}$ by (16) and (17).
5) Estimate the transmitted symbols $\hat{\mathbf{B}}$ by (18).

IV. Simulation Results

A set of computer simulation results are presented to illustrate the advantages of the proposed VCIR estimation scheme for the STBC-MC-CDMA system with multiple arbitrary antennas. We use the differential binary phase shift keying (DBPSK) modulation mode in our simulations.

In simulations, Gold codes with length $G = 31$ are assigned to different users. 200 Monte-Carlo trials are performed for each simulation. We use samples within 70 MC-CDMA symbols to estimate uplink channels for all active users. Moreover, we assume a rich scattering environment and the FIR coefficients are i.i.d. complex Gaussian random variables with zero-mean and variance $1/L$, where L is the length of the wireless channel and $L = 3$. θ_i are randomly chosen, but what do not satisfy theorem 1 are rejected. The signal-to-noise ratio (SNR) is defined as $10 \log 2/\sigma_v^2$. The performance is evaluated by the signal to bit error ratio (BER) in the symbol level.

Fig. 2 and Fig. 3 show the BER versus the SNR and the BER versus the numbers of users for the equivalent space-time MMSE multiuser detector as in (18), respectively. In the case of Fig. 2, the number of active users is fixed to 16. In the case of Fig. 3, the SNR is fixed to 0dB.

As shown in Fig. 2, three curves are associated with 1 antenna, 2 or 4 antennas, respectively. Clearly, as the number of antennas or the SNR increases, the performance increases significantly.

As shown in Fig. 3, three curves are associated with 1 antenna, 2 or 4 antennas, respectively. Clearly, as the number of users increases, the performance slightly decreases. Moreover, in contrast, the performance of the scheme proposed in [4] is also shown in Fig. 3. With the simulation, we observe that the performance of our multiuser detection is close to what proposed in [4], but its maximum active users is 14 only.

V. Conclusion

After interpreting the MC-CDMA system as a DS-CDMA system with specific spreading codes, we obtain the equivalent spatial-temporal system model of the proposed STBC-MC-CDMA scheme with arbitrary multiple antennas. Based on this model, we address an algorithm for blind estimation of the uplink VCIR.

Superior to the MC-CDMA system with one transmit and one recieve antenna, our scheme is more robust again noise and well mitigates the multiple access interference (MAI) in multiuser scenarios. Furthermore, the maximum number of the active users is significantly increased. The combination of the STBC, MC-CDMA and multiple-transmit multiple-receive antennas, can meet the needs of the high speed and huge capacity in the next generation (NextG) wireless communication systems.

References

[1] G. J. Foschini, Jr. And M. J. Gans, "On limits of wireless communication in a fading environment when using multiple antennas," Wireless Personal Commun., vol. 6, pp. 311-335, Mar. 1998.
[2] Vahid Tarokh, Hamid Jafarhani, and A. R. Calderbank, "Space-Time Block Codes from Orthogonal Designs," IEEE Trans. Inform. Theory, vol. 45, pp. 1456-1467, July 1999.
[3] S. M. Alamouti, "A simple transmitter diversity scheme for wireless communications," IEEE J. Select. Areas Commun., vol. 16, pp. 1451-1458, Oct. 1998.
[4] Hongbin Li, "Semi-blind multiuser receiver for space-time coded CDMA systems," Proc. IEEE ICASSP'2002, pp 13-17, May 2002
[5] Xiaojun Wu, Qinye Yin, Aigang Feng, Ke Deng, "Equivalently Blind Time-Domain Channel Estimation for MC-CDMA System over Frequency-Selective Fading Channels in Multiuser Scenario", Proc. IEEE VTC 2001/Fall, Atlantic, pp. 2687-2691.
[6] H. Liu and G. Xu, "Closed-form blind symbol estimation in digital communications," IEEE Trans. SP, Vol. 43, pp. 2714-2723, Nov. 1995.

Appendix: the Proof of Theorem 1

Because $\tilde{\mathbf{U}}$, $\mathbf{A}_k$, and $\mathbf{h}_{k,m}^i$ in (14) have some periodicity. Then we first check the condition that $\mathbf{h}_{k,m}^i$ can be uniquely determined in a "slice" of (14) (Because $\mathbf{h}_{k,m}^i$ is unknown in (14), we substitute it with $\mathbf{x}_{k,m}^i$)

$$\begin{bmatrix} \theta_1 \mathbf{U}_{2m-1}\mathbf{C}_k & \theta_4^* \mathbf{U}_{2m}\mathbf{C}_k^* \\ -\theta_3 \mathbf{U}_{2m}\mathbf{C}_k & \theta_2^* \mathbf{U}_{2m-1}^*\mathbf{C}_k^* \end{bmatrix} \begin{bmatrix} \mathbf{x}_{k,m}^1 \\ \mathbf{x}_{k,m}^2 \end{bmatrix} = 0 \quad (20)$$

(20) contains two sets of equations: the top set and the bottom set. According to [6] and the property of the EVD, the dimension of the solution space of every set is 2, and the solution of the top set can be written as

$$\xi = a_1 \begin{bmatrix} \mathbf{h}_{k,m}^1 \\ \mathbf{h}_{k,m}^{2*} \end{bmatrix} + a_2 \begin{bmatrix} \theta_2/\theta_1 \mathbf{h}_{k,m}^{2*} \\ -\theta_3^*/\theta_4^* \mathbf{h}_{k,m}^1 \end{bmatrix} \quad (21)$$

Similarly, the solution of the bottom one is

$$\xi = a_3 \begin{bmatrix} \mathbf{h}_{k,m}^1 \\ \mathbf{h}_{k,m}^{2*} \end{bmatrix} + a_4 \begin{bmatrix} \theta_4^*/\theta_3^* \mathbf{h}_{k,m}^{2*} \\ -\theta_1/\theta_2 \mathbf{h}_{k,m}^1 \end{bmatrix} \quad (22)$$

Becasue of (20), ξ in (21) and (22) are identical, therefore the following equation holds by subtracting (22) from (21)

$$(a_1 - a_3) \begin{bmatrix} \mathbf{h}_{k,m}^1 \\ \mathbf{h}_{k,m}^{2*} \end{bmatrix} = \begin{bmatrix} (-a_2\theta_2\theta_3 + a_4\theta_1\theta_4)/(\theta_1\theta_3)\mathbf{h}_{k,m}^2 \\ (a_2\theta_2^*\theta_3^* - a_4\theta_1^*\theta_4^*)/(\theta_2^*\theta_4^*)\mathbf{h}_{k,m}^{1*} \end{bmatrix} \quad (23)$$

$\mathbf{h}_{k,m}^1$ and $\mathbf{h}_{k,m}^2$ are linearly independent according to the assumption. Therefore, the fact that (23) holds implies the both sides of the equal mark are 0, i.e.

$$a_1 = a_3 \quad (24)$$
$$a_2\theta_2\theta_3 - a_4\theta_1\theta_4 = 0 \quad (25)$$
$$a_2^*\theta_2\theta_3 - a_4^*\theta_1\theta_4 = 0 \quad (26)$$

When a_2 and a_4 are not all 0, to guarantee that (25) and (26) hold simultaneously, it requires that $\frac{\theta_2\theta_3}{\theta_1\theta_4}$ is a real number. Then we can draw the conclusions:

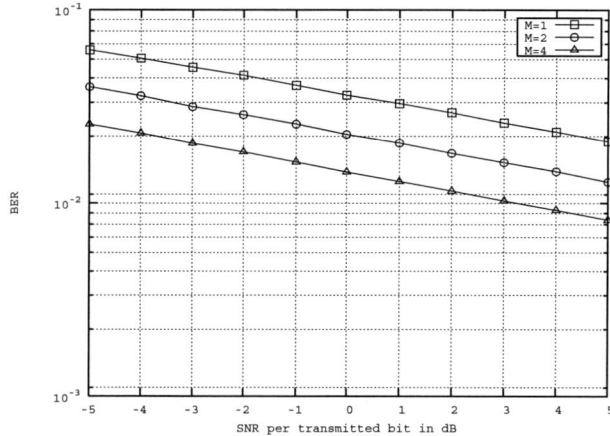

Fig. 2. The BER versus the SNR

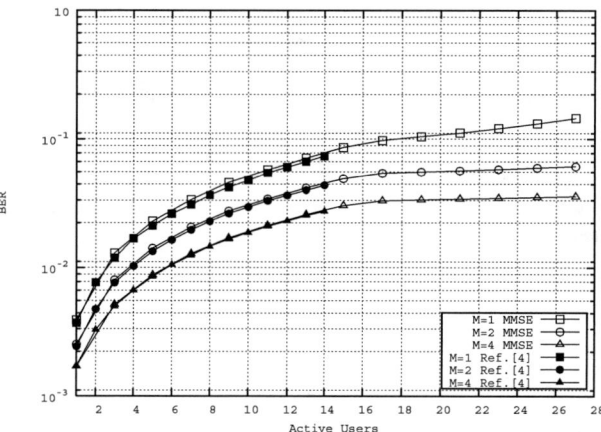

Fig. 3. The BER versus the number of users

1) When $\frac{\theta_2\theta_3}{\theta_1\theta_4}$ is a real number, denoted by k. So long as $a_4 = ka_2$, (23) holds. Because a_2, a_4 need not be all 0, the dimension of the solution of (20) is 2, and then $\mathbf{h}_{k,m}^i$ can not be estimated. This was be illustrated by (19).

2) Otherwise, it requires that $a_2 = a_4 = 0$ when $\frac{\theta_2\theta_3}{\theta_1\theta_4}$ is not a real number. Then the dimension of the solution of (20) is 1, and we can obtain the channel matrix.

To every slice in (14), the θ_i are the same and the condition is the same too. That is, when $\frac{\theta_2\theta_3}{\theta_1\theta_4}$ is not a real number, the solution space of (14) is with dimension 1. ∎

An alternative Natural Gradient approach for Multichannel Blind Deconvolution

Massimo Tomassoni, Stefano Squartini and Francesco Piazza
Dipartimento di Elettronica, Intelligenza Artificiale e Telecomunicazioni
Università Politecnica delle Marche
Ancona 60131, Italy
m.tomassoni@deit.univpm.it, sts@deit.univpm.it, upf@deit.univpm.it

Abstract— This paper presents an alternative natural gradient based solution to the multichannel blind deconvolution (MBD) problem. The derived learning rule comes from the definition of a new Riemannian metric in the space of linear system, rather than the one relative to the algorithm already existing in the literature. Moreover, it is proved that the novel approach satisfies the equivariance property if the MBD problem is formulated in a certain way. Experimental results have shown that the two gradients lead to the the same deconvolution performances.

I. INTRODUCTION

The convergence speed is a relevant issue to take care of when we deal with algorithms in blind deconvolution. Many approaches existing in literature allow to get a satisfying level of signal recovering but they turn out to be inefficient from the perspective of number of iterations they need to converge. The ones based on gradient descent techniques try to minimize a suitable cost function $l(\mathbf{w})$, defined in the generic space S of parameters $\mathbf{w}$ where we are searching our optimal solution. As already proved [1,2], the introduction of concepts related to Riemannian geometry can help to improve the performances usually achieved. This consists in defining a new gradient, namely the natural gradient, that is shown to be the steepest descent direction to follow in the minimization process of $l(\mathbf{w})$, assuming that S is Riemannian. This is valid also in the case of multichannel blind deconvolution (MBD) problem where S is the space of linear systems [1]. Up to authors' knowledge, only one metric has been considered to define such a natural gradient so far. The present work proposes a new metric which will lead to an alternative natural gradient based learning rule for the algorithm usually implemented in MBD. This will require to formulate the original problem in a different way. As already done in the case of BSS (blind source separation) by some of the authors [3], we will show that the new approach yields as good deconvolution performances as the existing ones.

II. MULTICHANNEL BLIND DECONVOLUTION

The multichannel blind deconvolution [4] is the problem of recovering the n sources mixed by a convolutive system into m mixture signals. The deconvolution procedure has to be carried out without knowing the sources and the system. However, some assumptions are ussually made to render the problem tractable: sources are i.i.d. and not gaussian, and $m \geq n$. From this perspective we can describe the problem as follows:

$$\mathbf{x}(k) = \sum_{p=-\infty}^{+\infty} H_p \mathbf{s}(k-p) \qquad \sum_{p=-\infty}^{+\infty} \|H_p\| < \infty \qquad (1)$$

where $\mathbf{s}(k)$ is the source (column) vector and $\mathbf{x}(k)$ the sensor one. $\mathbf{H}_p$ is the mixing matrix or convolutive system: it has dimensions equal to $m \times n$ and each entry is a FIR filter h_{ijp} having p taps. Let $*$ denote the convolution operator; then we can write in the time domain $\mathbf{x}(k) = H_p * \mathbf{s}(k)$ and in the Z-domain $\mathbf{x}(k) = [\mathbf{H}(z)]\mathbf{s}(k)$.

The deconvolution algorithm shoul yield the source recovered, apart from some acceptable indeterminancies: gain, delay and/or permutation. Our goal is nothing but applying a multichannel equalizer to the mixture signals in order to get:

$$\mathbf{y}(k) = \sum_{p=-\infty}^{+\infty} W_p(k) \mathbf{x}(k-p) = \mathbf{W}(k) * \mathbf{x}(k) \qquad (2)$$

where $\mathbf{y}(k) = [y_1(k), y_2(k), ..., y_n(k)]^T$ is the output vector and $\mathbf{W}(k) = \{W_p(k), -\infty \leq p \leq +\infty\}$ is the demixing system at time k. Similarly, in the Z-domain: $\mathbf{y}(k) = [\mathbf{W}(z,k)]\mathbf{x}(k)$. The global system becomes:

$$\mathbf{y}(k) = [\mathbf{C}(z,k)]\mathbf{s}(k), \qquad \mathbf{C}(z,k) = \mathbf{W}(z,k)\mathbf{H}(z).$$

Therefore, MBD can be seen as the problem of finding out a system $\mathbf{W}(z,k)$ such that:

$$\lim_{k\to+\infty} \mathbf{C}(z,k) = \mathbf{W}(z,k)\mathbf{H}(z) = PAD(z)$$

where $PAD(z)$ is the system describing the effects of permutation, amplification and delay respectively.

Several approaches have been proposed in the literature so far. Among them, we shall consider that one based on the Kullback-Leibler divergence, that lead to the following cost function to minimize [4,5]:

$$l(\mathbf{W}(z,k)) = -\log\left|\det(W_0(k))\right| - \sum_{i=1}^{n} \log p_{s_i}(y_i(k)). \quad (3)$$

The gradient is equal to:

$$\frac{\partial l(\mathbf{W}(z,k))}{\partial W_p(k)} = -W_p(k)\delta_p + g(\mathbf{y}(k))\mathbf{x}^T(k-p)$$

where $g(y_i) = -\dfrac{d(\log(p(y_i)))}{dy_i}$ and $p(y_i)$ is the output pdf. Therefore, the on-line learning rule (which is gradient descent based) turns out to be:

$$\Delta W_p(k) = \mu(k)\left\{W_p^{-T}(k)\delta_n - g(\mathbf{y}(k))\mathbf{x}^T(k-p)\right\}. \quad (4)$$

III. NATURAL GRADIENTS IN SPACE OF LINEAR SYSTEMS

As aforementioned, the steepest descent direction of the cost function $l(\mathbf{w})$ in a Riemannian space is given by the natural gradient, which is related to the standard (Euclidean) one by [1,2]: $\widetilde{\nabla}l(\mathbf{w}) = G^{-1}(\mathbf{w})\cdot\nabla l(\mathbf{w})$, where $G^{-1}(\mathbf{w})$ is the inverse of the metric tensor $G(\mathbf{w}) = \{g_{ij}(\mathbf{w})\}$ and $\nabla l(\mathbf{w})$ is the standard gradient made of the n derivatives respect with the $\mathbf{w}$ entries. The metric tensor can be easily obtained by exploiting some interesting properties of the parameter space we are dealing with. In our case, the space of linear systems $\mathcal{M}(L)$ is a Lie group [4], together with proper multiplication and inversion operations, which are:

$$\mathbf{B}^\dagger(z) = \sum_{p=0}^{L} B_p^\dagger z^{-p}, \quad \mathbf{B}(z)\otimes\mathbf{C}(z) = \sum_{p=0}^{L}\sum_{q=0}^{p} B_q C_{p-q} z^{-p},$$

where $\mathbf{B}(z), \mathbf{C}(z) \in \mathcal{M}(L)$, and $B_p^\dagger$ defined as follows:

$$B_0^\dagger = B_0^{-1}, \quad B_p^\dagger = -\sum_{q=1}^{p} B_{p-q}^\dagger B_q B_0^\dagger.$$

The neutral element is $\mathbf{E}(z) = \mathbf{I}$. We shall introduce the metric tensors we are focusing on by accordingly defining the inner products in the tangent space. We will follow the usual two-step procedure to do that:

1. Definition of the inner product on the tangent space at a point in the Lie group (usually the neutral element of the group).
2. Imposing that the inner product be invariant under translations in the Lie group.

The inner product at the neutral element is assumed to be:

$$\langle \mathbf{X}(z), \mathbf{Y}(z)\rangle_{\mathbf{E}(z)} = \sum_{p=0}^{L} tr(X_p Y_p^T) = \\ = \frac{1}{2\pi j} \oint tr\left[\mathbf{X}(z^{-1})\mathbf{Y}^T(z^{-1})\right] z^{-1} dz \quad (5)$$

where $\mathbf{X}(z), \mathbf{Y}(z) \in T_{\mathbf{E}(z)}$, being $T_{\mathbf{E}(z)}$ the tangent space at the neutral element.

A. The right translation

Such an operator ($\forall \mathbf{V}(z) \in \mathcal{M}(L)$) is described by:

$$\mathfrak{T}_{\mathbf{V}(z)}: \mathcal{M}(L) \to \mathcal{M}(L) \quad \mathfrak{T}_\mathbf{V}(\mathbf{A}(z)) = \mathbf{A}(z)\cdot\mathbf{V}(z)$$

For the claimed invariance of the inner product we can write (being $\mathbf{X}(z), \mathbf{Y}(z) \in T_{\mathbf{W}(z)}$):

$$\langle \mathbf{X}(z), \mathbf{Y}(z)\rangle_{\mathbf{W}(z)} = \langle \mathbf{X}(z)\otimes\mathbf{B}(z), \mathbf{Y}(z)\otimes\mathbf{B}(z)\rangle_{\mathbf{W}(z)\otimes\mathbf{B}(z)}.$$

As shown in [4,6], the (nonholonomic) differential variables in the Riemannian space are:

$$d\mathbf{X}(z,k) = d\mathbf{W}(z,k)\otimes\mathbf{W}^\dagger(z,k) = \left[d\mathbf{W}(z,k)\mathbf{W}^{-1}(z,k)\right]_L;$$

while the natural gradient based learning rule is:

$$\Delta W_p(k) = \mu(k)\left\{W_p(k) + g(\mathbf{y}(k))U_p^T(k)\right\} \\ U_p(k) = \sum_{q=0}^{L} W_q^T(k)\mathbf{y}(k-p+q) \quad (6)$$

We cannot use (6) as it appears, since it is not causal. Therefore we have to apply a kind of causalization technique. As proposed in [8], we can introduce a delay $d = L$ in (6) so that $\Delta W_p(k-d)$ is the weight update to consider at time instant k. It follows:

$$\Delta W_p(k) = \mu(k)\left\{W_p(k-d) - g(Y(k-d))U_p^T(k-d)\right\} \\ U_p(k) = \sum_{q=0}^{L} W_q^T(k) Y(k-p+q) \quad (7)$$

B. The equivariance property

As stated in [8], such a property is surely relevant. Indeed, if it satisfied by the learning rule under study, then we can say that the algorithm performances do not depend on the parameter values of the mixing and/or demixing model. In other words, the final solution is not related to the choice of the initial conditions. That is what happens in the case of the right natural gradient based algorithm: in fact (6) does not depend on $\mathbf{W}(z,k)$ or $\mathbf{H}(z)$ but only on $\mathbf{C}(z,k) = \mathbf{W}(z,k)\mathbf{H}(z)$. To prove that, it suffices to write the learning rule in the Z-domain and multiply it by the mixing matrix.

C. The left translation

Such an operator ($\forall \mathbf{V}(z) \in \mathcal{M}(L)$) is described by:

$$_{\mathbf{V}(z)}\mathfrak{T}: \mathcal{M}(L) \to \mathcal{M}(L) \qquad \mathfrak{T}_{\mathbf{V}}(\mathbf{A}(z)) = \mathbf{V}(z) \cdot \mathbf{A}(z)$$

that is different from the right translation for the non-commutativity property of $\mathcal{M}(L)$. Following a procedure similar to that one considered to derive (6) in [4], we have:

$$\langle \mathbf{X}(z), \mathbf{Y}(z) \rangle_{\mathbf{W}(z)} = \langle \mathbf{B}(z) \otimes \mathbf{X}(z), \mathbf{B}(z) \otimes \mathbf{Y}(z) \rangle_{\mathbf{B}(z) \otimes \mathbf{W}(z)}$$

that becomes equal to

$$\langle \mathbf{X}(z), \mathbf{Y}(z) \rangle_{\mathbf{W}(z)} = \langle \mathbf{W}(z)^{\dagger} \otimes \mathbf{X}(z), \mathbf{W}(z)^{\dagger} \otimes \mathbf{Y}(z) \rangle_{\mathbf{E}(z)}. \quad (8)$$

We use the following notation for the involved gradients (standard and natural respectively):

$$\nabla l(\mathbf{W}(z)) = \sum_{p=0}^{L} \partial_p l(\mathbf{W}(z)) z^{-p}$$

$$\widetilde{\nabla} l(\mathbf{W}(z)) = \sum_{p=0}^{L} \widetilde{\partial}_p l(\mathbf{W}(z)) z^{-p}$$

Through simple algebraic calculations, we can derive the following from (8):

$$\partial_p l = \sum_{k=0}^{L-p} \sum_{q=0}^{k} W_k^{\dagger T} W_q^{\dagger} \widetilde{\partial}_{k-q} l \quad (9)$$

The new (nonholonomic) differential variables in the Riemannian space now considered are:

$$d\mathbf{X}(z,k) = \mathbf{W}^{\dagger}(z,k) \otimes d\mathbf{W}(z,k) = \left[\mathbf{W}^{-1}(z,k) d\mathbf{W}(z,k) \right]_L \quad (10)$$

They allow us to derive the following, as done in the case of right translation [4]:

$$\partial_p l = \sum_{k=0}^{L-p} W_k^{\dagger T} \frac{\partial l(\mathbf{W}(z))}{\partial X_{k-p}} \quad (11)$$

Comparing (9) and (11) we can easily prove that holds:

$$\widetilde{\nabla} l(\mathbf{W}(z)) = \mathbf{W}(z) \otimes \frac{\partial l(\mathbf{W}(z))}{\partial \mathbf{X}(z)} \quad (12)$$

Such equation, together with the cost function defined in (3), will help us out to obtain the final formula for the left natural gradient. However, if we use the same term $(\partial l(\mathbf{W}(z))/\partial \mathbf{X}(z))$ used in the case of right natural gradient, we will get a learning rule that does not satisfy the equivariance property, as it can be intuitively understood looking at (12). That is why it has been decided to describe the deconvolution problem in terms of row vectors (rather than column ones) as follows:

$$\mathbf{x}(k) = \mathbf{s}(k) * H_p = \mathbf{s}(k)\left[\mathbf{H}(z)\right]$$
$$\mathbf{y}(k) = \mathbf{x}(k) * W_p(k) = \mathbf{x}(k)\left[\mathbf{W}(z,k)\right] \quad . \quad (13)$$
$$\mathbf{C}(z,k) = \mathbf{H}(z)\mathbf{W}(z,k).$$

We shall call this the "transpose formulation" of MBD problem, in contrast to the "common" one considered so far. Now we can calculate the infinitesimal increment $dl(\mathbf{W}(z,k)) = l(\mathbf{W}(z,k) - d\mathbf{W}(z,k)) - l(\mathbf{W}(z,k))$ of the cost function, as well as done in [4,6]. Such an increment corresponds to an increment $d\mathbf{W}(z,k)$, that is related to the increment $d\mathbf{X}(z,k)$ through (10). Simple algebraic and differential calculus yields:

$$dl(\mathbf{W}(z,k)) = -tr(dX_0(k)) + \\ + \left\{ \sum_{p=0}^{L} \mathbf{y}(k-p) dX_p(k) \right\} g(\mathbf{y}(k))$$

which immediately leads to:

$$\frac{\partial l(\mathbf{W}(z))}{\partial X_p(z)} = \left\{ \mathbf{I}\delta_p - \mathbf{y}^T(k-p)g(\mathbf{y}(k)) \right\} \quad . \quad (14)$$

Taking into account that:

$$W(z,k)\left[-\mathbf{I}\delta_p\right] = -W_p(k)$$
$$W(z,k)\left[Y^T(k-p)\right]g(Y(k)) = \\ = \left(\sum_{q=0}^{L} W_q(k) Y^T(k-p+q) \right) g(Y(k))$$

and substituting (14) into (12) we finally get:

$$\Delta W_p(k) = \mu(k)\left\{ W_p(k) - U_p^T g(\mathbf{y}(k)) \right\}$$
$$U_p = \left(\sum_{q=0}^{L} \mathbf{y}(k-p+q) W_q^T(k) \right) \quad . \quad (15)$$

Also in this case, we can apply the same considerations as above to get our causal version of (15):

$$\Delta W_p(k) = \mu(k)\left\{ W_p(k-d) - U_p^T(k-d) g(\mathbf{y}(k-d)) \right\}$$
$$U_p(k) = \sum_{q=0}^{L} Y(k-p+q) W_q^T(k). \quad (16)$$

The equivariance property is satisfied also in this case.

IV. EXPERIMENTAL RESULTS

In this section the deconvolution performances of the two algorithms dealt with (in their forms (7) and (16), respectively) are considered. The criterium addressed to evaluate the convergence status and the achieved deconvolution quality is the mISI (multichannel InterSymbol Interference) of the global system $\mathbf{C}(z,k)$. The sources are random signals with Laplacian distribution with an user-defined number of samples. The chosen non-linear functions g_i s are those typically used for supergaussian variables. The sources are mixed through a known convolutive system $\mathbf{H}(z)$ (whose ISI values for the two mixed channels are: 2.9537 dB and 2.6734 dB). The initial parameter values of the demixing system $\mathbf{W}(z,k)$ are the following:

$$\begin{cases} w_{ij}(k) = [0,...,0,...,0], \text{ for } i \neq j \\ w_{ij}(k) = [0,...,1,...,0], \text{ for } i = j. \end{cases}$$

Fig.1 depict the ISI curves in both natural gradient case studies: the performances are essentially identical, corresponding to a good deconvolution level achieved (the differences in shape are due to randomness of involved signals). The choice of the stepsize is fundamental to get the convergence. In these tests it decreases as the number of iterations k (30000 in total) grows up. It has been chosen as follows: $\mu = 10^{-3}$ for $k \leq 15000$, $\mu = 10^{-4}$ for $15000 < k \leq 25000$ and $\mu = 10^{-5}$ for $k > 25000$.

It must be highlighted that further simulations have been carried out in those cases where the equivariance property is not satisfied: the right natural gradient in the transposed formulation and the left natural gradient in the common formulation. Fig.2,3 depict the ISI curves in both natural gradient case studies in the Z-domain. We can see that the performances are better in the cases in which the equivariance property is satisfied. So, the right natural gradient is better in the "common formulation" of the problem while the left natural gradient is better in the "transpose formulation" of the problem.

V. CONCLUSIONS

In this paper a novel natural gradient approach to the multichannel blind deconvolution problem has been presented. The new method is based on the definition of a different metric tensor rather than the one used in the conventional natural gradient. The outcomes of the algorithm developed are identical to those already got in the literature. In particular, this occurs when the equivariance property is satisfied, that requires to rewrite the problem in its transposed version. Finally, this work could have relevant implications in the future research, since we can assume to derive other natural gradients based on different Riemannian metrics, and possibly rate their deconvolution performances.

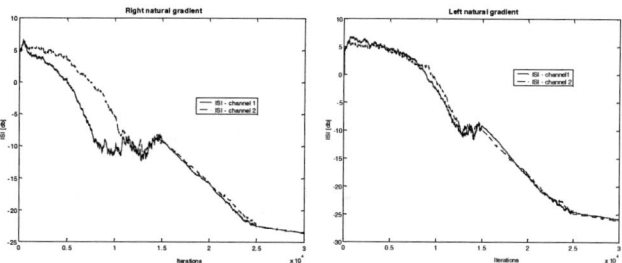

Figure 1. ISI curves in right and left natural gradient case study.

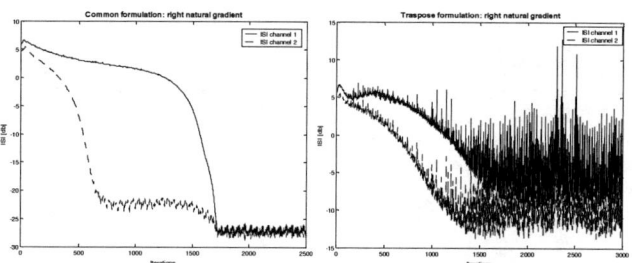

Figure 2. ISI curves in right natural gradient case study.

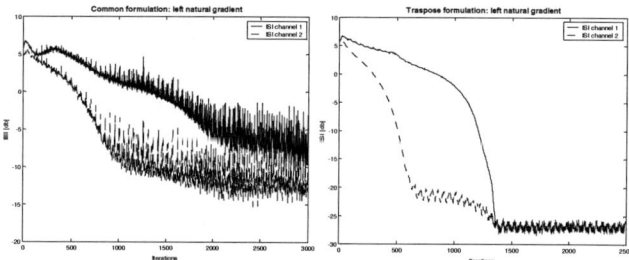

Figure 3. ISI curves in left natural gradient case study.

REFERENCES

[1] S. Amari, *"Natural Gradient works efficiently in learning"*. Neural computation, vol. 10, pp. 271-276,1998.

[2] S.Amari, and S.C.Douglas, *"Why natural gradient?"*. Proc. of Int. Conf. on Acoustics speech and signal processing. - ICASSP, vol. 2, pp. 1213-1216, Seattle, USA, May 1998.

[3] A. Arcangeli, S. Squartini, and F. Piazza, "An alternative Natural Gradient Approach for ICA based learning algorithms in Blind Source Separation", *EUSIPCO 2004*, Vienna, Austria, 2004.

[4] A.Cichocki, and S.Amari, *"Adaptive Blind Signal and Image Processing"*. Wiley&Sons, West Sussex, England, 2002.

[5] A.Bell, T.Sejnowski, *"An information-maximization approach to blind separation and blind deconvolution"*, Neural Computation, 7(6): 1129-1159, 1995.

[6] S.Amari, S.C.Douglas, A.Cichocki and H.H.Yang, *"Multichannel blind deconvolution and equalization using the natural gradient"*. In The First Signal Processing Workshop on Signal Processing Advances in Wireless Communications, pag. 101-104, France, 1997.

[7] M.Cohen and G.Cauwenberghs, *"Blind separation of linear convolutive mixtures through parallel stochastic optimization"*, in Proceedings of ISCAS, 1998.

[8] J.F.Cardoso and B.H.Laheld, *"Equivariant adaptive source separation"*. IEEE Trans. Signal Processing, vol.44(12), pp. 3017-3030, December 1996.

# Decision Feedback Equalizer With The Blind Matched Filter Estimation

İzzet Özçelik, İzzet Kale
Department of Electronic Systems
Applied DSP and VLSI Group, University of Westminster, London, UK
Telephone: +(44) 20 7911 5083
Email: { i.ozcelik, kalei } @wmin.ac.uk

Buyurman Baykal
Electrical and Electronics Engineering Department
Middle East Technical University, Ankara, Turkiye
Telephone: (90) 312 210 4582
Email: buyurman@metu.edu.tr

Abstract— The optimal receiver front-end is the Matched Filter (MF) for a maximum Signal-to-Noise Ratio (SNR) in a communication channel with an Additive White Gaussian Noise (AWGN). The sampled MF output can be processed with the suboptimal yet computationally effective Decision Feedback Equalizer (DFE). In this paper, the filters of the DFE are derived in terms of the channel autocorrelation sequence which is estimated blindly using the sampled channel output signal. Initially, the MF followed by the DFE receiver the structure is operated in the Linear Equalizer (LE) mode (without the Decision Device (DD)) with the symbol-rate sampled MF placed at the end of the receiver structure. During this LE mode, the MF is estimated blindly using the CMA. After the CMA converges, the structure is put back into its original form and the DD is included in the FeedBack Filter (FBF). The MF is adapted in the system identification structure using the detected symbols from the DD. In this way, the MF estimate from the CMA is improved further and an all-blind DFE is realized for a Single Input Single Output (SISO) system. The Bit Error Rate (BER) values obtained from this structure are very close to those obtained with the true values of the MF and hence they proved that the structure is efficient and effective.

I. Introduction

Practical communication channels introduce memory in the received signals causing a spreading on the transmitted symbols over time. This effect created by the channel is known as the InterSymbol Interference (ISI) and must be substantially eliminated for the correct detection of the transmitted symbols. Basically, there are two main methods to cancel the ISI at the channel output: equalization and the Maximum Likelihood Sequence Estimation (MLSE) [1]. Even though the equalization and the MLSE perform the same ISI canceling operation, they work on completely different operation principles. The equalizers are composed of a simple filter (or filters) and they invert the Channel Impulse Response (CIR) obtaining an approximate impulse response as the effective equivalent channel. The MLSE can only be applied to Finite Impulse Response (FIR) channels and they are based on the finite state machine model of the channel and provides the optimal solution by searching all possible state transitions based on the received signal samples theoretically. The biggest drawback of the MLSE is the computational complexity which grows exponentially as the channel memory and the symbol alphabet size increases. Due to this fact, the equalizers are of utmost importance for the practical systems since they are easy to implement. The main problem with the equalizers is their poor performance. Although an equalizer composed of a single linear filter exhibits a performance far from the one obtained from the optimal MLSE, the DFE provides a good compromise between the equalizers and the MLSE from the efficiency and effectiveness points of view. Most of today's practical communication systems rely on the trained methods obtained by the transmission of a pilot symbol sequence before the actual data transmission starts or if the channel has time-varying characteristics the pilot symbol sequence needs to be transmitted periodically. The training period is a waste of the communication channel bandwidth and must be avoided for fast communication applications. Therefore, blind methods which do not need any access to the transmitted symbols are required. The Constant Modulus Algorithm (CMA) is the main blind approach, which is very simple to implement on a sample by sample basis or a block of samples, and it can operate both in symbol-rate spaced or fractionally spaced scenarios [2]. Some blind algorithms based on the fractionally spaced channel output signal have been studied for an optimal blind equalization [3], [4]. The fractionally spaced channel can be considered as a Single Input Multiple Output (SIMO) system. By the fractional sampling a diversity is obtained and this diversity provides the blind estimation and the equalization of the channel. These methods involve some kind of matrix decomposition which can be quite computationally expensive e.g. the Singular Value Decomposition (SVD). In the work presented in this paper, we study the Single Input Single Output (SISO) DFE and offer a solution based on an all-blind structure. The DFE is based on the spectral decomposition of the channel output autocorrelation sequence [1]. An optimal receiver front-end involves the Matched Filter (MF) which provides the maximum Signal-to-Noise Ratio (SNR) and the output of the MF represents the channel autocorrelation sequence with an additive noise. The feedforward and the feedback filters of the DFE realize the inverses of the decomposed components of the MF output and these filters can be obtained from the Channel Impulse Response (CIR) autocorrelation sequence. Since the DFE filters can be computed from the sampled channel output signal, it is possible to estimate the MF operating the DFE without the DD by the use of the CMA if it is placed at the end of the receiver chain. Once the CMA has converged, the structure is switched into the DFE mode with the MF placed as the first component processing the channel output signal. In this mode, the DFE provides an improved Bit Error Rate (BER) provided that the MF estimate from the CMA is good enough to produce an equalized output and hence the symbols from the DD are used to adapt the MF in the system identification mode since the MF is the time-reversed CIR. This provides further improvement in the MF estimation and the BER. The CMA has been shown to have a performance close to the Minimum Mean Squared Error (MMSE) solution [5]. In our results for the blind MF estimation in the DFE with the CMA, it is observed that the CMA works adequately paralleling with earlier findings [5]. In our paper the basic novelties introduced are the estimation of the symbol-rate sampled MF using the CMA and an all-blind DFE with the MF.

A. System Model

The output signal $r(t)$ of a typical communication channel can be represented with the convolution $r(t) = \sum_{n=-\infty}^{\infty} a(n)h(t-nT) + v(t)$, where $a(n)$ is the transmitted symbol, $h(t)$ is the continuous-

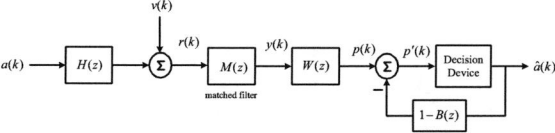

Fig. 1. DFE With The Matched Filter

time channel impulse response and $v(t)$ represents the Additive White Gaussian Noise (AWGN). The symbols are transmitted with a period of T seconds and hence the channel output can be sampled at a frequency of $1/T$ resulting in a discrete-time model. The sampling frequency can be chosen equal to $1/T$ or P/T with P being an integer denoting the oversampling ratio. The continuous-time model is converted to an equivalent discrete-time model by replacing t with $k\frac{T}{P}$ as shown in (1).

$$r(t)|_{t=k\frac{T}{P}} = \sum_{i=-\infty}^{\infty} a(i)h(k\frac{T}{P} - iT) + v(k\frac{T}{P}) \quad (1)$$

The discrete-time model will be as in $r(k) = \sum_{i=-\infty}^{\infty} a(i)h_{k-iP} + v(k)$ by simply removing $\frac{T}{P}$. In the rest of the paper, it is assumed that the symbols $a(k)$ are independent and identically distributed with a variance of σ_a^2, the noise $v(k)$ is uncorrelated with the input symbols having a variance σ_v^2. Moreover, the channel is assumed to be causal and have a finite memory of length L.

II. DERIVATION OF THE FINITE DFE FILTERS IN TERMS OF THE CHANNEL AUTOCORRELATION SEQUENCE

Consider the channel output signal sampled at the symbol-rate, i.e. $P = 1$, which is written in (2).

$$r(k) = \sum_{i=0}^{L} a(i)h_{k-i} + v(k) \quad (2)$$

Figure 1 gives the discrete-time DFE where $M(z)$ shows the transfer function of the MF, $W(z)$ and $B(z)$ denote the FeedForward Filter (FFF) and the FBF respectively. The symbol-rate sampled MF m_n will be the conjugated time-reversed version of the CIR, i.e. $m_n = h_{-n}^*$ The channel is assumed to be causal therefore for the MF to be causal h_{-n}^* must be delayed by L. The delayed MF expression is given as $m_n = h_{L-n}^*$. The equivalent system between the output of the MF and the transmitted symbol is the convolution of the channel with the MF when the noise is ignored (the noise variance will be estimated and subtracted from the 0th lag of the autocorrelation sequence). Let $x_n \triangleq h_n * m_n$ and the mathematical expression for x_n is given as $x_n = \sum_{i=0}^{L} h_i m_{n-i}$. The MF impulse response equals to h_{L-n}^* and hence if m_{n-i} is replaced by h_{L+i-n}^*, (3) will be obtained.

$$x_n = \sum_{i=0}^{L} h_i h_{L+i-n}^*, \quad 0 \le n \le 2L \quad (3)$$

The limiting values for the index of x_n are found by considering the values of n for which h_{L+i-n}^* is not 0. There are $2L+1$ nonzero values for x_n as expected since it is the convolution of h_n and m_n both of which are of length $L+1$. It is obvious that x_n is the autocorrelation sequence of the CIR since it describes $E\{h_i h_{i+n}^*\}$. After having noted this, we can formulate the DFE in terms of x_n.

For $N+1$ values of $y(k)$ the MF output can be expressed as a matrix equation

$$\mathbf{y}(k) = \mathbf{M}\mathbf{r}(k) + \mathbf{M}\mathbf{v}(k) \quad (4)$$

where $\mathbf{M}$ is the convolution matrix of size $(N+1) \times (N+L+1)$, and the vectors are defined as $\mathbf{y}(k) = [y(k) \cdots y(k-N)]^T$, $\mathbf{r}(k) = [r(k) \cdots r(k-N-L)]^T$, $\mathbf{v}(k) = [v(k) \cdots v(k-N-L)]^T$. Similarly the channel output $r(k)$ can be written as a vector for $N+L+1$ values $\mathbf{r}(k) = \mathbf{H}\mathbf{a}(k) + \mathbf{v}(k)$ where $\mathbf{H}$ is the convolution matrix of size $(N+L+1) \times (N+2L+1)$, and $\mathbf{a}(k) = [a(k) \cdots a(k-N-2L)]$. If $\mathbf{r}(k)$ is substituted into (4), the equation for $\mathbf{y}(k)$ will be obtained in terms of the MF and the CIR convolution matrices added with the noise vector term multiplied by $\mathbf{M}$.

$$\mathbf{y}(k) = \mathbf{MH}\mathbf{a}(k) + \mathbf{M}\mathbf{v}(k) \quad (5)$$

$\mathbf{MH}$ is the equivalent channel convolution matrix between $a(k)$ and $y(k)$ and it can be formed from the autocorrelation sequence since the autocorrelation sequence is actually the effective channel impulse response between $a(k)$ and $y(k)$. Let $\mathbf{X} \triangleq \mathbf{MH}$ and has a size of $(N+1) \times (N+2L+1)$, with the indices calculated as $x_n = \sum_{i=0}^{L} h_i h_{L+i-n}^*$ ($n = 0 \cdots 2L$) using the CIR. This reveals the fact that the effective channel is described by the autocorrelation sequence of the actual channel $H(z)$.

Having established the mathematical background, the DFE filters can be solved using the MMSE criterion. In the structure shown in Figure 1 the error signal is defined as the difference between the transmitted symbols and the input to the DD. The symbols must be delayed since the filtering operations will in introduce a delay. Hence, the error signal is written as in (6) with Δ denoting the delay.

$$e(k) = a(k-\Delta) - p'(k) \quad (6)$$

The length of the FFF is given by N and that of the FBF is assumed to be equal to the channel memory L. The FFF, $W(z)$, and the FBF, $B(z)$, can be put into vector forms as $\mathbf{w}(k) = [w_0(k) \cdots w_N(k)]^T$ and $\mathbf{b}(k) = [b_0(k) \cdots b_L(k)]^T$ respectively. The symbols entering to the FBF are assumed to be true and stacked into the vector $\mathbf{a}(k-\Delta) = [a(k-\Delta-1) \cdots a(k-\Delta-L)]^T$. The orthogonality principle will provide the MMSE solution [6]. The solutions obtained from the orthogonality principle for the FFF and the FBF coefficients vector will be as in (7) and (8) respectively,

$$\mathbf{w}(k) = (\mathbf{R}_{yy} - \mathbf{R}_{ay}\mathbf{R}_{ay}^T)^{-1}\sigma_a^2 \mathbf{g} \quad (7)$$
$$\mathbf{b}(k) = -(\mathbf{R}_{yy} - \mathbf{R}_{ay}\mathbf{R}_{ay}^T)^{-1}\sigma_a^2 \mathbf{g}\mathbf{R}_{ay} \quad (8)$$

where $\mathbf{R}_{yy} \triangleq E\{\mathbf{y}(k)\mathbf{y}(k)^T\}$, $\mathbf{R}_{ay} \triangleq E\{\mathbf{a}(k)\mathbf{y}(k)^T\}$ and $\mathbf{g} \triangleq [x_{-L+\Delta} \cdots x_{-L+\Delta+N}]$.

The Zero Forcing (ZF) solution of w_n and b_n is obtained by making the noise variance zero in $\mathbf{R}_{yy}$. With the noise eliminated, the autocorrelation matrix for $\mathbf{y}(b)$ is defined as $\mathbf{R}'_{yy} = \sigma_a^2 \mathbf{X}\mathbf{X}^T$. By replacing $\mathbf{R}_{yy}$ by $\mathbf{R}'_{yy}$ in (7) and (8), the ZF $\mathbf{w}(k)$ and $\mathbf{b}(k)$ will be solved. The ZF solution will provide a better estimate if the channel does not have deep nulls in its spectrum. From the above derivations, it is seen that $\mathbf{R}'_{yy}$, $\mathbf{R}_{ay}$ and $\mathbf{g}$ only depend on the sequence x_n which is the CIR autocorrelation sequence.

A. Channel Autocorrelation Estimation

It was shown by equations (7) and (8) that if the channel autocorrelation sequence can be estimated, it is possible to design the DFE. Blind calculation of the channel autocorrelation sequence is possible from the channel output $r(k)$ using the simple formula given in (9)

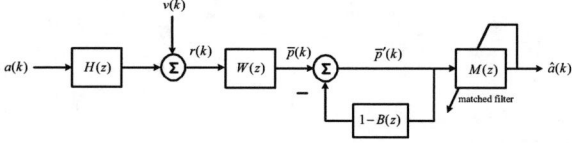

Fig. 2. The Novel Structure With The Blindly Adapted MF

since the transmitted symbols $a(k)$ are independent and identically distributed.

$$\hat{x}_k = \sum_{i=0}^{M} r(i)r(i+k)/(M\sigma_a^2), \ k \neq 0 \quad (9)$$

For $\hat{x}_0$ the AWGN variance σ_v^2 appears as a term added to the estimate. Due to this, it must be estimated and subtracted from the 0th channel autocorrelation estimate resulting in $\hat{x}_0 = \sum_{i=0}^{M} r(i)r(i)/(M\sigma_a^2) - \hat{\sigma}_v^2$. For the AWGN variance estimation, we need to form the channel output autocorrelation matrix of size greater than $L+1$ from the oversampled channel output samples ($P > 1$). The smallest eigenvalue of this autocorrelation matrix corresponds to the noise variance [4]. In the sequel, it will be assumed that oversampled channel output is available for the noise variance estimation.

III. NOVEL STRUCTURE FOR THE BLIND ESTIMATION OF THE MATCHED FILTER

The DFE shown in Figure 1, has a nonlinear structure due to the DD. If the DD is not involved, the Linear Equalizer (LE) will be obtained and it allows the reorganization of the receiver as in Figure 2 where the MF is placed at the end of the whole structure. The structure will be reffered to as the Blind Matched Filter Receiver (BMFR) [7]. Since the FFF and FBF are calculated from the channel autocorrelation sequence, in the receiver the MF is the only unknown and by placing it at the end of the receiver the application of the CMA is made possible. In this way, the equalizer filters are calculated using the channel autocorrelation sequence and the MF is estimated with the CMA rather than trying to estimate all parts of the receiver by using only the CMA. The DFE is actually the spectral decomposition of the channel autocorrelation sequence. As pointed out before, the channel autocorrelation sequence is equal to $h_n * m_h$. Provided the Paley-Wiener condition [8] is satisfied, any autocorrelation sequence can be decomposed uniquely into a causal, monic, stable minimum phase part with a noncausal monic maximum phase part [8]. Hence, the z-transform of the autocorrelation sequence of h_n can be decomposed as $\sigma_a^2 X(z) = \chi G(z) G^*(1/z^*)$ where χ denotes the gain factor. The FFF is realized as $1/\sqrt{\chi} G^*(1/z^*)$ while the FBF equals to $1/\sqrt{\chi} G(z)$. Let the transfer function from $a(k)$ to $\overline{p}'(k)$ in Figure 2 be $Q(z)$. If the noise is eliminated to address the ZF case, $Q(z)$ will be equal to $Q(z) = \frac{\sigma_a^2 H(z)}{\chi G(z) G^*(1/z^*)}$. Since $\chi G(z) G^*(1/z^*)$ equals to $\sigma_a^2 H(z) M(z)$ therefore the equivalent transfer function until $\overline{p}'(k)$, is $Q(z) = \frac{H(z)}{G(z)G^*(1/z^*)} = \frac{H(z)}{H(z)M(z)} = \frac{1}{M(z)}$. After these observations we show that the effective channel up until the MF input is $1/M(z)$. The CMA will inverse this equivalent channel resulting in the estimation for the MF $M(z)$ in addition to providing the equalized signal $\hat{a}(k)$ which is an estimate for $a(k)$. Based on the structure given in Figure 2 and the facts established, the following sequence of operations define a way to estimate the components $M(z)$, $W(z)$ and $B(z)$ blindly.

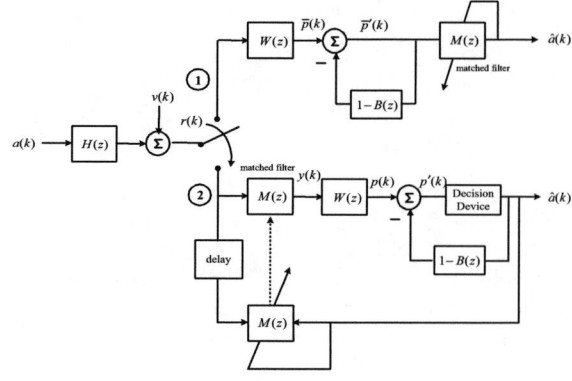

Fig. 3. Novel Blindly Adapted MF Receiver In The DFE Mode

1) Estimate the channel autocorrelation sequence from the channel output $r(k)$ with the noise variance estimated from the oversampled channel output autocorrelation matrix and subtracted from the 0th lag autocorrelation value.
2) Calculate w_n and b_n using (7) and (8).
3) Process the channel output to obtain $\overline{p}'(k)$ with w_n and b_n.
4) Apply the CMA using the signal $\overline{p}'(k)$ for the estimation of m_n which is equal to the time-reversed CIR i.e. h_{-n}.

The CMA update algorithm is given in (10) [2] where $R_p = \frac{E\{|a(k)|^{2p}\}}{|a(k)|^p}$, $\overline{p}'(k)$ is the input to the MF as indicated in Figure 2.

$$\hat{\mathbf{m}}(k+1) = \hat{\mathbf{m}}(k) + \mu \overline{p}'(k)(R_p - |\overline{p}'(k)|^p)^2 \overline{p}'(k). \quad (10)$$

IV. ALL-BLIND DFE

A novel way of using the BMFR is revealed in Figure 3. Initially, the switch is moved to position 1, and the MF (CIR) is estimated using the CMA in the LE mode. The CMA is run and when the decision error is reduced to a predefined value, the switch is moved to position 2. The switching can be performed based on the distance between the input of the DD and the output of the DD. The square of this distance can be averaged and tracked as adaptation goes on. Let the switching variable be $S(k)$ and it can be written as $S(k) = \gamma S(k-1) + (1-\gamma)|DD(\hat{a}(k)) - \hat{a}(k)|^2$ where γ is the forgetting factor for the simple lowpass filtering [9] for switch position 1. When $S(k)$ gets smaller than the predefined value say S_0, switching to position 2 is done. In this structure, the MF is put after the channel and performs the usual matched filtering operation and the FBF is operated with a DD, realizing the DFE. Since the BER of the DFE will be much better than the LE provided that the MF is estimated properly using the CMA, the estimated symbols from the DD are used to adapt the CIR estimate from the MF estimate of the CMA. In other words, the coefficients of the CIR (MF) are updated in the system identification mode. The CIR estimate will improve compared to the one obtained from the CMA due to the fact that the DFE delivers a lower BER. In this adaptation process, the symbol at the output of the DD is delayed by $N + L$ and the channel output used in the CIR (MF) adaptation needs to be delayed by N for the correct update of the MF. The updated CIR coefficients are loaded to the MF. These operations are shown in Figure 3 for the switch position 2. If the decision errors increase, the switch is shifted to position 1 operating the blind CMA in the LE mode otherwise tracking will be lost. The main novelty with this blind SISO DFE is that the MF is estimated blindly using the CMA and it does not involve any matrix decomposition techniques as in SIMO methods [3].

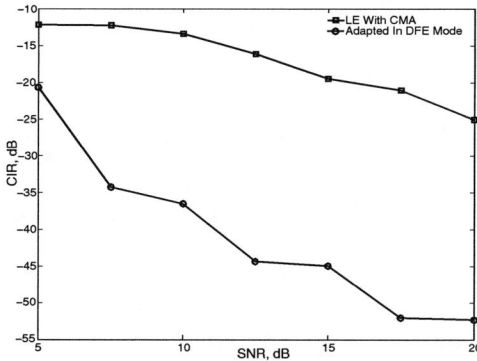

Fig. 4. CIR Estimation Errors For The Blind DFE

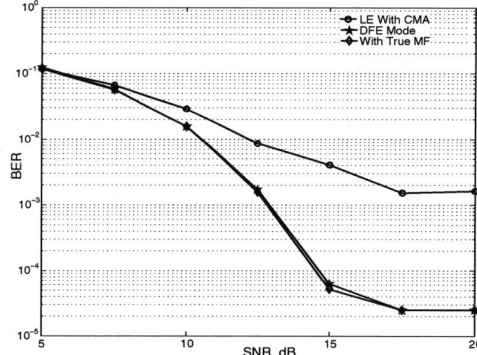

Fig. 5. BER Results Of The Blind DFE With The BER Values Obtained From The True Matched Filter

V. SIMULATIONS

Simulations were performed on the channel $\mathbf{h} = [1.0\; 0.57\; 1.12\; -0.12\; 0.95]$. BPSK symbols were transmitted through the channel. The autocorrelation sequence was estimated using 5000 samples for each case and the noise variance σ_v^2 was found from the eigendecomposition of the channel output autocorrelation matrix calculated using 5000 fractionally ($T/2$) spaced samples assuming ($T/2$) spaced samples are available. The ZF solutions of the WF and FBF were derived using (7) and (8) respectively by eliminating the noise term in autocorrelation matrix with $N = 14$. When the switch was in position 1, the normalized CMA was applied with the variable step size $\frac{\mu}{20+\|\mathbf{p}'(k)\|^2}$ where μ was 0.01 and the input vector to the MF is defined as $\overline{\mathbf{p}}'(k) = [\overline{\mathbf{p}}'(k) \cdots \overline{\mathbf{p}}'(k-L)]$. For the DFE mode, the Least Mean Square (LMS) [6] algorithm was applied with a fixed step-size $\mu = 0.01$ to adapt the MF in the system identification mode. The structure shown in Figure 3 was operated in the LE mode with the CMA, when the decision error decreased to a certain level ($S_0 = 0.2$, S_0 should be chosen larger for low SNR cases), the structure was turned into the DFE mode and the adaptation of the MF (CIR) continues in the system identification mode using the symbols detected by the DD. Figure 4 shows the estimation errors in the CIR values for the CMA and the DFE mode. The improvement in the CIR estimation is obvious, this is due to the low BER of the DFE once the filter settings are adjusted properly. Figure 5 confirms the effectiveness of the our novel approach offering a BER performance close to the one obtained with the true MF. The MF coefficient trajectories during the adaptation are shown in Figure 6 with the plussed dotted lines (+++) corresponding to the true values. The abrupt correction in the DFE mode is also observed clearly.

VI. CONCLUSIONS

A novel all-blind DFE has been introduced using the blind CMA which is a simple adaptive algorithm for SISO system. The structure is derived from the fact that the DFE is the spectral decomposition of the CIR autocorrelation sequence which is possible to calculate blindly from the channel output samples by sample averaging. The autocorrelation sequence estimate provides the solutions for the FFF and the FBF of the DFE and the MF is obtained from the CMA. The usual DFE structure with the MF is run in LE mode first for the MF estimate with the CMA and then the nonlinear DFE structure is used resulting in a better estimate for the transmitted symbols. These estimated transmitted symbols are used to improve the MF estimate further. As a whole the blind DFE introduced relies on simple well established methods for the blind estimation of the components and

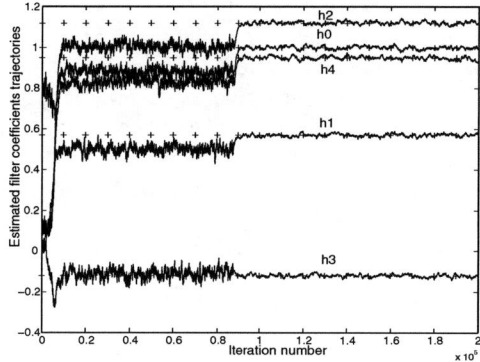

Fig. 6. Matched Filter Coefficient Trajectories For The Proposed Blind DFE

the BER results obtained are close to those obtained with the true MF.

REFERENCES

[1] G. D. Forney, "Maximum-likelihood sequence estimation of digital sequences in the presence of intersymbol interference," *IEEE Trans. Inf. Theory*, vol. IT18, pp. 363-378, 1972.

[2] C. R. Johnson, Jr., P. Schniter, T. J. Endres, J. D. Behm, D. R. Brown, and R. A. Casas, "Blind equalization using the constant modulus criterion: A review," *Proc. IEEE, Special Issue on Blind System Identification and Estimation*, vol. 86, pp. 1927-1950, Oct. 1998.

[3] L.Tong, G. H. Xu, and T. Kailath, "Blind identification and equalization based on second-order statistics: A time domain approach", *IEEE Trans. Inf. Theory*, vol. 40, pp. 340-349, 1994.

[4] E. Moulines, P. Duhamel, J.-F. Cardoso, and S. Mayrargue, "Subspace methods for the blind identification of multichannel FIR filters," *IEEE Trans. Signal Proc.*, vol. 45. pp. 516-525, Feb. 1995.

[5] H. H. Zeng, L. Tong, C. R. Johnson, Jr., "An analysis of Constant Modulus Receiver," *IEEE transac. on signal processing*, vol. 47, no. 11, pp. 2990-2999, November 1999.

[6] S. Haykin, *Adaptive Filter Theory*. Englewood Cliffs, NJ: Prenctice-Hall, 1991.

[7] I. Ozcelik, B. Baykal, I. Kale, "Blind adaptation of a matched filter using the constant modulus algorithm coupled with an optional correction method", *7th International Symp. on Signal Proces. and Its Applications Proc.*, vol. 2, pp. 291-294, July 2003.

[8] A. H. Sayed, T. Kailath, "A Survey of spectral factorization methods", *Numerical Linear Algebra With Applications*, vol. 08, pp. 467-496, 2001.

[9] J. Labat, O. Macchi and C. Laot, "Adaptive Decision Feedback Equalization : Can you Skip the Training Period ? ", *IEEE Trans. on. Com.*, vol. 46, No 7, pp. 921-930, July 1998.

INTEGRATED BLIND ELECTRONIC EQUALIZER FOR FIBER DISPERSION COMPENSATION

Foster F. Dai, Senior Member, IEEE, Shengfang Wei, Richard Jaeger, Fellow, IEEE
Electrical and Computer Engineering Dept., 200 Broun Hall, Auburn University, Auburn, AL 36849-5201
Tel. 334-844-1863, Fax. 334-844-1809, Email: Daifa01@eng.auburn.edu

ABSTRACT

This paper presents an adaptive blind electronic equalization technique for fiber dispersion compensation. Constant-modulus blind adaptive algorithm is proposed to provide reference-free feedback. The proposed 10Gb/s 7-tap electronic equalizer is designed in a 50GHz SiGe technology and the blind feed back algorithm can be implemented in an FPGA with 100MHz operation frequency.

Keywords: fiber dispersion, chromatic dispersion, PMD, blind equalization, CMA

1. INTRODUCTION

Modal, chromatic and polarization mode dispersions are the major sources of transmission impairments in high data rate fiber communications. Without proper compensation, the performance of the fiber communication systems will be severely limited. The available dispersion compensation fiber is static in nature, therefore does not support agile optical networks. Other optical solutions are only capable of compensating one form of the dispersions with very high cost, high insertion loss and slow tuning speed if they are tunable at all. This paper presents an adaptive electronic equalizer for the compensation of all forms of fiber dispersions, including modal, chromatic and polarization mode dispersions. The electronic equalizer can be dynamically tuned at high speed and has much smaller form factor and much lower cost. An electronic equalizer can be integrated into a single chip using the high-speed Silicon Germanium (SiGe) or Indium Phosphorus (InP) technology [1]. Further cost reduction is possible if electronic equalizer and other circuits on the receiver are integrated on the same die. An electronic equalizer module is superior considering its cost, size, reliability, flexibility and speed.

2. ELECTRONIC EQUALIZERS IN ADAPTIVE FIBER RECEIVER

Fig. 1 shows an adaptive receiver with feed forward equalizer (FFE) and decision feedback equalizer (DFE). The received optical signal is converted to electrical signal using a photo-detector (PD). Due to the square-law-detection, the amplitude of the electrical signal is proportional to the intensity of the optical signal. A trans-impedance amplifier (TIA) is used to convert the electrical current mode signal to voltage mode signal. Then, the voltage signal is fed to the data input of the feed-forward equalizer (FFE), which performs the dispersion compensation for the distorted voltage signal. The output of the FFE is experienced additional compensation at a decision feedback equalizer (DFE). The compensated signal is then fed into a clock-data recovery block (CDR) to recover the clock.

FFE, a linear equalizer, is used to compensate for moderate amplitude distortion. The equalizer is said to be linear in the sense that the estimate of a desired response is linearly related to the input vector when steady state conditions are established. On the other hand, a decision feedback equalizer is used to improve receiver performance in the presence of moderate to severe amplitude distortion. A DFE is intrinsically nonlinear as it feeds back to the input a sequence of hard decisions made at the equalizer output. Herein lie both the strength and weakness of the DFE. The feedback section removes the lagging inter-symbol interference that is beyond the reach of the forward section. However, the DFE suffers from the problem of error propagation since hard decisions may destroy information: once a wrong decision is made, errors are fed back to the equalizer input, which in turn leads to a higher probability of error on subsequent decisions. The end result could be a burst of errors. Moreover, like any feedback system, a DFE design involves stability analysis, while a FFE is always stable.

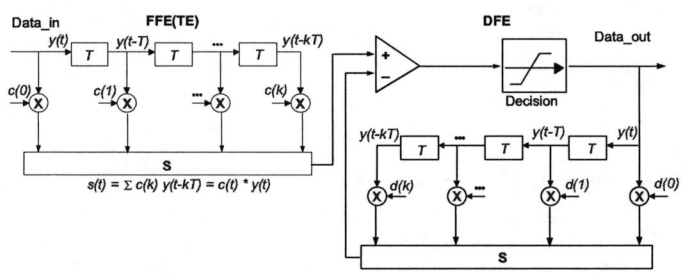

Fig.1 Block diagram of an electronic equalizer including TE and DFE.

3. ADAPTIVE EQUALIZER FEEDBACK ALGORITHM

Various adaptive equalizer algorithms have been developed for digital signal processing. LMS maximum likelihood algorithms are widely used in equalizer designs. *Recursive least-squares* (RLS) algorithm achieves faster convergence, but is computationally more complex than LMS since matrix inversion is required. The reference signal required by the LMS algorithm can be either provided by a transmitted training sequence or estimated by the detected bits. Decision-directed tap adaptation uses the output of the slicer/decision circuit as an estimation of the transmitted sequence. LMS is simple and can be implemented for FFE and DFE at 10Gb/s data rate. For higher data rate, steepest descend algorithm can be used, which consecutively dithers the equalizer taps to achieve the optimized signal quality. As a simplified variant of LMS, sign-sign LMS algorithm evaluates only the signal signs, which has greatly reduced computation time, yet with very "noisy" gradient estimate. The biggest advantages of sign-sign LMS algorithm is that it does not require any high-speed ADC and thus can be implemented at high speed.

The drawback of conventional equalization algorithms lies on the fact that they need the assistance of a training sequence, which is not suitable for applications with varying channel environment. On the other hand, blind equalization provides a reference-free equalization algorithm, which meets the current fiber communication transmission standard. Blind equalization is thus critical for agile optical networks, where the channel environment varies constantly. Since binary ON OFF key modulation with constant envelope is used in fiber communication system, *constant-modulus adaptive algorithm* (CMA) is proposed for fiber dispersion compensation using blind equalization.

3.1 LMS equalizer with training sequence

If a reference training sequence is transmitted during a pre-assigned time slot, the equalizer coefficients can be adapted by using LMS adaptive algorithm so that the output of the equalizer closely matches the training sequence. The LMS reference can also be estimated by the detected bits. In this case, a blind equalization is implemented with the goal of widening the eye-open for improved BER. If r_k is the reference signal and y_k is the equalizer output at time instance k, respectively, the error signal of LMS is defined as

$$e_k = r_k - y_k = r_k - X_k^T W_k \quad (1)$$

where r_k is the reference (training) signal, y_k is the equalizer output at time instance k. The cost function to be minimized is defined as

$$J_k = e_k^2 = [r_k - X_k^T W_k]^2 \quad (2)$$

In each iteration of the adaptive process, we have a gradient estimate of

$$\nabla J_k = \frac{\partial J_k}{\partial W_k} = 2e_k \frac{\partial e_k}{\partial W_k} = -2e_k X_k \quad (3)$$

With this simple estimate of the gradient we can specify a steepest descent adaptive algorithm that is described by the following equations

$$W_{k+1} = W_k - \mu \nabla J_k = W_k + 2\mu e_k X_k \quad (4)$$

3.2 CMA blind equalizer

Inclusion of the training sequence with the transmitted information adds an overhead and thus reduces the throughput of the system. Therefore, to reduce the system overhead, adaptation schemes are preferred that do not require training. In blind equalization, instead of using the training sequence, one or more properties of the transmitted signal are used to estimate the inverse of the dispersive channel. Unlike LMS that assumes a reference r_k, the error e_k in Eq. (1) cannot be used to define the cost function in blind equalization. Since the transmitted signal in fiber communication is of the constant amplitude, we can assume constant amplitude A in the absence of signal degradations. Then, the error signal in blind equalization can be defined as [3]

$$e_k = |y_k|^2 - A^2 \quad (5)$$

Further define the cost function as

$$J_k = E\{e_k^2\} = E\{(y_k^2 - A^2)^2\} \quad (6)$$

The true gradient of J at time k could be approximated by its instantaneous value

$$\nabla J_k = \frac{\partial J_k}{\partial W_k} = 2e_k \frac{\partial e_k}{\partial W_k} = 4e_k y_k X_k = 4(y_k^2 - A^2) y_k X_k \quad (7)$$

And the CMA recursion expression becomes

$$W_{k+1} = W_k - \mu \nabla J_k = W_k - \mu (y_k^2 - A^2) y_k X_k \quad (8)$$

where $y_k = W_k^T * X_k$ \quad (9)

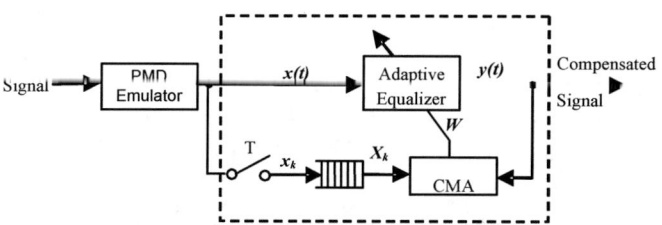

Fig. 2. Block Diagram of CMA adaptive equalizer

4. SIMULATION RESULTS

4.1 LMS equalizer

To demonstrate the effect of electronic equalizer on fiber dispersion compensation, we have conducted system level simulations in MATLAB. We simulated a 7-tap analog feed forward equalizer (FFE) with LMS feedback algorithm for PMD compensation (Fig. 3) and chromatic dispersion compensation (Fig. 4). Note that the severely distorted signal can be compensated using just 7 taps in an analog transversal equalizer. With the adaptive feedback, the tap coefficients are constantly adjusted based on the error signal. Even without feedback, the fixed coefficient equalizer can still open the distorted eye. For fixed network, we can thus predetermine the equalizer tap coefficients and load them into the equalizer at the field, which is very attractive from a cost reduction point of view. For chromatic dispersion, tap coefficients can be programmed based on the length and type of the fiber, providing a much more economic way than the current solution using dispersion compensation fiber (DCF), which is static and also wavelength dependent.

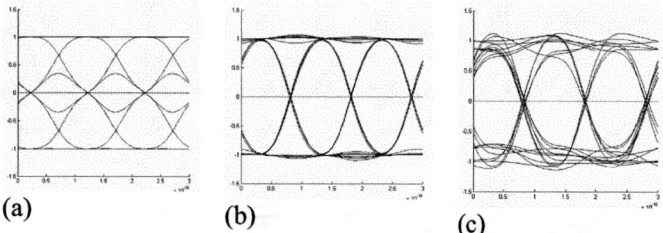

(a) (b) (c)

Fig.3 PMD compensation using a 7 tap transversal equalizer. (a) Eye diagram distorted by 1st order PMD with power splitting ratio = 0.5 and DGD = 0.8 period. (b) Compensated eye diagram with LMS adaptive feedback. (c) Compensated eye diagram with fixed tap coefficients.

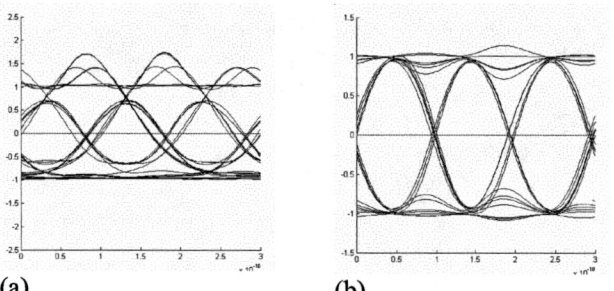

(a) (b)

Fig. 4 Chromatic dispersion compensation using a 7 tap analog FFE equalizer. (a) Eye diagram distorted by chromatic dispersion after transmission in a 75km SMF-28 fiber at wavelength 1550nm. (b) Compensated eye diagram with LMS adaptive feedback.

4.2 CMA blind equalizer

In our simulation for blind equalization, we set signal amplitude A to 1 and normalized the input signal to the range of [-1, 1]. Baud-Spaced CMA (sampling period equals to the symbol period) is used in our equalizer, and W_0 is initialized to be "Single-Spike" [2] (one non-zero tap coefficient, and all other to 0).

Three scenarios with different setting of DGD are investigated. Symbol period is set to be 0.025ns (25e-12 seconds). Simulation results when DGD is set to 48% symbol period, 64% the symbol period, and 80% the symbol period, are shown in Fig. 5, Fig. 6, and Fig. 7 respectively. In all scenarios, initial W_0 is chosen to be $[0\ 0\ 0\ 1\ 0\ 0]^T$, and feedback gain μ is set to 0.005. Assuming the eye is constant along the horizontal direction and the eye-open is then measured as the vertical height between the mean of peak amplitudes and the mean of valley amplitudes, which converges to value 2 ideally.

From the results, we see that when DGD is less than half (48%) symbol period (Fig. 5), the eye could be fully opened quite fast (in about 10 ns, or 400 symbol periods); when DGD increases to 64% of symbol period (Fig. 6), the eye can be almost fully opened in less than 50ns or 2000 symbol periods; when DGD is further increased to 80% of symbol period (Fig. 7), it takes more than 100ns for the coefficients to get stable, and the cost function J can not converge to 0 and stays around 0.1 (Fig. 7-d), but the eye is opened much wider (Fig. 7-b) than the distorted one (Fig. 7-a).

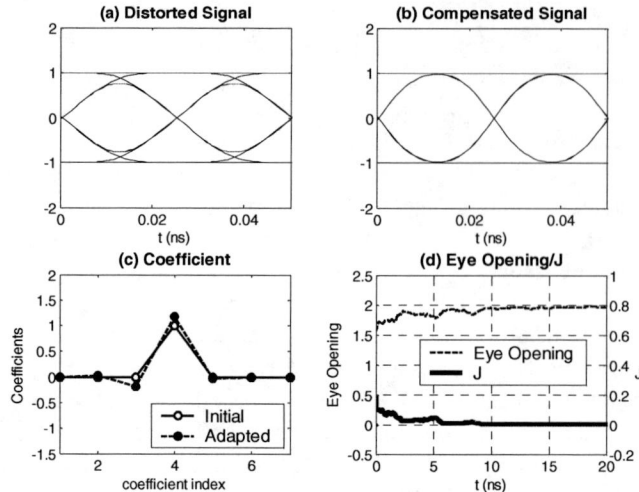

Fig. 5 DGD=48% symbol period. (a), (b) Eye diagrams before/after the compensation, and they are plotted within a time window of two symbol periods and trace number of 20; (c) Initial and adapted coefficients; (d) Eye Opening and cost function J. J is defined in (6) and ideally converges to 0.

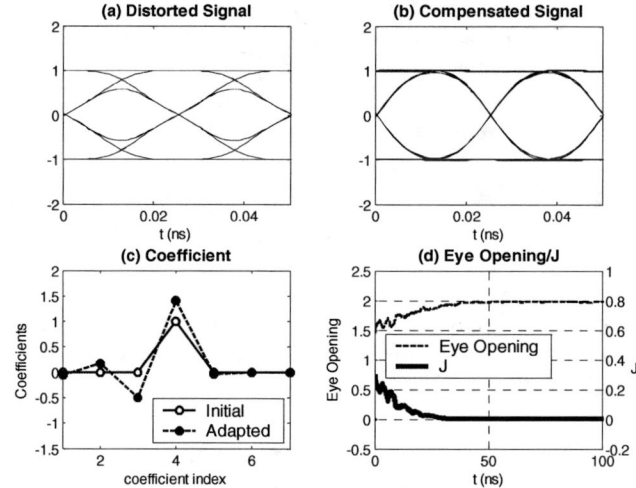

Fig. 6: DGD=64% symbol period. (Refer to the notes on Fig 5 for the meaning of each measure)

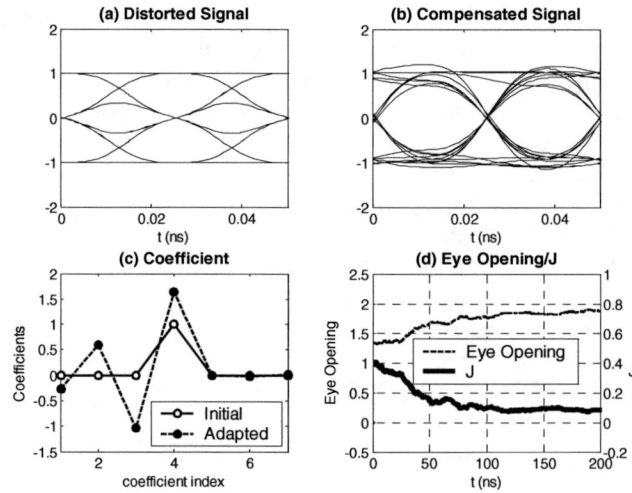

Fig. 7: DGD = 80% symbol period. (Refer to the notes on Fig 5 for the meaning of each measure)

5. IMPLEMENTATION OF HIGH SPEED FEED FORWARD EQUALIZER IN SIGE TECHNOLOGY

The proposed 7-tap electronic equalizer was implemented in a 50GHz SiGe technology. The critical block of the equalizer is the variable gain stage, which requires 10GHz bandwidth with small phase group delay for OC-192 fiber networks. The maximum gain of the variable gain stage is set such that its output would not saturate the following stage under any circumstance. A Gilbert gain-cell is used for this purpose and the simplified circuit is given in Fig. 8. The simulated magnitude of the equalizer tap gain under various gain settings is shown in Fig. 9, demonstrating linear gain tuning with larger than 10GHz bandwidth. The tap delay line is implemented using the cascaded unity gain stages with total delay of half data period (50ps). The equalizer circuit design is given in details in [5].

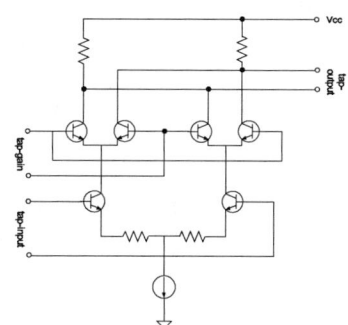

Fig. 8 Simplified equalizer tap gain stage.

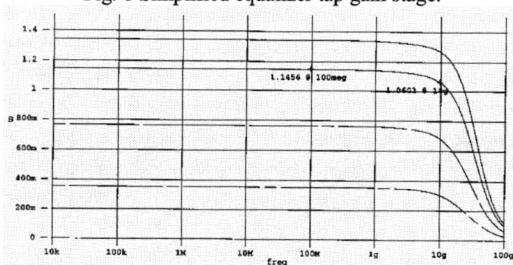

Fig.9 Magnitude of equalizer tap gain under various gain settings.

6. IMPLEMENTATION OF CMA BLIND EQUALIZION FEEDBACK IN A 100MHZ FPGA

The current CMA algorithm is very easy to implement. Based on (9) and (8), the implementation is shown in Fig. 10. If the filter size is L, we see that only L accumulators, 3L+1 multipliers, and L adders/subtractors are needed for CMA algorithm (Fig. 10). Note that

i) The small gradient gain μ in (8) doesn't really need a multiplier. If μ is chosen to be 1/256 ($\approx$0.004), we actually only need to truncate (discard) the lowest 8 bits (A=8 in Fig. 10).

ii) Although there are 3L+1 multipliers, most of them can work in parallel, for example, the left most L multipliers in the 1st stage, middle (L+1) multipliers in the 2nd stage, and right most L multipliers in the 3rd stage, therefore the delay due to multiplier is only $3T_{mul}$, where T_{mul} is the average multiplication time.

iii) We have different clock signals for the sampling (CLK1 in Fig. 10), and the coefficient updating (CLK2 in Fig. 10) which is much slower than CLK1. Note that multiplier has extra CLK which is not shown in the block diagram.

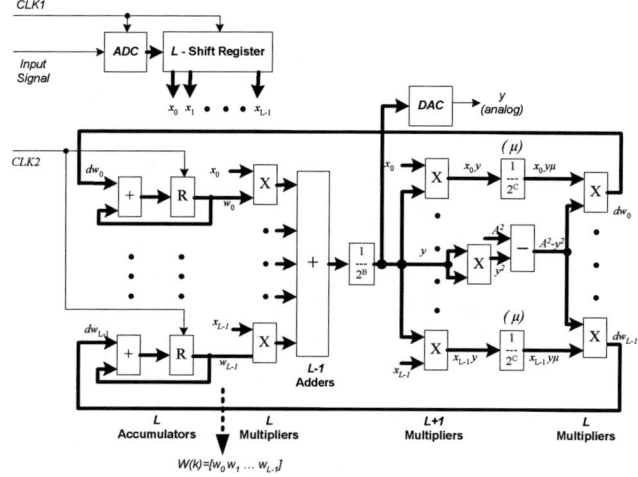

Fig. 10 Block Diagram of hardware implementation.
CLK1: Sample Clock (equal to symbol rate);
CLK2: Coefficient Updating Clock
L: Length of Equalizer;
A: Constant Amplitude
$1/2^B$ and $1/2^C$: bit "Truncator" (takes off the lowest B/C bits), (where C=[$\log_2\mu$], B= number of fractional bits in fixed point format of coefficient w)
μ: step size (gradient gain)
Bolded lines represent multiple bits (bus) signals and thin lines for analog and control signals

7. CONCLUSIONS

This paper presents a novel CMA based blind electronic equalization technique for fiber dispersion compensation. Simulation demonstrated effective compensation means for chromatic and PMD dispersions using the LMS and CMA blind equation techniques.

8. REFERENCES

[1] Azadet, K.; Haratsch, E.F.; Kim, H.; Saibi, F.; Saunders, J.H.; Shaffer, M.; Song, L.; Meng-Lin Yu, "Equalization and FEC techniques for optical transceivers," *IEEE Journal of Solid-State Circuits*, Vol. 37, 317 -327, (2002).

[2] C. Richard Johnson, Jr., Philip Schniter, Thomas J. Endres, James D.Behm, Donald R. Brown, Raúl A. Casas, "Blind equalization using the Constant Modulus Criterion: a review," *Proceeding of the IEEE* - 86(10):1927--1950, October 1998.

[3] Michael G. Larimore, C. Richard Johnson, John R. Treichler, "Theory & design of adaptive filters," ISBN 0130402656, Published March 2001.

[4] J.R. Treichler and B.G. Agee, "A new approach to multipath correction of constant modulus signals," *IEEE Trans. On Acoustics, Speech, and Signal Processing*, vol. ASSP-31, 459-472, (1983).

[5] V. Kakani, F. F. Dai, and R. C. Jaeger, "An High Speed Integrated Equalizer for Dispersion Compensation in 10Gb/s Fiber Networks", *IEEE International Symposium on Circuits and Systems (ISCAS)*, Kobe, Japan, May 2005.

Linear and Nonlinear Macromodels for Power/Signal Integrity

I. S. Stievano, S. Grivet-Talocia, I. A. Maio, F. G. Canavero

Politecnico di Torino, Dipartimento di Elettronica, Torino, Italy
{stievano,grivet,maio,canavero}@polito.it

Abstract—In this paper, a systematic methodology for the assessment of Power/Signal Integrity effects in high-speed communication and information systems is presented. The proposed methodology leads to accurate and efficient macromodels for logic devices, transmission-line interconnects and discontinuities that can be easily implemented in any commercial tool and can be combined for the simulation of the whole system. The SPICE implementation of macromodels is used for the prediction of power/ground noise and signals propagating on interconnects. An assessment of the impact that possible simplifications in the macromodel generation phase have on the simulation of a realistic application example, is discussed.

I. INTRODUCTION

In today's digital and mixed-signal circuit design, a serious concern is represented by interference noise, caused by crosstalk on signal busses and by current transients due to the synchronized switching of a huge number of digital gates. To assess and circumvent such problems prior to fabrication, a reliable analysis needs to be conducted during the design and verification process.

Such an assessment amounts to predicting the power and ground noise fluctuations and the signals propagating on the interconnects via numerical simulation. The combination of propagation effects with possibly very complex geometry and with nonlinear/dynamic behavior of logic devices makes a direct full-wave approach not feasible. Therefore, the feasible strategy that we present here amounts to subdividing the propagation path into separate and well-defined sub-structures, each of which is separately characterized by a macromodel, *i.e.*, a set of equations that are able to reproduce with sufficient accuracy the port behavior. Different macromodeling strategies are needed for the different structures: the nonlinear drivers and receivers terminating the interconnects, the interconnection elements characterized by a complex geometry, *i.e.*, discontinuities such as packages or via holes, and segments of uniform transmission-line structures at chip, module or board level. The proposed macromodels offer good accuracy and improved efficiency and can be easily implemented in any commercial tool as SPICE subcircuits and be used in any conventional circuit solver for the simulation of the whole system.

The following Sections provide short descriptions of the separate modeling methodologies adopted for the three classes of devices that we use to build the entire communication system. The paper is then concluded by a discussion of the impact that different simplifications in the generation of macromodels can produce on predicted signals.

II. DISCONTINUITIES

Discontinuities in the transmission paths represent a critical bottleneck for the integrity of high-speed signals in electronic systems, since their presence typically imposes serious bandwidth limitations. Such discontinuities can be vias and via arrays, bends, junctions, connectors, etc. A careful assessment of their effects requires a proper modeling of such structures, whose geometry and material properties must be taken into account, thus requiring a complex full-wave electromagnetic analysis.

The modeling approach is based on the characterization of discontinuities via full-wave analysis and/or by direct measurement. As a result of the characterization, a set of port responses in time or frequency domain are available, and constitute a large amount of data to be processed to obtain a global macromodel. For this reason, a partial macromodeling of several disjoint subsets of P_k port responses each is done first; the partial macromodels, which are processed independently, are then easily assembled into a global macromodel for the entire structure.

The Vector Fitting algorithm, either in its conventional frequency-domain formulation [1] or in its more recent time-domain extension [2] is the main macromodeling engine providing a rational approximation for each subset in the following form

$$\boldsymbol{H}_k(s) \simeq \boldsymbol{H}_{k,\infty} + \sum_{n=1}^{N_k} \frac{\boldsymbol{R}_{k,n}}{s - p_{k,n}}, \quad (1)$$

with diagonal $\boldsymbol{H}_{k,\infty}$ and $\boldsymbol{R}_{k,n}$. The global macromodel for the entire structure is recovered by tiling the partial macromodels according to

$$\boldsymbol{H}(s) = \sum_{k=1}^{K} \boldsymbol{Q}_k^T \boldsymbol{H}_k(s) \boldsymbol{P}_k, \quad (2)$$

where matrices $\boldsymbol{P}_k$ and $\boldsymbol{Q}_k$ are $P_k \times P$ selectors having a single unitary entry in each row with all the other entries vanishing. As a final step, the poles/residues representation (1) is translated by standard techniques into an equivalent state-space representation, which is suitable for implementation into standard industrial simulation tools like SPICE and VHDL-AMS.

The resulting macromodel is usually characterized by excellent accuracy, and it is stable by construction. However, it might not be passive since it was identified by a sequence of least squares solutions that do not guarantee passivity a priori. Since a non-passive macromodel can lead to unstable solutions when its terminations are changed, passivity is checked and enforced using spectral perturbation of an associated Hamiltonian matrix [3].

III. TRANSMISSION-LINE INTERCONNECTS

At each scale (chip, board or system level), the information is exchanged via busses, that often can be represented as transmission lines. Their main effects on the conducted signals are delays due to their physical length, shape distortion due to losses in the conductors and substrate, and crosstalk among adjacent traces. An accurate and efficient transient analysis of the information propagation on busses is still a challenging task, due to the intrinsic difficulties in the design of stable algorithms for the time-domain analysis of structures with frequency-dependent parameters.

As a simulation strategy, we employ a generalized version of the well-known Method of Characteristics, that allows to deal with multiconductor lines whose per-unit-length parameter matrices have a possibly complex dependence on frequency. In such case, both the characteristic admittance and the propagation operators of the line are frequency-dependent. In our approach [4], the line equations are projected onto their high-frequency asymptotic modes. This leads to a particular form of the propagation operator that allows an easy extraction of the asymptotic line modal delays, as follows:

$$\mathbf{P}(s) = \mathrm{diag}\{e^{sT_k}\} \mathbf{M}_\infty^{-1} e^{-\mathcal{L}\Gamma(s)} \mathbf{M}_\infty, \quad (3)$$

where $\Gamma^2(s) = \mathbf{Y}(s)\mathbf{Z}(s)$, and $\mathbf{Z}(s)$, $\mathbf{Y}(s)$ are the per-unit-length matrix impedance and admittance of the line, respectively; T_k represent the line modal delays and $\mathbf{M}_\infty$ is a matrix of their corresponding eigenvectors.

A second step is the generation of a rational approximation of the above-defined delayless propagation operator,

$$\mathbf{P}(s) \simeq \sum_n \frac{\mathbf{R}_n^P}{s - q_n} + \mathbf{P}_\infty$$

and of the characteristic admittance

$$\mathbf{Y}_c(s) = \Gamma^{-1}(s)\mathbf{Y}(s) \simeq \sum_n \frac{\mathbf{R}_n^Y}{s - p_n} + \mathbf{Y}_\infty \quad (4)$$

The final step of the macromodeling procedure is the generation of a lumped equivalent that can be used in a system-level simulation environment. The synthesis of lumped equivalents corresponding to rational expansions of characteristic admittance and delayless propagation operators is a standard task: SPICE-like implementation uses the basic elements, combined with dependent sources and ideal lossless lines, while the VHDL-AMS model implements the corresponding delayed differential equations.

IV. LOGIC DEVICES

In the distribution networks of digital signals, the terminations represented by the IC drivers and receivers are critical elements, since their intrinsic nonlinear and dynamic behavior can significantly affect the capacity of the propagation paths. Therefore, accurate and efficient macromodels are required.

In this Section, we briefly review the Mπlog approach [5], providing an effective methodology for the construction of accurate and efficient behavioral models of logic devices. In this approach, output buffer constitutive relations are sought as dynamic nonlinear parametric two-piece models of the form

$$i(t) = w_1(t)i_1(v(t)) + w_2(t)i_2(v(t)) \quad (5)$$

where $i_n(t)$, $n = 1, 2$ are submodels describing the port behavior in the HIGH and LOW logic states, respectively, and $w_n(t)$, $n = 1, 2$ are weighting coefficients describing state transitions.

Parametric nonlinear relations and system identification methods like those involving the identification of mechanical systems, economic trends, etc. allow us to obtain improved nonlinear dynamic models for submodels $i_{1,2}$ in (5), see [7]. Parametric models are usually expressed as sums of sigmoidal functions of the involved variables and their parameters are estimated by fitting the model responses to suitable transient responses of the input and output variables related by the model. In this case, the related variables are the voltage and current of the output port in fixed logic state and the model parameters are computed by minimizing a suitable error function between voltage and current waveforms of the model and real device. Specific algorithms are available to solve this problem, that depend on the choice of the family of basis functions used to define the parametric models. Parametric models offer rigorous mathematical foundations, identifiability from external observations, good performances for the problem at hand as well as preserving the ability to hide the internal structure of the modeled devices. Finally, parametric models can be readily implemented according to standard industrial simulation tools like SPICE and VHDL-AMS. In addition, such already-mentioned SPICE and VHDL-AMS implementations are completely compatible with the multilingual extension of IBIS (Input/output Buffer Information Specification), which is the most established standard for the behavioral description of IC ports. In fact, ver. 4.1 of IBIS specification [6] is an extension, recently devised to overcome some limitations of the original standard, allowing for more general models not necessarily based on simplified circuit interpretations.

Details on parametric modeling of single-ended CMOS devices can be found in [7], where the parametric approach is applied to the modeling of input and output ports of commercial devices by means of the transient responses of their transistor level models. The estimation of parametric models from measured transient responses is demonstrated in [8]. The extension of the methodology in order to take into account the device temperature, the power supply voltage, the

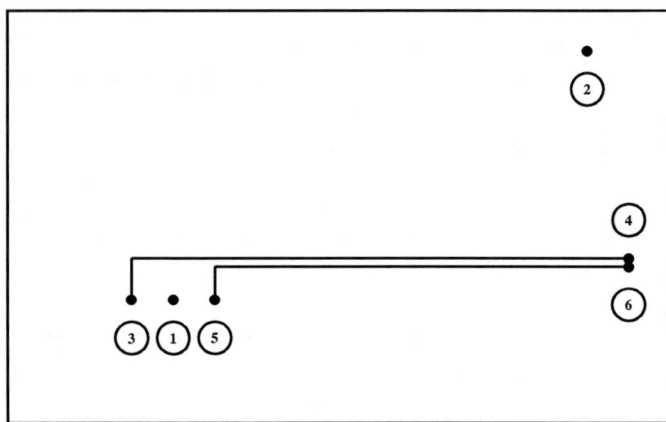

Fig. 1. PCB structure used for illustration of the proposed methodology. The board size is 16×10 cm, with a power-signal-ground configuration ($\sigma = 5.9 \times 10^7$ S/m for all conductors). Each layer ($\varepsilon_r = 4.2$, $\tan \delta = 0.001$) is 0.7 mm high. The stripline conductors are 0.2 mm wide with a separation in the coupled segment of 0.5 mm. Port locations are, in mm units from the bottom-left corner, 1:(40,30), 2:(140,90), 3:(29,31), 4:(150,41.1), 5:(49,31), and 6:(150,40.6).

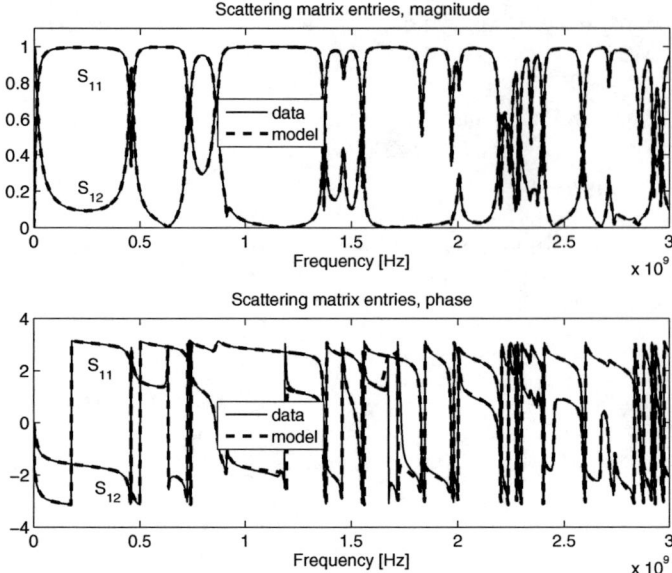

Fig. 2. Macromodel generation for the power/ground structure depicted in Fig. 1.

power supply current drawn by buffers as well as to model tri-state devices is addressed in [5]. Finally, some preliminary results on the modeling of differential Low Voltage Differential Signaling (LVDS) devices has appeared in [9].

V. EXAMPLES

In this section, we apply the proposed methodology to a simple PCB test case, depicted in Fig. 1. Ports 1 and 2 are located between the power and ground conductors, while ports 3–6 provide the termination to a coupled stripline structure. A full-wave transient solver based on the Finite Integration technique [10] was used to generate the frequency-dependent scattering responses of the structure up to a maximum frequency of 3 GHz. Two different simulations were performed. The first simulation included both power/ground planes and stripline conductors, and resulted in a 6-port scattering matrix. In the second simulation, we removed the signal conductors and we obtained a simplified unpopulated power/ground structure with only ports 1 and 2 left. This second run is aimed at the assessment of the power-ground model variations due to the presence of additional signal conductors in the structure.

Passive macromodels were generated for both populated and unpopulated structures. First, a rational fit based on the Vector Fitting algorithm was generated. Then, the correction scheme of Section II was used to enforce the macromodel passivity. As an example, we compare in Fig. 2 some macromodel responses to the original scattering responses for the populated board case. The plots show that the accuracy that can be achieved with the proposed technique is excellent, throughout the entire modeling bandwidth.

The above macromodels were synthesized into equivalent circuits in SPICE form, and three different SPICE decks were run. In all cases, the two stripline conductors were connected at port 3 and 5 to the output ports of two identical switching drivers forcing a bit pattern of 01001 and 00100, respectively. The driver is a Philips LVC244 3.3 V CMOS device, whose reference transistor-level description is available from the official site www.semiconductors.philips.com. A macromodel of the driver has been generated from the responses of the reference model by means of the Mπlog procedure of Sec. IV and the SPICE implementation of the macromodel is used for the simulation experiment. Also, the striplines were terminated by 50Ω loads at ports 4 and 6. Port 2 was connected by a 3.3 V battery in series with a 1Ω resistance, while port 1 was directly connected to the power supply ports of the two switching drivers.

The difference in the three SPICE runs is only on the models that were used for the power, ground, and signal conductors, as itemized below.

(i) The first setup used a fully rational macromodel for the complete 6-port structure representing power, ground, and signal conductors.

(ii) The second setup used a rational macromodel only for the 2-port power/ground structure obtained for the populated board case. The stripline conductors were modeled as lossy transmission lines using the procedure outlined in Section III. The two lines were considered coupled only in the parallel segment (see Fig. 1).

(iii) The third setup used the simplest rational macromodel of the unpopulated 2-port power/ground structure. The stripline conductors were modeled as lossy transmission lines as for the second setup.

Note that, in order to produce a fair comparison between the three cases, a correction for the DC resistance of the stripline conductors of case (i) was necessary. In fact, the employed full-wave solver allowed only a $\sqrt{f}$ type of conductor loss,

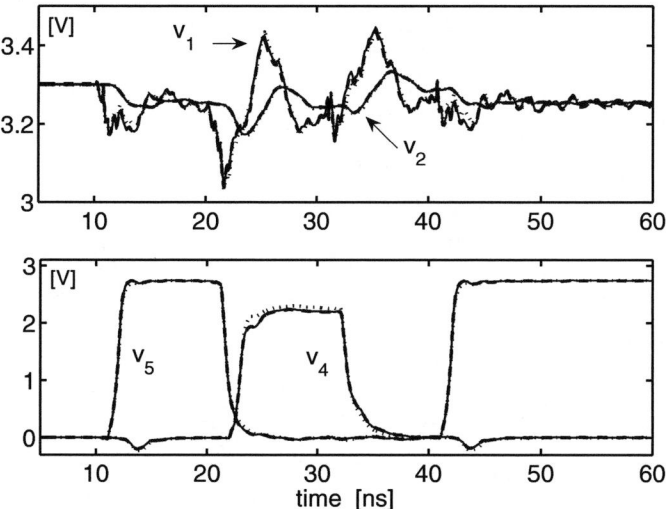

Fig. 3. Set of port voltage responses for the three SPICE runs of the example test case of Sec. V: (i) full macromodel for the structure of Fig. 1 (dotted lines); (ii) two-port macromodel for the populated power net and the line macromodel (dashed lines); (iii) simplified two-port macromodel for the power net and the line macromodel (solid lines).

resulting in no DC losses for all conductors. Consequently, no DC loss was present in the corresponding 6-port rational macromodel. On the other hand, an accurate set of frequency-dependent per-unit-length line parameters was generated for cases (ii) and (iii) via 2D transverse electromagnetic simulation, including the DC resistance. This resistance (about 5Ω) was therefore inserted as a lumped component in series to each stripline port in the SPICE deck of case (i).

Figure 3 shows the transient responses of voltages $v_1(t)$, $v_2(t)$, $v_4(t)$ and $v_5(t)$ for the above SPICE simulations. For the example board of Fig. 1, the above curves highlight a very good agreement between the three different predictions of transmitted signals on conductors, on power/ground noise and on sensitive effects like crosstalk. Even if simplified (uncoupled) macromodels for the power/ground part are used and combined to conventional transmission line macromodels for the signals conductors, only negligible differences can be registered for the steady state values of voltage responses v_4 and v_5 due to the correction of DC resistance, as already mentioned above.

As an additional index of performance, Table I summarizes the results on the efficiency of macromodels for the computation of the curves of Fig. 3, thus highlighting the speed-up improvement introduced by the simplified macromodels used in the SPICE runs (ii) and (iii).

TABLE I
CPU TIME COMPARISON FOR THE PREDICTION OF THE CURVES OF FIG. 3.

SPICE run	Sim. time
(i) fully rational macromodel	426 s
(ii) 2-port macromodel	129 s
(iii) simplified 2-port macrom.	119 s

VI. CONCLUSIONS

This paper presents different macromodeling strategies devised for the various structures that can be found along high-speed propagation paths. Macromodels of logic devices, that are nonlinear dynamic circuits, are expressed in terms of parametric equations reproducing the device behavior; signal busses, that can be represented as transmission lines with significant propagation delays together with dispersion and losses, are macromodeled as lumped equivalents combined with the line modal delays; finally, macromodels of discontinuities are linear, lumped equivalents fitting the port characteristics of the structure. All macromodels can be cast in terms of equivalent circuits, ready for system-level analyses using SPICE-like circuit solvers. The application example developed in this work shows the efficiency of using the advocated decomposition approach for the power and signal integrity characterization of high-performance digital systems. The considerable simulation speed-up and the readiness for trying several different configurations by simply linking macromodels is appealing for a designer who must perform what-if analyses during each new product development. Also, the excellent accuracy achieved in this paper, although for a simplified study case, is very promising and makes more complex structures of common use in real systems worth of further investigations.

ACKNOWLEDGMENTS

This work was supported in part by the Italian Ministry of University (MIUR) under a Program for the Development of Research of National Interest (PRIN grant #2004093025).

REFERENCES

[1] B. Gustavsen, A. Semlyen, "Rational approximation of frequency responses by vector fitting", IEEE Trans. Power Delivery, Vol. 14, July 1999, pp. 1052–1061.
[2] S. Grivet-Talocia, "Package Macromodeling via Time-Domain Vector Fitting", IEEE Microwave and Wireless Components Letters, pp. 472-474, vol. 13, n. 11, November, 2003.
[3] S. Grivet-Talocia, "Passivity enforcement via perturbation of Hamiltonian matrices", IEEE Trans. CAS-I, pp. 1755-1769, vol. 51, n. 9, September 2004.
[4] S. Grivet-Talocia, H-M. Huang, A. E. Ruehli, F. Canavero, I. M. Elfadel, "Transient Analysis of Lossy Transmission Lines: an Effective Approach Based on the Method of Characteristics", IEEE Trans. Adv. Packaging, pp. 45-56, vol. 27, n. 1, Feb. 2004.
[5] I. S. Stievano, I. A. Maio, F. G. Canavero, "Mπlog Macromodeling via Parametric Identification of Logic Gates," IEEE Trans. Adv. Packaging, Vol. 27, No. 1, pp. 15–23, Feb. 2004.
[6] "I/O Buffer Information Specification (IBIS) Ver. 4.1," on the web at http://www.eigroup.org/ibis/ibis.htm.
[7] I. S. Stievano, F. G. Canavero, I. A. Maio, "Parametric Macromodels of Digital I/O Ports," IEEE Trans. Adv. Packaging, Vol. 25, No. 2, pp. 255–264, May 2002.
[8] I. S. Stievano, I. A. Maio, F. G. Canavero, "Behavioral Models of I/O Ports from Measured Transient Waveforms," IEEE Trans. on Instrumentation and Measurement, Dec. 2002.
[9] I. S. Stievano, C. Siviero, I.A. Maio, F. Canavero, "Behavioral Macromodels of Differential Drivers" , *8th IEEE Workshop on Signal Propagation on Interconnects (SPI), Heildelberg (Germany)*, pp. 131-134, May 9-12, 2004
[10] *CST Microwave Studio Manual, version 4*, Computer Simulation Technology GmbH, Germany, 2003 (www.cst.de).

Delay Extraction from Frequency Domain Data for Causal Macro-modeling of Passive Networks

Rohan Mandrekar[1], Madhavan Swaminathan[2]
School of Electrical and Computer Engineering
Georgia Institute of Technology
Atlanta, Georgia 30332–0250
Email: rohan[1],madhavan.swaminthan[2]@ece.gatech.edu

Abstract— Causality, which deals with the precise timing of signal propagation through passive structures like interconnects, is an important problem in the time domain simulation of distributed passive networks. If unaccounted for, it can lead to significant error in the signal integrity analysis of high-speed digital systems. Distributed passive systems are characterized by multiple causality conditions in their time-domain response. The port-to-port delay, which is the time taken by the electrical signal to reach from one port to the other, forms the basis for these causality conditions. This paper describes a technique to extract the port-to-port delay in passive networks directly from their frequency domain parameters. The technique can be applied to either S, Y or Z parameters of passive systems. It can also be extended to multi-port networks. The delay thus determined has been used to enforce causality on the S parameters of a passive system.

I. INTRODUCTION

The ever-increasing complexity and operating frequencies in state-of-the-art digital systems present the need for more accurate and faster, simulation and modeling techniques. One of the problems being researched in this domain is the modeling of passive structures in an electronics system and the co-simulation of these passive structures with active devices in a time domain simulator like SPICE. Passive structures have traditionally been analyzed in the frequency domain. To understand their behavior in conjunction with active devices, which are analyzed in the time domain, the Inverse Fourier Transform can been used. According to the Inverse Fourier Transform, a passive frequency domain response can be transformed into a stable time domain response subject to the causality condition which forces the time domain response to be zero for $t < 0$ [1]. However with increasing clock frequencies, the size of the passive structures is comparable to the signal wavelength at the operating frequency, leading to distributed effects like delay playing an important role in the time domain analysis. These distributed effects imply that there are many causality conditions that need to be satisfied, to generate the correct signal response in the time domain [2]. Figure 1 shows the multiple causality conditions due to the finite velocity of the electromagnetic waves propagating on a transmission line. Present frequency domain macro-modeling techniques analyze a passive structure in the frequency domain and then approximate the frequency response using lumped elements which can be incorporated into SPICE. However such lumped element representations of passive structures are unable to accurately capture the distributed effects since the data obtained from frequency domain analysis of passive systems is bandlimited and without explicit information on the delay embedded in the systems. This often leads to violation of the causality conditions as shown in Figure 2 where the frequency response was approximated with rational functions using the method described in [3].

Such violations of causality can considerably affect the signal integrity analysis in high-speed systems [4]. Other existing techniques like the W-element models for transmission lines in HSPICE simulate causality conditions well. However these models are not applicable to arbitrary passive structures and have some accuracy issues [2]. Hence there is a clear need for new macro-modeling techniques which accurately account for

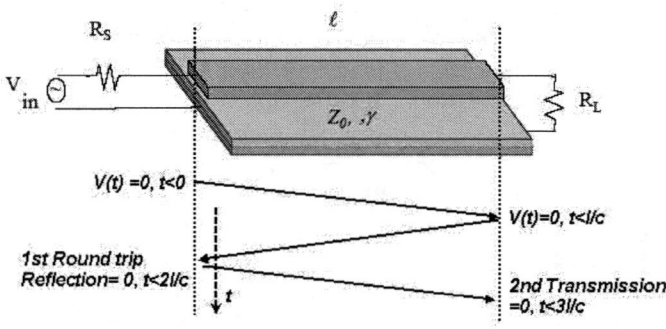

Fig. 1. Multiple causality conditions on a transmission line response

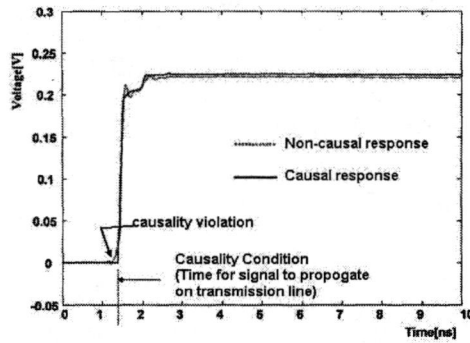

Fig. 2. Violation of causality conditions in macro-modeling

the multiple causality conditions in passive structures. Since the port-to-port delay in a passive system is the basis for the causality conditions (Figure 1), determination of this delay from the frequency domain response is an important step towards causal macro-modeling.

II. CAUSALITY IN MACRO-MODELING OF PASSIVE NETWORKS

Macro-modeling of a passive network involves development of a black-box representation of the network which approximates it's port-to-port behavior [3]. Such a representation is generated by approximating the frequency response of the network using complex poles and residues in the form

$$H(s) = \sum_{n=1}^{N} \frac{\alpha_n}{s - \beta_n} + k_d + k_l s \qquad (1)$$

where β_n are the complex poles, α_n are the complex residues and $s = j\omega$ where ω is the angular frequency. Once the poles and residues are known, they can be represented in a lumped-element-circuit form to be used in SPICE. $H(s)$ generated this way is stable if all the poles β_n lie in the left half of the complex s-plane. To ensure passivity of $H(s)$ several methods have been proposed in literature. For instance, the one described in [3] imposes a set of conditions on the residues α_n, k_d and k_l, to ensure that the developed macro-model is passive. Macro-models developed using such techniques satisfy the stability and passivity criteria, but not causality. This is because distributed passive systems like transmission lines have infinite poles, and equation 1 approximates their response using only a finite number of poles N, obtained using bandlimited frequency response data. This prevents $H(s)$ from accurately capturing the delay in the network, since capturing delay using a function in the pole-residue form shown in equation 1 would require an infinite number of poles N. This problem is illustrated in Figure 2 where the technique described in [3] was used to macro-model a transmission line using it's bandlimited frequency response. The time domain response obtained using the macro-model was non-causal.

III. DELAY EXTRACTION IN PASSIVE NETWORKS

In electronic systems, passive structures absorb, transfer and dissipate the electrical energy provided to them by the active devices. They are limited by their inability to amplify signals or perform switching fundamentals in their circuit performance. Such characteristics of passive networks result in interesting properties of their responses which have been used in this paper to determine the delay inherent in those networks. The property being used is that passive responses show minimum phase [1]. To understand the concept of minimum phase consider a one-port passive network with impedance parameter $Z11(s)$ where $s = j\omega$ and ω is the angular frequency. If the system is stable then all the poles of $Z11(s)$ lie in the left half of the complex s-plane. Now the same system can also be represented using admittance parameter $Y11(s)$ where $Y11(s) = 1/Z11(s)$. Since the system is stable, all the poles of $Y11(s)$ also lie in the left half of the complex s-plane. However, since the poles of $Y11(s)$ are the zeros of $Z11(s)$ and vice-versa, all the poles and zeros of $Z11(s)$ and $Y11(s)$ lie in the left half of the complex s-plane. This property constrains the phase response of the system such that $-\pi < \angle Z11(s) < \pi$ and $-\pi < \angle Y11(s) < \pi$. Such a system is called a minimum phase system and $Z11(s)$ and $Y11(s)$ are called minimum phase functions. The phase response of such functions does not show any phase transition.

In multi-port passive networks, this property of minimum phase is observed only for the self-responses i.e., only for the diagonal elements of the system matrix. Consider a 2-port passive network represented using impedance parameters

$$Z(s) = \begin{bmatrix} Z11(s) & Z12(s) \\ Z21(s) & Z22(s) \end{bmatrix} \qquad (2)$$

In this system only $Z11(s)$ and $Z22(s)$ are minimum phase functions. The transfer impedances $Z12(s)$ and $Z21(s)$ are stable but do not exhibit minimum phase. This is because of the port-to-port delay embedded in these transfer impedance responses. Let Td be the delay between ports 1 and 2 in the above system. Then $Z12(s)$ can be written as

$$Z12(s) = Z12'(s)e^{-sTd} \qquad (3)$$

According to linear system theory [1] any stable system function can be represented as a product of a minimum phase function and an all-pass function, where an all-pass function is one whose magnitude is unity over the entire frequency range. Therefore

$$Z12(s) = H_{\min}(s).H_{AP}(s) \qquad (4)$$

Comparing equations 3 and 4 and noting that e^{-sTd} has unity magnitude, it can be seen that if $Z12(s)$ is separated into a product of a minimum phase function and an all-pass function, the all-pass function will represent the delay between the two ports. This separation can be performed using the Hilbert Transform [1].

The Hilbert Transform relates the magnitude and phase of a minimum phase function $H_{\min}(j\omega)$ through the equation

$$\arg[H_{\min}(j\omega)] = -\frac{1}{2\pi}P\int_{-\pi}^{\pi} \log|H_{\min}(j\theta)|\cot\left(\frac{\omega - \theta}{2}\right)d\theta \qquad (5)$$

where P is the Cauchy Principal value. Since an all-pass function has unity magnitude, the magnitude response of the minimum phase function $Z12'(s)$ in equation 3 is the same as that of $Z12(s)$. Therefore the port-to-port delay Td embedded in the transfer impedance parameter $Z12(s)$ can be determined as follows

$$|Z12'(j\omega)| = |Z12(j\omega)| \qquad (6)$$

$$\arg[Z12'(j\omega)] = -\frac{1}{2\pi}P\int_{-\pi}^{\pi} \log|Z12(j\theta)|\cot\left(\frac{\omega - \theta}{2}\right)d\theta \qquad (7)$$

$$e^{-j\omega Td} = \frac{Z12(j\omega)}{Z12'(j\omega)} = Z12_{AP}(j\omega) \qquad (8)$$

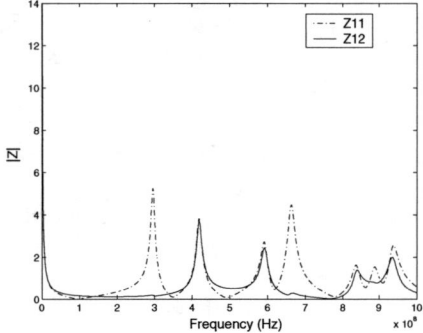

Fig. 3. Z-parameter magnitude response for the plane

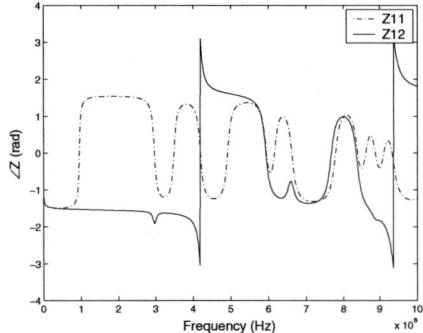

Fig. 4. Z-parameter phase response for the plane

$$Td = -\frac{\arg(Z12_{AP}(j\omega))}{\omega} \qquad (9)$$

This technique can be used to determine the delay from the S, Y or Z parameter representation of a passive system.

To demonstrate the proposed technique, a power/ground PCB plane pair was analyzed using the cavity resonator method [5] to obtain the Z-parameter representation. The plane pair was 25cm x 25cm with 8mil separation and the two ports under consideration were located at (1.67,2.33)cm and (22.67,2.33)cm respectively. Using the velocity of propagation of electromagnetic waves in a dielectric medium, the delay between the two ports was found to be about 1.5ns. Next, the technique previously described in this section was used to determine the delay between the two ports. Starting with the Z-parameters, Figures 3 and 4 show the comparison between the magnitude and phase responses of $Z11$ and $Z12$. From the phase response, it can be easily inferred that $Z11$ is a minimum phase response as against $Z12$ which has 2 phase transitions. Using equations 6 through 8 $Z12$ was separated into a minimum phase function $Z12'_{min}$ and an all-pass function $Z12'_{AP}$. The magnitude and phase responses for $Z12'_{min}$ and $Z12'_{AP}$ are shown in Figures 5 and 6. Since $Z12'_{AP}$ is of the form e^{-sTd} the port-to-port delay Td can be computed as the negative gradient of the phase of $Z12'_{AP}$. The minor deviations from the ideal magnitude and phase responses of $Z12'_{AP}$ seen in Figures 5 and 6 can be eliminated by averaging. The delay thus determined was found to be 1.517ns, which is in good agreement with the expected value.

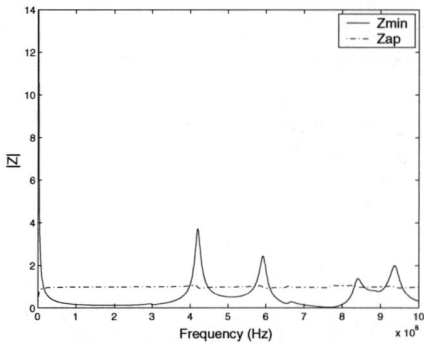

Fig. 5. Magnitude response for $Z12'_{min}$ and $Z12'_{AP}$

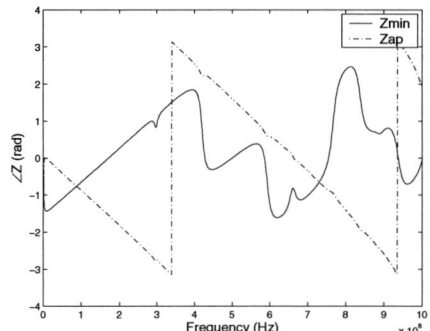

Fig. 6. Phase response for $Z12'_{min}$ and $Z12'_{AP}$

	Measured		Extracted	
	Even	Odd	Even	**Odd**
Microstrip	230ps	239ps	230.5ps	236.8ps
Stripline	247ps	247ps	243.3ps	242.6ps
Buried microstrip	229ps	240ps	227.7ps	237ps

TABLE I

DELAY EXTRACTION FOR DIFFERENTIAL TRANSMISSION LINES

A. Delay extraction in mixed-mode systems

Mixed-mode passive structures like differential transmission lines are characterized by two types of delays: differential mode (odd mode) delay and common mode (even mode) delay. The technique discussed in the previous section can be easily extended to extract the even and odd mode delays in mixed-mode systems. The process involves transforming the given system parameters into mixed-mode parameters [4] and applying the above technique on the new set of mixed-mode parameters. This technique was used to extract delay from 4-port S-parameter measurements done on differential transmission lines using Agilent's PLTS system. The extracted delay was compared with the measured values and the results are shown in table I.

IV. CAUSALITY ENFORCEMENT USING DELAY EXTRACTION

The technique described in section II to determine the port-to-port delay in a passive system can be used to enforce causality on the system response. This is demonstrated through

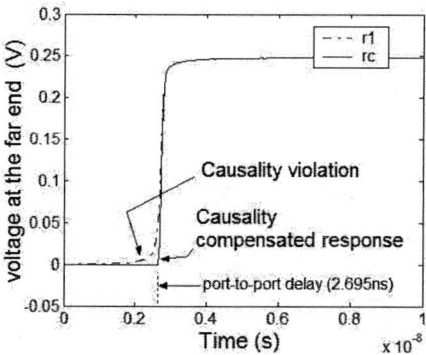

Fig. 7. Response at the far end of the transmission line

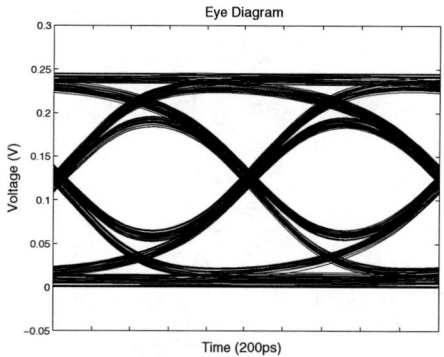

Fig. 8. Eye diagram using uncompensated s-parameters

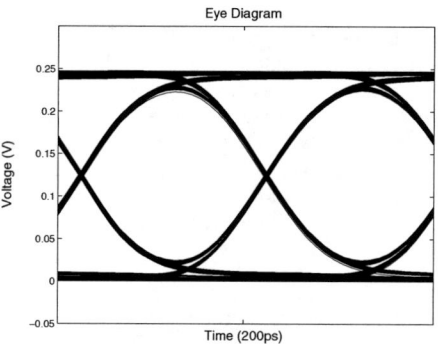

Fig. 9. Eye diagram using causality compensated s-parameters

a signal integrity analysis on a transmission line structure like the one shown in Figure 1. A microstrip transmission line 50cm in length and with a characteristic impedance of 50Ω was constructed in Agilent's Advanced Design System (ADS). The dielectric used had a relative permittivity (ϵ_r) of 3.4 and a loss tangent ($tan\ \delta$) of 0.03. The line was terminated with a 50Ω load at the far end and was excited at the near end with a 250mV/100ps step source having 50Ω input impedance. The response seen at the far end of the line is shown as $r1$ in Figure 7. The simulation was carried out by the applying the inverse Fourier Transform on the frequency response (ADS S-parameter sweep upto 5GHz) of the transmission line to get its impulse response and then using convolution in the time domain to get the far end response. A similar response was obtained using the time domain simulator in ADS. For simplicity the transmission line was perfectly matched at both ends so that there were no reflections in the system. The delay estimation technique described previously was applied to $S12$ of the given structure and the delay associated with the line was found to be 2.695ns. Hence the response $r1$ violates causality because the signal is seen to reach the far end before the 2.695ns delay inherent in the transmission line due to finite velocity of signal propagation. To understand the impact such a non-causal response can have on the signal integrity analysis in high-speed digital systems, the same transmission line structure was driven by a 2.5GHz random bit pattern with a rise time of 100ps to obtain an eye-diagram. The results are shown in Figure 8. The $x-axis$ in the eye-diagrams spans a time period of 200ps. Next, the delay determined from the transfer response $S12$ of the transmission line was used to enforce causality on the system by compensating the S-parameters. This compensation was done in the time domain on $S12$ and $S21$ such that they are forced to remain zero over the time period corresponding to the port-to-port delay. These compensated S-parameters result in a response at the far end of the transmission line given by rc in Figure 7. It is clearly seen that rc satisfies the causality requirements. Additionally, the compensated S-parameters do not violate the passivity conditions in the frequency domain [3]. Next the random bit pattern that was used to generate the eye-diagram in Figure 8 was applied on the compensated S-parameters to observe the effects of causality compensation on the eye-diagram. The results are shown in figure 9. From figures 8 and 9 considerable difference is observed in the eye-spreads obtained using the two different sets of S-parameters.

V. Conclusion

Causality is an important problem when dealing with time domain simulation of high-speed distributed passive systems. This paper describes a technique to extract the port-to-port delay in passive networks directly from their frequency domain parameters. The technique can be applied to either S, Y or Z parameters of passive systems. It can also be extended to mixed-mode systems to extract the even and odd mode delays. The extracted delay has been used to enforce causality on the S parameters of a transmission line system. The modified causal parameters have been shown to have considerable impact on the signal integrity analysis of the transmission line.

References

[1] A. Oppenheim and R. Schaffer, *Discrete-time Signal Processing*, 2nd ed. Prentice Hall, 1999, Ch 5,11.
[2] W. Kim and M. Swaminathan, "*Validity of non-physical RLGC models for simulating lossy transmission lines*", Proceedings of the ISAP, Vol.3, 2002. pp 786-789.
[3] S. Min and M. Swaminathan, "*Efficient construction of passive macro-models for resonant networks*", Proceedings of 10th EPEP, Aug, 2001.
[4] W. Kim , "*Development of measurement based time-domain models and its application to wafer level packaging*", Ph.D Thesis, ECE, Georgia Tech, 2004 Ch 1,3.
[5] S. Chun , "*Methodologies for modeling simultaneous switching noise in multi-layered packages and boards*", Ph.D Thesis, ECE, Georgia Tech.

Passive Approximation of Tabulated Frequency-Data by Fourier Expansion Method

Yuichi Tanji
Dept. of RISE
Kagawa University
2217-20, Hayashi-cho, Takamatsu, 761-0396 Japan
E-mail: tanji@eng.kagawa-u.ac.jp

Hidemasa Kubota
Graduate School of Electronic Science and Technology
Shizuoka University
Johoku3-5-1, Hamamatsu, 432-8561, Japan
E-mail: hidem@tzasai7.sys.eng.shizuoka.ac.jp

Abstract—The passive approximation of tabulated frequency-data for modeling package, PCB, and integrated circuits is presented. The macromodels in the Laplace-domain are directly obtained by the Fourier series approximation of the frequency-data and are represented using a Laguerre basis. The rigorous proof of passivity preservation associated with the proposed models is provided.

I. INTRODUCTION

Analysis of mixed electromagnetic and lumped circuit systems becomes indispensable for power/signal integrity of package, PCB, and integrated circuits. Then, the package, PCB, and integrated circuits are measured/calculated in the frequency-domain [1]-[2], and the transient analysis with other lumped circuits is carried out on the SPICE-like simulators requiring macromodeling of the tabulated frequency-data measured/calculated. In this case, the macromodels should be guaranteed to be passive, in order to carry out the timing analysis stably.

Some authors have provided the passive modeling algorithms [3]-[6]. The works [3], [4] apply optimization methods to the state space representation of scattering matrix of multiport networks to enforce the passivity. The passivity check algorithm is presented in [5]. A Laguerre basis is used for positive real (PR) and bounded real approximations of distributed systems [6].

In this paper, passive approximation of tabulated frequency-data is presented, where it is assumed that the tabulated frequency-data of impedance or admittance matrix are obtained from measurement or numerical analysis. This method is based on the Hosono's discussion and approximation methods about positive real function (PRF) and positive real matrix (PRM) [7], [8]. As a result, the frequency-data on the imaginary axis are approximated by the Fourier series with period $[-\infty, \infty]$, and the macromodels are represented using a Laguerre basis. In the previous work [6], the distributed system is approximated by the almost same Laguerre basis, but the proof of passivity is not completed and the macromodels obtained from this method may not be always guaranteed. Here, preserving the passivity of original frequency-data, we propose a new model. The rigorous proof of the passivity preservation is presented, based on the theory of PRM [9].

II. 1-PORT NETWORKS

The passive macromodeling of tabulated frequency-data networks in this paper is based on the Hosono's Fourier expansion method for approximating 1-port networks by a PRF [7], [8]. Before providing our method, we review the Hosono's method briefly.

An arbitrary PRF is expressed using the Herglotz-Caurer theorem [8].

Theorem 1: An arbitrary PRF is represented by the Stieltjes integral:

$$W(s) = h_\infty s + \frac{s}{\pi}\int_{-\infty}^{\infty}\frac{1+\omega^2}{s^2+\omega^2}dM(\omega) \qquad (1)$$

where $h_\infty > 0$ and $M(\omega)$ is non-decreasing odd function.

Also, the following theorem holds [8].
Theorem 2: The relation

$$R(\omega) = (1+\omega^2)\frac{dM(\omega)}{d\omega} \quad \text{a.e.} \qquad (2)$$

holds for the arbitrary RBF (1), where $R(\omega)$ is the real part of $W(j\omega)$ and a.e. means almost everywhere.

The non-decreasing odd function is decomposed into

$$M(s) = M_{sd}(s) + M_{sc}(s) + M_{ac}(s) \qquad (3)$$

where $M_{sd}(s)$, $M_{sc}(s)$, and $M_{ac}(s)$ are discontinuous singular, continuous singular, and absolute continuous non-decreasing functions, respectively. The absolute continuous non-decreasing function is defined by

$$dM_{ac}(j\omega)/d\omega = N(j\omega) \quad \text{a.e.}$$

and

$$dM_{ac}(j\omega) = \int_0^\omega N(j\omega)d\omega.$$

Using the three functions of (3) and Theorem 1, we can

consider three types of PRF's as

$$W_r(s) = h_\infty s + \frac{s}{\pi}\int_{-\infty}^{\infty}\frac{1+\omega^2}{s^2+\omega^2}dM_{sd}(\omega) \quad (4)$$

$$W_q(s) = \frac{s}{\pi}\int_{-\infty}^{\infty}\frac{1+\omega^2}{s^2+\omega^2}dM_{sc}(\omega) \quad (5)$$

$$W_m(s) = \frac{s}{\pi}\int_{-\infty}^{\infty}\frac{1+\omega^2}{s^2+\omega^2}dM_{ac}(\omega) \quad (6)$$

where $W_r(s)$, $W_q(s)$, and $W_m(s)$ are called reactance, quasi-reactance, and minimum reactance functions, respectively.

The reactance function is described by

$$W_r(s) = \frac{h_0}{s} + \sum_{i=1}^{\infty-1}\frac{h_i s}{s^2+\omega_i^2} + h_\infty s. \quad (7)$$

From Theorems 1 and 2, the minimum reactance function is described by

$$W_m(s) = \frac{s}{\pi}\int_{-\infty}^{\infty}\frac{R(\omega)}{s^2+\omega^2}d\omega. \quad (8)$$

In the Fourier expansion method [7], [8], it is assumed that each element of the the reactance part (7) is obtained in advance or ignored, and the minimum reactance part (8) is only considered.

Consider the $s-z$ transform:

$$z = \frac{1-s}{1+s} \quad (9)$$

Then, the right-half s-plane is mapped within the unit circle on the z-plane. Especially, the imaginary axis $j\omega$ on the s plane is corresponding to the unit circle on the z-plane as

$$e^{j\tau} = \frac{1-j\omega}{1+j\omega}. \quad (10)$$

From (8)-(10), The mapping of (8) into the z-plane[1] becomes

$$w_m(z) = \frac{1}{2\pi}\int_{-\pi}^{\pi}\frac{e^{j\tau}+z}{e^{j\tau}-z}R(\tan\frac{\tau}{2})\,d\tau. \quad (11)$$

Since a PRF is non-singular on the right-half s-plane, the PRF is also non-singular within the unit circle on the z-plane. Therefore, the binomial expansion gives

$$\frac{e^{j\tau}+z}{e^{j\tau}-z} = 1 + 2\sum_{n=1}^{\infty}e^{-jn\tau}z^n \quad (12)$$

Then, the PRF (11) is approximated by the Fourier series:

$$w_m(z) = \sum_{n=0}^{\infty}b_n z^n \quad (13)$$

where

$$b_0 = \frac{1}{2\pi}\int_{-\pi}^{\pi}R(\tan\frac{\tau}{2})\,d\tau$$

$$b_n = \frac{1}{\pi}\int_{-\pi}^{\pi}R(\tan\frac{\tau}{2})e^{-jn\tau}\,d\tau.$$

[1]The upper and lower letters imply the functions in the s- and z-planes, respectively.

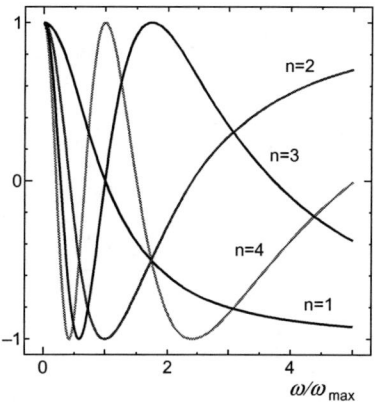

Fig. 1. Basis functions for PR approximation.

The PRF on the s-plane is obtained from the $s-z$ transform of (13). However, the function truncated until finite degree is not guaranteed to be positive real.

Fortunately, the positive realness is guaranteed by taking account into the Cesàro average which is defined by

$$\mathcal{K}_N(x) = \frac{1}{N+1}\sum_{k=0}^{N}\phi_n(x) \quad (14)$$

where $\phi_n(x) = \sum_{k=-n}^{n}c_k e^{jkx}$ is the Fourier series of a function $\xi(x)$. It is known that if the function $\xi(x)$ is periodic, the N-degree Cesàro average converges $\xi(x)$ itself, that is,

$$\xi(x) = \lim_{N\to\infty}\mathcal{K}_N(x). \quad (15)$$

Eventually, the PR approximation of 1-port networks is represented by

$$W_m^N(s) = \sum_{n=0}^{N}\left(1 - \frac{n}{N+1}\right)b_n\left(\frac{1-s}{1+s}\right)^n. \quad (16)$$

III. N-PORT NETWORKS

A. PR approximation

When a passive multi-port network is characterized by impedance or admittance matrix with sampled date, the PR approximation of 1-port networks in the previous section is also applied to the multi-port networks.

The PR approximation is derived from the following PRM expression by Herglotz and Yula [9]:

Theorem 3: An arbitrary PRM is represented by the Stieltjes integral:

$$\boldsymbol{W}(s) = \boldsymbol{Q} + \boldsymbol{A}_\infty s + \frac{s}{\pi}\int_{-\infty}^{\infty}\frac{1+\omega^2}{s^2+\omega^2}d\boldsymbol{K}(\omega)$$

$$+ \frac{1}{\pi}\int_{-\infty}^{\infty}\frac{s^2-1}{s^2+\omega^2}\omega d\boldsymbol{L}(\omega) \quad (17)$$

where $\boldsymbol{Q}$ and $\boldsymbol{L}(\omega)$ are real anti-symmetric matrices, $\boldsymbol{A}_\infty$ is a real symmetric and positive definite matrix, and $\boldsymbol{K}(\omega)$ is a real symmetric matrix. Further, $\boldsymbol{K}(\omega) = -\boldsymbol{K}(-\omega)$, $\boldsymbol{L}(\omega) = \boldsymbol{L}(-\omega)$, and if $\boldsymbol{y}$ is a complex vector, $\boldsymbol{y}^*[\boldsymbol{K}(\omega) - j\boldsymbol{L}(-\omega)]\boldsymbol{y}$ becomes a non-decreasing bounded function.

Since the matrices Q and $L(\omega)$ are real anti-symmetric matrices, an arbitrary symmetric PRM is expressed by

$$W_s(s) = A_\infty s + \frac{s}{\pi} \int_{-\infty}^{\infty} \frac{1+\omega^2}{s^2+\omega^2} dK(\omega). \quad (18)$$

Theorem 2 holds for PRM. Then, the matrix $K(s)$ satisfies

$$R(\omega) = (1+\omega^2)\frac{dK(\omega)}{d\omega} \quad \text{a.e.} \quad (19)$$

where $R(\omega)$ is the real part of $W(j\omega)$. Further, the matrix $K(s)$ is decomposed into

$$K(s) = K_{sd}(s) + K_{sc}(s) + K_{ac}(s). \quad (20)$$

The relation (20) is corresponding to (3) for PRM. Therefore, we can consider three types of PRM as PRF's (4)-(6) and describe the reactance matrix $W_r(s)$ and the minimum reactance one $W_m(s)$ by

$$W_r(s) = \frac{H_0}{s} + \sum_{i=1}^{\infty-1} \frac{H_i s}{s^2+\omega_i^2} + H_\infty s \quad (21)$$

$$W_m(s) = \frac{s}{\pi} \int_{-\infty}^{\infty} \frac{R(\omega)}{s^2+\omega^2} d\omega. \quad (22)$$

As the same with PR approximation of 1-port networks, the minimum reactance matrix $W_m(s)$ is only considered. As a result, the PR approximation of N-port networks is represented by

$$W_m^N(s) = \sum_{n=0}^{N}\left(1-\frac{n}{N+1}\right) B_n \left(\frac{1-s}{1+s}\right)^n \quad (23)$$

where

$$B_0 = \frac{1}{2\pi} \int_{-\pi}^{\pi} R(\tan\frac{\tau}{2}) d\tau \quad (24)$$

$$B_n = \frac{1}{\pi} \int_{-\pi}^{\pi} R(\tan\frac{\tau}{2})e^{-jn\tau} d\tau. \quad (25)$$

B. Implementation

The coefficient matrices (24) and (25) can be calculated by the discrete Fourier transform [6], based on the the collocation point orthogonality. Then, the PR approximation (23) provides the interpolated matrix at the tabulated data. However, since the collocation points are not uniformly distributed in $\omega \in [0, \omega_{max}]$, the important information of the tabulate data may be lost [6]. Alternatively, we apply the trapezoidal rule for calculating the integral values.

First, the variable τ of the coefficients matrices (24) and (25) is converted into ω.

$$B_0 = \frac{2}{\pi} \int_0^{\infty} \frac{1}{1+\omega^2} R(\omega) d\omega \quad (26)$$

$$B_n = \frac{4}{\pi} \int_0^{\infty} \frac{1}{1+\omega^2} R(\omega) \cos\left(2n \tan^{-1} \omega\right) d\omega \quad (27)$$

Assuming the sample points $\omega_0, \omega_1, ..., \omega_M$ and using the trapezoidal rule, we calculate the coefficient matrices by

$$\tilde{B}_0 = \frac{1}{\pi}\left\{F_0(\omega_0) + F_0(\omega_M) + 2\sum_{k=1}^{M-1} \delta\omega_k F_0(\omega_k)\right\} \quad (28)$$

$$\tilde{B}_n = \frac{2}{\pi}\left\{F_n(\omega_0) + F_n(\omega_M) + 2\sum_{k=1}^{M-1} \delta\omega_k F_n(\omega_k)\right\} \quad (29)$$

where

$$F_k(\omega) = \frac{1}{1+\omega^2} R(\omega) \cos\left(2n \tan^{-1} \omega\right)$$

and $\delta\omega_i = \omega_i - \omega_{i-1}$. Then, the PR approximation is obtained as

$$\tilde{W}_m^N(s) = \sum_{n=0}^{N}\left(1-\frac{n}{N+1}\right) \tilde{B}_n \left(\frac{1-s}{1+s}\right)^n. \quad (30)$$

Moreover, the angular frequency ω is normalized as $[0, a]$, where $a > 1$ is a positive number. When $a = 1$, the tabulated frequency-data on the imaginary axis are approximated by the basis functions shown in Fig. 1. The macromodel in the Laplace-domain is represented by the Laguerre basis $(1-s)/(1+s)$ in (30) as [6]. However, the coefficient matrices $\tilde{B}_n (n = 0, 1, ...)$ are calculated by the real part of frequency-data only, which is important for the passivity preservation.

C. Poof of Passivity

Theorem 4: The rational matrix (30) is positive real, thus, the multi-port networks whose impedance or admittance matrices are approximated by (30), is passive.

Proof) Equation (23) is described on the imaginary axis by

$$W_m^N(j\omega) = \frac{1}{2\pi} \int_{-\pi/2}^{\pi} R(\tan\frac{\tau}{2}) \left\{1 + \sum_{n=1}^{N} 2\left(1-\frac{n}{N+1}\right) e^{-jn(\tau-\beta)}\right\} d\tau$$

$$= \frac{1}{\pi} \int_0^{\pi} R(\tan\frac{\tau}{2}) \mathcal{K}_N(\tau-\beta) d\tau. \quad (31)$$

where $\beta = -2\tan^{-1}\omega$ and $\mathcal{K}_N(x)$ is the Cesàro average associated with the Fourier series $Q_N(x) = \sum_{k=0}^{N} c_N e^{-jNx}$ ($c_0 = 1$, $c_n = 2$ $(n \neq 0)$).

The Cesàro average $\mathcal{K}_N(x)$ satisfies the following relations:

$$(N+1)\mathcal{K}_N(x) - N\mathcal{K}_{N-1}(x) = Q_N(x). \quad (32)$$

The Fourier series $Q_N(x)$ can be written by

$$Q_N(x) = \frac{\cos Nx - \cos(N+1)x}{1-\cos x} + j\frac{\sin(N+1)x - \sin Nx - \sin x}{1-\cos x}. \quad (33)$$

From (32) and (33), it is shown that the real part of the Cesàro average is equal to Fejér kernel $\mathcal{F}(x)$

$$\mathcal{F}(x) = \frac{1}{N+1}\left(\frac{\sin\frac{(N+1)x}{2}}{\sin\frac{x}{2}}\right)^2. \tag{34}$$

Since $\operatorname{Re}\mathcal{K}_N(\tau-\beta) = \mathcal{F}(\tau-\beta) \geq 0$ and $\boldsymbol{R}(\cdot) \geq 0$, the real part of $\boldsymbol{W}_m^N(j\omega)$ satisfies $\operatorname{Re}\boldsymbol{W}_m^N(j\omega) \geq 0$.

The real part of (30) on the imaginary axis is obtained from the application of trapezoidal rule to (31) as follows:

$$\begin{aligned}\operatorname{Re}\tilde{\boldsymbol{W}}_m^N(j\omega) &= \frac{1}{2\pi}\{\boldsymbol{H}_N(\tau_0) + \boldsymbol{H}_N(\tau_M)\} \\ &\quad + \frac{1}{\pi}\sum_{k=1}^{M-1}\delta\tau_k \boldsymbol{H}_N(\tau_k)\end{aligned} \tag{35}$$

where $\boldsymbol{H}_N(\tau) = \boldsymbol{R}(\tan\frac{\tau}{2})\operatorname{Re}\mathcal{K}_N(\tau-\beta)$, $\tau_i = 2\tan^{-1}\omega_i$, and $\delta\tau_i = \tau_i - \tau_{i-1}$. Since $\boldsymbol{H}_N(\tau) \geq 0$ and $\delta\tau_i > 0$, $\operatorname{Re}\tilde{\boldsymbol{W}}_m^N(j\omega) \geq 0$, thus, the multi-port networks whose impedance or admittance matrices are approximated by (30), is passive.

IV. SIMULATION

To evaluate the proposed method, the admittance matrix of RLCG distributed interconnect was treated as sampled data, where the interconnect parameters are 0.5 [Ω/cm], 10.0 [nH/cm], 0.004 [nF/cm], and 0.0005 [S/cm], and the length is 1 [cm]. The admittance matrix was approximated by (30). The result was shown in Fig. 2, where N = 1000 and a = 2 are used. Although the proposed method needs a large number of terms, we can reduce the order of macromodels by the introduction of passivity guaranteed model order reduction algorithms [10], [11] without loss of accuracy.

V. CONCLUSIONS

The passive approximation of tabulated frequency-data for modeling package, PCB, and integrated circuits has been presented. The macromodels in the Laplace-domain are represented by a Laguerre basis, where the coefficient matrices are directly obtained by the Fourier series approximation of the frequency-data. The rigorous proof of passivity preservation associated with the proposed models is also presented. Although the proposed macromodels are guaranteed to be passive, the macromodels are not computationally efficient. However, introducing passivity guaranteed model order reduction algorithms [10], [11], we can provide the compact macromodels. We will report it in near future.

REFERENCES

[1] P. P. Sylvester, "*Finite Elements for Wave Electromagnetics: Methods and Techniques*", IEEE Press, New York, 1994.
[2] R. F. Harrington, "*Field Computation by Moment Methods*", IEEE Press, New York, 1993.
[3] S. Grivet-Talocia, "Passivity enforcement via perturbation of hamiltonian matrices," *IEEE Trans. Circuit Syst. I*, vol. 51, no. 9, pp. 1755-1769, 2004.
[4] C. P. Coelho, J. Phillips, and L. M. Silveira, "A convex programming approach for generating guaranteed passive approximation to tabulated frequency-data," *IEEE Trans. CAD*, pp. 293-301, Feb. 2004.

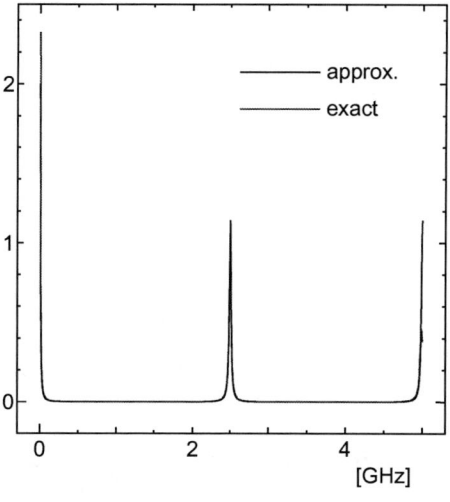

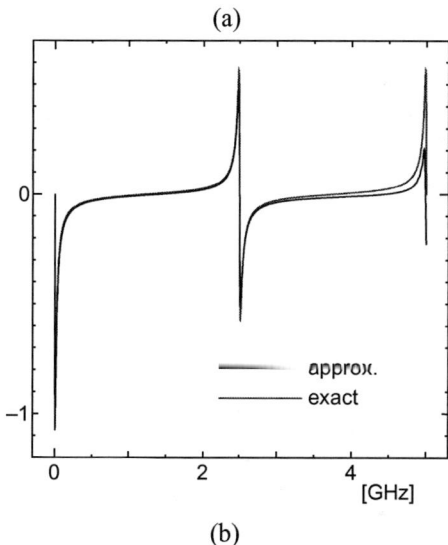

Fig. 2. PR approximation of admittance matrix of RLCG distributed interconnect. (a)real part of $Y_{11}(j\omega)$. (b)imaginary part of $Y_{11}(j\omega)$.

[5] D. Saraswat, R. Achar, and M. S. Nakhla, "A fast algorithm and practical considerations for passive macromodeling of measured/simulated data," *IEEE Trans. Advanced Pack.*, vol. 27, no. 1, pp. 57-69, Feb. 2004.
[6] L. Daniel and J. Phillips, "Model order reduction for strictly passive and causal distributed systems," *Proc. 39th Design Automation Conf.*, 2002.
[7] T, Hosono, "On general positive real function," *IEICE Trans.*, vol. 58-A, no. 10, pp. 641-648, 1975 (in Japanese).
[8] T, Hosono, *Senkei-black-box-no-kiso*, Corona, 1980 (in Japanese).
[9] W. R. Wholer, *Lumped and distributed passive networks*, Academic Press, New York and London, 1969.
[10] A. Odabasioglu, M. Celik, and L. T. Pilleggi, "PRIMA: passive reduced-order interconnect macromodeling algorithm," *IEEE Trans. CAD*, vol. 17, no. 8, pp. 645-654, Aug. 1998.
[11] J. R. Phillips, L. Daniel, and L. M. Silveira, "Guaranteed passive balancing transformations for model order reduction," *IEEE Trans. CAD*, vol. 22, no. 8, pp. 1027-1041, Aug. 2003.

Noise Generation, Coupling, Isolation, and EM Raidaiton in High-speed Package and PCB

Joungho Kim, Junso Pak, Jongbae Park, and Hyungsoo Kim
Terahertz Interconnection and Package Lab.,
EECS Dept., KAIST, Daejon, Korea
joungho@ee.kaist.ac.kr, http://tera.kaist.ac.kr

Abstract— Return current path is the most critical part of high-speed interconnection design in package and PCB. When the return current path is disturbed, significant amount of noise generation, coupling, and radiated emission problems occur. Signal via and power/ground via are producing the return current path disconnect problem. In this paper, we demonstrate that the via is a major source of the SSN generation, coupling, and edge radiated emission in multi-layer package and PCB

I. INTRODUCTION

High-speed digital multi-layer package and PCB have countless closely spaced metallic interconnection structures such as trace, via, pad, lead, partial plane, and plane cavity. Especially, when return current path is disturbed by such interconnection structures, these densely spaced interconnection structures become main sources of high frequency noise generation, noise coupling, and radiated emission, imposing serious signal and power integrity issues as well as EMI/EMC problems. These noises worsen noise and timing margin of digital and analog circuits, resulting in reduction of maximally achievable jitter, BER, and system reliability. Also, phase noise and SNR performance in RF and wireless communication circuits suffer the coupled noises from the fast switching digital devices.

In the high-speed and high-density package and PCB, major element of the high frequency noise is Simultaneous Switching Noise (SSN) from the fast switching digital circuits, as clock frequency and amount of the switching current are significantly increased. [1] The SSN could be generated by power/ground return current path disconnection combined with the fast switching current. [2] Accordingly, power/ground via should be carefully designed to minimize the generation of the SSN. The SSN could be coupled to adjacent interconnections or be radiated into the free space. The generated SSN constitutes standing waves inside the power/ground cavity at the cavity resonance frequencies, producing huge amount of the SSN spectrum and radiated emission spectrum from the PCB edge.[3] Furthermore, though-hole signal via structure is a considerable noise coupling structure, especially when the signal return current path is exchanging its reference planes. [2] In this paper, we presents major design cases where the return current path discontinuity at the signal and the power/ground network causes significant amount of noise generation, coupling and edge radiation problem.

II. SSN GENERATION AND EDGE RADIATION

A voltage fluctuation is induced across the power/ground cavity by the signal return current path discontinuity at the through-hole signal via. Frequency spectrum of the power/ground voltage fluctuation is depending on the position of the via as well as the signal current and the rise time. Also, the voltage fluctuation spectrum is determined by the resonance frequencies of the power/ground plane cavity of the package substrate or the PCB. The resonance frequencies of the cavity are decided by the dielectric constant and the cavity dimensions. Accordingly, the signal suffers huge amount of the signal loss at the cavity resonance frequencies when the signal via exchanges the reference plane, causing the return current discontinuity. Figure 1 illustrates the excitation mechanism of the cavity resonance by the through-hole signal via, and the edge radiation from the package or PCB edge. As noted in Figure 2, it is found that significant insertion loss is produced at the resonance frequencies at the though-hole signal via.

As a result, power/ground voltage fluctuation is generated and propagated toward the edge of the power/ground plane cavity, producing standing waves inside the cavity at the resonance frequencies. Consequently, as shown in Figure 3, the standing electromagnetic waves are distributed with a same behavior as impedance curve, and produce maximum magnetic current at the open edges of the power/ground plane cavity. Consequently, the magnetic current at the open-ended power/ground plane cavity edge becomes the source of the radiated field emission. The generated SSN voltage waveforms are presented depending on the decoupling capacitor design in Figure 4.

This is the mechanism of the radiated field emission from the power/ground plane cavity edge excited by the through-hole signal via. The measured spectrum of the radiated emission is shown in Figure 5 and Figure 6. Especially, when the via is excited by a clock signal, the radiated emission spectrum becomes the harmonic frequencies of the fundamental clock frequency, and the peak radiation frequency is coincident with the resonance frequency of the power/ground cavity, as verified in Figure 6..

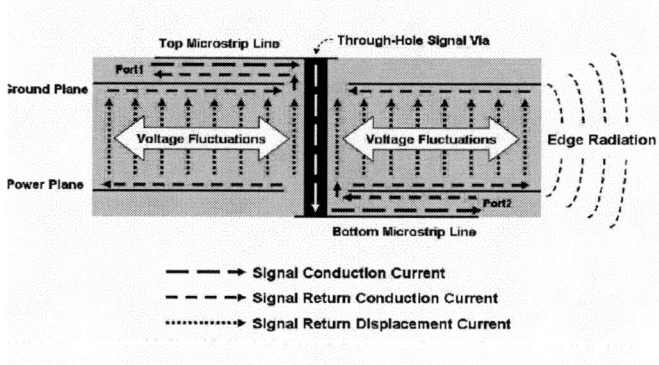

FIGURE 1. Radiated emission mechanism excited by SSN from via transition.

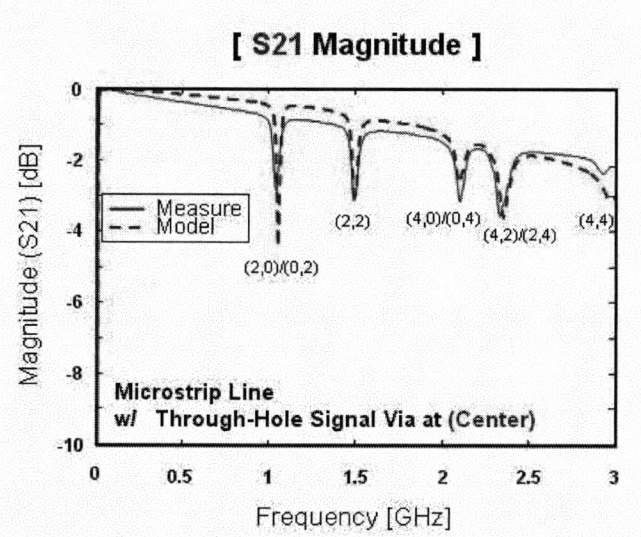

FIGURE 2. Insertion loss by the signal via with reference plane exchange.

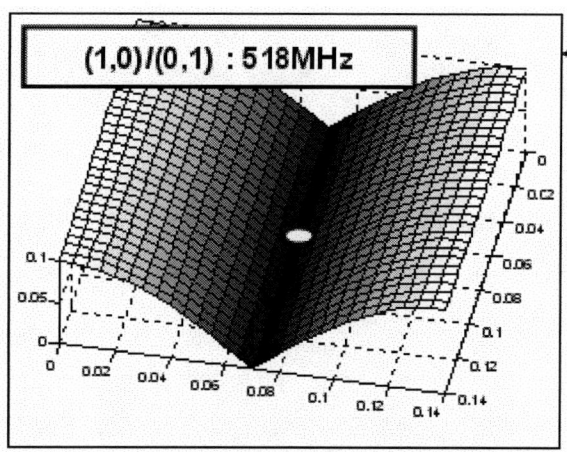

FIGURE 3. Power/Ground cavity resonance mode(1,0) and (2,2) and field distribution of the voltage fluctuations.

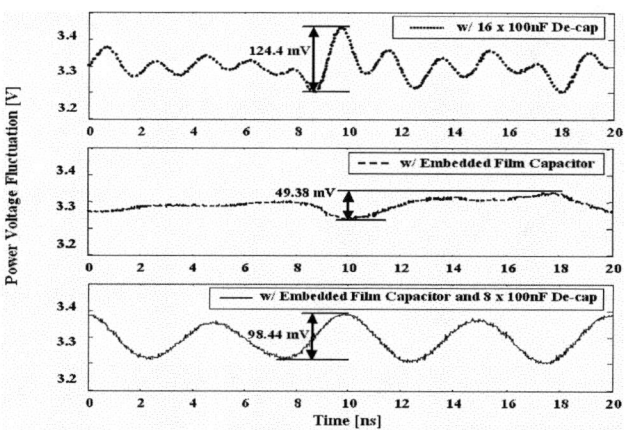

FIGURE 4. Measured SSN depending on decoupling capacitor design, and dielectric thickness with the embedded film capacitor.

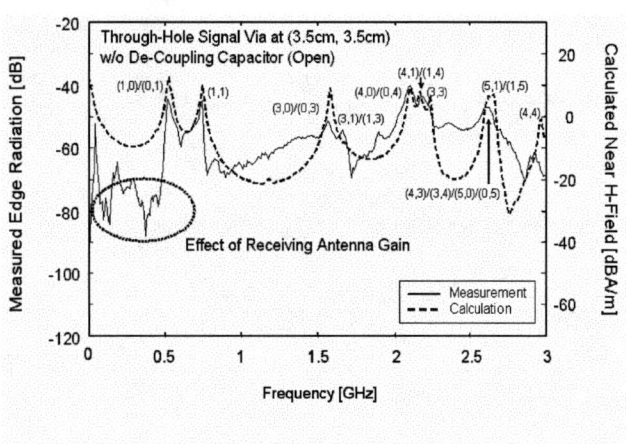

FIGURE 5. Measured radiated emission spectrum excited by SSN from via transition.

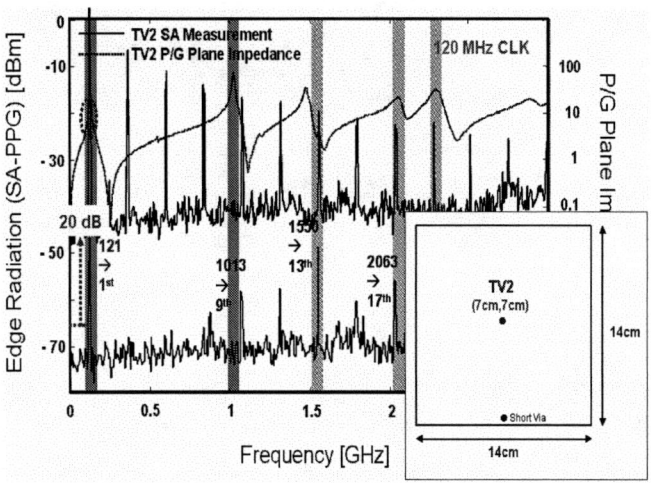

FIGURE 6. Measured radiated emission spectrum excited by a clock driver with via transition.

II. SSN COUPLING THROUH VIA TRANSITION

When signal trace exchanges its reference plane though via, return current path can be severely interrupted and the interrupted return current can cause considerable signal integrity problems. As a result, the via could be a receptor of the SSN from the power and ground plane cavity.

We have designed and tested a series of test PCB vehicles. The size of the test vehicles is 80 mm x 190 mm, and has a 6-layer stack-up. The size and the stack-up of the test vehicles are commonly used in commercial DDR memory module. The layer structures of the test PCB are shown in Figure 7. Then, we have executed a series of time-domain and frequency domain analysis and measurement using these test vehicles.

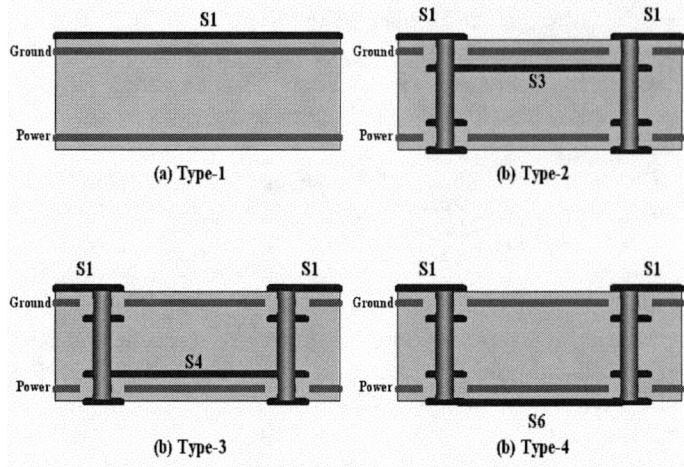

FIGURE 7. Four different interconnection structures in 6-layer package or PCB to study the effect of the signal transition through the via. (a) microstrip line with reference to ground plane (Type-1). (b) microstrip line to strip line transition without reference plane exchange (Type-2), (c) microstrip line to strip line transition with reference plane exchange to power plane (Type-3), and (d) microstrip line with reference plane exchange to power plane (Type-4).

Figure 8 show the measured SSN coupling coefficient, S21, obtained to evaluate the SSN coupling effect. The coupling coefficient of the Type-1 layer structure is neglected since the signal trace does not have the via transition in the Type-1 layer structure. It is noted that the Type-3 and the Type-4 layer structures have much higher SSN coupling compared to that of the Type-2 layer structure. It is understood that the Type-2 layer structure maintains a same reference plane(layer 2) even after the via transition. On the other hand, the Type-3 and the Type-4 layer structures suffer the reference plane exchange and the interrupted return current path by the via transitions. From these simulation and measurement, it is well confirmed that the signal via becomes a significant receptor of the SSN coupling when its return current path is disconnected.

To generate the SSN, we mounted a clock driver chip(TI, CDCVF2310) at the port 1 location, and the output drivers consumes totally 200 mA simultaneous output driver current with a 200 MHz clock frequency. And a precisely controlled 200 MHz clock source signal is supplied to the clock driver chip from a port of a two-channel pulse pattern generator. In order to complete the series of the time domain measurements, we have monitored distorted clock waveforms and eye patterns interfered by the SSN coupling at the via transitions and the strip line inside the cavity for the case of the four different layer structures of Type-1, Type-2, Type-3, and Type-4 as illustrated in Figure 7.

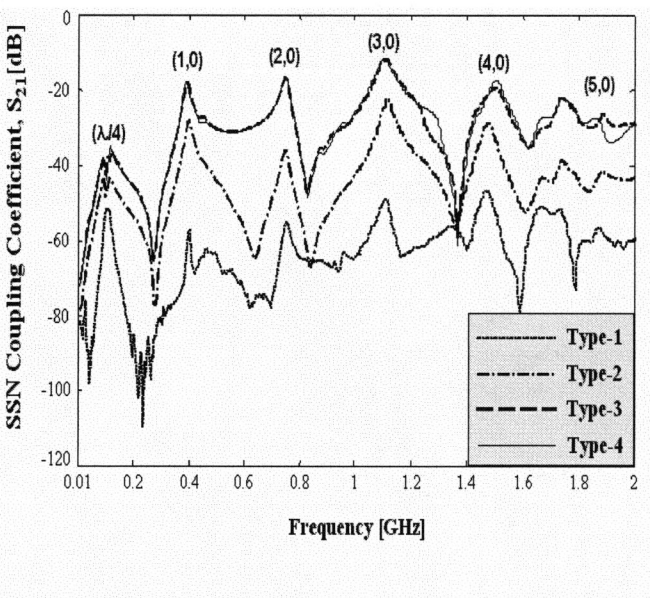

FIGURE 8. Measured coupling coefficient, S21, to evaluate the coupling of power/ground noise to the signal trace. Port1 is placed between power plane and ground plane at port 1 and port2 is located at the end of the signal trace. Type-1 (dotted line), Type-2 (dash-dotted line), Type-3 (dashed line), and Type-4 (solid line)

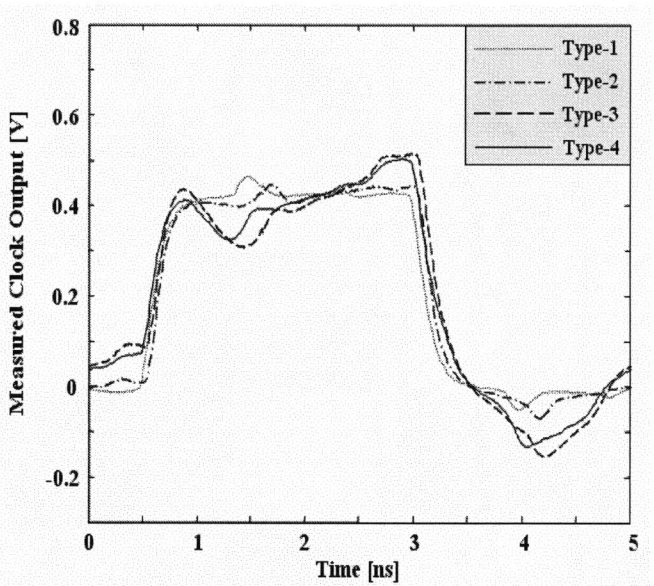

FIGURE 9. Measured output clock waveforms at the port 2 with the 200MHz input clock of 500 mV peak to peak voltage, depending on the layer structure when a 600 mV SSN exists at the power/ground cavity, excited at the port 1.

The SSN of a 600 mV peak-to-peak voltage was produced at the port 1 location by the clock driver chip and the distorted waveforms were measured at the port 2. The measured clock waveforms are plotted in Figure 9. It is manifest again that the Type-3 layer structure produces the worst waveform distortion, in which the strip line has a reference plane of the power plane(layer 5) inside the cavity, and has a via transition with the reference plane exchange from the ground plane(layer 2) to the power plane(layer 5). The SSN coupling occurs not only at the via transition with the reference plane exchange, but also at the strip line inside the cavity. However, the SSN coupling at the via transition is the major coupling mechanism compared to that at the strip line inside the cavity.

III. CONCLUSION

High-speed digital multi-layer package and PCB has countless closely spaced metallic interconnection structures such as trace, via, pad, lead, partial plane, and plane cavity. It is well demonstrated that these densely spaced interconnection structures become sources of high frequency noise generation, noise coupling, and radiated emission, imposing serious signal and power integrity issues as well as EMI/EMC problems, especially when the signal or power/ground return current is disconnected. Furthermore it is confirmed the SSN is heavily generated, coupled, and radiated caused by the via structure in the multi-layer package and PCB through a series of experimental demonstrations..

REFERENCES

[1] Hyungsoo Kim, et al, "High Dielectric Constant Thin Film Embedded Capacitor for Suppression of Simultaneous Switching Noise and Radiated Emission," *2004 IEEE International Symposium on Electromagnetic Compatibility*, Aug 2004.

[2] Jongbae Park, et al, "Noise Coupling to Signal Trace and Via from Power/Ground Simultaneous Switching Noise in High Speed DDR Memory Module", *2004 IEEE International Symposium on Electromagnetic Compatibility*, Aug, 2004.

[3] Jun So Pak, et al, "PCB Power/Ground Plane Edge Radiation excited by High-Frequency Clock," *2004 IEEE International Symposium on Electromagnetic Compatibility*, Aug 2004.

ACCURATE AND CLOSED-FORM SPICE COMPATIBLE PASSIVE MACROMODELS FOR DISTRIBUTED INTERCONNECTS WITH FREQUENCY DEPENDENT PARAMETERS

Natalie Nakhla, Ram Achar and Michel Nakhla

Department of Electronics, Carleton University, Ottawa, Ontario, K1S 5B6
Tel (613) 520-5651; Fax: (613) 520-5708; Email: achar@doe.carleton.ca

ABSTRACT: *Fast and accurate signal integrity analysis is becoming a major requirement in validation of high-speed designs. This demands efficient and accurate models for high-speed distributed interconnects. Also it has been demonstrated that, preserving passivity of the macromodel is essential to guarantee a stable global transient simulation. In this paper, an efficient method for the analysis of high-speed distributed interconnects with frequency-dependent parameters is presented. The proposed method enables representation of the distributed stamp in terms of simple delay and resistive elements. The new method while guaranteeing the passivity of the macromodel, provides significant speedup, and enables easy implementation. Necessary formulation and validation examples are given.*

I. INTRODUCTION

RECENT trend in the VLSI industry toward higher operating speeds, sharper rise times and low-power requirements has made the signal integrity analysis a challenging task. The high-speed interconnect effects such as ringing, delay, distortion, crosstalk, attenuation and reflections, if not predicted accurately at early design stages, can severely degrade the system performance. Interconnects can be found at various levels of design hierarchy, such as on-chip, packaging, MCM's and PCB's [1]–[7].

At higher frequencies, lumped models fail to accurately characterize the interconnects [2]–[8]. As the operating frequency increases, due to the skin, edge and proximity effects, the current distribution becomes uneven across the cross section. To address these high-frequency effects, distributed models based on Telegrapher's equations with frequency-dependent RLCG parameters become necessary.

However, the major difficulty usually encountered while linking the distributed transmission lines with frequency-dependent parameters (TLwFDP) and nonlinear simulators is the problem of mixed frequency/time. This is because TLwFDP elements are characterized in the frequency-domain whereas nonlinear components such as drivers and receivers are represented only in time-domain. Hence it is essential to model TLwFDP elements such that they can be directly incorporated in SPICE-like circuit simulators.

Several publications can be found in the literature to address this issue [2]-[7]. Approaches based on conventional lumped segmentation provide a brute force solution to the problem of mixed frequency/time simulation. However, these methods lead to large circuit matrices, rendering the simulation inefficient [2]. Algorithms based on method of characteristics (MoC) [2], [4] provide fast solutions for long low-loss lines. Although they ensure the transmission-line causality [5] (which implies that an excitation signal entering one end of a transmission line segment will appear at the other end only after the time-of-flight delay), they do not guarantee the passivity of resulting macromodel. However, passivity of the macromodel is important since non-passive but stable models when coupled with arbitrary nonlinear elements can lead to unstable systems [3].

On the other hand, algorithms based on matrix rational approximations (MRA) [6]-[7] guarantee the macromodel passivity. The MRA technique is based on using pre-determined coefficients, which are computed as function of the per-unit-length parameters of the line. However, in the presence of large delay lines (e.g. long lines with small losses), MRA requires high-order approximations (to accurately capture the flat delay portion) leading to inefficient transient simulation. In order to address the above problem, a passive macromodel based on modified Lie's formula [8] was recently suggested [9] in terms of multiplicative approximation of the exponential stamp of the MTL stamp. The circuit implementation of the each product term consists of cascade of lossless lines represented by delay elements and lossy lines approximated via MRA.

In this paper, a new closed-form SPICE compatible passive macromodel for practical class of distributed transmission lines with parameters described by frequency-dependent resistance $R(s)$ and inductance $L(s)$, constant C and $G = 0$ matrices. The method is based on the modified Lie formula. A significant advantage of the proposed method is that, it doesn't require MRA to represent the lossy part of the transmission line. The new macromodel can be computed analytically (closed-form) in terms of only delay and resistive elements, based on the information of p.u.l. line parameters. The new method while guaranteeing the passivity, provides significant speedup and an easy means to implement MTL macromodels in SPICE-like simulators.

II. REVIEW OF DELAY EXTRACTION BASED PASSIVE TL MACROMODELS

It is apparent from the introduction that, although MoC based algorithms employ delay extraction to generate compact macromodels, they do not guarantee the passivity of the macromodel. On the other hand, delay extraction is essential to ensure the compactness (lower-orders) of the macromodel. In order to address this difficulty, a delay extraction and MRA based algorithm was recently suggested [9]. A brief review of the algorithm as relevant to this paper is given in this section. Consider the matrix exponential form of Telegrapher's equations [2]

$$\begin{bmatrix} V(l,s) \\ -I(l,s) \end{bmatrix} = e^{(A+sB)} \begin{bmatrix} V(0,s) \\ I(0,s) \end{bmatrix}; \quad (1)$$

$$A(s) = \begin{bmatrix} 0 & -R \\ -G & 0 \end{bmatrix} l; \quad B(s) = \begin{bmatrix} 0 & -L \\ -C & 0 \end{bmatrix} l \quad (2)$$

where R, L, G, C are the per-unit-length (PUL) parameter matrices of the transmission-line, $V(s), I(s)$ are the terminal voltage and current vectors of the transmission line and l is the length of the line. Using perturbation and assuming that $\|A\| \ll \|s_{max} B\|$ (where s_{max} corresponds to the maximum frequency of interest), (1) can be approximated as [8]

$$e^{(A+sB)} \approx e^{\frac{sB}{m}} \prod_{k=1}^{m} e^{C_k}; \quad (C_k \equiv f(A,B)) \quad (3)$$

where $\|C_1\| \gg \|C_2\| \gg \dots \|C_m\|$. It was shown that the product [7]

$$\prod_{k=1}^{m} Q_k + \varepsilon_m; \quad Q_k \equiv e^{\frac{A}{2m}} e^{\frac{sB}{m}} e^{\frac{A}{2m}} \quad (4)$$

converges asymptotically to $e^{(A+sB)}$ as $m \to \infty$. The associated error (ε_m) in this case is given by

$$\|\varepsilon_m\| = \max_{0 \leq s \leq s_{max}} \left\| e^{(A+sB)} - \prod_{k=1}^{m} Q_k \right\| \cong O\left(\frac{1}{m^2}\right) \quad (5)$$

If $\|A\| \ll \|s_{max} B\|$ (which is the case for long low lossy lines), then an alternative form for (4) can be used with a better error-bound, and is given by:

$$\prod_{k=1}^{m} Q_k + \varepsilon_m; \quad Q_k \equiv e^{\frac{sB}{2m}} e^{\frac{A}{m}} e^{\frac{sB}{2m}} \quad (6)$$

which further reduces the error. The products represented by (4) or (6) can be viewed as a cascade of m transmission lines. In addition, each of the kth product term can be viewed as a cascade of lossy and lossless transmission lines. The lossy terms are macromodeled using the passive matrix rational approximation [6]. The resulting macro-

models are of lower orders (since a significant delay portion is already extracted from these terms). They are later combined with the lossless terms using the method of characteristics approach [4]-[5]. For example, each Q_k in (6) can be realized as shown in Fig. 1, where

$$\hat{B}(s) = B(s) - B(\infty).$$

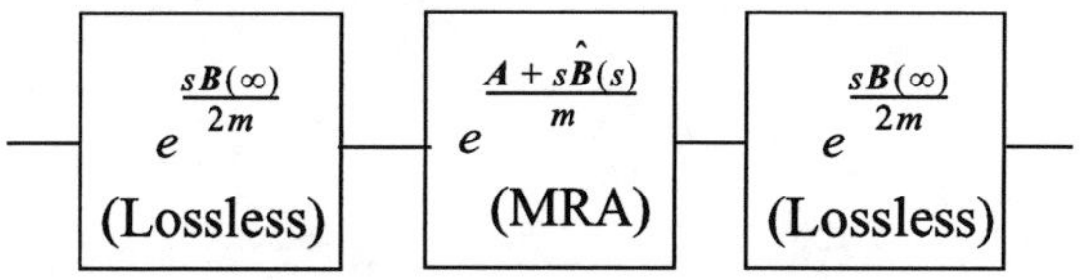

─── $e^{\frac{sB(\infty)}{2m}}$ (Lossless) ─── $e^{\frac{A+s\hat{B}(s)}{m}}$ (MRA) ─── $e^{\frac{sB(\infty)}{2m}}$ (Lossless) ───

Fig. 1: Macromodel Realization of the Product Terms in (6).

It is to be noted that the passivity of the entire macromodel is now guaranteed as the passivity of each sub-line in (6) (see Fig. 1) is preserved.

III. PROPOSED CLOSED-FORM PASSIVE MACROMODEL

In this section, we present a new passive macromodel for distributed transmission lines. The new model employs delay extraction based on modified Lie formula. One of the important advantages of the proposed algorithm is that, it doesn't require MRA based approximation in the resulting macromodel. The final macromodel is realized as an exact implementation (without any approximation) of the multiplicative product terms as represented by the modified Lie formula (6). The resulting macromodel consists of only delay and lumped elements, which are computed analytically based on the information of p.u.l parameters, leading to an easy implementation in SPICE like circuit simulators.

Two practical cases are considered here:

III-1 Frequency-Independent Parameters

For the case of distributed transmission lines with frequency independent RLGC parameters, the delay extracted exponential matrix representing a lossy module can be expressed as

$$e^{Z(s)} = e^{\frac{A}{m}} \quad (7)$$

which represents the middle term in (6) (modified Lie formula). In the following section, we will show that (6) can be implemented without any numerical approximation (exact), using only resistive networks.

a) Realization of Single Lossy Sections

For the case of single lines, (7) can be implemented in terms of a purely resistive network as shown in Fig. 2:

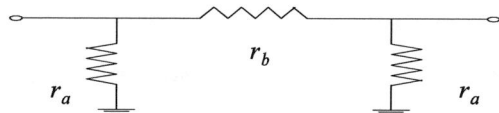

Fig. 2. Resistive network realization for single lossy sections

where

$$r_a = \frac{z_0(1-e^{-2\gamma d})}{(1-e^{-\gamma d})^2}; \qquad r_b = \frac{z_0(1-e^{-2\gamma d})}{2e^{-\gamma d}} \qquad (8)$$

$$\gamma = \sqrt{ZY} = \sqrt{RG}$$
$$z_0 = \sqrt{Z/Y} = \sqrt{R/G}$$
$$d = (l/m) \qquad (9)$$

As seen from (8) and (9), they represent an exact implementation (without any approximation). Also, the macromodel is realized in terms of pure resistive elements, whose values are computed analytically solely based on the information of the line parameters. This virtue of the algorithm greatly simplifies its implementation in SPICE like simulators.

b) Realization of Coupled Lossy Sections

For coupled lossy sections, a decoupling method is employed (similar to the one used for representing lossless lines) prior to realization in terms of resistive networks similar to the one described by Fig. 2.

Note that, the terminal voltages/currents of an n-port lossy section are related by Telegrapher's equations

$$\frac{\partial}{\partial x}v(x,t) = -Ri(x,t)$$
$$\frac{\partial}{\partial x}i(x,t) = -Gv(x,t) \qquad (10)$$

Next, (10) is expressed in terms of auxiliary voltages $\hat{v}(x,t)$ and currents $\hat{i}(x,t)$ as [2]:

$$v(x,t) = E_V\hat{v}(x,t); \qquad i(x,t) = E_I\hat{i}(x,t) \qquad (11)$$

The transformation matrices E_V and E_I can be found such that they simultaneously diagonalize both R and G as follows

$$E_V^{-1}RE_I = \hat{R} = diag\{\hat{r}_1, \hat{r}_2, \ldots, \hat{r}_n\}$$
$$E_I^{-1}GE_V = \hat{G} = diag\{\hat{g}_1, \hat{g}_2, \ldots, \hat{g}_n\} \qquad (12)$$

Using (11)-(12), (10) can be expressed as:

$$\frac{\partial}{\partial x}\hat{v}(x,t) = -\hat{R}\hat{i}(x,t)$$
$$\frac{\partial}{\partial x}\hat{i}(x,t) = -\hat{G}\hat{v}(x,t) \qquad (13)$$

It is to be noted that (13) represents a set of decoupled equations (since $\hat{R}$ and $\hat{G}$ are diagonal matrices). This implies that the solutions for the auxiliary voltages and currents have a general form similar to that of the single line case. Once $\hat{R}$ and $\hat{G}$ are known, the solution of (10) can be represented by an equivalent circuit of resistive elements [10] (details are not given here due to the lack of space), giving an exact implementation (without any approximation) of coupled lossy section given by Modified Lie formula. Also, the macromodel is realized in terms of purely resistive elements, whose values are computed analytically solely based on the information of the line parameters.

III-2 **Frequency-Dependent Parameters**

We consider here the practically important case for lines with frequency-dependent resistance $R(s)$ and inductance $L(s)$, constant C and $G = 0$ matrices.

In this case, the delay extracted exponential matrix representing a lossy module can be expressed as

$$e^{Z(s)} = e^{\frac{A(s)+s\hat{B}(s)}{m}} \qquad (14)$$

which represents the middle term in (6) (modified Lie Formula), where

$$A(s) = -\begin{bmatrix} 0 & R(s) \\ 0 & 0 \end{bmatrix} l$$
$$\hat{B}(s) = -\begin{bmatrix} 0 & L(s)-L(\infty) \\ 0 & 0 \end{bmatrix} l \qquad (15)$$

Using (15), it can be easily seen that (14) can be implemented using lumped resistive and inductive components. For example, for the case of single lines, (14) can be represented as a frequency-dependent resistor $\hat{R}(s) = R(s)/m$, in series with a frequency-dependent inductor $\hat{L}(s) = (L(s)-L(\infty))/m$.

IV. NUMERICAL EXAMPLES

In this section, an example is presented to demonstrate the accuracy and efficiency of the proposed method. The example deals with a 40cm PCB interconnect network referred to as Line-6 in [11], however with constant C and G=0. The input voltage is a unit step response with a rise time of 0.035ns. Fig. 4 shows the transient response of the proposed method and the conventional MRA. As seen from the figure, the results from both methods are in excellent agreement. The transient simulation using the proposed model took 3 seconds of CPU time, while MRA took 74 seconds and the conventional lumped segmentation required 4641 seconds.

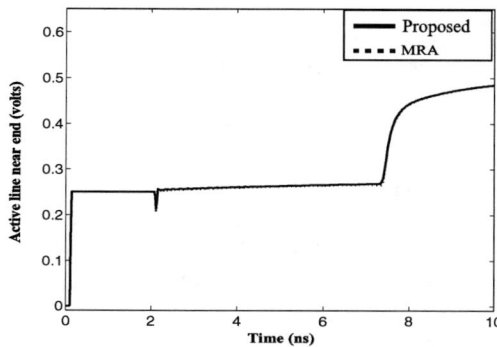

Fig. 3. Time-domain response for active line near end

V. CONCLUSIONS

In this paper, a new closed-form passive macromodel for distributed transmission lines with frequency-dependent parameters is presented. The macromodel is based on modified Lie formula and results in significantly lower orders. An important advantage of the proposed method is that, it doesn't require any numerical approximation to express the lossy sections of the modified Lie formula. The new macromodel can be computed analytically (closed-form) in terms of only delay sources and resistive elements, based on the information of per-unit-length line parameters. The proposed method while guaranteeing the passivity, provides significant speedup and an easy means to handle TLwFDP elements in SPICE-like simulators.

References

[1] A. Deustsch, "Electrical characteristics of interconnections for high-performance systems," *Proc. of IEEE*, vol. 86, pp. 315-355, Feb. 1998.

[2] C. R. Paul, Analysis of Multiconductor Transmission Lines. New York: Wiley, 1994.

[3] A. Odabasioglu, M. Celik and L. T. Pilleggi, "PRIMA: Passive Reduced-Order Interconnect Macromodeling Algorithm, "*IEEE Trans. on CAD*, pp. 645-653, August 1998.

[4] F. Y. Chang, "The generalized method of characteristics for waveform relaxation analysis of lossy coupled transmission lines," *IEEE Trans. MTT*, pp. 2028-2038, Dec. 1989.

[5] S. G. Talocia, F. Canavero, "Topline: a delay pole-residue method for simulation of dispersive interconnects, *Proc. EPEP-2002*, pp. 359-362. Monterey, CA. Nov. 2003.

[6] A. Dounavis, R. Achar and M. Nakhla "A General Class of Passive Macromodels for Lossy Multiconductor Transmission Lines," *IEEE Transactions on MTT*, pp. 1686 -1696, October. 2001.

[7] A. Dounavis, R. Achar and M. Nakhla. "Efficient passive circuit models for distributed networks with frequency-dependent parameters," *IEEE Trans. Adv. Packaging*, vol. 23, pp. 382-392, Aug. 2000.

[8] F. Fer, "Resolution de l'equation matricielle dU/dt = pU par produit infini d'exponentielles matricielles", *Acad. Roy. Belg. Cl. Sci.*, vol. 44, no. 5, pp. 818-829, 1958.

[9] N. Nakhla, A. Dounavis, R. Achar and M. Nakhla, "Delay Extraction and Passive Macromodeling of Lossy Coupled Transmission Lines", *IEEE Trans. on Advanced Packaging*, (accepted for publication).

[10] N. Nakhla, R. Achar and M. Nakhla, "Delay extraction based closed-form SPICE compatible passive macromodels for distributed transmission line interconnects", *Proc. ASP-DAC*, pp.1082-1085, Shanghai, China, Jan. 2005.

[11] A. Ruehli, A.C. Cangellaris and H-M Huang, "Three test problems for the comparison of lossy transmission line algorithms," *Proc. EPEP*, pp. 347-350, Oct. 2002.

Modeling of Power Distribution Networks with Signal Lines for SPICE Simulators

Takayuki Watanabe[†] and Hideki Asai[††]

[†]School of Administration and Informatics, University of Shizuoka, 52-1 Yada, Shizuoka, 422-8526, Japan
(TEL/FAX: +81-54-264-5444, E-mail: watanat@u-shizuoka-ken.ac.jp)
[††]Dept. of Systems Engineering, Faculty of Engineering, Shizuoka University, 3-5-1 Johoku, Hamamatsu, 432-8561, Japan
(TEL/FAX: +81-53-278-1237, E-mail: hideasai@sys.eng.shizuoka.ac.jp)

Abstract: *This paper describes the modeling of multilayered power distribution networks that include signal lines and vias. The signal lines are modeled as transmission lines, and vias are represented as self and mutual inductances. The structures can be analyzed using the conventional SPICE simulators in the time- and the frequency-domain. The accuracy of the simulation using our model is comparable to the full-wave electromagnetic simulations.*

1. Introduction

For the design of high-speed digital circuits, it is important to model the power distribution networks (PDN) in order to estimate and analyze unwanted noises, such as ground bounce, delta-I noise, and simultaneous switching noise (SSN). In the gigahertz (GHz) packages and boards, the PDN is designed using multilayered power/ground plane pairs. Due to the transient switching current of the CMOS transistors, voltage fluctuations are generated by the parasitic inductances/capacitances of the planes, and the power plane resonance causes a performance degradation of the system [1].

Detailed analysis of the PDN using full-wave electromagnetic (EM) simulators provides more accurate results. However it takes enormous CPU time and huge memory capacity. Instead of full-wave simulators, it is more convenient that the power plane is modeled by using lumped circuit elements, such as the cavity resonator model [2] and the unit cell model [3][4]. Considering the modeling of power planes having irregular shape, the unit cell model is useful because it is based on the finite deference approximation. However the method based on the unit cell model could not include effects of currents flowing though the signal lines and signal vias.

In this paper, a new feature has been added to the method of [3]. Signal lines have been modeled as transmission lines, and have been integrated in the PDN model. Through this modification, SSN generated when signal lines are referenced to the power/ground plane can be easily analyzed using the SPICE simulators. The accuracy of the simulation using our model is comparable to the full-wave electromagnetic simulations.

2. Modeling of the Power Distribution Networks

2.1 Multilayered Power/Ground Planes

In this section, the modeling of the power delivery system as illustrated in Fig. 1 is described. In the case of on-chip power distribution grids, the PDN can be modeled as power and ground (P/G) lines. On the other hand, in the case of packages and boards, the PDN can be modeled as two-dimensional P/G planes in many cases. As is well known, each P/G plane pair can be discretized spatially into $(M-1) \times (N-1)$ unit cells as shown in Fig. 2. Each RLGC parameter of the equivalent circuit of the unit cell is derived by dimensions and medium coefficients [3][4]:

$$R = R_{dc} + R_{ac}, \qquad R_{dc} = \frac{2}{\sigma_c t}, \qquad R_{ac} = 2\sqrt{\frac{\pi f \mu_0}{\sigma_c}} \qquad (1)$$

$$L = \mu_0 d \qquad (2)$$

$$C = \varepsilon_0 \varepsilon_r \frac{w^2}{d} \qquad (3)$$

$$G = 2\pi f C \tan(\delta) \qquad (4)$$

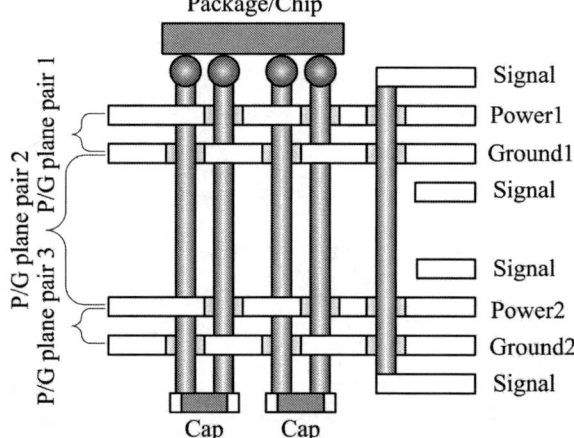

Fig. 1: Typical 8 layer PCB Stack-up.

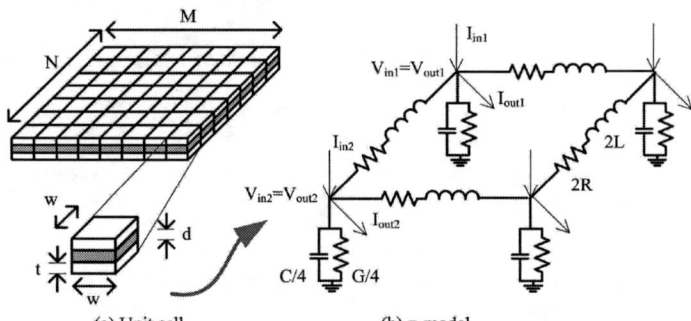

Fig. 2: Unit cell and equivalent circuit.

2.2 Via Interactions

The multilayered power delivery system can be represented as P/G plane pairs connected by many vias. Also there are thousands of signal vias through the P/G plane pairs in realistic boards and packages as shown in Fig. 3. In the past, several kinds of via model have been proposed [5]. Basically, vias can be modeled as partial self/mutual inductances and capacitances to the planes. To simplify the problem, only partial self/mutual inductances of vias have been included in this paper.

The self and mutual inductances of vias can be calculated by several electromagnetic field solvers, such as FastHenry, HFSS and so on. In [4], assuming that RF-currents flowing the P/G via pair have an equal value and an alternate direction (i.e. differential-mode currents), the P/G via pair can be modeled as one loop inductance. However, common-mode currents might flow through into the P/G via pair in realistic situation. Therefore the P/G vias are modeled separately in our method as shown in Fig. 4. In Fig. 4, in order to model voltage fluctuations between ground planes, two inductors are connected between node 'a' and node 'b' in series. Additionally, the power via and the ground via have a mutual coupling $M12$.

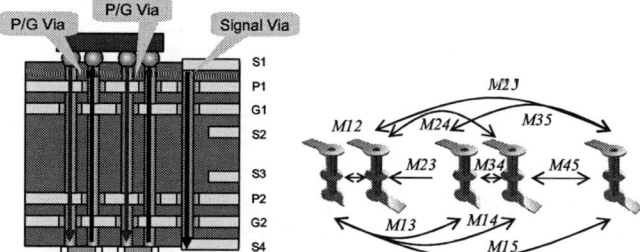

Fig.3: Modeling of via interaction.

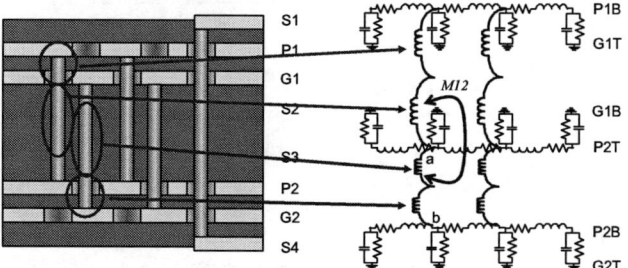

Fig.4: Modeling of P/G via pair.

3. Modeling of the Signal Lines

In this paper, a new feature is added to the PDN model. Signal lines are modeled as transmission lines, and are integrated in the PDN model. Signal lines can be represented as single and coupled transmission lines.

From the telegrapher's equation [6], the relation of voltages and currents between the near-end and the far-end in the Laplace domain is given by

$$\begin{bmatrix} \mathbf{V}_F \\ \mathbf{I}_F \end{bmatrix} = \exp\left\{\left(\begin{bmatrix} 0 & -\mathbf{R} \\ -\mathbf{G} & 0 \end{bmatrix} + s \begin{bmatrix} 0 & -\mathbf{L} \\ -\mathbf{C} & 0 \end{bmatrix}\right) \cdot d\right\} \begin{bmatrix} \mathbf{V}_N \\ \mathbf{I}_N \end{bmatrix}, \quad (5)$$

where $\mathbf{R}$, $\mathbf{L}$, $\mathbf{G}$ and $\mathbf{C}$ are per-unit-length transmission line parameters, s is $j\omega$, and d is the length of the transmission line. In order to solve Eq. (5), several transmission line models can be used in the conventional SPICE simulators, such as T-element of the Berkeley SPICE, W-Element of the H-Spice and so on.

Next, the signal via is modeled as the via inductance. The return currents of the signals are flowing through the plane pairs. These currents cause voltage fluctuations between planes. Finally, reference voltages of the signal line might swing. To model this effect, CCCS (Current Controlled Current Source) and VCVS (Voltage Controlled Voltage Source) are used in the lumped model as shown in Fig. 5.

The path of the return currents of the signal lines can not be identified on the P/G planes. However, the return current, which has an equal value of the signal current, would flows on the surface of the anti-pad to the opposite direction of the signal current. Therefore, this return current can be modeled as the CCCS in Fig. 6.

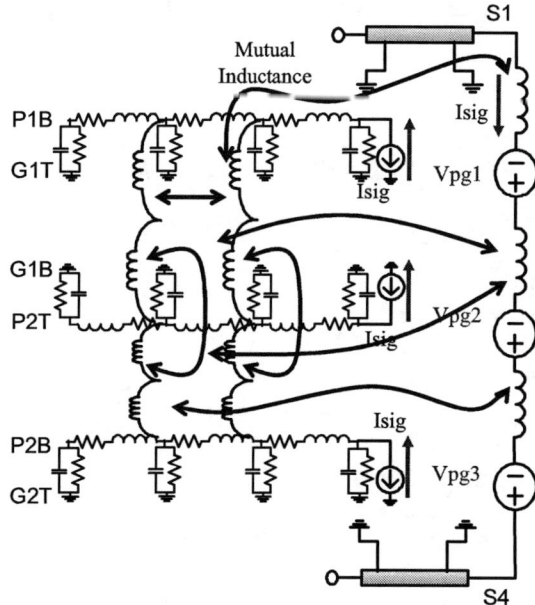

Fig. 5: The lumped model of Fig.1.

In Fig. 7(a), if the voltages (Vpg1, Vpg2, and Vpg3) between the P/G planes fluctuate, the reference voltages of signal lines will become to swing. Assuming that the reference node of the top transmission line is a unique reference point, the reference node of the bottom transmission line has the potential difference (Vpg1 + Vpg2 + Vpg3) to the unique reference point in Fig. 7(b). In order to model this potential difference, three VCVSs are inserted between the reference nodes of the transmission lines in Fig. 7(c). Also, Fig. 7(c) and Fig. 7(d) are equivalent.

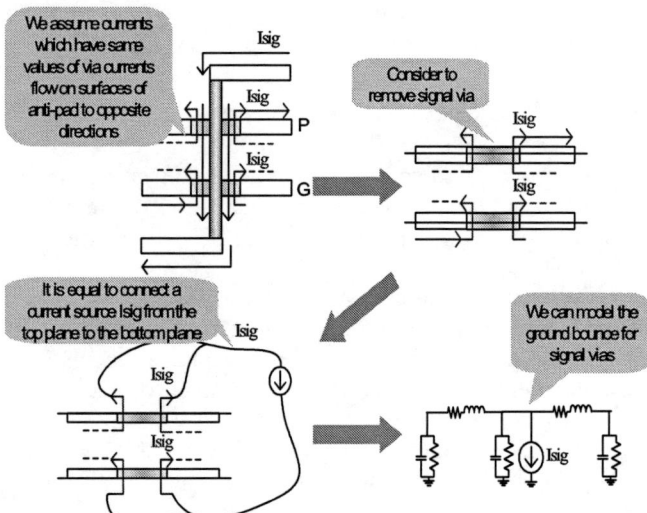

Fig. 6: Modeling Signal Return Currents.

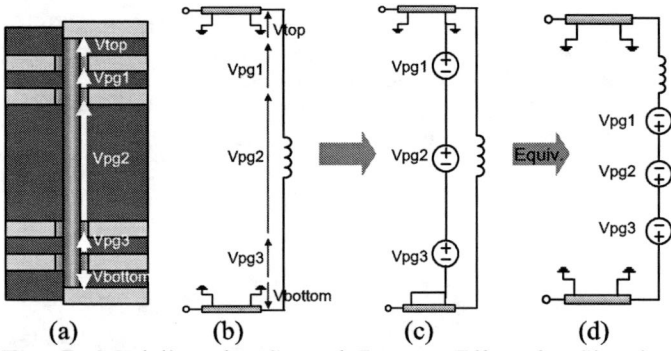

Fig. 7: Modeling the Ground Bounce Effect for Signal Lines.

4. Numerical Results

An irregular shaped board shown in Fig. 8 was analyzed. In order to verify the validity of the lumped model in Fig. 5, we compared the transient responses calculated using the SPICE (modified version of Berkeley Spice 3f5) and the FDTD (Finite Difference Time Domain) simulator. In the SPICE simulation, each plane pair was modeled by the unit cells (Fig. 9) without frequency dependent effects, such as the skin effect. Via inductance were calculated by the FastHenry program [7]. In the FDTD simulation, nonuniform cells and the PML (Perfect Matched Layer) absorb boundary condition were used.

As a first example, P1 (25mm, 75mm) was excited with a Gaussian pulse. The currents flowing in the left hand side power and ground vias in Fig. 8 is plotted in Fig. 10 (a). Fig. 10 (a) shows the difference of the currents flowing in the power via and the ground via. Our model can accurately simulate this phenomenal because the P/G vias are modeled separately in our method. Also the currents flowing in the input port is plotted in Fig. 10 (b). They indicate that there is a good correlation between the SPICE and the FDTD results.

As a second example, the transmission line was excited with a Gaussian pulse as shown in Fig. 11. The transmission line was modeled as a T-element in the SPICE, terminated with a 25.87Ω load resistor. The voltage waveforms at the terminals of the signal lines are shown in Fig. 12 (a) and (b). In Table 1, we compared the CPU time of the SPICE and the FDTD simulations. From these results, our lumped model including the vias and the signal lines is effective to verify not only the power integrity but also the signal integrity.

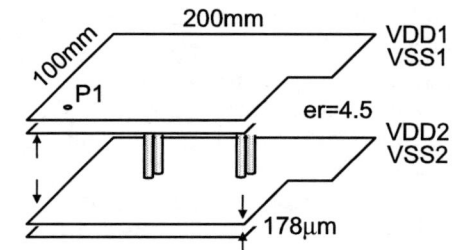

Fig. 8: Example board 1.

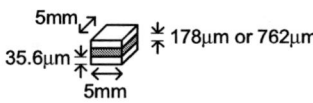

Fig. 9: Unit cell size.

Table 1: The CPU time comparisons.

	Problem Size	CPU Time
SPICE(Our model)	11,065 nodes	303.3 sec
FDTD	188,811 cells	953.3 sec

5. Conclusions

In this paper, the modeling method of the power distribution network including transmission lines has been presented. In this method, all plane pairs and transmission lines are connected by vias, which are modeled as self and mutual inductors. From the numerical results, the accuracy of the simulation using our model is comparable to the full-wave FDTD simulation, and our method is effective to verify not only the power integrity but also the signal integrity.

Acknowledgment

The authors would like to acknowledge Prof. M. Swaminathan and Mr. K. Srinvasan, School of Electrical and Computer Engineering, Georgia Institute of Technology, for valuable discussion on the unit cell modeling.

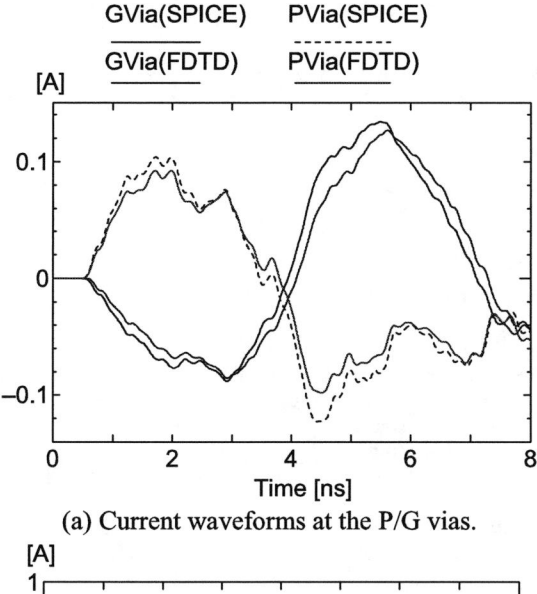

(a) Current waveforms at the P/G vias.

(b) Current waveforms at the input port.

Fig. 10: Transient responses of the example board 1 calculated by SPICE and FDTD simulator.

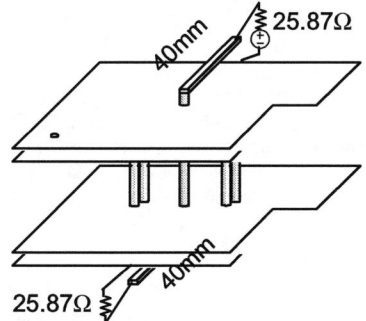

Fig. 11: Example board 2.

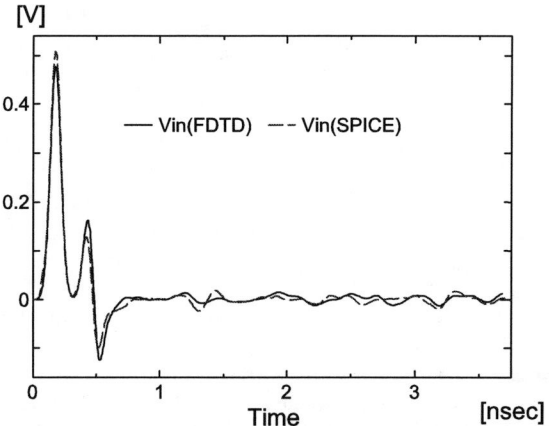

(a) Voltage waveforms at the near-end of T-Line.

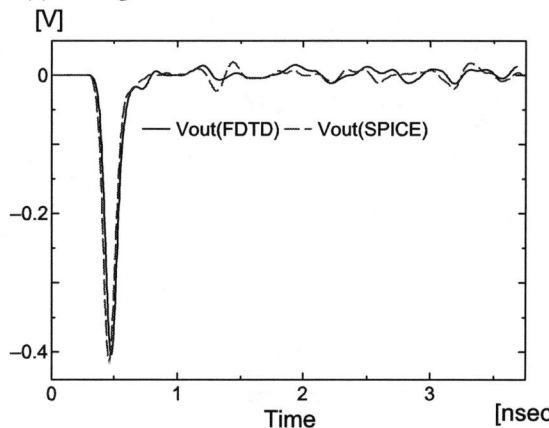

(b) Voltage waveforms at the far-end of T-Line.

Fig. 12: Transient responses of the example board 2 calculated by SPICE and FDTD simulator.

References

[1] L. Polka, S. Chickamenahalli, C.Y. Chung, D.G. Figueroa, Y.L. Li, K. Merley, D. Wood and L. Zu, "Package-Level Interconnect Design for Optimum Electrical Performance," *Intel Technology Journal Q3*, 2000.

[2] N. Na, J. Choi, S. Chun, M. Swaminathan, and J. Srinivasan, "Modeling and transient simulation of planes in electromagnetic packages," *IEEE Trans. Comp., Packag., Manufact. technol. B*, vol. 21, pp. 157–163, May. 1998.

[3] L. Smith, R. Raymond, and T. Roy, "Power plane spice models and simulated performance for materials and geometries, " *IEEE Trans. Adv. Packag.* vol. 24, pp. 277–287, Aug. 2001.

[4] J.H. Kim, and M. Swaminathan, "Modeling of Multilayered Power Distribution Planes Using Transmission Matrix Method," *IEEE Trans. Adv. Packag.* vol. 25, pp. 189–199, May. 2002.

[5] K.S. Oh, J.E. Schutt-Aine, R. Mittra, and B. Wang, "Computation of the equivalent capacitance of a via in a multilayered board using the closed-form Greens function," *IEEE Trans. Microwave Theory Tech*, vol. 44, pp. 347–349, Feb. 1996.

[6] R. Achar, M.S. Nakhla, "Simulation of high-speed interconnects, " *Proceedings of the IEEE*, vol. 89, No. 5, pp.693-728, May 2001.

[7] M. Kamon, M.J. Tsuk, and J. White, "Fasthenry: A multipole-accelerated 3-d inductance extraction program," *IEEE Transactions on Microwave Theory and Techniques*, vol. 42, pp. 1750–1758, Sept. 1994.

Scheduling Algorithm for Partially Parallel Architecture of LDPC Decoder by Matrix Permutation

In-Cheol Park and Se-Hyeon Kang
Department of Electrical Engineering and Computer Science, KAIST
373-1 Guseong-dong, Yuseong-gu, Daejeon 305-701, Republic of Korea
{icpark, shkang}@ics.kaist.ac.kr

Abstract— The fully parallel LDPC decoding architecture can achieve high decoding throughput, but it suffers from large hardware complexity caused by a large set of processing units and complex interconnections. A practical solution of area-efficient decoders is to use the partially parallel architecture in which a PU is shared for a several rows or columns. It is important in the partially parallel architecture to determine the rows or columns to be processed in a PU and their processing order. The dependencies between rows and columns should be considered to minimize the overall processing time by overlapping the decoding operations. This paper proposes an efficient scheduling algorithm that can be applied to general LDPC codes, which is based on the concept of the matrix permutation. Experimental results show that the proposed scheduling achieves a higher decoding rate, leading to a reduction of 25% processing time on the average. A 1024-bit rate-1/2 LDPC decoder employing the proposed scheduling algorithm provides almost 1Gbps decoding throughput and occupies one-fifth area compared to the fully parallel decoder.

I. INTRODUCTION

Low density parity check (LDPC) codes are originally devised to exploit low decoding complexity by constructing sparse parity check matrices. Though the LDPC code does not have a maximized minimum distance due to the randomly generated sparse parity check matrix, the typical minimum distance increases linearly as the block length increases. Moreover, the error probability decreases exponentially for a sufficiently long block length, whereas the decoding complexity is linearly proportional to the code length. Recent simulation results show that the LDPC code can achieve a performance that is within 0.04 dB of Shannon limit and the performance is close to that of the turbo code if the block length is larger than 1000 bits [2].

Despite of these advantages, when the LDPC code was first introduced, it made a little impact on the information theory community because of the storage requirements in encoding and the computational complexity in decoding. Modern VLSI technology is so advanced that it enables parallel architectures exploiting the benefit of inherently parallel LDPC decoding algorithms. Blanksby et al. implemented an 1-Gb/s fully parallel decoder in which the message passing algorithm is directly mapped [3]. This architecture, however, requires a large number of complex routings between concurrent processing units (PUs) each of which corresponds to a node of the message passing algorithm, leading to the average net length of 3mm and the total die size of 52.5mm^2. On the other side, Yeo et al. proposed an area-efficient architecture that serializes the computations by sharing PUs [4]. Consequently, one iteration takes 9612 cycles and wide-input multiplexers are required to select one of 18432 intermediate values to be fed into the shared PUs. These two counter examples show that high throughput LDPC decoding architectures should exploit the benefit of parallel decoding algorithms while reducing the interconnection complexity.

The partially parallel architecture is a good trade-off between throughput and hardware cost. Since a PU is shared for a number of rows or columns, the number of PUs becomes much smaller than that of the fully parallel architecture. As decoding operations are parallel in nature, it is important to determine which rows or columns are processed in a PU. In the grouping, the dependencies between rows and columns should be considered to minimize the overall cycles by overlapping the decoding operations. There has been a heuristic scheduling algorithm proposed for quasi-cyclic LDPC codes [9], but it cannot be applied to general LDPC codes. This paper proposes an efficient scheduling algorithm that can be applied to general LDPC codes. The proposed algorithm is based on the concept of the matrix permutation.

The rest of this paper is organized as follows. Section II explains briefly about the low density parity check code and the decoding algorithm. Section III proposes a new scheduling algorithm which helps partially parallel LDPC decoder to achieve high throughput using matrix permutation. Experimental results of the proposed algorithm are shown in Section IV. Finally, Section V addresses some concluding remarks.

II. LDPC DECODER ARCHITECTURE

The LDPC code, which was first introduced by Gallager in 1962 [1], is defined by a binary linear block code of length n and a parity check matrix H with a column weight γ and a row weight ρ: (n, γ, ρ). The parity check matrix H has n columns and J rows that correspond to the block length and the total number of parity check equations of the code, respectively. The column weight γ and the row weight ρ represent the number of 1's in a column and a row, respectively, and they are much smaller than n to achieve the sparse matrix H. Fig. 1 (a) shows an example matrix H of (12, 3, 6) LDPC code with indicating the code parameters.

A. Message Passing Algorithm

The LDPC code is often represented by a factor graph to make it easy to understand the message passing decoding algorithm, which is a bipartite graph that expresses how a global function of many variables is factored into a product of local functions [7]. Fig. 1 (b) shows the factor graph of a (12, 3, 6) LDPC code, which consists of two sets of nodes: i.e. variable nodes, $\{v_j\}$, and check nodes, $\{c_i\}$. The edge between a variable node v_j and a check node c_i is constructed if there is 1 at (i, j) in the parity check matrix. Therefore each check node represents a check equation used in generating parity check bits and each variable node represents one bit in the codeword. The variable nodes that are connected to a check node are called the neighbor variable nodes of the check node. For a variable node, the neighbor check nodes are defined similarly.

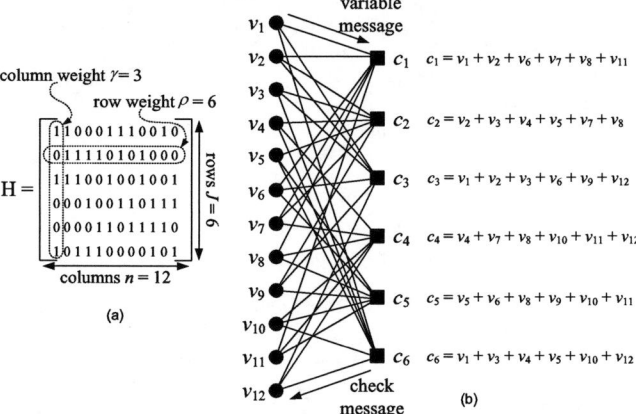

Fig. 1. A (12, 3, 6) LDPC code. (a) Parity check matrix. (b) Factor graph.

Unlike the general parity check codes, a LDPC code cannot be optimally decoded since its decoding is NP-complete. An iterative approximate algorithm called a message-passing algorithm is used instead, which is also known as sum-product [8] or belief propagation [5]. The message-passing algorithm can be expressed easily based on the factor graph as below.

1) Initialize each variable node v_j to the probability ratio of the corresponding received bit. The probability ratio is the first variable message to be sent to the neighbor check nodes.

$$\Delta_{ij} = \Delta_j = \frac{p(r_j | x_j = +1)}{P(r_j | x_j = -1)} = \exp(\frac{2r_j}{\sigma^2}) \quad (1)$$

2) Variable-to-check (VTC) step: Each check node c_i computes the likelihood ratio, Λ_{ij}, by using variable messages of its neighbor variable nodes except from v_j. This is the check message to be sent to variable node v_j.

$$\Lambda_{ij} = \prod_{j' \in N(i) \setminus j} \frac{1 - \Delta_{ij'}}{1 + \Delta_{ij'}} \quad (2)$$

3) Check-to-variable (CTV) step: Each variable node v_j computes probability ratios, denoted by Δ_{ij}, by using check messages of its neighbor check nodes except from c_i. This is the variable message to be sent again to check node c_i.

$$\Delta_{ij} = \frac{p(r_j | +1)}{P(r_j | -1)} \prod_{i' \in M(j) \setminus i} \frac{1 - \Lambda_{i'j}}{1 + \Lambda_{i'j}} \quad (3)$$

4) For each variable node, create a tentative bit-by-bit decoding x_j using the pseudo-posteriori probability Δ_j.

$$\Delta_j = \frac{p(r_j | +1)}{P(r_j | -1)} \prod_{i \in M(j)} \frac{1 - \Lambda_{ij}}{1 + \Lambda_{ij}} \quad (4)$$

$$x_j = \begin{cases} 0 & \text{when } \Delta_j \leq 1 \\ 1 & \text{when } \Delta_j > 1 \end{cases} \quad (5)$$

5) Check if $x \cdot H^T = 0$ is satisfied. If it is satisfied or the maximum number of iterations is reached, the decoding algorithm finishes. Otherwise, the algorithm repeats from step 2).

The symbols, Δ and Λ, correspond to the outgoing messages of the variable and check nodes, respectively, and $N(i)$ and $M(j)$ represent the set of neighbor nodes of check node c_i and variable node v_j, respectively. The not-notation (') marked as superscript means that the product is calculated over the indexes excluding its own index. The VTC operation calculates the check messages $\{\Lambda_{ij}\}$ of check node c_i using the variable messages $\{\Delta_{ij}\}$ for variable nodes identified by the i-th row of the matrix H. A detailed explanation of the algorithm can be found in [6].

B. Partially Parallel Architecture

The fully parallel architecture is inherited from the message-passing algorithm and all the operations required in each node are implemented as a PU. The number of PUs is as many as the number of the variable and check nodes. Although the fully parallel implementation of the message-passing algorithm is straightforward and results in a high throughput decoder, it faces with complex interconnections required to sum up the Λ_{ij} and Δ_{ij} messages that are calculated in the PUs spread over the chip area.

In the partially parallel architecture, each PU takes in charge of several numbers of rows or columns as shown in Fig. 2. As a PU is shared for a number of rows or columns, the number of PUs becomes much smaller than that of the fully parallel architecture. Thus the number of VTC (CTV) operations processed in a cycle is as many as the number of check (variable) PUs. The check messages calculated row-by-row by a check PU can be grouped and stored into a local memory to save area, and variable PUs access them later.

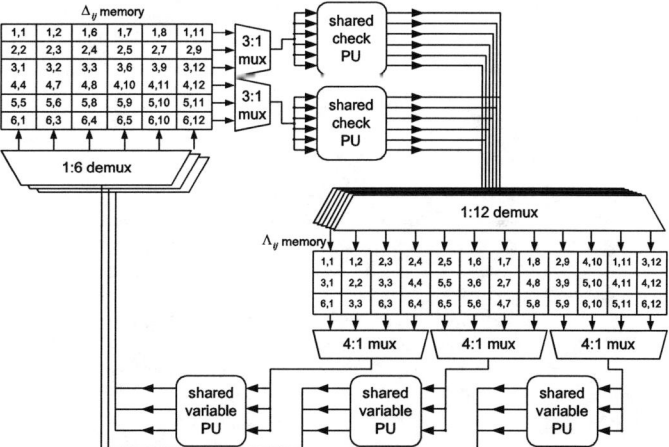

Fig. 2. Partially parallel LDPC decoding architecture.

The VTC and CTV operations are parallel in nature and thus a PU can process any rows or columns regardless of their order in matrix H. Since a PU takes in charge of several rows or columns in the partially parallel architecture, we have to determine which rows or columns are processed in the PU. In the grouping, the dependencies between rows and columns should be considered to minimize overall cycles by overlapping the VTC and CTV operations. For example, the CTV operation for v_1 in the (12, 3, 6) LDPC code requires the VTC operations for c_1, c_3 and c_6 to be completed beforehand. In addition, it is important to determine the processing order of rows or columns for a PU, because the dependence of the Λ_{ij} and Δ_{ij} can hinder the overlapped processing.

III. SCHEDULING ALGORITHM

Traditionally, the CTV operations can start only after the entire VTC step finishes completely, and vice versa. A well-scheduled sequence of the message calculations can reduce the latency, because a CTV operation can start in the course of the VTC step if the corresponding row summation values are available, and vice versa. The overlapped processing of the VTC and CTV steps results in a reduced number of cycles for an iteration.

A. Overlapped Processing

Fig. 3 shows an example of overlapped processing for the (12, 3, 6) LDPC code, assuming that the numbers of check PUs and variable PUs are two and three, respectively. Each arrow denotes a clock cycle and the numbers above an arrow indicate the row or column indices processed at that cycle. If the VTC and CTV steps are performed according to the increasing order of indices without scheduling, no overlapped processing is possible. Though two CTV operations for v_2 and v_7 can start at the last cycle of the VTC step, the CTV step does not start until three CTV operations are enabled for easy control design. If the VTC operations are scheduled as shown in Fig. 4 (b), three CTV operations for v_2, v_6 and v_9 can start processing at the last cycle of the VTC step. Furthermore, the CTV operations can be scheduled to enable the VTC operations of the next iteration to start earlier. In this example, the scheduling of the VTC and CTV operations saves two cycles.

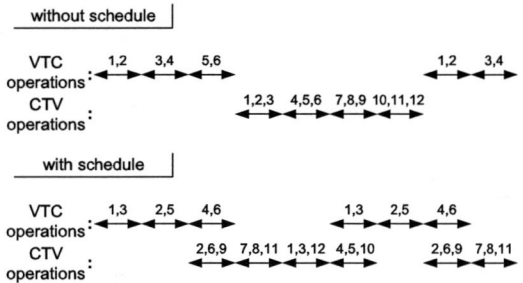

Fig. 3. An example of overlapped processing

The scheduling process can be viewed as the row and column permutation of the parity check matrix H. Fig. 4 (a) is the original parity check matrix H whose size is $M \times N$. In this case, the size of H is 6×12 and we assume 2 check PUs and 3 variable PUs. If there are m check PUs, m rows at the top of the matrix are processed in the first cycle, the next m rows in the second cycle, and so on. Thus the VTC step takes M/m cycles. Similarly, the CTV step takes N/n cycles if there are n variable PUs. The permutation changes the sequence of the VTC and CTV operations to enlarge the empty slots at the lower left and upper right corners as shaded in Fig. 4 (b).

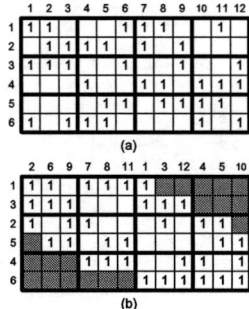

Fig. 4. Scheduling by the permutation of matrix H. (a) Matrix H. (b) Permuted matrix.

The empty slots in a column mean that the processing of the column can start as long as the rows above the empty slots are processed beforehand. If there are k empty slots in a column, the column processing can start $\lfloor k/m \rfloor$ cycles earlier, and if the first n columns have more than m empty slots, all the variable PUs can start their processing one cycle earlier. The simultaneous early starting is possible if the $m \times n$ slots that are at the lower left part surrounded by thick lines are all empty, and it is required to achieve an easy control design. For example, as three columns have empty slots larger than two in Fig. 4 (b), they can start after two cycles of the VTC step. Similarly, the empty slots in a row mean that the processing of the row can start as long as the columns at the left hand side of the empty slots are processed beforehand, leading to two rows that can start after three cycles of the CTV step.

In fact, the earliest cycle found by the permutation is not the real starting time of the CTV (VTC) step. For the sake of easy control design, we assume that the CTV (VTC) step performs continuously without skipping cycles once it starts. If there is a cycle in which any column (row) cannot be processed during the CTV (VTC) step, the former operations are delayed to achieve continuous CTV (VTC) operations.

B. Scheduling Algorithm

It takes a considerable amount of time to find an optimum schedule that minimizes the overall cycles because of the large number of rows and columns. There has been a heuristic scheduling algorithm proposed for quasi-cyclic LDPC codes [9], but it cannot be applied to general LDPC codes. We describe a new scheduling algorithm developed for the partially parallel LDPC decoding architecture. The proposed algorithm described below is based on the concept of the matrix permutation. The algorithm makes the row sequence and column sequence that can result in empty spaces in the lower left and upper right corners of the permuted matrix when the matrix H is rearranged according to the sequences.

Step 1: row sequence = null, column sequence = null.
Step 2: Select a row r_k randomly.
The row sequence is enlarged by appending r_k, and the column sequence is appended by inserting $\{c_y | c_y \in r_k\}$. Go to Step 4.
Step 3: If all the rows contained in a column are selected, such a column is called a completed column. If a column contained in a row is also in the column sequence, it is called a common column of the row. For each unselected row, count the number of common columns (CC) and the number of columns to be completed (CTC) when the row is selected. Select a row from unselected rows, r_i, that has the maximal CTC. If there are more than two rows associated with the maximal CTC, we select the one that has the largest CC.
The row sequence is appended by r_i, and the column sequence is enlarged by appending the columns of r_i that are not in the current column sequence.
Step 4: Find completed columns, and move them to the front of the first uncompleted column in the column sequence.
Step 5: If there are unselected rows, go to Step 3.
Step 6: Permute the matrix H according to the row sequence and column sequence.
Step 7: If there are m check PUs, the first m rows at the top of the permuted matrix are assigned to the first cycle of the VTC step, the next m rows to the second cycle, and so on. Similarly, for n variable PUs, the first n columns at the left of the permuted matrix are assigned to the first cycle of the CTV step, the next n columns to the second cycle, and so on.

The random row selection in Step 2 is to explore more search space by starting from a different row. Step 3 to 5 are the main loop for matrix permutation and Step 6 to 7 assign rows and columns to PUs considering easy control design after scheduling. In the algorithm, rows are selected to make the upper right part of the permuted matrix as empty as possible and completed columns are moved to the left in order to make enlarged empty space in the lower left part. An example of the algorithm is illustrated in Fig. 5.

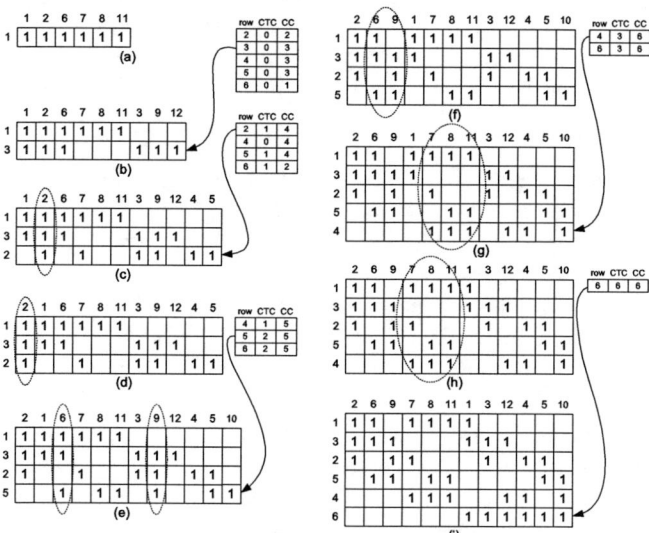

Fig. 5. An illustrative example of the proposed scheduling. (a) Select r_1. (b) Append r_3. (c) Append r_2. (d) Move c_2. (e) Append r_5. (f) Move c_6, c_9. (g) Append r_4. (h) Move c_7, c_8, c_{11}. (i) Append r_6.

IV. EXPERIMENTAL RESULTS

By applying the scheduling algorithm to a (1024, 3, 6) LDPC code, we can save the number of processing cycles per iteration as summarized in Table I, where parallelism (m, n) stands for an implementation associated with m check PUs and n variable PUs.

TABLE I
CYCLES SAVED BY SCHEDULING PER ITERATION

Parallelism	Cycles w/o schedule	Cycles w/ schedule	Saved cycles	Saved cycles (%)
(4, 8)	256	195	61	23.8
(8, 8)	192	139	53	27.6
(16, 8)	160	129	31	19.4
(8, 16)	128	93	35	27.3
(16, 16)	96	70	26	27.1
(32, 16)	80	60	20	25.0
(16, 32)	64	47	17	26.6
(32, 32)	48	36	12	25.0
(64, 32)	40	30	10	25.0

We designed two (1024, 3, 6) LDPC decoders using a 0.18 um CMOS process for parallelism (16, 32) and (32, 64). The performances of the decoders are summarized in Table II. The first decoder corresponding to parallelism (16, 32) occupies an area of 6.29 mm2 and provides more than 500 Mbps decoding throughput at the frequency of 200 MHz, and the second decoder corresponding to parallelism (32, 64) takes 10.08 mm2 and provides almost 1Gbps decoding throughput at the same frequency. The resulting bit rate of the second decoder is comparable to the fully parallel architecture proposed by Blanksby et al., but significant area reduction is achieved. Although the iteration number in Blanksby's implementation is fixed to 64 due to the scan-chain like I/O mechanism, there is no restriction in the proposed architecture. We chose 8 iterations to provide sufficient BER performance.

TABLE II
COMPARISON OF LDPC DECODERS

	Blanksby [3]	Proposed	
		Parallelism (16,32)	Parallelism (32,64)
Technology	0.16 um	0.18 um	0.18 um
Bit rate	1 Gbps	582 Mbps	985 Mbps
Area	52.5 mm²	6.29 mm² MEM: 3.46 mm² Other: 2.83 mm²	10.08 mm² 6.45 mm² 3.63 mm²

V. CONCLUSIONS

This paper has presented an efficient scheduling algorithm for partially parallel LDPC decoders to minimize the overall processing time by overlapping CTV and VTC steps. As the VTC and CTV operations are parallel in nature, a PU can process any rows or columns regardless of their order in matrix H, and thus it is very important to determine which rows or columns are processed in a PU. In the grouping, the dependencies between rows and columns should be considered to overlap the VTC and CTV operations. Based on the concept of the matrix permutation, the proposed scheduling algorithm generates the row sequence and column sequence that can result in empty spaces in the lower left and upper right corners of the permuted matrix when matrix H is rearranged according to the sequences. Experimental results show that the proposed scheduling achieves a higher decoding rate, leading to a reduction of 25% processing time on the average.

ACKNOWLEDGMENT

This work was supported by Institute of Information Technology Assessment through the ITRC, by the Korea Science and Engineering Foundation through MICROS center and by IC Design Education Center (IDEC).

REFERENCES

[1] R. G. Gallager, "Low density parity check codes," *IRE Trans. Info. Theory*, vol. IT-8, pp. 533-547, Jan. 1962.
[2] S. Chung, D. Forney, T. Richardson, and R. Urbanke, "On the design of low-density parity-check codes within 0.0045 db of the Shannon limit," *IEEE Comm. Letters*, vol. 5, pp. 58-60, Feb. 2001.
[3] A. Blanksby and C. Howland, "A 690-mW 1-Gb/s, rate-1/2 low-density parity-check code decoder," *IEEE J. of Solid-State Circuits*, vol. 37, pp. 404-412, Mar. 2002.
[4] E. Yeo, P. Pakzad, B. Nikolić and V. Anantharam, "VLSI architectures for iterative decoders in magnetic recording channels," *IEEE Trans. Magnetics*, vol. 37, pp. 748-755, Mar. 2001.
[5] J. Pearl, *Probabilistic Reasoning in Intelligent Systems: Networks of Plausible Inference*, Morgan Kaufmann, 1988.
[6] K. Lo, *Layered space time structures with low density parity check and convolutional codes*, MS Thesis, School of EIE, Univ. of Sydney, 2001.
[7] F. R. Kschischang, B. J. Frey, and H. Loeliger, "Factor graphs and the sum-product algorithm," *IEEE Trans. Info. Theory*, vol. 47, pp. 498-519, Feb. 2001.
[8] N. Wiberg, *Codes and decoding on general graphs*, PhD thesis, Dept. of EE, Linköping Univ. Sweden, 1996.
[9] Y. Chen, K. K. Parhi, "Overlapped message passing for quasi-cyclic low-density parity check codes," *IEEE Trans. Circuits and Syst. I*, vol. 51, pp. 1106-1113, Jun. 2004.

Quantized LDPC Decoder Design for Binary Symmetric Channels

Rohit Singhal
Computer Science
Texas A & M University
College Station, TX 77843

Gwan S. Choi
Electrical Engineering
Texas A & M University
College Station, TX 77843

Rabi N. Mahapatra
Computer Science
Texas A & M University
College Station, TX 77843

Abstract—Binary Symmetric Channels (BSC) like the Interchip buses and the Intra-chip buses are gaining a lot of attention due to their widespread use with multimedia storage devices and on system-on-chips (SoC) respectively. While the audio and video traffic between systems has increased many-fold over the years, SoC is a reality due to the advances in technology as predicted by the Moore's Law. These buses are prone to error arising from crosstalk between wires, propagation delay etc. Due to low latency requirements, re-transmission is undesirable in the event of an error and forward error correction (FEC) becomes more and more desirable is a necessity. This paper focuses on the low density parity check (LDPC) codes as a means of FEC. Several quantization schemes to reduce the size of the decoder, and the associated code performance, are presented herein. The reduction in size due to the quantization schemes is made apparent via implementation on a Xilinx Virtex FPGA.

I. INTRODUCTION AND BACKGROUND

The advances in the multimedia technology have resulted in widespread data communication between storage devices and user interfaces. The increasing resolution in the audio and video formats results in higher data-rates through the interconnecting buses. If these applications are on a real-time basis, re-transmission of the data received in error is prohibited in most cases. The onus of delivering data reliably to the receiver then lies on the forward error correction (FEC) mechanisms [1].

More use of fast and efficient error correcting mechanisms can be found on Intra-chip interconnects, where phenomenon like crosstalk etc. impose a significant data-reliability concern [2]. Moreover, the wires on the system-on-chips (SoC) have a large end-to-end delay and consume a significant amount of energy in data transmission. These power and delay concerns make re-transmission of data very difficult.

It is imperative to design FEC schemes for these Binary Symmetric Channels (BSC) that have the following characteristics in addition to exceptional error correction capabilities.

- Low latency in end-to-end transmission.
- Low power consumption.
- Small implementation size.

Turbo codes [3] are perhaps the most popular code for error correction and come very close to the Shannon's limit. They, however, are very bulky to implement and have a high latency of decoding. In 1962, Gallager proposed a class of linear block codes called the Low Density Parity Check (LDPC) codes [4], [5] that rival the performance of turbo codes. Several implementation designs for LDPC decoder are presented in [6], [7], [8], [9], [10], [11], [12], [13], [14]. A finite precision scheme based on uniform quantization, to reduce the decoder implementation size for additive white gaussain noise channels, is presented in [15]. This paper focuses on improving the LDPC decoder design for FEC on Binary Symmetric Channels through empirical analysis of the data flow in the decoder.

The rest of this paper is organized as follows. Section II describes the LDPC code construction and the decoding procedure. Section III describes reduction in data precision as a means of trimming the implementation size of a decoder. The effect on the code performance and the implementation size is also quantified. Section IV states the conclusion, while Section V contains the acknowledgements.

II. LOW DENSITY PARITY CHECK CODES

LDPC codes are linear block codes, where n bits of a code word are formed with the linear combination of the k bits of an input word. The number of information bits for every transmitted bit is therefore $R = k/n$. The LDPC codes can be uniquely defined by a $(n-k) \times n$ parity check matrix H such that if C be the set of all valid n bit long code words, then all vectors $x \in C$ satisfy the relation $H.x^T = \mathcal{O}$, where $\mathcal{O}$ is a vector having $n - k$ zeros. The H of a regular LDPC code can be constructed randomly such that each column contains exactly $j \geq 3$ ones and each row contains $\rho > j$ ones. This code is called an (n, j, ρ) LDPC code.

Alternate representation of an LDPC code is a bipartite graph which can be constructed from an H matrix uniquely. One side of the bipartite graph contains n Variable Nodes (VN) corresponding to the n columns of H matrix, while the other side contains $n - k$ Check Nodes (CN) representing the rows. A 1 on the rth row and the cth column of the H matrix shows the presence of a bidirectional edge between the rth VN and the cth CN.

A. LDPC Decoding Process

The LDPC decoding process follows an *iterative message passing* algorithm, where messages flow back and forth between the two sides of the bipartite graph. At each iteration, the passed messages are better estimates of the transmitted bits than the last iteration. These messages, also known as the extrinsics, are expressed as a Log Likelihood ratio (LLR) defined as

$$L(m) = L(c_m|r_m) = \log\left[\frac{p(c_m = 1|r_m)}{p(c_m = 0|r_m)}\right], \quad (1)$$

where, r_m is the value received from the communication medium and c_m is binary estimate of it.

On each side of the bipartite graph, updates are calculated and are called the VN update (VNU) or the CN update (CNU) respectively. Each node takes the *a priori* probabilities of the bits as inputs, and based on them, calculates the *a posteriori* probabilities [4].

Every VN has j LLR inputs $(L_{cb,1}, L_{cb,2}, \ldots, L_{cb,j})$ from the CN in addition to the channel output L_{ch}. The VN calculates j outputs $(L_{bc,1}, L_{bc,2}, \ldots, L_{bc,j})$ according to the following equation.

$$L_{bc,i} = L_{ch} + \sum_{k \neq i} L_{cb,k}. \quad (2)$$

Every CN has ρ LLR inputs $(L_{bc,1}, L_{bc,2}, \ldots, L_{bc,\rho})$ from the VN, and calculates ρ outputs $(L_{cb,1}, L_{cb,2}, \ldots, L_{cb,\rho})$ according to the following equation.

$$L_{cb,i} = \left(-\prod_{k \neq i} \text{Sign}(L_{bc,k})\right) \times \psi^{-1}\left(\sum_{k \neq i} \psi(L_{bc,k})\right) \quad (3)$$

where, $\psi(x) = \log\left(\tanh\left(\left|\frac{x}{2}\right|\right)\right)$, and $\psi^{-1}(x) = -\psi(x)$.

An expanded view of the bipartite graph to facilitate unidirectional flow of information is shown in fig. 1. Each successive iteration produces a better estimate of the input bits. The last VN stage takes a decision about the information bit. The output L_O is computed at each VN by summing all $j + 1$ inputs. If the resulting $L_O > 0$ then the final binary estimate $c_o = 1$ otherwise $c_o = 0$.

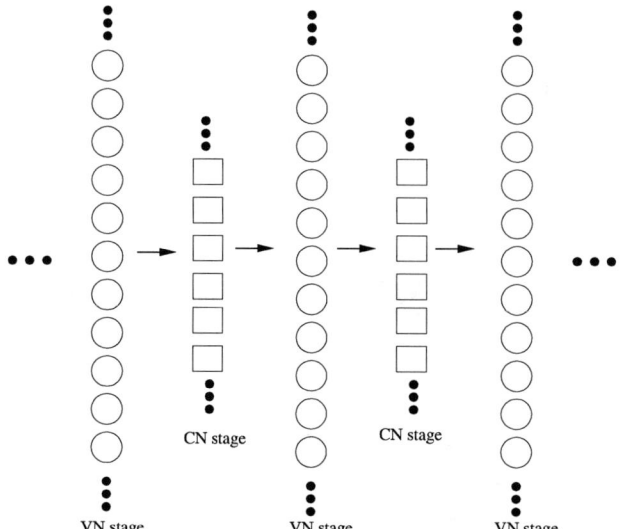

Fig. 1. Schematic of the Decoder.

Several papers showing the implementation of the LDPC decoder have been presented recently. These range from a completely serial architecture [6] to a fully parallel architecture [7]. The serial architecture has the advantage of lower hardware requirement, but it suffers from low throughput. On the other hand, although the parallel architecture permits high throughput, the hardware resource requirements may be prohibitive. Several other approaches that combine advantages of both serial and parallel designs have also been developed [8], [9], [10], [11], [12], [13], [14].

B. The Need for Efficient Quantization

This work studies the application of LDPC codes in BSC, which usually dictate a low latency decoder implementation. This can be ensured by a fully parallel LDPC decoder architecture. However, as discussed above, the fully parallel architecture suffers from a high hardware-resource requirement.

The parity check matrix H is a low density matrix, implying a sparse bipartite graph. There are $n \times j$ signal buses running between the VN stage to the CN stage and from the CN stage to the next VN stage. Even though, this many wires may be prohibitive for parallel implementation for large frame lengths (e.g., like those in AWGN channels), the frame size n in typical binary communications is much smaller resulting in a trivial routing complexity.

The problem that still remains, however, is that the parallel architecture has a massive hardware resource requirement. For just a single iteration, there are n VN and $n - k$ CN to be implemented. For additional iterations, more such blocks are required. The implementation complexity of each VN and CN is proportional to the precision of the LLR values in them. Moreover, the widths of the buses connecting the VN to the CN are also proportional to the precisions of the LLR values. Therefore, the minimization of the decoder size requires minimization of the information precision at the nodes. The highest precision (and the best performance) comes from floating point implementation of the LLR values. The next section discusses quantization in the LLR values at the VN and the CN without degrading the decoder performance noticeably.

III. THE CASE FOR QUANTIZATION

This section studies the effects of quantization on the LDPC code performance, as well as the reduction in the circuit size due to it. Even though, this discussion considers a $(264, 3, 12)$ code with a rate, $R = 0.75$, it can be extended to any general LDPC code for BSC channel.

A. Uniform Quantization

Uniform quantization means mapping of a number to the closest integer multiple of a small number ϵ. In the context of the LDPC decoder, the LLR values at the VN and the CN are mapped to integer multiples of ϵ_{vn} and ϵ_{cn} respectively. These values are known to the decoder and only the multiplicative integer is passed on as the message between stages. To minimize the complexity of the architecture, it is required that the number of multiplicative integers is minimized. This can be done by maximizing ϵ_{vn} and ϵ_{cn}.

The remaining part of this section studies the effect of changing ϵ_{vn} and ϵ_{cn} on the performance of the decoder. The aim of this experiment is to choose the maximum ϵ_{vn} and ϵ_{cn} that result in a satisfactory, "near floating point" performance of the decoder.

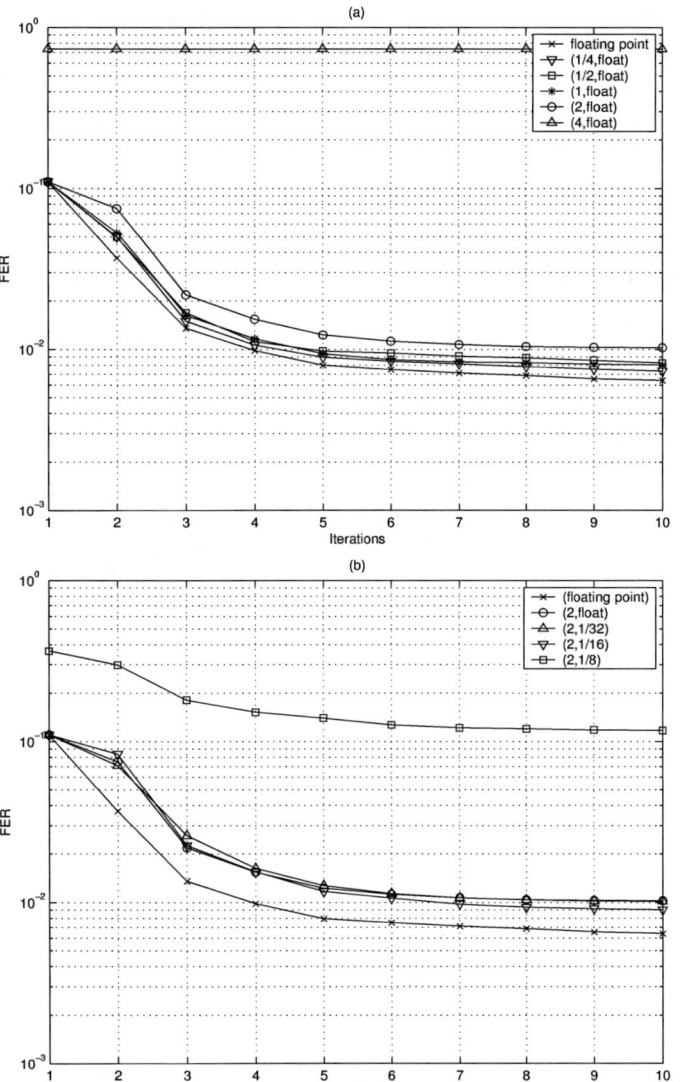

Fig. 2. The Comparison of Various Implementations of the LDPC decoder (a) with varying ϵ_{vn} (b) with varying ϵ_{cn}.

The floating point precision design is implemented for reference by keeping ϵ_{vn} and ϵ_{cn} infinitesimally small. The first step is to increase ϵ_{vn}, keeping ϵ_{cn} constant. Fig. 2(a) shows the performance of different implementations of the decoder as a function of the number of iterations. The simulations are done at channel BER of 10^{-2} and the different implementations are noted as $(\epsilon_{vn}, \epsilon_{cn})$. It is seen that there is no significant performance degradation even when ϵ_{vn} is as high as 2. However, there is a significant performance loss when $\epsilon_{vn} = 4$.

The next step is to increase ϵ_{cn}, selecting $\epsilon_{vn} = 2$. Fig. 2(b) shows the performance of several implementations of a $(2, \epsilon_{cn})$ design. Simulations at 10^{-2} channel BER show that satisfactory performance levels are achieved for $\epsilon_{cn} < \frac{1}{16}$. Any increase in ϵ_{cn} results in unacceptable level of performance. The slight differences in performance of the various designs shown in Fig. 2 (a) and (b) however, mean that the choice of the design will vary depending on the design parameters like correction capability, latency tolerated, power dissipation etc.

The VN and the CN circuit modules were implemented on a Xilinx Virtex FPGA for different values of ϵ_{vn} and ϵ_{cn}. Table I summarizes the hardware resource requirements of these implementations in terms of the LUTs utilized. These implementations assume a maximum LLR value of 7 at either node.

ϵ	$\frac{1}{64}$	$\frac{1}{32}$	$\frac{1}{16}$	$\frac{1}{8}$	$\frac{1}{4}$	$\frac{1}{2}$	1	2
LUTs for VN	55	50	45	40	35	30	25	20
LUTs for CN	767	684	601	-	-	-	-	-

TABLE I
THE LUT UTILIZATION FOR VN AND CN

The above experiments were done at a channel BER of 10^{-2}. The chosen design should be able to perform satisfactorily compared to the floating point implementation for different channel conditions. Fig. 3 shows the performance of the $(2, 1/16)$ implementation against the floating point implementation of the LDPC decoder as a function of BER.

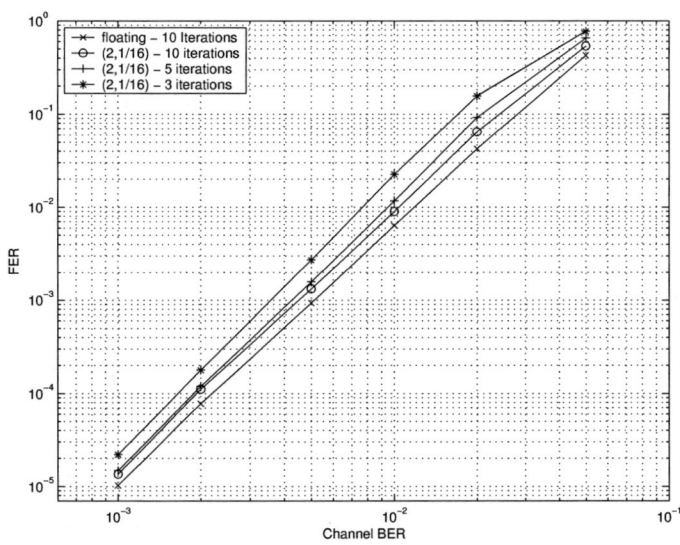

Fig. 3. The Error Performance of the Quantized LDPC decoder.

B. Variable Quantization

As discussed above, the LLR values at the CN are mapped into integer multiples of ϵ_{cn}. In other words LLR $= k \times \epsilon_{cn}$, where k is an integer. Fig. 4 plots the histograms of the values that k assumes in successive iterations while simulating several frames on the $(2, 1/16)$ decoder implementation. Let K_i be the set of values that k assumes in the ith iteration, then from fig. 4, $K_1 = \{\pm 1\}$, $K_2 = \{0, \pm 1\}$, $K_3 = \{0, \pm 1, \pm 4\}$ and $K_4 = \{0, \pm 1, \pm 4\}$. Even though the histograms for the remaining

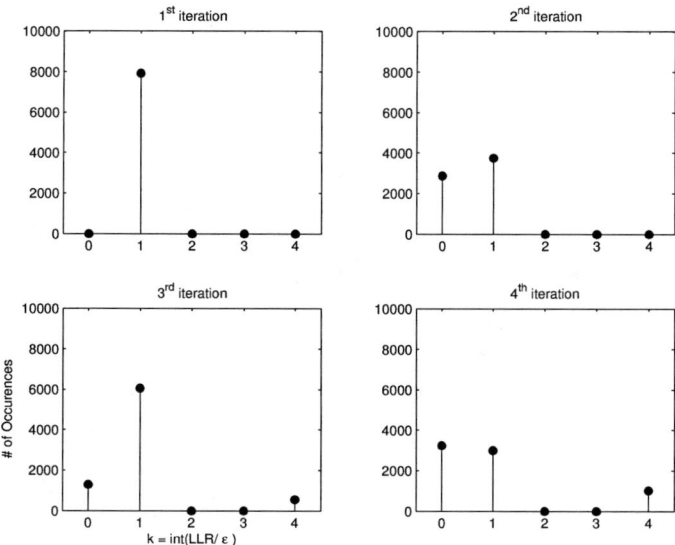

Fig. 4. The Histogram for LLR values at the CN.

iterations are not plotted, simulation results show that $K_i = K_4$, for all $i > 4$.

Following from the results above, a better quantization method can be devised to implement the CN stage than a simple uniform $\frac{1}{16}$ resolution. For the original uniform quantization implementation, as given in table I, the CN was implemented with 12 8-bit long inputs and 12 8-bit long outputs. The insight gained from fig. 4 permits CN implementation with much smaller input/output word widths than before as shown in table II

Iteration	Input Width	Output Width	CN Size
1	1	1	24 LUTs
2	2	6	193 LUTs
> 3	4	8	388 LUTs

TABLE II

THE CN IMPLEMENTATION SIZE FOR EACH ITERATION

A similar analysis can be done for quantization of the LLR values in the VN. Based on the implementations of the VN and the CN in adjacent iterations, the interfacing ψ and ψ^{-1} function modules can be designed accordingly. If the hardware resource limitations do not allow multiple iterations to be implemented sequentially, then a loop-back circuit can be implemented with just one VN stage and one CN stage, taking care that the resolution of this implementation is the highest among all the iterations.

IV. CONCLUSION

Floating point implementation results in best precision and dynamic range of the LLR values. As a result, this implementation gives the best performance of the LDPC decoder, albeit using up expensive hardware resources. A decoder based on uniform quantization of the LLR values is presented herein. This uniform quantization results in significant hardware reduction, while preserving a "near floating point" performance of the code. For example, a floating point implementation gives 6.5×10^{-3} frame error rate while a quantized decoder produces 9.6×10^{-3} FER in a 10^{-2} BER channel.

A variable quantization scheme is also presented in this paper. The variable quantization of the LLR values further reduces the decoder implementation size compared to the uniform quantization scheme. Moreover, the mathematical precision followed in this variable quantization scheme is exactly the same as the one followed by the uniform quantization scheme. Therefore, there is no performance degradation.

V. ACKNOWLEDGEMENT

This work benefited greatly from the interactions with the members of the Computer Engineering Laboratory at Texas A&M University. In particular, the help of Mr. Praveen Bhojwani, Mr. Euncheol Kim, Mr. Pankaj Bhagawat, Ms. Rajeshwary Tayade, Mr. Anand Mannivan, Mr. Shanghoan Chang, and Mr. Kiran Gunnam is greatly appreciated.

REFERENCES

[1] K.A.S. Abdel-Ghaffar, M. Hassner, "Multilevel error-control codes for data storage channels," IEEE Trans. Information Theory, vol. 37, pp. 735-741, May 1991.
[2] Kwang-Hyun Baek, Ki-Wook Kim, Sung-Mo Kang, "A low energy encoding technique for reduction of coupling effects in SoC interconnects," in Proc. Midwest Symposium on Circuits and Systems, vol. 1, pp. 80-83, Aug. 2000.
[3] C. Berrou, A. Glavieux, and P. Thitimajshima, "Near Shannon limit error-correcting codes and decoding," in Proc. Int. Conf. Comm. '93, May 1993, pp 1064-1070.
[4] R. Gallager, "Low-density parity-check codes," IRE Trans. Inform. Theory, Jan. 1962.
[5] T. J. Richardson, M. A. Shokrollahi, and R. L. Urbanke, "Design of capacity-approaching irregular low-density parity-check codes," IEEE Trans. Inform. Theory, vol. 47, pp. 619-637, Feb. 2001.
[6] E. Yeo, B. Nikolic, V. Anantharam, "Architectures and implementations of low-density parity check decoding algorithms," Circuits and Systems, The Midwest Symposium on, vol. 3 , pp. 437-440, Aug. 2002.
[7] A.J. Blanksby, C.J. Howland, "A 690-mW 1-Gb/s 1024-b, rate-1/2 low-density parity-check code decoder," Solid State Circuits, IEEE Journal of, March 2002. pp. 404 -412, vol:3, Issue:3.
[8] R. Singhal, G.S. Choi, N. Mickler, P. Koteeswaran, "Scaleable check node centric architecture for LDPC decoder," in Proc. Int. Symposium on Circuits and Systems, vol. 4, pp. 189-192, May 2004 .
[9] M. Karkooti, J.R. Cavallaro, " Semi-parallel reconfigurable architectures for real-time LDPC decoding," in Proc. Int. Conf. on Information Technology: Coding and Computing, Vol. 1, pp. 579-585, Apr. 2004.
[10] A. Selvarathiam, G. Choi, K. Narayanan, A. Prabakhar, E. Kim, "A Massively Scalable LDPC Decoder Architecture," Circuits and Systems, IEEE International Symposium on , 2003. Volume: 3, pp. 93 -96.
[11] Tong Zhang, K.K. Parhi, "Joint code and decoder design for implementation-oriented (3, k)-regular LDPC codes," Signals, Systems and Computers, Conference Record of the Thirty-Fifth Asilomar Conference on, 2001. Volume: 2, Page(s): 1232 -1236.
[12] G. Al-Rawi, J. Cioffi, M. Horowitz, "Optimizing the mapping of low-density parity check codes on parallel decoding architectures," Information Technology, Coding and Computing, Proceedings of International Conference on, 2001, Page(s): 578 -586
[13] M. M. Mansour and N. R. Shanbag, "Low-Power VLSI Decoder Architectures for LDPC Codes," Low Power Electronics and Design, Proceedings of the 2002 International Symposium on, Page(s): 284-289.
[14] S. Sivakumar,"VLSI implementation of encoder and decoder for low density parity check codes." MS Thesis, Texas A&M University, December 2001.
[15] T. Zhang, Z. Wang and K. K. Parhi, "On Finite Precision Implementation of Low-Density Parity-Check Codes Decoder," in Proc. Int. Symp. on Circuits and Systems, vol. 4, pp. 202-205, May 2001.

Low Complexity, High Speed Decoder Architecture for Quasi-Cyclic LDPC Codes

Zhongfeng Wang
School of EECS, Oregon State Univ.
Corvallis, CA 97331, USA
Email: zwang@eecs.orst.edu

Qing-wei Jia
Seagate Technology International
Singapore 118249
Email: qingwei.jia@seagate.com

Abstract

This paper presents a low complexity, very high speed decoder architecture for quasi-cyclic Low Density Parity Check (QC-LDPC) codes, specifically Euclidian Geometry (EG) based QC-LDPC codes. Algorithmic transformation and architectural level optimizations are employed to increase the clock speed. Enhanced partially parallel decoding architectures are proposed to linearly increase the overall throughput with the introduction of a small percentage of extra hardware. Based on the proposed architecture, a FPGA implementation of a (8176, 7154) EG-LDPC decoder can achieve a worst-case throughput of 169 Mbps.

1. Introduction

Recently a class of structured LDPC codes, namely quasi-cyclic LDPC codes [1], which can achieve comparable performance to random codes, have been proposed Further works include irregular cirlulant-based QC-LDPC codes [2] and Euclidean Geometry-based QC-LDPC codes [3][4]. QC-LDPC codes are well suited for hardware implementation. The encoder of a QC-LDPC code can be easily built with shift-registers [5] while random codes usually require complex encoding circuitry to perform complex matrix and vector multiplications. In addition, QC-LDPC codes also facilitate efficient high-speed decoding because of the regularity in its parity check matrix. A memory-based partially parallel decoding architecture has been presented in [6] to obtain a good trade-off between hardware complexity and decoding speed.

The paper is organized as follows: after a brief review of the conventional BP algorithm [7], a modified version based on algorithmic transformation is discussed. VLSI architectures and optimizations for the Variable node Processing Units (VPU) and Check node Processing Units (CPU) will be presented for the new algorithm. An enhanced memory-based partially parallel decoding architecture is proposed to linearly increase the throughput with small percentage of hardware overhead. A FPGA implementation of a (8176, 7154) EG-LDPC code is shown to achieve a maximum throughput of 169 Mbps, which is significantly higher than existing works such as [6] and [9].

2. Belief Propagation Algorithm and Modification

The conventional BP algorithm is composed of two phases of message passing, i.e., variable-to-check node message passing and check-to-variable node message passing. Let R_{cv} denote the check-to-variable message conveyed from the check node c to variable node v, and L_{cv} represent the variable-to-check message conveyed from the variable node v to the check node c, then R_{cv} can be computed as follows:

$$R_{cv} = -S_{cv}\Psi\{\sum_{n \in N(c)\backslash v}\Psi(L_{cn})\}, \quad (1)$$

where

$$S_{cv} = \prod_{n \in N(c)\backslash v} sign(L_{cn}). \quad (2)$$

S_{cv} is the sign part of R_{cv}, and $N(c)\backslash v$ denotes the set of variable nodes connected to the check node c excluding the variable node v. The nonlinear function $\Psi(x) = \log(\tanh(|x|/2))$ is generally implemented with a look-up table (LUT) in hardware. On the other hand, the variable-to-check message L_{cv} can be computed with the following equation:

$$L_{cv} = \sum_{n \in M(v)\backslash c} R_{nv} - 2*r_v/\sigma^2 \quad (3)$$

where $M(v)\backslash c$ denotes the set of check nodes connected to the variable node v excluding the variable node c, and $2*r_v/\sigma^2$ is the intrinsic information related to the received soft symbol r_v and the estimated standard deviation of the channel noise. The log likelihood ratio for the variable node v, denoted as L_v, is computed as follows.

$$L_v = \sum_{c \in M(v)} R_{cv} \quad (4)$$

The sign of L_v is taken as the estimated information bit (+1 or -1).

It can be observed that the conventional BP algorithm has unbalanced computation complexity between the two decoding phases. This leads to unbalanced date-paths between VPU's and CPU's. Actually, the critical path of a CPU consists of a summation operation and two LUT operations while that of a VPU consists of only a summation operation. As the clock speed will be upper-bounded by the longest data-path, the throughput of a LDPC decoder employing the BP algorithm will be limited. In [8], a modified version based on algorithmic transformation was proposed in order to balance the computation load between the two decoding phases. The new algorithm is expressed as the follows.

$$R_{cv} = -S_{cv} \sum_{n \in N(c) \backslash v} \Psi(L_{cn})\} \quad (5)$$

$$L_{cv} = \sum_{n \in M(v) \backslash c} -sign(R_{nc})\Psi(R_{nc}) - 2*r_v/\sigma^2 \quad (6)$$

where S_{cv} is computed as before. It should be noted that the real value of R_{cv} computed here is different from what is obtained with the original algorithm. The major benefit with the modified algorithm is that the computation complexity and thus computation delay are balanced between two decoding phases. As shown in [8], this modification not only helps reduce the clock cycle time, but also facilitate 100% hardware utilization efficiency.

3. Architectures for Node Processing Units and Optimizations

Consider a (3, 5) quasi-cyclic LDPC codes. The architectures of VPU and CPU with the original BP algorithm are shown in Figure 1 and 2 respectively. The input signal z_v in Fig. 2 stands for the intrinsic information. As can be observed from Fig. 1, two LUT operations are involved in the critical path of each CPU. With the new algorithm, we move one LUT operation to the critical path of every VPU. We can further eliminate the sign-magnitude to 2's complement conversion block [SM-2's] and the 2's complement to sign-magnitude [2's-SM] conversion block in order to shorten the critical path further. The compensation for the removal of these blocks includes adding the sign bit to the input of a normal LUT and forcing the output of LUT to have different formats at different decoding phases.

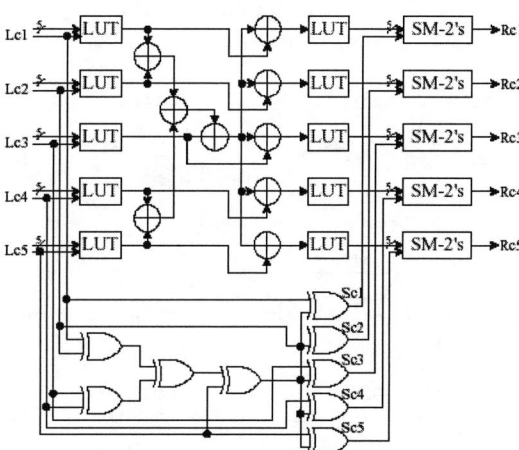

Figure 1. The architecture of CPU with original BP algorithm

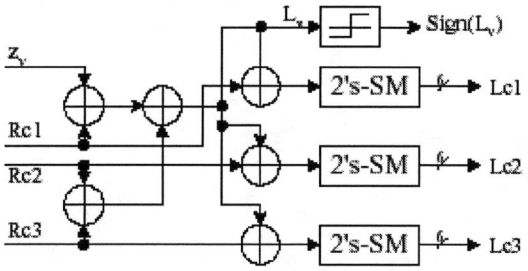

Figure 2. The architecture of VPU with original BP algorithm

In addition, we propose to utilize architecture-level optimization techniques to further reduce the total computation delay for the summation part. The new architectures for VPU and CPU are shown in Figure 3 and Figure 4, respectively. The dashed lines indicate possible positions for inserting pipeline stages. Actually with 3-stage pipelining, the critical path for either type of node processing units is reduced to about 1 multi-bit addition, which leads to nearly 6 times speed-up over the design presented in [8].

4. Partially Parallel Decoding Architecture and Enhancement

Several papers have addressed partially parallel decoding architecture for regular LDPC codes such as [6] and [9]. This kind of architecture generally achieves a good trade-off between hardware complexity and decoding throughput. A partially parallel decoder architecture for general (3, 5) QC-LDPC codes is shown in Figure 5, where totally 3*5=15 memory banks are used to store the soft message symbols conveyed at both decoding phases, memory bank Z's are used to store the intrinsic information, and the memory bank C's are used to store the estimated data bits.

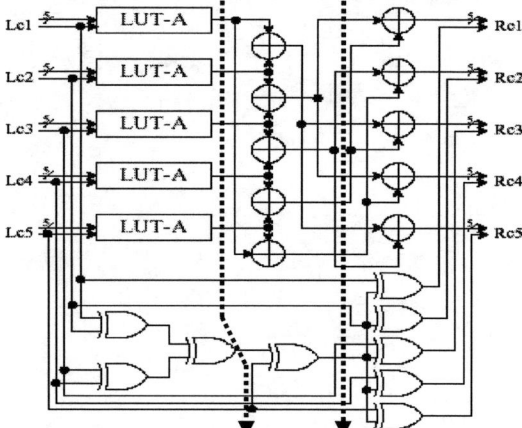

Figure 3. An optimal architecture for CPU.

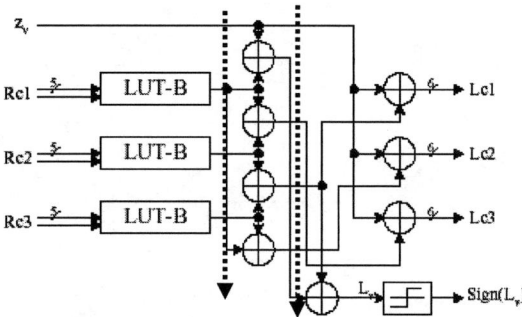

Figure 4. An optimal architecture for VPU.

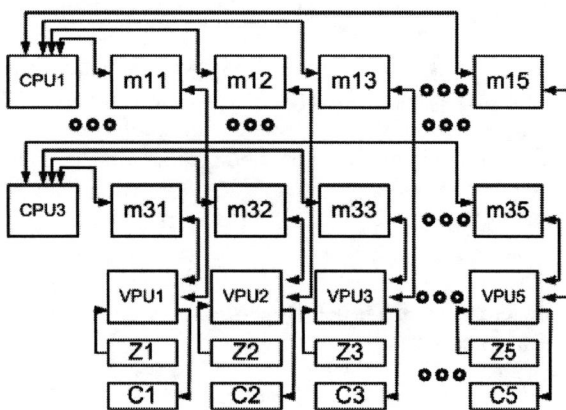

Figure 5. The structure of a partially parallel decoder for (3, 5) QC-LDPC codes.

For QC-LDPC codes, the address generator for each memory bank can be built with a simple counter, which not only simplifies the hardware design, but also improves the circuit speed. In general, each node processing unit takes 1 clock cycle (assuming dual-port memories are used, otherwise 2 cycles are needed) to complete message updating for each row (or column). To increase the parallelism, we can force each node processing unit to process multiple rows (or columns) at the same time. However, this will generally cause memory access problems. Figure 6 shows a small sub-matrix of a QC-LDPC parity-check matrix.

$$\begin{matrix} 0 & 0 & 0 & 1 & 0 & 0 & 0 & 0 & 0 & 0 \\ 0 & 0 & 0 & 0 & 1 & 0 & 0 & 0 & 0 & 0 \\ 0 & 0 & 0 & 0 & 0 & 1 & 0 & 0 & 0 & 0 \\ 0 & 0 & 0 & 0 & 0 & 0 & 1 & 0 & 0 & 0 \\ 0 & 0 & 0 & 0 & 0 & 0 & 1 & 0 & 0 & 0 \\ 0 & 0 & 0 & 0 & 0 & 0 & 0 & 1 & 0 & 0 \\ 0 & 0 & 0 & 0 & 0 & 0 & 0 & 0 & 1 & 0 \\ 0 & 0 & 0 & 0 & 0 & 0 & 0 & 0 & 0 & 1 \\ 1 & 0 & 0 & 0 & 0 & 0 & 0 & 0 & 0 & 0 \\ 0 & 1 & 0 & 0 & 0 & 0 & 0 & 0 & 0 & 0 \\ 0 & 0 & 1 & 0 & 0 & 0 & 0 & 0 & 0 & 0 \end{matrix}$$

Figure 6. A sub-matrix of a QC-LDPC parity-check matrix.

As discussed in [8], all soft message symbols corresponding to all the 1-components in the sub-matrix are stored in one memory bank. Thus, a straightforward approach to enable each process unit to process multiple rows/columns per cycle is to store multiple soft symbols at each memory entry. For example, we can store 2 soft symbols corresponding to two adjacent 1-components at one memory entry. This design easily solves the problem of processing 2 rows at each cycle. However, it is generally not applicable to column processing. Assume we start column processing from the first column for the same example. In the first cycle, no memory access conflict happens, as both soft symbols are stored at the same memory entry. In the second cycle, we are supposed to process the third and the fourth columns at the same time. However, the required two soft symbols are located in different memory entries. This situation would become even worse when there are multiple sub-blocks in one (block) column of the parity check matrix, e.g., there are 3 sub-blocks in one column for a (3, 5)-regular QC-LDPC parity-check matrix. Using multi-port memories is a possible solution, but not an efficient one as the overall hardware will be linearly increased. To fix this problem, the authors of [10] added one more constraint on each circulant matrix, i.e., the shift value of each shifted identity matrix must be multiples of δ, where δ denotes the number of soft symbols stored in one memory entry, e.g., $\delta = 2$ in the previous example. The added constraints unavoidably limit the performance of the LDPC codes.

Euclidian geometry-based QC-LDPC codes (EG-LDPC) were first introduced in [11] and late improved in [12][4]. This class of codes has comparable performance to random codes. Particularly, they have very low error floor. Consider a (8176, 7156) (4, 32)-regular EG-LDPC code discussed in [4]. The coding gain is only 1 dB away from the Shannon limit at BER=10^{-7}. Our simulations have shown that the error floor for this code is at least blow 10^{-10}. A key feature of the parity-check matrix of this code is that each sub-matrix consists of two overlapped cyclic-shifted identity matrices. There is no other constraint other than that the size of the sub-block must be a prime number. A more general case would be to have multiple (e.g., m>2) independently cyclic-shifted identity matrices overlapped in one sub-block. To efficiently decoder this class of codes, we propose to have one separate memory bank for each independently (cyclic) shifted identity matrix. Hence, we need 2 memory banks for each sub-block. To double the parallelism of a partially parallel decoder, we propose three solutions as follows:

I. Store two adjacent soft symbols at one memory entry while utilizing extra buffers (c.f. [13]) to solve the memory access problem.
II. Partition each memory bank into two sub-banks with one contains even-numbered entries and the other for odd-numbered entries.
III. Combine Approach I and Approach II.

Due to the limited space, we will only elaborate on Approach II in this paper. For any shifted identity matrix, it is guaranteed that one of two adjacent soft symbols (whether from row processing point of view or from column processing point of view) belongs to the even-numbered sub-bank and the other belongs to the odd-numbered sub-bank. Therefore, this method successfully solves the memory access problem, though extra multiplexers will be needed and the control circuitry will be slightly more complex. It can be observed that the proposed approach can be extended to the case with m>2. Most recently, we have implemented a LDPC decoder with Xilinx Virtex II 6000 for a (8176, 7156) EG-LDPC code, where all the techniques discussed in the above are employed. Our design has shown a worst-case decoding throughput of 169 Mbps (with 15 iterations), which is significantly higher than any existing work such as [6] and [9].

5. Conclusion

In this paper, we have discussed modified BP algorithm and presented optimized architectures for both type of node processing units based on the new algorithm. We have further presented enhanced partially parallel decoder architectures for QC-LDPC codes.

References

[1] *D. Sridhara, T. Fuja, and R. M. Tanner*, "Low density parity check codes from permutation matrices," Conf. on Info. Science and Systems. The John Hopkins University, March, 2001.

[2] *D. Hocevar*, "Efficient encoding for a family of quasi-cyclic LDPC codes," IEEE GLOBECOM'03. Vol. 7. Pages: 3996 - 4000 vol.7.

[3] *Y. Kou, J. Xu, H. Tang, S. Lin, and K. Abdel-Ghaffar*, "On circulant low density parity check codes," ISIT'2002. Pages:200.

[4] *L. Chen, J. Xun, I. Djurdjevic, and S. Lin*, "Near Shannon Limit Quasi-Cyclic Low Density Parity-Check Codes," to appear in IEEE Trans. on Communications, July, 2004.

[5] *Z. Li, L chen, and S. Lin, W Fong and P Yeh*, "Efficient Encoding of Quasi-Cyclic Low Density Parity-Check Codes," to appear in IEEE trans. on communications, 2004.

[6] *Y. Chen and D. Hocevar*, "A FPGA and ASIC implementation of rate 1/2, 8088-b irregular low density parity check decoder," IEEE GLOBECOM '03. Volume: 1, 1-5 Dec. 2003. Pages:113 – 117.

[7] *D. J. MacKay*, "Good error-correcting codes based on very sparse matrices," IEEE Trans. Infor. Theory, vol. 45, pp. 399-431, Mar. 1999.

[8] *Z. Wang, Y Chen, and K Parhi*, "Area-efficient quasi-cyclic LDPC code decoder architecture," in ICASSP 2004.

[9] *Tong Zhang and Keshab Parhi*, "A 54 Mbps (3,6)-regular FPGA LDPC decoder," in Proc. IEEE SiPS'2002, pp 127-132, 2003.

[10] *M. Karkooti and J. R. Cavallaro*, "Semi-parallel reconfigurable architectures for real-time LDPC decoding," ITCC'2004., Volume: 1, 5-7 April 2004. Pages:579 – 585, Vol.1.

[11] *Y. Kou, J. Xu, H. Tang, S. Lin, and K. Abdel-Ghaffar*, "On circulant low density parity check codes," in Proc. of 2002 IEEE ISIT, Pages:200.

[12] *S. Lin, L. Chen, J. Xu, and I. Djurdjevic*, "Near Shannon limit quasi-cyclic low-density parity-check codes," GLOBECOM '03, 2003. Volume: 4, Dec. 2003. Pages:2030 – 2035, vol.4.

[13] *Z Wang, Y Tan, and Y. Wang*, "Low Hardware Complexity Parallel Turbo Decoder Architecture", in ISCAS'2003.

An Analog/Digital Mode-Switching LDPC Codec

David Haley*, Chris Winstead†, Alex Grant‡, Vincent Gaudet§ and Christian Schlegel§

*Cohda Wireless, AUSTRALIA, Email: david.haley@cohdawireless.com
†Department of Electrical and Computer Engineering, Utah State University, USA
‡Institute for Telecommunications Research, University of South Australia, AUSTRALIA
§Department of Electrical and Computer Engineering, University of Alberta, CANADA

Abstract—We present a novel time-multiplexed codec for a class of low-density parity-check codes, which switches between analog decode and digital encode modes. In order to achieve this behaviour from a single circuit we have developed mode-switching gates. These logic gates are able to switch between analog (soft) and digital (hard) computation. Only a small overhead in circuit area is required to transform the analog decoder into a full codec. The encode operation can be performed two orders of magnitude faster than the decode operation, making the circuit suitable for full-duplex applications. Throughput of the codec scales linearly with block size, for both encode and decode operations. The low power and small area requirements of the circuit make it an attractive option for small portable devices.

I. INTRODUCTION

Low-density parity-check (LDPC) codes [1] are becoming increasingly popular for error control in modern communication systems. An iterative message passing decoder circuit can be built using check and variable node building blocks, with routing mapped according to the parity-check matrix of the code [2]. Messages representing binary probability mass functions are passed between nodes. The check and variable nodes are built using soft-logic gates which calculate the output message $\mathsf{p}_Z(z) = (\mathsf{p}_Z(0), \mathsf{p}_Z(1))$ using input messages $\mathsf{p}_X(x)$ and $\mathsf{p}_Y(y)$. LDPC codes offer very good error correction performance, and can outperform turbo codes. However, in general, they require more complex encoder circuit implementations than turbo encoders. In previous work [3] we introduced a class of *reversible* LDPC codes. These codes allow the decoder to be reused to perform iterative encoding via the Jacobi method for iterative matrix inversion, as follows.

Consider a binary systematic (n, k) code with codewords arranged as row vectors $\mathbf{x} = [\mathbf{x}_p \mid \mathbf{x}_u]$, where $\mathbf{x}_u$ are the information bits and $\mathbf{x}_p$ are the parity bits. Likewise partition the parity check matrix, $\mathbf{H} = [\mathbf{H}_p \mid \mathbf{H}_u]$. Defining $\mathbf{b} = \mathbf{H}_u \mathbf{x}_u^\top$, encoding becomes equivalent to solving $\mathbf{H}_p \mathbf{x}_p^\top = \mathbf{b}$. For $m \times m$ non-singular $\mathbf{H}_p$, we have $\mathbf{x}_p^\top = \mathbf{H}_p^{-1} \mathbf{b}$. The Jacobi method for iterative matrix inversion is applied over $\mathbb{F}_2$ by performing the iteration

$$\mathbf{x}_{p:k+1} = (\mathbf{H}_p \oplus \mathbf{I}) \mathbf{x}_{p:k}^\top + \mathbf{b} \qquad (1)$$

The parity check matrix shown in (2) represents a reversible

*This work was done when D. Haley was with the University of South Australia, and was supported by the Australian Government under ARC SPIRT C00002232.

LDPC code which is encodable using 4 iterations of (1).

$$\mathbf{H} = \begin{bmatrix} 1 & 1 & 0 & 1 & 0 & 0 & 0 & 0 & 1 & 0 & 0 & 1 & 1 & 0 & 0 & 0 \\ 0 & 1 & 1 & 0 & 1 & 0 & 0 & 0 & 1 & 1 & 0 & 0 & 0 & 0 & 1 & 0 \\ 0 & 0 & 1 & 1 & 0 & 1 & 0 & 0 & 0 & 1 & 1 & 0 & 0 & 0 & 0 & 1 \\ 0 & 0 & 0 & 1 & 1 & 0 & 1 & 0 & 0 & 1 & 1 & 0 & 1 & 0 & 0 & 0 \\ 0 & 0 & 0 & 0 & 1 & 1 & 0 & 1 & 0 & 0 & 1 & 1 & 1 & 1 & 0 \\ 1 & 0 & 0 & 0 & 0 & 1 & 1 & 0 & 1 & 0 & 1 & 0 & 0 & 0 & 0 & 1 \\ 0 & 1 & 0 & 0 & 0 & 0 & 1 & 1 & 0 & 0 & 0 & 1 & 0 & 0 & 1 & 1 \\ 1 & 0 & 1 & 0 & 0 & 0 & 0 & 1 & 0 & 0 & 1 & 1 & 0 & 1 & 0 & 0 \end{bmatrix} \qquad (2)$$

Analog circuit implementations for iterative decoding have been shown to offer power, size and speed advantages in comparison to digital implementations. Several small analog decoders have been fabricated (see [4] and the references therein) and more recently a larger decoder has been built [5].

Reusing the decoder circuit to also perform encoding reduces the circuit area of the codec. However, investigations into reusing the analog decoder to perform iterative encoding (1), have identified two issues [6]. Firstly, the output of the soft-XOR gates decays over time during the iterative process, thus requiring amplification. Furthermore, the asynchronous arrival of messages at a node makes it possible for the circuit to stray from the iterative Jacobi path, before settling to the correct steady state solution. These problems both arise from the fact that we are trying to map the discrete time digital encoding algorithm onto a continuous time analog circuit. The algorithm passes hard messages $\in \{0, 1\}$ yet the analog circuit passes soft messages $\in (0, 1)$. Motivated by this we present an alternative design here which switches between analog decode and digital encode modes.

II. MODE-SWITCHING GATES

A fundamental circuit contribution of this work is the *mode-switching gate*. A mode-switching XOR gate is shown in Figure 1. During decoding these gates operate as analog sub-threshold CMOS soft-XOR gates [2]. They are then switched to operate as digital *resistor-transistor-logic* (RTL) XOR gates during the encode operation. Mode switching between encode (*enc*) and decode ($\overline{enc}$) modes is accomplished by the addition of four transmission gates to the analog soft-XOR gate. Minimum sized FETs may be used for these transmission gates, presenting very little overhead. The cell may be switched between encode (hard) and decode (soft) modes according to Table I.

We now explain how the circuit generates the output $\mathsf{p}_Z(1)$. The output $\mathsf{p}_Z(0)$ is generated in a similar manner.

In decode mode ($enc = Gnd, \overline{enc} = V_{dd}$), the circuit is wired as a soft-XOR gate [2]. The voltage V_u represents an external connection to the driving FET of a p-type current

Mode	enc	$\overline{enc}$	V_u	V_{refN}	V_{refP}
Decode	Gnd	V_{dd}	V_u	V_{refN}	V_{refP}
Encode	V_{dd}	Gnd	Gnd	Gnd	V_{dd}

TABLE I
MODE-SWITCHING GATE SETTINGS.

operation is performed digitally, using the mode-switching XOR gates presented above. We design a mode-switching codec for the small reversible LDPC code described by (2).

The analog sum-product decoder circuit [2] consists of soft-equal and soft-XOR gates, and wires routed according to (2). By replacing the soft-XOR gates with mode-switching XOR gates this circuit can also be used to encode reversible LDPC codes. Figure 2 shows the codec circuit for (2), separating the components into information and parity sections.

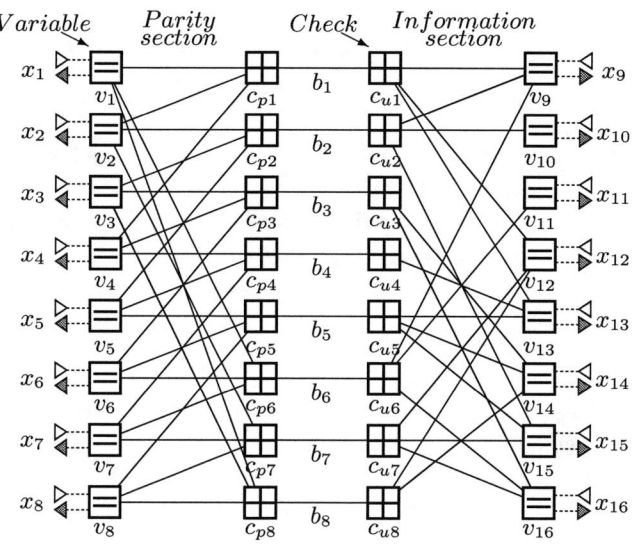

Fig. 2. Codec circuit for **H**.

The information variable nodes ($x_9 \ldots x_{16}$) are extended to allow encoding as shown in Figure 3 for the case of x_9. Similarly, the parity variables ($x_1 \ldots x_8$) are extended as shown in Figure 4 for the case of x_1. Transmission gates and a multiplexer (MUX) are used to switch each variable node between encode (enc) and decode ($\overline{enc}$) modes. The multiplexer is also built from transmission gates. Here 2-wire probability vector buses (thick lines) carry messages representing ($p_X(0), p_X(1)$). Where necessary these have been expanded into single wires (thin lines).

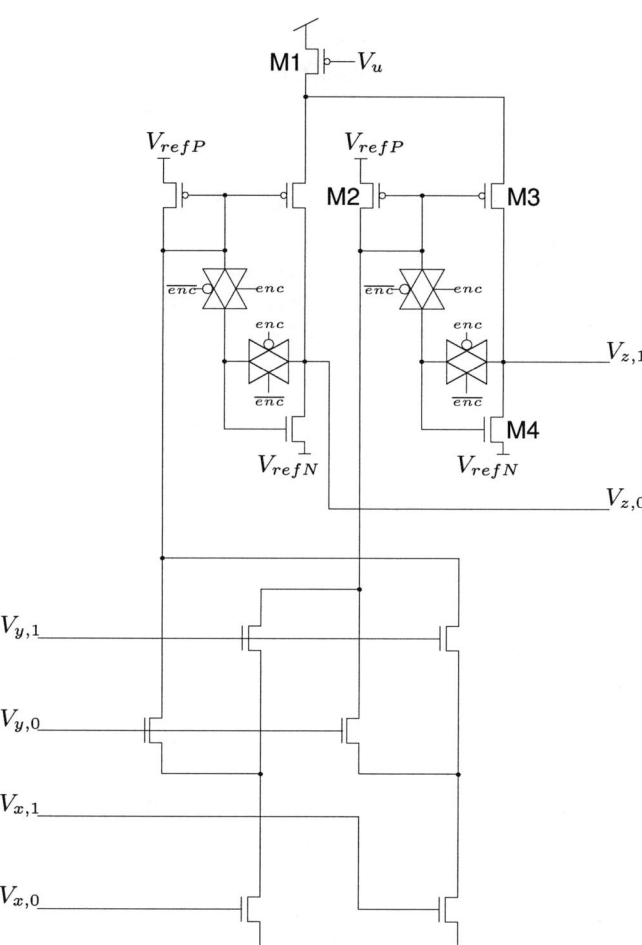

Fig. 1. Mode-switching XOR gate.

mirror, such that transistor M1 is biased to supply the unit current I_u. Transistors M2 and M3 form part of the normalisation network, and M4 is diode connected.

In encode mode ($enc = V_{dd}, \overline{enc} = Gnd$), the circuit is wired as a digital resistor-transistor-logic XOR gate. We assume that hard decision $\{Gnd, V_{dd}\}$ voltages drive the inputs. We set $V_{refP} = V_{dd}$ and use transistor M2 as a resistive load to the multiplication matrix. The gate voltage of M2 represents the inverted result for $p_Z(1)$. We set $V_u = Gnd$ to turn on M1, and $V_{refN} = Gnd$ so that transistors M3 and M4 form an inverter. The true form of $p_Z(1)$ is then presented at the output.

III. CODEC CORE ARCHITECTURE

We now describe how the analog sum-product decoder core can be extended to allow iterative encoding. The encode

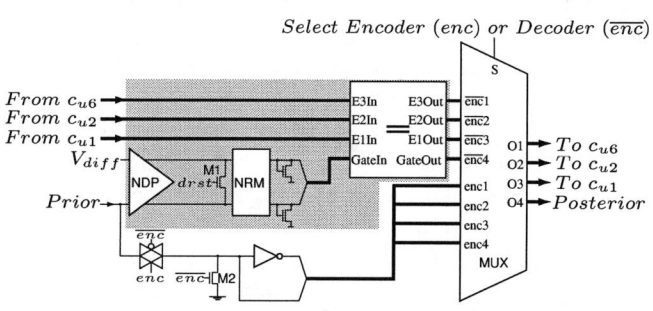

Fig. 3. Information variable node structure for x_9

In decode mode ($enc = Gnd, \overline{enc} = V_{dd}$) the circuit operates as a subthreshold CMOS analog sum-product decoder [2]. All checks (mode-switching XOR gates) are set to operate as

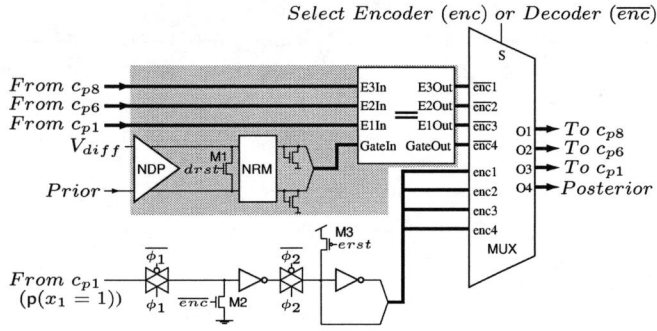

Fig. 4. Parity variable node structure for x_1.

Function	Components	FETs
Mode-switching	Transmission gates	192
Edge multiplexing	Transmission gates	512
Input multiplexing	Transmission gates	32
Feedback shift registers	Transmission Gates/Inverters	80

TABLE II

TRANSISTOR OVERHEAD FOR ENCODING.

soft-XOR gates. Information and parity variables both perform the same function. Each variable is based upon a soft-equal gate, as shown in the shaded region of Figure 3 and Figure 4. The equal gate operates as a voltage mode cell, however the output result (posterior) is presented as a current-mode pair. Gate edge connections E1Out, E2Out and E3Out are routed to adjacent check nodes. Edge connections E1In, E2In and E3In receive messages routed from adjacent check nodes. We assume that the received channel observation (prior) is represented as a voltage, in log-likelihood form. This voltage is converted into a current-mode probability vector, using the n-type differential pair (NDP). The reference voltage (V_{diff}) represents a log-likelihood value of zero. A vector normalisation circuit (NRM) [2] then feeds two diode connected FETs. These FETs provide a voltage-mode representation of the vector, for input (via GateIn) to the equal gate. The decoded soft output (posterior) is then taken from the output (GateOut port) of the equal gate.

A reset FET, M1, is connected across the input to the normalisation circuit. To reset the decoder, i.e. clear the previous result, we briefly set $drst = V_{dd}$. This causes M1 to turn on and sets a uniform distribution, $\mathsf{p}(x_s = 0) = \mathsf{p}(x_s = 1) = 0.5$, for all variables.

Once released from reset we allow the decoder circuit some fixed time, t_{dec}, to settle to its steady state. At this point an interface circuit [7] may be used to sample the posterior core output, and present this result as a hard decision at the output of the chip.

In encode mode ($enc = V_{dd}, \overline{enc} = Gnd$) the circuit operates as a digital message-passing Jacobi encoder [3]. All checks (mode-switching XOR gates) are set to operate as digital RTL XOR gates.

The first step toward encoding is to generate the vector $\mathbf{b} = \mathbf{H}_u \mathbf{x}_u^\top$ from the information variables, as shown in Figure 2. Using a multiplexer we bypass the equal gate that is used in decode mode (see Figure 3). The ($\mathsf{p}(x_s = 1)$) information bit value is presented as a hard decision voltage at the (prior) node input. An inverter is then used to generate $\mathsf{p}(x_s = 0)$, and both values are sent to each outgoing edge of the variable node. Transistor M2 at the input of this inverter connects it to ground during the decode mode ($\overline{enc} = V_{dd}$). This connection is made to prevent the input of the inverter from floating, which can create a resistive path between the power rails.

Each check node includes the three adjacent incoming information bits in the hard XOR operation, thus producing $\mathbf{b}$. For example, the XOR operation at check c_{u1} includes information bits x_9, x_{12} and x_{13} to produce b_1.

To implement the Jacobi iteration (1) we require only the check node output passed along the path $\mu_{c_s \to v_s}$ for each v_s representing the parity bit x_s. This value for v_s is then fed back into the checks $c \in \Gamma(v_s) \setminus c_s$, and also forms the final decision for x_s. We use a two phase shift register, consisting of two transmission gates and two inverters, to latch the feedback (see Figure 4). The input to this shift register is taken from the voltage representing $\mathsf{p}(x_s = 1)$. We initially ignore the value representing $\mathsf{p}(x_s = 0)$, and generate it later at the output stage of the shift register. Non-overlapping signals ϕ_1 and ϕ_2 are used to clock the shift register. Upon completion, at time t_{enc}, the result is latched through to the chip interface.

Initialisation is performed using transistor M3, which ties the input of the second inverter to V_{dd}. We hold the encoder in the reset state by setting $erst = Gnd$. During the decode mode ($\overline{enc} = V_{dd}$) we set $erst = Gnd$. Transistors M2 and M3 then turn on, and connect the input of each inverter to a supply rail, in order to prevent these nodes from floating.

A. *Estimate of Encoder Implementation Overhead*

An approximate count of the number of FETs used to add encoder functionality to this ($n = 16$) analog decoder core is summarised in Table II. We note that it is not necessary to transform all soft-XOR gates into mode-switching gates. Only those gates which route messages in the path of the iterative encoding algorithm require conversion, thus saving some transistor overhead.

The total number of FETs required to transform the above decoder core into a codec is 816. This represents approximately 15% of the total number of transistors used to build the core. The average overhead is 51 FETs per bit, and this count can be used as a linear guide to predict the overhead requirement for a larger codec. Most of these transistors are used to build transmission gates, and hence they can have minimum dimensions. The routing overhead is negligible, as only a small number of control signals have been added, and data paths have been reused from the decoder.

IV. CIRCUIT SIMULATIONS

We have built a T-SPICE description of the complete codec core described above, for the TSMC 0.18μm CMOS technology, using BSIM3v3 simulation device models obtained through Canadian Microelectronics Corporation. Circuit simulation parameters are provided in Table III. The voltages

Parameter	Symbol	Value
Unit bias current	I_u	100nA
Voltage supply	V_{dd}	1.8V
Reference voltage (n-type)	V_{refN}	0.4V
Reference voltage (p-type)	V_{refP}	1.4V
Reference voltage (differential)	V_{diff}	0.8V

TABLE III

CODEC CORE SIMULATION PARAMETERS.

Parameter	Decode	Encode
Time (per block)	$10\mu s$	50ns
Power (per block)	$110\mu W$	21.6mW
Energy	138pJ/decoded info bit	135pJ/encoded parity bit
Throughput	800kbit/sec (info bits)	160Mbit/sec (parity bits)

TABLE IV

CODEC CORE SPECIFICATION.

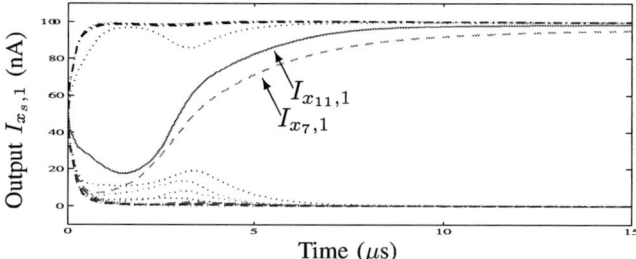

Fig. 5. Decoder output with bits x_7 and x_{11} corrected.

representing log-likelihood prior input to the decoder variables have a maximum deflection of ± 0.2V about V_{diff}.

Figure 5 shows the simulation of an example block decode. The circuit output current $I_{x_s,1}$, representing $\mathrm{p}(x_s=1)$, is shown for each symbol. In this example the decoder is released from reset at time zero, and the channel has flipped bits x_7 and x_{11}. The decoder successfully corrects these two errors, to arrive at the codeword $x = [0011001000101010]$. The settled decoding may be sampled by the interface at $t_{dec} = 10\mu s$.

To demonstrate the operation of the encoder we apply the information vector $x_{9...16} = [00110111]$ and expect the parity vector $x_{1...8} = [11111011]$. The voltage $V_{x_s,1}$ representing $\mathrm{p}(x_s=1)$ for each parity output bit is shown in Figure 6. The information vector is applied at time zero and the reset latch released 5ns later. The circuit is then clocked for $\kappa = 4$ iterations, to arrive at the correct codeword, i.e. x_6 is the only output bit having $\mathrm{p}(x_s=1) = 0$. The encoded result may be latched through to the output interface at $t_{enc} = 50$ns.

A summary of specifications for the codec core, measured from circuit simulation, is provided in Table IV. We expect the power requirements and throughput of a larger core to grow linearly with block length, for both encode and decode modes.

The digital encoder draws significantly more power than the analog decoder but operates 200 times faster. Hence the per block energy requirements of the two modes balance. The average power consumption, assuming an equal number of encode and decode operations, is $217\mu W$ per block. From these results, and considering a reversible code that is encodable in κ iterations, we expect the time taken to encode a block to be approximately 0.125κ% of that taken to decode a block.

V. CONCLUSIONS

The analog decoder architecture is well suited to continuous time soft decision decoding. However, the discrete time hard decision Jacobi encoder is more appropriately implemented with a digital circuit. In order to achieve these seemingly contradictory roles for a circuit that is based upon architecture re-use, we have introduced mode-switching logic gates. These gates also have other potential applications, e.g. for built-in-self-test circuit verification [8].

Analog decoders are robust to design and fabrication errors, making their behaviour difficult to verify, both at design time and after fabrication. However, encoding is a deterministic process in which errors are exposed quickly. Since we are reusing the decoder for encoding, verification of the encode operation also provides implicit verification for components of the decoder circuit.

The additional area required to convert the analog sum-product decoder into a full codec circuit is very low. Most of the additional transistors may have minimum dimensions and the routing overhead is negligible.

The time taken for the codec to perform an encode operation is insignificant in comparison to that taken to decode, thus offering full-duplex communications. The efficient power consumption and small area requirements of the codec make it a good candidate for portable and biomedical applications.

REFERENCES

[1] R. G. Gallager, *Low-density parity-check codes.* Cambridge, MA: MIT Press, 1963.

[2] F. Lustenberger, "On the design of analog iterative VLSI decoders," Ph.D. dissertation, ETH, Zürich, Switzerland, 2000.

[3] D. Haley, A. Grant, and J. Buetefuer, "Iterative encoding of low-density parity-check codes," in *Proc. GLOBECOM 2002*, vol. 2, Taipei, Taiwan, 2002, pp. 1289–1293.

[4] C. Winstead, V. Gaudet, and C. Schlegel, "Analog iterative decoding of error control codes," in *Canadian Conf. on Electrical and Computer Eng.*, vol. 3, Montréal, Canada, 2003, pp. 1539–1542.

[5] C. Winstead et al., "A CMOS analog $(16, 11)^2$ turbo product decoder," in *3rd Analog Decoder Workshop*, Banff, Canada, 2004, http://www.analogdecoding.org/docs/Winstead_ADW04.pdf.

[6] D. Haley, C. Winstead, A. Grant, and C. Schlegel, "An analog LDPC codec core," in *Proc. Int. Symp. on Turbo Codes and Related Topics*, Brest, France, 2003, pp. 391–394.

[7] M. Helfenstein, F. Lustenberger, H.-A. Loeliger, F. Tarköy, and G. S. Moschytz, "High-speed interfaces for analog, iterative decoders," in *Proc. IEEE Int. Symp. on Circuits and Systems, ISCAS '99*, vol. II, Orlando, Florida, 1999, pp. 424–427.

[8] M. Yiu et al., "A digital built-in self-test approach for analog iterative decoders," in *3rd Analog Decoder Workshop*, Banff, Canada, 2004.

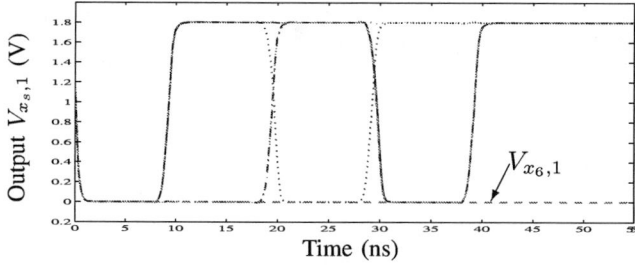

Fig. 6. Encoder parity bit outputs.

Digital VLSI OFDM Transceiver Architecture for Wireless SoC Design

Wei-Hsiang Tseng, Ching-Chi Chang, and Chorng-Kuang Wang
Graduate Institute of Electronic Engineering and Department of Electrical Engineering
National Taiwan University
Taipei 106, Taiwan R.O.C.
ckwang@cc.ee.ntu.edu.tw

Abstract—This paper presents the VLSI architecture of an OFDM baseband transceiver for wireless communications. The open-/closed-loop carrier recovery achieves the stepping frequency acquisition for high-band RF systems, and the proposed timing recovery cooperated with the self-correcting interpolation realizes an OFDM baseband digital IP design. Hardware sharing and power-of-2 coefficients fulfill this compact transceiver system chip. Simulations show that the receiver can deliver 10% packet error rate (PER) requirement under all specified SNRs for IEEE 802.11a. Using the typical 0.25μm CMOS technology, the chip occupies 3.5×3.5 mm^2 area and consumes 109 mW under 2.5 V power supply.

I. INTRODUCTION

The demand of wide-band communications leads to a tremendous growth of wireless LAN. In 1997, the IEEE 802.11 standard was first established for the wireless LAN. It supports the data rates of 1 and 2Mbps which are not sufficient for many multimedia applications. The IEEE 802.11a standard was proposed for efficient bandwidth utilization, which selects orthogonal frequency division multiplexing (OFDM) as the basis for the physical layer and supports the data rates from 6 up to 54 Mbps in the 5 GHz band. OFDM has been recognized as an optimal solution to combat burst noises and the multi-path fading.

Since the transmission data format is clearly specified in communication systems [1], how to utilize the limited signaling resources to design a practical and robust receiver becomes a tough challenge. Whereas the significant drawback of OFDM systems is the sensitivity to synchronization errors, the multi-step recovery and precise adaptation techniques represent nowaday wide-band VLSI system design methodologies. In order to realize a digital transceiver IP, the digitally self-compensating solution of the timing recovery is preferable [8]. Using an interpolator with a long buffer can solve the problem of sample stuffed or rubbed when the timing frequency offset (TFO) occurs [8]. However, the buffer size may be infinite in non-stopping transmissions and it is impractical. We propose the interpolator architecture with a sliding window to realize the self-adjustment of sample skipping or duplicating, which prevents the conventional digital timing recovery from fatal jitter problem caused by illegal indexing.

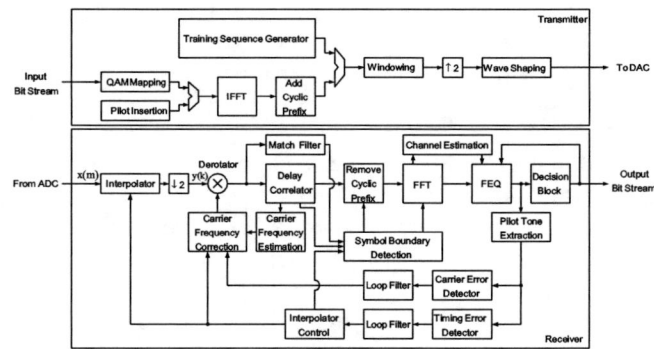

Fig. 1. The proposed transceiver architecture.

This paper organizes as follows. Section II presents the proposed baseband transceiver architecture. Section III shows the circuits of the functional blocks in the receiver. The simulation environment and the evaluation of the designed system are delivered in Section IV. Section V shows the chip layout and the post-layout summary. Finally, the conclusions are given in Section VI.

II. TRANSCEIVER ARCHITECTURE

The proposed baseband transceiver architecture is shown in Fig. 1 where the upper part is the transmitter and the lower one is the receiver. The transmitter mainly consists of training sequence generation, QAM mapping, pilot insertion, IFFT, cyclic prefix (CP) extension, windowing, and wave shaping functional blocks. The wave shaping filter adopts a raised cosine filter with a roll-off factor 0.22. The transmitted signal completely complies with the power spectrum mask requirement [1].

In the receiver, other than an interpolator, a symbol boundary detector, a de-rotator, a delay correlator, a match filter, and a FFT unit, the receiving chain also includes a multi-step carrier recovery loop, a self-correcting timing recovery loop, and an adaptation aided frequency-domain equalizer (FEQ). FFT and IFFT functions share the resource in a single architecture.

Wei-Hsiang Tseng is currently with MediaTek Inc., Taiwan, R.O.C..
This program is supported by the National Science Council (NSC), Taiwan, R.O.C..

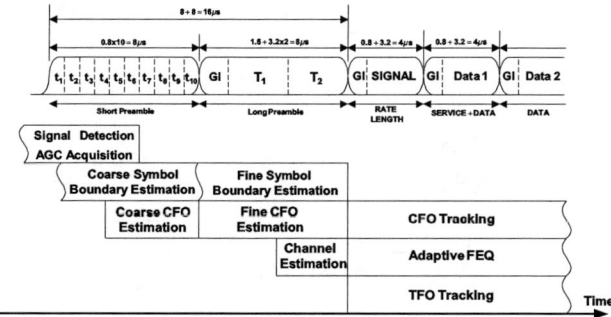

Fig. 2. The training process of the receiver design.

TABLE I. SUMMARY OF THE CFO ESTIMATION

CFO Estimation Type	Coarse	Fine
Estimation Range	± 625 KHz	± 156 KHz
Maximum Estimation Error	± 32.6 KHz	± 6.22 KHz

TABLE II. DESIGN PARAMETERS OF THE CARRIER RECOVERY LOOP

Parameter	Value
Lock-in Range $\Delta\omega_L$	7.5 KHz
Damping Factor ξ	0.707
Natural Frequency ω_n	3.3×10^4 rad/s
Phase Detector Gain K_d	1 unit/rad
NCO Gain K_o	20 MHz/unit
PI Filter Coefficient C_1	2^{-11}
PI Filter Coefficient C_2	2^{-14}

III. RECEIVER DESIGN

A. Training Process

The training process designed in the receiver is shown in Fig. 2. The first part of the short preamble, about 4.8 μs, is reserved for signal arriving detection and AGC acquisition. The remaining short preamble is utilized for the coarse symbol boundary detection and the carrier frequency offset (CFO) compensation. Then, the fine symbol boundary detection, the fine CFO estimation, and the channel estimation are finished in the duration of the long preamble. The residual CFO, the TFO, and the coefficients of the FEQ are adaptive tracked after the preamble.

B. Coarse and Fine Symbol Boundary Detections

The receiver has to extract the FFT interval within the consecutive received samples. In the short preamble period, the delay correlator moves over the ten repetitive intervals to detect the coarse symbol boundary. A mechanism basing on the binary search algorithm is proposed to track the maximum value of the correlation. Following the short preamble, the long-preamble matched filter is exploited to squeeze the exact symbol boundary. When the symbol boundary detection is finished, the receiver removes the CP and the FFT demodulates the remaining samples.

C. Multi-step Carrier Recovery

The OFDM system is sensitive to the CFO. Both amplitude reduction caused by phase shift and the inter-carrier inference (ICI) severely worsen the packet error rate (PER) [2]. The proposed carrier recovery has three steps. Based on the maximum likelihood estimation [3], the 16-tap delay correlator devised in the symbol boundary detector also achieves the coarse CFO estimation in the short preamble period. Then, a 64-tap delay correlator further completes the fine CFO compensation in the long preamble. Table I lists the performance summaries of the open-loop CFO estimations. By means of the pilots imbedded in the OFDM symbols, the remaining CFO is adaptive tracked by a phase-locked loop (PLL) [4]. The estimated carrier frequency error, $\varepsilon_{c,f}$, is

$$\varepsilon_{c,f}(n) = \frac{1}{2\pi T_{Symbol}} \cdot \frac{1}{N_i} \sum_i (\angle Y_{n,i} - \angle Y_{n-1,i}), \quad (1)$$

where $i \in [-21, -7, 7, 21]$ [5]. $Y_{n,i}$ is the equalized constellation point in the ith sub-carrier of the nth OFDM symbol. T_{Symbol} is the symbol duration and N_i denotes the number of the pilots which is 4 in this design. Correspondingly, the estimated carrier phase error, $\varepsilon_{c,phase}$, is

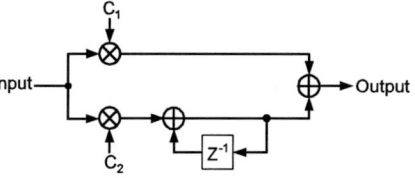

Fig. 3. PI filter

$$\varepsilon_{c,phase}(n) = \sum_{k=0}^{n} 2\pi \varepsilon_{c,f}(n) \cdot T_{Symbol} = \frac{1}{N_i} \sum_i \angle Y_{n,i}. \quad (2)$$

The proportional-and-integral (PI) filter shown in Fig. 3 stands the loop filter in the PLL. Table II lists the design parameters of the adaptive carrier recovery loop. To eliminate multiplications, the coefficients of the PI filter are truncated to power-of-2 numbers. In our design, the performance degradation caused by residual CFO can be suppressed down to 0.1 dB.

D. Self-correcting Timing Recovery

The design of the timing recovery adopts an all-digital solution with a Farrow interpolator [6] mounted after the free running analog-to-digital (A/D) converter. The conventional all-digital timing recovery designs ignored that the drifting interpolation phase due to the TFO would jump beyond sample boundaries to induce either sample stuffing or rubbing [7]. The large sampling jitters could crash the signal detection. Therefore, skipping or duplicating samples in the CP is required to reset the interpolating phase [8]. In order to combine the Farrow interpolator with the skip/duplicate process in the view of VLSI, we propose the architecture of the interpolator with a sliding window as shown in Fig. 4 which adds two additional delay registers for the original Farrow structure. The self-correcting timing recovery automatically slides the window, $s_k=1+\mu_k$ or $s_k=-1+\mu_k$, to fit the boundary-crossed interpolating phases in the reception of the samples within the FFT period and skips/duplicates a sample of the CP at the start of the next received OFDM symbol to reset the windowing position, $s_k=\mu_k$. The hardware implementation is given in Fig. 5. A PLL is also used to adaptively track the TFO, and the PI filter is chosen as the loop filter in the timing recovery loop. The estimated sampling frequency error, $\varepsilon_{s,f}$, is

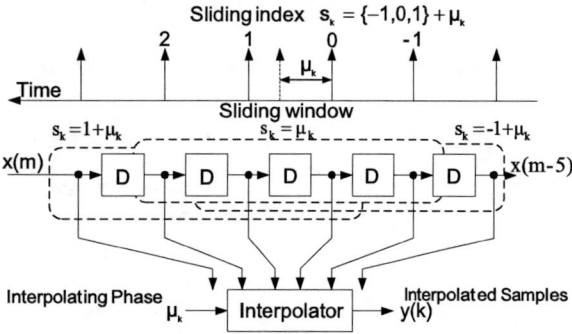

Fig. 4. The proposed interpolator with the sliding window.

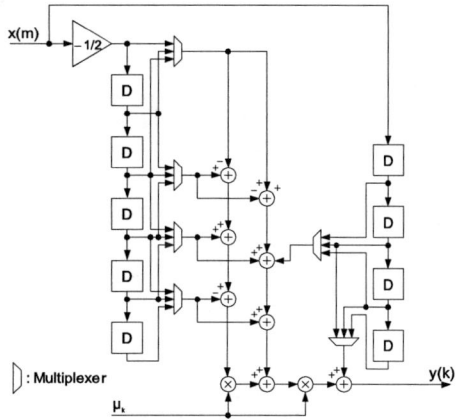

Fig. 5. Implementation of the interpolator with the sliding window.

$$\varepsilon_{s,f}(n) = \frac{1}{2\pi T_{sample}} \cdot \frac{1}{N_{ij}} \sum_{i,j} \frac{1}{i-j} \left[(\angle Y_{n,i} - \angle Y_{n,j}) - (\angle Y_{n-1,i} - \angle Y_{n-1,j}) \right], \quad (3)$$

where $i, j \in [-21, -7, 7, 21]$ [5] and T_{sample} is the sample period. The sampling phase error, $\varepsilon_{s,phase}$, is

$$\varepsilon_{s,phase}(n) = \sum_{k=0}^{n} 2\pi \varepsilon_{s,f}(n) \cdot T_{Symbol} = \frac{N_s}{N_{ij}} \sum_{i,j} \frac{1}{i-j} \left[(\angle \hat{Y}_{n,i} - \angle \hat{Y}_{n,j}) \right], \quad (4)$$

where N_s denotes the number of the samples in an OFDM symbol. Table III lists the parameters of the timing recovery loop.

E. FFT/IFFT

The 64-point pipeline FFT/IFFT processor is implemented with the radix-2/4/8 algorithm [9] in the single-path delay feedback (SDF) architecture. For a 64-point FFT/IFFT, it can be divided into two pipeline stages, and only one complex multiplier is required. The SQNR of the FFT/IFFT is 46.8 dB for this transceiver design.

F. Frequency-domain Equalizer

The FEQ exploits the minimum mean square error (MMSE) criterion [10] to perform the channel estimation in the end of the long preamble,

TABLE III. DESIGN PARAMETERS OF THE TIMING RECOVERY LOOP

Parameter	Value
Lock-in Range $\Delta\omega_L$	1.9 KHz
Damping Factor ξ	0.707
Natural Frequency ω_n	8.5×10^3 rad/s
Phase Detector Gain K_d	0.18 unit/rad
NCO Gain K_o	40 MHz/unit
PI Filter Coefficient C_1	2^{-12}
PI Filter Coefficient C_2	2^{-17}

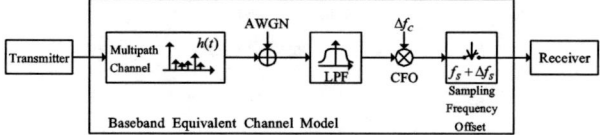

Fig. 6. The simulation environments.

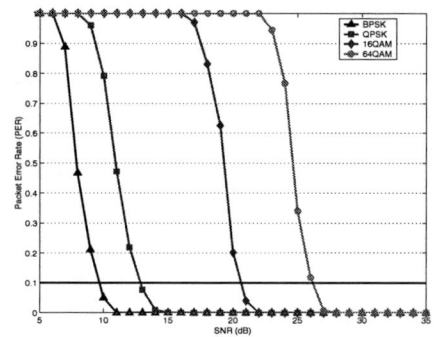

Fig. 7. PER performance for the channel delay spread 0 ns.

$$C_K = H_K^* / \{|H_K|^2 + \sigma_n^2 / \sigma_s^2\}, \quad (5)$$

where C_K and H_K are the equalization coefficient and the channel frequency response at the Kth sub-carrier, respectively. And σ_n^2 is the variance of the AWGN while σ_s^2 is the variance of the transmitted symbols. The least mean square (LMS) algorithm is adopted to adaptively update the coefficients [11],

$$C_K^{(j+1)} = C_K^{(j)} + \Delta \varepsilon_K Y_K^*, \quad (6)$$

where $C_K^{(j)}$ is the jth coefficient at the Kth sub-carrier, Y_K is the FFT demodulated output, ε_K is the decision error, and Δ is the scale factor. The additional LMS algorithm improves the equalization performance about 3 dB SNR.

IV. SIMULATION RESULTS

Multipath fading, CFO, TFO, and AWGN are taken into account in the simulation environments as shown in Fig. 6. The indoor multipath channel model is Saleh's one [12]. In the IEEE 802.11a standard, a PER less than 10% is required for 1000 bytes transmissions [1]. Fig. 7, 8, and 9 show the simulation results of the PER performances, which contain the cases of the delay spread equal to 0 ns, 50 ns, and 100 ns with the maximal tolerable CFO and TFO. The maximal tolerable CFO is ± 40 ppm of the local oscillator (LO) frequency, 5.8 GHz, in the zero-IF architecture and the maximal tolerable TFO is ± 40 ppm of the A/D sampling rate, 40 MHz, in this design.

V. Chip Layout and Performance Summary

Fig. 10 shows the chip layout. Total gate counts are about 302K. The bit numbers of both interfaces of the A/D and D/A converters are 10 bits. Using cell-based design methodologies, this chip is implemented in $0.25\mu m$ CMOS technology and occupies 3.5×3.5 mm^2 chip area. With 2.5 V supply voltage, the power dissipation is 109 mW. The performance summary is listed in Table IV.

VI. Conclusions

This paper presents the VLSI architecture of an OFDM digital baseband transceiver for wireless LAN systems. The 3-step carrier recovery acquires the coarse carrier frequency with the preliminary two open-loop estimations in the training period and maintains good jitter performances with the pilot aided PLL in the data transmission interval. The self-correcting timing recovery realizes the OFDM transceiver digital IP which can be ideally integrated with the analog-front-end and RF circuits for the SoC design. Because the proposed interpolator with the sliding window adjusts the rational interpolation phases in the cyclic prefix, it can be applied to discrete multi-tone (DMT) and single carrier modulation using frequency domain equalizer (SC-FEQ) systems. The OFDM transceiver realized in the IEEE 802.11a system can provide 10% PER requirement under specified SNRs in the indoor transmission environments. Using 0.25μm CMOS process, this transceiver is implemented in 3.5×3.5 mm^2 chip area. The power consumption is 109 mW under 2.5 V power supply.

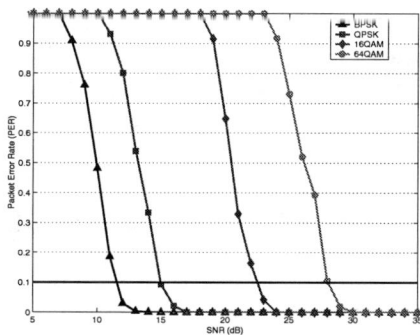

Fig. 8: PER performance for the channel delay spread 50 ns.

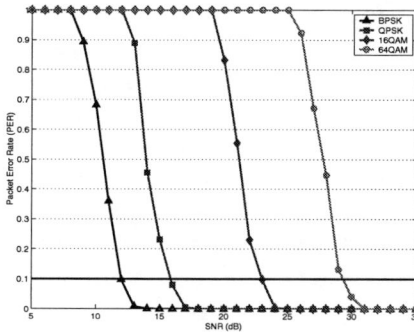

Fig. 9. PER performance for the channel delay spread 100 ns.

References

[1] Part 11: Wireless LAN medium access control (MAC) and physical layer (PHY) specifications for high-speed physical layer in the 5 GHz band. *IEEE std. 802.11a*, Nov. 1999.

[2] P. H. Moose, "A technique for orthogonal frequency division multiplexing frequency offset correction," *IEEE Trans. on Comm.*, Vol. 42, No. 10, pp. 2908-2914, Oct. 1994.

[3] J. J. van de Beek et al., "ML estimation of time and frequency offset in OFDM systems," *IEEE Trans. on Signal Processing*, Vol. 45, No. 7, pp. 1800-1805, July 1997.

[4] R.E. Best, *Phase-Locked Loops*, 3rd ed., McGraw-Hill, 1998.

[5] M. Sliskovic, "Carrier and sampling frequency offset estimation and correction in multicarrier systems," *IEEE GLOBECOM*, pp. 285-289, Nov. 2001.

[6] L. Erup et al., "Interpolation in digital modems-part ii: implementation and performance," *IEEE Trans. on Comm.*, Vol. 41, No. 6, pp. 998-1008, June 1993.

[7] C-F Hsu, Y-H Huang and T-D Chiueh, "Design of an OFDM receiver for high-speed wireless LAN," *ISCAS*, Vol. 4, pp. 558-561, May 2001.

[8] T. Pollet and M. Peeters, "Synchronization with DMT modulation," *IEEE Comm. Mag.*, Vol. 37, No. 4, pp. 80-86, Apr. 1999.

[9] L. Jia et al., "A new VLSI-oriented FFT algorithm and implementation," *IEEE ASIC Conf.*, pp. 337-341, Sept. 1998.

[10] H. Sari et al., "Transmission techniques for digital terrestrial TV broadcasting," *IEEE Comm. Mag.*, Vol. 33, No. 2, pp. 100-109, Feb. 1995.

[11] J. G. Proakis, *Digital Communication*, 4th ed., McGraw Hill, 2001

[12] A. A. M. Saleh and R. A. Valenzuela, "A statistical model for indoor multipath propagation," *IEEE J. Selected Areas in Comm.*, Vol. SAC-5, No. 2, pp. 128-137, Feb. 1987

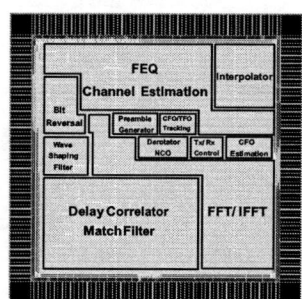

Fig. 10. Chip layout

TABLE IV. The Performance Summary of This Chip

Data Rate	6, 9, 12, 18, 24, 36, 48, 54 Mbps
Modulation	OFDM (BPSK, QPSK, 16QAM, 64QAM)
Tolerance of CFO	± 232.2 KHz
Tolerance of TFO	± 1.6 KHz
Technology	0.25μm CMOS
Chip Area	3.5 x 3.5 mm^2
Gate Counts	302189
Clock Rate	40 MHz
Power Dissipation	109 mW
Supply Voltage	2.5 V
DA/AD Interface	10 Bits

A One-Quadrant Discrete-Time Cellular Neural Network CMOS Chip for Pixel-Level Snakes

V.M. Brea, M. Laiho*,**, D.L. Vilariño, A. Paasio*, D. Cabello
Department of Electronics and Computer Science
University of Santiago de Compostela
Santiago de Compostela, Spain
Phone:+34981563100, Ext. 13572. Fax:+34981528012. Email:victor@dec.usc.es
*University of Turku, Department of Information Technology, Finland
**Electronic Circuit Design Laboratory, Helsinki University of Technology, Finland

Abstract— This paper introduces a CMOS on-chip implementation intended for Pixel-Level Snakes (PLS). The resultant architecture follows the SIMD paradigm. The B/W processing is executed on a Discrete-Time Cellular Neural Network (DTCNN) array with a one-quadrant (1Q) model. The gray-scale processing is also run locally on a dedicated hardware. Electrical simulations on a proof-of-concept chip with a resolution of 9×9 pixels in a $0.18 \mu m$ CMOS technology process, (ST Microelectronics), give some estimation of the figures of merit expected with the future chip measurements.

I. INTRODUCTION

Pixel-Level Snakes (PLS) have been widely discussed in the literature [1], [2]. In the image processing field, PLS are placed midway between energy and level-set based models [3]–[5]. This makes this technique very efficient when dealing with complex applications like medical image processing with a low S/N content, or applications with several contours on the scene, like moving object segmentation [2], [6].

From the hardware point of view, the pixel-level discretazation would lead to its natural projection onto an SIMD chip architecture. In line with this, the synergy of PLS and CNN seems to be a very promising approach to solve applications demanding a fast-time response [6], [7].

The PLS technique has been run on what it might be called a gray-scale architecture, as well as on a binary CNN-based model. In the former solution, the SIMD paradigm is implemented with a general purpose CNN circuit, where the cells (pixels) are designed for gray-scale processing, (B/W is a particular case of the more general gray-scale processing). The binary CNN-based approach splits the cell into a B/W and a gray-scale sub-cell [8]. In this case, the design efforts can be emphasized on the most demanding operation: either B/W or gray-scale.

The PLS CMOS on-chip implementation presented here follows this approach. The B/W sub-cell is designed to meet hard requirements on speed in operations like hole filling. The one-quadrant (1Q) CNN model with 1-bit of programmability for the templates is not only suitable for satisfying such speed demands, but also to have a dense chip (cells/mm^2). In the proof-of-concept circuit presented here, the aforementioned 1Q model is employed with a DTCNN architecture. For the time being, the gray-scale sub-cell is thought as being compatible with the B/W part.

II. THE PLS TECHNIQUE

The PLS technique comprises gray-scale and B/W operations [2]. The gray-scale operations are aimed at extracting the guiding forces for the contours to be shifted/moved towards the target. The B/W processing entails the evolution of contours. These are represented as sets of eight-connected pixels on a binary image.

Fig. 1 shows the operations performed in the PLS technique implemented here. The algorithm is iteratively run along the four cardinal directions. The OR gate depicted in Fig. 1 is just a way of indicating which one of the contour images is being processed, either the initial one or the result of the former iteration in the algorithm.

The algorithm displayed on Fig. 1 can be split into three major parts. The Guiding Force Extraction (GFE) block extracts the guiding information, a B/W image displaying in black the locations towards the contours can be moved/deformed. This information is extracted from the B/W contour image (containing the contours) and the so-called external potential image. This is a gray-scale image provided with the most relevant information from the scene [3]. It is fundamental in guiding the contours towards the target. The Active Contour Evolution (ACE) block performs the contour movement according to the GFE outcome. This is accomplished in two steps: Directional Contour Expansion (DCE) and Directional Contour Thinning (DCT). These are the only B/W operations specific to the PLS technique. The Topologic Transformations (TPT) block deals with several contours when needed. The latter encompasses morphological operations of erosion and dilation, as well as a propagating task, hole filling, and the binary edge detection. All of them are deeply discussed in the literature [9]. In order to get a better understanding of the PLS technique implemented here, the reader is addressed to [10], where an extensive set of examples with active contour applications like contour tracking or image segmentation can be found.

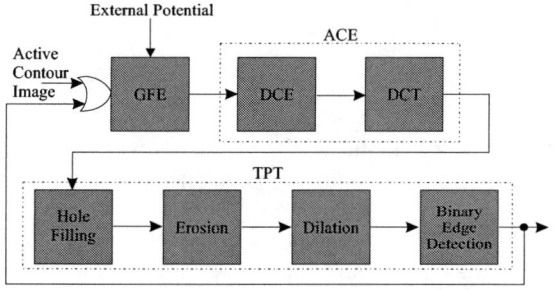

Fig. 1. Operations performed in the PLS technique.

III. THE CELL ARCHITECTURE

As can be seen in Fig. 1, the PLS technique is fed with two input images: the external potential and the active contour image. The external potential is a gray-scale image, which along with the internal and the balloon potentials, real-valued arrays derived from the contour themselves, guide their evolution. The external potential is calculated from the intensity image (the image to be processed). This can be done either on- or off-chip. The latter is the solution adopted in this implementation.

The PLS algorithm is projected onto a multilayer 2D-DTCNN network with cyclic time-variant clonning templates controled by a clock signal ck. The gray-scale processing is performed in a dedicated hardware. All the templates for B/W processing are designed following the 1Q model and with 1 bit of programmability. The bias term has 2 bits of programmability. In the 1Q DTCNN PLS architecture presented here all the operations involving only one image are performed with only one B-template. Operations with two images are realized with two B-templates. Nevertheless, for the current PLS technique it is always possible to design one of the B-templates with only one coefficient. The other B-template might have the 9 coefficient values different from zero. The result is an approach with only 10 coefficient circuits for the template execution. Concerning the notation adopted henceforth, normal B-templates are labeled as A. The additional B-templates with only one non-zero coefficient at the center of the template are marked with A^+. As both A and A^+ scale the outputs of a previous DTCNN step (set to either 0 or 1), A and A^+ are interchangeable, which might significantly reduce hardware [6]. The template design procedure followed here consists of splitting the task under study into as many sub-tasks as needed, as well as to account for the inversion of the images and to fix the evolution of either black or white pixels when needed [11].

The guiding information image (B/W) is extracted during the gray-scale processing. This is accomplished with a threshold function on the gradient of the potential image. The potential image is the result of summing three real values: the external, the internal and the balloon potentials [7].

Fig. 2 displays the architecture of the PLS cell irrespective of the mathematical domain of the implementation, either analog or digital. All the labels are given in the framework of the CNN-Universal Machine (UM) [9]. Also, the B/W and the gray-scale parts are clearly seen. In between, an interface circuitry starts the processing and selects which processing is taking place, B/W or gray-scale. Within the B/W circuitry, a set of digital memories, LLM and SRAM, stores the outcomes of DTCNN steps (clock cycles, ck). These values are properly steered to the coefficient circuits. The box labeled FSM (Fixed State Map) freezes the evolution of either black or white pixels when needed. It is also worth mentioning that in this implementation the FSM does not require an external input (image) to work properly. In the gray-scale cell, the external potential is led to a memory bank. The potential images are represented by IP (Internal Potential), BP (Balloon Potential) and EP (External Potential) [10]. They are summed and compared (threshold function) to the potential value of the neighboring cell under the current processing direction.

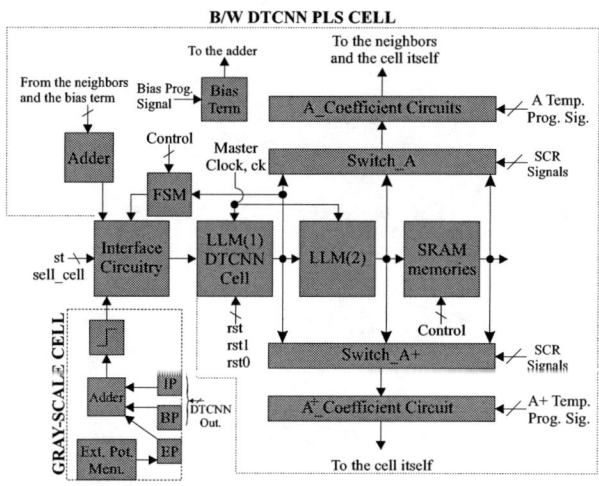

Fig. 2. Architecture of the PLS cell.

IV. CIRCUIT DESIGN

In our design, the main aim is to optimize area and processing speed. The former leads to chips with more resolution. The latter might tackle applications needing a fast time response. Power dissipation is also a major concern. Architectures with a selectable "stand-by" mode assist in solving this issue, alongside low bias curents and voltages. In line with this, the power supply is set to the minimum value, i.e. 1.8V. The robustness (accuracy) is also fundamental in achieving a dense realization. Its value is always enlarged as high as possible, particulary in the 1Q template design. Concerning the mathematical domain, the PLS cell contains both analog and digital circuits. Following, we go through the main constituent parts of the cell, outlining the applied principles of design.

A. Coefficient Circuits

The coefficient circuits are implemented with the load transistor of a current mirror. Their particular realization is determined by the 1Q and 1-bit programmable templates. The solution adopted here is that addressed in [12] and here recalled in Fig. 3. This solution fixes the gate to source voltage drop in the current mirror. The programming voltage drives a digital transistor (switch), so the DTCNN outcomes do (memory contents in Fig. 2). All the switches are realized with NMOS transistors and sized to the minimum area (0.28/0.18) (in μms). The size and bias current in the current mirror is a trade-off between area and power consumption in order for the mismatching effects to be within the limits imposed by the robustness of the templates [6]. The least robustness in the 1Q templates is around 14% [11]. In the coefficient circuits, only analog PMOS transistors are employed for the current mirrors in order to achieve high robust circuits, due to their superior V_{gs} compared to their NMOS counterpart when both draw the same current [13]. The resultant numbers are (1 μA) as unity current, and an area of (0.28/1) (μm) for the analog PMOS transistors. The bias term is implemented with NMOS transistors. Finally, the adder depicted in Fig. 2 is a simple node, as all the neighboring contributions are summed by KCL.

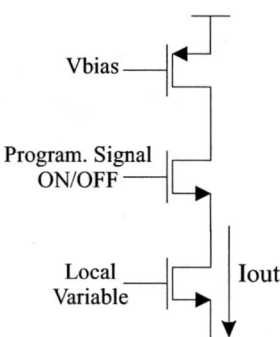

Fig. 3. The coefficient circuit.

B. B/W Memories and Switches

All the memory cells and switches employed here are sized to the minimum dimension (0.28/0.18) (μm). The Local Logic Memories (LLMs) are dynamic memories. LLM(1) also solves the DTCNN equation [6]. In order to deal with charge injection and feed-through effects, transmission gates are used. Also, four SRAM cells are included, as the outcome of DTCNN steps (ck cycles) must be stored for several cycles. The FSM is another SRAM cell with a logic gate to decide whether to freeze black or white pixels in their evolution. Concerning the switches driving the coefficient circuits, (A or A^+), it should be noted that the major concern is to avoid having two switches driving the same type of coefficient circuit in conduction mode at the same time, otherwise a charge sharing phenomenon would take place, leading to misprocessed data. This is controlled by the programming template sequence and the SCR (Switch Configuration Register) signals. The interface circuitry is also implemented with transmission gates.

C. Gray-Scale Implementation

In the present work, the gray-scale cell implementation has been kept as simple as possible. The use of current steering D/A converters is the most straightforward approach. The digital bits are always local signals, (low fan-out), preserving a high speed processing. In terms of area and power dissipation, again there is a trade-off between size and bias current to comply with the accuracy constraints. Here, four bits for the external potential are assumed. These are locally stored in a bank of SRAM cells (Fig. 2). The internal potential is properly weighted, and implemented with 2 bits of programmability. The balloon potential, also properly weighted, is estimated with 1-bit. These values are summed (KCL, single node), providing an analog signal equivalent to a five-bit digital word in resolution. In the D/A circuit implementation, again PMOS transistors are employed in the current mirrors, searching for a high accuracy. The unity current has been set to 0.5 μA, and the unity PMOS transistor is (1/3) (μm) in size. Additional switches selecting a "stand-by" mode (no bias current) within the three D/A converters (EP, IP and BP in Fig. 2) are also included. Finally, the comparison between a cell and its neighbor along the current processing direction is done by means of a current comparator provided with a low input impedance node.

D. Chip Data

The work presented here is a proof-of-concept chip. As such, the most significant data concern the cell itself. In terms of area, every cell amounts to $40 \times 32 \mu m^2$, leading to a cell density greater than 700 cells/mm^2. Every processing step, ck cycle, can be set to less than 50 ns. This would lead to a slowest processing time of $6\mu s$ in a hole filling operation with a 128×128 image (assuming a chip of 128×128 cells). The bottleneck in speed, however, is caused by the uploading/downloading tasks. In the future, embedded photosensors will speed up the uploading process. The downloading (B/W contours) will follow a row-wise scheme. High values of power dissipation are not expected, as there are few cells (pixels), only the contours, with active coefficient circuits (drawing current). Also, the bias currents and the voltage supply are low. The floorplan of the chip encompasses a digital control unit for an array of 9×9 cells. The control unit contains a digital memory of 128×12 SRAM cells for template storage, and also decoders for getting access to every cell in the array. The chip has been designed in a full-custom style. Fig. 4 displays the layout view of a cell (Cadence). The bigger size of the transistors in the gray-scale cell is plain. Also, it can be observed part of the routing from the cell to its neighbors. The chip has been submitted to fabrication. Experimental results might be shown in the conference.

V. CONCLUSION

The PLS technique has been implemented on a proof-of-concept chip following the so-called 1Q model for the B/W processing. The gray-scale processing is executed with dedicated D/A converters. The result is a high dense chip, 700 cells/mm^2 in 0.18μm, having an accuracy equivalent to that of a 5-bit D/A converter. The power dissipation is also expected low, due to the low voltages and bias currents, as well as a "stand-by" architecture, and a reduced number of cells active at the same time. The speed, although here is set by the master

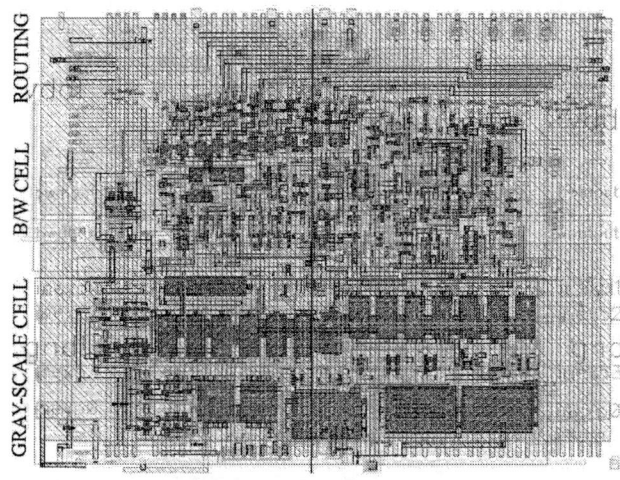

Fig. 4. Cell layout.

clock cycle, ck, around 50ns, in a practical application, the use of photosensors is required to achieve a fast uploading of the image to be processed. This, alongside the on-chip external potential calculation are some of the steps to be done in the subsequent PLS on-chip implementation.

VI. ACKNOWLEDGMENTS

This work was funded by Ministerio de Ciencia y Tecnologia (Spain) under the Project TIC2003-09521.

REFERENCES

[1] D.L. Vilariño, et al., "Pixel-Level Snakes", 15th International Conference on Pattern Recognition, vol. 1, pp. 640–643, 2000.
[2] D.L. Vilariño, et al., "Cellular Neural Networks and Active Contours: A Tool for Image Segmentation", Image and Vision Computing, vol. 21, no. 2, pp. 189–204, 2003.
[3] M. Kass et al,. "Snakes: Active Contours Models", International Journal of Computer Vision, vol. 1, pp. 321–331, 1988.
[4] V. Caselles, et al., "A Geometric Model of Active Contours in Image Processing", Num. Math., vol. 66, no. 3, 1993.
[5] R. Malladi, J.A. Sethian, B.C. Vemuri, "Shape Modelling with Front Propagation: A Level Set Approach", IEEE Transactions on Pattern Analysis and Machine Intelligence, vol. 17, no. 2, pp. 158–174, 1995.
[6] V.M. Brea, D.L. Vilariño, A. Paasio, D. Cabello, "Design of the Processing Core of a Mixed-Signal CMOS DTCNN Chip for Pixel-Level Snakes", IEEE Transactions on Circuits and Systems-I, vol. 51, no. 5, pp. 997-1013, May 2004.
[7] D.L. Vilariño, Cs. Rekeczky, "Implementation of a Pixel-Level Snake Algorithm on a CNNUM-based Chip Set Architecture", IEEE Transactions on Circuits and Systems-I, vol. 51, no. 5, pp. 885–891, May 2004.
[8] A. Paasio et al., "Different Approaches for CNN VLSI Implementation", ECCTD'99, vol. 2, pp. 1347-1350, 1999.
[9] L.O. Chua, T. Roska, " Cellular Neural and Vision Computing. Foundation and Applications", Cambridge University Press, 2002.
[10] D.L. Vilariño and Cs. Rekeczky: "Pixel-Level Snakes on the CNNUM: Algorithm Design, On-Chip Implementation and Applications", International Journal of Circuit Theory and Applications, 2004 (In Press).
[11] V.M. Brea, M. Laiho, D.L. Vilariño, A. Paasio, D. Cabello, "A One-Quadrant Discrete-Time Cellular Neural Network Architecture for Pixel-Level Snakes: B/W Processing", ISCAS2005.
[12] J. Flak, M. Laiho, A. Paasio, K. Halonen, "VLSI Implementation of a Binary CNN: First Measurement Results", Proceedings CNNA2004, pp. 129-134, 2004.
[13] P.G. Drennan and C.C. McAndrew, "Understanding MOSFET Mismatch for Analog Design", IEEE J. Solid-State Circuits, vol. 38, pp. 450–456, Mar. 2003.

Various implementations of topographic, sensory, cellular wave computers

Ákos Zarándy, Péter Földesy, Péter Szolgay,
Szabolcs Tőkés, Csaba Rekeczky, and Tamás Roska

Analogical and Neural Computing Laboratory
MTA-SZTAKI
Budapest, Hungary
<zarandy, foldesy, szolgay, tokes, Rekeczky, roska>@sztaki.hu

Abstract—The Cellular Wave Computer architecture, based on the CNN Universal Machine principle, has been implemented recently in many different physical forms. The mixed mode CMOS, the emulated digital (cell wise or as aggregated arrays), FPGA, DSP, as well as optical implementations are the main examples. In many cases, the sensory array is integrated as well. The new self contained unit, called Bi-i, winning the product of the year title at the Vision 2003 in Stuttgart as the fastest camera-computer, shows the application interest and impact being capable of sensing-computing with 50, 000 frame per second. In this paper a clear and concise comparison will be presented between the various implementation modes.

I. INTRODUCTION

Topographic sensory, cellular wave computers (derivatives of CNN Universal Machine principle [1,2]) are locally interconnected processor arrays, with optical or other type of sensors in each processor. The architecture of these devices is very similar to the biological vision systems or other topographic neuro computers. These computers can very efficiently process 2D discrete data sets, because in most cases, the data set and the function can be directly mapped to the processor array.

There are numerous advantages of this computer architecture. Since each processor is directly interconnected with its sensor(s), the sensor readout is extremely fast (few microsecond only). Moreover, the induvidual sensor characteristics can be locally tuned by the processors for each sensor in the large sensor array. This is an absolutely unique feature in artifitial sensor arrays, only natural sensory arrays used this special trick so far.

In this paper, we introduce a large number of implementation methods of this computer architecture, compare their performance. Among these implementations methods, one can find analog VLSI; two types of emulated digital silicon; FPGA; software simulation, running on the highest speed Texas DSP; and even optical implementation. It is a very exciting experiment, to compare this exceptional wide range of implementation methods of the same computer architecture.

II. ANALOG VLSI IMPLEMENTATION

At least a dozen of analog cellular computers was developed by various universities or research institutes in the last decade. The most complex of those is the ACE16k chip [8], which can calculate feedback and feed-forward convolution, accurate diffusion, and averaging in the analog domain. It has 8 internal analog memories and fast optical sensor. Its typical operation execution time is in the 5-10 microsecond for 128x128 sized images. Its high power consumption, large silicon size, and relatively slow IO makes its industrial usage difficult. However, it can be used very well in research, as specially in retina modeling [7].

Its successor, the eye-RIS chip [9], which is already an industrial design will hopefully the first analog cellular processor array, which can be used by the industry.

III. EMULATED DIGITAL IMPLEMENTATION

The accuracy, the difficult realization of the non-linear template operations and the multi-layer structures are the weak points of the analog implementation. To overcome these problems, but still having high performance, some laboratories has started to implement the digital array processors. Here we introduce two different concepts.

A. Cell wise implementation

In the emulation of the functionality of the analog CNN-UM the first evident step is the cell wise replication by means of digital resources. One of the crucial decisions is to use standard cell based synthesis (see fig.) or to use full-custom design (see fig.). Applying further advanced circuit solutions, such as dynamic ram mass storage, dynamic gates, low transistor count pass-transistor based logics, even two-three orders of magnitude more dense cell array can be built than the classic RTL (Register Transfer Level) description based digital design style allows.

From the technology point of view, as the optical sensitivity drops drastically with technology generations, the standard CMOS technology does not enable optical sensors below the 0.18 micron feature size. This limits the digital replication with embedded light sensing, apart from exotic solutions (e.g. amorphous silicon sensor grow over the die passivation).

Due to the most probable large count of control wires, large array size, and power consumption, it is difficult to increase the nominal clock frequency over one or two hundred megahertz.

Finally, taking into account the above considerations. the full custom option with average 80-90 kgate/mm^2 at the 0.18 micron technology, a specialized cell architecture of 300-400 equivalent gates and 10-20 clock cycle per operation, 100 MHz clock frequency, the 200-250 cell per mm^2 density can be achieved with about 1-2 GOPS (@8-bit) per mm^2. This system integrated into a 1x1 cm^2 die, results in a flexible, precise, CNN-UM emulator with app. 100 GOPS (@8-bit) and 256x256 optical resolution.

A 128x96 sized experimental version, called C-TON, of this architecture is under development in our laboratory. We expect to reach 30x30 micron cell size. Its speed in binary operations (like erosion, dilution) is expected to be 100 times faster, while it will be the same in feed-forward convolutions.

B. Aggregated array implementation

The other way of emulated digital implementation is the aggregated array implementation. This means, that the array of the simple tiny cellular processors is divided to stripes, and each stripe is processed with a high performance digital custom designed processor. An array of 2*3 aggregated processors (CASTLE ARRAY processor array) was designed and implemented by using a 0.35µ CMOS technology with a 1 ns/ virtual CNN cell/iteration speed supposing 12-bit-accuracy. The accuracy of computation can be set to 1 bit, 6 bits or 12 bits. The processing speed is inversely proportional with the selected accuracy. The templates can be space variant.

The basic architecture of core arithmetic unit was derived from state equation of a CNN cell discretized in time (by forward Euler method). To accelerate the operation a *4-level pipeline was used with three multipliers in arithmetic unit* to calculate the sum of the products of 9 operand-pairs and the constant, and to execute the limitation.

Except for the left and right processors, the interconnections are obvious. If the horizontal communication between the subarrays was solved in the same way as at the inner edges of the processors, the number of I/O ports would be extremely high. Consequently, two bi-directional I/O buses serve this communication in time-multiplex mode.

Using a two phase clock signal, the *Timing and Control Unit* generates the register-transfer signals. The *2x3* array of **CASTLE** processors is shown in Figure 1. .

The processor array was interfaced to Aladdin Pro system [6] where the hardware-software environment provides similar high-level support to develop analogic CNN algorithms to the analog VLSI CNN-UMs. The logic processor was tested by some black&white test-images. The size of these images is 120*80 pixels because the width of the input register array is 120 pixels. The length can be arbitrary. Approximately 5780 images can be processed by a logic processor in a second.

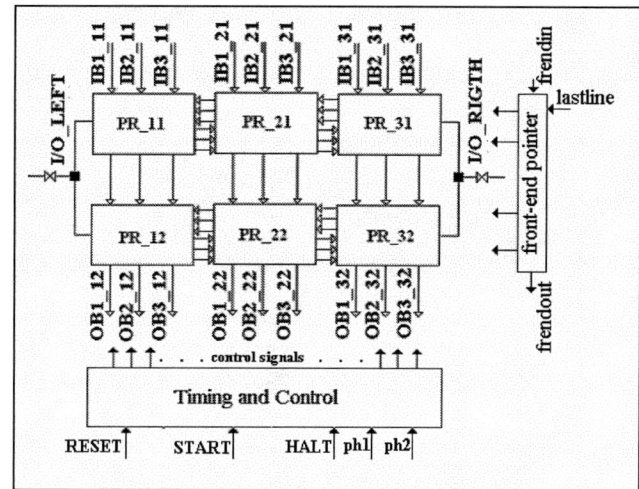

Figure 1. The Block diagram of 2*3 CASTLE array

IV. FPGA IMPLEMENTATION

Today the high end FPGAs have 6-8 million equivalent gates. Using one of these FPGA-s, we have designed a flexible emulated digital CNN-UM processor array, called Falcon [5]. The FPGA implementation is an aggregated array implementation. Though it is slower than the CASTLE, the processors can be reconfigured easily, to allow the implementation of special functions.

Let us examine a standard emulated digital CNN cell with neighborhood size r. The arithmetic unit can be built up from 2r+1 multipliers, an adder tree to sum the multiplied values and an accumulator register to store the partial results, the template operation is computed row-wise in 2r+1 clock cycles.

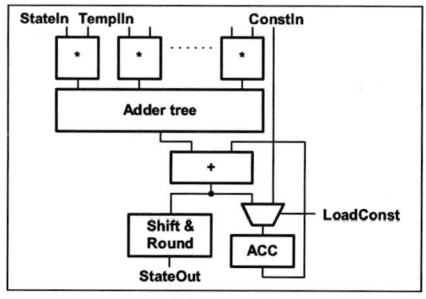

Figure 2. FPGA arithmetic unit

Since each state value is required in the update of (2r+1)*(2r+1), these values are read into the FPGA array only once, and distributed by the mixers to the arithmetic units in the required order. This solution significantly reduces the IO requirements of the FPGA. The mixer unit contains 3 groups of shift registers the main, the left and the right shift register (r=1).

Beside the single layer CNN arrays, we have implemented multi-layer CNN processor also. In this implementation, every CNN array layer was connected together in all possible ways. This means that the arithmetic unit must do r^2 times more work than in the single layer case. Templates in the multi-layer case can be treated as r*r pieces of single-layer templates.

These single-layer templates can be grouped by layers and the template operation can be computed row-wise by a single layer arithmetic unit. It is possible in high precision cases that this arithmetic unit requires larger area. In this case the parallelism is reduced, and one multiplier is used for every single-layer template.

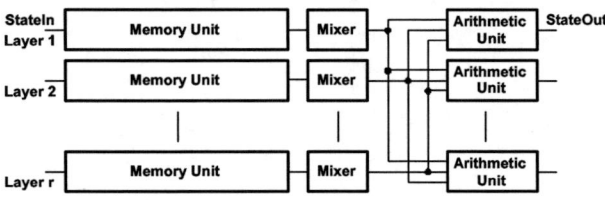

Figure 3. Multilayer emuletd digital architecture

After these design considerations we were able to make a synthesizable RTL-level VHDL description of our single- and multi-layer processor architectures. We used Synopsis FPGA-Express to synthesize our processors. The processors can be configured before the synthesis. The configurable parameters are the following:

- the bit width and displacement of the radix point for the state, constant and template values, possible values for width are between 2 and 64
- the number of templates
- the neighbourhood value of the templates
- the width of the cell array slice
- the number of processor core rows and columns
- the number of layers in the multi-layer case

The large number of configuration parameters makes easy to synthesize the Falcon architecture, which is optimal for our requirements, if our requirements have changed the same FPGA can be used but with differently configured Falcon processors.

Typical emulated digital FPGA applications are the neuromorph retina model implementations and PDE based solutions of elasticity or flowing fluids.

V. DSP IMPLEMENTATION

We have implemented a CNN simulator on a 1GHz Texas fixed point DSP implemented with 90nm technology. This is the most powerful DSP today. The resolution of the values is 8 bit. Its is very interesting, that the binary operations (like shift or erosion) for 128x128 sized images stored already in internal memory, takes roughly the same as on the ACE16k chip, and a feed-forward convolution takes roughly 3 times longer.

VI. OPTICAL IMPLEMENTATION

Our optical implementation is called as Programmable Optical Analogic Array Computer (POAC). Its kernel is a novel type of holographic correlator using bacteriorhodopsin as a dynamic holographic material. Presently POAC can execute feed-forward-only (B-template operations). Its main advantage is the massive parallelism both for the input and for the template. A time sequence of these operations builds up sophisticated algorithms. An Optical Template and Algorithm Library has been developed for POAC. The architecture of the POAC is shown in Figure 4. .

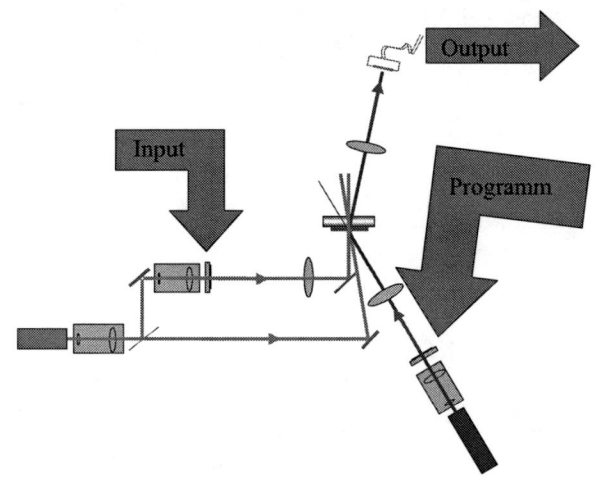

Figure 4. Schematics of the two-wavelength POAC.

The most important application of POAC is adaptive pattern recognition and target tracking.

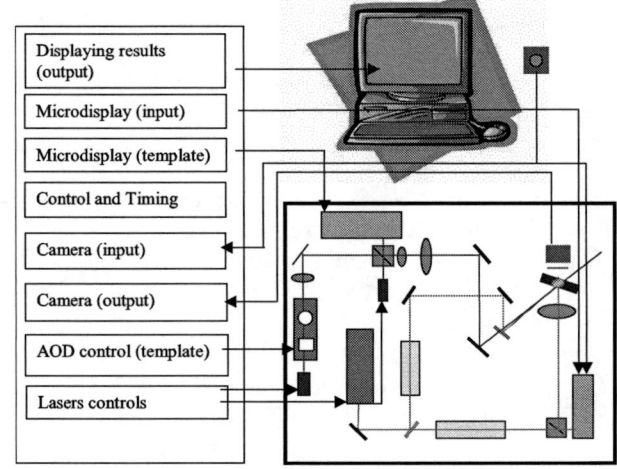

Figure 5. System of the Laptop-POAC

VII. Comparison of the Implementations

After the introduction of the different implementation methods, here we compare their performance in TABLE I. All the running time figures are corresponding to equivalent 128x128 sized images. There is no execution time information of the eye-RIS, however, one can expect roughly 2 times speed advantage compared to ACE16k. The POAC is not shown in the large comparison table, because it is very difficult to compare its parameters to the other devices. On the other hand, its parameters are summarized in TABLE II.

TABLE I. COMPARISON TABLE OF THE VARIOUS IMPLEMENTATIONS (RUN TIME FIGURES NORMALIZED TO EQUIVALENT 128X128 SIZED IMAGES)

	ACE16k	eye-RIS	C-TON	CASTLE	FALCON	DSP
implementation type	analog/mixed signal VLSI	analog/mixed signal VLSI	emulated digital cell-wise	emulated digital aggregated	FPGA	software
date	2001	2005	2005	2004	2003	2004
implementation features	0.35μ, technology; ~70x70μ cell size; 128x128 array; 6W;	0.18μ, technology; ~25mm^2; 176x144 array; 0.1W;	0.18μ, technology; ~30x30μ cell size; 128x96 array; 0.5W;	0.35μ, technology; 2x3 aggregated proc.;	XC2V1000 FPGA (1 million gate)	TMS320C6215 DSP @1GHz.
binary operations (e.g. erosion)	8 μs	-	220ns	1,6 μs	800ns	9,5 μs
grayscale convolution (B template)	8 μs	-	20 μs	16 μs	16 μs	26 μs
grayscale feedback convolution (A template)	15 μs	-	100 μs	90 μs	105 μs	130 μs

Acknowledgment

This work was sponsored by Info-Communications Technologies Applications, Ministry of Education, Hungary (IKTA-00002/2001).

TABLE II. THE MOST IMPORTANT PARAMETERS OF THE POAC

	Comparison of present and future POAC versions		
	Current setup	Next version	Considerably enhanced
Input resolution	500 x 500	1000 x 1000	2000 x 2000
Input frame rate	30	50	200
Template resolution	64 x 64 or 32 x 32	256 x 256	256 x 256
Template frame rate [Hz]	1000 (32 x32) or 256 (64 x6 4)	1000 (16 x 16) or 300 (256 x 256)	1000 (256x256) 10^5 (32x32)
Output frame rate [Hz]	30	100	1000
Performance [GOPS]	250-500	2 10^3	20 10^3
Size m^2	0.1	0.05	0.01

References

[1] T.Roska and L.O.Chua, "The CNN Universal Machine: an analogic array computer", IEEE Transactions on Circuits and Systems-II Vol.40, pp. 163-173, March,1993.

[2] L.O.Chua and L.Yang, "Cellular neural networks: Theory and Applications ", IEEE Trans. on Circuits and Systems, Vol.35, pp. 1257-1290, 1988.

[3] P. Keresztes, Á. Zarándy, T. Roska, P. Szolgay, T. Bezák, T. Hidvégi, P. Jónás, A. Katona "An emulated digital CNN implementation" Journal of VLSI Signal Processing Systems Kluwer Academic Publishers, Vol. 23. pp. 291-303, 1999.

[4] T.Hidvégi, P. Keresztes, P. Szolgay "Enhanced Modified Analized Emulated Digital CNN-UM (CASTLE) Arithmetic Cores" Journal of Circuits, Systems, and Computers, special issue on "CNN Technology and Visual Microprocessors" Vol. 12, No. 6

[5] Z. Nagy, P. Szolgay "Configurable Multi-Layer CNN-UM Emulator on FPGA" IEEE Transactions on Circuits and Systems I: Fundamental Theory and Applications, Vol. 50, pp. 774-778, 2003 Szabolcs Tőkés, László Orzó, Ahmed Ayoub and Tamás Roska: Laptop POAC: a Compact Optical Implementation of CNN-UM, proceedings of IEEE CNNA2004, pp. 70-75, Budapest, 2004

[6] Á. Zarándy, Cs. Rekeczky, I. Szatmári: "Vision Systems based on the 128x128 focal plane Cellular Visual Microprocessor Chips", ISCAS 2003 Bangkok.

[7] B. Roska, "Mammalian Retinal Vision: From Living Cells to Silicon" IEEE International Workshop on Cellular Neural Networks and their Applications (CNNA-2004), Budapest, Hungary 2004

[8] G. Liñán, R. Domínguez-Castro, S. Espejo, A. Rodríguez-Vázquez, "ACE16k: A Programmable Focal Plane Vision Processor with 128 x 128 Resolution". ECCTD '01 -European Conference on Circuit Theory and Design, pp.: 345-348, August 28-31,2001, Espoo, Finland

[9] eye-RIS 1.0 Leaflet of AnaFocus Ltd, Seville Spain, 2004.

IMPLEMENTATION OF SIMD VISION CHIP WITH 128x128 ARRAY OF ANALOGUE PROCESSING ELEMENTS

Piotr Dudek

School of Electrical and Electronic Engineering, University of Manchester
PO Box 88, Manchester M60 1QD, United Kingdom
p.dudek@manchester.ac.uk

ABSTRACT

This paper presents the latest implementation of the SIMD Current-mode Analogue Matrix Processor architecture. The SCAMP-3 vision chip has been fabricated in a 0.35μm CMOS technology and comprises a 128×128 general-purpose programmable processor-per-pixel array. The architecture of the chip is overviewed and implementation issues are considered. The circuit design of the analogue register is presented, the layout of the Analogue Processing Element is discussed and the design of control-signal drivers and readout circuitry is overviewed.

1. INTRODUCTION

Programmable vision chips, which combine image sensors and pixel-parallel processors, can offer significant advantages in many computer vision applications, providing image pre-processing capability with low power consumption and high computational performance. Inherently parallel low-level image processing algorithms map naturally to fine-grained pixel-parallel processor arrays, while the ability to perform computations right next to the sensors reduces the power consumption and I/O bandwidth requirements. However, the requirement of physically co-locating photosensor array and processor array, in a processor-per-pixel manner, introduces severe constraints in terms of the physical design of the device, circuit area, and power consumption. This is especially true if a low-cost requirement implies the use of a standard CMOS technology. To meet the implementation constraints, mixed-mode and analogue circuits are usually employed in the design of the processor. It has been demonstrated [1-3] that use of analogue processors can offer significant benefits in terms of cost, processing speed and power consumption, when limited accuracy of processing is acceptable. The implementation of a large analogue VLSI circuit, however, is a challenging task. The trade-offs between cell area, power dissipation and processing speed have to be suitably resolved. A full-custom design is obviously required, and the constraints on the physical geometry of the circuit topology are an important consideration affecting the circuit and system design. This implies a co-design of processor architecture, circuitry and layout. Furthermore, to achieve a robust implementation, the issues of noise and mismatch have to be carefully controlled. Again, the architecture and physical design have to be considered together.

In this paper the design of a 128×128 pixel-per-processor vision chip (SCAMP-3, shown in Figure 1) is presented. The chip has been fabricated in a 0.35μm CMOS technology and comprises over 1.8 million transistors (most of which are working in analogue mode). The design is based on the design used in our previous 39×48 array chip, which has been reported in [4]. Some details of the processing element design and readout architecture were presented in [5] and [6]. In this paper, several aspects of the chip implementation are elaborated. In particular, detailed schematic diagram of the analogue register circuit is presented, the layout of the processing element is discussed and the control/readout circuitry is overviewed.

2. CHIP ARCHITECTURE

The architecture of the chip is depicted in Figure 2. The pixel-parallel image processing capability is provided by a 128×128 array of analogue processing elements (APEs). Each APE is a simple processor, which includes nine registers (eight general-purpose ones and one used for exchanging data with nearest neighbours), a comparator with an activity-flag latch, and I/O circuits. Each APE also contains a photodetector circuit, so that a 128×128 image sensor array is embedded within the processor array. The array operates in SIMD

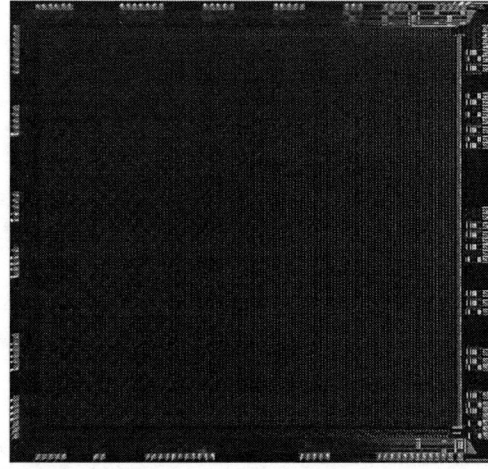

Fig.1. The SCAMP-3 chip

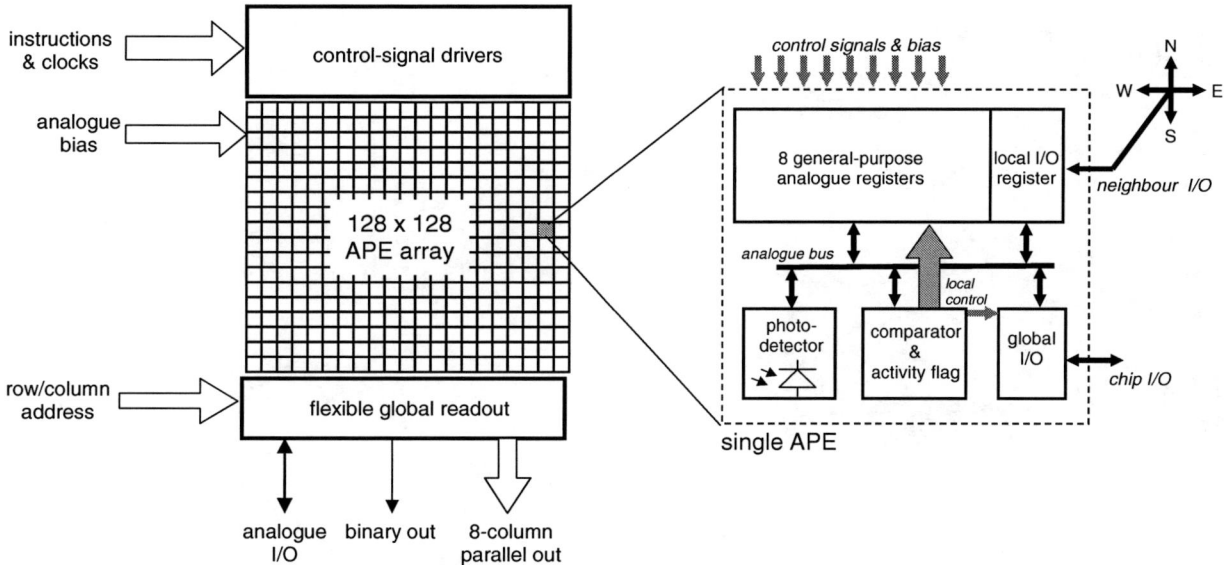

Fig.2. Architecture of the SCAMP-3 vision chip and its Analogue Processing Element (APE)

(Single Instruction Multiple Data) mode, the micro-instructions are issued by a single (external) controller and distributed to all APEs in the array via drivers located at the periphery of the array. The processing results are read-out from the array via flexible global readout circuitry, which enables pixel-addressing for analogue, binary and column parallel readout as well as global summation and logic OR operations.

3. IMPLEMENTATION

The most critical aspect of the overall chip design is the design of the processing element. This is where the constraints on the cell size and power consumption have to be carefully weighted against desired functionality, performance, and accuracy levels. Fortunately, small silicon area available to a single processor implies a simple cell structure, which makes a co-design of architecture, circuitry and layout of the device a manageable task. It has to be said, that when designing a general purpose processor, the trade-offs between speed, power, area and accuracy are resolved in a somewhat arbitrary way, guided to some extent by the envisaged applications: in this design we decided to minimize the size of the APE, so that reasonably high-resolution array (128×128) was feasible on a 50 mm^2 chip. At the same time, low power consumption was a priority (peak power below 250 mW per chip), and accuracy had to be kept at an acceptable level to perform low-level image processing. As a compromise, a moderate speed of operation, 1.25 MIPS (million instructions per second) per cell was achieved. Nevertheless, the massively parallel operation results in 20 GIPS (giga-instructions per second) per chip, which compares quite favourably with conventional digital signal processors. This speed is sufficient for performing, for example, 1250 operations per pixel at 1000 frames per second, which should be sufficient for majority of computer vision applications. On the other hand, when executing simpler algorithms, e.g. working at 20 frames per second with 200 instructions per pixel, the power consumption of the chip could be reduced to below one milliwatt. Therefore, the chip should be particularly suited to low-power requirements of battery-powered systems.

The APEs are implemented as "analogue micro-processors" [1], they execute software instructions, but achieve this using analogue circuits, and store data in a form of analogue sampled signals. Switched-current signal processing techniques are employed to achieve "ALU-free" design, i.e. no dedicated hardware exists to implement arithmetic operations. Instead, the basic operations of addition, inversion and division are executed directly in the registers/analogue bus system [5]. This results in significant silicon area savings and consequently also improves the accuracy of processing. Further discussion of benefits of this approach was presented in [7].

The layout of the APE is shown in Figure 3. The APE occupies silicon area of 49.35μm×49.35μm, most of this area is occupied by registers. The overall accuracy is ensured by a mixture of hardware and software techniques. The registers were designed as S^2I memory cells [8], as depicted in Figure 4. Large-area transistors M_{MEM} and M_{REF} are used to improve accuracy - using long and narrow transistors leads to a nominal biasing current of 1.7 μA with good voltage swing. The analogue power supply voltage is 2.5 V. The cells are laid out to minimize the capacitive coupling onto

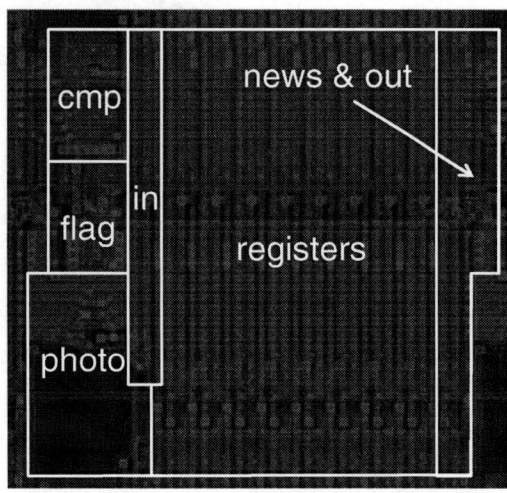

Fig.3. The APE layout. The cell size is 49.35μm×49.35μm

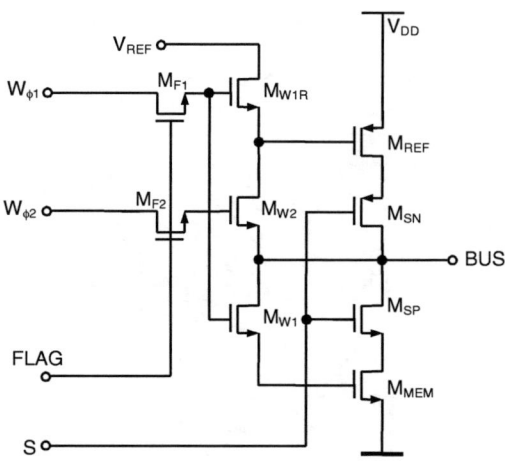

Fig.4. Schematic diagram of the register cell.

sensitive nodes from control signals (which are routed over the register area), and adjacent registers. As can be seen in Figure 3, each register extends for the full height of the APE cell. The register is connected to the analogue bus using transistors M_{SN} and M_{SP}. Consequently, the biasing current is switched off when register is not used, this minimizes power consumption. The storage is enabled by switches M_{W1} and M_{W1R} (operating during phase 1) and M_{W2} (operating during phase 2), according to the S^2I technique [8]. Conditional opening of these switches (controlled by a state of the local activity flag register) is implemented using transistors M_{F1} and M_{F2}.

The register associated with the neighbour communication is laid out on the right of the register bank (see Figure 3) to minimize the differences in errors when storing data in this register, as compared with other registers. It connects to analogue buses of the neighbours via four transistor switches (in North, East, West and South direction). The input circuit, placed to the left of the register bank, uses a circuit similar to the register cell, with transistors M_{MEM} and M_{REF}, and thus provides layout-level matching of the surrounding of the first register in the register bank.

The current comparator is implemented using simple voltage differential amplifier, with gate of one transistor in the pair connected to the analogue bus and the gate of the second transistor biased by a dc voltage (this voltage is equal to the nominal analogue bus voltage on phase 2 of the storage cycle, its value is close to V_{REF}). The comparator is followed by a latch, which serves as the activity flag register.

Photodetector circuit uses an n-type diffusion diode and can operate in integration mode, with close-to-linear characteristic, or in continuous-time log-compression mode. The pixel fill-factor is approximately 5.6%.

Control signals, analogue bias voltages and power supply lines have been routed over the APE area. The power supply and ground planes should ideally be used, but these have not been possible in the 3-layer metal technology. Instead, a power supply grid has been implemented. The registers have been shielded to prevent signal coupling from the digital control signals. The metal layers have been also used as an optical shield, so that only the photodiode area was exposed to light.

Apart from a careful design, to minimize errors related to clock-feedthrough, charge injection and output conductance of the S^2I cells and fixed-pattern noise errors caused by mismatch, a number of techniques (e.g. correlated double sampling, cancellation of signal-independent errors, algorithmic compensation of division error) can be implemented in software to improve the overall accuracy of processing.

The simple architecture, algorithmic error cancellation techniques, and careful circuit design result in a robust operation of the processor, which has been confirmed by our experiences with the previous 39×48 chip. Control signals are digital, except one analogue voltage that provides input value to the APE. No adjustments of instruction parameters are required – the programs which are designed in a simulator can be directly transferred to the hardware and execute as expected. The accuracy is, of course, limited, but we have successfully implemented linear and non-linear filters (e.g median filter, gray-level morphology), orientation-selective filters, edge detectors, motion detectors, active contours, hole filling, skeletons, models of autowave propagation in excitable medium, etc. The measurement results from the 128×128 chip are not yet available, but the experimental results from the 39×48 chip, with identical APE design, have been reported in [4] and it expected that the overall performance of this implementation will be similar to that of the previous design.

4. CONTROL AND READOUT

Each APE requires 40 control signals and 10 analogue voltages (bias and power supply). Digital control signals are derived from instruction-code-words that are provided to the chip from an external controller. The relatively low frequency of control signals (1.25 MHz) simplifies the design of the signal distribution network. Control signals are routed horizontally and vertically, with one buffer/driver for each row and column of the array. All analogue bias voltages are driving high-impedance nodes, providing either fixed dc bias to gates of transistors, or at most they are required to provide relatively small charge supplement at low frequency. The bias voltages are provided externally and routed globally.

The output interface has been designed to allow binary image readout at 80 Mpixels/second, via 8-bit column-parallel port. During binary read-out the state of the activity flag register is read. A grey-scale image read-out is designed to operate at 1 Mpixel/second (it is expected that in typical high-frame-rate applications reading-out of full grey-scale images will not be necessary). During grey-scale read-out the value of the output current from an analogue register is routed to the chip I/O pin, it is converted to a digital value using an external A/D converter.

A flexible global readout architecture [6] permits addressing groups of APEs in the array, to facilitate global operations. This is simply implemented using address decoder similar to the one shown in Figure 5. In a usual address decoder, the N-bits of the address bus are hard-wired to the inputs of the N-input 'AND' gate, corresponding to the physical address of the respective row/column, using either straight or inverted bit of the address signal A_i for each input i. If, instead of using inverted bits of the address A_i, an independent signal B_i is used, then the usual addressing is extended by the capability of an equivalent *don't care* bit in the address word, as illustrated in the table in Figure 5. Consequently, a number of rows/columns in the array can be addressed at the same time. In binary read-out mode, a logic OR operation is performed on the selected group. In analogue read-out mode, a summation of the currents from selected APEs is performed (high dynamic range of the output currents, resulting from the possibility of addressing one, many, or all 16k APEs at the same time, has to be handled by an external variable-gain amplifier). This simple scheme allows rapid calculation of global image descriptors (e.g. pixel counts, histograms), control if iterative procedures (e.g. loop until none of the pixels have changed in one iteration), multi-resolution read-out with pixel binning, extraction of object coordinates, etc.

5. CONCLUSIONS

The design of a 128×128 general-purpose pixel-per processor array vision chip has been presented in this paper. The chip has been fabricated in a 0.35µm technology. At the time of writing this paper the chip has not been tested yet. However, the scaled-down version of this chip, containing a 39×48 array, is operating successfully. Experimental results from the 128×128 chip are expected to be available for presentation at the conference in May 2005.

ACKNOWLEDGEMENT

This work has been supported by the EPSRC, under grant no. GR/R52688/01

REFERENCES

[1] P.Dudek and P.J.Hicks, "A CMOS General-Purpose Sampled-Data Analogue Processing Element", in IEEE Transactions on Circuits and Systems - II: Analog and Digital Signal Processing, vol. 47, no. 5, pp. 467-473, May 2000

[2] A.Dupret, J.O.Klein and A.Nshare, "A DSP-like analogue processing unit for smart image sensors", in Int. Journal of Circuit Theory and Applications, vol.30, pp.595-609, 2002.

[3] G. Liñán-Cembrano et al. "A Processing Element Architecture for High-Density Focal-Plane Analog Programmable Array Processor Arrays", Proc. ISCAS 2002, vol.III, pp.341-344, May 2002.

[4] P.Dudek, "A 39x48 General-Purpose Focal-Plane Processor Array Integrated Circuit", Proc. ISCAS 2004, vol.V, pp.449-452, Vancouver, May 2004

[5] P.Dudek, "A Processing Element for an Analogue SIMD Vision Chip", Proc. European Conference on Circuit Theory and Design, ECCTD'03, vol.III, pp.221-224, September 2003.

[6] P.Dudek, "A Flexible Global Readout Architecture for an Analogue SIMD Vision Chip", Proc. ISCAS 2003, Bangkok, vol.III, pp.782-785, May 2003.

[7] P.Dudek, "Accuracy and Efficiency of Grey-level Image Filtering on VLSI Cellular Processor Arrays", Proc. CNNA 2004, pp.123-128, Budapest, July 2004

[8] J. B. Hughes and K. W. Moulding, "S^{2}I: A Switched-Current Technique for High Performance", in Electronics Letters, vol.29, no.16, pp.1400-1401, August 1993.

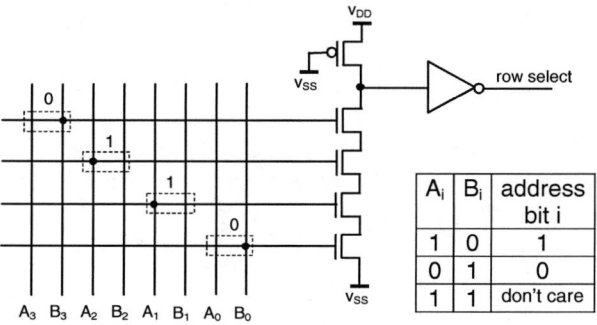

Fig.5. Address decoder. In this simplified example, a 4-bit decoder for row no.6 (i.e. "0110") is shown.

Dynamically Coupled Multi-Layer Mixed-Mode CNN

Mika Laiho, Ari Paasio
University of Turku
Microelectronics Laboratory
Lemminkaisenkatu 14-18
20540 Turku, Finland
Email: mlaiho@ecdl.hut.fi

Abstract— This paper shows a multi-layer cellular nonlinear network with feedback interconnections between layers. The multi-layer approach allows for imitation of some complex spatio-temporal visual information processing tasks that are motivated by recent biological studies. Mixed-mode architecture with discrete time integration is used, which makes it possible to process the data layer by layer and still achieve inter-layer feedback interaction. It is shown how the integration step can be programmed differently for each layer with the mixed-mode cell. Also, the in-cell analog-to-digital converter ADC is described and the selected multiplication scheme, the use of binary weighted multipliers, is described.

I. INTRODUCTION

The cellular nonlinear network (CNN) [1] theory offers a convenient model for describing many phenomena that rely on local interactions of regularly arranged data. A class of CNNs is the multilayer approach which can model complex spatiotemporal phenomena [2]. In a multilayer CNN, inter-layer feedback connections exist in addition to intra layer connections. The multi-layer structure can model e.g. some aspects of the way the retina performs pre-processing of visual data [3]. This has inspired a hardware realization of a continuous-time multi-layer analog CNN with 32x32 programmable cells [4]. A similar type of implementation is the 32x32x4 CNN chip in which two pairs of complementary layers represent two arrays of variables with both positive and negative values [5]. The chip is intended to model reaction-diffusion-type dynamics. Both of these realizations operate in continuous-time and in analog mode and the full parallel structure provides a high speed. A possible downside when implementing many layers is that each additional layer requires another set of multipliers (which require a lot of area). The approach of [6] is to use separate chips with dedicated functionalities operating in parallel. Through effective communication methods inter-chip feedback connections are made possible.

This paper proposes another candidate for consideration when choosing hardware for a multi-layer CNN. A multi-layer CNN, with feedback interconnections between the layers, is considered. A mixed-mode approach is followed, namely the multiplication and summing of the multiplication results (weighted summing) are done in analog domain, whereas integration and storage are digital. Since discrete-time integration is performed, it is possible to process the next value of each layer one after the other and update the data after all layers have been processed. With this time-multiplexed approach, the hardware used for intra- and inter-layer interconnections can be re-used for all layers, which saves in the number of multipliers. By increasing the amount of memory per cell, more layers can be processed. Reference [7] describes a mixed-mode polynomial CNN in which single-layer data was folded and processed with a 36-layer architecture.

The ideas of [7] are here applied to process multi-layer data. Especially, this paper shows the cell structure of the multi-layer mixed mode CNN and demonstrates how the CNN time constant can be made programmable so that each layer may integrate at a different pace. Also, the multiplication scheme is explained and the analog-to-digital converter structure is shown.

II. MULTI-LAYER MIXED-MODE CNN

A. Operating Principle

Figure 1 shows a multi-layer mixed-mode CNN cell at conceptual level. The sum of currents I_{sum} that enters the cell at the sum node describes the rate of change of cell state. I_{sum} is converted to digital using an analog-to-digital converter (ADC). Euler's integration method is used to integrate the digitized rate of change of cell state $r(n)$. The multipliers are composed of binary weighted current sources that are directly controlled with the digital cell state. Global weight signals provide the current references to the current sources. The digital cell state of the layer that is under processing, namely $x^{(CL)}(n)$, controls multipliers $x_1 - x_8$. Digital cell states $x^{(1)}(n) - x^{(L)}(n)$ (L is the number of layers) control the inter-layer multipliers $xi_1 - xi_L$. Register REG_B stores the cell input u and controls the corresponding multiplier x_B (all layers share the same cell input).

Since the full signal range (FSR) model [8] is used, the cell output equals the digitally limited cell state. The next state of layer CL, marked with $x^{(CL)}(n+1)$, is written to layer CL of register REG_A1. Register REG_A2 contains the current state data of the layers. After all layers of REG_A1 have been updated, their contents are moved to REG_A2. Two sets of registers are needed in the multi-layer cell core so that the state data of all layers stays synchronized. Here one vertical connection from each layer to the layer under processing was

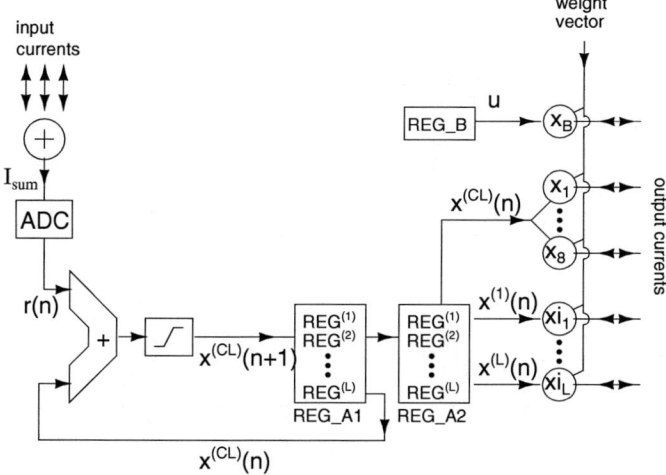

Fig. 1. Conceptual structure of a multi-layer mixed-mode CNN cell that uses Euler's integration method.

assumed. The bias current source (I-template) is not shown in the figure.

B. State Evolution

The sum of cell input currents (when processing layer CL) in i^{th} row and j^{th} column ($i \in [1, M]$, $j \in [1, N]$) is

$$I^{(CL)}_{sum,i,j}(n) = \sum_{k,l} \left[x^{(CL)}_{m,p}(n) \cdot A^{(CL)}_{k,l} \right] + \sum_{z=1}^{L} \left[x^{(z)}_{2,2}(n) \cdot Ai^{(CL)}_z \right] + u_{i,j} \cdot B^{(CL)} + I^{(CL)} \quad (1)$$

where $k, l \in [1, 3]$, $m = i + k - 2$ and $p = j + l - 2$. Furthermore, $A^{(CL)}$ defines the horizontal (intra layer) feedback interconnections of layer CL and $Ai^{(CL)}$ defines the inter layer interconnections when processing layer CL. Notice that $A^{(CL)}_{2,2}$ is equivalent to $Ai^{(CL)}_z$ when $z = CL$ (both of these are in-cell feedback weights). Therefore, either one can be kept zero (in Fig. 1 $A^{(CL)}_{2,2}$ is zero). $B^{(CL)}$ and $I^{(CL)}$ are the B-templates and the bias templates of layer CL. The input current of (1) is digitized to $r^{(CL)}_{i,j}(n)$. This digital rate of change of state is used to define the next state of a layer according to

$$x^{(CL)}_{i,j}(n+1) = h^{(CL)} \cdot r^{(CL)}_{i,j}(n) \quad (2)$$

where $h^{(CL)}$ is the layer-dependent integration step defined as

$$h^{(CL)} = \frac{1}{k^{(CL)}_A \cdot 2^{k^{(CL)}_D}}, \quad (3)$$

where $k^{(CL)}_A$ is a programmable scaling factor (real number, physically it is defined by the reference current of the ADC) and $k^{(CL)}_D$ is an integer that defines how many bit positions $r_{i,j}(n)$ is shifted towards lsb before it is fed to the integrator. Therefore, the integration step does not need to be an integer power of two and can be programmed differently for different templates/layers.

C. The Digitizing Process

The digitizing of $I^{(CL)}_{sum,i,j}(n)$ with a limited number (b_A) of bits should be done so that the loss of significant information is minimized. The absolute value of the maximum of the sum current, namely I_{max}, is upper bounded by the sum of all weights. The digitized rate of change of state is represented with b_A bits so that $r^{(CL)}_{i,j}(n) \in]-2^{b_A-1}-1, 2^{b_A-1}-1[$ (in sign and magnitude form). The maximum value is assigned to $r^{(CL)}_{i,j}(n)$ according to

$$r^{(CL)}_{i,j}(n) = 2^{b_A-1} - 1 \text{ if } I^{(CL)}_{sum,i,j}(n) \geq k_A \cdot I_{unit}, \quad (4)$$

where k_A together with I_{unit} (absolute value of current produced with unity weight and saturated state) defines the maximum rate of change of state that is represented digitally. Similarly, the minimum value of $r^{(CL)}_{i,j}(n)$ saturates to

$$r^{(CL)}_{i,j}(n) = -2^{b_A-1} - 1 \text{ if } I^{(CL)}_{sum,i,j}(n) \leq -k_A \cdot I_{unit}. \quad (5)$$

Therefore, if $k_A \cdot I_{unit} < I_{max}$ the maximum amplitude of the digital rate of change of state is susceptible to being bounded by the digitizing process. Also, since

$$r^{(CL)}_{i,j}(n) = 0 \text{ if } \left| I^{(CL)}_{sum,i,j}(n) \right| < \frac{k_A \cdot I_{unit}}{2^{b_A}}, \quad (6)$$

the choice of k_A is a tradeoff between representing small currents accurately and upper bounding the digital rate of change of state. Therefore, care must be taken when using k_A for programming the integration step.

In the following of this paper, the number of bits in the ADC, digital integrator and the multipliers (DACs) are $b_A = 8$, $b_i = 12$ and $b_D = 7$, respectively. Therefore, k_D can be assigned values 0, 1, 2, 3 and 4. Consequently, the integration step with $k_D = 0$ is 16 times shorter than with $k_D = 4$ assuming a constant k_A. More tuning range for the integration step is available with k_A. The integrator uses 12 bits to represent the state because it has to be able to count with $b_A + k_D$-bit data. When the 8-bit $r^{(CL)}_{i,j}(n)$ is added to the 12-bit $x^{(CL)}_{i,j}(n)$, the empty bits are replaced by zeros for positive numbers and ones for negative numbers since the integrator computes with two's complement numbers.

III. CELL STRUCTURE

The mixed-mode CNN realization of [7] was relatively complex because polynomial terms (linear, quadratic and cubic A-templates) were included, a second order integration method was used (Heun's method) and autozeroing was necessary before each iteration. Since the proposed multi-layer cell uses binary weighted current sources for multiplication, no autozeroing is needed. Euler's integration method was chosen to keep the cell as simple as possible.

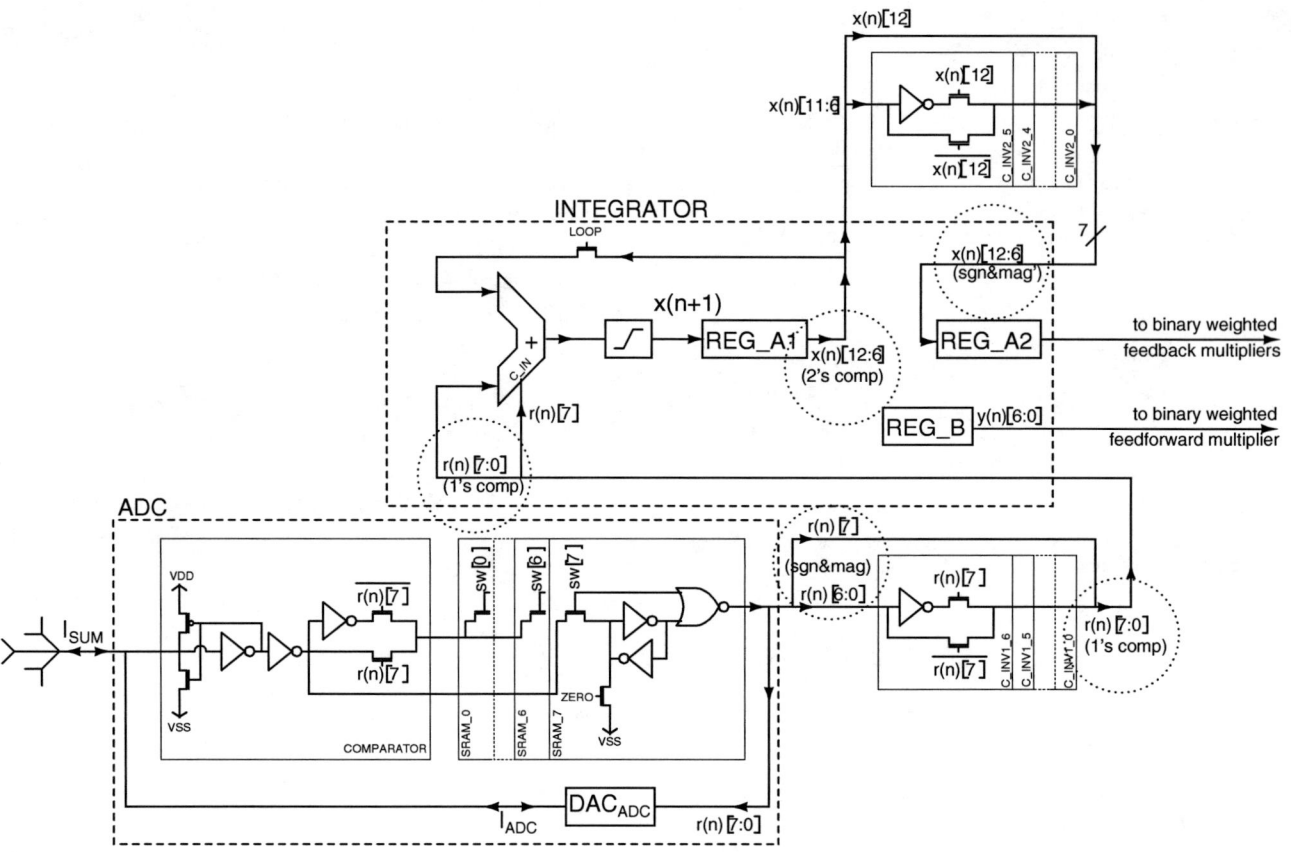

Fig. 2. Mixed-mode cell design (multipliers are not shown).

A. Data and Number Conversions

Figure 2 presents another view of the proposed multi-layer mixed-mode cell. The ADC, a current mode successive approximation register (SAR) [9], is modified from that of [7] so that the autozeroing circuitry is not included. The ADC is composed of a comparator, SRAMS 7-0 and binary weighted current sources (DAC_{ADC} in the figure). The outputs of the SRAMs are determined by a NOR function with $sw[7:0]$. Starting with $sw[7]$, the conversion proceeds bit by bit towards lsb as control signals corresponding to lower bits activate [7].

Between the ADC and integrator, a conversion from sign and magnitude to 1's complement is made with C_INV1_6 - C_INV1_0. The integrator is composed of a 13-bit full adder (in order to be able to compute with 12-bit inputs), a digital nonlinear circuit and two registers. The digital nonlinear block truncates the output of the adder to twelve bits. The carry in bit of the lsb of the adder (C_IN) is controlled by $r(n)[7]$ which effectively turns $r(n)[7:0]$ to two's complement. Registers REG_A1 and REG_A2 have L layers as shown in Fig. 1 but the layers are not shown here to simplify the picture. Control signal $LOOP$ is made LO when REG_A1 is written by the nonlinear block in order to prevent a feedback loop from taking place. REG_A1 stores 12-bit words, while REG_A2 can cope with storing 7-bit words since it controls the 7-bit multipliers.

B. Multipliers

The multipliers are binary weighted current sources. Therefore, the cell output $x(n)[12:6]$ can be used to directly control the current sources. Consequently, no separate DAC is needed. The output of a multiplier is conditionally mirrored by taking an XOR function of $x(n)[12]$ and the msb of the weight. The output of the integrator is in 2's complement form. The conversion to sign and magnitude is made using C_INV2_6 - C_INV2_0 and by having an extra lsb bit in each multiplier that is controlled by $x(n)[12]$.

Figure 3 shows the binary weighted current sources of the selected multiplier. The output current of this block goes either directly or through an NMOS current mirror to the sum node of a cell (the mirror and the switches are not shown in the figure. The largest current source ($8x$) is composed of eight 0.32/3 sized transistors in parallel and the smallest current source $0.25x$ has 4 of these transistors in series. The layout of the structure of Fig. 3 was drawn in an area of 20x7um^2 with ST 0.18um, 6-metal CMOS process. By optimizing the layout and integrating the multipliers tightly together, the area could be significantly reduced.

The circuit of Fig. 3 was Monte Carlo simulated with Eldo on level 59 parameters. All the magnitude control bits $x(n)[11:6]$ were kept LO so that all the current sources contributed to the total current. The upper part of Fig. 4

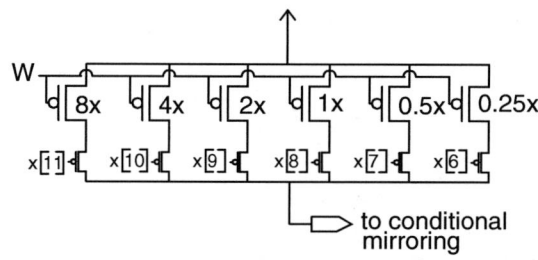

Fig. 3. Binary weighted current sources.

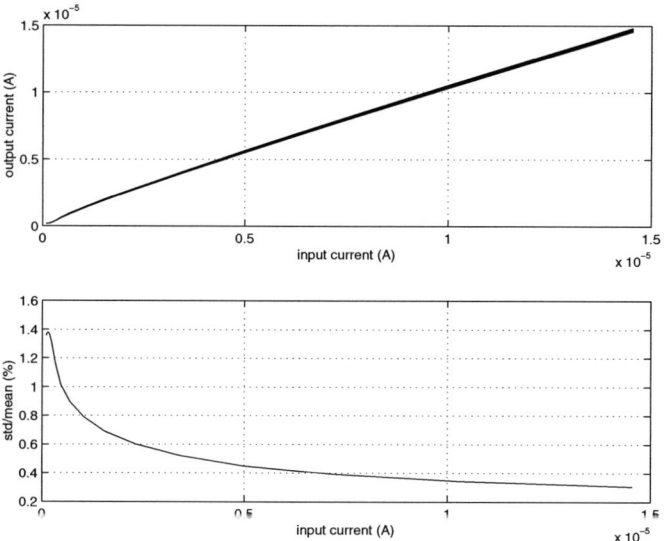

Fig. 4. Monte Carlo simulation of multiplier output current and the standard deviation divided by the mean of the output current.

shows the output currents of 100 Monte Carlo runs while sweeping the reference current value in transient simulation. The lower part of the figure shows how the standard deviation divided by the mean of the Monte Carlo runs varies with the reference current value. Since mismatch of currents relative to the current level is worse with small currents, large coefficient values should be favored in the algorithms.

IV. COMPUTING SPEED

The computing speed of the mixed-mode CNN is limited by the speed of the ADC. The conversion speed depends e.g. on the converter type, input current level and resolution. The speed of the ADC shown in Figure 2 is limited by the speed of the comparator. The comparator used has a small input voltage swing [10], which makes it suitable for high speed operation. An input current that equals half of the lsb current has to be able to drive the input node from HI to LO (or LO to HI) in a time that is reserved for determining one bit. The parasitic capacitances of the multipliers and comparator determine the maximum slew rate that is possible with a given lsb current. The comparator of Figure 2 was simulated with Eldo (with level 59 parameters of a 0.18μm CMOS process) so that the total width of multiplier transistors connected to the input node was 65μm. The width (multiplied by overlap capacitance) determines the capacitance since the current source multipliers are in saturation. The comparison speed with $I_{unit} = 5\mu A$, $k_A = 4$ and $b_A = 8$ was 50ns which corresponds to a total eight-bit conversion time of 400ns. By allowing 100ns for analog computation and digital integration, the processing of one layer would take 500ns. However, due to parasitic capacitances of the contacts and wires that are connected to the input node, the real comparison speed is somewhat slower.

V. CONCLUSIONS

This paper presented a multi-layer cellular nonlinear network with feedback interconnections between layers. The approach is a candidate for building hardware for imitation of some complex spatio-temporal visual information processing tasks that are motivated by recent biological studies. The proposed design uses a mixed-mode architecture with discrete time integration, which makes it possible to process the data layer by layer and still achieve inter-layer feedback interaction. It was shown how the integration step can be programmed differently for each layer and the cell description was complemented with selected simulations.

ACKNOWLEDGMENTS

This work was funded by the Academy of Finland (project 106451).

REFERENCES

[1] L. O. Chua, L. Yang, 'Cellular Neural Networks: Theory', IEEE Transactions on Circuits and Systems, Vol. 35, pp. 1257-1272, 1988.
[2] Cs. Rekeczky, T. Serrano-Gotarredona, T. Roska, A. Rodriguez-Vazquez, "A Stored Program 2nd Order/3-Layer Complex Cell CNN-UM", 6th IEEE International Workshop on Cellular Neural Networks and Their Applications, Catania, Italy, pp. 213- 217, 2000.
[3] Cs. Rekeczky, B. Roska, Erik Nemeth, Frank Werblin, "Neuromorphic CNN Models for Spatio-Temporal Effects Measured in the Inner and Outer Retina of Tiger Salamander", 6th IEEE International Workshop on Cellular Neural Networks and Their Applications, Catania, Italy, pp. 15- 20, 2000.
[4] R. Carmona, et al., "Second-Order Neural Core for Bioinspired Focal-Plane Dynamic Image Processing in CMOS", IEEE Trans. on Circuits and Systems-I, Vol. 51, No. 4, pp. 913-925, 2004.
[5] B. Shi, T. Luo, "Spatial Pattern Formation via Reaction-Diffusion Dynamics in 32x32x4 CNN Chip", IEEE Trans. on Circuits and Systems-I, Vol. 51, No. 4, pp. 939-947, 2004.
[6] B. Shi, "A CNN Model of Multi-dimensional Stimulus Selectivity in Primary Visual Cortex", International Joint Conference on Neural Networks, Budapest, pp. 1741-1747, 2004.
[7] M. Laiho, A. Paasio, A. Kananen, K. Halonen, 'A Mixed-Mode Polynomial Cellular Array Processor Hardware Realization', IEEE Trans. on Circuits and Systems-I, Vol. 51, No. 2, pp. 286-297, 2004.
[8] S. Espejo, R. Carmona, R. Dominguez-Castro, A. Rodriguez-Vazquez, "A VLSI-oriented Continuous-time CNN Model", International Journal of Circuit Theory and Applications, Vol. 24, no. 3, pp. 341-356, 1996.
[9] K. Chen, C. Svensson, J-R. Yuan, 'A CMOS Implementation of a Video-Rate Successive Approximation A/D converter', Proceedings of the International Symposium on Circuits and Systems, Espoo, Finland, pp. 2577-2580, 1988.
[10] A. Rodriguez-Vazquez, et.al., 'High Resolution CMOS Current Comparators: Design and Applications to Current-Mode Function Generation', Analog Integrated Circuits and Signal Processing, No. 7, pp. 149-165, 1995.

Spatiotemporal pattern formation in the ACE16k CNN Chip

Müştak E. Yalçın
Istanbul Technical University
Faculty of Electrical and Electronic Eng.
80626, Maslak, İstanbul, Turkey
Email:mey@ieee.org

Johan A.K. Suykens, Joos Vandewalle
Katholieke Universiteit Leuven
Department of Electrical Eng., SCD/SISTA
Kasteelpark Arenberg 10,
B-3001, Leuven, Belgium
Email:Johan.Suykens@esat.kuleuven.ac.be

Abstract—In this paper, pattern formation occurring on the ACE16k CNN chip is presented. The CNN chip can be programmed with a cloning template in order to generate spiral waves and autowaves. The waves diffract from the *internal sources* which cannot be relocated on the network. However, by using initial and/or input images, sources (*external sources*) can be located at any place on the network. Furthermore a competition between autowaves generated by external and internal sources is observed. Propagation of autowaves on the inhomogeneous CNN array, formed by the fixed-state map, is presented.

I. INTRODUCTION

Autowaves are a particular class of nonlinear waves resulting from strongly nonlinear active media [1]. The fundamental properties of autowaves are

- the shape and amplitude of autowaves remain constant during the propagation,
- they do not reflect at the medium boundaries,
- two colliding autowaves annihilate.

The CNN framework provides a very useful tool for studying spatial-temporal pattern formation [2], [3], [4]. Autowaves are well described by reaction-diffusion equations. As a result the reaction-diffusion CNN model has been widely used to generate such waves. In [1], Munuzuri *et al.* have presented a review on the study of spatial-temporal behaviors on an array of Chua's circuits, where Chua's circuit is used as the reaction term in the model. In [3], Manganaro *et al.* have used a second-order system for the reaction term and have observed autowaves, spiral waves and Turing patterns. In fact, the model used by Manganaro *et al.* [3] can be thought of as a two layer CNN with a Chua-Yang model. In [5], a VLSI implementation of a two layer CNN with Full-range CNN model (CACE1k) has been experimentally verified spatial-temporal phenomena on CNNs. An experimental observation of autowaves on a ACE16k chip is shown here. The first experimental observation of these phenomena on a ACE16k chip has been given by Yalçın *et al.* in [6] and [7].

This paper is organized as follows. Section 2 shortly describes the ACE16k chip. Section 3 presents pattern formation on an ACE16k chip. Section 4 presents propagation of autowaves on the inhomogeneous CNN array which is formed by the fixed-state map.

II. ACE16 CNN CHIP

Analogic Cellular Engines (ACEs) are designed based on the CNN-Universal Machine (CNN-UM) architecture. They are capable of realizing a very large variety of image related spatial-temporal operations and algorithms through the execution of a suitable sequence of instructions (or templates). The ACE16k chip is the third generation of ACE chips and it consists of an array of 128×128 identical analog Full-range CNN cells. Complex dynamic behaviors in Full-range model have been studied in [8] and a strange attractor similar to the chaotic attractor observed in a CNN with Chua-Yang model has been observed. An experimental observation of similar chaotic attractors in ACE4k [9] has been reported for an asymmetric template class in [10]. For information on the ACE16k chip, we refer to [11] and [12].

III. EXPERIMENTS

For the discussed experiments, the vector of the recommended settings of the internal references (<hwparams>) is set to [0 0 0 229 200 172 145 109 78 37 23] [13]. The internal references include optical, weight and signal references which are setting currents and voltages for the specific blocks in the ACE16k. The current and bias which define the threshold z are set to 0. Furthermore, a fixed boundary condition with 0 was used during our experiments.

A. Experimental results: Autowaves

In the first experiment, the following templates

$$A = \begin{bmatrix} 0 & 0 & 0 \\ 0 & -3 & 0 \\ 0 & 0 & 0 \end{bmatrix}, B = \begin{bmatrix} 0 & 0 & 0 \\ 0 & -3 & 0 \\ 0 & 0 & 0 \end{bmatrix} \quad (1)$$

were chosen and a full white image was used for initial and for input. Figure 1 shows several snapshots depicting the obtained autowave during the time evolution for the cloning template (1). There are four wave sources in Figure 1 (a) which are located at the corners. These are called *internal sources* because it has been observed that these sources stay active during the different experiments. The waves in Figure 1 (a) show the fundamental properties of autowaves: two waves spreading in opposite directions do not pass each other (as is usual for the classical conservative waves) but mutually annihilate. Also the shape and amplitude of the waves remain constant during propagation and the waves do not reflect at the boundaries of the network.

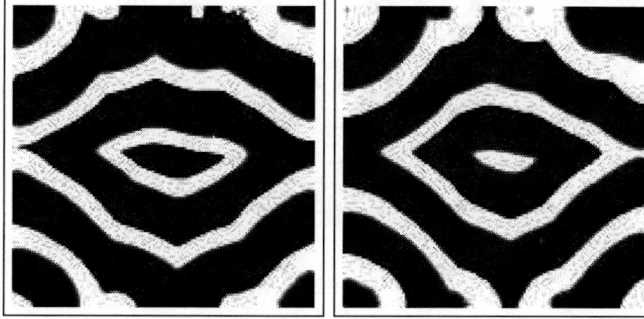

Fig. 1. (a-b) The time evolution for the cloning template (1). Each snapshot has been obtained by executing the same template with the same input and initial state but running time was increased as much as time stamp between the snapshots. The waves diffract form the corners. The shape and amplitude of the waves remain constant during propagation. There are no interferences between waves but the colliding waves annihilate. Furthermore the waves do not reflect from the boundaries of the network.

In a second experiment, it is shown that new autowave sources which is called here as *external sources* can be located in any place on the network. A source can be located with black dots on the initial and input images. In Figure 2(a) (the box around the figure is not a part of the image) a source is located on the left-bottom corner. Two consecutive snapshots depicting the obtained autowave in time. One can clearly see two autowave sources in Figure 2(b). While one of them is the *internal source* of the network, the other is located where the black dot is placed on the initial and input images. This experiment has been also repeated for four sources which are placed at the corners (see Figure 2(d)). The result is shown in Figure 2(e) and a consecutive frame is shown in Figure 2(f). Figures 2(e) and (f) clearly shows that the waves diffract from the defined sources. They interact approximately in the middle of the network and then annihilate each others.

Figure 3 shows the competition [1] between the autowaves which are generated by four internal sources and one external source. Initially the waves from the external source and internal sources start interacting and then annihilate (see Figure 3(a)). The region where they interact and annihilate is drifting towards the right-top corner of the network. Figures 3(a-f) show successive snapshots on the network during the competition. A movie file for whole time evolution can be seen in http://www.esat.kuleuven.ac.be/~mey/NLab/autowaves/test2.avi. This competition because of the autowaves from internal sources emits with a frequency higher than the frequency of the autowaves from the external sources.

B. Experimental results: Spiral waves

In a next experiment, we use the following templates

$$A = \begin{bmatrix} 0 & 0 & 0 \\ 0 & -3 & 0 \\ 0 & 0 & 0 \end{bmatrix}, B = \begin{bmatrix} 0 & 0 & 0 \\ 0 & 0 & 0 \\ 0 & 0 & 0 \end{bmatrix} \quad (2)$$

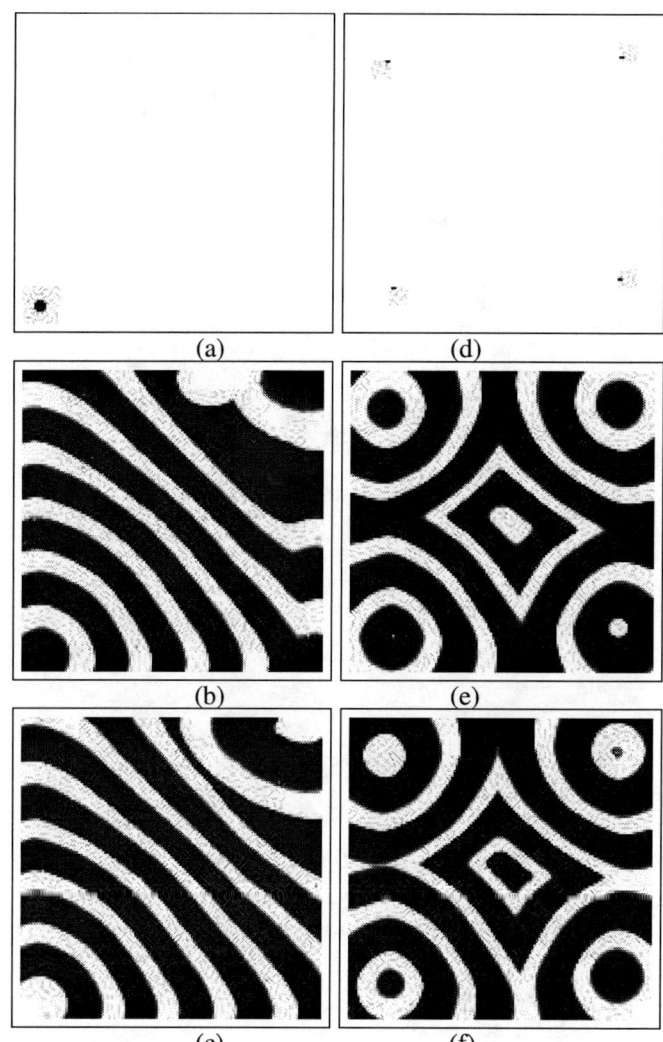

Fig. 2. (a,d) Initial and input image (the box around the figure is not a part of the image). (b-c) and (e-f) evolutions in time for the cloning template (1) corresponding to the initial and input images from (a) and (d), respectively.

which has the same feedback template as the previous experiment. However it includes a zero central element in the control template. Figure 4(a) shows the obtained result with a full white image as initial condition. The network has two internal spiral wave sources at the left and right side of the image. Using the given initial conditions in Figures 4(b) and 4(c), external sources for spiral wave can be located on the network as shown in Figure 4(d) and (e).

IV. PROPAGATION OF AUTOWAVES ON THE INHOMOGENEOUS CNN ARRAYS

The fixed-state map of a CNN is a binary image *e.g.* Figure 5(c-e), specifies which CNN cells are in an active or inactive state for all time. The state variables of these cells are frozen to fixed values and do not change in time. Therefore, the fixed state option offers an inhomogeneous structure for the CNN array. The ACE16k CNN chips allow this fixed-state map.

In our first experiment, the network has been divided into two sub-networks using the fixed-state which is given in

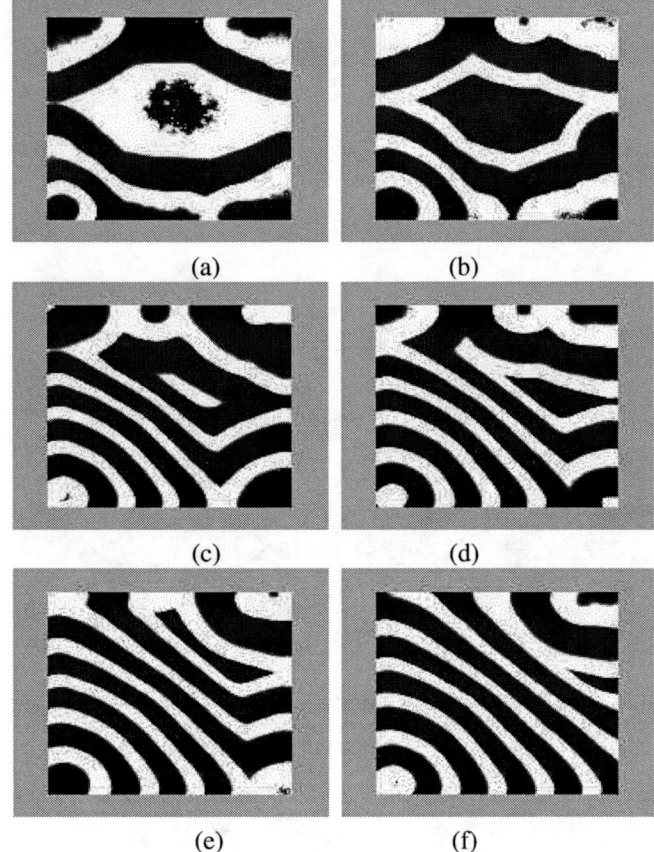

Fig. 3. Competition between autowaves resulting from the chip-internal sources and an autowave resulting from an externally imposed source. The interaction region of the autowaves moves to the right-top corner of the network.

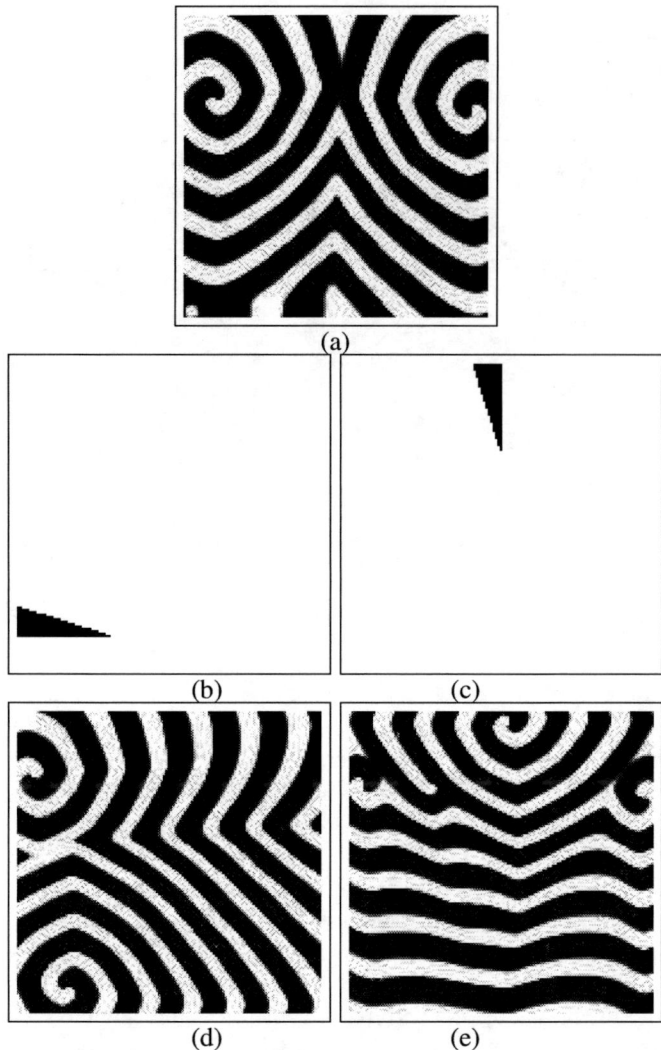

Fig. 4. (a) Spiral wave obtained on the ACE16k chip for a full white image as initial image. (b)(c) two different initial images. (d)(e) spiral waves obtained for the initial states (a) and (b), respectively.

Figure 5(c). In order to see the propagation of autowaves on these two sub-networks, Figure 5(a) has been chosen as an initial and input images and template (1) has been used. As expected, one external autowave source is located on the each sub-networks. There is also no interference between the autowaves propagated from external sources because of the isolation by the fixed-state (see Figure 5(f)). However there is an interaction between the waves propagated from the external source and the internal sources on the same sub-network which is at the right hand side of the network.

In a second experiment, Figure 5(d) has been used as a fixed-state map and Figure 5(b) has been taked as an initial and input images. In this experiment we have observed that the waves propagated from the external source can pass through the opening on the array (see Figure 5(g)). Experimentally it has been verified that if there are at least three cells active (or not fixed), waves can pass through the opening.

The experiment has been repeated with the fixed-state for Figure 5(e). Figure 5(h) shows that the waves propagate to the right on the inhomogeneous CNN arrays. The waves from the internal source propagate from the opposite direction. We can also see a competition between external and internal sources in the inhomogeneous network.

V. Conclusions

The erroneous behavior observed in VLSI implementations of CNN causes unavoidable and undesirable features which make CNN chips loose reliability for certain template operations [14]. Here we have taken advantage of these undesirable features in order to generate complex phenomena on the VLSI implementation of CNN *i.e.* ACE16k. Although the chip has internal spiral and autowave sources which cannot be relocated on the array, we have shown that external sources can be defined and relocated anywhere on the array. Furthermore we have presented propagation of autowaves on the inhomogeneous CNN array formed by the fixed-state map. These phenomena can be further employed towards applications such as e.g. path finding in labyrinths.

Acknowledgment

This research work was carried out at the ESAT laboratory and the Interdisciplinary Center of Neural Networks ICNN

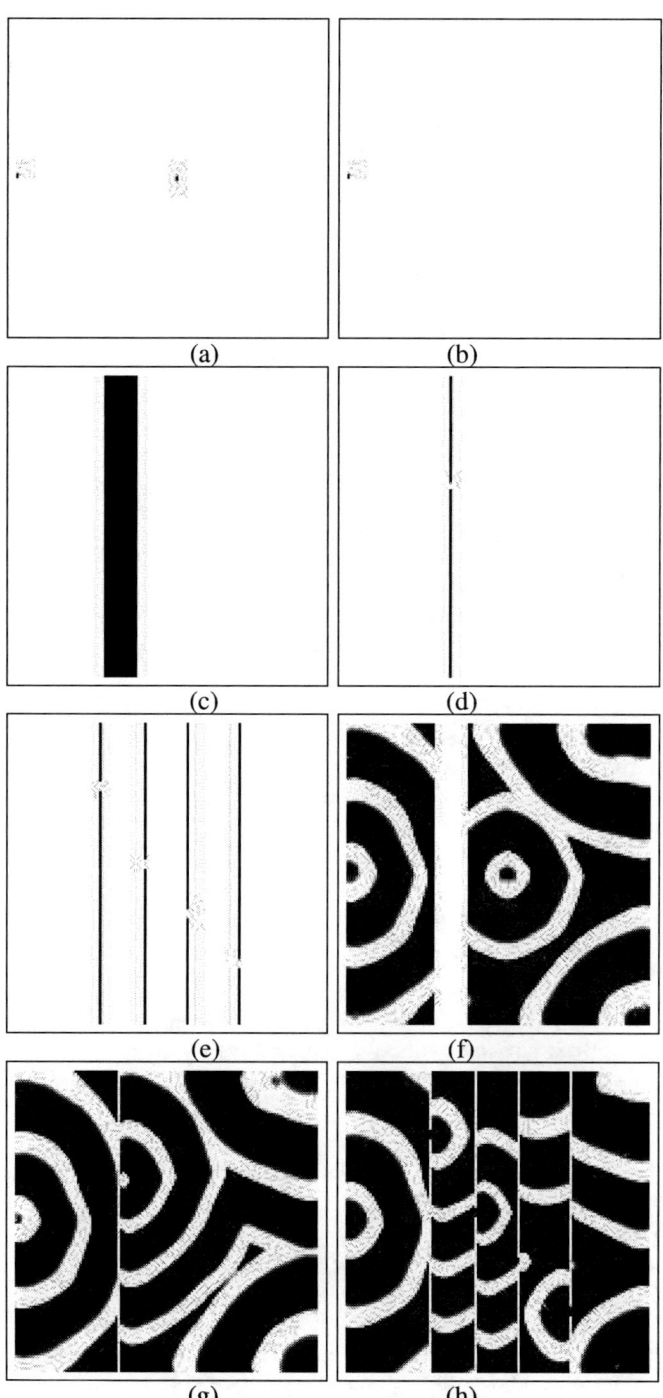

Fig. 5. (a,b) Images which are used as initial and input images. (c-e) examples of different fixed-state maps. The black regions have state variable values that are fixed in time. (f-h) Propagation of autowaves on the CNN corresponding fixed-state (c-e) using the same initial and input images of (a) for (f) and (b) for (g-h), respectively.

Collective Behavior and Optimization: an Interdisciplinary Approach.

REFERENCES

[1] A. P. Munuzuri, V. P. Munuzuri, M. G. Gesteria, L. O. Chua, and V. P. Villar, "Spatiotemporal structures in discretely-coupled arrays of nonlinear circuits: A review," *Int. J. Bifurcation and Chaos*, vol. 5, no. 1, pp. 17–50, 1995.

[2] L. O. Chua, *CNN: a Paradigm for Complexity*, World Scientific, Singapore, 1998.

[3] G. Manganaro, P. Arena, and L. Fortuna, *Cellular Neural Networks Chaos, Complexity and VLSI Processing*, Springer-Verlag, Berlin, Heidelberg, 1999.

[4] L. O. Chua, M. Hasler, G.S. Moschytz, and J. Neirynck, "Autonomous cellular neural networks: a unified paradigm for pattern formation and active wave propagation," *IEEE Trans. Circuits and Systems-I*, vol. 42, no. 10, pp. 559–577, 1995.

[5] R. Carmona, F. Jimenez-Garrido, R. Dominguez-Castro, S. Espejo, T. Roska, C. Rekecky, I. Petras, and A. Rodriguez-Vazquez, "A bio-inspired 2-layer mixed-signal mixed-signal flexible programmable chip for early vision," *IEEE Trans. Neural Networks*, vol. 14, no. 5, pp. 1313–1336, 2003.

[6] M. E. Yalçın, J. A. K. Suykens, and J. Vandewalle, "Experimental observation of autowaves on the ACE16k CNN chip," in *Proceedings of the 8th IEEE International Workshop on Cellular Neural Networks and their Applications*, Budapest, Hungary, July 2004, pp. 172–177.

[7] M. E. Yalçın, J. A. K. Suykens, and J. Vandewalle, *Cellular neural networks, multi-scroll chaos and synchronization*, World Scientific, Singapore, 2005, to appear.

[8] M. Biey, M. Gilli, and P. Checco, "Complex dynamic phenomena in space-invariant cellular neural networks," *IEEE Trans. Circuits and Systems-I*, vol. 49, no. 3, pp. 340–345, 2002.

[9] G. Linan, S. Espejo, R. Dominguez-Castro, and A. Rodriguez-Vazquez, "ACE4k: an analog I/O 64x64 visual microprocessor chip with 7-bit analog accuracy," *Int. J. Circuit Theory and Applications*, vol. 30, pp. 89–116, 2002.

[10] I. Petras, T. Roska, and L. O. Chua, "New spatial-temporal patterns and the first programmable on-chip bifurcation test bed," *IEEE Trans. Circuits and Systems-I*, vol. 50, no. 5, pp. 619–633, 2003.

[11] A. Rodriguez-Vazquez, G. Linan-Cembrano, L. Carranza, E. Roca-Moreno, R. Carmona-Galan, F. Jimenez-Garrido, R. Dominguez-Castro, and S. E. Meana, "ACE16k: the third generation of mixed-signal SIMD-CNN ACE chips toward VSoCs," *IEEE Trans. Circuits and Systems-I*, vol. 51, no. 5, pp. 851–863, 2004.

[12] G. Linan, S. Espejo, R. Dominguez-Castro, and A. Rodriguez-Vazquez, "Architectural and basic circuit considerations for a flexible 128 × 128 mixed-signal SIMD vision chip," *Analog Integrated Circuits and Signal Processing*, vol. 33, pp. 179–190, Nov. 2002.

[13] Analogic Computers Ltd., Budapest, *Aladdin Professional: hardware manual*, version 3.0 edition, 2003.

[14] S. Xavier de Souza, M. E. Yalçın, J. A. K. Suykens, and J. Vandewalle, "Toward CNN chip-specific robustness," *IEEE Trans. Circuits and Systems-I*, vol. 51, no. 5, pp. 892–902, 2004.

of the Katholieke Universiteit Leuven, in the framework of the Belgian Programme on Interuniversity Poles of Attraction, initiated by the Belgian State, Prime Minister's Office for Science, Technology and Culture (IUAP P4-02, IUAP P4-24, IUAP-V), the Concerted Action Project MEFISTO of the Flemish Community and the FWO project G.0080.01

CNN WAVE BASED COMPUTATION FOR ROBOT NAVIGATION ON ACE16K

P. Arena, L. Fortuna, M. Frasca, G. Vagliasindi
DIEES – Dipartimento di Ingegneria Elettrica, Elettronica e dei Sistemi
Facoltà di Ingegneria - Università degli Studi di Catania
v.le A. Doria 6, 95125 Catania, Italy
gvaglia@diees.unict.it

A. Basile
Automation and Robotics Team
STMicroelettronics
Stradale Primosole, 50, 95121 Catania, Italy
adriano.basile@st.com

M. E. Yalçın
Faculty of Electrical and Electronic Eng
Istanbul Technical University
80626, Maslak, Istanbul, Turkey
mey@ieee.org

J. A. K. Suykens
Department of Electrical Engineering, ESAT-SCD-SISTA
Katholieke Univesiteit Leuven
Kasteelpark Arenberg 10, B-3001, Heverlee (Leuven), Belgium

Abstract—**The CNN wave based computation is an approach for real time robot navigation in a complex environment based on the idea of considering the environment in which the robot moves as an excitable medium. Obstacles represent the sources of autowave generation. The waveform propagating in the CNN medium provide to the robot all the information to achieve an adaptive motion avoiding the obstacles. In this paper we implement entirely this strategy on the ACE16K CNN-chip.**

I. INTRODUCTION

Robot navigation in complex and dynamically changing environment is an actual and challenging problem of robotics. A very common approach for the robot navigation problem is the use of artificial potential fields [1], where the whole experimental arena is mapped into the computational architecture of the robot navigator: obstacles generate repulsive fields and target attractive fields. However the computational resources required by this approach are quite large and therefore realtime adaptation to moving obstacles is difficult to achieve. In [2], [3] a new approach for robot navigation is introduced. It is based on the application of wave computation implemented on Cellular Nonlinear Networks (CNNs) [4] in order to generate autowave fronts that, interacting with the robot, provide the information to achieve an adaptive motion to avoid obstacles and reach the target. In the approach described in [3] the whole robot arena is mapped on a CNN, in order to exploit its parallel and real-time processing capability to the identification of the obstacles and the target in the arena. Subsequently a Reaction-Diffusion (RD) CNN generates autowave fronts. This approach was targeted for the ACE4K (Analogic Cellular Engine), used for the obstacles-target recognition, and for the RD-CNN chip CACE1K (Complex Analogic Computing Engine), for the autowave generation. Although the methodology has revealed really useful, its hardware implementation needed the design of a complicated software-hardware routines for optimize the interface between the ACE4K and the CACE1K. Recently an experimental observation of autowaves on the ACE16K [5] has opened the way to a new compact implementation of the wavebased approach using the same platform both for implementing the complex image filtering routines, needed for processing the robot arena, and for propagating the autowaves used for the generation of the adaptive trajectories. In this way an on-board implementation is more easily achievable, leading to the realization of a fully autonomous robot with the capability of adapting its trajectories in real-time to a dynamically changing environment.

II. AUTOWAVES ON ACE16K CNN CHIP

The first experimental observation of autowaves on ACE16K has been given in [5]. Autowaves are a particular class of nonlinear waves and they result from strongly nonlinear active media [6]. The shape and amplitude of autowaves remain constant during the propagation and they do not reflect from the boundaries of the network. Furthermore two colliding autowaves annihilate rather than penetrate one another. The ACE16K elementary cell is depicted in Fig. 1 [7]. This circuitry includes only two functional blocks, which are multipliers and processing blocks, of eight functional blocks. These two blocks basically define the dynamics of the cell which can be programmed using the switches (see Fig. 1), $w_{0,0}^{(1)}$, $w_{0,0}^{(2)}$, $w_{0,0}^{(3)}$ and the synaptic law of the model which is given by

$$I_{i,j}^s = \sum_{k,l \in S_{i,j}} w_{k-i,l-j} v_{C1_{k,l}} \qquad (1)$$

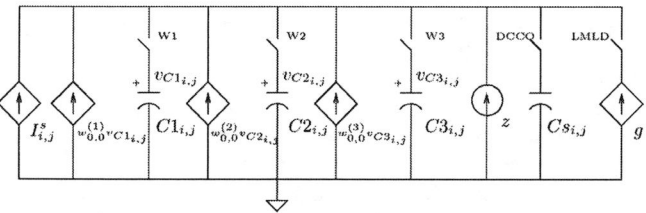

Figure 1. A cell circuitry of the ACE16K chip

with sphere of influence $S_{i,j}(1)$ where

$$S_{i,j}(1)=\{C_{k,l}:\max(|k-i|,|l-j|)\leq 1, 1\leq k,l \leq 128, k\neq i, l\neq j\}.$$

In [1], as in this application, the ACE16K chip is hosted by the Aladdin Professional [8] system which provides user friendly tools and hardware interfacing elements. In the Aladdin system the templates are defined in the template file. For ACE16K the template file configures the switches $w_{0,0}^{(1)}$, $w_{0,0}^{(2)}$, $w_{0,0}^{(3)}$ and the synaptic law according to the execution mode. The execution mode is chosen by the cloning template structures and ten modes have been defined in [9]. In this paper only the mode 1 is considered which allows for the following cloning template structure

$$A=\begin{bmatrix} a_{-1,-1} & a_{-1,0} & a_{-1,1} \\ a_{0,-1} & 2a_{0,0} & a_{0,1} \\ a_{1,-1} & a_{1,0} & a_{1,1} \end{bmatrix}, B\begin{bmatrix} 0 & 0 & 0 \\ 0 & 2b_{0,0} & 0 \\ 0 & 0 & 0 \end{bmatrix}, z=0 \quad (2)$$

The following mathematical model describes the cell dynamics of the ACE16K chip for mode 1

$$(C1_{i,j}+Cs_{i,j})\dot{v}_{C1_{i,j}}=-g(v_{C1_{i,j}})+2w_{0,0}^{(1)}v_{C1_{i,j}}+2w_{0,0}^{(2)}v_{C2_{i,j}}+I_{i,j}^{s} \quad (3)$$

with state variable $v_{C1_{i,j}}$ and

$$g(v_{C1_{i,j}})=\begin{cases} m(v_{C1_{i,j}}-1)+1 & v_{C1_{i,j}}<-1 \\ v_{C1_{i,j}} & |v_{C1_{i,j}}|\leq 1 \\ m(v_{C1_{i,j}}+1)-1 & v_{C1_{i,j}}>1 \end{cases} \quad (4)$$

The control and feedback templates are set by choosing $w_{i-k,j-l}^{(1)}=a_{i-k,j-l}$, $w_{0,0}^{(1)}=a_{0,0}$, $w_{0,0}^{(2)}=b_{0,0}$ and the input values is written to $v_{C2_{i,j}}$. In [5] the cloning template used for autowave propagation were

$$A=\begin{bmatrix} 0 & 0 & 0 \\ 0 & -3 & 0 \\ 0 & 0 & 0 \end{bmatrix}, B=\begin{bmatrix} 0 & 0 & 0 \\ 0 & -1 & 0 \\ 0 & 0 & 0 \end{bmatrix} \quad (5)$$

Using full white images as input and initial condition, it was observed that the chip has four wave sources, called internal sources, located at the corners of the images (see Fig. 2(a)). When an external source (see Fig. 2 (b)), represented by black dots on the initial and input images, is present a

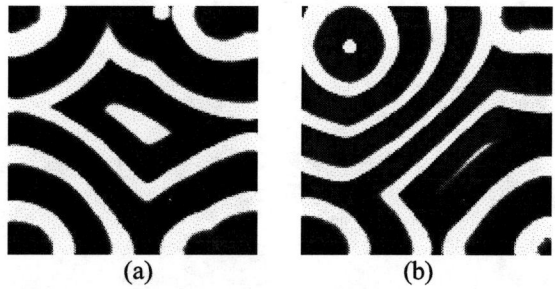

Figure 2. (a) The autowaves diffract from the internal sources. (b) Autowaves obtained from external and internal sources. An external source is located at left-top corner.

competition is established with the autowaves produced by the internal sources. This competition leads to the annihilation of the autowaves produced by the internal sources after an evolution time depending on the position of the external source in the image. Examining the effect of the external source position in the initial and input images, it raised out that with the templates used in [5] the evolution of the autowaves was influenced by this position in the sense that the autowaves tend to evolve in spirals when the source is in the middle of the image. This effect depends also on the dimensions of the source. Moreover the wave based navigation algorithm works better if the wave fronts spread a smaller number of pixels. In order to avoid the birth of spiral waves and to have a more compact wave front, experimental tests have led to the definition of new templates as follows:

$$A=\begin{bmatrix} 0 & 0 & 0 \\ 0 & -4.5 & 0 \\ 0 & 0 & 0 \end{bmatrix}, B=\begin{bmatrix} 0 & 0 & 0 \\ 0 & -3 & 0 \\ 0 & 0 & 0 \end{bmatrix} \quad (6)$$

III. THE ROBOT NAVIGATION CONTROL

A flow diagram of the whole algorithm is reported in Fig. 3. In the picture is also highlighted the part of the algorithm which is executed by the ACE16K. This part of the algorithm has been implemented in Analogic Macro Code (AMC) [9] which is a specific language developed to manage the Aladdin Professional System. At a higher level, the AMC program is compiled through a console program called CNNRun [9]. This is the graphical interface which allows the interaction with the ACE16K device. The remainder part is executed in a Matlab environment. This solution was selected to rapidly implement the algorithm. Once acquired the image reproducing the robot arena, it is resized in order to be processed by the ACE16K chip, which is able to host images till 128x128 pixel in size. A future improvement will be the utilization of the photo-sensors located in the focal plane of the chip to acquire directly the image of the arena: in this case the resize of the image will be no more necessary. Then the chip executes a set of instructions in order to filter the image from the noise. If the black pixels present in the image after a threshold operation represent only noise then the algorithm jumps directly to the control direction part of it as, in this case, the robot arena is free of

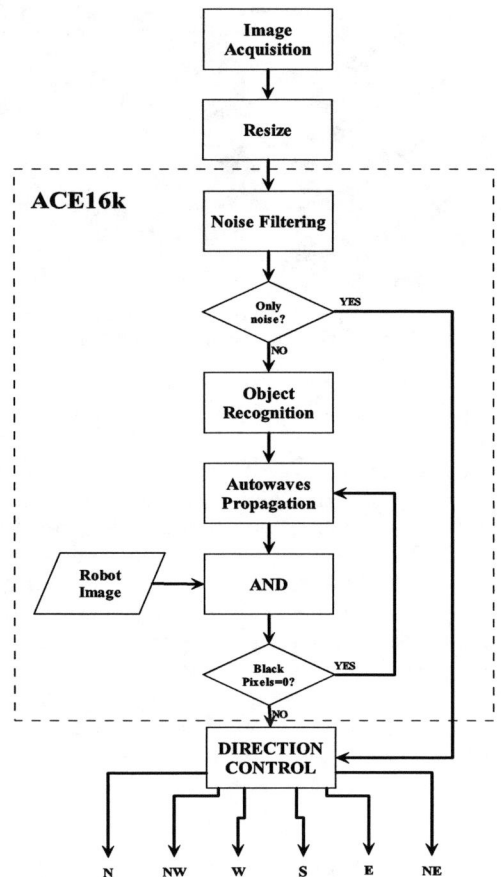

Figure 3. Flow diagram of the control algorithm. It is highlighted also the portion of the algorithm which is implemented on chip.

obstacles/target and the robot can continue moving in the previous direction. When other objects are present in the image, then an object recognition instruction is done, in order to distinguish between obstacles and target. An important hypothesis in this step is that the obstacles and the target can be detected by using a priori information. In particular the recognition is based on feature extraction. For instance the obstacles may be squared black objects while the target may have a circular shape. The resulting image is then used as input and initial condition of the autowaves generation, that makes use of templates (6). A logic AND operation is then performed between the snapshot of wave propagation and the image containing the robot that is viewed as a eighteen active pixel object divided in three sectors, six pixels each, called top, left and right. Depending on which sector of the robot's active pixels is activated by the wave front, a particular motion instruction is given to the robot.

IV. THE EXPERIMENTAL SET-UP

The experimental set-up is provided by a real environment, where a roving robot is required to move in a pre-specified direction. In this case no target recognition is required. A robot-centered perception strategy is implemented in our experimental setup [2]. The camera is placed on the robot through a pan-tilt, oriented in such a way that it frames the environment in the forward motion direction of the robot. The camera is connected to the personal computer in which is hosted the ACE16K chip housed in the Aladdin Professional System. At each step the robot camera takes a picture of the environment. The obstacles are then used as initial condition for the autowaves propagation template execution. The waveforms generated reach the robot and, depending on the direction they come from, a proper command is sent to the robot by the pc. In order to reduce the boundary distortion of the image produced by the camera, a Fixed Mask is used during the application of the template in order to block the dynamics of the cells in the border of the 128x128 image. In this condition (i.e. in absence of a target) we observed that the only instruction needed for noise filtering is a threshold operation. This operation is able to clearly identify the obstacles in good contrast with the background. In Fig. 4 several snapshots of an experimental test are reported. The robot while moving among some obstacles placed in the arena is shown. In Fig. 5 the results of the algorithm executed on chip are reported. The images are screenshots of the CNNRun program. In each image, from the top-left to the bottom-right, are reported in sequence: the image acquired by the camera, the results of the noise filtering and the obstacle identification, the autowave propagation and its logical not, the robot image and the results of the AND operation between the last image and the autowave fronts. In particular, the frames reported are related to the frames 10, 11 and 12 of Fig. 4. It can be seen the robot view while approaching the obstacle and, in the last frame, the wave fronts leading to a steering command to the left, as it can be observed on frame 13 of Fig. 4. The robot here used is a Lego Mindstorm® robotic rover chosen to overcome useless mechanical and interface complications. On the other hand the Lego robot infrastructure is very time consuming as regards command handling. Considering only the part of the algorithm running on the ACE16K and outlined in the square of fig.3, the execution time is about 100 ms. This time can be further reduced to about 50 ms, discarding the operations of loading images in memory, that can be avoided if the image is acquired directly through the focal plane. This would allow a control rate of about 20 commands per second, that represents a real breakthrough with respect to any other algorithm for navigation control based on visual feedback, ready for real-time applications. Future improvement in the algorithm as well as in the hardware-software implementation of the whole system could lead to higher rates and to a further improvement.

V. CONCLUSIONS

A definite improvement of the robot navigation strategy based on the paradigm of wave-based computation has been reported. This algorithm was originally implemented on two CNN chips: the ACE4K for image processing and the CACE1K for the generation of the wave fronts. In this paper the analog and massively parallel computing power of the

ACE16K chip, joined to the possibility to generate complex wave fronts, provides a self-contained platform and allows the real-time control of the robot trajectory in a complex environment. Some experimental results are proposed confirming the feasibility of the approach described. Interested people can find some videos of the experimental results in the following web page http://www.scg.dees.unict.it/activities/biorobotics/movie3.htm An on-going development is the realization of a fully autonomous robot, with the ACE chip embedded on it. It will use the photo-sensors located at the focal plane of the CNN chip to acquire directly the image of the arena, improving in this way the execution time of the whole algorithm.

REFERENCES

[1] C.J. Fourie, "Intelligent path planning for a mobile robot using a potential field algorithm", in *Proc. 29th Int. Symp. Robotics. Advanced Robotics: Beyond 2000, DMG Business Media,* Redhill, U.K., 1998, pp 221-224.

[2] A. Adamatzky, P.Arena, A. Basile, R.Carmona-Galan, B. De Lacy Costello, L. Fortuna, M. Frasca, A. Rodriguez-Vazquez, "Reaction-Diffusion Navigation Robot Control: from Chemical to VLSI Analogic Processors", *IEEE Trans. CAS-I: Regular Papers,* vol. 51, no. 5, May 2004, pp. 926-938.

[3] P. Arena, A. Basile, L. Fortuna, M. Frasca, "CNN Wave Based Computation for Robot Navigation Planning", *in Proc. of the 2004 IEEE ISCAS,* May 2004, Toronto, Canada.

[4] L.O.Chua, L. Yang, "Cellular Neural Networks: Theory and Applications", *IEEE Trans. CAS-I,* 35, pp. 1257-1290, 1988.

[5] M.E. Yalcin, J.A.K. Suykens, J. Vanderwalle, "Experimental Observation of Autowaves on the ACE16k CNN Chip", *Proc. of the 8th IEEE Int. Workshop on CNNA,* Budapest, Hungary, July 2004.

[6] A. P. Munuzuri, V.P. Munuzuri, M.G. Gesteria, L.O. Chua and V.P. Villar, "Spatiotemporal structures in discretely-coupled arrays of nonlinear circuits: A review," Int. J. Bifurcation and Chaos, Vol.5, no. 1, pp. 17-50, 1995.

[7] G. Linan, S. Espejo, R. Dominguez-Castro, A. Rodriguez-Vazquez, "Architectural and basic circuit considerations for a flexible 128x128 mixed signal SIMD vision chip", *Analog Integrated Circuits and Signal Processing,* 33, 179-190, Nov. 2002.

[8] A, Zarandy, C. Rekeczky, P. Foldesy, I. Szatmari, "The new framework of applications – the Aladdin system", *J. Circuits, Systems and Computers,* 12(6): 769-781, 2003.

[9] http://www.analogic-computers.com

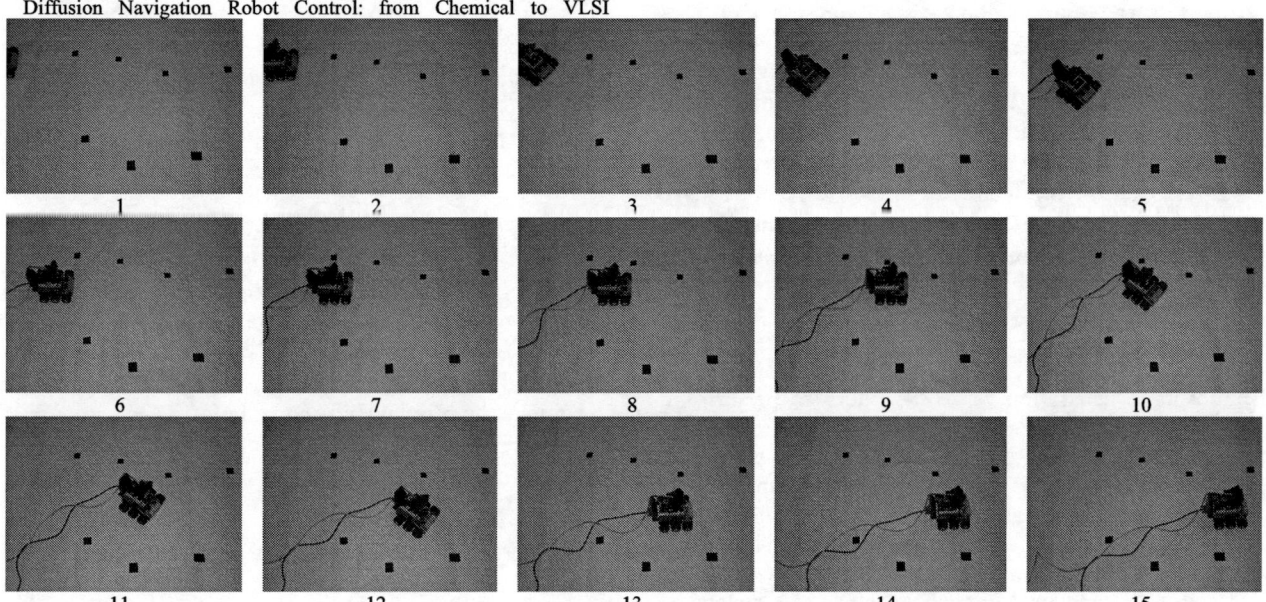

Figure 4. Snapshots of an experimental test. The image sequence in from left to right, from top to bottom. It can be seen how the robot is able to avoid the obstacles and to move in the path created.

Figure 5. Some CNNRun screenshots. In each one, are reported in sequence: the image acquired by the camera, the results of the noise filtering and the obstacle identification, the autowave propagation and its logical not, the robot image and the results of the AND operation between the last image and the autowave fronts.

A Single-Chip FPGA Design for Real-Time ICA-Based Blind Source Separation Algorithm

Charayaphan Charoensak, MIEEE and Farook Sattar, MIEEE

School of Electrical and Electronic Engineering, Nanyang Avenue,
Nanyang Technological University, Singapore 639798.
e-mail: {ecchara, efsattar}@ntu.edu.sg

Abstract— Blind source separation (BSS) of independent sources from their mixtures is a common problem in real world multi-sensor applications. In this paper, we propose efficient hardware architecture for the implementation of real-time BSS that can be implemented using a low-cost FPGA. The architecture offers a good balance between hardware requirement (gate count and minimal clock speed) and separation performance. The FPGA design implements the modified Torkkola's BSS algorithm for audio signals based on ICA (Independent Component Analysis) technique. The separation is performed by implementing noncausal filters, instead of the typical causal filters, within the feedback network. Architecture of the hardware is described. Results of various FPGA simulations and real-time testing of the final hardware design in real environment are given.

I. INTRODUCTION

Blind source separation, or BSS, refers to performing inverse channel estimation despite having no knowledge about the true channel (or mixing filter) [1]. BSS technique has been found to be very useful in many real-world multi sensor applications such as blind equalization, fetal ECG detection, and hearing aid. BSS method based on ICA (independent component analysis) technique has been found to be effective. A noncausal filters [2] can be used, instead of the typical causal, to shorten the filter length and a variable step-size learning process can be applied to provide a fast and stable convergence.

FPGA (Field Programmable Gate Array) architecture allows optimal parallelism needed to handle the high computation load of DSP algorithms in real-time. Being fully custom-programmable, FPGA offers rapid hardware prototyping and algorithm verification. Advances in IC processing technology and innovations in the architecture have made FPGA highly suitable alternative to using the powerful but expensive computing platform.

In spite of its potential applications, there have been very few published papers on real-time hardware implementation of the BSS algorithm. Many of the works such as in [6] focus on the VLSI implementation and do not provide a detailed set of specifications of the BSS implementation that offer a good balance between hardware requirement (gate count and minimal clock speed) and separation performance. Here, we propose an efficient hardware architecture that can be implemented using a low-cost FPGA and yet offers a good blind source separation performance. An extensive set of experimentations are presented.

Section 2 briefs the BSS algorithm. Section 3 describes the hardware architecture of our FPGA design. Section 4 presents the system level design of the FPGA followed by simulations and real-time experimentation.

II. BACKGROUND OF BSS ALGORITHM

As shown in Fig. 1, minimizing the mutual information between outputs u_1 and u_2 can be achieved by maximizing the total entropy at the output [3]. The architecture can be simplified by forcing W_{11} and W_{22} to be a scaling coefficient to achieve the relationship shown below [2, 4]:

$$u_1(t) = x_1(t) + \sum_{k=0}^{L_{12}} w_{12}(k) u_2(t-k) \quad (1)$$

$$u_2(t) = x_2(t) + \sum_{k=0}^{L_{21}} w_{21}(k) u_1(t-k) \quad (2)$$

And the learning rules for the separation matrix:

$$\Delta w^{ij} \propto (1 - 2y_i) u_j(t-k) \quad (3)$$

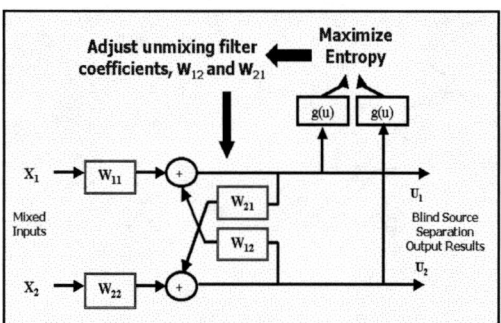

Fig. 1. Torkkola's feedback network for BSS.

The Torkkola's algorithm described above may be modified to use noncausal unmixing FIR filters for W_{12} and W_{21} to realize a non-minimum phase system [2]. The relationships between the signals are now changed to:

$$u_1(t) = x_1(t+M) + \sum_{k=-M}^{M-1} w_{12}(k) u_2(t-k) \quad (4)$$

$$u_2(t) = x_2(t+M) + \sum_{k=-M}^{M-1} w_{21}(k) u_1(t-k) \quad (5)$$

M is half of the filter length, L, ($L = 2M+1$) and :

$$\Delta w^{ij}_{(t_1-p_1+M)} = \Delta w^{ij}_{(t_0-p_0+M)} + K(u_i(t_0)u_j(p_0)) \quad (6)$$

where
$$K(u_i(t_0)) = \lambda * (1 - 2 y_i(t_0)) \quad (7)$$

$$y_i(t_0) = \frac{1}{1+e^{-u_i(t_0)}} \quad (8)$$

and $\quad t_1 = t_0 + 1 \quad p_0 = t_0 - k$ for $k = -M, -M+1, \ldots, M$

$p_1 = t_1 - k$ for $k = -M, -M+1, \ldots, M$

λ in (7) represents the variable learning step size.

III. ARCHITECTURE OF HARDWARE FOR BSS ALGORITHM

The top-level block diagram of the FPGA architecture for the BSS algorithm is shown in Fig. 2.

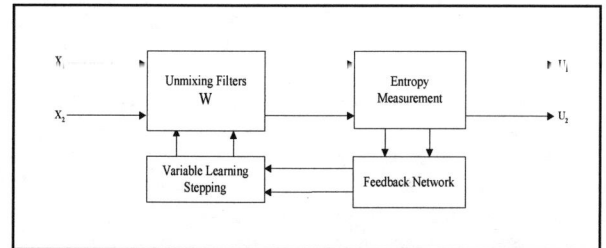

Fig. 2. Top-level block diagram of hardware architecture for BSS algorithm based on Torkkola's Network.

A. Practical Implementation of the Modified Torkkola's Network for FPGA Realization

In order to understand the effect of filter parameters on the separation performance, a number of MATLAB programs were written and a set of compromised specification proposed for practical hardware realization [4]. We propose that in order to reduce the FPGA resource needed, as well as to ensure real-time BSS separation given the limited maximum FPGA clock speed, the specifications shown below are to be used.

- Filter length, $L = 161$ taps,
- Buffer size for iterative convolution, $N = 2,500$,
- Maximum number of iterations, $I = 50$,
- Approximation of the exponential learning step size using linear piecewise approximation.

The linear piecewise approximation is used to avoid complex circuitry needed to implement the exponential function. We have chosen a small filter length considering the case of echo in small room.

B. Implementation of feedback network mechanism

The block diagram shown Fig. 3 depicts the hardware implementation of the feedback network. Note that the implementation of the FIR filtering of w_{12} is done through multiply-accumulate unit (MAC) which significantly reduces the numbers of multipliers and adders needed compared to direct parallel implementation. The tradeoff is that the FPGA has to operate at oversampling frequency.

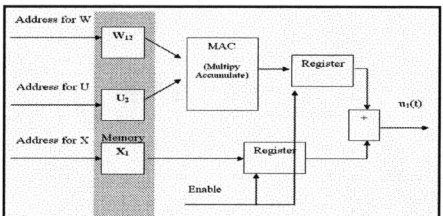

Fig. 3 Implementation of the modified feedback network.

C. Implementation of variable learning step size

In order to speed up the learning of the filter coefficients shown in Equations 6, we implement a simplified variable step size technique. In our application, the variable learning step size in Equation 7, λ, is implemented using Equation 10 below where n is the iteration level, λ_0 is the initial step size, and I is the maximum number of iterations, i.e. 50.

$$\lambda = \exp(-u_0 - \frac{n}{I}) \quad (10)$$

where $\quad u_0 = -\log_2(\lambda_0) - \frac{1}{I} \quad (11)$

The exponential term is difficult to implement in digital hardware. CORDIC algorithm (COordinate Rotation DIgital Computer) may be used but its circuitry will impose a very long latency, if not heavily pipelined, and will result in the need for even higher FPGA clock. Instead, we used a linearly decreasing variable step size in Equation 12.

$$\lambda = 0.0006 - 0.000012n \quad (12)$$

D. Calculation of the required FPGA clock speed

The required FPGA clock frequency can be calculated using Equation 13. Fs is the sampling frequency of the input signals, L is the tap length of the FIR filter, and I is the number of iteration.

$$FPGA\ Clock\ Frequency = L * I * Fs \quad (13)$$

In our FPGA design, filter tap $L = 161$, iterations $I = 50$, and input sampling frequency $Fs = 8,000$ Hz, the FPGA clock frequency is thus 161*50*8,000 = 64.4 MHz. This means that the final FPGA design must operate properly 64.4 MHz.

IV. SYSTEM LEVEL DESIGN AND TESTING OF THE FPGA FOR BSS ALGORITHM

A. System-level design of FPGA for BSS algorithm

System Generator provides a bit-true and cycle-true FPGA blocksets for functional simulation under MATLAB Simulink™ environment. The FPGA design using System Generator is different from the more typical approach of using HDL (Hardware Description Language). The FPGA is designed by means of Simulink environment. The FPGA functional simulation can be carried out easily right inside Simulink environment. After the successful simulation, the synthesizable VHDL (VHSIC HDL) code is automatically generated from the models.

B. Using fixed-point arithmetic in FPGA implementation

When simulating DSP algorithms using C or MATLAB, double precision floating point is commonly used. In hardware, however, fixed-point numeric is more practical. Although several groups have implemented floating-point adders and multipliers using FPGAs, very few practical systems exist. The main disadvantages using floating-point in hardware are high resource requirements and high clock frequency. We use 16-bit fixed-point numeric as shown in Fig. 4. One bit is used for sign, 2 bits for integer part, and 13 bits for the fractional part. When performing arithmetic operations, normalization was performed to avoid overflow and make sure that the data path was always best utilized.

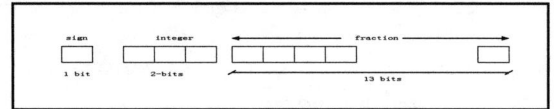

Fig. 4 Fixed-point numeric format used in the FPGA design

C. Analysis of critical path in FPGA implementation of BSS algorithm

Referring to figures 1 and 2, one can see that the maximum operating speed of the FPGA is determined by the critical path in the multiply-accumulate blocks. The propagation delay in this critical path is the summation of the combinational delay made up of one multiplier and one adder as shown in Fig. 5. In Xilinx data sheet for Virtex-E, the delay of the dedicated multiplier and adder units are:

Delay of 16-bits adder: 4.3 nsec
Delay of pipelined 16-bit multiplier: 6.3 nsec

Thus, the approximated total delay in the critical path is 4.3+6.3=10.6 nsec, or 94 MHz. This is higher than required 64.4 MHz. Accurate speed can also easily be extracted from the FPGA synthesis results given in later section.

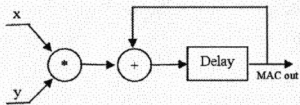

Fig. 5 Simplified block diagram of MAC operation.

D. Simulation Results of the FPGA for ICA-Based BSS

1) FPGA simulation using two female voices

In this simulation, we use wave files of two female voices. Each file is approximately one second long, with sampling frequency of 8,000 Hz. The two inputs were mixed using the instantaneous mixing program called *'instamix.m'* downloaded from the web site http://www.ele.tue.nl/ica99 [5]. The mixing matrix is [0.6 1.0, 1.0 0.6]. The FPGA simulation result showed that the separation was successful.

Fig. 6(a) compares the linear piecewise function and the exponential function (Equation (8)), in the circuit for adaptive step size. Fig. 6(b) compares the averaged changes of filter coefficients Δw_{12} and Δw_{21}, for all the 161 taps, plotted against number of iterations. It can be seen that for the maximum number of iterations used (=50), the learning step size using both exponential function and linear piecewise function converge to zeros.

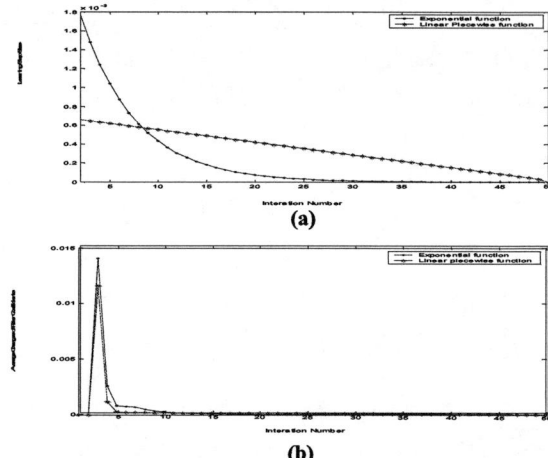

Fig. 6 (a) Learning step sizes using exponential and linear piecewise functions, (b) averaged changes of filter coefficients Δw_{12} and Δw_{21} using the two functions.

2) FPGA simulation using one female voice mixed with Gaussian noise

To measure the robustness of the BSS under noisy environment, the first female voice was mixed with white Gaussian noise and the Signal-to-Noise ratios (*SNRs*), before and after the operation, measured. By adjusting the variance of the noise source, σ^2, the input *SNR* varies in the range -9 dB to 10 dB. The input *SNR*, *SNRi*, is defined as (see Fig. 7):

$$SNRi(dB) = 10\log\left[\sum_{i=1}^{T} x_1^2(i) / \sum_{i=1}^{T} x_2^2(i)\right] \quad (14)$$

T is the total number of samples in the time period of measurement, $x_1(i)$ and $x_2(i)$ are the original female voice and the white Gaussian noise respectively. Similarly, the output Signal-to-Noise ratio *SNRo*, is defined in Equation 15. $e_1(i)$ represents the overall noise left in the first separated output u_1 and defined as u_1-x_1. The improvement in *SNR* after BSS operation is defined in Equation 16.

$$SNRo(dB) = 10\log\left[\sum_{i=1}^{T} x_1^2(i) / \sum_{i=1}^{T} e_1^2(i)\right] \quad (15)$$

$$SNRimp(dB) = 10\log\left[\sum_{i=1}^{T} x_2^2(i) / \sum_{i=1}^{T} e_1^2(i)\right] \quad (16)$$

Fig. 8 shows the average output *SNRs* and improvement of *SNRs* plotted against input *SNRs*. It can be seen that as the input *SNRs* varies, the output *SNRs* is almost constant at approximately 35 dB. The maximum achievable output *SNRs* is limited by the width of the datapath implemented inside the FPGA. Due to this reason, the amount of improvement of *SNRs* decreases with the increasing input *SNRs*.

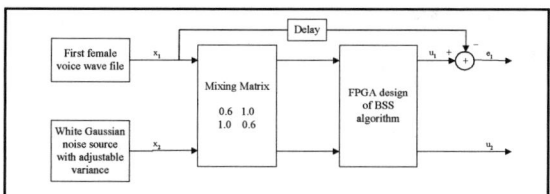

Fig. 7 FPGA simulation to test BSS using one female voice mixed with white Gaussian noise with adjustable variance, σ^2.

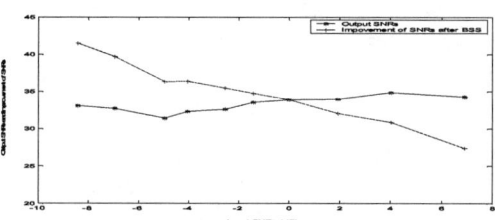

Fig. 8 Results of *SNR* measurements using white Gaussian noise. The figure shows output *SNRs* and improvement of *SNRs*

3) FPGA simulation using two female voices mixed using simulated room environment

To measure separation performance under realistic room environment, we used the program *'simroommix.m'* from the web site mentioned earlier. The coordinates (in meter) of the first and second signal sources are (2, 2, 2) and (3, 4, 2) respectively. The locations of the first and second microphone are (3.5, 2, 2) and (4, 3, 2). The room size is 5x5x5. The simulated room arrangement is shown in Fig. 9. The measurement of separation is based on Equation 17 [5]. Here, S_j is the separation performance of the jth separated output where $u_{j,xj}$ is the jth output of the cascaded mixing/unmixing system when only input x_j is active. It was found that the BSS hardware improves the separation by 3 dB and 5.9 dB for channel 1 and 2 respectively.

$$S_j = 10 \log \left(\frac{E\{(u_{j,x_j})^2\}}{E\{(\sum_{i \neq j} u_{j,x_i})^2\}} \right) \quad (17)$$

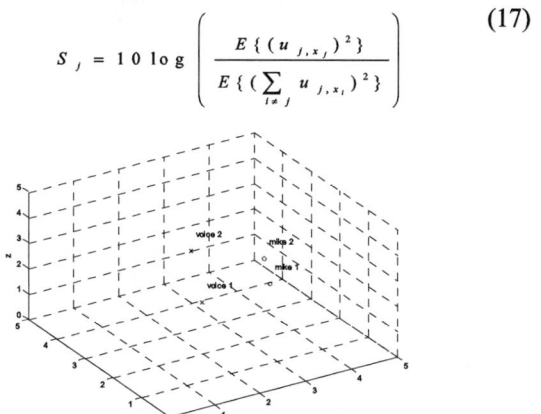

Fig. 9 Arrangement of female voice sources and microphones in simulated room environment using *'simroommix.m'*.

E. FPGA synthesis and Real-time Experimentation

The VHDL codes were generated and synthesized using Xilinx ISE 5.2i; targeted for Virtex-E family. The design requires 100,213 gates and maximum frequency is 71.2 MHz.The real-time testing of the FPGA was done using a prototype board equipped with a Virtex-E FPGA device and A/D and D/A converters (Fig. 10). The convolutive mixtures used in previous subsection were encoded into the left and right channels of a stereo MP3 file which is then played back using a portable MP3 player. The FPGA streamed out the separated outputs which were then converted into analog signals and played back on speaker. By listening to the sound, it was concluded that the same level of separation as in earlier simulations is achieved.

Fig. 10 Real-time testing of the FPGA for BSS.

V. CONCLUSIONS

In this paper, the hardware implementation of the modified BSS algorithm was realized using FPGA. A set of compromised parameters was proposed taking into consideration the separation performance and hardware resource requirements. FPGA functional simulations were carried out using additive mixtures, convolutive mixtures, and room environment. The final FPGA hardware achieves the real-time operation using minimal FPGA resource.

In the case where the echo is more serious, a longer filter tap may be needed. We propose that, in order to keep the FPGA clock frequency the same, multiple MAC engines be used together with the implementation of polyphase decomposition. For example, for the tap length of 161*16 = 2,576, 16 MACs are to be implemented with the 16-band polyphase decomposition. Since one MAC engine is made up of only one multiplier and one accumulator, the additional MAC engines will not lead to significant gates increase.

REFERENCES

[1] T-W Lee, "Independent Component Analysis - Theory and Applications", Kluwer Academic Publishers, 1998.

[2] F. Sattar, M. Y. Siyal, L. C. Wee, and L. C. Yen, "Blind Source Separation of Audio Signals Using Improved ICA Method", IEEE SSP'01 Workshop, Singapore.

[3] K. Torkkola, "Blind separation of convolved sources based on information maximization", IEEE Workshop Neural Networks for Signal Processing, Kyoto, Japan, Sept 4-6,1996.

[4] Charayaphan Charoensak and Farook Sattar, "Hardware for real-time ICA-based blind source separation," in Proc. 15th IEEE Int. Conf. SOCC, Sept. 12-15, 2004, pp. 139-140.

[5] Daniël Schobben, Kari Torkkola, and Paris Smaragdis, "Evaluation of Blind Signal Separation Methods", International Workshop on Independent Component Analysis and Blind Signal Separation, Aussois, France, Jan. 1999, pp. 261-266.

[6] Abdullah, C., Milutin, S. and Gert, C. "Mixed-signal Real-Time Adaptive Blind Source Separation", International Symposium on Circuits and Systems, ISCAS 2004, Vancouver, Canada, May 2004, pp. 760-763.

Design and FPGA Implementation of Finite Ridgelet Transform

Isa Servan Uzun* and Abbes Amira
School of Computer Science
Institute for Electronics, Communications and Information Technologies (ECIT)
The Queen's University of Belfast
Belfast, BT7 1EN, United Kingdom
Email: isu@ieee.org

Abstract—The Ridgelet transform was recently introduced to overcome the weakness of wavelets in higher dimensions. In this paper, we present design and FPGA implementation of the Finite Ridgelet Transform (FRIT) for image processing applications. The proposed architecture uses the Finite Radon Transform (FRAT) and 1-D Discrete Biorthogonal Wavelet Transform (DBWT) as building blocks. A detailed evaluation of FPGA implementation for the proposed architectures targeting Xilinx Virtex-II device family has been reported based on maximum system frequency, chip area and image size. The implementation results show that the core speed for the proposed FRIT architecture is around 100MHz and it occupies 491 Slices for an input image size of $7x7$.

I. INTRODUCTION

Many image processing tasks take advantage of sparse representations of image data where most information is packed into a small number of samples. Typically, these representations are achieved via invertible and nonredundant transforms. Currently, the most popular choices for this purpose are the wavelet transform [1], [2] and the discrete cosine transform [3].

The success of wavelets is mainly due to the good performance for piecewise smooth functions in one dimension. However, the wavelet transform has many limitations when it comes to representing straight lines and edges in images. To overcome the weakness of wavelets in higher dimensions, Cands[4] recently proposed the Ridgelet transform which deals effectively with line singularities in 2-D. It has become a very popular tool in image processing applications [5], [6], [7].

Although impressive image processing performance has been achieved with Ridgelet transform, the complexity of its implementation still remains as a heavy burden on standard microprocessors where large amounts of data have to be processed. Therefore, either the design of high-performance dedicated circuits or parallel systems is strategic for applications that require real-time performances. Surveying the literature, only one FPGA-based implementation has been found [8] for Ridgelet transform. In this work, two FRIT architectures have been proposed based on Radon transform and Haar wavelet for a fixed input image size which is 17x17.

In this paper, a parametrisable architecture for the Finite Ridgelet Transform (FRIT)and its efficient FPGA implementation have been proposed. Performance results for the hardware implementations of the FRIT filter architecture on the Xilinx Virtex-II FPGA chip are presented based on maximum system frequency, chip area and image size.

The outline of this paper is as follows. In the next section, the FRIT is shortly introduced. The proposed architecture and its computing blocks are described in Section 3. FPGA implementations with performance results are presented in Section 4. Conclusions are summarised in Section 5.

II. THE FINITE RIDGELET TRANSFORM (FRIT)

The FRIT was introduced in [6] in order to provide a sparse representation for functions defined on the continuum plane. The transform allows representing edges and other singularities along curves in a more efficient way, in terms of compactness of the representation, than traditional transformations, such as the wavelet transform, for a given accuracy of reconstruction. The basic idea is to map a line singularity in the two-dimensional (2-D) domain into a point by means of the Radon transform. Then, a one-dimensional (1-D) wavelet is performed to deal with the point singularity in the Radon domain.

A. The Finite Radon Transform (FRAT)

In recent years the Radon transform have received much attention. It is able to transform two dimensional images with lines into a domain of possible line parameters, where each line in the image will give a peak positioned at the corresponding line parameters. This have lead to many applications within image processing, computer vision, and seismics [9].

Numerous discretizations of the Radon transforms have been devised to approximate the continuous formulae [6]. However, most of them were not designed to be invertible transforms for digital images. Alternatively, the finite Radon transform theory (which means transform for finite length signals) originated from combinatorics, provides an interesting solution.

The FRAT [10] is defined as summations of image pixels over a certain set of lines. Those lines are defined in a finite geometry in a similar way as the lines for the continuous Radon transform in the Euclidean geometry. Denote $Z_p = 0, 1, ..., p-1$, where p is a prime number.

The FRAT of a real function on the finite grid Z_p^2 is defined as:

$$r_k[l] = FRAT_f(k,l) = \frac{1}{\sqrt{p}} \sum_{(i,j) \in L_{k,l}} f(i,j) \quad (1)$$

Here, $L_{k,l}$ denotes the set of points that make up a line on the lattice Z_p^2 as follows:

$$L_{k,l} = \{(i,j) : j = ki + l(mod\, p), i \in Z_p\}, 0 \leq k < p, \quad (2)$$
$$L_{p,l} = \{(l,j) : j \in Z_p\} \quad (3)$$

Computing the kth Radon projection, i.e., the kth row of the array, we need to pass all pixels of the original image once and use p histogrammers: one for every pixel in the row. At the end, all p histogrammed values are divided by p to get the average values. The computation of FRAT is illustrated by the pseudo-code given below [10]:

```
1   for k = 0 to p − 1 do
2       FRAT(k, l) = 0
3       n = k for j = 0 to p − 1 do
4           n = n − k
5           if (n < 0) then
6               n=n+p
7           end
8           l = n-1
9           for i = 0 to p − 1 do
10              l = l+1
11              if (l ≥ p) then
12                  l=l-p
13              end
14              FRAT(k, l) = FRAT(k, l) + f(i, j)
15          end
16      end
17      FRAT(k, l) = FRAT(k, l)/p
18  end
19  FRAT(p, j) = 0
20  for j = 0 to p − 1 do
21      for i = 0 to p − 1 do
22          FRAT(p, j) = FRAT(p, j) + f(i, j)
23      end
24  end
25  FRAT(p, j) = FRAT(p, j)/p
```

Algorithm 1: Finite Radon Transform Pseudo-code

B. The Discrete Wavelet Transform

A wavelet transform breaks a signal into shifts and translations of a basis function called the 'mother wavelet'. This is mathematically represented in equation 8. The direct implementation of this equation is computationally very intensive as the time shift and the scaling factor a can assume any real value.

$$CWT_x(\tau, a) = \frac{1}{\sqrt{a}} \int x(t) \cdot h * (\frac{t-\tau}{a}) dt \quad (4)$$

The work by Mallat [1] and Daubechies [2] led to the discrete filter based interpretation of wavelets. Through this, wavelets can be implemented as a set of filter banks comprising a high-pass and a low-pass filter, each followed by downsampling by two. The low-pass filtered and decimated output $a^j(n)$ having $N^j = N/2^j$ samples is recursively passed through similar filter banks to add the dimension of varying resolution at every stage. This is mathematically expressed in equation ?? and schematically shown in figure 1.

$$a^j(n) = \sum_{i=0}^{L-1} l(i) \cdot a^{j-1}(2n-i) \quad 0 \leq n < N_j \quad (5)$$

$$d^j(n) = \sum_{i=0}^{L-1} h(i) \cdot d^{j-1}(2n-i) \quad 0 \leq n < N_j \quad (6)$$

The coefficients $a^j(n)$ and $d^j(n)$ refer to approximation and detailed components in the signal at decomposition level j, respectively. The $l(i)$ and $h(i)$ represent the coefficients, respectively, of low-pass and high-pass L-tap filters.

In this paper, biorthogonal wavelet filters have been considered. These filters are very attractive for implementing pyramidal structures since they do not require phase compensation decomposition levels. Biorthogonal wavelet filters possess a linear-phase property and they have a symmetric (or anti-symmetric) impulse response. Their filter coefficients can thus be written as follows:

$$l(n) = \pm l(L-1-n), \quad n = 0, 1, .., \lceil L/2 \rceil - 1 \quad (7)$$

where L is the filter length and $\lceil \rceil$ represents maximum integer.

For the sake of illustration, the case of a (9-tap) 1-D biorthogonal filter ($L=9$) will be considered in the rest of the paper. For an (9)-tap biorthogonal filter, the symmetry along with the filter coefficients can be written as:

$$l(4-n) = \pm l(4+n), \quad n = 0, 1, .., 4 \quad (8)$$

By taking into account the symmetry given in Equation 7, biorthogonalised version of DWT (which is abbreviated as DBWT in this study) given in Equation 8.a can be described as follows:

$$a^j(n) = l(4) \cdot a^{j-1}(2n-4) + \sum_{i=0}^{3} \{a^{j-1}(2n-8-i) + a^{j-1}(2n-i)\} \quad (9)$$

where the case of filter length $L = 9$ is considered.

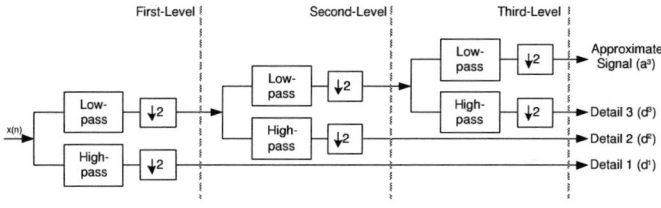

Fig. 1. A three-level wavelet decomposition system.

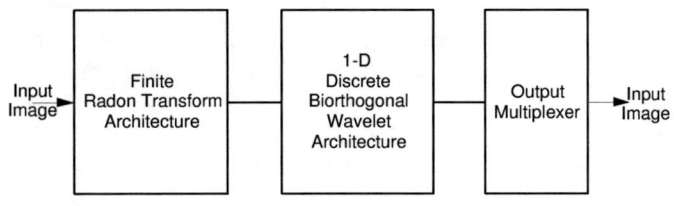

Fig. 2. Finite Ridgelet Transform Architecture

III. ARCHITECTURES

Once the Radon and wavelet transform have been implemented, the ridgelet transform is straightforward. Each output of the radon projection, i.e, each row of radon transformed image, is simply passed through the wavelet transform before it reaches the output multiplier. The block diagram of Ridgelet transform is shown in Figure 2

A. FRAT Implementation

Figure 3 shows the block diagram of the FRAT implementation. The proposed architecture of FRAT is a straightforward implementation of the pseudo-code given in the previous section. The address logic initializer along with controller block constitute the address generator that generates addresses, i.e., $L_{k,l}$ for memory blocks. The accumulator is a L_O-bit accumulator that accumulates the lth pixel value for the kth Radon projection. The controller block orginazes the flow of this process with input and output data flow.

B. 1-D DBWT Architecture

A generic 1-D DBWT architecture has been designed by combining the symmetry property of biorthogonal filters with polyphase decomposition [11]. The architecture for 9-tap biorthogonal filter is depicted in figure 4-a. It employs 3 multipliers, where each multiplier performs the computations related to two coefficients of the filter.

As shown in the functional analysis diagram (figure 4-b), the input to DBWT block is a sequential data stream, from Radon transform block, at a rate of 1 sample per cc. The computations are periodic with 2. The periods are

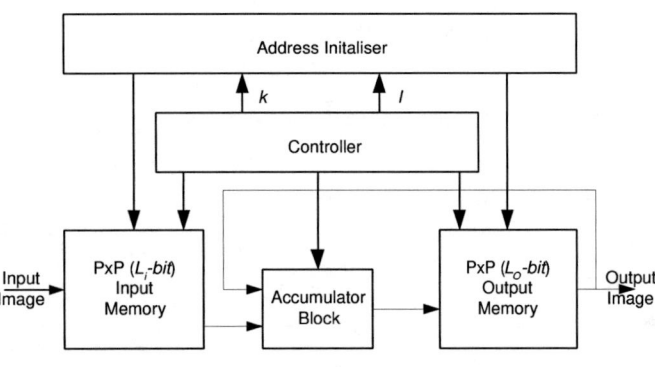

Fig. 3. Block diagram of proposed FRAT implementation

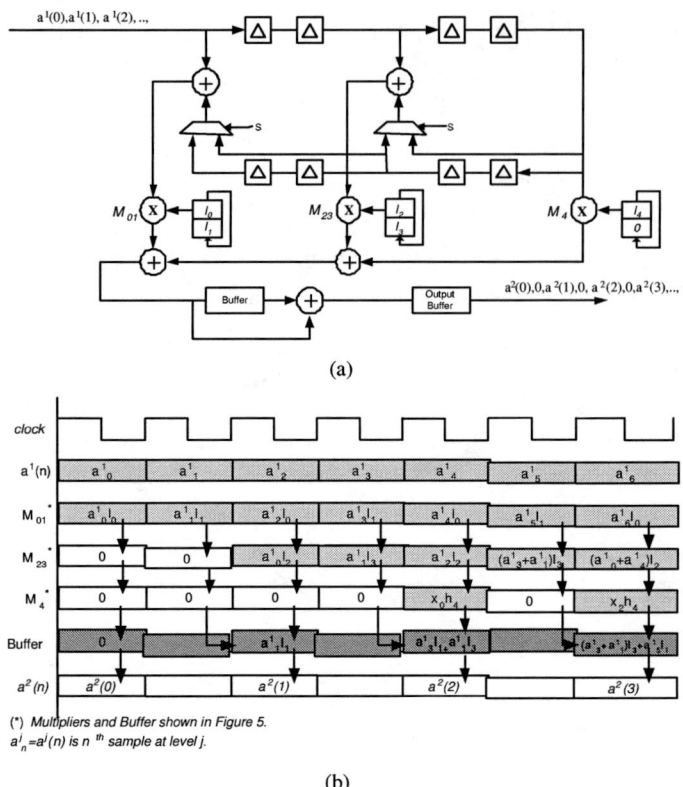

Fig. 4. (a) 1-D DBWT architecture for 9 tap low-pass analysis filter, (b) Functional analysis diagram.

divided in two subperiods identified by a 1-bit select signal S. In the subperiod $S=0$, the multiplier M_k performs the multiplication of input sample by the filter coefficients h_{2k} and adds these products to the data stored in the buffer during the previous subperiod. In the subperiod $S=1$, M_k performs the multiplication of the input sample by the filter coefficients h_{2k+1} and stores the addition of these products into buffer. By this way, output pixels are produced on the even-numbered ccs.

IV. FPGA IMPLEMENTATION

In order to verify the performance of the proposed architectures, designs have been ported to a Xilinx Virtex-II FPGA chip using Handel-C [12]. The designs of the architectures have been parameterised in terms of:

- Input image size (N)
- Number of wavelet decomposition levels (J)
- Input and output data wordlength (L_i and L_o)
- Wavelet filter coefficients wordlength (W_c)

A. FRAT Architecture

The FRAT implementation has been carried out for three different input image sizes where $N=7, 17$ and 31. 8-bits of data wordlength has been used in order to represent the input image while $log_2(p*255)$-bits of output data wordlength is used for the output of FRAT block.

Table I shows the FPGA implementation performance

results of FRAT for different input image size. In the implementations, two types of approaches - Distributed RAM and Block RAM- have been used in order to store input image data and output radon projections. The results indicate that higher f_{max} rates can be achived with the use of distributed RAM while Block RAM usage reduces the occupied FPGA area in terms of slices.

TABLE I

FPGA IMPLEMENTATION RESULTS FOR FINITE RADON TRANSFORM.

Image Size	Distributed RAM		Block RAM	
	Area(Slices)	f_{max}(MHz)	Area(Slices)	f_{max}(MHz)
7	198	112.867	131	67.295
17	636	87.443	300	49.319
31	1118	80.295	500	42.384

B. 1-D DWT Architecture

In this paper, biorthogonal (9,7) and (9,3) 1-D wavelet filters have been implemented on FPGA. Since the FRAT output coefficient are not normalised to reduce overall computational cost, input data length of $log2(p*255)$-bits has been used for wavelet architecture block while filter coefficients have been represented using 9-bits. The output data wordlength is selected as 16-bits since the wavelet coefficients from all levels are bounded by 2^{10}. FPGA implementation performances for 1-D DBWTs are reported in table II in terms of FPGA area (Slices occupied) and maximum clock frequency (f_{max}).

TABLE II

PERFORMANCE RESULTS FOR FPGA IMPLEMENTATIONS OF THE PROPOSED 1-D DBWT ARCHITECTURES.

Design	Wavelet Type	Decomposition Levels	Area (Slices)	f_{max} (MHz)
Wavelet	Bior(9,7)	1	293	129
	Bior(9,3)	1	224	135

In addition to individual blocks performance results, Table III shows the FPGA implementation results for the whole FRIT architecture. It also shows the comparison with the only existing FPGA-based FRIT implementation. It can be seen from the table that our implementation outperforms existing work in terms of f_{max}, FPGA area occupied and input image size.

V. CONCLUSION

The Ridgelet transform [4] was recently introduced to overcome the weakness of wavelet transforms. In this paper, an architecture and its efficient FPGA implementation for the Finite Ridgelet transform have been proposed. The implementations have been carried out for different input image size where N=7,17 and 31. A detailed evaluation of FPGA implementation for the proposed architectures targeting

TABLE III

PERFORMANCE RESULTS COMPARISON FOR FPGA-BASED FINITE RIDGELET TRANSFORM IMPLEMENTATION.

Design	Wavelet Type	Input Size	Area Slices)	f_{max} (MHz)	Speed/Area Ratio
Wisinger-Generic [8]	Haar	17x17	n/a	33	n/a
Wisinger-Serial	Haar	17x17	828	18	0.021
Proposed-I	Bior(9,7)	7x7	476	101	0.212
Proposed-I	Bior(9,7)	17x17	911	82	0.090

Xilinx Virtex-II device family has been reported based on maximum system frequency, chip area and image size. The implementation results show that proposed implementation outperforms existing work in terms of both area and system speed.

Our future work will be concerned with the desing and FPGA implementation of the Curvelet transform that uses "a'torus" wavelet and the Ridgelet transform as building blocks. The final Curvelet implementation will target the development of an FPGA-based adaptable compression system for HDTV.

REFERENCES

[1] S. G. Mallat, "A theory for multiresolution signal decomposition: The wavelet representation," *IEEE Trans. Pattern Anal. Mach. Intell.*, vol. 11, no. 7, pp. 674–693, 1989.

[2] I. Daubechies, "The wavelet transform, time-frequency localization and signal analysis," *IEEE Trans. Inform. Theory*, vol. 36, pp. 961–1005, 1990.

[3] K. Rao and P. Yip, *Discrete Cosine Transform: Algorithms, Advantages, Applications*. New York: Academix, 1990.

[4] E. J. Cands, "Ridgelets: Theory and applications," Ph.D. dissertation, Department of Statistics, Stanford University, U.S.A, 1998. [Online]. Available: http://www.acm.caltech.edu/ emmanuel/publications.html

[5] P. Campisi, D. Kundur, and A. Neri, "Robust digital watermarking in the ridgelet domain," *IEEE Signal Processing Letters*, vol. 11, no. 10, pp. 826–830, Oct. 2004.

[6] M. Do and M. Vetterli, "The finite ridgelet transform for image representation," *IEEE Transactions on Image Processing*, vol. 12, no. 11, pp. 16–28, Jan. 2003.

[7] H. Biao, J. Li-cheng, and L. Fang;, "Image denoising based on ridgelet," in *6th International Conference on Signal Processing*, vol. 1, 2002, pp. 780–783.

[8] J. Wisinger and R. Mahapatra, "Fpga based image processing with the curvelet transform," Department of Computer Science, Texas A&M University, TX, Tech. Rep. TR-CS-2003-01-0, 2003. [Online]. Available: http://faculty.cs.tamu.edu/rabi/recent-work.html

[9] P. Toft, "The radon transform - theory and implementation," Ph.D. dissertation, Department of Mathematical Modelling, Technical University of Denmark, Denmark, June 1996. [Online]. Available: http://pto.linux.dk/PhD/

[10] F. Matus and J. Flusser, "Image representation via a finite radon transform," *IEEE Transactions on Pattern Analysis and Machine Intelligence*, vol. 15, no. 10, pp. 996–1006, Oct. 1993.

[11] I.S.Uzun, A. Amira, and A. Bouridane, "An efficient architecture for 1-D discrete biorthogonal wavelet transform," in *IEEE International Symposium on Circuits and Systems (ISCAS 2004)*, vol. 2, 2004, pp. 697–700.

[12] *Handel-C Language Reference Manual*, Celoxica. [Online]. Available: www.celoxica.com

A NOVEL METRIC REPRESENTATION FOR LOW-COMPLEXITY LOG-MAP DECODER

Byonghyo Shim

Dept. of Mathematics,
University of Illinois at Urbana-Champaign,
1409 West Green Street, Urbana, IL 61801.
bshim@uiuc.edu

Hyung G. Myung

Dept. of Electrical & Computer Engineering
Polytechnic University,
Brooklyn, NY 11201.
hmyung01@poly.edu

ABSTRACT

In this paper, we propose a novel state metric representation of log-MAP decoding which does not require any rescaling in both forward and backward path metrics and LLR. In order to guarantee the metric values to be within the range of precision, rescaling has been performed both for forward and backward metric computation, which requires considerable arithmetic operations and decoding delay. In this paper, by applying the homomorphism in a finite abelian group Z_b associated with modulo 2^b addition, we show that the proposed metric representation does not need any rescaling in metric and LLR computation. In this general observation, we show that the Hekstra's scheme [5] is a special case for the path metric rescaling. Besides the fact that proposed technique saves design time considerably, we show through the complexity analysis that proposed technique saves the ACSU (Add-compare-select unit) complexity and reduces the critical path delay of decoder significantly.

1. INTRODUCTION

Recently, maximum a posteriori (MAP) decoding algorithm [1] has been paid considerable attention as an effective decoding method of Turbo codes, an optimum equalization scheme for space-time codes, and a joint decoding and equalization scheme for ISI channel [2]-[3]. Contrary to the Viterbi algorithm, which performs maximum likelihood sequence estimation (MLSE), the MAP algorithm finds an estimate of the probability that the information bit is 0 (or equivalently a 1) from received sequence [1]. Thus, MAP decoding method naturally lends itself to provide the soft estimates that can be used in iterative decoding.

In order to compute the probability of each bit, usually called *log-likelihood ratio* (LLR), both forward and backward state metrics are employed. Even though the process of each state metric computation is similar to that of Viterbi decoding, it is more complicated and hence judicious consideration should be given in many practical aspects such as metric representation, precision assignment, and metric normalization. In particular, due to the addition of forward state metric, backward state metric, and branch metric for computing LLR information, Hekstra's two's complement based metric representation scheme [5] popularly employed in Viterbi decoding cannot be applied directly to MAP decoding.

In this paper, we propose a novel state metric representation scheme for MAP decoding which does not require any rescaling in both forward/backward path metrics and LLR. Specifically, by exploiting the homomorphism in a finite abelian group Z_b associated with modulo 2^b addition operation, we show that proper precision assignment and modulo arithmetic operation lead to no rescaling in LLR as well as path metric computation. In this general observation, we show that the Hekstra's scheme [5] is a special case for path metric rescaling.

The organization of this paper is as follows: After briefly reviewing the Log-MAP algorithm in the next section, we present modulo 2^b based state metric representation in section III. In the section IV, we discuss the complexity reduction and conclude.

2. LOG-MAP ALGORITHM

In this section, we start from the resulting equation of Log-MAP decoding algorithm. For the detailed description, refer to [1].

Let $T = \{1, 2, \cdots\}$ be a time step and S be a set of all possible states of encoder state machine. The forward path metric of a state s at time k is given by

$$\alpha_k(s) = \max_{s' \in S}{}^* \alpha_{k-1}(s') + \gamma_k(s', s) \tag{1}$$

where $\gamma_k(s', s)$ is the branch metric of state from s' to s

$$\gamma_k(s', s) = x_k^s(L_{in}^e + y_k^s) + x_k^p \cdot y_k^p \tag{2}$$

and $\max^*(\delta_1, \delta_2) = \max(\delta_1, \delta_2) + \ln(1 + e^{-|\delta_1 - \delta_2|})$. Similarly, the backward state metric is given by

$$\beta_k(s) = \max_{s' \in S}{}^* \beta_{k+1}(s') + \gamma_{k+1}(s, s'). \tag{3}$$

With these metrics, the log-likelihood ratio (LLR) of information bit d_k is computed:

$$L(u_k) = \max_{s', s: u_k = +1}{}^* l_k(s', s) - \max_{s', s: u_k = -1}{}^* l_k(s', s) \tag{4}$$

where $l_k(s', s) = \alpha_{k-1}(s') + \gamma_k(s', s) + \beta_k(s)$. Note that since $\alpha_k(s)$ and $\beta_k(s)$ are chosen by $\max^*$ operation, an input shift will result in an output shift. However, this shift does not affect the eventual LLR information.

Lemma 1 (Shift invariance) *The forward state metric update process given in (1) depends only on differences of metrics.*

Proof) Suppose we add some constant K to all the forward metrics, $\forall \alpha_{k-1}(s)$, $s \in S$. Because forward update is performed via $\max^*$ operation, an output will be shifted by the same amount, i.e, $\max^*(a + K, b + K) = \max^*(a, b) + K$. This shift, however, will be removed in a LLR computation of (4) □

In the same way, lemma 1 can be applied to the backward state metric $\beta_{k+1}(s')$.

Lemma 2 (Boundedness in difference) *Let $B_{\max}$ be the maximum absolute value of branch metrics, and*

$$|\gamma_k(s, s')| \leq B_{\max}, \quad \forall k \in T, \forall s, s' \in S \quad (5)$$

We define the difference between metrics as

$$\Delta_k(s_1, s_2) = |\alpha_k(s_1) - \alpha_k(s_2)|, \quad k \in T, s_1, s_2 \in S. \quad (6)$$

Then, the bound of difference between metrics is given by

$$\max \Delta_k(s_1, s_2) \leq \Delta_{max} = (2B_{\max} + \ln(2))m \quad (7)$$

where m is memory order of the convolution code.

Proof) see [4], [5]. □

Using Lemma 1 and 2, we can express $\alpha_k(s)$ as a finite precision along with a rescaling function. This property also holds for $\beta_k(s)$.

3. MODULO 2^B BASED STATE METRIC REPRESENTATION

In this section, we show the invariance of the log-likelihood computation in mod based operation, which is key factor for preventing rescaling. In order to do this, we first define the finite group Z_b associated with modulo addition for representing the metric. After that, we present the invariance in the metric and LLR computation.

3.1. mod 2^b based metric representation

In order to express the modulo representation that is exactly matched with two's complement number system, we slightly modify the definition of modulo operation.

Definition 3 *Let $u' \in \{-2^{b-1}, \cdots, 2^{b-1} - 1\}$ be a mod 2^b representation of $u \in Z$, i.e., $u' \equiv u \mod 2^b$, then*

$$u = u' - k \cdot 2^b \quad (8)$$

where $k = \min\{i \mid \arg\min_i |u + i \cdot 2^b|, i \in Z\}$.

Example 4 *If $b = 3$ and $u = -9$, then $-9 \mod 2^3 = -1$ as $k = \arg\min_j |-9 - j \cdot 2^3| = -1$. If $b = 3$ and $u = 4$, then both 0 and -1 satisfy $\arg\min_j |4 + j \cdot 2^3| = 1$. Clearly, $-1 = \min\{0, -1\}$, so $4 \mod 2^3 = 4 + -1 \cdot 2^3 = -4$.*

Note that $u \mod 2^b$ represents b-bit two's complement representation. We denote the modulo addition operation as $\oplus_b$.

Lemma 5 *A set $Z_b = \{-2^{b-1}, \cdots, 2^{b-1} - 1\}$ equipped with modulo 2^b addition $\oplus_b$ is an abelian group.*

Proof) It is easy to check that 0 is an identity and $u \in Z_b$ has an inverse in Z_b. In addition, for any $u, v, w \in Z_b$, it is also easy to show the closure property $u \oplus_b v \in Z_b$ and associativity $u \oplus_b (v \oplus_b w) = (u \oplus_b v) \oplus_b w$. Since it satisfies all group axioms, and is also commutative, it is an abelian group. □

Lemma 6 *For a given b, a map $\phi : Z \to Z_b$ defined by $\phi(u) = u \mod 2^b$ is a homomorphism.*

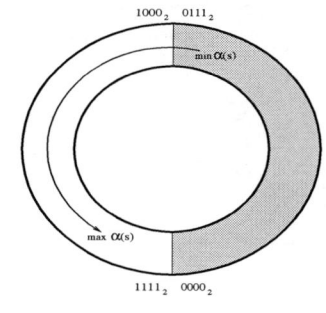

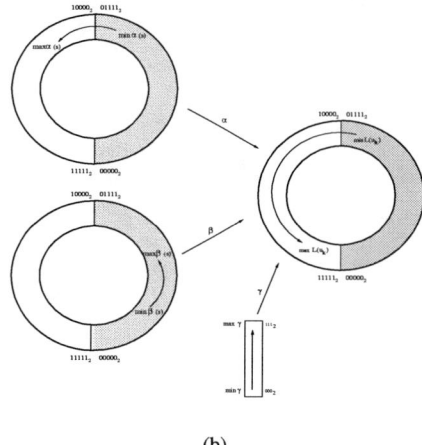

Fig. 1. Principle of modulo metric representation; (a) Path metric, and (b) LLR.

Proof) For any $u, v \in Z$, we can express them as $u = m \cdot 2^b + \phi(u)$ and $v = n \cdot 2^b + \phi(v)$. We then have $u + v = (m + n) \cdot 2^b + (\phi(u) + \phi(v))$, and $\phi(u + v)$ is the addition of $\phi(u)$ and $\phi(v)$ in Z_b ($\phi(u) \oplus_b \phi(v)$). That is,

$$\phi(u + v) = \phi(u) \oplus_b \phi(v). \quad (9)$$

Thus, ϕ is a homomorphism. □

We skip the rigorous definition of inverse and denote the modulo subtraction as $\ominus_b$.

Corollary 7 $u - v = \phi(u) \ominus_b \phi(v).$

For convenience, we express $\tilde{u} = \phi(u)$ from here.

Theorem 8 (Difference preservation) *Let u and v be given integer values satisfying $|u - v| < \Delta$. In addition, suppose $\tilde{u}$ and $\tilde{v}$ are mod 2^b representation of u and v in group Z_b. If we set $b = \lceil \log_2 \Delta \rceil + 1$, then the result is invariant and also preserves the range, i.e.,*

$$(i) \quad \tilde{u} \ominus_b \tilde{v} = u - v \quad (10)$$

$$(ii) \quad |\tilde{u} \ominus_b \tilde{v}| < 2^{b-1} \quad (11)$$

Table 1. ACSU complexity comparison.

metric	operation	conventional (add, select)	proposed (add, select)
α	ACS	$3N_s, N_s$	$3N_s, N_s$
	Min. metric	N_s-1, N_s-1	$0, 0$
	Rescaling	$N_s, 0$	$0, 0$
β	ACS	$3N_s, N_s$	$3N_s, N_s$
	Min. metric	N_s-1, N_s-1	$0, 0$
	Rescaling	$N_s, 0$	$0, 0$
total		$10N_s-2, 4N_s-1$	$6N_s, 2N_s$

Table 2. ACSU delay comparison.

metric	conventional	proposed
serial	$4T_a + 2T_s$	$2T_a + T_s$
parallel	$(3+\lceil \log_2 N_s \rceil)T_a + (1+\lceil \log_2 N_s \rceil)T_s$	$2T_a + T_s$

Proof) From the definition 3, it is clear that

$$\phi(x) = x, \text{ if } -2^{b-1} \leq x \leq 2^{b-1} - 1. \quad (12)$$

Note also that $\Delta = 2^{\log_2 \Delta} \leq 2^{\lceil \log_2 \Delta \rceil} = 2^{b-1}$ and thus $|u - v| < 2^{b-1}$. Hence,

$$u - v \underbrace{=}_{(a)} \phi(u-v) \underbrace{=}_{(b)} \phi(u) \ominus_b \phi(v) = \tilde{u} \ominus_b \tilde{v}. \quad (13)$$

where (a) is by (12) and (b) is by corollary 7. Immediately from (13) and given condition, we have $|u-v| = |\tilde{u} \ominus_b \tilde{v}| < 2^{b-1}$. □

Corollary 9 $\tilde{\alpha}_k(s)$ *(or* $\tilde{\beta}_k(s)$*), a mod 2^b representation of $\alpha_k(s)$ (or $\beta_k(s)$) in Z_b with $b = \lceil \log_2 \Delta_{\max} \rceil + 1$, is invariant in the metric update process in (1) and (3).*

Proof) It is immediate by replacing u, v with $\alpha_k(s_1)$ and $\alpha_k(s_2)$. Same for β. □

The difference preservation property can be illustrated in a circle as shown in Fig. 1(a). In principle, one additional bit is required for resolving an ambiguity in the metric computation. As an example, suppose $\Delta_{\max} = 6$ (hence $b = 4$) with the maximum and minimum metric at time k to be $\alpha_k(s_1) = 29$ and $\alpha_k(s_2) = 23$, respectively. Then their modulo representation will be $\tilde{\alpha}_k(s_1) = -3(= (1101)_2)$ and $\tilde{\alpha}_k(s_2) = 7(= (0111)_2)$, respectively. So the original and modulo difference are equal:

$$\alpha_k(s_1) - \alpha_k(s_2) = 6$$
$$\tilde{\alpha}_k(s_1) \ominus_b \tilde{\alpha}_k(s_2) = (-3 - 7) \bmod 16 = 6.$$

The mod 2^b based metric representation discussed so far works well in a single metric update process such as Viterbi decoding [5]. However, log-likelihood computation in (4) requires the sum of forward, backward, and branch metrics and hence it may corrupt the LLR results when the different signed metric is added together. In the next subsection, we discuss the methods to avoid an ambiguity in arithmetic operations.

3.2. Modulo 2^b based LLR Representation

Here, we show the main property that log-likelihood computation is invariant in mod 2^b operation by the proper precision allocation.

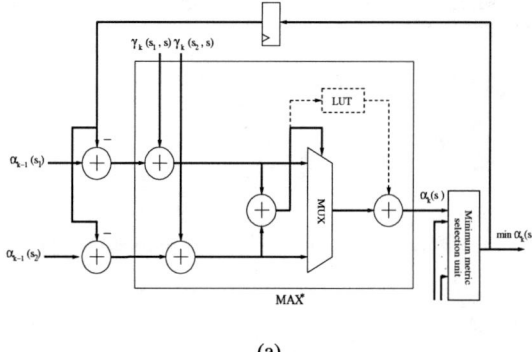

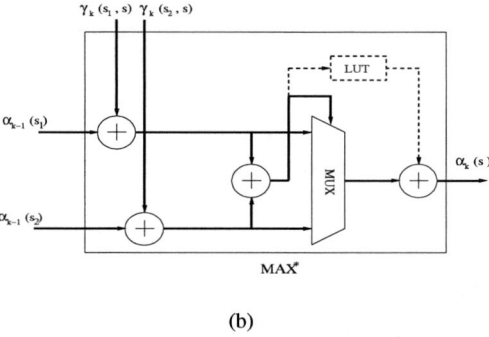

Fig. 2. ACSU architecture; (a) Rescaling based, and (b) Proposed.

Theorem 10 *Let $u_1, \cdots, u_N$ and $v_1, \cdots, v_N$ be given integer values satisfying $|u_j - v_j| < \Delta_j$, $\forall j \in \{1, ..., N\}$. Further, let η_1 and η_2 be the sum of integers u_i and v_i, $i \in \{1, \cdots N\}$,*

$$\eta_1 = \sum_{i=1}^{N} u_i, \; \eta_2 = \sum_{i=1}^{N} v_i. \quad (14)$$

Suppose $\tilde{\eta}_1$ and $\tilde{\eta}_2$ be the elements of Z_b with $b = \lceil \log_2 \sum_j \Delta_j \rceil + 1$, then

$$\tilde{\eta}_1 \ominus_b \tilde{\eta}_2 = \eta_1 - \eta_2 b, \text{ and } |\tilde{\eta}_1 \ominus_b \tilde{\eta}_2| < 2^{b-1}$$

Proof) From condition,

$$|\eta_1 - \eta_2| = |\sum_{i=1}^{N}(u_i - v_i)| \leq \sum_{i=1}^{N}|u_i - v_i| < \sum_{i=1}^{N}\Delta_j \quad (15)$$

and also $\sum_j \Delta_j = 2^{\log_2 \sum_j \Delta_j} \leq 2^{\lceil \log_2 \sum_j \Delta_j \rceil} = 2^{b-1}$. From (15), $|\eta_1 - \eta_2| < 2^{b-1}$, by applying similar step to theorem 8, we get

$$\eta_1 - \eta_2 = \phi(\eta_1 - \eta_2) = \phi(\eta_1) \ominus_b \phi(\eta_2) = \tilde{\eta}_1 \ominus_b \tilde{\eta}_2 \quad (16)$$

and $|\tilde{\eta}_1 \ominus_b \tilde{\eta}_2| < 2^{b-1}$. □

By applying Theorem 8, we can show that difference preservation property holds for LLR computation.

Corollary 11 (Difference preservation in LLR) *The computation of $L(u_k)$ is invariant in mod 2^b operation in Z_b if b is set to*

$$b = \lceil \log_2(2\Delta_{\max} + B_{\max}) \rceil + 1. \quad (17)$$

Proof) Recall the LLR computation in (4) is the difference between two component LLR values and LLR is the sum of α, β, and γ. By Lemma 2, α and β are bounded by $\Delta_{\max}$ and the maximum of γ is $B_{\max}$, so this case exactly corresponds to $N=3$ of Theorem 11. Therefore, for any $\tilde{l}_k(s'_1, s_1)$ and $\tilde{l}_k(s'_2, s_2)$, we have

$$l_k(s'_1, s_1) - l_k(s'_2, s_2) = \tilde{l}_k(s'_1, s_1) \ominus_b \tilde{l}_k(s'_2, s_2). \quad (18)$$

This condition also holds for any specific $l_k(s', s)$, so

$$L(u_k) = \max_{s',s:u_k=+1} l_k(s'_1, s_1) - \max_{s',s:u_k=-1} l_k(s'_2, s_2)$$
$$= \max_{s',s:u_k=+1} \tilde{l}_k(s'_1, s_1) \ominus_b \max_{s',s:u_k=-1} \tilde{l}_k(s'_2, s_2). \quad (19)$$

This proves corollary. □

4. COMPLEXITY ANALYSIS

Figure 2(a) shows the architecture of conventional add-compare-select unit (ACSU). Suppose the number of states in the decoder to be $N_s = 2^m$, then the path metric update unit requires $3N_s$ addition ($4N_s$ if we consider the log term) and N_s selection operations. Besides this, minimum metric selection unit (MMSU) requires $N_s - 1$ addition and selection operations to produce $\min \alpha_k(s)$. Note that adder in conventional scheme needs to include saturation function to avoid overflow when the signal reaches the maximum. Once $\min \alpha_k(s)$ is obtained, subtraction for metric rescaling is performed in the next cycle to get the rescaled outputs, which also requires N_s additions. Whereas, neither rescaling nor adjustment of precision are required in path metric update and LLR computation of the proposed scheme (see Fig. 2(b)). Hence, the proposed scheme only requires $3N_s$ addition and N_s selection operations for forward and backward metric computation (see Table I). In case of $m=3$ ($K=4$) which is a standard for WCDMA [6], the complexity reduction in add and select operation of ACSU are 38% and 48%, respectively.

Besides this, the speed increase of proposed scheme is also considerable. Table II compares the critical path delay of the proposed and conventional scheme, where T_a and T_s are the delay of adder and mux, respectively. While more than 50% of the delay is due to the MMSU and rescale operation in the conventional method, the proposed method does not require them and hence can reduce significant amount of delay. For a 0.18 μm standard cell technology, the speed increase of the proposed scheme ($m=3$) given by $\Delta T = (T_{conv} - T_{prop})/T_{conv} \times 100$ is 50% (serial) and 68% (parallel), respectively.

4.1. Precision comparison

In order to observe the performance for the precision variation, simulation of the proposed rescalingless log-MAP algorithm and conventional one is performed. The polynomial we used in our simulations are $(13, 15)_8$ which has 8-states [6] and the block size is set to $N = 1000$. For all our simulations, we used 4 bit ADC model in the receiver and AWGN noise model is assumed. In case of conventional fixed point simulations, saturation scheme is used to avoid overflow when the internal value reaches the maximum value. In the proposed scheme, the precision is set to 9 bits by (17). Figure 3 shows the simulation results of of proposed scheme and that of conventional method (both floating point and fixed point). We can see that the performance of the proposed scheme is nearly same as that of infinite precision. Besides, we can also observe that

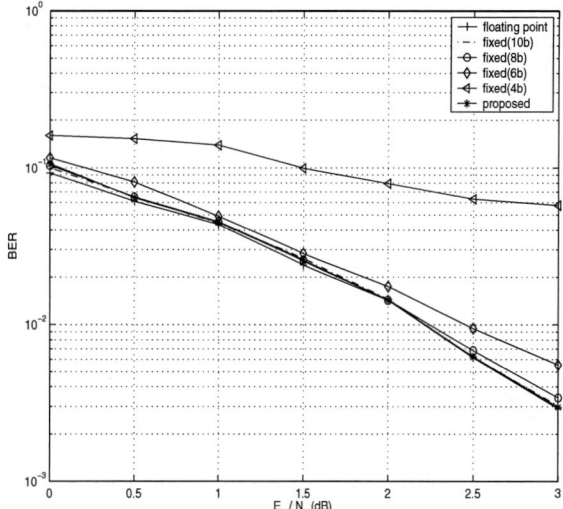

Fig. 3. BER performance of proposed scheme.

the results of proposed scheme show exactly same performance (no algorithmic difference) as conventional fixed point having precision $b > 9$ where no saturation is occurred.

5. CONCLUSION

In this paper, we presented a novel state metric representation scheme for Log-MAP decoding which does not require any rescaling in both forward and backward path metrics and LLR. Specifically, proposed scheme exploits the homomorphism in a finite abelian group Z_b associated with modulo 2^b addition operation for computing both path metrics and LLR without rescaling. Further research is being directed towards the application of proposed scheme into Turbo decoding to achieve improvement in decoding speed, complexity, and design time.

6. REFERENCES

[1] L. R. Bahl, J. Cocke, F. Jelinek, and J. Raviv, "Optimal decoding of linear codes for minimizing symbol error rate," *IEEE Trans. on Info. Theory.*, vol. 42, pp. 284-287, March. 1974.

[2] C. Berrou, A. Glavieux, and P. Thitimajshima, "Near Shannon limit error-correcting coding: Turbo codes," in *Proc. IEEE Int. Conf. Commun.*, pp. 1064-1070, May 1993.

[3] C. Douillard, A. Picart, P. Didier, M. Jezequel, C. Berrou, and A. Glavieux, "Iterative correction of intersymbol interference: Turbo equalization," *European Trans. on Telecomm.*, pp. 507-511, Sept. 1995

[4] E. Boutillon, W. J. Gross, P. G. Gulak, "VLSI architectures for the MAP algorithm," *IEEE Trans. Comm.*, vol. 51, pp. 175-185, Feb. 2003.

[5] A. P. Hekstra, "An alternative to metric rescaling in Viterbi decoders," *IEEE Trans. Commun.*, vol. 37, pp. 1220-1221, Nov. 1989.

[6] 3G TS 3G TS 25.212, "3rd Generation Partnership Project; Technical Specification Group Radio Access Network; Multiplexing and channel coding (FDD),".

Gradient Pile up for Edge Detection on Hardware

André Soares, Altamiro Susin
Instituto de Informática
Universidade Federal do Rio Grande do Sul - UFRGS,
Porto Alegre, Brazil
borin@inf.ufrgs.br

Leticia Guimarães, Viviane Cordeiro
Departamento de Engenharia Elétrica
Universidade Federal do Rio Grande do Sul – UFRGS
Porto Alegre, Brazil
leticia@eletro.ufrgs.br

Abstract - Edge detection plays a fundamental role on image processing Many edge detectors have been proposed. Most of them are based on step edges model and applying smooth filters to minimize the noise and the image derivative or gradient to enhance the edges. However, image sensors have a limited spatial bandwidth producing ramp edges with the same gradient magnitude as those produced by noise. This work presents an hardware implementation of the proposed edge detection algorithm. Our approach enhances the gradient correspondent to ramp edges without amplifying the noisy ones, for real time computer vision applications. The experimental results show that the proposed hardware implementation enhances the gradient of ramp edges, improving the gradient magnitude without shifting the edge location. Moreover, the execution time is appropriated for real time applications.

I. INTRODUCTION

The estimation of the boundary location of an object is an important subject on image processing. Since edge detection have been studied, several techniques for edge enhancement were proposed. Most of the edge enhancement are based on step edges model and apply the derivative to enhance the edge of the objects on an image. Usually, the derivative of an image is produced by the convolution with a gradient operators [1], as Sobel, Roberts, Prewitt, for example.

However, there are three problems on the use of the derivative as edge detector: a) false edges produced by noise are enhanced; b) ramp edges produce lower derivative magnitude; c) the location of the ramp edges are not accurate. Several edge detection methods as proposed by Marr-Hildreth [2], Canny [3], Perona-Malik [4], have been proposed in order to provide tools to overcome the edge detection problems by applying the derivative to edge detection. Most of these methods apply a smooth gaussian operator in order to suppress the noisy edges, then, estimate the location of the edges by detection of the zero-crossing on the second derivative of an image. Edge detectors proposed by Petrou-Kitter[5] and Zang et. al. [6] are based on ramp edges model. Petrou-Kitter[5] propose a ramp edge model and an optimal detector. Zang et. al. [6] applies correlation matching to enhance the gradient at ramp edge location. In [7] D. Wang proposed a multiscale gradient algorithm that enhances the morphological gradient of ramp edges.

Our approach aims to enhance the gradient value of ramp edges without enhances noisy gradient. In order to achieve our objective the enhancement occurs over the two component vector of the gradient, the horizontal and vertical gradient map separately. First, a smooth filter is applied to the original image. Second, the horizontal and vertical components of the gradient of an image are calculated. Third, each component of the gradient generates a connected component map, the gradient enhancement map (GEM), that guides the gradient enhancement process. Finally, the horizontal and vertical gradient map are enhanced by a process similar to a pile up, guided by their respective GEM. The proposed method presents many processing steps to be efficiently calculated in real time. This work presents a hardware implementation to use the method in real machine vision applications.

II. GRADIENT MAP CALCULATION

There are two components of the gradient, the horizontal and vertical component. The two-dimensional function gh(x,y) represents the horizontal gradient map and gv(x,y) represents the vertical gradient map of an image f(x,y), respectively. They are calculated by the convolution of the image f(x,y) with a horizontal and a vertical gradient operator. The gradient magnitude map g(x,y) is the magnitude of the vector composed by two components.

III. GRADIENT ENHANCEMENT MAP GENERATION

The gradient enhancement map GEM guides the pile up process. The pile up process occurs on two gradient maps separately following the same rules. Therefore, two gradient enhancement maps are generated, $GEM_H(x,y)$ that guides the enhancement of the horizontal gradient gh(x,y) and $GEM_V(x,y)$ that guides the enhancement of the vertical gradient gv(x,y).

The GEM is based on connected components map proposed by Bieniek and Moga [8]. The GEM is composed by seven elements: a) zero (Z); b)right arrow (→); c) left arrow (←); d) up arrow (↑); e) down arrow(↓); f) plateau and (P); g) maximum (M). In [8] the connected components map is used to locate the minima, in our approach we intend to locate the maxima. Therefore, the arrows on GEM point to the maximum neighbor in magnitude. Since the direction of the arrows are related to the gradient direction, GEM_H is composed only by up and down arrow and GEM_V is composed by only left and right arrow.

The GEM is generated in two steps: first, the initial map is generated, then, the final map is produced eliminating false maximum plateaus and detecting hidden zeros of the initial map. The false maximum plateaus are labeled with arrows and the true maximum plateaus with M.

The plateaus P are the candidates to maximum M and are present only on the initial map. Considering that the origin of an image coordinates is located at the upper left of an image, the labeling for the initial GEM is as follows:

$$gem_h(x,y) = \begin{cases} Z & if\ gv(x,y)=0 \vee (|gv(x,y)|<|gv(x-1,y)| \wedge |gv(x,y)|<|gv(x+1,y)|), \\ \leftarrow & if\ |gv(x,y)|<|gv(x-1,y)| \wedge |gv(x,y)|\geq|gv(x+1,y)|, \\ \rightarrow & if\ |gv(x,y)|<|gv(x+1,y)| \wedge |gv(x,y)|\geq|gv(x-1,y)|, \\ P & otherwise. \end{cases}$$

where $gem_h(x,y)$ represents the initial horizontal enhancement map.

The final GEM is obtained by the detection of noisy zeros and false maximum plateaus detection and substitution. This step is performed as follows:

$$GEM_H(x,y) = \begin{cases} \downarrow & if\ gem_h(x,y)=P \wedge gem_h(x,y-1)=\downarrow \wedge gem_h(x,y+1)=P, \\ \uparrow & if\ gem_h(x,y)=P \wedge gem_h(x,y+1)=\uparrow \wedge gem_h(x,y-1)=P, \\ M & if\ gem_h(x,y)=P \wedge gem_h(x,y-1)=\downarrow \wedge gem_h(x,y+1)=\uparrow, \\ Z & if\ (gem_h(x,y)=\uparrow \wedge gem_h(x,y-1)=\downarrow) \vee (gem_h(x,y)=\downarrow \wedge gem_h(x,y+1)=\uparrow). \end{cases}$$

where $GEM_H(x,y)$ represents the final horizontal enhancement map. The same process is applied in the vertical direction.

Fig. 1 (a) shows an one-dimensional ramp edge, the corresponding gradient, the initial and final GEM. Notice the gradient value is only one, even though, the difference between **a** and **b** level is five. Furthermore, the gradient location is not evident, in this case.

IV. GRADIENT ENHANCEMENT PROCESS (PILE UP PROCESS)

The gradient piled-up process is guided by the gradient enhancement maps GEM of the gradient maps, the horizontal and vertical map. The GEM indicate the direction, the start and the end points of the gradient pile up process. The arrows indicate the pile up direction. The start point of the pile up process is an arrow that the neighbor is zero (Z) and points to another arrow or maximum. The gradient value of a point at an arrow is piled up over the neighbor gradient value that presents an arrow of same direction or a maximum. The process ends at a maximum. In Fig. 1, observe, the arrows on final GEM in Fig.1(a) are substituted by zeros (Z) as the pile up occurs, in Fig. 1(b). At the end of the pile up process, the GEM presents only maxima (M) and zeros (Z) and the gradient is enhanced from one, the original gradient value, to five, the difference between **a** and **b** level

in Fig. 1(a). Fig. 2 shows the flowchart for the left and right pile up process to the vertical gradient enhancement.

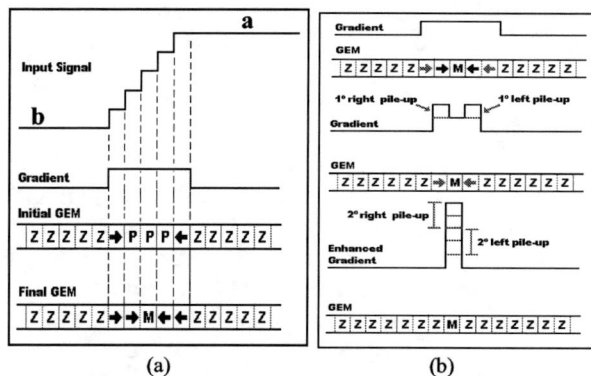

(a) (b)
Fig. 1. Example of gradient enhancement process for one dimensional signal; (a) gradient enhancement map; (b) Gradient enhancement process (pile up process) for the one dimensional ramp edge of the Fig. 1(a).

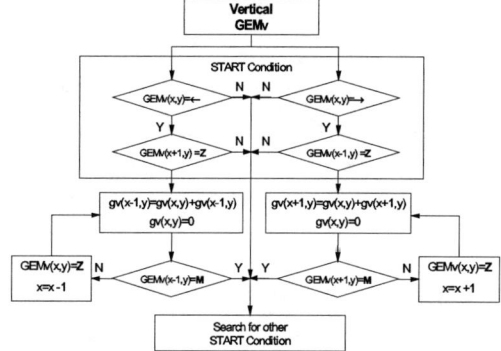

Fig. 2. Schema of the gradient enhancement process (pile up algorithm) for the vertical gradient enhancement.

More details and a performance evaluation of the method can be obtained in[10].

V. IMPLEMENTATION

Hardware presents a possibility to execute many operations in parallel, but there is a bottleneck in the communication between several processors and a main memory. The approach used here is to reduce the number of clock cycles per pixel without divide the image among a giver number of processors.

This idea can be exploited employing the amount of parallelism available in the neighborhood of each pixel for each operation. A pixel and its neighbors are stored in registers like an architecture of a filter, but different operators for neighbor relationship are developed in a combinational form.

All operations are executed first over the lines and then over the columns, with exception of the filtering operation. Operations are executed using line-buffers to store intermediate results. The horizontal piled-up gradient must be stored in a separate memory region because it is added to the vertical piled-up gradient, and even having completed the first phase (horizontal gradient) the original image must be stored to calculate the second phase. The final system is composed by the main processor, a memory and an IO manager. The complete pile-up method is executed in hardware in the follow steps:

A. Step 1: filtering

The 3x3 filtering operation was executed in pipeline by storing the filter input values in 9 registers preserving the neighborhood positions and the pixel values are shifted one pixel position each clock. Three lines of the image are stored to reduce the memory access. The output of the filter is stored in a line buffer to the next step. The filter is a combinational circuit that uses 9 multipliers and a tree of adders. There is one output of the filter in each clock cycle. The output of the filter is thresholded with a value to eliminate gradients of very small value.

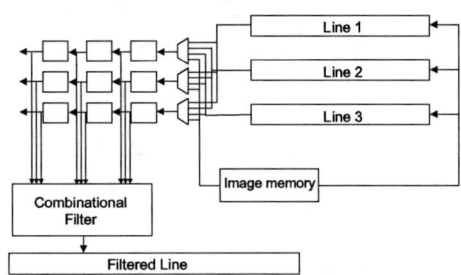
Fig. 3: Schematic representation of the filtering block

B. Step 2: map construction

In this step the output of the filter stored in the line buffer is analyzed to mount the map according to the rules presented in previous section. The line feeds three registers in a pipeline, to permit the evaluation of the relative to the neighborhood of each pixel, in parallel. This permit to generate each map element in a clock cycle. The condition evaluator block consists of comparators to verify the neighborhood of each pixel and the output of the comparators are connected to a LUT with the possible components of the map.

C. Step 3: plateau processing

In this step, the pixels indicated as maximum points are assessed to verify if they are a real maximum or if they point to a maximum. In this step, each pixel and its neighborhood must be evaluated with fixed conditions in each line assessment in order to find the correct position of the maximum. In order to maintain fixed conditions, the result of the assessment is stored in a second line buffer(fig. 5). As several assessments can be necessary to each image line, this two line buffers are used in an alternate way, until no pixels change its condition. The conditions are evaluated in one clock using three registers as in the last step. The maximum assessment block is a LUT with the outputs to the possible neighborhood inputs. Two map lines are necessary due to the recursive verification of the plateau. Each time the step is repeated one of the buffers is used.

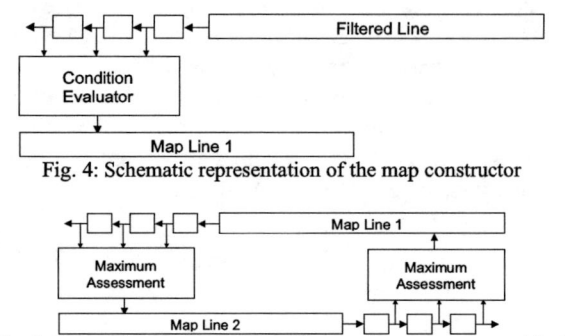
Fig. 4: Schematic representation of the map constructor

Fig. 5: Schematic representation of the maximum assessment block

D. Step 4: pile-up

In this step, the pile-up operation is executed. In a first phase the map line buffer is sweept in one direction, verifying which elements must be piled-up. Intermediate results are stored in an accumulator and when a maximum is reached or a junction of two opposite directions, the value of the accumulator is stored in the output line buffer. The same procedure is executed in the opposite direction. In fig. 6 a schematic representation of the circuit is shown. As the previous step can store the processed line in one of the two map line buffers, a multiplexer selects which line will be processed. The Pile-Up assessment block is a LUT that is used to verify if the gradient is or is not accumulated to be piled-up and stored in the maximum value.

E. Step 5: Transfer content to the memory

In this step, the content of the piled-up line is transfered to the memory. If the architecture is processing the lines, the result is stored directly in the memory. When the columns are being processed the output is added to the result of the previous operation.

Steps 1-4 are also executed for the columns, in the same way. All steps have finite duration with exception of step 3. We can estimate (and adjust) a maximum duration of step 3 by combining the threshold value with the limits of the image values. For example, assuming that the threshold value is 10 and the value of the pixel is in the range 0-255 we well have no more than 25 adjacent pixels in a plateau. If more than 25 adjacent pixels could have the same value the image will be out of the range 0-255 and if the pixels have different gradient values, then there is a smaller plateau. In this case, the line is verified a maximum of 25 times.

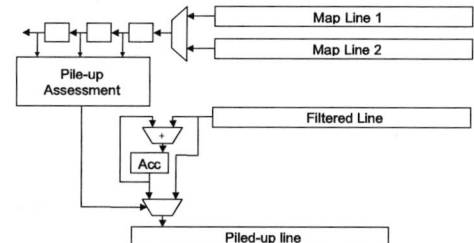

Fig. 6: Schematic representation of the pile-up assessment block.

VI. RESULTS

The entire architecture was developed in VHDL and prototyped in a FPGA running at 50MHz with a single communication channel to a PC. A C++ program controls transfers from the PC to the board and back to the PC. In Fig. 8 can be observed a picture of the experimental setup.
Each pixel of the image needs a minimum of 12 and a maximum of 60 cycles to be processed(horizontal and vertical). A 3.2 megapixel image, for example, can be processed in 1 second (worst case) if the architecture runs at 200MHz. Typical images in machine vision applications (640x480) can be processed in 90ms (worst case) at the same frequency. The cost of the architecture is 2743 4 input LUTs for the prototype in FPGA, 6 8x256 SRAM and 3 2x256bit SRAM. The prototype can run until 67.8 MHz (estimated). We can observe in fig 7 the original Lena image (grayscale, 256x256) and the result of the pile-up process. The processing is executed in 986085 cycles (15 clocks/pixel in this case). Performance can be enhanced if we consider that the image is stored in a memory bank to permit more than one memory access simultaneously. Each phase (line processing or column processing) can be processed simultaneously due to the way that the algorithm is executed.

In order to compare the results obtained with the custom architecture,a version of the pile-up was coded for the ADSP21XX family of DSP processors. In the DSP, the same Lena image (grayscale,256x256) is processed in 105228666 cycles. The difference is due to the number of instructions needed to execute each operation which needs only one cycle in the developed architecture.

VII. CONCLUSIONS

The pile-up method for edge detection was presented. An architecture to perform in hardware the pile-up edge detection was developed and prototyped. The performance of the architecture was maximized considering the conventional memory communication bottleneck. The communication between the memory and the processor was reduced to the image transfer. One advantage is that the architecture is scalable, permitting to process images with different sizes with a processing time proportional to the number of pixels of the image. The limitation is that line buffers must have the size of the larger image to be processed.

Obtained results showed that the developed architecture presents a high performance once that each processing step is executed in only cycle for each pixel of the image.

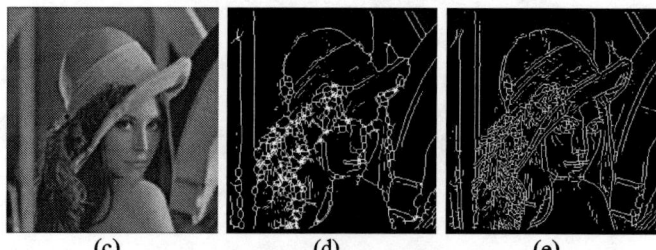

Fig.7: Edge detection on Lena; (a) original image; (b) ideal edge map of Lena. (c) edge map obtained using the pile-up method architecture

Fig. 8: Experimental setup used to evaluate the architecture.

ACKNOWLEDGEMENTS

This work was supported by Conselho Nacional de Pesquisa CNPq – BRAZIL.

REFERENCES

[1] Gonzalez, R., Woods, R., Digital Image Processing, Addison-Wesley, 1992.
[2] Marr, D. and Hildreth, E. C., Theory of Edge Detection, Proc. of the Royal Society of London B207, pp. 187-217, 1980.
[3] Canny, J., A Computational Approach to Edge Detection, PAMI, V. 8, No.6, pp. 679-698, 1986.
[4] Perona, P. and Malik, J., Scale-Space and Edge Detection Using Anisotropic Diffusion, PAMI ,V.12 , No. 7, 1990.
[5] Petrou, M. and Kitter, J., Optimal Edge Detectors for Ramp Edges, PAMI, V. 13, No. 5, pp. 483-491, 1991.
[6] Wang, Z., Rao, K. R. and Ben-Arie, J., Optimal Ramp Edge Detection Using Expansion Matching, PAMI, V. 18, No. 11, pp. 1092-1097, 1996.
[7] Wang, D., A Multiscale Gradient Algorithm for Image Segmentation using Watersheds, Pattern Recognition, V. 30, No. 12, pp. 2043-2052, 1997.
[8] Bieniek, A and Moga, A., An efficient watershed algorithm based on connected components. Pattern Recog., V. 33 , No. 6 , pp. 907-916, 2000.
[9] Ballard, D.H., Brown, C.M., Computer Vision, Prentice Hall Inc., 1982.
[10] Guimarães,L.V., Soares, A.B., Cordeiro,V., Susin,A., Gradient Pile Up Algorithm for Edge Enhancement and Detection. ICIAR, 2004

Area-Efficient Systolic Architectures for Inversions over $GF(2^m)$

Zhiyuan Yan
Department of ECE
Lehigh University
Bethlehem, PA 18015
Email: yan@lehigh.edu

Dilip V. Sarwate
Department of ECE
University of Illinois
Urbana, IL 61801
Email: sarwate@uiuc.edu

Zhongzhi Liu
Department of ECE
Lehigh University
Bethlehem, PA 18015
Email: liz4@lehigh.edu

Abstract—Based on a *new* reformulation of the extended Euclidean algorithm, we propose two types of in-place systolic architectures for inversion in $GF(2^m)$: *bit-parallel* and *folded bit-parallel* architectures. Our bit-parallel architectures have throughput of either $1/(2m-1)$ with interleaving or $1/(4m-2)$ without interleaving. Compared with the best previously proposed bit-parallel architectures of which we are aware, our new architectures require *less* hardware and achieve *shorter* critical path delays with approximately the *same* throughput and latency. We also propose a folded version of our bit-parallel architectures which achieves the $1/(4m-2)$ non-interleaved throughput with even less hardware; To our best knowledge, no comparable architecture has been proposed before.

I. INTRODUCTION

Finite fields of characteristic 2, $GF(2^m)$, have been widely used in cryptography, digital signal processing, and error-control codes. Among the arithmetic operations in $GF(2^m)$, inversion and division are the most complicated, and various algorithms and architectures for inversions/divisions have been proposed (see, for example, [1], [3]–[6] and the references therein). The extended Euclidean algorithm (EEA) (see, e.g., [1], [5], [6]) and the extended Stein's algorithm (ESA) (see, for example, [3]) have attracted some attention since systolic architectures based on the EEA or ESA can achieve $O(m^2)$ area-time (AT) complexity and $O(m)$ latency. In this paper, we focus on inversion architectures based on the EEA.

Many high-speed architectures based on the EEA (see, for example, [1], [5], [6]) consist of two-dimensional arrays of cells with area complexity $O(m^2)$. With continuous inputs, these architectures achieve throughput 1, producing one output per clock cycle. These architectures are suitable for applications with high throughput requirements, but are not practical for applications that require high hardware efficiency and only moderate throughputs. Instead, architectures designed for applications of the latter type (see, for example, [1], [4]) trade throughput for lower area complexity and achieve a different balance between area complexity and throughput: they have area complexity $O(m)$ and throughput $O(1/m)$. These more hardware-efficient architectures are referred to as *in-place* architectures henceforth.

In this paper, we design two types of new in-place architectures for inversion based on a variant of the EEA we recently proposed [5]. Our bit-parallel architectures have throughput of either $1/(2m-1)$ with interleaving or $1/(4m-2)$ without interleaving. Compared with the best previously proposed bit-parallel architectures of which we are aware, our new architectures require *less* hardware and achieve *shorter* critical path delays (CPDs) with approximately the *same* throughput and latency. We also propose a folded version of our bit-parallel architectures which achieves the $1/(4m-2)$ non-interleaved throughput with even less hardware. The folded bit-parallel architecture is suitable for applications where either interleaving is not possible or a throughput of $1/(4m-2)$ is sufficient. To our best knowledge, no comparable scheme has been proposed previously. These two types of architectures provide different balances between throughput and hardware cost, and offer designers a range of options.

We treat only inversion architectures in this paper, which suffices for those applications wherein only inversions are needed. However, divisions are commonly needed, and many architectures (for example, [1] and [4]) have been designed to compute divisions B/A directly. The architectures in [1] can be simplified if only inversion is needed while the architectures in [4] cannot. Other architectures (for example, [6]) compute divisions B/A as an inversion followed by a multiplication: $A^{-1} \cdot B$. We use this approach for the same reason as in [6]. Since hardware implementations of finite field multiplications have been well studied, we do not include the details here. To ensure fairness, we compare our inversion architectures only with the simplified versions of the previously proposed inversion/division architectures in our comparisons in Section III.

II. NEW AREA-EFFICIENT SYSTOLIC INVERSION ARCHITECTURES

A. Bit-Parallel Systolic Architectures

Our new bit-parallel systolic inversion architecture based on Algorithm III of [5], as illustrated in Figure 1(a), consists of a linear array of m type-6 computing cells controlled by one type-7 control cell. The parallel bits of the inputs enter the circuit in a skew fashion, and are then loaded into the cells using a control signal LOAD that propagates across the array. If the inputs to this architecture are interleaved (see, for example, [1]), two sets of inputs are loaded in successive clock cycles. The type-7 cell generates control signals $\text{ADD}^{(j)}$ and $\text{SWAP}^{(j)}$, and passes them to the adjacent type-6 cell. Each

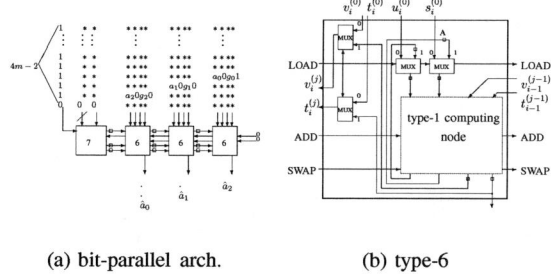

(a) bit-parallel arch. (b) type-6

Fig. 1. New bit-parallel architecture for inversion in $GF(2^3)$ (the initial values shown here are for the adder-based control) and the circuitry for type-6 computing and cell

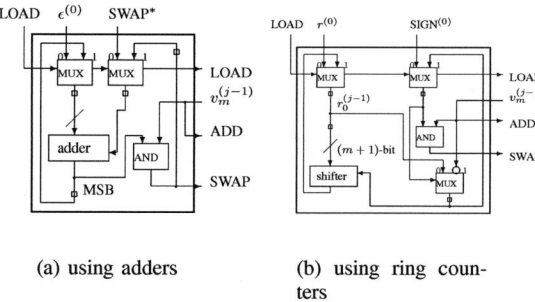

(a) using adders (b) using ring counters

Fig. 2. The circuitry for type-7 control cells

of the m type-6 cells updates the bits in the cell as described in Eqs. (1) and (2) of [5], and passes $\text{ADD}^{(j)}$ and $\text{SWAP}^{(j)}$ onto the next type-6 cell. Also, the i-th type-6 cell passes $v_i^{(j)}$ and $t_i^{(j)}$ to the cell on its left. After $4m-2$ clock cycles, the results are read out of the cells in skew fashion while the next set(s) of inputs are being loaded.

The type-6 computing cell is simply the type-1 computing cell of [5] (not shown here) plus four additional MUXes for initialization. Two implementations of the type-7 control cell are illustrated in Figures 2(a) and 2(b). Since the implementation of the control cell is similar to those in [6] and [5], detailed descriptions are omitted. Note that $\epsilon^{(0)}$ and SWAP* are both initialized to zeros and the $(m+1)$-bit shift register and $\text{SIGN}^{(0)}$ are initialized to $100\cdots000$ and 1.

The area complexity of the bit-parallel architectures is $O(m)$ with a latency of $(5m-2)$ clock cycles. The throughput and the hardware utilization efficiency (HUE) both depend on whether the inputs are interleaved. For interleaved inputs, this architecture produces two inverses every $(4m-2)$ clock cycles and the HUE is 100%; With non-interleaved inputs, this architecture produces only one inverse every $(4m-2)$ clock cycles and the HUE is only 50%. The CPD of the architecture in Figure 1(a) is given by $t_{\text{bp}} = \max\{t_6, t_7\}$, where t_6 and t_7 are the CPDs of the type-6 and type-7 cells respectively. It can be shown that t_{bpa}, the CPD of the bit-parallel architecture in Figure 1(a) using the adder-based control cell, is estimated as $t_{\text{XOR}} + t_{\text{MUX}} + t_{\text{AND}} \leq t_{\text{bpa}} < 2t_{\text{XOR}} + t_{\text{MUX}} + t_{\text{AND}}$, where t_{XOR},

t_{MUX}, and t_{AND} denote the delays of 2-input XOR, MUX, and AND gates respectively. For the control cell in Figure 2(b), the shifter is simply a bank of $(m+1)$ 2-input MUX gates. With this modification, t_{bpr}, the overall CPD of the bit-parallel architecture in Figure 1(a) when using the ring counter-based control cell, is $t_{\text{MUX}} + \max\{t_{\text{XOR}} + t_{\text{AND}}, t_{\text{MUX}}\}$.

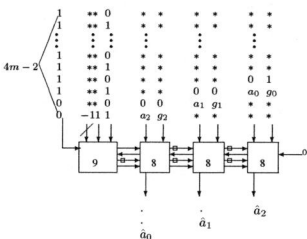

Fig. 3. New folded bit-parallel architecture for inversion in $GF(2^3)$ (the initial values shown here are for the adder-based control cell)

B. Folded Bit-Parallel Systolic Architectures

The HUE of the bit-parallel systolic architecture is only 50% and its throughput is only $1/(4m-2)$. Two techniques are usually used to improve the HUE: data interleaving and folding (processor sharing). Interleaving data has been used in some architectures (see, for example, [1]) to improve the HUE to 100% and the throughput to $1/(2m-1)$. However, interleaving is not practical for some applications. For example, the point multiplication over affine coordinates in elliptic curve cryptography [2] requires field divisions; however, the division operations are not consecutive because a division operation uses the result of the previous division as input. Another possible scenario is that the throughput of $1/(4m-2)$ may be sufficient for some applications, and hence it is not necessary to increase the throughput while improving the HUE. In both cases, folding is a better option since it improves the HUE while maintaining the throughput without interleaving.

As a standard architectural technique, folding is often used to reduce the hardware costs at the expense of throughput. In this case, folding is used to improve the HUE while maintaining the throughput $1/(4m-2)$ (without interleaving) since the HUE of the bit-parallel systolic architecture is 50% to start with. One straightforward option is to fold two processing elements into one. However, this approach needs extra control mechanisms and hence leads to overhead and potentially a longer critical path delay. Instead, we take advantage of the similarity of the operations between the two pairs—u_i and v_{i-1}, s_i and t_{i-1}—and fold these two sets of operations into one. Since the latter approach results in little hardware overhead and only a small increase in the critical path delay, we have adopted it and obtained the following bit-level algorithm for our new folded bit-parallel architecture.

Algorithm IV:

IV.1 Initialization: $P_i^{(0)} = a_i$, $Q_i^{(0)} = g_i$, $P_i^{(1)} = 0$, $Q_i^{(1)} = 0$, and $R_i^{(1)} = Q_i^{(0)}$ for $i = 0, 1, \cdots, m-1$. $\text{UV}^{(0)} = 0$, $\text{UV}^{(1)} = 1$, and $\delta^{(0)} = \delta^{(1)} = -1$.

IV.2 For $j = 1, 2, \cdots, 4m-2$ **do**

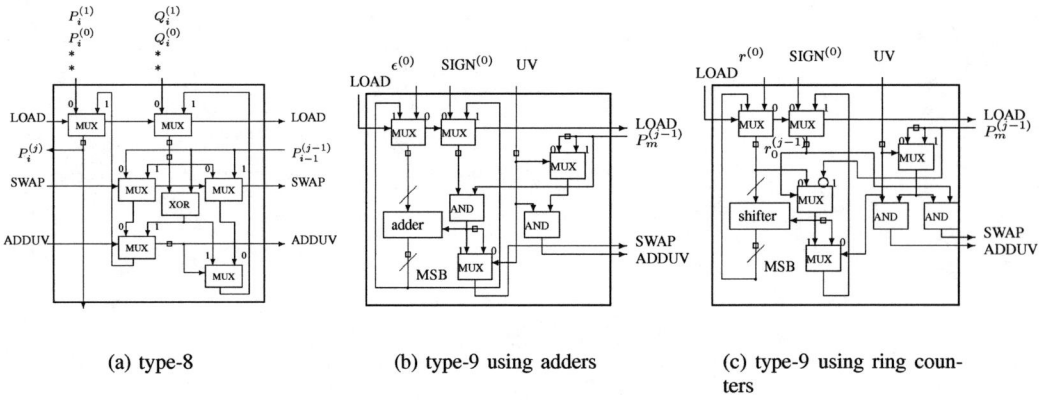

Fig. 4. The circuitry for type-8 computing and type-9 control cells

If $UV^{(j)} = 1$, **then set:** $ADD^{(j)} \stackrel{\text{def}}{=} P_m^{(j-1)}$,
$SWAP^{(j)} \stackrel{\text{def}}{=} P_m^{(j-1)} \wedge (\epsilon^{(j)} < 0)$, and $\epsilon^{(j+1)} = \begin{cases} -1 + \epsilon^{(j)}, & \text{if } SWAP^{(j)} = 0 \\ -1 - \epsilon^{(j)}, & \text{if } SWAP^{(j)} = 1 \end{cases}$
else set: $ADD^{(j)} = ADD^{(j-1)}$, $SWAP^{(j)} = SWAP^{(j-1)}$, $\epsilon^{(j+1)} = \epsilon^{(j)}$.
Set: $ADDUV^{(j)} = ADD^{(j)} \wedge UV^{(j)}$, $R_i^{(j+1)} = Q_i^{(j)}$, $ADDST^{(j)} = ADDUV^{(j-1)}$, and $UV^{(j+1)} = \neg UV^{(j)}$.
and **For** $i = m-1, \cdots, 1, 0$, **set:**
$$\left(X_i^{(j)}, Y_i^{(j)}\right) = \begin{cases} \left(P_{i-1}^{(j-1)}, R_i^{(j)}\right), & \text{if } SWAP^{(j)} = 0, \\ \left(R_i^{(j)}, P_{i-1}^{(j-1)}\right), & \text{if } SWAP^{(j)} = 1, \end{cases}$$
$$P_i^{(j+1)} = \begin{cases} X_i^{(j)}, & \text{if } ADDUV^{(j)} = 0, \\ P_{i-1}^{(j-1)} + R_i^{(j)}, & \text{if } ADDUV^{(j)} = 1, \end{cases}$$
$$Q_i^{(j+1)} = \begin{cases} Y_i^{(j)}, & \text{if } ADDST^{(j)} = 0, \\ P_{i-1}^{(j-1)} + R_i^{(j)}, & \text{if } ADDST^{(j)} = 1, \end{cases}$$

IV.3 Output $\hat{a}_i = P_{m-1-i}^{(4m-1)}$ for $i = 0, \cdots, m-1$.

Algorithm IV is implemented by the folded bit-parallel systolic architecture shown in Figure 3, which consists of m type-8 computing cells and one type-9 control cell. The circuitry for the type-8 computing cell is shown in Figure 4(a), and two possible implementations using adders and ring counters of the type-9 control cell are shown in Figures 4(b) and 4(c) respectively. The area complexity of the folded bit-parallel architecture in Figure 3 is $O(m)$, and the latency of the architecture is $(5m-2)$ clock cycles. Since the computations of each iteration are carried out in two phases in Algorithm IV, the folded bit-parallel architecture computes one inverse every $4m-2$ clock cycles, achieving a throughput of $1/(4m-2)$. Due to folding, the HUE is 100% without interleaving.

The CPD of the architecture in Figure 3 is given by $t_{\text{fbp}} = \max\{t_8, t_9\}$, where t_8 and t_9 are the CPDs of the type-8 and type-9 cells respectively. Clearly, t_8 is $\max\{t_{\text{MUX}}, t_{\text{XOR}}\} + 2t_{\text{MUX}}$. The two type-9 cells shown in Figures 4(b) and 4(c) use adders and ring counters respectively, and let us denote their CPDs as t_{9a} and t_{9r} respectively. t_{9a} evidently depends on the implementation of the adder unit, and so does t_{fbpa}, the CPD of our folded bit-parallel architecture using adders. The analysis of t_{9a} and t_{fbpa} is hence similar to that of t_{7a} and t_{bpa} respectively, and is not repeated here. Since t_{9r} is clearly $3t_{\text{MUX}}$, the CPD of our folded bit-parallel architecture using ring counters, t_{fbpr}, is hence the same as t_8.

TABLE I

NUMBERS OF GATES, TRANSISTORS, AND INTER-CELL CONNECTIONS FOR VARIOUS ARCHITECTURES

hardware costs	folded bit-parallel		bit-parallel			
	Fig. 3 (adder)	Fig. 3 (ring)	[4] (ring)	Fig. 1(a) (ring)	Fig. 1(a) (adder)	[1] (adder)
OR gates	0	0	2	0	0	0
NOT gates	1	1	1	0	0	0
adder	1	0	0	0	1	1
zero check	0	0	0	0	0	1
AND gates	3	3	$3m$	$2m$	$2m$	$2m$
XOR gates	m	m	m	$2m$	$2m$	$3m$
XOR3 gates	0	0	m	0	0	0
MUX	$6m$	$7m$	$8m$	$7m$	$6m$	$9m$
latches	$6m$	$8m$	$14m$	$11m$	$9m$	$14m$
trans.	$90m$	$132m$	$190m$	$150m$	$128m$	$192m$
I/O	4/4	4/4	6/6	5/5	5/5	6/6

III. PERFORMANCE COMPARISONS AND IMPLEMENTATION

In this section, we compare the two types of architectures discussed above with the best previously proposed bit-parallel in-place architectures [1], [4] of which we are aware. The key parameters of all the architectures are summarized in Tables II and I. In our comparisons, we have used the simplified form of the inversion/division architecture in [1] when only inversion is needed. The inversion/division architectures in [4] do not seem to simplify when only inversion is needed. Also note that the architectures shown in Figures 1(a) and 3 both have two choices for their control cells: using adders or using ring counters. Since the choice of control mechanism affects the hardware costs, we use "adder" and "ring" to differentiate them when ambiguity might arise in the discussions below.

Table I gives a detailed comparison of the areas of all architectures by showing both the gate and transistor counts

TABLE II

COMPARISON OF DIFFERENT INVERSION ARCHITECTURES. ALL ARCHITECTURES HAVE AREA COMPLEXITY $O(m)$ AND LATENCY APPROXIMATELY $5m$.

Type	Arch.	Control	CPD	interleaved throughput	non-interleaved throughput
folded bit parallel	Fig. 3	adder	t_{fbpa}	N/A	$1/(4m-2)$
	Fig. 3	ring	$\max\{t_{\text{MUX}}, t_{\text{XOR}}\} + 2t_{\text{MUX}}$	N/A	$1/(4m-2)$
bit parallel	[4]	ring	$t_{\text{XOR3}} + t_{\text{AND}} + t_{\text{MUX}}$	$1/(2m-1)$	$1/(4m-2)$
	Fig. 1(a)	ring	t_{bpr}	$1/(2m-1)$	$1/(4m-2)$
	[1]	adder	$\max\{t_{ref}, 2t_{\text{XOR}} + t_{\text{MUX}} + t_{\text{AND}}\}$	$1/(2m-2)$	$1/(4m-4)$
	Fig. 1(a)	adder	t_{bpa}	$1/(2m-1)$	$1/(4m-2)$

and the inter-cell connections. Note that only the dominant terms of gate counts of all the architectures are compared; the estimated number of transistors for each architecture is obtained assuming that a 2-input AND gate, a 2-to-1 multiplexer, and a 2-input XOR gate all require 6 transistors whereas a latch uses 8 transistors. For each architecture, the numbers of input and output lines for the computing cell, as a measure of the inter-cell connections [5], are given on the last line (marked I/O) of Table I.

TABLE III

THE CRITICAL PATH DELAY AND TOTAL AREA FOR THE IMPLEMENTATION OF OUR NEW ARCHITECTURES USING THE TSMC AVANTI $0.18\mu m$ CMOS STANDARD CELL LIBRARY

m	Total Core Area (μm^2) bit-parallel/folded bit-parallel	Critical Path Delay (ns) bit-parallel/folded bit-parallel
8	12,097/9,758	1.43/1.46
16	24,575/18,716	1.43/1.51
32	48,568/36,518	1.43/1.51
64	92,836/70,841	1.42/1.52
128	184,437/139,346	1.46/1.52

In comparison to our bit-parallel architecture, our folded bit-parallel architecture has a smaller area at the expenses of interleaved throughput and (possibly) a slightly longer CPD. For the purpose of verification, we implemented our new bit-parallel and folded bit-parallel architectures with ring counter control for a variety of field sizes using the TSMC Avanti $0.18\mu m$ CMOS standard cell library. Table III lists the total core area and the CPD of the core circuitry reported by the synthesis tool for various values of m. The implementation results in Table III are consistent with the comparisons between our two types of architectures given above.

We are not aware any architecture comparable to our folded bit-parallel architecture in the literature. Hence, we compare only our bit-parallel architecture, shown in Figure 1(a), with those in [1] and [4].

Gate/Transistor Counts and Input/Output: As shown in Table I, our new bit-parallel architecture shown in Figure 1(a), with either adders or ring counters, uses fewer gates and latches than the inversion architectures in [1] (which uses adders) and [4] (which uses a centralized ring counter). The transistor counts of our bit-parallel architectures with adder and ring counter implementation are approximately 33% and 21% smaller than that of [1] and [4] respectively. Finally, each computing cell in our bit-parallel architectures has *ten* inputs and outputs as opposed to *twelve* inputs/outputs in [1] and [4].

Latency, Throughput, and Critical Path Delay: All architectures have latencies of approximately $5m$ clock cycles. The throughput of our bit-parallel architectures is slightly worse than that in [1] and the same as that in [4]. As noted above, the CPD of the inversion architecture in [1] is given by $\max\{t_{\text{ref}}, 2t_{\text{XOR}} + t_{\text{MUX}} + t_{\text{AND}}\}$, and t_{bpa}, the CPD of the inversion architecture shown in Figure 1(a) (with adders), is clearly shorter than or the same as the CPD in [1]. When m is small enough that t_{ref} is less than the path delays of the computing cells, t_{bpa} is a 3-gate delay as opposed to the 4-gate delay in [1]. The CPD of the inversion architecture in [4] (with ring counter) is $t_{\text{AND}} + t_{\text{MUX}} + t_{\text{XOR3}}$. In comparison, the critical path delay of our architecture with centralized ring counters is t_{bpr}. Under the assumption that all 2-input gates have roughly the same delay and $t_{\text{XOR3}} > t_{\text{XOR}}$, which is true for most cell libraries, the CPD of [4] is larger than t_{bpr}. In fact, if we further assume that t_{XOR3} is nearly twice as much as t_{XOR}, as is the case in some cell libraries, the CPD of [4] could be as much as 33% more than t_{bpr}.

In summary, our bit-parallel inversion architectures with adders and ring counter achieve smaller CPDs and require less hardware than the architectures in [1] and [4] while maintaining approximately the same throughput and latency.

REFERENCES

[1] J.-H. Guo and C.-L. Wang, "Hardware-efficient Systolic Architecture for Inversion and Division in GF(2^m)," in *IEE Proceedings on Computers and Digital Techniques*, 1998, pp. 272–278.

[2] D. Hankerson, J. L. Hernandez, and A. Menezes, "Software Implementation of Elliptic Curve Cryptography over Binary Fields," in *Proceedings of Second International Workshop on Cryptographic Hardware and Embedded Systems, CHES 2000*, 2000, pp. 1–24.

[3] Y. Watanabe, N. Takagi, and K. Takagi, "A VLSI Algorithm for Division in GF(2^m) Based on Extended Binary GCD Algorithm," *IEICE Transactions on Fundamentals of Electronics, Communications and Computer Sciences*, vol. E85-A, no. 5, pp. 994–999, May 2002.

[4] C. H. Wu, C. M. Wu, M. D. Shieh, and Y. T. Hwang, "An Area-Efficient Systolic Division Circuit over GF(2^m) for Secure Communication," in *Proceedings of ISCAS'02*, 2002, pp. 733–736.

[5] Z. Yan, D. V. Sarwate, and Z. Liu, "High-Speed Systolic Architectures for Finite Field Inversion," *Integration: the VLSI Journal*, vol. 38, no. 3, pp. 383–398, January 2005.

[6] Z. Yan and D. V. Sarwate, "New systolic architectures for inversion and division in $GF(2^m)$," *IEEE Transactions on Computers*, vol. 52, pp. 1514–1519, November 2003.

A Speech Recognizer With Selectable Model Parameters

Wei HAN, Cheong-Fat CHAN, Chiu-Sing CHOY and Kong-Pang PUN
Department of Electronic Engineering,
The Chinese University of Hong Kong, Shatin, Hong Kong
Email: whan@ee.cuhk.edu.hk

Abstract — This paper presents the design and simulation results of a Hidden Markov Model (HMM) based isolated word recognizer IC. The new design can handle any combination of states and mixtures (up to 16 states and 8 mixtures). The speech IC has been verified with 353 test speech data. The recognition accuracy is 93.8% (48-bit) with no truncation and 88.9% with truncation (16-bit).

I. INTRODUCTION

Hidden Markov Model (HMM) approach is a well-known and widely used speech recognition method. It statistically characterizes the spectral properties of the frames of a pattern and provides a natural and highly reliable way of recognizing speech for a wide range of applications [1, 2]. People have done researches concentrated on the implementation of the HMM-based speech recognizer for more than ten years [3—6]. In an HMM-based speech recognition system, the word model has great effect on the recognition accuracy. If the model represents more detailed properties, the recognizer will be more accurate. The information contained in the word model is determined by the training process and the model parameters, e.g., the number of states S and the mixture number M in one state. These two parameters are determined by recognition task and the available training patterns. Different combination of S and M will affect the structure of the recognizer.

In this paper a speech recognizer which can work with variable combinations of S and M is presented. To reduce the amount of computations, a look-up table is used to implement the add-log function involved in a multi-mixture HMM based recognition system. This new architecture can perform different recognition tasks without any vital modifications and produces acceptable recognition accuracy.

II. HMM BASED SPEECH RECOGNITION

A. Viterbi Algorithms

HMM method uses the statistical information inherent in speech pattern (i.e., mean and covariances) to do speech recognition. Given an HMM function $\lambda = (A, B, \pi)$, here A is the state-transition probability distribution, B is the observation symbol probability distribution and π is the initial state distribution, and a observation sequence O, we can choose a best match sequence by the Viterbi algorithm [1], which can be described as:

a. Initialization
$$P_1(1) = b_1(o_1) \quad (1)$$

b. Recursion
$$P_t(j) = \max\{P_{t-1}(i)a_{ij}\}b_j(o_t) \quad (2)$$

c. Termination
$$P_{final} = P_T(N) \quad (3)$$

where $b_j(o_t)$ is the output probability calculated from the probability density function (pdf) at time t and the jth state, a_{ij} is the state-transition probability from the ith state to the jth state.

Figure 1 illustrates the Viterbi algorithm by a simplified Trellis state diagram. Here each state can only be reached by itself or the previous one, and the state sequence must begin with the state a and end in the state c. The initialization sets the point a_1 as the starting point. From a_1 we can get to a_2 and b_2, and so on. The point b_3 has two possible paths from a_2 and b_2 respectively. The algorithm will calculate the probabilities of these two paths and only keeps the one with the higher probability. The same comparison will be carried out at the point c_4. When the searching process reaches the last point c_4, the whole calculation is terminated. The length of the search path is related to the number of states of word model and the number of frames of input speech.

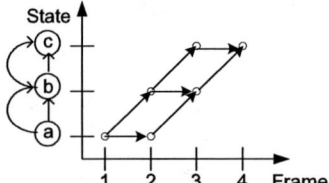

Figure 1. A simplified Trellis diagram

B. Probability Density Function

With a continuous observation density Hidden Markov Model, the most general representation of pdf is

$$b_j(o_t) = \sum_{k=1}^{M} c_{jk} N(o_t, \mu_{jk}, U_{jk}), \quad 1 \leq j \leq N \quad (4)$$

where o_t is the observation vector with the dimensionality of n at time t, c_{jk} is the mixture coefficient for the kth mixture in the state j and N is a multivariate Gaussian with the mean vector μ_{jk} and the covariance matrix U_{jk} for the kth mixture component in the state j. N is in the format of

$$N(o,\mu,U) = \frac{1}{\sqrt{(2\pi)^n |U|}} e^{-\frac{1}{2}(o-\mu)'U^{-1}(o-\mu)} \quad (5)$$

Since in the hardware implementation hardware cost of a multiplier is expensive and the result of multiplication is easily overflowed, equation (2) will be implemented in logarithmic domain:

$$\ln[P_j(o_t|M)] = \ln[b_j(o_t)] + \ln\{\max[P_i(o_{t-1}|M)\cdot a_{ij}]\} \\ = \ln[b_j(o_t)] + \max\{\ln P_i(o_{t-1}|M) + \ln a_{ij}\} \quad (6)$$

Equation (6) involves only addition and comparison if we calculated $\ln[b_j(o_t)]$ ahead. Thus the main task here is to compute $\ln[b_j(o_t)]$.

Express equation (5) as
$$N = Ae^X \quad (7)$$
Therefore equation (4) can be expressed as

$$b(o) = \sum_{k=1}^{M} c_k A_k e^{X_k} = \sum_{k=1}^{M} C_k e^{X_k}, \quad C_k = c_k A_k \quad (8)$$

$$\ln[b(o)] = \ln(\sum_{k=1}^{M} C_k e^{X_k})$$

$$C_k = \frac{c_k}{\sqrt{(2\pi)^n |U_k|}} \quad (9)$$

$$X_k = -\frac{1}{2}(o_k - \mu_k)'U_k^{-1}(o_k - \mu_k) \\ = -\frac{1}{2}[(o_k + (-\mu_k)]'U_k^{-1}[(o_k + (-\mu_k)] \quad (10)$$

Variables in equation (9) are constant, so C_k can be calculated in advance and stored in an external memory as model parameters. Equation (10) requires only addition and multiplication if we convert the frame vectors into negative values. Thus to calculate logged pdf as equation (8), the most difficult part is the exponential-add-log operation ($\ln\sum\exp$). A simple but efficient approach with acceptable recognition accuracy is to implement it using a look-up table as proposed in our previous design [7].

Divide each operand of the add operation of equation (8) by $C_{k\max}e^{X_{k\max}}$, here $C_{k\max}e^{X_{k\max}} = \max\{C_k e^{X_k}\}$.

$$\ln[b(o)] = \ln\left[C_{k\max}e^{X_{k\max}}(1 + \sum_{k=1,k\neq k\max}^{M} \frac{C_k}{C_{k\max}}e^{X_k - X_{k\max}})\right] \quad (11) \\ = \ln C_{k\max} + X_{k\max} + \ln(1 + \sum_{k=1,k\neq k\max}^{M} \frac{C_k}{C_{k\max}}e^{X_k - X_{k\max}})$$

In equation (11), $\frac{C_k}{C_{k\max}}e^{X_k - X_{k\max}}$ is always no more than 1, so that the value of $\ln(1 + \sum_{k=1,k\neq k\max}^{M} \frac{C_k}{C_{k\max}}e^{X_k - X_{k\max}})$ is finite and can be stored in a table. This approach can substantially reduce the complexity of the design with relatively high speed and acceptable accuracy.

III. VLSI IMPLEMETAION

A. Implemetation of Muli-mixture Pdf Using Look-up Table

In a multi-mixture HMM system, the logged pdf can be expressed by equation (11) as stated before:

$$\ln[b(o)] = \ln C_{k\max} + X_{k\max} + \ln(1 + \sum_{k=1,k\neq k\max}^{M} \frac{C_k}{C_{k\max}}e^{X_k - X_{k\max}}) \quad (11)$$

where C and X is defined in equations (9) and (10).

The term $\ln(1 + \sum_{k=1,k\neq k\max}^{M} \frac{C_k}{C_{k\max}}e^{X_k - X_{k\max}})$ is implemented by a look-up table in this project. Define $C_{k\min}e^{X_{k\min}} = \max\{C_k e^{X_k}, 1 \leq k \leq M, k \neq k_{\max}\}$, the second largest term of $C_k e^{X_k}$. Thus the log term can be re-written as below:

$$\ln(1 + \sum_{\substack{k=1,k\neq k\max}}^{M} \frac{C_k e^{X_k}}{C_{k\max}e^{X_{k\max}}}) \\ = \ln[(1 + \frac{C_{k\min}e^{X_{k\min}}}{C_{k\max}e^{X_{k\max}}}(1 + \sum_{\substack{k=1 \\ k\neq k\max \\ k\neq k\min}}^{M} \frac{C_k e^{X_k}}{C_{k\min}e^{X_{k\min}}})] \quad (12)$$

In this equation the value of $\frac{C_k e^{X_k}}{C_{k\min}e^{X_{k\min}}}$ is always between 0 and 1. Thus $\sum_{\substack{k=1 \\ k\neq k\max \\ k\neq k\min}}^{M} \frac{C_k e^{X_k}}{C_{k\min}e^{X_{k\min}}}$ is always between 0 and M-2.

We approximated this term to be (M-2)/2 to make our result close to the theoretical one while much less calculations is required. Now equation (12) is simplified to become

$$\ln[(1 + \frac{C_{k\min}e^{X_{k\min}}}{C_{k\max}e^{X_{k\max}}}(1 + \sum_{\substack{k=1 \\ k\neq k\max \\ k\neq k\min}}^{M} \frac{C_k e^{X_k}}{C_{k\min}e^{X_{k\min}}})] \\ \approx \ln[(1 + \frac{C_{k\min}e^{X_{k\min}}}{C_{k\max}e^{X_{k\max}}}(1 + \frac{M-2}{2})] \quad (13) \\ = \ln[(1 + e^{(\ln C_{k\min} + X_{k\min}) - (\ln C_{k\max} + X_{k\max}) + \ln(\frac{M}{2})})]$$

From equation (13) the index of the look-up table is $(\ln C_{k\min} + X_{k\min}) - (\ln C_{k\max} + X_{k\max}) + \ln(\frac{M}{2})$, which can be easily calculated for different value of M. Also because:

$$0 \leq \frac{C_{k\min} e^{X_{k\min}}}{C_{k\max} e^{X_{k\max}}} \leq 1$$

$$0 \leq \frac{C_{k\min} e^{X_{k\min}}}{C_{k\max} e^{X_{k\max}}} (1 + \frac{M-2}{2}) \leq \frac{M}{2}$$

$$1 \leq 1 + \frac{C_{k\min} e^{X_{k\min}}}{C_{k\max} e^{X_{k\max}}} (1 + \frac{M-2}{2}) \leq \frac{M}{2} + 1$$

the value of equation (13) is in the range of 0 and ln(M/2+1). In this design the maximum number of mixture M is set to 8 with an accuracy of 2 places after the decimal. Then the content of the look-up table is 162 digits, starts from 0 and ends at 1.61 (ln5) with a solution of 0.01.

B. Architecture of the Speech Recognizer

Figure 2 is a block diagram of the new speech recognizer. In the external memory the library of the recognizer is stored in the format of model parameters in the sequence of word index, and feature vectors are stored according to the frame sequence. These data are read into Temp Register sequentially before calculation: a frame vector is read into the Temp Register in the first clock cycle, in the second cycle one model parameter is read in and stored, and then the same process is repeated. The use of Temp Register reduces the number of input data pins, hence reduces the chip die area in this pad-limit design.

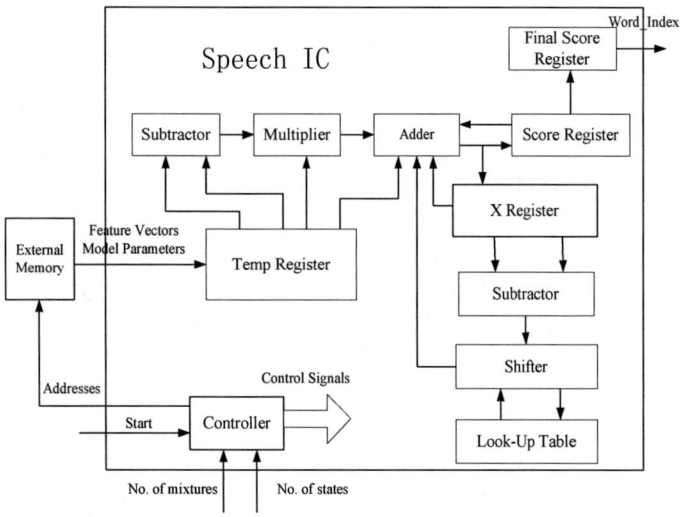

Figure 2. The block diagram of this new recognizer.

Subtractor, Multiplier and Adder blocks calculate X according to equation (10) for every frame and mixture at each state. The block X Register stores $\ln C_{k\max} + X_{k\max}$ (the largest term) and $\ln C_{k\min} + X_{k\min}$ (the second largest one), and then passes these two values to Subtractor. The difference between them will be shifted to the right scale of the given task to calculate the index of the look-up table. The output of the table is also shifted to obtain the exactly value of the term $\ln(1 + \sum_{k=1, k \neq k\max}^{M} \frac{C_k}{C_{k\max}} e^{X_k - X_{k\max}})$. Then Adder calculates the partial probability $\ln[P_j(o_t | M)] = \ln[b_j(o_t)] + \max\{\ln P_i(o_{t-1} | M) + \ln a_{ij}\}$, where $\ln P_i(o_{t-1} | M)$ is the previous partial probability stored in Sore Register. The new partial probability will replace the old one after addition and comparison. In the end of the probability calculation of one word, Score Register will give the final probability and the index of this word to Final Score Register. Final Score Register then compares this probability with the one stored inside before. The larger score and the corresponding word index will be kept in Final Score Register. The above process continues until the whole library has been searched. Finally the word index in Final Score Register is outputted as the recognition result.

The number of registers in Score Register is decided by the number of states S. The maximum value of S is set to 16 in this design, thus there are altogether 17 registers, where there are 16 registers to store the probabilities along the states, one register to carry out the select operation in equation (6). The feature vectors and model parameters are all 16-bit binary numbers. In a no truncation system, X will be a 48-bit binary number after two multiplications. Thus registers in Look-up Table, Score Register and final Score Register will be set to 50-bit wide to store all the efficient bits. This is not an area efficient design because the build-in look-up table occupies about half of the core area according to the area estimation during synthesis. If we perform a 32→16 truncation in the multiplication step, then 20-bit registers are wide enough to store the following computation results. We estimate a 50% reduction in core area with only minor reduction in recognition accuracy.

IV. RESULTS

A test chip is fabricated with an AMS 0.35-micron CMOS technology. Figure 3 is the micro photograph of the new IC. Table 1 is a brief specification of the proposed speech recognition chip.

This IC can handle up to 16 states and 8 mixtures. The maximum number of frames of input speech is 256. Generally each frame represents a 25ms speech segment with 10ms overlapping, thus 256 frames can represent a speech in the length of $(256-1)\times(25-10)+25 = 3851ms = 3.85s$, which is long enough for an isolated word.

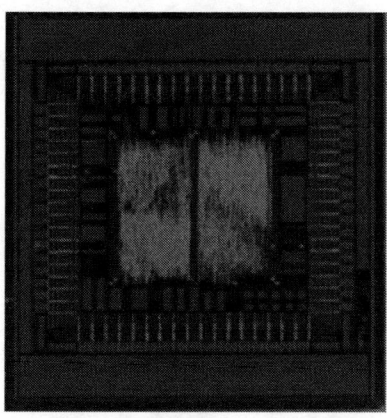

Figure 3. The micro photograph of the new speech IC

TABLE I. SPECIFICATION OF THE NEW SPEECH IC

Specification	Value
CMOS Technology	0.35um
Operating Voltage	3.3V
Core Area	1.75 sq.mm
Package	PGA68
Max. No. of State	16
Max. No. of Mixture	8
Max. No. of Frame	256

In the functional simulation, we verified the recognizer with the utterances from the AURORA 2 database [8]. The utterances in this database are noise-corrupted by adding noise to clear utterances. We used single-digit utterances with 10 dB SNR in both training and recognition. The single digits include "0", "1", "2", ..., "9", and for "0", there are two different pronunciations, namely "O" and "zero". As a result, there were totally 11 word models, each being an 8-state double-mixture HMM. They were trained by the HTK toolkit [9] with a set of the training utterances of 1144 speeches from 52 males and 52 females. The utterances to be recognized were 353 speeches from 55 males and 56 females, who were totally different from the ones giving the training utterances. The feature vectors of these speeches were also obtained by the HTK toolkit. The model parameters and feature vectors were converted into fix-point binary format from the initially floating-point data for the speech IC to work with. The simulation result is shown in Table 2. For comparison purpose, we also tablet the results from a software recognizer running on a desktop computer with floating-point calculations [9] and another 48-bit fix-point hardware recognizer with a similar architecture [7]. The software processor has a recognition accuracy of 94.3%, the 48-bit hardware recognizer reaches an accuracy of 93.8%, and the accuracy of the new 16-bit recognizer is 88.9%. We have reduced the core area by about 50% with an acceptable reduction in recognition accuracy.

TABLE II. SIMULATION RESULTS

	Software (floating-point)	Hardware (fix-point)	
		48-bit	16-bit
Word Accuracy (%)	94.3%	93.8%	88.9%
Core Area* (sq.mm)	-	2.66	1.18

*: The area estimation is performed during synthesis.

V. CONCLUSION

An HMM speech recognizer which can handle different combination of the number of state and the number of mixture has been presented in this paper. A simple look-up table is used to realize the computation of multi-mixtures probability density function. A 16-bit databus is employed in this design. Simulation results indicate we can reduce the core area by 50% with only 4.9% recognition accuracy reduction. A 0.35 micro CMOS test chip is fabricated. We will present the measurement result at the conference.

REFERENCES

[1] L. R. Rabiner and B. H. Juang, "Fundamentals of speech Recognition", Prentice Hall, 1993
[2] L. R. Rabiner, "A Tutorial on Hidden Markov Models and Selected Applications in speech Recognition", Proceedings of the IEEE, Volume: 77 Issue: 2, pp. 257-286 Feb. 1989
[3] J.M. Jou, Y.H. Shiau and C.J. Huang, "An Efficient VLSI Architecture for HMM-Based Speech Recognition", Electronic, Circuits and Systems, 2001. ICECS 2001, The 8th IEEE International Conference. Vol.1, pp. 469–472, 2001.
[4] J. Pihl, T. Svendsen and M.H. Johmsen, "A VLSI Implementation of Pdf Computations in HMM Based Speech Recognition", TENCON '96. Proceedings, 1996 IEEE TENCON, Digital Signal Processing Applications. Vol.1, pp. 241–246, 1996.
[5] G. Park, K.S. Cho and J.D. Cho, "Low Power VLSI Architecture of Viterbi Scorer for HMM-Based Isolated Word Recognition", Quality Electronic Design, 2002. Proceedings. International Symposium. pp. 235–239, 2002.
[6] S. Stölzle, K. Narayanaswamy, J. Kornegay, J. Rabaey and R.W. Brodersen, "A VLSI Wordprocessing Subsystem for a Real Time Large Vocabulary Continuous Speech Recognition System", Custom Integrated Circuits Conference, 1989. Proceedings of the IEEE 1989. pp. 20.7/1–20.7/5, 1989.
[7] Wei HAN, Kwok-Wai HON and Cheong-Fat CHAN, "An HMM-Based Speech Recognition IC", ISCAS 2003.
[8] H.-G. Hirsh and D. Pearce, "The AURORA Experimental Framework for the Performance Evaluation of Speech Recognition Systems under Noisy Conditions", Proceedings of ISCA ITRW ASR 2000. Paris, France, September 2000.
[9] S. J. Young, P.C. Woodland and W. J. Byrne, "The HTK BOOK (for HTK Version 2.2)", Entropic Ltd., Jan. 1999

Reconfigurable Multiple Scan-Chains for Reducing Test Application Time of SOCs

Jiann-Chyi Rau, Chih-Lung Chien, and Jia-Shing Ma
Department of Electrical Engineering, Tamkang University
151, Ying-Chuan Rd. Tamsui, Taipei Hsien 251, Taiwan, R.O.C
{jcrau, clchien, jsma}@ee.tku.edu.tw

Abstract

We propose an algorithm based on a framework of reconfigurable multiple scan chains for system-on-chip to minimize test application time. For the framework, the control signal combination causes the computing time increasing exponentially. The algorithm we proposed introduces a heuristic control signal selecting method to solve this problem. We also minimize the test application time by using the balancing method to assign registers into multiple scan chains. It could show significant reductions in test application times and computing times.

I. Introduction

For testing a System-on-Chip (SOC), it requires a test wrapper for each core, internal scan registers within each core, and a test access mechanism (TAM) [1]. The test wrapper is comprised of a standard cell at each core input and output that enables isolation of the core from the SOC for testing independently. The internal scan registers are designed for the necessary Design-For-Testablity (DFT) by the core providers. TAM is a mechanism to transport test data (test patterns as well as responses) and test control signals between SOC pins and core I/O and internal scan chains. The scan-based testing methodology needs high test application time because scan requires test data to be shifted in and out by one or more scan chains. The recent approaches to minimize test application time include [2], [3], and [4].

Our algorithm is based on a framework for scan chain design proposed in [5]. For the framework of Reconfigurable Multiple Scan Chains, the computing time is increasing exponentially with the number of control signals. So we propose an algorithm for the control signal selection to reduce the control signal space. Further the computing time can be reduced. We also modified the registers assignment to more balancing way to reduce the test application time.

The rest of the paper is organized as follows. In Section 2 we introduce the reconfigurable scan chain model and define the problem. The algorithm of control signal selection is presented in Section 3 and the modified registers assignment is presented in Section 4. Section 5 is the experimental results and Section 6 is Conclusion.

II. Model of Reconfigurable Scan Chain

Cores from providers are included necessary DFT to be integrated as a SOC by system integrators. The cores are prepared with internal scan chains and test vectors for each different core. The SOC integrators just saw the terminals of the I/O and the internal scan chains. That allowed integrators to insert a wrapper cell to each input and output. Further more, all the wrappers and internal scan chains would be assigned into one or several scan chains of TAM and the test vectors needed to be recombined based on the assignment. Reconfigurable Multiple Scan Chains are one kind of architectures to construct the scan chains. The following figure 1 is an example of SOC using Reconfigurable Multiple Scan Chains.

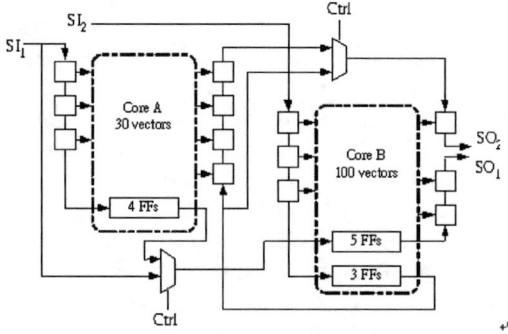

Figure 1: An example of reconfigurable scan chain design

In Figure 1, the SOC contents two cores, Core A and Core B. There are two scan chains. SC_1 contains 3 input wrappers, the internal scan chain of 4 flip-flops of Core A, the internal scan chain of 5 flip-flops and 2 output wrappers of Core B. SC_2 contains 3 input wrappers, 3 flip-flops internal scan chain of Core B, 4 output wrappers of Core A and a output wrapper of Core B. Both scan chains are reconfigurable by using the 2-to-1 multiplexers controlled by signal *Ctrl*. Two cores are tested concurrently. The scan chains are reconfigurable by the *Ctrl* signal and the multiplexers that are capable to bypass Core A.

A SOC contents many cores. Let n denote the number of cores in the SOC, each with a distinct test length, and let C = (C_1, C_2, ..., C_n) denote the cores ordered in terms of strictly increasing test lengths. If two cores have the equal test lengths, they can be treated as a single core. Let L = (L_1, L_2, ..., L_n) denote the test lengths in the set C. By the definition, $L_1 < L_2 < ... < L_n$.

In overlapped test application scheme, the test for a SOC consists a sequence of test sessions. In each session, test patterns are simultaneously applied to a subset of cores in the SOC until the test set for one core is exhausted. For an example in Figure 1, C = (C_1=core A, C_2=Core B), L = (30, 100). In the first test session, L_1 = 30 test patterns are applied to both cores. The test set for C_1 is exhausted at the end of TS_1. In the next test session, there are only $L_2 - L_1 = 70$ test patterns are applied to C_2. So, if there are n cores in the SOC, there are n test sessions as a test schedule (TS_1, TS_2, ..., TS_n).

Let CC_i denote the chain cycle under the test session TS_i which is the minimum number of clocks required to shift in bits of a test vector in and to shift out test responses captured in the chains. Because of the control signals *Ctrls* and the MUX, the every shift cycle CC_i for test session TS_i may not be the same. For an example in Figure 1, two cores mean two test sessions. For the TS_1, CC_1=12. After applying 30 test patterns, core A is exhausted, so next 70 test patterns would content the don't-care bits for core A if we ignore the MUXs. It would increase the test application time. If we active the Ctrl with the MUXs at the end of TS_1, all the wrappers and internal scan chains for the core A are bypass. The CC_2 would change to 6 for TS_2. That would decrease the test application time.

In the reconfigurable multiple scan chains, the control signals are defined that once a control signal is activated it remains active until the last test session and the signals could be activated at the end of test sessions only. Let $Ctrl_i$ denote the control signal activated at the end of TS_i. Once $Ctrl_i$ is activated, it is possible to bypass the registers in core C_1, C_2, ..., C_i. The ideal number of control signals is n-1 which means there is a control signal activated at the end of every test sessions besides the last test session. But the more *Ctrls* would increase the routing area since the MUXs is small. And replacing two *Ctrls* could have one more scan chain. So the number of control signals, say t, must be limited.

The total test time τ for a given multiple scan chain configuration is the sum of each test session. The total test time is given by

$$\tau = \sum (L_i - L_{i-1})(CC_i + 1) + CC_1,$$

where $L_0 = 0$.

CC_i means the shift cycle of TS_i. If there is not having a control signal activated at the end of TS_{i-1}, CC_i for TS_i is equal to CC_{i-1} for TS_{i-1}. Since the scan chains are not reconfigured.

III. Control Signal Selection

The number of control signals, say t, is limited so that we must choose which *Ctrl* would active to make total test time minimum. Trying each choice needs a lot of computing time and computing time would increase exponentially by t. We propose an algorithm for selecting the control signals.

First we initial the Control Signal Selected Table (CSST). CSST = (CS_1=0, CS_2=0, ..., CS_{n-1}=0), 0 denote not selected. For an example, the SOC with 4 cores would initial CSST = (0, 0, 0). Second we build a 1 × n matrix, named TSP, each element represent the number of test patterns for each test session which is $L_i - L_{i-1}$. Then we build another n × 1 matrix, named CSC, each element represent the minimum shift cycle for the single core with the TAM width w for the SOC. For the example, $TSP = \begin{matrix} 15 & 20 & 8 & 10 \end{matrix}$,

$$CSC = \begin{matrix} 12 \\ 8 \\ 16 \\ 20 \end{matrix}$$

We multiply the two matrixes as a data matrix, say M, for calculating which control signal would be chosen. In the data matrix, each element means the cycles for the session. For the example, CSC × TSP is showing in Figure 2. Based on the data matrix, we can build an array, named S, represent the cycles decreased if the *Ctrl* is chose. S = (S_1, S_2, ..., S_t). The element in S is calculated as following:

$$S_i = \sum_{j=1}^{i} \sum_{k=i+1}^{n} M_{(j,k)}$$

S_m is the maximum number in S and represent for choosing $Ctrl_m$. Then we update the elements summed by S_m in M to 0 and set the $CSST_m$ to 1. After choosing the first signal, we can repeat calculating S and updating M for choosing next signal until t signals are chose. For the example in Figure 2, we assume t=2. S = (456, 360, 360). So we set $CSST_1$=1 and updating M. M after updating is showing in Figure 3. Based M after updating, S is calculated again as S = (0, 144, 240). So we set $CSST_3$=1. Two control signals are chose.

$$M = \begin{array}{c} \\ Core_1 \\ Core_2 \\ Core_3 \\ Core_4 \end{array} \begin{array}{cccc} TS_1 & TS_2 & TS_3 & TS_4 \\ 180 & 240 & 96 & 120 \\ 120 & 160 & 64 & 80 \\ 240 & 320 & 128 & 160 \\ 300 & 400 & 160 & 200 \end{array}$$

Figure 2, the example of a data matrix.

$$M = \begin{array}{c} \\ Core_1 \\ Core_2 \\ Core_3 \\ Core_4 \end{array} \begin{array}{cccc} TS_1 & TS_2 & TS_3 & TS_4 \\ 180 & 0 & 0 & 0 \\ 120 & 160 & 64 & 80 \\ 240 & 320 & 128 & 160 \\ 300 & 400 & 160 & 200 \end{array}$$

Figure 3, the updating after choosing $Ctrl_1$.

The algorithm above for control signal selection is roughly approaching the best choice. Here we propose a parameter to increasing the accuracy, say p. we would choose t + p control signals as the new control signal space. Because the number of elements in control signal space is decreasing, the computing time for all the choices is decreasing by the user defined parameter p. The table contented every choice with t control signals in the new space is built for the registers assignment. Each choice would go though the registers assignment once to find the best solution. The registers assignment is presented in next Section.

IV. Registers Assignment

The registers would be reassigned and computed test cycles for each control signal choices. For the registers assignment, the cores in the later test session would be assigned first. Because the registers in the later test session would not be bypass by Ctrl early and the patterns would be applied to the registers to the end. So the test session order set for register assignment is (TS_n, TS_{n-1}, ..., TS_1). Considering the numbers of control signals, t controls would divide the test sessions into t+1 blocks. Each block may contain one or several test sessions. There is no control signal would be activated during the same block. In other words, the register assignment would not change for each test session in the same block. So we can treat the cores in the same block as a single core and assign the registers to the minimum shift cycles.

For each block, first we assign the internal scan chains into the given TAM width as decreasing orders. Next we assign bi-direct registers, inputs and output. The purpose is making the test process during the same block would be balanced. That makes the shift cycles for the block minimum and the total test application time can be decreased. The algorithm is presented in Algorithm 1. After calculate every choice, the solution for t control signals would be recorded in *BestAns*.

Algorithm 1 : the registers assignment
1 For every *choice* {
2   order the *TS* in decreasing order;
3   divide the *TS* to t+1 *blocks* by *t* control signals;
4   for every *block* (in the order above) {
5   sort the internal scan chains of the cores in the *block* in decreasing order;
6   assign the internal scan chains to *TAM* in the order above;
7   assign *Bidir*, *Input*, and *Output* to *TAM*;
8   *cycles* = calculate the test cycles for the *block*;
9   *TotalCycle* = *TotalCycle* + *cycles*; }
10   compare to *BestAns*
     if (*TotalCycle* < *BestAns* cycles)
11   copy and replace current *TAM* content and *TotalCycle* to *BestAns*;
12   Clear *TAM*, *TotalCycle*; }

V. Experimental Results

For serial test schedules, each Core in a SOC is needed a control signal for switching TAM to each different Cores. For TAM width is 16 to a SOC with 10 cores, it needs 16×2+10=42 pins for the testing process. Based on reconfigurable multiple scan chains of the parallel test schedules, the number of control signals is limited as a constraint. For the same SOC with 10 cores, set t, the number of control signals, to 6. Then the TAM width could increase to 18. It can be said as a trade-off between the TAM width and the number of control signals.

To evaluate the proposed method we have simulated the ITC02 SOC test benchmarks [6].

In Table 1 we compare the test times of four SOC benchmarks using different test scheduling approaches: (1) the Test Bus Architecture optimization method base on ILP and exhaustive enumeration in [7], (2) the generalized rectangle-packing-based optimization (GRP) in [8], (3) the cluster-based TestRail Architecture optimization in [9], (4) a test time reduction algorithm for TestRail Architecture in [10]. The numbers after the SOC names represent the number of cores each SOC included. For example, d695(10) means there are 10 cores in SOC d695. W represents TAM width and PINs represents the total pins for the test scheduling comparing to W. For the proposed part, t is the number of control signals that is used, SCs means the number of scan chains is used after t is decided and cycles represents the test application time for the choice of t and SCs.

In the experimental results for the four SOCs, we can find a common characteristic. Our method is performed well for the situations when the SOCs are tested with the few TAM width. With the less TAM width, our method could save the control signals and changed into more scan chains. As more cores embedded in the SOC, the better performance for our method.

SOC	W	PINs	ILP [7]	GRP [8]	Cluster [9]	TR [10]	Proposed t	Proposed SCs	Proposed cycles
d695 (10)	16	42	42644	43713	44330	44307	1	20	44689
							6	18	36122
							8	16	41528
	32	74	22268	23021	23488	21518	1	36	26548
							6	34	24697
							8	32	27767
p22810 (29)	16	61	468011	452639	(N/A)	458068	5	28	606795
							11	25	325837
							27	16	456963
	32	93	246322	246150	259975	222471	5	44	542203
							9	42	244989
							11	41	251277
							27	32	343044
p34392 (20)	16	52	1033210	1023820	(N/A)	1010821	1	25	908814
							4	24	841720
							17	16	1075617
	32	84	591027	544579	585309	551778	6	39	646062
							12	36	616186
							17	32	698426
p93791 (33)	16	65	1786200	1851135	(N/A)	1791638	3	31	969757
							5	30	962566
							12	26	1079224
							24	16	1711254
	32	97	894342	975016	(N/A)	912233	5	46	606060
							7	45	658576
							24	32	1121699

Table 1, Comparison of test time among different test scheduling methods

VI. Conclusions

In the paper, we have proposed an effective and efficient algorithm based on the framework of Reconfigurable Multiple Scan Chains to solve core-based SOC schedule problem. In our algorithm, the computing time is decreased by the Control Signal Selection and the Registers Assignment is simplified by the blocks divided by the control signals. The algorithm is performed well for the SOC with a large number of cores embedded and tested by few pins.

References

[1] Yervant Zorian, Erik J. Marinissen, and Sujit Dey, "Testing Embedded-Core Based System Chips", *In proceedings IEEE International Test Conference*, pp 130-134, 1998

[2] Vikram Iyegnar, Krishnendu Chakrabarty, and Erik Jan Marinissen, "Test Wrapper and Test Access Mechanism Co-Optimization for System-on-Chip", *In Proceedings IEEE International Test Conference*, pp 1023-1032, 2001.

[3] Vikram Iyegnar, Krishnendu Chakrabarty, and Erik Jan Marinissen, "Test Access Mechanism Optimization, Test Scheduling, and Tester Data Volume Reduction for System-on-Chip", *IEEE Transaction on Computers*, 52(12), pp. 1619-1631, 2003.

[4] Chih-pin Su, and Cheng-wen Wu, "A Graph-Based Approach to Power-Constrained SOC Test Scheduling", *Journal of Electronic Testing: Theory and Application* 20, 45-60, 2004.

[5] Md. Saffat Quasem, and Sandeep Gupta, "Designing Reconfigurable Multiple Scan Chains for System-on-Chip", *Proceeding of the 22nd IEEE VLSI Test Symposium (VTS 2004)*.

[6] Erik J. Marinissen, Vikram Iyegnar, and Krishnendu Chakrabarty, "ITC2002 SOC benchmarking initiative", http://www.extra.research.philips.com/itc02socbenchm.

[7] V. Iyengar, K. Chakrabarty, and E.J. Marinissen, "Efficient Wrapper/TAM Co-Optimization for Large SOCs", *in Proc. Design, Automation and Test in Europe (DATE)*, Paris, 2002, pp. 491-498.

[8] V. Iyengar, K. Chakrabarty et al. "On Using Rectangle Packing for SOC Wrapper/TAM Co-Optimization," *In Proceedings IEEE VLSI Test Symposium (VTS)*, 2002, pp. 253-258.

[9] S.K. Goel and E.J. Marinissen, "Cluster-based Test Architecture Design for System-on-Chip," *In Proceedings IEEE VLSI Test Symposium (VTS)*, 2002, pp. 259-264.

[10] S.K. Goel and E.J. Marinissen, "A Test Time Reduction Algorithm for Test Architecture Design for Core-Based System Chips," *Journal of Electronic Testing: Theory and Applications*, 2003, pp. 425-435.

The Improvement for Transaction Level Verification Functional Coverage

Wang Zhong-hai
Microelectronics Center
Harbin Institute of Technology
Harbin, P.O.B.313
China
oki_wang@hit.edu.cn

Ye Yi-zheng
Microelectronics Center
Harbin Institute of Technology
Harbin, P.O.B.313
China
yeyizhen@public.hr.hl.cn

Abstract—For hardware design, simulation is still the primary approach for functional verification of circuit descriptions written in hardware design language. The coverage metrics measure the process of validation and indicate the unexplored parts of the design. This paper describes a coverage-directed method that is suitable to transaction level verification. The approach is based on random test generation, and the coverage is increased by using fault insertion method. Using case studies, we show how to establish the testbech and how this approach was used to improve the quality of transaction level functional verification.

I. INTRODUCTION

As one of the most difficult problems associated with high-complexity and high performance circuit, functional verification comprises a large portion of the effort, and the cost in expert time and computer resources is big. In current design process, simulation-based verification is the most common approach [1][2], and most of the verification is done by using a large amount of tests. One of the key problems in behavioral validation is how to gauge the quality of test patterns. The code coverage is one popular metric to validate the design written in hardware design language (HDL), and there are also other coverage metrics proposed for HDL, such as line coverage, event coverage, etc [3]. These metrics are mostly based on traversing the HDL code structures, and can not find or show the functional bugs. The finite state machine (FSM) model can be used to verify the functionality, this approach is to apply input patterns to traverse the whole state transition graph (STG) during the simulation process. But the size of state space is too huge to traverse, using this approach is impractical. Now, functional coverage is popular used to indicate the actual coverage obtained by running the supplied tests. Functional coverage is based on design functionality, rather than the metrics that based on HDL code structure. Functional coverage measures the effectivity the tests exercised to verify the functionality implemented by the code.

In order to increase the design productivity, the design abstraction has been raised to the system level. And the higher level of abstraction takes benefit in transaction-level modeling, synthesis, and verification [3][4][5]. Transaction level model (TLM) is a higher level model of the hardware than RTL, and TLM is useful for architectural exploration, algorithmic evaluation, etc. Developing TLM takes lower effort, yet the model has higher simulation speed. Some research groups have applied TLMs in the design [6][7].

Here, we describe functional coverage as [8], using assertions express the design functionality is suitable to transaction level verification (TLV). The prior method of functional coverage closely tied to the input HDL, which is avoided in TLV. In this paper, we describe the approach to an accurate functional coverage in TLV, we also setup an environment which contains SystemC model and HDL model to validate the approach, and the approach led to good prediction results.

In Section II, we present an overview of the TLV and functional coverage. The approach and the refinement are described in Section III. The experimental results are discussed in Section IV. We conclude with Section V.

II. TRANSACTION LEVEL VERIFICATION AND FUNCTIONAL VERIFICATION

A. Transaction Level Verification and TLM

The transaction-based verification (TBV) is introduced in 1998 [9], and the purpose is to make it easy to create and reuse the testbenches, to debug and analyze coverage by introducing the concept of transactions. A transaction is defined as a single conceptual transfer of high-level data or control between the test bench and the design under verification (DUV) over an interface. A transaction can be as simple as a memory read, and also can be as complex as the transfer of an entire structured data packet, the transaction level is the level at which the design is designed.

Identify applicable sponsor/s here. *(sponsors)*

In [10] four types of TLM are defined: component-assembly model, bus-arbitration model, bus-functional model, and cycle-accurate computation model. In this paper we choose bus-functional model (BFM) as experimental testbench. TLM is the functional reference model and there are two ways to establish transaction level model, one is using HDL, one is using other new hardware languages, like SystemC, SystemVerilog. The top layer of TLM is the tests, which describe the system transaction-level activity without regard to the detail of the signal-level protocols. The bottom layer is to map transaction level requests made by the tests and the detailed signal-level protocols.

B. Functional Coverage

In current practice, verification consists of the generation and simulation of huge amounts of random tests. The main technique for checking the testing process is to create a list of tasks, then checking each task covered during verification.

The functional space consists of the complete set of possible bus transactions and other data-transfer abstractions, functional coverage (FC) is defined by the percentage of the testbench to traverse as much of the functional space.

$$FC = \frac{number\ of\ passed\ tests}{number\ of\ total\ tests}$$

A random testing environment can allow for the better simulation performance, so the BFM command and their arguments are separated to be randomized. There should be coverage-object to define the functional coverage goal, when the goal is not achieved the simulation process should be modified. Figure 1 depicts the testbench architecture for BFM, and the usage of FC.

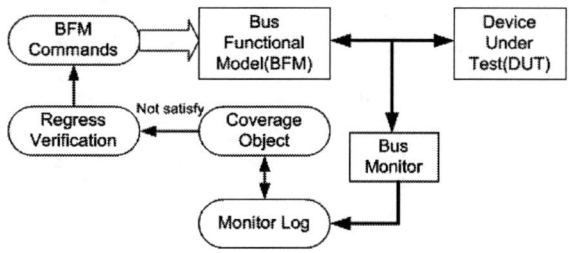

Figure 1. Architecture for BFM based on functional coverage

III. VERIFICATION WITH FUNCTIONAL COVERAGE

A. Problems in Functional Coverage

Functionalities are typically specific to the specification of design, and often have the form of carefully constructed lists of error-prone execution scenarios and the functionality fragments. Coverage tools report the number of times each such case is exercised, and a monitor should be constructed to identify each case. When develop the test plan of the design, functional coverage tasks can be specified.

The normal complete coverage of functional tasks only guarantees the behavior of the design, and the percentage is to indicate the passed functional tasks in the lists. So, this coverage is not the accurate coverage of the functionalities for the design specification. Typically, designers use a set of cases to measure the validation progress, beginning with some simple ones that require little effort. Gradually more complex test cases are introduced, and the higher coverage goal means more effort for stimulus generation.

B. Functional Test and Test Coverage

Typically, the functional test vector issue is formulated as follows. Suppose V is the input space such that members of V are states of the input vector v, S is the system state such that members of S are states of the system state vector s, and W is the output space such that members of W are states of the output vector w. The set of functional specifications that defines a mapping from the input space to the output space can be considered as a discrete function:

$$f: V \times S \to W \times S$$

where V, S, and W are finite sets. Using Cartesian products, the problem can be denoted by X and Y respectively. An element of X, x, is a state of vector (v, s), and an element of Y, y, is a state of vector (w, s). The probability that a test case exercises k new blocks which have not been exercised after covering n blocks more than once is given by [12] as hypergeometric distritribution:

$$prob\{k \mid n\} = \frac{_{N-n}C_k \times C_{p-k}}{_N C_p} \quad (1)$$

where N is the number of blocks in a program, and p is the average number of blocks covered by a test case during functional test. The expected number of blocks exercised by the i-th test case is given by:

$$e_i = \sum_{j=1}^{p} j \times prob\{j \mid n_{i-1}\} \quad (2)$$

where n_{i-1} is the number of blocks covered before running the i-th test case. [13] solves (2) and obtains the relationship between the coverage and the number of test cases executed:

$$c(x) = 1 - e^{-\frac{p}{N}x} \quad (3)$$

where $c(x)$ is the coverage after executing x test cases. (3) suggests that increasing test coverage beyond a certain point is not cost effective.

C. Generation of Random Test Vectors

The random tests should be simple and small at the beginning, and these tests can be aimed at single transaction. Then the complexity of test can be increased to match the scale of validation status.

There are two steps for the generation of random test vectors. First is to form random verification plan, second is to design test generator.

Random verification plan is the description of TLV goals, and consists of detailed lists of the verification tasks, forming verification plan is a very time-consuming job. Verification plan should include all functions and scenarios to be tested in the relevant module or unit. The proper tests amount for relevant transaction and its coverage goal can be calculated according to (3)., Designing the test generator should get started when verification plan is available. The system in Figure 2 shows verification with functional coverage.

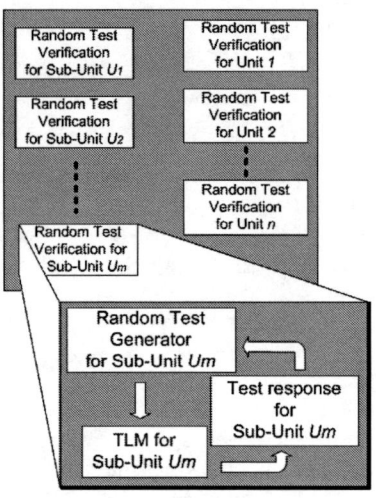

Figure 2. Illustrative example of verification which functional coverage

D. The Optimization Algorithm

The optimization algorithm is discussed in this section, and the algorithm assumes that the coverage goals set C_{goal} is set initially. Let N_{goal} be the amount of test cases set, which stores the relevant number for each element of C_{goal}. The algorithm is depicted in following can be divided into three parts for:

- Constraint checking.
- Test scheduling
- Result checking and regression

The algorithm operates according to the following steps:

a) Sort the sub-unit and unit according to TLM, and inject the verification sequence tag. The test amounts should be attached to relevant unit/sub-unit, here use set N_{goal} to store them.

b) Select one unit/sub-unit and check the constraints, if the constraints are ready, then enter loop body, the loop times can be found in set N_{goal}. The coverage goals set C_{goal} can be dynamically updated according to the different coverage precision. So is set N_{goal}.

c) Stimulating TLM by using the relevant random test generator, and there is a monitor in the loop body, when errors occur, the monitor creates a log file named with the relevant unit/sub-unit, and the related simulation message is stored in this file.

d) If the unit/sub-unit is an independent part of the design, we choose test parallelization to increase the test time. In this case, the unit/sub-unit in TLM will duplicate 3 copies and simulate with the generated tests, which are divided into 3 parts. This approach can accelerate the TLV process.

e) When the unit/sub-unit loop stop, if there is no error log, go to next unit/sub-unit verification loop body, and repeat step *b)* to step *d)*.

f) If there exists an error log file after step *e)*, use error-diagnosis program to analyze the error type and try to correct it. If the error can be corrected, then go to automatic regression program and re-simulate the test which brings error previously. If the error can not be corrected by the diagnosis program or the re-simulation still can not pass, this error should be recorded in regression file. Then go to step *b)* and start next unit/sub-unit verification loop body.

g) After the main loop is terminated, if there exists the regression file, go to regression process, which needs manual intervention. Modification for some of the TLM unit/sub-units may be taken in this step, if the TLM unit/sub-unit is modified, all the tests should be verified from beginning.

The computational complexity for the above algorithm comes mainly from the main loop and sub loops. The worst case for the algorithm occurs when step *f)* repeats in each sub loops and the automatic regression program executes in success.

E. Improvement of The Algorithm

As discussed before, increasing the test cases can bring limited effect when the amount of tests beyond a certain point. So, we focus on the generation and organization of the test cases, sophisticated tests can improve the coverage of functional verification. We use the concept from software engineering, fault insertion (FI), to improve the FC, this approach mainly contains two parts for:

- Fault type.
- The influence of the inserted fault for function.

The hardware signals can be divided into two groups: control signals and data signals, so the fault types belong to one of them. The control signals relate to functionality, the data signals relate to the transferred data. After inserting a fault into the TLM, an evaluation for it should be made. This is helpful to detect the efficiency of random test generator, if the results are not satisfied, none of the inserted faults were found in the relevant probability, the randomization need modifying. And if the verification result shows that only the inserted faults were found, the FC can be considered accurate. When the faults that are not correlative with the inserted fault are discovered, we suggest analyzing the new fault first, if no clues can be found, increasing the amount of tests maybe the best choice.

Through practice we find that inserting one fault each time is better than inserting two or more, because inserting more than one fault into the TLM may interact and cause the overlap fault functions, the influence of the faults can not be evaluated. This will effect on the estimation for FC.

IV. EXPERIMENTAL RESULTS

In Figure 3 the experimental testbench is shown. We use SystemC to establish the random test generator and the fault inserter, which can accelerate the simulation speed. DUT are modeled by HDL and SystemC, HDL model is suitable to the RTL details and SystemC is suitable to quick modeling.

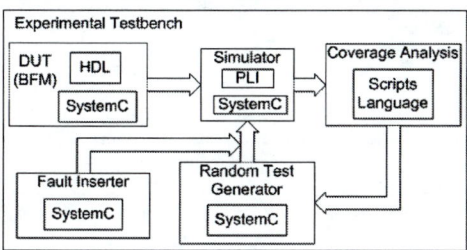

Figure 3. Experimental testbech

We use an AHB slave device to perform the experiment, this device consists of 7 different Verilog models and the results is shown in Table 1.

TABLE I. EXPERIMENTAL RESULT

Model	Normal tests			Tests with one fault insertion		
	Num	Ran time (s)	Coverage (%)	Num	Ran time(s)	Coverage (%)
M1	6100	3124	97.43	5230	2643	97.43
M2	4900	2972	99.25	4523	2758	99.25
M3	4600	3527	98.54	3962	2994	98.34
M4	5700	4745	96.72	5134	4316	96.72
M5	2300	1309	98.31	1985	1230	98.19
M6	10300	7381	99.16	8764	6185	99.16
M7	2900	973	99.17	2628	867	99.17

From Table 1, after using fault insertion the coverage of M3 and M5 were changed, the simulation time was reduced by using fault insertion. The reason for the changed coverage is: after inserting fault into TLM, the error found by using the original randomization is not enough, so the coverage was modified, and the new coverage is more accurate than that of the normal approach. The fault insertion approach can be regarded as the checker for the normal FC accuracy, and it also can be used to replace the normal method.

V. CONCLUSION

We have presented an approach to improve the accuracy of the functional coverage for TLV. This approach can be used to replace the normal functional coverage method, and it needs fewer tests and takes less time. This approach can also be used as a checker to collate the functional coverage with the original one.

REFERENCES

[1] F. Fallah, MIT, P. Ashar, CCRL and S. Devadas, Functional Vector Generation for Sequential HDL Models Under an Observability-Based Code Coverage Metric, IEEE Transactions on Very Large Scale Integration (VLSI) Systems 2002, pp.919–923.

[2] Qiushuang Zhang, and Ian G. Harris, A data flow fault coverage metric for validation of behavioral HDL descriptions, ICCAD'00. 2000, pp.369–373

[3] Surrendra Dudani, and Jayant Nagda, High Level Functional Verification Closure, ICCD2002. 2002, pp.91–96

[4] Rohit Jindal, and Kshitiz Jain, Verification of Transaction-Level SystemC models using RTL Testbenches, MEMOCODE 2003, pp.199–203

[5] Fotis Andritsopoulos, C. Charopoulos, Gregory Doumenis, Fotis Karoubalis, Yannis Mitsos, et al, Verification of a Complex SoC: The PRO3 Case-Study, DATE 2003, pp.20224–20231

[6] S. Pasricha, Transaction Level Modelling of SoC with SystemC 2.0. In Synopsys User Group Conference, 2002.

[7] P. Paulin et al. StepNP: A System-Level Exploration Platform for Network Processors. In IEEE Trans. On Design and Test, Nov-Dec 2002.

[8] Richard Goering, DAI introduces Test-Generation Tool, EE Times, November 17, 1998

[9] Steve Forde, Steve Bishop, and Ramnath S. Velu, Streamling HDL Code Coverage Analysis, Integrated Systems Design, Desember, 1998

[10] Lukai Cai, and Daniel Gajski, Transaction Level Modeling: An Overview, Proc. 1st IEEE Intl. Conf. on Hardware/Software Codesign & System Synthesis, 2003, pp. 19–24

[11] S. Chonnad, and B. Needamangalam, A Layered Approach to Behavioral Modeling of Bus Protocols, ASIC2000, pp170–173

[12] Y. Tohma, et. al., Structural approach to the estima- tions of the number of residual software fault based on hypergeometric distribution, IEEE Trans. Software Engineering, 1989, pp.345–355

[13] Piwowarski P., Ohba M.; and Caruso J. Coverage Measurement Experience During Function Test, Proceedings of the 15th International Conference on Software, 1993, pp.287–301

Parallely Testable Design for Detection of Neighborhood Pattern Sensitive Faults in High Density DRAMs

Ju Yeob Kim, Sung Je Hong and Jong Kim
Department of Electric and Electrical Engineering
Pohang University of Science and Technology
San -31, Hyoja-dong, Pohang, 790-784, Korea
{juyeob,sjhong,jkim}@postech.ac.kr

Abstract—The number of test patterns for DRAM increases at least linearly as the memory density increases. It affects the increase in total cost of memory test. In this paper, we only consider the Neighborhood Pattern Sensitive faults, which are the major and complicated faults in a high density DRAMs. Thus for a 1G DRAM the testing time may be several hours if test patterns are applied to memory cells one by one. In order to speed up the testing of the high density DRAMs, we propose a parallel accessible decoder, which allows multiple read/write operations at a time. With this scheme we can reduce the testing time roughly 500 times. This new decoder requires only 8 extra transistors per bit line.

I INTRODUCTION

Semiconductor memory is one of the most fundamental integrated circuit devices and cores in digital systems. Failure of memory fatally affects not only memory but also digital system. Therefore, testing memory is crucial to the reliability of product before shipping and in operating. In development of high density DRAMs, this is one of hard works in process of DRAM fabrication because minimized feature size increases cell capacity and unexpected faults are frequently presented. This is cause of which traditional models and detection methods of DRAM cell faults have a limitation of test time and fault coverage in evolution of high density memory. Therefore, efficient test solution needs to be introduced to reduce test time and improve of fault coverage.

Shrinking feature size is a main cause of faults through coupling and linking between multi cells. NPSF (Neighborhood Pattern Sensitive Fault) is the most complicated fault in these fault models. The fault complexity is practically difficult to deal with test cost which is test time and fault coverage. Since the application of test patterns has serially achieved, the burden of test time has especially grown heavier for NPSF detection as table 1. Perspectively, to detect NPSFs, serial access to memory array is not suitable for reduction of test time, even if area overhead is ignored. Also, Serial and sequential access method in testing memory have modified and added test method different from classic method, such as the transparent memory test [12].

Table 1. NPSF test time as a function of memory size

	Test method Complexity			
	Hayes Test [4] $n(2+3k)2^k$	Suk Test [5] $n(k+5)2^{k-1}$	March PS [6] 92n	March-12N[7] 96n
1k	0.036	0.011	0.006	0.006
16k	0.579	0.170	0.098	0.102
256k	9.269	2.726	1.568	1.636
4M	148.311	43.622	25.082	26.172
64M	2372.969	697.932	401.311	418.759
1G	37967.511	11166.915	6420.976	6700.149

Assumption: time unit =sec, memory cycle time=65ns, the number of neighborhood cell (k) = 5

Therefore, the test time for detecting NPSFs can be reduced by simultaneously accessible bit cells. Since it is a method that applies test patterns to several target bit cells in parallel, multi bit are tested every cycle time. But the parallel access requires the decoder circuit to allow selecting several cells at the same time.

This paper describes the parallely testable design for detection of NPSFs. It modifies circuit components which has functions related with cell activation and data I/O. So, we propose a parallel accessible decoder, which allows multiple read/write operations at a time. In two dimensional architecture of memory cell array, bit line decoder activates only one cell after selecting a word line. But when modified bit line decoder is accessible to several cells in a word line, simultaneously, total scanning time of cell array can be reduced. And read operation in parallel needs a function for distinguishing between fault BL and non fault BL. It should have the alternative which can be switched in test and normal mode. This is, DB and /DB line shared with several bit lines output result data of parallel test. The test mechanism provides the solution that reduces test time to

$O(N^{1/2})$ when row and column size is the same. From external device, such as test pattern generator (TPG), all combination of NPSFs is applied by Eulerian sequence [1] [2] [3].

To discuss these, this paper is organized as follows. Section 2 introduces definition and classification of NPSF model. Section 3 proposes several schemes in this paper. Section 4 discusses the improvement of proposed schemes in comparison with previous methods. Section 5 concludes the paper.

II NPSF MODEL

A pattern sensitive fault (PSF) is a conditional coupling fault in which the content of a memory cell or the ability to change its content is influenced by a certain bit pattern in other cells in the memory [1]. A NPSF model is a special case of PSF model, where the influencing cells are in the spatial neighborhood of the influenced cell. The influenced cell is called the base cell and the influencing cells are called the deleted neighborhood cells. NPSFs are detected by applying apply patterns, which activate faults, to the base cell and the deleted neighborhood cells.

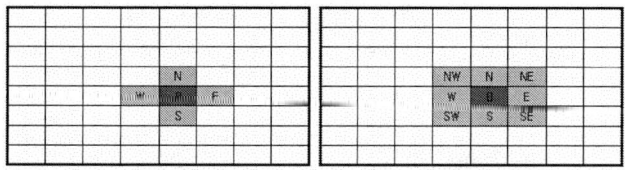

(a) 5-neighborhood cells   (b) 9-neighborhood cells
Fig 1. Two types of tiling group

A cell division of regular pattern generally is needed for detecting NPSFs through the division into the base cell and the deleted neighborhood cells. This method is called the tiling method. In the tiling method, memory is totally covered by a group of neighborhoods which do not overlap. Such a group will be called a 'tiling group' [1]. Each tiling group is independently to apply data to target memory cells because a tiling group has the assumption that is not influenced by other tiling group state. That is, a tiling group state is not influenced by other tiling group state. This means that memory can be tested in parallel. Fig 1 shows the two types of tiling group.

NPSF model is classified by three detail models. The definition of each model needs to state clearly for an appropriate pattern. This paper includes three models as follow.

- Active NPSF: The base cell changes its content due to a change in the deleted neighborhood pattern. For detecting this fault, it should verify whether the base cell has state 0 and state 1 for all transitions in the deleted neighborhood cells.
- Passive NPSF: The content of the base cell cannot be changed due to certain neighborhood pattern. For detecting this fault, it should verify whether the base cell has two transitions, from state 1 to state 0 and from state 0 to state 1, for all states in the deleted neighborhood cells.
- Static NPSF: the content of the base cell is forced to a certain state due to a certain deleted neighborhood pattern. For detecting this fault, it should verify whether the base cell has state 0 and state 1 for all possible combination of the deleted neighborhood pattern.

III PROPOSED SCHEMES

Each memory cell array is divided into same 5-neighborhood tiling group and bitmap is made by labeling the base cell and the deleted neighborhood cells. The bitmap is used to apply patterns to each label in the tiling group. Labeling cells consist of Base (B), West (W), East (E), North (N) and South (S). When a word line is selected, every cell in a word line is covered in five cycles in order to read and write operation, respectively. This tiling method reduces write operation from $n*2^k$ to $(n/k)*2^k$ for static NPSFs, and from $n*k*2^k$ to $n*2^k$ operations for active and passive NPSFs together, where k is the size of neighborhood and n is the total number of memory cells [1]. Therefore, the tiling method is efficient strategy for reducing the number of test patterns.

A Parallel Accessible Decoder

Parallel accessible decoder can select all the same label cells in a word line, simultaneously. We modify only bit line decoder in two decoders, word line decoder and bit line decoder. It minimizes the modified parts and the number of added transistor for the parallel access function. The type of 5-neighborhood cell is taken up for tessellating the cell array. So, the bit lines of modified decoder are coupled at intervals of 5 bits as Fig 2. This property is accessible to each tiling group in the same word line as Fig 1.

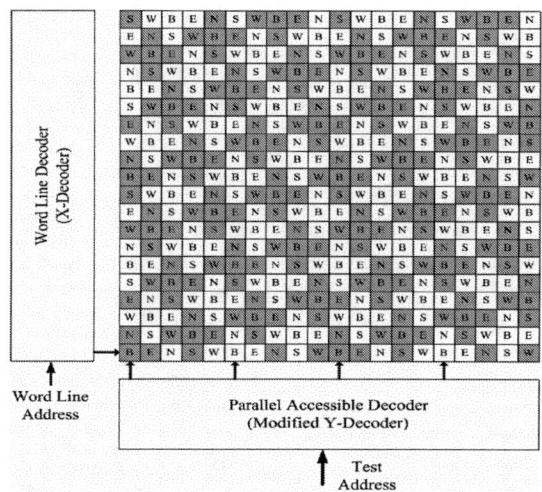

Fig 2. Bitmap of 5-neighborhood cell.

Parallel accessible decoder operates in two modes, the normal operation mode and the test mode. In the normal operation mode, *test_enable* signal is set to 0 and only one bit cell is activated. In test mode, *test_enable* signal is set to 1 and several bit cells can be activated at the same time. General address pins, A0~A3, are used to drive 5-neighborhood cells and another pins have state of high impedance (Hi-z) in the test mode. Each neighborhood cell address is decided by the bitmap and the configuration of tiling group in memory cell array. By each word line, the location of neighborhood cells is changed but if external device, such as TPG (Test Pattern Generator), catches a changed pattern of bitmap, it is possible to trace the same labeling cells.

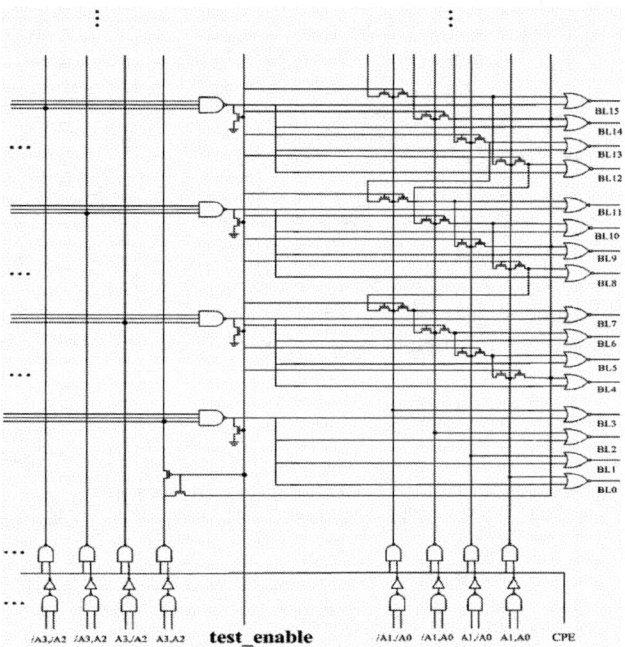

Fig 3. Parallel accessible decoder

Our parallel accessible decoder can reduce the test time by applying the several test patterns in one cycle in the test mode. Since this method is not related to increase in the number of bit line, it is merely affected by the capacity and the number of word line. Assume that row and column size are equal in general case of memory dimension. Using parallel accessible decoder may gain the reduction of test operation complexity $O(N^{1/2})$. It is test time that only depends on the number of word lines. In case of 256Kb memory cell array, 256*1024, the access to bit lines is completed in 5 cycle times. It needs originally access of 256 cycle times for writing and reading every bit cell in a word line. That is, this reduces test time from 256 cycle/word to 5 cycle/word.

B Parallel Comparison

Bit lines in cell array share the DB line and are selectively connected with DB line by column decoder (CD) signal. In parallel test operation, data of several bit line are loaded on same DB line. This connection has a difficulty for distinguishing between fault BL and non-fault BL. Parallel write operation is achieved but parallel read operation is not achieved for detecting faults if fault cells exist in selected BL. Logical two states, '0' and '1', have to be separated into distinguished line and medium. Lately, since almost all DRAM cell arrays are complementary BL and DB line, fault BL presents opposite of expected states in BL and /BL. Connection with BL and DB line has to be decided through options, normal operation mode and parallel read operation mode. So, we introduce /Test signal for a new connection with BL and DB line in Fig 4. The role of /Test signal is that BL or /BL of state '0' drives DB and /DB through added transistors, T1~T4. In the precharge state of DB and /DB, if parallel read operation is accomplished, one of DB and /DB certainly drops from state '1' to state '0'.

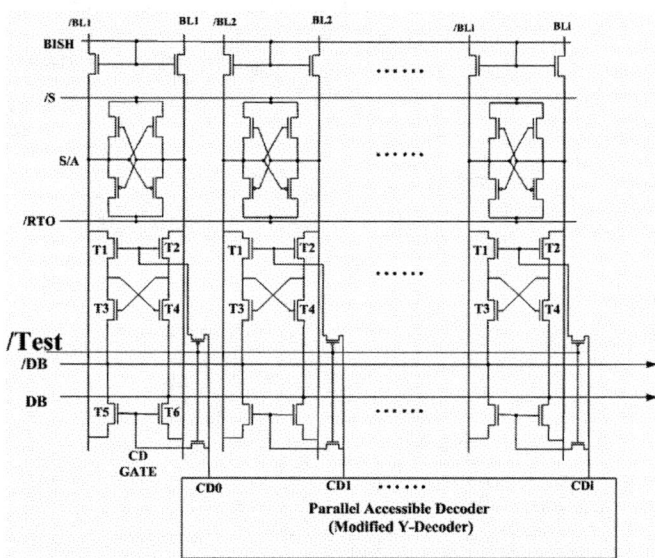

Fig 4. New connection with BL and DB lines

Suppose that there is a faulty cell, then the BL line corresponding to non-faulty cells drive DB line, while the BL line corresponding to other faulty cell should drive /DB line. therefore, both DB and /DB are in state '0'. Therefore, a fault is detected, if DB and /DB are in the same states.

IV Performance Comparison

When testing memory is tried with proposed schemes, Eulerian sequence is suitable to apply all test patterns for NPSFs. It generates 161 patterns for detecting all conditions of NPSFs in 5 neighborhood cells. To compare the proposed schemes with other test methods as Table 2, Eulerian sequence is applied to the sequential bit-by-bit test and the proposed scheme. [8] applies peculiar test patterns. Sequential bit-by-bit test method is serially applied to a target cell with normal architecture and operation of memory. This method does not almost have area overhead but it can not be practically accepted as a strategy which

Table 2. Performance Comparison

		Sequential Bit-by-Bit Test	P. Mazumder [8]	Proposed Schemes
Test cycles	Read	42,040,320	168,960	824,320
	Write	8,408,064	33,795	164,864
Type of patterns		All patterns	Partial patterns	All patterns

Target memory size=256kb (256×1024)

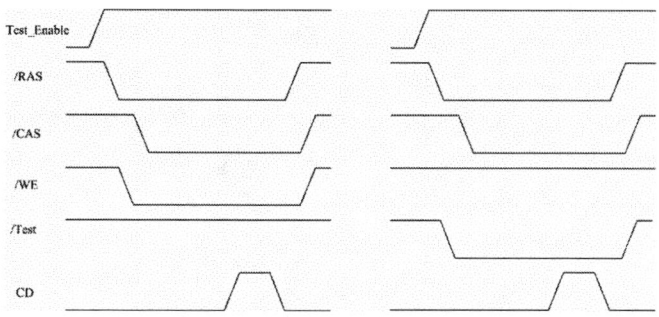

Fig 5. Read and write operation waveforms in parallel test

reduces test time because of the overall number of test cycle. The test algorithm of [8] has the test time of $195W$ (W: the number of word lines). This is, it completes the whole test procedure through accessing each word line 195 times. But this method has a limitation that it does not completely appear all combination of patterns for NPSFs. It affects fault coverage because odd bit line of CD is coupled with other odd bit line of CD in selecting a word line and vice versa. Thus, it lowers fault coverage of NPSFs in trend of high density DRAMs. So, [8] is comparatively difficult to conserve the reliability of memory test.

Using the proposed schemes, the parallel accessible decoder and the comparison method, all patterns for activating NPSFs are applied. The extra signals are *test_enable* and */Test*. Several existing control signals, /WE, /RAS, /CAS and etc, are used in the test mode. The number of extra transistors in order to allow parallel access is 8 per bit line as shown in Fig 3 and Fig 4.

V Conclusion

Increase in memory density goes with increase in test time of memory test. Testing memory is compromised with evaluation of memory performance through design for testability (DFT) and built-in self test (BIST) and etc. In this paper, we propose a DFT design for parallel access to memory cells. It focuses on reduction of test time for detecting NPSFs. Proposed scheme provides the function, parallel accessible decoder and parallel comparison, which is accessible to memory cell array in parallel. This method reduces repetitive access for reading/writing operation through multiple cell activation. Also, several patterns are applied to tiling group, simultaneously. This reduces test time to 500 times for 250 kb capacity of normal sub-cell array. And addition of test operation is minimized with almost conserving normal memory operation.

References

[1] A. J. Van De Goor, " Testing of Semiconductor Memories: Theory and Practice," John Wiley & Sons 1995.
[2] A. Chrisanthopoulos and G. Kamoulakos, "A Test Pattern Generation Unit For Memory NPSF Built-In Self Test," IEEE of the 7th Conference on a Electronics, Circuits and Systems, pp. 425, December. 2000.
[3] R. S. Sable and R. P. Saraf, "Built-in Self-test Technique for Selective Detection of Neighborhood Pattern Sensitive Faults in Memories," IEEE Proceedings of the 17th International Conference on VLSI Design (VLSID), pp 753, January 2004.
[4] J. P. Hayes, "Testing Memories For Single-Cell Pattern-Sensitive Faults," IEEE Transactions. Computers, Vol. C-29, No. 3, pp. 249, March. 1980.
[5] D. S. Suk and S. M. Reddy, "Test Procedures For a Class Of Pattern-Sensitive Faults in Semiconductor Random-Access Memories," IEEE Transactions. Computers, Vol. C-29, No. 6, pp. 419, June. 1980.
[6] V. Yarmolik and Yu. Kilmets, "March PS(23N) Test for DRAM Pattern-Sensitive Faults," IEEE Proceedings of Seventh Asian Test Symposium, pp. 354, December 1998.
[7] K. L. Cheng and M. F. Tsai, "Efficient Neighborhood Pattern-Sensitive Fault Test Algorithms for Semiconductor Memories," in Proceedings of nineteenth VLSI Test Symposium(VTS), pp 225, April 2001.
[8] P. Mazumder and J. K. Patel, "Parallel Testing for Pattern-Sensitive Faults in Semiconductor Random Access Memories," IEEE Trans Comput, Vol. C-29 no. 6, pp. 394, March 1989.
[9] Y. Morooka and S. Mori, "An Address Maskable Parallel Testing For Ultra High Density DRAMs," IEEE Proceedings of International Test Conference, pp 556, October 1991.
[10] R. David and A. Fuentes, "Random Pattern Testing Versus Deterministic Testing of RAM's" IEEE Transactions On Computer, Vol. 38, Np. 5, May 1989.
[11] T. Sakuta and M. Muranaka, "Circuit Techniques for Multi-Bit Parallel Testing of 64Mb DRAMs and Beyond," Symposium on VLSI Circuit Digest of Technical, 1992.
[12] M. G. Karpovsky and V. N. Yarmolik, "Transparent Memory Testing For Pattern Senstive Faults," IEEE Proceedings of International Test Conference, pp. 860, October. 2-6, 1994.
[13] H. J. Yoo, "DRAM Design" Hongpub. 1996.

Efficient Power Model for Crossbar Interconnects

B. Afzal and A. Afzali-Kusha
Low-Power High-Performance Nanosystems Laboratory
ECE Dept., University of Tehran
Tehran, Iran
b.afzal@ece.ut.ac.ir, afzali@ut.ac.ir

M. El Nokali
Department of Electrical and Computer Engineering,
University of Pittsburgh,
Pittsburgh, PA, U.S.A
elnokali@engr.pitt.edu

Abstract-In this paper, an efficient yet accurate model for determining the power consumption of mux-based crossbar interconnects is presented. First, an analytical model for the power consumption of a cell in the crossbar interconnect is introduced. The model is based on the modified n-th power law which is suitable for short channel MOSFET I-V characteristics. Using the cell model as well as input and control signals, an analytical expression for the power consumption of the whole interconnect is presented. The proposed model is compact and scalable making it suitable for power estimation CAD tools. The results of proposed model for a current and predictive technology of CMOS show an excellent accuracy in comparison with the results of HSPICE simulations.

I. INTRODUCTION

Power has become a critical parameter in the design of digital systems including battery-powered embedded systems, portable devices, and high performance systems [1][2]. To design and implement a low power system, one often needs to estimate the power consumption of the various components of the system very efficiently. A digital system consists of various components such as I/O drivers, interconnects, buses, and gates. Among these components is a crossbar interconnection network that configures which input is connected to which output of the network [3].

In this work, we present an analytical model for the power consumption of a crossbar interconnection network. In [3], the special requirements for a crossbar switch in a processor were studied without considering the power. Recently, an accurate model for the power dissipation of a transmission-gate crossbar has been proposed [2] which is based on the energy model for the evaluation of power dissipation. In this paper, we model the main building block of mux-based crossbar interconnects and estimate the power consumed by this block. The rest of the paper is organized as follows. Section II discusses the types of the crossbar interconnect. In Section III, we introduce the *I-V* characteristic model used in this work while the proposed crossbar interconnect power model is described in Section IV and V. The results of the model compared with those of the HSPICE simulations are given in Section VI with the summary and conclusion presented in Section VII.

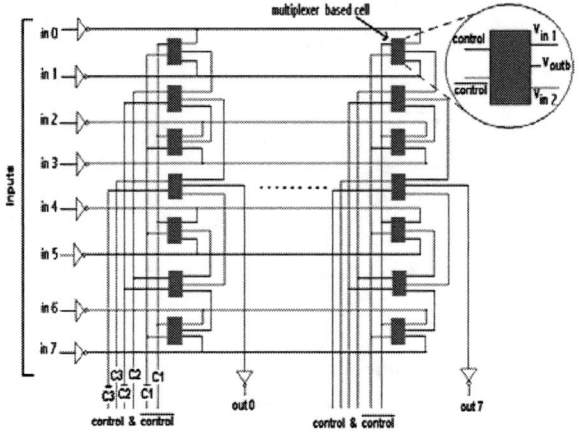

Figure 1. 8-port, 1-bit mux-based crossbar design [3]

II. CROSSBAR INTERCONNECTS

All basic crossbar designs rely on switching signals from horizontal input busses to vertical output busses. In the simplest scheme, a single transmission-gate connects the horizontal and vertical busses at each cross point [2]. A decoder for each port enables one switch at each output. The study in [3] shows that the transmission-gate based design is quite attractive for small crossbar interconnects with up to 16 ports, providing low latency and compact size. The design becomes impractically slow for high speed systems as the number of ports increase [3].

An alternate implementation makes use of a stage multiplexer at each output line. The multiplexer provides switching, decoding, and buffering with a better performance for large designs [3]. The structure of the multiplexer-based crossbar design is shown in Figure 1 where the main building block is depicted in the inset. The number of vertical control lines increases as $N \times \log(N)$ rather than N^2 in the former type. In this structure, the multiplexer buffers the signals such that driving each output port is not slower than driving a single output port. Furthermore, all long interconnect lines are driven by buffered gates. This results in a design which is more compact and much faster for large switches [3].

In this work, the model is derived for a mux-based crossbar interconnect.

III. I-V CHARACTERISTIC MODEL

To obtain an accurate model for the mux-based crossbar interconnects, an accurate submicron MOSFET model with a small number of parameters is required. The n-th power model, proposed in [4] offers a simple, yet accurate enough, empirical model for the drain current. The model is given by the following expressions:

Linear Region:

$$\begin{cases} V_{DS} \leq V_{DSsat} \\ I_D = I_{Dsat}(2\dfrac{V_{DS}}{V_{DSsat}} - \dfrac{V_{DS}^2}{V_{DSsat}^2})(1+\lambda V_{DS}) \end{cases} \quad (1)$$

Saturation Region:

$$\begin{cases} V_{DS} \geq V_{DSsat} \\ I_D = I_{Dsat}(1+\lambda V_{DS}) \end{cases} \quad (2)$$

where λ is the channel length modulation parameter, and I_{DSsat} and V_{DSsat} are defined in [4] as

$$I_{DSsat} = B\frac{W}{L}(V_{gs} - V_t)^n \quad (3)$$

$$V_{DSsat} = K(V_{gs} - V_t)^m \quad (4)$$

Here, the technology-dependent constants n, m, K and B, describe the short-channel effects and can be extracted by solving single variable equations.

IV. POWER FOR A CROSSBAR INTERCONNECT CELL

In CMOS circuits, the power dissipation consists of the static component and the dynamic (switching) power. In the case of conventional CMOS logic gates (like the inverter), the short circuit current also consumes some dynamic power which should be considered. For the static power components, there are analytical expressions as functions of the transistor terminal voltages which can be used for the power estimation. During the switching of a mux-based crossbar interconnect, the energy consumption can be expressed as [2]

$$E_{crossbar} = E_{reconfiguration} + E_{datatransfer} \quad (5)$$

where $E_{reconfiguration}$ is the energy consumed during the switch reconfiguration and $E_{datatransfer}$ is the energy dissipated due to the switching on the input and output lines during a data transfer from an input port to an output port. The first component may be calculated by the same energy model as that used in [2]. The second components include the dynamic power dissipated by the transmission gate (hereafter, refereed to as switch) and the inverter. The inverter power is calculated using the model presented in [5]. As explained here, the dynamic power consumption of the switch is calculated using the n-th power law model.

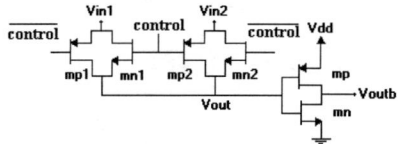

Figure 2. The circuit used for $E_{datatransfer}$ calculation.

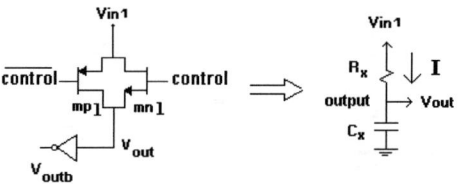

Figure 3. Equivalent circuit of the TG switch and its output.

To model the switch power, which is the main building block of the mux-based crossbar interconnect, we utilize the schematic circuit shown in Figure 2. When the control bit is high (low), mn1/mp1 are on (off) and mn2/mp2 are off (on) and hence V_{in1} (V_{in2}) signal is transferred to the output. If we assume that V_{in1} rises with a constant slope of S_R such that the output node has enough time for charge and discharge, then we can write

$$V_{out} \simeq V_{in1} \quad (6)$$

When the switch is on, we model the circuit as a series combination of a resistance (R_x) and a capacitance (C_x), as shown in Figure 3. Note that the values of R_x and C_x depend on the operation region of the transistors. For the RC circuit, assuming constant (time-invariant) R and C, the current $I(t)$ may be expressed as

$$I(t) = S_R C_x (1 - e^{-\frac{t-t_1}{R_x C_x}}) + I(t_1) e^{-\frac{t-t_1}{R_x C_x}} \quad (7)$$

where t_1 is the time that the transition occurs from one region to another. To simplify the evaluation of (7), in our model, we assume R_x to be voltage dependent and C_x constant in each region of operation. Having the voltage (V_{in1}) and current (I), one can find the instantaneous power and energy dissipation of the switch during the data transfer. When the switch energy is added to the inverter energy dissipation, the total $E_{datatransfer}$ can be calculated. Based on this argument, the evaluation of the energy dissipation relies on the availability of suitable piecewise models for the RC parameters during the data transfer.

We start our modeling by finding R_x in each operating region of mn1/mp1:

Region 1 ($0 \leq V_{in1} \leq V_{tp}$): mn1 is in the linear region and mp1 is off.

Region 2 ($V_{tp} \leq V_{in1} \leq V_{DD} - V_{tn}$): both transistors are in the linear region.

Region 3 ($V_{DD} - V_{tn} \leq V_{in1} \leq V_{DD}$): mn1 is off while mp1 is in the linear region.

In the first region for mn1, (1) may be rewritten as:

$$I_D = I_{DSsat}(1+\lambda V_{DS})(2-\frac{V_{DS}}{V_{DSsat}})\frac{V_{DS}}{V_{DSsat}} \quad (8)$$

Assuming $V_{in1} \cong V_{out}$ and, hence, $V_{DS} \approx 0$, we can write

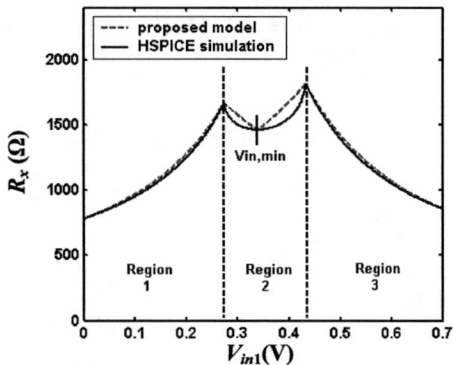

Figure 4. The resistance of the transmission gate as a function of V_{in1}.

$$R_x = \frac{1}{K_n(V_{DD} - V_{tn} - V_{in1})^{n_n - m_n}} \quad (9)$$

where

$$K_n \equiv (2W/L)(B/K)$$

A similar expression for mp1 in the linear region may be obtained. The transmission gate resistance obtained by HSPICE simulation, for a 0.07μm technology [6], as a function of the input voltage is shown in Figure 4. For mn1 in the first region, one may approximate (9) by

$$R_x = a/(1+bV_{in1}) \quad (10)$$

where a and b are constants determined by solving the following two equations:

$$a = \frac{1}{K_n(V_{DD} - V_{tn})^{n_n - m_n}} \quad (11)$$

$$b = \frac{\left(\frac{V_{DD} - V_{tn} - V_{tp}}{V_{DD} - V_{tn}}\right)^{n_n - m_n} - 1}{V_{tp}} \quad (12)$$

In the second region, R_x is a parallel combination of two resistances. To simplify R_x in the second region, we split it into two parts where the boundary is specified as the input voltage where the minimum of R_x occurs. It is given by the following approximation [7]:

$$V_{in,\min} \approx \frac{V_{DD} - V_{tn} + \gamma V_{tp}}{1+\gamma} \quad (13)$$

The approximation does not make a considerable error in calculating $V_{in,\min}$ and hence does not lead to a significant error in R_x. The second region, therefore, is split into two separate regions which are $V_{tp} \leq V_{in1} \leq V_{in,\min}$ and $V_{in,\min} \leq V_{in1} \leq V_{DD} - V_{tn}$. Now for each of these regions we can use expression (10). In the third region, R_x is the same as that of the first region. As Figure 4 reveals, the resistance obtained from the analytical model for the main building block of mux-based crossbar interconnect has a very good accuracy.

The total capacitance of the output node, C_x, is given by

$$C_x = C_{gs_{mn1}} + C_{bs_{mn1}} + C_{dg_{mp1}} + C_{db_{mp1}} + C_{gs_{mn2}} + C_{sb_{mn2}}$$
$$+ C_{dg_{mp2}} + C_{db_{mp2}} + C_{gs_{mp}} + C_{gb_{mp}} + C_{gs_{mn}} + C_{gb_{mn}} \quad (14)$$

where in C_{ij}, i and j refer to the corresponding terminals for the transistors of Figure 3. Since the components are defined for each region of transistor operation (e.g., see [8]), we define five operating regions for the inverter as a function of V_{in1}. These regions are given in Table I.

TABLE I. INVERTER REGIONS OF OPERATION

Region	Mn	Mp	Condition
1	Off	Linear	$0 \leq V_{in1} \leq V_{tn}$
2	Saturation	Linear	$V_{tn} \leq V_{in1} \leq V_{C1}$
3	Saturation	Saturation	$V_{C1} \leq V_{in1} \leq V_{C2}$
4	Linear	Saturation	$V_{C2} \leq V_{in1} \leq V_{DD} - V_{tp}$
5	Linear	Off	$V_{DD} - V_{tp} \leq V_{in1} \leq V_{DD}$

The main problem is to determine the boundaries of region 3 where mp and mn are both in saturation. The voltage, V_{outb}, may be approximated here by

$$V_{outb} \cong e + fV_{in1} \quad (15)$$

where e and f constants are determined using by (15) as

$$f = \frac{\partial V_{out}}{\partial V_{in}}\bigg|_{V_{in1} = V_{inb}} \quad (16)$$

$$e = V_{outb}\bigg|_{V_{in1} = V_{inb}} - fV_{in1} \quad (17)$$

where V_{inb} can be specified using any point in the region. Similar to [9], we select this point at the input voltage where both transistors are in saturation. For the left boundary, V_{C1}, note that mp is just at the boundary of the linear and saturation and, therefore, one can write [7]

$$V_{C1} \cong (e - V_{tp})/(1-f) \quad (18)$$

To calculate the right boundary, V_{C2}, notice that mn is at the boundary of saturation and linear. Therefore, [7]

$$V_{C2} \cong (e + V_{tn})/(1-f) \quad (19)$$

Having defined the inverter regions of operation, we can calculate the total capacitance. One should note that the interconnect capacitances, which are not voltage dependent, also are added to C_x using [6]. Using the resistance and capacitance obtained above, the power may be calculated for a cell in the crossbar interconnect.

V. TOTAL POWER FOR THE CROSSBAR INTERCONNECT

Having obtained the power of one cell, P_{cell}, one can determine the total power consumed for each cross-bar transition based on the current and previous input as well as the current control signals. As an example, consider the case if an eight bit crossbar with three control bits for each output bit shown in Figure 1. To calculate the power, when the input or the control signal changes from one clock cycle to the other, one should follow each input bit through the cells and count the number of cells whose output change compared to the previous cycle. Multiplying the number by the power consumed by the power of a cell yields the total power. Note that the output bit is selected by blocking some of the input bits in each of the three stages of the selection process and, hence, we should determine the number of stages (cells) that this input goes through. Showing the bit number of an input by a three bit binary number as $b_j = b_{j2}b_{j1}b_{j0}$, one may express the number of cells that this bit go through as the following summation

$$k_j = C_0 \oplus \overline{b_{j0}} + (C_0 \oplus \overline{b_{j0}}) \cdot (C_1 \oplus \overline{b_{j1}}) + (C_0 \oplus \overline{b_{j0}}) \cdot (C_1 \oplus \overline{b_{j1}}) \cdot (C_2 \oplus \overline{b_{j2}}) \quad (20)$$

Here, "$\oplus$" and "." represent the logical XOR and AND. The first, second, and third terms in the above equation shows that if this bit passes through the first, second, and third cell, respectively. If the input bit is not changed from one cycle to the other, the cell does not consume dynamic power. Therefore,

$$P_j = k_j |V_j(N) - V_j(N-1)| \times P_{cell} \quad (21)$$

where $V_j(N)$ and $V_j(N-1)$ are the values of input bit j in the current and previous clock cycles, respectively. The total power consumed is obtained from

$$P = \sum_j P_j \quad (22)$$

VI. RESULTS AND DISCUSSION

In order to validate the proposed analytical model, we have compared the model and HSPICE results. The results were obtained for a 0.18μm (V_{DD} of 2V) and 0.07μm (V_{DD} of 0.7V) technologies [6]. The transient current of the switch and its consumed energy for the 0.07μm technology is depicted in Figure 5. The error of the proposed analytical model for the switch is less 1.2% and 0.59% for the technologies of 0.07μm and 0.18μm technologies, respectively. The error of the analytical model for the total consumed energy of the crossbar interconnect, $E_{crossbar}$, which includes the reconfiguration energy, the switch energy, and the inverter energy, was 1.73% and 2.95% for the technologies respectively technologies which show an excellent agreement with HSPICE.

Figure 6 shows the error of using (22) compared to the results of HSPICE simulations for the standard CMOS technology 70nm. For 128 changes of the inputs where all the inputs have changed and the control bits are changed randomly. The error is less than 2.1% which is less than the error of a cell which was 3%.

VII. SUMMARY AND CONCLUSION

In this paper, we proposed an analytical model for the power estimation of the mux-based crossbar interconnects. The validity of the model was confirmed by the HSPICE simulations. The model, which was based on the n-th power law MOSFET model, was very accurate and scalable. Since the analytical model is very efficient, it is appropriate for the power estimation CAD tools.

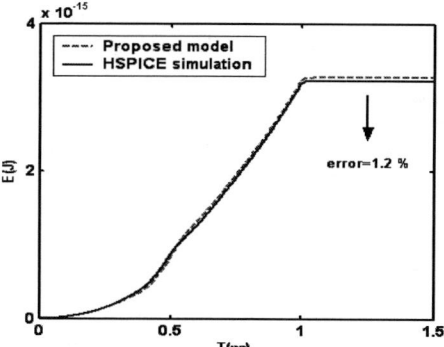

Figure 5. Energy as a function of time.

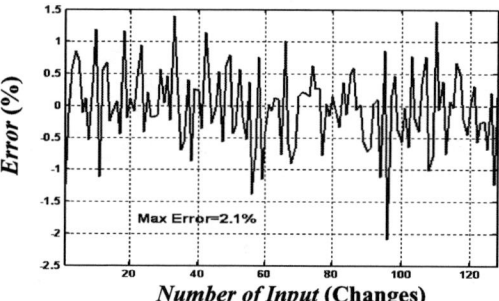

Figure 6. Error as a function of the number of input changes.

ACKNOWLEDGMENT

The authors would like to thank Mr. Ehasn Fathi for fruitful discussions and to Iran Telecom Research Center (ITRC) for the financial support during the course of this research.

REFERENCES

[1] A. P. Chandrakasan, S. Sheng, and R. W. Brodersen, "Low-Power CMOS Digital Design," *IEEE. J. Solid State Circuits.* Vol. 27, pp.473-484, Apr. 1992.

[2] E. Geethanjali, V. Narayana, M. J. Irwin, "An Analytical Power Estimation Model for Crossbar Interconnects," *ASIC/SOC Conference, 2002. 15th Annual IEEE International*, 25-28 Sept. 2002, pp. 119 -123

[3] S. Dutta, K. O'Connor, and A. Wolfe, "High Performance Crossbar Interconnect for a VLIW Video Signal Processor," in *IEEE ASIC Conference*, pp. 45-50, Sept. 1996.

[4] T. Sakurai and A. R. Newton, "A Simple MOSFET Model for Circuit Analysis," *IEEE Trans. Electron Device*, vol. 38, pp. 887-894, Apr.1991

[5] M. Taherzadeh Sani, B. Amelifard, H. Iman-Eini, A. Afzali-Kusha, M. Nourani and F. Farbiz, "Power and Delay Estimation of CMOS Inverters Using Fully Analytical Approach," in *Mixed-Signal Design Southwest Symposium*, pp. 116-120, Feb. 2003.

[6] http://www.device.eecs.berkeley.edu/~ptm/.

[7] B. Afzal and A. Afzali-Kusha, "Power Estimation of Crossbar Interconnects Using Fully Analytical Approach," in Proceedings of the 16th International Conference on Microelectronics, Tunis, Tunisia, December 6-8, pp. 219-222, 2004.

[8] B. Razavi, "Design of analog CMOS integrated circuits," McGRAW-HILL, 2001.

[9] A. P. Chandrakasan and R. W. Brodersen, "Low Power Digital CMOS Design," Kluwer Academic Publishers, 1995.

Coupling Reduction Analysis of Bus-Invert Coding

Rung-Bin Lin

Computer Science and Engineering
Yuan Ze University
Chung-Li, 320 Taiwan
csrlin@cs.yzu.edu.tw

Abstract—Theoretical analysis of bus-invert coding for reducing switching activity was previously investigated. In this paper we conduct a theoretical analysis of this method for coupling reduction. Closed-form formulas are derived to compute the number of couplings per bus transfer for a non-partitioned versus a partitioned bus. Our contribution complements the work done previously and helps establish a sound theoretical foundation for bus-invert coding.

I. INTRODUCTION

Due to significant capacitive coupling between two adjacent lines in deep submicron process technologies, many approaches have been proposed to alleviate bus energy loss and extra bus delay due to coupling. Among them, bus encoding has been widely investigated. In the past, bus encoding focused more on reducing switching activity [1-4]. Recently, many coupling reduction methods have been proposed to lower the energy dissipation or eliminate crosstalk-induced delay. They can be mainly classified into three categories. The methods in the first category [5-9] completely eliminate worst crosstalk coupling (a wire switching oppositely against its two neighboring wires) in order to get rid of crosstalk-induced delay. The methods in the second category [7, 10-17] re-arrange wiring layout for reducing coupling energy based on a priori knowledge about bus switching behavior. These methods can be best applied to an instruction address bus. Those in the third category [18-22] focus on coupling energy reduction without a priori information. In [22], the authors also derive a formula to compute energy dissipation per bus transfer for a given mapping of code words.

Many of the aforementioned methods have either included bus-invert coding [1,2] as a part of the methods or used it as a basis for performance comparison. Bus-invert coding was first proposed by Fletcher [1], later elaborated by Stan and Burleson [2, 3], and recently theoretically analyzed by Lin and Tsai [23]. The previous analysis performed by Lin and Tsai mainly focuses on switching activity and weight reduction. In this paper, an in-depth theoretical analysis was performed for coupling reduction. The main theoretical results included some closed-form formulas for computing the number of couplings on a partitioned and a non-partitioned bus. Using these results along with those presented in [23], we could obtain the following facts:

- Coupling reduction percentage with respect to an un-encoded bus first increases, culminates at 17.5% with a 6-bit bus, and then decreases with increasing bus width.
- Bus energy reduction percentage does not vary noticeably with the ratio of coupling capacitance to self-capacitance when bus width is larger than eight.
- Partitioning a bus into a number of smaller buses, each of which is encoded independently, also brings more coupling reduction.
- Partitioning a bus into many 2-bit buses results in most coupling reduction.
- Partitioning a bus into smaller buses, each of which has odd number of bits, is not a viable approach to coupling reduction.
- The expected number of couplings per bus transfer on a partitioned bus with clustered invert lines is equal to that with distributed invert lines.

The rest of the paper is organized as follows. Section II gives an overview of energy dissipation on coupled bus lines, bus-invert coding, and the limited-weight code. Section III describes how we derive closed-form formulas for computing couplings. Section IV discusses how energy reduction could vary with bus width and the ratio of coupling capacitance to self-capacitance and examines to what extent energy reduction could be brought about through partitioning a bus. The last section draws a conclusion.

II. PRELIMINARY

A. Bus Line Energy Dissipation

The energy dissipation on a non-terminated line i of an n-bit bus can be formulated as follows:

$$E_i = E_{is} + E_{ic} \qquad (1)$$

where E_{is} and E_{ic} are the energy dissipation due to self-capacitance and coupling capacitance of line i, respectively. To ease the problem formulation, we assume that E_{ic} for line i includes only the energy dissipation due to the coupling capacitance to line $i+1$. Although the energy dissipation due to the coupling capacitance to line $i-1$ is not counted in E_{ic}, it will be included in $E_{(i-1)c}$. Herein, the coupling for a particular line i always means the coupling between lines i and $i+1$. We assume that all bus lines have the same self-capacitance C_s, all-pairs of adjacent lines have the same coupling capacitance[1] C_c, and all signal changes appear on the bus lines at the same time. For $0 \leq i \leq n-2$,

a). $E_{ic} = 0$ if lines i and $i+1$ have no transitions or both have transitions in the same direction.
b). $E_{ic} = 0.5 C_c V^2$ if either line i or $i+1$ has a transition.
c). $E_{ic} = 2 C_c V^2$ if lines i and $i+1$ both have transitions in the opposite direction.

Here V is the voltage swing on the bus. We assume $E_{(n-1)c} = 0$ for line $n-1$. If we define a coupling as an occasion dissipating an

---

[1] In fact, the left-most (right-most) bus line would have larger self-capacitance than the lines in the middle if it does not have any close neighbors on its left (right) side. Our study also shows that for wires with a large height/width ratio, the coupling to the second nearest line from the line of interest is very small.

amount of energy equal to $0.5 C_c V^2$, then the number of couplings is 0 for case (a), 1 for case (b), and 4 for case (c). E_{is} can be 0 or $0.5 C_s V^2$ depending on whether there is a transition on line i. Then, the energy dissipation of a bus per bus transfer is

$$\xi = \sum_{i=0}^{n-1} E_i \qquad (2)$$

Assume $C_c = \lambda C_s$. Then, computing ξ is equivalent to counting the couplings per bus transfer. Although we can easily obtain ξ for a given datum, we are interested in the expected value of ξ.

B. Bus-Invert Methods and Limited-Weight Code

Here we investigate Hamming-distance-based (*HDB*) and weight-based (*WB*) bus-invert methods [23]. We assume that the invert line is numbered as line n and placed next to line $n-1$. The *HDB* approach computes the Hamming distance H_d between the next data value and the present bus value (*including the invert line*). The data value is inverted for transmission and the invert line is set to 1 if $H_d > \lfloor n/2 \rfloor$. Otherwise, the data value is unchanged and the invert line is set to zero. To facilitate our discussion, some results from [23] for *HDB* are given below.

For even bus width n:

$$p_v = p_b = \frac{1}{2} - 2^{-(n+1)} C\!\left(n, \frac{n}{2}\right). \qquad (3)$$

$$N(n) = (n+1)\left(\frac{1}{2} - 2^{-(n+1)} C\!\left(n, \frac{n}{2}\right)\right) \qquad (4)$$

For odd bus width n:

$$p_v = 0.5; \quad p_b = \frac{1}{2} - 2^{-n} C\!\left(n-1, \frac{n-1}{2}\right) \qquad (5)$$

$$N(n) = \frac{n+1}{2} - n\, 2^{-n} C\!\left(n-1, \frac{n-1}{2}\right) \qquad (6)$$

where $p_v > 0$ and $p_b > 0$ are the transition probability of the invert line and a bus line, respectively. $N(n)$ is the total number of signal transitions per bus transfer.

The *WB* approach [23] works essentially the same as the *HDB* approach except that the weight of the *next data value*, which is the number of 1's on the next data value, is used to decide whether a data value should be inverted before transmission.

Limited-weight codes (*LWCs*) [2,3] play an important role in deriving closed-form formulas for computing the number of couplings per bus transfer. The *HDB* approach generates code words in terms of *transition signaling*. The weight of a code word is the number of 1's in the code word. A perfect (*k*, *l*) *LWC* consists of all the code words whose length is *l* and whose weights are smaller than or equal to *k*. TABLE I gives the *LWC* generated by the coding process with $n = 4$. Note that the right most bit in a code word is generated by the invert line. The *WB* approach generates code words in terms of *level signaling*. The code words for even bus width are identical to those generated by the *HDB* approach. For example, the code words shown in Table 1 are also the ones generated by the *WB* approach for $n = 4$.

III. BUS COUPLING CHARACTERISTICS

A. Coupling Characteristics of an Un-encoded Bus

Herein, we assume that the un-encoded data are spatially and temporally independent and uniformly distributed. Based on this assumption, we have the following theorem.

Theorem 1: *The expected number of couplings on an*

TABLE I. A PERFECT (2,5) LWC WITH $n = 4$.

00000	00001	00010	00100	01000	10000	00011	00101
01001	10001	00110	01010	10010	01100	10100	11000

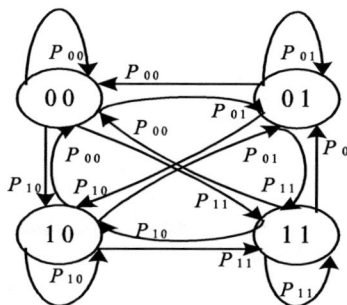

Figure 1. A four-state Markov chain.

un-encoded n-bit bus per bus transfer is
$$M(n) = n - 1. \qquad (7)$$

To prove this theorem, we first compute the probability $P(X_i)$, i.e., the probability of line i with X_i couplings. X_i can be either 0, 1, or 4. For any lines i and $i+1$ with $i < n-1$, we can create a four-state Markov chain shown in Fig. 1. States 00, 01, 10, and 11 represent the logic values of the two bus lines being 00, 01, 10, and 11, respectively. The transition probabilities from one state to another are $P_{00} = P_{01} = P_{10} = P_{11} = 0.25$. Hence, each state probability in steady state is 0.25. Therefore,

$$P(X_i = 0) = 0.375,\ P(X_i = 1) = 0.5,\ P(X_i = 4) = 0.125. \qquad (8)$$

Consequently, the expected number of couplings on line i per bus transfer is equal to 1 and Theorem 1 thus follows.

B. Coupling Characteristics of HDB Approach

Here, we want to find the *expected* number of couplings per bus transfer for even bus width. This can be done by summing up the expected number of couplings on all the lines. Hence, we have to compute $P(X_i)$ for $0 \le i \le n-1$. To compute $P(X_i = 1)$, we count the number of the code words whose bit values at the ith and $(i+1)^\text{th}$ positions are either 0 and 1 or 1 and 0, and then divide this number by the total number of code words. For example, in TABLE I, the number of distinct code words starting with 01 or 10 is 8 and thus $P(X_0 = 1) = 0.5$. Similarly, we can compute $P(X_i = 4)$ by considering only the code words whose bit values at the ith and $(i+1)^\text{th}$ positions are 1's. For example, $P(X_i = 4) = 1/32$ for the code words shown in TABLE I. Since the code words for even n form a *perfect* $(n/2, n+1)$ *LWC* and occur with equal probability, we have for each line $0 \le i \le n-1$

$$P(X_i = 1) = 2 \sum_{h=0}^{n/2-1} C(n-1, h) 2^{-n} = 0.5 \qquad (9)$$

$$P(X_i = 4) = 2^{-3} - 2^{-n-1} C\!\left(n-1, \frac{n}{2} - 1\right) \qquad (10)$$

$$P(X_i = 1) + 4 P(X_i = 4) = 1 - 2^{-n+1} C\!\left(n-1, \frac{n}{2} - 1\right) \qquad (11)$$

The hth term in the series of (9) counts the number of code words, each of which has 0 and 1 (or 1 and 0) respectively at the ith and $(i+1)^\text{th}$ positions and has h 1's at the other $n-1$ positions. Equation (10) is obtained in a similar way.

Theorem 2: *The expected numbers of couplings per bus transfer with even bus width n for the HDB approach is*

$$M(n) = n - n\,2^{-n+1} C\!\left(n-1, \frac{n}{2}-1\right) \quad (12)$$

Similarly, we have the following theorem for odd bus width.

Theorem 3: *The expected number of couplings per bus transfer with odd bus width n for the HDB approach is*

$$M(n) = n - (2n-1)2^{-n} C\!\left(n-1, \frac{n-1}{2}\right) \quad (13)$$

C. Coupling Characteristics of WB Approach

An approach similar to that for an un-encoded bus can be used to derive the number of couplings for the weight-based approach. Due to space limitation, only the results are presented.

Theorem 4: *The expected number of couplings per bus transfer with even bus width n for the WB approach is*

$$M(n) = n. \quad (14)$$

Theorem 5: *The expected number of couplings per bus transfer with odd bus width n for the WB approach is*

$$M(n) = n + 8\left(2^{-n-1} C\!\left(n-1, \frac{n-1}{2}\right)\right)^2. \quad (15)$$

D. Coupling Characteristics of a Partitioned Bus

Here, we would like to investigate to what extent partitioning would help reduce couplings. Suppose the bus lines are partitioned into k equal-sized groups, each of which has $m = n/k$ lines. We consider two possibilities of placing the invert lines. One is to place an invert line immediately next to the right most line in the underlying group as shown in Fig. 2 (left). In this case, the invert lines are said to be distributed among bus groups. The other is to cluster all the invert lines on the right of the kth groups as shown in Fig. 2 (right). We will deal with only the *HDB* approach for it being more effective in reducing couplings.

To obtain the expected number of couplings for a partitioned bus with distributed invert lines, it is yet to know the expected number of couplings between the invert line of group j and the first bus line of group $j+1$. Since groups are independent of each other, the expected number of couplings can be derived using (3) and (5), i.e., the transition probabilities of the invert line of group j and the first bus line of group $j+1$. A Markov chain as shown in Fig. 1 with the state transition probability matrix given in (16) can be made to model the transition behavior of the two lines.

$$\begin{array}{c} \\ 00 \\ 01 \\ 10 \\ 11 \end{array}
\begin{bmatrix}
(1-p_v)(1-p_b) & (1-p_v)p_b & p_v(1-p_b) & p_v p_b \\
(1-p_v)p_b & (1-p_v)(1-p_b) & p_v p_b & p_v(1-p_b) \\
p_v(1-p_b) & p_v p_b & (1-p_v)(1-p_b) & (1-p_v)p_b \\
p_v p_b & p_v(1-p_b) & (1-p_v)p_b & (1-p_v)(1-p_b)
\end{bmatrix} \quad (16)$$

Hence, the expected number of couplings between the invert line of group j and the first bus line of group $j+1$ is simply

$$M_p = p_b + p_v. \quad (17)$$

Theorem 6: *The expected number of couplings per bus transfer on an n-bit bus which is partitioned into k groups with distributed invert lines is*

$$M(n,k) = kM(m) + (k-1)M_p \quad (18)$$

Similarly, we can find out the expected number of couplings per bus transfer for a partitioned bus with clustered invert lines. It is surprising that we have the following theorem.

Theorem 7: *The expected number of couplings per bus transfer on a partitioned bus with clustered invert lines is equal to that on a partitioned bus with distributed invert lines.*

IV. DISCUSSIONS

A. Coupling Reduction

Fig. 3 gives the average number of couplings per pair of lines per bus transfer in terms of bus width. Because the *WB* approach can not reduce couplings, the discussion in the sequel will solely center around the *HDB* approach. Fig. 4 presents the percentage of transition reduction and coupling reduction on the whole bus per bus transfer with respect to an un-encoded bus. Coupling reduction first increases, reaches its maximum of 17.5% for $n=6$, and then decreases with increasing bus width. It is interesting to see that the percentage of coupling reduction is closer to that of transition reduction as bus width increases.

B. Total Energy Reduction

The average energy dissipation on the whole bus per bus transfer can be formulated as follows:

$$E(\xi) = (N(n) + \lambda M(n))E_s \quad (19)$$

where $C_c = \lambda C_s$, $E_s = 0.5 C_s V^2$, $M(n)$ is defined as above, and $N(n)$ is the average number of transitions per bus transfer given by (4) or (6). Fig. 5 plots the percentage of energy reduction for the *HDB* approach with different λ and bus width. The value of λ covers a wide range of choices in wire width and spacing. It is interesting to note that the variation in λ has little influence on

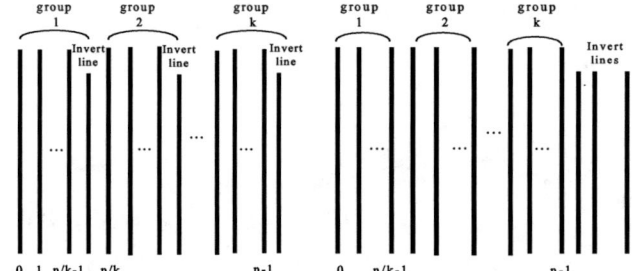

Figure 2. Distributed versus clustered invert lines.

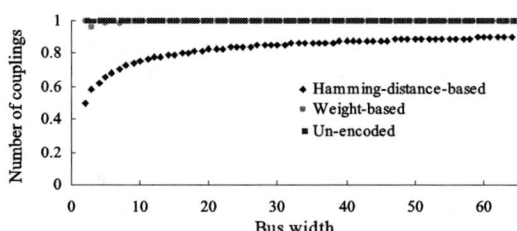

Figure 3. Average couplings per pair of lines per bus transfer.

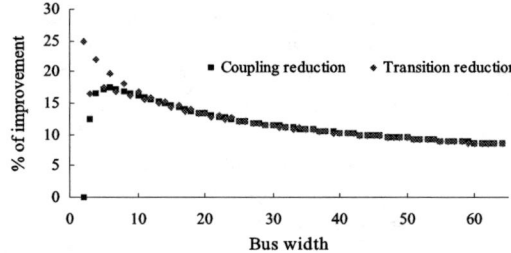

Figure 4. Coupling versus transition reduction.

the energy reduction percentage for a given bus with $n \geq 8$ as more clearly shown in Fig. 6. The reason for this is that, as shown in Fig. 4 when n is sufficiently large, the small difference between transition and coupling reduction percentages makes the total energy reduction percentage insensitive to λ.

C. Energy Reduction of a Partitioned Bus

It was shown that partitioning a bus into a number of smaller buses, each of which has odd number bits, is not a viable approach to switching activity reduction [23]. This property also holds for coupling reduction and can be justified in a way similar to that used in [23]. Its justification can also be observed from TABLE II that gives the results of partitioning a 24-bit bus into different number of groups. Also shown in the table is that partitioning a 24-bit bus into 12 two-bit groups would save 24% of couplings and 25% of transitions with respect to the un-encoded case.

V. CONCLUSIONS

We have performed an in-depth theoretical analysis of the bus-invert method for coupling reduction. The main results include some closed-form formulas for computing the number of couplings per bus transfer. These formulas and those presented in our previous work [23] together establish a sound theoretical foundation for bus-invert coding.

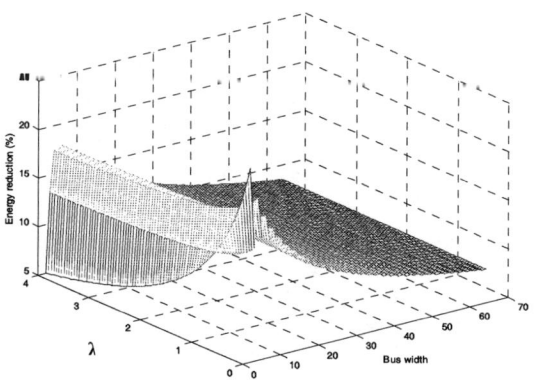

Figure 5. Energy reduction in terms of λ and bus width.

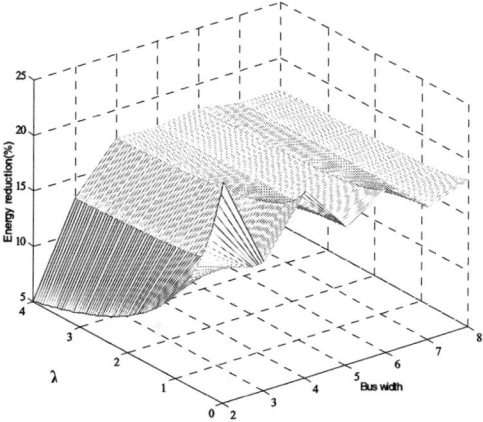

Figure 6. Energy reduction for small bus width.

TABLE II. BUS COUPLING CHARACTERISTICS WITH n=24.

# of bits per group	24	12	8	6	4	3	2
# of groups	0*	1	2	3	4	6	8	12
# of couplings	23	20.13	19.36	18.89	18.56	18.13	19.25	17.5
# of transitions	12	10.49	10.07	9.81	9.63	9.38	10	9

0* indicates bus transfer without coding

REFERENCES

[1] R. J. Fletcher, "Integrated circuit having outputs configured for reduced state changes," U.S. Patent 4,667,337, May 1987.

[2] M. R. Stan and W. P. Burleson, "Bus-invert coding for low power I/O," IEEE Trans. on VLSI, Vol. 3, No. 1, pp. 49-58, March 1995.

[3] M. R. Stan and W. P. Burleson, "Low-power encodings for global communication in CMOS VLSI," IEEE Trans. on VLSI Systems, Vol. 5, No.4, pp. 444-455, Dec. 1997.

[4] S. Ramprasad, N. R. Shanbhag, and I. N. Hajj, "A coding framework for low-power address and data buses," IEEE Trans. on VLSI Systems, Vol. 7, No. 2, pp. 212-221, June 1999.

[5] J. –S. Yim and C. –H. Kyung, "Reducing cross-coupling among interconnect wires in deep-submicron datapath design," DAC, pp. 485-490, June 1999.

[6] B. Victor and K. Keutzer, "Bus encoding to prevent crosstalk," ICCAD, pp. 57-64, Nov. 2001.

[7] K. –H. Baek, K. –W Kim, and S. –M. Kang, "EXODUS: Inter-module bus-encoding scheme for system-on-a-chip," Electronics Letters, pp. 615-617, No. 7, Vol. 36, March 2000.

[8] C. –G Lyuh and T. Kim, "Low power bus encoding with crosstalk delay elimination," ASIC/SOC, pp. 389-393, Sept. 2002.

[9] P. Subrahmanya, R. Manimegalai, V. Kamakoti, and M. Mutyam, "A bus encoding for power and crosstalk minimization," Intl. Conf. on VLSI Design, Jan. 2004.

[10] J. Henkel and H. Lekatsas, "A^{2}BC: adaptive address bus coding for low power deep sub-micron designs," DAC, pp. 744-749, June 2001.

[11] H. Lekatsas and J. Henkel, "ETAM++: extended transition activity measure for low power address bus designs," Intl. Conf. on VLSI Design, Jan. 2002.

[12] Y. Shin and T. Sakurai, "Coupling-driven bus design for low-power application-specific systems," DAC, pp. 750-753, June 2001.

[13] E. Macii, M. Poncino, and S. Salerno, "Combining wire swapping and spacing for low-power deep-submicrom buses," Great Lakes Symposium on VLSI, pp. 77-82, April 2003.

[14] Wong and Tsui, "Re-configurable bus encoding scheme for reducing power consumption of cross coupling capacitance for deep sub-micron instruction bus," DATE, 2004.

[15] Petrov and Orailoglu, "Power efficiency through application-specific instruction memory transformation," DATE, pp. 30-35, 2003.

[16] E. Naroska, S. –J. Ruan, F. Lai, U. Schwiegelshohn, and L. –C. Liu, "On optimizing power and crosstalk for bus coupling capacitance using genetic algorithms," ISCAS, pp. V-277-V-280, May 2003.

[17] C. –G Lyuh, T. Kim, and K. –W. Lim, "Coupling-aware high-level interconnect synthesis," IEEE Trans. on CAD, No. 1, Vol. 23, pp. 157-164, Jan. 2004.

[18] K. -W. Kim, K. -H, Baek, N. Shanbhag, C. L. Liu, and S. -M. Kang, "Coupling-driven signal encoding scheme for low-power interface design," ICCAD, pp. 318-321, Nov. 2000.

[19] Y. Zhang, J. Lach, K. Skadron, and M. R. Stan, "Odd/even bus invert with two-phase transfer for buses with coupling," ISLPED, pp. 80-83, 2002.

[20] M. Lampropoulos, B. Al-Hashimi, and P. Rosinger, "Minimization of crosstalk noise, delay and power using a modified bus invert technique," DATE, 2004.

[21] T. LV, J. Henkel, H. Lekatsas, and W. Wolf, "A dictionary-based en/decoding scheme for low-power data bus," IEEE Trans. on VLSI, No. 5, Vol. 11, pp. 943-951, Oct. 2003.

[22] P. P. Sotiriadis and A. P. Chandrakasan, "A bus energy model for deep submicron technology," IEEE Trans. on VLSI Systems, vol. 10, no. 3, pp. 341-350, 2002.

[23] Rung-Bin Lin and Chi-Ming Tsai, "Theoretical analysis of bus-invert coding," IEEE Trans. on VLSI Systems, Vol. 10, No. 6, pp. 929-935, 2002.

Energy and Latency Evaluation of NoC Topologies

Marcio Kreutz[1], Cesar Marcon[1,2], Luigi Carro[1], Ney Calazans[2] and Altamiro A. Susin[1]

GME – Informática – UFRGS[1] - Porto Alegre, RS, Brazil – PUCRS[2] - Porto Alegre, RS, Brazil

[kreutz, marcon,carro]@inf.ufrgs.br, calazans@inf.pucrs.br, susin@eletro.ufrgs.br

Abstract — Mapping applications onto different networks-on-chip (NoCs) topologies is done by mapping processing cores on local ports of routers considering requirements like latency and energy consumption. In this work, an algorithm devoted to evaluate different topologies is proposed. The evaluation starts with an application model called Application Communication Pattern (ACP), which specifies tasks with the computation load and communication profile. ACP focuses on communication aspects and is an appropriate model to obtain mappings that comply with application requirements. ACP allows fast analysis over many NoC topologies, helping the system designer to evaluate the communication performance of a NoC-based system; this performance strongly depends on the placement of the cores, and it is computationally hard to find the optimal placement.

I. INTRODUCTION

Future billion transistors Systems-on-Chip (SoCs) will allow the development of new applications, which will work in a distributed way and require reusable communication architectures offering scalable bandwidth and parallelism. Networks-on-Chip (NoCs) emerge as a potential tile-based architecture to meet such requirements. NoCs are communication infrastructures composed by a set of routers interconnected by communication channels, which can provide asynchronous communication between synchronous domains. An important issue for NoC-based system designers will be to find a solution of the communications-to-network mapping problem in order to satisfy the communication requirements.

In the upcoming years, a NoC is expected to accommodate more than 10 x 10 tiles [1]. The search for appropriate models and algorithms for mappings problems becomes mandatory. The efficient implementation of tile-based architectures requires efficient mapping strategies. This paper introduces the *Application Communication Pattern* (*ACP*), a model that enables to capture not only the communication capacity, but also the communication ordering. We use ACP to evaluate latency and energy consumption on different NoC topologies. The goal is to find, among regular and irregular NoC topologies, the one that better fits the application requirements. This will help on the design space exploration for NoCs at earlier stages of design.

II. RELATED WORK

Hu and Marculescu [2] showed that mapping algorithms reduce over 60% of energy consumption when compared to ad hoc mapping solutions. Murali and De Micheli, in [3], implement a similar solution. The focus of their papers is to present an algorithm that maps the cores onto mesh NoC architecture under bandwidth constraints, aiming to minimize the energy consumption and the average communication delay. We emphasize that application models, as the one presented in [2][3], omit essential information to estimate the latency of the application, since these models do not consider precisely the time where each communication take place. The knowledge of the communication *ordering* leads to the ACP model (presented in this paper), which aims for a better mapping solution with low extra computational effort, if compared to previous models.

Murali and De Micheli [4] extend the work presented in [3], by the introduction of a tool called SUNMAP. SUNMAP built inside a predefined library of topologies and uses a multi-objective function, which encompasses average communication delay, area and energy consumption. The main objective of the tool is to select automatically the best topology for a given application and to generate a mapping of cores onto that topology.

Hu and Marculescu [5] introduce a model that captures communication and computation scheduling, which is represented by communication task graph (CTG). CTG allows obtaining more accurate results than the one presented in [2][3], since it takes into account the effects of the traffic dynamics. However, while the input data of ACP is easily extracted through the application simulation, CTG implies an extra effort, since the designer has to describe the application and also its computation and communication scheduling.

Our approach uses ACP, which models the messages ordering and traffic load. Similarly to [4], this paper explores the design space for NoC topologies, however, we employ analytical energy models in the search for an optimized topology. In addition, we extend the work with the analysis of the effects of the tile size in the energy consumption.

III. PROBLEM FORMULATION

Given a distributed application, we can state our mapping problem as the task of minimizing the latency and energy consumption by determining the better place for cores on different NoC topologies. The solution for this problem relies in a very complex task, because more than one variable is considered concurrently in the search space: core place, topology and tile size.

It is assumed an application whose tasks were previously mapped onto a set of cores, so we envision a scenario where functions are distributed among a number of cores in a SoC.

We have defined an *application communication pattern* (ACP) to represent the semantics of communicating cores. Communications can be mainly described by the relationships among the cores, by the parallelism among messages exchanges and by the order in which the communications must occur. To model such behavior, it is possible to define a data structure like a "list of sets", where each element of a "time tag list" points to a set of messages.

Definition 1: An ACP is a list of sets. Let $C = \{c_1, c_2, ..., c_n\}$ be the set of application cores, and $b_q \in \mathbb{N}$ the number of bits of the q-th message. Then $m_{ijq} = (c_i, c_j, b_{ijq}) \mid c_i, c_j \in C$ is the q-th message from core c_i to core c_j with b_{ijq} bits. Let $M = \{m_{ijq} \mid c_i, c_j \in C\}$ be the set of all messages between application cores and m be a subset of M. ACP = $\{(t, m) \mid t \in \mathbb{N}, m \subset M\}$ represents an ordered list of message sets, such that t is a time tag that marks the start time of all messagens of m.

The communication architectures can also be modeled as a graph, whose vertices represent tiles and the set of oriented edges express all the links given by the network topology. This data structure is defined as the *communication resource graph* (CRG).

Definition 2: A $CRG = <\Gamma, L>$ is a directed graph, where $\Gamma = \{\tau_1, \tau_2, ..., \tau_p\}$ denotes the set of tiles, corresponding to the set of CRG vertices, and $L = \{(\tau_i, \tau_j) \mid \tau_i, \tau_j \in \Gamma\}$ designates the set of links from τ_i to τ_j, corresponding to the set of CRG edges. The way the edges are connected represents the network topology. Each directed edge $l_{ij} = (\tau_i, \tau_j)$ has associated a structure s_{ij} composed by parameters that expresses the link characteristics in a given topology, such as a link width.

Figure 1 illustrates the above definitions through a hypothetical application with four cores $C = \{a, b, c, d\}$, six messages, and a 2x2 NoC. Figure 1(a) depicts an ACP = $\{(t_1, \{(a, b, 15), (c, a, 20)\}), (t_2, \{(c, a, 15), (a, d, 15), (b, d, 40)\}), (t_3, \{(d, b, 15)\})\}$. Figure 1(b) depicts a CRG with an arbitrary valid mapping of C onto CRG, generating the following association: $\{(\tau_1, b), (\tau_2, a), (\tau_3, d), (\tau_4, c)\}$.

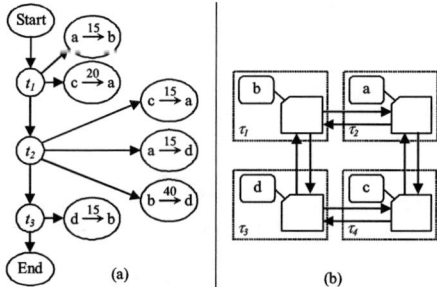

Figure 1 – ACP (a) and CRG (b) examples

For each application message, it is necessary to find in the CRG a path between its sender and receiver vertices in order to determine if the bandwidth offered by this path matches that one required by the application.

Definition 3: A path $p_{ij} = (\tau_i, l_{iy}, ..., \tau_w, l_{wj}, \tau_j)$ is an alternating sequence of CRG vertices and edges, to transport a message from core c_i to core c_j. A path is formed according to the routing strategy implemented in the network tiles the CRG represent. p_{ij} may also correspond to an indirect path in an irregular topology, for instance, a path that uses a router of level two in a fat-tree topology.

Finally, in order to find a valid mapping, one must map each message of the ACP to links and local ports associated to tiles of the CRG.

Definition 4: Given a CRG, for each $m_{ijq} \in$ ACP there exists a corresponding $p_{ij} \in$ CRG, i.e. there is a mapping function $F: \text{ACP} \to \text{CRG}$ such that $\forall m_{ijq} \in \text{ACP} \exists p_{ij} \in \text{CRG}$.

A. Energy Model

This work uses an energy model similar to the ones presented in [2][3], and extends those concepts to estimate energy for different topologies. We use the same concept of bit energy $Ebit$ to estimate the dynamic energy consumption for each bit. $Ebit$ is split into: bit dynamic energy consumed on the buffers ($EBbit$), on the logic gates of each switch ($ESbit$); and on the links between tiles ($ELbit$), which is directly proportional to tile dimension. Because of the strong topological dependence to compute $ESbit$, $EBbit$ and $ELbit$, we define α as a function to estimate the total sum of $ESbit$ and $EBbit$ and φ to estimate $ELbit$. Equation 1 computes the dynamic energy consumed by a single bit traversing the NoC, from tile τ_i to tile τ_j - where communicating cores were previously mapped - and TOP represents the topology selected for evaluation.

$$Ebit_{ij\,TOP} = \alpha_{TOP}(i, j, EBbit, ESbit) + \varphi_{TOP}(i, j, ELbit) \quad (1)$$

To estimate mesh, folded torus and fat-tree NoC topologies, we assume some considerations for all topologies: *(i)* the router area is insignificant front of the core area; and *(ii)* all tiles are regular and with the same square dimensions. These considerations allow estimating the energy consumption for each topology, permitting a comparative evaluation of each layout at earlier design stages.

Assuming *(ii)*, Figure 2(a) depicts that vertical and horizontal links have the same size, i.e. $l_{mV} = l_{mH} = l$. Since $ELbit$ is proportional to l size, for mesh topology function φ of equation 1 can be approximated to equation 2, where η corresponds to the number of routers through which the bit passes from tile i to tile j.

$$\varphi_{MESH}(i, j, ELbit) = (\eta\text{-}1)\,ELbit \quad (2)$$

Assuming *(ii)* Figure 2(b) depicts that $l_{tm} = l$, and as a result of *(i)* if $l_r << l_t$, than $l_{tM} \cong 2l$ making equation 3 an approximation of function φ for folded torus topology.

$$\varphi_{TORUS}(i, j, ELbit) = (2 \text{ or } 1) \times (\eta\text{-}1)\,ELbit \quad (3)$$

$ESbit$ and $EBbit$ of mesh and folded torus topologies are similarly estimated since routers are the same. Then, the function α of equation 1 can be approximated to equation 4.

$$\alpha_{MESH\_OR\_TORUS}(i, j, EBbit, ESbit) = \eta\,(ESbit + EBbit) \quad (4)$$

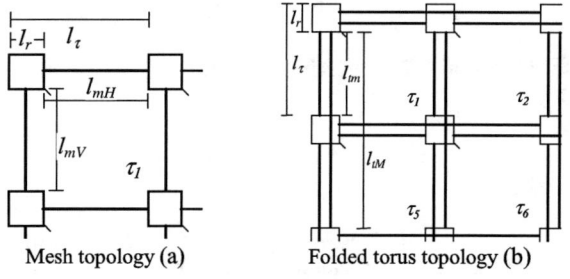

Mesh topology (a)   Folded torus topology (b)

Figure 2 – Partial view of NoCs

The irregularity of fat-tree topology implies paths with different sizes and consequently different $ELbit$. For instance, Figure 3 shows the minimum paths size among one router of level one and all routers of level-two of a fat-tree topology with sixteen tiles. Considering the depicted routing layout and assuming (i) and (ii) $la \cong l/2$, $lb \cong l$, $lc \cong l$ and $ld \cong 2l$. For instance, $ELbit$ is proportional to $3l$ ($2\,la$ and $2\,lb$), when the path uses a router of level one in the same tile of the router of level two. On the other hand, $ELbit$ is negligible, if the routing does not use routers of level two. The function α of equation 1 can be approximated to equation 5 for fat-tree topology.

$$\alpha_{TREE}(i, j, EBbit, ESbit) = (0, 3, 4 \text{ or } 5)\,ELbit \quad (5)$$

The routing inside a level one fat-tree router implies the use of a buffer and a switch that consumes – in average – double energy than the switches of mesh or folded torus topologies. This happen because switches and buffers of routers from level 1 must provide a path for packets coming from the upper level as well a form the same router. When a router of level two is used, there are three buffering and switching stages: one for each level, and for the target. In this case, equation 6 represents *ESbit* and *EBbit* computation.

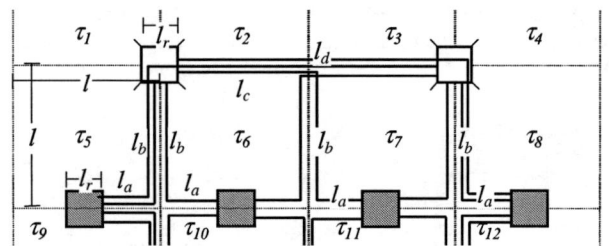

Figure 3 – Partial view of a NoC with fat-tree topology. It is depicted 2 level one and 4 level two routers (shaded)

$$\alpha_{TREE}(i, j, EBbit, ESbit) = (1 \text{ or } 3) \, ESbit + (2 \text{ or } 4) \, EBbit \quad (6)$$

The complete $Ebit_{ij}$ can be estimated by equations 7, 8 and 9, for mesh, folded torus and fat-tree topologies, respectively.

$$Ebit_{ij \, MESH} = \eta \, (ESbit + EBbit) + (\eta\text{-}1) \quad (7)$$

$$Ebit_{ij \, TORUS} = \eta \, (ESbit + EBbit) + (2 \text{ or } 1) \, (\eta\text{-}1) \, ELbit \quad (8)$$

$$Ebit_{ijTREE} = (1 \text{ or } 3) \, ESbit + (2 \text{ or } 4) \, EBbit + (0, 3, 4 \text{ or } 5) \, ELbit \quad (9)$$

Let n_{ijq} be the total amount of bits of a message $m_{ijq} \in M$ going from c_i to c_j. Then, for all topologies, $Ebit_q = n_{ijq} \times Ebit_{ij}$. Equation 10 gives the total amount of *NoC dynamic energy consumption* (*EDyNoC*), which computes all bit traffic of all k messages.

$$EDyNoC = \sum_{q=0}^{k} Ebit_q \quad (10)$$

Given this energy model, an evaluation algorithm can be developed to optimize NoC topologies regarding energy consumption. In the next section our strategy to optimization and the associated algorithm are detailed.

IV. Evaluating Mapping onto NoC Topologies

The problem of mapping cores onto NoC topologies is an instance of the *set covering problem* (SCP) which is proven to be NP-complete [9]. The search space grows exponentially with the system size: even for a 4x4 NoC there are 16! possible mappings. To solve the problem we adopt the *Tabu Search* (TS) algorithm [7], since it has been used to solve many covering problems in many different areas [8]. Figure 4 depicts the pseudo-code for the TS algorithm, where t stands for one step of the algorithm, k is the cost function for energy consumption and latency, Y is the set of all possible solutions and nt is the total number of iterations the algorithm should execute.

The general strategy of TS is to explore, from a set of p resources $S = (1, 2, ..., p)$, all possible moves from the current solution to a neighboring one. The move leading to a neighboring solution can be accepted, even if this results in a deterioration of the objective function. To prevent the search process from cycling, a Tabu list (T) is used to store the last moves for a certain number of iterations (nt). Thus, all solutions, which can be obtained by applying a move stored in T, are excluded from the search. An aspiration criterion allows overriding the Tabu status of a move, for instance, if the move leads to a new best solution. The OPTIMUM function selects a best move among a set of neighbors solutions.

```
Tabu_Search(resources S) {
  select an initial solution:
    y ∈ Y e y* = y; T = ∅
1 if (S(y)-T) == ∅
    stop;
  else
    t = t + 1;
  select best y = OPTIMUM(s(y): s ∈ S(y)-T);
  if k(y) < k(y*)   // y* → best solution
    y* = y;
  if t > nt
    stop;
  else
    update T;
  go to 1;
}
```

Figure 4 – Pseudo code for a Tabu Search algorithm

The S set is made equal to cores places, where each new solution (a move) in the search space means a core swapping, which changes the source and destination for communications; $S = \{c_2, c_4, c_8, ..., c_n\}$; where $S[0] = c_2$ means that core c_2 is placed on the tile 0.

The values for latency and energy consumption may vary for each application execution, due to network *contentions*. Due to its unpredictable behavior, contentions can only be obtained at execution time. We developed a simulator that simulates C++ models for generic components of NoC routers. By simulating a CRG for a target NoC, a system designer can observe how NoC components behave when they send and receive application messages. The NoC simulator is used to evaluate each mapping given by each step in the TS algorithm.

A *simulation step* stands for all messages sent concurrently by the application at each instant of time t, according to the Definition 1. Since the TS algorithm simulates the communications for each solution found by the OPTIMUM function, it was necessary to develop a simulator, which could run as fast as possible. This is the reason because all NoC components were described in C++, accompanying with a dedicated simulator.

We have modeled direct (torus and mesh) and indirect (fat-tree) topologies for evaluation. For direct topologies, it was employed the handshake flow control, a deterministic and source-based routing strategy (XY routing), wormhole packet switching, round-robin arbitration and input buffering. The fat-tree NoC differs in the routing policy, which is partially adaptive – to send messages to the second level of the tree structure – and partially static, the address of the destination node is conceived before a message is sent. These settings are commonly used for NoCs, so they can represent typical NoCs considered for SoCs design.

V. Experimental Results

To corroborate the efficiency of the proposed TS-based algorithm we have tested three applications running on three different NoC topologies: a 4x4 mesh, a 4x4 folded torus and a fat-tree with 16 terminals. Communication channels width was set to 32 bits and the length of messages vary stochastically between 100 and 800 bytes. The first application is a mathematical application that calculates the *Romberg* integration [9]; the second one is an 8-point Fast Fourier Transform (*FFT*) [10]; the third one is an "*image processing*" application for object recognition in a video frame.

The results for latency and energy Tabu-based optimization are given in Figure 5 and Figure 6, respectively. Figure 5 shows that the folded torus and fat-tree topologies could be better optimized for all applications due to higher level of parallelism offered in comparison to the mesh topology. The sequential accesses to a memory and to a central processor in the image application increase the latency of folded torus topology when compared to fat-tree topology. Its reduced number of hops justifies the best result of fat-tree topology.

When energy consumption is taken into account, the mesh structure presented best results, as outlined in Figure 6. The reason for that relies on the shorter wires length a mesh structure has, in comparison to others. For the image application, an optimal result could be achieved for the fat-tree topology, because the energy consumption is nearly as good as the one achieved for the mesh. The reason for that relies on the fact that for this structure the TS algorithm could find the best local minima, which implies in lower switching activities, as well as lower usage of the longer wires.

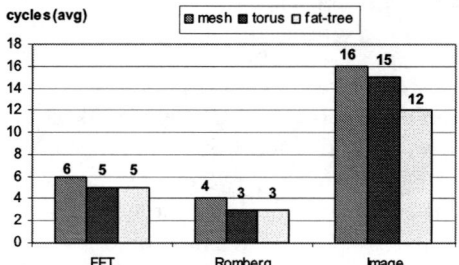

Figure 5 – Latency Optimization

We conclude that, by taking an intersection between latency and energy consumption results, the fat-tree is the better choice for executing the image application.

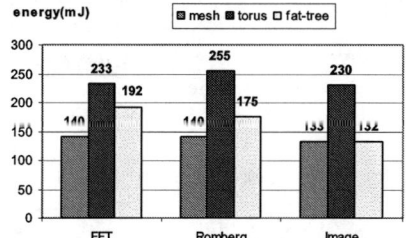

Figure 6 – Energy Optimization

Although the folded torus topology has presented good results regarding latency for all applications, it is not a good choice when energy consumption is critical. A good trade-off between latency and energy consumption could be found for the mesh topology, because it has – in average – about 10% lower performance, but 35% lower energy consumption. Another interesting experiment regards on tiles area overhead. The area of a tile can be enlarged to accommodate the area of the target core. This may increase the length (and capacity load) of communication wires used to connect them, which lead to new values for *ELbit* (defined in section III.A). In Figure 7, it is outlined the results achieved by running the Romberg application for tiles 2, 4 and 8 times larger than the ones previously tested.

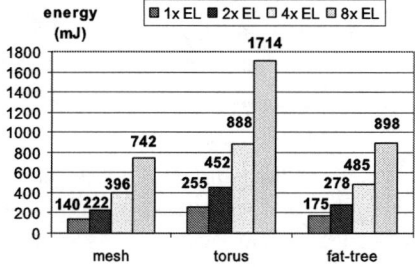

Figure 7 – Latency optimization for different tiles size

As previously shown in Figure 6, the mesh topology can achieve the best results for energy consumption, regardless of the area of tile it is mapped to. Another interesting point concerns the fact that all topologies tested, suffer *proportionally* the same penalty – in energy consumption – for longer wires in communication channels: the bigger *ELbit* is, the higher is the energy consumption at the same proportion for all topologies. Therefore, if bigger cores replace the original ones, the NoC optimized and selected to run the application will remain the same.

Finally, regarding simulation performance, for all Tabu executions, it needed not more than two or three seconds to find an optimal solution in a system composed by a 1 GHz Pentium IV machine with 512 MHz of memory. Therefore, the simulator allows larger applications to be explored in affordable time.

VI. CONCLUSIONS AND FUTURE WORK

In this work, we have presented an approach to evaluate NoC topologies, which is based on a simulation engine and on a Tabu-based heuristic optimization algorithm. Since NoCs architectural components exploration turn to a huge design space to be inspected, we have chosen to employ a heuristic algorithm to evaluate different network topologies. This was made with the objective of network concurrency exploration, in order to find a structure that could better optimize latency and energy consumption. To do that, an analytical energy model was developed, which is able to capture energy consumption estimations for regular and irregular topologies. The design exploration proposed encompasses NoC concurrency evaluation by simulating applications messages on different topologies and by mapping processor cores in a way that reduces NoC contentions.

We showed that the algorithm employed can find optimal solutions for regular and irregular topologies in an affordable time. In addition, this paper shows that fat-tree topology is a strong candidate to fulfill the latency constraints for many applications, while mesh topology achieves the less energy consumption, and for a image processing application the best trade-off between latency and energy consumption is obtained with fat-tree topology.

As future works, we plan to continuously improve the heuristic by adding processor cores execution times in order to evaluate not only the communication subsystem, but also a complete SoC.

REFERENCES

[1] S. Kumar et al. *A Network on Chip Architecture and Design Methodology*. **IEEE Computer Society Annual Symposium on VLSI**, pp.105-112, April 2002.

[2] J. Hu and R. Marculescu. Energy-Aware Mapping for Tile-based NoC Architectures under Performance Constraints. **ASP-DAC**, pp.233-239, January 2003.

[3] S. Murali and G. De Micheli. *Bandwidth-Constrained Mapping of Cores onto NoC Architectures*. **DATE**, pp.896-901, February 2004.

[4] S. Murali and G. De Micheli. SUNMAP: A Tool for Automatic Topology Selection and Generation for NoCs. **DAC**, pp.914-919, June 2004.

[5] J. Hu and R. Marculescu. Energy-aware communication and task scheduling for network-on-chip architectures under real-time constraints. **DATE**, pp.234-239, February 2004.

[6] M.R. Garey and D.S. Johnson. Computers and Intractability: A Guide to the Theory of NP-Completeness. **W.H. Freeman and Co**. 1979.

[7] F. Glover. *Tabu search – Part I*. **ORSA Journal on Computing**. vol. 1, pp.190-206, 1989.

[8] F. Glover. *Tabu search – Part II*. **ORSA Journal on Computing**. vol. 1, pp.4-32, 1990.

[9] R. Burden and J. D. Faires – **Study Guide for Numerical Analysis**, McGraw-Hill, New York, 2001.

[10] M. Quinn – **Parallel Computing- Theory and Practice**, McGraw-Hill, New York, 1994.

An Adaptive Super-Exponential Deflation Algorithm for Blind Deconvolution of MIMO Systems Using the Matrix Pseudo-Inversion Lemma

Kiyotaka Kohno*, Yujiro Inouye† and Mitsuru Kawamoto‡
*†‡Department of Electronic and Control Systems Engineering, Shimane University
1060 Nishikawatsu, Matsue, Shimane 690-8504, Japan
*kohno@yonago-k.ac.jp, †inouye@riko.shimane-u.ac.jp, ‡kawa@ecs.shimane-u.ac.jp

Abstract—The multichannel blind deconvolution of finite-impulse response (FIR) or infinite-impulse response (IIR) channels is investigated using the multichannel super-exponential deflation methods. We propose a new adaptive approach to the multichannel super-exponential deflation methods using the matrix pseudo-inversion lemma (which is extended from the matrix inversion lemma in the full-rank case to the rank-degenerate case) and the higher-order cross correlations of the channel and the equalizer outputs. In order to see the effectiveness of the proposed approach, many computer simulations are carried out for time-invariant MIMO channels along with time-variant MIMO channels. It is shown through computer simulations that the proposed approach is effective for even time-variant channels.

I. INTRODUCTION

Multichannel blind deconvolution has recently received attention in such fields as digital communications, image processing and neural information processing [1].

In the early 1990s, Shalvi and Weinstein proposed an attractive approach to single-channel blind deconvolution called the *super-exponential methods* (SEMs) [2]. One of the attractive properties of the SEM is that it converges iteratively to desired solution regardless of initialization at a super-exponential rate. Extensions of their idea to multichannel deconvolution were presented in [3], [4], [5], [6], [7], [9]. In particular, Inouye and Tanebe [3] proposed the multichannel super-exponential deflation method (MSEDM) using the second-order correlations. Moreover, Kawamoto *et al.* [7] and Kohno *et al.* [9] proposed MSEDMs using the higher-order correlations instead of the second-order correlations in order to reduce the computational complexity and accelerate the performance of deconvolution. The MSEDMs are to deconvolve sequentially the source signals one by one. The most important property of the MSEDMs is that it converges globally to desired solution except for pathological cases. However, it is not considered that the underlying channel exhibits change in time for almost all the conventional MSEDMs, because they do not have an adaptive algorithm which is capable of tracking the varying characteristics of the channel.

In the present paper, we propose an adaptive multichannel super-exponential deflation algorithm (AMSEDA) using the higher-order correlations for MIMO wide band channels (convolutive mixtures). We already proposed two type of adaptive multichannel super-exponential algorithms (AMSEAs), the one in covariance (correlation or Kalman-filter) form and the other in QR-factorization form, for the degenerate rank case of the correlations matrices [8]. We propose an AMSEDA using the matrix pseudo-inversion lemma (the covariance form) in this paper, and we show the effectiveness of the proposed algorithm by computer simulations in comparison with the AMSEDA using the QR-factorization [10].

The present paper uses the following notation: Let Z denote the set of all integers. Let $C^{m \times n}$ denote the set of all $m \times n$ matrices with complex components. The superscripts T, $*$, H and $\dagger$ denote, respectively, the transpose, the complex conjugate, the complex conjugate transpose (Hermitian) and the (Moore-Penrose) pseudoinverse operations of a matrix. Let $i = \overline{1, n}$ stand for $i = 1, 2, \cdots, n$.

II. ASSUMPTIONS AND PRELIMINARIES

We consider an MIMO channel with n inputs and m outputs as described by

$$\boldsymbol{y}(t) = \sum_{k=-\infty}^{\infty} \boldsymbol{H}^{(k)} \boldsymbol{s}(t-k), \quad t \in Z, \quad (1)$$

where
$\boldsymbol{s}(t)$ n-column vector of input (or source) signals,
$\boldsymbol{y}(t)$ m-column vector of channel outputs,
$\boldsymbol{H}^{(k)}$ $m \times n$ matrix of impulse responses.

The transfer function of the channel is defined by

$$\boldsymbol{H}(z) = \sum_{k=-\infty}^{\infty} \boldsymbol{H}^{(k)} z^k, \quad z \in C. \quad (2)$$

For the time being, it is assumed for theoretical analysis that the noise is absent in (1).

To recover the source signals, we process the output signals by an $n \times m$ equalizer (or deconvolver) $\boldsymbol{W}(z)$ described by

$$\boldsymbol{z}(t) = \sum_{k=-\infty}^{\infty} \boldsymbol{W}^{(k)} \boldsymbol{y}(t-k), \quad t \in Z. \quad (3)$$

The objective of multichannel blind deconvolution is to construct an equalizer that recovers the original source signals only from the measurements of the corresponding outputs.

We put the following assumptions on the channel and the source signals.

A1) The transfer function $\boldsymbol{H}(z)$ is stable and has full column rank on the unit circle $|z| = 1$ [this implies that the unknown system has less inputs than outputs, i.e., $n \leq m$, and there exists a left stable inverse of the unknown system].

A2) The input sequence $\{s(t)\}$ is a complex, zero-mean, non-Gaussian random vector process with element processes $\{s_i(t)\}$, $i = \overline{1,n}$ being mutually independent. Moreover, each element process $\{s_i(t)\}$ is an i.i.d. process with a nonzero variance σ_i^2 and a nonzero fourth-order cumulant γ_i. The variances σ_i^2's and the fourth-order cumulants γ_i's are unknown.

A3) The equalizer $W(z)$ is an FIR channel of sufficient length L so that the truncation effect can be ignored.

Remark 1: As to A1), if the channel $H(z)$ is FIR, then a condition of the existence of an FIR equalizer is rank $H(z) = n$ for all nonzero $z \in C$ [11]. Moreover, if $H(z)$ is irreducible, then there exists an equalizer $W(z)$ of length $L \leq n(K-1)$, where K is the length of the channel [11]. Besides, it is shown in [4] that there exists generically (or except for pathological cases) an equalizer $W(z)$ of length $L = \lceil \frac{n(K-1)}{m-n} \rceil$, where $\lceil x \rceil$ stands for the smallest integer that is greater than equal to x.

Let us consider an FIR equalizer with the transfer function $W(z)$ given by

$$W(z) = \sum_{k=L_1}^{L_2} W^{(k)} z^k, \quad (4)$$

where the length $L := L_2 - L_1 + 1$ is taken to be sufficiently large. Let $\tilde{w}_i$ be the Lm-column vector consisting of the tap coefficients (corresponding to the ith output) of the equalizer defined by

$$w_i := [w_{i,1}^T, w_{i,2}^T, \cdots, w_{i,m}^T]^T \in C^{mL}, \quad (5)$$

$$w_{i,j} = [w_{i,j}^{(L_1)}, w_{i,j}^{(L_1+1)}, \cdots, w_{i,j}^{(L_2)}]^T \in C^L, \quad (6)$$

where $w_{i,j}^{(k)}$ is the (i,j)th element of matrix $W^{(k)}$.

Inouye and Tanebe [3] proposed the *multichannel super-exponential algorithm* (MSEA) for finding the tap coefficient vectors $\tilde{w}_i$'s of the equalizer $W(z)$, of which each iteration consists of the following two steps:

$$\tilde{w}_i^{[1]} = \tilde{R}_L^\dagger \tilde{d}_i \quad \text{for } i = \overline{1,n}, \quad (7)$$

$$\tilde{w}_i^{[2]} = \frac{\tilde{w}_i^{[1]}}{\sqrt{\tilde{w}_i^{[1]H} \tilde{R}_L \tilde{w}_i^{[1]}}} \quad \text{for } i = \overline{1,n}, \quad (8)$$

where $(\cdot)^{[1]}$ and $(\cdot)^{[2]}$ stand respectively for the result of the first step and the result of the second step. Let $\tilde{y}(t)$ be the Lm-column vector consisting of the L consecutive inputs of the equalizer defined by

$$\tilde{y}(t) := [\bar{y}_1(t)^T, \bar{y}_2(t)^T, \cdots, \bar{y}_m(t)^T]^T \in C^{mL}, \quad (9)$$

$$\bar{y}_i(t) := [y_i(t-L_1), y_i(t-L_1-1), \cdots, y_i(t-L_2)]^T$$
$$\in C^L, \quad (10)$$

where $y_i(t)$ is the ith element of the output vector $y(t)$ of the channel in (1). Then the correlation matrix $\tilde{R}_L$ is represented as

$$\tilde{R}_L = E\left[\tilde{y}^*(t)\tilde{y}^T(t)\right] \in C^{mL \times mL}, \quad (11)$$

and the fourth-order cumulant vector $\tilde{d}_i$ is represented as

$$\tilde{d}_i = \text{cum}(z_i(t), z_i(t), z_i^*(t), \tilde{y}^*(t))$$
$$= E\left[|z_i(t)|^2 z_i(t) \tilde{y}^*(t)\right]$$
$$-2E\left[|z_i(t)|^2\right] E\left[z_i(t)\tilde{y}^*(t)\right]$$

$$-E\left[z_i^2(t)\right] E\left[z_i^*(t)\tilde{y}^*(t)\right] \in C^{mL}, \quad (12)$$

where $E[x]$ denotes the expectation of a random variable x. We note that the last term can be ignored in case of $E\left[s_i(t)^2\right] = 0$, in which case $E\left[z_i(t)^2\right] = 0$ for all $i = \overline{1,n}$.

III. AN ADAPTIVE SUPER-EXPONENTIAL ALGORITHM USING THE MATRIX PSEUDO-INVERSION LEMMA

Kohno et al. proposed two types of AMSEAs, the one in covariance (correlation or Kalman-filter) form and the other in QR-factorization form, for the degenerate rank case of the correlations matrices [8]. Except for the case when the number of outputs equals the number of inputs, i.e., $m = n$, the correlation matrix $\tilde{R}_L$ is not of full rank. Situations with the number of independent sources (or inputs) being strictly less than the number of sensors (or outputs) are often encountered in various applications such as digital communication, image processing and neural information processing. Moreover, if the underlying channel exhibits slow changes in time, processing all the available data jointly is not desirable, even if we can accommodate the computational and storage loads of the batch algorithm in (7) and (8), because different data segments correspond to different channel responses. In such a case, we want to have an adaptive algorithm which is capable of tracking the varying characteristics of the channel.

Consider the batch algorithm in (7) and (8). The equation (8) constraints a weighted norm of vector $\tilde{w}_i$ to equal one, and thus we assume this constraint is always satisfied using a normalization or an automatic gain control (AGC) of $\tilde{w}_i$ at each time t. To develop an adaptive version of (7), we must specify the dependency of each time t and rewrite (7) as

$$\tilde{w}_i(t) = \tilde{R}_L^\dagger(t)\tilde{d}_i(t), \quad i = \overline{1,n}. \quad (13)$$

Here the subscript L of $\tilde{R}_L(t)$ is omitted for simplicity hereafter.

In order to develop an adaptive version of the MSEA, we should obtain recursion formulas for time-updating of matrix $\tilde{R}(t)$, vector $\tilde{d}_i(t)$ and pseudoinverse $\tilde{R}^\dagger(t)$ in (13), respectively.

$$\tilde{R}(t) = \alpha \tilde{R}(t-1) + (1-\alpha)\tilde{y}^*(t)\tilde{y}^T(t), \quad (14)$$

$$\tilde{d}_i(t) = \alpha \tilde{d}_i(t-1) + (1-\alpha)\tilde{y}^*(t)\tilde{z}_i(t), \quad (15)$$

where

$$\tilde{z}_i(t) := (|z_i(t)|^2 - 2<|z_i(t)|^2>)z_i(t) - <z_i^2(t)>z_i^*(t). \quad (16)$$

Here $<|z_i(t)|^2>$ and $<z_i^2(t)>$ denote respectively the estimates of $E\left[|z_i(t)|^2\right]$ and $E\left[z_i(t)^2\right]$ at time t, α is a positive constant close to, but less than one, which accounts for some exponential weighting factor or forgetting factor [13].

By applying the pseudo-inversion lemma [8] to (14) for obtaining a recursion formula for time-updating of pseudoinverse $P(t) = \tilde{R}^\dagger(t)$, we obtain the following lemma.

Lemma 1: Let $b(t)$, $b_1(t)$ and $b_2(t)$ are defined as

$$b(t) = \sqrt{(1-\alpha)}\tilde{y}^*(t), \quad (17)$$

$$b_1(t) = P(t-1)\tilde{R}(t-1)b(t), \quad (18)$$

$$b_2(t) = \left\{I - P(t-1)\tilde{R}(t-1)\right\}b(t). \quad (19)$$

Then formula of the recursion for the pseudoinverse $P(t)$ from $P(t-1)$ by using the pseudo-inversion lemma is explicitly expressed, depending on the values of vectors $b_1(t)$ and $b_2(t)$ and matrix $\tilde{R}(t-1)$, as follows:

1) If $b_2(t) = 0$, then
$$P(t) = \frac{1}{\alpha}\left[P(t-1) - \frac{P(t-1)b_1(t)b_1^H(t)P(t-1)}{\alpha + b_1^H(t)P(t-1)b_1(t)}\right]. \quad (20)$$

2) If $b_2(t) \neq 0$ and $b_1(t) = 0$, then
$$P(t) = \frac{1}{\alpha}P(t-1) + \frac{b_2(t)b_2^H(t)}{\left\{b_2^H(t)b_2(t)\right\}^2}. \quad (21)$$

3) Let $l(t)$ be a non-negative number defined by
$$l(t) := |1 + b_1^H(t)P_b(t)b_2(t)|^2 - b_1^H(t)P_b(t)b_1(t)b_2^H(t)P_b(t)b_2(t), \quad (22)$$
where $P_b(t)$ is defined by
$$P_b(t) := \frac{1}{\alpha}\left[P(t-1) - \frac{P(t-1)b_1(t)b_1^H(t)P(t-1)}{\alpha + b_1^H(t)P(t-1)b_1(t)}\right] + \frac{b_2(t)b_2^H(t)}{\left\{b_2^H(t)b_2(t)\right\}^2}. \quad (23)$$

Then in the case when $b_2(t) \neq 0$ and $b_1(t) \neq 0$,
$$P(t) = P_b(t) - P_b(t)[b_1(t), b_2(t)]P_d(t)[b_1(t), b_2(t)]^H P_b(t), \quad (24)$$
where
$$P_d(t) := \frac{1}{l(t)}\begin{bmatrix} -b_2^H(t)P_b(t)b_2(t) & 1 + b_1^H(t)P_b(t)b_2(t) \\ 1 + b_2^H(t)P_b(t)b_1(t) & -b_1^H(t)P_b(t)b_1(t) \end{bmatrix}. \quad (25)$$

These equations are initialized by their values appropriately selected or calculated by the batch algorithm in (7) and (8) at initial time t_0 and used for $t = t_0 + 1, t_0 + 2, \cdots$.

Remark 2: In Lemma 1, in order to keep the stability of $P(t)$, the positive constant value α has to be chosen for the appropriate value which is very close to 1, or which is asymptotically approached to 1 with time t, for example $\alpha = 1 - 1/(1000 + t)$.

Based on Lemma 1 along with (15), we obtain following theorem with gives a recursion formula for time-updating of the tap vector $\tilde{w}_i(t)$ for $i = \overline{1, n}$.

Theorem 1: The recursion for $\tilde{w}_i(t)$ is
$$\tilde{w}_i(t) = P(t)\tilde{R}(t)\tilde{w}_i(t-1) + k(t)\left[\tilde{z}_i(t) - \tilde{y}^T(t)\tilde{w}_i(t-1)\right], \quad (26)$$
where
$$k(t) := (1-\alpha)P(t)\tilde{y}^*(t), \quad (27)$$
$$\tilde{z}_i(t) := (|z_i(t)|^2 - 2<|z_i(t)|^2>)z_i(t) - <z_i^2(t)>z_i^*(t), \quad (28)$$
$$<|z_i(t)|^2> := \beta<|z_i(t-1)|^2> + (1-\beta)|z_i(t)|^2, \quad (29)$$
$$<z_i^2(t)> := \beta<z_i^2(t-1)> + (1-\beta)z_i^2(t). \quad (30)$$
Here β is a positive constant less than α, and $P(t) = \tilde{R}^\dagger(t)$ is the pseudoinverse of $\tilde{R}(t)$.

IV. AN ADAPTIVE SUPER-EXPONENTIAL DEFLATION ALGORITHM USING THE MATRIX PSEUDO-INVERSION LEMMA

The MSEDM proposed by Inouye and Tanebe [3] uses the second-order correlations to estimate the contributions of an extracted source signal to the channel outputs. Kohno et al. [9] proposed an MSEDM using the higher-order correlations instead of the second-order correlations to reduce the computational complexity in terms of multiplications and to accelerate the performance of equalization. For the details of the MSEDM using the higher-order correlations, see the equations from (13) through (30) in [9]. In the present paper, we proposed a new AMSEDA which is an adaptive version of the MSEDM using the higher-order correlations and the matrix pseudo-inversion lemma described in the previous chapter.

In the new AMSEDA, the following procedures are carried out in each time when channel outputs are observed.

Before the following procedures are carried out, it is necessary that $\tilde{R}$, $\tilde{d}_i$, $\tilde{w}_i$ and P are initialized.

At first, set $t = t_0$, and set $l = 1$ where l denotes the number of channels (or the sources) equalized.

Then, $\tilde{R}(t)$ is calculated by (14), $\tilde{d}_1(t)$ is calculated by using (15), (28), (29) and (30), $P(t)$ is calculated by using from (17) to (25), and $\tilde{w}_1(t)$ is calculated by the two steps (26) and (8). By these procedures, the first equalized output $z_1(t)$ is obtained.

Next, the MSEDM using the higher-order correlations is carried out. We calculate the contribution signals by using the equalized output $z_1(t)$, and remove the contribution signals from the channel outputs in order to define the outputs of a multichannel with $n-1$ inputs and m outputs. The number of inputs becomes deflated by one. The procedures mentioned above are continued until $l = n$, where we obtain the last equalized output $z_n(t)$ for $t = t_0$. If $t < t_f$ (where t_f is a final time), then set $t = t_0 + 1$ and iterate the same procedures as the previous time t. If $t = t_f$, then stop here. The n equalized outputs $z_1(t), \cdots, z_n(t)$ are obtained for $t = t_0, t_0 + 1, \cdots, t_f$.

Therefore, the proposed algorithm is summarized as shown in Table 1.

Table 1. The proposed algorithm.

Step	Contents
1	Set $t = t_0$ (where t_0 is an initial time).
2	Set $l = 1$ (where l denotes the number of the channels equalized).
3	Calculate $\tilde{R}(t)$ using (14).
4	Calculate $\tilde{d}_l(t)$ using (15), (28), (29) and (30).
5	Calculate $P(t)$ using from (17) to (25).
6	Calculate $\tilde{w}_l(t)$ using (26) and (8).
7	Carry out the deflationary process using the MSEDM with the higher-order correlations [9].
8	If the subscript l is less than n, then set $l = l + 1$, and the procedures (from Step 3 through Step 7) are continued until $l = n$.
9	If $t < t_f$ (where t_f is a final time), then set $t = t + 1$ and iterate the procedures from Step 2 through Step 8. If $t = t_f$, then stop here.

V. SIMULATIONS

To demonstrate the effectiveness of proposed method, some computer simulations were conducted. We considered an MIMO channel with two inputs and three outputs, and assumed that the length of channel is three ($K = 3$), that is

$\boldsymbol{H}^{(k)}$s in (1) were set to be

$$\boldsymbol{H}(z) = \sum_{k=0}^{2} \boldsymbol{H}^{(k)} z^k = \begin{bmatrix} 1.00 + 0.15z + 0.10z^2 & 0.65 + 0.25z + 0.15z^2 \\ 0.50 - 0.10z + 0.20z^2 & 1.00 + 0.25z + 0.10z^2 \\ 0.60 + 0.10z + 0.40z^2 & 0.10 + 0.20z + 0.10z^2 \end{bmatrix}. \quad (31)$$

The length of the equalizer was chosen to be seven ($L = 7$). We set the values of the tap coefficients to be zero expect for $w_{12}^{(4)} = w_{21}^{(4)} = 1$. Two source signals were 4-PSK and 8-PSK signals, respectively. For recovering first source signal, the initial values of $\tilde{\boldsymbol{R}}$, $\tilde{\boldsymbol{d}}_i$ and $\boldsymbol{P}$ were estimated using 5,000 data samples. For recovering second source signal, the initial value of $\tilde{\boldsymbol{R}}$ and $\boldsymbol{P}$ were set the identity matrix $\boldsymbol{I}$. The values of α and β were chosen as $\alpha = 0.999$ and $\beta = 0.05$, respectively. Besides, we used the fourth-order correlation method for subtracting the contributions of an extracted source signal to the channel outputs. As a measure of performance, we use the multichannel intersymbol interference (M_{ISI}) [3],[9].

Fig. 1 and Fig. 2 show performance results of our proposed algorithm for the time-invariant and the time-invariant channel, respectively, compared with the existing AMSEDA using the QR-factorization [10] with same conditions obtained by using 100,000 data samples. In Fig. 2, the last matrix $\boldsymbol{H}^{(2)}$ of the impulse response of the channel was varied by adding 0.3 to all its elements at discrete time $t =30,000$ for the proposed algorithm and at $t =2,500$ for the existing AMSEDA.

It can be seen from Fig. 1 that the proposed algorithm deconvolved all source signals and it is effective for the time-invariant channel. Also it can be seen from Fig. 1 that in the proposed algorithm it takes much time (about $t =20,000$) until all source signals are deconvolved, however the accuracy is very high (about -22dB) and the stability with the time is good. In the meantime, in the existing AMSEDA all source signals are deconvolved quickly (about $t =2,500$), however the accuracy is not so high (about -3dB) and the stability with time is bad, because the value of M_{ISI} gradually increases after all source signals are deconvolved.

It can be seen from Fig. 2 that the proposed algorithm is effective for even the time-variant channel, however the existing AMSEDA is not effective for the time-variant channel.

We think that one of reasons why the proposed algorithm is superior to the existing AMSEDA for the time-variant channel is that the pseudo-inversion lemma gives an explicit recursion formula for time-updating of calculating the pseudoinverse $\tilde{\boldsymbol{R}}^{\dagger}(t)$. This fact holds also true in the full-rank case of the correlation matrices.

VI. CONCLUSIONS

We have considered the problem of adaptive multichannel blind deconvolution based on the super-exponential algorithms using deflation methods proposed by Inouye and Tanebe [3]. In this paper, we proposed a new approach to the adaptive multichannel deflationary blind deconvolution using the matrix pseudo-inversion lemma and the higher-order correlations. In order to see the effectiveness of the proposed approach, we

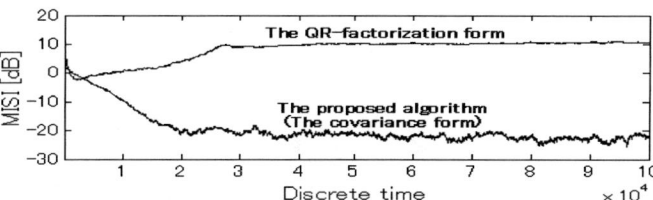

Fig. 1. Performance of the proposed algorithm for the non-adaptive model.

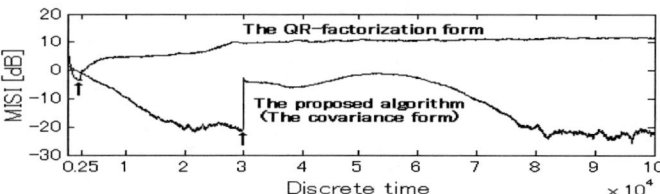

Fig. 2. Performance of the proposed algorithm for the adaptive model.

have considered computer simulations for two types of MIMO channels, that is, the first one is time-invariant and the second one is time-variant. It has been shown through computer simulations that the proposed approach is effective for time-invariant channels and even time-variant channels.

REFERENCES

[1] Special issue on blind system identification and estimation, Proc. IEEE, vol. 86, no. 10, pp. 1907-2089, Oct. 1998.
[2] O. Shalvi and E. Weinstein, "Super-exponential methods for blind deconvolution," IEEE Trans. Information Theory, vol. 39, no. 2, pp. 504-519, Mar. 1993.
[3] Y. Inouye and K. Tanebe, "Super-exponential algorithms for multichannel blind deconvolution," IEEE Trans. Signal Processing, vol. 48, no. 3, pp. 881-888, Mar. 2000.
[4] M. Martone, "An adaptive algorithm for antenna array low-rank processing in cellular TDMA base stations," IEEE Trans. Commun., vol. 46, no. 5, pp. 627-643, May 1998.
[5] M, Martone, "Fast adaptive super-exponential multistage beamforming for cellular base-station transceivers with antenna arrays," IEEE Trans. Vehicular Tech., vol. 48, no. 4, Jul. 1999.
[6] K. L. Yeung and S. F. Yau, "A cumulant-based super-exponential algorithm for blind deconvolution of multi-input multi-output systems," Signal Process., vol. 67, pp. 141-162, 1998.
[7] M. Kawamoto, K. Kohno and Y. Inouye, "Robust Super-Exponential Methods for Deflationary Blind Equalization of Instantaneous Mixtures," to appear in IEEE Trans. Signal Processing.
[8] K. Kohno, Y. Inouye, M. Kawamoto and T. Okamoto, "Adaptive Super-Exponential Algorithms for Blind Deconvolution of MIMO Systems," in Proc. ISCAS, vol.V, Vancouver, Canada, 2004, pp.680-683.
[9] K. Kohno, Y. Inouye and M. Kawamoto, "Super-Exponential Methods Incorporated with Higher-Order Correlations for Deflationary Blind Equalization of MIMO Linear Systems," in Proc. 5th Int. Conference on Independent Component Analysis and Blind Signal Separation (ICA2004), Granada, Spain, 2004, pp.685-693.
[10] K. Kohno, Y. Inouye, M. Kawamoto and T. Okamoto, "An Adaptive Super-Exponential Deflation Algorithm for Blind Deconvolution of MIMO Systems Using the QR-factorization of Matrix Algebra," in Proc. MWSCAS, vol.III, Hiroshima, Japan, 2004, pp.419-422.
[11] Y. Inouye and R-W. Liu, "A system-theoretic foundation for blind equalization of an FIR MIMO channel system," IEEE Trans. Circuits and Systems–I, Fundam. Theory Appl. vol. 49, no. 4, pp. 425-436, Apr. 2002.
[12] G. H. Groub and C. F. Van Loan. Matrix Computations 2nd Ed., Baltimore, MD: The Johns Hopkins University Press, 1989
[13] S. Haykin. Adaptive Filter Theory 3rd Ed., Upper Saddle River, NJ: Prentice-Hall, 1996.
[14] G. W. Stewart, "Methods of simultaneous iteration for calculating eigenvectors of matrices," in Topics in Numerical Analysis II, J. H. Miller, Ed. New York: Academic, pp. 169-185, 1975.

Filterbank-based Blind Signal Separation with Estimated Sound Direction

Hyung-Min Park, Chandra Shekhar Dhir, Do-Kwan Oh, and Soo-Young Lee

Brain Science Research Center and Department of Biosystems
Korea Advanced Institute of Science and Technology
Daejeon, Republic of Korea
{hmpark, sylee}@kaist.ac.kr

Abstract— Based on the mathematical model of human auditory pathway a binaural auditory system has been developed for sound localization and speech enhancement. The developed system includes a filterbank, and the blind signal separation is conducted at each filtered signals. The estimated directions of arrival are collected at each subband, and are utilized to localize the multiple sources from convolved mixtures. Imposing additional constraint based on the estimated sound directions shows faster convergence and also better signal separation is expected for noisy multi-source speeches.

I. INTRODUCTION

The blind source separation (BSS) is to estimate source signals from sensory observations that are mixtures of the sources [1]. Independent component analysis (ICA) has been a quite popular method for solving the BSS problem resorting to the statistical independence among the sources.

In real-world situations, due to reverberation the signals are also convolved with acoustic channels. Many researchers had proposed blind separation methods for convolved mixtures, and most of them can be categorized into the time-domain methods and the frequency-domain methods. The time-domain methods result in slow convergence because many parameters are to be adapted, and heavy computational load is required for real-time applications [2]. On the other hand, the frequency-domain methods usually result in poor performance mainly due to the frame-size trade-offs [3].

Recently, filterbank-based methods had been proposed [2,4]. Over-sampled filter bank approach proposed by Park et al. divides the complex separation problem into many simpler problems without the block effects of frequency domain approaches. With decimation on each filtered subband signals, this method shows faster convergence with lesser computational load [2].

BSS tries to find out demixing system which corresponds to the inverse of the mixing channels. For complex reverberation the demixing system should be constructed by very long adaptive filters. Although the filterbank-based method can reduce the required number of parameters in each subband, this number is still quite large. Thus, the convergence speed may be the limiting factor for real-time applications with practical length of the reverberation. Moreover, in strong noisy multisource environment the subband filters may not converge to the correct solutions.

In this paper, we propose a method to improve the performance of the filterbank-based ICA by imposing an additional physical constraint, i.e., all the subband signals should have the same direction of arrival. The proposed method improves the convergence speed, and also helps in removing the inherent permutation problem by exploiting the stationarity of the mixing environment [5].

II. FILTERBANK-BASED ICA

In human auditory pathway, the arriving sound signals initially go through the cochlear filterbank, which is an array of special subband filters at the cochlea. Then, the filtered subband neural signals from the left and right ears are combined at the superior olivery complexes for sound localization and speech enhancement [6]. Although the conventional models of binaural processing usually do not incorporate convolutive signals, it basically performs similar functions to the blind signal separation. Therefore, we believe in the need of extending binaural processing to convolutive signals as well, and the blind signal separation algorithms need to learn from binaural processing. The filterbank-based ICA for the blind signal separation had been developed in this context [2].

Fig. 1 shows the filterbank-based ICA method for 2 sources and 2 outputs. The input signals are mixtures of unknown independent sources, which are split into subband signals by analysis filters. Each subband signal is subsampled by factor M for computational efficiency. Usual time-domain ICA algorithm may be applicable for the convolved mixtures in each subband. Although binaural processing does not reconstruct the signals, we may need to reconstruct them from the subbands for the blind signal separation. For this case after fixing the permutation and

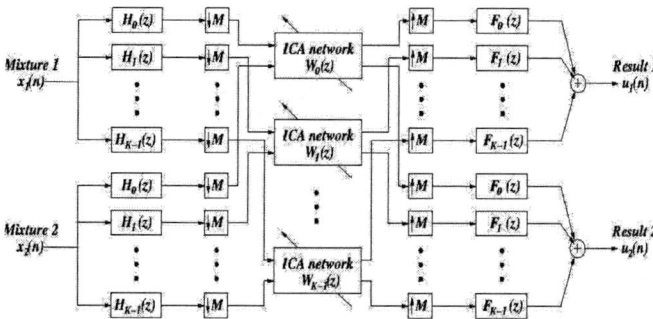

Figure 1. The oversampled filterbank-based ICA method.

scaling problems, each output signal from the ICA network is over-sampled by factor M and the estimated source signals are reconstructed from the subband output signals by the synthesis filters [7]. The filterbank-based ICA algorithm is easily extensible to more sources and more observed mixtures.

The aliasing effect can be neglected because of the sharp transitions between the pass and stop bands of the oversampled filterbank and also smaller decimation factor than the number of analysis filters. As a result, the adaptive filter coefficients in each subband can be adjusted without any information of the other subbands. To implement the oversampled filterbank, we make use of the uniform complex-valued filterbank obtained by the generalized discrete Fourier transform [8].

For the separation network in each subband, the feedback architecture is chosen as [2, 9]

$$u_i(n) = \sum_{m=0}^{L} w_{ii}(m) x_i(n-m) + \sum_{j=1, j \neq i}^{N} \sum_{m=1}^{L} w_{ij}(m) u_j(n-m), \quad (1)$$

where the adaptive filters $w_{ij}(m)$ generate the outputs $u_i(n)$ from the mixed observations $x_i(n)$ similar to the original source signals $s_i(n)$. Here, n and m denote the time indices, and L and N denote the maximum number of time delays for the acoustic reverberation and the number of sources, respectively. It is worth noticing that the complex-valued filterbank results in complex-valued $x_i(n)$, $u_i(n)$ and $w_{ij}(m)$. For the complex-valued BSS, the entropy maximization algorithm provides learning rules of the adaptive filter coefficients as [2,9]

$$\Delta w_{ii}(0) \propto 1/w_{ii}^*(0) - \varphi(u_i(n)) x_i^*(n),$$
$$\Delta w_{ii}(m) \propto -\varphi(u_i(n)) x_i^*(n-m), \quad m \neq 0, \quad (2)$$
$$\Delta w_{ij}(m) \propto -\varphi(u_i(n)) u_j^*(n-m), \quad i \neq j,$$

$$\varphi(u_i(n)) = -\frac{\frac{\partial p(|u_i(n)|)}{\partial |u_i(n)|}}{p(|u_i(n)|)} \exp(j \cdot \angle u_i). \quad (3)$$

where $\varphi(\cdot)$ is the score function.

The entropy maximization algorithm tries to make the outputs temporally whitened, which may degrade the performance in many applications such as separation of natural signals. The whitening of the recovered output signals can be avoided by forcing direct filters, $w_{ii}(m)$'s, to scales [9].

The filterbank-based ICA method has one demixing network for each subband and adapts the filter coefficients of the network independent of the other subbands. Also, it is well known that the standard ICA algorithm has the indeterminacy of the recovered independent signals up to the permutation and arbitrary filtering. Therefore, the filterbank-based method has the same permutation problem as the frequency-domain method. To solve the problem, many researchers had utilized the close correlation between subband signals and the stationarity of the mixing environment [5].

The filtered signals are band-limited and may be decimated for computational efficiency. Since ICA at each subband does not interfere with the others, filterbank-based ICA method essentially breaks down the complicated time-domain problem into several smaller time-domain problems, which results in much faster convergence. The filterbank-based ICA method does not have the performance limitation of frequency-domain approaches because ICA in each subband resembles the time-domain method. Since each subband ICA network can independently compute the subband output signals and adapt the filter coefficients of the demixing networks, this method is also appropriate for parallel processing.

III. ADDITIONAL CONSTRAINT: THE SOUND DIRECTION

Although the filterbank-based ICA algorithm results in better performance than frequency-domain approaches and much faster convergence with lesser computational complexity than the standard "full-band" time-domain approaches, there is still a room for improvement.

Especially the filterbank-based ICA is subject to additive noise. The colored noise may easily disturb the signals and hence the ICA results on a subband. This perturbs the learning of the demixing filters and results in much slower convergence. In addition, it may also give wrong converged results. To overcome this difficulty we noticed that the direction of sound arrival should be the same for all the subbands. So we try to put this as an additional constraint on the ICA paradigm.

The filterbank approach is also advantageous to estimate the sound directions from noisy multisource mixtures. At each filtered subband signals one estimates the time delays, which may be averaged with appropriate weights for better estimates. For the estimation of the sound direction at each subband, simple cross-correlation method was employed before the filtered subband signals are decimated. Moreover, the weight values for each subband were given as

$$w_{CC} = G_{x_1 x_2}(f_c), \quad (4)$$

where $G_{x_1 x_2}(f_c)$ is the cross spectral density between two observations $x_1(n)$ and $x_2(n)$ at the frequency f_c, the center frequency of the corresponding subband filters. Multiple sound localizations can be reliably estimated by picking up the indices of the maximum values as the time delays between the sources, which are directly related to the direction of arrival.

Although the new constraint may be incorporated into the ICA adaptation rules in (2) and (3), we chose the easiest one wherein the estimates of the direction of arrival are used to initialize the demixing filters at each subband before performing ICA. Decimation before ICA in each subband reduces the length of the demixing filter by a factor M. Considering this, the initial values of the cross filters for each subband were set to 0.5 at the nearest integer time delay, k, where k is given as

$$k = \text{nint}\left(abs\left(\frac{\text{delay estimate}}{M}\right)\right). \quad (5)$$

The *nint(.)* function returns the nearest integer to the real argument passed to it. All the other values of the cross filter taps were set to zero. Although this simple method does not guarantee right convergence, the ICA algorithm is very stable and usually results in solutions nearest to the initial values. However, the cross filters may not be able to adapt themselves to the correct sound direction in noisy environment and this may require incorporation of the new constraint within the ICA learning algorithm.

IV. EXPERIMENTAL RESULTS

Experiments were conducted on the BSS of speech signals to show the performance of the proposed algorithm. To construct a 2 × 2 mixing system, impulse responses were generated by the image method which simulates multipath acoustics between two points in a rectangular room [10]. Fig. 2 shows the virtual room to simulate impulse responses from 2 speaker points to 2 microphone points, and Fig. 3 shows the resulting room impulse response. The room reflection coefficients at the walls, ceiling and floor were set to 0.6. Here we had tried to make the simulated acoustic channels as realistic as possible.

Considering that the speakers and microphones have ideal characteristics, two speech signals were assumed to be played at the source points. Each signal had 5 seconds length at 16 *kHz* sampling rate. Since speech signal approximately follows the Laplacian distribution, *sgn(.)* was used as the score function.

The source directions, which are directly related to the time delays, were estimated using the cross correlation between each subband signals and is summarized in Table I. The time delays corresponding to the source locations are estimated by taking a weighted average of the index at which the maximum peak of the cross-correlation occurs. The presence of silence in the speech introduces errors in the delay estimates.

TABLE I. ESTIMATES OF DELAY FROM THE OBSERVATIONS

	Delay estimate (samples)
Source 1	-10.4267
Source 2	9.8958

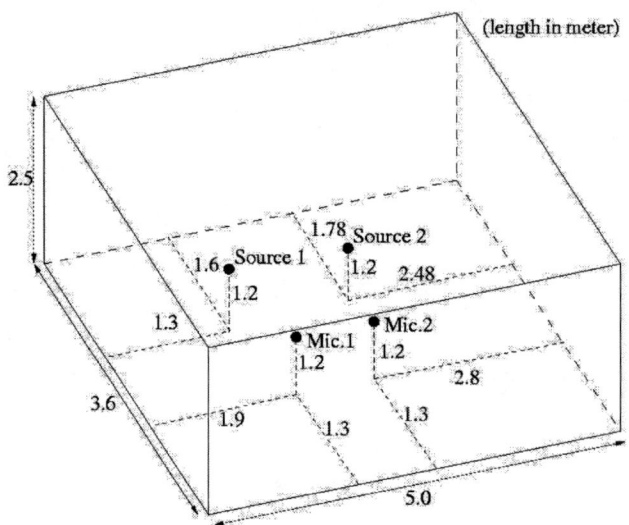

Figure 2. Virtual room to simulate impulse responses from 2 speaker points to 2 microphone points.

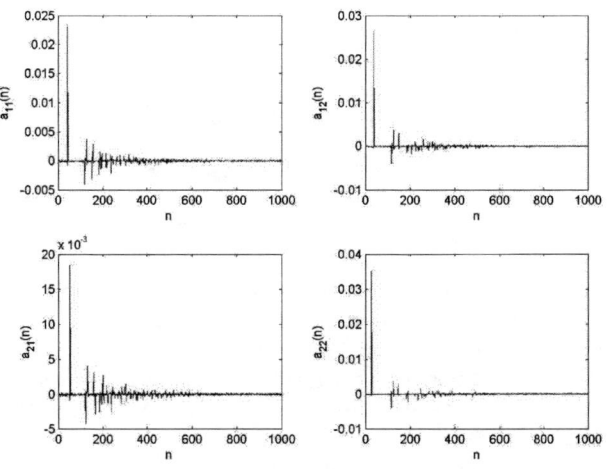

Figure 3. Impulse reponses of the mixing system for experiments on blind source separation.

Fig. 4 shows the frequency response of analysis filters of a uniform oversampled filterbank. The filterbank was designed for alias-free reconstruction with decimation factor $M = 10$ and it was constructed from a prototype filter with 220 taps.

Using the results in Table I and (5), the initial values of the demixing filters at each subband were set to 0.5 for the corresponding tap and zero for all the other filter taps. For comparison we also conduct experiments with all zero initial values of subband filters. From these initial values, the ICA adaptation went on with (2) and (3). Experimental results

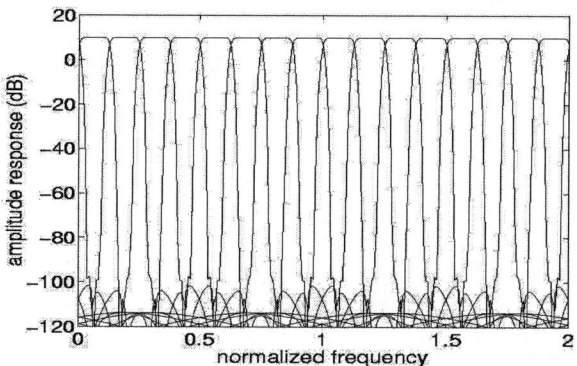

Figure 4. Frequency response of analysis filters of a uniform oversampled filterbank.

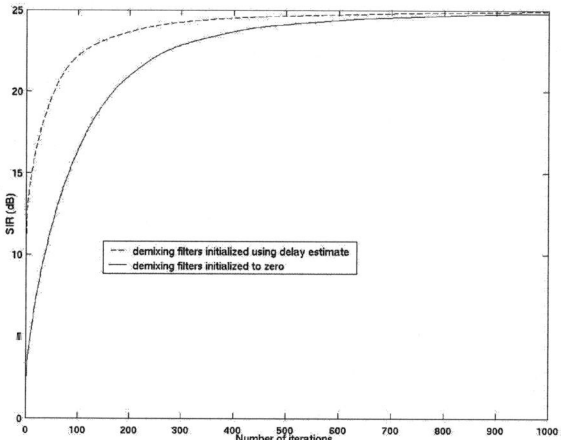

Figure 5. Learning curve of filterbank ICA using different initialization.

were compared in terms of signal-to-interference ratio (SIR). For a 2 × 2 mixing/demixing system, the SIR is defined as a ratio of the signal power to the interference power at the outputs as [11]

$$\text{SIR}(dB) = \frac{1}{2} \cdot \left| 10\log\left(\frac{\langle (u_{1,s_1}(n))^2 \rangle}{\langle (u_{1,s_2}(n))^2 \rangle} \cdot \frac{\langle (u_{2,s_2}(n))^2 \rangle}{\langle (u_{2,s_1}(n))^2 \rangle} \right) \right|. \quad (6)$$

Here, $u_{j,s_i}(n)$ denotes the jth output of the cascading mixing-demixing system only when $s_i(n)$ is active, i.e., the other $s_j(n)$ are set to zero.

Fig. 5 shows the learning curve when the estimated source direction is used as the initial condition for the demixing filters. The choice of demixing filters at every subband as an impulse at the integer time delay results in an initialization closer to the expected solution, it also accelerates the separation performance. Although the final SIRs show some improvement over the previous method without the constraint on source directions, the improvement is not significant. One reason may be that the previous method already produced excellent signal separation performance at all the subbands. By checking the demixing filters for all the subbands, we were able to confirm that the time delays of the maximum demixing filters were already the same for all the subbands without the direction information. However, in case of noisy mixtures such alignment of subband demixing filters was not observed at some subbands. This may be due to the fact that presence of noise may affect the correct solution by not letting it converge to the global minima.

CONCLUSION

In this paper, we proposed to utilize the directions of sound arrival for performance improvement of filterbank-based ICA for the BSS of acoustic convolved mixtures. For the easy problem of 2 sources and 2 observed mixtures, we demonstrated much faster convergence speed with only a slight improvement on the separation performance. In future, for real time applications a VLSI will be implemented based on the proposed algorithm.

ACKNOWLEDGMENT

This work was supported as the Brain Neuroinformatics Research Program sponsored by Korean Ministry of Science and Technology.

REFERENCES

[1] T.-W. Lee, Independent Component Analysis, Kluwer academic Publishers, Boston, MA, 1998.

[2] H.-M. Park, Sang-Hoon Oh, and S.-Y. Lee, "A uniform oversampled filter bank approach to independent component analysis", in Proc. Int. Conf. on Acoustics, Speech and Signal Processing, vol. 5, April. 2003, pp. 249-252.

[3] S. Araki, S. Makino, R. Mukai, T. Nishikawa, and H. Saruwatari, "Fundamental limitation of frequency domain blind source separation for convolved mixtures of speech," in Proc. Int. Conf. on ICA and BSS, Dec. 2001, pp. 132-137.

[4] S. Araki, S. Makino, R. Aichner, T. Nishikawa, and H. Saruwatari, "Subband based blind source separation with appropriate processing for each frequency band," in Proc. Int. Conf. on ICA and BSS, April 2003, pp. 499-504.

[5] C. S. Dhir, H.-M Park, and S.-Y Lee, "Permutation correction of filter bank ICA using static channel characteristics," Int. Conf. on Neural Information Processing, Nov. 2004.

[6] W. A. Yost, Fundamentals of Hearing – An Introduction, Academic Press, 2000.

[7] H.-M. Park, T. Kim, C.-M. Kim, and S.-Y. Lee, "Blind signal separation and adaptive noise control with filterbank-based independent component analysis," in Int. Congress on Acoustics, vol. I, April 2004, pp. 327-328.

[8] S. Weiβ, On Adaptive Filtering on Oversampled Subbands, Ph.D. thesis, Signal Processing Division, Univ. Strathclyde, Glasgow, May 1998.

[9] K. Torkkola, "Blind separation of convolved sources based on information maximization," in Proc. IEEE Int. Workshop on Neural Networks and Signal Processing, 1996.

[10] J. B. Allen and D. A. Berkley, "Image method for efficiently simulating small-room acoustics," Journal of the Acoustical Society of America, vol. 65, no. 4, pp. 943-950, April 1979.

[11] D. Schobben, K. Torkkola, and P. Smaragdis, "Evaluation of blind signal separation," in Proc. Int. Conf. on ICA and BSS, Jan. 1999, pp. 261-26.

Blind signal separation into groups of dependent signals using joint block diagonalization

Fabian J. Theis
Institute of Biophysics, University of Regensburg
93040 Regensburg, Germany, Email: fabian@theis.name

Abstract—Multidimensional or group independent component analysis describes the task of transforming a multivariate observed sensor signal such that groups of the transformed signal components are mutually independent - however dependencies within the groups are still allowed. This generalization of independent component analysis (ICA) allows for weakening the sometimes too strict assumption of independence in ICA. It has potential applications in various fields such as ECG, fMRI analysis or convolutive ICA. Recently we could calculate the indeterminacies of group ICA, which finally enables us, also theoretically, to apply group ICA to solve blind source separation (BSS) problems. In this paper we introduce and discuss various algorithms for separating signals into groups of dependent signals. The algorithms are based on joint block diagonalization of sets of matrices generated using several signal structures.

I. INTRODUCTION

In this work, we discuss *multidimensional blind source separation (MBSS)* i.e. the recovery of underlying sources s from an observed mixture x. As usual, s has to fulfill additional properties such as independence or diagonality of the autocovariances (if s possesses time structure). However in contrast to ordinary BSS, MBSS is more general as some source signals are allowed to possess common statistics. One possible solution for MBSS is *multidimensional independent component analysis (MICA)* — in section IV we will discuss other such conditions. The idea MICA is that we do not require full independence of the transform $\mathbf{y} := \mathbf{W}\mathbf{x}$ but only mutual independence of certain tuples $y_{i_1}, \ldots, y_{i_2}$. If the size of all tuples is restricted to one, this reduces to ordinary ICA. In general, of course the tuples could have different sizes, but for the sake of simplicity we assume that they all have the same length k.

Multidimensional ICA has first been introduced by Cardoso [1] using geometrical motivations. Hyvärinen and Hoyer then presented a special case of multidimensional ICA which they called independent subspace analysis [2]; there the dependence within a k-tuple is explicitly modelled enabling the authors to propose better algorithms without having to resort to the problematic multidimensional density estimation.

II. JOINT BLOCK DIAGONALIZATION

Joint diagonalization has become an important tool in ICA-based BSS (used for example in JADE [3]) or in BSS relying on second-order time-decorrelation (for example in SOBI [4]). The task of (real) *joint diagonalization* is, given a set of commuting symmetric $n \times n$ matrices $\mathbf{M}_i$, to find an orthogonal matrix $\mathbf{E}$ such that $\mathbf{E}^\top \mathbf{M}_i \mathbf{E}$ is diagonal for all i.

In the following we will use a generalization of this technique as algorithm to solve MBSS problems. Instead of fully diagonalizing $\mathbf{M}_i$ in *joint block diagonalization (JBD)* we want to determine $\mathbf{E}$ such that $\mathbf{E}^\top \mathbf{M}_i \mathbf{E}$ is block-diagonal (after fixing the block-structure).

Introducing some notation, let us define for $r, s = 1, \ldots, n$ the (r, s) *sub-k-matrix of* $\mathbf{W} = (w_{ij})$, denoted by $\mathbf{W}_{rs}^{(k)}$, to be the $k \times k$ submatrix of $\mathbf{W}$ ending at position (rk, sk). Denote $\mathrm{Gl}(n)$ the group of invertible $n \times n$ matrices. A matrix $\mathbf{W} \in \mathrm{Gl}(nk)$ is said to be a *k-scaling matrix* if $\mathbf{W}_{rs}^{(k)} = 0$ for $r \neq s$, and $\mathbf{W}$ is called a *k-permutation matrix* if for each $r = 1, \ldots, n$ there exists precisely one s such that $\mathbf{W}_{rs}^{(k)}$ equals the $k \times k$ unit matrix.

Hence, fixing the block-size to k, JBD tries to find $\mathbf{E}$ such that $\mathbf{E}^\top \mathbf{M}_i \mathbf{E}$ is a k-scaling matrix. In practice due to estimation errors, such $\mathbf{E}$ will not exist, so we speak of approximate JBD and imply minimizing some error-measure on non-block-diagonality.

Various algorithms to actually perform JBD have been proposed, see [5] and references therein. In the following we will simply perform joint diagonalization (using for example the Jacobi-like algorithm from [6]) and then permute the columns of $\mathbf{E}$ to achieve block-diagonality — in experiments this turns out to be an efficient solution to JBD [5].

III. MULTIDIMENSIONAL ICA (MICA)

Let $k, n \in \mathbb{N}$. We call an nk-dimensional random vector $\mathbf{y}$ k-*independent* if the k-dimensional random vectors $(y_1, \ldots, y_k)^\top, \ldots, (y_{nk-k+1}, \ldots, y_{nk})^\top$ are mutually independent. A matrix $\mathbf{W} \in \mathrm{Gl}(nk)$ is called a k-*multidimensional ICA* of an nk-dimensional random vector $\mathbf{x}$ if $\mathbf{W}\mathbf{x}$ is k-independent. If $k = 1$, this is the same as ordinary ICA.

Using MICA we want to solve the (noiseless) linear MBSS problem $\mathbf{x} = \mathbf{As}$, where the nk-dimensional random vector $\mathbf{x}$ is given, and $\mathbf{A} \in \mathrm{Gl}(nk)$ and $\mathbf{s}$ are unknown. In the case of MICA $\mathbf{s}$ is assumed to be k-independent.

A. Indeterminacies

Obvious indeterminacies are, similar to ordinary ICA, invertible transforms in $\mathrm{Gl}(k)$ in each tuple as well as the fact that the order of the independent k-tuples is not fixed. Indeed, if $\mathbf{A}$ is MBSS solution, then so is $\mathbf{ALP}$ with a k-scaling matrix $\mathbf{L}$ and a k-permutation $\mathbf{P}$, because independence is invariant under these transformations. In [7] we show that these are the only indeterminacies, given some additional weak restrictions to the model, namely that $\mathbf{A}$ has to be k-admissible and that $\mathbf{s}$ is not allowed to contain a Gaussian k-component.

As usual by preprocessing of the observations $\mathbf{x}$ by whitening we may also assume that $\mathrm{Cov}(\mathbf{x}) = \mathbf{I}$. Then $\mathbf{I} = \mathrm{Cov}(\mathbf{x}) = \mathbf{A}\,\mathrm{Cov}(\mathbf{s})\mathbf{A}^\top = \mathbf{A}\mathbf{A}^\top$ so $\mathbf{A}$ is orthogonal.

B. MICA using Hessian diagonalization (MHICA)

We assume that $\mathbf{s}$ admits a $\mathcal{C}^2$–density $p_\mathbf{s}$. Using orthogonality of $\mathbf{A}$ we get $p_\mathbf{s}(\mathbf{s}_0) = p_\mathbf{x}(\mathbf{A}\mathbf{s}_0)$ for $\mathbf{s}_0 \in \mathbb{R}^{nk}$. Let $\mathbf{H}_f(\mathbf{x}_0)$ denote the Hessian of f evaluated at $\mathbf{x}_0$. It transforms like a 2-tensor so locally at $\mathbf{s}_0$ with $p_\mathbf{s}(\mathbf{s}_0) > 0$ we get

$$\mathbf{H}_{\ln p_\mathbf{s}}(\mathbf{s}_0) = \mathbf{H}_{\ln p_\mathbf{x} \circ \mathbf{A}}(\mathbf{s}_0) = \mathbf{A}\mathbf{H}_{\ln p_\mathbf{x}}(\mathbf{A}\mathbf{s}_0)\mathbf{A}^\top \quad (1)$$

The key idea now lies in the fact that $\mathbf{s}$ is assumed to be k-independent, so $p_\mathbf{s}$ factorizes into n groups depending only on k separate variables each. So $\ln p_\mathbf{s}$ is a sum of functions depending on k separate variables hence $\mathbf{H}_{\ln p_\mathbf{s}}(\mathbf{s}_0)$ is block-diagonal i.e. a k-scaling.

The algorithm, *multidimensional Hessian ICA (MHICA)*, now simply uses the block-diagonality structure from equation 1 and performs JBD of estimates of a set of Hessians $\mathbf{H}_{\ln p_\mathbf{s}}(\mathbf{s}_i)$ evaluated at different points $\mathbf{s}_i \in \mathbb{R}^{nk}$. Given slight restrictions on the eigenvalues, the resulting block diagonalizer then equals $\mathbf{A}^\top$ except for k-scaling and permutation. The Hessians are estimated using kernel-density approximation with a sufficiently smooth kernel, but other methods such as approximation using finite differences are possible, too. Density approximation is problematic, but in this setting due to the fact that we can use many Hessians we only need rough estimates. For more details on the kernel approximation we refer to the one-dimensional Hessian ICA algorithm from [8].

MHICA generalizes one-dimensional ideas proposed in [8], [9]. More generally, we could have also used characteristic functions instead of densities, which leads to a related algorithm, see [10] for the single-dimensional ICA case.

IV. MULTIDIMENSIONAL TIME DECORRELATION

Instead of assuming k-independence of the sources in the MBSS problem, in this section we assume that $\mathbf{s}$ is a multivariate centered discrete WSS random process such that its *symmetrized autocovariances*

$$\bar{\mathbf{R}}_\mathbf{s}(\tau) := \frac{1}{2}\left(\mathbf{E}\left(\mathbf{s}(t+\tau)\mathbf{s}(t)^\top\right) + \mathbf{E}\left(\mathbf{s}(t)\mathbf{s}(t+\tau)^\top\right)\right) \quad (2)$$

are k-scalings for all τ. This models the fact that the sources are supposed to be block-decorrelated in the time-domain for all time-shifts τ.

A. Indeterminacies

Again $\mathbf{A}$ can only be found up to k-scaling and k-permutation because condition 2 is invariant under this transformation. One sufficient condition for identifiability is to have pairwise different eigenvalues of at least one $\mathbf{R}_\mathbf{s}(\tau)$, however generalizations are possible, see [4] for the case $k=1$. Using whitening, we can again assume orthogonal $\mathbf{A}$.

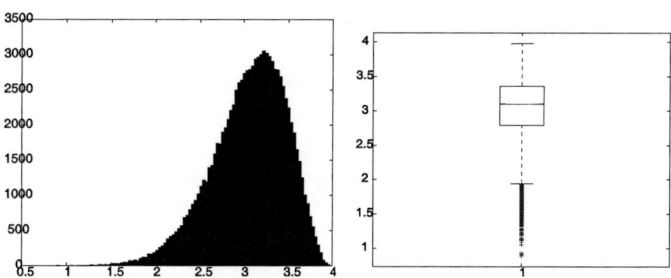

Fig. 1. Histogram and box plot of the multidimensional performance index $E^{(k)}(\mathbf{C})$ evaluated for $k=2$ and $n=2$. The statistics were calculated over 10^5 independent experiments using 4×4 matrices $\mathbf{C}$ with coefficients uniformly drawn out of $[-1, 1]$.

B. Multidimensional SOBI (MSOBI)

The idea of what we call *multidimensional second-order blind identification (MSOBI)* is now a direct extension of the usual SOBI algorithm [4]. Symmetrized autocovariances of $\mathbf{x}$ can easily be estimated from the data, and they transform as follows: $\bar{\mathbf{R}}_\mathbf{s}(\tau) = \mathbf{A}^\top \bar{\mathbf{R}}_\mathbf{x}(\tau)\mathbf{A}$. But $\bar{\mathbf{R}}_\mathbf{s}(\tau)$ is a k-scaling by assumption, so JBD of a set of such symmetrized autocovariance matrices yields $\mathbf{A}$ as diagonalizer (except for k-scaling and permutation).

Other researchers have worked on this problem in the setting of convolutive BSS — due to lack of space we want to refer to [11] and references therein.

V. EXPERIMENTAL RESULTS

In this section we demonstrate the validity of the proposed algorithms by applying them to both toy and real world data.

A. Multidimensional Amari-index

In order to analyze algorithm performance, we consider the index $E^{(k)}(\mathbf{C})$ defined for fixed n,k and $\mathbf{C} \in \mathrm{Gl}(nk)$ as

$$\begin{aligned}E^{(k)}(\mathbf{C}) &= \sum_{r=1}^{n}\left(\sum_{s=1}^{n}\frac{\|\mathbf{C}_{rs}^{(k)}\|}{\max_i \|\mathbf{C}_{ri}^{(k)}\|} - 1\right) \\ &+ \sum_{s=1}^{n}\left(\sum_{r=1}^{n}\frac{\|\mathbf{C}_{rs}^{(k)}\|}{\max_i \|\mathbf{C}_{is}^{(k)}\|} - 1\right).\end{aligned}$$

Here $\|.\|$ can be any matrix norm — we choose the operator norm $\|\mathbf{A}\| := \max_{|\mathbf{x}|=1}|\mathbf{A}\mathbf{x}|$. This *multidimensional performance index* of an $nk \times nk$-matrix $\mathbf{C}$ generalizes the one-dimensional performance index introduced by Amari et al. [12] to block-diagonal matrices. It measures how much $\mathbf{C}$ differs from a permutation and scaling matrix in the sense of k-blocks, so it can be used to analyze algorithm performance:

Lemma 5.1: Let $\mathbf{C} \in \mathrm{Gl}(nk)$. $E^{(k)}(\mathbf{C}) = 0$ if and only if $\mathbf{C}$ is the product of a k-scaling and a k-permutation matrix.

Corollary 5.2: Consider the MBSS problem $\mathbf{x} = \mathbf{A}\mathbf{s}$ from section III respectively IV. An estimate $\hat{\mathbf{A}}$ of the mixing matrix solves the MBSS problem if and only if $E^{(k)}(\hat{\mathbf{A}}^{-1}\mathbf{A}) = 0$.

In order to be able to determine the scale of this index, figure 1 gives statistics of $E^{(k)}$ over randomly chosen matrices in the case $k = n = 2$. The mean is 3.05 and the median 3.10.

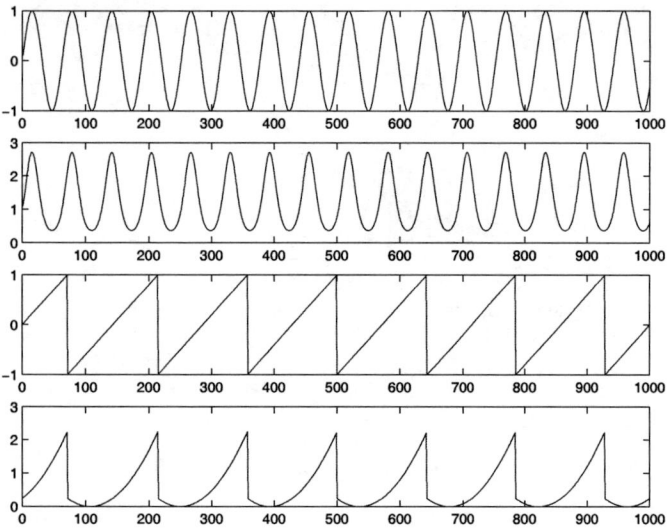

Fig. 2. Simulation, 4-dimensional 2-independent sources. Clearly the first and the second respectively the third and the fourth signal are dependent.

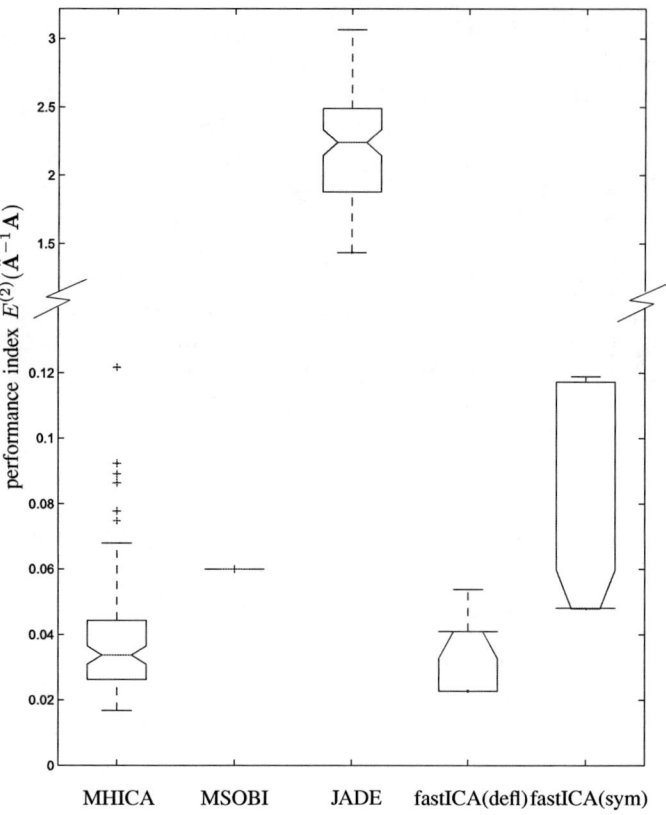

Fig. 3. Simulation, algorithm results. This notched boxed plot displays the performance index $E^{(2)}$ of the mixing-separating matrix $\hat{\mathbf{A}}^{-1}\mathbf{A}$ of each algorithm, sampled over 100 Monte-Carlo runs. The middle line of each column gives the mean, the boxes the 25th and 75th percentile. The deflationary fastICA algorithm only converged in 12% of all runs, the symmetric-approach based fastICA in 89% of all cases; the statistics are only given over successful runs. All other algorithms converged in all runs.

B. Simulations

We will discuss algorithm performance when applied to a 4-dimensional 2-independent toy signal. In order to see the performance of both MSOBI and MHICA we generate 2-independent sources with non-trivial autocorrelations. For this we use two independent generating signals, a sinusoid and a sawtooth given by

$$\mathbf{z}(t) := (\sin(0.1\,t), 2\lfloor 0.007\,t + 0.5 \rfloor - 1)^\top$$

for discrete time steps $t = 1, 2, \ldots, 1000$. We thus generated sources

$$\mathbf{s}(t) := (z_1(t), \exp(z_1(t)), z_2(t), (z_2(t) + 0.5)^2)^\top,$$

which are plotted in figure 2. Their covariance is

$$\mathrm{Cov}(\mathbf{s}) = \begin{pmatrix} 0.50 & 0.57 & 0.01 & 0.01 \\ 0.57 & 0.68 & 0.01 & 0.01 \\ 0.01 & 0.01 & 0.33 & 0.33 \\ 0.01 & 0.01 & 0.33 & 0.42 \end{pmatrix}$$

so indeed $\mathbf{s}$ is not fully independent.

$\mathbf{s}$ is mixed using a 4×4 matrix $\mathbf{A}$ with entries uniformly drawn out of $[-1, 1]$, and comparisons are made over 100 Monte-Carlo runs. We compare the two algorithms MSOBI (with 10 autocorrelation matrices) and MHICA (using 50 Hessians) with the ICA algorithms JADE and fastICA, where in the latter both the deflation and the symmetric approach was used. For each run we calculate the performance index $E^{(2)}(\hat{\mathbf{A}}^{-1}\mathbf{A})$ of the product of the mixing and the estimated separating matrix. Since the one-dimensional ICA algorithms are unable to use the group structure, for these we take the minimum of the index calculated over all row permutations of $\hat{\mathbf{A}}^{-1}\mathbf{A}$.

Figure 3 displays the result of the comparison. Clearly MHICA and MSOBI perform very well on this data, and MSOBI furthermore gives very robust estimates with the same error and negligibly small variance. JADE cannot separate the data at all — it performs not much better than random choice of matrix, see figure 1; this is due to the fact that the cumulants of k-independent sources are not block-diagonal. FastICA only converges in 12% (deflation approach) respectively 89% (symmetric approach) of all cases. However, in the cases it converges it gives results comparable with the multidimensional algorithms. Apparently, especially the symmetric method seems to be able to use the weakened statistics to still find directions in the data.

C. Application to ECG data

Finally we illustrate how to apply the proposed algorithms to real-world data set. Following [1], we will show how to separate fetal ECG (FECG) recordings from the mother's ECG (MECG). The data set [13] consists of eight recorded signals with 2500 observations; the sampling frequency is misleadingly specified as 500 Hz (which would mean around 168 mother heartbeats per minute), it should be closer to around 250 Hz. We select the first three sensors cutaneously recorded on the abdomen of the mother. In order to save space and to compare the results with [1] we plot only the first 1000 samples, see figure 4(a).

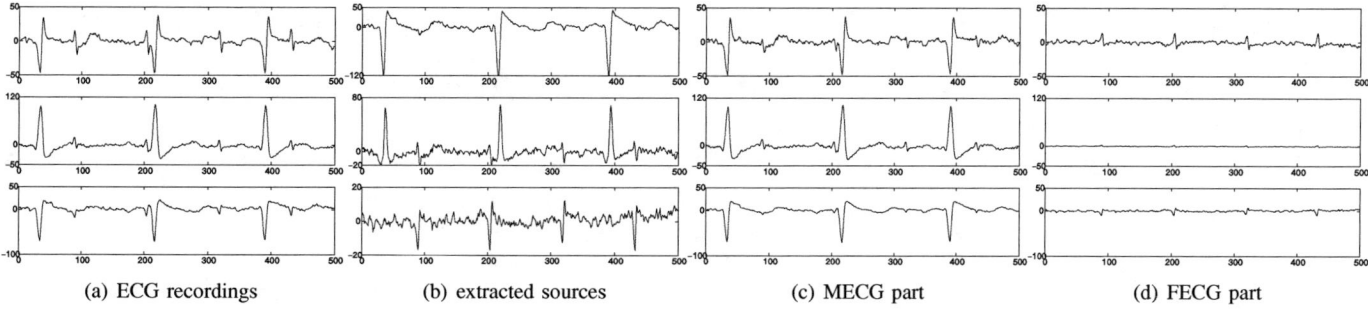

Fig. 4. Fetal ECG example. (a) shows the ECG recordings. The underlying FECG (4 heartbeats) is partially visible in the dominating MECG (3 heartbeats). Figure (b) gives the extracted sources using MHICA with $k=2$ and 500 Hessians. In (c) and (d) the projections of the mother sources (components 1 and 2 in (b)) respectively the fetal sources (component 3) onto the mixture space (a) are plotted.

Our goal is to extract an MECG and an FECG component; however it cannot be expected to find an only one-dimensional MECG due to the fact that projections of a three-dimensional vector (electric) field are measured. Hence modelling the data by a multidimensional BSS problem with $k=2$ (but allowing for an additional one-dimensional component) makes sense. Application of MHICA (with 500 Hessians) and MSOBI (with 50 autocorrelation matrices) extracts a two-dimensional MECG component and a one-dimensional FECG component. After block-permutation we get the following estimated mixing matrices ($\mathbf{A}$ using MHICA and $\mathbf{A}'$ using MSOBI)

$$\mathbf{A} = \begin{pmatrix} 0.37 & 0.42 & -0.81 \\ 0.75 & 0.80 & 0.16 \\ 0.55 & -0.16 & 0.57 \end{pmatrix} \quad \mathbf{A}' = \begin{pmatrix} 0.22 & 0.91 & -0.40 \\ -0.84 & 0.23 & -0.33 \\ 0.50 & 0.34 & 0.85 \end{pmatrix},$$

The thus estimated sources using MHICA are plotted in figure 4(b). In order to compare the two mixing matrices, calculation of

$$\mathbf{A}^{-1}\mathbf{A}' = \begin{pmatrix} 0.85 & 1.02 & 0.64 \\ -0.23 & 1.11 & 0.35 \\ -0.01 & -0.08 & 0.98 \end{pmatrix}.$$

yields a somewhat visible block structure; the performance index is $E^{(2)}(\mathbf{A}^{-1}\mathbf{A}') = 1.12$. The block structure is not very dominant, which indicates that the two models — block independence versus time-block-decorrelation — are not fully equivalent.

A (scaling invariant) decomposition of the observed ECG data can be achieved by composing the extracted sources using only the relevant mixing columns. For example for the MECG part this means applying the projection $\Pi_M := (\mathbf{a}_1, \mathbf{a}_2, 0)\mathbf{A}^{-1}$ to the observations. This yields the projection matrices

$$\Pi_M = \begin{pmatrix} 0.52 & 0.38 & 0.84 \\ -0.10 & 1.08 & 0.17 \\ 0.34 & -0.27 & 0.41 \end{pmatrix} \quad \Pi_F = \begin{pmatrix} 0.48 & -0.38 & -0.84 \\ 0.10 & -0.08 & -0.17 \\ -0.34 & 0.27 & 0.59 \end{pmatrix}$$

onto the mother respectively the fetal ECG using MHICA and

$$\Pi'_M = \begin{pmatrix} 0.78 & 0.21 & 0.45 \\ -0.18 & 1.17 & 0.36 \\ 0.47 & -0.44 & 0.05 \end{pmatrix} \quad \Pi'_F = \begin{pmatrix} 0.22 & -0.21 & -0.45 \\ 0.18 & -0.17 & -0.36 \\ -0.47 & 0.44 & 0.95 \end{pmatrix}.$$

using MSOBI. The results of the first algorithm are plotted in figures 4 (c) and (d). The fetal ECG is most active at sensor 1 (as visual inspection of the observation confirms). When comparing the projection matrices with the results from [1], we get quite high similarity of the ICA-based results, and a modest difference with the projections of the time-based algorithm. Other one-dimensional ICA-based results on this data set are reported for example in [14].

VI. CONCLUSION

We have shown how the idea of joint block diagonalization as extension of joint diagonalization helps us to generalize ICA and time-structure based algorithms such as HICA and SOBI to the multidimensional ICA case. The thus defined algorithms are able to robustly decompose signals into groups of independent signals. In future work, besides more extensive experiments and tests with noise and outliers, we want to extend this result to a version of JADE using moments instead of cumulants, which preserve the block structure.

REFERENCES

[1] J. Cardoso, "Multidimensional independent component analysis," in *Proc. of ICASSP '98*, Seattle, 1998.
[2] A. Hyvärinen and P. Hoyer, "Emergence of phase and shift invariant features by decomposition of natural images into independent feature subspaces," *Neural Computation*, vol. 12, no. 7, pp. 1705–1720, 2000.
[3] J.-F. Cardoso and A. Souloumiac, "Blind beamforming for non gaussian signals," *IEE Proceedings - F*, vol. 140, no. 6, pp. 362–370, 1993.
[4] A. Belouchrani, K. A. Meraim, J.-F. Cardoso, and E. Moulines, "A blind source separation technique based on second order statistics," *IEEE Transactions on Signal Processing*, vol. 45, no. 2, pp. 434–444, 1997.
[5] K. Abed-Meraim and A. Belouchrani, "Algorithms for joint block diagonalization," in *Proc. EUSIPCO 2004*, Vienna, Austria, 2004, pp. 209–212.
[6] J.-F. Cardoso and A. Souloumiac, "Jacobi angles for simultaneous diagonalization," *SIAM J. Mat. Anal. Appl.*, vol. 17, no. 1, pp. 161–164, Jan. 1995.
[7] F. Theis, "Uniqueness of complex and multidimensional independent component analysis," *Signal Processing*, vol. 84, no. 5, pp. 951–956, 2004.
[8] ——, "A new concept for separability problems in blind source separation," *Neural Computation*, vol. 16, pp. 1827–1850, 2004.
[9] J. Lin, "Factorizing multivariate function classes," in *Advances in Neural Information Processing Systems*, vol. 10, 1998, pp. 563–569.
[10] A. Yeredor, "Blind source separation via the second characteristic function," *Signal Processing*, vol. 80, no. 5, pp. 897–902, 2000.
[11] C. Févotte and C. Doncarli, "A unified presentation of blind separation methods for convolutive mixtures using block-diagonalization," in *Proc. ICA 2003*, Nara, Japan, 2003, pp. 349–354.
[12] S. Amari, A. Cichocki, and H. Yang, "A new learning algorithm for blind signal separation," *Advances in Neural Information Processing Systems*, vol. 8, pp. 757–763, 1996.
[13] B. D. M. (ed.), "DaISy: database for the identification of systems," *Department of Electrical Engineering, ESAT/SISTA, K.U.Leuven, Belgium*, Oct 2004. [Online]. Available: http://www.esat.kuleuven.ac.be/sista/daisy/
[14] L. D. Lathauwer, B. D. Moor, and J. Vandewalle, "Fetal electrocardiogram extraction by source subspace separation," in *Proc. IEEE SP / ATHOS Workshop on HOS*, Girona, Spain, 1995, pp. 134–138.

Blind extraction of a dominant source from mixtures of many sources using ICA and time-frequency masking

Hiroshi Sawada   Shoko Araki   Ryo Mukai   Shoji Makino

NTT Communication Science Laboratories, NTT Corporation
2-4 Hikaridai, Seika-cho, Soraku-gun, Kyoto 619-0237, Japan
Email: {sawada,shoko,ryo,maki}@cslab.kecl.ntt.co.jp

Abstract— This paper presents a method for enhancing a target source of interest and suppressing other interference sources. The target source is assumed to be close to sensors, to have dominant power at these sensors, and to have non-Gaussianity. The enhancement is performed blindly, i.e. without knowing the total number of sources or information about each source, such as position and active time. We consider a general case where the number of sources is larger than the number of sensors. We employ a two-stage process where independent component analysis (ICA) is first employed in each frequency bin and time-frequency masking is then used to improve the performance further. We propose a new sophisticated method for selecting the target source frequency components, and also a new criterion for specifying time-frequency masks. Experimental results for simulated cocktail party situations in a room (reverberation time was 130 ms) are presented to show the effectiveness and characteristics of the proposed method.

I. INTRODUCTION

The technique for estimating individual source components from their mixtures at sensors is known as blind source separation (BSS) [1], [2]. With some applications such as brain imaging or wireless communications, it makes sense to extract as many source components as possible, because many sources are equally important. However, with audio applications such as speech enhancement, the sources do not necessarily have equal significance. We often want to extract only one source that is close to sensors, has dominant power, and/or has interesting features.

This paper presents a method for extracting a source signal of interest and suppressing other interference sources blindly. Let us formulate the task. Suppose that a target source s_1 and other interference sources $s_2, \ldots, s_N$ are convolutively mixed and observed at M sensors

$$x_j(t) = \sum_{k=1}^{N} \sum_l h_{jk}(l) s_k(t-l), \; j=1,\ldots,M, \quad (1)$$

where $h_{jk}(l)$ represents the impulse response from source k to sensor j. The goal is to have an output signal $y_1(t)$ that is close to the component of s_1 measured at a selected sensor J:

$$x_{J1}(t) = \sum_l h_{J1}(l) s_1(t-l). \quad (2)$$

Note that $x_j(t) = \sum_{k=1}^{N} x_{jk}(t)$. The task should be performed only with the M observed signals. The number of sources N is unknown and may be larger than M.

The first problem is how to extract the target source s_1 blindly. Even if N could be larger than M, independent component analysis (ICA) [2] with an $N = M$ assumption produces M components that maximize an ICA criterion such as non-Gaussianity. We assume that the target source s_1 is non-Gaussian, close to sensors, and dominant in the mixtures. Therefore, we expect that one of the M components corresponds to s_1 whose ICA criterion is high.

We employ ICA in the time-frequency domain. The reason is that it is efficient and also fits time-frequency masking, which is discussed in the next paragraph. An additional operation that should be performed is the selection of the s_1 component in every frequency bin. This is considered to be the permutation problem of frequency-domain BSS. It has been reported that the selection of a component with maximum kurtosis works well when the target is speech and the interferences are babble sources [3]. However, this does not always work well for a case where the interferences are also speech. Thus, we exploit the information of basis vectors (8) produced by ICA. Our previously reported methods estimate the directions [4], [5] and/or the distances [5] of the sources from the basis vectors, and then cluster the estimations to solve the permutation problem. However, the system needs to know the locations of sensors to estimate such geometric information about the sources. In Sec. II-C, we propose a new method for solving the permutation problem. With this approach, we do not need to know the sensor locations, simply the maximum distance between a sensor and any other sensor. This relaxation makes it easy to use a non-uniform arrangement of sensors, and also eliminates the need for sensor calibration.

The next issue is that some interference still remains in the extracted frequency components when $N > M$. Post filtering [3], [6] can be used to reduce such residual interference. However, it needs additional adaptation where the step size should be controlled based on the short-term power of the target. Another approach is time-frequency masking [7]–[11], which is efficient for sources with sparseness in the time-frequency domain, such as speech. The performance of time-frequency masking depends on how well we can specify the time-frequency slots where the target source is active. A simple way to specify such slots is to calculate the phase and/or amplitude difference between the observations of different sensors [7], [8]. Another recently proposed approach involves calculating the power ratio between an input and outputs of a spatial filter (beamformer [9], [11] or ICA [11]). However, such a power-based criterion depends on the scaling ambiguity of ICA or beamformer outputs. We propose a new criterion for specifying masks in Sec. II-D. It is based on the cosine distance between a sample vector and the basis vector corresponding to the target. The closeness is calculated in a spatially whitened space where the target basis vector is expected to be almost orthogonal to those of interferences. Therefore, the new criterion does not suffer from the problem of scaling ambiguity.

The next section describes our proposed method. Section III shows experimental results, and Section IV concludes this paper.

II. THE PROPOSED METHOD

A. Frequency domain operations

Figure 1 shows the flow of the method discussed here. First, time-domain signals $x_j(t)$ sampled at frequency f_s are converted into frequency-domain time-series signals $x_j(f, \tau)$ with an L-point short-time Fourier transform (STFT):

$$x_j(f, \tau) \leftarrow \sum_{r=-L/2}^{L/2-1} x_j(\tau + r) \text{win}(r) e^{-j2\pi f r}, \quad (3)$$

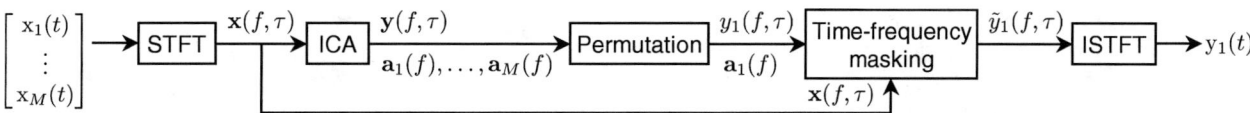

Fig. 1. Flow of proposed method

where $f \in \{0, \frac{1}{L}f_s, \ldots, \frac{L-1}{L}f_s\}$ is a frequency, $\mathrm{win}(r)$ is a window that tapers smoothly to zero at each end, such as a Hanning window $\frac{1}{2}(1 + \cos\frac{2\pi r}{L})$, and τ is a new index representing time.

The remaining operations are performed in the frequency domain. There are two advantages to this. First, the convolutive mixtures (1) can be approximated as instantaneous mixtures at each frequency:

$$x_j(f,\tau) \approx \sum_{k=1}^{N} h_{jk}(f) s_k(f,\tau), \quad (4)$$

where $h_{jk}(f)$ is the frequency response from source k to sensor j, and $s_k(f,\tau)$ is a frequency-domain time-series signal of $s_k(t)$ obtained by the same operation as (3). The frequency-domain counterpart of (2) is

$$x_{J1}(f,\tau) \approx h_{J1}(f) s_1(f,\tau), \quad (5)$$

where J should be the same for all frequency bins f. The second advantage is that the sparseness of a source signal becomes prominent in the time-frequency domain if the source is colored and non-stationary such as speech. The possibility of $s_k(f,\tau)$ being close to zero is much higher than that of $s_k(t)$.

Through several operations, which will be discussed in the following subsections, we have an output $\tilde{y}_1(f,\tau)$, which should be close to (5) in each frequency bin. At the end of the flow, we have an output $y_1(t)$ by an inverse STFT (ISTFT):

$$y_1(\tau + r) \leftarrow \frac{1}{L \cdot \mathrm{win}(r)} \sum_{f \in \{0, \frac{1}{L}f_s, \ldots, \frac{L-1}{L}f_s\}} \tilde{y}_1(f,\tau) e^{j 2\pi f r}.$$

B. Independent component analysis (ICA)

Let us have a vector notation of the mixing model (4):

$$\mathbf{x}(f,\tau) \approx \sum_{k=1}^{N} \mathbf{h}_k(f) s_k(f,\tau), \quad (6)$$

where $\mathbf{x} = [x_1, \ldots, x_M]^T$ is a sample vector and $\mathbf{h}_k = [h_{1k}, \ldots, h_{Mk}]^T$ is the vector of frequency responses from source s_k to all sensors. Independent component analysis (ICA) is used as a first step to identify the vector $\mathbf{h}_1$ of a dominant source s_1.

Even though the number of independent components N may be larger than the number of sensors M, we employ ICA by assuming that N is equal to M:

$$\mathbf{y}(f,\tau) = \mathbf{W}(f)\mathbf{x}(f,\tau), \quad (7)$$

where $\mathbf{y} = [y_1, \ldots, y_M]^T$ is a vector of independent components and $\mathbf{W} = [\mathbf{w}_1, \ldots, \mathbf{w}_M]^H$ is an $M \times M$ separation matrix. In the experiments shown in Sec. III, we calculated $\mathbf{W}$ by using a complex-valued version of FastICA [2], and improved it further by using InfoMax [12] combined with the natural gradient [13] whose nonlinear function is based on the polar coordinate [14].

Then, we calculate the inverse of $\mathbf{W}$ to obtain basis vectors

$$[\mathbf{a}_1, \cdots, \mathbf{a}_M] = \mathbf{W}^{-1}, \quad \mathbf{a}_i = [a_{1i}, \ldots, a_{Mi}]^T. \quad (8)$$

By multiplying both sides of (7) by $\mathbf{W}^{-1}$, the sample vector $\mathbf{x}(f,\tau)$ is represented by a linear combination of basis vectors $\mathbf{a}_1, \ldots, \mathbf{a}_M$:

$$\mathbf{x}(f,\tau) = \sum_{i=1}^{M} \mathbf{a}_i(f) y_i(f,\tau). \quad (9)$$

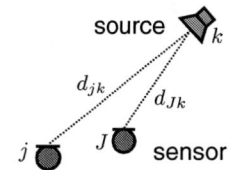

Fig. 2. Direct-path (nearfield) model

Since s_1 is assumed to be a dominant non-Gaussian source, it is strongly expected that one of $y_1, \ldots, y_M$ corresponds to s_1 and thus one of $\mathbf{a}_1, \ldots, \mathbf{a}_M$ corresponds to $\mathbf{h}_1$.

C. Permutation

The next operation is to find i for each frequency f such that $\mathbf{a}_i(f)$ corresponds to $\mathbf{h}_1(f)$. As shown in [4], integrating the basis vector $\mathbf{a}_i(f)$ and signal envelope $|y_i(f,\tau)|$ information solves the permutation problem robustly and precisely, and we also employ this approach here. In the rest of this subsection, we discuss a new method for exploiting the basis vector information.

The new method involves normalizing all basis vectors $\mathbf{a}_i(f)$, $i = 1, \ldots, M$, for all frequency bins $f = 0, \frac{1}{L}f_s, \ldots, \frac{L-1}{L}f_s$ such that they form clusters, each of which corresponds to an individual source. The normalization is performed by selecting a reference sensor J and calculating

$$\bar{a}_{ji}(f) \leftarrow |a_{ji}(f)| \exp\left[j\frac{\arg[a_{ji}(f)/a_{Ji}(f)]}{4 f c^{-1} d_{\max}}\right] \quad (10)$$

where c is the propagation velocity and $d_{\max}$ is the maximum distance between the reference sensor J and a sensor $^\forall j \in \{1, \ldots, M\}$. The rationale of this operation will be explained afterwards. Then, we apply unit-norm normalization

$$\bar{\mathbf{a}}_i(f) \leftarrow \bar{\mathbf{a}}_i(f) / \|\bar{\mathbf{a}}_i(f)\| \quad (11)$$

for $\bar{\mathbf{a}}_i(f) = [\bar{a}_{1i}(f), \ldots, \bar{a}_{Mi}(f)]^T$.

The next step is to find clusters $C_1, \ldots, C_M$ formed by normalized vectors $\bar{\mathbf{a}}_i(f)$. The centroid $\mathbf{c}_k$ of a cluster C_k is calculated by

$$\mathbf{c}_k \leftarrow \sum_{\bar{\mathbf{a}} \in C_k} \bar{\mathbf{a}} / |C_k|, \quad \mathbf{c}_k \leftarrow \mathbf{c}_k / \|\mathbf{c}_k\|,$$

where $|C_k|$ is the number of vectors in C_k. The clustering criterion is to minimize the total sum $\mathcal{J}$ of the squared distances between cluster members and their centroid

$$\mathcal{J} = \sum_{k=1}^{M} \mathcal{J}_k, \quad \mathcal{J}_k = \sum_{\bar{\mathbf{a}} \in C_k} \|\bar{\mathbf{a}} - \mathbf{c}_k\|^2. \quad (12)$$

This minimization can be performed efficiently with the k-means clustering algorithm [15].

This paragraph explains the reason why normalized basis vectors $\bar{\mathbf{a}}_i(f)$ form a cluster for a source. Let us approximate the multi-path mixing model (1) by using a direct-path (nearfield) model (Fig. 2)

$$h_{jk}(f) \approx \frac{q(f)}{d_{jk}} \exp\left[j 2\pi f c^{-1}(d_{jk} - d_{Jk})\right], \quad (13)$$

where $d_{jk} > 0$ is the distance between source k and sensor j. We assume that the phase $2\pi f c^{-1}(d_{jk} - d_{Jk})$ depends on the distance normalized with the distance to the reference sensor J. This assumption makes the phase zero at the reference sensor J. We also assume that the attenuation $q(f)/d_{jk}$ depends on both the distance and a frequency-dependent constant $q(f) > 0$. By considering the permutation and scaling ambiguity of ICA, a basis vector and its elements are represented as

$$\mathbf{a}_i \approx \alpha_i \mathbf{h}_k, \quad a_{ji} \approx \alpha_i h_{jk}, \qquad (14)$$

where α_i represents the scaling ambiguity, and index k, which may be different from index i, represents the permutation ambiguity. Substituting (13) and (14) into (10) and (11) yields

$$\bar{a}_{ji}(f) \approx \frac{1}{d_{jk}D} \exp\left[\jmath \frac{\pi}{2} \frac{(d_{jk}-d_{Jk})}{d_{\max}}\right], \quad D = \sqrt{\sum_{i=1}^M \frac{1}{d_{ik}^2}},$$

which is independent of frequency, and dependent only on the positions of the sources and sensors. From the fact that $\max_{j,k} |d_{jk} - d_{Jk}| \leq d_{\max}$, an inequality

$$-\pi/2 \leq \arg[\bar{a}_{ji}(f)] \leq \pi/2$$

holds. This property is important for the distance measure (12), since $|\bar{a} - \bar{a}'|$ increases monotonically as $|\arg(\bar{a}) - \arg(\bar{a}')|$ increases.

Once we have found M clusters $C_1, \ldots, C_M$, we need to identify a cluster that corresponds to the target source s_1. We decide that a cluster C_K with the minimum variance $K = \arg\min_k \mathcal{J}_k/|C_k|$ corresponds to s_1. The rationale behind this is that the mixing model (13) is more valid for s_1 than for the other sources. The direct-path components of impulse responses h_{j1} are distinct since s_1 is assumed to be close to the sensors. Finally, the output index i for each frequency f is selected by

$$I(f) = \arg\min_i ||\bar{\mathbf{a}}_i(f) - \mathbf{c}_K||^2.$$

This means that basis vector $\mathbf{a}_{I(f)}(f)$ corresponds to $\mathbf{h}_1(f)$.

After we align the index as $\mathbf{a}_1(f) \leftarrow \mathbf{a}_{I(f)}(f)$ and $y_1(f,\tau) \leftarrow y_{I(f)}(f,\tau)$, we solve the scaling ambiguity in (9):

$$\mathbf{a}_1(f) y_1(f,\tau) = (\alpha_1 \mathbf{a}_1(f))(y_1(f,\tau)/\alpha_1),$$

for any non-zero complex scalar α_1. This is easily solved by

$$y_1(f,\tau) \leftarrow a_{J1}(f) y_1(f,\tau),$$

where J is the index of the sensor specified in (5). The reason is as follows. The goal in each frequency bin is to make $y_1(f,\tau)$ as close to $x_{J1}(f,\tau)$ defined in (5) as possible. And we can derive relations

$$x_{J1}(f,\tau) \approx h_{J1}(f) s_1(f,\tau) \approx a_{J1}(f) y_1(f,\tau).$$

from (5), the $\mathbf{h}_1$ term in (6) and the $\mathbf{a}_1$ term in (9).

D. Time-frequency masking

Suppose that the permutation ambiguity of ICA is solved at this stage. The extraction of s_1 by ICA (7) is represented by

$$\begin{aligned} y_1(\tau) &= \mathbf{w}_1^H \mathbf{x}(\tau) \\ &= \mathbf{w}_1^H \mathbf{h}_1 s_1(\tau) + \sum_{k=2}^N \mathbf{w}_1^H \mathbf{h}_k s_k(\tau). \end{aligned}$$

If $N \leq M$, $\mathbf{w}_1$ satisfies $\mathbf{w}_1^H \mathbf{h}_k = 0, \forall k \in \{2,\ldots,N\}$ and makes the second term zero. However, we assume that the number of sources N is generally larger than M. In this case, there exists a set $\mathcal{K} \subseteq \{2,\ldots,N\}$ such that $\mathbf{w}_1^H \mathbf{h}_k \neq 0, \forall k \in \mathcal{K}$. Thus, $y_1(\tau)$ contains an unwanted residual $\sum_{k\in\mathcal{K}} \mathbf{w}_1^H \mathbf{h}_k s_k(\tau)$. The purpose of time-frequency masking is to obtain another output $\tilde{y}_1(\tau)$ that contains less power of the residual $\sum_{k\in\mathcal{K}} \mathbf{w}_1^H \mathbf{h}_k s_k(\tau)$ than $y_1(\tau)$.

Time-frequency masking is performed by

$$\tilde{y}_1(f,\tau) = \mathcal{M}(f,\tau) \cdot y_1(f,\tau),$$

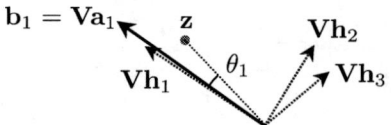

Fig. 3. Angle θ_1 calculated in whitened space

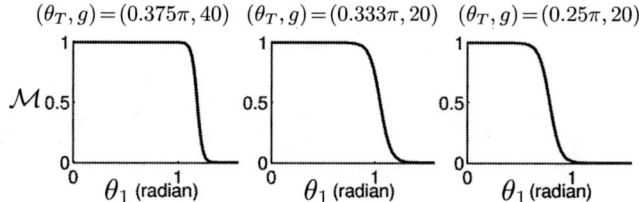

Fig. 4. Masking functions with three sets of parameters (θ_T, g)

where $0 \leq \mathcal{M}(f,\tau) \leq 1$ is a mask specified for each time-frequency slot (f,τ). We specify masks based on the angle $\theta_1(f,\tau)$ between $\mathbf{a}_1(f)$ and $\mathbf{x}(f,\tau)$ calculated in the space transformed by a whitening matrix $\mathbf{V}(f) = \mathbf{R}^{-1/2}$, $\mathbf{R} = \langle \mathbf{x}(\tau)\mathbf{x}(\tau)^H \rangle_\tau$. Let $\mathbf{z}(f,\tau) = \mathbf{V}(f)\mathbf{x}(f,\tau)$ be whitened samples and $\mathbf{b}_1(f) = \mathbf{V}(f)\mathbf{a}_1(f)$ be the basis vector in the whitened space. The angle is calculated by

$$\theta_1(f,\tau) = \arccos \frac{|\mathbf{b}_1^H(f) \cdot \mathbf{z}(f,\tau)|}{||\mathbf{b}_1(f)|| \cdot ||\mathbf{z}(f,\tau)||} \qquad (15)$$

for each time-frequency slot (Fig. 3). Then, we calculate a mask by using a logistic function (Fig. 4)

$$\mathcal{M}(\theta_1(f,\tau)) = \frac{1}{1 + e^{g(\theta_1 - \theta_T)}}, \qquad (16)$$

where θ_T and g are parameters specifying the transition point and its steepness, respectively. As θ_T becomes smaller, the residual power that appears in $\tilde{y}_1$ decreases but the musical noise in y_1 increases.

The effectiveness of the above operation depends on the sparseness of sources. If we assume that the possibility of $s_k(f,\tau)$ being close to zero is very high, (6) can be approximated as

$$\mathbf{x}(f,\tau) \approx \mathbf{h}_k(f) s_k(f,\tau), \quad k \in \{1,\ldots,N\}, \qquad (17)$$

where k depends on each time-frequency slot (f,τ). Let us consider the whitened-space counterpart of (17), while distinguishing between cases where s_1 is the only active source (18) and other cases (19):

$$\mathbf{z}(f,\tau) \approx \mathbf{V}(f)\mathbf{h}_1(f)s_1(f,\tau) \approx \mathbf{V}(f)\mathbf{a}_1(f)y_1(f,\tau) \quad (18)$$

$$\mathbf{z}(f,\tau) \approx \sum_{k=2}^N \mathbf{V}(f)\mathbf{h}_k(f)s_k(f,\tau). \qquad (19)$$

If the number of sources N is equal to or less than the number of sensors M, vectors $\mathbf{V}\mathbf{h}_1, \ldots, \mathbf{V}\mathbf{h}_N$ in the whitened space are orthogonal to each other. Even if $N > M$, the vector $\mathbf{b}_1 = \mathbf{V}\mathbf{a}_1$ of a dominant source s_1, which points in almost the same direction as $\mathbf{V}\mathbf{h}_1$, tends to have large angles with the other vectors $\mathbf{V}\mathbf{h}_2, \ldots, \mathbf{V}\mathbf{h}_N$. Figure 3 shows such a case. Therefore, calculating the angle (15) provides information about whether or not s_1 is the only active source at a time-frequency slot (f,τ).

III. EXPERIMENTS

We performed experiments to enhance a dominant speech that was close to microphones. We measured impulse responses $h_{jk}(l)$ under the conditions shown in Fig. 5. The speaker positions simulated a cocktail party situation. Mixtures at the microphones were made by convolving the impulse responses and 6-second English speeches sampled at 8 kHz. Microphone arrangement was 3-dimensional

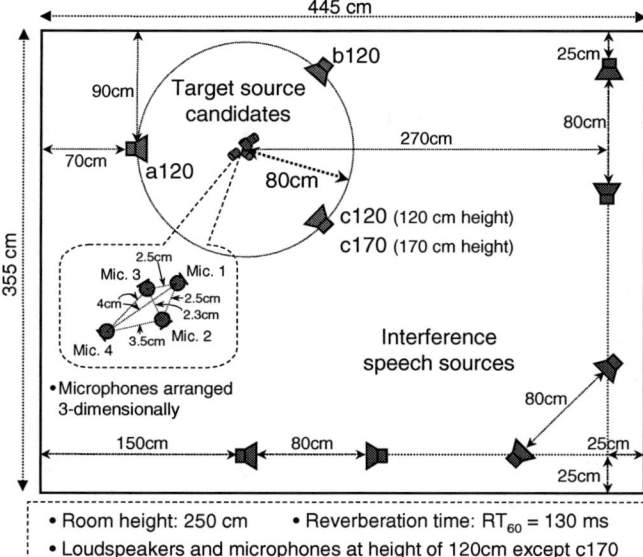

Fig. 5. Experimental conditions

TABLE I
AVERAGE SIR IMPROVEMENT FOR EACH POSITION (dB)

Target position	a120	b120	c120	c170
InputSIR	1.3	1.5	1.9	−0.0
Only ICA	11.7	11.8	9.0	13.0
ICA and T-F masking $(0.375\pi, 40)$	15.4	14.6	12.5	16.9
ICA and T-F masking $(0.333\pi, 20)$	16.8	15.8	14.1	18.3
ICA and T-F masking $(0.25\pi, 20)$	19.5	18.2	16.9	21.0

IV. CONCLUSION

We have presented a method for extracting a dominant target source and suppressing other interferences. The process of ICA and following permutation alignment extracts the target source, and estimates the corresponding basis vector. The new method for permutation alignment makes it easy to use a 3-dimensional non-uniform arrangement of sensors without exact measurement or calibration. Time-frequency masking in the second stage reduces the power of the residuals caused by ICA when $N > M$. It exploits the sparseness of sources. We have proposed a new criterion for specifying masks. It is based on the angle between the target basis vector and a sample vector, and gives information about whether or not the target source is active. The experiments showed good results for extracting a dominant source out from six interferences mixed in a real room.

and non-uniform. The system knew only the maximum distance (4 cm) between the reference microphone (Mic. 1) and others. For each setup, we selected one of the four speakers (a120, b120, c120, c170) as a dominant target source, and the others were kept silent. The six speakers away from the microphones were used as interferences for every setup. The frame size L of STFT (3) was 1024 (128 ms). The computational time was around 12 seconds for 6-second speech mixtures. The program was coded in Matlab and run on Athlon 64 FX-53. The performance was evaluated in terms of the signal-to-interference ratio (SIR) improvement, which is OutputSIR − InputSIR. These two types of SIRs are defined by

$$\text{InputSIR} = 10\log_{10}\frac{\langle |\mathbf{x}_{J1}(t)|^2\rangle_t}{\langle |\sum_{k\neq 1}\mathbf{x}_{Jk}(t)|^2\rangle_t} \quad (\text{dB}),$$

$$\text{OutputSIR} = 10\log_{10}\frac{\langle |\mathbf{y}_{11}(t)|^2\rangle_t}{\langle |\sum_{k\neq 1}\mathbf{y}_{1k}(t)|^2\rangle_t} \quad (\text{dB}),$$

where $\mathbf{x}_{Jk}(t)$ is defined in (2), and $\mathbf{y}_{1k}(t)$ is the component of $\mathbf{s}_k$ that appears at output $\mathbf{y}_1(t)$, i.e. $\mathbf{y}_1(t) = \sum_{k=1}^{N}\mathbf{y}_{1k}(t)$.

Experiments were conducted with 16 combinations of 7 speeches for each target position. Table I shows the average SIR improvements obtained only with ICA, and by the combination of ICA and time-frequency (T-F) masking. The SIR improvements depend on the target position. Positions a120 and b120 were fairly good for enhancement. This is because the interferences came from different directions. If we consider the speaker arrangement 2-dimensionally, positions c120 and c170 seems to be a hard position as many interferences came from similar directions. However, the result for position c170 was very good. This is because the height of c170 was different from those of interferences, and the 3-dimensionally arranged microphones enable the system to exploit the height difference.

We used three sets of parameters for function (16) specifying a mask for each time-frequency slot. The shapes of these functions are shown in Fig. 4. Table I shows that a smaller θ_T resulted in greater SIR improvements by T-F masking. However, some sounds with a small θ_T were unnatural. We observed that in many cases parameter $(\theta_T, g) = (0.333\pi, 20)$ produced natural sounds with sufficient interference suppression. Some sound examples can be found on our web site [16].

REFERENCES

[1] S. Haykin, Ed., *Unsupervised Adaptive Filtering (Volume I: Blind Source Separation)*. John Wiley & Sons, 2000.
[2] A. Hyvärinen, J. Karhunen, and E. Oja, *Independent Component Analysis*. John Wiley & Sons, 2001.
[3] S. Y. Low, R. Togneri, and S. Nordholm, "Spatio-temporal processing for distant speech recognition," in *Proc. ICASSP 2004*, vol. I, May 2004, pp. 1001–1004.
[4] H. Sawada, R. Mukai, S. Araki, and S. Makino, "A robust and precise method for solving the permutation problem of frequency-domain blind source separation," *IEEE Trans. Speech Audio Processing*, vol. 12, no. 5, pp. 530–538, Sept. 2004.
[5] R. Mukai, H. Sawada, S. Araki, and S. Makino, "Frequency domain blind source separation using small and large spacing sensor pairs," in *Proc. ISCAS 2004*, vol. V, May 2004, pp. 1–4.
[6] R. Mukai, S. Araki, H. Sawada, and S. Makino, "Removal of residual crosstalk components in blind source separation using LMS filters," in *Proc. NNSP 2002*, Sept. 2002, pp. 435–444.
[7] M. Aoki, M. Okamoto, S. Aoki, H. Matsui, T. Sakurai, and Y. Kaneda, "Sound source segregation based on estimating incident angle of each frequency component of input signals acquired by multiple microphones," *Acoustical Science and Technology*, vol. 22, no. 2, pp. 149–157, 2001.
[8] S. Rickard, R. Balan, and J. Rosca, "Real-time time-frequency based blind source separation," in *Proc. ICA2001*, Dec. 2001, pp. 651–656.
[9] N. Roman and D. Wang, "Binaural sound segregation for multisource reverberant environments," in *Proc. ICASSP 2004*, vol. II, May 2004, pp. 373–376.
[10] S. Araki, S. Makino, A. Blin, R. Mukai, and H. Sawada, "Underdetermined blind separation for speech in real environments with sparseness and ICA," in *Proc. ICASSP 2004*, vol. III, May 2004, pp. 881–884.
[11] D. Kolossa and R. Orglmeister, "Nonlinear postprocessing for blind speech separation," in *Proc. ICA 2004 (LNCS 3195)*, Sept. 2004, pp. 832–839.
[12] A. Bell and T. Sejnowski, "An information-maximization approach to blind separation and blind deconvolution," *Neural Computation*, vol. 7, no. 6, pp. 1129–1159, 1995.
[13] S. Amari, "Natural gradient works efficiently in learning," *Neural Computation*, vol. 10, no. 2, pp. 251–276, 1998.
[14] H. Sawada, R. Mukai, S. Araki, and S. Makino, "Polar coordinate based nonlinear function for frequency domain blind source separation," *IEICE Trans. Fundamentals*, vol. E86-A, no. 3, pp. 590–596, Mar. 2003.
[15] R. O. Duda, P. E. Hart, and D. G. Stork, *Pattern Classification*, 2nd ed. Wiley Interscience, 2000.
[16] [Online]. http://www.kecl.ntt.co.jp/icl/signal/sawada/demo/dominant/

Independent Arrays or Independent Time Courses for Gene Expression Time Series

Sookjeong Kim, Seungjin Choi
Department of Computer Science
Pohang University of Science and Technology
San 31 Hyoja-dong, Nam-gu
Pohang 790-784, Korea
Email: {koko,seungjin}@postech.ac.kr

Abstract—In this paper we apply three different independent component analysis (ICA) methods, including spatial ICA (sICA), temporal ICA (tICA), and spatiotemporal ICA (stICA), to gene expression time series data and compare their performance in clustering genes and in finding biologically meaningful modes. Only spatial ICA was applied to gene expression data [3], [4]. However, in the case of yeast cell cycle-related gene expression time series data, our comparative study reveals that tICA outperforms sICA and stICA in the task of gene clustering and stICA finds linear modes that best match the cell cycle.

I. INTRODUCTION

Microarray technology allows us to measure expression levels of thousands of genes simultaneously, producing gene expression profiles generated by gene interactions. For example, gene expression data analysis is useful in discriminating cancer tissues from healthy ones or in revealing biological functions of certain genes. Successive microarray experiments over time, produces gene expression time series data. Main issues in these experiments (over time), are to detect cellular processes underlying regulatory effects, to infer regulatory networks, and ultimately to match genes with associated biological functions.

Linear model-based methods explicitly describe expression levels of genes as linear functions of common hidden variables which are expected to be related to distinct biological causes of variations such as regulators of gene expression, cellular functions, or responses to experimental treatments. Such linear model-based methods include singular value decomposition (SVD), principal component analysis (PCA), independent component analysis (ICA), Bayes decomposition, and the plaid model. Standard clustering methods (such as k-means and hierarchical clustering) assign a gene (involving various biological functions) to one of clusters, however linear model-based methods allow the assignment of such a gene to null, single, or multiple clusters.

ICA is an exemplary linear model-based method that has been widely used in a variety of applications. Given a set of multivariate data, ICA aims at finding a linear decomposition where statistical independence is maximized over space (sICA) or over time (tICA). On one hand, tICA has been widely used in the context of blind source separation (for example, acoustic source separation, co-channel signal separation in digital communications, brain wave separation in EEG, and so on), since a set of temporally independent time courses is sought for in such applications. On the other hand, sICA was successfully applied to the field of medical image analysis (for example, fMRI and PET) where mutually independent source images and a corresponding dual set of unconstrained time courses, are of interest [5]. Spatiotemporal ICA (stICA) is a method which permits a trade-off between the mutual independence of spatial underlying variables (for example, images in fMRI) and the mutual independence of their corresponding time courses [8].

In the context of bioinformatics, Liebermeister [4] showed that expression modes and their influences, extracted by sICA, could be used to visualize the samples and genes in lower-dimensional space and a projection to expression modes could highlight particular biological functions. In addition, sICA was also used in gene clustering [3]. So far, only sICA has been considered as a tool for gene expression data analysis, because it seems to better fit in such a task. However, regarding gene expression time series data, tICA might be more suitable for gene clustering and temporal mode analysis, because it tries to maximize mutual independence over time. Numerical experimental study with several sets of yeast cell cycle-related gene expression time series data, shows that tICA outperforms sICA and stICA, which is an interesting result. Although sICA, tICA, and stICA are known methods, a main contribution of this paper, is to compare these three methods in the context of gene expression time series data analysis, showing that tICA is more suitable for gene clustering and stICA finds linear modes that best match the cell cycle.

II. LINEAR MODELS

Linear models assume that the data matrix $X = [X_{ij}] \in \mathbb{R}^{m \times N}$ (where the (i,j)-element, X_{ij} represents the expression level of the ith gene associated with the jth sample (time point), $i = 1, \ldots, m$, $j = 1, \ldots, N$.) is modelled as

$$X = SA, \qquad (1)$$

where $S \in \mathbb{R}^{m \times n}$ and $A \in \mathbb{R}^{n \times N}$ are the encoding variable and linear mode matrix, or vice versa, depending on constraints over time or over space.

We briefly overview of linear model-based methods including PCA, sICA, tICA, and stICA. We follow some notations used in [8].

A. PCA

PCA is a widely-used linear dimensionality reduction technique which decomposes high-dimensional data into low-dimensional subspace components. PCA is illustrated as a linear orthogonal transformation which captures maximal variations in data.

Suppose that the singular value decomposition (SVD) of X is given by

$$X \approx UDV^\top, \quad (2)$$

where $U \in \mathbb{R}^{m \times n}$ corresponds to eigenarrays, $V \in \mathbb{R}^{n \times N}$ is associated with eigengenes, and D is a diagonal matrix containing singular values. In order to choose an appropriate value of n, we use the method, *PCA-L* which is based on the Laplace approximation [6].

In this paper, we use PCA for two reasons: (1) in order to provide a comparison with ICA methods; (2) to provide a reduced rank data set as input to ICA. Following notations in [8], we define $\widetilde{X} \approx X$ as

$$X \approx \widetilde{X} = UDV^\top = \left(UD^{1/2}\right)\left(VD^{1/2}\right)^\top = \widetilde{U}\widetilde{V}^\top. \quad (3)$$

B. Spatial ICA

Spatial ICA seeks a set of independent arrays S_S and a corresponding set of dual unconstrained time courses A_S. It embodies the assumption that each eigenarray in $\widetilde{U}$ is composed of a linear combination of n independent arrays (associated with independent component patterns), i.e., $\widetilde{U} = S_S \widetilde{A}_S$, where $S_S \in \mathbb{R}^{m \times n}$ contains a set of n independent m-dimensional arrays and $\widetilde{A}_S \in \mathbb{R}^{n \times n}$ is an encoding variable matrix (mixing matrix).

Define $Y_S = \widetilde{U} W_S$ where W_S is a permuted version of $\widetilde{A}_S^{-1}$. That is $Y_S = S_S P$ where P is a generalized permutation matrix. With this definition, the n dual time courses $A_S \in \mathbb{R}^{n \times N}$ associated with the n independent arrays, is computed by $A_S = W_S^{-1} \widetilde{V}^\top$, since $\widetilde{X} = Y_S A_S = \widetilde{U}\widetilde{V}^\top = Y_S W_S^{-1} \widetilde{V}^\top$. Each row vector of A_S corresponds to a temporal mode.

C. Temporal ICA

Temporal ICA finds a set of independent time courses and a corresponding set of dual unconstrained arrays (spatial patterns). It embodies the assumption that each eigengene in $\widetilde{V}$ consists of a linear combination of n independent sequences, i.e., $\widetilde{V} = S_T \widetilde{A}_T$, where $S_T \in \mathbb{R}^{N \times n}$ has a set of n independent temporal sequences of length N and $\widetilde{A}_T \in \mathbb{R}^{n \times n}$ is an associated mixing matrix.

Unmixing by $Y_T = \widetilde{V} W_T$ where $W_T = \widetilde{A}_T^{-1} P$, leads us to recover the n dual arrays A_T associated with the n independent time courses, by calculating $A_T = W_T^{-1} \widetilde{U}^\top$, which is a consequence of $\widetilde{X}^\top = Y_T A_T = \widetilde{V}\widetilde{U}^\top = Y_T W_T^{-1} \widetilde{U}^\top$. Fig. 1 illustrates how a set of dual arrays are calculated when tICA is applied to gene expression time series data.

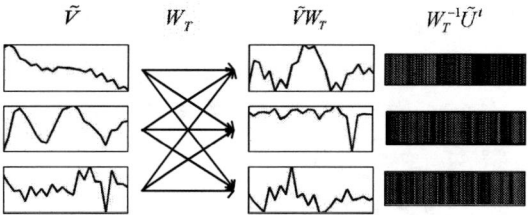

Fig. 1. An illustration of finding a set of dual arrays associated with n independent time courses, when tICA is applied to gene expression time series data. Temporal ICA first learn a mixing matrix A_T from the eigengene matrix V, in order to compute $W_T = A_T^{-1} P$. Eigengenes are linearly transformed by W_T, to produce $Y_T = V W_T$. Then n dual arrays A_T (each row of A_T corresponds to m-dimensional array) are computed by $A_T = W_T^{-1} U^\top$.

D. Spatiotemporal ICA

In linear decomposition, sICA enforces independence constraints over space, to find a set of independent arrays, whereas tICA embodies independence constraints over time, to seek a set of independent time courses. Spatiotemporal ICA finds a linear decomposition, by maximizing the degree of independence over space as well as over time, without necessarily producing independence in either space or time. In fact it allows a trade-off between the independence of arrays and the independence of time courses.

Given $\widetilde{X} = \widetilde{U}\widetilde{V}^\top$, stICA finds the following decomposition:

$$\widetilde{X} = S_S \Lambda S_T^\top, \quad (4)$$

where $S_S \in \mathbb{R}^{m \times n}$ contains a set of n independent m-dimensional arrays, $S_T \in \mathbb{R}^{N \times n}$ has a set of n independent temporal sequences of length N, and Λ is a diagonal scaling matrix. There exist two $n \times n$ mixing matrices, W_S and W_T such that $S_S = \widetilde{U} W_S$ and $S_T = \widetilde{V} W_T$. The following relation

$$\begin{aligned}\widetilde{X} &= S_S \Lambda S_T^\top = \widetilde{U} W_S \Lambda (\widetilde{V} W_T)^\top \\ &= \widetilde{U} W_S \Lambda W_T^\top \widetilde{V}^\top = \widetilde{U}\widetilde{V}^\top,\end{aligned} \quad (5)$$

implies that $W_S \Lambda W_T^\top = I$, which leads to

$$W_T = (W_S^{-1})^\top (\Lambda^{-1})^\top. \quad (6)$$

Linear transforms, W_S and W_T, are found by jointly optimizing objective functions associated with sICA and tICA. That is, the objective function for stICA has the form

$$\mathcal{J}_{stICA} = \alpha \mathcal{J}_{sICA} + (1-\alpha)\mathcal{J}_{tICA}, \quad (7)$$

where $\mathcal{J}_{sICA}$ and $\mathcal{J}_{tICA}$ could be infomax criteria or log-likelihood functions and α defines the relative weighting for spatial independence and temporal independence. More details on stICA can be found in [8].

III. NUMERICAL EXPERIMENTS

We applied sICA, tICA, and stICA to 3 sets of yeast cell cycle-related data [1], [7] (see Table I). Procedures that we took from preprocessing till statistical significance test, are summarized below.

1) Preprocessing: The gene expression data matrix X was preprocessed such that each element is associated with $X_{ij} =$

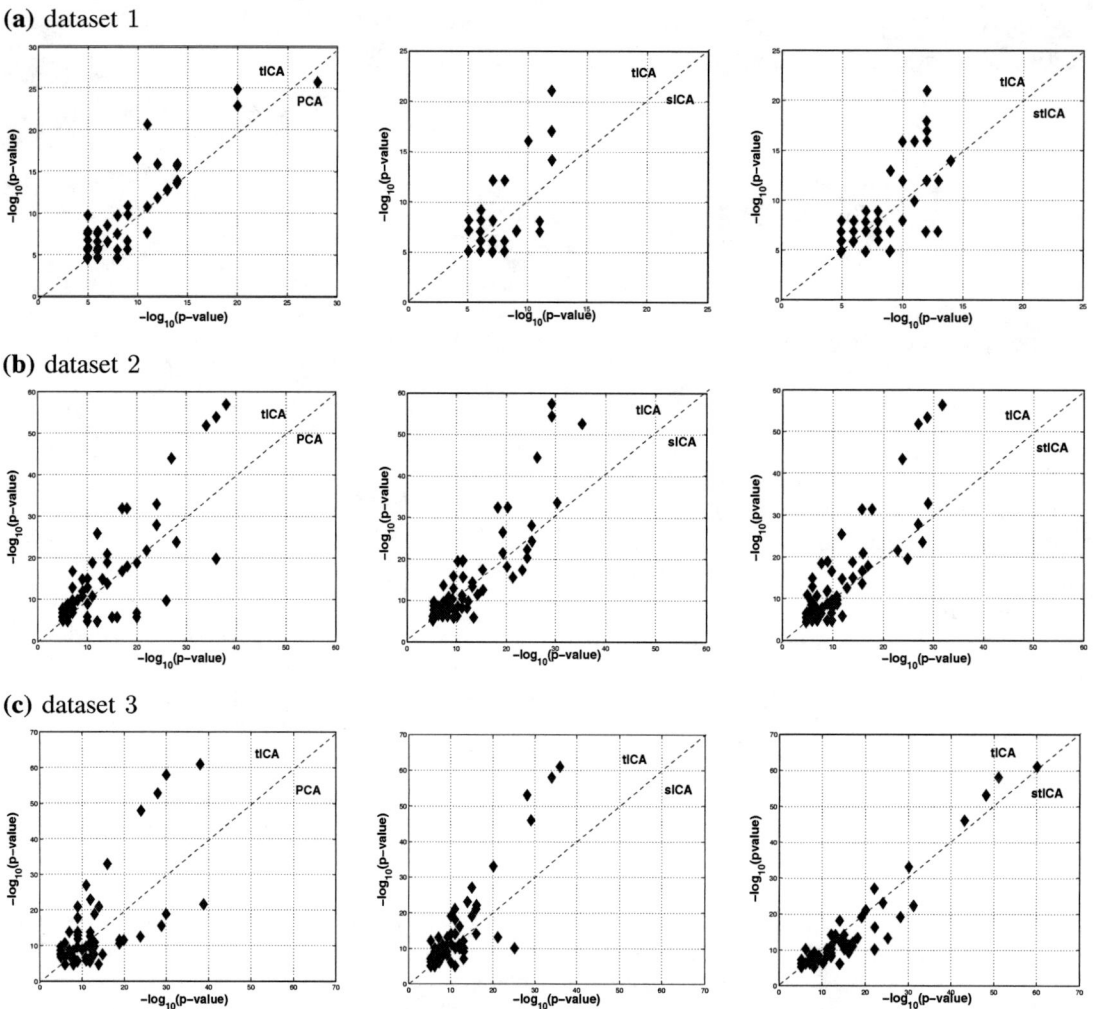

Fig. 2. Performance comparison of three different ICA methods (such as sICA, tICA and stICA) and PCA on **(a)** dataset 1 (alpha) **(b)** dataset 2 (cdc15) and **(c)** dataset 3 (elutriation). Each point corresponds to $-\log_{10}(p-value)$ of a GO annotation (biological function).

TABLE I
DATASETS AND THEIR PROPERTIES.

no	experiment	# of ORFs	# time points	# of eigenvectors
1	alpha	4579	18	6
2	cdc15	5490	24	7
3	elutriation	5981	14	4

$\log R_{ij} - \log G_{ij}$ where R_{ij} and G_{ij} represent red and green light intensity, respectively. We removed genes whose profiles have missing values more than 10%. Then we applied the *KNNimput* method [10], in order to fill in missing values. The data matrix was centered such that each row and each column have zero mean.

2) *Dimensionality Selection:* We chose the dimension n using the *PCA-L* method [6] that is based on the Laplace approximation. Then SVD was applied to find the decomposition in (3).

3) *Decomposition by ICA:* A conjugate gradient method was used to find independent components for sICA, tICA, and stICA. The initial condition for the unmixing matrix were randomly chosen, but were identical for three ICA methods. The hypothesized density for ICA algorithms were chosen as a super-Gaussian distribution.

4) *Gene Clustering:* Column vectors of S_S in sICA are independent arrays. For each column vector, genes with strong positive and negative values are grouped, which leads to two clusters related to induced and repressed genes. For tICA, the same method is applied to the row vectors of A_T containing unconstrained dual arrays.

5) *Statistical Significance Test:* For each cluster, we measured the enrichment with genes of known functional annotations. Using the Gene Ontology (GO) annotation databases [2], we calculated the p-value for each cluster with every annotated genes. The hypergeometric distribution was used to obtain the chance probability of observing the number of genes from a particular GO functional category within each cluster [9].

IV. RESULTS

Our analysis was carried out in accordance with procedures mentioned above on the publicly available yeast cell cycle-related data. The expression levels of genes were preprocessed to be log-ratios $X_{ij} = \log R_{ij} - \log G_{ij}$ where R_{ij} and G_{ij} denote red and green intensities, respectively. Then we applied ICA algorithms (including tICA, sICA, stICA) and PCA to the gene expression matrix X. All results were based on the analysis of a reduced rank data $\widetilde{X}$ of rank n. We computed n independent components and n principal components by ICA and PCA, respectively. Each eigenarray or independent array (or dual array) went through grouping into two clusters, each of which contains 10% genes with significantly high or low influences. Statistical significance for each cluster was evaluated by computing p-values which tell us how well the genes in a cluster match a certain functional category. Only p-values less than 10^{-5} were considered. Scatter plots of the negative logarithm of the best p-value for each cluster are shown in Fig. 2. Among three ICA methods, tICA was the best in all cases. In addition, ICA methods produced significantly lower p-value than PCA did.

The ICA decomposition of the gene expression data matrix of rank n, leads to n temporal modes (or dual time courses). These modes were already shown to be related to cell cycle behavior when those modes were calculated by sICA [4]. Here we calculated (dual) temporal modes by three ICA methods. An interesting point that we found in our numerical experiments, was that temporal modes calculated by stICA exhibited the cell cycle behavior more clearly, compared to other two ICA methods. Fig. 3 shows the temporal behavior of mode 2, 3, 4 when stICA calculated 6 temporal modes. The result is also summarized in Table II).

TABLE II
THE THREE MOST SIGNIFICANT TEMPORAL MODES ON DATASET 1
(ALPHA). THE TEMPORAL MODES WERE CHARACTERIZED ACCORDING TO
FUNCTIONALLY RELATED CATEGORIES.

mode	Induced functions	Repressed functions
2	sexual reproduction, cell wall, bud	protein amino acid glycosylation, nucleosome
3	cell cycle, cell proliferation, DNA replication, response to stress, chromosome, replication fork,	ribosome biogenesis, rRNA processing, nucleolus, RNA helicase activity
4	cell proliferation, cell cycle, DNA repair, chromosome, cell wall	

V. DISCUSSION

We have applied three different ICA methods, including sICA, tICA, and stICA, to a problem of gene expression time series analysis. We compared three ICA methods in the context of gene expression time series data analysis, showing that tICA is more suitable for gene clustering in all datasets and stICA finds linear modes that best match the cell cycle. tICA and stICA would be expected to reflect the specific characteristics

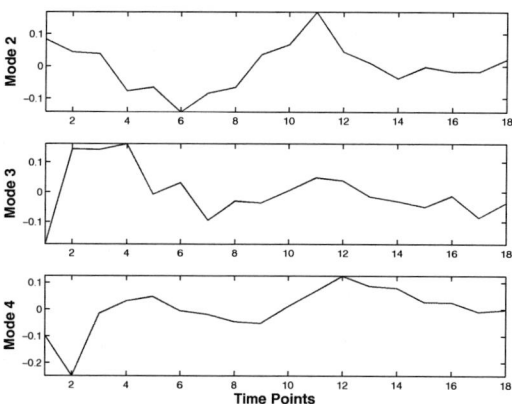

Fig. 3. Temporal modes that best match the cell cycle on dataset 1 (alpha).

of gene expression time series data. For gene clustering, ICA methods performed well than PCA did. Consequently, the linear temporal modes and independent arrays (spatial patterns) will help to highlight particular biological functions in gene expression time series data.

ACKNOWLEDGMENT

We thank J.V. Stone *et al.* for sharing their stICA MATLAB code with us. This work was supported by Systems Bio-Dynamics Research Center and POSTECH Basic Research Fund.

REFERENCES

[1] R. J. Cho, M. J. Campbell, E. A. Winzeler, L. Steinmetz, A. Conway, L. Wodicka, T. G. Wolfsberg, A. E. Gabrielian, D. Landsman, D. J. Lcokhart, and R. W. Davis, "A genome-wide transcriptional analysis of the mitotic cell cycle," *Mol. Cell*, vol. 2, pp. 65–73, 1998.
[2] Gene Ontology Consortium, "Creating the gene ontology resource: Design and implementation," *Genome Research*, vol. 11, pp. 1425–1433, 2001.
[3] S. Lee and S. Batzoglou, "ICA-based clustering of genes from microarray expression data," in *Advances in Neural Information Processing Systems*, vol. 16. MIT Press, 2004.
[4] W. Liebermeister, "Linear modes of gene expression determined by independent component analysis," *Bioinformatics*, vol. 18, no. 1, pp. 51–60, 2002.
[5] M. J. McKeown, T. Jung, S. Makeig, G. Brown, S. S. Kindermann, T. W. Lee, and T. J. Sejnowski, "Spatially independent activity patterns in functional magnetic resonance imaging data during the stroop color-naming task," *Proc. Natl. Acad. Sci. USA*, vol. 95, pp. 803–810, 1998.
[6] T. P. Minka, "Automatic choice of dimensionality for PCA," in *Advances in Neural Information Processing Systems*, vol. 13. MIT Press, 2001.
[7] P. T. Spellman, G. Sherlock, M. Q. Zhang, V. R. Iyer, K. Anders, M. B. Eisen, P. O. Brown, D. Botstein, and B. Futcher, "Comprehensive identification of cell cycle-regulated genes of the yeast *saccharomyces cerevisiae* by microarray hybridization," *Molecular Biology of the Cell*, vol. 9, pp. 3273–3297, Dec. 1998.
[8] J. V. Stone, J. Porrill, N. R. Porter, and I. W. Wilkinson, "Spatiotemporal independent component analysis of event-related fmri data using skewed probability density functions," *NeuroImage*, vol. 15, no. 2, pp. 407–421, 2002.
[9] S. Tavazoie, J. D. Hughes, M. J. Campbell, R. J. Cho, and G. M. Church, "Systematic determination of generic network architecture," *Nature Genetics*, vol. 22, pp. 281–285, 1999.
[10] O. Troyanskaya, M. Cantor, G. Sherlock, P. Brown, T. Hastie, R. Tibshirani, D. Botstein, and R. B. Altman, "Missing value estimation methods for DNA microarrays," *Bioinformatics*, vol. 17, no. 6, pp. 520–525, 2001.

Blind Separation of a Class of Nonlinear ICA Models

Jan Eriksson and Visa Koivunen
SMARAD CoE, Signal Processing Laboratory
Department of Electrical Engineering
Helsinki University of Technology, FIN-02015 HUT, Finland
Tel: +358-9-451 5827, Fax: +358-9-452 3614
Email: {jan.eriksson,visa.koivunen}@hut.fi

Abstract—In this paper we consider a class of nonlinear ICA models that may be described using the Addition Theorem (AT) [1]. Such models cover a wide variety of nonlinear systems of interest in engineering applications. In general, some nonlinear distortions always remain after performing signal separation using such models. In this paper we find a class of AT models, i.e. nonlinear mixing systems, that may be separated up to conventional scaling ambiguity. A theorem proving the separability is provided as well. A connection between AT models and commonly-used post-nonlinear (PNL) models is established. Furthermore, we extend the proposed AT models to a more general case where the functional form of nonlinearity is parameterized and consider the separability of such systems as well.

I. INTRODUCTION

The general *instantaneous time-invariant noiseless ICA* model is described by the equation

$$\bar{x} = \mathcal{F}(\bar{s}), \quad (1)$$

where $(\bar{s}_1, \ldots, \bar{s}_m)^T = \bar{s}$ are unknown mutually independent non-degenerate random variables (r.v.'s) called *sources*, $\mathcal{F}(\cdot)$ is an unknown *mixing function*, and $\bar{x} = (\bar{x}_1, \ldots, \bar{x}_p)^T$ are *mixtures*, i.e., the observed r.vc. (sensor array output). In the blind source separation (BSS) problem, the idea is to recover the original sources $\bar{s}$, i.e. to *separate* the sources, by transforming the mixtures $\bar{x}$ using only the knowledge that the original sources are independent. In a conventional ICA problem, the mixing function $\mathcal{F}$ in eq. (1) is assumed to be linear. In this case, it is now well-known [2], [3] that the separation can be achieved essentially if there is at least as many mixtures as sources and at most one source is Gaussian. Recent textbooks provide excellent tutorial material and extensive review on linear ICA methods [4], [5].

It is now well-known [6] that the general model of eq. (1) is not separable in the sense [7] that transforming the mixture $\bar{x}$ such that the marginal r.v.'s are independent would imply that each marginal is a transformation of one source only. That is, given a mixture $\bar{x}$ from the general model there exist infinitely many transformations that produce independent marginal r.v.'s but the marginal r.v.'s are mappings of two or more source r.v.'s. Therefore, some restrictions are needed either for the allowed source r.v.'s or the allowed mixing functions. An interesting class of separable nonlinear ICA models were introduced in [8]. We continue the study of such nonlinear systems by defining a class of models that have the desirable property that the sources can be recovered without any nonlinear distortions. We also proof a theorem associated with such models. It allows relaxing the conditions on the separability of the post-nonlinear (PNL) [9] model for bounded sources. Also, these special addition theorem models can be combined to form a single generic nonlinear model that has all the desirable properties.

This paper is organized as follows. The models obeying the addition theorem, introduced in [8], are reviewed in Section II. An important subclass of these models possessing a desirable scaling distortion property are introduced in Section III. An illustrative simulation example is also presented. The connection between the addition theorem models and the post-nonlinear model [9] is established in Section IV. Also an enhanced version of the theorem addressing the separability of post-nonlinear model for bounded sources is proved. The proof removes the unnecessary assumption on analytic mixing functions. A novel nonlinear ICA model possessing the desirable properties in introduced in Section V. Finally, some concluding remarks are given.

II. ADDITION THEOREM MODELS

Symmetry is a fundamental property of the physics of nature. Mathematically symmetry is described by the group theory [10]. Therefore, from the application point of view, it is natural to require that the component mixing functions in eq. (1) satisfy the group axioms with respect to source r.v.'s. This motivates the construction presented next.

A closed operation $\circ$ on a set forms a *group* [10], if the following conditions hold: the operation is associative, there exists a unit element, and every element has an inverse element. Moreover, a group is called *Abelian* if the group operation is also commutative. It is straightforward to check that any continuous and strictly monotonic (i.e. invertible) function $\mathcal{G} : \mathbb{R} \to \mathbb{U}$ gives an Abelian group on an open interval $\mathbb{U} \subseteq \mathbb{R}$ by defining for all $u, v \in \mathbb{U}$,

$$u \circ v \triangleq \mathcal{G}(\mathcal{G}^{-1}(u) + \mathcal{G}^{-1}(v)). \quad (2)$$

By denoting $x = \mathcal{G}^{-1}(u)$, $y = \mathcal{G}^{-1}(v)$, and $\mathcal{H}_2(u,v) = u \circ v$, eq. (2) can be written as

$$\mathcal{G}(x+y) = \mathcal{H}_2(\mathcal{G}(x), \mathcal{G}(y)). \quad (3)$$

This type of equations are called *addition theorems* (AT's) [1]. It can be shown [1] that given an Abelian operation $\circ$ on an open interval $\mathbb{U}$, eq. (3) is satisfied for a unique (up to constant multiplication of its argument) continuous strictly monotonic function $\mathcal{G}$. The converse is also true. For a fixed continuous function $\mathcal{G}$ satisfying (3) with some function $\mathcal{H}_2(\cdot,\cdot)$, the operation $\mathcal{H}_2(\cdot,\cdot)$ necessarily defines an Abelian group and $\mathcal{G}$ is strictly monotonic.

Since $\mathcal{G}$ has the inverse function $\mathcal{G}^{-1}$, it follows by (3)

$$\mathcal{G}(kx) = \mathcal{G}(x+x+\cdots+x) = \mathcal{G}(x) \circ \mathcal{G}(x) \circ \cdots \circ \mathcal{G}(x)$$
$$\triangleq k \star \mathcal{G}(x) = k \star u,$$

which defines a new multiplication operation $\star$ for an integer k. This extends uniquely by continuity to all reals a by defining

$$a \star u \triangleq \mathcal{G}(ax) = \mathcal{G}(a\mathcal{G}^{-1}(u)). \quad (4)$$

Using operators defined by expressions. (2) and (4) one can define a nonlinear function $\mathcal{F}: \mathbb{U}^m \to \mathbb{R}$,

$$\mathcal{F}(u_1,\ldots,u_m) = (a_1 \star u_1) \circ (a_2 \star u_2) \circ \cdots \circ (a_m \star u_m), \quad (5)$$

where the parenthesis could be dropped since $\circ$ is associative. Using these type of functions as mixing functions, a nonlinear ICA model of eq. (1) can be rewritten as

$$\boldsymbol{x} = \boldsymbol{\mathcal{F}}(\bar{\boldsymbol{s}}) = \begin{bmatrix} \mathcal{F}_1(\bar{s}_1,\ldots,\bar{s}_m) \\ \mathcal{F}_2(\bar{s}_1,\ldots,\bar{s}_m) \\ \vdots \\ \mathcal{F}_p(\bar{s}_1,\ldots,\bar{s}_m) \end{bmatrix}$$
$$= \begin{bmatrix} a_{11} \star \bar{s}_1 \circ a_{12} \star \bar{s}_2 \circ \cdots \circ a_{1m} \star \bar{s}_m \\ a_{21} \star \bar{s}_1 \circ a_{22} \star \bar{s}_2 \circ \cdots \circ a_{2m} \star \bar{s}_m \\ \vdots \\ a_{p1} \star \bar{s}_1 \circ a_{p2} \star \bar{s}_2 \circ \cdots \circ a_{pm} \star \bar{s}_m \end{bmatrix} \triangleq \boldsymbol{A} \otimes \bar{\boldsymbol{s}}, \quad (6)$$

where the matrix $\boldsymbol{A} = (a_{kl})$, $k = 1,\ldots,p$, $l = 1,\ldots,m$ and $\otimes$ is a shorthand notation for the nonlinear mixing above. These models were introduced in [8], and they are called *addition theorem models*. It should be emphasized that *any* continuous strictly monotonic function introduces a model of the form (6) by eq.'s (2) and (4). Notice also that $\mathcal{G}(x) = x$ gives the standard linear ICA model.

Since

$$\boldsymbol{\mathcal{G}}^{-1}(\bar{\boldsymbol{x}}) \triangleq \begin{bmatrix} \mathcal{G}^{-1}(\bar{x}_1) \\ \mathcal{G}^{-1}(\bar{x}_2) \\ \vdots \\ \mathcal{G}^{-1}(\bar{x}_p) \end{bmatrix} = \boldsymbol{A} \begin{bmatrix} \mathcal{G}^{-1}(\bar{s}_1) \\ \mathcal{G}^{-1}(\bar{s}_2) \\ \vdots \\ \mathcal{G}^{-1}(\bar{s}_m) \end{bmatrix} \quad (7)$$

describes a standard linear ICA model with source r.v.'s $\mathcal{G}^{-1}(\bar{s}_k)$, $k = 1,\ldots,m$, the separability of an AT model of eq. (6) is reduced to the separability [3] of the standard linear model. Hence, we have the following [8]:

Theorem 1 (Separability of the AT models): Suppose that in the model (6) at most one of the r.v.'s $\mathcal{G}^{-1}(\bar{s}_k)$, $k = 1, 2, \ldots, m$, where $\mathcal{G}$ is the function defined by the operator $\circ$, is normal and the matrix $\boldsymbol{A}$ is of full column rank. Then

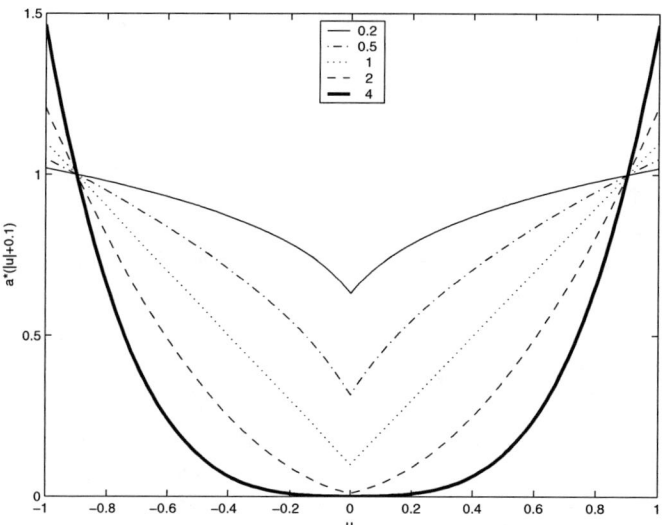

Fig. 1. Distortion $a \star (|u| + 0.1)$ plotted for the different values of a in the AT model described by $\mathcal{G}(u) = \exp(u)$.

the model (6) is separable up to arbitrary permutation and a constant $\star$-multiplication as defined in eq. (4) of each source.

An important consequence of Theorem 1 is that if $\mathcal{G}$ does not preserve normality, it is possible to separate Gaussian r.v.'s unlike in the case of the linear model.

Eq. (7) suggests also a generic method of separating sources from an AT model:
1) Apply $\boldsymbol{\mathcal{G}}^{-1}$ to the mixture $\bar{\boldsymbol{x}}$, i.e. $\bar{\boldsymbol{y}} = \boldsymbol{\mathcal{G}}^{-1}(\bar{\boldsymbol{x}})$.
2) Use a linear ICA algorithm to separate $\bar{\boldsymbol{y}}$.
3) Transform the obtained separation result with $\boldsymbol{\mathcal{G}}$.

III. SEPARABLE ADDITION THEOREM MODELS WITH CONVENTIONAL SCALING AMBIGUITY

The sources can be recovered up to constant $\star$-multiplication in the AT models by Theorem 1. Although mathematically this is almost equivalent to the usual scaling distortion in the linear model, there may be significant, noticeable differences for a human observer in the recovered sources depending on the specific AT model. This is due to the fact that the constant $\star$-multiplication may distort the *shape* of the underlying signal waveform. This is illustrated in Figure 1 for the signal $|u|+0.1$, $-1 \le u \le 1$, in the model described by $\mathcal{G}(u) = \exp(u)$, i.e. the component mixing functions of the eq. (5) are of the form $\mathcal{F}(u_1,\ldots,u_m) = u_1^{a_1} u_2^{a_2} \cdots u_m^{a_m}$. This model arises in the image processing context, see [8] for an example. Notice also that due to the above shape distortion, the quantitative error measures such as the mean square error (MSE), the mean absolute error (MAE), and the signal-to-interference ratio (SIR) may not adequately describe the quality of separation results for the AT models.

Since the nonlinear distortion described above is unavoidable, it would be interesting to find models which are separable up to the conventional scaling. Therefore, we characterize all AT models such that $a \star u = b(a)u$ for all constants $a \in \mathbb{R}$, for some constant $b(a) \in \mathbb{R}$ that depends only on a, and for all $u \in \mathbb{U}$. Using the eq. (4) this can be written as

$$\mathcal{G}(ax) = b(a)\mathcal{G}(x) \tag{8}$$

for all $x, a \in \mathbb{R}$ and for a real-valued function $b(\cdot)$, i.e. we need to find all continuous strictly monotonic solutions $\mathcal{G}$ of the functional equation (8).

Theorem 2: The functional equation (8) has a strictly monotonic solution $\mathcal{G}$ continuous in a point if and only if

$$\mathcal{G}(x) = c \, \text{sign}(x) |x|^\gamma, \tag{9}$$

where $c \neq 0$ and $\gamma > 0$.

Proof: Suppose $\mathcal{G}(x)$ is a strictly monotonic solution for (8). If $b(a) = 0$ for any $a \neq 0$ in eq. (8), then $\mathcal{G}(x) \equiv 0$, which does not qualify as a strictly monotonic solution. Hence $b(a) \neq 0$ for all $a \neq 0$, and from $\mathcal{G}(0) = b(a)\mathcal{G}(0)$ it follows that $\mathcal{G}$ equals zero only at zero by the strict monotonicity.

From $\mathcal{G}(1) = b(1)\mathcal{G}(1) = \mathcal{G}((-1)(-1)) = b(-1)^2\mathcal{G}(1)$ it follows $b(1) = b(-1)^2 = 1$. Now $\mathcal{G}(-x) = b(-1)\mathcal{G}(x)$, and since $\mathcal{G}$ is strictly monotonic, $b(-1) = -1$ and $\mathcal{G}(-x) = -\mathcal{G}(x)$. Since $\mathcal{G}(-x) = b(x)\mathcal{G}(-1) = -\mathcal{G}(x) = -b(-x)\mathcal{G}(-1)$, also $b(-x) = -b(x)$. Therefore, it is enough to find the solution for the eq. (8) with positive x and a. However, this is now a special case of one of the Cauchy-Pexider equations [1], which admits, assuming $\mathcal{G}$ is continuous in a point, the solution ([1], p. 144) $\mathcal{G}(x) = cx^\gamma$ for some constants $\gamma, c \neq 0$. Therefore $\mathcal{G}(x) = -c|x|^\gamma$ for negative x, and by the strict monotonicity $\gamma > 0$. Hence we have the solution (9).

On the other hand, the function $\mathcal{G}(x)$ of the eq. (9) is strictly monotonic, continuous and satisfies the eq. (8) with $b(a) = \text{sign}(a)|a|^\gamma$. ∎

The strict monotonicity was only needed in Theorem 2 for combining the solutions for negative and positive arguments. If this assumption is dropped we simply get two solutions of the form (9) ($\gamma \neq 0$) valid separately for negative and positive arguments. Furthermore, the assumption of continuity at a point can be further weakened [1], e.g. by only requiring that $\mathcal{G}$ is bounded on an interval. Actually, *without any* requirements on $\mathcal{G}$ and assuming the eq. (8) holds for a fixed a, one still obtains [11] the solution of the type (9) multiplied by a 1-*periodic* function whose argument depends on both x and a.

Since the scaling is an unavoidable ambiguity anyway, we can take without a loss of generality $c = 1$ in the eq. (9) when generating our models. For such AT models with only the conventional scaling ambiguity we use the shorthand notation $\otimes_\gamma$ for the function $\mathcal{F}$ defined by the equations (6) and (9), i.e. the model generated with $\mathcal{G}$ of eq. (9) is written as

$$\bar{x} = A \otimes_\gamma \bar{s}. \tag{10}$$

Assuming A is invertible, the inverse $\mathcal{F}^{-1}$ is given as $A^{-1} \otimes_{1/\gamma}$, i.e. $A^{-1} \otimes_{1/\gamma} (A \otimes_\gamma x) = x$. As an example of a model with only the scaling ambiguity, consider a two dimensional case with $\mathcal{G}(x) = \text{sign}(x)|x|^3 = x^3$. The

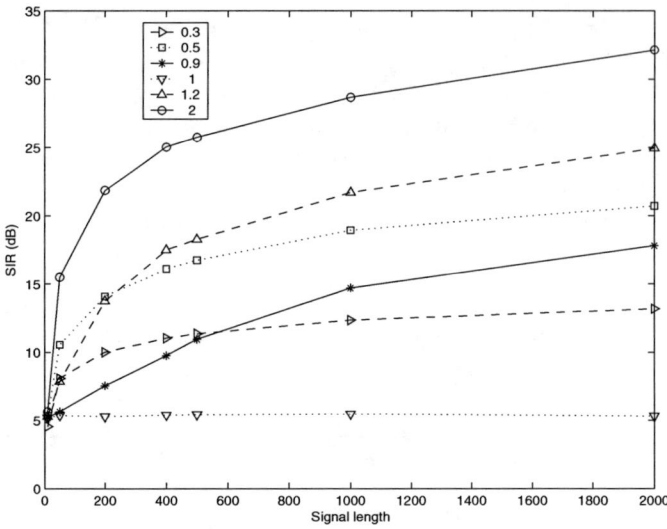

Fig. 2. Median separation performance for mixture of three standard normal signals in scale separable AT models with different γ's.

corresponding model is described as

$$\bar{x} = \begin{bmatrix} a_{11} & a_{12} \\ a_{21} & a_{22} \end{bmatrix} \otimes_3 \begin{bmatrix} \bar{s}_1 \\ \bar{s}_2 \end{bmatrix} = \begin{bmatrix} (a_{11}\sqrt[3]{\bar{s}_1} + a_{12}\sqrt[3]{\bar{s}_2})^3 \\ (a_{21}\sqrt[3]{\bar{s}_1} + a_{22}\sqrt[3]{\bar{s}_2})^3 \end{bmatrix}$$
$$= \begin{bmatrix} a_{11}^3 \bar{s}_1 + \frac{1}{3}a_{11}^2 a_{12} \sqrt[3]{\bar{s}_1^2 \bar{s}_2} + \frac{1}{3}a_{11}a_{12}^2 \sqrt[3]{\bar{s}_1 \bar{s}_2^2} + a_{12}^3 \bar{s}_2 \\ a_{21}^3 \bar{s}_1 + \frac{1}{3}a_{21}^2 a_{22} \sqrt[3]{\bar{s}_1^2 \bar{s}_2} + \frac{1}{3}a_{21}a_{22}^2 \sqrt[3]{\bar{s}_1 \bar{s}_2^2} + a_{22}^3 \bar{s}_2 \end{bmatrix}.$$

As with all AT models, it is possible to separate normal signals in the AT models with only the conventional scaling ambiguity ($\gamma \neq 1$). This is illustrated in Figure 2, where the median separation performance in terms of signal-to-interference ratio (SIR) for the mixture of three standard normal distributed signals is presented for different γ-values with varying signal length. The mixing coefficients (A) were randomly generated, the performance was calculated over 1001 realizations, and the linear separation part was done with the standard JADE algorithm [12]. It is worth noticing the relatively good results with the model with a minor nonlinearity, i.e., $\gamma = 0.1.2$. The signals are in our subjective view clearly separated when the SIR is close to 15dB.

IV. SEPARABILITY OF POST-NONLINEAR MODEL

A well-known nonlinear ICA model (1) is described by

$$\bar{x} = \mathcal{F}(A\bar{s}), \tag{11}$$

where $\mathcal{F}(x) = (\mathcal{F}_1(x_1), \ldots, \mathcal{F}_p(x_p))^T$ is a *component-wise* nonlinear function. The model is known as *post-nonlinear* (PNL) model [9], and can be viewed as linear mixing followed by a nonlinear distortion at each sensor. Since $\mathcal{F}(\cdot)$ is also unknown, the PNL model provides a wide variety of nonlinear ICA systems.

It should be emphasized that the PNL model, unlike the AT models, does not describe truly nonlinear mixing but rather *nonlinear distortion after linear mixing*. However, if the components $\mathcal{F}_k$ of $\mathcal{F}$ are the same, i.e. $\mathcal{F}(x) = (\mathcal{F}(x_1), \ldots, \mathcal{F}(x_m))^T$, then by writing $\bar{s} = \mathcal{F}^{-1}(\bar{r})$ and using the addition theorem (3), the model of the eq. (11) reduces to an addition theorem model (6) in source r.v.'s $\bar{r}_k$, $k = 1, \ldots, m$. Therefore, in such a case PNL system can be viewed as an outcome of a truly nonlinear mixing. If the mixing functions of the eq. (6) satisfy conditions for PNL mixture, the statement may be reformulated in an alternate way: the addition theorem models belong to the class of PNL models with post-processing.

The separability of the PNL model means that the sources can be recovered up to permutation, scaling, and an additive constant [9]. The separability was proved under rather restrictive conditions in [9]. Another proof for bounded sources was given in [13], [14]. The proof for bounded sources is based on the geometrical interpretation of the statistical independence, and it is additionally assumed that the functions $\mathcal{F}(x_k)$, $k = 1, \ldots, p$, are analytical. Coincidently, the assumption on analytic functions is only needed for finding (the analytic) solutions (Lemma 3.1. in [14]) of a functional equation, which is equivalent to the functional eq. (8) with a fixed a. Interestingly the distribution function the argument on "borders of parallelogram" can be extended to other "lines" in the sample space, and consequently the eq. (8) has to be satisfied for all a. Now using Theorem 2 the assumption of analyticity to can be removed, and we have the following.

Theorem 3: Suppose all component functions in the eq. (11) are continuous and strictly monotonic, and all sources are bounded. Assume further that the matrix A is of full column rank, and every column and every row of A has at least two non-zero entries. Then the PNL model of the eq. (11) is separable.

V. PARAMETRIC SEPARABLE ADDITION THEOREM MODEL WITH THE SCALING AMBIGUITY

Although the class of AT models seems to describe most reasonable nonlinear mixing models one has to know the structure of the system in order to apply them. That is, one has to know the function $\mathcal{G}$. In some applications the system structure may be unknown, and there may be no way of learning the system structure. For such situations, it would be desirable to have a single model that could capture a vast variety of models. The PNL model is such a general model but it does not in general describe a truly nonlinear mixing as pointed out earlier.

A natural way of defining such generic models is to take a parametrized class of AT models, and letting also the parameters to be unknown. A very appealing model is therefore obtained by considering all AT models with the scaling ambiguity as a *single model* with the unknown mixing parametrized by the matrix A and the structure describing parameter γ. The model is simply described by the eq. (10) with *also γ as an unknown*. It should be emphasized that this model is the most general independence-based instantaneous blind separation model, which satisfies the natural group axioms for the mixing, and moreover it has the property that if it is separable, the sources can be recovered up to the normal scaling.

VI. CONCLUSION

We have introduced a class of nonlinear ICA models that allow the signal separation up to the conventional scaling. These models cover a wide variety of nonlinear settings, and they have some interesting properties. It was demonstrated that the models allow the separation of Gaussian mixtures without nonlinear distortion, which has not been possible with previously described ICA models. The theorem associated with these models also allowed enhancing the theorem [13], [14] for the separability of the post-nonlinear model for the bounded sources.

The parametric nonlinearity model introduced in Section V is intended as a generic nonlinear model with all the desirable properties. However, in order to prove that the parametric model is indeed separable, one should show that the r.vc $B \otimes_{\gamma_1} (A \otimes_{\gamma_2} \bar{s})$ has independent marginal r.v.'s if and only if $\gamma_2 \gamma_1 = 1$ and BA is a permutated version of a diagonal matrix for any allowed source r.vc. $\bar{s}$. This seems to be a nontrivial task beyond a rather restricted classes of source r.vc.'s $\bar{s}$. However, due to the relationship described in Section IV, the conditions proved for the PNL model apply also for this model. Thus, the parametric model is separable at least for the bounded sources. However, we firmly believe that the class of allowed sources is much more general, but this needs further studying. Also, it would be interesting to develop parameter estimation and consequently separation algorithms for the model.

REFERENCES

[1] J. Aczél, *Lectures on Functional Equations and Their Applications*. New York, NY: Academic Press, 1966.

[2] P. Comon, "Independent component analysis, a new concept?" *Signal Processing*, vol. 36, no. 3, pp. 287–314, Apr. 1994.

[3] J. Eriksson and V. Koivunen, "Identifiability, separability, and uniqueness of linear ICA models," *IEEE Signal Processing Lett.*, vol. 11, no. 7, pp. 601–604, July 2004.

[4] A. Cichocki and S. Amari, *Adaptive Blind Signal and Image Processing: Learning Algorithms and Applications*. John Wiley & Sons, 2002.

[5] A. Hyvärinen, J. Karhunen, and E. Oja, *Independent Component Analysis*. John Wiley & Sons, 2001.

[6] C. Jutten, M. Babaie-Zadeh, and S. Hosseini, "Three easy ways for separating nonlinear mixtures?" *Signal Processing*, vol. 84, no. 2, pp. 217–229, Feb. 2004.

[7] A. Taleb, "A generic framework for blind source separation in structured nonlinear models," *IEEE Trans. Signal Processing*, vol. 50, no. 8, pp. 1819–1830, Aug. 2002.

[8] J. Eriksson and V. Koivunen, "Blind identifiability of class of nonlinear instantaneous ICA models," in *Proc. of XI European Signal Processing Conference*, Toulouse, France, Sept. 2002, pp. 7–10.

[9] A. Taleb and C. Jutten, "Source separation in post-nonlinear mixtures," *IEEE Trans. Signal Processing*, vol. 47, no. 10, pp. 2807–2820, Oct. 1999.

[10] W. Scott, *Group Theory*, reprint ed. New York, NY: Dover Publications, Inc., 1987.

[11] S.-M. Jung and K.-S. Lee, "Stability of generalized additive Cauchy equations," *Internat. J. Math. & Math. Sci.*, vol. 24, no. 11, pp. 721–727, Nov. 2000.

[12] J.-F. Cardoso, "JADE algorithm," Matlab code with references, http://www.tsi.enst.fr/~cardoso/stuff.html.

[13] M. Babaie-Zadeh, C. Jutten, and K. Nayebi, "A geometric approach for separating post non-linear mixtures," in *Proc. of XI European Signal Processing Conference*, Toulouse, France, Sept. 2002, pp. 11–14.

[14] M. Babaie-Zadeh, "On blind source separation in convolutive and nonlinear mixtures," Ph.D. dissertation, INPG, Grenoble, France, 2002.

A Novel Application Specific Network Protocol for Wireless Sensor Networks

Jichuan Zhao[1], Ahmet T. Erdogan[1,2] and Tughrul Arslan[1,2]
[1]University of Edinburgh, School of Engineering and Electronics
Edinburgh, EH9 3JL, Scotland, United Kingdom
[2]Institute for System Level Integration, The ALBA Campus
Livingston, EH54 7EG, Scotland, United Kingdom
J.Zhao@ed.ac.uk, Ahmet.Erdogan@ee.ed.ac.uk, Tughrul.Arslan@ee.ed.ac.uk

Abstract—The recent interest in Wireless Sensor Networks (WSN) has led to emergence of many application specific communication protocols which must be energy-efficient. Among those protocols developed for WSN, LEACH (Low Energy Adaptive Clustering Hierarchy) protocol is one of the most popular protocols. In this paper, a novel application specific energy efficient protocol based on LEACH is presented, combining cluster based architecture and multiple-hop routing. Multi-hop routing is utilized for inter-cluster communication between cluster heads and the base station, instead of direct transmission in order to minimize transmission energy. Besides, this protocol adds some mechanisms to CSMA/CD (Carrier Sense Multiple Access with Collision Detection) so as to avoid collisions, instead of using other more complicated MAC protocols. Simulation results, compared with LEACH, demonstrate that our novel protocol can reduce energy consumption and hence prolong the lifetime of WSN.

I. INTRODUCTION

Distributed sensing and computing have been made possible and practical by the advance of wireless communication technology and the availability of integrated miniature sensors and many lightweight, compact, portable computing devices. Wireless Sensor Networks (WSN) can achieve data collection, aggregation and communication from a remote environment through many distributed individual sensor nodes, called microsensors, which can be connected by radio link. WSN can be used to monitor a variety of environments for applications such as surveillance, machine failure diagnosis, and chemical/biological detection [1].

There are many strict constraints in the design of WSN such as small size, light weight, ultra-low energy consumption and low cost [2]. Among these energy efficiency should be considered one of the most critical issues since it is impractical to replace batteries on thousands of microsensors. Furthermore, in some cases the microsensors may not be accessible for battery replacement. Therefore designing power-efficient protocols is crucial for prolonging the lifetime of WSN.

The recent interest in WSN has led to a number of network protocols. Because of high correlation of data from the neighboring nodes, some protocols adopted cluster based network architectures. The researchers in [3] proposed a new chain-based protocol called PEGASIS that minimizes the energy consumption of each sensor node. In [4], the authors presented a new minimum spanning tree-based protocol, called PEDAP and its power-aware version PEDAP-PA. In [5], a protocol called APTEEN is introduced that uses an enhanced TDMA schedule for efficiently incorporating query handling.

Besides the protocols above, the LEACH protocol presented in [6] provides an elegant solution to the data aggregation problem where clusters are formed in a self-organized manner to fuse data before transmitting to the end user. In LEACH, a designated node in every cluster, called the cluster head, is responsible for collecting and aggregating the data from sensors in its cluster and eventually transmitting the result to the end user.

This paper builds on the LEACH protocol described in [6] by integrating a multiple-hop routing scheme for inter-cluster communication between cluster heads and the base station, instead of direct transmission, for prolonging the lifetime of WSN. Furthermore, CSMA incorporating some collision avoidance mechanisms is utilized as media access scheme during the period of cluster formation.

II. LEACH PROTOCOL ARCHITECTURE

In LEACH, the operation of the whole network is divided into many rounds. Every round includes set-up phase and steady-state phase. The latter is divided into many frames. During the period of set-up phase all nodes are organized into some clusters through communicating with short messages and one node becomes cluster head. Every cluster head sets up a TDMA schedule for all member nodes of its cluster. All nodes broadcast short messages using carrier-sense multiple access (CSMA) MAC protocol [7]. Following the set-up phase, the data are transferred from member nodes to cluster heads according to the TDMA schedule during a frame, aggregated to reduce redundant data and then passed on to the base station (BS) at the end of each frame. Fig.1. shows the time line of one stretched round.

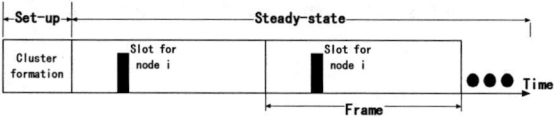

Figure 1. Time line in one round

A potential problem with LEACH protocol is that all cluster heads send the compressed data to the BS directly. If all sensor nodes are pervasive in a large area, some clusters are far from the BS and others are close to the BS. This can lead to great difference on the transmission energy dissipations that the nodes use to transmit data to the BS. According to the free space channel model, the minimum required amplifier energy which dominates the transmission energy is proportional to the square of the distance from the transmitter to the destined receiver ($E_{Tx-amp} \propto d^2$) [8]. Therefore after the network operates for some rounds there will be considerable difference on energy consumption between the nodes near the BS and those far from the BS. If all nodes begin at the same energy storage, the nodes far from the BS will use up their energy before those near the BS. As a result the network will be partitioned into regions with live nodes and dead nodes and hence the performance of the network will decline.

Besides there can be high probability of collisions among the short messages at the set-up phase in LEACH because all nodes are within communication range of each other and utilize the same frequency band. This may lead to the result that some important short messages fail to be received by their expected destination nodes. As a result some nodes cannot be organized into clusters and lose connectivity with the network during a round. Thus the network loses its function partly.

III. OUR NOVEL PROTOCOL

To solve the preceding problems, we propose a novel hybrid protocol. This protocol combines cluster architecture with multi-hop routing to reduce transmission energy.

In many WSN and Ad hoc wireless networks multi-hop routing is adopted. This approach makes a node that wants to transmit data to a destination find one or multiple intermediate nodes. The data packets from the source node are relayed among the intermediate nodes until it reaches the destination [9]. The main advantage of this approach is that the transmission energy consumption can be reduced.

In our protocol after clusters are organized, the cluster heads could form a multi-hop routing backbone. For the communication within a cluster every member node sends data to the cluster head directly. While for the communication between the cluster heads and the BS, a multi-hop routing is adopted to reduce the transmission energy and minimize the difference of energy consumption among all nodes. Our protocol uses "minimum transmission energy" (MTE) routing [10] as the routing algorithm.

In order to reduce the probability of collisions at set-up phase, we add some collision avoidance mechanism to CSMA/CD. In contrast to many developed MAC protocols capable of avoiding collisions effectively, such as 802.11 [11] which is applied to most Ad hoc wireless networks, our approach is much simpler and more energy efficient. Thus it is more suitable for WSN.

For the development of our protocol, we make the same assumptions as LEACH about the network model as follows. Firstly, every node in the network has the same infrastructure and is homogenous. Besides, every node has limited energy. Secondly, all nodes in the network have enough power to directly communicate with any node in the network including BS. This means that all nodes can use power control to vary their transmission power and range. At the same time, each node has enough processing power to support different protocols and signal processing tasks. These assumptions have been made possible because of the availability of many advanced radio frequency devices and low power computing devices. Finally, we assume that those nodes nearby have highly correlated data that is redundant for BS. The following section describes our protocol in more detail.

A.   Period of Cluster Formation

At the beginning of this phase the whole network sets a timer to control when this phase would end and steady phase would begin and keep every node synchronously. Every node in the network must decide whether it will become a cluster head or not. Each node decides to become a head with probability P_i. P_i is calculated according to a cluster head selection algorithm given in [6]. This algorithm ensures that each node can take its turn to act as cluster head. After some nodes have become cluster heads, every head broadcasts an announcement (ANNOU) message through CSMA/CD. This message is a short message including the node's ID and message content indicating that the message is ANNOU. ANNOU also contains the unique spreading code of the node for the intending direct-sequence spread spectrum (DSSS) within a cluster and coordinate of the node's location for multi-hop routing among cluster heads at steady phase.

At this stage all cluster heads load a random delay time t_1 after they decide to become heads. After t_1 they broadcast ANNOU messages. The random delay time t_1 is uniformly distributed between zero and T_1. The value of T_1 should be set appropriately according to the number of cluster heads and end-to-end delay of the network to make sure that there can be enough interval time between transmissions of two random heads. This approach ensures that the ANNOU messages can reach a member node at different times. Thus the probability of collisions is minimized and nearly all messages can be received successfully by member nodes.

A constant period of time later when every member node receives several ANNOU messages from different heads it will choose the closest cluster head to join the cluster according to the received signal strength of the messages. Then each member node would load a random delay time t_2. After t_2 each member node sends a join request (JOIN) message to the chosen cluster head and adds the unique spreading code of the head to the message. From then on all signals in the network would be transmitted and received with appropriate spreading codes. Therefore, in our protocol spreading codes are used from the set-up phase, and not from steady-state as in LEACH. After the cluster head node receives the JOIN message it will send an acknowledgement (ACK) message to the member from which the JOIN comes so as to set up a time division multiple access (TDMA) schedule and allocate a time-slot number to the member. In order to ensure that every node can be organized into a cluster, we adopt a retransmission mechanism. After a member sends a JOIN, it would wait a fixed period of time t_3 for the ACK message. If it does not receive the message after t_3, it would retransmit the JOIN message. This process will be executed iteratively before the steady-state phase comes till it receives the ACK message. This mechanism can prevent any node from not being organized into a cluster. If a member node receives the ACK message it would wait for the beginning of the steady-state phase. Then all nodes in the network would enter the steady-state phase synchronously.

B. Steady-state Phase

The steady-state is made up of many frames. Each frame includes many time-slots during which every member node can send its data to the cluster head only once during its unique time-slot. This TDMA schedule avoids collisions among data messages in a cluster and allows the radio devices of each member to be turned off when it is not its time-slot. Thus the energy consumption is reduced. At the end of each frame the cluster head aggregates the data from its all member nodes, reduce redundant data and then send the data packet to the BS.

With our protocol every cluster head that has a data packet ready to be transmitted would select a route to relay the packet to the BS indirectly instead of transmitting the packet directly to the BS as in LEACH. The route is chosen according to MTE routing algorithm. Therefore, the algorithm chooses one or some intermediate nodes so that the sum of squared distances is minimized. As described in the preceding section, the dominant part of the transmission energy is proportional to the square of the distance. Thus the total transmission energy is minimized.

In our protocol, if there is a head node A that wants to send a packet, it would calculate the function D(X) of all other heads which is defined as below:

$$D(X) = d_{A-X}^2 + d_{X-BS}^2 \quad X \in \{\text{All other heads}\} \quad (1)$$

Then the minimum of these would be picked and compared to the square of the distance from head node A to the BS. Only if

$$Min(D(X)) < d_{A-BS}^2 \quad (2)$$

the node that makes the function minimal (we name the node B) would be selected as intermediate node. Otherwise the node A would still transmit to the BS directly. When the packet arrives at node B, the above algorithm will be repeated to decide whether node B should select an intermediate node or transmit to the BS directly. This process would be iterated till the packet reaches the BS.

All nodes within a cluster transmit data using a unique spreading code that was assigned in set-up phase as mentioned. When a cluster head wants to communicate with another head or the BS, it would use the appropriate spreading code of the destination. At the inter-cluster level CSMA/CA (Collision Avoidance) [12] is used as MAC protocol.

IV. SIMULATION AND RESULTS

We use the network simulator OPNET to model our protocol. OPNET [13] provides a fairly realistic simulation environment for WSN among the available network simulators. Especially it takes into account the effect of noise on the performance of networks. In order to compare with original LEACH we have also built a model for LEACH using OPNET and used the same power model as in [6] for both models to evaluate and compare their energy efficiency. Our simulation is based on a network with 40 nodes distributed in a 1km*1km area.

Fig.2. shows the energy consumptions over time of three nodes with different distances to the BS using the original LEACH protocol. Clearly, there is great difference on the energy consumption between the node far from the BS and that of the closest one as expected. The farthest node consumes almost eight times more energy than the closest node, after 300 minutes of simulation time.

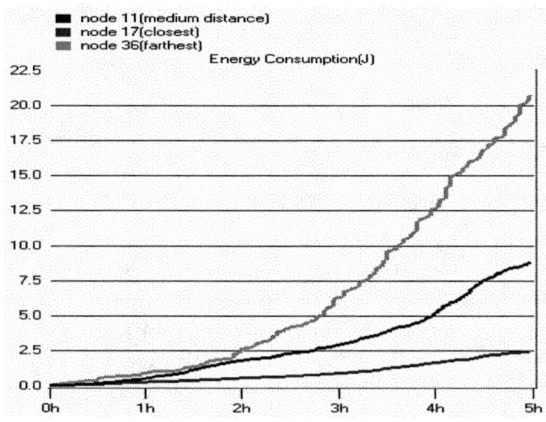

Figure 2. Energy consumption over time of three nodes with LEACH

Fig.3. shows the same profile with our novel protocol as in Fig.2. However, as evident from the graph, the energy consumptions of the same two nodes (node 36 and 11) are reduced significantly for the same duration of time compared to using LEACH. With our protocol, the maximum energy consumption is 4.4J and the minimum is a little more than 2J, compared to 20.7J and 2.5J respectively for LEACH protocol. Clearly the difference of energy consumption is also reduced significantly. However, the energy consumption of the closest node (node 17) rises up a little (from 2.5J to 4.3J). This is because the nodes near the BS take the responsibility of intermediate nodes in multi-hop routing and hence consume a little more energy.

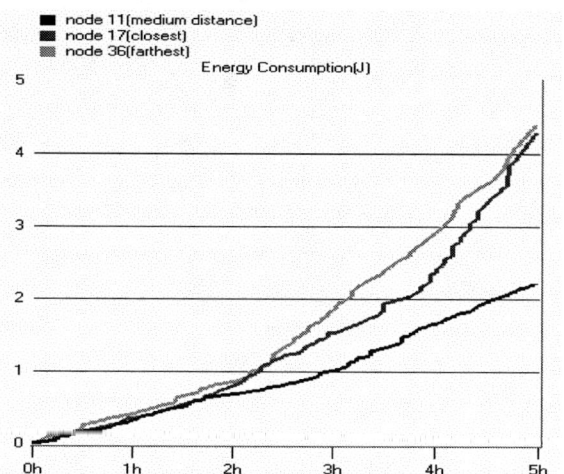

Figure 3. Energy consumption over time of three nodes with our protocol

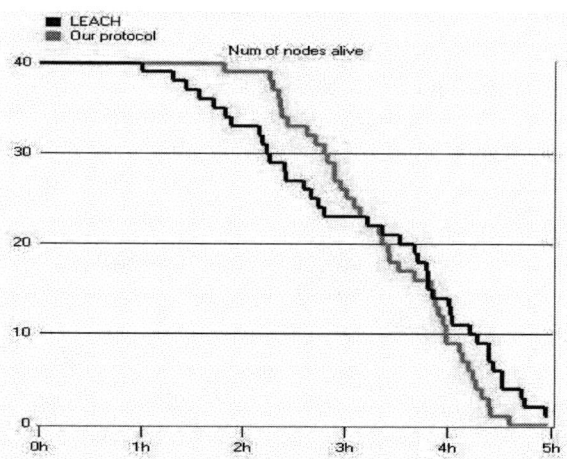

Figure 4. Number of live nodes with both protocol

If we limit the storage energy of every node to 2J in the simulation, we can get the number of live nodes over time for LEACH and our novel protocol as shown in Fig.4. The number of live nodes in the network using LEACH begins to fall after 1 hour but that in the network using our protocol begins to fall after nearly 2 hours. Thus the lifetime of the network is prolonged greatly. However, the number of live nodes in the network using our protocol falls more sharply than that of using LEACH. Therefore, the network using our protocol stays alive as a whole longer, and declines a little faster.

V. CONCLUSIONS AND DISCUSSIONS

In this paper, we have presented a novel hybrid network protocol for WSN and compared it to the LEACH protocol. Results from our simulations show that our protocol provides better performance for energy efficiency and network lifetime.

Our protocol can still be improved further. For example, multi-hop routing algorithm can be implemented for all nodes in the network. This means that when a cluster head has a packet to send to the BS, it would route the packet using all nodes including both cluster heads and members to find the optimal route.

REFERENCES

[1] J. M. Rabaey, M. J. Ammer, J. L. da Silva, D. Patel, and S. Roundy, "PicoRadio Supports Ad Hoc Ultra-Low Power Wireless Networking", *IEEE Computer,* 33(7), July 2000, pp. 42-48

[2] Anantha Chandrakasan, Rajeevan Amirtharajah, SeongHwan Cho, James Goodman, Gangadhar Konduri, Joanna Kulik, Wendi Rabiner, Alice Wang, "Design Considerations for Distributed Microsensor Systems", *Proc. CICC 1999,* pp. 279-286

[3] S. Lindsey, C. Raghavendra, K. M. Sivalingam, "Data Gathering Algorithms in Sensor Networks using Energy Metrics", *IEEE Trans. Parallel and Distributed Systems,* vol. 13, no. 9, pp. 924-935, Sept, 2002

[4] H. O. Tan, I. Korpeoglu, "Power Efficient Data Gathering and Aggregation in Wireless Sensor Networks", *Proc. ACM Int. Conf. Management of Data (ACM SIGMOD),* vol. 32, no. 4, pp. 66-71, Dec. 2003

[5] A. Manjeshwar, Q-A. Zeng, D. P. Agarwal, "An Analytical Model for Information Retrieval in Wireless Sensor Networks using Enhanced APTEEN Protocol", *IEEE Trans. Parallel and Distributed Systems,* vol. 13, no. 12, pp. 1290-1302, Dec. 2002

[6] Wendi B. Heinzelman, Anathan P. Chandraskan, and Hari Blakrisshnan, "An Application-Specific Protocol Architecture for Wireless Microsensor Networks", *IEEE Trans. on Wireless Communications*, 1 (4): 660-670, OCT 2002

[7] K. Pahlavan and A. Levesque, *Wireless Information Networks.* New York: Wiley, 1995

[8] T. Rappaport, *Wireless Communications: Principles & Practice.* Englewood Cliffs, NJ: Prentice-Hall, 1996.

[9] Rahul C. Shah and Jan Rabaey, "Energy Aware Routing for Low Energy Ad Hoc Sensor Networks", *IEEE Wireless Communications and Networking Conf. (WCNC),* March 17-21, 2002, Orlando, FL.

[10] M. Ettus, "System capacity, latency, and power consumption in multihop-routed SS-CDMA wireless networks," in *Proc. Radio and Wireless Conf. (RAWCON),* Colorado Springs, CO, Aug. 1998, pp. 55-58.

[11] LAN MAN Standards Committee of the IEEE Computer Society, *Wireless LAN medium access control (MAC) and physical layer (PHY) specification,* IEEE, New York, NY, USA, IEEE Std 802.11-1997 edition, 1997.

[12] V. Bharghavan, A. Demers, S. Shenker, and L. Zhang, "Macaw: A media access protocol for wireless lans," in Proceedings of the ACM SIGCOMM Conference, 1994

[13] OPNET Technologies, Inc., "OPNET MODELER", http://www.opnet.com/products/modeler/home.html, 2004

High-Efficiency Power Amplifier for Wireless Sensor Networks

Devrim Aksin
Department of Electrical Engineering
University of Texas at Dallas
Dallas, Texas, USA
devrim.aksin@ieee.org

Stefano Gregori
School of Engineering
University of Guelph
Guelph, Ontario, Canada
sgregori@uoguelph.ca

Franco Maloberti
Department of Electronics
University of Pavia
Pavia, Italy
franco.maloberti@unipv.it

Abstract— We designed a high-efficiency class-E switched-mode power amplifier for a wireless networked micro-sensors system. In this system, where each sensor operates using a micro-battery, has local processing capability, and contains on the same chip integrated sensing elements and a RF transmitter, most of the power is dissipated by the transmitter. The proposed amplifier achieves 92.4% maximum drain efficiency and can vary the transmitted power between -4.2 to -0.2 dBm with almost constant efficiency. This last feature is obtained by controlling the modulation duty cycle and by switching the capacitors' values in the parallel circuit. The possibility of choosing the transmitted power depending on the distance from a base station or other sensors and on the charge level of the battery, combined with power aware network protocols, improves network lifetime, reliability, and adaptability. The amplifier is designed in 0.18 µm CMOS process and operates with nominal 1.2 V supply.

I. INTRODUCTION

The integrated circuit technology, supporting the ongoing miniaturization of sensing, processing, storage, and transmission, enables new kinds of embedded systems. These systems are distributed and deployed in environments where conditions change dynamically. They comprise a collection of devices, each acting as a node of a physical communication network collecting and disseminating a wide range of environmental data.

Low data rate sensor networks are rapidly emerging as a major new player in the information technology arena. Currently many research efforts focus on the design of small, low-cost, lightweight, fully integrated, wireless communication sensor nodes. The network lifeblood, fueling range, versatility, and reliability, comes from the available power sources and from the efficiency in their usage. Each node has a battery or an energy storage device and may have means of extracting energy from the environment (solar, vibration, temperature difference) [1] or receiving wirelessly transmitted energy from a base station [2].

This work has been supported by the National Science Foundation grant ECS-0225528.

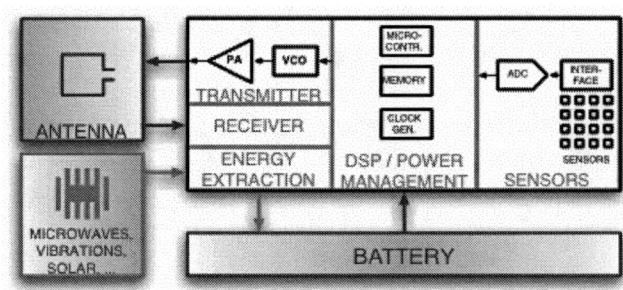

Figure 1. Block diagram of the micro-sensor.

Physical dimensions and cost of energy extraction/storage devices, limit the node power budget to a level somewhere below 100 µW to have a reasonable network lifetime [2], [3]. In nodes organized as in Fig. 1, the most power-hungry block is the transmitter, which generally uses more than 50% of the available power.

Considering power dissipation, the key component in the whole sensor node is the power amplifier (PA). The challenges faced in designing a power amplifier for wireless sensor networks are quite different with respect to its counterpart of conventional wireless transmitters (e.g. Bluetooth, Wi-Fi). Here the power constraints typically limit transmission duty cycle to less than 1%, so that occasional bursts of data pockets are sent with a low data rate (a few tens of Kbit/s) to neighboring nodes or to a base station in a few tens of meters range. Power efficiency is the most important requirement, which directly affects network range, lifetime, and reliability. In addition a power control circuit must be designed in order to change the output power for various transmission distances, variable fading conditions, and charge level of the battery. Finally, since the power amplifier must be off when idle and turned on just for short transmission bursts, the turn-on transient must be very fast to reduce the overhead power dissipation.

In this paper we present a high-efficiency class-E switched-mode power amplifier suitable for wireless networked micro-sensors systems. Switched-mode power amplifiers have a better efficiency than linear amplifiers (A, B, A/B), thus a non-linear modulation scheme (i.e. FSK) is the best choice for low-power consumption. In addition the proposed transmitter uses the 433 MHz ISM band, which gives a shorter path-loss than higher frequencies, allows more power-efficient transmitter circuits, and still provides enough bandwidth for the transmitted signal and allows the design of a relatively small antenna. The amplifier, together with the power control circuit, has been designed in 0.18μm CMOS process and operates with nominal 1.2 V supply.

The first part of the paper describes design issues and trade-offs. The following section discusses simulation results and the last section offers our concluding remarks.

II. Design Considerations

A. Approach

The schematic and circuit waveforms of a class E tuned amplifier are shown in Fig. 2. It consists of a transistor (M1), operated as a switch, an inductor L_0, large enough that the current through it can be considered almost constant, and a parallel load network (L-C-R_A-C_1). The signal V_{IN} is a square waveform generated by the preamplifier, which turns on and off M1 at the operating frequency (433 MHz). The circuit is characterized by two resonant frequencies: when M1 is on, the series-resonant circuit consists of C, L, and R_A; when M1 is off, consists of C_1, C, L, and R_A. For optimum operation (i.e. minimum power dissipated by M1), the drain voltage V_D and its derivative dV_D / dt must be zero when M1 turns on (at $\omega t = 2\pi$).

Fig. 2 shows possible current and voltage waveforms for optimum operation. This condition, for a given operating frequency, is met by setting the values of C_1, C, L, and R_A, and by choosing a proper duty cycle. A detailed analytical model for the circuit and the derivation of the steady-state waveforms are described in [4].

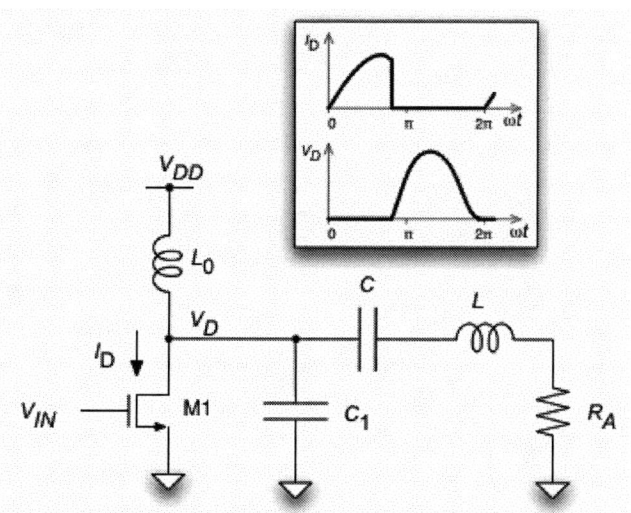

Figure 2. Power amplifier and circuit waveforms.

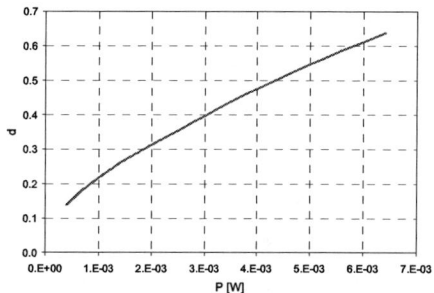

Figure 3. Duty cycle vs. output power.

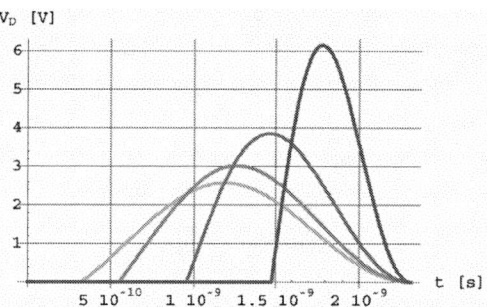

Figure 4. Voltage waveforms for variable output power levels.

The output power of the circuit is related to the waveforms, which in turn depend on V_{DD}, the operating frequency, the duty cycle, and the values of C_1, C, L, and R_A. The variation of any of these parameters affects the output power, but, in general, drives the working conditions away from the maximum efficiency point.

Changing the output power, while keeping the conditions for maximum efficiency, requires the simultaneous variation of more than one of the mentioned parameters. Parameters that can actually be changed in an integrated circuit are the duty cycle (that can be accomplished by the preamplifier, Sec. II.B), the capacitor C_1 (additional capacitors can be connected in parallel using nMOS switches), the resistor R_A (which in practical applications results in a matching network, whose input impedance depends on the values of capacitors that can be varied the same way as C_1, Sec. II.C).

The values of the parameters corresponding to any desired value of output power can be calculated starting from the equation describing the steady-state waveform for V_D and applying the optimum turn-on conditions ($V_D = 0$ and $dV_D / dt = 0$). Figs. 3, 4, and 5 present results from numerical simulations. Fig. 3 shows the required duty cycle for maximum efficiency when the output power ranges from 0.4 to 6.4 mW. Drain voltage waveforms corresponding to different output power levels (0.4, 1.2, 3.2, 6.4 mW) are plotted in Fig. 4.

The same parameters variation can also be used to maintain a constant output power while V_{DD} changes (for example because of battery discharge). Fig. 5 presents voltage waveforms for a constant output power (1.2 mW), and V_{DD} = 0.8, 1.2, and 1.6 V.

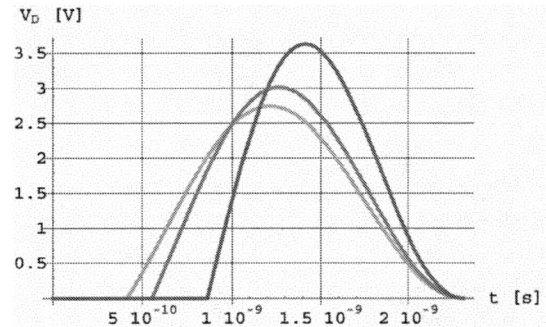

Figure 5. Voltage waveforms for constant output power and variable V_{DD}.

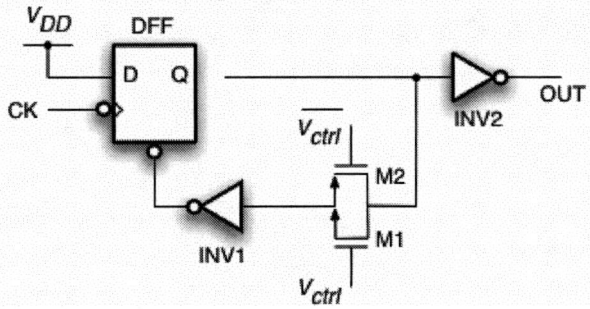

Figure 6. Duty cycle modulator (DCM).

B. Preamplifier with Duty Cycle Adjustment

The duty cycle adjustment of carrier frequency is obtained using the circuit shown in Fig. 6. The pulse is adjusted with a variable resistor created by pass transistors M1-M2 and control voltages. The circuit works as follows, whenever a rising edge occurs at the input, the output of DFF starts rising, this voltage is applied to the input of the inverter INV1 with an RC delay, determined by the on resistance of the pass transistors and the input capacitance of the inverter. Then, INV1 resets DFF. Hence the pulse width is determined by the propagation delay from DFF's Q terminal to INV1's output terminal plus DFF's reset terminal to Q terminal. Fig. 7 shows the variation of the duty cycle with respect to the applied control voltage. The transient simulation results showing the output of the duty cycle modulator is given in Fig. 8. The circuit dissipates 170 µW at 433 MHz. Although, it is not shown here, digital programming of the RC delay line is also possible. For the digital programming scheme, transistors M1 and M2 are designed as multi finger pass transistors that can be turn on or off digitally so that effective channel width of the pass transistors can be programmed.

C. Capacitively Programmable Load

In order to change the equivalent resistive load with respect to different output power (or equivalently duty cycle setting), a capacitively programmable load is designed. The whole class E power amplifier schematic is given in Fig. 9, and capacitively programmable load consists of the components C_2, C_3, L_1, and R_A. The real part of the power amplifier's load in proximity to the resonance frequency can be expressed as:

$$R_L = \frac{R_A}{1 + \left(R_A^2 L_1^2 - 2L_1 C_3\right)\omega^2 + L_1^2 C_3^2 \omega^4} \quad (1)$$

As it is clear from (1), effective load near the resonance frequency can be modified using C_3. Since, changing the value of C_3 changes also equivalent resonance frequency of the load, the value of another capacitor, i.e. C_2, is changed in correlation with the changes of C_3 so that overall resonance frequency can be set at the desired 433 MHz.

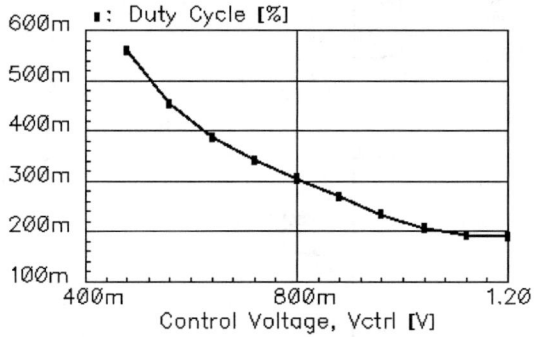

Figure 7. The variation of the duty cycle with control voltage.

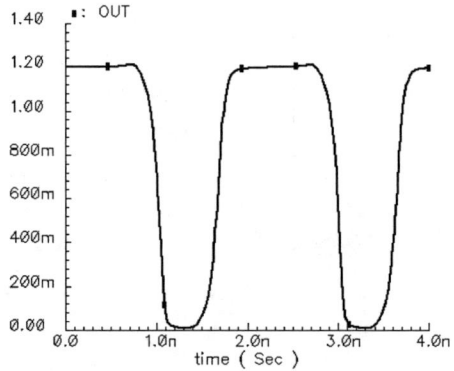

Figure 8. Transient simulation result of the duty cycle modulator.

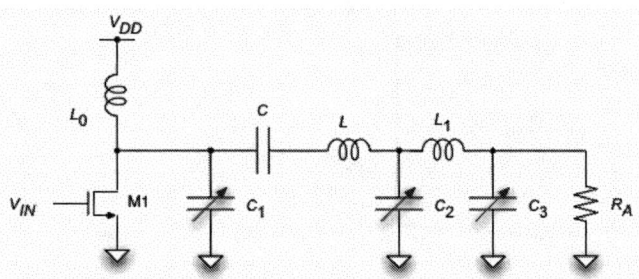

Figure 9. Circuit schematic of the power amplifier.

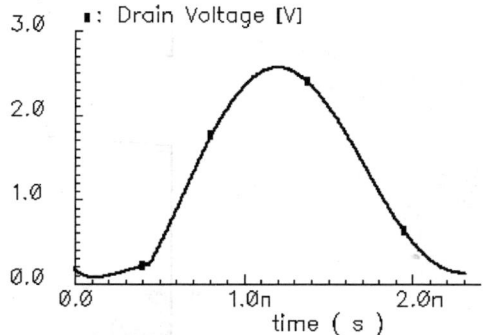

Figure 10. Drain voltage of switch transistor M1 (PSS analysis).

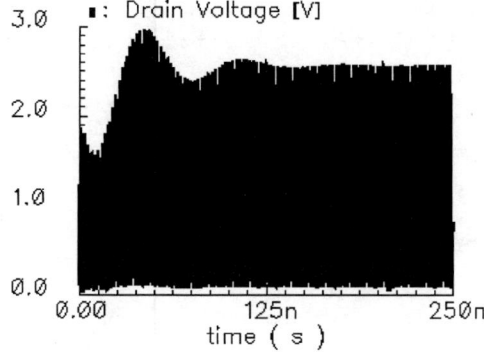

Figure 11. Drain voltage of switch transistor M1 (transient analysis).

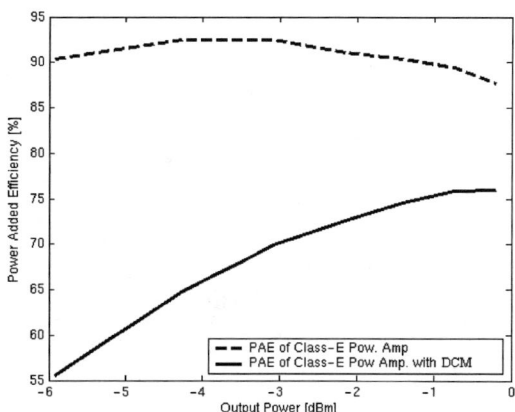

Figure 12. Power Added Efficiency (PAE) of the class-E power amplifier: a) alone; b) with preamplifier, i.e. duty cycle modulator.

III. RESULTS

To prove the validity of the proposed scheme, the class-E power amplifier shown in Fig. 9 has been extensively simulated. Table 1 shows the obtained output power and power added efficiency of the class-E power amplifier for different C_1, C_2, C_3 and duty cycle values. Fig. 10 shows the drain voltage of the switch transistor M1 for the output power level of -0.211 dBm obtained using periodic steady state analysis (PSS). Fig. 11 shows the initial transient at the switch drain terminal. The power added efficiency of the class-E amplifier, together the PAE of the whole power amplifier including the duty cycle modulator with respect to the output power is shown in Fig. 12. Class-E power amplifier shows more than 87% throughout the whole output power range and of course overall PAE decrease with the decrease of the output power because of the constant power consumption of the duty cycle modulator (DCM).

IV. CONCLUSION

Wireless sensor networks set new challenges in designing low-power communication circuits. The power available to each micro-sensor node is a vital resource for the functioning a sensor network and the transmitted power is a substantial part of it. This paper presents a high-efficiency power amplifier, which is particularly well suited for wireless networked micro-sensors systems. Simulation results shows that it keeps optimum efficiency conditions, with PAE over 87%, while delivering variable output power, from -4.262 to -0.211 dBm, without needing for additional power control mechanisms.

REFERENCES

[1] J.M. Rabaey, J. Ammer, T. Karalar, S. Li, B. Otis, M. Sheets, T. Tuan, "Picoradios for wireless sensor networks: the next challenge in ultra-low-power design," Digest of Technical Papers of the 2002 Solid-State Circuit Conference, Vol. 1, pp. 201-202.

[2] S. Gregori, Y. Li, H. Li, J. Liu and F. Maloberti, "2.45 GHz power and data transmission for a low-power autonomous sensor platform," Proceedings of the 2004 International Symposium on Low Power Electronics and Design, pp. 269-273, Aug. 2004.

[3] Y.-H. Chee, J. Rabaey, A.M. Niknejad, "A class A/B low power amplifier for wireless sensor networks," Proceedings of the 2004 International Symposium on Circuits and Systems, Vol. 4, pp. 409-412, May 2004.

[4] M. Kazimierczuk and K. Puczko, "Exact analysis of class E tuned power amplifier at any Q and switch duty cycle," IEEE Trans. on Circuits and Systems, Vol. 34, Issue 2, pp. 149-159, Feb. 1987.

TABLE I. OUTPUT POWER AND POWER ADDED EFFICIENCY FOR DIFFERENT COMPONENT VALUES AND DUTY CYCLE

Duty Cycle	L [H]	C [F]	C_1 [F]	L_1 [H]	C_2 [F]	C_3 [F]	Output Power [dBm]	Power Added Efficiency [%]
0.178	1.00E-07	6.50E-11	1.25E-12	9.19E-09	2.16E-11	2.48E-11	-4.262	92.46
0.205	1.00E-07	6.50E-11	1.20E-12	9.19E-09	2.20E-11	2.81E-11	-3.075	92.44
0.230	1.00E-07	6.50E-11	1.15E-12	9.19E-09	2.20E-11	3.05E-11	-2.153	91.10
0.253	1.00E-07	6.50E-11	1.10E-12	9.19E-09	2.19E-11	3.24E-11	-1.407	90.40
0.275	1.00E-07	6.50E-11	1.05E-12	9.19E-09	2.18E-11	3.41E-11	-0.741	89.42
0.296	1.00E-07	6.50E-11	1.00E-12	9.19E-09	2.16E-11	3.56E-11	-0.211	87.94

A new switched capacitor circuit for parallel-pixel image processing

N.Massari, N.Viarani
ITC-irst, via Sommarive 16,
I-38050 Trento, Italy, Tel. (+39) 0461 314510
massari@itc.it; nviarani@itc.it

M.Gottardi
ITC-irst, via Sommarive 16,
I-38050 Trento, Italy, Tel. (+39) 0461 314535
gottardi@itc.it

Abstract— A new Switched Capacitor (SC) circuit for image processing is presented. Optimized area occupancy and power consumption allow the integration of the proposed circuit into each pixel, in order to implement high speed pixel-parallel processing. The operation performed by the circuit consists of the accumulation of the absolute value of the difference between two voltage inputs, which is the base of a large set of image processing algorithms. Two circuit configurations have been developed and compared in this paper: a basic implementation and a second improved solution with higher sensitivity and linearity over the full output range. The improved pixel consists of 10 transistors and 3 capacitances and provides an output voltage range of 1.8V with 3.3V power supply. A test chip was fabricated in 0.35μm double-poly triple-metal CMOS process. The resulting 35μm square pixel has a power consumption 3μW @ 3.3V. The single absolute difference and accumulation is executed in 2μs, which turns into a computing figure of *1.2 GOPS/mm^2* and *1.3 TOPS/W*.

I. INTRODUCTION

The demand to extract real-time information from images has led towards the design of vision sensors with integrated signal processing. In some cases analogue processing, embedded in the focal plane array, can perform many operations with lower power consumption and better area efficiency than its digital counterpart. Moreover, many low-level tasks do not require high accuracy but need massive parallel processing to be efficiently implemented. The advantages mentioned above are of great importance for image filters that require pixel-parallel operations over small kernels. In particular, pixel-level architectures can drastically reduce both the computational time and the amount of data to be read out of the sensor array. Analog pixel-level approach has been deeply investigated recently [1]-[7] and has brought to two main types of pixel topologies. The first one is oriented toward the execution of few simple operations [1], [2] or very specialized algorithms with poor programmability [3]-[5]; the second is based on a general purpose processing block, called Analog Processing Elements (APE) [6],[7]. This latter is typically implemented in current-mode and requires a quite complicated hardware, related to the level of programmability.

The present work describes a novel SC circuit to be used as analog processing block embedded in each pixel of a digitally programmable vision sensor with filtering capabilities. A large class of signal processing algorithms, where only two operations are used (accumulation and absolute difference), can be executed. The assignment of the filter coefficients is accomplished by means of a pulse-based technique, which can be easily implemented with a proper timing control.

In the next section the types of programmable image filtering, executed by the presented circuit, are explained. In Section III circuit description and simulation results are presented. Two different implementations have been considered: a first basic realization of the main circuit and an improved version of it, provided with better sensitivity. In Section IV, experimental results are reported for both versions of the circuit and compared with the simulated behaviors.

II. ALGORITHMS

Some image filters can be realized by implementing simple algorithms, with a good tradeoff between computational ability and filter complexity. In particular, it is possible to define a specific class of image processing algorithms suitable to be integrated at sensor level. They can be implemented in analog domain with a small number of transistors and low power consumption. The class of algorithms proposed in this work is based on the following two main operations: 1) extraction of the absolute value of the voltage difference between some signals; 2) accumulation of the result provided by operation 1). These operations are implemented at pixel level with small area occupancy. A proper switched capacitor (SC) circuit with two analog memories (MA and MB) placed into each pixel, performs both image acquisition and signal processing. Moreover, the SC circuit is suitable for implementing pixels interconnections. For a given 3x3 pixel kernel, (1) shows the pre-processing computation executed by the circuit of pixel P_0, based on the two operations mentioned above:

$$V_{G0} = c_0 |V_{P0B}(T_{-1}) - V_{P0A}(T_0)| + \sum_{k=1}^{8} c_k |V_{Pk}(T_0) - V_{P0}(T_0)| \quad (1)$$

where the voltages $V_{P0A}(T_0)$ and $V_{P0B}(T_{-1})$ represent the values of the pixel P_0 stored in the two memories at times T_0 and T_{-1}, respectively; V_{Pk} is the k-th value among of the 8 neighboring pixels and c_0 and c_k are integer coefficients.

A. Spatial Gradient

A kernel of 3x3 pixels constitutes a good tradeoff between sensor architecture complexity and filter resolution. The image spatial gradient at time T_0 is extracted from (1), letting $c_0=0$:

$$V_{G0} = \sum_{k=1}^{8} c_k \cdot |V_{Pk}(T_0) - V_{P0}(T_0)|. \quad (2)$$

A great flexibility in gradient computation can be obtained by choosing proper values of the integer coefficients c_k. In fact, in (2), the k-th gradient component can be neglected, by setting $c_k=0$, or can be emphasized, by choosing $c_k>1$. By doing this, directional spatial filtering can be carried out over all of the four main directions allowed by the 3x3 pixel kernel (0°, 90°, ±45°).

B. Temporal Gradient

Using the two memories, temporal gradient can be computed between two successive frames:

$$V_{G0} = c_0 \cdot |V_{P0B}(T_{-1}) - V_{P0A}(T_0)|; \quad (3)$$

where $V_{P0A}(T_0)$ and $V_{P0B}(T_{-1})$ are the values stored in the two memories at time T_0 and T_{-1} respectively, with $c_0 \geq 1$. Exploiting (3), real-time motion detection is computed by image subtraction. In this case, no information is provided related to the direction of motion, due to the absolute value operator. If (3) is used sampling the pixel value right after the reset (T_{-1}) and at the end of integration time (T_0), the readout with kTC noise reduction of the photosensor is performed.

III. CIRCUIT DESCRIPTION

The pixel (shown in Fig. 1) consists of a photodiode, two embedded memories and a SC circuit whose input can be connected to one of the 8 neighbor pixels, one of the 2 memories, or the photodiode itself. This last configuration is used to integrate the light. The voltage of MA is available to the 8 neighboring pixels, implementing the 3x3 pixel kernel connectivity. The SC circuit is intended to work in two modes: a) photodiode readout mode, where the photogenerated voltage (V_{out}) is amplified and sampled in MA or MB; b) signal processing mode, where V_{in} is switched between two inputs as many times as needed, performing the operations as shown in (1). Core of the pixel is the SC circuit. Two versions of it have been developed: the first one is the basic implementation while the second one is an improved version of the former and uses a few more transistors to achieve better performance in terms of circuit sensitivity and linearity. In both versions, the inverting amplifier consists of a class AB amplifier [8] with a DC voltage gain of 57 dB and a power consumption of 2 µW.

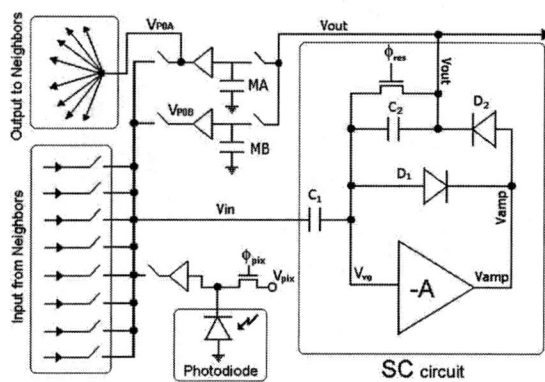

Figure 1. Pixel schematic.

A. Basic circuit

The schematic of the basic circuit is shown in Fig. 2. The inverting SC gain stage uses diode D_2 as blocking component, implementing the absolute value and the signal accumulation operations. For a negative input voltage variation ($\Delta V_{in} < 0$), V_{out} increases supplied by the amplifier output V_{amp}. During positive input voltage variations ($\Delta V_{in} > 0$), the signal path (C_2, D_2) is turned off, due to the blocking diode (D_2), while D_1 maintains the feedback path, avoiding the amplifier to saturate.

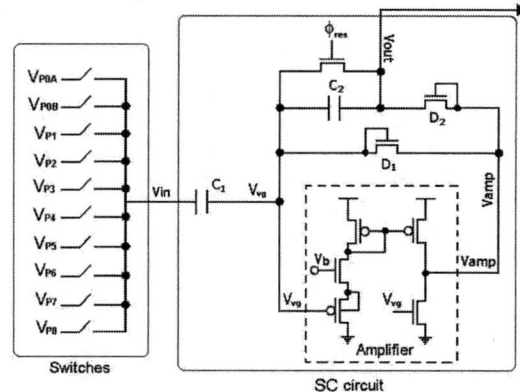

Figure 2. Schematic of basic configuration.

The computation of (3), for example, is executed by changing the input of the SC circuit (V_{in}) over the two voltages stored in MA (V_{P0A}) and MB (V_{P0B}). The proper operation is computed by switching V_{in} from MA to MB and to MA again on order to guarantee a negative input voltage variation. Only in this case the absolute value is accumulated on V_{out}. Repeating the same procedure N times, the assigned integer coefficient c_0 is set to N:

$$V_{G0} = N\frac{C_2}{C_1}|V_{P0A} - V_{P0B}|. \quad (4)$$

The circuit was simulated using a sinusoidal voltage input (V_{in}) with 75mV amplitude and 400kHz frequency. The simulation results are reported in Fig. 3.

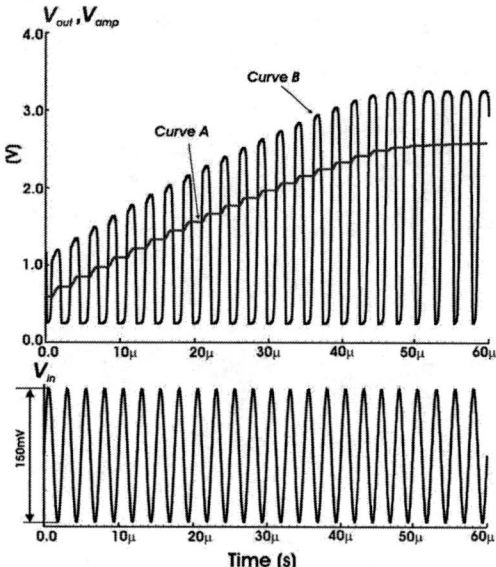

Figure 3. Basic circuit simulation results.

Curve A represents the output node of the circuit (V_{out}). During the negative input voltage variations, the circuit works as a pure inverting amplifier ($\Delta V_{out} \approx -\Delta V_{in}$), being $C_1 = C_2$. During the positive input transition, the feedback (C_2, D_2) is interrupted so that the output voltage (V_{out}) is held on C_2. Repeating the operation more than once, the related output voltage will be accumulated to the output node (V_{out}), thanks to the blocking diode (D_2) which prevents C_2 from discharging, until saturation level. Curve B, representing the output of the amplifier (V_{amp}), changes between two values: the lowest one is always the same because it is set by the feedback D_1 during positive input variations; the second strongly depends on the output signal, held on C_2, in order to guarantee the signal path through feedback (C_2, D_2). It can be noted that, the larger the output signal (V_{out}), the larger the V_{amp} range the amplifier has to provide to lock the two different feedback paths. This turns into a signal loss, due to the limited amplifier slew-rate and gain, which is directly proportional to the output level, as shown in (5) and (6). Indeed, during a negative input transition, the amplifier output voltage is forced to:

$$V'_{amp} \cong V_{out} + V_{T2}; \quad (5)$$

where V_{T2} is the D_2 threshold voltage. During a positive input variation the feedback (D_1) is guaranteed setting V_{amp} down to:

$$V''_{amp} \cong V_{vg} - V_{T1}; \quad (6)$$

where V_{vg} is the amplifier virtual ground.

B. Improved circuit

The second circuit, shown in Fig.4, uses an additional capacitor (C_3) and two more transistors with respect to the original version.

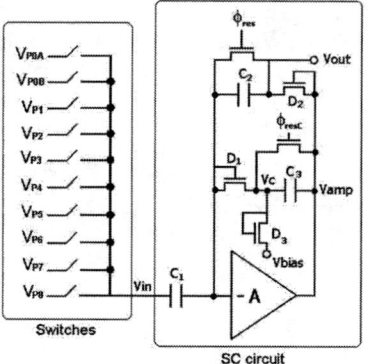

Figure 4. Schematic of the improved circuit.

The new topology forces the V_{amp} variations to be independent with respect to the output signal (V_{out}). In this way, the signal loss is much smaller and better circuit sensitivity is obtained with the same amplifier (A). C_3 is inserted in the feedback path of D_1, while the diode-connected transistor D_3 works as a voltage clamp for V_c. During a negative input transition, V_{amp} increases, turning on the feedback (C_2, D_2). At the same time, V_c tries to follow V_{amp}, but it is clamped at ($V_{bias} + V_{gs3}$) by D_3. The voltage range of V_c is set by the voltage drop between D_1 and D_3. By doing this, during a positive input variation, the feedback path (D_1, C_3) is immediately activated, thanks to the voltage difference pre-charged at C_3 in the previous phase. V_{bias} has to be set carefully in order to force D_1 into small conduction so that the feedback can be inserted almost instantaneously, minimizing the signal loss.

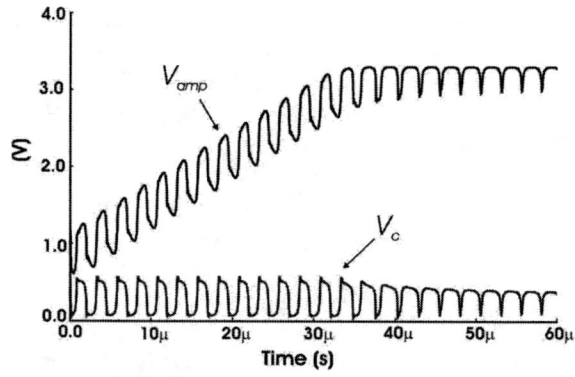

Figure 5. Simulated V_{amp} and V_c for a 400 KHz sinusoidal input voltage with 75 mV amplitude.

Fig. 5 shows the simulation result of the circuit with respect to a 400 kHz sinusoidal input voltage with 75 mV amplitude. It can be noted that V_{amp} follows the output signal V_{out} with a constant negative voltage variation:

$$\Delta V_{amp} \cong (V_{bias} + V_{gs3}) - (V_{vg} - V_{gs1}).$$

Fig. 6 shows a simulated comparison between the outputs of the two implemented circuits for the same input signal. It can be observed that *Curve A* performs a much better linearity with respect to *Curve B*, representing the output of the original circuit. This means that, in the improved version, the signal loss is greatly reduced within the whole signal range.

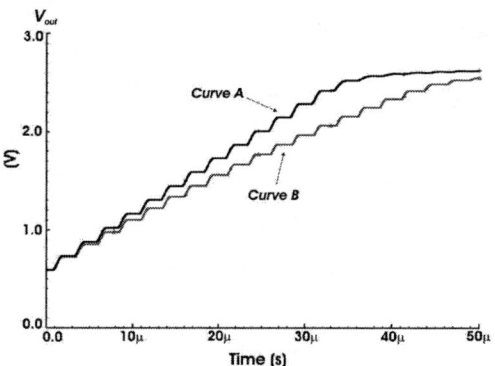

Figure 6. Comparison of simulated output V_{out} between the basic (*Curve B*) and improved circuit (*Curve A*).

IV. EXPERIMENTAL RESULTS

The two test circuits have been fabricated in 0.35μm double-poly triple-metal CMOS process and electrically tested. They provide an output voltage range of 1.8V with a power supply of 3.3V with a total power consumption of 3μW @ 3.3 V, which is mainly due to the amplifier bias current. The pixel pitch is 35 μm and fill factor is 20%, which can be considered a good tradeoff between resolution and elaboration's capability.

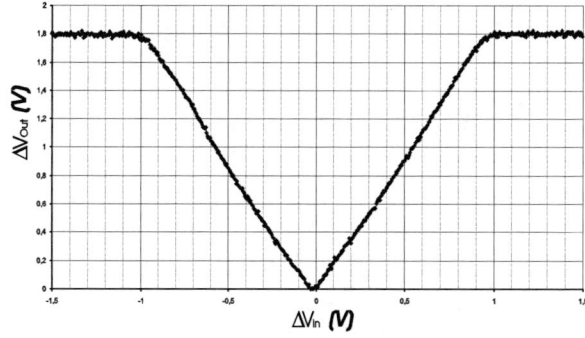

Figure 7. Absolute value of the input voltage difference.

Fig. 7 shows the DC characteristic of the original circuit, implementing the absolute value of the input voltage difference. Fig. 8 shows the outputs of both circuits for a square input signal of 50mV amplitude. The signal polarities are inverted with respect to the simulation due to the inverting read-out circuit. It can be noted that the improved circuit (voltage ramp on the right) performs a better linearity along the entire output voltage range, while the basic version (voltage ramp on the left) provides a visible signal loss which is strongly dependent on the output signal (V_{out}), in agreement with the simulation results.

The single operation of absolute value of difference is executed in 2μs, which turns into a computing figure of *1.2 GOPS/mm^2* and *1.3 TOPS/W*.

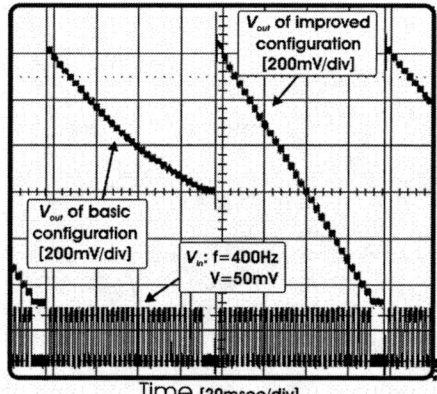

Figure 8. Comparison between original (left voltage ramp) and improved (right voltage ramp) circuit.

V. CONCLUSIONS

Two types of circuits, to be used as signal processing blocks suitable for CMOS image sensors, have been realized and electrically tested. They implement the accumulation of the absolute value of the input voltage difference, which is the basic operation for a large class of image processing algorithms. The two devices have been integrated in a 0.35μm CMOS technology. Both circuits have been successfully tested and the electrical measurements are in agreement with the simulation results.

REFERENCES

[1] Yoshinori Muramatsu, Susumu Kurosawa et al., "A signal-processing CMOS image sensor using a simple analog operation", *IEEE Journal of Solid State Circuits*, Vol. 38, No. 1, pp. 101-106, January 2003.

[2] Shyh-Yih Ma and Liang-Gee Chen, "A single-chip CMOS APS with direct frame difference output," *IEEE Journal of Solid State Circuits*, Vol. 34, No. 10, pp. 1415-1418, October 1999.

[3] D.Stoppa, A.Simoni, D.Gonzo, M.Gottardi and G.F.Dalla Betta, "Novel CMOS Image Sensor with a 138-dB Dynamic Range", *IEEE J. of Solid State Circuit*, Vol. 37, No 12, December 2002.

[4] P.-F. Ruedi, "A 128x128 pixel 120 dB dynamic range vision sensor chip for image contrast and orientation extraction" *IEEE Journal of Solid State Circuits*, vol.38, No 12, 00. 2325-2333, December 2003.

[5] Massimo Barbaro et al., "A 100x100 pixel silicon retina for gradient extraction with steering filter capabilities and temporal output coding", *IEEE Journal of Solid State Circuits*, Vol. 37, No. 2, February 2002.

[6] G.Linan Cembrano, et al., "A 1000 FPS at 128x128 Vision Processor with 8-bit digitez I/O" *IEEE Journal of Solid State Circuits*, Vol. 39, No. 7, pp. 1044-1055, July 2004.

[7] P Dudek and P.J. Hicks, "A general-purpose CMOS Vision chip with a Processor-per-Pixel SIMD array" Proceedings of ESSCIRC 2001, Villach, Austria, pp. 782-785, September 2001.

[8] J.N. Babanezahad, and R. Gregorian "A programmable Gain/Loss Circuit", *IEEE Journal of Solid State Circuits*, vol.22, No 6, pp 1082-1090, December 1987.

Broadband Dielectric Spectroscopy CMOS Readout Circuit for Molecular Sensing

Youngbok Kim[*], Anuj Agarwal[*], S. R. Sonkusale[+]

[*]Department of Electrical Engineering, Texas A&M University, College Station, TX
[+]Department of Electrical and Computer Engineering, Tufts University, Medford, MA

ABSTRACT

The readout circuit of Broadband Dielectric Spectroscopy (BDS) for molecular sensing is described. The circuit performs a high-precision capacitance-to-current conversion and amplification of the difference of signals generated by two very small capacitors in the conventional CMOS 0.18um technology. We make use of the continuous-time common mode error amplification and offset–canceling low-noise lock-in architecture to measure the admittance as a function of frequency. The power consumption of the front-end is under 30uW/channel. It has quasi-linear conversion ratio of 164pA/aF.

1. Introduction

Nowadays, BDS is widely used for investigating the characteristics of a great number of materials. If the BDS devices can be implemented in small size and with low power consumption, it can work as a chemical detector in a portable environment for biomedical and remote sensing applications. However, previously published BDS devices are too bulky and immobile to operate for these applications [1].

BDS determines the complex dielectric constant (permittivity) versus frequency. If the size of the probe-capacitor is at the nano scale, BDS study of single organic molecules becomes possible. Thus single-molecule detection and recognition can be achieved. However, the available capacitive measuring techniques [2]-[7] are not appropriate for sensing the change of the dielectric permittivity in the high frequency range for BDS measurement. Even though C. Baltes [8] reported a single-chip gas detection system using the dielectric constant change, it was not adaptable for high frequency application, such as the Ghz range, which is the important regime of BDS. Recently, Sarpeshkar [9] used synchronous demodulation technique adapted in a low noise lock-in architecture for capacitive sensing. His technique has shown some potential for implementing the BDS on the silicon chip because it can vary the frequency of the probing signal. But, the achieved modulating frequency is not high enough for the BDS frequency range and is not designed for low power application. Also, for small-size sensing devices, the sensed capacitor should be minimized and integrated with the sensing circuitry. Even though the parasitic capacitance and mismatch of the circuitry can cause a serious offset in the output signal, if we integrate a great number of sensing nodes, we can achieve an accurate sensing result through averaging and redundancy. Therefore, in this paper, wide-band, low-power, and low-voltage capacitive sensing circuit for measuring the permittivity variations at the nano-size [10] capacitor with tens of femto farads is demonstrated with the conventional 0.18 μm CMOS technology. The current-mode output of the device is designed to be fed to a low-power, current-mode ADC.

2. Review of Broadband Dielectric Spectroscopy

Excitations of single molecules with electromagnetic fields of the proper strength and frequency can result in transitions between electronic, vibrational and rotational energy states. These kinds of dielectric responses can be observed by high frequency spectroscopy. These phenomena are often due to dipole relaxation effects arising from the reorientation motions of molecular dipoles and electrical conduction arising from the translational motions of electrical charges.

$$\varepsilon = \varepsilon^* - j\varepsilon^{**} \qquad (2\text{-}1)$$

$$\varepsilon^* = \varepsilon_\infty + \frac{\varepsilon_s - \varepsilon_\infty}{1 + (\omega\tau)^2} \qquad (2\text{-}2)$$

$$\varepsilon^{**} = \frac{\omega\tau(\varepsilon_s - \varepsilon_\infty)}{1 + \omega^2\tau^2} + \frac{\sigma_{dc}}{\omega\varepsilon_0} \qquad (2\text{-}3)$$

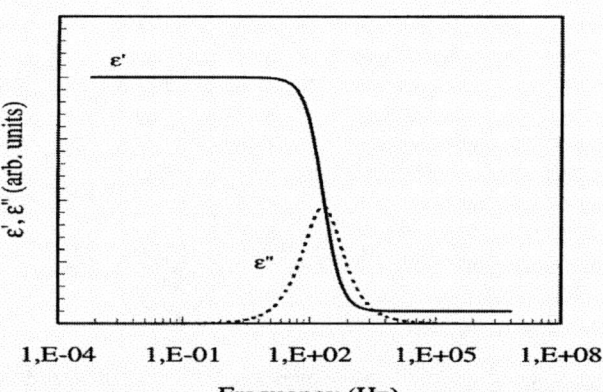

Fig 1. Sketch of the frequency variation of the real and imaginary parts of the complex relative permittivity of the material [14]. (e' : real part e'' : imaginary part)

Therefore, these effects can dominantly influence the dielectric properties of the material in the relevant frequency range. At the microelectronics side, the dielectric properties of the material are expressed by the complex electrical impedance. The underlying effects mentioned above and their impact on the impedance can be modeled by a couple of standard models such as Debye, Cole-Cole and Cole-Davidson models [1].

The classical physical behavior of the frequency dependence of the real and imaginary parts of the permittivity was described by Debye [1]. His model can be summarized by (2-1), (2-2) and (2-3), where ε_∞ is the absolute value of the permittivity at infinite frequency, ε_s is the absolute value of the permittivity at zero frequency, ε_0 is the vacuums permittivity, τ is a characteristic relaxation time and σ_{dc} is the DC conductivity of the system. Fig 1 shows us the graphical way of expression of the frequency-dependence of the permittivity of the measured material according to (2-1), (2-2) and (2-3) [1].

3. Review of the lock-in capacitive sensing

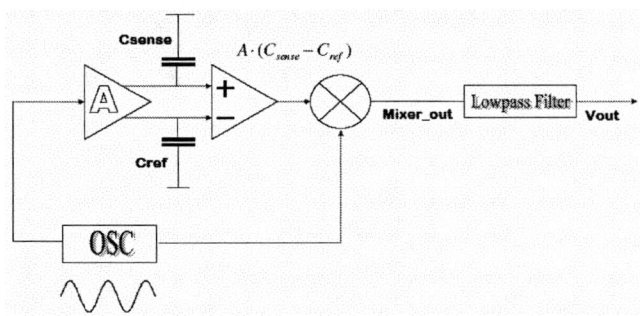

Fig 2. The conceptual diagram on the lock-in capacitive sensing

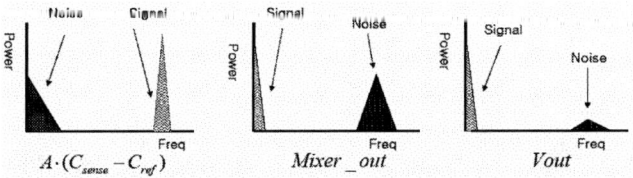

Fig 3. The spectral representation of the lock-in capacitive sensing system

The lock-in technique is very useful for sensing the small amplitude signal. It is because it can attenuate the low frequency error and noise signal such as 1/f noise and other offset signal generated in the device [9].

The Fig 2 and Fig 3 show us a novel common mode lock-in technique for the capacitive sensing. Fig 2 demonstrates its systematic implementation and Fig 3 shows the spectral representation of the system per one stage. First of all, the carrier signal from the same sources applied to two capacitors, the reference capacitor C_{ref} and the sense capacitor to be measured, C_{sense}. The mismatch between two capacitors is modulated by the applied carrier signal, generating a signal quasi-linearly proportional to C_{ref}-C_{sense}. After this step, the signal is demodulated by the same carrier signal in the mixer. Mixing causes the low frequency noise and offset signal to be transferred to the high frequency range and the high frequency information signal at the carrier frequency is moved into the low frequency range. Mixer_out signal is created at the output of the mixer. Finally, by making use of a low pass filter on Mixer_out, we can dramatically attenuate the originally low frequency noise and capture the mismatch information signal now available at the low frequency range. Changing the frequency of the carrier enables us to perform the measurement at different frequencies which is a central need for BDS measurements.

4. Analog System and Building Blocks

4.1 Conceptual System

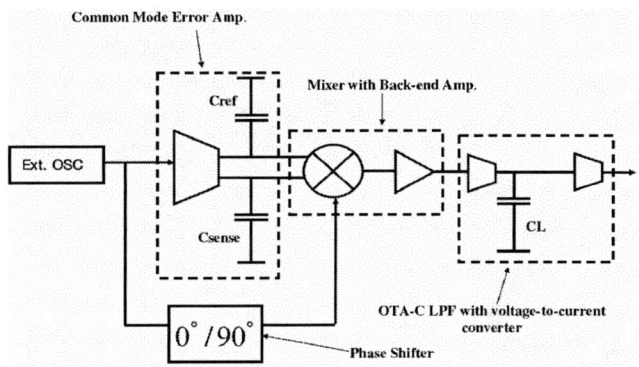

Fig 4. The conceptual diagram of the front-end circuit system

The conceptual block diagram of the sensor front-end for BDS measurement is shown in Fig 4. First of all, the mismatch between the reference capacitor and the sensed capacitor is modulated and amplified using a common mode error amplifier as a capacitance detector in the first stage. Next, the modulated signal is demodulated at the mixer stage. The phase shifter should be applied between the oscillator and the mixer input to sense both of the amplitude change and the phase change by use of quadrature modulation. Since the demodulated signal has very weak signal power, we need an additional gain stage of a conventional OTA. Finally, the demodulated signal is low-pass filtered with the OTA-C filter and converted into current. The current domain output can then be fed to a low-power current-mode ADC for digitization [11, 15].

4.2 Common Mode Error Amplifier

Fig 5 shows the implementation of the Common Mode Error Amplifier (CMEA). To amplify the common mode error, the conventional differential gain stage is used. The input V_{RF} to the CMEA is the modulating carrier signal of high frequency. Since each common mode output caused by the common mode input has the different value due to the capacitive mismatch, CMEA can be used as a capacitance detector to detect the difference between a small reference capacitor C_{ref} and the sense capacitor C_{sense} (with molecules to be detected as dielectric) as modulated common mode error signal at a non-zero frequency. Also, because the effects of the parasitic capacitance are cancelled at the output stage, the circuit can work at the high frequency range. In addition, the simple structure of the circuit has the merit of low power implementation. The additional common mode feedback circuit is attached to stabilize the output common mode signal. The output function of the amplifier is formulated in (4-1) with the capacitive mismatch value in its amplitude. Therefore, it is shown that the capacitance mismatch value is included in the amplitude value. By sensing the value, we can detect the change of the capacitance.

$$V_{out}(s) = \frac{g_{m1,2}R_{out}^2 \Delta C \cdot s}{(1+2g_{m1,2}r_{o3})(1+s/\omega_1)(1+s/\omega_2)} \quad (4\text{-}1)$$

$$\omega_1 = \frac{1}{R_{out}(C_{sense}+C_p)}, \quad \omega_2 = \frac{1}{R_{out}(C_{ref}+C_p)}$$

R_{out} value is the output resistance of CMEA and $gm_{1,2}$ is the transconductance value of the driving transistors M1 and M2. Also, r_{o3} means the output resistance of M3 and C_p is the symbol for the parasitic capacitance around the output port. Finally, ΔC means C_{sense}-C_{ref} value.

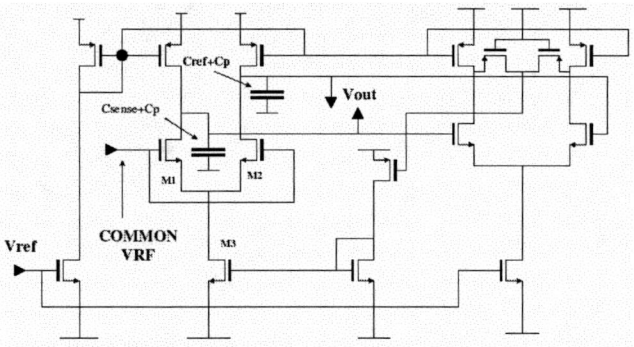

Fig 5. The common mode error amplifier with CMFB

4.3 Mixer with the back-end amplifier

The output signal from CMEA is demodulated at the single balanced mixer and amplified at the back-end amplifier.

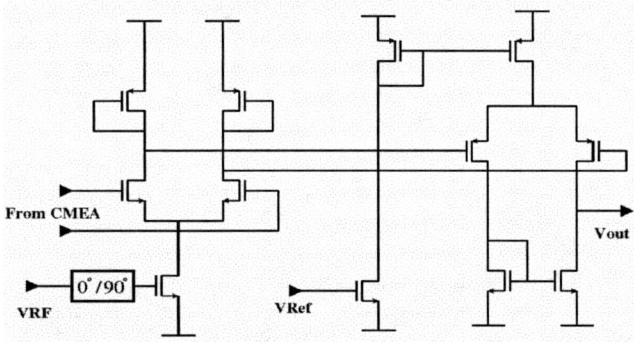

Fig 6. The Mixer stage with the back-end amplifier

Through this stage, the low frequency noise signal is moved into the high frequency range around the carrier frequency. Also, the modulated signal of the capacitance mismatch is transferred into the low frequency range for future low-pass filtering. Finally, the conventional OTA amplifies the demodulated signal.

4.4 OTA-C LPF with the voltage-current converter

Since the major signal power to be sensed is concentrated in the low frequency range, we need a very low cut-off frequency low-pass filter to capture the signal power and cancel the noise component present in the high frequency range.

The current splitting technique [12] is used to realize the OTA with a very small transconductance to implement OTA-C low pass filter with very low cut-off frequency. The low pass filter is made up as a simple 1st order OTA-C filter since the noise signal resides at high frequency range and there is no need for higher order filters. Fig 7 shows the circuit implementation of the filter and voltage-to-current converter. Additionally, the conventional input stage of the current mode ADC [13] is shown in the figure to understand the output interface requirement of the BDS front-end.

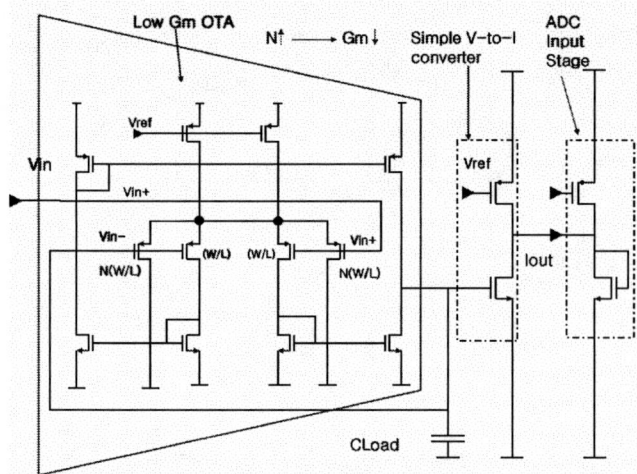

Fig 7. OTA-C filter and voltage-to-current converter(N=10)

5. Simulation Results

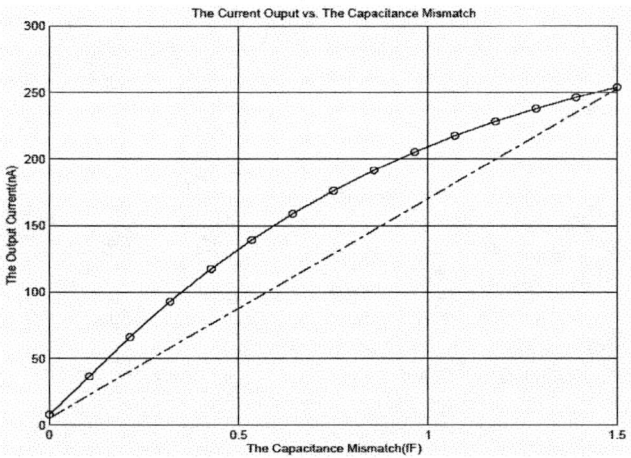

Fig 8. The output current vs. the capacitance mismatch plot for 1 GHz 20mV amplitude input V_{RF} signal

The overall circuit was simulated with the conventional current-mode ADC input section [11, 13, 15] shown in Fig 7. The reference capacitor and the sensed capacitor have very small capacitor value around few femto farads. It is because this circuit was designed for sensing the very small capacitance of a nano-size silicon structure used for chemical sensing using innovative dense integration technique of the sensed capacitor and the signal sensing circuitry together on the same chip for portable and medical applications. Due to the recent advancement in semiconductor technology, such a small capacitor can be realized on the silicon with the conventional lithography [10]. The circuit was designed with the conventional 0.18um CMOS technology and 0.65V power supply. Fig. 8 demonstrates that this circuit has the quasi-linear relationship between the output current and the capacitance mismatch change. The dotted line shows that the quasi-linear conversion ratio is about 164pA/aF and the overall power consumption is about 30uW with 1 GHz modulating signal.

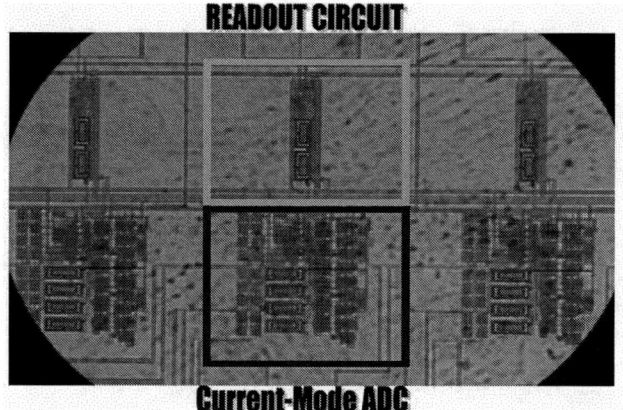

Fig 9. The fabricated chip layout (Upper box means the readout circuit)

6. CONCLUSION

In this paper, the low voltage and low power capacitance sensor readout circuit has been designed for BDS application in the portable and medical environment. It has the current-mode output as the input signal to the low voltage, low power current-mode ADC. Also, to get rid of the low frequency error noise, the lock-in architecture has been used. To realize the architecture, CMEA, mixer and low cut-off frequency OTA-C filter have been applied. Also, to minimize the power consumption, the circuits have been designed as simple as possible. In the simulation results, it shows the quasi-linear conversion ratio of 164pA/aF. Fig 9 shows the chip layout fabricated with TSMC 0.18um CMOS technology. After post-fabrication process which connects the nano-size capacitor to the circuitry on the wafer, the chip will be tested soon.

7. ACKNOWLEDGEMENT

The authors would like to thank Dr. Kish for his special support and helpful discussion about Broadband Dielectric Spectroscopy.

8. REFERENCES

[1] Friedrich Kremer and Andreas Schönhals *Broadband Dielectric Spectroscopy*. Springer, Berlin, Germany, 2003
[2] Bramani, M., "A high sensitivity measuring technique for capacitive sensor transducer", *IEEE, Trans. On Industrial electronics*, vol.37, N0.6. pp 584-586, Dec. 1990.
[3] Toth, F.N., Meijer, G.C.M; and Kerkvliet, H.M.M., "A very accurate system for multielectrode capacitive sensors", IEEE. Trans. On Instrumentation and Measurement, vol.45, No.2, pp531-535, April 1996.
[4] Goes, F.M.L.V., and Meijer, G.C.M., "A novel low-cost capacitive sensor interface", IEEE. Trans. On Instrumentation and Measurement, vol.45, No.2, pp 536-540, April, 1996.
[5] Mochizuki, K., Masuda, T., and Watanabe, K.,"An Interface circuit for high-accuracy signal processing of differential capacitance transducer". IEEE Instrumentation and Measurement Technology Conference, Brussels, Belgium, pp. 1200-1204, June 4-6, 1996.
[6] Kung, J.T., Mills, R.N., Lee, H.S., "Digital cancellation of noise and offset for capacitive sensors", IEEE Trans. On Instrumentation and Measurement, vol. 42, No.5, pp. 939-942, Oct. 1993.
[7] S. Ranganathan, M. Inerfield, S. Roy, and S.L Garverick, "Sub-Femtofarad Capacitive Sensing for Microfabricated Transducers Using Correlated Double Sampling and Delta Modulation", IEEE Trans. Circuits Sys. –II, vol. 47(11), pp. 1170-76, Nov. 2000.
[8] Christoph Hagleitner, Dirk Lange, Andreas Hierlemann, Oliver Brand and Henry Baltes, "CMOS Single-Chip Gas Detection System Comprising Capacitive, Calorimetric and Mass-Sensitive Microsensors", IEEE J. Solid-State Circuits, vol. 37, No. 12, Dec. 2002.
[9] M. Tavakoli and R. Sarpeshkar, "An offset-canceling low-noise lock-in architecture for capacitive sensing", J. Solid-State Circuits, Vol.38, No.2, Feb. 2003, pp. 244-253.
[10] S. Hashioka, H. Matsumura, "10 nm size fabrication of semiconductor substrate and metal thin lines by conventional photolithography", Microprocess and Nanotechnology Conference, 2000 International, 11-13 July,2000, pp. 184-185.
[11] Chih-Cheng Chen, Chung-Yu Wu, "Design Techniques for 1.5-V Low-Power CMOS Current-Mode Cyclic Analog-to-Digital Converters", IEEE Trans. On Circuits and Systems—II: Analog and Digital Signal Processing, Vol. 45, No.1, Jan. 1998.
[12] A. Veeravalli, E. Sanchez-Sinencio and J. Silva-Martinez," Transconductance Amplifiers with Very Small Transconductance: A Comparative Design Approach" Solid-State Circuits, IEEE Journal of , Volume: 37 , Issue: 6, June,2002 Pages:770 – 775
[13] David G. Nairn, C. Andre T. Salama, "Current-Mode Algorithmic Analog-to-Digital Converters", J. Solid-State Circuits, Vol. 25, No.4, August 1990.
[14] P.Q. Mantas, "Dielectric Response of Materials: Extension to the Debye Model", J. the European Ceramic Society 19 (1999) 2079-2086.
[15] Sameer Sonkusale, Mosong Cheng, Laszlo Kish, Anuj Agarwal, Youngbok Kim, Chao Liu, Sungkyu Seo, "A CMOS Sensor Array IC of Nanowell Devices for Molecular Sensing", Third IEEE conference on Sensors, IEEE Sensors 2004, Oct.24-27, 2004

Read-Out circuit in RT-Fluxgate

S. Baglio, V. Sacco
DIEES
Università degli studi di Catania
Catania, Italy
vincenzo.sacco@diees.unict.it

A. Bulsara
SPAWAR
San Diego, U.S.
bulsara@spawar.navy.mil

P. Nouet
LIRMM/UMR CNRS
Université Montpellier II
Montpellier, France
nouet@lirmm.fr

Abstract— The Residence Time Fluxgate magnetometer approach is based on the characterization of the time spent by the core magnetization in each of the two stable equilibrium states of the potential function, that underpins the dynamic of a typical ferromagnetic core, when a (carefully configured) periodic excitation signal is applied. This time domain technique presents several interesting features and, in particular, it can be implemented by using bias signals having lower amplitude and frequency than those used in conventional fluxgate processing schemes. In this paper we present the RT-Fluxgate magnetometer and the design of an Analog Digital conditioning circuit for reading out the magnetic filed information carried on spikes time position in the output waveform.

I. INTRODUCTION

Fluxgate magnetometers have always been of interest to the technical and scientific communities as practical and convenient sensors for weak magnetic field measurements [1,2] at room temperature, finding applicability in fields such as non destructive testing, geophysical exploration and mapping, as well as assorted military applications. Recently, however, the possibilities offered by new technologies, materials, and readout schemes in realizing devices with improved performance, as well as miniaturization, and enhanced noise-tolerance, have lead to a renewed interest in fluxgate magnetometer technology [2] for a variety of applications involving room temperature magnetic field measurements. In this sense Fluxgates fill the gap between SQUIDs (see [3] for a good overview), that offer far greater sensitivity but with significantly greater cost as well as difficult operating conditions, and the relatively inexpensive (though less accurate) Hall Effect sensors [4]. Alternative technologies, based on the anisotropic magneto resistance, giant magneto-resistance and magneto-impedance effects have been also recently studied in the context of precise magnetometers. Fluxgate systems prevail over these competitive technologies not only because of their higher sensitivity but also because of lower noise level, robustness and remarkable thermal and long-term stability.

Residence Times Difference (RTD) Fluxgate magnetometer [5] is based on a time domain read-out; i.e. the magnetic field information is carried out on the different position in time between spikes of the output waveform. This is the important point in the RTD approach: "events" are to be detected rather than "signal amplitudes". The paper will explain the basic operating principle of the sensors, and will present the design of a new Analog-digital integrated circuit for Residence-times reading.

II. THE RT-FLUXGATE

A. Working Principle

An RTD FluxGate is based on a two-coils structure (a primary coil and a secondary coil) wound around a suitable ferromagnetic core having a hysteretic input-output characteristic (see Fig. 1). A periodic driving current, I_e, is forced in the primary coil and generates a periodic magnetic field, H_e parallel to the geometry of the core. This geometry is adopted to guarantee uniformity of magnetic field along the ferromagnetic core. A target field H_x is applied in the same direction of H_e; the secondary coil is used as pick-up coil and the output voltage V_{out} is used to detect the target according to the follow working principle.

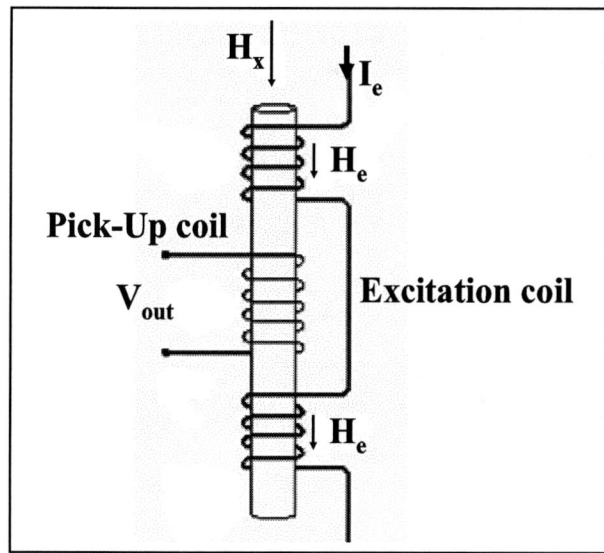

Figure 1. Single core RTD based sensor design

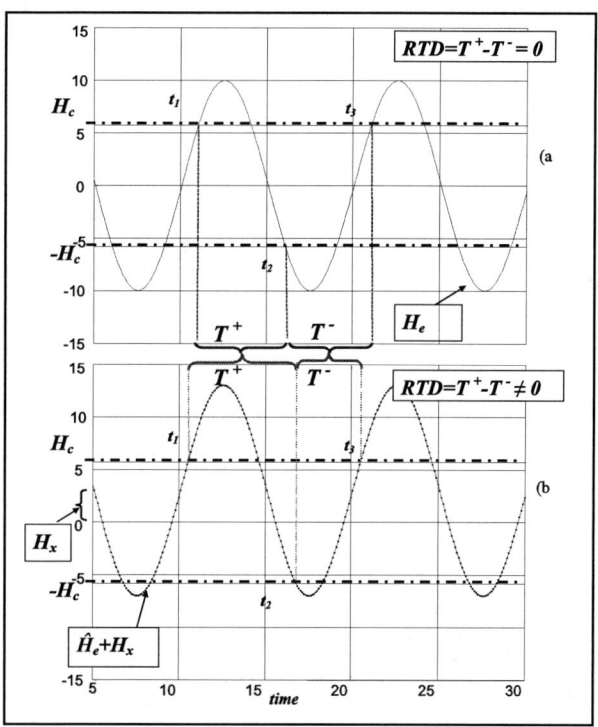

Figure 2. Working principle of the RTD Fluxgate

The basic idea is that the magnetic core has two commutation thresholds and a two state output, whose behaviour can be described via bistable dynamics governing a double-well potential energy function $U(m)$ [6]. The magnetization m of the magnetic core is governed by the excitation field, H_e, produced in the primary coil, and the bistable potential energy function $U(m)$, which underpins the crossing mechanism between the two steady magnetization states of the magnetic core [6]. In order to reverse the core magnetization (from one steady state to the other one), the driving field (H_e) must cross the switching thresholds of the magnetic core. In the case of a time-periodic excitation having amplitude large enough to cause switching between the steady states and in the absence of any target field, the hysteresis loop (or the underlying potential energy function $U(m)$) is symmetric and two identical Residence Times are obtained. The presence of a target dc signal (H_x) leads to a skewing of the loop with a direct effect on the Residence Times, which are no longer the same [7]. Finally, we assume a sharp hysteretic characteristic for the magnetic core; in turn, this allows us to infer that switching between the two stable states of the magnetization occurs instantaneously when the applied magnetic field exceeds the coercive field level H_c. Under these conditions, the device operates almost like a static hysteretic nonlinearity, e.g. a Schmitt Trigger. Mathematically, this amounts to the assumption of a very small time-constant ζ; in fact, for calculation purposes, it may be assumed that the signal frequency is smaller than ζ^{-1}. The above considerations are schematized in Fig. 2. Under these assumptions, the residence time, T^+, in the right well of the potential $U(x)$ can be defined as the time interval elapsed between the crossing of the upper H_c level (at time t_1 in Fig. 2) and the successive crossing of the lower H_c level (at time t_2);

the residence time in the left well, T^-, is defined as the time interval between the lower threshold crossing (time t_2) and the upper threshold crossing in the next period of the bias signal (at time t_3). The Residence Times Difference, $RTD = T^+ - T^-$, is zero in the absence of the external dc magnetic field ($H_x = 0$); for nonzero target signal, it affords a quantification of the target signal amplitude (H_x). We want to point out that the bias signal is forced to the core through a sinusoidal current in the primary coil and that the residence time differences can be computed from the voltage at the pick-up coil V_{out}; this voltage is in fact proportional to the first derivative of the magnetic flux (whose form is a two state waveform i.e. a square-wave) and hence to a spike train. Fig. 3 represents a typical set of signals [8].

B. Technology issue on RT-Fluxgate

In the following some notes about a RTD-FluxGate sensor developed in Printed Circuit Board (PCB) technology are given. The sensor has been realized in PCB that has already been successfully used for other magnetic sensors [9,10] and is made up of three layers. An Amorphous Metal (also known as metallic glass alloys) has been chosen for the magnetic core layer due to its suitable hysteretic characteristic which encourages the use of a readout strategy based on the estimation of the RTD. In particular, an as cast Magnetic Alloy 2705 As Cast (Cobalt-based), by Metglas® has been adopted for its very sharp characteristic [11].

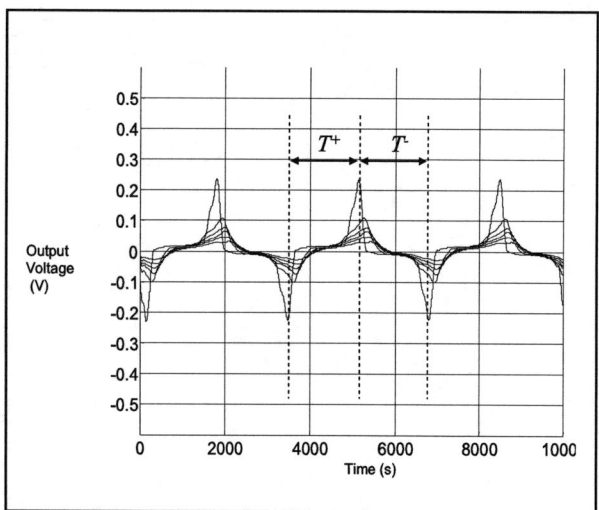

Figure 3. Experimental signals for different values of H_c with a strongly suprathreshold bias signal

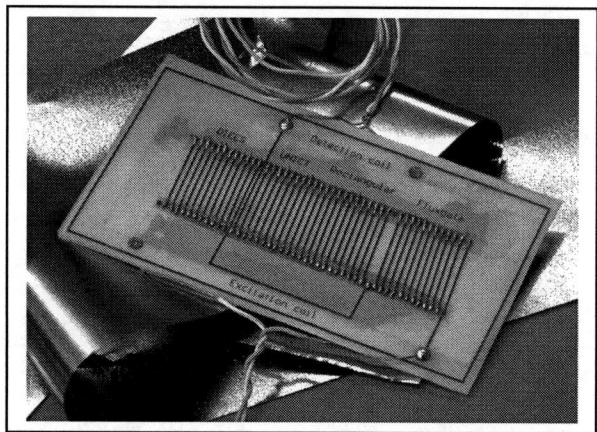

Figure 4. Experimental prototype of *RTD* Fluxgate

The sides of the PCB two PCB layer that face away from the core material are printed with copper wirings to form the windings for the driving coil and the sensing coil. Solder is used to fuse the two sheets together to complete the circuit for the windings. Fig. 4 shows the RTD-Fluxgate prototype and a sample of the magnetic material. The signals in Fig. 3 has been obtained in the prototype of Fig. 4. Other important parameters of the adopted core are its coercive field, $H_c=1.6A/m$, and the D.C. permeability $\mu = 290000$.

C. Performance

The set of experiments performed in our laboratories [8] has showed very good agreement between expected and actual behaviour; the maximum value of experimental sensitivity obtained for the prototype is, $S = 0.02s/(A/m)$. We remark that *0.02* seconds is the variation of the residence times differences when a field of *1 (A/m)* is applied; this implies that by using a *100 Mhz* counter to estimate the RTD the digit uncertainty would produce a an intrinsic resolution of about 0.5×10^{-6} *(A/m)* in terms of target field. This remarkable result, however must be compared with the fluctuation of RTD produced by magnetic and electric noise; the actual resolution of the device ,in fact can be indeed limited by the noise floor and further studies are actually in progress to characterize the device in terms of noise.

III. RTD EXTRACTION CIRCUIT

A. Working principle

In this section we present the design of a new Analog-digital integrated circuit for Residence times reading. The analog part of the circuit consists in an instrumentation amplifier (amplifies the differential voltage at the pick-up coil to get a suitable signal level to activate the next stage, the trigger) cascade with a Schmitt trigger that modifies the spike train into a square-wave i.e. a digital signal containing the RTD (see Fig. 5).

The working principle is based on an Up/Down *n* bit digital counter; the circuit counts up during T^+ and down during T^-, at the end of each period the output of the counter matches exactly the RTD. If we suppose to have a frequency clock of f_c and using a two's complement representation of the output bits (for measurement of negative field), the absolute value of RTD must relay within the value:

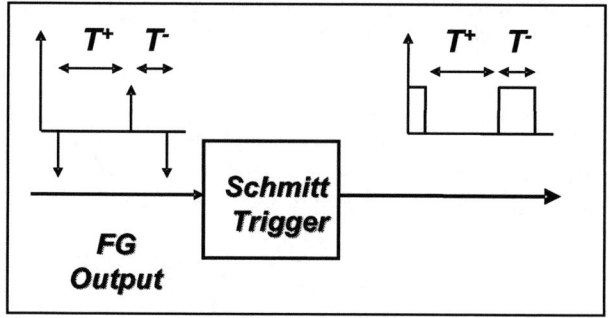

Figure 5. Analog to digital conversion

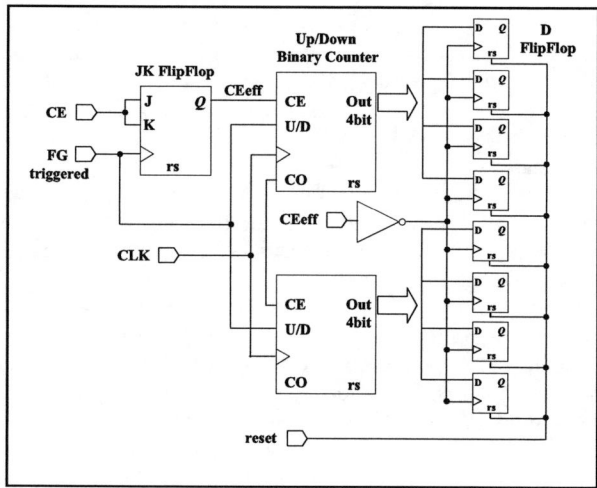

Figure 6. Schematic of RTD extraction circuit

$$(2^{n-1}-1)/f_c \ (s) \quad (1)$$

that leads to an operating range defined by the expression:

$$(2^{n-1}-1)/(f_c \times S) \ (A/m) \quad (2)$$

with an intrinsic resolution in target field of:

$$1/(f_c \times S) \ (A/m). \quad (3)$$

Fig. 6 shows the schematic of the RTD counter; the JK Flip-Flop at the input of the system synchronize the beginning of the counting with the rising edge of the inverted square-wave and hence with the falling edge of the triggered output signal (where the T^+ starts; see Fig. 5). In the example the cascade of two *4-bits* accomplishes the Up/Down counting and eight D-Type Flip-Flop latch the output for consistency reading.

B. Technology and realization

The Cells used in the design of the circuit belong the Austriamicrosystems [12] Standard Cell families available for *0.8 µm* double metal CMOS process technologies and address logic and system designers developing Application Specific Integrated Circuits (ASICs). Lowest power consumption, high speed, noise immunity and applicability to a wide range of automotive, communications and industrial design requirements are the key benefits of this advanced process. The whole circuit has been designed and simulated (See Fig. 7) in a CADENCE HIT – KIT environment using logic port and operational amplifiers of the *0.8 µm* Standard Cell family, these last presents an open loop gain of *103 db* and a cut-off frequency of *20 Mhz*. The circuit contains also a *8-bit* Up/Down counter whose technology is compatible with the *20 Mhz* frequency clock; the expected value of operating range and resolution are calculated according to (2,3) and leads to values *317,5 × 10⁻⁶ (A/m)* and *2,5 × 10⁻⁶ (A/m)*;

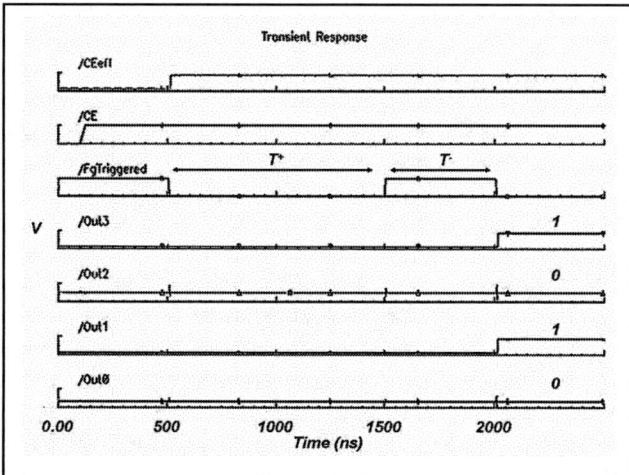

Figure 7. Signals during the counting operation for a 20Mhz 4-bit RTD counter; at the end of T⁻ the ouput is updated and the value 1010 (10) i.e. the ratio between RTD and the clock period.

This value satisfies the specific of the circuit for the PCB prototype, whose fluctuation in terms of magnetic field is in the order of *50 × 10⁻⁶ (A/m)*. Fabrication is expected in the second half of February.

IV. CONCLUSIONS

The paper presents the operating principle of the sensor, based on a time domain reading, and the design of an analog-digital integrated circuit for read-out. Experimental results has showed the suitability, in terms of sensitivity, for applications requiring measurement with a resolution of *2,5 × 10⁻⁶ (A/m)*.

However further research and measurement on the realized circuit will address the objective to estimate the noise-floor level in the system and confirm the expected value of resolution.

ACKNOWLEDGMENT

The work has been carried out on behalf of a research contract with CDI and SPAWAR.

REFERENCES

[1] W. Geyger, "Nonlinear Magnetic Control Devices" (McGraw Hill, New York, 1964), for a good review of fluxgate magnetometer technology in its early stages.

[2] W. Bornhofft, G. Trenkler in Sensors, "A Comprehensive Survey" Vol. 5 eds. W. Gopel, J. Hesse, and J. Zemel (VCH, New York 1989), and P. Ripka: Magnetic Sensors and Magnetometers, (Artech House, Boston, 2001), for good reviews.

[3] R. Barone, G. Paterno: Physics and Applications of the Josephson Effect, (Wiley, New York, 1982).

[4] R. Popovic, "Hall Effect Devices", (IOP Publishing, Bristol, 1991).

[5] B. Andò, S. Baglio, A.R. Bulgara, V. Caruso, S. Castorina, "A new readout strategy for fluxgate sensor". Proceedings of Inst and Meas tech Conf. IEEE IMTC 2003, Vail, Colorado, 2003

[6] G. Bertotti, Hysteresis in Magnetism, Academic Press, San Diego, 1998.

[7] A. Bulsara, C. Seberino, L. Gammaitoni, M. Karlsson, B. Lundqvist, J. Robinson, "Signal Detection via Residence Times Asymmetry in Noisy Bistable Devices". Phys. Rev. E, Vol. 67,No 1, art 016120, Jan 2003).

[8] B. Andò, S. Baglio, A.R. Bulsara, V. Sacco "Effects of driving mode on RTD-FluxGate performances". Inst and Meas tech Conf. IEEE IMTC 2004, Como,Italy 2004.

[9] O. Dezuari, E. Belloy, S. E. Gilbert, M. A. M. Gijs; New Hybrid Technology for Planar Fluxgate Sensor Fabrication, IEEE Trans. Magn. 35, 2111-2117 (1999).

[10] O. Dezuari, E. Belloy, S. E. Gilbert, M. A. M. Gijs; High Inductance Planar Transformers, Sensors and Actuators A81, 200-204 (2000).

[11] Honeywell METGLASS Solution, magnetic materials, METGLASS Magnetic Alloy 2114A datasheet, http://www.metglass.com

[12] http://asic.austriamicrosystems.com/

Integrated Interface Circuits for Chemiresistor arrays

Carina K. Leung and Denise M. Wilson
Department of Electrical Engineering, University of Washington,
Seattle WA 98195-2500
denisew@u.washington.edu

Abstract -- Integrated circuits customized to transduce resistance information from small footprint composite polymer chemiresistor arrays are presented. The circuits are designed to fit underneath the sensor platform itself, therefore enabling 100% fill factor of the chemically active area as a percentage of total sensor surface area. Two approaches to transducing very small changes in resistance on top of a large baseline resistance are presented. Both seek to transduce small changes in resistance, typical to composite polymer chemiresistors in practical applications, to a usable output with high sensitivity, detection limit, and resolution. The circuits have been simulated, fabricated, and tested in an integrated array underneath an array of sensor platforms in a CMOS process. Resolution for the resistance-to-digital circuit exceeds the noise characteristics of the chemiresistor while the resolution for the resistance-to-frequency circuit allows for the detection of changes in resistance as low as 0.02% of the baseline resistance.

I. Introduction

Chemiresistors are a good choice for the fabrication of high resolution, general purpose electronic noses because of their broad selectivity and small size. Metal-oxide chemiresistors have been studied extensively for a wide range of chemical sensors sizes and types since the 1970's [1]. Conducting polymers and organic films[2] are also popular for miniaturized sensor applications. In the past decade, the composite polymer film, where a chemically sensitive insulating polymer is combined with a conducting particle, such as carbon black, has gained popularity for high resolution chemical sensor arrays. These films can be fabricated with a wide range of materials and compositions, thereby providing a broad range of widely selective sensors for high resolution sensor arrays. Broad selectivity and high resolution are two necessary components for furthering the development of a true electronic nose. The electronic nose, like olfaction, relies on large numbers of sensors to overcome limitations of individual sensors to provide general purpose, aggregate odor sensing capability. High resolution arrays demand on-board dedicated processing circuits to extract and amplify small signals before they are transferred off the sensing plane. This paper presents two possible approaches for (a) conforming measurement circuit size to sensor platform size and (b) extracting the small changes in resistance on top of large and broadly varying baseline resistances that are typical of chemiresistor response to vapor phase analytes. Experimental results for resistance-to-digital and resistance-to-frequency conversion circuits in terms of resolution, detection limit, and sensitivity of the resulting chemiresistor pixel are presented.

II. Sensor Description

The chemiresistors for which the interface circuits have been developed are made of composite polymer films which consist of an insulating polymer matrix implanted with conductive particles of carbon-black. Details of the fabrication of these chemiresistors are described by Lewis *et al* in [3][4]. As an analyte is selectively adsorbed onto the sensor film, the polymer swells. This swelling causes the conductive carbon-black particles to move farther apart, thereby changing the conductivity of the sensor. Large amounts of swelling produce a non-linear relationship with concentration, as determined by percolation theory. For typical use, however, in small concentrations, the response of these sensors can be described as linear:

$$\frac{\Delta R}{R_o} = k[C] \qquad (1)$$

R_o is the baseline resistance of the sensor; ΔR is the change in resistance from the baseline value; $[C]$ is the concentration of the analyte, typically in ppm; and k is a sensor constant. The value k depends in part on the analyte to which the sensor is responding and the type of chemically sensitive polymer used. Data used in this analysis is supplied by the Lewis laboratory at Cal Tech, using an array of 40 total sensors

By their very nature, the composite polymer chemiresistor does not scale cleanly during miniaturization. Miniaturization increases the part of the chemiresistor that is chemically sensitive, as the surface becomes a larger part of the overall sensor volume; this increase in signal, however, is often offset by the noise, inherent in the resistor itself (thermal noise) as well as surface fluctuations. Typical composite polymer chemiresistors have baseline resistances (no analyte present) that vary widely from 3kΩ to over a 1MΩ in typical fabrication. This variation, in miniaturized chemiresistors, is a direct result of the fact that the (inherently) poorly controlled surface fabrication is a larger percentage of the overall sensor volume. Changes in resistance representative of many applications for the electronic nose are typically 1% maximum. Interface circuits for the composite polymer chemiresistor, therefore, must (a) accommodate a wide range of baseline conditions; (b) detect and amplify small changes in the baseline while filtering out the baseline condition itself; and (c) remain small enough to fit underneath the sensor platform on a CMOS substrate, of typical area 0.04mm^2. A ring-type electrode platform, designed to minimize noise at small sensor sizes, is shown in Figure 1. The electrode geometry and size have been optimized for reducing noise in previous efforts by Lewis et al at Cal Tech [5].

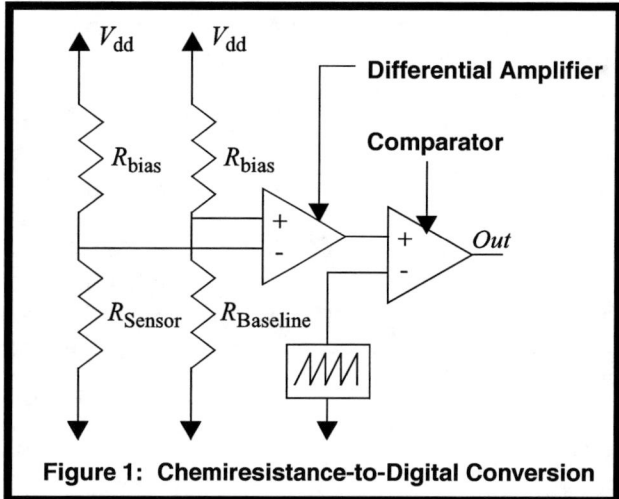

Figure 1: Chemiresistance-to-Digital Conversion

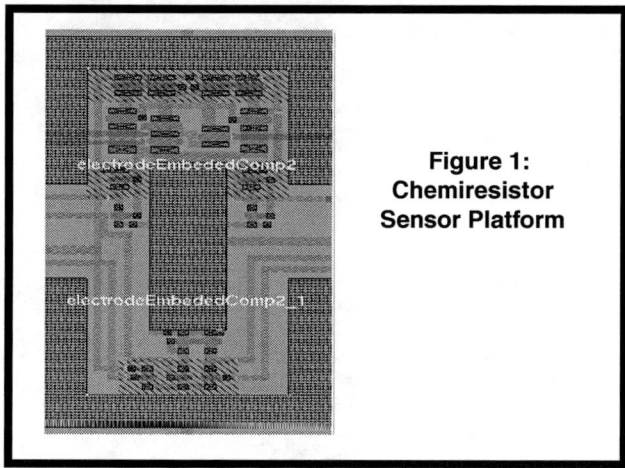

Figure 1: Chemiresistor Sensor Platform

III. Circuit description

The two transduction circuits are shown in Figure 1 and Figure 2 for resistance-to-digital and resistance-to-frequency conversions respectively. The first circuit relies on the division of the chemiresistor sensor platform into a three terminal device as shown in Figure 1. The polymer film deposited on this platform is designed to be a combination of reference and measurement sensors. The reference film is deposited on one side and is either (a) shielded from the sensing layer through an overlying passivation layer or (b) adjusted to be significantly less sensitive to the sensing environment than the measurement sensor. The measurement side of the platform is designed to be highly sensitive and is fully exposed to the sensing environment. The result is an inherently differential signal between baseline (reference) and sensing films that is transferred, first through a differential amplifier and then a comparator for a ramping analog-to-digital conversion. The bias resistors (or transistors) are chosen so that the differential amplifier is set to the lower end of the amplifier's linear operating range; as the chemiresistor responds to the environment, the (sensor) input voltage increases through the full linear range of the amplifier.

The second transduction circuit relies on a sensor platform that is two rather than three terminals (the outer rings of Figure 1 are simply connected together). The sensor resistance is connected to a current source that charges the capacitor C_o. As the capacitor charges, the input to the Schmitt trigger increases; as the Schmitt trigger surpasses its upward threshold, the control signals C' and C go high and low respectively, causing the sensor current to cease charging the capacitor C_o and begin discharging it instead (through the C' branch of the circuit). As the capacitor discharges, the input to the Schmitt trigger decreases; when it surpasses its downward threshold, the discharge signal turns off and the charge signal C turns on again, restarting the charge/discharge signal. The frequency of the charging cycle (and the output signal *Out*) is dependent on the hysteresis of the Schmitt trigger, the relative sizes of the current source transistors, and the charging current sourced by the sensor resistance R_{sensor}.

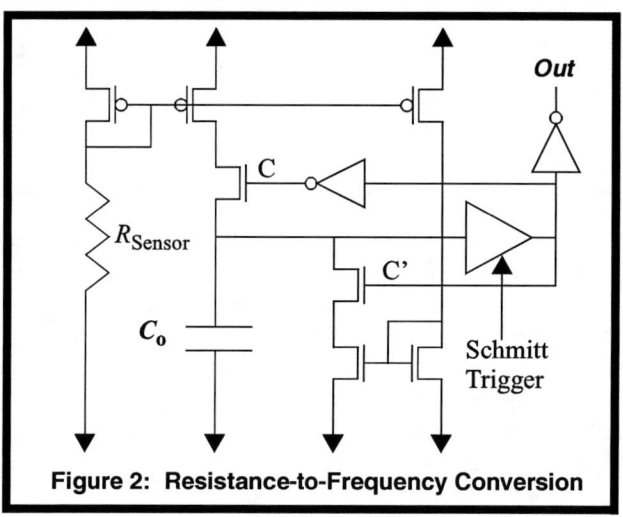

Figure 2: Resistance-to-Frequency Conversion

Both the resistance-to-digital and resistance-to-frequency conversion circuits have been fabricated on a tiny chip (2mm X 2mm) in an AMI 1.5 micron, CMOS process through the MOSIS prototyping service. Both circuits are optimized within the constraint of the sensor platform footprint, laid out underneath the sensor platform. Results from characterization and testing of these circuits at the circuit and application level follow. The circuits are first characterized experimentally using electronically controlled inputs and their transfer curves extracted. Then, they are tested in an electronic nose context using actual composite polymer chemiresistor data.

IV. Experimental results

The resistance to digital conversion circuit produces a digital output whose trip point (on the ramping reference voltage into the comparator) is linearly proportional to the input sensor resistance. The differential amplifier is fabricated to have the highest gain possible while still remaining within the sensor electrode footprint. The gain is calculated in the linear range of the amplifier; experimentally, the gain is 20. The comparator can digitize a minimum change of 500μV between the input voltage (= output voltage of the differential amplifier) and the reference voltage (Figure 3). Trans-

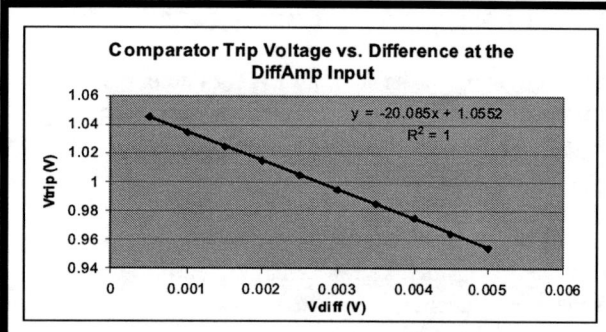

Figure 3: Experimental Results
for the resistance-to-digital transduction approach are shown for a range of baseline resistances from 730kΩ to 9.26kΩ. Gain characteristics do not change across the range of baseline resistances due to the proper selection of bias and the differential nature of the input signal.

lated to the input stage, the circuit can then detect a minimum change of 500μV/20 or 25μV. This resolution is much better than the noise level of the chemiresistor, therefore making the circuit itself capable of detecting changes in the input chemiresistor signal at the noise floor of the chemical sensor itself. The comparator outputs a digital signal in thermometer code at an overall sensitivity of 148kbits/V (output over input). Converting thermometer code to regular binary code gives a sensitivity of 17 bits/V.

The resistance to frequency conversion circuit produces a output frequency that is linearly proportional to the input resistance (chemiresistor) for typical response ranges of a composite polymer resistor. The sensitivity of the circuit ranges from 0.12%/Ω at large baseline resistances (730kΩ - -Figure 4a) to 4.15%/Ω at small baseline resistances (9.26kΩ--Figure 4b).

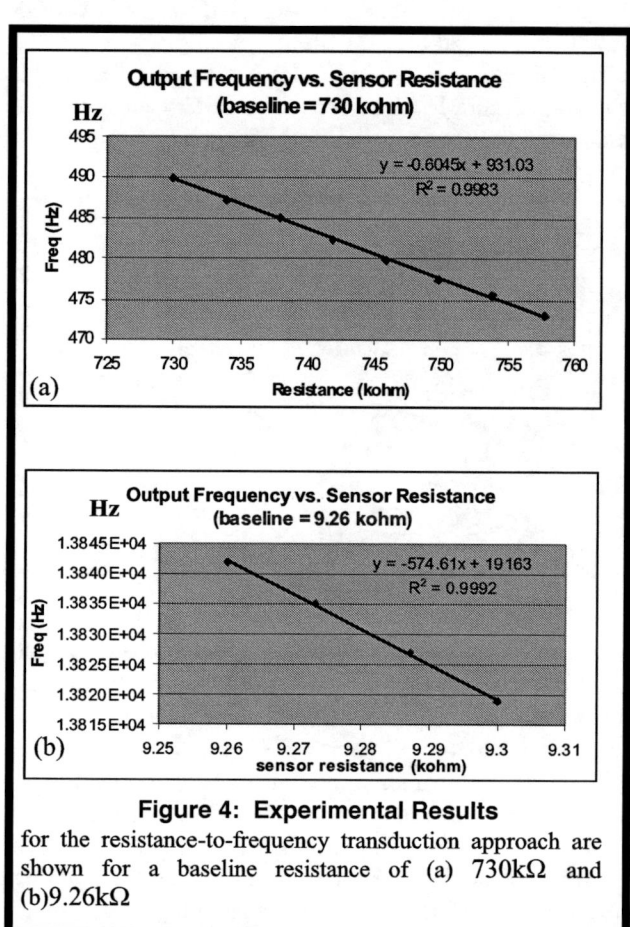

Figure 4: Experimental Results
for the resistance-to-frequency transduction approach are shown for a baseline resistance of (a) 730kΩ and (b) 9.26kΩ

The sensitivity of the circuit, in an application context, is the ratio of output/input relative to the baseline conditions at both input and output. For example, at a baseline resistance of 730kΩ, the frequency decreases 0.12% from the baseline frequency for every ohm increase from the baseline resistance. The resolution (and detection limit) are identical for these linear sensors and in circuit testing, based on noise measurements, are determined to be at least 0.07% for a baseline resistance of 730kΩ and 0.02% for a baseline resistance of 9.26kΩ. To put these numbers in perspective, a typical chemiresistor experiences only a 1% change in resistance from baseline conditions for practical concentration ranges; the resolutions of 0.02% and 0.07% for these circuits correspond to typical concentrations in the sub-ppm range.

V. Conclusions

Two interface circuits have been presented that are compatible with processing the outputs of composite polymer chemiresistors. A resistance-to-digital circuit demonstrates sensitivity and resolution that are better than the noise floor of the chemiresistors themselves (Refer to Figure 5 for a typical response curve of a composite polymer resistor). Theoretically, then, these circuits do not limit the transduction of chemical information to electrical output. However, due to the noisy input signals generated by typical chemiresistors, it is expected that this circuit will show greater instability (bit fluctuation) than is theoretically obtained from circuit characterization results. The resistance-to-frequency conversion circuit, on the other hand, has built in resilience to hysteresis (through the use of a Schmitt trigger) and offers greater stability to the sensor signal transduction process at a sensitivity of 0.12%/Ω and resolution of 0.02% change in chemical sensor resistance from baseline conditions. The performance of both circuits is suitable for chemiresistor signal measurement and their size has been adapted to a sensor electrode footprint that is on the order necessary for high resolution, electronic nose style arrays.

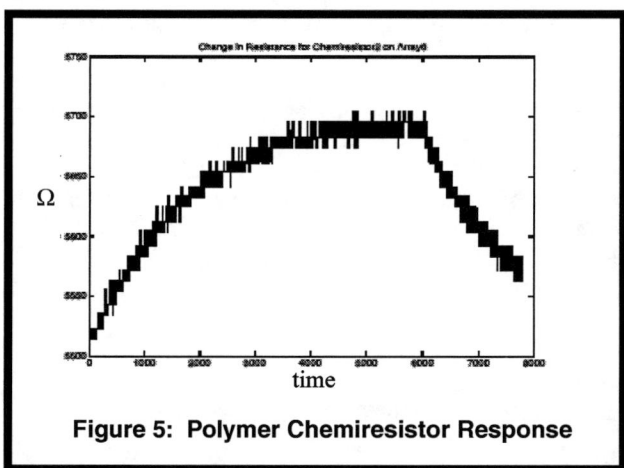

Figure 5: Polymer Chemiresistor Response

VI. ACKNOWLEDGEMENTs

The authors would like to thank Nathan Lewis and his graduate group at the California Institute of Technology for data and technical assistance, as well as a subcontract through CalTech on ARO Grant DAAG55-98-1-0266.

VII. REFERENCES

[1] Kousuke Ihokura and Joseph Watson, <u>The Stannic Oxide Gas Sensor, Principles and Applications</u>, CRC Press: 1994.

[2] D.M. Wilson, Sean Hoyt, Jiri Janata, Louis Abando, and Karl Booksh, "Chemical Sensors for Portable, Handheld Field Instruments," *IEEE Sensors Journal*, vol. 1, no. 4, December 2001, pp. 256-276.

[3] Mark C. Longergan, Michael S. Freund, Erik J. Severin, Brett J. Doleman, Robert H. Grubs, and Nathan S.Lewis, "Array-based vapor sensing using chemically sensitive, polymer composite resistors," *IEEE Aerospace Applications Conference Proceedings*, vol. 3, 1997, pp. 583-631.

[4] Adam Matzger, Thomas P. Vaid, and Nathan S. Lewis, "Vapor sensing with arrays of carbon black polymer composites," *Proceedings of SPIE - The International Society for Optical Engineering:* vol. 3710, 1999, pp.315-320.

[5] Shawn M. Briglin, Michael S. Freund, Phil Tokumaru, and Nathan Lewis, "Exploitation of spatiotemporal information and geometric optimization of signal/noise performance using arrays of carbon black-polymer composite vapor detectors," *Sensors and Actuators, B: Chemical*, vol. 82, no, 1, Feb 1, 2002, p 54-74.

SHD Movie Distribution System Using Image Container with 4096×2160 Pixel Resolution and 36 Bit Color

Takahiro Yamaguchi, Mitsuru Nomura, Kazuhiro Shirakawa, Tetsuro Fujii
NTT Network Innovation Laboratories
Nippon Telegraph and Telephone Corp.
Yokosuka-Shi, Kanagawa, 239-0847, Japan
{yamaguchi.takahiro, mitsuru.nomura, kazuhiro.shirakawa, tetsuro.fujii}@lab.ntt.co.jp

Abstract—The super-high-definition (SHD) movie distribution system described in this paper can transmit and display exceptionally high quality movies using an image container with 4096×2160 pixel resolution and 36-bit color. The total bit rate of an SHD movie to be shown at 24 frames per second is 7.6 Gbps, and should be compressed by 15:1 for transmission via Gigabit IP networks. The system is based on JPEG2000 coding technology. Coded streams of 500 Mbps can be continuously transmitted from the server to the real-time decoder, from which decompressed SHD movies are projected onto a screen.

I. INTRODUCTION

Advances in computer and telecommunication technology and the growth of broadband networks have stimulated the development of applications that use extra high quality image communications. To satisfy professional users in the printing, medicine, and image archiving industries, a precision color imaging system is required to achieve digital images of excellent quality. An image category, called Super High Definition (SHD), has been defined to have a resolution of at least 2000 scanning lines with 24-bit color separation [1]. SHD images surpass the quality of 35-mm film in terms of spatial resolution and approach the quality of 60-mm film. SHD movies are also very promising in many applications, such as medical or educational video archives, and especially in the commercial movie industry as possible replacements for conventional 35-mm cinefilm (Figure 1). We have already developed an SHD movie system for digital cinema that uses a 3840×2048 pixel image format and 24 frame per second (fps) motion [4,5,6]. The image quality of this movie system is quadruple that of HDTV in terms of resolution. We have proposed that a next-generation digital cinema format be defined in which a system would have 2000 or more scanning lines, be progressively scanned, and be shown at 24 fps. This concept has been endorsed by professionals working in Hollywood and has been drawn up as a draft "Digital Cinema System Specification" written by Digital Cinema Initiatives, LLC (DCI). DCI opened the new specification on December 6th, 2004 [7]. The main characteristics of the Digital Cinema Distribution Master (DCDM) Image Structure are listed in Table I.

TABLE I. MAIN CHARACTERISTICS OF IMAGE STRUCTURE OF DCI'S "DIGITAL CINEMA SYSTEM SPECIFICATION V4.3"

Size	4K: 4096 × 2160 pixel 2K: 2048 × 1080 pixel
Sampling	Aspect ratio 1:1
Pixel	$\gamma = 2.6$ CIE XYZ 12 bit/each component
Coding	JPEG2000 image coding system

The standardized 4K digital cinema format is slightly larger than the size of the image container able to be handled by our previous system (Motion picture professionals working in Hollywood measure the cinema resolution in terms of the number of horizontal pixels.). Thus, we have developed a system that conforms to this image structure.

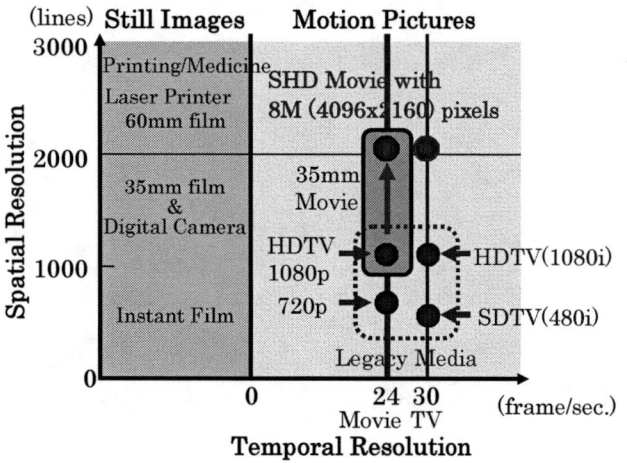

Figure 1. Resolution and Frame Rate of SHD Movie

In this paper, we introduce our latest SHD movie distribution system that supports an image container with 4096×2160 pixel resolution and 36-bit color. Section 2 describes the design of the system. Performance evaluations are described in section 3.

II. SHD MOVIE DISTRIBUTION SYSTEM

A. System Design Concepts

Figure 2 shows the configuration of the SHD movie distribution system. SHD movie/video cameras are not yet commercially available, and an enormous number of film-based movies exist. An archiving system that can preserve all film contents as digital media could be tied to a digital cinema system that works as a video on demand (VOD) service. Film-based movie contents are digitized by scanning 35-mm or 65-mm cinefilms, after which the resulting digital data are compressed and stored. Because the total bit rate of raw SHD movie data can be as high as 7.6 Gbps (4096×2160 pixel resolution, 24fps motion and 36 bit/pixel color mode), the movie should be compressed by 15:1 (or more) by using JPEG2000 coding algorithms [8], in order to transmit it by the envisioned wide-area optical IP networks using gigabit Ethernet (GbE).

To transmit SHD movie contents on public optical networks, an exceptionally high-performance decoder and imaging system will be required if the movies are also to be displayed in real-time as they are streamed. Additionally, an SHD real-time decoder and an SHD projection device will be required to present SHD movies. The high-speed/gigabit optical networks would distribute the contents of VOD servers to wherever the SHD real-time decoders and SHD projectors have been installed. The real-time decoder would then decompress the video streams transmitted from the server and output the digital video data to the projector.

We have developed a prototype digital cinema system that can store, transmit, and display SHD movies with 8 million pixel resolution. The original resolution was set as 3840×2048 pixels, and it has been extended to up to 4096×2160 pixels. The system consists of three main devices, a video server, a real-time decoder, and a projector, described below.

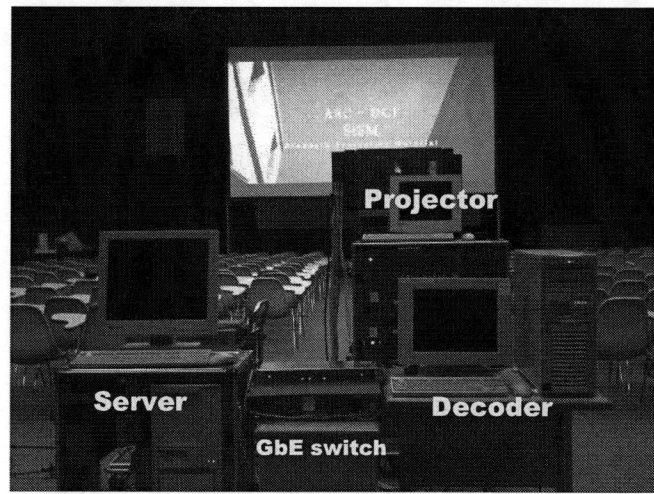

Figure 3. Appearance of the new SHD movie delivery system

B. Real-time Decoder

The decoder can perform the real-time decompression at a speed of 250M pixels per second, using JPEG2000 processors in parallel, and it can handle an input flow of up to 500 Mbps. It has two circuit blocks, a PC/Linux part with a GbE network interface, and newly developed JPEG2000 decoder boards (Figure 4). The JPEG2000 processing chips are Analog Devices Inc. ADV202 ES3s. Four chips are installed in one board and four boards are installed on the PCI-X-bus in order to process up to 24 frames of 4k×2k pixels with up to RGB (4:4:4) 36-bit color images in a second. The decoder works with the following frame sizes: 3840×2048 pixel (previous mode), 3840×2160 pixel (Quad HDTV mode) and 4096×2160 pixel (4K DCDM mode). It supports frame rates of 24 fps (cinema mode) and 30 fps (TV mode).

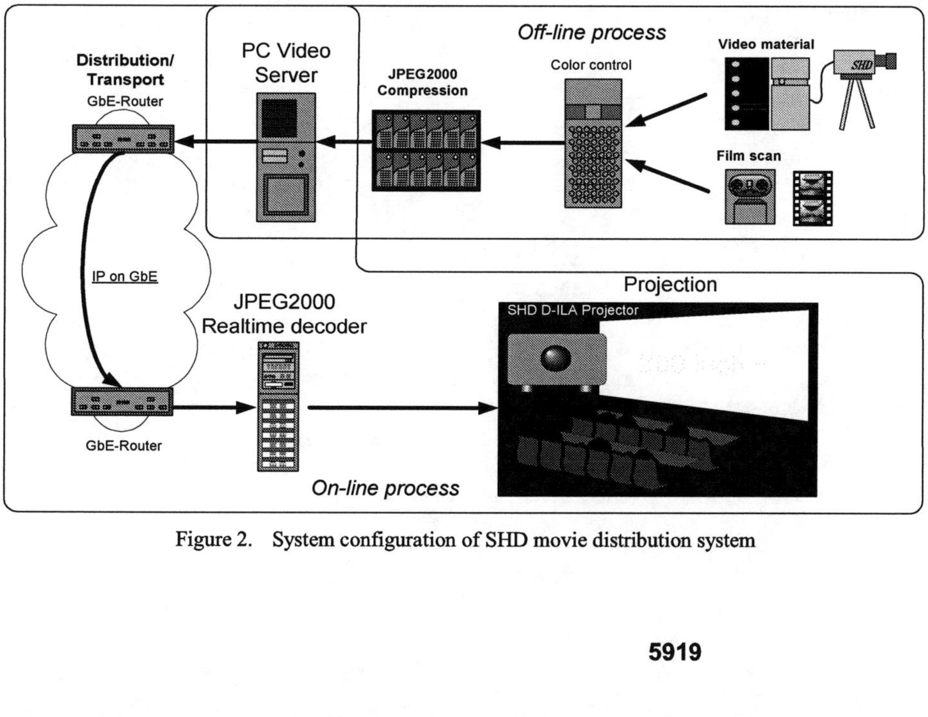

The PC part also manages the accompanying audio data tracks via a PCI digital sound card (RME, HDSP9652) that can output 6-channel, 48-kHz, 24-bit sampling digital audio based on ADAT digital I/O. The audio data are not compressed, because no degradation in sound quality is permitted in high-picture-quality large-screen digital-imagery applications. The total bit rate for 6-channel, 48-kHz, 24-bit sampling audio signals is only 7 Mbps, which is negligible compared to the data rate of the video signal.

Figure 2. System configuration of SHD movie distribution system

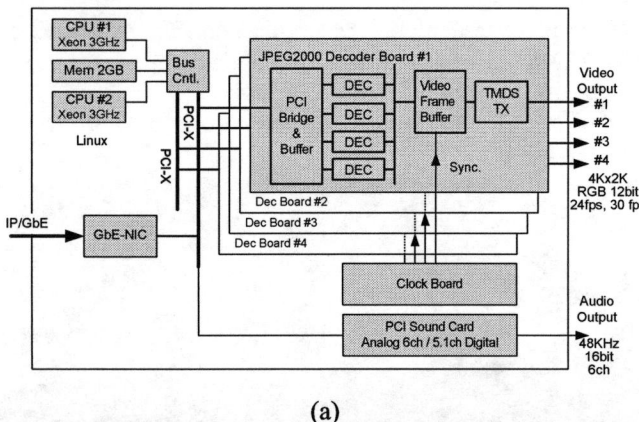

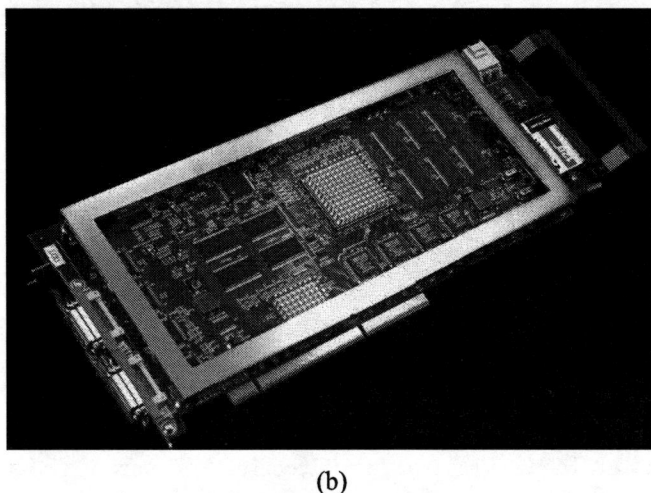

Figure 4. Block diagram of JPEG2000 real-time decoder (a) and photograph of JPEG2000 decoder boad (b).

C. Projector

The prototype SHD projector itself was developed by JVC using D-ILA (Direct-Drive Image Light Amplifier) technology. The high picture quality of D-ILA is derived from the advanced LCoS (Liquid Crystal on Silicon) technology and a high-precision optical system. The major features of D-ILA are high brightness, high resolution, high contrast ratio, analog gradation, and high-speed response. The prototype SHD projector uses three 1.7 inch, 3840×2048 pixel D-ILA devices for RGB, and it represent 30-bit color images. The effective brightness exceeds 5000 ANSI lumens by using a 1600-W xenon lamp, which is bright enough to show images on as large as 300-inch diagonal screens.

Because DCI has opened the digital cinema system specification, new 4K projectors based on the specification were recently announced from JVC and Sony, respectively. JVC has developed their 4K2K D-ILA projector that uses three newly developed 4096×2160 pixel D-ILA devices and a 2000-W xenon lamp [9]. Sony has also developed their 4K SXRD (Silicon X-tal Reflective Display) projector [10]. Our decoder can be connected to these projectors by using four HD-SDI Dual Links via a TMDS/HD-SDI signal converter. However, a Dual Link digital interface for 2048×1080 picture format is not an SMPTE standard as of yet. Therefore, the output from our decoder is limited to 3840×2160 pixels, which is the same as Quad HDTV resolution. The decoder can work in 12-bit processing mode, because at present, we don't have any device to handle 36-bit color images. This will be a topic for future study.

D. Server

The PC video server consists of Linux running on dual CPUs (Xeon, 3.0GHz), a high-performance Software-RAID (Serial ATA disk 250GB×7, RAID0, XFS file format), and a GbE network interface. A large set of still images is compressed in advance. The 4096×2160 pixel images in 16-bit TIFF files are divided into 2048×135 pixel image tiles, compressed/encoded with a suitable data-format for the decoder, and made into one big file bringing 1000 frames together as a unit that is stored in the RAID of the server. A server program that receives the data transfer command reads the data from the RAID and writes them periodically to the GbE network interface. The server has a large shared buffer in order to enhance the maximum transmission rate by averaging the disk read speed.

E. Transmission

UDP and TCP functionalities were implemented in the server and the decoder. Usually we use TCP as the connection protocol from the server to the decoder. TCP is adequate for a stably connected transmission and is the best method to share the bandwidth of IP-router-based networks. Thus, it would be very easy to use in a multiplex cinema. However, for large RTT (Round trip time) networks, it is very difficult to realize the full performance of TCP. To overcome this problem, we have used (1) a large TCP window, (2) multiple TCP connections, and (3) a shaping control function. As a result, we have succeeded in transmitting an SHD movie over 3000 km at 300 Mbps [11]. To implement adequate functions, TCP can be applied to long-distance transmissions.

III. PERFORMANCE EVALUATIONS

The image quality of SHD movies and the performance of the distribution/transmission system were evaluated using the standard evaluation material (StEM) of digital cinema produced by DCI and ASC (American Society of Cinematographers) [12].

As mentioned before, our current system uses ADV-202 ES3s. As far as using this ES3 chip, the tile size (2048×135 pixel), code block size (128×32 pixel), and 4-level frequency components with a 9/7-wavelet filter are ineludible specifications for enough real-time decoding performance. To extract higher coding performance without any restriction, we should use the whole image area as one tile and 6-level frequency components with the 9/7-wavelet filter. These are the decoder requirements of the DCI's specification [7]. We have evaluated the coding performance of our current system in comparison with ideal conditions (DCI's spec. decoder) and our previous decoder (128×128 pixel tile and 3-level with the 5/3-wavelet filter). We assumed a bit rate from 100 Mbps to 700 Mbps. The test image was one of the frames showing the landscape wedding party of the Warm Night scene in the StEM. (Figure 5) The results are shown in Figure 6. Our current system had a coding performance equivalent the ideal conditions and showed about a 1.0-dB improvement over that of the previous system. Thus, its coding scheme is good enough for digital cinema. We would like to increase the number of test images as part of a future study.

Figure 5. Landscape Wedding Party of the Warm Night scene in the StEM

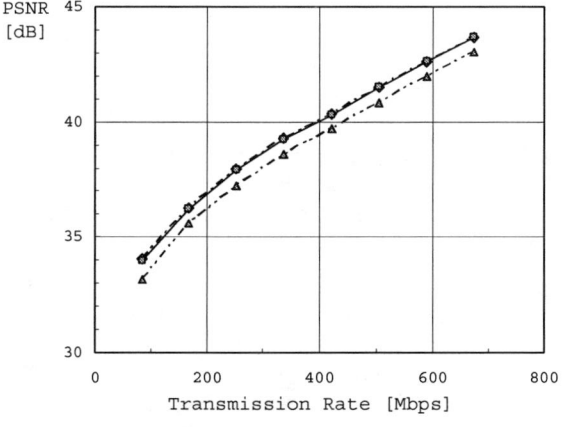

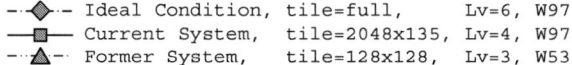

Figure 6. Comparison of JPEG2000 coding performances. Ideal condition vs. current system vs. former system.

The full data of "StEM" (16-bit TIFF file, 16605 frames) that occupy 700 GBytes are compressed to 32 GBytes of JPEG2000 encoded stream at the rate of 500 Mbps. Currently the file format is our own one, not based on the Motion-JPEG2000 standard [13] or digital cinema packaging format of the DCI's specification [7]. The time required for JPEG2000 compression was about 10 seconds per frame using a Linux PC. We can use 10 PCs in parallel to shorten the off-line encoding process.

IV. CONCLUSIONS

The SHD movie distribution system has been designed for a contents delivery network that provides a 500-Mbps video streaming service via a gigabit network. The network transmission capability is so important that the prototype SHD decoder uses JPEG2000 with IP on GbE transmission mode in order to show 4k×2k pixel movies on the SHD projector. The image container has 4096×2160 pixel resolution and 36-bit color. The data path from the video server to the display devices is completely digital, which means that the SHD movie system achieves actual 2k scanning lines and 4k-pixel wide resolution. Everybody agrees that cinema will become digital in the future. We have concerted our efforts towards realizing a complete 4K digital cinema system that fully matches the quality of 35-mm cinefilm. Now that we have achieved this goal, a worthy target would be development of a business model and an adequate a security scheme for digital cinema production and distribution.

REFERENCES

[1] S. Ono, N. Ohta, and T. Aoyama, "All-Digital Super High Definition Images", Signal Processing: Image Communication 4, pp. 429-444, 1992.

[2] S. Ono and J. Suzuki, "Perspective for Super High Definition Image Systems", IEEE Communication Magazine, pp. 114-118, Jun, 1996.

[3] T. Fujii, T. Fujii and K. Ishimaru, "Performance Evaluation of Protocols for ATM Transmission of Super High Definition Images", ICC'96, pp. 155-159, Jun., 1996.

[4] T. Fujii, M. Nomura, J. Suzuki, I. Furukawa, and S. Ono, "Super High Definition Digital Movie System", SPIE VCIP'99 Vol. 3653, pp. 1412-1419, Jan. 1999.

[5] T. Fujii, M. Nomura, D. Shirai, T. Yamaguchi, T. Fujii, K. Hagimoto, and S. Ono, "IP Transmission System for Digital Cinema Using 2048 Scanning Line Resolution", IEEE GLOBECOMM2002, pp. 1643-1647, Nov. 2002.

[6] T. Fujii, M. Nomura, D. Shirai, T. Yamaguchi, T. Fujii, and S. Ono, "Digital Cinema System Using JPEG2000 Movie of 8 Million Pixel Resolution", IS&T/SPIE Electronic Imaging IVCP2003, 5022-07, Jan. 2003.

[7] Digital Cinema Initiatives, LLC Technology Committee, "Digital Cinema System Specification v4.3", December 6th, 2004.

[8] ISO/IEC 15444-1, Information technology - JPEG2000 image coding system - Part 1: Core coding system, 2002.

[9] JVC press release, http://www.jvc-victor.co.jp/press/2004/4k2k_d-ila.html (in Japanese), September 24th, 2004.

[10] SONY press release, http://news.sel.sony.com/pressrelease/4864, June 4th, 2004.

[11] T. Yamaguchi, D. Shirai, T. Fujii, M. Nomura, T. Fujii, S. Ono, "SHD Digital Cinema Distribution Over a Long Distance Network of Internet2", SPIE VCIP2003, Vol.5150, pp.1760-1769, July, 2003.

[12] DCI press release, "ASC And DCI Creating Digital Cinema Test Film", http://www.dcimovies.com/press/09-24-03.tt2, September 24th, 2003.

[13] ISO/IEC 15444-3, Information technology - JPEG2000 image coding system - Part 3: Motion JPEG 2000, 2002.

An Implementation of JPEG 2000 Interactive Image Communication System

Junichi Hara

Software Research and Development Group
RICOH Company LTD.
1-1-17 Koishikawa, Bunkyo-ku, Tokyo 112-0002, Japan
jun@src.ricoh.co.jp

Abstract—An implementation of a JPEG 2000 interactive communication system is described, which is based on JPIP image communication protocol. JPIP specifies an image communication scheme, which utilizes JPEG 2000 resolution/ quality/position scalability so as to enhance transmission efficiency. Two transmission modes are defined in JPIP, tile-based JPT-stream mode and precinct-based JPP-stream mode. In the proposed system, transmission efficiency of JPT-stream is raised in such a way to re-construct tile data by using a set of packets at the server end. Comparison between transmission modes and details of system implementation are also explained.

I. INTRODUCTION

JPEG 2000 is the latest image compression standard. This standard enables excellent quality codestreams at any compression ratios, and has properties of resolution/quality /position scalability features, which are distinct from previous standards such as JPEG [1]. These properties enable creations of a scaling image, reduction of codestream quality and quantity, and decoding a part of image without decoding whole codestream. JPEG 2000 Part 9, called JPIP, is an interactive protocol standard for viewing of JPEG 2000 images in a client-sever system, and utilizes these scalable features of JPEG 2000 codestreams. The JPIP protocol allows the client to fetch the region of interest image without directly accessing the compressed target file, instead the client formulates requests using a simple descriptive syntax that identifies the current "focus window" of the client-side application [3]. Put another way, the client identifies a view-window region, image components and spatial resolution of interest, and requests it from the server. Then the server divides a JPEG 2000 codestream into data-bins and responds by rate-distortion optimally sequencing the codestream while considering previous responses.

JPIP defines two transmission schemes for partitioning JPEG 2000 codestreams, based on either tiles or precincts as the predominant data-bins. In both cases, a main header of the codestream is assigned its own data-bin. The tile-based method assigns each tile as a single data-bin, which represents the stream of data formed by concatenating all tile-parts of the tile including all tile-part headers. The precinct-based method assigns each tile header and precinct as its own data-bins. The tile-based approach allows simple servers and clients, but it reduces flexibility in the order of data that can be transmitted, and also reduces spatial accessibility. The precinct-based method allows efficient interactive spatial accessibility even if the entire image is compressed as a single tile [4], however, it requires extra implementation cost at both client and server. The proposed method achieves the JPP-stream capability using JPT-stream. This method is based on re-constructing tile's codestream data, which is made by selecting packet data that is related to the requested view-window region and then by merging these packet data into a new tile-part's codestreams at the server side. This method needs the same system implementation cost as that of the precinct-based system, however no complexity change to implement the client.

This paper is organized as follows. In Section 2, two JPIP transmission modes, tile and precinct based, are explained, and then in Section 3 we describe the proposed tile-based method. The system implementation detail is explained in Section 4, and comparison between proposed method and the precinct-based method are shown in Section 5. Finally, concluding remarks are made in Section 6.

II. OVERVIEW OF JPIP TWO TRANSMISSION MODES

JPIP has two methods for transmitting JPEG 2000 codestream, "JPT-stream" and "JPP-stream" media types. Each media type consists of a concatenated sequence of messages, where each message contains a portion of a single data-bin preceded by a message header. Data-bins contain portions of a JPEG 2000 compressed image representation, such that it is possible to construct a stream that completely represents the information present in a JPEG 2000 codestream. Each message is completely self-describing, so that the sequence

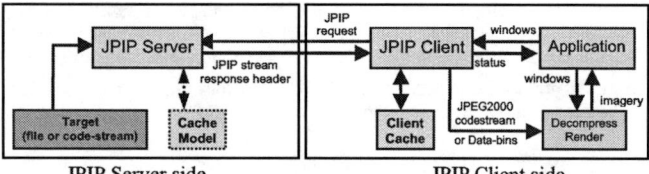

Figure 1. The JPIP interactive image communication system.

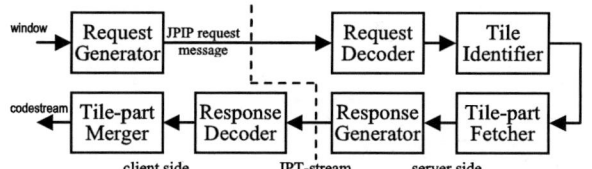

(a) tile-based system

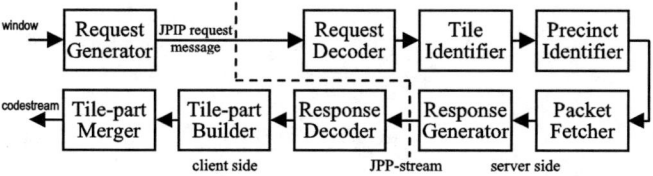

(b) precinct-based system

Figure 2. JPIP clieant and server architecture

of messages may be terminated at any point and messages may be re-ordered subject to minimal constraints without losing their meaning. A simplified architecture of both tile and precinct based systems are illustrated in Figure 2. This figure shows the precinct-based system has two extra functional blocks that is "Precinct Identifier" and "Tile-parts Builder," and it also shows the precinct-based server has "Packet Fetcher" instead of "Tile-part Fetcher."

A. JPT-Stream – Tile-based transmission mode

In this mode, a "Request Generator" first translates the target image and view-window information into JPIP requested messages, and a "Request Decoder" gets the requested information from those messages that are made by the client. A "Tile Identifier" selects tiles that belong to the requested view-window area. The selected tile's upper left and lower right corners on the reference grid are (tx_0, ty_0) and (tx_1, ty_1) respectively, where

$$tx_0 = \lfloor (wx_0 - T_{ox})/T_{sx} \rfloor \quad ty_0 = \lfloor (wy_0 - T_{oy})/T_{sy} \rfloor \quad (1)$$

$$tx_1 = \lfloor (wx_1 - T_{ox})/T_{sx} \rfloor \quad ty_1 = \lfloor (wy_1 - T_{oy})/T_{sy} \rfloor \quad (2)$$

where (wx_0, wy_0) and (wx_1, wy_1) are normalized upper left and lower right corners of the view-window on the reference grid of the requested subband domain respectively, and (T_{ox}, T_{oy}) and (T_{sx}, T_{sy}) are tile's offset and size on the reference grid of the requested domain respectively. "Tile-part Fetcher" gets the tile-parts and "Response Generator" makes the response messages. The client receives these messages and concatenates them into a codestream for a decoder by "Tile-part Builder" and "Tile-part Merger." The JPEG 2000 decoder makes an image from this codestream, and the client application displays this image onto a screen.

The tile is a spatial unit of transmitting codestream, so this spatial accessibility depends on the tile size. Since the JPEG 2000 standard has the starting positioning and tile length marker, Start-Of-Tile-part (SOT), it is easy to retrieve and concatenate tile-parts. Therefore, there is very low complexity for implementing of JPT-stream based JPIP server and client system. However, this approach reduces flexibility in spatial accessibility.

B. JPP-Stream – Precinct-based transmission mode

In this mode, the target image name and focus view-window information are also transmitted to the server, and the server select the tiles that belong to the request area, like tile-based mode. A "Precinct Identifier" selects precincts that belong to the requested view-window area and processed tile area. The selected precinct's upper left and lower right corners on the reference grid in the subband domain are (px_0, py_0) and (px_1, py_1) respectively, where

$$px_0 = \lfloor \max(tx(n)_0, wx'_0)/2^{PPx} \rfloor$$
$$py_0 = \lfloor \max(ty(n)_0, wy'_0)/2^{PPy} \rfloor \quad (3)$$

$$px_1 = \lfloor \min(tx(n)_1, wx'_1)/2^{PPx} \rfloor$$
$$py_1 = \lfloor \min(ty(n)_1, wy'_1)/2^{PPy} \rfloor \quad (4)$$

where $(tx(n)_0, tx(n)_0)$ and $(tx(n)_1, tx(n)_1)$ are the upper left and lower right corners of the n-th tile on the reference grid in the subband domain, and PPx and PPy is the precinct width and height exponents respectively, and (wx'_0, wy'_0) and (wx'_1, wy'_1) are a view-window region, which is enlarged according as the wavelet filters, on the reference grid in the subband domain. A "Packet Fetcher" picks up packets associated with the selected precincts. Then, the server creates and sends response message, and the client gets that and concatenates these messages into a codestream for a decoder. The client application might decode it and display an image.

For creating the packet header, the tag tree-coding scheme is used for coding of code-block's inclusion and the coding passes information, and it uses previous packet header's coding state of code-blocks within the precinct. Therefore, identifying packets is not as simple as identifying tiles, and JPEG 2000 standard offers simple tools, identifying or storing length markers, that identify packet or packet length. Since a codestream merely has these markers and the system contains two more extra blocks, the precincts-based system has more complexity than that of the tile-based system. However, since the precinct size might be smaller than the tile's one, this approach enables more effective spatial interactive accessibility than that of JPT-stream based especially if the entire image is compressed as a single tile.

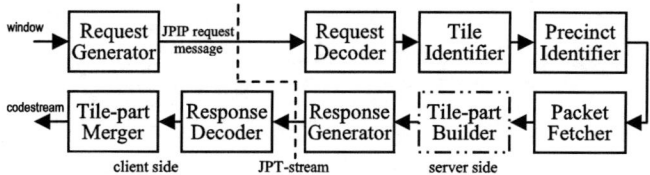

Figure 3. The proposed tile-based system architecture

III. PROPOSED TILE-BASED SCHEME

As shown in previous section, the tile-based system is low complexity but it only has mediocre spatial interactive accessibility, while the precinct-based system has excellent accessibility but requires additional implementation cost. This section describes a method that offers excellent spatial accessibility even if the system is based on the tile-part transmitting. The idea of this method is to reconstruct tile-part data at the server side. Figure 3 shows the architecture of this method. This architecture uses the same blocks in the precinct-based method and the "Tile-part Builder" block is placed in the server instead of in the client. Therefore, the complexity of this system is the same as that of precinct-based system, but this somewhat increases server complexity. Using selected packet data and its' position, the "Tile-part Builder" makes new tile-parts which have the tile header and only packet data that associated with the precincts, which belong to the requested view-window. For progressive or resolution level accessing to a full image, the selected packets are complete to build new JPEG 2000 tile-parts data. However, a compliant JPEG 2000 tile-part requires information for every precinct even if the client only wishes to examine a portion of the image. To rebuild the compliant JPEG 2000 codestream, the "Tile-part Builder" block fills the packets without packet body, called "Zero Length Packet" (ZLP) that size is one byte only, to the unselected precinct areas. Figure 4 shows an example of ZLP filling at some subband level. In this example, this subband level tile has 25 precincts, and the requested view-window region covers a precinct's region of (iii, b)-(v, e). The "Precinct Identifier" selects these precincts, and the "Packet Fetcher" takes packet data that belong to the selected precinct areas. The "Tile-part Builder" fills ZLP to these unselected precincts area and makes a new tile-part. The precinct-based server only sends the selected packet data, but this server also transmits the ZLP.

IV. IMPLEMENTATION OF SYSTEM

The proposed tile-based method and precinct-based transmitting method are implemented in our JPIP client and server system. Figure 5 is the block diagram of our system. This system can send an image via HTTP and HTTP-TCP, and transmit three media types; jpeg, JPP-stream and JPT-stream. The jpeg media type is for compatibility with legacy

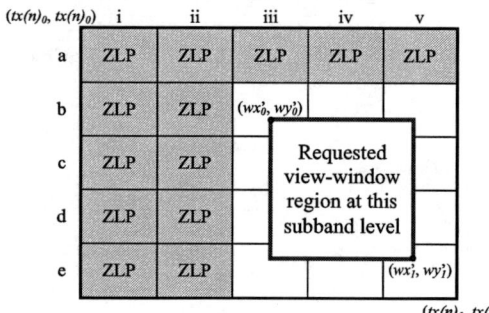

Figure 4. Example of filling ZLP into a tile at certain subband level.

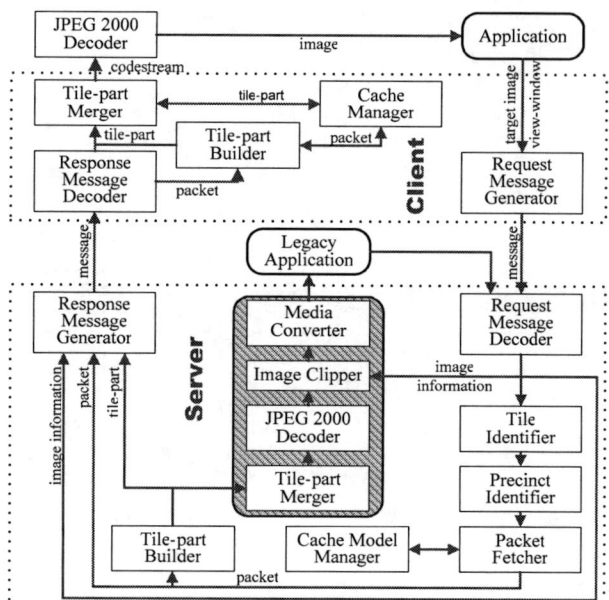

Figure 5. The proposed tile-based system architecture

clients, such as web browsers. Hence, the browser can obtain an arbitrary region of a target image with a HTML page that contains embedded JPIP requests as static targets. The gray area of this figure is for this function. In this area, a reconstructed codestream is decoded and image is clipped according with the view-window, and finally the image type is converted to image/jpeg media type. The server manages server's cache model, which has the packet status of already transposed or not. The client has cache for storing JPIP data-bins and stores data-bins for each media types. This system builds complete codestreams for decoding with unspecialized JPEG 2000 decoders. The system is developed in Java, and we evaluated the performance of the proposed method, in the next section.

V. PERFORMANCE OF PROPOESED SCHEME

The spatial and progressive accessibilities performance is evaluated in both the proposed method and precinct-based method. The input data is the "lena" image (grayscale, 512×512, and 8 bits/pixel) compressed by a JPEG 2000 encoder. The wavelet decomposition level values of 3 and 6 are selected for the spatial and progressive evaluations respectively, and the 256×256 tile size and 16×16 precinct size are used in both evaluations. In the Table I and Table II, the client has no data in the cache initially.

A. Spactial Accessibility Performance

The proposed method allows the client to show a portion of the image without receiving the whole tile data. Figure 6 shows an example of this accessibility of both the normal tile-based method and the proposed method. This figure shows the proposed method can transmit tile's packet data related to the requested view-window only, whereas the normal one sends the entire tile's packet data. Table I shows

 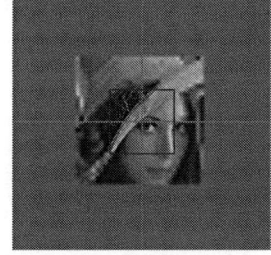

(a) the normal tile-based system   (b) the proposed tile-based system

Figure 6. The spactial accessibility performance of proposed method. The center rectangle area and the crossed lines are regested view-window area and tile borders respectively.

(a) first resolution   (b) second resolution   (c) third resolution

Figure 7. The progressive display avirity of the proposed method.

TABLE I. THE PERFOEMANCE OF SPATIAL ACCESIVIRITY

Requested View-window Area	Transmitted Bytes		Advantage of proposed JPT-stream
	Proposed JPT-stream	JPP-stream	
R1: (192, 192) - (320, 320)	14,206	13,275	-931
R2: (152, 152) - (280, 280)	19,623	18,892	-731
R3: (0, 0) - (256, 256)	42,696	44,211	1,515
R4: (0, 0) - (128,128)	10,394	10,481	87
R5: (0, 0) - (30, 30)	1,851	1,513	-338

Note1: the codestream size is 166,582 bytes.
Note2: the frame size, fsiz, is (512, 512).

TABLE II. THE PERFORMANCE OF RESOLUTION ACCESIVIRITY

Requested Frame Size	Transmitted Bytes		Advantage of proposed JPT-stream
	Proposed JPT-stream	JPP-stream	
16 × 16	1,049	994	-55
32 × 32	1,778	1,750	-28
64 × 64	4,390	4,424	34
128 × 128	13,629	13,974	345
256 × 256	47,187	48,741	1,554
512 × 512	168,626	174,794	6,168

Note1: the codestream size is 164,962 bytes.
Note2: the transmitted byte of normal JPT-stream is 168,626 bytes.

this performance between the both methods. This table uses three view-window sizes which are 30×30 ("R5"), 128×128 ("R1", "R2," and "R4") and 256×256 ("R3"), and the only upper left tile is accessed in the case of R3, R4 and R5. Table I shows the performance of the proposed method is close to that of the precinct-based method, even though our method requires several bytes for storing ZLP while the tile-based method requires SOT and SOD (Start-Of-Data) markers that indicate the starting position of tile and its data respectively. The reason is that each packet or tile-part, which consists of packets, is divided and wrapped with message headers and the last precinct messages are smaller. Accordingly, the precinct-based method requires more message headers than the tile-based method, and this increases transmitting data size. Therefore the data sent by the precinct-based method and our method is identical, the precinct method requires a message header consisting of multiple bytes for each precinct, whereas our method uses ZLP to indicate which precincts are missing, and the position of the remaining precincts is implicit from their position in the codestream.

B. Progressive Accesibility Performance

The proposed method can also access each resolution image without transmitting all tile data like the precinct-based method. Figure 7 shows this accessing example and also demonstrates the progressive resolution accessing of the proposed method. Table II shows the performance of resolution level image accessing both the precinct-based method and the proposed method. Since there is no missing precinct in each resolution level, our method requires no ZLP in this case. Therefore, the proposed method has an advantage over the precinct-based method if the accessing size is greater than 64×64. This is caused by the same reason discussed in section *V-A*.

VI. CONCLUSIONS

A tile-based JPEG 2000 image communication scheme has been proposed to enhance transmission efficiency like a precinct-based system, and an implementation of this system has been described. The scheme exploits ZLP and the selected packet data, which are associated with the requested view-window region, and reconstructs tile data that might contain no packet body. This scheme requires some extra cost to implement the server like a precinct-based system, but this offers good accessibility performance and no extra cost on the client side.

REFERENCES

[1] M. Boliek, C. Christopoulos and E. Majani, "J JPEG 2000 Part 1 020719 (Final Publication Draft)," ISO/IEC JTC 1/SC 29/WG1 N2678, July 2002.

[2] M. Gormish, D. Lee and M Marcellin, "JPEG 2000: Overview, Architecture, and Applications," Int. Conf. on Image Processing, Vancouver, Canada, September 2000.

[3] R. Prandolini, G. Colyer, and S. Houchin, "15444-9:2004 JPEG 2000 image coding system - Part 9: Interactivity tools, APIs and protocols – JPIP," Final Publication Draft Revision 3, ISO/IEC JTC 1/SC 29/WG 1 N3463, November 2004.

[4] D. Taubmana and R. Prandolini, "Architecture, Philosophy and Performance of JPIP: Internet Protocol Standard for JPEG2000," VCIP2003 SPIE volume 5150 pp. 649-663, July 2003.

[5] M. Gormish and S. Banerjee, "Tile-based transport of JPEG 2000 images," VLVB03, Madrid, Spain, September 2003.

Designing and Packaging Technology of Renesas SIP

Noriaki Sakamoto, Norihiko Sugita, Takafumi Kikuchi, Hideki Tanaka, and Takashi Akazawa
System Solution Integrated Product Design Dept., Custom LSI Business Unit, System Solution Business Group
Renesas Technology Corp.
Tokyo, JAPAN
sakamoto.noriaki/sugita.norihiko/kikuchi.takafumi/tanaka.hideki/akazawa.takashi@renesas.com

Abstract—Renesas Technology Corp. has started SIP (Solution Integrated Product) Project, since April in 1999 aiming at promotion of SiP (System in Package) business. SiP can achieve 1/10-1/6 design TAT (Turn Around Time) in comparison with SoC (System on a Chip). SiP has also advantages of EMI noise reduction and customer's substrate area reduction by using signal integrity analysis technology and packaging technology of a planar and a stack structure. On the basis of these technologies, we can enlarge SIP for digital consumer field, analog included digital field, and other fields.

I. INTRODUCTION

Semiconductor suppliers and customers had designed mainly SoC (System on a Chip) solutions, which can realize one application system only by a single chip. However, recent 0.13 μm, 90 nm process and more advanced technologies bring huge investment and long TAT (turn around time) issues. Therefore, customers require shorter design cycle and lower design investment especially for digital consumer market segment; Digital Still Camera (DSC), Digital Video Camera (DVC), mobile phone, and other consumer electronics products. On the other hand, SiP, which packs a few chips in a single package, can resolve these issues, since it is basically composed of well-known chips. Thus, SiP business is enhanced, according to requirement of reduction of TAT and investment. In many cases, SiP is composed of CPU/ASIC and one/several memories including SDRAM and Flash, etc. We can realize one application system by using it.

We have noted these advantages of SiP and have already started "SIP (Solution Integrated Product)" business since April 1999.

This paper describes characteristics of SiP and new technologies concerning it in the field of designing, packaging, and testing, including application examples. In chapter II, we describe main characteristics of SiP. In chapter III-V, designing, packaging, and testing technology are shown, respectively. In chapter VI, technology of SiP implemented in DSC made by CASIO Computer Co., Ltd. (CASIO) is introduced. Finally, we describe SiP technical issues and conclude this paper.

II. CHARACTERISTICS OF SiP

In this chapter, we describe characteristics of SiP in comparison with SoC. Advantages of SoC are described below.

1) High integration with advanced process technology.

2) High performance including low power consumption and high-speed operation.

However, SoC has weaknesses below;

3) Long design Turn Around Time (TAT)

4) High cost for designing and photomasks

5) High potential of design failure

On the other hand, SiP can cover these defects by using well-known chips [1].

1) Short development TAT

2) Low cost for Non-recurring Engineering (NRE)

3) Low potential of design failure because of relatively easy design between well-known chips.

Furthermore, SiP has other merits described below.

4) High flexibility of components in an application system

5) Ability to implement large volume of memory with high speed logic

6) Easy implementation and realization of small-area mounting on customer's board

7) Reduction of noises such as Electromagnetic Interference (EMI)

For 1)-2), we examined by designing a SoC and a SiP, respectively, which have about 1M gates. As a result, design TAT of SiP in case of using well-known chips was 1/10-1/6 and development cost was 1/10-1/5 in comparison with SoC. Furthermore, for 3), we can design a SiP with less mistakes since we use well-known chips and connect signals between several chips in a SiP with prevention of degeneration of electrical characteristic and noise interference.

SiP has unique characteristics because of combining several chips. Firstly, we can change combinations of chips for each application according to customer's needs, as shown in 4). Furthermore, we can derive optimal performance from selection of CPU and memory as shown in [2], [3]. Next, for 5), many customers demand larger volume of memory in implementation of application, for example, image processing, sound processing. These applications need to deal with large amount of data and to have large volume of memory. Thus a single chip with small volume of memory cannot realize such data processing. On the other hand, SiP can have large volume of memory within the package size which customers request. For 6), customers have to procure components, for example, CPU, SDRAM and Flash memory, respectively. SiP solves this issue since it is all in one. In order to minimize application system size, mounting area in customers' board should be necessarily small. Therefore, SiP, which can minimize mounting area by implementation of several chips in one package, satisfies customers' demand.

For 7), we monitored EMI noise for boards with SoCs and with a SiP. As shown in Fig. 1, we prepared a four-layer print circuit board (75 mm x 120 mm), which includes a Renesas SuperH™ CPU core, SH-4 (SH7750), and four SDRAM LSIs, and a board with a SiP (27 mm x 27 mm), which includes the same chips. Fig. 2 illustrates the results of EMI noise at 81 MHz operation with using the same program. As can be seen in Fig. 2, the SiP can achieve 30 % noise reduction, on which high frequency current influences in comparison with SoCs. This result shows effects of the reduction of voltage source layers and voltage source pins.

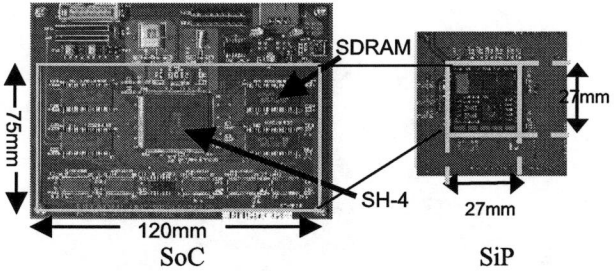

Figure 1. Outline of print board composed of SoCs and that composed of a SiP substrate which includes the same chips.

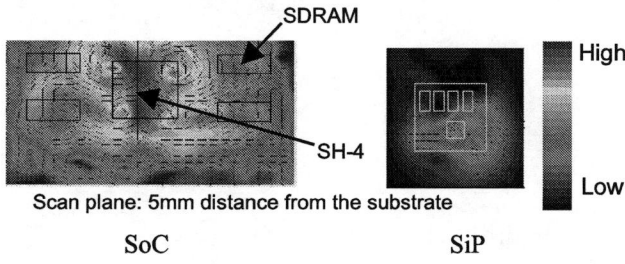

Figure 2. High frequency current distribution in operation in Fig.1.

We can strongly promote SiPs on the basis of plenty of IPs, such as Super-H, M32R, M16C, H8 series with high performance CPU core from 4-bit to 32-bit, and many memory lineup, such as SDRAM, SRAM, Flash memory, etc.

III. Design Technology of SiP

SiP needs different and advanced technologies in designing, packaging, and testing in comparison with SoC.

In system designing, we define system specification and connection between embedded chips. After that, we design SiP substrate and simulate SiP operation. Fig. 3 shows a substrate design flow. After inputting circuits by using design entry tool, we perform functional verification, thermal analysis simulation, and transmission line analysis to develop high quality SiPs.

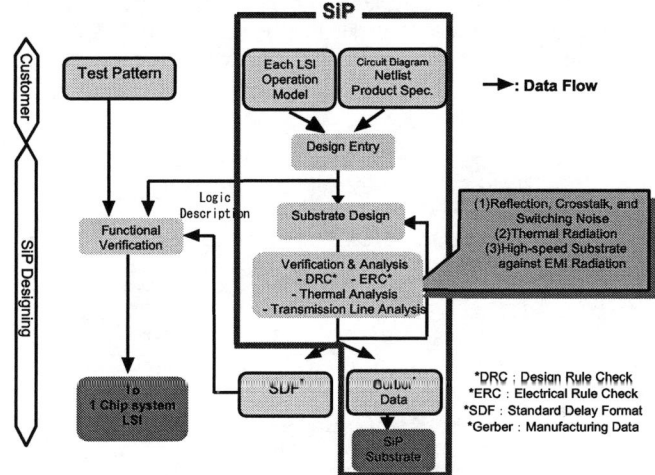

Figure 3. Outline of print board composed of SoCs and that composed of a SiP substrate which includes the same chips.

IV. Packaging Technology of SiP

A. Wire and Flip Chip Bonding Method

Packaging technology is one of the most important technologies for SiPs.

For chip connecting methods, wire bonding method and flip chip bonding method are generally adopted. Wire bonding method uses gold wire to connect signal pads of a die with a substrate of SiP at low cost, while packaging area is relatively large. On the other hand, flip chip bonding is a method by which a chip is electrically connected with substrate face to face after putting connecting metals on the circuit face. Although the flip chip bonding technology is relatively difficult and its cost is higher, this method can realize almost the same package size as the chip size. We can also realize high electrical performance because of short wire length as shown in [4]. Furthermore, thermal radiation becomes better, since a back surface of a chip is upper side and exothermic influence is decreased for shorter wire length.

B. Packaging Method

There are generally two types of wire bonding methods; one is a planar type and the other is a stack type, as can be seen in Table 1. The planar type uses easier bonding technology, while the package size is relatively large. This type is suitable for a request of a low cost and wide ball pitch package.

On the other hand, the stack type can realize small packaging size, which is mainly used for mobile phones and consumer electronics. The stack type packaging by wire bonding methods is influenced on the size of each upper and lower die, since it is difficult if two chips are almost the same size. On the contrary, flip chip attachment is not influenced on the chip sizes and can make package size smaller than wire bonding method because of connecting die pads directly with a SiP substrate. In order to make SiPs small packages, advanced SiP technologies such as narrow wiring technology for substrate and thinning technology for stacked dies are requested.

TABLE I.   SiP PACKAGING TECHNOLOGY

	Bonding Method	Cross Section	Characteristc
Planar Type	Wire-Bonding		- Low cost - Larger package size
	FCA (Flip chip Attach)		- Better electrical characteristics - Chip size package
Stack Type	Stack FBGA Wire Bonding+Wire		- Low cost - Small package size - Lower chip limited by constraints
	Stack FBGA FCA + Wire Bonding		

C. Advanced SiP Packaging Technology

Back grinding (BG) technology is very important for one of advanced SiP technologies of package minimization, high-speed operation, and a few kinds of memory implementation. BG technology is developed for the purpose of thin stack packaging and chip thickness reaches 90 μm for 200 mm (8 inch) wafer and 300 mm (12 inch) wafer in SiP in recent days. As a next step, under 90 μm back grinding technology should be developed. In addition, thickness of substrate influences on the total thickness. In order to solve this issue, build-up substrate is used. This substrate can achieve high density of wiring by decreasing penetration via, which affects on flexibility of wiring in substrate.

In addition, die bonding technology, which is used for the stack type packaging, changes from paste bonding to tape bonding method, and in the near future, thin tape bonding technology will become a serious issue.

For a new SiP technology, Wafer level μBGA (WLBGA), which Renesas calls Wafer Process Packaging (WPP) in [5], has been investigated. This technology, as depicted in Fig. 4, is a wafer-level CSP (chip scale package), which can reduce packaging, inspection, and substrate costs in comparison with flip chip packaging. Furthermore, it does not require underfill assembly technology, which is applied between a lower die and a SiP substrate in order to keep reliability for testing such as Temperature Cycle Testing (TCT). This technology can reduce loads of solder bump bonding points by adding stress moderation layer and rewiring lines.

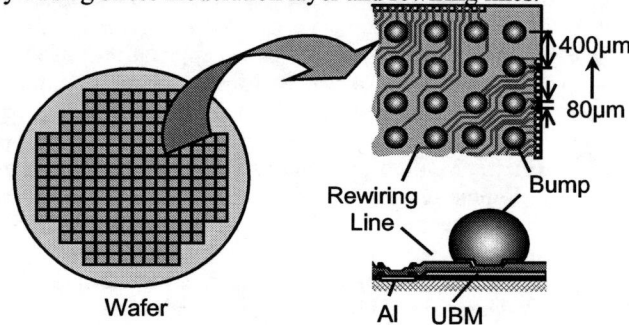

Figure 4.   Wafer Process Packaging (WPP).

V. TESTING TECHNOLOGY OF SiP

Testing of SiP is difficult and complex, since we must verify not only operation of each chip but also system operation of SiP. If SiP is composed of bare chips, chip-level testing before assembly becomes stricter not to throw all dies in SiP away after assembly. Therefore, testing each chip before SiP assembly is very important. After assembly, we can provide high quality products by testing all of chips in SiP again, as can be seen in Fig 7.

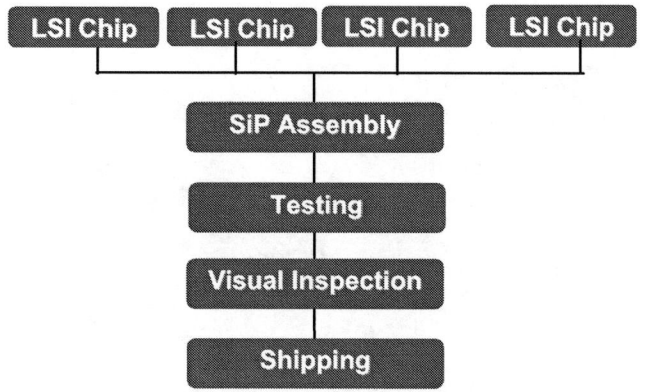

Figure 5.   SiP test flow.

VI. ADVANCED PACKAGING TECHNOLOGY OF SiP EXAMPLE

Renesas develops various packaging types of SIP. Ball Grid Array (BGA) type SiP is mainly investigated, while Quad Flat Package (QFP) type, Package on Package (PoP) are also developed to satisfy customers' demand.

One of the most impressive Renesas SIPs is embedded on DSC "EXILIM (EX-S1/S2/M1/M2)", which was presented by CASIO on May 13th, 2002 and was thinnest card-size DSC in the world. The size of the camera is as large as a credit card and its thickness is 11.3 mm, whose thickness was required and was realized by SIP technology in cooperation with CASIO. This SIP is composed of 4 chips; a microprocessor, an ASIC, a Flash memory, and a SDRAM and all chips are mounted in planar type by flip chip bonding. This technology realized 70 % reduction of mounting area in comparison with in a past mounting method and achieved 11.3 mm thickness, which was thinnest for DSC in the world.

Next, CASIO presented "EXILIM-Zoom (EX-Z3)" with 3x optical zoom function in January 2003 in the same size. Accordingly, we changed from a planar SIP type to a stack type. Flash memory is stacked on microprocessor and SDRAM, and ASIC is mounted apart from these chips. This stack type SIP can reduce 40 % area size in comparison with the planar type.

In recent days, Renesas has developed advanced stacked SIPs. These SIPs are composed of ASIC, SDRAM which used for work and image processing, and Flash memory for data storage. These SIPs have same pin assignment and package outline of 13 x 13 mm, while these have three different capacities of memory. These advanced stacked SIPs are 35 % reduction of density of the stacked SIP for EX-Z3.

Cross section of this five-stacked SIP is shown in Fig. 6. This height is 1.6 mm typical. In order to satisfy package size specification, volume of SDRAM is divided in two and two SDRAM dies are stacked through a spacer chip.

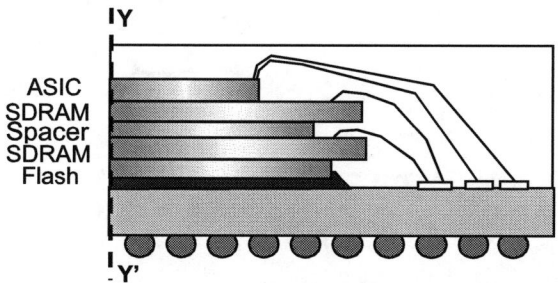

Figure 6. Cross section of five-stacked SIP.

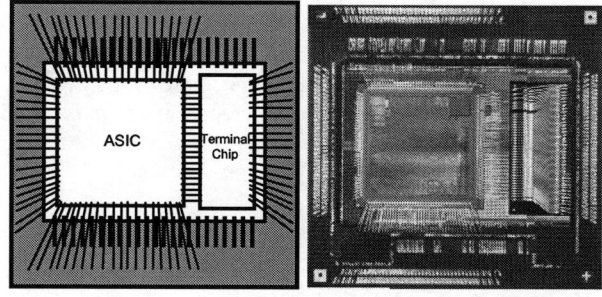

Figure 7. Top view of five-stacked SIP for CASIO DSC.

Fig. 7 illustrates top view of this SIP with photograph. Although the top die of this SIP is ASIC, the long wire length from ASIC pads to substrate is critical issue for package height, electrical characteristics, and stable mass production. Therefore, terminal chip, on which only wiring lines are arranged, is used for short wire length.

VII. ADVANCED TECHNICAL ISSUE FOR SiP

According as process generation changes from 130 nm to 90 nm, SiP bonding technology for narrower pad pitch is requested. Therefore, development for under 50 μm pad pitch bonding technology is necessary, considering gold wire diameter. WPP technology is also one of answers for this problem.

As semiconductor process technology advances, internal noise from voltage sources and signals influences directly on SiP, since it integrates more chips. Therefore, verification of SiP is very important before assembly, and thus we must develop total simulation environment of signal integrity (SI), power integrity (PI), and EMI for SiP, etc.

In the near future, the new technology will be needed, such as (i) analog integrated SiP, (ii) higher frequency SiP for Bluetooth and wireless LAN, and (iii) high voltage SiP, for instance, motor driver. Additionally, (iv) wide temperature range (WTR) SiP, which requires high performance and high quality, will be developed. In developing these technologies, critical system circuit designing and simulation is more important.

VIII. CONCLUSION

SiP, which has advantages of short design TAT and low investment, has advanced step by step for small digital consumer equipments because of low risk of design failure, embedded high volume memory, and minimization of packaging area. This progress has promoted packaging technology from the planar type to the stack type and narrow pad pitch bonding technology, and so on. On the basis of these effective technologies, SiP business is expected to enlarge in combination with SoC business.

ACKNOWLEDGMENT

The authors would like to thank CASIO Computer Co., Ltd. for its valuable supports contributed to SiP investment.

REFERENCES

[1] T. Akazawa, Nikkei Electronics, vol. 9-1, pp. 158-159, Sep. 2003 (in Japanese).

[2] W. Garg, S. Lacy, D. E. Schimmel, D. Stogner, C. Ulmer, D. S. Wills, and S. Yalamanchili, "Incorporating Multi-Chip Module Packaging Constraints into System Design," Proc. 1996 European Design and Test Conference, pp. 508-513, 1996.

[3] P. Dehkordi, K. Ramamurthi, D. Bouldin, and H. Davidson, "Early Cost/Performance Cache Analysis of a Split MCM-Based MicroSparc CPU," Proc. the MCM Conference, Feb. 1996.

[4] A. I. Kayssi, K. A. Sakallah, R. B. Brown, R. J. Lomax, T. N. Mudge, and T. R. Huff, "Impact of MCMs on System Performance Optimization," Proc. of Multichip Module Workshop, Santa Cruz, pp. 919-922, March 1991.

[5] A. Kazama, T. Satoh, Y. Yamaguchi, I. Anjoh, and A. Nishimura, "Development of Low-Cost and Highly Reliable Wafer Process Journal of Japan Institute of Electronics Packaging, vol. 5, No. 3, pp. 264-271, May, 2002 (in Japanese).

System LSI Design with C-based Behavioral Synthesis and Verification

Kazutoshi Wakabayashi
System Devices Research Laboratories
NEC corp.
Kawasaki, Japan

Abstract— This paper presents the effects of System LSI design with C language-based behavioral synthesis. The proposed C-based tool flow, and then how to synthesize and verify an entire system LSI will be explained. The merits of behavioral synthesizable modules and configurable processors are then discussed.

I. INTRODUCTION

The design productivity gap problem is becoming more and more serious according to the increase of SoC size. In the mid-1980s, gate-level design shifted to register transfer level (RTL) design when the gate size exceeded the 100K gates. We assume a hundred thousand is the upper limit for directly handled parts to design in several months with suitable number of people which can be practically paid for. Currently, more than one million gates circuits are commonly used just for random logic portion, which require several hundreds thousand lines of RTL description. Therefore, we are already in a stage to shift to a higher level of abstraction, namely the behavior level, which allows designers "less number of description" and "higher reusability".

A higher level description requires less number of codes and gives higher simulation performance. These two facts are the main effects of higher-level shifting. They are "code size" and "simulation performance". For our typical example, one million logic-gates-circuit requires about 300KL RTL (e.g. verilog) codes, and 40KL C code with our statistical data on several designs. Also, RTL simulation for 300KL sometimes takes 10 or 100 times slower than 40KL behavioral simulation. This statistical analyses is discussed in our paper[2]. It is important that implementing SoC once at a behavior level is mandatory to enjoy those two merits. Since these merits are recognized through some commercial chip successes, Behavior Synthesis or High Level Synthesis is gaining acceptance within the design community, especially in Japanese Industries. As a behavior description, C language is preferred because embedded software is often described in C, and its design tools, such as debugger and libraries, editors, are easily available. In this paper, we first

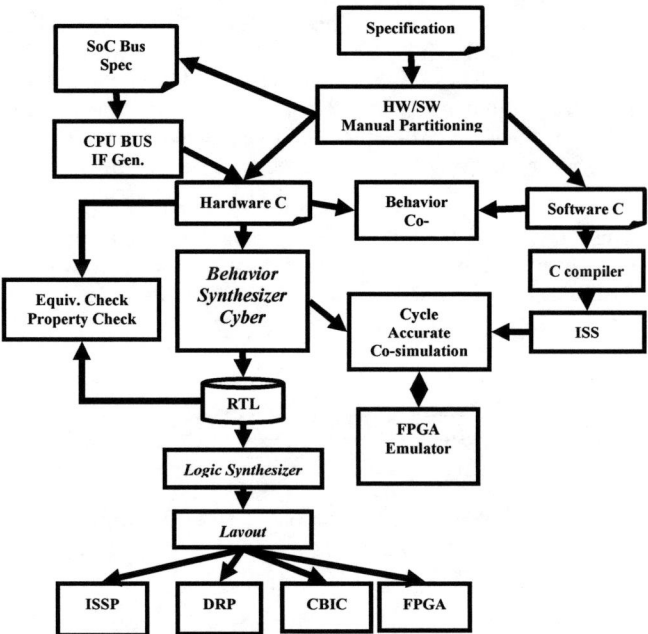

Fig.1 C-based Design Flow

describe our C-based design flow with tool chain, then discuss on the statistical data for code efficiency and simulation performance. Using some industrial designs, we show how our behavior synthesis creates high-quality chips and can be used for wide application areas, not only for some specific areas like filters.

II. C-BASED DESIGN FLOW

We assume that a typical system LSI contains several CPUs or DSPs and dedicated hardware modules and some pre-designed or fixed IP modules, and they are connected through buses. Initially, each dedicated hardware module such as ECC, encryption module, is described in C, and after its functionality is verified in C simulator and debugger, the hardware module is synthesized with our behavioral synthesizer. Also, configurable processors are synthesized from C description in our environment. The bus interface circuits to a CPU bus are automatically generated according to a library of the CPU bus specifications. After synthesizing and verifying all hardware modules, our design environment allow designers to create an cycles-accurate simulation model for an entire system LSI including CPUs, DSPs and such hardware modules. With this simulation model, designers verify both functionality and performance of their hardware and of their embedded software run on the CPU, DSP and the generated configurable processors. Behavioral synthesis is very quick, so designer can modify the hardware modules and embedded software until the last stage of the logic design stage. Note that currently, we have to synthesize each hardware module one by one, and that we do not use any automatic partitioning tool which partitions one huge C source code into various hardware modules and software modules. This means our current behavioral synthesizer just exploits operational parallelisms, and architectural level parallelization is defined by a designer.

Our design flow is shown in Fig.1. A hardware C is transferred into synthesizable RTL description with our "Cyber" behavioral synthesizer[1] according to design constraints such as clock frequencies, number and kind of function units, memories. Functionality of hardware C can be verified at behavior level, and performance and timing should be verified at cycle-accurate level (or RTL). The generated RTL is so optimized by the behavior synthesizer that debugging the RTL is not so easy. Therefore, so we provide a behavioral C source code debugger linked to cycle-accurate or RTL simulation.

After verifying each hardware module, designer simulates the entire SoC to analyze the performance and/or to find inter-modules problems such as low performance by bus collision, or inconsistent bit orders between modules. This entire chip performance simulation takes huge period with RTL-based HW-SW co-simulation. The Cyber behavior synthesizer generates cycle accurate C++ simulation models which can run hundred times faster than RTL model. Our HW-SW co-simulator called "ClassMate" takes the generated cycle-accurate model. The ClassMate simulator allows designers to simulate and debug both hardware and software at the C source code level at the same time. If some performance problems are found, designers should change the hardware software partitioning or algorithm in C description, and then can repeat the entire chip simulation in short turn-around-time. The C description is the only final implementation in this environment. This entire chip simulation can be accelerated by FPGA emulation board[3]. Formal verification tools are tightly linked to our behavior synthesizer. The generated-circuit information, such as distinction of control-datapath registers and state transitions, are used by our property model checker, and the input-behavior-description information, such as loop boundaries and array expansions, are used by our C-RTL formal equivalency checker. Such information enables formal tools to handle larger circuits. Designer can specify assertions or properties in the behavioral C description. A test bench generator is available, which inputs test patterns for a C

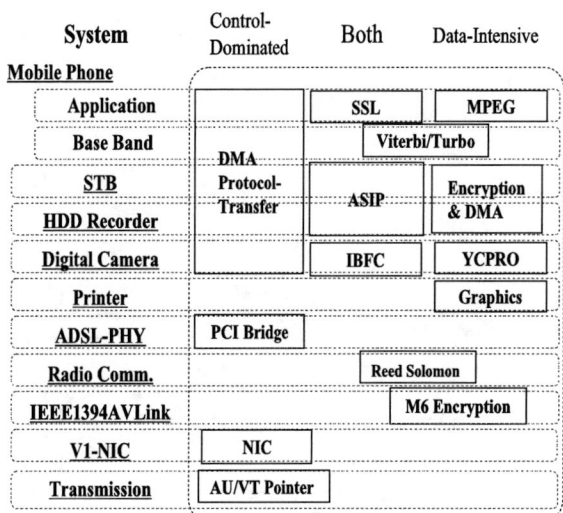

Fig.2 Applied Applications for Behavior Synthesis

simulation and outputs stimulus to an RTL simulator at appropriate cycles. Analysis data on generated circuits, such as number of wires, areas, execution cycles, are used for quick micro-architectural exploration.

III. APPLICABLE AREAS FOR C-BASED SYNTHESIS.

It is often said that behavior synthesis is applicable just for digital signal processing areas. DSP area is actually a favorite area for behavior synthesis, because DSP application often has a simple data flow, and inter-module communication often uses simple mechanism such as FIFO where complicated protocol sequence is not needed. For such designes, a behavior synthesizer has a large freedom to schedule interface operations, and this is rather easily handled by the automatic scheduler of a conventional behavioral synthesizer. On the contrary, a complex controller communicating with other modules at almost every cycle is not easily handled by a simple automatic scheduler, but this type of circuits also can be behaviorally synthesized. For this type of circuit, cycle-by-cycle behavior should be described in the "behavioral domain", not in the "structural domain" like RTL language, and it should be synthesized to realize the complex communication as exactly described. This constrained communication problem often occurs in designing bus or memory controllers. However, recently, many hardware modules are communicate through a fixed on-chip CPU buses instead of special hardware interface. The interface for the prefixed CPU bus, such as AMBA bus, are easily generated automatically according to the pre-designed interface library.

Some behavior synthesis tool might work well only for some favorite application areas, but behavior synthesis technique itself can be applied for wide range of applications including control dominated and data-intensive circuits. It is important to use single language for entire design. If some modules are described in C, and some others should be described in RTL, designers rather like to use only RTL.

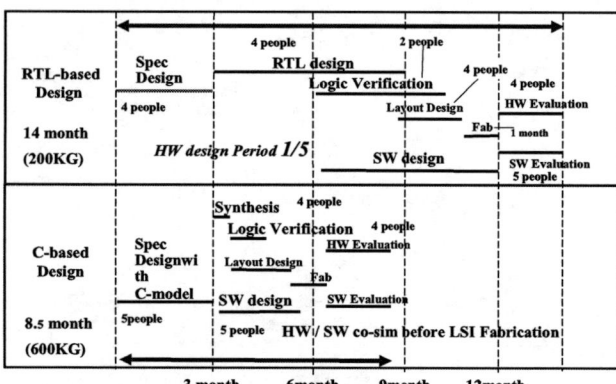

Fig.3 Design Period Comparison

Fig.2 shows our design examples with our behavioral synthesizer. Wide ranges of application including control dominated circuits and data-intensive circuits are successfully designed.

IV. HARDWARE-SOFTWARE CONCURRENT DESIGN

Since C-based behavior synthesizer automates functional design of "hardware", hardware design period is naturally shortened as described in the previous sections. On the other hand, we would like to emphasize that it enables us to shorten the design period for embedded software. We successfully designed an entire 2.5Generation-mobile – phone-application-chip whose functions are motion CODEC, audio recorder, auto graphic resizer, etc. with our C-based design. It was reported that all logic bugs are eliminated before chip fabrication using the entire chip simulation with our "ClassMate" HW-SW co-simulator[2] and our FPGA emulator[3]. Fig.3 shows design period comparison for the two different mobile phone base-band chips. Smaller and simpler circuit (200KG) with the RTL design takes longer time to production than larger circuit (600K) with the C-based design. The first reason is that we cannot use hardware-software co-simulation for the RTL-besed design since RTL co-simulator is too slow to debug embedded software. Second reason is that software designer has to wait long time for hardware designer to accomplish the RTL design. Moreover, in many cases, software designer has to wait an evaluation chip to evaluate their software. C-based design allows very quick creation of simulation model for embedded software in very early stage. This means C-based design enables true hardware software concurrent design. Moreover, it is not always possible to fabricate several evaluation chips until all hardware bugs are fixed, so some remaining hardware bugs have to be escaped by its embedded software in the RTL-based design. Our fast HW-SW co-simulation allows us to eliminate the hardware bugs using real embedded software before chip fabrication.

V. REUSABILITY OF BEHAVIOR DESCRIPTIONS

The importance of C-based behavioral design exists in high-reusability of behavior model (behavioral IP). An RT level reusable module, called "RTL synthesizable IP", is successfully used just for circuits of fixed performance such as bus interface circuits. However, RTL-IP for general functional circuits such as encryption is of use only for short term, since RTL-IP's "performance" is hard to change drastically. For instance, an RTL-IP for encryption for 200Mbps is difficult to be applied to encryption for 400Mbps, because RTL-IP structure is fixed and a logic synthesis tool is not able to reduce its delay to the half. On the contrary, a behavioral IP is more flexible and more reusable than RTL-IP, since it can change its structure and behavior synthesis tool can generate various performance circuits by changing synthesis constraints such as number of function units and clock frequencies. Table-1 shows that various circuits of different "clock-frequency" can be generated from a single behavior IP. This IP is a BS broadcast descramblers (Multi2). All generated circuits satisfy the required performance (more than 80Mbps) with various frequencies. Note that the high speed circuit (108MHz) uses smaller gate size than the slow speed circuit (33MHz). This is never happened to an RTL-IP, which follows the area-delay tradeoff relation of logic synthesis. However, it is natural that a behavior synthesizer generates a smaller circuit of higher clock frequency for the same performance, since less parallel operations are necessary to achieve the same performance at higher clock frequency. Lastly, RTL-IP in VHDL cannot be changed to that in verilog so easily. Behavior IP also solves this language problem.

More substantially, a behavior IP is easy to modify its "functionality" and "interface" than an RTL-IP. We designed two types of "Viterbi" decoders for mobile phone and communication satellite. The two requires different Bit Error Rate, and so several parameters like encode rate and constraint bit length affecting BER have to be changed. Changing these parameters requires significant modification of RTL-IP description, however, only slight modification is necessary for a behavior IP.

VI. CONFIGURABLE PROCESSOR SYNTHESIS

Since chip fabrication cost becomes very expensive, system LSI should have more customizability. For this sake, recent system LSI usually has several configurable processors besides a main CPU. These configurable processors are in the hardware part from the embedded software designers' point of view. Therefore, these configurable processor should be small and of high performance for some specific application. Such a configurable processor is also called as an Application Specific Instruction set Processor (ASIP). ASIP employs

Table-1 Multi2 Descrambler for Various Clock Frequencies

Clock Frequency	Generated Gate size	Generated RTL size	Performance
33MHz	57KG	7.0KL	80Mbps
54MHz	42KG	5.9KL	80Mbps
108MHz	26KG	2.5KL	80Mbps

Table.2 BEHAVIRORAL CONFIGURABLE PROCESSOR SYNTHESIS

	STB stream	Base-band DSP	Application DSP
MIPS(Clock)	72(108MHz)	15(15MHz)	60(60MHz)
#.of Inst.	Base: 81 +Adding: 24	Base: 17 +Adding: 17	Base: 65 +Adding: 21
Gate Size	43K	20K	120K
Behavior	2.1KL	1.3KL	2.5KL
Generated RTL	13.0KL	11.4KL	26.0KL
Man-power	1.5 m-m	0.5 m-m	0.8 m-m

Table.3 BEHAVIRORAL BASE-BAND DSP SYNTHESIS RESULTS

	Behavior	RTL
Code Size	1.3KL (1/7.6)	9.2KL (1)
Simulation	61.0 Kc/s(203x) Pentium3@1GHz	0.3Kc/s(1) UltraSparc-II@450MHz
Gate Size	19KG	18KG

some adding instruction-sets to accelerate some application specific procedure. There are several commercial ASIPs, such as Xtensa[6] from Tensilica and Mep[7] from Toshiba. Their base-processor and co-processors for adding instructions are described in RTL, and they are logic synthesisized. On the other hand, our ASIP's base processor and supplementary instructions are described fully in behavioral C description, and they are behavioral synthesized. Therefore, the base-processors and the adding instructions are allowed to share function units. This sharing leads to smaller circuits than the conventional RTL-based ASIP generation. For an ASIP base-processor, we added 24 instructions suitable for stream processing, such as CRC calculation, with only 25% area increase (34KG to 42KG) because of FU sharing. Also, C-based ASIPs are more flexible than RTL-based ones in terms of public register number, pipeline stage number or interrupt policy. In Table-2, three ASIPs synthesis results are summarized[4]. All ASIPs were relatively small, but had enough performance for a specific application area by using the adding instructions. All C-based ASIP designs required only as one tenth man-power as RTL-based. The Base-band DSPs were designed both with C-based and manual RTL design. The results are shown in Table-3[5]. The two designs had comparable gate size and delay. The code efficiency of C-based is 7.6 times and simulation speed-up of that is 200 times. These values for ASIPs are similar to those for other circuits in Section III, which means behavior design has similar efficiency also for ASIPs.

VII. SUMMARY AND CONCLUSIONS

We demonstrated that our C-based design environment is able to increases design productivity in terms of code efficiency and simulation performance, with statistical analysis. Currently, we are using our behavior synthesis for most of newly designed modules and more system LSIs are verified with our C-based simulation. High reusability of behavioral module, such as changing performance, clock frequency, functionality and communication protocols much easier than RTL, is discussed with a commercial chip example. Currently, our behavior synthesis tool is obviously as mature as logic synthesis tools in the late 80's, when designers started to use them widely. However, it is not so easy for some RTL designers to shift from RTL "structural" domain thinking to "behavioral" domain thinking, and it shields conventional RTL designers from such paradigm shifting. Therefore, education on the behavioral thinking for RTL designers is a crucial and difficult task. We already trained several hundreds designers for a few years, and somehow succeeded to persuade them to higher level of abstraction, but not completely succeeded to understand them how behavioral description should be. Verification and debugging capability at behavior level is essential for this behavior design environment. As a future plan, timing validation and wirebility checking, ECO at layout level should be considered in our environment to treat with large and fast operating circuits. Our behavioral synthesizer fits for low power design, but not so good for very high clock frequency design like 1GHz logic. We plan to implement several tricks for high speed circuits in our behavioral synthesis. Also, we plan to develop system level tools.

ACKNOWLEDGEMENTS

The author would like to acknowledge his many colleagues in System CAD Technology Group at NEC corpration, and at NEC Information Systems Ltd., at NEC Electronics Corp. for their effort to develop the tools in this C-based environment. The author works for research on behavior synthesis and the entire tool flow organization, and he introduced the concept of the total C-based design environment on behalf of the whole development team.

REFERENCES

[1] K.Wakabayashi and T.Okamoto, "C-based SoC Design Flow and EDA Tools: An ASIC and System Vendor Perspective," IEEE Trans. on Computer-Aided Design of Integrated Circuits and Systems, Vol. 19, No.12, pp.1507-1522, Dec. 2000

[2] H.Kurokawa, et al., "Study and Analysis of System LSI Design Methodologies Using C-Based Behavioral Synthesis", IEICE Trans. On Fundamentals, Vol.E85-A,. Jan. 2002

[3] Yuichi Nakamura, Kouhei Hosokawa, Ichiro Kuroda, Ko Yoshikawa, Takeshi Yoshimura, " A Fast Hardware/Software Co-Verification Method for System-On-a-Chip by Using a C/C++ Simulatorand FPGA Emulator with Shared Register Communication", DAC,2004, pp.299-304

[4] S.Takahashi, H.Kurokawa and M.Otsubo, "behavior synthesis of processor for STB streami data", A-3-16, No.7476, Proc. of IEICE annual meeting. March,2002

[5] H.Nakajima,"Semi-custom processor design experiences using C-based behavioral synthesis", A-3-15, No.7475, Proc. of IEICE annual meeting. March, 2002

[6] Xtensa, http://www.tensilica.com

[7] Mep,. http://www.mepcore.com/english/

2005 IEEE International Symposium on Circuits and Systems (ISCAS)
Approach for Physical Design in Sub-100nm Era

Hiroo Masuda and Shinichi Okawa
Semiconductor Technology Academic Research Center
(STARC)
Yokohama, Japan

Masakazu Aoki
Electronic Systems Engineering
Tokyo University of Science
Suwa, in Nagano, 391-0292 Japan

Abstract— In sub-100nm processes, various physical phenomena come up as critical red-brick in designing circuits and LSIs. We focus on design for variability (DFV) for LSI-chip design, taking the within-die variations into consideration. Main approach for the purpose is a new Test Structure (TEG: Test Element Group) to measure the within-die variation of elements (MOS, R, C) and ring-oscillators. The precise measurement has been achieved with careful TEG design including on-chip circuit, such as CBCM, Kelvin pattern. Reliable measurement data were analyzed statistically. Variation caused systematic and random physical sources have been successfully decomposed with newly developed extraction strategy.
The data exhibits an extremely large variation in N/PMOS drain current (Ids) and threshold voltage (Vth). The main sources of the random variation are doping fluctuation and line edge roughness (LER) in small size MOS transistors. Ids variation is affected with the doping fluctuation. On the other hand, Vth variation is sensitive to LER. In interconnect variation is essentially small compared with the Ids/Vth variation of MOS transistor, however its variation is systematic component dominant. Ring oscillator Tpd variation found to be closely related to the Ids variations, showing the correlation coefficient of 0.9. In summary, Design for variability is one of the most difficult challenges in 65-90nm processes. Statistical design in early stage of design will be necessary.

Agenda

- Introduction
- Precise measurement of Variability
  - DMA-TEG
- Analysis and Design Variations
  - Die to Die: Systematic
  - Within Die: Systematic and Random
- Summary

Overview: DFM (Design for Manufacturability)

- **Reduction of relative variability margin**
- **65-90nm: sources of complicated variability design**
      Increase of WID (Within-die) random variation
      Density effects, Dummy fill, OPC ...
      Interconnect rich LSI: RC systematic variation
- **Product-level yield and variability prediction**
      High-volume of data and their mining
      Defect analysis
- **Variability design methodology: Statistical!**
      Systematic variation: Realistic worst(Least Conservative)
      Random variation: Statistical treatment

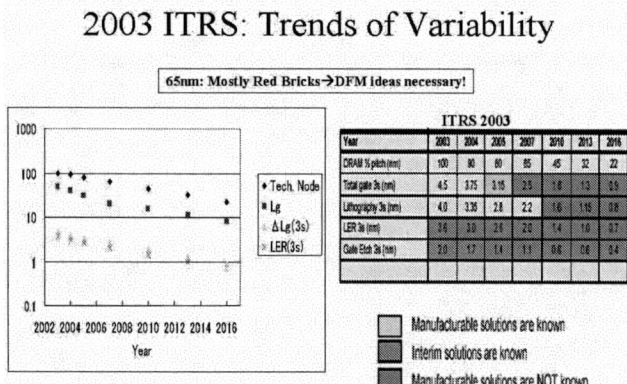

2003 ITRS: Trends of Variability

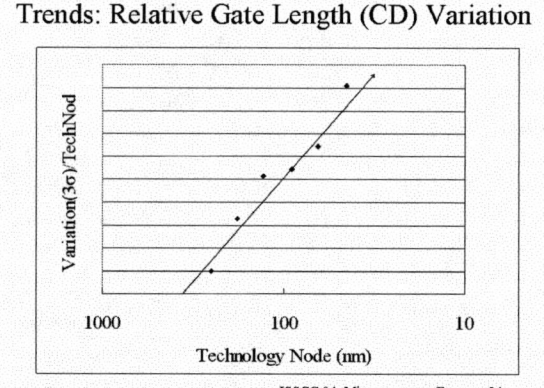

Trends: Relative Gate Length (CD) Variation

ISSCC 04, Microprocessor Forum, p24

Trends: NMOS Vth Variation (Relative)

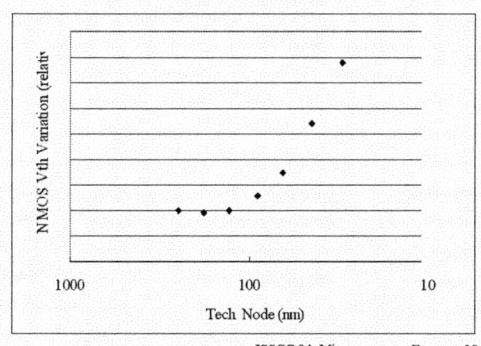

ISSCC 04, Microprocessor Forum, p29

Performance Trend for Future Technology Node

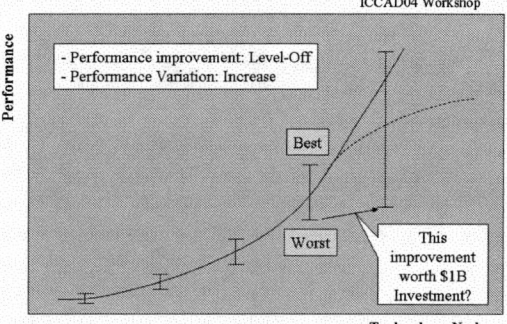

ICCAD04 Workshop

STARC Approach for Variability
(TEG: from components to CKTs)

- Within-Die Variation TEG(DMA-TEG)
    - 16*16 Cell Matrix Array
    - TRS, R, C, Ring-OSC/Cell

- Characterization into Systematic & Random components of Variation
    - D2D: Systematic
    - Within-Die: Systematic & Random

- => Get Statistical !

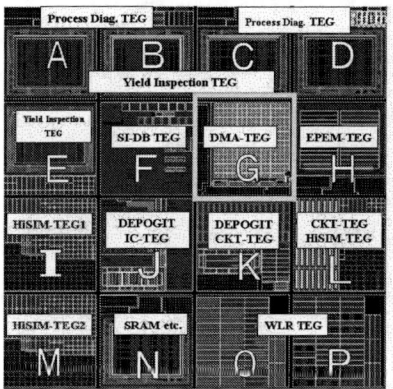

- Chip Size
  20.5X20.5 mm2
  (5X5mm2 16Sub-chips)
- Design Rule
  90nm ASPLA-DRM
- Contents
  -DMA-TEG
  WID variation of components
  (MOS, R, C, Ring-osc)
- Yield Inspection TEG
  Yield, Variability
  Electrical measure dimensions
- Process Diagnosis TEG
- WLR TEG
  (Wafer Level Reliability)

Features: 90nm DMA-TEG
(Device Matrix Array)

- WID (Within Die): Accurate variation measure of devices
- All the devices used in CKT-design:
    - NMOS, PMOS, R, C, Ring-osc
- High Precision measurement with on-chip circuit and low-leakage jig
- Etc.

90nm DMA-TEG

Integrity: 350 devicesX16² = 90K devices@chip
TEG area: 4mmX4mm

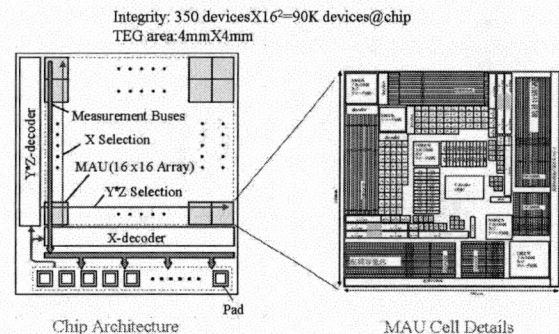

Chip Architecture   MAU Cell Details

- 90nm DMA(Device Matrix Array) : Features in detail

2002 DMA (STARC 90nm) vs. 2001 DMA (STARC 130nm) #1

Items	90nm DMA chip	130nmDMA chip
Chip size	5.0mmX5.0mm	ibid.
DMA area	4.2mmX4.4mm	4.0mmX4.0mm
MAU size	240umX240um	ibid.
Array size	16X16	14X14
Samples/chip	256	196
Devices/MAU	350*1, *2	148*1
Total # devices/chip	89600	29008

Components	90nm DMA chip	130nm DMA chip
C-measure structures	41X2 *2	30
R-measure structures	62X2 *2	51
NMOS structures	26X2 *2	26
PMOS structures	26X2 *2	26
Ring-osc types	20X2 *2	15
Total	350	148

*1 Including size-variation
*2 0/90 degree dependency

Capacitor Measurement ($3\sigma = 23aF$)

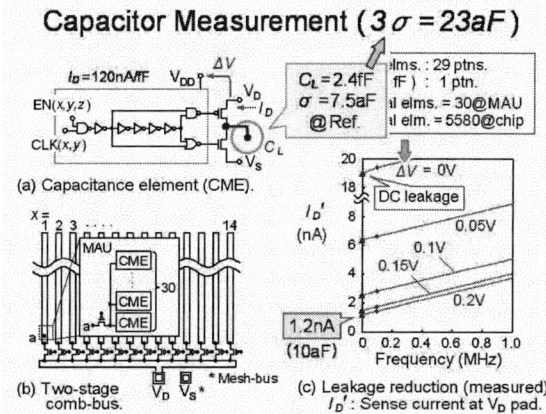

(a) Capacitance element (CME).
(b) Two-stage comb-bus.
(c) Leakage reduction (measured).
I_D': Sense current at V_D pad.

Typical Library Cell NMOS Transistor:

* Large WID variation!

  Idsat 3σ=19%
  Vth 3σ= 67mV

* Random>>Systematic Compo.

  NMOS Gate area(L*W) minimization causes impurity fluctuation under the gate

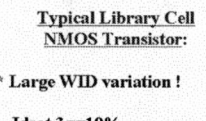

Extraction technique of Systematic & Random components of variation

◆ Application of a 4th-order polynomial fitting for date decomposition.

$$z(x,y) = a_0 + a_1x + a_2y + a_3x^2 + a_4xy + a_5y^2 \\ + a_6x^3 + a_7x^2y + a_8xy^2 + a_9y^3 + a_{10}x^4 \\ + a_{11}x^3y + a_{12}x^2y^2 + a_{13}xy^3 + a_{14}y^4.$$

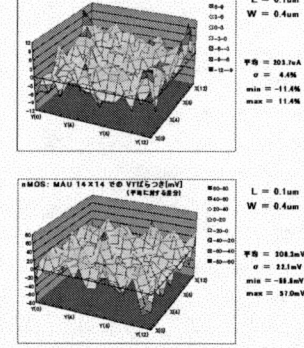

(a) Raw data as measured. (b) Fit data as a systematic part. (c) Residual as a random part.

WID Random-variation: "Strictly" Normal distribution!

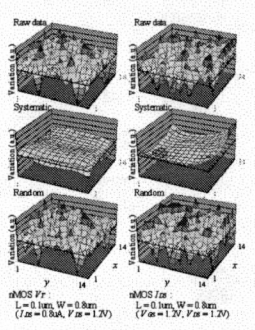

WID Systematic-variation: not unique among die-to-die in Wafer

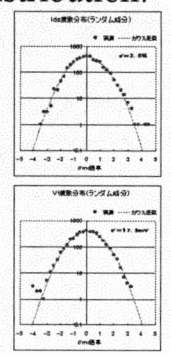

WID Systematic variation: Correlation length (mm)

Correlation Length=1-3mm: Correlation is High within ~1mm distant
→ Small Chip <1mm : "Purely" Systematic (Correlation=1)

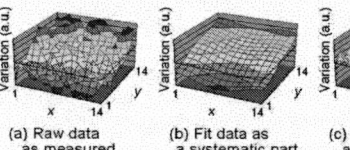

Random Variation σ; Gate area dependency

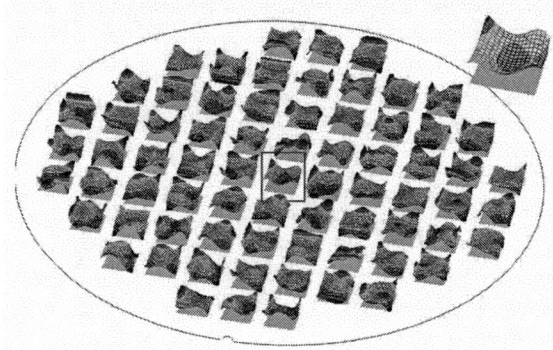

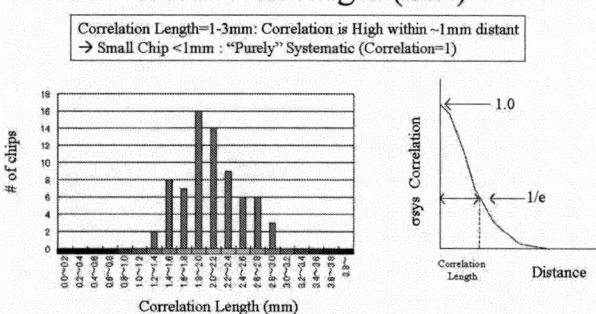

90nm DMA(Device Matrix Array):
NMOS Ids WID(Within Die) Variation vs. size

Sample Data #1 : nMOS Vt

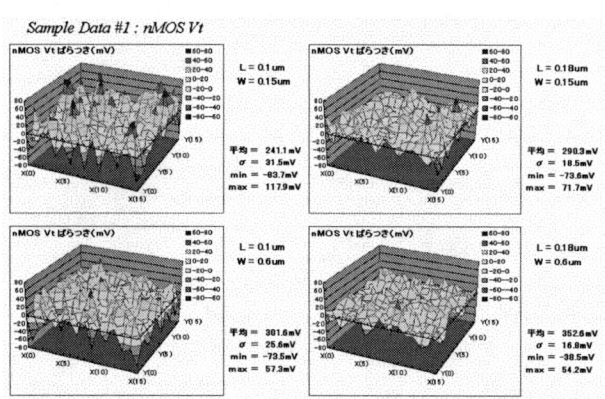

Sample Data #3 : nMOS Ids

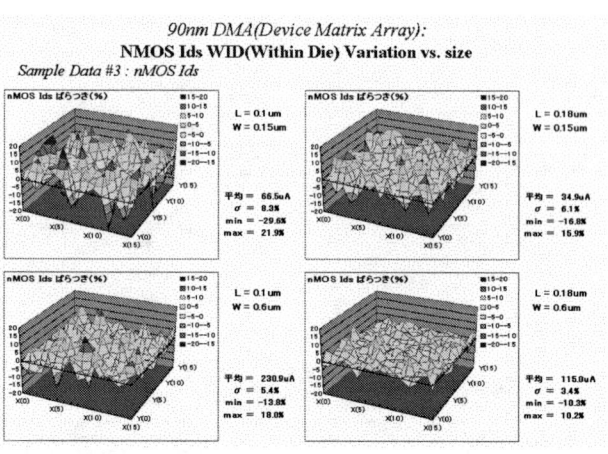

90nm DMA(Device Matrix Array):
M2-R WID(Within Die) Variation vs. dimension
Sample Data #5 : Metal2 Interconnect Resistance

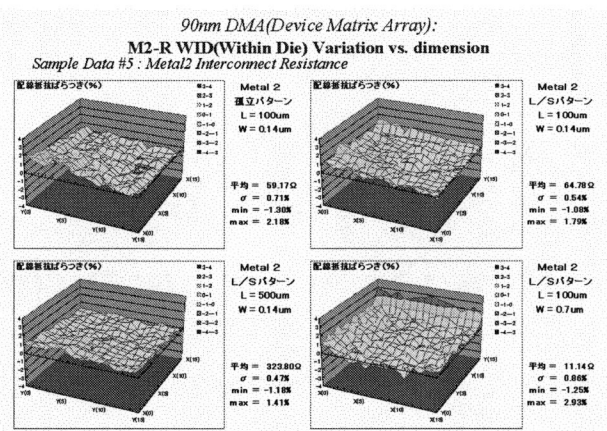

N/PMOS Ids Variation vs. gate-area
Comparison between 130nm/90nm

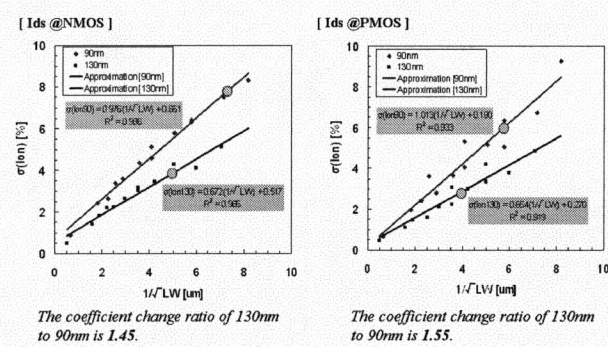

The coefficient change ratio of 130nm to 90nm is **1.45**.

The coefficient change ratio of 130nm to 90nm is **1.55**.

WID & WIW Variation(90nm Process)
(X & Y Axis:Relative)

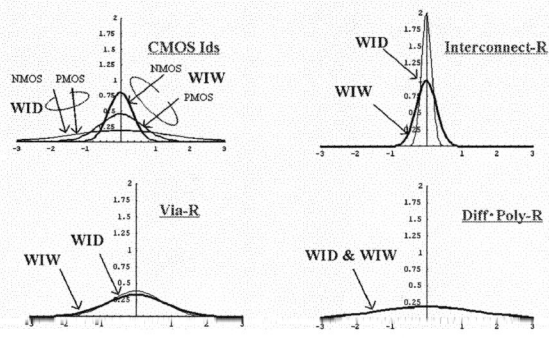

Ring-OSC: Tpd-Variation (Experiment)

Type:Inv, 2NAND, 2NOR, 3NAND, 3NOR
Loading:FO=1&4, M2&M3 Wires(FO=4 level)

(1) Tpd(WID) Variation:
  * σ=3.74 - 7.48%, (3σ=11-22%)

(2) Other measurement results:
  * N/PMOS size (area) dependency:
    large area -> smaller σ
  * Loading effects(@M2, FO=4): change in negligible

→ MOS Ids variation = Tpd variation/Stage

Strict correlation between
NMOS Ids vs. Ring-osc Frequency

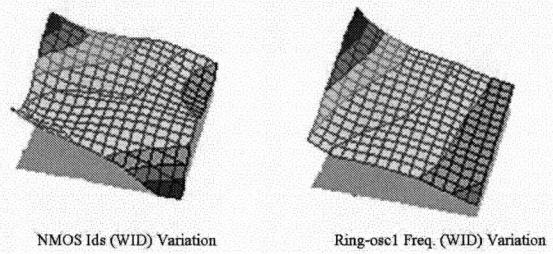

NMOS Ids (WID) Variation   Ring-osc1 Freq. (WID) Variation

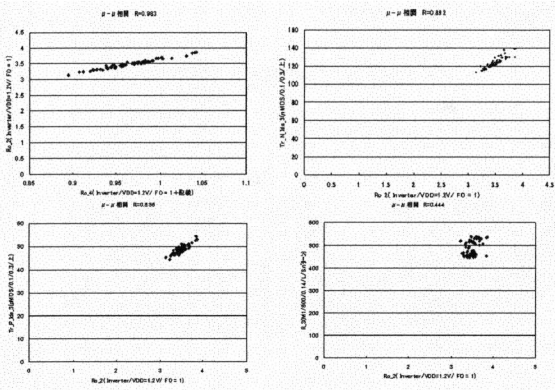

Summary on 90-130nm DMA-TEG

- CMOS Ids variation used in Library found to be un acceptably large (3σ :ΔIds/Ids=25-30%)
- New process generation (90nm against 130nm) causes 1.5-2.0X Increase in Ids-random variation
- Wire R/C(100-500um long): small WID variations compared with Ids ones(3σ:ΔR/R=1.5-4.0%)*
- Tpd/Stage variation strongly correlate with N&P MOS Ids & Tpd variation dominated by MOS Ids variations

(*) Variation of Wire R/C: Systematic variation dominant

Conclusion

- DFV (Design for Variability) necessary in 65-90nm process era
  - Relative increase of variation, A new WID variation
  - Timing Closure will be more difficult
- A new TEG for DFV has been established, with DMA-TEG and efficient analysis software
- Future effects on circuit design:
  - Statistical treatment of Random variation of Circuits
  - Minimize Systematic variation with sophisticated process control
- Get Statistical! In early stage of design

An Advance RTLtoGDS2 Design Methodology for 90nm and below System LSI's to Solve Timing Closure, Signal Integrity and Design for Manufacturing

Nobuyuki Nishiguchi
Design Technology Development Department
STARC
Yokohama, Japan
nishiguchi@starc.or.jp

Abstract— An advance design methodology for 90nm and below system LSI is described. The methodology provides the total solution to solve timing closure, signal integrity issues and design for manufacturing in RTLtoGDS2 silicon implementation. It also focuses on hierarchical and low power design implementation. Sign-off criteria to guarantee the first silicon success. And also the methodology is adapted for actual CPU RISC core design in 90nm process technology and the first silicon work well as designed.

I. DESIGN REQUIREMENTS IN 90NM AND BELOW PROCESS TECHNOLOGY

Recent system LSI's are using advance process technology such as 90nm and below and are complicated due to integrating much more functionalities. In implementing the system LSI from RTL to GDS2 [1], the most critical design issues are along with timing closure, signal integrity and design for manufacturing (DFM). The timing closure issues include accuracy of delay calculation, on-chip and off-chip delay variation due to the process variations and interpolate method for voltage and temperature. The signal integrity issues consist of cross talk noise, cross talk induced delay and IR drop in the both of static and dynamic phenomena. In 90nm and below process technology, design should consider manufacturing such as very complicated design rules, metal density rule and antenna effects. Those issues are not independent but dependent each other. Eventually it takes long design time with a lot of iterations in design converging and closure and then design cost increases drastically. Therefore design methodology is more important to solve the issues simultaneously and effectively and to guarantee design which makes the first silicon work well on EDA design environment. This paper describes RTLtoGDS2 design implementation methodology, called STARCAD-21, to solve the issues and to achieve first silicon success.

II. CONCEPTS OF THE STARCAD-21

The STARCAD-21 introduces two concepts to reduce design iterations and to realize strait forward design flow to solve design issues simultaneously such as timing closure, signal integrity and DFM. One is "Check and go" and another is "Prediction and Prevention". The RTLtoGDS2 design implementation is split to three design phases. They are "Estimation phase", "Refinement phase" and "Final implementation phase". Among the phases the hand-off criteria are set up and design should be satisfied with them and then goes to the next design phase. Finally there are the sign-off criteria before manufacturing. The sign-off criteria are defined to guarantee design which makes the first silicon success. Each design phase introduces "Prediction and prevention" method for design issues such as area, timing, signal integrity, power and DFM. The Fig. 1 shows the concept.

III. FEATURES OF THE STARCAD-21

The STARCAD-21 is developed as three release versions and one future release version. The first version was released in April, 2004 and the second in October, 2004 and is released the third in April 2005. The other future version will be release in 2006. The first version provides the basic functions in 90nm system LSI design methodology. The second supports the hierarchical design with several types of blocks and pushing down power and signal lines into lower hierarchy.

This project is supported by Incorporated Administrative Agency NEDO as a part of the Focus 21 program.

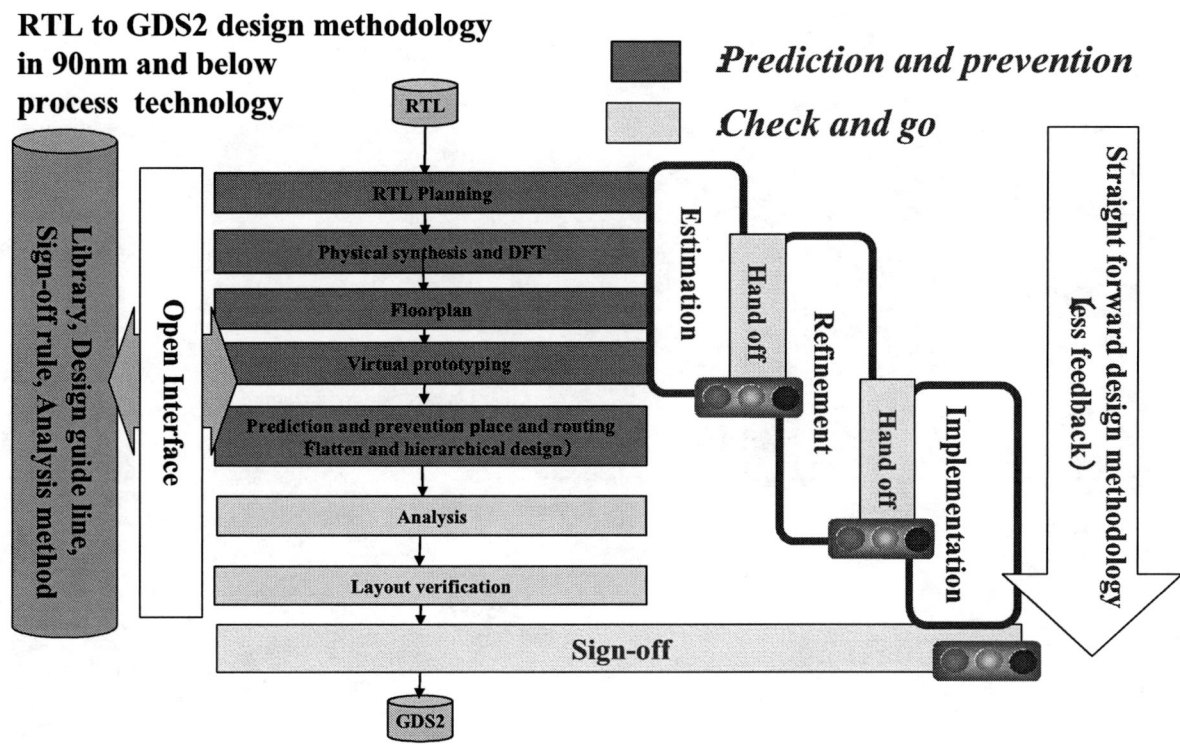

Figure 1. Concept of the STARCAD 21

The third supports low power design methodology including multi-power supply on a system LSI. The STARCAD-21 has two basic design flows based on EDA tools which are provided from two EDA vendors respectively as main stream flows.

A. The basic design methodology

The features are as follows.

a) Basic design methodology in 90nm technology

　1) Integrated methodology for timing, SI and DFM closures

　2) Higher efficiency in timing with pseudo hierarchical design

b) Consistent EDA library specification with EDA tools focusing on timing and SI

c) Clear sign-off criteria taken into account of process variation with consideration of process variation without any penalty in timing closure (Fig. 2)

d) Important tool evaluation for 90nm phenomena analysis focusing on Clock tree synthesis (CTS), Signal integrity (SI), Layout parasitic extraction (LPE) and Electron migration (EM)

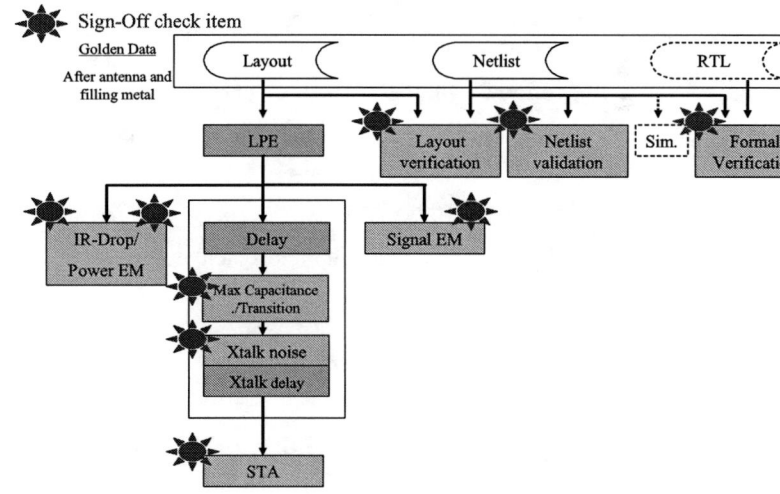

Figure 2. Sign-off criteria

B. The standard design methodology

Here are the features.

a) Hierarchical design methodology to take true benefit (Fig.3)

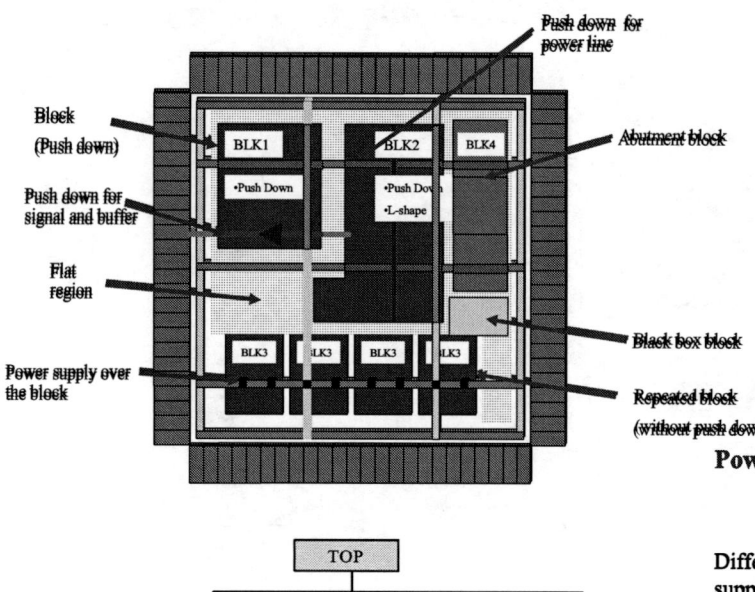

Figure 3. The feature in hierarchical design

1) Handling several types of hierarchical blocks such as abutment, black box, hard macro and repeated blocks

2) Pushing down power stripes, signal lines and buffers into lower hierarchy

b) Reduction for timing verification corners

c) Definition for verification corners in cross talk noise, IR drop and signal EM

d) Timing constrains checker

e) IP modeling specification in hierarchical design

C. *The low power design methodology*

The features described as flows.

a) Low power design (Fig. 4)

    1) Design flow for multi threshold cell

    2) Design flow for multi power

    3) Design rule check for multi power
    (Ex. Power island , Level shifter insertion)

    4) Cross talk analysis in multi power (among different powers)

    5) Clock tree synthesis with multi power island

    6) Gated clock

    7) Voltage variable library (Single library and accurate interpolate tool)

b) Hierarchical sign-off toward minimizing verification for whole chip

c) Enhance for timing constraints handling with multi corners, multi scenario and golden timing constraint flow

d) Sophisticated hand-off criteria

e) Sign-off criteria with less design margin for on-chip variation

f) Robust IP modeling

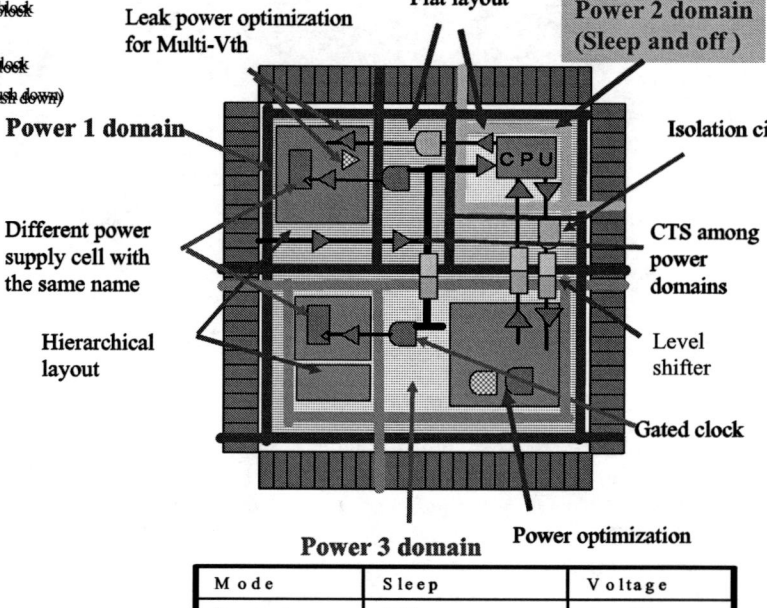

Mode	Sleep	Voltage
Power1	ON	1.2V
Power2	OFF	1.2V
Power3	ON	1.0V

Figure 4. The feature in low power design

D. *The efficient design methodology in future*

As a future plan in the next year the STRACD-21 will has high effective design methodology. More efficient timing closure method such as multi corners and scenarios handling in timing verification, golden timing constraints generation and reduction of design margin will be developed. In DFM design intents in the implementation should go to OPC (Optical Proximity Correction) and mask data preparation to shorten design and operation turn around time and improve yielding in making mask and chip fabrication. Moreover physical phenomena such as considerations of on-chip variation will be more critical in system LSI design therefore the STARCAD-21 will be enhanced with them.

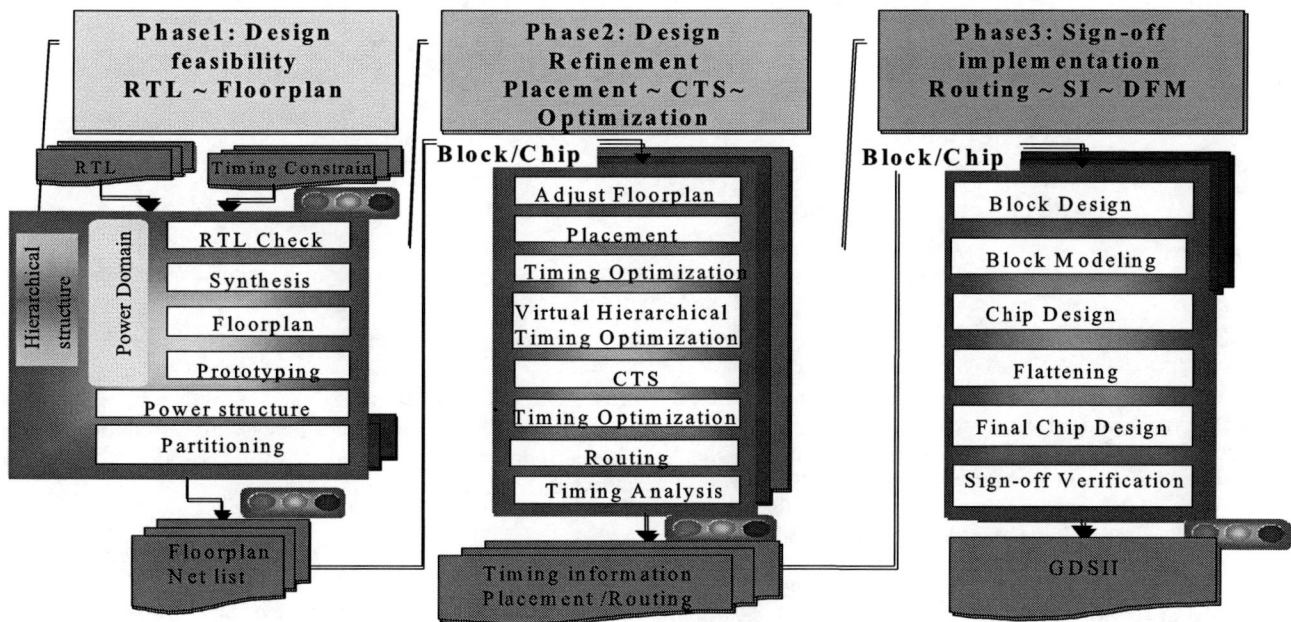

Figure 5. Design flow

IV. DESIGN FLOW

The Fig. 5 shows the design flow in the third version which is supporting low power design. When the design flow is developed, each design tool in every phase is very carefully evaluated in its function, performance and accuracy and then selected. And also according to accuracy in analysis tools, hand-off and sign-off criteria are set up with proper design margins take into account process technology including variation. And design libraries are also important because they depend on tool performance and accuracy. The design libraries are also generated under investigation to keep accuracy in the tools and the libraries. The STARCAD-21 also provides how to generate the libraries to keep consistencies from early to end stage of design.

On the basic design flow, the hierarchical design flow is established with budgeting in the early stage of design and modeling for each block which is used in top level design and verification.

In the third release, multi power supply planning capability and check functions are added on the top of the second release.

V. EXPERIENCE

The first version of STARCAD-21 is able to design 32bits RISC CPU cores. Chip images in shown in Fig. 6 (a), (b) based on the two basic flows. The first silicon works well in the both case. And it is to realize similar design turn around time in 90nm to 130nm. It means 33% up in design productivity. Complete solutions in 90nm technology difficulty such as timing, signal integrity and design for manufacturing are provided.

VI. CONCLUSION

The STARCAD-21 is state of the art design methodology for 90nm and below system LSI and is applicable to real system LSI now. The designers in Japanese semi-conductor vendors are use it and the STARCAD-21 is the de-facto standard design methodology

REFERENCES

[1] "TSMC Reference flow," URL, http://www.tsmc.com/

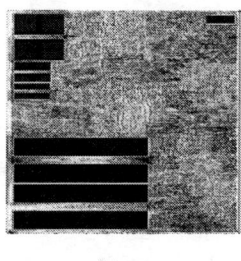

Figure 6. Chip image

CMOS High-Linear Wide-Dynamic Range RF On-Chip Filters Using Q-enhanced LC Filters

Shengyuan Li, Susanta Sengupta, Huseyin Dinc, and Phillip E. Allen
School of Electrical and Computer Engineering
Georgia Institute of Technology
Atlanta, Georgia 30332 USA
gte582w@prism.gatech.edu, ssgupta@ece.gatech.edu, dinc@ece.gatech.edu, pallen@ece.gatech.edu

Abstract—An RF on-chip filter using Q-enhanced LC filter with digitally synchronized gain, center frequency (f0), and quality factor (Q) tuning is implemented in a standard 0.18um CMOS process. The circuit consists of a new high-speed, wide-range, low-distortion, constant-gm OTA design and a new constant tunable discrete capacitor design that maximizes the linearity. The concept is illustrated through the design of a front-end RF filter operating at f0 of 5.775GHz. Simulation results show that the filtering Q can be digitally tuned from 6 to 65 while maintaining a f0 shifting within +-0.2%, gain variation 0.03 dB, 1-dB compression point (P_{1dB}) from –0.9 to 2.69 dBm, and dynamic range from 57 to 80 dB over the 100 MHz passband.

I. INTRODUCTION

Since the first use of on-chip inductor for RF LC filter design [1], consistent effort on improving LC filter performance can be found [2-5]. Theoretically [6], an LC filter can achieve a comparable dynamic range (DR) performance as the current off-chip ceramic and/or SAW filter solution. The upper limit of LC filter's selectivity is mainly limited by the achievable inductor quality factor (Q). In reality, due the nonlinearity introduced by the negative resistance for loss compensation, the varactor for frequency tuning, and the input OTA for gain control, the P_{1dB} is usually limited to only –10 dBm or so with deep-submicron technologies. In this paper, a divide-and-conquer method is presented to improve the overall system linearity while keeping the high dynamic range giving the overall system as shown in Fig. 1. Ideal parallel LC tank loss compensation methods without center frequency shifting are introduced in section II.A, the idea of synchronous tuning is explained in section II.B, tunable discrete capacitor (TDC) for frequency adjusting is present in II.C followed by a new gm-cell design

present in II.D. In section III, a 5.775GHz filter design is present to illustrate the above ideas with simulation results. The simulation results for a 900MHz design are also included. Section IV concludes this paper.

II. FILTER WITH Q-ENHANCED LC TANK

A. Ideal Parallel Compensation of LC Tank Without Center Frequency Shifting

The simple LC tank with parallel loss compensation by R_N is shown in Fig. 2. Inductor L's loss is modeled by a simple series resistor R_L. The objective here is to achieve a quality factor of Q at resonant frequency ω_0. The voltage transfer function is given in (1). There are two requirements or knowns (the desired resonant frequency and Q) and two control variables or unknowns (negative resistance R_N for Q and either reduced inductance or capacitance for ω_0), so the solution is unique. The two methods are illustrated in Fig. 3 and the corresponding solutions are given in Table 1.

B. Synchronous Tuning of f0, Gain, and Q

Synchronous tuning of f0, gain and Q means that when the Q is digitally tuned to meet different selectivity/DR [6] requirements by changing -R_N (= -1/g_N), the center frequency shouldn't drift and the gain should be kept constant also.

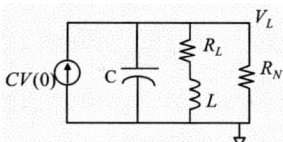

Figure 2. Simple LC tank with parallel compensation

$$V_L = \frac{[(R_L+SL)/L]V(0)}{S^2+S\left(\frac{R_L}{L}+\frac{1}{R_N C}\right)+\left(1+\frac{R_L}{R_N}\right)\frac{1}{LC}} = \frac{[(R_L+SL)/L]V(0)}{S^2+S\left(\frac{\omega_0}{Q}\right)+\omega_0^2} \quad (1)$$

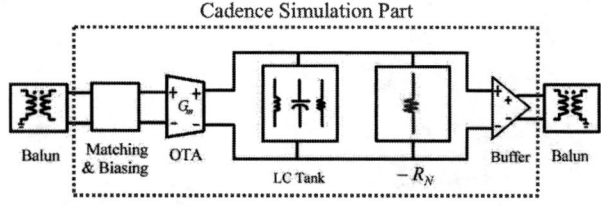

Figure 1. Overall filter system with Q-enhanced LC tank

Thanks to Agilent Lab at Palo Alto, CA for funding

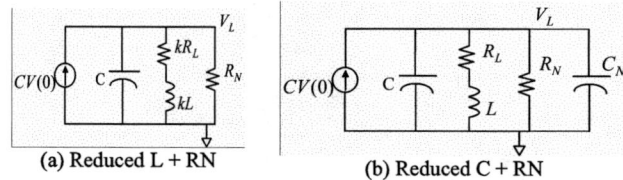

(a) Reduced L + RN (b) Reduced C + RN

Figure 3. Two parallel compensation methods for the LC tank

By observing some simple relations (2), (3) between Q, gain, center frequency f0 and tunable variables such as G_m of input OTA, g_{mP} of equivalent LC tank loss, and $-g_{mN}$ of negative resistance, assuming 0dB gain, the synchronous tuning can be achieved. Notice that there is an inverse relation between G_m and g_{mN} that can be realized via some inverters between their control signals. I.e. for example, suppose that g_{mN} is decreased for lower Q, then G_m should be increased to keep the gain constant. The center frequency shifting mainly due to switching out some g_{mN} cells can be compensated through adding a corresponding amount of capacitance that is implemented through the new wide-range high-linear TDC circuit.

TABLE I. TWO SOLUTIONS FOR PARALLEL COMPENSATION

Method I: Reduced L+R_N	Method II: Reduced C+R_N
$R_N = 1 \Big/ \left[\left(\dfrac{\omega_0}{Q} - \dfrac{R_L}{L} \right) C \right]$	$R_N = -R_L + 1 \Big/ \left[\left(\dfrac{\omega_0}{Q} - \dfrac{R_L}{L} \right) C \right]$
$k = R_N/(R_N - R_L)$	$C_N = C(R_N - R_L)/R_N - C = CR_L/R_N$

$$Q = \frac{\omega_0 C}{g_{mP} - g_{mN}} \quad (2)$$

$$G_m = g_{mP} - g_{mN} \quad (3)$$

C. Digital Tunable Discrete Capacitor (TDC) Design

One copy of TDC is illustrated in Fig. 4. The key idea is instead of operating inversion-mode-only pMOS based varactor (I-mode varactor) in the middle sharp continuous range, we bias it in a way that only uses the on- and off-state hence a two-level capacitance where the capacitance variations are much smaller over a much wider range [7].

The new designs here are: (1) Two DC biasing voltage VS1 and VS2 are connected through four 1 MΩ n-well resistors to the four I-mode varactors to compensate the threshold voltages needed to form inversion layers. I.e. VS1 and VS2 shift the sharp transition region to the middle of rail-rail voltage. (2) Four high-linearity, equal-value, fixed capacitors are connected in series with the I-mode varactors to further expand the allowable signal swing. (3) Q_ctl1 and Q_ctl1Bar are two complementary digital control signals that are either 0V or 1.8V. They are converted to smaller two-level digital voltages that are either 0.35V or 1.45V via the inverter+diode-connected active loads. That ensures that TDC is biased at the middle voltage of each capacitance level, hence allows the maximum ac swing. (4) We use two complementary copies by flipping the G/DS connections of I-mode varactors to compensate the small capacitance variation slope mainly due to the gate depletion width changes, hence an improved linearity is achieved during ac signal swing.

Fig. 5 shows the simulation results of 3-bit TDC design. Both ac and transient simulation reveals a similar behavior.

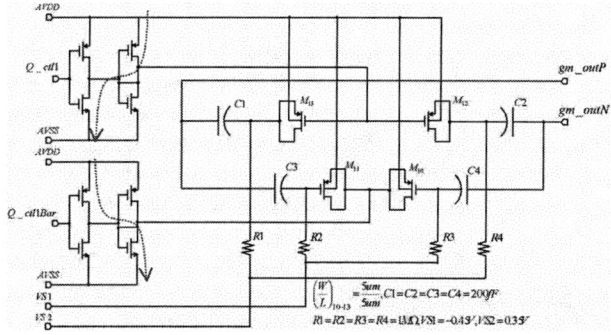

Figure 4. One copy of Tunable Discrete Capacitor

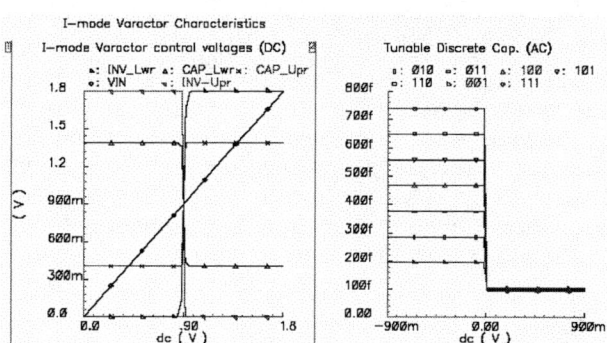

Figure 5. Simulation Results of TDC

In reality, due to finite Q of each element, there will be some variation.

D. Input Digital Tuned OTA Design via Pseudo-Differential Pair (PDP) [8]

One copy of PDP is illustrated in Fig. 6. The key idea is to enhance the linearity without using feedback for high speed. The details are beyond the scope of this paper [8]. But the operating principles highlighted here are: (1) Two cascode transistors MC1 and MC2 are used for I/O isolation. (2) Two additional transistors MR1 and MR2 are biased such that they absorb most of the nonlinearity of the two input transistors MF1 and MF2, the resulted difference linear currents are taken from output transistors MC1 and MC2. (3) MF1 and MF2 are biased at the middle of rail-rail voltages and gates of MC1 and MC2 are tied to VDD that

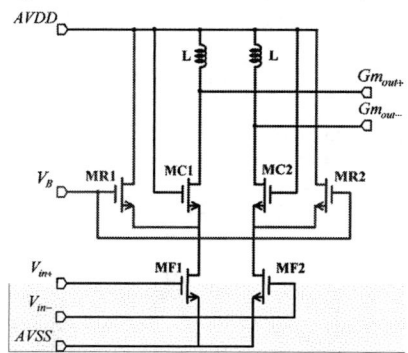

Figure 6. One Copy of PDP Design

allows the maximum overdrive for minimum HD3. The simulation results of comparing the PDP and the conventional cascode (without MR1,2) are shown in Fig. 7. For a typical 0.18um process with a power supply of 1.8V, the total harmonic distortion (THD) can be as low as −64/48/39 dB with 1.1VPP at 0.1/1/10 GHz and maintain a gm variation within 0.15% and a gm drop at 10 GHz for about 10.75%. It is therefore called a high-speed, low-distortion, wide-range, constant-gm cell design. Another advantage of this cell is a linear relationship between power, available g_m and the W/L while the biasing voltage remains constant. Hence it is linearly scalable with W/L.

III. A 5.775GHZ BANDPASS FILTER DESIGN AND THE SIMULATION RESULTS

To illustrate the above ideas, a 3-bit synchronous tuned filter using Q-enhanced LC tank has been designed to operating at center frequency of 5.775GHz. The total in-band noise (both flicker and thermal) is integrated over 100MHz passband (5.725~5.825GHz) for a HiperLAN/2 application. The circuit is implemented in a 0.18um standard digital CMOS process with 1.8V voltage supply and simulated with Cadence specterS.

The input OTA is shown in Fig. 8. Part 1 is for matching and biasing. A simple linear poly 50Ω resistor is used instead of inductive source degeneration (ISD) for matching for the following reasons: (1) more area cost of ISD matching, (2) strong frequency dependent matching of ISD, (3) huge gate inductance if using ISD for low frequency application such as GSM-900MHz, (4) 50Ω resistor simply contribute negligible noise considering that its noise floor is only −168dBm/Hz and (5) more complicated design for ISD. The DC biasing part provides some extra high-pass filtering to assist the overall low-frequency attenuation as the LC tank alone has a low-pass peaking response. Part 2 generates some fixed amount of G_m. Part 3 provides a small amount of continuously tuned G_m. It is meant to be small as the linearity is not as good as PDP. Part 4,5, and 6 are 3-bit digitally tuned G_m blocks. Instead of tuning the current, the W/L is tuned for linear G_m changes.

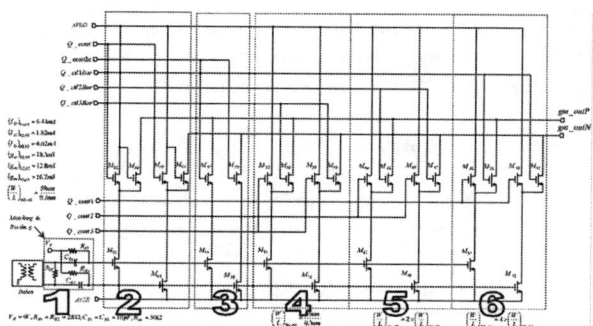

Figure 8. 3-bit Input OTA (Parts 1-6 Shown on Figure)

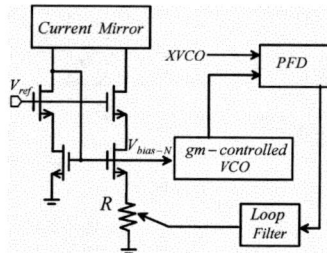

Figure 7. Constant-gm Biasing Circuits for Input OTA

To make sure that the G_m is invariant to process, voltage, temperature (PVT) variations, a constant-gm biasing circuit [9] is used as illustrated in Fig. 9.

The negative resistance circuit is shown in Fig. 10. It is based on cross-coupled transistors. Part 1 generates some fixed amount of $-g_m$. Part 2 provides a small amount of continuously tuned g_m. Part 3,4, and 5 are 3-bit digitally tuned $-g_m$ blocks [6]. Instead of using level shifter, maximum overdrive is used to minimize its HD3. Also, by tying the G/D of cross-coupled transistors to the LC tank terminals, the G/D capacitance can be absorbed into the tank enabling fastest speed. Again, instead of tuning the current, the W/L is tuned for linear $-g_m$ changes.

The AC simulation results with and without synchronous tuning (only Q tuning) are shown in Fig. 11. Without synchronous tuning, there is a gain variation of about 20dB and f0 shifting of 154MHz when Q is tuned from 6 to 65. With synchronous tuning, gain variation is only 0.03dB and f0 shifting is 17MHz. The detailed performance is listed in Table II where the total noise is integrated through the 100

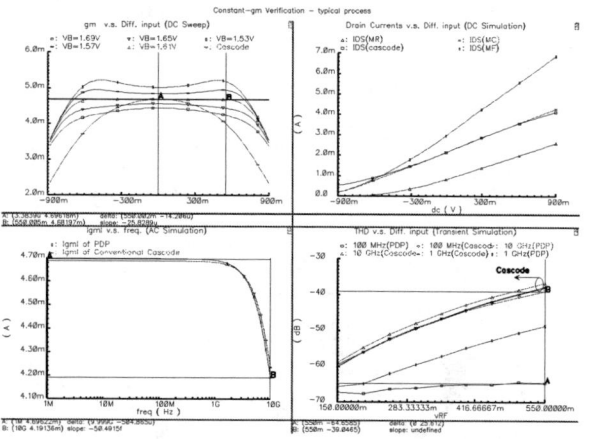

Figure 9. Simulation Results of PDP and Conventional Cascode

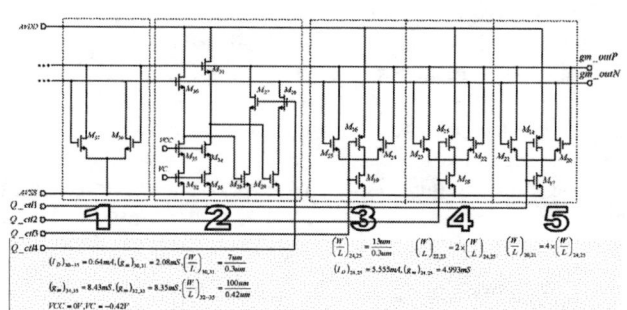

Figure 10. 3-bit Negative Resistor (Parts 1-5 Shown on Figure)

MHz passband (5.725~5.825GHz) and the DR is determined by the difference between the P1dB and inband total noise power all referred to 50 Ohms with which the input and output are matched. A performance comparison between this work and recently published work is detailed in Table III. In Table III, design 1 and design 2 represent this work with design 1 at center frequency 5.775GHz for HiperLAN/2 application and design 2 at center frequency at 900MHz for GSM application as they represent the highest and lowest end frequency of the wireless communication standards [10]. Exact same structure and techniques is used for both designs.

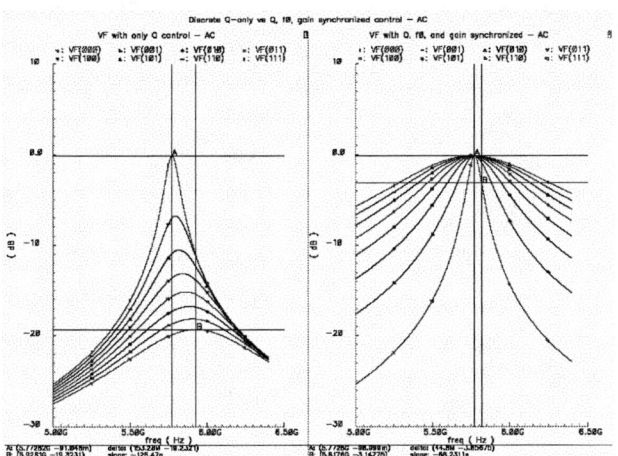

Figure 11. AC Simulation Results with/without Synchronous Tuning

TABLE II. 5.775GHz BPF PERFORMANCE

Digital Word	111	110	101	100	011	010	001	000
F0 (GHz)	5.773	5.771	5.769	5.768	5.764	5.762	5.759	5.756
BW (MHz)	88.9	219.3	348.48	476.29	603.4	728.34	852.6	975.1
Filter Q	64.9	26.32	16.55	12.11	9.55	7.91	6.75	5.9
IL (dB)	0.091	0.175	0.19	0.18	0.176	0.158	0.142	0.123
I_gm (mA)	2.598	6.355	10.11	13.87	17.62	21.38	25.14	28.89
I_RN (mA)	23.48	21.08	18.69	16.3	13.91	11.53	9.145	6.765
I_sum (mA)	26.08	27.44	28.8	30.17	31.54	32.91	34.28	35.66
Power (mW)	46.94	49.39	51.84	54.3	56.77	59.24	61.71	64.19
P1dB (dBm)	-0.91	1.36	2.03	2.35	2.48	2.57	2.63	2.69
VIN2 (dBm)	-58.72	-66.31	-70.13	-72.65	-74.50	-75.93	-77.08	-78.03
DR (dB)	57.81	67.67	72.16	75.00	76.98	78.50	79.71	80.72

TABLE III. RF FILTER PERFORMANCE COMPARISONS

	[2]	[3]	[4]	[5]	[6]	Design 1	Design 2
Tech.	0.25um CMOS	0.25um BiCMOS	0.35um CMOS	0.5um CMOS	0.5um SOS	0.18um CMOS	0.18um CMOS
f_c	2.14GHz	1.88GHz	2.19GHz	1.84GHz	900MHz	5.773~5.756GHz	900~899.65MHz
BW	60MHz	150MHz	53.8MHz	80MHz	20MHz	88.9~975.1MHz	20.1~163.9MHz
I.L.	0dB	2dB	5dB	9dB Gain	11dB Gain	0.09~0.12dB	0.23~0.04dB
Power	7mA	18mA	4mA	16mA	13 mA	26.1~35.7mA	41.9~30.2mA
VDD	2.5V	2.7~3.3V	1.3V	2.7V	3.0V	1.8V	1.8V
P_{1dB}	-13.4dBm	-9.5dBm	-30dBm	-16dBm	-5.5dBm	-0.91~2.69dBm	8.22~9.36dBm
Noise Floor	-155 dBm/Hz	-156 dBm/Hz	-152 dBm/Hz	-137 dBm/Hz	-142 dBm/Hz	-139~158 dBm/Hz	-135~154 dBm/Hz
DR	64dB Over BW	65dB Over BW	38dB Over BW	42dB Over BW	63dB Over BW	58~81dB over 5.725~5.825GHz	70~91dB over 890~910MHz

IV. CONCLUTIONS

A high-linear, wide-dynamic range RF filter using Q-enhanced LC filter is presented in this paper. The contributions of this work are: (1) the new high-speed, low-distortion, wide-range, constant-gm cell design for input OTA, (2) the new high-linear two-level TDC design for center frequency fixing, (3) maximum overdrive and direct connecting to LC tank of cross-coupled negative resistor design to ensure the high-speed and low-distortion, (4) the idea of digitally synchronous tuning inspired by the results of [6]. The resulting filter can have a comparable DR performance and a much better P_{1dB} with the recent published work. Since design 1 and design 2 represent the two frequency ends of wireless communication standards, it can be inferred that the filter designed with the structure and the techniques presented in this paper can be applied to any wireless communication standards with a center frequency in between. However due to its relatively large current consumption, it might not be ideal for portable devices but would be more suitable to fixed applications.

ACKNOWLEDGMENT

The first author thanks Zhijie Xiong from Analog IC Design Group at Georgia Tech for technical discussions.

REFERENCES

[1] N.M. Nguyen and R.G. Meyer, "Si IC-compatible inductors and LC passive filters," IEEE Journal of Solid-State Circuits, vol. 27, No. 10, pp. 1028–1031, August 1990.

[2] T. Soorapanth and S.S. Wong, "A 0-dB IL 2140+-30MHz bandpass filter utilizing Q-enhanced spiral inductors in standard CMOS," IEEE Journal of Solid-State Circuits, vol. 37, No. 5, pp. 579–586, May 2002.

[3] D. Li and Y. Tsividis, "Design techniques for automatically tuned integrated gigahertz-range active LC filters," IEEE Journal of Solid-State Circuits, vol. 37, No. 8, pp. 967–977, August 2002.

[4] F. Dulger, E. Sanchez-Sinencio and J. Silva-Martinez, "A 1.3-V 5-mW fully integrated tunable bandpass filter at 2.1 GHz in 0.35-um CMOS," IEEE Journal of Solid-State Circuits, vol. 38, No. 6, pp. 918–928, June 2003.

[5] A.N. Mohieldin, E. Sanchez-Sinencio and J. Silva-Martinez, "A 2.7-V 1.8-GHz fourth-order tunable LC bandpass filter based on emulation of magnetically coupled resonators," IEEE Journal of Solid-State Circuits, vol. 38, No. 7, pp. 1172–1181, July 2003.

[6] W.B. Kuhn, D. Nobbe, D. Kelly and A.W. Orsborn, "Dynamic range performance of on-chip RF bandpass filters," IEEE Transactions on Circuits and Systems-II: Analog and Digital Signal Processing, vol. 50, No. 10, pp. 685–694, October 2003.

[7] R.B. Staszewski, C.M. Hung, D. Leipold, and P.T. Balsara, "A first multi-gigahertz digital controlled oscillator for wireless applications," IEEE Transactions on Microwave Theory and Techniques, vol. 51, No. 11, pp. 2154–2164, November 2003.

[8] S. Li and P.E. Allen, "A high-speed wide-range low-distortion and constant-gm cell design for GHz applications," Accepted for publication in ISCAS-05.

[9] A. McLaren and K. Martin, "Generation of accurate on-chip time constants and stable transconductances," IEEE Journal of Solid-State Circuits, vol. 36, No. 4, pp. 691–695, April 2001.

[10] 2003 Worldwide Wireless Communication Standards: www.semiconductors.phillips.com/comms

70MHz CMOS Gm-C IF FILTER

Muhammad S. Qureshi and Phillip E. Allen

School of Electrical and Computer Engineering
Georgia Institute of Technology
Atlanta, Georgia 30332, USA

ABSTRACT

Implementation of high dynamic range on-chip 70MHz gm-C filter for UMTS (WCDMA) super-heterodyne receiver is described in this paper. The filter is designed in National Semiconductor's 0.18 micron standard CMOS process. Chebyshev approximation is used to implement 6-th order filter in active gm-C. Positive feedback is used to realize large quality factor (Q). This filter provides blocker attenuation with dynamic range (DR) of 42 dB and with power consumption of 21.78mW.

1. INTRODUCTION

The demand for high performance cellular phones with low cost and small size is increasing. Lowering the complexity, cost, power, and number of external components is the primary criterion in various cellular phone design approaches [1]. Currently, most of the commercially available mobile receivers are designed using super-heterodyne architectures [2]. In super-heterodyne receivers (shown in Figure 1), proper filtering is mandatory and is done by external surface acoustic wave (SAW) filters. These external filters are large and expensive, but they are unavoidable in super-heterodyne architectures. These filters are the major impediment to increasing the level of integration of wireless radio because they are not implemented monolithically [3]. To overcome the problems of those external filters, other receiver architectures such as the homodyne and the low-IF architecture were developed. Unfortunately, these techniques require complicated correcting circuitries [4].

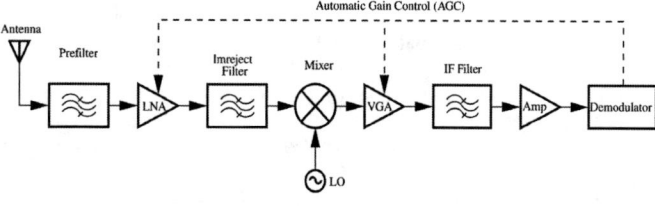

Figure 1. Super-heterodyne receiver

IF-stage filtering can be done by making use of active filters. The gm-C filter offers many advantages in terms of low-power and high frequency capability. Noise and linearity performance of a gm-C filter is strongly related to the filter specifications as well as the total capacitance used in the filter [5]. However, the most prominent issue with the gm-c active filter implementation is the dynamic range.

This work was supported by Georgia Tech Analog Consortium (GTAC). Fabrication services are provided by National Semiconductor Corporation (NSC).

Dynamic range is defined as the ratio of the maximum signal power that the circuit can tolerate, to the noise power that is present at the same time [6]. Dynamic range characterizes the signal handling capability and the overall robustness of a filter. Noise limits the lower end of the signal range in active filter, and nonlinear distortion limits the higher end of the signal. Table 1 shows the normalized dynamic range of some the recently published CMOS filters along with this work. Dynamic range is normalized as

$$DR_{norm}(dB) = DR(dB) + 10\log f_o(MHz) - 10\log P(mW) + 10\log Q \quad (1)$$

where f_o is the center frequency and Q is the quality factor.

Table 1. Recently published CMOS filters

Ref.	Freq. [MHz]	Power [mW]	DR [dB]	DR$_{norm}$ [dB]
1	25	90	60	54.4
2	4	10	57.6	53.6
3	60-350	70	54	58.6
4	5.5	12	62	58.6
5	7	30	40	33.7
6	15	96	55	46.9
7	4	16	70	64
8	63	77	68	67.1
9	10	12	62	61.2
10	50	12.5	34	57
11	1	5.8	64	56.4
12	10.7	220	68	71.2
This work	70	21.78	42	119.3

The filter designed in this paper is intended to replace the bulky and expensive SAW filters in super-heterodyne receivers. A simple low-Q bandpass block is placed in positive feedback loop for Q enhancement [14]. This technique improves dynamic range and reduces the power requirements.

2. FILTER DESIGN

2.1 Linear Transconductor

Gm-c is the most popular technique used to implement integrated high frequency continuous-time filters. The performance of a gm-c biquad relies heavily upon the characteristics of the transconductor employed. The main focus of the design is on the linearity and the noise of the transconductor. We implement active biasing [13] with the transconductor block for linearization as shown in Figure 2.

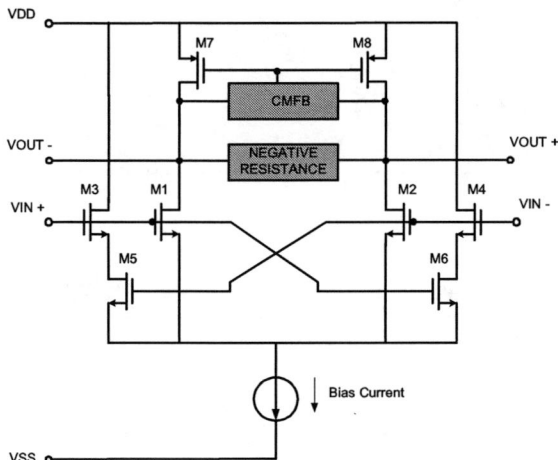

Figure 2. Transconductance block with active biasing

The differential pair M3-M6 acts as an active biasing circuit. It shares the same bias current source with the differential pair M1-M2. When the signal amplitude is small, M3 and M4 are in saturation while M5 and M6 are in triode region. The M3-M6 branches remove some of the bias current. When the signal amplitude is positive and large, M6 will go into saturation and M5 will be cut off. Smaller sum of the bias current will be taken out by M3-M6. That causes more current to be supplied to M1 and M2 to compensate for the drop of the transconductance. The sizing of the transistor has to be optimized for maximum linearity. Following equation is used for this purpose [13].

$$K_3 : K_1 = \frac{4\sqrt{C+1} + 3C - 5}{1} \quad (2)$$
$$K_3 : K_5 = C$$

The value of variable C is chosen equal to one. This will give the remaining ratios for transistor sizing that will give an optimal value of linearity. All the transconductor are designed in the similar manner. This active biasing technique has high power efficiency both in the linear range and the equivalent transconductance, low noise, and small transconductance variation over a wide range. However, good matching of biasing transistors is required in physical design. Negative resistance block is used at the output of each transconductor to boost the output impedance and decrease the effects of output impedance on gain and Q of the filter [14]. Negative resistance blocks are made using the same transconductor structure with cross coupled inputs. Common mode feedback (CMFB) blocks are used to fix the output drain voltages.

2.2 Gm-C Biquad

Figure 3 shows the biquad structure used in this filter.

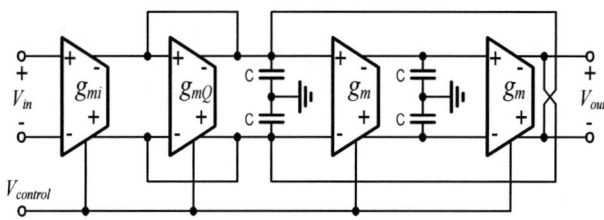

Figure 3. Filter biquad structure

This biquad implement's the bandpass function. The transfer function of this biquad is given by

$$\frac{V_{out}}{V_{in}} = \frac{g_{mi}}{g_{mQ}} \frac{\frac{g_{mQ}}{C}s}{s^2 + \frac{g_{mQ}}{C}s + \frac{g_m^2}{C^2}} = H_o \frac{\frac{\omega_o}{Q}s}{s^2 + \frac{\omega_o}{Q}s + \omega_o^2} \quad (3)$$

where

$$H_o = \frac{g_{mi}}{g_{mQ}} \qquad \omega_o = \frac{g_m}{C} \qquad Q = \frac{g_m}{g_{mQ}}$$

All the transconductor blocks are designed making use of active biasing. These cells are designed to be fully differential to reject even order distortion. CMFB block in gmi and gmq is unnecessary as the common mode voltage is provided by the CMFB block in gm cell. The negative resistance block is also not necessary for gmi and gmq blocks. The bias voltage in gmi, gmq and gm are used to control the gain, Q and the frequency of the filter respectively.

2.3 Sixth Order Filter

The design specifications for the filter are shown in Table 2. It is apparent from Matlab system level simulations that 6-th order filter is needed for achieving the required blocker attenuations. Q Multiplication technique is used in this filter design. We make use of low-Q bandpass network in a positive feedback to achieve low power and low noise. Figure 4 shows the cascading of three Q enhanced blocks. The resulting biquad transfer function after Q enhancement is given by [14]

$$H_{BP} \Rightarrow H_o \frac{\frac{\omega}{Q}(1-AH_o)s}{s^2 + \frac{\omega}{Q}(1-AH_o)s + \omega^2} \quad (4)$$

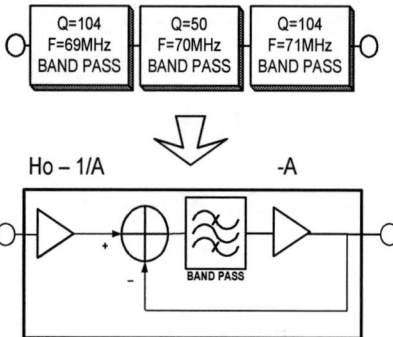

Figure 4. Gm-C biquad

"A" is the gain of the amplifier used in this positive feedback structure. The gain of the preamplifier in this structure has a tradeoff between noise and linearity of the biquad. This gain is selected for an optimal value of 0.25 V/V. The design value of the transfer function of this 6'Th order filter is

$$H(s) = \frac{(2.93 \times 10^{20})s}{(s^2 + 4.29 \times 10^6 s + 1.99 \times 10^{17})} \times$$
$$\frac{(2.93 \times 10^{20})s}{(s^2 + 4.16 \times 10^6 s + 1.88 \times 10^{17})} \times \quad (5)$$
$$\frac{(2.93 \times 10^{20})s}{(s^2 + 8.45 \times 10^6 s + 1.93 \times 10^{17})}$$

The gain of the second amplifier inside the loop is combined with the biquad inside the loop. Similarly, the preamplifier is combined with the summing block. Summing circuit is shown in Figure 5. The governing equations for this circuit is

$$Vout = \frac{Gm_1}{Gm_L}V_1 - \frac{Gm_2}{Gm_L}V_2 \quad (6)$$

A buffer stage is necessary to isolate the filter from the parasitic capacitance of the pad, socket, and PCB, because the capacitance at the output node determines the AC response of the filter. The tuning of this filter is done externally by controlling the bias currents of the transconductor blocks.

2.4 Chip Layout

The layout of the filter is shown in Figure 6. The chip area is 1.5mm by 1.9mm. All the capacitors are integrated on chip and they take the majority of the chip area. Capacitors are kept big to reduce kT/C noise and effectively increasing the dynamic range of the filter. All the current mirrors and differential pairs are drawn using inter digitized layout methods to improve matching. Non-ESD pads are used along with GSG pads for probe testing. The process used is a 0.18 micron digital CMOS twin poly process, provided by National Semiconductor Inc. Test measurements on this chip will be made once this chip is fabricated.

Table 2. Filter design specifications

Application	UMTS
Filter Order / Type	6-th / Active Gm-C
Center Frequency	70MHz
Channel Spacing	2MHz
Quality Factor (Q)	35
Pass Band Gain	0 dB
Pass Band Ripple	±0.5 dB
Dynamic Range (DR)	58 dB
Selectivity @ ± 5MHz	- 33 dB
Selectivity @ ± 10MHz	- 58 dB
Technology	0.18μ CMOS
Power Supply	1.8V

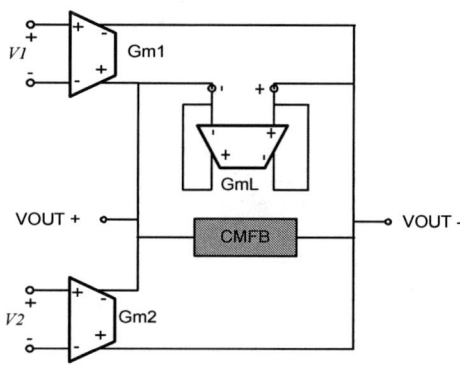

Figure 5. Summing circuit

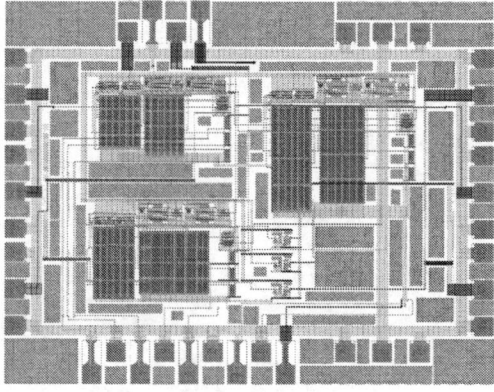

Figure 6. Chip layout

3. RESULTS

Table 3. Simulation results

Center Frequency	70MHz
Pass Band Gain	-0.5 dB
Pass Band Ripple	± 0.369 dB
1 dB Compression	-28.41 dB Volt
Noise	-71.35 dB Volt
Dynamic Range (DR)	42.94 dB
Selectivity @ -5MHz	- 43.35 dB
Selectivity @ +5MHz	- 41.97 dB
Selectivity @ -10MHz	- 62.7 dB
Selectivity @ +10MHz	- 59.9 dB
Power Consumption	21.78 mW

Table 3 shows the simulation results of this filter in detail. The tunable center frequency is 70MHz with a bandwidth of 2MHz. The pass band gain is -0.5 dB with ripple of ±0.365 dB. The signal dynamic range is 42.94 dB and it provides blocker rejection at 5/10 MHz offsets from the center frequency. Figure 7 shows the AC response of this 6'Th order filter.

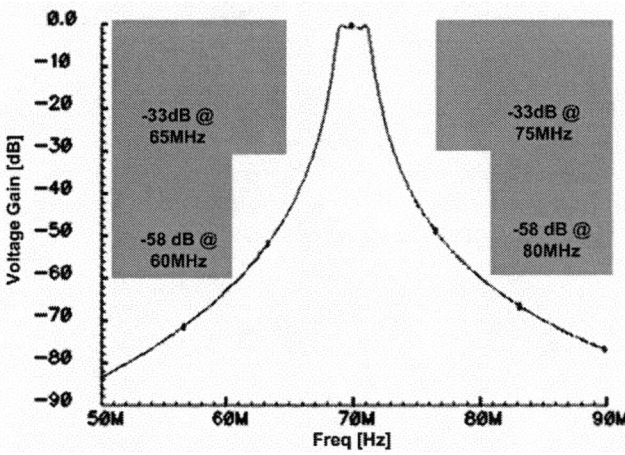

Figure 7. AC response

4. SUMMARY

Implementation of on-chip 70MHz Gm-C filter for UMTS (WCDMA) super heterodyne receiver architecture is described in this paper. It makes use of positive feedback for Q enhancement. The filter is designed in NSC 0.18 micron standard CMOS process. This filter provides blocker attenuation with dynamic range (DR) of 42 dB and power consumption of 21.78mW.

5. ACKNOWLEDGMENTS

The authors would like to thank Fang Lin, LSI Logic Inc. and Zhijie Xiong, Mindspeed Inc. for their valuable assistance and discussions on this project.

6. REFERENCES

[1] I. Mehr and D.R. Welland, "A CMOS continuous-time Gm-C filter for PRML read channel applications at 150Mb/s and beyond," *IEEE J. Solid-State Circuits,* vol. 32, no. 4, pp. 499-513, April 1997.

[2] C. Yoo, S. Lee and W. Kim, "A ± 1.5V, 4MHz CMOS continuous-time filter with a single-integrator based tuning," *IEEE J. Solid-State Circuits,* vol. 33, no. 1, pp. 18-27, January 1998.

[3] S. Pavan, Y. P. Tsividis and K. Nagaraj, "Widely programmable high-frequency continuous-time filters in digital CMOS technology," *IEEE J. Solid-State Circuits,* vol. 35, no. 4, pp. 503-511, April 2000.

[4] C. H. Mensink, B. Nauta and H. Wallinga, "A CMOS soft-switched transconductor and its application in gain control and filters," *IEEE J. Solid-State Circuits,* vol. 32, no. 7, pp. 989-998, July 1997.

[5] B. Stefanelli and A. Kaiser, "A 2um CMOS fifth-order low-pass continuous-time filter for video-frequency applications," *IEEE J. Solid-State Circuits,* vol. 28, no. 7, pp. 713-718, July 1993.

[6] J.M Khoury, "Design of a 15MHz CMOS continuous-time filter with on-chip tuning," *IEEE J. Solid-State Circuits,* vol. 26, no. 12, pp. 1988-1997, December 1991.

[7] F. Krummenacher and N. Joehl, "A 4MHz CMOS continuous-time filter with on-chip automatic tuning," *IEEE J. Solid-State Circuits,* vol. 23, no. 3, pp. 750-758, June 1988.

[8] B. Nauta, "A CMOS transconductance-C filter technique for very high frequencies," *IEEE J. Solid-State Circuit,* vol. 27, no. 2, pp. 142-153, February 1992.

[9] U. Gatti, F. Maloberti, G. Palmisano and G. Torelli, "CMOS triode-transistor transconductor for high-frequency continuous-time filters," *IEE Proc. –Circuit Devices Systems,* vol. 141, no. 6, December 1994.

[10] K. Manetakis and C. Toumazou, "A 50MHz high-Q bandpass CMOS filter," *IEEE International Symposium on Circuit and Systems,* pp. 309-312, June, 1997, Hong Kong.

[11] S. Lindfors, K. Halonen and M. Ismail, "A 2.7V elliptical MOSFET-only GM-C-OTA filter," *IEEE Trans. Circuits and Systems-II: Analog and Digital Signal Processing,* vol. 47, no. 2, pp. 89-95, February 2000.

[12] J. Silva-Martinez, M. Steyaert and W. Sansen, "A 10.7MHz 68dB SNR CMOS continuous-time filter with on-chip automatic tuning," *IEEE J. Solid-State Circuits,* vol. 27, no. 12, pp. 1843-1853, December 1992.

[13] C. S. Kim, Y. H. Kim, and S. B. Park, "New CMOS linear transconductor," *Electron. Letter,* vol. 28, no. 21, pp. 1962-1964, October 1992.

[14] F. Lin, "High-Q High-Frequency CMOS Bandpass Filters for Wireless Applications," Ph.D. thesis, Department of Electrical and Computer Engineering, Georgia Institute of Technology, Atlanta, Georgia, July. 2003.

Voltage-Mode High-Order OTA-Only-Without-C Low-Pass (from 215 M to 705M Hz) and Band-Pass (from 214 M to 724M Hz) Filter Structure

Chun-Ming Chang

Dept. of Electrical Engineering, Chung Yuan Christian University, Chung-Li, Taiwan 32023, R. O. C.
E-mail: chunming@dec.ee.cycu.edu.tw

Abstract: Only using single-ended-input OTAs and grounded capacitors, a low sensitivity voltage-mode high-order OTA-C low-pass and band-pass filter structure with the least number of components is presented by using analytical synthesis approach. Since each internal node has a grounded capacitor to absorb its parasitic capacitance, after replacing the traditional capacitor with the parasitic capacitance, a novel voltage-mode high-order OTA-only-without-C low-pass and band-pass filter structure is then proposed. H-Spice simulations show that the operational frequencies are from 215M to 705M Hz and from 214M to 724M Hz, respectively, for the proposed tenth-order OTA-only-without-C low-pass and band- pass filter.

I. INTRODUCTION

Replacing the traditional capacitors with the parasitic capacitances to do the circuit design has been proposed in [1]. This idea makes the parasitic capacitance become an element useful for analogue circuit synthesis. Using a traditional capacitor with proper magnitude to cover the effect of the parasitic capacitance is no need which restricts the maximum operational frequency from the de-normalization point of view: the smaller the capacitance is the higher the operational frequency of an active-RC circuit provided that other conditions remain the same. Therefore, the limitation of the operational frequency of an active circuit is only relevant to (i) active elements, and (ii) process parameters for integration.

In [1], an nth-order current-mode low-pass and high-pass OTA (Operational Transconductance Amplifier)-only-without-C (Capacitor) filter structure was obtained by deleting all of the grounded capacitors from its corresponding OTA-C filter structure presented in [2]. The successful replacement with parasitic capacitances is constructed by the special topology that all the parasitic capacitances are just located at the same positions of all the given capacitors. Moreover, the OTA-C filter structure presented in [2] employs only single-ended-input OTAs (for covering the feed-through effects associated with differential-input OTAs), grounded capacitors (for absorbing all the shunt parasitic capacitances), and the least number of components (for reducing the parasitic capacitance......, etc.), the three important criteria [2] for the design of OTA-C circuits, which are good for high-frequency operation.

Analytical synthesis methods presented recently [3-5] by a succession of algebra manipulation operations to decompose a complicated nth-order transfer function into n or n+1 simple and realizable equations have been verified to easily achieve the three important criteria stated above. Therefore, the analytical synthesis approach is also used in this paper to find an nth-order voltage-mode OTA-C low-pass and band-pass filter structure. The proposed new single-ended-input OTA-grounded C low-pass and band-pass filter structure is simpler by one single-ended-input OTA than that presented recently [5]. Hence, it also meets the three important criteria stated above. That there is a grounded capacitor located at each internal node of the proposed filter structure makes it possible again to use the parasitic capacitance, like an element, instead of the given capacitor. H-Spice simulation validates the theoretical prediction using tenth-order examples, and demonstrates that the OTA-only-without-C filter achieves lower sensitivities than those of the OTA-C filter as some sensitivities have the same variation tendency.

II. HIGH-ORDER OTA-C LOW-PASS AND BAND-PASS FILTER STRUCTURE

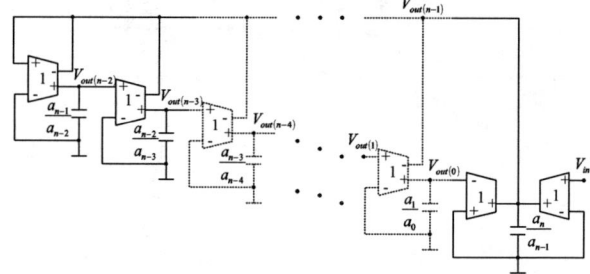

Fig. 1 Nth-order OTA-C LP and BP filter structure

Fig. 1 shows the proposed nth-order filter structure where V_{in} is the filter input voltage, and $V_{out(0)}$, $V_{out(1)}$, $V_{out(2)}$......, $V_{out(n-2)}$, and $V_{out(n-1)}$ are the n filter voltage outputs. The settings of the n filter output voltages determine the filter functions (low-pass and band-pass). It can be seen that the structure employs only single-ended-input OTAs and grounded capacitors. The general transfer function of an nth-order low-pass and band-pass filter with different outputs can be written as

$$V_{out(i)} = V_{in}\left(\frac{a_i s^i}{\Delta}\right), i = 0, 1, 2......, n-1, \text{ where}$$

$$\Delta = a_n s^n + a_{n-1} s^{n-1} + a_{n-2} s^{n-2} + ... + a_2 s^2 + a_1 s + a_0 \quad (1)$$

If $i = 0$ in Eq. (1), then

$$V_{out(i)} = V_{in}\left(\frac{a_0}{a_n s^n + a_{n-1} s^{n-1} + a_{n-2} s^{n-2} + ... + a_2 s^2 + a_1 s + a_0}\right) \quad (2)$$

Cross multiply Eq. (2), divide it by a_0, and re-arrange,

$$V_{in} - V_{out(0)} = \sum_{i=1}^{n}\left(V_{out(i)}\frac{a_i s^i}{a_0}\right) \quad (3)$$

From Eq. (1), we have

$$V_{out(i)} = V_{out(0)}\left(\frac{a_i s^i}{a_0}\right) \text{ for } i = 1, 2......, n-1 \quad (4)$$

$$V_{out(i)} = V_{out(i-1)}\left(\frac{a_i s}{a_{i-1}}\right) \text{ for } i = 1, 2......, n-1 \quad (5)$$

Substituting Eqs. (4) and (5) into Eq. (3) yields

$$V_{in} - V_{out(0)} = \sum_{i=1}^{n-1}\left(V_{out(i-1)}\frac{a_i s}{a_{i-1}}\right) + \left(V_{out(0)}\frac{a_n s^n}{a_0}\right) \quad (6)$$

$$= \sum_{i=1}^{n-1}\left(V_{out(i-1)}\frac{a_i s}{a_{i-1}}\right) + \left(V_{out(0)}\frac{a_{n-1} s^{n-1}}{a_0}\right)\left(\frac{a_n s}{a_{n-1}}\right)$$

$$= \sum_{i=1}^{n-1}\left(V_{out(i-1)}\frac{a_i s}{a_{i-1}}\right) + V_{out(n-1)}\left(\frac{a_n s}{a_{n-1}}\right)$$

Eq. (6) is a voltage relationship. In order to be consistent with the input-and-output current relationship of an OTA, i.e., $(V_+ - V_-)g_m = I_{out}$, we multiply each side of Eq. (6) by a unity trans-conductance, leading to

$$(V_{in} - V_{out(0)})(1) = \sum_{i=1}^{n-1}\left[V_{out(i-1)}\left(\frac{a_i s}{a_{i-1}}/1\right)(1)\right] + V_{out(n-1)}\left(\frac{a_n s}{a_{n-1}}\right) \quad (7)$$

Then, Eq. (5) can be derived as

$$V_{out(i)} = V_{out(i-1)}\left[\frac{a_i s}{a_{i-1}}/1\right] \text{ for } i = 1, 2, \ldots, n-1 \quad (8)$$

namely, $V_{out(i)}(1) = V_{out(i-1)}\left(\frac{a_i s}{a_{i-1}}\right)$ for $i = 1, 2, \ldots, n-1$ (9)

Substituting Eq. (9) into Eq. (7) yields

$$(V_{in} - V_{out(0)})(1) = \sum_{i=1}^{n-1}[V_{out(i)}(1)] + V_{out(n-1)}\left(\frac{a_n s}{a_{n-1}}\right) \quad (10)$$

Eq. (9) can be easily realized by an OTA with unity transconductance followed by a grounded capacitor with capacitance (a_i/a_{i-1}) shown in Fig. 2.

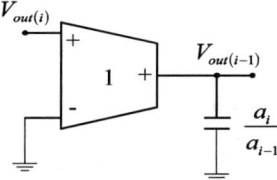

Fig. 2 OTA-C Implementation of Eq. (9)

In Eq. (10), the unity represents the value of transconductance and a_n/a_{n-1} represents the value of capacitance. Eq. (10) is a current relationship for the node with voltage $V_{out(n-1)}$ shown in Fig. 1. The combination of the n sub-circuits realized from Eqs. (9) and (10) is the proposed nth-order OTA-C low-pass and band-pass filter structure shown in Fig. 1. Note that the filter structure employs n+1 single-ended-input OTAs, which is less than n+2 single-ended-input OTAs used in the very recent analytical synthesis method [5], namely, the least number of active components, and n (the minimum number) grounded capacitors, and furthermore has a grounded capacitor at each internal node. The above advantages lead to a new OTA-only-without-C filter structure which will be presented in Sec. IV.

III. LOW SENSITIVITIES FOR OTA-C FILTERS

The sensitivities of the transfer function of the new third-order filter structure shown in Fig. 3 to individual capacitance and transconductance are shown as follows.

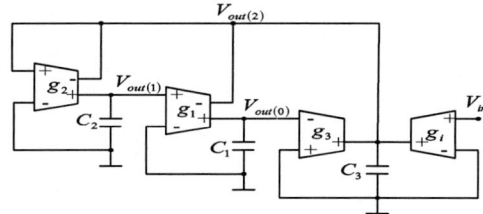

Fig. 3 Third-order OTA-C LP and BP filter

Circuit analysis for Fig. 3 yields the following three transfer functions

$$H_L \equiv \frac{V_{out(0)}}{V_{in}} = \frac{g_i g_1 g_2}{s^3 C_1 C_2 C_3 + s^2 C_1 C_2 g_2 + s C_1 g_1 g_2 + g_1 g_2 g_3} \quad (11)$$

$$\equiv \frac{b_0}{a_3 s^3 + a_2 s^2 + a_1 s + a_0} \equiv \frac{b_0}{\Delta}$$

$$H_{B1} \equiv \frac{V_{out(1)}}{V_{in}} = \frac{s C_1 g_i g_2}{s^3 C_1 C_2 C_3 + s^2 C_1 C_2 g_2 + s C_1 g_1 g_2 + g_1 g_2 g_3} \quad (12)$$

$$H_{B2} \equiv \frac{V_{out(2)}}{V_{in}} = \frac{s^2 C_1 C_2 g_i}{s^3 C_1 C_2 C_3 + s^2 C_1 C_2 g_2 + s C_1 g_1 g_2 + g_1 g_2 g_3} \quad (13)$$

Then, the sensitivities of H_L to the components are

$$S_{g_i}^{H_L} = 1,\ S_{g_1}^{H_L} = (a_3 s^3 + a_2 s^2)/\Delta,\ S_{g_2}^{H_L} = a_3 s^3/\Delta,\ S_{g_3}^{H_L} = -a_0/\Delta,$$

$$S_{C_1}^{H_L} = -(a_3 s^3 + a_2 s^2 + a_1 s)/\Delta,\ S_{C_2}^{H_L} = -(a_3 s^3 + a_2 s^2)/\Delta,\ S_{C_3}^{H_L} = -a_3 s^3/\Delta. \quad (14)$$

Obviously, all sensitivities are not larger than unity. In other words, it enjoys low sensitivity performance. Moreover, it is very interesting that the sum of some sensitivities in Eq. (14) has a null total. They are

$$S_{C_3}^{H_L} + S_{g_2}^{H_L} = 0,\ S_{C_2}^{H_L} + S_{g_1}^{H_L} = 0,\ S_{C_1}^{H_L} + S_{g_3}^{H_L} + S_{g_i}^{H_L} = 0,$$

$$S_{C_3}^{H_L} + S_{g_2}^{H_L} + S_{C_2}^{H_L} + S_{g_1}^{H_L} + S_{C_1}^{H_L} + S_{g_3}^{H_L} + S_{g_i}^{H_L} = 0. \quad (15)$$

The above null relationships lead to zero sensitivities if the variations of some sensitivities concerned have the same increment. Obviously, the above sensitivity analysis offers the advantage of low sensitivities which has been achieved by the well-known doubly terminated LC ladder monolithic integrated circuit. So do the sensitivities of H_{B1} and H_{B2}. And so do the nth-order OTA-C low-pass and band-pass filter structure by using deduction approach.

IV. OTA-ONLY-WITHOUT-C LOW-PASS AND BAND-PASS FILTER STRUCTURE

In Fig. 1, all of the positions of parasitic capacitances are exactly at the same positions of all of the given capacitors. It becomes possible to replace the traditional capacitors with the parasitic capacitances without changing the output responses. The proof is shown as below.

Numerous publications have reported non-ideal effects of OTA-C filters. One of the most recent is [2]. The non-ideal effects [2] of the CMOS OTA include (i) frequency dependent transconductances $A_j(s)$ which can be reasonably represented as $g_j(1-sT_j)$, (ii) input parasitic capacitances C_{ipj}, (iii) output parasitic capacitances C_{opj}, and (iv) output parasitic conductances G_{opj}. If we insert the frequency dependent transconductances and the parasitic capacitances and conductances into the new filter structure shown in Fig. 1, the

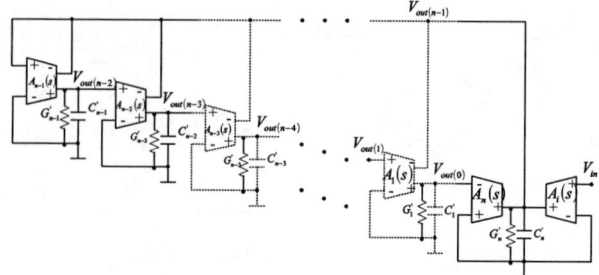

Fig. 4 Non-ideal nth-order OTA-C LP and BP filter structure

non-ideal filter structure of the nth-order voltage-mode OTA-C low-pass and band-pass filter structure is shown in Fig. 4 in which C_j' represent the sum of capacitances of the given capacitor and the parasitic capacitances presenting at that position, and G_j' represent the sum of all parasitic conductances presenting at that position. The non-ideal output voltage signals of the proposed third-order OTA-C filter structure are given by

$$\frac{V_{out(0)}}{V_{in}} = \frac{\rho_0'(s)}{\Delta}, \frac{V_{out(1)}}{V_{in}} = \frac{\rho_1'(s)}{\Delta}, \frac{V_{out(2)}}{V_{in}} = \frac{\rho_2'(s)}{\Delta},$$

where $\Delta = \rho_3'(s) + \rho_2(s) + \rho_1(s) + \rho_0(s)$ (16)

in which

$\rho_3'(s) = s^3 C_1' C_2' C_3' + s^2 C_1' C_2' G_3' + s^2 C_1' C_3' G_2' + s^2 C_2' C_3' G_1'$
$+ sC_1' G_2' G_3' + sC_2' G_1' G_3' + sC_3' G_1' G_2' + G_1' G_2' G_3'$

$\rho_2(s) = s^2 C_1' C_2' g_2(1 - sT_2) + sC_1' G_2' g_2(1 - sT_2)$
$+ sC_2' G_1' g_2(1 - sT_2) + G_1' G_2' g_2(1 - sT_2)$

$\rho_1(s) = sC_1' g_1 g_2 (1 - sT_1)(1 - sT_2) + G_1' g_1 g_2 (1 - sT_1)(1 - sT_2)$

$\rho_0(s) = g_1 g_2 g_3 (1 - sT_1)(1 - sT_2)(1 - sT_3)$

$\rho_2'(s) = s^2 C_1' C_2' g_i (1 - sT_i) + sC_1' G_2' g_i (1 - sT_i)$
$+ sC_2' G_1' g_i (1 - sT_i) + G_1' G_2' g_i (1 - sT_i)$

$\rho_1'(s) = sC_1' g_1 g_2 (1 - sT_i)(1 - sT_2) + G_1' g_1 g_2 (1 - sT_i)(1 - sT_2)$

$\rho_0'(s) = g_1 g_2 g_i (1 - sT_1)(1 - sT_2)(1 - sT_i),$

and $C_1' = C_1 + C_{ip3} + C_{op1}, C_2' = C_2 + C_{ip1} + C_{op2}$,

$C_3' = C_3 + C_{opi} + C_{op1} + C_{op2} + C_{ip2}$,

$G_1' = G_{op1}, G_2' = G_{op2}, G_3' = G_3 + G_{opi} + G_{op1} + G_{op2} + G_{op3}$.

If we get rid of all the given capacitors shown In Fig. 1, we obtain an nth-order voltage-mode OTA-only-without-C low-pass and band-pass filter structure shown in Fig. 5 in which all the parasitic capacitances, which have not been shown in Fig. 5, replace all the given capacitors. The transfer function of the proposed third-order OTA-only-without-C filter is the same as Eq. (16) but $C_1 = C_2 = C_3 = 0$.
Based on Eq. (16) with $C_1 = C_2 = C_3 = 0$, the output responses are depended upon the following requirements. For $\rho_3'(s), \rho_2(s), \rho_2'(s), \rho_1(s),$ and $\rho_1'(s), G_j'$ must be rather smaller than sC_j'. And for $\rho_0(s)$ and $\rho_0'(s), sT_j$ must be smaller than unity. If $C_j' = 0.025\text{pF}$ (referring to Sec. VI),

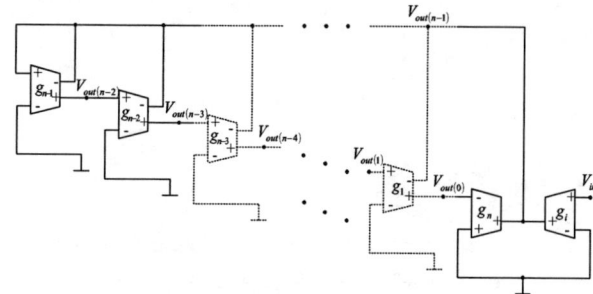

Fig. 5 Nth-order OTA-only-without-C LP and BP filter structure

$G_j' = 954\text{nS}$ [2], and $T_j = 1/(2\pi \times 370 MHz)$ [6], then G_j' are rather smaller than sC_j' and sT_j are smaller than unity when the proposed filter operates at frequencies from 214M Hz to 724M Hz (referring to section VI). This should lead to a successful low-pass and band-pass output responses.

V. SENSITIVITY PREDICTION FOR OTA-ONLY-WITHOUT-C FILTERS

The sensitivity of an OTA-only-without-C filter is not the same as that of an OTA-C filter. The main reason is that the existed parasitic capacitance doesn't keep equal as the frequency varies. In Sec. VI, it is shown that there is an increasing parasitic capacitance region from 215M to 515M Hz and from 229M to 692M Hz, shown in Figs. 7 and 8, respectively, for the proposed third-order OTA-only-without -C filter. And there is an unstable or oscillation region from 515M to 705M Hz in Fig. 7. As operating in the increasing region, the sensitivity to individual transconductance will be affected by the existed inner total variation of all parasitic capacitances. Since the sensitivities of H_L presented in Sec. III, to individual $g_i, g_1,$ and g_2 are "positive" but the sensitivities of H_L to individual parasitic $C_{p1}, C_{p2},$ and C_{p3} are "negative", it can be evaluated that the simulation sensitivities of H_L to individual $g_i, g_1,$ and g_2 will be reduced by the addition to total negative sensitivities relevant to $C_{p1}, C_{p2},$ and C_{p3}, which have existed in the filter already, if some relative components vary up or down together. Therefore, the sensitivities of the transfer function of the proposed OTA-only-without-C filter to individual transconductance are much lower than those of the proposed OTA-C filter if some relative transconductances and capacitances increase or decrease, in magnitude, together. This theoretical prediction will be verified in the next section.

VI. H-SPICE SIMULATIONS

A tenth-order voltage-mode OTA-only-without-C low-pass and band-pass filter is the example used to verify the feasibility of the filter shown in Fig. 5 by using H-Spice simulation with UMC05 level-49 process parameters. The CMOS implementtation of an OTA proposed in [6] is employed in the simulation with ± 2.5 V supply voltages and W/L=5μ/0.5μ and 10μ/0.5μ for NMOS and PMOS transistors, respectively. The feasible case with the minimum operational frequency having the bias currents: $I_{bi} = I_{b10} = 250\mu A$ ($g_i = g_{10} = 301\mu S$), $I_{b3} = 50\mu A$ ($g_3 = 144\mu S$), and $I_{b2} = I_{b4} = \ldots = I_{b9} = 10\mu A$ ($g_{b2} = g_{b4} = \ldots = g_{b9} = 18.9\mu S$)

leads to 215M Hz (f_{3dB}) low-pass, and $I_{bi}=I_{b10}$=10.3μA ($g_i=g_{10}$=29.8μS) and $I_{b2}=I_{b3}=...=I_{b9}$=15μA ($g_{b2}=g_{b3}=...=g_{b9}$= 36.9μS) leads to 214M Hz (central frequency) band-pass filtering signals, shown in Fig. 6, with the parasitic capacitances 0.022607pF and 0.026863pF, respectively. The feasible case with the maximum operational frequency having the bias currents: $I_{bi}=I_{b10}$=250μA ($g_i=g_{10}$=190μS), $I_{b2}=I_{b4}$=50μA ($g_2=g_4$=60.3μS), I_{b3}=150μA (g_3=154μS), and $I_{b1}=I_{b5}=...=I_{b9}$=240μA ($g_{b1}=g_{b5}=...=g_{b9}$=191.3μS) leads to 705M Hz (f_{3dB}) low-pass, and $I_{bi}=I_{b10}$=74.9μA ($g_i=g_{10}$=58.1μS) and $I_{b2}=I_{b3}=...=I_{b9}$=136μA ($g_{b2}=g_{b3}=...=g_{b9}$=210μS) leads to 724M Hz (central frequency) band-pass filtering signals, shown in Fig. 6, with the parasitic capacitances 0.024790pF and 0.028797pF, respectively. The simulation results show that the OTA-only-without-C low-pass and band-pass biquad can be operated well from 215 M to 705 MHz and from 214 M to 724 MHz, respectively. The simulated parasitic capacitances have a straight increment about 0.002904pF from 0.0022607pF to 0.025511pF, varying with frequency from 215M to 515 M Hz, shown in Fig. 7, for the low-pass biquad, and a parabolic increment about 0.003277pF from 0.026489pF to 0.029766pF, varying with frequency from 229M to 692 M Hz, shown in Fig. 8, for the band-pass biquad. In addition to the tenth-order example, the sixth-order, third-order, and second-order simulation results have shown very good low-pass and band-pass amplitude frequency responses.

On the other hand, the sensitivity simulations have also been done for the proposed third-order OTA-only-without-C low-pass filter, i.e., the residue of Fig. 3 after replacing C_1, C_2, and C_3, with parasitic capacitances, denoted by C_{p1}, C_{p2}, and C_{p3}, respectively. The sensitivity simulations with +5% g_i, +5% g_1, and +5% g_2 tolerances have f_{3db}=473M Hz with 1.285% error (compared with the nominal f_{3db}=467M Hz as $g_i=g_3$=236.8μS, g_1=38.8μS, and g_2=98.8μS), f_{3db}=471M Hz with 0.857% error, and f_{3db}=471M Hz with 0.857% error, respectively, all of which are much lower than +5% component tolerance, shown in Figs. 9 and 10, and meet the theoretical prediction in Sec. V.

VII. CONCLUSIONS

In addition to present a new voltage-mode high-order OTA-only-without-C low-pass and band-pass filter structure, which can operate well from 215M to 705M Hz and from 214M to 724M Hz, respectively, for a tenth-order example, very low sensitivity merit is achieved from sensitivity prediction and verified by sensitivity simulation, as the relative components vary up or down together. Two curves with straight or parabolic increasing part show how parasitic capacitances vary with frequency.

REFERENCES

[1] C. M. Chang, "6.61M to 317M Hz nth-order current-mode low-pass and high-pass OTA-only -without-C filter", Proc. IEEE/ISCAS, Vol. I, pp. 37-40, Vancouver, Canada, May 2004.

[2] Y. Sun, and J. K. Fidler, "Synthesis and performance analysis of universal minimum component integrator-based IFLF OTA-grounded capacitor filter", IEE Proc.-Circuits Devices Syst, vol. 143, no. 2, pp. 107-114, Apr. 1996.

[3] C. M. Chang, and B. M. Al-Hashimi, "Analytical synthesis of high-order current-mode OTA-C filters", IEEE Trans. Circuits Syst.-I, vol. 50, no. 9, pp. 1188-1192, Sep. 2003.

[4] C. M. Chang, and B. M. Al-Hashimi, "Analytical synthesis of voltage-mode OTA-C all-pass filters for high frequency operation", Proc. IEEE/ISCAS, vol. I, pp. 461-464, Bangkok, Thailand, May 2003.

[5] C. M. Chang, B. M. Al-Hashimi, Y. Sun, and J. N. Ross, "New high-order filter structures using single-ended-input OTAs and grounded capacitors", IEEE Trans. Circuits Syst.-I, vol. 51, no. 9, pp. 458-463, Sep. 2004.

[6] S. Szczepanski, A. Wyszynski, and R. Schaumann, "Highly linear voltage-controlled CMOS transconductors", IEEE Trans. Circuits Syst.-I, vol. 40, no. 4, pp. 258-262, Apr. 1993.

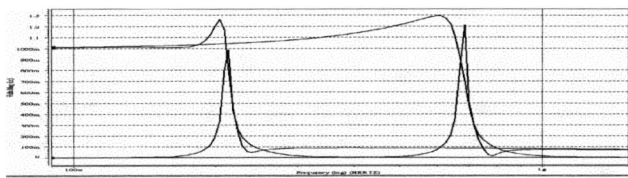

Fig. 6 Tenth-order filter amplitude responses with the min. and max. operational frequencies

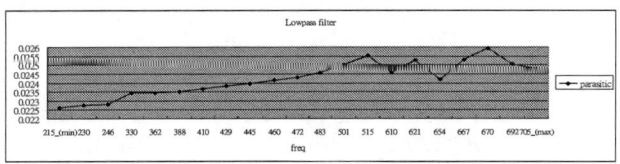

Fig. 7 Variation of parasitic capacitances with frequency (low-pass)

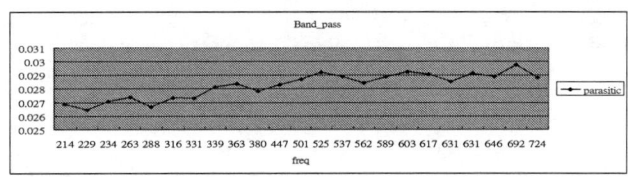

Fig. 8 Variation of parasitic capacitances with frequency (band-pass)

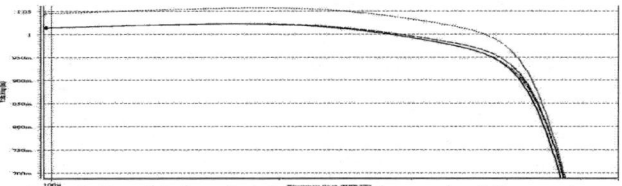

Fig. 9 Sensitivity simulations with nominal (lowest), +5%g_i (highest), +5%g_1, and +5%g_2 tolerance lines

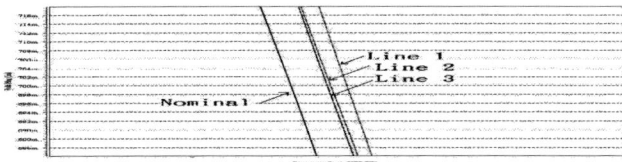

Fig.10 Amplification of Fig.9 around f_{3dB} (nominal: real line, +5%g_i: line 1, +5%g_1: line 2, and +5%g_2: line 3)

Application of Reverse-Active *npn*s for Compact, Wide-Tuning f_T-Integration-based Filters in SiGe HBT BiCMOS Technology

Phanumas Khumsat
Department of Electrical Engineering
Faculty of Engineering, Prince of Songkhla University
Hat-Yai, Thailand e-mail: phanumas.k@psu.ac.th

Apisak Worapishet
Mahanakorn Microelectronics Research Centre
Mahanakorn University of Technology
Bangkok, Thailand e-mail: apisak@mut.ac.th

Abstract— The f_T–integration principle is investigated in SiGe HBT 0.8μm BiCMOS process through an implementation of wide frequency tuning 5-GHz biquadratic filters. The techniques formerly employed in Si f_T–integrators, including the circuit structures and the use of lateral *pnp* as well as the vertical *npn* operating in a reverse-active mode for wide tuning, are successfully demonstrated in SiGe HBT process. The prototype filters achieves a center frequency tuning from 1.9GHz to 6.7GHz (over 300%) and SFDR of 30dB and 33dB for the corresponding Q of 10 and 5 respectively at a 1.8-V supply. The vertical reverse-active tuning *npn*'s help save chip area by four times over the use of lateral *pnp*'s, while overall performance is still preserved.

I. INTRODUCTION

The rapid growth of today's wireless multi-gigabit communication market has placed an ever increasing demand on integrated filter performance particularly in terms of very high operating frequencies with acceptable dynamic range and low power consumption. Among conventional active filtering techniques such as G_m–C and log–domain filters, f_T–integration technique has portrayed itself as a viable alternative for very high frequency applications [1], [2], [3], [4].

A mature SiGe HBT BiCMOS technology is an excellent match for realisation of cost-effective, highly integrated high-speed data communication and wide-bandwidth digital wireless architectures at microwave frequencies. This is because high performance analog circuitry (provided by heterojunction bipolar transistor –HBT) can co-exist with sophisticated digital blocks (CMOS). Bandgap engineering enables process designers to achieve a SiGe HBT with all-around superior performances compared to a conventional Si BJT device [5], [6], [7], [8].

This work demonstrates the feasibility in applying HBT devices in SiGe BiCMOS technology to f_T–integration technique. Specifically, the HBT's will be employed to implement a low-voltage, wide-tuning 5-GHz f_T-integration-based bandpass filter. It will be shown that a large frequency tuning range at microwave operating frequencies is accomplished in the SiGe filter via the existing structure using lateral *pnp*'s as large base transit time tuning devices. The improved structure using reversed active vertical *npn*'s with a significant reduction in chip area is proposed.

II. f_T-INTEGRATION PRINCIPLE

f_T–integration technique has been thoroughly discussed in previous literatures [1], [2], [3], [4] and only a brief overview of the principle is addressed here for better circuit understanding in subsequent sections.

Base charging dynamic of a bipolar junction transistor results in a charge control equation which relates *large-signal* base current I_b and collector current I_c as

$$I_b = \tau_b \frac{dI_c}{dt} + \frac{I_c}{\beta_0} \quad (1)$$

where τ_b is the transit time in the base. This mathematical function of integration lies as a basis of f_T–integration. The operation of SiGe *heterojunction* bipolar junction transistor (HBT) is also governed by the base charging dynamic similar to the conventional homojunction BJT. Therefore it is conceptually feasible to apply f_T–integration technique to HBT's for filter implementation.

III. WIDE-TUNING f_T-INTEGRATOR DESIGN IN SiGe HBT BiCMOS TECHNOLOGY

Based upon the f_T–integration principle presented in the previous section, the first version of the f_T–integrator designed within SiGe HBT 0.8μm BiCMOS process is shown in Fig. 1 which has been migrated from the low-voltage f_T–integrator presented in [4]. Four lateral *pnp*'s Q_F connected in parallel have been used on each side of the integrator in order to achieve the required nominal operating frequency of 5GHz and wide frequency tuning range. Note that the lateral *pnp* symbol has an extra node indicating a gate terminal which always has to tied to the most positive supply. This is because this lateral *pnp* is simply constructed

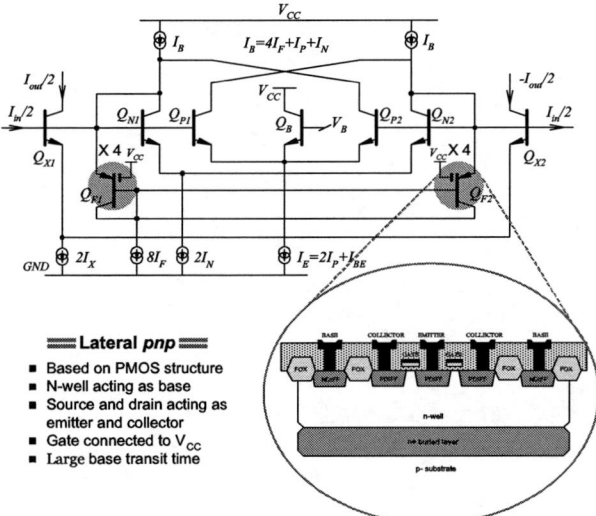

Figure 1. f_T-integrator in SiGe BiCMOS with lateral pnp as tuning device (migrated from [4])

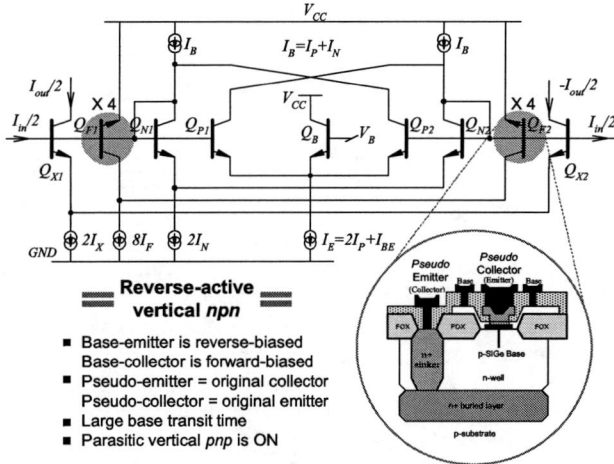

Figure 2. The proposed f_T-integrator with vertical reverse-active npn as frequency tuning device

from the existing PMOS structure, its gate thus has to be biased positively so that channel inversion never occurs.

The unity-gain frequency f_U of the f_T–integrator presented in Fig. 1 can be expressed as

$$f_U = \frac{1}{2\pi\tau_b \left\{ 1 + m \dfrac{\tau_{bf}}{\tau_b} \dfrac{I_F}{I_X} + \dfrac{I_P}{I_X} + \dfrac{I_N}{I_X} \right\}} \quad (2)$$

where τ_b is the base transit time of the forward-active npn and τ_{bf} is that of the frequency-tuning device Q_F with m being the number of Q_F in parallel. Since typical values for I_P and I_N are negligibly small compared to I_X and I_F, the tuning range can be estimated from (2) to be

$$\frac{f_{U\max}}{f_{U\min}} \approx \frac{f_U|_{I_F=0}}{f_U|_{I_F=I_{F\max}}} = 1 + m \frac{\tau_{bf}}{\tau_b} \frac{I_{F\max}}{I_X}$$

$$= 1 + m \frac{\tau_{bf}}{\tau_b} \frac{p \cdot I_{F0}}{I_X} \quad (3)$$

where $m=4$ in Fig. 1. I_{Fmax} is the maximum possible frequency-tuning current which may be written as $I_{Fmax}=p \cdot I_{F0}$, where p is the ratio of the maximum possible I_F relative to the nominal bias value I_{F0}. Normally the ratio p is the same for any type of HBT or BJT. Equation (3) suggests that the term on the right hand side has to be large in order to achieve a wide tuning range. An advantage in utilizing lateral pnp, commonly available in standard BiCMOS process, for Q_F is a considerable tuning range improvement while keeping power consumption at minimum. That is, only a small m is required to accomplish a large tuning range, owing to the pnp's inherently large base transit time. Moreover, the lateral pnp normally requires low bias current I_{F0}. In other words, with lateral pnp, the second term in (3) can be made large without sacrificing excessive extra current consumption. Small nominal bias current also helps minimise collector and base shot noise.

However, the employment of lateral pnp results in one major disadvantage of occupying huge silicon area and so this might not be an ultimate solution for tuning range extension. It would be ideal if there is a more compact device but with an inherently large base transit time and low nominal bias current in a similar order to that of the lateral pnp. Our proposal is to employ a vertical npn operating in a reverse-active region, i.e., with the reverse biased base-emitter and forward biased base-collector junctions. Because of its asymmetry (both in doping profile and physical structure, see Fig. 2), a reverse-active npn possesses an inferior characteristic compared to the forward-active mode. The large base transit time of a reverse-active npn, which is normally considered inferior, can be turned into an advantage by performing a significant role of frequency tuning range enhancement. In a typical BiCMOS, bipolar or SiGe BiCMOS process including the one employed in this work, it is found that lateral pnp and reverse-active npn possess a similar large value of $\tau_{bf}I_{F0}$, which means that these two types of devices require the same number of transistors to obtain the same tuning range. Therefore with a benefit of saving chip area, there is no penalty on either current consumption or number of devices in replacing lateral pnp with a reverse-active vertical npn. In this way, we are able to extend f_T–integrator's frequency tuning range without sacrificing neither power consumption nor chip area by simply utilising npn operation in a region which has been widely regarded worthless.

The f_T–integrator structure employing a reverse-active vertical npn as frequency-tuning device is illustrated in Fig. 2. Similar to the lateral pnp case, four reverse-active npn are employed on each side of the integrator in order to achieve a 5GHz nominal operating frequency. Both circuits in Fig. 1

and Fig. 2 use a supply voltage of $V_{BE} + 2V_D$, where V_D is the voltage drop across each current source and the typical supply voltage could be estimated to be $0.8 + 2(0.5) = 1.8V$.

There are various issues that have to be taken care of when designing circuit using reverse-active vertical *npn*. Firstly, voltage across emitter-base junction has to be kept lower than a relatively low emitter-base breakdown voltage. The supply voltage of circuit in Fig. 2 is thus restricted to a low value. If it is necessary to have a high supply voltage (e.g. 3V), the reverse-active *npn* Q_F in Fig. 2 can be modified by disconnecting its emitter (pseudo collector) from V_{CC} and tying this to the base. In this way, the Q_F's has formed a diode-connected configuration and base-emitter breakdown damage can be avoided. Secondly, when operating the transistor in a reverse mode, the parasitic vertical *pnp* formed by base-collector-substrate structure is inevitably turned ON. Consequently, there will always be a small leakage current wasted into substrate.

IV. SIMULATION RESULTS

The resonators based on the proposed integrators of Fig. 1 and Fig. 2 have been simulated with $V_{CC}=1.8V$ using SpectreRF within Cadence design platform. The devices are taken directly from the library of the foundry's design kits of HBT SiGe 0.8μm BiCMOS process without any modification. A double-base *npn* occupies a total area of 60μm^2 (effective *emitter* area = 2μm×0.8μm) which is much smaller than 400μm^2 of lateral *pnp* (effective emitter area = 3.6μm×3.6μm). Therefore, in this particular technology, the reverse-active vertical *npn* occupies silicon area less than one-sixth of the lateral *pnp*'s area which significantly help optimise circuit compactness and reduce fabrication cost. For rough estimation, if a single vertical *npn* represents one unit chip area, it would mean that the proposed integrator of Fig. 2 consumes only one-fourth of the area occupied by the integrator in Fig. 1. Regarding device characteristics, the *npn* HBT has a peak f_T of 30GHz at bias collector current = 750μA with corresponding $β_0$ of 80 and base resistance = 300Ω. In a reverse-active mode, the same device possesses a peak f_T= 800MHz with $β_0$ = 10 at collector current of 80μA. Comparing to a reverse-active *npn*, lateral *pnp* have a slightly superior device characteristic with peak f_T =3GHz at collector current of 90μA with the corresponding $β_0$ =20.

Simulation results show that the centre frequency (f_c) can be widely tuned from 1.9GHz to 6.7GHz by varying I_F from 600μA to 0.25μA while fixing I_X at 750μA as shown in Fig. 3. Such a wide frequency tuning of over 300% is a solid confirmation of reverse-active *npn* and lateral *pnp*'s expected functionalities. Resonator frequency responses are illustrated in Fig. 4 indicating centre-frequency (f_c) and quality factor (Q) tuning.

The spurious-free dynamic range, SFDR (ratio between a wanted signal and an unwanted third-order intermodulation at the point where the third-order intermodulation power is equal to noise power) is plotted against centre frequencies is shown in Fig. 5 for two values of Q = 5 and 10. Note that

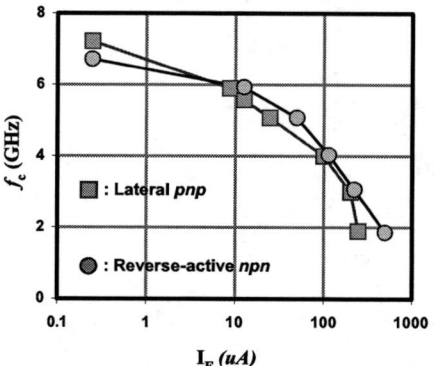

Figure 3 Centre-frequency vs I_F

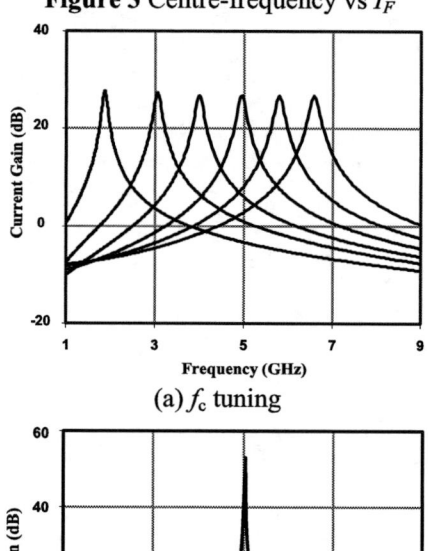

(a) f_c tuning

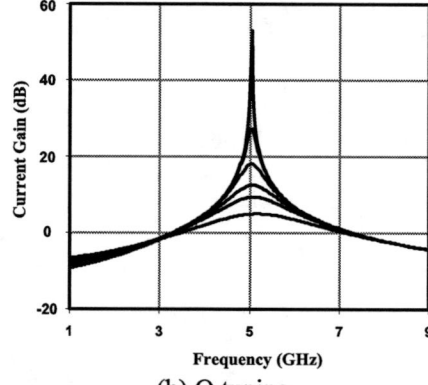

(b) Q tuning

Figure 4 Frequency response (reverse-active)

SFDR is simulated with input two-tone signals of 0.5% frequency separation and the noise power is measured by integrating ac-simulated noise within -3dB signal bandwidth. At nominal centre frequency of 5GHz, the SFDR's are found to be 30dB and 33dB for Q = 10 and 5 respectively.

Fig. 6 illustrates how SFDR varies with Q for the proposed circuits at f_c=5GHz and they are compared with the prototype in Si BiCMOS at f_c =1GHz. It can be seen that the proposed SiGe circuits render SFDR similar to those reported in [4] but with an operating frequency higher by fivefold. A fair comparison of filter overall performance is normally obtained by means of a figure of merit (FoM) which is defined here as

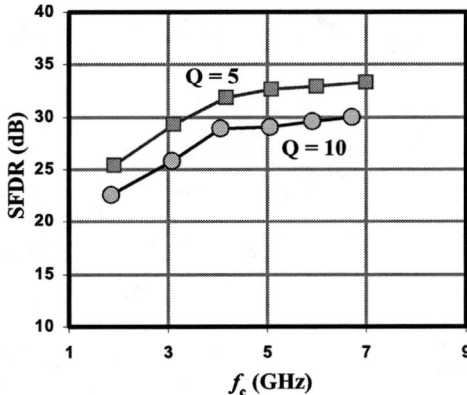

Figure 5 SFDR vs f_c (reverse-active)

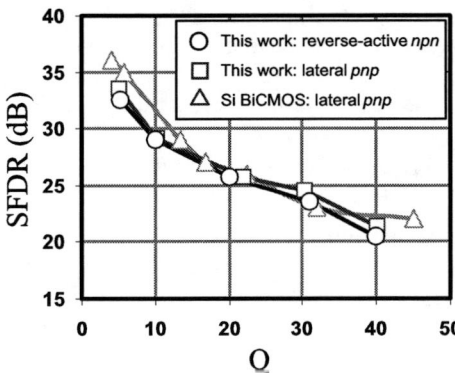

Figure 6 SFDR vs Q (reverse-active)

$$\text{FoM} = \frac{f_c \times \text{SFDR}}{\text{Power consumption per pole}} \quad (4)$$

According to FoM definition, the filter with a better performance possesses a higher FoM. FoM comparisons are plotted against f_c in Fig. 7 for $Q = 10$. It is clear that at the region around nominal operating frequencies, the proposed circuits (both with lateral *pnp* and reverse-active *npn* as a tuning device) give better FoM over the filter implemented in Si BiCMOS where an improvement by a factor as large as four can be observed.

V. CONCLUSION

f_T-integration technique has been successfully demonstrated via simulations in HBT SiGe 0.8μm BiCMOS technology. Addition to existing technique in deployment of a lateral *pnp* for frequency tuning range enhancement, a reverse-active vertical *npn* has been proposed to achieve the same task while also keeping current consumption and chip area at minimum. The resonators can be widely tuned from about 2GHz to more than 6GHz, i.e., tuning range is greater than

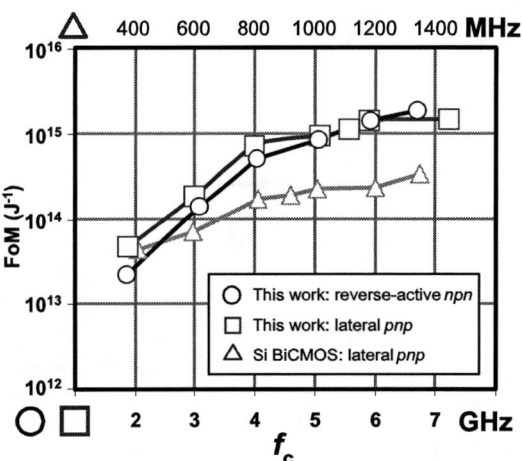

Figure 7 Figure of merit comparison

three-fold. At nominal, the resonators render SFDR of 30dB and 33dB for $Q = 10$ and 5, respectively. FoM comparison between the SiGe filters and the previous prototype implemented in Si enjoys improved performances as large as four times.

ACKNOWLEDGEMENT

This work is financially supported by Thailand Research Fund (TRF) under grant number MRG4780085.

REFERENCES

[1] J. Mahattanakul, C. Toumazou and S. Pookaiyaudom, "Low-distortion Current-mode Companding Integrator Operating at f_T of BJT," *Electronics Letters*, Vol. 32, No. 21, October 1996.

[2] A. Worapishet and C. Toumazou, "f_T Integrator – A New Class of Tuneable Low-Distortion Instantaneous Companding Integrators of Very High-Frequency Applications," *IEEE Transactions on Circuits and Systems - II: Analog and Digital Signal Processing*, Vol. 45, No. 9, pp. 1212-1219, September 1998.

[3] P. Khumsat, A. Worapishet, and A.J. Payne, "f_T integration employing lateral *pnp*'s in BiCMOS/CMOS technologies," *Electronics Letters*, Vol. 35, No. 24, pp. 2138-2140, November 1999.

[4] P. Khumsat, A. Worapishet, and A.J. Payne, "Wide-tuning, low-voltage 1GHz bandpass filter based on f_T-integration," in Proceedings of the 45th IEEE Midwest Symposium on Circuits and Systems, August 2002.

[5] M. Feng, S.C. Shen, D.C. Caruth, and J.J. Hunag, "Device Technologies for RF Front-End Circuits in Next-Generation Wireless Communications," *Proceedings of the IEEE*, Vol. 92, No. 2, Feb. 2004.

[6] K. Washio, "SiGe HBT and BiCMOS Technologies for Optical Transmission and Wireless Communication Systems," *IEEE Transactions on Electron Devices*, Vol. 50, No. 3, March 2003.

[7] J.D. Cressler, "SiGe HBT Technology: A New Contender for Si-Based RF and Microwave Circuit Applications," *IEEE Transactions on Microwave Theory and Techniques*, Vol. 46, No. 5, May 1998.

[8] M. Soyuer, H.A. Ainspan, M. Meghelli, and J.-O. Plouchart, "Low-Power Multi-GHz and Multi-Gb/s SiGe BiCMOS Circuits," *Proceedings of the IEEE*, Vol. 88, No. 10, October 2000.

# A 2V 0.25μm CMOS 250MHz Fully-differential Seventh-order Equiripple Linear Phase LF Filter

Masood-ul-Hasan
School of ECEE
University of Hertfordshire,
Hatfield, Herts, AL10 9AB, UK
m.hasan@herts.ac.uk

Yichuang Sun
School of ECEE
University of Hertfordshire,
Hatfield, Herts, AL10 9AB, UK
y.sun@herts.ac.uk

Abstract— A fully-differential seventh-order 0.05° equiripple linear phase low-pass filter based on multiple loop feedback (MLF) leapfrog (LF) topology is presented for read/write channels. The filter is designed and simulated with the proposed fully balanced, highly linear operational transconductance amplifier (OTA). This OTA contains two complementary differential cross-coupled input pairs and a pair of regulated cascode output in order to achieve both low-distortion and wide dynamic rang in high-frequency operation. Simulations in 0.25μm CMOS show that the cutoff frequency of the low pass filter without and with gain boost ranges from 50 to 150MHz and 65 to 250MHz respectively, dynamic-range is over 65dB and total harmonic distortion is less than 40dB. The group delay ripple is less than 5% for frequencies up to 1.5 times of the cutoff frequency, and for a 2-volt power supply, the maximum power consumption is 216mW.

I. INTRODUCTION

Filters are an important signal processing building block in mixed-signal integrated circuits (ICs). Modern computer and communication systems demand for high performance filters with attractive features. For example, high-performance high-frequency filters are required for read/write channel equalization in computer hard-disk drive systems and intermediate-frequency (IF) filtering for communication systems. Continuous-time operational transconductance amplifier and capacitor (OTA-C) filters have been shown to perform well in high frequency applications [1]. Unlike switched-capacitor filters, OTA-C filters do not suffer from the switching noise problem and doe not need pre- or post-filtering. For these reasons OTA-C filters have become the preferred choice for filtering tasks in mixed-signal ICs. In recent years, several integrated continuous-time filters with high frequency range and low power consumption have been realized successfully and used for read/write channels in hard disk drive systems [2-9]. Most of the filters are based on the cascade structure. As is well known, the ladder-based topology has very low sensitivity, but can implement transmission zeros only on imaginary axis. The cascade topology can be used to realize a filter with arbitrary transfer functions, but sensitivity is higher, in particular, as filter order increases. One of the solutions is to use the multiple loop feedback (MLF) structure [1-3]. The single-ended MLF OTA-C filter based on the leapfrog (LF) configuration for read channel application has been discussed in [3]. However, in mixed-signal SoC environment, digital switching noise signals will be injected into the analog circuitry through chip parasitics, degrading the performance of the analog circuits. To combat this problem, fully-differential, rather than signal-ended, circuitry is normally used [10].

A fully-balanced OTA-C filter design can ensure a completely symmetrical layout such that parasitic injections will couple perfectly equally into both the signal paths and appear as a common-mode signal. Thus, very high common-mode rejection ratio can be achieved and both even-order harmonic distortion components and effects of power supply noise can be reduced. In this paper we therefore present the design of a fully-balanced read/write channel equiripple linear phase OTA-C filter based on the MLF LF architecture. For this a new high performance fully-differential OTA is proposed. The designed filter has potential commercial application in hard disk drive systems. Its most important specification is the linearity on its phase response. Our design is targeted for a cut-off frequency of 250MHz with accurate linear phase and programmable gain boost.

The fully-differential OTA is discussed in Section II. The seventh-order equiripple group delay low-pass filter with finite real-axis zeros is presented in Section III. Filter simulation results are given in Section IV, and finally conclusions are summarized in Section V.

II. OPERATION TRANSCONDUCTANCE AMPLIFIER

For high-frequency operation, large tranconductances are needed for the implementation of filter poles. BiCMOS technology based filters have been reported in [5, 6] but they have parasitic poles and high power consumption. To fulfill filter requirements in CMOS technology both parasitic poles and power consumption must be reduced. In this paper, a fully-balanced OTA shown in Fig. 1 is used.

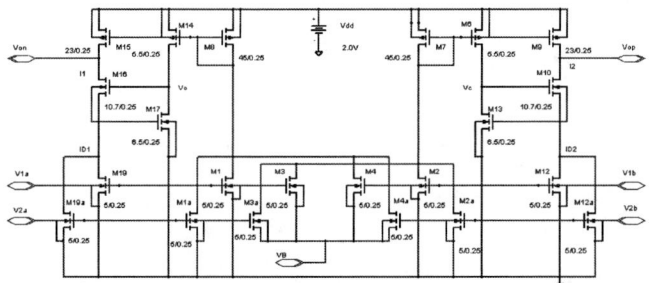

Fig.1 Fully-balanced OTA unit cell with two differential inputs and regulated cascode output stage (RGC).

The OTA is based on dual cross-coupled differential pairs with current tracking regulated cascode output stage. The cross-coupled input pairs enhance the OTA transconductance linearity and overcome the conflict between tuning and common-mode input range as occurred in OTA based on source-coupled pairs. The regulated output stage increases dynamic range and output voltage swing and the cascode stage compensates the GBW product and frequency behavior. Assume that all the input transistor pairs M_1–M_4 and M_{1a}–M_{4a} are identical and operate in saturation region. Neglecting second-order effects, currents I_1 and I_2 are given by

$$I_1 = I_{1d} + I_{4d} + I_{1da} + I_{4da} \quad (1)$$

$$I_1 = k\left[\begin{array}{l}(V_{gs1}-V_t)^2 + (V_{gs4}-V_t)^2 \\ + (V_{gs1a}-V_t)^2 + (V_{gs4a}-V_t)^2\end{array}\right]$$

$$I_1 = k\left[\begin{array}{l}(V_{1a}-V_t)^2 + (V_{1b}-V_B-V_t)^2 + \\ (V_{2a}-V_t)^2 + (V_{2b}-V_B-V_t)^2\end{array}\right] \quad (2)$$

Similarly,

$$I_2 = k\left[\begin{array}{l}(V_{1b}-V_t)^2 + (V_{1a}-V_B-V_t)^2 \\ + (V_{2b}-V_t)^2 + (V_{2a}-V_B-V_t)^2\end{array}\right] \quad (3)$$

where $k = 0.5\mu C_{ox}(W/L)_n$ is the transconductance parameter, and μ, C_{ox}, W and L are the mobility, oxide capacitance per unit area, and channel width and length, respectively.

The final output is taken from the drain of biasing transconductance elements. The current mirror output of I_1 and I_2 is set by the drain voltages of transistor M_{12} and M_{19}. The value of the mirrored current is small enough so that for the range of input gate voltage considered both transistors always operate in their triode region, thus ensuring a reasonably large linear range. The voltage-current relationship is given by

$$I_{D1} = I_{d19} + I_{d19a}$$

$$I_{D1} = k\left[(V_{1a}-V_t)^2 V_C - \frac{1}{2}V_C^2 + (V_{2a}-V_t)^2 V_C - \frac{1}{2}V_C^2\right] \quad (4)$$

and similarly,

$$I_{D2} = k\left[(V_{1b}-V_t)^2 V_C - \frac{1}{2}V_C^2 + (V_{2b}-V_t)^2 V_C - \frac{1}{2}V_C^2\right] \quad (5)$$

Note that the positive and negative output currents are given by

$$I_{op} = I_2 - I_{D2} \quad (6)$$
$$I_{on} = I_1 - I_{D1} \quad (7)$$

Using Equations (2)-(7) we can derive the differential output current of the OTA as

$$I_{out} = I_{op} - I_{on}$$
$$I_{out} = k[(V_{1a}-V_{1b}) + (V_{2a}-V_{2b})](V_C - 2V_B) \quad (8)$$
$$I_{out} = kV_A V_{id} = g_m V_{id} \quad (9)$$

where $V_{id} = (V_{1a}-V_{1b}) + (V_{2a}-V_{2b})$ is the differential input voltage, $V_A = V_C - 2V_B$ where V_B is the bias voltage and $g_m = kV_A$ is the transconductance. Equation (9) shows that the transconductance of the proposed OTA is linear and tunable by varying bias voltage.

III. FILTER ARCHITECTURE AND SYNTHESIS

Design of all-pole and transmission zero MLF OTA-C filters have been well investigated in [1, 10]. The seventh-order fully-balanced MLF LF filter structure with input distribution OTA network to realize gain boost zeros is shown in Fig.2.

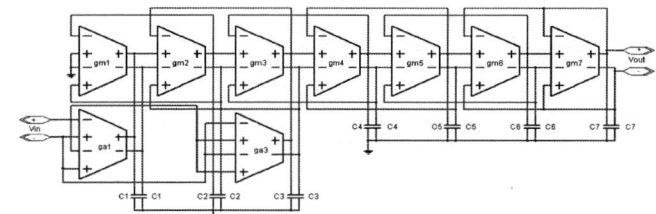

Fig.2 Seventh-order low-pass MLF LF filter structure with gain boost zeros.

The transfer function for the seventh-order LF filter with gain boost in Fig.2 can be derived using nodal equation as

$$H(s) = \frac{V_{out}}{V_{in}} = \frac{N(s)}{D(s)} \quad (10)$$

where

$$N(s) = \beta_3 \tau_1 \tau_2 s^2 + (\beta_3 - \beta_1)$$

and

$$D(s) = \tau_1\tau_2\tau_3\tau_4\tau_5\tau_6\tau_7 s^7 + \tau_1\tau_2\tau_3\tau_4\tau_5\tau_6 s^6 +$$
$$(\tau_1\tau_2\tau_3\tau_4\tau_5 + \tau_1\tau_2\tau_3\tau_4\tau_7 + \tau_1\tau_2\tau_3\tau_6\tau_7 +$$
$$\tau_1\tau_2\tau_5\tau_6\tau_7 + \tau_1\tau_4\tau_5\tau_6\tau_7 + \tau_3\tau_4\tau_5\tau_6\tau_7)s^5 +$$
$$(\tau_1\tau_2\tau_3\tau_4 + \tau_1\tau_2\tau_3\tau_6 + \tau_1\tau_2\tau_5\tau_6 + \tau_1\tau_4\tau_5\tau_6 + \tau_3\tau_4\tau_5\tau_6)s^4$$
$$+ (\tau_1\tau_2\tau_3 + \tau_1\tau_2\tau_5 + \tau_1\tau_2\tau_7 + \tau_1\tau_4\tau_7 + \tau_1\tau_6\tau_7$$
$$+ \tau_3\tau_4\tau_5 + \tau_3\tau_4\tau_7 + \tau_3\tau_6\tau_7 + \tau_5\tau_6\tau_7)s^3 +$$
$$(\tau_1\tau_2 + \tau_1\tau_4 + \tau_1\tau_6 + \tau_3\tau_4 + \tau_3\tau_6 + \tau_5\tau_6)s +$$
$$(\tau_1 + \tau_3 + \tau_5 + \tau_7)s + 1$$

where $\beta_j = g_{aj}/g_{mj}$ and $\tau_j = C_j/g_{mj}$ are the zero and pole parameters respectively. The values of time constants τ_j and gain boost coefficients β_j can be determined by comparing the coefficients in Equation (10) with those of the desired transfer function in Equation (11).

$$H_d(s) = \frac{A_2 s^2 - A_0}{B_7 s^7 + B_6 s^6 + B_5 s^5 + B_4 s^4 + B_3 s^3 + B_2 s^2 + B_1 s + 1} \quad (11)$$

The calculation formulas are derived as

$$\tau_7 = \frac{B_7}{B_6} \qquad \tau_6 = \frac{B_6}{B_5 - B_4\tau_7}$$

$$\tau_5 = \frac{B_5 - B_4\tau_7}{B_4 - (B_3 - B_2\tau_7)\tau_6}$$

$$\tau_4 = \frac{B_4 - (B_3 - B_2\tau_7)\tau_6}{B_3 - B_2\tau_7 - [B_2 - (B_1 - \tau_7)\tau_6]\tau_5}$$

$$\tau_3 = \frac{B_3 - B_2\tau_7 - [B_2 - (B_1 - \tau_7)\tau_6]\tau_5}{B_2 - (B_1 - \tau_5 - \tau_7)\tau_4 - (B_1 - \tau_7)\tau_6} \quad (12)$$

$$\tau_2 = \frac{B_2 - (B_1 - \tau_3 - \tau_5)\tau_7 - (B_1 - \tau_7)\tau_6}{B_1 - \tau_3 - \tau_5 - \tau_7}$$

$$\tau_1 = B_1 - \tau_3 - \tau_5 - \tau_7$$

$$\beta_3 = \frac{A_2}{\tau_1\tau_2} \qquad \beta_1 = A_0 + \beta_3$$

The normalized characteristic of a seventh-order low-pass filter with 3-dB gain-boost real zeros at the cut-off frequency is given by [3]

$$H(s) = \frac{(s^2 - 1)}{D(s)} \quad (13)$$

where

$$D(s) = 0.055617s^7 + 0.291094s^6 +$$
$$1.095656s^5 + 2.554179s^4 + 4.255922s^3$$
$$+ 4.676709s^2 + 3.176156s + 1$$

To realize the function in Equation (13) for the cut-off frequency of 150MHz, the fully-balanced OTA in Fig.2 are designed with identical transconductance g_m of value of 980μS to improve the tracking and performance of transconductance with respect to the tuning voltage. Matching Equations (11) and (13) to find A_j and B_j and using Equation (12), the values of capacitances from C_1 to C_7 for the seventh-order low-pass equiripple linear phase filter with real zeros are calculated as

$C_1 = 2.25$pF $\qquad C_2 = 1.49$pF

$C_3 = 6.49$pF $\qquad C_4 = 1.07$pF $\qquad (14)$

$C_5 = 0.93$pF $\qquad C_6 = 2.91$pF

$C_7 = 1.11$pF

$g_{m1,2,4,5} = 980$μS $\qquad g_{m3} = 5 \times 980$μS

$g_{m6,7} = 4 \times 980$μS $\qquad g_{a1} = 328$μS $\qquad g_{a3} = 672$μS.

IV. SIMULATION RESULTS

The proposed OTA in Fig.1 was used for simulation of the seventh-order linear phase filter shown in Fig.2. The OTA structure is very simple, therefore there is no need for using common-mode feedback circuitry. Due to the small dimensions of the transistors, parasitic node capacitances are subtracted from the values of C_j's in Equation (14). The OTA and the filter were simulated with the TSMC 0.25μm CMOS process model using the following key model parameters, $V_{tno} = 0.407736$V, $V_{tpo} = -0.587912$V, $K_n = 121.9$μA/V^2 and $K_p = -23.2$μA/V^2 provided by the MOSIS. The total power consumption (PC) of the seventh-order low pass filter is only 216mW for a single 2V supply. Fig.3 shows the tuning ranges of the cutoff frequencies without and with gain boost are 50-150 MHz and 65 – 250 MHz, respectively.

The boosted gain has been maintained throughout the tuning range with the tolerance of ±0.25dB. Fig.3 also indicates that the magnitude response of the filter decreases as bias voltage V_B decreases. This is due to the finite output conductance of the OTA.

The filter phase response is fairly linear, as can be seen from Fig.4, the group delay has a very small variation up to about twice the cut-off frequency. The group delay ripple (GDR) is calculated as 4.5% by using the definition given in [7] and this is well within the limit of read/write channel filter specification.

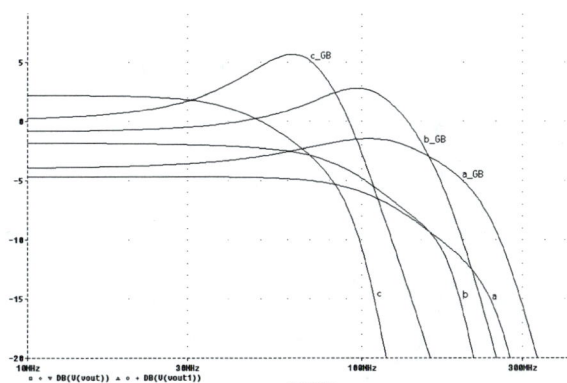

Fig.3 Simulated magnitude response of the filter without and with gain boost. GB stands for gain boost. Curves a, a_GB, b, b_GB and c, c_GB correspond to tuning voltages of V_B = 50mV, 110mV and 130mV, respectively.

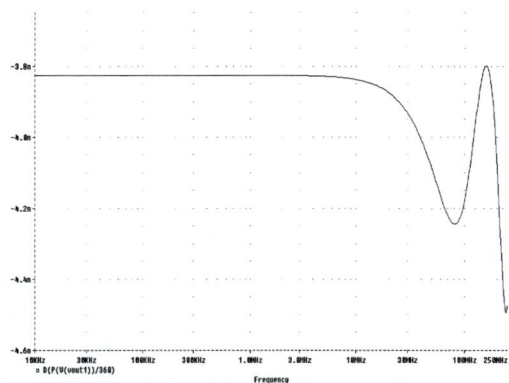

Fig.4 Filter group delay response, the ripple is less than 700ps up to 300 MHz.

Simulation results of the filter have shown a total harmonic distortion (THD) of less than 1% for a single tone of 250mVpp at 10MHz. The overall filter input-referred spectral noise density is around 9.14×10^{-18} V^2/Hz.

TABLE I. COMPARISON WITH OTHER 7TH-ORDER FILTER DESIGNS

Parameters	Specifications				
Design By	[2]	[8]	[9]	[3]	This work
Filter_Type	Eq.Rip	Eq.Rip	L.Phase	Eq.Rip	Eq.Rip
Filter_Conf.	IFLF	-	Cascade	LF	LF
Range MHz	5 -20	0 -100	0-200	8-32	50-150
GBR_MHz	10-47	-	-	15-100	65-250
Fc Stability	±5%	±5%	±5%	±5%	±5%
GDR	3-6dB	<4.6%	<4%	±7%	±4.5%
NBW MHz	-	-	200	500	1GHz
Input	1Vpp	450mV	500mV	500mV	250mV
SNR	-	-	>40dB	>60dB	>65dB
THD	<-46dB	<-40dB	<-44dB	<-40dB	<-40dB
DR	-	-	50dB	55dB	65dB
PC mW	120	210	60	322	216
P.Supply	5V	-	±1.5V	5V	2V
CMOSTech	2μ	0.25μ	0.35μ	0.25μ	0.25μ

GBR=gain boost range; NBW=noise bandwidth

Results from this work and some previous designs are summarized in Table I. The performances of group delay ripple (GDR), signal to noise ratio (SNR), total harmonic distortion (THD), and dynamic range (DR) of our filter are equal or better than all other realizations. The equalizer described in this work is the only MLF LF equalizer known so far that realizes a cut-off frequency of up to 250MHz. This fully-balanced design has achieved performances all better than the single-ended design [3].

V. CONCLUSIONS

This paper has described a fully-balanced continuous-time seventh-order linear phase filter as required in a hard disk drive read/write channel. The filter is based on a multiple-loop feedback leapfrog structure for lower sensitivities. A dual differential pair OTA with cross-coupled input and a typically larger transconductance has been used. Simulation results in 2V 0.25μm CMOS show that group delay variation is around 4.5 %, in-band harmonic distortion is less than -40dB and power consumption of the entire filter system is 216mW. These results have shown that the MLF LF filter is well suitable for hard disk drive read/write channel applications.

REFERENCES

[1] T. Deliyannis, Y. Sun and J. K. Fidler. Continuous-time Active Filter Design, CRC Press, Florida, USA, 1999.

[2] D. H. Chiang and R. Schaumann, "A CMOS fully-balanced continuous-time IFLF filter design for read/write channels", Proc. IEEE Int. Symp. Circuits and Systems, pp. 167-170, 1996.

[3] H. W. Su and Y. Sun, "A CMOS 100MHz continuous-time seventh order 0.05° equiripple linear phase leapfrog multiple loop feedback G_m-C filter", IEEE Int. Symp. on Circuits and Systems, vol. 2, pp 17-20, May 2002.

[4] G. A. De Veirman and R. Yamasaki, "A 27MHz programmable bipolar 0.05° equiripple linear phase low pass filter." in IEEE ISSCC Dig. Tech. paper, pp. 64-65, Feb. 1992.

[5] B. E. Bloodworth, P. P. Siniscalchi, G. A. De Veirman, A. Jezdic, R. Pierson, and R. Sundararaman, "A 450-Mb/s analog front end for PRML read channels," IEEE Custom Integrated Circuits Conf., pp. 309-316, 1998.

[6] N. Rao, V. Balan, and R. Contreras, "A 3V 10-100MHz continuous-time seventh-order equiripple linear phase filter," in IEEE Int. Solid-State Circuits Conf. (ISSCC) Dig. Tech. papers, pp.44-46, Feb. 1999.

[7] W. Dehaene, M. S. J. Steyaert, and W. Sansen, "A 50-MHz standard CMOS pulse equalizer for hard disk read channels", IEEE J. Solid-State Circuits, vol. 32, No.7, pp.977-988, July 1997.

[8] V. Gopinathan, M. Tarsia and D. Choi, "Design considerations and implementation of a programmable high-frequency continuous-time filter and variable-gain amplifier in sub micrometer CMOS", IEEE J. Solid-State Circuits, vol. 34, No.12, pp.1698-1707, Dec. 1999.

[9] J. Silva-Martinez, J. Adut, J.M. Rocha-Perez, M. Robinson and S. Rokhsaz, "A 60-mW 200-MHz continuous-time seventh-order linear phase filter with on-chip automatic tuning system", IEEE J. Solid-State Circuits, vol. 38, No.2, pp.216-225, Feb. 2003.

[10] Y. Sun and J. K. Filder, "Fully-balanced structures of continuous-time MLF OTA-C filters", Proc. IEEE Int. Conf. Electronics, Circuits and Systems, pp. 157-160, 1998.

ANALYSIS OF TRAVELING WAVE & TRANSVERSAL ANALOG ADAPTIVE EQUALIZERS

Shanthi Pavan & Shankar Shivappa

Department of Electrical Engineering
Indian Institute of Technology, Madras
Chennai 600 036, India

ABSTRACT

We present a general framework for the analysis and simulation of traveling wave and transversal adaptive equalizers in the context of a real communication channel. Simulation results for adaptive equalizers operating at a data rate of 10 Gbps are given. Intuition is derived for the superior performance of traveling wave FIR filters.

1. INTRODUCTION

The performance of high speed optical communication links is limited by intersymbol interference due to fiber impairments like polarization mode and modal dispersion. When compared to optical mitigation methods, electronic signal processing solutions are attractive due to their potentially reduced cost [1]. Adaptive electrical equalization using a finite impulse response (FIR) filter is one such low complexity technique.

Equalization can be accomplished in the analog or digital domain. At several Gbps, analog adaptive filters are promising low power alternatives to digital equalizers. Recently, there has been a resurgence of the traveling wave FIR filter architecture, as evidenced in several publications ([2], [3] & references therein). The first paper describes a seven tap fractionally spaced equalizer dissipating only 40 mW while attempting to equalize 10 Gbps NRZ data.

The idea of using a traveling wave amplifier as an FIR filter by using the "anti-sync" end as the output is not new [4]. A more apparent FIR architecture is the transversal filter [1] [5]. The common reason to prefer a traveling wave approach seems to be the higher potential bandwidth achievable. It has only recently been pointed out [6] that a traveling wave (TW) filter is also topologically superior to a transversal structure when reflections and series losses in the transmission lines are considered. Most of the work on TW and transversal FIR filters has been concentrated realizing programmable filtering in the analog domain. To our knowledge, there appears to be no study of the performance of such filters in the context of a real communication channel and practical circuit nonidealities. In this work, we present a framework that enables us to answer questions like the following-

a. How do termination impedance mismatch and transmission line loss impact system performance?

b. How do the input and output capacitances of the transconductors effect the FIR filter performance ? How much capacitance can we tolerate ?

Analyzing FIR filters in isolation will result in unrealistically tight specifications for individual circuit elements (like transmission line loss, or transconductor bandwidth). As we will see later in this paper, TW FIR filters are *topologically* remarkably tolerant to

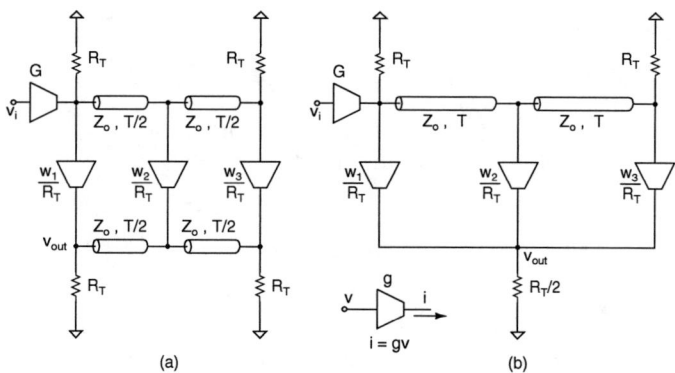

Fig. 1. Schematics of traveling wave & transversal FIR filters.

circuit nonidealities. Other important issues, like timing recovery and adaptive algorithms are not addressed here.

The rest of the paper is organized as follows. In Section 2 we derive the minimum mean square solution for the equalizer weights. Nonidealities like series loss, termination impedance mismatch and transconductor load capacitances are included in the analysis. Section 3 discusses methodologies used for simulation and gives representative results. Conclusions are presented in Section 4.

2. MINIMUM MEAN SQUARE ERROR (MMSE) SOLUTION FOR TRAVELING-WAVE AND TRANSVERSAL EQUALIZERS

Figure 1 shows the schematics of a three tap fractionally spaced traveling-wave equalizer and transversal filter for reference. Tap weights are realized using programmable transconductors. w_1, w_2, w_3 are dimensionless. Ideally, the transmission lines are lossless, $R_T = Z_o$ and $T = T_b/2$, where T_b is the symbol period. Note that the output resistor in the transversal filter is $R_T/2$. This ensures that under ideal circumstances, both filters have the same impulse response. We assume that all transconductors have zero reverse transmission. In practice, reverse transmission can (and should) be made very small with careful circuit design and layout techniques. The end-to-end system considered in this paper is shown in Figure 2(a). While the optical channel is considered here as an example, a similar diagram could be drawn for a variety of channels. The transmitted sequence is represented as $\sum_n b_n \delta(t - nT_b)$, where b_n takes on values of 1 or 0 with equal probability. The transmit filter is an NRZ pulse, followed by a band limiting filter. The optical fiber is modeled as a Polarization Mode Dispersion (PMD) channel, along with its propagation loss. At the receiver, the

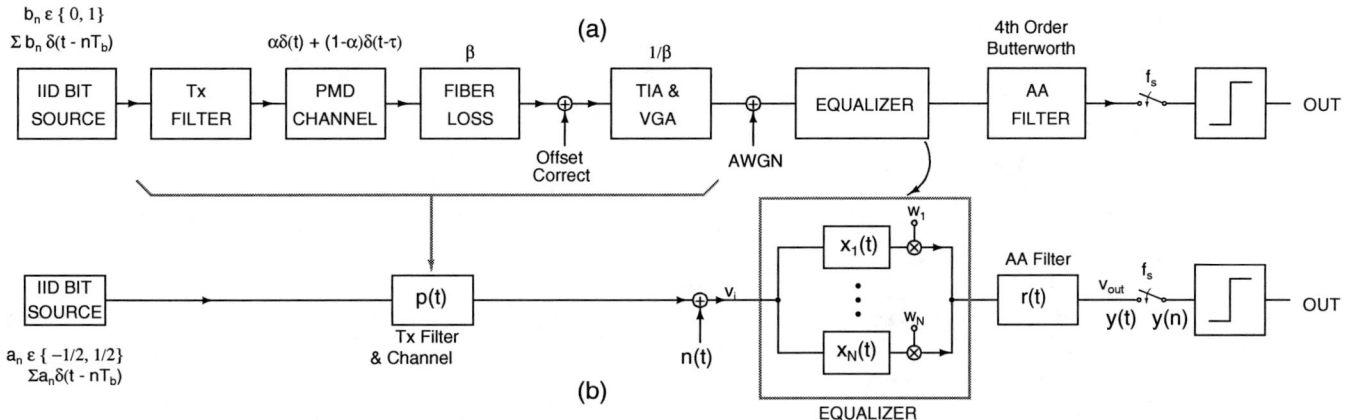

Fig. 2. System model.

transimpedance amplifier (TIA) and variable gain amplifier (VGA) cancel DC offset and compensate for fiber loss. In this process, they add noise which is represented as an equivalent noise source at the output. The block diagram of Figure 2(a) can be simplified according to the following. We replace the on-off bit source b_n and the offset correction mechanism by an offset free source a_n. The TIA-VGA gain compensates for the fiber loss and hence both may be removed. The transmit pulse shape and the PMD channel are combined into a single filter. The resulting block diagram is shown in Figure 2(b). Here a_n takes on values $\pm 1/2$. The various quantities of interest are as follows-

1. The transmitted data is $a(n)$. $p(t)$ represents the complete channel response prior to the receiver - it is a cascade of the transmit pulse shape (NRZ), the transmit filter and the PMD channel.

2. Equalizer input noise (which is the output noise of the TIA-VGA combination), considered white and having a double sided spectral density of $N_o/2$, is denoted as $n(t)$.

3. $x_i(t)$ denotes the impulse response of the filter (either traveling wave or transversal) with $w_i = 1$ and all other weights being 0. An N-tap equalizer consists of a bank of N such responses. $\mathbf{w}$ is the $N \times 1$ tap weight vector.

4. $r(t)$ represents the impulse response of the receive "anti-aliasing" filter. Without this, out-of-band noise could significantly degrade receiver performance due to the periodic nature of the equalizer frequency response.

5. $c_i(t) = p(t) * x_i(t) * r(t)$ is the impulse response from the transmitter to the detector due to the ith equalizer tap. $f_i(t) = x_i(t) * r(t)$.

6. The output of the equalizer is $y(t)$. t_o represents the phase difference between the transmit and receive clocks. We assume that the receive and transmit clocks have the same frequency. $y(n)$ is the nth output sample.

We derive an expression for filter tap weights that minimize the mean squared error at the equalizer output. In what follows, $*$ represents the convolution operation. The output of the equalizer is

$$y(t) = \sum_{i=1}^{N} \sum_{k=-\infty}^{\infty} w_i a(k) c_i(t - kT_b) + \sum_{i=1}^{N} w_i n(t) * f_i(t) \quad (1)$$

$y(n)$ can be written in matrix form as follows - the samples of $c_i(t)$ are represented such that $C_{rs} = c_s((r-1)T_b + t_o)$. Assuming that $c_i(t)$ is negligible for $t > LT_b$, we see that the size of $\mathbf{C}$ is $(L+1) \times N$. Further, let $\mathbf{a}^T(n) = [a(n) \ a(n-1) \ \cdots \ a(n-L)]$ and $\mathbf{\eta}^T(n) = [(n(t) * f_1(t)) \ \cdots \ (n(t) * f_N(t))]|_{t=nT_b+t_o}$. We then have,

$$y(n) = \mathbf{a}^T(n)\mathbf{C}\mathbf{w} + \mathbf{\eta}^T(n)\mathbf{w} \quad (2)$$

Defining $\mathbf{h}_\delta^T = [\ 0 \ \cdots \ 0 \ 1 \ 0 \ \cdots \ 0\ ]$ (δ zeros preceding the 1 and $(L - \delta)$ zeros after it), we see that the desired equalizer output is $a(n - \delta) = \mathbf{a}^T(n)\mathbf{h}_\delta$ - we want the transmitted symbol, with a delay of δ symbols. The error at the equalizer output is

$$\begin{aligned}
e(n) &= y(n) - \mathbf{a}^T(n)\mathbf{h}_\delta \\
&= \mathbf{a}^T(n)(\mathbf{C}\mathbf{w} - \mathbf{h}_\delta) + \mathbf{\eta}^T(n)\mathbf{w} \\
E[e^2(n)] &= (\mathbf{C}\mathbf{w} - \mathbf{h}_\delta)^T E[\mathbf{a}^T(n)\mathbf{a}(n)](\mathbf{C}\mathbf{w} - \mathbf{h}_\delta) \\
&\quad + \mathbf{w}^T E[\mathbf{\eta}^T(n)\mathbf{\eta}(n)]\mathbf{w}
\end{aligned} \quad (3)$$

Since the transmitted sequence is i.i.d, $E[\mathbf{a}^T(n)\mathbf{a}(n)] = \sigma_a^2 \mathbf{I}$ where $\sigma_a^2 = E[|a(n)|^2]$. $E[\mathbf{\eta}^T(n)\mathbf{\eta}(n)]$ is given by

$$\mathbf{M} = \frac{N_o}{2} \begin{bmatrix} \int_0^\infty f_1(t)f_1(t)dt & \cdots & \int_0^\infty f_1(t)f_N(t)dt \\ \vdots & \ddots & \vdots \\ \int_0^\infty f_N(t)f_1(t)dt & \cdots & \int_0^\infty f_N(t)f_N(t)dt \end{bmatrix}$$

so that

$$MSE = \sigma_a^2 (\mathbf{C}\mathbf{w} - \mathbf{h}_\delta)^T (\mathbf{C}\mathbf{w} - \mathbf{h}_\delta) + \mathbf{w}^T \mathbf{M} \mathbf{w} \quad (4)$$

Our aim is to find $\mathbf{w}$ which minimizes the expected squared error at the output of the equalizer. In terms of $\mathbf{A} = \mathbf{C}^T\mathbf{C} + (1/\sigma_a^2)\mathbf{M}$, the technique of "completing the squares" yields

$$\mathbf{w_{opt}} = \mathbf{A}^{-1}\mathbf{C}^T\mathbf{h}_\delta \quad (5)$$

and the minimum mean square error (which is a function of δ) is

$$MSE_{min} = \sigma_a^2 \mathbf{h}_\delta^T (\mathbf{I} - \mathbf{C}\mathbf{A}^{-1}\mathbf{C}^T)\mathbf{h}_\delta \quad (6)$$

The optimal delay corresponds to the minimum diagonal element of $\mathbf{I} - \mathbf{C}\mathbf{A}^{-1}\mathbf{C}^T$ [7]. Therefore,

$$\delta_{opt} = \operatorname{argmin}_\delta \left\{ \left[\mathbf{I} - \mathbf{C}\left(\mathbf{C}^T\mathbf{C} + (1/\sigma_a^2)\mathbf{M}\right)^{-1}\mathbf{C}^T \right]_{\delta,\delta} \right\} \quad (7)$$

Notice that the MSE is still a function of the timing offset t_o between the receive and transmit clocks.

3. SIMULATION METHODOLOGY

In this section, we present a methodology for simulating the performance of traveling-wave and transversal filters. Referring to Figure 2, the crux of the problem here is to evaluate $x_1(t), \cdots, x_N(t)$. Once these impulse responses are known, knowledge of the channel can be used to determine **C**, **M** and **A**. Using these in (5), (6) and (7) yields all the information that we are after.

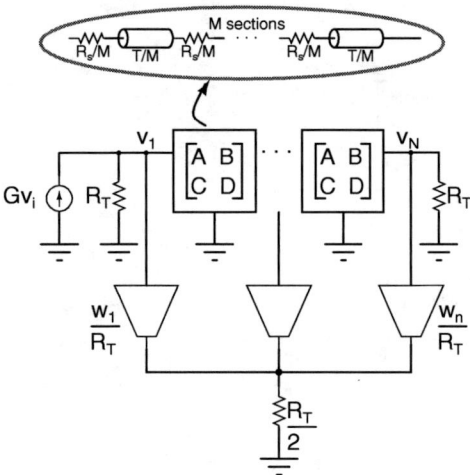

Fig. 3. Simulation methodology for a transversal FIR filter.

Figure 3 shows the simulation framework for a transversal filter. The use of ABCD parameters makes the methodology very general and can take into account all the important practical nonidealities in such filters. To simulate uniform series loss, each transmission line section of delay T is broken up into M equal sections with delay T/M and resistance R_s/M. Working in the frequency domain, the ABCD matrix for a single section is seen to be

$$\begin{bmatrix} A & B \\ C & D \end{bmatrix} = \left\{ \begin{bmatrix} 1 & \frac{R_s}{M} \\ 0 & 1 \end{bmatrix} \begin{bmatrix} \cos(\omega T/M) & jZ_0 \sin(\omega \frac{T}{M}) \\ \frac{j}{Z_0} \sin(\omega T/M) & \cos(\omega \frac{T}{M}) \end{bmatrix} \right\}^M$$

We denote

$$\begin{bmatrix} A_{tot} & B_{tot} \\ C_{tot} & D_{tot} \end{bmatrix} = \left\{ \begin{bmatrix} 1 & 0 \\ \frac{1}{R_T} & 1 \end{bmatrix} \begin{bmatrix} A & B \\ C & D \end{bmatrix}^{(N-1)} \begin{bmatrix} 1 & 0 \\ \frac{1}{R_T} & 1 \end{bmatrix} \right\}$$

$$\begin{bmatrix} A_i & B_i \\ C_i & D_i \end{bmatrix} = \begin{bmatrix} A & B \\ C & D \end{bmatrix}^{(N-i)} \begin{bmatrix} 1 & 0 \\ \frac{1}{R_T} & 1 \end{bmatrix} \quad (8)$$

With a little algebra, it can be shown that

$$X_i(j\omega) = A_i \frac{G}{2C_{total}} \quad i = 1, \cdots, N \quad (9)$$

where $X_1, \cdots X_N$ denote the Fourier transforms of the corresponding time-domain impulse responses. Input admittance Y_{in} of the transconductors can be easily incorporated into the above setup by modifying the ABCD matrix of a single section. Finite bandwidth of the transconductors can be modeled by making G frequency dependent. The time domain responses are obtained by using the inverse Fourier transform. A similar strategy is used for a TW-FIR filter. The simulation setup is implemented in MATLAB. To build intuition, we compare the responses of both kinds of filters to a

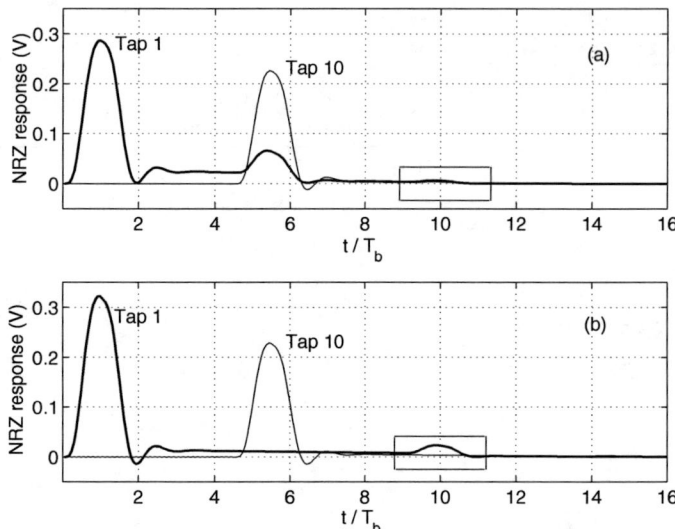

Fig. 4. The first and tenth tap responses to a 10 GHz fourth order Butterworth filtered NRZ pulse of (a) A 10 tap traveling wave FIR filter & (b) A 10 tap transversal FIR filter. Both are $T_b/2$ spaced equalizers, with $T_b = 100$ ps, $G = 1/50\,\Omega$, $Z_0 = 50\,\Omega$, $R_T = 65\,\Omega$, $R_s = 5\,\Omega$.

filtered NRZ pulse. Figure 4 shows the first and tenth tap responses of traveling-wave and transversal $T_b/2$ spaced equalizers, with $T_b = 100$ ps, $G = 1/50\,\Omega$, $Z_0 = 50\,\Omega$, $R_T = 65\,\Omega$ and $R_s = 5\,\Omega$. Notice that in both cases, the responses have slowly decaying tails (due to series loss) and reflections (due to terminal impedance mismatch). The delay of the tenth tap corresponds to the span of the equalizers. Comparing the responses of Tap 1 in both filter implementations, we see that the traveling-wave equalizer has a larger deviation than the transversal filter within the equalizer span. Outside of this, the transversal filter has a much larger deviation. Notice, in particular the portion shown in the box around time $10\,T_b$. For the transversal case, it is the first reflection, that occurs at twice the equalizer span and is proportional to the reflection coefficient at the termination (Γ_T). In a traveling-wave structure, the first reflection occurs at time $5\,T_b$. Its strength is twice that of the first reflection of the transversal filter - this is because of reflections occurring on both input and output lines. The second reflection occurs at $10\,T_b$ and is proportional to Γ_T^2, and is much smaller than the *first* reflection of the transversal filter. Since the filter coefficients will be adapted, it is apparent that any deviation from ideal behavior within the equalizer span will be compensated for by the adaptive process, but reflections outside this span cannot be corrected for [6]. This intuitively explains why the traveling wave architecture is a *fundamental* improvement over the transversal filter.

The simulation framework presented in this paper allows a circuit designer to examine trade-offs involved in the design of TW and transversal equalizers, so that specifications for individual blocks (like delay lines and transconductors) can be arrived at, given the desired system performance.

We now present representative simulation results for ten tap TW and transversal filters equalizing a channel impaired by polarization mode dispersion (PMD). 10 Gbps NRZ data ($T_b = 100$ ps) with a 7.5 GHz transmitter bandwidth is assumed. The channel impulse response is of the form $\alpha\delta(t) + (1-\alpha)\delta(t-\tau)$. α and τ are referred to as power-split and differential group delay respectively. We show simulations for $\alpha = 0.5$, and $\tau = 0.8 T_b$. Figure 5 shows the post-equalized SNR for a TW-FIR equalizer for different values of series loss as R_T is varied. It is seen a TW-equalizer is robust with varying terminations and series loss. The performance of a transversal equalizer is shown in Figure 6. Notice that when the lines exhibit series loss, the SNR peaks when $R_T \approx Z_o - R_s$, indicating that termination impedance should be deliberately made smaller than Z_o. This makes sense due to the following : series loss gives rise to positive tails in the tap impulse responses. When $R_T < Z_o$, Γ is negative, and these reflections cancel some of the effects of series loss.

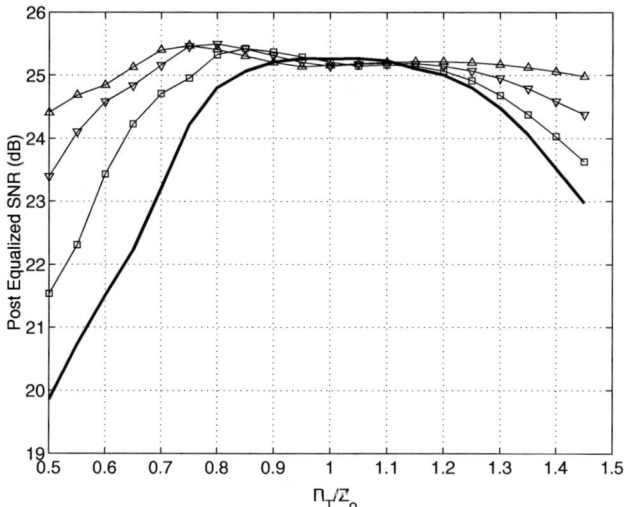

Fig. 5. Effect of mistermination and series loss on the performance of a 10 tap traveling wave filter : (–) $R_s = 0$, ($\square$) $R_s = 0.075 Z_o$, ($\triangledown$) $R_s = 0.150 Z_o$ and ($\triangle$) $R_s = 0.225 Z_o$. $SNR_{in} = 30$dB.

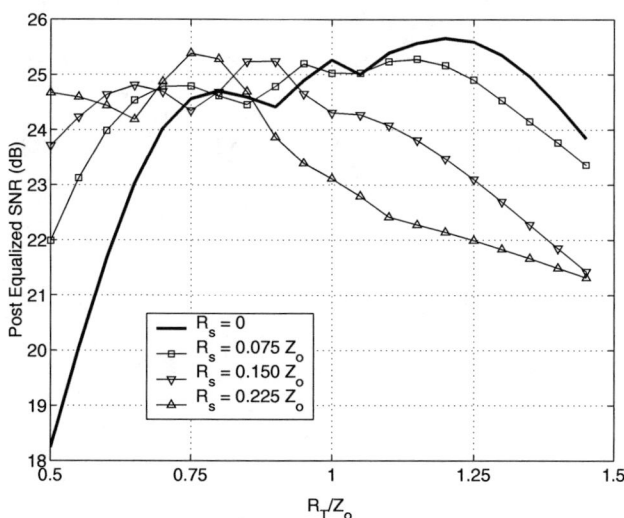

Fig. 6. Effect of mistermination and series loss on the performance of a 10 tap transversal filter : (–) $R_s = 0$, ($\square$) $R_s = 0.075 Z_o$, ($\triangledown$) $R_s = 0.150 Z_o$ and ($\triangle$) $R_s = 0.225 Z_o$. $SNR_{in} = 30$dB.

4. DISCUSSION & CONCLUSIONS

In this section we present some key points from our simulation studies of TW and transversal FIR filters. Exhaustive results are omitted here due to lack of space.

When practical circuit nonidealities are considered, *fractionally spaced* transversal filters exhibit significant performance sensitivity to timing offset between transmit and receive clocks, while *fractionally spaced* TWA-filters are relatively insensitive to timing offset.

Transmission line loss reduces the effects of terminal impedance mismatch on the performance of a traveling wave and transversal filters. One should deliberately choose a nominal terminating impedance *lower* than Z_o for better performance, because the effect of reflections can be made to partially cancel the effect of series loss. While the benefit of a lower R_T is seen to be marginal in a traveling wave design, a transversal filter benefits significantly. In such filters, contrary to popular belief (and practice), a terminating impedance $R_t = Z_0 - R_s$ is a better choice than $R_t = Z_0$.

Unexpectedly, the performance of both topologies (traveling-wave & transversal) are similar when parasitic capacitances of the transconductors are considered.

In this work, we presented a framework that enables a circuit designer to examine the impact of transmission line and transconductor nonidealities on the performance of a communication link employing TW or transversal equalizers. Representative simulation results were shown.

5. REFERENCES

[1] J. Winters and R. Gitlin, "Electrical signal processing techniques in long-haul fiber-optic systems," *IEEE Transactions on Communications*, vol 38, no. 9, pp. 1439-53, September 1990.

[2] H. Wu, J. A. Tierno, P. Pepeljugoski, J. Schaub, S. Gowda, J. A. Kash and A. Hajimiri, "Integrated transversal equalizers in high-speed fiber optic systems," *IEEE Journal of Solid State Circuits*, vol 38, no. 12, pp. 2131-37, Dec 2003.

[3] A. Freundorfer, D. Choi and Y. Jamani, "Adaptive transversal preamplifier for high speed light wave systems," *IEEE Microwave and Guided Wave Letters*, vol 11, no. 7, pp. 293-5, Jul 2001.

[4] C. Raucher, "Microwave active filters based on traveling wave and recursive principles," *IEEE Transactions on Microwave Theory and Techniques*, vol. 33, no. 12, pp. 1350-60, Dec 1985.

[5] C. Rogers and N. Ahmed, "Implementation of a VHF broadband FIR adaptive filter," *IEEE Transactions on Instrumentation and Measurement*, Vol. 38, No. 6, pp. 1074-9, Dec 1989.

[6] S. Pavan, "Continuous-time integrated FIR filters at microwave frequencies," *IEEE Transactions on Circuits and Systems II - Express Briefs*, Vol. 51, No. 1, pp. 15-20, Jan 2004.

[7] C. R. Johnson, P . Schniter, T .J . Endres, J .D . Behm, D .R . Brown and R .A . Casas, "Blind equalization using the constant modulus criterion: a review," *Proceedings of the IEEE*, Vol.86, No. 10, October 1998, pp. 1927-50.

Rules for Systematic Synthesis of All-Transistor Analogue Circuits by Admittance Matrix Expansion

Phil Corbishley and David G Haigh
Dept of Electrical & Electronic Engineering
Imperial College London
London SW7 2BT, UK
d.haigh@ee.ic.ac.uk

Abstract—A previous method for synthesis of all-transistor circuits suffered from the disadvantage that it generated a large number of possible circuits which would require computer generation and evaluation. We propose restrictions on the transformations used in the synthesis such that the number of circuits generated is much reduced while retaining circuits which have desirable performance characteristics. Finally, we consider reduction of the trial-and-error element in the synthesis process by exploring carrying out the synthesis entirely in the admittance matrix domain.

I. INTRODUCTION

In analogue VLSI, all-transistor circuits play a very important role, especially for implementing building blocks such as operational amplifiers, buffers and transconductors [1, 2, 3]. Such circuits are traditionally synthesised using a bottom-up approach whereby sub-blocks, such as differential pairs, gain stages and current mirrors are assembled to realise the complete circuit. As an alternative to such an approach, a method of systematic synthesis of all-transistor circuits has been proposed [4] which starts from a port admittance matrix description of the wanted circuit function and expands this to a nodal admittance matrix describing an interconnection of transistors. Ideal transistors are represented in the circuit by nullors consisting of a nullator-norator pair, and linked infinity parameters [5] are used to describe the ideal transistors and in many cases the wanted port matrix too. A problem with the approach in [4] is that it results in a large number of possible circuits. Thus a computer approach would be needed to generate and assess the circuits, thus leaving the designer out of a critical part of the design.

The purpose of this paper is to propose restricting the transformations used in [4] in order that (a) the number of circuits produced is considerably reduced, (b) the circuits produced include known useful ones and (c) the non-ideal performance changes during the synthesis are incremental so that good performance may be consolidated and enhanced throughout the process. We begin by reviewing key elements of the synthesis method in [4].

II. BACKGROUND

We assume that the element for the synthesis is the field effect transistor (FET). An ideal FET M_i may be represented in an admittance matrix using linked infinity parameters [5]:

$$\begin{array}{c} \quad\quad g \quad\quad s \\ d \begin{bmatrix} \infty_i & \infty_i \\ \infty_i & \infty_i \end{bmatrix} \end{array} \quad (1)$$

where g, d and s are the nodes to which the gate, drain and source are connected and ∞_i represents the transconductance of the ideal FET; note that one of the elements is always on the main diagonal in a position corresponding to the source node and that this element is always positive on account of the voltage and current polarity conventions.

The ideal FET can be represented in a circuit by means of a nullator element to represent the gate-source port and a norator element to represent the drain-source port as shown in Fig 1a; the nullator is defined by $V = I = 0$ and the norator by V and I arbitrary. A nullator-norator pair is referred to as a nullor. The nullator and norator in a nullor representing a transistor must have a common node, which is the source node. As an alternative to the admittance matrix representation in eqn (1) we can use a bracket between columns to indicate the nodes between which a nullator is connected and a bracket between rows to indicate the nodes between which a norator is connected [5].

$$\begin{array}{c} \overline{\quad g \quad\ s\ } \\ \begin{bmatrix} d & 0 & 0 \\ s & 0 & 0 \end{bmatrix} \end{array} \quad (2)$$

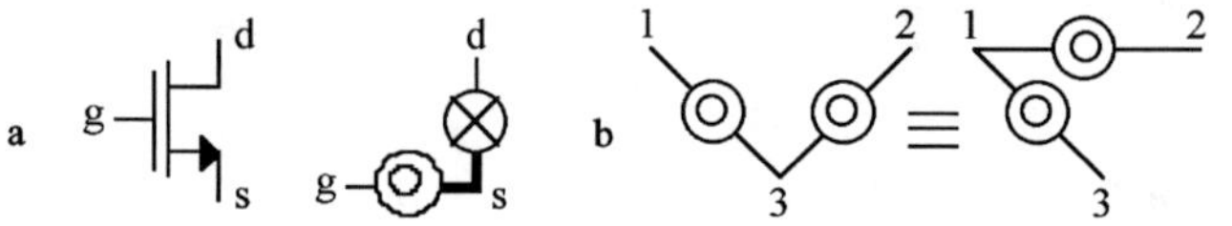

Figure 1. a Nullor equivalent for the ideal FET (thick wire denotes source end of nullator and norator) b V-7 transformation (shown for nullators)

The non-ideal degradation behaviour for the device description in eqn (1) is such that the device becomes a voltage-controlled current source with finite transconductance ($\Box_i \Box G_i$), which is an appropriate model for a non-ideal FET; admittance terms representing finite input and feedback capacitance, y_i and y_f, and output conductance y_o, may be introduced in eqn (1). The nodal admittance matrix of a circuit containing ideal or non-ideal FETs may be reduced by conventional methods to obtain the port matrix and desired network functions [5].

A series combination of a nullator and a norator is equivalent to an open-circuit and a parallel combination of the same elements is equivalent to a short-circuit; these will be referred to as nullor open- and short-circuits, respectively. The synthesis method for all-transistor circuits in [4] is based on the introduction into a circuit of one or more nullor open- and short-circuits and the application of subsequent equivalence transformations. These include the V-7 transformation, illustrated for nullators in Fig 1b, although it is also valid for norators. Another equivalence transformation used in [4] is the re-pairing of nullators and norators, which does not affect the properties of the ideal circuit [1]. The synthesis method in [4] can also be interpreted as a process of expansion of the circuit admittance matrix.

Both the admittance matrix and the nullor representations for the FET are small-signal linear representations. So the biasing arrangements for transistors will have to be considered in addition to the small signal synthesis. Having summarised the synthesis method of [4], we now consider introducing some additional rules.

III. Rules for Transformation

The admittances presented by a nullor open- and short-circuit with non-ideal FET parameters of finite transconductance, G, and finite input, output and feedback admittances, y_i, y_o and y_f, are as follows:

$$Y_{oc} = y_f + \frac{y_i y_o}{G + y_i + y_o} \quad (3)$$
$$Y_{sc} = G + y_i + y_o$$

For the case of $y_i = y_o = y_f = 0$, the nullor open-circuit has $Y_{oc} = 0$ and the nullor short-circuit has $Y_{sc} = G$; hence, the nullor open-circuit retains ideal behaviour even with finite G whereas the short-circuit is significantly degraded. Even with practical values for y_i, y_o and y_f, the nullor open-circuit is likely to be significantly more ideal than the nullor short-circuit. Since in circuit synthesis we are looking for refinement of circuit performance rather than dramatic changes, we adopt the policy of introducing during synthesis nullor open-circuits between any nodes but restrict the introduction of a nullor short-circuit to the case where it is in series with the gate terminal of a nullator or the drain terminal of a norator. Then, the addition of a nullor open- or short-circuit to an existing circuit has little or no effect.

Subsequent to the addition of a nullor open- or short-circuit, further transformations are carried out in [4], one of which is re-pairing of the nullator and norator in the added nullor with those in the existing circuit. It is clear that such nullator-norator re-pairing represents a substantial modification to the existing circuit in such a way that its non-ideal performance is likely to be significantly affected. In order to allow beneficial performance characteristics to be consolidated at each stage of the synthesis process, we state that nullator-norator re-pairing should be avoided. We adopt the nullor notation in eqn (1), which embodies defined nullator-norator pairing, in preference to that in eqn (2).

Having eliminated nullator-norator re-pairing from the synthesis toolbox, the process that is mainly responsible for change in circuit non-ideal performance is the V-7 transformation of Fig 1b. Avoidance of nullator-norator re-pairing places a number of restrictions on where nullor open- and short-circuits may be introduced and how subsequent V-7 transformations may be applied. In particular, a nullor open- or short-circuit must not be introduced in such a way that it forces a subsequent V-7 transformation to separate the nullator from the norator in the added or in any existing nullor. The ways in which it is permitted to add a nullor open- or short-circuit to an existing nullor are illustrated in Figs 2 and 3, respectively, together with the results of the V-7 transformation. The effect of avoiding nullator-norator re-pairing is to preserve the integrity of existing and added FETs, just as if the circuit was being constructed in the lab by adding discrete transistors one by one to the existing circuit and modifying connections according to the V-7 rule.

We now illustrate the synthesis of transconductor and current mirror examples of [4], together with some further ones, using the reduced set of transformations.

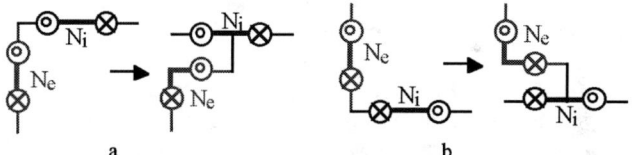

Figure 2. Permissible V-7 transformations using nullor open-circuit; existing nullor –N_e; introduced nullor – N_i

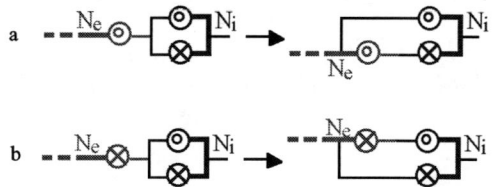

Figure 3. Permissible V-7 transformations using nullor short-circuit; existing nullor –N_e; introduced nullor – N_i

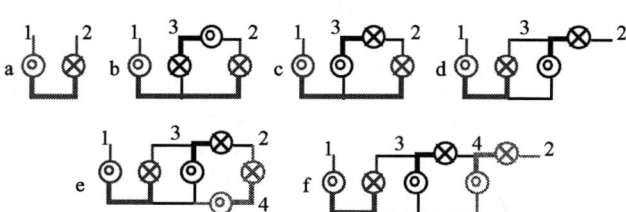

Figure 4. Transformation of a transconductor to obtain low output conductance

IV. GENERATION OF TRANSCONDUCTOR CIRCUITS

Consider the 2-port nullor circuit (nullor a) in Fig 4a. It has the following ideal and non-ideal admittance matrices:

$$\begin{bmatrix} 0 & 0 \\ a & 0 \end{bmatrix} \quad \begin{bmatrix} 0 & 0 \\ G & g \end{bmatrix} \quad (4)$$

where G is transconductance and g is output conductance. The open circuit voltage gain is $A_V = -y_{21}/y_{22} = -G/g$. We wish to reduce g in order to increase A_V. Since it is the output conductance that we wish to reduce, let us add a nullor open-circuit (nullor b) across the output port as shown in Fig 4c. The alternative orientation of the nullor open-circuit in Fig 4b does not allow any permitted V-7 transformations to be applied. Consider the V-7 transformations that may be applied to Fig 4c. The only permitted transformation is to move the drain terminal of norator a from node 2 to node 3 as shown in Fig 4d; this corresponds to the well known cascode transconductor which has output conductance $g_o \approx g^2/G$ and $A_V \approx (G/g)^2$. Since this operation has reduced the output conductance of the transconductance, it is reasonable to suppose that repetition of the process would give a further reduction. This is the case and the transformations are shown in Figs 4e and f, leading to an output conductance $g_o \approx g^3/G^2$ and $A_V \approx (G/g)^3$. A permitted variant of this circuit is obtained by moving the gate terminal of nullator c from node 0 to node 3, which gives a circuit with similar good properties.

An alternative to the double cascode circuit may be derived as follows starting from the single cascode circuit of Fig 4d. We add a nullor short-circuit (nullor c) in series with the gate terminal of nullator b, as shown in Fig 5a. Using the V-7 transformation, we move the gate terminal of nullator c from node 4 to node 3, as shown in Fig 5b. The circuit is the familiar regulated cascode circuit [1]. This example shows how the above rules reduce the number of possible circuits generated, retain circuits with good performance and allow beneficial performance features to be further enhanced.

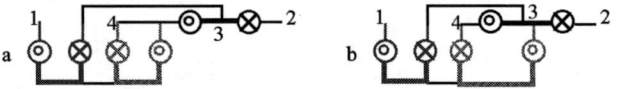

Figure 5. Transformation of a transconductor to obtain very low output conductance

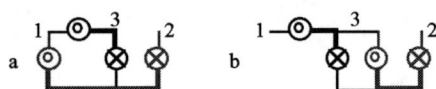

Figure 6. Transformation of a transconductor to obtain low input conductance

We now wish to increase the input admittance of the transconductor in Fig 4a. We therefore connect a nullor open-circuit (nullor b) across the input port, as shown in Fig 6a. The alternate orientation for the nullor open-circuit does not allow any permissible V-7 transformations. The only permissible V-7 transformation that can be applied to the circuit of Fig 6a is to move the gate terminal of nullator a from node 1 to node 3 as shown in Fig 6b. This circuit represents a common-drain FET in cascade with a common-source FET and clearly has a lower input admittance than the original single nullor.

In order to enhance the transfer characteristics of the simple transconductor, we can add a nullor open-circuit between the input and output nodes as shown in Fig 7a. The alternate orientation for the nullor open-circuit does not allow any permissible V-7 transformations. Two possible V-7 transformed circuits are shown in Figs 7b and c, where the gate and drain terminal, respectively, of nullor a have been moved. If nullors represent BJTs with current gain β, then the short-circuit current gain of the circuit in Fig 7b is about β^2. If the nullors have open-circuit voltage gain μ, then the open-circuit voltage gain of the circuit in Fig 7c is about μ^2. Thus both of the circuits in Figs 7b and c have enhanced transfer characteristics.

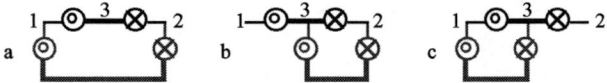

Figure 7. Transformation of a transconductor to obtain enhanced transfer characteristics

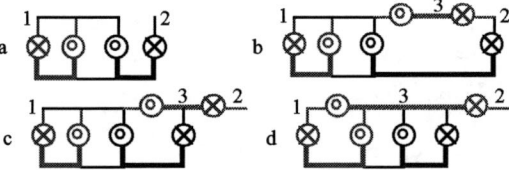

Figure 8. Transformation of current mirror circuit

V. GENERATION OF CURRENT MIRRORS

It has been shown that the admittance matrix representation of the current-controlled current source may be expressed in the following form [4]:

$$\begin{bmatrix} a & 0 \\ b & 0 \end{bmatrix} \quad (5)$$

where the current gain is given by $-\beta_b/\beta_a$. This matrix corresponds to the nullor implementation in Fig 8a which represents the basic 2-FET current mirror circuit [4]. In order to improve performance, we add a nullor open-circuit (nullor c) between nodes 1 and 2, as shown in Fig 8b. This allows us to move norator b connection from node 2 to node 3 as shown in Fig 8c. This circuit corresponds to the cascode current mirror in [4] (where an additional short-circuit nullor is introduced in order to facilitate the biasing). It is then possible to move nullators a and b from node 1 to node 3, as shown in Fig 8d. This circuit corresponds to the Wilson current mirror [3].

We have shown that the adoption of a restricted set of synthesis transformations reduces the number of circuits generated while retaining those with good performance properties. However, the exploration of the alternatives still involves trial-and-error. We now explore the idea of a more systematic and targeted synthesis approach which is based on carrying out the synthesis within the admittance matrix.

VI. Synthesis Using the Admittance Matrix

Consider the single FET transconductor of Fig 4a which has the following ideal and non-ideal admittance matrices:

$$\begin{bmatrix} 0 & 0 \\ G_a & 0 \end{bmatrix} \quad \begin{bmatrix} 0 & 0 \\ G_a & g_a \end{bmatrix} \quad (6)$$

where G_a and g_a are FET finite transconductance and output conductance, respectively. We aim to reduce the y_{22}' term in eqn (6) and hence increase open-circuit voltage gain $A_V = -y_{21}'/y_{22}'$. The properties of a circuit are unchanged if we add an additional row and column to the admittance matrix corresponding to an isolated node (3) in the circuit; we should then introduce linked infinity parameters ($\pm\infty_b$) in order to link the new row (3) and column (3) to another row and to another column [6]. When the 3 × 3 matrix is reduced to the 2 × 2 matrix, the y_{22}' term in eqn (6) will depend on the y_{23}, y_{32} and y_{33} elements in the 3 × 3 matrix. Let us add the new $\pm\infty_b$ terms between rows 2,3 and columns 3,0 respectively in order to obtain the following:

$$\begin{bmatrix} 0 & 0 & 0 & 0 \\ G_a & 0 & \infty_b & 0 \\ 0 & 0 & \infty_b & 0 \end{bmatrix} \begin{bmatrix} 0 & 0 & 0 & 0 \\ G_a & g_a + g_b & G_b & g_b \\ 0 & g_b & G_b + g_b \end{bmatrix} \quad (7)$$

Consider reducing eqn (7) in the non-ideal case by Gaussian elimination; since the y_{23} and y_{33} terms are exactly equal and opposite, the $-g_b$ term at y_{32} is effectively shifted without change to y_{22}, where it cancels the g_b term and leaves just the g_a term. So circuit performance is unchanged with respect to eqn (6); eqn (7) describes the circuit in Fig 4c which has a nullor open-circuit connected between nodes 2 and 0.

By virtue of the $\pm\infty_b$ terms in eqn (7), we can move the ∞_a term from y_{21} to y_{31} [5]:

$$\begin{bmatrix} 0 & 0 & 0 & 0 \\ 0 & 0 & \infty_b & 0 \\ \infty_a & 0 & \infty_b & G_a \end{bmatrix} \begin{bmatrix} 0 & 0 & 0 & 0 \\ 0 & g_b & G_b & g_b \\ G_a & g_b & G_b + g_a + g_b \end{bmatrix} \quad (8)$$

Since norator a is now connected to node 3 instead of to node 2, the g_a term moves from y_{22} in eqn (7) to y_{33} in eqn (8). Now the y_{32} and y_{33} terms are only approximately equal and opposite; hence the $-g_b$ term at y_{32} only approximately cancels the g_b term at y_{22}. But since now g_b is the sole term in y_{22}, y_{22} falls to a very low value. y_{21}' and y_{22}' are now take the following forms:

$$\begin{aligned} y_{22}' &= g_a g_b / (G_b + g_a + g_b) \approx g_a g_b / G_b \\ y_{21}' &= G_a (G_b + g_b)/(G_b + g_a + g_b) \approx G_a \end{aligned} \quad (9)$$

Eqn (8) describes the cascode transconductor of Fig 4d.

We now consider the requirement that y_{22}' is to be further reduced. Our strategy will be to augment the y_{23} and y_{33} terms in eqn (8) such that they become more equal and provide better cancellation of the $\pm g_b$ terms in y_{22} and y_{32} when the matrix is reduced. In order to dominate over the existing terms, the additional terms in y_{23} and y_{33} must involve a product of G terms. This requires expansion of eqn (8) to a 4 × 4 matrix and introduction of a nullor. In order to obtain dominant terms in y_{23} and y_{33} in the 3 × 3 matrix, we will need the 4 × 4 matrix to have G terms in y_{24}, y_{34}, y_{43} and a g term in y_{44}. We introduce a nullor short circuit between node 4 and the reference node which allows $\pm\infty_b$ terms to move from column 0 to column 4:

$$\begin{bmatrix} 0 & 0 & 0 & 0 \\ 0 & 0 & \infty_b & 0 \\ \infty_a & 0 & \infty_b & 0 \\ 0 & 0 & 0 & \infty_c \end{bmatrix} \rightarrow \begin{bmatrix} 0 & 0 & 0 & 0 \\ 0 & 0 & \infty_b & \infty_b \\ \infty_a & 0 & \infty_b & \infty_b \\ 0 & 0 & 0 & \infty_c \end{bmatrix} \quad (10)$$

By virtue of the $\pm\infty_b$ terms, we can move the ∞_c term to y_{43}.

$$\begin{bmatrix} 0 & 0 & 0 & 0 \\ 0 & 0 & \infty_b & \infty_b \\ \infty_a & 0 & \infty_b & \infty_b \\ 0 & 0 & \infty_c & 0 \end{bmatrix} \begin{bmatrix} 0 & 0 & 0 & 0 \\ 0 & 0 & g_b & G_b, g_b & G_b \\ G_a & g_b & G_b + g_a + g_b & G_b \\ 0 & 0 & G_c & g_c \end{bmatrix}$$

We now have G terms in y_{24}, y_{34}, y_{43} and a g term in y_{44} as required. Hence, when the matrix is reduced, equal and opposite terms $\pm G_b G_c / g_c$ will be introduced in y_{23} and y_{33}. Gaussian reduction to a 2 × 2 port matrix gives:

$$y_{22}' = g_a g_b g_c / [G_b G_c + (G_b + g_a + g_b) g_c] \approx g_a g_b g_c / G_b G_c$$

$$y_{21}' \approx G_a$$

The matrix describes the regulated cascode circuit of Fig 5b.

VII. Conclusions

We have shown that for a previously proposed method of synthesizing all-transistor circuits, we can reduce the excessive number of circuits produced by restricting the set of synthesis transformations. We still retain circuits with good performance synthesized previously and good performance characteristics can be consolidated and enhanced. We have also explored the synthesis of transconductor circuits to have high gain by means of mathematical operations within the admittance matrix.

References

[1] W-K. Chen (editor), *The circuits and filters handbook*, CRC-IEEE Press, 1995.

[2] P E Allen and D R Holberg, *CMOS analog circuit design*, Oxford University Press, 2002.

[3] P R Gray et al, *Analysis and design of analogue integrated circuits*, John Wiley, 4th edition, 2001.

[4] D G Haigh, F Q Tan and C Papavassiliou, "Systematic synthesis of analogue circuits - Part III All-transistor circuit synthesis", *Procs 2004 IEEE ISCAS*, Volume I, pp. 709-712, 2004.

[5] D G Haigh and P M Radmore, "Systematic synthesis of analogue circuits - Part I Notation and synthesis toolbox", *Procs 2004 IEEE ISCAS*, Volume I, pp. 701-704, 2004

[6] D. G. Haigh, F-Q. Tan and C. Papavassiliou, "Systematic synthesis of active-RC circuit building blocks", accepted for *Analog Integrated Circuits and Signal Processing*, Kluwer, April/May 2005.

Fast Iterative Method Package for High Frequency Circuits Analysis

Somsak Akatimagool

Faculty of Technical Education
King Mongut's Institute of Technology North Bangkok
1518, Pibulsongkram Road, Bangsue, Bangkok 10800, Thailand,
E-mail : ssa@kmitnb.ac.th

Abstract - *We present the fast iterative method for MMIC's and iris circuit's analysis. A numerical method, based on electro-magnetic wave diffraction and an iterative formulation, allows the analysis of characteristic iris circuits and complex multilayer integrated circuits with fast computation time. The obtained results are good agreement with both theoretically analysis and CAD simulations.*

Keywords: waves, iterative method, simulation, iris, MMIC

I. INTRODUCTION

MIC and MMIC technology present many design of passive elements [5-8]. For simulations in the spatial domain, thin multilayer stacks are responsible of highly increased calculation complexity. Certain methods of electromagnetic analysis available on the market present the simulated circuit parameters. Furthermore, it can't be directly presented to the description of electromagnetic parameters and the lower speed of computation time.

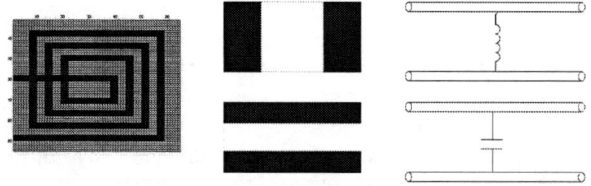

Fig. 1 MMIC and iris structures

The general principle, in this analysis [1], concerns the iterative methods of the conception of the waves with fast iterative algorithm. This method presents the alternating between the space domain (pixels) and the modal domain. In the electromagnetic scattering, on the surface of the circuit, the boundary conditions are presented in several sub-domains; the metal, the insulating and the source. For a speed of the resolution, we benefit the fast Fourier Transform (FFT) expression [7-8].

In this work, we will present the 2.5D electromagnetic wave analysis of the planar circuits with the lossy conductor and iris circuits using the fast iterative method of the electromagnetic waves. We will focus the example of high frequency circuits; microstrip line, couple line, inductive and capacitive iris circuits.

II. THEORY

A. Fast iterative method

The fast iterative method is mainly result of the building of an iterative procedure that avoids the inversion of the integral operator to solve the boundary conditions problem owing to its waves based formulation [1-4]. On the printed surfaces, the boundary conditions are expressed in term of waves rather than in terms of tangential fields.

As a result, the set of boundary conditions defines an integral diffraction of operator $\hat{S}$ built in the spatial domain. The numerical representation of $\hat{S}$ is achieved with a basis of pixel-like functions. Then, the iterative procedure rests on the balance conditions between the diffracted waves by the printed surfaces and the reflected waves by the circuit environment: substrate and enclosure. The advantage of this 2.5D environment is preserved by using a reflection operator defined in the spectral domain from the admittance operator.

Substrate resistively and layer stacking are included in these eigen values. A fast Modal Transform based on a fast Fourier Transform insures the toggling between the spectral and spatial domains.

The whole printed surface is discredited by pixel-like functions [2]. The collection for each sub-domain the S-Matrix, the scheme of the successive iterations is then given by :

$$B = \hat{S} A + B_0 \quad \text{in the spatial domain} \quad (1)$$
$$A = \hat{\Gamma} B \quad \text{in the spectral domain}$$

The toggling between the spatial and the spectral representation of the waves A and B is ensured by a fast modal transform quite similar to the fast Fourier transform which takes advantages in the pixel-like discretization of the printed surface.

The spectral formulation of the waves is based on the box with electric wall defined in term of the electric field as:

$$\vec{E}^{\alpha}_{x(x,y)} = K^{\alpha}_x Cos\left(\frac{m\pi x}{a}\right) Sin\left(\frac{n\pi y}{b}\right) \quad (2)$$

$$\vec{E}^{\alpha}_{y(x,y)} = -K^{\alpha}_y Sin\left(\frac{m\pi x}{a}\right) Cos\left(\frac{n\pi y}{b}\right) \quad (3)$$

which α refers to the TE and TM mode, a and b are the dimension of the box and K name as the value constant respect to the x and y direction.

The formulation spectral of the wave is based on the box with magnetic wall defined in term of the electric field as:

$$\vec{E}^{\alpha}_{x(x,y)} = C^{\alpha}_x Sin\left(\frac{m\pi x}{a}\right) Cos\left(\frac{n\pi y}{b}\right) \quad (4)$$

$$\vec{E}^{\alpha}_{y(x,y)} = -C^{\alpha}_y Cos\left(\frac{m\pi x}{a}\right) Sin\left(\frac{n\pi y}{b}\right) \quad (5)$$

which α refers to the TE and TM mode, a and b are the dimension of the box and C name as the value constant respect to the x and y direction.

The definition of the fast iterative method is extended two formulations, a fast Fourier transform and inverse fast Fourier transform of the cosine and sine function [9-10]. If M and N are mapped into the matrix of the two dimensions, a fast Fourier transform defined as:

$$F^{\alpha}_{x(m,n)} = \beta_{(m)}\gamma_{(n)} \sum_{x=0}^{M-1}\sum_{y=0}^{N-1} e^{\alpha}_{x(x,y)} \cos\left(\frac{\pi(2x+1)m}{2M}\right) \sin\left(\frac{\pi(y+1)(n+1)}{N+1}\right) \quad (6)$$

and a inverse fast Fourier transform defined as:

$$e^{\alpha}_{x(x,y)} = \sum_{m=0}^{M-1}\sum_{n=0}^{N-1} \beta_{(m)}\gamma_{(n)} F^{\alpha}_{x(m,n)} \cos\left(\frac{\pi(2x+1)m}{2M}\right) \sin\left(\frac{\pi(y+1)(n+1)}{N+1}\right) \quad (7)$$

with
$\begin{cases} \text{for } 0 \leq x \leq M-1 \\ \text{for } 0 \leq y \leq N-1 \end{cases}$

while $\beta_{(0)} = \sqrt{\frac{1}{M}}$, $\beta_{(m)} = \sqrt{\frac{2}{M}}$ for $1 \leq m \leq M-1$

and $\gamma_{(n)} = \sqrt{\frac{2}{N+1}}$ for $0 \leq n \leq N-1$.

Since the pixel description is adopted, fields and currents are available on the whole printed surface at each step of the iteration procedure. Increasing the circuit complexity does not penalize iterative method computation time or memory storage while it usually happens in method of Moments.

B. Iris simulation

This simulation is a 2.5D electromagnetic wave analysis of the inductive and capacitive iris circuits, show in fig. 3. The fast iterative method can be used to calculate the impedance and scattering characteristics of iris circuits inside a rectangular metallic waveguide.

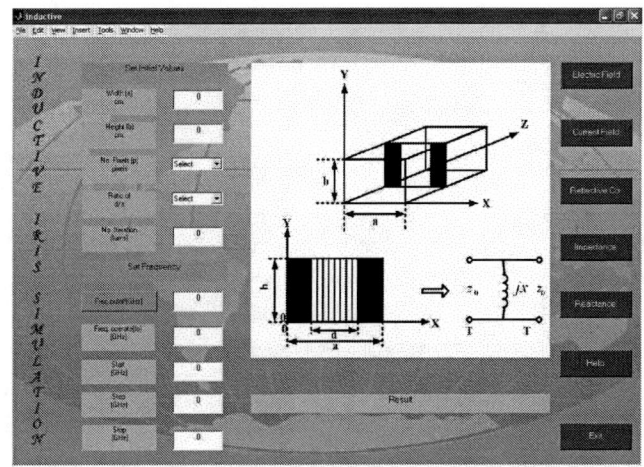

Fig. 3 Iris simulation

C. MMIC's simulation

This simulation is a 2.5D electromagnetic wave analysis of the complex structure of MMIC and components. The iterative method can be used to calculate the impedance and electromagnetic fields of planar and coplanar circuits witch is enclosed inside a rectangular box with perfect electric or magnetic wall.

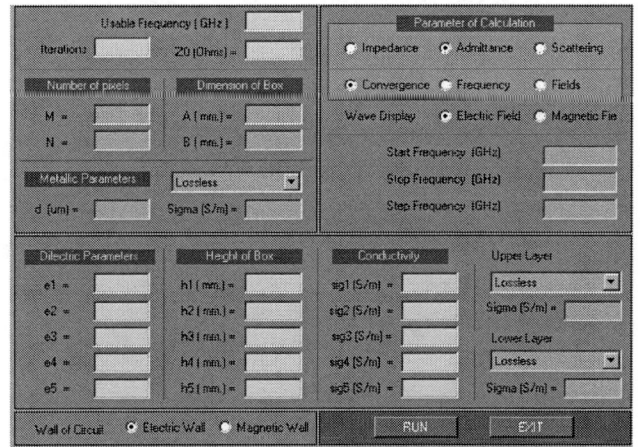

(a)

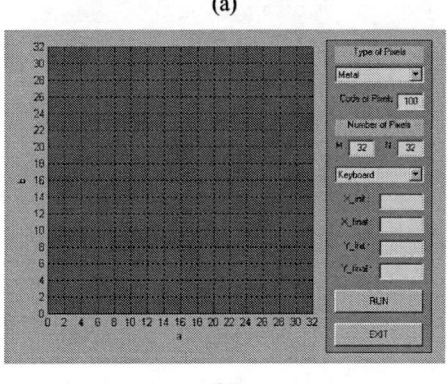

(b)

Fig. 4 MMIC's simulation, (a) main menu calculation and (b) circuit surface design

III. NUMERICAL RESULTS

In this work, we have been presented the electromagnetic wave analysis tools for the planar circuits and microwave components with the fast iterative method (WCIP : Wave Concept Iterative Procedure), the microstrip line, the couple lines, the inductive and capacitive iris are presented. These examples presented in this paper include perfect and loss conductors [8] and electromagnetic field distribution on the surface of the circuits. This calculating is checked by comparison with the theory results and the SONNET simulation results.

A. Inductive circuits

Fig.5 presents a thin metallic iris of a waveguide structure and the comparison of inductance of inductive iris between WCIP and theoretically analysis (a=b=10 cm, x2-x1=5 cm at 1.8-3.6 GHz).

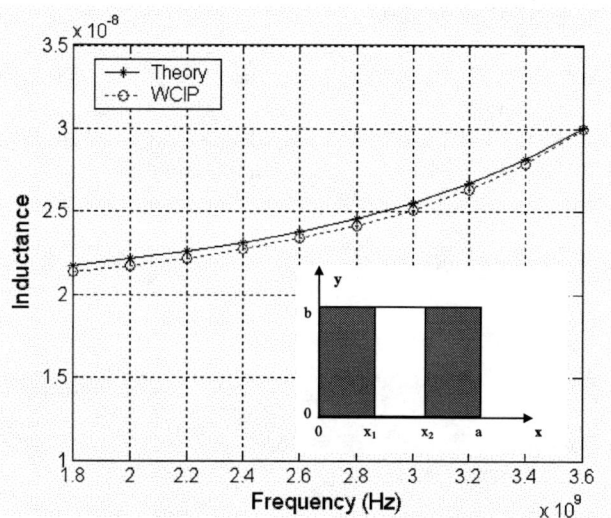

Fig. 5 Comparison of inductance of inductive iris between WCIP and theoretically analysis

After calculating, current density and electric field are in readly obtained. The magnitude of the electric field distribution is displayed in fig.6(a) and the magnitude of the current density is displayed in fig.6(b).

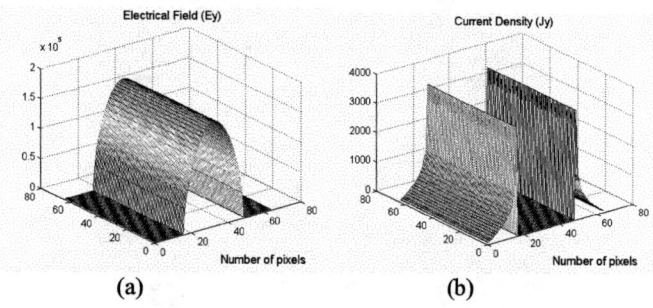

Fig. 6 (a) magnitude of the electric field and (b) magnitude of the current density on the inductive iris circuit

B. Capacitive circuits

Fig.7 presents a thin metallic iris of a waveguide structure and the capacitance of capacitive iris by WCIP simulation. (a=b=10 cm, x2-x1=5 cm at 1.8-3.6 GHz).

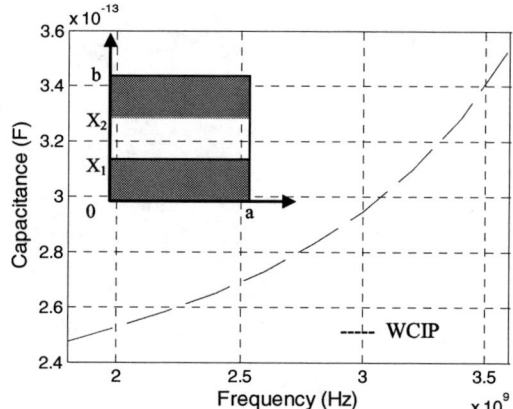

Fig. 7 Capacitance of capacitive iris by WCIP

The magnitude of the electric field distribution is displayed in fig.8 (a) and the magnitude of the current density at x2-x1=2.5 cm is displayed in fig.8(b).

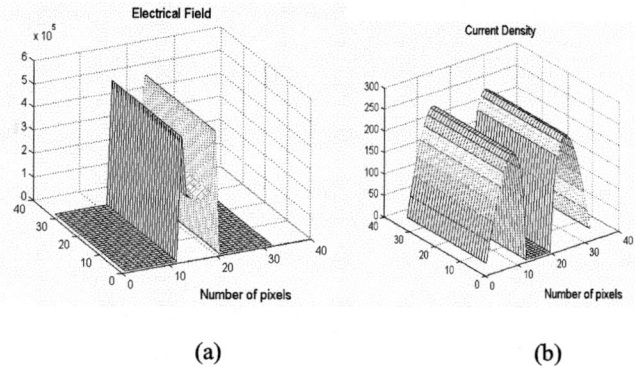

Fig. 8 (a) magnitude of the electric field and (b) magnitude of the current density on the capacitive iris

C. Microstrip line

We presented a new iterative method package for the characteristic analysis of microstip line circuits. Fig.9 presents a microstrip line structure (B=6 mm, W=0.375 mm, $\varepsilon_{r1}=1, \varepsilon_{r2}=9.6$, h1=4 mm and h2=0.635 mm). The results of scattering parameters (S11 and S21) of microsrip line circuit is showed in fig.10.

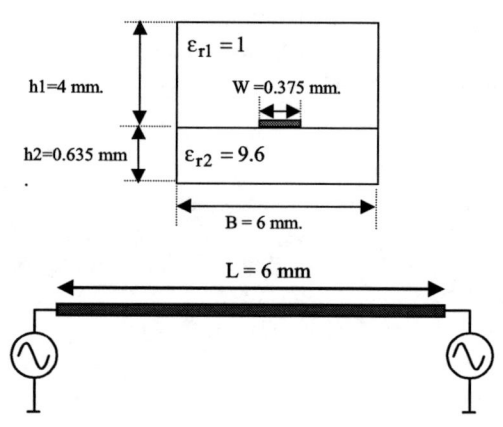

Fig. 9 Microstrip line structure
Fig. 10 Scattering parameters S11 and S21 of microstrip line by using WCIP simulation

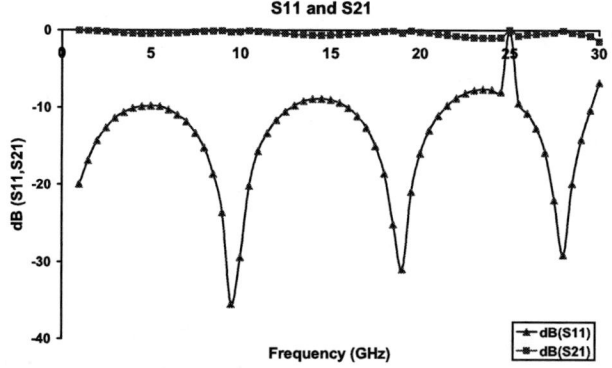

The result of scattering parameters of microstrip line circuit in function of frequency between 1 to 30 GHz is presented by using WCIP method is good agreement in comparing with theory.

D. Couple lines

Fig.11 presents a couple lines structure. The comparison of the coupling scattering parameters (S_{31}) of two microstrip couple lines between WCIP and SONNET simulation is showed in fig.12.

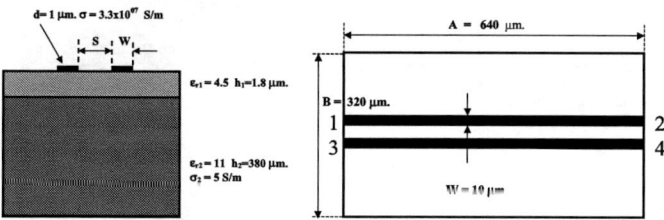

Fig. 11 Couple lines structure

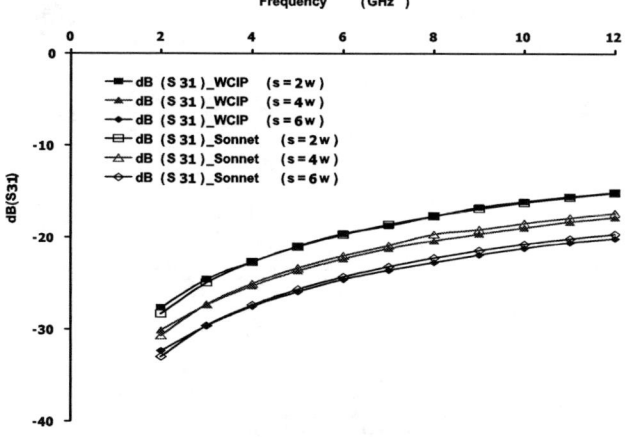

Fig. 12 Comparison of coupling scattering parameter (S31) between WCIP and SONNET simulation

Fig.12 shows the coupling S-parameters (S_{31}) by WCIP and SONNET with various distances (s) of two lines at 2-12 GHz. The observed differences, the coupling S-parameter decreases as the distances increased.

IV. CONCLUSION

We presented a new iterative method package for the characteristic analysis of high frequency circuits. This method has been presented to the complex description of electromagnetic waves and the parameter analysis of MMIC and iris circuits. The advantage of this method is computationally efficient because the alternation between the space and modal domain is used by fast Fourier transform expression.

The results were favorably compared to theoretically analysis and another CAD simulation. The fast iterative method presented in this paper is expected to have wide applications in multi-layer MMIC and waveguide circuit structures.

V. ACKNOWLEDGEMENTS

The author would like to thank Prof. Bajon, SUPAERO, Prof. Baudrand, ENSEEIHT, Toulouse, France and Ms. Boonsanit for helpful technical and data support. This research was supported by the KMITNB Foundation for Innovation.

REFERENCES

[1] S.Akatimagool, D.Bajon, H.Baudrand, "Modelisation of Microwave Interconnects IC's with Iterative Procedure using Fast Modal Algorithm", *Progress in Electromagnetics Research Symp.*, JAPAN, July 2001.

[2] S.Akatimagool, D.Bajon, H.Baudrand, "Analysis of Multi-layer Integrated Inductor with Wave Concept Iterative Procesdure (WCIP)", *IEEE MTT-S Intern. Microwave Symp. Digest.*, Arizona, USA, May 2001.

[3] F.Bouzidi, H.Aubert, D.Bajon, H.Baudrand. "Equivalent network representation of boundary conditions involving generalized trial quantities : Application to lossy transmission lines with finite metallization thickness ". *IEEE Trans. on Microwave Theory and Techniques*, Vol. 45, N° 6, pp. 869–876, June 1997.

[4] H.Baudrand "The Wave Concept in Electromagnetic Problems : Application in Integral Methods", *Asia Pacific Microwave Conference APMC'96*, New Dehli, 1996.

[5] C.Patrick Yue, S.Simon Wong; "On-Chip Spiral Inductors with Patterned Ground Shields for Si-Based RF IC's", *IEEE Journal of Solid-State Circuit*, Vol. 33, No. 5, may 1998.

[6] J.N.Burghartz, Keith A.Jenkins; "Multilevel-Spiral Inductors Using VLSI Interconnect Technology", *IEEE Electron Device Letters*, Vol. 17, No.9, September 1996.

[7] T.Becks, I.Wolff; "Analysis of 3-D Metallization Structures by a Full-Wave Spectral Domain Technology", *IEEE Transactions on Microwave Theory and Techniques*, Vol.40, No.12, December 1992.

[8] Ferenc Mernyei, Franz Darrer, Matthijs Pardoen, "Reducing the Substrate Losses of RF Integrated Inductors", *IEEE microwave and guided wave letters*, Vol. 8 No. 9, September 1998.

[9] Anil K., *"Fundamental of Digital Image Processing"*, Prentice-Hall, Englewood Cliffs, NJ, 1989.

[10] Ronald N. Bracewell, *"The Fourier Transform and Its Applications"*, McGraw Hill, Third edition, 2001.

Mixed signal and SoC design flow requirements

Tuna B. Tarim

Abstract— This paper addresses today's mixed signal and SoC design flow and methodology issues, and discusses the requirements for a successful design flow and methodology from industry point of view. The importance of a strong collaboration between IP owner and customer, maximum re-use, and top level verification (TLV) is discussed in detail. An example is given to emphasize the importance of using behavioral models during TLV to reduce simulation time in large designs.

Index Terms—Design flow, mixed signal, SoC, TLV

I. INTRODUCTION

THE development of mixed signal design flows has been the focus of much attention over the past few years [1]. In traditional IC designs, different parts of the design have been on different chips, reducing the need to simulate at the top level and verifying the interfaces between the chips has been considered to be enough for traditional IC designs.

Today's system-on-chip (SoC) designs, however, combine analog, digital, and RF designs on the same chip [2-3]; this places absolute requirements on the efficiency of the design process all the way from specification (spec) definition to sample test, without sacrificing design time, design performance, chip area, reliability and cost.

As SoC designs become the trend in today's industry, it is of utmost importance that mixed signal design flows and simulation methodologies catch up with the requirements of SoC simulations. Traditional methodologies need to be modified to help manage the complexity of the designs, to allow collaboration with customers and all groups involved in the design of SoC. The flows supporting all of the above also need to be flexible to address custom problems.

This paper describes a top-down design flow and methodology that will allow close collaboration between all involved parties and allow maximum re-use of existing design blocks as well as the top level verification (TLV) of SoC designs. Section II explains today's mixed signal and SoC design flow and methodology problems in detail. Section III shows the mixed signal and SoC design flow and methodology recommended for use. The usage of this flow is explained with an example in Section IV. Section V summarizes the discussion, and describes future work.

Manuscript received 30 September 2004.
Tuna B. Tarim is the EDA manager in the Wireless Analog Technology Center, Texas Instruments, Dallas, TX 75243 USA (Phone: +1 (214) 480 3384; fax: +1 (214) 480 7014; e-mail: tuna@ti.com).

II. TODAY'S MIXED SIGNAL AND SoC DESIGN FLOW AND METHODOLOGY PROBLEMS

Today's problems in mixed signal and SoC designs can be grouped and not limited to the five items listed below:

A. Strong collaboration between IP owner and customer

Today's mixed signal and SoC designs require a great amount of interaction with customers, from spec definition, all the way to in-system evaluation, therefore, it is very important that the flow and methodology in place allows customer interaction starting early stages of the project.

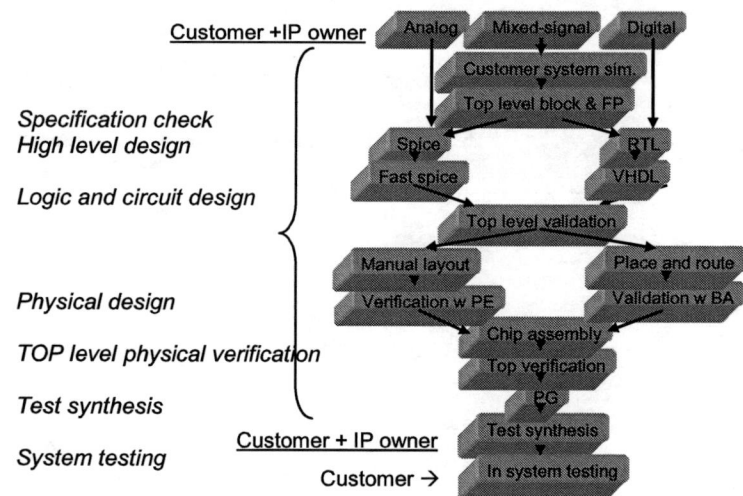

Figure 1: Today's mixed signal and SoC design flows should allow customer interaction starting early stages of the project

Earlier designs were less complicated and would allow parts of the design to be done by different groups, sometimes without communicating with each other until the very last step. Today's complicated SoC designs, however, will not tolerate this format of collaboration between the customer and intellectual property (IP) owner. A methodology and flow which allows for a strong collaboration can save tremendous time during the project.

B. Maximum re-use

One important concept to consider for a successful mixed signal and SoC methodology is "re-use" [4]. The key idea of re-use is to pre-create modules aligned to the customers needs for re-use across many projects. While re-using blocks from one project to another, it is important to have a system/process wherein the issues pertaining to a block found on silicon are passed onto the project where the block is being re-used. It is almost inevitable that there will be a few new blocks needed, however, minimizing the number of new blocks and maximizing re-use will save a great amount of time in project delivery schedules (Figure 2(a) and 2(b)).

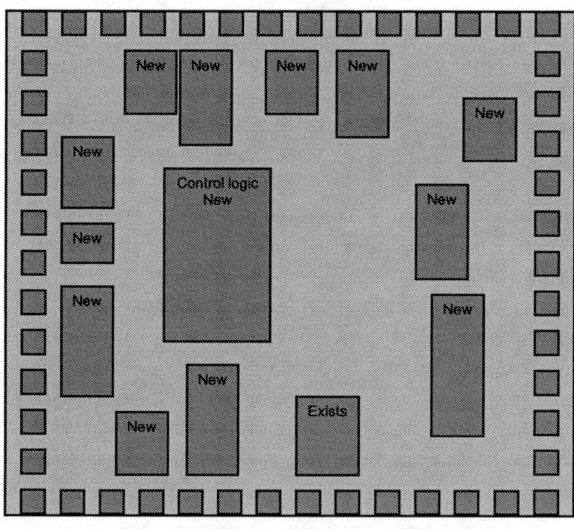

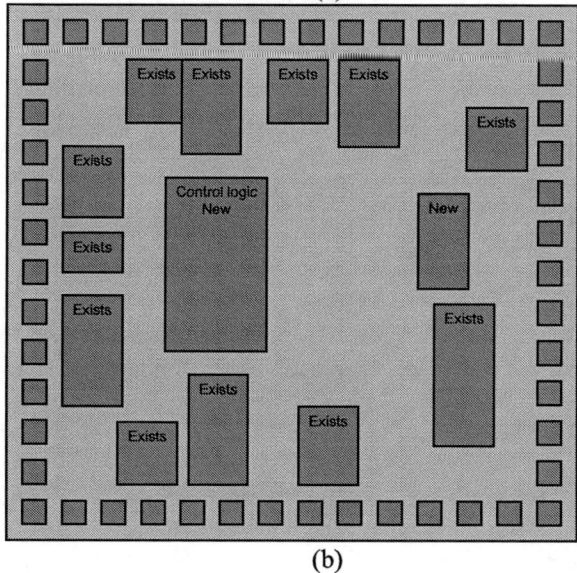

Figure 2: Limiting number of new blocks (a) and maximizing amount of re-use (b) will result save a great amount of time in the project

C. Mixed mode hierarchical simulator

A mixed mode hierarchical simulator that supports top-down design methods and behavioral system simulation is needed for a successful SoC design flow. The simulator will have to handle VHDL(-AMS), Verilog(-AMS), RTL, MATLAB, C models and symbols, SDF, DSPF and similar formats as inputs to tools within the flow (Figure 3) [3-5].

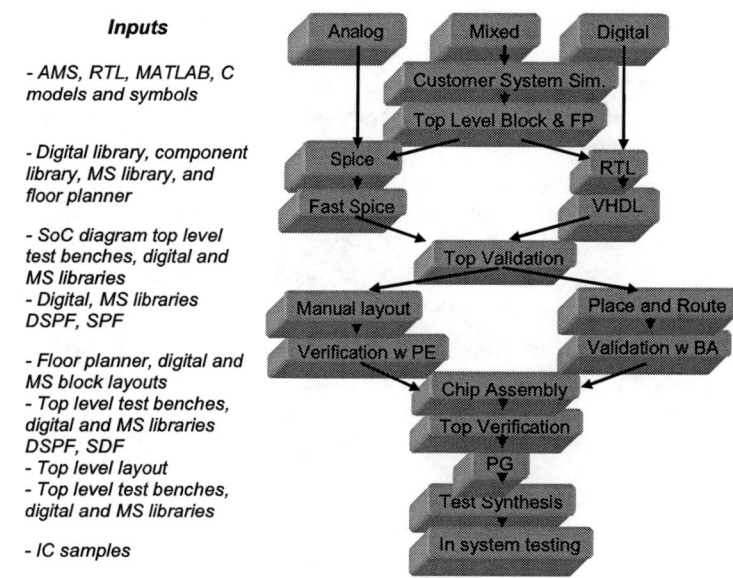

Figure 3: The mixed mode simulator should be able to handle VHDL(-AMS), Verilog(-AMS), RTL, mathematical inputs, C models and symbols, SDF, DSPF and similar formats as inputs to tools within the flow

A hierarchical simulator will allow the user to simulate different configurations, such as transistor level analog blocks and gate level digital blocks, or behavioral level analog blocks and RTL level digital blocks, etc. This will allow the user to concentrate on the details of a specific block based on the goal of simulation, as well save time during top level simulations.

D. Top level verification

The process of spec validation and/or checking whether the design matches the spec is called top level simulation. The *flow* in which many top level simulation runs are executed in parallel by an entire design team within a short time-period is called TLV. TLV is a critical path activity for all projects just before tape-out. The definitions of top level simulation and TLV may differ from one publication to another; please refer to the above definitions while reading this article.

TLV requires a significant amount of time and therefore needs to start as early as possible, preferably as soon as the pin definitions for all blocks are complete and the design of the control block is finished. During TLV, the top level design will be verified, top level hook-ups will be checked, power up sequences will be run and checked. It is possible to run more than 3000 simulations will during the course of TLV. The configuration may be different from one run to another, based on the block of focus. Needless to say, if there is any change in any of the blocks, this will need to be reflected back to the TLV scenario. In this case, it is very important that only the simulations related to the block of focus need to be re-run and the rest of the simulations do not have to be repeated; this requires an extremely effective TLV methodology and flow. If the design methodology and design flow does not support quick regression of new netlist/models, TLV will become the bottleneck for meeting project schedules and delivery dates.

In today's designs, TLV continues to be a crucial step in validating the spec and making sure that the design and spec are matching. With this many number of simulations to run and verify the results, the process being able to simulate long runs and being able to view more than 3000 simulation results continues to be a challenge.

III. Requirements for a successful SoC design flow and methodology

While considering the requirements for a successful SoC design flow and methodology, one needs to consider the IP block library requirements, simulator requirements, floorplan and routing tool, test synthesis requirements, and functional vs. parasitic netlists.

A. IP block library requirements

Within the mixed signal and ASIC IP block library, one would need the specifications of the blocks, schematics, functional and parasitic netlists. The IP block library should also include mixed signal behavioral models with functional and parametric simulation views (only for mixed signal IP blocks), layout bounding box with pins. The ASIC library should include placement and routing support files. Finally, both mixed signal and ASIC libraries should include test benches for the IP blocks.

The ideal SoC design flow would be where the IP owner provides the customer with the IP block library. The IP owner will also define missing IP block models and create estimated bounding boxes. The customer and IP owner will work together to approve top level simulation and floorplan. The IP owner will then execute SoC design by designing missing blocks, validating missing blocks in top level test bench and synthesizing and validating the RTL control logic. The customer and IP owner will together approve the final top level simulation and floorplan and the IP owner will complete the placement, top level hook-up and interblock parasitic checks. Finally with the customer's approval, the design will be released to the fab. The test engineers will use top level design simulation to generate initial test setups and vectors.

B. Simulator requirements

Simulator requirements are one of the most important challenges the mixed signal design world is facing [6]. The mixed signal simulator not only needs to manage the complexity of mixed signal and SoC designs, it also needs to allow collaboration with customers and all groups involved in the design of SoC. The flows supporting all of the above also need to be flexible to address custom problems. The following list contains all features necessary in a mixed signal and SoC design flow:

a) Accurate SPICE + RTL + fast-spice + gate level + C + AMS (in one simulator)
 - Fast spice accuracy and speed comparable to HSIM
 - (VHDL or VERILOG) + C.
 - AMS industry standard VHDL(-AMS) and VERILOG-(AMS) compatible
 - Spice accuracy and speed comparable
b) Common test benches possible in all cases
c) Available analog and digital behavioral models
d) Supply domain sensitive d/a a/d interface elements
e) Able to change hierarchy without re-netlisting
f) Simulation result I/O for standard viewers and
 - Validation to spice simulator
 - Post processing
g) Hierarchical backannotation
h) Fast functional simulation for hookup validation
i) Fast top level simulation to drive test synthesis

All of the above challenges need to be met to accomplish a successful SoC design.

Today's typical collaborative SoC designs usually start with the customer developing SoC specifications from system simulations using their own models and a mathematical simulator. The customer then provides SoC specification to the IP owner. If the design methodology and design flow is broken, and each design project is treated as a new project and maximum re-use is not possible, the struggle to validate design blocks and verify top level simulation continue to play a significant role in the failure of timely deliveries.

C. Floorplan and routing tools

The floorplan and routing tools need to be compatible with the existing tool standards. These tools need to handle hierarchical mixed signal and analog simulation data. Sensitive and high current density routing features, and pad and ESD place and route are essential features that these tools should be able to handle.

D. Test synthesis requirements

The test synthesis requirements can be listed as a model library of tester resources, mixed signal simulator that can quickly simulate test benches and SoC top level test bench netlist in tester format, and a program to convert simulation output to analog and digital test "patterns".

These requirements can be pursued in two categories: Simulation to test generation integration and test, and floorplanner, and P&R tool integration and test. The former includes the task of writing and adapting a test description language (TDL) tool, scripting this flow from simulation to tester simulator, and testing this flow at the top level. The latter includes scripting the flow into the P&R tool, scripting the flow to view the outputs, and testing this flow at the top level.

E. Functional vs. parasitic netlists

Today's parasitic netlisting is done by netlisting all component parasitics. This quadruples the netlist size and increases run time. The suggested netlisting can be done in a separate way for functional simulations and parametric simulations: The functional simulation will use the minimal component subcircuits. This minimal component subcircuit can be substituted with a full parasitic subcircuit only for parametric simulations. This will help spice and fast spice cope with large functional simulations.

IV. EXAMPLE

The example mixed signal design discussed in this section is a relatively small test case with about 1200 analog components and 800 digital components. The design is first partitioned to a top level block and test fixture (Figure 4(a)). The top level block is partitioned to an analog and digital block (Figure 4(b)). The main intention for using this design is to be able to prove the use of hierarchical simulator and a standard behavioral language to be able to simulate different configurations in a much shorter time than a full transistor simulation. The top level blocks are shown in Figure 4.

VHDL-AMS models were written for the top level analog and digital blocks (TOP_ANALOG and TOP_DIGITAL), and the test bench (test_fixture). While running a 27ms transient simulation at the top level design at transistor level using a spice simulator takes approximately 6 hours to complete, replacing all three top level blocks with VHDL-AMS models reduced the simulation time to 10 seconds. The accuracy of the models can be questioned, however, validated when the blocks were designed and behavioral models created. This example is given to emphasize the time savings in TLV by using transistor level for the block of focus and using behavioral models for the rest of the blocks.

This example can also be used for all the requirements listed in Section III. Obviously, a larger design will be more appropriate for more accurate results for SoC designs. A design with around 900K components is being tested for a SoC flow development. This design is not included in this paper due to proprietary reasons.

V. CONCLUSION AND FUTURE WORK

This paper addresses today's mixed signal and SoC design flow and methodology issues and discusses the requirements for a successful design flow and methodology from industry point of view. The mixed signal flow and TLV flow was developed based on customer feedback and internal requirements, to be able to accomplish first pass success. Section IV briefly goes through an example to emphasize the importance of using behavioral models during TLV to reduce simulation time in large designs.

The team at TI will continue to work on the mixed signal and SoC design flow development and automation based on the requirements of Wireless Analog Technology Center (WATC) at Texas Instruments, Inc., and based on the discussions in Section III.

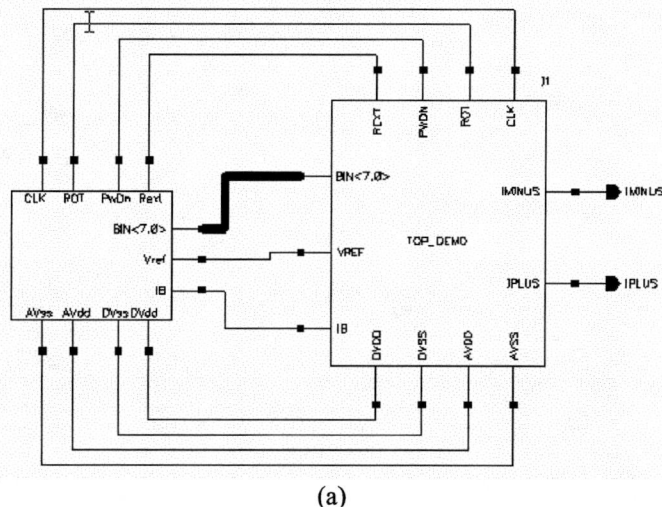

(a)

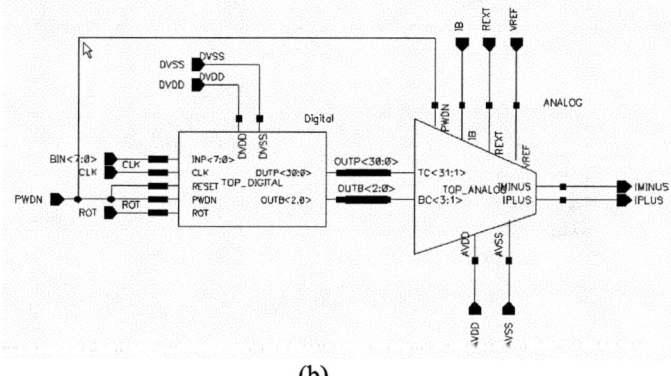

(b)

Figure 4: Example mixed signal design: a) Top level design partitioning: Top level block + test fixture, b) Top level block partitioning: Analog block + digital block

REFERENCES

[1] C. Ajluni, "Automated Analog/Mixed-Signal Design Flows Draw Closer To Reality, Electronic Design , online, http://www.elecdesign.com/Articles/Print.cfm?ArticleID=4690, September 2000.

[2] R. Ahola, "Mixed-signal design flow enables RF CMOS chip", EE Design, online, http://www.eedesign.com/article/printableArticle.jhtml?articleID=16506315, December 2002.

[3] R. B. Staszewski and S. Kiriaki, "Top-down simulation methodology of a 500 MHz mixed-signal magnetic recording read channel using standard VHDL," Proceedings of Behavioral Modeling and Simulation Conference., sec. 3.2, October 1999.

[4] E. Sperling, "Who Benefits from SoCs?", Electronic News, online, http://www.reed-electronics.com/electronicnews/article/CA455657, September 2004

[5] J.A. Lear, "High Speed Mixed-signal Models in VHDL", Proceedings of Mentor Graphcs User2User Conference, April 2004.

[6] Robert Bogdan Staszewski, Chan Fernando, and Poras T. Balsara, "Event-Driven Simulation and Modeling of an RF Oscillator", Proceedings of the 2004 IEEE International Symposium on Circuits and Systems, Vol.. 4, pp. 641-644, May 2004.

Analog VLSI Circuit-Level Synthesis using Multi-Placement Structures

Raoul F. Badaoui and Ranga Vemuri
University of Cincinnati
{rbadaoui,ranga}@ececs.uc.edu

Abstract—[1] This paper contributes to the circuit-level design field the novel idea of multi-placement structures. They enable a fast and optimized placement instantiation in analog circuit-level synthesis. A multi-placement structure needs to be generated only once for a specific circuit topology. We propose its use in synthesis. This pre-generated structure instantiates different layout floor-plans for different sizes and parameters of a circuit. It offers a multitude of high-quality variants for placements of a circuit along with a fast execution time. Speed results from the efficiency of the structure. The optimality of placements variants derives from the intelligent search process used to build it. The target benchmarks of these structures are analog ones that need to be synthesized. They are in the vicinity of 25 modules. This paper presents a circuit synthesis approach for analog circuits using Multi-Placement Structures.

I. INTRODUCTION

Layout generation information is needed within analog circuit synthesis to reach convergence due to parasitics-induced effects. Major placement algorithms follow two main synthesis approaches [1]: optimization based and template based.

Optimization based methods rely on heuristic algorithms such as simulated annealing to meet specific performance constraints [2], [3], [4], [5], [6]. Highly optimized placements are attained though high convergence time on each run makes it impossible to use in a layout-aware synthesis process.

Template based methods mostly use procedural module generators describing layout templates like BALLISTIC [7], MOGLAN [8], [9] or MSL [10]. Fixed placements on variable sized layouts considerably reduce generation time. It makes them highly suitable for layout-inclusive synthesis processes. Nevertheless, possible higher performance circuits might be achieved for certain sizes if the layout was to be placed differently.

We propose the usage of an approach aimed at retaining benefits from both techniques: A multitude of placement variants not restricted to a fixed template along with fast instantiation times. This approach has been proposed in [11]. We present the efficiency of such an approach in several experimental analog circuit synthesis processes. The target circuits of our methodology are analog circuits in the range of 25 modules.

The proposed approach relies on a one-time generation of a multi-placement structure for a circuit topology as depicted in Figure 1. The generated structure would be used in a layout-inclusive synthesis process : numerical sizes are provided to it from a sizing engine, then, a specific floor-plan for the circuit is returned. Results will show that the best floor-plan possible for the various specified sizes of the circuit shall be chosen, performance being the major constraint for the process.

The rest of the paper is organized as follows: Section II defines briefly the multi-placement structure and its handling of the sizing search space along with the algorithm used to generate it [11]. Section III proposes the analog circuit synthesis flow proposed.

[1] This work is sponsored in part by the DARPA/MTO NEOCAD program under contract number F33615-01-C-1977 monitored by the Air Force Research Laboratory.

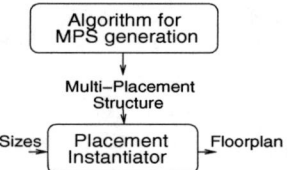

Fig. 1. *Generation of the multi-placement structure and its use in synthesis*

Finally, Section IV presents experimental results of analog circuit synthesis experiments supporting the feasibility, the effectiveness, and the applicabitlity to circuit-level synthesis of the method.

II. THE MULTI-PLACEMENT STRUCTURE AND ITS GENERATION

A multi-placement structure[11] is a mapping of placements variants to intervals of sizes and parameters for a given circuit. It is generated **once** for a specific topology. In a synthesis process, a multi-placement structure is used iteratively. The most suitable placement of modules with respect to their input widths and heights is instantiated as input parameters are fed to the structure. This section outlines the structures and functions comprising the multi-placement structure.

A. Overview of the Multi-Placement Structure

A circuit is defined as a set of N blocks. A block is any module defined by its module generator functions. The variables w_i and h_i represent the width and height of block i while constants w_{m_i}, h_{m_i}, w_{M_i} and h_{M_i} are set by the designer of the blocks as the minimum and maximum widths and heights of block i. A specific placement of the set B of blocks would be defined as a set of x_i and y_i values. They represent the coordinates of Block i on the floor-plan with the constraint of not having overlaps among blocks.

The generation of a multi-placement structure takes as input a circuit topology with the following variables: widths (w_i), heights (h_i), x_i and y_i coordinates of the N blocks.

Each set of w_i and h_i of all blocks is to be mapped to a set of x_i and y_i coordinates representing the best placement to use for the specified widths and heights. A mathematical function representing the multi-placement structure $\mathcal{M}$ could be outlined as follows.

Set Π is defined as a set of placements stored in a multi-placement structure while vector V and its w_i and h_i values consist of the possible dimensions of the various blocks. V is an input to function $\mathcal{M}$. Placement p represents the best placement to use for those specific widths and heights of the blocks and is thus the output of the said function.

For specific intervals of the w and h values for all block, one corresponding placement exists. Conversely, each placement maps to a specific set of intervals for w and h of all blocks.

Traditionally, layout templates such as the ones used in the template-based synthesis flow described in the introduction have a fixed placement that is set by an expert designer while designing the

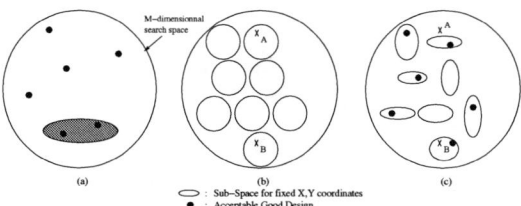

Fig. 2. *Representation of the M-dimensional search space and the sub-space a Multi-Placement Structure is supposed to cover*

Circuit	Blocks	Nets	Terminals
TwoStage op-amp	5	9	22
TwoStage op-amp 2	7	12	26
SingleEnded op-amp	9	14	32
SingleEnded op-amp 2	13	18	42
Mixer	8	6	15
Cascode Two-Stage Opamp	21	36	46

TABLE I
TEST BENCHMARKS

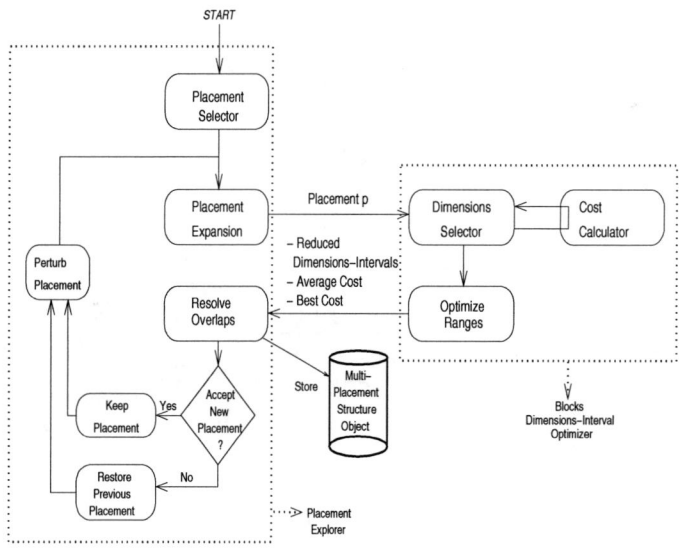

Fig. 4. *Algorithm Flow Detailed Representation* [11]

module. They are fast to instantiate which makes them suitable to use in a circuit synthesis loop. No placement algorithm is executed during layout generation due to the time consumption of such an algorithm.

Good attainable design solutions that would exist in another location in the block placement search space could not be explored using this approach. There lies its drawback. The placement chosen by the expert designer who developed the modules yields most probably good solutions in a slice of the search space of variables. For other sets of parameters and sizes for the circuit, other placements might yield better results.

To illustrate this idea, a M-dimensional search space for some arbitrary circuit is represented as a two-dimensional circle in Figure 2.a . M is equal to the number of parameters in the circuit added to the (x,y) coordinates variables of the blocks. The black dots in the figure represent potential good solutions of the design problem. When using templates to generate the layouts, the placement is set to a fixed set of (x,y) coordinates. The sizing algorithm is hence constrained to a sub-space of the M-dimensional search space. In Figure 2.a, a shaded elliptic area illustrates conceptually what such a constraint imposes on the sizing algorithm. As shown, numerous *good* solutions are hidden in the non-shaded area. Thus, the synthesis process is not able to explore solutions outside its grey shaded area and find potential sub-optimal solutions. An exhaustive search of the whole M-dimensional space is practically impossible for time constraints.

The Multi-Placement Structure tries to include most possible good solutions of the search space in a reduced search space of the synthesis algorithm. Looking at Figure 2.a, if a fixed placement with specific (x,y) coordinates for the blocks is represented as one grey shaded area, then a multi-placement structure and its set of placements Π would be represented as a set of grey shaded areas such as the ones shown in Figure 2.b. These elliptic areas can be overlapping in the search subspace of the synthesis tool (The synthesis tool does not include (x,y) coordinates in its search parameters). For example, points A and B represent one solution with the same values for all the parameters of the circuit. The only difference is in the (x,y) coordinates values of the blocks. Based on the definition of the Multi-Placement Structure, the latter should return one specific placement for each unique set of circuit parameter values: the best one. In Figure 2.b, both points A and B are inside one shaded area of the search space. Thus, to make them comply with the main condition of returning only one placement , the placements stored in the Multi-Placement Structure are shrunk in the circuit parameters range search space as conceptually shown in Figure 2.c. Each placement p_j along with its reduced widths and heights space shall then have a quadruple $[w_{start_{i,j}}, w_{end_{i,j}}, h_{start_{i,j}}, h_{end_{i,j}}]$ attached to each of its blocks representing the range in which that placement is acceptable.

Ultimately, function $\mathcal{M}$ returns one and only one placement for each set of input widths and heights presented to it.

B. Multi-Placement Structure Generation

An algorithm aimed at generating the multi-placement structure has been presented in [11]. The main steps and modules involved in the generation of the multi-placement structure are depicted in Figure 4. A brief overview of the major parts of the algorithm and its flow are presented below and in Figure 4.

The **Placement Explorer** is a search-like tool that intelligently chooses various placements by selecting values for the (x_i,y_i) coordinates. It then finds out which range of values of w's and h's yields best performance for those specific (x_i,y_i) values, and sets the value of the $(w_{start}, w_{end}, h_{start}, h_{end})$ 4-tuple accordingly. Finally, it stores this placement in a multi-placement structure such as the one described previously. The placement explorer obtains a *cost* value for this placement using the other part of the tool, the Block Dimensions-Interval Optimizer.

The **Block Dimensions-Intervals Optimizer** takes a placement with fixed (x_i,y_i) values as input along with the 4-tuple (w_{start}, w_{end}, h_{start}, h_{end}). It runs a search algorithm (with the w and h dimensions of the blocks as variables) to try and reduce those w and h intervals around the values that result in the lowest wiring lengths and area for the circuit. This tool returns to the placement explorer the 4-tuple representing the reduced dimensions interval fed in along with an average value of the cost induced by the various wire lengths and areas encountered during the search. The best attained value of that cost is also returned. The said average value returned would be used as the cost indicator of the placement explorer as stated above.

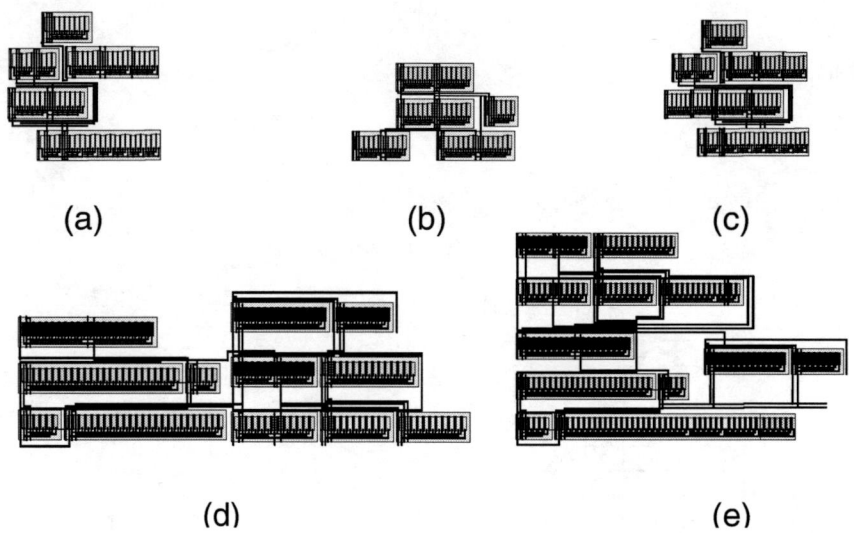

Fig. 3. Various sizes floor-plan instantiations using Multi-Placement Structures for the two-stage and the single-ended opamp circuits.

Circuit	CPU Generation Time	Placements	Instantiation
TwoStage op-amp	52m45s	82	**0.09s**
SingleEnded op-amp	1h55m	115	**0.12s**
TwoStage op-amp 2	55m02s	79	**0.09s**
SingleEnded op-amp 2	1h35m	97	**0.11s**
Cascode-TwoStage op-amp	2h36m35s	124	**0.14s**
Mixer	1h42m	97	**0.11s**

TABLE II
MULTI-PLACEMENT STRUCTURES GENERATED

Circuit	Attribute	Constraints	Prop.App.	Plac. Explored
SEO1	$DC\ Gain$	$\geq$25 dB	25.2581dB	76
	F_{-3dB}	$\geq$5.0e+05	4.2566e+6	
	UGF	$\geq$2.0e+07	5.4950e+7	
	PM	$\geq 50°$	141.91°	
SEO2	$DC\ Gain$	$\geq$25 dB	28.234 dB	58
	F_{-3dB}	$\geq$5.0e+05	6.815e+05	
	UGF	$\geq$2.0e+07	2.7845e+07	
	PM	$\geq 50°$	137.97°	
TSO1	$DC\ Gain$	$\geq$50 dB	60.0451dB	67
	F_{-3dB}	$\geq$1.0e+05	1.52015e+5	
	UGF	$\geq$1.0e+07	1.58773e+8	
	PM	$\geq 60°$	172.45°	
TSO2	$DC\ Gain$	$\geq$50 dB	51.1031 dB	72
	F_{-3dB}	$\geq$1.0e+05	2.003e+05	
	UGF	$\geq$1.0e+07	1.298e+07	
	PM	$\geq 60°$	173.285°	
TSO-cascode	$DC\ Gain$	$\geq$25 dB	25.8912dB	87
	F_{-3dB}	$\geq$1.0e+06	5.2552e+6	
	UGF	$\geq$1.0e+07	7.0304e+7	
	PM	$\geq 40°$	49.2873°	

TABLE III
CIRCUIT SYNTHESIS RESULTS

Finally, the multi-placement structure would be filled with a multitude of placements, mapping to widths and heights of the blocks present in the circuit. The next section will show through some example benchmarks the usage of the proposed method and its efficiency.

III. CIRCUIT SYNTHESIS USING MULTI-PLACEMENT STRUCTURES

We propose the use of the Multi-Placement Structure as shown in Figure 5. The search engine (Optimization Algorithm) is based on a Simulated Annealing approach. It produces sizes and parameters for the circuit being synthesized. These sizes are used by the placement

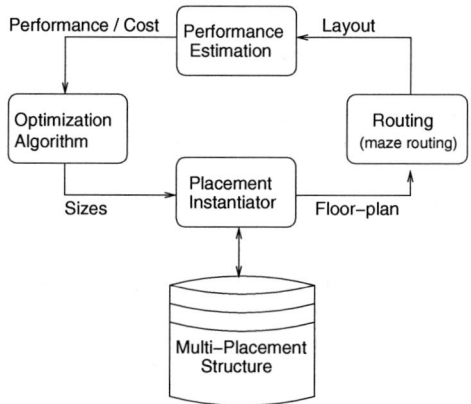

Fig. 5. *Circuit Synthesis flow using Multi-Placement Structures*

instantiator to instantly return the corresponding placement based on the Multi-Placement Structure that would have been pre-generated. A maze router routes the interconnections present in the circuit to produce an exact physical layout. The latter enables the performance estimator (HSPICE in our case) to accurately assess the circuit's behaviour including all parasitics effects.

This method allows a fast instantiation time for the circuit enabling its use in an iterative search process as shown. At the same time, the usage of a Multi-Placement structure provides the search process with the possibility of exploring a multitude of variants for the possible placements of the layout. This ability shall cover parts of the search space previously not coverable using layout templates in such processes.

Section IV presents a multitude of fairly-sized analog circuits synthesized using the proposed approach using their corresponding pre-generated Multi-Placement Structure.

IV. Experimental Results

Each of the circuits presented in Table I has been used as a benchmark to test the circuit synthesis approach proposed. A multi-placement structure has been generated for each. The synthesis flow algorithm has been written in C++, and run on a SUN-Blade-1000 workstation with 2GB of RAM.

A. Generation and Instantiation

Table II shows the details of the multi-placement structures generation. The *placements* column shows the number of possible template placements modeled in each multi-placement structure. The *instantiation* column reveals the time it takes to instantiate one placement when the structure is fed with sizes for the circuit. Those instantiation times prove to be short enough for use in a layout-inclusive synthesis process.

B. Placements Variants and Coverage

Figures 3.a,3.b and 3.c show examples of layout placement variants for the TwoStage opamp circuit. Depending on the input sizes and parameters of the circuit, the multi-placement structure has generated distinct placements that would work best with the input sizes at hand. Figures 3.d and 3.e present further examples of placement variants for the Single-ended opamp circuit.

C. Circuit Synthesis Results

The circuits presented above along with their pre-generated multi-placement structures have been synthesized using the proposed method shown in Figure 5. Table III displays the results obtained from the synthesis process. The *plac. explored* column shows the number of variants of placements used during the search process.

V. Conclusion

This paper presented a circuit-level synthesis approach using the novel concept of multi-placement structures. These structures provide optimized placements in a synthesis loop without including the time consuming placement algorithms within the process. This method is comparable to fixed template-based approaches with regards to speed while it instantiates various floor-plan variants depending on the parameters and sizes of the circuit, as if optimization-based methods were being used.It instantiates placements within milliseconds and has been shown of being capable of providing the right means for optimized placement generation during an iterative layout-inclusive circuit synthesis.

References

[1] Georges G.E. Gielen and Rob A. Rutenbar. Computer-aided design of analog and mixed-signal integrated circuits. *Proceedings of the IEEE*, 88(12):1825–1852, December 2000.
[2] J.M Cohn, D. J. Garrod, R. A. Rutenbar, and L. R. Carley. *Analog Device-Level Layout Generation*. Kluwer, 1994.
[3] John M. Cohn, David J. Garrod, Rob A. Rutenbar, and L.Richard Carley. Koan/anagram 2: New tools for device-level analog placement and routing. *IEEE Journal of Solid-State Circuits*, 26(3):330–342, March 1991.
[4] L.Zhang et.al. A genetic approach to analog module placement with simulated annealing. In *IEEE International Symposium on Circuits and Systems*, volume 1, pages 345–348, ISCAS 2002.
[5] K. Lampaert, G. Gielen, and W. Sansen. *Analog Layout Generation for Performance and Manufacturability*. Kluwer, 1999.
[6] K. Lampaert, G. Gielen, and W.M. Sansen. A performance driven placement tool for analog integrated circuits. *IEEE Journal of Solid-State Circuits*, 30(7):773–780, July 1995.
[7] B.R.Owen, R.Duncan, S.Jantzi, C.Ouslis, S.Rezania, and K.Martin. Ballistic: An analog layout language. In *IEEE 1995 Custom Integrated Circuits Conference*, pages 41–44, 1995.
[8] M. Wolf, U. Kleine, and B. J. Hosticka. A novel analog module generator environment. In *Proceeding of the European Design and Test Conference*, pages 388–392, March 1996 1996.
[9] M. Wolf and U. Kleine. Automatic topology optimization for analog module generators. In *Design Automation and Test in Europe*, pages 961–962, February 1998.
[10] H.Sampath and Ranga Vemuri. Msl: A high-level language for parameterized analog and mixed-signal layout generators. In *IFIP 12th International Conference on VLSI*, 2003.
[11] Raoul F. Badaoui and Ranga Vemuri. Multi-placement structures for fast and optimized placement in analog circuit synthesis. In *Proceedings of the Design, Automation and Test in Europe, Munich-Germany*, March 7-11 2005.

Improved Modeling of Sigma-Delta Modulator Non-Idealities in SIMULINK

A. Fornasari, P. Malcovati and F. Maloberti

Department of Electrical Engineering, University of Pavia
Via Ferrata 1, 27100 Pavia, Italy
E-mail: {andrea.fornasari, piero.malcovati, franco.maloberti}@unipv.it

Abstract—The goal of this paper is to present an extension of the behavioral models, implemented in the Matlab/Simulink™ environment, previously presented in [1, 2] and available in [3]. This toolbox allows us to simulate at behavioral level most of the switched-capacitor (SC) sigma-delta ($\Sigma\Delta$) modulator non-idealities, such as sampling jitter, kT/C noise and operational amplifier limitations (finite bandwidth, finite DC gain, slew rate and saturation). Although very effective in simulating wide-band, medium-resolution $\Sigma\Delta$ converters the lack of a model for flicker noise and multi-bit quantizers makes this toolbox less attractive for simulating narrow band high resolution converters. The proposed extension not only fixes this limitation, but introduces a predictive model of the effect of capacitor mismatch in the internal multi-bit D/A converter.

I. INTRODUCTION

Due to the inherent non-linearity of the sigma-delta ($\Sigma\Delta$) modulator loop, the optimization of the basic building blocks has to be carried out with behavioral time-domain simulations [1]. The Matlab/Simulink® toolbox (SD Toolbox) presented in [1, 2] is a good trade-off between accuracy and speed of the simulations. In this paper two additional blocks are presented. The first allows us to include in the Matlab environment data about the noise power spectral density (PSD) of the operational amplifiers obtained by a circuit simulator (e.g. Spectre or Eldo), including flicker noise. The second one allows us to estimate the impact of the mismatches among capacitors in the feedback DAC of a multi-bit $\Sigma\Delta$ modulator on the signal-to-noise and distortion ratio (SNDR). In order to validate the proposed models, we simulated both at behavioral and transistor level the second-order switched-capacitor (SC) $\Sigma\Delta$ modulator architecture shown in Fig. 1 [4], which features a 12 levels internal DAC (11 comparators in the ADC).

II. NOISE MODEL

In the original toolbox (SD Toolbox) all the possible noise sources (mainly the contributions of the operational amplifiers and of the voltage references) were supposed to be white. The parameter V_n of the noise model (i.e. the noise rms voltage) had to be evaluated using a transistor-level noise simulation in the proper clock phase and including all the load capacitors. The output referred noise PSD obtained from the simulation had then to be integrated over the whole frequency spectrum, thus obtaining the total noise power V_n^2.

The square root of this value, V_n, was finally used in the model to scale the output of a Gaussian distributed random signal. This model allows very fast simulations and can be used without any worry, if one of the two following considerations is satisfied:

- the flicker noise (1/f) can be neglected in the specific field of application (wide band converters);
- the noise spectrum is folded, due to sampling operation, a number of time sufficient to be considered white.

If these two conditions are not satisfied a more accurate model has to be used.

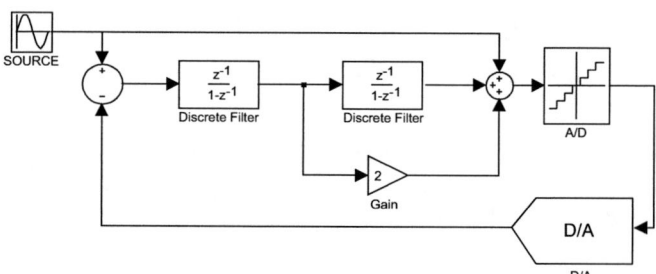

Figure 1. $\Sigma\Delta$ modulator topology porposed in [4] and used to test behavioral blocks.

III. COLORED NOISE MODEL

The noise PSD (expressed in V^2/Hz), provided by most transistor-level simulator, can be considered as the spectrum of the sum of N sine-waves with arbitrary phase (Fourier theorem), each having a power equal to the area of a slide of the PSD as large as F_{MAX} divided by N (i.e. as large as a bin)

$$V_{Noise} = \sum_{i=1}^{N} a_i \sin\left(2\pi \frac{F_{MAX}}{N} it + \varphi_i\right), \quad (1)$$

This simple consideration is the basis for the proposed noise model, whose flow chart is shown in Fig. 2. Basically, we pass in the Matlab environment a detailed description of the noise PSD, elaborate it (basically folding it around the sampling frequency F_s) and calculate the value of V_{Noise} at the end of each clock period (T_s). The first possibility to pass the PSD in the Matlab environment is to sample the waveform (Fig. 3) provided by the transistor-level simulator (e.g. using the Ocean commands in the Cadence environment) and to reconstruct the function in Matlab. A simpler possibility is to take advantage of the knowledge about the shape of the noise PSD. Considering that the noise power is additive, the PSD can be considered as the sum of a term due to flicker noise and one due to thermal noise, low-pass filtered by the circuit transfer function:

$$S_N = \left(c + \frac{k_1}{f}\right) \frac{1}{1 + f^2/f_P^2}. \quad (2)$$

In this way, by providing the coordinates of only two points, the corner frequency (f_C, y_C) and the pole frequency (f_P, y_P), it is possible to estimate the parameters k_1 and c, according to

$$c = -\frac{-f_C^3 y_C - f_C f_P^2 y_C + 2 f_P^3 y_P}{(f_C - f_P) f_P^2}$$

$$k_1 = -\frac{f_C \left(f_C^2 y_C + f_P^2 y_C + 2 f_P^2 y_P\right)}{(f_C - f_P) f_P} \quad (3)$$

as shown in Fig. 3.

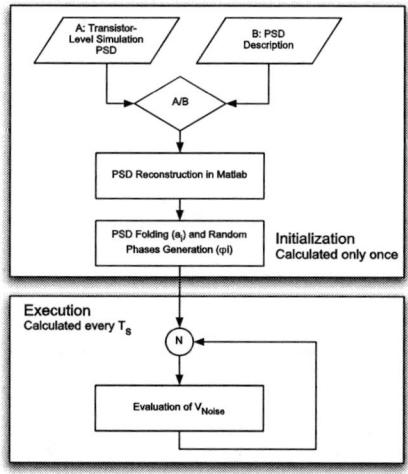

Figure 2. Flow chart of the proposed colored noise model.

Once obtained the analytic function of the noise PSD, it is possible to define its Fourier series. In order to reduce the complexity of the model, we calculate the impact of the sampling operation on the noise spectrum in the initialization phase, defining an equivalent envelope limited in the frequency range [0; F_s/2]. In this way it is possible to use a smaller number of sine-waves or, alternatively, to have a better frequency resolution with the same number of sine-waves (Fig. 4). All the code is written using vectorizing algorithms, i.e. carefully avoiding the use of "for" and "while" loops and replacing them with the equivalent vector or matrix operation. This allows to speed-up simulations [6].

Since with this model we can include the data coming form the circuit simulator, it is also possible to evaluate the impact of techniques as auto-zero or correlated double sampling (CDS) on the performance of the whole converter [5].

This can simply be accomplished by connecting at the output of the noise model block the proper z-domain transfer function.

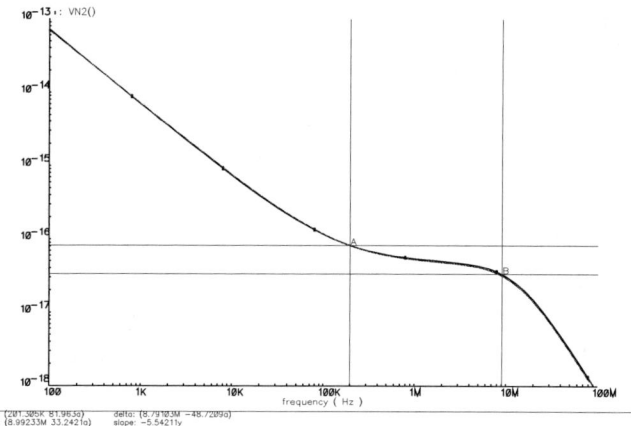

Figure 3. Noise output PSD of an operational amplifier obtained from a circuit simulator (black) and reconstructed by Matlab (red). The PSD integral on 1 GHz bandwidth is 1.089 nV^2, which means V_n=33 µV.

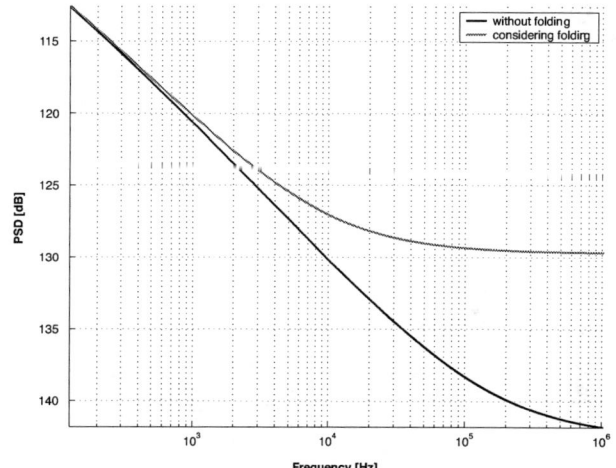

Figure 4. Noise output PSD in the band [0 F_S] before and after having considered the folding due to the sampling operation.

IV. SIMULATION RESULTS

To validate the proposed model, we performed several simulations with Simulink using the model shown in Fig. 7 of the 2nd order modulator of Fig. 1. A sampling capacitor C_S of 12 pF was chosen. The circuit was simulated for two different values of the sampling frequency F_s (2.5 MHz and 1.25 MHz, assuming a jitter of 1‰ of the clock period) with different values of the oversampling ratio (OSR) to highlight the impact of 1/f noise in different operating conditions (Table 1 and Fig. 6). The different versions of the toolbox were compared initializing random generators with the same seeds to better evaluate the algorithms. As operational amplifier we used a simple differential pair with active load and a bias current of 40 µA, having a dc gain of 40 dB and a gain-bandwidth (GBW) product of 6 MHz. Its output noise PSD has already been reported in Fig. 2. A transistor level noise simulation (in time domain) of the ΣΔ modulator with F_s=1.25 MHz and OSR=256 was made to verify the improvement in simulation accuracy of the proposed colored noise block.

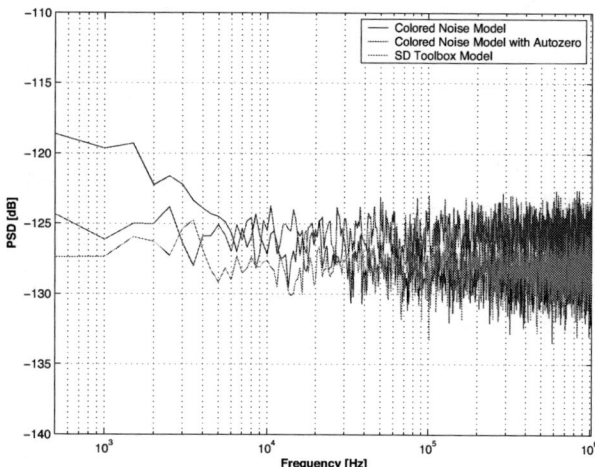

Figure 5. Noise output PSD modeled by the original (black) and the proposed block with (red) and without (green) autozero. The PSD was obtained averaging ten FFTs on a window of 2^{15} points.

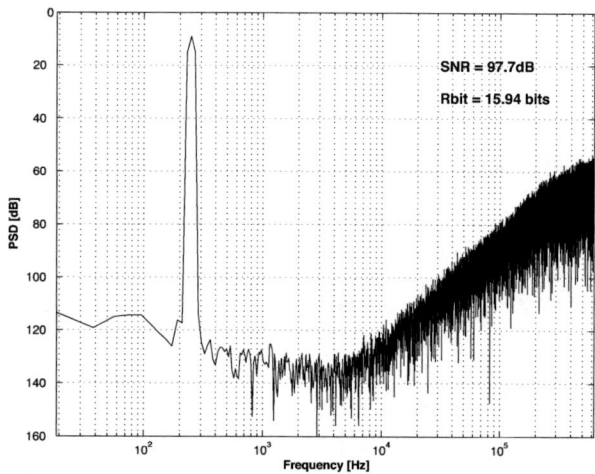

Figure 6. PSD of the 2nd order ΣΔ modulator with F_s=1.25 MHz and OSR=256. It is clearly visible the colored noise floor.

V. MULTI-BIT QUANTIZER MODEL

It is well known [5] that a mismatch among capacitors in the internal DAC of a multi-bit ΣΔ modulator causes an increase in the noise floor and in the harmonic distortion. Performance degradation is proportional to capacitor standard deviation (σ), given by:

$$\sigma\left(\frac{\Delta C}{C}\right) = \frac{k}{\sqrt{W \cdot L}} \ [\%/\mu m] \qquad (4)$$

TABLE I. SIMULATED PERFORMANCES

F_s=2.5 MHz		
OSR=512 Bandwidth=2.4 kHz		
Model	SNR [dB]	Bit
Ideal	142.6	23.4
SD Toolbox	109.6	17.9
SD Toolbox update	99.4	16.2
SD Toolbox update w/ autozero	110.5	18.1
OSR=256 Bandwidth=4.9 kHz		
Model	SNR [dB]	Bit
Ideal	129.1	21.2
SD Toolbox	108.3	17.7
SD Toolbox update	94.4	15.4
SD Toolbox update w/ autozero	105.4	17.2
OSR=64 Bandwidth=19.5 kHz		
Model	SNR [dB]	Bit
Ideal	97.4	15.9
SD Toolbox	94.4	15.4
SD Toolbox update	94.4	15.4
SD Toolbox update w/ autozero	95.2	15.5
F_s=1.25 MHz		
OSR=512 Bandwidth=1.2 kHz		
Model	SNR [dB]	Bit
Ideal	138.9	22.7
SD Toolbox	111.3	18.2
SD Toolbox update	97.7	15.9
SD Toolbox update w/ autozero	108.0	17.6
OSR=256 Bandwidth=2.4 kHz		
Model	SNR [dB]	Bit
Ideal	129.1	21.2
SD Toolbox	106.4	17.4
SD Toolbox update	97.8	15.9
SD Toolbox update w/ autozero	105.6	17.3
Transistor-level noise simulation	96.2	15.7
OSR=64 Bandwidth=9.7 kHz		
Model	SNR [dB]	Bit
Ideal	98.6	16.1
SD Toolbox	95.9	15.6
SD Toolbox update	93.5	15.2
SD Toolbox update w/ autozero	95.3	15.5

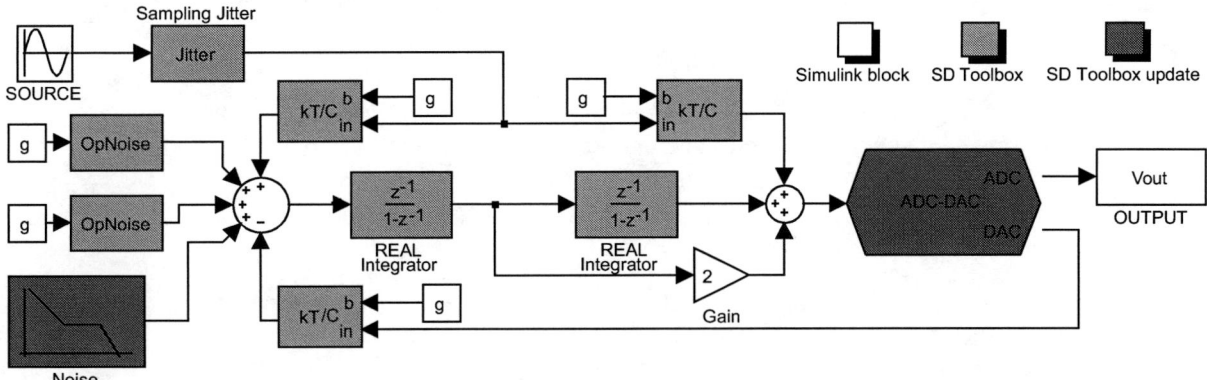

Figure 7. Simulink model of the ΣΔ modulator simulated with blocks introduced in the previous version of SD Toolbox and those proposed in this paper

Therefore, σ is inversely proportional to the square root of the capacitor size (the constant k depends on the technology and is usually provided by the silicon foundry). Considering that the sampling capacitor value impacts the constraints of almost all basic building blocks (e.g. operational amplifiers and voltage references) it has to be determined at the very beginning of the design phase. This makes approaches based on circuit simulator (e.g. Montecarlo simulation) ineffective (not only time consuming).

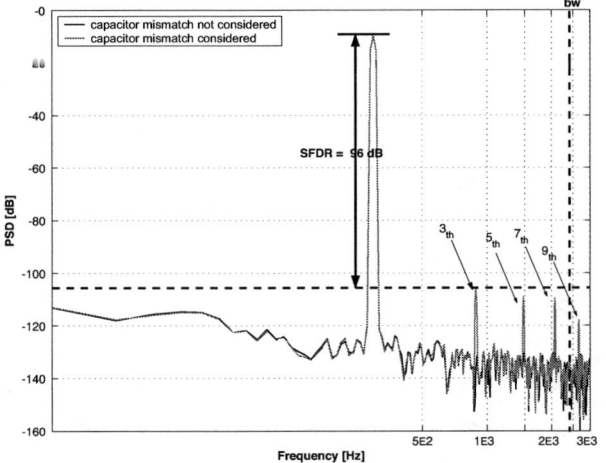

Figure 8. ΣΔ modulator output spectrum with and without capacitor mismatch. The FFT is performed on a window of 2^{17} points.

Therefore, we developed a block which models the ADC and DAC of a ΣΔ modulator, including the mismatch effects. This block can be used to evaluate if, given a sampling capacitor size, the performance degradation due to mismatch can be considered negligible with respect to thermal noise, or if some correction technique, e.g. dynamic element matching (DEM), has to be applied (Fig. 8). The internal DAC was supposed to have an odd symmetry (as it happens in reality in all fully differential circuits and in all single ended circuits carefully designed), which means that the same elements are used to construct both positive and negative values. Under this assumption no even distortion can be introduced by DAC. This shrewdness is fundamental to avoid overrating mismatch effect on the output spectrum.

CONCLUSIONS

In this paper we presented an extension of the SD Toolbox, which includes a more general noise model with also flicker noise and a multibit quantizer model considering capacitor mismatches. Transistor level simulation have demonstrated that, under specific conditions, the proposed noise model achieves results by far more accurate than the original one. Moreover, the multibit quantizer model allows us to accurately estimate the sampling capacitance value required for achieving a given harmonic distortion at the very early stage of the design.

REFERENCES

[1] S. Brigati, F. Francesconi, P. Malcovati, D. Tonietto, A. Baschirotto, F. Maloberti, "Modeling sigma-delta modulator non-idealities in SIMULINK", Proceedings of ISCAS '99, Vol. 2, , pp. 384-387, 1999.

[2] P. Malcovati, S. Brigati, F. Francesconi, F. Maloberti, F. Cusinato, A. Baschirotto, "Behavioral modeling of switched-capacitor sigma-delta modulators", IEEE Trans. on Circuits and Systems I, Vol. 50, No. 3, pp. 352-364, 2003.

[3] Category: Control Systems, File: SD Toolbox [Online]. Available: http://www.mathworks.com/matlabcentral/fileexchange

[4] J. Silva, U. Moon, J. Steensgaard, G. Temes; "Wideband low-distortion delta-sigma ADC topology", Electronics Letters, Vol. 37, pp. 737-738, June 2001.

[5] C. Enz and G. Temes, "Circuit techniques for reducing the effects of op-amp imperfections: autozeroing, correlated double sampling and chopper stabilization", Proc. IEEE, Vol. 84, pp. 584-614, Nov. 1996.

[6] SIMULINK and MATLAB Users Guides, The MathWorks, Inc., Natick, MA, 1997.

[7] S. Norsworthy, R. Schreier, G. Temes, "Delta-sigma data converters: Theory, design and simulation", IEEE Press, Piscataway, NJ, 1997.

Analysis of Supply and Ground Noise Sensitivity in Ring and LC Oscillators

Volodymyr Kratyuk, Igor Vytyaz, Un-Ku Moon, Kartikeya Mayaram

School of Electrical Engineering & Computer Science
Oregon State University, Corvallis, OR 97331, USA

Abstract— Supply and ground noise sensitivity of a wide variety of ring and LC oscillators has been analyzed based on the perturbation projection vector (PPV) technique. The resulting PPV provides an understanding of how specific frequency content of supply/ground noise is converted to oscillator phase noise. Based on this analysis oscillators that are tolerant to supply/ground noise can be identified and used for low noise oscillator design.

Keywords: supply noise, ground noise, oscillators, noise in oscillators, noise analysis, perturbation projection vector.

I. INTRODUCTION

Phase-locked loops (PLLs) are key building blocks of frequency synthesizers and clock generators and they are used in nearly all analog, digital, and RF ICs. Within the PLL the voltage controlled oscillator (VCO) is a critical block since its noise performance determines the performance of the overall PLL. Given the drive for high levels of integration in current and future generations of integrated systems-on-a-chip (SoC), the design of VCOs is critical to ensuring first-pass silicon of future SoCs. The major components that contribute noise in VCOs are the *device thermal* and *flicker* noise, *substrate coupling noise*, and noise coupling through the *power supply* [1]. Although techniques are available for analyzing the phase noise of an oscillator due to the intrinsic thermal and flicker noise sources [2, 3], analysis of supply and substrate noise has been very limited [4, 5].

In this paper, we use our version of SPICE3 with the perturbation projection vector (PPV) based noise analysis technique [3] to evaluate the supply noise sensitivity of a wide range of ring and LC oscillators. The paper is organized as follows. In Section II, the PPV based noise analysis method is briefly described. Simulations of the PPV for a wide range of oscillators are presented in Section III followed by a discussion of the various structures in Section IV. Finally conclusions are provided in Section V.

II. SIMULATION ENVIRONMENT

For the accurate simulation of phase noise in VCOs, the circuit simulator SPICE3 has been extended to handle phase noise analysis [6] based on a non-linear perturbation analysis for oscillators [3, 7, 8]. The implemented technique allows for an accurate simulation of phase noise.

For $0 \leq f_m \ll f_0$, the single-sideband phase noise spectrum $L(f_m)$ in dBc/Hz can be approximated as [8]:

$$L(f_m) = 10\log_{10}\left(\frac{f_0^2 c(f_m)}{\pi^2 f_0^4 c^2(f_m) + f_m^2}\right) \quad (1)$$

where f_0 is the frequency of oscillation, f_m is the offset frequency and the scalar constant $c(f_m)$ is frequency dependent in the general case and is given by:

$$c(f) = c_w + \sum_{m=1}^{M}|c_{cm}(f)|^2 \quad (2)$$

where c_w is a contribution to the scalar c from white noise sources and is given by:

$$c_w = \frac{1}{T}\int_0^T v_1^T(\tau) B_w(x_s(\tau)) B_w^T(x_s(\tau)) v_1(\tau) d\tau \quad (3)$$

and c_{cm}, is the contribution to the scalar c from the m-th colored noise source and is given by:

$$c_{cm} = \frac{1}{T}\int_0^T v_1^T(\tau) B_{cm}(x_s(\tau)) d\tau \quad (4)$$

where $v_1(\tau)$ is the perturbation projection vector (PPV) [8]. The PPV is a periodic vector which serves as a transfer function from the noise sources to the scalar c, and hence to the overall phase noise power spectral density. The PPV scales the amount of noise transferred to the scalar c at each point of time. B_w is a state-dependent matrix that maps white noise sources with unity PSD to the system of differential algebraic equations (DAEs) which describe a circuit. B_{cm} is a state-dependent vector that maps the m-th colored noise source to the system of DAEs. Thus, in order to obtain the single-sideband phase noise spectrum $L(f_m)$ in dBc/Hz, first a scalar c (Eq. (2)) needs to be calculated using (3) and (4) followed by computation of the phase noise spectrum using (1). The method to obtain the PPV and matrices B_w and B_{cm} is described in [6, 8]. The PPVs can be calculated based on a time-domain periodic steady-state analysis, or using the frequency-domain harmonic balance method. In this paper, both approaches have been used and the resulting PPVs are in good agreement.

Since the PPV serves as a transfer function from a noise source to the scalar c, it is similar to the use of the impulse sensitivity function (ISF) [1, 2] and can be used to identify the sensitivity of a node to noise. For this reason, one can compare the PPVs from the supply/ground for two different oscillators and evaluate their sensitivity to supply/ground noise. The oscillator with a higher magnitude for the PPV will be more sensitive to supply/ground noise. In the next section we compare a variety of ring and LC oscillators for supply/ground noise tolerance based on the PPV analysis.

III. PPV SIMULATION RESULTS

All the oscillators in this section were designed for an oscillation frequency of approximately 1GHz in a 0.18 μm CMOS process. BSIM3 MOSFET models were used for the transistor models.

A. Single-ended Ring Oscillators

Consider a three-stage ring oscillator with a simple inverter delay cell and a current starved delay cell as shown in Fig. 1. The PPVs for these two types of oscillator cells are shown in this figure as well where it is seen that the simple delay cell is less sensitive to both supply and ground noise. Also note that the PPV has a dc component as well as a 3rd harmonic content due to the switching nature of the three stage single-ended oscillator. Even though there are six switching/inverting occurrences in a period of this oscillator, only three of them are switched to supply and the other three switch to ground, resulting in a 3rd harmonic PPV content. For an N-stage single-ended oscillator one will see N^{th} harmonic content.

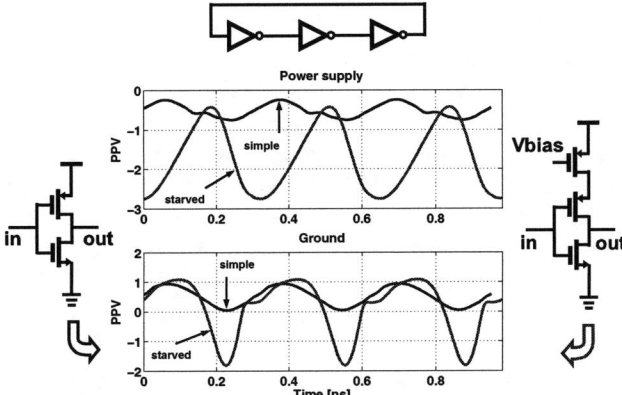

Figure 1: PPV (center) for three-stage single-ended ring oscillator with simple inverter delay cell (left) and current starved delay cell (right).

B. Differential Ring Oscillators

We consider three different ring oscillator delay cells and analyze the supply/ground noise sensitivity for a three-stage differential ring oscillator.

First we consider two commonly used delay cells: the Maneatis delay cell [9] and the Lee-Kim delay cell [10]. The PPV plots for these oscillator cells are shown in Figs. 2 and 3. From these figures one can conclude that the Lee-Kim cell is more sensitive to ground noise compared with the Maneatis delay cell in agreement with previous results [11]. Another observation is that the PPV has a dc component as well as a 6th harmonic content due to the differential switching of the circuit. Differential switching creates paths to both supply and ground at each of the switching/inverting occurrences, resulting in $2*N^{th}$ harmonic content in the PPV (twice the harmonic in comparison to the singled-ended case) for an N-stage oscillator.

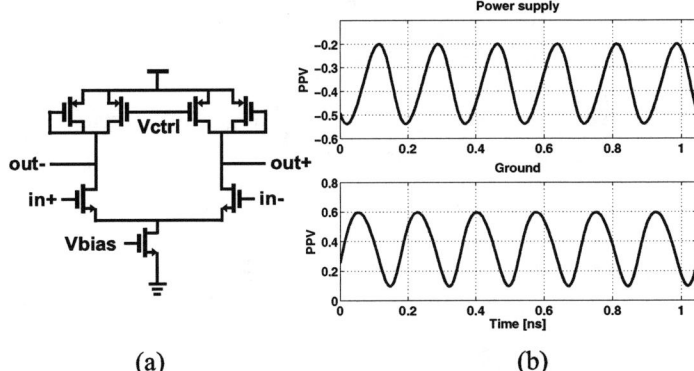

Figure 2: (a) Maneatis delay cell of [9], (b) power supply PPV (top) and ground PPV (bottom).

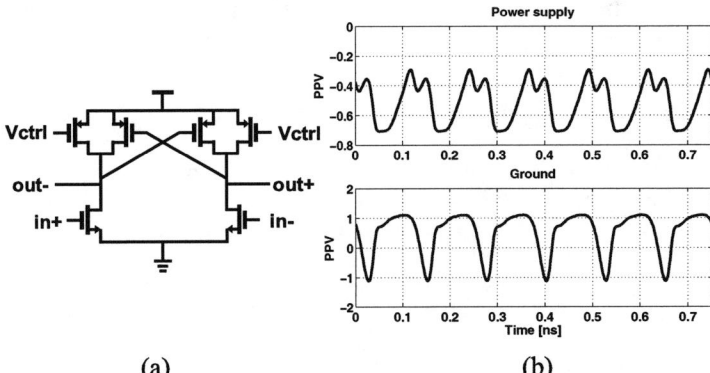

Figure 3: (a) Lee-Kim delay cell of [10], (b) power supply PPV (top) and ground PPV (bottom).

Next we consider the oscillator based on the delay cell shown in Fig. 4 (a) [12] and the PPV for the power supply and ground are shown in Fig. 4(b). There are two different PPV analyses shown in Fig. 4(b) one with an ideal ground connection (short) and one with a small resistance (R). The PPV analysis provides insight into the noise sensitivity of the supply and ground nodes. By comparing the PPV magnitudes we see that the delay cell with an ideal ground connection is more sensitive to ground noise and not as sensitive to power supply noise. The current sources at the top make this cell more tolerant to supply noise. For this reason we expect that the delay cell with a small resistance to ground would provide an improved noise immunity. This result is observed in Fig. 4(b) where the ground PPV reduces in magnitude. However, an adverse affect is observed on the supply as that PPV increases in magnitude.

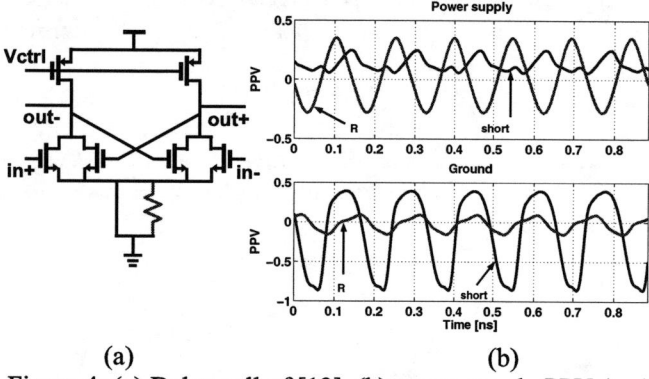

Figure 4: (a) Delay cell of [12], (b) power supply PPV (top) and ground PPV (bottom).

The oscillators analyzed above and the resulting PPV provides additional insight. For example, when the PPV results in a dc component plus a 6th harmonic, it indicates an up-conversion of dc (low frequency) supply noise to the carrier's phase noise as well as down-conversion of supply noise at the 6th harmonic frequency to the carrier's phase noise. This insight provides the designer some control in choosing the number of stages (which dictates the harmonics of the PPV) used in the oscillator if reasonable information about the supply/ground noise spectral content is available.

C. *Single-ended LC (Colpitts) Oscillator*

A simple Colpitts oscillator is included here as an example of a single-ended LC oscillator. The circuit and the PPV for supply and ground nodes are shown in Figs.

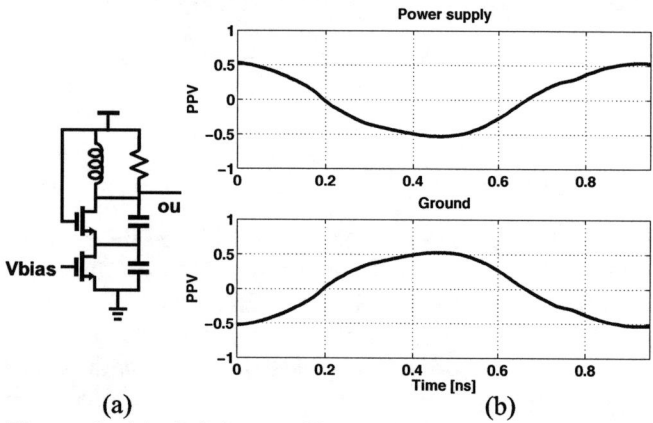

Figure 5: (a) Colpitts oscillator, (b) power supply PPV (top) and ground PPV (bottom).

5(a) and 5(b), respectively. From the PPV we see that this circuit has a similar sensitivity to supply and ground noise since the PPVs are identical in magnitude. One difference compared with the single-ended ring oscillator is the absence of the 3rd harmonic in the PPV. This is because there is no switching of individual stages as in the 3 stage ring oscillator. So called "switching" occurs only once to supply and once to ground in a period of oscillation.

The PPVs for a NMOS cross-coupled oscillator (Fig. 6(a)) are shown in Fig. 6(b). In this case we see that the PPVs for both the supply and ground are of similar magnitude. The PPVs display a 2nd harmonic (and 4th, 6th, etc.) due to the fact that the *differential* LC oscillator "switches" to both supply and ground *two times* per period of oscillation.

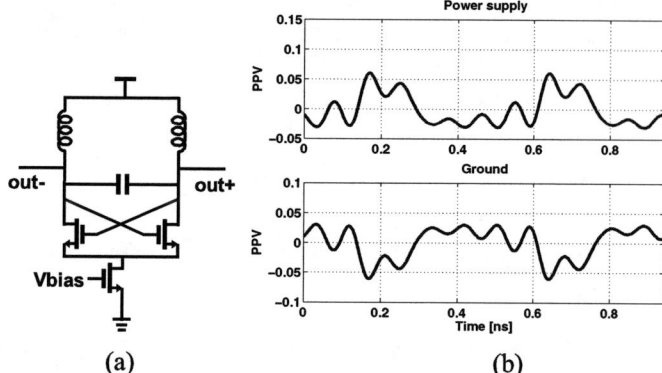

Figure 6: (a) NMOS cross-coupled oscillator, (b) power supply PPV (top) and ground PPV (bottom).

A complementary cross-coupled oscillator of Fig. 7(a) is analyzed with different options, such as a NMOS tail current source before the ground connection (tail CS), a small resistance (R), and an LC filter that is tuned to the second harmonic of the oscillation frequency (LC). PPVs for these cases together with PPVs of the oscillator with no modifications (short) are shown in Fig. 7(b). From this figure we see that a direct ground connection results in the most sensitivity to supply/ground noise. The use of a tail current source, resistor, or the LC filter are effective techniques to reduce the magnitude of the PPV and hence the sensitivity to supply/ground noise. It is interesting to note that when an LC tank isolation is used to suppress the down-conversion of supply/ground noise at the 2nd harmonic, the PPV changes from having the 2nd to 4th harmonic component. While one normally does not have the choice to change the number of stages in the design of an LC oscillator, such a supply/ground filtering technique effectively changes the PPV characteristic that can represent a multi-stage oscillator behavior. Thus a similar design flexibility for LC oscillators (with respect to supply sensitivity) is achieved as in the case of ring oscillators.

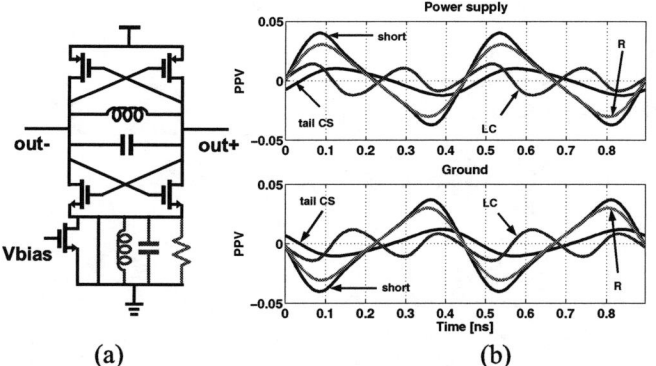

Figure 7: (a) Complementary cross-coupled oscillator, (b) power supply PPV (top) and ground PPV (bottom).

The complementary cross-coupled oscillator cell can be further modified to use resistors or LC filters in both the supply and ground connections. This modified oscillator is shown in Fig. 8(a) and the corresponding PPVs are shown in Fig. 8(b). Once again we see that the LC filter and the resistors are effective in reducing the PPV magnitude for supply and ground as in the previous design.

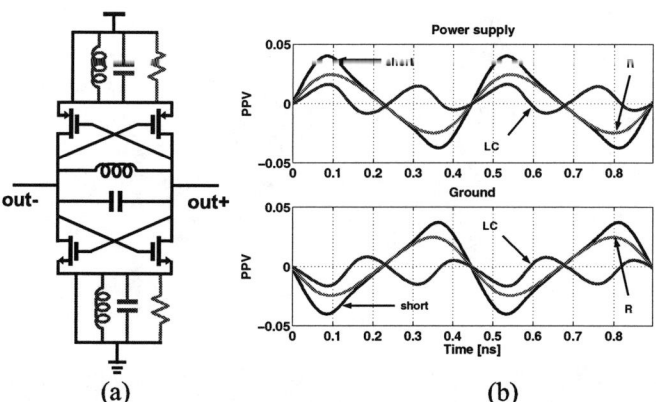

Figure 8: (a) Complementary cross-coupled oscillator, (b) power supply PPV (top) and ground PPV (bottom).

IV. DISCUSSION

From the various simulation results the following conclusions can be derived. For ring VCOs with single ended delay cells, the phase noise is affected by supply and ground noise at baseband and (# of stages)*N harmonics. In the differential ring oscillators the phase noise is affected by supply and ground noise at baseband and 2*(# of stages)*N harmonics. In a single-ended LC oscillator, the phase noise is affected by supply and ground noise at the baseband and harmonics of the oscillation frequency. However, for a differential LC oscillator, supply and ground noise at the baseband and even harmonics of the oscillation frequency are important. Furthermore, use of LC tanks to isolate supply/ground noise in LC oscillators results in the modification of PPVs in a similar manner to that for ring oscillators with different number of stages.

V. CONCLUSION

A simulation technique based on the perturbation projection vector (PPV) has been used to accurately simulate the supply/ground noise sensitivity of a wide variety of ring and LC oscillators. Based on this analysis insight is gained into the supply/ground noise sensitivity of the various oscillators. This work provides the foundation for identifying oscillators that are tolerant to supply and ground noise.

ACKNOWLEDGMENTS

This work is supported in part by the SRC under contracts 2003-HJ-1076 and 2001-TJ-922, and by NSF under grant CCR-0120275.

REFERENCES

[1] A. Hajimiri and T. H. Lee, *The Design of Low Noise Oscillators*, Kluwer Academic Publishers, Boston, 2000.

[2] A. Hajimiri and T. Lee, "A general theory of phase noise in electrical oscillators," *IEEE J. Solid-State Circuits*, vol. 33, pp. 179-194, Feb. 1998.

[3] A. Demir, A. Mehrotra, and J. Roychowdhury, "Phase noise in oscillators: a unifying theory and numerical methods for characterization," *IEEE Trans. Circuit Syst.-I*, vol. 47, pp. 655-674, May 2000.

[4] F. Herzel and B. Razavi, "A study of oscillator jitter due to supply and substrate noise," *IEEE Trans. Circuits Syst.-II*, vol. 46, pp. 56-62, Jan. 1999.

[5] N. Barton, D. Ozis, T. Fiez, and K. Mayaram, "The effect of supply and substrate noise on jitter in ring oscillators," CICC'02, pp. 505-508, May 2002.

[6] V. Kratyuk, "Algorithms and tools for optimization of integrated RF VCOs," *M.S. Thesis*, Oregon State University, June 2003.

[7] A. Demir, "Phase noise in oscillators: DAEs and colored noise sources," in *Proc. ICCAD*, pp. 170-177, Nov. 1998.

[8] A. Demir, "Floquent theory and non-linear perturbation analysis for oscillators with differential-algebraic equations," *International Journal of Circuit Theory and Applications*, pp. 163-185, March-April 2000.

[9] J. Maneatis and M. Horowitz, "Precise delay generation using coupled oscillators," *IEEE J. Solid-State Circuits*, vol. 28, pp. 1273-1282, Dec. 1993.

[10] J. Lee and B. Kim, "A low-noise fast-lock phase-looked loop with adaptive bandwidth control," *IEEE J. Solid-State Circuits*, vol. 35, pp. 1137-1145, 2000.

[11] M. Brownlee, P. K. Hanumolu, U. Moon, and K. Mayaram, "The effect of power supply noise on ring oscillator phase noise," *Proc. NEWCAS*, pp. 225-228, June 2004.

[12] R. Prasun, "A 0.6-1.2V low-power configurable PLL architecture for 6GHz-300MHz applications in a 90nm CMOS process," *Symposium on VLSI circuits*, pp. 232-235, June 2004.

Multichannel SVD-Based Image De-Noising

Y. Wongsawat, K.R. Rao, and S. Oraintara
Department of Electrical Engineering
University of Texas at Arlington
Box 19016, 416 Yates Street, Arlington, TX 76019, USA
Email: yxw1769@exchange.uta.edu
rao@uta.edu
oraintar@uta.edu

Abstract—In this paper, we propose a multichannel SVD-based image de-noising algorithm. The IntDCT is employed to decorrelate the image into sixteen subbands. The SVD is then applied to each of the subbands and the additive noise is reduced by truncating the eigenvalues. The simulation results illustrate that this technique can effectively filter the noisy images without assuming any statistics of the image by using data compression technique.

I. Introduction

Preprocessing is a very important process for image and video compression. Once we import the inputs from the analog sources, additive noise usually appears in the raw images or video sequences. De-noising before encoding the images or video frames yields the advantages for the compression and the complexity. Since 1995, Donoho has proposed the famous soft-thresholding using dyadic wavelets to optimally de-noise smooth signals based on the statistics of the data [3]. However, this de-noising algorithm yields superior results only for signals with slow variations. Recently, many works have been done to improve Donoho's de-noising algorithm. Yang and Nguyen proposed the image de-noising using lapped transform instead of dyadic wavelets, and then rearranged the coefficients into octave-based model [6]. They set up a maximum a posteriori (MAP) estimation problem to find the estimate of the original DCT coefficients before the degradation to improve the performance. Portilla *et al* [7] proposed the de-noising algorithm based on the statistical model of the coefficients of an over-complete multiscale oriented basis. They developed a model for neighborhoods of oriented pyramid coefficients based on a Gaussian scale mixture. However, all of these techniques still require the statistics of the noise and the input signal.

In 1997, Konstantinides *et al* proposed a block-based singular value decomposition (SVD) filter to reduce the noise without using the statistics of the noise and the input image [1]. Although, this system shows superior results over Donoho's algorithm, it requires high complexity in order to calculate the SVD in each image subblock and lacks robustness to adjust the thresholds to discard the eigenvalues of the SVD that represent the noise.

In this paper, we propose the multichannel SVD-based image de-noising. This algorithm can improve the de-noising performance of the conventional SVD algorithm [1], while reducing the complexity without using the statistics of the noise and the input image. The integer discrete cosine transform (IntDCT) [8] is employed to decompose the image into the desired subbands before applying the SVD.

In section II, the SVD-based de-noising algorithm is briefly discussed. In section III, the proposed multichannel SVD-based image de-noising system is presented. Simulation results and discussions are addressed in section IV. Section V concludes the paper and discusses some future works.

II. SVD-Based De-Noising

The SVD-based de-noising algorithm can be summarized as follows. First, the image is divided into non-overlapping 8×8 subblocks. Each subblock is treated as a square matrix which is decomposed using SVD. The eigenvalues from the SVD of each 8×8 subblock are discarded by an optimal threshold using the method proposed in [2]. Finally, the new eigenvalues and the eigenvector matrices are recovered back as the de-noised image. Let $\mathbf{A}$ be an $n \times n$ matrix. The SVD of $\mathbf{A}$ can be given by

$$\mathbf{A} = \mathbf{U}\Sigma\mathbf{V}^T, \quad (1)$$

where $\mathbf{U}$ is an $n \times n$ orthogonal matrix, $\mathbf{V}$ is an $n \times n$ orthogonal matrix, and Σ is an $n \times n$ matrix whose off-diagonal entries are all 0s. The diagonal elements of Σ, λ_is, satisfy such that $\lambda_1 \geq \lambda_2 \geq \ldots \geq \lambda_n \geq 0$. The λ_is determined by this factorization are unique and are called the eigenvalues of $\mathbf{A}$. The kth column of $\mathbf{V}$, $\mathbf{v}_k$, can be calculated from

$$\left(\acute{\mathbf{A}} - \lambda_k \mathbf{I}\right)\mathbf{v}_k = 0, \quad k = 1, 2, \ldots, r, \quad (2)$$

where r is the rank of $\mathbf{A}$ and $\acute{\mathbf{A}} = \mathbf{A}^T\mathbf{A}$. The kth column of $\mathbf{U}$, $\mathbf{u}_k$, can be obtained from

$$\mathbf{u}_k = \frac{1}{\sqrt{\lambda_k}}\mathbf{A}\mathbf{v}_k, \quad k = 1, 2, \ldots, r. \quad (3)$$

In reality, $\mathbf{A}$ appears together with the additive noise resulting in the noisy input, $\mathbf{B} = \mathbf{A}+\mathbf{E}$, to the system, where $\mathbf{E}$ is a random noise perturbation matrix of full rank [1]. In this case,

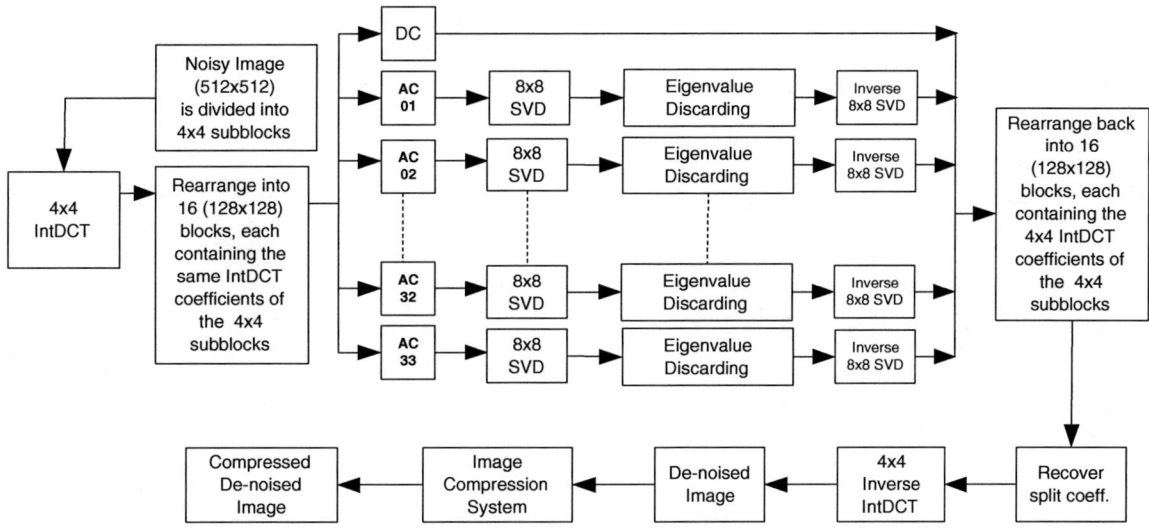

Fig. 1. The proposed multichannel SVD-based image

the rank of **B** can be greater than r. Thus, we can define the effective rank of **B** as r if

$$\acute{\lambda}_r \geq \epsilon_1 \geq \acute{\lambda}_{r+1}, \quad (4)$$

where $\acute{\lambda}_i$ are the sorted eigenvalues of **B**, $1 \leq r \leq n$, and $\epsilon_1 = \|E\|_2$ is the norm-2 of **E**. If the elements of **E** are independent and identically distributed (i.i.d.) random Gaussian variables with zero mean and σ^2 variance, the upper bound of ϵ_1 is

$$\epsilon_1 \leq n\sigma. \quad (5)$$

According to this, we can adjust the threshold ϵ_1 in order to filter out the noise and retain most of the information using the knowledge of data compression [1]. In terms of the eigenvalues, it is easy to select the required energy. Therefore, it is also easy to determine which eigenvalues and corresponding eigenvectors can be discarded.

III. MULTICHANNEL SVD-BASED IMAGE DE-NOISING

The proposed multichannel SVD-based image de-noising can be summarized as follows. We partition the $n \times n$ image into 4×4 subblocks. Each of the subblocks is transformed by the 4-point IntDCT [8] instead of the floating point DCT in order to reduce the computational complexity while maintaining the similar results. The transform coefficients are then regrouped into sixteen subbands by grouping all subblocks' coefficients of the same band together. Thus, the size of each subband is $(\frac{n}{4}) \times (\frac{n}{4})$. Each of the subbands except for the DC one, is then partitioned into block of size 8×8. Each of the 8×8 blocks is treated as a noisy matrix, which is then decomposed by the SVD. The de-noising process is applied by discarding some of the small eigenvalues according to a preset threshold. Fig. 1 illustrates the de-noising algorithm using multichannel SVD.

In order to find the optimal thresholds, as an example, we apply the same threshold, ϵ, between 45 and 75 (using gray scale image, pixel values from 0 to 255), to all 15 channels, except the DC. After that, we encode each image with JPEG-LS [9], [11] in order to see the nature of the compressed size (bit rate) of the de-noised images versus ϵ [1] (see Fig. 3). From [1], we claim that in order to find the knee point which is the point that Fig. 3 has the maximum second derivative with respect to $\log \epsilon$, we can find the point that threshold, ϵ, yields the maximum PSNR (Peak Signal to Noise Ratio) (see Fig. 3). In fact, any lossless compression algorithms can be used instead of JPEG-LS, for example Lempel-Ziv algorithm [1]. We do not use the lossy compression algorithm because we want to use all information to determine the thresholds.

After getting the knee point, we can approximate the thresholds as follows. First, we divide 16 subbands into 7 levels according to the zigzag scan lines (Fig. 2). The threshold for the 4th level (ϵ) is the knee point. From the 3rd to 7th level, we can represent the thresholds as the linear model, while the 2nd level is different. It rapidly decreases from the 3rd level, because of the level of decay of DCT which keeps most of the information in the low frequency subbands [4]. This method is equivalent to compressing the image using the lossy compression (In this case, SVD is used). We try to find the optimal thresholds in order to discard the high frequency information (noise) while retaining most of the original information.

IV. SIMULATION RESULTS AND DISCUSSION

In this section, we compare the simulation results of the various de-noising algorithms including the conventional SVD de-noising algorithm [1], the proposed multichannel SVD-based image de-noising and Donoho's statistical based de-

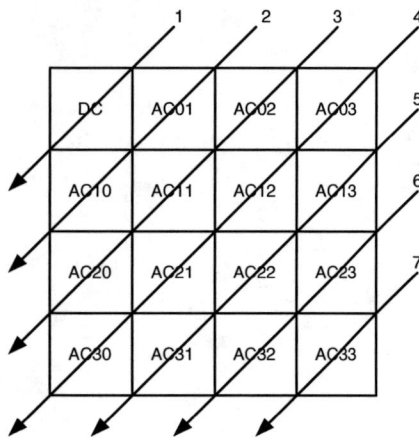

Fig. 2. IntDCT coefficients of a 4×4 subblock are divided

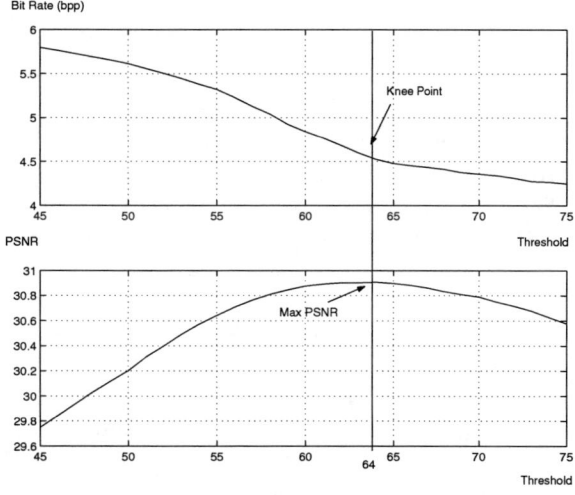

Fig. 3. Bit rate and PSNR versus threshold, ϵ

noising algorithm [3]. Figs. 4 (a) and (b) show the noisy Barbara and Lena images with additive noise at standard deviation (STD) of $\sigma=10$.

In Table I, for Barbara image, which contains both high and low frequency information, the multichannel SVD outperforms both the conventional and Donoho's algorithms. Figs. 5 (a) and (b) show the multichannel SVD and Donoho's de-noised Barbara images, with PSNR equal to 31.0053 dB and 25.0365 dB, respectively. Figs. 5 (c) and (d) illustrate the error images between Figs.5 (a) and (b), respectively, and the original Barbara image. It is evident that Donoho's algorithm still has some residual structure in the error image. This has been reduced in the case of the proposed multichannel SVD.

In Table II, for Lena image, which contains low frequency information, Donoho's algorithm [3] outperforms the conventional SVD algorithm while multichannel SVD system outperforms both of them. Figs.6 (a) and (b) show the multichannel SVD and Donoho's de-noised Lena images, with PSNR equal to 32.2275 dB and 25.0365 dB, respectively. Figs. 6 (c) and (d)

illustrate the error images of Figs. 6 (a) and (b), respectively, compared with the original Lena image. Table III shows that the proposed algorithm outperforms (achieves higher PSNR while using lower bit rate) the conventional algorithm [1] for different noisy environment using JPEG-LS [11].

TABLE I
SIMULATION RESULTS USING BARBARA IMAGE

σ(STD)	Noisy	SVD	16-channel SVD	Donoho*
10	28.1433	30.7162	31.0053	25.0365
15	24.6141	27.6910	28.4174	24.2158
20	22.1154	25.4677	26.5900	23.7573

*Note: for Donoho, we use 2 level DWT (Daubechies-4 filters),σ=10, 15, 20: Threshold= 66.6537, 90.5338, 114.5676

TABLE II
SIMULATION RESULTS USING LENA IMAGE

σ(STD)	Noisy	SVD	16-channel SVD	Donoho**
10	28.1433	31.4154	32.2275	30.1072
15	24.6141	28.6876	30.0657	29.1281
20	22.1154	26.6577	28.5429	28.3432

**Note: for Donoho, we use 2 level DWT (Daubechies-4 filters),σ=10, 15, 20: Threshold= 56.6189, 80.8025, 105.202

In the conventional algorithm, all the 8×8 subblocks have the same priority, so we have to use the same threshold in order to discard the eigenvalues that contain the noise. In the multichannel SVD-based de-noising algorithm, we split the coefficients into several channels using the IntDCT, which can be implemented using just additions and shifts [8]. The priority of each channel is assigned using the zigzag scan, which is optimal for DCT. Using this fact, we can adjust the threshold for each level of the subbands, which yields the improvement from the conventional algorithm [1]. Moreover, for the conventional algorithm [1], we have to calculate SVD for all 4,096 subblocks. However, by skipping the SVD calculation for the most important subband (DC), only 3,840 subblocks have to be calculated. Thus, the number of SVD-calculated blocks can be reduced by 6.25 percent (24.66 percent in time complexity).

V. CONCLUSION

In this paper, we proposed a novel multichannel SVD-based image de-noising. In order to reduce the computational complexity in calculating SVD, we introduce a multichannel SVD and rearrange the priority of each channel. Therefore, calculating the SVD for the DC channel can be neglected. Consequently, using multichannel, we can adjust the thresholds to discard the eigenvalues according to the decay level of DCT. We can improve the performance by using 8-point IntDCT [12] to achieve 64-channel SVD-based algorithm. We can also extend this algorithm to video de-noising.

TABLE III

RELATION BETWEEN PSNR AND BIT RATE OF EACH ALGORITHM USING JPEG-LS FOR BARBARA IMAGE

Algorithm	PSNR (σ=10)	Bit Rate (σ=10)	PSNR (σ=10)	Bit Rate (σ=10)
Noisy	28.1433	6.1741	22.1154	6.9355
SVD	30.7162	5.0000	25.4677	5.3904
16-channel SVD	31.0053	4.7224	26.5900	4.8050

ACKNOWLEDGMENT

The authors would like to thank the group members of the Multimedia Processing Lab (MPL) and the Multirate Signal Processing Lab (MSP), the University of Texas at Arlington, for their suggestions, and comments.

REFERENCES

[1] K. Konstantinides, B. K. Natarajan, and G. S. Yovanof, " Noise Estimation and Filtering Using Block-based Singular Value Decomposition," *IEEE Trans. IP*, vol.6, pp.497-483, Mar. 1997.
[2] B. K. Natarajan, " Filtering Random Noise from Deterministic Signals via Data Compression," *IEEE Trans. SP*, vol.43, pp.2595-2605, Nov. 1995.
[3] D. L. Donoho, " De-noising by Soft Thresholding," *IEEE Trans. Info. Theory*, vol.41, pp.613-627, May 1995.
[4] K. R. Rao and P. C. Yip, *The Transform and Data Compression Handbook*, Boca Raton: CRC Press, 2001.
[5] K. Konstantinides and K. Yao, " Statistical Analysis of Effective Singular Values in Matrix Rank Determination," *IEEE Trans. ASSP*, vol.36, pp.757-763, May 1988.
[6] S. Yang and T. Nguyen, " Denoising in the Lapped Transform Domain," *IEEE ICASSP*, pp.173-176, April 2003.
[7] J. Portilla, V. Strela, M. J. Wainwright, and E. P. Simoncelli, " Image Denoising using Scale Mixtures of Gaussians in the Wavelet Domain," *IEEE Trans. IP*, vol.12, pp.1338-1351, Nov. 2003.
[8] I. E. G. Richardson *H.264 and MPEG-4 Video Compression*, Chichester, West Sussex: Wiley, 2002.
[9] K. R. Rao and J. J. Hwang, *Techniques and Standards for Image, Video and Audio Coding*, Upper Saddle River, NJ: Prentice Hall, 1996.
[10] E. F. Deprenttere, *SVD and Signal Processing: Algorithms, Applications and Architectures*, Amsterdam: North-Holland, 1988.
[11] M. J. Weinberger, G. Seroussi and G. Sapiro, "The LOCO-I Lossless Image Compression Algorithm: Principles and Standardization into JPEG-LS," *IEEE Trans. IP*, vol. 9, pp. 1309-1324, Aug. 2000 (http://www.hpl.hp.com/loco).
[12] G. J. Sullivan, P. Topiwala, A. Luthra, "The H.264/AVC Advanced Video Coding Standard: Overview and Introduction to the Fidelity Range Extensions," *SPIE Conference on Application of Digital Image Processing XXVII*, pp. 53-74, Aug. 2004.

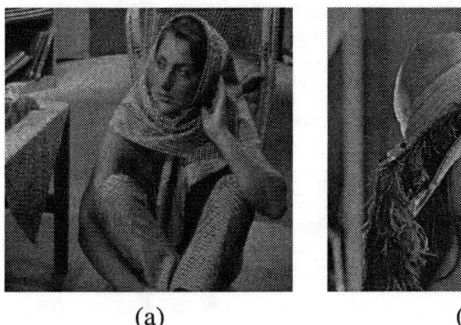

(a)   (b)

Fig. 4. Noisy images of (a) Barbara ($\sigma = 10$) and (b) Lena ($\sigma = 10$).

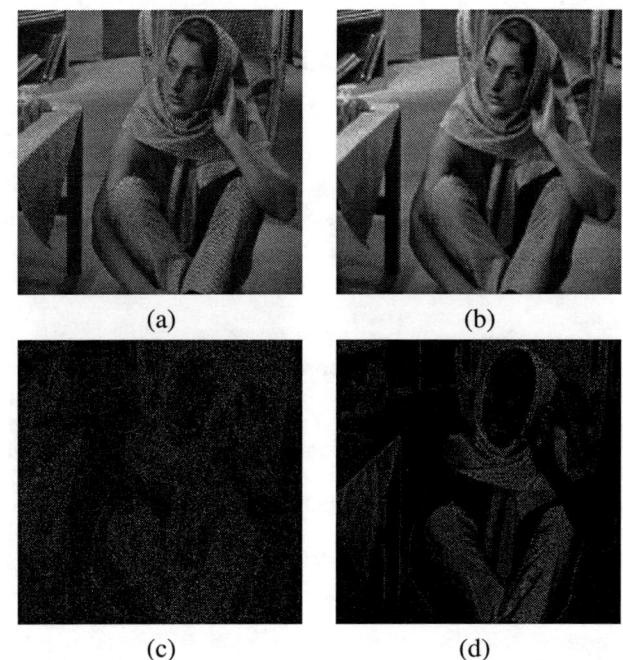

Fig. 5. Denoised Barbara images using (a) multichannel SVD (PSNR = 31.01dB) and (b) Donoho's algorithm (PSNR = 25.04dB), and error images using (c) multichannel SVD and (d) Donoho's algorithm

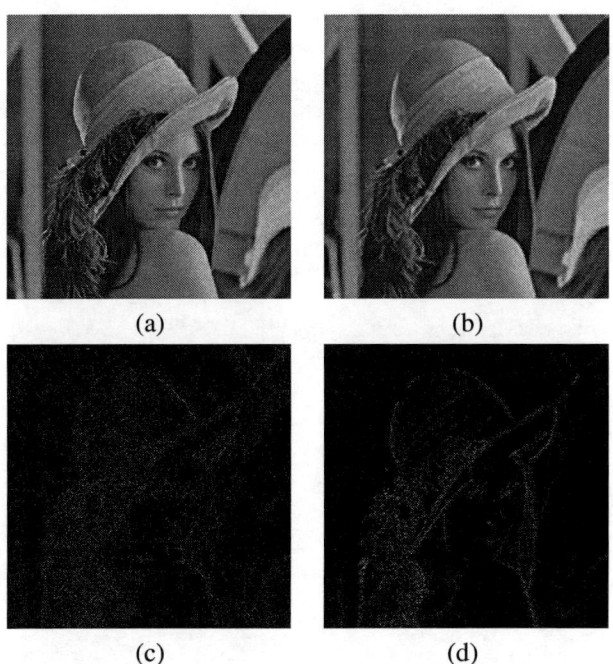

Fig. 6. Denoised Lena images using (a) multichannel SVD (PSNR = 32.23dB) and (b) Donoho's algorithm (PSNR = 30.11dB), and error images using (c) multichannel SVD and (d) Donoho's algorithm

Analysis of Nonlinear Residual Echo Suppressors for Telecommunications

Sen M. Kuo[1] and Woon S. Gan[2]

[1]Department of Electrical Engineering, Northern Illinois University, DeKalb, IL, 60115, USA
[2]School of Electrical & Electronic Engineering, Nanyang Technological University, Singapore

Abstract—This paper analyzes the performance of nonlinear processors (NLP) that are commonly used for attenuating residual echoes in adaptive echo cancellers for voice communications. The NLP (or residual echo suppressor) implemented as a center clipper will produce undesired nonlinear distortions. In this paper, the center-clipping effects are analyzed using Fourier analysis, which decomposes the clipped signal as the summation of a set of rectangular waves and a pulse-amplitude modulated signal. The theoretical analysis shows that the clipping of a sinusoidal signal produces extra odd harmonics, which can be reduced by using the uncompressed center clippers. The analysis results can be extended to speech signals since they can be modeled as the summation of multiple sinusoidal components.

I. INTRODUCTION

Adaptive echo cancellation (AEC) is an important application of adaptive filtering to the attenuation of undesired echoes in the telecommunication networks [1, 2]. As illustrated in Figure 1, AEC is accomplished by modeling an unknown echo path $P(z)$ using the adaptive filter $W(z)$, and subtracting the estimated echo $\hat{d}(n)$ from the undesired echo $d(n)$, which is generated by leaking the far-end speech $s(n)$ through the echo path. AEC can also be applied to cancel acoustic echoes in speakerphones and hands-free telephones, which is also called acoustic echo cancellation. The development of echo canceling chips and especially advances in digital signal processors [3] have made the implementation of adaptive echo cancellers a commercially acceptable cost.

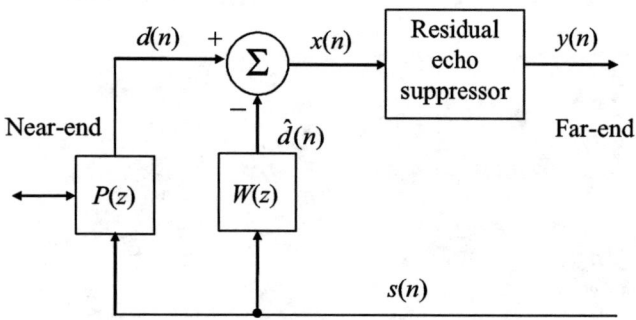

Figure 1 Adaptive echo canceller with residual echo suppressor

Nonlinearities and noises in the telephone circuit limit the amount of achievable echo cancellation. Therefore, the echo canceller output $x(n)$ still consists of residual echo, which is more noticeable and annoying when the echo delay is long. The residual echo suppressor (or NLP) shown in Figure 1 is thus applied to remove the remaining echo. This NLP based on center clipping also effectively removes undesired echo before the convergence of the adaptive echo canceller. There are two types of center clippers [4] designed to remove the low-level echoes which cannot be canceled by the AEC.

As illustrated in Figure 2(a), a compressed center clipper clips the signal component of the near-zero region (central region) and thus reduces (compresses) the signal amplitudes of the non-central region with the predetermined values. The function of this clipper can be described as

$$y(n) = \begin{cases} 0, & \text{if } |x(n)| \leq C_{th} \\ x(n) - C_{th}, & \text{if } x(n) > C_{th} \\ x(n) + C_{th}, & \text{if } x(n) < -C_{th} \end{cases}, \quad (1)$$

where $x(n)$ and $y(n)$ are the input and output signals of the clipper, respectively, and $C_{th} > 0$ is the clipping threshold. The clipping threshold C_{th} determines how "choppy" the speech will sound with respect to the echo level. A large value of C_{th} suppresses all the residual echoes, but also deteriorates the quality of the near-end speech. In this paper, we will show that this type of clipper produces more undesired distortions.

The better and more widely used residual echo suppressor is the uncompressed center clipper with an input-output characteristic illustrated in Figure 2(b). The nonlinear operation of generating output signal can be expressed as

$$y(n) = \begin{cases} 0, & \text{if } |x(n)| \leq C_{th} \\ x(n), & \text{if } |x(n)| > C_{th} \end{cases}. \quad (2)$$

This center clipper eliminates signals below the clipping threshold, but leaves instantaneous signal values greater than the clipping threshold unaffected. Thus, large signals go through unchanged, but small signals are eliminated. Since small signals are consistent with echo, this device suppresses residual echoes. By comparing equations (1) and (2), we note that the uncompressed center clipper not only requires less computation for implementation, but also produces fewer distortions to the desired near-end speech.

The nonlinear suppression of signals as shown in Figure 2 has undesired effects. In this paper, we developed a simple model to analyze the effects of both compressed and uncompressed center clippers used for adaptive echo cancellers.

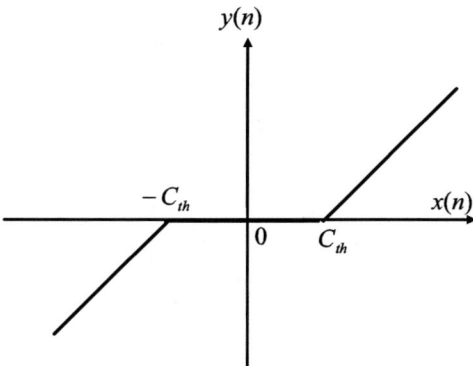

(a) Compressed center clipper

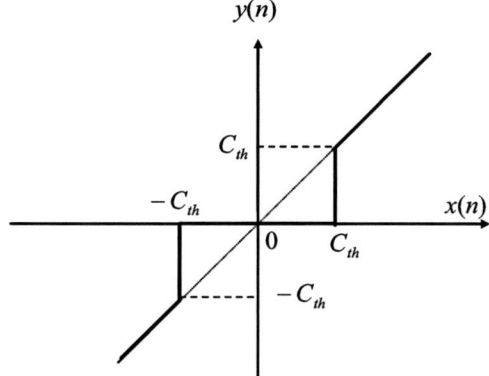

(b) Uncompressed center clipper

Figure 2 Input-output relationships of (a) compressed and (b) uncompressed center clippers

II. ANALYSIS MODEL

Speech signal may be modeled as the summation of multiple sinusoidal components with different amplitude and phase [5]. To simplify analysis, we use a single sine function as the input signal for the center clipper, which can be extended to signals that are composed of multiple sinewaves.

A center-clipped signal can be represented by a set of functions whose spectra are already known, thus the characteristics of the center-clipped signal can be derived in a closed form. The clipped signal can be constructed by the original signal in conjunction with a set of rectangular signals [6]. Because the spectrum characteristics of a rectangular function can be formularized in a closed form, the center-clipped sinewave can be analyzed.

Suppose $x(t)$ is the original signal and C_{th} is the clipping threshold, the center-clipped sinewave $y(t)$ can be represented by

$$y(t) = x(t)f_1(t) - C_{th}f_2(t) - C_{th}f_3(t) \quad \text{(Compressed)} \quad (3)$$

or

$$y(t) = x(t)f_1(t) \quad \text{(Uncompressed)}, \quad (4)$$

where $f_1(t)$ and $f_2(t)$ are rectangular functions with positive unit magnitude, and $f_3(t)$ is a rectangular function with negative unit magnitude. The spectrum components of a center clipped signal can be calculated by terms in the right side of equation (3) or (4). As an example, Figure 3 illustrates this concept when the magnitude of the original signal is 2 and the clipping threshold $C_{th} = 1$. Figure 3(a) shows the original sinewave $x(t)$ and the rectangular function $f_1(t)$. Figure 3(b) shows the uncompressed center-clipper output $x(t)f_1(t)$ given in equation (4), which is the multiplication result of the two waveforms shown in Figure 3(a). Figures 3(c) and 3(d) show $C_{th}f_2(t)$ and $C_{th}f_3(t)$, respectively. Figure 3(e) shows the compressed center-clipped sinewave $y(t)$ given in equation (3).

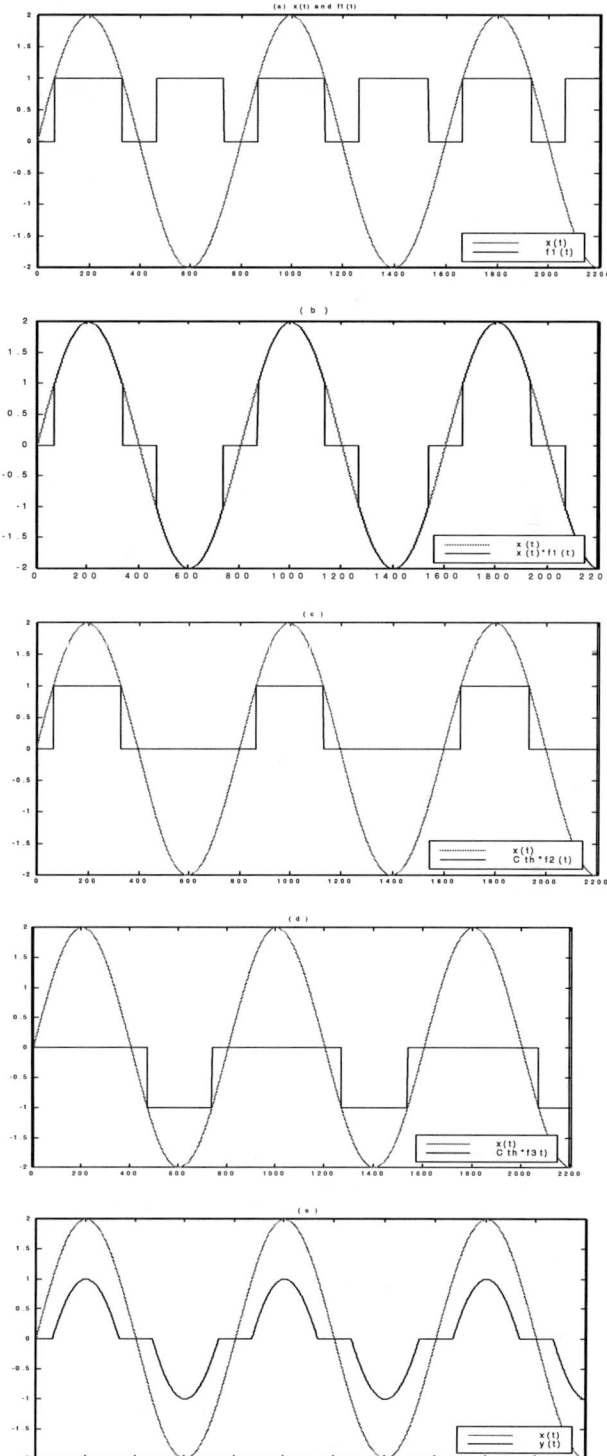

Figure 3 Construction of a clipped sinusoidal signal by a set of rectangular waveforms. (a) sinewave $x(t)$ (dashed line) and square wave $f_1(t)$ (solid line). (b) $x(t)$ and the uncompressed center-clipped signal $x(t)f_1(t)$ (solid line). (c) $x(t)$ and $C_{th}f_2(t)$ (solid line). (d) $x(t)$ and $C_{th}f_3(t)$ (solid line). (e) $x(t)$ and the compressed center-clipped signal $y(t)$ (solid line).

III. Analysis of Center Clipping

Any periodic signal $f(t)$ can be formed by the summation of time-shifted generating function $g(t)$, which is equal to $f(t)$ in one period. The relationship between $f(t)$ and $g(t)$ can be expressed as

$$g(t) = \begin{cases} f(t), & -\frac{T_0}{2} \le t \le \frac{T_0}{2} \\ 0, & \text{elsewhere} \end{cases} \quad (5)$$

and

$$f(t) = \sum_{m=-\infty}^{m=\infty} g(t - mT_0), \quad (6)$$

where T_0 is the period of $f(t)$ in seconds. The Fourier transform of $f(t)$ is expressed as

$$F(\omega) = \omega_0 G(\omega) \sum_{m=-\infty}^{\infty} \delta(\omega - m\omega_0), \quad (7)$$

where $G(\omega)$ is Fourier transform of $g(t)$ and $\omega_0 = 2\pi/T_0$ is the fundamental frequency in radians.

In equation (3) or (4), $f_1(t)$ is formed by the generating function

$$g_1(t) = rect\left(\frac{t}{\tau}\right), \quad (8)$$

and its Fourier transform can be expressed as

$$G_1(\omega) = \tau \text{sinc}\left(\frac{\tau\omega}{2}\right), \quad (9)$$

where τ is determined by the clipping threshold C_{th}, and the period of $f_1(t)$ is $T_1 = T_0/2$. From equation (6), we have

$$f_1(t) = \sum_{m=-\infty}^{m=+\infty} g_1(t - mT_1) \quad (10)$$

and

$$F_1(\omega) = \sum_{m=-\infty}^{m=+\infty} \omega_1 G(m\omega_1)\delta(\omega - m\omega_1)$$
$$= \sum_{m=-\infty}^{m=+\infty} C_1(m)\delta(\omega - 2m\omega_0), \quad (11)$$

where

$$C_1(m) = 2\omega_0 G(2m\omega_0) = 2\omega_0 \tau \text{sinc}(\tau\omega_0 m). \quad (12)$$

The sinusoidal signal $x(t)$ can be expressed as

$$x(t) = A_0 \exp(j\omega_0 t) \quad (13)$$

and its Fourier transform is given as

$$X(\omega) = A_0 \delta(\omega - \omega_0). \quad (14)$$

Thus, the Fourier transform of $f_1(t)x(t)$ is

$$FT[f_1(t)x(t)] = \frac{1}{2\pi}F_1(\omega) * X(\omega)$$
$$= \frac{A_0}{2\pi}\sum_{m=-\infty}^{m=+\infty} C_1(m)\delta(\omega - (2m+1)\omega_0). \quad (15)$$

In the last step, we use the fact that $h(t) * \delta(t - t_0) = h(t - t_0)$.

In equation (3), $f_2(t)$ is formed by the generating function $g_2(t)$, which is the same as $g_1(t)$. However, the period of $f_2(t)$ is $T_2 = T_0$. Therefore, we have

$$f_2(t) = \sum_{m=-\infty}^{m=+\infty} g_2(t - mT_2) \quad (16)$$

and

$$F_2(\omega) = \sum_{m=-\infty}^{m=+\infty} \omega_2 G(m\omega_2)\delta(\omega - m\omega_2) = \sum_{m=-\infty}^{m=+\infty} C_2(m)\delta(\omega - m\omega_0) \quad (17)$$

where

$$C_2(m) = \omega_0 \tau \text{sinc}\left(\frac{\tau\omega_0 m}{2}\right). \quad (18)$$

The generating function $g_3(t)$ is defined as

$$g_3(t) = -rect\left(\frac{t - \Delta}{\tau}\right), \quad (19)$$

and its Fourier transform is given as

$$G_3(\omega) = -\tau \text{sinc}\frac{\tau\omega}{2}\exp(-j\omega\Delta), \quad (20)$$

where $\Delta = T_0/2$. In equation (3), $f_3(t)$ is formed by the generating function $g_3(t)$, and the period of $f_3(t)$ is $T_3 = T_0$. Therefore,

$$f_3(t) = \sum_{m=-\infty}^{m=+\infty} g_3(t - mT_3), \quad (21)$$

and its Fourier transform can be computed as

$$F_3(\omega) = \sum_{m=-\infty}^{m=+\infty} \omega_3 G(m\omega_3)\delta(\omega - m\omega_3)$$
$$= \sum_{m=-\infty}^{m=+\infty} C_3(m)\delta(\omega - m\omega_0), \quad (22)$$

where

$$C_3(m) = -\omega_0 \tau \text{sinc}\left(\frac{\tau\omega_0 m}{2}\right)\exp(-jm\omega_0\Delta)$$
$$= -C_2(m)\exp(-jm\pi)$$
$$= \begin{cases} -C_2(m), & \text{if } m \text{ is even} \\ C_2(m), & \text{if } m \text{ is odd} \end{cases}. \quad (23)$$

We analyze the more complicated compressed center clipper defined in (3) first. The result can be simplified for the uncompressed center clipper given in (4), which only consists of the first term in the right side of equation (3). Taking Fourier transform of equation (3) and substituting the results obtained in equations (15), (17), and (22), we have

$$Y(\omega) = \frac{1}{2\pi}X(\omega) * F_1(\omega) - C_{th}F_2(\omega) - C_{th}F_3(\omega)$$
$$= \frac{A_0}{2\pi}\sum_{m=-\infty}^{m=+\infty} C_1(m)\delta[\omega - (2m+1)\omega_0] - C_{th}\sum_{m=-\infty}^{m=+\infty}[C_2(m) + C_3(m)]\delta(\omega - m\omega_0) \quad (24)$$

Therefore, $Y(\omega)$ only have odd frequency components. From equation (23), we can simplify equation (24) as

$$Y(\omega) = \frac{A_0}{2\pi}\sum_{m=-\infty}^{m=+\infty} C_1(m)\delta[\omega - (2m+1)\omega_0]$$
$$- 2C_{th}\sum_{m=-\infty}^{m=+\infty} C_2(2m+1)\delta[\omega - (2m+1)\omega_0] \quad (25)$$

for the compressed center-clipper defined in equation (3). Note that the uncompressed center-clipper given in (4) only has the first term in (25). Therefore, the uncompressed center-clipper produces fewer distortions and is widely used in practical applications.

Substitute $C_1(m)$ and $C_2(m)$ into equation (25), we have

$$Y(\omega) = Y_1(\omega) + Y_2(\omega), \qquad (26)$$

where

$$Y_1(\omega) = \frac{A_0 \omega_0 \tau}{\pi} \sum_{m=-\infty}^{m=\infty} \operatorname{sinc}(\tau \omega_0 m) \delta[\omega - (2m+1)\omega_0] \qquad (27)$$

and

$$Y_2(\omega) = -2 C_{th} \omega_0 \tau \sum_{m=-\infty}^{m=\infty} \operatorname{sinc}\left(\frac{\tau \omega_0 (2m+1)}{2}\right) \delta[\omega - (2m+1)\omega_0]. \qquad (28)$$

Therefore, the clipped signal has extra odd harmonics. The amplitude of the harmonic is determined by the function $\operatorname{sinc}(x)$ whose amplitude decreases dramatically when x increases. Hence, it is expected that higher-order harmonics do not contain significant energy. The amplitudes of the harmonics also vary according to τ, which depends on the clipping threshold C_{th}. If the clipping threshold is small, τ is large, and the amplitudes of harmonics are small. In other words, the larger clipping threshold will generate more distortions with extra strong odd harmonics components.

IV. COMPUTER SIMULATIONS

As discussed in Section 3, nonlinear center clipping of sinusoidal signals generates extra odd harmonics. The amplitudes of these harmonics decrease dramatically when the order increases. Amplitudes of odd harmonics also vary according to the clipping threshold. The higher threshold value results in the larger amplitude of harmonics. The following simulations verify these analysis results based on the worst case compressed center clipper. In these simulations, the original 10 Hz sinewave has magnitude 2.

In Figure 4, the center-clipping threshold is 0.5, 25% of peak amplitude (2) of the original signal. Figure 4(a) shows the comparisons of original sinewave and its center-clipped signals in time domain. Figure 4(b) shows the comparisons of the same signals in frequency domain. It shows that only odd harmonics (30 Hz, 50 Hz, etc.) are generated. Also, the amplitudes of these odd harmonics decrease as the order m increase.

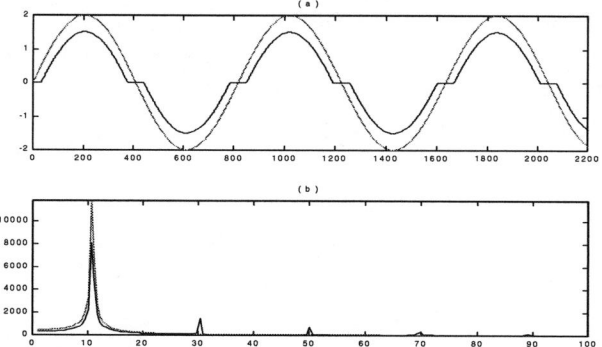

Figure 4 Comparison of the original and center-clipped sinewaves. (a) Original signal (dashed line) and center-clipped signal in time domain. (b) Original signal (dashed line) and center-clipped signal in frequency domain.

The analysis results can be further verified by the comparison of the spectrum calculated by equation (26) and the spectrum calculated by FFT of the clipped signal. The amplitude of the original signal is 2 and the clipping threshold is 1, thus $\tau \approx 0.4$ is used for calculation. Figure 5(a) shows the spectrum calculated by the FFT of the center-clipped signal and Figure 5(b) shows the spectrum calculated by equation (26). To compare the relative amplitude, the data are normalized in Figure 5. Also, by comparing Figure 4(b) with Figure 5(a), we show that using a larger clipping threshold (1) in Figure 5(a) produces larger old harmonics.

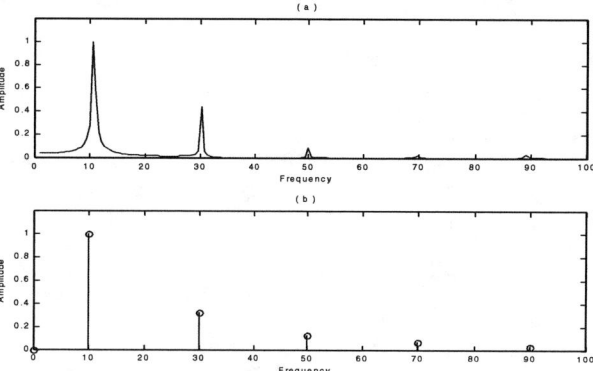

Figure 5 Comparisons of spectra. (a) The spectrum calculated using FFT of the clipped signal. (b) The spectrum calculated from equation (26).

V. CONCLUSIONS

In this paper, two center clipping operations are analyzed. Fourier analysis shows that extra odd harmonics are generated in the frequency domain for a center-clipped signal. The uncompressed center clipper produces fewer nonlinear distortions. The amplitudes of harmonics are related to the clipping threshold C_{th}. Computer simulations show that the spectrums calculated by derived equations fit the spectrum calculated by FFT of the center-clipped signal, thus verify the analysis results.

REFERENCES

[1] S. L. Gay and J. Benesty, *Acoustic Signal Processing for Telecommunication*, Boston MA: Kluwer Academic Publishers, 2000.

[2] S. M. Kuo and B. H. Lee, *Real-Time Digital Signal Processing*, New York, NY: Wiley, 2001.

[3] S. M. Kuo and W. S. Gan, *Digital Signal Processors*, Upper Saddle River, NJ: Prentice Hall, 2005.

[4] ITU-T Recommendation G.168, Digital Network Echo Cancellers, International Telecommunication Union, 1997.

[5] T. F. Quatieri and R. J. McAulay, "Speech Transformations Based on a Sinusoidal Representations," *IEEE Trans. on Acoustics, Speech, Signal Processing*, vol. ASSP-34, pp. 1449-1464, Dec. 1986.

[6] S. M. Kuo, H. T. Wu, F. K. Chen, and M. R. Gunnala, "Saturation Effects in Active Noise Control Systems," *IEEE Trans. on Circuits and Systems-I: Regular Papers*, vol. 51, no. 6, pp. 1163-1171, June 2004.

SMF Robust Filtering In Impulsive Noise

Li Guo and Yih-Fang Huang
Department of Electrical Engineering
University of Notre Dame, Notre Dame, IN 46556
Email: <lguo1, huang>@nd.edu

Abstract— An adaptive M-estimation algorithm based on set-membership filtering (SMF) is presented for robust filtering in impulsive noise. The proposed algorithm has unique features of data-dependent weights and selective update. It is derived from the general M-estimation and a SMF-type cost function. Simulation results show that the proposed algorithm performs much better than conventional recursive least-squares algorithms and conventional SMF algorithms in impulsive noise. Simulation results also demonstrate that the proposed algorithm has tracking capability superior to the least M-estimation approach, and it is more resistant to outliers.

I. INTRODUCTION

The performance of conventional linear least-squares (LS) and least-mean-square (LMS) adaptive algorithms can degrade significantly in impulsive noise, which is often encountered in communications. Impulsive noise, which can be caused by both man-made and natural interferences, consists of random occurrences of energy spikes with random amplitude and spectral content. In broadband indoor wireless communications, the principal sources of impulsive noise are devices with electromechanical switches, e.g., copy machines and printers [1]. A number of filtering methods have been proposed to deal with impulsive noise in the literature, see, e.g., [2]–[4]. Many of those methods are nonlinear, e.g., nonlinear recursive least-squares (NRLS) [2], robust mixed-norm (RMN) [3] algorithms, etc. In the context of linear estimation, recursive least M-estimation (RLM) [4] and least mean M-estimation (LMM) [5] algorithms are among the more notable ones.

In addition to the conventional LS and LMS methods, there exists a different class of estimation algorithms, referred to as set-membership filtering (SMF), see, e.g., [6]–[10]. With a pre-specified error bound, SMF algorithms seek filter weights such that the worst-case error is bounded by this error bound [10]. It has been shown that adaptive SMF algorithms have excellent convergence and tracking performance thanks to the data-dependent *selective* updates and 'optimal' weight at each update. However, simulation experiences have shown that performance of SMF algorithms also degrades in impulsive noise. This paper investigates such performance degradation of SMF algorithms. It also presents an adaptive set-membership M-estimation algorithm (SM-AM) that yields robust filtering in impulsive noise. The proposed algorithm retains the unique features of conventional SMF algorithms [6], [10], namely, data-dependent selective update and optimal forgetting factor. It will be shown that SM-AM algorithm outperforms conventional exponential weighted recursive least-squares (WRLS) algorithm, and conventional adaptive SMF algorithms. Furthermore, SM-AM algorithm has superior tracking capabilities over least M-estimation algorithm (like RLM) for time-varying systems in impulsive noise.

II. SET-MEMBERSHIP FILTERING AND M-ESTIMATION

Consider a general linear-in-parameter filter with input $\mathbf{x}$, weight θ and output y. Let d be the desired output, and $e(\theta)=d-\theta^T\mathbf{x}$ be the corresponding filter error. To estimate the parameter vector θ, SMF employs a bounded error criterion. With an *a priori* error-bound specification γ, SMF seeks to find θ such that

$$|e(\theta)|^2 \leq \gamma^2, \quad \forall (\mathbf{x}, d) \in \mathcal{S} \quad (1)$$

where $\mathcal{S}$ is the design space comprising of all input vector-desired output pairs $(\mathbf{x}, d)$. The set of feasible solutions, termed *feasibility set* Θ, is given by

$$\Theta = \bigcap_{(\mathbf{x},d)\in\mathcal{S}} \{\theta \in \mathbf{C}^L : |d - \mathbf{x}^T\theta| \leq \gamma\} \quad (2)$$

where L is the dimension of input vector $\mathbf{x}$. Many adaptive algorithms have been derived in the above framework, including the well-known family of optimal bounding ellipsoid (OBE) algorithm [6], [7], and SM-NLMS algorithm [9], [10].

The set-membership filtering problem can be formulated in the framework of the generalized M-estimation method of Huber [11], which seeks parameter estimates that minimize a weighted cost defined as:

$$J(\theta) = \sum_{i=1}^{n} w(\mathbf{x}_i)\rho(e_i(\theta)), \quad (3)$$

or, equivalently, solving the following equation

$$\sum_{i=1}^{n} w(\mathbf{x}_i)\psi(e_i(\theta))\mathbf{x}_i = 0 \quad (4)$$

where $\rho(.)$ is the M-estimation cost function, $\psi = d\rho/de$ and $w(.)$ is the weight. If the cost function is defined as

$$\rho(e) = \begin{cases} \gamma^2 & |e| \leq \gamma \\ \infty & |e| > \gamma \end{cases} \quad (5)$$

and $w(\mathbf{x}_i)=1$, then the SMF formulated by (1) and (2) follows. The existing adaptive SMF algorithms, like DH-OBE [6] and

SM-NLMS [10] are, in essence, the general M-estimation with cost function defined as

$$\rho(e) = \begin{cases} \gamma^2 & |e| \leq \gamma \\ e^2/2 & |e| > \gamma. \end{cases} \quad (6)$$

In the above definitions, $\rho(e) = \gamma^2$ for $|e| \leq \gamma$ is reflective of SMF that requires bounded filter output error. Outside the error bound, we assign quadratic cost as in (6). In adaptive M-estimation, the contribution of incoming data depends on the value of $\psi(e)$. As seen in (6), $\psi(e) = 0$ for $|e| \leq \gamma$, thus the contribution of incoming data is zero and no update is needed. This is simply the data-dependent selective update feature of SMF, see, e.g., [6], [10]. Outside the error bound, $\psi(e) = e$ and the contribution of incoming data on update is proportional to the value of prediction error. In impulsive noise, the filter's performance degradation is mainly caused by the false updates due to outliers. However, from (6), we can see that the false updates of outliers can not be avoided in conventional adaptive SMF algorithms since a contribution of $\psi(e) = e$ is assigned to the outliers, and even a larger contribution is usually assumed for the outliers in OBE and SM-NLMS algorithms because the value of $|e|$ resulting from outliers is generally very large. Thus the performance can degrade significantly. This result is also supported by our simulation experiences.

III. SET-MEMBERSHIP ADAPTIVE M-ESTIMATION ALGORITHM

An algorithm is presented here that employs the general M-estimation (4) with the following SMF-type cost function for robust filtering in impulsive noise. With an *a priori* error bound specification γ, the cost function is defined as

$$\rho(e) = \begin{cases} \gamma^\alpha/\alpha & |e| \leq \gamma \\ |e|^\alpha/\alpha & \gamma < |e| \leq a \\ |e|^2/2 + a^\alpha/\alpha - a^2/2 & a < |e| \leq b \\ b|e| + a^\alpha/\alpha - \frac{a^2+b^2}{2} & b < |e| \leq c \\ bc + a^\alpha/\alpha - \frac{a^2+b^2}{2} & \text{otherwise} \end{cases} \quad (7)$$

with $\alpha \geq 2$. Based on this cost function, the resulting algorithm can provide robust filtering in impulsive noise and retain the feature of data-dependent selective updates. Once again, in (7), a constant $\rho(e)$ inside the error bound retains the bounded error property of SMF. The false update of outliers is suppressed by the cost with a less steep gradient in the error region, (c, ∞) and $(b, c]$. Whenever $|e_n| > c$, it is determined that an outlier has occurred and the false update is avoided by setting $\rho(e)$ a constant (with respect to e), thus $\psi(e)=0$ and the contribution of incoming data is zero. In the region of $b < |e_n| \leq c$, it is still possible that an outlier has occurred, and the impact of the outlier is mitigated by the reduced contribution, $\psi(e) = b$. Within the region of $(\gamma, a]$ and $(a, b]$, it is unlikely that an outlier has occurred, thus quadratic or higher order cost is assigned for normal update. Between the error bound γ and threshold a, a cost with an order of $\alpha > 2$ is employed to compensate the impaired estimator efficiency caused by the reduced contribution of error outside the threshold b.

Using simple weight $w(\mathbf{x}_i) = \lambda_i$, a general update equation for adaptive solution of (4) is given by [12]:

$$\theta_n = \begin{cases} \theta_{n-1} & \text{if } |e| \leq \gamma \text{ or } |e| > c \\ \theta_{n-1} + \frac{\lambda_n P_{n-1}\mathbf{x}_n\psi(e_n)}{1+\lambda_n\dot{\psi}(e_n)G_n} & \text{otherwise} \end{cases} \quad (8)$$

where

$$P_n = P_{n-1} - \frac{\lambda_n \dot{\psi}(e_n) P_n \mathbf{x}_n \mathbf{x}_n^T P_{n-1}}{1 + \lambda_n \dot{\psi}(e_n) G_n} \quad (9)$$

with $G_n \triangleq \mathbf{x}_n^T P_{n-1} \mathbf{x}_n$. In (8), no update is needed whenever the error is smaller than γ or larger than c. For $|e_n| \leq \gamma$, no update is needed because the information contained in the received data is insufficient to require an update. On the other hand, whenever $|e_n| > c$, the estimator determines that an outlier has occurred and an update should be avoided. If the error falls within $(\gamma, c]$, an important issue is the determination of the optimal weight λ_n in (8) and (9). Here, an *a posteriori* error bound amplitude criterion is employed to calculate λ_n, i.e., the amplitude of the *a posteriori* error is γ at each update step. Based on the value of e_n, there are three different cases in the determination of λ_n:

Case 1: $\gamma < |e_n| \leq a$. In this error region, $\psi(e_n)=e_n|e_n|^{\alpha-2}$ and $\dot{\psi}(e_n)=(\alpha-1)|e_n|^{\alpha-2}$. So λ_n is obtained by solving

$$\left| \frac{|e_n| + \lambda(\alpha-2)G_n|e_n|^{\alpha-1}}{1+\lambda(\alpha-1)G_n|e_n|^{\alpha-2}} \right| = \gamma \quad (10)$$

and the solution is

$$\lambda_n = \frac{|e_n| - \gamma}{G_n |e_n|^{\alpha-2}((\alpha-1)\gamma - (\alpha-2)|e_n|)}. \quad (11)$$

Case 2: $a < |e_n| \leq b$. In this case, the cost function is quadratic, and $\psi(e_n)=e_n$ and $\dot{\psi}(e_n)=0$. The solution of λ_n is obtained by letting $\alpha=2$ in (8):

$$\lambda_n = \frac{1}{G_n}\left(\frac{|e_n|}{\gamma} - 1\right). \quad (12)$$

Case 3: $b < |e_n| \leq c$. In this case, the cost function is linear and $\psi(e_n)=\text{sign}(e_n)b$ and $\dot{\psi}(e_n)=0$. γ_n is obtained by solving the following equation

$$\left|1 - \lambda G_n \frac{b}{|e_n|}\right| = \frac{\gamma}{|e_n|}. \quad (13)$$

The above equation has two positive solutions: $\gamma^{(1)}=\frac{|e_n|+\gamma}{G_n b}$ and $\gamma^{(2)}=\frac{|e_n|-\gamma}{G_n b}$, while λ_n takes the smaller value for stability.

One should note that the weight λ_n in SM-AM algorithm is data-dependent and is optimized at each update step. This is in contrast to the fixed forgetting factor used in least-squares and least M-estimation algorithms. This contributes to the superior tracking abilities of SM-AM algorithm in impulsive noise.

The thresholds (a, b, c) in the cost function (7) need to be determined, while γ is a performance specification and is determined *a priori*. Others are generally unknown *a priori* and should be estimated continuously during the adaptation. One approach is to set the value of (a, b, c) based on the estimated variance of output errors, $\hat{\sigma}_n^2$, which can be estimated by weighted averaging: $\hat{\sigma}_n^2 = \lambda_\sigma \hat{\sigma}_{n-1}^2 + (1-\lambda_\sigma)e_n^2$, where λ_σ is the smoothing factor for $\hat{\sigma}_n^2$. Generally, there are no constraints on b and c. However, the value of a should be smaller than $\frac{\alpha-1}{\alpha-2}\gamma$ for the case of $\alpha>2$ to ensure λ_n to be positive which guarantees the stability of the algorithm.

IV. SIMULATION RESULTS

Results of simulation studies on the proposed SM-AM algorithm are presented here, and a comparison is made with the conventional algorithm like WRLS, DH-OBE and RLM. Superior tracking performance of the SM-AM in impulsive noise for time-variant systems has been observed. In all the simulations, the following mixture model is used for impulsive noise, $v(n)=v_b(n)+b(n)v_I(n)$ where $v_b(n)$ is the zero-mean background noise with variance σ_b^2, $v_I(n)$ is zero-mean Gaussian process with variance σ_I^2 and $b(n)$ is a Bernoulli process with occurrence probability $Pr(b(n)=1)=p_r$. Furthermore, $v_I(n)$ and $b(n)$ are independent and, respectively, represent the random amplitude and position of the impulses. The extent of impulsiveness of $v(n)$ is characterized by the ratio between the power of impulsive and background noise, in particular, $R_{im}=p_r\sigma_I^2/\sigma_b^2$. Further details of the simulation are omitted here for the sake of brevity. In all the following simulation examples, $\alpha=3$ (as in (7)) and $\gamma=3\sigma_b$.

Simulation Example 1: Identification of an unknown system. Here, SM-AM algorithm is employed to identify an unknown fourth-order FIR filter in impulsive noise. Background noise is uniformly distributed with variance 0.01. Impulse parameters are $p_r=0.01$ and $R_{im}=300$. To visualize clearly the effect of impulsive noise, the locations of impulses generated by $b(n)$ are fixed but their amplitude are generated randomly according to $v_I(n)$. The mean-squared error (MSE) results are shown in Fig.1. The robustness of SM-AM algorithm is clearly seen in Fig.1. The WRLS algorithm and DH-OBE algorithm, however, are not as robust, although their initial convergence rate is faster than SM-AM algorithm. The SM-AM and RLM algorithms have almost identical initial convergence rate, steady-state error, and robustness to impulses. More interestingly, the number of updates needed by SM-AM algorithm is around 26 out of a total of 1500 samples, while achieving comparable performance to RLM.

Simulation Example 2: Adaptive Equalization in Impulsive noise. The performance of SM-AM algorithm applied to adaptive equalization in a raised cosine ISI channel is considered. The channel impulse response coefficients are $\{0.28, 1.0, 0.28\}$. A decision feedback equalizer with nine forward and one feedback taps is used with QPSK signaling. Background noise is zero-mean Gaussian with variance $\sigma_b^2=0.1$, $R_{im}=300$ and the $p_r=0.01$. Again, the positions of impulses are fixed but the amplitude of impulses is random according to $v_I(n)$. The positions of impulses are at $n=383, 850$ and marked in Fig.2. The results shown in Fig.2 illustrate that SM-AM can effectively suppress the impulsive noise in equalization. In contrast, the WRLS algorithm is rather vulnerable to impulsive noise. In the simulation, SM-AM algorithm had 260 updates out of 1500 samples.

Simulation Example 3: Tracking time-varying systems in

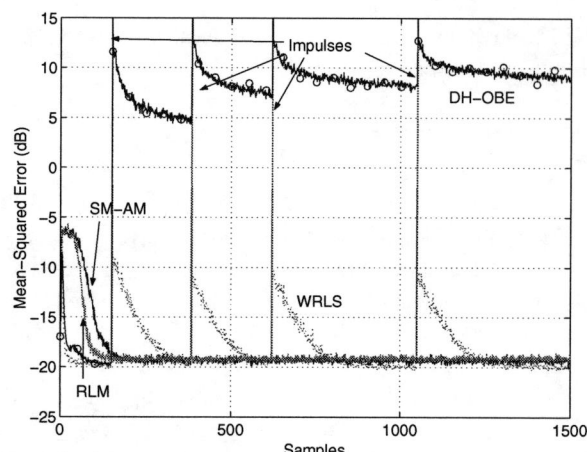

Fig. 1. Identification of a forth-order FIR filter

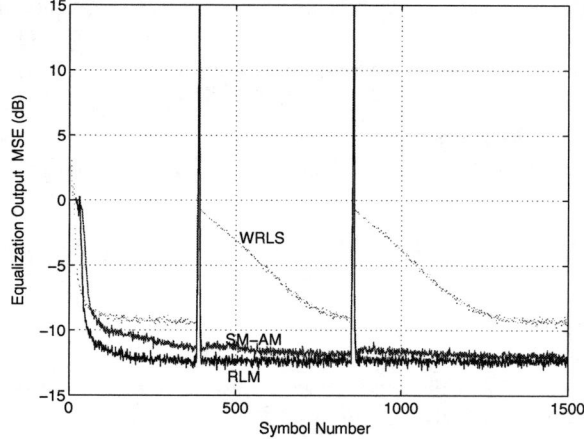

Fig. 2. Adaptive equalization with input impulsive noise

impulsive noise. Identification of a fourth-order FIR filter is simulated similarly to Example 1, except that the system coefficients vary randomly. The parameter variations are introduced by having random jumps at every 200th sample. Background noise is zero-mean uniformly distributed with variance $\sigma_b^2=0.0316$, $R_{im}=0.01$ and $p_r=0.01$. The position and amplitude of the impulses are generated randomly according to $b(n)$ and $v_I(n)$, respectively. The first and third coefficients are shown in Fig.3 and Fig.4 respectively. It is clear that the SM-AM algorithm has superior tracking capability over RLM (with an exponential forgetting factor of 0.9) while using only around 12% of the received data to update parameter estimates.

Simulation Example 4: Filtering with non-impulsive noise. SM-AM algorithm is used to identify the fourth-order FIR filter used in Example 1, but in zero-mean Gaussian noise with variance 0.1. In general, the LS approach will provide the best solution for this case. It is seen from Fig.5 that the steady-state performance of the SM-AM algorithm is comparable (almost the same) to WRLS, although the rate of convergence of SM-

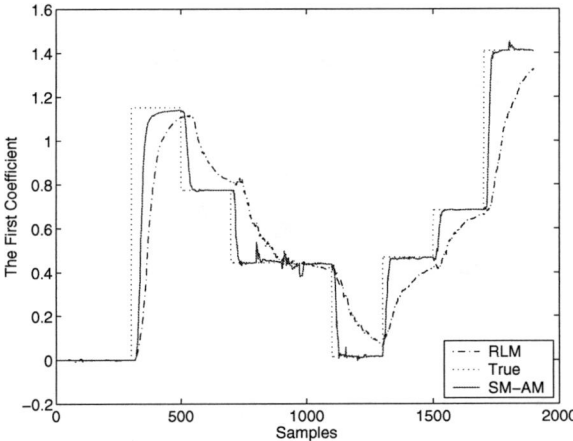

Fig. 3. Performance of tracking time-varying coefficients

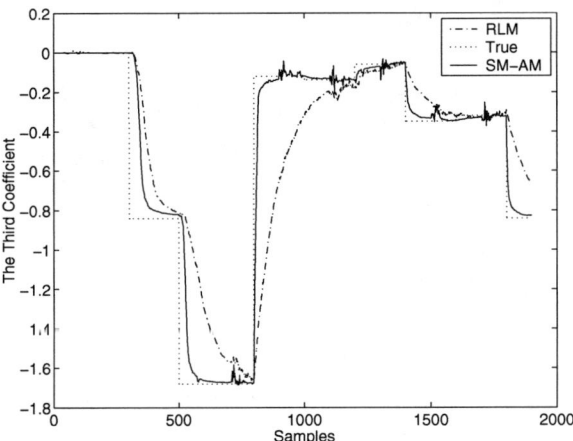

Fig. 4. Performance of tracking time-varying coefficients

AM is slightly slower. But SM-AM algorithm used less than 4% of the data to update parameter estimates!

V. CONCLUSION

In this paper, we first examine the robustness of conventional SMF algorithms in impulsive noise. We then present an SM-AM algorithm that explores general M-estimation of Huber along with an SMF-type cost function, which is defined in such a way that the false updates in the presence of outliers can be reduced while retaining the unique features of data-dependent weight and selective update mechanism. Simulation results show that the SM-AM algorithm is robust with respect to impulses and outperforms WRLS and OBE algorithms. The superior tracking ability of SM-AM algorithm over RLM algorithm has also been observed.

ACKNOWLEDGMENT

This work has been supported, in part, by the National Science Foundation under Grant EEC02-03366, by the U.S. Department of the Army under Contract DAAD 16-02-C-

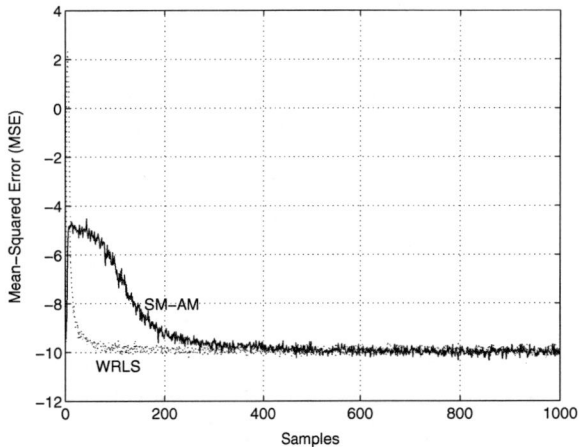

Fig. 5. Convergence performance with non-impulsive noise

0057-P1, and by the Indiana 21st Century Fund for Research and Technology.

REFERENCES

[1] K. Blackard, T. Rappaport, and C. Bostian, "Measurements and models of radio frequency impulsive noise for indoor wireless communications," *IEEE J. Select. Areas Commun.*, vol. 11, no. 7, pp. 991–1001, Sept. 1993.

[2] J. Weng and S. Leung, "Adaptive nonlinear rls algorithm for robust filtering in impulse noise," in *Proceedings of 1997 IEEE International Symposium on Circuits and Systems, 1997. ISCAS '97.*, 1997, pp. 2337–2340.

[3] J. Chambers and A. Avlonitis, "A robust mixed-norm adaptive filter algorithm," *IEEE Signal Processing Lett.*, vol. 4, no. 2, pp. 46–48, Feb. 1997.

[4] Y. Zou, S. Chan, and T. Ng, "A recursive least m-estimate (rlm) adaptive filter for robust filtering in impulse noise," *IEEE Signal Processing Lett.*, vol. 7, no. 11, pp. 324–326, Nov. 2000.

[5] ——, "Least mean m-estimate algorithms for robust adaptive filtering in impulse noise," *IEEE Trans. Circuits Syst. II*, vol. 47, no. 2, pp. 1564–1569, Dec. 2000.

[6] S. Dasgupta and Y. F. Huang, "Asymptotically Convergent Modified Recursive Least-Squares with Data-Dependent Updating and Forgetting Factor for Systems with Bounded Noise," *IEEE Trans. Inform. Theory*, vol. IT-33, no. 3, pp. 383–392, May 1987.

[7] J. R. Deller, Jr., M. Nayeri, and M. S. Liu, "Unifying the Landmark Developments of Optimal Bounding Ellipsoid Identification," *Int. J. Adaptive Control and Signal Processing*, vol. 8, pp. 48–63, Jan-Feb 1994.

[8] S. Nagaraj, S. Gollamudi, S. Kapoor, and Y. F. Huang, "BEACON: An Adaptive Set-Membership Filtering Technique with Sparse Updates," *IEEE Trans. Signal Processing*, vol. 47, no. 11, pp. 2928–2941, November 1999.

[9] P. S. Diniz and S. Werner, "Set-Membership Binormalized Data Reusing LMS Algorithms," *IEEE Trans. Signal Processing*, vol. 51, no. 1, pp. 124–134, Jan 2003.

[10] S. Gollamudi, S. Nagaraj, S. Kapoor, and Y. F. Huang, "Set-Membership Filtering and a Set-Membership Normalized LMS Algorithm with an Adaptive Step Size," *IEEE Signal Processing Lett.*, vol. 5, no. 5, pp. 111–114, May 1998.

[11] P. Huber, *Robust Statistics*. New York: Wiley, 1981.

[12] S. Nagaraj, *Multiuser Signal Estimation and Downlink Beamforming for Wireless Communications*. University of Notre Dame: Ph.D Dissertation, 2000.

Modelling of High-Order Mechanical Plate Vibration Systems by Multidimensional Wave Digital Filters

Chien-Hsun Tseng
Department of Electronics Engineering
Southern Taiwan University of Technology, Taiwan
Email: jct@mail.stut.edu.tw

Stuart Lawson
School of Engineering
University of Warwick, UK
Email: S.S.Lawson@eng.warwick.ac.uk

Abstract— The application of MDWDF technique derived from the Wave Digital Filter (WDF) paradigm to the modelling of Mindlin plate vibration system is presented. Initial and boundary conditions of the plate vibration system are properly embedded in the MDWDF algorithm in terms of state output quantities, which are fully implemented in the algorithm. More specifically, different types of boundary conditions encountered in practice can be easily fulfilled in a very simple and efficient manner without remodifying the whole algorithm, which is usually proceeded by finite elements based method with various boundary conditions. This is due to the local connectivity of the MDWDF model, in the sense that the behaviour at any point on the grid is directly influenced only by its nearest-neighbour grid points. Graphical results illustrating the plane wave propagation on a square plate are given for which free edges boundary conditions are fully considered.

I. INTRODUCTION

The modelling of vibration in mechanics is of importance to engineering and science, especially in the area of aerospace, marine and construction sectors. The vibration design aspect is even more increasingly important in micro-machines such as electronic packaging, micro-robots, etc. because of their enhanced sensitivities to vibration. The most popular kind of models for mechanical vibration system can be represented by sets of linear and/or non-linear partial differential equations (PDEs) with properly imposed initial and boundary conditions. Techniques used to solve the PDEs such as finite elements and finite differences have some success. A remarkable alternative approach to the PDE system simulation is recently proposed that builds on the Wave Digital Filter paradigm and analogies with electrical networks. The technique involves firstly finding a multidimensional lumped electrical network, which represents the behavior of the linear or linearized system. From this network, a discrete-time equivalent is developed that is a multidimensional wave digital filter (MDWDF) [1], [2].

The technique of MDWDF originally introduced for the implementation of digital filters has been successfully applied to a numerical integration of PDEs [2], [3] and many other contributions. These PDE systems usually involve low-order theories for governing equations. As a result, the systems can be directly handled by a standard PDE solver such as the MATLAB PDE Toolbox [5]. To further demonstrate the inherent capabilities of the method in handling the high-order governing equations, in this contribution, the MDWDF technique is applied to the simulation of plane wave propagation of Mindlin system.

We begin by considering a typical thick plate theory, Mindlin plate [6], [7], [8], the first development of an approximate theory for especially dealing with flexural waves in a relative thick plate. The equation of motion for a 2D Mindlin plate can be written as a 4th-order system of PDEs for the displacement w in space (x, y) and time t [6], [7]:

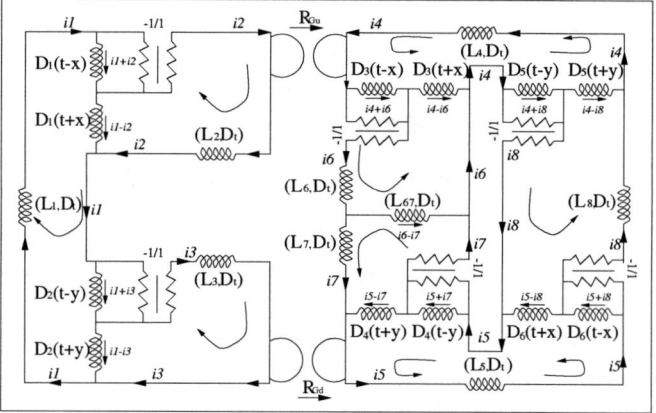

Fig. 1. MD Kirchhoff circuit for Mindlin system (1.1).

$$\left(D\nabla^2 - \frac{\rho h^3}{12}\frac{\partial^2}{\partial t^2}\right)\left(\nabla^2 - \frac{\rho}{\kappa^2 G}\frac{\partial^2}{\partial t^2}\right)w + \rho h \frac{\partial^2 w}{\partial t^2} = 0 \quad (1.1)$$

where h is the plate thickness, ρ is the material density, κ^2 is a shear correction factor, G is the modulus of elasticity in shear, and D is the flexural rigidity of the plate. Defining the transverse and bending rotation velocities (v, w_x, w_y), and the plate stress (M_x, M_y, M_{xy}), the 4th-order system (1.1) can be decomposed into 1st-order systems of eight PDEs [6]. These systems are mathematically manipulated and interpreted in graphical form of a 3D Kirchhoff circuit in Fig. 1 where the same physical dimension is preserved so that systems variables denoted as mesh voltages are replaced by mesh currents i_k with positive constants r_1, r_2 and r_3 to be determined:

$$\begin{array}{l}(v, Q_x, Q_y, w_x, w_y, M_x, M_y, M_{xy}) = \\ (r_1 i_1, i_2, i_3, r_2 i_4, r_3 i_5, i_6, i_7, i_8).\end{array} \quad (1.2)$$

Noting that (Q_x, Q_y) are the transverse shear forces per unit length of the plate. In view of the MDKC of Fig. 1, we have 3 loops in the left circuit and 5 loops in the right circuit, each loop consisting of the Jaumann structures and self-inductors. Furthermore, 2 gyrators with the gyration coefficients, $R_{Gu} = r_2$, and $R_{Gd} = r_3$ are connected in the middle of the circuit. The self-inductors with inductances L_j in the MDKC are given with arbitrary positive constants δ_j:

$$\begin{array}{l}L_1 = r_1^2 \rho h - \delta_1 - \delta_2, L_2 = \frac{1}{\kappa^2 G h} - \delta_1 \\ L_3 = \frac{1}{\kappa^2 G h} - \delta_2, L_8 = \frac{24(1+\nu)}{Eh^3} - \delta_5 - \delta_6 \\ L_4 = \frac{\rho h r_2^2}{12} - \delta_3 - \delta_5, L_5 = \frac{\rho h r_3^2}{12} - \delta_4 - \delta_6 \\ L_6 = \frac{12(1-\nu)}{Eh^3} - \delta_3, L_{67} = \frac{12\nu}{Eh^3}, L_7 = \frac{12(1-\nu)}{Eh^3} - \delta_4\end{array} \quad (1.3)$$

where E is the Young's modulus and ν stands for Poisson's ratio. In addition, the partial derivative operators $D_j(\cdot)$ with respect to the 2-port Jaumann structure are defined by

$$D_j(t \pm x) = 1/2(\delta_j D_t \pm r_k D_x), j = 1, 3, 6; k = 1, 2, 3$$
$$D_j(t \pm y) = 1/2(\delta_j D_t \pm r_k D_y), j = 2, 4, 5; k = 1, 3, 2$$

where equally placed signs correspond to one another.

II. MDWDF ALGORITHM BASED ON SPACE-TIME-DOMAIN ANALYSIS

Applying a generalized trapezoidal rule [2] for the linear discretization of the self-inductors and Jaumann structures involved in the continuous setting (x, t) or (y, t), we have arrived at an integral approximation for these structures. Accordingly, mesh voltages given in terms of $D_j(\cdot)$ and appeared in Fig. 1 are discretized in the form of $[x, y, t]' = [mT_x, nT_y, kT_t]'$, $(m, n, k \in N)$. Furthermore, the continuous settings of inductors are stated in terms of shift operators: $\mathbf{T} = [0, 0, T_t]'$, $\mathbf{T}_1 = [-T_x, 0, T_t]'$, $\mathbf{T}_2 = [T_x, 0, T_t]'$, $\mathbf{T}_3 = [0, -T_y, T_t]'$, $\mathbf{T}_4 = [0, T_y, T_t]'$. To develop a stable digital system directly from an analog network, the passive effect plays an important role in the lumped circuit. When the passivity requirements of inductances are held in the MDKC, this simply implies that $L_j \geq 0$ in (1.3). Clearly the simplest choice for these δ_j is thus given by

$$\begin{cases} \delta_1 = \delta_2 = \frac{1}{\kappa^2 Gh} \triangleq \bar{\delta}_1 \\ \delta_3 = \delta_4 = \delta_5 = \delta_6 = \frac{12(1-\nu)}{Eh^3} \triangleq \bar{\delta}_2. \end{cases} \quad (2.1)$$

This leads to a simplified MDKC with $T_x = T_y \triangleq \hbar$, $r_1 = \hbar/T_t\bar{\delta}_1$, $r_2 = r_3 = \hbar/T_t\bar{\delta}_2$, and

$$\begin{cases} L_1 = \bar{\delta}_1[(\hbar/T_t)^2\bar{\delta}_1\rho h - 2] \geq 0 \\ L_2 = L_3 = L_6 = L_7 = 0, L_8 = 48\nu/Eh^3 \geq 0 \\ L_4 = L_5 = \bar{\delta}_2[(\hbar/T_t)^2(\rho h^3/12)\bar{\delta}_2 - 2] \geq 0. \end{cases} \quad (2.2)$$

Due to the space samplings and following from $L_j \geq 0, j = 1, 2, 3$ in (2.2), we obtain a lower bound of a ratio between the density of sampling in space and that of sampling in time, i.e. $\frac{\hbar}{T_t} = \frac{r_1}{\sqrt{2}\bar{\delta}_1\bar{\delta}_2}\sqrt{\bar{\delta}_1^2 + \bar{\delta}_2^2} \geq \sqrt{2}\mu_0$ where $\mu_0 = \sqrt{\kappa^2 G/\rho}$ is the phase velocity of a wave [6] traveling freely according to the system (1.1). This is the Courant-Friedrichs-Levy (CFL) bound [9], a stability criterion for many approaches to the numerical solution of the wave equation. From the point of view of the minimal CFL bound, it imposes the least restriction on the density of the sampling in time for a given density of the sampling in space. Thus, the passivity requirements (and hence the numerical stability) of Mindlin systems are guaranteed.

Let the voltage waves [10] be adopted by means of a forward wave a and a backward wave b:

$$a \triangleq u + Ri, b \triangleq u - Ri.$$

Applying the voltage waves to the discrete terms of the self-inductors and Jaumann structures, it leads to $b(\mathbf{t}) = -a(\mathbf{t} - \mathbf{T}_j), j = 1, \ldots, 4$ and $b(\mathbf{t}) = -a(\mathbf{t} - \mathbf{T})$ where $\mathbf{t} \triangleq [x; y; t]$. Utilizing properties of the 2-port lattice structure, which is equivalent to the 2-port Jaumann structure, we may now transform the simplified MDKC into a MDWDF algorithm of Fig. 2 where the adaptors marked N'(1) and N''(1) are given in [1]. The port resistances in the MDWDF model are determined according to the generalized trapezoidal rule.

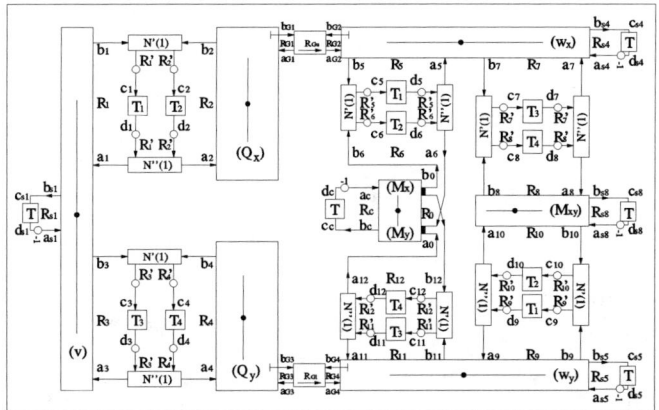

Fig. 2. A simplified MDWDF model for numerical simulation of Mindlin system (1.1).

Thus, they are all positive, which is inductances divided by shift operators. This has resulted in all port resistances being passive (and hence the stability of the MDWDF algorithm is ensured). Details of the port resistances can be viewed in [11]. Furthermore, by principles of wave digital filters [10], the structures of n-port adaptors $(n > 2)$ representing the discrete type of the system variables, the coupled 2-port series adaptor described by vector waves, and the 2-port series adaptors due to the gyrator couplings can be, respectively, described as

$$b_j = a_j - \gamma_j \sum_j^n a_j, \gamma_j = \frac{2R_j}{\sum_j^n R_j}, \begin{cases} j = s1, 1, 3 \\ j = s2, 2, G1 \\ j = s3, 4, G3 \\ j = s4, G2, 5, 7 \\ j = s5, G4, 9, 11 \\ j = s8, 8, 10 \end{cases} \quad (2.3)$$

$$\begin{cases} \mathbf{b}_0 = \mathbf{a}_0 - 2\mathbf{R}_0(\mathbf{R}_0 + \mathbf{R}_c)^{-1}(\mathbf{a}_0 + \mathbf{a}_c) \\ \mathbf{b}_c = \mathbf{a}_c - 2\mathbf{R}_c(\mathbf{R}_0 + \mathbf{R}_c)^{-1}(\mathbf{a}_0 + \mathbf{a}_c) \end{cases} \quad (2.4)$$

$$\begin{cases} a_{G1} = 1/\Delta_{u+}(\Delta_{u-}b_{G1} - 2R_{Gu}R_{G1}b_{G2}) \\ a_{G2} = 1/\Delta_{u+}(2R_{Gu}R_{G2}b_{G1} + \Delta_{u-}b_{G2}) \\ a_{G3} = 1/\Delta_{\ell+}(\Delta_{\ell-}b_{G3} - 2R_{Gd}R_{G3}b_{G4}) \\ a_{G4} = 1/\Delta_{\ell+}(2R_{Gd}R_{G4}b_{G3} + \Delta_{\ell-}b_{G4}). \end{cases} \quad (2.5)$$

where $\Delta_{u\pm} = R_{Gu}^2 \pm R_{G1}R_{G2}$ and $\Delta_{\ell\pm} = R_{Gd}^2 \pm R_{G3}R_{G4}$, and the vector waves $\mathbf{a}_0 = [a_6; a_{12}]$, $\mathbf{b}_0 = [b_6; b_{12}]$, $\mathbf{a}_c = [a_{13}; a_{14}]$, and $\mathbf{b}_c = [b_{13}; b_{14}]$ due to the mutually coupled inductances in the MDKC. The desired plate variables following from the above derivation can now be expressed in the form a_{sj} and a_j (vector index $\mathbf{t}$ is omitted) as:

$$v = \frac{r_1}{R_{s1} + 2\bar{R}_1}(a_{s1} + a_1 + a_3) \quad (2.6)$$

$$Q_x = \frac{1}{\Delta_{u+}}[R_{G2}a_2 + r_2(a_{s4} + a_5 + a_7)] \quad (2.7)$$

$$Q_y = \frac{1}{\Delta_{\ell+}}[R_{G4}a_4 + r_3(a_{s5} + a_9 + a_{11})] \quad (2.8)$$

$$w_x = \frac{r_2}{\Delta_{u+}}[R_{G1}(a_{s4} + a_5 + a_7) - r_2 a_2] \quad (2.9)$$

$$w_y = \frac{r_3}{\Delta_{\ell+}}[R_{G3}(a_{s5} + a_9 + a_{11}) - r_3 a_4] \quad (2.10)$$

$$[M_x; M_y] = (\mathbf{R}_0 + \mathbf{R}_c)^{-1}(\mathbf{a}_0 + \mathbf{a}_c) \quad (2.11)$$

$$M_{xy} = \frac{1}{R_{s8} + 2\bar{R}_2}(a_{s8} + a_8 + a_{10}). \quad (2.12)$$

III. INITIAL AND BOUNDARY CONDITIONS

A. Initial conditions

Let initial conditions of Mindlin plate variables given in (1.2) be imposed as follows:

$$\begin{cases} i_k = \frac{f_k}{r_j}, D_t(i_k) = \frac{g_k}{r_j}, j = 1,2,3; k = 1,4,5 \\ i_k = f_k, D_t(i_k) = g_k, k = o.w. \end{cases} \quad (3.1)$$

where $i_k = i_k(\mathbf{x}, 0), f_k = f_k(\mathbf{x}), g_k = g_k(\mathbf{x})$ and $\mathbf{x} = (x,y)$. We first consider the first derivative initial conditions $D_t(i_k(\mathbf{x},0))$. Replaced the plate variables in (2.6)-(2.12) by their initial values in (3.1) and following from the results obtained in [4], we have the incoming waves a_{sj} across each self-inductor with inductances L_i, and vector waves across a coupled inductor with a vector of inductance $\mathbf{L}_c$ in the MDKC of Fig. 1:

$$\begin{cases} a_{s1} = \frac{R_{s1}}{r_1}\left[f_1 + \frac{T_t}{2}g_1\right], a_{s4} = \frac{R_{s4}}{r_2}\left[f_4 + \frac{T_t}{2}g_4\right] \\ a_{s5} = \frac{R_{s5}}{r_3}\left[f_5 + \frac{T_t}{2}g_5\right] \\ a_{sj} = R_{sj}\left[f_j + \frac{T_t}{2}g_j\right], j = 2,3,8 \\ \mathbf{a}_0 = \mathbf{R}_0\mathbf{f}_{67} - \mathbf{L}_c\mathbf{g}_{67}, \quad \mathbf{a}_c = \mathbf{R}_c\mathbf{f}_{67} + \mathbf{L}_c\mathbf{g}_{67} \end{cases} \quad (3.2)$$

where $\mathbf{f}_{67} = [f_6; f_7]$ and $\mathbf{g}_{67} = [g_6; g_7]$. Taking sums and differences of the plate variables replaced by their initial values in (2.7)-(2.10), and then combining together with (2.6), (2.12), and (3.2), we finally obtain two sets of state outputs, respectively, corresponding to initial conditions of the plate variables in two separate subsystems of the MDWDF algorithm by defining $d_1(\mathbf{x},0) = \eta_1(\mathbf{x})$, $d_5(\mathbf{x},0) = \eta_2(\mathbf{x})$, $d_7(\mathbf{x},0) = \eta_3(\mathbf{x})$, and $d_{11}(\mathbf{x},0) = \eta_4(\mathbf{x})$ where indices $(\mathbf{x},0)$ and $(\mathbf{x})$ are omitted:

$$\begin{cases} d_2 = R_{G1}f_2 - f_4 + d_1 \\ d_3 = \frac{1}{2}\left[\frac{R_{s1} + 2\bar{R}_1}{r_1}f_1 - R_{G3}f_3 + f_5 - a_{s1} - d_1 - d_2\right] \\ d_4 = R_{G3}f_3 - f_5 + d_3 \\ d_6 = \bar{R}_2 f_6 - \frac{T_t}{2}R_{67}(g_6 - g_7) + d_5 \\ d_8 = r_2 f_2 + \frac{R_{G2}}{r_2}f_4 - a_{s4} - d_5 - d_6 - d_7 \\ d_9 = \frac{1}{2}[r_3 f_3 + \frac{R_{G4}}{r_3}f_5 - (R_{s8} + 2\bar{R}_2)f_8 - a_{s5} \\ \quad + a_{s8} - d_7 + d_8 - d_{11} - d_{12}] \\ d_{10} = (R_{s8} + 2\bar{R}_2)f_8 - a_{s8} + d_7 - d_8 + d_9 \\ d_{12} = \bar{R}_2 f_7 + \frac{T_t}{2}R_{67}(g_6 - g_7) + d_{11}. \end{cases}$$

B. Boundary conditions

There are four common types of boundary condition applied to the plate edge of Mindlin plates [7], [8]: Free edge, Hard-type simply-supported edge, Soft-type simply-supported edge, and Clamped edge. Furthermore, there are many types of mixed edges boundary conditions based on the above common types of boundary (see [11] for more details). Due to limited space, we only consider the free edge case in this section. Let the free edge boundaries be given by $Q_y = M_y = M_{xy} = 0$ according to [6]. Considering a rectangular Mindlin plate whose grid points are calculated by state outputs, it seems clear from the MDWDF model that the plate can be decomposed into three separate subplates depicted in Fig. 3.

Starting with the straight plate boundaries illustrated in Fig. 3, clearly one of four neighbourhood points instituted by the number of shift operators in the MDWDF algorithm is outside the boundary of each subplate. This results in four combinations of state outputs: (d_4, d_8, d_{12}), (d_2, d_6, d_{10}), (d_3, d_7, d_{11}), and (d_1, d_5, d_9). These sets of state outputs, respectively, corresponding to the southern boundary (SB), western boundary (WB), northern boundary (NB), and eastern boundary (EB) are accommodated by the given boundary conditions provided at most an additional condition. Similar to the straight line edge grid points, two of four grid points are outside the vertex v_j for each subplate. As a result, at most two additional boundary conditions have to be imposed for providing vertices of these subplates. By assembling these three subplates into one unit of Mindlin plate, we, therefore, can obtain vertices of the rectangular plate (v_4, v_8, v_{12}), (v_2, v_6, v_{10}), (v_3, v_7, v_{11}), (v_1, v_5, v_9), which are included in the directions of SB, WB, NB, and EB, respectively.

Let the plate variables described by the state equations in (2.6)-(2.12) be given in terms of state output quantities. We then obtain the following state output equations by setting zero to all variables:

$$d_1 + d_2 + d_3 + d_4 = d_{s1} \quad (3.3)$$
$$r_2(d_5 + d_6 + d_7 + d_8 - d_{s4}) = R_{G2}(d_1 - d_2) \quad (3.4)$$
$$r_3(d_9 + d_{10} + d_{11} + d_{12} - d_{s5}) = R_{G4}(d_3 - d_4) \quad (3.5)$$
$$R_{G1}(d_5 + d_6 + d_7 + d_8 - d_{s4}) = r_2(d_2 - d_1) \quad (3.6)$$
$$R_{G3}(d_9 + d_{10} + d_{11} + d_{12} - d_{s5}) = r_3(d_4 - d_3) \quad (3.7)$$
$$\left.\begin{array}{c} d_6 - d_5 - d_{13} = 0 \\ d_{12} - d_{11} - d_{14} = 0 \end{array}\right\} \quad (3.8)$$
$$d_8 - d_7 + d_{10} - d_9 = d_{s8}. \quad (3.9)$$

Following the conditions $Q_y = M_y = M_{xy} = 0$ from (3.5), (3.8), and (3.9), the state outputs outside the straight line edge exclusive the vertices are obtained for each subplate by providing the additional conditions $Q_x = 0$. It is worth noting that the additional condition $Q_x = 0$ is given to provide the state outputs d_1 and d_2 for EB and WB, respectively, in the subplate I of Fig. 3. We now consider vertices of these three subplates by following the above results obtained. Due to the gyrator coupling between the loops with currents $\bar{i}_4$ and $\bar{i}_{11}$ that, respectively, determine the plate variables Q_y and w_y, taking sums and differences of (3.5) and (3.7) yields

$$d_4 - d_3 = 0, \quad d_9 + d_{10} + d_{11} + d_{12} - d_{s5} = 0. \quad (3.10)$$

Combining the given boundary conditions (3.8) and (3.9) with two additional conditions $Q_x = w_y = 0$, which result in (3.4) and (3.7), one is able to assemble these three subplates into a unit of Mindlin plate so that the state outputs appeared at each set of vertices (v_i, v_j, v_k) for the plate are obtained.

IV. NUMERICAL RESULTS

In this section, the implementation of the MDWDF algorithm is used to demonstrate the plane wave propagation of Mindlin system with free edges boundary conditions. The material body used for the simulation is an isotropic square elastic plate made of steel with side lengths $l_x = l_y = 1m$ and the plate thickness is fixed by $h = 10cm$. The

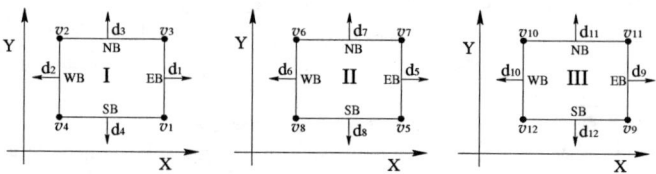

Fig. 3. Decomposed rectangular Mindlin plates with boundaries calculated by state output on edges and vertices.

material parameters are thus given by $\rho = 7.85 \times 10^4 kg/m^3$, $E = 190 GPa (GPa = 10^9 N/m^2)$, and $\nu = 0.3$. The shear correction factor κ^2 is 5/6, and the grid spacing is set to $\hbar = 4cm$. The transverse velocity is initially centered at $(l_x/3, l_y/3)$ and takes the form

$$v(x,y,0) = \cos(\sqrt{x^2+y^2})/\sqrt{x^2+y^2}.$$

Using the golden section search method, the results of minimal ratio $\hbar/T_t = 2630.2$ as compared to the CFL bound $\sqrt{2}\mu_0 = 1245.6$ according to the material body used, are applied to the MDWDF algorithm for the optimal performance of maintaining the passivity and stability of Mindlin systems. Considering different time slots t_k, the graphic illustrations shown in Fig. 4 have given a clear picture of the plane wave propagation in terms of the displacement including energy reflection and transmission from and across the boundary. Similar results are also to be found in [12] where the digital waveguide network approach was used.

V. CONCLUSIONS

Utilizing the principles of the MDWDF technique, we presented the development of a model for Mindlin plate vibration, described by a 4th order system of PDEs, and its subsequent simulation in the plane wave propagation. In particular, the resulting algorithm required only local interconnections for which the behaviour at any point in space is directly related to the points in its nearest neighbourhood. This property is thus very suited to develop different types of boundary conditions encountered in practice. The 4th order system we have investigated is of sufficient complexity to enable the effective consideration of a suitable parallel implementation. This is the subject of ongoing research.

ACKNOWLEDGMENT

The authors would like to acknowledge support for this work in part by EPSRC(UK) under grant no. GR/N22298 and in part by NSC(Taiwan) under grant no. NSC 93-2213-E-218-046.

REFERENCES

[1] A. Fettweis and G. Nitsche, "Transformation approach to numerical integrating PDEs by means of WDF principles", Multidimensional Sys. Sig. Proc. Vol. 2, pp. 127-159, 1991.
[2] A. Fettweis and G. Nitsche, "Numerical integration of partial differential equations using principles of multidimensional wave digital filters", J. of VLSI Sig. Proc., Vol. 3, pp. 7-24, Jun., 1991.
[3] H. Krauss, R. Rabenstein, and M. Gerken, "Simulation of wave propagation by multidimensional digital filters", J. of Sim. Prac. Theory, Vol. 4, pp. 361-382, 1996.
[4] C. H. Tseng, S. S. Lawson, "Incorporating initial and boundary conditions into the multidimensional wave digital filter", IEEE ICASSP 2002, May 2002, Orlando, USA.
[5] Computer Solutions Europe AB, Partial Differential Equation Toolbox, The Math Works, Inc., 1996.
[6] K. F. Graff, Wave Motion in Elastic Solids. Dover, New York, 1975.
[7] H. Huang, Static and Dynamic Analyses of Plates and Shells. Springer-Verlag, London, U.K., 1989.
[8] K. Liew, C. Wang, Y. Xiang, and S. Kitipornchai, Vibration of Mindlin plates: Programming the p-version Ritz Method. Elsevier, Amsterdam, The Netherlands, 1st. edition, 1998.
[9] A. A. Samarskij, Theorie der Differenzenverfahren, Leipzig: Akademische Verlagsgesellschaft, 1984 in German.
[10] A. Fettweis, "Wave digital filters: Theory and practice", Proceedings of IEEE, Vol. 74, pp. 270-327, 1986.
[11] C. H. Tseng, S. S. Lawson, "Initial and boundary conditions in multidimensional wave digital filter algorithms for plate vibration", IEEE Trans. Cirs. Sys.-I: Regular papers, Vol. 51, No. 8, pp. 1648-1663, 2004.
[12] S. D. Bilbao, Wave and Scattering Methods for the Numerical Integration of Partial Differential Equations, Ph.D thesis, Stanford University, 2001.

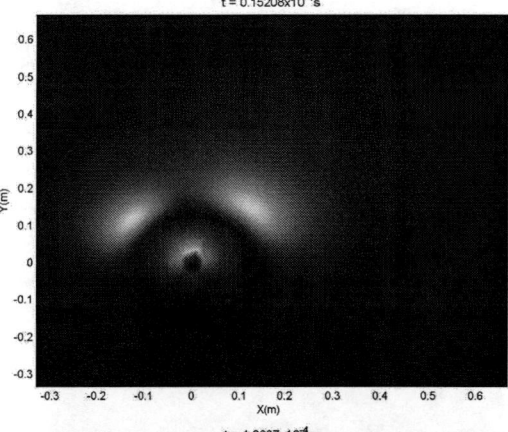
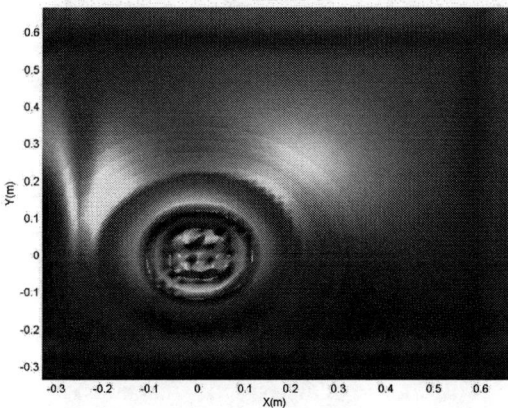
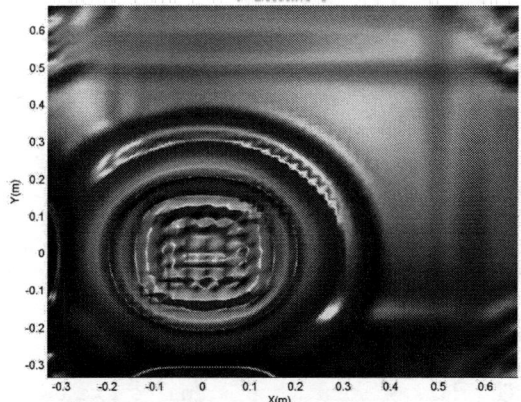
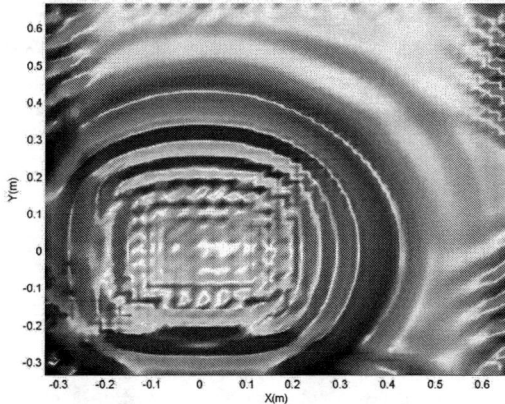

Fig. 4. Plane wave propagation in terms of the displacement for an elastic thick square plate made of steel at different time points.

A COST-EFFECTIVE MEMORY-BASED REAL-VALUED FFT AND HERMITIAN SYMMETRIC IFFT PROCESSOR FOR DMT-BASED WIRE-LINE TRANSMISSION SYSTEMS

Hsiang-Feng Chi
Department of Communication Engineering
National Chiao-Tung University
Hsin-Chu, Taiwan
hfchi@faculty.nctu.edu.tw

Zhao-Hong Lai
Department of Communication Engineering
National Chiao-Tung University
Hsin-Chu, Taiwan
jhlai.cm91g@nctu.edu.tw

Abstract—This paper presents an efficient computation scheme for the memory-based FFT/IFFT processor used in DMT (discrete multi-tone) systems. Only half-size FFT/IFFT is required to transform real-valued data and Hermitain symmetric data. That is, the cost in processing elements and memory can be reduced by two. Finally, a variable-size radix-4 memory-based FFT/IFFT processor with block scaling scheme is designed for DMT systems.

I. INTRODUCTION

As the multi-carrier modulation scheme becomes popular, Fast Fourier transform (FFT) and Inverse Fast Fourier transform (IFFT) are indispensable functions in OFDM-based and DMT-based transmission systems. To provide high-bandwidth internet access, the broadband FDD-based multi-carrier systems require the large-size FFT/IFFT processors, which usually consume most of the silicon area in the integrated circuit. In the DMT-based system, like VDSL, the Hermitian symmetric property of real-valued FFT enables us to develop a reduced-complexity FFT/IFFT processor architecture. As a result, a half-size complex FFT/IFFT can be used to conduct the transform between a real-valued sequence and a Hermitian symmetric sequence at the minor extra expense on the post-processing/pre-processing. In this paper, we start with the real-valued FFT (RFFT) and Hermitian symmetric IFFT (HSIFFT) algorithms in Section II. The radix-4 memory-based FFT/IFFT processor architecture is applied to the half-size FFT/IFFT and the post-processing/pre-processing in Section III. In Section IV, the hardware cost and the timing requirement are analyzed. A conclusion is made in Section V.

II. REAL-VALUED FFT AND HERMITIAN SYMMETRIC IFFT ALGORITHMS

It has been known that the real-valued FFT can be completed by using a half-size FFT [1, 2, 3]. Different from the previously proposed real-FFT algorithm, our RFFT and HSIFFT computation schemes are aimed for the memory-based FFT processor. Representing all operations in the form of butterfly and shuffle, the resulting RFFT and HSIFFT algorithms can easily operate upon a butterfly-based processing element (PE) and shared memory banks.

A. Real-valued FFT (RFFT)

The FFT of a N-point sequence $x(n)$ is

$$X(k) = \sum_{n=0}^{N-1} x(n) W_N^{nk}, \quad k = 0, 1, \cdots, N-1 \quad (1)$$

where $W_N = \exp(-j2\pi)$.

From the property of discrete Fourier transform, it follows that $x(n)$ is real if and only if $X(k)=X^*(-k)=X^*(N-k)$. In this paper, we consider the radix-4 FFT/IFFT processor. Two cases of transform sizes should be considered, separately.

Case-I: $N = 2 \times 4^M$:

Similar to [1, 2, 3], we can calculate the RFFT of $x(n)$ as follows.

1) Form an ($N/2$)-long complex sequence $y(n)$, in which the real part and the imaginary part are the even-indexed $x(n)$ and the odd-indexed $x(n)$, respectively
$$y(n) = x(2n) + j \cdot x(2n+1), \quad 0 \leq n \leq N/2-1 \quad (2)$$
2) Perform size-$N/2$ FFT of $y(n)$ to obtain to obtain $Y(k)$.
3) Calculate $H(k)$ and $G(k)$ as, $0 \leq k \leq N/2$
$$H(k) = (Y(k) + Y^*(N/2 - k))/2 \quad (3a)$$
$$G(k) = (Y(k) - Y^*(N/2 - k))/(2j) \quad (3b)$$
4) The FFT of $x(n)$ is calculated as the butterfly, $0 \leq k \leq N/2$
$$\begin{pmatrix} X(k) \\ X^*(N/2-k) \end{pmatrix} = \begin{pmatrix} 1 & 1 \\ 1 & -1 \end{pmatrix} \cdot \begin{pmatrix} H(k) \\ W_N^k G(k) \end{pmatrix} \quad (4)$$

Case-II: $N = 4^M$:

Likewise, we can calculate RFFT as follows.
1) Form two ($N/4$)-long complex sequence $y_0(n)$ and $y_1(n)$ as, $0 \leq n \leq N/4-1$
$$y_0(n) = x(4n) + j \cdot x(4n+1) \quad (5a)$$
$$y_1(n) = x(4n+2) + j \cdot x(4n+3) \quad (5b)$$
2) Perform two size-$N/4$ FFT of $y_0(n)$ and $y_1(n)$ to obtain $Y_0(k)$ and $Y_1(k)$, respectively.
3) Calculate $H_0(k)$, $H_0(k)$, $G_0(k)$ and $G_1(k)$ as, $0 \leq k \leq N/4$

$$G_0(k) = (Y_0(k) + Y_0^*(N/4-k))/2 \quad (6a)$$
$$G_1(k) = (Y_0(k) - Y_0^*(N/4-k))/(2j) \quad (6b)$$
$$H_0(k) = (Y_1(k) + Y_1^*(N/4-k))/2 \quad (6c)$$
$$H_1(k) = (Y_1(k) - Y_1^*(N/4-k))/(2j) \quad (6d)$$

4) The FFT of $x(n)$ is calculated as the butterfly, $0 \le k < N/4$

$$\begin{pmatrix} X(k) \\ X^*(N/4-k) \\ X(N/4+k) \\ X^*(N/2-k) \end{pmatrix} = \begin{pmatrix} 1 & 1 & 1 & 1 \\ 1 & j & -1 & j \\ 1 & -j & -1 & j \\ 1 & -1 & 1 & -1 \end{pmatrix} \cdot \begin{pmatrix} G_0(k) \\ W_N^k G_1(k) \\ W_N^{2k} H_0(k) \\ W_N^{3k} H_1(k) \end{pmatrix} \quad (7)$$

In summary, when N is of the form 2×4^M, the RFFT can be completed by an $N/2$-point complex-valued FFT followed by the post-processing described by (3) and (4). When N is of the form 4^M, the RFFT can be completed by two $N/4$-point complex-valued FFT followed by the post-processing described by (6) and (7).

B. Hermitian Symmetric IFFT (HSIFFT)

The IFFT of a N-point sequence $X(k)$ is

$$x(n) = \sum_{k=0}^{N-1} X(k) W_N^{-nk}, \quad n = 0,1,\cdots,N-1 \quad (8)$$

Likewise, the HSIFFT can be completed by reversing the computation of RFFT. Thus, when N is of the form 2×4^M HSIFFT can be completed by an $N/2$-point complex-valued IFFT preceded by a pre-processing. When N is of the form 4^M, the HSIFFT can be completed by two $N/4$-point complex-valued IFFT preceded by a pre-processing.

III. RADIX-4 MEMORY-BASED RFFT/HSIFFT PROCESSOR ARCHITECTURE

Figure 1 shows the signal flow graphs (SFG) of RFFT and HSIFFT. Note that both pre-processing and post-processing should have linear order input and linear order output. In RFFT, the input of the half-size FFT should be digit-reversed order and the output should be linear order. Contrarily, in HSIFFT, the input of the half-size FFT should be linear order and the output should be digit-reversed order. The proposed radix-4 memory-based RFFT/HSIFFT processor would conduct the computation serially based on the SFG shown in Figure 1. Figures 1(a) and 1(b) shows that the RFFT and HSIFFT are completed by the half-size FFT/IFFT followed and preceded by post-processing and pre-processing, respectively, when N is of the form 2×4^M. The similar SFG can be obtained for the case $N = 4^M$.

Figures 2(a) and 2(b) are the SFGs of size-4^M radix-4 FFT and IFFT. The shuffle interconnection networks are represented as permutation matrix as follows:

$$\mathbf{I}_{4^{M-1-i}} \otimes \mathbf{S}_{4^i,4} \quad i = 0,1,\cdots,M-1$$

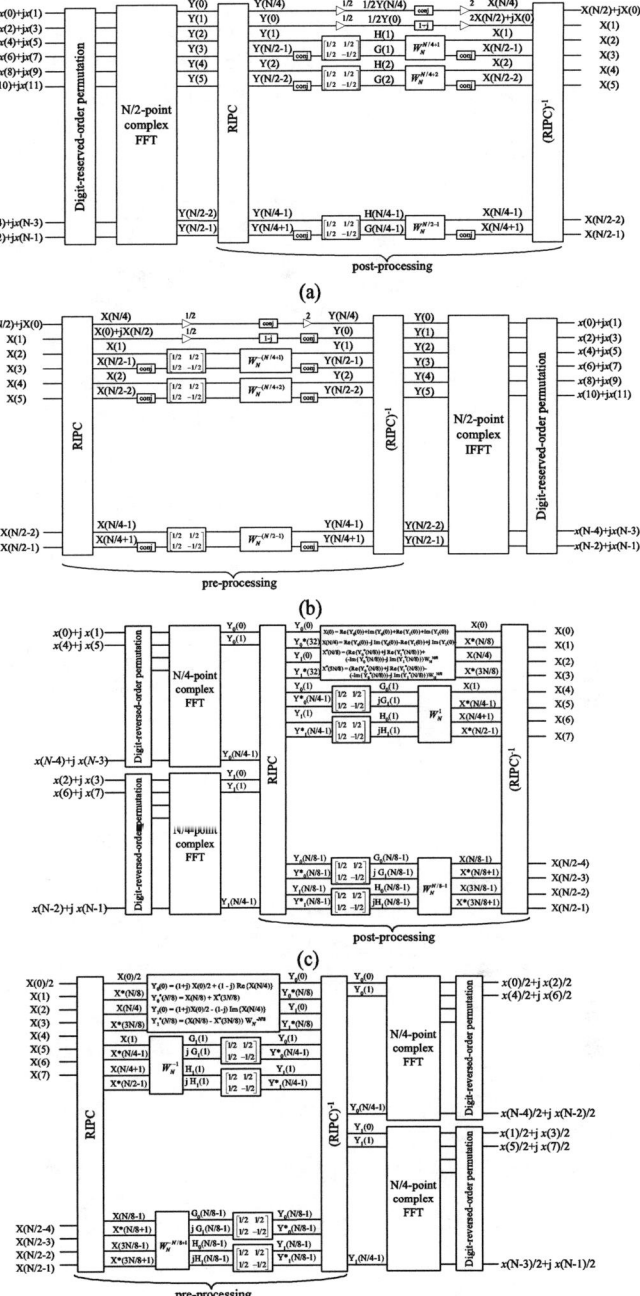

Figure 1: (a) SFG of RFFT for $N=4^M$. (b) SFG of HSIFFT for $N=4^M$. (c) SFG of RFFT for $N=2\times 4^M$. (d) SFG of HSIFFT for $N=2\times 4^M$.

where '$\otimes$' is Kronecker product, and $\mathbf{S}_{4^i,4}$ is a size-4^{i+1} permutation matrix of radix-4 perfect shuffle.

The block RIPC performs "Reverse-Interleave Permutation and Conjugation" described as the permutation from a size-$N/2$ array $\{A(m) : 0 \le m \le N/2-1\}$ to a size-$N/2$ array $\{B(m) : 0 \le m \le N/2-1\}$ as

$$B(m) = \begin{cases} A(i) & ,m=2i \text{ for } 0 \le i \le N/4-1 \\ A^*(N/2-i) & ,m=2i+1 \text{ for } 1 \le i \le N/4-1 \\ A^*(N/4) & ,m=1 \end{cases}$$

Note that the permutations on the both sides of butterfly PEs are mutually inversed. It results in the in-place computation, which is essential for the memory-based processing. The shuffle permutation can be achieved by memory access at addresses produced from a shuffle address generator, which will be described later.

Two address generators are designed for accessing data and coefficients. The data address generators DAG and R_DAG produce the addresses for shuffle permutation and RIPC, respectively, used on data memory access. The coefficient address generator (CAG) produces the addresses for coefficient memory, where the twiddle factors are stored.

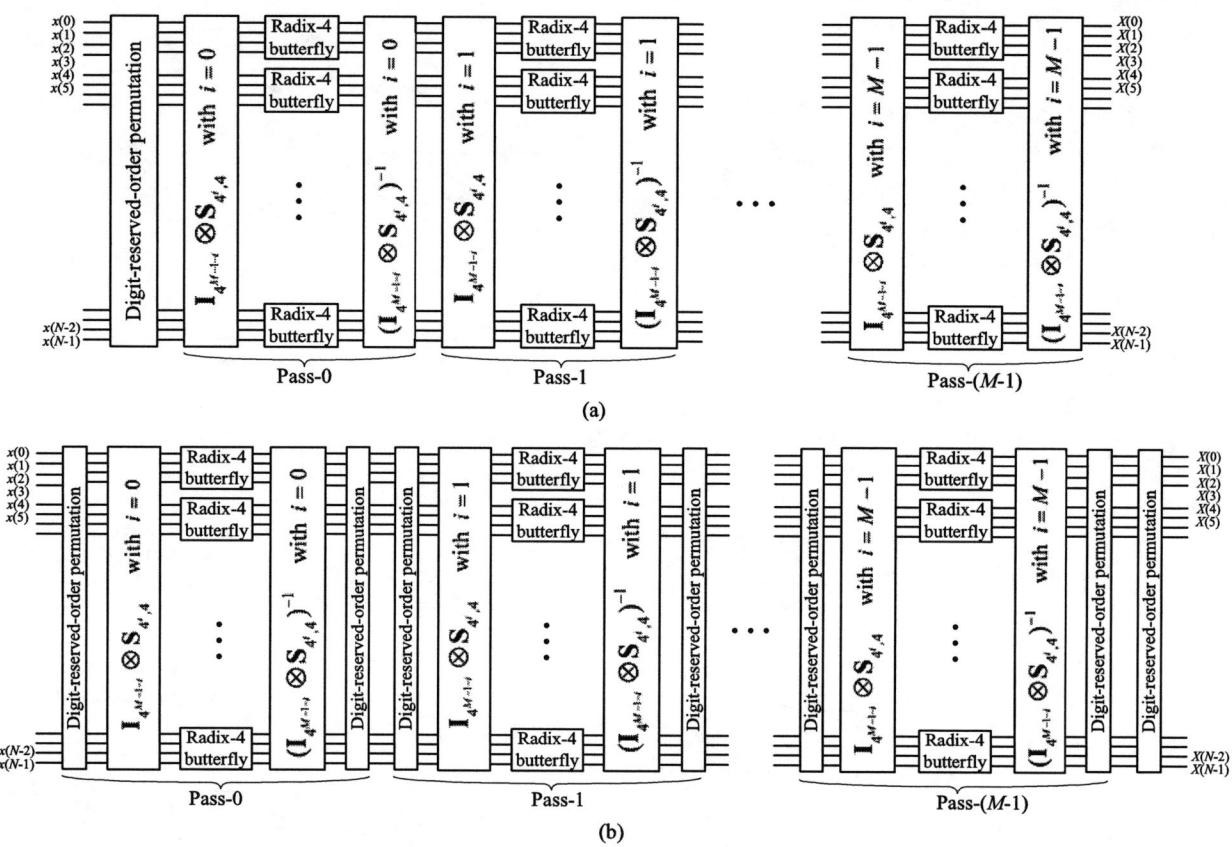

Figure 2: (a) Radix-4 decimation-in-time FFT SFG. (b) Radix-4 decimation-in-time IFFT SFG.

Figure 3 shows the proposed radix-4 memory-based RFFT/HSIFFT processor architecture. We use the block floating point (BFP) method [4] to improve the precision.

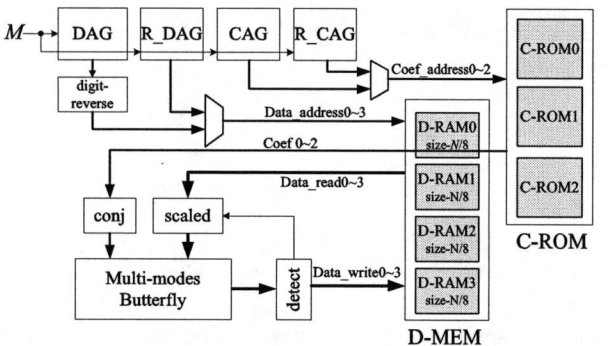

Figure 3: The proposed radix-4 memory-based RFFT/HSIFFT architecture with BFP

In the memory-based processor architecture, the address generators are designed to produce addresses for memory modules in order to retrieve and store data for butterfly PEs.

When being used for the shuffle permutation [5], the DAG can be implemented as a digit-rotated counter.

Three twiddle factors are used for a radix-4 butterfly. Three addresses are required to access coefficients from three coefficient ROMs. During the normal FFT/IFFT computation, the three addresses generated by CAG are described in (9). Figure 4 shows an example of C_address0 of radix-4 FFT for $N = 64$. To make the RFFT/HSIFFT processor operate with a variable transform size. The parameter M should be programmable in the address generation.

$$C\_address0 = \{00, \text{linear\_counter} \ll 2M - 2i - 2\} \quad (9a)$$
$$C\_address1 = \{00, \text{linear\_counter} \ll 2M - 2i - 2\} \ll 1 \quad (9b)$$
$$C\_address2 = C\_address0 + C\_address1 \quad (9c)$$
$$i = 0, 1, \cdots, M - 1$$

In order to increase the memory access bandwidth, it is necessary to partition the memory into four banks. However, the memory-conflict problem [6] occurs when the data of the butterfly PE are accessed from the same memory bank.

We can solve the memory-conflict problem by arranging the data address $(d_{M-1}, d_{M-2}, ..., d_2, d_1, d_0)_4$ as follows.

$$\text{bank\_id} = (d_{M-1} + d_{M-2} + ... + d_2 + d_1 + d_0) \bmod 4$$
$$\text{address(in bank)} = (d_{M-2}, d_{M-3}, ..., d_1, d_0)_4$$

where, $M = \log_4 N$

address	C-ROM	Stage-0	Stage-1	Stage-2	extend	liner counter
000000	W_{64}^0	000000	000000	000000	00	0000
000001	W_{64}^1	000000	000100	000001	00	0001
000010	W_{64}^2	000000	001000	000010	00	0010
000011	W_{64}^3	000000	001100	000011	00	0011
000100	W_{64}^4	000000	000000	000100	00	0100
000101	W_{64}^5	000000	000100	000101	00	0101
000110	W_{64}^6	000000	001000	000110	00	0110
000111	W_{64}^7	000000	001100	000111	00	0111
001000	W_{64}^8	000000	000000	001000	00	1000
001001	W_{64}^9	000000	000100	001001	00	1001
001010	W_{64}^{10}	000000	001000	001010	00	1010
001011	W_{64}^{11}	000000	001100	001011	00	1011
001100	W_{64}^{12}	000000	000000	001100	00	1100
001101	W_{64}^{13}	000000	000100	001101	00	1101
001110	W_{64}^{14}	000000	001000	001110	00	1110
001111	W_{64}^{15}	000000	001100	001111	00	1111

Figure 4: An example of *C-address*0 for *N*=64.

To use a single radix-4 butterfly PE in the RFFT/HSIFFT processor, we need a multi-modes butterfly PE. The operation mode of the PE includes the normal radix-4 butterfly operation for regular FFT/IFFT, the dual radix-2 butterfly operations, and the radix-4 operations for the post-processing/pre-processing.

IV. MEMORY AND TIMING BUDGETING

Based on the aforementioned in-place computation schemes and memory partition, we need only single-port memory in the RFFT/HSIFFT processor. We summarize the memory requirement of different transform sizes in Table 1. W is the word length used as internal precision. 2W bits are required for a complex word.

In the proposed memory-based RFFT/HS-IFFT processor, *N*/2 data are partitioned and stored into four size-*N*/8 single-port memory modules. Data-reading and data-writing upon single-port memory cannot be executed at the same time. It follows that two clock cycles are consumed to complete a butterfly computation. The three complex multiplications can be operated upon three real multipliers within two clock cycles. Table 1 also shows the requirement of the clock cycles for different transform sizes.

Table 2 shows FFT/IFFT sizes and sampling rates for different DMT-based wire transmission systems. According to Table 1, the listed operating frequency must be fulfilled for the proposed memory-based RFFT/HS-IFFT processor in order to provide adequate throughputs the operating frequency is calculated as follows.

$$\text{operating frequency} = \frac{\text{sampling rate} \times \text{require\_clock\_cycles}}{(\text{transform\_size} + \text{cyclic\_prefix})}$$

Transform size	Radix-4 RFFT/HS-IFFT			
	# of real multipliers	Memory spec. (depth x width)	# of memory modules	Required cycles
8192	3	1Kx2W	x4	16384
4096	3	512x2W	x4	12288
2048	3	256x2W	x4	3854
1024	3	128x2W	x4	2560
512	3	64x2W	x4	768
256	3	32x2W	x4	512
128	3	16x2W	x4	160

Table 1: Memory and timing requirement of the RFFT/HSIFFT processor when different transform sizes are applied

Transmission System	RFFT/HSIFFT Size	Sampling Rate	Radix-4 processor clock rate
HomePlug 1.0	256	50 MHz	60 MHz
ADSL (CPE)	512	2.2 MHz	3.3 MHz
VDSL	8192	34.5 MHz	69 MHz
	4096	17.3 MHz	51.9 MHz
	2048	8.6 MHz	16.2 MHz
	1024	4.3 MHz	10.75 MHz
	512	2.2 MHz	3.3 MHz

Table 2: FFT/IFFT size and sampling rate for different DMT systems

V. CONCLUSION

To meet the broadband requirement in DMT-based systems, large-size FFT and IFFT are inescapable functions in high bandwidth transmission systems. The short symbol period makes FFT/IFFT timing-critical calculation and results in expensive hardware in the integrated circuit design. Fortunately, the real and the Hermitian symmetric properties can be utilized to reduce the complexity. We have designed a variable-size RFFT/HS-IFFT processor, in which almost a half number of clock cycles and a half memory are saved. The proposed RFFT/HS-IFFT processor can be applied to most of the current DMT-based broadband wire-line transmission systems without using high operation clock rates. The processor with the maximal transform size 8192 and 14-bit internal precision (W=14) has been implemented on an Altera Stratix EP1S25 FPGA board running at 40 MHz clock rate for verification purpose.

VI. REFERENCES

[1] Henrik V. Sorensen, Douglas L. Jones, C. Sidney Burrus, and Michael T. Heideman, "Real-Valued Fast Fourier Transform Algorithm," *IEEE Trans. Acoust, Speech, and Sig. Proc.*, vol. ASSP-35, no. 6, June 1987.

[2] J. B. Martens, "Discrete Fourier transform algorithms for real valued sequence," *IEEE Trans. Acoust., Speech, Sig. Proc.*, vol. ASSP-32, pp. 390-396, Apr. 1984.

[3] H. Ziegler, "A fast Fourier transform algorithm for symmetric real-valued series," *IEEE Trans. Audio Electroacoust.* vol. AU-20, pp. 353-356, Dec. 1972.

[4] E. Bidet, D. Castelain, C. Joanblanq, P. Senn, "A fast single-chip implementation of 8192 complex point FFT", *IEEE J. Solid-State Circuits*, Vol. 30, pp. 300 – 305, Mar. 1995

[5] Marc Davio, "Kronecker Products and Shuffle Algebra," *IEEE Trans. Computers*, vol. C-30, no. 2, February 1981.

[6] L. G. Jonhson, "Conflict Free Memory Addressing for Dedicated FFT Hardware," *IEEE Trans. Circuit and System-II: Analog and Digital Sig. Proc.*, vol. 39 no. 5, pp. 312-316, May 1992.

Defining Correlation Functions and Power Spectra for Multirate Random Processes

Charles W. Therrien
Department of Electrical and Computer Engineering
Naval Postgraduate School
Monterey, California 93943–5000
Email: therrien@nps.edu

Abstract—This paper proposes a representation for the time-lag cross-correlation function of two random processes sampled at different rates and its related cross-power density spectrum, using the theory of lattices. The representation of autocorrelation and the (auto)power spectral density function then follows directly.

I. INTRODUCTION

The time-lag cross-correlation function for two (continuous-time) random processes is defined as

$$R_{xy}(t;\tau) = E\left\{x(t)y^*(t-\tau)\right\} \quad (1)$$

while what will be called the "traditional" cross-correlation function is defined as

$$\mathrm{R}_{xy}(t,t') = E\left\{x(t)y^*(t')\right\} \quad (2)$$

The time-lag form has a number of advantages. Among these are that it is easily related to concepts of the spectrogram, various stochastic time-frequency distributions (Wigner, Rihaczek) as well as the time-frequency correlation or "ambiguity" function and cyclic correlation and power density for cyclostationary (or "periodically correlated") random processes [1], [2]. The time-lag form also leads to a clear distinction between random processes that are wide-sense stationary (wss), those that are wide-sense cyclostationary (wscs), and those that are periodic. Stationary processes are independent of t, cyclostationary processes are periodic in t but not in τ, and periodic processes are periodic in both variables.

The analysis of random processes sampled at different rates from a statistical point of view leads to the concept of joint cyclostationarity between these processes. The representation for the time-lag cross-correlation function and cross-spectra of two such random processes, seems to have eluded researchers in signal processing, however, and has impeded research for statistical multirate signal processing. We believe that such representation is fundamental to the statistical analysis of multirate processes and is necessary for further progress in the field.

II. CROSS-CORRELATION

Consider two continuous-time random processes $x(t)$ and $y(t')$ as illustrated in Fig. 1. Beginning at $t = t' = 0$, these two processes are sampled at rates of F_x and F_y samples/second with sampling intervals of $T_x = 1/F_x$ and $T_y = 1/F_y$. It

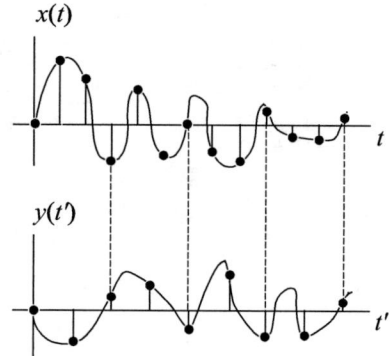

Fig. 1. Continuous random processes sampled at different rates. (Note that the samples do not line up in time except at points indicated by the dashed lines.)

is reasonable to assume that the sampling rates are integer-valued. We further assume that the samples "line up" at multiples of T, where T is called the *common period* and is defined by the smallest integers M_x and M_y satisfying

$$T = M_x T_x = M_y T_y \quad (3)$$

With some abuse of notation, we denote the resulting discrete-time signals as $x[m_x] = x(m_x T_x)$ and $y[m_y] = x(m_y T_y)$ where m_x and $m_y \in \mathbb{Z}$ (the set of integers).

The traditional and time-lag forms of the cross-correlation function for the continuous-time random processes are defined by (1) and (2) above with the relation

$$\tau = t - t' \quad (4)$$

The definitions apply for all $t, t', \tau \in \mathbb{R}$ (the real numbers).

It is straightforward and intuitive to define a discrete cross-correlation function for the two sampled random processes in the traditional form as

$$\mathrm{R}_{xy}[m_x, m_y] = E\left\{x[m_x]y^*[m_y]\right\} = \mathrm{R}_{xy}(m_x T_x, m_y T_y)$$

This cross-correlation function is defined on a rectangular lattice [3] denoted by Λ_R and depicted in Fig. 2. The generating matrix for the lattice is given by

$$\mathbf{V}_\mathrm{R} = \begin{bmatrix} \mathbf{v}_x & \mathbf{v}_y \end{bmatrix} = \begin{bmatrix} T_x & 0 \\ 0 & T_y \end{bmatrix}$$

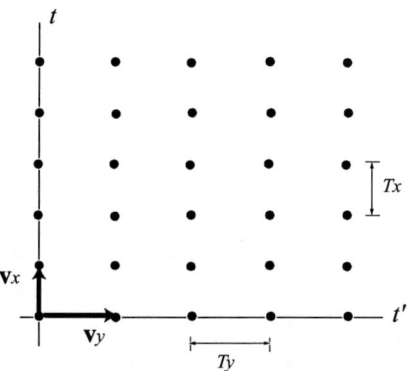

Fig. 2. Cross-correlation function $R_{xy}[m_x, m_y]$ defined on rectangular lattice Λ_R.

where $\mathbf{v}_x$ and $\mathbf{v}_y$ are the two basis vectors shown in the figure. A point on the lattice is thus represented by

$$\begin{bmatrix} t \\ t' \end{bmatrix} = \mathbf{V}_R \begin{bmatrix} m_x \\ m_y \end{bmatrix} = m_x \mathbf{v}_x + m_y \mathbf{v}_y \quad (5)$$

Now consider the representation of the time-lag correlation function $R_{xy}(t;\tau)$ in discrete time. The variable t is of the form $t = m_x T_x$. Also, τ can only take on values corresponding to the actual samples available. To see what values are allowed, we write (4) as

$$\tau = t - t' = m_x T_x - m_y T_y$$

Thus for any fixed value of m_x, the allowable values of τ are separated by integer multiples of T_y. The possible values for (t, τ) can be seen to fall on the (non-rectangular) lattice Λ_L depicted in Fig. 3. The generating matrix for this lattice

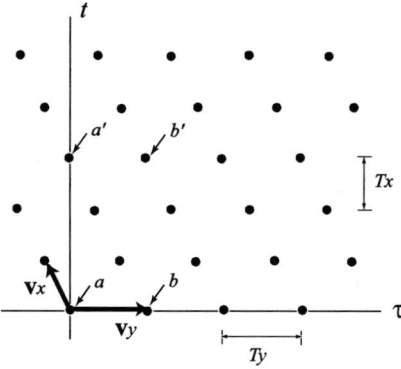

Fig. 3. Time-lag cross-correlation function $R_{xy}[m; l]$ represented on a lattice Λ_L.

is given by

$$\mathbf{V}_L = \begin{bmatrix} \mathbf{v}_x & \mathbf{v}_y \end{bmatrix} = \begin{bmatrix} T_x & 0 \\ T_x - T_y & T_y \end{bmatrix} \quad (6)$$

Any point on the lattice can be represented by coordinates

$$\begin{bmatrix} t \\ \tau \end{bmatrix} = \mathbf{V}_L \begin{bmatrix} m_x \\ l_y \end{bmatrix} = m_x \mathbf{v}_x + l_y \mathbf{v}_y \quad (7)$$

where $m_x, l_y \in \mathbb{Z}$.

Now, in order to represent the discrete, time-lag correlation function algebraically, observe that the variables t, t' and τ are related by

$$\begin{bmatrix} t \\ t' \end{bmatrix} = \mathbf{T} \begin{bmatrix} t \\ \tau \end{bmatrix} = \begin{bmatrix} 1 & 0 \\ 1 & -1 \end{bmatrix} \begin{bmatrix} t \\ \tau \end{bmatrix} \quad (8)$$

The matrix $\mathbf{T}$ is an *involutory* matrix ($\mathbf{T}^{-1} = \mathbf{T}$). Using (5), (8), and (7) in that order, produces

$$\begin{bmatrix} m_x \\ m_y \end{bmatrix} = \mathbf{V}_R^{-1} \mathbf{T} \mathbf{V}_L \begin{bmatrix} m_x \\ l_y \end{bmatrix}$$

It is easily shown by carrying out the matrix multiplication that

$$\mathbf{V}_R^{-1} \mathbf{T} \mathbf{V}_L = \begin{bmatrix} 1 & 0 \\ 1 & -1 \end{bmatrix} = \mathbf{T}$$

Thus it is seen from the previous equation that $m_y = m_x - l_y$. We can therefore define the discrete, time-lag cross-correlation function as

$$\boxed{R_{xy}[m_x; l] \stackrel{\text{def}}{=} E\{x[m_x] y^*[m_x - l]\}} \quad (9)$$

where we have dropped the subscript on the lag variable to simplify notation.

The algebraic definition of cross-correlation in the multirate case is seen to be the same as would be used for processes with a single sampling rate, but *the interpretation is different*. As seen above, the variables m_x and l specify a position on a lattice Λ_L. This lattice degenerates to a rectangular lattice only in the case of identical sampling rates ($T_x = T_y$), where the generating matrix (6) becomes diagonal.

Taking a similar approach, one can define the cross-correlation function R_{yx} as

$$R_{yx}[m_y; l'] \stackrel{\text{def}}{=} E\{y[m_y] x^*[m_y - l']\}$$

Observe that this function is defined on its own lattice $\Lambda_{L'}$ and that the lag variable l' is distinct from the lag variable l in R_{xy}. The relation between these two cross-correlation functions and the traditional forms of cross-correlation is given by

$$\begin{align} R_{xy}[m_x; l] &= R_{xy}[m_x, m_x - l] \quad (10) \\ &= R_{yx}^*[m_x - l, m_x] = R_{yx}^*[m_x - l; -l] \end{align}$$

All four of the forms are equivalent in the sense that any one of them can be used to specify all the possible correlation values that exist between the two sampled random processes.

III. INDEXED CORRELATION FUNCTION

While the representation (9) is quite general, an important property is not directly apparent. That property has to do with *stationarity* (or more generally *cyclostationarity*, if it exists) of the underlying continuous random processes. Refer to Fig. 3 again, and assume that the random processes $x(t)$ and $y(t')$ are jointly wide-sense stationary (wss). In this case the continuous-time cross-correlation function $R_{xy}(t;\tau)$ is a function of τ alone. Thus $R_{xy}(t;\tau)$ is a surface that resembles a tunnel with its axis parallel to the t axis. The contours of equal correlation are straight lines parallel to the t axis. Because the function is constant in t, the periodic sampling of the lattice induces a periodicity in correlation. For example, the points a' and b' shown in Fig. 3 have the same value of correlation as the points a and b because they are at the same distance from the t axis. Further, observe that this periodicity is *not* two-dimensional. The periodicity occurs along only one direction, namely that of the t axis.

The periodicity of the cross-correlation function $R_{xy}[m_x; l_y]$ is not immediately evident from (9) because the lattice basis vectors do not align with the t axis. It can also be shown algebraically, however, as follows. If $x(t)$ and $y(t')$ are jointly wss then (9) can be written as

$$\begin{aligned} R_{xy}[m_x; l_y] &= E\{x[m_x]y^*[m_x - l_y]\} \\ &= R^c_{xy}(m_x T_x - (m_x - l_y)T_y) \end{aligned}$$

where R^c_{xy} is written here with a single argument (τ) which represents the difference of the sample times. To demonstrate the periodicity, let $m'_x = m_x + M_x$ and $l'_y = l_y + M_x - M_y$ where M_x and M_y are defined by (3). Then

$$\begin{aligned} R_{xy}[m'_x; l'_y] = &R^c_{xy}(m_x T_x - m_x T_y + l_y T_y \\ &+ \underline{M_x T_x - M_x T_y + (M_x - M_y)T_y}) \\ &\text{added terms} \end{aligned}$$

By virtue of (3), the added terms sum to 0 and the periodicity is demonstrated.

For the case of wss processes, the variables m_x and l_y can therefore be confined to the first period of the lattice (i.e., the period containing the point $m_x = l_y = 0$). That is, given arbitrary values for the variables m_x and l_y, we can find equivalent values m and l in the first period of the lattice in the sense that $R_{xy}[m; l] = R_{xy}[m_x; l_y]$. To confine the indices to the first period, choose m as

$$m = m_x - PM_x \quad (11)$$

where $P = \lfloor m_x/M_x \rfloor$. In other words, P is the number of full periods covered by m_x and $m \equiv m_x \mod M_x$. Then it follows from the arguments in the previous paragraph that choosing

$$l = l_y - P(M_x - M_y) \quad (12)$$

forces $R_{xy}[m; l]$ and $R_{xy}[m_x; l_y]$ to have the same values.

The equivalent correlation function defined by

$$\boxed{\begin{aligned} R_{xy}[m; l] &= R^c_{xy}(m(T_x - T_y) + l\,T_y) \\ &0 \leq m \leq M_x - 1, \quad -\infty < l < \infty \end{aligned}} \quad (13)$$

will be called an "indexed" correlation function because it can be viewed as a family of correlation functions of a single variable (l) indexed by m.

IV. Autocorrelation, Mean, and Covariance

For a single random process $x[n]$, it is natural to define the discrete, time-lag *auto*correlation function as

$$R_x[m_x; l] \stackrel{\text{def}}{=} E\{x[m_x]x^*[m_x - l]\} \quad (14)$$

Although this function is also defined on a lattice, the lattice is rectangular, since there is no diversity in the sampling rates. The second-moment statistics then are as listed in Table I. The

TABLE I
Listing of time-domain statistics for multi-rate signal processing.

Definition	Rate
$\mu_x[m] = E\{x[m]\}$, $\mu_y[m] = E\{y[m]\}$	F_x, F_y
$R_x[m; l] = E\{x[m]x^*[m - l]\}$	F_x
$R_{xy}[m; l] = E\{x[m]y^*[m - l]\}$	mixed

covariance can be defined in an analogous way and is related to the other quantities as

$$\begin{aligned} C_x[m; l] &= R_x[m; l] - \mu_x[m]\mu_x^*[m - l] &\text{(a)} \\ C_{xy}[m; l] &= R_{xy}[m; l] - \mu_x[m]\mu_y^*[m - l] &\text{(b)} \end{aligned} \quad (15)$$

Discrete time processes with constant or periodic mean, and with correlation functions of the form (13) are said to be *jointly wide-sense cyclostationary* (wscs).

V. Spectral Representation for Jointly WSS Processes

A. Cross-Power Spectral Density

The two-dimensional (2-D) cross-power spectral density function of the continuous signals $x(t)$ and $y(t)$ can be defined as

$$S^c_{xy}(f', f) = \int_{-\infty}^{\infty} \int_{-\infty}^{\infty} R_{xy}(t; \tau) e^{-j2\pi f' t} e^{-j2\pi f \tau} dt\, d\tau \quad (16)$$

The following observations can be made:
1) If $x(t)$ and $y(t)$ are jointly wss, then R_{xy} is constant with respect to t. This implies that $S^c_{xy}(f', f)$ has support *only* on the f axis, where it is *singular*.
2) If $x(t)$ and $y(t)$ are jointly wscs with period P, then $R_{xy}(t; \tau)$ is periodic in t. This implies that $S^c_{xy}(f', f)$ has support on an infinite set of lines $f' = k \cdot 2\pi/P$ for $k \in \mathbb{Z}$.

These are the two cases of primary interest since wss or wscs processes sampled at different rates become jointly wscs.

It follows from the theory of sampling on a lattice (e.g., [4]) that the corresponding sampled version of $R_{xy}(t; \tau)$ has its spectrum repeated at points on a lattice in frequency space with generating matrix $\mathbf{U} = (\mathbf{V}_L^T)^{-1}$. This lattice is known as the *reciprocal* lattice of Λ_L. From (6), the generating matrix of the reciprocal lattice is given by

$$\mathbf{U} = \frac{1}{T_x T_y} \begin{bmatrix} T_y & T_y - T_x \\ 0 & T_x \end{bmatrix} = \begin{bmatrix} F_x & F_x - F_y \\ 0 & F_y \end{bmatrix} \quad (17)$$

Let us consider the case where the underlying processes $x(t)$ and $y(t)$ are jointly wss. In this case the spectrum, before sampling, has support only on the f axis, where it is singular (impulsive). The spectrum corresponding to the sampled correlation function thus has the spectrum repeated along lines parallel to the f axis with centers at points on the reciprocal lattice. This is illustrated in Fig. 4(a) where the basis vectors corresponding to (17) are depicted as $\mathbf{u}_x$ and $\mathbf{u}_y$. The parallelogram defines one period of the spectrum. Notice that the aliasing that normally occurs when the correlation function is sampled in one dimension is less severe because the spectral components are spread out in the plane. This is to be expected because the sampling in two dimensions effectively increases the sampling rate of $R_{xy}^c(\tau)$.

For the sampled signals the cross-power density spectrum can be defined as

$$S_{xy}^{(k)}(\omega) \stackrel{\text{def}}{=} \sum_{m=0}^{M_x-1} \sum_{l=-\infty}^{\infty} R_{xy}[m;l] e^{-j\omega l} e^{-j2\pi km/M_x} \quad (18)$$

and used in place of (16). This spectral density function is depicted in Fig. 4(b).

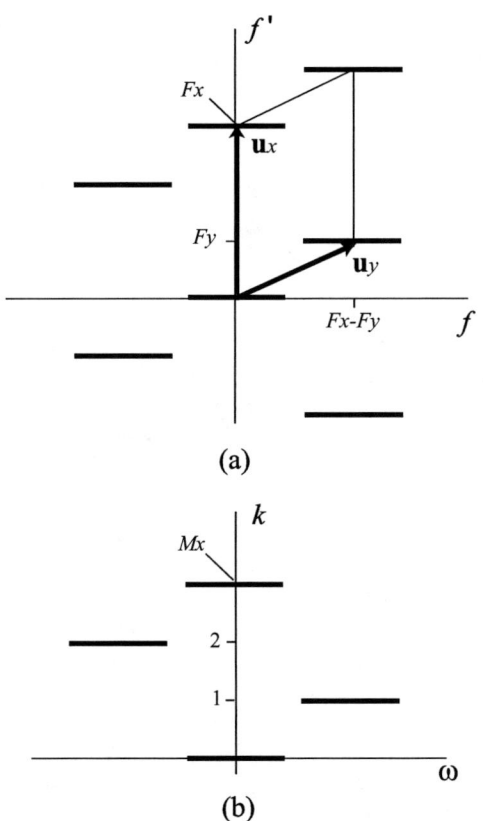

Fig. 4. Two-dimensional cross-spectral density function.

An alternative representation for jointly wss random processes with different rates is the time-dependent cross-power spectrum defined as

$$S_{xy}[m,\omega] = \sum_{l=-\infty}^{\infty} R_{xy}[m;l] e^{-j2\pi km/M_x}$$

This can be thought of as a set of one-dimensional spectra defined by transforming the indexed correlation function, and is analogous to the spectrogram.

B. Auto-Power Spectral Density

When the random process is wss, the autocorrelation function defined by Table I is independent of m. The power spectral density function is then defined in the usual way as the (one-dimensional) Fourier transform of the autocorrelation function. In our two-dimensional k,ω representation, this function has support only on the ω axis, where it is singular.

If x is wscs, then a two-dimensional cyclic spectrum $S_x^{(k)}(\omega)$ can be defined in analogy with the definition for $S_{xy}^{(k)}(\omega)$. In this definition R_{xy} is replaced by R_x, and M_x becomes the cyclic period. This function is periodic in two dimensions with centers occuring at points on a rectangular lattice defined by the sampling frequency F_x. In addition, this function has a positivity constraint which can be written as

$$S_x^{(k)}(\omega) \geq 0; \quad k=0$$

The term $S_x^{(0)}(\omega)$ can be thought of as the "stationary" component of the spectrum.

VI. DISCUSSION

The representation of the correlation function in the time-lag form (as opposed to the traditional time-time form) has important advantages for the analysis of random processes sampled at different rates. While the latter representation is straightforward, the former has been elusive. The correct definition is clear however, if one considers correlation as represented on a two-dimensional lattice. The corresponding spectral representation fits well with established descriptors for processes exhibiting cyclostationarity.

In future publications we will demonstrate how this representation naturally fits in to the extension of linear optimal filtering to the multirate case both in time and frequency domains. Some previous publications [5], [6] have addressed these problems without the benefit of the results discussed here.

REFERENCES

[1] L. L. Scharf, B. Friedlander, P. Flandrin, and A. Hanssen, "The Hilbert space geometry of the stochastic Rihaczek distribution," in *Proceedings of the 35th Asilomar Conference on Signals, Systems, and Computers*, November 2001, pp. 720–725, (Pacific Grove, CA).

[2] C. W. Therrien, "Some considerations for statistical characterization of nonstationary random processes," in *Proceedings of the 36th Asilomar Conference on Signals, Systems, and Computers*, November 2002, pp. 1554–1558, (Pacific Grove, CA).

[3] J. W. S. Cassels, *An Introduction to the Geometry of Numbers*. New York: Springer-Verlag, 1959.

[4] D. E. Dudgeon and R. M. Mersereau, *Multidimensional Digital Signal Processing*. Upper Saddle River, New Jersey: Prentice Hall, Inc., 1984.

[5] R. Cristi, D. A. Koupatsiaris, and C. W. Therrien, "Multirate filtering and estimation: The multirate Wiener filter," in *Proceedings of the 34th Asilomar Conference on Signals, Systems, and Computers*, November 2000, pp. 450–454, (Pacific Grove, CA).

[6] R. J. Kuchler and C. W. Therrien, "Optimal filtering with multirate observations," in *Proceedings of the 37th Asilomar Conference on Signals, Systems, and Computers*, November 2002, pp. 1208–1212, (Pacific Grove, CA).

Pilot Tone Design for Peak-to-Average Power ratio Reduction in OFDM

Shinji Hosokawa, Shuichi Ohno, Kok ann Donny Teo and Takao Hinamoto
Dept. of Artificial Complex Systems Engineering, Hiroshima University
Higashi-Hiroshima, 739-8527 Japan

Abstract—In Orthogonal Frequency Division Multiplexing (OFDM), the composite time signal exhibits a high Peak-to-Average Power ratio (PAPR). Due to non-linearities of the transmit power amplifiers, this high PAPR generates spectral spreading, in-band distortion and out of band noise (OBN) which degrades the bit-error rate (BER). In this paper, we propose a way to combat this problem without sacrificing channel estimation and frequency-offset tracking accuracy, by designing a sub-optimal configuration of the pilot tones. The effectiveness of the proposed method to reduce PAPR and improve BER will be validated by computer simulations based on the conditions in IEEE 802.11a standard.

I. Introduction

The class of multicarrier modulation systems which makes use of orthogonal frequency channels, known as Orthogonal Frequency Division Multiplexing (OFDM), has been the subject of numerous dissertations in recent years, mainly due to its fulfillment of three criterion in the arena of mobile wireless communications: high data rate transmission with high bandwidth efficiency, adaptability to channels with nonuniform gain/noise characteristics versus frequency, and exceptional robustness to multi-path fading. It is applied widely in wireless Local Area Networks (WLAN) standards, for example IEEE 802.11 and HIPERLAN/2 [1]. The finalization of IEEE 802.11a standard permits data rate transmissions of up to 54Mbps, providing an extensive range of multimedia communication services in indoor as well as outdoor wireless environments [2].

However, OFDM systems are not without its limitations [3], [4]. The composite time signal in an OFDM system is formed from the linear addition of the independently modulated subcarriers. As a result, when the number of subcarriers is large, spurious high amplitude peaks appear in the OFDM signal which induces a large Peak-to-Average Power ratio (PAPR). Usually, the OFDM signal pass through a non-linear power amplifier before it is transmitted over the channel. When the input signal exceeds a certain value, the output of the amplifier becomes saturated, causing distortion in the signal, and hence a deterioration in the BER performance. Various approaches [3] to alleviate this problem have been proposed. For instance, coding in the form of special forward error correcting codes (FEC) excludes OFDM symbols with a large PAPR while probabilistic techniques like selected mapping (SLM), partial transmit sequences (PTS) and tone injection (TI) reduce the probability of the occurrence of peak values. On the other hand, clipping effect transform involves signal distortion techniques which minimize the peak amplitudes by nonlinearly distorting the OFDM signal at or around the peaks.

The OFDM symbol is composed of the information symbols and the pilot tones. The pilot tones, known at the receiver prior to transmission, are used mainly for tracking the channel frequency-offset in order to remove its influence from the signal. If the amplitudes of the pilot tones are fixed, the frequency-offset tracking accuracy remains unchanged, leaving the phases of the pilot tones the only factors free for manipulation. In this paper, we propose a novel method to reduce the PAPR by exploiting the freedom in phase of the pilot tones, in an OFDM system adopting IEEE 802.11a standard. Since it is difficult to obtain an analytical expression for the probability density function of the amplitude, an exhaustive search for the phases of the pilot tones which minimize its maximum will be performed. PAPR reduction and BER improvement owing to this sub-optimal designation will then be substantiated by computer simulations.

The remainder of the paper is organized as follows. After shortly reviewing OFDM system in Section 2, we give a brief description of PAPR in Section 3. In addition, we develop our method to reduce PAPR in detail in Section 4. Section 5 presents the simulation results and the corresponding analysis, while conclusions are drawn in Section 6.

II. OFDM System

Fig. 1 depicts a simplified OFDM transmitter. At the transmitter, the serial data sequence in the frequency domain undergoes serial-to-parallel (S/P) conversion to be stacked into one OFDM symbol. Suppose that the number of subcarriers is N. After S/P conversion, N-points inverse fast Fourier transform (IFFT) follows to produce the N dimensional signal in the time domain. Next, this N dimensional signal is parallel-to-serial (P/S) converted, and a cyclic prefix (CP) of length greater than or equal to the channel length is appended to introduce redundancy to mitigate interference between adjacent OFDM symbols and to preserve the orthogonality between the subcarriers. The resultant symbol is then applied to a power amplifier and subsequently sent over the channel.

The detailed structure of the OFDM packet is shown in Fig. 2. In the case of WLAN standard IEEE 802.11a, 64 slots are available in the OFDM symbol, out of which 48 are assigned for information symbols, 4 for pilot tones, while the rest serve as spectral nulls so as to mitigate the interferences from/to OFDM symbols in adjacent bands [5]. Long training

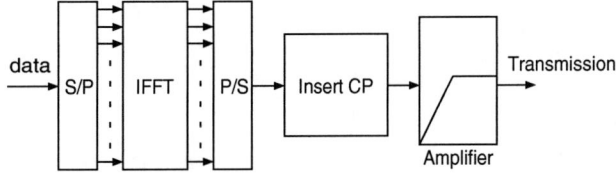

Fig. 1. A baseband OFDM system transmitter

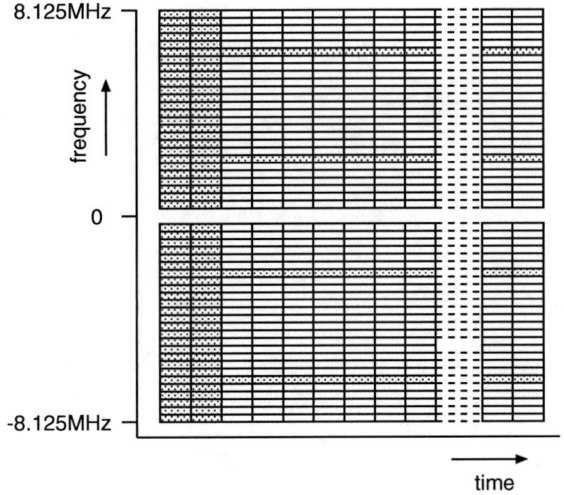

Fig. 2. The time-frequency structure of an OFDM packet. Shaded subcarriers contain pilots

sequences are sent at the beginning of the transmission process to estimate the channel, followed by pilot subcarriers at fixed interval to track the remaining frequency-offset [6].

We consider one OFDM signal. Let the symbol at the kth subcarrier be s_k, the serial signal in the time domain that is applied to the amplifier is then expressed as

$$s(t) = \frac{1}{\sqrt{N}} \sum_{k=0}^{N-1} s_k \exp j(\frac{2\pi kt}{N}), t \in [-N_{cp}, N], \quad (1)$$

where N_{cp} is the length of CP. Note that this serial signal $s(t)$, in turn, can be factored into two components, the pilot signal $p(t)$ and the information signal $d(t)$, given by

$$s(t) = d(t) + p(t). \quad (2)$$

III. PEAK-TO-AVERAGE POWER RATIO (PAPR)

In an OFDM system, the samples of the discrete time baseband signal exhibit large peaks caused by the coherent addition of several independently modulated tones. Consequently, the continuous time signal, obtained by passing the discrete time samples to the D/A converter and filtering the outputs using a pulse-shaping function, exhibits even larger peaks. When the peak power, defined as the power of the sine wave with an amplitude equal to the maximum envelope value, is high, the corresponding PAPR, expressed by [2]

$$\text{PAPR} = \frac{max|s(t)|^2}{E\{|s(t)|^2\}}, \quad (3)$$

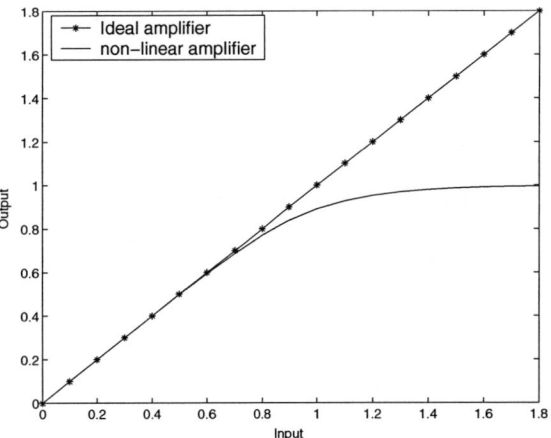

Fig. 3. Power amplifier

where $E\{\cdot\}$ denotes expectation, is also high. This is true especially under the condition that the number of subcarriers N is large. Under this condition, as apparent from (3), PAPR $\simeq N$, we know that PAPR increases proportionally with the number of subcarriers. An exceedingly large PAPR results in intractable problems such as in-band distortion, spectral spreading and OBN when the signal is applied to a non-linear power amplifier [7].

Fig. 3 illustrates the input-output characteristics of the linear and non-linear power amplifier which is based on the following function [3]:

$$g(A) = \frac{A}{(1+A^6)^{\frac{1}{6}}}, \quad (4)$$

where A denotes the normalized input to the amplifier.

In Fig. 3, saturation occurs for the non-linear amplifier. A large backoff can be used to set the operating point of the non-linear amplifier far away from saturation but this leads to severe power inefficiency and expensive transmitter. Hence, the alternative to reduce PAPR is desirable. Moreover, the reduction of PAPR allows a higher average power to be transmitted for a fixed peak power, thus improving the overall signal-to-noise ratio (SNR) at the receiver.

IV. PILOT TONE DESIGN

In Fig. 2, the pilot tones are deterministic, whereas the information symbols are stochastic random values. Existing techniques focus on lowering the peak amplitude of the information signal $d(t)$ to minimize the PAPR but in our novel approach, we modify the configuration of the pilot tones. One main advantage of this pilot tone design is that it does not incur additional loss in channel estimation and frequency-offset tracking accuracy.

Pilot tones having the same amplitude are optimal for channel estimation. This is vindicated in [8] which stipulates that the optimal pilot tones to be embedded in the OFDM symbol minimizes the channel mean square error (MSE) provided they are equidistant and equipowered. The freedom in phase available in pilot tones can be exploited to reduce PAPR.

TABLE I
SIMULATION PARAMETERS

Parameters	Specifications
number of subcarriers (N)	64
number of information symbols	48
number of pilot tones (K)	4
first pilot position (d)	7
signal source modulation	QPSK
noise	AWGN
channel model order (L) profile	Rayleigh 3(typical urban scenario) with power [0,-3,-6,-9] (in dB)

However, it is difficult to obtain an analytical expression for the probability density function of the amplitude which is essential in the derivation of the optimal phase of the pilot tones. Therefore, we conduct an exhaustive search for the phases of the pilot tones that can minimize the maximum value of the amplitude in the time domain, in other words, the sub-optimal value.

Pilot tones perform the dual function of frequency-offset synchronization and channel estimation, both of which are dependent on the amplitude of the pilot tones. The pilot tones $\{p_0, p_1, \ldots, p_{K-1}\}$ are set to have amplitude 1 and are defined as

$$p_k = \exp(j\theta_k), k \in [0, K-1], \quad (5)$$

where θ_k is the phase of the kth pilot tone. For simplicity, we assume N is an integer multiple of K so that the pilot tones are inserted into each OFDM signal uniformly at $i \in I$, where I is represented by

$$I = \{0, J, \ldots, (K-1)J\}, \quad (6)$$

and $J = N/K$. The pilot signal $p(t)$ in the time domain can then be written as

$$p(t) = \sum_{k=0}^{K-1} p_k \exp(j\frac{2\pi Jkt}{N}) = \sum_{k=0}^{K-1} \exp j(\theta_{k-1} + \frac{2\pi Jkt}{N}). \quad (7)$$

OFDM signal is formed from the linear addition of the information signal $d(t)$ and pilot signal $p(t)$. For a given $d(t)$, phases of the pilot tones $(\theta_0, \theta_1, \ldots, \theta_{K-1})$ can be readily obtained such that the amplitude of the OFDM signal is reduced. However, the amplitude of the OFDM signal may not be reduced for varying values of $d(t)$ using the same pilot tones. Considering that the phases of the pilot tones are invariably fixed, we know that $p(t)$ cannot be varied for different $d(t)$ for the purpose of reducing the amplitude of the OFDM signal. Hence, for the same $p(t)$, we reduce the PAPR of the OFDM signal by suppressing the peak amplitude of $p(t)$ through our pilot tone design, i.e., we find the sub-optimal phases of the pilot tones such that

$$(\theta_0, \theta_1, \ldots, \theta_{K-1}) = \arg \min_{(\theta_0, \theta_1, \ldots, \theta_{K-1})} \max_t |p(t)|^2, \quad (8)$$

where $|p(t)|^2$ denotes the power of the pilot signal. In view of the fact that it is theoretically prohibitive to compute the value

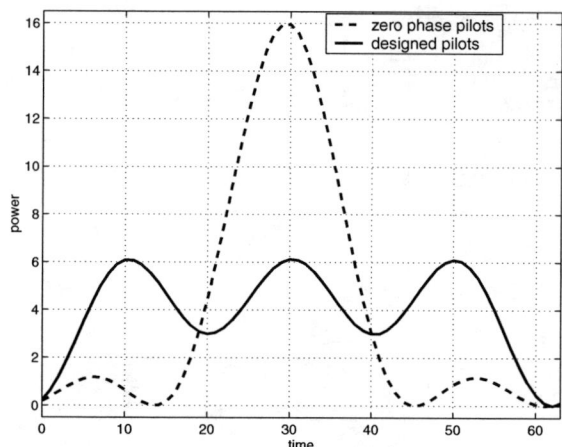

Fig. 4. Power of the pilot signal, $a = 7$

θ_k, we adopt the alternative method of manually deriving θ_k, through an exhaustive search via computer calculations.

The effectiveness of our design, based on the sub-optimal phases of the pilots acquired through exhaustive search, on PAPR and BER will be tested via Monte-Carlo simulations in the following section.

V. NUMERICAL SIMULATIONS

In this section, we illustrate the merits of our approach to reduce the PAPR and BER through realistic simulations which adhere to IEEE 802.11a standard by adopting the system parameters listed in Table I. Since we also need to observe channel estimation performance, we assume perfect timing and carrier synchronization. Interleaving at the transmitter as well as deinterleaving at the receiver, which serve the function of randomizing the bit errors, are not employed. Coding will be omitted in the simulation as it affects the bandwidth efficiency and data rate adversely. The results for 10^4 independent channels are averaged.

Fixing $\theta_0 = 0$, exhaustive search for the phases of the remaining pilot tones that minimize the maximum amplitude yields

$$\theta_1 = \frac{\pi a}{20}, \theta_2 = \frac{\pi a}{10} + \frac{23\pi}{40}, \theta_3 = \frac{3\pi a}{20} + \frac{69\pi}{40}, \quad (9)$$

for an arbitrary $a = 1, \ldots, 20$, in IEEE 802.11a standard where the number of pilot tones is 4.

Using the equidistant pilots equipped with these phases, we try to minimize the symbol MSE by apportioning power to the information and pilot symbols optimally [9]. For a given transmit power budget, let 52α and $52(1-\alpha)$ be the power allocated to the information and pilot symbols, respectively. According to [8], α_{opt} is found to be approximately

$$\alpha_{opt} = \frac{1}{1 + \sqrt{\frac{L+1}{M}}}, \quad (10)$$

where M is the number of information symbols in an OFDM symbol. In IEEE 802.11a standard, since $M = 48$ and the number of channel taps $L + 1 = K = 4$, we obtain $\alpha_{opt} =$

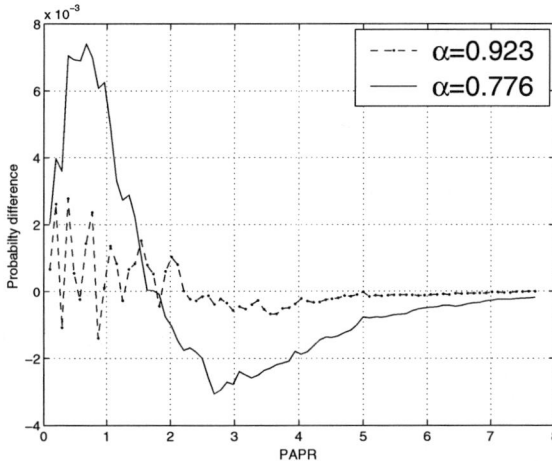

Fig. 5. Probability difference for $\alpha = 48/52 \simeq 0.923$ and $\alpha = 0.776$

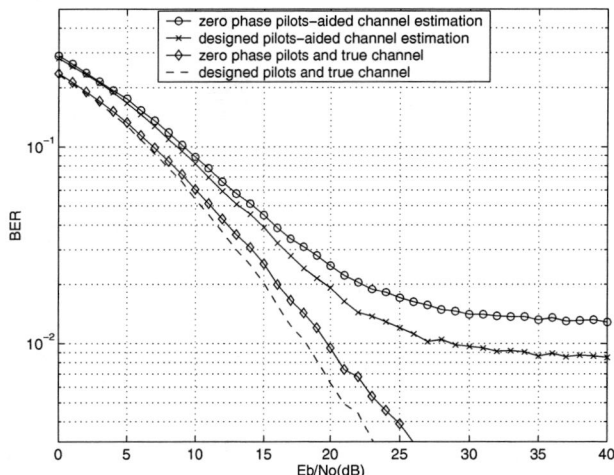

Fig. 6. BER performance

0.776. This α_{opt} is also optimal for channel estimation and data detection.

We choose $a = 7$ in (9), such that the designed pilot tones have phases $\theta_0 = 0, \theta_1 = 7\pi/20, \theta_2 = 51\pi/40$ and $\theta_3 = 111\pi/40$. Fig. 4 shows the power of the pilot symbols for both the zero phase pilot tones and the designed pilot tones in the time domain. Comparing the two, we verify that the peak power is indeed reduced by using the designed pilot tones.

The probability that the PAPR falls below a certain value are found for both the designed pilots and the zero phase pilots. Fig. 5 depicts the probability difference between the designed pilots and the zero phase pilots w.r.t. PAPR, for $\alpha = 48/52 \simeq 0.923$ and $\alpha = 0.776$. In the former, we put the same power to every subcarriers. For equal power allocation, i.e., $\alpha = 0.923$, the probability difference hovers at 0. In contrast, for the optimal power allocation, i.e., $\alpha = 0.776$, the probability difference is positive at low PAPR and negative at high PAPR. This is because at $\alpha = 0.923$, the number of pilot tones is small which implies a low power contribution to the pilot signal $p(t)$ and the benefits of the designed pilots are negligible. The peak value of the OFDM signal is then solely dependent on the information signal $d(t)$ which has the bulk of the power allocation. On the other hand, at $\alpha = 0.776$, more power is allocated to the pilots and the benefits of the designed pilots become evident. The number of OFDM signals with high peak is reduced.

Assigning α_{opt} as 0.776 for the designed pilots, we insert these equidistant, equipowered pilot tones in the OFDM symbol together with the information symbol and sent them over the channel. Pilot-aided channel estimation and coherent symbol detection are also performed. BER performance w.r.t. SNR for the range 0 to 40dB is simulated. Simulation results pertaining to the designed pilots and zero phase pilots using the estimated channel and the true channel for coherent detection are displayed in Fig. 6. From Fig. 6, it is obvious that BER performance for the designed pilots is better than that for the pilots set at zero phases. This demonstrates that the designed pilots contributes to a reduction in PAPR and hence a corresponding decrease in spectral spreading, in-band distortion and OBN which arises due to non-linearities of the transmit power amplifier, thus accounting for the improvement in BER compared to the zero phase pilots.

VI. CONCLUSIONS

In an OFDM system adopting IEEE 802.11a standard, the method of designing the sub-optimal pilots, in terms of selecting an appropriate power distribution and an exhaustive search for the sub-optimal phases of the pilots, proves effective in reducing PAPR. This exploitation of the phase does not incur any additional burden on the resource used for carrier frequency-offset and channel estimation. The reduction in PAPR translates into an amelioration in the BER as corroborated by the simulation results.

In future, we hope to accomplish two tasks. First, to derive the theoretical expression for the optimal phases of the pilots in reducing the PAPR, and second, to further develop the proposed method for a more significant BER improvement.

REFERENCES

[1] H. Harada and R. Prasad, *Simulation and Software radio for Mobile Communications*, Artech House, 2002.
[2] A. D. S. Jayalath and C. Tellambura, "Peak-to-Average Power ratio of IEEE 802.11a PHY layer Signals, " in *DSPCS'02*, pp. 31–36, 2002.
[3] R. V. Nee and R. Prasad, *OFDM for Wireless Multimedia Communications*, Artech House, 2000.
[4] M. J. F. Garcia, O. Edfors and J. M. Paez-Borrallo, "Joint Channel Estimation and Peak-to-Average Power Reduction in Coherent OFDM: A Novel Approach, " in *Proc. VTC 2001 Spring*, pp. 815–819, 2001.
[5] B. Sklar, *Digital Communications*, Prentice Hall, 2nd ed., 2001.
[6] R. V. Nee, G. Awater, M. Morikura, H. Takanashi, M. Webster and K. W. Halford, "New High-Rate Wireless LAN Standards, " *IEEE Commun.*, vol. 37, no. 12, pp. 82–88, 1999.
[7] E. Lawrey and C. J. Kikkert, "Peak to Average Power ratio Reduction of OFDM signals using Peak Reduction Carriers, " in *Proc. ISSPA '99*, pp. 737–740, 1999.
[8] S. Ohno and G. B. Giannakis, "Optimal Training and Redundant Precoding for Block Transmissions with Application to Wireless OFDM, " *IEEE Trans. Commun.*, vol. 50, no. 12, pp. 2113–2123, 2002.
[9] S. Ohno and G. B. Giannakis, "Average-Rate Optimal PSAM Transmissions Over Time-Selective fading Channels, " *IEEE Trans. Wireless Commun.*, vol. 1, no. 4, pp. 712–720, 2002.

HIGH-SPEED AND LOW-POWER DESIGN OF PARALLEL TURBO DECODER

Zhiyong He, Sébastien Roy, and Paul Fortier

Department of Electrical and Computer Engineering, Laval University,
Sainte-Foy, Quebec, Canada, G1K 7P4

ABSTRACT

This paper presents the high speed and low power design of a turbo decoder with parallel architecture. To solve the memory conflict problem of extrinsic information in such parallel architectures, a two-level mapping approach is proposed for designing a collision-free parallel interleaver. Since the warm-up process in the parallel architecture increases the decoding delay, a new parallel architecture without warm-up is proposed for high speed applications. The proposed parallel architecture increases decoding speed by 6-50% for a 16-parallel decoder. To reduce the power consumption of the decoder with parallel architecture, a simple truncation approach is proposed to reduce the storage requirement of the extrinsic information and path metrics without any extra hardware cost. The proposed truncation approach reduces the power consumption with little performance degradation.

I. INTRODUCTION

Turbo codes were invented in 1993 and have received a lot of interest since they offer the potential of near-Shannon limit communication [1]. Third generation (3G) wireless system standards have adopted turbo codes as the channel coding scheme. To apply turbo codes in 4G wireless systems [2], high throughput of decoders for turbo codes is a critical issue. A parallel decoding architecture is usually favored to increase the throughput and reduce the latency. In a decoder with a parallel architecture, an information block is divided into M sub-blocks and is assigned M component decoders. Then, M component decoders process in parallel.

Several implementations of parallel decoders have already emerged such as [3] – [6]. To initialize the path metrics in each sub-block, these parallel decoders have to warm up for several clock cycles at each iteration. With the increasing degree of parallelism, the warm-up process reduces the decoding throughput substantially. Another decoding delay in parallel architectures is due to collisions in memory access [7, 8], i.e., the M component decoders will likely collide when reading (writing) the extrinsic information from (into) M storage elements. Collision-free parallel interleavers are required for high-speed parallel decoders.

In this paper, a two-level mapping approach is proposed to design a collision-free parallel interleaver. To maximize the decoding speed of the parallel decoder, new architectures without warm-up are presented. Finally, a simple truncation approach for reducing the bit width of the storage is proposed for low power design of parallel decoders.

II. COLLISION-FREE TWO-LEVEL MAPPING INTERLEAVER

Let an information block of length $N = M \times L$ be written row-by-row into an $M \times L$ matrix. In the bottom-level mapping, L information bits in each row are interleaved and then swapped. In the top-level mapping, N information bits in the whole block are interleaved by reading diagonal-wise from the $M \times L$ matrix. The simulation results show that the two-level mapping plus swapping interleaver outperforms the collision-prone S-random interleaver. To illustrate how the collision-free interleaver is designed in the top-level mapping, Fig. 1 gives an example for an information block of length $N = 18$. By reading diagonal-wise, the data at a given position relative to the beginning of a sub-block is mapped to the same position within another sub-block. The solid arrows in Fig. 1 show the mapping of the second bits in 3 sub-blocks.

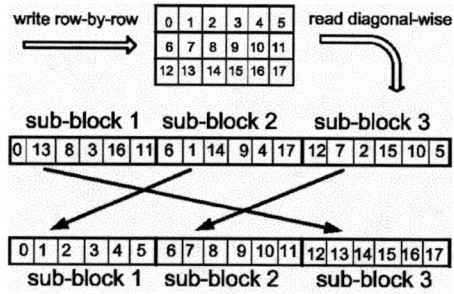

Fig. 1. Design of collision-free parallel interleaver.

III. PARALLEL DECODING ARCHITECTURES

A. Sliding window and parallel window

There are two classes of parallel architectures, the parallel window architecture (e.g., [4]) and the sliding window architecture (e.g., [5, 6]). The parallel decoder with the sliding window architecture operates slower but requires less memory, while the parallel window architecture operates faster but requires more memory.

Shown in Fig. 2(a) is an example of a parallel window scheme with a speed-up factor of 3. An information block of length N is divided into 3 windows, where each window consists of $W = N/3$ information bits. In order to initialize the boundary distributions of the path metrics for each window, overlapping of L information bits between adjacent windows is added for the warm-up of decoding. If each component decoder decodes one bit each clock cycle, the decoding delay is $2(L+W)$ cycles each iteration. The main drawback with the parallel window version is that it requires a large amount of memory to store the path metrics from the starting to the ending trellis stage. With an increase in block size, its extraordinary hardware complexity will become proportionally prohibitive for many practical purposes.

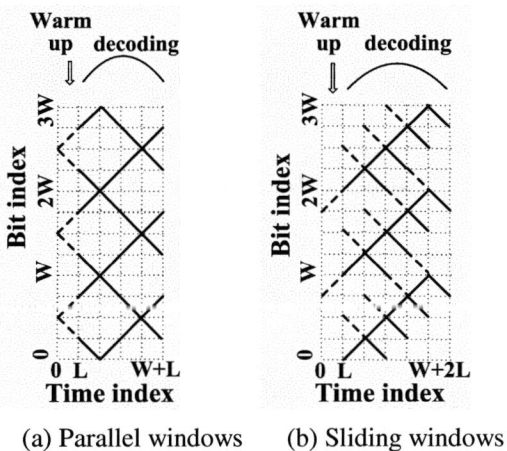

(a) Parallel windows   (b) Sliding windows

Fig. 2. Parallel architectures with warm-up.

The parallel architecture with the sliding window approach was proposed to reduce the memory requirements of the path metrics. Shown in Fig. 2(b) is an example of a three-parallel decoding scheme with the sliding window approach. Each sub-block of length W is divided into several sliding windows, where each sliding window consists of L information bits. Only the forward path metrics of L information bits need to be stored in each sub-block. The decoding delay for the sliding window approach is $2(2L+W)$ clock cycles each iteration. In the sliding window architecture in Fig. 2(b), three processors are used to calculate the path metrics. In [9], by storing the backward path metrics computed in the previous iteration, only two processors are used for computation of forward and backward path metrics.

The number of bit L for the warm-up phase is normally chosen to be 5 to 6 times the constraint length of the code. Since the length W of each sub-block decreases linearly with an increasing degree of parallelism, the warm-up process represents a large portion of the decoding delay in a high speed parallel implementation.

For example, for a decoder with 16-parallel scheme, an information block of 256 bits is divided into 16 sub-blocks of length 16 bits. Half of the decoding time is used to warm up the path metrics.

B. New parallel architectures

We propose two new parallel decoding architectures for high speed applications: warm-up-free parallel window and warm-up-free sliding window. Instead of the warm-up process using two overlapped sliding windows, the new architectures use the path metrics that were computed in the previous iteration of the adjacent sub-block to initialize the path metrics for each sub-block in the next iteration. Fig. 3 shows an example of three-parallel schemes with the new architectures. Without the warm-up process, the decoding delay of the new architectures decreases to $2I \times W$ and $2I(W + L)$ for the parallel window and the sliding window versions, respectively.

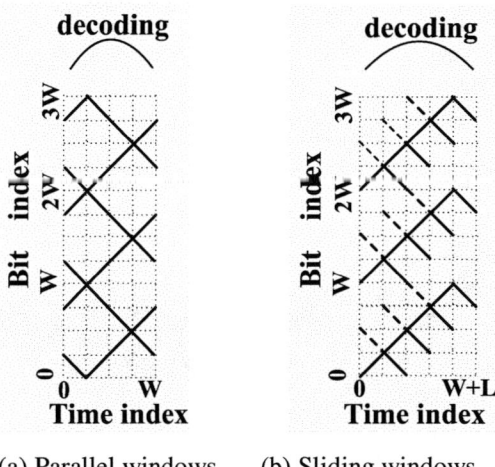

(a) Parallel windows   (b) Sliding windows

Fig. 3. Parallel architectures without warm-up.

For an M-parallel decoder of the 4-state turbo code, assuming 10 bits are used to represent each path metric, where 7 bits are used for the integer part and 3 bits for the fractional part, the storage requirement for initialization of the path metrics would be a total of $4 \times 10 \times (M-1) = 40(M-1)$ bits. Since only the path metrics at the boundary of each sub-block are stored at each iteration, the writing (reading) period to (from) memory is W clock cycles, where W is the number of bits in each sub-block. The power consumption incurred by the storage of these path metrics is negligible compared to the storage of the extrinsic information.

To validate the new architectures, we performed simulations for a 4-state turbo code with a generator [111,101], and a code rate $k/n = 1/2$, assuming an AWGN channel and BPSK modulation. The bit error rate

(BER) at the 6th iteration versus the signal-to-noise ratio per bit E_b/N_0 are compared in Fig. 4(a) for the 16-parallel architectures with and without warm-up process. No performance degradation is observed. To further reduce the storage requirement, we propose that only the integer part of the path metric, i.e., 7 bits, be stored for the initialization in the next iteration. Fig. 4(b) compares the BER of the 16-parallel decoders with the proposed architecture when 10 bits and 7 bits of path metric are stored. The performance degradation is negligible when the fractional part of the path metrics is truncated.

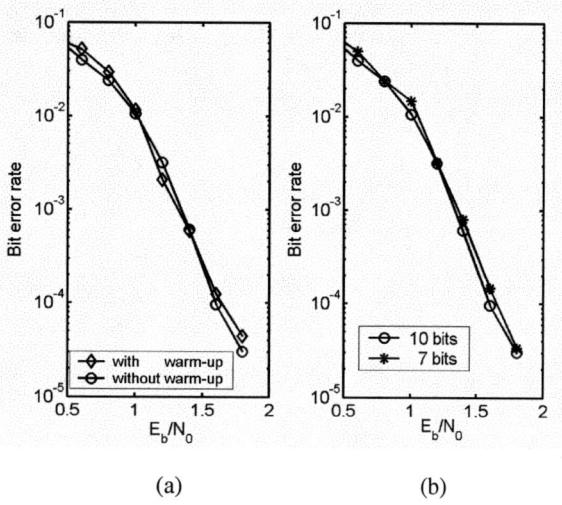

(a)　　　　　　　　(b)

Fig. 4. (a) Performance comparison of parallel architecture with and without warm-up.
(b) Performance comparison of proposed architecture with storage of 10 bits and 7 bits of path metric.

C. FPGA implementation results

To compare the computation complexities of various parallel architectures, we implemented a series of parallel decoders for 4-state turbo codes into Xilinx FPGA devices. To achieve good performance, a precision of 3 bits was chosen, i.e., 3 bits were used to represent the fractional part of each quantity within the algorithm. The extrinsic information was represented by a vector of 7 bits. The representable range was in [-8, 7.875]. The path metric was represented by a bit vector of 10 bits by employing modulo normalization.

The hardware resource utilization statistics as a function of the decoding speed are shown in Fig. 5 for the parallel decoders with the sliding window version for 4-state turbo codes with a block length of 2048 bits. The decoding speeds are calculated by assuming a decoding clock frequency $F = 100$ Hz and a number of decoding iterations $I = 6$. The proposed parallel architectures without warm-up increase largely the decoding speed, compared to the parallel architecture with warm-up. For example, by employing 0.7 million gates, the parallel architectures with and without warm-up can attain decoding speeds of 180 and 210 Mbps, respectively. The parallel architecture without warm-up thus increases the speed by 15%. The decoding speed of the 16-parallel decoders with the sliding window version is shown in Fig. 6 as a function of the block size. Compared to the architecture with warm-up, the proposed architecture without warm-up increases the speed by 6-50% for block sizes ranging from 256 bits to 4096 bits.

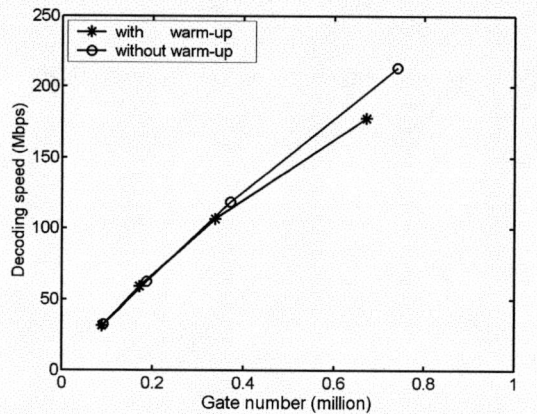

Fig. 5. Decoding speed versus hardware cost for parallel decoders with sliding window. Block size = 2048 bits.

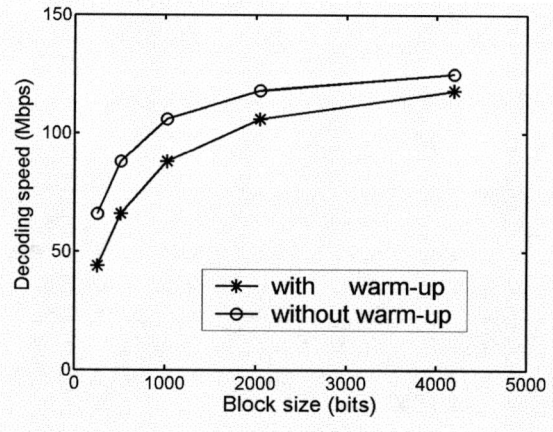

Fig. 6. Decoding speed of 16-parallel decoder with sliding window for various block sizes.

IV. POWER REDUCTION TECHNIQUE

Parallel architectures increase the decoding speed but also increase the power consumption of the decoders. For a serial decoder, the memories of both the extrinsic information and the path metrics are written and read once per decoding clock cycle. 72% of the overall power consumption comes from the storage of the path metrics for a serial decoder with the sliding window approach [10]. For an *M*-parallel decoder, the memories of both the extrinsic information and the path metrics are written and read *M* times per clock cycle. A very large part of the overall power consumption is therefore imputable to the memory accesses of the extrinsic information and the path metrics. In [11], a nonlinear quantization approach was proposed to reduce the bit width of the extrinsic information to 3 bits by adding a transformation module and an inverse transformation module into the decoder. For an *M*-parallel decoder for a 4-state turbo code, *5M* transformation modules and *5M* inverse transformation modules are thus needed.

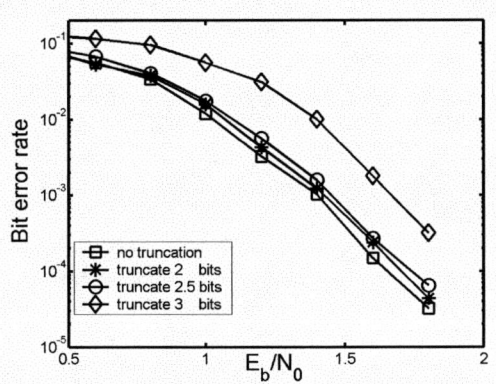

Fig. 7. Performance comparison of 16-parallel decoders after truncating.

We propose a simple truncation approach to reduce the bit width of the storage without extra hardware cost. The basic idea is to truncate several LSB bits after calculations of the extrinsic information and the path metrics and only write their MSB bits into the memories. Fig. 7 shows the performances of 16-parallel decoders for a 4-state turbo code with a block size of 2048 bits after employing the truncation approach. The truncation operation of 2.5 bits means that 2 bits were cut from the extrinsic information and 3 bits were cut from the path metric. It is shown clearly in Fig. 7 that the truncation operation of 2.5 bits is feasible with a very small performance loss. Since 7 and 10 bits are used to represent the extrinsic information and the path metrics, the proposed truncation scheme reduces the power consumption of the extrinsic memory by 30%.

V. CONCLUSION

A two-level mapping approach has been proposed to design a collision-free interleaver for high speed decoders with parallel architectures. To increase decoding speed, we have proposed two classes of warm-up-free parallel architectures. By storing path metrics at the boundary of each sub-block for initialization of the path metrics, the proposed warm-up-free parallel architectures achieves a significant increase in decoding speed without any BER performance degradation. To reduce the power consumption of decoders, a simple truncation approach is proposed to reduce the storage requirement of the extrinsic information and path metrics without any extra hardware cost.

This work has been supported in part by the Natural Sciences and Engineering Research Council (NSERC) of Canada and the Canadian Microelectronics Corporation (CMC) under its System-on-Chip Research Network (SOCRN).

REFERENCES

[1] C. Berrou, A. Glavieux, and P. Thitimajshima, "Near Shannon limit error-correcting coding and decoding: Turbo-codes," *Proc. Int. Conf. Communications (ICC'93)*, Vol. 2, pp. 1064-1070, May 1993.

[2] W. W. Lu, B. H. Walke, and X. Shen, "4G Mobile Communications: Toward Open Wireless Architecture," *IEEE Wireless Communications*, Vol. 3, No. 2, pp. 6-8, April 2004.

[3] J. Hsu, and C. Wang, "A parallel decoding scheme for turbo codes," *IEEE International Symposium on Circuits and Systems*, Vol. 4, pp. 445-448, June 1998.

[4] A. Worm, H. Lamm, and N. Wehn, "VLSI architectures for high-speed MAP decoders," *14th International Conference on VLSI Design*, pp. 446-453, 2001.

[5] Z. Wang, Z. Chi, and K. K. Parhi, "Area-efficient high-speed decoding schemes for turbo decoders," *IEEE Transaction on very large scale integration (VLSI) systems*, Vol. 10, No. 6, pp. 902-912, Dec. 2002.

[6] Y. Zhang and K. K. Parhi, "Parallel turbo decoding," *IEEE International Symposium on Circuits and Systems*, Vol. 2, pp. 509-512, May 2004.

[7] A. Giulietti, L. van der Perre, and A. Strum, "Parallel turbo coding interleavers: avoiding collisions in accesses to storage elements," *IEE Electronics Letters*, Vol. 38, No. 5, pp. 232-234, May 2002.

[8] Jaeyoung Kwak, Sook Min Park, Sang-Sic Yoon, and Kwyro Lee, "Implementation of a parallel turbo decoder with dividable interleaver," *IEEE International Symposium on Circuits and Systems*, Vol. 2, pp. 65-68, May 2003.

[9] F. Raouafi, A. Dingninou, C. Berrou, "Saving memory in turbo-decoders using the max-log-MAP algorithm," *IEE Colloquium on Turbo Codes in Digital Broadcasting - Could It Double Capacity?* (Ref. No. 1999/165), pp. 14/1-14/4, 1999.

[10] C. Schurgers, F. Catthoor, M. Engels, "Memory optimization of MAP turbo decoder algorithms," *IEEE Transactions on VLSI systems*, Vol. 9, No. 2, pp. 305-312, April. 2001.

[11] J. Vogt, J. Ertel, A. Finger, "Reducing bit width of extrinsic memory in turbo decoder realizations," *IEE Electronics Letters*, Vol. 36, No. 20, pp. 1714-1716, Sept. 2000.

Efficient View Maintenance in Wireless Networks

Huaizhong Lin[1], Bo Zhou[1], Zengwei Zheng[1, 2], Chun Chen[1]
1. College of Computer Science, Zhejiang University, Hangzhou 310027, China
{linhz, bzhou, zhengzw, chenc}@zju.edu.cn
2. City College, Zhejiang University, Hangzhou 310015, China

Abstract—In mobile databases, data caching in mobile hosts often takes the form of materialized views. Data is updated by communicating only the incremental part to reduce the demand for network bandwidth. But previous incremental update algorithms may be inefficient or even cannot accomplish view maintenance tasks when the wireless bandwidth drops dramatically. In the paper, a priority-based incremental update algorithm is proposed for real-time view maintenance in mobile databases. The algorithm orders data items by dynamic freshness and transmits data to mobile hosts according to the priority. The algorithm improves communication efficiency and data freshness when the wireless bandwidth is low and tends to vary widely.

I. INTRODUCTION

Due to the rapid development of mobile device and wireless network technologies, more and more interests are paid on the research areas of mobile databases. Mobile computing environments are characterized by low bandwidth, excessive latency, instability of connection, and constant disconnection [1,2]. Data caching, which often takes the form of materialized views regarding mobile database applications, is the most common approach to support data operations in such weakly connected network. Views are updated by communicating only the incremental part to reduce the demand for wireless network bandwidth. Many mobile database applications, such as dissemination of stock exchange information, statistic analysis of gathered data from varied data sources etc, have strong requirements on timeliness of data. But previous incremental view update algorithms don't pay much attention to the characteristics of wireless network and may be inefficient or even cannot accomplish view maintenance tasks when the wireless bandwidth drops dramatically [3]. These algorithms are not suitable for real-time mobile database applications.

In the paper, the notion of data freshness, which is based on time measurement, and the model of incremental update with real-time constraints are proposed. Through a quantitative analysis, we suggest a priority-based incremental update algorithm for real-time view maintenance in mobile databases. The algorithm orders data items by dynamic freshness and transmits data to mobile hosts according to the priority. The algorithm will transmit the data updates with high priority and thus improve communication efficiency and data freshness when the wireless bandwidth is low and tends to vary widely.

The rest of the paper is organized as follows. In section 2, we describe the incremental view update problem with real-time constraints. In section 3, we propose the algorithm. We give the performance evaluation in section 4, and conclude the paper in section 5.

II. INCREMENTAL UPDATE WITH REAL-TIME CONSTRAINTS

A. View Agent

The architecture of mobile database is usually composed of three layers, which are mobile host layer, synchronization server layer, and database server layer [4]. Mobile hosts (MH for short), which include notebook, palm PC, and PDA etc, connect with synchronization servers by wireless communication. The synchronization servers reside in fixed network, and play the role of bridge between MHs and database servers. In MHs, data caching often takes the form of materialized views. User queries are served quickly by search in materialized views without accessing remote database. View is updated by communicating only the incremental part to reduce the demand for wireless bandwidth.

Because most database servers don't provide incremental view maintenance for clients, each MH has a corresponding view agent in the synchronization server [3]. The functions of a view agent include: (1) acts as a proxy for MH in fixed network. It gathers information from varied data sources, establishes and maintains index structure, and transmits only incremental part to MH, and (2) data prefetching and buffer, to overcome the resource scarcity problem in MH, such as limited memory and short battery life. A view agent can be proxy for several MHs. When MHs move between base stations, the view agents can communicate each other to handle the hand-over process of MHs easily.

B. Data Freshness

Let V be the view to be maintained. $r_1, r_2, ..., r_m$ denote rows of V, and $c_1, c_2, ..., c_n$ denote columns of V. r_{ij} represents the intersection data item in row r_i and column c_j of V. s_{ij} is the size of r_{ij}.

This work is supported by the Natural Science Fundation of Zhejiang Province, China (Grant no. M603230) and the Research Fund for Doctoral Program of Higher Education from Ministry of Education of China (Grant no. 20020335020).

There are many ways to measure data freshness, e.g. the number of uncommitted updates or transactions on a data item [5]. We adopt the measurement based on time. Fr, denoted for data freshness, is a real number between 0 and 1 with Fr=1 presenting that the data item is up to date.

Definition 1 Suppose the data item r_{ij} is up to date at time t_0, i.e. Fr=1. With time going on, the freshness will go down gradually. The freshness of item at time t ($t \geq t_0$) can be expressed by a function $f(\Delta t)$ ($\Delta t = t - t_0$), i.e. $Fr(r_{ij})=f(\Delta t)$. The freshness function $f(\Delta t)$ satisfies the following conditions:

(1) The definition field of $f(\Delta t)$ is $[0,+\infty)$ and the range of $f(\Delta t)$ is $[0,1]$;

(2) $f(\Delta t)$ is a monotonically decreasing function;

(3) $f(0)=1$, $f(+\infty)=0$.

$f(\Delta t)$ is a function representing the decreasing data freshness of a stale data item when time goes on. In practice, $f(\Delta t)$ can be a linear function for simplicity:

$$f(\Delta t) = \begin{cases} 1 - \Delta t / T, & 0 \leq \Delta t \leq T \\ 0, & \Delta t \geq T \end{cases}.$$

T in the above equation is the active period of the data item, which means that the data item will be no value if it hasn't been updated over a period of T.

In many applications, users will pay different attention to varied data items. Some key items will be much more important than the others. So, we give weights to different columns to model user interests in the measurement of data freshness.

We give weight w_j to every column c_j in V, which satisfies:

$$\sum_{j=1}^{n} w_j = 1.$$

The weight is a measurement of user's demand of data timeliness. In our algorithm, columns with higher weight will have higher priority.

The size of data item should be taken into consideration in the definition of row freshness. So, we revise the weights of columns according to the size:

$$\overline{w_j} = \frac{w_j s_{ij}}{\sum_{j=1}^{n} w_j s_{ij}}.$$

Definition 2 Suppose t_{ij} be the last update time for r_{ij}. $f_j()$ is the freshness function for c_j (We suppose that data items in one column share the same freshness function for simplicity.) The freshness of row r_i at time t is the weighted arithmetic average of each data item ($\Delta t_{ij}=t-t_{ij}$):

$$Fr(r_i) = \sum_{j=1}^{n} \overline{w_j} Fr(r_{ij}) = \frac{\sum_{j=1}^{n} w_j s_{ij} f_j(\Delta t_{ij})}{\sum_{j=1}^{n} w_j s_{ij}}.$$

Similarly, we can define the freshness of view, which is the arithmetic average of all rows.

C. Description of Problem

Base on the above definition of freshness, we present a quantitative description of the incremental view update problem with real-time constraints. Let ΔV be the incremental part of view. It is the duty for view agent to compute and transmit ΔV to MH. When the wireless bandwidth is too low to transmit all information in ΔV, there is a problem how to maximize the freshness of view $Fr(V)$ by selecting an appropriate part of ΔV to be transmitted to MH.

Definition 3 Suppose t_{max} be the maximum transmission time required by user. B is the network bandwidth. The incremental view update problem with real-time constraints can be defined as selecting a subset Ω of ΔV under the constraints

$$\sum_{i,j \in \Omega} s_{ij} \leq B \cdot t_{max}, \quad (1)$$

so that

$$Fr_{sum} = \frac{\sum_{i,j \notin \Omega} w_j s_{ij} f_j(\Delta t_{ij}) + \sum_{i,j \in \Omega} w_j s_{ij}}{\sum_{i=1}^{m} \sum_{j=1}^{n} w_j s_{ij}} \quad (2)$$

is maximized.

In the above definition, inequation (1) represents the real-time constraints, and equation (2) represents the freshness after transmitting the selected part Ω of ΔV, which is the target of the problem.

Now, we simplify the target of the problem as follows. Based on (2), the increase of freshness can be computed as

$$\Delta Fr = Fr_{sum} - Fr(V) = \frac{\sum_{i,j \in \Omega} w_j s_{ij} - \sum_{i,j \in \Omega} w_j s_{ij} f_j(\Delta t_{ij})}{\sum_{i=1}^{m} \sum_{j=1}^{n} w_j s_{ij}}$$

$$=\frac{\sum_{i,j\in\Omega}w_j(1-f_j(\Delta t_{ij}))s_{ij}}{\sum_{i=1}^{m}\sum_{j=1}^{n}w_j s_{ij}}$$

Let $p_{ij}=w_j(1-f_j(\Delta t_{ij}))$, then

$$\Delta Fr=\frac{\sum_{i,j\in\Omega}p_{ij}s_{ij}}{\sum_{i=1}^{m}\sum_{j=1}^{n}w_j s_{ij}}$$

As the freshness Fr(V) at time t and the denominator of the above equation are fixed, the problem is equal to selecting a subset Ω under constraints (2), so that

$$H=\sum_{i,j\in\Omega}p_{ij}s_{ij} \qquad (3)$$

is maximized.

From (3), it is obvious that the subset Ω should include data times with higher value p_{ij}, i.e. items with larger weight and longer period of not being updated.

III. ALGORITHM

In our algorithm, a structure Tuple_Info (Fig. 1) is used to represent a row. The incremental part ΔV of view is expressed by a list of Tuple_Info, denoted by Tuple_List. Columns in Tuple_Info can be classified as fixed and variable value columns. The Tuple_Info records last update time of data items in database server and MHs, denoted by u_s and u_{mh} respectively. u_s is always greater or equal than u_{mh}.

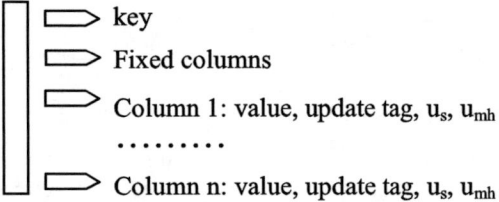

Figure 1. Structure of Tuple_Info

Here are some other data structures used in the algorithm:

- Column_Queue$_j$ (j=1,...,n): Queues with each element pointing to jth column in a Tuple_Info structure. The elements are ordered by decreasing update period (i.e. $u_s - u_{mh}$) of items in column j;
- Delete_Queue: A queue of deleted rows, each element of which points to a Tuple_Info structure.

The insertion, deletion, and update operations of view are shown in algorithm 1-3. The priority-based incremental update algorithm is shown in algorithm 4. The priority value of items in algorithm are computed according to the following equation (T_j is the active period of data items in column j, and $\Delta t=u_s-u_{mh}$.) The items with larger value have higher priority. Deleted rows are processed firstly in algorithm 4, for it needn't transmit data.

$$\text{priority}=\begin{cases}0, & \Delta t\leq 0\\ w_j\cdot\Delta t/T_j, & 0<\Delta t<T_j\\ w_j, & \Delta t\geq T_j\end{cases}$$

Algorithm 1 Insertion

1. new a Tuple_Info structure to record row values
2. update tag='Y', u_s=current clock, u_{mh}=$-\infty$
3. add Tuple_Info to Tuple_List
4. add each item to head of corresponding Column_Queue$_j$

Algorithm 2 Deletion

1. add the row into Delete_Queue
2. delete items from corresponding Column_Queue$_j$

Algorithm 3 Update

1. update data value in Tuple_Info
2. update tag='Y', u_s=current clock
3. reorder updated item in Column_Queue$_j$ according to its modified priority

Algorithm 4 Transmission

1. empty the transmission queue
2. for every row in Delete_Queue
3.   append deletion message to tail of transmission queue
4.   delete row from Tuple_List
5. end for
6. repeat
7.   select the highest priority item from head elements of all n Column_Queue$_j$ do {
8.   update tag='N', u_s=current clock
9.   add the item to tail of transmission queue
10. move the element to tail of corresponding Column_Queue$_j$ }
11. until received stop message or all updated items have been transmitted

Compared with previous incremental update algorithms, our algorithm introduces limited additional data structures Column_Queue$_j$, Delete_Queue, and update time of data items, for MHs usually have small capacity of memory and views to be maintained are small. Moreover, the algorithm is executed in synchronization server and the additional computation is limited. Therefore, the space and time overhead of our algorithm is limited and it is practical, especially when the numbers of columns and rows of views are small and the size of data items is large. In practice, we can combine several columns into one column to reduce overhead.

IV. EXPERIMENTAL RESULTS

We perform experiments to analyze the performance of the proposed algorithm by simulation. Additionally, we adopt the incremental update algorithm view holder (VH for short) [3] in the literature for comparison. View holder is a proxy to be used in fixed network gathering data from database or data warehouse, and transmitting the incremental part to MH. The experiment sample is a view with 1000 rows and 10 columns, 8 of which are variable. The view only occupies 2M space, for the memory in MH is relatively small. The wireless network bandwidth varies above or below the average bandwidth by 20~30% to model the instability of wireless connection. A simulator generates insertion, deletion, and update operations and the sampling interval of data is 1s. Other settings of experimental parameters are shown in table 1.

TABLE I. EXPERIMENTAL PARAMETERS

Parameter description	Values
Average bandwidth of wireless network	100K-1Mbps
Size of data items	0.05-0.5Kbytes
Update probability of each item	0.01-0.1/s
Deletion probability of rows	3 rows/s
Insertion probability of rows	3 rows/s
Active period of data items	5-50s
Weight of columns	0.05-0.2

Fig. 2 shows the freshness comparison of our algorithm and VH algorithm under different bandwidth. We omit the initial phase of execution when computing freshness. In fact, the freshness of setup time of our algorithm is higher than VH algorithm. From the figure, it is obvious that our algorithm is superior to VH algorithm. Two algorithms perform the same good work when the wireless bandwidth is abundant. But when the bandwidth is low, i.e. not all update information can be transmitted to MH, our algorithm outperform VH algorithm. The advantages of our algorithm become more and more obvious with the drop of bandwidth.

Fig. 3 shows the restoration of data freshness after MH disconnects from synchronization server for a period of 30s. From the figure, it is obvious that the freshness of our algorithm is higher than VH algorithm in restoration phase. The average improvement is about 16.1%. It can be easily inferred that our algorithm will be even better than VH algorithm when bandwidth drops several orders of magnitude in a dynamic varying wireless connection.

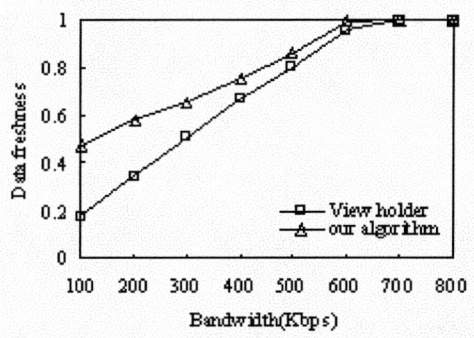

Figure 2. Comparison of data freshness

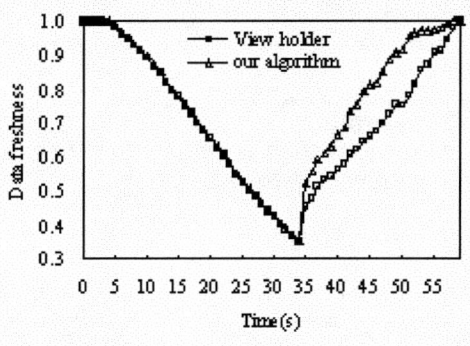

Figure 3. restoration of data freshness after 30s disconnection from fixed network (bandwidth=700Kbps)

V. CONCLUSION

We proposed a priority-based incremental view update algorithm to address the real-time view maintenance problem in wireless network. The proposed algorithm introduces limited space and time overhead and outperforms previous algorithms under low bandwidth, improving data freshness evidently. The algorithm is practical and appropriate when the bandwidth varies dynamically in mobile environments.

REFERENCES

[1] E. Pitoura and B. Bhargava, "Data consistency in intermittently connected distributed systems," IEEE Transactions on Knowledge and Data Engineering, vol. 11, no. 6, 1999, pp. 896-915.

[2] Huaizhong Lin and Chun Chen, "Optimistic voting for managing replicated data," Journal of Computer Science and Technology, vol. 17, no. 6, 2002, pp. 874-881.

[3] S. W. Lauzac and P. K. Chrysanthis, "Programming views for mobile database clients," Proceedings of the 9th DEXA, Vienna, Austria, 1998, pp. 408-413.

[4] M. H. Dunham and A. Helal, "Mobile computing and databases: anything new," SIGMOD Record, vol. 24, no. 4, 1995, pp. 5-9.

[5] E. Pacitti, E. Simon, and R. Melo, "Improving data freshness in lazy master schemes," Proceedings of the 18th ICDCS, Amsterdam, Netherlands, 1998, pp. 164-171.

Digital Signal Processing Engine Design for Polar Transmitter in Wireless Communication Systems

Hung-Yang Ko, Yi-Chiuan Wang and An-Yeu (Andy) Wu

Graduate Institute of Electronics Engineering, and
Department of Electrical Engineering,
National Taiwan University, Taipei, 106, Taiwan, R.O.C.

Abstract-Polar modulation techniques offer the capability of multimode wireless system and the potential for the high efficiency *Power Amplifier* (PA). This paper describes a new design of *Digital Signal Processing* (DSP) engine for the polar transmitter. The digital part includes rectangular-to-polar converter and digital phase modulator, and the engine is designed for EDGE (2.5G) system. We employ the *Coordinate Rotation Digital Computer* (CORDIC) and *Direct Digital Frequency Synthesizer* (DDFS) techniques in our design. A prototype chip has been designed and fabricated in *UMC 0.18 um* CMOS process with *1P6M* technology.

1. INTRODUCTION

Polar modulation offers the capability of achieving high linearity and high efficiency simultaneously in a wireless transmitter. Improved efficiency is achieved by using a highly efficient and non-linear PA to work at its peak efficiency. Linear transmission is achieved by modulating the envelope of the signal through the voltage supply of the PA.

Polar transmission utilizes envelope and phase component to represent the digital symbols instead of the conventional I/Q format [1]. The baseband signal $V(t)$ is split into the phase signal $\theta(t)$ and the envelope signal $A(t)$.

$$V(t) = x(t) + j \cdot y(t). \quad (1)$$

$$A(t) = \sqrt{x(t)^2 + y(t)^2},$$
$$\theta(t) = \tan^{-1}\left(\frac{y(t)}{x(t)}\right). \quad (2)$$

It is clear that from Eq. (2) we can have a phase-only signal through phase modulator and multiplied with its envelope at the PA to recreate the original complex signal $V(t)$. This polar modulation process is like the *Envelope Elimination and Restoration* (EER) [2] architecture. In the conventional design, one part goes through a limiter to remove the envelope and keeps the phase information only. And the other part is detected by an envelope detector to extract the envelope information.

This work is supported by the MediaTek Inc., under NTU-MTK wireless research project.

But both circuits would suffer from the non-linearity and distortion of the analog devices and would cause mismatch problem through the two paths. In this paper we proposed a DSP engine which includes rectangular-to-polar converter and digital *Phase Modulator* (PM). The design does not have the distortion problem caused by the analog components and the phase modulation process can be precisely controlled by the digital phase modulator. The baseband phase signal is modulated through digital phase modulator at the specific frequency range. The phase modulated signal is represented as $S_{IF-PM}(t)$.

$$S_{IF-PM}(t) = \cos(w_c t + \phi(t)). \quad (3)$$

The PA stage of *amplitude modulator* (AM) operates in principle as a multiplier in our design model. This gives the output signal in the specific frequency band as follows:

$$\begin{aligned} S_{IF}(t) &= A(t) \cdot S_{IF-PM}(t), \\ &= A(t) \cdot \operatorname{Re}\{e^{j\phi(t)} \cdot e^{jw_c t}\}, \\ &= x(t)\cos(w_c t) + j \cdot y(t)\sin(w_c t). \end{aligned} \quad (4)$$

For convenience of the simulation model [2], the gain of the PA is set to one. Thus the Eq. (4) is equal to the signal of EDGE, which is up-converted at *Intermediated Frequency* (IF) band. The non-linearity of PA and analysis of up-converter to *Radio Frequency* (RF) stage are beyond the scope of this paper.

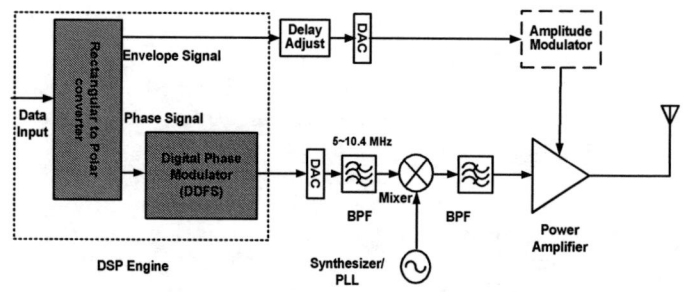

Fig. 1. Architecture of polar transmitter.

2. POLAR TRANSMITTER ARCHITECTURE

The architecture of the polar transmitter is shown in Fig. 1. The rectangular-to-polar converter extracts the symbol phase and envelope information in the digital domain. Then the phase information is modulated through digital phase modulator to create a constant envelope and phase modulated signal. The phase modulated precision and channel selection can be well controlled in the digital part first. In this paper we use the concept from [3] to realize the digital phase modulator design. The digital fine-tune frequencies are generated by the DDFS. The DDFS interpolates the carrier frequencies between the coarse frequencies generated by the integer-N PLL. The main design considerations of the DSP engine include: (a) the bandwidth of the envelope and the phase signal; (b) the numbers of the fine-tune frequencies generated by the DDFS would affect the clock rate of DDFS and rectangular-to-polar converter; (c) the quantization effect in digital domain will cause phase noise and frequency spurs. And this effect also influences the *Error Vector Magnitude* (EVM) performance and the signal spectrum. Typically the bandwidths of envelope and phase signal are equal to 1~2 MHz and larger than the EDGE signal bandwidth 200k Hz. The clock rate of the DDFS can be derived [3] as below:

$$f_{clk} = S \cdot f_{sym} > \frac{1}{0.4} \cdot (f_{cs} \times (N+1) + \frac{f_{tb}}{2}) \quad (5)$$

Where f_{clk} is the clock rate of DDFS, S is the number of samples per symbol and f_{sym} is the symbol rate of the EDGE signal. The maximum output frequency of DDFS is limited to 0.4 times the clock frequency. The parameter f_{cs} is the carrier spacing (200 kHz) in EDGE system, N is the number of digital fine tune frequency and f_{tb} is the transition BW of the filter which is located after up-converter stage. In our design, we choose N=25, f_{tb}=10 MHz, f_{cs}=200 kHz and S=96. Thus the clock rate of DDFS should be operated at 26 MHz. The digital fine tuning frequencies are generated by the DDFS and locating at 5 MHz~10.4 MHz. Each interpolated frequency (channel) is stored in the *fine-tune Frequency Control Word* (FCW) table.

3. RECTANGULAR-TO-POLAR CONVERTER

For a coordinate axis converter, we adopt the CORDIC algorithm in our design since the CORDIC algorithm is very simple and low hardware cost. In order to further reduce the complexity, we also apply the technique in [4] to our rectangular-to-polar converter. For the first iteration we move the input vector into 1_{-th} and 4_{-th} quadrant with simply sign inversing and data exchanging. Second we replace y_i by $y_i \cdot 2^{-i}$ as compared with conventional CORDIC algorithm. This modification can save once iteration and one barrier shifter in the rectangular-to-polar converter. This can save more area in our design. For i=1 and input vector is (x_1, y_1) from the EDGE signal:

$$\begin{aligned} x_2 &= d_1 \cdot y_1, \\ y_2 &= -d_1 \cdot x_1, \\ z_2 &= 0.5 \cdot d_1. \end{aligned} \quad (6)$$

$$d_i = sign(y_i) = \begin{cases} -1, & y_i < 0 \\ 1, & y_i \geq 0 \end{cases}.$$

And the remaining iterations (for i=2~n) are shown in Eq. (7).

$$\begin{aligned} x_{i+1} &= x_i + d_i \cdot 2^{-2(i-2)} \cdot y_i, \\ y_{i+1} &= 2 \cdot [y_i - d_i x_i], \\ z_{i+1} &= z_i + d_i \cdot p_i. \end{aligned} \quad (7)$$

$$p_i = \frac{1}{\pi} \cdot \tan^{-1}(2^{-(i-2)}),$$

$$K = \frac{1}{\sqrt{\sum_i 1 + 2^{-2(i-2)}}}.$$

The desired phase is z_{i+1} and the desired envelope value is x_{i+1} multiplied by a constant scaling factor K. Due to the iterative feature of CORDIC algorithm, the clock rate of this module is n*f_{clk}, and n is iteration number. It is hard for the module to operate at such high clock rate. A compromise is to use unfolded technique and the architecture is shown in Fig. 2.

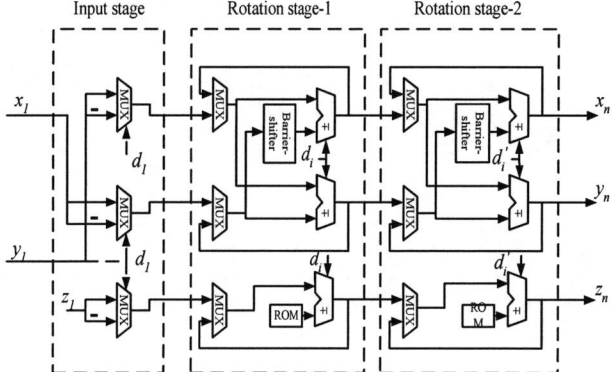

Fig. 2. Architecture of rectangular-to-polar converter.

4. DIGITAL PHASE MODULATOR

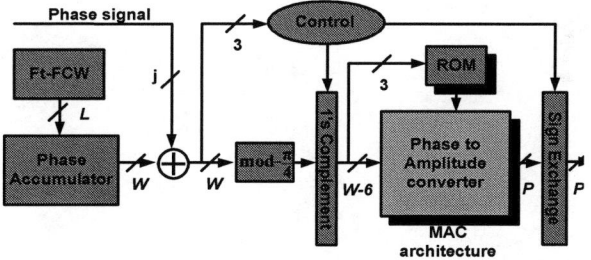

Fig. 3. Architecture of DDFS

The DDFS architecture is shown in Fig. 3. The DDFS has three basic blocks: FCW table, phase accumulator and phase-to-amplitude converter. The FCW table stores the desired fine-tune frequency control words and can be derived from Eq. (8).

$$f_c = \frac{FCW \cdot f_{clk}}{2^L}, \quad \forall \quad FCW < 2^{L-1} \quad (8)$$

In our design we focus on the phase-to-amplitude converter design and propose an architecture which is based on *Least Squared* (LS) algorithm [5] and *Merged-Multiply Accumulator* (MAC) technique [6]. The input phase is first truncated by 3-bit according to the $\pi/4$ *symmetry* and the amplitude of the sine function can be express by the polynomial. The approximated polynomial is generated according to the LS algorithm. In this paper we compare the *Spurious Free Dynamic Range* (SFDR) performance with the other approximation algorithm such as Taylor and Chebyshev [9]. The comparison method is set the input phase from 0 to $\pi/2$. The phase word-length is 15-bit and amplitude output is 15-bit. From the simulation result in Fig. 4, we can easily see that the LS-based polynomial can achieve better performance than Taylor and Chebyshev approximation algorithm with less polynomial order. The less order of polynomial means that low hardware complexity can also be achieved.

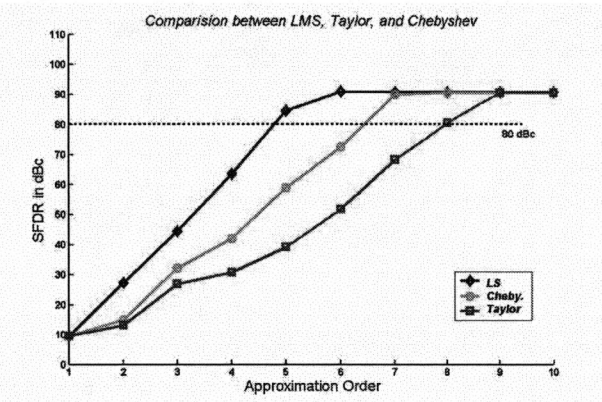

Fig. 4. SFDR comparison between LS, Taylor and Chebyshev.

In order to reduce the polynomial order we further divide the approximated region into eight segments. In each segment, the approximated polynomial $p(X)$ can be represented as in Eq. (9).

$$p(X) = c_2 \cdot X^2 + c_1 \cdot X + c_0$$
$$= \sum_{i=0}^{n_1-1} R_i \cdot 2^i + [c_1]_{n_2} \cdot [X]_{n_3} + \sum_{k=0}^{n_4-1} C_k \cdot 2^k \quad (9)$$
$$= MAC([rom\_1]_{n_1} + \sum_{j=0}^{n_2/2} Q_j \cdot [X]_{n_3} \cdot 4^j + [rom\_2]_{n_4}).$$

$$Q_j = -2c_{1,2j+1} + c_{1,2j} + c_{1,2j-1},$$
$$c_{1,j} = 0,1 \quad \text{and} \quad c_{1,-1} = 0, \quad (10)$$
$$[\ ]_n : \text{denote the truncation with } n-\text{bit}.$$

Where c_i represents the coefficient, and X is the phase of each divided region. In Eq. (9) we store the first term and third term in the look-up table. The size of rom_1 and rom_2 are 1,536 bits and 232 bits respectively. The operations in Eq. (9-10) now become one booth multiplication and two constant additions. These can be merged into a modified-MAC (Fig. 4).

First the binary phase X is inputted to the booth decoder circuit and the partial product term is generated in each row of MAC. The partial product terms are summed through *Carry-Save-Adder* (CSA) tree. As compared with the direct implementation of 2th order polynomial, the CSA tree can prevent the carry ripple problem in the early stages, and the carry ripple only occurs at the final stage. Due to the EDGE spectral requirement we target the desired SFDR over 80dBc. From Matlab simulation, we set the truncated accumulated phase word-length to W=15 bits and amplitude word-length to P=14 bits. These hardware parameters can achieve SFDR=86dBc. The other parameter is the word-length of the phase of the EDGE signal. This will also introduce phase noise and spurs in the output spectrum and we will discuss in section 5. The proposed DDFS circuit is simulated by the *NANOSIM* tool and compares with state of the art in Table 1. It is obvious that the proposed DDFS can achieve high SFDR performance. The power efficiency is also superior to the other designs.

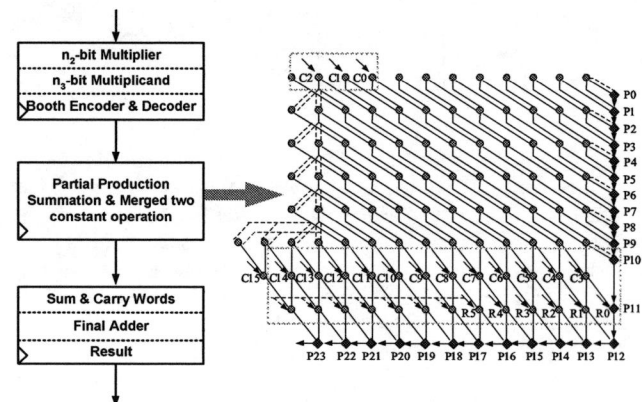

Fig. 4. Architecture of Modified-MAC.

Table 1. Comparison with the existing DDFS designs.

DDFS	CMOS tech.	SFDR	Latency	Power efficiency (mW/MHz)
Ours	0.18	86	5	0.15
Ref [7]	0.18	84	-	0.22
Ref [8]	0.25	90.3	13	0.66
Ref [9] (Taylor)	0.35	82.5	9	0.26
Ref [9] (Chebyshev)	0.35	73	7	0.35
Ref [10]	0.35	80	2	0.44

5. SIMULATION RESULT

For *Mobile Station* (MS), the requirements of EVM-rms and EVM-peak are below 9% and 30%. For *Base-Tranceiver Station* (BTS) EVM-rms and EVM-peak are below 7% and 22%. The SFDR performance of the digital frequency synthesizer is suitable for the up-link and down-link spectral requirement. But the phase signal word-length also contributes spurs and phase noise. And the wordlength also affects the EVM and the signal spectrum. In this paper we simulate the finite word-length (J) effect of the phase signal with the EVM measurement and spectral mask requirement. The performance summary is in Table 2.

Table 2. Simulation result and EVM measurement.

J-bits	EVM-rms	EVM-peak	Spectral requirement
9-bits	0.028%	0.094%	No (Spurs at -66dBc)
10-bits	0.014%	0.046%	No (Spurs at -74dBc)
11-bits	0.007%	0.018%	No (Spurs at -79dBc)

| 12-bits | 0.003% | 0.011% | Yes (Spurs at -81dBc) |

| Power consumption@26MHz | 3.92 mW |

From the Table 2, we can see that the errors produced by the phase quantization are very small for the word-length higher than 9-bits. And the errors introduced by the entire digital phase modulator can be eliminated. But the spectrum of the $S_{IF-PM}(t)$ signal is not exactly below the spectral mask. Especially for BTS-mask, the requirement of the mask is more stringent than MS-mask. Since the quantization phase error will degrade the synthesizer SFDR performance. It is conservative to choose J=12-bit in our design. The signal spectrum with J=12-bit at the carrier which equals to 8 MHz is shown in Fig. 5. The digital phase modulated signal generated by the DDFS can meet the spectral requirement for BTS-mask and MS-mask.

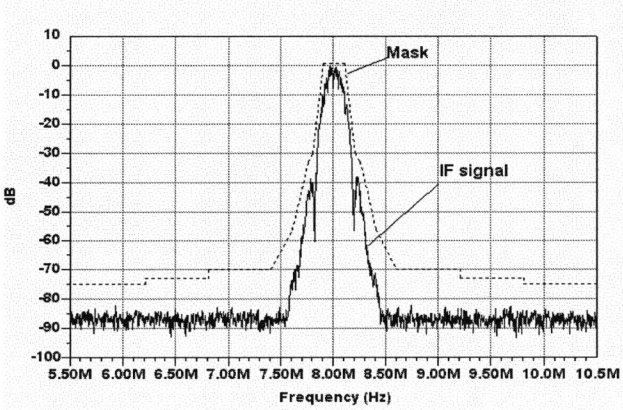

Fig. 5. The spectrum of the EDGE signal through DSP engine.

6. IMPLEMENTATION RESULT

The proposed DSP engine was implemented in *UMC 0.18 um* CMOS process with *1P6M* technology. The layout of the DSP engine is shown in Fig. 6. The summary of the circuit is list in Table 3.

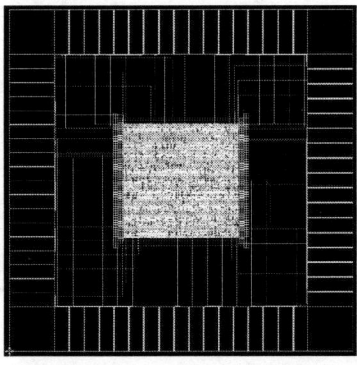

Fig. 6. layout of the proposed DSP engine.

Table 3. Implement summary of the DSP engine.

Technology	UMC 0.18 um 1P6M CMOS
Voltage	1.8 V
Core layout area	0.51x0.51 mm^2
Chip layout area	1.114x1.114 mm^2
System clock Frequency	26MHz

7. CONCLUSION

In this paper, we proposed the DSP engine for the polar transmitter. The engine is realized by the CORDIC and DDFS techniques. In the digital phase modulator we adopt the LS algorithm. We also apply MAC technique in our DDFS architecture to reduce the hardware complexity and decrease the carry ripple problem of the direct polynomial implementation. The chip implementation with *UMC 0.18 um* CMOS process with *1P6M* technology is also presented in this paper.

8. REFERENCES

[1] Nagle, P.; Burton, P.; Heaney, E.; McGrath, F., "A wide-band linear amplitude modulator for polar transmitters based on the concept of interleaving delta modulation," *IEEE Journal of Solid-State Circuits*, vol. 37, pp. 1748-1756, Dec. 2002.

[2] Rudolph, D.; "Out-of-band emissions of digital transmissions using Kahn EER technique," *IEEE Trans., Microwave Theory and Techniques*, vol. 50, pp. 1979-1983. Aug. 2002.

[3] Vankka, J.; "Digital frequency synthesizer/modulator for continuous-phase modulations with slow frequency hopping," *IEEE Trans., Vehicular Technology*, vol. 46, pp. 933-940, Nov. 1997.

[4] Chen, A.; Yang, S.; "Reduced complexity CORDIC demodulator implementation for D-AMPS and digital IF-sampled receiver," in *Proc. Globecom '98*, vol.3, pp.1491-1496, Nov. 1998.

[5] M. Flickner, J. Hafner, and E.J. Rodriguez, and J.L.C. Sanz; "Fast least-squares curve fitting using quasi-orthogonal splines," *IEEE Int. Image Processing*, vol. 1, pp. 686-690, Nov. 1994.

[6] Elguibaly, F.; "A fast parallel multiplier-accumulator using the modified Booth algorithm," *IEEE Trans., Circuits and Systems*, vol. 47, pp. 902-908, Sept. 2000

[7] Langlois, J.M.P.; Al-Khalili, D.; "Low power direct digital frequency synthesizers in 0.18 /spl mu/m CMOS," *IEEE CICC Proceedings*, pp. 21-24, Sept. 2003.

[8] A. Torosyan, Dengrwei Fu, Jr. Willson A. N., "A 300 MHz quadrature direct digital synthesizer/mixer in 0.25 μm CMOS," in *IEEE Solid-State Circuits Conference*, vol. 1, pp. 132-133, 2002.

[9] Kalle I. Palomaki and Jarkko Niittylahti; "Phase-to-Amplitude Mapping in Direct Digital Frequency Synthesizers Using Series Approximation," *EURASIP Journal on Applied Signal Processing*, 2001.

[10] D. De Caro, E. Napoli, A. G. M. Strollo, "Direct digital frequency synthesizers using high-order polynomial approximation," *IEEE* Solid-State Circuits Conference, vol. 1, pp. 134-135, 2002.

A Novel Technique for I/Q Imbalance and CFO Compensation in OFDM Systems[*]

Jui-Yuan Yu, Ming-Fu Sun, Terng-Yin Hsu, and Chen-Yi Lee
Department of Electronics Engineering, National Chiao-Tung University, Taiwan
E-mail: blues@ieee.org

Abstract—In this paper, a novel baseband IQ estimation and compensation technique is proposed to overcome the joint effects of CFO and IQ imbalance. The proposed method uses a one-shot algorithm which is able to estimate and compensate the non-ideal effect up to gain error 1 dB, phase error 10 degree, and CFO 50 ppm at carrier frequency 2.5GHz in 64QAM OFDM system. Under this condition, the system performance has less than 0.8 dB design loss in terms of PER and is shown to have more than 10 dB improvement compared with the reference design.

Index Terms—CFO, Direct Down Conversion, IQ Imbalance, OFDM, Zero IF

I. INTRODUCTION

Orthogonal Frequency Division Multiplexing (OFDM) [1][2] is an effective and spectrally efficient signaling technique for communications over frequency selective fading channels. Unfortunately, OFDM is also sensitive to non-ideal front-end effect and non-perfect synchronization, which lead to serious system performance degradation. It also causes heavy front-end specifications and results in an expensive front-end circuit. One of the key effects coming from non-ideal RF circuit is IQ imbalance, which is due to the gain and phase mismatch between in-phase (I) and quadrature-phase (Q) paths. More specifically, it occurs when the difference of the phase in I and Q channels from local oscillator is not exactly 90 degree and the gain is not the same.

Also, OFDM is sensitive to *Carrier Frequency Offset* (CFO) between transmitter and receiver [3]. In real front-end circuit, however, IQ imbalance and CFO will jointly occur and severely degrade the system performance. An I/Q correction scheme under a noisy Rayleigh fading channel is proposed in [4], but it does not take CFO effect into consideration. The correction method discussed in [5] can only compensate the mismatch with gain error up to 0.414 dB and phase error to 10 degree, which does not reach the conventional tolerance [8], say gain error 1dB and phase error 10 degree. Although, the algorithm in [6] consider jointly IQ and CFO phenomenon, the CFO tolerance has only the range from -47 kHz to 47 kHz, which does not even reach the half of conventional specification value, -125 kHz to 125 kHz [1]. Another approach [6], however, uses the iteration algorithm, which does not guarantee the convergence of estimation when the packet based WLAN standards are considered. Another approach [7] is to do the work by analog circuit.

In this paper, an all-digital IQ imbalance algorithm is proposed. We first describe the IQ mismatch model with presence of CFO. Then a novel scheme for IQ estimation is developed, including a robust algorithm for IQ correction in the presence of CFO, which can tolerate CFO up to 125 kHz, gain error 1dB, and phase error 10 degree.

II. SYSTEM MODEL

IQ imbalance arises when the signal in I and Q channels do not meet the orthogonality and the power balance. This effect can be modeled in two parameters: gain error ε and phase error φ. Based on the direct conversion architecture in [6][8][9], the system block can be depicted as Fig. 1.

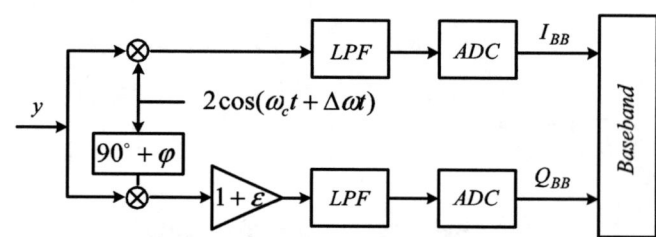

Fig. 1. OFDM receiver with IQ imbalance and CFO

The mathematical modeling can be derived from

$$\begin{aligned} r_{BB} &= I_{BB} + jQ_{BB} \\ &= \cos(\Delta\omega t)\Re\{y\} - \sin(\Delta\omega t)\Im\{y\} \\ &\quad + j(1-\varepsilon)\sin(\Delta\omega t + \varphi)\Re\{y\} \\ &\quad + j(1-\varepsilon)\cos(\Delta\omega t + \varphi)\Im\{y\} \end{aligned} \quad (1)$$

where I_{BB} and Q_{BB} is the baseband data in *I* and *Q* channel respectively which are distorted by IQ imbalance with CFO $\Delta\omega$ induced from the front-end, and *y* is the coming signal in the receiver. Also, (1) can be further summarized as

$$r_{BB} = \xi \cdot y \cdot e^{j\Delta\omega t} + \sigma \cdot (y \cdot e^{j\Delta\omega t})^* \quad (2)$$

---
[*] Work supported by MOEA of Taiwan, ROC, under Grant 93-EC-17-A-03-S1-0005

Thus the received signal can be regarded as a gain, say *Signal Gain (SG)* ξ, of its original one added by the conjugate multiplied by a delta value, say *Mirror Gain (MG)* σ, which are given as

$$SG : \xi = \frac{1}{2}[1+(1+\varepsilon)e^{-j\varphi}]$$
$$MG : \sigma = \frac{1}{2}[1-(1+\varepsilon)e^{j\varphi}] \quad (3)$$

If neither gain nor phase error exists, *MG* will reduce to zero, and *SG* remains unit. Note that the phase rotation is inversed in the direction between the original signals and its conjugate if the CFO is present. This means the conventional compensation algorithm [3] which simply multiplies the data with an exponential term has to be modified in accordance with gain and phase errors. This will be discussed in section IV. In OFDM systems, r_{BB} is further transformed into frequency domain in the receiver,

$$\begin{aligned}R_{BB,n} &= DFT\{r_{BB}\}_N \\ &= \xi \cdot DFT\{x \cdot e^{j\Delta\omega t}\} + \sigma \cdot DFT\{(x \cdot e^{j\Delta\omega t})^*\} \\ &= \xi \cdot Y_n \cdot (D_n + ICI_n) + \sigma \cdot Y_{-n}^* \cdot (D_n^* + ICI_n^*) \\ &= \xi \cdot Y_n \cdot C_n + \sigma \cdot Y_{-n}^* \cdot C_n^* \end{aligned} \quad (4)$$

Y_n is the sub-carrier of received signals in one DFT block, N represents the DFT point number, and $-N$ in the footnote makes Y mirrored in block size N. D_n is the distortion due to CFO which is accompanied with inter-carrier interference ICI_n [3]. If y is convolved with a multipath channel, the data in receiver baseband is $Y_n = X_n \cdot H_n$, where X_n and H_n denote the original baseband signals and channel response respectively.

III. PARAMETER ESTIMATION

A. Estimation for CFO

CFO exists in any wireless communication systems. This is due to the imperfect PVT variation of the front-end circuit which results in the mismatch of the carrier frequency between transmitter and receiver. The effects on OFDM systems are revealed in [3]. In OFDM WLAN systems, there are two fields, short preamble and long preamble, for CFO estimation. For example [1], the short preamble consists of 10 periodic short symbols which lasts 0.8 us for coarse estimation, and the long preamble has only two symbols but it has longer period for fine estimation. Therefore CFO can be calculated by dividing two consecutive symbols which have the same value but a phase rotation exists as well. Then take its argument.

$$\Delta\hat{\omega} = \arg\{\frac{y \cdot e^{j\Delta\omega t_2}}{y \cdot e^{j\Delta\omega t_1}}\} \quad (5)$$

With its negative value, the coming signal in the receiver can be compensated.

B. Estimation of Gain and Phase Error

When estimating the IQ imbalance effect, there are several parameters have to be considered jointly: estimated channel response $\hat{H}$, SG, MG, and $\Delta\omega$. Some WLAN standards for example [1] have fixed pilot symbols for channel parameter estimation, and the subcarrier pattern is pre-known. Thus, the channel response can be estimated by dividing the received subcarrier by the preamble. The result, however, will be different when IQ imbalance is involved, and is interfered by a conjugate and mirror of itself. Therefore, the resulting estimated channel response is then

$$\begin{aligned}\hat{H}_n &= \frac{R_{BB,n}}{X_n} \\ &= \xi \cdot H_n \cdot C_n + \sigma \cdot \frac{X_{-n,N}^*}{X_n} \cdot H_{-n}^* \cdot C_n^* \end{aligned} \quad (6)$$

with $X_n \in \{-1,+1\}$ the preamble subcarrier. When $X_n \neq X_{-n,N}$, there is a large transition in the received channel response $\hat{H}$, and $\hat{\sigma}$ can be calculated with this property as illustrated in Fig. 2. As can be seen, there are several sharp transitions, and we take the average of $\hat{\sigma}$, which is estimated from those transitions, to be the *Mirror Gain*. The transition locations keep the same in every preamble since the pattern is pre-defined, and the relationship can be derived from (6).

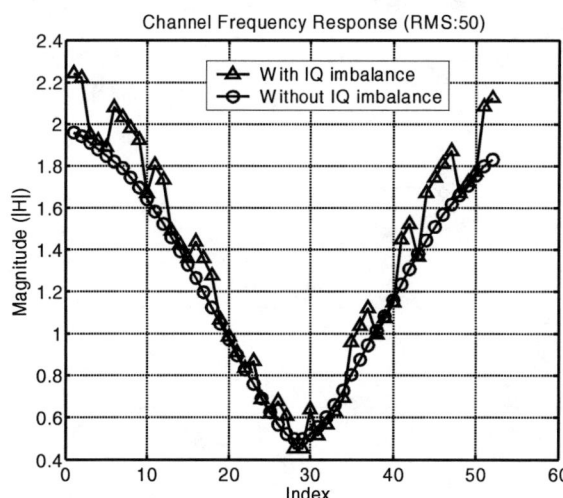

Fig. 2. The estimated channel response under IQ imbalance

In the following discussion, we take $n = 1$ and 2 to be the reference index for algorithm derivation. Thus,

$$\hat{H}_1 = \xi \cdot H_1 \cdot C_1 + \sigma \cdot H_{63}^* \cdot C_1^*$$
$$\hat{H}_2 = \xi \cdot H_2 \cdot C_2 - \sigma \cdot H_{62}^* \cdot C_2^* \tag{7}$$

with n = 0 the DC value. Subtracting $\hat{H}_1$ from $\hat{H}_2$, we'll have an estimation of MG,

$$\sigma = \frac{(\hat{H}_1 - \hat{H}_2) - \xi(H_1 C_1 - H_2 C_2)}{(H_{62} C_2 + H_{63} C_1)^*} \tag{8}$$

Similarly, we can find out $H_{62}C_2$ and $H_{63}C_1$ from (6), which are shown as

$$H_{62}C_2 = \frac{\hat{H}_{62} \cdot C_2 \cdot C_{62}^{-1} + \sigma \cdot H_2^* \cdot C_{62}^* \cdot C_{62}^{-1} \cdot C_2}{\xi}$$
$$H_{63}C_1 = \frac{\hat{H}_{63} \cdot C_1 \cdot C_{63}^{-1} - \sigma \cdot H_1^* \cdot C_{63}^* \cdot C_{63}^{-1} \cdot C_1}{\xi} \tag{9}$$

Substituting (9) into (8), and we can find that

$$\sigma = \frac{(\hat{H}_1 - \hat{H}_2) \cdot \xi^* - (H_1 C_1 - H_2 C_2) \cdot |\xi|^2}{\left[(\hat{H}_{62} C_2 C_{62}^{-1} + \hat{H}_{63} C_1 C_{63}^{-1}) - \sigma(H_1^* C_{63}^* C_{63}^{-1} C_1 - H_2^* C_{62}^* C_{62}^{-1} C_2)\right]^*} \tag{10}$$

From the above derivation, the MG changes its original behavior due to the existence of CFO. Before solving the equation for MG, we have first to reduce and simplify some terms in (10). To estimate MG and SG, we follow the conventional CFO compensation scheme to multiply the received signal with an inversed rotation term. Thus equation (4) is then reduced to

$$R_{BB,n} = \xi \cdot Y_n + \sigma \cdot Y_{-n}^* \cdot (C_n^*)^2 \tag{11}$$

Use (11) and follow the derivation in (6) ~ (10), we can find that (10) becomes

$$\sigma = \frac{(\hat{H}_n - \hat{H}_{n+1})\xi^* - (H_n C_n - H_{n+1} C_{n+1}) \cdot |\xi|^2}{(\frac{X_{-n}}{X_n} \hat{H}_{-n} C_n^* - \frac{X_{-n-1}}{X_{n+1}} \hat{H}_{-n-1}^* C_{n+1}^*) - \sigma^*(H_n C_{-n} C_n^* - H_{n+1} C_{-n-1} C_{n+1}^*)}$$
$$\cong \frac{(\hat{H}_n - \hat{H}_{n+1})\xi^*}{(\frac{X_{-n}}{X_n} \hat{H}_{-n} C_n^* - \frac{X_{-n-1}}{X_{n+1}} \hat{H}_{-n-1}^* C_{n+1}^*) - \sigma^*(H_n C_{-n} C_n^* - H_{n+1} C_{-n-1} C_{n+1}^*)} \tag{12}$$

with the reasonable assumption that the channel changes slowly in its frequency response. We also define some terms as

$$\begin{cases} m = (\hat{H}_{-n-1} \cdot C_{n+1} + \hat{H}_{-n} \cdot C_n)^* \\ n = H_n \cdot C_n^* \cdot C_{-n} - H_{n+1} \cdot C_{n+1}^* \cdot C_{-n-1} \\ p = (\hat{H}_n - \hat{H}_{n+1}) \cdot \xi^* \end{cases} \tag{13}$$

Thus we can solve MG and SG from the equation below.

$$\begin{cases} n|\sigma|^2 - m\sigma + p = 0 \\ \xi = 1 - \sigma^* \end{cases} \tag{14}$$

All the parameters in (14) are complex numbers, and solution can be found by setting the real part and imaginary part to be equal to zero. Each complex number is denoted separately as $x = (x_R, x_I)$. Therefore we'll have the equation as

$$\begin{cases} (\sigma_R^2 + \sigma_I^2)n_R - (\sigma_R m_R - \sigma_I m_I) + p_R = 0 \\ (\sigma_R^2 + \sigma_I^2)n_I - (\sigma_R m_I - \sigma_I m_R) + p_I = 0 \end{cases} \tag{15}$$

Some parameters u, v, w are also defined first.

$$\begin{cases} u = [1 + (\frac{m_R n_I - n_R m_I}{m_R n_R + m_I n_I})^2] n_R \\ v = \frac{2(m_R n_I - n_R m_I)(n_R p_I - n_I p_R) n_R}{(m_R n_R + m_I n_I)^2} + \frac{(m_R n_I - n_R m_I) m_I}{m_R n_R + m_I n_I} - m_R \\ w = \frac{(n_R p_I - n_I p_R)^2 n_R}{(m_R n_R + m_I n_I)^2} + \frac{(n_R p_I - n_I p_R) m_I}{m_R n_R + m_I n_I} + p_R \end{cases} \tag{16}$$

Rearrange the equation (15) with the notation in (16), it can be easily found

$$u\sigma_R^2 + v\sigma_R + w = 0 \tag{17}$$

Thus, σ_R can be solved by

$$\sigma_R = \frac{-v \pm \sqrt{v^2 - 4uw}}{2u} \tag{18}$$

Replace equation (16) with (19), and σ_I can be written as

$$\sigma_I = \frac{-m_I \pm \sqrt{m_I^2 - 4n_R(\sigma_R^2 n_R - \sigma_R m_R + p_R)}}{2n_R} \tag{19}$$

With the one-shot computation, MG is found from (18) and (19), and SG can be solved from taking the conjugate of MG and subtracting it from one as shown in (14). Note that, however, we cannot know SG and MG in the beginning of calculation since the parameter p involves the initial SG value. According to the simulation, we can follow equation (13)~(19) by first setting SG to be unit. After the first calculation, we put the results into (14)~(19) and solve them again, then accurate SG and MG can be always found in five iterations. Those calculations will be done in receiving the guard interval of an OFDM symbol, and will be ready for the compensation of the coming data.

IV. DATA COMPENSATION

With the information of estimated channel response and the existing CFO amount, we can calculate the necessary MG and SG for data compensation. With the mathematical derivation, the original data can be solved from

$$\hat{y} = \frac{\hat{\xi}^* \cdot r_{BB} - \hat{\sigma} \cdot r_{BB}^*}{|\hat{\xi}|^2 - |\hat{\sigma}|^2} \quad (20)$$

To illustrate the overall algorithm for estimation and compensation algorithm, we summary them as the block diagram in Fig. 3.

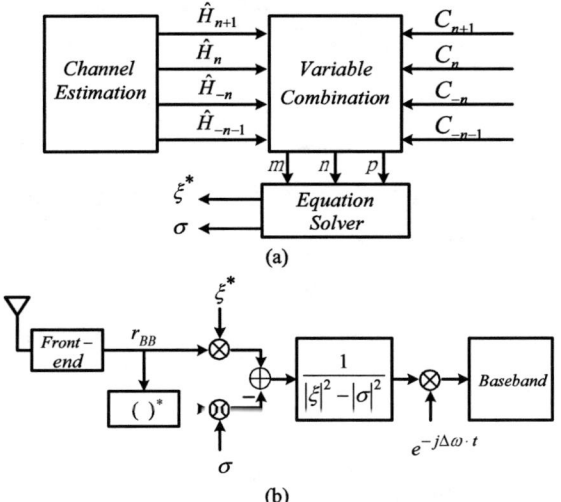

Fig. 3. (a) The block diagram for the IQ estimation (b) The compensation blocks for IQ imbalance and CFO

V. SIMULATION RESULTS

To evaluate the proposed algorithm, an OFDM system based on IEEE 802.11g [1] for WLAN is considered. With the presence of IQ imbalance, the gain error and phase error are set to be 1dB and 10° as the design target [8]. The CFO amount is simulated at the maximum value 50 ppm and 0 ppm respectively at a carrier frequency 2.5GHz. Also, the proposed algorithm is verified under the worst transmission condition, i.e., the highest data rate 54Mbps and use 64QAM with the OFDM modulation. This QAM order is most sensitive to any non-ideal effect in the simulation platform, especially the discussed IQ imbalance.

Simulation result shows that the compensated PER curve without CFO effect approaches the reference one in which only AWGN is applied, whereas the design loss is less than 0.8 dB with CFO 50 ppm. Both estimation and compensation are completed in a one-shot manner for each packet. Also, it is shown to have more than 10 dB system improvement compared with the reference design [5] as illustrated in Fig. 4.

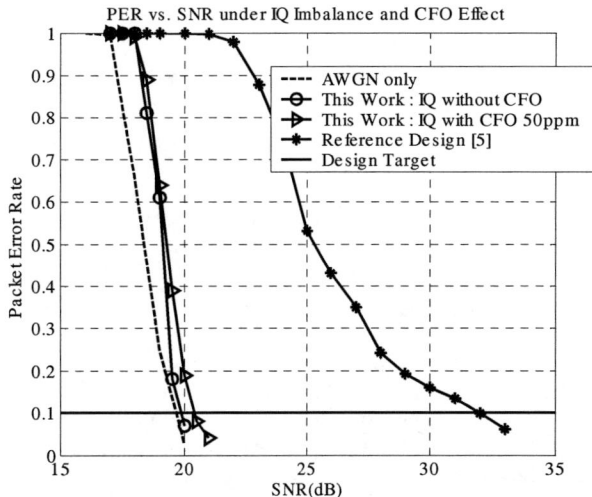

Fig. 4. The compensation result with gain error 1dB and phase 10°

VI. CONCLUSION

In this paper, an IQ imbalance model with CFO effect is investigated. It is also found that the joint effects of IQ imbalance and CFO degrade the system performance dramatically. Thus we propose a novel estimation algorithm and a robust compensation scheme which is shown to work well under gain error 1dB, phase error 10°, and maximum CFO tolerance 50 ppm at 2.5GHz carrier frequency. All of the estimations are done in the known preambles, and the estimated parameters converge fast. Therefore, our estimation and compensation algorithm makes a high performance digital OFDM receiver.

REFERENCES

[1] Supplement to standard for LAN/MAN part 11: Wireless MAC and PHY specifications: Further Higher Data Rate Extension in the 2.4GHz band, IEEE 802.11g, 2003.
[2] DVB-T – Framing structure, channel coding, and modulation for digital terrestrial television, Jan 2001.
[3] P. H. Moose, "A technique for orthogonal frequency division multiplexing frequency offset correction", IEEE Transactions on Communications, vol. 45, no. 12, December 1997.
[4] Ediz Cetin, Izzet Kale, Richard C. S. Morling, "On the structure, convergence, and performance of an adaptive I/Q mismatch corrector", Vehicular Technology Conference, 2002 Proceedings, Volume 4.
[5] Jan Tubbax, Andrew Fort, Liesbet Van der Perre, Stephane Donnay, Marc Engels, Hugo De Man, "Joint compensation of IQ imbalance and frequency offset in OFDM systems", Global Telecommunications Conference, 2003. GLOBECOM '03. IEEE , Volume: 4 , 1-5 Dec. 2003
[6] S. Fouladifard, H. Shafiee, "Frequency offset estimation in OFDM systems in presence of IQ imbalance", IEEE International Conference on Communications, Volume: 3, 11-15., 2003
[7] Sebastien Simoens, Marc de Courville, Francois Bourzeix, Paul de Champs, "New I/Q imbalance modeling and compensation in OFDM systems with frequency offset", Personal, Indoor and Mobile Radio Communications, Volume: 2 , 15-18 Sept. 2002.
[8] Behzad Razavi, "RF Microelectronics", Prentice Hall, 1998.
[9] Pengfei Zhang; Nguyen, T.; Lam, C.; Gambetta, D.; Soorapanth, T.; Baohong Cheng; Hart, S.; Sever, I.; Bourdi, T.; Tham, A.; Razavi, B., "A 5-GHz Direct Conversion CMOS Transceiver", Solid-State Circuits, IEEE Journal of , Volume: 38 , Issue: 12 , Dec. 2003, Pages:2232 - 2238

Architectural Issues in Base-Station Frequency Synthesizers

Sankaran Aniruddhan and David J. Allstot[1]

Dept. of Electrical Engineering, Box 352500, University of Washington, Seattle, WA 98195-2500

Abstract — Base station frequency synthesizers have extremely stringent specifications in terms of low integrated RMS phase error and low lock time. Satisfying both these conflicting specifications demands the selection of the right architecture. At the same time, other significant issues such as spur suppression and tuning range necessitate the use of allied techniques. In this paper, the different architectural choices available for this application are compared vis-à-vis their respective benefits and drawbacks. A Dual-Loop-PLL-based architecture that meets very strict specifications is designed and simulated at 2GHz. This synthesizer has an integrated RMS phase error of 1° while having a phase noise of -120dBc/Hz @ 600kHz offset. The lock time is 40μs, and the tuning range is 100MHz.

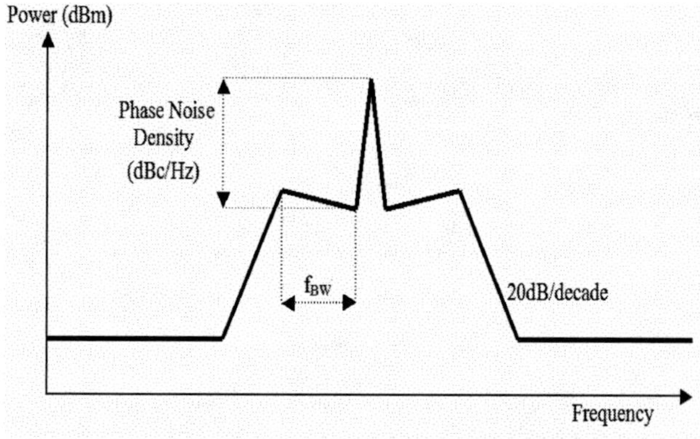

Fig. 1. PLL Output Spectrum

I. INTRODUCTION

The most significant trend in the semiconductor industry in last decade has been the emergence of CMOS system-on-chip (SoC) solutions in wireless communications. With traditionally "discrete" wireless circuits quickly moving into the SoC domain, new challenges are surfacing. Usually specific to a particular application, they require a combination of established solutions and unconventional techniques to achieve the state-of-art performances.

Base-stations are central radio transceivers that channel signals between mobile subscribers within a given range. Handling wireless traffic from multiple subscribers simultaneously, they have extremely stringent specifications for the RF front-end, especially for the frequency synthesizer. Traditionally, these have been implemented using high-quality off-chip components like ceramic resonators and hyper-abrupt junction varactors.

In this paper, we discuss the design issues of fully integrated base-station synthesizers, which are critical in meeting the demanding receiver specifications. Section II details the RF specifications of the synthesizer, while Section III compares and contrasts the different PLL architectures suitable for this application. The design and optimization of a 2GHz dual-loop synthesizer meeting the strict specifications are presented in Section IV. Finally, the paper is concluded in Section V with preliminary simulation results.

II. BASE-STATION SPECIFICATIONS

The typical phase noise spectrum of a PLL, as shown in Fig. 1, has two major regions: inside the loop bandwidth, where it depends on the noise characteristics of the reference oscillator, charge pump and loop filter and has a fairly constant value; and outside the loop bandwidth, where the phase noise has a slope that corresponds to the VCO phase noise [1].

Base-station synthesizers have a number of strict specifications on the phase noise, lock time and other performance metrics of the PLL. In particular, the following parameters are quite crucial to its overall performance: RMS Phase Error, Lock Time, In-Band Noise Floor, Out-of-Band VCO Phase Noise, Spur Level and Channel Frequency.

RMS phase error can be determined from the VCO spectrum by integrating both phase noise {L(Δf)} sidebands:

$$\text{RMS Phase Error (in deg.)} = \frac{180}{\pi} \cdot \sqrt{2 \cdot \int_a^b L(\Delta f) d\Delta f} \quad (1)$$

where L(Δf) is the phase noise and *a* & *b* are frequency offsets whose value depends on the particular standard targeted. The RMS phase error and Lock-time specifications are conflicting ones. Low RMS phase error implies a small PLL bandwidth, which in turn, degrades the transient response. Therefore, other techniques (discussed in section IV) must be employed to reduce the lock time of the synthesizer.

The *order* of the PLL is usually chosen such that the phase margin can be fixed independent of other PLL parameters such as loop bandwidth, noise floor etc. To achieve stable operations, phase margins of 45-60° are common.

III. LOOP ARCHITECTURES

There are three major classes of PLLs – Integer-N, Fractional-N and Dual-Loop [2] – each offering its advantages and drawbacks depending upon the application.

Integer-N PLL

An Integer-N PLL consists of a phase detector, loop filter, a voltage-controlled oscillator (VCO) and a pulse-swallow divider, as shown in Fig. 2. It has been extremely popular due to its simple architecture, easy implementation and low-cost in terms of design time. However, Integer-N PLLs do have quite a few

---
[1] Research supported by NSF ITR Grant # CCR - 0086032

disadvantages. An integral division ratio constrains the reference frequency to be exactly equal to the channel frequency. This results in major drawbacks including: (a) large division ratios - increased area and power, and (b) low-bandwidth – large lock times.

To maintain good stability, the loop bandwidth is restricted to ~1/10th the reference frequency. This constrains the bandwidth to be very small, and thus significantly increases the lock time of the PLL. For example, a synthesizer with 2GHz carrier frequency and 100kHz channel frequency requires a division ratio of 20,000 operating at full-rate (2GHz) or close to full-rate, and a bandwidth < 10kHz. This would entail a 15-bit divider, and result in long switching times of hundreds of μs. A further disadvantage is that Integer PLLs exhibit spurs at multiples of the reference (channel) frequency, lying inside the band of interest. These spurs increase the RMS phase error, while acting as interferers on the LO. Clearly, the Integer-N PLL is not a suitable choice for an integrated solution, especially for a high-performance system.

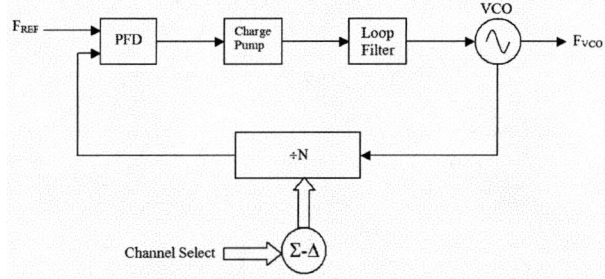

Fig. 3.　Fractional-N PLL Architecture (with SDM)

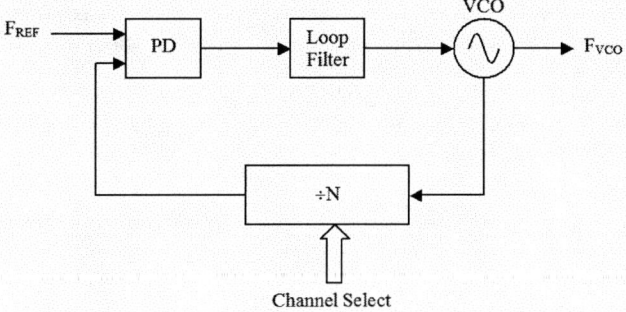

Fig. 2.　Integer-N PLL Architecture

Fractional-N PLL

Fractional-N PLLs, as shown in Fig. 3, use Sigma-Delta Modulators (SDMs) to adjust the instantaneous division ratio while keeping the average value fixed, thus enabling non-integer division ratios [3]. Effectively, the reference frequency and the channel frequency are decoupled from each other, allowing design of synthesizers with higher reference frequencies and bandwidths and lower lock times, while still having small channel frequencies [2]. This also eases the phase noise requirements of the VCO. The reference frequency is an integral multiple of the channel frequency, and this pushes the reference spurs further away from the carrier where the loop filter can filter them out more effectively [1].

The principal drawback of Fractional-N synthesizers is the presence of fractional spurs in the output at multiples of the channel frequency that are caused by periodic disturbances on the control voltage node. SDMs reduce the fractional spur levels considerably, by high-pass shaping of the noise spectrum. However, these are still significant in some high-performance systems.

Dual-Loop PLL

In the Dual-Loop PLL (Fig. 4), a wideband primary loop that operates at a fairly high reference frequency produces the overall output of the synthesizer, and is referred to here as the Main PLL (M-PLL). An RF mixer is used in the feedback path to perform part of the frequency translation [2], [4], [5], whose output is then divided down to the reference frequency using a fixed-ratio divider. The other input of the mixer comes from a second narrowband PLL that is known as the Reference PLL (R-PLL). This PLL can be an Integer-N PLL, or a Fractional-N PLL with a simple 1st or 2nd order SDM.

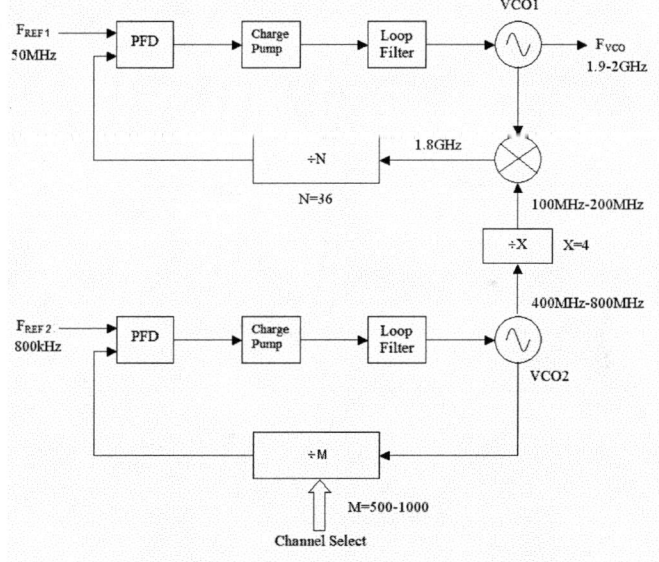

Fig. 4.　Block Diagram of the Dual-Loop Architecture

The synthesizer switches channels through a programmable divider in the R-PLL. If the R-PLL has an Integer-N topology, its reference frequency is equal to the channel frequency, while its bandwidth is much smaller. The spurious harmonics and sidebands produced by the mixer lie far away from the carrier, and since the mixer is placed inside the loop, get attenuated by the loop filter [2]. A divider can be placed immediately after the R-PLL to suppress its phase noise [4]. However, this increases the tuning range of the R-PLL by the same amount, and therefore requires careful frequency planning.

IV. DESIGN AND OPTIMIZATION

As shown in the previous section, the dual-loop PLL architecture is a reasonable architectural choice for integrated realization of high-performance synthesizers. For validation, a 2GHz Dual-Loop PLL was implemented in a 0.18μm IBM RF-CMOS process. System-level simulations of stability and phase noise were performed using MATLAB, while circuit level simulations were done in *SpectreRF*. Initial analysis shows that a main loop bandwidth of around 50kHz achieves an integrated RMS phase error of 1°, and this was verified using MATLAB.

Frequency Planning

The overall frequency plan for the system is depicted in Fig. 4. The reference frequency of the M-PLL was chosen to be 50MHz, and the IF frequency was fixed at 1800MHz. A divide-by-36 circuit divides the IF down to the reference frequency. The tuning range of the main VCO is 1900-2000MHz, while that of the auxiliary R-PLL is 100MHz-200MHz. A divide-by-4 block is placed between the R-PLL and the M-PLL to suppresses the noise of the R-PLL by 12dB. This ensures minimum contribution of the R-PLL to the output phase noise. However, it increases the actual tuning range of the R-PLL fourfold to 400MHz-800MHz. The R-PLL is implemented as an Integer-N loop, with a divide-ratio of 500-1000. An 800kHz reference frequency is used for the R-PLL to achieve a channel frequency of 200kHz.

Building Blocks

A dead-zone-free digital phase-frequency detector (PFD) and a differential offset-cancelled charge-pump with symmetric loads were implemented as in [6]. These two blocks can be replicated in both loops with minimal adjustments, thus saving valuable design time. Passive loop filters were chosen over active ones to preserve simplicity. A third-order loop filter was implemented in the main loop to achieve maximum out-of-band suppression of noise and spurs while preserving loop stability. A Gilbert-cell mixer can be used to combine the outputs of the two loops. Any undesired frequency components at the output of the mixer can produce spurs in the PLL output. Therefore, an image-reject architecture has to be utilized to suppress the image component in the mixer output [4]. LC-tuned loads will further suppress spurs to the maximum extent. Device matching influences RF and LO feedthrough to the output. Therefore great care is required in layout and use of techniques such as fingering, inter-digitation etc may be necessary.

The VCO in the M-PLL was implemented as an LC-VCO due to its demanding phase noise requirement. A cross-coupled NMOS-only architecture was chosen because it can potentially have better phase noise than a complementary structure when slightly more power is dissipated [7]. To meet the exceedingly demanding specifications without external inductors, bondwire inductors with a Q of ~35 used to realize the LC-tank. Frequency tuning was accomplished using Hyper-abrupt Junction Varactors (HAV) available in this particular process. This ensures an exceedingly linear tuning range, as well as a low VCO gain. The VCO also employed digital switched-capacitor coarse-tuning to obtain low VCO gain and large tuning range. A series resistor-inductor combination was used to bias the VCO, which eliminates any flicker noise up-conversion from the bias circuitry, while filtering any unwanted ground noise [8]. The R-PLL VCO requires a 67% tuning range, which indicates that it may probably be realized as a voltage controlled ring oscillator (VCRO). However, since the phase noise requirements are quite stringent, a better option is to design an LC-VCO with band-switching capacitors. Further, since the HAV diodes can operate at higher power supplies with linear tuning, higher-voltage IO devices may be used to implement the charge pump circuit, thereby increasing the control voltage range to boost the tuning range.

Lock-time Issues

As the bandwidth of the main loop is approximately 50kHz, the lock time is greater than 250μs. To accelerate this locking, two techniques were explored. First, the acquisition process of the PLL can be divided into a coarse and fine locking sequence. In the coarse process, a frequency discriminator/detector and the VCO coarse tuning achieve a frequency lock to the required channel [9]. The PLL then enters the fine sequence, where phase locking takes place. In the second technique, the loop filter component values are modulated such that during the locking process, the bandwidth of the PLL is increased, while in steady state they return to their original values. The charge-pump current also needs to be modified appropriately such that PLL stability and phase margin are maintained during the locking process [10]. The latter technique gave better results for this architectural scheme and hence, was incorporated in the final implementation.

Noise Issues

A qualitative comparison may be made between conventional synthesizers and the dual-loop one based on phase noise. At first sight, the dual-loop architecture has more number of components, including an additional PFD, CP, LF and VCO and an RF mixer in the M-PLL. Considering the overall phase noise, at low frequency offsets, the in-band noise of the R-PLL is the dominant factor due its amplification by the R-PLL loop transfer function. This makes the design more flexible by decoupling the noise from other parameters of the RF M-PLL. Second, it will be relatively easier to achieve a lower noise floor in the low-frequency R-PLL. As in the case of the regular PLL, the phase noise at larger offsets is dominated by that of the M-PLL RF VCO due to the inherent high-pass-filtering action of the loop.

Two important points are to be noted regarding the noise of the mixer. First, any mixer phase noise is attenuated via the succeeding fixed-ratio divider in the main loop. Second, the inherent output phase noise of the mixer is actually quite low. In its usual application in transceiver systems, the mixer is preceded by an LNA precisely because it is a significant source of noise when referred to its RF input. However, viewed as an externally driven system, the mixer actually has a rather low noise floor as compared to other noise sources. Unlike autonomous systems like oscillators whose phase is free to drift, the phase of the mixer is well determined by the inputs. In the present design, phase noise simulations on Spectre show that the noise floor at the output of the mixer is below –140dBc/Hz at 10kHz offset from 1800MHz.

V. SIMULATION RESULTS

System level simulations for phase noise were performed in MATLAB, while PLL lock-time and other circuit level simulations were carried out in Spectre. The 2GHz LC-VCO in the M-PLL has a simulated phase noise of -125dBc/Hz@600kHz while consuming 10mW from a 1.8V power supply. It has a tuning range of 200MHz (1.9GHz to 2.1GHz) and a K_{VCO} (VCO Gain) of 35MHz/V. A bondwire inductance of 2nH was used in the LC-tank.

The charge pump sources and sinks a current of 10uA through the loop filter. The loop filter occupies an area of $0.3mm^2$, the majority of which is taken up by dual-MIM capacitors. The loop was designed for a phase margin of at least 60° to ensure good stability in the frequency range of interest.

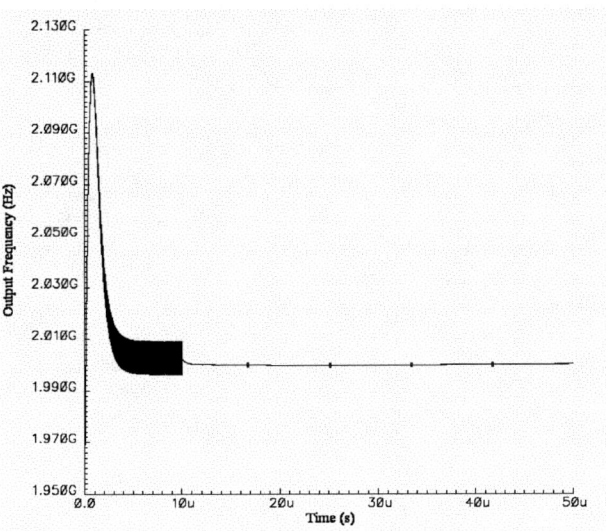

Fig. 5. Transient Response of the Dual-Loop PLL

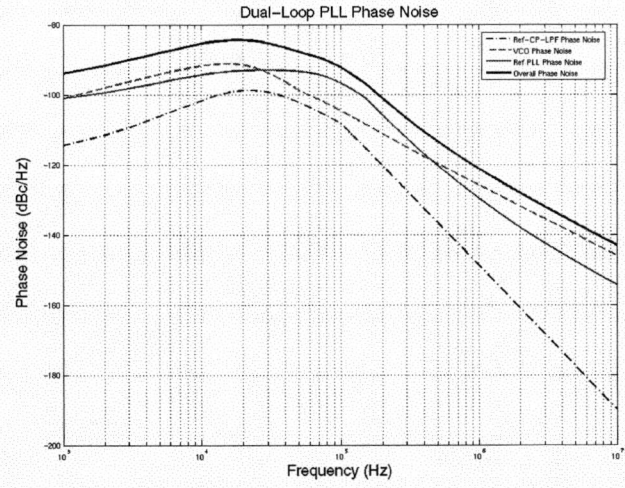

Fig. 6. Phase Noise of the Dual-Loop PLL

The transient response of the PLL to a frequency step is shown in Fig. 5. The lock time is 40µs. As previously discussed, capacitors were switched out of the loop filter to increase the bandwidth during the first phase (10µs) of the locking sequence. This can be implemented using a digitally-controlled counter that switches the capacitors after a specified time. The overall phase noise of the PLL with the important individual contributions is shown in Fig. 6. The mixer output phase noise, which is below −140dB, is not shown in the figure to preserve a reasonable scale on the plot. The integrated RMS phase error calculated as in [1] is 1°. The PLL performance is summarized in Table 1.

Table 1. Simulated PLL Performance Summary

PLL Parameter	Simulated Result
Center Frequency and Tuning range	1.9-2GHz
K_{VCO} (VCO Gain)	35MHz/V
PN@10kHz offset (flat region)	<−85dBc/Hz
PN@600kHz offset	<−120dBc/Hz
Lock Time	40µs
Channel Frequency	200kHz
RMS Phase Error [1]	1°

REFERENCES

[1] D. Banerjee, "PLL performance, simulation and design," *Dean Banerjee Pubns*, 3rd edition, Oct. 2003.

[2] B. Razavi, "Challenges in the design of frequency synthesizers for wireless applications," *IEEE Custom IC Conference*, pp. 395-402, May 1997.

[3] T.A.D. Riley, et al., "Delta-sigma modulation in fractional-N frequency synthesis," *IEEE J. Solid-State Circuits*, vol. 28, pp. 553-559, May 1993.

[4] T. Aytur, et al., "Advantages of dual-loop frequency synthesizers for GSM applications," *IEEE International Symposium on Circuits and Systems*, vol. 1, pp. 17-20, June 1997.

[5] T.K.K. Kan, et al, "A 2-V 1.8-GHz fully integrated CMOS dual-loop frequency synthesizer," *IEEE J. Solid-State Circuits*, vol. 37, pp. 1012-1020, Aug. 2002.

[6] J.G. Maneatis, "Low-jitter process-independent DLL and PLL based on self-biased techniques," *IEEE J. Solid-State Circuits*, vol. 31, pp.1723-1732, Nov. 1996.

[7] A. Hajimiri, et al., "Design issues in CMOS differential LC oscillators," *IEEE J. Solid-State Circuits*, vol. 34, pp. 717-724, May 1999.

[8] E. Hegazi, et al., "A filtering technique to lower LC oscillator phase noise," *IEEE J. Solid-State Circuits*, vol. 36, pp. 1921-1930, Dec. 2001.

[9] C.Y. Yang, et al., "Fast-switching frequency synthesizer with a discriminator-aided phase detector," *IEEE J. Solid-State Circuits*, vol. 35, pp. 1445-1452, Oct. 2000.

[10] National Semiconductor Corporation Application Note AN-1000, "A Fast Locking Scheme for PLL Frequency Synthesizers," © 1995 National Semiconductor Corporation.

INTEGRAL OBSERVER APPROACH FOR CHAOS SYNCHRONIZATIONN WITH TRANSMISSION DISTURBANCES

Guo-Ping Jiang[1], Wei Xing Zheng[2], Wallace Kit-Sang Tang[3], Guanrong Chen[3]

[1] Department of Electronic Engineering, Nanjing University of Posts & Telecommunications, Nanjing, 210003, P.R. China (e-mail: jianggp@njupt.edu.cn)

[2] School of Quantitative Methods and Mathematical Sciences, University of Western Sydney, Penrith South DC NSW 1797, Australia

[3] Department of Electronic Engineering, City University of Hong Kong, Kowloon, Hong Kong SAR, P.R. China

ABSTRACT

This paper addresses the issue of chaos synchronization with disturbances in the transmission channel. Using an integral observer approach, a new scheme for chaos synchronization is developed for a class of chaotic systems. Based on the Lyapunov stability theory, a sufficient condition is derived for chaos synchronization in the aforementioned setting. By using the Schur theorem and some matrix operation techniques, this criterion is then transformed into a Linear Matrix Inequality form, which can be easily verified and solved by using the MATLAB LMI Toolbox. It is then shown that under the proposed scheme and derived criterion, the effect of the transmission disturbances can be greatly reduced, and consequently chaos synchronization is achieved satisfactorily. The chaotic Murali-Lakshmanan-Chua system is simulated to verify the effectiveness of the scheme and to validate the criterion suggested in this paper.

1. INTRODUCTION

During the past ten years or so, chaos synchronization has attracted increasing interests [2-6,8-11], and found many applications in secure communications [2,4,12], based on chaotic masking, chaotic shifting keying, chaotic modulation, etc. Chaos synchronization can be viewed from a state observer perspective, in the sense that the response system can be considered as the state observer of the drive system [2,4,8,9].

Recently, it has been reported that there is a limitation in the traditional proportional state observer, where the output or measurement disturbance is amplified by the proportional gain [1]. To overcome this limitation, the so-called integral observer has been proposed for linear systems, and extended to a class of nonlinear systems lately. It has been shown that the integral observer has a better performance in reducing measurement or output disturbances as compared to the traditional proportional observer [1].

In this paper, we apply the integral observer approach to chaos synchronization with disturbance in the transmission channel. Firstly, a new scheme of chaos synchronization is proposed for a class of chaotic systems. Then, a new sufficient condition is derived for chaos synchronization based on the Lyapunov stability theory. By using the Schur theorem and some matrix operation techniques [7], this criterion is transformed into the Linear Matrix Inequality (LMI) form, which can be easily verified and solved by using the MATLAB LMI Toolbox. It is then shown that under the proposed scheme and derived criterion, the effect of transmission disturbances can be greatly reduced, consequently chaos synchronization is achieved satisfactorily. Finally, the proposed scheme and derived criterion are applied to the chaotic Murali-Lakshmanan-Chua (MLC) system [5] for illustration.

2. THE PROPORTIONAL OBSERVER APPROACH AND THE PROBLEM STATEMENT

Consider a chaotic system described by

$$\dot{\mathbf{x}} = \mathbf{A}\mathbf{x} + \mathbf{g}(\mathbf{x}) + \mathbf{h}(t)$$
$$y = \mathbf{C}\mathbf{x} + d \qquad (1)$$

where $\mathbf{A} \in \mathbf{R}^{n \times n}$, $\mathbf{C} \in \mathbf{R}^{n \times 1}$, $(\mathbf{A}, \mathbf{C})$ is observable, $\mathbf{x} \in \mathbf{R}^{n \times 1}$ is the state variables, $\mathbf{h}(t) \in R^{n \times 1}$ is the external input signal, d is the bounded disturbance or noise in the transmission channel, y is a scalar output chaotic signal, which is interfered with the disturbance, d, and will be transmitted to the response system, and $\mathbf{g}(\mathbf{x})$ is a continuous nonlinear function satisfying Lipschitz condition, namely,

$$\|\mathbf{g}(\mathbf{x}) - \mathbf{g}(\tilde{\mathbf{x}})\| \le \rho \|\mathbf{x} - \tilde{\mathbf{x}}\| \qquad (2)$$

where $\|\cdot\|$ denotes the Euclidean norm and ρ is the Lipschitz constant.

Based on the traditional proportional state observer approach, the response system can be constructed as follows [2,4,8,9]:

$$\dot{\tilde{\mathbf{x}}} = \mathbf{A}\tilde{\mathbf{x}} + \mathbf{g}(\tilde{\mathbf{x}}) + \mathbf{h}(t) + \mathbf{L}(y - \tilde{y})$$
$$\tilde{y} = \mathbf{C}\tilde{\mathbf{x}} \qquad (3)$$

where $\mathbf{L} = [l_1, l_2, \cdots, l_n]^T$ is the observer gain to be designed.

From (1) ~ (3), one can get the following error dynamical system:

$$\dot{\mathbf{e}} = \dot{\mathbf{x}} - \dot{\tilde{\mathbf{x}}} = \mathbf{Ax} + \mathbf{g}(\mathbf{x}) + \mathbf{h}(t) - (\mathbf{A}\tilde{\mathbf{x}} + \mathbf{g}(\tilde{\mathbf{x}}) + \mathbf{h}(t) + \mathbf{L}(y - \tilde{y}))$$
$$= (\mathbf{A} - \mathbf{LC})\mathbf{e} + \mathbf{g}(\mathbf{x}) - \mathbf{g}(\tilde{\mathbf{x}}) - \mathbf{L}d \quad (4)$$

where $\mathbf{e} = \mathbf{x} - \tilde{\mathbf{x}}$ denotes the error vector.

From the error dynamical system (4), one can see that the term, $\mathbf{L}d(t)$, caused by the disturbance $d(t)$, will lead the error vector $\mathbf{e}$ not to tend to zero asymptotically. In particular, if a high proportional gain $\mathbf{L}$ is chosen, then the disturbance will be amplified, and consequently $\mathbf{e}$ will be far away from zero. As is known, this is an inherent disadvantage of the traditional proportional observer [1].

3. INTEGRAL OBSERVER APPROACH AND SYNCHRONIZATION CRITERION

To overcome the above limitation, for chaos synchronization based on the traditional proportional observer, we propose a new scheme using the integral observer here.

Consider system (1). Set $x_0 = \int_0^t y(\tau)d\tau$ so that $\dot{x}_0 = y = \mathbf{Cx} + d$. Then, we can obtain the following augmented system:

$$\begin{bmatrix} \dot{\mathbf{x}} \\ \dot{x}_0 \end{bmatrix} = \begin{bmatrix} \mathbf{A} & \mathbf{O} \\ \mathbf{C} & 0 \end{bmatrix} \begin{bmatrix} \mathbf{x} \\ x_0 \end{bmatrix} + \begin{bmatrix} \mathbf{g}(\mathbf{x}) \\ 0 \end{bmatrix} + \begin{bmatrix} \mathbf{h}(t) \\ 0 \end{bmatrix} + \begin{bmatrix} \mathbf{O} \\ 1 \end{bmatrix} d \quad (5)$$
$$y_0 = x_0$$

Letting $\mathbf{z} = \begin{bmatrix} \mathbf{x} \\ x_0 \end{bmatrix}$, $\mathbf{F}(\mathbf{z}) = \begin{bmatrix} \mathbf{g}(\mathbf{x}) \\ 0 \end{bmatrix}$, $\mathbf{H}(t) = \begin{bmatrix} \mathbf{h}(t) \\ 0 \end{bmatrix}$, $\mathbf{E} = \begin{bmatrix} \mathbf{O} \\ 1 \end{bmatrix}$, and $\mathbf{A}_1 = \begin{bmatrix} \mathbf{A} & \mathbf{O} \\ \mathbf{C} & 0 \end{bmatrix}$, we can rewrite system (5) as

$$\dot{\mathbf{z}} = \mathbf{A}_1 \mathbf{z} + \mathbf{F}(\mathbf{z}) + \mathbf{H}(t) + \mathbf{E}d \quad (6)$$
$$y_0 = \mathbf{C}_1 \mathbf{z}$$

where $\mathbf{C}_1 = [\mathbf{O} \quad 1]$. Notice that the pair $(\mathbf{A}_1, \mathbf{C}_1)$ is observable if $(\mathbf{A}, \mathbf{C})$ is observable.

The response system, or observer for (6), can be reconstructed by

$$\dot{\tilde{\mathbf{z}}} = \mathbf{A}_1 \tilde{\mathbf{z}} + \mathbf{F}(\tilde{\mathbf{z}}) + \mathbf{H}(t) + \mathbf{L}_1 (y_0 - \tilde{y}_0) \quad (7)$$
$$\tilde{y}_0 = \mathbf{C}_1 \tilde{\mathbf{z}}$$

where $\mathbf{L}_1 = [l_1 \quad l_2 \quad \cdots \quad l_{n+1}]^T$ is the observer gain vector.

From (6) and (7), we get the error dynamical system,

$$\dot{\mathbf{e}}_1 = \mathbf{A}_1 \mathbf{e}_1 + \mathbf{F}(\mathbf{z}) - \mathbf{F}(\tilde{\mathbf{z}}) - \mathbf{L}_1(y_0 - \tilde{y}_0) + \mathbf{E}d$$
$$= (\mathbf{A}_1 - \mathbf{L}_1\mathbf{C}_1)\mathbf{e}_1 + \mathbf{F}(\mathbf{z}) - \mathbf{F}(\tilde{\mathbf{z}}) + \mathbf{E}d \quad (8)$$

where $\mathbf{e}_1 = \mathbf{z} - \tilde{\mathbf{z}}$.

Remark 1: From the above, we can see that the variable, $y_0 = x_0 = \int_0^t y(\tau)d\tau$, is chosen as a new output signal (drive signal), and used to design the state observer for the augmented system (5) or (6). This leads to no disturbance term in the output equation of the augmented system (6). Comparing (4) and (8), we see that the term $\mathbf{E}d = [\mathbf{O} \quad 1]^T d$ is independent of the feedback gain $\mathbf{L}_1$, which will provide some flexibility for choosing the feedback gain $\mathbf{L}_1$. Therefore, we can choose a reasonably high gain $\mathbf{L}_1$ such that the stabilizing term prevails over the perturbation term $\mathbf{E}d$, and consequently, it will result in a very small error between the two systems (6) and (7), as will be made clearer in the following. On the other hand, the integral term, $\int_0^t y(\tau)d\tau$, acts as the function of a low-pass filter, which can greatly reduce the effect of high frequency disturbances contained in the output signal.

Theorem 1: If a suitable feedback gain $\mathbf{L}_1$ is selected such that

$$\mathbf{A}_1^T \mathbf{P} + \mathbf{P}\mathbf{A}_1 - \mathbf{C}_1^T \mathbf{L}_1^T \mathbf{P} - \mathbf{P}\mathbf{L}_1\mathbf{C}_1 + \rho^2 \mathbf{P}\mathbf{P} + \mathbf{I} + 2\delta\mathbf{P} < 0 \quad (9)$$

where $\mathbf{P}$ is a positive definite symmetric matrix, $\mathbf{I}$ is the identity matrix, and δ is a positive constant, then the error dynamical system (8) will converge to the following neighborhood of the origin,

$$\mathbf{D}_0 = \{\mathbf{e}_1 : \|\mathbf{e}_1\| \le (\gamma/\delta) \cdot |d| + \mu\} \quad (10)$$

where $\gamma = \|\mathbf{E}^T\mathbf{P}\| / \lambda_{\min}(\mathbf{P})$, $\lambda_{\min}(\mathbf{P})$ denotes the minimum eigenvalue of matrix $\mathbf{P}$, and μ is a small positive constant. Consequently, the coupled systems (6) and (7) are globally synchronized with a small bounded error.

Proof: Define a Lyapunov function, $V = \mathbf{e}_1^T \mathbf{P}\mathbf{e}_1$. Differentiating V along the error dynamical trajectory (8) and using (2) yield

$$\dot{V} = \dot{\mathbf{e}}_1^T \mathbf{P}\mathbf{e}_1 + \mathbf{e}_1^T \mathbf{P}\dot{\mathbf{e}}_1 = [(\mathbf{A}_1 - \mathbf{L}_1\mathbf{C}_1)\mathbf{e}_1 + \mathbf{F}(\mathbf{z}) - \mathbf{F}(\tilde{\mathbf{z}}) + \mathbf{E}d]^T \mathbf{P}\mathbf{e}_1$$
$$+ \mathbf{e}_1^T \mathbf{P}[(\mathbf{A}_1 - \mathbf{L}_1\mathbf{C}_1)\mathbf{e}_1 + \mathbf{F}(\mathbf{z}) - \mathbf{F}(\tilde{\mathbf{z}}) + \mathbf{E}d]$$
$$= \mathbf{e}_1^T ((\mathbf{A}_1 - \mathbf{L}_1\mathbf{C}_1)^T \mathbf{P} + \mathbf{P}(\mathbf{A}_1 - \mathbf{L}_1\mathbf{C}_1))\mathbf{e}_1 + 2[\mathbf{F}(\mathbf{z}) - \mathbf{F}(\tilde{\mathbf{z}})]^T \mathbf{P}\mathbf{e}_1 + 2\mathbf{E}^T \mathbf{P}\mathbf{e}_1 d$$
$$\le \mathbf{e}_1^T ((\mathbf{A}_1 - \mathbf{L}_1\mathbf{C}_1)^T \mathbf{P} + \mathbf{P}(\mathbf{A}_1 - \mathbf{L}_1\mathbf{C}_1))\mathbf{e}_1 + 2\rho\|\mathbf{e}_1\| \cdot \|\mathbf{P}\mathbf{e}_1\| + 2\|\mathbf{E}^T\mathbf{P}\|\|\mathbf{e}_1\||d|$$

Since $2\|\mathbf{e}_1\| \cdot \rho\|\mathbf{P}\mathbf{e}_1\| \le \rho^2\|\mathbf{P}\mathbf{e}_1\|^2 + \|\mathbf{e}_1\|^2$, using (9) we further have

$$\dot{V} \le \mathbf{e}_1^T ((\mathbf{A}_1 - \mathbf{L}_1\mathbf{C}_1)^T \mathbf{P} + \mathbf{P}(\mathbf{A}_1 - \mathbf{L}_1\mathbf{C}_1))\mathbf{e}_1 + \rho^2\|\mathbf{P}\mathbf{e}_1\|^2 + \|\mathbf{e}_1\|^2 + 2\|\mathbf{E}^T\mathbf{P}\|\|\mathbf{e}_1\||d|$$
$$= \mathbf{e}_1^T (\mathbf{A}_1^T\mathbf{P} + \mathbf{P}\mathbf{A}_1 - \mathbf{P}\mathbf{L}_1\mathbf{C}_1 - \mathbf{C}_1^T\mathbf{L}_1^T\mathbf{P} + \rho^2\mathbf{P}\mathbf{P} + \mathbf{I})\mathbf{e}_1 + 2\|\mathbf{E}^T\mathbf{P}\|\|\mathbf{e}_1\||d|$$
$$\le -2\delta\mathbf{e}_1^T \mathbf{P}\mathbf{e}_1 + 2\|\mathbf{E}^T\mathbf{P}\|\|\mathbf{e}_1\||d| \le -2\delta\lambda_{\min}(\mathbf{P})\|\mathbf{e}_1\|^2 + 2\|\mathbf{E}^T\mathbf{P}\|\|\mathbf{e}_1\||d|$$
$$= -2\delta\lambda_{\min}(\mathbf{P})\|\mathbf{e}_1\|(\|\mathbf{e}_1\| - (\|\mathbf{E}^T\mathbf{P}\|/(\delta\lambda_{\min}(\mathbf{P})) \cdot |d|)$$
$$= -2\delta\lambda_{\min}(\mathbf{P})\|\mathbf{e}_1\|(\|\mathbf{e}_1\| - (\gamma/\delta) \cdot |d|)$$

If the error variable $\mathbf{e}_1$ belongs to $\mathbf{D}_0$, i.e., $\|\mathbf{e}_1\| > (\gamma/\delta)|d| + \mu$, then we have

$$\dot{V} \le -2\mu\delta\lambda_{\min}(\mathbf{P})\|\mathbf{e}_1\|, \quad \forall \|\mathbf{e}_1\| > (\gamma/\delta)|d| + \mu.$$

Based on the Lyapunov stability theory, the error dynamical system (10) will converge to the neighborhood, $\mathbf{D}_0$. ∎

Lemma 1: (Schur Complements [7]) For a given symmetric matrix, $\mathbf{S} = \begin{bmatrix} \mathbf{S}_{11} & \mathbf{S}_{12} \\ \mathbf{S}_{21} & \mathbf{S}_{22} \end{bmatrix}$, where $\mathbf{S}_{11} = \mathbf{S}_{11}^T$, $\mathbf{S}_{12} = \mathbf{S}_{21}^T$, $\mathbf{S}_{22} = \mathbf{S}_{22}^T$, the condition $\mathbf{S} < \mathbf{0}$ is equivalent to

$$\mathbf{S}_{22} < \mathbf{0}, \text{ and } \mathbf{S}_{11} - \mathbf{S}_{12}\mathbf{S}_{22}^{-1}\mathbf{S}_{12}^T < \mathbf{0}. \tag{11}$$

Using Lemma 1, the condition (9) can be easily transformed to be

$$\begin{bmatrix} \mathbf{PA}_1 + \mathbf{A}_1^T\mathbf{P} + 2(\delta - \rho)\mathbf{P} - \mathbf{PL}_1\mathbf{C}_1 - \mathbf{C}_1^T\mathbf{L}_1^T\mathbf{P} & \rho\mathbf{P} + \mathbf{I} \\ \rho\mathbf{P} + \mathbf{I} & -\mathbf{I} \end{bmatrix} < 0 \tag{12}$$

Let $\mathbf{X} = \mathbf{PL}_1$. Then, (12) can be easily transformed into the following LMI form.

Theorem 2: If suitable matrices $\mathbf{X}$ and $\mathbf{P}$ are selected such that the following LMI

$$\begin{bmatrix} \mathbf{PA}_1 + \mathbf{A}_1^T\mathbf{P} + 2(\delta - \rho)\mathbf{P} - \mathbf{XC}_1 - \mathbf{C}_1^T\mathbf{X}^T & \rho\mathbf{P} + \mathbf{I} \\ \rho\mathbf{P} + \mathbf{I} & -\mathbf{I} \end{bmatrix} < 0 \tag{13}$$

is satisfied, then the errors of system (8) with the feedback gain $\mathbf{L}_1 = \mathbf{P}^{-1}\mathbf{X}$ will converge to a small neighborhood of the origin, $\mathbf{D}_0$.

4. AN ILLUSTRATIVE EXAMPLE

The chaotic MLC circuit is simulated to illustrate the above-derived criterion for chaos synchronization. The chaotic MLC circuit is described by [5]

$$\begin{bmatrix} \dot{x}_1 \\ \dot{x}_2 \end{bmatrix} = \begin{bmatrix} 0 & 1 \\ -\beta & -\sigma \end{bmatrix}\begin{bmatrix} x_1 \\ x_2 \end{bmatrix} + \begin{bmatrix} -f(x_1) \\ 0 \end{bmatrix} + \begin{bmatrix} 0 \\ F\sin(\omega t) \end{bmatrix} \tag{14}$$

where $\sigma > 0$, $\beta > 0$, $F > 0$, $\omega > 0$, and $f(.)$ is a piecewise linear function,

$$f(x_1) = bx_1 + 0.5(a-b)(|x_1 + 1| - |x_1 - 1|) \tag{15}$$

with $a < b < 0$. From (15), one easily has [3]

$$f(x_1) - f(\tilde{x}_1) = k_{x1,\tilde{x}1}(x_1 - \tilde{x}_1) \tag{16}$$

where $k_{x1,\tilde{x}1}$ depends on x_1 and $\tilde{x}_1$, and varies within the interval $[a, b]$ for $t \geq 0$. So, $k_{x1,\tilde{x}1}$ is bounded, $a \leq k_{x1,\tilde{x}1} \leq b < 0$.

From (14), we get $\mathbf{A} = \begin{bmatrix} 0 & 1 \\ -\beta & -\sigma \end{bmatrix}$, $\mathbf{h}(t) = \begin{bmatrix} 0 \\ F\sin(wt) \end{bmatrix}$ and $\mathbf{g}(\mathbf{x}) = \begin{bmatrix} -f(x_1) \\ 0 \end{bmatrix}$. Choose $\mathbf{C} = \begin{bmatrix} c_1 & 0 \end{bmatrix}$, where $c_1 \neq 0$. Obviously, $(\mathbf{A}, \mathbf{C})$ is observable.

Referring to (5) and (6), we can get the augmented drive system as

$$\begin{aligned} \dot{\mathbf{z}} &= \mathbf{A}_1\mathbf{z} + \mathbf{F}(\mathbf{z}) + \mathbf{H}(t) + \mathbf{E}d \\ y_0 &= \mathbf{C}_1\mathbf{z} \end{aligned} \tag{17}$$

where $\mathbf{z} = \begin{bmatrix} x_1 \\ x_2 \\ x_0 \end{bmatrix}$, $\mathbf{F}(\mathbf{z}) = \begin{bmatrix} -f(x_1) \\ 0 \\ 0 \end{bmatrix}$, $\mathbf{H}(t) = \begin{bmatrix} 0 \\ F\sin(\omega t) \\ 0 \end{bmatrix}$, $\mathbf{E} = \begin{bmatrix} 0 \\ 0 \\ 1 \end{bmatrix}$,

$\mathbf{A}_1 = \begin{bmatrix} 0 & 1 & 0 \\ -\beta & -\sigma & 0 \\ 1 & 0 & 0 \end{bmatrix}$, $\mathbf{C}_1 = \begin{bmatrix} 0 \\ 0 \\ 1 \end{bmatrix}^T$.

From (7), the response system can be constructed as

$$\begin{aligned} \dot{\tilde{\mathbf{z}}} &= \mathbf{A}_1\tilde{\mathbf{z}} + \mathbf{F}(\tilde{\mathbf{z}}) + \mathbf{H}(t) + \mathbf{L}_1(y_0 - \tilde{y}_0) \\ \tilde{y}_0 &= \mathbf{C}_1\tilde{\mathbf{z}} \end{aligned} \tag{18}$$

where $\mathbf{L}_1 = \begin{bmatrix} l_1 & l_2 & l_3 \end{bmatrix}^T$.

Subtracting (18) from (17), we get the error dynamical system

$$\dot{\mathbf{e}}_1 = (\mathbf{A}_1 - \mathbf{L}_1\mathbf{C}_1)\mathbf{e}_1 + \mathbf{F}_1(\mathbf{z}) - \mathbf{F}_1(\tilde{\mathbf{z}}) + \mathbf{E}d \tag{19}$$

where $\mathbf{e}_1 = \begin{bmatrix} x_1 - \tilde{x}_1 & x_2 - \tilde{x}_2 & x_0 - \tilde{x}_0 \end{bmatrix}^T$. Now, consider

$$\mathbf{F}_1(\mathbf{z}) - \mathbf{F}_1(\tilde{\mathbf{z}}) = \begin{bmatrix} -(f(x_1) - f(\tilde{x}_1)) \\ 0 \\ 0 \end{bmatrix} = \begin{bmatrix} -k_{x1,\tilde{x}1}(x_1 - \tilde{x}_1) \\ 0 \\ 0 \end{bmatrix}$$

$$= \begin{bmatrix} -k_{x1,\tilde{x}1} & 0 & 0 \\ 0 & 0 & 0 \\ 0 & 0 & 0 \end{bmatrix}\mathbf{e}_1 = \mathbf{M}_{z,\tilde{z}}\mathbf{e}_1 \tag{20}$$

where $\mathbf{M}_{z,\tilde{z}} = \begin{bmatrix} -k_{x1,\tilde{x}1} & 0 & 0 \\ 0 & 0 & 0 \\ 0 & 0 & 0 \end{bmatrix}$. Since $a \leq k_{x1,\tilde{x}1} \leq b < 0$, letting $k_{x1,\tilde{x}1} = (a+b)/2 + \bar{k}_{x1,\tilde{x}1}$ with $(a-b)/2 \leq \bar{k}_{x1,\tilde{x}1} \leq -(a-b)/2$ yields

$$\mathbf{F}_1(\mathbf{z}) - \mathbf{F}_1(\tilde{\mathbf{z}}) = \begin{bmatrix} -(a+b)/2 & 0 & 0 \\ 0 & 0 & 0 \\ 0 & 0 & 0 \end{bmatrix}\mathbf{e}_1 + \begin{bmatrix} -\bar{k}_{x1,\tilde{x}1} & 0 & 0 \\ 0 & 0 & 0 \\ 0 & 0 & 0 \end{bmatrix}\mathbf{e}_1$$

$$= \overline{\mathbf{A}}_1\mathbf{e}_1 + \overline{\mathbf{F}}_1(\mathbf{z}) - \overline{\mathbf{F}}_1(\tilde{\mathbf{z}}) = \overline{\mathbf{A}}_1\mathbf{e}_1 + \overline{\mathbf{M}}_{z,\tilde{z}}\mathbf{e}_1 \tag{21}$$

where $\overline{\mathbf{A}}_1 = \begin{bmatrix} -(a+b)/2 & 0 & 0 \\ 0 & 0 & 0 \\ 0 & 0 & 0 \end{bmatrix}$, $\overline{\mathbf{M}}_{z,\tilde{z}} = \begin{bmatrix} -\bar{k}_{x1,\tilde{x}1} & 0 & 0 \\ 0 & 0 & 0 \\ 0 & 0 & 0 \end{bmatrix}$.

Then, (19) can be rewritten as

$$\begin{aligned} \dot{\mathbf{e}}_1 &= (\mathbf{A}_1 - \mathbf{L}_1\mathbf{C}_1)\mathbf{e}_1 + \mathbf{F}_1(\mathbf{z}) - \mathbf{F}_1(\tilde{\mathbf{z}}) + \mathbf{E}d \\ &= (\mathbf{A}_1 + \overline{\mathbf{A}}_1 - \mathbf{L}_1\mathbf{C}_1)\mathbf{e}_1 + \overline{\mathbf{F}}_1(\mathbf{z}) - \overline{\mathbf{F}}_1(\tilde{\mathbf{z}}) + \mathbf{E}d \\ &= (\mathbf{A}_2 - \mathbf{L}_1\mathbf{C}_1)\mathbf{e}_1 + \overline{\mathbf{F}}_1(\mathbf{z}) - \overline{\mathbf{F}}_1(\tilde{\mathbf{z}}) + \mathbf{E}d \end{aligned} \tag{22}$$

where $\mathbf{A}_2 = \mathbf{A}_1 + \overline{\mathbf{A}}_1 = \begin{bmatrix} -(a+b)/2 & 1 & 0 \\ -\beta & -\sigma & 0 \\ 1 & 0 & 0 \end{bmatrix}$, and

$$\left\|\overline{\mathbf{F}}_1(\mathbf{z}) - \overline{\mathbf{F}}_1(\tilde{\mathbf{z}})\right\| = \left\|\overline{\mathbf{M}}_{z,\tilde{z}}(\mathbf{z}-\tilde{\mathbf{z}})\right\|$$
$$\leq \left|-\overline{k}_{x1,\tilde{x}1}\right| \cdot \|\mathbf{e}_1\| = |(a-b)/2|\|\mathbf{e}_1\| = \rho\|\mathbf{e}_1\| \quad (23)$$

where $\rho = |(a-b)/2|$.

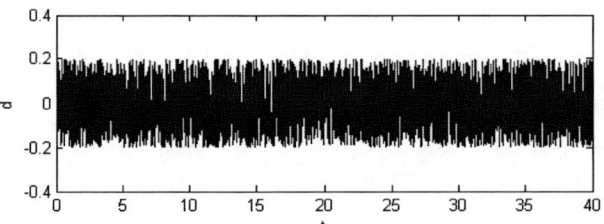

Figure 1. The transmission disturbance d.

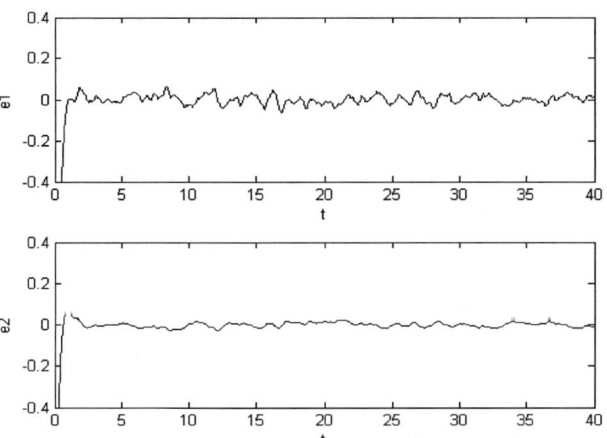

Figure 2. Chaos synchronization of chaotic MLC circuit with transmission disturbance.

The parameters of the circuit used are $\sigma = 1.015$, $\beta = 1.0$, $F = 0.15$, $\omega = 0.75$, $a = -1.02$, and $b = -0.55$, for which the system exhibits chaotic behavior [5]. We have $\rho = 0.235$ from (23). Choose $c_1 = 1$ and $\delta = 1.0$. Then, we easily obtain $\mathbf{L}_1 = [38.1328 \quad 7.9196 \quad 10.3535]^T$ from Theorem 2 by using the MATLAB LMI Toolbox. Figure 1 shows the profile of the disturbance d, which is a uniformly distributed random noise. Figure 2 shows the behavior of chaos synchronization. From Figure 2, we can see that the effect of the transmission disturbance on chaos synchronization has been greatly suppressed, showing good performance, as expected.

5. CONCLUSIONS

A new scheme of chaos synchronization has been developed for a class of chaotic systems from the integral observer approach. Based on the Lyapunov stability theory, a sufficient condition has been derived for chaos synchronization. This criterion has been transformed into the LMI form. It has been shown that under the proposed scheme and derived criterion, the effect of transmission disturbances can be greatly reduced, and consequently chaos synchronization is achieved satisfactorily. The scheme and criterion have also been successfully applied to the chaotic MLC circuit with simulation verification.

It is noted that the scheme and criterion proposed herein are applicable to a large class of chaotic systems. A similar approach can be applied to Chua's circuit, Rössler system, Lorenz system, Chen's system, and so on.

6. ACKNOWLEDGEMENTS

This work is supported in part by the 'Qing Lan Project' Program of Jiangsu Province [JS200407], China, in part by the Key Project of Natural Science Foundation of Universities of Jiangsu Province [04KJA510092], China, in part by a research grant from the Australian Research Council [A00102928], and in part by the City University of Hong Kong [SRG 7001174], China.

7. REFERENCES

[1] K.K. Busawon and P. Kabore, "Disturbance attenuation Using Proportional Integral Observers," *Int. J. Control*, 74(6):618-627, 2001.

[2] G. Chen and X. Dong. *From Chaos to Order – Methodologies, Perspectives and Applications*. World Scientific, Singapore, 1998.

[3] G.-P. Jiang, G. Chen and K.S. Tang, "A new criterion for chaos synchronization using linear state feedback control," *Int. J. Bifurcation and Chaos*, 13(9):2343-2351, 2003.

[4] T.-L. Liao and N.-S. Huang, "An observer-based approach for chaotic synchronization with application to secure communications," *IEEE Trans. on Circuits Syst., I*, 46:1144 –1150, 1999.

[5] K. Murali and M. Lakshmanan, "Synchronization through Compound Chaotic Signal in Chua's Circuit and Murali-Lakshmanan-Chua Circuit," *Int. J. Bifurcation and Chaos*, 7:415-421, 1997.

[6] L.M. Pecora and T. L. Carroll, "Synchronization in chaotic systems," *Phys. Rev. Lett.*, 64:821-824, 1990.

[7] S. Boyd, L.E. Ghaoui, E. Feron and V. Balakrishnan. *Linear Matrix Inequalities in System and Control Theory*, SIAM, Philadelphia, 1994

[8] H. Nijmeijer and I.M.Y. Mareels, "An observer looks at synchronization," *IEEE Trans. Circuits Syst. I*, 44:882-890, 1997

[9] T. Ushio, "Synthesis of synchronized chaotic systems based on observers," *Int. J. Bifurcation and Chaos*, 9:541-546, 1999.

[10] J.A.K. Suykens and J. Vandewalle, "Master-slave synchronization of Lur'e systems," *Int. J. Bifurcation and Chaos,* 7:665-669, 1997.

[11] C.W. Wu and L. O. Chua, "A unified framework for synchronization and control of dynamical systems," *Int. J. Bifurcation and Chaos,* 4:979-998, 1994.

[12] F.C.M. Lau and C.K. Tse, *Chaos-Based Digital Communication Systems: Operating Principles, Analytical Methods and Performance Evaluation*, Springer-Verlag, Germany, 2003.

SYNCHRONIZING CHAOTIC COLPITTS CIRCUITS ADAPTIVELY WITH PARAMETER MISMATCHES AND CHANNEL DISTORTIONS

Cheng Shen, Zhiguo Shi and Lixin Ran

Department of Information and Electronic Engineering, Zhejiang University, Hangzhou 310027, China

ABSTRACT

This paper presents an approach to achieve synchronization between chaotic Colpitts circuits in practical communications. By applying the adaptive controller to compensate for the parameter mismatches between transmitter and receiver and to offset the channel distortions, the synchronization performance is markedly improved. Both time-constant and time-varying circuit parameter mismatches and channel distortions are considered in simulations.

1. INTRODUCTION

Research into applications of chaos in communications has been motivated by the observation that chaotic systems can be synchronized [1], which offers a potential advantage over non-coherent detection in terms of noise performance and data rate when the information is recovered from a noisy distorted received signal [2]. The applications of synchronized chaotic systems for communication largely rely on the robustness of the synchronization within the transmitter-receiver pair.

Synchronization of chaotic Colpitts circuits [3,4] has been discussed in much literature [5-7]. But in the most of previous works the drive and response systems were assumed to be identical, the circuit parameters were assumed to be time-constant, and the channel between the two systems was assumed to be ideal. However, this is not the practical circumstance at all [7].

To achieve the synchronization in the presence of channel distortions and parameter mismatches, one good approach is to employ adaptive controllers to compensate for the transmitted signals. This approach has been successfully applied in the Chua's circuits [8] and has obtained enhancement in synchronization performance in both simulations and experiments.

In this paper, we study the adaptive controller and its capability of synchronizing chaotic Colpitts circuits with channel distortions and parameter mismatches. We compare the respective influence exerted by the time-constant mismatch of each of the three main circuit parameters on synchronization, and uncover the most pivotal parameter whose mismatch has the most negative influence on synchronization. Then we use the adaptive controller to offset de-synchronization caused by time-varying parameters of the transmitter. We also study the synchronization with time-constant and time-varying channel distortions respectively, and employ adaptive controllers to offset these distortions.

2. SYSTEM CONFIGURATION

In this system, we use two identical chaotic Colpitts circuits as the transmitter and the receiver and adopt the error feedback synchronization scheme, with resistor R_1 coupling the received signal into the receiver. The signal V_{C2}, serving as the transmitted signal, passes through a time-varying channel and becomes a distorted version defined as $\tilde{V}_{Tr}$. To implement synchronization adaptively between transmitter and receiver, we construct an adaptive controller before resistor R_1. The received signal $\tilde{V}_{Tr}$ is processed by the adaptive controller and is then coupled into the receiver through R_1. The system configuration is shown in Fig. 1.

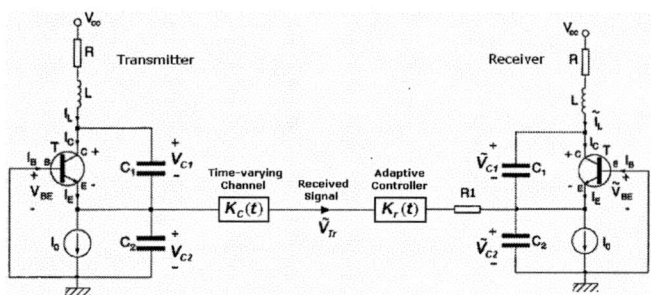

Figure 1. Synchronization scheme of two chaotic Colpitts circuits with an adaptive controller over a time-varying channel

In our simulations, the parameters of Colpitts circuits are: $C1 = C2 = 237nf$, $L = 2.1mH$, $R = 74.5\Omega$, $V_{CC} = 5V$, $I_0 = 2.5mA$.

With these parameters, both the transmitter and the receiver exhibit chaotic oscillation separately. While the circuit parameters are time-constant and the channel has no distortion (i.e. $\tilde{V}_{Tr} = V_{C2}$), synchronization between two circuits can be achieved by properly selecting the value of R_1. This has been demonstrated by both numerical simulations and experimental investigation [5].

In our simulations, we set initial conditions of the transmitter and the receiver as $(V_{C1}(0), V_{C2}(0), I_L(0)) = (4.9V, -0.4V, 0mA)$ and $(\tilde{V}_{C1}(0), \tilde{V}_{C2}(0), \tilde{I}_L(0)) = (5.9V, -0.4V, 0mA)$, respectively.

The fourth order Runge-Kutta method with fixed step-size $h = 10^{-6}s$ is used to simulate the system.

With the coupling resistor $R_1 = 100\Omega$, perfect synchronization between transmitter and receiver can be easily achieved in simulations. However, when the parameter mismatches or channel distortions are exerted, the system will be desynchronized as indicated by our simulation results below. To compare the synchronization performance, we define the average attractor distance between the transmitter and the receiver as:

$$D = \lim_{t_s \to \infty} \frac{\int_{t_0}^{t_s} \sqrt{e_1^2 + e_2^2 + e_3^2}\, dt}{t_s - t_0} \quad (1)$$

where $e_1 = V_{C1} - \tilde{V}_{C1}$, $e_2 = V_{C2} - \tilde{V}_{C2}$, $e_3 = I_L - \tilde{I}_L$, and t_0 denotes the settling time when the transient part of the signal has passed. A bigger value of D means a worse synchronization performance.

3. ADAPTIVE CONTROL FOR TIME-VARYING PARAMETER COMPENSATION

The three main passive circuit components, namely the inductor L, the capacitors C_1 and C_2, compose the Colpitts circuit together with BJT and generate chaotic waveforms [3,5]. Therefore, it is necessary to examine the effect of a mismatch between these parameters of the transmitter and the receiver on the recovery of transmitted signals in chaotic communication systems.

3.1 Comparison between Synchronization Performance on Each Time-Constant Parameter Mismatch

In this simulation, we set a certain mismatch of one parameter, taking the inductor L for example, between the drive and response system and fix the values of all other circuit parameters in the drive system to the corresponding values in the response system in each simulation step. To compare effects of parameter mismatches in different degrees on synchronization, we vary the value of L in the drive system in the range from 0.9 to 1.1 times that of its corresponding value in the response system, and record the average attractor distance D for each value of L. Then we employ the same method on parameters C_1 and C_2 respectively and record the average attractor distance D for each.

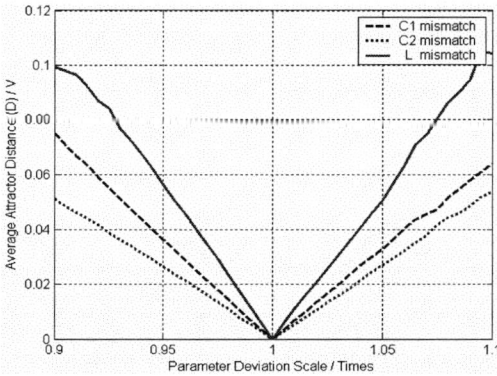

Figure 2. Average attractor distance between the transmitter and the receiver when three transmitter parameters L, C_1 and C_2 deviate from 0.9 to 1.1 times that of their corresponding values in the receiver, respectively

Fig. 2 shows effects of parameter mismatches of these three main circuit components on synchronization of chaotic Colpitts system. Comparing the average attractor distance D of the three curves, one can see that the mismatch of inductor L (solid line) causes the greatest degradation on synchronization performance. Thus, we conclude that the synchronization of Colpitts system is more sensitive to the mismatch of L than that of C_1 (dashed line) and C_2 (dotted line). Further, since only 10% parameter mismatch of either L, C_1 or C_2 will lead to great de-synchronization (D is more than 0.05), we conclude that the tolerance degree of chaotic Colpitts system on parameter mismatches is very low.

3.2 Compensating for Time-varying Parameter Mismatches

In this simulation, we consider the parameter mismatch being time-varying. We write the drive system as follows:

$$\begin{cases} \dfrac{dV_{C1}}{dt} = \dfrac{K_{C1}(t)}{C_1}(-f(-V_{C2})+I_L) \\ \dfrac{dV_{C2}}{dt} = \dfrac{K_{C2}(t)}{C_2}(I_L - I_0) \\ \dfrac{dI_L}{dt} = \dfrac{K_L(t)}{L}(-V_{C1}-V_{C2}-I_L R + V_{CC}) \end{cases} \quad (2)$$

where $K_{C1}(t)$, $K_{C2}(t)$ and $K_L(t)$ are the time-varying factors of circuit parameters C_1, C_2 and L, respectively.

The response system is as follows:

$$\begin{cases} \dfrac{d\tilde{V}_{C1}}{dt} = \dfrac{\tilde{K}_{C1}(t)}{C_1}(-f(-\tilde{V}_{C2})+\tilde{I}_L) \\ \dfrac{d\tilde{V}_{C2}}{dt} = \dfrac{\tilde{K}_{C2}(t)}{C_2}(\tilde{I}_L - I_0 + \dfrac{\tilde{V}_{Tr} - \tilde{V}_{C2}}{\tilde{R}_1}) \\ \dfrac{d\tilde{I}_L}{dt} = \dfrac{\tilde{K}_L(t)}{L}(-\tilde{V}_{C1}-\tilde{V}_{C2}-\tilde{I}_L R + V_{CC}) \end{cases} \quad (3)$$

where $\tilde{K}_{C1}(t)$, $\tilde{K}_{C2}(t)$ and $\tilde{K}_L(t)$ are compensating adjustments of circuit parameters C_1, C_2 and L, respectively, which are adaptively modified by using adaptive controllers.

Here, we consider that only parameter L of the transmitter mismatches with that of the receiver. We make the value of inductor L time-varying at the transmitter, with a time-varying factor $K_L(t)$ as defined by the following sinusoidal function:

$$K_L = 1 - 0.1\sin(10\pi t) \quad (4)$$

The dynamics of the compensating adjustment $\tilde{K}_L(t)$ is given as

$$\begin{aligned} \dot{\tilde{K}}_L(t) &= k_1 \operatorname{sgn}\left(\dfrac{\partial \dot{\tilde{I}}_L}{\partial \tilde{K}_L}\right)(V_{C2} - \tilde{V}_{C2}) \\ &= k_1 \operatorname{sgn}\left(\dfrac{1}{L}\left[-\tilde{V}_{C1} - \tilde{V}_{C2} - \tilde{I}_L R + V_{CC}\right]\right) \times (V_{C2} - \tilde{V}_{C2}) \end{aligned} \quad (5)$$

where $k_1 = 5 \times 10^4$. The simulation time is 0.3s.

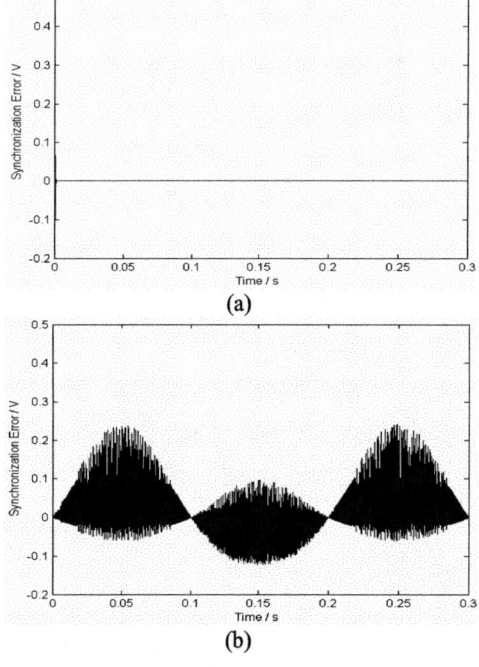

Figure 3. Synchronization of chaotic Colpitts circuits with time-varying inductor L in the transmitter
(a) The synchronization error $(V_{C2} - \tilde{V}_{C2})$ when the adaptive controller is used. (b) The synchronization error $(V_{C2} - \tilde{V}_{C2})$ when no adaptive controller is used.

For comparison, the synchronization error $(V_{C2}-\tilde{V}_{C2})$ with and without using the adaptive controller are plotted in Fig. 3(a) and Fig. 3(b), respectively. One can see that the adaptive controller markedly offset the de-synchronization caused by the time-varying parameter mismatch and successfully recovers the transmitted signal at the receiver. Furthermore, since the synchronization of chaotic Colpitts circuits has more immunity on mismatches of capacitors C_1 and C_2, it is safe to say that the adaptive controller can also compensate for mismatches of them, which has also been verified by our simulations.

4. ADAPTIVE CONTROL FOR TIME-VARYING CHANNEL COMPENSATION

When the system parameters keep identical and time-constant, the channel distortions between the transmitter and the receiver pose the major problem on synchronization, which are unavoidable in practical chaotic communications.

First, we introduce the time-varying channel gain $K_c(t)$ into the system (shown in Fig. 1), hence $\tilde{V}_{Tr}=K_c(t)V_{C2}$. Constant unit-gain channel corresponds to $K_c(t)=1$ and $\tilde{V}_{Tr}=V_{C2}$. In the receiver, we construct an adaptive gain $K_r(t)$ such that $K_c(t)K_r(t)\to 1$ as $t\to\infty$ to maintain the synchronization. Then the receiver should be rewritten as:

$$\begin{cases} \frac{d\tilde{V}_{C1}}{dt}=\frac{1}{C_1}(-f(-\tilde{V}_{C2})+\tilde{I}_L) \\ \frac{d\tilde{V}_{C2}}{dt}=\frac{1}{C_2}(\tilde{I}_L-I_0)+\frac{1}{R_1C_2}(K_r(t)K_c(t)V_{C1}-\tilde{V}_{C2}) \\ \frac{d\tilde{I}_L}{dt}=\frac{1}{L}(-\tilde{V}_{C1}-\tilde{V}_{C2}-\tilde{I}_L R+V_{CC}) \end{cases} \quad (6)$$

The dynamics of $K_r(t)$ is given by one of the following adaptive controllers:

Controller #1:
$$\dot{K}_r(t)=-k_1(K_r(t)|\tilde{V}_{Tr}|-|\tilde{V}_{C2}|) \quad (7)$$

Controller #2:
$$\dot{K}_r(t)=-k_1(K_r(t)\tilde{V}_{Tr}^2-\tilde{V}_{Tr}\tilde{V}_{C2}) \quad (8)$$

Controller #3:
$$\dot{K}_r(t)=-k_1\operatorname{sgn}\left(\frac{\partial\dot{\tilde{V}}_{C2}}{\partial K_r}\right)(K_r(t)\tilde{V}_{Tr}(t)-\tilde{V}_{C2}) \quad (9)$$

Note that **NO** parameter mismatch is considered in this section.

4.1 Compensating for Time-Constant Channel Fading

In this simulation, we consider the time-constant fading as the only channel distortion. The channel gain $K_c(t)$ is set as a constant in each simulation step. For comparison, we vary the channel gain $K_c(t)$ in the range from 0.1 to 1 and record the average attractor distance D for each value of $K_c(t)$. The second controller (8) with $k_1=5\times 10^4$ is used here.

Fig. 4(a) shows that the synchronization performance sharply declines as the fading factor decreases. While only less than 5% of the transmitted signal is attenuated, the D rises up to 0.1, which means a bad synchronization state. So it can be inferred that the chaotic Colpitts system can tolerate a very low degree of channel fading in practical communications.

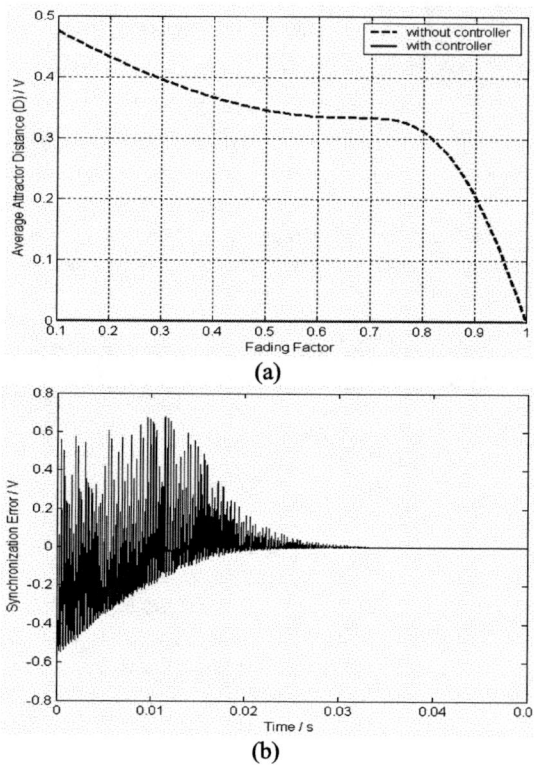

Figure 4. Synchronization of chaotic Colpitts circuits with time-constant fading channel
(a) Average attractor distance between the transmitter and the receiver when the fading factor varies from 0.1 to 1. (b) The synchronization error $(V_{C2}-\tilde{V}_{C2})$ with adaptive controller is used when the fading factor is 0.1.

Fortunately, the adaptive controller can be employed to remarkably compensate for the time-constant channel fading, represented by the solid line in Fig. 4(a). Even as 90% of the transmitted signal is attenuated, the D still keeps almost zero, which means the system is perfectly synchronized.

Fig. 4(b) reveals the process of compensating for the time-constant channel fading by the adaptive controller, with the fading factor being only 0.1. As time proceeds, the adaptive controller calculates appropriate adaptive gain $K_r(t)$ to offset the channel attenuation, gradually reduces the synchronization error, and finally achieves the synchronization. It can be inferred that as long as the transmitted signal does not submerge in the channel, the adaptive controller can finally recover it at the receiver. Certainly, as the channel attenuation getting worse, the settling time will be longer. So the adaptive controller is powerful in terms of compensating for time-constant channel fading.

4.2 Compensating for Sinusoidal Channel Fading

In this simulation, we consider the channel fading being time-varying. We set the channel gain $K_c(t)$ as a sinusoidal function given by, $K_c(t)=0.5-0.1\sin(x\pi t)$ (13) with a certain frequency in each simulation step. Since different frequencies reflect different channel varying rates, we vary the frequency of sinusoidal function in the range from 0Hz to 300 Hz and record the average attractor distance D for each frequency. The second controller (8) with $k_1=10^5$ is used here.

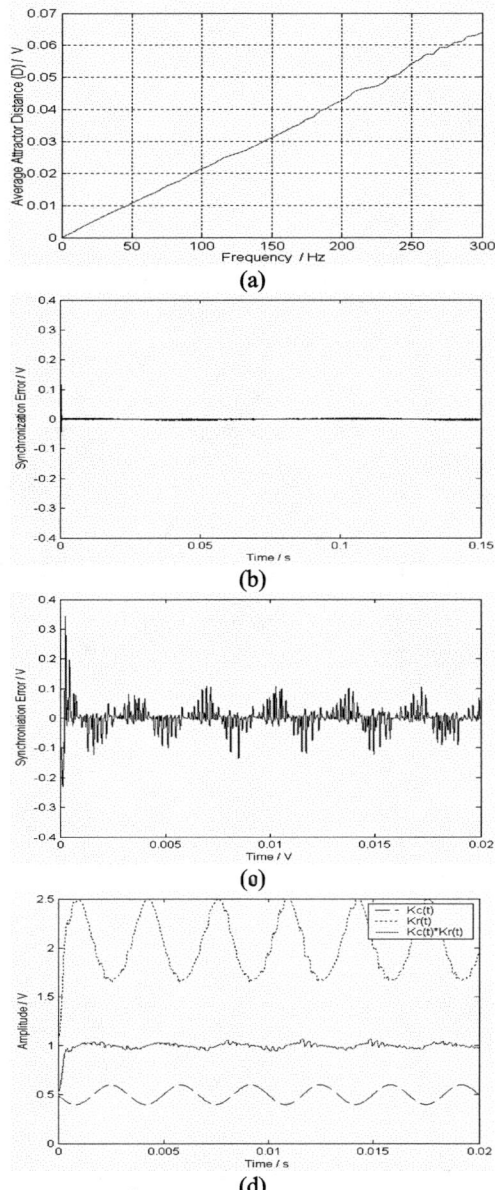

Figure 5. Synchronization of chaotic Colpitts circuits with sinusoidal fading channel
(a) Average attractor distance between the transmitter and the receiver when the frequency of channel function varies from 0 to 300Hz. (b) The synchronization error $(V_{C2} - \tilde{V}_{C2})$ with frequency being 10Hz. (c) The synchronization error $(V_{C2} - \tilde{V}_{C2})$ with frequency being 300Hz. (d) $K_c(t)$, $K_r(t)$ and $K_c(t)K_r(t)$ with frequency being 300Hz.

Fig. 5(a) plots the effects of increasing frequency of sinusoidal channel function on the synchronization performance. As the frequency increases, the value of D rise almost linearly, which means the synchronization performance linearly declines. As high as the frequency being 250Hz, the value of D rises above 0.05, corresponding to a bad synchronization state.

To have a deeper insight, we compare two cases with the frequency of channel function being 10Hz (Fig. 5(b)) and 300Hz (Fig. 5(c)). Obviously, the synchronization performance differs considerably between these two cases. The synchronization error presents periodically corresponding with the varying of sinusoidal fading channel. When the channel varying rate goes faster, the synchronization error rises accordingly and the adaptive controller becomes incapable of following the fast changing of channel fading. Fig. 5(d) illustrates that the $K_c(t)K_r(t)$ (solid line) can no longer keep time-constant unit gain as the channel varying rate increases to 300Hz. From the simulation results, we conclude that the adaptive controller is capable of compensating for sinusoidal channel fading. But such capability declines with the increasing frequency of sinusoidal channel function. Fortunately, since the channel fading is of low-varying rate in most practical environment, the Colpitts system will definitely benefit from the adoption of adaptive controllers for practical communications.

5. CONCLUSION

We have demonstrated that two chaotic Colpitts circuits can be properly synchronized with employment of adaptive controllers while the circuit parameters and the channel are time-varying. The synchronization performance of chaotic Colpitts circuits is markedly improved by applying the adaptive controller to offset time-constant and time-varying circuit parameter mismatches and channel distortions. Simulation results indicate that this approach has tremendous potential for developing practical chaotic spread-spectrum communication systems.

ACKNOWLEDGMENTS

The authors thank the National Natural Science Foundation of China (NSFC) for the financial support.

6. REFERENCES

[1] L.M. Pecora and T.L. Carroll, "Synchronization in chaotic systems". *Phys. Rev. Lett.*, vol. 64, no. 8, pp. 821–824, 1990

[2] G. Kolumban, M.P. Kennedy and L.O. Chua, "The roll of Synchronization in digital communication using chaos – part II". *IEEE Trans. Circuits Syst. I*, vol. 45, no. 11, pp. 1129–1140, Nov. 1998

[3] M.P. Kennedy, "Chaos in the Colpitts oscillator". *IEEE Trans. Circuits Syst. I*, vol. 41, no. 11, pp. 771–774, Nov. 1994

[4] Z.G. Shi and L.X. Ran, "Design of Chaotic Colpitts Circuit with Prescribed Frequency Distribution". *Int. J. Nonlinear Science and Numeric Simulations*, vol. 5, no. 4, pp. 89–94, 2004

[5] A. Baziliauskas, A. Tamaševičius, S. Bumelienė and E. Lindberg, "Synchronization of chaotic Colpitts oscillators". in *Scientific Proc. of Riga Tech. Univ.*, Ser. 7, no. 1, pp. 55–58, 2001

[6] V. Rubezic and R. Ostojic, "Synchronization of chaotic Colpitts oscillators with applications to binary communications". in *Proc. IEEE ICECS '99*, vol. 1, pp. 153–156, Sep. 1999

[7] Z.G. Shi, J.T. Huangfu and L.X. Ran, "Performance comparison of two synchronization schemes for Colpitts circuits based chaotic communication system over noisy channel". presented at the 5th World Congress on Intelligent Control and Automation, 6, pp. 1276–1279, 2004

[8] L.O. Chua, T. Yang, G.Q. Zhong and C.W. Wu, "Synchronization of Chua's circuits with time-varying channels and parameters". *IEEE Trans. Circuits Syst. I*, vol. 3, no. 10, pp. 862–868, Oct. 1996

2005 IEEE International Symposium on Circuits and Systems (ISCAS)

Synchronization in an array of chaotic systems coupled via a directed graph

Chai Wah Wu

IBM T. J. Watson Research Center, P. O. Box 218, Yorktown Heights, NY 10598, U. S. A. (e-mail: chaiwahwu@ieee.org)

Abstract—Most analytical results on the synchronization of coupled chaotic systems consider the case of reciprocal coupling, i.e. the coupling matrix is symmetric and the underlying topology is an undirected graph. In this paper we study synchronization in arrays of systems where the coupling is nonreciprocal. This corresponds to the case where the underlying topology can be expressed as a weighted directed graph. We show that several recently proposed definitions of the algebraic connectivity of directed graphs are useful in deriving sufficient conditions for synchronization. In particular, we show that an array synchronizes for sufficiently strong cooperative coupling if the coupling topology includes a spanning directed tree. This is an intuitive result since the existence of such a tree implies that there is a system which influences directed or indirectly all other systems and thus it is possible to make every system synchronize to it.

I. INTRODUCTION

Recently, there have been many analytical results on the synchronization in arrays of chaotic systems with arbitrary coupling [1]–[4]. Many of these results focus on the case of reciprocal coupling which corresponds to the case where the underlying coupling graph is undirected. The sufficient conditions are usually given in terms of the algebraic connectivity or other property of the underlying undirected graph. The purpose of this paper is to consider synchronization in arrays with nonreciprocal coupling, in which the underlying graph is a weighted directed graph. In particular, we show that several recent definitions of the algebraic connectivity of directed graphs are useful in deriving sufficient conditions for synchronization in this case. Finally, we show that sufficiently strong cooperative coupling will synchronize an array of chaotic systems if and only if the underlying interaction graph contains a spanning directed tree.

II. PROBLEM DEFINITION

Consider an array of coupled dynamical systems

$$\dot{x} = (f(x_1,t),\ldots,f(x_n,t))^T + (G\otimes D)x + u(t) \quad (1)$$

where $x = (x_1,\ldots,x_n)^T$, $u = (u_1,\ldots,u_n)^T$ and $G\otimes D$ is the Kronecker product of the matrices G and D. We assume that $\lim_{t\to\infty}\|u_i - u_j\| = 0$ for all i, j. The individual systems which are all identical are given by $\dot{x}_i = f(x_i,t)$. The coupling matrix G describes the coupling topology of the array, i.e. $G_{ij} \neq 0$ if x_j influences x_i, i.e. x_j appear in the equation for $\frac{dx_i}{dt}$. The matrix D describes the coupling between a pair of systems and is usually chosen to be negative semidefinite[1]. If $G_{ij} < 0$ ($G_{ij} > 0$), we call such a coupling

---
[1]We say a (not necessarily symmetric) matrix A is positive definite (semidefinite) if for all $x \neq 0$, $x^T A x > 0$ ($x^T A x \geq 0$).

element *cooperative* (*competitive*).

Definition 1: The class of zero row sums real matrices with nonpositive off-diagonal elements is denoted as **W**. The class of matrices in **W** which are symmetric and irreducible is denoted as **W**$_\mathbf{s}$.

In the rest of this paper we assume that $G \in \mathbf{W}$. This means that all coupling is cooperative. For a matrix G, we define its weighted directed graph as the graph with an edge from vertex i to vertex j with weight G_{ij} if and only if $G_{ij} \neq 0$. We define the *interaction graph* of G as the graph with an edge from vertex i to vertex j with weight G_{ji} if and only if $G_{ji} \neq 0$. In other words, the interaction graph is obtained from the graph of G by reversing the orientation of all the edges, i.e. the adjacency matrix of the interaction graph of G is the transpose of the adjacency matrix of the graph of G. In particular, the interaction graph and the graph of G are identical for undirected graphs (i.e. G is symmetric).

Definition 2: A function $f(y,t)$ is V-uniformly decreasing if $(y-z)^T V(f(y,t)-f(z,t)) \leq -\mu\|y-z\|^2$ for some $\mu > 0$ and all y, z, t.

$f(y,t)$ being V-uniformly decreasing implies that $\dot{y} = f(y,t)$ is globally asymptotically stable in the sense that for any two trajectories $y_1(t)$ and $y_2(t)$, $\lim_{t\to\infty}\|y_1(t)-y_2(t)\| = 0$. This can be seen by using the quadratic Lyapunov function $y^T V y$. This property is also called quadratically stable in the control theory literature. For differentiable f, $f(y,t)$ is V-uniformly decreasing if and only if $V D_1 f(y,t)+\delta I$ is negative semidefinite for some $\delta > 0$ and all y,t [5].

III. SYNCHRONIZATION CRITERION VIA LYAPUNOV'S DIRECT METHOD

A typical synchronization result using the Lyapunov function approach is given by:

Theorem 1 ([1], [6]–[8]): Let Y be a matrix and V be a symmetric positive definite matrix such that $f(x,t) + Yx$ is V-uniformly decreasing. Then the array of coupled dynamical systems in Eq. (1) synchronizes in the sense that $\|x_i - x_j\| \to 0$ as $t \to \infty$ for all i, j if the following condition is satisfied: There exists a matrix $U \in \mathbf{W_s}$ such that $(U \otimes V)(G \otimes D - I \otimes Y)$ is negative semidefinite.

Suppose that we assume that VD is symmetric negative semidefinite. This is a reasonable assumption as it occurs for the following scenarios which are found in practice:

1) V and D are both diagonal and D is negative semidefinite;
2) D is a nonpositive multiple of the identity matrix;

3) V is a positive multiple of the identity matrix and D is symmetric negative semidefinite.

If in addition, we make the assumption that $Y = \alpha D$, then the condition in Theorem 1 reduces to the existence of a matrix $U \in \mathbf{W_s}$ such that $U(G - \alpha I)$ is positive semidefinite. Thus we have reduced the synchronization condition to a matrix inequality. Next we show that existence of such a matrix U is closely related to the algebraic connectivity of the graph of G. For simplicity, for a given α we say in the sequel that the matrix G satisfies *synchronization condition* $A(\alpha)$ if there exists $U \in \mathbf{W_s}$ such that $U(G - \alpha I)$ is positive semidefinite.

IV. ALGEBRAIC CONNECTIVITY OF UNDIRECTED GRAPHS

Suppose $G \in \mathbf{W}$ is symmetric, i.e. the interaction graph of G is undirected and all eigenvalues of G are real. It was shown that G satisfies synchronization condition $A(\alpha)$ if $\lambda_2(G) \geq \alpha$, where $\lambda_2(G)$ is the second smallest eigenvalue of G. To prove this, U can be chosen to be either $U = G$ [1] or $U = I - \frac{1}{n}ee^T$ [7] where $e = (1,\ldots 1)^T$. Note that the choice $U = I - \frac{1}{n}ee^T$ is independent of G which will be useful when we look at the case of time-varying coupling (Section VII).

The quantity $\lambda_2(G)$ is also called the *algebraic connectivity* of the graph of G and is related to many graph-theoretical properties [9], [10]. Since $\lambda_2(G) > 0$ if and only if the graph is connected, a consequence is that the array synchronizes if the graph is connected and the matrix D is sufficiently large[2] [1] (which results in α to be small in the equation $Y = \alpha D$).

V. ALGEBRAIC CONNECTIVITY OF DIRECTED GRAPHS

When G is not symmetric, the corresponding graph is not undirected and there are several ways to extend the notion of algebraic connectivity to directed graphs. Consider a matrix $G \in \mathbf{W}$ written in Frobenius normal form [11]:

$$G = P \begin{pmatrix} B_1 & B_{12} & \cdots & B_{1k} \\ & B_2 & \cdots & B_{2k} \\ & & \ddots & \vdots \\ & & & B_k \end{pmatrix} P^T \quad (2)$$

where P is a permutation matrix and B_i are square irreducible matrices. Each B_i can be decomposed as $B_i = L_i + D_i$ where L_i is a zero row sums matrix and D_i is diagonal. Let w_i be the unique positive vector such that $\|w_i\|_\infty = 1$ and $w_i^T L_i = 0$ and W_i be the diagonal matrix with w_i on the diagonal. Let w be the unique nonnegative vector such that $\|w\|_\infty = 1$ and $w^T G = 0$ and W be the diagonal matrix with w on the diagonal. We can then define the following generalizations of algebraic connectivity to directed graphs:

- $a_1(G) = \min_{x \perp e, \|x\|=1} x^T G x$;
- $a_2(G) = \min_{x \perp e, \|x\|=1} x^T W G x$;
- $a_3(G) = \min_{x \neq 0, x \perp e} \frac{x^T W G x}{x^T \left(W - \frac{ww^T}{\|w\|_1}\right)x}$;
- $a_4(G) = \min_{1 \leq i \leq k} \eta_i$ where $\eta_i = \min_{x \neq 0} \frac{x^T W_i B_i x}{x^T W_i x}$ for $1 \leq i \leq k-1$ and $\eta_k = \min_{x \neq 0, x \perp e} \frac{x^T W_k B_k x}{x^T \left(W_k - \frac{w_k w_k^T}{\|w_k\|_1}\right)x}$.

[2]corresponding to sufficiently strong cooperative coupling.

It is easy to see that $a_1(G) = a_2(G) = a_3(G) = a_4(G) = \lambda_2(\frac{1}{2}(G + G^T))$ when the graph is undirected or balanced[3]. a_1 and a_2 can be expressed as an eigenvalue. For instance $a_2(G) = \lambda_{\min}\left(\frac{1}{2}K^T(WG + G^T W)K\right)$ where K is an n by $n-1$ matrix whose columns form an orthonormal basis of the linear subspace orthogonal to e.

The following can be shown about these quantities which generalizes the corresponding statements about the algebraic connectivity of undirected graphs [8], [12], [13]:

1) $a_3(G) > 0$ if the graph is strongly connected;
2) $a_i(G) \leq 0$, $1 \leq i \leq 4$ if the graph is not weakly connected;
3) If the graph is strongly connected, then $a_3(G) \geq a_2(G)$;
4) For a strongly connected graph with local coupling, $a_2(G) \to 0$ as $n \to \infty$;
5) $a_4(G) > 0$ if and only if the interaction graph of G contains a spanning directed tree;
6) (Super-additivity) For $G, H \in \mathbf{W}$, $a_1(G + H) \geq a_1(G) + a_1(H)$;
7) If the off-diagonal elements of G are random variables chosen independently according to $P(G_{ij} = 1) = p$, $P(G_{ij} = 0) = 1 - p$, then $a_1(G) \approx pn$ in probability as $n \to \infty$.

Similar to the algebraic connectivity in the undirected case, these quantities also provide a sufficient condition for the array to synchronize:

Theorem 2 ([8], [12], [13]): G satisfies synchronization condition $A(\alpha)$ if one of the following conditions is satisfied:

1) $a_1(G) \geq \alpha$;
2) $a_4(G) \geq \alpha$;
3) The graph of G is strongly connected and $a_3(G) \geq \alpha$.

The sufficient condition for synchronization $a_4(G) \geq \alpha$ along with the fact that $a_4(G)$ is positive if and only if the interaction graph contains a spanning directed tree implies the following:

Proposition 1: The array synchronizes if the interaction graph contains a spanning directed tree and the coupling matrix D is large enough.

This generalizes the corresponding statement for undirected graphs and is intuitive since the existence of a spanning directed tree in the interaction graph implies the existence of a system (located at the root of the tree) which influences directly or indirectly all other systems. The converse is also true under certain conditions for arrays of chaotic systems. If there does not exist a spanning directed tree, then there are two groups of systems which are not influenced by any other system [8]. If these two coupled groups remain chaotic, they will not synchronize to each other.

One interpretation for the synchronization occurring in coupled systems where the interaction graph contains a spanning directed tree is that all the systems are synchronized to the system at the root of the spanning directed tree (which we

[3]We define a directed graph to be balanced if for each vertex, the sum of the weights of incoming edges is equal to the sum of the weights of outgoing edges. This is equivalent to G having zero row and column sums.

call the *root system*). This suggests that any control to the system should be applied to the root system for maximum efficiency. Of course, in general graphs, the spanning directed tree is not unique and neither is the root system. It is clear that a root system is unique if and only if there is coupling from the root system to other systems and not vice versa (Fig. 1).

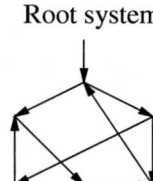

Fig. 1. Interaction graph with a unique root system.

If the matrix G is not irreducible, it is m-reducible for some $1 \leq m \leq n$ [14]. This implies that there are m directed trees that span the interaction graph. The m strongly connected components (SCC) of the graph which contain the m roots of these trees will not have coupling from other systems and thus they will synchronize within themselves (when the coupling is sufficiently cooperative). Thus there are at least m groups or clusters of systems which synchronize among themselves.

There exists a lower bound on $a_4(G)$ which is related to the structure of the graph. In particular, in [14] a lower bound was given in terms of quantities a_2 and a_3 of the SCC's of the graph and the number of edges between them. The larger the values of a_2 of the SCC's, the larger the value of this bound. Similarly, the more coupling between the SCC, the larger the value of this bound. This is somewhat intuitive as the a_i's are measures of "connectivity" of the graph.

More precisely, the SCC's of the graph correspond to the matrices B_i in the Frobenius normal form (Eq. (2)) with decomposition $B_i = L_i + D_i$. D_i describes the coupling to the i-th SCC from other systems. If we remove this coupling, we obtain L_i which corresponds to the coupling within the i-th SCC. Then η_i, $1 \leq i \leq k-1$ in the definition of a_4 can be bounded as

$$\eta_i \geq \frac{a_2(L_i)}{\left(1 + \sqrt{1 + \frac{a_2(L_i)}{w_i^T D_i e}}\right)^2 n + 1}$$

whereas $\eta_k = a_3(B_k) \geq a_2(B_k)$.

The facts that $a_2(G)$ vanishes for local coupling and that $a_1(G) \approx pn$ (which is the largest possible for the given number of 1's in the matrix [12]) for random coupling indicate that local coupling and random coupling form two extremes in the corresponding array's ability to synchronize, the same as in the case of symmetric G [7], [15]. This also indicate that the distribution of the vertex degrees is itself not enough to ensure high synchronizability. A locally connected k-regular graph and a random k-regular graph have the same degree distribution, yet their abilities to synchronize are at opposite ends of the range.

For the case of reciprocal coupling (symmetric G), because of the super-additivity of λ_2, adding additional reciprocal coupling does not decrease λ_2. In other words, adding cooperative coupling can only help achieve synchronization in the array. An important difference between reciprocal and nonreciprocal coupling is that this property is not longer true for nonreciprocal coupling. In particular, consider the following directed graph

with corresponding coupling matrix

$$G = \begin{pmatrix} 1 & -1 & & & \\ & 1 & -1 & & \\ & & \ddots & \ddots & \\ & & & & 0 \end{pmatrix}.$$

In this case $a_4(G) = 1$. By adding an additional coupling, we obtained the following directed graph which is balanced

with corresponding coupling matrix

$$G = \begin{pmatrix} 1 & -1 & & & \\ & 1 & -1 & & \\ & & \ddots & \ddots & \\ -1 & & & & 1 \end{pmatrix}$$

For this graph, $a_4(G) = 1 - \cos(\frac{2\pi}{n})$ which decreases to 0 for large n. Thus by *adding* a single coupling element, we have *reduced* its synchronizability significantly. This can be explained by noting that the SCC's in the first graph are small; they are single vertices, whereas the SCC is large in the second graph; it is the entire graph with a large diameter which makes the array harder to synchronize.

VI. OPTIMAL α AND SYNCHRONIZABILITY OF THE COUPLING TOPOLOGY

For a fixed coupling matrix G, what is the largest α such that G satisfies synchronization condition $A(\alpha)$, i.e. there is a matrix $U \in \mathbf{W_s}$ for which $U(G - \alpha I)$ is positive semidefinite? This value of α, which we denote as $\mu(G)$ gives yet another measure on how easy it is to synchronize an array with the given coupling topology (expressed as G). The following theorem gives some upper and lower bounds on $\mu(G)$.

Theorem 3 ([16]): For a matrix $G \in \mathbf{W}$,
1) $\lambda_{\min}\left(\frac{1}{2}(G + G^T)\right) \leq \mu(G)$;
2) $a_1(G) \leq \mu(G)$ and $a_4(G) \leq \mu(G)$;
3) If the graph of G is strongly connected, $a_2(G) \leq a_3(G) \leq \mu(G)$;
4) $\mu(G) \leq Re(\lambda)$ for all eigenvalues λ of G not corresponding to the eigenvector e;
5) If G has both zero rows and zero column sums, then $\lambda_2^s(G) \leq \mu(G)$, where $\lambda_2^s(G)$ is the smallest eigenvalue of $\frac{1}{2}(G + G^T)$ not belonging to the eigenvector e;
6) If G is a normal matrix, then $a_1(G) = \mu(G) = \mu_2(G)$, where $\mu_2(G) = \min_{\lambda \in L(G)} Re(\lambda)$, and $L(G)$ are the eigenvalues of G not belonging to the eigenvector e;

7) If G is a triangular matrix after simultaneous permutation of its rows and columns (i.e. the graph of G is acyclic), then $\mu(G) = \mu_2(G)$.

The value of $\mu(G)$ can be computed efficiently by solving a sequence of semidefinite programming problems [16] which are known to be solvable in polynomial time.

VII. TIME-VARYING COUPLING

In this case the state equations are given by:

$$\dot{x} = (f(x_1,t), \ldots, f(x_n,t))^T + (G(t) \otimes D(t))x + u(t) \quad (3)$$

and the corresponding synchronization theorem is given by:

Theorem 4: Let $Y(t)$ be a time-varying matrix and V be a symmetric positive definite matrix such that $f(x,t) + Y(t)x$ is V-uniformly decreasing. Then the network of coupled dynamical systems in Eq. (3) synchronizes if the following condition is satisfied: There exists a matrix $U \in \mathbf{W_s}$ such that $(U \otimes V)(G(t) \otimes D(t) - I \otimes Y(t))$ is negative semidefinite for all t.

Similarly to Section III, if $Y(t) = \alpha(t)D(t)$ and $VD(t)$ is symmetric negative semidefinite for all t, then this condition is reduced to checking whether $U(G(t) - \alpha(t)I)$ is positive semidefinite for some $U \in \mathbf{W_s}$ and all time t. In the constant coupling case, to prove that $a_3(G) \geq \alpha$ implies synchronization, we choose $U \in \mathbf{W_s}$ which depends on G. Therefore, in contrast to the constant coupling case, the synchronization condition in the time-varying coupling case is not necessarily satisfied if $a_3(G(t)) \geq \alpha(t)$ for all t since the matrix U is not allowed to change for different t. By choosing $U = I - \frac{ee^T}{n}$, it was shown in [12] that $U(G(t) - \alpha(t)I)$ is positive semidefinite for all t and thus the array synchronizes if $a_1(G(t)) \geq \alpha(t)$ for all t. This synchronization criterion does not depend on the rate of change of $G(t)$ and $D(t)$.

On the other hand, if there exists a single positive vector w such that $w^T G(t) = 0$ for all t, then using $U = W - \frac{ww^T}{\|w\|_1}$ shows that Eq. (3) synchronizes if $a_3(G(t)) \geq \alpha(t)$ for all t.

VIII. LYAPUNOV EXPONENTS APPROACH TO SYNCHRONIZATION

In [17], local synchronization criteria are derived based on numerical estimates of Lyapunov exponents. In this case, for appropriate D (e.g., $D = I$), the array synchronizes if the nonzero eigenvalues of G have real parts which are large enough. In [13], [14] it was shown that all nonzero eigenvalues of G have nonzero real parts if and only if the interaction graph contains a spanning directed tree. Thus the Lyapunov exponents based synchronization criterion shows that the array synchronizes locally if and only if the interaction graph contains a spanning directed tree and the cooperative coupling is large enough and this statement is qualitatively the same as Proposition 1.

IX. CONCLUSIONS

We study synchronization in arrays of coupled systems where the coupling is nonreciprocal and the coupling graph is directed. We show synchronization criteria which depend on various quantities of the graph, including various generalizations of algebraic connectivities to directed graphs. These results can be summarized succinctly as follows. For constant coupling, the array synchronizes if a_4 of the coupling topology is large enough[4]. This occurs if the cooperative coupling is large enough and there is a system which influences directly or indirectly all other systems. For time-varying coupling the array synchronizes if a_1 of the coupling topology is large enough at each time. These results can be extended to the case where some of the coupling contains delay terms [18]. Finally, recall that if the interaction graph is undirected (coupling is reciprocal and G is symmetric) or balanced then $a_1 = a_4$. Thus these bounds for constant and time-varying coupling are the same in these cases.

REFERENCES

[1] C. W. Wu and L. O. Chua, "Synchronization in an array of linearly coupled dynamical systems," *IEEE Trans. on Circ. and Syst.–I*, vol. 42, no. 8, pp. 430–447, 1995.

[2] C. W. Wu, *Synchronization in coupled chaotic circuits and systems*. World Scientific, 2002.

[3] X. F. Wang and G. Chen, "Synchronization in small-world dynamical networks," *Int. J. of Bif. and Chaos*, vol. 12, no. 1, pp. 187–192, 2002.

[4] V. N. Belykh, I. V. Belykh, and M. Hasler, "Connection graph stability method for synchronized coupled chaotic systems," *Physica D*, vol. 195, pp. 159–187, 2004.

[5] L. O. Chua and D. N. Green, "Graph-theoretical properties of dynamic nonlinear networks," Tech. Rep. Memo ERL-M507, College of Engineering, University of California, Berkeley, 1975.

[6] C. W. Wu and L. O. Chua, "Application of Kronecker products to the analysis of systems with uniform linear coupling," *IEEE Trans. on Circ. and Syst.–I*, vol. 42, no. 10, pp. 775–778, 1995.

[7] C. W. Wu, "Perturbation of coupling matrices and its effect on the synchronizability in arrays of coupled chaotic circuits," *Physics Letters A*, vol. 319, pp. 495–503, 2003.

[8] C. W. Wu, "Synchronization in networks of nonlinear dynamical systems coupled via a directed graph," *Nonlinearity*, in press.

[9] M. Fiedler, "Algebraic connectivity of graphs," *Czechoslovak Mathematical Journal*, vol. 23, no. 98, pp. 298–305, 1973.

[10] B. Mohar, "Some applications of Laplace eigenvalues of graphs," in *Graph Symmetry: Algebraic Methods and Applications* (G. Hahn and G. Sabidussi, eds.), pp. 225–275, Kluwer, 1997.

[11] R. A. Brualdi and H. J. Ryser, *Combinatorial Matrix Theory*. Cambridge University Press, 1991.

[12] C. W. Wu, "Algebraic connectivity of directed graphs," *Linear and Multilinear Algebra*, in press.

[13] C. W. Wu, "On Rayleigh-Ritz ratios of a generalized Laplacian matrix of directed graphs," *Linear Algebra and Its Applications*, in press.

[14] C. W. Wu, "On bounds of extremal eigenvalues of irreducible and m-reducible matrices," *Linear Algebra and Its Applications*, in press.

[15] C. W. Wu, "Synchronization in systems coupled via complex networks," in *Proceedings of IEEE ISCAS 2004*, pp. IV-724–727, 2004.

[16] C. W. Wu, "On a matrix inequality and its application to the synchronization in coupled chaotic systems," IBM Research Report RC23155, 2004, submitted for publication.

[17] L. M. Pecora and T. L. Carroll, "Master stability functions for synchronized chaos in arrays of oscillators," in *Proceedings of the 1998 IEEE Int. Symp. Circ. Syst.*, vol. 4, pp. IV-562–567, IEEE, 1998.

[18] C. W. Wu, "Synchronization in arrays of coupled nonlinear systems with delay and nonreciprocal time-varying coupling," *IEEE Trans. Circuits and Systems-II*, in press.

[19] M. Vidyasagar, *Nonlinear Systems Analysis, 2nd Ed.*, Prentice-Hall, 1993.

---

[4]This is also true if the coupling changes slow enough. This can be shown using stability results for slowly varying nonlinear systems [19].

A Subtle Link in Switched Dynamical Systems: Saddle-Node Bifurcation Meets Border Collision

Yue Ma*, Chi K. Tse[†], Takuji Kousaka[‡] and Hiroshi Kawakami*

*Dept. of Electrical and Electronic Engineering Department, The University of Tokushima, Tokushima, Japan
[†]Dept. of Electronic and Information Engineering, Hong Kong Polytechnic University, Hong Kong
[‡]Dept. of Electronic and Electrical Engineering, Fukuyama University, Japan
Emails: *mayue(kawakami)@ee.tokushima-u.ac.jp; [†]encktse@polyu.edu.hk; [‡]kousaka@fuee.fukuyama-u.ac.jp

Abstract—Switched dynamical systems are known to exhibit border collision, in which a particular operation is terminated and a new operation is assumed as one or more parameters are varied. In this paper, we report a subtle relation between border collision and saddle-node bifurcation in such systems. Our main finding is that the border collision and the saddle-node bifurcation are actually linked together by unstable solutions which have been generated from the same saddle-node bifurcation. Since unstable solutions are not observable directly, such a subtle relation has not been known. In this paper, we describe an effective method to track solutions regardless of their stability, allowing the subtle phenomenon to be uncovered. Two typical DC-DC converters are observed to verify our finding.[1]

I. INTRODUCTION

Being a commonly observed phenomenon in switched dynamical systems, border collision has attracted much attention in recent years [1]–[3]. Like other typical bifurcation scenarios, border collision manifests itself as a sudden change of qualitative behavior of a system as one or more parameters are varied, and hence can be regarded as a kind of *bifurcation* phenomenon. However, border collision has always been considered separately from such traditional bifurcations as saddle-node bifurcation and period-doubling bifurcation. This can be attributed to the fundamental difference in the mechanisms underlying border collision and traditional bifurcations. Specifically, traditional bifurcations are caused by a loss of stability of an operating orbit and the assumption of a new stable orbit, whereas border collision is resulted from operational change in which an operating orbit fails to maintain itself due to some inherent structural property of the system [4]. The system thus typically jumps to another stable operating orbit at border collision. Moreover, no loss of stability is required for border collision. Up to now, because of the apparent lack of commonality between border collision and traditional bifurcations, no connection has been known that links the two types of coexisting bifurcations in a given switched dynamical system.

In our previous work [5], the relation between border collision and period-doubling bifurcation has been discussed briefly. In this paper, we will find a strong link that connects border collision with saddle-node bifurcation. Specifically, by tracking the border collision that occurs in unstable periodic solutions, we will show how border collision is connected to a coexisting saddle-node bifurcation via unstable solutions. Based on this finding, we are able to explain why and when a "jump" or "continuous transition" occurs during border collision.

II. SADDLE-NODE BIFURCATION AND BORDER COLLISION

According to traditional bifurcation theory, if a saddle-node bifurcation occurs, one of the characteristic multipliers of the characteristic equation for a fixed point equals to 1. Such a bifurcation can produce or destroy a pair of solutions: one is a node and the other is a 1-dimensional unstable saddle. Meanwhile, a border collision is a bifurcation phenomenon which is often observed in switched dynamical systems. Unlike traditional bifurcations, border collision is independent of characteristic multipliers. Instead, it occurs when the system experiences a structural change which causes a stable operation to cease, as a result of the system state hitting a spatial or temporal "border" [4]. Specifically, in our previous work [5], a method has been proposed to locate the occurrence of border collision. However, the manifestation of the transition at border collision (i.e., the way in which the system jumps from one orbit to another) is still a complicated problem which is not generally solved.

In much of the previous study, only stable solutions are considered in detail, as limited by the way in which codimension-1 bifurcation diagrams are obtained through deriving the steady-state solutions of the system. In the following we will introduce a method to compute the fixed points of switched dynamical systems regarding of their stability. Then, by tracking the bifurcation of unstable solutions, we are able to uncover an important role that unstable solutions play in connecting border collision and saddle-node bifurcation.

III. PERIODIC SOLUTIONS AND STABILITY

A switched dynamical system can be described briefly as follows. Consider a simple but general case where a switched dynamical system consists of two sub-systems: $S_1 : \dot{x} = f(x, \lambda_1)$ and $S_2 : \dot{x} = g(x, \lambda_2)$, where $x \in R^n$ and λ

---
[1]This work was supported in part by the Research Grant Council of Hong Kong under a competitive-bid earmarked research grant (No. PolyU 5241/03E).

denote the state variables and system parameters, respectively. Solution flows of the sub-systems are represented by $\varphi(t, x_{f0})$ and $\psi(t, x_{g0})$, where x_{f0} and x_{g0} are initial points. Switching is modulated by borders for each sub-system; that is, whenever φ or ψ hits the border specifically defined for one sub-system, switching occurs. Suppose borders for S_1 and S_2 are given by $B_1 = \{x \in R^n : \beta_1(x,t) = 0\}$ and $B_2 = \{x \in R^n : \beta_2(x,t) = 0\}$, respectively. Then, the simplest solution flow can be formulated as

$$\begin{aligned} x_1 &= \varphi(\tau_1, x_0) \\ x_2 &= \psi(\tau_2 - \tau_1, x_1) \\ \beta_1(x_1, \tau_1) &= 0 \\ \beta_2(x_2, \tau_2) &= 0 \end{aligned} \quad (1)$$

where x_0 is the starting point and x_2 is the ending point of the flow, and τ_1 and τ_2 are switching times. Thus, if $x_2 = x_0$, it becomes a periodic solution. Moreover, x_1 is the ending point of the first interval as well as the starting point of the second interval. Clearly, switching points must satisfy the border functions $\beta_1 = 0$ and $\beta_2 = 0$. Finally, as (1) contains $2n + 2$ scalar equations with $2n + 2$ scalar unknowns, i.e., $\{x_0, x_1, \tau_1, \tau_2\}$, we can solve the periodic solution using an appropriate numerical method. Unstable solutions can be also found since our method does not discriminate the stability status. To determine the stability of the computed periodic solution, one needs to find the Jacobian of the map $F : R^n \to R^n$; $x_0 \mapsto x_2$, for the solution flow given in (1), and preferably in terms of $\partial x_2 / \partial x_0$.

From the first two equations of (1), we get

$$\frac{\partial x_1}{\partial x_0} = \frac{\partial \varphi}{\partial t} \frac{\partial \tau_1}{\partial x_0} + \frac{\partial \varphi}{\partial x_0} \quad (2)$$

$$\frac{\partial x_2}{\partial x_0} = \frac{\partial \psi}{\partial t} \frac{\partial \tau_2}{\partial x_0} - \frac{\partial \psi}{\partial t} \frac{\partial \tau_1}{\partial x_0} + \frac{\partial \psi}{\partial x_1} \frac{\partial x_1}{\partial x_0} \quad (3)$$

where φ and ψ denote $\varphi(\tau_1, x_0)$ and $\psi(\tau_2 - \tau_1, x_1)$, respectively. Moreover, from the border functions (i.e., last two equations in (1)), we have

$$\frac{\partial \beta_1}{\partial x_0} = \frac{\partial \beta_1}{\partial x} \frac{\partial x_1}{\partial x_0} + \frac{\partial \beta_1}{\partial t} \frac{\partial \tau_1}{\partial x_0} = 0 \quad (4)$$

$$\frac{\partial \beta_2}{\partial x_0} = \frac{\partial \beta_2}{\partial x} \frac{\partial x_2}{\partial x_0} + \frac{\partial \beta_2}{\partial t} \frac{\partial \tau_2}{\partial x_0} = 0. \quad (5)$$

Hence, $\partial \tau_2 / \partial x_0$, $\partial \tau_1 / \partial x_0$ and $\partial x_1 / \partial x_0$ can be obtained easily. Direct substitution yields

$$\frac{\partial x_2}{\partial x_0} = \frac{\left(\frac{\partial \psi}{\partial x_1} \frac{\partial \beta_1}{\partial t} + \frac{\partial \psi}{\partial t} \frac{\partial \beta_1}{\partial x}\right) \frac{\partial \varphi}{\partial x_0} \frac{\partial \beta_2}{\partial t}}{\left(\frac{\partial \beta_2}{\partial t} + \frac{\partial \psi}{\partial t} \frac{\partial \beta_2}{\partial x}\right) \left(\frac{\partial \beta_1}{\partial t} + \frac{\partial \varphi}{\partial t} \frac{\partial \beta_1}{\partial x}\right)}. \quad (6)$$

All terms in the above equation can be calculated by an appropriate numerical method. Thus, by finding the roots of the characteristic equation, we can determine the stability of any periodic solution.

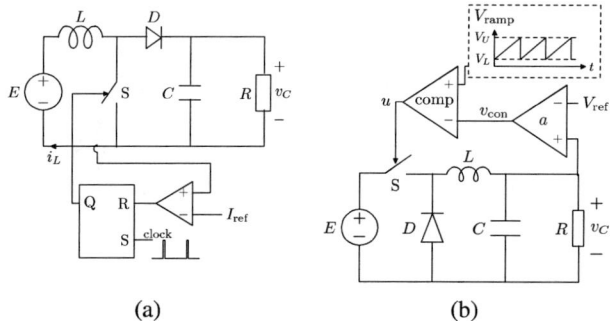

Fig. 1. (a) Current-mode controlled boost converter; (b) Voltage-mode controlled buck converter.

IV. ILLUSTRATION OF MAIN FINDINGS

In much of the previous study, border collision has been shown to exhibit a "jump" in the bifurcation diagram, as a stable operating orbit suddenly gives way to another stable orbit in a discontinuous fashion. Border collision "terminates" a specific solution and the system "jumps" to another attractor. The resulting solution is entirely new and seems to have little relation with the original solution assumed before the onset of border collision. In the following, we will illustrate that the new solution is often generated by a saddle-node bifurcation. In fact, saddle-node bifurcation gives birth to a node and a saddle. While the node can be directly observed, the saddle (or the unstable solution) is invisible and much less considered.

The problem is best illustrated with examples. We will consider the current-mode controlled boost dc-dc converter and the voltage-mode controlled buck dc-dc converter. They provide very accessible examples on explaining the subtle connection between the coexisting border collision and saddle-node bifurcation.

Example 1: Current-Mode Controlled Boost Converter

In the current-mode controlled boost converter circuit shown in Fig. 1(a) [7], the switch is turned on periodically and turned off whenever the current of the inductor reaches a reference value. Thus, the two sub-systems are given by

$$S_1 : \dot{x} = \begin{bmatrix} -1/RC & 0 \\ 0 & 0 \end{bmatrix} x + \begin{bmatrix} 0 \\ 1/L \end{bmatrix} E \quad (7)$$

$$S_2 : \dot{x} = \begin{bmatrix} -1/RC & 1/C \\ -1/L & 0 \end{bmatrix} x + \begin{bmatrix} 0 \\ 1/L \end{bmatrix} E. \quad (8)$$

When it works in continuous conduction mode, the two borders can be described by

$$B_1 = \{(x,t) \in R^2 \times R : \beta_1 = i_L - I_{\text{ref}} = 0\} \quad (9)$$
$$B_2 = \{(x,t) \in R^2 \times R : \beta_2 = t - kT = 0\} \quad (10)$$

where $x = [v_C \; i_L]^T$, $k = 1, 2, \cdots$. A typical one-parameter bifurcation diagram is shown in Fig. 2, with the reference current I_{ref} being the bifurcation parameter and the sampled i_L being the variable. Here, as revealed from the blow-up views, as I_{ref} increases, border collision takes place to terminate the period-2 solution, and the system jumps to a period-4 solution. Moreover, if we move backward (decreasing I_{ref}), we see that

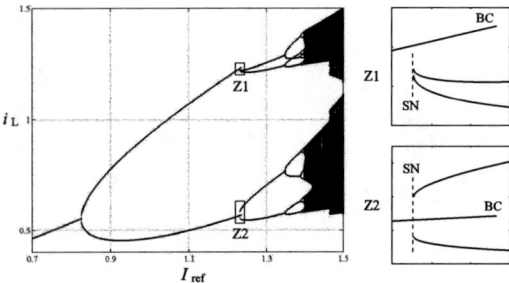

Fig. 2. One-parameter bifurcation diagram of current-mode controlled boost converter, with two blow-up views of regions z1 and z2. Parameters are $R = 40\ \Omega$, $L = 1.5$ mH, $T = 100\ \mu$s, $\tau_C = RC/T = 2$ and $E = 10$ V.

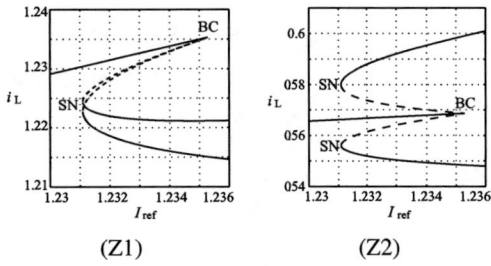

Fig. 3. Blow-up views of the one-parameter bifurcation diagram showing bifurcations of unstable solutions. Parameters are same as those in Fig. 2.

the period-4 solution ends up at a saddle-node bifurcation, and the solution jumps back to the original period-2 solution.

Although only a stable period-4 solution is observed, an unstable period-4 solution exists. This unstable period-4 solution, being unobservable, has been rarely noticed, let alone discussed. Using the method introduced previously, however, we can track the position of the unstable solutions easily. The results are shown in Fig. 3, where unstable period-4 solutions are traced and shown as dashed curves.

The key message of Fig. 3 is that unstable solutions are also terminated by the same border collision. In other words, the stable period-2 solution and the unstable period-4 solution are terminated by the same border collision, but in opposite parametric directions. Thus, we see that the border collision is actually being linked to the saddle-node bifurcation by the unstable period-4 solution.

Example 2: Voltage-Mode Controlled Buck Converter

We now consider a voltage-mode controlled buck converter, as shown in Fig. 1(b). A feedback signal, V_{con}, is compared with a ramp voltage signal, V_{ramp}, to control the switch. Therefore, we can write the two sub-systems as

$$S_1\ :\ \dot{x} = \begin{bmatrix} -1/RC & 1/C \\ -1/L & 0 \end{bmatrix} x \quad (11)$$

$$S_2\ :\ \dot{x} = \begin{bmatrix} -1/RC & 1/C \\ -1/L & 0 \end{bmatrix} x + \begin{bmatrix} 0 \\ 1/L \end{bmatrix} E \quad (12)$$

where $x = [v_C\ i_L]^T$. The border of this system is defined as

$$\beta_1(x,t) = \beta_2(x,t)$$
$$= a(v_C - V_{\text{ref}}) - \frac{(V_U - V_L)t}{T} - V_L = 0 \quad (13)$$

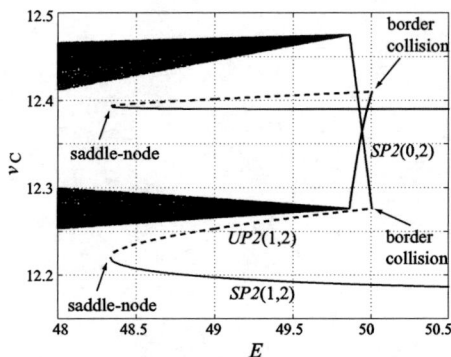

Fig. 4. One-parameter bifurcation diagram from voltage-mode controlled buck converter. Parameters are $T = 400\mu$s, $L = 20$mH, $a = 8.4$, $V_{\text{ref}} = 11.3$V, $V_L = 3.8$V, $V_U = 8.2$V and $C = 47\mu$F.

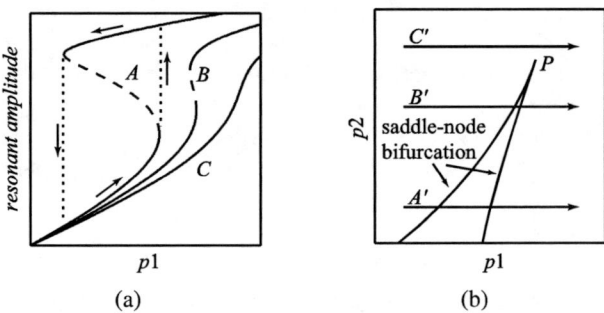

Fig. 5. Illustration of jump and hysteresis phenomena. (a) One-parameter bifurcation diagram; (b) two-parameter bifurcation diagram. Varying parameter $p1$ along lines A, B and C in (b), we get three curves of resonant amplitude characteristics in (a).

for $t \in [0, T]$, which is a common border for both two sub-systems.

Our previous studies [5] have shown that a period-2 solution can jump to another stable period-2 solution via a border collision, as shown in Fig. 4. Here, we track the unstable period-2 solution generated from saddle-node bifurcation, and observe a similar phenomenon. As denoted by dashed curves in Fig. 4, the unstable solution conjoins with the stable solution at the border collision point.

V. Details of the Bifurcation Connection and Interaction

In the foregoing, we have described how stable solutions connected to different bifurcations are tied together by unstable solutions. This is similar to the phenomenon of *hysteresis*, which is often found in systems with resonating states associated with saddle-node bifurcation. At this point, it is instructive to recall the familiar behavior of the Duffing's equation shown simplistically in Fig. 5. Here, curves A, B and C indicate the resonant amplitude characteristics corresponding to movements of parameters along the horizontal lines A', B' and C'. In the case of parameter movement along A', solution moves along the solid curves in the arrow direction in Fig. 5 (a). When it comes to the point with vertical tangency, a slight change of $p1$ will cause a discontinuous jump of the amplitude to the upper or lower portion of the curve. Between

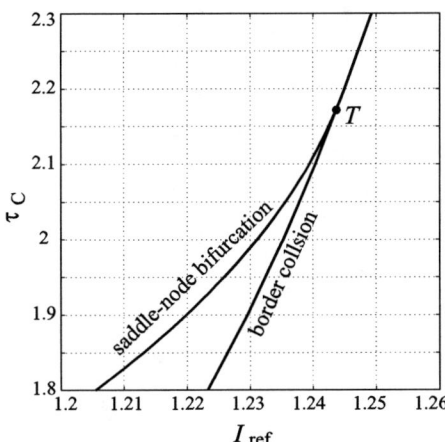

Fig. 6. Bifurcation diagram of boost converter with same set of parameters as Fig. 2.

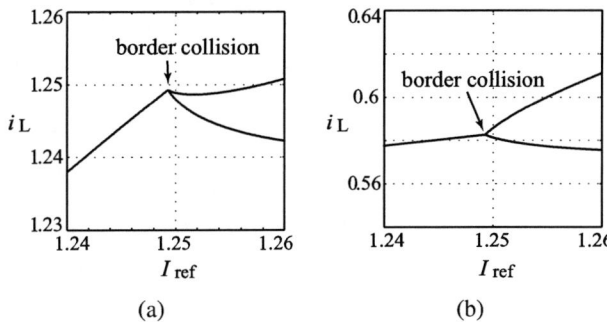

Fig. 7. One-parameter bifurcation diagrams showing no hysteresis jump, with $\tau_C = 2.3$. This corresponds to the case above the tangent point T in Fig. 2.

the two bifurcation curves in Fig. 5 (b), coexistence of the two solutions is possible. Moreover, as we increase $p2$, the two bifurcation curves in Fig. 5 (b) will move closer to each other. For example, along line B', the coexistence region is much narrower than that along line A. Finally, for the case of movement along line C', i.e., beyond the cusp point P of the bifurcation curves, no bifurcation occurs for the solution and coexistence is no longer observed.

Returning to our switched systems, from Figs. 3 and 4, we see that saddle-node bifurcation and border collision are articulated in a similar manner as the aforementioned jump and hysteresis phenomenon.

We now take a detailed look at the bifurcation behavior of the current-mode controlled boost converter around the region where border collision and saddle-node bifurcation occur. Fig. 6 shows the bifurcation diagram in the (I_{ref}, τ_C) plane. Here, we see that the saddle-node bifurcation curve merges tangentially with the border collision curve at point T. Stable period-2 solutions exist on the left-hand side of the border collision curve and stable period-4 solutions exist on the right-hand side of the saddle-node bifurcation curve. Thus, between the two bifurcation curves, coexistence and hysteresis can be observed, as shown in Fig. 3, where $\tau_C = 2$. Similar to the situation illustrated by the Duffing's system (Fig. 5), increasing τ_C above the tangent point T will make saddle-node bifurcation vanish. However, the border collision still remains. If we move parameter I_{ref} along the line $\tau_C = 2.3$, we get a one-parameter bifurcation diagram, as shown in Fig. 7. Comparing with Fig. 3, we see that unstable solutions disappear due to the tangential merger of border collision and saddle-node bifurcation. Finally, the scenes shown in Fig. 7 look deceptively like a period-doubling bifurcation. Here, it is clear that they are the results of the interaction between saddle-node bifurcation and border collision, corresponding to the choice of parameters above the tangent point T.

Similar phenomena can also be observed in the buck converter. As R is decreased, the saddle-node bifurcation will merge tangentially with the border collision (see Fig. 4), and the two period-2 solutions will be connected together. This clearly explains why border collision may manifest itself as a jump (hysteresis) or as a continuous turning point, depending upon the parameter values being below or above the tangent (merging) point.

VI. CONCLUSION

Being fundamentally different in the underlying bifurcation mechanisms, border collision and saddle-node bifurcation have rarely been considered in a unified way. In this paper we have discussed the relation between border collision and saddle-node bifurcation, and identified the way in which the two bifurcations are connected. The key finding is that border collision is connected to saddle-node bifurcation via unstable solutions that have been generated by the same saddle-node bifurcation. This explains the jump phenomenon that occurs at border collision. Moreover, as parameters vary, the border collision and the saddle-node bifurcation points merge tangentially. Beyond this merger point, border collision no longer manifests as jumps, but rather as continuous transitions which typically resemble turning points with possible period-multiplying in one-parameter bifurcation diagrams.

REFERENCES

[1] H. Nusse, E.Ott, and J. Yorke, "Border-collision bifurcations: an explanation for observed bifurcation phenomena," *Physical Review E*, vol. 49, pp. 1073–1076, 1994.
[2] S. Banerjee, P. Ranjan, and C. Grebogi, "Bifurcation in two-dimensional piecewise smooth maps – theory and applications in switching circuits," *IEEE Transactions on Circuits and Systems Part I*, vol. 47, no. 5, pp. 633–643, May 2000.
[3] M. di Bernardo, C. J. Budd, and A. R. Champneys, "Grazing and border-collision in piecewise-smooth systems: a unified analytical framework," *Physical Review Letters*, vol. 86, no. 12, pp. 2553–2556, March 2001.
[4] C. K. Tse, *Complex Behavior of Switching Power Converters*. Boca Raton: CRC Press, 2003.
[5] Y. Ma, H. Kawakami, and C. K. Tse, "Analysis of bifurcation in switched dynamical systmes with periodically moving borders," *IEEE Transactions on Circuit ans Systems: Part I*, vol. 51, no. 6, pp. 1184–1193, June 2004.
[6] C. K. Tse and M. di Bernardo, "Complex behavior of switching power converters," *Proc. IEEE*, vol. 90, no. 5, pp. 768–781, 2002.
[7] W. C. Y. Chan and C. K. Tse, "Study of bifurcations in current-programmed dc/dc boost converter: From quasi-periodicity to period-doubling," *IEEE Transactions on Circuits and Systems Part I*, vol. 44, no. 12, pp. 1129–1142, 1997.

Bifurcation and Transitional Dynamics in Asymmetrical Two-Coupled Oscillators with Hard Type Nonlinearity

Takuya Yoshimura, Kuniyasu Shimizu, and Tetsuro Endo
Department of Electronics and Communications
Meiji University
Kawasaki 214-8571, JAPAN
TELFAX: +81-44-934-7344, Email: endoh@isc.meiji.ac.jp

Abstract- This paper investigates various global bifurcations and related change of dynamics including the transitional phenomenon in asymmetrical two-coupled oscillators with hard type nonlinearity. There exist periodic attractors for comparatively large ε (=a parameter showing the degree of nonlinearity), and they disappear for small ε via saddle-node bifurcation. We draw nodes and saddles with their unstable manifold near the bifurcation point on Poincare section for various values of k^2 (=deviation between two oscillators' intrinsic frequencies). It is recognized that the unstable manifold changes a lot with parameter variation around the bifurcation point, and hence the associated dynamics changes drastically.

I. INTRODUCTION

In our previous papers [1][2], we investigated the transitional phenomenon of a switching attractor in identical (hence symmetric) two-coupled oscillator systems with hard type nonlinearity. We elucidated that the reason was a heteroclinic cycle connecting two degenerate saddles and nodes which was formed at the bifurcation point. In this paper, we investigate various dynamics including transitional phenomenon in asymmetrical two-coupled oscillators with hard type nonlinearity. In particular, we investigate how the heteroclinic cycle in symmetrical system disappears and new connections of nodes and saddles appear when the system becomes asymmetric. Namely, we draw various diagrams of nodes, saddles and their unstable manifolds (UM's)[1] in terms of frequency deviation k^2 to observe change of the behavior of UM's with the value of k^2. The assumption that each oscillator has a slightly different intrinsic oscillation frequency is very natural and realistic, because there exist unavoidable errors between two oscillators in practice. We focus our attention to the connection between nodes and saddles by UM's to elucidate the total dynamics for such an asymmetric system. As a result, we confirm that the system presents various connections associated with the degree of frequency deviation.

---

[1] To draw unstable manifold, we fix the initial point on the unstable eigenvector associated with a saddle fixed point and continue Poincare mapping.

II. FUNDAMENTAL EQUATION AND THE SWITCHING PHENOMENON

The asymmetric two inductively-coupled oscillators with hard type nonlinearity can be written by the following 4th-order autonomous system:

$$\begin{cases} \dot{x}_1 = x_2 \\ \dot{x}_2 = -\varepsilon(1-\beta x_1^2 + x_1^4)x_2 - x_1 + \alpha x_3 \\ \dot{x}_3 = x_4 \\ \dot{x}_4 = -\varepsilon k^2(1-\beta x_3^2 + x_3^4)x_4 - k^2 x_3 + \alpha k^2 x_1 \end{cases} \quad (1)$$

where x_1 denotes the normalized output voltage of one oscillator, x_2 is its derivative, and where x_3 denotes the normalized output voltage of the other oscillator, x_4 is its derivative. The parameter $\varepsilon > 0$ shows the degree of nonlinearity. The parameter $0 < \alpha < 1$ is a coupling factor; namely, $\alpha = 1$ means maximum coupling, and $\alpha = 0$ means no coupling. The parameter β controls amplitude of oscillation. The parameter k^2 presents the frequency deviation of two oscillators; namely, $k^2 = 1$ means that two oscillators have an equal intrinsic frequency, and $k^2 \neq 1$ means that they have some frequency deviation. This system has three attractors in general for small ε; namely, the same-phase (periodic) attractor, the reverse-phase (periodic) attractor, and the double-mode (quasi-periodic) attractor. When ε becomes large, the double-mode attractor becomes two periodic attractors as shown later. In particular, for ε smaller than, but close to the bifurcation point, one can observe the switching phenomenon of these attractors[1] which is the ultimate form of the double-mode attractor. The reason for this switching solution is a formation of a heteroclinic cycle associated with degenerate saddles. In this paper, we investigate global bifurcation of this heteroclinic cycle for $k^2 \neq 1$.

III. GLOBAL BIFURCATION OF THE HETEROCLINIC CYCLE

Taking a Poincare section at $x_2 = 0$ in (1), flows in four dimensional phase space become discrete maps in three dimensional phase space (x_1, x_3, x_4). First of all, we will present the result of $k^2 = 1$ for review. Figure 1 presents a bifurcation diagram for $k^2 = 1.0$, $\alpha = 0.1$ and $\beta = 3.1$ of two periodic solutions which bifurcate to be a quasi-periodic oscillation for small ε. Equation (1) is invariant by replacing x_1 by x_3 and x_2 by x_4 for $k^2 = 1$, therefore, if there exists a periodic solution P: ($x_1(t)$, $x_2(t)$, $x_3(t)$, $x_4(t)$), then there exists another periodic solution P': ($x_3(t), x_4(t), x_1(t), x_2(t)$) by the above replacement. This means that two saddle-node bifurcations occur at the same value of $\varepsilon = 0.449$ in Fig.1[2]. In this case, the same-phase and the reverse-phase periodic solutions are stable, and no bifurcation occurs for $0 < \varepsilon < 0.6$, therefore they are omitted. Fig.2 shows nodes (corresponding to the stable periodic attractors) and the associated saddles, of which UM's connect two nodes and saddles for $\varepsilon = 0.449$. Note a cycle connecting two (almost degenerate) saddles, which is a heteroclinic cycle. When ε becomes a bit smaller, the nodes and the saddles disappear but their "locus" exist. The same is said for the UM's. Therefore, the flow for example, for $\varepsilon = 0.447$ behaves as follows. 1) Flow stays around the locus of one node for a long time, and 2) quickly moves along the locus of UM, and 3) it stays again around the locus of the other node for a long time, and 4) moves quickly along the locus of UM, vice versa. We call this the switching phenomenon. This flow behavior is verified by the Poincare mapped points of the switching attractor in Fig.2. The triangular mark and the square mark show the reverse and the same phase periodic attractors, respectively. In this case, they have no relation to the switching solution.

Fig.3 shows the bifurcation diagram of the periodic solution for the asymmetric case for $k^2 = 1.033, \alpha = 0.1$ and $\beta = 3.1$. The S-N bifurcation point for the upper periodic solution is $\varepsilon = 0.445$, and that for the lower one is $\varepsilon = 0.438$. Fig.4 shows two nodes, two saddles and so on. In this case there is no heteroclinic cycle but a homoclinic cycle associated with the SN-pair of B. It passes through very close to the "locus" of the SN-pair of A (note that the saddle and node in A are already disappeared for $\varepsilon = 0.438$). Therefore, it looks like a heteroclinic cycle in practice, and hence one can observe the switching phenomenon for smaller ε. The switching phenomenon is verified by Poincare mapped points in Fig.4 for $\varepsilon = 0.437$. The reason why there are more mapped points on the lower UM than those on the upper UM, is that the deviation from the bifurcation point is smaller for the lower UM ($\Delta\varepsilon = 0.001$) than for the upper UM ($\Delta\varepsilon = 0.008$). When $0.438 < \varepsilon < 0.445$, the saddle-node pair denoted by A is lost, but that denoted by B still exists. Therefore, if one gives an initial condition around A, the flow follows the locus of the upper UM and converges to

---
[2] The precise value of S-N bifurcation is a bit smaller than this value. Same is said for Fig.3 and Fig.5 and so on.

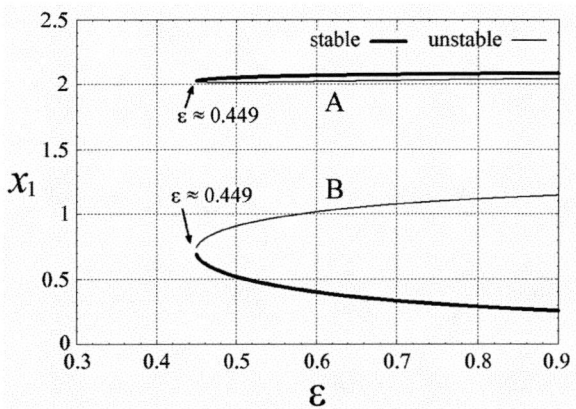

Figure 1: Bifurcation diagram of two symmetric periodic solutions for the symmetric system : $k^2 = 1.0$, $\alpha = 0.1$ and $\beta = 3.1$. The upper trace corresponds to the periodic attractor associated with the initial condition ($x_1(0), x_2(0), x_3(0), x_4(0)$)=(2,0,0,0). The lower trace corresponds to the periodic attractor associated with (0,0,2,0).

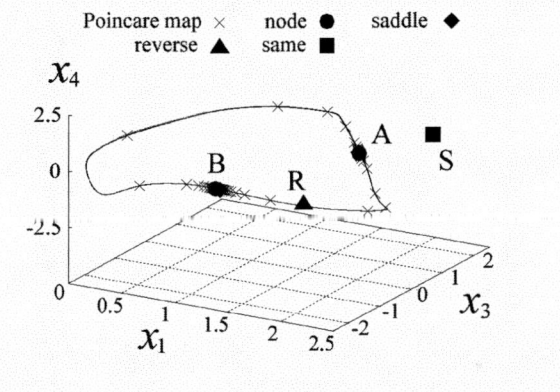

Figure 2: Nodes and saddles with their UM's for $k^2 = 1.0$, $\alpha = 0.1$ and $\beta = 3.1$. The UM is drawn for $\varepsilon = 0.449$. The cross marks ($\times$) present the Poincare mapped points of the switching attractor at $\varepsilon = 0.447$. Saddles are overlapped by nodes, hence invisible in this figure. Same is said for Fig.4 and Fig.6.

the node in B. For $\varepsilon < 0.438$, the upper and lower unstable manifolds both disappear, therefore the flow switches between A and B.

Fig.5 shows the bifurcation diagram of the periodic solution for the asymmetric case for $k^2 = 1.035$, $\alpha = 0.1$ and $\beta = 3.1$. The S-N bifurcation point for the upper solution is $\varepsilon = 0.444$, and that for the lower one is $\varepsilon = 0.437$. Fig.6 shows the relationship between nodes and saddles and so on. Note that there is no homoclinic cycle; namely, the upper UM connects two nodes in A and B, while the lower UM connects the node in B and R (=stable reverse-phase solution). Therefore starting around the initial condition in A, the flow reaches the node in B for $0.437 < \varepsilon < 0.444$ and stays there forever. Further, for $\varepsilon < 0.437$ the flow stays in B for some time but eventually moves toward R. In the same way, the resulting transitional dynamics change with the value of k^2 due to

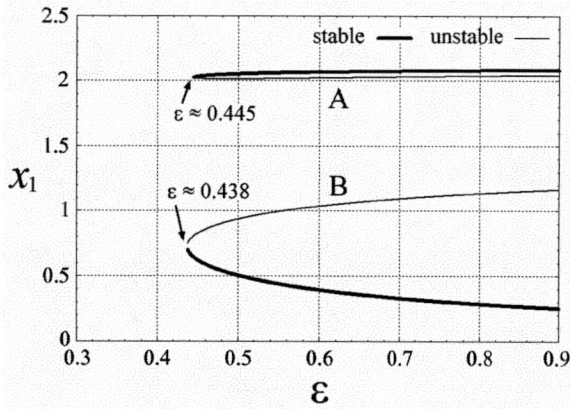

Figure 3: Bifurcation diagram of two periodic solutions for the asymmetric system : $k^2 = 1.033$, $\alpha = 0.1$ and $\beta = 3.1$. The periodic solutions corresponding to the upper and lower traces are explained in Fig.1.

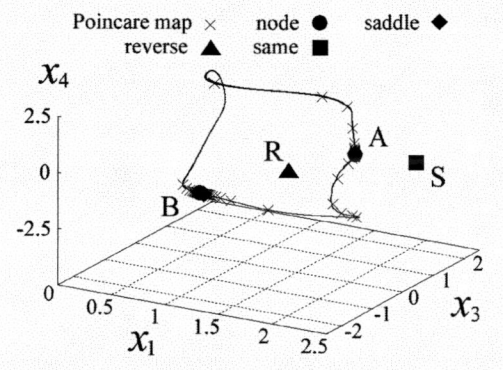

Figure 4: Relationship between nodes and saddles for $k^2 = 1.033$, $\alpha = 0.1$ and $\beta = 3.1$. The homoclinic cycle associated with the SN-pair of B is drawn for $\varepsilon = 0.438$. It passes through very close to a locus of the SN-pair of A. The cross marks (×) present the Poincaré mapped points for the switching attractor at $\varepsilon = 0.437$.

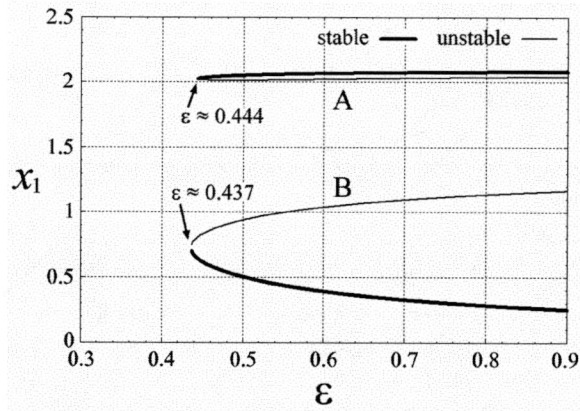

Figure 5: Bifurcation diagram of two periodic solutions for the asymmetric system : $k^2 = 1.035$, $\alpha = 0.1$ and $\beta = 3.1$. The periodic solutions corresponding to upper and lower traces are explained in Fig.1.

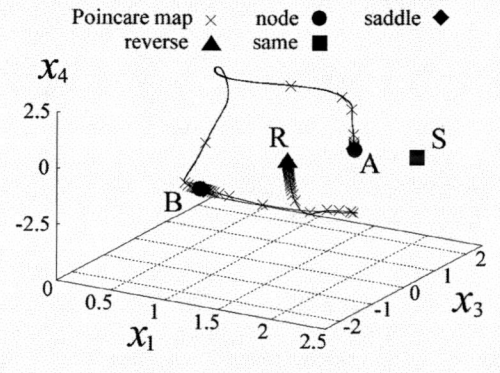

Figure 6: Relationship between nodes and saddles for $k^2 = 1.035$, $\alpha = 0.1$ and $\beta = 3.1$. The upper UM is drawn for $\varepsilon = 0.444$ and the lower UM is drawn for $\varepsilon = 0.437$. The cross marks (×) present the Poincaré mapped points for the initial condition around A at $\varepsilon = 0.436$. The S is unstable and the R is stable.

different behavior of UM's. The behavior of the flow starting around A at $\varepsilon = 0.436$ is shown by cross marks in Fig.6. Observe that the points move along the upper UM quickly ($\Delta\varepsilon = 0.008$) and stay on the locus of node in B for a long time and again move along the lower UM relatively slowly ($\Delta\varepsilon = 0.001$) and converge to R.

Fig.7 shows the relationship between nodes and saddles and so on at $k^2 = 1.048$. For the S_1-N_1 pair A, the S-N bifurcation occurs approximately for $\varepsilon = 0.440$ and for the S_2-N_2 pair B, it occurs approximately for $\varepsilon = 0.431$. For $0.431 < \varepsilon < 0.440$, the flow starting around A follows the locus of UM and converges to N_2 in B. For $\varepsilon < 0.431$, the flow stays around the locus of S_2-N_2 in B and around R (=unstable reverse-phase solution) for a long time and moves quickly along the locus of the UM.

Fig.8 shows the relationship between nodes and saddles and so on at $k^2 = 1.052$. For the S_1-N_1 pair A, the S-N bifurcation occurs approximately for $\varepsilon = 0.439$ and for the S_2-N_2 pair B, it occurs approximately for $\varepsilon = 0.428$. In this case the heteroclinic cycle in Fig.1 is divided in two homoclinic cycles. Therefore, for the initial condition around A at $\varepsilon = 0.438$, the mapped points draw the homoclinic cycle associated with A, namely a quasi-periodic oscillation can be observed. Further, for the initial condition around B at $\varepsilon = 0.427$, the mapped points draw the homoclinic cycle associated with B, namely another quasi-periodic oscillation can be observed.

Fig.9 summarizes the relationship between nodes, saddles and UM's for various value of k^2. In Fig.9 (a) for $k^2 = 1$, the S_1-N_1 and S_2-N_2 bifurcation points are the same, and hence, a heteroclinic connecting S_1 and S_2 exists at the bifurcation point. In Fig.9 (a') for $1 < k^2 < 1.034$, however, there is a homoclinic cycle associated with S_2 instead of the heteroclinic cycle, because the bifurcation points of the S_1-N_1 pair and the S_2-N_2 pair are different.

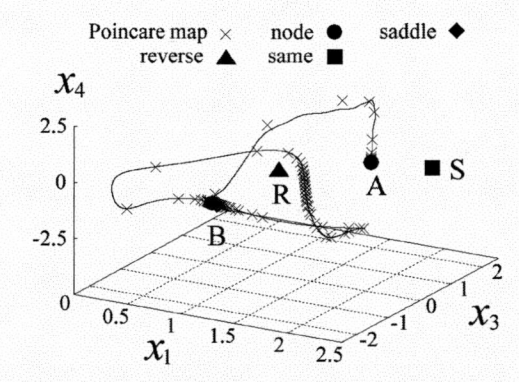

Figure 7: Relationship between nodes and saddles for $k^2 = 1.048$, $\alpha = 0.1$ and $\beta = 3.1$. The upper UM is drawn for $\varepsilon = 0.440$ and the lower UM is drawn for $\varepsilon = 0.431$. The cross marks ($\times$) present the Poincare mapped points for the initial condition around A at $\varepsilon = 0.429$. The S and R are unstable.

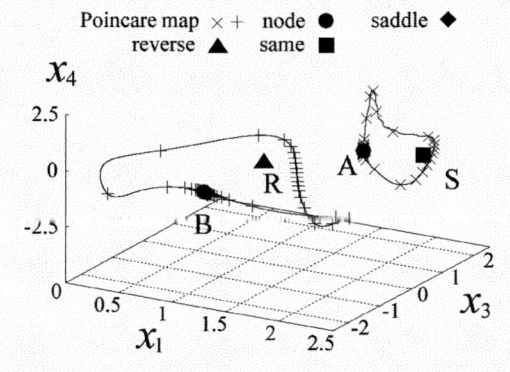

Figure 8: Relationship between nodes and saddles for $k^2 = 1.052$, $\alpha = 0.1$ and $\beta = 3.1$. The right-hand side UM is drawn for $\varepsilon = 0.439$ and the left-hand side UM is drawn for $\varepsilon = 0.428$. The cross marks ($\times$) present the Poincare mapped points for the initial condition around A at $\varepsilon = 0.438$ and the other cross marks ($+$) present them for the initial condition around B at $\varepsilon = 0.427$. The S and R are unstable.

Since the homoclinic cycle passes through very near the "locus" of the S_1-N_1 pair and stays there for a long time, it looks like a heteroclinic cycle. In Fig.9 (b) for $1.034 < k^2 < 1.039$, the homoclinic cycle is broken; namely one of the UM's of S_2 goes to the reverse-phase solution R. In Fig.9 (c) for $1.039 < k^2 < 1.050$, a homoclinic cycle associated with S_2 is formed (this is not close to the S_1-N_1 pair). In Fig.9 (d) for $k^2 > 1.050$, there are two homoclinic cycles associated with S_1 and S_2. It is interesting that manner of connections of nodes and saddles by UM's change a lot with k^2, and hence transitional dynamics change with k^2.

IV. Conclusions

In this paper, we elucidate the global bifurcation

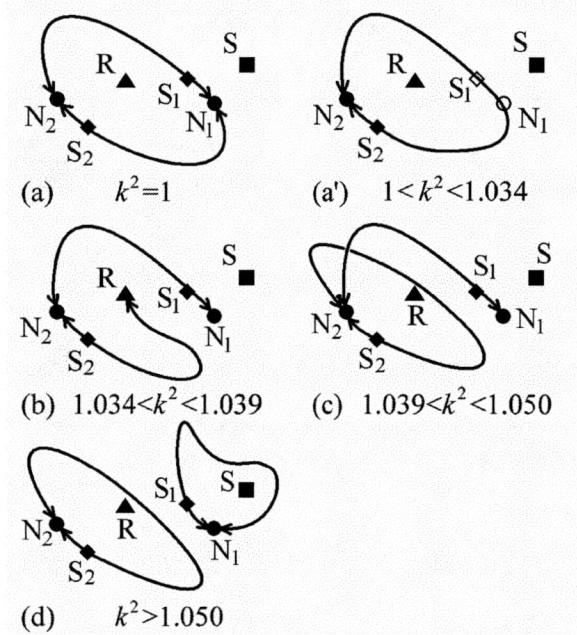

Figure 9: Schematic diagrams for various values of k^2. $N_1(N_2)$, $S_1(S_2)$, R and S denote nodes, saddles, the reverse-phase periodic solution and the same-phase periodic solution, respectively.

associated with the unstable manifold of saddles for the periodic solution bifurcated from a quasi-periodic (double-mode) attractor in asymmetric two-coupled oscillators with hard type nonlinearity. Mathematical background of such a global bifurcation may be found partly in [3]. In the future, we will investigate more thorough bifurcation of this system including the same and reverse phase attractors.

References

[1] Y. Aruga and T. Endo, "Transient dynamics observed in strongly nonlinear mutually-coupled oscillators", NOLTA2002, pp.135-138, 2002.

[2] Y. Aruga and T. Endo, "Transitional dynamics and chaos in coupled oscillator systems", Tranc. IEICE (A), vol. J86-A, No.5, pp.559-568, 2003.

[3] Y.A.Kuznetsov, *Elements of Applied Bifurcation Theory*, Springer-Verlag, New York, 1998.

Bifurcations in Modified BVP Neurons Connected by Inhibitory and Electrical Coupling

Shigeki Tsuji*, Tetsushi Ueta†, Hiroshi Kawakami* and Kazuyuki Aihara‡

*Faculty of Engineering, Tokushima University, Tokushima, 770-8506 Japan
Email: {shigeki@is, kawakami@ee}.tokushima-u.ac.jp
†Center for Advanced Information Technology, Tokushima University, Tokushima, 770-8506 Japan
Email: tetsushi@ait.tokushima-u.ac.jp
‡Institute of Industrial Science, University of Tokyo, Tokyo, 153-8505, Japan,
and ERATO Aihara Complexity Modeling Project, Japan
Email: aihara@sat.t.u-tokyo.ac.jp

Abstract—Electrical coupling among inhibitory interneurons has been discovered in various regions of the brain. Since these local networks indicate the sophisticated activity by the interplay between electrical and inhibitory synapses, the emergence of synchronous phenomena have been investigated by many researchers. We propose coupled neuronal models connected by both inhibitory and electrical synapses and investigate the emergence of various synchronous phenomena and its bifurcational structures by the bifurcation analysis. Although it is generally known that strong gap junction synchronize coupled neurons, we show in-phase and anti-phase phenomena in coupled neurons connected by both inhibitory and electrical synapses. They change to chaotic solutions via period-doubling and pitchfork bifurcations.

I. INTRODUCTION

Recently, gap-junctionally coupling has been discovered extensively in local interneuronal networks, and it is suggested that the important roles are played by gap junctions about detection and transmission of synchronous activity in the neocortex [1], [2], [3]. Moreover, they often coupled by bi-directional or uni-directional inhibitory synapses. The interplay of two different functions, voltage dependency and short-term plasticity, can develop advanced information processing in inhibitory interneuronal networks.

To investigate various synchronous firing by the different dynamics of these coupling, some neuronal models connected by both electrical and inhibitory synapses have been proposed [4], [5]. In these reports, although gap junction can generate generally the in-phase synchronous phenomena, anti-phase solution is generated by gap junction and bi-directional inhibitory coupling. In certain ratio of each coupling coefficient, these synchronous phenomena coexist. On the other hand, nearly in-phase solutions are shown in the case of uni-directional inhibitory coupling. However, coupled interneurons are analyzed in the limit of weakly coupling, and therefore uninvestigated parameter values (or regions) may provide novel phenomena and features.

In this paper, to clarify the relationship between various synchronous phenomena and parameter value of each coupling coefficient, we investigate coupled neuron models connected by gap junctions and inhibitory synapses by using the bifurcation analysis. This method has no restriction to any parameter values. It is necessary to choose single neuron model before analyzing the bifurcation phenomena in coupled neurons, because the characteristics of single neuron model greatly affect the firing patterns in coupled systems. Therefore, it is important to understand the characteristic of the single neuron model applied to the coupled system. In fact, we clarified that spiking phenomena of coupled neurons differ widely in Class 1 and Class 2 excitability [14]. We firstly adopt modified Bonhöffer-van der Pol (abbr. BVP) as single model. It exhibits some basic excitability by the different parameter sets, although it is a low dimensional and very simplest model. We set the parameter values of each single model exhibiting Class 1 excitability, because many interneurons show its excitability in the neocortex [6], [7]. Nextly, we investigate the bifurcation phenomena in coupled neurons connected by both inhibitory and electrical synapses, and show the emergence of in-phase and anti-phase phenomena via some bifurcations. Moreover, they coexist in certain region of parameter space as shown in previous study [4], [5]. We clarify that these solutions change to the chaotic phenomena via period-doubling and pitchfork bifurcations, although two neurons are connected by fully and symmetrically coupling. These results may indicate the additional variety of information representation by two different couplings in asymmetrically local interneuronal networks.

II. MODIFIED BVP NEURON MODEL

Hodgkin suggested that there are two different classes of neurons according to their frequency response characteristics when a constant current is injected to the cell body [8]: Class 1 spiking is concerned with saddle-node bifurcations in which the bifurcating state shows zero frequency response. On the other hand, Class 2 spiking is related to the Andronov-Hopf bifurcation, characterized by the fact that the bifurcating state holds a certain non-zero frequency. Many mathematical models have been proposed to describe neural activities [9], which may exhibit either or both of two firing modes.

Despite their high ability in the description of nerve dynamics, high-dimensional neuron models have potential degeneracy in bifurcations with high codimensions as well as the difficulty caused by multiplicity of parameters. Notwithstanding of their simplicity, single neuron models reduced to two- or three variable systems, such as BVP [10], [11] and Morris-Lecar models [12], may have advantages not only in their practicality in simulations with large scale coupled neuronal systems but also in its clarity of mathematical essence of bifurcation structure, i.e., the spike generation mechanism in terms of parameters.

Hence, we firstly consider a simplified neuron model described as follows:

$$\begin{cases} \dfrac{dx}{dt} &= c\left(x - \dfrac{x^3}{3} - y + z\right) \\ \dfrac{dy}{dt} &= \dfrac{x^2 + dx - by + a}{c} \end{cases} \quad (1)$$

where x and y denote the cell membrane potential and a recovery variable, respectively. z represents the external stimulus.

We construct this model based on the original BVP neuron model. Since the original BVP neuron model that $dy/dt = 0$ forms a linearity, it exhibits only Class 2 excitability and spiking for all the range of a injected constant current. Namely, it only shows the firing state via the subcritical Hopf Bifurcation within the limits of valid parameter values. However, most of interneurons seem to exhibit Class 1 excitability. (see e.g., Ref. [13]). Hence, we modify the BVP neuron model as a simplest neuronal model with Class 1 excitability, the form of $dy/dt = 0$ is changed to a quadratic nullcline. By this modification, in fact, our model shows Class 1 excitability via the saddle-node bifurcation [14]. It is well known that its state is generated by this bifurcation phenomena [15]. Moreover, Equ. (1) also shows the Class 2 excitability and the subtype of each excitability.

III. COUPLED NEURONS INTERCONNECTED ELECTRICAL AND CHEMICAL SYNAPSES

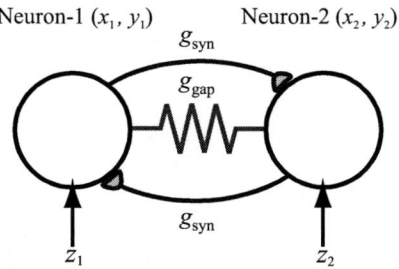

Fig. 1. Coupled system connected by both electrical and bi-directional inhibitory synapses.

In some recent reports, interneurons of the same type are extensively connected by gap junction, and they often coupled by bi-directional or uni-directional inhibitory synapses [1], [2], [3]. The interplay of two different functions, voltage dependency and short-term plasticity, can develop advanced information processing in inhibitory interneuronal networks. So, in such systems, it is important to elucidate the synchronous and asynchronous phenomena generated by interaction between electrical and inhibitory synapses. To clarify these phenomena, we investigate bifurcation phenomena in a fundamental system interconnected by electrical and bi-directional inhibitory synapses as shown in Fig.1. The equations are

$$\begin{cases} \dot{x}_1 &= c_1(x_1 - x_1^3/3 - y_1 + z_1 \\ &\quad + g_{\text{gap}}(x_2 - x_1) + g_{\text{syn}}s_1(x_{\text{syn}} - x_1)) \\ \dot{y}_1 &= (x_1^2 + d_1 x_1 - b_1 y_1 + a_1)/c_1 \\ \dot{s}_1 &= \alpha(1 - s_1)/(1 + \exp(-x_2/0.1)) - \beta s_1 \\ \dot{x}_2 &= c_2(x_2 - x_2^3/3 - y_2 + z_2 \\ &\quad + g_{\text{gap}}(x_1 - x_2) + g_{\text{syn}}s_2(x_{\text{syn}} - x_2)) \\ \dot{y}_2 &= (x_2^2 + d_2 x_2 - b_2 y_2 + a_2)/c_2 \\ \dot{s}_2 &= \alpha(1 - s_2)/(1 + \exp(-x_1/0.1)) - \beta s_2 \end{cases} \quad (2)$$

where g_{gap} and g_{syn} are the maximal electrical and synaptic conductances, respectively. For the inhibitory synaptic transmission, we adopt the first-order kinetics equation, $s_{1,2}$, with a sigmoid function of pre-synaptic membrane potential. The synaptic reversal potential, the rise and decay time constant are set to $x_{\text{syn}} = -2.5$, $\alpha = 1.0$ and $\beta = 0.05$, respectively. Although this model is not equipped with biophysically meaning, we can investigate various synchronous phenomena and the effects generated by some important parameters in coupled system, qualitatively, e.g., the effects of any intensity of each coupling or variation of firing frequency by the decay time constant of synapse transmission.

Here, we suppose that both single neurons have the same internal parameters, i.e., they exhibit the same excitatory type. We fix the parameter $a_1 = a_2 = 0.42$, $b_1 = b_2 = 1.0$, $c_1 = c_2 = 3.0$, $d_1 = d_2 = 1.8$ as each neuron shows the Class 1 excitability [14]. Additionally, we assume that the same constant current $z_1 = z_2 = 0.5$ is injected to the individual neurons. This situation indicates that each neuron shows a firing state. To investigate the relationship between the value of each coupling coefficient and the emergence of firing state, we compute the bifurcation diagram in the g_{syn}-g_{gap} plane. In order to classify each neuron, we label (x_1, y_1, s_1) as Neuron-1, and (x_2, y_2, s_2) as Neuron-2, and we assume that each neuron exhibits the firing state when $x \geq 0$.

IV. BIFURCATIONS

We compute bifurcation diagram of limit cycles in the g_{syn}-g_{gap} plane as shown in Fig. 2A. In this bifurcation diagram, G_i, I_i and Pf_i represent the tangent, period-doubling and pitchfork bifurcation, respectively. i is a nominal number. S_{in} and S_{anti} indicates the region existing a stable in-phase and anti-phase solution, respectively.

Stable in-phase synchronous solution is observed extensively in this diagram expect the region surrounded by the x-axis and I_2. In addition, stable anti-phase solution generates by inhibitory synapses, and it is observed in the region surrounded by x-axis, Pf_1 and G_2. In the overlapping region, in-phase and anti-phase solutions coexist. On the y-axis (it means only gap-junctional coupling), in-phase solution is also observed by the function which decreases the difference of each membrane potential. Moreover, on the x-axis (it means only inhibitory

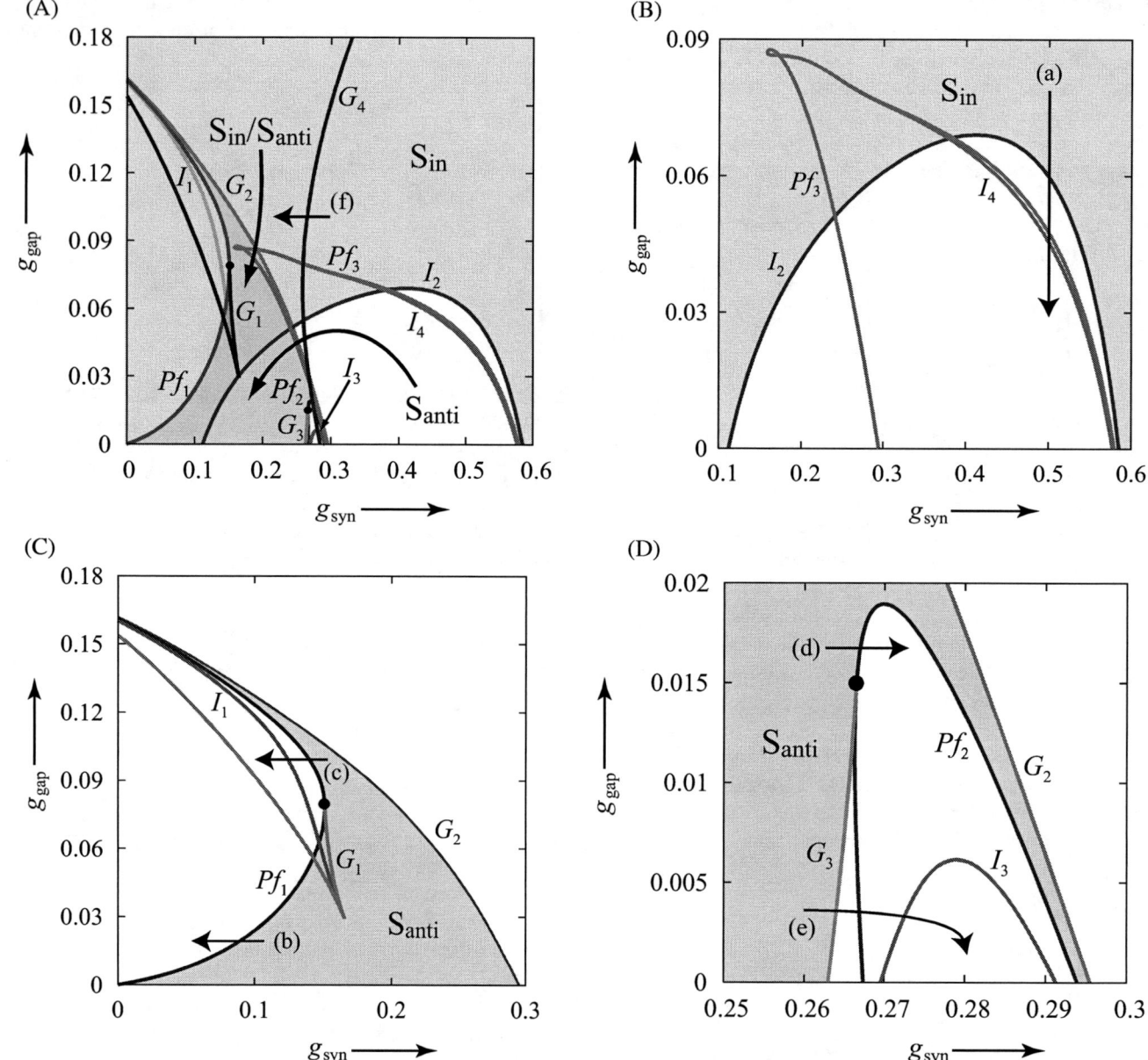

Fig. 2. (A): Bifurcation diagram of limit cycles in the g_{syn}-g_{gap} plane. (B), (C), (D): The enlargements of bifurcation diagram (A).

coupling), anti-phase and in-phase solutions coexist, but in-phase solution is destabilized via I_2 when $g_{syn} \gtrsim 0.1$.

These solutions bifurcate via some bifurcation curves in the diagram 2A. In Fig. 2B, in-phase solution exists in the shaded region as shown in Fig. 3-1. By decreasing g_{gap} along the line (a), this solution changes to period-2 solution (Fig. 3-2) via I_2 Moreover, its period-2 solution changes to two period-2 solutions via Pf_3, and they change to two chaotic solutions (Fig. 3-3) by period-doubling cascade (I_4 and more), simultaneously. Finally, these chaotic solutions are incorporated by more decreasing g_{gap}, and then it seems that one chaotic solution appears suddenly as shown in Fig. 3-4. On the other hand, by changing g_{syn} along the line (b) and (c) in Fig. 2C, anti-phase solution (Fig. 3-5) bifurcates two period-1 solutions (Fig. 3-6) via Pf_1, whereas anti-phase solution disappears via G_2 by increasing g_{syn}. On the line (c), two period-1 solutions change to two chaotic solutions (Fig. 3-7) by period-doubling cascade (I_1 and more), simultaneously. The same results are observed in Fig. 2D. Hence, by changing parameter such as a line (d), anti-phase solution changes to two period-1 solutions via Pf_2, and we observe that these solutions bifurcate chaotic solution, on the line (e). In Fig. 2A, we show two solutions (Fig. 3-8) that only one of two neurons fires by the function of inhibitory synapses. these solutions are observed in the right-side region divided by G_4, and they disappear via G_4 by decreasing g_{syn} along the line (f).

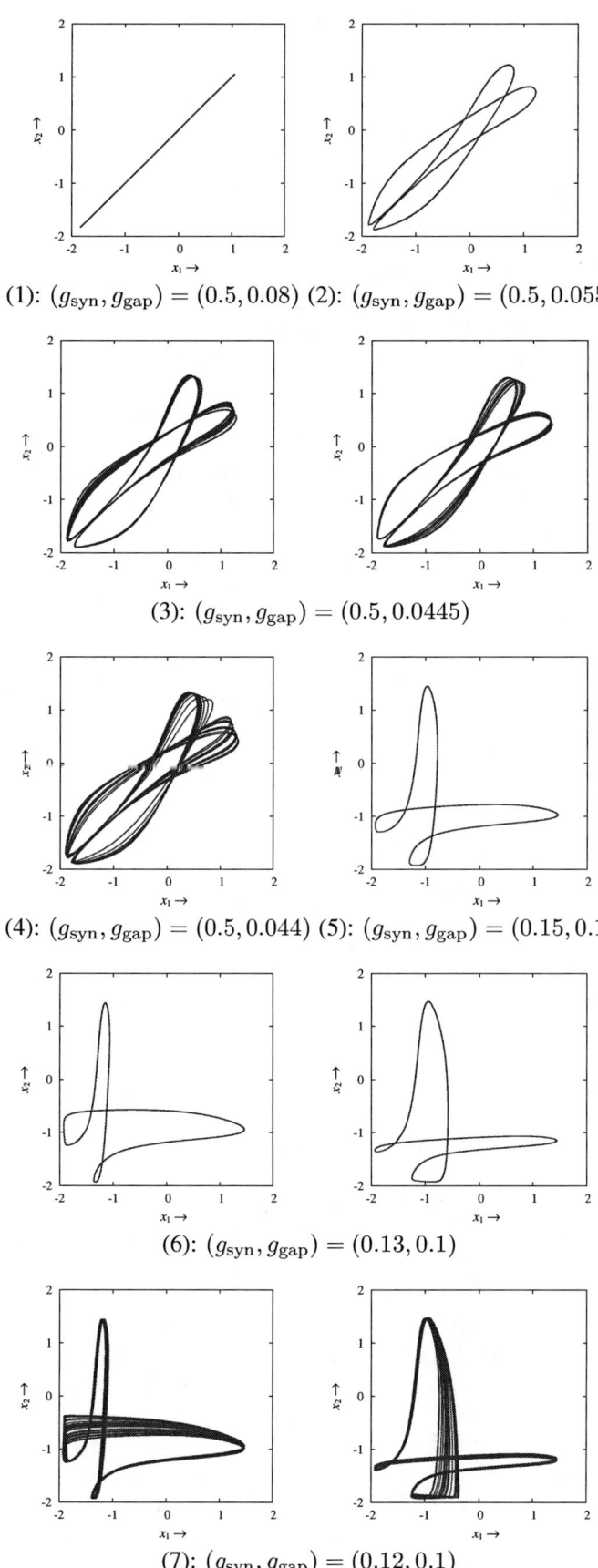

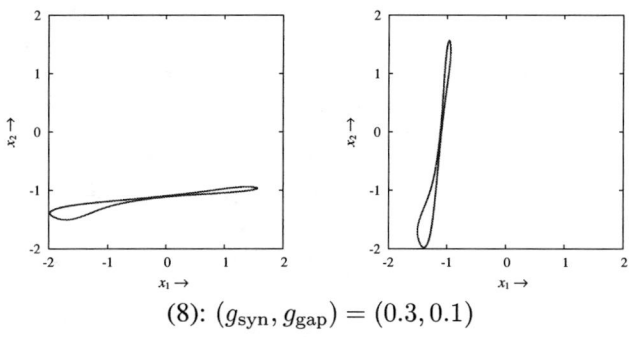

(8): $(g_\text{syn}, g_\text{gap}) = (0.3, 0.1)$

Fig. 3. (Continued)

V. CONCLUSIONS

We investigated the bifurcation phenomena in coupled neurons connected by both inhibitory and electrical synapses, and show the emergence of in-phase and anti-phase phenomena via some bifurcations. Moreover, we clarified that these solutions change to the chaotic phenomena via period-doubling and pitchfork bifurcations, although two neurons are connected by fully and symmetrically coupling. These results may indicate that symmetrically or asymmetrically large-scale networks exhibit spatial and temporal complicated dynamics by the interplay of two functionally different connections.

ACKNOWLEDGMENT

This work was partly supported by Grant-in-Aid for JSPS Fellows (No. 1800).

REFERENCES

[1] M. Galarreta, and S. Hestrin, "A network of fast-spiking cells in the neocortex connected by electrical synapses," *Nature*, **402**, 1999.
[2] M. Galarreta, and S. Hestrin, "Spike transmission and synchrony detection in networks of GABAergic interneurons," *Science*, **292**, 2001.
[3] J. R. Gibson, M. Beierlein, and B. W. Connors, "Two networks of electrically coupled inhibitory neurons in neocortex," *Nature*, **402**, 1999.
[4] T. J. Lewis & J. Rinzel, "Dynamics of spiking neurons connected by both inhibitory and electrical coupling," *J. Comput. Neurosci.*, **14**, 2003.
[5] M. Nomura, T. Fukai, and T. Aoyagi, "Synchrony of fast-spiking interneurons interconnected by GABAergic and electrical synapses," *Neural Comput.*, **15**, 2003.
[6] M. A. Rogawski, "The A-current: how ubiquitous a feature of excitable cells is it?," *TINS*, **8**, 1985.
[7] J. L. Massengill, M. A. Smith, D. I. Son, and D. K. O'Dowd, "Differential expression of $K_{4\text{-}AP}$ currents and Kv3.1 potassium channel transcripts in cortical neurons that develop distinct firing phenotypes," *J. Neurosci.*, **17**, 1997.
[8] A. L. Hodgkin, "The local Electric changes associated with repetitive action in a non-medullated axon," *J. Physiol.*, **107**, 1948.
[9] E. M. Izhikevich, "Neural Excitability, Spiking and Bursting," *International Journal of Bifurcation and Chaos*, **10**, 2000.
[10] R. FitzHugh, "Impulses and physiological state in theoretical models of nerve membrane," *Biophy. J.*, **1**, 1961.
[11] J. Nagumo, S. Arimoto & S. Yoshizawa, "An active pulse transmission line simulating nerve axon," In *Pro. of IRE*, **50**, 1962.
[12] C. Morris & H. Lecar, "Voltage oscillations in the Barnacle giant muscle fiber," *Biophys.*, **35**, 1981.
[13] B. Cauli, E. Audinat, B. Lambolez, M. C. Angulo, N. Ropert, K. Tsuzuki, S. Hestrin, and J. Rossier, "Molecular and physiological diversity of cortical nonpyramidal cells," *J. Neurosci.*, **17**, 1997.
[14] S. Tsuji, T. Ueta, H. Kawakami, and K. Aihara, "Bifurcations in Modified BVP Neurons Coupled by Gap-Junctions," Proc. NCSP'04, 2004.
[15] J. Rinzel & G. B. Ermentrout, "Analysis of neural excitability and oscillations," *Methods in Neuronal Modeling*, 1989.

Fig. 3. Phase portraits $(x_1\text{-}x_2)$ in Fig. 2.

Error sensitivity testing for the MC-EZBC scalable wavelet video coder

Tamer Shanableh
Department of Computer Science
American University of Sharjah
P.O. Box 26666, Sharjah, UAE
Email: tshanableh@aus.ac.ae

Tony May
Motorola Labs
Jays Close,
Basingstoke, Hampshire RG22 4PD UK
Email: Tony.May@motorola.com

Abstract - This work looks at the error resiliency of the Motion-Compensated Embedded Zero Block Coder (MC-EZBC). The MC-EZBC is gaining momentum as a motion compensated temporal filtering solution. We propose to demultiplex the MC-EZBC bitstream into seven bit categories and examine the effect of bit errors on each bit category. It is shown that different bit categories have different immunity to bit errors. The significance map at quadtree level zero, the motion vectors and the coefficients sign bits are shown to be the most sensitive. These categories exceed 60% of the total bitrate. The paper concludes that employing unequal error protection for the motion fields is sufficient. The coding algorithm need not be modified, hence preserving its bitstream embedding and universal scalability feature.

1. Introduction

Bitstream error resiliency is reported in the literature for various scalable wavelet video/image codecs. For instance error resiliency in the EZW (embedded zerotree wavelet) algorithm is realized through partitioning and processing the wavelet transform coefficients into groups belonging to different spatial regions [1].

However, it was shown that higher immunity to channel errors is achieved through utilizing Forward Error Correction (FEC) [2]. Subsequently it was proposed in [3] to divide a SPIHT (Set Partitioning into Hierarchical Trees) bitstream into sub streams with various error immunities. As such unequal error protection can be applied to different sub bitstreams. In their method, Alatan *et al* divided a SPIHT bitstream into two categories. The first category is the Value Bit Class (VBC), which contains the refinement pass and sign bits. The second category is the Location Bit Class (LBC), which contains the sorting pass or the coefficient location information, or in other words the significance map. Based on the bit-plane index the LBC is further divided into high bit-plane LBC- and low bit-plane LBC+. These two categories are protected with different channel coding rates. In summary the SPIHT stream is demultiplexed into three sub streams, LBC-, LBC+ and VBC. The LBCs are protected by rate-compatible punctured convolutional codes (RCPC codes) and Cyclic Redundancy Coding (CRC).

The attractive feature of Alatan's work is that it did not alter or sacrifice the bitstream embedding property. Additionally, the reviewed work was extended to SPIHT-based video coding in [4]. There, the stream was divided into LBC and VBC without any further stream demultiplexing. The FEC of choice was BCH (Bose-Chaudhuri-Hochquenghem code).

Likewise, the work of [1] was extended to 3D-SPIHT in [5]. In their work Cho and Pearlman modified the 3D-SPIHT algorithm to work independently in a number of spatial-temporal blocks resulting in error suppressing and localization. FEC is then applied to each block.

Clearly the rearrangement of the bitstreams into independent spatial-temporal blocks defeats the purpose of universal scalability. Nevertheless the authors claim that scalability can still be achieved by interleaving the coded blocks. Similar idea of spatial block localization appeared in [6].

Bajic and Woods reported their work on error resiliency and MC concealment in MC-EZBC codecs in [7]. Full information regarding the MC-EZBC algorithm are found in [8] Error resiliency is provided through multiple description FEC (MD-FEC) [9]. It was proposed to dedicate higher protection to the motion vectors; the rest of the bitstream was not demultiplexed and protected as a whole. The heavily protected motion vectors were then used for MC error concealment. It was shown that the low temporal frequency band of a corrupted GOP (group of pictures) could be concealed by predicting it from the scaled version of the last frame of the previous GOP. The last motion vector field of the latter GOP is used for that concealment purpose.

2. Proposed testing technique

A common drawback to the above reviewed techniques is the degradation in coding efficiency under error-free channel communications. This is due to either algorithmic modifications, which can be avoided, or the excessive overhead used for channel coding, which is inevitable. Hence in this work the error sensitivity of the MC-EZBC bitstream is tested by classifying it into seven bit categories. These categories are derived from the embedded bitstream organization of the MC-EZBC coding algorithm, thus

avoiding the need for algorithmic modifications. The bit categories are:
1. Refinement bits. Bits of the coefficients found significant in the previous bit-plane scanning pass (category 0).
2. Significance map at quadtree level 0. This map contains two important pieces of information; the most significant bit of a wavelet coefficient and its location (category 1).
3. Significance map at quadtree level 1 (category 2).
4. Significance map at quadtree level 2 and above (category 3).
5. Sign bits of the significant coefficients (category 4).
6. Motion vector bits.
7. Block modes.

3. Experimental results and empirical observations

In the experimental setup the pull function (a.k.a. transcoder) is first used to extract a MC-EZBC bitstream at a given spatio-temporal resolution and a given bitrate. The bitstream is then exposed to a Binary Symmetric Channel (BSC) with error rates of 10^{-2} and 5×10^{-3}. For generating the errors, three parameters are employed; the Bit Error Rate (BER), the bit category and the spatial subband. A publicly available software implementation of the MC-EZBC package including the pull function is available at [10].

A. Textural information testing

In Figure 1, bit errors are introduced to one bit category belonging to one spatial subband at a time. The spatial subband number versus the average PSNR per bit category is shown for two different BERs. The experiments are carried out for two cases; encoded only bitstreams and pulled bitstreams at various spatio-temporal resolutions and bitrates. The reason behind this is to test the effect of bit errors aside from any effects that might be associated with lower rates and spatio-temporal resolutions. The legends of the bit categories are given in Table 1.

Category	Name
0	Refinement Pixels
1	Significance Map Level 0
2	Significance Map Level 1
3	Significance Map Level 2 and up
4	Sign bits

Table 1. Category legends for Figure 1.

The plots show that the worst affected bit category is the significance map at quadtree level 0 followed by the sign bit. In most but not all the experiments, the spatial subband 0 was the most sensitive to bit errors.

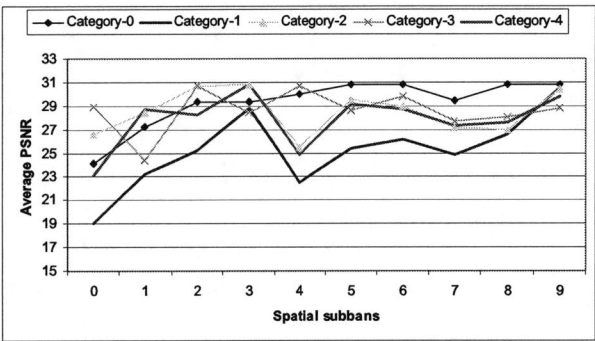

a. FOREMAN. QCIF, 4 temporal levels at 128 kbit/s. BER 0.005.

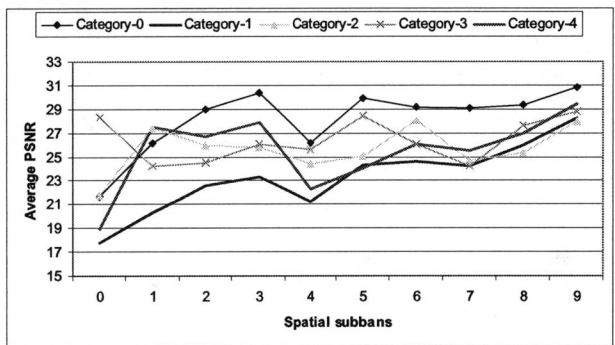

b. FOREMAN. QCIF, 4 temporal levels at 128 kbit/s. BER 0.01.
Figure 1. Error sensitivity results, the spatial subband number versus the average PSNR per bit category are shown. CIF sequences with a GOP size of 16 and a total of 200 frames.

The significance map at quadtree level 0 represents two important pieces of visual information namely, the location of the significant coefficient and its most significant bit. Thus this category is the most sensitive. The sign bit proved to be error sensitive as well. However, the MC-EZBC coding algorithm interleaves the sign bits with the significance map at quadtree level 0. Hence both categories need not be separated for error protection, but can be protected equally.

That said, additional experiments relating to the contribution of each bit category to the resultant bitrate reveals that the significance map quadtree level 0 and the sign bit comprise around 50% of the bitrate. These experiments are shown in Figure 2 below.

This 50% is quite a high ratio. Hence in an attempt to reduce the size of the bits that need to be protected, the experiments are conducted at one spatial subband at a time. Intuitively, one would think that the coarsest spatial subband is the most sensitive to bit errors. However this was not always the case. In some experiments, the subbands representing either the horizontal or vertical frequencies were the most sensitive. The sensitivity of the spatial subbands depends on the coding parameters and the features of the underlying image. Thus the low frequency spatial subbands did not assist in the error protection prioritization.

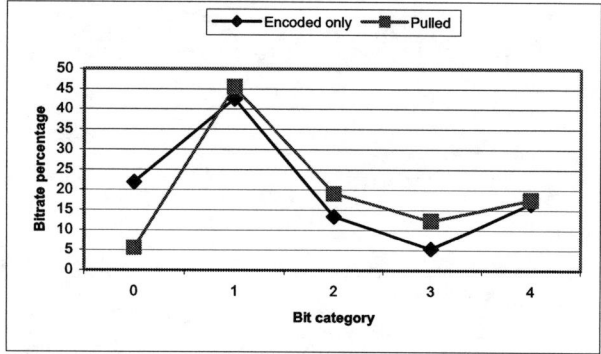

Figure 2. Bitstream proportion for each bit category. Encoded only: FOREMAN sequence versus transcoded to QCIF, 4 temporal levels at 128 kbit/s.

The experiments in Figure 2 are repeated for each test sequence with and without the use of the pull function. The figure shows that the proportion of bit category 1 bits is similar across both experiments.

B. Motion parameter testing

In Figure 3, the effect of bit errors on motion vectors is examined. Three bit error rates are employed: 0.001, 0.005 and 0.01. The effect of motion vector bit errors on the Container and Busman sequences are severe. The average PSNR drops to under 15 dB. Overall, the motion vector bits are very sensitive and need to be protected the most. To illustrate the importance of the motion vector bits in comparison to the texture information, Figure 4 reconstructs a GOP using the motion vectors, block modes and the coarsest temporal subband only. All texture information belonging to the remaining frames are discarded. For reasonable reconstruction a small GOP size of 4 is used in the figure. It is shown that the whole GOP can be reasonably reconstructed whilst discarding most of the texture information. Thus in the MC-EZBC algorithm, protecting the motion vectors and block modes should be given the highest priority.

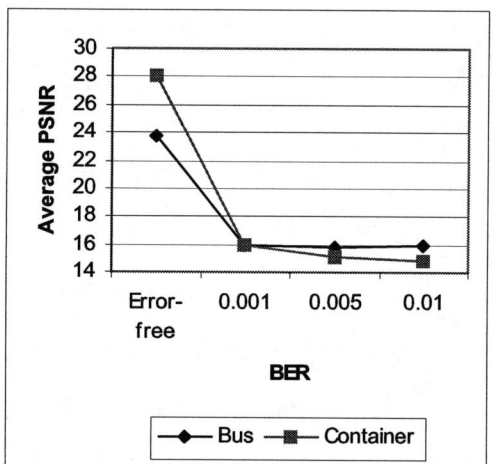

Figure 3. Effect of bit errors on coded motion vectors

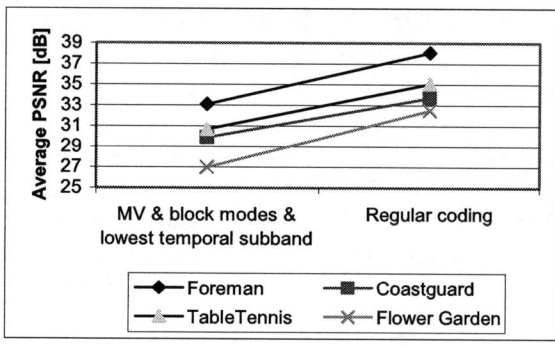

Figure 4. Reconstructing a GOP without texture information of high frequency temporal bands. CIF sequences, 100 frame / sequence. Transcoded to 1 Mbit/s. Flower sequence pulled at 2Mbit/s.

Conclusion

This paper introduced a novel technique of testing the error sensitivity of the MC-EZBC wavelet video coder. The embedded bitstream is divided into seven categories for testing.

The textural information experimental results revealed that the significance map at quadtree level 0 is the most sensitive to bit errors. The coefficient sign bits proved to be sensitive to bit errors as well. According to the MC-EZBC coding algorithm, the aforementioned categories are interleaved in the embedded bitstream and therefore can be protected equally without the need to rearrange the bitstream structure.

On the other hand, it was shown that bit errors on the differentially coded motion vectors result in severe degradations. Nevertheless since motion information is coded at the beginning of the MC-EZBC bit stream then it can be protected against channel errors more heavily as was proposed by Bajic and Woods in [7].

In summary, since each spatial subband is encoded as an independent arithmetic message, the MC-EZBC textural information should provide adequate error resiliency with the attractive feature of the ability to apply error protection to the motion fields without bitstream reorganization.

Acknowledgment

This work was carried out at Motorola UK research lab as part of the author's research visit during Summer 2003.

References

[1] C. D. Creusere, "A new method of robust image compression based
on the embedded zerotree wavelet algorithm," *IEEE Transactions on Image Processing*, v 6, n 10, pp. 1436-1446, October, 1997

[2] H. Man, F. Kossentini, and M. J. T. Smith, "Robust EZW image coding for noisy channels," *IEEE Signal Processing Letters*, vol. 4, pp. 227–229, August 1997.

[3] A. A. Alatan, M. Zhao, and A. N. Akansu, "Unequal error protection of SPIHT encoded image bit streams," IEEE J. Select. Areas Commun., vol. 18, pp. 814–818, June 2000.

[4] E. Khan and M. Ghanbari, "Error Resilient Virtual SPIHT For Image Transmission Over Noisy Channels ", Proc. of European Association for Signal, Speech, and Image Processing (EUSIPCO-2002), September 3-6, 2002 Toulouse, France, Volume II pp. 369-372

[5] S. Cho and W. A. Pearlman, "A Full-Featured, Error-Resilient, Scalable Wavelet Video Codec Based on the Set Partitioning in Hierarchical Trees (SPIHT) Algorithm," IEEE transactions on circuits and systems for video technology, 12(3), pp. 157-171, March, 2002

[6] E. Khan, "Efficient and Robust wavelet based image/video coding techniques," PhD thesis, University of Essex, England, October 2002

[7] I. V. Bajic and J. W. Woods, "EZBC video streaming with channel coding and error concealment," *Proc. SPIE VCIP 2003*, v 5150 I, p 512-522, Lugano, Switzerland, July 2003.

[8] Peisong Chen, "FULLY SCALABLE SUBBAND/WAVELET CODING," PhD thesis, Rensselaer Polytechnic Institute, New York, May, 2003

[9]. R. Puri and K. Ramchandran, "Multiple description source coding using forward error correction codes," Proc. 33rd Asilomar Conf. on Signals, Systems and Computers, p 342-346, Pacific Groove, CA, October 1999.

[10] Peisong Chen, Software package of MC-EZBC wavelet coder is publicly available at ftp://ftp.cipr.rpi.edu/personal/chen.

A Model-based Rate Allocation Mechanism for Wavelet-based Embedded Image and Video Coding

Ya-Hui Yu and Chun-Jen Tsai
Department of Computer Science and Information Engineering,
National Chiao Tung University,
Hsinchu, 300, Taiwan, R.O.C
E-mail: {yhyu, cjtsai}@csie.nctu.edu.tw

Abstract—In wavelet-based embedded coding for still images and/or videos, a trade-off between scalability and coding efficiency is achieved via layered-packetization of the embedded bitstreams optimized for several target operating bitrate points. This process is called rate allocation (tier-2 coding). The typical rate allocation mechanism is formulated as a rate-distortion optimization problem and a simple searching method with slow convergence rate is used. In this paper, a highly efficient model-based rate allocation mechanism is proposed. The algorithm is base on the adaptive analysis of the relationship between rate and distortion of the image/video data. Experiments conducted on a scalable video codec show that the time to obtain the optimal solution is greatly reduced. The techniques can be applied to various wavelet-based embedded image and video coding schemes, such as JPEG2000 and scalable video coding.

I. INTRODUCTION

The most unique characteristic of wavelet-based embedded coding is that it readily provides full-dimensional scalability. In another words, the spatial resolution, bitrate (i.e. SNR), and frame rate (for video coding) can be changed on-the-fly after the encoding process. For different applications on various devices or under different network conditions, the available bandwidth and resource may be highly divergent. Therefore, a rate allocation mechanism which can extract embedded bitstream with accurate target bit-rate in tolerable time is very important.

Several literatures formulated the rate control scheme for embedded image coding as a rate-distortion (R-D) optimization problem [1][2]. Each truncated point of the bitstream represents an R-D point, and the Lagrange multiplier is used to find the optimal point for different quality layer by minimizing the cost function. The search algorithm for the Lagrange multiplier is typically done by traditional iterative methods which are time consuming. Early termination of the iterations to save time results in lost of precision. Another possibility is to take a large number of evenly-spaced operating points in the Lagrange multiplier domain so that the operating points (logarithmically-spaced) in rate domain can hit all major target bitrates [1]. However, coding efficiency suffers in this case. Similar rate-allocation processes are used in 3-D embedded wavelet video coding [3][4] while the bisection method[5] is adopted to find the Lagrange multiplier for specific target bit-rate.

In this paper, we propose an approach to adaptively model the rate and distortion relationship based on the image/video data for rate allocation mechanism. In [11], a proposal of modeling simplified R-D curve has been made. Here, we propose to use an R-λ model instead of an R-D model for fast λ determination. With the model, the relation between bitrate and Lagrange multiplier can be established accurately. Therefore, the Lagrange multiplier can be inferred from the model to speed up the optimization process.

The paper is organized as follows. In section II, the general algorithm of wavelet-based embedded coding is introduced, and the rate allocation approach suitable for both embedded image and video coding is described. In section III, the novel rate allocation model is proposed. Some experimental results are shown in section IV. Finally, the conclusion of the paper is given in section V.

II. WAVELET-BASED EMBEDDED CODING

A. Wavelet-based Coding Framework

A general framework for wavelet-based embedded image and/or video coding [1][3][6] first transforms the input YC_BC_R frame data into frequency domain via temporal (for video coding only) and spatial subband decomposition. The transform process is followed by the quantizer and, finally, the entropy coder with rate allocation mechanism. Arithmetic coding with adaptive context modeling is often used as the entropy coder. The rate allocation procedure explores bitrate (quality) scalability of the embedded bitstream. The main differences between embedded image coding and video coding are in the application of temporal subband decomposition and the context models for each subband in the entropy coder. The rate allocation mechanism can be applied to either wavelet image or video coding in a similar manner.

B. Rate Allocation Algorithm

In rate allocation process, each subbands are coded independently, and the pixels in a subband are split into coding blocks. After entropy coding, the R-D curve of each coding block is formed. The rate allocation process tries to find the optimal embedded points on the R-D curve that meets the target bit-rates for the truncated bitstreams (Fig. 1).

The optimization problem for rate allocation is described as follows. For the constraint $R \leq R_{target}$, the "achievable" R-D point on the R-D curve with minimum distortion is the optimal solution. To solve the constrained optimization problem, it is easier to transforming the problem into an unconstrained problem, as Eq. (1). As a result, for a given Lagrange multiplier λ, a rate-distortion point with minimum $J(R)$ in Eq. (1) is the optimal solution.

$$J(R) = D + \lambda R. \quad (1)$$

The key issue of allocating the correct amount of bits to form a quality layer that matches a target bitrate is about how to find the Lagrange multiplier λ. As mentioned in section I, most existing techniques either uses an iterative process with slow convergence rate or creating a large number of evenly-distributed quality layers so that many prospective target bitrates can be supported. The drawbacks of these approaches are that either high coding complexity or low coding performance.

III. PROPOSED RATE ALLOCATION ALGORITHM

In this section, the R-λ model for the relationship between rate and Lagrange multiplier is described first. With the R-λ model, the correct Lagrange multiplier can be computed quickly by solving the model for the target bitrate.

A. Adaptive R-λ Model

Although the true rate-distortion model is data dependent and complicated, a simpler model has been derived and used for video coding analysis [7].

$$R(D) = a \ln\left(\frac{\sigma^2}{D}\right). \quad (2)$$

Eq. (2) has been used for source modeling [9] and coding mode decision for transform-quantization based video compression techniques [9]. An example of applying this model to R-D analysis of DCT-based codec is described in [10]. Bit-plane coding used in the wavelet formulation is essentially a non-uniform quantization process and does not change the form of Eq. (2) [8]. Truncation of some of the significant subband coefficients (based on the results of R-D analysis) may cause deviation from the model. However, since these cases should not happen often, Eq. (2) should still be a reasonable representation of the R-D relationship.

In Eq. (2), the parameter a depends on the distribution of the source, and the parameter σ is also source dependent. For

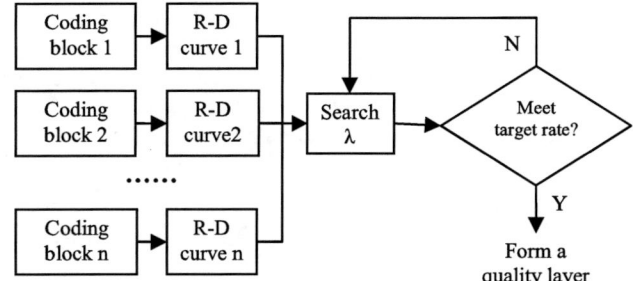

Figure 1. Algorithm for quality layer formation.

a given value λ, the minimization of $J(R)$ in Eq. (1) can be obtained when the first derivative $dJ(R)/dR = 0$, that is,

$$\frac{dJ(R)}{dR} = \frac{dD(R)}{dR} + \lambda = 0, \quad (3)$$

and

$$\lambda = -\frac{dD(R)}{dR}. \quad (4)$$

Solve Eq. (2) for D and substitute it into Eq. (4), the relationship between the Lagrange multiplier and the bitrate can be derived as follows,

$$\lambda = -\frac{dD(R)}{dR} = -\frac{d(R^{-1}(D))}{dR}$$
$$= -\frac{d(\sigma^2 e^{-R/a})}{dR} = \left(\frac{1}{a}\right)\sigma^2 e^{-R/a} \quad (5)$$

In summary, the R-λ model is established as Eq. (6) where the parameters α and β are source dependent:

$$\lambda = \alpha e^{\beta R}. \quad (6)$$

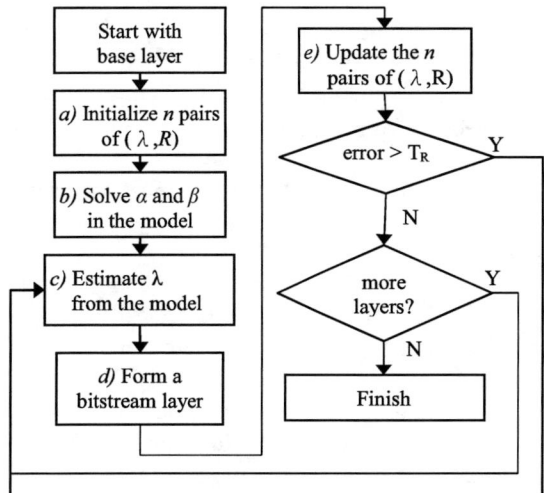

Figure 2. Flow chart of the proposed algorithm.

B. Proposed Rate Allocation Algorithm

The proposed algorithm uses the R-λ model (6) for two purposes. One is to search for the optimal λ for a quality layer in GOP level, and the other is to describe the R-D characteristics of a single block in coding block level. Eq. (6) is used as an adaptive model since the source dependent parameters α and β are estimated causally based on the input data. Given n pairs of numerical data (λ_i, R_i), $i = 0, \ldots n-1$, the parameter α and β can be calculated as follows. First, Eq. (6) can be rewritten as $\ln \lambda = \ln \alpha + \beta \cdot R$. Therefore, for $n > 2$ we have an over-determined system of equations,

$$\begin{pmatrix} \ln \lambda_0 \\ \ln \lambda_1 \\ \vdots \\ \ln \lambda_{n-1} \end{pmatrix} = \begin{pmatrix} 1 & R_0 \\ 1 & R_1 \\ 1 & \vdots \\ 1 & R_{n-1} \end{pmatrix} \begin{pmatrix} \ln \alpha \\ \beta \end{pmatrix}. \quad (7)$$

The system can be solved by least-squares estimation. Once the parameters α and β are determined, the relationship between the Lagrange multiplier and rate is directly established. The proposed algorithms are as follows.

Case 1) Search for optimal Lagrange multiplier: The R-λ model is adopted in this step to simulate the behavior of data in a GOP. The aim is to speed up optimal Lagrange multiplier search by better understanding of the R-λ relationship rather than blind iterative search with bisection method. Detail flow chart of the algorithm is shown in Fig.2 and described step by step as follows:

a) Find the first n pairs of (λ, R) in the base layer, n is typically 3 (using max, min, and midpoint λs).

b) Solve for the parameter (α, β) using Eq. (7).

c) Given target bitrate, estimate λ using Eq. (6).

d) Use the estimated λ to form the bitstream layer and obtain another *(λ, R)* data point.

e) Update the *n (λ, R)* pairs with the pairs which the *R* value is near to the target rate.

f) Iteratively doing b)-e) until the R value is close enough to the target bitrate within a tolerable error range T_R.

g) Repeat the procedure for the enhancement layers.

In the proposed algorithm, no additional memory storage is needed. Furthermore, the smaller the n is, the lower the computational overhead will be. The experimental results in next section show that even for a small n value, the accuracy of the algorithm is still good.

Case 2) Represent RD property of a coding block: In block d) of Fig. 2, a bitstream layer is formed by giving a Lagrange multiplier value. The truncated point of each coding block is determined at the fractional bitplane pass with the nearest Lagrange multiplier value. To achieve the typical coding block level rate allocation, the Lagrange multiplier value of each fractional bitplane pass in all coding blocks should be stored during entropy coding tier 1. In order to reduce the memory storage of the information and distribute the rate among all coding blocks with theoretical basis, the R-λ model is applied to describe the property of each coding block. Therefore, only the parameters α and β should be stored for a single coding block, and the coding block level rate allocation can be easily done by adopting the inverse R-λ model with a given Lagrange multiplier. In the proposed method, the truncated point would be the fractional bitplane pass with the nearest rate.

II. EXPERIMENTS

In this section, some experiments on the proposed algorithm are conducted using the MSRA scalable video codec [4], with the MPEG test sequences, STEFAN, FOREMAN, and FOOTBALL in CIF resolution. The coding parameters used in the experiments are as follows. The GOP size is 64 frames, and the frame rate is 30 fps. There are 9 quality layers which are optimized for the following target bitrates, 256 kbps, 320 kbps, 400 kbps, 512 kbps, 750 kbps, 1024 kbps, 1350 kbps, 1750 kbps and 2048 kbps. The parameter n in the algorithm is set to 3, and the bitrate error threshold T_R is set to 1% of the target bitrate. The number of iterations required before the solution converges for the proposed method and the commonly used bisection searching method is shown in TABLE I. And the average result is also shown in the table that when the error range is limited to 1%, the saving ratio is up to 64.39%.

As an example, a comparison of λ search points between a conventional method and the proposed method is illustrated in Fig. 3. In conventional bisection method, the search range of λ value is half-eliminated by each iterative operation. On the other hand, the proposed method relates the λ value and the target bitrate by an adaptive R-λ model. Each iterative step adapts and fine tunes the parameters of the model to new input data. Therefore, the model can represent the characteristic of the content well and the accuracy of the λ predicted value is higher.

Since the proposed mechanism allocates rate for each coding block in a model based method, the rate distribution in a GOP is rearranged. The coding efficiency graphs are shown in Fig. 4 and Fig. 5. The figures show that the proposed method achieves similar PSNR performance at any range of the rate.

Even though the experiments are conducted on a wavelet-based scalable video codec, similar results should apply to wavelet-based still image codec due to the similarity of these techniques.

TABLE I. COMPARASION BETWEEN PROPOSED ALGORITHM AND ORIGINAL BISECTION METHOD. THE ERROR RANGE OF THE RATE IS UNDER 1% AND THE TOTAL LAYBER NUMBER IS 9.

Sequence	Bisection	R-λ Model	Saving Ratio
Stefan	10.43	3.33	68.07%
Foreman	10.55	3.38	63.22%
Football	10.23	3.9	61.88%
Average	**10.40**	**3.70**	**64.39%**

Unit: iterative times/layer

III. CONCLUSION

In this paper, a novel adaptive model-based rate allocation mechanism for embedded wavelet image/video coding is proposed. According to the initial experiments, the algorithm reduces the search time for the Lagrange multiplier by up to 64.39% during the optimal truncation point search process for a given target bit-rate. We are currently refining the model, analyzing the rate allocation effects, and simplify the operations in the model in order to develop real-time hardware solution. Further improvements can be expected with these efforts.

IV. ACKNOWLEDGEMENT

This research is partly funded by National Science Council, Taiwan, R.O.C., under grant number NSC 93-2220-E-009-008.

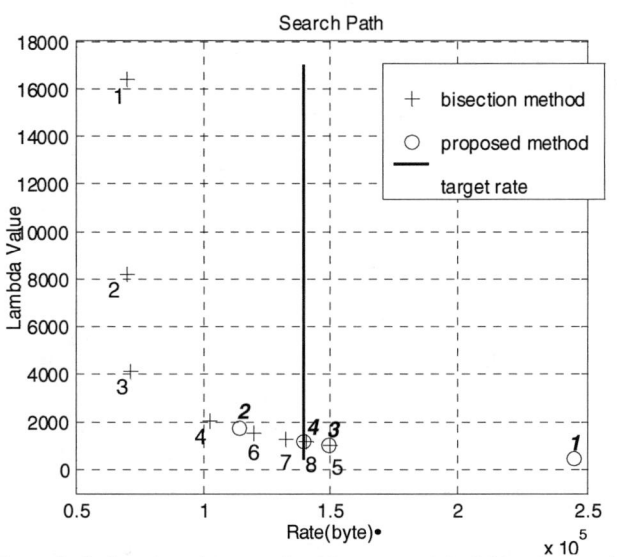

Figure 3. λ Search path comparison between original bisection method (step 1-8) and proposed model-based method (step 1-4 in bold).

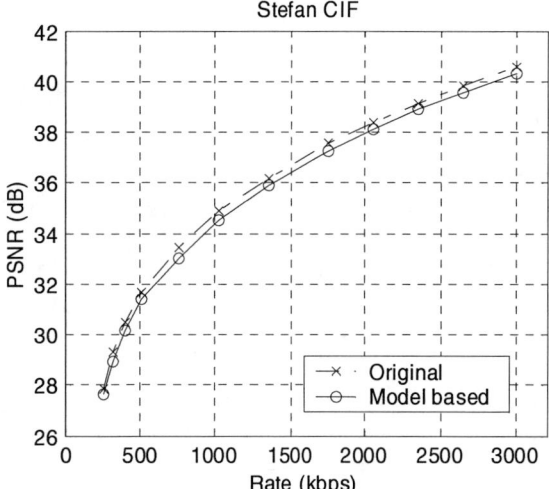

Figure 4. Coding efficiency comparison of sequence Stefan at CIF resolution

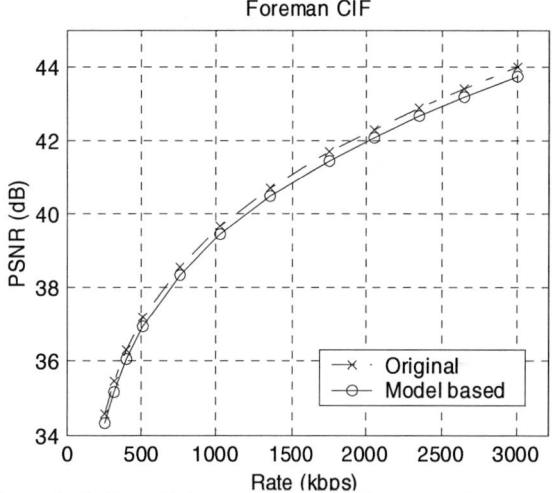

Figure 5. Coding efficiency comparison of sequence Foreman at CIF resolution

REFERENCES

[1] D. Taubman, "High performance scalable image compression with EBCOT," Image Processing, IEEE transactions on, vol. 9, pp. 1158-1170, July 2000.

[2] D. Taubman, E.Ordentlich, M. Weinberger, G. Seroussi, I. Ueno, F. Ono, "Embedded Block Coding in JPEG2000," Proc. of IEEE International Conference on Image Processing, vol. 2, pp. 33-36, Sep. 2000.

[3] J. Xu, Z. Xiong, S. Li and Y.Q. Zhang, "Three-Dimentional embedded subband coding with optimal truncation (3-D ESCOT)," Applied and Computational Harmonic Analysis: Special Issue on Wavelet Applications in Engineering, vol. 10, pp. 290-315, May 2001.

[4] J. Xu, R. Xiong, B. Feng, G. Sullivan, M. Lee, F. Wu, and S. Li, "3D Sub-band Video Coding using Barbell lifting," MPEG meeting input document M10569, Munich, Mar. 2004.

[5] Y. Shoham and A. Gersho, "Efficient bit allocation for an arbitrary set of quantizers," IEEE transactions on Acoustics Speech Signal Process.36 (1998), 1445-1453.

[6] JPEG 2000 Part I Final Committee Draft Version 1.0, ISO/IEC JTC1/SC29/WG1 N1646R, Mar. 2000.

[7] H. Gish and J.N. Pierce, "Asymptotically efficient quantizing," IEEE Transactions on Information Theory, vol. 14, pp. 676-683, Sep. 1968.

[8] N. S. Jayant and P. Noll, *Digital Coding of Waveforms*, Prentice-Hall, 1984.

[9] H. M. Hang and J. J. Chen, "Source Model for Transform Video Coder and Its Application – Part I: Fundamental Theory," IEEE transactions on Circuits and Systems for Video Technology, vol. 7, pp. 289-298, Apr. 1997.

[10] G. J. Sullivan and T. Wiegand, "Rate-Distortion Optimization for Video Compression," IEEE Signal Processing Magazine, vol. 15, pp. 74-90, Nov. 1998

[11] A. Aminlou and O. Fatemi, "Very Fast Bit Allocation Algorithm, Based on Simplified R-D Curve Modeling," *Proceedings of 10th IEEE International Conferences on Electronics, Circuits, and Systems 2003*, pp. 112-115, Dec. 2003.

OPTIMAL RESYNCHRONIZATION FOR LAYERED VIDEO OVER WIRELESS CHANNEL

Tao Fang, Lap-Pui Chau

School of Electrical & Electronic Engineering
Nanyang Technological University
50 Nanyang Avenue, 639798, Singapore

ABSTRACT

A scalable video codec can produce bitstreams with different rates and reconstructed quality. This paper focuses on Fine Granularity Scalable (FGS) video coding scheme although the proposed framework can work with all the layered video and image coding methods. FGS, which has been adopted in the MPEG-4 streaming video profile, provides a layered and fine granularity scalable bitstream with different importance at different bitplanes. A framework is proposed for the positioning of resynchronization markers for FGS video such that the resynchronization markers can be optimally inserted in each bitplane according to the importance of the bitplane and the time-varying channel condition. The effectiveness of this scheme lies in reducing loss of information based on significance of each bitplane of FGS video streaming. Experimental results demonstrate that this scheme significantly improves video quality for wireless video transmission.

1. INTRODUCTION

Recent advances in technology have led to a significant growth in wireless communications and the widespread access to information via the Internet, which have resulted in a strong demand for reliable transmission of video data. The video applications have to provide sufficient robustness to ensure that the quality of the decoded video is not overly affected by the channel unreliability.

Much effort has been invested in building error resilience into the compressed bitstream [1]. Among the state-of-art error-resilient techniques, resynchronization has been proved to be a very effective tool. Previous video coding standards such as H.261 and H.263 (Version 1) logically partition each of the images into rows of macroblocks (MBs) called Group Of Blocks and resynchronization markers are allowed to occur only at the left edge of the images. Since the resynchronization markers are likely to unevenly spaced in the bitstream, some areas of the picture will be more susceptible to errors. This drawback is overcome in MPEG-4 encoder, where resynchronization markers are inserted in the bitstream at approximately constant intervals. Thus, in the presence of a short burst of errors, the decoder can quickly localize the error to within a few MBs in the high activity areas of the image and preserve the image quality in these important areas.

Besides the conventional approaches mentioned above, there are still many literatures considering the insertion of resynchronization markers [2]-[4]. For instance, in [2], Jeong, Kang and Kim propose an optimal resynchronization marker positioning technique using a novel information measure, i.e., the difference between the image recovered by error concealment and the image without error. Methods in [2]-[4] share the same nature that they are all designed at *picture level*, i.e., discussing positioning of markers in the bitstream of one picture. On the other hand, Luis and Han [5] develop resynchronization approaches to *group-of-picture (GOP) level*, assigning different slice sizes to the video frames based on their type and order of appearance within a GOP. However, all the above methods [2]-[5] are discussed only for single-layer video. In [6], Yan, Wu, Li and Tao design a hierarchical enhancement layer bitstream structure with resynchronization markers and Header Extension Code (HEC) for FGS video. In this scheme, unfortunately, the resynchronization markers are inserted in a fixed mode (same number of resynchronization markers in each bitplane) without considering the significance of different bitplanes and different channel condition. In this paper, we introduce an optimal resynchronization method for layered video, which can optimally insert resynchronization markers within the bitplanes of the enhancement-layer based on the significance of the bitplane and the time-varying channel condition to perform a graceful degradation of video streaming when transmitting over wireless channels. Note that the proposed framework can be applied to all the layered video and image coding methods, e.g., SNR scalability coding method, although this paper focuses on FGS video coding scheme.

The rest of this paper is organized as follows. In Section 2, we propose a resynchronization approach based on the different importance of each bitplane and channel condition. Section 3 gives the simulation results and Section 4 concludes the paper.

2. OPTIMAL RESYNCHRONIZATION APPROACH FOR FGS VIDEO

In this section, we first analyze the factors that affect the video quality, based upon which, the optimization problem will be formulated thereafter. Finally, a fast and effective algorithm will be described to solve this seemingly prohibitive problem.

2.1 Factors Affecting Video Quality

The bitplanes in the enhancement-layer of FGS video are not equally important. For each bitplane, the amount of bits included and the importance to the constructed video quality is different from other bitplanes. Summarily, there are three factors to be considered when we want to insert resynchronization markers in each bitplane:

1. The channel condition. With more resynchronization markers, the bitstream will be more robust to channel errors while the coding efficiency will drop. It is supposed that there must exist a tradeoff between the bits used for source coding and those used for resynchronization markers according to certain channel condition.

2. The amount of bits in the bitplane. Generally, with the level of bitplane getting higher, the amount of bits in this bitplane soars. If we insert same number of resynchronization markers in all the bitplanes, the slices in the higher bitplanes will be more prone to be lost due to the large slice size when error happens.

3. The significance of the bitplane. For FGS video, the bitplanes in the enhancement-layer are not equally important. This is not only due to the different amount of bits in each bitplane, but also due to the position of the bitplane in the enhancement layer. Intuitively, if we only consider the latter factor, we should put more resynchronization markers in the lower bitplanes. Another important thing is how to evaluate the significance of each bitplane. In this paper, the criterion to evaluate the caused distortion is defined as *decreased PSNR* due to the loss of the slices in each bitplane without consideration of the dependencies of distortion among bitplanes.

2.2 Optimization Problem

Below, we discuss the problem of optimally inserting resynchronization markers subject to an overall target bit rate R_{Budget}. In the following formulation, we do not involve channel coding, but we should mention that the video quality can be further enhanced using unequal loss protection [7]. We firstly assume that there are totally L bitplanes to be transmitted. Given an overall coding rate, $R_{Overall}$, we want to optimally insert resynchronization markers such that the total distortion is minimized, that is,

$$\text{Min } D(\vec{N}), \text{ subject to } R_{overall} \leq R_{Budget} \quad (1)$$

with $\vec{N} = (N_1, N_2, ..., N_L)$ and N_l is the number of slices (also resynchronization markers) in bitplane l. $R_{Overall}$ is defined as

$$R_{Overall} = R_S + R_{RM} \quad (2)$$

with R_S being the source rate and R_{RM} being the rate consumed by resynchronization markers. The objective therefore of the optimization is to find $\vec{N}^*$, where

$$\vec{N}^* = \arg \min D(\vec{N}) \quad (3)$$

so that the overall rate does not exceed R_{Budget} while minimizing the overall expected distortion, which can be expressed as

$$D(\vec{N}) = \sum_{l=1}^{L} \sum_{s_l=1}^{N_l} \Delta PSNR_l(s_l) P(s_l) + D_{RM}(\vec{N}) \quad (4)$$

There are two parts of distortion in the right side of the above equation. The first part of distortion is due to the loss of source bits because of channel errors and the second part depends on the number of bits consumed by inserted resynchronization markers.

In the first part, $\Delta PSNR_l(s_l)$ is defined as *decreased PSNR* due to loss of s_l slices in bitplane l and $P(s_l)$ is the probability that s_l slices are lost in bitplane l. $\Delta PSNR_l(s_l)$ can be obtained at the encoder by calculating the difference between the image with loss and the original image. We calculate $P(s_l)$ in the following way:

$$P(s_l) = \binom{N_l}{s_l} SER_l^{s_l} (1 - SER_l)^{1-s_l} \quad (5)$$

where SER_l is the slice error probability, i.e., the probability that one slice in bitplane l will be lost. If we use BER as bit error rate, then we can calculate SER_l as

$$SER_l = 1 - (1 - BER)^{B_l} \quad (6)$$

with B_l being the average number of bits in each slice in bitplane l.

The second part is the distortion caused by the consumption of bits by inserting resynchronization markers in all the L bitplanes. Note that for synchronization markers that we add to a stream, the same

Table 1. Pseudo-code of resynchronization marker assignment algorithm. M is the number of MBs in each bitplane and T_l is the number of bits in bitplane l. The variables $\vec{N}_{opt}$, $\vec{N}_{last}$, and $\vec{N}_{temp}$ are vectors that store redundancy assignments.

$\vec{N}_{opt} = (1,1,1,...,1);$
while ($\vec{N}_{opt} \neq \vec{N}_{last}$){
  $\vec{N}_{last} = \vec{N}_{opt};$
  for ($l = 1; l \leq L; l++$){
    for ($q = -Q; q \leq Q; q++$){
      $\vec{N}_{temp} = \vec{N}_{last};$
      $\vec{N}_{temp,l} = \vec{N}_{temp,l} + q;$
      if ($\vec{N}_{temp,l} < 0$ or $\vec{N}_{temp,l} > M$)
        continue to next q;
      if ($q > 0$)
        for ($i = l+1; i \leq L; i++$)
          $\vec{N}_{temp,i} = \left[T_i / \max(T_l/\vec{N}_{temp,l}, T_i/\vec{N}_{temp,i}) \right]$
      else
        for ($i = 1; i \leq l-1; i++$)
          $\vec{N}_{temp,i} = \left[T_i / \min(T_l/\vec{N}_{temp,l}, T_i/\vec{N}_{temp,i}) \right]$
      calculate $D(\vec{N}_{temp});$
      if ($D(\vec{N}_{temp}) < D(\vec{N}_{opt})$){
        $\vec{N}_{opt} = \vec{N}_{temp};$
        $D(\vec{N}_{opt}) = D(\vec{N}_{temp});$
      }
    }
  }
}
$\vec{N}^* = \vec{N}_{opt};$

---

amount of source data needs to be removed to cater for the bandwidth requirement $R_{overall} \leq R_{Budget}$. We will not transmit this portion of source data in the highest bitplane L since it is the least important. Thus, we can express the second part of distortion as

$$D_{RM}(\vec{N}) = \Delta PSNR_L \left(r \sum_{l=1}^{L} N_l / B_L \right) \quad (7)$$

with r being the number of bits for one resynchronization marker.

Now, by solving the equations (1), (2) and (4), we can optimally insert resynchronization markers in each bitplane according to the amount of bits in each bitplane, the importance of each bitplane and the time-varying channel condition. However, this seems like an extremely difficult task especially for a large value of L. For example, if each layer has M MBs, the computational complexity will be $O(M^L)$, which is highly prohibitive. To circumvent this problem, we develop a simple and effective local hill-climbing algorithm, which is also adopted to assign FEC codes for layered image in [8]. We will give a brief description of it in the following.

2.3 Hill-climbing Algorithm

First, we initialize the number of resynchronization markers as $(1,1,1,...,1)$. For each iteration, we examine $2QL$ possible assignments to find certain $\vec{N}$ that can reduce $D(\vec{N})$, where Q is the search distance (maximal number of resynchronization markers that can be added to or subtracted from each bitplane). We calculate $D(\vec{N})$ after adding or subtracting 1 to Q resynchronization markers to each bitplane while satisfying our constraint $B_l < B_{l+1}$. The searching process is repeated until the best $\vec{N}$ minimizing $D(\vec{N})$ is found. The algorithm pseudo-code is given in Table 1. Our algorithm finds a local minimum, which is quite close to the global minimum with tolerable computation.

3. EXPERIMENTAL RESULTS

Simulations have been carried to test the performance of the proposed scheme. To clearly show the effect of the proposed scheme, channel coding is not used in our experiment although it can be jointly considered in real applications. In the rest of this work, we will use Foreman (CIF) to show our two sets of experiment results: two- and three-bitplane scalability. In the experiment, every I frame was followed by 29 P frames. The frame rate is 10 frames/s. The base-layer was compressed using the same QP 20. For comparison, we also show the results obtained with the scheme of fixed synchronization insertion method proposed in [6]. We denote it as the conventional method in the rest of the paper. In the conventional method, one slice consists of three rows of MBs, i.e., 66 MBs, for the first bitplane. For other bitplanes, one slice consists of one row of MBs, i.e., 22 MBs.

The end-to-end performance (actual decoded bits and average PSNR) for two- and three-bitplane scalability is illustrated in Fig. 1 and Fig. 2, respectively. In contrast, the proposed solution exhibits excellent performance and always performs above the performance of the conventional solution over a wide range of bit error rate. For two-bitplane scalability, PSNR value is improved around 1 dB while for three-bitplane case, up to 2 dB PSNR improvement is achieved.

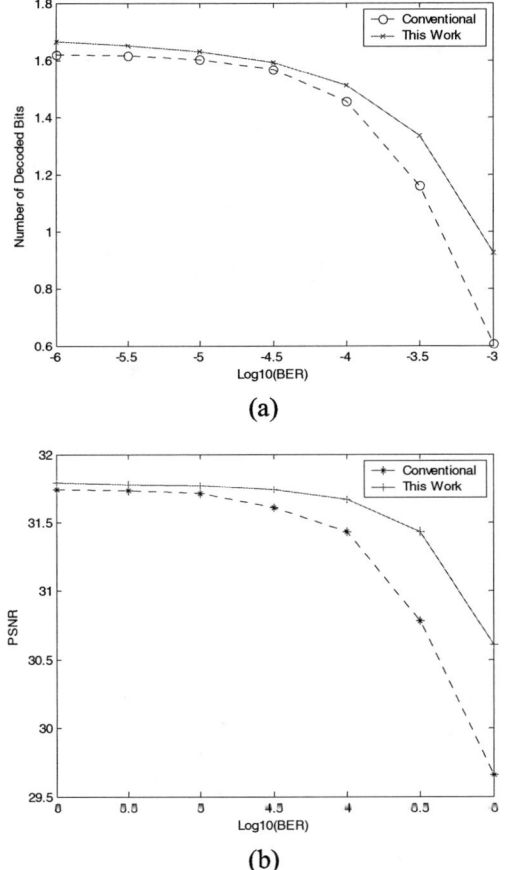

(a)

(b)

Fig. 1. Two-bitplane scalability results for Foreman. (a) The average actual decoded bits and (b) Average PSNR.

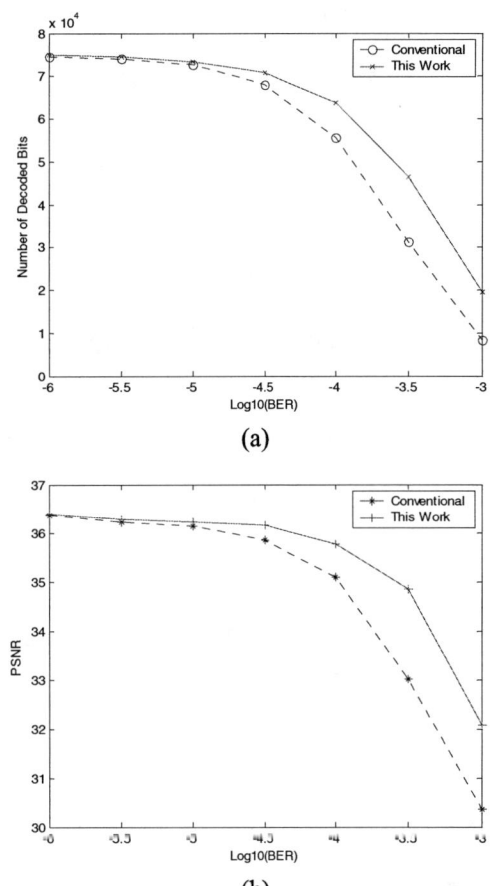

(a)

(b)

Fig. 2. Three-bitplane scalability results for Foreman. (a) The average actual decoded bits and (b) Average PSNR.

4. CONCLUDING REMARKS

In this paper, we proposed a resynchronization scheme for layered video so that we can optimally insert resynchronization markers in each bitplane according to three factors: the amount of bits in each bitplane, the importance of each bitplane and the time-varying channel condition. The simulation results indicated that our proposed scheme encodes the video sequences with higher quality as compared to that of the conventional approach over error-prone channels.

5. REFERENCES

[1] I. Moccagatta, S. Soudagar, J. Liang and H. Chen, "Error-resilient coding in JPEG-2000 and MPEG-4", *IEEE J. Select. Areas Commun*, vol. 18, pp. 899-914, Jun. 2000.

[2] J. -H. Jeong, H. -S. Kang, J. -K. Kim, "Optimal resynchronization marker positioning method using a novel information measure," *Signal Processing: Image Communication* 17, pp. 799-806, 2002.

[3] G. Cote, S. Shirani and F. Kossentini, "Optimal mode selection and synchronization for robust video communication over error-prone networks," *IEEE J. Select. Areas Commun*, vol. 18, pp. 952-965, Jun. 2000.

[4] K. H. Yang, D. W. Kang and A. F. Faryar, "Efficient intra refreshment and synchronization algorithms for robust transmission of video over wireless networks," *IEEE ICIP*, Oct. 2001, vol. 1, pp. 938–941.

[5] L. O. -Barbosa and T. Han, "On the use of frame-based slice size for the robust transmission of MPEG video over ATM networks," *IEEE Trans. Broadcasting*, vol. 46, pp. 134-143, Jun. 2000.

[6] R. Yan, F. Wu, S. Li and R. Tao, "Error resilience methods for FGS video enhancement bitstream," *IEEE PCM*, Dec. 2000.

[7] M. van der Schaar and H. Radha, "Unequal packet loss resilience for fine-granular-scalability video," *IEEE Trans. Multimedia*, vol. 3, pp. 381-393, Dec. 2001. [8] E. Mohr, E. A. Riskin and R. E. Ladner, "Unequal loss protection: graceful degradation of image quality over packet erasure channels through forward error correction," *IEEE J. Select. Areas Comm.*, vol. 18, pp. 819-828, Jun. 2000.

Sub-Sequence Video Coding For Improved Temporal Scalability

Dong Tian
Tampere International Center for Signal Processing
Tampere, Finland
dong.tian@tut.fi

Miska M. Hannuksela
Nokia Research Center
Tampere, Finland
miska.hannuksela@nokia.com

Moncef Gabbouj
Tampere University of Technology
Tampere, Finland
moncef.gabbouj@tut.fi

Abstract—Compression efficiency and bitrate scalability are among the key factors in video coding. The paper introduces novel sub-sequence coding techniques for temporal scalability. The presented coding schemes provide a wider range for bitrate scaling than conventional temporal scalability methods and maintain high coding efficiency at the same time. The proposed sub-sequence techniques are adopted into the latest video coding standard H.264, making it easy to identify sub-sequences and possible to discard them intentionally. As shown by the extensive simulations, a wide range of applications, from mobile messaging to consumer electronics such as digital TV can benefit from sub-sequences.

I. INTRODUCTION

In recent years, scalable video coding has been one of the key challenges in the field of video coding. Scalable bitstreams can be used for various purposes, such as adjustment of the transmitted bitrate according to the prevailing network throughput in streaming applications and scaling the complexity of the decoding process according to the available computational resources. Scalable coding also partitions the coded bitstream into sections with different impact on decoded video quality. These sections can be used in the transport layer to implement unequal error protection. Scalable video coding methods can be classified into temporal, spatial, and SNR techniques, as well as any combination of them.

Two general categories exist for interframe coding in temporal scalable video coding algorithms: predictive coding and subband coding [1]. All prevailing video coding standards, such as H.263, H.264 (aka MPEG-4 AVC), MPEG-2 Visual, and MPEG-4 Visual, deploy motion compensation predictive techniques, and hence this paper focuses on the temporal scalability for predictive coding.

The paper introduces a novel sub-sequence coding technique, which is an enhancement of the known temporal scalability methods. It is shown that the range for bitrate scaling is wider and the compression efficiency is the same or better compared to earlier methods. Thus, the proposed method gives more flexibility in applications utilizing bitrate scalability, such as rate scaling in streaming servers.

Modern video coding techniques often utilize multiple reference pictures for motion compensation to improve compression efficiency and error resilience. The sub-sequence technique also makes use of multiple reference pictures. A typical mode for reference pictures operation is "sliding window", which removes the oldest reference frame from the buffer when a new reference frame is decoded and the buffer is full.

This paper is organized as follows. Section II reviews the conventional temporal scalable coding. The proposed sub-sequence technique and coding schemes for improved temporal scalability are given in Section III. Section IV discusses the simulation results. Finally, we conclude the work in Section V.

II. CONVENTIONAL TEMPORAL SCALABILITY

A. Individually Disposable Pictures

In other video coding standards than H.264, bi-predictive (B) pictures are not used as prediction references. Consequently, they provide a way to achieve temporal scalability.

The enhanced reference picture selection mode (Annex U) of H.263 allows signaling whether a particular picture is a reference picture for any inter prediction of any other picture. Consequently, a picture not used for prediction (a non-reference picture) can be safely disposed. The H.264 syntax includes similar signaling to distinguish between reference and non-reference pictures.

B. Disposal of Picture Chains

A known method in today's streaming systems to cope with drastically dropped channel throughput is to transmit Intra pictures only. When the network throughput is restored, Inter pictures can be transmitted again from the beginning of the next Group of Pictures (GOP).

Generally, any chain of Inter pictures can be safely disposed, if no other picture is predicted from them. This fact can be utilized to treat Inter pictures at the end of a prediction chain as less important than other Inter pictures. The known layered coding techniques put some pictures into enhancement layers for temporal scalability, but do not identify the dependencies of pictures. In addition, multiple prediction chains are often maintained to achieve temporal scalability. In the conventional solutions, it is hard for the server or gateway to discard pictures intentionally without affecting the decoder behavior.

III. SUB-SEQUENCES AND H.264

A. Sub-Sequence and Sub-Sequence Layer

The proposed sub-sequence represents a number of inter-dependent pictures that can be disposed without affecting the decoding of any other sub-sequence in the same sub-sequence layer or any sub-sequence in any lower sub-sequence layer. The sub-sequence technique enables easy identification of disposable chains of pictures when processing pre-coded bitstreams.

Disposal of a sub-sequence on which there are no dependencies in the bitstream maintains a valid bitstream. Thus, the decoding process for the remaining bitstream and the reference picture buffer handling in particular has to be such that it does not depend on the presence or absence of any disposable sub-sequences. Subsection III.C describes the fundamentals how the decoding process of H.264 takes sub-sequences into consideration.

Pictures in a coded bitstream can be organized into sub-sequences and sub-sequence layers in multiple ways provided that the structure fulfills the requirements for dependencies between sub-sequences and sub-sequence layers. In most applications, a single structure of sub-sequences and sub-sequence layers is sufficient. Each picture belongs to exactly one sub-sequence, and each sub-sequence belongs to exactly one sub-sequence layer in any sub-sequence structure.

Sub-sequence layers are arranged hierarchically based on their dependency on each other. The base layer (layer 0) is independently decodable. Sub-sequence layer 1 depends on some of the data in layer 0, i.e., correct decoding of all pictures in sub-sequence layer 1 requires decoding of all the previous (in decoding order) pictures in layer 0. In general, correct decoding of sub-sequence layer N requires decoding of layers from 0 to N-1. It is recommended to organize sub-

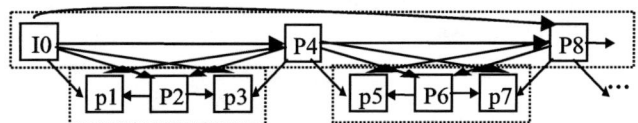

Figure 1. Example of sub-sequences: coding pattern "IpPpP" (*The numbers in the figure indicates the output order and the number of reference frames is 3.*)

sequences into sub-sequence layers in such a way that discarding of enhanced layers results in a constant or nearly constant picture rate. Picture rate and therefore subjective quality increase along with the number of decoded sub-sequence layers.

Compared with the conventional layered scalability, sub-sequences can be a non-layered (i.e. one-layer) bitstream with no added complexity on handling multiple layers. Sub-sequence technique enables easy identification of independent sub-sequences within the layers, making the bitrate shaping more efficient.

Since a sub-sequence in the base layer can be decoded independently of any other sub-sequences, the beginning of a base layer sub-sequence can be used as a random access position.

B. Use of Sub-Sequences

Sub-sequences can be used for improved bitrate scalability and error resiliency. Improved bitrate scalability can be achieved without sacrificing compression efficiency. In this sub-section, we present the sub-sequence coding scheme for improved bitrate scalability. We also discuss how the fast forward operation can be improved with the proposed sub-sequence scheme. The use of sub-sequences in error resilience has been demonstrated at least in [4] and [5] and we do not discuss the topic here.

Fig. 1 illustrates an example of a sub-sequence coding scheme referred to as IpPpP within H.264 codec. 'P' and 'p' denote reference picture and non-reference picture, respectively. The decoding order of pictures is as follows: I0 P4 P2 p1 p3 P8 P6 p5 p7. The midmost P picture in IpPpP is not used as a reference picture for pictures other than the two p pictures in the same sub-sequence. Any non-reference picture (p picture) can be safely discarded. Any sub-sequence pPp can be discarded without affecting the decoding of other sub-sequences pPp. A modification of the sub-sequence coding scheme IpPpP is to replace the P and p in sub-sequence layer 1 to B and b, respectively. Noting that B pictures can also be used as references in H.264 (see subsection III.C).

There are at least two methods that are often used with the conventional GOP structure (referred to as IbbP in this paper) for the fast forward operation: decoding only the I pictures of each GOP and decoding only the I and P pictures. The proposed sub-sequence scheme (IbBbP) provides an additional method for the fast forward operation: decoding only the reference pictures in layer 0. In other words, the IbBbP scheme enables one additional fast forward speed in player implementations.

C. Sub-Sequences in H.264

1) Overview

H.264 includes three main differences in the concept of P and B pictures and their relation to reference picture buffering when compared to previous video standards such as H.263. First, both a P slice and a B slice allow using multiple reference pictures to predict sample values. However, each block in P slices can only use at most one motion vector, whereas each block in B slices can use at most two motion vectors. Second, whether a picture is a reference picture is indicated independently from the slices types, which implies that a B picture can be stored as a reference picture as well. Third, the decoding order of pictures is totally decoupled with their output (presenting) order. Thus, the decoded picture buffer is not only for buffering reference pictures but also for storing such non-reference pictures that are output with a delay.

2) Gaps in frame number

Frame number (the frame_num syntax element in the slice header) is used to identify different reference frames. By monitoring the continuity of frame numbers, decoders can detect losses of reference frames. Further actions can be invoked upon the founding of gaps in frame numbers. However, when a streaming server or a gateway disposes a sub-sequence intentionally, an H.264 decoder should not infer any frames losses. Instead, the decoder inserts "non-existing" frames into the decoded picture buffer as if the frames with absent frame numbers were decoded normally. Only when any "non-existing" frames are referred in the following decoding process, unexpected frame losses can be deduced.

3) Sub-sequences related SEI messages

Supplemental enhancement information (SEI) is data embedded in the coded bitstream that is not required for correct decoding of the sample values. However, SEI messages may help the decoder at least in displaying the decoded pictures or concealing transmission errors. Three types of SEI messages are defined for sub-sequences. The sub-sequence information SEI message maps a coded picture to a certain sub-sequence and sub-sequence layer. The sub-sequence layer characteristics SEI message and the sub-sequence characteristics SEI message give statistical information, such as bitrate, on the indicated sub-sequence layer and sub-sequence respectively. Furthermore, the dependencies between sub-sequences are indicated in the sub-sequence characteristics SEI message. Decoders can use these messages to scale the decoding process computationally in case of lack of computational resources and to detect in which sub-sequences and sub-sequence layers accidentally lost pictures (during transmission) resided, and thus improve error resilience.

4) File format

Information on sub-sequences and sub-sequence layering can be included in the file format specified for H.264 [3]. The file format is based on the ISO base media file format and can be used as an extension of the MP4 file format, for example. As consequences, streaming servers can easily adapt the bitrate of the transmitted streams by deciding which sub-sequence layers and sub-sequences are transmitted. File players can use the sub-sequence information for the implementation of the fast forward operation.

IV. SIMULATIONS

A. Simulation Environment

To evaluate the coding performance of IpPpP and IbBbP, they were compared with IPPP, IppP and IbbP within H.264 codec. In IPPP, all the Inter pictures are P pictures. In IppP, the two p pictures are non-reference pictures predicted from both the previous frames and the subsequent frame in output order. In IbbP, the two b pictures are non-reference pictures.

To demonstrate the usefulness of the proposed technique to a variety of applications, such as mobile messenger and digital TV, we carried out simulations for the following picture sizes and frame rates: QCIF 15 Hz, QCIF 30 Hz, CIF 30 Hz, and 525SD 25 Hz. The size of the decoded picture buffer was selected according to level 1 (QCIF), level 2 (CIF) and level 3 (525SD) of H.264. As the decoded picture buffer stores also the non-reference frames whose output is delayed, the number of reference frames (the size of the "sliding window" for reference pictures) for IpPpP and IbBbP is one less than that for IPPP, IppP and IbbP. The number of reference frames in each case is listed in Table I.

We used a constant quantization parameter (QP) value for all pictures in sub-sequence layer 0. In sub-sequence layer 1, we used a constant QP value that is 2 units larger than the QP value in the base layer. We coded each original sequence six times, QP values for layer 0 pictures being 20, 24, 28, 32, 36 and 40.

B. Marking Reference Pictures

The midmost P picture in IpPpP was not used as a reference picture after the decoding of the second p picture. Memory management control operation (MMCO) command in H.264 allows marking a reference picture to be unused for reference. Since MMCO commands can only be associated to reference pictures, we assigned a MMCO command to P8 to mark P2 to be unused for reference (when the notation as of Fig. 1 is used). P6 was marked to be unused for reference at P12, and so on. Similar MMCO commands were used in IbBbP too.

C. Simulation Results

We ran simulations to compare the rate-distortion performance of different coding schemes at full frame rate. The rate-distortion curve of Paris in CIF at 30Hz is shown in Fig. 2 as an example. Bjontegaard delta PSNR [6] was used to evaluate the average differences between rate-distortion curves. Table II contains the Bjontegaard delta PSNR values of the three competitive pairs: IpPpP vs. IPPP, IpPpP vs. IppP and IbBbP vs. IbbP. A positive value implies the former scheme outperforms the latter. It can be found that the compression performance of IpPpP is very close to that of IppP and IbBbP even outperforms IbbP a little in most cases.

The comparisons of H.264 Main/Extended profile with the Baseline profile, i.e., IbbP vs. IppP or IbBbP vs. IpPpP, are also presented in Table II. We can easily see the superiority of B and b pictures over P and p pictures regarding the compression efficiency.

The share of bits allocated for sub-sequence layer 0 and all reference pictures is shown in Table III. It can be seen that the proposed sub-sequence schemes provide a larger range to adapt the bitrate of a transmitted or decoded bitstream. Moreover, the proposed sub-sequence schemes provide two steps of bitrate scalability that result into a constant picture rate, whereas the IbbP and IppP schemes provide only one such step. On the average, the IpPpP coding scheme provides bitrate steps at constant frame rate at about 50% and about 70% of the full bitrate, whereas the IppP coding scheme can be scaled down to an average of 60% of the full bitrate while maintaining constant frame rate. Similarly, the IbBbP coding scheme provides bitrate steps of about 60% and 80% of the full bitrate, whereas decoding of the reference frames in the IbbP coding scheme results into an average of 70% of the full bitrate.

V. CONCLUSIONS

This paper proposes a novel sub-sequence coding technique which can be applied to any video coding standards with multiple reference pictures buffer. IpPpP and IbBbP are proposed to provide more scalability compared to IPPP, IppP, and IbbP patterns while maintaining at least as high coding efficiency. We presented how sub-sequences are adopted in H.264, including the decoding process on gaps of frame number, sub-sequence related SEI messages and file format for H.264. Finally, the extensive simulations show the improvement in performance compared to conventional schemes.

REFERENCES

[1] G. J. Conklin, S. S. Hemami, "A Comparison of Temporal Scalability Techniques," IEEE Trans on CSVT, vol. 9, no. 6, pp. 909-919, Sept 1999.

[2] T. Wiegand, G. Sullivan and A. Luthra, "Draft ITU-T Recommendation and Final Draft International Standard of Joint Video Specification," document JVT-G050r1, May 2003

[3] D. Singer and T. Walker, "Study Text of Amd7 of ISO/IEC 14496-1 PDAM7," MPEG-4 Systems, N5096, Aug 2002

[4] S. Wenger, "Video Redundancy Coding in H.263+," PV 1997

[5] M. M. Hannuksela, "Simple Packet Loss Recovery Method for Video Streaming," PV2001, South Korea, May 2001

[6] G. Bjontegaard, "Calculation of average PSNR differences between RD-curves," ITU-T VCEG-M33, March 2001

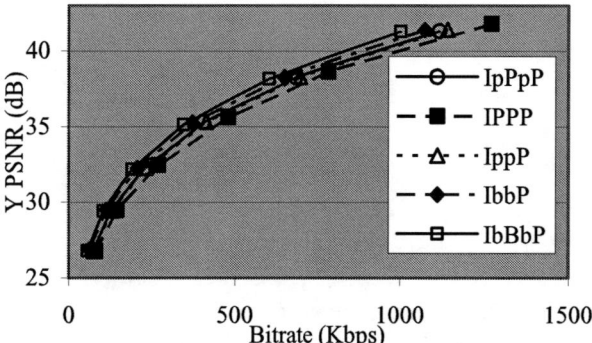

Figure 2. Rate-distortion curves for Pairs (*CIF @ 30 Hz*)

TABLE I. NUMBER OF REFERENCE FRAMES

-	IpPpP, IbBbP	IPPP, IppP, IbbP
QCIF	3	4
CIF	5	6
SD	4	5

TABLE II. AVERAGE RATE-DISTORTION DIFFERENCES (DB) AT FULL FRAME RATE (a: IpPpP vs. IPPP, b: IpPpP vs. IppP, c: IbBbP vs. IbbP, d: IppP vs. IbbP, e: IpPpP vs. IbBbP. A positive value implies the former scheme outperforms the latter)

Sequences		a	b	c	d	e
QCIF 15Hz	Foreman	0.07	-0.11	-0.01	-0.27	-0.37
	Paris	0.55	-0.08	0.08	-0.46	-0.63
	Tempete	0.32	-0.09	0.10	-0.45	-0.63
QCIF 30Hz	Foreman	0.27	-0.15	-0.00	-0.43	-0.58
	Paris	0.69	-0.06	0.29	-0.65	-1.02
	Container	1.03	0.16	0.25	-0.79	-0.90
CIF 30Hz	Mobile	0.62	-0.00	0.21	-0.74	-0.95
	Paris	0.63	0.09	0.84	-0.54	-0.74
	Tempete	0.47	0.05	0.21	-0.45	-0.61
SD 25Hz	Mobile	0.05	-0.12	0.20	-0.85	-1.18
	Parkrunner	0.03	-0.10	0.23	-0.59	-0.92

TABLE III. BITRATE PERCENTAGES AT LOWER FRAME RATES (%). (The fraction in the column titles (1/2, 1/3, 1/4) indicates the picture rate compared to the full picture rate.)

Sequences		IpPpP		IppP	IbBbP		IbbP
		1/4	1/2	1/3	1/4	1/2	1/3
QCIF 15Hz	Foreman	47.0	68.1	54.6	56.5	75.7	63.5
	Paris	49.5	69.5	57.4	59.1	77.0	65.6
	Tempete	44.9	65.3	53.8	58.2	75.5	64.6
QCIF 30Hz	Foreman	50.9	70.7	59.0	64.6	81.0	70.6
	Paris	52.6	71.7	61.1	66.0	82.0	71.4
	Container	65.6	78.1	73.0	79.9	87.2	85.2
CIF 30Hz	Mobile	46.6	66.1	55.9	62.2	78.2	68.5
	Paris	50.2	70.2	59.0	61.6	79.2	67.4
	Tempete	44.2	65.1	53.0	59.2	76.4	65.2
SD 25Hz	Mobile	46.3	65.8	54.2	69.2	81.7	73.3
	Parkrunner	47.0	68.1	57.0	64.0	79.7	70.0
Average Percent		49.5	68.9	58.0	63.6	79.4	69.5

Scalable Multiview Video Coding Using Wavelet

[1],[*]WenxianYang, [2]Feng Wu, [2]Yan Lu, [1]Jianfei Cai, [3]King Ngi Ngan, [2]Shipeng Li

[1]Nanyang Technological University, Singapore
wxyang@pmail.ntu.edu.sg
[2]Microsoft Research Asia, Beijing, China
[3]The Chinese University of Hong Kong, Shatin, N.T., Hong Kong

Abstract—Most existing multiview video coding (MVC) techniques are based on the traditional hybrid DCT-based video coding schemes, e.g., MPEG-2, MPEG-4 and H.264. They neither fully exploit the redundancy among different views nor provide an easy way of implementation for scalabilities. In this paper, we propose an MVC scheme based on wavelet, which can provide temporal, spatial, SNR as well as view scalabilities. To the best of our knowledge, wavelets have not been used for MVC in the literature before. In particular, we consider a multiview video as a 2D matrix with 1D along the temporal axis, 1D along the view axis, and each element in the 2D matrix represents a video frame. We first apply 1D wavelet decomposition to the 2D matrix along the temporal axis with motion compensation, and then along the view axis with disparity compensation. After that, 2D spatial wavelet decomposition is applied to each element in the matrix. Finally, all the subbands are encoded by 3D-ESCOT with rate-distortion optimization. Compared with traditional 3D wavelet coding, our proposed wavelet-based MVC scheme applies one more dimensional wavelet transform along the view direction to exploit the redundancy between adjacent views. Some preliminary experiments are carried out using standard multiview video sequences and the efficiency of the proposed system is confirmed.

I. INTRODUCTION

It has been recognized that multiview video coding is a key technology for a wide variety of future applications including FVV (free-viewpoint video) or FTV (free-viewpoint television), 3DTV (3D television), immersive teleconference and surveillance etc [1]. Due to the huge data volume of multiview video, highly efficient compression is very demanding. In addition to the redundancy exploited in 2D video for compression, the common idea for MVC is to further exploit the redundancy between adjacent views. This is because multiview video is captured by multiple cameras at different positions and significant correlations exist between neighbor views.

In the past, multiview video coding has been included in several video coding standards, e.g., MPEG-2 MVP (multiview profile) and MPEG-4 MAC (multiple auxiliary component). Based on the framework provided by the video standards, recently, some standard-compatible improved MVC schemes have been proposed, e.g., the scheme based on MPEG-2 [2] and the scheme based on MPEG-4 [3]. Some non-standard techniques [4, 5] have also been proposed. In [4], the authors considered the issues of structure and motion estimation in teleconferencing type multiview sequences. In [5], a sprite generation algorithm in multiview sequence was proposed to improve the coding efficiency. More recently, an H.264-based multiview video coding scheme was proposed in [6], which utilizes the multiple reference structure in H.264. Although this method does exploit the correlations between adjacent views through inter-view prediction, it has some constraints for practical applications. Firstly, simply adding a reference from another view may not be efficient since it costs the overhead for disparity vectors, which are usually larger than motion vectors. Secondly, the GGOP structure in [6], which contains all the GOPs in different views at the same time interval, is required to be predefined. This is hard to achieve for an arbitrary number of views.

In addition, all the existing MVC schemes mentioned above use DCT-based coding. A fundamental problem for DCT-based block coding is that it is not convenient to achieve scalability, which has become a more and more important feature for video coding and communications. On the other hand, wavelet-based image and video coding has been proved to be an excellent way to achieve both good coding performance and full scalabilities including spatial, temporal and SNR scalabilities. Therefore, in this paper, we propose to use wavelet for scalable multiview video coding. To the best of our knowledge, wavelets have not been used for MVC in the literature before. In particular, we consider a multiview video as a 2D matrix with 1D along the temporal axis, 1D along the view axis, and each element in the 2D matrix represents a video frame. We first apply 1D wavelet decomposition to the 2D matrix along the temporal axis with motion compensation, and then along the view axis with disparity compensation. After that, 2D spatial wavelet decomposition is applied to each element in the matrix. Finally, all the subbands are encoded by 3D-ESCOT with rate-distortion optimization. Compared with traditional 3D

---

[*]This work has been done while the author is with Microsoft Research Asia.

wavelet coding, our proposed wavelet-based MVC scheme applies one more dimensional wavelet transform along the view direction to exploit the redundancy between adjacent views.

The rest of this paper is organized as follows. A previous related work on scalable video coding (SVC) using 3D wavelet is described in Section II, which provides the fundamental techniques for our proposed MVC scheme. We then introduce the proposed scheme in details in Section III. The experimental results are given in Section IV, and finally Section V concludes the paper with some discussions.

II. SVC USING 3D WAVELET

As mentioned in Section I, 3D wavelet is widely used in SVC. In 3D wavelet decomposition, typically, 1D wavelet transform is first applied along the temporal axis, followed by the traditional 2D spatial wavelet decomposition. The temporal wavelet decomposition is often performed on a motion-aligned basis. This is because co-located pixels in different frames are usually out of alignment due to global and/or local motions across frames. Figure 1 shows a typical temporal wavelet decomposition along with motion trajectory, where each column represents a frame. Pixels in one frame are not linked with the co-located pixels in adjacent frames but with the pixels along the motion trajectory, which is specified by forward (or both forward and backward) motion vectors. Either global or local motion models can be applied to the motion-aligned temporal wavelet decomposition.

In this paper, we adopt the SVC scheme proposed in [7] to develop the multiview video codec. Any other 3D wavelet based SVC scheme can also be used. In particular, in [7], the raw video data is first input to the motion estimation module,

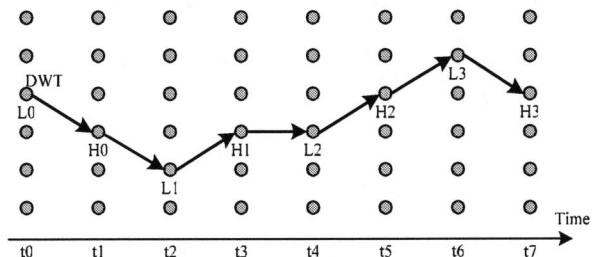

Figure 1: The wavelet decomposition along with motion trajectory for an image sequence.

which performs motion estimation on variable block sizes from 16x16 to 4x4 with quarter pixel precision. Then, the Barbell lifting is employed for the temporal wavelet decomposition. The Barbell lifting contains three steps: split, predict and update. In the first step, *Split*, video data is split into two subsets, X and Y. Then, in the second step, *Predict*, the subset X is used to predict the subset Y and the high-pass wavelet coefficients H is calculated as the prediction error. The overlap motion alignment is enabled in the predicting step. The third step, *Update*, uses the high-pass wavelet coefficients H to update subset X to ensure the preservation of moments in the low-pass L. After the temporal wavelet decomposition, the resulting high-pass and low-pass frames are further decomposed by 2D spatial wavelet transform with the 9/7 filters. Finally, all subbands are coded with the 3D ESCOT [8].

For multiview video coding, a straightforward extension is to align pixels across different views and then apply another 1D wavelet decomposition along the view axis. In other words, we consider a multiview video sequence as a 2D matrix with 1D along the temporal axis, 1D along the view axis and each element as a video frame. We will describe our proposed MVC scheme in the next section in details.

III. SCALABLE MULTIVIEW VIDEO CODING

A. System structure

Figure 2 shows the proposed multiview video coding scheme. In particular, we first apply the motion-compensated temporal-directional digital wavelet transform (MC-TDWT) to the multiview video matrix. Then, the low-pass frames are further decomposed along the view axis by disparity compensated view-directional digital wavelet transform (DC-VDWT) while the high-pass frames are directly sent to the 2D spatial wavelet decomposition module. Finally, after the temporal, view and spatial wavelet decomposition, we employ the 3D-ESCOT algorithm in [8] for entropy coding of all the generated subbands.

Note that, in the current implementation, for simplicity we only apply the view direction wavelet decomposition to the t-LLLL band, as shown in Figure 3(a). Other decomposition structures such as the one shown in Figure 3(b) may be more efficient in exploiting the correlations

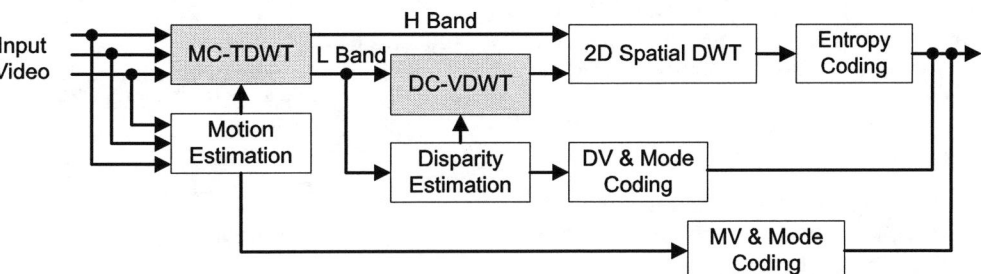

Figure 2: The block diagram of the proposed scalable multiview video encoding using wavelet.

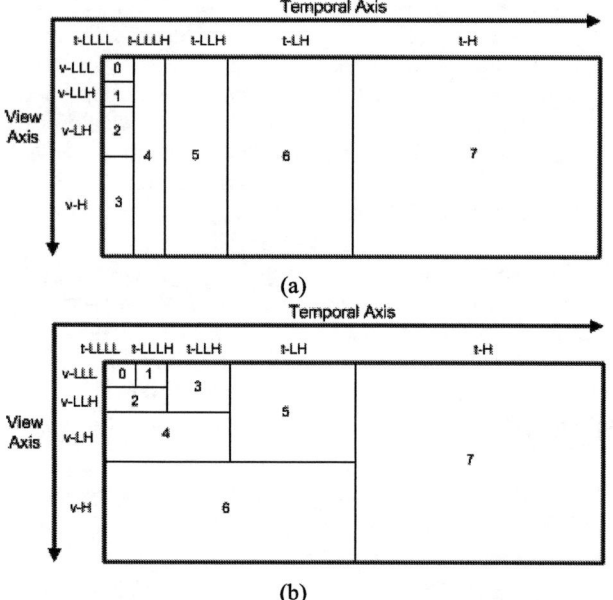

Figure 3: The decomposition structure along the temporal axis and the view axis.

among adjacent frames in either the temporal or view direction, which needs further investigation in our future work.

B. Entropy coding and rate allocation

As stated earlier, in our proposed system, the 3D-ESCOT algorithm is adopted for entropy coding. In particular, each subband is coded independently in order to provide flexible spatial, temporal and view scalabilities. A rate-distortion (R-D) optimized truncation scheme is used to multiplex different subband bitstreams into different layers to further provide SNR scalability. The basic idea is to cut the R-D curves of different subbands at the same slope for a particular layer.

C. Disparity estimation

Disparity is the name of motion along the view axis. Globe disparity always exists in multiview video since different views are captured by different cameras. In the DC-VDWT module, an affine model representing the global disparity between adjacent views is combined with the block-based local disparity model. We first calculate the model parameters using the FFRGMET (Feature-based Fast and Robust Global Motion Estimation Technique) in the MPEG-2 Optimization Model ver 2.0 [9], and then the reference frame is warped according to the model for the prediction of the current frame. The pixels in the occlusion area after warping are padded by using the low-pass extra-interpolation (LPE) scheme.

During the disparity estimation, each macroblock (MB) is predicted by either the global disparity or the local disparity, which is selected based on R-D optimization. In particular, besides the MB modes defined in [7], three additional MB modes, forward, backward and bi-directional global disparity modes, are introduced in our proposed multiview video coding system. If a MB is selected with a local disparity mode, we encode the local disparity vectors. If a MB is selected with a global disparity model, no vector needs to be encoded since three pairs of global motion vectors only need to be encoded once for an entire frame.

IV. SIMULATION RESULTS

Two standard video sequences for multiview video coding [9], *golf2* and *race1*, are used for testing the proposed system. Both sequences are dense 8-view video captured by 1D parallel cameras at 30 fps, and the frame size is 320 pixels by 240 pixels. We encode *golf2* and *race1* at average 128 kbps and 256 kbps per view, respectively. The results are compared with the simulcast video coding schemes, where each view is independently coded by either H.264/AVC or the SVC. Table 1 shows the parameters for the H.264 simulcast coding [10].

The PSNR curves of the three methods are given in Figure 4. The PSNR is the average PSNR over 8 views and the simulation is conducted over 90 frames. For golf2, the proposed multiview video coding scheme performs saliently better than the two simulcast coding schemes. The average PSNR of our proposed scheme is 2 dB higher than the simulcast coding using H.264, and 1.3 dB higher than the simulcast coding using the SVC. For race1, the average PSNR gain is 1.5 dB and 0.87 dB over H.264 and the SVC, respectively.

In addition, we compare the proposed DC-VDWT with the MC-TDWT for encoding the view-directional sequences in Figure 5. With a combined global disparity model, DC-VDWT is more efficient for removing view-directional redundancy. We observe from the experimental results that nearly 28% MBs are predicted by one of the three global disparity modes. This proves that the global disparity compensation is very efficient in multiview video coding.

V. CONCLUSIONS

In this paper, a wavelet-based multiview video coding scheme has been presented, which can provide temporal, spatial, SNR as well as view scalabilities. The preliminary simulation results demonstrated that the proposed scheme not only provides a variety of scalabilities but also improves

Table 1. The AVC Parameters.

Feature / Tool / Setting	AVC Parameters
Rate control	Yes
RD optimization	Yes
Specific settings	Loop filter, CABAC
Search range	±32 for CIF/VGA
# Reference picture	5
I-frame period	1 sec
GOP Structure	IBBP...

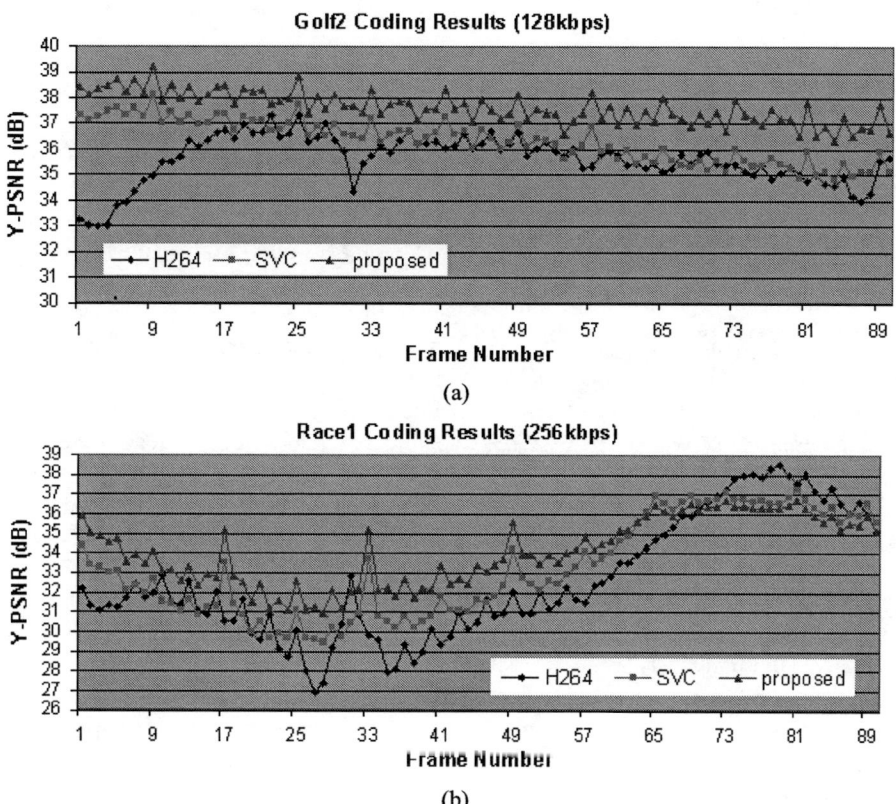

Figure 4: The coding results comparison with simulcast coding using H.264 and the SVC. (a) *golf2* at 128 kbps per view, (b) *race1* at 256 kbps per view.

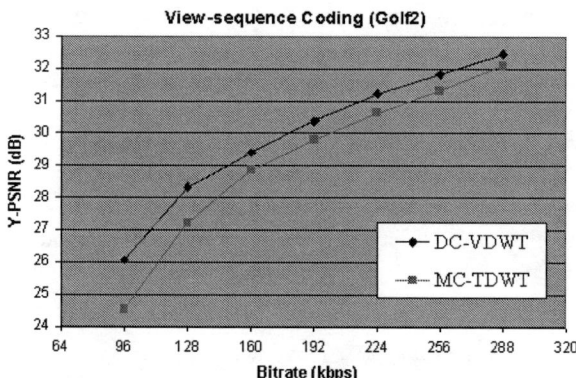

Figure 5: Comparison of DC-VDWT and MC-TDWT for view-sequence coding (*golf2*).

the coding efficiency and the visual quality of the decoded images. In the future work, we will further study the wavelet decomposition structure and the global disparity issues in order to obtain the optimal performance.

REFERENCES

[1] ISO/IEC JTC1/SC29/WG11 N6501, "Requirements on Multi-view Video Coding", Redmond, USA, July 2004.

[2] J. Lim, K. Ngan, W. Yang and K. Sohn, "Multiview Sequence CODEC with View Scalability", Signal Processing: Image Communication, vol. 19, no. 3, pp. 239-256, March 2004.

[3] W. Yang and K. Ngan, "MPEG-4 based Stereoscopic Video Sequences Encoder", IEEE International Conference on Acoustics, Speech, and Signal Processing (ICASSP 04), vol. 3, pp. 741-744, 17-21 May 2004.

[4] R. S Wang, Y. Wang, "Multiview Video Sequence Analysis, Compression, and Virtual Viewpoint Synthesis," IEEE Trans. Circuits Syst. Video Technol. Vol.10, pp.397-410, April 2000.

[5] N. Grammalidis, D. Beletsiotis, and M. G. Strintzis, " Sprite Generation and Coding in Multiview Image Sequences," IEEE Trans. Circuits Syst. Video Technol., vol.10, pp.302-311, Mar. 2000.

[6] Guoping Li, Yun He, "A Novel Multi-View Video Coding Scheme Based on H.264", ICICS-PCM 2003, pp. 493-497, Dec. 2003.

[7] Ruiqin Xiong, Feng Wu, Jizheng Xu, Shipeng Li and Ya-Qin Zhang, "Barbell Lifting Wavelet Transform for Highly Scalable Video Coding", Picture Coding Symposium (PCS'04), Dec. 2004.

[8] J. Xu, Z. Xiong, S. Li, and Y.-Q. Zhang, "3-D embedded subband coding with optimal truncation (3-D ESCOT)," J. Applied and Computational Harmonic Analysis: Special Issue on Wavelet Applications in Engineering, vol. 10, pp. 290-315, May 2001.

[9] ISO/IEC JTC1/SC29/WG11 N3675 (MPEG-2 OM), "Optimization model version 2.0," LaBaule, October 2000.

[10] ISO/IEC JTC1/SC29/WG11 MPEG2004/N6494, "Preliminary Call for Evidence on Multi-View Video Coding", Redmond, USA, July 2004.

Stereo Video Coding System with Hybrid Coding Based on Joint Prediction Scheme

Li-Fu Ding, Shao-Yi Chien, Yu-Wen Huang, Yu-Lin Chang, and Liang-Gee Chen

DSP/IC Design Lab, Graduate Institute of Electronics Engineering and Department of Electrical Engineering,
National Taiwan University, Taipei, Taiwan, Email: {lifu, shaoyi, yuwen, ylchang, lgchen}@video.ee.ntu.edu.tw

Abstract— Stereo video systems require double bandwidth and more than twice computational complexity relative to mono-video systems. Thus, An efficient coding scheme is necessary for transmitting stereo video. We propose a novel stereo video coding system by exploiting joint prediction scheme which combines three prediction schemes to achieve high coding efficiency and low computational complexity. Joint block compensation improves the visual quality. Motion vector prediction and mode pre-decision utilize the features of stereo video to reduce the computational complexity with up to 8–9 times acceleration. Experiments show that the proposed joint prediction scheme is 2dB better than MPEG-4 TS and 3dB better than MPEG-4 SP.

I. INTRODUCTION

Stereo video can make users have 3D scene perception by showing two frames to each eye simultaneously. With the technologies of 3D-TV getting more and more mature [5], stereo and multi-view video coding draw more and more attention. In recent years, MPEG 3D auido/video (3DAV) Group has worked toward the standardization for multi-view video coding [2], which also makes advancement of stereoscopic video applications. Although stereo video is attractive, the amount of video data and the computational complexity is doubled. A good coding system is required to solve the problem of huge data with limited bandwidth. Besides, in a mono-video coding system, motion estimation (ME) requires the most computational complexity [3]. By comparison, computational loading is heavier in stereo video coding systems due to additional ME and disparity estimation (DE). Thus, an efficient prediction scheme is required to overcome these problems. Finally, it is preferred that the proposed video encoding system is easily integrated by existing video standards.

In the past years, some stereo video coding systems are proposed. Stereo video coding can be supported by temporal scalability tools of existing standards, such as MPEG-2 multiview profile [1]. However, it cannot achieve good coding efficiency. I3D [7] is a famous approach, where the texture information is contained in a synthetic view, and the depth information is contained in a disparity map. It has good coding efficiency and compatibility with existing standard. However, additional operations for extracting disparity maps and synthesizing stereo views are required in the encoder and the decoder, respectively, which are not building blocks of conventional video coding systems. A mesh-based and block-based hybrid approach is proposed [9]. It has good compatibility and achieves acceptable coding efficiency. It needs additional preprocessing for segmentation to prevent matching failure around object boundary. In addition, the computational complexity is very high.

In this paper, we propose a new stereo video coding system with joint prediction scheme for general stereo video applications. The joint prediction scheme contains three coding tools. First, The joint block compensation is employed for better subjective and objective quality. Second, a new motion vector prediction algorithm is proposed according to the features of stereo video. Then, a mode pre-decision scheme is adopted to reduce computational complexity. The rest sections are organized as follows. Section II describes the proposed stereo video coding system. The experimental results are shown in Section III. Finally, Section IV gives the conclusion.

II. PROPOSED STEREO VIDEO CODING SYSTEM

For the purpose of compatibility, the coding system adopts a base-layer-enhancement-layer scheme, as shown in Fig. 1. The left view is set as the base layer, and the right view is set as the enhancement layer. The base layer is encoded with MPEG-4 video encoder. The proposed stereo video coding system is based on three concepts. First, in the compensation step, a block is not only compensated by the block of left or right reference frames, but also the combination of them due to different types of content in the current block. Second, in order to reduce the computational complexity of ME, the properties of stereo video, which is introduced later, are considered. It is adopted for accurate motion vector prediction. Third, the computational complexity of DE can be reduced for the similar reason with mode pre-decision scheme. Based on these three concepts, in this section, the encoding flow is introduced first. Next, the details of joint prediction scheme are shown in the rest subsections.

A. Encoding Flow

The block diagram of the proposed stereo video encoder is shown in Fig. 2. The main differences between the left channel and the right channel are DE and mode pre-decision, which are introduced later. Note that reference frames from left and right channels are both reconstructed. After encoding, the left compressed data, M and L, and the right compressed data of a small amount, N and R, are transmitted.

B. Joint Block Compensation Scheme

In ME and DE steps of the right channel, the current block has two reference frames, as shown in Fig. 3. Grey region is

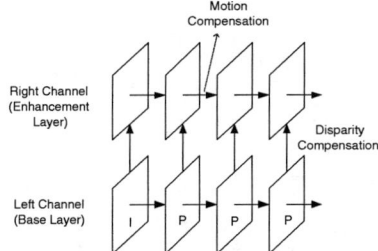

Fig. 1. Base-layer-enhancement-layer scheme of the proposed system.

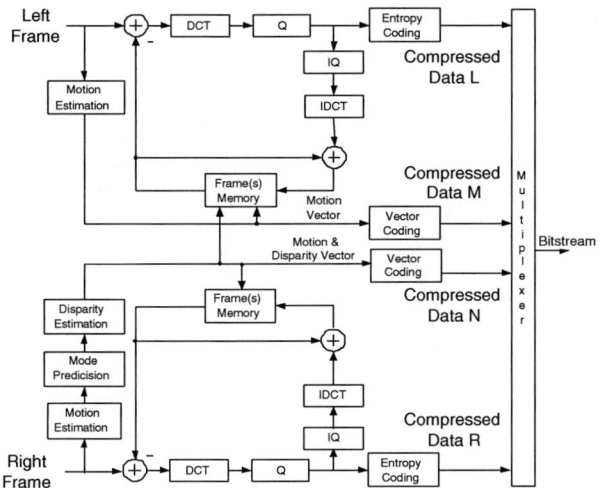

Fig. 2. Block diagram of the proposed stereo video encoder.

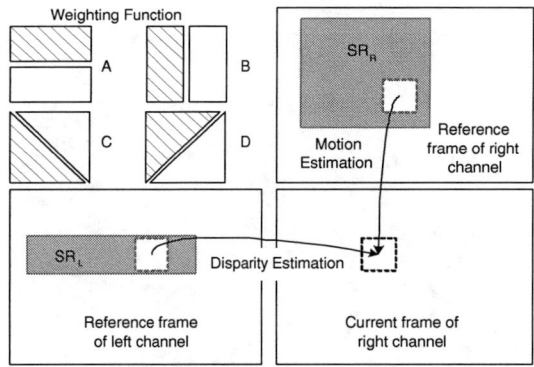

Fig. 3. The illustration of prediction directions and search range of two reference frames.

tively. $I_r(\mathbf{B})$ is the right current block, $I_{r-1}(\mathbf{B_R})$ is the right reference block, and $I_l(\mathbf{B_L})$ is the left reference block. $SR_R(B)$ and $SR_L(B)$ are the search ranges in right and left reference frames of block B, respectively. DE and ME result in the best matching blocks, $I_l(B'_L)$ and $I_{r-1}(B'_R)$, in left and right reference frames. Then the proposed joint block is composed of the weighted sum of the two blocks, $I_l(B'_L)$ and $I_{r-1}(B'_R)$. W_n and $W_{n'}$ are complementary weighting functions that describe a weighting parameter or some example patterns such as A to D, as shown in Fig. 3. In (3), the SAD value D_{j_n} is derived. I indicates that weighting parameters all equals one after weighted sum. Finally, the mode decision is described as follows,

$$Mode = \arg\min_{mode}\{D_{motion}, D_{disparity}, D_{j_1}, ..., D_{j_n}\}. \quad (4)$$

C. Fast Algorithm of Motion Estimation in the Right Channel

In general stereo video systems, ME and DE are the key operations. However, compared with mono-video systems, additional ME and DE of the right channel greatly increase computational burdens. Therefore, in this and next sections, accurate motion vector prediction scheme and mode pre-decision scheme before DE are proposed. First, the correlation between motion vectors (MVs) and disparity vectors (DVs) is shown.

1) The correlation between DVs and MVs: The correlation between these four vectors Fig. 4 can be described as the following equation:

$$DV_{k-1} + MV_R = MV_L + DV_k \quad (5)$$

In general, because the difference between two frames in the temporal domain is tiny, we have the relation below,

$$DV_{k-1} \approx DV_k, MV_R \approx MV_L \quad (6)$$

According to the correlation, MVs derived in the left channel (MV_L) are set to be predictors of MVs in the right channel (MV_R). Because of the parallel camera structure, there is an global horizontal displacement between left and right channels. This displacement is called "global disparity." In order to find the predictors, the global disparity should be derived first due to the relation between MVs and DVs introduced above. Here, we use a simple way to find the global disparity rather than global motion estimation (GME) scheme [6].

the search range of a reference frame. Note that search range of the left reference frame is not square because cameras are parallel-structured, so the candidate blocks are only on a belt of region.

There are three types of compensated blocks in the proposed stereo video encoder. 1) Motion-compensated block: it often occurs in the background due to its zero or slow motion. Occlusions between left and right frames will also compensated by this type of blocks. 2) Disparity-compensated block: it often occurs in the moving objects because of their deformation during motion. In this case, disparity-compensated blocks usually have better prediction capability. 3) Joint block: it often occurs in the block which contains both foreground and background in it because the foreground and the background may be suitably predicted by different types of blocks.

According to the criterion of sum of absolute difference (SAD), the best matching block mode is selected. For each macro block of the current frame, the distortion of three types of blocks are computed as follows,

$$D_{motion} = \min \sum_{\mathbf{B_R} \in SR_R(B)} |I_r(\mathbf{B}) - I_{r-1}(\mathbf{B_R})|, \quad (1)$$

$$D_{disparity} = \min \sum_{\mathbf{B_L} \in SR_L(B)} |I_r(\mathbf{B}) - I_l(\mathbf{B_L})|, \quad (2)$$

$$D_{j_n} = \sum |I_r(B) - [W_n \cdot I_l(B'_L) + W_{n'} \cdot I_{r-1}(B'_R)]| \\ W_n + W_{n'} = I \quad (3)$$

where D_{motion} and $D_{disparity}$ are the minimum SAD values of motion- and disparity-compensated blocks, respec-

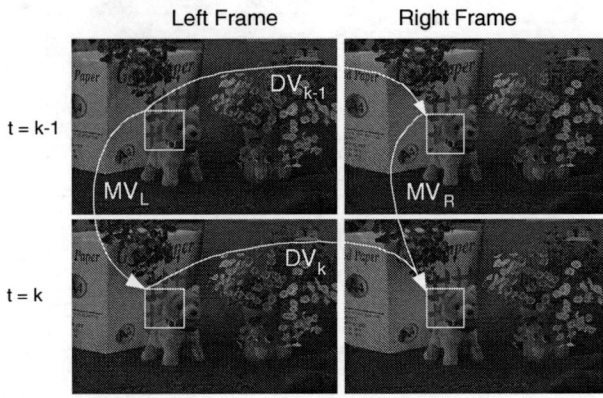

Fig. 4. The relation between DVs and MVs.

2) Background detection and global DE of the right channel: The background detection is determined as below,

$$F_{diff}(N) = \sum_B |I_r(\mathbf{B}) - I_{r-1}(\mathbf{B})|, \quad (7)$$

$$Background(N) = \begin{cases} true, & \text{if } MV_N(x,y) = (0,0) \text{ OR} \\ & F_{diff}(N) < Threshold \\ false, & \text{otherwise.} \end{cases} \quad (8)$$

where $Background(N)$ is the state of the Nth block in the left frame. $MV_N(x,y)$ is the MV of the Nth block. If the MV is zero or $F_{diff}(N)$ is smaller than a threshold, this block is probably belongs to background, so $Background(N)$ is set to *true*. After this step, the global disparity vector, GD, is derived as follows,

$$GD = \arg\max_{DV} \{ Num(DV) \} \quad (9)$$

The DVs of these zero motion blocks are gathered for statistics. The DV with the highest appearing frequency, $Num(DV)$, is regarded as GD. For the first P-frame in the left channel, background detection scheme is used to find GD. Before ME of the right channel, the corresponding block (in the left frame) of the current block (in the right frame) can be found by using GD. Then the MV of this related block is regarded as the predictor of the current block in the right frame, and MV_R can be derived correctly within only small search range to reduce computation. However, the DVs of background are usually smaller than those of foreground. If SAD is over a threshold, we adaptively extend the search range to find a better MV. Then, a more precise global disparity vector is fed back to the system. To avoid error propagation, GD can be updated every M frames, where M is a flexible parameter.

D. Mode Pre-decision Before Disparity Estimation

In our experiments shown in Fig. 5, statistics show that in a frame of the right channel, 40%–70% blocks are motion-compensated, 25%–60% blocks are joint-compensated, only about 5% blocks are disparity-compensated. Because a joint block results from both ME and DE, it means 25%–60% blocks need to perform DE and ME. From above analysis, over 95% blocks must perform ME, while only 40%–60%

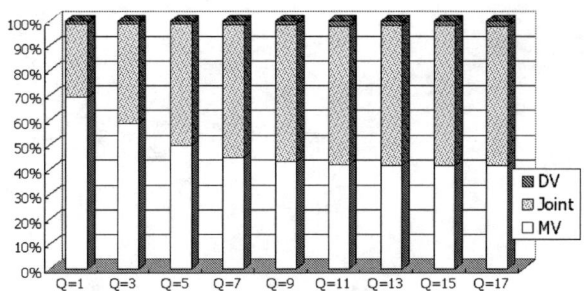

Fig. 5. Percentage of block prediction types of sequence "Race2."

blocks must perform DE. Thus, unnecessary DE step should be skipped to reduce computational complexity. From our analysis, the MV-predicted blocks are often have zero motion, such as blocks in the background or blocks with slow motion caused by moving cameras. Therefore, by utilizing these features, mode pre-decision scheme is proposed by following equations,

$$SAD_{ME} = \min \sum_{\mathbf{B}' \in SR(B)} |I_r(\mathbf{B}) - I_{r-1}(\mathbf{B}')|, \quad (10)$$

$$Skip = \begin{cases} true, & \text{if } F_{diff} < Threshold_1 \text{ AND} \\ & SAD_{ME} < Threshold_2 \\ false, & \text{otherwise.} \end{cases} \quad (11)$$

If $F_{diff} < Threshold_1$ and $SAD_{ME} < Threshold_2$ are simultaneously established, the mode of this block is usually MV-predicted. Then DE of this block is skipped, and the computational burdens is decreased.

III. EXPERIMENTAL RESULTS AND ANALYSIS

The proposed system is compared with MPEG-4 Simple Profile (SP) and Temporal Scalibility (TS) encoder [4]. Rate-distortion performance of only right channels (enhancement layer) are compared because the left channels are all encoded by MPEG-4 SP. Sequence "Race2" (320×240, 30 fps) and "Soccer2" (720×480, 30 fps) are taken as test sequences.

Figure 6 shows the comparison between the proposed algorithm, MPEG-4 TS, and MPEG-4 SP. Proposed joint prediction scheme is 2–3dB better than other two MPEG-4 profiles. Figure 7 shows the performance of different coding tools. It is shown that without joint block compensation, the PSNR degradation is serious. The proposed algorithms (curve "Joint Block" and "Joint Prediction") is over 3dB better than MPEG-4 SP at low bit-rate, which is the target bit-rate for the right channel according to the "asymmetrical spatial resolution property" [8]. Besides, after applying MV-predictor prediction and mode pre-decision scheme (curve "Joint Prediction"), most of the computational complexity can be reduced, as shown in Fig. 8. In our experiments, the search range of ME is set to ±16 horizontal and vertical, and the search range of DE is set to ±32 horizontal and ±4 vertical. Without our algorithm, the overhead (additional search points relative to mono-video systems) is 150%. The proposed algorithm reduce the overhead under 20% in some cases, that is, it can speed up 8–9 times. From Fig. 9, we can see that if the search range is

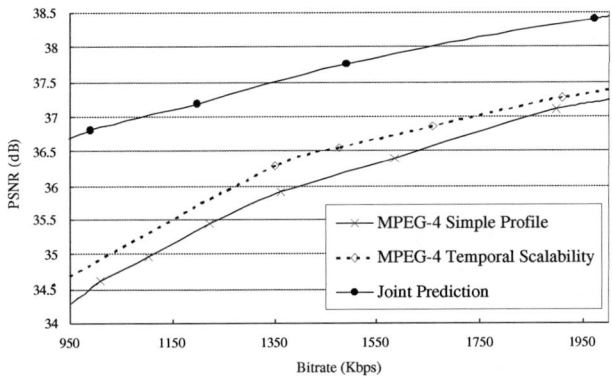

Fig. 6. Rate-distortion curve of sequence "Soccer2."

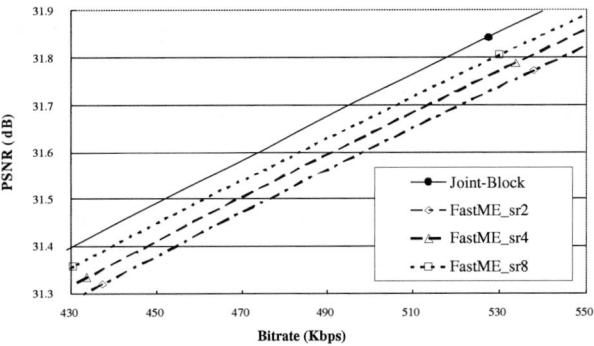

Fig. 9. Rate-distortion curve of Fast algorithm with various search range of sequence "Race2."

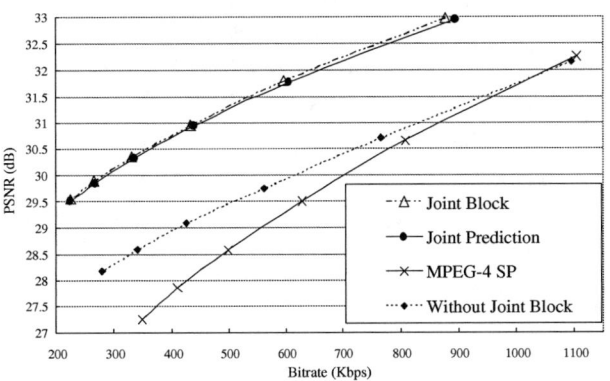

Fig. 7. Rate-distortion curve of sequence "Race2."

Fig. 10. Reconstructed frames of frame 46 of the sequence "Race2." (a) Proposed Joint prediction scheme@334.47 Kbps. (b) Joint block compensation@330.16 Kbps. (c) MPEG-4 SP@348.84 Kbps. (d), (e), and (f) are zoom-in views of (a), (b), and (c), respectively

reduced from ±16 to ±2, the PSNR degradation is only about 0.1dB, while the computational complexity is greatly reduced. Figure 10 demonstrates the subjective quality of the proposed algorithm. The reconstructed frames of the proposed coding system with the bitrate 334.47 Kbps are shown in Fig. 10(a)(d). Figure 10(b)(e) is the proposed system without MV predictor prediction and mode pre-decision scheme. It shows that it is hard to recognize the difference between them. However, block artifacts can be easily observed in Fig. 10(c) and (f), which are encoded by MPEG-4 SP.

IV. CONCLUSION

In this paper, we propose a stereo video coding system with joint prediction scheme which combines three coding tools. Joint block compensation utilizes the weighted sum of motion- and disparity-compensated blocks. Our system outperforms MPEG-4 TS and SP by 2–3 dB in rate-distortion performance. Moreover, MV-prediction and mode pre-decision

Algorithm	Race2	Flamenco	Soccer2	Puppy	Golf
Full Search (L)	1024	1024	1024	1024	1024
Full Search (R)	1536	1536	1536	1536	1536
Joint Prediction	381.42	200.45	415.75	177.62	234.89
Search point reduction	75.20%	86.95%	73.92%	88.44%	84.71%
Overhead (original)	150.00%	150.00%	150.00%	150.00%	150.00%
Overhead (proposed)	37.24%	19.58%	40.60%	17.35%	22.94%
Speed Up	4.03	7.66	3.69	8.50	8.65

Fig. 8. Search points and reduction rate of proposed scheme with and without complexity-reduction scheme.

utilize the correlation between MVs and DVs, which reduce the computational burdens with up to 8–9 times acceleration. This system can be easily integrated by the existing standard and extend to the multi-view video coding system.

REFERENCES

[1] F. Isgrò, E. Trucco, P. Kauff, and O. Schreer, "Three-dimensional image processing in the future of immersive media," *IEEE Transations on Circuits and Systems for Video Technology*, vol. 14, no. 3, pp. 388–303, Mar. 2003.

[2] *Requirements on multi-view video coding.* ISO/IEC JTC1/SC29/WG11 N6501, 2004.

[3] H.-C. Chang, L.-G. Chen, M.-Y. Hsu, and Y.-C. Chang, "Performance analysis and architecture evaluation of MPEG-4 video codec system," in *Proceedings of 2000 IEEE International Symposium on Circuits and Systems (ISCAS 2000)*, 2000.

[4] *Proposed draft amendment No. 3 to 13818-2 (multi-view profile).* ISO/IEC JTC 1/SC 29/WG11 N1088, 1995.

[5] J.-R. Ohm and K. Müller, "Incomplete 3-D multiview representation of video objects," *IEEE Transations on Circuits and Systems for Video Technology*, vol. 9, no. 2, pp. 389–400, Mar. 1999.

[6] R.-S. Wang and Y. Wang, "Multiview video sequence analysis, compression, and virtual viewpoint synthesis," *IEEE Transactions on Circuits and Systems for Video Technology*, vol. 10, no. 3, pp. 397–410, Apr. 2000.

[7] H.-Z. Jia, W. Gao, and Y. Lu, "Stereoscopic video coding based on global displacement compensated prediction," in *International Conference on Information and Communications Security*, 2003, pp. 61–65.

[8] S. Pastoor, "3D-television: a survey of recent research results on subjective requirements," *Signal Processing: Image Communication*, vol. 4, no. 1, pp. 21–32, 1991.

[9] S. Cho, K. Yun, B. Bae, Y. Hahm, C. Ahn, Y. Kim, K. Sohn, and Y. h. Kim, *Report for EE3 in MPEG 3DAV.* ISO/IEC JTC1/SC29/WG11 M9186, December 2002.

A Generalized Semi-blind Channel Estimation for Pilot-aided OFDM Systems

K.Y. Ho and S.H. Leung

Department of Electronic Engineering, City University of Hong Kong
Tat Chee Avenue, Kowloon Tong, Kowloon, Hong Kong
email: EEEUGSHL@cityu.edu.hk

Abstract— In this paper, a new semi-blind channel estimation method for orthogonal frequency division multiplexing (OFDM) is presented. A deterministic frequency domain objective function for an OFDM frame is defined, which contains a sum of squared error for the pilot signals and a sum of weighted squared generalized magnitude error for data signals. It is found that the sum of squared absolute magnitude error for data signals, as a special case of the generalized term, provides an excellent block mean square error. An efficient iterative algorithm for obtaining an optimal solution is derived, that takes several iterations to converge to the steady state solution. The new algorithm is shown to provide a better performance gain than other pilot-based channel estimation methods and semi-blind method.

I. INTRODUCTION

Orthogonal frequency division multiplexing (OFDM) is an attractive transmission method for high data rate transmission because of its excellent bandwidth efficiency and good immunity to multi-path fading and impulse noise. Owing to its desirable advantages, it has been adopted as standard for digital audio broadcasting, digital video broadcasting and indoor wireless LAN.

Accurate channel estimation is essential for coherent OFDM detection otherwise there is degradation in signal-to-noise ratio (SNR). Pilot-aided schemes are commonly used for channel estimation and tracking in fast fading channels. The maximum likelihood estimator (MLE) [1] and the minimum mean square error estimator (MMSEE) [2] are two popular pilot-aided methods. The advantage of MLE is its simplicity that requires no information on the channel statistics and SNR. On the other hand, the MMSEE method with the information of the SNR and channel information gives better channel estimate than MLE.

Blind and semi-blind channel estimation methods for OFDM have been developed in the literature [3] [4] [5]. Most of these methods are subspace-based approach. However these methods need to use many OFDM blocks in order to obtain a good statistical estimate. Therefore, their performances are generally degraded in fast multipath fading channels. Besides their computational complexity is generally high because of the computationally expensive singular value decomposition.

In this paper, we present a semi-blind channel estimation method for OFDM using PSK signaling. This method is a generalization of the semi-blind method in [6]. Its objective function contains a sum of squared error for the pilot signals and a sum of weighted squared generalized magnitude error for data signals. The new method is a single frame-based deterministic algorithm which does not require any statistical information. As the method takes only one OFDM-frame signal for performing channel estimation, it is therefore useful for fast fading channel. The method uses both the pilot signals and data signals for estimation, that gives not only the performance gain in block mean square error and also in bit error rate over pilot-based methods.

The notations of the paper are summarized as follows. The letters in bold denotes vectors or matrices. The superscripts $(\cdot)^H$ and $(\cdot)^T$ denote the Hermitian operation and matrix transposition, respectively.

II. OFDM SIGNAL MODELLING

Consider an OFDM system using N-point Inverse Fast Fourier Transform (IFFT) for modulation with M ($M < N$) subcarriers used for data transmission and N-M virtual carriers on the edge of the spectrum. The data is a phase-shift-keyed (PSK) signal. The OFDM symbol is finally formed by adding a cyclic prefix (CP) to the IFFT output. The input data including the uniformly spaced pilot signals for the lth OFDM block is expressed as:

$$\mathbf{a}^{(l)} = [a_0^{(l)}, a_1^{(l)}, ..., a_{M_1}^{(l)}, 0, ..., 0, a_{N-M_2}^{(l)}, ..., a_{N-1}^{(l)}]^T \quad (1)$$

where $M_1 + M_2 + 1 = M$ and $|M_1 - M_2| \le 1$

We assume the channel to be a multipath channel with additive white Gaussian noise. The impulse response of the channel for the l-th block is defined as:

$$\mathbf{h}^{(l)} = [h_0^l, h_1^l, h_2^l, ..., h_{L-1}^l]^T \quad (2)$$

where L is the length of the impulse response.

The frequency response of the channel, $\mathbf{H}^{(l)}$, is an $N \times 1$ vector expressed as

$$\mathbf{H}^{(l)} = \mathbf{G}\mathbf{h}^{(l)} \quad (3)$$

where $\mathbf{G}$ is the $N \times L$ FFT matrix with entries $\exp\{-j2\pi mn/N\}$ for $0 \leq m \leq N-1$ and $0 \leq n \leq L-1$.

In the following, the superscript l is dropped for the sake of brevity unless there is ambiguity occurred.

The received signal in frequency domain, $\mathbf{Y}$, is an $N \times 1$ vector expressed as

$$\mathbf{Y} = [Y(0), \cdots, Y(N-1)]^T = diag\{\mathbf{a}\}\mathbf{H} + \mathbf{w} \quad (4)$$

where $\mathbf{w}$ is the independent white Gaussian channel noise vector with covariance vector $\sigma_n \mathbf{I}$, where $\mathbf{I}$ is an $N \times N$ identity matrix and $diag\{\mathbf{a}\}$ transforms the vector $\mathbf{a}$ to a diagonal matrix.

III. Pilot aided channel estimation

For pilot-aided OFDM systems, N_p known symbols are placed uniformly in the vector $\mathbf{a}$. Let $\{1_m\}$ for $1 \leq m \leq N_p$ denote the location indices of the pilots. The received pilot signal, $\mathbf{Y}_p$, in frequency domain, is an $N_p \times 1$ vector expressed as

$$\mathbf{Y}_p = diag\{\mathbf{a}_p\}\mathbf{B}_p\mathbf{h} + \mathbf{w}_p \quad (5)$$

where $\mathbf{a}_p$ is the $N_p \times 1$ vector containing the known pilot signals, $\mathbf{w}_p$ is the $N_p \times 1$ noise vector, and $\mathbf{B}_p$ is the $N_p \times L$ FFT matrix for the pilot subcarriers with entries $\exp\{-j2\pi i_m k/N\}$ for $1 \leq m \leq N_p$ and $0 \leq k \leq L-1$.

The maximum likelihood estimator (MLE) for (5) is given by [1].

$$\hat{\mathbf{h}}_{MLE} = (\mathbf{B}_P^H \mathbf{B}_P)^{-1} \mathbf{B}_P^H \mathbf{Z}_p \quad (6)$$

where $\mathbf{Z}_p = diag\{\mathbf{a}_p\}^H \mathbf{Y}_p$.

The MLE method is a deterministic estimation method that does not need any statistical information.

IV. New Semi-blind Channel Estimator

For PSK signaling, not only the pilots provide the channel information, the data subcarriers also give the information about the magnitude response of the channel. The application of data-subcarriers for channel estimation has been reported in [6]. Let N_d denote the number of data subcarriers used for channel estimation and $\{g_m\}$ for $1 \leq m \leq N_d$ denote the location indices of the data subcarriers. The received data signal, $\mathbf{Y}_d$, in frequency domain is an $N_d \times 1$ vector expressed as

$$\begin{aligned}\mathbf{Y}_d &= [Y(g_1), \ldots, Y(g_{N_d})]^T \\ &= diag\{\mathbf{a}_d\}\mathbf{B}_d \mathbf{h} + \mathbf{w}_d = diag\{\mathbf{a}_d\}\mathbf{H}_d + \mathbf{w}_d\end{aligned} \quad (7)$$

where $\mathbf{a}_d$ is the $N_d \times 1$ vector containing the unknown PSK signals, $\mathbf{w}_d$ is the $N_d \times 1$ noise vector, $\mathbf{B}_d$ is the $N_d \times L$ FFT matrix for the data subcarriers with entries $\exp\{-j2\pi g_m k/N\}$ for $1 \leq m \leq N_d$ and $0 \leq k \leq L-1$, and $\mathbf{H}_d = \mathbf{B}_d \mathbf{h} = [H(g_1), \ldots, H(g_{N_d}))]^T$.

We define an objective function, which is composed of a sum squared error for pilot signals and a sum of weighted squared generalized magnitude error for data signals in frequency domain, for the estimating the channel impulse response as follows:

$$\begin{aligned}F &= \lambda \|\mathbf{Z}_p - \mathbf{B}_p \mathbf{h}\|_2^2 \\ &\quad + (1-\lambda)(|\mathbf{Y}_d|^\rho - |\mathbf{H}_d|^\rho)^T \mathbf{K}_d (|\mathbf{Y}_d|^\rho - |\mathbf{H}_d|^\rho) \\ &= \lambda \|\mathbf{Z}_p - \mathbf{B}_p \mathbf{h}\|_2^2 \\ &\quad + (1-\lambda) \sum_{m=0}^{N_d-1} \frac{(|Y(g_m)|^\rho - |H(g_m)|^\rho)^2}{|H(g_m)|^q}\end{aligned} \quad (8)$$

where λ is a weighting parameter, $\{\rho, q\}$ are the parameters used to maximize the performance, and

$$|\mathbf{Y}_d^\rho| = [|Y(g_1)|^\rho, \ldots, |Y(g_{N_d})|^\rho]$$
$$|\mathbf{H}_d^\rho| = [|H(g_1)|^\rho, \ldots, |H(g_{N_d})|^\rho]$$
$$\mathbf{K}_d = diag\{|H(g_1)|^{-q}, \ldots, |H(g_{N_d})|^{-q}\}$$

When $\rho = 2$ and $q = 0$, the objective function is the same as the one in [6].

Using the definitions of matrix derivative in [7], we have:

$$\begin{aligned}\frac{\partial F}{\partial \mathbf{h}^*} &= -\lambda \mathbf{B}_p^H (\mathbf{Z}_p - \mathbf{B}_p \mathbf{h}) - \frac{(1-\lambda)}{2} \sum_{m=1}^{N_d} (|Y(g_m)|^\rho \\ &\quad -|H(g_m)|^\rho)[q|Y(g_m)|^\rho + (2\rho - q)|H(g_m)|^\rho] \\ &\quad |H(g_m)|^{-(q+2)} \mathbf{B}^H(g_m) H(g_m) \\ &= -\lambda \mathbf{B}_p^H (\mathbf{Z}_p - \mathbf{B}_p \mathbf{h}) - \frac{(1-\lambda)}{2} \sum_{m=1}^{N_d} [q|Y(g_m)|^{2\rho} \\ &\quad +2(\rho - q)|Y(g_m)|^\rho |H(g_m)|^\rho - (2\rho - q)|H(g_m)|^{2\rho}] \\ &\quad |H(g_m)|^{-(q+2)} \mathbf{B}^H(g_m) \mathbf{B}(g_m) \mathbf{h}\end{aligned} \quad (9)$$

where $\mathbf{B}^H(g_m)$ is a column vector with entries $\exp\{j2\pi g_m k/N\}$ for $0 \leq k \leq L-1$.

To obtain efficient iterative algorithms for finding an optimum solution $\hat{\mathbf{h}}$ of $\frac{\partial F}{\partial \mathbf{h}^*} = 0$, we set the following condition in order to make $|H(g_m)|^{2\rho - q - 2}$ in (9) equal one for simplifying the iteration,

$$2\rho - q - 2 = 0 \quad (10)$$

Under this condition, an iterative relation for finding the estimate $\hat{\mathbf{h}}_n$ for $\frac{\partial F}{\partial \mathbf{h}^*} = 0$ is described as:

$$(\lambda \mathbf{B}_p^H \mathbf{B}_p + (1-\lambda) \mathbf{B}_d^H \mathbf{B}_d) \hat{\mathbf{h}}_{n+1} =$$

$$\lambda \mathbf{B}_p^H \mathbf{Z}_p + \frac{(1-\lambda)}{2} \mathbf{B}_d^H \mathbf{M}_n \mathbf{B}_d \hat{\mathbf{h}}_n \quad (11)$$

where $\mathbf{M}_n = diag\{M_n(g_1), \ldots, M_n(g_{N_d})\}$ with

$$M_n(g_m) = [q|Y(g_m)|^{2\rho} + 2(\rho - q)|Y(g_m)|^{\rho}|\hat{H}_n(g_m)|^{\rho}]$$
$$|\hat{H}_n(g_m)|^{-(q+2)} \quad (12)$$
$$\mathbf{B}_d = [\mathbf{B}^H(g_1), \ldots, \mathbf{B}^H(g_{N_d})]^H \quad (13)$$

and

$$\hat{H}_n(g_m) = \mathbf{B}(g_m)\hat{\mathbf{h}}_n \quad (14)$$

Equation (11) can be expressed as

$$\hat{\mathbf{h}}_{n+1} = \mathbf{D}^{-1}(\lambda \mathbf{B}_p^H \mathbf{Z}_p + \frac{(1-\lambda)}{2} \mathbf{B}_d^H \mathbf{M}_n \mathbf{B}_d \hat{\mathbf{h}}_n)$$
$$= \mathbf{h}_p + \bar{\mathbf{h}}_{n+1} \quad (15)$$

where $\mathbf{D} = \lambda \mathbf{B}_p^H \mathbf{B}_p + (1-\lambda)\mathbf{B}_d^H \mathbf{B}_d$ is a constant matrix in terms of λ, $\mathbf{h}_p = \lambda \mathbf{D}^{-1}\mathbf{B}_p^H \mathbf{Z}_p$ is the part of the estimate related to the pilot signals, and $\bar{\mathbf{h}}_{n+1}$ is the part of the estimate refined by the data signals.

By the condition $2\rho - q - 2 = 0$, the matrix $\mathbf{D}$ becomes a constant matrix and its inverse can be precomputed; otherwise its inverse needs to be computed at each iteration. In (15), we only compute the terms $\mathbf{h}_p$ at the beginning of the iterative algorithm. In regard to the term $\bar{\mathbf{h}}_{n+1}$, its computation can be simplified by setting $q = 0$ to remove the first term in (12), that gives $\rho = 1$ in accordant to the condition (10). For this setting, the iterative algorithm is expressed as:

$$\hat{\mathbf{h}}_{n+1} = \mathbf{h}_p + (1-\lambda)\mathbf{D}^{-1}\mathbf{B}_d^H \mathbf{Y}_d \mathbf{P}_n \quad (16)$$

where $\mathbf{Y}_d = diag\{|Y(g_1)|, \ldots, |Y(g_{N_d})|\}$
$\mathbf{P}_n = [\exp\{j\angle \hat{H}_n(g_1)\}, \ldots, exp\{j\angle \hat{H}_n(g_{N_d})\}]^T$

In (16), the term $(1-\lambda)\mathbf{D}^{-1}\mathbf{B}_d^H \mathbf{Y}_d$ is computed at the beginning of the algorithm. Therefore the update is simply done by multiplying the phase vector $\mathbf{P}_n$ to a fixed matrix and then compute FFT to obtain the frequency response to update the phase vector.

The iterative algorithm is summarized as follow:
1. compute $\mathbf{h}_p = \lambda \mathbf{D}^{-1}\mathbf{B}_p^H \mathbf{Z}_p$
2. compute $\mathbf{C}_d = (1-\lambda)\mathbf{D}^{-1}\mathbf{B}_d^H \mathbf{Y}_d$
3. compute $\hat{\mathbf{h}}_o = \hat{\mathbf{h}}_{MLE} = (\mathbf{B}_p^H \mathbf{B}_p)^{-1}\mathbf{B}_p^H \mathbf{Z}_p$
4. compute $\mathbf{P}_o = \exp\{j\angle \mathbf{B}_d \hat{\mathbf{h}}_o\}$
$= [\exp\{j\angle \mathbf{B}(g_1)\hat{\mathbf{h}}_n\}, \cdots, \exp\{j\angle \mathbf{B}(g_{N_d})\hat{\mathbf{h}}_n\}]^T$
($\mathbf{B}_d \hat{\mathbf{h}}_o$ can be computed by FFT)
5. iteration: $\hat{\mathbf{h}}_{n+1} = \hat{\mathbf{h}}_p + \mathbf{C}_d \mathbf{P}_n$
6. compute $\mathbf{P}_{n+1} = \exp\{j\angle \mathbf{B}_d \hat{\mathbf{h}}_{n+1}\}$
7. repeat steps 5 and 6.

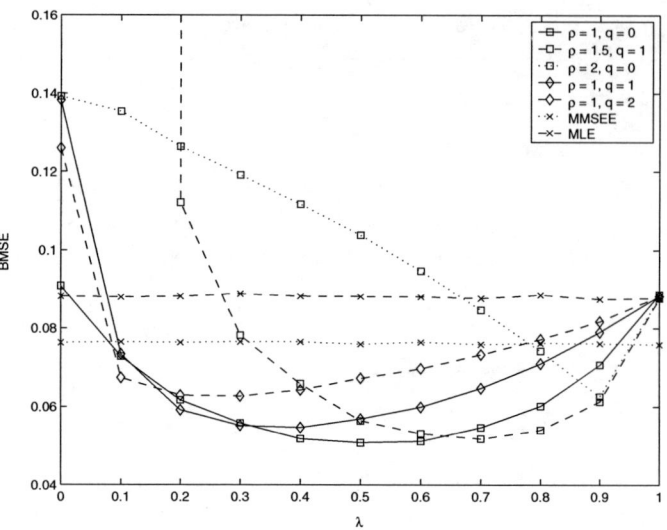

Fig. 1. BMSE over λ for different parameters.

V. SIMULATION RESULT

In our experiments, the OFDM system has $N = 512$ subcarriers of which $M = M_1 + M_2 + 1 = 433$ subcarriers are set for data and pilots. The number of pilots for channel estimation is 20. The signal-to-noise ratio (SNR) is defined as $SNR = -10 \log \sigma_n^2$, where σ_n^2 is the variance of the zeros mean white Gaussian channel noise. The multipath channel has length of $L = 16$ and its coefficients are independent complex Gaussian random variables with the variance defined by an exponential power profile, i.e. $\sigma_k^2 = e^{-k/10}$ for $k = 0, 1, \ldots, L-1$.

We use block mean square error to measure the error performance of estimator over Q OFDM blocks. It is defined as

$$BMSE = E\{\frac{1}{QM}\sum_{l=0}^{Q-1}[\sum_{m=0}^{M_1}(|\hat{\mathbf{H}}^{(l)}(m)| - |\mathbf{H}^{(l)}(m)|)^2$$
$$+ \sum_{m=N-M_2}^{N-1}(|\hat{\mathbf{H}}^{(l)}(m)| - |\mathbf{H}^{(l)}(m)|)^2]\} \quad (17)$$

We used Q=80 for evaluation and averaged the BMSE over 50 different noise samples.

First we compare the BMSE of the semi-blind method versus the weighting parameter λ for different ρ and q in Fig.1. The semi-blind method with $\rho = 2$ and $q = 0$ is the same method as described in [6]. The results show that the semi-blind method with $\rho = 1$ and $q = 0$ gives the minimum BMSE of 0.0508 at $\lambda = 0.5$. In comparing with that of $\rho = 2$ and $q = 0$, the BMSE of $\rho = 1$ and $q = 0$ is improved by 23%. The optimum results of different parameter sets are summarized in the following table.

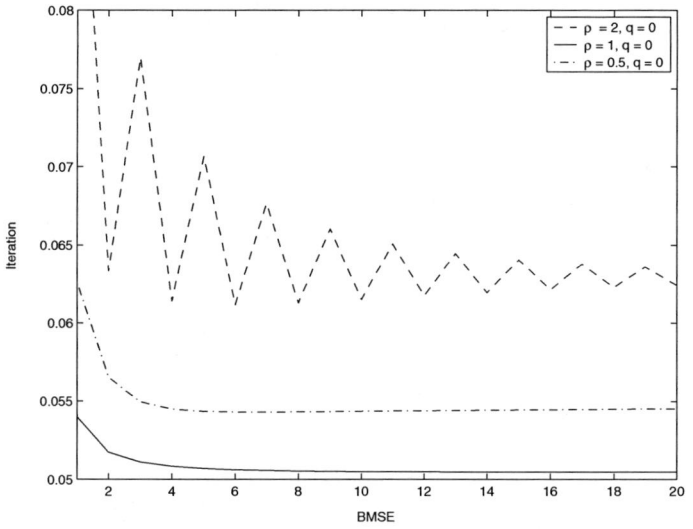

Fig. 2. Convergence of the algorithm for different ρ and q=0.

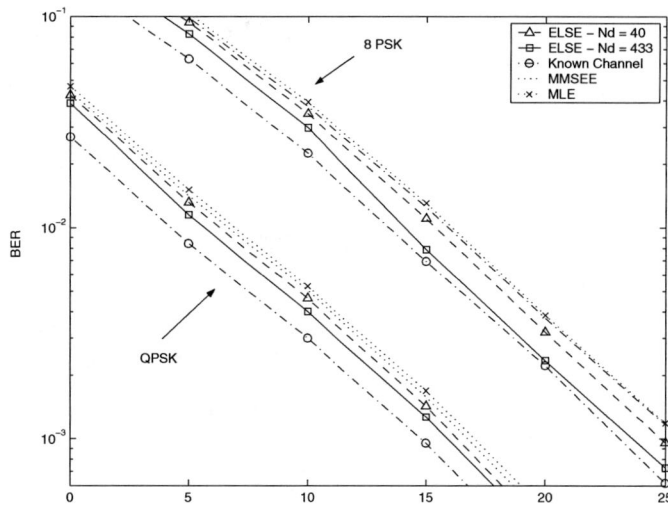

Fig. 3. BER versus SNR for different channel estimators.

ρ	q	optimum λ	optimum BMSE
1	0	0.5	0.0508
1.5	1	0.7	0.0518
2	0	0.9	0.0626
1	1	0.3	0.0551
1	2	0.3	0.0627

We plot the BMSE of the semi-blind method versus iteration for $\rho = 0.5, 1, 2$ and $q = 0$ in Fig.2. It is found that the semi-blind algorithm with $\rho = 0.5$ and the method with $\rho = 1$ take about 6 iterations to converge, while the method with $\rho = 2$ (the same method as [6]) oscillates and takes many iterations to converge.

The bit error rate (BER) of the semi-blind method with $\rho = 1$ and $q = 0$ versus SNR for $N_d = 40, 433$ for QPSK and 8PSK and plotted in Fig.V. The BERs of known channel impulse response for QPSK and 8PSK, maximum likelihood estimator (MLE) [1] and minimum mean square error estimator (MMSEE) [2] and also plotted in the figure for comparison. The results show that the BERs of MLE and MMSEE are close to each other even though their BMSEs are different. However, the semi-blind method provides better BER than MLE and MMSEE. For QPSK, the semi-blind method of $N_d = 40$ and $N_d = 433$ respectively gives 0.5dB and 1dB gain over MLE and MMSEE. For 8PSK case, the performance gains are respectively 0.75dB and 2dB for $N_d = 40$ and $N_d = 433$. For high SNR, it is found that the semi-blind method for $N_d = 433$ gives a performance close to that of known impulse response case.

VI. CONCLUSION

In this paper, a new single frame-based semi-blind channel estimation method has been presented. For this method, the generalized magnitude error term for data signals incorporated with pilot signals is proved to yield a channel estimate giving better BMSE than MLE and MMSEE. The semi-blind method with $\rho = 1$ and $q = 0$ is shown to give good performance in regard to BMSE and BER and also have an efficient implementation. Unlike most of subspace-based semi-blind methods that needs many OFDM blocks for channel estimation, the new method just takes one single OFDM block for carrying out estimation, that enables it to perform in fast fading channel.

Acknowledgment: The work described in this paper was fully supported by a grant from CityU (Project No. 7001421).

REFERENCES

[1] R. Negi and J. Cioffi, "Pilot Tone Selection for Channel Estimation in a Mobile OFDM System," *IEEE Trans. Consum. Electron.*, vol. 44, pp. 1122-1128, Aug. 1998.
[2] Michele Morelli and Umberto Mengali, "A Comparision of Pilot-Aided Channel Estimation Methods for OFDM Systems," *IEEE Trans. Signal Processing*, vol. 49, pp. 3065-3073, Dec. 2001.
[3] Muquet, B.; de Courville, M.; Duhamel, P. "Subspace-based blind and semi-blind channel estimation for OFDM systems," *IEEE Trans. Signal Processing*, Vol. 50, pp. 1699 - 1712, July 2002.
[4] Chengyang Li; Roy, S. "Subspace-based blind channel estimation for OFDM by exploiting virtual carriers," *IEEE Trans. Wireless Commun.*, Vol. 2, pp. 141 - 150, Jan. 2003.
[5] Shengli Zhou; Giannakis, G.B. "Finite-alphabet based channel estimation for OFDM and related multicarrier systems," *IEEE Trans. Commun*, Vol. 49, pp.1402 - 1414, Aug. 2001.
[6] Thomas, T.A.; Vook, F.W.; Baum, K.L. "Semi-blind channel identification in OFDM ," *Proc. 55th IEEE Vehicular Technology Conference*, May 2002, Vol. 4, pp. 1747 - 1750.
[7] S. Haykin, Adaptive Filter Theory, 3rd edition, Prentice Hall, 1996.

Jitter Limitations on Multi-Carrier Modulation

J.H.R. Schrader, E.A.M. Klumperink, B. Nauta
IC Design Group, MESA+ Research Inst., Univ. of Twente
Enschede, The Netherlands
E-mail: j.h.r.schrader@utwente.nl

J.L. Visschers
NIKHEF
Amsterdam, The Netherlands

Abstract—A feasibility study is made of an OFDM system based on analog multipliers and integrate-and-dump blocks, targeted at Gb/s copper interconnects. The effective amplitude variation of the integrator output caused by jitter is explained in an intuitive way by introducing correlation plots. For a given rms jitter and error rate, high frequency carriers allow for less modulation depth than low frequency carriers. A jitter limit on the total system bit rate is calculated, which is a function of rms jitter, bandwidth, and specified system symbol error rate. It is concluded that, because of the high sensitivity to timing errors inherent to OFDM, traditional PAM systems with equal bandwidth and error rate are more feasible.

I. INTRODUCTION

There is a continuous demand for higher bit rate in short-range wireline and PCB communication. The traditional way of transmitting data over short length copper wires is to use Pulse Amplitude Modulation (PAM), e.g. [1]. However, the spectral efficiency (bps/Hz) of such systems is much lower than that of Digital Subscriber Line (DSL) modem techniques. Furthermore, these techniques might be interesting when the channel transfer contains spectral nulls (PCB tracks and connectors). We investigate the possibility of extending the bandwidth of Orthogonal Frequency Division Multiplexing (OFDM) techniques to the GHz order. In common Discrete Multi-Tone (DMT) implementations the ADCs and DACs and the complex digital processing put limits on the maximum bandwidth. A possible way to overcome the bandwidth limitation is to use analog multipliers and integrate-and-dump blocks. We will study the system shown in fig. 1.

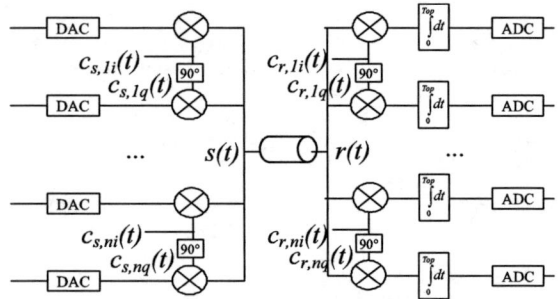

Fig. 1. Multi-carrier system using analog correlation

Parallelization is used for the AD/DA-converters and integrate-and-dump blocks, which relaxes bandwidth requirements. As with any OFDM system, this system transmits and receives symbols that contain multiple tones (where 'tone' is defined as a single carrier frequency). Both the in-phase and quadrature component of a tone are modulated with data. At the transmitter side the in-phase and quadrature carrier signals $c_{s,xi}(t)$ and $c_{s,xq}(t)$ ($x=\{1,n\}$) are multiplied by the data and the result added in $s(t)$. At the receiver, correlation is implemented with a multiplier and an integrate-and-dump block. The signal $r(t)$ is demodulated using locally generated carriers $c_{r,xq}(t)$ and $c_{r,xi}(t)$ ($x=\{1,n\}$). The system choice may seem arbitrary but is well suited for analysis of jitter effects.

Jitter is expected to have a large impact on the system. In this paper the jitter limitation on the system bit rate will be calculated for a certain specified symbol error rate, and compared to a PAM system. Jitter analyses have been performed in e.g. [2], which is a thorough mathematical analysis of jitter effects on DMT systems. However, [2] is based on a standard DMT system and furthermore we feel that the purely mathematical approach lacks intuitive insight to help designers analyze and improve their systems. The correlation plot based analysis in this paper presents an intuitive way of understanding the mechanism of effective amplitude variation of the integrator output caused by jitter. It serves to improve understanding of jitter impact on system bit rate limits and to draw conclusions about the feasibility of such a system.

II. JITTER ANALYSIS

A. Orthogonality loss caused by jitter

The transmitted data is modulated on several orthogonal carriers. In order to avoid interference these have to comply with the orthogonality constraint, which is defined as

$$\int_0^{Top} c_n(t) \cdot c_m(t) dt = 0 \quad n \neq m. \quad (1)$$

where $c_{n,m}(t)$ are carriers and T_{op} is the length of the receiver integration period ('orthogonality period'). Candidates for tone frequencies f_c are harmonic frequencies

n/T_{op}. Integration over exactly T_{op} delivers perfectly orthogonal carriers.

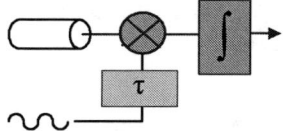

Fig. 2. Influence of delay τ on correlator receiver output

The analysis goal is to determine the change in the integrator output as a function of variations in τ (time shift between transmitter and receiver), as illustrated in fig. 2. Time offsets over the symbol, caused by jitter, lead to imperfect separation of the in-phase and quadrature component of the carrier. We will analyze crosstalk between the in-phase and quadrature components of a carrier at a given frequency.

B. Definitions

The receiver generated in-phase carrier $c_{r,i}(t)$ is defined as

$$c_{r,i}(t) = A_r \sin(2\pi f_c t). \quad (2)$$

The transmitted symbols $s_i(t)$ and $s_q(t)$ (resp. in-phase and quadrature component) are defined as

$$s_i(t) = \{A_i \sin(2\pi f_c t), \quad t = \{-T_{gt}/2, T_{op} + T_{gt}/2\}\} \quad (3)$$

$$s_q(t) = \{A_q \cos(2\pi f_c t), \quad t = \{-T_{gt}/2, T_{op} + T_{gt}/2\}\} \quad (4)$$

where A_i and A_q are chosen from interval $\{-A_{max}, A_{max}\}$. A guard time T_{gt} is assumed to be included in the symbol to improve robustness against symbol transition effects [3]. The following analysis is valid for $\tau = \{-T_{gt}/2, T_{gt}/2\}$.

C. Auto- and Cross-correlations

The receiver integrates over the interval $t = \{0, T_{op}\}$. We calculate the (normalized) autocorrelation $z_i(\tau)$ between $c_{r,i}(t)$ and $s_i(t)$:

$$z_i(\tau) = \frac{2}{T_{op}} \int_0^{T_{op}} s_i(t-\tau) \cdot c_{r,i}(t) dt = A_i A_r \cos(2\pi f_c \tau), \quad (5)$$

and the crosscorrelation $z_q(\tau)$ between $c_{r,i}(t)$ and $s_q(t)$:

$$z_q(\tau) = \frac{2}{T_{op}} \int_0^{T_{op}} s_q(t-\tau) \cdot c_{r,i}(t) dt = A_q A_r \sin(2\pi f_c \tau). \quad (6)$$

Mutatis mutandis these calculations (for the in-phase carrier) deliver similar results for the quadrature carrier $c_{r,q}(t)$. In fig. 3, example correlations are shown ($A_i = A_q = A_{max}$). The units on the x-axis are τ/T_c, where $T_c = 1/f_c$. It can be noticed that there are only few points where the auto-correlation is exactly maximum and the cross-correlation is exactly zero, and vice versa. The maximum auto-correlation point is (by definition) found at $\tau = 0$ (optimum match between transmitter and receiver).

However, at $\tau = T_c/4$ the *cross*-correlation is maximum. It can be seen that the time shift between the local carrier and the received signal is very critical for optimum reception.

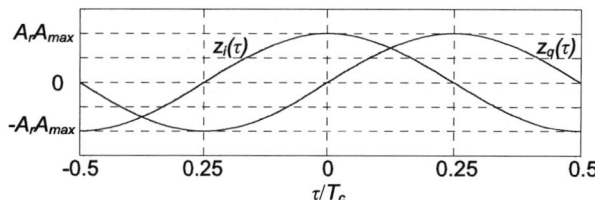

Fig. 3. Autocorrelation $z_i(\tau)$ and crosscorrelation $z_q(\tau)$ for $A_i = A_q = A_{max}$

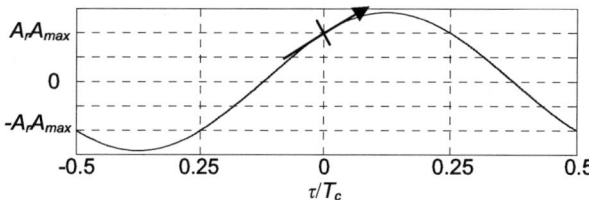

Fig. 4. Correlation $z_{iq}(\tau)$ with summed signals, and derivative at optimum detection point (arrow) for $A_i = A_q = A_{max}$

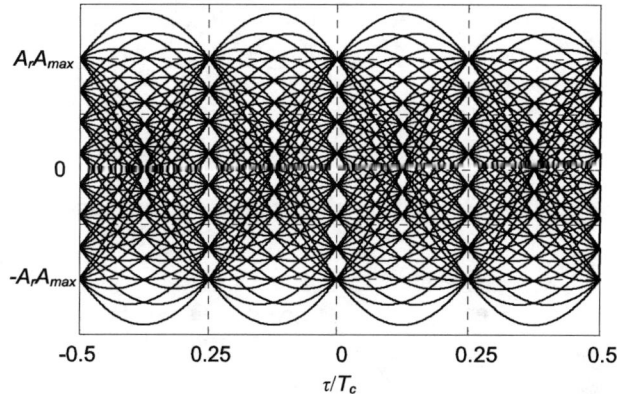

Fig. 5. Correlations $z_{iq}(\tau)$ for all possible combinations of A_i and A_q

Focusing on detection of the in-phase component, we calculate the correlation $z_{iq}(\tau)$ of $c_{r,i}(t)$ with the summed transmitted signal as

$$z_{iq}(\tau) = \frac{2}{T_{op}} \int_0^{T_{op}} (s_i(t-\tau) + s_q(t-\tau)) \cdot c_{r,i}(t) dt$$
$$= A_r (A_i \cos(2\pi f_c \tau) + A_q \sin(2\pi f_c \tau)). \quad (7)$$

This is shown in fig. 4 (for $A_i = A_q = A_{max}$) together with the derivative at the optimum detection point.

Next, we calculate $z_{iq}(\tau)$ for all possible combinations of $A_i = \{-A_{max}, A_{max}\}$ and $A_q = \{-A_{max}, A_{max}\}$. Plotting all these correlations on top of each other looks a bit like a normal eye diagram. In fig. 5 an example is shown where 3 bits are modulated on both the in-phase and quadrature carrier, resulting in 8 possible levels.

Fig. 5 resembles an eye diagram but it is not the same. Like an eye diagram, these plots can actually be used, in a very similar way, to find the optimum detection moment and to analyze the effect of amplitude and time errors on symbol error rate. However, note that the x-axis is not time but relative *time shift* between $r(t)$ and $c_{r,i}(t)$, in units of τ/T_c. The figure shows the effect of a time shift (away from the optimum detection point) on the integrator output.

The impact of a time shift depends on the steepness $y(\tau)$ of the lines around the optimum detection point. We need to calculate this steepness to be able to translate jitter into effective amplitude variation. The steepness is calculated as:

$$y(\tau) = \frac{d}{dt}\left(z_{iq}(\tau)\right) = 2\pi f_c A_r \left(A_q \cos(2\pi f_c \tau) - A_i \sin(2\pi f_c \tau)\right) \quad (8)$$

For $\tau=0$, $y(\tau)$ is completely determined by $A_r A_q$, so it can take on l discrete values, where l is the number of levels used in modulation. To be able to translate from time jitter into worst-case amplitude deviation, we calculate the maximum absolute steepness of these lines y_{max} as

$$y_{\max} = \max(|y(\tau)|_{\tau=0}|) = 2\pi f_c A_r A_{\max}. \quad (9)$$

(Using [4] it can be shown that the jitter accumulation *during* the integration period is negligible for $\kappa\sqrt{T_{op}} \ll T_c$, where κ is an oscillator figure of merit.)

D. Probability of symbol error

In this section, a 'per-tone symbol error rate' P_e is calculated. A 'tone error' occurs when either the in-phase component or the quadrature component of that specific tone is detected incorrectly. The methodology is as follows:

- calculate effective standard deviation of amplitude of integrator output (σ_{Aeq}) as a function of jitter standard deviation (σ_t),
- calculate SNR per symbol from σ_{Aeq} and distance between levels,
- calculate P_e using cumulative normal distribution function.

The total system error rate will be limited by the worst performing tone, so the system should be designed to have an equal error rate for each tone. It is assumed that the jitter coming from the PLL has a Gaussian time distribution with an rms standard deviation of σ_t. Its size is determined by the PLL noise and loop bandwidth. For a well-designed LC-based PLL in the GHz range, currently the rms jitter can be as low as σ_t=1ps.

The receiver will compare the integrator output to a number of $(l-1)$ thresholds that are placed in between the amplitude levels. In order to calculate the error rate, we need to calculate the probability that the received signal crosses the threshold between two amplitude levels. In fig. 6, this is illustrated; t_x are the thresholds and s_x the signal points.

The worst-case effective amplitude standard deviation σ_{Aeq} as a function of the jitter standard deviation is

$$\sigma_{A_{eq}} = y_{\max}\sigma_t = 2\pi f_c A_r A_{\max}\sigma_t. \quad (10)$$

We can express the distance between levels $2d$ as a function of $A_r A_{max}$ as

$$2d = \frac{2A_r A_{\max}}{l-1}. \quad (11)$$

The error rate is a function of d/σ_{Aeq}. Expressing d/σ_{Aeq} in terms of f_c, σ_t and l gives

$$\frac{d}{\sigma_{A_{eq}}} = \frac{1}{2\pi\sigma_t f_c(l-1)}. \quad (12)$$

Next we calculate P_i, the error rate for the in-phase component, taking into account a factor $(l-1)/l$ because the uppermost and lowermost levels have only one neighbor:

$$P_i = \frac{(l-1)}{l}\frac{1}{\sqrt{2\pi}}\int_{\frac{d}{\sigma_A}}^{\infty} e^{\left(-\frac{y^2}{2}\right)} dy = \frac{(l-1)}{l}Q\left(\frac{d}{\sigma_{A_{eq}}}\right) \quad (13)$$

where $Q(x)$ is the cumulative normal distribution function. Substituting (12) into (13) leads to

$$P_i = \frac{(l-1)}{l}Q\left(\frac{1}{2\pi\sigma_t f_c(l-1)}\right) \quad (14)$$

The probability of error P_q for the quadrature component is equal to P_i. The (total) probability of a tone error P_e is

$$P_e = 1 - (1 - P_i)^2. \quad (15)$$

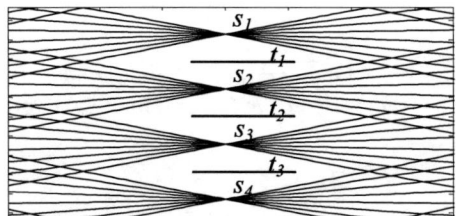

Fig. 6. Amplitude levels and thresholds

Now we can plot P_e (at a given σ_t) as a function of f_c for a number of different modulation depths n_b (= $^2\log(l)$, where n_b is expressed in bits). This is shown in fig. 7 for σ_t=1ps. If necessary, we can convert from symbol to bit errors, e.g. assuming the use of Gray code, so that one symbol error will imply one bit error.

The error rate caused by jitter is a function of modulation depth n_b and carrier frequency f_c. The number of bits that can be modulated onto a carrier (for a given error rate) is limited by jitter, with higher frequency carriers being able to carry fewer bits. In an optimum multi-carrier system, higher frequency carriers should have fewer constellation points to achieve the same error rate. This corresponds with results in [2]. In addition, statistical simulations have been done which confirm SNR accuracy of the analysis to within 4dB margin.

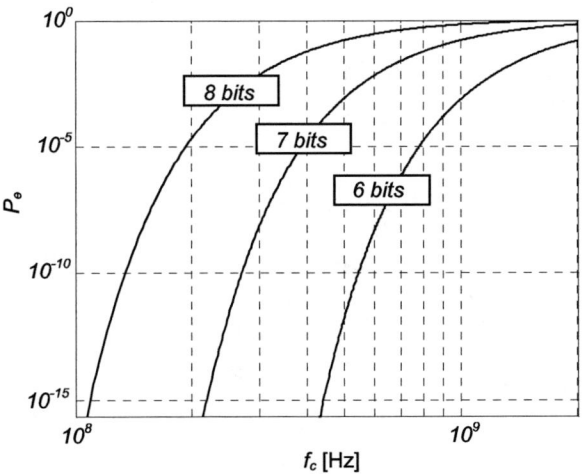

Fig. 7. Probability of error vs. carrier frequency for σ_t=1ps

E. Bit rate limits

The [max. number of bits that can be modulated] $n_{b,max}$ has been calculated as a function of σ_t, f_c and P_e, using a numeric solver on (14). Fig. 8 shows the outcome for three different values of σ_t = {0.1ps, 1ps, 10ps}, which corresponds to {excellent, good, fair}, for P_e=1·10^{-12}.

It is interesting to see what the jitter limited maximum bit rate of such a multi-carrier system could be. This is then compared to a PAM system with an equal bandwidth and error rate. In [1] a PAM system is described that can achieve a bit rate of ~7Gb/s for an error probability of ~1·10^{-12}, with a bandwidth of 2GHz and an rms jitter of 4ps. In our analysis, the upper bound on the multi-carrier system's bit rate is found by integration of $n_{b,max}$ over a 2GHz bandwidth and multiplying by two (because both in-phase and quadrature component are used). This delivers a bit rate limit of 14 Gb/s (for σ_t=4ps and P_e=1·10^{-12}).

The bit rate limit calculated for the multi-carrier system is higher than for the PAM system in [1], but it will have to be corrected downwards for practical implementations due to insertion of a guard time in the symbol. Furthermore, in an implementation with simple switching mixers, a sine wave on the local oscillator port will generate a square wave on the output. This will produce harmonics that fall onto other tone frequencies, creating unusable areas in the spectrum. Next, in case of frequency- or duty-cycle mismatch, inter-carrier interference with carriers at other frequencies will arise. It can be proven that duty-cycle deviations of >5% in the receiver generated carrier already cause P_e>1·10^{-6}. From the above it is clear that there is no spectacular improvement in bit rate to be expected from a GHz multi-carrier system at current state-of-the-art rms jitter figures. Unless properties like robustness against spectral nulls are an important issue, the timing sensitivities make such a system unattractive.

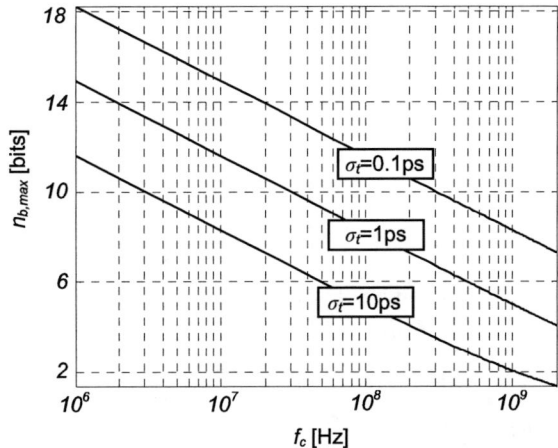

Fig. 8. Max. no. of bits that can be modulated vs. car. freq. for P_e=1·10^{-12}

III. CONCLUSION

The feasibility of a Gb/s analog OFDM system is analyzed. Correlation plots are introduced to analyze the impact of jitter (coming e.g. from the PLL) on such a multi-carrier system. Jitter causes crosstalk between the in-phase and quadrature channels. The maximum bit rate which can be achieved, given a certain specification for the symbol error rate, is limited by this jitter. Assuming a certain rms jitter specification, where the jitter has a Gaussian distribution, it is concluded that low frequency tones can carry more bits than higher frequency tones for the same error rate. Using correlation plots this can be understood in an intuitive way. A jitter limit on the system bit rate is calculated by integrating the area under the plot of [maximum number of bits that can be modulated] versus carrier frequency. The expectations of high spectral efficiencies will not be fulfilled because of the system's high sensitivity to timing errors. It seems that traditional Pulse Amplitude Modulation (PAM) systems with a comparable bandwidth still have the better cards.

ACKNOWLEDGEMENTS

The authors would like to thank Wim van Etten for helpful discussions.

REFERENCES

[1] R. Farjad-Rad, C. K. Yang, M. Horowitz, and T. H. Lee, "A 0.3-μm CMOS 8-Gb/s 4-PAM serial link transceiver," *IEEE J. Solid-State Circuits*, vol. 35, pp. 757–764, May 2000.

[2] T. N. Zogakis and J. M. Cioffi, "The effect of timing jitter on the performance of a discrete multitone system," *IEEE Trans. Commun.*, vol. 44, pp. 799–808, July 1996.

[3] J. A. C. Bingham, "Multicarrier modulation for data transmission: An idea whose time has come," *IEEE Commun. Mag.*, vol. 28, pp. 5–14, May 1990.

[4] J. McNeill, "Jitter in ring oscillators," *IEEE J. Solid-State Circuits*, vol. 32, pp. 870–879, June 1997.

Estimating the fading coefficient in mobile OFDM systems using state-space model

Mihai Enescu and Visa Koivunen

Signal Processing Laboratory, Helsinki University of Technology
P.O. Box 3000, FIN-02015 HUT, Finland

Abstract—In this paper the problem of estimating the fading coefficient in OFDM systems is addressed. Our approach is based on building a state-space model for the OFDM transmission that allows the estimation of the fading coefficient from the received pilot data. Kalman filter is then applied to estimate and track the time-varying channels in frequency domain. Our simulations show that reliable channel estimation can be performed under realistic conditions.

I. INTRODUCTION

Radio spectrum is a scarce resource in wireless communication. High spectral efficiency is therefore a major goal of future mobile wireless communications systems. Multicarrier systems, including orthogonal frequency division multiplexing (OFDM), play an important role in future beyond 3G systems. A key benefit of OFDM is its ability to turn a frequency selective channel into a set of parallel narrowband channels, which leads to very simple equalization since the transmission becomes free of Intersymbol Interference (ISI).

Accurate channel estimation is required in achieving the benefits promised by OFDM systems. It has been investigated in Single-Input Single-Output (SISO) systems [3], [2], [6]. Due to the high mobility that is considered for future systems [8], channel tracking over time is required. Hence, dynamic models that describe the channel time evolution are needed. Recursive estimators, such as Kalman Filter, that are suitable for tracking time-varying channels have been used in OFDM, to estimate the wireless channel in frequency domain [3], [4] or in time domain [13].

Using pilot tones in OFDM is a widely used approach to achieve reliable performance at high speeds. The optimal placement of pilots on the time-frequency grid in the case of block fading channels has been found for maximizing the channel capacity [1], or to minimize the MSE of least-squares channel estimator in [12]. The selection of pilots is optimized with respect to the MSE of the channel estimator for time-invariant channels in [11].

The main contribution of this paper is the estimation of the fading correlation coefficient in mobile OFDM systems. The fading coefficient is characterizing the degree of time variation and can be related to the Doppler frequency. Hence, for various speeds of the mobile station we have different values of the fading coefficient. When using a state-space model in channel tracking it is typical that the fading correlation coefficient is assumed to be known. The importance of this parameter has been highlighted in [5]. It has been recognized that in many practical implementations the erroneous values of this parameter degrade significantly the performance of the estimator [14]. Using an accurate fading coefficient improves the tracking performance of the time varying channel.

When modeling the wireless channel as an autoregressive (AR) process, it is worth mentioning that we would expect that the higher the AR model order, the more precise the model would be. However, this can lead to over fitting and ultimately result in degraded performance and increased complexity. Information-theoretic results [16] have demonstrated that implementing a first-order Markovian model offers sufficient accuracy to model the Rayleigh narrowband time-varying channel. Hence, throughout this paper we will consider an AR(1) model to describe the temporal correlation [9]. Estimating AR parameters is a well researched topic in signal processing [10]. Typical techniques process directly the AR signal [10], they are based on more simple state-space models [7] or use higher order statistics [15]. Our approach is tailored on the OFDM model and hence leads to a low complexity estimation technique for the tracking of time-varying fading coefficients.

The rest of the paper is organized as follows. In the next section, we briefly present the OFDM system model. In Section 3 we introduce the time and frequency domain state space representation, while in section 4 we present the fading coefficient estimation algorithm. In Section 5 simulation results are presented. Finally, Section 6 concludes the paper.

II. OFDM SYSTEM

The OFDM transmission model used in this paper is presented in Figure 1. The OFDM modulated block at time k

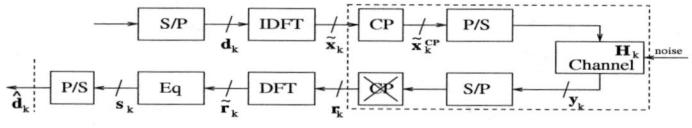

Fig. 1. OFDM transmission.

after cyclic prefix insertion is $\tilde{\mathbf{x}}_k = \mathbf{T}_{CP}\mathbf{F}_N\mathbf{d}_k$ where $\mathbf{T}_{CP}$ is the cyclic prefix insertion matrix of dimension $(L+N) \times N$, $\mathbf{F}_N$ is the $N \times N$ inverse discrete Fourier transform (IDFT) matrix, and N is the total number of subcarriers [13]. The vector $\mathbf{d}_k$ is the $N \times 1$ symbol vector and L is the length of the cyclic prefix. As a result, the total length of each OFDM symbol is $P = N + L$.

The received $N \times 1$ signal block after cyclic prefix insertion followed by transmission on the wireless channel and cyclic prefix removal is then expressed as:

$$\mathbf{r}_k = \mathbf{R}_{CP}\mathbf{H}_k\mathbf{T}_{CP}\mathbf{F}_N \mathbf{d}_k + \mathbf{w}_k, \quad (1)$$

where $\mathbf{R}_{CP}$ performs the cyclic prefix removal, $\mathbf{H}_k$ is the Toeplitz channel convolution matrix of dimension $(N+L) \times (N+L)$ and $\mathbf{w}_k$ is the noise term, assumed to be circular white complex Gaussian. Due to cyclic prefix insertion and removal operations, the $N \times N$ channel matrix $\widetilde{\mathbf{H}}_k = \mathbf{R}_{CP}\mathbf{H}_k\mathbf{T}_{CP}$ becomes circulant, with the (k,l)th entry given by $h_{(k-l) \mod N}$.

In our state-space model, the state vector contains the channel taps to be estimated. We start from equation (1) which can be rewritten as:

$$\mathbf{r}_k = \widetilde{\mathbf{X}}_k \mathbf{h}_k + \mathbf{w}_k, \quad (2)$$

where $\widetilde{\mathbf{X}}_k$ is a $N \times L_h$ circulant matrix formed with the modulated block $\tilde{\mathbf{x}}_k$, and $\mathbf{h}_k = [h_0(k), h_1(k), \ldots, h_{L_h-1}(k)]^T$, L_h being the channel length. In this paper the channel length is considered equal with the length of the CP ($L_h = L$). The time evolution of the channel taps can be described as:

$$\mathbf{h}_k = \mathbf{A}\, \mathbf{h}_{k-1} + \mathbf{v}_k, \quad (3)$$

where $\mathbf{h}_k$ is the $L_h \times 1$ state vector containing the channel taps, $\mathbf{A}$ is the state transition matrix and $\mathbf{v}_k$ is the white Gaussian state noise. Equations (2)-(3) form the time domain state-space model that will be used later in the paper.

After performing the discrete Fourier transform (DFT) we get:

$$\tilde{\mathbf{r}}_k = \mathbf{F}_N^H \widetilde{\mathbf{H}}_k \mathbf{F}_N \mathbf{d}_k + \tilde{\mathbf{w}}_k, \quad (4)$$

where $\tilde{\mathbf{w}}_k = \mathbf{F}_N^H \mathbf{w}_k$ and $\mathbf{F}_N^H$ is unitary DFT matrix.

Circulant matrices implement circular convolutions, they are diagonalized by DFT and IDFT operations, and thus equation (4) can be rewritten as:

$$\tilde{\mathbf{r}}_k = \bar{\mathbf{H}}_k \mathbf{d}_k + \tilde{\mathbf{w}}_k, \quad (5)$$

where the diagonal matrix $\bar{\mathbf{H}}_k = \mathbf{F}_N^H \widetilde{\mathbf{H}}_k \mathbf{F}_N$ contains the frequency response of the channel, evaluated at the subcarrier frequencies. Hence, the initial wideband frequency selective channel has been turned into a set of N narrowband flat fading channels.

Finally, single-tap MMSE equalization can be performed in the frequency domain as follows:

$$\mathbf{s}_k = \left[\bar{\mathbf{H}}_k + \sigma_{\tilde{n}}^2 \mathbf{I}\right]^{-1} \tilde{\mathbf{r}}_k, \quad (6)$$

where $\sigma_{\tilde{w}}^2$ is the variance of the noise. Then, decisions are carried out on $\mathbf{s}_k$ in order to obtain the symbol estimate $\hat{\mathbf{d}}_k$.

So far we have seen the OFDM transmission model. In the next section we will present the frequency domain state-space model. Both channel estimation and equalization, as well as the parameter estimation, are performed in frequency domain.

III. Time and Frequency domain OFDM state-space models

We start from the time domain (TD) independent state-space model used in OFDM. Considering the fact that all the taps have the same fading coefficients, i.e. $\mathbf{A}_d = a\mathbf{I}$, the following TD state-space model can be written:

$$\mathbf{h}_{k+1} = a\mathbf{h}_k + \mathbf{v}_k \quad (7)$$
$$\mathbf{r}_k = \widetilde{\mathbf{X}}\mathbf{h}_k + \mathbf{w}_k \quad (8)$$
$$\mathbf{r}_k = \widetilde{\mathbf{H}}_k \mathbf{x}_k + \mathbf{w}_k. \quad (9)$$

To obtain the frequency domain (FD) representation we apply a Fourier transform on the previous equations. The measurement equation is obtained as follows:

$$\tilde{\mathbf{r}}_k = \mathbf{F}^H \widetilde{\mathbf{H}}_k \mathbf{F} \mathbf{d}_k + \mathbf{F}^H \mathbf{w}_k \quad (10)$$
$$= \bar{\mathbf{H}}_k \mathbf{d}_k + \tilde{\mathbf{w}}_k. \quad (11)$$

On the other hand we have:

$$\tilde{\mathbf{r}}_k = \mathbf{D}_k \widetilde{\mathbf{h}}_k + \mathbf{F}^H \mathbf{w}_k \quad (12)$$

where $\mathbf{D}_k = \sqrt{N} diag\{d_1, \ldots, d_N\}$ contains the transmitted data on the N subcarriers at time index k and frequency response $\widetilde{\mathbf{h}}_k = \mathbf{F}_{tr} \mathbf{h}_k$, with $\widetilde{\mathbf{h}}_k$ of dimension $N \times 1$. $\mathbf{F}_{tr}$ is a $N \times L$ sub-matrix of the Fourier matrix, more precisely it contains the first L columns, and $\mathbf{h}_k$ is the channel in TD.

Now, we investigate the connection between the state equation in TD and the state equation in FD. We use the fact that $\widetilde{\mathbf{h}}_k = \mathbf{F}_{tr}\mathbf{h}_k$ and starting from the state equation (7) in TD we get:

$$\mathbf{F}_{tr}\mathbf{h}_{k+1} = \mathbf{F}_{tr} a\mathbf{h}_k + \mathbf{F}_{tr}\mathbf{v}_k \quad (13)$$

which leads to: $\widetilde{\mathbf{h}}_{k+1} = a\widetilde{\mathbf{h}}_k + \tilde{\mathbf{v}}_k$. To summarize, the FD state-space equations are:

$$\widetilde{\mathbf{h}}_{k+1} = a\widetilde{\mathbf{h}}_k + \tilde{\mathbf{v}}_k \quad (14)$$
$$\tilde{\mathbf{r}}_k = \bar{\mathbf{H}}_k \mathbf{d}_k + \tilde{\mathbf{w}}_k \quad (15)$$
$$\tilde{\mathbf{r}}_k = \mathbf{D}_k \widetilde{\mathbf{h}}_k + \tilde{\mathbf{w}}_k. \quad (16)$$

The fact that all channels are assumed to have the same fading correlation coefficient, a, decouples the state-space equations. This property allows one to select any of the model equations in order to estimate the state. In order to perform channel estimation using pilots [11], [5] one can do the following. Perform channel estimation using the model:

$$\widetilde{\mathbf{h}}_{k+1,p} = a\widetilde{\mathbf{h}}_{k,p} + \mathbf{F}_p \tilde{\mathbf{v}}_{k,p} \quad (17)$$
$$\tilde{\mathbf{r}}_{k,p} = \mathbf{D}_{k,p} \widetilde{\mathbf{h}}_{k,p} + \tilde{\mathbf{w}}_{k,p}, \quad (18)$$

where the index p represents the indices where pilot symbols are transmitted. For this purpose we have to select $p \geq L+1$. Once we have the p channel estimates we have to build the LN frequency response that is needed in equalization. This is done as:

$$\widetilde{\mathbf{h}}_{k+1} = \mathbf{F}_{N-p} \mathbf{F}_p^+ \widetilde{\mathbf{h}}_{k+1}, \quad (19)$$

where, having the Fourier matrix $\mathbf{F}$ of dimension $N \times N$, $\mathbf{F}_{tr}$ is the truncated Fourier matrix of dimension $N \times L$, $\mathbf{F}_p$ contains the p rows of the truncated matrix and $\mathbf{F}_{N-p}$ the remaining $N-p$ rows, $\mathbf{F}_p^+$ is the pseudo inverse.

Our goal is to estimate the fading coefficient from the received data. A novel estimation method is presented in the next section. Since we have the diagonal symbol matrix and independent channels, channel estimation can be performed on assigned pilot channels. For this purpose we can use Kalman Filter.

IV. ESTIMATING THE FADING CORRELATION COEFFICIENT

Due to the decoupled nature of the state-space model, we need only a scalar state-space model in order to estimate the fading coefficient. The coherence time of the channel is at least double of the symbol duration. We considered the following pilot model in FD:

$$\widetilde{h}^i_{k+1,p} = a\widetilde{h}^i_{k,p} + \widetilde{v}^i_{k,p} \quad (20)$$
$$\widetilde{r}^i_{k,p} = d^i_{k,p}\widetilde{h}^i_{k,p} + \widetilde{w}^i_{k,p}, \quad (21)$$

where index i refers to the selected channel on which we perform the estimation. We investigate $E\left[\widetilde{r}^i_{k,p}\widetilde{r}^{i*}_{k-2,p}\right]$ and $E\left[\widetilde{r}^i_{k-1,p}\widetilde{r}^{i*}_{k-2,p}\right]$, where $E[\cdot]$ is the expectation operation and $*$ denotes the complex conjugate. Computing the first expectation using the previous state space model, we have:

$$\begin{aligned} E\left[\widetilde{r}^i_{k,p}\widetilde{r}^{i*}_{k-2,p}\right] &= E\left[\left(d^i_{k,p}\widetilde{h}^i_{k,p} + \widetilde{w}^i_{k,p}\right)\left(d^i_{k-2,p}\widetilde{h}^i_{k-2,p} + \widetilde{w}^i_{k-2,p}\right)^*\right] \\ &= d^i_{k,p} E\left[\widetilde{h}^i_{k,p}\widetilde{h}^{i*}_{k-2,p}\right] d^{i*}_{k-2,p}. \end{aligned} \quad (22)$$

The expectation $E\left[\widetilde{h}^i_{k,p}\widetilde{h}^{i*}_{k-2,p}\right]$ can be further written as:

$$\begin{aligned} E\left[\widetilde{h}^i_{k,p}\widetilde{h}^{i*}_{k-2,p}\right] &= E\left[\left(a\widetilde{h}^i_{k-1,p} + \widetilde{v}^i_{k-1,p}\right)\widetilde{h}^{i*}_{k-2,p}\right] \\ &= aE\left[\widetilde{h}^i_{k-1,p}\widetilde{h}^{i*}_{k-2,p}\right]. \end{aligned} \quad (23)$$

We obtain the expression:

$$E\left[\widetilde{r}^i_{k,p}\widetilde{r}^{i*}_{k-2,p}\right] = d^i_{k,p} a E\left[\widetilde{h}^i_{k-1,p}\widetilde{h}^{i*}_{k-2,p}\right] d^{i*}_{k-2,p}. \quad (24)$$

Considering now the expectation $E\left[\widetilde{r}^i_{k-1,p}\widetilde{r}^{i*}_{k-2,p}\right]$ and following the same steps as in the previous case we obtain:

$$E\left[\widetilde{r}^i_{k-1,p}\widetilde{r}^{i*}_{k-2,p}\right] = d^i_{k-1,p} E\left[\widetilde{h}^i_{k-1,p}\widetilde{h}^{i*}_{k-2,p}\right] d^{i*}_{k-2,p} \quad (25)$$

Combining equations (24) and (25) we get:

$$a = (d^i_{k,p})^{-1} \left(E\left[\widetilde{r}^i_{k,p}\widetilde{r}^{i*}_{k-2,p}\right]\right) \left(E\left[\widetilde{r}^i_{k-1,p}\widetilde{r}^{i*}_{k-2,p}\right]\right)^{-1} d^i_{k-1,p}.$$

The above equation can be further written as:

$$\hat{a} = (d^i_{k,p})^{-1} \left(\sum_{n=3}^{k} \widetilde{r}^i_{k,p}\widetilde{r}^{i*}_{k-2,p}\right) \left(\sum_{n=3}^{k} \widetilde{r}^i_{k-1,p}\widetilde{r}^{i*}_{k-2,p}\right)^{-1} d^i_{k-1,p}. \quad (26)$$

The recursive structure of (26) allows also the tracking of the correlation coefficient. However, an adaptive update with forgetting factor may be needed.

V. SIMULATIONS

In this section simulation results are presented. Our OFDM setup is the following: the carrier frequency is $f_0 = 2.4$ GHz and the number of subcarriers is set to $N = 128$. The available bandwidth is chosen equal to $B = 1$ MHz. The subcarrier symbol rate is of 7.8 KHz and the modulation employed is QPSK. The channel is Rayleigh fading with independent propagation paths. The Doppler spectrum is Jake's and the power loss and delay profiles are: $[0, -1, -3, -9]$ [dB] and $[0, 1, 2, 3]$ [μs] which correspond to an Urban type of scenario. The channel is considered to be quasi-stationary during the OFDM symbol. The number of pilots in FD is equal to $p = 5$ per OFDM symbol. This is the minimum required number, as $p \geq L + 1$, and occupies around 4% of the OFDM tones. For comparison purposes we perform channel estimation also in TD, where an equal number of pilots (as in FD) is sent. In TD when referring to a pilot we consider that the whole OFDM block is known. Our pilot approach in TD is just for illustration purposes, and it is by no means optimized.

The estimation of the fading coefficient, using equation (26), is shown in Figure 2. Different values for the fading coefficient have been estimated 0.98, 0.94 and 0.88 which correspond to different speeds of the mobile station that have been used in this simulation: 3, 80 and 160 km/h. The results have been averaged over 10 realizations.

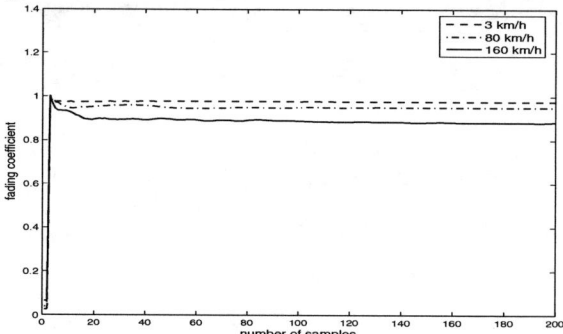

Fig. 2. Estimation of the fading coefficient (SNR = 15 dB).

Since the equalization stage operates in the frequency domain, accuracy in estimating frequency responses of the channels at the subcarrier frequencies needs to be investigated. Figure 3 and 4 respectively show amplitude and phase responses, for the true and estimated channel, at a given OFDM block time. In this case Kalman filter has been used to estimate the channel.

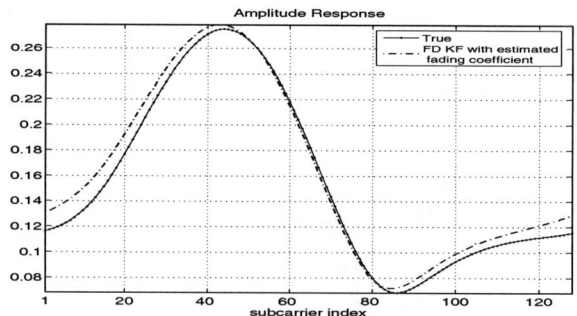

Fig. 3. Amplitude responses (SNR = 15 dB).

The performance criterion is the bit error rate as a function of noise variance, presented in Figure 6. A lower bound for the performance of the tracking algorithm is given by using the ideal channel state information (CSI), i.e. perfectly known channel at the receiver side. Since in the algorithm loop the decoding stage is using the channel estimate obtained at previous step, we have performed also another decoding after the channel estimation stage. This is depicted in the figure as '2 Dec'. As shown by simulation curves, tracking in FD with

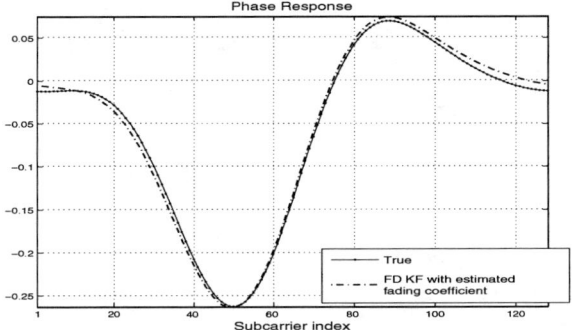

Fig. 4. Phase responses (SNR = 15 dB).

second decoding provides us with results close to the ones obtained with known channel.

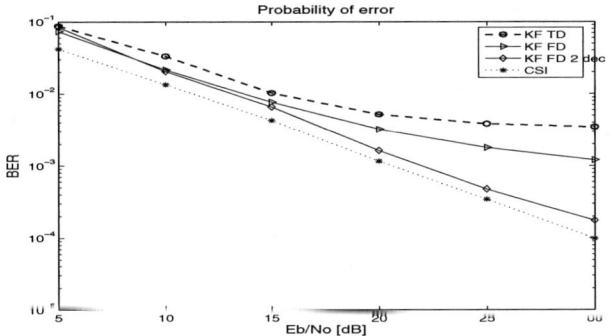

Fig. 5. Bit error rate performance (over 5000 OFDM blocks) for a mobile station speed of 40 km/h, 4% pilot.

Finally, simulations for variable terminal velocities, ranging from 3 up to 200 km/h, have been performed at 20 dB SNR. We have transmitted 1000 OFDM symbols and the results are averaged over 50 realizations. A 4% pilot rate has been used, for both TD and FD channel estimation.

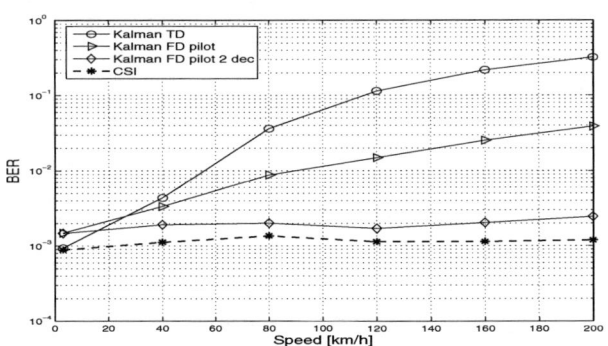

Fig. 6. Bit error rate performance vs. speed at 20 dB SNR(over 5000 OFDM blocks), 4% pilot.

VI. CONCLUSION

In this paper a method for estimating the fading coefficient in state-space OFDM has been proposed. By building a state-space model in frequency domain, using pilot sequences we have been able to estimate the fading coefficient from the received data. This has been incorporated in a Kalman filter estimator in order to get the channel frequency response. The reliable estimation led to performant bit error rates of our receiver even at high speeds.

REFERENCES

[1] S. Adireddy, L. Tong, and H. Viswanahtan, "Optimal placement of training for frequency-selective block-fading channels", IEEE Trans. on Information Theory, Vol. 48, no. 8, pp. 2338 - 2353, Aug. 2002.

[2] J.-J. van de Beek, O. Edfors, M. Sandell, S. K. Wilson, and P. O. Borjesson, "On channel estimation in OFDM systems," Proc. 45th IEEE Vehicular Technology Conf., pp. 815–819, 1995.

[3] S. B. Bulumulla, S. A. Kassam, and S. S. Venkatesh, "An adaptive diversity receiver for OFDM in fading channels," International Conference on Communications (ICC 1998), vol. 3, pp. 1325–1329, 1998.

[4] S. B. Bulumulla, S. A. Kassam, and S. S. Venkatesh, "A systematic approach to detecting OFDM signals in a fading channel," IEEE Transactions on Communications, vol. 48, no. 5, pp. 725–728, 2000.

[5] M. Dong, L. Tong, B. M. Sadler, "Optimal pilot placement for time-varying channels", IEEE Workshop on Signal Processing Advances in Wireless Communications, SPAWC 2003, pp. 219 - 223, 2003.

[6] O. Edfors, M. Sandell, J.-J. van de Beek, S. K. Wilson, and P. O. Borjesson, "OFDM channel estimation by singular value decomposition," IEEE Transactions on Communications, vol. 46, pp. 931–939, 1998.

[7] W. Gersch, "Estimation of the autoregressive parameters of a mixed autoregressive moving-average time series", IEEE Trans. on Automatic Control, Vol. 15, pp. 583 - 588, 1970.

[8] G. D. Golden, I. Sohn, H. Lee, J. Y. Ahn, S. Kapoor, "Channel Models and Performance Implications for OFDM-based MBWA", IEEE 802.20 Working Group, available at http://www.ieee802.org/20/.

[9] S. Haykin, K. Huber, Z. Chen, "Bayesian Sequential State Estimation for MIMO Wireless Communications", Proc. of the IEEE, Vol. 92, no. 3, pp. 439 - 454, 2004.

[10] P. Stoica, R. L. Moses, "Introduction to Spectral Analysis", Prentice Hall, 1997.

[11] R. Negi, J. Cioffi, "Pilot tone selection for channel estimation in a mobile OFDM system", IEEE Trans. on Consumer Electronics, vol. 44, no. 3, pp. 1122 - 1128, 1998.

[12] S. Ohno; G. B. Giannakis, "Optimal training and redundant precoding for block transmissions with application to wireless OFDM", IEEE Transactions on Communications, Vol. 50, no. 12, pp. 2113 - 2123, Dec. 2002.

[13] T. Roman, M. Enescu, V. Koivunen, "Time-domain method for estimating dispersive channels in OFDM systems", IEEE VTC spring, pp. 1318 -1321, 2003.

[14] Z. S. Roth, H. Xu, "Effects of Modeling Errors on the Stability and Innovations Moments of Kalman Filters", IEEE International Conference on Control & Applications, pp. 643-644, 1989.

[15] M. K. Tsatsanis, G. B. Giannakis, G. Zhou, "Estimation and Equalization of Fading Channels with Random Coefficients", Signal Processing, vol. 53, pp. 211-229, 1996.

[16] H. Wang, P. Chang, "On Verifying the first-order Markovian assumption for a Rayleigh fading channel model", IEEE Tran. on Vehicular Technology, Vol. 45, pp. 353 - 357, 1996.

Block-wise Adaptive Modulation for OFDM WLAN Systems

Yin-Tsung Hwang, Chen-Yu Tsai and Cheng-Chen Lin

Department of Electronic Engineering,
National Yunlin University of Science and Technology
Touliu, Yunlin 640, Taiwan, ROC

Abstract—Adaptive modulation has been shown to be effective in performance enhancement for multi-carrier wireless communication. In this paper we present an adaptive modulation framework for the OFDM based 802.11a wireless LAN system. The goal is to increase the data transmission rate subject to fixed transmission power and the upper bound constraint of packet error rate. Two adaptive modulation schemes are proposed and both use the estimated subband SNR values to adjust the modulation scheme in a block wise manner. The simulation results indicate both schemes can effectively increase the average data rate up to 147% under 802.11a model. The performance improvement is even prominent for severe frequency selective fading environments.

I. INTRODUCTION

OFDM (Orthogonal Frequency Division Multiplexing) technique has been widely adopted in various communication and broadcasting systems. It is more robust to frequency selective fading environment and the orthogonality among sub-carriers leads to better channel capacity than the conventional FDM schemes. An OFDM system is inherently convenient in adaptive modulation and power allocation. In the former case, due to the channel condition discrepancy among subbands (subcarriers), adaptive bit loading [1,2] can be applied by using different modulation scheme per subband to maximize the data transmission rate. In the latter case, instead of data, power is portioned among subbands to achieve a better quality of communication, i.e. lower bit error rate. In this paper, aiming at the wireless LAN applications, we will investigate novel adaptive modulation schemes for OFDM systems to maximize data transmission rate under the condition of fixed transmission power and packet error rate constraint.

To achieve adaptive modulation, some system information must be obtained as the guideline for modulation switching. These may include received signal strength, channel impulse response, SNR or BER. In [3], the modulation is switched depending either on the received signal strength or on the error detection of a systematic BCH codec. In [4], an MMSE channel prediction scheme is used. An iterative algorithm is then employed to increase the bit rate in each subband until the target bit rate is accomplished. In [5], a Mean Least Square (MLS) algorithm is first used for power estimation. Pilot assisted channel and SNR estimation schemes then follow for modulation scheme selection. In [6], prior channel information is assumed and the bit loading in each subband is calculated from the channel information subject to the BER constraint. In most of the previous works, the modulation or bit loading in each subband is determined solely by comparing the estimated SNR or BER values with predefined thresholds. The overall channel condition, however, is not taken into account. In addition, most of the systems addressed above do not include channel coding and are evaluated based on the OFDM kernel only. This deviates from most communication systems in practice. In this paper, two schemes of block wise adaptive modulation are developed. One of them is based on simple SNR threshold selection and the other one additionally takes the overall SNR information into account. In particular, we adopt an IEEE 802.11a standard compliant system as the test bed of the proposed scheme to verify the performance. The system features pulse shaping (higher frequency subbands are not used), data interleaving and Viterbi coding. The standard suggested channel environment is also employed in system simulation. Since 802.11a is mainly for indoor wireless communication, a slow fading (varying on the basis of packet length) system is assumed. For simplicity, we will assume the channel information is given in advance and focus on the adaptive modulation scheme. In practice, various methods such as [7] can be readily applied for channel estimation with some possible performance degradation.

II. ADAPTIVE OFDM SYSTEM MODEL

Since our adaptive modulation scheme is tailored to 802.11a wireless LAN environment, our system consists of all 802.11a blocks and additional adaptive modulation modules as shown in Fig. 1. Adaptive modulation module changes modulation in each subband subject to the feedback information from the receiving end. This information can be coded in the signal field of a PPDU frame from the receiver. The feedback channel is assumed to be ideal (error free). Because the interleaving scheme for 802.11a is based on fixed modulation scheme, it must be modified subject to bit rate change in adaptive modulation. At the receiving end, for simplicity, we assume ideal synchronization and channel estimation. The adaptive algorithm module uses channel information to determine the modulation in each subband and signals the transmitter through a return packet. A time varying complex channel impulse response (CIR) can be expressed as follows

$$h(t) = \sum_{k=0}^{N-1} \alpha_k e^{j\theta_k} \delta(t - \tau_k), \quad (1)$$

where k is the index of the path, α_k is the associated gain, θ_k is the phase rotation, and τ_k is the path delay. Two channel models are adopted in this paper: The first one is JTC model [8]

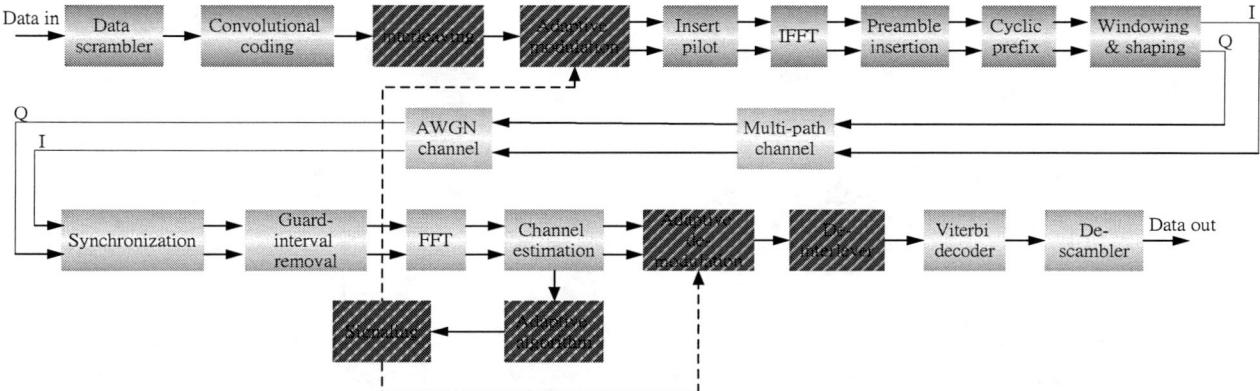

Figure 1. Adaptive modulation system architecture for multi-carrier WLAN

with 3 delay paths and normalized power. Its CIR, representing a moderate fading channel model, is illustrated in Fig. 2.

The second one is suggested in IEEE 802.11 standard for indoor wireless communication and features a much more complicated Rayleigh fading channel. The mean power of each path decays exponentially with the path delay. The kth path can be expressed mathematically as follows:

$$h_k = N\left(0, \tfrac{1}{2}\sigma_k^2\right) + j \cdot N\left(0, \tfrac{1}{2}\sigma_k^2\right), \quad (2)$$

where $\sigma_k^2 = \sigma_0^2 \cdot e^{-kT_s/T_{RMS}}$, $\sigma_0^2 = 1 - e^{-T_s/T_{RMS}}$ and $N\left(0, \tfrac{1}{2}\sigma_k^2\right)$ is a random variable with zero mean and a variance of $\sigma_k^2/2$. T_s is the sampling period and T_{RMS} is the RMS value of the delay spread. In this paper, T_{RMS}'s used in simulation are 18ns and 50ns for JTC and 802.11 channel model respectively. After channel estimation, the essence of adaptive modulation lies in the SNR estimation. The modulation level or the bit loading depends on the channel quality, i.e. SNR value, of each subband. The selection is by comparing the estimated SNR versus a set of pre-determined SNR thresholds. The SNR value of each subband can be estimated as follows

$$SNR(n) = \frac{H(n) \cdot \sigma_s^2}{\hat{\sigma}_n^2}, \quad (3)$$

where $H(n)$ and σ_s^2 are the channel gain and signal power of the nth subband, respectively. $\hat{\sigma}_n^2$ is the AWGN noise power estimated by using the information of preamble and pilot sub-carrier in a 802.11a packet (PPDU frame).

III. ADAPTIVE MODULATION SCHEMES

The development of an adaptive modulation scheme starts with the profiling of BER versus SNR values through system simulation. Because 802.11a adopts packet error rate (PER), therefore, the simulation in Fig. 3 is measured on PER using size 1kB packets. Via simulation, we try to characterize the relationship between the modulation scheme and PER under different SNR conditions. Since each subband is considered as a narrow band channel, a flat fading AWGN channel model suffices for the simulation. The system depicted in Fig. 1 is used in the simulation. The bit rates in Fig. 3 correspond to different modulation and convolution code puncturing schemes as suggested in 802.11a standard. The SNR values where PERs reach 10^{-1}, the 802.11a constraint, are considered as the SNR threshold in modulation mode selection. The adaptive modulation process is accomplished by the following steps. Note that block wise adaptive modulation means the subbands are grouped into blocks using the same modulation. This is to reduce the information needed to be sent back from the receiver to the transmitter.

(1) The long preamble defined in 802.11a is considered as the training symbol for the receiver to perform channel estimation. (In this paper, this step is omitted by assuming prior channel information)

(2) The estimated channel frequency response, along with the estimated AWGN power (the received signal power before

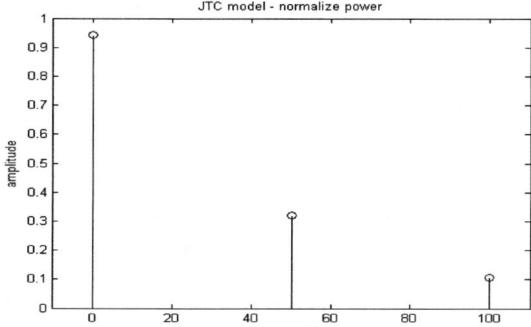

Figure 2. the normalized power CIR of JTC channel model

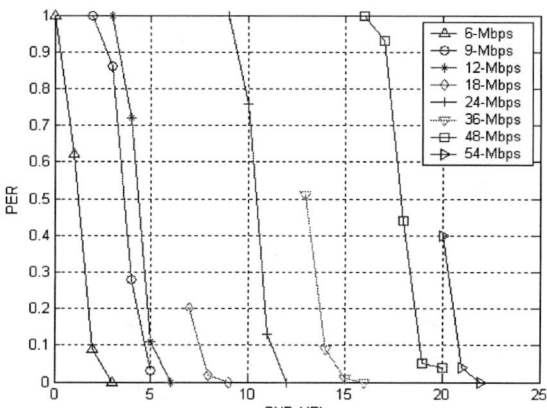

Figure 3. simulation result for our 802.11a system

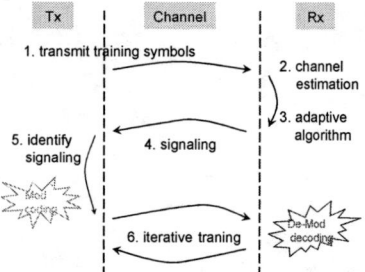

Figure 4. Adaptive modulation process

the packet arrival) is used to calculate the estimated SNR according to Eq (3).

(3) The estimated SNR is compared with the thresholds derived from Fig. 3 to determine the modulation scheme in each block – scheme 1.

(4) The modulation selection information is then coded in the signal field of a PPDU frame and returned to the transmitter (signaling).

This process is illustrated in Fig. 4. Fortunately, such signaling process does not occur often in a slowly varying communication environment. We will next examine two adaptive modulation schemes.

Adaptive modulation scheme 1:

We call the simple SNR threshold scheme as suggested in step (3) as scheme 1. The SNR threshold values are compiled in Table I. As mentioned, this scheme only considers the channel state information of each subband and does not take the entire channel condition into consideration.

TABLE I. SNR THRESHOLDS OF SCHEME 1

Modulation / Target	BPSK (T_1)	QPSK (T_2)	16-QAM (T_3)	64-QAM (T_4)
PER < 10^{-1}	2	5.1	11.3	18.9

Adaptive modulation scheme 2:

In this scheme, an adaptation range is first selected and different SNR thresholds are defined for each range. Table II summarizes the SNR thresholds for each subband to switch the modulation level. The range is selected subject to the average of all estimated subband SNR values. A cross out entry in the table means the subband should be discarded due to ill channel condition. In other words, the 10^{-1} PER constraint cannot be met using that modulation scheme. A negative infinity ($-\infty$) means the corresponding modulation can always be applied without jeopardizing the PER constraint regardless of the subband condition. The rationale behind this scheme is that: Due to the uncertainty in subband SNR estimation, those subbands in a worse channel should adopt a more conservative modulation scheme than those in a better channel even though they both have the same estimated subband SNR value. The corresponding flow chart of the scheme is shown in Fig. 5. Note that the ranges and the threshold values presented in

TABLE II. RANGES AND SNR THRESHOLDS OF SCHEME 2

Modulation	BPSK (T_1)	QPSK (T_2)	16-QAM (T_3)	64-QAM (T_4)
Range 1 (< 4dB)	0.5	5.2	×	×
Range 2 (4 ~ 7.3dB)	1	3.5	×	×
Range 3 (7.4 ~ 13.3dB)	1	5.4	9.7	×
Range 4 (> 13.4dB)	×	$-\infty$	7.9	16.4

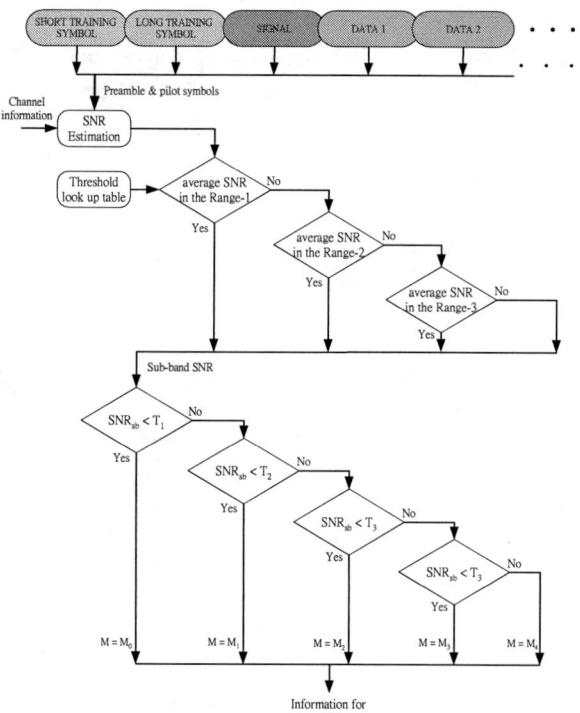

Figure 5. Flow chart of scheme 2

scheme 2 are derived using an iterative algorithm and using the threshold values in scheme 1 as initial values.

IV. SIMULATION RESULTS

Both schemes are simulated under the Fig. 1 environment to verify the performance. The simulation conditions are listed in table III where parameters are compliant to the 802.11a standard. The block size is chosen to be 2. We also set the convolution coding rate as 1/2 (mandatory in 802.11a) so that only 6, 12 and 24Mbps transmission rate are supported. The proposed schemes can be equally applied to higher coding rates. Since higher coding rates require higher SNR condition to meet the PER constraint, the benefit of adaptive modulation scheme will be less significant due to the reduction of SNR headroom for each subband to adopt bigger constellation. Fig. 6 illustrates the simulation results under JTC channel model. The staircase like curve represents the result of using fixed modulation. The modulations of all subbands are switched manually at optimal SNR thresholds. Both schemes achieve nearly linear data rate improvements as SNR value increases. Scheme 2 performs apparently better than scheme 1. In addition, under most SNR conditions, it outperforms the fixed modulation scheme. Scheme 1, while becoming inferior in certain cases due to conservative SNR thresholds, still has a better average data rate than that of the fixed modulation scheme. The benefits of adaptive modulation are not fully exploited in this case because JTC model does not have large channel gain fluctuations among subbands. Fig. 7 shows the simulation results under the 802.11 channel model. Both schemes now outperform the fixed modulation scheme overwhelmingly. The selection probabilities for each modulation scheme using scheme 2 are demonstrated in Fig. 8. It is evident that large SNR value leads to higher probability of using high level modulation. We also calculate the average

PERs of the two schemes. They are 0.007 and 0.051 for scheme 1 and 2, respectively. Since the PER constraint is 0.1 in 802.11a, scheme 1 apparently has much more conservative SNR thresholds. The average data transmission improvements of the two schemes are summarized in Table IV. Under IEEE channel model, the improvement can be as high as 146.8% using scheme 2. We also include the pure AWGN channel condition into our comparison. The information revealed is – our adaptive modulation scheme can achieve better data transmission rates under frequency selective fading environment compared with what a fixed modulation scheme can achieve under a flat fading AWGN environment. It shows the proposed schemes are effective in mitigating inferior channel environment for higher data rate transmission.

TABLE III. SIMULATION PARAMETERS

Selectable Modulation	BPSK, QPSK, 16-QAM, 64-QAM
Channel	JTC & IEEE 802.11 model
Channel estimation	ideal
Channel decoder	Viterbi hard-decision
FFT size	64
# of data/pilot subcarriers	48/4
# of blocks	24
Guard interval	16
# of bytes per packet	1000
# of packet per simulation	100

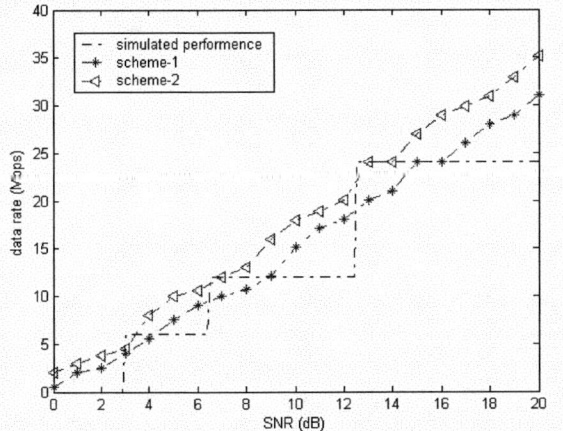

Figure 6. Performance of 2 schemes comparison, JTC model

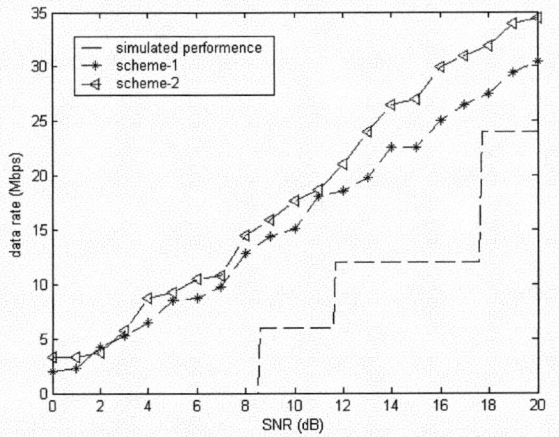

Figure 7. Performance comparison of 2 schemes, 802.11 model

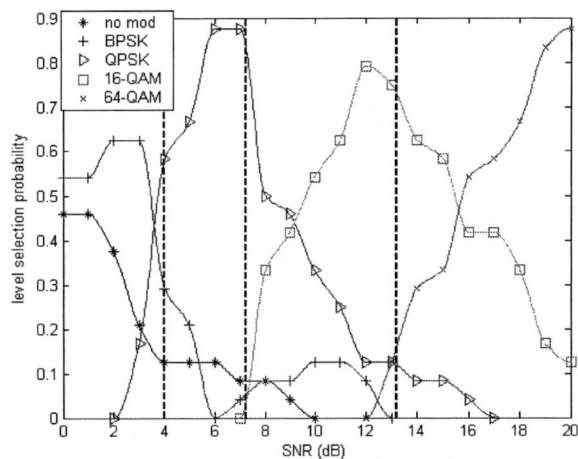

Figure 8. The modulation selected probability versus SNR

TABLE IV. TRANSMISSION RATE COMPARISON

Adaptive methods Non-adaptive	Scheme-1		Scheme-2	
	IEEE	JTC	IEEE	JTC
AWGN	4.51%	0.56%	20.23%	19.26%
AWGN + JTC		11.24%		29.65%
AWGN + IEEE	117.9%		146.8%	

V. CONCLUSIONS

In conclusion, in this paper we present an adaptive modulation framework for 802.11a wireless LAN system to increase the data transmission rate subject to fixed transmission power and PER upper bound constraint. Two adaptive modulation schemes are proposed and the simulation results indicate both schemes can effectively increase the data rate. The performance is even significant for severe frequency selective fading environment.

VI. REFERENCES

[1] I.Kalet, "The Multitone Channel," *IEEE Transactions on Communications*, vol.37, no.2, Feb. 1989.

[2] J.A.C. Bingham, "*ADSL, VDSL, and Multicarrier Modulation,*" John Wiley and Sons, 2000.

[3] L.Hanzo, W.Webb and T.Keller, "*Single- and Multi-carrier Quadrature Amplitude Modulation,*" John Wiley and Sons, 2000.

[4] M.Munster and L.Hanzo, "MMSE Channel Prediction Assisted Symbol-by-symbol Adaptive OFDM," *ICC*, vol.28, pp.416-420, 2002.

[5] C.-J.Ahn and L.Sasase, "The effects of modulation combination, target BER, doppler frequency, and adaptation interval on the performance of adaptive OFDM in broadband mobile channel," *IEEE trans. on Consumer Electronics*, vol.48, no.1 , pp.167-174, 2002.

[6] S.Ye, R.S.Blum and L.J.Cimini, "Adaptive Modulation for Variable-Rate OFDM Systems with Imperfect Channel Information," *VTC*, vol.2, pp.767-771, May 2002.

[7] Luc Deneire, and others," A Low-Complexity ML Channel Estimator for OFDM", *IEEE Trans. Comm.*, pp.135-140, Feb. 2003.

[8] P.Krishnamurthy and others"Modeling of the Wideband Indoor Radio Channel for Geolocation Applications in Residential Areas," *VTC*, vol.1, pp.175~179, May, 1999.

A Hybrid Space-time and Collaborative Coding Scheme for Wireless Communications

M. Ma, E. Masoud
School of Electronic, Communication and Electrical Engineering
University of Hertfordshire
Hatfield Herts AL10 9AB, UK

Y. Sun and J. M. Senior
School of Electronic, Communication and Electrical Engineering
University of Hertfordshire
Hatfield Herts AL10 9AB, UK

Abstract: This paper addresses at space-time coding techniques for broadband wireless communications. A brief overview of the space-time block coding and Collaborative Coding Multiple Access (CCMA) techniques is presented. A new coding scheme which combines the CCMA and space-time block coding techniques is proposed. The new coding technique with transmit diversity is simulated. Results are presented, which show that the hybrid coding technique is advantageous over the space-time block coding and CCMA techniques.

I Introduction

Space-time coding has received much interest for broadband wireless and mobile communications [1]. Several interesting coding approaches have been suggested to combat the impairments in mobile fading channels. One interesting approach is space-time Trellis coding [2] which combines signal processing at the receiver with coding techniques appropriate to multiple transmit antennas and provides significant gain. The cost for this scheme is additional processing, which increases exponentially as a function of bandwidth efficiency and the required diversity order. For the simplicity of decoding, Alamouti provided a remarkable scheme for transmission using two transmit antennas [3]. Tarokh et al [4] introduced space-time block coding which generalizes the transmission scheme of Alamouti to an arbitrary number of transmit antennas and is able to achieve the full diversity promised by the transmit and receive antennas. These codes have a very simple maximum likelihood decoding algorithm based only on linear processing at the receiver.

In another research direction, a collaborative coding multiple access technique allows simultaneous communications by several users in the same bandwidth by means of special codes, known as collaborative codes, without subdivision in time, frequency or orthogonal codes [5-8]. This technique has theoretically been shown to achieve higher transmission rate than conventional multiple access techniques. The combining of signals to implement the multiple access channel (MAC) is reasonably simple to achieve at baseband, in which signals can be represented as voltages or currents which can add or combine appropriately. However, this combining of signals over mobile radio channel will introduce distortion due to channel fading and it then become less practical should no measures be taken to combat the effect of fading.

A new hybrid CCMA and space-time coding scheme is outlined in this paper to combine the advantages of these two kinds of coding. The principle of this novel scheme is that considering a T-user multiple access communication system with T independent users communicating simultaneously over a common MAC using a T-user collaborative code, the output of the T-user CCMA is phase mapped and space-time block code encoded, and the encoded symbol is divided into streams which are simultaneously transmitted using m transmit antennas. In the receiver side, the received signals are first space-time block decoded using a maximum likelihood algorithm, the result is then CCMA decoded to obtain the information from the T independent users.

II Space-Time Block Coding

Considering a mobile communication system with n antennas at the transmitter and m antennas at the receiver. At each time slot t, signals c_t^i, $i=1,2,...,n$ are transmitted simultaneously from the n transmit antennas. The channel is assumed to be a flat Rayleigh fading channel and the path gain from transmit antenna i to receive antenna j is defined as $g_{i,j}$. The channel is assumed to be quasi-static so that the path gains are constant over a frame of length l and vary from one frame to another. At time t, the signal r_t^j, received at antenna j, is given by

$$r_t^j = \sum_{i=1}^{n} g_{i,j} c_t^i + \eta_t^j \qquad (1)$$

where the noise samples η_t^j are independent samples of a zero-mean complex Gaussian random variable with variance $n/2$ per complex dimension. Assuming a perfect channel estimation, the receiver computes the decision metric and decides in favor of the code word that minimizes the sum.

$$\sum_{t=1}^{l}\sum_{j=1}^{m}\left| r_t^j - \sum_{i=1}^{n} g_{i,j} c_t^i \right|^2 \qquad (2)$$

A space-time block code is defined by a $p \times n$ transmission matrix Y. The entries of the matrix Y are linear combinations of the variables $x_1, x_2, ..., x_k$ and their conjugates. The number of transmission antennas is n, and we usually use them to separate different codes from each other. For example, for two transmit antennas, the transmission matrix is defined by

$$Y_2 = \begin{pmatrix} x_1 & x_2 \\ -x_2^* & x_1^* \end{pmatrix} \quad (3)$$

Assume that transmission at the baseband employs a signal constellation with 2^b elements. At time slot 1, kb bits arrive at the encoder and select constellation signals $s_1, s_2, ..., s_k$. Let $x_i = s_i$, $i = 1, 2, ..., k$ in Y, we arrive at a matrix C with entries of linear combinations of $s_1, s_2, ..., s_k$ and their conjugates. The entry c_t^i represents the element in the tth row and the ith column of C. The entries c_t^i, $i = 1, 2, ..., n$ are transmitted simultaneously from transmit antennas $1, 2, ..., n$ at each time slot $t = 1, 2, ..., p$. So the ith column of C represents the transmitted symbols from the ith antenna and the tth row of C represents the transmitted symbols at time slot t.

Figure 1 shows the baseband representation of the two branch transmit diversity scheme. The scheme uses two transmit antennas and one receive antenna.

At a given symbol period t, two signals are simultaneously transmitted from the two antennas. The signal transmitted from antenna one is denoted as s_1, and s_2 from antenna two. During the next symbol period $(t+T)$, signal $(-s_2^*)$ is transmitted from antenna one and s_1^* transmitted from antenna two. The channel at time t can be modeled by a complex multiplicative distortion $g_1(t)$ for transmit antenna one and $g_2(t)$ for transmit antenna two. Assuming that fading is constant across two consecutive symbols, we can obtain:

$$g_1(t) = g_1(t+T) = a_1 e^{j\theta_1} \quad (4)$$
$$g_2(t) = g_2(t+T) = a_2 e^{j\theta_2}$$

where T is the symbol duration. The received signals can be expressed as

$$r_1 = r(t) = g_1 s_1 + g_2 s_2 + \eta_1 \quad (5)$$
$$r_2 = r(t+T) = -g_1 s_2^* + g_2 s_1^* + \eta_2$$

or in the matrix form

$$\vec{r} = \begin{bmatrix} r_1 \\ r_2 \end{bmatrix} = \begin{bmatrix} s_1 & s_2 \\ -s_2^* & s_1^* \end{bmatrix} \begin{bmatrix} g_1 \\ g_2 \end{bmatrix} + \begin{bmatrix} \eta_1 \\ \eta_2 \end{bmatrix}$$

$$\vec{r} = Y_{2, x_i = s_i} \vec{g} + \vec{\eta} \quad (6)$$

where r_1 and r_2 are the received signals at time t and $t+T$ and η_1 and η_2 are complex random variables representing receiver noise and interference.

The combiner shown in Figure 1 creates the following two combined signals that are sent to the maximum likelihood detector:

$$\tilde{s}_1 = g_1^* r_1 + g_2 r_2^* \quad (7)$$
$$\tilde{s}_2 = g_2^* r_1 + g_1 r_2^*$$

or by matrix

$$\vec{\tilde{s}} = \begin{bmatrix} \tilde{s}_1 \\ \tilde{s}_2 \end{bmatrix} = \begin{bmatrix} g_1^* & g_2 \\ g_2^* & -g_1 \end{bmatrix} \begin{bmatrix} r_1 \\ r_2^* \end{bmatrix} \quad (8)$$

$$= (|g_1|^2 + |g_2|^2) \begin{bmatrix} s_1 \\ s_2 \end{bmatrix} + \begin{bmatrix} g_1^* & g_2 \\ g_2^* & -g_1 \end{bmatrix} \begin{bmatrix} \eta_1 \\ \eta_2^* \end{bmatrix}$$

These combined signals are then sent to the maximum likelihood detector to make a decision on which signal is dispatched.

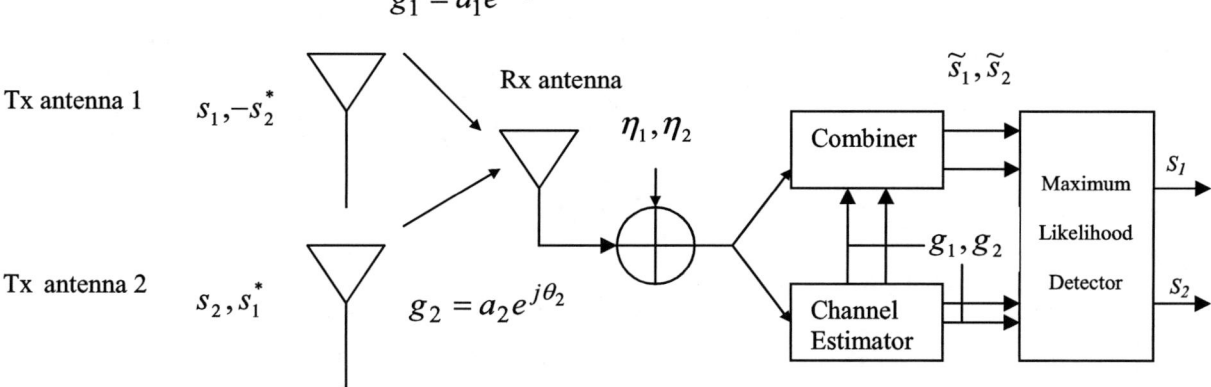

Figure 1 The two branch transmit diversity scheme with one receiver

III Collaborative Coding Multiple Access

In a situation where the bandwidth is a restricted resource such as radio frequency bands, it is necessary to study efficient ways of sharing it between as many users as possible. Furthermore, it is of considerable importance to use a simple and effective multiple access coding technique capable of error control. CCMA is an attractive proposition since it allows a substantial increase in the number of users that can access the channel simultaneously leading to a higher combined information rate. CCMA techniques exist which lie between the two extreme cases of TDMA and CDMA and offer in certain circumstances the possibility of rate sums higher than unity with modest synchronisation requirements [6]. There are two main approaches for the CCMA code design for the discrete adder channel. The first one focuses on achieving the bounds promised by multiple access information theory where all the users are active [5, 6]; the second approach aims at code construction for T active users out of M multiple access systems where the primary goal of code construction is not to achieve channel capacity [9]. Previous work has covered both approaches to the CCMA coding [5, 7, 8].

Code constructions for CCMA schemes are restricted since the composite code resulting from the individual user's code combinations have to be uniquely decodable. A composite code is said to be uniquely decodable if it can decode each of the component codes uniquely and deliver the corresponding sink information reliably to their intended destinations. Various block codes have been designed to meet the unique decodability criteria. It was found that the best rate sum would be achieved if block length N is kept to a minimum [10]. The rate sum decreases with increase in N tending to unity. Code constructions in this instance are based on the multiple access information theory (MAIT) approach which began with a coding theorem developed in [11]. The search for codes in this case is complicated by the fact that at least one of the component codes must be non-linear in order to achieve a rate point near the boundary of the capacity region of the MA adder channel. A similar approach based on achieving channel capacity asymptotically as the number of users (M) goes to infinity is also described. Here, each user gets two codewords and the overall rate sum is M/N (bits/channel use). The original model of such a scheme was proposed by [10] and represents a uniquely decodable code pair of block length $N=2$, as is shown in Table 1.

Table 1: 2-user Block Code

User1 \ User 2	00	11
00	00	11
01	01	12
10	10	21

User one has two code words $C_1=(00,11)$ and User two has three code words $C_2=(00,01,10)$. The individual rates for User one and User two are $R_1=0.5$ and $R_2=0.792$ respectively. The composite coding scheme, shown in Table 1, has a total rate sum $R_T=R_1+R_2=1.292$ (bits/channel use).

The rate of a component code is expressed as

$$R_i = \frac{\log_2 W_i}{N} \quad \text{(bits/channel use)} \quad (9)$$

where W_i is the number of distinct codewords in component code C_i and N is the block length. The rate sum $R_T(M)$ of an M-user code $(C_1, C_2, ..., C_M)$ is defined as:

$$R_T(M) = R_1 + R_2 + ... + R_M \ (bits/channel use) \quad (10)$$

The simple coding scheme above can be extended to length N, where C_1 is the two N-tuples $(000... 0)$ and $(111... 1)$, and C_2 is the N-tuples $(000... 0)$ and all the other N-tuples except the all one vector. The omission of the all one vector from C_2 is made in order to maintain unique decodability. It is clear that User one code is a repetition code that has one message symbol which is repeated N times. The total rate sum of a 2-user scheme based on this construction can be seen to decrease with increase in N tending to unity.

IV Hybrid Coding Scheme Combining CCMA and STC

In this section, a hybrid CCMA/space-time coding is introduced to combine the advantages of these two kinds of coding. The principle of this new scheme is that considering a T-user multiple access communication system with T independent users communicating simultaneously over a common multiple access channel using a T-user collaborative code, the output of the T-user CCMA is phase mapped and space-time block code encoded. The encoded symbol is divided into streams which are simultaneously transmitted using m transmit antennas. On the receiver side, the received signals are first combined using the method described in Section 2, and the combined signals are then sent to the maximum likelihood decoder to obtain the sink signals of the T-users. Figure 2 shows the proposed hybrid coding scheme corresponding to a two-user CCMA, two transmit antennas and one receive antenna system.

For example, assuming $C_1 = (00, 11)$ and $C_2 = (01, 10)$, the output of the adder of C_1 and C_2 is uniquely decodable. The BPSK (Binary Phase Shift Keying) output of C_1 is $S_{11}S_{12} = (1 1, -1-1)$, the BPSK output of C_2 is $S_{21}S_{22} = (1-1, -1\ 1)$, and the BPSK output of both C_1 and C_2 has four possibilites, that is $C=(C_{11}C_{12}, C_{21}C_{22}, C_{31}C_{32}, C_{41}C_{42})= (2\ 0, 0-2, 0\ 2, -2\ 0)$. Let $S_1=S_{11}+S_{21}$ and $S_2=S_{12}+S_{22}$. Following the same process as in Section 2, we can obtain decoded $\widetilde{S}_1$ and $\widetilde{S}_2$. We then compute the following distances:

$$d_k^2 = \sum_{i=1}^{2} \left| \widetilde{S}_i - C_{k,i} \right|^2, k=1,2,3,4 \quad (11)$$

Selecting the smallest distance and getting thecorresponding $C_{k,i}$, we can follow the look-up table to sink information to complete the decoding process.

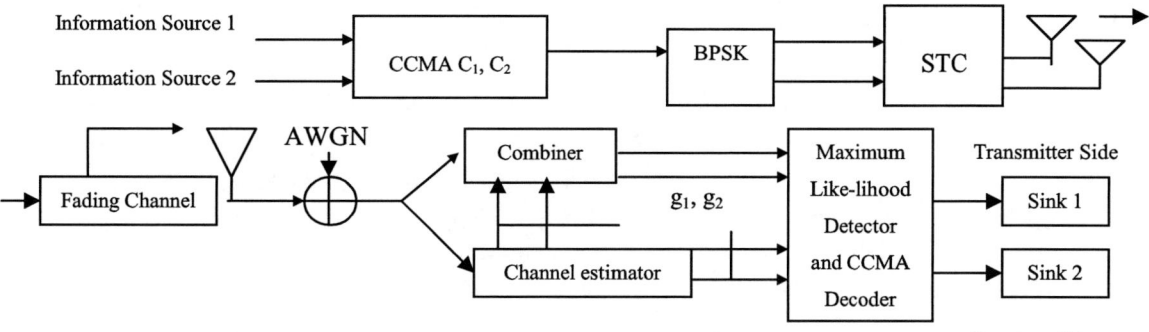

Figure 2: Hybrid CCMA and STC Scheme

V Simulation Results

In this section we provide simulation results for the performance of the hybrid CCMA/space-time coding scheme described in Section 4 (Figure 2 illustrates a block diagram of the proposed system). The main simulation parameters which we used were a data rate 9.6 kbit/s and a maximum Doppler frequency 100 Hz. Figure 3 shows the bit error performance of CCMA coding with and without space-time coding. The results demonstrate that significant gain can be achieved by transmit antenna diversity.

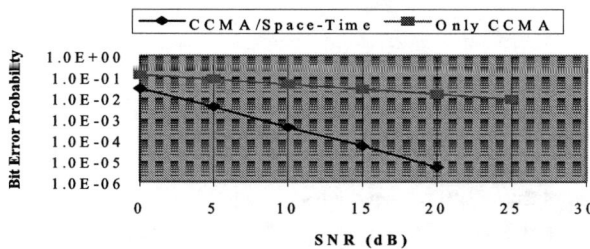

Figure 3: Performance of Hybrid Coding Compared with CCMA

Figure 4 shows the bit error rate of the proposed hybrid 2-user CCMA/space-time block coding with two transmit antennas and one receive antenna. From Figure 4 it can be seen that the combined CCMA/space-time coding system can give about 3dB improvement at bit error probability 10^{-5} with very little additional coding and decoding complexity compared with the space-time block coding with coherent BPSK.

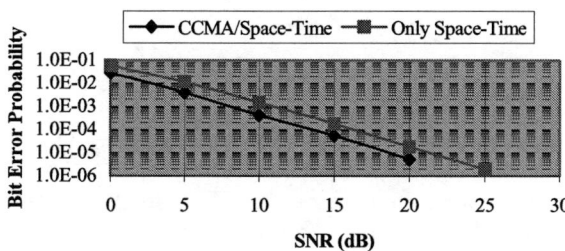

Figure 4: Performance of Hybrid Coding Compared with STC

VI Conclusions

Space-time block coding and collaborative coding multiple access techniques have been briefly discussed and a new coding scheme which combines the collaborative coding multiple access and space-time block coding techniques has been presented. The new coding technique with transmit diversity has been simulated. Initial results have been presented which show that the hybrid coding technique is advantageous over the space-time block coding and CCMA techniques.

References

[1] D. Gesbert, M. Shafi, D. Shiu, P. J. Smith and A. Naguib, "From theory to practice: an overview of MIMO space-time coded wireless systems," IEEE J. Selected Areas in Communications, Vol.21, No.3, pp. 281-302, April 2003.

[2] V. Tarokh, N. Seshadri, and A. R. Calderbank, "Space-time codes for high data rate wireless communication: Performance criterion and code construction." IEEE Trans. Information Theory, Vol. 44, no. 2, pp 744-765, March 1998.

[3] S. M. Alamouti, "A simple transmit diversity technique for wireless communications", IEEE Journal on Selected Areas in Communications, Vol. 16, No. 8, October 1998.

[4] V. Tarokh, H. Jafarkhani and A.R. Calderbank, "Space-time block codes from orthogonal designs", IEEE Transactions on Information Theory, Vol. 45, No. 5, July 1999.

[5] F. Ali and B. Honary, "Collaborative coding and decoding techniques for multiple access channel", IEE Proceedings-Communications, Vol.141, No.2, pp.56-62, April 1994.

[6] P. G. Farrell, "Survey of channel coding for multi-user systems" in SKWIRZYNSKI, J.K. (Ed.): New Concept in Multi-user Communication, Sijthoff and Noordhoff, pp. 133-159, 1981.

[7] F. H. Ali and B. Honary, "Low complexity soft decision decoding technique for T-user collaborative coding multiple access channels," Electronic Letters, 27, (13), pp. 1167-1169, 1991.

[8] B. Honary, F. H. Ali and M. Darnell, "Capacity of T-user collaborative coding multiple access scheme operating over a noisy channel," Electron. Lett., Vol. 25, No.11, pp. 742-744, May 1989.

[9] P. Mathys, "A class of codes for T active users out of N, for a multiple access communication system," IEEE Trans. Inform. Theory, Vol. 36, pp.1206-1219, Nov.1990.

[10] T. Kasami and S. Lin, "Coding for a multiple access channel", IEEE Trans. Inform. Theory. Vol. IT-22, No.2, pp. 129-137, March 1976.

[11] R. Ahlswede, "Multi-way communications channels," in Proc. of the 2nd Int. Symp. Inform. Theory, Armenian S.S.R, pp. 23-52, 1971.

An Analog Modulator/Demodulator using a Programmable Arbitrary Waveform Generator

Ravi Chawla*, Christopher M. Twigg[†], and Paul Hasler[‡]
School of Electrical and Computer Engineering
Georgia Institute of Technology, Atlanta, Georgia 30332–0250
Email: *ravic@ece.gatech.edu; [†]ctwigg@ece.gatech.edu; [‡]phasler@ece.gatech.edu

Abstract— In this paper, we present an analog implementation of a modulator/demodulator (PAMD) for analog signal processing applications. The PAMD architecture is fully programmable enabling it to be used for a variety of communication schemes and not specific to a particular application. PAMD uses a direct digital frequency synthesis architecture to generate the waveforms. We use our programmable floating-gate MOS transistors as analog memory cells to store easily programmed samples of the waveforms. Experimental results showing the programmability along with both modulation and demodulation using the presented architecture are presented. The IC prototype generating eight programmable waveforms was fabricated in a 0.5μm n-well CMOS process and occupies 0.95mm^2 of area.

I. OVERVIEW

The explosive growth of wireless and signal processing applications has resulted in an increasing demand for such systems with low cost, low power consumption, and small die area. To meet this demand, much work is focused on, and has recently demonstrated, fully integrated single-chip systems in low–cost CMOS processes. With the integration of these systems, power consumption becomes an important performance parameter. An IF band signal processing system requires the use of an array of DSPs operating in parallel to meet the speed requirements [1], [2]. This is a power intensive approach and makes use of certain communication schemes impractical in portable applications. The front-end ADC and back-end DAC converters required in these systems become expensive when the signal is of wideband nature and greater resolution is required [3], [4]. Recent focus has been in processing signals as much as possible in the analog domain before converting them digital.

One of the building blocks that would enable a lot of analog signal processing applications is a programmable waveform generator [5]. We propose an analog programmable arbitrary waveform generator that can be used for a variety of signal processing applications. The waveform generator is fully programmable through use of floating-gate MOS transistors. Floating-gate MOS transistors are modified EEPROM elements working as analog storage in a standard CMOS process. A more detailed discussion on floating-gate elements and how they are programmed accurately is discussed in [6].

An important applications where the programmable arbitrary waveform generator can be used is as a building block for a Programmable Analog Modulator/Demodulator (PAMD). PAMD can be one of the fundamental blocks in the transceiver

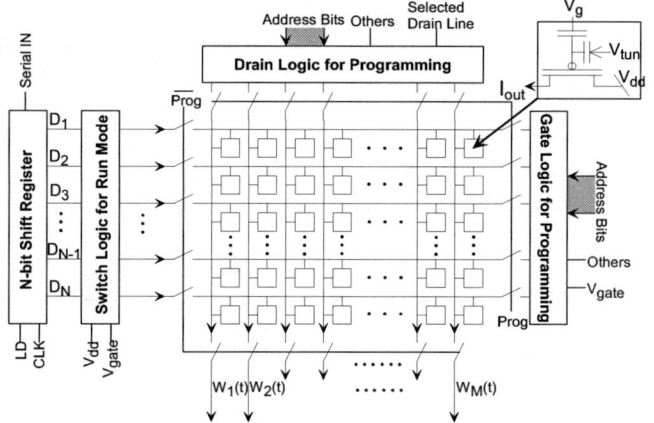

Fig. 1. **Programmable waveform generator using floating-gate transistors**: Analog implementation of a waveform generator using floating-gate devices. This architecture is similar to a direct digital frequency synthesizer implementation with floating-gates acting as analog memory cells.

enabling a lot of other signal processing functions in the analog domain. This approach is power and area efficient as compared to complex DSPs and relaxes the requirement on the design of converter specifications.

The proposed PAMD implementation can be used in various communication schemes such as Orthogonal Frequency Division Multiplexing (OFDM) and radar signal processing [3]. The biggest advantage comes from the fact that the waveforms generated can be arbitrary and are programmable. In section II, we discuss the programmable analog waveform generator. Section III presents the PAMD implementation using the programmable waveform generator along with measured experimental results for normal operation. We conclude in section IV with possible applications of the presented architecture.

II. WAVEFORM GENERATOR

Figure 1 shows the block diagram of the waveform generator that is used in the presented modulator architecture. This architecture is similar to that of a direct digital frequency synthesizer implementation. All rows in the waveform generator consist of floating-gate MOS transistors that can be programmed to any analog value. Each floating-gate in the row can be individually programmed to store a precise analog value. Using this architecture, any arbitrary waveform can be programmed when needed. W_1 to W_M can be any

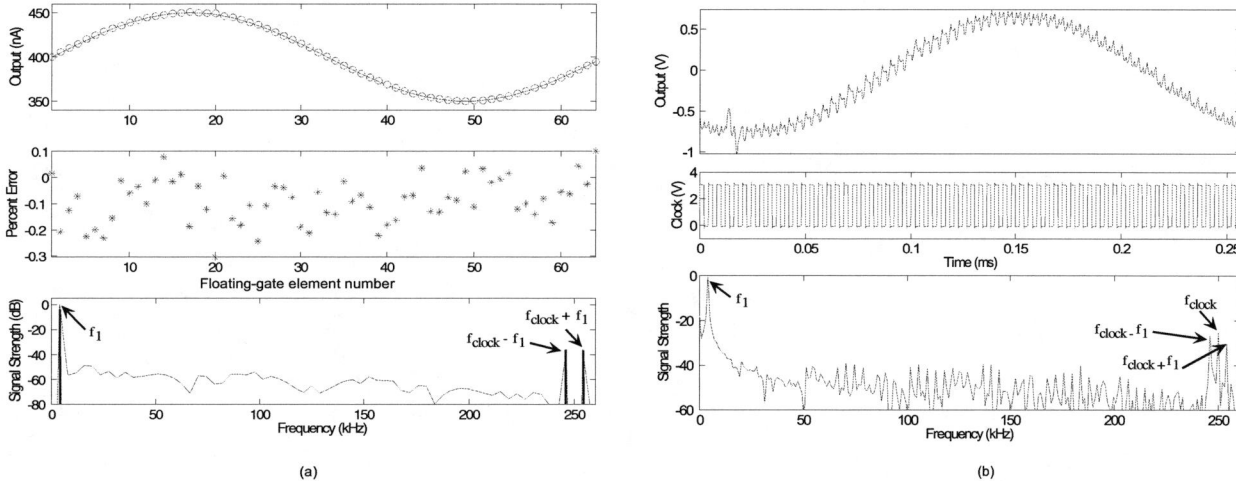

Fig. 2. **Waveform generator measurements**: (a) Measurement showing the output waveform when a $100nA_{pp}$ sine wave is programmed riding on a $300nA$ DC current. Each row has 64 floating-gate elements.. (b) Measurement showing the output waveform when a clock of $250kHz$ is applied to the waveform generator programmed with the sine-wave as shown in (a). The output frequency of the waveform was $250kHz/64$ or $3.9kHz$. Comparing FFTs of the two waveforms, they are very similar apart from the noise floor. FFT of the programmed waveform does not have any frequency component at clock frequency as there was no physical clock present in that measurement.

arbitrary set of waveforms that are programmed and can be used to modulate or demodulate any input signal. Details of the programming scheme such as speed and accuracy can be found in [6].

During the normal operation, a shift register scans through the entire row of programmed floating-gates and generates a sampled waveform at the output. The frequency of this output signal can be changed by changing the clock frequency. The shift register that scans through the row of floating-gate transistors during normal operation is designed for appropriate frequency performance and uses dynamic logic for fast response. To reduce the effect of the line capacitance on the frequency response, a cascode is added at the end of each column. The cascode lowers the impedance at the drain node of the floating-gates and isolates it from any variations at the output. Adding the cascode also helps improve the distortion due to the isolation from the output signal variations. This analog implementation eliminates the need for an adder at the output as the addition of currents can be simply done by connecting the outputs together. The number of floating-gate transistors in each row determines the quantization noise of the generated waveform.

$$I = I_o e^{-\kappa V_{charge}/U_T} \quad (1)$$

where I_o is the DC bias current in the entire row and V_{charge} depends on the charge offset programmed on each floating-gate MOS. Figure 2(a) shows a measurement of a programmed $100nA_{pp}$ sine wave riding on a $300nA$ DC current. As evident from (1), programmed current shown in Fig. 2(a) is proportional to the charge stored on each floating-gate node. We obtained a worst case programming error of 0.2% and it takes about 10 pulses of $100us$ to programmed each floating-gate [6]. The FFT of this waveform is also shown and is clearly limited by the quantization noise. The FFT was performed assuming a $256us$ time-period for the entire programmed sine-wave. This was done in order to compare the results directly with the measured data when a clock of $250kHz$ is applied to the PAMD system. Figure 2(b) illustrates the output waveform as it looks when the clock of the shift register is turned ON. As can be seen from the FFT of the programmed charges and the output waveform, a clean frequency can be generated without any observable higher-order harmonics. The measurement is limited by the noise of the measurement setup. The FFT also shows the clock frequency and images of the signal around clock frequency. Thus, the system requires a clean clock signal and a programmable lowpass filter at the output to filter out anything outside the bandwidth of the desired output waveform. Figure 3 illustrates the measurement of the waveform generator block when programmed to ω and 2ω. Figure 3 shows that this waveform generator can clearly be used to synthesize any arbitrary waveform such as chirp or any other modulating waveform.

III. Modulator Architecture

Figure 4 shows the block diagram of the programmable analog modulator/demodulator (PAMD) system using the floating-gate waveform generator. PAMD system has differential gilbert-cell mixers at the output to modulate or demodulate the differential input signal. Figure 5 shows the output when a $15.9kHz$ input signal is modulated with the $3.9kHz$ signal generated by the modulator. The input signal, $15.9kHz$, is generated using a Stanford Research System (SRS) function generator. This input signal source has a limited phase noise performance. The $3.9kHz$ signal is generated with a sine wave programmed on a row of 64 floating-gates and using a clock speed of $250kHz$. Figure 5 illustrates the basic modulation operation and shows the FFT of the output spectrum. The

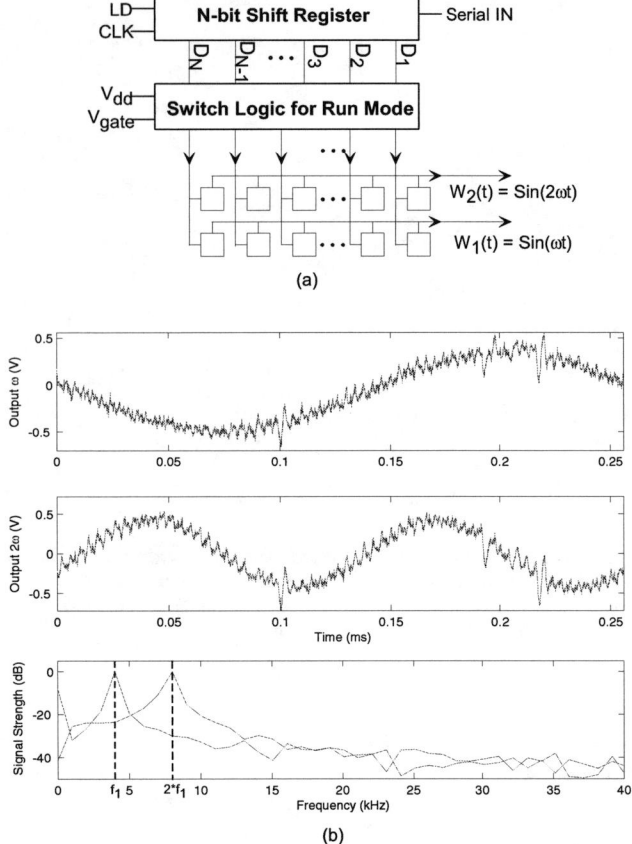

Fig. 3. **Generated output waveform at ω and 2ω**: Measurement showing the output waveforms when two rows (one cycle and two cycles) were programmed with a $100 nA_{pp}$ sine wave riding on a $300 nA$ DC current. The clock speed is $250 kHz$ and the number of elements in a row are 64. The output signal frequency generated from the two rows is $3.9 kHz$ and $7.8 kHz$, respectively. As is clear, waveform generator can be used to generate arbitrary waveform with varying frequencies.

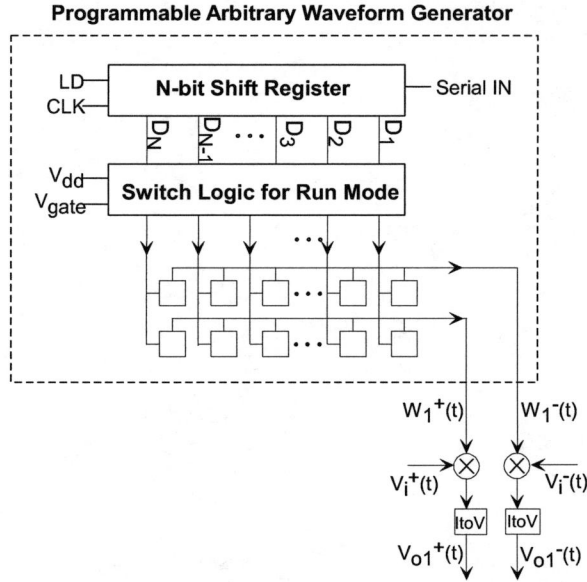

Fig. 4. **Block diagram showing modulation/demodulation**: Block diagram for the analog modulator/demodulator system. It can be easily extended for multi-channel system by adding more rows to the waveform generator.

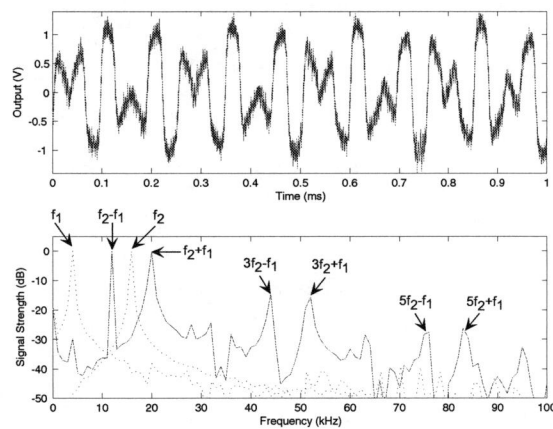

Fig. 5. **Measurement showing modulation**: Output waveform and spectrum when a $15.9 KHz$ signal is modulated with a $3.9 KHz$ signal.

output spectrum signal can be appropriately filtered to select the desired signal.

Figures 6 and 7 show the demodulation operation to near DC and at DC for the input signal, respectively. Figure 6 shows when a $3.4 KHz$ input signal is demodulated to $500 Hz$ using the generated waveform, $3.9 KHz$. This signal can be easily filtered from the spectrum to reject the high frequency spurious signal at $7.3 KHz$. Figure 7 shows the demodulation to extract the DC signal strength of the input signal by demodulating it to DC. In the current experiment, the input signal was left running and output of the modulator was turned ON after some time to see the transition in the DC level of the output signal. The output waveform still has a very slow AC component of approximately $1.5 Hz$. This is attributed to the limited precision of the function generator used to provide the input signal. As is clearly evident, this can be used to extract the spectral content of an input signal at desired frequencies by demodulating them with the desired frequencies and filtering the DC signal out.

IV. APPLICATIONS

The proposed architecture can be used for a variety of other applications along with the described modulator/demodulator. The key advantage for the presented architecture is due to the ability to generate programmable arbitrary waveforms. One such application can be generating arbitrary waveforms to perform on-chip testing of other circuits and systems. This can be easily made as part of a Built-in Self Test with a control loop to test various designs. Figure 3 illustrates how PAMD can be used for OFDM-based communication schemes, where multiples of fundamental frequency are used to orthogonally modulate different channels. The same waveforms can be used to demodulate the input signal for different channels at the receiver end. Traditionally, these operations are performed as

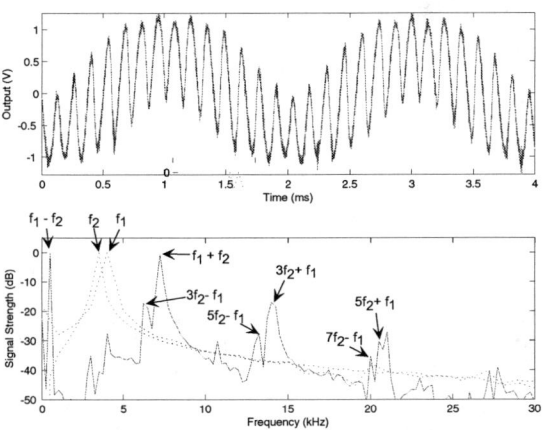

Fig. 6. **Measurement showing demodulation to near DC**: Output waveform and spectrum when a $3.4KHz$ signal is demodulated with a $3.9KHz$ signal. The output signal at $500Hz$ can be filtered to reject the high frequency components.

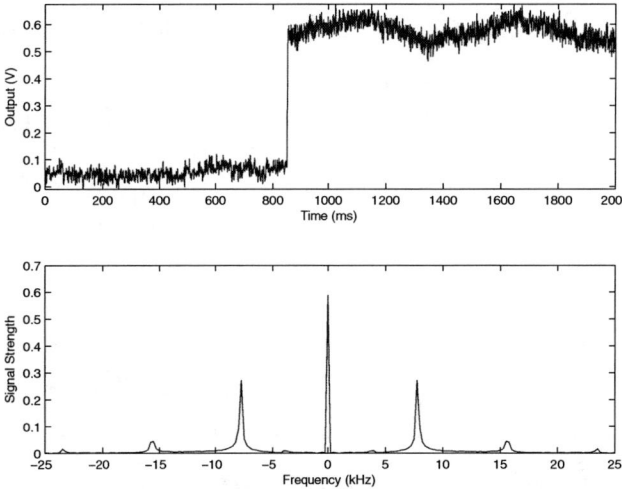

Fig. 7. **Measurement showing demodulation of input signal to DC**: Output waveform and spectrum when a $3.9KHz$ signal is demodulated with a $3.9KHz$ signal. The output signal at DC can be filtered to reject the high frequency component. This approach can be used to extract the spectral content of the input signal at desired frequencies. In the current experiment, the input signal was left running and output of the modulator was turned ON after some time to see the transition in the DC level of the output signal and was filtered to extract the low-frequency information. The output waveform still has a very slow AC component of approximately $1.5Hz$. This is attributed to the limited precision of the function generator used to provide the input signal.

FFT/IFFT in digital domain that are computationally area and power intensive [3]. The presented waveform generator can be used as a part of an adaptive equalizer system. It can be programmed to generate any waveform that can be used to then perform equalization. The compact nature of the architecture and low power consumption makes it suitable for multiple-channel processing and array signal processing.

Intermediate Frequency (IF) band applications, with reasonable accuracy, for such a system depends on the speed of shift register (shown in Fig. 1) along with the frequency performance of floating-gate. To increase accuracy of the output waveform, the number of elements in each row of PAMD should be increased that further makes the performance of the shift register important. We use dynamic logic to improve the frequency performance of the shift register. The frequency response of the row of floating-gate response depends on the total line capacitance. As briefly discussed earlier, adding a regular or active cascode at the end of each row helps in alleviating the effect of the line capacitance for high frequency applications. The performance will now depend on the input stage, which is a function of the number of parallel rows being driven and resistance of switches. High speed operation of the complete system puts a design constraint on the clock speed as well. The quality of clock in terms of rise-time, fall-time, and jitter along with coupling of the clock will affect the quality of the generated signal. Thus, generating a clean clock signal for high frequency applications becomes crucial.

V. CONCLUSION

This paper presented an analog modulator/demodulator that can be used for various communication schemes and array signal processing applications. This approach can be both power and area efficient compared to existing implementations using DSPs [2], [7]. PAMD consists of a programmable arbitrary waveform generator using floating-gate MOS devices. We presented results for the programmable waveform generator along with the spectral energy plot. We showed results with the basic operation of a modulator and demodulator operation. We also discussed and presented how to extract spectral content of an incoming signal at specific frequencies by performing auto-correlation using the proposed structure. The presented structure with proper design can be used for a variety of other applications as discussed and is being explored along those lines.

REFERENCES

[1] W. Shao, J. Xie, and G. Wang;, "Structure and implementation of smart antennas based on sortware radio," in *IEEE International Conference on Man and Cybernatics*, 2003, pp. 1938–1943.
[2] S. Berner and P. Leon;, "Parallel digital architectures for high-speed adaptive dsss receivers," in *Asilomar Conference on Signals, Systems and Computers*, 2000, pp. 1298–1302.
[3] Y.-P. Lin and S.-M. Phoong, "Analog representation and digital implementation of ofdm systems," in *Proceedings of the International Symposium on Circuits and Systems*, vol. 4, May 2003, pp. 9–12.
[4] J. Vankka, J. Ketola, J. Sommarek, O. Vaananen, M. Kosunen, and K. Halonen, "A gsm/edge/wcdma modulator with on-chip d/a converter for base stations," *IEEE Transactions on Circuits and Systems II*, vol. 49, pp. 645 – 655, October 2002.
[5] H. V. Tran, T. Blyth, D. Sowards, L. Engh, and B. Nataraj, "A 2.5v, 256-level non-volatile analog storage device using eeprom technology," in *Proceedings of ISSCC,*, 1996, pp. 270–271.
[6] G. Serrano, P. Smith, H. Lo, R. Chawla, T. Hall, C. Twigg, and P. Hasler, "Automated Rapid Programming of Large Arrays of Floating-gate Elements," in *Proceedings of the International Symposium on Circuits and Systems*, vol. I, May 2004, pp. 373–376.
[7] A. Ghazel, M. Zhili, and N. Youssef;, "Optimized dsp implementation of gmsk software modem for gsm transceiver," in *Proceedings of Vehicular Technology Conference*, 2000, pp. 2573 – 2577.

A Novel Low-Cost High-Performance VLSI Architecture for MPEG-4 AVC/H.264 CAVLC Decoding

Hsiu-Cheng Chang, Chien-Chang Lin, and Jiun-In Guo

Dept. of Computer Science and Information Engineering
National Chung Cheng University
Chia-Yi 621, Taiwan, ROC
e-mail: {chc92, lcc, jiguo}@cs.ccu.edu.tw

Abstract—The demand of high quality video and high data compression enables the MPEG-4 AVC/H.264 adopting the Context-based Adaptive Variable Length Code (CAVLC) technique as contrary to the traditional MPEG-4 VLC techniques. This paper presents a novel low-cost, high-performance VLSI architecture design for MPEG-4 AVC/H.264 CAVLC decoding. In the proposed design, we exploit five different techniques to reduce both the hardware cost and power consumption, as well as increase the data throughput rate. They are *PCCF* (Partial combinational component Freezing), *HLLT* (Hierarchical logic for Look-up tables), *ZTEBA* (Zero-left table elimination by arithmetic), *IDS* (Interleaved Double Stacks), and *ZCS* (Zero Codeword Skip). As a result, the proposed design can decode every syntax element per cycle. The synthesis result shows that the design achieves the maximum speed at 175 MHz. When we synthesize the proposed design at clock constraint of 125MHz, the hardware cost is about 4720 gates under a 0.18um CMOS technology, which achieves the real-time processing requirement for H.264 video decoding on HD1080i format video. *(Abstract)*

I. INTRODUCTION

Variable Length Coding (VLC) plays an important role in nowadays MPEG video and image coding applications. It is often applied together with other lossy image compression techniques to further increase the data compression rate. The main idea for variable length coding is to minimize the average codeword length. Shorter codewords are assigned to frequently occurring data and longer codewords are assigned to infrequently occurring data. In the typical MPEG video coding standards like MPEG-1/2/4, the VLC technique has been used to reduce the amount of video data streams according to a fixed statistical model. In order to further increase the data compression ratio, MPEG-4 AVC/H.264 has adopted the context-based adaptive variable length coding (CAVLC) technique to encode the residual data organized as 4x4/2x2 blocks of transform coefficients. Due to the context-adaptive feature in the H.264 CAVLC, its coding efficiency is much higher than that of the traditional MPEG entropy coding at the cost of much higher complexity in algorithmic point of view. Apart from high algorithmic complexity, the real-time processing demand in the H.264 video coding applications also make the dedicated hardware implementation of CAVLC decoding inevitable. These considerations motivate the proposed low-cost, high-performance VLSI architecture for MPEG-4 AVC/H.264 video decoding.

In recent years, there are many researches involved in designing the traditional MPEG VLC decoding hardware [8-12]. The methods they adopted include the bit-serial method and the bit-parallel method. Though the bit-serial method is simple and straightforward, but it is not suitable for real-time processing applications. Thus, most VLC decoding hardware designs are based on the bit-parallel method for achieving high-performance in speed. However, considering the distinctive algorithm possessed by the CAVLC decoding, the traditional bit-parallel method cannot be directly applied in implementing the H.264 CAVLC decoder. Up to now, the researches on implementing the CAVLC coding hardware design are few [4-6]. The design [4] used decoding hardware to parse CAVLC codeword. However, they don't describe their design in detail. The design [5] used HW/SW synchronization method, and parsed the CAVLC codeword by CPU directly. The design [6] proposed by Wu Di is a CAVLC decoding hardware. It presented a VLSI implementation of CAVLC decoder for H.264/AVC, which includes a coeff_token decoder, level decoder, total_zeros decoder, and run_before decoder. Together with a barrel shifter and controller, the pipeline architecture in [6] can decode every syntax element in one clock cycle. The maximum work frequency of their implementation is 125 MHz based on a 0.25um CMOS technology. Its area consumes about 6100 gates.

In this paper, we propose a novel low-cost, high-performance VLSI architecture for the MPEG-4 AVC/H.264 CAVLC decoding. In order to achieve the goals of efficient hardware implementation on the CAVLC decoding with high algorithmic complexity, we first analyze the data decoding flow in the CAVLC and propose our hardware implementation methods on the CAVLC decoding. The proposed implementation methods include PCCF (Partial combinational component Freezing), HLLT (Hierarchical logic for Look-up tables), ZTEBA (Zero-left table elimination by arithmetic), IDS (Interleaved Double Stacks), and ZCS (Zero Codeword Skip). Applying these implementation methods can achieve low hardware cost, low power consumption, and high data throughput rate. As a result, the proposed design can decode every syntax element per cycle. The synthesis result shows that the design achieves the maximum speed at 175 MHz. When we synthesize the proposed design at clock constraint of 125MHz, the hardware cost is about 4720 gates under a 0.18um CMOS technology, which achieves the real-time processing requirement for H.264 video decoding on HD1080i format video. As compared to the existing design [6] for CAVLC decoding, the proposed design outperforms the design [6] in terms of 23 % reduction in the hardware cost and 40% improvement in speed.

The rest of this paper is organized as follows. We will illustrate the decoding flow in the proposed CAVLC decoder in Section II.

Section III introduces the hardware architecture design, and implementation of the proposed design. The performance evaluation and comparison on the proposed design is included in Section IV. Finally, we conclude this paper in Section V.

II. CAVLC DECODING FLOW

Since the CAVLC coding algorithm is quite different from that used in the traditional MPEG VLC, we first describe the decoding flow shown in Fig. 1 in the proposed CAVLC decoding hardware architecture in the following steps

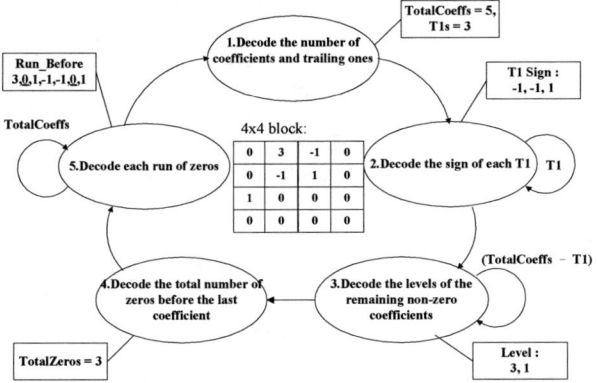

Figure 1: The decoding flow of the proposed CAVLC decoder.

(1) Decoding the number of coefficients and trailing ones:
The first step in CAVLC decoding is to decode the total number of non-zero coefficients (TotalCoeffs) and the number of trailing +/-1 values (T1). The TotalCoeffs is ranged from 0 to 16, while T1 is ranged from 0 to 3. There are five choices of look-up tables in this part, i.e. Num-VLC0 (nC=-1), Num-VLC1 ($0<=nC<2$), Num-VLC2 ($2<=nC<4$), Num-VLC3 ($4<=nC<8$), and FLC ($8<=nC$). The variable nC is calculated as follows: If upper (nU) and left-handed (nL) coded blocks are available, $nC = (nU + nL +1)/2$. If only block nU is available, $nC = nU$. If only block nL is available, $nC = nL$. Otherwise, $nC = 0$.

(2) Decoding the sign of each T1:
From the above coeff_token process, we can know the value of T1. Thus, a single bit is used to decode the sign (0=+1, 1=-1) of each T1 in reverse order.

(3) Decoding the levels of the remaining non-zero coefficients:
The level of each non-zero coefficient is decoded in reverse order. The choice of VLC tables to decode each level is decided according to the magnitude of each successive coded level, i.e. the context adaptive feature in CAVLC. There are seven choices of the tables in this part, i.e. tables Level_VLC0 to Level_VLC6. In additional, the Level_VLC0 table has two escape conditions, and all the others only have one escape condition.

(4) Decoding the total number of zeros before the last coefficient:
The total_zeros tables are applied for decoding AC 4x4 blocks or DC 2x2 blocks. The VLC tables to decode the total zeros is decided according to the total number of the non-zero coefficients in the current AC 4x4 or DC 2x2 blocks.

(5) Decoding each run of zeros:
The number of zeros preceding each non-zero coefficient is decoded in reverse order. The VLC table for each run of zeros is chosen depending on the previous the number of zeros (zeroleft) not been decoded.

III. PROPOSED CAVLC DECODER

Based on the above-mentioned CAVLC decoding flow, the proposed VLSI architecture for H.264 CAVLC decoding is illustrated in Fig. 2. The proposed design contains the following components, i.e. Coeff_token decoder, T1 decoder, Level decoder, Totalzero decoder, Run_before decoder, Flush-unit, Parameter interface, Prediction data R/W module, and IDS (Interleave double stacks). The flush unit is the most important component in the VLC decoding hardware, because we have to flush the previous codeword in the bitstream and provide the aligned bitstream for the next decoding process. The flush unit includes two registers, a barrel shifter, and an accumulator, which is similar to the traditional MPEG VLC decoding hardware designs.

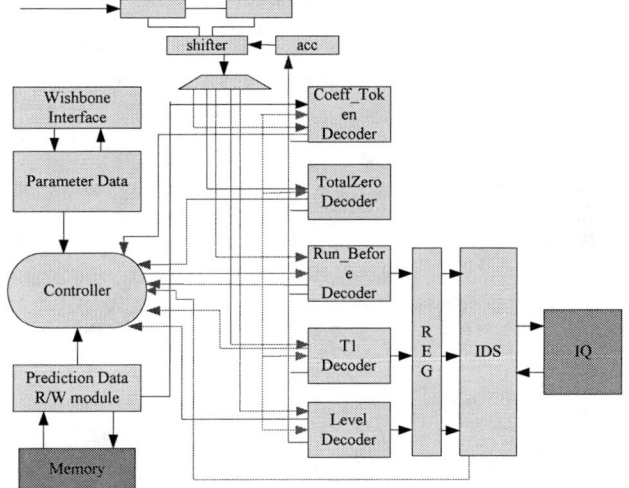

Figure 2: The architecture of the proposed CAVLC decoder

The role of the controller in the proposed CAVLC decoder is to assign the decoding tasks to the different hardware components. First, the controller decides which steps in the CALVC decoding flow to be selected. Second, in Coeff_token decoding process, the number of the previous non-zero coefficients must be already prepared. After finishing Coeff_token decoding, the controller would handle the counts of the T1 decoder and Level decoder. Then, according to the number of non-zero coefficients in the same decoding block, it starts to decode the value of total-zero. Finally, the controller handles the last process that decodes the number of zeros preceding each non-zero coefficient. In addition, the controller will write the value of Level and Run_before into IDS and skip decoding on all the zero codewords (i.e. ZCS, Zero Codeword Skip) when all of coefficients in 4x4 or 2x2 block are zeros. The function of the ZCS not only reduces the computation time, but also achieves low-power consumption. In addition, each component in the proposed CAVLC decoder likes Coeff_token decoder, Trailing one decoder, Level decoder, Totalzero decoder, and Run_before decoder equips one enable signal to freeze the non-operated component of combinational circuits to achieve low-power consumption. This technique is called *PCCF* (Partial combinational component Freezing) that has been used in our previous work for achieving low-power consumption [8]. In the following, we will describe the architecture of each the components in the proposed CAVLC decoder in more details, and illustrate another three design techniques, i.e. HLLT (Hierarchical logic for Look-up tables), ZTEBA (Zero-left table elimination by arithmetic), and IDS (Interleaved Double Stacks) used in the proposed design.

3-1: Architecture for Coeff_Token decoder

The maximum code length needed in Coeff_Token decoder in Num-VLC1 is 16, which stands for the table will have 65536 entries in maximum. But the Num-VLC1 has only 62 entries. Thus, we partition this look-up table (LUT) into multiple smaller ones. The method we used in partitioning Num-VLC1 LUT is denoted as *HLLT* (Hierarchical logic for Look-up tables). The implementation of *HLLT* partitions the original big LUT into many small LUTs. We put the codewords with higher occurring frequency on Table I as illustrated in Fig. 3. While, the codewords with lower occurring frequency are put in the LUTs with higher number of indices, and the codeword with higher occurring frequency stands for the shorter length. When no codeword is found in the current LUT, the next-layer LUT will be enabled to find the current codeword. Otherwise, the right-hand side LUTs are disabled in advance for reducing power consumption, which exploits the essence of PCCF technique mentioned before. The entries for each Num-VLCN table are 62. The value specified in the LUTs in Fig. 3 denotes the number of entries needed for each look-up TableN. In addition, the implementation of the condition (nC >= 8) in the proposed design is using a ROM-based architecture. In summary, combining the proposed *HLLT and PCCF techniques together can* make the proposed design possess the features of low-cost, low-power, and high performance for achieving the real-time processing requirement needed for CAVLC decoding on HD1080i video when it is operated at 125MHz.

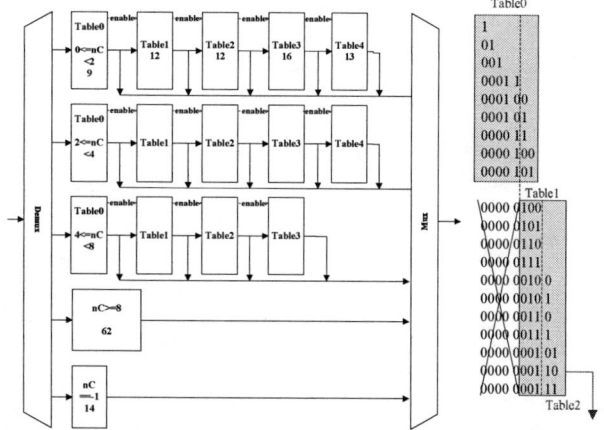

Figure 3: The architecture for Coeff_Token decoder

3-2: Architecture for T1 decoder

In the proposed design, we implement the T1 decoder through straight approach, which results in the associated hardware consisting of an inverter and comparator.

3-3: Architecture for Level decoder

The Level decoder is used to decode the level of non-zero coefficients. There are seven VLC tables to be selected for the non-zero levels. The syntax of the level codeword is organized by Golomb code, so we need the *First One detector* to find the first one bit position in the bitstream. Then, we can produce the code length and level with the information of the suffix. Besides, the Level_VLC0 has two conditions of escapes and other Level_VLCs only have one condition of escape. The calculated code length and level magnitude of the Level_VLC0 are different from those in Level_VLC1~Level_VLC6. Therefore, we separate the Level decoder tables into two parts as illustrated in Fig. 4. Also, the parameter VLCN is used to select table, so that the level decoder will generate the correct code length and level magnitude. Although Level_VLC0~Level_VLC6 are denoted as "tables", the tables are implemented by arithmetic calculation and logic circuits instead of using ROM LUTs for reducing hardware cost.

3-4: Architecture for TotalZero decoder

The TotalZero decoder is used to decode the total number of zeros before the last non-zero coefficient with the information of TotalCoeff at the same 4x4 or 2x2 blocks. This TotalZero decoder seems like the Coeff_Token decoder. We partition the tables by the TotalCoeff and the attributes of the block. Then, PCCF is also adapted by the TotalZero decoder. Besides, there are no more entries in each condition for TotalCoeff, so we don't use the *HLLT* design method.

3-5: Architecture for Run_Before decoder

The total entries of the LUT used in Run_Before decoder are only 45. It may not be suitable for look-up tables implemented by ROM even if you divide them into multiple smaller ones, since the area cost will be expensive. In the proposed design, we adopt a *ZTEBA* (Zero-left table elimination by arithmetic) method in realizing these LUTs by finding out the rules among these tables. Fig. 5 shows the architecture for the Run-Before decoder.

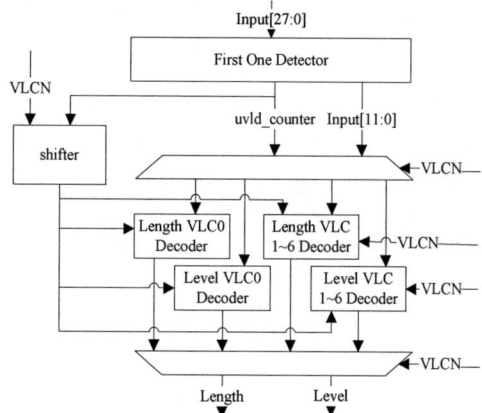

Figure 4: The architecture for Level decoder

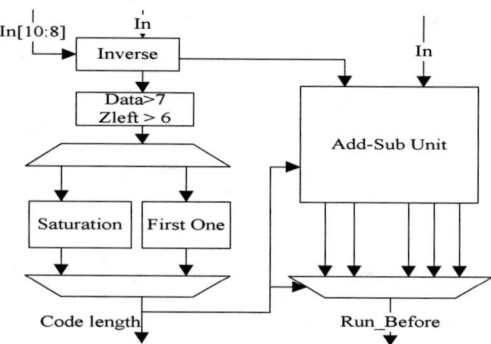

Figure 5: The architecture for the Run_Before decoder

3-6: IDS (Interleave Double Stacks) buffering

The target of the IDS buffering is to communicate the transmission between CAVLC decoder and Inverse Quantization

(IQ). In the proposed design, we use the two stacks with 16-entry in depth and 4-bit in width, two stacks with 16-entry in depth and 16-bit in width, and two single-port RAM. If we use one-stack system, the next decoded block processes will stop and wait for the current block been read by the single-port RAM. If we use double-stack system, we can eliminate the chance of data waiting mentioned above and speed-up the data transition between CAVLC decoder and IQ. The proposed IDS buffering will change run_before into the number of the zeros before each non-zero coefficient and write the rest of 0 values automatically. Moreover, the proposed IDS buffering also provides the function of zigzag scan and mb-field scan.

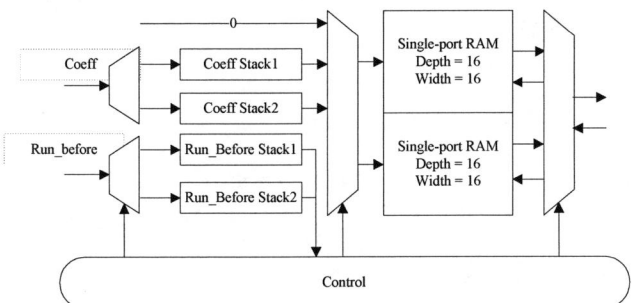

Figure 6: The architecture of the IDS buffering

IV. PERFORMANCE EVALUATON AND COMPARISON

We use Verilog language to implement the proposed design and adopt 0.18um CMOS standard cell-based library. In addition, we build a H.264 video decoder system to verify the correctness and analyze the performance of the proposed design for system verification. Because the design [4,5] don't describe detail CAVLD performance, the Table I only summarizes the performance evaluation and comparison of the proposed design with the design [6] and the Table II shows hardware cost analysis of the proposed design. The synthesis result shows that the design achieves the maximum speed at 175 MHz. When we synthesize the proposed design at clock constraint of 125MHz, the hardware cost is about 4720 gates under a 0.18um CMOS technology, which is more efficient as compared to the design [6]. In order to achieve the real-time processing requirement for H.264 video decoding on HD1080i video, the proposed CAVLD should run over 94MHz when one pixel decodes per cycle. That is, the average number of cycles needed to decode one macroblock in the proposed design is 384 cycles for meeting this real-time processing requirement. The Table III and Table IV respectively show the throughput of proposed CAVLD and CAVLD+IDS running different sequences. All the sequences in our simulation are composed of only I-frames, because decoding I-frames needs more working cycles per MB than decoding B-frames or P-frames in the CAVLD processing.

TABLE I. Hardware cost evaluation of the proposed CAVLD design

	Design[6]	*Proposed Design*
Gate-count	6100	4720
Target Spec.	*Baseline Profile*	*Main Profile@L4.1*
RAM	*Not implemented yet*	*1152 bits (for IDS)*

TABLE II. Hardware cost analysis of the proposed design

Module	*CAVLD*	*IDS*	*Prediction Registers*	*ALL*
Gate-count	4720	2793	2430	9943
RAM	-	1152 bits	-	1152 bits

V. CONCLUSIONS

In this paper, a novel low-cost, high-performance VLSI architecture for MPEG-4 AVC/H.264 CAVLC decoding has been proposed. We have analyzed the decoding flow for CAVLC and exploited the algorithmic properties to propose five design techniques denoted as *PCCF, HLLT, ZTEBA, IDS,* and *ZCS* for reducing both the hardware cost and power consumption as well as achieving high data throughput rate. As a result, the proposed design can decode every syntax element per cycle. The proposed design has been verified through C behavioral simulation, Verilog RTL modeling, and SYNOPSYS gate-level synthesis and simulation. The maximum operating frequency of the proposed CAVLC decoder achieves 175 MHz under a 0.18um CMOS technology. This performance is fast enough for meeting the real-time processing requirement of CAVLC decoding on HD1080i video. As compared to the existing design [6], the proposed design possesses 23 % reduction in the hardware cost and 40 % improvement in speed.

TABLE III. The average cycles per macroblock for the CAVLD design

Pattern	*Mobile*	*Foreman*	*NL1_Sony_D*	*BA1_Sony_D*	*SVA_BA1_B*
Average Cycle/MB	*345*	*184*	*187*	*187*	*140*

TABLE IV. The average cycles per macroblock for CAVLD+IDS design

Pattern	*Mobile*	*Foreman*	*NL1_Sony_D*	*BA1_Sony_D*	*SVA_BA1_B*
Average Cycle/MB	*434*	*294*	*297*	*297*	*240*

REFERENCES

[1] Joint Video Team (JVT) of ISO/IEC MPEG&ITU-T VCEG, "ISO/IEC 14496-10", 2003
[2] Joint Video Team (JVT) reference software JM8.2a
[3] S. K. Chio, J. G. Jeon, W. S. Shim, W. K. Jang, and Victor H. S. Ha, "Design and Implementation of H.264-based Video Decoder for Digital Multimedia Broadcasting," Proc. ICME 2004.
[4] H. Y. Kang, K. A. Jeong, J. Y. Bae, Y. S. Lee, and S. H. Lee, "MPEG4 AVC/H.264 decoder with scalable bus architecture and dual memory controller," Proc. ISCAS 2004.
[5] S. H. Wang, W. H. Peng, Y. He, G. Y. Lin, C. Y. Lin, S. C. Chang, C. N. Wang, and P. Chiang, "A platform-based MPEG-4 advanced video coding (AVC) decoder with block level pipelining," Information, Communications and Signal Processing, pp. 15-18 Dec. 2003.
[6] Wu Di, Gao Wen, Hu Mingzeng and Ji Zhenzhou, "A VLSI architecture design of CAVLC decoder", Proc. 5th International Conference on ASIC, Vol.2 pp. 962-965, 21-24 Oct. 2003.
[7] Qiang Peng and Jin Jing, "H.264 codec system-on-chip design and verification", Proc. 5th International Conference on ASIC, pp. 922 - 925 Vol.2,'21-24 Oct. 2003.
[8] T. L. Chang, Y. M. Tsai, H. C. Chang, C. D. Chien and J. I. Guo, "A High-Performance Reconfigurable Multi-MPEG Bitstream Processing IP Core", Proc. 2004 VLSI DESIGN/CAD symposium, Aug. 12-15, 2004.
[9] T. L. Chang, Y. M. Tsai, C. D. Chien, C. C. Lin, and J. I. Guo, "A High-performance MPEG4 bitstream processing core", Proc. ICME'2004, June 2004.
[10] Ming-T.Sun, "VLSI Architecture and Implementation of A High-Speed Entropy Decoder", Proc. ISCAS-91, pp. 200-203, 1991.
[11] T-H Tsai, W-C Chen and C-N Liu, "A Low Power VLSI for Variable Length Decoder in MPEG-1 Layer III", Proc. ICME2003, Vol. 1, no. 6-9, pp. 133-136, 2003.
[12] Sung-Won Lee and In-Cheol Park, "A Low-Power Variable Length Decoder for MPEG-2 Based on Successive Decoding of Short Codewords," IEEE Trans. on CAS-II., vol. 50, no. 2, p.73~82, February 2003.
[13] Michael Keating and Pierre Bricaud, "*Reuse Methodology Manual*" 3rd edition by *Kluwer Academic Publishers*, 2002.

A Hardware-based Predictive Motion Estimation Algorithm

Saku Hamalainen, Lauri Koskinen, Kari Halonen
Electronic Circuit Design Laboratory
Helsinki University of Technology
P.O.Box 3000
02015 HUT, Finland
Email: sha@ecdl.hut.fi

Abstract—For low-power motion estimation hardware solutions, advanced hardware aimed algorithms and architectures are a necessity. A hardware solution for the modern emerging standard H.264 is researched in this work. Design trade-offs, including search patterns, predictors, and memory architecture, have been made with power consumption and hardware complexity in mind. The memory architecture has been optimised for variable block sizes used in H.264 motion estimation. All predictors are refined with the small search pattern. An average PSNR of $-0.0438dB$ compared to full-search algorithm was simulated. A $2.793mm^2$ layout using $0.18\mu m$ CMOS technology of the design running at $7.61MHz$ is presented.

I. INTRODUCTION

Real-time streaming video will be among the next innovations to be included to mobile hand-held devices. A camera is already installed in many mobile phones, but for streaming video, real-time low-power encoding is necessary. One of the largest issues is the computational complexity of motion estimation within the encoders. As new motion estimation algorithms emerge, it is necessary to research their suitability for low-power hardware solutions. A new hardware-based algorithm was designed after inspecting the EPZS algorithm [1].

The H.264 [2] standard variable block sizes are introduced. The seven block sizes range from 4x4 to 16x16. The standard does not define how to partition the macro-blocks into smaller block sizes. In this work partitioning of the blocks was calculated separately using a CNN algorithm [3] but, for example, Lagrange optimisation [4] could be used instead.

Previous work in hardware-based motion estimation has been mostly done using the *full-search* (FS) algorithm, such as in [5]. Due to differences in memory architectures, the results are not easily comparable to this implementation.

II. BACKGROUND THEORY

A. Power Consumption

To achieve low-power consumption it is necessary to try to minimize the dynamic power consumption of the chip. Equation for this is

$$P_{dyn} = CV_{DD}^2 K_D f_{ck}, \quad (1)$$

where K_D is the duty factor and f_{ck} is the clock frequency. It is worth noticing that CV_{DD}^2 is seen as a constant here. Therefore f_{ck} should be minimized for lower power consumption. The difficulty in lowering f_{ck} is that in some cases K_D has tendency to rise when f_{ck} decreases. Nevertheless a smaller clock frequency leads to smaller power consumption.

B. Motion Estimation Fundamentals

Motion estimation is the single component causing most of the power consumption in current video coding standards. To achieve good image quality with low bit rate it is necessary to find a good motion vector. This is achieved by comparing current image with previous image and trying to find the block with the smallest difference or *distortion*. There are many formulas for comparing the difference, but in this work the simplest formula, Sum of Absolute Differences (SAD) is used:

$$SAD(dx, dy) = \sum_{m=x}^{x+N-1} \sum_{n=y}^{y+N-1} |I_k(m,n) - I_{k-1}(m+dx, n+dy)|. \quad (2)$$

Commonly $I_k(m,n)$ is referred as *current block* and $I_{k-1}(m+dx, n+dy)$ is referred as *reference block*. The spatial distance (dx, dy) between minimum distortion blocks is referred as *motion vector* (MV). The minimum SAD solution is searched within a *search area* (SA) to reduce the complexity of the calculation. In FS algorithm all points are calculated in a search area but other algorithms use only certain points within the SA.

In predictive motion estimation, previously calculated MVs are used to help the calculation by predicting the likely area of the minimum SAD. Most commonly used prediction vector is the *median vector predictor* (MVP), this is also specified as the reference vector for coding of the MVs in H.264. The MVP is formed calculating a median value from selected previously calculated MVs referred as A (left), B (top), C (right-top) and D (left-top). D is used if C is unavailable and with some block shapes special rules are applied.

In H.264 a *skip block* is defined so that the MVP is used as the MV and no distortion values are encoded. The algorithm simulations presented in this document all implement the skip block.

III. HARDWARE DESIGN ASPECTS

As this algorithm is based on hardware implementation, it is necessary to be able to calculate the search patterns in the most efficient way. There are basically two different methods to solve the calculations; a parallel solution and a sequential solution. Most of the software algorithms are based on a sequential solution. Hardware solutions are in most cases fixed and calculating variable block sizes is complicated with good efficiency and clock rate.

In this work it is assumed that calculation of the motion vector for the smallest block shape, the 4x4 block is the worst case calculation. The worst case term here refers to the usage percentage of the memory bandwidth. Therefore the computation should be as efficient as possible in the worst case. All of the other block shapes are constructed from 4x4 blocks so calculating the distortion in 4x4 partial solutions and then adding the results will solve all of the other block shapes.

A. Search Pattern Analysis

It is possible to solve all points in any search pattern for a 4x4 block in one cycle. This is based on the fact that almost all of the needed pixels for the points in the search pattern are either same as in the previous pattern or can be assumed as being very close to each other in the memory. The size of the search pattern then only determines the required hardware resources and memory bandwidth. This kind of calculation assumes that it is possible to open in parallel a memory area which is related to the size of the search pattern. If the search pattern would have points within range $(-1, 1)$, a pipelined solution would require a 6x6 parallel memory bandwidth for the reference block and only small search patterns would be possible to calculate. In this work range $(-2, 2)$ was selected as a good trade-off, so a 8x8 parallel memory bandwidth for reference block must be implemented. This sets an upper limit to the size of some patterns described in software solutions.

First the small diamond and large diamond patterns from the MVFAST [6] algorithm and the small square pattern from the EPZS [7] algorithm were analyzed for suitability in a parallel hardware solution. These patterns are shown in Fig.1. The largest pattern in EPZS was not considered, because the number of calculation points was too large for the selected amount of calculation pipelines. If both the diamond patterns were used the size of the large pattern would determine the amount of needed SAD calculation pipelines as there are 9 needed calculation points in the large diamond pattern but only 5 in the small. In this case the small diamond would use the 9 pipelines inefficiently. As a minimum of 9 pipelines was needed the small square was chosen instead of the small diamond. The advantage of the small diamond is the smaller number of computed SAD values but this also has a disadvantage of a greater tendency of getting stuck at local minimum. Also, in low-power hardware design, the optimisation of the memory bandwidth is of greater significance that the optimisation of the calculation [7].

It was also noticed if a large square pattern would be selected as the large pattern instead of the large diamond

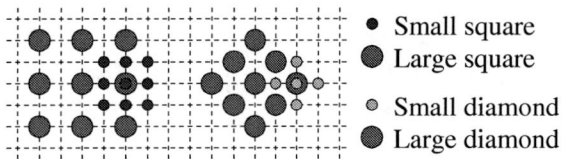

Fig. 1. The referred search patterns.

pattern, the small pattern would fill the gaps of the pattern in more efficient manner (Fig. 1). The large square pattern also has 9 computation points and also uses the needed memory bandwidth in more efficient manner than the large diamond pattern. The only issue with the large square pattern is that it calculates only every other SAD point. In some cases, where the best solution lies between the large square points, the large diamond pattern would be better.

B. Predictor Calculation

The predictors described in the EPZS algorithm were also analyzed for suitability. Predictors using other frames were not considered. As the target of this implementation is mobile applications that encode streaming video, the memory size restrictions of such applications restrict the use of only the previous frame as the reference frame in the near future.

As this implementation is able to calculate all points in a search pattern in same amount of clock cycles as it calculates only one point, it is possible to refine each prediction vector using a search pattern. All vectors are refined with the small square pattern. The predictors used in this work are (denominations from [1]):

- Median. Median of vectors A-D (Chap.II-B)
- Zero. Collocated block of reference frame
- Spatial. Vectors A-D (Chap.II-B)
- 4 Adjacent temporal. Positions $\pm 1x$, $\pm 1y$ of current position in reference frame.
- Previous best. The best predictor of the reference frame's collocated block.

The additional fixed grid in EPZS is also not used the number of clock cycles needed is too large and also if each of the 16 points would be calculated using small square pattern, the number of calculation points would be close to FS algorithm.

The level based pattern selection scheme from MVFAST is implemented in this algorithm which decides the start search pattern based on the city-block length of the A,B,C and D MVs.

IV. IMPLEMENTATION SOLUTIONS

For each calculation point in a search pattern, one pipeline is needed. One pipeline must be able to calculate the distortion value for 4x4 pixel area. The pixel data is spread from the 8x8 memory output to 9 pipelines based on the shape of the search pattern. In Fig.2 the distribution of the memory output to the 9 SAD pipelines is portrayed. The distribution for the small and the large patterns is also shown. Only left-top, center-top and right-bottom points of the search pattern are portrayed

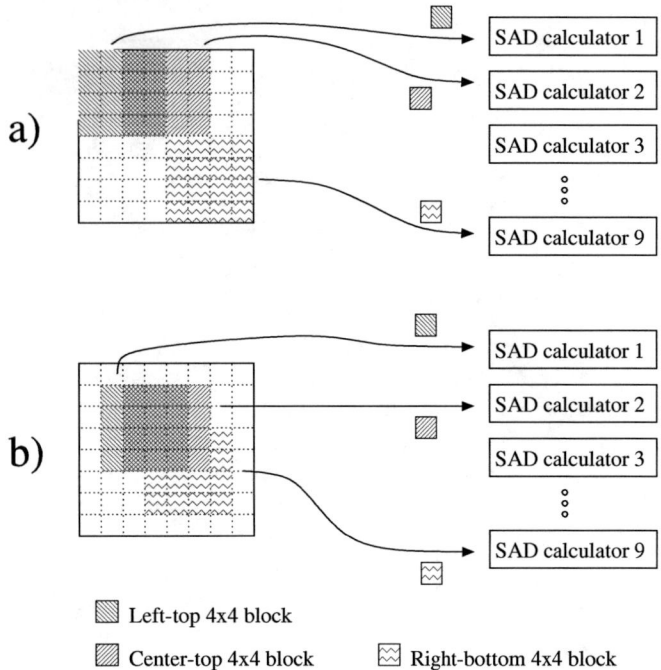

Fig. 2. Distribution of the 8x8 pixel area to the pipelines. a) Large square pattern, b) Small square pattern

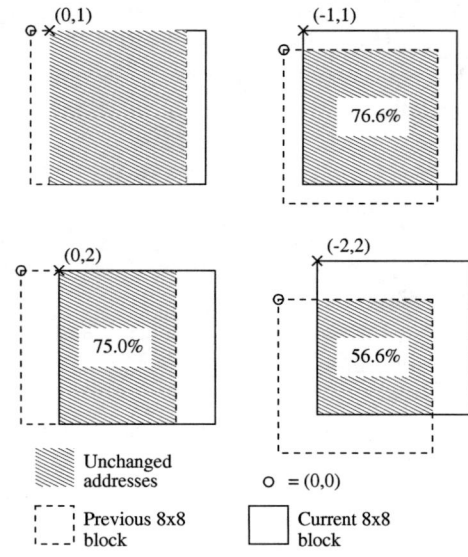

Fig. 3. As the search pattern moves, many of the memory addresses remain unchanged resulting to decreased power consumption. This happens for 4x4 blocks only however.

for the sake of simplicity. The center-center point would be distributed to SAD pipeline 5.

The current block data is the same for all of the pipelines, so only 4x4 bandwidth is needed for the current block.

A. Memory Architecture

In ME applications it is advantageous to split the frame memory into sub-blocks. This memory partition has a large effect on functionality of the memory. In this solution it was selected so that each memory block has a one pixel bandwidth and there are 8x8 memory blocks.

The pixels are spread to the memory blocks in such manner that pixels 1,8,16,... are in the first block and pixels 2,9,17,... are in the second block and so on. This spreading is done in 2 dimensions so that the first memory block holds pixels from picture coordinates $(1,1)$, $(1,8)$, $(8,1)$, etc. In other words rows and columns are both spread with modulo 8.

Each of these memory blocks has unique address bus. This enables to fetch each pixel with different address. Only 4 different addresses are needed for the operation, however. When the search pattern moves, because it can only move maximum of 2 pixels, many of the memory addresses remain unchanged (Fig. 3).

Because the address changes were minimized, the parallel memory architecture doesn't output the data in linear form. Therefore after being fetched from the memory the 8x8 data array must be 2-dimensionally rotated before it is ready for use.

With the described pipeline and memory architecture the data re-use problem is somewhat solved also, as now many of the pixels addresses in the search pattern remain unchanged. For example, with a 8x4 block 50% of the addresses for the respective 4x4 block stay unchanged.

V. Performance

One search pattern for a 4x4 block is calculated in 4 clock cycles. For larger blocks only 1 cycle is added for each 4x4 block the larger block contains because they can be subsequently calculated in the pipelines during a clock cycle. Therefore a 8x4 block takes 5 clock cycles. The reason why search patterns cannot be calculated each cycle is because the results of previous search pattern must be ready before it is known what direction the pattern must be moved. The implementation has total of 144 processing elements, which compute the SAD values.

To be able to calculate the worst case performance for a search algorithm like this, it is needed to know how much the search pattern is able to move. Many implementations have a search area limited calculation. Search area could be for example $(-16, 15)$. In this implementation a 'hop'-counter is used. This limits the search by limiting directly the amount of search pattern movement or 'hops'. Hop number is not constant, so it could even be different for each search.

Worst case performance for a CIF image with 396 macro-blocks, containing 16 4x4 blocks and with 10 hops is $4*10*16*396 = 253440$ clock cycles per frame. For real-time at 30 frames per second, 7.61MHz is the needed clock rate. For a 30 fps VGA-sized image with 1200 macro-blocks the respective figure is 23.04MHz. The layout in shown Fig.6.

As the hop count directly affects the needed clock rate of the chip and the hop count isn't constant, it is possible to optimise the hop count so that it will adaptively change within one frame. This can be used to decrease the clock rate or to improve the results of the motion estimation.

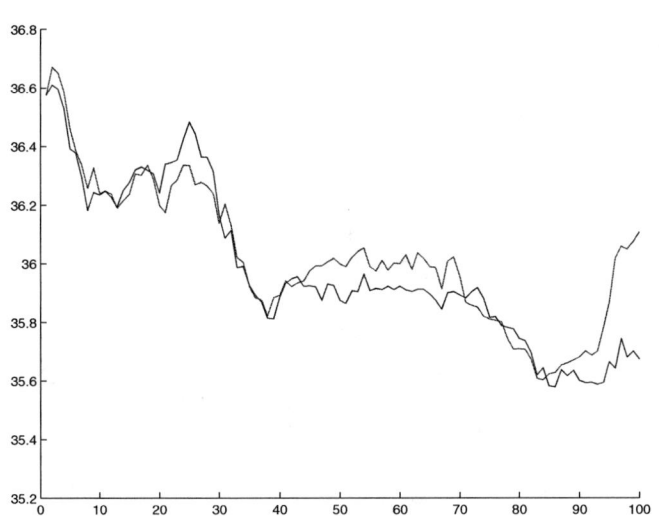

Fig. 4. Simulation results. Foreman QCIF frames 151-250. Red is FS. Blue is this algorithm.

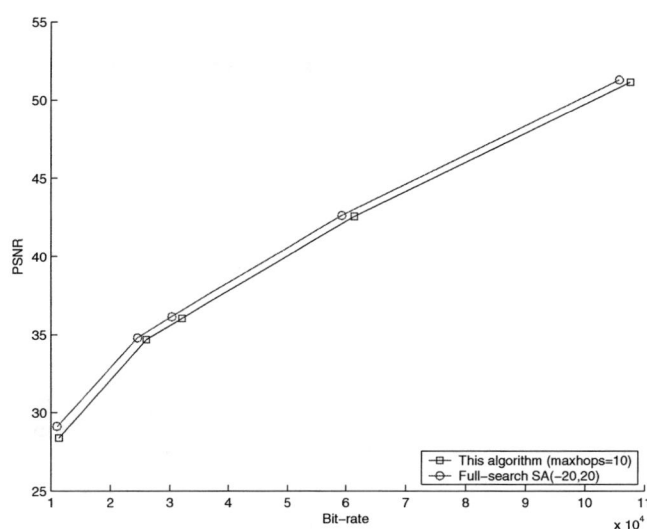

Fig. 5. Rate-distortion figure. Foreman QCIF frames 151-250.

VI. SIMULATION RESULTS VERSUS THE FS ALGORITHM

In this work only the pattern search engine is implemented in hardware due to lack of resources. Other parts of the algorithm, like the prediction vectors and motion compensation are simulated using Matlab.

The simulation results for the high motion part (frames 151-250) of the QCIF Foreman sequence (Q_p=28) are portrayed in Fig.4. Such a high Q_p value would be likely to be used in low bit-rate mobile applications. The average PSNR difference in this sequence is $-0.0438dB$ compared to the FS (SA ±32) algorithm.

The rate-distortion graph of Fig.5 is computed using Q_p values 10,20,28,30, and 38. The maxhops value indicates the maximum amount of moves the search pattern can perform before the search is terminated. Maxhops=10 makes the maximum search area (-20,20).

VII. CONCLUSIONS

A predictive algorithm for variable block size motion estimation is introduced. The algorithm is designed for a efficient hardware implementation. The search patterns and predictors are hardware optimised. The memory architecture is optimized for variable block size calculation. The results show a low-frequency hardware is possible with comparable image quality to the FS. To improve the design the worst case (4x4 block) usage of the memory bandwidth would have to be increased. The clock frequency of the design could then be reduced proportionally to the increase in worst case memory bandwidth.

REFERENCES

[1] H.-Y.C.Tourapis et al., *"Fast Motion Estimation within the JVT codec"*, Proc. of the 2003 Int. Conf. on Multimedia and Expo, ICME03, pp. III 517 - III 520

[2] Joint Video Team of ITU-T and ISO/IEC JTC 1, *"Draft ITU-T Recommendation and Final Draft International Standard of Joint Video Specification (ITU-T Rec. H.264 and ISO/IEC 14496-10 AVC)"*, doc. JVT-G050r1, May 2003

[3] L.Koskinen et al., *"CNN Algorithm for H.264 Partitions"*, Proc. of the Cellurral Neural Networks and their Applications, 2004. (CNNA 2004), pp. 346-351

[4] Wiegand, T., et al. *"Rate Constrained Coder Control and Comparison of Video Coding Standards"*. IEEE Trans. on Circ. and Sys. for Video Tech., Vol. 13, July 2003 pp. 688 -70

[5] T.Enomoto et al., *"Low-Power CMOS Circuit Techniques for Motion Estimators"*, Proc. of the 2003 Int. Symp. on Circuits and Systems, Vol. 5, pp. V409-V412

[6] ISO/IEC JTC1/SC29/WG11, *"Optimised Reference Software for Coding of Audio-Visual Objects Version 4.0"*, ISO/IEC DTR 14496-7, Pattaya, December 2001

[7] Kuhn P. *"Algorithms, Complexity Analysis and VLSI Architectures for MPEG-4 Motion Estimation"*. Kluwer Academic Publishers, 1999, 239 p.

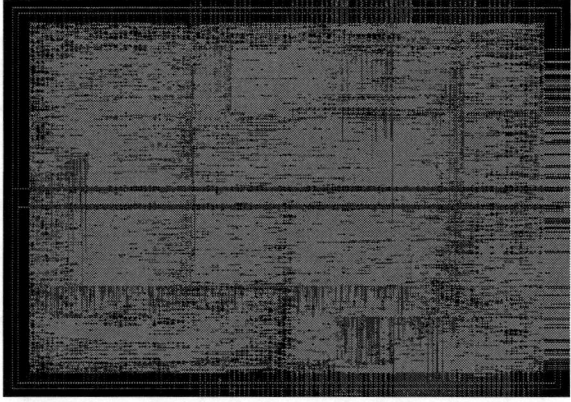

Fig. 6. Image of the layout. Size of the layout is $2.793mm^2$.

A MULTIPLICATION-ACCUMULATION COMPUTATION UNIT WITH OPTIMIZED COMPRESSORS AND MINIMIZED SWITCHING ACTIVITIES

Li-Hsun Chen, Oscal T.-C. Chen

Signal and Media Laboratories,
Department of Electrical Engineering,
National Chung Cheng University,
Chia-Yi, 621 Taiwan

Teng-Yi Wang, Yung-Cheng Ma

Computer System Development Dept.,
Computer & Communications Research Labs.,
Industrial Technology Research Institute,
Hsinchu, 310 Taiwan

Abstract—A low-power Multiplication-Accumulation Computation (MAC) unit using the radix-4 Booth algorithm is proposed in this work, by reducing its architectural complexity and minimizing the switching activities. However, to maintain a high performance, the critical delays and hardware complexities of MAC units are explored to derive at a MAC unit with a high performance and a low hardware complexity. In addition, a carry-save addition operation with optimized compressors is proposed to omit the use of half adders to further reduce the hardware complexity. Furthermore, the scheme to reduce the switching activity is proposed in this work to lower the power consumption of the proposed MAC unit. In performing a MAC for $X \times Y + Z$, the effective dynamic ranges of X and Y are detected, the one with the smaller effective dynamic range is processed for Booth decoding so as to increase the probability of the partial products being zero, and thus the switching activity of the MAC unit is reduced. Moreover, the effective dynamic range of the result from this multiplication is also estimated and compared with the effective dynamic range of the datum, Z. The larger effective dynamic range of the two data is considered as the effective word length for an addition operation. Pipelined latches are used to make the non-effective operation maintaining the status of the previous operation so to reduce the switching activities from the addition performed in MAC. After the addition operation, sign extension is performed on the result from the effective sign bit copied to non-effective bits to derive at a correct output datum. When comparing to the conventional MAC units, the proposed MAC unit is able to reduce 21.09% to 43.74% of power consumption. Additionally, the proposed MAC unit outperforms the conventional ones in comparing the product of critical delay, area, and power consumption.

I. INTRODUCTION

The multiplication-accumulation computation is considered as one of the fundamental operations in digital signal processing. Such operation is applicable in digital filtering, speech processing, video coding or communication, etc. The MAC unit becomes a basic component in many Digital Signal Processors (DSP) and Application-Specific-Integrated-Circuits (ASIC). Moreover, with the ever-increasing demand for portable electronic products, an electronic component with low power consumption would surely lead the market trend. Therefore, it is needed to design a low-power MAC unit.

The majority of power consumption in the circuit comes from the switching activities, as shown with the following equation:

$$P_{switching} = \alpha C V_{dd}^2 f_{clk} \qquad (1)$$

of which, α is the switching activity parameter, C is the loading capacitor, V_{dd} is the operating voltage, and f_{clk} is the operating frequency. By reducing the hardware complexity, the loading capacitor can be lowered, which in turn lower its power dissipation. In addition, αC could be viewed as the effective loading capacitor during switching operations so that power consumption can also be lowered from reducing circuit's switching activities.

Many researchers have attempted in designing a MAC architecture with high computational performance and low power consumption. Elguibaly proposed a fast pipelined implementation to lower the MAC architecture's critical delay [1]. Murakami *et al.* adopted the half array implementation to design a high-speed and area-effective MAC architecture [2]. Raghuneth *et al.* made use of a carry-save multiplier that can simplify sign extension and saturation, and further applies it on a MAC architecture to reduce the unit's area and power consumption [3]. While the abovementioned schemes may increase the computational performance or lower the power dissipation of MAC architectures, these schemes are applied on one type of MAC architecture instead of exploring the other types, which implies that there still is room to improve on a conventional MAC architecture.

To realize the low-power MAC unit proposed in this work, first we examine the critical delays and hardware complexities of conventional MAC architectures to derive at a unit with low critical delay and low hardware complexity. Next, a carry-save addition operation is proposed in the unit to further reduce its hardware complexity. Finally, a design to reduce the number of switching activities is applied to lower power consumption of the proposed MAC unit.

II. CONVENTIONAL MAC ARCHITECTURES

With $X \times Y + Z$ as the MAC, X and Y are two input data with k bits, and Z is the input datum with $2k$ bits. A $k \times k$-bit multiplier and a $2k$-bit Carry Propagation Adder (CPA) can be applied to perform this computation. As shown in Fig. 1, the $k \times k$-bit multiplier is applied to perform the multiplication of X and Y, while the $2k$-bit CPA carries out the addition of the multiplication result and a datum, Z, via a carry lookahead adder, a carry select adder, a carry skip adder and so on to increase computational performance. The $k \times k$-bit multiplier can also be broken down into three parts: Partial-Product Generation (PPG), Partial-Product Summation (PPS) and $2k$-bit carry propagation adder denoted as mul_CPA. The PPG unit generates many partial products based on the algorithm of the multiplication such as Baugh-Wooley, Booth, and modified Booth algorithms. The PPS unit performs addition on these partial products, and to speed up its computation, the carry-save addition technique is used to generate the sum and carry values, as shown in Fig. 1. Then, the $2k$-bit mul_CPA unit adds up two data to derive at the result of this multiplication. Additionally, the PPS unit can be implemented using a carry-save adder array, a carry-save adder tree or a hardware platform composed of compressors, as illustrated in Fig. 1. Both of the carry-save adder array and carry-save adder tree allocate many carry save adders (CSAs) to perform PPS. With allocating CSAs in a serial form, the carry-save adder array has a long critical delay. The carry-save adder tree, on the other hand, has a high operating speed by parallel processing;

however, this would increase the number of CSAs so that the carry-save adder tree has a higher hardware complexity than the carry-save adder array. In working with a hardware platform composed of compressors, the addition of every bit position is completed using a compressor. Each compressor takes the bit values from the PPG unit in its bit position and performs addition via a tree form with the output bits from the neighboring compressor, then using the compressors would have the same computational performance as that of the carry-save adder tree. Furthermore, by integrating the addition of input bits in the same bit position, it would have a hardware complexity close to that of the carry-save adder array. Taking both computational performance and hardware complexity in consideration, operating using compressors would outperform the other two hardware platforms.

To reduce the hardware complexity, the addition of the MAC can be merged into PPS. By performing the carry-save addition on the Z datum and the partial products produced from the multiplication, sum and carry values are generated. A $2k$-bit CPA then performs the addition on the two values to derive at an output datum of MAC. Based on the above operation, the MAC architectures can be classified into tree types as shown in Fig. 2 [1-3]. Type I may have a less hardware complexity than Type II and III, but it has a long critical delay due to its pipelined design. In comparing Type II and Type III, Type II using one more $2k$-bit CSA has a higher hardware complexity and longer critical delay than Type III. Based on its high speed and low hardware complexity, Type III is considered more favorably in the design of the proposed MAC architecture.

III. PROPOSED LOW-POWER MAC UNIT

Based on the architecture of Type III shown in Fig. 2, the proposed MAC unit is designed. A design scheme to reduce its switching activities is also included to lower the power consumption, while a carry-save addition operation is proposed to reduce its hardware complexity. The proposed MAC architecture is shown in Fig. 3. With X, Y and Z as the three input data for the proposed architecture, it is able to perform the multiplication of a 16-bit X and a 16-bit Y, and to carry out the addition of the multiplication result and a 32-bit Z datum. The radix-4 Booth algorithm is applied to perform the multiplication to design a low-complexity and high-speed architecture. The functional blocks of proposed low-power MAC unit are illustrated bellow.

A. Proposed low-power design

In Fig. 3, a Dynamic-Range Determination (DRD) unit for multiplication (MUL_DRD unit), a dynamic-range determination unit for addition (ADD_DRD unit), and a Sign Extension (SE) unit are used to perform the low-power operation. A MUL_DRD unit can detect the effective dynamic ranges of two multiplication input data. The one with smaller effective dynamic range is processed for Booth decoding so that the probability of the partial products becoming zeros is increased to reduce the number of switching activities. In addition to detecting the effective dynamic ranges of the two input data, MUL_DRD unit can also estimate the effective dynamic range of the multiplication result and pass on the control signals, which represent the effective dynamic range of the product result, to the ADD_DRD unit. With these control signals and datum, Z, the ADD_DRD unit is able to detect the non-effective and effective addition of the multiplication result and the datum, Z. Control signals are generated from the ADD_DRD unit to control pipelined latches in carry-save addition unit so that the non-effective addition can maintain the status of the previous operation to reduce the switching activities. With the output from non-effective addition maintaining the status of the previous operation, to attain a correct result, the SE unit has to perform the sign extension on the CPA's output [4, 5, 6].

The two functions of the MUL_DRD unit is one, to detect the effective dynamic ranges of two input data for multiplication. The one with smaller effective dynamic range is perceived as the multiplier while the one with larger effective dynamic range is the multiplicand. The second function is to estimate the effective dynamic range of the multiplication result. A diagram of the MUL_DRD unit is shown in Fig. 4. For the first function, this MUL_DRD unit is based on detection performed on 3 bits per group, since radix-4 is used for Booth decoding. When all three bits are either all zero or 1, the control signals would be 1, or otherwise 0. The control signals would pass through logic gates to obtain the effective dynamic ranges of two input data. These two effective dynamic ranges are compared to see which one is smaller. After controlling the input data paths, the input datum with a smaller effective dynamic range is used for the Booth decoding. As for the second function, six AND gates, one inverter, and six 1-bit latches would be used to generate six control signals (m_ctl0, m_ctl1, ..., m_ctl5) to represent the effective dynamic range of the multiplication result and to perform pipelining [4].

As shown in Fig. 5, the ADD_DRD unit determines the effective dynamic range of an input datum, Z, based on the resolution of 4 bits where the 5-bit comparators are utilized. The effective dynamic range of the other input datum is generated from the MUL_DRD unit. Which one of two input data has the larger effective dynamic range is then determined in this unit. Additionally, the control signals are produced to control the operations of pipelined latches and the SE unit to lower power consumption and to obtain a correct output, as shown in Fig. 3 [5].

With the inclusion of the ADD_DRD unit, some non-effective bits in the input data for a 29-bit CSA and a 32-bit CPA are kept in previous states, so to minimize the switching activities and to reduce power consumption. However with these bits for addition in the previous states, their added results must be corrected. The SE unit, realized by multiplexers, has the responsibility of restoring the correct result from the output of the CPA. Fig. 3 shows the SE unit that is composed of multiplexers based on the control signals of a_ctl0', a_ctl1', ..., a_ctl5' generated from the ADD_DRD unit where these multiplexers determine if their output values are taken from the sign or value of the CPA's output result.

B. Carry-save addition unit

As shown in Fig. 3, carry-save addition unit is required to perform carry-save addition on partial products and carry-in bits produced by the Booth decoders, the Z datum, and a sign correction term. Control signals (a_ctl0, a_ctl1, a_ctl2, ..., a_ctl5) and clock signal (clk) are responsible for the control of pipelined latches so that the non-effective bits can maintain the status of the previous operation to further reduce the switching activities. Additionally, the pipelined latches can be used to reduce the critical delay of the MAC unit. Carry-save addition is normally implemented using a full adder and a half adder. However, half adder is only able to take two bits in the same bit position and add them up to produce two bits in different bit positions so that the number of bits can't be reduced. Therefore, the use of half adders would lead to hardware waste. In the proposed carry-save addition unit, a compressor is used in each bit position to perform the carry-save addition operation. In design of each compressor, if the number of input bits is odd, only the full adder is required to perform the carry-save addition. But if number of the input bits is even, then a half adder is needed to design a compressor. Take Fig. 6 as an example, when the compressor is performing carry-save addition in the ith bit position, there are two kinds of input data involved. One is the bit values of the partial products, carry-in bits or sign correction term in the ith bit position, which number of bits is assumed as d. The other kind is the output bits from the compressor of the $i-1^{th}$ bit position, and its number of bits is assumed as l. Thus the total number of the input bits of this compressor is $d+l$. If the result of $d+l$ is an odd number, then the compressor shown in Fig. 6 would not require a half adder, otherwise yes.

Based on this principle, two methods are undertaken to ensure that the number of input bits of the compressor in each bit position is odd to omit the use of the half adder. The first method is to insert the carry-in bit ($Cin0$) of the Booth decoder in the least significant bit of the 32-bit CPA, so that the number of input bits of the compressors in the least significant bit positions would be odd. The second method takes the fact that sign correction term is a constant so that it can use some inverters to perform its addition. For example, if a 1-bit x is added to a 1-bit with the value of 1, the output would be a 2-bit xx' value, where x' is the inversion of x. Such operation would allow the number of input bits of the compressors in the most significant bit positions to become odd too. The operation of the proposed carry-save addition unit reflecting these two methods is displayed in Fig. 7. In this diagram, twenty-five compressors are applied to perform the carry-save addition of eight 17-bit partial products, eight carry-in bits and a sign correction term. Here, the twenty-five compressors consist of four 3:2 compressors, four 4:2 compressors, four 5:2 compressors, four 6:2 compressors, four 7:2 compressors and five 8:2 compressors, and as shown in Fig. 7, the numbers of input bits of these compressors are odd so that only full adder is used. In addition, a 29-bit CSA is used to complete the addition process with the datum, Z. Lastly, the 32-bit sum and carry values are produced to input into the 32-bit CPA along with $CinO$ for one addition operation to obtain a MAC result.

In the hardware complexity, we further assume the input data, X, Y and Z, of MAC are k, k and $2k$ bits, according to the above two methods, the proposed carry-save addition unit would need $((k^2-k-2)/2)$ full adders and $(k/2)$ inverters. As for computation delay, due to the different arrival time of input bits in a compressor, some are inputted earlier than the others. Of a compressor, some computation paths have longer delays, while others have shorter delays. To reduce the critical delay of the compressor, we can write the delayed input bits into the computation paths with a short delay, and the bits inputted earlier into the computation paths with a long delay.

To take numbers in comparison, the proposed unit is compared with the conventional carry-save adder array, carry-save adder tree, and a hardware platform composed of compressors. Table 1 compares the hardware complexities and critical delays of the three conventional and the proposed designs, with X and Y at 16 bits and Z at 32 bits. It is evident that, by changing the number of input bits of every compressor, the proposed operation can omit the use of the half adder without increasing to the number of full adders and critical delay. The proposed operation does prove to have a lower hardware complexity than that of the conventional designs.

IV. POWER ANALYSES

In analyzing and comparing power consumption, critical delay, and hardware complexity between the proposed and conventional MAC units, the TSMC 0.25μm standard cell is utilized to realize all the MAC units discussed in this work. The Cadence tool is applied to place and route to get the layouts of these MAC units. Power-mill and Time-mill tools are used to estimate their power consumption and critical delays. Here, the clock frequency of the MAC units is set at 50MHz for fairly comparing their power consumption. Input data are taken from the practical signals going through the adaptive differential pulse code modulation (ADPCM) audio, G.723.1 speech, and wavelet-based image coder. The detail to take these input data is described in [5].

Based on the above analysis, Table 2 compares the power consumption, areas and critical delays between the proposed and conventional units. With using both multiplication and addition to reduce switching activities, the proposed unit can be further broken down into three architectures for its power consumption analysis: 1. PMA, with the use of both multiplication and addition to reduce switching activities; 2. PM, with only the use of multiplication to reduce switching activities; 3. P, without any use to reduce switching activities. An architecture with only the use of addition to reduce switching activities is not considered here because to do so would still require the control signals from the MUL_DRD unit that is used for multiplication at low-power operations. With the proposed MAC unit, the hardware complexity for multiplication is much higher than that for the accumulation, thus adopting the method on multiplication to reduce switching activities, rather than that on addition, would reduce more power consumption. As Table 2 clearly shows, the proposed MAC unit using PM can reduce more than 6% of power consumption compared to that of the proposed MAC unit using P, while the proposed MAC unit using PMA can at best lower only another 2.32% of power consumption from the one using PM. In comparing the proposed MAC units using PMA, PM, and P, they all have the same critical delay. However, the proposed MAC unit using PMA having both multiplication and addition to reduce switching activities, it would have a larger area, and lower power dissipation than the ones using PM and P. However, if the products of areas, critical delays and power consumption all need to be taken into consideration, the proposed MAC unit using PM would outperform the other ones.

In conventional architectures, Elguibly used the Type I architecture in Fig. 2 so that it has a smaller area and longer critical delay [1]. With the Type III architecture in Fig. 2, Murakami's design has an area larger than that of Elguibly, but its critical delay is smaller than Elguibly's [2]. Raghunath employed the Type II architecture in Fig. 2 so that its hardware complexity is increased [3]. However Raghunath was able to decrease the area by simplifying the additions of sign extension and carry-in bits. Moreover, since the carry-save adder array is used in the three architectures for carry-save additions, the three conventional architectures have a longer critical delay than the proposed ones. The proposed MAC unit using P could reduce 14.31% to 35.60% of power consumption from the conventional ones. With the use of only multiplication to reduce switching activities, the proposed MAC unit using PM could reduce 19.47% to 42.41% of power consumption. And lastly, with the use of both multiplication and addition to reduce switching activities, the proposed MAC unit using PMA reduces 21.09% to 43.74% of power consumption. Furthermore, the three proposed MAC units also outperform the conventional ones in comparing the products of critical delays, areas, and power dissipation.

V. CONCLUSION

Critical delays and hardware complexities of the conventional MAC architectures is looked at in this work to propose a MAC unit with high computational performance and low power consumption. In addition, a carry-save addition operation using compressors is proposed. The proposed operation would change the number of input bits of every compressor so that the use of half adders is omitted to further reduce hardware complexity. With detecting the effective dynamic ranges of input data for the MAC unit and to manipulate data flows in multiplication and addition, the number of switching activities is reduced to lower the power consumption of the proposed unit.

REFERENCES

[1] F. Elguibaly, "A fast parallel multiplier-accumulator using the modified Booth algorithm," *IEEE Trans. Circuits and Systems II: Analog and Digital Signal Processing*, vol. 47, pp. 902-098, Sept. 2000.

[2] H. Murakami, et al." A multiplier-accumulator macro for a 45 MIPS embedded RISC processor," *IEEE J. Solid-State Circuits*, vol. 31, pp. 1067-1071, July 1996.

[3] R. K. J. Raghunath, et al. "A compact carry-save multiplier architecture and its applications," *Proc. IEEE 40th Midwest Symp. Circuits and Systems*, vol. 2, pp. 794-797, Aug. 1997.

[4] O. T.-C. Chen, S. Wang, Y. W. Wu, "Minimization of switching activities of partial products for designing low-power multipliers," *IEEE Trans. VLSI Systems*, vol. 11, pp. 418-433, June 2003.

[5] O. T.-C. Chen, R. R.-B. Sheen, s. Wang, "A low-power adder operating on effective dynamic data ranges," *IEEE Trans. VLSI Systems*, vol. 10, pp. 435-453, Aug. 2002.

[6] O. T.-C. Chen, N. Y. Shen, C. C. Shen, "A low-power multiplication accumulation calculation unit for multimedia applications," *Proc. IEEE ICASSP*, vol. 2, pp. 645-648, April 2003.

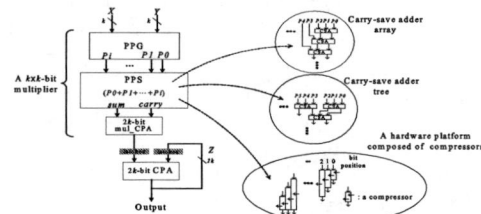

Fig. 1 The conventional MAC architecture.

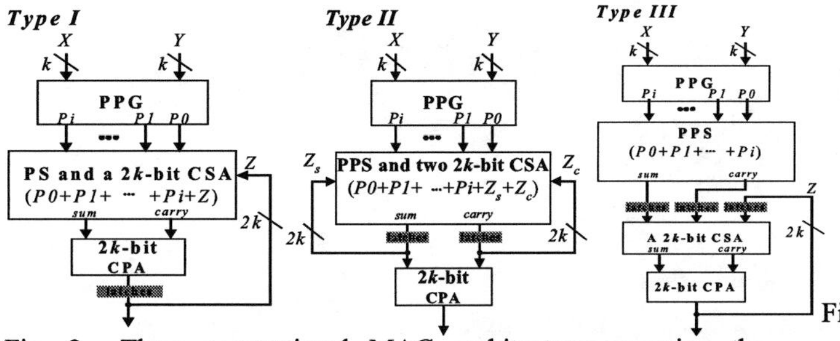

Fig. 2 Three conventional MAC architectures merging the addition into PPS.

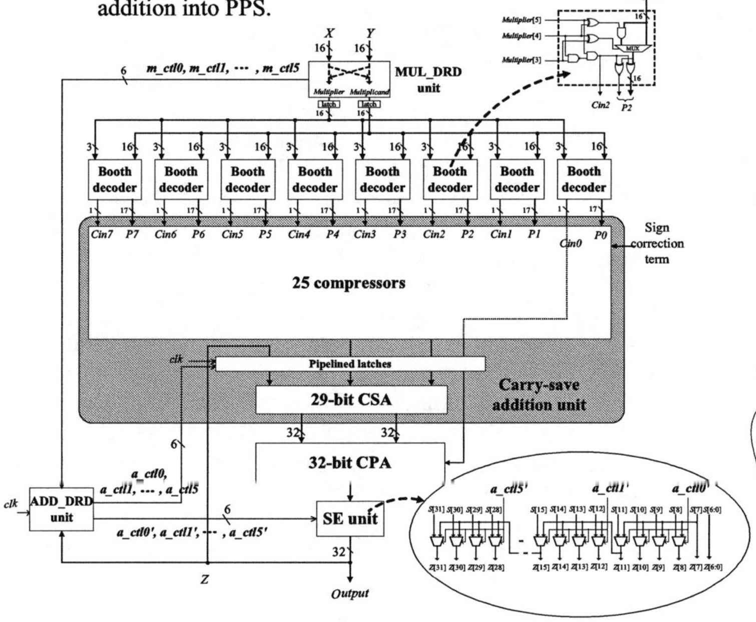

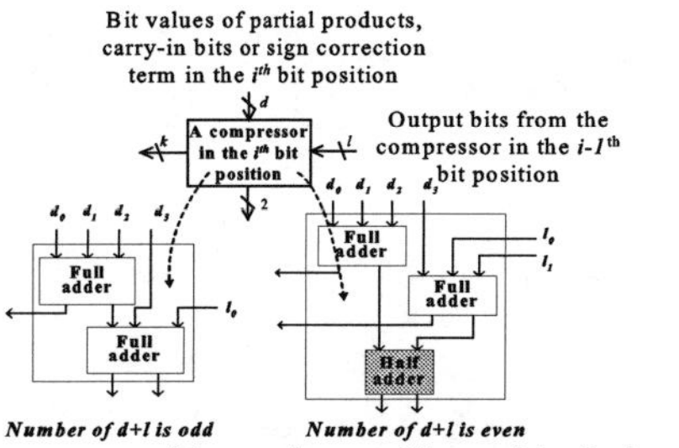

Fig. 6 The compressors with numbers of input bits being odd and even.

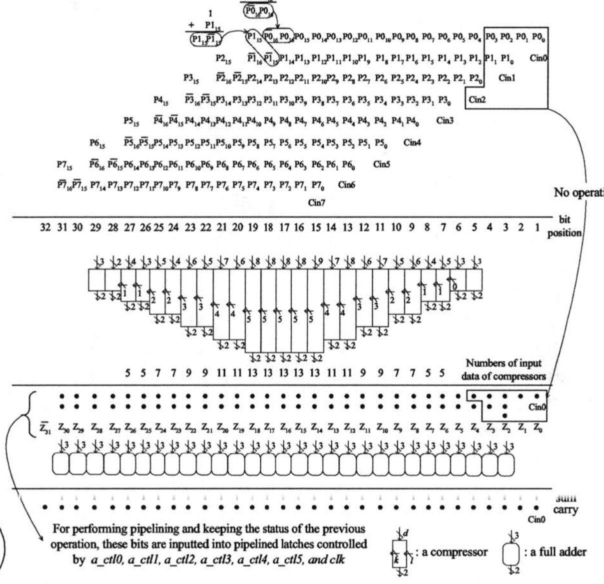

Fig. 3 The proposed MAC architecture.

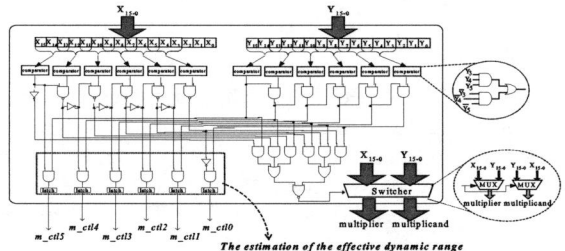

Fig. 4 The dynamic-range determination unit for multiplication.

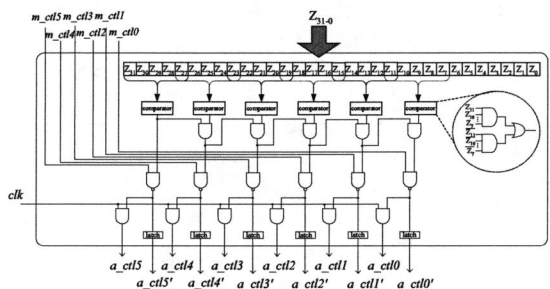

Fig. 5 The dynamic-range determination unit for addition.

Fig. 7 The proposed carry-save addition operation.

Table 1 Hardware complexities and delays of the proposed and conventional designs.

Designs \ Features		Hardware complexities	delays (delay of a full adder)
Conventional carry-save adder array		121 full adders 23 half adders	7
Conventional carry-save adder tree		120 full adders 36 half adders	4
Hardware platform composed of compressors	Conventional operation	121 full adders 22 half adders	4
	Proposed operation	119 full adders 8 inverters	4

Table 2 Comparison between the proposed and conventional MAC units.

MACs \ Features		Critical delay (ns)	Area (µm²)	ADPCM		G.723.1		WAVELET	
				Power (mW)	Critical delay× Area× Power (×10³)	Power (mW)	Critical delay× Area× Power (×10³)	Power (mW)	Critical delay× Area× Power (×10³)
Proposed MAC	P	9.33	260.84	5.81	14.14	5.47	13.31	5.30	12.90
	PM		272.36	5.46	13.81	5.06	12.86	4.74	12.04
	PMA		280.61	5.35	14.01	5.02	13.14	4.63	12.12
Elguibly's MAC		14.95	254.55	8.80	33.49	8.20	31.21	8.23	31.30
Murakami's MAC		11.47	269.10	7.25	22.38	6.48	19.99	6.80	20.99
Raghunath's MAC		13.17	269.01	6.78	24.02	6.38	22.60	6.23	22.07

A Cost-Effective Media Processor for Embedded Applications

Wen-Kai Huang, I-Ting Lin, Shi-Wei Chen and Ing-Jer Huang
Department of Computer Science and Engineering
National Sun Yat-sen University
Kaohsiung, Taiwan
wkhuang@eslab.cse.nsysu.edu.tw; ijhuang@cse.nsysu.edu.tw

Abstract— this paper presents a low-cost media processor which was specially designed for mid-range embedded applications. The key feature is to extend the media processing capabilities of the traditional low-end microcontrollers. Main components include a PIC16C-compatible 8-bit CPU with multimedia extensions, which are comprised of a single-instruction-multiple-data instruction-set, and a reformed memory interface to improve the efficiencies of DSP-oriented memory accesses. An MP3 audio decoder has been implemented as an example to demonstrate the feasibility of the proposed processor, while the circuit gate count is kept smaller than the designs which based on the high-end microprocessors or the digital signal processors.

I. INTRODUCTION

Due to the evolution of system-on-chip technologies, portable devices equipped with multimedia functionalities, such as audio, image, video are growing up rapidly. A sensible evolution trend of these devices is that the price is getting lower, and the size is getting smaller. For designing such cost-economic systems, the chip area might be the most concern of the hardware designers, while the performance is just enough to finish the application-specific workloads. Chips with small area are not only to possess economical die cost, but also to imply the reduced energy dissipation. Although the adoptions of application-specific hardware accelerators can derive a satisfying performance/cost ratio for specific workloads, however, it usually results in poor flexibility. Designs based on the general-purpose processors (GPPs) might be the preferred solutions in order to meet the requirements for fast time-to-markets. Concluding the considerations for low cost and higher flexibility, the low-end microcontrollers (MCUs), such as ones usually adopted narrow data-path (8-bit for example), and simple micro-architectures with a low operating frequency, would be the better candidates in comparison with most of high-end processor based designs [1][2].

Even so, despite the simple-core advantages of such low-end microcontrollers, the computation capabilities are very unacceptable for multimedia workloads. There are two major weaknesses for multimedia applications: the insufficient computing performance and the insufficient memory accessing capabilities. In order to overcome the insufficient, this paper introduces the multimedia extensions to low-end MCU architectures for lost-cost, high-volume embedded applications. The main components contain a PIC16C-compitable 8-bit integer CPU with multimedia extensions. The extensions involve a single-instruction-multiple-data (SIMD) based instruction-set for 32-bit data packet to exploit sub-word parallelism, and an adapted memory interface that improves the efficiencies of DSP-oriented memory accesses. The processor chip also integrates several media-dedicated peripherals to further reduce the overall system cost. Our goal is not only to improve the media performance of PIC16C, but also to keep the small-area benefit of the low-end microcontrollers.

Furthermore, we have successful implemented an MP3 audio decoder based on the proposed media processor. The result shows the feasibility of doing the high-end media applications under the proposed low-cost design.

Related Works

The PIC16C 8-bit MCU cores have been widely used in many embedded applications [3]. It employs an RISC instruction-set-architecture (ISA) with two-stage execution pipeline. The Harvard memory interfaces allow a 14-bit wide instruction with the separate 8-bit data. The reduced chip area is achieved by the simplify accumulate-architecture with the 8-bit data-path. For improving the performance for such general-purpose processors under the restricted overhead, the multimedia-specific instruction-set extensions to existent ISA is a popular method in CPU design fields [4][5]. In our previous work, we have presented a *MultiMedia Extension* (MME) module [6] for GPPs. Base on MME; a low-cost implementation of the multimedia extensions to ARM7 microprocessors has been introduced [7]. These previous efforts provide good design guidelines in adjusting the low-end MCUs for multimedia applications.

The rest of this paper is organized as follows: Section 2 introduces the multimedia extensions in proposed processor.

Section III discusses the chip implementation. Section 4 presents an MP3 application design based on the proposed processor. Finally, section 5 concludes the paper with a discussion on future works.

II. MULTIMEDIA EXTENSIONS

A. Multimedia Instructions

In order to extend the abilities for handling media workloads, we added 60 new multimedia instructions to the standard PIC16C instruction-set architecture according to the functionalities provided by MME module. The main essence of these new instructions is to exploit data parallelism by the SIMD mechanism. It is the fundamental reason for the speedups achieved by the MME SIP. To achieve the best coordinating with SIMD, data-path width of PIC16C must be stretched from 8-bit to certain wider. We decided to adopt 32-bit width because the previous investigation in [7] reveals the satisfied performance/cost ration for 32-bit width with SIMD architecture. Base on the 32-bit SIMD structure, the MME SIP can operate on groups of four bytes or two half-words in a single instruction. These groups of 32 bits are referred to as packed data. For instance, one MME packed-bytes addition can handle the workload with one execution cycle that is operated by PIC16C addition instructions with more than four cycles. Table I shows the summary of the instruction-set architecture.

TABLE I.   INSTRUCTION SUMMARY

Instructions	# of instr.
General-Purpose PIC16C Instruction-Set (8-bit accumulator architecture with *W* register)	
Byte-oriented operations	18
Bit-oriented operations	4
Literal-oriented & control operations	13
Total	**35**
Multimedia Instruction-Set (32-bit load-store architecture with 8 G.P.R.)	
SIMD arithmetic with saturation	14
SIMD barrel shift operations	18
SIMD data packing/unpacking	3
SIMD compare and set	6
SIMD multiply-accumulation	4
Bit-wise logical operations	6
32-bit load/store operations	2
Register data transfer operations	7
Total	**60**

B. Instruction-Set Architecture

The standard PIC16C adopts accumulator-architecture which employs only one working register *W* for instruction execution. The major advantage of this architecture is that instruction format tends to be easy to encode and yields good density. The other benefit acquired by the reduced gate counts. However, from points of view at performance, it may not be qualified to do the multimedia jobs such as sub-word operations and localized recurring operations. In the new 32-bit data-path of multimedia instructions, the load-store architecture has been applied to reduce the latency of data memory accesses when performing the parallel operations in SIMD. Eight 32-bit *Multimedia-Extension-Registers* (MER) had been provided as the general-purpose registers in this model. In other words, two different type of data-path, i.e. the accumulator-architecture and the load-store-architecture, are possessed. However, programmers are allowed to use the PIC16C instructions to access the eight general purpose MERs via the memory mapped mechanism.

C. Enhanced Memory Interface

The insufficient memory interface is the other shortcoming for media workloads of PIC16C. We extended the addressing space for both program memory and data memory to 64K bytes. The length of *offset* field in instruction encoding has been increased and new special-purpose registers has been added as the base registers for addressing the extended space indirectly. Additionally, the bit-reversal addressing mode and the circular addressing mode have been supported for the DSP-oriented operations such as Fast Fourier Transform (FFT). Reforming the memory interface discussed above simultaneously changes the encoding of instructions, and causing the un-compatible with standard PIC16C. Face to this situation, we designed a software translation tool that can translate the program with encoding in standard PIC16C format to the new format, so that the existent PIC16C applications can be correctly executed in the new architecture.

III. CHIP IMPLEMENTATION

The design was developed as a synthesizable silicon intellectual property (SIP) by using Verilog HDL. Figure 1 shows the pipeline microarchitecture of the CPU core. It comprises four separate modules of two pipeline stages each. Stage-one fetches instructions (IF module), generates the control signals and obtains the operands from the data memory or MER bank (DE module). Stage-two performs the corresponding execution (EX module) and writes back the result to the data memory or MER bank (WB module). A clocking method similarly to the dual-edge-triggered were applied to whole pipeline, i.e., the IF, EX modules update its states on rising edge of the clock, while the DE, WB modules updated on falling edge. Using both edges of the clock to update pipelines allows the 50% reduction in clock distribution power, and it allows all instructions to execute in a single clock cycle, except for branches and 32-bit data movements. Therefore the average cycles-per-instruction (CPI) has been decreased significantly.

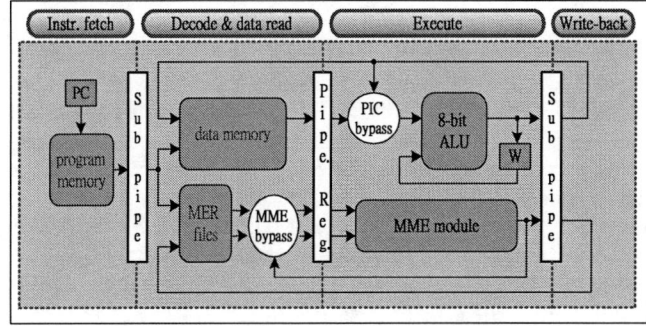

Figure 1.   Microarchitecture of CPU Core

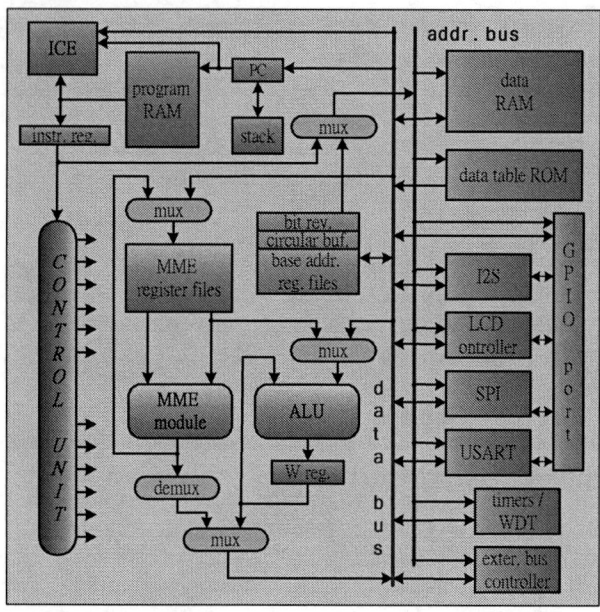

Figure 2. Block Diagram of the Processor Test Chip

Bypass logic are provided to eliminate the pipeline stalls caused by data hazards. In order to balance the circuit delay between each pipeline stage, we adopted different bypass policy for each of two data-paths individually. In the load-store path, the bypass unit was placed in the DE module to reduce the delay of MME operations. On the other hand, the bypass unit in accumulator path are placed in the EX module to reduce the delay for accessing the data memory. Timing simulation indicated that we achieved a balance of delays between the pipelines.

Figure 2 shows the block diagram of the test chip for the proposed media processor. It employs 32KB data memory (includes 16KB ROM for table-lookup applications), and a 16K words program memory (RAM) with a boot-up loader. In addition to on-chip memory, the external bus controller has been provided. Peripherals include a inter-IC sound (I2S) interface, a serial-parallel interface (SPI), a universal asynchronous receiver transmitter (UART), timers/watchdog, eight general-purpose I/O ports (totally 40-bit), and an programmable interrupt controller. The standard JTAG port has been support for chip testing, and an embedded in-circuit emulator (ICE) which provides a breakpoint register and a watchpoint register has also been integrated for software debugging.

We use Artisan cell library with UMC 0.18um 1P6M process technology to implement the test chip via National Chip Implementation Center (CIC). In our measurement, the maximum operating frequency of the test chip is 70MHz. However, in our analysis for MP3 decoding, we derive a conclusion that the decoder can be executed in real time if the clock frequency is above 50MHz. Thus, the final test chip was implemented at 50MHz to reduce the chip area and power consumptions. Detail information of the MP3 decoder design is described in next section. Figure 3 shows the chip plot. Chip characteristics are listed in Table II.

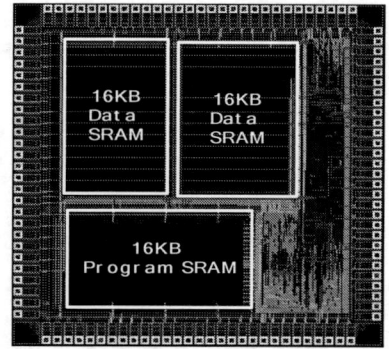

Figure 3. GDSII-based plot of test chip

TABLE II. CHIP CHARACTERISTICS (INCLUDE SRAM CELL)

Process	0.18 um	Vdd	1.8 V
Clocks	50 MHz	Area	3.1 x 3 mm²
Package	128 CQFP	mW/MHz	0.68

IV. APPLICATION EXAMPLE: MP3 AUDIO DECODER

In order to prove the feasibility of using the proposed processor to perform high-end multimedia workloads, we have implemented an MP3 (MPEG-1 audio layer III) audio decoding system based on the proposed processor. MP3 is the most popular decoding format for playback of high quality compressed audio for portable devices. A large number of previous researches which make efforts in improving the efficiencies have been introduced. These works can be briefly classified into two strategies, i.e. hardware optimization [8] and software optimization [9]. Figure 4 shows the operating flow of the MP3 decoding application.

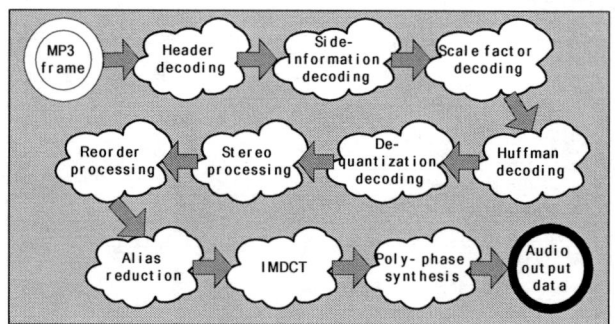

Figure 4. Illustration of MP3 decoding flow

In our implementation, although the multimedia extensions in our media processor can provide a certain level of performance enhancements, however, the MP3-specific fine-tuning of codes would still be required if the best performance/cost ratio is expected. Following optimization methods have been implemented in each module if possible:

1) Algorithm optimization: main efforts are paid to reduce control hazards of the pipeline microarchitecture. Loop unrooling and instruction reordering are performed to exploit the instruction-level parallelism. Macros are timely used to reduce the frequency of sub-routine calls.

Simultaneously, register re-allocations according to the the chromatic number are performed to reduce the requirement of multimedia-extension registers for operations.

2) Data optimization: a fixed-point implementation has been adopted instead of floating-point design. Although the fixed-point implementation suffers from the risk of precision loss, it allows the area- and power-efficient design at lower frequency. Besides, data reordering for exploiting sub-word parallelisms can further improve the efficiency for the SIMD instructions. Figure 5 illustrates the partition of data memory. There are 12K bytes address spaces for table-lookup operations to simplify the computation complexity.

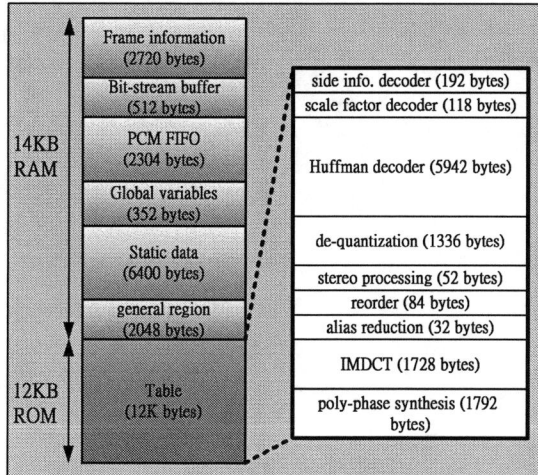

Figure 5. Usage of Data Memory

For achieving a high-degree of programming efficiency, we developed most codes under hand-coded in assembly level. The application is running at 50MHz for 44.1 kHz of sample rate with 128 kbps bit rate. Table III gives a summary of static code size of each module.

TABLE III. STATIC INSTRUCTION COUNT

Module	Instructions
header and side information decoding	794
scale factor decoding	310
Huffman decoding	594
de-quantization	368
stereo processing	1,585
reorder processing	134
alias reduction	159
IMDCT	266
poly-phase synthesis	607
Total instruction count	**4,817 (9.5 KB)**
PIC instructions	2,160 (44.83 %)
Multimedia instructions	2,657 (55.17%)

For the requirement of real time decoding, the upper-bound of clock cycles spent for decoding one frame can be derived as equation (1):

$$\text{clock cycles/frame} \leq \frac{\text{max. number of samples per frame}}{\text{sample rate}} \times \text{clock rate}$$

$$\leq \frac{1152}{44.1k} \times 50M = 1,306,122 \quad (1)$$

That is, the decoder must be able to decode 38.28 frames per second at 50MHz. The symphony Tchaikovsky 1056-6 is used as the benchmark. Table IV shows the cycles for decoding the worst frame. Except the cycles for the proposed processor, the ones for standard PIC16C are also provided for comparison. Even operating in such worst case, our media processor still achieves the throughput requirement of 38.28 frames per second for real time decoding.

TABLE IV. CYCLES FOR THE WORST FRAME DECODING

	Module	Cycles
Proposed Media Processor	header and side information decoding	24,405
	scale factor decoding	3,906
	Huffman decoding	308,916
	de-quantization	231,370
	stereo processing	29,210
	reorder processing	29,898
	alias reduction	3,972
	IMDCT	236,350
	poly-phase synthesis	424,448
	Total Clock Cycles	**1,292,477**
	Throughput: 38.69 frames/second	**(met)**
Standard PIC16C	Total Clock Cycles	2,318,827
	Throughput: 21.56 frame/second	**(violated)**

V. CONCLUSION

This paper introduces a media processor which extends the multimedia capabilities of the low-end microcontrollers for low-cost embedded applications. The extensions contain a SIMD-based instruction-set for exploiting the sub-word parallelism, and an adapted memory interface for DSP-oriented accesses. An MP3 decoder has been implemented based on the test chip of the proposed processor. The result shows that the decoder can operate in real time mode when the proposed processor is working at 50MHz. In future works, we are interested in developing the JPEG encode/decode applications in the proposed media processor.

REFERENCES

[1] S. Purcell, "The impact of Mpact 2", IEEE Signal Processing Magazine, Vol. 15, Issue 2, pp. 102-107, Mar. 1998

[2] J.T.J. van Eijndhoven, et al., "TriMedia CPU64 Architecture," Proc. International Conference on Computer Design, pp. 586-592, Oct. 1999

[3] PIC16Cxx Data Sheet, Microchip Technology Inc., 2000

[4] A. Peleg and U. Weiser, "MMX Technology Extension to the Intel Architecture," IEEE Micro, pp. 42-50, July 1996

[5] R.B. Lee, "Multimedia Extensions for General-Purpose Processors," Proc. IEEE Workshop on Signal Processing Systems, pp. 9-23, Nov. 1997

[6] I.J. Huang, W.K. Huang and C.F. Kao, "A Parameterized MMX IP Module for RISC Microprocessors", Proc. the 12th VLSI Design/CAD Symposium, p. 35, Aug. 2001

[7] I.J. Huang, W.K. Huang, R.T. Gu and C.F. Kao, "A Cost Effective Multimedia Extension to ARM7 Microprocessors", Proc. IEEE International Symposium on Circuit and Systems, pp. 304-307, Mar. 2002

[8] G. Maturi, "Single Chip Audio Decoder," IEEE Trans. on Consumer Electronics, Vol. 38, Issue 3, pp. 348-356, Aug. 1992

[9] V. Gurkhe, "Optimization of an MP3 Decoder on the ARM Processor," Conference on Convergent Technologies for Asia-Pacific Region, Vol. 4, pp. 1475-1478, Oct. 2003

Design and Implementation of A New Cryptographic System for Multimedia Transmission

Jui-Cheng Yen
Department of Electronic Engineering
National United University
Miaoli, Taiwan, R.O.C
jcyen@nuu.edu.tw

Hun-Chen Chen
Department of Electronic Engineering
National United University
Miaoli, Taiwan, R.O.C
hcchen@nuu.edu.tw

Shu-Meng Wu
Department of Electronic Engineering
National United University
Miaoli, Taiwan, R.O.C
M9222505@nuu.edu.tw

Abstract – In this paper, we propose a new cryptographic system based on the Shannon product theory. It is analyzed to be high security. Regarding the real-time applications, a five-stage pipeline architecture has been designed, implemented, and verified based on a UMC 0.18 *um* CMOS technology. The simulation result shows that the proposed design achieves the throughput rates of encryption part up to 2.79 Gbps at the cost of silicon area of 225360 um^2 and the power consumption of 16.93 *mW*. The comparison shows that the performance of the proposed design is much better than the existing designs.

I. INTRODUCTION

Recently, owing to rapid increase in bandwidth, it is very prevalent to transmit multimedia signal over Internet. Meanwhile, illegal data access has become more prevalent and the losses have been much serious. Hence, data security [1]-[8] has become a critical and imperative issue. In this paper, a new cryptographic system composed of 4-pipelined functions is proposed. The 4 functions are as follows. i) Every 15 bytes of input data are randomly expanded to 16 bytes. ii) Random swapping on the expanded data is made in four levels. iii) The 8 bit-planes are randomly XORed or XNORed to two random operands Seed1 or Seed2. iv) Two rounds of 2D 64-bit rotation operation are performed. All the operations are under the control of a binary sequence generated from a chaos-based pseudorandom bit generator.

Regarding the implementation of the proposed system in real-time, a five-stage pipeline architecture has been designed, implemented, and verified by using Verilog and high-level synthesis tool (Design Compiler) based on a UMC 0.18 *um* CMOS technology. The proposed design verified in front-end possesses the throughput rates of encryption part up to 2.79 Gbps at the cost of silicon area of 225360 um^2 and the power consumption of 16.93 *mW*. To compare with other designs [4]-[8], the simulation results show that the proposed design possesses much better performance in terms of the evaluation index of data rate per area.

II. THE NEW CRYPTOGRAPHIC SYSTEM

Based on the Shannon product theory, a new multimedia cryptography system (MCS) is proposed. The block diagram of the encryption part of MCS is shown in Fig. 1.

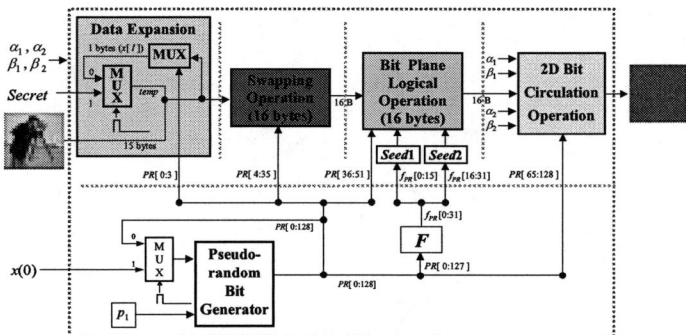

Fig. 1. The block diagram of the encryption part of MCS.

A detailed description on Fig. 1 is given in the following.

Pseudorandom Bit Generators in MCS

Kocarev and Jakimoski construct a class of chaos-based pseudorandom bit generators (PBGs) in [9]. We adopt the second PBG with p=419, m=8, M=64, and k=64 as the pseudorandom bit generator in MCS.

The initial state of PBG is the 129-bit $x(0)$ and the output is denoted as 129-bit $PR[0:128]$, where 129 comes from $M+k+1$. The output $PR[0:128]$ of PBG is used to be operands and control signal. To be operands, $PR[0:127]$ is transformed to $f_{PR}[0:31]$ by the map defined as

$$f_{PR}[k] = (PR[4k] \oplus PR[4k+1]) \oplus (PR[4k+2] \oplus PR[4k+3])$$

for $k = 0, 1, \ldots, 31$. The 16-bit operand Seed1 is composed of $f_{PR}[0:15]$ and Seed2 is composed of $f_{PR}[16:31]$.

Data Expansion in MCS

Every 15 bytes of input data are expanded to 16 bytes in the data expansion unit. At first, the appended byte is the parameter *Secret*. Afterward, the appended one is randomly selected from the previous batch of 16 bytes according to $PR[0:3]$.

Swapping Operation in MCS

Definition 1: The operation $Swap_w(g(m), g(n))$ is defined to swap $g(m)$ and $g(n)$ if w is equal to 1 or preserve their original positions if w is equal to 0.

The swapping operation for a batch of 16-byte data $g(n)$s, $0 \le n \le 15$, contains four levels. They are defined as follows:

#1) $Swap_{PR[k]}(g[i], g[j])$ for $(i, j, k) \in \{(0, 8, 4), (1, 9, 5), (2, 10, 6), (3, 11, 7), (4, 12, 8), (5, 13, 9), (6, 14, 10), (7, 15, 11)\}$,

#2) $Swap_{PR[k]}(g[i], g[j])$ for $(i, j, k) \in \{(0, 4, 12), (1, 5, 13), (2, 6, 14), (3, 7, 15), (8, 12, 16), (9, 13, 17), (10, 14, 18), (11, 15, 19)\}$,

#3) $Swap_{PR[k]}(g[i], g[j])$ for $(i, j, k) \in \{(0, 2, 20), (1, 3, 21), (4, 6, 22), (5, 7, 23), (8, 10, 24), (9, 11, 25), (12, 14, 26), (13, 15, 27)\}$,

#4) $Swap_{PR[k]}(g[i], g[j])$ for $(i, j, k) \in \{(0, 1, 28), (2, 3, 29), (4, 5, 30), (6, 7, 31), (8, 9, 32), (10, 11, 33), (12, 13, 34), (14, 15, 35)\}$.

Bit Plane Logical Operation in MCS

Definition 2: The ith bit plane BP_i, $0 \leq i \leq 7$ of $g(n)$s, $0 \leq n \leq 15$ is defined to be the set of all the ith bits of $g(n)$s from least significant bit.

The logical operation on the bit planes is defined as follows:

FOR $i = 0$ TO 7 DO
    Switch ($2 \times PR[36+2i] + PR[37+2i]$)
        Case 3: BP_i XOR $Seed1$;
        Case 2: BP_i XNOR $Seed1$;
        Case 1: BP_i XOR $Seed2$;
        Case 0: BP_i XNOR $Seed2$;
End

2D Bit Rotation Operation in MCS

Let M be an 8×8 binary matrix. Two mappings are defined.

Definition 3: The mapping $RotateX_i^{p_i, r_i} : M \to M'$ is defined to rotate each bit in the ith row of M, $0 \leq i \leq 7$ in the left direction r_i bits if p_i equals 1 or in the right direction r_i bits if p_i equals 0, where $0 \leq r \leq 7$.

Definition 4: The mapping $RotateY_j^{q_j, s_j} : M \to M'$ is defined to rotate each bit in the jth column of M, $0 \leq j \leq 7$ in the up direction s_j bits if q_j equals 1 or in the down direction s_j bits if q_j equals 0, where $0 \leq s \leq 7$.

Regard the first 8 bytes of the batch of data after the logical operation as the 8×8 binary matrix M_1 and the second 8 bytes as M_2. The operation sequentially performs

$$\left(\prod_{j=0}^{7} RotateY_j^{q_j, s_j}\right) \bullet \left(\prod_{i=0}^{7} RotateX_i^{p_i, r_i}\right)(M_1) \quad \text{and}$$

$$\left(\prod_{j=0}^{7} RotateY_j^{q_j, s_j}\right) \bullet \left(\prod_{i=0}^{7} RotateX_i^{p_i, r_i}\right)(M_2).$$

For $0 \leq i, j \leq 7$ in M_1, $p_i = PR(65+2i)$, $q_j = PR(81+2j)$, $r_i = \alpha_1 + \beta_1 \times PR(66+2i)$, and $s_j = \alpha_1 + \beta_1 \times PR(82+2j)$. For $0 \leq i, j \leq 7$ in M_2, $p_i = PR(97+2i)$, $q_j = PR(113+2j)$, $r_i = \alpha_2 + \beta_2 \times PR(98+2i)$, and $s_j = \alpha_2 + \beta_2 \times PR(114+2j)$.

Now, we consider the security problem. We regard MCS be a composite function MCS under the control of $PR[0:128]$. The MCS is composed of the following four cascaded operations: the data expansion operation (DE), swapping operation (SP), bit plane logical operation (BP), and 2D bit-rotation operation (BR). Hence,

$$MCS = BR \bullet BP \bullet SP \bullet DE$$

So, it belongs to the Shannon product cipher. The input data will be diffused and confused by the randomly cascaded operations under the control of $PR[0:128]$. Moreover, in the chaotic systems [10], it is well-known that i) it has sensitive dependence on initial conditions; ii) the trajectories are dense, bounded, but nonperiodic in the state space; and iii) it has noise-like spectrum. Especially, Kocarev and Jakimoski have proved that the adopted PBG is cryptographically secure.

To demonstrate the parameter sensitivity of MCS by MATLAB simulation, the root mean square difference (RMSD) is computed. Let f'_A and f'_B be the encryption results of the image f of size $L \times P$ pixels under $x_A(0)$ and $x_B(0)$, respectively. The RMSD is defined as

$$RMSD \equiv \left(\frac{1}{L \times P}\sum_{i=0}^{L-1}\sum_{j=0}^{P-1}(f'_A(i,j) - f'_B(i,j))^2\right)^{\frac{1}{2}}.$$

Randomly generate $x(0)$. Take the complement of the jth bit of $x(0)$ and denote it as $x[cj]$ $0 \leq j \leq 128$. Let $\alpha_1 = 2$, $\beta_1 = 2$, $\alpha_2 = 1$, $\beta_2 = 2$, and $Secret = 123$. After applying MCS to "Lena" under $x(0)$ and $x[cj]$, the RMSDs between the results of $x(0)$ and $x[cj]$ are listed in Table 1. Table 2 shows the RMSDs between $x[ci]$ and $x[cj]$. Moreover, according to [11], compute the fractal dimensions (fds) of the encryption results under $x(0)$ and $x[cj]$s. The fds are listed in Table 3.

From Tables 1 and 2, the encryption results are quite different under just 1-bit or 2-bit variation. From Table 3, the encryption results of MCS are completely disorderly.

Table 1. The RMSDs between the results of $x(0)$ and $x[cj]$.

j	20	40	60	80	100	120
RMSD	104.68	104.59	104.73	104.89	104.29	104.16

Table 2. The RMSDs between the results of $x[ci]$ and $x[cj]$.

(i,j)	(10, 70)	(20, 80)	(30, 90)	(40, 100)	(50, 110)	(60, 120)
RMSD	104.84	104.28	105.05	104.31	104.71	104.48

Table 3. The fds under $x(0)$ and $x[cj]$s.

j	*	20	40	60	80	100
fd	2.9999	2.9996	2.9998	2.9998	2.9994	2.9997

Note: * denotes $x(0)$.

III. THE VLSI ARCHITECTURE DESIGN

Fig. 2 shows the VLSI architecture of the proposed MCS. It includes a signal encryption unit (SEU), a signal decryption unit (SDU) and the pseudorandom bit generators (PBGs). Fig. 3 shows the five-stage pipeline architecture of the SEU including three-level data swapping for position permutation, and one-level XOR/XNORed operation with random keys of $Seed1$ or $Seed2$ and 2D circulation for bit-recirculation with random direction and random number of bits for value transformation. The detail of the building blocks in the MCS design is shown as Fig. 4. For balancing the pipeline architecture, we further split the 2D circulation stage to two stages. Fig. 5 shows the architecture of the PBG, where 129 bits of control signal are generated as the key of the proposed MCS to randomly decide the operations in each pipeline stage. Observing the formulation of the chaos-based pseudorandom bit sequence generation, we concatenate the result of multiplication by the way of wiring rather than the complex operations of modular operation and truncation operation to minimize the hardware cost of PBG.

For the clocking issue of the proposed design, since the computation time for generating the random sequence is much longer than that needed in the SEU's, we exploit the design concept of multiple clock sources in the proposed design. That means we can use a slower clock source in the PBG design by dividing the original clock source by a certain factor. The value of the dividing factor should be determined by the consumption rate of data processing stages. Regarding the issue of data flowing, with the proposed algorithm, each encryption data packet consists of fifteen data elements. They are sampled serially, and concatenated an additional data as secret at first stage such that sixteen data elements are issued to be encrypted. Thus it makes two clock frequencies are needed. One is used for data sampling. The other is used as the pipeline clock. For maintaining the data flowing continuously in this system, the two clocks must be synchronized with each other. Namely, the period of fifteen cycles for data sampling must be same as the period of sixteen cycles for sending out the sixteen encrypted data elements. However, synchronization of the two clocks is not so easy. We adopt the manner that stalls one cycle to sample data every sixteen clock cycles to synchronize the data sampling and sending out.

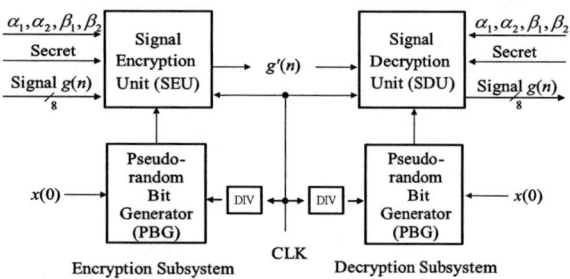

Fig. 2. The VLSI architecture of the proposed MCS.

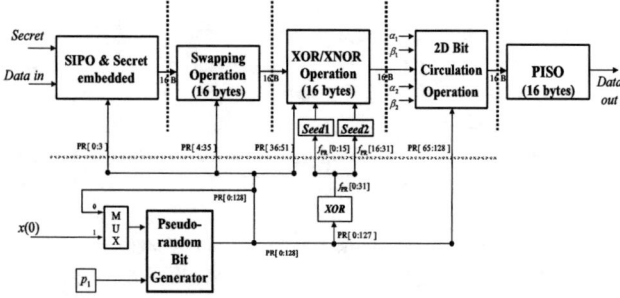

Fig. 3. The pipeline architecture for the SEU in MCS.

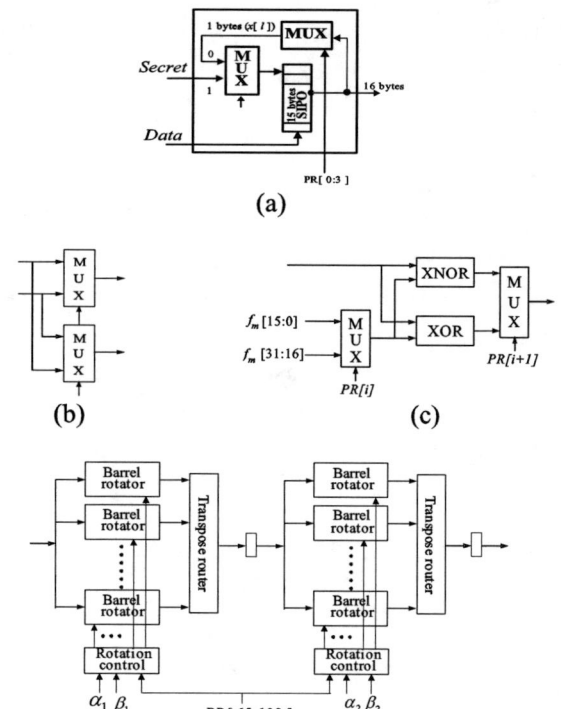

Fig. 4. Detailed architecture in MCS: (a) SIPO & secret embedded unit, (b) swapping operation unit, (c) bit-plane logical operation unit, and (d) 2D bit-circulation unit.

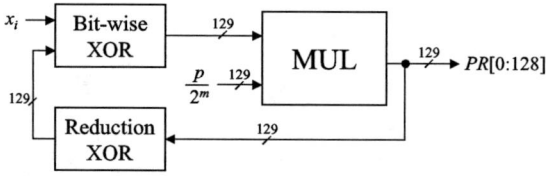

Fig. 5. The architecture of the PBG.

IV. PERFORMANCE EVALUATION AND COMPARISONS

For a fair verification, we verify the proposed design of SEU with cell-based design flow by using Verilog HDL and high-level synthesis tool (Design Compiler) based on UMC 0.18 um VST cell-library. Table 4 shows the detail of the front-end implementation on the SEU of the proposed MCS. According to the simulation result, the throughput rates of the proposed MCS are larger than 2.79 Gbps with the area of 225360 um^2. This performance is suitable for most of the requirement in high quality multimedia applications.

Table 4. The detail of the front-end implementation on the SEU of the proposed MCS

Item	Simulation result
Technology	UMC 0.18 um
Area cost	225360 um^2
Critical path delay	2.69 ns
Power consumption	45.5 uW/MHz

This section also provides the performance evaluation of the proposed design with other existing designs [4]-[8]. In order to eliminate the factor of different fabrication technologies, we define an index of normalized area (denoted as NArea), which is the silicon area normalized to a 0.35 um technology. It is defined as

$$NArea = \frac{Area}{(Technology/0.35)^2}. \quad (1)$$

Besides, we also define an index of data rate per area (DRPA), i.e. Data rate/NArea, to reflect the efficiency of the hardware design for data encryption and decryption. It is defined as

$$DRPA = \frac{Data\ rate}{NArea}(\frac{Mbps}{mm^2}). \quad (2)$$

Based on the above two indices, we have summarized the comparison results of the proposed design with the existing ones [4]-[8] in Table 5. From Table 5, we find that the proposed design is better than the design [4] in providing higher data processing rate at lower hardware cost weighted using gate count based on a 0.35 um CMOS technology. Also, the proposed design achieves higher data processing rate at lower hardware cost in gate count as compared with the design [8]. As for the comparison to the other popular data encryption/decryption approaches [5]-[8], we use the normalized index of DRPA defined in equation (2) in the evaluation of the efficiency in these approaches. We find that the proposed design is better than the existing designs [5]-[7] in possessing higher DRPA. That means the proposed design can provide higher efficiency considering the data processing rate and required hardware cost. Though the above comparison cannot provide the absolute comparison in every aspect (including the efficiency as well as other factors like security and so on) between the proposed design and others, it can reflect the efficiency of the proposed design under the situation that the security of the proposed design and the existing ones [4]-[8] is good enough for the applications.

V. CONCLUSIONS

In this paper, a new signal cryptographic system has been proposed. It has been analyzed to be high security. Simulations have indicated the encryption results are completely disorderly and very sensitive to parameter variation. For the real-time requirement, its five-stage pipeline architecture has been designed, implemented, and verified. The proposed design verified in front-end possesses the throughput rates of SEU up to 2.79 Gbps at the cost of silicon area of 225360 um^2 and the power consumption of 16.93 mW. To compare with other designs, the simulation results show that the proposed design possesses much better performance in terms of the evaluation index of DRPA. Finally, it is believed that many real-time multimedia transmission can benefit from the proposed system.

Table 5. Comparison of the proposed design with the existing designs [4]-[8].

Design	Technology (um)	Area (mm^2)	Data rate (Mbps)	NArea (mm^2)	DRPA (Mbps/mm^2)	Encryption algorithm	Cipher block length (bit)
Design [4]	0.35	---	395	---	---	FEA-M	64
Design [5]	0.7	29	251.8	7.25	34.73	SAFER K-128	64
Design [6]	0.35	13.69	600	13.69	43.82	AES	128
Design [7]	1.2	107.8	177.8	9.17	19.39	IDEA	64
Design [8]	0.25	---	609	---	---	AES	128
Proposed	0.18	0.22536	2790	0.852	3274.65	MCS	120

ACKNOWLEDGMENTS

This work is supported by the National Science Council, Republic of China, under Grant NSC 93-2215-E-239-002.

REFERENCES

[1] "Data encryption standard," FIPS PUB 46, National Bureau of Standards, Washington, D. C., Jan. 1997.

[2] J. Daemen and V. Rijmen, "AES Proposal: Rijndael, AES Algorithm submission," Sep. 1999.

[3] X. Yi, C. H. Tan, C. K. Kheong, and M. R. Syed, "Fast Encryption for Multimedia," *IEEE Trans. on Consumer Electronics*, Vol. 47, No. 1, pp. 101-107, Feb. 2001.

[4] X. Yi, C. H. Tan, C. K. Siew, and M. R. Syed, "Fast encryption for multimedia" *IEEE Trans. on Consumer Electronics*, vol. 47, no. 1, pp. 101-107, Feb. 2001.

[5] A. Schubert, V. Meyer, and W. Anheier, "Reusable cryptographic VLSI core based on the SAFER K-128 algorithm with 251.8 Mbit/s throughput," *Proc. 1998 IEEE Workshop on Signal Proc. Sys.*, pp. 437-446, 1998.

[6] Y. Mitsuyama, etc., "VLSI implementation of high performance burst mode for 128-bit block ciphers," *Proc. ISCAS'2001*, vol. 2, pp. 344-347, 2002.

[7] A. Curiger, etc., "VINCI: VLSI implementation of the new secret-key block cipher IDEA," *Proc. IEEE 1993 Custom Integrated Circuits Conf.*, pp. 15.5.1~15.5.4, 1993.

[8] C. C. Lu etc., "Integrated design of AES Encrypter and Decrypter," *Proc. ASAP'2002*, pp. 277-285, 2002.

[9] L. Kocarev and G. Jakimoki, "Pseudorandom bits generated by chaotic maps," *IEEE Trans. On Circuits and Systems – Part I*, vol. 50, pp. 123-126, Jan. 2003.

[10] T. S. Parker and L. O. Chua, "Chaos - A tutorial for engineers," *Proc. IEEE*, vol. 75, pp. 982-1008, 1987.

[11] C. C. Chen, etc., "Fractal feature analysis and classification in medical imaging," *IEEE Trans. on Medical Imaging*, vol. 8, no. 2, pp. 133-142, 1989.

Low-Cost Implementation of a Super-Resolution Algorithm for Real-Time Video Applications

Gustavo M. Callicó, Sebastián López, José Fco. López, Roberto Sarmiento and Antonio Núñez
Research Institute for Applied Microelectronics (IUMA)
Electronic and Automatic Engineering Department (DIEA)
University of Las Palmas de Gran Canaria (ULPGC), E-35017, Spain
{gustavo, seblopez, lopez, roberto, nunez}@iuma.ulpgc.es

Rafael Peset Llopis
Philips Consumer Electronics, ICLab
Eindhoven, The Netherlands
rafael.peset.llopis@philips.com

Ramanathan Sethuraman
Philips Research Laboratories, NatLab
Eindhoven, The Netherlands
ramanathan.sethuraman@philips.com

Abstract — In this paper, a novel algorithm based on super-resolution (SR) techniques for increasing the quality of video sequences, is presented together with its mapping into the Philips Research proprietary hardware-software platform. As result, a low-cost real-time implementation is obtained, suitable for personal multimedia applications. Low cost constraints are accomplished by re-using a video codec and introducing some changes in order to avoid the use of specific SR hardware. As a consequence, a drastic reduction in the memory requirements is obtained at the expense of less than a 7% of quality loss with respect to previous works.

I. INTRODUCTION

The straightforward way to increase the resolution of an image is to use higher resolution sensors, at the expense of higher costs. However, this results in a decrease of the size of the active pixel area where the integration of light is performed. In this case, lower amounts of light will reach the sensor, making it more sensitive to shot noise. It has been estimated that the minimum photo-sensor size is around $50\mu m^2$ [1], a limit that has already been reached by the CCD technology. One solution to this drawback is to increase the resolution of the video sequence by using SR techniques, wherein high-resolution images are obtained from low-resolution sensors at low costs. SR represents also a smart strategy to perform image zooming without using mechanical parts to move the lenses, giving as result an important reduction in the power dissipation.

This approach to SR consists on gathering information from a shifted image set in order to integrate all the available information in a new super-resolved image. The algorithm presented in this paper is based on [2], although modifications in the iterative scheme are introduced aimed at allowing real time applications. Furthermore, memory requirements are minimized by combining in a single step the new incoming information contained in the frame and delivering a new SR image per every incoming frame.

II. THE HYBRID VIDEO ENCODER PLATFORM

This work is included in the frame of a 'SOC platform oriented design', in the sense that the hardware/software platform is already established by Philips [3] and the objective is to introduce minor modifications in order to map the new SR algorithm into it. It has to be mentioned, however, that the SR developed algorithm is easily translated into any other hybrid video encoder.

The architecture used in Philips is shown in Fig. 1. Software and hardware tasks are executed on an ARM processor and on four VLIW processors respectively. The pixel processor (PP) communicates with the pixel-domain (image sensor or display) and performs line to stripe (16 lines) conversion and vice-versa for a video/image encode and decode operation respectively. The motion estimator processor (MEP) evaluates a set of candidate vectors received from the ARM processor and selects the best vector for full, half and quarter pixel refinements. The output of the MEP consists of motion vectors, sum-of-absolute-difference (SAD) values, and intra metrics. This information is used by the software processor to determine the encoding approach for the current macro-block (MB) (skipped, intra coded, inter coded with zero motion vector, inter coded with motion vector, or inter coded with four 8x8 motion vectors). The texture processor (TP) performs the encoding of MBs and stores the decoded MBs in the loop memory. The output of the TP consists of variable length encode (VLE) codes for the DCT coefficients of the current MB. The TP also performs the core functionality for H.263 decoding and JPEG encode/decode. Finally, the stream processor (SP) packs the VLE codes for coefficients and headers generated

---

This work has been partially funded by Philips Electronics Netherlands B.V. in the frame of a research collaboration between NatLab (Eindhoven) and IUMA (Univ. Las Palmas de Gran Canaria). The support provided by the European Community through the *CAMELLIA* project (IST-2001-34410) as well as the one given by the Spanish Government through the ARTEMI project (TIC2003-09687-C02-02.) is also acknowledged.

by software on the ARM. In 0.18μm CMOS the area is 20 mm² and the power dissipation is 30mW.

III. THE SUPER-RESOLUTION ALGORITHM FOR VIDEO

After an exhaustive study of previous work in SR, were it could be highlighted references [4], [5] and 0, the first Super-Resolution Algorithm (SRA) developed during this research was described in [7], where a static iterative SRA was successfully mapped onto the previously described platform. Nevertheless the algorithm exhibited an iterative behavior which prevents real-time execution. In pursuit of a real-time implementation, the algorithm was firstly modified to avoid the iterative behavioral obtaining a non-iterative version for the static SR described in [8], where the mapping details are detailed in [9]. Although this version works properly for real time applications, it exhibits a major drawback in the memory requirements, which results to be very high for a low-cost single-chip implementation.

A. Preliminary considerations

When applying SR techniques to real sequences, it must be taken into account two main limitations. Firstly, an object inside the image is able to move very fast, producing occlusions of other objects in the scene. Secondly, fast movements of the camera produce changes among non-correlated scenes. This is known as 'context changes', and under this conditions, no information can be inferred from previous images. The only solution in these cases is to interpolate those pixels that can not being improved through SR techniques. In this sense, the information provided by the MEP to help the system to decide the type of compression for every MB is used in the proposed SRA. Such information consists in three parameters. The first one is the motion vectors set, which can be local or global; the second one is the SAD_inter, and the third one is the SAD_intra. Depending on the used motion vectors, the SAD_inter can be global or local. The SAD_intra is a measure of the variations inside the image, being computed as the absolute differences between every MB pixel and the same MB average value. From the motion vectors, a new parameter is computed, called 'mv_sad', obtained as the absolute differences of every motion vector with the global motion vector. If the 'mv_sad' is low, that means that the image is dominated by global movement. On the contrary, the image is dominated by local movements. The 'mv_sad' per MB is computed as the summation of the absolute value of the differences of their components in both directions. Based upon the previous MB indicators, the control system must decide the action to be taken for each MB: applying SR using local vectors, applying SR using the global vector or interpolating the present MB.

B. Proposed Dynamic SRA description

The algorithm starts by obtaining the values of the SAD_inter for every motion vector and the SAD_intra, together with the local motion vectors. The global motion vector is computed from the local motion vectors. Once the

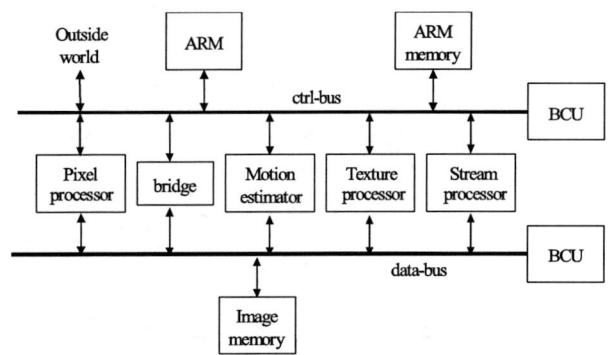

Fig. 1. Architecture for the multi-standard video/image codec developed in Philips Research.

global vector is known, the 'mv_sad' is computed. Moreover, by passing the global vector to the MEP, the SAD associated to the global vector can be obtained. If the 'mv_sad' is higher than the MV_SAD_threshold, then the global motion must be promoted. If the 'sad_global' is lower than the 'sad_local' then the global motion vector is chosen. In any other case, the local motion vector is selected. Once local or global movements have been discriminated, it is necessary to decide between applying super-resolution using the chosen motion vector or performing an interpolation. In case it is the first frame of a scene, the whole image is interpolated. If it is not, it must be found out if a context change situation is taking place. This fact is determined by applying the following empiric equation:

$$\text{context\_change} = 32 \cdot \frac{\text{sad\_intra} - \text{sad\_inter\_local}}{\text{sad\_intra} + \text{sad\_inter\_local}} \quad (1)$$

Equation (1) has demonstrated to be a quite stable estimator to determine the context change, taken into account that all the input values are directly given by the MEP. If the value given by this equation is lower than a threshold named CONTEXT_CHANGE_threshold, then the MB is interpolated. If not, SR is applied using local or global motion vectors depending on the value obtained for mv_sad. These thresholds are empirically determined from a detailed study of several image sequences, and offer good quality in all the tested conditions. This methodology increases even more the algorithm robustness, assuring that in no case the quality does not drop below the interpolation level in any case. For this particular case, the two thresholds previously introduced are set to a value of 5.

In Fig. 2 the block diagram of this SRA is presented. Although the final super-resolved image is always stored, it has been necessary to define a new memory, HR_B, where the SR image with holes (zeroes) [8] is re-introduced. Even though it is necessary to fill the holes prior to display the image, it is better to keep such holes inside the algorithm loop, because in that way it is possible to fill the missing information without mixing it with spoiled values coming from the interpolation with the neighbour pixels. This feedback scheme largely decreases the memory used by the algorithm, where the last SR image is reused to increase the resolution of the present image. In the proposed scheme, the new low-resolution image is interpolated to high-resolution, leaving holes instead of interpolated pixels to increase the image size. The pixels from the new low-resolution image are considered valid and are all kept. The algorithm tries to fill the generated holes with the information of the previous super-resolution image, applying a similitude criterion per MBs between the previous low-resolution image and the new obtained image. If the criterion is not satisfied, the holes are interpolated with the surrounding information.

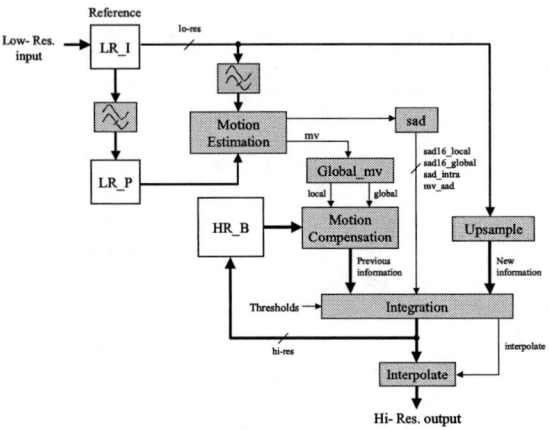

Fig. 2. Block-diagram of the SRA for video applications.

IV. RESULTS

In order to evaluate the image quality achieved by the SRA, several tests have been carried out with synthetic and with real movement and aliasing sequences. Synthetic sequences have the advantage of obtaining quantitative measures. For real sequences only qualitative measures are possible. In this stage only global movement is introduced.

A. Tests with synthetic sequences

The KRANT sequence with context change in frame number 8 is used in this test experiment. The PSNR reached by the luminance signal for the SRA, always above the interpolations levels, is depicted in Fig. 3. In Fig. 4 the SR frame number 12 (b.1) together with the bilinear (c.1) and nearest neighbour (a.1) interpolations are shown, while the errors with respect to the original reference image are located in the bottom of the same Figure. It can be seen clearly that the super-resolved image outperforms the quality of the interpolated images, exhibiting sharper edges and lower overall errors.

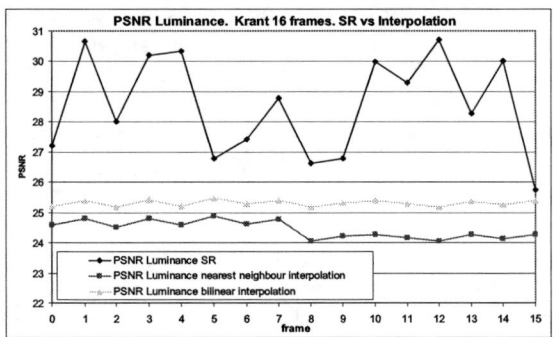

Fig. 3. PSNR of the luminnace for the KRANT secuence of 16 frames with context change in frame number 8.

B. Tests with real movement and aliasing sequences

With respect to applying this SRA to real movement sequences, the sentence *'Let's make things better'* has been recorded with different font sizes, indicating the size on the left, in vertical and horizontal orientations. This text was recorded using a low-grade camera that introduced substantial aliasing. As the camera was hand held, it

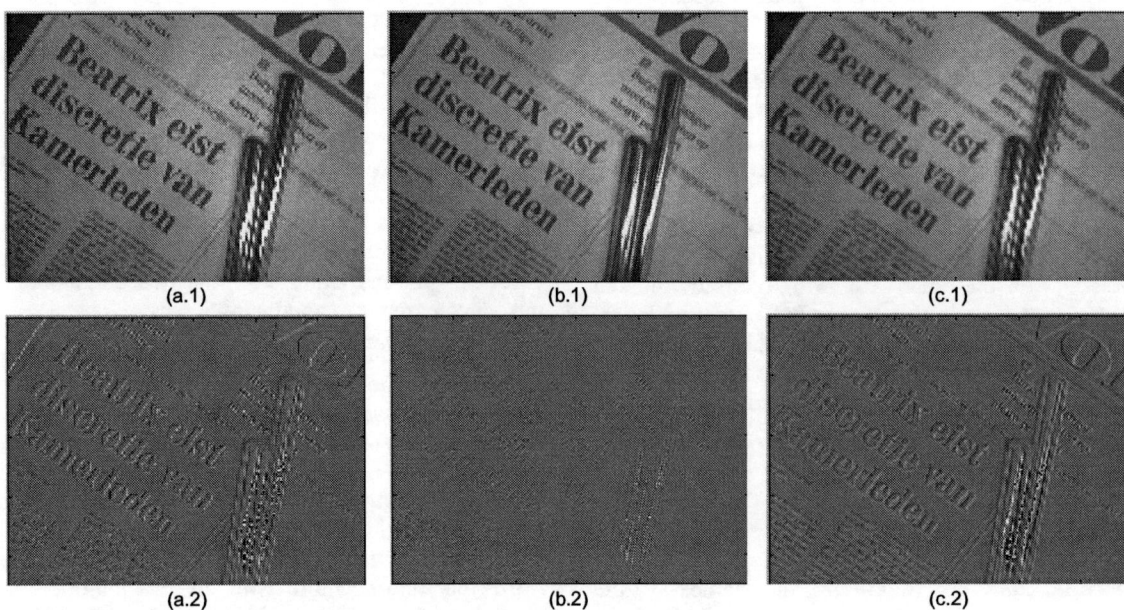

Fig. 4. Frame number 12 of the super-resolved sequence (b.1), toghether with the nearest neighbour interpolation (a.1) and the bilinear interpolation (c.1). In the botton side are shown the errors when compared with the reference image.

produced the necessary shifts to increase the images resolution. In Fig. 5 (a) an enlarged detail of frame number 7 of the input sequence can be observed, and in Fig. 5 (b) the super-resolved image is shown. In general it can be appreciated a recovery in the character edges (specially for the lower font size of 24 points) and a decrease in the background noise, which is obvious as every resulting image comes from the average of some other related images.

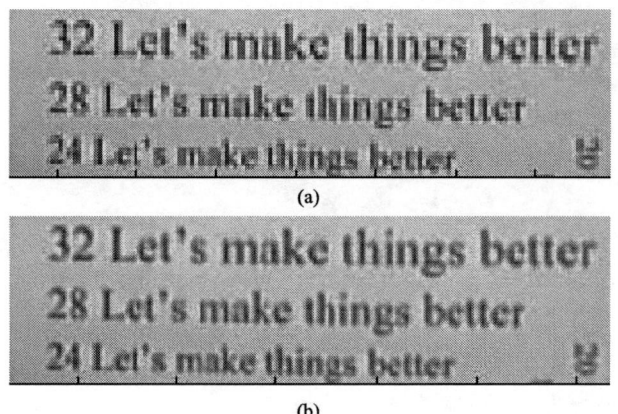

Fig. 5. Enlarged detail of the frame number 7 of the input sequence (a) and of the output sequence (b) of the SRA.

C. Memory requirements versus image quality

Comparisons among different previously reported versions of the SRA are shown in Fig. 6. It can be seen that the use of a feedback memory drastically reduces the memory requirements when compared with the static iterative SRA in [7], and the static non-iterative SRA in [8] and [9]. As it is depicted in Fig. 7, moving from static iterative to static non-iterative, produces a 40.32% increase in the memory requirements while the quality increases in 31.39%. However, moving from the static to the dynamic SRA (both non-iterative) decreases the quality in 1.92 dB (only a 6.3 %) while the memory requirements decrease in 94.84%.

V. CONCLUSIONS

A novel dynamic SRA implemented on a hardware/software multi-standard video codec has been presented in this paper. The results reveal important improvements in the image quality compared to the low-resolution input sequence, obtaining a drastic reduction in the memory requirements with respect to the previous SRA and enabling the algorithm to be implemented at the expenses of a slight quality loss. As result, the codec can be used either in compression mode or in SR mode by including minimal changes on the original platform. The algorithm, although implemented on a Philips Research proprietary codec architecture, can be easily mapped upon any other hybrid video encoder platforms.

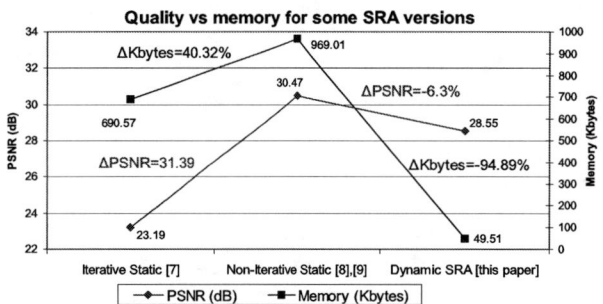

Fig. 7. Average quality versus memory used in different published versions of the SRA.

REFERENCES

[1] T. Komatsu, T. Igarashi, K. Aizawa, T. Saito, "Very high resolution imaging scheme with multiple different-aperture cameras," Signal Processing: Image Communication vol. 5, pp. 511-526, Dec. 1993.

[2] M. J. Op De Beeck & R. P. Kleihorst, "Super-Resolution of Regions of Interest in a Hybrid Video Encoder," Philips Conf. on DSP, 1999.

[3] R. Peset Llopis, M. Oosterhuis, R. Sethuraman, P. Lippens, A. van der Werf, S. Maul and J. Lin, "HW-SW Codesign and Verification of a Multi-Standard Video and Image Codec," IEEE ISQED, San Jose, California, March 2001, pp. 393-398.

[4] T. S. Huang and R. Y. Tsay, "Multiple Frame Image Restoration and Registration," Advances In Computer Vision and Image Processing (Ed. -T. S. Huang), vol. 1, JAI Press Inc., Greenwich, CT, 1984, pp. 317-339.

[5] Lucas J. Van Vliet and Cris L. Luengo Hendriks, "Improving spatial resolution in exchange of temporal resolution in aliased image sequences," in proc. of 11th Scandinavian Conf. On Image Analysis, Kaugerlussauaq, Greenland, pp. 493-499, 1999.

[6] H. Ur and D. Gross, "Improved Resolution from Sub-pixel Shifted Pictures," CVGIP: Graphical Models and image Processing, vol. 54, pp. 181-186, March 1992.

[7] Gustavo M. Callicó, Rafael P. Llopis, Antonio Núñez, Ramanathan Sethuraman, Marc Op de Beeck. "A Low-Cost Implementation of Super-Resolution based on a Video Encoder," IEEE IECON, vol. 2, pp. 1439-1444, Seville, Spain, Nov. 2002.

[8] Gustavo M. Callicó, Rafael P. Llopis, Antonio Núñez, Ramanathan Sethuraman. "Low-Cost and Real-Time Super-Resolution over a Video Encoder IP," IEEE ISQED, pp. 79-84, San Jose, California, USA, March 2003.

[9] Gustavo M. Callicó, Rafael P. Llopis, Antonio Núñez, Ramanathan Sethuraman, "Mapping of Real-Time and Low-Cost Super-Resolution Algorithms on a Hybrid Video Encoder", SPIE, vol. 5117, pp. 42-52, Maspalomas, Spain, May 2003.

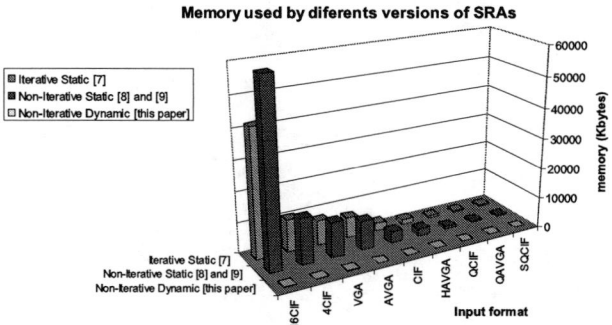

Fig. 6. Comparative of the memory used by several published versions of SRAs for different input formats.

A 0.35μm CMOS Comparator Circuit For High-Speed ADC Applications

Samad Sheikhaei, Shahriar Mirabbasi, and Andre Ivanov
Department of Electrical and Computer Engineering
University of British Columbia
2356 Main Mall, Vancouver, BC, V6T 1Z4, Canada
{samad, shahriar, ivanov}@ece.ubc.ca

Abstract—A high-speed differential clocked comparator circuit is presented. The comparator consists of a preamplifier and a latch stage followed by a dynamic latch that operates as an output sampler. The output sampler circuit consists of a full transmission gate (TG) and two inverters. The use of this sampling stage results in a reduction in the power consumption of this high-speed comparator. Simulations show that charge injection of the TG adds constructively to the sampled signal value, therefore amplifying the sampled signal with a modest gain of 1.15. Combined with the high gain of the inverters, the sampled signals are amplified toward the rail voltages. This comparator is designed and fabricated in a 0.35μm standard digital CMOS technology. Measurement results show a sampling frequency of 1GHz with 16mV resolution for a 1V input signal range and 2mW power consumption from a 3.3V supply. The architecture can be scaled down to smaller feature sizes and lower supply voltages.

I. INTRODUCTION

High-speed comparators are essential building blocks in high-speed flash analog-to-digital converters (ADCs). Such ADCs are widely used in many applications including data storage systems, fast serial links and high-speed measurement instruments. In these applications, typically a low-to-medium resolution ADC (i.e., 4 to 6 bits of resolution) and speeds of the order of GHz are desired [1].

In this paper, an architecture for a high-speed comparator is presented. A block diagram of this comparator is shown in Fig. 1. It consists of three main stages: a *preamplifier*, a *latch* and an *output sampler*.

In high speed ADCs, the conventional comparator architecture consists of several stages of (pre)amplifiers and latches [1, 2]. In such architectures, in the last latching stage, during the regeneration process, two cross-connected inverters create a positive feedback and generate the rail-to-rail voltages at the output of the comparator. During this process, a low resistance path is formed from V_{DD} to ground through the cross-connected inverters. As a result, the last latching stage usually consumes a significant power. The proposed architecture reduces this power consumption by using an output sampler as the last latching stage, while maintaining the speed performance.

A TG is used to implement the output sampler circuit. As explained in Section II, the proposed architecture takes advantage of the typically problematic charge injection phenomenon of the transistors in the transmission gate.

A test circuit based on the proposed architecture was fabricated in a 0.35μm standard digital CMOS technology, and was successfully tested.

This paper is organized into 4 sections. The circuit blocks of the proposed comparator are described in Section II. Measurement results are presented in Section III. Finally, conclusions are drawn in Section IV.

II. CIRCUIT BLOCKS OF THE PROPOSED COMPARATOR

The circuit-level schematics of the main building blocks of the proposed comparator are shown in Fig. 2. These blocks include: a preamplifier (Fig. 2.a), a latch and output samplers (Fig. 2.b) and sampling clock generators (Fig. 2.c). A timing diagram and the relation between comparator clock and output sampler clock are also shown in Fig. 2.d.

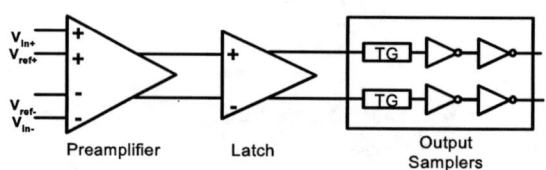

Figure 1. Architecture of the proposed comparator

This research is supported by Natural Sciences and Engineering Research Council of Canada (NSERC) and Micronet.

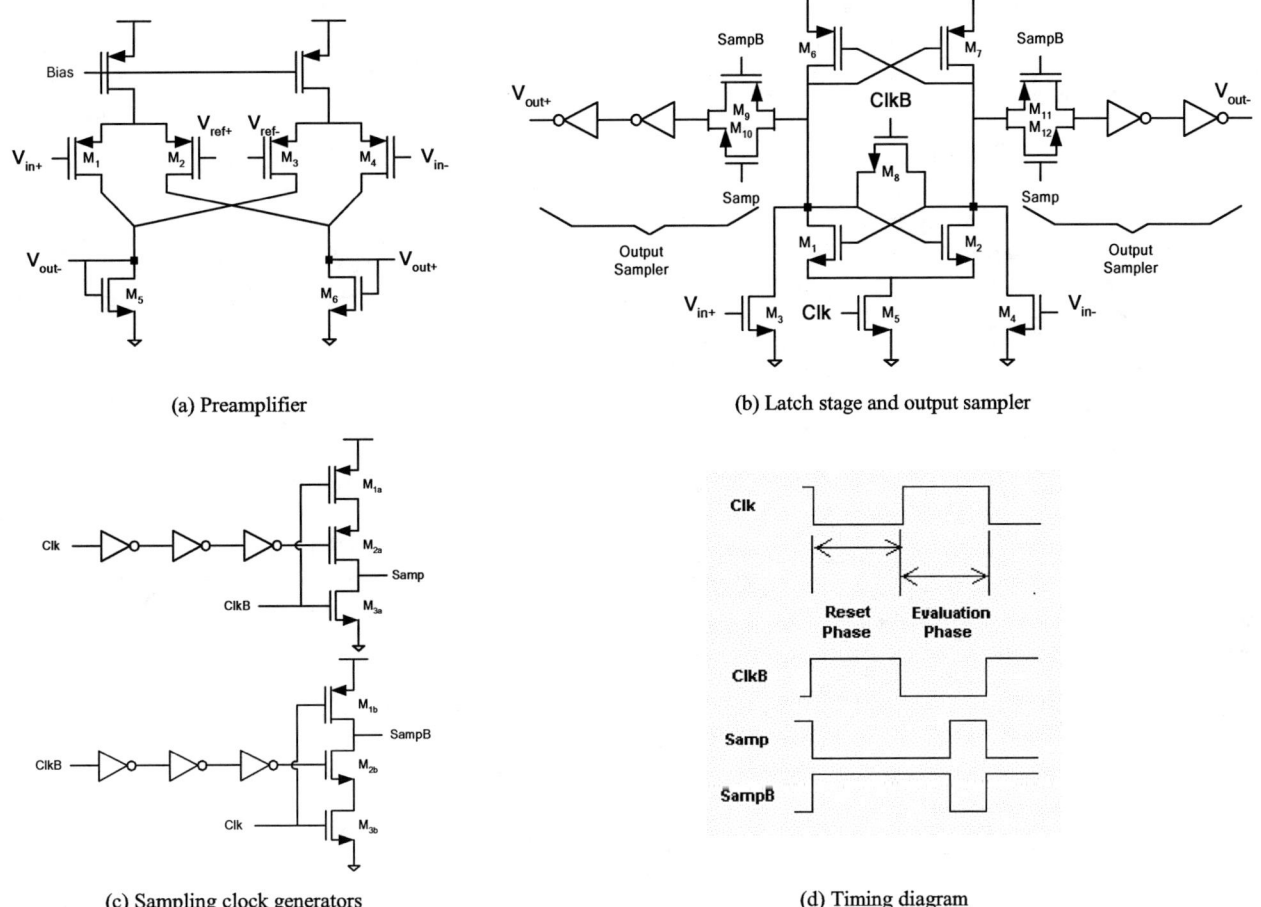

Figure 2. The proposed comparator

The detailed operation of each building block is described in the following subsections.

A. Preamplifier Circuit

The preamplifier is shown in Fig. 2.a. The input differential pairs are PMOS rather than NMOS transistors. This is partly because the range of the input voltage is assumed to be below 1V.

The output currents of the preamplifier are mirrored into the latch stage through transistors M5-M6 in Fig. 2.a and M3-M4 in Fig. 2.b.

B. Latch Stage

The latch stage (Fig. 2.b) consists of a cross-coupled pair of NMOS and PMOS transistors, which are connected to the ground through the clock enabled transistor M5. When the clock (Clk) is low (i.e., ClkB is high), the latch is in its reset state (Fig. 2.d). In this state, output voltages of the latch are at the mid-point of the rails. This yields a faster regeneration time in comparison with the case of starting from the rail voltages [3]. During the reset phase, the preamplifier translates the voltage difference between the inputs of the comparator into an unbalanced state in the latch stage. Then, in the evaluation phase (when Clk goes high), the latch stage is activated. Because of the positive feedback this unbalanced state is amplified towards the rail voltages. At the end of this phase, there is still a gap between latch outputs and rail voltages. The outputs of the latch are sampled and then the latch is reset through the reset switch, M8.

Deferring the sampling time of the output sampler to the end of the evaluation phase decouples the latch stage from the output sampler (refer to Figure 2). This decoupling allows for a faster operation of the latch circuit at the beginning of its evaluation phase. Furthermore, it minimizes the kickback effects of the output sampler and the error due to charge injection of the TG on the output of the latch..

C. Output Samplers

The output sampling circuit is also shown in Fig. 2.b and consists of a full TG and two inverter buffers. This configuration is in fact a dynamic latch. The combination of the input capacitance of the inverter gate and the output capacitance of the TG acts as a holding capacitor. The outputs are sampled at the end of evaluation phase. A short pulse signal is needed as the sampling clock (Fig. 2.d). The samples are amplified (and buffered) using the output inverters. The final output samples remain constant for the whole clock period, which relaxes the timing requirements for the following stages (e.g., encoder).

D. Slight Amplification by the Transmission Gate

Typically, charge injection is a major problem when a TG is utilized as a sampling switch. However, in the proposed architecture, voltage change due to charge injection adds constructively with the sampled signal and helps push the sampled signals towards the rail voltages. To further clarify this property, consider the TG shown in Fig. 3. A full TG passes rail-to-rail voltages. The NMOS transistor acts as a closed switch for low-to-medium voltages while the PMOS transistor operates as a low resistance path for medium-to-high (near supply) voltages.

When the TG is in track mode, i.e., the "Samp" signal is high, the M_n transistor acts as a closed switch for low input voltages. At the end of the track phase, "Samp" changes to low. This opens the M_n switch and causes half of its channel charge to be injected into the sampling capacitor. Since the channel charge is negative, it shifts the sampled signal towards lower voltages. A similar complementary process happens for M_p. Thus, the differential sampled signal is amplified.

Simulations show a modest gain of 1.15, as depicted by the input-output characteristic of the TG in Fig. 4.

The gain of the TG can be estimated using the following equations. In these equations, ΔV_{NMOS} and ΔV_{PMOS} are the voltage changes at the output of the TG due to the charge injection of NMOS and PMOS transistors, respectively [3]. C_H is the total capacitance at the output of the TG that acts as the holding capacitor. This capacitor has two parts, C_{TG} and C_{inv}, which are the output capacitance of the TG and the input capacitance of the inverter, respectively. G is the small-signal gain of the TG.

$$\Delta V_{NMOS} = -\frac{W_N LC_{ox}(V_{DD} - V_{in} - V_{TN})}{2C_H} \quad (1)$$

$$\Delta V_{PMOS} = \frac{W_P LC_{ox}(V_{in} - |V_{TP}|)}{2C_H} \quad (2)$$

$$V_{charge-inj} = \Delta V_{PMOS} \text{ or } \Delta V_{NMOS} \quad (3)$$

For $W_N = W_P = W$

$$G = \frac{\partial V_{out}}{\partial V_{in}} = \frac{\partial (V_{in} + V_{charge-inj})}{\partial V_{in}} = 1 + \frac{WLC_{ox}}{2C_H} \quad (4)$$

$$C_H = C_{TG} + C_{inv}$$
$$\cong \frac{1}{2}WLC_{ox} + (W+2W)LC_{ox} = \frac{7}{2}WLC_{ox} \quad (5)$$

$$G \cong 1 + \frac{1}{7} = 1.14 \quad (6)$$

E. Sampling Clock Generators

Fig. 2.c shows the output sampling clock generators. Two delay lines, each consisting of three inverters, delay the rising edge of the "Samp" and falling edge of the "SampB" signals. As a result, the required sampling signals of Fig. 2.d are generated.

F. Offset of the Comparator

In a flash ADC, the offset of comparators is an important issue, which results in the degradation of the accuracy of the ADC. There are various methods reported in the literature to reduce this offset; including input offset storage [2], offset averaging [1], and digital calibration [4]. The input offset storage method is used for the ADCs which permit idle times (i.e., not continuously working). Therefore, this method is not applicable for the potential applications of the proposed comparator. The offset averaging method is a popular technique, which is commonly used in flash ADCs. The last method, digital calibration, is obtained at the expense of added complexity in the digital domain. However, it can be used for precise offset removal of a flash ADC.

The offset averaging and digital calibration methods are appropriate for use in conjunction with the proposed comparator when the comparators are employed in a flash ADC. However, none of these methods were implemented in the fabricated stand-alone comparator. The offset averaging method is not applicable for a single comparator, while the digital calibration method adds considerable complexity to the circuit of a single comparator. Although no offset cancellation technique was used in the fabricated comparator, the offset of the comparator was measured in view of enhancing future designs.

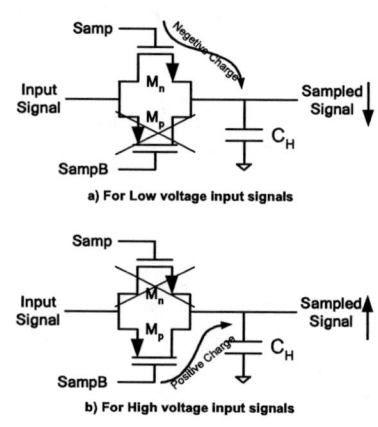

Figure 3. Charge injection effect in a full TG

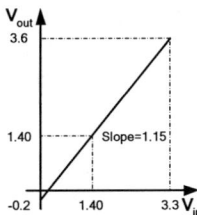

Figure 4. Amplification by a full TG

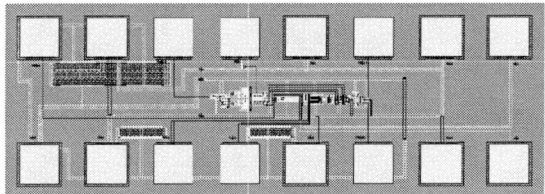

Figure 5. Layout of the test circuit

III. MEASUREMENT RESULTS

The layout of the circuit and a micrograph of the chip under test are shown in Figures 5 and 6, respectively. It should be noted that 50 ohm internal terminating resistors were used for the clock and the differential input ports, and open drain NMOS amplifiers were used to facilitate driving the measurement equipment. Fig. 7 shows the test setup of the chip. Reference voltages were generated using a DC power supply. A single signal generator was employed to generate synchronous clock and input signals. Attenuators were added to the input path for precise resolution and offset measurements. Table 1 shows the frequency measurement results for three sample chips. Table 2 shows the performance measurement results averaged for three chips, compared with the fastest reported comparator in 0.35µm CMOS [1]. The ratio of the comparator speed to its power consumption can be used as a performance metric of the comparator. From the data in the table, the $\frac{Speed}{Power}$ metric for the proposed comparator shows a major improvement in comparison with the comparator in [1].

IV. CONCLUSIONS

In this paper, a comparator architecture suitable for high-speed applications is presented. Using an output sampler consisting of a TG and two inverters as the last latching stage reduces the power consumption. The TG's charge injection enhances the voltage level of the sampled signal. For proper operation of this architecture, a short-pulse sampling signal is required to activate the output sampling circuitry. A simple pulse-generator which constructs such short pulses from the comparator's clock, was also presented. The comparator was fabricated in 0.35µm standard digital CMOS process. The circuit works at 1GHz while consuming 2mW from a 3.3V supply.

REFERENCES

[1] Choi, Abidi, "A 6b 1.3Gsample/s A/D converter in 0.35µm CMOS", *IEEE J. Solid-State Circuits*, vol. 36, pp. 1847–1858, Dec. 2001.

[2] Mehr, Dalton, "A 500-MSample/s, 6-b nyquist-rate ADC for disk-drive read-channel application", *IEEE J. Solid-State Circuits*, vol. 34, pp. 912-920, July 1999.

[3] D. Johns and K. Martin, *Analog Integrated Circuit Design*, John Wiley, 1997.

[4] C. Donovan and M. P. Flynn, "A 'digital' 6-bit ADC in 0.25µm CMOS," *IEEE J. Solid-State Circuits*, vol. 37, pp. 432-437, March 2002.

Figure 6. A micrograph of the chip under test

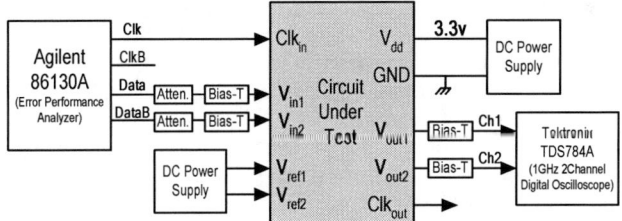

Figure 7. The test setup

TABLE I. SAMPLING FREQUENCY MEASUREMENT RESULTS FOR 3 SAMPLE CHIPS

Chip Number	Max Freq
#1	960M
#2	1.00G
#3	1.03G

TABLE II. MEASUREMENT RESULTS AVERAGED FOR 3 CHIPS COMPARED WITH THE COMPARATOR IN [1]

Performance metrics	The proposed comparator	Comparator in [1]
Technology	0.35µm CMOS	0.35µm CMOS
Supply Voltage	3.3V	3.3V
Input signal range	1V	1V
Resolution	16mV = 6bit	16mV = 6bit
Sampling frequency	1.0GHz	1.3 GHz
Power consumption	2mW	4.3mW
$\frac{Speed}{Power}$	500 GS/J	300 GS/J
Input referred offset voltage	50mV	Cancelled by offset averaging

A 4-Bit 5GS/s Flash A/D Converter in 0.18μm CMOS

Samad Sheikhaei, Shahriar Mirabbasi, and Andre Ivanov
Department of Electrical and Computer Engineering
University of British Columbia
2356 Main Mall, Vancouver, BC, V6T 1Z4, Canada
{samad, shahriar, ivanov}@ece.ubc.ca

Abstract—A 4-bit 5GS/s flash analog-to-digital converter (ADC) is designed and simulated in a 0.18μm CMOS technology. Low-swing operation both in the analog and the digital circuitry results in high-speed low power operation. The ADC dissipates 70mW power from a 1.8V supply while operating at 5GHz. Offset averaging is used to minimize the effect of comparator offsets. Simulation results show that offset voltages with 67mV standard deviation (i.e., 1 LSB) can be tolerated. Static INL and DNL errors are 0.34 LSB and 0.24 LSB respectively, and the ENOB is 3.65 bits. The simulation results of this non-time-interleaved flash ADC demonstrates a significant improvement in terms of power and area compared to those of previously reported ADCs.

I. INTRODUCTION

Flash ADCs are the key building blocks in many applications including the read channels of magnetic and optical data storage systems, high data rate digital communications, high-speed instruments, wideband radar and optical communications. These applications require 4 to 6 bits of resolution at conversion rates of 1GHz or beyond.

Several works were published earlier in the area of 4-bit flash ADCs [1]-[4]. In these ADCs, multi-GHz sampling rates are achieved using time-interleaved architectures. Because of the gain and offset mismatches among the different ADC channels, time-interleaved architectures usually require digital calibration methods [1], [4]. These calibration circuits significantly increase the power and / or area of the flash ADC [1].

In the proposed architecture, a speed of 5GS/s is achieved based on low swing operation in the complete ADC. Two-stage resistor offset averaging gives an ENOB of 3.65. Therefore, no digital calibration is required, resulting in substantial saving in power and area.

This paper is organized into 6 sections. The architecture of the ADC is presented in Section II. Circuit descriptions are presented in Section III. Advantages of the proposed ADC over the previously published works are described in Section IV. Simulation results are demonstrated in Section V. Finally, conclusions are drawn in Section VI.

II. ARCHITECTURE OF THE ADC

Fig. 1 shows the architecture of the proposed ADC. The resistor ladder generates 21 tap voltages from two clean reference voltages, 0.9V and 1.6V. 21 multi-stage comparators, including 15 main and 3 over-range comparators at each end of the array, compare the input signal with the tap voltages and generate a thermometer code. Finally, a current-mode logic (CML) encoder translates the thermometer code to a binary code through an intermediate Gray code.

No single front-end track-and-hold (T/H) is used in this ADC. Instead, a distributed sampling is performed in the first latch of the comparator array.

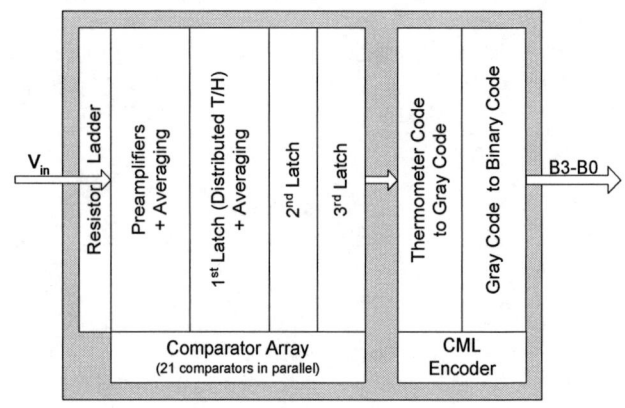

Figure 1. Architecture of the proposed 4-bit flash ADC

This research is supported by Natural Sciences and Engineering Research Council of Canada (NSERC) and Micronet.

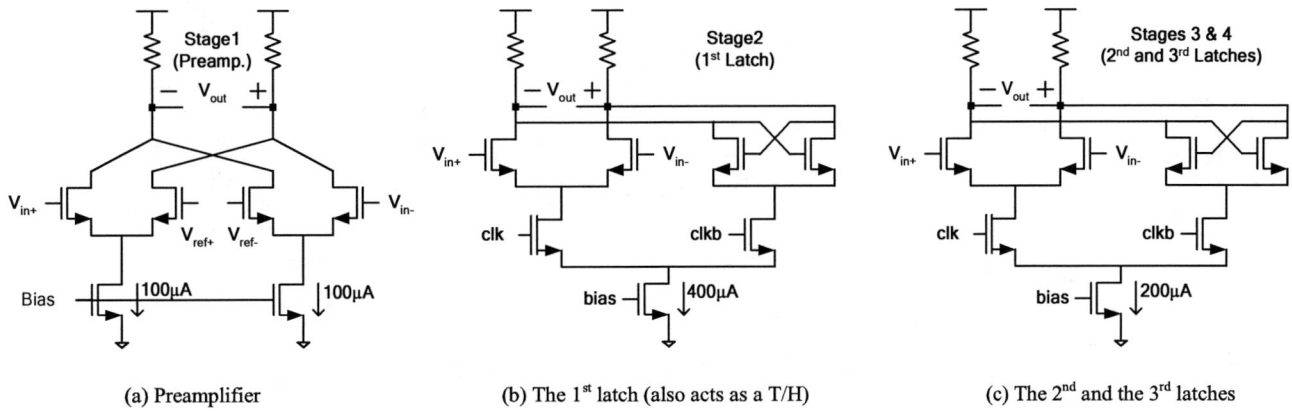

(a) Preamplifier   (b) The 1st latch (also acts as a T/H)   (c) The 2nd and the 3rd latches

Figure 2. Circuit for the comparator

III. CIRCUIT DESCRIPTIONS

A. Circuit for the Comparator

As shown in Fig. 2, the comparator is comprised of 4 stages. The first stage is a low gain preamplifier. The next 3 stages are latching stages, which gradually amplify the differential output voltages to generate +/-0.4V swing at the comparator's outputs. In comparison with the comparator used in the 6-bit ADC of [5], reset switches in the preamplifier and the latching stages can be removed in this design, because there is enough overdrive voltage for each comparator in a 4-bit ADC to override the previous decision.

In the first latch, when the clock is high, outputs follow the inputs. As the clock goes down, the amplified signal is sampled on the output. This process acts as a distributed sampling scheme for the proposed ADC.

Due to amplifying-and-latching through the remaining stages the required swing is achieved at the output of the comparator. These three latching stages largely decrease the meta-stability errors in the ADC.

B. Circuit for the Encoder

Due to low swing nature (i.e. +/- 0.4V) of the output of comparators, the encoder is implemented using CML blocks. Using CML circuits ensures high-speed low-power operation on low-swing signals.

In order to design the CML encoder, a gate-based encoder is preferred as compared to a ROM-based one. The structure of the encoder in [6] was chosen which is shown in Fig. 3. In this structure, the encoder first converts the thermometer code to an intermediate Gray code and then to a binary code. The intermediate Gray code is intended to reduce the effect of bubble errors in the thermometer code.

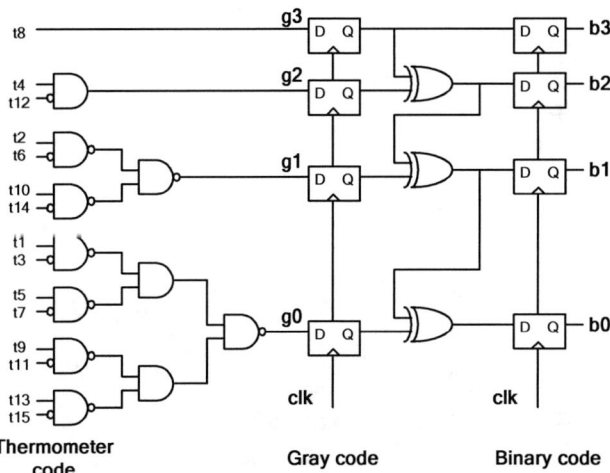

Figure 3. Circuit for the encoder [6]

C. Layout

Fig. 4 shows the layout of the ADC.

Distributed T/Hs suffer from skews in the clock and input signals distributed across the array [7]. To reduce this effect, ring paths are used to deliver the input signals to the preamplifiers and the clock signals to the first latch stages (which act as the T/H). Fig. 5 shows the path for the input signals.

IV. SOME ADVANTAGES OF THE PROPOSED ADC

This ADC has a number of advantages over the previously published works. (1) All the signals in this ADC both in the analog part and the digital part are differential, which ensures the immunity of the system to the common-mode noise. This is important, especially when the noise generating digital processing circuits are implemented in the vicinity of the ADC core. (2) All the signals in the circuit are low swing signals. As a result, a lower noise is generated at the high-frequency of operation of the circuit. (3) As all the clocked transistors in the circuit are in differential pairs, low-swing clock signals are also applicable to the ADC. Simulations show that the complete ADC circuit works with a differential low-swing 'sinusoidal' clock. (4) Because all the stages of this ADC consist of differential pairs, all the tail currents in the circuit are reused and none of the current sources are turned off. As a result, supply noise due to switching is minimized.

V. SIMULATION RESULTS

Fig. 6 shows the effect of offset averaging of the preamplifiers. Fig 6a shows the normal trip points (or threshold voltages) for all the 21 comparators when there is no offset. Fig. 6b shows the absolute value of the voltage offsets (with a σ_{offset} of 1 LSB) randomly applied to each comparator in the array. Fig. 6c shows the ADC trip points affected by the offsets. Simulations show that the offsets degrade the ENOB to 2.51. Fig. 6d shows the trip points for all the 15 main comparators in ADC after resistor averaging. As a result, an ENOB of 3.65 is achieved. Fig. 6e and Fig. 6f show the static INL and DNL performance of the flash ADC.

Table 1 demonstrates the performance summary in comparison with the two other ADCs in [1] and [4]. As shown in this table, simulations predict a speed of 5GS/s with a power consumption of 70mW from a 1.8V supply.

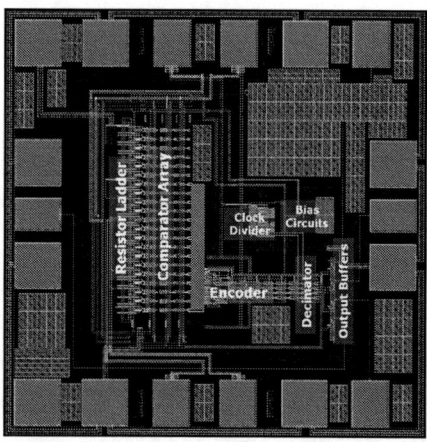

Figure 4. Layout of the ADC

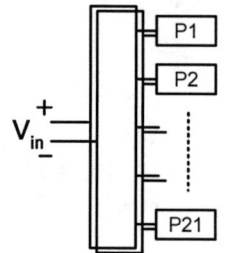

Figure 5. Using a ring path for delivery of the input signal to the preamplifier array

VI. CONCLUSIONS

In this paper, a 4-bit flash ADC is presented. Low-swing operation allows achieving the speed of 5GS/s with a single non-time-interleaved ADC. Thus, with no digital calibration technique the proposed 4-bit ADC achieves a static ENOB of 3.65, while consuming low power and area. This circuit is submitted for fabrication in TSMC 0.18μm CMOS technology. For future designs, digital calibration can be added to implement ultra-high-speed time-interleaved ADCs.

ACKNOWLEDGMENTS

The authors would like to thank Roberto Rosales for technical discussions.

REFERENCES

[1] W. Ellersick, C. K. Yang, M. Horowitz, and W. Dally; "GAD: A 12GS/s CMOS 4-bit A/D converter for an equalized multilevel link," *IEEE Symp. VLSI Circuits, Dig.Tech. Papers*, June 1999, pp. 49-52.

[2] C. K. Yang, V. Stojanovic, S. Mojtahedi, M. Horowitz, W. Ellersick, "A serial-link transceiver based on 8-G samples/s A/D and D/A converters in 0.25μm CMOS," *IEEE J. Solid-State Circuits*, vol. 36, pp. 293–301, Nov. 2001.

[3] Nathawad, Urata, Wooley and Miller, "A 40-GHz-bandwidth, 4-bit, time-interleaved A/D converter using photoconductive sampling", *IEEE J. Solid-State Circuits*, vol. 38, pp 2021 - 2030, Dec. 2003.

[4] S. Naraghi and D. Johns, "A 4-bit analog-to-digital converter for high-speed serial links", Micronet Annual Workshop, April 26-27, 2004, Aylmer, Quebec, Canada, pp. 33-34.

[5] Choi, Abidi, "A 6b 1.3Gsample/s A/D converter in 0.35μm CMOS", *IEEE J. Solid-State Circuits*, vol. 36, pp. 1847 -1858, Dec. 2001.

[6] P. Scholtens and M. Vertregt, "A 6-b 1.6-Gsample/s flash ADC in 0.18μm CMOS using averaging termination," *IEEE J. Solid-State Circuits*, Vol. 37, pp. 1599 -1609, Dec. 2002.

[7] B. Razavi, "Design of sample-and-hold amplifiers for high-speed low-voltage A/D converters," *Custom Integrated Circuits Conf.*, May 1997, pp. 59 -66.

[8] K. Uyttenhove and M. Steyaert, "Speed-power-accuracy tradeoff in high-speed CMOS ADCs" *IEEE Trans. Circuits and Systems II: Analog and Digital Signal Processing*, Vol. 49, pp. 280-287, April 2002.

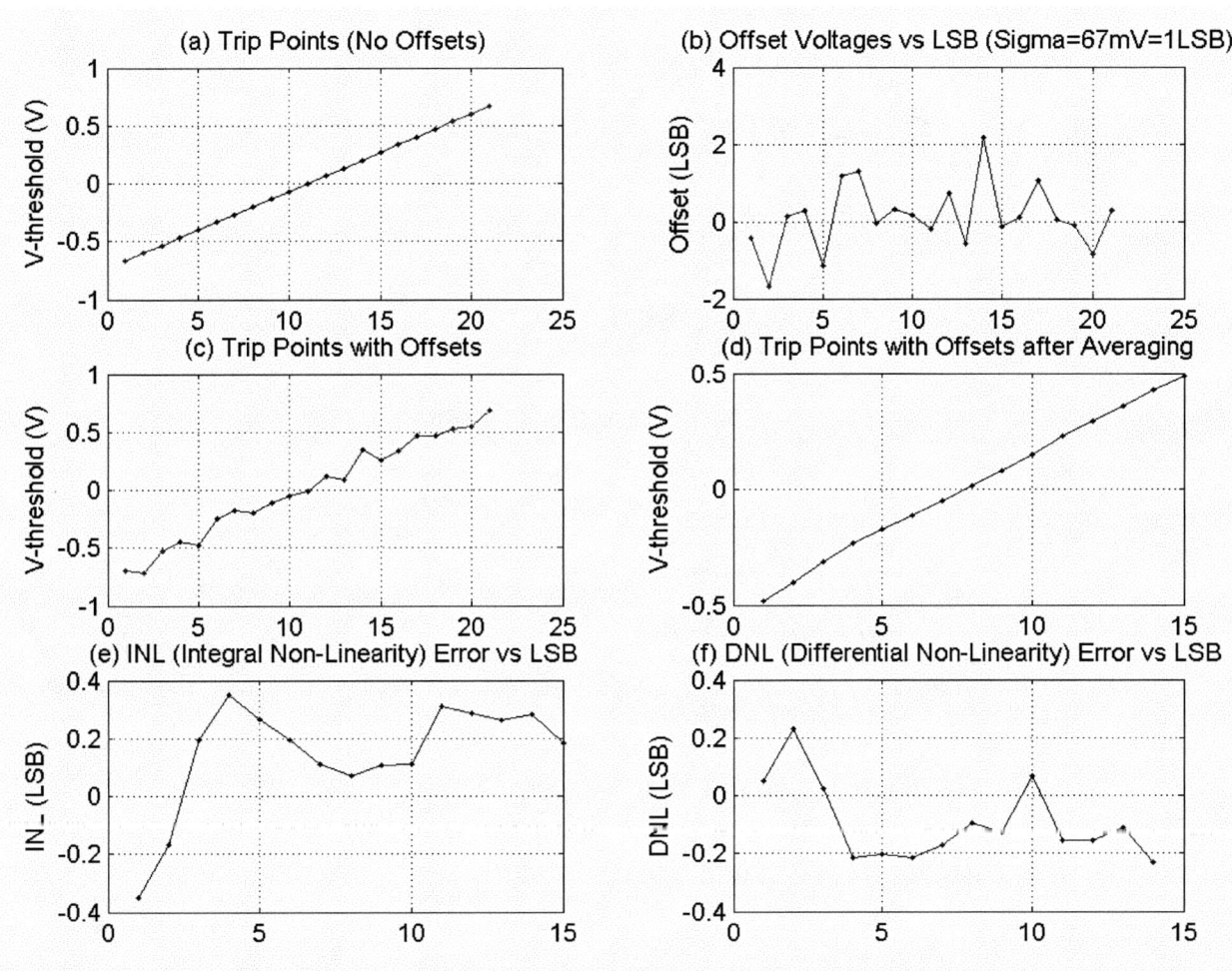

Figure 6. Effect of the offsets in the proposed flash ADC
Notes: (1) Horizontal axes are the comparator number in the array (2) 1LSB = 67mV

TABLE I. COMPARISON OF THE PERFORMANCE

	The proposed ADC	ADC in [4]	ADC in [1]
Technology	Standard 0.18μm CMOS	Standard 0.18μm CMOS	Standard 0.25μm CMOS
Published in	Proposed in 2004	2004	1999
Supply Voltage	1.8V	1.8V	2.5V
Input Signal Range	1V_{p-p} differential	0.88V_{p-p} differential	2 V_{p-p} differential
Resolution	4 bits = 67mV	4 bits	4 bits
Max Sampling freq.	5.0GS/s @TT70	3.2GS/s	12GS/s
Power Consumption	70mW	131mW	1.0 W
Figure of Merit [1]	950 GS/J	316 GS/J	133 GS/J
Active Area	0.2 mm^2	0.4 mm^2	> 9 mm^2
Total Area	0.9 mm^2	1.92 mm^2	10.6 mm^2
INL	23mV=0.34 LSB @ DC	0.6 LSB @ DC	0.4 LSB @ DC
DNL	16mV= 0.24 LSB @ DC	0.4 LSB @ DC	0.4 LSB @ DC
ENOB	3.65@DC	3.6@DC, 2.4@1.3GHz	3.34 @ DC
Architecture	Non-time-interleaved	2-way time-interleaved	8-way time-interleaved
Digital calibration	No	Yes, Off-chip	Yes, On-chip

Note (1): Figure of merit is defined as [8]:

$$F.M. = \frac{Speed \times Accuracy^2}{Power}$$

A Low-Power 4-b 2.5 Gsample/s Pipelined Flash Analog-to-Digital Converter Using Differential Comparator and DCVSPG Encoder

Shailesh Radhakrishnan, Mingzhen Wang and Chien-In Henry Chen
Department of Electrical Engineering
Wright State University
Dayton, USA
{sradhakr, mwang, cihchen}@cs.wright.edu

Abstract - This paper presents a 4-bit high-speed, low-power, pipelined flash analog-to-digital converter (ADC). The proposed ADC is pipelined and mainly consists of three stages: 1) track-and-hold (T/H), 2) differential comparator, and 3) Differential Cascode Voltage Switch with Pass Gates (DCVSPG) encoder. The T/H uses a current mode, dual-array structure to reduce the aperture jitter for high input frequency. The differential comparator eliminates the use of the resistor ladder circuit by generating the reference voltages internally. The DCVSPG encoder has a full output signal swing and compact logic design style of pass gate circuits which makes it suitable for high sampling frequency. The DCVSPG encoder reduces the power consumption by a factor of 88% as compared to the conventional ROM encoder. The ADC is designed in 130 nanometer CMOS technology. FFT tests prove proper operation of the ADC sampled at 2.5 GHz for input signal frequency up to 1 GHz.

I. INTRODUCTION

Data conversion between analog and digital signals is important in signal processing. In many System-on-a-Chip (SoC) applications, it is necessary to have a low-power, low-voltage, fast data conversion which has led to the development of new architecture for ADCs. Recent high-speed ADCs [1-3] have been reported to improve conversion by time interleaving, averaging termination, interpolating, folding, etc., but conversion rate of 2.5 Giga samples per second (GSPS) is still a challenge in CMOS technology. Meanwhile designing ADCs that operate at a low-voltage faces great challenge because of the relatively high threshold voltage of transistor.

A flash ADC usually consists of three components: track-and-hold, comparators and encoder. The performance of flash ADCs depends on the ability of sampling the input without jitter. One approach is adopting T/H circuitry at the input and the sample time of T/H needs to be small enough for fast data conversion. The other is using the clocked comparator and it is necessary that all comparators be clocked simultaneously to avoid jitter. This limitation reduces the resolution of ADCs at high speed. T/H circuitry can be implemented using dual-array [1] or single-array [4] structure. Dual-array T/H achieves higher data throughput. The proposed ADC uses current mode, dual-array T/H.

Conventional voltage comparators use resistor ladder circuit to generate 2^n-1 reference voltages (V_{ref}) for 2^n-1 comparators, ranging from $V_{ref}(max)$ to $V_{ref}(min)$. Generating equally spaced reference voltages determines quantization performance of ADCs. Threshold inverter quantization (TIQ) comparator [5] and Quantum voltage (QV) comparator [6] eliminate the use of resistor ladder circuit by generating the reference voltages internally. TIQ comparator uses two cascaded inverters to generate the range of internal reference voltages. At the first inverter the analog input quantization is set by adjusting the W/L ratio of the transistors. The second inverter is used to increase the voltage gain and prevent unbalanced propagation delay. QV Comparator is a type of differential voltage comparator and is a cascaded voltage comparator. The idea of generating the internal reference voltages is similar to TIQ comparators. Differential comparator has less susceptible to noise than TIQ comparator. Power consumption of the differential comparator is much less as compared to the TIQ comparator.

Encoder is a bottleneck for achieving high-speed ADCs. ROM encoder and fat-tree encoder [7] convert the thermometer code to binary code in two steps. The thermometer code is first converted into 1-out-of-n code using 0-1 generators or XOR gates. The 1-out-of-n code is then converted into the binary code by using either the ROM encoder or the fat-tree encoder. This paper presents a DCVSPG [8] encoder which directly converts the thermometer code to binary code in one step.

II. PIPELINED FLASH ADC

The proposed ADC is a 4-bit pipelined flash ADC. The ADC is designed to digitize input signal frequency up to 1 GHz at a sampling rate of 2.5 GHz. The ADC is pipelined at three stages: 1) the input T/H circuitry, 2) the input of DCVSPG encoder, and 3) the output of DCVSPG encoder. The ADC takes one clock cycle for data conversion (i.e., the digitized output is produced in one clock cycle after the analog input is clocked at input of T/H circuit). The timing of the ADC is shown in Fig. 1.

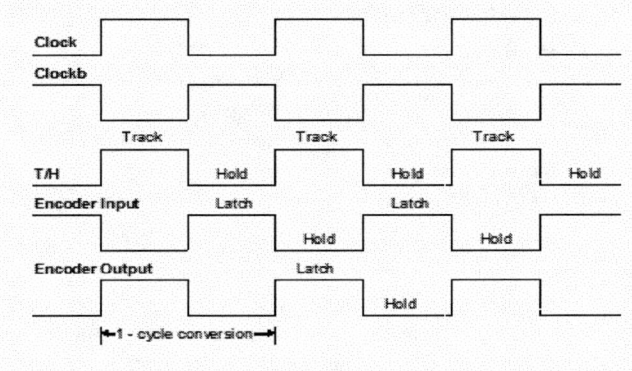

Figure 1. Timing of the 4-bit flash ADC

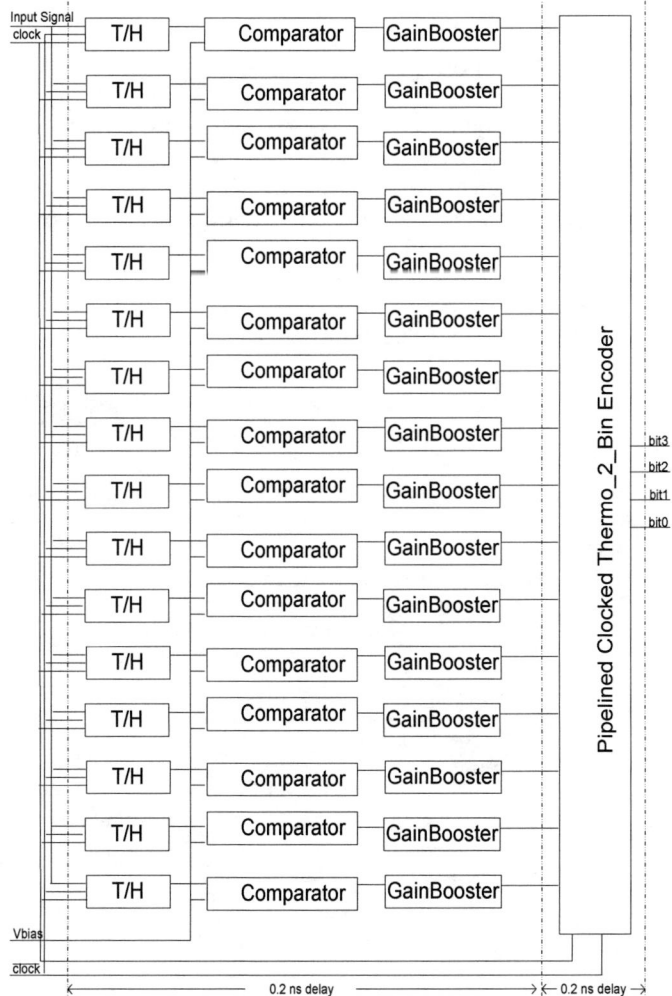

Figure 2. Architecture of the 4-bit flash ADC

The architecture of 4-bit pipelined flash ADC is shown in Fig. 2. It consists of fifteen clocked dual-array T/H, fifteen cascaded differential comparators, fifteen gain boosters and one clocked DCVSPG encoder. The T/H circuit reduces the aperture jitter for high input frequency. The 15 differential comparators compare the T/H output with internal reference voltages. The output of the 15 comparators is a 15-b thermometer code (t15,..,t1) which is then converted to 4-b binary code (b3,..,b0) using the DCVSPG encoder.

A. Track-and-hold (T/H) Circuit

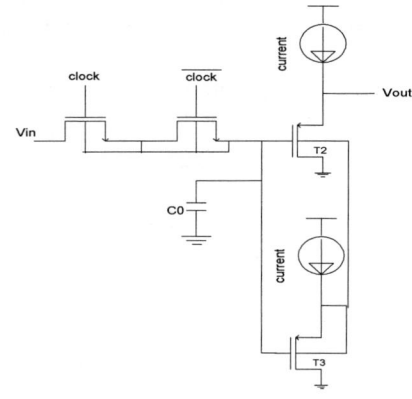

Figure 3. Schematic of the dual-array T/H circuit

The T/H circuit shown in Fig. 3 uses a current mode, dual-array structure. The T/H is clocked for pipelining in ADC. Dynamic range at the output of the T/H is controlled by the capacitor (C0), the transistor sizes of PMOS T2 and T3, and the current source. The output of the T/H circuit has a dynamic range of 0.4 V, ranging from 0.4 to 0.8 V. The T/H determines the signal offset voltage and signal amplitude to the comparator. The dynamic range is also determined to ensure proper operating points of the transistors in the comparator. Utmost care has been taken to ensure the dynamic range of T/H is nearly independent of the input signal frequency. Fig. 4 presents the dynamic performance of T/H for input frequency up to 1 GHz. The harmonics increases with the frequencies. The maximal spurious free dynamic range (SFDR) for 100 and 900 MHz is 45.17 and 25.6 dB, respectively.

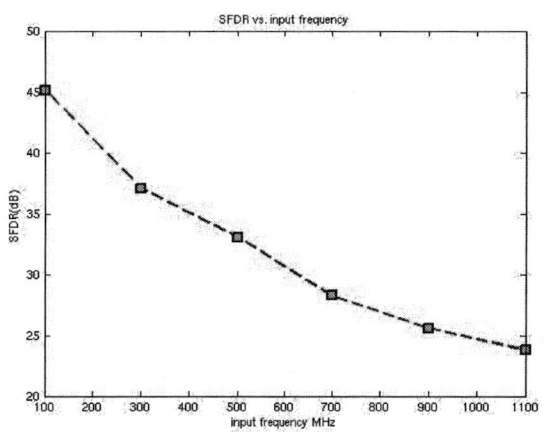

Figure 4. Simulated dynamic performance of T/H

B. Differential Comparator

Comparator of Fig. 5 is a type of differential voltage comparator and has good common-mode noise rejection. The current mirror is designed to produce constant channel currents in Q1 and Q2. The differential pair consists of Q3 and Q4, where the current is determined by the voltage difference between V3 and V4. Transistor Q5 is a current source and its current is determined by V5. The 15 internal reference voltages generated by 15 comparators are nearly equally spaced and are almost identical to the ideal values. The output of the cascaded comparator is fed to the gain booster which operates as a signal buffer to provide sharp transition of the comparator output and fully digital rail to rail swing.

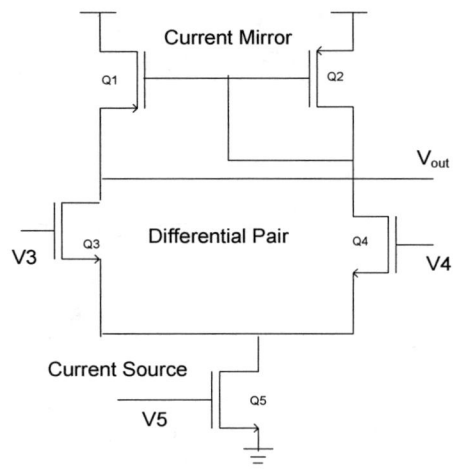

Figure 5. Schematic of the Differential Comparator

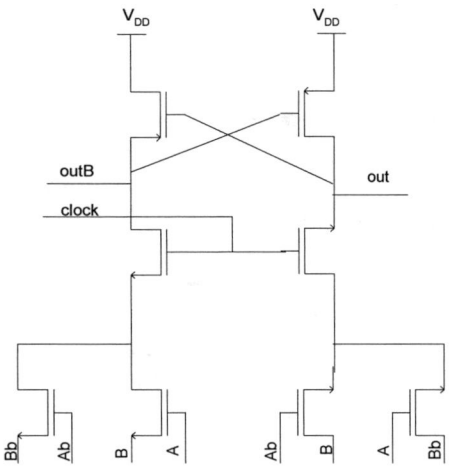

Figure 6. Clocked DCVSPG XOR gate

C. DCVSPG Encoder

The pipelined encoder is designed using clocked DCVSPG logic. An example of clocked XOR gate is presented in Fig. 6. Due to the cross-coupled PMOS device load, the DCVSPG logic is designed to have a built-in latch structure. The output is latched at the previous output value when the clock is inactive. Both out and outb are nearly produced at the same instance which avoids extra inverter delay for generating the outb. DCVSPG encoder is a ratioless logic and the output obtained has no glitches. It has superior performance with power and area.

The DCVSPG encoder converts the thermometer code to binary code in one step. The Boolean expression of thermometer code to binary code is shown below.

$b3 = t8$
$b2 = t4 \oplus t8 \oplus t12$
$b1 = (t2 \oplus t4 \oplus t6) \oplus (t8) \oplus (t10 \oplus t12 \oplus t14)$
$b0 = (t1 \oplus t2) + (t3 \oplus t4) + (t5 \oplus t6) + (t7 \oplus t8) + (t9 \oplus t10) + (t11 \oplus t12) + (t13 \oplus t14) + (t15)$

The thermometer to gray encoder is also implemented using clocked DCVSPG logic. The Boolean expression of thermometer code to gray code is shown below.

$b3 = t8$
$b2 = t4 \oplus t12$
$b1 = (t2 \oplus t6) + (t10 \oplus t14)$
$b0 = (t1 \oplus t3) + (t5 \oplus t7) + (t9 \oplus t11) + (t13 \oplus t15)$

The DCVSPG encoder has less propagation delay and power consumption as compared to the ROM encoder. The thermometer code to gray code encoder has even superior delay and power, as shown in Table I. The power consumption of the DCVSPG encoder is 88 % less than the conventional ROM encoder.

TABLE I. ENCODER PERFORMANCE SUMMARY

Encoder Design	Number of Transistors	Propagation Delay (ns)	Power Dissipation (mW)
ROM	314	0.220	0.1934
DCVSPG (Ther – Bin)	296	0.186	0.02229
DCVSPG (Ther – Gray)	170	0.060	0.01295

TABLE II. PERFORMANCE OF 4-BIT ADC

ADC Type	Flash
Resolution	4 bit
Power Supply	1.2 V
Sampling Rate	2.5 GSPS
Power	23.78 mW
Area	0.0572 mm^2
V_{REF}	0.425– 0.775 V
V_{LSB}	0.025 V
DNL	0.296 LSB
INL	0.260 LSB

III. EXPERIMENTAL RESULTS

The 4-bit flash ADC is designed and simulated in 130 nanometer CMOS. The ADC is tested for input frequency ranging up to 1 GHz. Table II summarizes the results. Both INL and DNL as shown in the table are less than 0.3 LSB. The average power consumption is 23.78 mW. The simulation of the ADC with 250 MHz input signal is shown in Fig. 7.

Fast Fourier transform (FFT) test was used to evaluate the performance and the dynamic parameters of the ADC [9]. Figure 8 shows the FFT test results with input frequency of 250 MHz. The FFT test exhibits harmonics below the fundamental frequency (250 MHz). The dynamic parameters for the 4-bit ADC are calculated by using the ideal FFT plot. Fig. 9 presents the dynamic performance of 4-b ADC for input frequency up to 1 GHz. The harmonics increases with the frequencies. The maximal SFDR for 125, 400, 700, 875 MHz and 1 GHz is 21.6, 19.83, 15.55, 11.2 and 9.08 dB, respectively.

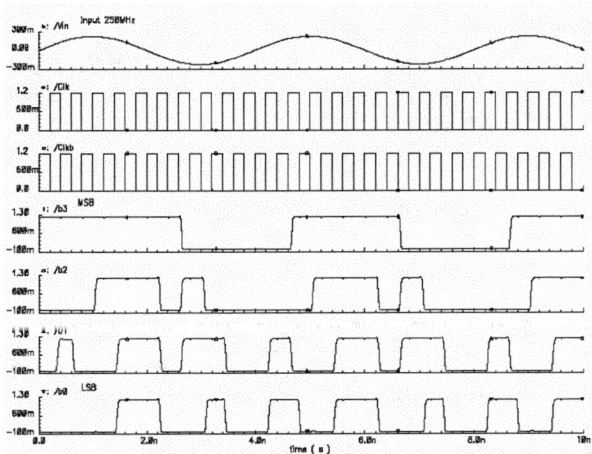

Figure 7. Output of the ADC with 250 MHz input

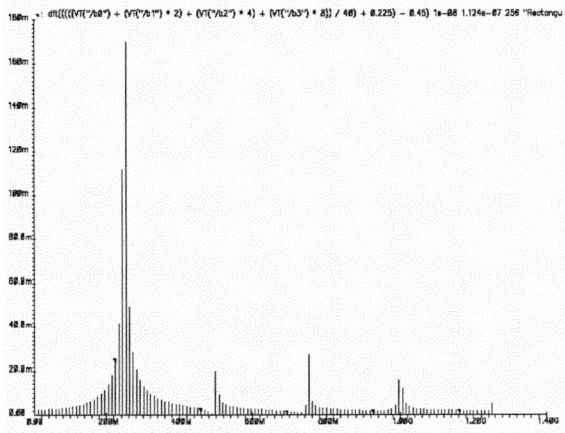

Figure 8. FFT test with input frequency of 250 MHz

IV. Conclusion

This paper presents a design technique (pipelined-flash architecture) and an encoder (DCVSPG encoder) for high-speed flash CMOS ADCs. The pipelined architecture achieves high data throughput and high speed by incorporating pipelined clocked track-and-hold and clocked DCVSPG encoder. The pipelined CMOS ADC offers a data conversion rate of 2.5 GSPS while maintaining low power consumption, which makes it suitable for SoC applications. FFT tests with input frequencies ranging up to 1 GHz prove proper operation over the broad-band frequency range. The DCVSPG encoder overcomes the speed limitation of the ROM encoder which has been a bottleneck of high-speed ADCs.

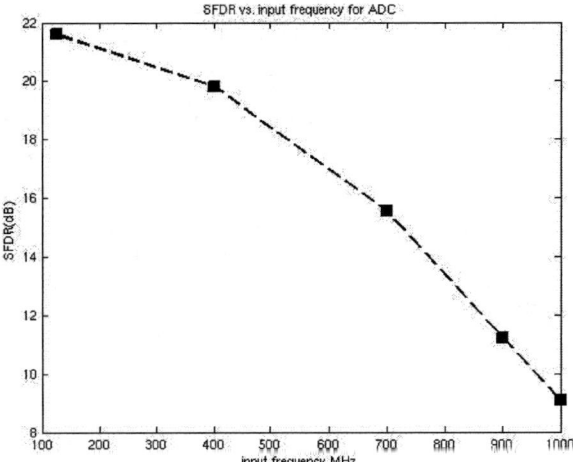

Figure 9. Simulated dynamic performance of ADC

References

[1] X. Jiang, Z. Wang and M.F. Chang, "A 2GS/s 6-b ADC in 0.18 µm CMOS," IEEE International Solid-State Circuits Conference, vol. 1, pp. 9-13, Feb. 2003

[2] M. Choi and A. Abidi, "A 6-b 1.3 Gsamples/s A/D converter in 0.35 µm CMOS," IEEE Journal of Solid-State Circuits, Vol. 36, pp. 1847-1858, Dec. 2001.

[3] P.C.S. Scholtens and M. Bertregt, "A 6-bit 1.6 Gsamples/s Flash ADC in 0.18 µm CMOS using Averaging Termination," IEEE Journal of Solid-State Circuits, Vol. 37, No. 12, December 2002.

[4] M. Wang and C.-I. H Chen, "Design Synthesis and Performance Measurement of Pipelined Flash ADC for SoC Applications," in press.

[5] J. Yoo, K. Choi, D. Lee. "Comparator Generation and Selection for Highly Linear CMOS Flash Analog-to-Digital Converter," Journal of Analog Integrated Circuits and Signal Processing, 35(2-3), pp. 179-187, 2003.

[6] J. Yoo, K. Choi, and J. Ghaznavi, "Quantum voltage comparator for 0.07 µm flash A/D converters," Proceedings of the IEEE Computer Society Annual Symposium on VLSI, pp. 20-21, Feb. 2003

[7] D. Lee, J. Yoo, K. Choi and J. Ghaznavi, "Fat-tree encoder design for ultra-high speed flash analog-to-digital converters," IEEE Midwest Symposium on Circuits and Systems, 2002.

[8] F. Lai and W. Hwang, "Design and implementaion of differential cascode voltage switch with pass-gate (DCVSPG) logic for high-performance digital systems," IEEE Journal of Solid-State Circuits, vol. 32, April 1997.

[9] Nicholas Gray, "ABCs of ADCs", National Semiconductor Corporation, August 2004.

A 1.2GHz Adaptive Floating Gate Comparator with 13-bit Resolution

Yanyi Liu Wong, Marc H. Cohen and Pamela A. Abshire
yanyi.wong/marc.cohen/pamela.abshire@ieee.org
Institute for Systems Research, University of Maryland, College Park, MD 20742, U.S.A.

Abstract—We present a high-speed voltage comparator that uses floating gate adaptation to achieve high comparison resolution. The comparator uses nonvolatile charge storage for either offset nulling or automatic programming of a desired offset. We exploit the negative feedback functionality of pFET hot-electron injection to achieve fully automatic offset cancellation. The design has been fabricated in a commercially available 0.35μm process. Experimental results confirm the ability to reduce the variance of the comparator offset 3600× and to accurately program a desired offset with maximum observed residue offset of **469μV** and standard deviation **199μV**. We achieve controlled injection to accurately program the input offset to voltages uniformly distributed from -1V to 1V. The comparator operates at 1.2GHz with a power consumption of 2.97mW.

I. INTRODUCTION

Comparators are decision-making circuits that interface between analog and digital signals. Comparators are the core element for A/D converters, and oftentimes their performance directly affects that of the resulting A/D converters. Mismatches in the pre-amplifier and regenerative stage due to process variations cause offset that directly affects resolution of a comparator. A common approach used to cancel offset is dynamic switching which adds switches and multiple non-overlapping clocks, and excellent results have been reported [1]. Another approach for high speed operation without switching is on-chip averaging, which has been shown to reduce mismatch and boost resolution [2]. Since offset is a constant value, it is natural to store it using nonvolatile storage on a floating gate. Floating gate circuits have been used to cancel offsets in imagers [3], to trim current sources [4]–[6], and to autozero amplifiers [7], [8]. We previously introduced a simple floating gate comparator [9] with the ability to accurately trim and store desired zero or nonzero offsets, a feature that is not readily available using existing offset cancellation techniques. In this paper, we present the design and testing of a high-speed high-resolution comparator using floating gate adaptation.

II. BACKGROUND

A floating gate MOSFET uses an electrically isolated material to store charge indefinitely. In our comparator, the circuit offset is stored in this high-retention charge form, and altered by means of differential injection and tunneling.

The injection mechanism has been extensively described in the literature [8], [10]. Impact-ionized hot-carrier injection occurs in p-type MOSFETs when two conditions are satisfied: a high lateral electric field E_L across the channel to increase the likelihood of impact-ionization and produce high energy electrons, and a high vertical electric field E_V across the gate oxide to sweep the hot electrons across the oxide barrier. For an ordinary pFET, it is relatively easy to achieve both conditions under normal operation. An accurate semi-empirical model in [11] suggests that injection current scales as an exponential function of source-to-drain voltage V_{sd}.

Using current sources and sinks, injection mechanisms have been exploited to satisfy different needs [3]–[10]. Our design approach is to use a current source at the source of the injection pFET to form a stable negative feedback loop that enables automatic and accurate adaptation [9].

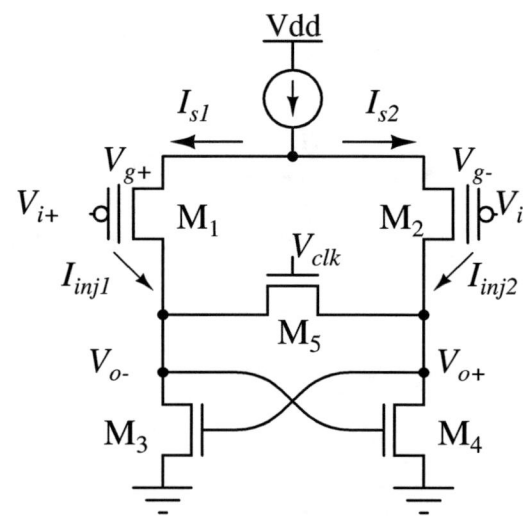

Fig. 1. Simple five-transistor AFGC in [9].

III. ADAPTIVE FLOATING GATE COMPARATOR

Figure 1 shows a 5-transistor implementation of differential-mode hot-electron injection that uses local control and adapts the charge on the input pFETS' floating gates [9]. Transistors M1,2 form the differential pair that compares the input voltages $V_{i+,-}$ and at the same time are responsible for controlling injection. The adaptive element is integrated within the comparator itself.

In this paper we present an AFGC that separates the adaptive elements from the comparator core. Because the comparator itself does not carry out adaptation, its design can be much more flexible. We based our design on a 3-stage-pipelined high performance CMOS comparator [2] as the comparator

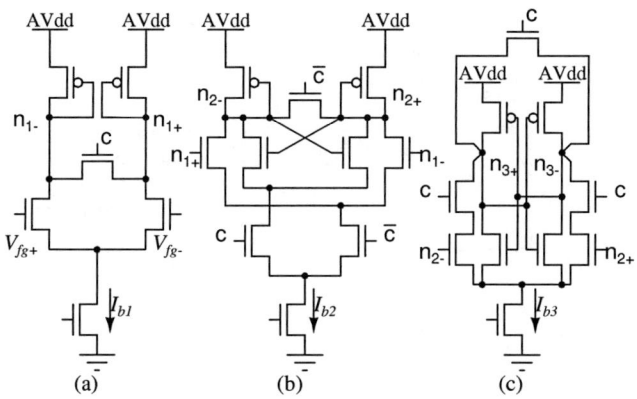

Fig. 2. The comparator core consists of 3 stages.

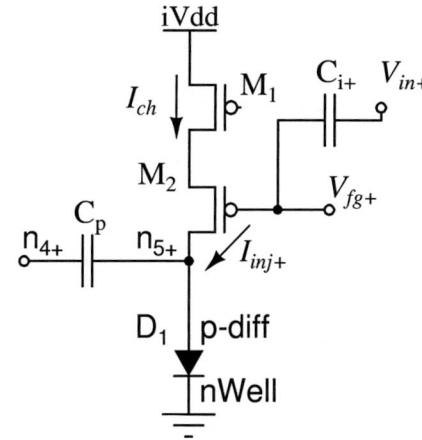

Fig. 3. The adaptive element for AFGC.

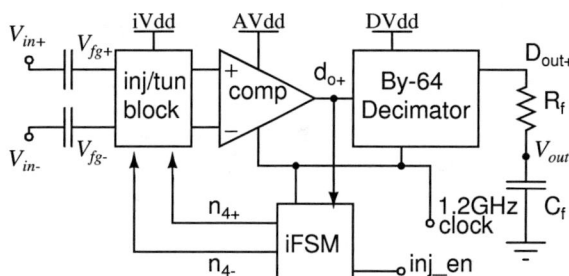

Fig. 4. The AFGC system with filtered output.

core for our AFGC. Figure 2 (a), (b) and (c) show the three stage comparator core. Currents $I_{b1,2,3}$ provide tail currents for the 1st, 2nd and 3rd stages, respectively. The inputs to the 1st stage $V_{fg+,-}$ are supplied by the floating gates of the nFET differential pair. Nodes $n_{1+,-}$ are the outputs of the 1st stage and the inputs to the 2nd stage. Nodes $n_{2+,-}$ are the outputs of the 2nd stage and the inputs to the 3rd stage. Nodes $n_{3+,-}$ are then fed to a latch that produces a rail-to-rail digital output d_{o+}. Clock c and its complement $\bar{c}$ are supplied to the comparator. A separate analog Vdd (AVdd) of 3.3V is supplied to the comparator core.

Figure 3 shows one of the adaptive elements that is used to change the charge on one of the input floating gates. The inputs $V_{in+,-}$ are the AFGC differential inputs, and they are coupled to comparator inputs $V_{fg+,-}$ through capacitors $C_{i+,-}$. M2 is the injection transistor, with channel current supplied by M1. M1 forms a current mirror with another diode-connected transistor and sets the channel current I_{ch}. Capacitor C_p and diode D_1 form a negative charge pump. During normal operation, node n_{4+} sits at the digital Vdd (DVdd), and n_{5+} sits around 0.65V since D_1 is forward biased. C_p holds a voltage of DVdd − 0.65V. The maximum source-to-drain voltage V_{sd} on M2 is 2.65V, which is insufficient to produce impact-ionized hot-electron injection. When n_{4+} goes to ground, n_{5+} immediately goes to −DVdd + 0.65V. This increase in the V_{sd} across M2 causes injection to occur, and a small amount of charge is transferred onto the floating gate so that V_{fg+} decreases by a small amount ΔV.

Suppose that injection occurs with inputs $V_{in+,-}$ connected to some desired voltage $V_{in+} - V_{in-} = V_d$, and there exists unknown charge on the floating gates $V_{fg+,-}$; since $V_{fg+,-}$ are capacitively coupled to clamped inputs $V_{in+,-}$, they are held constant. We pulse n_{4+} to ground when the comparison outcome is positive (plus side greater than minus side), and pulse n_{4-} to ground when the comparison outcome is negative. Differential injection occurs in the direction that makes the outcome reverse. By pulsing n_{4+} to ground, we decrease the gate voltage V_{fg+}, and we move in the direction of a negative outcome. Eventually, the system reaches an equilibrium and the comparison outcome alternates for each cycle, causing injection on the corresponding side of the floating gate. After equilibrium has been established the residual offset left on the floating gates after each injection is smaller than ΔV. As injection proceeds, the maximum V_{sd} for M2 decreases, so the incremental voltage change ΔV gradually diminishes, and we successfully program a precise desired offset V_d onto the differential floating gates of the AFGC. Setting $V_d = 0$, we achieve offset cancellation. This method of differential injection uses the outcome of comparison to correct offset; therefore the adaptation feedback loop encompasses all mismatch and offset within the circuit, and accurate offset adaptation can be achieved.

In the above description, injection is performed after every comparator outcome. In practice, we perform injection every 3 clock cycles. This is because the outcome immediately after an injection cycle is the old comparison result in the pipeline and should not be used to determine the update direction. After 3 clock cycles the pipeline is flushed and the outcome and injection are correctly aligned in time. We implement this delay with a finite state machine (shown as iFSM in Fig.4).

Injection is a one way process that lowers the floating gate voltages. To raise these floating gate voltages, we perform tunneling by grounding inputs $V_{in+,-}$ and raising the power supply voltage on the adaptive element iVdd, to 9.16V. The back gate (nWell) of M2 is also connected to iVdd and therefore also raised to 9.16V. A large electric field now exists across the gate oxide at the side edge of M2, from nWell to the floating gates. This large electric field tunnels electrons off the floating gate, raising the floating gate voltages.

IV. HARDWARE, EXPERIMENTS AND RESULTS

The AFGC was fabricated in a commercially available 0.35μm CMOS process and packaged in a DIP-40 ceramic package. As in all high-speed digital chips, several pins are dedicated to DVdd and GND to reduce ground bounce [12]. These pins are located near pin numbers 10 and 30, the center pins on the left and right side of the DIP-40 package, where the parasitic inductance in the package is minimum. Large area sandwiched layers of metal1, metal2, metal3, poly and poly2 form on-chip decoupling capacitors and at the same time satisfy the chemical-mechanical polishing (CMP) requirements for planarized processes. A 4-layer PCB was made with two inner layers dedicated to power and ground. We use surface mount ceramic capacitors to decouple the power rails. A PC-based DAC card supplies the common-mode voltage to the negative input V_{in-} and a precision voltage source (having 5μV steps) supplies the differential input $V_i = V_{in+} - V_{in-}$.

We use a fast, low-noise, fully differential voltage controlled oscillator (VCO) as the on-chip clock generator [13]. True single-phase clocked (TSPC) D-flip-flops are used as the building block for all synchronous logic circuits to enable a sampling rate above 1GHz [14].

Figure 4 shows the system architecture of the AFGC. We use a decimator to subsample the digital output by 64X to relax requirements on the output buffer that drives the pin D_{out+}. The 1.2GHz clock feeds the comparator core, the decimator and the iFSM. We enable adaptation by turning on inj_en. DVdd is supplied with 4.3V to boost the clock voltage to 4.3V because the clocked switches in the comparator core (Fig.2) are too small to provide sufficient reset conductance with a 3.3V clock, which causes hysteresis. Future designs will remedy this by making all clocked switches bigger. We observe a maximum operating frequency of 1.2GHz with DVdd=3.3V. Therefore, we confine the operating frequency to 1.2GHz even though we have the ability to operate faster for higher values of DVdd.

We low-pass-filter the subsampled digital output D_{out+} with $R_f = 25.6$kΩ and $C_f = 0.1$nF to get a filtered output V_{out}. The cut-off frequency is 62.2kHz. We then sample V_{out} with PC-based data acquisition at 1kHz. We reduce noise by $2\sqrt{10}$ in the measured voltage by averaging over 40 samples.

Using the lowpass filter followed by averaging, the measured V_{out} approaches the average value for the comparator output D_{out+}. Therefore, the normalized quantity $\frac{V_{out}}{DVdd}$ approaches the mean m for the comparison outcome, which includes the effects of deterministic offset as well as random noise. Let X be the random variable representing the actual input offset, and suppose that the outcome of a comparison is zero ($D_0 = 0$) when the differential input signal $V_i = (V_{in+} - V_{in-}) < X$, and one ($D_1 = 1$) when $V_i > X$. Then, m is equivalent to the cumulative distribution function (cdf) $p_1 = P[X < V_i]$ since $m = \sum p_i D_i = p_0 \cdot 0 + p_1 \cdot 1 = p_1$, where $p_0 = P[X > V_i]$. Empirically, we find that the distribution $P[X < V_i]$ is Gaussian. Figure 5 shows

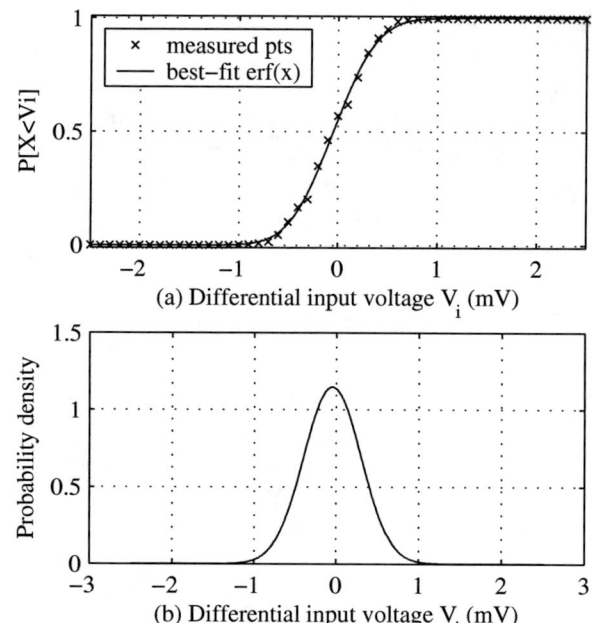

Fig. 5. (a) The normalized V_{out} plotted against differential input voltage V_i and (b) the corresponding probability density function.

the normalized measurement points V_{out} along with a least-square-fit Gaussian cdf curve. Here the measurement is taken after adaptation and has an offset of $E[X] = -46$μV.

In addition to the floating gate comparator, there is also a comparator core with its inputs connected directly to external pins. We were able to measure the offset performance for this "bare" comparator and compare it to the AFGC. In the following, we present offset statistics for the bare comparator and for the AFGC before and after injection. We erase random charges on the floating gates by tunneling prior to adaptation. We measure 21 available chips and plot their offset distribution in Fig.6. The mean μ and standard deviation σ statistics for the 21 measured offsets are; $\mu_0 = 7.081$mV and $\sigma_0 = 11.942$mV for the bare comparator, $\mu_b = 33.685$mV and $\sigma_b = 25.246$mV for the AFGC before injection, and $\mu_a = -6$μV and $\sigma_a = 199$μV for the AFGC after injection. These results clearly demonstrate the AFGC's ability to reduce the mean offset by a factor of 1000 and to reduce the standard deviation by a factor of 60. Note that the mean approaching zero for a large number of samples simply means that the injection mechanism is perfectly balanced. The magnitude of the residual offset on the AFGC after adaptation is limited by the standard deviation. Thus, by reducing σ by 60, we gain approximately 6 bits in resolution compared to the bare comparator. An input offset with $\sigma_a = 199$μV for a 3.3V peak-to-peak sine wave input signal corresponds to an SNR of 81.4dB, or 13 effective bits.

We programmed one AFGC to 21 offset values evenly distributed from -1V to 1V. Figure 7 shows the residual offset (measured−programmed) versus the programmed offset. The magnitude of the residue is under 0.5mV over all programming voltages. The standard deviation for the residue is 178μV, which is comparable to the standard deviation obtained with

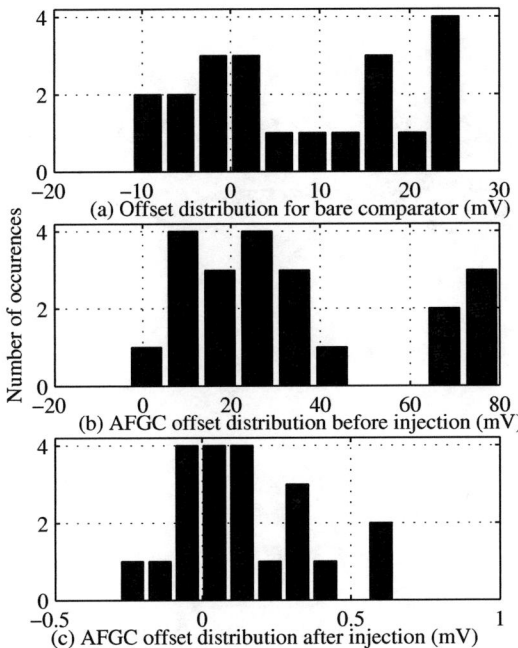

Fig. 6. The input offset distribution over 21 chips for (a) the bare comparator, (b) AFGC before injection and (c) AFGC after injection.

21 different chips. This demonstrates that the AFGC's performance after adaptation holds for chip-to-chip variations, and is independent of programming voltage.

For the above experiments, we inject with $I_{ch} = 8\mu A$. When we apply one short pulse ($10\mu s$) on the inj_en pin in Fig.4, the offset shifts by 21.3mV. For higher I_{ch}, we observe larger shifts. The offset shift is not linear in I_{ch} because the negative voltage on node n_{5+} in Fig.3 holds for less time the higher the I_{ch}. The offset stops shifting if we apply $32\mu A$ or more.

To quantify the ability of the AFGC to retain its post-adaptation stored offset voltage, we programmed three chips with 150mV, 0V and -150mV offset and, after three days observed a 3.935mV, 0.838mV and 0.219mV offset drift.

V. CONCLUSION

A "bare" three stage comparator architecture was redesigned to include floating gate pFET adaptive circuit elements at the input stage. The output decision of this new floating gate comparator is used as the feedback control signal to guide the adaptation process so as to virtually eliminate all inherent circuit mismatches. The AFGC uses the mechanism of hot electron injection to adjust the voltages on the floating nodes of each of the input pFET transistors. When enabled, hot electron injection occurs during normal comparator operation. The AFGC was fabricated in a $0.35\mu m$ technology and when supplied with 3.3V, it operates at 1.2GHz and consumes 2.97mW. Standard deviation of initial offset is reduced by a factor of 60 which translates into a 6 bit gain in resolution when compared with the "bare" comparator. An A/D converter using multiple AFGCs with input offset standard deviation of $199\mu V$ can achieve an SNR of about 80dB which corresponds to 13 effective bits, when converting a 3.3V peak-to-peak sine wave. The performance of the adaptation is independent of chip-to-chip process variations and programming voltage. In addition to canceling offset, the AFGC can accurately store an arbitrary input offset, a feature not readily available in

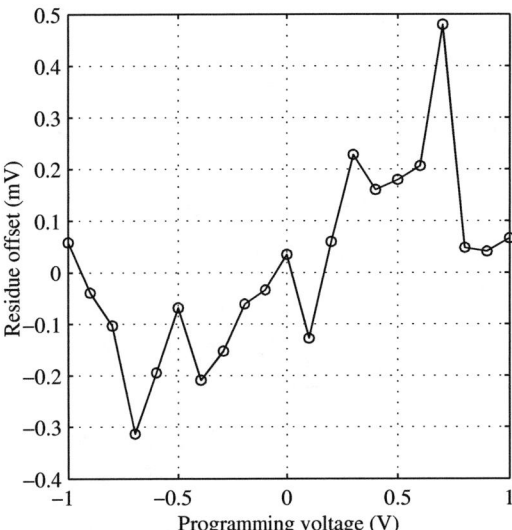

Fig. 7. The residue offset versus programming voltage.

other offset cancellation schemes. Ongoing and future work will use our AFGC to implement new classes of adaptive data converters.

VI. ACKNOWLEDGEMENTS

We thank the MOSIS service for providing chip fabrication through their Educational Research Program. Y.W. is supported by Johns Hopkins University Applied Physics Laboratory. P.A. is supported by an NSF CAREER Award (NSF-EIA-0238061).

REFERENCES

[1] B. Razavi and B. Wooley, "Design techniques for high-speed, high-resolution comparators," *IEEE JSSC*, vol. 27, no. 12, pp. 1916–1926, December 1992.
[2] M. Choi and A. Abidi, "A 6-b 1.3-Gsample/s A/D converter in 0.35-μm CMOS," *IEEE JSSC*, vol. 36, no. 12, pp. 1847–1858, December 2001.
[3] M. Cohen and G. Cauwenberghs, "Floating-gate adaptation for focal-plane online nonuniformity correction," *IEEE TCAS.II*, vol. 48, no. 1, pp. 83–89, January 2001.
[4] S. Shah and S. Collins, "A temperature independent trimmable current source," in *IEEE ISCAS*, vol. 1, May 2002, pp. I713–I716.
[5] S. Jackson, J. Killens, and B. Blalock, "A programmable current mirror for analog trimming using single poly floating-gate devices in standard CMOS technology," *IEEE TCAS.II*, vol. 48, no. 1, pp. 100–102, Jan. 01.
[6] J. Hyde, T. Humes, C. Diorio, M. Thomas, and M. Figueroa, "A 300-MS/s 14-bit digital-to-analog converter in logic CMOS," *IEEE JSSC*, vol. 38, no. 5, pp. 734–740, May 2003.
[7] P. Hasler, B. Minch, and C. Diorio, "An autozeroing floating-gate amplifier," *IEEE TCAS.II*, vol. 48, no. 1, pp. 74–82, January 2001.
[8] T. Constandinou, J. Georgiou, and C. Toumazou, "An auto-input-offset removing floating gate pseudo-differential transconductor," in *IEEE ISCAS*, vol. 1, May 2003, pp. 169–172.
[9] E. Wong, P. Abshire, and M. Cohen, "Floating gate comparator with automatic offset manipulation capability," in *IEEE ISCAS*, vol. 1, May 2004, pp. I–529–532.
[10] P. Hasler and J. Dugger, "Correlation learning rule in floating-gate pFET synapses," *IEEE TCAS.II*, vol. 48, no. 1, pp. 65–73, January 2001.
[11] K. Rahimi, C. Diorio, C. Hernandez, and M. Brockhausen, "A simulation model for floating-gate MOS synapse transistors," in *IEEE ISCAS*, vol. 2, May 2002, pp. 532–535.
[12] H. W. Johnson and M. Graham, *High-speed digital design: a handbook of black magic*. Englewood Cliffs, NJ: Prentice Hall, 1993.
[13] L. Dai and R. Harjani, *Design of high performance CMOS voltage-controlled oscillators*. Boston, MA: Kluwer, 2003.
[14] J. Yuan and C. Svensson, "New single-clock CMOS latches and flipflops with improved speed and power savings," *IEEE JSSC*, vol. 32, no. 1, pp. 62–69, January 1997.

An 8-bit 160 MS/s Folding-Interpolating ADC with Optimizied Active Averaging/Interpolating Network

Meysam Azin
Department Of Electrical Engineering
Sharif University of Technology
Tehran, Iran
m_azin@mehr.sharif.edu

Hamid Movahedian, Mehrdad Sharif Bakhtiar
Department Of Electrical Engineering
Sharif University of Technology
Tehran, Iran

Abstract— An 8-bit CMOS folding-interpolating analog-to-digital converter is presented. A new method for designing optimized averaging circuit is also described. Careful circuit design and layout leads to a high-speed (160 MSPS) and low power (70 mW in 2.5 V supply voltage) ADC. The ADC is successfully implemented in 0.25um CMOS digital process and it takes 1x1.4 mm2 silicon area.

I. INTRODUCTION

High-speed, low-power, low-voltage ADC's are the key elements in many digital signal processing systems such as high-speed digital communication and flat-panel displays which need at least 8-bits resolution and over 100 MHz sampling rate. CMOS flash ADC's with auto-zero comparators have been reported in [1] and [2] with around 500 MHz sampling rate. On the other hand, the complexity of flash ADC's grows exponentially as resolution increases. This is why flash converters are only used for low-resolution (6-7 bits) applications. Pipeline ADC's are used for many high-speed applications, which require higher resolution. Pipeline ADC's, however, introduce a large latency which makes them unsuitable for some applications such as control and data storage. Folding-Interpolating technique is another alternative to avoid the complexity of flash ADC's. Folding ADC's can also achieve high-speed, high-resolution and low power characteristics as well as low latency.

One of the problems in designing high performance folding-Interpolating ADC's is the offset of input devices, which directly affects ADC performance. Averaging is a method to reduce the effect of input offsets. In this work, a new structure for offset averaging is presented. Unlike the resistive offset averaging structures [3], this structure can be used in current mode circuits and results in greater error correction factor (ECF).

System level considerations of converter and ADC structure are investigated in section 2. In section 3, ADC circuits are described. Post layout simulation results are presented in section 4.

II. ADC STRUCTURE

The block diagram of designed ADC is shown in figure 1. The folding factor is selected to be 8. 4 parallel folding blocks are used to generate 4 main folded signals. 32 additional signals are generated by the interpolating network. In this work, main folded signals and 32 interpolated signals are generated differentially. Because of the non-ideal characteristics of folding blocks, Interpolation makes a systematic error called interpolation inherent error [5]. The value of this error in the proposed ADC is 0.1 LSB for integral nonlinearity error (INL) and 0.05 LSB for differential nonlinearity error (DNL).

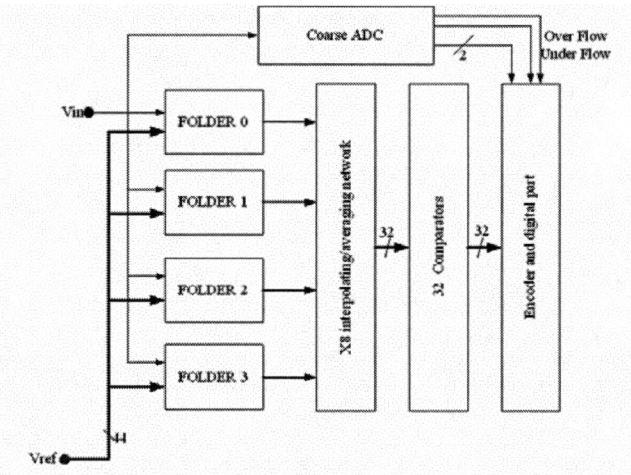

Figure 1. ADC's block diagram

A coarse converter is used to make 2 significant bits of the output code. Other bits are extracted from the fine ADC which has 32 fine comparators. The number of comparators in the coarse ADC is 10 including over-flow and under-flow detectors.

As mentioned earlier, averaging is an effective way to reduce offset induced error. This is done by using redundant information in main signals. First order resistive networks are typically used as offset averaging. In this kind of networks, a number of resistors, which are also used as interpolating resistors, connect adjacent folding blocks.

In an offset averaging network, each output is a linear combination of input signals. If Sin_i and $Sout_i$ present input and output signal to the network respectively, it can be shown that:

$$Sout_0 = \sum_{i=0}^{M-1} h_i \cdot Sin_i \qquad (1)$$

Where, M is the number of total input signals to the averaging network and h_i's are offset averaging factors. These factors determine network characteristics such as ECF (error correction factor). Errors in the input blocks of folding-interpolating ADC's can be divided into vertical and horizontal errors [3]. Vertical errors are caused by deviation of folding characteristics curve from desired curve in the Y (output) direction such as mismatch of tail current sources. Horizontal errors are caused by deviation of the characteristics curve in the X (input) direction such as offset of folder pair. It should be notified that vertical error is not input dependent. However, horizontal error depends on the input value. It can be shown that the highest value of horizontal error occurs at zero crossings of a folding characteristic while its lowest value occurs when signal is at minimums and maximums. This is shown in figure 2 which e is the horizontal error normalized to its maximum.

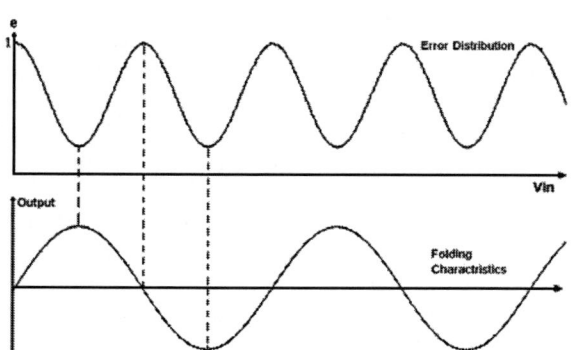

Figure 2. Horizontal error distribution

If the second half of input signals are in negative polarity of the first half, it is proved that ECF's for an averaging network are:

$$ECF_{VE} = Ga \cdot \left[\sum_{i=0}^{\frac{M}{2}-1} (h_i - h_{\frac{M}{2}+i})^2 \right]^{-1/2} \quad \text{for vertical errors} \qquad (2)$$

$$ECF_{HE} = Ga \cdot \left[\sum_{i=0}^{\frac{M}{2}-1} ((h_i - h_{\frac{M}{2}+i})^2 \cdot e_i^2) \right]^{-1/2} \quad \text{for horizontal errors} \qquad (3)$$

Where **Ga** is the gain of the offset averaging network and can be obtained by differentiating equation.1. And e_i is the value of e for Sin_i when Sin_0 has a zero value. It can be shown that $e_0=1$ and e_i is less than 1 for $i > 0$.

Equation.1 can be implemented by use of current mirrors, as shown in figure 3. This network can be implemented together with the active interpolating network to save components and area. Optimization of equation.3 results in the addition of $\frac{1}{3}$ of each input signal to $\frac{1}{3}$ of two neighbor signals to make an output signal. Using Monte Carlo simulation, it is shown that the optimized averaging circuit improves vertical errors by a factor of 1.3 and horizontal errors by a factor of 1.5. Using previous methods, (first order resistive offset averaging [3]) the above improvement would be limited to 1.2 for vertical errors and 1.3 for horizontal errors.

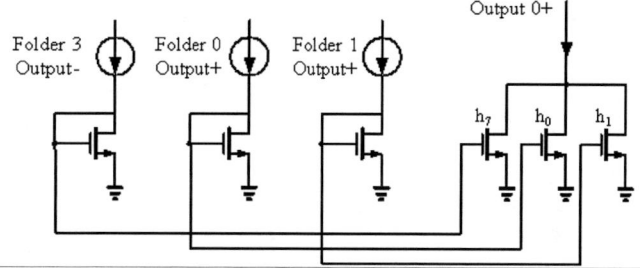

Figure 3. Current mode offset averaging

III. ADC CIRCUIT DESIGN CONSIDERATIONS

A. Folders

Folding block architecture is shown in figure 4. Among 11 used differential pairs 8 ($3^{rd} \sim 10^{th}$) are the main folders, two (2nd, 11th) are to enhance the range for interpolation and one (1st) is added to make a DC balance. Output currents of the 11 differential pairs and a dummy pair are added together to make two folded currents. These currents are subtracted from two constant currents and then flow to the Mr1, Mr2 transistors. These two transistors lower the resistance seen by nodes V1 and V2, resulting in low time constant for these nodes, which are loaded by large capacitances. Two output currents are fed in two current mirror transistors (ML1, ML2) which drive the interpolating/averaging network. This network heavily loads the output nodes of folding blocks. In order to increase the speed of output nodes, an equalizing switch shorts these nodes for a certain clock period (around $\frac{1}{3}$).

For large rising input pulses, common mode voltage of V1 and V2 falls, turning off Mr1 and Mr2. In order to improve the common mode voltage settling, two NMOS transistors (Msp1 and Msp2) are added to the circuit. The value of Vb2 is one threshold voltage higher than common mode voltage of V1 and V2 nodes, so when voltage of these nodes decreases, Msp1 and Msp2 turn on speeding up the folding block settling.

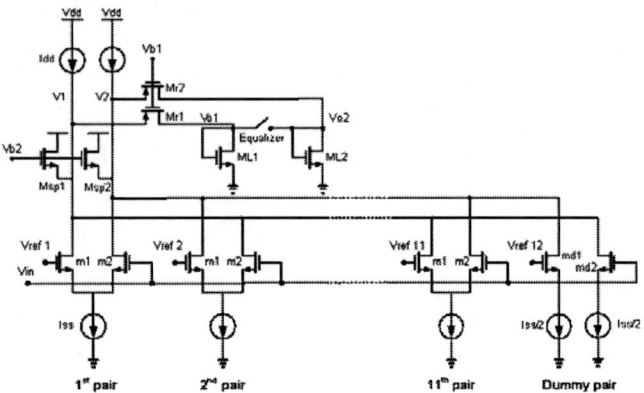

Figure 4. Folding block circuit

The finite output resistance of tail current sources results in an input-dependent current offset in the output signals of a folding block. The input dependency of this offset is due to this fact that for a differential pair with $V_{in} < V_{ref}$, the current source voltage is constant, and for $V_{in} > V_{ref}$ the source voltage follows input signals. The variation of output offset with input voltage is shown in figure 5 (dotted line). V_{FS} is full input range voltage and R_{CS} is current source output resistance.

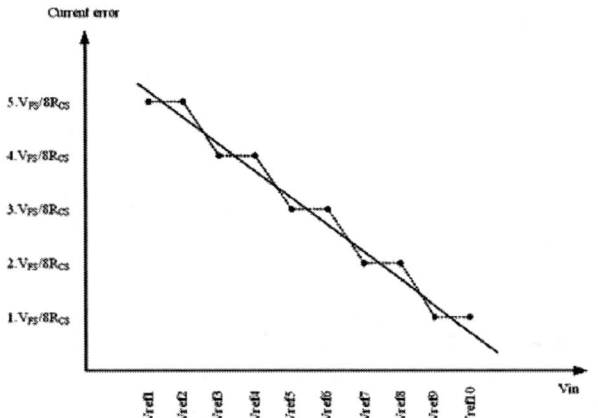

Figure 5. Error induced by current source resistance

The input voltage is applied to a dummy pair with

$$Rcs\_dummy = 2.Rcs \tag{4}$$

and

$$Vref12 = Vref11 + \frac{3}{16}V_{FS} \tag{5}$$

The output currents of this pair will also have input dependent variation as shown in figure 5 (solid line). If the output currents of this dummy pair are subtracted from the output currents of the folding block, the current offset will be limited to the difference of the two curves of figure 5. Using this method, the current error is reduced by a factor of 16. In practice, because of nonlinear behavior of current source resistance, this improvement is not better than 5.

B. Comparators

Current comparators are used to detect zero crossing point of 32 folded signals. Because of pre-amplification of signals in the folding blocks, the functionality of comparators in folding-interpolating ADC's is not so critical. The designed current comparator is shown in figure 6.

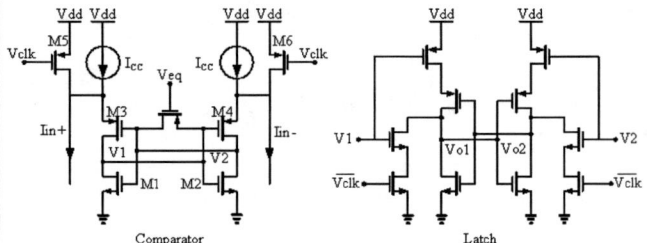

Figure 6. Fine comparator and latch

Two input signals are subtracted from two constant current and then flow to the bistable circuit (M1~ M4). When **Veq** signal goes high, the bistable circuit goes to one of two stable states according to the difference of input currents. It should be considered that the current source value (I_{CC}) should be greater than the common mode current of input signals. It can be shown that if I_{CC} has a value close to the common mode value of input currents, mismatch of M1 and M2 has a negligible effect on the comparator offset. As I_{CC} increases, the effect of M1 and M2 mismatch becomes more dominant. Hence, in order to have a precise comparator, I_{CC} should be selected as close to the common mode value of inputs as possible. On the other hand, small value of difference between I_{CC} and common mode value results in a large settling time for the bistable circuit. In order to have precision and speed simultaneously, two transistors (M5 and M6) turn on shortly (1 nsec) after Veq goes high. This will cause much current flow to the bistable circuit and it reaches its final state quickly. This technique and careful transistor sizing leads to a 1uA current offset and 1 nsec comparison time for these comparators.

IV. PHYSICAL IMPLEMENTION

The ADC is implemented in the 0.25um digital CMOS process with one poly and five metal interconnections. The designed layout measured 1x1.4 mm^2 excluding reference ladder. The chip layout is shown in figure 7.

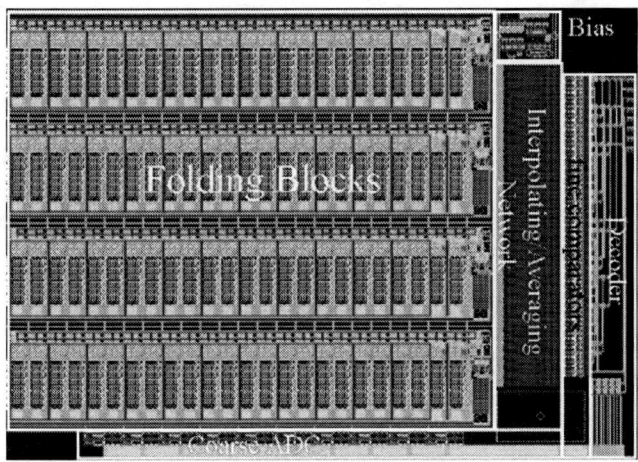

Figure 7. the proposed chip layout

Table 1. Simulation results summary

Technology	CMOS 0.25um
Supply Voltage	2.5 volt
Full scale input voltage	1.3 volt
INL, DNL	< 0.7, < 0.5
Maximum clock frequency	166 MSPS
ENOB (low frequency)	7.5
ENOB (Nyquist frequency)	7.3
SFDR	> 55dB
Power Dissipation (reference ladder included)	70 mW

V. SIMULATION RESULTS AND CONCLUSION

Monte Carlo parameters are extracted using matching rules indicated in the process design kit. Monte Carlo simulation results show a maximum INL and DNL value of 0.7 LSB and 0.5 LSB respectively (figure 9). The dynamic performance of ADC was simulated based on the assumption of 10-bit resolution and 2.5 nsec settling time for Track and Hold circuit. The output spectrum for a full-scale sinusoidal input signal is shown in figure 8. ADC characteristics are depicted at table 1.

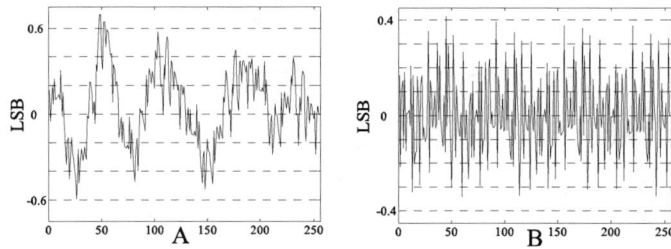

Figure 9. A: INL and B: DNL

REFERENCES

[1] S. Tsukamoto, T. Endo, and W. Schofield, "A CMOS 6-b 400-MSample/s ADC with error correction," *IEEE J. Solid-State Circuits*, vol. 33, pp. 1939–1947, Dec. 1998.

[2] Y. Tamba and K. Yamakido, "A CMOS 6-b 500-MSample/s ADC for a hard disk driver read channel," in *ISSCC Dig. Tech. Papers*, pp. 324–325, Feb. 1999.

[3] H. Pan, M. Segami, M. Choi, J. Cao, F. Hatori and A. Abidi, "A 3.3V, 12b, 50MS/ s A/D converter in 0.6-um CMOS with over 80dB SFDR," in *ISSCC Dig. Tech Papers*, pp. 40- 41, Feb. 2000

[4] B. Nauta and A. G. W. Venes, "A 70- MS/ s 100- mW 8- b CMOS folding and interpolating A/ D converter," *IEEE J. Solid- State Circuits*, vol. 30, pp. 1302- 1308, Dec. 1995.

[5] Razavi, Behzad, "Principles of Data Conversion System Design," New York, IEEE Press, c1995.

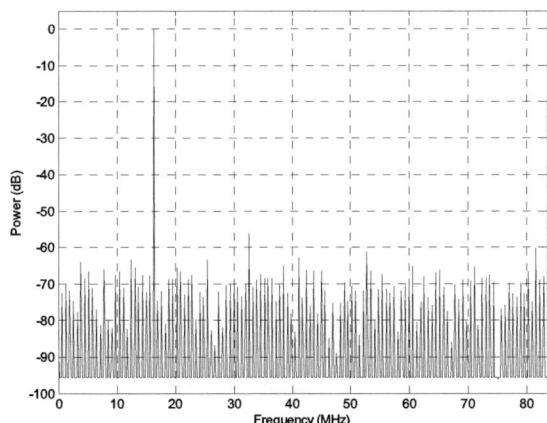

Figure 8. Output spectrum, input frequency=16.27 MHz, Clock frequency=166.6 MHz

OFFSET COMPENSATION IN FLASH ADCS USING FLOATING-GATE CIRCUITS

Philomena Brady and Paul Hasler

Georgia Institute of Technology
Department of Electrical and Computer Engineering
Atlanta, GA 30332

ABSTRACT

Traditionally, high speed Analog to Digital converters (ADCs) are built using various forms of the Flash converter architecture. However, all of these structures suffer from one major flaw, offsets. Whether these offsets are introduced from the resistive biasing network, mismatches in the comparators, or from other sources, extensive care must be taken to compensate for them. There are several methods used to correct for these offsets, but many of these methods require complicated compensation circuitry or greatly increase the cost and area of the ADC. This paper presents an alternative to these methods, implementing the high speed data converters using floating-gate circuits. The floating-gate circuits provide a method of accurately programming the reference voltages of the Flash converter, instead of obtaining them from a resistive biasing network. As a result, any form of offset that is introduced can be compensated for by adjusting the reference voltage.

Today, many applications require the use of high-speed data converters. To obtain the shortest possible conversion time, basic Flash architectures process the input signal using 2^N-1 parallel comparators, shown in Fig.1a. Each comparator is connected to two signals, V_{in}, and a reference voltage that is tapped off from a resistor string. The comparators create a thermometer code that is then decoded by the Schmitt trigger, INV-NOR-INV stage, and the encoder which outputs the appropriate N bit digital word.

One of the most important issues to consider when designing a Flash ADC is the effect of offsets. Variations in the comparator offsets as well as variations in the biasing network and other additional circuitry can cause several problems, including a decrease in the resolution of the ADC. To correct these problems Flash designs are often implemented using complex compensation schemes, or increased resistor and transistor sizes [1, 2, 3]. Our design approaches this problem quite differently. We have built a 6-bit Flash ADC and 6-bit DAC in 0.35μm process using floating-gate circuits. The floating-gate circuits are used to compensate for any offsets that could hinder the accuracy of the Flash structures.

1. FLOATING-GATE FLASH ADC STRUCTURE

One of the basic structures used in our Flash ADC is the electric potentiometer. E-pots are a type of floating-gate circuit which can provide an on-chip, non-volatile analog memory cell that can be configured in an addressable array and programmed easily [4]. E-pots are capable of providing precisely controlled voltages for long periods of time with very little noise or drift. The e-pot's atomic storage element is the floating-gate of a MOS transistor, which can

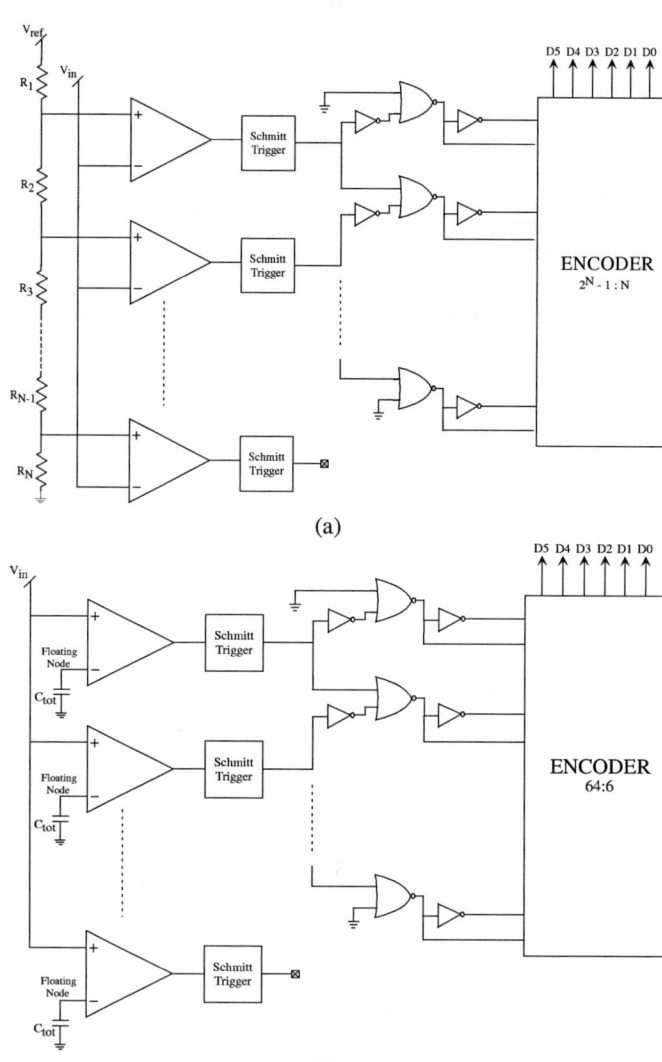

Fig. 1: Two N-Bit Flash ADCs(a) Block diagram of a simple Flash ADC. The bias voltages are provided by the resistor string. (b) Block Diagram of our Floating-Gate Flash ADC. We biased the comparators by programming analog voltages onto the floating nodes. The charge is stored on C_{tot}, the total capacitances present at that floating node.

This work was partially supported by grants National Science Foundation (CISE-1068549) and by corporate donations to the Georgia Tech Analog Consortium by Texas Instruments Inc.

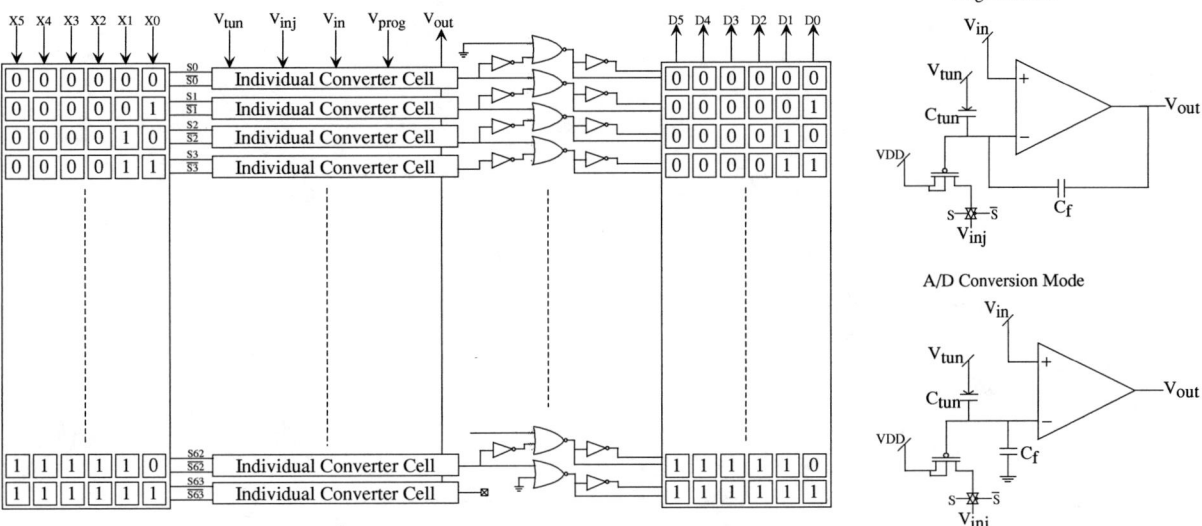

Fig. 2: Overall Structure of the Floating Gate, 6-Bit, Flash Converter. The flash converter uses an e-pot based structure that allows us to program the trip points of the comparators. The e-pots operate in two modes, programming mode and A/D conversion mode. During programming, inputs X0 to X5 are the digital inputs used to select individual converter cells for programming. V_{tun} is the tunneling voltage, and V_{inj} is the injection voltage. V_{in} has a dual role, as the reference voltage during programming, and the analog input voltage during the A/D conversion. V_{prog} is a digital select that sets operating mode for the ADC. V_{out} is the analog voltage output from the e-pot/comparator structures. D0 to D5 are the digital output signals. The output of the final comparator cell is not connected to any digital circuitry. It is not used for analog to digital conversion. However, it is necessary if it is desired to use the same Flash structure as a DAC. When operating in A/D conversion mode, the individual cells are configured as comparators. In this mode C_f has been removed from the feedback path, and connected in parallel with the other capacitances.

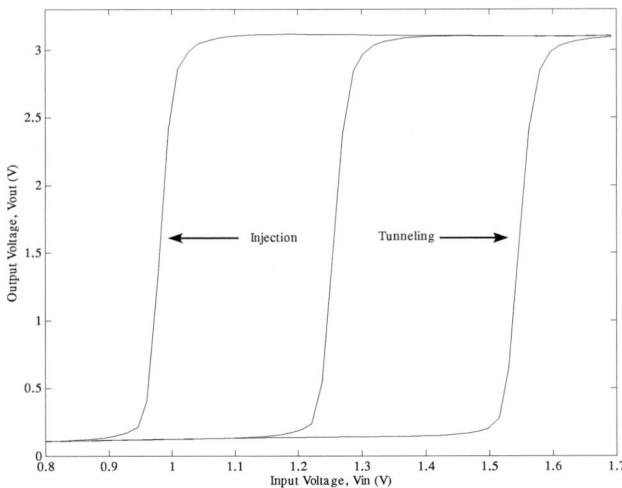

Fig. 3: Affect of tunneling and injection on the comparator trip points. Tunnelling increases the comparator trip points where as injection decreases the comparator trip points.

hold charge indefinitely and with essentially no leakage. To program a desired value to the memory, we tunnel and inject electrons through thin SiO_2 [5, 6]. The e-pots in our Flash have a dual function. In programming mode they create the biasing network for the Flash, and in A/D conversion mode the act as the comparators needed for the conversion process.

Our floating-gate Flash ADC operates in two different modes, but it is typically configured for analog to digital (A/D) conversion, as shown in Fig. 1b and Fig. 2. During this time the Flash converter is arranged similarly to a standard Flash converter. The floating-gate charge is stored on the total parallel capacitance at the negative terminal of each comparator. The Schmitt trigger adds hysteresis to the comparator output, and acts similarly to the internal latch used in many comparator designs. The INV-NOR-INV combined with the encoder is used as the thermometer decode circuit.

To create the reference voltage for each comparator we configure each comparator as an e-pot, Fig. 2. We use a 6-bit decoder to select individual e-pots for programming. Once a particular row is selected we are able to tunnel or inject the e-pot in order to change the floating-node voltage/comparator trip point, allowing us to compensate for any offsets.

When programming the Flash, we can not measure the floating-gate voltage directly. Instead we measure the output voltage of the e-pot which is capacitively coupled to the floating-gate voltage. In [7] we demonstrated that the following relationship holds.

$$V_{out} = (1 + C_{tot}/C_f) \cdot (V_{in} - V_{trip}). \quad (1)$$

Additionally, the 6-bit decoder allows us to program each converter cell, but it also gives our Flash added versatility. When the Flash is configured in programming mode it operates as a Flash DAC. Once the converter cells are programmed, a particular digital word can be fed into the decoder. The decoder selects the appropriate converter cell, and outputs a preprogrammed analog voltage. For this design we implemented a separate 6-bit DAC in order to eventually combine it with the 6-bit ADC in a 12-bit, 2-step Flash ADC.

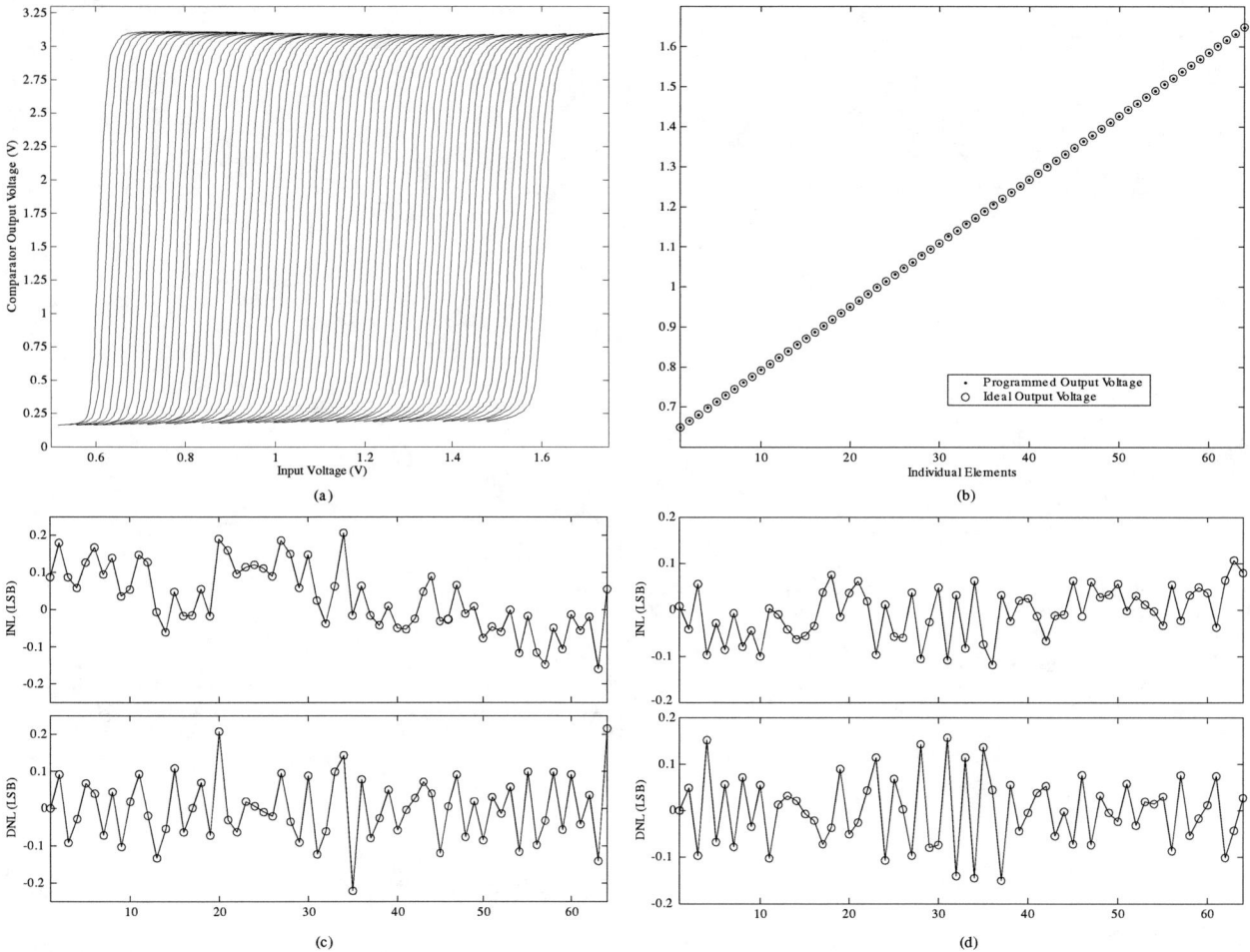

Fig. 4: Several plots showing experimental results obtained after programming the Flash Converter. (a) Plot of the comparator outputs during A/D conversion mode. (b) Plot of the output voltages for each comparator when the Flash is configured as a DAC. The output voltages of the DAC can be programmed to match the trip points of the ADC exactly. (c) Plot of the INL and DNL of our 6-bit Flash ADC. (d) Plot of the INL and DNL for our 6-bit Flash DAC

2. CHARACTERIZATION OF THE FLASH ADC/DAC

To characterize our 6-bit Flash we performed several experiments. We first determined the capacitive gain for each e-pot circuit in order to calculate the necessary output voltages for each trip point. We then programmed the ADC and DAC and examined the stored output voltages and corresponding trip points. Figure 4 shows the programmed trip points and output voltages as well as the INL and DNL for both the ADC and DAC. The INL and DNL for the ADC was found to be within ± 0.2LSB, and for the DAC they were within ± 0.15LSB.

Figure 5 shows the results for our speed tests for the Flash ADC and DAC combined. We began our experiments by providing the circuit with an input sine wave and determining the frequency at which we began to see distortion. Figure 5a shows that we began to see increased distortion around 1MHz. The amplitude of the output waveform is attenuated, and slightly delayed with respect to the input. The phase delay can be easily explained as the effect of the latency of ADC and DAC combined. Therefore, we also applied a 100kHz sine wave to better see the ability of the floating-gate Flash ADC and DAC. As shown in Fig. 5b, the full input range was converted. Figure 5c shows the resulting SNR information. We found the SNR to be 41.55dB which is equivalent to 6.6 ENOB for the ADC and DAC combined. Therefore as shown from the INL and DNL plots, a 7 or 8-bit resolution for the individual circuits seems reasonable. Figure 5d shows the combined response to a large input step. We measured the latency to be 60.81ns.

The magnitude attenuation shown in Fig. 5a, can also be explained as a result of our comparator being slew rate and bandwidth limited. During the conversion process the 63 comparators will output the thermometer code, which is then decoded by the INV-NOR-INV circuit and the encoder. As each encoder row is selected it turns on the corresponding DAC row. As a result, if one encoder row is not selected then the corresponding DAC output will not appear. Therefore, at higher speeds we see some magnitude distortion due to the fact that the edge comparators, for example comparators 1 and 63, do not transition fast enough to latch the Schmitt triggers. Once these two problems are resolved, our Flash ADC should be able to operate at much higher speeds.

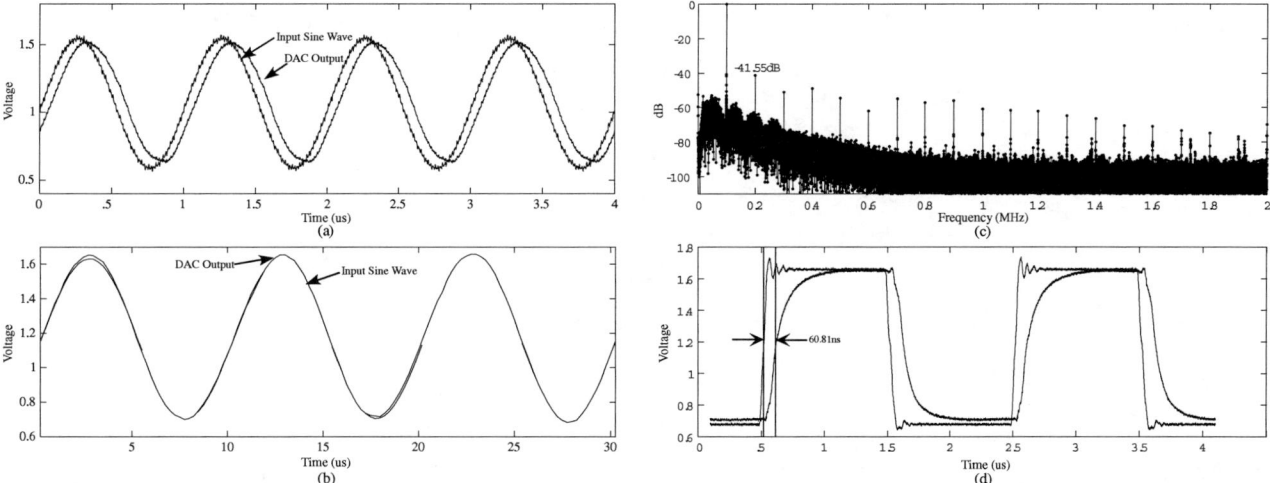

Fig. 5: Plots showing the dynamic characteristics of the 0.35μm 6-bit Flash ADC and DAC. Our Flash ADC design and layout was not optimized to minimize any offset. We relied solely on our ability to accurately program the different trip points of the comparators to compensate for any offsets. (a) Results obtained from a 1MHz input sinusoid. The output waveform experiences some slight distortion, including a loss in magnitude and phase delay. The phase delay is due to the latency of the ADC and DAC combined. The magnitude distortion is due to the comparators in the ADC. At high frequencies the edge comparators are not able to switch fast enough in order to latch the Schmitt triggers. (b) Results obtained from a 100kHz waveform. These results are much cleaner than the results from the 1MHz signal, and better show the ability of the floating-gate Flash ADC and DAC. (c) Plot of the SNR for a 100kHz input sinusoid. From this we obtained that our SNR for the ADC and DAC combined was 41.55dB, giving an overall ENOB of 6.6 bits. (d) Plot showing the response to a 500kHz square wave. The plot demonstrates the overall latency of the ADC and DAC, which was found to be 60.81ns.

Table 1: Summary of Experimental Parameters and Results

Resolution	6-bits
INL for ADC	0.2062/-0.1601
DNL for ADC	0.2155/-0.2216
INL for DAC	0.1069/-0.1187
DNL for DAC	0.1569/-0.1505
Latency	60.81ns
SNR	41.88dB
THD (%)	0.00718
Voltage Supply	3.3V
Area of ADC	0.2mm^2
Area of DAC	0.09mm^2
Technology	0.35μm CMOS

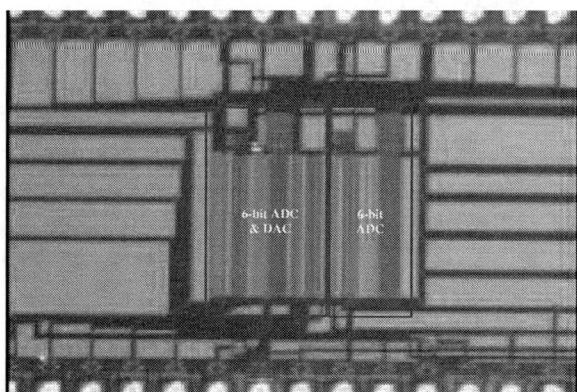

Fig. 6: Photomicrograph of the 0.35μm Floating-Gate Flash ADC and DAC. On this chip we implemented a 6-bit ADC and 6-bit DAC plus another 6-bit ADC. The size of the 6-bit ADC and DAC combined was 0.29mm^2. The size of the second ADC, which would be used to implement a 12-bit, 2-step Flash, is 0.19mm^2.

3. REFERENCES

[1] Y. Tamba and K. Yamakido, "A CMOS 6b 500MSample/s ADC for a Hard Drive Read Channel," in *IEEE Internation Solid-State Circuits Conference*, 1999.

[2] M. Choi and A. A. Abidi, " A 6-b 1.3-Gsample/s A/D Converter in 0.35-μm CMOS ," *IEEE Journal of Solid-State Circuits*, vol. 36, no. 12, pp. 1847–1858, 2001.

[3] C. Donovan and M. P. Flynn, " A "Digital" 6-bit ADC in 0.25-μm CMOS ," *IEEE Journal of Solid-State Circuits*, vol. 37, no. 3, pp. 432–437, 2002.

[4] P. Hasler and T. Lande, "Special issue on floating-gate devices, circuits, and systems," *IEEE Journal of Circuits and Systems*, vol. 48, no. 1, 2001.

[5] R. Harrison, J. Bragg, P. Hasler, B. Minch, and S. Deweerth, "A CMOS programmable analog memory cell array using floating-gate circuits," *IEEE Transactions on Circuits and Systems*, vol. 48, no. 1, 2001.

[6] P. Hasler, B. A. Minch, J. Dugger, and C. Diorio, "Adaptive circuits and synapses using pFET floating-gate devices," in *Learning in Silicon*, G. Cauwenbergs, Ed. Kluwer Academic, 1999, pp. 33–65.

[7] P. Brady and P. Hasler, "Investigations Using Floating - Gate Circuits for Flash ADCs," in *IEEE Midwest Symposium on Circuits and Systems*, Tulsa, USA, 2002.

A Novel Approach for Implementing Ultra-High Speed Flash ADC Using MCML Circuits

H. Dang, M. Sawan and Y. Savaria

PolySTIM Neurotechnology Laboratory,
École Polytechnique de Montréal, QC, Canada
dang@grm.polymtl.ca

Abstract—We propose in this paper a new MOS current-mode logic (MCML) approach for implementing ultra-rapid and accurate comparators for flash ADC applications. The speed-to-power ratio of the MCML comparator is improved by a factor of 3 compared to CMOS latch counterparts. MCML is also used to build up the subsequent 6-bit flash decoder block, which exhibits better performance than the fastest published 6-bit CMOS decoder. A 6-bit full-flash ADC has been implemented and sent to TSMC for fabrication in 0.18-μm CMOS. Post-layout simulations show that the ADC operates at 1.25-GHz, with an effective resolution bandwidth (ERBW) near 600-MHz.

I. INTRODUCTION

Recently, the concept of Software-Defined-Radio (SDR) has received considerable attention, as pushing digital processing closer to the antenna enables implementing more flexible RF communication links. This aspect requires wideband ADCs with a sampling rate close to the RF frequencies. At this moment, only flash ADCs can come close to this requirement. Although SDR requires at least 10-bit of resolution, it is possible to increase the effective number of bits (ENOB) through oversampling [1]. An ADC with ultra-high conversion rate facilitates design of agile SDRs with sufficient ENOB that cover wide frequency bands. However, the drawback of the flash architecture is its high power dissipation, resulting from the large number of power hungry CMOS comparators.

In the field of high-speed flash ADC, most efforts in the past have focused on analog design techniques such as interpolation [2] and folding for reducing the power consumption. An interesting work relies on a frequency comparator to construct ultra high-speed ADCs with a fraction of the power consumption of conventional flash utilizing CMOS comparators [3]. Nevertheless, the linearity of the ADC is strongly bound to the precision of the R-C network and the fine-tuning of the inverter's transition voltages, which are difficult to preserve due to process variations.

We propose in this paper a novel MOS current mode logic (MCML) approach for implementing ultra-high speed flash ADCs. We explain in section II the motivation for MCML circuits. Section III describes the ADC architecture and defines the *MCML comparator* and the MCML decoder. Section IV reports the post-layout simulation results of the ADC, and section V summarizes our conclusions.

II. MOTIVATION FOR MCML

The basic structure of an MCML circuit, as depicted in Fig-1(a), features pull-up and pull-down networks and a constant bias

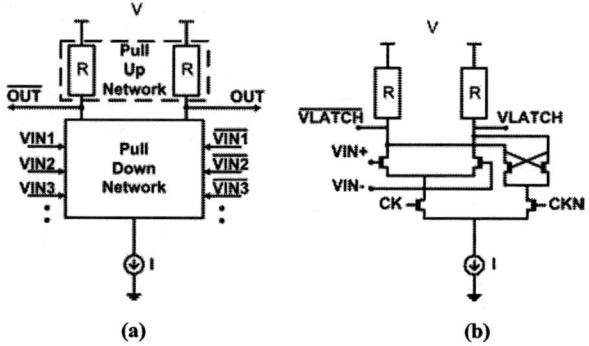

Fig-1. Circuit diagram featuring (a) the basic structure of an MCML circuit, and (b) the structure of an MCML latch.

current source [4, 5]. MCML does not require a rail-to-rail output swing, as opposed to the CMOS logic family. Indeed, $\Delta V = IR$ gives the maximum voltage swing on both output nodes, where I is the bias current and R is the pull-up resistor. Fig-1(b) shows an MCML latch configuration. The clock signal (CK) activates the input pair, while the negated clock (CKN) activates the cross-coupled transistors that regenerate the output levels. The differential topology of MCML makes them resistant to common-mode noise.

A first motivation to use MCML is the fact that its switching speed is independent of the supply voltage, as opposed to CMOS circuits. Thus, there is an operating range where it is possible to lower power dissipation without affecting performances. The delay of MCML gates is proportional to:

$$\tau = \frac{\Delta V}{I} \cdot C = RC \qquad (1)$$

where C is the output capacitive load. CMOS has a delay time constant expressed as [4]:

$$\tau = RC = C\frac{2V}{k \times (V - V_t)^\alpha} \qquad (2)$$

where V and V_t are the supply and threshold voltages. The constants k and α depend on the process and transistor sizes. From (1) and (2), MCML could maintain the same performance at 1.8V or 1.2V, while CMOS is guaranteed to suffer from performance degradation.

The second motivation to use MCML relates to its low dissipation at sub-GHz/GHz frequency ranges. MCML dissipates static power. However, as each gate tail current is approximately constant, power dissipation does not depend on the operating frequency (*f*). By contrast, CMOS circuits dissipate dynamic power according to a fCV^2 relationship. Even though CMOS power dissipation is low at low frequencies, due to the dependency on f, it may consume more than MCML at frequencies close to a GHz [5].

III. ADC DESIGN

A. ADC Architecture

A preliminary version of a 6-bit flash ADC using MCML circuits was implemented with the 0.18-μm Taiwan Semiconductor Manufacturing Company (TSMC) CMOS technology. The simplified block-diagram of the 6-bit ADC, including a front-end sample-and-hold (S/H) [6] made of PMOS switches and NMOS buffers, is shown in Fig-2. The resistive ladder provides a reference voltage for 83 preamplifiers, where 20 dummy preamplifiers are used to fulfill the averaging requirement [7]. The preamplifiers, along with the 63 MCML comparators, convert the analog signal coming from the S/H into thermometer code (TC). At the MCML decoder stage, the TC code gets translated into binary code following a quasi-Gray encoding scheme. The decoder also removes first-order bubbles occurring in the TC through a parallel bank of 63 MCML AND latches. The differential nature of the S/H, preamplifiers and comparators greatly reduces second-order distortions. Except for the S/H, all other modules have an MCML topology, and work under 1.2-V supply. The preamplifier also uses an MCML-like topology. Only the S/H works at 1.8-V.

B. MCML Comparator

The power dissipation of the flash ADC is dominated by the comparators. This section motivates our choice of MCML, instead of static CMOS, to implement ultra-rapid and low-power comparators. Our comparison is based on metastability and hysteresis properties, which are further discussed below.

The probability of observing metastability (P_M) in a latch is given by:

$$P_M = \frac{1}{A_{eff}} = \exp\frac{-(A-1)\cdot T}{\tau} \qquad (3)$$

where A is the regenerative gain of the positive feedback loop, τ is the regeneration time constant, T is the time spent in the latch-mode, and A_{eff} is the effective gain. Since a CMOS latch has two pairs of cross-coupled transistors (NMOS and PMOS), as compared to a single pair for the MCML counterpart (NMOS), it has a higher regenerative gain (A). However, pipelining latches is also an efficient alternative for increasing the gain A_{eff}, as it raises T. This option is not attractive to CMOS latch comparators, as they dissipate a lot of power. The opposite is true for MCML latches. In this work, each *MCML comparator* takes the form of an MCML latch cascaded with a D flip-flop (DFF), as shown in Fig-2. Since the DFF is, in fact, composed of two MCML latches arranged in a master-slave configuration, each MCML comparator can be thought of as a cascade of three MCML latch stages.

CMOS class AB latch comparators are widely used in the design of ultra-high speed flash ADCs [8,9], as it stands out compared to other CMOS comparators in terms of speed and accuracy. Overdrive-recovery test [10], as depicted in Fig-3, shows that the MCML comparator has a lower probability of causing metastability, and it dissipates less power, compared to the CMOS class AB latch comparator. Here, the test consists of toggling ΔV_{in} repeatedly between full-scale value and +/− 0.5-V_{LSB} and observing

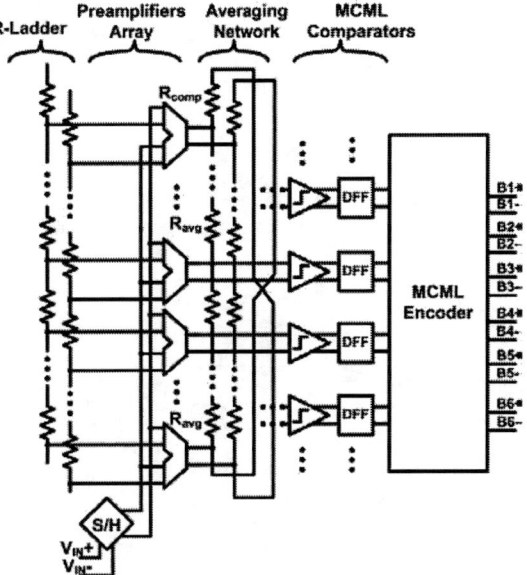

Fig-2. General block diagram of a 6-bit flash ADC.

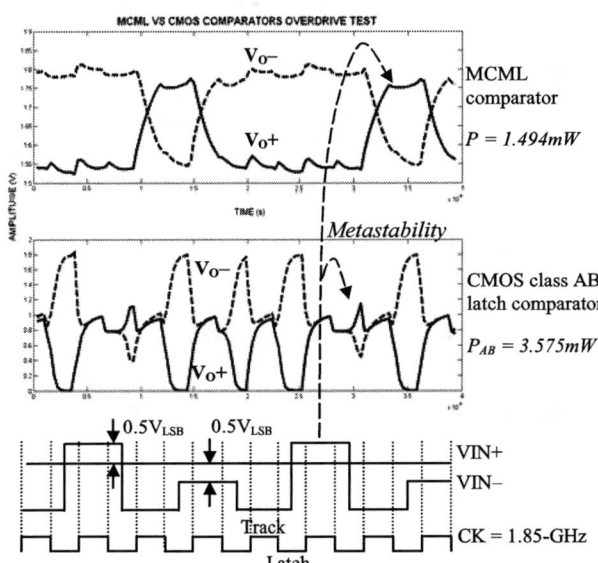

Fig-3. Overdrive-recovery test results of an MCML comparator versus a CMOS class AB latch comparator under a clock rate of 1.85GHz, displaying metastable behavior for the class AB

the output changes (V_O+ and V_O-). Both comparators have been optimized for operating under the lowest power level.

Test conditions are as follow: the clock (CK) frequency is set to 1.86-GHz, the supply voltages are 1.8-V, and both comparators have the same capacitive load. The resulting power consumption of MCML (P) and CMOS (P_{AB}) are 1.494-mW and 3.575-mW respectively. Simulation results indicate that the CMOS comparator begins to show signs of metastability, while the MCML comparator still switches correctly. This suggests that cascading MCML latches is an effective way to augment the gain A_{eff}. Moreover, the dissipation of the MCML comparator is further reduced to 0.98-mW (less than $P_{AB}/3$), at a supply voltage of 1.2-V. This power reduction leaves the performance intact.

Regarding the *hysteresis*, as illustrated in Fig-4, the phenomenon occurs when the comparator makes erroneous decisions, usually after sensing very small input changes. It results from insufficient time allowed to overcome the previous voltage level.

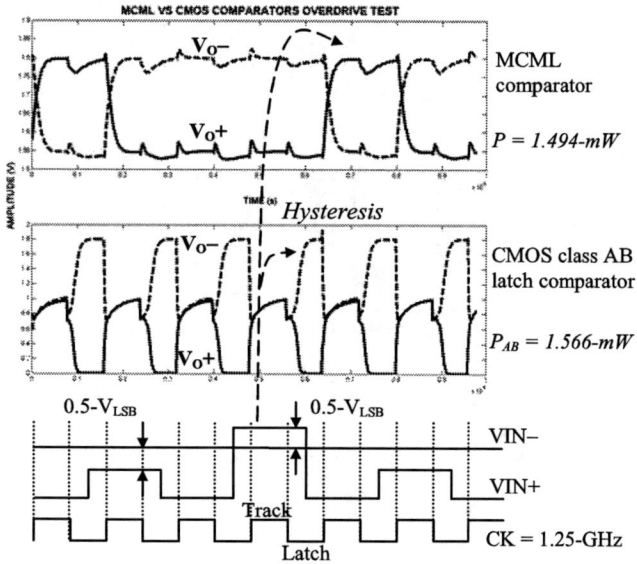

Fig-4. Overdrive-recovery test results of an MCML comparator versus a CMOS class AB latch comparator under a clock rate of 1.25-GHz, displaying hysteresis for the class AB counterpart.

Once again, overdrive-recovery test shows that MCML comparator is more immune to hysteresis than the CMOS class AB latch counterpart, at the same power level. Due to its small voltage swing, it has a better *recovery time* when compared to the CMOS latch class AB comparator with a rail-to-rail output voltage. Test conditions are as follow: the operating frequency is set to 1.25-GHz, the supply voltages are 1.8-V in both cases, the comparators have the same capacitive load and the power consumptions are adjusted to a comparable level, MCML (1.494-mW) and CMOS (1.566-mW). Clearly, the CMOS comparator exhibits hysteresis. However, the CMOS comparator does not exhibit hysteresis at a power consumption level of 2.289-mW. More interestingly, at 1.6-GHz, the CMOS comparator necessitates around 3.3-mW to overcome hysteresis, while its MCML counterpart still tracks properly without increasing power. Finally, the MCML comparator is capable of meeting the hysteresis requirement at 1.86-GHz, with a consumption level of 0.98-mW under a supply voltage of 1.2-V.

C. MCML Flash Decoder

MCML comparators outperform CMOS latch comparators in the various aspects presented above, and dissipate only a fraction of their power. However, MCML circuits do not have rail-to-rail output voltages, which make MCML comparators incompatible with existing CMOS decoders. Therefore, an MCML decoder is needed. It follows that the structure of the fat-tree decoder is very suitable for MCML circuits. This encoding scheme makes use of pipelined OR gates to convert the 1-out-of-N-code into binary code [11]. It has been shown that the fat-tree decoder beats the conventional ROM decoder in terms of speed-to-power ratio. According to their simulations, the 6-bit CMOS decoder can process incoming data at a rate of 2.0-GSPS. The same decoder implemented with MCML OR gates, as depicted in Fig-5, can handle data rate as fast as 2.6-GSPS, using the same power level.

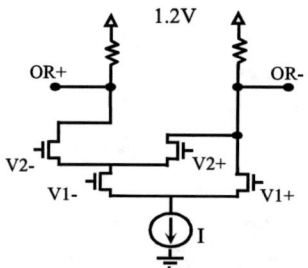

Fig-5. Structure of the 2-input MCML OR gate.

The average power of MCML decoder corresponds to their maximum power dissipated, as opposed to the CMOS decoder. Table-1 gives more details.

Table-1. Comparison of 0.18-μm CMOS and MCML 6-bit fat-tree decoders.

Design Method	DC Supply (V)	Max. Speed (GHz)	Power Dissipation (mW)	
			Avg.	Max.
CMOS [11]	1.8	2.0	22.70	38.65
MCML	1.2	2.6	22.09	22.09

IV. SIMULATION RESULTS

Post-layout simulations with parasitic extraction show that the proposed ADC achieves 6-bit resolution at a maximum sampling frequency (f_s) of 1.25-GHz. The ADC dissipates around 260-mW with the clock buffer included. The spectrum analysis of the ADC in response to a sinewave input frequency (f_{in}) of 9.77-MHz and 585-MHz – near the ADC Nyquist frequency – are illustrated in Fig-6(a) and Fig-6(b) respectively. The quality of the response is quantified by the signal to noise and distortion ratio (SNDR) and the spurious free dynamic range (SFDR). By definition, the effective resolution bandwidth (ERBW) is the input-signal frequency where the SNDR of the A/D converter has fallen by 3-dB (0.5 bit). This definition suggests that ERBW is greater than 585-MHz, but lower than 600-MHz for which the SNDR is 34.34-dB. Dynamic input offset voltage resulting from transistor mismatches in the preamplifier is the main source of non-linearity in flash ADCs. The averaging technique is used to reduce this effect, and simulations have shown that averaging improves the

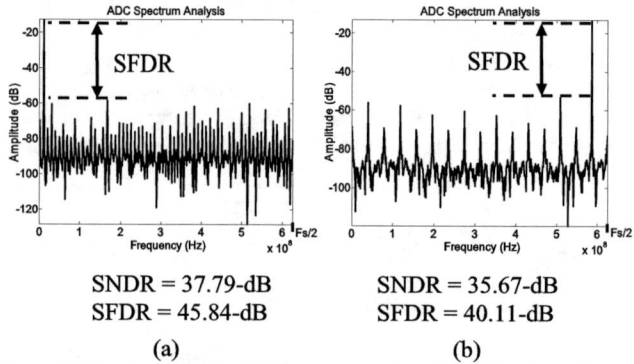

SNDR = 37.79-dB   SNDR = 35.67-dB
SFDR = 45.84-dB   SFDR = 40.11-dB
(a)   (b)

Fig-6. FFT analyses of the ADC response at f_s = 1.25-GHz, (a) with f_{in} = 9.77-MHz, and (b) with f_{in} = 585-MHz.

differential non-linearity (DNL) and integral non-linearity (INL) by a factor of 9.66 and 3.75 respectively. Moreover, the simulated DNL_{max} and INL_{max} are 0.13-LSB (least significant bit) and 0.27-LSB, given a standard deviation for the dynamic input offset of 0.5-LSB. The layout of the entire ADC, as depicted in Fig-7, was submitted for fabrication. Table-2 summarizes our results.

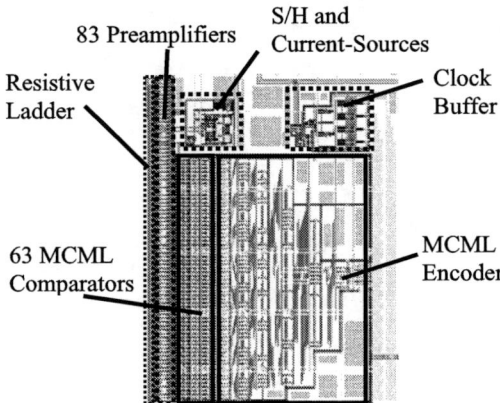

Fig-7. Layout view of the entire ADC 1200×1600-μm^2.

The clock buffer alone dissipates around 83.0-mW. Half of this power is used to drive the MCML latches, which are used to pipeline the decoder. It is possible to cut off half of this power by removing pipelining latches and utilizing only MCML gates to build the decoder, such as the one in Section III-C. Latches were initially employed to facilitate data synchronization with the clock, but they are unnecessary in this case.

Table-2. Post-simulation results of the full-flash ADC under CMOS 0.18-μm technology.

Resolution	6-bit
Sampling Rate	1.25-GHz
SNDR/SFDR @ f_{in} = 9.77-MHz	37.79-dB/45.84-dB
Power (with clock buffer)	260-mW
Input Range	460 mV_{p-p}
Common Mode Input	1.5 V
585-MHz $\leq$ ERBW < 600-MHz	

Moreover, the interpolation technique may be applied in this work to further reduce the power dissipation, as in [2]. Accordingly, we foresee that it is possible to decrease power dissipation below the 200-mW level, by fixing the decoder and interpolating the preamplifiers. The 6-bit interpolating flash ADC fabricated under CMOS 0.18-μm in [2] dissipates 340-mW at 1.6-GSPS. Although the sampling rate is very high, we estimate its ERWB to be roughly 400-MHz [2, Fig. 23(b)], which is much less than the 800-MHz ADC Nyquist frequency. It is hypothesized that their preamplifiers are not fast enough, and their comparators may suffer from hysteresis when the input frequency is close to the ADC Nyquist frequency. In fact, it is harder for the CMOS comparator to track high-frequency signals, as it has to recover from the previous latch-mode rail-to-rail output voltages. Our MCML comparator is more advantageous regarding this matter, as its output swing is much smaller.

V. CONCLUSION

This study shows the various aspects of designing ultra high-speed ADC with MCML circuits. MCML has better power-speed tradeoffs than CMOS circuits at GHz-range frequency. It has also been shown that cascading MCML latches is a good alternative for implementing high-speed and highly accurate comparators, while dissipating less than 1/3 of the CMOS latch comparator power. Moreover, the MCML decoder exhibits better performance than the fastest published 6-bit CMOS decoder. A 6-bit 1.25-GSPS full-flash ADC was implemented and submitted for fabrication at TSMC and consumes about 260-mW. It is important to note that this power consumption level can be improved by removing latches from the decoder. Moreover, an improved floorplanning of the decoder combined with careful layout can improve the speed of the ADC. Finally, the interpolating technique could be used to further trim down the power consumption and die area.

ACKNOWLEDGMENTS

The authors would like to acknowledge the financial support from the Fonds de recherche sur la nature et les technologies du Québec and design facilities from CMC. The authors also thank Mr. Abdelouahab Djemouai, Philippe Ménard Beaudoin and Amine Mounaim for their contribution to this work.

REFERENCES

[1] J.A. Wepman, "Analog-to-Digital Converters and Their Applications in Radio Receivers," IEEE Communications Magazine, Vol. 33, no. 5, pp. 39-45, May 1995.

[2] P.C.S. Scholtens, M. Vertregt, "A 6-b 1.6-Gsample/s flash ADC in 0.18-/spl mu/m CMOS using averaging termination," IEEE Journal of Solid-State Circuits, Vol. 31, no. 12, pp.1599-1609, Dec. 2002.

[3] Q. Diduck, M. Margala, "6-Bit low power low area frequency modulation based flash ADC," IEEE ISCAS, Vol. 1, pp. 137-140, May 2004.

[4] J.M. Musicer, J. Rabaey, "MOS current mode logic for low power, low noise cordic computation in mixed-signal environment," IEEE ISLPED, pp. 102-107, 2000.

[5] M. Yamashina and H. Yamada, "An MOS current mode logic (MCML) circuit for low-power sub-gigahertz processors," IEICE Trans. Electron., vol. E75-C, pp. 1181–1187, Oct. 1992.

[6] D. Wei, D. Sun, and A. Abidi, "A 300-MHz mixed-signal FDTS/DFE disk read channel in 0.6-μm CMOS," IEEE Solid-State Circuits Conf., pp. 186–187, Feb. 2001.

[7] P. Hui, A.A. Abidi, "Spatial Filtering in Flash A/D converters," IEEE Transactions on Circuits and Systems, Vol. 50, no. 8, pp. 424-436, Aug. 2003.

[8] K. Uyttenhove, M.S.J. Steyaert, "A 1.8-V 6-bit 1.3-GHz flash ADC in 0.25-/spl mu/m CMOS," IEEE Journal of Solid-State Circuits, Vol. 38, no. 7, pp. 1115-1122, 2003.

[9] P.M. Figueiredo, J.C. Vital, "Low kickback noise techniques for CMOS latched comparators," IEEE ISCAS, Vol. 1, pp.537-540, May 2004.

[10] B. Razavi, Principle of Data Conversion System Design, New York: IEEE Press, 1995, p.183.

[11] Daegyu Lee, Jincheol Yoo, Kyusun Choi, J. Ghaznavi, "Fat tree encoder design for ultra-high speed flash A/D converters," IEEE MWSCAS, Vol. 2, pp. II-87-II-90, Aug. 2002.

A 1.8V 3.2µW Comparator for Use in a CMOS Imager Column-Level Single-Slope ADC

M.F. Snoeij, A.J.P. Theuwissen and J.H. Huijsing
Electronic Instrumentation Laboratory
Delft University of Technology
Delft, The Netherlands
e-mail: M.F.Snoeij@ewi.tudelft.nl

Abstract—In this paper, a 1.8V 3.2µW comparator is presented. It features a hybrid offset compensation scheme and achieves over 60dB gain with an input offset below 150µV. The comparator is designed in a 0.18µm CMOS process and is specifically designed to be used as the key component of a column-level single-slope ADC of a CMOS imager. This ADC architecture is attractive because of its low noise, but so far this has come at the price of a relatively high power consumption. Using this comparator design, the power consumption of column-level single-slope ADCs can be reduced significantly.

I. INTRODUCTION

CMOS imagers have evolved in the last decade into the imaging technology of choice for many applications. Compared to conventional CCD imagers, they offer lower power consumption at lower supply voltages. Their most important advantage is the possibility to realize both sensor and readout electronics on a single substrate, thus creating a camera-on-a-chip. However, in spite of the advantages, some challenges still remain. Firstly, the imaging quality of a CMOS imager is still lower than that of most CCDs due to higher noise levels. Secondly, there is a continuing demand for higher resolution imagers. Thirdly, a further decrease in power consumption is highly desirable in some applications, particularly in cell phones. These three design challenges have an important impact on the imager's readout circuit.

First of all, the readout circuitry significantly contributes to the total noise of the imager. Therefore, to increase the imaging quality, the noise in the readout circuitry needs to be decreased. Furthermore, the speed of the readout circuitry has to be increased in order to be able to increase the resolution of the imager. Finally, since the readout circuitry consumes most of the power in a CMOS imager, its power consumption must be reduced in order to decrease the overall power consumption of the imager. For a given readout architecture, these demands conflict with each other, as increasing the readout speed means either a higher power consumption or higher noise, and decreasing the noise means either increasing power consumption or decreasing bandwidth. Therefore, a change on the architectural level is necessary to create a readout circuit that has less noise and a lower power consumption but a higher speed.

The most important part of the readout circuit of a CMOS imager is the ADC. Most of today's image sensors use a chip-level ADC, i.e. all pixel outputs are input to a single ADC. In a typical VGA imager for mobile applications, this ADC would consume in the order of 15mW at 30frames/second. Although a chip-level ADC has several advantages, it has an important drawback: the bandwidth of the ADC has to be high (>10MSPS). This has a negative impact on the noise performance. In contrast, it has been shown [1-3] that a column-parallel ADC can offer a significantly lower noise, as each ADC channel has a much lower bandwidth. Moreover, as the ADC is in the column, the analog signal path can be shorter, which also decreases noise.

In Fig. 1, a possible approach for a column-level ADC is depicted [2]. Here, the well-known single-slope ADC architecture is used. One central ramp generator is used for all the column circuits, and therefore only a comparator and a latch have to be implemented in every column. The advantage of this approach is that, compared to other ADC architectures, it is relatively easy to compensate for circuit non-uniformities, since these non-uniformities only consist of comparator offset. However, so far imagers with such ADCs have a higher power consumption than imagers with a chip-level ADC, e.g. 60mW in [1] and 20mW in [2] (the latter number is recalculated to VGA resolution at 30 frames/second). Most of this power is probably consumed by

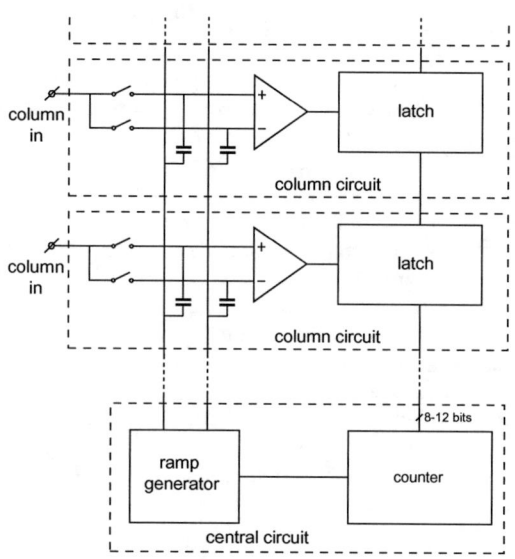

Fig. 1. Column-level single-slope ADC architecture

the column circuitry, in particular by the comparator. Therefore, to decrease the ADC power consumption, a low power comparator design is needed.

In this paper, a compact low-power comparator is presented that can be used in a column-level single-slope ADC. It is designed for a 0.18µm CMOS process and features a hybrid offset compensation scheme. It achieves over 60dB gain at a power consumption of only 3.2µW. Therefore, this design will enable a 10 bits ADC for a VGA resolution imager that has a power consumption below 5mW.

This paper is organized as follows. In section II, the system design of the comparator is presented, as well as its operation inside the single-slope ADC. In section III, the circuit design of the comparator is presented. In section IV, simulation results of the circuitry are presented. Finally, conclusions are drawn in section V.

II. SYSTEM DESIGN AND OPERATION

A. ADC Target Specifications

The comparator is intended to be used in a VGA imager that can output 30 frames/second, and therefore has a line time of about 65µs, of which 50µs can be used for the A/D conversion. The target resolution is 10 bits, which means that a clock frequency of 20MHz is necessary, i.e. the ramp increases with one LSB every 50ns. The pixel design dictates that the maximum signal input of the ADC is 500mV, thus the LSB voltage is 500µV. As the comparator has to discriminate between two LSBs, the comparator gain should therefore be at least 1000.

B. Circuit Topology

There are two classes of circuits that can be used to create a comparator: gain stages and regenerative latches. At a first glance, designing a comparator using only gain stages might seem more attractive, as they do not 'pollute' the supply rails and ramp voltage with current peaks due to switching. However, creating a gain of over 1000 in every column can easily cause instability. As a gain stage is continuously switched on, there is a high-gain forward signal path constantly present. The comparator inputs have a high impedance, as they are only connected to sampling capacitors. Therefore, if a small parasitic capacitance is present from output to input, it can create a positive feedback loop that may cause instability. Therefore, a regenerative latch is preferable, as it only has a high gain at the clock edge. To reduce the kick-back effect, i.e. the charge injection into the input of the regenerative latch, a preamp has to be added in front of the latch.

C. Delay

In a classical single-slope ADC design, the delay of the comparator has to be less than one clock period, as a higher delay creates an offset in the digital output. However, a comparator with a delay less than 50ns, as required in our case, would consume a lot of power and have a relatively high noise due to its high bandwidth. Therefore, the comparator is allowed to have a higher delay and compensate for it on system level. A delay measurement is performed after every A/D conversion and the measured delay is subtracted from the signal in the digital domain. A typical delay of 600ns is accepted, to enable the design of a low-power low-noise circuit, as will be shown in section III and IV.

D. Offset

Comparator offset can create column non-uniformities in the image, to which the human vision system is very sensitive. Therefore, the comparator offset should be well below ½ LSB; our target is 125µV. A preliminary analysis of the intended circuit blocks shows that the preamp can have a minimum input offset of 9mV (4σ). The optimal input offset specification for the regenerative latch is at least 30mV (4σ); although lower offsets are possible, this would increase power consumption quite significantly. Therefore, a dynamic offset cancellation technique will be necessary. Since the input signal is already sampled, it is most advantageous to apply auto-zeroing.

There are two methods of implementing the auto-zero. On one hand, a separate A/D conversion for offset could be performed, which could be conveniently combined with the delay measurement mentioned in the previous section. The main disadvantage is the relatively long conversion time this takes; measuring an offset of ±9mV in 125µV increments takes 144 clock cycles.

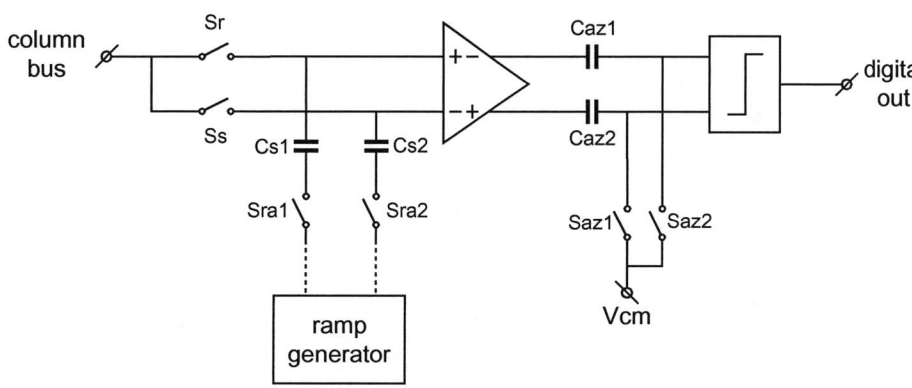

Fig. 2. Column circuit topology

On the other hand, it is also possible to implement an auto-zero at circuit level. The offset of the preamp can be corrected by storing the offset on capacitors, either at the input or the output of the preamp. We have chosen to consider only output storage, as input storage can interfere with the sampling capacitors at the preamp input. This limits the gain of the preamp to about 50, as a higher gain would cause the comparator outputs to saturate during offset sampling.

The input referred offset of the latch is at minimum 30mV/50=600μV, which is too high for our application. We therefore use a *hybrid* offset compensation scheme: first, we reduce the offset by auto-zeroing the preamp; second, we perform a combined delay/offset measurement at the system level. This approach reduces the required conversion time needed for the system auto-zero, while it does not require a low-offset regenerative latch that consumes a lot of power.

The resulting system topology is depicted in Fig. 2.

III. CIRCUIT DESIGN

In Fig. 3, the preamp circuit is depicted. Only transistors M1-M7 have to be implemented in every column; the biasing circuitry can be implemented centrally. Differential pair M2-M3 is loaded with current sources M4 and M5. The gain is determined by the combined output resistance of M2-M5 and the gm of M2-M3. The output resistance of the current sources is negligibly high; the voltage gain of the preamp is therefore determined by the length of the input transistors. As a result, the gain will vary considerably over temperature and process variations; however, as the preamp is part of a comparator, an accurately defined gain is not necessary. The output common-mode voltage is regulated to voltage V_{cm} by transistors M6-M7 that operate in triode region.

In Fig. 4, the regenerative latch circuit is depicted. This latch is based on the circuit presented in [4] and consists of two stages. The first stage, consisting of transistors M1-M7 provides most of the dynamic gain and is reset every clock cycle in order to eliminate any memory effect. The second stage amplifies the output of the first stage to digital levels. When the clock is high, transistor M5 resets the first latch. During this phase, transistors M8-M9 will be switched on, but transistors M8-M11 are dimensioned in such a way that the output of the latch does not change in this phase. When the clock goes low, a positive feedback loop will be created around M6-M7. Although this loop will not deliver a rail-to-rail output, it will be enough to drive transistors M8-M9. These will change the state of the second stage latch if the input value has changed sign since the last clock period; therefore, a digital output value will be continuously available at the output of the circuit.

The advantage of this latch circuit is that, unlike many other regenerative latch circuits, the first stage draws a constant current. Although the second stage does have a current peak at the moment of switching, this only happens when the latch output changes; moreover, as the operation of the second stage is not very critical, it can be connected to the digital supply (vddd). This greatly reduces the chance that one comparator can trigger the next in our parallel ADC structure.

IV. SIMULATION RESULTS

In Fig. 5, simulation results of the comparator are given. In this simulation, we apply a 500mV ramp voltage to the comparator. The comparator input crosses 0V at t=25μs. As can be seen in Fig. 5a, the preamp output has a delay of about 600ns. As can be seen in Fig. 5b, the latch has a negligible delay. As described earlier, one of the design goals for the comparator was to have an analog supply current that is as constant as possible. In Fig. 5c, the simulated analog and digital supply current are depicted. As can be clearly seen in the figure, there is a high peak current in the digital supply current (shown in black). This is caused by the switching of the second stage of the regenerative latch, which only occurs when the output of the comparator

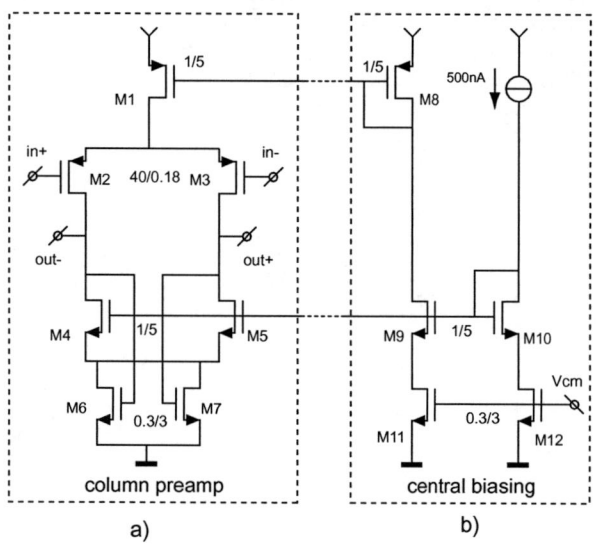

Fig. 3. a) Column preamp circuit b) Central biasing circuit

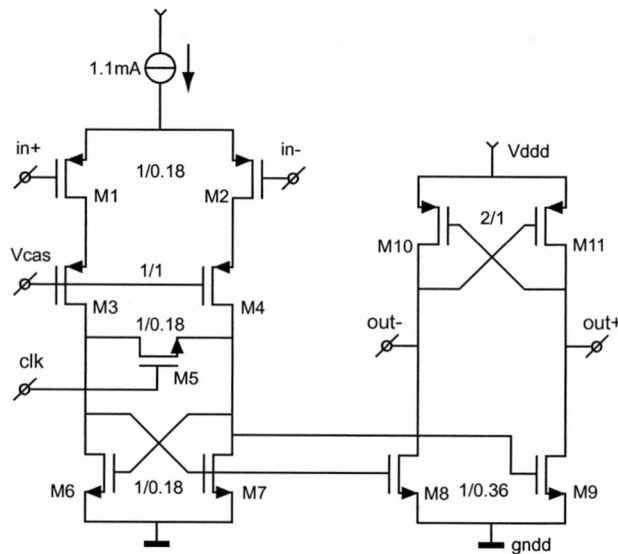

Fig. 4. Regenerative latch circuit

changes value. However, the analog supply current (shown in gray), is nearly constant. The total average supply current for the comparator is about 1.8µA, which equals 3.2µW. Finally, in Fig. 5d, the settling of the output of the preamp is shown during the auto-zero phase.

Ensuring correct operation over process corners is obviously very important. We have therefore checked the spread by doing corner simulations with the following parameters:

- Transistor spread (slow/fast corners)
- Temperature (-15 to +100°C)
- Bias current (±20%)

By exhaustively testing for all combinations of these parameters, the worst-case spread can be found. Results of these simulations for the preamp are given in table I.

TABLE I. SIMULATED WORST-CASE PREAMP PARAMETER SPREAD

	Min.	Typ.	Max.
GBW (MHz)	7	11,9	17,2
-3dB freq. (kHz)	174	354	653
DC voltage gain	26.3x	33.1x	40.3x
Delay (ns)	250	450	820

As can be seen in the table, there is a large spread in both AC frequency response and in DC gain. The spread in frequency response results in a varying delay. This will be corrected by delay compensation; since the worst-case delay is 820ns here, we will take a delay measurement on system level with a maximum of 1µs to have design margin. The spread on DC gain should stay within limits posed by the offset compensation scheme: the gain should stay below 50 to prevent clipping of the preamp output, and above 20 to limit the input-referred offset of the latch. As can be seen from the table, this is the case.

Corner simulations of the regenerative latch show that it is insensitive to changes in the described process parameters. Its main parameter of interest is its delay, which should be shorter than one clock period. This is the case for all corners.

V. CONCLUSION

A 1.8V comparator for use in a column-level single-slope ADC of a CMOS imager was presented. By using delay compensation it achieves over 60dB gain at a power consumption of only 3.2µW. The comparator features a hybrid offset compensation that decreases its offset to below 150µV. Using these techniques, the comparator can be realized in a compact circuit of only 18 transistors, which makes it very suitable for implementation in the column of a CMOS imager. Simulation results show that the comparator performs correctly over process corners. Work on a silicon implementation of the comparator is ongoing.

VI. ACKNOWLEDGEMENT

The authors would like to thank Philips Semiconductors for their financial and practical support of this project.

REFERENCES

[1] T. Sugiki et al., "A 60 mW 10b CMOS image sensor with column-to-column FPN reduction", *IEEE International Solid-State Circuits Conference*, vol. XLIII, pp. 108-109, February 2000

[2] K. Findlater et al., "SXGA pinned photodiode CMOS image sensor in 0.35µm technology", *IEEE International Solid-State Circuits Conference*, vol. XLVI, pp. 218-219, February 2003

[3] A. Krymski, N. Khaliullin and H. Rhodes, "A 2e noise 1.3 megapixel CMOS sensor", *IEEE workshop on CCDs and advanced imager sensors*, May 2003

[4] I. Mehr and D. Dalton, "A 500-MSample/s, 6-bit Nyquist-rate ADC for disk-drive read-channel applications", *IEEE Journal of Solid-State Circuits*, Vol. 34, No. 7, July 1999

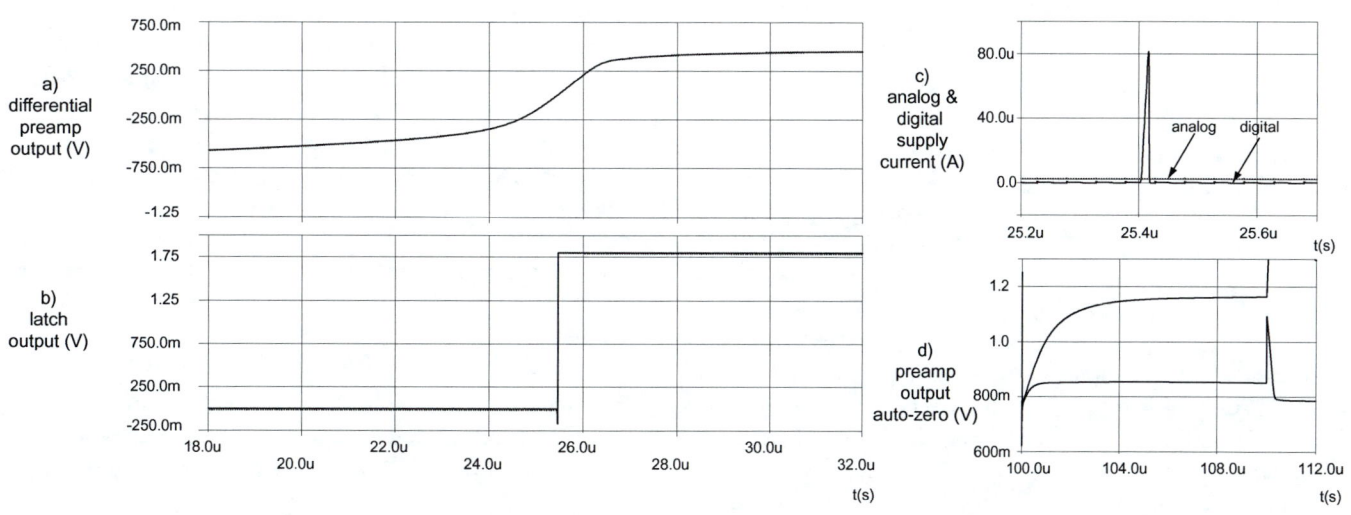

Fig. 5. Comparator simulation: a) differential preamp output b) latch output c) analog & digital supply current d) preamp output during auto-zero

Design Considerations of a Floating-Point ADC with Embedded S/H

Johan Piper and Jiren Yuan

johan.piper@es.lth.se, jiren.yuan@es.lth.se

CCCD, Department of Electroscience, Lund University, 211 00 Lund, Sweden

Abstract—This paper presents the implementation and test results of a 10+5 bit 50MS/s floating-point ADC, along with the design considerations. The combination of resistive weighting with identical chopped gain stages proved successful in gain, delay and offset matching. It demonstrated that the input referred thermal noise of the gain stages needs to aim for 15 bits, while the rest requirements such as channel matching (gain, delay, offset) and settling time need only 10 bits. The channel selecting logic gives a serious impact on the ADC distortion, especially at high frequencies. For this reason, a robust channel selecting logic is suggested.

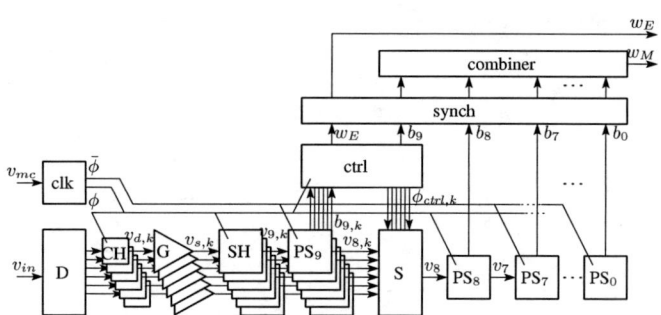

Fig. 1. The architecture of the 10+5 bit FP-ADC chip. The pipeline stages 6–1 are not shown in the figure.

I. INTRODUCTION

To extend the dynamic range of an ADC, without increasing the effective number of bits, a floating-point technique can be used. The basic technique used to convert an electrical signal using a floating-point ADC (FP-ADC) includes; (i) determine the exponent, (ii) normalize the input signal, and (iii) convert the normalized signal. In this article a 10+5 bit FP-ADC with embedded S/H is presented [1–4]. In the literature other variants of FP-ADC are also reported [5–15].

The block level architecture of the 10+5 bit FP-ADC is given in Figure 1. In the divider D the signal is attenuated successively by a factor of two for every divider output. Each divider output is amplified by a factor of G_G in the gain stage column G and then sampled in the sample-and-hold stage column SH. The sample-and-hold stages have a gain of G_{SH}. The total gain is 2^m in each row, which means that the output of the k-th sampling stage equals

$$v_{9,k} = v_{in} 2^{m-k} \qquad (1)$$

where k ranges from 0 to m. Here m is the extended dynamic range of the FP-ADC, expressed in bits. This way the binary weighted gain is implemented. The signals $v_{9,k}$ are sampled by the pipeline stage column (PS_9). The column corresponds to the first pipeline stage in the n-bit ADC. The controller uses the data from the PS_9 to decide which of the rows shall be used by setting the switch S accordingly. The controller also calculates the exponent and forwards the digital signal from the selected row to the synchronizer.

After PS_9, the signal is converted through a single row of pipeline stages PS_{8-0}. All the digital outputs from the pipeline stages and the controller are synchronized. The data from the pipeline stages are then combined into the mantissa.

The clock phases are generated in the phase generator. The phases used for the pipeline stages are toggled when going down the pipeline. It ensures that when one pipeline stage is amplifying, the next is sampling.

II. DESIGN CONSIDERATIONS

When a parallel structure is used it is very important that all the rows match each other. Between two successive rows the amplitudes should differ exactly by a factor of two. The gain, the offset and the delay should be identical for all the rows, to guarantee that no spurious signals appear when the signal is reconstructed. If the rows are not identical mismatch errors will come into effect. This is similar to time-interleaved converter arrays [16, 17]. But unlike in time-interleaved converters the distortion of the matching errors in an FP-ADC appears as harmonic distortion.

To get the best matching for gain and delay the active blocks (gain stages and SH-stages) are made identical and the binary weighted gains are provided by a passive voltage divider. To remove the offset chopping is used. The efforts to match the FP-ADC rows with regards to gain, delay and offset are discussed in more detail in [2, 3].

III. RESULTS OF CHIP MEASUREMENT

A 10+5 bit FP-ADC was designed and manufactured in a 0.35μm CMOS process with 4 metal and 2 poly-silicon layers. A chip photo of the 10+5-bit FP-ADC is presented in Figure 7.

The FP-ADC was measured to acquire the performance in terms of dynamic range, nonlinearity and speed. The measurement results are summarized in table I. A plot of the

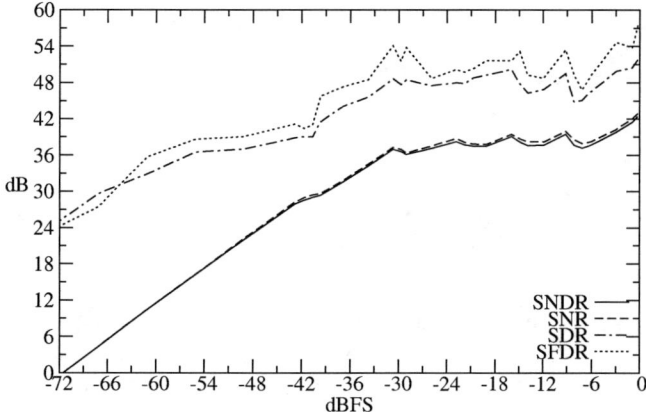

Fig. 2. The signal to noise and distortion ratio when sweeping the input amplitude, $f_s = 10$ MHz, $f_{in} = 610$ kHz.

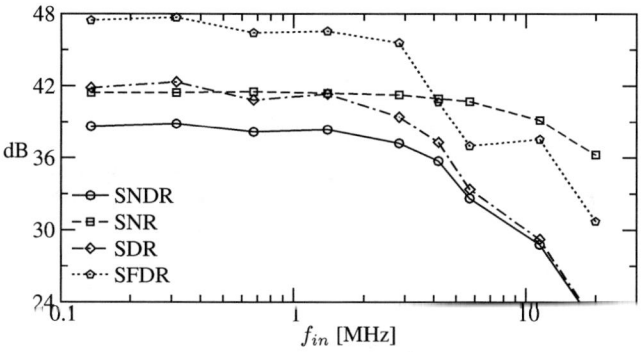

Fig. 3. The signal to noise and distortion ratio when sweeping the input frequency, $f_s = 46$ MHz, $A = -0.6$ dBFS.

dynamic range versus the amplitude of a single tone input is shown in Figure 2 and in Figure 3 a plot of the dynamic range versus the input frequency for a full-scale input signal is shown.

The measurements show that the maximal SNDR is 42.5 dB – corresponding to 6.8 bits. The amplitude sweep in Figure 2 shows that the total dynamic range is 71.5 dB – corresponding to 11.6 bits. Thus the dynamic range has been extended by 29 dB – corresponding to 4.8 bits. The measurements proves that this idea is technically feasible and a distributed FP-ADC with embedded SH can well extend the dynamic range by 5 bits.

The measurements also reveal two design limits; the noise limits the resolution to 6.8 bits and control algorithm limits the analog input bandwidth to 4.2 MHz for a full-scale input signal. The causes behind the limits of the performance are discussed in the following sections.

IV. THE CONTROL ALGORITHM

The purpose of the controller is to select the row where the signal has been amplified just within the range of the ADC. The inputs to the controller are the data from the comparators in each row, and the outputs are the control signals to the CMOS switches. The timing is controlled by the system clock phases.

The selection is done in such a way that the row with the highest index k, which satisfies the equation

$$V_{ctrl} = \frac{V_R}{2} - V_{safe} \leq |v_{9,k}| \quad (2)$$

will be selected for conversion. V_{ctrl} is half the reference voltage, V_R, minus a safety margin, V_{safe}, to ensure that no clipping occurs due to offsets in the comparators. This way the controller can reject gain stages that are saturated. The selection algorithm of the controller is illustrated in Figure 4.

A. The comparators of the controller

The reason why the row selecting switch is moved to after the first pipeline stage is to avoid using comparators before the signal has been sampled. The signals are continuous time and are likely to have a fast slope. The comparators need to be very fast and accurate timing-wise. If they are not fast enough the delay time in the comparator will introduce a large offset error for high frequency signals. It is also troublesome that the number of comparators is multiplied by the number of rows, so the use of high power and fast comparators is not feasible.

The solution is to move one pipeline stage before the row selecting switch. In this way half a clock period is gained and the signals are discrete time. Then the rate of change is almost zero. This has the benefit that comparators with a long delay can be used. Still the comparators need to be very fast but now in terms of latency, i.e. the comparators need to decide quickly when probed.

The increase in power consumption will not be dramatic as the pipeline stages are low-power compared to the gain stage. Also the comparators inside the first pipeline stage can be merged with the controller's comparators. Since we accept a $\pm V_{FS}/8$ error to the threshold value in an RSD 1-bit pipeline stages [18], this operation is permitted. The same thresholds are used for the first pipeline stage and for the controller. In this way the required number of comparators is reduced.

B. Gain stage saturation

The fact that the gain stages are always in operation means that at some point one or more of the gain stages will be saturated to the supply voltage or ground. This is not a problem as long as the saturated signal is definitely outside the input range of the ADC. These rows will be rejected by the controller and their outputs will not be used. The gain stages are reset to zero during the AZ phase. The AZ makes sure that no gain stage is saturated at the beginning of each amplifying phase.

However this scheme will not alleviate the problem of the dynamic settling behavior when the gain stage goes out of saturation. The problem arises when the input amplitude goes from high amplitude and returns to zero, so the gain stages come out of saturation. The recovery from saturation is not

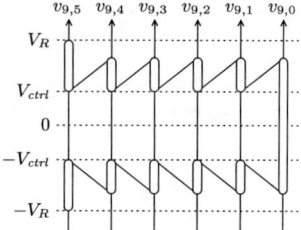

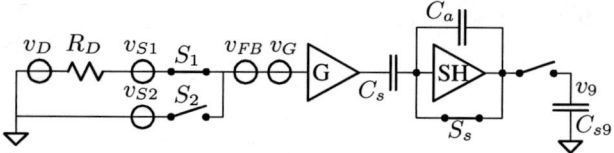

Fig. 4. Illustration on how the controller selects the appropriate row for conversion ($v_{9,k} = v_{in}2^{5-k}$).

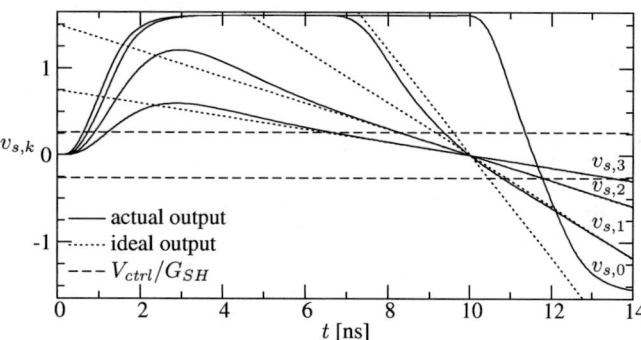

Fig. 5. A simulation of saturation in the gain stages. The slope corresponds to the zero-crossing of a 2 MHz full-scale signal. The corresponding V_{ctrl} at the input of the sample-and-hold stage is plotted as well.

instantaneous. It takes some time for the output of the gain stage to settle to its correct value. During the recovery time (which can be long) the signal from the gain stage may be sampled.

A simulation of the transient response of the gain stages, where the input signal is a slope corresponding to a zero-crossing of a 2 MHz signal, is presented in Figure 5. Initially the gain stages are in the AZ phase. At $t = 0$ they are switched to the amplifying phase and start to amplify the signal, $v_k = v_i G_G 2^{-k}$. Initially the output of gain stage number 0 goes into saturation but recovers from the saturation when the signal returns to zero. If the signal is sampled between 9–11 ns then the controller will select row 0 which will give rise to a large error as the output of gain stage 0 is still saturated.

In the measurements this effect was found to be the dominant source of the distortion at high frequencies.

C. New control algorithms

A solution to this problem is to detect whether a gain stage has been saturated and not yet recovered with a mono-stable latch. The latch should be active during the longest recovery time before sampling and be probed at the sampling instant. If $v_{s,k} > V_{ctrl}$ during the amplifying phase the latch will detect that the gain stage has, at one point before the sampling instant, been saturated. Just to be sure this row will be rejected.

The solution reported in [12], is to insert a memory (low-pass filter) in the logic to ensure that if a sample from a row is saturated, the row will not be used for some time. This ensures that the row has enough time to recover before it is used again.

V. THE NOISE IN THE FP-ADC

The noise from the divider and the gain stage will dominate the total noise performance of the FP-ADC. The model used for calculating the noise is shown in Figure 6. Here are v_D the voltage noise from the divider, v_{FB} the voltage noise from the gain stage feedback resistors, and v_G the voltage noise of the gain stage amplifier. It is evident that the noise contribution is dominant from the circuits prior to and within the gain stage. After the gain stage the noise power contributions are a factor $1/G_G^2$ smaller when referred to the input of the FP-ADC.

The noise energies, from both the amplifying and the sample phase, appear at v_9 due to the AZ. The noise at the input of the n-bit ADC is

$$v_9 = 2^m \left(v_D + v_{S1} + v_{S2} + \sqrt{2} v_{FB} + \sqrt{2} v_G \right) \quad (3)$$

where $G \cdot C_a/C_s = 2^m$. The noise energies of the divider and the two chopper switches just appear once during a clock cycle. The noise energies of v_{FB} and v_G appear both in the amplifying and sampling phase (correlated double sampling [19]) which doubles the thermal noise energy, hence the $\sqrt{2}$ factor.

By using Wiener-Kintchine's theorem [20] and matching every spectrum to an equivalent thermal noise resistor $S_x = 4kTr_x$, the equivalent noise resistor is obtained

$$r_9 = 2^{2m} \cdot (r_D + 2r_{S1} + 2r_{FB} + 2r_G) = 2^{2m} r_{eq} \quad (4)$$

where r_{eq} is the equivalent noise resistor of the noise sources at the input of the gain stage.

If we want the thermal noise to be lower than the quantization noise of the n-bit ADC [21]

$$4kTr_9 \cdot B < \frac{V_{FS}^2}{12 \cdot 2^{2n}} \quad (5)$$

then the maximal r_{eq} can be solved. Replacing the bandwidth B by the noise bandwidth [19], which is related to n and the sampling frequency f_s, gives the expression [4]

$$B = (n+1) \ln 2 \frac{f_s}{2} \quad (6)$$

Fig. 6. Block level figure of the noise from the divider (v_D), the chopper switches (v_{S1} and v_{S2}) and the gain stage (v_{FB} and v_G).

Dynamic range	71.5 dB
SNDR	42.5 dB
SDR	51.7 dB
SFDR	57.4 dB
Sampling rate	53 MHz
Input bandwidth	4.2 MHz
Input range	2.8 V_{FS}
Voltage supply	3.3 V
Current supply	100 mA
Core size	1.4×1.2 mm

TABLE I
THE MEASURED OVERALL CHARACTERISTICS OF THE 10+5 BIT FP-ADC CHIP.

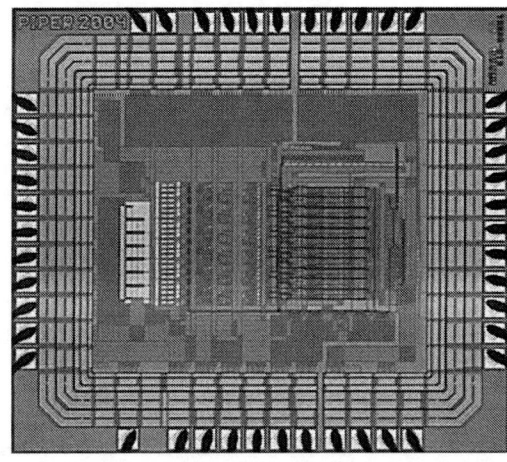

Fig. 7. Chip photo of the 10+5-bit FP-ADC.

when assuming a first order settling behavior. Solving for r_{eq} yields

$$r_{eq} < \frac{V_{FS}^2}{4kT\frac{(n+1)\ln 2}{2}f_s \cdot 12 \cdot 2^{2n+2m}} \quad (7)$$

The noise requirement on the $n + m$ bit FP-ADC is high, but more relaxed compared to a linear $(n + m)$-bit ADC, especially for high values of m, due to the settling time.

VI. CONCLUSIONS

When designing an FP-ADC with embedded SH in CMOS, there are some considerations that need to be addressed.

It is important that the signal channels until and including the SH-circuit are matched in gain, delay and offset. To get the best matching for gain and delay the active blocks (gain stages and SH-stages) are suggested to be identical and the binary weighted gain can be provided by a passive voltage divider. To remove the offset chopping is used.

The noise requirements on the divider and gain stage are set by the total dynamic range. But compared to a linear $(n+m)$-bit ADC the noise requirements are lower. This is because the settling of the first pipeline stage needs not $(n + m)$-bit, but only n-bit settling.

In addition, for the control logic to detect the gain stage saturation it is not sufficient to only detect the sampled signal. Instead the saturation should be detected continuously on the gain stage output by a mono-stable latch, within the recovery time before the sampling instant.

REFERENCES

[1] J. Piper and J. Yuan, "A delay-balanced binary-weighted CMOS amplifier tree for a floating point A/D converter," in *Proc. IEEE NORCHIP Conference*, pp. 131–138, Nov. 1998.
[2] J. Yuan and J. Piper, "Floating-point analog-to-digital converter," in *IEEE International Conference on Electronics, Circuits and Systems*, vol. 3, pp. 1385–1388, 1999.
[3] J. Piper and J. Yuan, "Realization of a floating-point A/D converter," in *IEEE International Symposium on Circuits and Systems*, vol. 1, pp. 404–407, 2001.
[4] J. Piper, *Floating-Point Analog-to-Digital Converter*. PhD thesis, Lund University, Sweden, 2004.
[5] F. Chen and C. S. Chen, "A 20-b dynamic-range floating-point data acquisition system," *Trans. Industrial Electronics*, vol. 38, pp. 10–14, February 1991.
[6] L. Grisoni, A. Heubi, P. Balsinger, and F. Pellandini, "Implementation of a micro-power 15-bit 'floating-point' A/D converter," in *Int. Symp. Low Power Electronic and Design*, pp. 247–252, August 1996.
[7] V. Z. Groza, "High-resolution floating-point ADC," *IEEE Trans. Instr. and Meas.*, vol. 50, pp. 1822–1829, December 2001.
[8] V. Groza, "Floating-point ADC optimized for acquisition of deterministic signals," in *IEE Instr. and Meas. Tech. Conference*, vol. 1, pp. 707–711, May 2002.
[9] G. M. Haller and D. R. Freytag, "Analog floating-point BiCMOS sampling chip and architecture of the BaBar CsI caliometer front-end electronics system at the SLAC B-factory," *IEEE Trans. Nuclear Science*, vol. 43, pp. 1610–1614, June 1996.
[10] J. Heath and T. Nagle, "Design a floating-point A/D converter," *Electronic design*, vol. 22, pp. 80–84, may 1974.
[11] G. Ootomo, K. Tsukamoto, T. Watahiki, and T. Miyata, "A floating-point A/D converter with self-calibration," in *Trans. of the inst. of Electronics, Information and Communication Eng.*, vol. E71, pp. 1303–1308, 1988.
[12] K. E. Prada, K. von der Heydt, and T. F. O'Brien, "A versatile multi-channel data acquisition system for seismic and acoustic applications," *OCEANS*, vol. 13, pp. 43–47, September 1981.
[13] S. Sharma, G. Otomo, K. Tsukamoto, and T. Miyata, "A floating-point A/D converter uses low resolution DAC to get wide dynamic range," *Int. J. of Electronics*, vol. 64, pp. 787–794, May 1988.
[14] D. U. Thompson and B. A. Wooley, "A 15-b pipelined CMOS floating-point A/D converter," *IEEE J. Solid-State Circuits*, vol. 36, pp. 299–303, Feb. 2001.
[15] T. Zimmerman and J. R. Hoff, "The design of a charge-integrating modified floating-point ADC chip," *IEEE J. Solid-State Circuits*, vol. 39, pp. 895–905, May 2004.
[16] W. C. Black, JR and D. A. Hodges, "Time interleaved converter arrays," *IEEE J. Solid-State Circuits*, vol. SC-15, pp. 1022–1029, Dec. 1980.
[17] J. Yuan and C. Svensson, "A 10-bit 5-MS/s successive approximation ADC cell used in a 70 MS/s ADC array in 1.5μm CMOS," *IEEE J. Solid-State Circuits*, vol. 29, pp. 866–872, August 1994.
[18] T. B. Cho and P. R. Gray, "A 10 b, 20 Msample/s, 35mW pipeline A/D converter," *IEEE J. Solid-State Circuits*, vol. 30, pp. 166–177, March 1995.
[19] R. Gregorian and G. Temes, *Analog MOS Integrated Circuits for signal processing*. John Wiley & Sons, 1986.
[20] C. J. M. Verhoeven, A. van Staveren, G. L. E. Monna, M. H. L. Kouwenhoven, and E. Yildiz, *Structured Electronic Design: Negative-feedback amplifiers*. Delft University of Technology, Delft, the Netherlands, 2002 ed., July 2002.
[21] R. van de Plassche, *Integrated analog-to-digital and digital-to-analog converters*. Kluwer, 1994.

A New CCII-Based Pipelined Analog to Digital Converter

Yuh-Shyan Hwang, Lu-Po Liao, Chia-Chun Tsai, Wen-Ta Lee, Trong-Yen Lee, Jiann-Jong Chen

Department of Electronic Engineering

National Taipei University of Technology

Taipei, Taiwan, R.O.C.

Abstract—This paper proposes a new pipelined analog to digital converter (ADC) based on second-generation current conveyors (CCIIs). Two main building blocks of the pipelined ADC, sample-and-hold (S/H) circuit and multiplying digital-to-analog converter (MDAC) are constructed of CCIIs instead of operational amplifiers (OAs). Simulation results show that the proposed CCII-based pipelined ADC can work at 10MHz with an 8-bit resolution. The DNL is within -0.4LSB and 0.5LSB and INL is within -0.4LSB and 0.7LSB, respectively. The ADC is realized in TSMC 0.35μm CMOS technology and consumed 29mW under a 3.3V power supply. The core size is 0.85x0.85mm^2.

I. INTRODUCTION

It has been proven that the digital signal processing (DSP) shows better accuracy than analog signal processing. The ADC is the main link between the analog input and DSP part. The ADC based on pipelined architectures [1-3] can offer both high resolution and high speed instead of large size and power consumption which make the architectures well suited for many applications. The switched-capacitor (SC) circuits are usually used in pipelined ADCs not only they can construct the S/H circuit but also for the MDAC. The SC circuit is usually made by OAs, for this reason, the efficacy of the OA always determines the SC circuit performance. Compared with traditional OAs, CCIIs [4-5] have higher accuracy and wider frequency ranges, hence the CCII-based circuits are popularly used in many applications. But no paper discussed the CCII-based ADC. This paper presents a new CCII-based pipelined ADC instead of using traditional OA-based SC techniques which can employ the advantages of CCII. The digital error correction technique is also applied to increase the performance of our pipelined ADC.

II. ARCHITECTURE OF PIPELINED ADC

The ADC is realized with the 1.5-bit/stage pipelined architecture where the 0.5-bit redundancy in each stage is used for digital correction to relax the requirement for the comparators. The block diagram of the 8-bit pipelined ADC is shown in Fig. 1. It is composed of one S/H stage and eight sub-ADC stages. Each pipelined stage performs a low resolution sub-ADC with 1.5-bit output and a MDAC.

The operation of the pipelined ADC consists of two phases. The even stages and odd stages operate in opposite, thus the conversion of a sample traverses two stages in a clock cycle. In the first phase, the S/H stage samples the input signal while the MDAC of even stages sample the output of odd stages and the sub-ADC dose the analog-to-digital conversion. In the second phase, the MDAC even stages generate and amplify the residue yielding the input signals for the even stages. The operation principle of every stage can be expressed as below

$$D = \begin{cases} 00, & -Vref < Vi < -\frac{1}{4}Vref \\ 01, & -\frac{1}{4}Vref < Vi < +\frac{1}{4}Vref \\ 10, & +\frac{1}{4}Vref < Vi < +Vref \end{cases} \quad (1)$$

where D is the digital output of every stage.

$$Vo = \begin{cases} 2 \times Vi + Vref, & -Vref < Vi < -\frac{1}{4}Vref \\ 2 \times Vi, & -\frac{1}{4}Vref < Vi < +\frac{1}{4}Vref \\ 2 \times Vi - Vref, & +\frac{1}{4}Vref < Vi < +Vref \end{cases} \quad (2)$$

where Vo is the analog output of every stage.

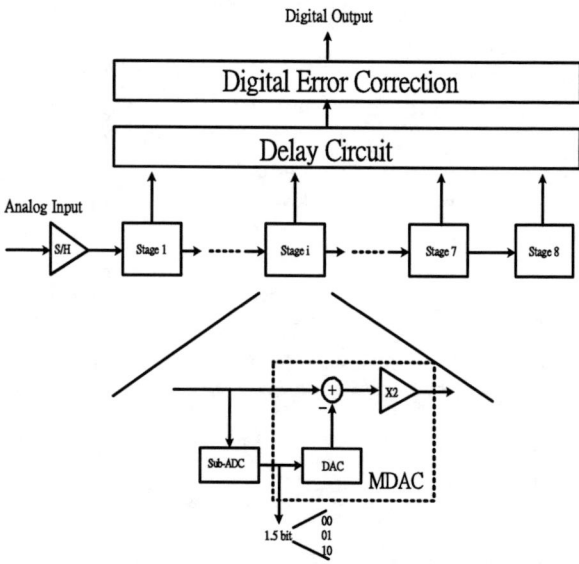

Fig. 1 1.5-bit/stage pipelined ADC architecture

III. BUILDING BLOCKS OF PIPELINED ADC

A. Second-generation current conveyor

Second-generation current conveyors (CCIIs) [4-5] are investigated for applications to filters, oscillators and so on. The CCII is a three-terminal device usually labeled with X, Y and Z as shown in Fig. 2.

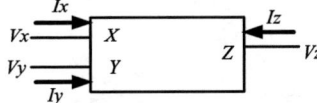

Fig. 2 The symbol of CCII

If we consider the non-ideal effect of CCII, the related equations can be represented by the following matrix

$$\begin{bmatrix} I_y \\ V_x \\ I_z \end{bmatrix} = \begin{bmatrix} 0 & 0 & 0 \\ \alpha_v & 0 & 0 \\ 0 & \alpha_i & 0 \end{bmatrix} \cdot \begin{bmatrix} V_y \\ I_x \\ V_z \end{bmatrix} \quad (3)$$

where $\alpha_v = 1 - \varepsilon_v$, $\alpha_i = 1 - \varepsilon_i$, and ε_v, ε_i express the voltage and current tracking errors, respectively.

The output current of the terminal Z only depends on the input terminal X. The current in terminal X can be directly supplied, or produced through a copy of the voltage of the terminal Y acting across the impedance connected at the terminal X.

The CMOS CCII circuit is designed as shown in Fig. 3 and the supply voltage is 3.3V. Transistors M1~M7 are composed as a voltage buffer and the current through M8 and M9 is as same as that through M6 and M7. M10 and C_c are used for frequency compensation and V_b is the bias voltage. Among all the other realizations, this realization of CCII exhibits good performance in terms of noise, linearity and voltage tracking error between terminals X and Y which can achieve the requirement in the S/H and MDAC design.

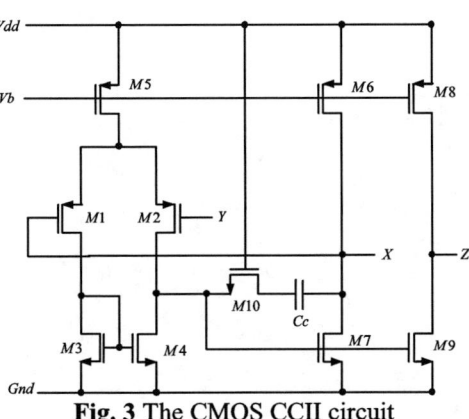

Fig. 3 The CMOS CCII circuit

B. Proposed CCII-based S/H and MDAC

With the above derived characteristics, the new CCII-based S/H and MDAC circuits are proposed. Figs. 4(a) and 4(b) show such two circuits.

Fig. 4(a) is the S/H circuit, when $\phi_1=1$, the S/H is in the sample mode and the input signal is sampled on the capacitor C1. When $\phi_1=0$, the S/H is in the hold mode and the output V_{out} will be copied from the voltage sampled in C1. The output voltage can be expressed as

$$V_{out} = \alpha_v V_{in} \quad (4)$$

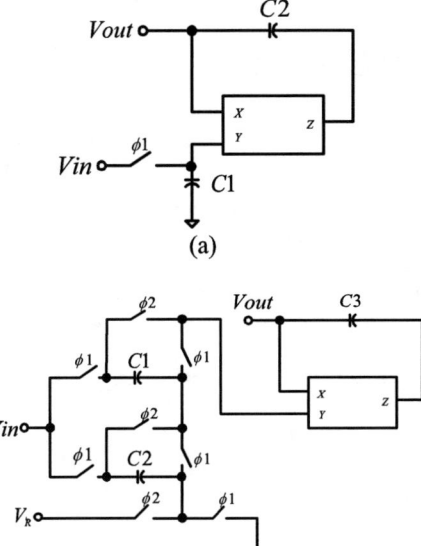

Fig. 4 CCII-based (a) S/H and (b) MDAC circuits

The MDAC circuit shown in Fig. 4(b) operates similar to the S/H circuit. When $\phi_1=1$, the input is sampled on both capacitors C1 and C2, and the two capacitors are series together in the next phase ϕ_2 and the bottom plate of capacitor C2 connects to V_R ($V_R = \pm V_{ref}$ or 0). The output voltage V_{out} is

$$V_{out} = 2\alpha_v V_{in} + V_R \quad (5)$$

To correspond 1.5-bit/stage architecture for the pipelined ADC, the V_{out} can be expressed as below by simplifying $\alpha_v = 1$.

$$V_{out} = 2V_{in} + V_R \quad (6)$$

The output V_{out} is determined by two terminals X and Z with the same current which makes the output voltage settling faster. The feedback capacitor C3 between terminals X and Z plays the major role to determine the output voltage simultaneously. The settling time t_s of the MDAC circuit can be derived as

$$t_s = \frac{\xi \times C_L (C_1 + C_2)}{I_x (1 + \alpha_i) C_3} \quad (7)$$

Where ξ is the factor that effects the settling time and C_L is the loading capacitance. Fig. 5 shows the simulation result of settling time for the MDAC circuit. The circuit with the Z-terminal capacitor feedback can decrease the settling time. With the Z-terminal capacitor feedback, the circuit is up to 38% in settling time faster than that of without the Z-terminal capacitor feedback.

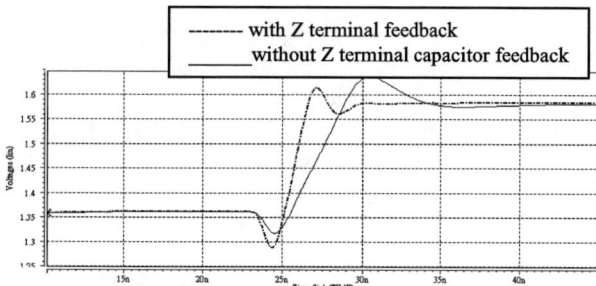

Fig. 5 Comparison of settling time for the MDAC circuit

C. Comparator

Comparing to the CCII, another important circuit in the pipelined ADC is the comparator which can be composed with resistors for use as the flash ADC (i.e. sub-ADC) in every stage. Since there are error correction circuits in the proposed pipelined ADC, the required accuracy of the comparator can be tolerated. Therefore, it can employ the lower accuracy comparators rather than the pre-amplifier ones, ±1/4Vref still can be achieved.

A low power transconductance latched comparator is employed in the pipelined ADC design and the circuit is shown in Fig. 6 [6]. This comparator is constructed of a power switch (M15), reset transistors (M13, M14), input transistors (M9, M11), a feedback latch circuit (M1~M4), and the cutting transistors (M10, M12) with feedback inverters (M5~M6, M7~M8). When the CLK is high, the comparator is in a reset period. At this period, the power switch M15 is turned off and the reset transistors are turn on which make the terminals Vout+ and Vout- at the GND level. When the CLK is low, the comparator is in the operation period. At this period, the power switch is turn on and reset transistors are turned off. The feedback latch circuit (M1~M4) amplifies the voltage gap at terminal Vout+ and Vout- to the full range which due to the difference in the input transconductance. When the voltage at terminal Vout+ or Vout- achieves to the voltage level of VDD, the cutting transistors (M10, M12) are turned off by the feedback inverter. Because the feedback inverter cut off the DC current path, the power consumption is very low.

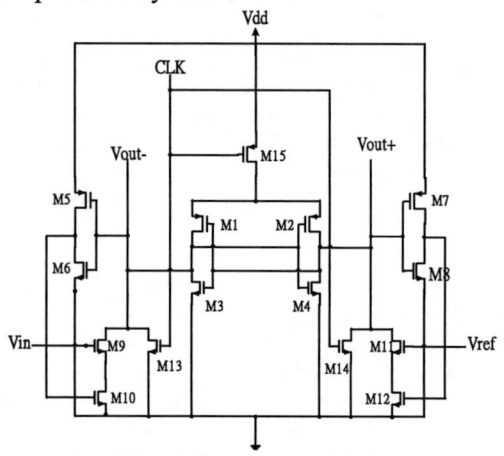

Fig. 6 Comparator circuit

D. Clock generator and delay circuit

There are many switches in the entire design. The circuit of the control signal generator is shown in Fig. 7, and it generates two control signals, CLK1 and CLK2. Under the 10 MHz operation frequency, the time interval between these two clocks is 1ns. The overall delay circuit used in proposed pipelined ADC is shown in Fig. 8. Here, we employ D-type flip-flops to complete the delay function. The digital error correction circuit is also realized in our design.

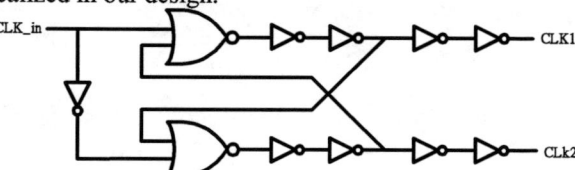

Fig. 7 Non-overlap clock generator

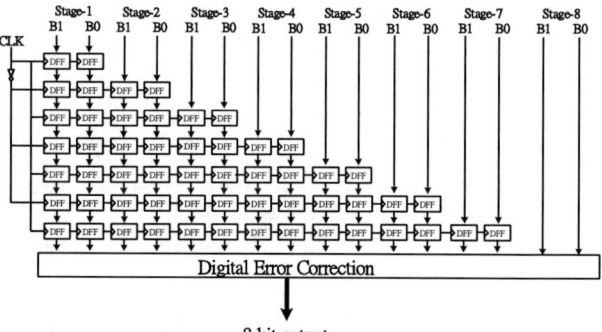

Fig. 8 Overall delay circuit

IV. SIMULATION RESULTS

The proposed pipelined ADC is implemented in TSMC 0.35μm double-poly four-metal process with poly-poly capacitors. It occupied a total area of 1.65x1.65mm² with I/O pad and the active area is 0.85x0.85mm². The physical layout of this chip is shown in Fig. 9. The input signal enters the chip from the upper left corner and is processed by a cascade of eight pipelined stages.

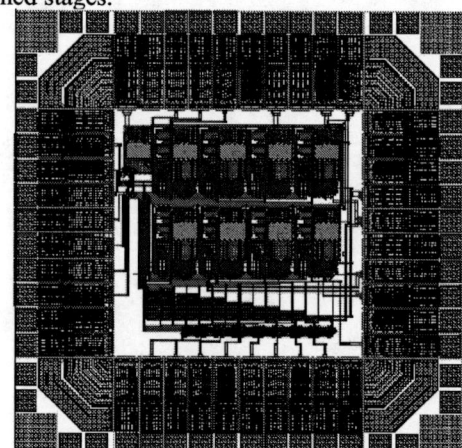

Fig. 9 Physical layout of CCII-based pipelined ADC

The dynamic linearity of the pipelined ADC was analyzed by the fast Fourier transform (FFT). Fig. 10 shows the FFT plot with 100 kHz sine-wave input sampled at 10MHz conversion rate. The signal to noise and distortion ratio (SNDR) for 100 kHz sine-wave input is 45.5dB.

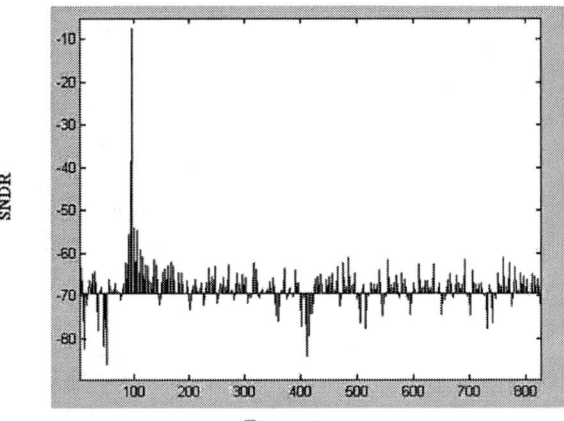

Fig. 10 FFT plot at fin=100 kHz, fs=10 MHz

The differential nonlinearity (DNL) and integral nonlinearity (INL) for static linearity simulation results are shown in Fig. 11 and Fig. 12. The DNL is within 0.4LSB and 0.5LSB and INL is within -0.4LSB and 0.7LSB, respectively.

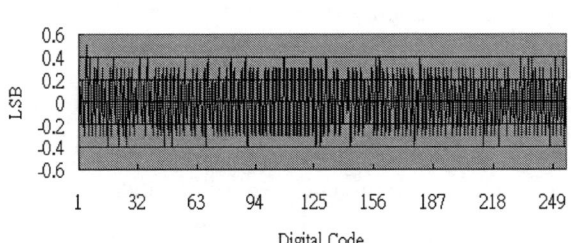

Fig.11 Differential nonlinearity

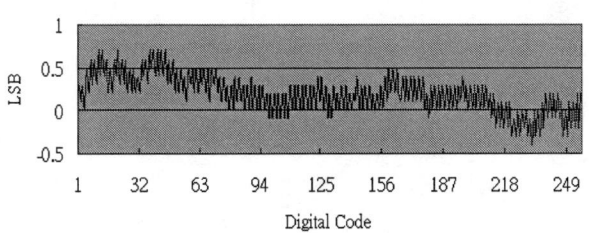

Fig. 12 Integral nonlinearity

V. CONCLUSION

A new CCII-based pipelined ADC is proposed and implemented in this paper. Two main building blocks of the pipelined ADC, S/H and MDAC circuits are constructed of CCIIs instead of OAs. The Z-terminal capacitor feedback in CCIIs can shorten the settling time. Because of the 1.5 bit/stage architecture with error correction circuits in the pipelined ADC, the required accuracy of the comparator can be tolerated. The CCII-based pipelined ADC is realized in a 0.35μm CMOS technology and consumes 29mW under a 3.3V power supply. A summary table is listed in Table I.

Table I Simulation results for proposed ADC

Technology	0.35μm CMOS
Power Supply	3.3V
Sampling Frequency	10MHz
Resolution	8-bit
SNDR	45.5dB @100kHz
DNL	-0.4LSB~0.5LSB
INL	-0.4LSB~0.7LSB
ENOB	7.3 bits
Power Dissipation	29mW
Active Area	0.85x0.85mm^2
Chip Area	1.65x1.65mm^2

ACKNOWLEDGMENT

The authors would like to thank the National Science Council of Taiwan for the financial supporting. The work was sponsored by NSC-92-2220-E-027-001 and NSC-93-2213-E-027-049.

REFERENCES

[1] Shang-Yuan Chuang and Terry L. Sculley, "A Digitally Self-Calibrating 14-bit 10-MHz CMOS Pipelined A/D Converter," *IEEE Journal of Solid-State Circuits,* vol. 37, no. 6, pp. 683-674, June 2002.

[2] Andrew M. Abo and Paul R. Gray, "A 1.5-V, 10-bit, 14.3-MS/s CMOS Pipeline Analog-to-Digital Converter," *IEEE Journal of Solid-State Circuits,* vol. 34, no. 5, pp. 599-606, May 1999.

[3] Hsin-Shu Chen, Bang-Sup Song, and Kantilal Bacrania, "A 14-b 20-MSamples/s CMOS Pipelined ADC," *IEEE Journal of Solid-State Circuits,* vol. 36, no. 6, pp. 997-1001, June 2001.

[4] Yuh-Shyan Hwang, Pei-Tzu Hung, Wei Chen, and Shen-Iuan Liu, "Systematic generation of current-mode linear transformation filters based on multiple output CCIIs," *Analog Integrated Circuits and Signal Processing,* vol. 32, pp. 123-134, 2002.

[5] Dong-Shiuh Wu, Shen-Iuan Liu, Yuh-Shyan Hwang, and Yan Pei Wu, "Multiphe-phase sinusoidal oscillator using second-generation current conveyors," *International Journal of Electronics,* vol. 78, no.4, pp. 645-651, 1995.

[6] Terada J., Matsuya Y., Morisawa F. and Kado Y., "8-mW, 1-V, 100-Msps, 6-bit A/D converter using a transconductance latched comparator," *IEEE Asia Pacific Conference on ASIC, AP-ASIC,* pp. 53-56, Aug. 2000.

A 12bits/200MHz Resolution/Sampling/Power-Optimized ADC In 0.25μm SiGe BiCMOS

Q. Wu and A. Wang
Department of Electrical and Computer Engineering
Illinois Institute of Technology
Chicago, IL 60616, U.S.A
awang@ece.iit.edu

Abstract— Design of a resolution/sampling-rate/power-optimized analog-to-digital converter (ADC) in a foundry 0.25μm SiGe BiCMOS for digital communications is presented. It is based on a combined four-stage multi-bit pipelined, subranging and interpolating architecture with novel digital correction technique. The expanded subranging structure along with the proposed multiple MDAC (multiplying DAC) and relative comparison method used in sub ADC simplifies the 2nd and 3rd stages. The ADC features 12bit resolution, 200MHz sampling rate, differential nonlinearity (DNL) of ± 0.6LSB, integral nonlinearity (INL) of ± 1.1LSB, power dissipation of 380mW and a die size of 6mm^2.

I. INTRODUCTION

Demand for high-performance ADC ICs with optimal combined specifications of resolution, sampling rate and power consumption (not uni-directional improvement) becomes obvious due to emerging applications in wireless communication, broad band transceivers, digital-IF receivers and countless of digital devices [1-4]. The required level of accuracy for ADC exceeds 10 bits at conversion speed of several hundreds MHz [5]. Conventional ADCs for high-speed applications have employed flash, folding and interpolating, subranging, and pipelined architectures [6]. Flash ADCs in SiGe HBT are reported to achieve an 8-GSample/s conversion rate [7]. However, flash ADC is not suitable for beyond 6-bit due to its complexity. Although, folding and interpolating ADC, using reduced number of comparators, is advantageous to wide bandwidth digitization, its resolution (6-10 bits) is typically limited by distortion associated with the strongly nonlinear folders. The filter effect at the output due to the bandwidth limitation displaces the zero-cross points and introduces distortion when referred back to the input of the ADC [8]. Subranging and pipelined architectures have been used for high speed, high resolution and low power ADCs with low complexity for video/image processing and communications, where the speed of a subranging ADC is limited by the settling time of the reference voltages in the resistor DAC and the bandwidth of preamplifiers, whereas a pipelined ADC is slowed down by the feedback factor in the closed-loop residue amplifiers [9].

In this paper, a novel ADC architecture is reported, which inherits the merits of both subranging and pipelined structures, while achieves a speed of 200MS/s. Fresh ideas used in this design include, substituting resistor DAC with multiple MDACs, employing two types of interpolation techniques and using relative comparison method in sub flash ADC to reduce the need for gain accuracy of the residue amplifier so as to make the open-loop amplifier structure possible for beyond 10-bit resolution.

II. ADC ARCHITECTURE

A. Overall Architecture and Timing

The proposed four-stage ADC diagram, as illustrated in Figure 1, consists of a track-and-hold circuit followed by four pipelined sub stages (3.5-bit, 3.5-bit, 3.5-bit and 3-bit). However, different from a conventional pipelined structure, a sub-ranging structure and two types of interpolation topologies are employed in this ADC to take advantage of the good amplifier gain matching and excellent capacitor matching offered by the commercial 0.25μm SiGe BiCMOS process used to achieve higher resolution and linearity at the high speed operation. In order to prevent the offset generated in the previous stage from being coupled into the following ones, capacitor-coupling circuits are adopted at stage interfaces. Digital correction technique is used to suppress the error introduced by the shift of decision level and to relax the offset requirement on the comparators. Bandgap voltage references and current sources are embedded into the ADC chip.

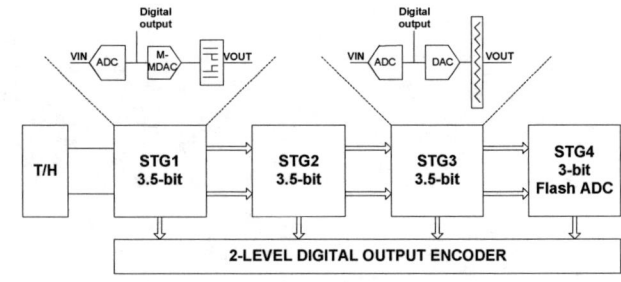

Figure 1. Block diagram of the 12bit/200MHz ADC.

A set of four-phase non-overlap clocks is internally generated for concurrent operations of all stages to convert analog input signals to digital output codes. The non-overlap condition of these clock signals not only ensures that the charge stored in the channels of transistor switches only introduce input-independent offset to the signal path, but also creates time margin for previous residue amplifiers settling.

B. Sub-Range Structure

Stage-1 consists of a sub-ADC and multiple MDACs (capacitor DAC, subtractor and residue amplifier). The M-MDAC sets up the sub-ranging structure, which divides each reference range in the reference ladder of the sub flash ADC into five reference voltage levels as illustrated in Figure 2, where "N" is the index of the reference ladder. The overlap of the adjacent sub-reference division provides redundancy for digital correction. The paralleled working MDACs used, different from a traditional subranging network, not only avoid the RC delay originated from resistor ladder DACs [10] and make higher speed ADC possible, but also, combined with the interpolation technique to be discussed in Section C, re-construct the multi residues around the fine reference levels. The smaller residues, the closer the fine sub-divided reference point to the analog input. Consequently, the input analog signal will fall into the range formed by the two residues around zero. The sub flash ADC in the second stage will compare the adjacent residue values V_{res1} and V_{res2} as following:

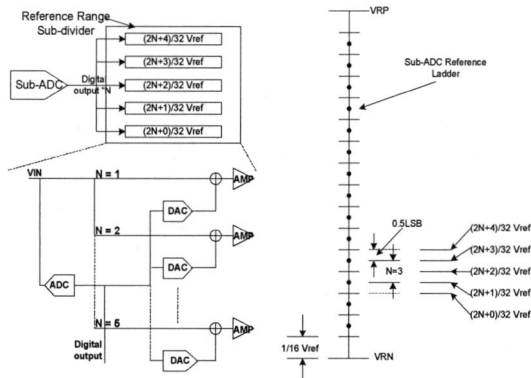

- If $V_{res1}>0$ and $V_{res2}>0$, D=1;
- If $V_{res1}>0$, $V_{res2}<0$, and $|V_{res1}| > N|V_{res2}|$, D=1;
- If $V_{res1}>0$, $V_{res2}<0$, and $|V_{res1}| < N|V_{res2}|$, D=0;
- If $V_{res1}<0$ and $V_{res2}<0$, D=0;

N is the ratio of the two coupling capacitors at the input of comparators. Therefore, the $V_{res1}-V_{res2}$, corresponding to the transition bit of the second stage sub-ADC, will define the fine reference range of the third stage. The $V_{res1}-V_{res2}$ is further divided using resistive interpolation as to be described in Section C. The same mechanism applies to Stage-3 until the final 12-bit resolution is achieved.

Since the sub flash ADCs used in the second, third and fourth stages only care about the relative values of the two neighboring residues around zero, the requirement on the output range of each stage is much relaxed. Hence, the gain matching, not the absolute value, becomes dominant to decrease the linearity error. An important error source in the pipelined ADC is therefore eliminated.

C. Interpolation Technique

Two types of interpolation techniques, 1) cross-connecting the differential inputs of the residue amplifiers, and 2) resistive interpolation, are used in this ADC to reduce the number of pre-amplifiers needed, hence reduce the large input capacitance, input power dissipation, stringent timing requirements and large area [10].

The first interpolation technique used in the Stage-1 output is shown in Figure 3(a), where the adjacent differential outputs of the MDAC blocks are cross-connected to generate an additional zero-crossing point, V_m. Combining the five zero-crossings generated from the M-MDAC sub-ranging configuration, the overall reference diagram is given in Figure 3(b).

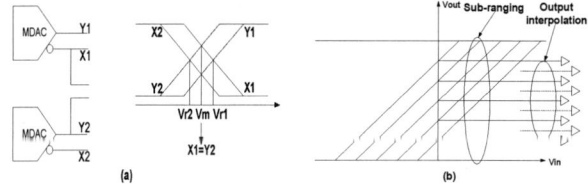

Figure 3. (a) One-level interpolation scheme. (b) Overall subranging reference defininaton.

The resistive interpolating employed in the Stages-2/3 extends the normal interpolation concept, hence, produces more quantization levels between every two consecutive reference voltages [10]. The unit resistor value in the interpolating network should be decided with care, not only considering the RC delay introduced by the resistor strings and the input gate capacitance of the following amplifier, but also taking into account the linearity errors. Too small a resistor value will narrow the output reference range, which causes saturation of digital output of the next stage and lead to the DNL errors around the reference level transition point showing as repeating codes. In this design, we separate the

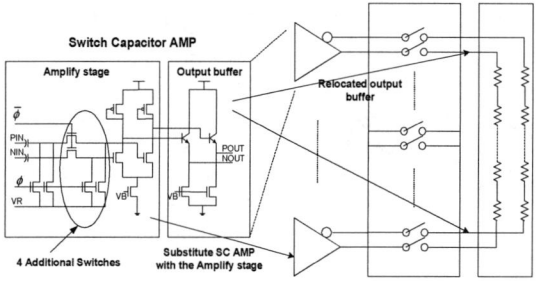

Figure 4. Stage-2 and Stage-3 diagram and SC circuit.

two-stage (amplifier stage and output buffer) switch capacitor amplifier used at the stage interface as illustrated in Figure 4 by replacing the original one with a amplifier stage and putting only two additional buffers at the two-end of the resistor string to increase the drivability for the resistor network. This approach helps to maintain the bandwidth without increasing the differential non-linearity and to decrease the overall power dissipation.

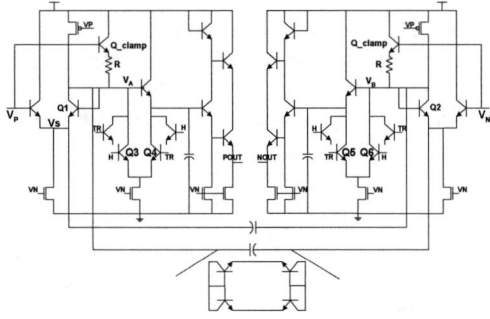

Figure 5. Track and Hold circuit diagram.

III. CRITICAL CIRCUIT BLOCKS

A. Track and Hold Circuit

A full bipolar SEF (Switch Emitter Follower) structure, with CMOS current biasing (shown in Figure 5), is used to take advantage of high frequency, low power dissipation and low noise characteristics of the SiGe HBT process [11-13]. The differential unit gain input buffer carrying the same current isolates the held voltage from input signal, therefore cancels the even-order distortion and reduces the distortion generated from input feedthrough effect. Compared with its CMOS counterpart, the SEF sampler is superior because of no channel discharging, no $R_{ds} \sim V_{gs}$ nonlinear effect and reduced clock-feedthrough behavior. The Darlington type output buffer structure is used to increase the drivability, hence, to speed up the operation. In addition, clamping transistors (Q_clamp) are employed to prevent transistor Q1 and Q2 from being pushed into deep depletion during the hold mode. Hence, the points V_A and V_B are pulled down due to the low impedance path provided by Q3 and Q6 until the Q_clamps turn on and fix V_A and V_B at V_p-V_{be} and V_n-V_{be} (w/o resistors) or $V_p-(V_{be}+V_R)$ and $V_n-(V_{be}+V_R)$ (with resistors), respectively, as illustrated in Figure 6b-1. The resistors used here adjust V_A and V_B to meet the requirement: $V_{diff} = V_{Tr}(N+1) - V_H(N) \leq V_{be}$ (refer to Figure 6b-2), otherwise if the V_{PP} is large enough, at the Nyquist sampling frequency, the input signals, especially the minimum value, will be fed to the hold capacitor and introduce distortion (refer to Figure 6b-3). In order to compensate the charge injected from the collector-base junction capacitor C_{bc} of the current switch Q3-Q6, the dummy switches driven with opposite phase clocks are used. The pedestal error, therefore, is limited to the nonlinear and mismatch contributions only. Meanwhile, the crossed feed-forward capacitors lower hold mode feed-through errors [14-15].

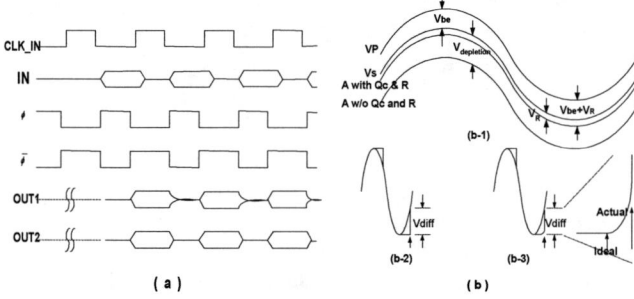

Figure 6. (a) SC timing diagram (b) T/H waveforms in hold mode

B. Inter-Stage Switch Capacitor Circuit

The switch capacitor circuit (Figure 4) used to remove the offset generated from the proceeding stages meets the requirements that the linear range is no smaller than 1V with at least unit gain and the output common voltage is appropriate to provide the common bias for the following amplifiers without being disturbed by the process variation. The additional switches employed in the circuit reduce the timing requirement and increase the delay tolerance between the control signals and the input signal, which is hard to estimate and adjust under the time-variant environment and with various process corners. Figure 6a shows the timing analysis. The phase relationship between CLK_IN (controlling the input), ϕ and $\overline{\phi}$ is determined by the system clock scheme. "OUT1" is the output without additional switches and "OUT2" is the output with additional switches. From this timing diagram, it can be seen that the introduction of four additional switches has eliminated the output dispersion generated from the uncertainty delay between ϕ and the input of the SC circuit. Meanwhile, since there is a 180° phase difference existing between ϕ and $\overline{\phi}$, and all of the switches have the same dimension, the clock feedthrough effect and channel discharging effect on the input gate capacitors have been partially reduced.

IV. ADC PERFORMANCE DATA

This ADC chip was designed and implemented in a commercial 0.25μm SiGe BiCMOS. Figure 7 shows the die photo of the design.

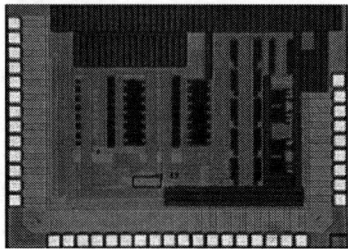

Figure 7. Die photo of the 12-bit 200MHz ADC.

A. Track and Hold Circuit Result

The track-and-hold circuit designed in SiGe achieves higher sampling rate, lower distortion and larger differential input range while maintaining lower noise as compared to comparable state-of-the-art T/H designed in other processes published [1-2]. It operates at 200MHz sampling frequency with 1.5Vpp input signals. Figure 8 Shows DFT analysis results for this T/H circuit operating at 200MHz.

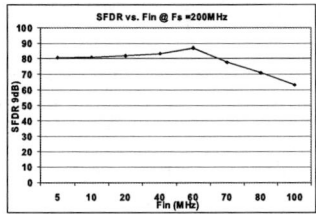 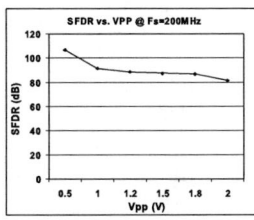

Figure 8. DFT results for SFDR vs. Fin and Vpp.

B. Noise Performance

There typically exist three noise sources in all types of ADC's, which affect the overall SNR. They are the quantization noises, the input-referred circuit noise (equivalent thermal noise and uncertainty, such as jitter, in track and hold circuit and residue amplifier), and kT/C noise. Since the overall gain of the residue amplifiers in first stage and the following capacitor coupling amplifiers is high enough, the overall noise performance will be dominated by the noise generated in track-and-hold circuit and Stage-1 only. Table I summarizes the noise performance results.

TABLE I.   ADC noise results

Input Equivalent Noise (μV)			
T/H	Capacitor	Residue AMP	Total
76.38	101.40	138.86	188.14 (-74.5dB)

C. Overall Performance

The die size of this ADC is 6mm^2. It achieves a 12b resolution at 200MHz sampling rate. Its critical performance specs, as summarized in Table II, are a DNL of ±0.6LSB, an INL of ±1.1LSB, and power consumption of 380mW.

V. CONCLUSION

This paper presents a 4-stage resolution/sampling/power-optimized ADC designed in a 0.25μm SiGe BiCMOS. The design uses a combined pipelined/subranging/interpolation architecture for optimal overall performance. Several unique circuit techniques are used including, capacitor coupling at the stage interface to remove offset, a digital correction technique that generates the redundancy by introducing overlap in reference division, adding resistors at emitter of the clamping transistors for the T/H circuit to achieve higher input range without extra distortion, using additional switches in SC circuit to avoid process corner effect on timing, etc. The ADC performance specs are: 12b resolution, 200MHz sampling rate, DNL of ±0.6LSB, INL of ±1.1LSB, power dissipation of 380mW and a die size of ~6mm^2.

TABLE II.   ADC performance

BiCMOS Technology	1-poly 3-metal 0.25μm SiGe
Resolution (bits)	12
Sampling Rate (MHz)	200
Supply Voltages (V)	4 V (TH), 2.5V
Input Range (V)	1.5 Vpp
Power (mW)	380
DNL (LSB)	+/-0.6
INL (LSB)	+/-1.1

REFERENCES

[1] C.Fiocchi, U.Gatti and F.Maloberti "Design issues on high-speed high-resolution track-and-holds in BiCMOS technology", *IEE Proc. Circuits Devices Syst.*, Vol.147. No.2, April 2000, pp.100-106.

[2] B. Razavi "A 200-MHz 15-mW BiCMOS sample-and-hold Amplifier with 3V supply", *IEEE J. Solid-State Cir.*, Vol.30, No.12, December 1995, pp.1326-1332.

[3] R. Walden "Analog-to-digital converter survey and analysis", *IEEE J. Selected Areas in Comm.*, Vol.17, No.4, April 1999, pp.539-550.

[4] B. Yu "A 900MS/s 6b interleaved CMOS flash ADC", *Digest VLSI Circuits Symp.*, 2001, pp.149-152

[5] S.Yoo, J. Park and S. Lee, "A 2.5-V 10-b 120-MSample/s CMOS pipelined ADC based on merged-capacitor switching", *IEEE Trans Cir. and Sys. II*, Vol. 51, No. 5, May 2004, pp.269-275.

[6] S. Yoo, J. Park1 and H. Yang, "A 10b 150MS/s 123mW 0.18 μm CMOS pipelined ADC", *Proc. IEEE ISSCC*, 2003, pp.326 – 497.

[7] P. Xiao, K. Jenkins, M. Soyuer and H. Ainspan, "A 4 b 8 GSample/s A/D converter in SiGe bipolar technology", *Proc. IEEE ISSCC*, 1997, pp.124-125.

[8] S. Limotyrakis, K. Nam, and B. Wooley," Analysis and simulation of distortion in folding and interpolating A/D converters", *IEEE Trans. Cir. and Sys. II*, Vol. 49, No. 3, March 2002, pp.161-169.

[9] M. Choe, B. Song, and K. Bacrania, "An 8-b 100-MSample/s CMOS pipelined folding ADC", *IEEE J Solid-State Cir.*, Vol. 36, No. 2, Feb. 2001, pp.184-194.

[10] B. Razavi, *Principles of data conversion system design*, IEEE Press, 1995.

[11] H. Kobayashi, T. Mizuta and Uchida, "Design consideration for folding/interpolation ADC with SiGe HBT", *Proc. IEEE Inst. and Meas. Tech. Conf.*, 1997, pp.1142-1147.

[12] D. Harame, et.al., "Si/SiGe epitaxial-base transistors – Part I: materials, physics and circuits", *IEEE Trans. Elec. Dev.*, Vol.42, No.3, March 1995, pp.435-468.

[13] D. Harame, et.al., "Si/SiGe epitaxial-base transistors – Part II: process integration and analog applications", *IEEE Trans. Elec. Dev.*, Vol.42, No.3, March 1995, pp.469-482.

[14] G. Caiulo, C.Fiocchi and U.Gatti, "On the design of high-speed high-resolution track and holds", *Proc. IEEE ISCAS*, v1, 1996, pp.73-76.

[15] C.Fiocchi, U.Gatti and F.Maloberti "Design issues on high-speed high-resolution track-and-holds in BiCMOS technology", *IEE Proc. Cir. Dev. Syst.*, Vol.147, No.2, April 2000, pp.100-106.

A Low Power Pipelined Analog-to-Digital Converter using Series Sampling Capacitors

SeongHwan Cho
Korea Advanced Institute of Science and Technology
(KAIST)
Daejon, Republic of Korea
chosta@ee.kaist.ac.kr

Sungmin Ock, Sang-Hoon Lee, and Joon-Suk Lee
Future Communications IC Inc.
(FCI)
Sungnam, Republic of Korea
{caruso,hooni,jlee}@fci.co.kr

Abstract— A low power pipelined analog-to-digital converter(ADC) that employs sampling capacitors connected in series is presented. The series sampling capacitors minimize the size of the sampling capacitors to the KT/C limit without degrading the ADC's performance due to mismatch. Using this technique, a 10-bit 100MHz pipelined ADC is designed and simulated. The ADC achieves 60dB of signal-to-noise-and-distortion ratio(SNDR) at 100MHz while consuming 47mW from 1.8-V supply in 0.18μm CMOS technology.

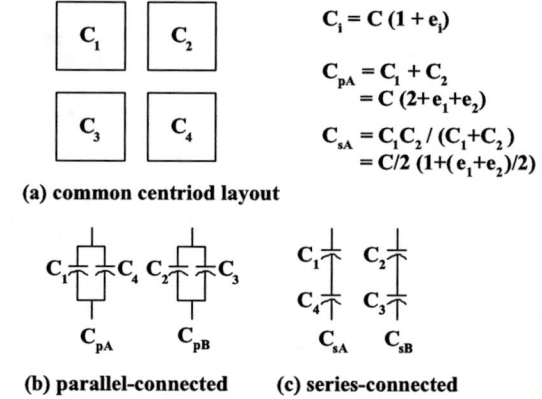

Fig. 1. Capacitors connected in parallel and in series.

I. INTRODUCTION

Pipelined analog-to-digital(ADC) architecture has gained great popularity in data communication and video processing applications where high sampling rates and medium-to-high resolutions are necessary [1], [2]. Due to the increasing demand for portability in these applications, reducing the power consumption of the ADC has become one of the key design criteria. While the pipelined ADC architecture is well suited for high sampling rates, reducing the power consumption is not an easy task mainly due to the opamps in the multiplying digital-to-analog converter(MDAC). In order to reduce the power consumption in the MDACs, several techniques have been proposed such as opamp sharing [3], and resolution per stage optimization [4], [5]. While these methods reduce the number of opamps in the ADC, the power consumption of the individual opamp is still determined by the performance requirements set by the sampling capacitors in the MDAC.

The value of sampling capacitors in the pipelined ADCs is determined by several factors such as mismatch, KT/C noise, charge leakage and linearity. Of these factors, mismatch and KT/C noise are the decisive constraints in todays submicron CMOS process technology. Charge leakage is on the order of several fA/(pF-V) and voltage dependent linearities are typically less than 100ppm [6], [7], all providing more than 14bits of resolution at sampling rates of hundreds of MHz. Thus, capacitance determined by the thermal noise and mismatch is often of critical concern. For high resolution ADCs near 10-bits, the size of the capacitor can be limited by the mismatch requirements in a standard digital CMOS process [8]. Although the capacitance requirement from mismatch can be relaxed using various layout [9] and circuit techniques, including calibration [10] and error averaging methods [11], [12], these techniques increase the system complexity. Hence designers are often lead to implementing a standard pipelined ADC with large capacitors to remove the effect of mismatch, which results in large power consumption. In this paper, a simple way to reduce the size of the sampling capacitors is presented.

II. SERIES SAMPLING CAPACITOR TECHNIQUE

In a typical process, the mismatch(σ) is a function of the area. That is, $\sigma = a/W$, where a is coefficent that depends on process technology and W is the width of the capacitor [13]. If a reliable 10-bit ADC performance is to be achieved in a 1.5-bit/stage pipelined ADC, then the 3σ capacitor mismatch needs to be less than 2^{-10} for the first stage of the pipelined ADC. This results in a capacitor size of $60\mu m \times 60\mu m$ for process with $a \simeq 2\%\mu m$ [13], [14]. With the capacitor density of $1fF/\mu m^2$, this leads to 3.6pF of capacitance, which is much larger than what is required by the kT/C noise, which is less than 1pF. Considering that the opamp's power consumption is proportional to the capacitance value, more than 70% of power is wasted due to capacitor mismatch.

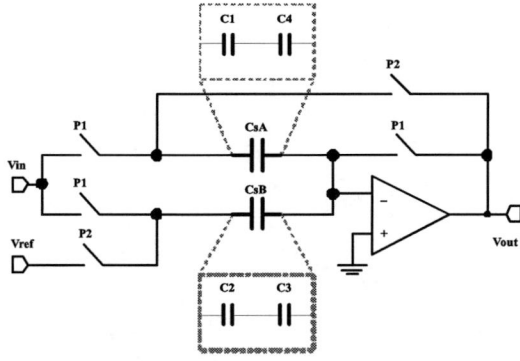

Fig. 2. MDAC with series sampling capacitors

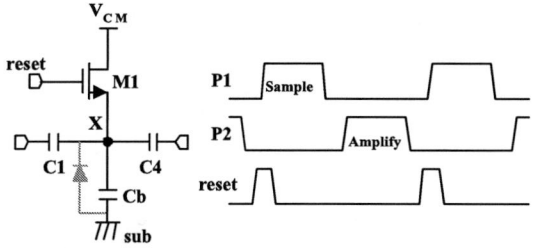

Fig. 3. Resetting the intermediate node.

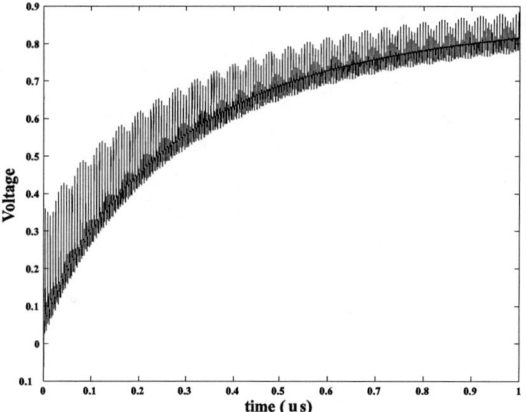

Fig. 4. Intermediate node voltage

A. Mismatch in Capacitors

To cancel mismatch from first order process gradients, common centroid layout using unit capacitors is commonly employed as shown in Fig. 1, where (C_1, C_4) forms one capacitor and (C_2, C_3) forms the other capacitor. Fig. 1 depicts two cases, where the capacitors are connected in parallel (C_{pA}, C_{pB}) and in series (C_{sA}, C_{sB}). Representing the capacitance of C_i as $C \cdot (1 + e_i)$, where e_i is a random error that is normally distributed with zero mean and standard deviation σ, (i.e., $e_i \sim N(0, \sigma)$), the mismatch between the capacitors can be represented as the following,

$$\frac{C_{pA} - C_{pB}}{C_{pA}} = \frac{C(2 + e_1 + e_4) - C(2 + e_2 + e_3)}{C(2 + e_1 + e_4)}$$
$$\simeq \frac{e_1 + e_4 - e_2 - e_3}{2 + e_1 + e_4} \simeq e_p \quad (1)$$

$$\frac{C_{sA} - C_{sB}}{C_{sA}} = \frac{\frac{C}{2}(1 + \frac{e_1 + e_4}{2}) - \frac{C}{2}(1 + \frac{e_2 + e_3}{2})}{\frac{C}{2}(1 + \frac{e_1 + e_4}{2})}$$
$$\simeq \frac{e_1 + e_4 - e_2 - e_3}{2 + e_1 + e_4} \simeq e_s \quad (2)$$

where e_s and e_p are both normally distributed with $N(0, \sigma)$. From Eq. 1 and Eq. 2, it can be seen that both schemes provide nearly identical matching performance. This is quite intuitive since both layouts use the same area and the value of mismatch should depend on area rather than the capacitance value. It should also be noted that in this example, the capacitance of series capacitors is four times smaller than that of the parallel capacitors.

B. Principles of Operation

A multiplying digital-to-analog converter(MDAC) of gain 2 is shown in Fig. 2. As can be seen, the sampling capacitors C_{sA} and C_{sB} are implemented using two series capacitors as opposed to a single capacitor in a conventional approach. Assuming single-ended operation for simplicity, the voltage transfer function of this MDAC can be represented by the following equation,

$$V_{out} = (2 + e_s)V_i - (1 + e_s)V_r \quad (3)$$

where e_s, is given in Eq. 2.

For a fair comparison of this result to a conventional MDAC that use single capacitors C_{pA} and C_{pB}, let's set the value of the effective sampling capacitance in both methods to be equal (i.e., $C_{sA} = C_{sB} = C_{pA} = C_{pB}$). Since the series connection reduces the effective capacitance by a factor of two, the series capacitors need to be twice as large as the capacitors in the conventional approach (i.e., $C_i = C_{pA}/2$). Hence the mismatch in the series capacitors is smaller by a factor of $\sqrt{2}$. In general, if the capacitor is formed by n series capacitors, the mismatch will be reduced by $\sqrt{n}$. Hence the effective sampling capacitance can be reduced down to the limit set by the KT/C noise.

C. Non-ideal effects

There are a few potential problems with the series sampling capacitors, which are mainly associated with the parasitic capacitance at the intermediate node X between the two series capacitors, as shown Fig. 3.

The first problem occurs from the fact that the intermediate node is a high impedance node and that it is bootstrapped during the ADC operation. For reliability purposes, many process technologies require an antenna rule so that the thin oxide is protected from stress that results from charge accumulation during the manufacturing process. The antenna rule usually requires a reverse biased diode to be placed at the intermediate node. Due

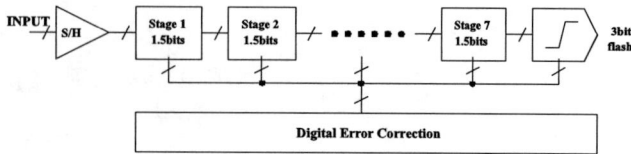

Fig. 5. ADC block diagram.

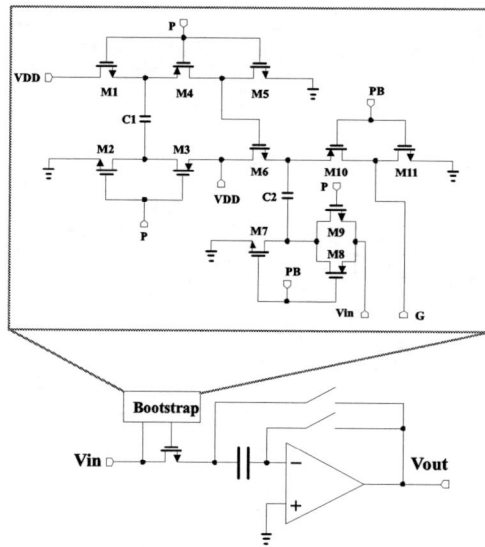

Fig. 6. Schematic of the bootstrapped switch in S/H (single-ended shown).

to bootstrapping, the voltage on the intermediate node may reach above or below the power supply rails, and can turn on the diode, malfunctioning the ADC. To avoid such problem, the intermediate node is reset to a common-mode voltage through a small transistor M1. The transistor M1 is turned on by a reset clock which is asserted high right before the sampling phase. The reset clock may overlap with the sampling clock by a little amount, but it does not degrade the ADC's performance since the settling of the input voltage on the sampling capacitors is dominated by the opamp's settling time. The size of M1 is made very small to minimize the amount of non-linear junction capacitance. Since the transistor M1 is not large enough to reset the intermediate node to the common-mode voltage in one reset clock cycle, the node slowly reaches the common-mode voltage level after many cycles, as shown in Fig. 4. Note that M1 also serves as the antenna diode.

Another problem with the series sampling capacitors is due to the parasitic bottom plate capacitance. Whether the capacitor is formed by a poly-to-poly or a metal-insulator-metal(MIM) capacitor, there is bottom plate capacitance(C_b) to the substrate and the junction capacitance(C_j) formed by the reset transistor. In addition to the increase in total sampling capacitance due to these parasitic capacitors, the mismatch of C_b and C_j can affect the output given in Eq. 3. However, this effect is negligible if C_b and C_j is small compared to the sampling capacitors.

III. ADC Implementation Details

The pipelined ADC is composed of seven 1.5-bit/stage, a 3-bit flash and a digital error correction logic as shown in Fig. 5. All analog paths are implemented differentially to minimize the effect of noise.

Bottom plate sampling technique is used throughout the stages to minimize the effect of charge injection. For the sample-and-hold (S/H) stage, bootstrapped switches are used to further minimize signal-dependent charge injection as shown in Fig. 6. It is formed of two bootstrapping circuits, where transistors M1–M5 form the first bootstrap circuit which provides boosted clock to the main bootstrap circuit formed by M6–M11.

The opamp is based on a telescopic topology with gain-boosting as shown in Fig. 7. The main opamp's common-mode feedback is generated by a switch capacitor circuit. The opamp has open-loop gain of 80dB with phase margin of 70 degrees and unity gain frequency of more than 1GHz.

The comparator is a dynamic comparator using cross-coupled inverter latches as shown in Fig. 8 [15]. Although

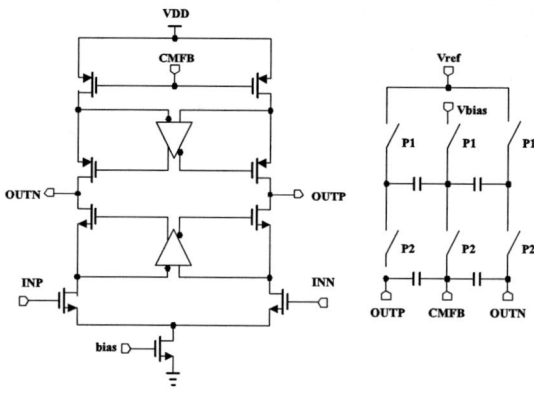

Fig. 7. Schematic of the opamp and common-mode feedback circuit.

there has recently been a variation of this circuit [2], where the node X,Y is pre-charged to VDD before the evaluation phase, the opamp output can be damaged through the parasitic capacitance C_p, as the pre-charged nodes falls to ground during the evaluation phase. This phenomenon becomes significant as the size of the sampling capacitance gets smaller and become comparable to the parasitic capacitance.

IV. Simulation Results

The proposed ADC has been designed using $0.18\mu m$ CMOS technology. The SNDR and SFDR is shown in Fig. 9 as the sampling frequency is varied. It can be observed that the ADC achieves more than 9.5 effective-number-of-bits (ENOB). The ADC performance falls below 8 ENOB as the sampling frequency is increased to above 150MHz. The simulation takes into account the bottom plate capacitance. The summary of proposed ADC is shown in Table I.

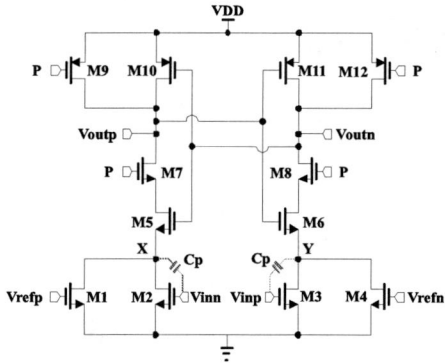

Fig. 8. Comparator schematic.

TABLE I
SIMULATION RESULTS SUMMARY.

Resolution	10 bits
Technology	0.18μm CMOS
V_{DD}	1.8V
Sampling frequency (f_s)	100MHz
SNDR	60dB (@ Nyquist)
SFDR	67dB (@ Nyquist)
INL	0.8 bit
DNL	0.4 bit
Power consumption	26mA from 1.8V

V. CONCLUSION

A low power analog-to-digital converter is implemented using series sampling capacitor technique. This method can ultimately remove the effect of mismatch in ADC's performance and reduce the sampling capacitors to the kT/C limit. A proof-of-concept ADC has been designed using a 0.18μm CMOS technology and the validity of this technique has been verified.

VI. ACKNOWLEDGMENT

The authors would like to S.H. Baik and M.W. Hwang of Future Communications IC Inc. for their invaluable contributions and K. Gulati of Engim Inc. for his comments.

REFERENCES

[1] K. Gulati et al., "A highly integrated analog baseband transceiver featuring a 12-bit 180MSPS pipelined A/D converter for multi-channel wireless LAN," in *Int. Symposium on VLSI Circuits*, June 2004.
[2] B.-M. Min, P. Kim, F. B. III, D. Boisvert, and A. Aude, "A 69mW 10-bit 80MSamples/s pipelined CMOS ADC," *IEEE J. Solid-State Circuits*, vol. 38, pp. 2031–2039, Dec. 2003.
[3] P. Yu and H. Lee, "A 2.5-V, 12-b, 5-Msample/s, pipelined CMOS ADC," *IEEE J. Solid-State Circuits*, vol. 31, pp. 1854–1861, Dec. 1996.
[4] S. Lewis, "Optimizing the stage resolution in pipelined, multistage, analog-to-digital converters for video-rate applications," *IEEE Trans. Circuits Syst. II*, vol. 39, pp. 516–523, June 1992.
[5] P. Kwok and H. Luong, "Power optimization for pipeline analog-to-digital converters," *IEEE Trans. Circuits Syst. II*, vol. 46, pp. 549–553, May 1999.

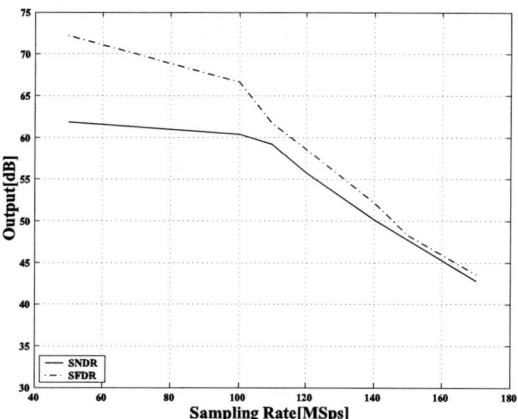

Fig. 9. ENOB vs. sampling frequency.

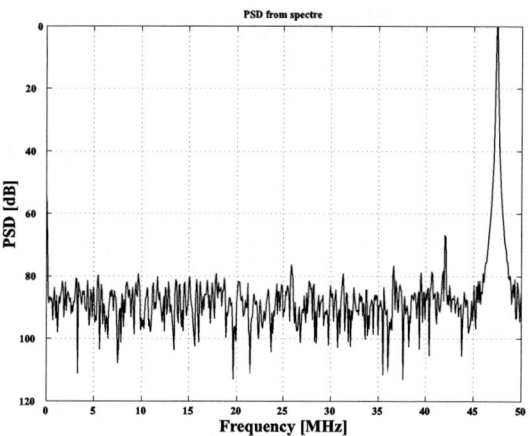

Fig. 10. FFT result of the ADC

[6] *RF and Analog/Mixed-Signal Technologies for Wireless Communications*, International Technology Roadmap for Semiconductors, 2003. [Online]. Available: http://public.itrs.net
[7] J. Babcock et al., "Analog characteristics of metal-insulator-metal capacitors using PECVD nitride dielectrics," *IEEE J. Solid-State Circuits*, vol. 23, pp. 1324–1333, Dec. 1988.
[8] A. Abo and P. Gray, "A 1.5-V, 10-bit 14.3-MS/s CMOS pipeline analog-to-digital converter," *IEEE J. Solid-State Circuits*, vol. 34, pp. 599–606, May 1999.
[9] M. McNutt, S. LeMarquis, and J. Dunkley, "Systematic capacitance matching errors and corrective layout procedures," *IEEE J. Solid-State Circuits*, vol. 29, pp. 611–616, May 1994.
[10] A. Karanicolas, H. Lee, and K. Barcrania, "A 15b 1MS/s digitally self-calibrated pipeline ADC," in *ISSCC Digest of Technical Papers*, Feb. 1993, pp. 60–61.
[11] B.-S. Song, M. Tompsett, and K. Lakshmikumar, "A 12-bit 1-Msample/s capacitor error-averaging pipelined A/D converter," *IEEE Electron Device Lett.*, vol. 22, pp. 230–232, May 2001.
[12] Y. Chiu, "Inherently linear capacitor error-averaging techniques for pipelined A/D conversion," *IEEE Trans. Circuits Syst. II*, vol. 47, pp. 229–232, Mar. 2000.
[13] Z. Ning, L. D. Schepper, R. Gillon, and M. Tack, "A floating gate ac nulling technique for characterisation of matching properties of MOS capacitors," in *Proceedings of ESSCIRC*, 2003.
[14] A. Abo, "Design for reliability of low-voltage switched capacitor circuits," Ph.D. dissertation, U.C. Berkeley, 1999.
[15] T. Cho and P. Gray, "A 10bit, 20MS/s 35mW pipeline A/D converter," in *Proceedings of IEEE Custom Integrated Cicuits Conference*, 1994, pp. 499–502.

A Generic Multilevel Multiplying D/A Converter for Pipelined ADCs

Vivek Sharma, Un-Ku Moon, Gabor C. Temes
Electrical Engg. & Computer Sci.
Oregon State University
Corvallis, USA
Vivek.Sharma@austriamicrosystems.com, moon@ece.orst.edu, temes@ece.orst.edu

Abstract—State-of-art implementations of pipelined ADCs can only realize a multiplying DAC (MDAC) with $(2^n - 1)$ levels. However, the number of levels needed to optimize the performance may differ from this number. A novel scheme is proposed allowing for realization of an arbitrary number of MDAC levels, while allowing for 1 bit of digital redundancy and digital error correction without any overhead.

I. INTRODUCTION

Pipelined ADCs find wide use in a variety of applications requiring high-speed, high-resolution operation at a reasonable cost. These benefits are derived from the design flexibility offered by these ADCs which allows the designer to choose the number of levels resolved in each stage. Previous research has shown the importance of choosing the number of levels *m* and also suggested that this number may assume arbitrary values depending on the specifications [1],[2],[3],[5]. However, the practical implementation of these ADCs is possible only if digital redundancy and error correction are included in the design. Digital redundancy allows for more robust design at the cost of a few extra comparators. The use of digital redundancy along with digital error correction makes the overall system tolerant to large comparator offsets and simplifies the design to such an extent that the use of extra comparators is not a cost-limiting factor. In the absence of these desirable features, the advantages of a pipelined ADC are lost. State-of-art implementations only allow for digital redundancy and error correction when the MDAC resolves $(2^n - 1)$ levels. Consequently, designers have had to work with a sub-optimal number of MDAC levels and incur a loss in the power vs. speed tradeoff offered by pipelined ADCs. This paper presents a technique to overcome this problem, thereby allowing designers greater freedom in choosing the number of MDAC levels.

A novel scheme for implementing 1 bit of digital redundancy in an *m*-level MDAC is presented in Section II and illustrated with the example of a 5-level MDAC. The relevance of this configuration has been discussed in [4]. Section III covers the technique for implementing digital error correction and presents a simple overlap-and-add scheme for the 5-level MDAC. Finally, the implementation details are briefly reviewed in Section IV and the discussion is concluded in Section V.

II. *m*-LEVEL MDAC WITH 1 BIT OF DIGITAL REDUNDANCY

A. Basic Definitions

Fig. 1 shows the block diagram of a typical MDAC used in a pipelined ADC stage. The ADC in each pipeline stage performs a coarse quantization of its input, and the error resulting from this quantization is then computed and scaled by the DAC and sample-and-hold amplifier (SHA), which are collectively known as the MDAC. This ADC is realized using a few comparators, while the MDAC is typically implemented using a switched-capacitor sample-and-hold amplifier. 1 bit of digital redundancy is usually enough and is included in each pipeline stage. The gain *G* of the SHA can then be written in terms of the number of MDAC levels *m* as

$$2G = m + 1. \qquad (1)$$

B. Example: 5-Level MDAC

The design methodology is now illustrated with the example of a 5-level MDAC with 1 bit of digital redundancy. The signal range of the ADC is assumed to be $-V_{REF}$ to $+V_{REF}$. We use the input-output transfer characteristic of the MDAC, also known as the *residue transfer curve* for our discussion. The scheme presented here is similar to that discussed in [6],[7] for the design of a 3-level MDAC with 1 bit of digital redundancy. The scheme involves halving the SHA gain *G* and leaving out 1 comparator threshold. Thus, we begin with an MDAC stage with *m*+1 levels and with a SHA gain of *2G*. For a 5-level MDAC, this corresponds to a stage with an SHA gain of 6 and with resolving 6 levels with 5 comparators. The residue transfer curve for such a stage is

This work was supported by the Catalyst Foundation, New York, NY.

shown in Fig. 2. This stage has almost no tolerance to comparator offsets and any offset forces the residue out of range introducing nonlinear distortion. Thus, the design of comparators becomes a great challenge.

The next step involves cutting the SHA gain in half to G while retaining the old comparator thresholds. The residue transfer curve for this scheme is shown in Fig. 3. Now, comparator errors as large as $+/-V_{REF}/6$ can be tolerated. This is a significant improvement over the previous case. However, the digital error correction for this case is more complicated. In order to understand this, we need to visualize a 2-stage pipeline where both stages have been designed for a residue transfer curve as shown in Fig. 3. The first stage suffers from comparator offsets which cause deviation from the ideal residue tranfer curve, but are not so large as to overload the next stage.

Now, the errors introduced in the digital output code of the first stage can be corrected by looking at the digital output code of the second stage. In the ideal scenario, the residue output of the first stage is bounded between $+/- 0.5\ V_{REF}$. If the residue output exceeds this range, then it can be viewed as an *overrange error*. Thus, the output codes (000) and (101) never occur in the ideal case. However, a positive or negative comparator offset in the first stage introduces a positive or negative overrange error respectively. Such an error can be detected by looking for the output codes (000) and (101) in the digital output of the second stage, and can be corrected for by subtracting or adding 1 LSB (least significant bit) to the output of the first stage. However, this process involves two steps: (a) detection of the nature of overrange error, and (b) addition or subtraction of 1 LSB depending on the nature of the error. It is desirable to reduce the number of steps involved in the digital error correction.

If we deliberately introduce an offset of $V_{REF}/6$ (0.5 LSB) in the residue transfer curve using the arrangement of Fig. 4, then the residue transfer curve is altered to that shown in Fig. 5. Now, there can only be a positive overrange error and the digital error correction scheme is simplified. The last comparator threshold is not needed as the overrange error can be easily detected even without it. Thus, it is discarded to yield the residue transfer curve of Fig. 6. Now, the MDAC resolves only m levels and has $m-1$ comparators. This is similar in appearance to the residue transfer curves of the well-known 1.5-bit/stage and 2.5-bit/stage MDACs.

The same technique can be applied to realize MDACs with any arbitrary number of levels and 1 bit of digital redundancy.

III. Digital Error Correction

The digital error correction schemes for conventional MDAC implementations are based on an overlap-and-add technique for combining the digital output code from different pipeline stages. This scheme merely accounts for the SHA gain G while adding the bit outputs from successive stages of the pipeline. This suggests that digital error correction might be realized similarly for any arbitrary number of MDAC levels. For the example of a 5-level MDAC, digital error correction is easily realized if the final digital output B_{ADC} of a k-stage pipeline is computed as:

$$B_{ADC} = 3^{k-1}D_1 + 3^{k-2}D_2 + \ldots + D_k \quad (2)$$

Thus, the digital error correction simply accounts for the SHA gain G in the digital domain, which seems reasonable. Eqns. (3)-(5) illustrate the manner in which the bit outputs are combined in 3 different 2-stage pipelined ADCs. Eqn. (3) shows the case $m = 4$ and no digital redundancy, (4) represents the case $m = 3$ and 1 bit of redundancy while (5) represents correction for $m = 5$ and 1 bit of digital redundancy.

$$
\begin{aligned}
&D_1 = 01, D_2 = 10;\ Output = 4.D_1 + D_2 \\
&\quad\ \ 0\ \ 1 \\
&+\ \ \underline{1\ \ 0\ \ \ \ } \\
&\quad 0\ \ 1\ \ 1\ \ 0\ \ ;\ Case(i) - m = 4, no\ redundancy
\end{aligned}
\quad (3)
$$

$$
\begin{aligned}
&D_1 = 01, D_2 = 10;\ Output = 2.D_1 + D_2 \\
&\quad\ \ 0\ \ 1 \\
&+\ \ \underline{1\ \ 0\ \ } \\
&\quad 1\ \ 0\ \ 0\ \ ;\ Case(ii) - m = 3, 1-bit\ redundancy
\end{aligned}
\quad (4)
$$

$$
\begin{aligned}
&D_1 = 011, D_2 = 100;\ Output = 3.D_1 + D_2 \\
&\quad\ \ 0\ \ 1\ \ 1 \\
&\quad\ \ 0\ \ 1\ \ 1 \\
&+\ \ \underline{1\ \ 0\ \ 0\ \ } \\
&\quad 1\ \ 1\ \ 0\ \ 1\ ;\ Case(iii) - m = 5, 1-bit\ redundancy
\end{aligned}
\quad (5)
$$

The proposed scheme for digital error correction has been found to be effective in eliminating the effect of comparator offsets, and detailed discussions may be found in [4]. The validity of this scheme has been verified through behavioral simulations of pipelined ADCs using MATLAB©. An example result (Fig. 7) shows the simulated INL & DNL of a 10-bit ADC with the 6-level MDAC described in the previous sections, in the presence of random comparator offsets within the correction range. The nonlinearity is found to be within the requisite bounds, proving the validity of the proposed scheme for digital error correction.

IV. Implementation Issues

The proposed scheme requires very little change in terms of implementation. The only difference is in the digital error correction where extra adders are required for realization of the SHA gain G in the digital domain. In conventional implementations, G assumes values of the form 2^n and this can be realized easily by left-shifting the digital code from each stage by a certain number of bits. In the proposed scheme, G may assume arbitrary integer values, and these need to be decomposed into bit-shifts. However, the practical values of G required for most m-level MDACs require only 2 terms and hence the cost of the extra digital hardware is minimal. The simplest implementation of G may require use of subtraction instead of addition. For instance, $G = 7$ may be represented as $(4 + 2 + 1)$ or as $(8 - 1)$. The latter representation requires just 2 terms, although it requires the realization of a negative number. This can be done easily, especially with 2's complement code, which is often used for encoding the output of pipelined ADCs.

Another point of interest is the practical values of m used for the MDAC. The proposed scheme may be used to realize MDACs with an arbitrary integer number of levels. However, for even-valued m, the SHA gain G is not an integer, but may assume values such as 1.5, 2.5, 3.5 etc. Although this can still be implemented correctly at the cost of some more digital hardware, it is desirable to avoid this and use only odd-valued m for the MDAC. The reason can be found by observing the residue transfer curve for such a case. The problem can be explained with the residue transfer curve of Fig. 2. This has an even number of MDAC levels, and hence an odd number of comparator thresholds, the middle one of which lies at zero. Although, the A/D conversion is not affected much by this, the arrangement is not preferred if the digitized signal is to be converted back to the analog form later as the presence of a comparator decision level at zero can cause *en masse* switching when there is a transition from 011...1 to 100...0, which can be a potential source of distortion [1]. Thus, it is desirable to avoid such an arrangement and hence it is preferable to use odd-valued m for the MDAC design.

Other than the slight increase in digital hardware, the proposed scheme does not require any special considerations or calibration when compared with conventional schemes.

V. CONCLUSION

A novel scheme was proposed for implementation of generic m-level MDACs with 1 bit of digital redundancy and error correction. The methodology for incorporating digital redundancy and error correction was presented with an example. The scheme was shown to blend seamlessly with conventional schemes, which are shown to be special cases of the proposed algorithm. The proposed scheme entails almost no design overhead. The scheme allows designers considerably greater freedom in choosing the number of MDAC levels in order to optimize the design tradeoffs. Thus, it makes it realistic to design ADCs which could not possibly be designed using existing techniques. As the design of cyclic ADCs is very similar to that of pipelined ADCs, the proposed scheme is also relevant to this category of ADCs.

ACKNOWLEDGMENT

This research was supported by the Catalyst Foundation, New York, NY.

REFERENCES

[1] D.W. Cline, "Noise, speed and power trade-offs in pipelined analog-to-digital converters," Ph.D. dissertation, Dept. Elect. Eng. & Comp. Sci., U.C. Berkeley, 1995.
[2] S.H. Lewis, "Video-rate analog-to-digital conversion using pipelined architectures," Ph.D. dissertation, Dept. Elect. Eng. & Comp. Sci., U.C. Berkeley, 1987.
[3] D.W. Cline, P.R. Gray, "A power optimized 13-b 5 Msamples/s pipelined analog-to-digital converter in 1.2 μm CMOS," *IEEE J. Solid-State Circuits*, vol. 35, no. 3, pp. 294-303, Mar. 1996.
[4] V. Sharma, "Generalized radix design techniques for low-power, low-voltage pipelined & cyclic ADCs," M.S. thesis, Dept. Elect. Eng. & Comp. Sci., Oregon State Univ., Corvallis, 2004.
[5] S.H. Lewis, "Optimizing the stage resolution in pipelined, multistage, analog-to-digital converters for video-rate applications," *IEEE Trans. Circuits & Systems II*, vol. 39, no. 8, pp. 516-523, Aug. 1992.
[6] K. Nagaraj *et al.*, "A 250-mW, 8-b, 52-Msamples/s parallel pipelined A/D Converter with Reduced Number of Amplifiers," *IEEE J. Solid-State Circuits*, vol. 32, no. 3, pp. 312-320, Mar. 1997.
[7] S.H. Lewis *et al.*, "A 10-b 20-Msample/s Analog-to-Digital Converter," *IEEE J. Solid-State Circuits*, vol. 27, no. 3, pp. 351-358, Mar. 1992.

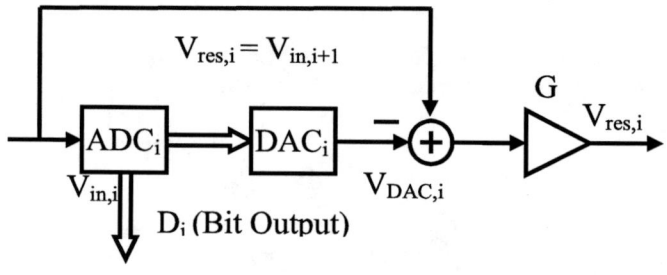

Fig. 1 Simplified representation of a pipeline stage (Stage i).

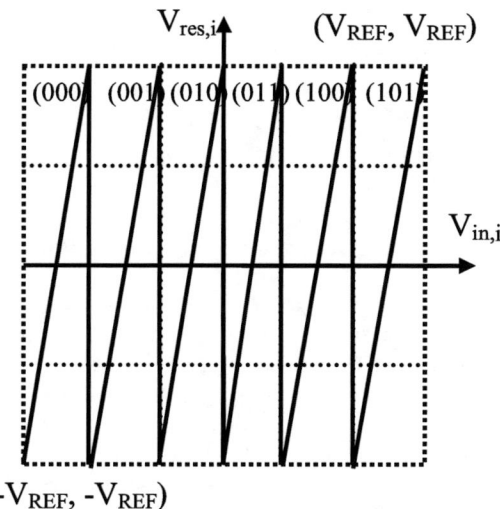

Fig. 2 Residue transfer curve of a 6-level MDAC with no digital redundancy.

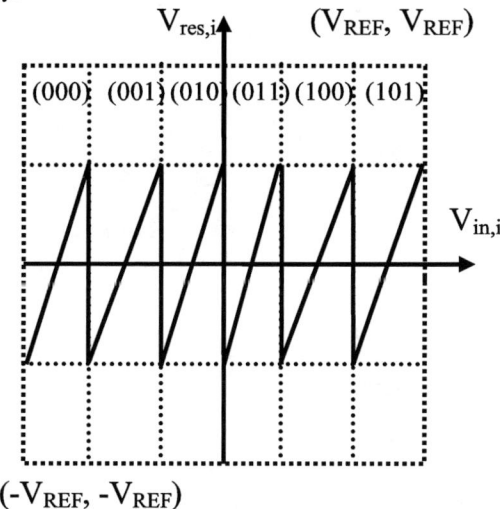

Fig. 3 Residue transfer curve of a 6-level MDAC with 1 bit of digital redundancy.

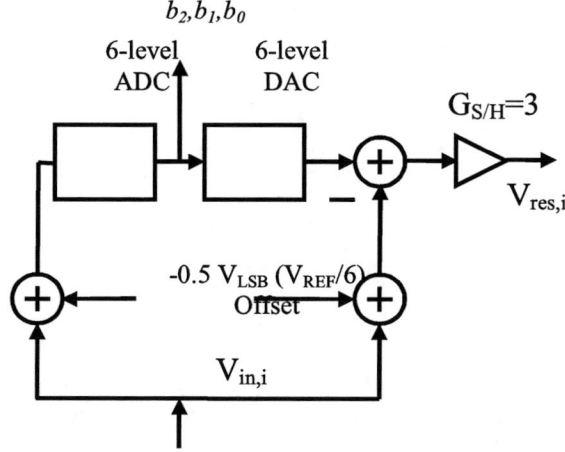

Fig. 4 Modification to the 6-level MDAC to simplify digital error correction.

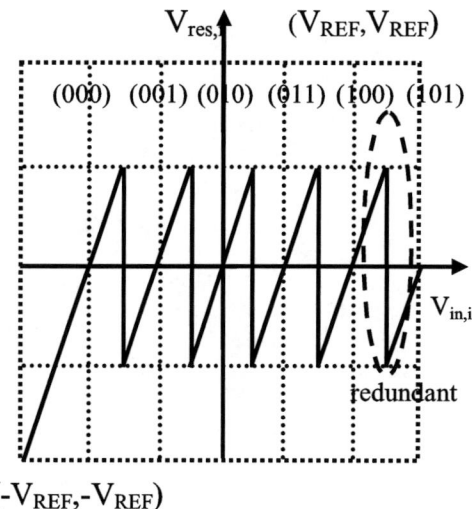

Fig. 5 Modified residue transfer characteristic with redundant comparator threshold (circled).

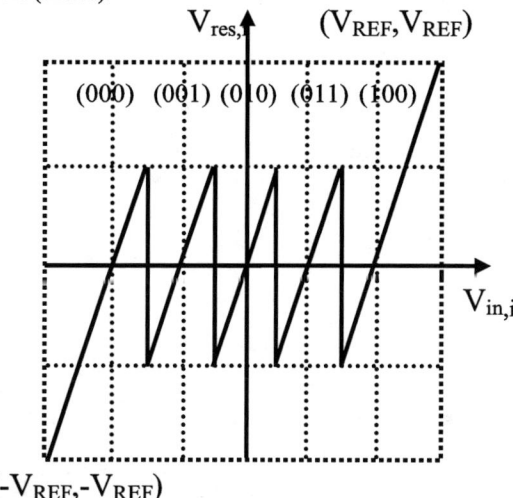

Fig. 6 Final residue transfer curve of a 5-level MDAC with 1 bit of digital redundancy.

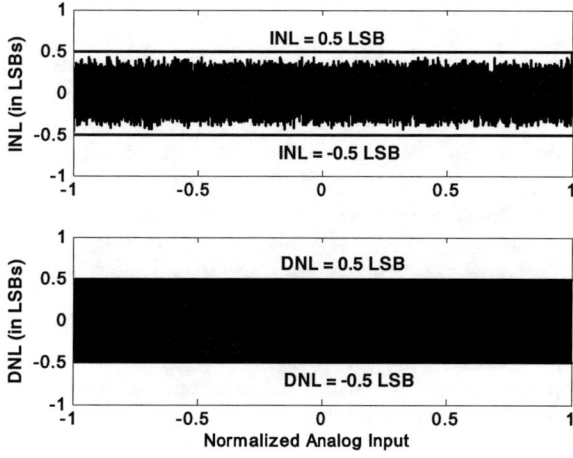

Fig. 7 Simulated INL & DNL of a 10-bit ADC with a 6-level MDAC.

A 10-Bit Algorithmic A/D Converter for Cytosensor Application

Thirumalai Rengachari, Vivek Sharma, Gabor C. Temes, Un-Ku Moon
School of Electrical Engineering and Computer Science, Oregon State University, USA

Abstract

A novel 10-bit algorithmic A/D converter for cytosensor applications is described in this paper. The converter is capable of a conversion rate of 1.5-bits/phase. It has advantages compared with conventional architectures with respect to nonideal effects.

I. Introduction

Analog-to-digital converters have been used in many sensor and instrumentation applications to analyse and store data. There is considerable interest in developing a low-cost, portable and generic cytosensor capable of serving as an ecological canary in various applications in order to provide an early warning regarding the presence of life-threatening agents in the environment. Previous research done on this rely on monitoring various responses of living organisms[1]. Most of the existing state-of-the-art sensors are too selective and require extensive use of pattern recognition and skilled human observers that reduce the efficiency. A simple and robust cytosensor system using a novel A/D converter was developed. The embedded converter is described in this paper. The block diagram of the sensor system is shown in Fig.1.

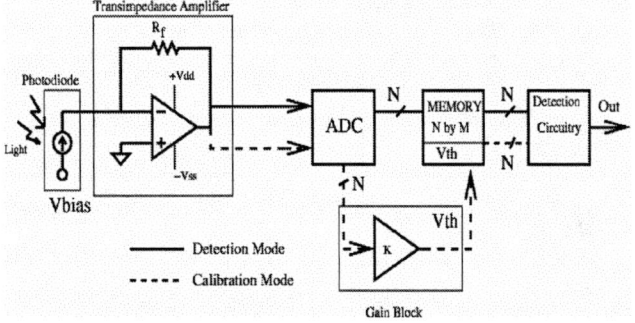

Figure 1. Sensor block diagram

The aggregation or the dispersion behaviour (which controls the intensity of light falling on the photodiode) of fish chromatophores (*Betta Splendens*-siamese fighting fish) is converted into current using a photodiode. The transimpedance amplifier converts the current thus generated into a voltage that is digitized by the A/D converter and stored in memory. Since the cytosensor is designed to be a low-cost, portable sensor, the A/D converter should consume low power and occupy low area. Section II describes the choice of converter and some of the existing algorithmic converter architectures. Section III introduces the novel algorithmic A/D converter. Section IV describes non-ideal effects in the converter, and compares it with the conventional approach and Section V describes needed components and simulation result.

II. Choice of Converter Architecture

A. Choice of Converter

The signal from the photodiode and the transimpedance amplifier are slow and can be considered as DC signals. Its resolution was determined by experimentation. Some of the low and medium speed converters can be used for the design. Successive approximation converters need a 10-bit accurate DAC. Charge redistribution converters have binary weighted capacitances that make the area large. Also, load on the amplifier is high which leads to high power consumption. First-order incremental converters are very slow, and higher-order incremental converters need complex digital and analog circuitry. Algorithmic converters have the simplest circuitry and consume the least power among all the converters.

B. Existing Algorithmic Converters

Most conventional algorithmic converters[3][4][5] employ a 1-bit pipeline stage or a 1.5-bit pipeline stage as their core circuitry. The multiply-by-2 circuit in the architectures is shown in figure 2.

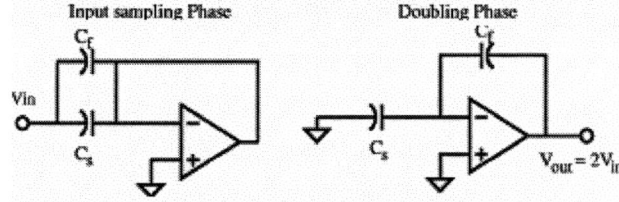

Figure 2. Multiply-by-2 circuit

The architecture in [2] is capable of giving 1.5-bits per phase. The operation of this architecture is as follows: during the reset phase (figure 3a), capacitors C_1 and C_2 are charged to the input voltage and the op-amp is connected in unity gain feedback (op-amp is in reset). Capacitors C_3 and C_4 are the load to the amplifier. The DAC voltage for the next conversion is also obtained during this phase using the 2 comparators.

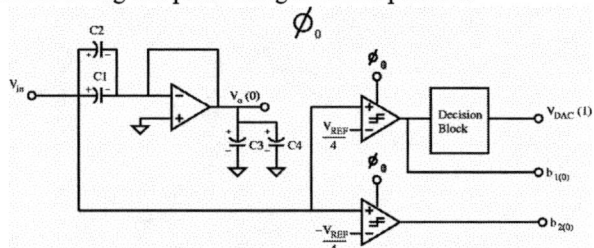

Figure 3a. Reset

During phase 1 (figure 3b), Capacitor C_2 is connected across the op-amp and C_1 discharges into C_2 to produce the desired residue, while C_3 and C_4 act as load capacitors and store the residual output. During phase 2 (figure 3c), C_1 and C_2 are connected as the load to the amplifier. C_4 is connected across the amplifier, and C_3 discharges into C_4 to produce the residue.

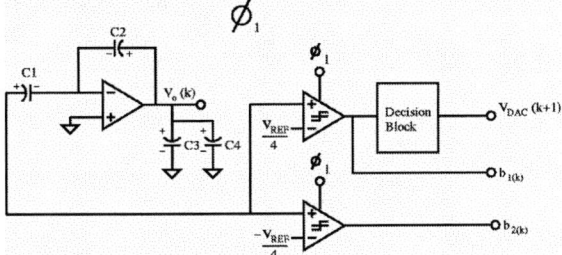

Figure 3b. Phase 1

Phases 1 and 2 are repeated until the required resolution is obtained.

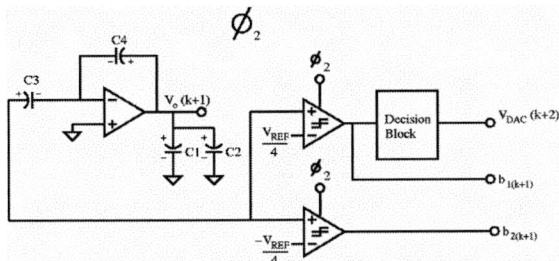

Figure 3c. Phase 2

III. A Novel Algorithmic Converter

The proposed new algorithmic converter contains 3 capacitors. The operation is explained below.

A. Reset Phase (Fig.4a):
Capacitors C_1 and C_2 are charged to the input voltage V_{in}. C_3 is the load capacitor to the amplifier which is being reset. Also during this phase, V_{in} is compared with references $\pm \frac{V_{REF}}{4}$ to give the MSB of the digital code.

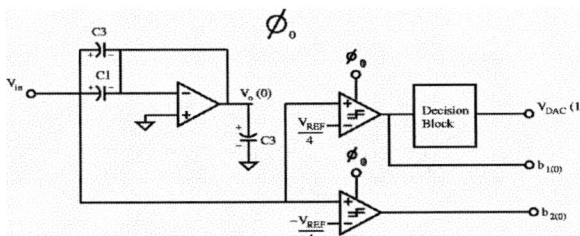

Figure 4a. Reset Phase

B. Phase 1 (Fig.4b):
Capacitor C_1 is connected as the feedback to the amplifier, and connecting one of capacitor C_2 to the DAC voltage does the reference subtraction. Hence C_2 discharges to the DAC voltage, and the residual output of the amplifier is given by

$$V_o(k+1) = 2 \cdot V_o(k) - V_{DAC}(k+1)$$

$$V_{DAC}(k+1) = V_{ref} \text{ for } V_o(k) < V_{ref}$$

$$V_{DAC}(k+1) = -V_{ref} \text{ for } V_o(k) > V_{ref}$$

C_3 and C_1 are charged to the residual voltage. The residual voltage is compared with the reference to obtain the next bit in the conversion and this decides the next DAC voltage to be subtracted from the next residue.

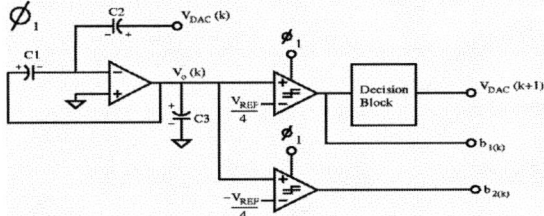

Figure 4b. Phase 1

C. Phase 2 (Fig.4c):

During this phase capacitors C_2 and C_3 are interchanged (C_1 and C_3 have the same potential stored on them), C_3 is connected to DAC voltage and the next residue is obtained. The next bit is obtained by comparing the residue to the reference.

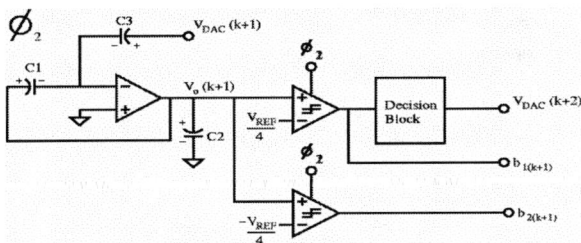

Figure 4c. Phase 2

The outputs of both comparators are used in the decision block to determine successive DAC voltages.

IV. Performance Limitations

A. Comparator Offsets
The use of a 1.5-bit per phase architecture removes the problem of comparator offsets. Offsets up to Vref/4 can be tolerated. The design of the comparator is easier and high-speed low power comparators can be used.

B. Amplifier Offsets
In the new architecture, the amplifier offsets affects the residues exponentially in each cycle. The residue equation is given as follows:

$$V_o(k) = 2^k V_{in} - 2^{k-1} V_{dac}(1) - 2^{k-2} V_{dac}(2) - \ldots - V_{dac}(k) + (2^{k-1} - 1) V_{os}$$

Since the input and the first residue are not affected by the offset, the first and the second MSBs of the converter output are offset free. The offset effect of the converter is illustrated in Figure 5. Redundancy error correction makes the offset of the amplifier give rise only to an effective offset in the system. The transfer curve of the ADC shifts left or right depending on the sign of the offset. Simulations show that the final offset of the system is twice the amplifier offset.

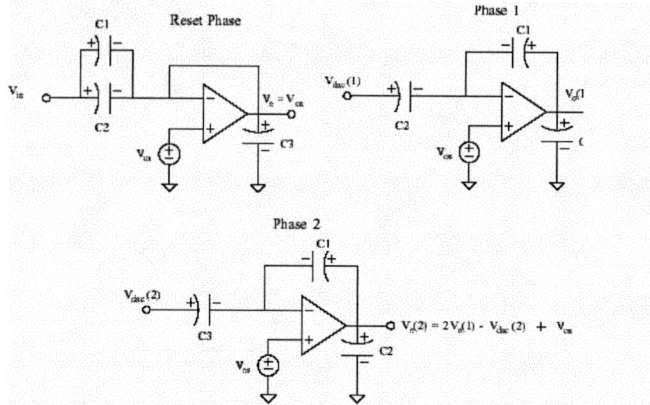

Figure 5 Effect of amplifier offset in the new converter

In the conventional architecture, the residues are affected by the offset effect as given by

$$V_o(k) = 2^k V_{in} - 2^{k-1} V_{dac}(1) - 2^{k-2} V_{dac}(2) - \ldots - V_{dac}(k) + 2(2^{k-1}-1)V_{os}$$

The final offset is twice that of the proposed new converter. With Vin = 0, the output of the converter represents the offset that can be stored and subtracted from subsequent output codes.

C. Finite Op-Amp Gain

Finite op-amp gain reduces the gain of the stage from 2 and thus causes INL and DNL. For the multiply-by-2 circuit shown in figure 4 the stage gain is given by

$$V_o = \frac{C_s + C_f}{C_f(1 + \frac{1}{A} + \frac{C_s}{C_f A})} V_{in} \approx 2(1 - \frac{2}{A}) V_{in}$$

In the proposed approach the residue with op-amp finite gain is given by

$$V_o(k) = \left[2 - \frac{1}{A_o}\left(3 + \frac{C_p}{C}\right)\right] V_o(n-1) - \left[1 - \frac{1}{A_o}\left(2 + \frac{C_p}{C}\right)\right] V_{dac}(n)$$

Cp is the parasitic capacitance at the virtual ground node of the op-amp. From simulations, an op-amp DC gain of 75 dB is required for the INL to be less than 0.5 LSBs. Simulation was performed with a parasitic capacitance of about 1.2pF at the virtual ground node of the op-amp ($C_1 = C_2 = C_3 = C = 2$ pF).

D. Capacitor Mismatch

Good layout techniques can give 0.1% mismatch, which corresponds to 10 bits of accuracy. For $C_2 = C_1(1+\alpha)$ and $C_3 = C_1(1+\beta)$, the residue equations during phase 1 and phase 2 are given respectively by

$$V_o(n) = V_{in} \cdot (2+\alpha)^{\frac{n+1}{2}} (2+\beta)^{\frac{n-1}{2}} - V_{dac}(1) \cdot (1+\alpha)(2+\alpha)^{\frac{n-1}{2}} (2+\beta)^{\frac{n-1}{2}}$$
$$- V_{dac}(2) \cdot (1+\beta)(2+\alpha)^{\frac{n-1}{2}} (2+\beta)^{\frac{n-3}{2}} \ldots - V_{dac}(n) \cdot (1+\alpha)$$

and

$$V_o(n) = V_{in} \cdot (2+\alpha)^{\frac{n-2}{2}} (2+\beta)^{\frac{n}{2}} - V_{dac}(1) \cdot (1+\alpha)(2+\alpha)^{\frac{n}{2}} (2+\beta)^{\frac{n-2}{2}}$$
$$- V_{dac}(2) \cdot (1+\beta)(2+\alpha)^{\frac{n-2}{2}} (2+\beta)^{\frac{n-2}{2}} \ldots - V_{dac}(n) \cdot (1+\beta)$$

The INL and DNL curves for 0.1% mismatch is shown in Figure 6. Mismatch > 0.1% makes the INL > 0.5 LSBs. MATLAB was used for simulations.

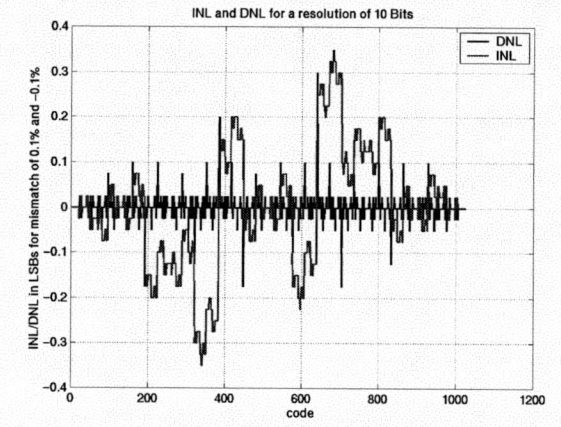

Figure 6 INL and DNL curves for a capacitor mismatch of 0.1%

E. Thermal Noise

The output referred kT-C noise of the multiply-by-2 circuit is given by

$$\frac{2kT}{C_2}\left(1 + \frac{C_1}{C_2}\right) \approx \frac{4kT}{C}$$

The signal gain is 2 and assuming C1 = C2, the input referred noise is given by kT/C, where C1 = C2 = C.

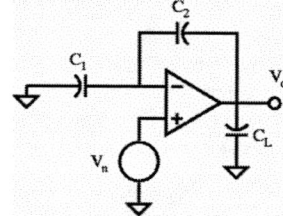

Figure 7. Op-amp in feedback configuration

For the feedback op-amp shown in Figure7, the input-referred noise is given by

$$\frac{4kT(1+N_f)}{3C_{L,eff}} \cdot \frac{1 + \frac{C_1 + C_p}{C_2}}{\left(\frac{C_1}{C_2}\right)^2}$$

Here,

$$C_{L,eff} = C_L + \frac{C_2(C_1 + C_p)}{C_1 + C_2 + C_p}$$

Nf refers to the contribution of the load transistors, and Cp represents the parasitic capacitance at the virtual ground node of the opamp. For $C_1 = C_2$, the total input referred noise of the multiply-by-2 circuit is given by

$$\overline{V_{n,stg}^2} = \frac{kT}{C} + \frac{8kT(1+N_f)}{3C_{L,eff}}$$

The total noise of an algorithmic stage using the above circuit is given by

$$V_{n,tot}^2 = V_{n,1}^2 + \frac{V_{n,2}^2}{4} + \frac{V_{n,3}^2}{16} + \ldots$$

where $V_{n,i}$ represents the noise from the ith cycle.

In the new converter, C1 is switched only in the reset phase and the noise in this phase is given by

$$\overline{V_{n,stg}^2} = \frac{kT}{C} + \frac{8kT(1+N_f)}{3C_{L,eff}}$$

In the subsequent phases C1 is never switched and hence does not contribute to the total noise. The noise in the subsequent cycles is given by

$$V_{n,i}^2 = \frac{kT}{2C} + \frac{8kT(1+N_f)}{3C_{L,eff}}$$

The total input referred noise is given by
$$\frac{7}{6} \cdot \frac{kT}{C} + \frac{4}{3} \cdot \frac{8kT(1+N_f)}{3C_{L,eff}}.$$

For a single stage op-amp, $C_{L,eff} = \frac{3C}{2}$, whereas for a 2 stage op-amp $C_{L,eff} = C_c$, where C_C is the compensation capacitor of the 2-stage op-amp.

F. Reduced Area and Power Consumption

Let C_{new} and C_{old} be the switching capacitors in the proposed and existing architectures respectively. Assuming single-stage op-amps for both the architectures, the respective load on the op-amps is ($3C_{new}/2$) and ($5C_{old}/2$). Assuming the input pair is the dominant source of noise in both the converters, the total input referred noise is given by $\frac{191kT}{54C_{new}}$ and $\frac{132kT}{45C_{old}}$ respectively. For equal SNR $\frac{C_{new}}{C_{old}} = 1.2$. Thus, there is a 10% reduction of capacitor area in the proposed architecture.

(1) For equal slew rate the current ratio is given by $\frac{I_{old}}{I_{new}} = \frac{5C_{old}}{3C_{new}} = 1.3$. The existing architecture requires 30% more current.

(2) For equal bandwidth, $\frac{gm_{old}}{gm_{new}} = \frac{5C_{old}}{3C_{new}}$. Since $gm \alpha \sqrt{I\frac{w}{l}}$,

$\frac{I_{old}(w/l)_{old}}{I_{new}(w/l)_{new}} = 1.69$. Considering equal slew rate, a 30 % bigger input transistors are required which increases the parasitic at the op-amp input. As mentioned earlier, virtual ground node parasitic reduces the interstage gain, which introduces non-linearity. Avoiding this will lead to a 69% higher current in the existing architecture.

Thus the proposed architecture has atleast 30% savings in power and 10% savings in capacitor area.

V. Circuit Components and Simulations

The proposed algorithmic converter requires an op-amp, 2 comparators, 3 capacitors and MOS switches. System level simulation was done using C programs and MATLAB. The following plot shows the FFT of the converter output simulated using MATLAB. The FFT plot is for an oversampling ratio of 64 and a mismatch of 0.2 % in the sampling capacitances. The obtained THD is -60.17 dB.

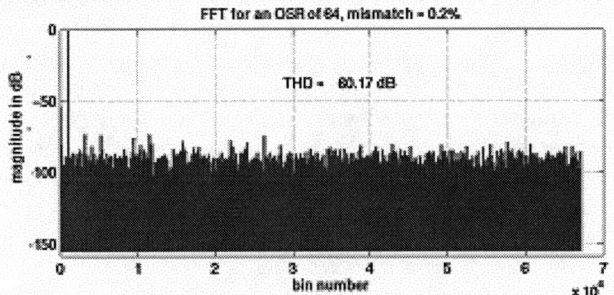

Figure 8. FFT of converter output

VI. Conclusions

An algorithmic converter with a novel switching scheme has been proposed. The conversion rate is 1.5-bits/phase. The non-ideal effects in the converter have been compared with existing architecture with the same conversion speed. The proposed converter was found to be economical in terms of both power and area.

VII. Acknowledgement

This research was funded by the Catalyst Foundation.

References

[1] EILATox-Oregon Workshop, September 9-13, 2002, Oregon State University.
[2] K. Nagaraj et al., " A 250 –mW, 8-b, 52-Msamples/s Parallel-Pipelined A/D Converter with Reduced Number of Amplifiers," IEEE J. of Solid-State Circuits, Vol. 32, No. 3, pp. 312-320, March 1997.
[3] Shang-Yuan et al.,"A Digitally Self-Calibrating 14-bit 10-MHz CMOS Pipelined A/D Converter," IEEE J. Solid-State Circuits, Vol 37, pp. 674-683, June 2002.
[4] Hae-Seung Lee, "A 12-bit 600 ks/s Digitally Self-Calibrated Pipelined Algorithmic ADC," IEEE J. Solid-State Circuits, Vol 29, No. 4, pp. 509-515, April 1994.
[5] Christian Enz et al., "Circuit techniques for Reducing the Effects of Op-Amp Imperfections: Autozeroing, Correlated Double Sampling, and Chopper Stabilization," Proceedings of the IEEE, Vol. 84, no. 11, pp. 1584-1614, November 1996
[6] Roubik Gregorian & Gabor C Temes, "Analog MOS Integrated Circuits for Signal Processing", Wiley-Interscience, 1986.

An Adaptive, Truly Background Calibration Method for High Speed Pipeline ADC Design

Degang Chen, Zhongjun Yu, Randy Geiger

Abstract: This paper presents a self-calibration method for designing high speed pipeline ADCs. Unlike all existing calibration algorithms, the proposed calibration does not insert any test signal or dithering signals to the pipeline signal path and it does not take any measurements at any internal nodes. It simply observes the ADC output digital codes during the normal operation of the ADC and extracts needed information about the ADC to generate the correction codes. This process is done adaptively and the correction codes are improved gradually as the ADC is being used for a longer time. Simulation results show that a 14-bit ADC with 7 bit original performance was gradually improved to close to 14 bit performance.

Index Terms— ADC Design, Background Calibration, Adaptive Calibration

I. INTRODUCTION

PIPELINE ADCs continue to be the architecture of choice for high speed and high resolution analog to digital conversion in communications, signal processing, and other demanding applications. To achieve moderate to high effective resolutions and high speed operation, calibration of one form or another or laser trimming is normally required. For example, 12 to 14 bit resolution may be achievable without calibration or trimming at sampling rate around a few MSPS. However, achieving even 10 – 12 bit effective resolution at 100 to 200 MSPS clock rates is a very challenging task without calibration or trimming. The higher clock rates necessarily require small capacitive loads and small parasitics which require small device sizes. The reduced device sizes inevitably lead to poor matching which limits the achievable effective resolution without calibration. On the other hand, the trend in the market place is clearly pushing towards very high-speed, high-volume, low-cost ADCs in both stand alone and embedded applications. The low cost requirement rules out laser trimming and it also makes on-chip fuses or ROM unattractive. Consequently, feasible candidates of calibration algorithms should require minimal area overhead and can perform real time background calibration.

Numerous techniques have been developed to improve the linearity of high-speed ADCs. Among them, the error averaging [1], reference feed forward [2], walking reference [3], and ratio-independent methods [4] are analog approaches, while calibration [5–8] and over sampling [9] are digital approaches. The analog approaches tend to be simpler, but the digital approaches are more flexible. Most Nyquist-rate high-resolution ADC works are based on variations of the pipelined architecture. However, all existing calibration techniques relies on applying certain stimulus signals to the ADC, measuring the ADC's output response, and comparing the response to its expected counter part to generate calibration codes. In this paper, we introduce a novel calibration approach for building high speed pipeline ADCs. The novelty is embodied in the following features: 1) it uses no stimulus input signals, 2) it take no internal measurements, 3) it never interferes with the ADC's normal operation, 4) its calibration accuracy gets better as the ADC is being used for longer time.

Due to space limitations, we won't be able to completely describe the calibration algorithm and the ADC design procedure in general terms. In the next section we will illustrate the basic principles by walking through a conceptual ADC design and calibration example. In section 3, we will present simulation results to demonstrate how the proposed calibration approach adaptively improves the ADC performance as the ADC is being used.

II. ADAPTIVE BACKGROUND CALIBRATION

The following steps illustrate a practical implementation of a truly background, self adapting, self-calibration method for high speed ADC design with moderate resolution. If parasitic nonlinearities are small, the major source of nonlinearity in the transfer characteristics of a pipelined ADC is attributable to incorrect interpretation of the digital output codes. This is caused by gain errors in the inter-stage amplifiers as well as by offset errors and DAC errors. These errors are all completely correctable if over-range protection is provided. If sufficient over-range protection is provided, these errors cause discontinuities in the output codes provided by the ADC. The ADC will be calibrated if these discontinuities can be removed.

We now use the following example to illustrate an efficient method for eliminating these discontinuities in the background without requiring training sequences or elaborate calibration hardware. For simplicity, let us consider a 10 bit pipeline ADC as an example.

First we design the pipeline with achieving the highest clock rate as the dominant focus, with minimal regard to matching accuracy. Suppose the process can provide 7 bit matching accuracy for small devices. The first 3 bits (MSB) of the ADC are pipelined with 1 comparator per stage. Size the nominal values of the capacitor ratios so that nominally the 3rd stage has 1 missing code when it transitions from 0 to 1, the 2nd stage has 2 missing codes, and the 1st stage has 4 missing codes. This gives 7 discontinuities in the ADC transfer curve. These discontinuities are created by making the inter-stage amplifier gains intentionally a little less than 2 to provide over-range protection.

Use 8 RAM cells to store the error correction codes for the 8 continuous segments of the transfer curve. Each cell is 4 bits wide. It is increased beyond the nominal 3 bits to allow for compensation of process variations. The 3 MSBs of the ADC raw code will be used to address the 8 cells.

At initial power up, the 8 cells will be set to equal to the nominal values of the 8 correction codes based on the 7 nominal discontinuities in the transfer curve. For the 7 gap sizes mentioned above, the 8 initial correction codes would be +6, +5, +3, +2, -2, -3, -5, -6 respectively.

During ADC operation, the 3 MSB of the ADC raw code

are used to fetch the corresponding correction code which is then added to the raw code to form the corrected code as the ADC output code. For the above example, this process nominally shifts the first 1/8 segment of the DC transfer curve up by 6 LSB, the second 1/8 segment up by 5 LSB, the third 1/8 segment by 3 LSB, and so on. This will make the nominal transfer curve continuous.

Seven pairs of "compare and store" circuits will be provided, one pair for each of the 7 expected gaps in the actual transfer curve. For example, the first gap happens when the ADC code transitions from 000xxxxxxx to 001xxxxxxx. In each clock cycle, the first "compare and store" circuit compares the ADC raw code (if it is of the form 000xxxxxxx) against its stored value and updates the stored value to be the larger of the two. Hence, at any time point, the first "compare and store" circuit is holding the largest observed code in the first 1/8 (of the form 000xxxxxxx) of the transfer curve. Similarly, the second "compare and store" circuit is holding the smallest observed code in the second 1/8 (of the form 001xxxxxxx) of the transfer curve, the third is holding the largest observed code in the second 1/8 (of the form 001xxxxxxx) of the transfer curve, the fourth holding the smallest observed code in the third 1/8 (of the form 010xxxxxxx) of the transfer curve, and so on. These "compare and store" circuits are essentially determining the number of missing codes at the corresponding major transition points. This information will later be used to remove the missing codes.

After a significant number of clock cycles (say one million), it is assumed that the actual largest and smallest codes in every 1/8 segment of the transfer curve will have all been hit. This should be a reasonable assumption if the ADC is used in a communications or signal processing circuit in which the ADC input is an AC signal ranging at least +-3/4 of the ADC input range. Therefore, the "compare and store" circuits are holding the true largest and smallest codes of each 1/8 segment of the transfer curve, and the differences (for example, the smallest code in 001xxxxxxx minus the largest code in 000xxxxxxx, and the smallest code in 010xxxxxxx minus the largest code in 001xxxxxxx, etc) are the true gap sizes in the actual transfer curve of the ADC in operation.

The measured true gap sizes will be used to adaptively update the correction codes. For example, a simple adaptation law could be of the form: (new code) = (old code)*(1-lamda) + (observation based code)*lamda, where (old code) is the correction code that is currently in effect, (new code) is the updated correction code to be used starting now, the (observation based code) is the code computed based on the observed gap sizes, and lamda (selected by the designer) is < 1 but > 0 that controls the adaptation speed.

A simple binary counter can be used to decide when a new update is to be performed. For example, if an update of every one million clock cycles is desired, then a 20 bit counter can be used. The overflow signal can act as the trigger for code update and the over flow also resets the counter to zero.

In the actual design, the nominal gap sizes and the corresponding nominal correction codes may be different from the ones given as examples above. These nominal values for obtaining a desired yield can be obtained from circuit level simulation after the amplifiers and capacitors have been designed.

If only 6 bit matching (rather than the 7 bits used in the above example) can be guaranteed, then the first 4 bits should have over-range protection with nominal gains less than two. Then there will be 15 nominal gaps in the transfer curve, dividing the curve into 16 segments. Then, 16 correction codes will be used, one code for each of the segments. Alternatively, if more than 10 bit resolution is needed, the number of gaps and the number of segments will be similarly increased. If the comparator offsets are not significant so that each segment is of approximately the same length, then the algorithm can be modified so that only one gap size associated with each comparator needs to be determined. This will significantly reduce the number of "compare and store" circuits and simplify the calibration logic circuit.

This is a truly background calibration and the ADC operation is never interrupted. This is an input-output based calibration, calibrating the true ADC signal path. There is no insertion of a test signal or measurement at any internal node. In contrast, most existing algorithms insert a test signal into the pipeline or take measurement at internal nodes. By doing so, these existing algorithms are actually calibrating an altered pipeline due to signal insertion and measurement. There is no need for a precision signal generator or pseudorandom signal generator since the suggested method is based on observing the actual operation of the ADC. The algorithm is easy to implement, requiring small hardware and software overhead.

III. SIMULATION RESULTS

For the simulation results presented here, the ADC has 14 bits of raw resolution. Its architecture contains 10 single-bit subradix-2 stages followed by a 4-bit flash stage. The flash stage has random errors in transition levels, but still has better than 4-bit linearity. Capacitor matching accuracy is at the 9 bit level. Comparator offsets are at the 7 bit level. Over range protection is set at 1%. Amplifier gains are 74 – 94 dBs. Amplifier linearity is at 11-bit linear or better.

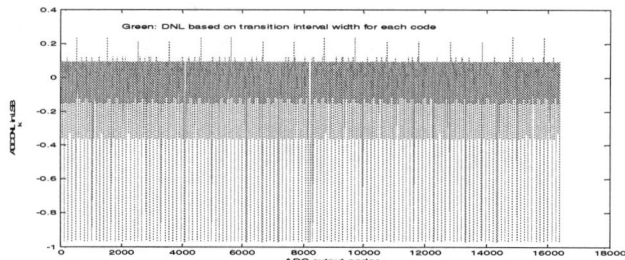

Figure 1 ADC DNL before calibration

As expected, Figure 1 shows that many codes have $DNL_k = -1$, which means code width = 0, which means that a code is actually missing. Hence, there are many missing codes. Groups of missing codes manifest as jump-ups in the ADC transfer curve and as vertical jump-downs in the INL curve as seen in Figure 2. These happen near the major transitions as expected. Also notice that there are no jump-ups in the INL curve, which means that there are no jump-downs in the ADC transfer curve. Hence, the ADC has monotonic transfer characteristics. This is the benefit of using subradix-2 gain stages for over range protection. The vertical jump-ups or gaps in the ADC transfer curve do not cause information loss.

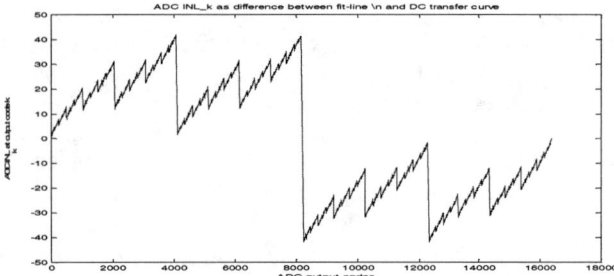

Figure 2 ADC INL before calibration

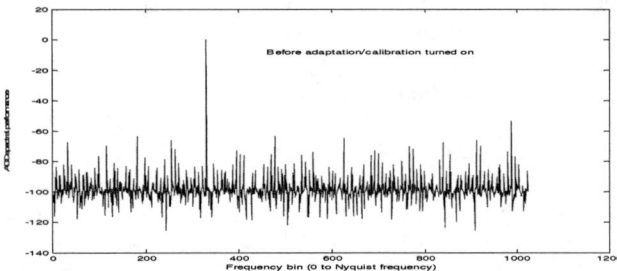

Figure 3 ADC spectral performance before calibration

Because of the 1% over range protection and because of the 9-bit level capacitor matching, the ADC before calibration has only about 7 bit INL performance or the transfer curve is only about 7 bit linear. We would expect the total harmonic distortion of the ADC to be somewhere near the 7 bit level or at the − 44 dB level. Figure 3 indicates that the ADC has about 54 dB SFDR performance, which is consistent with what we would expect based on component matching accuracy.

Then the self-adaptive self calibration capability was turned on. No testing signal or internal measurement is used for the generation of the calibration code. The self-adaptive self calibration algorithm simply watches and analyzes the ADC output codes while the ADC is being used for regular conversion. To reduce the waiting time, the algorithm in this simulation is set to update the adaptive calibration codes every 2048 clock cycles. In real implementation, the update time can be set to be much longer to better average out noise effect. In this simulation, a sine wave is input to the ADC, which is unknown to the algorithm. Only the raw ADC output codes are available to the algorithm for analysis.

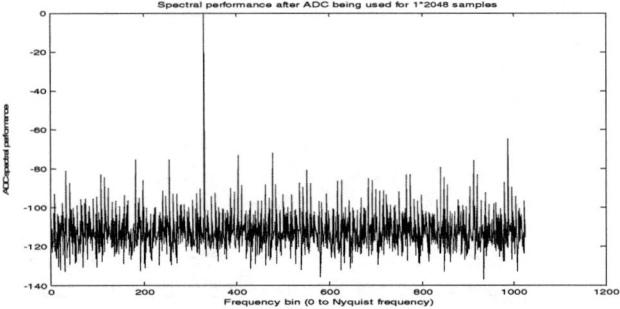

Figure 4 ADC spectral performance after 2048 samples

After the ADC is being used for 2048 clock cycles and the calibration code is updated for the first time, the calibration codes are frozen and the ADC's spectral performance is tested again as shown in Figure 4. Notice that the ADC SFDR performance has been improve to about 64 dB. That is a 10 dB improvement in SFDR after 2048 clock cycles of normal ADC use.

Then the adaptive self-calibration algorithm is turned back on and the ADC is set to resume its normal conversion.

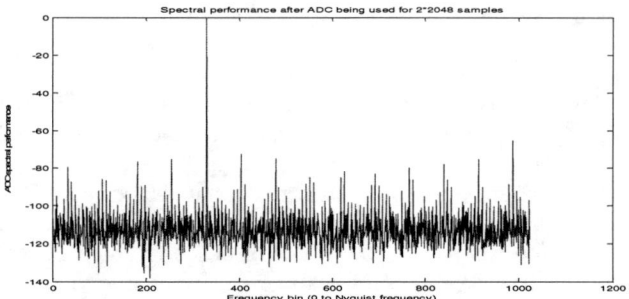

Figure 5 ADC spectral performance after 4096 samples

After another 2048 clock cycles, the calibration codes are updated for the second time. Then the calibration codes are frozen again and the ADC's spectral performance is tested again as shown in Figure 5. Notice that the ADC SFDR performance is still at about 64 dB. Hence, between the first and second updates, the ADC spectral performance does not exhibit any improvements, even though the actual calibration codes may have been changed. This should not be surprising since a 2048 point FFT does not hit all ADC codes.

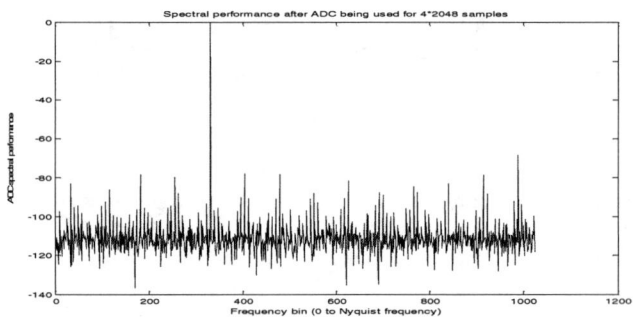

Figure 6 ADC spectral performance after 8196 samples

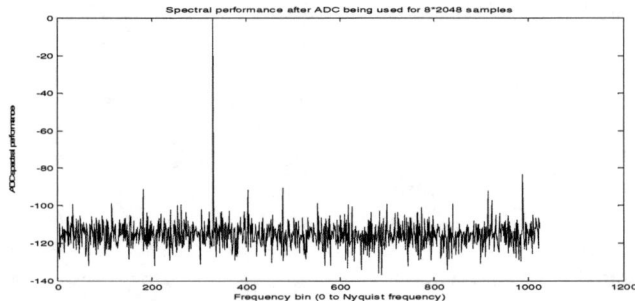

Figure 7 ADC spectral performance after 8*2048 samples

After another 4096 clock cycles and after the calibration codes are updated for the forth time, the calibration codes are frozen again and the ADC's spectral performance is tested again, as shown in Figure 6. Notice that the ADC SFDR performance is now at about 68 dB and the other harmonic components are also reduced to lower levels. Therefore the adaptation is seen to be working.

Another 4*2048 clock cycles later, the calibration codes are updated for the eighth time. The calibration codes are frozen again and the ADC's spectral performance is tested again. Figure 7 shows that the ADC SFDR performance is now at about 83 dB and the other harmonic components are also reduced to significantly lower levels. This further confirms that the self-adapting self-calibration algorithm is in deed working effectively.

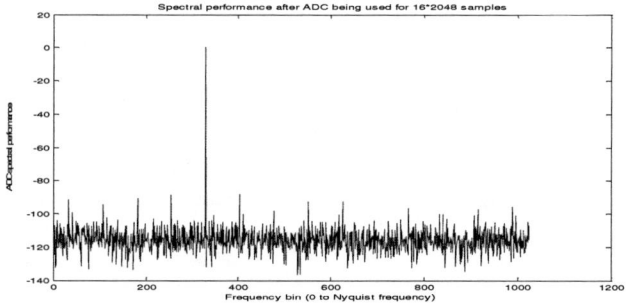

Figure 8 ADC spectral performance after 16*2048 samples

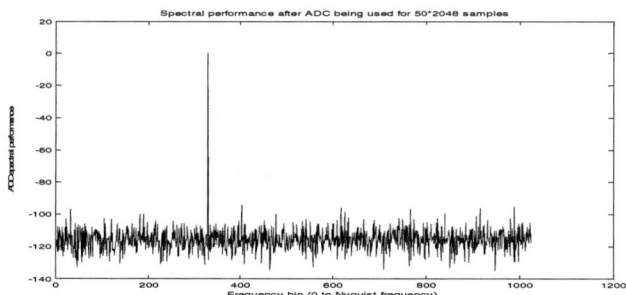

Figure 9 ADC spectral performance after 50*2048 samples

At the end after 16*2048 and 50*2048 clock cycles of regular ADC normal conversion, the calibration codes are updated for the 16th and 50th times respectively. The calibration codes are frozen at those times and the ADC's spectral performance is tested again. Figures 8 and 9 show that the ADC SFDR performance has been improved to about 88 dB and 94 dB respectively. The other harmonic distortion components are also reduced to significantly lower levels.

In fact, after about 30*2048 clock cycles of ADC conversion, the ADC linearity performance has been improved to a level that is similar to a 14-bit linear ADC. Further adaptation is in-effective in further improving the ADC linearity performance. However, if the ADC has aged, or the operating environment has changed, or something else has caused the ADC to change, we would expect the self-adaptation mechanism to kick in and adaptively converge to the correct calibration codes.

Figures 10 and 11 show the DNL and INL performance after the ADC has been calibrated. Notice that there are no codes with DNL=$-$1 and therefore no missing codes.

IV. CONCLUSION

In this paper, we have introduced a novel calibration method for high speed pipeline ADC design. The calibration algorithm is a truly background self-calibration algorithm. It distinguishes itself from all existing algorithms by not needing any testing signal or internal measurements and by having the capability of gradually improving its own linearity performance as the ADC is being used for longer time. Simulation results demonstrated that DNL, INL, as well as spectral performance can be improved 7 bit level to 14 bit level. Both hardware and software overhead is relatively small. Simple implementation schemes have been illustrated.

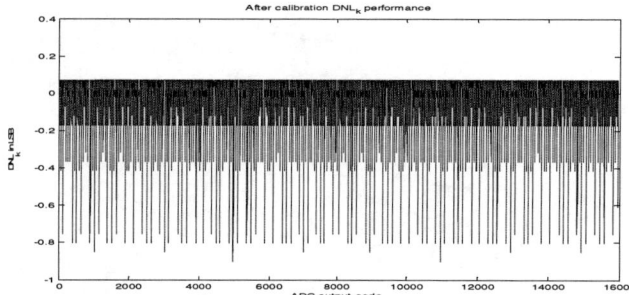

Figure 10 ADC DNL after sufficiently long use

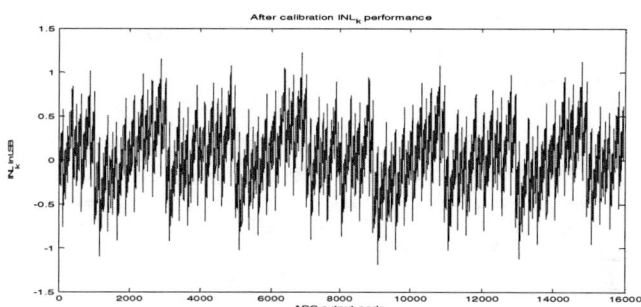

Figure 11 ADC INL after sufficiently long use

REFERENCES

[1] B.-S. Song, M. Tompsett, and K. Lakshmikumar, "A 12-bit 1-MSample/s capacitor error-averaging pipelined A/D converter," *IEEE J. Solid-State Circuits*, vol. 23, pp. 1324–1333, Dec. 1988.

[2] S. Sutarja and P. R. Gray, "A pipelined 13-bit 250-ks/s 5-V analog-to digital converter," *IEEE J. Solid-State Circuits*, vol. 23, pp. 1316–1323, Dec. 1988.

[3] D. A. Kerth, N. S. Sooch, and E. J. Swanson, "A 12-bit 1-MHz two-step flash ADC," *IEEE J. Solid-State Circuits*, vol. 24, pp. 250–255, Apr. 1989.

[4] J. Wu, B. Jeung, and S. Sutarja, "A mismatch independent DNL-pipelined analog to digital converter," in *Proc. IEEE Int. Symp. Circuits and Systems*, vol. 5, 1994, pp. 461–464.

[5] T.-H. Shu, B.-S. Song, and K. Bacrania, "A 13-b 10-MSample/s ADC digitally calibrated with over sampling delta–sigma converter," *IEEE J. Solid-State Circuits*, vol. 30, pp. 443–452, Apr. 1995.

[6] S.-U. Kwak, B.-S. Song, and K. Bacrania, "A 15-b 5-MSample/s low-spurious CMOS ADC," *IEEE J. Solid-State Circuits*, vol. 32, pp. 1866–1875, Dec. 1997.

[7] M.-J. Choe, B.-S. Song, and K. Bacrania, "A 13-b 40-MSample/sCMOS pipelined folding ADC with background offset trimming," in *ISSCC Dig. Tech. Papers*, 2000, pp. 36–37.

[8] C. Moreland, M. Elliott, F. Murden, J. Young, M. Hensley, and R. Stop, "A 14-b 100-MSample/s 3-stage A/D converter," in *ISSCC Dig. Tech. Papers*, 2000, pp. 34–35.

[9] S. A. Paul, H.-S. Lee, J. Goodrich, T. F. Alailima, and D. D. Santiago, "A Nyquist-rate pipelined over sampling A/D converter," *IEEE J. Solid-State Circuits*, vol. 34, pp. 1777–1787, Dec. 1999.

The Realization of a Mismatch-free and 1.5-bit Over-sampling Pipelined ADC

Shigeto Tanaka
Depertment of E.E.C.E.
Chuo University
Tokyo, Japan 112-8551
Email: tanaka@sugi.elect.chuo-u.ac.jp

Yuji Ghoda
Graduate School of E.E.C.E.
Chuo University
Tokyo, Japan 112-8551
Email: goda@sugi.elect.chuo-u.ac.jp

Yasuhiro Sugimoto
Depertment of E.E.C.E.
Chuo University
Tokyo, Japan 112-8551
Email: sugimoto@sugi.elect.chuo-u.ac.jp

Abstract—This paper proposes a simple method to realize an over-sampling pipelined analog-to-digital converter (ADC) with 1.5-bit bit-blocks. The ADC performs conversion by permuting internal capacitors in alternate clocks of the upper 1.5-bit bit-blocks in the analog domain, then averaging the data from bit-blocks in the digital domain. The behavioral simulation of a 14-bit ADC verified that this over-sampling pipelined ADC with 1.5-bit bit-blocks has more than 70 dB of spurious-free dynamic range (SFDR) for up to an 8 MHz input signal when each of the upper three bit-blocks has a gain error of $+0.8$ %. Using a S/H circuit in front improves the SFDR to 95 dB up to the signal frequency bandwidth of 25.6 MHz when the clock frequency is 102.4 MHz.

I. INTRODUCTION

Although the pipelined ADC is very popular, its accuracy is generally limited to 10- to 12-bit equivalents, due to device mismatches such as capacitors that cause gain and offset errors [1].

Various ways to eliminate these errors have been demonstrated. Performing a calibration, either in the fore-ground [2] or in the back-ground, [3] is very effective. However, the calibration method requires precise error measurements, additional hardware and complicated algorithms.

Among the methods, those that are most promising include the over-sampling method [4] and the capacitor averaging method [5],[6]. The over-sampling method inverts the input signal and multiplies the output digital code by -1 at one of the alternate clocks. This rearranges the errors due to device mismatch in the opposite direction of the frequency region compared to the errors produced in the normal operation. However, this method's application are limited to a pipelined ADC with a 1-bit bit-block, even though most pipelined ADC adopts a 1.5-bit architecture. The capacitor averaging method is considered one of the derivatives of the over-sampling method. It reduces the capacitor mismatch error by taking the average of the size equivalence of the positive and negative errors that are produced by permuting capacitors in an error amplifier. However, until now the averaging has been done in the analog domain. It would be natural for the ADC to produce digital data with positive and negative errors alternately, so that the averaging could be done in digital domain.

The purpose of this paper, therefore, is to offer a simple and effective over-sampling method for a pipelined ADC with 1.5-bit bit-blocks, with an averaging method performed in the digital domain.

II. AN OVER-SAMPLING PIPELINED ADC WITH 1.5-BIT BIT-BLOCKS

In the over-sampling pipelined ADC reported in reference 4, the error produced in the bit-block is an even function around the center of the input-versus-output voltage relationship. However, in a pipelined ADC with 1.5-bit architecture, the error becomes an odd function. Figure 1 shows transfer curves between the input signal voltage and the output signal voltage of a 1.5-bit bit-block.

In Figure 1, the input and output voltage values have the same range and are normalized to be ± 1. When the input voltage exceeds $+1/4$, digital code "1" is generated, and digital code "0" is generated when the input voltage is less than $-1/4$. When the input voltage is between $-1/4$ to $+1/4$, the digital code is not fixed in the current bit-block, and the decision is postponed until the digital codes of the succeeding bit-blocks are fixed. A solid line in the figure represents the ideal transfer curve. The chain line shows the one with the positive gain error caused by the capacitor mismatch, and the dotted line with the negative gain error. It is evident that the difference between the ideal transfer curve and the error curve becomes the odd function around the center. It is also seen in Figure 1 that the gain error is cancelled when both curves with positive and negative errors are added because they are

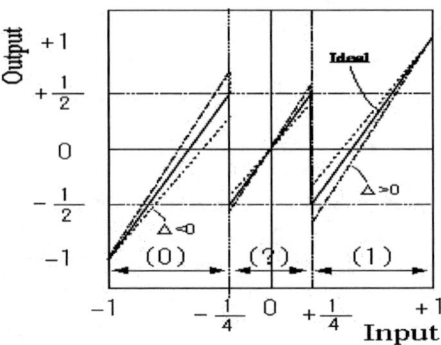

Fig. 1. The input vs. output voltage relationship of a 1.5-bit bit-block.

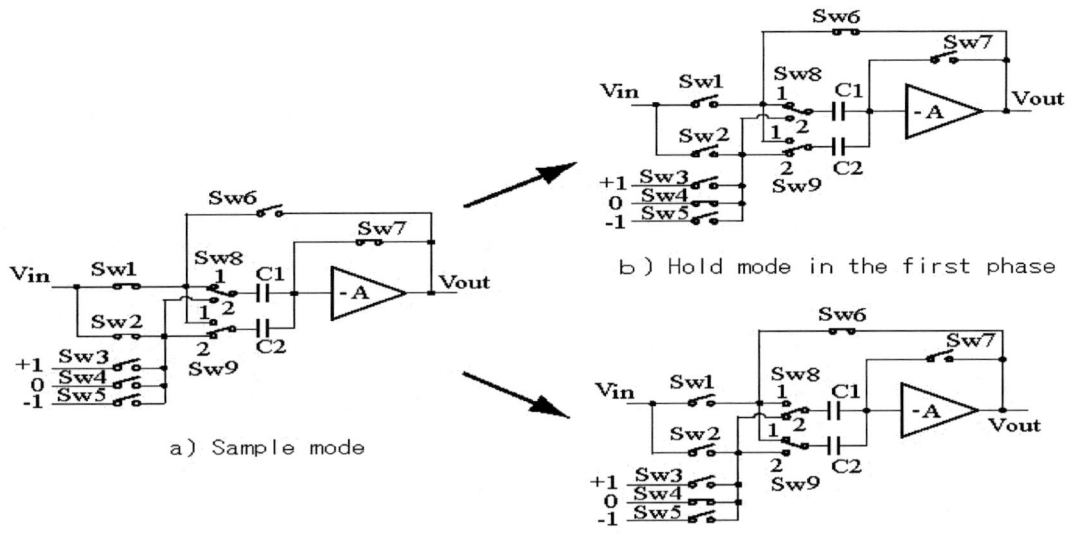

Fig. 2. The mismatch error reduction scheme.

of equal amounts but in different polarities.

This observation leads us to form a 1.5-bit bit-block in which the permutation of capacitors is performed at alternate sampling clocks as shown in Figure 2 as a circuit of an MOS pipelined ADC. The circuit resembles that of the capacitor averaging method in reference 5; in our case, however, the averaging is performed in the digital domain, and the control sequence to produce the digital code from the bit-block is different. The circuits consist of a sample-and-hold circuit, a sub-DAC, and a residue amplifier. Figure 2(a) shows the circuit in sample mode, with the input signal stored in capacitors C1 and C2. We assume a capacitor mismatch between C1 and C2, that is, $C_1 = C$, and $C_2 = C(1+\Delta)$, where Δ is the relative mismatch error. Because the charge stored in the capacitors is equal in both the sample and hold modes, equation (1) is derived for the circuit shown in Figure 2(b). V_{REF} in equation (1) becomes +1, 0, or −1, depending on which switch (SW3, SW4, or SW5) turns on.

$$V_{out} = \left(1 + \frac{C_2}{C_1}\right) V_{in} - \frac{C_2}{C_1} V_{REF}$$
$$= \{1 + (1+\Delta)\} V_{in} - \{(1+\Delta), 0, or - (1+\Delta)\} \quad (1)$$

Figure 2(c) has a configuration such that the feedback capacitor of the amplifier is C2 instead of C1, in contrast to that shown in Figure 2(b). In this case, the V_{out}, with one of the three switches Sw3, Sw4, and Sw5 on and with other switches in the positions shown in Figure 2(c), becomes

$$V_{out} = \left(1 + \frac{1}{1+\Delta}\right) V_{in} - \left\{\frac{1}{1+\Delta}, 0, or \frac{-1}{1+\Delta}\right\}$$
$$\approx \{1 + (1-\Delta)\} V_{in} - \{(1-\Delta), 0, or - (1-\Delta)\} \quad (2)$$

It is evident that the capacitor mismatch error Δ is cancelled out by summing equations (1) and (2). In the proposed ADC, the circuits shown in Figure 2(a) and 2(b) are used for the sample and hold mode operation in one clock period of alternate sampling clocks, while the circuits shown in Figure 2(a) and 2(c) are used for the sample and hold mode operation in the other clock period. Figure 2(a) is common to the sample mode operation in both clock periods. As a result, we obtain a digital output code from the bit-block at each of the sampling clocks, but the digital output code which is the A-D conversion results from either of the circuits shown in Figure 2(b) and 2(c) at alternate sampling clocks. The error obtained from the bit-block in digital form in one clock period has a polarity opposite to that of the other clock period, as long as the input signal frequency is low. It therefore becomes possible to eliminate the error at the output of the ADC in digital form.

The whole ADC is constructed in such a way as shown in Figure 3. The 1.5-bit bit-block consists of a sample-and-hold circuit (S/H), a sub-ADC, a sub-DAC, a subtraction circuit, and an amplifier producing a voltage gain of 2. As described previously, the capacitors in the bit-block are permutated in hold mode at alternate sampling clocks, and the digital code from a bit-block contains the positive and negative errors in turn. Therefore, the output of the digital correction logic shown in Figure 3 contains the positive and the negative errors in turn, as well. In order to eliminate the offset error and gain error that are included in the alternative digital outputs in different polarities, the averaging function using a 1-clock delay and an adder is introduced. The absolute value of the transfer function of the averaging function by using Z-transform is expressed as,

$$|H(j\omega)| = |1 + e^{-j\omega T}| = \sqrt{2\{1 + \cos(\omega T)\}} \quad (3)$$

where T is the time interval of the sampling clock periods. As

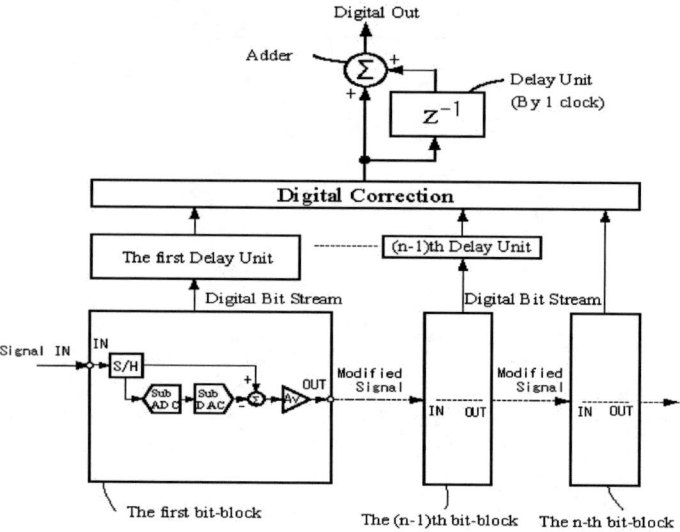

Fig. 3. Block diagram of the proposed ADC with 1.5-bit bit-blocks and an averaging function at the output.

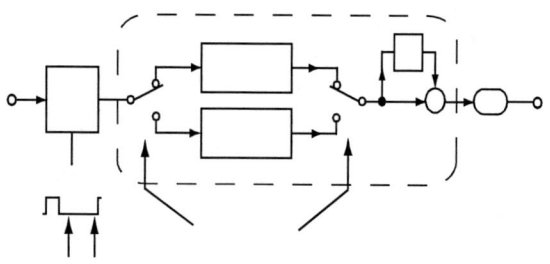

Fig. 4. The ADC with a sample-and-hold function in front of the ADC in Figure 3.

equation 3 becomes zero when

$$f = \frac{\omega T}{2\pi T} = \frac{(2n-1)}{2T} \quad (4)$$

where n is the integer, the frequency characteristics of the ADC in Figure 3 have the null at odd integer multiples of half the sampling clock frequency. This means that the offset error is completely eliminated because its frequency is half the sampling clock frequency, and that the gain error at low frequency is minimized also because its frequency is in the vicinity of half the sampling frequency.

III. THE OVER-SAMPLING PIPELINED ADC WITH AN INPUT SAMPLE-AND-HOLD

The proposed ADC produces output data by summing two consecutive conversion results from digital correction logic as shown in Figure 3. The configuration eliminates the constant offset error; however, the error caused by the gain error remains when the input signal level changes between consecutive sampling periods.

In order to solve the problem, a S/H circuit was placed in front of the over-sampling pipelined ADC, as shown in Figure 4. Half of a clock period is used for sampling the input signal, and the following one-and-a-half clock periods are used for the hold operation. Two clock periods are needed to complete the operation in this S/H circuit. The over-sampling pipelined ADC shown in Figure 3 performs A-D conversions with errors in opposite polarities at alternate clocks, and is considered to have two different ADCs (though in reality they comprise just one ADC) with a positive error and a negative error, which are combined at the output as shown in Figure 4. In the hold period of the sample-and-hold circuit, the ADC converts its input signal twice, first by ADC1 and then by ADC2. As the output data of ADC1 and ADC2 have errors that have been analyzed in equations (1) and (2), the errors of these two output data are cancelled out by taking their average, and the non-linearity disappears. In this configuration, shown in Figure 4, the sampling frequency and the signal frequency bandwidth become 1/2 and 1/4 of the clock frequency, respectively.

IV. THE SIMULATION RESULTS

In order to demonstrate the effectiveness of the proposed ADCs shown in Figures 3 and 4, the behavioral simulations with a C program were performed. The program consisted of 14 1.5-bit bit-blocks, delay units and the digital correction logic to form a 14-bit ADC. It imitates the ADC shown in Figure 3. Gain errors at each of the first three bit-blocks were set at +0.8 %. Clock timing was assumed to be aligned and the timing jitter was not taken into account.

Figure 5 shows the Fast-Fourier-Transform (FFT) result of the output data. The rest of bit-blocks are assumed ideal. The input signal voltage and frequency are ±2 V and 1.00625 MHz, respectively, and the sampling clock frequency is 102.4 MHz. It does not have the function to permutate capacitors or to take the average of the output data at alternate clocks. The number of collected data points is 16,384. As seen in Figure 5, gain errors produce harmonic components in the frequency domein and odd multiple of harmonics become eminent because the gain error in the bit-block is the odd function. The signal frequency bandwidth is 51.2 MHz and the SFDR is measured 57.5 dB.

When the permutation of capacitors in the first three bit-blocks at alternative clocks is done, the frequency spectrum changes such that as shown in Figure 6. The averaging function at the output has not been taken yet. Harmonic components become aligned in the reverse order, in contrast to the Figure 5, in the frequency domain starting from 51.2 MHz. The SFDR doesn't change much from that of Figure 5 and is measured 56.5 dB with the signal frequency bandwidth of 51.2 MHz.

The improvement of the SFDR is further achieved by taking the sum of the ADC outputs at alternate clocks by using a unit delay function and an adder as shown in Figure 3. As equation (4) shows, the averaging function introduces zero at multiples of half the clock frequency. The offset error which is caused by permuting capacitors has the frequency component at half the clock frequency, and is eliminated by

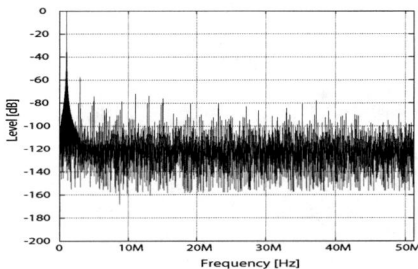

Fig. 5. The frequency spectrum of the simulated conventional 14-bit ADC.

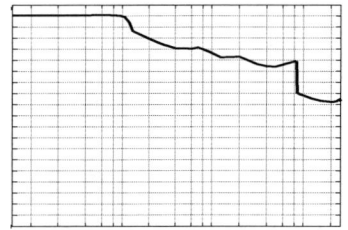

Fig. 7. The input signal frequency vs. SFDR characteristics of the ADC in Figure 3.

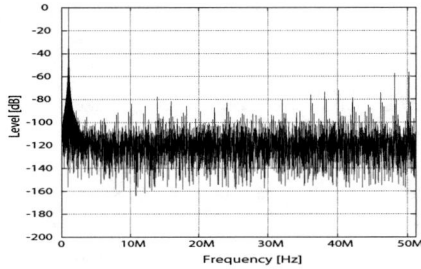

Fig. 6. The frequency spectrum of the simulated 14-bit ADC with permuting capacitors.

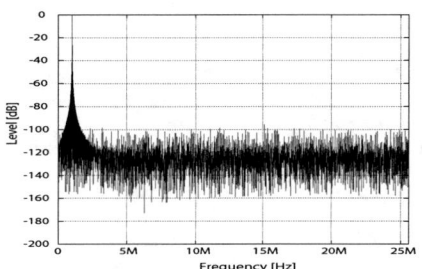

Fig. 8. The frequency spectrum of the simulated 14-bit ADC in Figure 4.

the zero. However, the introduction of the zero reduces the signal frequency bandwidth to half the clock frequency.

When the input signal frequency increases, however, the cancellation of the non-linearity becomes less effective. Figure 7 shows the input signal frequency vs. SFDR characteristics. Due to the complex error function, the SFDR degradation does not become linear, and the SFDR goes down to approximately 55 dB at one-fourth the clock frequency of 25.6 MHz.

The complete cancellation of the harmonics or distortion components was achieved by A-D converting the held output value of the S/H circuit twice as shown in Figure 4. Figure 8 shows the output frequency spectrum. The frequency range in the figure is limitted to 25.2 MHz because the signal frequency bandwidth remains 25.2 MHz. The input signal was sampled by an input S/H circuit at an equivalency rate of 51.2 MHz when the clock frequency was 102.4 MHz. In the hold mode of the S/H circuit, two rounds of A-D conversions were performed by permuting capacitors. As the input signal was constant, the offset error and the gain error contained in the two conversion digital data cancelled each other out. With an ideal S/H circuit and with aligned clocks, the SFDR reached 95.6 dB. The SFDR value does not depend on the input signal frequencies.

Table-1 summarizes the relationships of the input signal frequency bandwidth, the sampling clock frequency, and the SFDR among various types of ADCs discussed in this paper.

TABLE 1
ADC performance comparison

ADCs	BW	fs	SFDR (1.00625 MHz Input)
Conventional	$51.2 MHz$	$102.4 MHz$	$57.5 dB$
Permuting	$51.2 MHz$	$102.4 MHz$	$56.5 dB$
ADC in Fig. 3	$25.6 MHz$	$102.4 MHz$	$78.7 dB$
ADC in Fig. 4	$25.6 MHz$	$51.2 MHz$	$95.6 dB$

V. Conclusion

This paper offers a very simple implementation of an over-sampling ADC with 1.5-bit bit-blocks. Using permuting capacitors in a bit-block with alternate clocks in the analog domain, offset and gain error cancellation became possible by taking their average in the digital domain.

References

[1] B.Razavi, "Principles of Data Conversion System Design", IEEE Press, 1995.
[2] S-Y.Chuang and T.L.Sculley, "A Digitally Self-Calibrating 14-bit 10-MHz CMOS Pipelined A/D Converter", IEEE J. of Solid State Circuits, Vol. 37, no. 6, pp.674-683, June 2002.
[3] Y-M.Lin, B.Kim and P.R.Gray, "A 13-b 2.5-MHz Self-Calibrated Pipelined A/D Converter in 3-μm CMOS", IEEE J. of Solid State Circuits, Vol. 26, no. 4, pp.628-636, Appril 1991.
[4] A.Shabra and H.S.Lee, "Oversampled Pipeline A/D Converters With Mismatch Shaping", IEEE J. of Solid State Circuits, Vol. 37, no. 5, pp.566-578, May 2002.
[5] Y.Chiu, "Inherently Linear Capacitor Error-Averaging Techniques for Pipelined A/D Conversion", IEEE Trans. on Circuits and Systems-II: Analog and Digital Signal Processing, Vol. 47, no. 3, pp.229-232, March 2000.
[6] R.J.Baker, "CMOS Mixed-Signal Circuit Design", IEEE Press, ISBN 0-471-22754-4, pp.368-372.

An 800Mbps System Interconnect Modeling and Simulation for High Speed Computing

Mohammad S. Sharawi and Daniel N. Aloi

Electrical and System Engineering,
Oakland University
Rochester, MI 48309
USA

msharawi@oakland.edu , aloi@oakland.edu

Abstract— System interconnect modeling for high speed systems is a vital bottleneck for high speed data transfer. We demonstrate the modeling process on a high speed computer differential net running at 400MHz (800Mpbs) with IBM I/O cells. The modeling of the traces on the boards was done using a field solver. The transmission line matrices were used in a SPICE model, and 3-simulation scenarios were tested for this model. The obtained EYE opening of the modeled interconnect simulation was $705mV$ while the measured EYE opening for the same net topology in the Lab was $710mV$. This shows a close match between the actual behavior and the model generated. Careful modeling can be very beneficial to get a design running at first time operation.

Keywords—High Speed, PCB, Interconnect, Signal Integrity.

I. INTRODUCTION

Today's high speed electronic devices and gadgets are considered a vital element in the life of professionals. Ranging from Personal Computers (PC), to Pagers, Cell phones, hand held devices, Personal Digital Assistants (PDA), and ending with Pocket PCs. The necessity to process and transfer large amounts of data in relatively short periods of time, is the driving factor in high speed industry, where the competition is based on how much data can you transfer and how fast?

In order to accomplish this task, special care should be taken when designing high speed circuits and interconnects that combine these circuits and the functionality of a system. As the bit rate of the data increases, and the frequency of switching increases, the interconnect behavior should be looked at carefully. If the design or model of the interconnect is not done properly, the data transfer between various devices might fail, and your product will be doomed.

High speed interconnect modeling, design and analysis is not an easy task. As a matter of fact, a new discipline in electrical engineering has evolved that specifically studies, analysis and models such behavior to ensure proper high speed data transfer and operation. This field is called *Signal Integrity* (SI). SI combines Analog and Digital circuit design methodologies with Electromagnetic field theory to analyze and characterize the behavior of various interconnects ranging from Printed Circuit Board (PCB) traces (wires), to connector models that connect boards or devices together, to board and cable irregularities that has an effect that cannot be ignored at high speeds on signal transmission.

Special attention and care should be taken when designing high speed boards, or investigating high speed interconnects. This is why we touch upon this important area in electronic design, and investigate the stages that a designer should follow to get an accurate and reliable interconnect model to be used in his high speed designs.

The paper is structured as follows: Section II considers modeling of high speed interconnects like PCB traces, and connectors. Section III presents an example of a high speed data net that travels from one machine through a number of boards to another via an external cable. Section IV illustrates the design results and models. The paper is concluded in section V.

II. MODELING OF HIGH SPEED INTERCONNECTS

As the rise time of the signal use gets lower, the transition speed of the signal becomes higher, and the wave is considered as a TEM (Transverse Electromagnetic) wave travelling from driver to receiver. The wire or PCB trace (or etch) connecting the driver to receiver circuits will no longer be treated as a lumped Resistive-Capacitive network (RC), but it will be considered as a transmission line (distributed network, Fig. 1) if the length of the wire/etch is greater than 1/6 of the ratio between the Rise time T_r and Propagation delay D, i.e. $l > \frac{T_r}{6D}$, or if $T_r < 6TD$, where TD is the time delay of the interconnect. In this case the PCB trace will be treated and modelled as a TL, and its characteristics will be determined based on the geometry, dielectric constant, frequency of operation, and the stackup of the board.

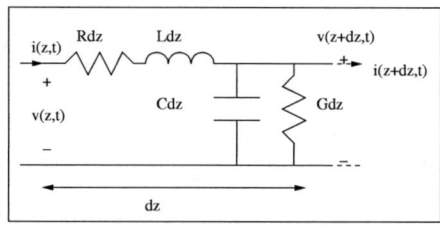

Fig. 1. Transmission Line Distributed Network.

Maxwell's equations for the TEM waves on multiconductor transmission lines reduce to the telegraphers equations. The general form of the telegrapher's equations in the frequency domain are given by [3]:

$$-\frac{\partial}{\partial z}\mathbf{v}(\mathbf{z},\omega) = [\mathbf{R}(\omega)+j\omega\mathbf{L}]\,\mathbf{i}(\mathbf{z},\omega) \quad (1)$$

$$-\frac{\partial}{\partial z}\mathbf{i}(\mathbf{z},\omega) = [\mathbf{G}(\omega)+j\omega\mathbf{C}]\,\mathbf{v}(\mathbf{z},\omega) \quad (2)$$

where, boldface lower-case and upper-case symbols denote vectors and matrices, respectively. **v** is the voltage vector across the lines and **i** is the current vector along the lines.

There are various rules of thumb to evaluate and calculate the parameters of the PCB trace, but usually if we are looking for an accurate model at a certain bandwidth of operation, we rely on a field solver (2D or 3D) to give us the parameters of the PCB trace structure for a certain geometry and stack-up. One of the most widely used field solvers is Maxwell (from Ansoft Inc.) [10]. Lossy Transmission lines that we usually incorporate in our PCB trace designs are usually characterized by the following,

- Capacitive Matrix: that includes self and mutual capacitance of the trace with adjacent traces, (**C**) in Farads/meter.
- Inductive Matrix: that includes self and mutual inductance of the trace with adjacent traces, (**L**) in Henrys/meter.
- DC Resistance Matrix, (**R**) in Ohms/meter.
- DC Conductance Matrix, (**G**) in Siemens/meter.
- AC Resistance Matrix (due to Skin effect), (**R$_s$**) in Ohms/meter$\sqrt{Hz}$.
- Conductance losses Matrix (due to dielectric losses), (**G$_d$**) in Siemens/meter Hz.

$$\mathbf{C} = \begin{bmatrix} C_{11} & & \\ C_{21} & C_{22} & \\ C_{31} & C_{32} & C_{33} \end{bmatrix} \quad \mathbf{L} = \begin{bmatrix} L_{11} & & \\ L_{21} & L_{22} & \\ L_{31} & L_{32} & L_{33} \end{bmatrix}$$

$$\mathbf{R} = \begin{bmatrix} R_{11} & & \\ 0 & R_{22} & \\ 0 & 0 & R_{33} \end{bmatrix} \quad \mathbf{G} = \begin{bmatrix} G_{11} & & \\ G_{21} & G_{22} & \\ G_{31} & G_{32} & G_{33} \end{bmatrix}$$

$$\mathbf{R_s} = \begin{bmatrix} R_{s11} & & \\ 0 & R_{s22} & \\ 0 & 0 & R_{s33} \end{bmatrix}$$

$$\mathbf{G_d} = \begin{bmatrix} G_{d11} & & \\ G_{d21} & G_{d22} & \\ G_{d31} & G_{d32} & G_{d33} \end{bmatrix} \quad (3)$$

These matrices include information about the value of the parameter for the trace itself, and the effect of adjacent traces on it. The matrices are lower triangular (3 × 3 since we have a three conductor stripline trace, see Fig.2) since the effect of a trace 1 on trace 2 is same as the effect of 2 on 1 (symmetric). Note that the **R** and **R$_s$** do not have any mutual effect (the DC resistance and skin effect are properties of traces themselves and there is no mutual effect). The frequency dependent relationships of these matrices are [3], [4], [6],

$$\mathbf{R(f)} = \mathbf{R_o} + \sqrt{f}\mathbf{R_s} \quad \frac{\Omega}{m} \quad (4)$$

$$\mathbf{G(f)} = \mathbf{G_o} + f\mathbf{G_d} \quad \frac{S}{m} \quad (5)$$

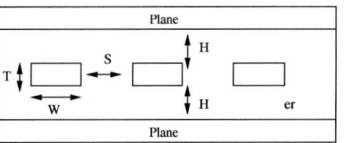

Fig. 2. A three conductor PCB Cross section.

$$\mathbf{C(f)} = \mathbf{C_o} \quad \frac{F}{m} \quad (6)$$

$$\mathbf{L(f)} = \mathbf{L_o} \quad \frac{H}{m} \quad (7)$$

where $\mathbf{L_o}, \mathbf{C_o}, \mathbf{R_o}$ and $\mathbf{G_o}$ are the DC valued matrices characterizing the TL. The characteristic impedance of every conductor can be found using its respective self inductance and capacitance using the equation

$$\mathbf{Z_o} = \sqrt{\frac{\mathbf{L}}{\mathbf{C}}} \quad (8)$$

i.e. for conductor 1, its $\mathbf{Z_o} = \sqrt{\frac{L_{11}}{C_{11}}}$ Ω. The Propagation delay of such a stripline is given by,

$$TD = \sqrt{LC} \quad (9)$$

i.e. for conductor 2, its $TD = \sqrt{L_{22}C_{22}}$ inches/sec. After creating the geometry and stackup of the layers, we run the field solver on the traces drawn to get these matrices. After getting the value of these matrices, which is usually saved in a file, we pass this generated file to the transistor level simulator, which is HSPICE is our case. A file for the high speed system should be specified in SPICE, and we have to incorporate as close to real life models in the design as possible, to get close to lab measurements values. This spice-deck should have the correct Driver/Reciever models taken from (generated by) the vendor of the IC (Integrated Circuit) used, the correct interconnect models, i.e. PCB traces, via models, and connector models, and any other parameters that might affect the operation of the system. Then, once we gather this file, we simulate to see the behavior of the circuit and net with such models. The driver/receiver models, and connector models, usually are obtained from the vendor of the device. But PCB and via modelling are done by the designer. And this is not an easy job, since it needs alot of experience, and patience in running long runs of simulation scenarios.

In HSPICE, the PCB trace is modelled via the $W-line$ TL model, because it has faster convergence times than the older $U-line$ model, no spurious ringing, frequency dependent loss is well and accurately modeled, and no limit on coupled conductor numbers [4], [6].

HSPICE based modeling is considered the most accurate in industry, unfortunately it is also time consuming, and consumes long simulation times (i.e. this net took about an hour to simulate for a couple of 100's of nsec on a multiprocessor machine). IBIS modeling on the other hand is not as accurate, but can get relatively good results with 0.1 times the simulation time of an HSPICE based one [9]. This will not be discussed further in this paper.

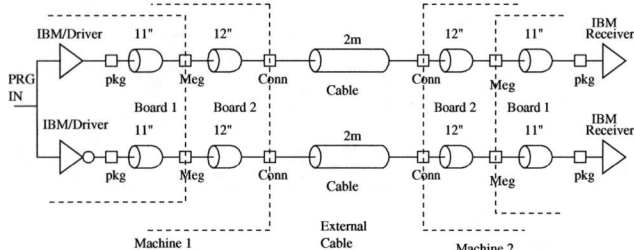

Fig. 3. The net topology simulated. Copper etch width is 4mils, thickness is 0.5Oz., and separation is 6mils.

III. A DESIGN EXAMPLE

In this section we will present a design example to demonstrate the modelling/simulation of various interconnects in a driver receiver path. We will consider a 2 conductor stripline connecting a driver chip (IBM 400MHz driver) on one computer machine, through a high speed low latency cable, to another machine that has the receiver (IBM 400MHz receiver), as shown in Fig. 3. The driver/reciever are differential components (operating with a 2.5V supply, and differential common mode voltage of 1.25V), i.e. the signal is driven with its complement on two conductor lines. Such a signal suffers less noise than single ended signals, because the generated mutual effects are opposite to each other, and their effect is minimized (almost cancelled) relative each other. The traces are being modeled using Maxwell (Fig. 4), with width of 4mils, separation of 6mils, distance from reference planes was 5mils, and thickness of 0.5oz. (0.7mils). The IC package model was taken from a typical 400MHz component[1], so is the Meg-Array connector model, that connects the two boards in each machine together, and the Cable connector model, that connects the two machines together. Such connector models are a combination of an RLGC values that represent the behavior of the connector to frequency changes and DC and AC losses[2]. The cable connecting the two machines is 2 meters long and has a DC loss of about 1Ω. Its a low loss, low latency, high speed cable. The etch length on machine 1 was 23inches (board 1 has 11inches, and board 2 has 12inches), and on machine 2 was the same (a scenario that might be encountered while designing high speed boards for high performance computers.) The PCB, connector and driver models are then combined in a single HSPICE file for simulation. The simulation was performed by applying a Pseudo Random bit generator (PRG) with a frequency of 400MHz (800 Mbps) at the driver inputs, and the EYE pattern was then observed at the receiver end. The EYE pattern is very important in Signal Integrity, and is very helpful in identifying problems, like Intersymbol Inteference (ISI), low Swing Levels, timing problems, and noise issues. We always need to see an EYE wide open to get good signal timing (setup and hold times), and have a good sampling margin. Also, the levels should be sufficient for the technology used. Finally, the crossing point should be as thin as possible to have less ISI.

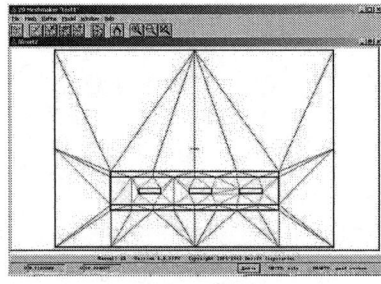

Fig. 4. Analyzing the PCB traces via Maxwell, mesh sections.

[1]such models are either generated by companies for their custom designed chips, or obtained from the IC vendor who sells the chip.
[2]The BGA (Ball Grid Array) Meg-Array model can be obtained from http://www.fciconnect.com. [11]

IV. SIMULATION AND MODELLING RESULTS

The simulation results of our differential net shown in Fig. 3 are presented in this section. The matrices obtained from the Maxwell simulation for the two conductor stripline operating at 800MHz[3], with $T = 0.7 mils$, $H = 5 mils$, $w = 4 mils$, $s = 6 mils$, $\epsilon_r = 4.3$, where,

$$\mathbf{C} = \begin{bmatrix} 1.118267e-10 & \\ -1.266590e-11 & 1.117687e-10 \end{bmatrix} \mathbf{F/m}$$

$$\mathbf{L} = \begin{bmatrix} 3.830068e-07 & \\ 4.340322e-08 & 3.832060e-07 \end{bmatrix} \mathbf{H/m}$$

$$\mathbf{R} = \begin{bmatrix} 9.880000e+00 & \\ 0.000000e+00 & 9.880000e+00 \end{bmatrix} \mathbf{\Omega/m}$$

$$\mathbf{G} = \begin{bmatrix} 0.000000e+00 & \\ 0.000000e+00 & 0.000000e+00 \end{bmatrix} \mathbf{S/m}$$

$$\mathbf{R_s} = \begin{bmatrix} 1.187494e-03 & \\ 5.245024e-05 & 1.161468e-03 \end{bmatrix} \frac{\mathbf{\Omega}}{\mathbf{m}\sqrt{\mathbf{Hz}}}$$

$$\mathbf{G_d} = \begin{bmatrix} 1.755680e-11 & \\ -1.988546e-12 & 1.754769e-11 \end{bmatrix} \frac{\mathbf{S}}{\mathbf{mHz}}$$

The single ended $\mathbf{Z_{o1}} = \mathbf{56.9\Omega}$, $\mathbf{Z_{o2}} = \mathbf{57.1\Omega}$, and the differential impedance $\mathbf{Z_{oDiff}} = \mathbf{113.9\Omega}$. The propagation delay for each trace $TD_1 = 6.544 nsec/m$, and $TD_2 = 6.545 nsec/m$. The simulation of the net was performed using a PRG with a bit width of $1.25 nsec$, rising edge of $T_r = 200 psec$, falling edge of $T_f = 200 psec$. The resulted EYE pattern at the receiver end looks like Fig. 5. To see the effect of etch length and width on the observed EYE parameters, we recorded some simulation scenarios in Table I. In Table I, the simulation of the net in Fig. 3 shows a differential level of $743 mV$ with no discontinuity in the path, represented by the via, while the existence of such a via degrades the voltage swing by about $40 mV$, which is a drop of about 5%. In the third scenario, we split the 12 inches of trace of board 2 in each machine into two parts, one of $4 mils$ width and $3 inches$ of length, and then a via is used to have this trace routed on another layer, and have a thickness of $8 mils$, twice as much, and length of $7 inches$. This will reduce the signal loss since the thickness of the long trace is $8 mils$ instead of $4 mils$, not to mention that the total trace was reduced by $2 inches$ on each machine,

[3]We simulated the traces at 800MHz because the data rate was 800 Mbps.

TABLE I
SIMULATION RESULTS TABLE.

Trace width	Trace thickness	Trace length/machine	Differential Eye Opening
4 mils	0.5 Oz.	23 in.	743 mV
4 mils	0.5 Oz.	23 in. with via	705 mV
4 mils	0.5 Oz.	(3+11) in. + via	842 mV
8 mils		7 in + via.[a]	

[a] the trace on board 2 in machines 1 and 2 (12 in.) was redesigned and split into two traces with different lengths to reduce signal loss and increase the EYE width.

with a total of $4inches$ in the signal path. The signal level increased by about $100mV$ or about 14%. In high speed signaling and with this amount of loss (signal at driver is 2.5V) on a long interconnect path (as a worse case, specially in huge systems and supercomputers that incorporate the interconnection of many boards, and then connecting subsystems together) such an improvement in signal levels and swing is highly appreciated by system and circuit designers. Fig. 5 shows the EYE pattern for the second case in

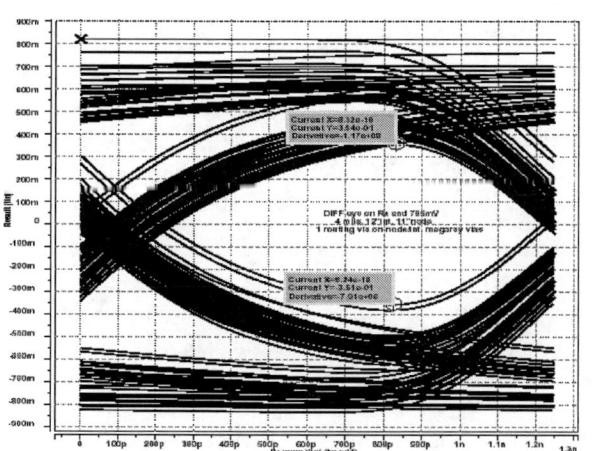

Fig. 5. Eye Pattern simulation. Trace width is $4mils$, thickness $0.5Oz.$, length $23inches$, and a via is present in the signal path.

Table I. As you can see the EYE voltage opening is about $705mV$ taken differentially, and has about $180psec$ of ISI. The width between the crossings of the smallest EYE is about $950psec$. Fig. 6 shows the measure EYE pattern of the topology assumed in the design example[4]. The EYE opening was $710mV$ for the second case in Table I. This shows that the simulation and modelling of such interconnects, and the high speed circuit does actually give a very good estimation about real life signal behavior, given the modeling was conducted in the correct fashion.

V. CONCLUSIONS

As the speed of data transfer is increased, careful modeling and design of system interconnects should be performed. The transmission line behavior of this high speed

[4]This test and measurement was performed in one of Silicon Graphics Inc. Lab, CA-USA.

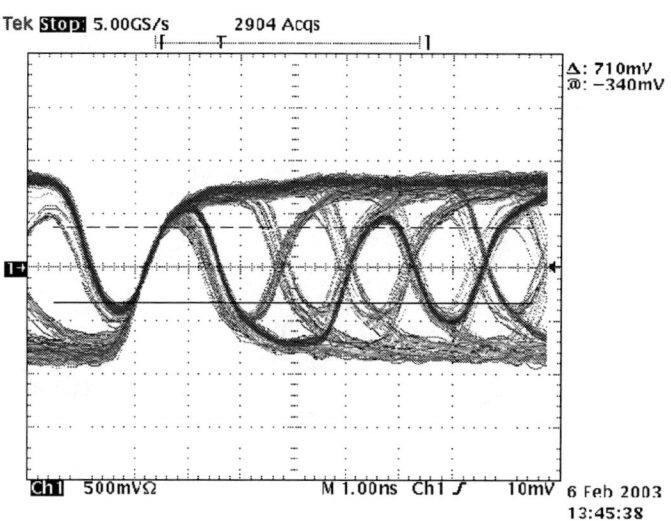

Fig. 6. Measured EYE on the receiving end at machine 2.

operation of such interconnects affects the quality and speed of the signal. The best way to model such interconnects is by using electromagnetic field solvers that can give us accurate high speed modeling. We have modeled a high speed differential net operating at 400MHz (800Mbps) using a transistor level simulator (SPICE), and a field solver called Maxwell (Ansoft Inc.), and compared the behavior of the net with that of a net with the same topology that was measured in the Lab. The interconnect model had a behavior as the one used in the real system. We have obtained an eye opening of $705mV$ using the simulated net, and we measured an eye opening of $710mV$ for the same net topology in the Lab. This means that by accurate and careful modeling of such system interconnects, we can get a very close to real life behavior of such critical nets.

ACKNOWLEDGEMENTS

The authors would like to thank Silicon Graphics Inc. (SGI) and Brad Juskiewicz for their help in accessing their labs and help in testing the modules.

REFERENCES

[1] H. Johnson and Martin Graham, *High Speed Digital Design: A Handbook of Black Magic*, Prentice Hall, 2^{nd}Edition, 1996.
[2] H. B. Bakoglu, *Circuits, Interconnections, and Packaging for VLSI*, Addison-Wesley, 1990.
[3] "Star HSPICE, User's Manual, Volume III", Avanti Inc., 1996.
[4] "Star HSPICE, User's Manual Supplement", Avanti Inc., 1997.
[5] S. Hall, J. Hall, and J. McCall, *High Speed Digital Design: A Handbook of interconnect theory and Design Practices*, John Wiley & Sons, Reading, 2000.
[6] "Star HSPICE, User's Manual", Avanti Inc. 2000.
[7] Dmitri Borisovich and Jose Schutt-Aine, "Optimal Transient Simulation of Transmission Lines", *IEEE Transactions on Circuits and Systems-I: Fundamental Theory and Applications*, vol. 43, no.2, pp. 110-121, February 1996.
[8] Nannapaneni Narayana Rao, *Elements of Engineering Electromagnetics*, Prentice Hall, 4th Edition, 1994.
[9] Mohammad S. Sharawi,"Modeling and Simulation of High Speed Digital Circuits and Interconnects ",*Middle East Conference on Simulation and Modeling* (MESM), Amman-Jordan, Sep. 14-16, 2004.
[10] http://www.ansoft.com
[11] http://www.fciconnect.com

Ternary Walsh Transform

Bogdan J. Falkowski and Shixing Yan

School of Electrical and Electronic Engineering, Nanyang Technological University
Block S1, 50 Nanyang Avenue, Singapore 639798

Abstract—The new ternary Walsh transform is introduced in this article. It is based on Kronecker product as well as known Galois Field (3) (GF(3)) and new ternary operations and its big advantage is that the same hardware implementation can be used for both forward and inverse ternary Walsh transforms.

I. INTRODUCTION

Walsh transforms are orthogonal, normal and complete [1]–[3]. They are important spectral representations of binary logic functions as the spectral Walsh domain with its global information provides much deeper insight into logical structure of combinational networks than logic domain [2], [3]. Spectral representation based on the Walsh transforms have been used in the classification of logic functions, functional decompositions, logic synthesis, analysis of logic complexity, analysis of balanced functions, detections of symmetries, linearization of decision diagrams, state assignment, cascade realizations, testing, and technology mapping [2]–[4].

The renewed interest in applications of spectral methods in VLSI circuits is caused by their excellent design for testability and the recent development of efficient methods to operate on spectra of logic functions directly from reduced representations such as arrays of disjoint cubes or decision diagrams (DDs) [2]. VLSI testing based on signatures, that are, the correlations between output functions and test vectors, use the Walsh spectral coefficients.

In this article, the new ternary Walsh transform based on Kronecker product is introduced. The big advantage of the new transform is that the same hardware can be used to calculate its forward and inverse butterfly diagrams. In addition, this transform can be used in the same applications for ternary logic design where Walsh transform is used for binary case.

II. WALSH TRANSFORM

The Walsh transform in Hadamard order which is also called Hadamard-Walsh transform is most widely used since it has recursive Kronecker product structure and its inverse transform is its own matrix of order 2^n with scaling factor of $\frac{1}{2^n}$ [1], [3].

Definition 1: The forward Walsh transform matrix of order 2^n is defined as:

$$\mathbf{W}(n) = \bigotimes_{i=1}^{n} \mathbf{W}(1) \quad (1)$$

where $\mathbf{W}(1) = \begin{bmatrix} 1 & 1 \\ 1 & -1 \end{bmatrix}$ and the symbol $\otimes$ represents the Kronecker product [1]–[4].

The inverse Walsh transform matrix $\mathbf{W}^{-1}(n)$ of order 2^n is defined as:

$$\mathbf{W}^{-1}(n) = \frac{1}{2^n}\mathbf{W}(n). \quad (2)$$

Property 1: The Walsh matrix can be factorized using fast algorithms as follows:

$$\mathbf{W}(n) = \prod_{i=1}^{n} \mathbf{C}_i(n) \quad (3)$$

where $\mathbf{C}_i(n) = \bigotimes_{j=1}^{n} \mathbf{C}_i^j(1)$,

$$\mathbf{C}_i^j(1) = \begin{cases} \mathbf{I}_2, & \text{for } j \neq i, \\ \mathbf{W}(1), & \text{for } j = i, \end{cases}$$

and $\mathbf{I}_2$ is the identity matrix of order 2.

Example 1: By using the fast algorithms shown in Property 1, the fast Walsh transforms of $\mathbf{W}(3)$ can be derived. Due to the factorized matrices of $\mathbf{W}(3)$, the corresponding butterfly diagram is shown in Fig. 1 where the solid lines and dotted lines represent addition and subtraction, respectively.

Since the forward and inverse Walsh transforms have the same recursive Kronecker product structure, the same fast Walsh transform hardware structure can be used to both forward and inverse Walsh transforms [1], [3]. Such fast algorithms will greatly reduce the number of arithmetic operations as compared to the computation of Walsh transform by the whole matrix. The computational costs mean the number of additions and subtractions required for the generation of Walsh transform. The computational costs for the Walsh transform by using whole matrix are 2^{2n}, while the computational costs using fast Walsh transform are $n2^n$.

III. TERNARY WALSH TRANSFORM

The Walsh transform has been used in many areas including logic design and binary coding [2]–[4]. As the Walsh transform is binary valued (± 1), it is limited in multiple-valued logic applications and higher valued codings. A new ternary Walsh transform which uniquely map ternary logic functions into ternary transform space is presented in this section. The operations in the new ternary Walsh transform are all over GF(3), which provides the unique and invertible properties of the transform.

A. Operations over GF(3)

There are two ternary operations used in the forward ternary Walsh transform and two ternary operations used in the inverse ternary Walsh transform.

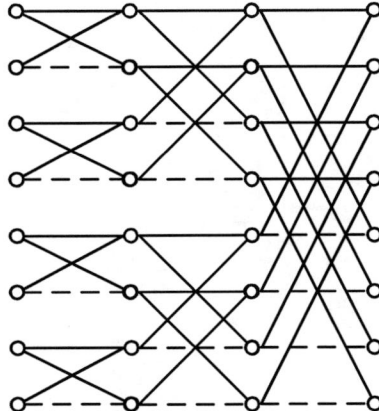

Fig. 1. Fast butterfly diagram for Walsh transform where $n = 3$.

TABLE I
FORWARD TERNARY OPERATIONS $\oplus$ AND $\ominus$

$a \oplus b$		b			$a \ominus b$		b		
		0	1	2			0	1	2
a	0	0	1	2	a	0	0	2	1
	1	1	2	0		1	1	0	2
	2	2	0	1		2	2	1	0

TABLE II
INVERSE TERNARY OPERATIONS $\boxplus$ AND $\boxminus$

$a \boxplus b$		b			$a \boxminus b$		b		
		0	1	2			0	1	2
a	0	0	2	1	a	0	0	1	2
	1	2	1	0		1	2	0	1
	2	1	0	2		2	1	2	0

Let a and b be ternary variables.

Definition 2: The following ternary operations $\oplus$ and $\ominus$ are used in forward ternary Walsh transform. The ternary operation $\oplus$ replaces the additions in fast Walsh transform structure, and it is defined as the modular 3 addition: $a \oplus b = (a + b)$ mod 3. The ternary operation $\ominus$ replaces the subtractions in fast Walsh transform structure, and it is defined as the modular 3 subtraction: $a \ominus b = (a - b)$ mod 3.

Table I shows the ternary operations $\oplus$ and $\ominus$ over GF(3).

Definition 3: The ternary operations $\boxplus$ and $\boxminus$ are used in inverse ternary Walsh transform. The ternary operation $\boxplus$ replaces the additions in fast Walsh transform structure, and it is defined as: $a \boxplus b = 2(a + b)$ mod 3. The ternary operation $\boxminus$ replaces the subtractions in fast Walsh transform structure, and it is defined as: $a \boxminus b = 2(a - b)$ mod 3.

Table II shows the ternary operations $\boxplus$ and $\boxminus$ over GF(3).

B. Definitions and Properties of Ternary Walsh Transform

The ternary Walsh transform is based on the fast Walsh transform structure while the four types of ternary operations are used in the butterfly structure instead of the additions and subtractions.

Definition 4: The forward ternary Walsh transform can be computed by replacing the additions and subtractions by ternary operations $\oplus$ and $\ominus$ in the butterfly structure of fast Walsh transform, respectively. The vector $\mathbf{S}$ computed by using ternary Walsh transform is called as ternary Walsh spectrum which is n-tuples over GF(3).

Definition 5: The inverse ternary Walsh transform can be computed by replacing the additions and subtractions by ternary operations $\boxplus$ and $\boxminus$ in the butterfly structure of fast Walsh transform, respectively.

Property 2: Let $R_1 = \{0, 1\}$, $R_2 = \{0, 1, 2\}$, and R_p^q means q-space Cartesian product of a set R_p where $p = 1, 2$. If the input vectors are in domain $R_2 = \{0, 1, 2\}$, the forward and inverse ternary Walsh transforms can be seen as the mappings: $(R_2^N)_F \to (R_2^N)_S$ and $(R_2^N)_S \to (R_2^N)_F$ where the subscripts F and S correspond to functional and spectral domains, respectively. When the input vectors are the truth vectors of binary logic functions, the forward and inverse ternary Walsh transforms can be seen as the mappings: $(R_1^N)_F \to (R_2^N)_S$ and $(R_2^N)_S \to (R_1^N)_F$, respectively.

The ternary Walsh transform is a non-linear transform, and it is also unique and invertible.

Example 2: For $n = 3$, let the ternary vector $\mathbf{F}_1 = [0, 2, 1, 0, 1, 0, 1, 2]^T$ where the superscript T is the matrix transpose operator. We can use ternary Walsh transform to compute its ternary Walsh spectrum $\mathbf{S}_1$ as shown in Fig. 2. Due to the invertibility of ternary Walsh transform, the vector $\mathbf{F}_1$ can be also recovered from $\mathbf{S}_1$ by using inverse Walsh transform as shown in Fig. 3.

Although the definitions of ternary Walsh transform are based on replacing the additions and subtractions in fast Walsh transform structure by the ternary operations, we can also use the matrix representation to express the ternary Walsh transform where both the elements of the matrix and the arithmetic operations in the matrix are also over GF(3). The alternative definitions of ternary Walsh transform by modifying standard Walsh matrix are obtained as shown below.

Definition 6: The basic matrix of ternary Walsh transform is modified by using modular arithmetic over GF(3), so that the forward ternary Walsh transform matrix $\mathbf{TW}(n)$ of order 2^n can be defined as follows:

$$\mathbf{TW}(n) = \bigotimes_{i=1}^{n} \mathbf{TW}(1) \quad (4)$$

where $\mathbf{TW}(1) = \begin{bmatrix} 1 & 1 \\ 1 & 2 \end{bmatrix}$.

The multiplications over GF(3) are used in the computations of Kronecker products to construct the matrix. Table III shows the multiplication operations over GF(3).

Definition 7: From the definitions of inverse ternary operations $\boxplus$ and $\boxminus$, the relationships between forward and inverse

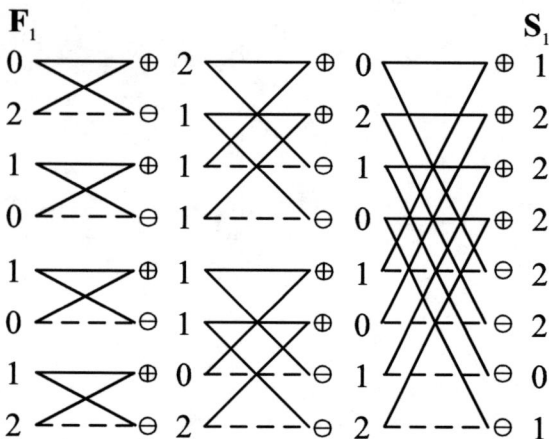

Fig. 2. Computation of forward ternary Walsh transform where $n = 3$.

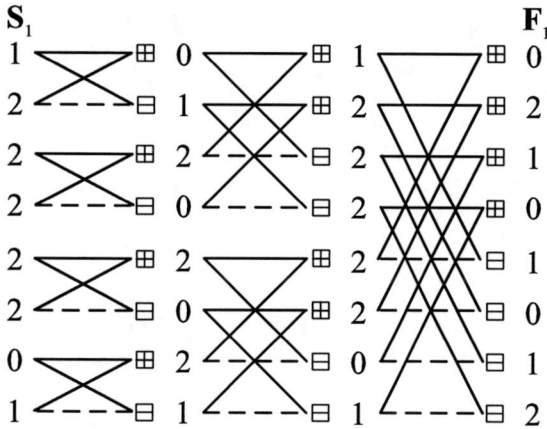

Fig. 3. Computation of inverse ternary Walsh transform where $n = 3$.

ternary Walsh transforms can be obtained. The inverse ternary Walsh transform matrix $\mathbf{TW}^{-1}(n)$ of order 2^n can be defined as follows:

$$\mathbf{TW}^{-1}(n) = \bigotimes_{i=1}^{n} \mathbf{TW}^{-1}(1) \quad (5)$$

where $\mathbf{W}(1) = \begin{bmatrix} 2 & 2 \\ 2 & 1 \end{bmatrix}$

or

$$\mathbf{TW}^{-1}(n) = 2 \times \mathbf{TW}(n). \quad (6)$$

where the operations are performed over GF(3).

Property 3: The matrices of forward and inverse ternary Walsh transforms are defined in Definitions 6 and 7, respectively. From (6) and Table III, it can be observed that the inverse ternary Walsh transform matrix can be constructed by simply replacing the elements 1 and 2 by 2 and 1, respectively, in the corresponding forward ternary Walsh transform matrix. The forward ternary Walsh transform matrix can be also generated from the inverse ternary Walsh transform matrix in the same way.

Example 3: The forward ternary Walsh matrix for $n = 3$

TABLE III
MULTIPLICATION RULES OVER GF(3)

×	0	1	2
0	0	0	0
1	0	1	2
2	0	2	1

can be constructed using (4) as:

$$\mathbf{TW}(3) = \bigotimes_{i=1}^{3} \begin{bmatrix} 1 & 1 \\ 1 & 2 \end{bmatrix} = \begin{bmatrix} 1 & 1 & 1 & 1 & 1 & 1 & 1 & 1 \\ 1 & 2 & 1 & 2 & 1 & 2 & 1 & 2 \\ 1 & 1 & 2 & 2 & 1 & 1 & 2 & 2 \\ 1 & 2 & 2 & 1 & 1 & 2 & 2 & 1 \\ 1 & 1 & 1 & 1 & 2 & 2 & 2 & 2 \\ 1 & 2 & 1 & 2 & 2 & 1 & 2 & 1 \\ 1 & 1 & 2 & 2 & 2 & 2 & 1 & 1 \\ 1 & 2 & 2 & 1 & 2 & 1 & 1 & 2 \end{bmatrix}.$$

The corresponding inverse ternary Walsh transform matrix can be calculated using (5) as:

$$\mathbf{TW}^{-1}(3) = \bigotimes_{i=1}^{3} \begin{bmatrix} 2 & 2 \\ 2 & 1 \end{bmatrix} = \begin{bmatrix} 2 & 2 & 2 & 2 & 2 & 2 & 2 & 2 \\ 2 & 1 & 2 & 1 & 2 & 1 & 2 & 1 \\ 2 & 2 & 1 & 1 & 2 & 2 & 1 & 1 \\ 2 & 1 & 1 & 2 & 2 & 1 & 1 & 2 \\ 2 & 2 & 2 & 2 & 1 & 1 & 1 & 1 \\ 2 & 1 & 2 & 1 & 1 & 2 & 1 & 2 \\ 2 & 2 & 1 & 1 & 1 & 1 & 2 & 2 \\ 2 & 1 & 1 & 2 & 1 & 2 & 2 & 1 \end{bmatrix}.$$

By comparing $\mathbf{TW}(3)$ with $\mathbf{TW}^{-1}(3)$, it can be seen that $\mathbf{TW}^{-1}(3)$ can be generated from $\mathbf{TW}(3)$ by replacing the elements 1 and 2 by 1 and 2, respectively, and vice versa. This relationship has been described in Property 3.

Definition 8: The ternary Walsh spectrum $\mathbf{S}$ of vector $\mathbf{F}$ can be calculated by using forward ternary Walsh transform matrix as follows:

$$\mathbf{S} = \mathbf{TW}(n) \times \mathbf{F} \quad (7)$$

Similarly, the input vector $\mathbf{F}$ can be also obtained as follows:

$$\mathbf{F} = \mathbf{TW}^{-1}(n) \times \mathbf{S} \quad (8)$$

The additions and multiplications in (7) and (8) are over GF(3) that correspond to the additions and multiplications over GF(3) shown in Table I and III.

The definitions of ternary Walsh transform based on the matrix representations have been proposed here, but the method using the ternary Walsh transform matrix requires more computational costs than the one through the fast Walsh transform structure shown in Definitions 4 and 5. Therefore, the matrix representations of ternary Walsh transform provide more theory than practicality. Hence, the butterfly structure of ternary Walsh transform should be used to get lower computational costs and more convenient hardware implementation.

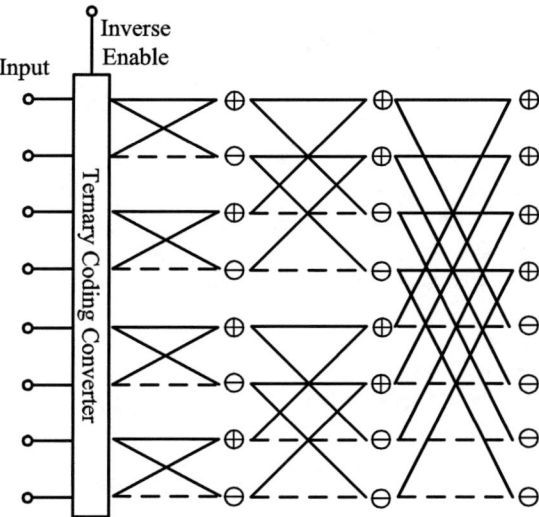

Fig. 4. Diagram of ternary Walsh transform where $n = 3$.

TABLE IV
RULES OF TERNARY CODING CONVERTER

Input	Output of Ternary Coding Converter	
	Inverse Enable=0	Inverse Enable=1
0	0	0
1	1	2
2	2	1

Property 4: The calculation of input vector **F** from its ternary Walsh spectrum by using (8) is shown in Definition 8. Let us rewrite (8) by using (6) as:

$$\begin{aligned} \mathbf{F} &= \mathbf{TW}^{-1}(n) \times \mathbf{S} = 2 \times \mathbf{TW}(n) \times \mathbf{S} \\ &= \mathbf{TW}(n) \times (2 \times \mathbf{S}). \end{aligned} \quad (9)$$

From (9), the vector **F** can be obtained by using forward instead of inverse ternary Walsh transform. Hence, we can use forward ternary Walsh transform to complete the functions of both forward and inverse ternary Walsh transforms and the inverse ternary Walsh transform can be ignored in the computations. In additions, we can modify **S** by replacing its elements 1 and 2 by 2 and 1, respectively, to use only the forward ternary Walsh transform.

From Property 4, the fast Walsh transform structure with only ternary operations ⊕ and ⊖ can be used in the design of butterfly structure, while the other ternary operations from Table II can be ignored by adding a ternary coding converter between the input and the butterfly which can change the input values 1 and 2 to 2 and 1 or bypass all the input values. The diagram of this structure for $n = 3$ is shown in Fig. 4 as an example, where the value of Inverse Enable decide whether to apply the ternary coding conversion or not for the input. Table IV shows the rules for the ternary coding converter where the Inverse Enable is set to 0 for the forward transform and to 1 for the inverse transform.

IV. CONCLUSION

Most of the digital VLSI/ULSI circuits that are used presently are based on binary logic which has been found to be reliable and compact. However, lately there is an increasing interest in circuits based on multiple-valued logic. This interest is fueled by their potential advantages over the binary ones, such as increased data processing capability per unit area, reduced number and complexity of interconnections, as well as reduced number of active devices inside a chip. It should be noticed that among the papers written on multiple-valued logic, quite a large number of them are on the design and implementation techniques as well as applications of 3-valued logic or ternary logic [5]. In [6], it is written that ternary logic has some inherent advantages in an environment where a 'middle' state between two outer ones can be found in which the outer devices are either both on or both off, as well as in the environments where two binary elements are combined at an upper and lower signal levels. Proposed applications of ternary logic include fail-safe logic and detection of hazard in binary logic circuits as well as evaluation of binary logic functions in the presence of unknown inputs [5]. It is also useful for designs of highly parallel circuits for some class of k-ary specifications composed of non-permutation unary operations and high-speed area-efficient multiplier for unsigned binary integers.

In this paper, the introduction of ternary Walsh functions is considered. Such a ternary Walsh transform can be used in a similar manner as Walsh transform for binary logic functions as shown in the introduction. The ternary Walsh functions can be also applied to digital signal and image processing and pattern recognition that involves the manipulation of multi-dimensional signals and operations on large number of data values. Since such a process generally requires high computational costs and substantial processing time, many attempts have been made to find efficient computation methods. Due to the properties of ternary Walsh transform described in this article, one of them is to speed-up the data computation by using dedicated and the same hardware as shown here.

REFERENCES

[1] P. S. Moharir, *Pattern-Recognition Transforms*. New York: John Wiley, 1992.

[2] T. Sasao and M. Fujita, Eds., *Representation of Discrete Functions*. Boston: Kluwer Academic, 1996.

[3] R. S. Stankovic, M. Stankovic, and D. Jankovic, *Spectral Transforms in Switching Theory, Definitions and Calculations*. Belgrade: Science Publisher, 1998.

[4] B. J. Falkowski, "Recursive relationships, fast transforms, generalizations and VLSI iterative architecture for Gray code ordered Walsh functions," *IEE Proceedings on Digital Techniques and Computers*, vol. 142, no. 5, pp. 325–331, Sept. 1995.

[5] Y. Iguchi, M. Matsuura, T. Sasao, and A. Iseno, "Realization of regular ternary logic functions using double-rail logic," in *Proceedings of Asia and South Pacific Design Automation Conference*, vol. 1, Hong Kong, Jan. 1999, pp. 331–334.

[6] K. C. Smith, "The prospects for multivalued logic: a technology and applications view," *IEEE Transactions on Computers*, vol. 30, no. 9, pp. 619–634, Sept. 1981.

A Post Layout Watermarking Method for IP Protection

Tingyuan Nie   *Tomoo Kisaka*   *Masahiko Toyonaga*

Dep. of Mathematics and Information Science,
Graduate School of Science, Kochi University
2-5-1 Akebono-cho, Kochi 780-8520, Japan
{nieteien, 00ss019, toyonaga}@is.kochi-u.ac.jp

Abstract—We propose a new watermarking system for VLSI layout design intellectual property protection (IPP) that will not damage circuit properties. The previous studies for layout design IPP are mainly restricted to pre-layout design, i.e. they would increase layout size and vary signal timing. The idea of our system is to use a special incremental router that removes wires of target nets and re-routes them by inserting redundant bends. We can distinguish the marked net from other. This redundant insertion is not always possible according to wire density around the wire, thus we use it iteratively. We evaluated the success possibilities of our watermarking system in various wire density benchmark circuits experimentally and found more than three iterations are enough for the practical post layout design to achieve successful watermarking.

I. INTRODUCTION

The progress of semiconductor manufacture process technology makes it possible to mount huge numbers of transistors on one chip. A new design methodology that can handle such large number of devices, called "system-on-chip" (SOC), becomes more and more important. Since large and complex circuit design requires long time and many designers, an integration design method with existing designs, so-called Intellectual Properties, can be one of the solutions. Reusable IP block is effective in reducing the SOC design cost and development TAT. However, it is difficult to provide such important IP without any security assurance which protects from the use of third person. Especially layout design IP blocks, that cannot hide its contents, are very easy to copy. Therefore, the author or owner of IP designs desire some guarantees that IPs are not illegally re-distributed, and the IP users desire guarantee that IP which they use are legal. There are already some feasible IP protection (IPP) techniques for such visible contents like layout design based on the watermarking [3-11]. They embed author's information to the layout design by adding extra restrictions than the usual design on pre-layout stage. Those approaches could be difficult to guarantee its performance of original IP design.

Each watermarking method has its own advantages and disadvantages, as well as will be attacked by different schemes. Watermarking attacks are categorized in four main classes [11]: unauthorized removal, unauthorized embedding, unauthorized detection, and system attacks. Unauthorized detection is not considered a high risk for IPs. Removal attacks aim at the removal of watermark information without breaking the security of watermark. The intruder tries to eliminate the watermark completely in the elimination attacks and aim at distorting the watermark detector in masking attack. Embedding attacks aim at embedding fabricated information in the design that can be done either by ghost searching or by re-embedding the watermark if he/she has the tools necessary to do this. System attacks aim at attacking the cryptographic base of the watermarking. This kind of attacks can be avoided by using appropriate protection techniques.

In this paper, we propose a new watermarking method at the post-layout design which confirms original IP performance. The copyright information is coded by DES (Data Encryption Standard) and watermarked on layout design using an incremental router [1, 2]. To evaluate our watermarking system, we applied it to several benchmark circuits with various wire density, and the results show our proposed method can watermark the post-layout design without increasing area and signal delay. We can also expect the DES system makes it safe from any fabrications of copyright. The following section explains the overview of our watermarking system for IPP. The third section explains the special Incremental Router algorithm applied after a basic layout design. The fourth section shows the result in experiments. The final section describes the conclusion of this paper.

II. PROPOSED WATERMARKING METHOD

A. System Overview

The outline of the layout design IP protection system is shown in Fig.1. The system is constructed of the following steps.

i) The copyright information is encoded into a bit string by DES system with the user's private key for each IP user.

ii) The nets allocated to '1' in the bit string are rerouted for watermarking by our incremental router.

This work is supported by Regional Science Promotion Program of Japan Science and Technology Agency

iii) The IP user reuses the watermarked physical IP and then integrates it into his/her LSI layout design.

iv) The IP checker, such as the entrusted silicon foundry or the IP owner, extracts the bit string from the LSI layout design by using our watermark decoder with the incremental router, and then decode into the copyright information by DES. The entrusted silicon foundry or the owner can easily confirm the validity of the integrated layout design for IP.

In our watermarking system, the each bit of the DES decoded bit string is corresponding to the net at corresponding sequence in the cross-reference of IP net-list. The wire of the net corresponding to '1' will be inserted an extra bend by our incremental router to distinguish from the original wire shape. Since the incremental router changes the shape of wires of the layout design slightly, it does not damage the area-size at all. Furthermore, the incremental router is not applied if it may damage the layout design properties of the original layout design IP. The comparison of approaches between the pre-design watermarking and the post-design watermarking are summarized in Table 1.

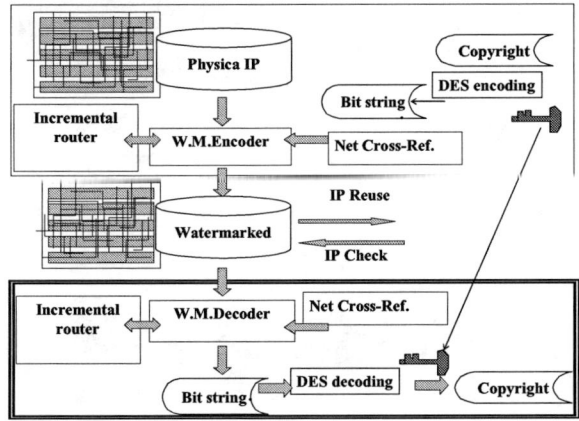

Figure 1 Over-view of IPP for Post-Layout Design

Table 1 Comparison of different watermarking approaches

	Pre-design	Post-design
Area size	Varies	No changes
Parasitic-capacity	Varies	No changes
Wire resistance	Varies	No changes

If the layout design includes congested area, the incremental router would fail to embed the external bends into wires. Thus our system allocates all bit information to more than two nets respectively after the cross-reference file as shown in Fig.2, i.e. the copyright bit string will be embedded into the layout design as more times as possible. This iterative embedding makes the system robust and enhances the success possibility of embedding. We assume the copyright information is constructed of a few dozen characters and the modern layout design IP includes more than thousands of nets, thus we expect the system can embed the information more than hundreds times.

To identify IP reuse validity, the watermarking information is decoded by the reverse order operation of the system. The embedded information will be decoded by the same incremental router, i.e. the incremental router generates exactly the same shape of wire if the net allocated '1'.

The final copyright information will be obtained by DES decoder with user's private key from the bit string. We consider this verification process should be managed by IP owner or entrusted silicon foundries.

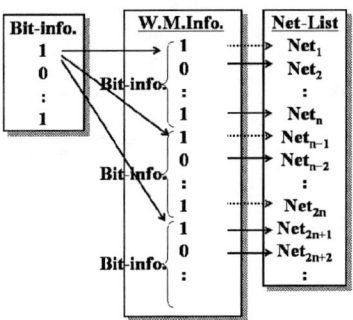

Figure 2 Cross-Reference of Net versus Bit information

B. Copyright Encoding by DES System

We employed DATA ENCRYPTION STANDARD (DES) system to encode copyright into the bit string. As the limit of this paper, we will not describe it in detail here.

C. Incremental Rip-Up and Re-Router

We developed a special incremental router which embeds bit information into the net in post-layout design stage. There are two kinds of requirements for the router as follows:

1) With no influences to performance and area size,

2) The obtained layout must be almost similar to the original to hide the embedded information.

Our incremental router embeds bit '1' by inserting extra bends at the corners of the wire of corresponding net. Since the general router has no meaning to insert such redundant bends, it is possible to distinguish from the original. Furthermore the bends around the corner does not change the wire length and the global shape of wire, therefore it does not damage circuit performance and it is difficult to detect by eyes. Our incremental router is implemented by following steps based on the multilayer maze routing method.

Incremental Routing Algorithm

Step1. Obtain the corner points in the initial wire.

Step2. Rip the original wire up.

Step3. Set obstacles around the possible corner point.

Step4. Re-route corner points using maze router.

Step5. If obtained wire includes more number of corners than that of the original, then go to Step7

Step6. If re-router failed to interconnect corners, then output "fail" message.

Step7. Stop.

The incremental router rips a wire up, and then searches the insertion region for redundant bends around the corners of the wire. In the Fig.3 (1), we show an example of wire of a net in layout design in "L" shape. Generally a router for layout design seeks the shortest path shape of a net including all corresponding terminals, while our incremental router seeks the redundant area around the corners of the original path. By setting barriers nearby the corners, the incremental router generates redundant path shape.

This will not change the wire length between of terminals and the global shape of the wire, thus the incremental wire has almost the same physical design properties. Those operations also do not change the area size at all. Furthermore, these extra bends are difficult to distinguish from the original corners, in other words, they are transparent.

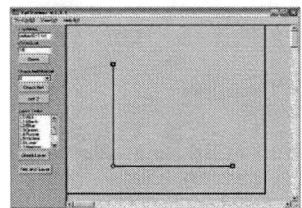 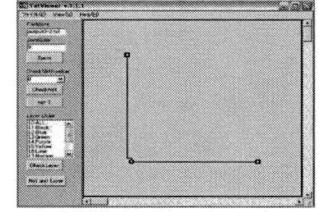

(1) Original wire shape (2) Embedding an extra bend
Figure3 Increment router for Watermarking

When the incremental router cannot find the new path shape, it returns 'fail' message. Thus our system embeds the bit string using incremental router into the layout design iteratively to enhance success possibility. This iteration makes sense for the crowded wiring area design.

D. Copyright Decoding

To ensure the copyright of IP reuse, the watermarking information will be decoded by the reverse order of embedding watermarks, if a net is allocated as '1', then the incremental router provide the same shape of a original wire, while it provides the different shape if the net is "0". Then the decoded bit string will be show the copyright information with user's private key and DES decoder.

III. EXPERIMENTAL RESULTS

We evaluated our post-layout watermarking system by hundreds of experiments. The system is implemented by C language of Cygwin. CPU was 2GHz Celeron with 512MB main memory.

The benchmark circuits and their layout designs are generated by using randomly distributed triplet terminals and a multilayer maze router. All of generated net-lists consist of 192 nets with 576 terminals that afford a few iteratively embedding the minimum unit of DES, i.e. "64 bit string". There are five categories for the generated circuits with the 150 grids x 150 grids area.

Fig. 4 shows the terminal distributions of benchmark circuits including R1 "local" networks and R2 "global" networks.

The benchmark circuits are categorized into five types according to the ratios of R1 and R2 respectively. We prepared thirty benchmark circuits which the numbers of terminals is in the ratio R1:R2. (R1:R2) are set (5:1), (4:1), (3:1), (2:1) and (1:1). Therefore 150 benchmark circuits are prepared.

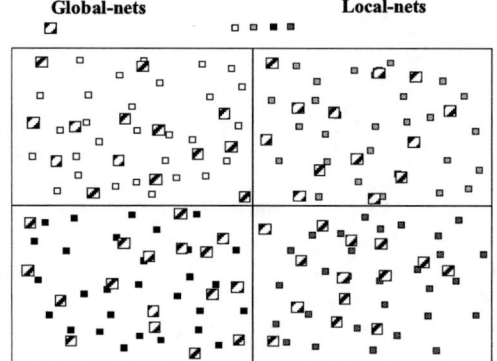

Figure4 Local nets and Global nets

We performed three iterative watermarking embedding to each benchmark circuit. The experimental results are shown in Table 2. The success probability per bit under one embedding is about 81 to 87%, however the success probability for the whole bit string is less than 5% in any cases. Contrarily success probability after three times' embedding is up to 90%~96%.

Table 2 Experimental results for benchmark circuits

Benchmark Type	Test Type	SUCC/TOTAL	SUCC Pro.	Avg. Succ. Pro. For bit
5:1	1 time's WM	1/30	3.33%	81.49%
	2 times' WM	21/30	70%	83.97%
	3 times' WM	29/30	96.67%	81.34%
4:1	1 time's WM	1/30	3.33%	87.55%
	2 times' WM	21/30	70%	86.08%
	3 times' WM	29/30	96.67%	88.86%
3:1	1 time's WM	0/30	0.00%	85.59%
	2 times' WM	16/30	53.33%	85.54%
	3 times' WM	28/30	93.33%	86.50%
2:1	1 time's WM	0/30	0.00%	85.98%
	2 times' WM	20/30	66.67%	87.78%
	3 times' WM	29/30	96.67%	87.83%
1:1	1 time's WM	1/30	3.33%	86.57%
	2 times' WM	19/30	63.33%	85.10%
	3 times' WM	27/30	90.00%	85.49%

We estimated the iterate number and the success possibility in Figure 5. It shows that if we can embed the bit string more than 5 or 6 times, the system will achieve 100% success probability, i.e. the 64 bit string can be embedded by our system if the number of net is more than 384(= 6x 64). We consider that the amount of nets is not so little for the layout design IP today.

To the several general ways of attacking watermarking scheme, our approach has a certain resistance. For the embedding attacks, the intruder has to choose to have a ghost searching because he/she has no our incremental router tool. It is difficult to break the encryption protocol, i.e. the breaking probability P will be less than 2^{-56} dues to DES with 56bit key, which is considered to be a practically impossible task. In Fig.6, we can see that the watermarking by our approach is indistinguishable.

IV. Conclusion

We propose a post-layout design watermarking system in this paper. The copyright information of IP owner is DES coded and then repeatedly embedded to the layout design IP using incremental router. This system is expected to preserve the circuit performance of layout design IP. We evaluated various benchmark circuits in several wire density and found that the system is in enough success possibility for practical layout design IP.

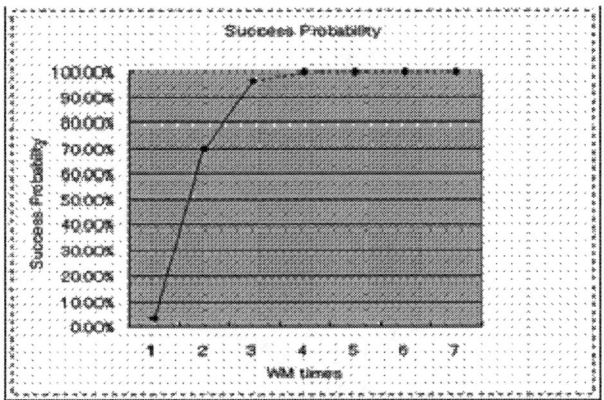

Figure5 Success possibilities to the number of iterations

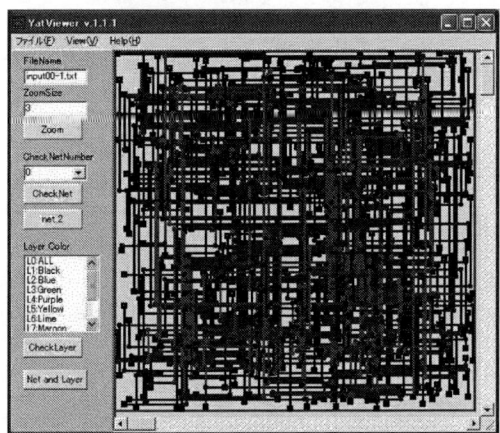

(1) Original physical layout design

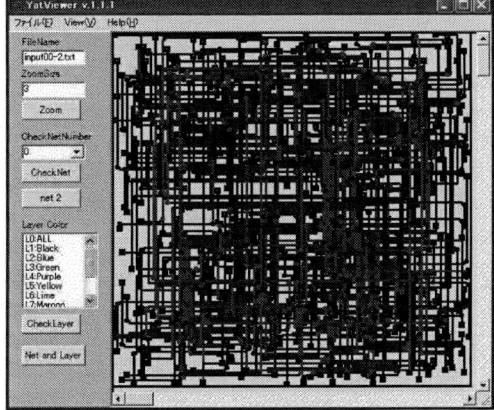

(2) Copyright Embedded physical layout design

Figure 6 Watermarking Results

Acknowledgment

Authors would like to thank to RSP (Regional Science Promotion Program) of JST (Japan Science and Technology Agency) for their supports. Authors would also like to thank to Prof. Shigeo Kuninobu and Prof. Ken-ichi Shiota of Kochi University for their technical suggestions.

References

[1] Nie Ting-yuan, Masahiko Toyonaga, Ken-ichi Shiota, "A watermarking system for VLSI layout using a special router," 2003 Shikoku-section joint convention record of the institutes of electrical and related engineers (in Japanese), 9-11 2003.

[2] Masato Kisaka, Nie Ting-yuan, Masahiko Toyonaga, "A special knob-router for layout water-marking", 2003 Shikoku-section joint convention record of the institutes of electrical and related engineers (in Japanese), 1-11,2003.

[3] D. Kirovski, Y.-Y. Hwang, M. Potkonjak and J. Cong "Intellectual Property Protection by Water-marking Combinational Logic Synthesis Solutions," Proc. ACM/IEEE International Conference on Computer Aided Design, San Jose, California, pp. 194-198, November 1998.

[4] Andrew B. Kahng, John Lach, William. H. Mangione-Smith, Stefanus Mantik, Igor L. Markov, Miodrag Potkonjak, Paul Tucker, Huijuan Wang, and Gregory Wolfe,"Constraint-Based Water-marking Techniques for Design IP Protection," IEEE Transactions on Computer-Aided Design of Integrated circuits and Systems, VOL. 20, NO. 10, pp. 1236-1252, October 2001.

[5] S.H. Kwok, C.C. Yang, K.Y. Tam, "Water-mark Design Pattern for Intellectual Property Protection in Electronic Commerce Applications", 33rd Hawaii International Conference on System Sciences-Volume 6, p.6038, January 04 - 07, 2000.

[6] John Lach, William H. Mangione-Smith, Miodrag Potkonjak, "Fingerprinting Techniques for Field-Programmable Gate Array Intellectual Property Protection," IEEE Transactions on Computer-Aided Design of Integrated circuits and Systems, VOL. 20, NO. 10, pp. 1253-1261, October 2001.

[7] A. B. Kahng, J. Lach, W. H. Mangione-Smith, S. Mantik, I. L. Markov, M. Potkonjak, P. Tucker, H. Wang and G. Wolfe, "Water-marking Techniques for Intellectual Property Protection", 35th Design Automation Conference, pp.776-781, June 1998.

[8] Andrew B. Kahng, Stefanus Mantik, Igor L. Markov, Miodrag Potkonjak, Paul Tucker†, Huijuan Wang and Gregory Wolfe, "Robust IP Water-marking Methodologies for Physical design", 35th Conference on Design Automation Conference (DAC'98), pp. 782-787,June 15 - 19, 1998.

[9] Greg Wolfe, Jennifer L. Wong, and Miodrag Potkonjak, "Water-marking Graph Partitioning Solutions," 38th Conference on Design Automation (DAC'01), pp. 486-489, June 18 - 22, 2001.

[10] Miodrag Potkonjak, "Water-marking While Preserving The Critical Path", 37th Conference on Design Automation (DAC'00), pp. 108-111, June 05 - 09, 2000.

[11] Amr T. Abdel-Hamid, Sofi`ene Tahar, El Mostapha Aboulhamid, "IP Water-marking Techniques: Survey and Comparison", The 3rd IEEE International Workshop on System-on-Chip for Real-Time Applications (IWSOC'03), p. 60,June 30 – July 02,2003.

Rapid and Precise Instruction Set Evaluation for Application Specific Processor Design

Masayuki Masuda   Kazuhito Ito

Department of Elecrical and Electronic Systems
Saitama University
255 Shimookubo, Sakura-ku, Saitama 338–8570, Japan
Email: {masuda,kazuhito}@elc.ees.saitama-u.ac.jp

Abstract— The selection of instruction set of a processor greatly influences the processor hardware and execution of software in speed, area, and power. Evaluation of instruction set is an important task in designing a processor specific to a given application. In this paper, a technique to rapidly and precisely evaluate instruction sets for the given application is proposed. It uses efficient branch and bound to explore the combination of instructions and evaluates the execution steps by task scheduling. The results show the proposed technique efficiently evaluates instruction sets for assumed processor hardware.

I. Introduction

With the advances in EDA tools for designing and synthesizing custom processors [1-3] and the advances in automated generation techniques of software development system for the processor [1,4,5], it is becoming practical to design a processor which is specific to a particular application.

In general, there exist many possible configurations in designing a processor, which is the combination of the types and the number of functional units, the number of registers, interconnection topology of these elements, the number of pipeline stages, and so on. The instruction set is also an important aspect of processors. For the specific application, each configuration results in different performances of execution speed, chip area, power consumption, etc. Among these alternative configurations, one configuration is selected to implement the processor which best meets the requirements of speed, area, and power.

Since it is very difficult to directly generate the best configuration for the given application, several candidates are generated and evaluated, and then the best one is selected. The evaluation consists of hardware evaluation and software evaluation. The hardware evaluation implies synthesis of the processor circuits and physical design (placement and routing) to determine the clock speed and chip area of the processor. The software evaluation implies scheduling the tasks within the given application to determine the application execution speed and chip area for program memory and work memory. The software evaluation may also include simulation to determine the power consumption based on the processor hardware and schedule of tasks.

In software evaluation, an instruction set is assumed for the given application and the assumed hardware configuration. Then the tasks (operations, conditional branches, etc.) within the application are scheduled to achieve the shortest execution steps with respect to the restriction of the instruction set. The assumption of an instruction set imposes restrictions on task scheduling. For example, the number of simultaneous invocation of operations is limited by the definition of instruction set. Therefore, selection of the instruction set greatly influences the final performance of the processor. The software evaluation implies the evaluation of instruction set. For the precise evaluation of instruction set, it is crucial to determine the number of execution steps of the application as precisely as possible by scheduling the tasks with respect to the restriction imposed by the instruction set.

In this paper, we propose a technique to rapidly and precisely determine the execution steps of the given application with respect to the assumed hardware configuration and assumed instruction set.

II. Related Works

Paying attention to the software evaluation, the existing works can be categorized into two types. One is to compile the given application with a general compiler customized to the designed processor [1]. The optimality strongly depends on the general compiler. The other is to first schedule tasks in given application and then generate instruction set to implement tasks as scheduled [4], [5]. This tends to result in large instruction set.

In [6] covering problem is used for optimal compilation for processors with a special architecture like DSP. This approach can be used for precise instruction set evaluation. One drawback is the CPU time. The technique to speed up the solution of covering problem is necessary.

III. Formulation of Problem

A. DAG

The application processing is given as a data-flow graph (DFG). A DFG consists of nodes and arcs which represent operations and data dependencies, respectively. A DFG implicitly expresses that the process is iteratively executed indefinitely. Arcs may have delays which represent data dependency between iterations.

The DFG can be obtained by data-flow analysis of software written in high-level programming language or directly from digital signal processing area. The DFG must be acyclic, that is, directed acyclic graph (DAG). If the given DFG is cyclic, it is converted to DAG by eliminating arcs with delay. Figure 1 (a) shows an example of DAG. In Fig. 1 (a), any primary input, primary output, and data dependency between iterations are removed for simplicity.

B. DAG Covering

An instruction executes one or more nodes. The set of the nodes executed by an instruction is called a *match*. Each node in the DAG must be executed at least once. *DAG covering* [6] is defined as the combination of matches where every node is included in at least one match. The *DAG covering problem* is to determine the optimal DAG covering.

C. Scheduling Constraints

Once a legal DAG covering is obtained, matches are scheduled so that every data dependency is maintained. In scheduling matches of a DAG covering, the following constraints must be satisfied.

1) Serialized instruction execution
   In each execution step, at most one instruction is initiated. This is the fundamental restriction on instruction execution of scalar processors.
2) Precedence constraint caused by data dependency
   Every node n must be executed no earlier than the end of nodes which generate data consumed by the node n.
3) Hardware constraint
   The number of nodes executed at the same time is no more than the number of functional units.

D. Cost of DAG Covering

In the problem of instruction set evaluation, the cost of DAG covering is defined as the required number of steps (clock cycles) to execute the matches in the DAG covering. This is illustrated by using a simple example.

The processor configuration is assumed to include one multiplier and two adders. The possible matches are induced by the assumed instruction set. For the DAG shown in Fig. 1 (a) and the following instructions, ADD (single addition), MUL (single multiplication), MUL·ADD (execute multiplication and then addition), and ADD+ADD (execute two additions in parallel), Fig. 1 (b), (c), (d) show the possible matches. The matches m1, m2, m4 correspond to ADD instruction, m3 to MUL, m5 and m6 to MUL·ADD, and m7 and m8 to ADD+ADD. The match corresponding to the execution of one node is called *single match*. The match corresponding to successive execution of nodes is called *serial match*. The match corresponding to parallel execution of nodes is called *parallel match*. The matches m1, m2, m3, and m4 are single matches, m5 and m6 are serial matches, and m7 and m8 are parallel matches.

Two DAG coverings for Fig. 1 are shown in Fig. 2. All the nodes n1, n2, n3, and n4 are included in at least one match. Therefore, both DAG coverings of Fig. 2 (a) and (b) are legal. Each DAG covering of Fig. 2 results in the program consisting of three instructions.

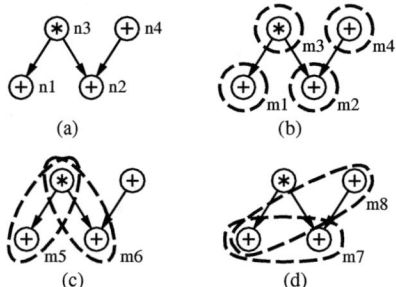

Fig. 1. Matches. (a) DAG (b)(c)(d) possible matches.

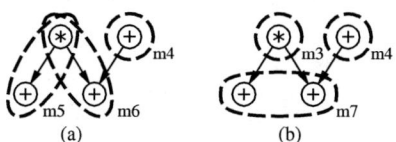

Fig. 2. DAG coverings.

Step	Match schedule	Node schedule			Step	Match schedule	Node schedule		
		M	A1	A2			M	A1	A2
0	m5	n3			0	m3	n3	n4	
1					1	m4			
2	m6	n3	n1		2	m7		n1	n2
3	m4		n4		3				
4			n2		4				
(a)					(b)				

Fig. 3. Schedule of DAG covering of Fig. 2

For each DAG covering, list scheduling [7] is applied to schedule the matches. The optimized schedules for these two DAG coverings are shown in Fig.3 (a) and (b), respectively. In the schedule shown in Fig. 3 (a), the instruction m5 is scheduled to start at step 0. Since the given hardware contains only one multiplier, the instruction m6, which uses the multiplier, is scheduled at step 2 to wait until the multiplier becomes available (here we assume the multiplier is not pipelined). The required number of steps for the DAG covering results in 5. In the schedule Fig. 3 (b), the instruction m3 is scheduled at step 0. The addition of m4 is delayed to step 1 to meet the scheduling constraint 1. The instruction m7 starts at step 2 to satisfy the scheduling constraint 2, i.e., the data dependency between node n3 and n4 to n1 and n2. The required number of steps for the DAG covering results in 3. Thus the DAG covering of Fig. 2 (b) is better than that of Fig. 2 (a).

IV. SOLUTION OF DAG COVERING

DAG covering problem can be solved by the depth first search of the search tree described below. Let the *DAG node* denote the node in DAG and *search node* denote the node in the search tree, respectively.

Fig. 4 shows the search tree for the DAG covering of the DAG and possible matches shown in Fig. 1. The root search node N_0 corresponds to the DAG nodes which does not have outgoing arcs. At the root search node, no match is selected for DAG covering. As shown in Fig. 4, DAG nodes to be included by matches are indicated in the lower half of search nodes. Except for the root node, each search node is assigned one or more matches. The matches are indicated in the upper half of search nodes. Instead of selecting matches one by one, a combination of more than one matches may be selected in a search node. The combination is determined so that it is feasible and avoids useless overlaps of DAG nodes. This minimizes the number of search nodes and thus speeds up the exploration. When the search reaches at search node N_x, it implies that the matches assigned to search node N_x and all the ancestor search nodes are selected for DAG covering. In other words, each search node corresponds to a partial DAG covering. Let $M(N_x)$ denote the set of matches of the partial DAG covering corresponding to search node N_x.

It must be noted that search node N_8 is assigned the match m3. The DAG node n3 is already included by the match m6 at the search node N_2. The DAG node n3 is executed as the MUL part of m6 and its result is sent only to the DAG node n2. The execution of DAG node n1 also requires the result of n3, but it is not obtained by the match m6. Therefore n3 needs to be executed again by the instruction corresponding to m3 to obtain n3 result for execution of n1. This is the reason why the match m3 is assigned to search node N_8. Alternatively, DAG nodes n3 and n1 can be executed by a single MUL·ADD instruction. This corresponds to the path (N_0, N_4, N_{11}).

Parallel match is selected only when all the parallel nodes are not yet included or one of the parallel nodes is included by a single match. When the parallel match is selected, that single match is no longer necessary and removed from the DAG covering. For example in Fig. 4, a parallel match m8 is assigned to search node N_7. If the search reaches N_7, the matches m2, m3, and m8 are selected and the match m1 is removed. Removal of the single match is indicated as '−m1' nearby the arc between two search nodes as shown in Fig. 4.

If the search reaches the node with the empty lower half, search node N_6 for example, all the DAG nodes have been included by any of the matches, that is, a legal DAG covering is found. As shown in Fig. 4, there exist 7 legal DAG coverings (N_6 to N_{12}) to the DAG of Fig. 1(a).

V. Branch and Bound for DAG Covering

The search tree for finding the optimal DAG covering easily becomes very large since there exist many feasible combinations of matches. Branch and bound technique can be employed to minimize the search effort. The branch and bound is to omit unnecessary exploration of the search tree by pruning arcs which lead to non-promising solution candidates.

In addition to considering D-cycle [6], the following criteria are used to bound the search by taking the characteristics of the problem of DAG covering into account.

Let S denote the tentative smallest number of execution steps of the DAG. The initial value of S is infinite. S is updated when the search reaches a legal DAG covering and its execution steps are less than before. At a search node N_x, if

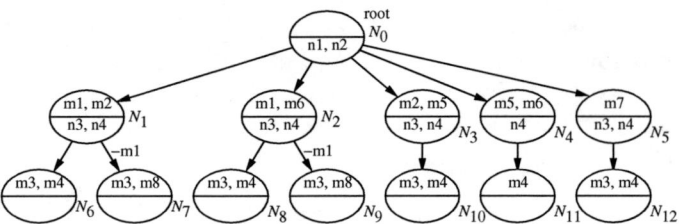

Fig. 4. DAG covering search tree of the DFG in Fig. 1(a)

one of the values L_x, $|M(N_x)|$, and P_x described below exceeds S, then the search does not proceed beyond N_x and tracks back. This corresponds to pruning the branch outgoing from N_x.

A. Estimated Lower Bound

At a search node N_x in DAG covering, DAG nodes are divided in to two subset Q_C are Q_N. A node in Q_C is included in DAG covering by one or more matches in $M(N_x)$. A node in Q_N are not yet included. Let the matches in $M(N_x)$ and the nodes in Q_N be scheduled so as to satisfy all the constraints 1, 2, and 3 described in III-C for nodes in $M(N_x)$ and the constraints 2 and 3 for DAG nodes in Q_N. Since the constraint 1 is not considered for future matches for Q_N, the number of execution steps, L_x, of this schedule is estimated generally smaller than or equal to the actual value. Thus L_x is the lower bound for any legal DAG covering which contains $M(N_x)$.

B. The Number of Matches

Since a match corresponds to an instruction and at most one instruction can be initiated in a step, the execution of the matches requires K or more steps if a DAG contains K matches. When the search reaches a search node N_x, some DAG nodes are included by matches $M(N_x)$. In that case at least $|M(N_x)|$ steps are needed to execute the DAG.

C. The Path Length

Let P_x denote the longest path length at a search node N_x. The required number of execution steps of a schedule is larger than or equal to P_x. Thun P_x is the lower bound of the execution steps of DAG.

VI. Fast Solution of Optimized DAG Covering

The above mentioned criteria guarantee the optimality of the solution, but the speeding up the exploration of the search tree is limited. To further speeding up the exploration, the following techniques are proposed at some possibility of degrading the solution optimality.

A. Reduction of Parallel Matches

Basically arbitrary parallel match can be used for DAG covering provided that no D-cycle is introduced. For example if ADD+ADD instruction exists in the instruction set, any pair of add nodes can be considered as a parallel match.

Selecting a parallel match introduces imaginary arcs to the DAG to constrain parallel execution of DAG nodes. Therefore, selecting a parallel match may derive long directed paths and hence results in long execution steps of the DAG. Parallel matches leading to such long execution steps would never be selected and could be excluded in advance of the search.

Let $l(n)$ denote the length of the longest path from the primary input of the DAG to DAG node n. For any pair of nodes n_1 and n_2, a parallel match corresponding to this pair is considered for DAG covering only if $|l(n_1) - l(n_2)| \leq C$ where C is a constraint parameter.

B. Dividing the DAG

One problem of DAG covering is that many parallel matches are generated if the instruction set includes the instruction to execute two or more nodes in parallel. Every combination of nodes corresponds to a parallel match. However, only small subset of these many parallel matches are adopted in the DAG covering. To limit parallel match, the technique is proposed where the given DAG is divided into sub-DAGs and these are DAG-covered in turn. Only parallel matches within the sub-DAGs are considered. This greatly reduces the solution time at the cost of possible degradation of the solution optimality.

VII. Experimental Results

The proposed software evaluation method is implemented by using C language and run on a PC with a 2GHz processor. DAGs are 5th-order wave elliptic filter (WEF) [8] and 8-point DCT (8DCT) [9]. WEF consists of 26 additions and 8 multiplications. 8DCT consists of 29 additions and 11 multiplications. From 8DCT, 4DCT is extracted which consists of 9 additions and 3 multiplications. For these DAGs, four instruction sets shown in Table I are selected for experiments.

Table II shows the optimal cost of DAG covering. These results are obtained without DAG division. The columns #M and #S show the number of matches and the shortest execution steps, respectively. The column #S under 'Basic B & B' shows the true optimal solutions. 'NA' indicates that the optimal DAG covering is not obtained within a practical time. The constraint parameter C is 2 in this experiment. The results show that the proposed parallel match reduction shortens CPU time for optimal DAG covering without degrading the solution optimality.

Table III shows the effect of DAG division. #M shows the number of matches of original (undivided) DAG. For each DAG, 10 patterns of dividing the DAG into two sub-DAGs are used. The search is terminated if CPU time exceeds 90 minutes for each pattern. The CPU times in Table III are the sum of CPU times of 10 patterns. In these experiments, the optimal (or near-optimal in a few cases) DAG coverings are obtained within a practical time by DAG division.

The proposed technique can derive the fact in a practical time, that instruction sets 3 and 4 are better than instruction set 2 for the given DFGs.

VIII. Conclusions

In this paper, a technique to rapidly and precisely evaluate instruction sets for the given application is proposed. By selecting more than one matches in DAG covering, using proposed lower bounds, and DAG division, the exploration of DAG covering is done efficiently. The proposed technique can be embedded in the application specific processor design flow. The further speeding up of the search process, sophisticated choice of DAG division, and consideration for the register count limit remain as future work.

TABLE I
Instruction Sets

ID	Instructions
1	MUL, ADD
2	MUL, ADD, MUL·ADD, ADD·ADD
3	MUL, ADD, MUL·ADD, ADD+ADD
4	MUL, ADD, MUL+ADD, ADD+ADD

TABLE II
Results of DAG Covering

DAG	IS	Basic B & B			P-Match Reduction		
		#M	#S	CPU	#M	#S	CPU
WEF	1	12	12	0sec	12	12	0sec
	2	22	10	0sec	22	10	0sec
	3	40	8	0sec	32	8	0sec
	4	54	8	0sec	44	8	0sec
4DCT	1	34	35	0sec	34	35	0sec
	2	73	24	54sec	73	24	54sec
	3	190	NA	NA	98	22	15sec
	4	284	NA	NA	132	22	27sec

TABLE III
Results of DAG Covering with DAG Division

DAG	IS	#M	w/o DAG div		with DAG div	
			#S	CPU	#S	CPU
WEF	1	34	35	0sec	35	0sec
	2	73	24	54sec	25	28sec
	3	98	22	15sec	22	14sec
	4	132	22	27sec	22	342sec
8DCT	1	40	10	0sec	10	0sec
	2	82	32	381min	33	103min
	3	216	NA	NA	25	548min
	4	368	NA	NA	23	671min

References

[1] B. Shackleford, et al., "Satsuki: An Integrated Processor Synthesis and Compiler Generation System," IEICE Trans. Inf. & Syst., vol. E79-D, No. 10, pp. 1373-1381, 1996.

[2] M. Itoh, et al., "Processor Generation Method for Pipelined Processors in Consideration with Pipeline Hazards," IPSJ Journal, vol. 41, No. 4, pp. 851-862, 2000.

[3] T. Sasaki, et al. "Rapid Prototyping of Complex Instructions for Embedded Processors using PEAS-III," Proc. SASIMI 2000, pp. 61-66, 2000.

[4] N. Ishiura, T. Watanabe, and M. Yamaguchi, "A Code Generation Method for Datapath Oriented Application Specific Processor Design," Proc. SASIMI 2000, pp. 71-78, 2000.

[5] O. Wahlen, et al., "Instruction Scheduler Generation for Retargetable compilation," IEEE Design & Test, vol. 20, No. 1, pp. 34-41, 2003.

[6] Stan Liao, et al., "Instruction Selection Using Binate Covering for Code Size Optimization," Proc. Int. Conf. Computer-Aided Design, pp. 393-399, 1995.

[7] M. C. McFarland, et al., "The High-Level Synthesis of Digital Systems," Proc. IEEE, vol. 78, No. 2, pp. 301-318, 1990.

[8] Sonia M. Heemstra de Groot et al., "Range-Chart-Guided Iterative Data-Flow Graph Scheduling," IEEE Trans. Circuits Syst.-I: Fund. Theory & Appl., vol. CAS-39, No. 5, pp. 351-364, 1992.

[9] C. Loeffler et al., "Practical Fast 1-D DCT Algorithms with 11 Multiplications," Proc. IEEE ICASSP, pp. 988-991, 1989.

Modern Floorplanning with Abutment and Fixed-Outline Constraints

Chang-Tzu Lin, De-Sheng Chen, Yi-Wen Wang, Hsin-Hsien Ho
Department of Information Engineering and Computer Science
Feng Chia University, Taichung, Taiwan

Abstract

Typical floorplanning problem concerns a series of objectives, such as *area, wirelength* and *routability*, etc., without any specific constraint in a *free-outline* style. Entering SOC era, however, modern floorplanning takes more care of providing extra options to place dedicated modules in the hierarchical designs, such as abutment, boundary and fixed-outline constraints, etc. It has been empirically shown that any of the modern constraints extremely restricts the solution space, that is, a large number of randomly generated floorplans might violate the constraint. This paper addresses modern floorplanning with abutment and fixed-outline constraints. In order to search the drastically limited solution space, we first investigate the feasible properties of a slicing floorplan with abutment constraint. The properties, coupled with an efficient evolutionary search algorithm, provide the way to produce floorplans with abutment constraint. We then extend the algorithm with minor modification to enable the abutment floorplans to be gradually fit into the desirable fixed outline. The methods are verified by using the MCNC and GSRC benchmarks, and the empirical results show that our methods can obtain promising solutions using short time.

1. Introduction

Floorplanning plays an important role in physical design of VLSI circuits and the problem was shown to be NP-complete [1]. Classical floorplanning formulation roughly determines the layout of a given set of modules, such that no modules overlap, and the enclosing layout region has minimum area and interconnection [2-5]. In this floorplanning step, it is common that designers will put additional placement constraints on the final packing to get meaningful designs for different purposes. For instance, abutment constraint is to have several dedicated modules in a circuit abut one after another to favor the transmission of data between them; boundary constraint is to arrange some I/O involved modules along the chip boundary to minimize off-chip connections. Therefore, in addition to the conventional area and interconnection requirement, modern floorplanning must handle different placement constraints efficiently. Meanwhile, in a hierarchical design flow, classical floorplanning formulation may generate a floorplan that is completely useless for a situation where its outline is dissatisfied. As a result, modern floorplanning must handle the fixed-outline constraint, which often occurs in the hierarchical design style, effectively as well.

Several works were proposed to handle floorplanning with either abutment or fixed-outline constraint, separately. Young et al. [6] first proposed an algorithm to handle abutment constraint based on slicing structure. Afterward, Ma et al. [7] proposed another algorithm to handle abutment constraint based on well-known nonslicing representation, Corner Block List [8]. The common strategy they used was to inspect the intermediate solutions during the simulated annealing process and repair the solutions by heuristics in case the constraints are violated. If violations still exist during annealing process, a penalty function is used. Adya et al. [9-10] first suggested new objective functions to drive simulated annealing and new types of moves to help exploiting fixed-outline satisfied floorplans. The new move may be helpful to reduce critical path. However, it does not guarantee to produce a floorplan with fixed-outline constraint.

In this paper, we handle the modern floorplanning with the abutment and fixed-outline constraints using the slicing representation. We first investigate the feasible properties of a slicing floorplan with abutment constraints. The properties, coupled with an efficient evolutionary search algorithm, provide the way to produce floorplans with abutment constraint. We then extend the algorithm with minor modification to enable the abutment floorplans to be gradually fit into the desirable fixed outline. The methods are verified by MCNC and GSRC benchmarks, ami49 and n100, which have 49 and 100 modules in totally different style of circuits, respectively. The empirical results show that our approach can successfully achieve the goal of efficiently generating floorplans with abutment and fixed-outline constraints.

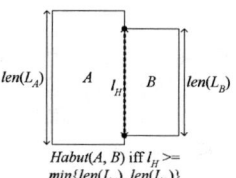

Figure 1. Definition of the horizontal abutment constraint.

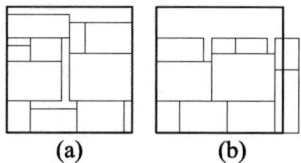

Figure 2. Fixed-outline floorplan (a) satisfied, (b) dissatisfied. (The bolder rectangle is the desirable fixed outline.)

2. Preliminaries

2.1. Abutment Constraint

As defined in [6], two modules A and B are said to be abutting horizontally (see Figure 1), denoted by $Habut(A, B)$, if a vertical boundary L_A of module A and a vertical boundary L_B of module B abut such that L_A lays immediately on the left of L_B and the length of the abutment l_H is at least $min\{len(L_A), len(L_B)\}$, where $len(L_A)$ is the length of L_A and $len(L_B)$ is the length of L_B. The abutment in the vertical direction is defined similarly. Notice that an abutment constraint can be composed of multiple modules. For instance, modules X_1, X_2 and X_3 are said to be abutting horizontally, denoted by $Habut(X_1, X_2, X_3)$, if the vertical distances between these modules are zero (i.e. they will align horizontally), and they are horizontally adjacent. In addition, one module cannot simultaneously belong to both horizontal and vertical abutment constraints.

2.2. Fixed-Outline Constraint

The modern floorplanning formulation defined in [9] is an "inside-out" version of the classical outline-free floorplanning formulation — the aspect ratio of the floorplan is fixed, but the aspect ratios of the modules can vary. That is, modern floorplanning formulation should be cast as a fixed-outline problem, and the packing must simultaneously achieve zero whitespace and zero overlap for the given choice of fixed die. Figure 2 shows the layout of a fixed-outline chip.

2.3. Problem Definition

A module M is a rectangle of height $h(M)$, width $w(M)$, and area $A(M)$. The *aspect ratio* of M is defined as $h(M)/w(M)$. The modules can be either *hard* or *soft*. The height and width of a hard module are fixed, but the module is free to rotate. The shape of a soft module can be

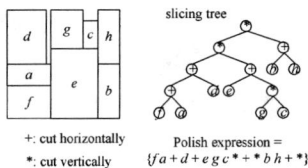

Figure 3. Slicing tree representation and its corresponding Polish expression of a slicing floorplan.

changed as long as the area remains a constant and the aspect ratio is within a given range. A *super-module* consists of several modules, also called a *sub-floorplan*. A floorplan for *n* modules consists of an enveloping rectangle *R* subdivided by horizontal lines and vertical lines into *n* non-overlapping rectangles such that each rectangle must be large enough to accommodate the module assigned to it. In our problem, we are given two kinds of soft modules $M_S = M_F \cup M_C$. The free modules in M_F have freedom to arbitrarily move while the modules in M_C are restricted to be abutted in the final floorplan, i.e. abutment constraints.

A *feasible* floorplan, *R*, is a floorplan that all abutment constraints are satisfied. Our objective is thus to construct a feasible floorplan *R* subject to fixed outline constraint.

2.4. Floorplan Representations
Floorplans can be divided into two categories, the *slicing structure* [11] and the *nonslicing structure* [12-16]. A slicing floorplan can be obtained by recursively cutting itself into two parts by either a vertical line or a horizontal line; however, a nonslicing floorplan is contrary. As shown in Figure 3, a slicing structure can be represented by a binary tree whose leaves denote modules, and internal nodes specify horizontal or vertical cut lines. Traversing the binary tree in postorder, we obtain a Polish expression of length $2n - 1$ for the slicing floorplan. There are several advantages of using slicing floorplan. First, focusing only on slicing floorplan significantly reduces the search space, which in turn leads to a faster runtime. Second, the shape flexibility of the soft modules can be fully exploited to give a tight packing. It has been proved mathematically it is achievable for packing slicing floorplans tightly [17] if the modules are flexible in shape.

3. Algorithm Description
3.1. Handling Abutment Constraint
Consider an abutment constraint consisting of a set of modules *C*. Let $\mathcal{N}$ be a subset of *C*. $\mathcal{N}$ contains *n* modules, b_i, $i = 1, 2, ..., n$, and coordinates (x_i, y_i) referring to the lower-left corner, respectively. Some definitions are given as follows,

Definition 1: A *free sub-floorplan* (F_f) is a sub-floorplan consisting of free modules or *completely Abut sub-floorplans*, which will be defined below.

Definition 2: A *strictly Habut sub-floorplan* (F_h) is a sub-floorplan consisting of $\mathcal{N}$ and free sub-floorplans such that
(1) modules in $\mathcal{N}$ are abutted one-by-one horizontally, $x_i + w(b_i) = x_{i+1}$, $1 \leq i \leq n-1$, and (x_1, y_1) is placed at the lower-left corner of the sub-floorplan.
(2) the width of the sub-floorplan is $\Sigma w(b_i)$, $1 \leq i \leq n$,

Definition 3: A *strictly Vabut sub-floorplan* (F_v) is a sub-floorplan consisting of $\mathcal{N}$ and free sub-floorplans such that
(1) modules in $\mathcal{N}$ are abutted one-by-one vertically, $y_i + h(b_i) = y_{i+1}$, $1 \leq i \leq n-1$, and (x_1, y_1) is placed at the lower-left corner of the sub-floorplan.
(2) the height of the sub-floorplan is $\Sigma h(b_i)$, $1 \leq i \leq n$,

Definition 4: An F_h or an F_v is a *completely Abut sub-floorplan* (F_c), if $\mathcal{N}$ is equal to *C*.

When consider combining two modules A and B vertically to obtain AB+, we have the following two lemmas.

Lemma 1: $\underline{A\,B\,+}$ is an F_h if A is an F_h, B is an F_f, and $w(A) \geq w(B)$.
Lemma 2: $\underline{A\,B\,+}$ is an F_v if A is an F_v, and B is an F_v.

The proofs for the lemmas are trivial, and thus are omitted. The cases where we combine two modules horizontally to obtain AB* can be considered similarly.

Lemma 3: $\underline{A\,B\,*}$ is an F_h if A is an F_h, and B is an F_h.
Lemma 4: $\underline{A\,B\,*}$ is an F_v if A is an F_v, B is an F_f, and $h(A) \geq h(B)$.

Input: A set of modules $\{m_1, m_2, ..., m_n\}$
Output: An area-efficient floorplan that satisfy abutment constraint
1. Randomly generating a set of floorplans to form the *initial population* Π ;
2. Evaluate area and wirelength of each floorplan;
3. Initialize the threshold value T_s;
4. **WHILE** (**NOT** exceeding the set # of generation) {
5. $CP1 = CP2 = \{\varnothing\}$;
6. **FOR** *i* **IN** 1 **TO** *cr*popsize* **LOOP**
7. Randomly choose P_1 and P_2 from Π ;
8. $CP1 = CP1 \cup \text{Crossover}(P_1, P_2)$;
9. **end LOOP**;
10. **FOR** *i* **IN** 1 **TO** *mr*popsize* **LOOP**
11. Randomly choose *P* from Π ;
12. $CP2 = CP2 \cup \text{Mutation}(P)$;
13. **end LOOP**;
14. Selection(Π, CP1, CP2);
15. **IF** (*period*)
16. Update T_s with a little value Δ; }

Figure 4. Overall procedure of the evolutionary algorithm.

To obtain a floorplan satisfying the given abutment constraints, we have the following Theorem.

Theorem 1: *The final floorplan satisfies the given set of abutment constraints, if the composition of modules follows Lemma1 through Lemma4.*

3.2. Floorplanning with Abutment Constraint
Based on Theorem 1, we develop a robust evolutionary algorithm to do solution exploration. The algorithm guarantees to generate slicing floorplans with abutment constraint. The overall procedure of the algorithm is described in Figure 4. Threshold value T_s is used in extracting area-efficient sub-floorplans. Two *Child Pools*, denoted as *CP*1 and *CP*2, are used in recording new floorplans generated from crossover and mutation. At the beginning, a set of floorplans is randomly generated to compose a *population* Π. The area and total wirelength of each floorplan is evaluated. Thereafter, the evolutional phases start. First, crossover randomly chooses two floorplans from Π to generate a new floorplan that satisfies the abutment constraint. The generated floorplan is put into *CP*1. Parameters *cr* and *popsize* correspond to *crossover rate* and *population size*, respectively. Note that a floorplan cannot be chosen twice in the same generation to prevent premature convergence. Thereafter, mutation proceeds to avoid local optimum and make diversity during the evolutionary process. The generated floorplan is put into *CP*2. Afterward, a new population whose population size is the same as $|\Pi|$ is selected. If the population is not updated within a given period of time, the *threshold value* T_s will be increased by a little value Δ. Finally, when the stop condition is met, the algorithm is completed.

3.2.1. Crossover
Figure 5 elaborates the process of crossover. The main compositions of crossover are *extract* and *merge*. During the evolutionary progress, the *extract* will choose the area-efficient sub-floorplans that are **strictly Habut** or **strictly Vabut**. The *merge* will combine the extracted sub-floorplans and the rest of modules, based on **Theorem 1**, to produce an area-efficient floorplan that satisfies the abutment constraint.

The input of crossover is a pair of floorplan P_1 and P_2. The output is also a floorplan that as far as possible inherits area-efficient sub-floorplans from P_1 and P_2 subject to the abutment constraint.

The fitness for each sub-floorplan SF_i of a floorplan is calculated by equation (1) to decide whether the SF_i should be inherited or not.

$$fitness(SF_i) = \frac{A(R) - \sum_{B_i \in R} A(B_i)}{A(R)} \quad (1)$$

,where $A(B_i)$ is the area of each module and $A(R)$ is the area of the floorplan *R*, consisted of modules $\{B_i| B_i \in R\}$. The fitness calculates the proportion of the dead area of sub-floorplans, that is, the more the fitness closes to zero, the less the dead area is. The threshold value, T_s, is initially set close to zero, which is used to differentiate the quality of sub-floorplans. If a sub-floorplan satisfies equation (2), we identify

Crossover1 (P_1, P_2) {
1. Extract all good sub-floorplans that are F_h or F_v from P_1, denoted as gs_1.
2. Extract all good sub-floorplans that are F_h or F_v from P_2, denoted as gs_2, subject to no modules in the gs_2 appeared in gs_1.
$GsP = gs_1 \cup gs_2$.
Regard the rest of modules as *degenerated good sub-floorplan*, denoted as *dgs*.
$GsP = GsP \cup dgs$.
3. Continue merging good sub-floorplans in *GsP*, based on **Theorem 1**, until $|GsP|=1$.
4. Evaluate the total area and interconnection of the generated floorplan. }

Figure 5. Process of the Crossover1.

Crossover2 (P_1, P_2) {
1. Extract all fixed-outline satisfiable good sub-floorplans that are F_h or F_v from P_1, denoted as gs_1.
2. Extract all fixed-outline satisfiable good sub-floorplans that are F_h and F_v from P_2, denoted as gs_2, subject to no modules in the gs_2 appeared in gs_1.
$GsP = gs_1 \cup gs_2$.
Regard the rest of modules as *degenerated good sub-floorplan*, denoted as *dgs*.
$GsP = GsP \cup dgs$.
3. Continue merging good sub-floorplans based on the Theorem 1 in *GsP* subject to fixed outline until there is no possible combination satisfying fixed-outline constraint. Afterward, the fixed-outline constraint is released to avoid infinite loop and repeat the process of merging.
4. Evaluate the total area and interconnection of the generated floorplan. }

Figure 7. Process of the Crossover2.

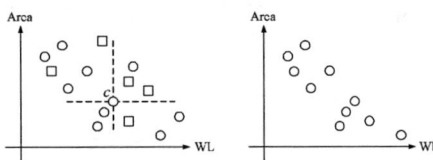

Figure 6. (a) Illustration for calculating *score(c)*, (b) The new developed *population*.

it as a *good sub-floorplan*.

$$fitness(SF_i) \leq T_s \quad (2)$$

Suppose that the number of good sub-floorplans of P_1 is larger than or equal to the number of good sub-floorplans of P_2. There are four main steps for crossover.

Step 1: Extract all good sub-floorplans that are *strictly Habut* or *strictly Vabut* from P_1, denoted as gs_1.

Step 2: Extract good sub-floorplans that are *strictly Habut* or *strictly Vabut* from P_2, denoted as gs_2, subject to no modules in the gs_2 appeared in gs_1. Then gs_1 and gs_2 are put into a *Good sub-floorplan Pool*, denoted *GsP*, i.e. $GsP = gs_1 \cup gs_2$. The rest of modules that are not appeared in *GsP* are viewed as *degenerated good sub-floorplans*, denoted as *dgs*. These modules are also put into *GsP*, i.e. $GsP = GsP \cup dgs$.

Step 3: This step is to enlarge sub-floorplans of *GsP* based on **Theorem 1**. First, a sub-floorplan is randomly selected from *GsP*. Next, the unselected sub-floorplans are sequentially chosen to combine with the selected sub-floorplan. Whether vertical or horizontal composition of the two sub-floorplans is randomly decided. Among the new generated sub-floorplans, the one with minimum fitness value is put into *GsP* and the corresponding two sub-floorplans are removed. This step continues until there exists only one floorplan in *GsP*.

Step 4: The final floorplan will be evaluated in terms of total area and interconnection.

3.2.2. Mutation

There are four mutation operators to be randomly performed in order to avoid local optimum and to refine the floorplan for reducing interconnection between modules, respectively.

Op1: *complement*: the complement operator is to change an originally relational operator to the other one in the Polish expression, such as + to * or * to +.

Op2: *swap*: the operator is to randomly swap a pair of sub-expressions.

Op3: *rotate*: the operator is to randomly rotate a module.

Op4: *exchange*: the exchange operator is to swap a pair of sibling sub-expressions.

3.2.3. Selection

After performing crossover and mutation, each floorplan in *CP1*, *CP2*, and Π is given a score, according to its area and total wirelength. The score is calculated based on the concept of *dominance relation* [18]. That is, suppose p and q are two solutions along with k objectives, denoted as p_i and q_i, p is said to dominate q, if and only if 1) $\forall i \in \{1, 2, ..., k\}: p_i \leq q_i$, and 2) $\exists j \in \{1, 2, ..., k\}: p_j < q_j$. For example, let p and q be two floorplans whose total area and wirelength correspond to *area(p)*, *wirelength(p)*, *area(q)*, and *wirelength(q)*, respectively.

Then p is said to dominate q, denoted as $p \square q$, if and only if 1) *area(p)* $\leq$ *area(q)* and *wirelength(p)* $\leq$ *wirelength(q)*, and 2) *area(p)* $<$ *area(q)* or *wirelength(p)* $<$ *wirelength(q)*.

Let f, g, and $c \in R^k$, and assume $f \square c$, and $c \square g$. We count the number of solutions that dominate c, denoted by $\sum |f \square c|$. We also count the number of solutions that dominated by c, denoted by $\sum |c \square g|$. Then the score for a solution, c, is defined as follows,

$$score(c) = \sum |c \angle g| - \sum |f \angle c| + popsize$$

Figure 6 illustrates the calculation for score of floorplan c, where objectives are total area and total wirelength. Let original *population* = {○}, *popsize* = 10, and $CP1 \cup CP2$ = {□}. Then $\sum |c \square g| = 3$ and $\sum |f \square c| = 2$. Hence, $score(c) = 3 - 2 + 10 = 11$. The scores for all floorplans are calculated and the ones with larger scores will have bigger probability to survive in the population.

3.3. Fixed-Outline Floorplanning with Abutment Constraint

In order to simultaneously satisfy the abutment and fixed-outline constraints, we extend the algorithm described in Section 3.2 with minor modification. The overall procedure of the algorithm is shown in Figure 7. The minor modifications focus on *extract* and *merge*, and are described as follows.

Steps 1 and 2 extract fixed-outline good sub-floorplans that are *strictly Habut* or *strictly Vabut* from P_1 and P_2, respectively. Notice that a fixed-outline good sub-floorplan is a sub-floorplan whose size is within the dimensions of the prescribed outline and the fitness satisfies equation (2).

Step 3 tries to enlarge sub-floorplans of *GsP* based on Theorem 1, subject to fixed-outline constraint, until a new floorplan is generated. However, if we cannot find any floorplan that satisfies both constraints, we will release the fixed-outline constraint to avoid the occurrence of infinite loop. Then, repeat this step until there exists only one floorplan in *GsP*.

4. Experimental Results

Our proposed method was applied to the larger cases of the MCNC and GSRC benchmarks (Table I). We implemented our algorithm in C++/STL on a PC with P4 1.8GHz CPU.

We will compare our method with the state-of-the-art approach for handling classical floorplanning with abutment constraint in Section 4.1. Subsequently, we will present the efficiency and effectiveness of our algorithm to tackle modern floorplanning with abutment and fixed-outline constraints. The following assumptions are adopted from [9] and will be used in our fixed-outline floorplanning formulation. For a given collection of modules with total area A and given *maximum percent of dead-space* γ, a fixed outline is constructed with expected fixed-outline ratio α, i.e. height/width. The *maximum percent of dead-space*, γ, is set to 15% in the final solutions. Hence, we have the height, H_*, and width, W_*, of outline of floorplan by the following equations.

$$H_* = \sqrt{(1+\gamma)A\alpha}, \quad W_* = \sqrt{(1+\gamma)A/\alpha}$$

If the current outline of found floorplan is smaller than H_* and W_*,

TABLE I
The MCNC and GSRC Benchmark Circuits.

Circuit	# modules	# nets	Total area (mm^2)
ami33	33	123	1.156
ami49	49	408	35.445
n100	100	885	0.1795

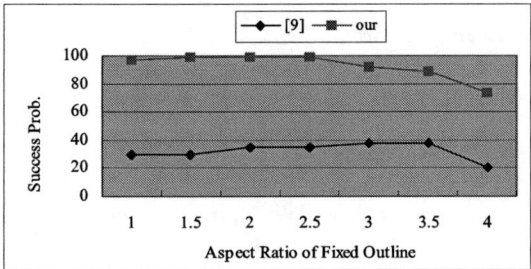

Figure 8. Successful Probability vs Aspect Ratio.

TABLE II
The Results of Floorplanning with Abutment Constraints.

Method	Examples	No. of constraints	Area (mm^2)	Deadspace (%)	Time (sec)	No. of violations
[6]	ami33	12	1.20 (1.16)	3.62 (0.75)	84.98 (79.45)	0 (0)
[6]	ami49	12	36.16 (35.79)	2.02 (0.94)	164.51 (136.36)	0.2 (0)
[Our]	ami33	12	1.157 (1.156)	0.08 (0)	17.6 (14)	0 (0)
[Our]	ami49	12	35.81 (35.445)	1.02 (0)	46.6 (39)	0 (0)

TABLE III
The Results of Fixed-Die Floorplanning With Abutment Constraints.

Examples	No. of abutment constraints	Fixed-outline Ratio	Avg Area (mm^2)	Avg Time (sec)	Best Area (mm^2)	Success Rate
ami49	8	1	35.7756	42.30	35.4454	100%
ami49	8	2	35.8304	40.32	35.4454	100%
ami49	8	3	35.6887	34.06	35.4454	100%
ami49	8	4	35.6295	33.20	35.4454	100%
ami49	10	1	35.7624	43.34	35.4454	100%
ami49	10	2	35.7286	38.82	35.4454	100%
ami49	10	3	35.6708	36.02	35.4454	100%
ami49	10	4	35.6102	32.92	35.4454	100%
n100	10	1	0.1913	84.10	0.1838	78%
n100	10	2	0.1909	68.39	0.1840	88%
n100	10	3	0.1905	67.89	0.1831	94%
n100	10	4	0.1888	65.38	0.1829	96%
n100	12	1	0.1931	61.05	0.1851	86%
n100	12	2	0.1910	50.88	0.1839	86%
n100	12	3	0.1894	55.82	0.1832	90%
n100	12	4	0.1887	51.28	0.1828	94%

its aspect ratio can be different from the aspect ratio of the fixed outline. We terminate the algorithm as soon as it finds a solution satisfying a given fixed outline. If the algorithm runs out of time and no solution is satisfied, we claim this evolution as failure. Figure 8 compares our method with [9] and shows plot of *the probability of success* of satisfying the fixed outline constraints vs desired aspect ratio of the fixed outline for the benchmark ami49. This plot reflects our method is near 100% success to satisfy the given outline. Even if the extreme case is tested, α = 4, our method still succeeds over 70%. This accordant rate of success confirms the practicability of this strategy.

4.1. Floorplanning with Abutment Constraint
We conducted five experiments for each benchmark data with aspect ratio of each module lying between 0.25 and 4.0. We compared our results with [6], which the experiments were carried out on a 143-MHz UltraSPARC workstation. The results for each benchmark in Table II (The best values are shown in brackets.) show that our method can effectively handle abutment constraints. Note that the strategy of penalty used in [6] cannot ensure that all the constraints are satisfied in the final floorplan in Table II. However, our method always produces feasible solutions, i.e. all constraints are satisfied.

4.2. Fixed-Die Floorplanning with Abutment Constraint
In this section, we tested the benchmarks, ami49 and n100, 50 times for different kind of fixed-outline ratio. The expected fixed-outline ratio α is set to different cases: $1 \leq \alpha \leq 4$. Table III lists the areas, runtimes and *probabilities of success* of simultaneously satisfying the fixed-outline and the abutment constraints. It is clear that the abutment constraints are always satisfied by using our method, and the fixed-outline floorplanning is much more difficult to be solved than the classical floorplanning. Without doubt, the solution space of the fixed-outline floorplanning with abutment constraints is seriously restricted. Finding feasible solutions in this kind of restricted solution space is very arduous. However, our method still can find promising solutions with highly successful probabilities to simultaneously satisfy abutment and fixed-outline constraints.

5. Conclusions
We have presented an efficient and effective algorithm to deal with the modern floorplan with the abutment and fixed-outline constraints. The algorithm is based on the slicing floorplan and evolutionary algorithm. We have investigated the feasible properties with the abutment constraint, and then have proposed an algorithm that can guarantee a feasible floorplan with abutment constraint to be gradually suitable in the desirable outline during evolutionary progress. The experimental results have shown the effectiveness and efficiency of our algorithm.

6. References
[1] Stockmeyer L., "Optimal Orientations Of Cells In Slicing Floorplan Designs," *Information and Control*, 91-101, 1983.
[2] Chang-Tzu Lin, De-Sheng Chen, Yiwen.Wang, "An Efficient Genetic Algorithm for Slicing Floorplan Area Optimization," *Proceedings of the International Symposium on Circuits And Systems*, pp. 879-882, 2002.
[3] Valenzuela C.L. and Wang P.Y., "VLSI placement and area optimization using a genetic algorithm to breed normalized postfix expressions," *IEEE Transactions on Evolutionary Computation*, pp. 390–401, 2002.
[4] S. Nakaya, T. Koide, and S. Wakabayashi, "An adaptive genetic algorithm for VLSI floorplanning based on sequence-pair," *Proceedings of the International Symposium on Circuits And Systems*, pp. 65–68, 2000.
[5] M. Rebaudengo, and M.S. Reorda, "GALLO: A genetic algorithm for floorplan area optimization," *IEEE Transactions on Computer-Aided Design of Integrated Circuits and Systems*, pp. 943–951, 1996.
[6] F.Y. Young, H.H. Yang, D.F. Wong, "On extending slicing floorplans to handle L/T-shaped blocks and abutment constraints," *IEEE Transactions on Computer-aided Design of Integrated Circuits and Systems*, pp. 800–807, 2001.
[7] Yuchun Ma, Xianlong Hong, Sheqin Dong, Yici Cai, Chung-Kuan Cheng and Jun Gu, "Floorplanning with abutment constraints based on corner block list," *Integration VLSI Journal*, pp. 65–77, 2001.
[8] Xianlong Hong, et al., "Corner Block List: An Effective And Efficient Topological Representation of Non-Slicing Floorplan," *Proceedings of the International Conference on Computer Aided Design*, pp. 8-12, 2000.
[9] Saurabh N. Adya and Igor L. Markov, "Fixed-outline Floorplanning Through Better Local Search," *Proceedings of the International Conference on Computer Design*, pp. 328-334, 2001.
[10] Saurabh N. Adya and Igor L. Markov, "Fixed-Outline Floorplanning: Enabling Hierarchical Design," *IEEE Transactions on Very Large Scale Integration Systems*, pp. 1120–1135, 2003.
[11] D. F. Wong, and C. L. Liu, "A New Algorithm for Floorplan Design," *Proceedings of the Design Automation Conference*, pp.101–107, 1986.
[12] P.-N. Guo, C.-K. Cheng, and T. Yoshimura, "An O-Tree Representation of Non-Slicing Floorplan and Its Applications," *Proceedings of the Design Automation Conference*, pp. 268–273, 1999.
[13] H. Murata, K. Fujiyoshi, S. Nakatake, and Y. Kajitani, "Rectangle-Packing Based Module Placement," *Proceedings of the International Conference on Computer Aided Design*, pp. 472–479, 1995.
[14] S. Nakatake, K. Fujiyoshi, H. Murata, and Y. Kajitani, "Module Placement on BSG-Structure and IC Layout Applications," *Proceedings of the International Conference on Computer Aided Design*, pp. 484–491, 1996.
[15] T. C. Wang, and D. F. Wong, "An Optimal Algorithm for Floorplan and Area Optimization," *Proceedings of the Design Automation Conference*, pp.180–186, 1990.
[16] Yun-Chih Chang; Yao-Wen Chang; Guang-Ming Wu; Shu-Wei Wu, "B*-trees: A New Representation for Non-slicing Floorplans," *Proceedings of the Design Automation Conference*, pp. 458 –463, 2000.
[17] Young, F.Y., and Wong, D.F., "How Good Are Slicing Floorplans," *Integration VLSI Journal*, pp. 61–73, 1997.
[18] S.-Y. Ho, and X.-I. Chang, "An efficient generalized multiobjective evolutionary algorithm," *Proceedings of the Genetic and Evolutionary Computation Conference*, pp. 871-878, 1999.

Modeling of MOS Transistors Based on Genetic Algorithm and Simulated Annealing

A. Abbasian, M. Taherzadeh-Sani, B. Amelifard*, and A. Afzali-Kusha

Low-Power High-Performance Nanosystems Laboratory
ECE Dept., University of Tehran, Tehran, Iran
{abbasian, taherzadeh, amelifard }@ece.ut.ac.ir, afzali@ut.ac.ir

Abstract— **A novel method to extract the efficient model for Metal-Oxide-Semiconductor (MOS) transistors in order to satisfy a specific accuracy is presented. The approach presented here utilizes a Genetic Algorithm (GA) to choose the necessary physical and heuristic elements in order to define a compact yet accurate model for MOS I-V characteristic. Then the values of the free parameters related to each element are determined using Simulated Annealing (SA). For a desired accuracy considered here, the accuracy of the results predicted by our model were within 3.1%, for PMOS, and 1.3%, for NMOS, of the results of BSIM3 model while having much less complexity compared to the BSIM3 model. When this model with a variable accuracy is implemented in a circuit simulator, it provides the freedom of making a selection between the time and the accuracy of the simulation.**

I. INTRODUCTION

The proposed analytical models describing the I-V characteristic of CMOS transistors have a trade-off between the complexity and the accuracy [1][2]. When these models are incorporated in a circuit simulator such as SPICE, the higher the accuracy is, the more time is needed for the CAD tool to simulate the circuit. The models with higher accuracies use large numbers of parameters while models with smaller accuracy have small numbers of parameters making them also appropriate for hand-calculations.

In order to define more accurate expression for I-V modeling of MOS transistors, several physical and heuristic approaches have been presented [2]-[7]. All of them claimed they are leading to more accuracy for MOS modeling. Although in a perfect model for MOS, all of them have to be considered, the resulting model would not suitable for developing CAD tools as well as hand calculations. Hence, developing a most compact model with a specific accuracy seems fascinating.

In this paper, in order to choose the necessary elements including physical and heuristic elements for an MOS model, a genetic algorithm (GA) based approach is utilized. The goal of the algorithm is to find the elements of the model so as the resulted model has a specific accuracy and the most compactness. After choosing building block elements, the value related to each element is determined using a simulated annealing (SA) method to prevent the local optimization for them. In order to reveal the effectiveness of the proposed approach, we present a new compact model with a closed form expression for the I-V characteristics of MOSFET's which is more accurate compared to the *n*-th power model [4] due to the incorporation of more physical effects. Using the genetic algorithm approach, the effects of various model parameters on the accuracy of the model are determined. The model, which has few parameters, can be easily used in CAD tools as well as for hand calculation of circuit analysis.

In Section II, the compact I-V model used in this work is presented. The use of the genetic algorithm and simulated annealing in determining the model parameters for a given accuracy are discussed in Section III. The results and discussion are given in Section IV while we present the summary and conclusion in Section V.

II. COMPACT MODEL

The α-power model [3] takes into account the velocity saturation effect while neglects the channel length modulation. Therefore, the behavior of short channel MOSFET's in strong inversion is not described properly in this model. To enhance the accuracy of the α-power model for short channel devices, the *n*-th power model has been proposed by Sakurai [4]. In this model, the effects of the channel length modulation and the velocity saturation have been taken into account. This model can be expressed by the following expressions:

$$V_T = V_{T0} + \gamma(\sqrt{2\varphi_f - V_{BS}} - \sqrt{2\varphi_f}) \quad (1)$$

$$V_{DSAT} = K \cdot (V_{GS} - V_T)^m \quad (2)$$

$$I_{DSAT} = \frac{W}{L_{eff}} \cdot B \cdot (V_{GS} - V_T)^n \quad (3)$$

$$I_D = I_{DSAT}(1 + \lambda V_{DS}); \quad \lambda = \lambda_0 - \lambda_1 V_{BS}$$
$$(V_{DS} \geq V_{DSAT}: \text{saturated region}) \quad (4)$$

$$I_D = I_{DSAT}(2 - \frac{V_{DS}}{V_{DSAT}})\frac{V_{DS}}{V_{DSAT}}(1 + \lambda V_{DS})$$
$$(V_{DS} < V_{DSAT}: \text{linear region}) \quad (5)$$

---

* Currently with System Power Optimization and Regulation Technology Lab, University of Southern California, Los Angeles, CA.

where W is the channel width and L_{eff} is the effective channel length, V_{T0} is the threshold voltage at zero bulk-source voltage, γ is the body factor, and $2\varphi_F$ is the band bending of the semiconductor at the onset of strong inversion, I_{DSAT} is the drain saturation current, V_{GS}, V_{DS}, V_{BS}, V_T, and V_{DSAT} are the gate-source, drain-source, bulk-source, threshold voltage, drain saturation voltages, respectively. Parameters K and m control the linear region characteristics while B and n determine the saturated region characteristics. Finally, the finite drain conductance in the saturation region has been modeled by λ_0 and λ_1 [4].

In [8], comparisons between the results of the I-V characteristics predicted by n-th power model and the results of complicated BSIM3 model [2] were presented. The comparisons show that the n-th model may not have enough accuracy for many applications. To improve the accuracy of the compact model, we consider several physical effects that have been ignored in the n-th power model. These effects which include velocity saturation, channel length modulation, static feedback, effective channel width, and deep inversion have been described in [8]. They may be included in the model based on the desired accuracy.

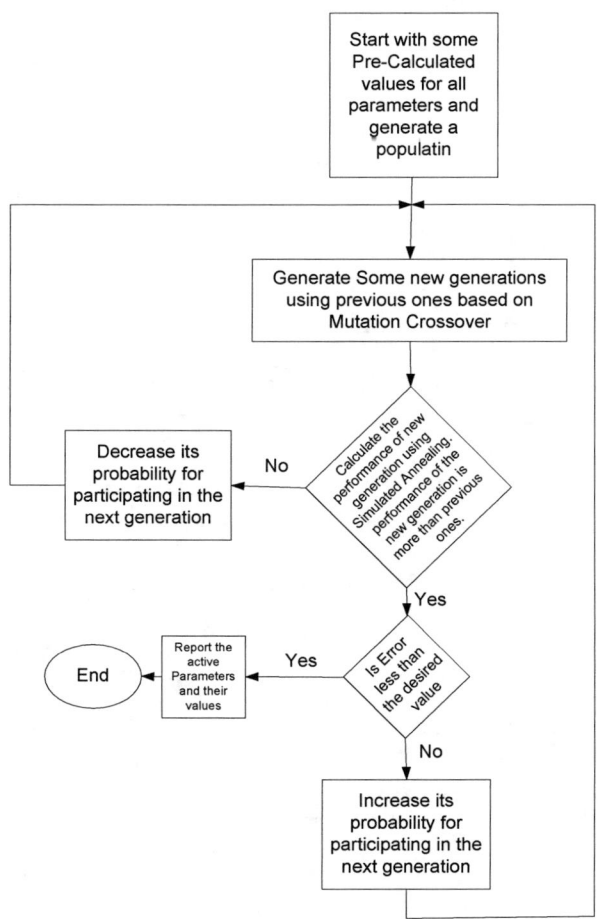

Figure 1. The flowchart of the proposed method.

III. CHOOSING THE PARAMETERS FOR A GIVEN ACCURACY

The accuracy of the model discussed in the previous section is higher compared to the n-th power model. This increase in the accuracy was achieved at the price of a bit more complex model compared to the initial model, though still much simpler compared to the BSIM3 model. Some of the parameters added to the compact model may not improve the accuracy considerably, and hence their omission leads to the simplification of the model.

The flowchart of the proposed model for choosing parameters and their values is illustrated in Figure 3 which can be described briefly with the following stages

- GA determines which terms is used in the model,
- SA determines the corresponding coefficient to each term,
- Fitness function is calculated,
- If the value of Fitness function is not below a desired value, the process is repeated.

As is understood from the figure, we use the genetic algorithm to determine these less important parameters, and simulated annealing to determine the best values for the parameters.

A. Genetic Algorithm

The Genetic Algorithm (GA) utilizes a non-gradient-based random search and is used in the optimization of complex systems [9]. The algorithm models the process of biological evolution and optimizes the parameters of problem. In the algorithm, each unknown parameter is called *gene* and each vector of these parameters is called a *chromosome* [9]. The purpose of the genetic algorithm is to determine the elements of the unknown vector (chromosome) which maximizes or minimizes the defined *fitness* function. The algorithm starts with a population of chromosomes. In each generation, new population of the chromosomes is enhanced in the fitness function by means of some operators such as *cross over* and *mutation*. The initial population is chosen randomly. More details about this method can be found in [9]. In our work, this method is used to determine existing parameters.

B. Simulated Annealing

The Simulated Annealing (SA) is a combinational optimization technique [10]. The principle behind it is analogous to what happen when metals are cooled at a controlled rate. A typical simulated annealing process starts with a very high temperature (T), where the system state is generated at random. The cost function is analogous to the energy $E(S)$ of a system in state S which should be minimized for the system stability. In [10], more details about this method are given. We use this method in our approach to determine the value of existing parameters.

C. Formulation of the Problem

There are thirteen parameters in our proposed compact model defined in [8]. In the GA, the chromosome is defined to be a bit vector with 13 elements where each bit relates to the model parameters as follows:

$$[K\ m\ n\ V_T\ \beta\ \kappa\ \lambda_C\ \lambda_V\ \sigma\ W_W\ W_{VGS}\ p\ q] \quad (6)$$

If a bit becomes zero, the corresponding parameter is eliminated from the model i.e. n would be replaced by 2, p and q by 1, β, κ, λ_V, σ, W_W, and W_{VGS} by 0 and K, λ_C and V_T by their physical values.

Using a simple GA approach, the bit-vector chromosome can be easily determined in order to minimize the value of a goal function. In order to define the goal function, two major objectives have been considered which are the accuracy and the complexity of the model. To consider the first objective in our GA, we define the error of the model by

$$\varepsilon = \frac{1}{N_W \cdot N_{V_{GS}}} \sum_{W=W_{min}}^{W_{max}} \sum_{V_{GS}=V_T}^{V_{DD}} \left[\frac{\sum_{V_{DS}=0}^{V_{DD}} (I_{D,Model} - I_{D,BSIM3})^2}{I_{D,Max}} \right]^{1/2} \quad (7)$$

where ε is the error of the model, $I_{D,Model}$ and $I_{D,BSIM3}$ are the drain currents predicted by the compact model and the BSIM3 model (HSPICE) for the same V_{DS}, V_{GS}, and W, and $I_{D,Max}$ is the drain current at $V_{DS} = V_{DD}$, N_W and $N_{V_{GS}}$ are the number of sampled widths and V_{GS} used for calculating error. For the second objective in the GA, the complexity factor is defined as

$$\theta = \frac{\delta}{\eta} \quad (8)$$

where δ is the number of nonzero elements in the chromosome, and η is the number of bits in each chromosome. This definition for complexity guarantees that the GA program considers minimizing the number of elements in the model to compact it. The fitness of the problem, therefore, is defined as

$$\varphi = \varepsilon + \chi \cdot \theta \quad (9)$$

where φ is the fitness, and χ is a weight coefficient which specifies the importance of the complexity versus the error. Having a very small χ means that the accuracy of the model is very important for the user. This leads to a model with more parameters. If χ is large enough, the model would be reduced to the α-power and n-th power models. We choose $\chi=1/50$ to attain an average error of about 2.5%.

In the SA optimization method, utilized to determine the values of the model parameters, the accepting and generating functions are the Boltzmann probability distribution and a Gaussian probability density function, respectively. The algorithm involves the following four steps. First, the objective function corresponding to the energy function must be identified. Second, one must select a proper annealing scheme consisting of decreasing temperature with increasing of iterations. Third, a method of generating a neighbor near the current search position is needed. Fourth, after a new point has been evaluated, SA decides whether to accept or reject it based on value of an acceptance function. In our optimization problem, the aforementioned functions define in (7).

Annealing schemes are selected as follow:

$$T(k) = \frac{T_0}{\ln k} \quad (10)$$

where k is iteration number and T_0 is a high starting temperature. Generating function defines as follow.

$$X_{i+1}(k) = X_i(k) + \frac{1}{2}\lambda\left\{\left[(1+\text{sgn}(\lambda-0.5))\left(X_i^{max}(k) - X_i(k)\right)\right] \right.$$
$$\left. + \left[(\text{sgn}(\lambda-0.5)-1)(X_i(k) - X_i^{min}(k))\right]\right\}; k=1:m$$
$$(11)$$

where sgn is the sign function, m is the number of characteristics to fit, X_i^{max} and X_i^{min} are the maximum and minimum of the ith dimension, and $\lambda \in [0,1]$. The generating function for λ has a Gaussian probability density function of

$$g(\Delta x, T) = (2\pi T)^{-n/2} \exp\left[-\|\Delta x\|^2 / (2T)\right] \quad (12)$$

The acceptance function has a Boltzmann probability distribution of

$$h(x) = \frac{1}{1 + \exp(\Delta\varepsilon / cT_k)} \quad (13)$$

where $\Delta\varepsilon = \varepsilon(X_{i+1}) - \varepsilon(X_i)$ and c is a system dependent constant and T_k is the temperature in kth iteration.

IV. RESULTS AND DISCUSSION

We have implemented the models discussed above in MATLAB. Our simulations show that for attaining 2.5% error, nine of the thirteen model parameters are required. These parameters include K, m, n, V_T, β, κ, λ_C, W_{VGS}, and p. The first five parameters belong to the n-th power model whereas the last four parameters have been added by the improvements that we have made to the model. The eliminated parameters were λ_V, σ, W_W, and q. It is obvious that choosing a different value for χ could have led to another set of the model parameters. In Figures 4 and 5, the I_D-V_{DS} and I_D-V_{GS} characteristics of a NMOS transistor obtained by the proposed model and the SPICE simulations are compared which show a very good accuracy for the proposed model. The errors and the CPU time (for a 500MHz Pentium III processor) of the new and the n-th power models for NMOS and PMOS transistors ($W_n = W_p = 1\mu m$, $L = 0.35\mu m$ and $V_{DD} = 3.3V$) are given in Table I. As a reference, the required time of BSIM3 is also given in this table. The times required for the compact models are much smaller compared to the BSIM3 model while having few percents of error.

V. SUMMARY AND CONCLUSION

We presented a compact MOSFET I-V model with the ability to modify the accuracy of the model using the genetic algorithm and simulated annealing. Using GA, one can select between the thirteen model elements based on the accuracy that needed for a specific application. For the accuracy considered in this paper, only nine elements were necessary and hence considered in the simulations. The value of each model parameter was found using simulated annealing. The results predicted by the model were compared to the results of the very accurate but complex model of BSIM3, predicting 1.3% and 3.1% of error. The very good accuracy obtained for this model suggests that for many (digital) applications, the model presented in this paper may be used in place of very complicated models like the one currently implemented in SPICE to speed up the circuit simulation.

REFERENCES

[1] A.P. Chandrakasan, W.J. Bowhill and F. Fox, *Design of High-performance Microprocessor Circuits*, IEEE Press. 2001.

[2] D.P. Foty, *MOSFET modeling with SPICE, principles and practice*, Prentice Hall, 1997.

[3] T. Sakurai and R. Newton, "Alpha-power low MOSFET model and its application to CMOS inverter delay and other formulas," *IEEE J.l of Solid-State Circuits*, vol. 25, pp. 584-594, Apr. 1990.

[4] T. Sakurai and A.R. Newton, "A simple MOSFET model for circuit analysis," *IEEE Trans. Electron Device*, vol. 38, pp. 887-894, 1991.

[5] W. Shockley, "A unipolar filed effect transistor," *Proc. IRE*, vol 40, pp. 1365-1376, Nov. 1952.

[6] R. Van Langevelde and F.M. Klaassen, "Accurate drain conductance modeling for distortion analysis in MOSFETs," *IEDM 1997 Tech. Digest*, pp. 313-316, 1997.

[7] Y. Tsividis, *Operation and modeling of the MOS transistor*, McGraw-Hill, 1999.

[8] M. Taherzadeh-Sani, A. Abbasian, B. Amelifard, and A. Afzali-Kusha, "MOS Compact I-V Modeling with Variable Accuracy Based on Genetic Algorithm and Simulated Annealing," in Proceedings of the 16th International Conference on Microelectronics, Tunis, Tunisia, December 6-8, pp. 364-367, 2004.

[9] D. E. Goldenberg, *Genetic algorithm in search, optimization and machine learning*, Addison Wesley, Reading MA, 1989.

[10] J. Roger Jang, C. Sun, and E. Mizutani, *Neuro-fuzzy and soft computing*, Prentice Hall, 1997.

TABLE I. COMPARISON OF BSIM3, N-TH POWER AND PROPOSED MODEL

Model	Error (PMOS)	Error (NMOS)	CPU time (ms)
This Model	3.1%	1.3%	0.3
n-th power	7.5%	3.7%	0.2
BSIM3	-	-	10

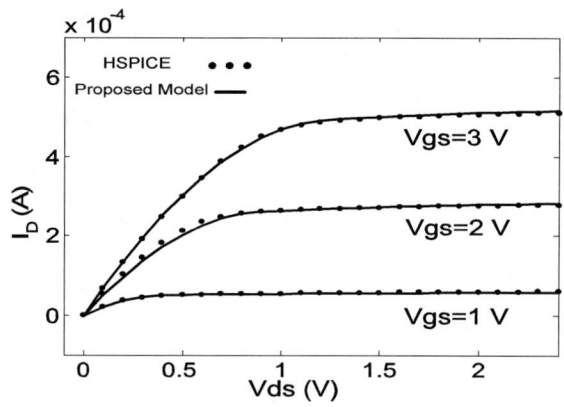

Figure 2. The I_D-V_{DS} characteristic of a NMOS transistor (W=1μm, L=0.35 μ and V_{DD}=3.3V).

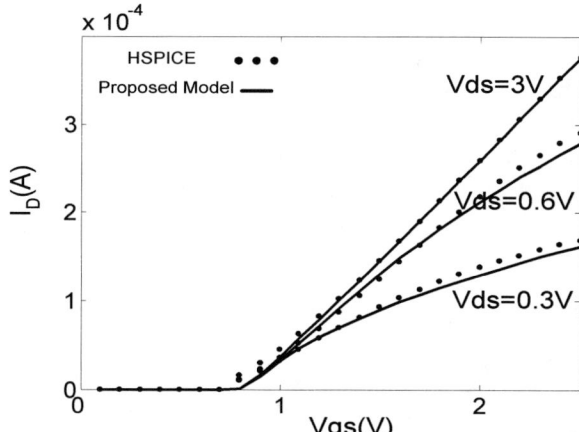

Figure 3. The I_D-V_{GS} characteristic of a NMOS transistor (W=1μm, L=0.35 μm and V_{DD}=3.3V).

VLSI Block Placement with Alignment Constraints based on Corner Block List [1]

Song Chen, Xianlong Hong, Sheqin Dong [2], Yuchun Ma
Computer Science & Technology
Tsinghua University
Beijing, China
chens00@mails.tsinghua.edu.cn

Chung-Kuan Cheng
Department of Computer Science and Engineering
University of California
San Diego, USA
kuan@cs.ucsd.edu

Abstract—Corner Block List (CBL) is an excellent representation of block floorplan/placement. In this paper, we give a sufficient and necessary condition for the feasibility of a CBL and alignment constraints in CBL are also dealt with. A method is proposed to identify topological relation between any two blocks in CBL. It is also found that a sufficient and necessary condition to judge that whether a CBL is feasible or not under alignment constraints. The experimental results have demonstrated the efficiency and effectiveness of the proposed method.

I. INTRODUCTION

A dramatic increase in the complexity of integrated circuits has taken place because of rapid advances in integrated circuit technology. Hierarchical design and IP reuse become very important. Therefore, the placement/floorplan has received vast attention in the latest decade. As you know, even the simple rectangle-packing problem is NP-hard [3]. As the application of simulated annealing on this problem, many placement/floorplan representations have been proposed, such as Corner Block List (CBL)[1][2], Sequence Pair (SP)[3], Bounded Slice-line Grid (BSG)[4], O-tree[5], B*-tree[7], Q--sequence[8], and etc.

Based on various floorplan/placment representations, People also proposed many methods to process placement constraints, such as boundary, abutment, preplaced, range, clustering and so on. There are also some previous work dealt with alignment constraints [12][13][14], but there are not the similar works under room-based representation (section II). Some room-based representations, such as CBL and Q-sequence, have linear time complexity of transformation into placement/floorplan, but it is difficult for these representations to deal with some placement constraints, such as symmetry, alignment and etc.

CBL is an excellent block floorplan/placement representation because of its linear complexity of the construction of floorplan/placement. Alignment constraint gets applications either in analog circuit layout or digital layout (bus-structure). In this paper, we give a sufficient and necessary condition for the feasibility of a CBL and alignment constraints in CBL are also dealt

with, which is the first work about alignment constraints in room-based floorplan representation (section II). A method is proposed to identify topological relation between any two blocks in CBL. It is also found that a sufficient and necessary condition to judge that whether a CBL is feasible or not under alignment constraints. The experimental results have demonstrated the efficiency and effectiveness of the proposed method.

The rest of this paper is organized as follows: Section II revisits the CBL representation and gives a sufficient and necessary condition for the feasibility of CBL. The alignment constraint in CBL is discussed in section III. Section IV is the experimental results and conclusions.

II. CORNER BLOCK LIST REPRESENTATION

Definition 1 Topological relation: *right of, left of, above, below* are four kinds of relations between any two blocks in a floorplan/placement. Block b_i is said to be *right of* block b_j if the left side of b_i is right of the right side of b_j. Similarly, *left of, above, below* relations between blocks are defined. And these relations are defined as *topological relations* between blocks.

Definition 2 A floorplan representation is *topological* if the topological relations between blocks in this representation are independent of the physical dimensions of blocks. And a set of all the topological relations is defined as a *floorplan topology*.

Described as in the introduction, many floorplan/placement representations have been proposed in latest decade. These representations can be classified into three categories. The first category is *non-topological* representations, including O-tree, B*-Tree that are representative, and so on, in which the topological relations between blocks depend on physical dimensions of blocks [2]. The second one denoted by us as *packing-based* representation, such as Transitive Closure Graph [7], SP and so on, and this type of floorplan representation is topological. Finally, the last one is *room-based* representation, such as BSG, Q-sequence, Selected Sequence Pair [12], CBL and so forth, which are also topological.

The difference between *packing-based* floorplan representations and *room-based* floorplan representations is that the former indicate only topological relations between blocks, while the later indicate not only floorplan topology but also a dissection of the chip, which satisfy that the chip is dissected into rooms and there is at most one block in each room (Dashed lines in Fig.1 (b) shows an example). It is also conclude that packing-based floorplan representations have at least $O(nlogn)$ complexity (thus far) of transformation into

---
[1] This work is supported by the National Natural Science Foundation of China (NSFC) 60473126, NSFC and Hongkong RGC joint Project 60218004, and Hi-Tech Research & Development (863) Program of China 2004AA1Z1050.
[2] The author is also with University of Kitakyushu, Japan.

floorplan/placement, where n is the number of blocks, while that of room-based floorplan representations are linear with the number of rooms. Moreover, the room-based floorplan representations are good for interconnect planning, especially buffer planning [15].

A. Corner Block List

CBL [1][2] is a room-based representation, which represents floorplan by a triple list of (S, L, T). The sequence L and T represent a dissection of the chip into rectangular areas, denoted as *room*, and the sequence S is an assignment of blocks to rooms subjecting to the rule of that there is one and only one block in each room. Dong [9] extended the Corner Block List by introducing dummy blocks.

S is the *Corner Block* sequence and is also the packing order of blocks. *Corner Block* means the upper-right block, i.e., no block is above or right of the corner block in the floorplan. Let S_i be the i-th block in S. S_n is the *corner block* in the floorplan of all n blocks and S_i is the *corner block* in the floorplan of blocks S_j, $j=1\ldots i$.

The list L records the orientation of each block except the block in the bottom-left corner of the floorplan/placement. The left and bottom boundary of each block compose a T-junction to dissect the chip into rooms. The T-junction has two alternative orientations: T rotated counterclockwise by 90° ('⊢') if the block is vertical and by 180° ('⊥') if the block is horizontal. Fig.1.(a) shows an example of 90° T-junction generated while inserting block d.

The binary sequence T records the number of T-junctions each *corner block* covers except the bottom-left block. A binary list T_i records the number of T-junctions the corner block b_i covers. T_i is comprised of consecutive '1's ended by one '0' and the number of '1's is the same as the number of T-junctions covered by b_i. In Fig.1.(a), *corner block d* covers one T-junction. In sequence T, there is a binary sequence '10' to record the number of T-junctions the block d covers. The maximum length of sequence T is $2n-3$[1].

Fig.1.(b) shows a placement and its CBL representation. The dissection of chip is also pointed out. A dummy block 0 is assigned to an empty room.

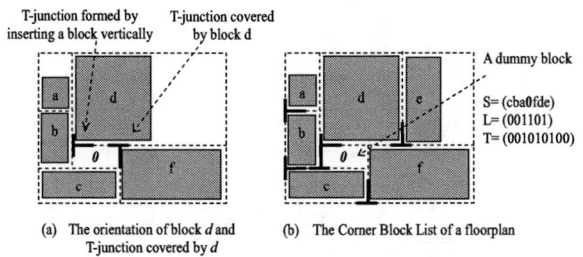

(a) The orientation of block d and T-junction covered by d

(b) The Corner Block List of a floorplan

Figure 1. The Corner Block List Representation

Property 1 The sequence S in the CBL observes that:

A block b_i is left of or below another block b_j in the floorplan/placement (or the block b_j is right of or above the block b_i) if b_i is before b_j in the sequence S.

B. A sufficient and necessary condition for feasible CBL

A CBL (S L T) is *feasible* if it can be transformed into a foorplan/placement. A feasible CBL must satisfy certain condition. In this sub-section, a sufficient and necessary condition of feasible CBL is proposed. Let b_i be the i-th block in S, l_i be the orientation of b_i, $l_i=0$ means vertical and $l_i=1$ means horizontal. Let t_i be the number of T-junction block b_i covers.

In the CBL, the t_1 and l_1 are not defined. Without loss of generality, we assume that $t_1 = 0$ and $l_1 = l_2$. The following theorem is concluded.

Theorem 1 A CBL (S, L, T) is feasible if and only if L and T satisfy the following condition.

$$\begin{cases} 0 \le t_i \le |P_i| - \sum_{j=1}^{i-1}(1-l_j)\cdot t_j, & if\ l_i = 0 \\ 0 \le t_i \le |Q_i| - \sum_{j=1}^{i-1} l_j\cdot t_j, & if\ l_i = 1 \end{cases}$$

where $2 \le i \le n$ and P_i is a set of blocks that can be covered vertically and Q_i is a set of blocks that can be covered horizontally. They are defined as follows.

$$\begin{cases} P_i = \{b_j\ |\ 1 < j < i \wedge l_j = 1\} \\ Q_i = \{b_j\ |\ 1 < j < i \wedge l_j = 0\} \end{cases}$$

The theorem gives the upper bound of the number of T-junction each block can cover.

III. ALIGNMENT CONSTRAINTS IN CBL

Alignment constraints get applications either in analog circuit layout (signal flow) or digital layout (bus-structure). In this research, we concentrate on the alignment constraints problem: Given a block set S, dimensions of each block and two subsets of S H and V, we try to find a floorplan/placement such that the area and some other objectives get optimal value and blocks in H and V are aligned vertically and horizontally at block center, respectively. It is obvious that $|H \cap V| \le 1$ since we cannot align two blocks vertically and horizontally, simultaneously.

A. Recogonition of topological relations between blocks

Under CBL representation, It is not straightforward that the topological relations between two blocks. In this sub-section, a simple method is developed to recognize these relations. Similar to [10], we give following definitions.

Definition 3 In a floorplan, except the block in the bottom-left corner, the T-junction at the bottom-left corner of a block has two alternative orientations: *90-T-junction* and *180-T-junction* (Section II.A). Given a T-junction, one of the two segments who combined the T-junction must have an end point on the other segment. The former is called *non-crossing* segment, and the later is called *crossing* segment. Fig.2 illustrates the concepts.

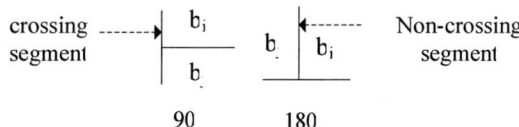

Figure 2. Segments and T-junction

Definition 4 C^{bl}-neighbor: Given a floorplan, assume that block b_i is not the bottom-left corner block. The T-junction at the bottom-left corner of b_i is either a *90-T-junction* or *180-T-junction*. Let b_j be the block adjacent to b_i by the non-crossing segment at the corner of that T-junction. b_j is called the C^{bl}-neighbor of b_i. In Fig.2, b_i is the C^{bl}-neighbor of b_i.

Lemma 1 Except the bottom-left block of the floorplan, each block has exactly one C^{bl}-neighbor. And all the blocks can be connected to their C^{bl}-neighbor to construct a binary tree, denoted as t^{bl}, whose root is the bottom-left block in the floorplan.

The binary tree t^{bl} is constructed as follows.

If the neighbor relation defined by a 90-T-junction, let the block be left child of its C^{bl}-neighbor. Otherwise (180-T-junction), let the block be right child of its C^{bl}-neighbor. An example of the binary tree t^{bl} is shown in fig.3.

Lemma 2 An in-order traversal on t^{bl} with left sub-tree visited first results in a sequence S_{IO}, which satisfies that

A block b_i is left of or above another block b_j in the floorplan/placement (or the block b_j is right of or below the block b_i) if $S_{IO}[b_i] < S_{IO}[b_j]$). $S_{IO}[b]$ represents the position of block b in S_{IO}.

According to *property 1* and *Lemma 2*, theorem 2 is concluded.

Theorem 2 Let $S[b]$ be the position of block b in S. For a block b_i, any other block b_j is uniquely one of the following four cases.

1. $L(b_i) = \{b_j \mid S[b_j] < S[b_i] \wedge S_{IO}[b_j] < S_{IO}[b_i]\}$,
2. $R(b_i) = \{b_j \mid S[b_j] > S[b_i] \wedge S_{IO}[b_j] > S_{IO}[b_i]\}$,
3. $A(b_i) = \{b_j \mid S[b_j] > S[b_i] \wedge S_{IO}[b_j] < S_{IO}[b_i]\}$,
4. $B(b_i) = \{b_j \mid S[b_j] < S[b_i] \wedge S_{IO}[b_j] > S_{IO}[b_i]\}$

b_j is *left of* b_i if $b_j \in L(b_i)$, b_j is *right of* b_i if $b_j \in R(b_i)$, b_j is *above* b_i if $b_j \in A(b_i)$ and b_j is *below* b_i if $b_j \in B(b_i)$.

B. *A sufficient and necessary condition for alignment constraint*

All of feasible CBL can not be transformed into a floorlan/placement with alignment constraint met. We give a sufficient condition to judge that whether a CBL is feasible under alignment constraints.

Assume that there are m blocks b_i ($i=1 \ldots m$) need to be aligned in the floorplan. Let $S[b_i]$ and $S_{IO}[b_i]$ record the position of block b_i in sequence S and S_{IO}, respectively. Without loss of generality, we assume that $S[b_i] < S[b_{i+1}]$, $i=1, \ldots, m-1$.

Theorem 3 The m blocks can be aligned in the floorplan if and only if one of the following conditions is met.

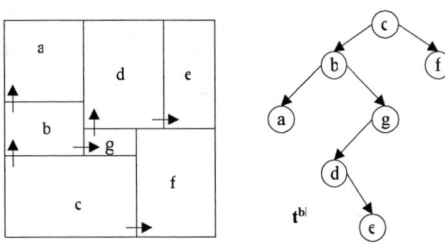

Figure 3. The C^{bl}-neighbor of the rooms and corresponding binary tree t^{bl}

$$\begin{cases} S_{IO}[b_i] < S_{IO}[b_{i+1}], & i = 1, \ldots, m-1. \\ S_{IO}[b_i] > S_{IO}[b_{i+1}], & i = 1, \ldots, m-1. \end{cases}$$

In the first case, the blocks are aligned based on y-coordinates (horizontally), while in the second case, we align the blocks based on x-coordinates (vertically).

C. *Packing blocks with alignment constraints met*

First, the coordinates of the blocks are computed according to the given CBL without consideration of alignment constraints, the binary tree t^{bl} is constructed simultaneously. And then we determine the base of alignment and move the blocks such that the blocks have alignment constraints among each other are aligned at center.

The following is the main steps of the packing algorithms.

PACKING_WITH_ALIGNMENT
1. Construct t^{bl} from CBL and calculate block coordinates with no consideration of alignment constraints;
2. Compute S_{IO} by performing a left sub-tree first in-order traversal on t_{bl};
3. Go to step 4 if the CBL is feasible under given alignment constraint; (Based on the sufficient condition in section III.B)
4. Calculate the alignment base according to the block coordinates computed in step 1.
5. Adjust block coordinates to meet alignment constraints

The above procedure is embedded a simulated annealing, in step 3, we will generate another CBL if the current CBL is unfeasible under alignment constraints. Hereafter, we take the horizontal case as the example to show the details of the algorithms and the vertical alignment can be dealt with similarly. Let S_i be the i-th block in the sequence S. $(b.x, b.y)$ represents the coordinates of the upper-right corner of block b. To avoid applying the complicated sufficient and necessary condition to judging the feasibility of a CBL, a feasible CBL is extracted from an unfeasible CBL by scanning the CBL once.

Procedure construct_TBL_ and_cal_coordinates
Input: CBL (S L T), a group of blocks that need to be aligned
Output: t^{bl}, coordinates of blocks (alignment constraint is not met)

/*the array Left and Low record the blocks adjacent to the top boundary and the right boundary of chip currently, respectively.*/
localMaxW = 0; localMaxH = 0; $x_i = 0$, $y_i=0$
k = 1; t = T[k]; $S_1.x = S_1$.width; $S_1.y = S_1$.height;
for i =2 to n;
  if b_i is vertical
    b_l = Left[S_i]; $b_{tmp} = S_{i-1}$;
    Low[S_i] = S_{i-1};   localMaxY = $S_{i-1}.y$.
    while (b_l exists) && (t = 1)) $^\alpha$
      localMaxY = max($b_l.y$, localMaxY);
      $b_{tmp} = b_l$; b_l = Left[b_l];
      t = T[k]; k++;
    Left[S_i] = b_l;
    if b_{tmp} exists
      Add the edge (S_i, b_{tmp}) to t^{bl} by letting S_i be the left child of b_{tmp}
    If Left[S(i)] does not exits
      $S_i.x = S_i$.width;
    else $S_i.x = b_{tmp}.x - b_{tmp}$.width +$S_i$.width;
    $S_i.y = S_i$.height + localMaxY;
  else if b_i is horizontal
    $b_{tmp} = S_{i-1}$;   Left[S_i] = S_{i-1};
    b_l = Low[S_{i-1}]; localMaxX = $S_{i-1}.x$;
    while (b_l exists) && (t is equal to 1)) $^\beta$
      localMaxX = max($b_l.x$, localMaxX);
      $b_{tmp} = b_l$; b_l = Low[b_l];
      t = T[k];   k++;
    Low[S_i] = b_l;
    if b_{tmp} exits
      Add edge (S_i, b_{tmp}) to t^{bl} having S_i as right child of b_{tmp}.
    if Low[S_i] does not exits
      $S_i.y = S_i$.height;
    else $S_i.y = b_{tmp}.y - b_{tmp}$.height + S_i.width;
if T[k]=0 k++; /* Skip 0 in sequence T*/

The above algorithm only scans (S, L, T) one time. Therefore, its complexity is linear with the number of blocks. For a given CBL, we need not to use the condition proposed in section II.B to decide the feasibility of the CBL. The above procedure can extract automatically a unique feasible CBL from the given un-feasible CBL by modifying the T_i that exceeds the upper bound given in theorem 1 to the upper bound of the number of T-junction the block S_i can cover. Steps α and β ensure that the number of T-junction the current block cover does not exceed the upper bound defined in theorem 1 when the CBL is unfeasible.

The alignment base of an alignment group b_i (i =1...m) is computed by the formula $A_b = \max_{i=1,...,m}(b_i.y)$.

The following algorithm moves blocks to satisfy the alignment constraints among blocks.

Let $S[b_i]$ and $S_{IO}[b_i]$ record the position of block b_i in sequence S and S_{IO}, respectively, and S_i be the i-th block of S. S_i.gap records the distance that the block S_i has to be moved to meet alignment constraints.

Procedure Adjust_Coordinates;
Input: S, S_{IO} and block coordinates calculated in Cal_S_{IO}_and_Cal_Base
Output: Coordinates of blocks (Alignment constraint is met)

for i =1 to n
  if block S_i belongs to the alignment group
    gap = A_b - S_i.y + S_i.height/2; // Calculate the space S_i has to be moved upward.
    S_i.gap = gap;
    for j=i+1 to n // Calculate the space each of blocks that are above of block S_i has to be moved upward.
      if $S_{IO}[S_j]$ < $S_{IO}[S_i]$
        if gap>S_j.gap S_j.gap = gap;
for i =1 to n
  S_i.y = S_i.y + S_i.gap.

The algorithm for adjusting the blocks coordinates has a complexity linear with n^2, where n is the number of block.

IV. EXPERIMENTAL RESULTS AND CONCLUSTION

We have implemented our algorithms in c++ language on Sun's v880 workstation which has a CPU speed of 750M. The algorithm *PACKING_WITH_ALIGNMENT* is embedded in a standard implementation of simulated annealing. For each new generated CBL, the sufficient condition proposed in section III.B is used to judge that whether it is feasible under the alignment constraints or not. Another new CBL is generated if the current CBL is not feasible under alignment constraint.

Two test cases ami33_t and ami49_t are generated from MCNC benchmark ami33 and ami49, respectively. Fig.4.(a) shows the packing results of ami33_t with five blocks (1, 3, 5, 7, 12) aligned horizontally at center. An experimental result from ami49_t is shown in fig.4.(b), where 4 blocks (3, 29, 33, 34) get horizontal alignment and another 4 blocks (21, 28, 38, 48) get vertical alignment. All the blocks are hard blocks. Table I shows the run time and usage in the situation of with/without alignment constraint, respectively. The experimental results are promising.

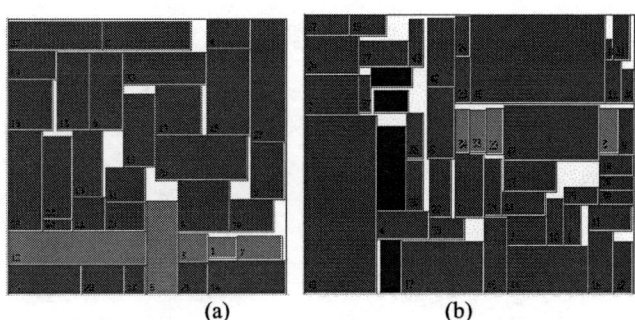

Figure 4. (a) A packing result of ami33. (b) A packing Results of ami49

TABLE I. EXPERIMENTAL RESULTS

Cicuit	#blocks	#aligned	With Constraint		Without Constraint	
			area usage	time(s)	area usage	time(s)
ami33_t	33	5	92.80	22.00	94.50	18.00
ami49_t	49	6	91.60	32.00	94.60	26.00

REFERENCES

[1] X. Hong, G. Huang, S. Dong, Y. Cai, C.K. Cheng, J. Gu, "Corner block list: an effective and efficient topological representation of non-slicing floorplan," Proc. ICCAD, pp.8-12, Nov.2000.

[2] X. Hong, S. Dong, G. Huang, Y. Cai, C.K. Cheng, "Corner Block List representation and its application to floorplan optimization", Circuits and Systems-II: Express Briefs, IEEE Trans. On, Vol. 51, No.5, 2004, p228-233.

[3] H. Murata, K. Fujiyoshi, S. Nakatake, Y, Kajitani, "VLSI module placement based on rectangle-packing by the sequence-pair," Computer-Aided Design of Integrated Circuits and Systems, IEEE Trans. on, Vol.15, No.12,1996, pp.1518 – 1524.

[4] S Nakatake, K Fujiyoshi, H Murata, Y Kajitani "Module packing based on the BSG-structure and IC layout applications", IEEE Trans. on Computer-Aided Design of Integrated Circuits and Systems, Vol.17, No.6, June 1998, pp.261-267.

[5] P. Guo, C.K. Cheng, T. Yoshimura, "An O-tree representation of non-slicing floorplan and its applications", Proc. DAC, 1999, pp.268-273.

[6] J-M.Lin, Y-W. Chang, "TCG: a transitive closure graph-based representation for non-slicing floorplans," Proc. DAC, 2001.

[7] Y. Chang, G. Wu, S. Wu, "B*-tree: A new representation for non-slicing floorplans", Proc. DAC, 2000, pp.458-463.

[8] K. Sakanushi, Y. Kajitani, Dinesh P. Mehta, 'The quarter-state-sequence floorplan representation", Circuits and systems-I: Fund-amental theory and applications, IEEE Trans. On, vol.50, no.3, 2003.

[9] S. Dong, S. Zhou, X. Hong, C.K. Cheng, J. Gu, Y. Cai, "An optimum placement search algorithm based on extended corner block list", J. Comput. Sci. & Technol., Vol.17, No.6, 2002, pp. 699-707.

[10] B. Yao, H. Chen, C.K. Cheng, R. Graham, "Floorplan representations: complexity and connections", Design Automation of Elecronic Systems, ACM Trans. on, Vol.8, No.1, 2003, p.55-80.

[11] C. Kodama, K. Fujiyoshi, "Selected sequence-pair: an efficient decodable packing representation in linear time using sequence-pair", Proc. ASP-DAC 2003, p331-335.

[12] F.Y. Young, Chris C. N. Chu, M. L. Ho, "Placement Constraints in floorplan design", Very Large Scale Integration Systems, IEEE Trans. On, Vol.12, No.7, 2004, p735-745.

[13] H. Xiang, X. Tang, D.F. Wong, "Bus-driven floorplanning", Proc. IEEE/ACM ICCAD, 2003. p. 66-73.

[14] R. Liu, X. Hong, S. Dong, J. Gu, "A block placement algorithm with predeined coordinate alignment constraint based on sequence pair representation", Journal of Software, Vol.14, No.8, 2003 p1418-1423.

[15] S. Chen, X. Hong, S. Dong, Y. Ma, Y. Cai, C. K. Cheng, J. Gu, "A buffer planning algorithm based on dead space redistribution", Proc. IEEE/ACM ASP-DAC 2003, p435-438.

# Multiobjective VLSI Cell Placement Using Distributed Simulated Evolution Algorithm

Sadiq M. Sait   Mustafa I. Ali   Ali Mustafa Zaidi

King Fahd University of Petroleum & Minerals
Computer Engineering Department
Dhahran 31261, Saudi Arabia
E-mail: {sadiq,mustafa,alizaidi}@ccse.kfupm.edu.sa

Abstract—Simulated Evolution (SimE) is a sound stochastic approximation algorithm based on the principles of adaptation. If properly engineered it is possible for SimE to reach near-optimal solutions in lesser time then Simulated Annealing [1], [2]. Nevertheless, depending on the size of the problem, it may have large run-time requirements. One practical approach to speed up the execution of SimE algorithm is to parallelize it. This is all the more true for multi-objective cell placement, where the need to optimize conflicting objectives (interconnect wirelength, power dissipation, and timing performance) adds another level of difficulty [3]. In this paper a distributed parallel SimE algorithm is presented for multiobjective VLSI standard cell placement. Fuzzy logic is used to integrate the costs of these objectives. The algorithm presented is based on random distribution of rows to individual processors in order to partition the problem and distribute computationally intensive tasks, while also efficiently traversing the complex search space. A series of experiments are performed on ISCAS-85/89 benchmarks to compare speedup with serial implementation and other earlier proposals. Discussion on comparison with parallel implementations of other iterative heuristics is included.

I. INTRODUCTION

Simulated Evolution algorithm (SimE) is a general search strategy for solving a variety of combinatorial optimization problems [2]. It operates on a single solution, termed as *population*. Each population consists of elements. In case of the placement problem, these elements are cells to be moved. The algorithm has one main loop consisting of three basic steps, Evaluation, Selection and Allocation.

In the *Evaluation* step, *goodness* of each element is measured as a single number between '0' and '1', which is an indicator of how near the element is from its optimal location.

Then comes *Selection*, which is the process of selecting elements which are unfit (badly placed) in the current solution. An individual having high goodness measure still has a non-zero probability of being *selected*. It is this element of non-determinism that gives SimE the capability of escaping local minima. The last step, *Allocation*, has the most impact on the quality of solution. Its main function is to mutate the population by altering the location of selected cells.

The above three steps are executed in sequence until no noticeable improvement to the population goodness is observed after a number of iterations, or a fixed number of iterations are completed.

The pseudo-code of SimE is similar to that given in Figure 1 [1]. Although the illustration depicts the slave process to be discussed later, if the entire set of rows is allocated to a single processor, then the execution of the algorithm is the same as that of the serial SimE.

For large test cases, SimE has large runtime requirements. The reason is that, like other stochastic iterative algorithms, SimE is blind. It has to be told when to stop. Depending on which stopping criteria are used, as well as the size of the problem, SimE may consume hours of CPU time before it stops. The most practical approach to speed up the execution of SimE algorithm is to parallelize it. Unlike Simulated Annealing [4], [5] Genetic Algorithms [6] and Tabu Search [7] the parallelization of SimE has not been the subject of much research. Kling and Banerjee suggested three ways of speeding up the SimE algorithm [2], [8].

A parallelization strategy for VLSI cell placement for a single objective (wirelength) was attempted on a network of workstations [2], where each station is assigned a number of rows of the placement problem, in a pre-determined order. The stations executes one iteration of the SimE algorithm on the cells of the rows assigned to it. In each iteration, the rows are redistributed among the processors in a predetermined order [2].

In this paper we are addressing the problem of parallelizing SimE to solve the multiobjective VLSI standard cell placement by using a cluster of low cost PCs. The goal is to achieve a placement quality very near or equal to that achieved by serial algorithm, but with run times that decrease linearly (or super-linearly) with increasing number of processors.

In the next section we present the details of our NP-hard, multiobjective, VLSI cell placement problem. Problem formulation and models for estimating the costs for the various objectives to be optimized are presented. In Section III, the distributed algorithm is detailed. Experimental setup, results obtained on ISCAS benchmark circuits, and other observations are given in Section IV, followed by Conclusion in Section V.

II. MULTIOBJECTIVE FUZZY COST FUNCTION

In this section, we formulate our multiobjective fuzzy cost function used in the optimization process.

Algorithm Slave_Process($CurS, \Phi_s$)
Notation
 (* B is the bias value. *)
 (* $CurS$ is the current solution. *)
 (* Φ_s are the rows assigned to slave s. *)
 (* m_i is module i in Φ_s. *)
 (* g_i is the goodness of m_i. *)
Begin
  Receive Placement_ And_ Indices
 Evaluation:
  ForEach $m_i \in \Phi_s$ evaluate g_i;
 Selection:
  ForEach $m_i \in \Phi_s$ **DO**
   Begin
    If $Random > Min(g_i + B, 1)$
    Then
     Begin
      $S = S \cup m_i$; Remove m_i from Φ_s
     End
   End
  Sort the elements of S
 Allocation:
  ForEach $m_i \in S$ **Do**
   Begin
    Allocate(m_i, Φ_s)
    (* Allocate m_i in local partial solution rows Φ_s. *)
   End
  Send_Partial_Placement_Rows
End. (*Slave_Process*)

Fig. 1. Structure of the Distributed Simulated Evolution Algorithm.

Algorithm Parallel_Simulated_Evolution
 Read_User_Input_Parameters
 Read_Input_Files
Begin
  Construct_ Initial_ Placement
  Repeat
   Generate Random_ Row-Indices
   ParFor
    Slave_ Process($CurS, \Phi_s$)
   (* Broadcast Cur Placement And Row-Indices. *)
   EndParFor
   ParFor
    Receive_Partial_Placement_Rows
   EndParFor
   Construct_Complete_Solution
   Calculate_Cost
  Until (Stopping Criteria is Satisfied)
  Return Best_Solution.
End. (*Parallel_Simulated_Evolution*)

Fig. 2. Outline of Overall Parallel Algorithm.

The objectives considered in our problem include: optimizing power consumption, improving timing performance (delay), and reducing overall wirelength, while, considering layout width as a constraint. A semi-formal description of the placement problem can be found in [3]. The multiobjective cost function is similar to the one formulated in [9]. The first objective, wirelength cost ($Cost_{wire}$) is estimated using an approximate Steiner tree algorithm.

The power consumption cost p_i is computed for each net i. Assuming a a fix supply voltage and clock frequency, the estimate can be obtained by $p_i \simeq C_i \cdot S_i$, (where S_i is the switching probability and C_i the total capacitance, of net i). This can be further improved to $p_i \simeq l_i \cdot S_i$ (since interconnect capacitances are a function of the interconnect lengths, and the input capacitances of the gates are constant). The total estimate of the power dissipation reduces to $Cost_{power} = \sum_{i \in M} p_i = \sum_{i \in M} (l_i \cdot S_i)$.

The delay cost is taken as the delay along the longest path in a circuit. The delay T_π of a path π consisting of nets $\{v_1, v_2, ..., v_k\}$, is expressed as: $T_\pi = \sum_{i=1}^{k-1}(CD_i + ID_i)$ where CD_i is the switching delay of the cell driving net v_i and ID_i is the interconnect delay of net vi. The placement phase affects ID_i because CD_i is technology dependent parameter and is independent of placement: $Cost_{delay} = max\{T_\pi\}$.

The layout width is constrained not to exceed a certain positive ratio α to the average row width w_{avg}.

Since we are optimizing three objectives simultaneously, we need to have a cost function that represents the effect of all three objectives in the form of a single quantity. We use fuzzy logic to integrate these multiple, possibly conflicting objectives into a scalar cost function. Fuzzy logic allows us to describe the objectives in terms of linguistic variables. Then, fuzzy rules are used to find the overall cost of a placement solution. In this work, we have used following fuzzy rule:

IF a solution has *SMALL wirelength* **AND** *LOW power consumption* **AND** *SHORT delay* **THEN** it is an *GOOD* solution.

The above rule is translated to *and-like* OWA fuzzy operator [10] and the membership $\mu(x)$ of a solution x in fuzzy set *GOOD solution* is obtained by:

$$\mu(x) = \begin{cases} \beta \cdot \min_{j=p,d,l}\{\mu_j(x)\} + (1-\beta) \cdot \frac{1}{3}\sum_{j=p,d,l}\mu_j(x); \\ \qquad \text{if } Width - w_{avg} \leq \alpha \cdot w_{avg}, \\ 0; \qquad \text{otherwise.} \end{cases} \quad (1)$$

Here $\mu_j(x)$ for $j = p, d, l, width$ are the membership values in the fuzzy sets *LOW power consumption, SHORT delay*, and *SMALL wirelength* respectively. β is the constant in the range $[0, 1]$. The solution that results in maximum value of $\mu(x)$ is reported as the best solution found by the search heuristic. The membership functions for fuzzy sets *LOW power consumption, SHORT delay*, and *SMALL wirelength* and the lower bounds for different objectives can be found in [9].

III. DISTRIBUTED SIMULATED EVOLUTION ALGORITHM

The parallelization of the SimE algorithm is carried out by partitioning the workload among available processors. The partitioning is done according to rows. The workload for each slave in the cell placement problem is the computation of SimE operations of Evaluation, Selection, and Allocation on it's assigned rows [2].

The row allocation pattern that was proposed in [2] is made up of two alternating sets of rows. In the even iterations, each

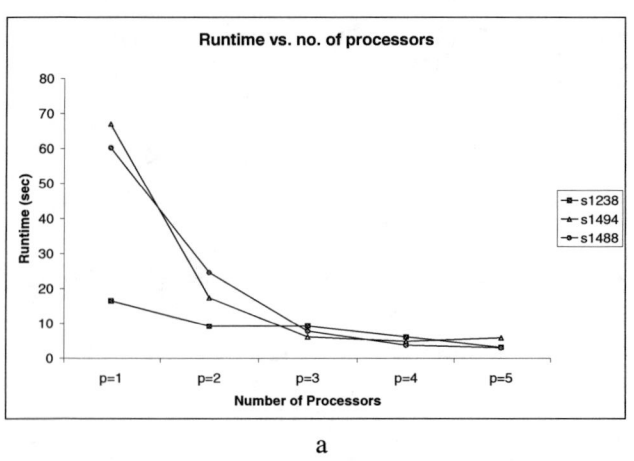
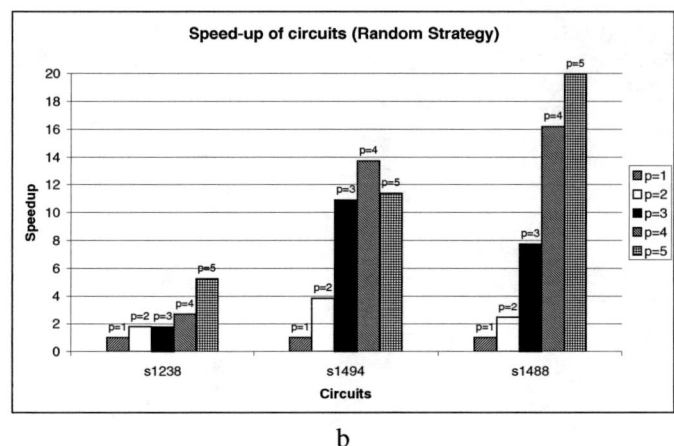

Fig. 3. (a) The decrease in runtime to reach a pre-defined fitness objective with increasing number of processors; (b) Speedup versus number of machines.

slave gets a slice of $\lceil \frac{K}{m} \rceil$ rows, (where m is the number of slaves, and K is the total number of rows in the placement) while in the odd iterations the jth slave gets the set of rows j, $j + m$, $j + 2m$, and so on. It has been shown that with the above fixed pattern of assigning rows to slaves in alternate steps, each cell can move to any position on the grid in at most two steps [2]. The consequence of row partitioning however is that the each processors has only a partial view of the placement. This hinders free cell movement, making it more difficult for cells to reach their optimal locations. Results from implementing this strategy on our multiobjective optimization problem revealed that even when given a large amount of time, the best solution obtained was poorer than one achieved by the serial implementation.

Though the lack of a global placement view will always exist in case of a distributed algorithm, the effects of restrictive cell movement can be alleviated by using a better row allocation pattern. Use of a pattern that facilitates a variety of combination among the rows sounds intuitively better. This lead us to experiment with a random row allocation.

The pseudo code of the parallel simulated evolution is illustrated in Figures 1 and 2. As can be seen, one of the processors (the master) is in-charge of running SimE on a particular partition as well as performing the following tasks periodically at the end of each iteration: (1) receive the partial placements from all other processors and combine them into a new solution and evaluate its fitness, (2) partition the new solution to obtain a new row allocation, and finally, (3) distribute the resulting sub-populations among the processors. The number of rows randomly assigned depends on the size of the placement and the number of processors. This is repeated for all iterations until the termination condition is met.

IV. RESULTS AND DISCUSSION

The parallel SimE strategies mentioned were implemented in C/C++ using MPICH Message Passing Interface implementation ver 1.2.4. for communication between nodes. The experimental environment used consists of a dedicated cluster of 8 Pentium IV 2 GHz PCs with 256 MB RAM, running RedHat Linux ver 7.3 connected with a fast Ethernet switch. ISCAS-89 circuits are used as performance benchmarks for evaluating the parallel SimE placement techniques. These circuits are of various sizes in terms of number of cells and paths, and thus offer a variety of test cases.

Table I shows the amount of time taken to reach a predefined fitness objective with increasing number of processors for both the proposed random row allocation strategy, and the fixed row allocation strategy. For the proposed strategy, as can be seen, there is a constant decrease in runtime for all circuits. Better trends are observed for medium to large circuits, than for smaller ones, as can be seen in Figure 3(a). Speedup is also illustrated in the bar-chart given in Figure 3(b). Due to space restrictions, and scaling factor limitations, not all results have been included in the same figure for sake of clarity.

The fitness values achieved with the proposed row allocation are consistently higher in all test cases when compared to the fixed row allocation scheme, as shown by the *Qual Fixed* column in Table I, the fixed row allocation never equals 100% of the solution quality obtained by the proposed scheme. Further, the run times are far better, and the speedup is super linear in most cases. This can be attributed to modified working space of the selection and allocation operators on each slave, as in each iteration different sets and combination of rows are addressed. This has resulted in even more reduced times to obtain desired solution quality than with workload partitioning alone.

A. Comparison With Other Iterative Heuristics

The runtimes and solution quality was also compared with those obtained from parallelizing simulated annealing [4], genetic algorithms (a distributed search space parallel strategy) [6], and tabu search [7]. For GAs, the time for completion to obtain solutions of a certain pre-specified quality were exorbitantly high. And in some cases, for the given run-time, acceptable solutions could not be obtained. For example, for the S1494, the serial GA implementation took 1883 seconds, and when the parallel version was executed on 7 processors the best time was 418 seconds (with 8% inferior quality than

TABLE I

TABLE DEPICTING THE RUN TIMES FOR A SPECIFIED FITNESS, FOR SERIAL, AND 2, 3, 4, AND 5 PROCESSORS, FOR BOTH RANDOM AND FIXED ROW ALLOCATION STRATEGIES. UH INDICATES UNREASONABLY HIGH TIMES.

Circuit Name	# of Cells	Random Row Distribution					Qual Fixed	Fixed Row Distribution			
		$N_p=1$	$N_p=2$	$N_p=3$	$N_p=4$	$N_p=5$		$N_p=2$	$N_p=3$	$N_p=4$	$N_p=5$
s641	433	UH	4.99	4.97	3.99	3.87	79.7%	9.14	1.08	0.76	0.55
s1238	540	16.5	9.24	9.29	6.12	3.14	95.8%	17.83	8.47	11.30	5.71
s1494	661	67	17.4	6.15	4.88	5.89	82.3%	2.77	1.85	1.76	4.34
s1488	667	60.23	24.6	7.78	3.72	3.02	96.6%	22.0	4.89	5.1	16
s3330	1961	UH	678.02	115	108.5	49.14	33.8%	316	215	4.6	3.4
s5378	2993	UH	1620	338.2	286.6	178.6	46.8%	UH	UH	124.3	95.0

that obatined by SimE).

Since cost computation of new generated solutions is very expensive in our problem, TS was parallelized by partitioning and distributing the candidate list (moves) to various slaves. While better quality was obtained in some cases at the cost of high computation time, for the same quality the run-time requirements for TS were over three times more than that required by parallel SimE. For example, for s1494, the time taken by serial TS was 268 seconds, and when parallel TS was run on 6 processors, the runtime was 57 Seconds, (compared to 5 Seconds by SimE) with slightly better quality, and TS took over 15 Seconds to obtain solutions of same quality as SimE. A similar trend was seen for all circuits.

For simulated annealing, the asynchronous multiple-Markov chain parallelization strategy was chosen [4]. Like TS, Parallel SA was also able to achieve slightly better quality solutions than SimE, given enough time. However, for a fixed quality, SimE was seen to be increasingly faster than SA as processors were increased. For instance, for s1494, with 2 processors SA took 86 seconds to achieve the desired quality, while SimE took only 17 seconds. With 5 processors, SA required 63 seconds on average, while SimE needed only 6. Similar trends are seen for most circuits.

For appreciable quality solutions, SimE has exhibited dramatic speedups with increase in number of processors, even when compared to other, more established heuristics. The results obtained suggest that in scenarios where placement quality considerations are overridden by design time constraints, the proposed parallel SimE algorithm should be favored.

V. FUTURE WORK, CONCLUSION & DISCUSSION

This paper presented the application of a modified Distributed SimE algorithm to a multi-objective VLSI cell placement problem. The algorithm focused on distributing the work load among processors. Random allocation of work load in each iteration resulted in better traversal of search for our complex multiobjective NP-hard design problem.

The results showed a significant reduction in runtime for all circuits, although the speedup was more obvious for larger ones. This speedup trend was compared to other established iterative and evolutionary heuristics from literature, and was shown to be more consistent with increasing number of processors.

This work can be extended along several lines. One would be to investigate suitable parameters for the SimE algorithm that will enable better quality and run-times. At the moment, the same parameters that have been set for serial execution are used. Another approach is to relieve computational resources during execution when the quality ceases to improve. This can be achieved by modifying the stopping criteria. If the quality does not improve for the last j iterations on k processors, then the number of processors can be reduced to $k-1$, and this can continue until all processors are relieved. The effects of this experiment will be, that while execution continues to improve the obtained best solution, the distribution of increased number of rows on reduced number of processors will enable exploring different regions of the search space in the same run, and will hopefully result in better quality with reduced resources. Our initial experiments on this idea have been encouraging.

ACKNOWLEDGMENT

The authors thank King Fahd University of Petroleum & Minerals (KFUPM), Dhahran, Saudi Arabia, for support under Project Code COE/CELLPLACE/263.

REFERENCES

[1] Sadiq M. Sait and Habib Youssef. *Iterative Computer Algorithms with Applications in Engineering: Solving Combinatorial Optimization Problems*. IEEE Computer Society Press, California, December 1999.
[2] Ralph M. Kling and Prithviraj Banerjee. ESP: A new standard cell placement package using simulated evolution. *Proceedings of 24th Design Automation Conference*, pages 60–66, 1987.
[3] Sadiq M. Sait and Habib Youssef. *VLSI Physical Design Automation: Theory and Practice*. World Scientific Pubishers, 2001.
[4] John A. Chandy, Sungho Kim, Balkrishna Ramkumar, Steven Parkes, and Prithviraj Banerjee. An evaluation of parallel simulated annealing strategies with application to standard cell placement. *IEEE Transactions on Computer-Aided Design of Integrated Circuits and Systems*, 16:398–410, April 1997.
[5] Robert Azencott, editor. *Simulated Annealing Parallelization Techniques*. John Wiley & Sons, 1992.
[6] Erick Cant-Paz. Designing efficient master-slave parallel genetic algorithms. *Genetic Programming*, 1998.
[7] E. Taillard. Robust tabu search for the quadratic assignment problem. *Parallel Computing*, 17:443–455, 1991.
[8] Prithviraj Banerjee. *Parallel Algorithms for VLSI Computer-Aided Design*. Prentice Hall International, 1994.
[9] Sadiq M. Sait, Mahmood R. Minhas, and Junaid Asim Khan. Performance and low-power driven VLSI standard cell placement using tabu search. *Proceedings of the 2002 Congress on Evolutionary Computation*, 1:372–377, May 2002.
[10] Ronald R. Yager. On ordered weighted averaging aggregation operators in multicriteria decision making. *IEEE Transaction on Systems, MAN, and Cybernetics*, 18(1), January 1988.

A Divide-and-Conquer 2.5-D Floorplanning Algorithm Based On Statistical Wirelength Estimation

Zhuoyuan Li, Xianlong Hong, Qiang Zhou,
Yici Cai, Jinian Bian
Department of Computer Science and
Technology, Tsinghua University, Beijing, China

Hannal Yang, Prashant Saxena,
Vijay Pitchumani
Strategic CAD Lab, Intel
Hillsboro, OR 97124

Abstract — **An efficient and effective divide-and-conquer 2.5-D floorplanning algorithm is proposed for wirelength optimization. Modules are pre-partitioned into different dies with respect to the statistical wirelength estimation result. Then floorplan is generated on each die for wirelength optimization. The new partitioning method successfully solves the confliction between wirelength minimization and inter-die via constraints. Experimental results show that our algorithm could provide noticeable improvement on the total wirelength compare to both 2-D design and previous 2.5-D floorplanning algorithm.**

1. INTRODUCTION

The 2.5-D and 3-D technology is being viewed as a potential alternative that could alleviate interconnect delay problem, increase transistor packing density and reduce chip area significantly [1]. However, until now not much work on circuit applications has been done due to lack of insight into 3-D circuit architecture and performance. Some work has been done on the knowledge about interconnection length and delay distribution of 2.5-D and 3-D circuits because that it is critical in determining device layer and interconnection layer distribution [2-4].

Floorplanning is one of the first steps of VLSI design. Common goals of floorplanning are minimization of the combination of chip area and interconnections length. Compared to 2-D design, there is another key problem of 2.5-D floorplanner: how to solve the confliction between wirelength reduction and vertical channel constraints [5]. In 2.5-D design, the modules with higher connectivity tend to be placed into different dies. Then the connections between them could utilize the vertical channel to eliminate global wires. However, a large number of vertical channels can be detrimental to the integration density due to additional area required by such channels and lead power and reliability issues [2]. There must be a careful planning to decide the tradeoff between vertical channels usage and manufacturing cost.

The wirelength-oriented floorplanning problem for 2.5-D design could be stated as:

$$minimize \quad Area + w \sum_{net} L_{net}$$
$$subject\ to\ N_{via} \leq V_{max}$$

where *Area* is the minimal packing area needed for stacked dies. L_{net} is the wirelength of net N. w is the weight. N_{via} is the number of vertical vias needed by interconnections between dies and V_{max} is the maximum number of vertical channels between dies.

There are commonly two types of approaches to realize 2.5-D floorplanner and solve this confliction: partition-based and un-partition-based. The partition-based methods pre-partition modules into different dies and generate floorplan on each die. With min-cut partitioning methods, the number of vertical vias could be minimized and the confliction is naturally solved. The results are more immune to inter-layer via capacitance variation. However, minimizing the number of vertical vias will lead an increase in the total wirelength so as to degrade the benefit of 2.5-D integration. The min-cut method should not be applied directly here [6].

For the un-partition-based method, a 2.5-D floorplanning algorithm for two active layers based on the array of 2-D BSG representation is proposed in [7]. The optimization is accomplished using simulated annealing engine with the cost function that is the weighted sum of total wire length, floorplan area, and total number of vertical vias. Given an initial solution, modules are moved inside or between upper and bottom dies to achieve smaller packing area and total wirelength. With this method, modules are assigned to different dies automatically during the SA process. The vertical channel constraints could be checked and validated in the SA process, or treated as a penalty term as [7]. As a preliminary successful trial, a significant reduction in both total wirelength and longest wire length is achieved.

However, the design complexity would be increased with this method. In [10] it is proven that the solution space of the un-partition-based method would be $N \times 2^N$ times larger than the partition-based method for N modules. It would cause much increase in the running time, while maintaining low design complexity is a key problem in 2.5-D IC design.

The above discussion could be summarized as the following statement: Treating the vertical channel constraints as the necessary condition, the main issue of any 2.5-D floorplanner is to determine the tradeoff between wirelength reduction and high efficiency.

This problem could be solved if we could calculate the possible wirelength of any inter-die partition without packing. It means we could predict the benefit of different partitions before floorplanning so that the 2.5-D floorplanning problem is transformed to two successive sub-problems:

1. Pre-partition modules to different dies with the

---

* The work was supported by the Hi-Tech Research & Development (863) Program of China 2002AA1Z1460 and The National Natural Science Foundation of China (NSFC) 90407005 minimal possible wirelength while satisfying the vertical channel constraints.

2. Generate floorplan on each die for packing area and wirelength optimization.

In this paper, we present a new divide-and-conquer wirelength-oriented floorplanning algorithm for two stacked dies. Based on pre-placement wirelength estimation result, modules are assigned to different dies before floorplanning. Then we perform wirelength optimization on each die. The rest of this paper is organized as follows: Section 2 formulates and solves the partitioning problem with the new wirelength estimation model. The overall algorithm is demonstrated in section 3. Experimental results are reported and compared in section 4. Section 5 is the conclusion.

2. INTER-DIE PARTITIONING

2.1 Problem Formulation

Given a set of rectangular modules $M = \{M_1, M_2,...,M_n\}$, each rectangular block M_i is defined by a tuple (h_i, w_i), where h_i and w_i are the height and the width of module M_i, respectively. The formulation of the inter-die partitioning problem could be expressed as follows:

$$minimize \sum_{net} PL_{net}$$
$$subject\ to\ N_{via} \leq V_{max}$$

where PL_{net} is the possible wirelength of net N for a inter-die partition scheme. The formulation of the inter-die partitioning problem is quite different from 2-D net-cuts oriented partitioning problem. The number of net-cuts is treated as a constraint, not the objective as these methods do. On the other hand, tightly interconnected modules could be placed close together on the same die or overlapped vertically on different dies to reduce the total wirelength in 2.5-D integration. The problem becomes much more difficult than 2-D design. Traditional partitioning methods are hard to tackle with it.

To solve this problem, the key factor is to calculate PL_{net} accurately. As the position of each module remains unknown during partitioning, it would be impossible to calculate the real wirelength without packing. Pre-placement wirelength estimation methods are needed to solve it. There is a vast literature on wirelength estimation during different design phase (refer to [8] for a brief survey). The *posteriori* and *on-line* wirelength estimation technique depend on the fixed or intermediate placement, which could not be achieved without packing. Previous *priori* methods are focused on the distribution of wires according to the number of gates. It is also hard to tackle with this problem. We propose a new method to estimate PL_{net}.

2.2 Partitioning Based on Statistical Wirelength Estimation

Wirelength depends on the positions of connected pins, which is decided by the positions and orientations of modules. With the probabilistic approach, we could predict the average distance between interconnected pins and calculate the statistical expectation of wirelength between them.

Assume that the upper and bottom dies are square. W_a and W_b denotes the width of the two dies, $W_a < W_b$ as shown in fig.1. Module M_i and M_j are assigned to the upper die and Module M_k is assigned to the bottom die. There are interconnections between them.

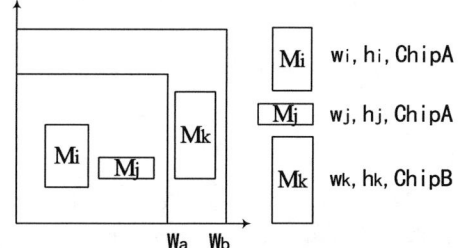

Fig.1 Two stacked dies

Suppose the coordinates of M_i and M_j are (x_i, y_i) and (x_j, y_j), respectively. They remain unknown without packing. Since x_i, x_j, y_i and y_j are uniformly distributed within the range of $[0, W_a]$, the density function is $P_a(x) = 2(W_a - x)/W_a^2$. As module M_i and M_j are on the same die, they could not overlap with each other. The statistical expectation of average wirelength between M_i and M_j is:

$$L_1 = \int_0^{W_a}\int_0^{W_a}(\int_0^{x_i-w_i} + \int_{x_i+w_i}^{W_a-w_i})\int_0^{W_a}(|x_i-x_j|+|y_i-y_j|)P_a(x_i)P_a(y_i)P_a(x_j)P_a(y_j)dy_jdx_jdy_idx_i$$
$$+ \int_0^{W_a}\int_{w_i}^{W_a}\int_{y_i-h_i}^{y_i+h_i}(\int + \int)(|x_i-x_j|+|y_i-y_j|)P_a(x_i)P_a(y_i)P_a(x_j)P_a(y_j)dy_jdx_jdy_idx_i$$
$$\underset{W_a \gg w_i,w_j,h_i,h_j}{\approx} \frac{2W_a}{3} + \frac{w_i+w_j+h_i+h_j}{2}$$

As M_i and M_k are located on different dies, they could overlap with each other. The statistical expectation of average wirelength between M_i and M_k is:

$$L_2 = \int_0^{W_a}\int_0^{W_a}\int_0^{W_b}\int_0^{W_b}(|x_i-x_k|+|y_i-y_k|)P_a(x_i)P_a(y_i)P_b(x_k)P_b(y_k)dy_kdx_kdy_idx_i$$
$$\underset{W_a,W_b \gg w_i,w_j,h_i,h_j}{\approx} \frac{2W_a}{3} + \frac{2(W_a-W_b)^2}{3W_b}$$

where (x_k, y_k) is the coordinate of M_k, $P_b(x) = 2(W_b - x)/W_b^2$ is the density function of a point located at x within the range of $[0, W_b]$.

We have had the pre-placement estimated wirelength for modules on the same or different dies. Given a inter-die partition, suppose the probability of any two modules connected by net N located on the vertex of the bounding-box is equal, the statistical expectation of the wirelength of N with n interconnected pins is as follows:

$$PL_{net} = \frac{2\alpha}{n(n-1)}(\sum_{M_i,M_j\ on\ the\ same\ die} L_1 + \sum_{M_i,M_j\ not\ on\ the\ same\ die} L_2) + N_{via}L_{via}$$

where N_{via} is the number of via and L_{via} is the required vertical wirelength between two dies.

The scaling factor α is used because all the floorplanning approaches tend to place modules with interconnections together to reduce wirelength. The actual wirelength would be smaller than the average distance between these modules. It is set to be 0.5 in our algorithm. As W_a and W_b remain unknown without packing, they are estimated as

$$W_a = \lambda \sqrt{\sum_{Mi\ on\ upper\ die} w_i \times h_i} \text{ and } W_b = \lambda \sqrt{\sum_{Mi\ on\ bottom\ die} w_i \times h_i}$$

where λ is the scaling factor for packing area. It is set to be 1.1 in our algorithm.

The inner meaning of the wirelength estimation model could be expressed as follows: Suppose the sizes of each die differ a little from each other, the difference by assigning two interconnected modules, M_i and M_j, to the same or different dies is

$$\Delta L = \sum_N \sum_{Mi, Mj \in N} (L_1 - L_2) \approx \sum_N \sum_{Mi, Mj \in N} \frac{w_i + w_j + h_i + h_j}{2}$$

It is because modules on the same die could not overlap with each other, so there would be an increase on the total wirelength to place them on the same die. Due to different orientations of modules, the increase on wirelength is also related to the size of modules. Furthermore, this difference caused by more tightly interconnected modules would be larger, so these modules would be assigned to different dies when minimizing the estimated wirelength. Then more long global wires could be changed to local wire through vertical vias. It corresponds fairly well with all the heuristic methods.

2.3 Partitioning Based on Interconnectivity

To show the advantages of above method, we also propose an interconnectivity-based partitioning method as traditional methods do. It is shown that tightly interconnected modules should be placed on different dies to reduce the total wirelength [6]. As a consequence, the partition with maximal interconnections between dies may achieve the minimal wirelength after floorplanning. With the vertical channel constraint, the evaluation criterion would be: inter-die partition with minimal interconnections inside each die while satisfying the constraint. For a net N with n interconnected pins, Suppose the probability of any two modules connected by net N located on the vertex of the bounding-box is equal, PL_{net} could be estimated as follows:

$$PL_{net} = \sum_{Mi, Mj\ on\ the\ same\ die} \frac{2}{n(n-1)} \Delta w$$

where M_i and M_j are connected by net N, Δw is the average wirelength needed to realize a horizontal interconnection. If M_i and M_j are located on different dies, a vertical via connects them and its length is assumed to be zero.

With above two estimation methods, the possible benefit on wirelength reduction could be calculated for different partitions. The optimization is accomplished with a SA algorithm. Then we get an inter-die partition with possible minimal wirelength.

3. OVERALL ALGORITHM

The overall 2.5-D floorplanning algorithm integrates the aforementioned estimation and partitioning approaches. Given a set of modules, they are assigned to different dies with the new partitioning method. The floorplans of each die after partitioning are generated using SA algorithm with CBL representation [9]. The objective is to minimize the weighted sum of total wire length and floorplan area. The design flow of our algorithm is shown in fig.2.

4. EXPERIMENTAL RESULTS

As there are two different methods to predict possible benefit on wirelength reduction for each partition, we implemented two 2.5-D floorplanning algorithms, A1 and A2, for hard modules in C language on a Sun V880 workstation with 4GB memory. A1 uses the estimation method by interconnectivity and A2 uses the statistical wirelength estimation method. To show the effectiveness of our methods on larger test cases, we create several synthetic circuits named ami33_x or ami49_x by duplicating the modules and nets in ami33 or ami49 for x times.[1] Experimental results are reported and compared from table 1 to table 3. The net length after floorplanning is estimated with half perimeter bounding box model.

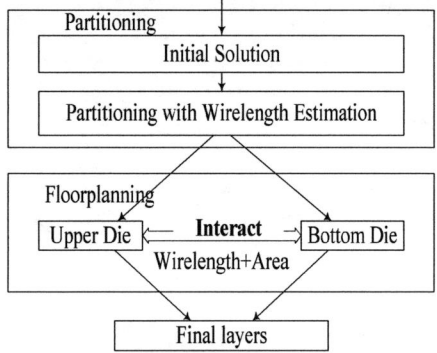

Fig.2 Design Flow of Our 2.5-D Floorplanner

A. Compare A2 With A1

As A1 and A2 are both based on divide-and-conquer technique, the computational efficiency of each algorithm differs little from each other. The key factor that would affect the floorplanning results of A1 and A2 is the threshold on maximum number of vertical channels. When the threshold is large enough, there is no constraint on the number of vertical vias. The first estimation method becomes a max-cut partitioning method. Previous heuristic methods are based on it. The second method still considers the influence on wirelength due to possible position and orientation of modules. If A2 achieves better results under different thresholds, our formulization and deduction is valuable, otherwise it is useless.

TABLE 1. COMPARE A2 WITH A1

	Wirelength result on ami33			
V_{max}	30	50	80	1000
A1	33089	31514	29344	27384
A2	30037	29176	27734	25900
Dec.	-9.2%	-7.4%	-5.5%	-5.4%
	Wirelength result on ami49			
V_{max}	150	180	250	1000
A1	881596	769426	641774	629482
A2	691502	648944	581210	546406
Dec.	-21.6%	-15.6%	-9.4%	-13.2%

Experimental results are reported and compared in table 1. As the constraint on vertical vias is relaxed, the total wirelength generated by A1 and A2 both decreases. It is shown that A2 could achieve much better results under different thresholds. The wirelength is reduced from 5% to 10% on ami33. Simultaneously it is reduced from 10% to 20% on ami49. As the maximum number of vertical vias becomes fewer, A2 would achieve more improvement on A1. It is because A2 gives out a more accurate vertical channel allocation scheme with respect to the size, possible position relation and possible orientation of modules. When there is no constraint on vertical via number ($V_{max} = 1000$), performance of A2 is still better than A1. This proves the effectiveness of the statistical wirelength estimation based partitioning method.

B. Compare A2 With Un-partition-based 2.5-D Floorplanning Algorithm

Experimental results compared with previous 2.5-D floorplanning algorithm are reported in Table 2. Since there are only results on ami33 and ami49 reported in [7], we compare our method with it on the two test cases. The total wirelength and longest wirelength are reduced by 30% and 21% on the average, while the total area of two dies increases slightly. The number of vertical channels, V_{net}, is also constrained below the threshold with our method. As the solution space of un-partition-based method is much larger than the divide-and-conquer method, our algorithm would be more efficient in addition to better results.

C. Compare A2 With 2-D Design

Table 3 shows the results compared with traditional 2-D design. We use the same parameters of the SA process for 2-D and 2.5-D floorplanning so the running time of our algorithm is almost the same with 2-D floorplanning for every benchmark. It is shown that the total wirelength and longest wirelength are reduced by 40% and 28% on the average without any increase on the design complexity. The packing area of the stacked-die is reduced by 44%, which is approximately half of the 2-D design. It means our algorithm does not cause much excessive sacrifice on the packing area. Moreover, the packing area of each die differs within 1% on every benchmark.

TABLE 2. COMPARE A2 WITH [7]

	ami33			
	T.W.*	L.W.**	Area	N_{via}/V_{max}
[7]	64713	2688	12.82	NA
A2	33622	1637	12.79	37/40
	ami49			
[7]	625769	8099	43.2	NA
A2	546372	7826	45.2	187/200
Dec.	-31.1%	-21.3%	+1.7%	

*T.W.: Total Wirelength; **L.W.: Longest Wirelength

TABLE 3. FLOORPLANNING RESULTS COMPARED WITH 2-D DESIGN

	2-D Floorplanning				A2						
	Area	T.W.	L.W.	Cpu(s)	Area*	Dec.%	T.W.	Dec.%	L.W.	Dec.%	Cpu(s)
Ami33	1309868	49168	2282	47	783118	**-40.2**	28189	**-43.2**	1701	**-25.5**	50
Ami33_2	2682848	109165	3276	96	1446480	**-46.1**	69692	**-36.1**	2408	**-26.5**	95
Ami33_4	5350212	273035	4627	200	3055444	**-42.3**	153461	**-43.8**	3500	**-24.4**	201
Ami33_6	8328726	363804	5775	295	4494868	**-46.0**	236264	**-35.1**	4039	**-30.1**	311
Ami49	41221152	1316756	12796	109	22581276	**-45.2**	546372	**-55.1**	7826	**-38.8**	106
Ami49_2	84863688	2532068	17570	208	49166992	**-42.1**	1744344	**-31.1**	13706	**-22.0**	206
Ami49_3	131625368	4107068	23058	298	73249512	**-44.4**	2430764	**-40.8**	16324	**-29.2**	304
Ami49_4	173009200	5481350	25970	399	100314368	**-42.1**	3473582	**-36.6**	19698	**-24.2**	396
Average						**-43.6**		**-40.1**		**-27.6**	

*Area: Packing area of the stacked-die

5. CONCLUSION

New wirelength-oriented floorplanning approach is proposed for 2.5-D integration. The new partitioning method successfully solves the confliction between wirelength reduction and inter-die via constraint. It also achieves area balancing between different dies automatically. Our divide-and-conquer technique is proven to be very efficient and offers a potential way for high performance 2.5-D design.

REFERENCE

[1] S.F. Al-Sarawi, D. Abbott, P.D. Franzon, "A Review of 3-D Packaging Technology", IEEE Transactions on Components, Packaging, and Manufacturing technology Part B, Vol. 21, No.1, 1998, pp. 2-14.

[2] R. Zhang, K. Roy, C.K. Koh, D.B. Janes, "Stochastic Interconnect Modeling, Power Trends, and Performance Characterization of 3-D Circuits", IEEE Transactions on Electron Devices, Vol. 48, No.4, 2001, pp. 638-652.

[3] A. Rahman, A. Fan, J. Chung, R.Reif, "Wire-Length Distribution of Three-Dimensional Integrated Circuits", IEEE International Interconnect Technology Conference Proceedings, 1999, pp. 233-235.

[4] K. Banerjee, S.J. Souri, P. Kapur, K.C. Saraswat, "3D-ICs: A Novel Chip Design for Improving Deep-Submicrometer Interconnect Performance and Systems on Chip Integration", Proceedings of the IEEE, Vol. 89, No.5, 2001, pp. 602-633.

[5] S. Das, et. al. "Performance, and Computer Aided Design of Three Dimensional Integrated Circuits", Proc. International Symposium on Physical Design, 2004

[6] S. Das, A. Chandrakasan and R. Reif, "Design Tools for 3-D Integrated Circuits", Proc. ASP-DAC, pp.53-58, Jan. 2003.

[7] Yangdong Deng and Wojciech P. Maly, "Interconnect Characteristics of 2.5-D System Integration Scheme", Proc. International Symposium on Physical Design, 2001

[8] Andrew E. Caldwell, Andrew B. Kahng, Stefanus Mantik, Igor L. Markov, and Alexander Zelikovsky, "On Wirelength Estimations for Row-Based Placement", IEEE trans. On CAD, Vol.18 No.9, 1999

[9] X.L. Hong, G. Huang, Y. Cai, J. Gu, S. Dong, C.-K. Cheng, and J. Gu, "Corner Block List: An effective and efficient topological representation of non-slicing floorplan," Proc. ICCAD, pp. 8–12, Nov. 2000.

[10] Zhuoyuan Li, "Wirelength-oriented partitioning method for two stacked dies", Technical report, Tsinghua Univ,, Dec. 2003

A New Congestion and Crosstalk Aware Router

Chin-Hui Wang, Yung-Ching Chen, Tsai-Ming Hsieh
Department of Information and Computer Engineering,
Chung Yuan Christian University,
Chung-Li, Taiwan, R.O.C.
{trf,ycchen}@fpga.ice.cycu.edu.tw; hsieh@cycu.edu.tw

Chih-Hung Lee
Department of Information Management,
Ling Tung College,
Taichung, Taiwan, R.O.C.
chlee@mail.ltc.edu.tw

Hsin-Hsiung Huang
Institute of Electronic Engineering,
Chung Yuan Christian University,
Chung-Li, Taiwan, R.O.C.
bear@fpga.ice.cycu.edu.tw

Abstract— In this paper, we study and implement a new congestion- and crosstalk-driven routing system. It first takes estimated congestion cost, crosstalk cost and the track utilization of each routing grid simultaneously into consideration to determine the global routing paths of all nets. Then a crosstalk-driven track assignment algorithm is applied to the global routing result to generate the corresponding detailed routing solution. The proposed approach can effectively disperse nets to lower congested regions. Compared with results of maze routing algorithm without consideration of crosstalk and the crosstalk aware routing solution [2], our algorithm archives 94% and 17% reduction of the overall effective coupling length in average, respectively.

I. Introduction

As the fabrication technology entered very deep sub-micron era, the complexity of routing problem is increasing dramatically along with the growth on demand of net density. Wire congestion can be regarded as an over-using of the finite routing resources on the chip and will directly affects the routability. Crosstalk not only makes the logic values of circuit nodes differ from the desired values but also makes unexpected timing change caused by switching behavior. Therefore, how to consider routability issue and crosstalk minimization in router design is a very important topic.

To cope with the increasing complexity of circuits, measuring congestion as earlier as possible in the design flow is necessary. A good congestion estimation model should be accurate enough to reflect the real post-routing result and fast enough to be embedded into the iterative algorithms for searching the better solution. Lou *et al.* [9] introduced a probabilistic analysis based method to estimate congestion. Hsieh *et al.* [3] proposed a new congestion model based on the probabilistic analysis and a new concept of irregular-size grids (IR-grids) instead of previous models with fixed-size grids. Sarkar et al. [7] proposed a congestion model that employed a 2-D rectangular grid based probabilistic map assuming L-shaped and Z-shaped routing for each net.

Coupling capacitance between wires is the main source of crosstalk. In fact, coupling capacitance between wires can account for over 70% of the total wiring capacitance in 0.25 μm processes [4]. There were several previous works addressing the crosstalk problem. Gao *et al.* [1] proposed track permutation on channel routing to reduce crosstalk. Zhou *et al.* [8] proposed a two stage sequential approach to solve the crosstalk-constrained global routing problem. Pan *et al.* [6] proposed track permutation and wire sizing after detailed routing to reduce crosstalk. Kastner *et al.* [5] minimized crosstalk using only one bend (L-shaped) routing pattern at global routing stage. However, the routability and the accuracy of estimated crosstalk seems to be dominated by the limitation of only one bend (L-shaped) routing pattern was considered. Ho *et al.* [4] embedded an intermediate stage of routing layer/track assignment into the multilevel routing framework for crosstalk minimization. This multilevel routing framework performed congestion-driven global routing at the coarsening stage, followed by crosstalk-driven layer/track assignment at the intermediate stage, and then detailed routing at the uncoarsening stage. Hsiao *et al.* [2] provided a crosstalk- aware two-pin net router which uses a constrained region graph which represents the necessary information.

In this paper, we propose an effective congestion and crosstalk-driven router (CCR). First, the entire chip area is divided into a set of grids. The wire congestion cost and crosstalk cost of each grid are evaluated at each grid, assuming that each wire is routed in either L-shaped or Z-shaped. An integrated cost related to congestion cost, crosstalk cost track utilization of each grid is then defined to determine the global routing paths of all nets. Finally, a modified crosstalk-driven track assignment algorithm of [4] is applied to the global routing result to get the detailed routing solution. Multi-bend routing patterns are allowed in the detailed routing solution if needed.

The rest of this paper is organized as follows: Section II presents our motivation. Section III briefly describes the congestion model, crosstalk model and problem formulation. In Section IV, the CCR algorithm is given in detail. Experimental results are shown in Section V. Finally, we conclude in Section VI.

II. Motivation

The routing problem is often solved by using the two-stage approach, global routing followed by detailed routing. In general, minimization on wire congestion and total wire length are mainly concerned at global routing stage. And the crosstalk minimization issue can be processed in more accurate at the detailed routing stage. However, the corresponding detailed routing solutions are restricted by the global routing solution.

Figure 1 illustrates the importance of considering the wire-coupling issue at the global routing stage. Assume the entire chip area are divided into 3×3 subregions, the horizontal/vertical tracks are marked in dotted lines, four nets N_1, N_2, N_3 and N_4 are distributed on the chip area. Assume that the coupling capacitance will occur if two parallel segments are assigned to adjacent tracks when the coupling length is larger or equal to 5 units. The feasible global routing solutions shown in Figure 1(a), (b) and (c) have no routing resource overflow problem, i.e. the number of nets passing through a subregion is not larger than the available tracks. Figure 1 (d), (e) and (f) are the corresponding track assigned solutions of (a), (b) and (c). From the viewpoint of wire congestion, Figure 1 (a) is better than both cases of (b) and (c). Under the presumption of no detoured routing are allowed and from the viewpoint of crosstalk issue at the detailed routing stage, both Figures 1 (d) and (e) have coupling capacitance. However, a coupling capacitance free solution in Figure 1(f) can be generated from the Figure 1(c). And

---

This work was supported by the National Science Council of Taiwan, R.O.C., under Grant NSC 93-2215-E-033-007.

we can say Figure 1(c) is a better global routing solution for congestion and crosstalk driven routing problem than Figure 1(b).

According to the observation on Figure 1, we have the motivation to find a new evaluation method that can help us to find a good global routing solution.

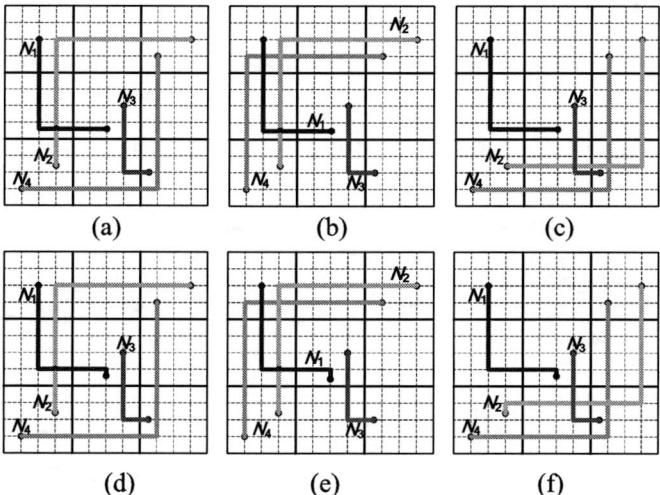

Figure 1: (a), (b) and (c) are three feasible global routing solutions. (d), (e) and (f) are the corresponding track assigned solutions of (a), (b) and (c).

III. Problem Formulation

A. Preliminaries

Assume that the crosstalk effect occurs when two parallel segments of signals with effective coupling length larger than or equal to the threshold value L_{max}, and the spacing of the two segments is smaller than the threshold spacing value S_{max}. Based on above assumption, we develop an algorithm to generate a set of horizontal panels $P_h = \{P_h(i) | i = 0, 1, .., p-1\}$ (vertical panels $P_v = \{P_v(j) | j = 0, 1, .., q-1\}$) to estimate the crosstalk cost by horizontal (vertical) cutting lines. Then the entire chip is divided into $p \times q$ irregular size subregions, $G_{i,j}$, $0 \leq i \leq p-1, 0 \leq j \leq q-1$, by these cutting lines. Every subregion $G_{i,j}$ consists a set of grid points, $G_{i,j} = \{(x,y) | x_i \leq x < x_{i+1}, y_j \leq y < y_{j+1}, 0 \leq i \leq p-1, 0 \leq j \leq q-1\}$. We then calculate the congestion cost for each subregion.

In Figure 2, there are three nets N_1, N_2 and N_3. Applying the Gen_Horizontal_Panel algorithm, we find four vertical cutting lines, $x = x_i$ ($i=0,1,2,3$) and four horizontal cutting lines, $y = y_j$, ($j=0,1,2,3$). Three horizontal panels and three vertical panels are formed. The entire chip is partitioned into 9 subregions, $G_{i,j}$, $i = 0,1,2,3$, $j = 0,1,2,3$. The following is procedure of generating horizontal panels, and the vertical panels can be generated in similar way.

B. Congestion Model

To simplify the complexity of routing path analysis, only L-shaped and Z-shaped routing patterns are considered for each net on the fixed size routing grids as [7]. The expected value of congestion on each grid can be figured out according to the probability table [7]. For a two-pin net N_u with the source point S located in grid $G_{0,0}$ and a sink point T located in grid $G_{2,2}$, $\delta C_h(i,j)$ and $\delta C_v(i,j)$ are the corresponding probabilities of N_u passing through grid $G_{i,j}$ ($0 \leq i \leq 2, 0 \leq j \leq 2$) in horizontal and vertical, respectively. The sum of probabilities of all nets acrossing grid $G_{i,j}$ horizontally is:

```
Algorithm Gen_Horizontal_Panel;
Input : A set of two-pin nets, L_max and S_max.
Output : A set of cuttinling lines.
Method :
    Initialization ; //*Generate bounding segments for all nets.*//
    Set bottom grid line y = 0 be a cutting line;
    Set bottom grid line y = 0 be the current_segment;
    while (the current_segment not the last grid line ) do
        begin
            scan up vertically and update the y-coordinate of the current_segment;
            if (a horizontal segment is found) then
                begin
                    find the y-coordinate of the found next segment;
                    find the x-coordinates of the two end points of the found next segment;
                    calculate the coupling length Δl of the current segment and the next segment;
                    calculate the spacing Δs between the current segment and the next segment;
                    if (Δl ≥ L_max) and (Δs ≤ S_max)
                        then add the next segment to the current processing panel;
                        else generate a new cutting line; //* a new panel is generated between
                                the new cutting line and the previous cutting line.*//
                end
        end
end
```

Figure 2: 9 subregions, $G_{i,j}$ are formed.

$$E_{i,j}^h = \sum_{u=1}^{n} P_u^h(i,j), \quad P_u^h(i,j) = \sum_{(k,l) \in G_{i,j}} \delta C_h(k,l)$$

The horizontal congestion cost $\mathcal{K}_{i,j}^h$ of $G_{i,j}$ is $\mathcal{K}_{i,j}^h = E_{i,j}^h / T_{i,j}^h$, where $T_{i,j}^h$ is the number of horizontal tracks of grid $G_{i,j}$. Of course, in the case of fixed grid size, the number of horizontal tracks on each grid is the same so that $T_{i,j}^h$ is a constant and we can define $\mathcal{K}_{i,j}^h$ as $E_{i,j}^h$. The vertical congestion cost $\mathcal{K}_{i,j}^v$ of $G_{i,j}$ can be defined in a similar way.

C. Crosstalk Model

In this paper, a statistical crosstalk estimation approach is embedded into the global routing stage so that the router was guided to search the solution with minimized crosstalk effect.

First, the horizontal (vertical) panels are generated by the algorithm mentioned in part A of Section III. Second, every routing path of all nets is divided into a set of horizontal and vertical segments with only consideration of L-shaped and Z-shaped. If the length of a segment is larger or equal to L_{max}, we say that it's a *long segment*. To simply the complexity of statistical crosstalk estimators, only the *long segment* of a panel is considered for crosstalk effect.

Suppose that the horizontal bounding segments of nets N_a and N_b have coupling length $L_{a,b}$, where $L_{a,b} \geq L_{max}$ and their spacing $S_{a,b}$ is closer than S_{max} in horizontal panel $P_h(i)$, the cross effect occurs. The expected horizontal crosstalk cost $\mathcal{X}_{i,j}^h$ of $G_{i,j}$ in panel $P_h(i)$ can be evaluated by :

$$\mathcal{X}_{i,j}^h = \sum_{\substack{\text{all nets } N_a, N_b \text{ in } P_h(i) \\ S_{a,b} \leq S_{max},\ L_{a,b} \geq L_{max}}} \frac{L_{a,b}}{R_a} + \frac{L_{a,b}}{R_b}, \quad j = 0,1,...,q-1$$

where R_a and R_b are the numbers of possible routing paths of net N_a and N_b, respectively. The expected vertical crosstalk cost of $G_{i,j}$ in panel $P_v(j)$ is defined in similar way.

The expected number of tracks demand by the nets in a panel which will cause crosstalk noise can be calculated by

$$\mathcal{N}_{i,j}^h = \sum_{\substack{\text{all nets } N_a, N_b \text{ in } P_h(i) \\ S_{a,b} \leq S_{max}, L_{a,b} \geq L_{max}}} \frac{1}{R_a} + \frac{1}{R_b}, \quad j = 0,1,...,q-1$$

We then define the horizontal integrated cost $C_{i,j}^h$ which integrates the congestion, the crosstalk and track utilization information. The value of $C_{i,j}^h$ is equal to $\mathcal{X}_{i,j}^h \times f(\mathcal{N}_{i,j}^h, \mathcal{T}_{i,j}^h)$ if $\mathcal{K}_{i,j}^h$ is greater than one, otherwise $C_{i,j}^h$ is infinite, where $f(\mathcal{N}_{i,j}^h, \mathcal{T}_{i,j}^h)$ is a track utilization function in a subregion $G_{i,j}$. In this paper, we use a product operation to balance the influence of $\mathcal{X}_{i,j}^h$ and $f(\mathcal{N}_{i,j}^h, \mathcal{T}_{i,j}^h)$, that is defined as $f(\mathcal{N}_{i,j}^h, \mathcal{T}_{i,j}^h) = \exp(\mathcal{N}_{i,j}^h / \mathcal{T}_{i,j}^h)$. Moreover, the vertical integrated cost $C_{i,j}^v$ of $G_{i,j}$ is defined in similar way.

D. Congestion and Crosstalk-Driven Routing Problem

The congestion and crosstalk-driven routing problem is defined as: Given a set of n two-pin nets $N_1, N_2, ..., N_n$, two layers l_1, l_2, and assume that the crosstalk effect occurs when two parallel segments of signals with effective coupling length larger than or equal to the threshold value L_{max}, and the spacing of the two segments is smaller than the threshold spacing value S_{max}, to find a solution with minimized congestion and crosstalk.

IV. CCR algorithm

Algorithm CCR
Input: A set of n two-pin nets $N_1, N_2, ..., N_n$, two layers $l = \{l_1, l_2\}$, S_{max} and L_{max}.
Output: The global routing paths of all nets.
Method:

1. Generate subregion $G_{i,j}$ according to net information.
2. Compute $\mathcal{K}_{i,j}^h$ for each $G_{i,j}$.
3. Compute $\mathcal{X}_{i,j}^h$, $\mathcal{T}_{i,j}^h$ and $C_{i,j}^h$ for each horizontal panel.
4. Compute $\mathcal{X}_{i,j}^v$, $\mathcal{T}_{i,j}^v$ and $C_{i,j}^v$ for each vertical panel.
5. Compute the sum of integrated cost of each subregion of all possible routing path passing through of a net.
6. Choose the minimum integrated cost.
7. Apply the modified congestion-driven track assignment algorithm to generate a solution with less congestion and crosstalk.

Figure 3 illustrates CCR flow briefly. In the first step, the entire chip area are partitioned into 9 subregions (Figure 3(a)) according to net information. In the step 2, the congestion model discussed above is applied to estimate horizontal (vertical) congestion cost $\mathcal{K}_{i,j}^h$ ($\mathcal{K}_{i,j}^v$) for each subregion, as shown in Figure 3(b). In step 3 and 4, the horizontal (vertical) integrated costs $C_{i,j}^h$ ($C_{i,j}^v$) of each subregion $G_{i,j}$ are figured out and listed in Figure 3(c) according to our crosstalk model mentioned in Section III. In step 5, CCR algorithm first figures out the routing cost of all possible routing paths for each net and choose the routing path with lowest cost in step 6, see Figure 3(d). For example, the cost of the lower-L routing path of net N_1 is computed as follows:

$$C_{0,0}^h + C_{0,1}^h + C_{0,2}^h + C_{0,2}^v + C_{1,2}^v + C_{2,2}^v$$
$$= 2.95 + 2.95 + 2.95 + 3.54 + 3.54 + 3.54 = 19.47$$

It is worthy to note that we use the penalty cost to avoid the bending of routing path by consideration of both the horizontal and vertical integrated cost of $G_{0,2}$. Among all possible routing paths of all nets, we choose the minimum integrated cost solution to guide the detailed router. The modified crosstalk-driven track assignment algorithm [4] embedded into CCR is applied to assign tracks but coupling crosstalk occurred, see Figure 3(e). CCR can detect this issue and perform step 7 to disperse two adjacent nets, N_1 and N_2 on the panel $P_h(0)$, to avoid crosstalk, see Figure 3(f) and 4.

V. Experimental Results

All of our experiments were implemented using C++ language on an Intel 2.4GHz processor with 256MB memory. The objective is to minimize wire congestion and crosstalk at global routing stage. We decompose each multi-pin net into a set of several two-pin nets by minimum spanning tree algorithm.

Three algorithms were implemented and preformed to test four benchmarks. First algorithm I is a maze router without considering the optimization of crosstalk and congestion. Second algorithm II is a crosstalk aware router [2]. Third algorithm III is our CCR algorithm. TABLE 1 describes parameters used in our experiments [2], 100 longest nets were selected and routed by the algorithms and the corresponding values of S_{max} and L_{max} were set as 10μm and 500μm according to the parameter of the 0.13μm CMOS process technology. The experimental results were shown in TABLE 2. Total effective coupling length represents the sum of all coupling length of the routing solution. The column labeled as "No. of Violation Nets" is the number of nets that can't be routed successfully. Compared with results of maze routing algorithm without consideration of crosstalk and the crosstalk aware routing solution [2], CCR archives 94% and 17% reduction of the overall effective coupling length in average, respectively. From the column of "run time", we discover that the efficiency of the CCR algorithm is good and it result in few additional resources after processing of track assignment, see TABLE 3.

Besides, we modify CCR algorithm by setting the value of $f(\mathcal{N}_{i,j}^h, \mathcal{T}_{i,j}^h)$ (or $f(\mathcal{N}_{i,j}^v, \mathcal{T}_{i,j}^v)$) is equal to one for each grid, called CR algorithm. We use four benchmarks to compare the CR and the CCR algorithm. All of the experimental results show that the CCR algorithm has the less sum of congestion cost than CR algorithm. Due to the limitation on publication space, we only report the result of ami49 in Figure 5. As the increasing on number of routed nets, we can find the corresponding congestion of solution generated by CCR is lower than CR. It means that CCR algorithm has more improvement than CR algorithm in routing optimization.

VI. Conclusions

In this paper, we study and implement a new congestion- and crosstalk-driven routing system. It takes estimated congestion cost, crosstalk cost and the track utilization of each routing grid simultaneously into consideration to determine the global routing paths of all nets. The proposed approach can effectively disperse nets to lower congested regions. Compared with results of maze routing algorithm without consideration of crosstalk and the crosstalk aware routing solution [2], our algorithm archives 94% and 17% reduction of the overall effective coupling length in average, respectively.

References

[1] T. Gao and C.L. Liu, "Minimum Crosstalk channel Routing," *IEEE Trans. Computer Aided Design*, pp. 465-474, 1996.

[2] M.F. Hsiao, M. Sadowska, M. and S.J. Chen, "A Crosstalk Aware Two-Pin Net Router," *Proc. of International Symposium on Circuits and Systems,* pp.25-28, 2003.
[3] Y.L. Hsieh and T.M. Hsieh, "A New Effective Congestion Model in Floorplan Design," *Proc. of Design, Automation and Test in Europe,* pp.1204-1209, 2004.
[4] T.Y. Ho, Y.W. Chang, S.J. Chen and D.T. Lee, "A Fast Crosstalk- and Performance-Driven Multilevel Routing System," *Proc. of IEEE International Conference on Computer Aided Design,* pp. 382-387, 2003.
[5] R. Kastner, E. Bozorgzadeh and M. Sarrafzadeh, "An Exact Algorithm for Coupling-Free Routing," *Porc. of ACM International Symposium on Physical Design,* pp. 10-15, 2001.
[6] S.R. Pan and Y.W. Chang, "Crosstalk-Constrained Performance Optimization by Using Wire Sizing and Perturbation," *Proc. of IEEE International Conference on Computer Design,* pp. 581-584. 2000.
[7] P. Sarkar and C.K. Koh, "Routability-Driven Repeater Block Planning for Interconnect-Centric Floorplanning," *IEEE Trans. Computer Aided Design,* pp.660-671, 2001.
[8] H. Zhou and D.F. Wong, "Global Routing with Crosstalk Constraints," *Proc. of Design Automation Conference,* pp.374-377, 1998.
[9] J. Lou, S. Thakur, S. Krishnamoorthy and Henry S.Sheng, "Estimating Routing Congestion Using Probabilistic Analysis," *Proc. of ACM International Symposium on Physical Design,* pp. 112-117, 2001.

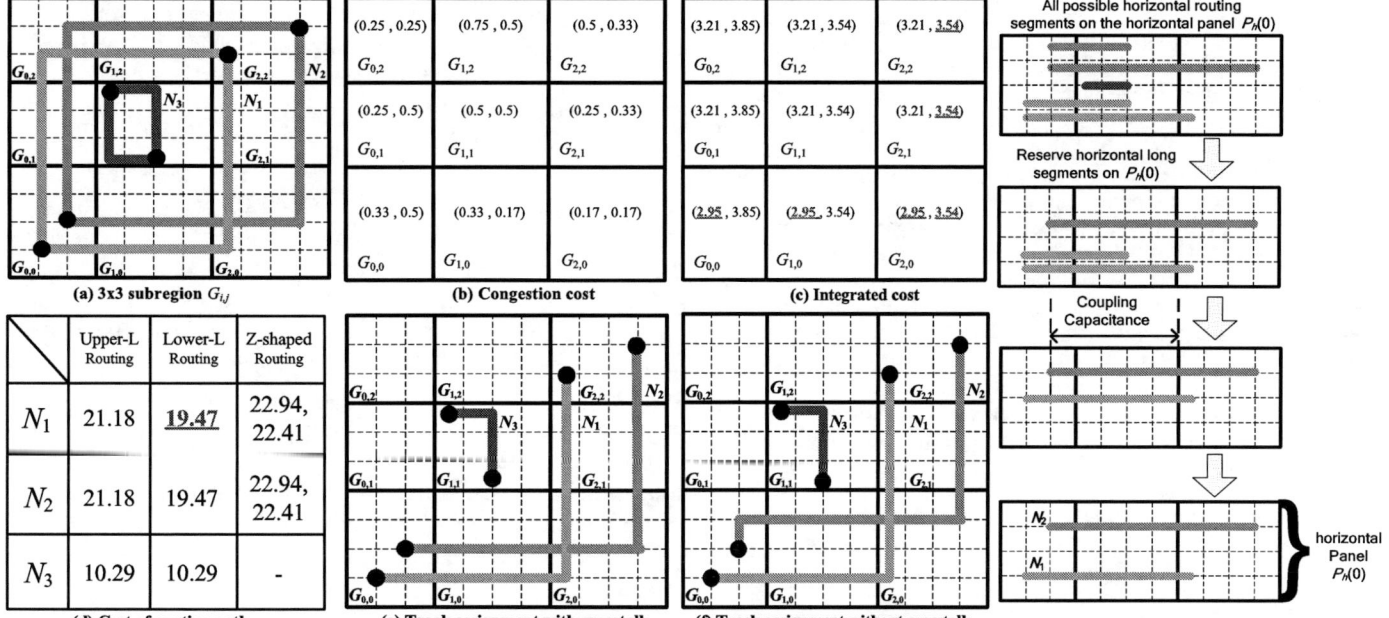

Figure 3: The flow of CCR algorithm (a)-(f)

Figure 4: Track assignment for $P_h(0)$

TABLE 1. Experimental parameters.

Circuit Name	Parameters used in [2]		
	No. of Critcal Nets	S_{max}	L_{max}
Primary1	100	10	500
Primary2	100	10	500
Avqsmall	100	10	500
Avqlarge	100	10	500

TABLE 2. Experimental results.

Circuit Name	Total Effective Coupling Length			Improvement (compared with I)		Improvement (compared with II)	No. of Violations Nets			Run Time(s)
	I	II	III	II	III	III	I	II	III	III
Primary1	6098	963	293	84.2%	95.2%	11%	2	0	0	0.202
Primary2	7152	1247	246	82.5%	96.6%	14.1%	2	0	0	0.218
Avqsmall	14498	5646	1065	61%	92.7%	31.7%	5	0	0	0.266
Avqlarge	16729	3264	952	80.5%	94.3%	13.8%	9	0	0	0.234
Average				77.0%	94.7%	17.6%				

TABLE 3. Additional routing resources caused by the track assignment.

Circuit Name	CCR without track assignment				CCR with track assignment				Extra Length (%)	Extra Bend (per net)
	Total wire length	Total bend	Max length	Avg. length	Total wire length	Total bend	Max length	Avg. length		
Primary1	29557	100	892	295.57	31457	277	896	314.57	6.43 %	1.77
Parmary2	30225	100	892	302.25	32201	277	896	322.01	6.54 %	1.77
Avqsmall	59555	101	1349	589.65	61369	263	1373	617.61	3.05 %	1.62
Avqlarge	52441	100	1589	524.41	54413	274	1597	544.13	3.76 %	1.74

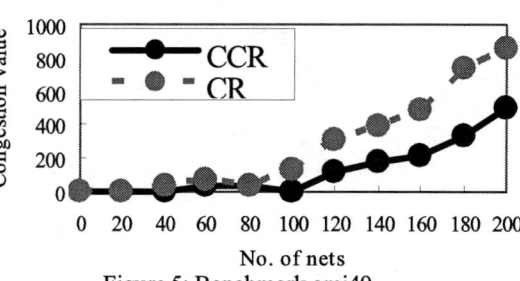

Figure 5: Benchmark ami49

Fast Integer Linear Programming Based Models for VLSI Global Routing

Laleh Behjat
and Andy Chiang
Department of Electrical & Computer Engineering
University of Calgary, Calgary, Alberta, Canada
Email: laleh, chianga@enel.ucalgary.ca

Abstract—Global routing is an essential part of VLSI physical design, and has been traditionally solved using sequential or concurrent methods. In the sequential techniques, routes are generated one at a time based on a predetermined ordering. These methods are very fast, but because of their sequential nature can result in sub-optimal solutions. The concurrent techniques attempt to solve the problem using global optimization techniques. These methods can provide a global view of the circuit's routing, but take a considerable amount of time. In this paper, a global router based on the concurrent techniques is presented. The proposed technique formulates the global routing problem as an integer linear programming (ILP) problem. This model combines the traditional wire length minimization model with the channel capacity minimization to obtain more accurate routings. In addition, the characteristics of the trees generated by our global router are investigated. A tree pruning technique based on the characteristics of the trees is developed to reduce the sizes of the ILP problem, and consequently reduce the solution time. The results show an average of 58% improvement in solving time without any loss in the quality of the results.

I. INTRODUCTION

In today's DSM (deep submicron) technology, a circuit may easily contain hundreds of thousands of interconnecting wires. This makes routing in VLSI physical design, an already NP-hard problem, difficult to manage. To simplify this problem, routing is usually done in two stages: global routing and detailed routing [1]. During the global routing, circuit components, i.e. modules and net, are represented as a graph. Then, the nets are routed without specifying the actual geometric layout of the wires. In detailed routing, specific layers and tracks are assigned to the nets.

The algorithms developed for global routing can be categorized as either sequential or concurrent based methods. The sequential algorithms [2], [3], [4], [5] route nets one at a time. In these algorithms, the nets are first ordered based on their importance. Then, each net is routed separately based on its order. Sequential routers are very fast compared to the concurrent routers. However, due to their sequential nature, the results are highly dependent on the ordering of the nets [5]. In addition, the lack of a global view in sequential routers makes predicting factors such as congestion and routability difficult. In [2], the geometric characteristics of Steiner trees generated for each net, as well as their correlation to the routability of the circuit are studied. In [3], a multilevel router which integrates global and detail routing is introduced. In this technique, routing is performed in two stages: coarsening and uncoarsening. The novelty of this router is in combining the global and detailed routing together. In [4] a multilayer global router based on a 3-D routing graphs is proposed.

In concurrent algorithms all the nets are routed at once [6], [7], [8], [9], [10]. In [10], the global routing is formulated as an integer linear programming (ILP) problem. The ILP is relaxed into a linear programming problem and solved using Karmarkar's [11] interior point method. Techniques to reduce the size of the ILP problem as well as determining the bounds for the cost function are proposed in [9]. In [6], an ILP objective function that incorporates congestion and via counts is proposed. An algorithm which includes timing and congestion into the global routing formulation is introduced in [8]. In [7], instead of using ILP formulation a wire length probability function is used to evaluate the cost of each tree. In this technique, multiple trees are generated for each net, then randomized rounding based [12] is performed to route the circuit.

Traditionally, concurrent routing techniques are used to solve small VLSI problems because they yield quality results in a reasonable amount of time. However, as technology advances, feature sizes are getting smaller, the ILP problems are getting larger which leads to a significant increase in solving time. In this paper, a concurrent routing technique based on the ILP model in [6] is presented. The proposed technique includes wire length and an estimate of congestion in the routing area to produce high quality routes. Furthermore, the routes that are likely to be in the congested areas are discarded to reduce the sizes of the ILP formulation. To increase the speed of the router while keeping the quality of results optimal, a new algorithm that examines the characteristics of the possible routes in the ILP formulation is presented. This algorithm reduces the time to solve the ILP problem by performing ILP in two stages. In the first stage, the trees that have low cost in the objective function are tagged as chosen. This set of chosen trees taken out of the ILP problem and the size of the ILP problems are reduced. In the second stage, the modified ILP problem is solved. By eliminating trees in the first stage, significant improvements in solving time are obtained. The novelty of the proposed model is in keeping the global view of the concurrent method while solving the global routing problem with speeds comparable to the sequential techniques.

The remainder of this paper is organized as follows: In Section II, the ILP formulation for global routing is given. In Section III, the proposed enhanced algorithm for solving the ILP model is introduced. Numerical results when applying the proposed model on the IBM-Place 2.0 benchmarks [13] are presented in Section IV. Finally, conclusions are given in Section V.

II. MATHEMATICAL MODELS FOR THE GLOBAL ROUTING PROBLEM

As interconnect dominates the overall system delay in the modern VLSI design, determining the optimal routes for a circuit becomes increasingly important. The input of the global routing is the position of the modules obtained at the placement stage. In a typical global routing problem, a circuit is modeled by a graph $G(V, E)$, where V is a set of nodes or vertices and E is a set of edges. When a graph is mapped directly to a circuit, the nodes in the graph represent the approximate position of the module terminals while the edges correspond to the routing areas. In order to route the nets, module terminals must be placed or "snapped" to the nearest node on the grid. An example of snap to the grid operation is illustrated in Figure 1. The edges of the graph represent the available routing areas or the routing tracks.

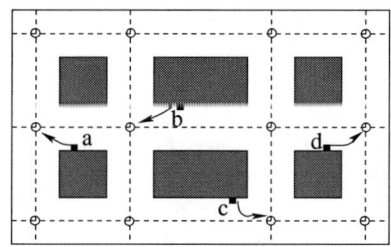

Fig. 1. Snap to grid example

A. Tree Generation and Congestion Estimation

The first step in the proposed global router is to generate a set of trees for each net. A minimum rectilinear Steiner tree (MRST) algorithm [14] is used to generate this set. For the two and three cell nets, only the minimum bend and minimum length trees are considered.

After producing a set of trees for each net, the congestion in the edges of the graph is estimated. The proposed congestion estimation [6] is based on the probability of a tree to be chosen. At this stage, the number of trees produced for each net and the edges that each tree passes through are known. The probability of $y_j \in N_k$ to be chosen in the final results is approximated by:

$$p(y_j) = \frac{1}{\text{number of trees produced for net } k},$$

where, the set of all trees produced for net k is represented by N_k. For example, if net k has two trees, the probability of each tree to be chosen is approximated equal to $\frac{1}{2}$. For each edge e_i, we introduce a measure called the *routing demand*, $r(e_i)$. The routing demand is an approximation of the number of trees that might pass through an edge and is equal to summation of the probabilities of each tree, $p(y_j)$, that passes through edge, e_i:

$$r(e_i) = \sum_{j=1}^{t} a_{ij} p(y_j),$$

here, a_{ij} is a binary value that is equal to one if tree y_j passed through edge e_i and zero otherwise. If the routing demand for an edge is higher than a predetermined percentage of the edge capacity, then that edge is considered congested. As the result, additional trees are made for all nets passing through that edge. These additional trees are produced such that they do not pass through any congested area.

After a suitable set of trees has been produced for all the nets, a binary variable y_j is assigned to each tree. If a tree is chosen to connect the nodes of a net in the final solution, the variable y_j is set to one; otherwise it is equal to zero.

A congestion measure associated with tree j, w_{c_j}, is calculated. This congestion is equal to the sum of the routing demands, $r(\cdot)$, of the edges that y_j passes through and is called the congestion factor.

$$w_{c_j} = \sum r(e_i), \quad e_i \in y_j$$

This congestion measure is used in the objective function of the global routing problem.

B. Mathematical Modeling

The traditional objective function for the global routing problem is based on the wire length minimization model where the total length of the wire is minimized. For a circuit with n nets and m modules, this problem can be formulated as [10] :

$$\begin{align}
\min \quad & \sum_{j=1}^{t} w_{l_j} y_j & (1)\\
\text{s.t.:} \quad & \sum_{y_j \in N_k} y_j = 1, \quad k = 1, 2, ..., n & (2)\\
& \sum_{j=1}^{t} a_{ij} y_j \leq c, \quad i = 1, 2, ..., p & (3)\\
& y_j \in \{0, 1\}, \quad j = 1, 2, ..., t & (4)
\end{align}$$

Where t is the total number of the trees, w_{l_j} is the length of the tree y_j, and n and p are the total number of nets and edges respectively. Constant c represents the capacity of the edges. The objective function of the problem is to minimize the total wire length of the circuit. The first set of constraints, (2), limits the final solution to have only one tree chosen for each net. The next set of constraints, (3), ensures that number of trees passing through an edge is within the routing area capacity limit, c. Finally, the last set of constraints, (4), confines the solution to be binary.

The wire length model function is not very effective in determining the "best" trees in the final solution. This is because a large number of trees can have equal lengths. In addition, the wire length model does not consider congestion in the solution. Congestion reduces the overall system performance due to detoured nets and can lead to infeasible solutions. Furthermore, congestion makes crosstalk and antenna effects more difficult to manage because of the lack of routing flexibility. Besides congestion, the number of bends in a net also plays an important role in global routing. A bend in the net means the insertion of a via is required to connect a vertical wire to a horizontal wire. The usage of vias should be minimized as each via adds more capacitive load to the wire and tends to be less reliable than a conventional wire. An improved objective function which takes congestion and via counts into consideration can be formulated as follows [6]:

$$\min \sum_{j=1}^{t} w_{l_j} y_j + w_{c_j} y_j + w_{v_j} y_j \quad (5)$$
$$\text{s.t.:} \quad \sum_{y_j \in N_k} y_j = 1, \quad k = 1, 2, ..., n \quad (6)$$
$$\sum_{j=1}^{t} a_{ij} y_j \leq c, \quad i = 1, 2, ..., p \quad (7)$$
$$y_j \in \{0, 1\}, \quad j = 1, 2, ..., t \quad (8)$$

where w_{l_j}, w_{c_j} and w_{v_j} are weighting factors for the wire length, congestion factor and via count for tree y_j, respectively. The problem formulated in equations (5) to (8) is an ILP problem. Because of the large size of the ILP problems encountered in VLSI circuits, solving the problem using traditional techniques such as branch and bound can be time consuming and impractical. Hence, this problem is usually relaxed into a linear programming (LP) problem by changing (8) to $0 \leq y_j \leq 1$.

In this research an interior point technique [15] is used to solve the LP problem. The solution of the LP problem contains fractional values that have to be rounded to obtain the integer solution. In [7], [9], a randomized rounding technique is used. The randomized rounding techniques [12] use the fractional values obtained in LP solution as probabilities to round trees. This rounding is performed for up to a hundred times and the best results are chosen to represent the final solution. In this paper we have used congestion estimation to avoid using randomized rounding techniques. Our results indicate that the solution of the ILP is very close to the LP solution and therefore the optimal results are obtained without needing to perform any iterations of rounding.

III. TIMING ENHANCED ILP SOLUTION

The interior point methods have been proven to be effective when solving linear programming problems with large number of constraints . Therefore, a primal-dual interior point (PD-IP) method [15] is used to solve the global routing problem formulated in Section II. The general formulation of the global routing problem can be written as:

$$\min \quad \mathbf{w}^T \mathbf{y} \quad (9)$$
$$\mathbf{Ay} \leq \mathbf{b}$$
$$\mathbf{y} \geq \mathbf{0},$$

where $\mathbf{y} \in \Re^t$ is a vector of unknown variables or trees, usually referred to as the primal variables. Matrix $\mathbf{A} \in \Re^{q \times t}$ is the constraint matrix consisting of constraints on the nets and capacity constraints and q is the total number of constraints. $\mathbf{b}$ is the right hand side vector containing the right hand side of equations (6) and (7). Note that the upper boundary constrains, $\mathbf{y} \leq \mathbf{1}$, are indirectly incorporated by the first set of equality constraints and are omitted from the final formulation to reduce the number of constraints in the problem.

The first step in the derivation of an interior point algorithm to solve (9) is to add non-negative slack variables $\mathbf{s}_1 \in \Re_+^q$ and $\mathbf{s}_2 \in \Re_+^t$ to transform all inequality constraints to equality constraints. The optimization problem then has the form:

$$\min \quad \mathbf{w}^T \mathbf{y}$$
$$\text{s.t.:} \quad -\mathbf{Ay} - \mathbf{s}_1 + \mathbf{b} = \mathbf{0}$$
$$\mathbf{y} - \mathbf{s}_2 = \mathbf{0}$$
$$\mathbf{s}_1, \mathbf{s}_2 \geq \mathbf{0}.$$

The non-negativity conditions $\mathbf{s}_1, \mathbf{s}_2 \geq \mathbf{0}$ are incorporated into the objective function using logarithmic barrier terms as follows:

$$\min \quad \mathbf{w}^T \mathbf{y} - \mu^k \sum_{i=1}^{q} \log(s_{1_i}) - \mu^k \sum_{i=1}^{t} \log(s_{2_i}) \quad (10)$$
$$\text{s.t.:} \quad -\mathbf{Ay} - \mathbf{s}_1 + \mathbf{b} = \mathbf{0}$$
$$\mathbf{y} - \mathbf{s}_2 = \mathbf{0},$$

where $\mu^k > 0$ is a monotonically decreasing series of *barrier parameters* and s_{1_i} and s_{2_i} represent the ith element of $\mathbf{s}_1$ and $\mathbf{s}_2$ respectively. The logarithmic terms impose the strict positivity conditions on the nonnegative variables. To solve the equality-constrained problem stated in (10), a Lagrange-Newton method is applied: First, the Lagrangian function associated with problem (10) is formed, then a local minimum of the Lagrangian function is obtained by finding a point that satisfies the Karush-Kuhn-Tucker (KKT) conditions. To simplify the presentation of the Lagrangian function, the vector $\mathbf{r} := (\mathbf{s}_1^T, \mathbf{s}_2^T, \boldsymbol{\lambda}^T, \boldsymbol{\pi}^T, \mathbf{y}^T)^T \in \Re^q \times \Re^t \times \Re^q \times \Re^t \times \Re^t$, is introduced, where $\boldsymbol{\lambda} \in \Re^q$ and $\boldsymbol{\pi} \in \Re^t$ are Lagrangian multipliers, usually referred to as the dual variables. The Lagrangian function $L_\mu(\mathbf{r})$ associated with the problem (10) is:

$$L_\mu(\mathbf{r}) := \mathbf{w}^T \mathbf{y} - \mu^k \sum_{i=1}^{q} \log(s_{1_i}) - \mu^k \sum_{i=1}^{t} \log(s_{2_i}) - \boldsymbol{\lambda}^T(-\mathbf{Ay} - \mathbf{s}_1 + \mathbf{b}) - \boldsymbol{\pi}^T(\mathbf{y} - \mathbf{s}_2) \quad (11)$$

To find a local minimum of the Lagrangian function $L_\mu(\mathbf{r})$, a stationary point of the Lagrangian function should be found. Assuming that an initial solution satisfying the strict positivity constraints (interior point) $\mathbf{r} > \mathbf{0}$ is provided, one step in the Newton's direction is taken to find an approximate solution of (11). The Newton step, $\Delta \mathbf{r}$, is computed by solving the following non-symmetric indefinite system of equations:

$$\nabla_\mathbf{r}^2 L_\mu(\mathbf{r}) \Delta \mathbf{r} = -\nabla_\mathbf{r} L(\mathbf{r}).$$

A suitable step size, $0 < \alpha^k \leq 1$, is calculated by performing a line search along the Newton direction and the new primal and dual variables are calculated by:

$$\mathbf{r}^{k+1} = \mathbf{r}^k + \alpha^k \Delta \mathbf{r},$$

At the end of each iteration, the updated variables are tested to determine if a local minimum has been obtained. The program terminates when the convergence criteria is satisfied; otherwise, k is incremented, a new barrier parameter $\mu^{k+1} < \mu^k$ is computed and another Newton direction is determined. Even though the interior point techniques are very efficient in solving LP problems, because of the large sizes of the problems encountered in today's VLSI problem, they can still be time consuming. This is because the interior point solver has a complexity of $O(n^3)$ [15]. It should be noted that the number of trees generated for benchmarks in our experiment has been already optimized by estimating the congestion in the routing areas and eliminating trees in the congested areas when possible. In this paper, the size of LP were further reduced by proposing a preprocessing technique that prunes the trees based on their values in the objective function in (9).

The optimization problem is presented as a minimization problem where the linear programming solver is trying to minimize the cost of a tree. This cost consists of three factors: length of a tree, congestion in the path of the tree and the number of vias or bends in a tree. Our algorithm is based on the special properties of two pin nets. The trees generated for two-pin nets have the same length and the same number of bends, but they pass through areas with different congestion; hence, they have different congestion factors. A tree with a lower congestion factor is preferred, since it minimizes to total objective function. Consequently, the optimizer tends to choose a tree with a lower congestion factor if none of the constraints of the problem are violated.

Because interior point solvers are global optimizers, the final solution of the problem can not be predicted. However, our experiments have shown that in cases where there is a significant difference between the congestion factors of the trees generated for a single net, the optimizer chooses the tree with the lower congestion factor. Based on the above observation, when formulating the ILP problem if the congestion between the trees produced for a net exceed a certain threshold, the trees with high congestion factors are eliminated from the ILP formulation or they are *pruned* as long as the capacity constraints are satisfied. The remaining tree with the low congestion factor becomes the solution for that particular net and the capacities of the routing areas are updated in the LP formulation. The proposed tree elimination algorithm for the ILP modeling of the global routing problem is summarized in Figure 2.

The basis of the preprocessing is that only the trees that are not likely to be chosen will be eliminated otherwise the trees are kept in the ILP problem. The accuracy of the pruning process is determined by the cut-off threshold, and it is preferred to pick a conserve value for the threshold since a large threshold may result in an inaccurate pruning process. The tree selection and pruning criteria used in this work was set as follows:

$$\frac{w_{c_i}}{\sum_{j \in N_k} w_{c_j}} \leq \theta$$

where, w_{c_i} is the congestion factor for tree i, N_k is the set of all trees made for net k, and θ is the cut-off threshold.

Proposed Global Routing Algorithm
1. Construct MRST for all nets
2. Estimate congestion factor for each tree
3. For all nets with two trees:
   if tree congestion factor is lower than threshold:
      select the tree with the lowest congestion factor
   else:
      Add all trees to LP formulation.
4. Add all other trees to LP.
5. Solve LP.
6. Extract integer solution from LP results

Fig. 2. Global routing algorithm

The tree elimination algorithm can be further extended to accommodate nets with more than two trees. The basic technique of comparing the properties of the trees is still the same; however, the tree elimination threshold value will be different. In addition, a net with more than two trees indicates that the MRSTs generated for the net may no longer have the same length; therefore, length also must be included as part of the elimination criteria.

IV. EXPERIMENTAL RESULTS

The proposed algorithm for the tree generation and pruning is implemented in C++. To solve the LP problem, the optimization package LOQO [15] is used. The fractional results obtained from LOQO are fed back to our program, where the tree with the lowest cost is chosen. The experimental circuits are chosen from IBM-Place 2.0 benchmark suits [13] and the placement results are generated by Dragon [16], a wire length driven placer package.

For the snap to grid operation, the row width was set to be equal to the standard cell height. The IBM benchmarks are derived from the TSMC $0.18\mu m$ standard-cell library and the capacities of the areas are determined by dividing the standard cell height by the sum of the minimum wire width, $0.28\mu m$, and minimum spacing parameters, $0.28\mu m$.

$$c = \lfloor \frac{\text{cell height}}{\text{minimum wire width} + \text{minimum wire spacing}} \rfloor,$$

where $\lfloor \cdot \rfloor$ indicated the floor function. Based on the above calculations, six routing layers, each with a capacity of nine were used. Table 1 shows the circuit information and the ILP formulation. Columns one to four in this table are the circuits' names, number of nets in each circuit, number of modules, number of horizontal and vertical column lines for the circuits' corresponding grids, respectively. In the last column, the number of variables for the ILP problem, i.e. total number of trees, for two cases without the pruning and after the pruning are given to show the reduction in the sizes of the problems.

The result for the benchmarks listed in Table I are given in Tables II and III. In our experiments various values for the cut off threshold were tested. In table II, the wirelength without and with tree pruning are listed in column 2 and 3 respectively.

circuit	# nets	# module	grid sizes	ILP Variables org.	ILP Variables pruned
ibm 01	11507	12274	132x132	7414	4738
ibm 02	18429	19321	153x153	10340	7112
ibm 07	44394	45098	223x223	34142	24704
ibm 08	47944	50958	243x243	24834	16148
ibm 10	64227	67506	321x321	35562	23588
ibm 11	67016	68452	273x273	44836	32106
ibm 12	67739	69372	338x338	44608	30690

TABLE I

IBM-PLACE 2.0 BENCHMARKS STATISTICS

As can be seen the wirelength are very similar between the two methods. Congestion and number of vias are included Table III. In Table III the total congestion is equal to the sum of the number of tracks used in all the edges. Similar to the wirelength, the pruning results are almost identical to the ones without pruning for both congestion and via counts. Since the total length, congestion and number of vias between the original and the pruned circuits are almost identical we can conclude that the solution quality is not affected by the pruning process.

circuit	Total Wire Length orginal	Total Wire Length pruned
ibm 01	432414	432423
ibm 02	2240186	2240186
ibm 07	4915125	4915214
ibm 08	2978800	2978449
ibm 10	6502378	6501863
ibm 11	5951209	5951025
ibm 12	9278638	9278661

TABLE II

GLOBAL ROUTING WIRE LENGTH RESULT

circuit	Tot. Congestion org.	Tot. Congestion pruned	Tot. of vias org.	Tot. of vias pruned
ibm 01	416264	416260	14819	14819
ibm 02	2269191	2269194	27719	27717
ibm 07	4808169	4808269	57261	57272
ibm 08	2876952	2876717	49458	49448
ibm 10	6292015	6289962	93208	93140
ibm 11	5668769	5668626	84057	84020
ibm 12	9009439	9009686	102669	102661

TABLE III

GLOBAL ROUTING CONGESTION AND VIA COUNT RESULT

Table IV shows the time to solve the LP problem for the two cases the original and pruned problems in seconds. The ILP solving time is dramatically reduced with our tree pruning technique. For example, the time to solve ibm01 which is a relatively small circuit is improved by almost 50%. For the larger circuits which take a considerable amount of time to solve, the solving time is improved by up to 70%. Furthermore, the routing quality is maintained since the wire length and via counts are essentially the same when the results are compared between the two algorithms.

circuit	ILP solution time (s) org.	ILP solution time (s) pruned	% improv.
ibm 01	3.094	1.65	**49.7**
ibm 02	7.797	4.03	**48.3**
ibm 07	305.06	106.383	**65.1**
ibm 08	55.509	20.9	**61.5**
ibm 10	110.099	35.875	**67.4**
ibm 11	674.689	387.72	**42.5**
ibm 12	405.406	119.763	**70.5**

TABLE IV

GLOBAL ROUTING SOLUTION TIME

V. CONCLUSION AND FUTURE WORK

In this paper, a fast and efficient algorithm to solve the global routing problem formulated as an ILP problem is introduced. In the ILP formulation stage, minimum bend Steiner trees are generated for routes and the problem is solved using an interior point method. Experimental results show our ILP formulation with pruning can significantly improve the solving time. In addition, the route quality is preserved compared to the ILP models without the tree pruning technique. Altogether with our algorithm, the router is able to gain global view of whole circuit with solving times that are comparable to global routers based on sequential techniques.

ACKNOWLEDGMENT

The authors would like to thank Natural Sciences and Engineering Research Council of Canada for Financial support of this work. We would also like to thanks Dr. Warme and GeoSteiner Inc. for providing a license for GeoSteiner.

REFERENCES

[1] N. Sherwani, *Algorithms for VLSI physical design automation.* The Netherlands: Kluwer Publishers, 1999.
[2] E. Bozorgzadeh, R. Kastner, and M. Sarrafzadeh, "Creating and exploiting flexibility in rectilinear steiner trees," *IEEE Transactions on Computer-Aided Design of Circuits and Systems*, vol. 22, pp. 605– 615, 2003.
[3] Y. Chang and S. Lin, "MR: a new framework for multilevel full-chip routing," *IEEE Transactions on Computer-Aided Design of Circuits and Systems*, vol. 23, pp. 793–800, 2004.
[4] L. E. Liu and C. Sechen, "Multilayer chip-level global routing using an efficient graph-based steiner tree heuristic," *IEEE Transactions on Computer-Aided Design of Circuits and Systems*, vol. 18, no. 10, pp. 1442–1451, 1999.
[5] R. Nair, "A simple yet effective technique for global wiring," *IEEE Transactions on Computer-Aided Design of Circuits and Systems*, vol. 6, pp. 165–172, 1987.
[6] L. Behjat and A. Vannelli, "Steiner tree construction based on congestion for the global routing problem," in *proceeding of 3rd IEEE WSoC*, 2003, pp. 28–31.
[7] R. C. Carden, J. Li, and C. Cheng, "Creating and exploiting flexibility in rectilinear Steiner trees," *IEEE Transactions on Computer-Aided Design of Circuits and Systems*, vol. 15, pp. 208– 216, 1996.
[8] T. Jing, X.-L. Hong, J.-Y. Xu, H.-Y. Bao, C.-K. Cheng, and J. Gu, "UTACO: a unified timing and congestion optimization algorithm for standard cell global routing," *IEEE Transactions on Computer-Aided Design of Circuits and Systems*, vol. 23, pp. 358–365, 2004.

[9] T. Lengauer and M. Lungering, "Provably good global routing of integrated circuits," *SIAM Journal of Optimization*, vol. 11, no. 1, pp. 1–30, 2000.

[10] A. Vannelli, "An adaptation of the interior point method for solving the global routing problem," *IEEE Transactions on Computer-Aided Design of Circuits and Systems*, vol. 10, no. 2, 1991.

[11] N. Karmarkar, "A new polynomial-time algorithm for linear programming," *Combinatorica*, vol. 4, no. 4, pp. 373–395, 1984.

[12] P. Raghavan and C. D. Thompson, "Multi-terminal global routing: A deterministic approximation scheme," *Algorithmica*, vol. 6, pp. 73–82, 1991.

[13] "Ibm-place 2.0 benchmark suits, http://er.cs.ucla.edu/benchmarks/ibm-place2/," Website, http://er.cs.ucla.edu/benchmarks/ibm-place2/.

[14] D. M. Warme, "A new exact algorithm for rectilinear steiner trees," in *International Symposium on Mathematical Programming*, 1997.

[15] R. Vanderbei, "LOQO: An interior point code for quadratic programming," Princeton University, Princeton, NJ, Tech. Rep., 1998.

[16] M. Wang, X. Yang, and M. Sarrafzadeh, "Dragon2000: Standard-cell placement tool for large industry circuits," in *IEEE/ACM International Conference on Compter-Aided Design*, 2000, pp. 260–263.

Floorplanning with Clock Tree Estimation

Chih-Hung Lee
Department of Information Management,
Ling Tung College,
Tai Chung, Taiwan, R.O.C.
chlee@mail.ltc.edu.tw

Chin-Hung Su, Shih-Hsu Huang
Department of Electronic Engineering,
Chung Yuan Christian University,
Chung Li, Taiwan, R.O.C.
{g9176020, shhuang}@cycu.edu.tw

Chih-Yuan Lin, Tsai-Ming Hsieh
Department of Information & Computer Engineering,
Chung Yuan Christian University,
Chung Li, Taiwan, R.O.C.
{g9277005, hsieh}@cycu.edu.tw

Abstract — Traditional floorplanners determine the module topology for the minimization of total chip area, total wirelength, and routing congestion. However, as the design complexity continues to increase, it is necessary to reduce the clock skew and estimate the clock latency during floorplanning. In this paper, we propose a two-stage simulated annealing (SA) floorplanner based on the sequence-pair representation. A clock tree estimation model is embedded into the floorplanner to guide the process of floorplan generation. Our clock tree model includes two parts. First, we use H-tree algorithm to predict an intra-module clock tree for each module. Secondly, the DME algorithm is applied for inter-module clock tree generation. Experimental data show that both the clock tree wirelength and the clock latency can be improved simultaneously, while the extra overhead on the chip area is small.
Index Terms — Clock Tree Planning, Floorplanning, Zero Clock Skew, Floorplan-Based Clock Tree Model.

1. Introduction

Clock distribution network plays a very important role for the operation and performance of a synchronous system. For example, the clock skew violation could cause system's functional operation error and the delay of clock signal may increase the probability of rise/fall time's violation. In traditional design flow, clock trees are synthesized in the post-placement stage. However, as the progress on design and manufacturing technology, the early clock planning becomes necessary and important. Especially, due to the dramatic increasing on the gate count of ICs, the traditional clock tree synthesis techniques cannot guarantee to generate the zero clock skew results in post-placement stage. Thus, Yim et.al [10] proposed a floorplan-based clock distribution planning methodology. However, this paper mainly discussed on solving the power/clock routing problem with the view of design flow.

In this paper, a two-stage simulated annealing (SA) based floorplanner integrated with a fast zero-skew clock tree estimation engine was proposed. We use the sequence pair [8][9] representation to implement this floorplanner. We use the chip area as the cost function in the first stage of the proposed floorplanner. During the second stage, the cost function of the floorplanner is in the form of a linear combination of chip area, total wirelength of general signals, estimated total wirelength of the clock tree, and estimated depth of the clock tree. The method to estimate the clock tree at the floorplanning stage is as below. First, we use a H-tree [6][7] generator to implement intra-module clock tree. Next, we use the DME algorithm [1][2][3][4][5] to complete the inter-module part of the whole clock tree.

The remainder of this paper is organized as the following: Motivations of this paper are shown in Section 2. The problem formulation is given in Section 3. Section 4 presents the clock tree plan model and the two stages SA algorithm. In Section 5, we present the numeral experimental results on MCNC benchmarks. Finally, conclusion is attached in Section 6.

2 Motivation

In traditional design flow, clock signal routing problem was processed at the so-called clock tree synthesis stage. It handles the clock tree generation after the entire chip floorplanning and cell placement are completed. However, to fix the clock skew and/or clock latency problem at the post-layout stage is usually very difficult and the cost is very expensive. As the increasing on design complexity and manufacturing technology, it is necessary to consider clock issues in the design cycle as early as possible. Therefore, we propose a methodology which integrate the clock distribution planning and estimation into the sequence-pair based floorplanner.

A simple example is given to show the motivation that why we should consider clock tree distribution problem in the floorplanning stage. Let's use Figure 1 as an example, which includes 6 modules. Assume that the root of each intra-module clock tree is located on the center point of each module and we use "module name/clock tree depth" to label the module name and its corresponding intra-module clock tree depth. Let's use two different module placement results shown in Figure 1 (a) and (b) as examples. After the DME algorithm is applied, we can find their differences on the inter-module clock tree length and the full clock tree depth. In the case of Figure 1(a), the clock tree length (without include intra-module part) is 87 and the estimated clock tree depth is 39.5. However, using Figure 1(a) as an example, if we exchange the relative position of module E and module F, as shown in Figure 1(b), the corresponding clock tree length (without include intra-module part) becomes 73 and the estimated clock tree depth becomes 38. This is the motivation that we integrate a clock tree estimation engine into the floorplanner.

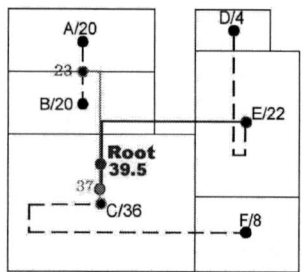

clock tree length (not include the intra-module part) = 87

(a)

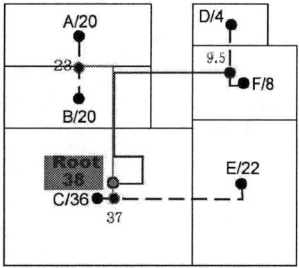

clock tree length (not include the intra-module part) = 73

(b)

Figure 1: Floorplan impact on clock tree

3 Problem Formulation

Let $M=\{m_1, m_2,..., m_n\}$ be a set of n rectangular modules. For module m_i, $1 \leq i \leq n$, the height h_i and width w_i are given. During the process of floorplanning, each module can be freely rotated in 90°, 180° and 270°, i.e. the value of each module's height and width can be exchanged. Besides, the number of synchronizing elements (e.g. flip-flops) in each module is pre-given. The clock tree plan driven floorplanning problem is formulated as follows:

Given n rectangular modules with their area sizes, width, height and the number of synchronous cells contained in each module, to find a floorplan and a corresponding zero skew clock tree with respect to the appropriate delay model. Besides, the chip area, the total wirelength and the depth of the generated clock tree are minimized.

4 Clock Tree Driven Floorplanning Algorithm

In this section, we propose the clock tree driven floorplanner. Section 4.1 shows the floorplan-based clock tree estimation model and Section 4.2 introduces the two-stage SA based floorplanner.

4.1 Floorplan-Based Clock Tree Model

We use the H-tree model to estimate the intra-module clock in the first step and then apply the DME algorithm to construct the inter-module clock tree in the second step.

4.1.1 H-tree Model for Intra-module Clock Tree

Assume that module m_i, $1 \leq i \leq n$, contains sp_i synchronous primitives. We use an H-tree to generate candidate locations for synchronous elements and to estimate the clock delay within this module. The hierarchical structure of H-tree can be decided not only by the synchronous elements number in module m_i but also by the H-tree's depth constraints that directly influence on the clock delay latency. However, the generated H-tree of module m_i must have at least k_i levels, where $k_i \geq \lceil \log_4 sp\_i \rceil$. And the distance from the H-tree root to each terminal is $(2^{k_i} - 1)(\Delta x_i + \Delta y_i)$ where $\Delta x_i = w_i / 2^{k_i+1}$ and $\Delta y_i = h_i / 2^{k_i+1}$. For the example shown in Figure 2, a 3-level H-tree structure (k_i=3) will be introduced when $17 \leq sp\_i \leq 64$. The corresponding distance from the H-tree root to each terminal is

$$(2^{k_i} - 1)(\Delta x_i + \Delta y_i) = (2^3 - 1)(\Delta x_i + \Delta y_i)$$
$$= (2^3 - 1)[(w_i / 2^{3+1}) + (h_i / 2^{3+1})] = (7/8)(w_i + h_i).$$

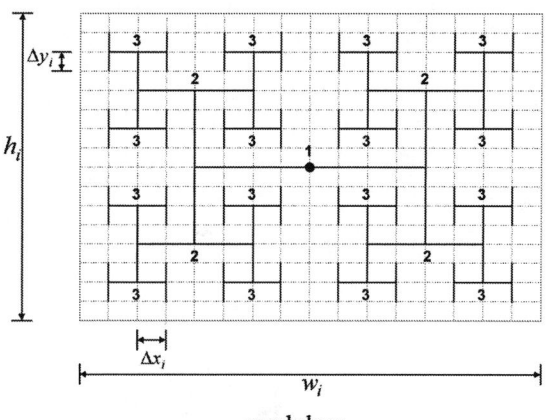

module m_i

Figure 2 H-tree generation for module m_i

Next, let's use Figure 3 as an example to illustrate our clock tree estimation model throughout this section. Note that Figure 3 has 4 modules. Assume that $17 \leq sp\_1 \leq 64$, $9 \leq sp\_2 \leq 16$, $2 \leq sp\_3 \leq 4$, $2 \leq sp\_4 \leq 4$ and $k_1 = 3$, $k_2 = 2$, $k_3 = 1$, $k_4 = 1$. Figure 3 gives the intra-module H-tree generation results of module 1, 2, 3 and 4. The corresponding H-tree depths are 21, 12.75, 6.5 and 4 for modules 1, 2, 3 and 4, respectively.

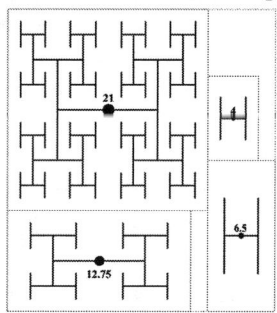

Figure 3 Intra-module H-trees of 4 modules placement result

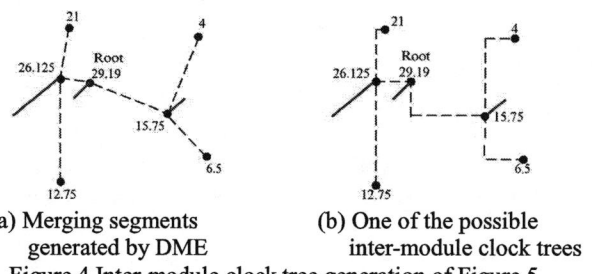

(a) Merging segments generated by DME

(b) One of the possible inter-module clock trees

Figure 4 Inter-module clock tree generation of Figure 5

4.1.2 DME Model for Inter-module Clock Tree

According to the initial clock topology generated by the H-tree model in the previous step, we then apply the DME algorithm to build the inter-module clock tree. For the case of Figure 3, the zero-skew clock tree was built as following: (1)By merging nodes with the weight 4 and 6.5, the DME algorithm generate the merging segment in the right side of Figure 4(a). (2) Merging nodes with weight 21 and 12.75 then generated the left merging segment. (3) Finally, the merging segment in the middle of Figure 4(a) was constructed by referring merging the left side and right side segments. Figure 4(b) is one of the possible inter-module clock trees.

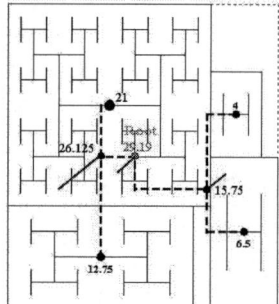

Figure 5 Integration of the intra-module clock tree and the inter-module clock tree

4.1.3 Integration of H-tree and DME

The Deferred-Merge Embedding (DME) algorithm [1][2][3][4][5] can be either applied to a given clock topology or combined with a clock topology generation algorithm to achieve zero skew with a smaller wirelength. In this paper, we use the H-tree model [6][7] to estimate the depth of intra-module clock tree for each module and we record this information with the location of H-tree root. Based on the clock topology constructed from clock roots of all modules, we can further generate the inter-module clock tree by the DME algorithm. Figure 5 shows the result that integrates the intra-module clock tree and the inter-module clock tree. In the case, the clock tree depth is 29.19 and the total clock tree wirelength (inter-module clock tree length plus intra-module clock tree length) is 481.005.

4.2 Two-Stage Floorplanning Algorithm

A two-stage simulated annealing (SA) floorplanner based on the sequence-pair representation is implemented. The clock tree estimation model given in Section 4.1 is embedded into the second stage of the 2-stage SA floorplanner. The first stage is mainly focus on the area minimization and the corresponding cost function is: $Cost1 = Area + \alpha \times Wire$, where $Area$ is the area size of the floorplan, $Wire$ is the estimated total wirelength of signals, and α is the weight. For the wirelenth calculation of general signals (not include clock signal), we first decompose multi-terminal net into a set of two-terminal nets by minimal spanning tree algorithm. Then the half-perimeter estimation method using the center of a module as the net terminal location is adopted. In the second stage, we further introduce the estimated clock routing wirelength and the depth of the estimated clock tree as parts of the second stage cost function. The cost function used in the second stage is:

$$Cost2 = Area + \alpha \times Wire + \beta \times clk\_tree\_depth$$
$$+ \gamma \times clk\_tree\_wirelength,$$

where $Area$ is the area size of the floorplan, $Wire$ is the estimated total wirelength of signals, clk_tree_depth is the clock tree depth that corresponds to the delay from root of clock tree to terminals, $clk\_tree\_wirelength$ is the the estimated clock routing wirelength and α, β, γ are the weights. Because we use two different cost functions in the two-stage SA optimization process, we should prevent the occurrence of unexpectedly drop or rise on worse solution's accept rate. When the stage transfer from stage 1 to stage 2, for maintaining the smooth property of cooling process, the new initial temperature of the second stage is adjusted by the following formula:

$$New\_Initial\_temperature = \frac{\Delta Cost2}{\Delta Cost1} \times Original\_temperature$$

where $\Delta Cost1$ is the mean value of cost differences from a number of random walks using the cost function of stage 1, and $\Delta Cost2$ is the mean value of cost differences from a number of random walks using the cost function of stage 2.

5 Experimental Results

We implement the proposed algorithm in C programming language and use MCNC benchmark circuits to test the effectiveness of our approach. The platform is on a SUN Blade workstation with 1GHz CPU and 8GB memory. Table 1 shows the number of modules, the number of nets, and the number of synchronous primitives used in our experiments. Since original benchmarks lack of synchronous primitive information, we randomly generate the synchronous primitive number according to the area size of each module. In Table 2, columns labeled " W/O CLK plan " are experimental results generated by the sequence-pair based floorplanner that did not introduce the clock tree optimization into the cost function. While columns labeled " W/CLK plan " are the results generated by our two-stage SA floorplanner that is also based on the sequence pair representation, and this algorithm introduces the consideration on the clock tree optimization in the second stage SA. In Table 2, Area represents the value of chip size. G_CLK_L is the inter-module clock wirelength. CLK_L is the total clock tree wirelength. Depth is the clock tree depth (in linear delay model, this value can direct reflect the clock latency). And Wire is the total signal wirelength. It's worthy to note that the total length of intra-module clock trees is independent to the floorplan, i.e., the H-tree length will not change when the module placement relation was modified. Thus we report both values of the inter-module clock wirelength and the total clock tree wire length for reader's reference. For each benchmark, the listed results on Table 2 were the average values of 20 runs that were randomly chosen from 100 runs. Besides, we use a real fuzzy logic controller design [11] to test the effectiveness of our approach on the clock tree optimization. Figure 9 (a) is the clock skew histograms of the result without introducing clock tree estimation into the design stage of floorplanning, and the clock skew is 0.1039. On the other hand, for another case using the design methodology proposed by this paper, the clock skew is 0.0867 and the corresponding clock skew histograms is shown as Figure 6 (b). Furthermore, we can find that the number of pins with clock skew larger then 0.047 in the case of Figure 6 (b) is obviously smaller then the case of Figure 6 (a).

6 Conclusions

This paper presents an approach to integrate clock tree planning/estimation into the floorplanning stage. The H-tree and DME techniques are combined for clock tree estimation. Our strategy is to use the total clock tree wirelength and the clock tree depth as a part of cost function in our sequence-pair based floorplanner. Experimental data consistently shows that our approach achieves good results. Besides, the zero skew property is guaranteed under the pathlength and Elmore delay models.

7 Acknowledgements

This work was supported in part by the National Science Council of R.O.C. under contract number NSC 92-2220-E-033-001.

References

[1] K.D. Boese and A. B. Kahng, "Zero-Skew Clock Routing Trees With Minimum Wirelength," in *Proc. of* IEEE ASIC Conference, pp. 1.1.1 - 1.1.5, 1992.

[2] T.H. Chao, Y.-C. Hsu, and J.-M. Ho, "Zero Skew Clock Net Routing,", in *Proc. of* ACM/IEEE Design Automation Conf., 1992, pp. 518-523.

[3] T.H. Chao, Y.-C. Hsu, J. M. Ho, K. D. Boese and A. B. Kahng, "Zero Skew Clock Routing With Minimum Wirelength," IEEE Trans. on Circuits and Systems, 39(11), Nov. 1992, pp. 799–814.

[4] M. Edahiro, "A Clustering-Based Optimization Algorithmin Zero-Skew Routing," in *Proc. of ACM/IEEE Design Automation Conference*, pp. 612–616, 1993.

[5] M. Edahiro, "Minimum Skew and Minimum Path Length Routing in VLSI Layout Design," NEC Research and Development, 32(4), Oct. 1991, pp. 569–575.

[6] A.L. Fisher and and H. T. Kung, "Synchronizing Large Systolic Arrays," in *Proc. of SPIE*, vol. 341, May 1982, pp.44-52.

[7] S.Y. Kung and R. J. Gal-Ezer, "Synchronous vs Asynchronous Computation in VLSI Array Processor," in *Proc. of SPIE*, vol. 341, May 1982, pp. 53–65.

[8] H. Murata and Ernest S. Kuh. "Sequence-Pair Based Placement Method for Hard/Soft/Pre-placed Modules", in *Proc. of International Symposium on Physical Design*, pp.167-172, 1998.

[9] H. Murata, K. Fujiyoshi, S. Nakatake and Y. Kajitani, "Rectangle-Packing-Based Module Placement," *Proc. ICCAD*, pp. 472-479, 1995.

[10] J.S. Yim, S.O. Bae and C.M. Kyung, "A Floorplan-Based Planning Methodology for Power and Clock Distribution in ASICs", in *Proc. of ACM/IEEE Design Automation Conference*, pp. 766-771, 1999.

[11] S.H. Huang and J.Y. Lai, "A High Speed VLSI Fuzzy Logic Controller with Pipeline Architecture", in the Proc. of IEEE International Conference on Fuzzy Systems, Vol. 2, pp.1054-1057, 2001.

Table 1 Benchmark information

	apte	xerox	hp	ami33	ami49
# of Modules	9	10	11	33	49
# of Nets	97	203	83	123	408
# of Syn. Cells	384	388	400	916	1756

Table 2 Experimental Results

	Area(μm^2)			G_CLK_L (μm)			CLK_L (μm)			Depth (μm)			Wire (μm)			Run Time (sec)	
	W/O CLK Plan	W/ CLK Plan	Improve %	W/O CLK Plan	W/ CLK Plan	Improve %	W/O CLK Plan	W/ CLK Plan	Improve %	W/O CLK Plan	W/ CLK Plan	Improve %	W/O CLK Plan	W/ CLK Plan	Improve %	W/O CLK Plan	W/ CLK Plan
apte	47055445	47643665	-1.25%	15373	12302	19.98%	210757	207686	1.46%	5492	4643	15.46%	851495	711069	16.49%	6	31
xerox	19983633	20053960	-0.35%	10527	9106	13.50%	163071	161650	0.87%	2861	2775	3.01%	797013	773842	2.91%	7	32
Hp	9196908	9202200	-0.06%	10346	8365	19.15%	113722	111741	1.74%	2921	2671	8.56%	335658	307678	8.34%	7	21
Ami33	1259361	1276108	-1.33%	5297	4967	6.23%	70560	70207	0.50%	674	647	4.01%	113465	114875	-1.24%	53	196
Ami49	39176000	39574080	-1.02%	42920	41430	3.47%	457936	456446	0.33%	3865	3795	1.81%	1896245	1855567	2.15%	233	798

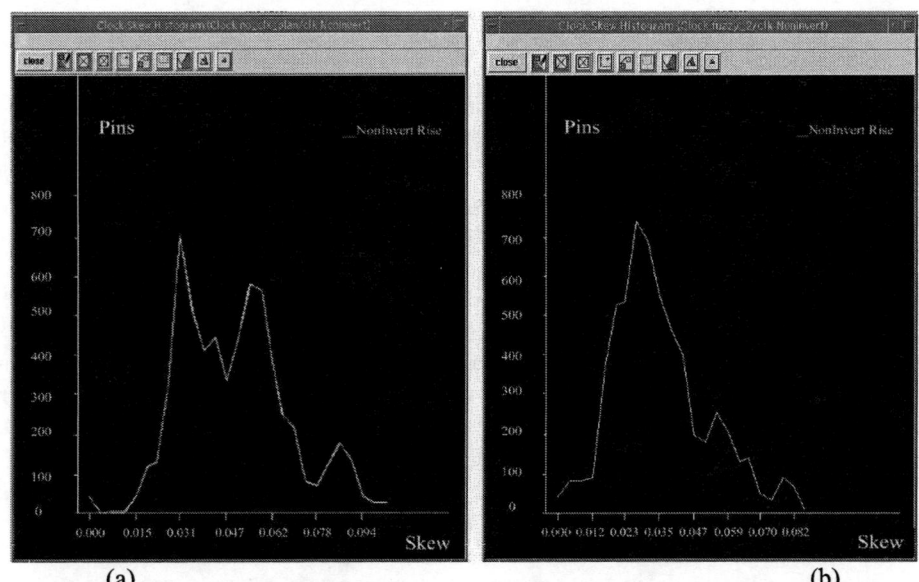

Figure 6 (a) Clock skew histogram of the fuzzy chip design without considering clock tree issue during floorplanning
(b) Clock skew histogram of the fuzzy chip design with considering clock tree issue during floorplanning

Segmented Channel Routing with Pin Rearrangements via Satisfiability

Fei He[*†], William N. N. Hung[‡], Xiaoyu Song[§], Ming Gu[†] and Jiaguang Sun[†]
[*] Department of Computer Science and Technology, Tsinghua University, Beijing, P.R. China
[†] School of Software, Tsinghua University, Beijing, P.R. China
[‡] Synplicity Inc., California, USA
[§] ECE Department, Portland State University, Oregon, USA

Abstract— We address a new segmented channel routing problem with pin rearrangements in FPGA Technology. In our routing model, the pins in each module have certain degree of freedom to be rearranged. With this flexibility, the wire routability can be improved in segmented channel routing. We present an efficient SAT-based approach to solve the problem. We use one of the best SAT-solvers, zChaff, to perform our experiments. Experimental results show the promising performance of the method.

I. INTRODUCTION

Field Programmable Gate Arrays (FPGA's) combine the flexibility of mask programmable gate arrays with the convenience of field programmability [1]. It is widely used as an important technology in hardware emulators for rapid prototyping of designs.

As discussed in [2], the FPGA architectures can be classified into two classes: island-style architecture and row-based architecture. The row-based architecture (segmented channel) [3] is consisted of rows of logic cells separated by segmented routing channels. Unlike the conventional channel architecture, the segmented routing channel contains predefined wiring segments of various lengths that are interconnected by programmable switches. These switches include vertical switches at the crossing of vertical segment and horizontal segment, and horizontal switches at the joint of two adjacent horizontal wiring segments. By programming these switches, we can build paths between terminals that belong to the same net. Due to manufacturing technologies, the switches have significant resistances and capacitances. So for nets should be routed using as little switches as possible.

Conventional channel routers have two assumptions: (i) the logic blocks that form the border of the channel are fixed; and (ii) the pins on the boundary of every block are fixed. Thus, if a routing instance cannot be routed by the conventional routing model, we have no remedial measure. The logic block locations are determined by the placement algorithm, so they are not easily changeable by the router alone. However, the pin positions within each block are determined by the block's internal logic configuration. Since there are usually many alternative logic configurations, the pin arrangements within each block can be changed. If we give some degree of freedom to rearrange the pins for each block, then some routing cases that are non-routable in conventional routers become routable. As a result, the routability is higher with the new formulation.

Even for the cases that are routable in conventional routers, we may also find better routing results with this flexibility in which less switches (segments) are used. As a result, the quality of the routing result is improved.

The idea of channel pin assignment was first proposed by Gopal [4] and have been studied in [5]–[8]. One proposal assumed that pin positions are fully permutational. Some proposed movable pins, where the either side of the channel may have empty (unused) slots such that pins can be moved to these positions. However all these studies are limited to conventional channel architectures. Their main objective was to minimize the conventional channel width (number of tracks) or to minimize the routing cost. In this paper, we consider channel pin assignment in segmented channel routing that arises in FPGA designs. We define this new problem of segmented channel routing with pin rearrangement. The segmented channel routing is different from conventional channel because the channel width (number of tracks) is already fixed by the FPGA architecture. Our objective is to increase routability and to minimize the routing cost (e.g. number of segments used by each net).

Our pin rearrangement is specified by the user as a set of possible permutation patterns (ordering of pins) for each block. The reason for our proposal is that not every permutational pattern may be feasible in hardware. So our router only considers feasible permutational patterns specified by the user. These permutation patterns may include unused pin positions. Hence the problems of permutational pins and movable pins are all subsets of our formulation.

We present a formal description for the problem of segmented channel routing with pins rearrangement. An approach based on satisfiability is proposed to solve this problem. We use a set of variables to designate the specific detailed route for each net and the adopted pins permutation pattern for each block. Based on these variables, both the horizontal constraint and vertical constraints for the problem are translated into Boolean equations. Any assignment of Boolean variables that satisfies the conjunction of all the equations specifies a valid routing.

The rest of this paper is organized as follows. In Section II, we introduce related work about the problem. In Section III, we present a formal problem description. Section IV develops the satisfiability formulations. Section V tests our approach for performance by benchmarks. Section VI concludes this paper.

II. RELATED WORK

In [4], the terminals are assumed to be movable and the channel routing is modeled as the process of finding the optimal wiring solution with some objective functions minimized. Different objective functions are considered in [4]. For some of which, they presented polynomial time algorithms, and for others, they proved they are NP-completeness. In [5], the authors consider the channel pin assignment problem subject to some constraints, such as the terminal position constraints, terminal order constraints and so on. In [6], a specified problem, named as performance-driven channel pin assignment problem that takes the performance-driven net into account is proposed. They prove this problem is NP-complete, and present a polynomial time algorithm for the sub problem which assume each module has at most 2 pin assignment scheme.

Many successful applications showed that the satisfiability-based approach is more efficient than other decision methods ([9], [10]) for many NP-complete problems occurred in layout. Davadas [11] developed a simple formulation of classical two-layer channel routing as Boolean satisfiability. For island-style FPGAs, Wood [12] and Nam [13] presented a satisfiablity-based approach respectively. For the row-based FPGAs, Hung [14] developed a set of Boolean equations whose conjunction determines the routability of this channel. In this paper, we extend the method shown in [11], [14] to our problem.

III. PRELIMINARIES AND DEFINITIONS

According to the grid-based model, a channel is divided into grids by some horizontal and vertical cut lines. Pins in top and bottom boundaries can only be connected along these cut lines. The horizontal cut line is known as track, and the vertical cut line is known as column.

As defined in [3], consider a channel with height $P+1$ and length $L+1$, there are P tracks and L columns, where the tracks are numbered from bottom to up, columns are numbered from left to right. Let Γ be the set of P tracks and Ξ be the set of L columns. Each track t is separated into a set G_t of g_t segments by $g_t - 1$ switches. We use $s_{t,i}$ to denote the segment i on track t, where the indexes of segments are numbered from left to right. Let $left(s)$, $right(s)$ be the leftmost and rightmost column in which the segment s is present respectively. Since each switch need occupy a column, then $left(s_{t,i+1}) = right(s_{t,i}) + 1$ for all $t = 0, \ldots, P-1$ and $i = 0, \ldots, g_t - 1$.

The top and bottom boundaries of the channel are composed of the boundaries of a set Ω of M blocks, as shown in Fig. 1. Let $left(b)$, $right(b)$ be the leftmost and rightmost columns in which the block b is present respectively. Blocks are indexed from left to right according to their leftmost column. The position for each block is fixed, and the terminals assigned to each block are known. For each block b to which p pins are assigned, we can arrange these pins by k_b different orders. Consider each arrangement for these pins as a permutation

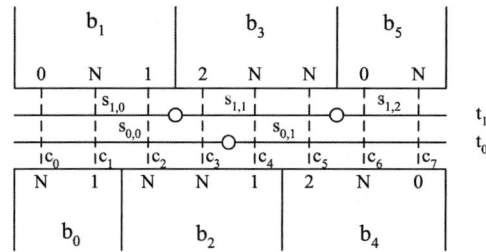

Fig. 1. A segmented channel example with P=2, L=8.

pattern, and then there are k_b permutation patterns for block b. If the pins in a block are rearranged by some order, we say this block is implemented by the corresponding permutation pattern.

If the pins can be rearranged with complete freedom, the number of permutation patterns for each block would be $N!$, where N is number of pins. But in real world, not every permutation is allowed. The pins rearrangement must be performed under the precondition that the structural intent of the functional block remains unchanged. In practical routing, the precise degree of the freedom for pins rearrangement relies on the designer, with the technology-dependent design rules perhaps been take into account. Different designers may have different constraints for the pins arrangement. They should specify a set of allowed permutation patterns for each block before the satisfiability check.

Let Φ be the set of n nets. A net is a connection of terminals to be connected. The span of a net n is defined by its leftmost terminal $(left(n))$ and rightmost terminal $(right(n))$. However, before every block is implemented, we cannot determinate the exact position of the leftmost and rightmost terminals of a net. Fortunately, from the problem specification we can predicate a set of blocks, in one of which the leftmost terminal (or rightmost terminal) is present in. The set for leftmost terminal of net n is known as *left critical modules* for net n, denoted as $LC(n)$; and the set for rightmost terminal of net n is known as *right critical modules* for net n, denoted as $RC(n)$. Consider the example shown in Fig. 1, the leftmost terminal of net n_0 can be present in only b_1, and the rightmost terminal of net n_0 may be present in either b_4 or b_5, so $LC(n_0) = \{b_1\}$, and $RC(n_0) = \{b_4, b_5\}$. Obviously, the sets $LC(n)$ and $RC(n)$ have at most two elements respectively.

The exact position for any pin on a module cannot be determined until the module is implemented. Assume the net n has a pin on the boundary of the module b, also assume the module b is implemented by the pin assignment k, then the index of column on which the pin of net n is assigned can be determined by the problem specification, we refer to this as $POS(b, k, n)$.

In this paper, we restrict our attention to the special case that a net is assigned to at most one track. We refer to this case as *dogleg-free segment channel routing*. It is reasonable for us to ignore the doglegged channel routing as the "doglegs" tend to add greatly to parasitic capacitances and can considerably degrade the circuit performance.

IV. ROUTING VIA BOOLEAN SATISFIABILITY

We define a *permutation variable* $\vec{x}(b)$ for each block b as the binary representation of the selected permutation pattern, i.e. $\vec{x}(b) = x_1(b)x_2(b)\cdots x_\alpha(b)$, where $x_i(b) \in \{0,1\}$ for $1 \leq i \leq \alpha$, and $\alpha = \lceil \log_2(k_b) \rceil$. As in [14], we also define a *track variable* $\vec{y}(n)$ for each net n as the binary representation of the track index where net n is assigned to, i.e. $\vec{y}(n) = y_1(n)y_2(n)\cdots y_\beta(n)$, where $y_j(n) \in \{0,1\}$ for $1 \leq j \leq \beta$, and $\beta = \lceil \log_2(P) \rceil$. Let $X = (\vec{x}(b_1), \vec{x}(b_2), \ldots, \vec{x}(b_M))$, and $Y = (\vec{y}(n_1), \vec{y}(n_2), \ldots, \vec{y}(n_n))$.

Since each net must be assigned to one track,

$$\forall n \in \Phi, 0 \leq \vec{y}(n) \leq P - 1 \quad (1)$$

Since each module must be implemented by a permutation pattern,

$$\forall b \in \Omega, 0 \leq \vec{x}(b) \leq k_b - 1 \quad (2)$$

A. Unlimited Segment Routing

For unlimited segment routing, there is no limit for the number of segments occupied by each net. We define a occupy function for net n to segment s on track t:

$$Occupy(n,s,t) = (right(s) \geq left(n))$$
$$\wedge (left(s) \leq right(n)) \wedge (\vec{y}(n) = t)$$

where $left(n)$ and $right(n)$ denotes the leftmost and rightmost column of net n respectively,

$$left(n) = \min\{POS(b, \vec{x}(b), n) | b \in LC(n)\} \quad (3)$$
$$right(n) = \max\{POS(b, \vec{x}(b), n) | b \in RC(n)\} \quad (4)$$

By combining above three formulas, we have

$$Occupy(n,s,t) = \exists b_l \in LC(n), \exists b_r \in RC(n),$$
$$(right(s) \geq POS(b_l, \vec{x}(b_l), n))$$
$$\wedge (left(s) \leq POS(b_r, \vec{x}(b_r), n))$$
$$\wedge (\vec{y}(n) = t)$$

Since each segment on each track cannot be occupied by more than one net,

$$\forall t \in \Gamma, \forall s \in G_t, \forall n_1, n_2 \in \Phi, n_1 \neq n_2 \rightarrow$$
$$\neg(Occupy(n_1, s, t) \wedge (Occupy(n_2, s, t)) \quad (5)$$

The overall routability check for dogleg-free unlimited segment routing with pins rearrangement is the conjunction of (1), (2), and (5).

B. K-Segment Routing

In K-segment routing, a net can occupy at most K segments on a track [3]. Thus there is:

$$\forall n \in \Phi, \forall t \in \Gamma, (\vec{y}(n) = t) \rightarrow \sum_{s \in G_t} Occupy(n,s,t) \leq K \quad (6)$$

The overall routability check for dogleg-free K-Segment routing with pins rearrangement is the conjunction of (1), (2), (5) and (6).

C. Performance Driven Nets

A performance driven net is a net satisfied following two conditions: 1) its span is shorter than a predetermined bound; 2) it is assigned to a set of predetermined tracks. These requirements sometimes originate due to phyical design issues [15], [16]. Let ϑ be the set of performance driven nets, Σ be the set of performance driven tracks, and l_n be the span upper bound of net n, then the constraint 1) for performance driven nets can be represented as:

$$\forall n \in \vartheta, right(n) - left(n) \leq l_n$$

Substitute (3) and (4) into the above formula, we have

$$\forall n \in \vartheta, \forall b_l \in LC(n), \forall b_r \in RC(n),$$
$$(POS(b_r, \vec{x}(b_r), n) - POS(b_l, \vec{x}(b_l), n) \leq l_n) \quad (7)$$

Constraint 2) for performance driven nets can also be represented as:

$$\forall n \in \vartheta, \exists t \in \Sigma, \vec{y}(n) = t \quad (8)$$

The overall routability check for dogleg-free performance driven nets routing with block pins reassigned is the conjunction of (1), (2), (5), (7) and (8).

V. EXPERIMENTAL RESULTS

We created a number of random benchmarks using a generator similar to [14]. All experiments are conducted on Linux with 850MHz Pentium III processor. We used zChaff [17] as our SAT solver. Table I shows the performance of our router on randomly generated benchmarks. For each benchmark, we randomly selected 10% of the tracks to be performance tracks and randomly selected 10% of the nets to be routed on these performance driven tracks. For each problem instance, with applied the router with fixed pins (i.e. the pin locations cannot be permuted), and compared with some movable pin patterns. As shown in Table I, the number of clauses and literals for the CNF increased due to pin rearrangement, which also lead to an increase in CPU seconds. But the problem has changed from unroutable to routable.

For more practical benchmarks, we used the segmentation design proposed by Zhu and Wong [18] and also used by Hung et al [14]. The benchmarks provide a set of channels for segmented channel routing with fixed pins. But not all channels and routing specifications are routable, as reported in [14]. We take the unroutable specifications and re-arrange their pins to make them routable. The results are shown in Table V.

The experimental results demonstrate that our approach is indeed more efficient than the traditional segmented channel router by improving the routability with an acceptable increase of CNF size and CPU time.

VI. CONCLUSION

In this paper, we consider segmented channel routing with pin rearrangement. In our routing model, the pins assigned to each module can be rearranged according to some user specified patterns. Utilizing this freedom, we improve the

TABLE I
RANDOM BENCHMARKS

Track	Column	Net	Fixed Pins				Movable Pins			
			Clause	Literal	CPU sec.	Routable	Clause	Literal	CPU sec.	Routable
20	60	50	40612	2121	0.01	No	71729	5995	4.55	Yes
30	85	75	131545	4613	0.02	No	204802	12991	1.32	Yes
40	100	70	150611	5374	0.01	No	235780	15358	1.14	Yes
40	150	80	192357	6153	0.02	No	296876	17616	1.60	Yes
40	500	75	170986	5803	0.02	No	265604	16631	1.08	Yes
50	150	75	213434	7094	0.03	No	349437	20762	0.26	Yes
50	200	80	244769	7654	0.03	No	376737	21688	1.02	Yes
50	300	100	354863	9423	0.05	No	543401	27583	5.31	Yes
80	450	150	1187729	21307	0.12	No	1743613	63393	4.33	Yes

TABLE II
ZHU AND WONG'S BENCHMARKS

Benchmark	Fixed Pins				Movable Pins			
	Clause	Literal	CPU sec.	Routable	Clause	Literal	CPU sec.	Routable
D1	56039	3007	0.01	No	119627	10179	0.10	Yes
D2	27980	2001	0.01	No	65919	6110	0.05	Yes
D3	74367	3744	0.02	No	183565	13614	0.65	Yes
D4	99010	4515	0.01	No	231097	15733	2.37	Yes
D5	9591	956	0.01	No	27177	3253	0.02	Yes
D6	24732	1901	0.01	No	65541	6367	0.05	Yes
Geo	58248	3284	0.01	No	165449	12926	0.27	Yes
Norm	10700	1089	0.01	No	39347	4210	0.03	Yes

routability of the segmented channel routing, and present a satisfiability based approach to solve this model. We used zChaff to perform our experiments. The experimental results show the improved routability of our approach to traditional segmented channel router. It should be noted that the efficiency of our satisfiability-based approach would directly benefit from the innovations in SAT solver research. Our approach is also incremental, it can be extended to take more constraints into account without changing other parts.

REFERENCES

[1] S. D. Brown, R. J. Francis, J. Rose, and Z. G. Vranesic, *Field Programmable Gate Arrays*. Norwell, MA: Kluwer, 1992.
[2] S. Trimberger, "Effects of FPGA architecture on FPGA routing," in *Proc. Design Automation Conference*, San Francisco, CA, June 1995.
[3] V. P. Roychowdhury, J. Greene, and A. E. Gamal, "Segmented channel routing," *IEEE Transactions on CAD*, vol. 12, no. 1, pp. 79–95, 1993.
[4] I. S. Gopal, D. Coppersmith, and C. K. Wong, "Optimal wiring of movable terminals," *IEEE Transactions on Computers*, vol. C-32, pp. 845–858, September 1983.
[5] Y. Cai and D. F. Wong, "Optimal channel pin assignment," *IEEE Transactions on Computer-Aided Design of Integrated Circuits and Systems*, vol. 10, pp. 1413–1424, 1991.
[6] T. W. Her, T. C. Wang, and D. F. Wong, "Performance-driven channel pin assignment algorithms," *IEEE Transactions on Computer-Aided Design of Integrated Circuits and Systems*, vol. 14, pp. 849–857, 1995.
[7] Y. Cai and D. F. Wong, "Minimizing channel density by shifting blocks and terminals," in *Proc. International Conference on Computer-Aided Design*, 1991, pp. 524–527.
[8] C. Y. R. Chen and C. Y. Hou, "A pin permutation algorithm for improving over-the-cell channel routing," *IEEE Transactions on Computer-Aided Design of Integrated Circuits and Systems*, vol. 14, pp. 1030–1037, 1995.
[9] Y. Wang et al., "Optimal Symmetry Detection for OKFDDs," in *Proc. IEEE Midwest Symposium on Circuits and Systems (MWSCAS 2000)*, August 2000.
[10] ——, "Single-faced Boolean Functions and their Minimization," *Computer Journal*, vol. 44, no. 4, pp. 280–291, 2001.
[11] S. Devadas, "Optimal layout via boolean satisfiability," in *Proc. International Conference on Computer-Aided Design*, 1989, pp. 294–297.
[12] R. G. Wood and R. A. Rutenbar, "FPGA routing and routability estimation via Boolean satisfiability," *IEEE Transactions on VLSI Systems*, vol. 6, no. 2, 1998.
[13] G. J. Nam, F. Aloul, K. A. Sakallah, and R. A. Rutenbar, "A comparative study of two Boolean formulations of FPGA detailed routing constraints," *IEEE Transactions on Computers*, vol. 53, no. 6, pp. 688–696, 2004.
[14] W. N. N. Hung, X. Song, E. M. Aboulhamid, A. Kennings, and A. Coppola, "Segmented channel routability via satisfiability," *ACM Transactions on Design Automation of Electronic Systems (TODAES)*, 2004, accepted for publication.
[15] S. Gao, K. Thulasiraman, et al., "Homotopic Routing of Multi-terminal Nets with Wire-length Minimization," *Journal of Circuits, Systems and Computers*, vol. 6, pp. 1–14, 1996.
[16] S. Gao and K. Thulasiraman, "Parallel Algorithm for Integrated Floor Planning and Routing," in *Proc. International Conference on High Performance Computing*, December 1995, pp. 457–462.
[17] L. Zhang, C. F. Madigan, M. W. Moskewicz, and S. Malik, "Efficient conflict driven learning in a boolean satisfiability solver," in *Proc. International Conference on Computer-Aided Design*, November 2001, pp. 279–285.
[18] K. Zhu and D. F. Wong, "On Channel Segmentation Design for Row-Based FPGAs," in *Proc. International Conference on Computer-Aided Design*, November 1992.

An Automatic Face Recognition System Based on Wavelet Transforms

A. Amira[*], P. Farrell
School of Computer Science
Institute of Electronics, Communications and Information Technology (ECIT)
Queen's University of Belfast
Belfast BT7 1NN United Kingdom
[*]a.amira@qub.ac.uk

Abstract- Face recognition is emerging as an active research area spanning several disciplines such as image processing, pattern recognition, computer vision and neural networks. Face recognition technology has numerous commercial and law enforcement applications. This paper presents an automatic system based on wavelet transforms for face recognition. A range of wavelet decompositions together with different threshold types and segmentation algorithms have been implemented in order to investigate the best performances. Haar, Gabor and 9/7 wavelet filters have been implemented as a part of the proposed algorithms due to their simplicity, suitability and regularity for face recognition using multiresolution approaches.

I. INTRODUCTION

In recent years, face recognition has been the subject of intensive research. With the current perceived world security situation, governments as well as businesses require reliable methods to accurately identify individuals, without overly infringing on rights to privacy or requiring significant compliance on the part of the individual being recognised. Face recognition provides an acceptable solution to this problem.

A multitude of techniques have been applied to face recognition and can be separated into two categories – geometric feature matching and template matching.

Geometric feature matching [1-6] involves segmenting the distinctive features of the face – eyes, nose, mouth, etc – and extracting descriptive information about them such as their widths and heights. Ratios between these measures can then be stored for each person and compared with those from known individuals.

Template matching is a non-segmentation approach to face recognition. Each face is treated as a two-dimensional array of intensity values, which is then compared with other facial arrays. Earliest methods treated faces as points in very high dimensional space and calculated the Euclidean distance between them. Dimensional reduction techniques including Principal Component Analysis (PCA) [2], [3], [9] have now been successfully applied to the problem, thus reducing complexity of the recognition process without negatively infringing on accuracy.

Another technique that has been applied to the field is Neural Network Models (NMM) [6]. With NMM, the system is supplied with a training set of images along with correct classification, thus allowing the neural network to ascertain a weighting system to determine which areas of an image are deemed most important. A further technique, Deformable Templates (DT) [10], involves the development of a template to identify the regions of interest in an image. The template is placed on an image and the computer automatically deforms the template to fit the object.

An HMM represents stochastic sequences as Markov chains where the states are not directly observed. HMM's have been successfully applied to face recognition, using one model for each class of face, where a class represents an individual. The Discrete Wavelet Transform (DWT) [7], [8] has also been used within the field of face recognition. The wavelet transform, which measures local frequency information in an image, allows features of an image, such as eyes or mouth, to be extracted.

This paper presents two techniques for face recognition based on a range of wavelet transforms using "*error tolerance*" and "*region tagging*" segmentation algorithms and their impact to the metric methodologies used for face recognition such as recognition error and accuracy (i.e. given a set of images, find the number that are matched to the correct person) of the system.

The outline of this paper is as follows. In the next section, the proposed system developed for face recognition is briefly described. Results and analysis are given in Section 3. Conclusions are presented in Section 4.

2. PROPOSED SYSTEM

This paper concerns face classification by geometric face matching using multiresolution analysis, namely wavelet decomposition. The images used in this paper have been taken from the FERET database [9].

Images are loaded from file, and may be pre-processed (i.e. application of low pass or high pass filters) before image transformation. The purpose of each transform is to identify the edges of the facial features. After transformation, the image is thresholded to remove any "background points".

A segmentation algorithm is then applied to the image to extract the eyes, nose and mouth from the image, and based on the dimensions of these features, a set of 14 ratios are stored. For accuracy purposes, the identity of the person in the image is also recorded.

A. Image Wavelet Transforms

Wavelets can be implemented as a set of filter banks comprising a high-pass and a low-pass filter, each followed by downsampling by two. The low-pass filtered and decimated output is recursively passed through similar filter banks to add the dimension of varying resolution at every stage. This is mathematically expressed in equation (1) [7].

$$DWT_{x(n)} = \begin{cases} d_{j,k} = \sum x(n) h_j^*(n - 2^j k) \\ a_{j,k} = \sum x(n) g_j^*(n - 2^j k) \end{cases} \quad (1)$$

The coefficients $d_{j,k}$ refer to the detailed components in the signal and correspond to the wavelet function whereas the coefficients $a_{j,k}$ refer to the approximation components in the signal. The functions $h(n)$ and $g(n)$ in this equation represent the coefficients of the high-pass and the low-pass filter respectively.

1) Haar Wavelet Transform (HWT)
HWT decomposition works on an averaging and differencing process as follows [7]:

Resolution	Averages	Detail Coefficients
4	[9 7 3 5]	
2	[8 4]	[1 -1]
1	[6]	[2]

It can be seen that the number of decomposition steps is $2^2 = 4$.

The decomposition on a two-dimensional array is given by these pseudo-codes

```
Haar Standard Decomposition (2DArray X)
pre X = 2^n*2^n pixels;
int i = Lg (number of rows)
while (i > 0)
        {
        for each r. r ε {rows of x}
                decompose r;
        for each c. c ε {columns of x}
                decompose c;
        i = i -1;
        }
post Haar Standard decomposition of x;
```

```
Haar Non-Standard Decomposition (2DArray X)
pre X = 2^n*2^n pixels;
int i = Lg (n)
while (i>0)
        {
        int j =n;
        while (j > 0)
                {
                decompose row j;
                decompose column j;
                }
        }
post Haar Non-Standard decomposition of x;
```

This paper addresses the first level decomposition of images. The differencing is the wavelet decomposition process highlights edges to give a coarse shape of the face.

Figure 1a. Subject Image   Figure 1b. First Level HWT Standard   Figure 1c. First Level HWT Non-Standard

First level Haar decomposition as shown in figure 1 divides an image into four clearly identifiable regions: LL, LH, HL and HH where L represents Low pass filtering, H represents High pass filtering and HL would mean Low pass filter, then High pass filter. It is dealt with the HL region for face segmentation algorithms presented in this paper.

2) Gabor Filters
Gabor filters allow local frequency information to be extracted from an image. They estimate the strength of certain frequency bands and orientations at each location in the image, giving a result in the spatial domain. Due to the high computational load associated, this paper employs the fast computation of Gabor transform [12]. A 2D Gabor wavelet kernel can be represented as a convolution of two orthogonal 1D components. These components are a Gaussian $g(x)$, and a wavelet $w(x)$, defined respectively by:

$$g(x) = \frac{1}{\sqrt{2\pi\sigma^2}} e^{\frac{-x^2}{2\sigma^2}} \quad (2)$$

and $w(x) = g(x) e^{j\omega x}$

Figure 2a. Real Gabor applied to figure 1a, λ=1, scale=2, filtersize=5, angle=45

Figure 2b. Imaginary Gabor applied to figure 1a, λ=1, scale=2, filtersize=5, angle=45

Figure 2c. Modular Gabor applied to figure 1a, λ=1, scale=2, filtersize=5, angle=45

where $j = \sqrt{-1}$ and ω is the frequency of the wavelet. The convolution of an image I with a Gabor kernel K is thus:

$$I * K = I * (g * w)$$

This paper deals with three types of Gabor wavelet, namely: *Real, Imaginary* and *Modular (Figure 2)*.

3) Biorthogonal 9/7
The Biorthogonal 9/7 filter is similar to the Haar transform in that it operates on the application of high and low pass filters [7]. In addition, it can be carried out in the standard and non-standard manners.

B. Threshold Algorithms
This paper deals with two types of Thresholds [7]:
1) Hard Thresholding
Given an image I, and a threshold value T_v, the thresholded image I^{HARD} is obtained by examining the value of each pixel $[x,y]$ in I and comparing it to the value T_v such that:
$$I^{HARD}_{[x,y]} = (I_{[x,y]} < T_v \;?\; 0 : I_{[x,y]})$$
2) Soft Thresholding
Given I and T_v as before, the resulting image I^{SOFT} is obtained thus: $I^{SOFT}_{[x,y]} = max(|I_{[x,y]}| - T_v, 0.0)$

C) Segmentation
This paper presents two novel segmentation algorithms for face features extraction as shown in figures 3 and 4.
1) Error Tolerance
This algorithm extracts the eyes from the transformed image, then the nose, and finally the mouth. The position of each feature is influenced by the position of features extracted previously i.e, the position of the nose is below that of the eyes, and the position of the mouth is below the nose and between the centre points of the eyes.

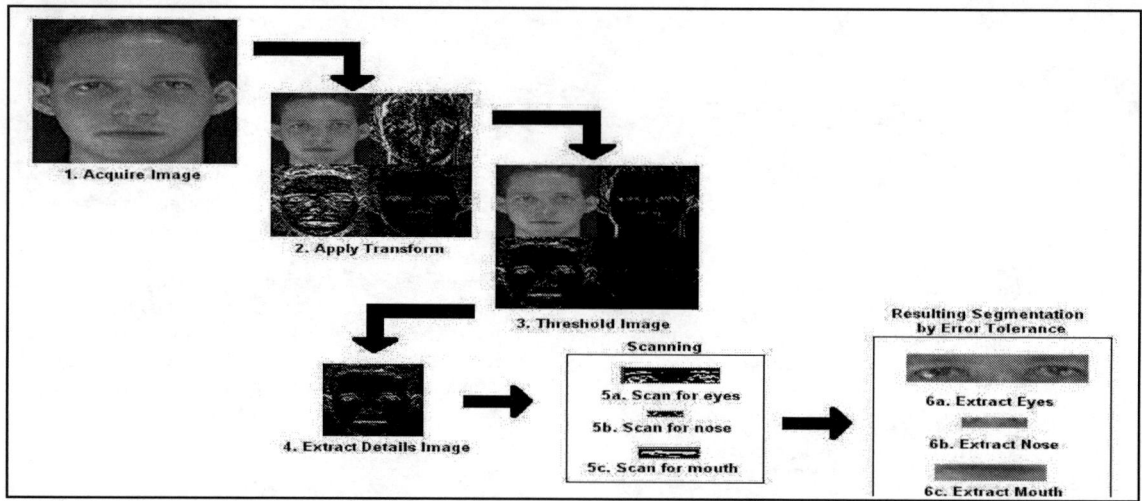

Figure 3. Error Tolerance Algorithm of figure 1a using Hard Threshold 3

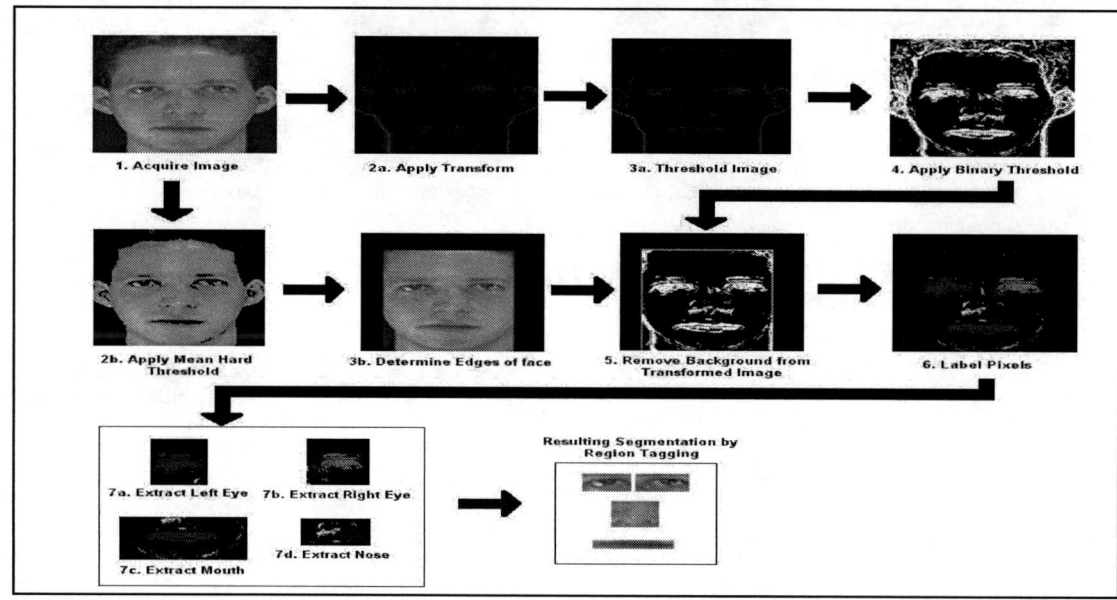

Figure 4. Region Tagging Algorithm of figure 1a using Hard Threshold 2

2) Region Tagging
This algorithm extracts the left eye from an image, then the right eye, then the mouth, and finally the nose.

3. RESULTS AND ANALYSIS
A) Recognition
1) Ratios Calculation
After segmentation, 14 ratios per image are stored. Example of ratios calculated:

1. Nose Height / Eyes External Distance (EED)
2. EED / Eyes Internal distance (EID)
3. Nose Height / Mouth Width
4. EED / Mouth Width
5. EED / Nose Width
6. Nose Height / Nose Width
7. Nose height / Face Width
8. Face Width / EID
9. Face Width / Nose Width....etc.

2) Classification of an Image
After ratio calculation, the identity of the person in the image is recorded – i.e. the name that goes with the face. This paper defines a person as a class, and thus images of that person are members of that class.

3) Accuracy
Having obtained image B for every image A, if A and B are members of the same class, a correct match is obtained. Thus accuracy is defined as:

Number of Correct Matches / Number of images for which ratios are stored

B) Results

Table 1 illustrates some results achieved in term of accuracy when applying the "Region Tagging Algorithm" with the soft threshold on a set of images such as those shown in figure 5.

Table 1. Accuracy obtained (Region Tagging algorithm)

Database Size	Haar		Gabor		Biorthogonal 9/7		Prewitt
	Standard	Non-Standard	Imaginary	Modular	Standard	Non-Standard	
15	93%	93%	93%	66%	100%	100%	93%
20	70%	70%	65%	60%	65%	65%	75%
25	76%	76%	64%	56%	68%	68%	76%

Figure 5. Face image samples

C) Discussion

A comparison with existing techniques for face recognition has been made to evaluate the proposed algorithms.

Work presented in [1] uses an eigenvalue based approach resulting in accuracy figures of around 50%. This figure is comparable to that of the proposed technique when few images are stored, yet an initial training set of images is required to initialise a weights matrix, which is not the case with the new algorithms.

Examining non-segmentation based approaches results in much greater accuracy, but typically at the expense of computation time or loss of automation.

Work presented in [11] involves Gabor transform and an eigenface approach, with an accuracy achieved of 91% using an image set with two images per person in the set. However, each image in the knowledge base requires the manual configuration of 34 facial landmarks, thus a high degree of automation has been lost.

Work presented in [12] also employs PCA approach using Gabor transform, but their work focuses in identifying characteristics (such as age, sex and race) rather than identification of individuals.

In [13], an algorithm for the problem quoting accuracy figures in excess of 99% by adopting another PCA technique is described. However, computation times are typically five times greater than those of the proposed algorithms.

The accuracy of a system operating under the proposed algorithms fell drastically as the number of images increased as shown in figure 6, but not without yielding some significant shortcomings that revised versions of the algorithms may cater for.

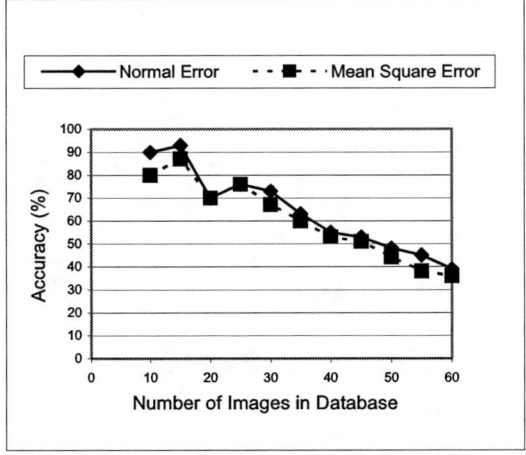

Figure 6. Accuracy Vs database size

4. CONCLUSIONS

Face recognition is becoming very important in many real world applications including security, robotics, banking…etc. This paper presents two techniques for face recognition based on a multiresolution approach using *"error tolerance"* and *"region tagging"* segmentation algorithms and their impact to the metric methodologies used for face recognition such as recognition accuracy. It is the aim of future work to investigate other novel, efficient, fast, robust and accurate algorithms for face recognition and features extractions based on hybrid approaches such DWT-HMM and DWT-PCA.

5. REFERENCES

[1] H. Demirel, T.J. Clarke and P.Y.K. Cheung "Adaptive Automatic Facial Feature Segmentation", 2nd International Conference on Automatic Face and Gesture Recognition, October 14-16, 1996.

[2] M. Turk and A. Pentland "Eigenfaces for Recognition", J. Cognitive NeuroScience, vol. 3, pp. 71-86, 1991.

[3] B. Moghaddam, W. Wahid and A. Pentland, "Beyond Eigenfaces: Probabilistic Matching for Face Recognition", Proceedings of Face and Gesture Recognition, pp 30-35, 1998.

[4] E. Kussul, T. Baidyk and M. Kussul "Neural Network System for Face Recognition", Proceedings of ISCAS'04, International Symposium on Circuits and Systems 2004.

[5] C. Sanderson "Face Processing & Frontal Face Verification", IDIAP-RR 03-20, Martigny, Switzerland, 2003.

[6] M. Bicego, U. Castellani and V. Murino "Using Hidden Markov Models and Wavelets for face recognition" ICIAP'03, 12th International Conference on Image Analysis and Processing, September 17-19, 2003.

[7] E. J. Stollnitz, T. D. DeRose and D. H. Salesin "Wavelets for computer graphics: a primer, part I," IEEE Computer Graphics and Applications, vol. 15, No. 3, pp. 76-84, May 1995.

[8] M. Harandi, M. Ahmadabadi and B. Araabi "Face Recognition Using Reinforcement Learning" Proceedings of ICIP'04, International Conference on Image Processing.

[9] The Facial Recognition Technology Database, URL:http://www.itl.nist.gov/iad/humanid/feret/

[10] A. L. Yuille "Deformable Templates for Face Recognition," J. Cognitive Neuroscience, vol. 3, no. 1, pp. 59--79, 1991.

[11] X. Wang and H. Qi "Face Recognition Using Optimal Non-Orthogonal Wavelet Basis Evaluated by Information Complexity" 16th International Conference on Pattern Recognition (ICPR'02) Volume 1, August 11 - 15, 2002, Quebec City, QC, Canada

[12] M. J. Lyons, J. Budynek, A. Plante and S. Akamatsu "Classifying Facial Attributes using a 2-D Gabor Wavelet Representation and Discriminant Analysis" Proceedings of the 4th International Conference on Automatic Face and Gesture Recognition, pp 202-207, 28-30 March, 2000, France.

[13] K. Lee, Y. Chung, "Hyeran Byun: Face Recognition Using Support Vector Machines with the Feature Set Extracted by Genetic Algorithms" AVBPA 2001, pp 32-37.

Hardware realization of panoramic camera with speaker-oriented face extraction for teleconferencing

Yukinori NAGASE, Takahiko YAMAMOTO, Takao KAWAMURA and Kazunori SUGAHARA

Faculty of Engineering, Tottori University
Koyama-Minami, Tottori 680-8552, JAPAN
Email: {kawamura,sugahara}@ike.tottori-u.ac.jp

Abstract—In this paper, a panoramic camera with speaker oriented face extraction function is proposed. For the face extraction, the Genetic Algorithm(GA) is implemented in a Field Prgramble Gate Array(FPGA) chip. Panoramic video signal generator circuit is also implemented in the FPGA for teleconferencing. Experimental results are included.

I. INTRODUCTION

The advancement of communication speed in digital networks makes it possible to exchange a large amount of information, such as movies, images or human voices, between distant places. Teleconferencing is one typical example of the practical applications in such fields.

In most of the current teleconferencing systems, a full view image of a conference room and voice of speakers are bidirectionally exchanged between distant places. However, in teleconferencing situations, face images of speakers are also requested to perceive facial expression. Fig.1 shows a typical example. To take both of full view images of conference rooms and face images of speakers, it is required to develop a multi camera system with human operations.

In this paper, a panoramic camera with speaker-oriented face extraction function is proposed. The proposed camera is constructed with three NTSC video camera elements, video decoder and encoder LSIs and the proposed video image processor. The video image processor has a panoramic image generation function from output video signals of three video camera elements and a face extraction function. The video image processor is implemented on one FPGA chip, and is applicable to develop stand alone systems.

II. HARDWARE CONFIGURATION

The hardware configuration of the proposed camera is shown in Fig.2. As shown in the figure, the camera is constructed with the following 5 blocks.

Fig. 1. Example of a preferred teleconferencing.

1) Image input block

Image input block is constructed with three NTSC video camera elements as shown in Fig.3. Each camera element in Fig.3 outputs NTSC video signal.

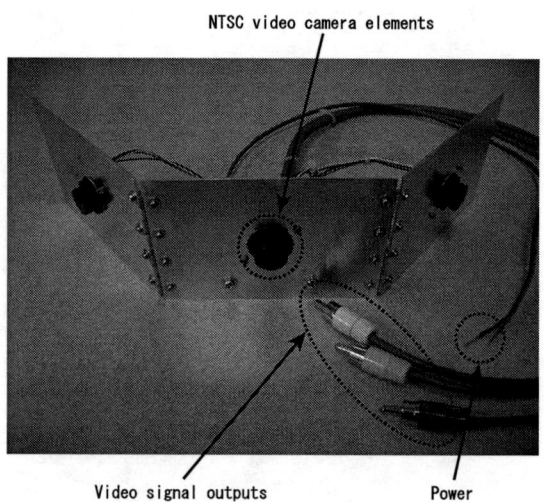

Fig. 3. Image input block.

2) Video signal decoder block

Video signal decoder LSI, MSM7664B[1] pro-

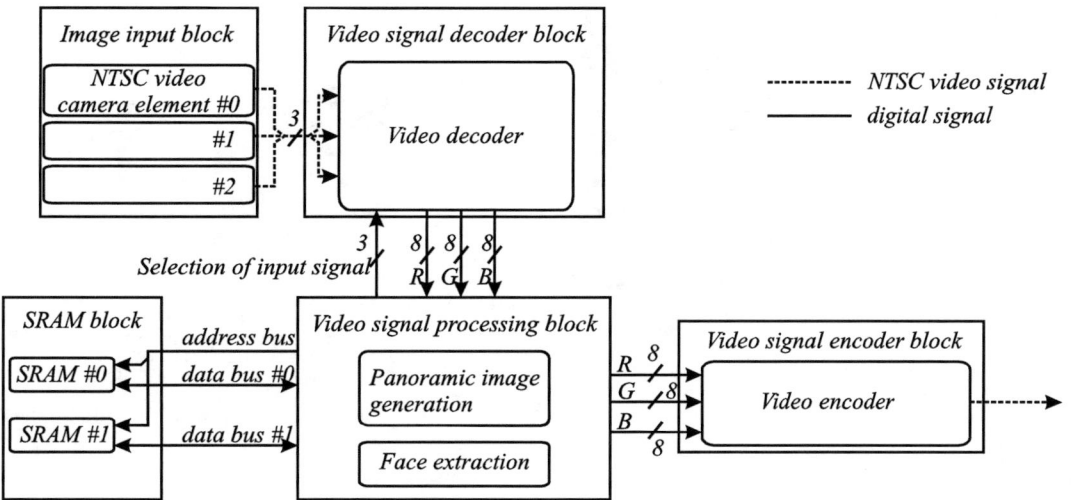

Fig. 2. Hardware configuration.

duced by Oki Electric Industry Co., Ltd., converts NTSC video signals of three video camera elements to 640 × 480 [pixels] size RGB digital image signals. The decoder chip has 4 input ports and 3 of them are utilized in this camera. The switching of these input ports is controlled by the address signals from the video signal processing block.

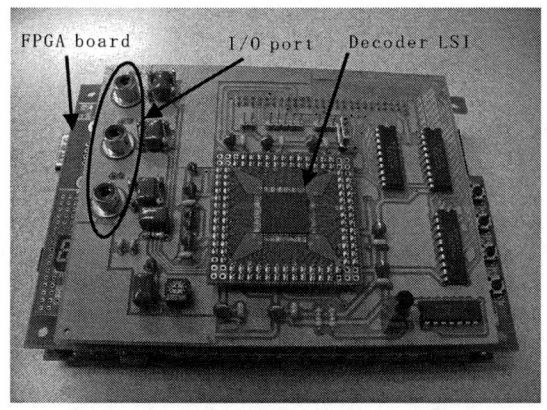

Fig. 4. FPGA, video decode and encode board.

3) Video signal processing block

In the video signal processing block, functions of panoramic image generation from three RGB digital images and of speaker face extraction are implemented in one FPGA. Input signal selection signals for the video signal decoder is also generated in this block. The details of the panoramic image generation function and of the face extraction function are described in the following chapters.

4) SRAM block

The input images and the generated panoramic image is stored in the SRAM block. The SRAM block is constructed with two SS-RAM(Synchronous Statical Random Access Memory) chips and they are able to be accessed independently.

5) Video signal encoder block

Video signal encoder LSI, MSM7654[2] produced by Oki Electric Industry Co., Ltd., generates a NTSC output signal according to the output digital image which the video signal processing block generated.

III. GENERATION OF PANORAMIC IMAGES

The 3 input images from the image input block are compliant with the NTSC specifications, they are 640 × 480 [pixels] size. The output image also should be compliant with the specifications, it should have the same size. Considering these points, the output image is designed as shown in Fig.5. The lower half area of the output image is arranged for the panoramic image and the extracted face image is enlarged and is represented on the upper half of it.

The panoramic image part in the output image is 640 × 180 [pixels] size and it is generated with 3/8 reduced 3 input images with two overlapped area in 40 pixel width.

IV. FACE EXTRACTION

The processing for the face extraction of the speaker is accomplished by the technique based on the GA (Genetic Algorithm)[3], [4]. The block diagram of the face extraction processing is shown in Fig.6. As shown

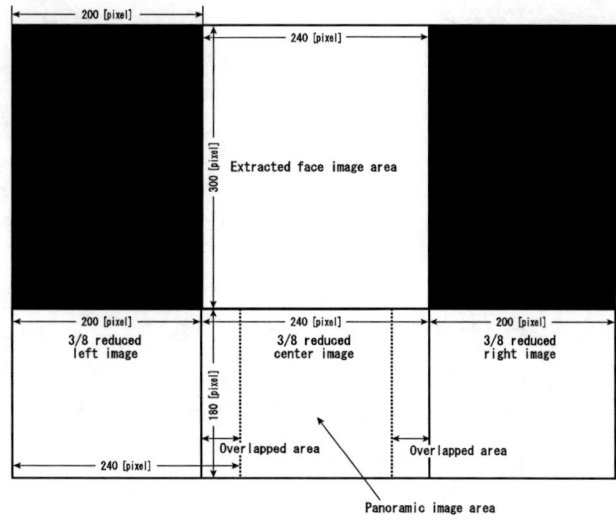

Fig. 5. Generation of panorama image.

in the figure, the GA process is constructed with the following functions,
1) The random number generator.
2) The evaluation of obtained intermediate results.
3) The selection of preferred results as parents genes.
4) The crossover and mutation of parent genes and generate child genes.

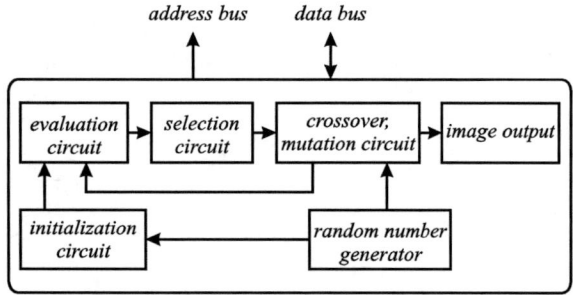

Fig. 6. Face extaction.

In this paper, the number of genes is set as 32 and the genetic code is defined as shown in Fig.7. The values of X-axis and Y-axis in this code represent the coordinates of the pixel in the bottom left corner of a sub-image, as shown in Fig.8.

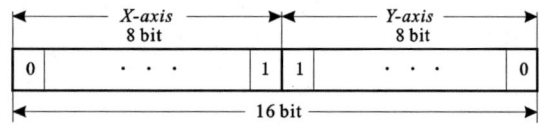

Fig. 7. Genetic code.

The evaluation of intermediate results are accomplished by the comparison of RGB values of each pixel

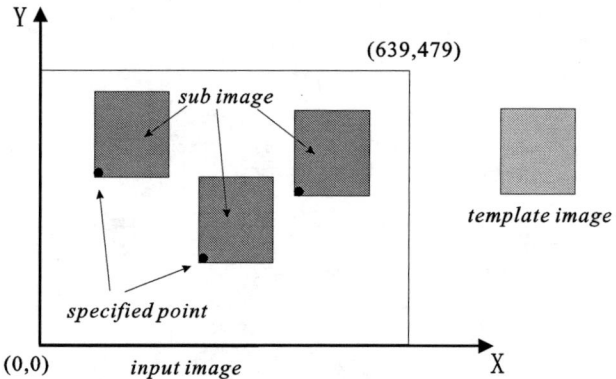

Fig. 8. Comparison of input and template image.

in sub image which addressed by the each genetic code with those in the given template image. The genes are classified into parents group and children group according to the evaluation results. The upper half of genes is classified into the parents group and the lower half is classified into the children group. And two genes selected randomly in the parents group generate two children by the crossover process. The crossover process is repeated 8 times and the generated 16 genes updates the children group. The evaluation and crossover processes are shown in Fig.8 and Fig.9, respectively.

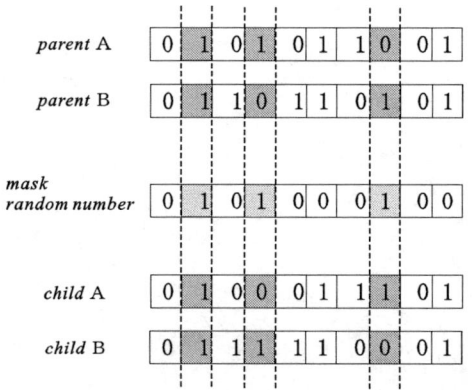

Fig. 9. Crossover.

After the crossover processes, each gene is updated by the mutation process as shown in Fig.10. In the mutation process, bit number of randomly selected genes are inverted according to the given mask data. Here, the mask data is also decided according to the random number.

V. EXPERIMENTAL RESULTS

A. Face extraction

To confirm the operation of the face extraction function in the produced FPGA, several images are examined.

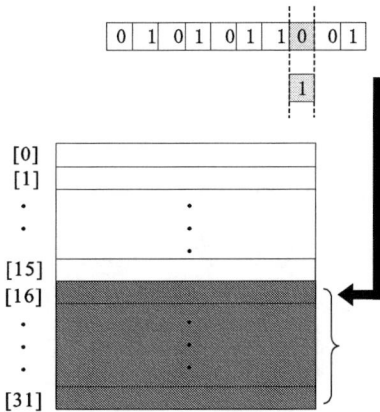

Fig. 10. Mutation.

Fig.11 shows one of the examined images and the locations of sub images corresponding to the initial genes. The template face image used here is shown in Fig.12.

The operation of the face extraction process is confirmed by the obtained final result shown in Fig.13.

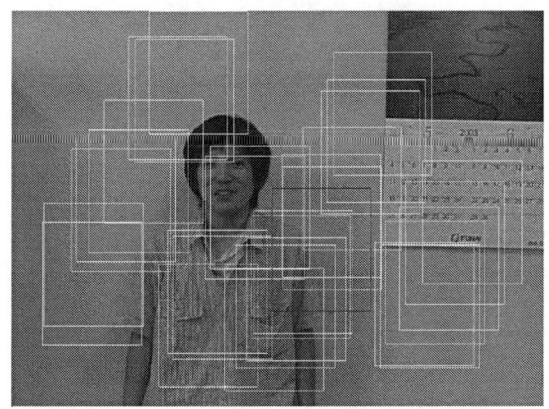

Fig. 11. Initial state.

Fig. 12. Template image.

B. Generation of panoramic images

The example of the generated panoramic images is shown in Fig.14. In Fig.14, the final result of face extraction of speaker is also included.

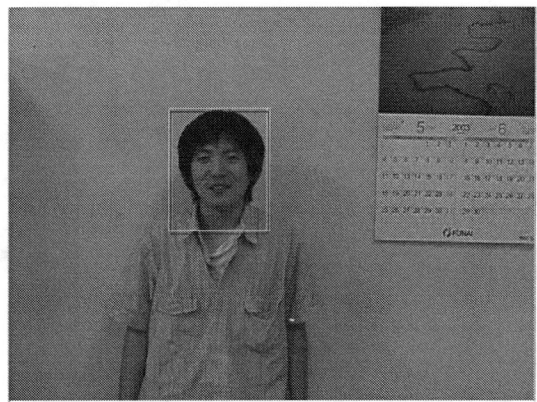

Fig. 13. Final result.

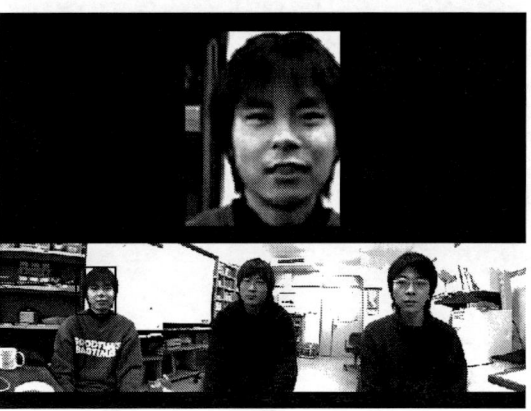

Fig. 14. Composite image.

VI. CONCLUSION

In this paper, the panoramic camera with speaker-oriented face extraction for teleconferencing is proposed. By using the proposed camera, not only full view of conference room but also facial expressions of speakers are able to be exchanged between distant places.

The functions of the face extraction and those of the panoramic image generation are implemented in one FPGA chip. The inputs and output images of the camera are compliant with the NTSC specifications, it is easily applicalbe to stand alone systems.

REFERENCES

[1] Oki Electric Industry: MSM7664B data sheets(in Japanese), (2001).
[2] Oki Electric Industry: MSM7654B data sheets(in Japanese), (1999).
[3] D.E.Goldberg:"Genetic Algorithms in Search, Optimization and Machine Learning", Addison Wesley (1989).
[4] Vose, Michael D.:"The Simple Genetic Algorithm", (1998).

A Novel Content-Adaptive Interpolation

Tai-Wai Chan, Oscar C. Au, Tak-Song Chong, Wing-San Chau
Department of Electrical and Electronic Engineering
Hong Kong University of Science and Technology
Hong Kong
eedavid@ust.hk, eeau@ust.hk, eects@ust.hk, eecws@ust.hk

Abstract— In this paper, we propose a content-adaptive interpolation algorithm for digital images. The proposed algorithm makes use of a non-linear spatial varying filter (SVF), which is simple but efficient in enlarging images. First, edge orientation statistics in an image are exploited, and then the information is used to determine suitable weights of neighboring pixels when interpolating the current pixel. Using the proposed approach, we could generate pixel values to replace the missing ones such that sharp edges could be accurately reconstructed. Simulation results show that the proposed algorithm obtains better results in both subjective and objective quality measurements over existing methods.

I. INTRODUCTION

Image interpolation addresses the problem of generating a larger, high-resolution image from a smaller, low-resolution version of the same image. Many conventional techniques like nearest neighbor, bilinear, cubic and spline interpolations have been developed in the literatures [1]-[7]. However, these methods are usually based on some spatial-invariant models and fail to capture the fast varying property around edges in images. Consequently, the interpolated images that use these methods are usually with blurred edges and contain annoying artifacts as a result. Some edge-preserving interpolation algorithms have also been developed, but the computational complexity of these algorithms might be too high. For example, the method in [6] makes use of the local covariance information to detect the edge direction in a low-resolution image and then use the information to interpolate the high-resolution image. That method requires around 1300 multiplications per pixel.

In the proposed algorithm, local edge orientation statistic in an image would be exploited in a simple but efficient way, and the computation and memory requirement is low. The rest of this paper is organized as follow. In section II, we would introduce the proposed algorithm, and the simulation results would be shown in section III. Lastly, conclusions are given in section IV.

II. PROPOSED ALGORITHM

The proposed algorithm is designed to interpolate a small, low-resolution image with size *NxM* pixels to a larger high-resolution image with size *(2N-1)x(2M-1)* pixels as illustrated in Fig.1.

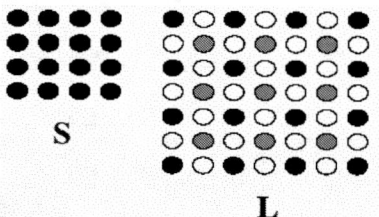

Fig.1 Small image **S** and enlarged image **L**

In Fig.1, the black dots are the pixels with original pixel value while the gray and white dots in L are the pixels that going to be interpolated. In the proposed algorithm, interpolation of an image is done in two phases. First, the missing pixel values located in gray dots are generated and the remained pixel values located in white dots are reconstructed in the second phase.

In the first phase, the pixels represented by the gray dots are interpolated. Referring to Fig.2(i), for each gray dot x in this situation, its pixel value $p(x)$ is reconstructed by taking the weighted average of its nearest four pixel values using a spatial varying filter (SVF) in (1). The weights w_1 and w_2 in the SVF are adaptively determined by (2). In case, both w_1 and w_2 are zero, $p(x)$ would then be set to the mean of its nearest pixel values using (3).

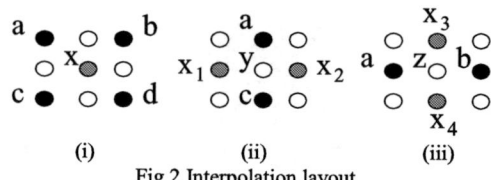

Fig.2 Interpolation layout

Let $p(x)$ be pixel value of pixel x:

$$p(x) = \frac{w_1(p(a)+p(d)) + w_2(p(b)+p(c))}{2(w_1+w_2)} \quad (1)$$

where

$$w_1 = f(|p(a)-p(d)|) \ \& \ w_2 = f(|p(b)-p(c)|)$$

$$f(i) = \begin{cases} k^{(\lfloor T_1/s \rfloor - \lfloor i/s \rfloor)} & i < T_1 \\ 0 & else \end{cases} \quad (2)$$

where k, s and T_1 are predefined values

$$p(x) = \frac{p(a)+p(b)+p(c)+p(d)}{4} \quad (3)$$

$$p(y) = \frac{p(a)+p(c)}{2} \quad (4)$$

$$p(y) = \frac{w_3(p(a)+p(c)) + w_4(p(x_1)+p(x_2))}{2(w_3+w_4)} \quad (5)$$

where

$$w_3 = f(|p(a)-p(c)|) \ \& \ w_4 = f(|p(x_1)-p(x_2)|)$$

$$p(z) = \frac{p(a)+p(b)}{2} \quad (6)$$

$$p(z) = \frac{w_5(p(a)+p(b)) + w_6(p(x_3)+p(x_4))}{2(w_5+w_6)} \quad (7)$$

where

$$w_5 = f(|p(a)-p(b)|) \ \& \ w_6 = f(|p(x_3)-p(x_4)|)$$

The idea behind the proposed approach is based on the fundamental property of an ideal edge that the intensity gradient along an edge is usually far less than that of across an edge. In the other words, if the pixel value difference between two pixels is small, then there is highly chance that they are on the same edge or same object, otherwise they should belong to two different objects. Base on this observation, we also noticed that if two pixels are on the same edge or same object, then a pixel between them (i.e. the current pixel) should also be on the same edge or same object, and their pixel values should be highly correlated. Therefore, in order to interpolate a pixel accurately, we would like to detect if there is any edge across the pixel that is to be interpolated, by utilizing the pixel value correlations of its neighboring pixels. Equation (2) is designed to exploit the pixel value correlations between the neighboring pixels of the current pixel. If the pixel value difference between two diagonal pixels around the current pixel is small, then there is highly chance that the current pixel is on the same edge or same object as that of the two diagonal pixels, therefore a large weight would be given to the pixel values of that diagonal pixels. Otherwise, the current pixel might not belong to the same object as that of the two diagonal pixels, and less weight would be given to the pixel values of that diagonal pixels. By using adaptive weights, the proposed algorithm would not blur the edges in object boundaries of an image.

It is worth to notice that SVF actually combine the advantages of low-pass filter and edge-preserving filter. In texture areas, SVF is able to reconstruct sharp edges by exploiting the corrections between the diagonal pixels of the current pixel, and giving suitable weights to them depending on their local characteristics. In smooth areas, as the pixel value differences between the diagonal pixels of the current pixel should be similar, and SVF would thus give similar weights to the neighboring pixels of the current pixel. As a result, the reconstructed pixel value would be similar to that of its neighboring pixels, and smoothness could be maintained in smooth areas.

After all the gray dots are interpolated, the pixel values located in while dots could be reconstructed using similar procedures. In the second phase, for each white dot y situated as shown in Fig.2(ii), we would first check if the pixel value difference between pixels *a* and *c* is smaller than a predefined threshold T_2. If yes, then there is probably an edge along pixels *a*, *c* and the current pixel, or they are belong to the same object, therefore the reconstructed pixel value *p(y)* would be set to the average pixel values of pixels *a* and *c* using (4). Otherwise, the reconstructed pixel value *p(y)* would be generated using (5).

Lastly, it is noticed that the geometric orientation of all white dots in Fig.2(iii) is exactly the same as that in Fig.2(ii) but with a 90-degree rotation. Therefore, all white dots in situation of Fig.2(iii) would be interpolated with similar procedures as that when interpolating white dots in Fig.2(ii). Referring to Fig.2(iii), the reconstructed pixel value *p(z)* would be set to the average pixel values of pixels *a* and *b* using (6) when the pixel value difference between them is smaller than a predefined threshold T_2. Otherwise, the reconstructed pixel value *p(z)* would be set by using (7).

III. SIMULATION RESULTS

The performance of the proposed algorithm has been compared to the conventional nearest neighbor, bilinear and bicubic interpolation methods. Images of size 352x288 pixels are first down-sampled by taking only the pixels with even-even coordinate, and then they are enlarged using different interpolation algorithms. For the proposed algorithm, the values of the constants are set as follow: T_1=88, T_2=12, k=2 and s=8. The average PSNR of the enlarged images are calculated with respect to the original high-resolution images and Table.1 shows some of the results. It could be seen from Table.1 that the proposed algorithm has higher PSNR than that of the nearest neighbor and bicubic algorithms, the proposed algorithm could gain about 2dB to 3dB in term of PSNR compared to bicubic interpolation. In addition, the correlation coefficient *C* between the original high-resolution image and the enlarged image is calculated using (8). In (8), *I* and *J* are the width and height of the enlarged image, *O(i,j)* and *L(i,j)* are the pixel values of the original high-resolution and enlarged images respectively, and *u* and *v* are the mean pixel values of the original high-resolution and enlarged images respectively. The value *C* is between 0 and 1. If C is more close to 1, the enlarged image is more similar to the original high-resolution image. The correlation coefficients of different sets of images are compared in Table.2, and it could be seen that the proposed algorithm obtains higher value than that of the nearest neighbor and bicubic algorithms. That means the enlarged image by using the proposed algorithm is more similar to the original high-resolution image. The perceptual quality of the enlarged images is also compared and some of the enlarged images are shown from Fig.3 to Fig.10. It could be seen that the proposed method results in clearer and sharper edges (e.g. the big edges on the wall of the "foreman" image) Although the bilinear interpolation obtains similar PSNR values and correlation coefficients compared to the proposed algorithm, it could be seen that the proposed algorithm could obtain the best perceptual quality of interpolated images.

$$C = \left| \frac{\sum_{i,j} O(i,j)L(i,j) - IJuv}{\sqrt{(\sum_{i,j} O^2(i,j) - IJu^2)(\sum_{i,j} L^2(i,j) - IJv^2)}} \right| \quad (8)$$

IV. CONCLUSIONS

A novel content-adaptive interpolation method is presented. The proposed algorithm is simple but efficient in enlarging images. Contrast to some convention interpolation algorithms, the proposed method able to reconstruct sharp edges accurately in the enlarged images, and with low computation and memory requirement. Both subjective and objective measurements show the proposed algorithm outperforms existing interpolation techniques.

Table.1
Average PSNR (dB) of enlarged images using different interpolation algorithms

Image sequence	Nearest neighbor	Bilinear	Bicubic	Proposed
akiyo	29.409	34.1933	30.4263	34.2961
coastguard	24.1249	28.2429	25.1985	27.7608
foreman	27.8355	32.2857	29.8320	32.9396
mother	31.1940	35.8296	32.5004	35.8325
silent	27.4490	31.6610	28.5435	31.4880

Table.2
Correlation coefficient of enlarged images using different interpolation algorithms

Image sequence	Nearest neighbor	Bilinear	Bicubic	Proposed
akiyo	0.9839	0.9947	0.9873	0.9947
coastguard	0.9512	0.9806	0.9596	0.9782
foreman	0.9828	0.9938	0.9894	0.9947
mother	0.9821	0.9954	0.9873	0.9954
silent	0.9744	0.9903	0.9800	0.9898

V. ACKNOWLEDGEMENT

This work has been supported in part by the RGC grant HKUST6203/02E, and the ITF grant ITS/122/03 of the Hong Kong Special Administrative Region, China.

REFERENCES

[1] A. K. Jain, Fundamentals of Digital Image Processing. Prentice-Hall, Inc., 1989.

[2] E. Maeland, "On the comparison of interpolation methods" IEEE Transactions on Medical Imaging, vol. 7, no. 3, September 1988

[3] H. Hou, and H. Andrews, "Cubic splines for image interpolation and digital filtering" IEEE Transactions on Acoustics, Speech, and Signal Processing, vol. 26, no. 6, pp. 508-517, 1978.

[4] N.A. Dodgson, "Quadratic interpolation for image resampling" IEEE Transactions on Image Processing, vol. 6, no. 9, September 1997

[5] R. Keys, "Cubic convolution interpolation for digital image processing" IEEE Transactions on Acoustics, Speech, and Signal Processing, vol. 29, no. 6, pp. 1153-1160, 1981.

[6] X. Li, and M. Orchard, "New edge directed interpolation" Proc. IEEE Int. Conf. Image Processing, vol. 2, 2000, pp. 311-314.

[7] Xin Li; M.T. Orchard, "Edge-directed prediction for lossless compression of natural images" IEEE Transactions on Image Processing, vol. 10 no. 6, June 2001.

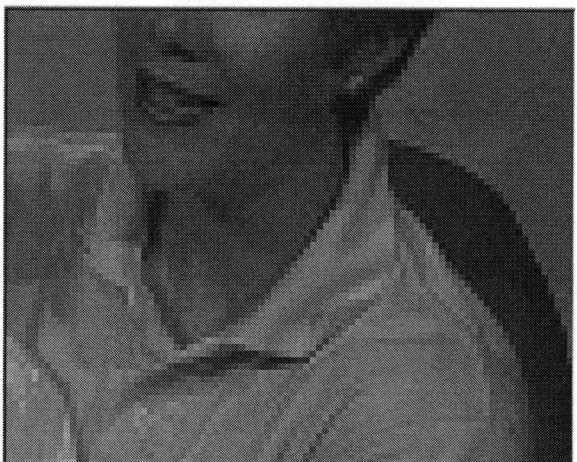

Fig.3 Part of enlarged "mother" image using nearest neighbor interpolation

Fig.4 Part of enlarged "foreman" image using nearest neighbor interpolation

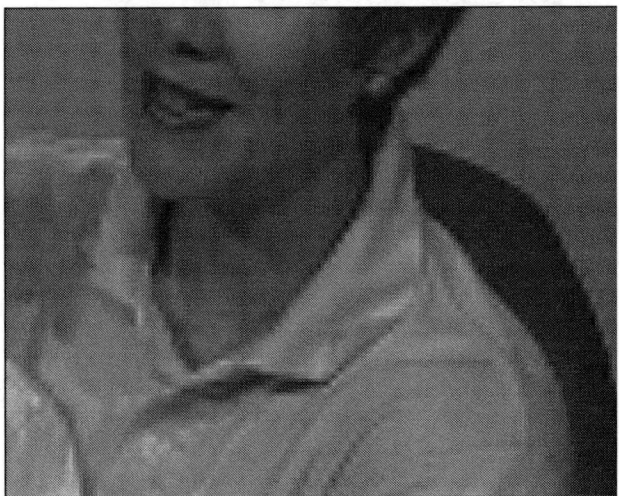
Fig.5 Part of enlarged "mother" image using bilinear interpolation

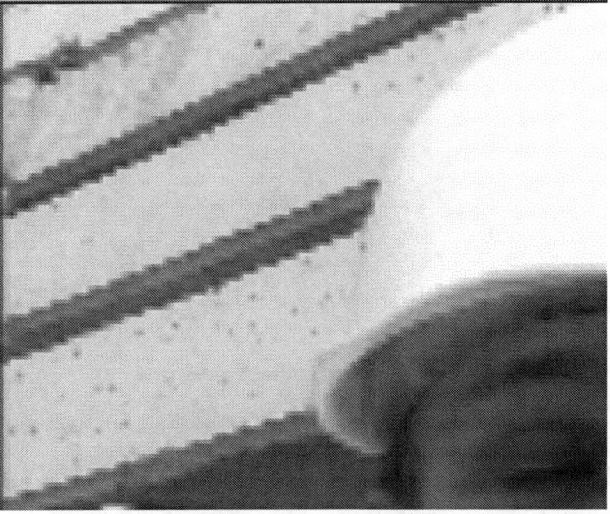
Fig.8 Part of enlarged "foreman" image using bilinear interpolation

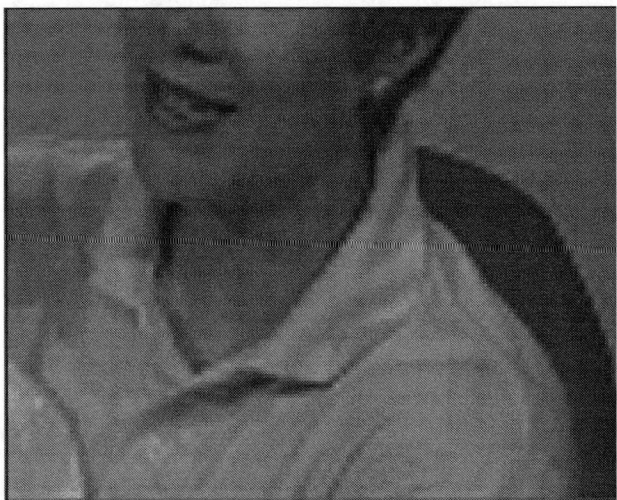
Fig.6 Part of enlarged "mother" image using bicubic interpolation

Fig.9 Part of enlarged "foreman" image using bicubic interpolation

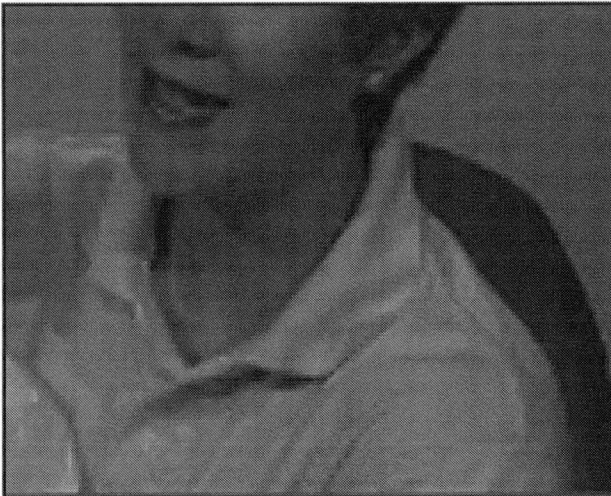
Fig.7 Part of enlarged "mother" image using proposed interpolation

Fig.10 Part of enlarged "foreman" image using proposed interpolation

Arbitrary Scale Image Enlargement with the Prediction of High Frequency Components

Shuai Yuan*, Akira Taguchi† and Masayuki Kawamata*

*Department of Electronic Engineering, Tohoku University, Sendai 980-8579, Japan
†Department of Electrical and Electronic Engineering, Musashi Institute of Technology, Tokyo 158-8557, Japan
Email: yuan@mk.ecei.tohoku.ac.jp; ataguchi@eng.musashi-tech.ac.jp; kawamata@ecei.tohoku.ac.jp

Abstract—In this paper, we propose an arbitrary scale enlargement method of digital images based on the Laplacian pyramid (LP) representation. Through the calculation of the arbitrary scale LP stages we obtain the ability of arbitrary scale enlargement, which solves the problem that the conventional LP enlargement methods are only capable of expanding an image up by a factor of two in size. Experimental results show that the proposed method can work effectively on arbitrary scale image enlargement.

I. INTRODUCTION

Enlargement is a prime technique in image processing. It is used in many important applications such as digital high-definition television, big screen display, copy and print machine, medical imaging and so on. Some typical enlargement methods, such as the bilinear, cubic and bicubic methods [1], are used widely.

However, the problem of typical enlargement methods is that the enlarged images will appear blurred because the high frequency components of the enlarged images are not enough. For solving this problem, some resolution enhancement (RE) methods were proposed. They can predict or find the unknown high frequency components and then to obtain high-resolution enlarged images. Most of the studies done in the resolution enhancement are from a group of low-resolution image frames to extract one high-resolution image, and they are called multi-image resolution enhancement (MIRE) methods [2]-[5]. MIRE methods often require a very heavy computational load. Moreover, they have not considered the condition that only one low-resolution image is supplied. Therefore, the resolution enhancement methods based on one low-resolution image were also proposed, and they are called one-image resolution enhancement (OIRE) methods [6]-[12]. Ref. [6] uses the DCT iteration method for RE enlargement, and [7] uses the orthogonal wavelet transform method for RE enlargement. OIRE methods [8]-[11] are based on the Laplacian pyramid (LP) representation. They are called LP methods. Utilizing the characteristic that the LP stages at different resolutions have the similar structure in the neighborhoods of the zero-crossing point, the LP methods can predict the unknown high frequency components to enhance the resolution of low-resolution images. Unfortunately, the LP methods of [8]-[11] are limited to the integer LP stages calculation, and thus they are only capable of expanding an image up by a factor of two in size (so called "zoom in"). In fact, we also need the enlargement scales as 1.41 times, 2.15 times and so on for many enlargement applications. Thus, in this paper we propose a new LP method that uses the calculation of the arbitrary scale stages of the Laplacian pyramid to obtain arbitrary scale enlargement ability. The proposed LP method has two parameters in the prediction of the unknown high frequency components. The parameters are evaluated via the response of the Laplacian pyramid to the unit step signal.

This paper is organized as follows. The Gaussian- and Laplacian-pyramid representations are described in Section II.A. The "zoom in" LP method is briefly reviewed in Section II.B. The proposed method is presented in Section II.C. In Sections III, the parameter evaluation is given. In Section IV, we illustrate the experimental results and compare the proposed method with the typical enlargement methods to show the effectiveness of the proposed method. Finally, the conclusion is included in Section V.

II. ARBITRARY SCALE ENLARGEMENT METHOD BASED ON THE LAPLACIAN PYRAMID REPRESENTATION

A. Gaussian- and Laplacian-Pyramid Representations

The Gaussian- and the Laplacian-pyramid representations, which were described by Burt and Adelson in [12], are one of the multi-resolution frameworks.

We refer to the Gaussian pyramid stages as $G_0, G_1, G_2,..., G_n$, in which the stage G_0 shows the original input image. The upward low-resolution stage G_{n+1} is given by

$$G_{n+1} = \text{DownSample}(\overline{G}_{n+1}) \qquad (1)$$
$$\overline{G}_{n+1} = W_{half} * G_n \qquad (2)$$

where $*$ is the convolution operator and W_{half} is the half-band Gaussian filter. The size of G_{n+1} is half of the size of G_n.

The Laplacian pyramid stages are composed of high-pass filtered versions (high frequency components) of the corresponding Gaussian pyramid stages. The stages $L_0, L_1, L_2,..., L_n$ of the Laplacian pyramid are constructed by

$$L_n = G_n - Expand(G_{n+1}) \qquad (3)$$
$$Expand(G_{n+1}) = W_{half} * G_{n+1}^0 \qquad (4)$$
$$G_{n+1}^0 = \text{UpSample}(G_{n+1}). \qquad (5)$$

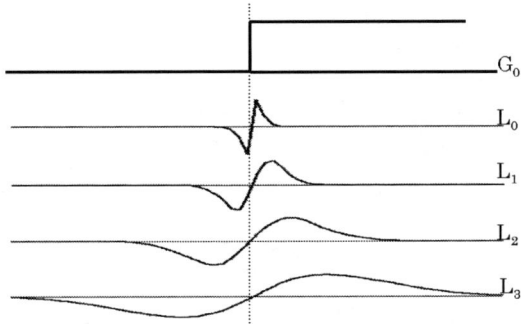

Fig. 1. Laplacian pyramid stages of a step signal.

The up-sampled version G_{n+1}^0 is obtained by interpolating zeros into G_{n+1}.

If we regard the Gaussian pyramid stage G_0 as a step signal, the forms of Laplacian pyramid stages are shown in Fig. 1.

B. "Zoom in" Method Based on the Laplacian Pyramid Representation

From the image pyramid algorithm we know that the size of G_{-1} is 2 times of the size of G_0. In (3) when $n = -1$, we have

$$G_{-1} = L_{-1} + Expand(G_0). \quad (6)$$

$Expand(G_0)$, which is the Gaussian interpolated version of G_0, has almost the same frequency components as G_0. $Expand(G_0)$ is calculated from (4) and (5), when $n = -1$. The stage L_{-1} is the unknown LP downward stage, of which frequency components are higher than the frequency components of the stage L_0.

From Fig. 1 we can find that the LP stages of a step signal have the similar structure, and the difference is only that the slope is different in the neighborhoods of the zero-crossing point. Therefore, we can predict the high frequency stage L_{-1} by changing the slope of the stage L_0 near the zero-crossing point [8]-[11]. However, [8]-[11] are limited to the representation of the integer pyramid stages and only have the ability of expanding an image up by a factor of two in size. In many applications we also need arbitrary enlargement scales, such as 1.41 times and so on. Thus, in next section we present a new enlargement method for arbitrary scale enlargement based on the Laplacian pyramid representation.

C. Proposed Arbitrary Scale LP Enlargement Method

We know that the relationship of the image pyramid representations exists not only among the integer stages but also among the non-integer stages. Then, we denote the stage index of an arbitrary scale pyramid stage as r. When $r < 0$, it means that the pyramid stages are high-resolution stages. In the proposed enlargement algorithm, we always have $r < 0$.

In (6) we can replace -1 with r, and then we have an enlarged high-resolution image of arbitrary scale as

$$G_r = L_r + Expand_r(G_0) \quad (r < 0), \quad (7)$$

in which G_r is 2^{-r} times high-resolution image of G_0. We denote the enlargement scale as S, and thus we have $S = 2^{-r}$. The image $Expand_r(G_0)$, which should keep the same frequency components as G_0, is the Gaussian interpolated version of G_0. The size of $Expand_r(G_0)$ is S times of the size of G_0. In the following, we use one-dimensional signal form to formulate the proposed algorithm.

At first, we define the Gaussian function as $W(x)$. Then, we have

$$W(x) = \frac{1}{2\sqrt{\rho\pi}} e^{-\frac{x^2}{4\rho}} \quad (8)$$

$$\rho = \frac{\sigma^2}{2}. \quad (9)$$

Since the half-band Gaussian filter has $\rho = 16/(9\pi)$, we would like to use the Gaussian filter W_{all}, which has $\rho = 4/(9\pi)$, to finish the Gaussian interpolation of G_0. Through this interpolation, we can approximately retain all of the G_0 frequency components in $Expand_r(G_0)$.

We then denote the pixel values of G_0 and $Expand_r(G_0)$ as $g_0(i)$ ($i = 1, 2, 3, ..., M$) and $\tilde{g}_r(j)$ ($j = 1, 2, 3, ..., (M \times S)$). We know that the size of $Expand_r(G_0)$ is S times of the size of G_0, and thus we can give $\tilde{g}_r(j)$ as

$$\tilde{g}_r(j) = \sum_{m=-2}^{2} W_{all}(m) \cdot g_0(\text{int}(\frac{j}{S}) + m) \quad (10)$$

$$W_{all}(m) = \frac{3}{4} e^{-\frac{9\pi}{16}(\triangle k + m)^2} \quad (11)$$

$$\triangle k = \text{int}(\frac{j}{S}) - (\frac{j}{S}). \quad (12)$$

The function $\text{int}(\bullet)$ takes only integer part.

The Laplacian pyramid stage L_0 is obtained from (3), when $n = 0$. The Gaussian interpolation for $Expand_r(L_0)$ is the same as the Gaussian interpolation for $Expand_r(G_0)$. Then, we have

$$\tilde{l}_r(j) = \sum_{m=-2}^{2} W_{all}(m) \cdot l_0(\text{int}(\frac{j}{S}) + m) \quad (13)$$

where $\tilde{l}_r(j)$ is the pixel value of $Expand_r(L_0)$ and $l_0(i)$ is the pixel value of L_0.

From the LP representation of a step signal in Fig. 1, we have known that changing the slope of known LP stage near zero crossing point can predict unknown LP stages. Therefore, we can give the equations for the prediction of L_r as

$$l_r(j) = \alpha_r \times \begin{cases} T & \text{if } \tilde{l}_r(j) > T \\ \tilde{l}_r(j) & \text{if } T \geq \tilde{l}_r(j) \geq -T \\ -T & \text{if } \tilde{l}_r(j) < -T \end{cases} \quad (14)$$

$$T = (1 - c_r) \times \tilde{l}_r(j)_{max}, \quad (15)$$

where $\tilde{l}_r(j)$ is the pixel value of $Expand_r(L_0)$, $\tilde{l}_r(j)_{max}$ is the maximum pixel value of $Expand_r(L_0)$ and $l_r(j)$ is the pixel value of the predicted L_r. The parameter T is a threshold, which is decided by the clipping parameter c_r. The parameter α_r is a constant, which fits the slope of $Expand_r(L_0)$ to the slope of L_r. Both of them should be pre-determined. In the next section, we present how to evaluate α_r and c_r.

III. PARAMETER EVALUATION

We have known that a step signal can show the relationship of the LP representation clearly. Therefore, the unit step signal is used as an ideal input for our evaluation. In [8] a similar thought was presented, but it is limited to the 2 times enlargement scale. In this section, we expand the evaluation of [8] to arbitrary enlargement scales.

It is known that the PSF (Point Spread Function), which is the optical system function, is the Gaussian function with standard deviation σ_0 (normally $\sigma_0 = 0.9$). Then, we have the equation of the PSF as

$$PSF = \frac{1}{\sqrt{2\pi}\sigma_0} e^{-\frac{x^2}{2\sigma_0^2}}.$$

The response of the PSF to the unit step signal is the convolution of the PSF and the unit step signal. It is equal to the integral probability distribution of the PSF from $-\infty$ to x. We denote the unit step signal as $U(x)$, and the convolution result $CR[U(x)]$ is given as

$$CR[U(x)] = \int_{-\infty}^{x} \frac{1}{\sqrt{2\pi}\sigma_0} e^{-\frac{x^2}{2\sigma_0^2}} dx.$$

We regard this response as the Gaussian pyramid stage G_0 of $U(x)$. Thus, we have

$$G_0[U(x)] = \int_{-\infty}^{x} \frac{1}{\sqrt{2\pi}\sigma_0} e^{-\frac{x^2}{2\sigma_0^2}} dx.$$

If we sample $G_0[U(x)]$ with distance 1, the Gaussian pyramid stage G_r, of which resolution is S times higher than the resolution of G_0, should be sampled with distance $1/S$. Thus, $G_r[U(x)]$ is equal to the integral probability distribution of the Gaussian function from $-\infty$ to x with $\sigma_r = \sigma_0/S$. Then, we have

$$G_r[U(x)] = \int_{-\infty}^{x} \frac{1}{\sqrt{2\pi}\sigma_r} e^{-\frac{x^2}{2\sigma_r^2}} dx.$$

In the algorithm of the Gaussian pyramid representation, the half-band Gaussian filter W_{half} is an approximation of the normalized Gaussian with $\sigma_w = 1.0$. The Gaussian pyramid stage G_1 is given by

$$G_1[U(x)] = \int_{-\infty}^{x} \frac{1}{\sqrt{2\pi}\sigma_1} e^{-\frac{x^2}{2\sigma_1^2}} dx.$$

where $\sigma_1 = \sqrt{\sigma_0^2 + \sigma_w^2}$. Moreover, we have the Laplacian pyramid stages $L_0[U(x)]$ and $L_r[U(x)]$ as

$$L_0[U(x)] = G_0[U(x)] - G_1[U(x)]$$

and

$$L_r[U(x)] = G_r[U(x)] - G_0[U(x)],$$

respectively.

Since the parameter α_r fits the slope of L_0 to the slope of L_r near the zero-crossing point, at the zero crossing point ($x = 0$) we have

$$\alpha_r \times \dot{L}_0[U(x)] = \dot{L}_r[U(x)], \quad (16)$$

where

$$\dot{L}_i[U(x)] = \frac{dL_i[U(x)]}{dx}.$$

The predicted L_r should have the same maximum value as the ideal L_r. We also have

$$\alpha_r \times (1 - c_r) \times L_0[U(x_{0max})] = L_r[U(x_{rmax})]. \quad (17)$$

The positions x_{0max} and x_{rmax} are the L_0 maximum value position and the L_r maximum value position, respectively.

From (16) and (17), we then obtain the two parameters in our algorithm as

$$\alpha_r = 3.022 \times (S - 1)$$

$$c_r = 1 - \frac{L_r[U(x_{rmax})]}{\alpha_r \times L_0[U(x_{0max})]},$$

where x_{0max} and x_{rmax} are calculated from

$$x_{0max} = \sqrt{\frac{2\log(\sigma_1/\sigma_0)}{1/\sigma_0^2 - 1/\sigma_1^2}}$$

and

$$x_{rmax} = \sqrt{\frac{2\log(\sigma_0/\sigma_r)}{1/\sigma_r^2 - 1/\sigma_0^2}},$$

respectively.

IV. EXPERIMENTAL RESULTS AND COMPARISON

We give some enlarged results at 1.41 times enlargement scale using the bilinear method, the bicubic method and the proposed method. For doing the comparison between the proposed method and the typical enlargement methods, we first use the Gaussian filter to low-pass the existing Lenna image (256×256). Then, we down-sample the low-passed Lenna image to 1/1.41 times of its original size. Next, we utilize the proposed method, the bilinear method and the bicubic method to enlarge the down-sampled Lenna image to its original size. Finally, we compare the enlarged Lenna images with the original Lenna image to get the MSE values. The smaller MSE value means that the enlarged result has the higher accuracy.

The enlarged results are shown in Fig. 2. In Fig. 2, the enlarged result using the proposed method is clearer than the enlarged results using the bilinear and bicubic methods. From the MSE values of the 3 enlarged results, we can also see that the enlarged result using the proposed method has the highest enlarged accuracy.

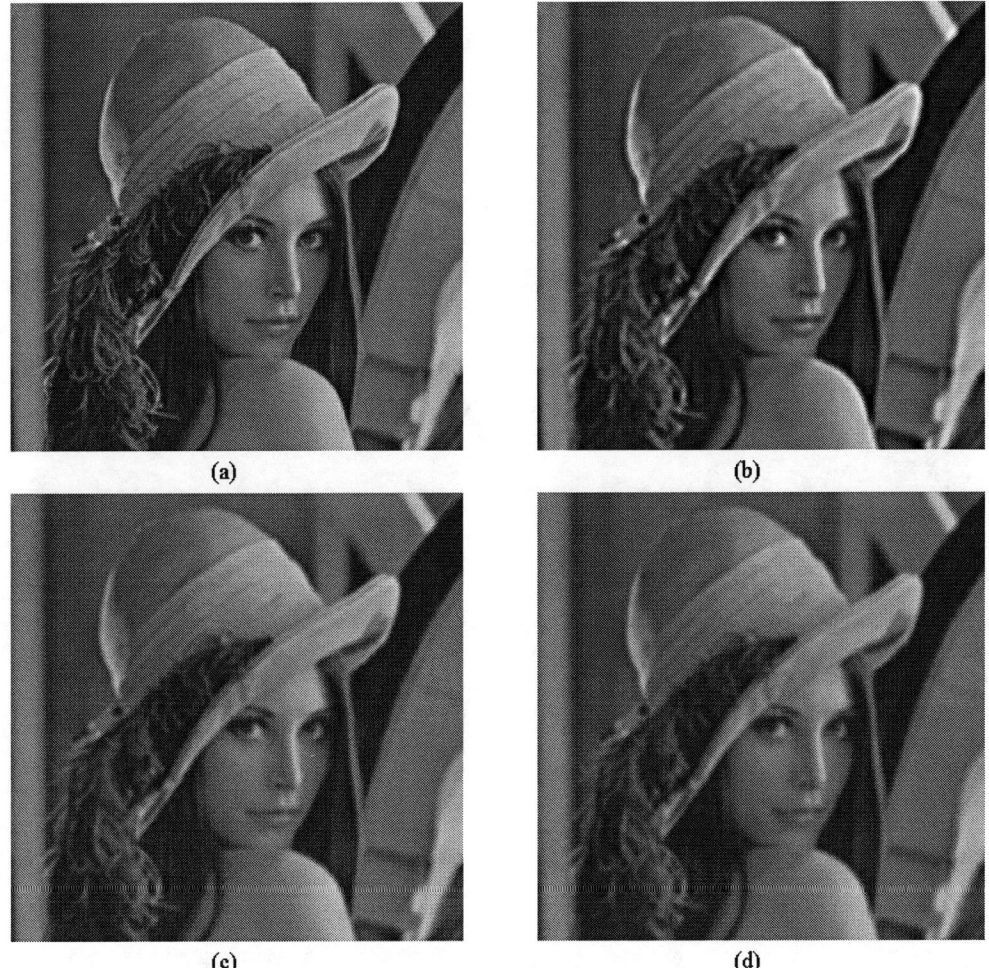

Fig. 2. Experimental results on 1.41 times enlargement scale. (a) Original Lenna image. (b) The proposed method result (MSE: 73.81). (c) The bicubic method result (MSE: 77.56). (d) The bilinear method result (MSE: 92.46).

V. CONCLUSION

The proposed method for image enlargement of arbitrary scale utilizes only one low-resolution image and its Laplacian pyramid representation to obtain the high-resolution enlarged image. Through the calculation of arbitrary scale stages of the image pyramids, we realize the ability of arbitrary scale enlargement. To evaluate the two parameters of the proposed algorithm, we analyze the responses of the unit step signal in the image pyramid representations of arbitrary scale.

REFERENCES

[1] W. K. Pratt, *Digital Image Processing*, New York: Wiley, 1991.
[2] S. P. Kim and W. Y. Su, "Recursive high resolution reconstruction of blurred multiframe images," *IEEE Trans. on Image Processing*, vol. 2, pp. 534-539, October 1993.
[3] R. Hardie, K. Barnard. and E. Armstrong, "Joint map registration and high resolution image estimation using a sequence of undersampled measured images," *IEEE Trans. on Image Processing*, vol. 6, pp. 1621-1632, December 1997.
[4] N. R. Shah and A. Zakhor, "Resolution enhancement of color video sequences," *IEEE Trans. on Image Processing*, vol. 8, no. 6, pp. 879-885, June 1999.
[5] B. K. Gunturk, Y. Altunbasak and R. M. Mersereau, "Super-resolution reconstruction of compressed video using transform-domain statistics," *IEEE Trans. on Image Processing*, vol. 13, no. 1, pp. 33-43, January 2004.
[6] R. W. Gerchberg, "Super-resolution through error energy reduction," *Opt, Acta*, vol. 21, no. 9, pp. 709-720, 1974
[7] S. Mallat, "A theory for multiresolution signal decomposition: The wavelet representation," *IEEE Trans. Pattern Anal. and Mach. Intell.*, vol. 11, no. 7, pp. 674-693, 1989.
[8] H. Greenspan, C. H. Anderson and S. Akber, "Image enhancement by nonlinear extrapolation in frequency space," *IEEE Trans. on Image Processing*, vol. 9, no. 6, pp. 1035-1048, June 2000.
[9] Y. Takahashi and A. Taguchi, "An enlargement method of digital image based on Laplacian pyramid representation," *SPIE vol. 3961: Nonlinear Image Processing XI*, pp. 163-169, 2000.
[10] D. Sekiwa, A. Taguchi, and Y. Murata, "Enlargement of digital image based on the Laplacian pyramid by using neural networks," *Electronics and Communications in Japan, PART 3, Script a Technical*, vol. 82, no. 1, pp. 1499-1508, January 1999.
[11] H. Greenpan and C. H. Anderson, "Image enhancement by non-linear extrapolation in frequency space," *SPIE vol. 2182: Image and Video Processing II*, pp. 1499-1508, September 1994.
[12] P. J. Burt and E. A. Andelson, "The Laplacian pyramid as a compact image code," *IEEE Trans. on Communications*, vol. 31, no. 4, pp. 532-540, April 1983.

// # Restoration from Image Degraded by White Noise Based on Iterative Spectral Subtraction Method

Tetsuya KOBAYASHI*, Tetsuya SHIMAMURA*, Tetsuo HOSOYA† and Yoshitake TAKAHASHI†
*Graduate School of Science and Engineering, Saitama University
Shimo-Okubo 255, Sakura-ku, Saitama, 338-8570, Japan
Email: {koba,shima}@sie.ics.saitama-u.ac.jp
†Daiichi Radioisotope Laboratories Ltd
Kyobashi 1-17-10, Chuo-ku, Tokyo, 104-0031, Japan
Email: {htetsuo,tyoshitake}@drl.co.jp

Abstract— In this paper, an application of the spectral subtraction method to restoration of image is investigated. Performance of the spectral subtraction method is compared with that of the Wiener filter in an ideal case, resulting in the spectral subtraction method providing an improvement. In order to improve further the performance of the spectral subtraction method, an iterative algorithm is utilized with a frequency-division based noise estimation technique.

I. INTRODUCTION

It is known that in the case where an image is degraded by white noise, the Wiener filter is more suitable for restoration than nonlinear filters such as the median filter and order-statistic filter [1]. The performance of the Wiener filter is, however, dependent on the image characteristics. Recently, division processing methods for the Wiener filter were addressed in [2] and [3]. In both papers, it was shown that a good performance is obtained by relying on division in the time or frequency domain.

On the other hand, in the area of speech processing, the spectral subtraction method has been successfully utilized for reducing the additive noise [4]. This motivates, in this paper, the use of the spectral subtraction method for the restoration from an image degraded by white noise.

In this paper, first the spectral subtraction method is extended from one-dimensional processing to two-dimensional processing. The two-dimensional spectral subtraction method was originally discussed for image processing in [5] and [6]. However, in this paper, a different type of the spectral subtraction method is considered. And then, the performance of the spectral subtraction method is investigated in an ideal case where the noise image is known. A noise estimation method is further considered and an iterative version of the spectral subtraction method is developed to improve the performance.

II. SPECTRAL SUBTRACTION

The model of an image degraded by white noise is expressed as

$$s(n_1, n_2) = x(n_1, n_2) + n(n_1, n_2) \tag{1}$$

where $s(n_1,n_2)$, $x(n_1,n_2)$ and $n(n_1,n_2)$ are the degraded image, original image and white noise, respectively. In the frequency domain, the degraded image in (1) is expressed as

$$S(\omega_1, \omega_2) = X(\omega_1, \omega_2) + N(\omega_1, \omega_2) \tag{2}$$

where $S(\omega_1,\omega_2)$, $X(\omega_1,\omega_2)$ and $N(\omega_1,\omega_2)$ are the Fourier transforms of $s(n_1,n_2)$, $x(n_1,n_2)$ and $n(n_1,n_2)$, respectively. From (2), the amplitude spectral subtraction method can be derived as

$$\hat{X}(\omega_1,\omega_2) = \begin{cases} (|S(\omega_1,\omega_2)| - \alpha|\hat{N}(\omega_1,\omega_2)|)e^{j\arg S(\omega_1,\omega_2)}, \\ \quad \cdots \alpha|\hat{N}(\omega_1,\omega_2)| \leq |S(\omega_1,\omega_2)| \\ 0, \\ \quad \cdots otherwise \end{cases} \tag{3}$$

where α is a parameter that controls the level of noise reduction and $|\hat{N}(\omega_1,\omega_2)|$ means an estimate of the noise amplitude spectrum. The resulting image, $\hat{x}(n_1,n_2)$, is obtained by the inverse Fourier transform of $\hat{X}(\omega_1,\omega_2)$.

III. RESTORATION RESULTS IN IDEAL CASE

We investigated first the performance of the spectral subtraction method in an ideal case where the noise image is known a priori. Under the same condition, the Median filter, Wiener filter and Averaging filter were implemented and compared. We used 8 images in SIDBA, each of which has a size of 256×256 with 256 gray scales. A white noise was generated and added to each image, resulting in the preparation of 0[dB] and 10[dB] noisy images for each image.

The results by each processing to the image degraded by white noise are shown in Tables I and II. In the spectral subtraction method, the noise amplitude spectrum was calculated from the true noise image and it was used as the estimate of the noise amplitude spectrum. The parameter α was set to 1.0. In the Wiener filter, the window size was set to 5×5. In the Median and Averaging filters, the window size was commonly set to 3×3. The SNR improvement in [dB] is defined as

$$10\log_{10}\frac{NMSE[x(n_1,n_2), s(n_1,n_2)]}{NMSE[x(n_1,n_2), \hat{x}(n_1,n_2)]} \tag{4}$$

TABLE I
SNR IMPROVEMENT (IDEAL CASE, 0[DB]).

Image	Spectral	Wiener	Median	Average
Airplane	7.931	7.503	5.690	7.095
Boat	9.588	7.792	6.416	7.495
Building	8.661	8.121	6.531	7.737
Cameraman	7.998	7.811	6.069	7.174
Girl	9.550	9.485	6.740	8.435
Lenna	9.671	8.958	6.631	8.187
Lighthouse	7.462	7.001	5.387	6.360
Woman	9.293	8.567	6.599	7.753

TABLE II
SNR IMPROVEMENT (IDEAL CASE, 10[DB]).

Image	Spectral	Wiener	Median	Average
Airplane	3.849	3.554	1.849	1.164
Boat	4.723	4.046	1.866	1.141
Building	4.850	4.566	3.740	3.507
Cameraman	4.264	3.913	2.580	2.272
Girl	5.509	5.168	4.577	4.902
Lenna	5.093	4.384	3.955	3.728
Lighthouse	2.967	2.945	0.107	-0.106
Woman	4.176	3.814	2.732	2.085

where

$$NMSE[x(n_1,n_2), s(n_1,n_2)] = 100 \times \frac{Var[x(n_1,n_2) - s(n_1,n_2)]}{Var[x(n_1,n_2)]} \quad (5)$$

$$NMSE[x(n_1,n_2), \hat{x}(n_1,n_2)] = 100 \times \frac{Var[x(n_1,n_2) - \hat{x}(n_1,n_2)]}{Var[x(n_1,n_2)]}. \quad (6)$$

From Tables I and II, it is observed that the spectral subtraction method is most effective among the 4 methods.

IV. PROPOSE METHOD

We cannot obtain the true noise amplitude spectrum directly. Thus, a method for noise amplitude spectrum estimation is proposed here. And, in order to improve the performance of the spectral subtraction method, its iterative version is proposed.

A. Noise Estimation by Frequency Band Division Processing

In general, the power spectrum of an image is concentrated in a low frequency region, while the power spectrum of white noise exists in all frequency regions. Therefore, by restoring a high frequency region of the degraded image, the noise image should be obtained. This is because the noise components are dominant in a high frequency region of the degraded image. Based on this principle, we derive the following noise estimation method, which is a kind of frequency band division processing.

(Step.1) We calculate $X(\omega_1, \omega_2)$ by the Fourier transform of the input image $x(n_1, n_2)$ and further calculate its logarithmic power spectrum as $P(\omega_1, \omega_2) = \log|X(\omega_1, \omega_2)|^2$.

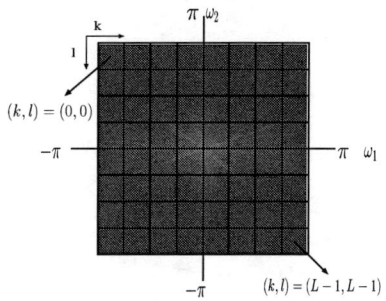

Fig. 1. Blocks in the frequency domain ($L = 8$).

(Step.2) Dividing $P(\omega_1, \omega_2)$ into $L \times L$ blocks as shown in Figure 1. And the average of each logarithmic power spectrum $P_{(k,l)}(\omega_1, \omega_2)$, $\bar{P}_{(k,l)}(\omega_1, \omega_2)$, is calculated for each block. $P_{(k,l)}(\omega_1, \omega_2)$ corresponds to the (k,l)-th block of $P(\omega_1, \omega_2)$.

(Step.3) A threshold decision is made for each $\bar{P}_{(k,l)}(\omega_1, \omega_2)$, and $X(\omega_1, \omega_2)$ is divided into its low and high frequency counterparts $X_{low}(\omega_1, \omega_2)$ and $X_{high}(\omega_1, \omega_2)$ as

$$X_{low}(\omega_1, \omega_2) = \begin{cases} X_{(k,l)}(\omega_1, \omega_2) & \bar{P}_{(k,l)}(\omega_1, \omega_2) \succ TH \\ 0 & otherwise \end{cases} \quad (7)$$

$$X_{high}(\omega_1, \omega_2) = \begin{cases} X_{(k,l)}(\omega_1, \omega_2) & \bar{P}_{(k,l)}(\omega_1, \omega_2) \leq TH \\ 0 & otherwise \end{cases} \quad (8)$$

$$TH = (P_{max} - P_{min})/100 * p + P_{min} \quad (9)$$

where P_{max}, P_{min} and p are the maximum and minimum values of $P(\omega_1, \omega_2)$, and the division ratio, respectively. $X_{(k,l)}(\omega_1, \omega_2)$ corresponds to $X(\omega_1, \omega_2)$ of the (k,l) block.

(Step.4) The noise image $n(n_1, n_2)$ is obtained by the inverse Fourier transform of $X_{high}(\omega_1, \omega_2)$. However, a negative value is set to 0 when $n(n_1, n_2)$ contains negative values.

B. Optimization of Parameters

Important parameters used for noise estimation by the frequency band division processing are L and p. We consider how the two parameters influence SNR improvement.

First, the relation between L and SNR improvement is examined. For images of SNR 0 and 10[dB], the relations are shown in Figures 2 (a) and (b), respectively. The division rate p has been chosen manually as the best value so that SNR becomes maximum. In Figures 2 (a) and (b), SNR is improved as the size of L becomes large. However, there are images which are degraded when $L = 64$ is set. Since the amount of calculations increases as the size of L becomes large, the parameter L is settled as 32 for below simulations.

Next, the relation between the division rate p and SNR improvement is examined. For images of SNR 0 and 10[dB], the relations are shown in Figures 5 (a) and (b), respectively. From Figures 3 (a) and (b), for values near the optimal value of the division rate p, there is no great difference in restoration accuracy. Since the optimal value of p often becomes around

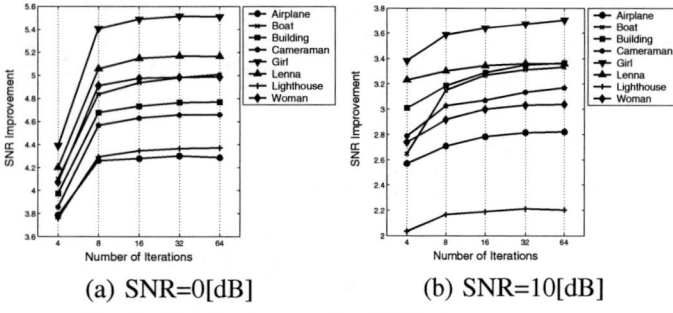

(a) SNR=0[dB] (b) SNR=10[dB]

Fig. 2. Relation between L and SNR improvement.

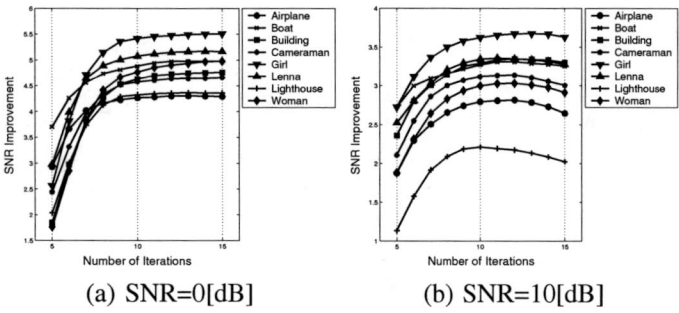

(a) SNR=0[dB] (b) SNR=10[dB]

Fig. 3. Relation between α and SNR improvement.

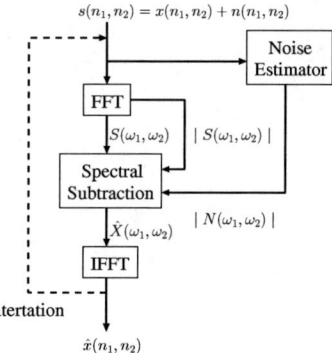

Fig. 4. Block diagram of the iterative spectral subtraction method.

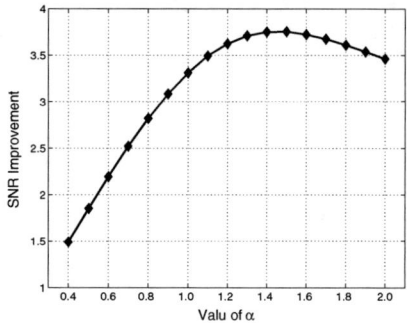

Fig. 5. Relation between α and SNR improvement.

12 in many images, the division rate p is settled as 12 for below simulations.

C. Iterative Spectral Subtraction

In order to improve the performance of the spectral subtraction method, its iterative version is proposed. The idea of the iterative spectral subtraction method is to repeat the spectral subtraction method by utilizing the image obtained by the spectrum subtraction method as the input image iteratively. If the number of iterations is t, then the result obtained by the t-th spectral subtraction method is expressed in the time domain as

$$\hat{x}_t(n_1, n_2) = s_t(n_1, n_2) - \hat{n}_t(n_1, n_2). \tag{10}$$

In this case, the input image of the $t+1$-th spectral subtraction method is expressed as

$$s_{t+1}(n_1, n_2) = \hat{x}_t(n_1, n_2). \tag{11}$$

Figure 4 shows a block diagram of the iterative spectral subtraction method.

D. Relations between α, Number of Iterations and SNR Improvement

The performance of the iterative spectral subtraction method is investigated. First, the relation between α and SNR improvement is examined. When α is set to a large integer, the amounts of noise to be removed is much. For the Boat image of SNR 10[dB], the relation is shown in Figure 5. In Figure 5, it is observed that basically the restoration accuracy is improved as α becomes large. We see, however, that the SNR improvement starts to deteriorate at a certain value of α.

Next, the relation between number of iterations and SNR improvement is examined. For the Boat image of SNR 10[dB], the relation is shown in Figures 6. In Figure 6, it is observed that basically the SNR improvement is improved when the spectral subtraction method is repeated. We see, however, that the SNR improvement starts to deteriorate at a certain number of iterations.

Next, the effect of the two parameters, α and number of iterations, on SNR improvement are examined. In the Boat image of SNR 10[dB], the effect on SNR improvement is shown in Table III. Setting of α and number of iterations which provides the best restoration accuracy is found in Table III. In order to improve the SNR improvement, it is effective to increase the number of iterations and to decrease the value of α. In order to determine the optimal parameter for image restoration, we restored all the prepared degraded images with various parameter sets. As a result, $\alpha = 0.4$ and number of iterations = 6 in SNR 0[dB] (=12 in SNR 10[dB]) brought a good result on the average. The number of iterations should be changed depending on the SNR like the above result. This is because the amount of noises increases as the SNR becomes low, which requires more iterations of spectral subtraction to reduce enough the additive noise.

E. Estimation of SNR of Degraded Image

In order to perform the iteration spectral subtracting method, it is necessary to decide the number of iterations. As men-

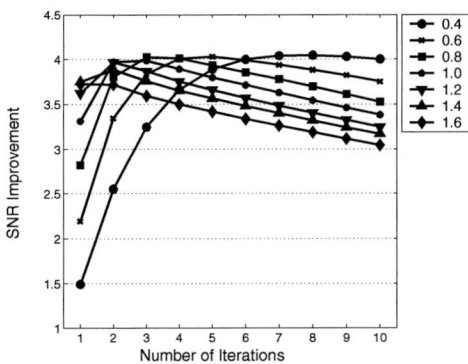

Fig. 6. SNR improvement for varying α.

TABLE III
NUMBER OF ITERATIONS AND α FOR THE BOAT IMAGE OF SNR 10[DB].

α	Number of Iterations	SNR Improvement
$\alpha = 0.4$	8	4.043
$\alpha = 0.6$	5	4.029
$\alpha = 0.8$	3	4.024
$\alpha = 1.0$	3	3.985
$\alpha = 1.2$	2	3.967
$\alpha = 1.4$	2	3.886
$\alpha = 1.6$	1	3.723

tioned above, the number of iterations depends on the SNR of image. However, in practice, we may have no knowledge about the SNR of the degraded image. In this case, we have to estimate the SNR. To do this, we found from the simulation experiment the fact that the SNR of a degraded image can be estimated from the degraded image as

$$SNR[dB] \doteq 10 \log_{10} \frac{Var[x_{low}^1(n_1, n_2)]}{Var[x_{high}^1(n_1, n_2)]} - 5 \quad (12)$$

where $x_{low}^1(n_1, n_2)$ and $x_{high}^1(n_1, n_2)$ are the low and high frequency images to be obtained by the first band division processing for the iterative method, respectively. The estimated result is shown by using images whose SNR is known in Table IV. Table IV shows that the estimated SNRs are very close to the original SNR. This means that (12) is effective to estimate the SNR, and based on the estimated result, we can determine the number of iterations.

V. RESTORATION RESULTS BY ITERATIVE SPECTRAL SUBTRACTION

The iterative spectral subtraction method was performed by using the noise estimation based on frequency band division processing, and all degraded images were restored. The parameters of the iterative spectral subtraction method used in the simulation were $\alpha = 0.4, L = 32$ and $p = 12$. The number of iterations was settled as 6 for the images of SNR 10[dB], while that was settled as 12 for the images of SNR 0[dB]. The resulting SNR improvement in the simulation is shown in Table V. Table V suggests that the degraded image of SNR 0[dB] and that of SNR 10[dB] were restored for 80

TABLE IV
SNR ESTIMATION.

Image	0[dB]	10[dB]
Airplane	0.654	10.100
Boat	-0.070	10.761
Building	-0.095	10.479
Cameraman	-0.383	9.810
Girl	-0.265	10.101
Lenna	-0.146	9.958
Lighthouse	-0.598	9.877
Woman	0.132	9.982

TABLE V
SNR IMPROVEMENT.

SNR [dB]	image	ideal	real	$real/ideal$
0	Airplane	7.931	7.031	0.886
	Boat	9.588	8.205	0.856
	Building	8.661	7.639	0.882
	Cameraman	7.998	6.332	0.792
	Girl	9.550	8.498	0.890
	Lenna	9.671	8.711	0.901
	Lighthouse	7.462	6.392	0.857
	Woman	9.293	8.103	0.872
10	Airplane	3.849	3.029	0.787
	Boat	4.723	3.999	0.847
	Building	4.850	4.061	0.837
	Cameraman	4.264	3.417	0.801
	Girl	5.509	4.409	0.800
	Lenna	5.093	3.917	0.769
	Lighthouse	2.967	2.048	0.690
	Woman	4.176	3.395	0.813

% and 86 % (on the average) relative to those in an ideal case respectively.

VI. CONCLUSION

In this paper, we investigated the method of spectral subtraction for restoration from an image degraded by white noise. In an ideal case, it was found that the spectral subtraction method is more effective. Then, in order to apply it in real cases, we proposed the noise estimation method based on frequency band division processing and the iterative spectral subtraction method, and investigated their combination. As a result, without relying on the use of true image, the degraded image of SNR 0[dB] and that of SNR 10[dB] were restored for 80 % and 86 % (on the average) relative to those in an ideal case respectively.

REFERENCES

[1] M.Muneyasu and A.Taguchi, "Nonlinear Digital Signal Processing", Asakura Pub., 2002.
[2] R.Nishimiya and A.Taguchi, "Image restoration by using multiple Wiener filters", IEICE Trans. Part A, Vol.J83-A, No.7, pp.892-902, 2000.
[3] A.Kakimoto and A.Taguchi, "A novel noisy image restoration method based on the frequency region", IEICE Trans. Part A, Vol.J85-A, No.7, pp.780-792, 2002.
[4] S. F. Boll, "Suppression of acoustic noise in speech using spectral subtraction", IEEE Trans. Acoustics, Speech and Signal Processing, Vol.ASSP-27, No.4, pp.113-120, 1979.
[5] J.S.Lim, "Image restoration by short space spectral subtraction", IEEE Trans. Acoustics, Speech and Signal Processing, Vol.ASSP-28, No.2, pp.191-197, 1980.
[6] J.S.Lim, "Two-Dimensional Signal and Image Processing", Prentice-Hall, 1990.

Compensation of Errors Generated by an Analog 2-D DCT

Kati Virtanen*, Mikko Pänkäälä*[†], Ari Paasio*
*University of Turku, Department of Information Technology, Microelectronics Laboratory
Lemminkäisenkatu 14-18 B, FIN-20014, Turku, Finland, Email: kapvirt@utu.fi
[†]Turku Centre for Computer Science, TUCS

Abstract— This paper proposes an error compensation method for cancelling the errors introduced in analog discrete cosine transform. Despite the simplicity of the method, significant improvements in image quality can be achieved. The operation was simulated with Matlab using standard test images and the results were examined with both visual and objective measures.

I. INTRODUCTION

The Discrete Cosine Transform (DCT) is a common transform used for example in video and image compression. Many video compression standards, such as MPEG-2, contain 8×8 DCT, and some recent standards like H.264 also support 4×4 DCT [1]. It is also used in some other real-time applications related to signal processing. Theoretically, the perfect representation of DCT coefficients would require infinite precision, because the basis functions for taking the DCT comprise irrational numbers. However, in real life fixed-point arithmetic is most often used instead of floating-point arithmetic for practical reasons. ISO/IEC standard 14496-2:2001 dictates accuracy requirement to be less than value 2 differences in the floating and fixed-point implementation. This can be achieved with careful design, e.g. [1] reports fixed-point implementation, where the differences are less than 0.875. Another error source in digital implementations is that input data coming from image sensor must be converted to digital bits before DCT. Thus, quantization errors are introduced before the transform.

If strict requirements are imposed on power consumption, we may consider performing the transform in the analog domain. Indeed, there exists many examples of analog DCT implementations, e.g. [2], [3], [4]. With an analog implementation there is no AD conversion, and thus no quantization noise before the transform. However, when designing analog circuits, there is always mismatch between transistors (or capacitors), and thus the accuracy requirements are sometimes very hard to fulfil. For example, excessive mismatch in the cosine coefficients causes an annoying checkerboard artifact in the reconstructed picture. Of course, we can always decrease the effects of the mismatch by increasing the size of the transistors, but this is not always desirable.

Process variations introduce errors that are not known in advance. For example, random doping fluctuations cause variation of the threshold voltage of transistors. Due to this, for example the output of a current mirror may differ from expected value. Mismatch is random in nature, but does not change after the chip has been fabricated. Consequently, the errors due to mismatch for a certain signal path on a certain chip are correlated regardless of the input. Thus, taking advantage of this correlation property, the errors due to mismatches can be decreased by using methods introduced in this paper. From another point of view, it can be stated that if error compensation is used, the DCT core does not have to be extremely accurate and still good results are achievable. This way area can be saved. In [4] it was discovered that the device mismatch is mostly an issue when I-frames

[1] In fact, an integer approximation of DCT

are processed. Hence, error compensation presented here is mainly targeted for correcting transformed I-frames. The error compensation was simulated with 4×4 blocks, but the same method can be applied also to 8×8 blocks. The compensation itself is easy to implement as will be explained in the remainder of this paper.

II. THE DCT

Using matrix notations, the forward 2-D DCT for a given $N \times N$ real signal can be written as:

$$Y = CXC^T \quad (1)$$

where X is the input block of spatial domain and C is matrix of cosine coefficients. For example, for 4×4 block size

$$\mathbf{C} = k \begin{pmatrix} b & b & b & b \\ a & c & -c & -a \\ b & -b & -b & b \\ c & -a & a & -c \end{pmatrix}$$

where $a = \cos(\pi/8)$, $b = \cos(2\pi/8)$, $c = \cos(3\pi/8)$, and k is a scaling factor.

III. COMPENSATION FOR ERRORS

Once the circuit is fabricated there is not much to do for mismatch errors, since usually there is no possibility of adjusting the circuit. Nevertheless, post-fabricate trimming of cosine coefficients is possible using floating gate transistors, though this approach requires some special arrangements like programming circuitry etc. One example of this is presented in [5]. Another possibility is to allow some inaccuracy due to analog computation and try to compensate for the errors afterwards. The unidealities of the DCT core can be measured, e.g. by using so called "calibration frame" and then comparing the output to the ideally transformed values. Armed with the knowlegde of the amount of errors we are able to take them into account. Basically, the error compensation can be performed before or after quantization and AD conversion, or even at the decoding phase. Because our analog DCT core [4] works in current-mode, the error compensation was chosen to be done right after DCT. This makes the implementation very compact and simple, since summing and subtracting currents can be easily realized.

If the nature of the pictures to be transformed is known, the calibration frame can be chosen to be similar to them. In practise, this information is not usually on hand and we are forced to seek other strategies. First possibility is to use some arbitrary picture for error compensation. In fact, quite good results can be achieved even using this method. However, the results are much better if the calibration frame is carefully chosen, i.e. its characteristics, for example intensity, are average. This method also produces best results with video data, since generally the sequential frames are highly correlated in video stream, but not necessarily when time between the frames is longer. When the calibration frame is not changed constantly according to input data, the complexity of the error compensation

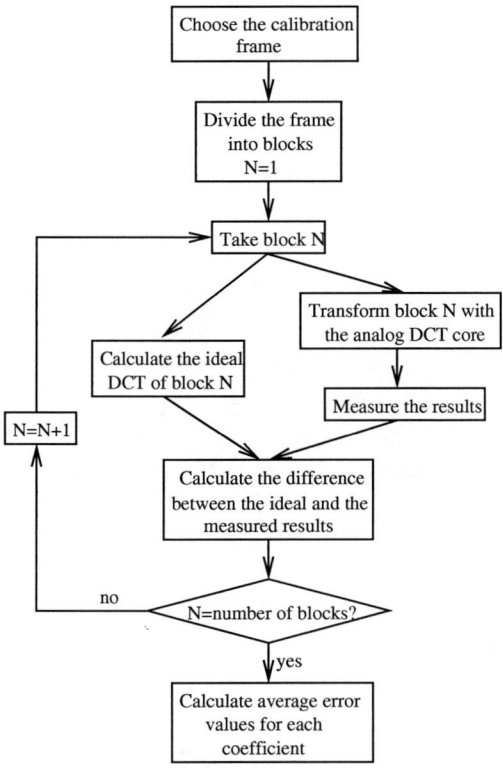

Fig. 1. Flowchart of the calibration algorithm

Fig. 2. Test image Peppers

Fig. 3. Test image Airplane

scheme is maintained low. It is apparent that the compensation is not worth using with the high frequency coefficients, because of number of reasons. Errors in the high frequency coefficients are usually very small and have little effect on the subjective image quality. Furthermore, the compensation of very small errors easily increases the magnitude of errors instead of decreasing. Usually, the errors caused by mismatch are larger in images with high intensity, because then the error is multiplied by large intensity values (see eq. 1).

After the calibration frame is chosen, the first block of it is fed to the DCT core. The results of analog DCT are measured. The same block is also transformed ideally, for example with Matlab. Every measured coefficient is subtracted from the corresponding ideal value to calculate the error. This is repeated for every block of the calibration frame. Finally, the average of errors is calculated for each DCT coefficient. This procedure is depicted in Fig. 1.

The average error, specific for each coefficient, can be used to compensate for errors. This is done by subtracting the average error from corresponding coefficient for each transformed block. Because the device mismatches introduced during fabrication remain the same, the average error calculation has to be done only once.

IV. SIMULATION RESULTS

The simulations were performed using Matlab. Block size 4×4 was used. Gaussian mismatch was randomly generated and applied to the cosine coefficients. Two cases, standard deviations 2% and 5%, were examined.

First, the effect of inaccurate cosine coefficients on DCT coefficients was examined. A whole image was transformed block by block. For every block, the absolute value of the error was calculated with and without compensation for every DCT coefficient. The results were compared by calculating average values for both cases.

Ten standard test images with 8 bits and a resolution of 512×512 pixels were used. Let us examine the results of images Peppers (Fig. 2), Airplane (Fig. 3) and Lax (Fig. 4) in more detail. It can be seen that the three pictures are diverse. Peppers was chosen to be the calibration frame, because it has quite average characteristics. Table I shows one example of the average of errors in all DCT coefficients for Peppers image with a standard deviation of 2%. These values were used in compensation.

Best results for a certain image are of course achieved when the image is compensated with itself. An example of this can be seen in Fig. 5. There, the average error of Peppers image has been used to compensate for the errors of Peppers image itself. The standard deviation was 2%. The upper picture shows the average absolute of error before the compensation and the lower picture after the compensation. It can be seen that for example the DC coefficient error, which is situated back on the left, is reduced from approximately 4 to 1.6, that is 60%.

Fig. 6 and Fig. 7 show the error before and after compensation for Airplane figure. The former is with 2% standard deviation and the latter with 5%. The effect of the standard deviation can be clearly seen. The smaller deviation gives a maximum average absolute of error of approximately 6 and the larger over 16. In Fig. 8 are the

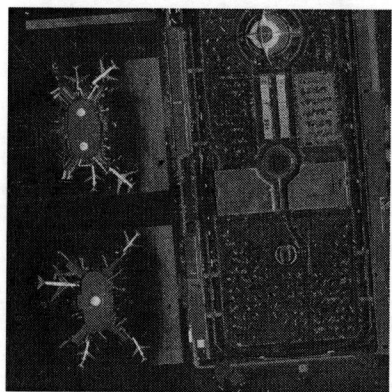

Fig. 4. Test image Lax

TABLE I
EXAMPLE OF AVERAGE OF ERRORS FOR PEPPERS

4.0217	0.2899	-1.5869	3.8020
-2.6399	0.0055	-0.0015	0.0158
-1.0194	0.0020	-0.0008	0.0052
-3.5920	0.0083	-0.0046	0.0286

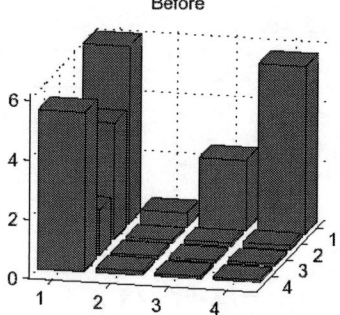

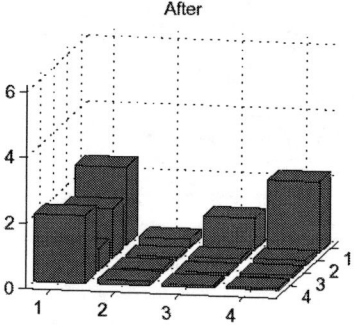

Fig. 6. Airplane error with 2% standard deviation

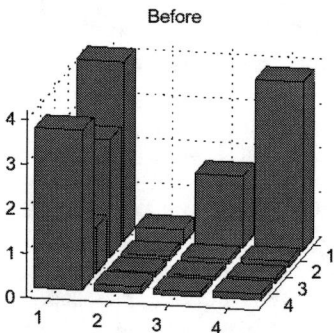

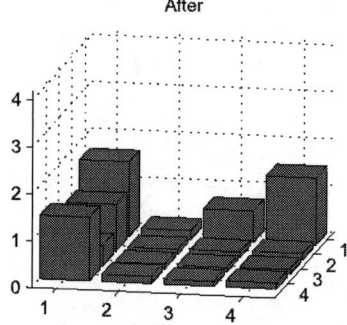

Fig. 5. Peppers error with 2% standard deviation

results for Lax image with 5% standard deviation. It has lower average intensity than Airplane, which affects the magnitude of the errors.

The second approach was to use reconstructed image in error estimation. The inverse DCT was taken ideally and no quantization was performed between transforms. The resulting image quality was evaluated by subjective means, i.e. visual inspection, and also with objective measures. Mean Squared Error (MSE) and Peak Signal to Noise Ratio (PSNR) were used, although they do not coincide with the visual quality of the images very well [6], but are still widely used and provide numerical values, which are easy to compare. MSE can be expressed as

$$MSE = \frac{\sum (f_{i,j} - F_{i,j})^2}{M \times N} \quad (2)$$

where $f_{i,j}$ is the original image, $F_{i,j}$ is the reconstructed image and $M \times N$ is the image size. The PSNR is usually expressed in decibels and calculated using formula

$$PSNR = 10 \cdot lg \frac{I_{max}^2}{MSE} \quad (3)$$

where I_{max} is the maximum intensity value.

A comparison of the PSNR with and without compensation for the three test images in the two mismatch cases is presented in Table II. It can be seen that the performance is always considerably better with error compensation. For all ten test images, the enhancement in the PSNR varied between 4.0 dB and 9.9 dB depending on the

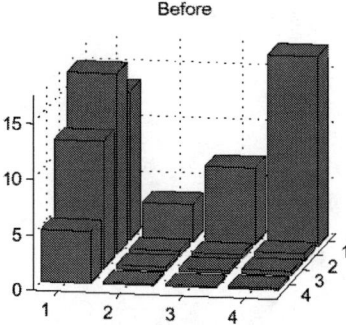

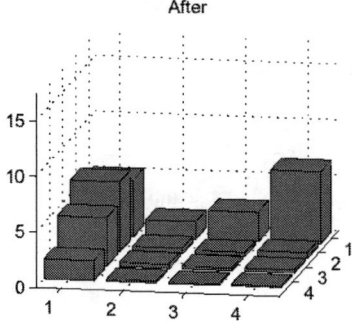

Fig. 7. Airplane error with 5% standard deviation

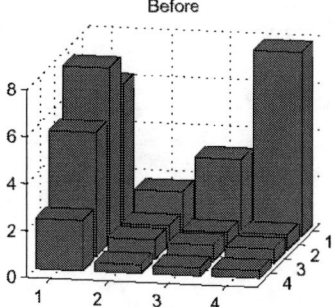

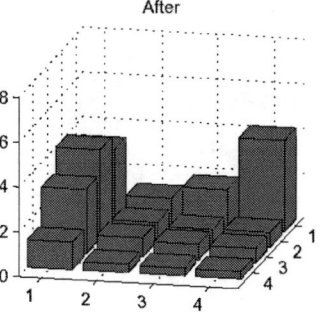

Fig. 8. Lax error with 5% standard deviation

TABLE II
PSNR COMPARISON

Image	Standard deviation	PSNR, not compensated	PSNR, compensated	PSNR difference
Peppers	2%	41.99 dB	49.50 dB	7.51 dB
Peppers	5%	33.10 dB	40.82 dB	7.72 dB
Airplane	2%	39.06 dB	46.74 dB	7.68 dB
Airplane	5%	30.14 dB	37.95 dB	7.81 dB
Lax	2%	45.14 dB	49.36 dB	4.22 dB
Lax	5%	36.39 dB	41.00 dB	4.61 dB

image with 2% standard deviation. The average enhancement was 7.4 dB. For 5% standard deviation, PSNR enhancement varied between 4.1 dB and 10.3 dB with an average of 7.6 dB. Visual inspection verified that the performance was improved. As can be expected, the errors were most visible in large uniform areas, e.g. the upper part of Airplane image.

V. CONCLUSION

In this paper, a simple but effective scheme for compensating errors generated by analog DCT core was proposed. This method is based on measuring the average error using a calibration frame. Implementation in current-mode designs is very straightforward and area effective. Correct operation and performance of the proposed method were verified by Matlab simulations. The results suggest that the image quality can be improved regardless of the input image. However, the amount of error reduction depends on the input and the calibration frame.

REFERENCES

[1] M Alam, W. Badawy, G. Jullien, 'Time distributed DCT architecture for multimedia applications', IEEE International Conference on Consumer Electronics, ICCE, pp. 166-167, 2003

[2] S. Kawahito, M. Yoshida, M. Sasaki, K. Umehara, D. Miyazaki, Y. Tadokoro, K. Murata, S. Doushou, A. Matsuzawa, 'A CMOS Image Sensor with Analog Two-Dimensional DCT-Based Compression Circuits for One-Chip Cameras', IEEE Journal of Solid-State Circuits, Vol. 32, No. 12, pp. 2030-2041, 1997

[3] A. Handkiewicz, M. Kropidlowski, M. Lukowiak, 'Switched-Current Technique for Video Compression and Quantization', ASIC/SOC Conference, pp. 299-303, 1999

[4] M. Pänkäälä, J. Poikonen, L. Vesalainen, A. Paasio, 'Realization of an Analog Current-Mode 2-Dimensional DCT', Proceedings of the 2004 International Symposium on Circuits and Systems, Vol. 1, pp. I 745-748, 2004

[5] P.D. Smith, M. Kucic, P. Hasler, 'Accurate Programming of Analog Floating-Gate Arrays', IEEE International Symposium on Circuits and Systems, Vol. 5, pp. V-489 -492, 2002

[6] A.M. Eskicioglu, P.S. Fisher, 'Image Quality Measures and Their Performance', IEEE Transactions on Communications, Vol. 43(12), pp. 2959-2965, 1995

An Optimal Tone Reproduction Curve Operator for the Display of High Dynamic Range Images

Guoping Qiu and Jiang Duan
School of Computer Science, The University of Nottingham
Jubilee Campus, Nottingham, NG8 1BB, UK
{qiu | jxd} @ cs.nott.ac.uk

Abstract We present a new tone mapping method for the display of high dynamic range images in low dynamic range devices. We formulate high dynamic range image tone mapping as an optimisation problem. We introduce a two-term cost function, the first term favours linear scaling mapping, the second term favours histogram equalisation mapping, and jointly optimising the two terms optimally maps a high dynamic range image to a low dynamic range image. We control the mapping results by adjusting the relative weightings of the two terms in the objective function. We also present a fast and simple implementation for solving the optimisation problem. We will present results to demonstrate that our method works very effectively.

I. INTRODUCTION

The ultimate goals of computer graphics and digital imaging are to develop systems that match or maybe even exceed the capabilities of the human visual system (HVS). One of the remarkable abilities of the HVS is that it can perceive real world scenes with brightness dynamic ranges of over 5 orders of magnitude and can distinguish even higher contrast through adaptation. Ordinary imaging sensors and reproduction media typically have dynamic ranges spanning a few orders of magnitude, much lower than those of the real world scenes.

High dynamic range (HDR) imaging technologies are designed to produce images that faithfully depict the full visual dynamics of real world scenes. There are two major technical challenges in HDR imaging, image capture and image reproduction. In the literature, several methods have been developed for creating high dynamic range still images [1, 2] and videos [3]. In order to display HDR images on monitors or print them on paper, we must compress the dynamic range of the HDR images to reproduce low dynamic range (LDR) images supported by these media. In recent years, several techniques have been developed by various researchers for mapping HDR images to LDR images for display. Please see the Related Work box for a brief review of these methods available in the open literature.

In this article, we present a new optimization method for mapping HDR images to LDR images. For a given HDR image, our method first defines an objective function. An optimal solution to the objective function corresponds to an optimal mapping of the HDR image to an LDR image for display. We will present a fast computational solution to solve the optimization problem. Our new method only manipulates the pixel distribution of an image and is a tone reproduction curve (TRC) based high dynamic range compression method. Our final algorithm has one single variable, which the users can change to control a reproduction to suit individual images and users' subjective preference. Compared with existing HDR image compression methods, our technique is more flexible and computational much more efficient. We will show that our new method has an excellent performance in mapping a variety of high dynamic range scenes.

II. RELATED METHODS

In the literature, a number of techniques have been developed for tone reproduction for high contrast images. There are two broad categories of technology [5]. Tone reproduction curve (TRC) based techniques manipulate the pixel distributions. Earlier pioneering work in this category include that of [6] which introduced a tone reproduction method that attempted to match display brightness with real world sensations. Recently, [7] presented a tone mapping method that modeled some aspects of human visual system. More recently, we have also developed a learning-based TRC tone mapping method [16] and a fast TRC tone mapping method [17].for high dynamic range compression. Perhaps the most comprehensive technique in this category is that of [10], which introduced a quite sophisticated tone reproduction curve technique that incorporated models of human contrast sensitivity, glare, spatial acuity and color sensitivity to exploit the limitations of human visual system.

Tone reproduction operator (TRO) based techniques involve the spatial manipulation of local neighboring pixel values, often at multiple scales. The scientific principle of this type of technique is based on the image formation model: $I(x, y) = L(x, y) R(x, y)$, which states that image intensity function $I(x, y)$ is the product of the luminance function $L(x, y)$ and the scene reflectance function $R(x, y)$. Because real world reflectance $R(x, y)$ has low dynamic range (normally not exceeding 100:1), reducing the dynamic range of $I(x, y)$ can be achieved by reducing the dynamic range of $L(x, y)$ if one could separate $L(x, y)$ from $R(x, y)$. Methods based on this principle include [9], [10] and [11]. They mainly differ in the way in which they attempted to separate the luminance component from the reflectance component. All TRO based methods can be regarded as related to the Retinex theory [14]. A direct use of the Retinex theory for high dynamic range compression was presented by Jobson and co-workers [15]. Recent development has also attempted to incorporate traditional photographic technology to the digital domain for the reproduction of high dynamic range images [12]. An impressive latest development in high dynamic range compression is that of [13]. Based on the observation that human visual system is only sensitive to relative local contrast, the authors developed a multiresolution gradient domain technique. This is also a TRO type technique and the authors reported very good results that are free from halo effects.

TRO based methods involve multiresolution spatial processing and are therefore computationally very expensive. Because TRO methods can reverse local contrast, they can sometimes cause "halo" effects in the reproduction. Another difficulty of these techniques is that there are too many parameters the users have to set and this makes them quite difficult to use. In many cases, the setting of these controlling parameters is rather ac hoc and involves many trial and errors. TRC based methods do not involve spatial processing, they are therefore computationally very simple. This is useful in real time applications such as high dynamic range video. TRC techniques also preserve the lightness orders of the original scenes and avoid artifacts such as halo that is often associated with TRO based methods. One of the weaknesses of TRC approaches as compared with TRO methods is that it may cause noticeable lost of spatial sharpness in some images. In many cases, this is not a serious problem if the TRC method is well designed and a simple standard image sharpening operation often suffices to bring back the sharpness. However, TRO methods sometimes could introduce too much (artificial) detail. Both types of techniques have their own merits and drawbacks in terms of computational complexity, easy implementation and practical application, and therefore are likely to co-exist for tone mapping in HDR imaging for the foreseeable future.

III. Optimal Tone Reproduction Curve Tone Mapping for HDR Images

For high dynamic range image tone mapping, there are at least two requirements. Firstly, it has to ensure that all features, from the darkest to the brightest, to be visible simultaneously. Secondly, it has to preserve the original scene's visual contrast impression to produce a visual sensation matching that of the original scene. To preserve the original scene's relative visual contrast impression, the simplest approach is to linearly map the pixels from a high dynamic range to a low dynamic range. However, since the dynamic range in the display devices is much narrower than that of the original scene, visibility will be lost due to compression. Also, linear scaling maps all values in the same way, some displayable values in the low dynamic range device may be empty or have too few pixels mapped onto them thus resulting in an under utilization of all displayable values. On the other extreme, one can render the low dynamic range image to have a maximum contrast, i.e., histogram equalized by distributing equal number of pixels to each display levels. However, this will alter the original scene's visual impression, because it exaggerates contrast in densely populated pixel value intervals while compresses too aggressively sparsely populated pixel value intervals.

The TRC based HDR image tone mapping problem can be conveniently explained by referencing to Figure 1. Essentially, there are more (discrete) levels in the input HDR axis than in the output LDR axis. If the HDR data is represented by 32-bit floating point number, then there are 4,294,967,296 possible levels in the HDR input. For the LDR output, typically, there will be 256 levels (can be any other numbers, but we will use this number for convenience). Clearly and straightforwardly, some HDR values will have to be merged and represented by the same LDR value. To do this, we cut the HDR dynamic range axis into 256 intervals at positions c1, c2, ..., c255. HDR values falling into the same interval are then assigned the same LDR display value. For convenience, we assume both the HDR range and the total number of pixels in the image are unity. If we divide the HDR input axis in such a way that pixel populations falling into each interval are identical and equal to 1/256, then this is histogram equalization mapping. On the other hand, if we make each interval to have the same length and equal to 1/256, then this is linear scaling mapping.

A TRC based tone mapping algorithm should assign relatively more display values to densely populated luminance intervals and relatively fewer display values to sparsely populated luminance intervals. A TRC based tone mapping algorithm should also maintain the relative visual contrast impressions of the original scene while simultaneously trying to fully utilize all available display levels. Therefore, with reference to Figure 1, c1, c2, ..., c255 must fall in between linear scaling and histogram equalization. For illustration purpose, Figure 3 shows examples of mapping 8bit/pixel images to 1bit/pixel images for display. In the left image, the majority of pixels have values greater than mid grey (128). The overall impression of this image is bright. The display image should also ensure that there are more bright pixels than dark pixels, which means that the cut should be below c_e. It should not be below c_l either for obvious reasons. In right image, there are more pixels have values below mid-grey, the image appears dark. The display should also have more dark pixels than bright pixels and this means that the cut should be above c_e. In this case, the cut should be below c_l. In these examples, we clearly see that if the mappings are outside the linear scaling or histogram equalization lines, then the mapped images will either have too many dark pixels or too many bright pixels. As a consequence, the overall visual impressions of the displays differ drastically from those of the originals. By cutting the image in between the linear scaling and histogram equalization lines, both the details and visual impressions of the original images are better preserved.

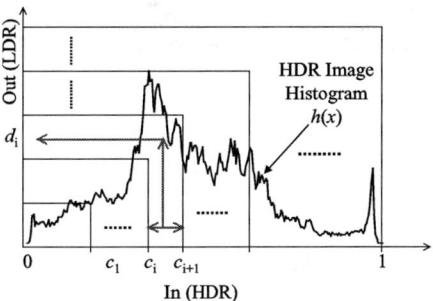

Figure 1 Tone reproduction curve based HDR image tone mapping problem can be solved by dividing the HDR range axis into segments, HDR pixels falling into the same segment are mapped to the same display value in the LDR image.

To formalize this mapping principle, again, with reference to Figure 1, we can define the following objective function

$$E = \sum_{i=1}^{255}\left(c_i - \frac{i}{256}\right)^2 + \lambda \sum_{i=1}^{255}\left(\int_0^{c_i} h(x)dx - \frac{i}{256}\right)^2 \quad (1)$$

where λ is the Lagrange multiplier. Setting $\lambda = \infty$, optimizing E becomes histogram equalization mapping, and $\lambda = 0$, optimizing E becomes linear scaling mapping. By choosing an appropriate l, we can strike a balance between the two extreme forms of mapping to suit individual images. An optimal solution to (1) can be found by solving following linear equations

$$\frac{\partial E}{\partial c_i} = 0 \quad i = 1, 2, \cdots, 255 \quad (2)$$

IV. A Fast Implementation

Equation (2) may be seriously ill conditioned, and a straightforward numerical solution to optimize E in (1) may be difficult to obtain. We here present an approximated fast solution. Instead of trying to find all cuts in one go, we use a recursive binary cut approach. We first divide the dynamic range of the HDR image into two intervals based on a modified objective function of (1). These resultant two intervals are then subsequently and independently divided into two intervals in a similar way. The schematic is illustrated in Figure 2. We first find c_0, which divides the full dynamic range into two intervals, we then find $c_{1,0}$ and $c_{1,1}$, which divide the resultant two intervals from previous level into 4 intervals. The process continues, until the desired numbers of intervals are found.

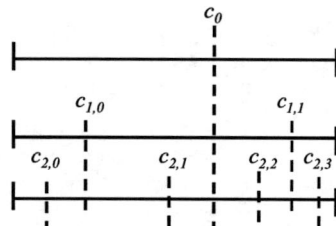

Figure 2 The full dynamic range of the HDR image is first divided into two intervals, each of which is then again independently divided into two intervals. The process is applied recursively until the desired numbers of intervals are created.

With the scheme in Figure 2, we can now reformulate (1) for each interval we want to divide into two. Assuming that the total length of the interval is L, the minimum is L_{min} and the maximum is L_{max}, the number of pixels falling within the interval is N, we find c that cuts the interval into two by optimizing the following objective function

$$E_b(c) = \frac{(c - 0.5(L_{max} + L_{min}))^2}{L^2} + \lambda \frac{\left(\left(\sum_{x=0}^{c} h(x)\right) - 0.5N\right)^2}{N^2} \quad (3)$$

Now to find a c that minimizes $E_b(c)$ in (3) is simple. We simply compute all possible values of E_b, $E_b(L_{min})$, $E_b(L_{min}+\Delta L)$, $E_b(L_{min}+2\Delta L)$, ..., $E_b(L_{min}+n\Delta L)$, ..., $E_b(L_{max})$ and find the minimum

$$c = \arg\left(\min_x (E_b(x))\right) \quad (4)$$

ΔL controls the precision of the division. We used the same ΔL for all the segments in the hierarchy of Figure 2 and we found that setting ΔL between 0.01% ~ 0.1% of the full dynamic range of the HDR image worked very well. Obviously, the smaller ΔL is, the more accurate is the division and the longer it will take to compute the optimization. Using a PC with a 2.66 GHz Pentium 4 Processor, calculating a 256-interval division according to (3) and (4) takes 0.016 to 0.672 second for ΔL ranging between 0.1% and 0.01% (codes written in C and not optimized for speed). For each intermediate segment, we apply (3) and (4) to divide it into two segments. Although it is possible to vary λ for different levels of the hierarchy in Figure 2, for simplicity, we use the same λ for all the segments.

V. RESULTS

Our technique has been tested on a variety of high dynamic range images. In our experiments, the luminance signal is calculated as: $L = 0.299*R+0.587*G+0.114*B$. Log(L) is computed to compile a histogram. After mapping, the LDR images are rendered for display using following formula

$$R_{out} = \left(\frac{R_{in}}{L_{in}}\right)^{\gamma} L_{out}, G_{out} = \left(\frac{G_{in}}{L_{in}}\right)^{\gamma} L_{out}, B_{out} = \left(\frac{B_{in}}{L_{in}}\right)^{\gamma} L_{out} \quad (5)$$

where L_{in} and L_{out} are luminance values before and after compression, γ controls display color (setting it between 0.4 and 0.6 worked well). How to compute the mapped luminance for display devices is a well-studied problem [4]. In our experiments, we simply give all pixels mapped to the first (lowest) interval a display value of 0, those mapped to the second interval a display value of 1, and so on. We did not carry out any color correction for the mapped images. Notice that for the best viewing effects, it is necessarily to carry out the correct calibration of the display devices. Because compression will inevitably loose some fine details, we found that using a standard sharpening filter can improve the spatial sharpness of the mapped images.

Our new method is very flexible. By controlling the value of λ, we can control the final appearance of the image in a simple and elegant way. A smaller λ renders the image with lower contrast whilst a larger λ renders the image with higher contrast. By choosing an appropriate λ, we can achieve a balanced reproduction. Figure 4 shows examples of HDR images mapped by our new method.

Our method is most similar to the method of Ward Larson [8]. Compared with Ward Larson's histogram adjustment method, our method is much more flexible and more general. For example, the histogram adjustment method only prevents the display contrast (produced by simple histogram equalization) from exceeding that of the real scene, but this will not prevent the display from having too low contrast. Whilst the histogram adjustment method is restrictive and inflexible, our method can adjust the contrast of the display by simply changing a single parameter. Figure 5 shows an informal comparison of our method and Larson's method.

Computationally, our method is very fast, using non optimized C code running on a PC with a 2.66 GHz Pentium 4 Processor, mapping a 512 x 768 pixel image takes 0.75 second, a 768 x 1024 image takes 1.28 seconds, and a 2048 x 1536 image takes 5.375 seconds (note that these times do not include file I/O operations but include all other necessary computations).

VI. CONCLUDING REMARKS

In this article, we have introduced an optimization method for HDR image tone mapping. We have introduced an objective function and developed a fast method to solve such an optimization problem. Experimental results demonstrated that our new method worked very well on a variety of HDR images. The merits of our new method includes that it is intuitive, simple, flexible and effective.

REFERENCES

[1] P. E. Debevec and J. Malik, "Recovering high dynamic range radiance maps from photographs", Proc. ACM SIGGRAPH'97, pp. 369 – 378, 1997

[2] T. Mitsunaga and S. K. Nayar, "High dynamic range imaging: Spatially varying pixel exposures", Proc. CVPR'2000, vol. 1, pp. 472-479, 2000

[3] S. B. Kang, M. Uyttendale, S. Winder and R. Szeliski, "High dynamic range video", ACM Transactions on Graphics, vol.22, no. 3, Pages: 319 – 325, July 2003

[4] R. Hall, Illumination and color in computer generated imagery, Spinger-Verlag, 1989

[5] J. DiCarlo and B. Wandell, "Rendering high dynamic range images", Proc. SPIE, vol.3965, pp. 392 – 401, 2001

[6] J. Tumblin and H. Rushmeier, "Tone reproduction for realistic images", IEEE Computer Graphics and Applications, vol. 13, pp. 42 – 48, 1993

[7] M. Ashikhmin, "A tone mapping algorithm for high contrast images", Proc. Eurographics Workshop on Rendering, P. Debevec and S. Gibson Eds., pp. 1 – 11, 2002

[8] G. W. Larson, H. Rushmeier and C. Piatko, "A visibility matching tone reproduction operator for high dynamic range scenes", IEEE Trans on Visualization and Computer Graphics, vol. 3, pp. 291 – 306, 1997

[9] K. Chiu, M. Herf, P. Shirley, S. Swamy, C. Wang and K. Zimmerman, "Spatially nonuniform scaling functions for high contrast images", Proc. graphics Interface'93, pp. 245 – 253, 1993

[10] J. Tumblin and G. Turk, "LCIS: A boundary hierarchy for detail preserving contrast reduction", ACM SIGGRAPH 1999

[11] F. Durand and J. Dorsey, "Fast bilateral filtering for the display of high-dynamic-range images", Proc. ACM SIGGRAPH'2002

[12] E. Reinhard, M. Stark, P. Shirley and J. Ferwerda, "Photographic tone reproduction for digital images", Proc. ACM SIGGRAPH'2002

[13] R. Fattal, D. Lischinski and M. Werman, "Gradient domain high dynamic range compression", Proc. ACM SIGGRAPH'2002

[14] E. H. Land and J. J. McCann, "Lightness and retinex theory", Journal of the Optical society of America, vol. 61, pp. 1-11, 1971

[15] D. J. jobson, Z. Rahman and G. A. Woodell, "A multiscale Retinex for dridging the gap between color images and the human observation of scenes", IEEE Transactions on Image processing, vol. 6, pp. 965-976, 1997

[16] J. Duan, G. Qiu and G. D. Finlayson, "Learning to display high dynamic range images", CGIV'2004, IS&T's Second European Conference on Color in Graphics, Imaging and Vision, Aachen, Germany, April 5-8, 2004J.

[17] J. Duan and G. Qiu, "Fast tone mapping for high dynamic range images", ICPR2004, 17th International Conference on Pattern Recognition, Cambridge, United Kingdom, 23 - 26 August 2004

Figure 4 Results of various HDR images mapped by our new method. Notice that no effort has been spent on selecting the best λ for individual images. In this example, $\lambda = 0.8$ for the top image and $\lambda = 0.5$ for the bottom image. Data courtesy of Paul Debevec and Sumant Pattanaik

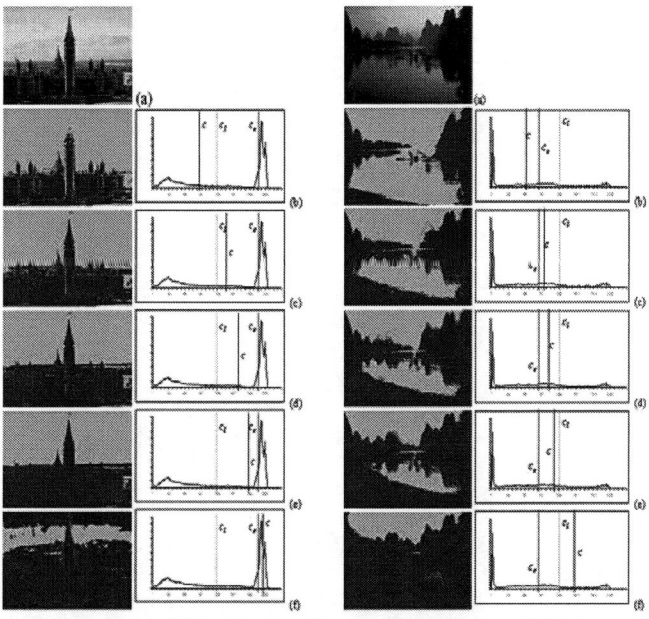

Figure 3 Left: (a) Original image, the overall impression of this image is bright. (b) Cut below liner mapping. (c), (d) and (e), Cuts between linear scaling and histogram equalization. (f) Cut outside histogram equalization. Outside c_l and c_e, there are either too many bright or too many dark pixels. Cuts in between c_l and c_e, visual impressions are more faithful to that of the original and also have more details. Right: (a) Original image, the overall impression of this image is dark. (b) Cut below liner mapping. (c), (d) and (e), Cuts between linear scaling and histogram equalization. (f) Cut outside histogram equalization. Outside c_l and c_e, there are either too many bright or too many dark pixels. Cuts in between cl and ce, visual impressions are more faithful to that of the original and also have more details

Figure 5 Memorial Church image. HDR data courtesy of Paul Debevec. Left: Result of Ward Larson's histogram adjustment technique, image courtesy of Dani Lischinski, printed with permision. Right: Results of our new optimal method, $\lambda = 0.7$.

The FIR Filter Bank with Given Analysis Filters That Minimizes Various Worst-Case Measures of Error at the Same Time

Yuichi Kida
School of Pharmaceutical Sciences
Ohu University
Koriyama-shi 963-8611, Japan
Email: ykida@pop12.odn.ne.jp

Takuro Kida
Department of EE
Nihon University
Koriyama-shi 963-8642, Japan
Email: kida@ee.ce.nihon-u.ac.jp
Tel/Fax: +81(24) 990-0576

Abstract—In this paper, we present the scan-type discrete approximation of FIR filter bank that minimizes various worst-case measures of error, including the long-range worst-case measures of error in the time-domain or the frequency-domain. The discrete interpolation functions presented in this paper vanish outside the prescribed domain in the integer time-axis and are realized by FIR filters.

I. INTRODUCTION

Recently, theory of filter banks becomes important in various fields of wireless and/or visual image processing [1]···[6]. The realization of FIR filter bank that minimizes various worst-case measures of error has been one of the difficult open-problems in the theory of filter banks and no exact solution has been obtained yet. In this paper, firstly, we show the modified version of the continuous interpolation functions with extended band-width that interpolate the proposed discrete interpolation functions and satisfy the condition called discrete orthogonality. This condition is one of the two conditions that constitute the necessary and sufficient condition for the above optimum approximation. Then, we present a definition of the set of signals that satisfy the other condition of those two conditions.

II. THE OPTIMUM APPROXIMATION

Consider the filter bank shown in Fig.1. Let Z be the set of integers. We denote by Λ the set of integers $\{0, 1, \ldots, M-1\}$. Let $H_m(\omega)$ $(m \in \Lambda)$ be the prescribed time-invariant linear systems. Suppose that $f_m(t)$ $(m \in \Lambda)$ are the output signals obtained by adding $f(t)$ to $H_m(\omega)$ $(m \in \Lambda)$. Then, $f_m(t) \longleftrightarrow H_m(\omega)F(\omega)$ $(m \in \Lambda)$ hold. $H_m(\omega)$ $(m \in \Lambda)$ are called analysis filters. We assume that $f_m(t)$ are continuous and contained in L_t^2. Let $\Delta(t)$ be the finite subset of (m, n) $(m \in \Lambda, n \in Z)$ defined at each t. Further, consider the sample points $t_{m,n} = nT + a_m$, where T, a_m are given positive integers and $(m, n) \in \Delta(t)$. When $\Delta(t)$ is independent of the time t, this set of sample points is corresponding to the approximation using the fixed finite set of sample points. If $\Delta(t)$ moves along the time axis sequentially, we consider that the above set of the sample points is corresponding to the sequence of the sample-points used in scanning type filter bank [7].

Now, let Ξ be the set of signals. In general, let $g(t) =$

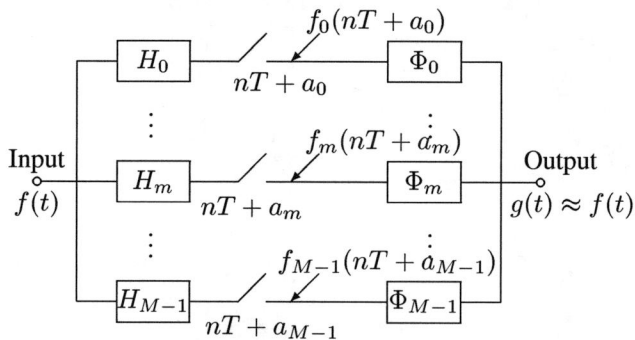

Fig. 1. Filter Bank System

$v_0[\{f_{m,n}\}, t]$ be a linear/nonlinear approximation for $f(t)$ in Ξ using the sample values $f_{m,n} = f_m(t_{m,n})$ $((m, n) \in \Delta(t))$. We assume that $v_0[\{f_{m,n}\}, t]$ is zero when all the $f_{m,n}$ $((m, n) \in \Delta(t))$ are zero. Let $g(t)$ be the optimum approximation that we consider in this discussion.

Further, let $\hat{g}(t) = \hat{v}_0[\{f_{m,n}\}, t]$ be another linear/nonlinear approximation for $f(t)$ in Ξ using the sample values $f_{m,n} = f_m(t_{m,n})$ $((m, n) \in \Delta(t))$. In this paper, we assume that $\hat{v}_0[\{f_{m,n}\}, t]$ is zero when all the $f_{m,n}$ $((m, n) \in \Delta(t))$ are zero. Let $\hat{g}(t)$ be the approximation that is compared with $g(t)$.

Suppose that $e(t) = f(t) - g(t)$ and $\hat{e}(t) = f(t) - \hat{g}(t)$ are the error of approximation for the corresponding approximations, respectively. Since the error of approximation $\hat{e}(t) = f(t) - \hat{g}(t)$ depends on the signal $f(t)$, we express the error of approximation $\hat{e}(t)$ such as $\hat{e}(t) = \hat{\varepsilon}[f(t)]$.

Let $\beta[\hat{e}(t)]$ be a function (or a functional/an operator) of $\hat{e}(t)$ and let $\beta[\hat{e}(t)]$ have non-negative value. We use similar notation $\beta[e(t)]$ for $e(t)$ also.

Further, let Θ be a subset in Ξ. Then, consider the following measure of error of approximation $E_\Theta(t)$ for $f(t)$ in Θ.

$$E_\Theta(t) = \sup_{f(t) \in \Theta} \{\beta[\hat{e}(t)]\}$$

With respect to $E_\Theta(t)$, it is natural to assume that $E_{\Theta_s}(t) \leq E_\Theta(t)$ holds for all the set of signals Θ_s satisfying $\Theta_s \subseteq \Theta$.

Now, let us pay attention to the following assumption. Let Ξ_e be the set of all the $e(t)$'s, where $e(t) = f(t) - g(t)$. Then, we assume that the following two conditions are satisfied with respect to $\hat{e}(t)$, $e(t)$, Ξ and Ξ_e.

Condition (1): $\hat{e}(t)|_{f(t)=e(t)} = \hat{\varepsilon}[e(t)] = e(t)$.

Condition (2): $\Xi_e \subseteq \Xi$.

Now, let $E(t) = E_\Xi(t) = \sup_{f(t) \in \Xi}\{\beta[\hat{e}(t)]\}$ be the (final) measure of error of approximation to be minimized. Then, as the consequence of $E_{\Theta_s}(t) \leq E_\Theta(t)$ and the above two conditions, we have

$$\sup_{f(t)\in\Xi}\{\beta[\hat{e}(t)]\} = \sup_{f(t)\in\Xi}\{\beta[\hat{\varepsilon}[f(t)]]\}$$
$$\geq \sup_{f(t)\in\Xi_e}\{\beta[\hat{\varepsilon}[f(t)]]\} = \sup_{e(t)\in\Xi_e}\{\beta[\hat{\varepsilon}[e(t)]]\}$$
$$= \sup_{e(t)\in\Xi_e}\{\beta[e(t)]\} = \sup_{f(t)\in\Xi}\{\beta[e(t)]\} \quad (1)$$

Hence, $g(t) = v_0[\{f_{m,n}\}, t]$ gives the minimum measure of error of approximation $\sup_{f(t)\in\Xi}\{\beta[e(t)]\}$ among all the measures of error of approximation $\sup_{f(t)\in\Xi}\{\beta[\hat{e}(t)]\}$ under consideration. In [7], we show that these two conditions are necessary.

For the sake of simplicity, if an approximation $g(t) = v_0[\{f_{m,n}\}, t]$ is the optimum approximation in this meaning, we call $g(t)$ is BAMIE (Best Approximation Having the Minimum Interpolation Error), or simply, MIE [7].

III. EXTENDED DISCRETE ORTHOGONALITY

Suppose that M is an arbitrary positive integer. Let R be a sufficiently large positive even integer satisfying $R \geq 2$ and let $\tau = 2/R$. Further, we define

$$\omega_\nu = \frac{M}{T}\pi, \quad \omega_\mu = \mu\pi \leq \frac{R\pi}{2}, \quad \omega_b = (R-1)\pi,$$
$$\omega_0 = R\pi, \quad \omega_1 = (R+1)\pi, \quad \frac{R\pi}{2} \leq \omega_b, \quad \omega_\nu < \omega_\mu \quad (2)$$

In the following discussion, we consider the weighting function $W(\omega)$ such that $W(\omega) > 0$ ($|\omega| \leq \frac{R\pi}{2}$), $W(\omega) = 0$ ($|\omega| > \frac{R\pi}{2}$), $W(\omega) \simeq 1$ ($|\omega| \leq \frac{M}{T}\pi$) and $W(\omega) \simeq \epsilon$ ($\frac{M}{T}\pi < |\omega| \leq \frac{R\pi}{2}$) where $\epsilon \simeq 0$. Further, we define analysis filter bank $\tilde{H}_m(\omega) = X_{in}(\omega)H_m(\omega)$, where $X_{in}(\omega) = 1$ ($|\omega| \leq \omega_\mu$), $X_{in}(\omega) = 0$ ($|\omega| > \omega_\mu$), $H_m(\omega) = \sum_{r=0}^{\nu_m} h_r^m e^{-jr\omega} \cdot e^{ja\omega}$ and $m \in \Lambda$. The quantity a is an integer and h_r^m are real constants. Moreover, we assume that $H_m(\omega)$ ($m \in \Lambda$) are well designed and satisfy $H_m(\omega) = \begin{cases} T & \left(\frac{m}{T}+2n\right)\pi \leq |\omega| < \left(\frac{(m+1)}{T}+2n\right)\pi \\ 0 & \text{(otherwise)} \end{cases}$ approximately. Although detail is omitted, we assume that $\tilde{H}_m(\omega)$ ($m \in \Lambda$) are independent of each other.

For these analysis filters $\tilde{H}_m(\omega)$ ($m \in \Lambda$), consider the optimum approximation

$$g(t) = \sum\sum_{(m,n)\in\Delta(t)} f_m(t_{m,n})v_{m,n}(t - a_m) \quad (3)$$

of $f(t)$ that uses the interpolation functions $v_{m,n}(t - a_m)$ ($m \in \Lambda, n \in Z$) having the infinite support provided in [7]. Because this optimum approximation uses the sample values at the fixed, that is, the infinite sample points, the corresponding interpolation functions $v_{m,n}(t)$ ($m \in \Lambda, n \in Z$) satisfy $v_{m,n}(t - a_m) = v_m(t - nT - a_m)$ and the generalized discrete orthogonality

$$v_m^k(nT + a_k - a_m) = \begin{cases} 1 & (k = m \text{ and } n = 0) \\ 0 & \text{(otherwise)} \end{cases} \quad (4)$$
$$(k \in \Lambda, \ m \in \Lambda, \ n \in Z)$$

where $v_m^k(t)$ is the output of the filter $\tilde{H}_k(\omega)$ ($k \in \Lambda$) for the input $v_m(t)$ as shown in [7]. Let $\Upsilon_m(\omega)$ be the Fourier transform of $v_m(t)$ for $m \in \Lambda$, then, as shown in [3], $\Upsilon_m(\omega)$ is band-limited in the same band-width of $X_{in}(\omega)$, that is, $|\omega| \leq \omega_\mu$.

Now, we turn our attention to the functions $\Theta_m(\omega)$ ($m =$

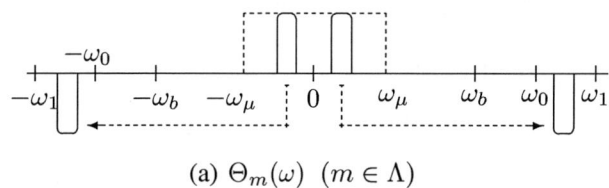

(a) $\Theta_m(\omega)$ ($m \in \Lambda$)

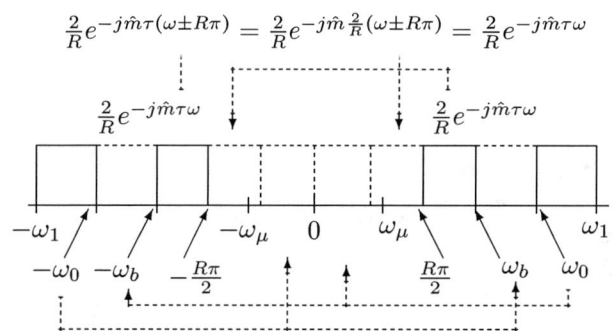

(b) $\Theta_m(\omega)$ ($m = M, M+1, \ldots, \frac{RT}{2} + M - 1$)

Fig. 2. $\Theta_m(\omega)$

$0, 1, \ldots, M-1, M, M+1, \ldots, \frac{RT}{2}+M-1$). Using $\Upsilon_m(\omega)$ for each m in Λ, we make the Fourier transforms $\Theta_m(\omega)$ of the functions $\theta_m(t)$ ($m \in \Lambda$) shown in Fig. 2(a) and Fig. 2(b). As is shown in Fig. 2(a), the left-hand side negative component of $\Theta_m(\omega)$ is made by transporting and upsetting the left-hand side component of $\Upsilon_m(\omega)$ and the right-hand side negative component of $\Theta_m(\omega)$ is made by transporting and upsetting the right-hand side component of $\Upsilon_m(\omega)$.

For $\hat{m} = m - M$, Fig. 2(b) shows the functions

$$\Theta_m(\omega) = \begin{cases} \frac{2}{R}e^{-j\hat{m}\tau\omega} & (\frac{R\pi}{2} < |\omega| \leq \omega_b) \\ \frac{2}{R}e^{-j\hat{m}\tau\omega} & (\omega_0 < |\omega| \leq \omega_1) \\ 0 & \text{(otherwise)} \end{cases} \quad (5)$$
$$(m = M, M+1, M+2, \ldots, \frac{RT}{2} + M - 1)$$

Now, let us consider new interpolation functions defined by

$$\phi_m(t-a_m) = \theta_m(t-a_m) + \sum_{p=M}^{\frac{RT}{2}+M-1} \sum_{s=-\infty}^{\infty}$$
$$\psi_m(sT+(p-M)\tau - a_m)\theta_p(t-sT) \quad (6)$$

for the prescribed $\psi_m(t)$ $(m \in \Lambda)$.

Then, as a direct consequence of Eq.(6) and $\Theta_m(\omega)$ shown in Fig.2, we can prove that these interpolation functions $\phi_m(t-a_m)$ $(m \in \Lambda)$ satisfy

$$\phi_m(q\tau - a_m) = \psi_m(q\tau - a_m) \quad (7)$$
$$\phi_m^k(pT + a_k - a_m) = \begin{cases} 1 & (k=m \text{ and } p=0) \\ 0 & (\text{otherwise}) \end{cases} \quad (8)$$
$$(k \in \Lambda, m \in \Lambda, p \in Z, q \in Z)$$

where $\phi_m^k(t)$ is the output of the filter $\tilde{H}_k(\omega)$ $(k \in \Lambda)$ for the input $\phi_m(t)$.

Now, we define new approximation $\tilde{g}(t)$ for $f(t)$ which is band-limited in $|\omega| \leq \omega_1$ by $\tilde{g}(t) = \sum\sum_{(m,n)\in\Delta(t)} f_m(nT+a_m)\phi_m(t-nT-a_m)$. Then, $\tilde{g}(q\tau) = \sum\sum_{(m,n)\in\Delta(t)} f_m(nT+a_m)\psi_m(q\tau-nT-a_m)$ holds for any $q \in Z$.

We define $\tilde{e}(t) = f(t) - \tilde{g}(t)$. Further, let $\tilde{g}_k(t)$ and $\tilde{e}_k(t)$ be the output of the filter $\tilde{H}_k(\omega)$ $(k \in \Lambda)$ for the input $\tilde{g}(t)$ and $\tilde{e}(t)$, respectively. Then, from the direct consequence of the discrete orthogonality Eq.(4) for the interpolation functions $\phi_m(t)$ $(m \in \Lambda)$, we can easily recognize that the sample values $\tilde{g}_k(t_{k,n})$ $(k \in \Lambda, n \in Z)$ are identical to the sample values $f_k(t_{k,n})$. Hence, all the sample values $\tilde{e}_k(t_{k,n}) = f_k(t_{k,n}) - \tilde{g}_k(t_{k,n})$ $(k \in \Lambda, n \in Z)$ of the error of approximation are equal to zero. Therefore, Condition (1) is satisfied by $\tilde{g}(t)$.

IV. An Example of Set of Signals

In this section, we assume that R is quite large and $\tau = 2/R$ is quite small. Let $F(\omega)$ be the Fourier transform of $f(t)$. Suppose that A is the prescribed positive number and let $X_0(\omega)$ be the ideal filter $X_0(\omega) = \begin{cases} 1 & (|\omega| \leq \omega_0/2) \\ 0 & (|\omega| > \omega_0/2) \end{cases}$. In this paper, we assume that each analysis filters have approximately independent sub-band with the band-width $\frac{2\pi}{T}$, respectively, and hence, there exists no signal $f(t)$ band-limited in $|\omega| \leq \frac{\omega_0}{2}$ that is not identically zero but all the corresponding sample values $f_m(nT+a_m)$ are zero. Then, we assume the following conditions hold.

Condition (a): Consider $f_B(t)$ such that the corresponding Fourier transform $F_B(\omega)$ is expressed by

$$F_B(\omega) = X_0(\omega) \sum_{q=-\lambda}^{\lambda} \sigma_q e^{-jq\tau\omega} \cdot e^{jc\omega} \quad (9)$$

where c, λ are the prescribed integers and σ_q $(q = 0, \pm 1, \pm 2, \ldots, \pm\lambda)$ are the prescribed real numbers. Let Ξ_B be the set of these $f_B(t)$.

Condition (b): Further, $F_B(\omega)$ satisfies

$$\int_{-\frac{\omega_0}{2}}^{\frac{\omega_0}{2}} \frac{|F_B(\omega)|^2}{W(\omega)} d\omega \leq A \quad (10)$$

Because the band-width $\omega_0 = R\pi$ of $X_0(\omega)$ is very large and $\tau = 2/R$ is quite small, if λ is sufficiently large, Ξ_B is quite similar to the set of continuous functions $F(\omega)$ satisfying $\int_{-\omega_0/2}^{\omega_0/2} \frac{|F(\omega)|^2}{W(\omega)} d\omega \leq A$, in good approximation.

Moreover, we consider that the following assumption holds.
Condition (c): Consider the functions $\Psi_m(\omega)e^{-ja_m\omega}$ $(m \in \Lambda)$ satisfy

$$\Psi_m(\omega)e^{-ja_m\omega} = X_0(\omega) \sum_{q=-\nu_m}^{\nu_m} \alpha_q^m e^{-jq\tau\omega} \cdot e^{jc_m\omega} \quad (11)$$

where α_q^m $(q = 0, \pm 1, \pm 2, \ldots, \pm\nu_m; \nu_m \leq \lambda)$ are prescribed real numbers and c_m $(m \in \Lambda)$ are prescribed integers. Let $\psi_m(t-a_m)$ $(m \in \Lambda)$ be the inverse Fourier transform of $\Psi_m(\omega)e^{-ja_m\omega}$. Since the band-width $\omega_0 = R\pi$ of $X_0(\omega)$ is very large and $\tau = 2/R$ is quite small, if ν_m is sufficiently large, $\psi_m(t-a_m)$ is favorable approximation of any continuous time-limited function. Besides, consider the approximation of $f_B(t)$ expressed by

$$\tilde{g}_B(t) = \sum\sum_{(m,n)\in\Delta(t)} f_m(nT+a_m)\psi_m(t-nT-a_m) \quad (12)$$

Further, let $\tilde{E}_B(\omega)$ be the Fourier transform of $\tilde{e}_B(t) = f_B(t) - \tilde{g}_B(t)$. Because $\psi_m(t-a_m)$ and $\tilde{g}_B(t)$ are band-limited in $|\omega| \leq \frac{\omega_0}{2}$, the error $\tilde{e}_B(t)$ is band-limited in $|\omega| \leq \frac{\omega_0}{2}$, also. Then, for the interpolation functions $\psi_m(t-a_m)$ $(m \in \Lambda)$, consider the following min-max value.

$$E_{\text{Min}}^{\text{Max}} = \min_{\text{all the } \alpha_q^m} \left[\max_{\text{all the } \sigma_q} \left\{ \int_{-\frac{\omega_0}{2}}^{\frac{\omega_0}{2}} \frac{|\tilde{E}_B(\omega)|^2}{W(\omega)} d\omega \right\} \right] \quad (13)$$

The integral $E_W = \int_{-\omega_0/2}^{\omega_0/2} \frac{|\tilde{E}_B(\omega)|^2}{W(\omega)} d\omega$ is smoothly continuous positive quadrature-form with respect to either set of the coefficients σ_q of $F(\omega)$ or set of the coefficients α_q^m of $\Psi_m(\omega)e^{-ja_m\omega}$ for the corresponding set of the suffix-integers q and m, respectively. Hence, if these coefficients are large, the value of E_W becomes large. Therefore, in the present discussion of minimizing E_W, we consider E_W for the bounded coefficients σ_q and α_q^m for the corresponding set of the suffix-integers q and m, respectively. Then, because E_W is smoothly continuous positive quadrature-form for either set of bounded coefficients σ_q or set of the bounded coefficients α_q^m, there exists a pair of the set of σ_q and the set of α_q^m realizing the above minimum value of the maximum value $E_{\text{Min}}^{\text{Max}}$.

When all the α_q^m $(m \in \Lambda, q = 0, \pm 1, \pm 2, \ldots, \pm\nu_m)$ are zero, we have $\int_{-\omega_0/2}^{\omega_0/2} \frac{|\tilde{E}_B(\omega)|^2}{W(\omega)} d\omega = \int_{-\omega_0/2}^{\omega_0/2} \frac{|F_B(\omega)|^2}{W(\omega)} d\omega \leq A$. Hence, obviously, $E_{\text{Min}}^{\text{Max}} \leq A$ holds and, from Eq.(13), we can recognize that $\int_{-\omega_0/2}^{\omega_0/2} \frac{|\tilde{E}_B(\omega)|^2}{W(\omega)} d\omega \leq A$ is valid. Therefore, the above Condition (b) is satisfied for the signal $f_B(t) = \tilde{e}_B(t)$, as well. Further, it is obvious that the signal $f_B(t) = \tilde{e}_B(t)$ satisfies the above Condition (a).

We consider the approximation $\tilde{g}(t)$ of $f_B(t)$ using the extended interpolation function $\phi_m(t)$ $(m \in \Lambda)$ that are made by the above $\psi_m(t)$ $(m \in \Lambda)$ and $\theta_m(t)$ $(m =$

$0, 1, 2, \ldots, \frac{RT}{2} + M - 1$) presented in this section.

We define $\tilde{e}(t) = f_B(t) - \tilde{g}(t)$. Further, let Ξ_e be the set of $\tilde{e}(t)$. Suppose that $\tilde{E}(\omega)$ is the Fourier transform of $\tilde{e}(t)$. Further, Let $\tilde{E}_b(t)$ be the band-limited version of $\tilde{E}(\omega)$ in $|\omega| \leq \frac{\omega_0}{2}$. Obviously, in the frequency band $|\omega| < \frac{\omega_0}{2}$, the approximation $\tilde{g}(t)$ is equal to the ideal approximation using the interpolation functions $v_{m,n}(t - a_m)$ ($m \in \Lambda$) with the infinite support only. Hence, in this case, the approximation uses the fixed infinite number of sample values, and hence, the condition $\int_{-\omega_0/2}^{\omega_0/2} \frac{|\tilde{E}_b(\omega)|^2}{W(\omega)} d\omega \leq A$ is satisfied [2].

From the consequence of Eq.(7), for any discrete time $q\tau$ ($q \in Z$), we have $\tilde{e}(q\tau) = \tilde{e}_B(q\tau)$. Hence, if we put $\tilde{e}(t) - \tilde{e}_B(t) = \tilde{e}_C(t)$, we have $\tilde{e}_C(q\tau) = 0$ for all the q in Z. Therefore, $\tilde{e}(t)$ is expressed by the sum of $\tilde{e}_B(t)$ band-limited in $|\omega| \leq \frac{\omega_0}{2}$ and $\tilde{e}_C(t)$ having zero sample values at $q\tau$ ($q \in Z$).

Further, we adopt the set of signals $\Xi = \Xi_B \cup \Xi_e$. Then, obviously, Condition (2) is satisfied. Since we adopt the interpolation functions $\phi_m(t)$ ($m \in \Lambda$) satisfying the discrete orthogonality, Condition (1) is satisfied, as well. Hence, $\tilde{g}(t)$ is MIE.

Now, we consider the maximum value of $\tilde{e}(q\tau)$ obtained by changing $f_B(t)$ in Ξ_B. Let $\tilde{e}_{max}(q\tau)$ be this maximum value. Then, $\tilde{e}_{max}(q\tau)$ is equal to the maximum value of $\tilde{e}(q\tau) = \tilde{e}_B(q\tau)$.

Since $\tilde{g}(t)$ is MIE, the approximation $\tilde{g}(q\tau)$ is equal to the approximation $\tilde{q}_{max}(q\tau)$ of $f_B(q\tau)$ minimizing $\tilde{e}_{max}(q\tau)$. Hence, the corresponding interpolation functions $\phi_m(q\tau) = \psi_m(q\tau)$ ($m \in \Lambda$) are equal to the interpolation functions $\hat{\psi}_m(q\tau)$ ($m \in \Lambda$) realizing the minimum $\tilde{e}_{max}(q\tau)$ for the signal in $\Xi = \Xi_B \cup \Xi_e$. However, for the signal $f(t) = \tilde{e}(t)$ in Ξ_e, the corresponding error is $\tilde{e}(t)$ itself. Hence, the corresponding interpolation functions $\phi_m(q\tau) = \psi_m(q\tau)$ ($m \in \Lambda$) are equal to the interpolation functions $\hat{\psi}_m(q\tau)$ ($m \in \Lambda$) realizing the minimum $\tilde{e}_{max}(q\tau)$ for the signal in Ξ_B.

Let Ξ_1 be the set of signals $f(t)$ such that the corresponding Fourier transform $F(\omega)$ is band-limited in $|\omega| \leq \frac{\omega_0}{2}$ and satisfies $\int_{-\omega_0/2}^{\omega_0/2} \frac{|F(\omega)|^2}{W(\omega)} d\omega \leq A$. Because signals in Ξ_B is good approximation of signals in Ξ_1, the above interpolation functions $\phi_m(q\tau) = \hat{\psi}_m(q\tau)$ ($m \in \Lambda$) are quite similar to the interpolation functions $\hat{\psi}_m(q\tau)$ ($m \in \Lambda$) realizing the minimum $\tilde{e}_{max}(q\tau)$ for the signal in Ξ_1, in good approximation.

Hence, the proposed discrete best approximation $\tilde{g}(q\tau)$ is equal to the discrete approximation

$$g_{\hat{\psi}}(q\tau) = \sum_{m=0}^{M-1} \sum_{n=-\infty}^{\infty} f_m(nT + a_m)\hat{\psi}_m(q\tau - nT - a_m) \quad (14)$$

realizing the minimum $\tilde{e}_{max}(q\tau)$ for the signals in Ξ_1, in actual. It should be noted that $g_{\hat{\psi}}(q\tau)$ is defined in the set of signals Ξ_1 only.

Indeed, let $\tilde{e}_{max}(q\tau)$ be the upper limit of $|\tilde{e}(q\tau)|$ obtained by changing $f(t)$ over all the signals $f(t)$ in Ξ_1. That is,

$$\tilde{e}_{max}(q\tau) = \sup_{f(t) \in \Xi_1} \{|\tilde{e}(q\tau)|\} \quad (15)$$

Then, from the discussion of [2], the worst-case measure of error $\tilde{e}_{max}(q\tau)$ is expressed by the following equation.

$$\tilde{e}_{max}(q\tau) = \frac{\sqrt{A}}{2\pi} \left\{ \int_{-\frac{\omega_0}{2}}^{\frac{\omega_0}{2}} W(\omega) \left| e^{j\omega(q\tau-d)} - \sum_{m=0}^{M-1} \sum_{n=-\infty}^{\infty} \hat{\psi}_m(q\tau - nT - a_m) H_m(\omega) e^{j\omega(nT+a_m)} \right|^2 d\omega \right\}^{\frac{1}{2}} \quad (16)$$

Further, we can obtain the optimum interpolation functions $\hat{\psi}_m(q\tau)$ by
(1) expanding $\tilde{e}_{max}(q\tau)^2$ with respect to $\hat{\psi}_m(q\tau - nT - a_m)$,
(2) differentiating $\tilde{e}_{max}(q\tau)^2$ with respect to $\hat{\psi}_m(q\tau - nT - a_m)$,
(3) making the resultant set of linear formulas to be zero, or shortly,

$$\frac{\partial \tilde{e}_{max}(q\tau)^2}{\partial \hat{\psi}_m(q\tau - nT - a_m)} = 0 \quad (17)$$

(4) solving the resultant set of linear equations with respect to $\hat{\psi}_m(q\tau - nT - a_m)$.

When the optimum interpolation functions are used, $\tilde{e}_{max}(q\tau)$ satisfies $\tilde{e}_{max}(q\tau) = \tilde{e}_{max}(q\tau + nT)$ (n is an arbitrary integer). As the direct consequence of this relation, the optimum interpolation functions obtained from Eq.(17) are expressed as $\hat{\psi}_m(q\tau - nT - a_m)$ and are realized by the shift-invariant FIR filters. This formulation is extended in the multi-dimensional case easily.

V. CONCLUSION

The optimum scan-type discrete approximation of FIR filter bank is presented that minimizes various worst-case measures of error, including the long-range worst-case measures of error in the time-domain or the frequency-domain.

This research was partially supported by a Grant from the Ministry of Education, Culture, Sports, Science, and Technology to promote multidisciplinary research projects at Nihon University, College of Engineering.

REFERENCES

[1] P. P. Vaidyanathan: "Multirate Systems and Filter Banks", Prentice Hall, 1993.
[2] T.Kida: "The Optimum Approximation of Multi-Dimensional Signals based on the Quantized Sample Values of Transformed Signals", IEICE Trans. Fundamentals, E78-A, No.2, pp.208-234, 1995.
[3] M. Vetterli and J. Kovacevic: "Wavelets and Subband Coding", Prentice Hall, 1995.
[4] Sankar Basu and Bernard Levy: "Multidimensional Filter Banks and Wavelets: Basic Theory and Cosine Modulated Filter Banks", Kluwer Academic Publishing, 1996.
[5] G. Strang and T. Nguyen: "Wavelets and Filter Banks" Wellesley-Cambridge Press, 1996.
[6] Ryan Prendergast: "Multirate Filter Bank Reconstruction of Signals from Bunched Samples", http://www.ece.ucdavis.edu/ hurst/papers/RyanP
[7] Yuichi Kida and Takuro Kida: "Theory of the optimum discrete approximation of FIR filter banks having the minimum worst-case measures of error", Proceedings of the 18th Workshop on Circuits and Systems in Karuizawa, to be published, 2005.

Directionally Weighted Color Interpolation for Digital Cameras

Hung-An Chang[1] and Homer Chen[1,2]

[1]Department of Electrical Engineering, [2]Graduate Institute of Communication Engineering
National Taiwan University, Taipei, Taiwan, R.O.C.
E-mail: hungana@ms56.hinet.net, homer@cc.ee.ntu.edu.tw

Abstract—Demosaicing is a color interpolation process that converts a raw image generated by a color filter array to a full color image by estimating the missing color components of each pixel from its neighbors. Most demosaicing approaches often introduce false colors or blur edges to areas of the image where there are dense edges. In this paper, we present a directionally weighted color interpolation method to address this problem. We consider finer edge orientations to improve the accuracy of edge indicators and exploit the inter-channel correlation of R, G, and B color planes in the interpolation process to reduce the undesirable artifacts and hence improve the perceptual image quality. Experimental results are shown to demonstrate the strengths of the proposed method.

I. INTRODUCTION

Most commercially available digital cameras use a color filter array (CFA) to capture images with a single CCD sensor array, resulting in a sub-sampled raw image with a single *R*, *G*, or *B* component for each pixel of the image. To reconstruct a full color image from the raw sensory data, a color interpolation scheme is required to estimate the two missing color components of each pixel. In this paper, we consider the interpolation for Bayer color arrays shown in Fig. 1.

The false colors and blur edges, often happening in image areas where there are dense edges, are due to the aliasing effects caused by averaging (low-pass filtering) image samples across the edges in the interpolation process. Many methods have been proposed [1]-[8] to solve the problem. These methods can be classified into two categories: non-adaptive interpolation and edge-directed adaptive interpolation.

For non-adaptive methods, the interpolation is carried out uniformly across the image. These methods reduce the aliasing by performing the interpolation in a different color space rather than the original *RGB* space. The smooth hue approach [1] and the color difference approach [4] belong to this category. Although such methods are relatively simple

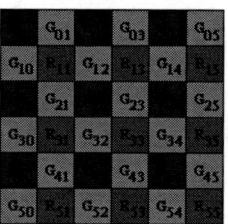
Figure 1. Bayer pattern.

to implement, their effectiveness is limited, for the lack of adaptivity near edge regions where aliasing is likely to occur.

The adaptive methods, on the other hand, adjust the weights of samples in the interpolation process by utilizing the edge information gathered by edge indicators. If an edge indicator senses a possible color intensity change across the direction of an edge, a small weight is assigned to the sample along that edge direction. Therefore, the effectiveness of an edge-adaptive method highly depends on its edge sensing capability.

In this paper, we propose a directionally weighted color interpolation method. By carefully designing the edge indicator for each edge direction, a good edge sensing performance is achieved. In our method, the color difference technique is also employed to exploit the inter-channel correlation between color planes. The details of this method are described in Sec. II. Experiments and performance comparisons are presented in Sec. III, followed by a conclusion in Sec. IV.

II. DIRECTIONALLY WEIGHTED COLOR ONTERPOLATION

A. Fine-Resolution Edge Indicators

To increase the edge sensitivity, we consider eight more samples in addition to the four commonly used samples for interpolation. This is illustrated in Fig. 2. Take the interpolation of the missing *G* at B_{44} as an example. The four nearby samples G_{43}, G_{34}, G_{45}, and G_{54} are commonly used by most edge-directed algorithms. The additional 8 samples G_{32}, G_{23}, G_{25}, G_{36}, G_{56}, G_{65}, G_{63}, and G_{52} considered in our approach are located one knight step away from B_{44}. Here, a knight step is a combination of one vertical step plus two horizontal steps, or vise versa, on the Bayer color array. For notational convenience, we number the directions from 1 to 12 as shown in Fig. 2. For a given sample, the nearest sample in the same color plane is either 2 or $2\sqrt{5}$ pixels away in each edge direction. For example, the nearest sample of B_{44} in direction 3 is B_{46}, which is 2 pixels away from B_{44}. Likewise, the nearest sample of B_{44} in direction 8 is B_{28}, which is $2\sqrt{5}$ pixels away.

In this multi-directional configuration, the difference of nearby samples (referred to as sample difference) along a direction is not a fair measure of the edge strength, because the nearest samples are not equally distanced in all directions. A proper adjustment on the sample difference is required to put the measure of edge strength along all edge directions on the same footing. There are two possible approaches.

The first approach is called linear adjustment. In this approach, a sample difference is normalized by the distance between the two samples based on which the sample difference is computed. Edge indicators of this kind are also referred to as directional derivatives [3].

The second approach, called stochastic adjustment, is based on the assumption that the image sample X can be modeled as a locally stationary Gaussian process with mean μ_X and standard deviation σ_X. Under this assumption, $\Omega(d)$, the correlation of two samples with distance d between them, can be approximated by $exp(-d^2/\rho^2)$ [2], where ρ is an image-dependent parameter, and the sample difference can be modeled as a Gaussian random variable with zero mean and standard deviation σ_d.

$$\sigma_d = \sqrt{E[(X(0)-X(d))^2]}$$
$$= \sqrt{E[(X'(0)-X'(d))^2]}$$
$$= \sqrt{E[(X'(0))^2] + E[(X'(d))^2] - 2E[X'(0)X'(d)]} \quad (1)$$
$$= \sigma_X \sqrt{2(1-\Omega(d))},$$

where $X(0)$ and $X(d)$, respectively, denote the current sample and a sample that is d pixels away from $X(0)$, $X'(0)=X(0)-\mu_X$, and $X'(d)=X(d)-\mu_X$. Normalizing the sample differences by σ_d is equivalent to multiplying the sample differences in directions 5-12 by a factor κ_n while keeping the sample differences in directions 1-4 unchanged.

$$\kappa_n = \sigma_X \sqrt{2(1-\Omega(2))} / (\sigma_X \sqrt{2(1-\Omega(2\sqrt{5}))})$$
$$\approx \sqrt{(1-\Omega(2))/(1-(\Omega(2))^5)}, \quad 5 \leq n \leq 12 \quad (2)$$

where the value of κ_n is computed based on the approximated $\Omega(d)$. κ_n is around 0.5 as long as $\Omega(2)$ is high enough. For example, $\kappa_n=0.519$ when $\Omega(2)=0.85$, $\kappa_n=0.5$ when $\Omega(2)=0.9$, and $\kappa_n=0.472$ when $\Omega(2)=0.95$.

For the two approaches, the edge indicator I_n for direction n can be obtained by:

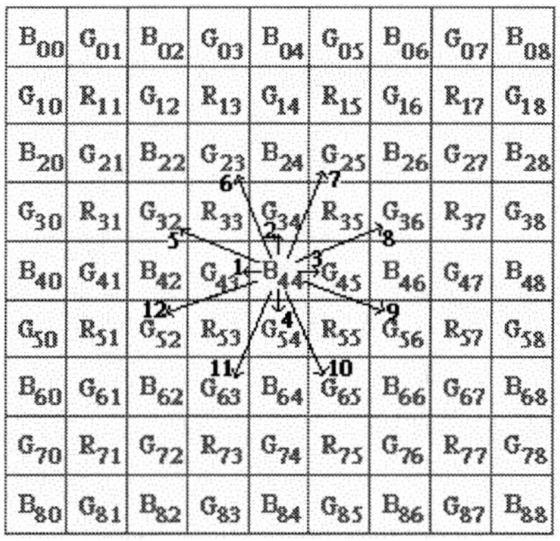

Figure 2. Directions of interpolation.

Table 1. POSITIONS OF NEARBY SAMPLES IN STEPS 1 AND 3.

n	v_n	h_n	n	v_n	h_n
1	0	-1	2	-1	0
3	0	+1	4	+1	0
5	-1	-2	6	-2	-1
7	-2	+1	8	-1	+2
9	+1	+2	10	+2	+1
11	+2	-1	12	+1	-2

$$I_n(i,j) = \kappa_n(abs(P(i+v_n, j+h_n) - P(i-v_n, j-h_n)) + abs(P(i+2v_n, j+2h_n) - P(i,j))) \quad (3)$$

where $P(i,j)$ denotes the sample at the position (i, j) and h_n and v_n, listed in Table 1, denote the horizontal and vertical positions, respectively, of a nearest sample relative to the sample to be interpolated. In our experiments, we set κ_n to the following values:

$$\kappa_n = \begin{cases} 1, & 1 \leq n \leq 4, \text{ for linear and stochastic adjustment} \\ 1/\sqrt{5}, & 5 \leq n \leq 12, \text{ for linear adjustment} \\ 0.5, & 5 \leq n \leq 12, \text{ for stochastic adjustment} \end{cases} \quad (4)$$

Experimental results of these two approaches are compared in Sec. 3.

B. Directionally Weighted Interpolation

Because of high correlation between the R, G, and B channels [4], the color difference values (G-R, G, G-B) are relatively smoother than the (R, G, B) values, making the aliasing effects less pronounced in the color difference space. Like some previous methods, our interpolation method is conducted in the color difference space.

Let $w_n(i,j)$, be the weight for direction n when interpolating a missing color at (i,j).

$$w_n(i,j) = (\frac{1}{1+I_n(i,j)}) / \sum_{n=1}^{12} \frac{1}{1+I_n(i,j)}, \quad (5)$$

where $I_n(i,j)$ is obtained from (3).

Our interpolation algorithm consists of four steps:

Step 1. Interpolate the missing green value of blue/red samples.

The missing green value $G(i,j)$ of a blue sample $B(i,j)$ is determined by:

$$G(i,j) = B(i,j) + \sum_{n=1}^{12} w_n(i,j) * K_{b,n}(i+v_n, j+h_n) \quad (6)$$

where

$$K_{b,n}(i+v_n, j+h_n) = G(i+v_n, j+h_n) - B(i+v_n, j+h_n) \quad (7)$$

is the color difference along direction n. The rule of calculating the color difference is as follows. If the value of B (or G) is available at the target location in the Bayer array, use it directly; otherwise, check to see if it has been interpolated previously. Use the interpolated value if it exists. If it does not, take the average of the two adjacent samples of the desired color in the Bayer array. Refer to Fig. 2 for an example. To interpolate the missing green value at $(4,4)$, $K_{b,6}(2,3)$ is needed and is obtained by calculating $G_{2,3} - (B_{2,2}+B_{2,4})/2$. Note again that the color difference, as opposed to green difference [3], is adopted here to enhance the accuracy of the interpolation. The missing green value of a red sample can be obtained in a way similar to (6) and (7).

Step 2. Interpolate the missing red/blue values of blue/red samples.

In this step, we consider only the four nearest samples of the same color in the diagonal directions because there are no similar samples in the other 8 directions (Fig. 2). The vertical and horizontal positions of these four samples (indexed from $n=1$ to $n=4$) relative to the sample to be interpolated are listed in Table 2. Let $I'_n(i,j)$ be the edge indicator for direction n.

$$I'_n(i,j) = abs(P(i+v'_n, j+h'_n) - P(i-v'_n, j-h'_n)) + \\ abs(P(i+2v'_n, j+2h'_n) - P(i,j)). \quad (8)$$

Denote the weight for direction n when interpolating a missing color at (i,j) by $w'_n(i,j)$,

$$w'_n(i,j) = \left(\frac{1}{1+I'_n(i,j)}\right) \Big/ \sum_{n=1}^{4} \frac{1}{1+I'_n(i,j)}. \quad (9)$$

The missing red value of a blue sample $B(i,j)$ is obtained by:

$$R(i,j) = G(i,j) - \sum_{n=1}^{4} w'_n(i,j) * K_{r,n}(i+v'_n, j+h'_n), \quad (10)$$

where

$$K_{r,n}(i+v'_n, j+h'_n) = G(i+v'_n, j+h'_n) - R(i+v'_n, j+h'_n). \quad (11)$$

is the color difference along direction n.

The interpolation of missing blue values of red samples is similar to (10) and (11).

TABLE 2. POSITIONS OF NEARBY SAMPLES IN STEP 2.

n	v'_n	h'_n	n	v'_n	h'_n
1	-1	-1	2	-1	+1
3	+1	+1	4	+1	-1

Step 3. Interpolate missing red/blue values of green samples.

The missing red value of a green sample $G(i,j)$ is obtained by:

$$R(i,j) = G(i,j) - \sum_{n=1}^{12} w_n(i,j) * K_{r,n}(i+v_n, j+h_n), \quad (12)$$

where $w_n(i,j)$ is computed by (5). The interpolation of the remaining missing blue values is similar to (12).

Step 4. Adjust the estimated green values of red/blue samples.

In step 1 the blue (red) values required to calculate the color differences are computed by averaging the values of adjacent samples. This average operation, however, will cause some aliasing. To reduce this aliasing, we re-compute the green values interpolated in step 1 by using the blue (red) values obtained in step 3 when calculating the color differences as in (7).

III. EXPERIMENTS AND COMPARISONS

The directionally weighted interpolation method is designed for reducing the artifacts around dense edges. To evaluate its effectiveness, the performance of the proposed method was compared with that of four existing color interpolation methods: the bilinear method, the color difference based method [4], the gradient-based method [1], and the C2D2 method [3].

The bilinear interpolation method simply averages the most nearby samples to interpolate missing color samples, and introduces serious artifacts near the edges. The color difference based method [4] considers same set of nearby samples as the bilinear interpolation, but utilizes advantages of the color differences. The gradient based method [1] is a hard decision, edge-directed method. It compares the horizontal and vertical gradients and chooses samples in the direction with smaller gradient for interpolation. The C2D2 (Color Correlations and Directional Derivatives) [3] method is an edge-directed method considering 8 directions for interpolation. The relative weights for these directions are calculated by the directional derivatives. Besides of these four methods, the performances of two adjustment approaches described in Sec. II are also compared.

The results of two test images are shown in Fig. 3-4 respectively. Fig. 3 is the down-right portion of a building, which contains very fine textures below each window. The image in Fig. 4 is cropped from the airplane image of Kodak image database [3]. The PSNR values of resulting images in Fig. 3-4 are listed in Table 3 and Table 4 respectively. Note that only the pixels shown in Fig. 3-4 are considered for the computation of PSNR to evaluate the performances of these methods near the edge regions. According to the experiment results, the artifacts are significantly reduced by our method. Besides, in terms of the PSNR results, the stochastic adjustment approach outperforms the linear adjustment approach. The results of the full images in Fig. 3-4 as well as results of other benchmark images are available on http://www2.ee.ntu.edu.tw/~b9901049/demosaicing.html.

IV. CONCLUSION

In this paper, we have described a directionally weighted color interpolation method for digital cameras using a color filter array to acquire the image. A general scheme for extracting edge information at finer orientation than previous approaches was also presented. By going for finer edge directions and exploiting inter-channel correlation between color planes, our method is able to achieve a better edge-sensing capability. The comparison results demonstrate the robustness of our method in dealing with densely spaced edges.

REFERENCES

[1] B. Eamanath, W.E. Synder, and G. L. Bilbro, "Demosiacking methods for Bayer color array," Journal of Electronic Imaging 11(3), pp. 306-315, Jul. 2002

[2] X. Li and M.T. Orchard, "New edge-directed interpolation," IEEE Trans. on Image Processing, vol.10, no. 10, pp. 1521-1527, Oct. 2001

[3] N. Kehtarnavaz, H. Oh, and Y. Yoo, "Color filter array interpolation using correlations and directional derivatives," Journal of Electronic Imaging 12(4), pp. 621-632, Oct. 2003

[4] S.C. Pei and I. K. Tam, "Effective color interpolation in CCD color filter arrays using signal correlation," IEEE Trans. on Circuit and System for Viedo Technology, vol. 13, no. 6, Jun. 2003

[5] R. Lukac, K.N. Plataniotis, D. Hatzinakos, and M. Aleksic, "A novel cost effective demosaicing approach," IEEE Trans. on Consumer Electronics, vol.50, no. 1, Feb. 2004.

[6] B. S Hur and M. G. Kang, "High definition color interpolation scheme for progressive scan CCD image sensor," IEEE Trans. on Consumer Electronics, vol. 47, no. 1 Feb. 2001.

[7] B. K. Gunturk, Y. Altunbasak, R. M. Mersereau, "Color plane interpolation using alternating projections," IEEE Trans. on Image Processing, vol. 11, no. 9, Sept. 2002.

[8] C.L. Hsu, *Demosaicking algorithm with noise removal based on adaptive spatial filter*, Master Thesis, GICE, National Taiwan University

TABLE 3. PSNR COMPARISON FOR FIGURE 3.

	PSNR_G	PSNR_R	PSNR_B
Bilinear	19.63	18.94	19.07
Color difference based	24.50	24.23	25.12
Gradient based	18.05	19.36	19.60
C2D2	23.39	23.48	24.51
Proposed (linear)	27.38	24.47	26.45
Proposed (stochastic)	27.49	24.58	26.59

TABLE 4. PSNR COMPARISON FOR FIGURE 4.

	PSNR_G	PSNR_R	PSNR_B
Bilinear	23.15	21.50	21.61
Color difference based	29.85	28.51	29.20
Gradient based	24.51	26.57	26.27
C2D2	30.32	31.45	31.00
Proposed (linear)	34.64	31.96	31.98
Proposed (stochastic)	34.77	32.08	32.08

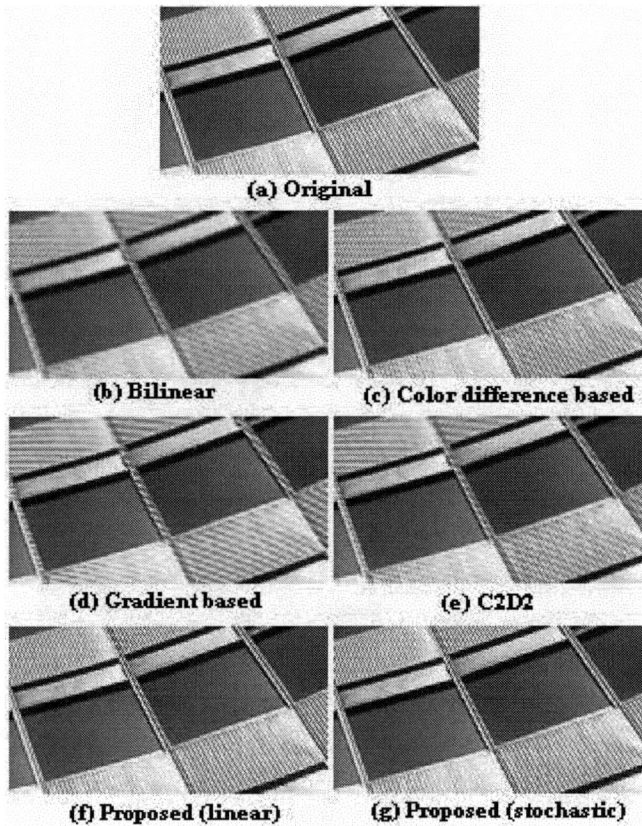

Figure 3. Comparison of the methods for the building image.

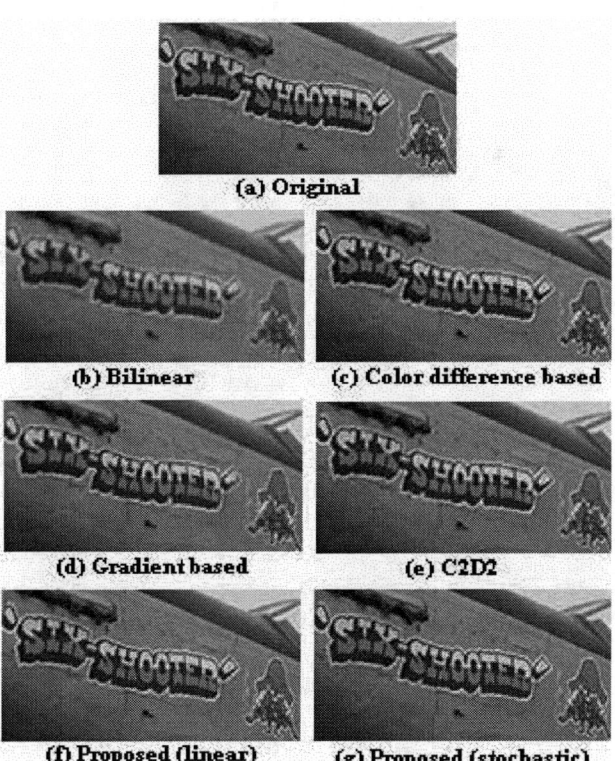

Figure 4. Comparison of the methods for the airplane images. Note the aliasing effects near the edges of the logo.

A Novel Color Interpolation Algorithm by Pre-estimating Minimum Square Error

Jhing-Fa Wang
Electrical Engineering
National Cheng Kung University
Tainan, Taiwan
wangjf@csie.ncku.edu.tw

Chien-Shun Wang
Electrical Engineering
National Cheng Kung University
Tainan, Taiwan
cs.wang@emc.com.tw

Han-Jen Hsu
Electrical Engineering
National Cheng Kung University
Tainan, Taiwan
hjhsu@icwang.ee.ncku.edu.tw

Abstract—In this paper, a novel color interpolation algorithm for Color Filter Array (CFA) in digital still cameras (DSCs) is presented. This work introduces pre-estimating the minimum square error to address the color interpolation for CFA. In order to estimate the missing pixels in Bayer CFA pattern, the weights of adjacent color pattern pairs are decided by the matrix computation. We adopt the color model (K_R, K_B) used in many color interpolation algorithms for CFA. The proposed algorithm can achieve better performance shown in the experimental results. Comparing the previous methods, the proposed color interpolation algorithm can provide high quality image in DSCs.

Index Terms—Color interpolation, color filter array, Bayer pattern

I. INTRODUCTION

Digital still cameras (DSCs) have been used in consumer electronics for wide applications. Many DSCs use single imagining sensor (CCD or CMOS) combined with color filter arrays (CFA) to reduce the hardware cost. The most popular CFA pattern is Bayer CFA pattern [1]. The pixels of *G* channel are designed to half of all, and the pixels of *B*, *R* channel are designed to share the others in Bayer CFA pattern. In order to estimate the lack colors exactly in CFA, many color interpolation algorithms [2]-[10] have been proposed. Sakamoto etc. [2] used the conventional interpolation techniques in image interpolation for CFA. For the consideration of integration into DSCs, they also provided the execution time of the implementation. However, the bilinear method can not be used to estimate "true" color in the edge region. The performance of different algorithms is similar in the flat image region. The main effort of the color interpolation is in the edge region. In order to reduce the zipper artifact in the edge region, an edge-sensing method is provided by Adams [3]. Kimmel [4] presented an adaptive edge-sensing method based on the edge indicators. S. C. Pei etc. [5] proposed an effective color interpolation (ECI) adopted the image model by Adams [6]. Gunturk etc. [7] proposed an alternating-projections (AP) method of frequency subbands in color planes. Y. P. Tan etc. [8] proposed the enhanced ECI and AP methods to improve the reconstructed image quality.

II. PROPOSED NOVEL COLOR INTERPOLATION ALGORITHM

This Section will introduce the proposed color interpolation algorithm. The algorithm is based on pre-estimating the minimum square error between the estimated value and original value of a pixel. The weight vector is decided by the matrix computation.

A. Basic Description of Proposed Color Interpolation Algorithm

The basic description builds the fundamentals of the proposed color interpolation algorithm as in (1).

$$K"(x) = \sum_{i=1}^{n} w_i K(x_i) \quad (1)$$

where $K"(x)$ is the estimated value of central pixel x, and $K(x_i)$ is the value of adjacent pixel x_i

The missing color value is calculated by linear estimation. Refer to [11], the minimum square error between the estimated value and the original true value $K(x)$ of a pixel is defined as in (2).

$$\begin{aligned}
&\{[K"(x)-K(x)]^2\} \\
&= \{[K"(x)]^2\} - 2\{K"(x) \cdot K(x)\} + \{[K(x)]^2\} \\
&= \sum_{i=1}^{n}\sum_{j=1}^{n} w_i w_j \{K'(x_i) \cdot K'(x_j)\} - 2 \cdot \sum_{i=1}^{n} w_i \{K'(x_i) \cdot K(x)\} + \{[K(x)]^2\} \\
&= \sum_{i=1}^{n}\sum_{j=1}^{n} w_i w_j C(x_i,x_j) - 2\sum_{i=1}^{n} w_i C(x,x_i) + C(0) \quad (2)
\end{aligned}$$

where $C(x_i,x_j) = \{K'(x_i) \cdot K'(x_j)\}$

In order to minimize the square error, the partial derivative of (2) is shown as in (3). The result of (3) must be zero to derivate (4).

$$\frac{\partial (2)}{\partial w_i} = 2\sum_{j=1}^{n} w_j C(x_i,x_j) - 2C(x,x_i), i=1,...,n \quad (3)$$

$$\sum_{j=1}^{n} w_j C(x_i,x_j) = C(x,x_i), i=1,...,n \quad (4)$$

From (4), we expand the equation as in (5)

$$\begin{bmatrix} C_{11} & C_{12} & C_{13} & C_{14} & 1 \\ C_{21} & C_{22} & C_{23} & C_{24} & 1 \\ C_{31} & C_{32} & C_{33} & C_{34} & 1 \\ C_{41} & C_{42} & C_{43} & C_{44} & 1 \\ 1 & 1 & 1 & 1 & 0 \end{bmatrix} \begin{bmatrix} W_1 \\ W_2 \\ W_3 \\ W_4 \\ \lambda \end{bmatrix} = \begin{bmatrix} C_{10} \\ C_{20} \\ C_{30} \\ C_{40} \\ 1 \end{bmatrix} \quad (5)$$

where C_{ij} is $C(x_i, x_j)$, C_{i0} is $C(x, x_i)$

$$\begin{bmatrix} W_1 \\ W_2 \\ W_3 \\ W_4 \\ \lambda \end{bmatrix} = \begin{bmatrix} C_{11} & C_{12} & C_{13} & C_{14} & 1 \\ C_{21} & C_{22} & C_{23} & C_{24} & 1 \\ C_{31} & C_{32} & C_{33} & C_{34} & 1 \\ C_{41} & C_{42} & C_{43} & C_{44} & 1 \\ 1 & 1 & 1 & 1 & 0 \end{bmatrix}^{-1} \begin{bmatrix} C_{10} \\ C_{20} \\ C_{30} \\ C_{40} \\ 1 \end{bmatrix} \quad (6)$$

where $W_1 + W_2 + W_3 + W_4 = 1$

B. Illustration of Proposed Color Interpolation algorithm

The G channel is the most important component in color interpolation. The better estimated quality will also improve the quality of R, B channel. The subscripts are used to describe the proposed algorithm in this paper. The R'', G'', or B'' are designed as estimated color value, and the R, G, or B are designed as original color value. The average color value are denoted as R', G', or B'. First, we will introduce how to estimate G channel in R, B planes. Fig. 1 is the reference Bayer CFA pattern. We consider the relationships of color pattern pairs in different directions. In G interpolation, we use the vertical and horizontal color pattern pairs. In R interpolation, the diagonal color pattern pairs are also used. The subscripts "1 to 8" denote the different relations of directions as shown in Fig. 2 to Fig. 4. And the index of central position is 0. The subscript with comma is used to define the actual position in Reference Bayer CFA pattern.

1) Step1: G channel interpolation at B pixel

The estimated value of central pixel can be calculated from adjacent pixels as in (7). We define the value in previous section as the color model ($K_R = G - R$, $K_B = G - B$). The sample pixel positions used to estimate G channel interpolation at B pixel are shown in Fig. 2.

$$K''_{B(i,j)} = W_1 \times K'_{B(i-1,j)} + W_2 \times K'_{B(i,j-1)} + W_3 \times K'_{B(i+1,j)} + W_4 \times K'_{B(i,j+1)} \quad (7)$$

Let
$$\begin{cases} K'_{B(i-1,j)} = G_{i-1,j} - ((B_{i-2,j} + B_{i,j})/2) \\ K'_{B(i,j-1)} = G_{i,j-1} - ((B_{i,j-2} + B_{i,j})/2) \\ K'_{B(i+1,j)} = G_{i+1,j} - ((B_{i+2,j} + B_{i,j})/2) \\ K'_{B(i,j+1)} = G_{i,j+1} - ((B_{i,j+2} + B_{i,j})/2) \end{cases} \quad (8)$$

The weights can be calculated from (6). In order to reduce computational complexity, C_{ij} is to simplify for hardware design.

The relation is a linear distance function combined with our empirical results. Refer to Fig.1 and Fig. 2 together, some of equations are shown below. ($C_{ij}, 1 \le i \le 4$ and $1 \le j \le 4$)

$$\begin{cases} C_{10} = (|B_{i-2,j} - B_{i,j}| + |G_{i-2,j}' - G_{i,j}'|)/256 + 1 \\ C_{20} = (|B_{i,j-2} - B_{i,j}| + |G_{i,j-2}' - G_{i,j}'|)/256 + 1 \\ C_{12} = (|B_{i-2,j} - B_{i,j-2}| + |G_{i-2,j}' - G_{i,j-2}'|)/512 + 1 \\ C_{34} = (|B_{i,j+2} - B_{i+2,j}| + |G_{i,j+2}' - G_{i+2,j}'|)/512 + 1 \\ C_{11} = C_{22} = C_{33} = C_{44} = 0 \end{cases} \quad (9)$$

where
$$\begin{cases} G_{i,j}' = (G_{i-1,j} + G_{i+1,j} + G_{i,j-1} + G_{i,j+1})/4 \\ G_{i-2,j}' = (G_{i-2,j-1} + G_{i-1,j} + G_{i-2,j+1})/3 \end{cases} \quad (10)$$

The average G' color values can be estimated by adjacent three or four pixels as in (10). The other G' color value pixels can be calculated by adjacent three pixels except for the central pixel. The final estimated color model can be used to estimate G channel interpolation as in (11).

$$G_{i,j}'' = K''_{B_{i,j}} + B_{i,j} \quad (11)$$

2) Step2: G channel interpolation at R pixel

The missing G channel at R pixels can be estimated in the same method. After the G channel interpolation at B, R pixels. Next, the R channel interpolation at B, R planes will be introduced. The main difference between G channel interpolation and R (or B) channel interpolation is the G color value will be estimated first. We can use the already estimated G channel to estimate the remainder colors.

3) Step3: R channel interpolation at B pixel

The sample pixel positions used to estimate R channel interpolation at B pixel are shown in Fig. 3(a). We only show part of equations in (12) to (15). The detailed process is omitted.

$$K''_{R(i,j)} = W_1 \times K'_{R(i-1,j-1)} + W_2 \times K'_{R(i+1,j-1)} + W_3 \times K'_{R(i-1,j+1)} + W_4 \times K'_{R(i+1,j+1)} \quad (12)$$

Let
$$\begin{cases} K'_{R(i-1,j-1)} = G_{i-1,j-1}'' - R_{i-1,j-1} \\ K'_{R(i+1,j-1)} = G_{i+1,j-1}'' - R_{i+1,j-1} \\ K'_{R(i-1,j+1)} = G_{i-1,j+1}'' - R_{i-1,j+1} \\ K'_{R(i+1,j+1)} = G_{i+1,j+1}'' - R_{i+1,j+1} \end{cases} \quad (13)$$

Refer to Fig. 2 and Fig. 3(a) together, some of equations are shown below. The index i, j of (14) is similar as $i+4, j+4$ in (9).

$$\begin{cases} C_{50} = (|B_{i-2,j-2} - B_{i,j}| + |G_{i-2,j-2}'' - G_{i,j}''|)/256 + 1 \\ C_{60} = (|B_{i+2,j-2} - B_{i,j}| + |G_{i+2,j-2}'' - G_{i,j}''|)/256 + 1 \\ C_{56} = (|B_{i-2,j-2} - B_{i+2,j-2}| + |G_{i-2,j-2}'' - G_{i+2,j-2}''|)/256 + 1 \\ C_{78} = (|B_{i+2,j+2} - B_{i-2,j+2}| + |G_{i+2,j+2}'' - G_{i-2,j+2}''|)/256 + 1 \\ C_{55} = C_{66} = C_{77} = C_{88} = 0 \end{cases} \quad (14)$$

$$R_{i,j}'' = G_{i,j}'' - K''_{R_{i,j}} \quad (15)$$

4) Step4: R channel interpolation at G pixel (mode 1)

There are two conditions to estimate R channel at G pixel shown in Fig. 4. The sample pixel positions used to estimate R channel interpolation at G pixel are shown in Fig. 3(b).

$$K''_{R(i,j)} = W_1 \times K'_{R(i-1,j)} + W_2 \times K'_{R(i,j-1)} + W_3 \times K'_{R(i+1,j)} + W_4 \times K'_{R(i,j+1)} \quad (16)$$

Let
$$\begin{cases} K'_{R(i-1,j)} = G_{i-1,j}" - R_{i-1,j} \\ K'_{R(i,j-1)} = G_{i,j-1}" - R_{i,j-1}" \\ K'_{R(i+1,j)} = G_{i+1,j}" - R_{i+1,j}" \\ K'_{R(i,j+1)} = G_{i,j+1}" - R_{i,j+1}" \end{cases} \quad (17)$$

Refer to Fig. 2 and Fig. 3(b) together, some of equations are shown below.

$$\begin{cases} C_{10} = \left(|G_{i-2,j} - G_{i,j}| + |R_{i-2,j}' - R_{i,j}'| \right)/256 + 1 \\ C_{20} = \left(|G_{i,j-2} - G_{i,j}| + |R_{i,j-2}' - R_{i,j}'| \right)/256 + 1 \\ C_{12} = \left(|G_{i-2,j} - G_{i,j-2}| + |R_{i-2,j}' - R_{i,j-2}'| \right)/256 + 1 \quad (18) \\ C_{34} = \left(|G_{i+2,j} - G_{i,j+2}| + |R_{i+2,j}' - R_{i,j+2}'| \right)/256 + 1 \\ C_{11} = C_{22} = C_{33} = C_{44} = 0 \end{cases}$$

where
$$\begin{cases} R_{i,j}' = (R_{i-1,j} + R_{i+1,j} + R_{i,j-1}" + R_{i,j+1}")/4 \\ R_{i-2,j}' = (R_{i-2,j-1}" + R_{i-1,j} + R_{i-2,j+1}")/3 \end{cases} \quad (19)$$

$$R_{i,j}" = G_{i,j} - K"_{R_{i,j}} \quad (20)$$

5) Step5: R channel interpolation at G pixel (mode 2)
The R channel interpolation at G pixels is similar as step 4.

6) Step6: B channel interpolation at R pixel
The B channel interpolation at R pixels is analogous to the step 3.

7) Step7: B channel interpolation at G pixel (mode 1)
The B channel interpolation at R pixels is analogous to the step 4.

8) Step8: B channel interpolation at G pixel (mode 2)
The B channel interpolation at R pixels is analogous to the step 5.

III. EXPERIMENTAL RESULT AND COMPLEXITY COMPARISON

We use Kodak images to evaluate the proposed algorithm. The subjective observation of small reconstructed image regions is also provided shown in Fig. 5. The complexity analysis of different algorithms is shown in Table. I. The PSNR results of different algorithms are shown in Table. II. The performance of the proposed algorithm is better than most of other algorithms.

IV. CONCLUSION

The contribution of this paper is to introduce pre-estimating the minimum square error in color interpolation. We provide a complete derivation of matrix computation in CFA. The proposed color interpolation algorithm can provide a solution to produce high quality images. The performance of B channel interpolation is better than previous algorithms.

REFERENCES

[1] B. Bayer, "Color imaging array," U. S. Patent No. 3,971,065, 1976.

[2] T. Sakamoto, C. Nakanishi, and T. Hase, "Software pixel interpolation for digital still camera suitable for a 32-Bit MCU," *IEEE Trans. Consumer Eletronics*, no. 4, 1998.

[3] J. E. Adams Jr., "Interactions between color plane interpolation and other image processing functions in electronic photography," in *Proc. SPIE*, vol. 2416, C. Anagnostopoulos and M. Lesser, Eds., Bellingham, WA, 1995, pp. 144–151.

[4] R. Kimmel, "Demosaicking: Image reconstruction from color CCD samples," *IEEE Trans. Image Processing*. vol.7, no. 3, pp. 1221-1228, 1999.

[5] S. C. Pei and I. K. Tam, "Effective color interpolation in CCD color filter arrays using signal correlation," *IEEE Trans. Circuits and Systems for Video Technology*, vol. 13, 2003.

[6] J. E. Adams Jr., "Design of practical color filter array interpolation algorithm for digital cameras," in *Proc. SPIE*, vol. 3028, D. Sinha, Ed., Bellingham, WA, 1997, pp. 117–125.

[7] B. K. Gunturk, Y. Altunbasak and R. M. Mersereau, "Color plane interpolation using alternating projections," *IEEE Trans. Image Processing*, vol.11, no.9, 2002.

[8] L. Chang and and Y. P. Tan, "Effective use of Spatial and Spectral Correlations for Color Filter Array Demosaicking," *IEEE Trans. Consumer Electronics*, 2004.

[9] T. W. Freeman, "Median filter for reconstructing missing color samples," U.S. Patent No. 4,724,395, 1998.

[10] X. Li and M. T. Orchard, "New edge directed interpolation," *IEEE Trans. Image Processing*, vol. 10, no. 10, 2001.

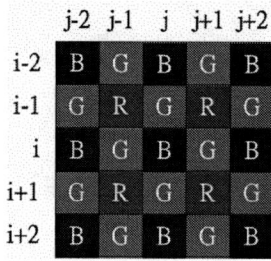

Figure 1. Reference Bayer CFA pattern

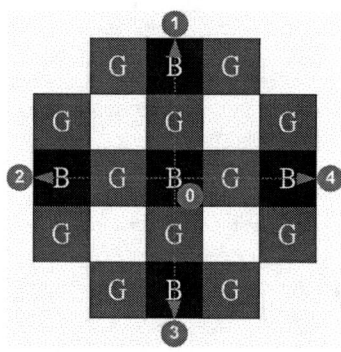

Figure 2. Reference samples of G channel interpolation

(a)

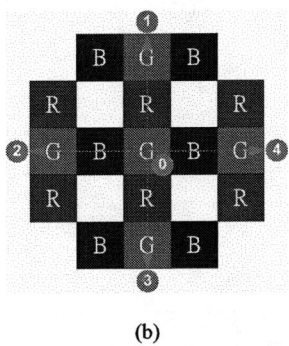

(b)

Figure 3. Reference samples of R channel interpolation

Figure 4. Two different conditions of R channel interpolation

TABLE. I COMPLEXITY ANALYSIS

	Original ECI [5]	Ours method	Enhanced ECI [8]	Original AP [7]	Enhanced AP [8]
Addition	10 MN	57.5 MN	58 MN	384 MN	510 MN
Multiplication	NO	8 MN	28 MN	384 MN	457 MN
Bit-shift	4 MN	11.5 MN	2 MN	NO	2 MN
Absolute Conversion	NO	16 MN	16 MN	NO	16 MN
Division	NO	1.5 MN	NO	NO	NO
Matrix	NO	2 MN	NO	NO	NO
Refinement Step	No needed	No needed	Needed	Needed	No needed
Complexity	Low	Middle	Middle	High	High
Hardware Implementation	Suitable	Suitable	Not Suitable	Not Suitable	Not Suitable
	Only addition and bit-shift, easy implementation	The matrix computation is regular	The weight computation is not regular	The hardware cost is too high	The hardware cost is too high

Image Size : $M \times N$

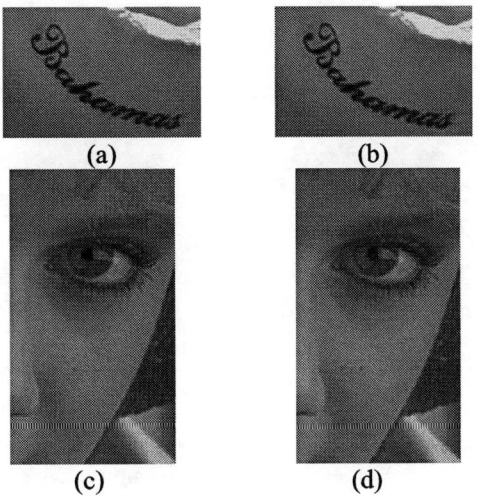

Figure 5. Small regions of experimental results. (a)original image of no.3 (b)reconstructed image of no.3. (c)original image of no.4 (d) reconstructed image of no.4

TABLE. II COMPARISON OF DIFFERENCE ALGORITHMS (PSNR RESULTS IN R, G, B PLANES)

	1	2	3	4	5	6	7	8	9	10	11	12	13	14	15	Avg.
Bilin-ear	25.31	31.89	33.45	32.61	25.75	26.70	32.57	22.53	31.56	31.84	28.17	32.67	23.09	28.15	28.15	28.96
	29.58	36.26	37.17	36.53	29.32	31.05	36.47	27.40	35.72	35.37	32.22	36.82	26.49	32.01	32.01	32.96
	35.35	32.36	33.83	32.91	25.92	27.00	32.58	22.48	31.38	31.23	28.34	32.35	23.00	28.60	28.60	29.73
Free-man [9]	30.75	35.95	39.02	37.07	32.88	32.14	39.10	27.40	36.86	38.02	33.70	37.77	29.52	33.54	35.13	34.59
	37.08	42.36	44.27	43.44	38.82	38.56	44.18	34.87	43.09	43.66	39.80	43.74	34.20	38.91	42.35	40.62
	30.90	37.77	39.08	39.44	33.02	32.30	38.84	27.40	37.09	37.51	34.20	37.70	29.14	34.24	38.17	35.12
NEDI [10]	33.77	37.24	39.57	36.89	31.45	35.59	38.58	30.58	39.53	38.14	34.98	40.17	29.52	34.41	34.76	35.68
	35.95	41.34	41.67	41.58	33.70	37.67	41.14	33.61	41.68	40.02	37.70	42.87	32.00	37.56	39.43	38.53
	34.33	39.87	39.20	40.09	31.13	35.17	38.83	30.27	40.07	38.10	35.32	40.40	29.37	34.70	37.48	36.29
ECI [5]	33.21	36.85	40.51	36.88	35.04	34.65	40.31	30.16	38.66	39.37	35.62	39.22	31.43	34.78	36.33	36.20
	35.65	41.28	43.11	42.16	36.84	36.98	42.16	33.09	41.37	42.07	38.06	42.40	32.64	37.87	41.11	39.12
	33.38	39.35	40.07	40.77	34.74	34.12	39.69	29.87	39.32	39.43	36.09	39.38	30.64	35.15	39.31	36.75
AP [7]	37.17	38.29	41.29	37.70	37.70	38.34	41.85	35.36	41.37	41.44	38.66	42.08	34.40	35.81	38.02	38.63
	40.02	40.20	43.23	42.15	39.42	41.21	43.32	37.99	44.00	44.30	40.90	44.30	36.79	37.75	40.09	41.04
	36.96	38.73	39.82	41.29	35.53	37.07	39.54	34.26	40.81	40.60	38.36	41.17	33.01	34.12	38.70	38.00
EECI [8]	37.00	38.01	41.98	37.71	37.60	37.57	42.08	34.37	41.62	41.28	38.24	41.76	34.24	36.10	36.97	38.44
	40.64	43.81	45.50	44.07	40.93	41.19	45.35	38.63	44.78	45.01	42.17	45.62	36.85	40.58	42.67	42.52
	37.44	41.44	41.68	42.16	37.01	37.12	41.60	34.43	41.56	41.20	39.39	41.95	33.24	36.86	40.34	39.16
EAP [8]	37.86	39.62	42.42	38.90	38.92	38.21	42.66	35.60	41.89	41.92	39.00	42.31	35.14	38.26	39.04	39.45
	40.76	41.62	44.64	43.42	40.54	41.42	44.28	38.40	44.26	44.77	41.41	45.00	37.66	39.86	41.47	41.97
	37.50	40.00	41.27	42.15	36.20	36.83	40.36	34.32	41.11	40.94	38.53	41.36	33.52	35.78	39.58	38.63
Ours	36.08	38.49	42.12	39.14	37.12	37.18	42.57	33.32	41.94	42.07	37.80	41.72	32.58	36.35	37.74	38.42
	37.25	42.71	44.13	42.51	37.76	38.50	43.76	35.24	43.15	43.31	39.32	43.77	32.87	39.24	41.73	40.35
	36.31	41.74	41.69	42.55	36.45	36.72	41.83	33.39	41.34	41.26	38.47	42.04	31.82	36.66	40.94	38.88

A Random-valued Impulse Noise Detector Using Level Detection

Noritaka YAMASHITA, Munenori OGURA, Jianming LU, Hiroo SEKIYA and Takashi YAHAGI

Graduate School of Science and Technology, Chiba University
1-33, Yayoi-cho, Inage-ku, Chiba, 263-8522 Japan
Email: n_yamashita@graduate.chiba-u.jp

Abstract—In this paper, we propose a new random-valued impulse noise detector from images using level detection. In our method, we use directional windows in order to search a level region in the images. One window whose variation is lowest is selected as a flat window from multi directional windows. In a flat window, random-valued impulse noise may move to the both ends of order statistics. Therefore, noise detection in selected window is easy. Consequently, the proposed method reduces undetected noise pixels without increasing mis-detections. Extensive simulations indicate that the proposed method performs significantly better than conventional methods.

I. INTRODUCTION

In the image processing, median filters have been widely used for removing impulse noise, since median filters are quite effective for the noise removal and the edge preserving. However, median filters tend to modify both noise pixels and undisturbed good pixels.

Recently, switching schemes have been studied for removal impulse noise in images [1]-[3]. These schemes detect whether the current pixel is corrupted by impulse noise at each pixel. Then, filtering is activated for pixels that is detected as noise pixels, while good pixels are kept. As a switching scheme, progressive switching median (PSM) filter [1] was proposed for removal impulse noise. With the PSM filter, both the impulse noise detector and the noise filter are applied progressively. The noise detector detects an impulse noise and outputs a binary flag image. The binary flag image denotes whether pixels are corrupted or not. According to a binary flag image, the noise filter processes to only noise pixels using neighborhood good pixels. Since the noise filter processes according to the binary flag image, the PSM filter performs satisfactorily in removing impulse noise. However, in the case of random-valued impulse noise, the performance of the noise detector is significantly reduced. If the random-valued impulse noise is located in the middle of order statistics, the noise detector cannot detect the noise. Therefore, the random-valued impulse noise detection is more difficult than the fixed-valued impulse noise detection.

In this paper, we propose a new random-valued impulse noise detector from images using level detection. In our method, we use directional windows in order to search level region in the images. One window whose variation is lowest is selected as a flat window from multi directional windows. In a flat window, random-valued impulse noise may move to the both ends of order statistics. Therefore, the noise detector can detect the noise in the selected window. Consequently, the proposed method reduces undetected noise pixels without increasing mis-detection. Extensive simulations indicate that the proposed method performs significantly better than conventional methods.

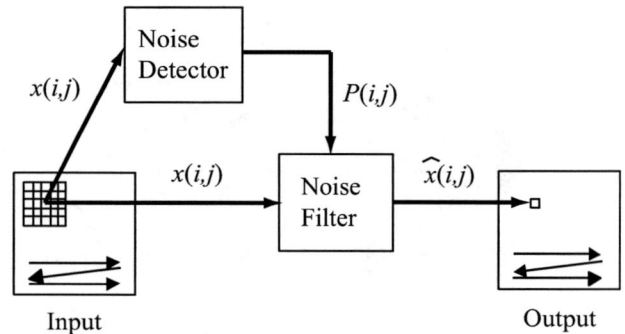

Fig. 1. Structure of the PSM filter.

II. PSM FILTER [1]

A structure of the PSM filter is shown in Fig.1. The PSM filter consists of the noise detector and the noise filter. In the noise detector, the current pixel $x(i,j)$ is judged whether it is corrupted by an impulse noise or not using neighborhood pixels. First, a median value of neighborhood pixels $m(i,j)$ is obtained. Next, a binary flag image $P(i,j)$ is given by

$$P(i,j) = \begin{cases} 1 & |x(i,j) - m(i,j)| \geq T_D \\ 0 & otherwise \end{cases} \quad (1)$$

where T_D is a threshold of the noise detection. $P(i,j) = 1$ means $x(i,j)$ is corrupted by an impulse noise. On the other hand, $P(i,j) = 0$ means $x(i,j)$ may be a good pixel. If T_D is small, almost noises are detected. However, good pixels are regarded as a noise pixel. On the other hand, in case that T_D is large, the mis-detection is decreased with increasing the undetection. In the noise filter, $x(i,j)$ is processed based on the binary flag image. The noise filter processes to only noise pixels using neighborhood good pixels. Since the noise filter processes based on the binary flag image, the PSM filter performs satisfactorily in removing impulse noise. Therefore, the performance of the noise filter depends on the binary flag image.

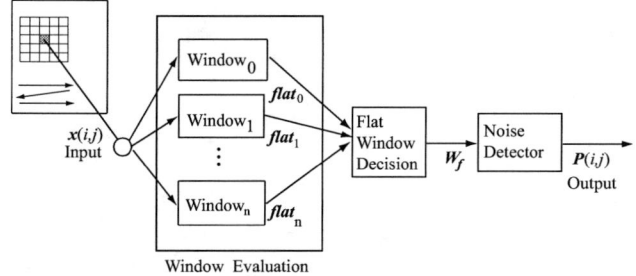

Fig. 2. Topology of proposed method by using level detection.

However, in the case of random-valued impulse noise, the performance of the noise detector is significantly reduced. If the random-valued impulse noise is located in the middle of order statistics, the noise detector cannot detect the noise and the undetection is increased. If T_D is set to small in order to decrease the undetection, the mis-detection is increased. Therefore, the random-valued impulse noise detection is more difficult than the fixed-valued impulse noise detection.

III. PROPOSED METHOD

In this paper, a new random-valued impulse noise detector from images is proposed. The input signal $x(i,j)$ is given by

$$x(i,j) = \begin{cases} x_0(i,j) & : \text{probability} \quad 1-q \\ h & : \text{probability} \quad q \end{cases} \quad (2)$$

where (i,j) denote the pixel coordinates of an image, $x_0(i,j)$ is the original image, q is impulse noise ratio, and h is uniformly distributed within [0,255].

A topology of the proposed impulse noise detector is shown in Fig.2. The detector consists of the window evaluation, the flat window decision and the noise detector.

A. Directional window

In our method, we use directional windows in order to search a level region in the images. One window whose variation is lowest selected as a flat window from multi directional windows includes a flat region. In the noise detector, an impulse noise is detected using the flat window. In a flat window, the random-valued impulse noise may move to the both ends of order statistics even if the noise is located in the middle of order statistics. Therefore, the noise detector can detect the random-valued impulse noise. In our method, we use some directional windows for searching the flat region. Directional windows $W_n (n = 1, 2, \cdots, 8)$ are shown in Fig.3. The size of the directional window is 2×5. Since we use multi windows with various direction, searching the flat region is easy at each pixel. Consequently, the performance of the noise detector is improved.

B. Flat window decision

In window evaluation, first, signals in the window W_n are sorted and $\mathbf{r}^n = \{r_1^n, r_2^n, \cdots, r_m^n\}$ ($r_i^n \leq r_{i+1}^n$) is obtained.

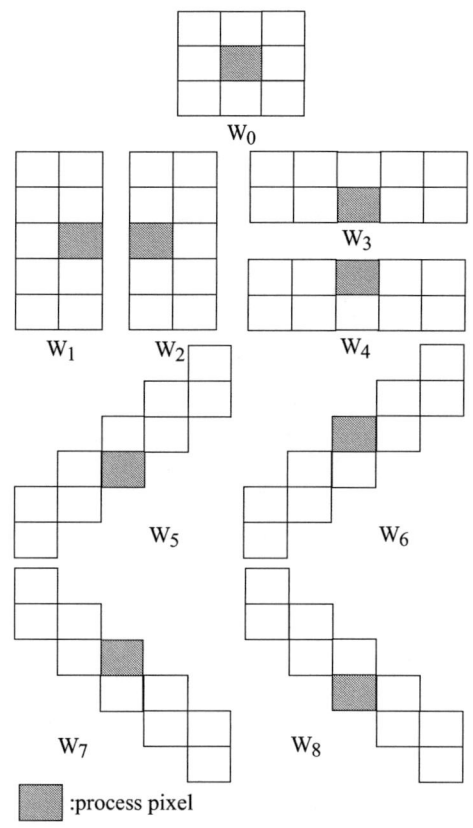

Fig. 3. Multi window used in this paper.

Next, the pair of k and l is searched according to

$$\begin{cases} r_l^n - r_k^n > Th \\ r_{l-1}^n - r_k^n \leq Th \end{cases} \quad (1 \leq k < l \leq m) \quad (3)$$

where Th is a threshold for flat decision. The k for maximum $(l-k)$ is searched from these pair. The intensity of flat is defined by

$$flat_n = l - k \quad (4)$$

where $flat_n$ is used to select the flat window. If $flat_n$ is small, W_n includes the edge or impulse noises. On the other hand, in case that $flat_n$ is large, W_n includes a flat region. Next, in the flat window decision, the window used to detect impulse noise is defined by

$$W_f = \begin{cases} W_n & \text{if} \quad flat_n \geq M \\ & \text{for maximum} \quad flat_n \\ W_0 & \text{if} \quad flat_n < M \\ & \text{for maximum} \quad flat_n \end{cases} \quad (5)$$

where M is threshold in order to allow that W_n is the flat window. If $flat_n < M$, there is no flat region in W_n. In this case, since we use pixel at the neighborhood of the current pixel, the noise detector uses the window W_0.

Here, if W_n includes the flat region, T_D can be set to small in order to reduce undetected noise pixels. Since the variation of W_n is low in the flat window, the noise detector generates few mis-detection for small T_D. On the other hand, in case

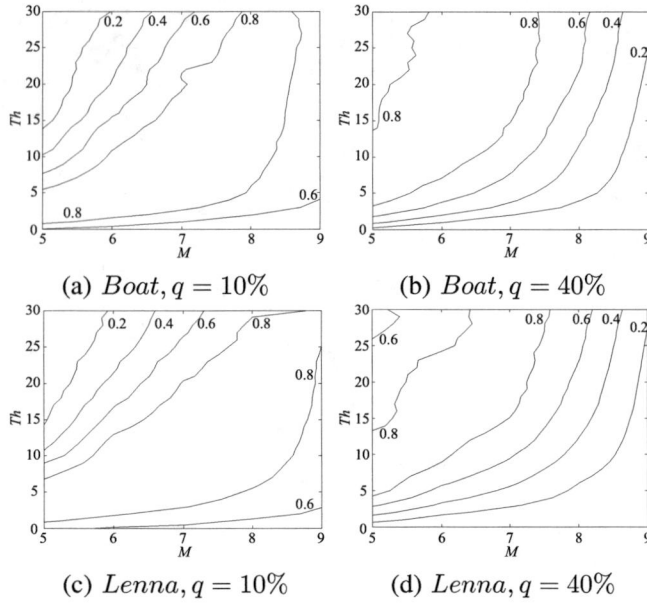

Fig. 4. Relation among M, Th and normalized PSNR

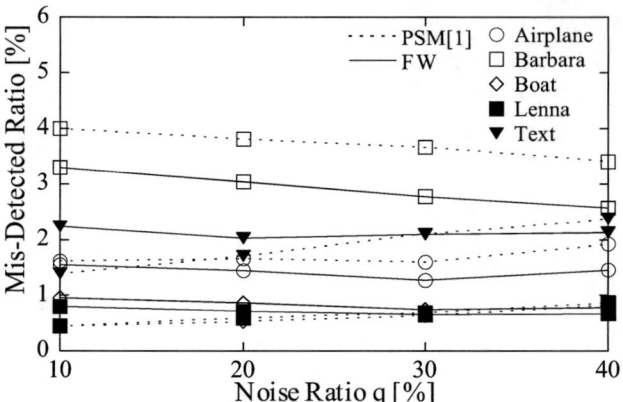

Fig. 5. The mis-detected ratio of the PSM and the FW.

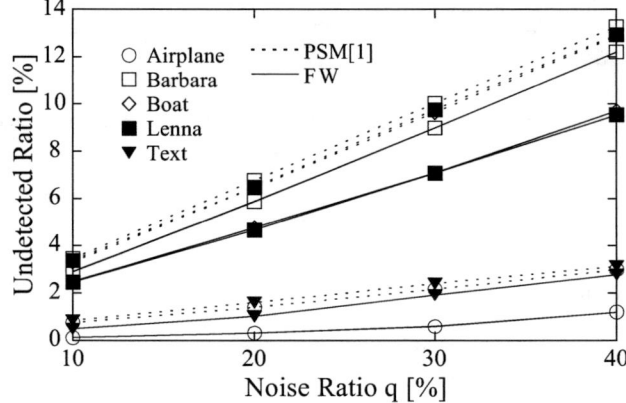

Fig. 6. The undetected ratio of the PSM and the FW.

that W_n has no flat region, T_D should be set to large in order to reduce mis-detection. If T_D is set to small in non-flat, mis-detection is increased.

IV. SIMULATION RESULTS

The performance of the proposed method has been evaluated by the simulations. In the simulation, windows of Fig.3 are used to search a flat region. And, the noise detector is carried out by the median-based impulse noise detector[1]. T_D is set to 20 for flat region and 50 for no flat region. *Boat*, *Lenna*, *Airplane*, *Barbara* and *Text* are used as processing images(256×256, $8bits$). This images is corrupted by random-valued impulse noise ($q = 10 \sim 40\%$).

The performance of noise detection is quantitatively measured by the mis-detected ratio and the undetected ratio. And the performance of restoration is quantitatively measured by the peak signal to noise ratio(PSNR).

In our method, two threshold Th and M are used to decide a flat window. We first study the effects of Th and M on the performance of impulse noise detection. Fig.4 shows the normalized PSNR of filtering *Boat* and *Lenna* corrupted by the random-valued impulse noise ($q = 10, 40\%$). In Fig.4(a)(c), the PSNR depends on Th for $M = 5, 6$. For low probability impulse noise ratio, the window includes few impulse noises. If M is set to small, non-flat is regarded as a flat region. Therefore, mid-detection is increased, and the performance of the restoration is reduced. In Fig.4(b)(d), the PSNR obtained for $M = 8, 9$ is low. Since the window includes many impulse noises for high impulse noise ratio, almost windows is judged to be $W_f = W_0$. Therefore, better results cannot be obtained for $M = 8, 9$. On the other hand, in case that M is small, Th obtaining better PSNR exists. If Th is smaller than optimal threshold, the window including flat region is judged as no flat window. Therefore, the performance of restoration is reduced. On the other hand, in case that Th is larger than optimal threshold, no flat region is regarded as flat region. Thus, the PSNR is declined. In Fig.4, we set $Th = 15$ and $M = 7$ in order to adapt various noise ratio.

The performance of the proposed method is compared with that of the PSM filter[1]. The parameter of the PSM filter is selected to obtain better performance for different cases. Here, the proposed method is named the *flat window* (FW) method. Fig.5 shows the mis-detected ratio of the PSM and the FW. And, Fig.6 shows the undetected ratio. In Fig.5, the mis-detected ratio of the FW and that of the PSM almost equivalent. In Fig.6, the undetected ratio of the FW is better than that of the PSM for various noise ratio and images. That is because random-valued impulse noise are detected using a flat window searched from multi directional windows. Therefore, the undetected ratio is reduced without increasing the mis-detected ratio.

Next, the FW is compared with the PSM in terms of filtering results. In this simulation, images are processed using each binary flag image with the noise filter[1]. Fig.7 shows the PSNR of filtering images corrupted by random-valued impulse noise with different noise ratios. In Fig.7, the FW exceeds

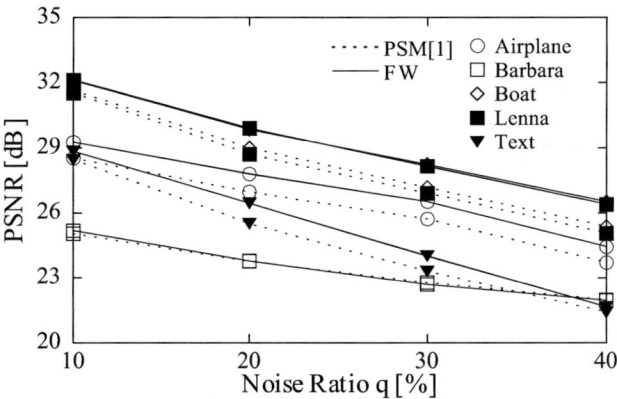

Fig. 7. Performance evaluation of the PSM and the FW in filtering images corrupted by random-valued impulse noise.

the PSM in PSNR. Since the FW reduces undetected noise pixel without increasing mis-detection, the performance of the noise filtering is improved. Fig.8 shows the results of filtering *Airplane* corrupted by the random-valued impulse noise ($q = 10\%$). And, Fig.9 shows the results of filtering *Boat* corrupted by the random-valued impulse noise ($q = 40\%$). In Fig.8(c) and Fig.9(c), impulse noises remain in resultant images by the PSM. Furthermore, in Fig.9(c), the PSM provides inferior performance in edge preservation. Since the detection of the random-valued impulse noise is difficult, impulse noise remains in resultant images. In Fig.8(d) and Fig.9(d), better results have been achieved by the FW with more effective noise rejection and edge preservation. That is because the performance of the noise detection is improved by searching flat region.

V. CONCLUSION

In this paper, we have proposed a new random-valued impulse noise detector from images using level detection. In our method, one window whose variation is lowest is selected as a flat window. In a flat window, random-valued impulse noise may move to the both ends of order statistics. Therefore, the performance of noise detection is increased. In the simulation, the proposed method has demonstrated superior performance in detecting impulse noise. Especially, it is note worthy that the proposed method has shown the noticeable difference from other methods in the noise detected rate as well as the image quality.

REFERENCES

[1] Z. Wang and D. Zhang, "Progressive switching median filter for the removal of impulse noise from highly corrupted images", IEEE Trans. Circuits Syst. II, Analog and Digit. Signal Process. , Vol. 46, No. 1, pp. 78-80, Jan. 1999.
[2] T. Sun and Y. Neuvo, "Detail-preserving median based filters in image processing," Pattern Recognit. Lett., Vol. 15, pp. 341-347, April 1994.
[3] T. Chen, K.-K. Ma and L.-H. Chen, "Tri-state median filter for image denoising," IEEE Trans. Image Processing, Vol. 8, pp. 1834-1838, Dec. 1999.

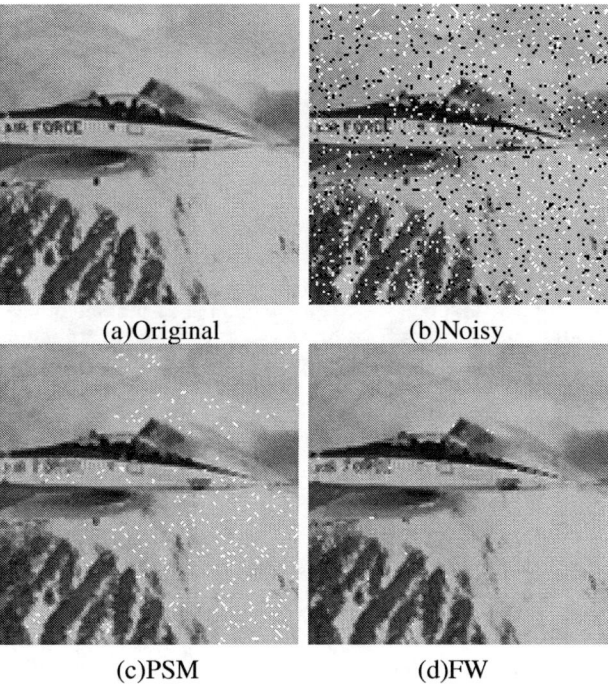

Fig. 8. Restoration images (*Airplane*, $q = 10\%$).

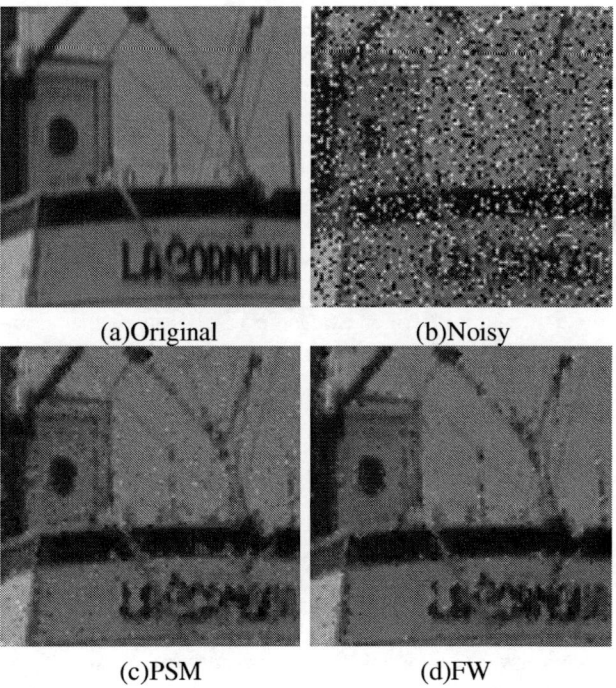

Fig. 9. Restoration images (*Boat*, $q = 40\%$).

Super-Resolution Image Restoration from Blurred Observations

Nirmal K. Bose   Michael K. Ng   Andy C. Yau

Abstract— In this paper, we study the problem of reconstruction of a high-resolution image from several blurred low-resolution image frames. The image frames consist of blurred, decimated and noisy versions of a high-resolution image. The high-resolution image is modeled as a Markov random field (MRF), and a maximum a posteriori (MAP) estimation technique is used for the restoration. We show that with the periodic boundary condition, a high-resolution image can be restored efficiently by using fast Fourier transforms. We also apply the preconditioned conjugate gradient method to restore high-resolution images in the aperiodic boundary condition.

Keywords: high-resolution, image restoration, regularization, preconditioned conjugate gradient method

I. INTRODUCTION

A very fertile arena for applications of some of the developed theory of multidimensional systems has been spatio-temporal processing following image acquisition by, say a single camera, mutiple cameras or an array of sensors. Due to hardware cost, size, and fabrication complexity limitations, imaging systems like CCD detector arrays or digital cameras often provide only multiple low-resolution (LR) degraded images. However, a high-resolution (HR) image is indispensable in applications including health diagnosis and monitoring, military surveillance, and terrain mapping by remote sensing. Other intriguing possibilities include substituting expensive high resolution instruments like scanning electron microscopes by their cruder, cheaper counterparts and then applying technical methods for increasing the resolution to that derivable with much more costly equipment. Resolution improvement by applying tools from digital signal processing technique has, therefore, been a topic of very great interest. This talk will provide in sufficient detail the various approaches to the attaining of superresolution following image acquisition by either a digital video camera or a multisensor array, see for instance [2], [3], [7], [8], [10], [11], [12], [14].

In this paper, we focus on the problem of reconstructing a high-resolution image from several blurred low-resolution image frames. The image frames consist of decimated, blurred and noisy versions of the high-resolution image [1], [13]. The high-resolution image is modeled as a Markov random field (MRF), and a maximum a posteriori (MAP) estimation technique is used for the restoration. We propose to use the preconditioned conjugate gradient method [4] instead to optimize the MAP objective function. We show that with the periodic boundary condition, the high-resolution image can be restored efficiently by using fast Fourier transforms (FFTs). In particular, an n-by-n high-resolution image can be restored by using two-dimensional FFTs in $O(n^2 \log n)$ operations. We remark such approach has been proposed and studied by Bose and Boo [2] for high-resolution image reconstruction. Here we consider a more general blurring matrix in the image reconstruction. By using our results, we construct a preconditioner for solving the linear system arising from the optimization of the MAP objective function when other boundary conditions are considered. Both theoretical and numerical results show that the preconditioned conjugate gradient method converges very quickly, and also the high-resolution image can be restored efficiently by the proposed method.

The outline of the paper is as follows. In §2, we briefly give a mathematical formulation of the problem. In §3, we study how to use fast Fourier transforms to restore high-resolution images efficiently. Finally, numerical results and concluding remarks are given in §4.

II. MATHEMATICAL FORMULATION

In this section, we give an introduction of the mathematical model for the high-resolution image restoration. Let us consider the low-resolution sensor plane with m-by-m sensors elements. Suppose that the downsampling parameter is q in both the horizontal and vertical direction. Then the high-resolution image is of size qm-by-qm. The high-resolution image Z has intensity values $Z = [z_{i,j}]$, for $i = 0, \cdots, qm-1$, $j = 0, \cdots, qm - 1$. The high-resolution image is first blurred by a different, but known linear space invariant blurring function. They have the following relation:

$$\widehat{z}_{i,j} = h(i,j) * z_{i,j}. \qquad (1)$$

where $h(i,j)$ is a blurring function and '$*$' denotes the discrete convolution.

The low-resolution image Y has intensity values $Y = [y_{i,j}]$, for $i = 0, \cdots, m-1$, $j = 0, \cdots, m-1$. The relationship between Y and $\widehat{Z}$ can be written as follows:

$$y_{i,j} = \frac{1}{q^2} \sum_{k=qi}^{(q+1)i-1} \sum_{l=qj}^{(q+1)j-1} \widehat{z}_{k,l}. \qquad (2)$$

E-mail: nkb@stspbkn.ee.psu.edu. Department of Electrical Engineering, The Spatial and Temporal Signal Processing Center, The Pennsylvania State University, University Park, PA 16802, U.S.A. Research supported in part by Research Office Grant DAAD 19-03-1-0261.

E-mail: mng@maths.hku.hk. Department of Mathematics, The University of Hong Kong, Pokfulam road, Hong Kong. Research supported in part by RGC Grant Nos. HKU 7130/02P, 7046/03P, 7035/04P

Department of Mathematics, The University of Hong Kong, Pokfulam Road.

We consider the low-resolution intensity is the average of the blurred high-resolution intensities over a neighborhood of q^2 pixels.

Let $\mathbf{z}$ be a vector of size q^2m^2-by-1 containing the intensity of the high-resolution image Z in lexicographically order. Let $\mathbf{y}_i$ is the m^2-by-1 lexicographically ordered vector containing the intensity value of the blurred, decimated and noisy image Y_i. Then the matrix form can be written as :

$$\mathbf{y}_i = \mathbf{DH}_i\mathbf{z} + \mathbf{n}_i \quad (3)$$

where $\mathbf{D}$ is a decimation matrix of size m^2-by-q^2m^2, $\mathbf{H}_i$ is a blurring matrix of size q^2m^2-by-q^2m^2 and $\mathbf{n}_i$ is a m^2-by-1 noise vector. The decimal matrix $\mathbf{D}$ consists of q^2 values of $\frac{1}{q^2}$ in each row and has the form

$$\mathbf{D} = \frac{1}{q^2}\begin{pmatrix} 1 & \cdots & 1 & & & & & 0 \\ & & & 1 & \cdots & 1 & & \\ & & & & & & \ddots & \\ 0 & & & & & 1 & \cdots & 1 \end{pmatrix}.$$

For the noise vector, $\mathbf{n}_i$, we assume that it is a zero mean independent identically distributed and given by

$$P(\mathbf{n}_i) = \frac{1}{(2\pi)^{\frac{m^2}{2}}\sigma^{m^2}} e^{-\frac{1}{2\sigma^2}\mathbf{n}_i^T \mathbf{n}_i}.$$

By using a MAP estimation technique [13], we find that the cost function of this model is given by

$$\min_{\mathbf{z}}\left\{\sum_{i=1}^{p}\|\mathbf{y}_i - \mathbf{DH}_i\mathbf{z}\|_2^2 + \alpha\|\mathbf{L}\mathbf{z}\|_2^2\right\} \quad (4)$$

where p is the number of observed low-resolution images, α is a regularization parameter and $\mathbf{L}$ is the first-order finite-difference matrix and $\mathbf{L}^T\mathbf{L}$ is the discrete Laplacian matrix. In the above formulation, the noise variance term is absorbed in the regularization parameter α. The minimization of the cost function (4) is equivalent to solving the following linear system:

$$\left(\sum_{i=1}^{p}\mathbf{H}_i^T\mathbf{D}^T\mathbf{DH}_i + \alpha\mathbf{L}^T\mathbf{L}\right)\mathbf{z} = \sum_{i=1}^{p}\mathbf{H}_i^T\mathbf{D}^T\mathbf{y}_i. \quad (5)$$

In the next section, we will discuss the coefficient matrix of the linear system (5) and suggest an algorithm to solve the above system efficiently.

III. ANALYSIS FOR PERIODIC BLURRING MATRICES

In this section, we discuss the linear system (5) for periodic blurring matrices, that is the blurring matrix $\mathbf{H}_i$ under the periodic boundary condition [9]. Then the linear system (5) becomes

$$\left(\sum_{i=1}^{p}\mathbf{C}_i^T\mathbf{D}^T\mathbf{DC}_i + \alpha\mathbf{L}_c^T\mathbf{L}_c\right)\mathbf{z} = \sum_{i=1}^{p}\mathbf{C}_i^T\mathbf{D}^T\mathbf{y}_i \quad (6)$$

where $\mathbf{C}_i$ is a block-circulant-circulant-block (BCCB) blurring matrix and $\mathbf{L}_c^T\mathbf{L}_c$ is a Laplacian matrix in BCCB structure.

Notice that $\mathbf{C}_i^T\mathbf{D}^T\mathbf{DC}_i$ is singular for all i, and $\mathbf{L}_c^T\mathbf{L}_c$ is positive semidefinite but it has only one zero eigenvalue. The corresponding eigenvector is equal to $\mathbf{1} = (1,\ldots,1)^T$, that is,

$$\left(\sum_{i=1}^{p}\mathbf{C}_i^T\mathbf{D}^T\mathbf{DC}_i + \alpha\mathbf{L}_c^T\mathbf{L}_c\right)\mathbf{1}$$
$$= \left(\sum_{i=1}^{p}\mathbf{C}_i^T\mathbf{D}^T\mathbf{DC}_i\right)\mathbf{1} \neq \mathbf{0}.$$

This shows that the coefficient matrix $\sum_{i=1}^{p}\mathbf{C}_i^T\mathbf{D}^T\mathbf{DC}_i + \alpha\mathbf{L}_c^T\mathbf{L}_c$ is nonsingular. Therefore, the system (6) can be solved and therefore the high-resolution image can be restored.

A. Decomposition of Coefficient Matrix

In this subsection, we discuss the coefficient matrix of the linear system (6). The coefficient matrix consists of two parts: the blurred down/upsampling matrix $\sum_{i=1}^{p}\mathbf{C}_i^T\mathbf{D}^T\mathbf{DC}_i$ and the regularization matrix $\alpha\mathbf{L}_c^T\mathbf{L}_c$.

For the regularization matrix $\alpha\mathbf{L}_c^T\mathbf{L}_c$, since it is a BCCB matrix, we can use the tensor product $\mathbf{R}_2 = \mathbf{F}_{mq} \otimes \mathbf{F}_{mq}$, i.e.,

$$\Lambda_{\mathbf{L}_c} = \mathbf{R}_2\mathbf{L}_c^T\mathbf{L}_c\mathbf{R}_2^*.$$

For the first part $\sum_{i=1}^{p}\mathbf{C}_i^T\mathbf{D}^T\mathbf{DC}_i$ of the coefficient matrix, it has a multilevel structure so that it cannot be diagonalized directly by $\mathbf{R}_2 = \mathbf{F}_{mq} \otimes \mathbf{F}_{mq}$. However, we can permute this matrix into the circulant-block matrix.

$$\mathbf{E} = \mathbf{P}_1\left(\sum_{i=1}^{p}\mathbf{C}_i^T\mathbf{D}^T\mathbf{DC}_i\right)\mathbf{P}_1^*$$
$$= \begin{pmatrix} \mathbf{A}_{1,1} & \mathbf{A}_{1,2} & \cdots & \mathbf{A}_{1,q} \\ \mathbf{A}_{2,1} & \mathbf{A}_{2,2} & \cdots & \mathbf{A}_{2,q} \\ \vdots & \vdots & \ddots & \vdots \\ \mathbf{A}_{q,1} & \mathbf{A}_{q,2} & \cdots & \mathbf{A}_{q,q} \end{pmatrix}$$

where $\mathbf{P}_1$ is a permutation matrix and $\mathbf{A}_{i,j}$ is of size qm^2-by-qm^2. For each $\mathbf{A}_{i,j}$, we can partition it into q-by-q BCCB matrices, that is,

$$\mathbf{A}_{i,j} = \begin{pmatrix} \mathbf{B}_{1,1} & \mathbf{B}_{1,2} & \cdots & \mathbf{B}_{1,q} \\ \mathbf{B}_{2,1} & \mathbf{B}_{2,2} & \cdots & \mathbf{B}_{2,q} \\ \vdots & \vdots & \ddots & \vdots \\ \mathbf{B}_{q,1} & \mathbf{B}_{q,2} & \cdots & \mathbf{B}_{q,q} \end{pmatrix}$$

where $\mathbf{B}_{i,j}$ is of size m^2-by-m^2. It follows that the matrix $\mathbf{E}$ can be block-diagonalized by the tensor product of the discrete Fourier matrix $\mathbf{R}_1 = \mathbf{I}_{q^2} \otimes \mathbf{F}_m \otimes \mathbf{F}_m$. Then we have the block-diagonal matrix $\mathbf{S} = \mathbf{R}_1\mathbf{E}\mathbf{R}_1^*$. Then the system (5) becomes

$$(\mathbf{RSR}^* + \alpha\Lambda_{\mathbf{L}_c})\mathbf{R}_2\mathbf{z} = \mathbf{R}_2\sum_{i=1}^{p}\mathbf{C}_i^T\mathbf{D}^T\mathbf{y}_i$$

where $\mathbf{R} = \mathbf{R}_2(\mathbf{R}_1\mathbf{P}_1)^*$. Next we note that the matrix $\mathbf{R}$ is a sparse matrix.

Theorem 1: Let $\mathbf{F}_n$ be the n-by-n discrete Fourier matrix and $\mathbf{I}_n$ be the identity matrix of size n-by-n. Then

$$\mathbf{R}_2\mathbf{P}_1^*\mathbf{R}_1^* \begin{cases} \neq 0, & a-l = 0 (\bmod\ m) \text{ and} \\ & x-y = 0(\bmod\ m), \\ = 0, & \text{otherwise.} \end{cases}$$

where $\mathbf{R}_1 = \mathbf{I}_{q^2} \otimes \mathbf{F}_m \otimes \mathbf{F}_m$, $\mathbf{R}_2 = \mathbf{F}_{mq} \otimes \mathbf{F}_{mq}$ and $\mathbf{P}_1$ is a permutation matrix. For those nonzero entries, they are given by

$$m^2 e^{\frac{-2\pi i[(a-1)(k-1)+(x-1)(t-1)]}{mq}}.$$

Here x and y are the row and column indices of the matrix $\mathbf{R}_2\mathbf{P}_1^*\mathbf{R}_1^*$ respectively, with $l = r(b, m+1) + 1$ with $b = y \bmod m^2$ for $y \neq m^2$ otherwise $b = m^2$, $a = r(x, qm+1) + 1$, $k = r(y, m^2+1) + 1$, $t = k \bmod q$ for $k = nq$ otherwise $t = q$ and $r(c, d)$ denotes the integral part of c/d.

According to Theorem 1, the structure of $\mathbf{R}$ can be described as follows. The matrix $\mathbf{R}$ can be considered as a q-by-q^2 block matrix and the size of each block matrix is qm^2-by-m. Each block matrix has the same structure. In particular, each block matrix can be again considered as a m-by-m block matrix and the size of each block is qm-by-m. In this level, all the blocks are just zero matrices except the main diagonal blocks. Such diagonal block matrices are q-by-1 block with block-diagonal matrix of size m-by-m. According to this nice structure, there are at most m nonzero entries in each row and each column of $\mathbf{R}$, and it implies that $\mathbf{R}$ is a sparse matrix. By using Theorem 1 and the fact that $\mathbf{S}$ is a block-diagonal matrix, it is clear that the matrix $\mathbf{R}^*\mathbf{S}\mathbf{R}$ is sparse, and therefore the matrix $\mathbf{R}^*\mathbf{S}\mathbf{R} + \alpha\Lambda_{\mathbf{L}_c}$ is also sparse.

IV. APERIODIC BLURRING MATRICES

For the aperiodic boundary condition, we denote that $\mathbf{T}_i$ be the block-Toeplitz-Toeplitz-block matrix [9], and $\mathbf{L}_e^T\mathbf{L}_e$ to be the discrete Laplacian matrix with the zero boundary condition. Then the system (5) becomes

$$\left(\sum_{i=1}^p \mathbf{T}_i^T\mathbf{D}^T\mathbf{D}\mathbf{T}_i + \alpha\mathbf{L}_e^T\mathbf{L}_e\right)\mathbf{z} = \sum_{i=1}^p \mathbf{T}_i^T\mathbf{D}^T\mathbf{y}_i. \quad (7)$$

In this case, we employ a circulant matrix $\mathbf{C}_i$ to approximate the Toeplitz matrix $\mathbf{T}_i$. Similarly, we use $\mathbf{L}_c^T\mathbf{L}_c$ to be the discrete Laplacian matrix with the periodic boundary condition to approximate $\mathbf{L}_e^T\mathbf{L}_e$. Then the system (7) becomes

$$\left(\sum_{i=1}^p \mathbf{C}_i^T\mathbf{D}^T\mathbf{D}\mathbf{C}_i + \alpha\mathbf{L}_c^T\mathbf{L}_c\right)\mathbf{z} = \sum_{i=1}^p \mathbf{C}_i^T\mathbf{D}^T\mathbf{y}_i. \quad (8)$$

which is exactly the linear system in (6). Therefore, we can use the same decomposition as before. Also, as the preconditioned matrix is symmetric positive definite, we can apply the preconditioned conjugate gradient method with the above preconditioner to solve for the system (7) efficiently.

V. EXPERIMENTAL RESULTS

In this section, we demonstrate our results. A 128-by-128 image is taken to be the original high-resolution image, and the desired high-resolution image is restored from several 64-by-64 noisy, blurred and undersampled images, i.e., we take the downsampling parameter $q = 2$. The original 128-by-128 images "Cameraman" is shown in Figure 1.

Fig. 1. The original 128-by-128 images "cameraman".

We assume the blur to be a Gaussian blur which is given by

$$H_{i,j} = e^{-D^2(i,j)/2\gamma} \quad (9)$$

The support of the blurring kernel for this model is 29.

A. Periodic Blurring Matrices

Figure 2 shows that the super-resolution image obtained by the single observed image. Table I shows further results for periodic blurring matrices. The results show that if we input the more low-resolution images, we can get the more accurate high-resolution image.

TABLE I
THE OPTIMAL REGULARIZATION PARAMETERS AND THE CORRESPONDING RELATIVE ERRORS.

Number of input images	Noise Level 40 dB	
	Optimal α	Relative Error
1	0.10	0.11903
3	0.04	0.11263
5	0.02	0.10943
7	0.02	0.10710
9	0.02	0.10536

Fig. 2. The blurred low-resolution image with $\gamma = 5$ (left) and its restored images with $\alpha = 0.1$ [rel. err. =0.11903] (middle) and $\alpha = 0.5$ [rel. err. =0.12252] (right).

B. Aperiodic Blurring Matrices

We have discussed to employ the preconditioned conjugate gradient method with circulant preconditioners to solve (7). Here we show the results for aperiodic blurring matrices.

Figure 3 shows the restored image from a single image. Figures 4 and 5 show other examples that the super-resolution image is obtained by seven low-resolution images. We find that the use of circulant preconditioner can speed up the conjugate gradient method and therefore the high-resolution restored image can be obtained more efficiently.

Fig. 3. The low-resolution image with $\gamma = 5$ (left) and its corresponding restored images with $\alpha = 0.09$ [rel. err. = 0.12448, CG iterations = 177, and PCG iterations = 96] (middle) and $\alpha = 0.15$ [rel. err. = 0.12535, CG iterations = 145, and PCG iterations = 75] (right).

Fig. 4. Seven blurred images with $\gamma = 3.8, 4.2, 4.6, 5, 5.4, 5.8, 6.2$.

Fig. 5. The restored images with $\alpha = 0.02$ [rel. err. = 0.11289, CG iterations = 301 and PCG iterations = 194] (left) and $\alpha = 0.1$ [rel. err. = 0.11838, CG iterations = 166 and PCG iterations = 89] (right).

C. Conclusions

It is shown that with the periodic boundary condition, a high-resolution image can be restored efficiently by using fast Fourier transforms. Also the preconditioned conjugate gradient method can be applied to restore high-resolution images with aperiodic boundary condition.

To improve the superresolution result, the point spread function (PSF) or blur of the acquisition system needs to be estimated. *Blind deconvolution* refers to the problem of restoring the original image from a degraded observation and incomplete blur information. Though some progress has been recently reported in the area of multiframe blind deconvolution, several problems remain to be tackled when the blur support is unknown and accurate estimate of the PSF is required directly from the low-resolution frames [5], [6].

REFERENCES

[1] B. Bascle, A. Blake, and A. Zissermann, *Motion deblurring and super-resolution from an image sequence*, in Proc. of European Conf. on Computer Vision, Cambridge, UK, 1996, Springer-Verlag: Berlin.

[2] N. K. Bose and K. J. Boo, *High-resolution image reconstruction with multisensors*, International Journal of Imaging Systems and Technology, 9 (1998), pp. 294–304.

[3] M. Elad and A. Feuer, *Resolution of Single Superresolution Image from Several Blurred, Noisy and Undersampled Measured Images*, IEEE Trans. on Image Proc., 6 (1997), pp. 1646–1658.

[4] G. Golub and C. Van Loan, *Matrix Computations*, 2nd ed., The Johns Hopkins University Press, Baltimore, MD, 1989.

[5] S. Lertrattanapanich and N. K. Bose. High resolution image formation from low resolution frames using Delaunay Triangulation. *IEEE Transactions on Image Processing*, 17(2):1427–1441, December 2002.

[6] S. Lertrattanapanich. Superresolution from Degraded Image Sequence Using Spatial Tessellations and Wavelets Ph. D. Dissertation, Department of Electrical Engineering, Pennsylvania State University, University Park, PA, USA, May 2003.

[7] M. Ng, N. K. Bose and J. Koo, *Constrained Total Least Squares Computations for High Resolution Image Reconstruction with Multisensors*, International Journal of Imaging Systems and Technology, 12 (2002), pp. 35–42.

[8] M. Ng, R. Chan and W. Tang, *A Fast Algorithm for Deblurring Models with Neumann Boundary Conditions*, SIAM J. Sci. Comput., 21 (1999), pp. 851–866.

[9] M. Ng, R. Chan, T. Chan and A. Yip, *Cosine Transform Preconditioners for High Resolution Image Reconstruction*, Linear Algebra & Its Appls., 316 (2000), pp. 89–104.

[10] M. Ng and A. Yip, *A Fast MAP Algorithm for High-Resolution Image Reconstruction with Multisensors*, Multidimensional Systems and Signal Processing, 12, 2 (2001), pp. 143–164.

[11] N. Nguyen and P. Milanfar, *A Wavelet-Based Interpolation-Restoration Method for Superresolution (Wavelet Superresolution)*, Circuits Systems Signal Processing, vol. 19, no. 4 (2000), pp. 321–338.

[12] N. Nguyen, P. Milanfar, and G. Golub, *Efficient Generalized Cross-Validation with Applications to Parametric Image Restoratioin and Resolution Enhancement*, IEEE Transactions on Image Processing, vol. 10, (2001), pp. 1299–1308.

[13] D. Rajan and S. Chaudhuri, *An MRF-based Approach to Generation of Super-Resoltuion Images from Blurred Observations*, Journal of Mathematical Imaging and Vision, 16 (2002), pp. 5–15.

[14] S. Rhee and M. Kang, *Discrete Cosine Transform Based Regularized High-Resolution Image Reconstruction Algorithm*, Optical Engineering, 38(8), August (1999), pp. 1348–1356.

A Nearest Neighbor Graph Based Watershed Algorithm

Wei-Chih Shen and Ruey-Feng Chang
Department of Computer Science and Information Engineering
National Chung Cheng University
Chiayi, Taiwan 621
wcshen@cs.ccu.edu.tw and rfchang@cs.ccu.edu.tw

Abstract— A novel watershed transformation defined on the nearest neighbor graph (*NNG*) is proposed in this paper. The *NNG* is considered as a relief map and some geographic features are defined. The proposed algorithm is not only defined for image segmentation but also utilized to refine the partition result. An image is transformed into the *NNG* and then partitioned by discovering the defined geographic features in the first step. The initial partition result is also transformed into the *NNG* again and then recursively distilled by the proposed algorithm. In the final result, the proposed algorithm is effective for capturing most of objects even though they are textured regions that are perceptually homogeneous.

I. INTRODUCTION

Watershed transformation is an important progress in fields of mathematical morphology for image segmentation. Watershed lines and river basins are both common geographic features in nature. The underlying concept of watershed algorithm is to consider an image as a relief and then to partition the image by discovering geographic features.

In the transformation from input image to a relief, the simplest method is mapping the gray scale to the altitude of the relief directly [1;2]. The gradient magnitude is also popularly used as the altitude due to that the local variations on the image can be reflected by the strengths [3;4]. For reducing the sensitivity of gradient, a Gaussian blurring is generally employed as a pre-processing before the computation of gradient [4]. Some more complex nonlinear blurring processes which preserving edge and blurring flat zone can be adopted to avoid the side effect of losing boundary positions in the linear blurring [3;5]. In [6], each tri-stimulus of a pixel in the *LUV* color space is integrated a gradient strength as the altitude. A texture gradient extracted using a nondecimated form of a complex wavelet transform is proposed in [7]. A same goal of these mentioned transformations is to reflect the local variations of an image on the undulation of formed relief.

On the formed relief, two major methods can be utilized to discover the geographic features. Firstly, the formed relief is progressively immersed into the water and watershed lines are discovered while the water comes from different basins is met [8]. The connected component on a graph produced from the input image is utilized to help the flooding process [9]. Secondly, a drainage network is built on the formed relief and then the river basins are explored by observing the formation [4].

The famous drawback of watershed algorithm is oversegmentation problem which means that a flat region is incorrectly subdivided into several small and niggling basins. A region adjacency graph (*RAG*) is commonly utilized to represent the relationships between formed basins to help the merging process [10]. Fast region merging method utilizes the *NNG* constructed from *RAG* to improve the performance of running time and data space [3]. In the merging process, each pair of vertices denoted by cycle on the *NNG* is individually compressed in to a new region.

Defining the geographic features on the *NNG*, a novel concept is proposed in the next section. The input image is transformed into the *NNG* firstly. Each pixel is mapped to a vertex and the edge set is utilized to represent the dissimilarities and spatial information between pixels. The formed edge set is directly considered as the drainage networks and the altitude of each vertex is ignored on the formed *NNG*. Intuitively, the formed *NNG* is huge and will consume a lot of execution time in the later works. In fact, the *NNG* can be represented by the 2D array with the same size of input image because of only a nearest neighbor must be recorded for each vertex. By discovering the defined geographic features, the image can be successfully partitioned as discussed in section III. Each formed basin on the initial partition result is mapped to a vertex and the adjacency relations associated with the dissimilarities between formed basins are represented by the edge set. The formed *NNG* is then recursively distilled by the proposed algorithm as discussed in section IV.

II. DEFINITIONS OF NNG BASED WATERSHED

The nearest neighbor graph is composed of a set of vertices V and a set of directional edges E [3;11]. For each vertex V_i, the nearest neighbor $V_j, V_i, V_j \in V$ and $V_i \neq V_j$, is one of adjacent neighbors and owns the minimum weight. This proximity relation of neighboring is represented by a directional edge $<V_i, V_j>$. Once two or more neighbors have the minimal weight, the nearest neighbor is decided by the index of vertex to make the unique.

For a given *NNG*, let $V = \{v_1, v_2, \Lambda, v_n\}$ be the vertex set and $E = \{e_1, e_2, \Lambda, e_n\}$ be the edge set. The vertex set V is directly considered as the relief map. An edge $e = \langle e_i, e_j \rangle$ denotes that drainage is built on vertex v_i and drains toward vertex v_j. The altitude of v_i is higher than that of v_j. However, the real altitude of v_i and v_j is ignored. That is, the edge set E forms the drainage networks on the relief. For any cycle $c = (\langle v_i, v_j \rangle, \langle v_j, v_i \rangle)$, the water makes a round trip between v_i and v_j by the definition of drainage. Hence, c is defined as minima to be the bottom of river basin. In the section IV, a threshold for disconnecting flow (*TDF*) will be proposed to prune the formed *NNG*. Some improper connections, edges, will be removed from the formed *NNG* by this threshold. These vertices which the removed edges originated from are individually defined as minima due to without any exports on these vertices when the water drains into them. Discovering the minima is very simple because they are either located in a cycle with length two or without edge connected to neighbor.

A river $R: D_1 \to D_2 \to \cdots \to D_n \to M$ from drainage D_1 to drainage D_n is a sequence of drainage $D_1, D_2, \ldots, D_n$ such that the direction of drainage D_i drains toward D_{i+1}. The river R originates from D_1 adjacent to watershed and ends in M the bottom of river basin. The river is important in the features discovering process. All the drainages which drain into the same minima are grouped into a river basin. Those isolated vertex acquired after the pruning process are individually formed river basins. Simply speaking, each connected component on the *NNG* corresponds to a river basin. The *TDF* is named from this definition due to the meanings of pruning some edges from the formed NNG is equivalent to disconnecting flow from relief map. The watershed line is utilized to separate the adjacent basins. Due to each vertex belongs one of formed basins, the watershed line doesn't really exist on the formed relief. While the partition result is shown, some vertices located in the contours of basins are temporarily set as the watershed lines.

III. IMAGE PARTITION USING *NNG* BASED WATERSHED

Although the RGB color data can be directly acquired from an input image, it is generally transformed into another color space for the lack of perceptual uniformity. The CIELAB color system is one of perceptual uniformity spaces which the distance almost reflects human perception [12]. For this property, it is employed in the proposed *NNG* based watershed algorithm to replace the RGB color space.

Two fundamental definitions are necessary in the transformation from an input image to a relief. Firstly, each pixel is assigned with a unique number as its identity and

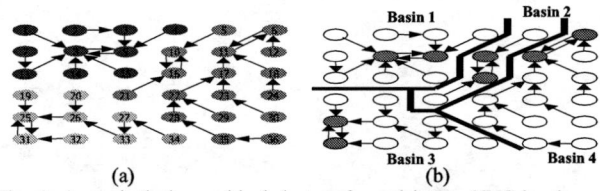

Fig. 1. A synthetic image block is transformed into a *NNG* by the color dissimilarity and the spatial information. The formed *NNG* is directly considered as a relief map and the drainage network is represented by the edge set. Each cycle colored by gray is defined as minima. All the pixels that drain into same minima are defined as a basin. Four basins are formed on the *NNG* and the watershed line is shown in bold.

then mapping to a vertex to form the relief. The identity assignment of a pixel $P_{i,j}$ in the image I with size $m \times n$ is defined as

$$Identity(P_{i,j}) = i \times n + j \quad where\ 0 \le i < m\ and\ 0 \le j < n \quad (1)$$

In this assignment, the identity of each pixel conforms to its element position of data array under the row major order. Secondly, the distance function $D(x,y)$ between two vertices x and y is defined as

$$D(x, y) = CD(x, y) \times w(x, y). \quad (2)$$

In this function, the $CD(x,y)$ is the Euclidean distance between pixel x and y in the CIELAB color space. The $w(x,y)$ is a weighted value to decide the candidates in the selecting nearest neighbor. The weighted value can be defined as:

$$w(x, y) = \begin{cases} 1 & \text{If } x \text{ and } y \text{ are adjacent} \\ \infty & \text{otherwise} \end{cases} \quad (3)$$

In this definition, only the adjacent pixel will be considered in the constructing *NNG* for each vertex. Two advantages, reducing the amount of distance computation and confirming the integrity of obtained regions, support this coefficient $w(x,y)$. In (3), only the adjacent pixels are required to compute the color distance. While the 8-connectivity is adopted, the computation needs four times at most because the distance is symmetric. On the other hand, the connected components in the *NNG* will correspond to the basins in the proposed definition. Once discontiguous pixels are connected in the *NNG*, a basin may be located in several isolated parts in the segmentation result. While fractional basins can be acceptable, the weight function can be utilized to control the maximum distance between isolated basins that belong to same basin by modifying the definition of neighborhood.

An *NNG* transformed from a synthetic image block with size 6×6 is shown in Fig. 1. Four cycles, minima, are firstly discovered and respectively assigned a unique basin identity on the formed *NNG*. Four basins are formed from these discovered minima by the definition of river. 89754 minima are discovered on a real partition result as shown in Fig. 2. Most of dissimilarities $D(x,y)$ between vertices are slightly but really reflected on numeric computation. This property is similar to the ripple on the relief formed by the gradient magnitudes and leads into the oversegmentation. In the traditional methods, this affect is reduced by a blurring processing.

Fig. 2. (a) The original image with size 505×756. (b) The initial partition result. Only minima are shown by red dots owing to the serious oversegmentation.

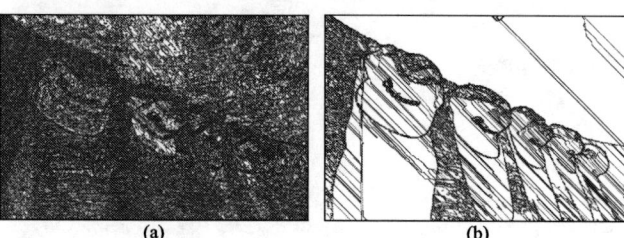

Fig. 3. The effects of BLF. (a) BLF = 1. (b) BLF = 4.

A Blurring-Like Factor (*BLF*) is proposed to reduce local variations on the relief. The *BLF* denotes a maximal tolerant of error in the distance computation and is defined as

$$CD(x,y) = \begin{cases} 0 & \text{If } CD(x,y) < BLF \\ CD(x,y) & \text{otherwise} \end{cases}. \quad (4)$$

All the color distances less than the *BLF* are set to 0. The *BLF* is quite different from the blurring process in essential but the effect is very similar. The original strength on each pixel is replaced by an integrated magnitude after the traditional blurring process. This replacement action may leads into losing the accuracy of boundary positions in the some blurring process, such as Gaussian. All the magnitudes on the image are not changed while the *BLF* is employed. The change only occurs on the formed *NNG*. As the partition result shown in Fig. 3(a), a lower *BLF* reduces the affection of local variations and the number of discovered minima is reduced from 89754 to 62945. For a larger *BLF*, the similar effects of blurring are clearly shown in Fig. 3(b) and the number of discovered minima is 7595. Some boundaries located between the sky and the clouds are lost in the result. Comparing two results, the local variations are eliminated in flat zones before losing boundary positions due to the dissimilarities between different regions are generally large. That is, the effects of *BLF* are similar to the nonlinear blurring process which is edge preserved and flat zone blurred while a proper value is assigned. However, the cost of *BLF* is very cheap.

The shape of formed basins is the side effect of *BLF* as shown in Fig. 3(b). The selection of nearest neighbor in the formation process of *NNG* is decided by the vertex identity when the minimal distance in case of ties. The smallest identity is assigned to the pixel $P_{0,0}$ and the largest identity is allocated to $P_{m-1,n-1}$ in (1). The neighbor with the smallest identity is always located in the northwest for each vertex.

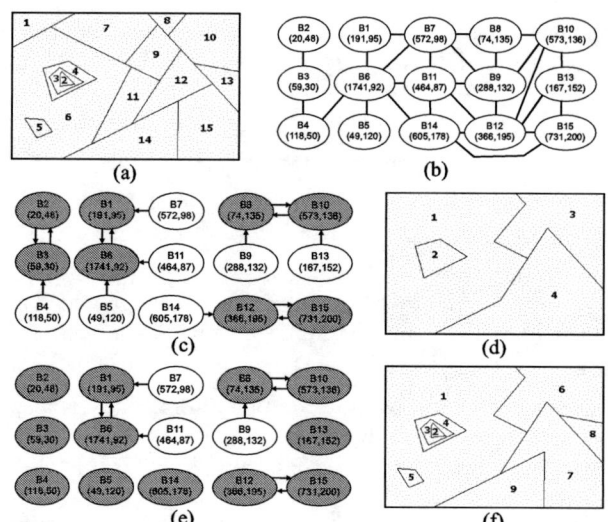

Fig. 4. (a) A synthetic partition result. (b) The corresponding RAG. For each vertex, the symbol $B_i(s,)$ denotes that the vertex corresponds to the i-th basin and the size and mean value are s and μ respectively. (c) The formed *NNG* of (b). (d) The integration result of (c). (e) The TDF is set to 15 and utilized to prune the formed *NNG*. (f) The integration result of (e).

That is, the northwest neighbor owns the first priority to be selected as the nearest neighbor when the minimal distance in a tie for each vertex. With the increasing of *BLF*, the drainages drained toward northwest are increased. At an extreme example, all the distances between pixels are adjusted to 0 when the *BLF* is ∞. The formed relief slopes toward the northwest and the minima is located in $P_{0,0}$ and $P_{0,1}$ and the whole relief forms a river basin in the result.

IV. INTEGRATION BASINS USING *NNG* BASED WATERSHED

A synthetic partition result shown in Fig. 4 illustrates the process of basins integration using the *NNG* based watershed transformation. In the synthetic result, image I is partitioned into 15 basins and the relationships between formed basins are represented by a RAG as shown in Fig. 4(a) and (b). Two additional information, basin size and mean value, are recorded on the each vertex to help the integration processes. Let $B = \{B_1, B_2, \cdots, B_n\}$ be the set of N-partition result for image I and $P_k = \{p_k^1, p_k^2, \cdots, p_k^m\}$ be the set of m pixels belong to B_k. The basin size $s(\cdot)$ and mean value $\mu(\cdot)$ are respectively defined as

$$s(B_i) = \|P_i\|$$

$$\mu(B_i) = (\frac{\sum_{m=1}^{\|P_i\|} L(P_i^m)}{\|P_i\|}, \frac{\sum_{m=1}^{\|P_i\|} a*(P_i^m)}{\|P_i\|}, \frac{\sum_{m=1}^{\|P_i\|} b*(P_i^m)}{\|P_i\|}) \quad (5)$$

where $\|P_i\|$ denotes the number of pixels in the B_i and $L(P_i^m)$, $a*(P_i^m)$ and $b*(P_i^m)$ denote the tri-stimulus (L, a,*, b*) of P_i^m in the LAB color space respectively.

By the mean value and identity recorded on the each vertex, the *RAG* is transformed into the *NNG* as shown in Fig. 4(c). The Euclidean distance is also adopted to evaluate the dissimilarities between adjacent basins. The nearest

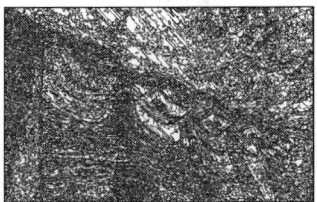 

Fig. 5. The integration results of NNG based watershed. (a) First iteration. (b) Five iterations.

Fig. 6. The integration results of *NNG* based watershed associated with *TDF*. (a) *TDF* = 6. (b) *TDF* = 12.

neighbor is decided by the basin identity in case of ties. For example, the distances between B_1 and its neighbors are both equal to 3. The B_6 is selected due to its identity is smaller. By the definitions of *NNG* based watershed, four minima are discovered on the formed *NNG*. That is, the original 15 basins in the initial partition result are integrated into four new basins as shown in Fig. 4(d).

An experiment result of basins integration by *NNG* based watershed is shown in Fig. 5. The initial partition result is acquired by the *NNG* based associated with *BLF*=1 as shown in Fig. 3(a). The basins are integrated into 15847 basins as shown in Fig. 5(a) after the first iteration. The integrated result is recursively refined by the *NNG* based watershed. After five iterations, the basins are integrated into 58 basins as shown in Fig. 5(b). The whole relief will be integrated into a basin after 9 iterations.

In the formation process of *NNG*, a nearest neighbor must be selected for each vertex even though the difference is quietly large. For example, the basins, B_2 and B_5, are located in the interior of adjacent basin as shown in Fig. 4(a). The only basin is selected to be the nearest neighbor in the formed *NNG* regardless. This property leads into an unreasonable integration result. A *TDF*, Threshold for Disconnecting Flow, is proposed for this problem and defined as

$$e(B_i,B_j)\begin{cases} \text{is removed} & \text{if } d(B_i,B_j) \geq TDF \\ \text{is remained} & \text{if } d(B_i,B_j) < TDF \end{cases}. \quad (6)$$

All the edges on the formed *NNG* are filtered by the *TDF*. The *TDF* is named due to the actions of removing edges from the formed *NNG* is similar to disconnect flow on the relief. In the illustration, the *TDF* is set to 15 to prune the formed *NNG* and the result is shown in Fig. 4(e).

In the experiments, the initial partition result is also acquired by the *NNG* based watershed associated with *BLF*=1. In the first experiment, the *TDF* is set to 6 and the integrated result is recursively distilled. The integration process is stopped after eight iterations due to that all distances between basins are larger than *TDF*. That is, all the edges on the formed *NNG* are pruned. The final result is shown in Fig. 6(a) and the number of remained basins is 2655. The cloud above the pink hat and most of grains on the wood wall are captured but serious oversegmentation occurs in a textured region located in the west. Although the texture region is perceptually homogeneous, the distances between basins in the area are large. While the *TDF* is set to 12 in the second experiment, the oversegmentation is eliminated as shown in Fig. 6(b). The number of remained basins is 508. Almost all the objects are successfully captured in the finale result.

V. CONCLUSIONS

A novel watershed algorithm defined on the *NNG* is proposed to simulate the drainage networks on the relief. The problem of image segmentation and refining result are both generalized by the *NNG*. Two parameters, *BLF* and *TDF*, are proposed to help the proposed approach by very cheap computations. By the evidence of final result, most of objects are successfully captured even though they are textured.

REFERENCES

[1] A. Bleau, J. Deguise, and A. R. Leblanc, "A New Set of Fast Algorithms for Mathematical Morphology .2. Identification of Topographic Features on Grayscale Images," *Cvgip-Image Understanding*, vol. 56, no. 2, pp. 210-229, Sept.1992.
[2] S. R. Sternberg, "Grayscale Morphology," *Computer Vision Graphics and Image Processing*, vol. 35, no. 3, pp. 333-355, Sept.1986.
[3] K. Haris, S. N. Efstratiadis, N. Maglaveras, and A. K. Katsaggelos, "Hybrid image segmentation using watersheds and fast region merging," *IEEE Transactions on Image Processing*, vol. 7, no. 12, pp. 1684-1699, Dec.1998.
[4] J. M. Gauch, "Image segmentation and analysis via multiscale gradient watershed hierarchies," *IEEE Transactions on Image Processing*, vol. 8, no. 1, pp. 69-79, Jan.1999.
[5] E. Dam and M. Nielsen, "Non-linear diffusion for interactive multiscale watershed segmentation," *Medical Image Computing and Computer-Assisted Intervention - Miccai 2000*, vol. 1935, pp. 216-225, 2000.
[6] L. Shafarenko, M. Petrou, and J. Kittler, "Automatic watershed segmentation of randomly textured color images," *IEEE Transactions on Image Processing*, vol. 6, no. 11, pp. 1530-1544, Nov.1997.
[7] P. R. Hill, C. N. Canagarajah, and D. R. Bull, "Image segmentation using a texture gradient based watershed transform," *IEEE Transactions on Image Processing*, vol. 12, no. 12, pp. 1618-1633, Dec.2003.
[8] L. Vincent and P. Soille, "Watersheds in Digital Spaces - An Efficient Algorithm Based on Immersion Simulations," *IEEE Transactions on Pattern Analysis and Machine Intelligence*, vol. 13, no. 6, pp. 583-598, June1991.
[9] A. Bieniek and A. Moga, "An efficient watershed algorithm based on connected components," *Pattern Recognition*, vol. 33, no. 6, pp. 907-916, June2000.
[10] A. Tremeau and P. Colantoni, "Regions adjacency graph applied to color image segmentation," *IEEE Transactions on Image Processing*, vol. 9, no. 4, pp. 735-744, Apr.2000.
[11] D. Eppstein, M. S. Paterson, and F. F. Yao, "On nearest-neighbor graphs," *Discrete & Computational Geometry*, vol. 17, no. 3, pp. 263-282, Apr.1997.
[12] Stephen Westland and Caterina Ripamonti, *Computational Colour Science Using MATLAB* John Wiley & Sons, 2004.

JPEG 2000 Encryption Enabling Fine Granularity Scalability without Decryption

Bin B. Zhu, Shipeng Li
Microsoft Research Asia, Beijing 100080, China
{binzhu, spli}@microsoft.com

Yang Yang
Dept. of Elec. Eng. & Info Sci., Univ. of Sci. & Technol.
of China, Hefei, Anhui 230027, China

Abstract—In this paper, we propose a novel encryption scheme for JPEG 2000 (J2K) and motion JPEG 2000. A block cipher in CBC mode is used to encrypt the bitstream of each J2K code-block. The encrypted J2K codestream preserves almost the same fine granularity scalability as the original J2K codestream yet with small or negligible overhead, and has fine and near RD-optimal truncations for a large range of bitrates. The proposed scheme enables desired transcoding directly on a single encrypted codestream without decryption to fit diverse capabilities of devices and heterogeneous networks with time-varying bandwidths. Any node, trusted or not, along the delivery path is able to perform desired transcoding without sacrificing the end-to-end security of the system.

I. INTRODUCTION

JPEG 2000 (J2K) is a new image coding standard with fine granularity scalability (FGS) [1]. A J2K codestream is organized in a hierarchical structure with structural elements tiles, components, resolution levels, precincts, and layers. A packet is the fundamental building block in a J2K codestream, and is uniquely identified by the five aforementioned structural parameters. A J2K codestream provides FGS: the codestream can be truncated to the preset layers (i.e. qualities), resolutions, components, or to break packets to truncate at coding passes to fit a large variety of applications with devices of diverse capabilities and heterogeneous networks of different characteristics. FGS of a J2K codestream allows near Rate-Distortion (RD)-optimal bitrate reduction for a large range of bitrates. JPEG 2000 has also defined motion JPEG 2000 which encodes each video frame independently [2].

Content should be protected against unauthorized usage. This is achieved typically by encrypting the content and ensuring that only authorized users can access the decryption keys. Protection can be further refined that authorized users can only consume protected content according to the acquired rights. This is done with a Digital Rights Management (DRM) system which provides persistent protection for content from creation to consumption [3][4]. In either simple or DRM protection, a J2K codestream should be encrypted such that the encrypted codestream still preserves certain level of scalabilities, preferably the original FGS. Such scalability enables desired transcoding directly on an encrypted stream without decryption. Otherwise each intermediate processing node, possibly untrusted, along the delivery path, needs to access the encryption secrets to decrypt the encrypted content first, transcode to a desired stream, and then re-encrypt the resulting stream, which may dramatically lower the end-to-end security of the system.

Many multimedia encryption schemes have been proposed in the literature. Some are designed specifically for scalable streams. A comprehensive review on scalable multimedia encryption schemes, i.e., the schemes that preserve certain level of scalabilities in the encrypted stream, is given in [5]. As for JPEG 2000 encryption, Grosbois et al. [6] proposed two encryption schemes to provide access control on either resolutions or layers. To provide access control on resolutions, signs of wavelet coefficients in high frequency subbands are pseudo-randomly flipped. The output of a pseudo-random sequence generator is used to determine if the sign of a coefficient is inverted or not. A different seed to the generator is used for each code-block. Each seed is encrypted and inserted into the codestream right after the last termination marker of the corresponding code-block by exploiting the fact that any byte appearing behind a termination marker is skipped by a J2K standard compliant decoder. The resulting encrypted codestream is J2K format compliant. To provide access control on J2K layers, the bitstream of coding passes belonging to last layers are pseudo-randomly flipped in the same way as that used for image resolution scrambling. One drawback of the scheme is that the two types of access control cannot be supported with a single encrypted stream. Another drawback is that a seed inserted after the last termination marker of a code-block may be lost during truncation or transmission, rendering the code-block undecryptible. Wee et al. [7] proposed an encryption scheme called *Secure Scalable Streaming* (SSS) that works with J2K. The scheme groups J2K packets into SSS packets. All data except header fields in each SSS packet are independently encrypted with a block cipher in Cipher Block Chaining (CBC) mode. The Initialization Vector (IV) used in the encryption is inserted into the header of each SSS packet, which may add significant overhead, esp. if FGS is needed to be supported in the encrypted stream. Scalable granularity is also reduced to a progressive SSS packet level. The supported adaptations in SSS are either to drop an entire SSS packet or to truncate trailing data in a SSS packet. To reduce encryption overhead, the number of SSS packets for each J2K compressed image is not high, resulting in very coarse granularity of scalability in an SSS-encrypted stream. The paper [7] gives an example of 9 SSS packets, in either 3RX3L or 1RX9L setting that supports 3 resolutions and 3 layers, or 1 resolution and 9 layers, respectively. Other scalabilities are not supported since individual J2K packets cannot be directly accessed after SSS encryption. For example, an SSS stream in the 1RX9L setting does not support multiple resolutions.

In this paper, we propose a novel encryption scheme for JPEG 2000 that enables FGS in an encrypted J2K codestream yet with very small or negligible overhead. In our scheme, the bitstream of coding passes of each code-block, possibly padded with stuffing bits to ensure the length is a multiple of the encryption block size if ciphertext stealing [8] is not used, is independently encrypted with a block cipher such as AES in CBC mode. A single "global" IV is randomly generated and inserted into the image's header fields. The IV used for encrypting each code-block is generated by hashing the global IV along with the parameters that uniquely identifies the code-block. The ciphertext of each code-block is then partitioned into smaller blocks, each block is aligned with the encryption block size, and put into J2K packets. Unlike SSS, we don't use our own packets. The granularity of scalability after the encryption with our proposed scheme is nearly the same as the original JPEG 2000: an encrypted codestream can be truncated at any J2K packet, or each individual J2K packet can be reshaped by truncating trailing ciphertext of one or more code-blocks inside the packet. Auxiliary data for RD-optimal cutoff points can be inserted into header fields for near RD-optimal truncations in a large range of bitrates. For motion JPEG 2000, an independent random IV is generated for each frame and inserted into the frame's header. The same encryption scheme is applied to encrypt each frame.

This paper is organized in the following way: In the next section, we briefly introduce JPEG 2000 which is the basis to describe our proposed scheme. Our scheme is described in detail in Section III, along with comparison with other proposed J2K encryption schemes. Experimental results are presented in Section IV. We conclude our paper in Section V. Before we go to the next section, we would like to note that unless explicitly mentioned otherwise, a packet means a J2K packet and a header is not encrypted in this paper.

II. JPEG 2000

JPEG 2000 (J2K) is a wavelet-based image coding standard [1]. In J2K, an image can be partitioned into smaller rectangular regions called *tiles*. Each tile is encoded independently. Data in a tile are divided into one or more components in a color space. A wavelet transform is applied to each tile-component to decompose into different resolution levels. The lowest frequency subband is referred to as the resolution level 0 subband, which is also resolution 0. The image at a resolution r ($r>0$) consists of the data of the image at resolution ($r-1$) with the subbands at resolution level r. Wavelet coefficients are quantized by a scalar quantization to reduce precision of the coefficients except in the case of lossless compression. Each subband is partitioned into smaller non-overlapping rectangular blocks called *code-blocks*. Each code-block is independently entropy-encoded. The coefficients in a code-block are encoded from the most significant bit-plane to the least significant bit-plane to generate an embedded bitstream. Each bit-plane is encoded within three sub-bitplane passes. In each coding pass, the bit-plane data and the contextual information are sent to an adaptive arithmetic encoder for encoding. The arithmetic coding is terminated at the end of the last bit-plane encoding for a code-block. For error resilience, J2K also allows for termination of the arithmetic coded bitstream as well as the re-initialization of the context probabilities at each coding pass boundary to enable independent decoding of the bitstream from each coding pass. The compressed bitstream from each code-block is distributed across one or more layers in the codestream. Each layer represents a quality increment. A layer consists of a number of consecutive bit-plane coding passes from each code-block in the tile, including all subbands of all components for that tile. J2K also provides an intermediate space-frequency structure known as a *precinct*. A precinct is a collection of spatially contiguous code-blocks from all subbands at a particular resolution level. The fundamental building block in a J2K codestream is called a *packet*. A packet is simply a continuous segment in the compressed codestream that consists of a number of bit-plane coding passes for each code-block in the precinct. Data length of each code-block in a packet is indicated in the packer header. Each packet can be uniquely identified by the five parameters: tile, component, resolution level, layer, and precinct. Each code-block can be uniquely identified by the six parameters: tile, component, resolution level, precinct, subband, code-block index. All packets of a tile can be ordered with different hierarchical ordering in a J2K codestream by varying the ordering of the parameters in nested "for loops", where each "for loop" is for one parameter from the above list. Details on J2K can be found in [1], and motion JPEG 2000 in [2].

III. FGS ENCRYPTION FOR JPEG 2000 & MOTION JPEG 2000

In our J2K FGS encryption scheme, a random IV is first generated and inserted into J2K header fields. This IV is referred to as a "global" IV for the image. The bitstream from each code-block is independently encrypted with a block cipher in CBC mode from the first coding pass of the most significant bit to the last coding pass of the least significant bit. A block cipher partitions a plaintext into blocks of the same length as the block size of the block cipher to be used, typically 64 or 128 bits. Such a block is referred to as *encryption block* in this paper. If the J2K bitstream of a code-block, referred to as plaintext, is not aligned with the encryption block size, the last partial block is padded with stuffing bits to a full block. These padding bits are overhead of our proposed encryption scheme, as well as the global IV which is of the same size as an encryption block.

The IV for encrypting a code-block is generated in the following way: a hash function such as SHA-1 [8] is applied to the global IV along with the parameters that uniquely identify the code-block. The resulting hash value is wrapped into blocks of the size of IV and XORed with each other. The result is used as the IV for encryption of the code-block. This code-block IV can be regenerated at decryption side and is *not* inserted into the codestream, which is very different from SSS proposed in [7].

After encryption, the ciphertext of each code-block is partitioned into smaller blocks which are all aligned with boundaries of encryption blocks and closest to the original partition if no encryption were used. These blocks of ciphertext are then placed to packets of different layers in a similar way as the original J2K packetization. Auxiliary data for RD-optimal cutoff points can be inserted into header fields for near RD-optimal truncations in a large range of bitrates if packets are needed to be rate-reshaped.

An image is usually encoded and encrypted at the highest rate of a range of supported bitrates. An encrypted J2K stream can be truncated at a preset resolution, layer, and/or component determined at the packetization time during encryption. For example, to truncate an encrypted J2K stream to a certain layer, all packets of higher layers are dropped. To truncate to a certain resolution, all packets of higher solution levels are dropped. Please note that packet headers are not encrypted in our scheme. Therefore each packet can be easily identified and directly accessed in an encrypted codestream. In addition to packet level truncations, a packet can also be rate-reshaped if finer granularity of scalability is needed. In this case, the trailing ciphertext of each code-block inside the packet can be independently truncated. Such a truncation should be aligned with encryption block boundaries.

At receiver side, the IV for each code-block is regenerated from the inserted global IV. The ciphertext of each code-block in each received packet can be fully decrypted, thanks to the aforementioned packetization and truncation methods which generate encryption block aligned ciphertext for each code-block inside a packet. Since the J2K packet header includes information for the length of bitstream from each code-block in the packet, the ciphertext for each code-block in a packet can be easily identified. After decryption, the compressed bitstream for each code-block in a packet is decoded with an arithmetic decoder. Since ciphertext of a code-block is aligned with the encryption block size, the bitstream of a coding pass may be partitioned into two packets. Therefore the decrypted bitstream of a code-block in a packet may end with a partial coding pass. In this case, the data corresponding to a partial coding pass are also input to the arithmetic coder for decoding. Decoding of the code-block pauses and waits for more data when the current decrypted bitstream is exhausted. When the next block of data of the code-block arrives, decoding is resumed. In this way, the data from each received packet is all used for decoding. The overhead is just those last bits that cannot generate a complete decodable symbol. This overhead is negligible. Therefore, even though our scheme has a very small overhead after encryption, the resulting codestream after truncations has negligible overhead. This is very different from SSS which has larger overhead in terms of percentage when a smaller number of SSS packets are used at decryption and decoding.

Due to limitation of paper length, performance of the proposed scheme over a lossy communication network will not be discussed in this paper. It will be discussed in detail in a separate and lengthy paper. We only mention the result here: our encryption scheme has the same error resilience performance as the original J2K codestream if the error resilience option is not used. When the error resilience option is turned on, which results in a significant overhead, our scheme has a little worse error resilience performance than the original unencrypted J2K codestream.

Our scheme has a few advantages over SSS. Our scheme has much finer granularity of scalability in the encrypted stream than SSS yet with smaller overhead, esp. when truncation occurs, as mentioned above. Each J2K packet and data of coding passes from individual code-blocks in a packet can be directly accessed. Therefore our scheme can generate a single encrypted stream for diverse applications without having to use different encrypted streams as used in SSS. As shown in the next section, our scheme always generates fine, near RD-optimal truncations even though quite a few resolutions are simultaneously supported with the same encrypted codestream. In SSS, to support resolutions, less quality layers are supported to maintain the overhead at a fixed level, which results in very coarse, stairstep-like RD curves. As compared to the scheme proposed in [6], our scheme uses a single encrypted stream to support both layer access and resolution access rather than two differently encrypted codestreams used in [6]. No overhead data is provided in [6]. We expect that it has a similar overhead as our scheme when no truncation is applied (one seed in [6] and padding bits in ours for each code-block) but our scheme has less overhead if truncation is applied. In addition, our scheme has much less information leaking and much higher security than the schemes proposed in [6].

An alternative scheme is to use the ciphertext stealing method [8] in CBC mode which generates ciphertext of exactly the same size as the plaintext for any plaintext of size larger than one encryption block. In this method, the last full block and the last partial block of plaintext are encrypted differently from the rest blocks. In this alternative scheme, if a plaintext is not aligned with the encryption block size, the ciphertext of the last two blocks must be packed and truncated together in a packet. All other blocks are not allowed to be truncated to a partial encryption block. In this way, the last two blocks encrypted with ciphertext stealing can be detected if the ciphertext of a code-block inside a packet is not aligned with encryption block size. In a rare case that the compressed bitstream of a code-block is less than a full block, the ciphertext stealing method cannot be used. In this case, the padding scheme mentioned above is used. If this rare case is ignored, this alternative scheme does not introduce any overhead other than the inserted global IV.

The proposed J2K FGS encryption scheme is equally applicable to motion JPEG 2000. For motion J2K, an independent random "frame" IV is generated for each frame. The frame IV which plays the same role as the global IV in J2K encryption is then inserted into the header fields of the frame. Each frame is encrypted in the same way as the image case. Auxiliary data can be inserted into frame header fields to allow near RD-optimal truncation for each frame.

IV. EXPERIMENTAL RESULTS

The proposed scheme has been implemented based on the publicly available J2K implementation JasPer (version 1.701.0) [9]. The block cipher and the hash function used in our implementation are Blowfish [8] and SHA-1 [8], respectively, from the publicly available Crypto++ library (version 5.2.1) [10]. Blowfish is a 64 bit block cipher. The following reported results are based on the experiments on a set of standard 8-bit grayscale images of 512 by 512 pixels with the proposed scheme which does *not* use the ciphertext stealing method. Each image is compressed to a nominal 1.0 bpp with 5 levels of wavelet decomposition (i.e., 6 resolution levels) and 10 layers. The nominal code-block size is set to 64 by 64. Table I shows the overhead of the proposed encryption scheme when *no* truncation is applied. The overhead is about 1.0% for all those images, which is smaller than SSS reported in [7], even though our scheme has much finer granularity of scalability. We mentioned that the overhead of our scheme would reduce with truncations. This is confirmed in our experiments to be described next.

TABLE I. Encryption overhead for 512 by 512 grayscale images compressed at 1.0 bpp without any truncation.

Images	Barbara	Boat	Goldhill	Lena	Peppers	Zelda
Overhead (%)	1.00	0.91	1.02	0.96	0.95	0.95

Figure 1. Images of the three highest resolutions from the encrypted codestream of the image Lena.

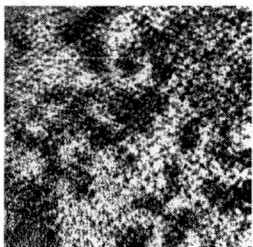

Figure 2. Encrypted Lena.

Figure 1 shows the images of the three highest resolutions supported with the encrypted codestream of Lena. Figure 2 shows the encrypted image Lena. It appears very random without any visual information leaking out. Figure 3 shows the RD curve truncated at each layer for the encrypted image Lena along with the original J2K RD curve without encryption. The two curves almost coincide with each other. This means that the RD performance when truncations are applied directly on an encrypted codestream produced by our scheme is almost the same as the original J2K coder for a large range of bitrates. Our scheme produces fine and smooth RD curves at each resolution, in great contrast to the stairstep-like RD curves when 3 resolutions are supported with SSS in the example given in [7]. Figure 3 also shows that our scheme's experimental points get closer to the corresponding points of non-encryption case when more layers are truncated, virtually overlapping each other at left side. This implies that the overhead of our scheme reduces when more data is truncated, and is negligible at low bitrates. SSS shows an opposite behavior that its overhead increases percentage-wise when more data is truncated.

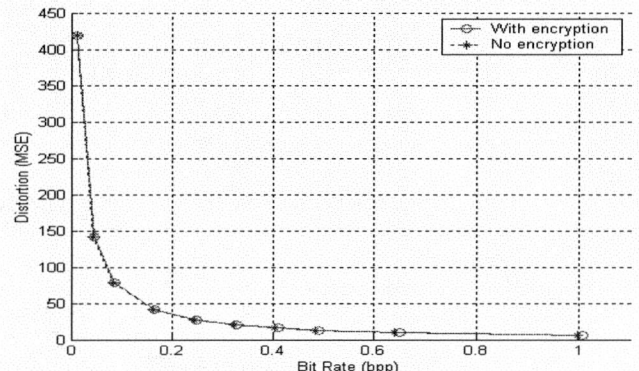

Figure 3. RD curves for both encrypted and unencrypted codestreams of the image Lena truncated at preset layers.

V. CONCLUSION

We have described a novel encryption scheme for JPEG 2000 which preserves in the encrypted codestream almost the same fine granularity scalability as the original unencrypted J2K codestream. The overhead is very small, about 1.0% for 8-bit 512 by 512 grayscale images at 1.0 bpp, when no truncation is applied. This overhead reduces to a negligible level when truncations are applied to code-blocks and more and more data is truncated. The described scheme produces encrypted codestreams with fine and smooth RD curves at each resolution for a large range of bitrates. We have also shown how the proposed scheme works with motion JPEG 2000.

REFERENCES

[1] *Information Technology – JPEG 2000 Image Coding System, Part 1: Core Coding System*, ISO/IEC 15444-1:2000.

[2] *Information Technology – JPEG 2000 Image Coding System, Part 3: Motion JPEG 2000*, ISO/IEC 15444-3:2002.

[3] R. Iannella, "Digital Rights Management (DRM) Architectures," *D-Lib Magazine*, vol. 7, no. 6, June 2001.

[4] A. M. Eskicioglu, J. Town, and E. J. Delp, "Security of Digital Entertainment Content from Creation to Consumption," *Signal Processing: Image Communication, Special Issue on Image Security*, vol. 18, no. 4, April 2003, pp. 237 – 262.

[5] B. B. Zhu, M. D. Swanson, and S. Li, "Encryption and Authentication for Scalable Multimedia: Current State of the Art and Challenges," *Proc. of SPIE Internet Multimedia Management Systems V*, vol. 5601, pp. 157-170, Philadelphia PA, Oct. 2004, (invited paper).

[6] R. Grosbois, P. Gerbelot, and T. Ebrahimi, "Authentication and Access Control in the JPEG 2000 Compressed Domain," *Proc. SPIE 46th Annual Meeting, Applications of Digital Image Processing XXIV*, San Diego, California, 2001.

[7] S. J. Wee and J. G. Apostolopoulos, "Secure Scalable Streaming and Secure Transcoding with JPEG-2000," *IEEE Int. Image Processing*, vol. 1, pp. I-205-208, Sept. 14-17, 2003.

[8] B. Schneier, *Applied Cryptography: Protocols, Algorithms, and Source Code in C*, 2nd ed., John Wiley & Sons, Inc. 1996.

[9] JasPer, http://www.ece.uvic.ca/~mdadams/jasper.

[10] Crypto++, http://www.eskimo.com/~weidai/cryptlib.html.

DSVD: A Tensor-Based Image Compression and Recognition Method

Kohei Inoue and Kiichi Urahama
Department of Visual Communication Design
Kyushu University
Fukuoka-shi, 815–8540 Japan
Email: k-inoue,urahama@design.kyushu-u.ac.jp

Abstract— Optimal dimensionality reduction of a single matrix is given by the truncated singular value decomposition, and optimal compression of a set of vector data is given by the principal component analysis. We present, in this paper, a dyadic singular value decomposition (DSVD) which gives a near-optimal dimensionality reduction of a set of matrix data and apply it to image compression and face recognition. The DSVD algorithm is derived from the higher-order singular value decomposition (HOSVD) of a third-order tensor and gives an analytical solution of a low-rank approximation problem for data matrices. The DSVD outperforms the other dimensionality reduction methods in the computational speed and accuracy in image compression. Its face recognition rate is higher than the eigenface method.

I. INTRODUCTION

While data-independent transformation such as the discrete cosine transform or the wavelet transform is popularly used for image compression, it is known that optimal compression is given by data-dependent projection such as the Karhunen-Loeve transform or the principal component analysis. An image is represented by a matrix of which optimal dimensionality reduction is given by the singular value decomposition (SVD). Pesquet-Popescu et al. have extended the SVD to joint decomposition of a set of matrices and have presented a greedy solution method called the joint singular value decomposition (JSVD)[1]. Shashua and Levin tackled the same problem and have presented a similar algorithm[2]. We have also presented a more sophisticated algorithm faster than them[3]. Ye formulated another extension of the SVD to multiple matrices and has presented its iterative solution method[4] which we abbreviate here to GLRA (generalized low rank approximation).

These dimensionality reduction methods are derived from the spectral decomposition of matrices, i.e. second order tensors. We present, in this paper, a third-order tensor-based dimensionality reduction method for a set of matrix data and apply it to image compression. We call this algorithm the DSVD (dyadic SVD) where the term "dyadic" comes from the dyadic product in the tensor algebra. This DSVD algorithm is derived from the higher-order singular value decomposition (HOSVD) of a third-order tensor. The HOSVD is an analytical solution method for low-rank approximation of a single tensor. The HOSVD itself is also usable for image compression.

The DSVD algorithm is a direct analytical solution method as same as the HOSVD in contrast to iterative solution methods such as the JSVD and the GLRA. We show by experiments that the DSVD outperforms the JSVD and the GLRA in computational speeds and outperforms the HOSVD in compression ratios.

We apply then the DSVD to image recognition and show its higher performance than the eigenface method[5] with experiments of classification of persons by their face images.

II. JOINT SVD

The truncated SVD is the optimal solution of the low-rank approximation of a single matrix A:

$$\min_{x_l, y_l, \sigma_l} \quad \|A - \sum_{l=1}^{r} \sigma_l x_l y_l\|_F^2$$
$$\text{subj.to} \quad x_k^T x_l = y_k^T y_l = \delta_{k,l} \quad (1)$$

where $\|\bullet\|_F$ is the Frobenius norm and $\delta_{k,l}$ is the Kronecker's delta. An extension of (1) to a set of matrices $A_k (k=1,...,N)$ is

$$\min_{x_l, y_l, \sigma_{kl}} \quad \sum_{k=1}^{N} \|A_k - \sum_{l=1}^{r} \sigma_{kl} x_l y_l\|_F^2$$
$$\text{subj.to} \quad x_l^T x_l = y_l^T y_l = 1 \quad (2)$$

where note that the orthogonality condition on x_l and y_l in (1) is discarded in (2). On the contrary to (1) which can be solved analytically, no analytical solution is known for (2). Pesquet-Popescu et al. have presented a greedy algorithm based on the Jacobi method for (2)[1], and Shashua and Levin have proposed an iterative solution method[2]. We[3] have also presented a similar iterative algorithm faster than them briefly written as follows:

Step 0: Set $\tilde{r} = 1$.
Step 1: Initialize $x_{\tilde{r}}$ and $y_{\tilde{r}}$.
Step 2: Compute

$$x_{\tilde{r}} = \frac{\sum_{k=1}^{N} (x_{\tilde{r}}^T A_k y_{\tilde{r}}) A_k y_{\tilde{r}}}{\|\sum_{k=1}^{N} (x_{\tilde{r}}^T A_k y_{\tilde{r}}) A_k y_{\tilde{r}}\|} \quad (3)$$

Step 3: Compute

$$y_{\tilde{r}} = \frac{\sum_{k=1}^{N}(x_{\tilde{r}}^T A_k y_{\tilde{r}}) A_k^T x_{\tilde{r}}}{\|\sum_{k=1}^{N}(x_{\tilde{r}}^T A_k y_{\tilde{r}}) A_k^T x_{\tilde{r}}\|} \quad (4)$$

Step 4: If $x_{\tilde{r}}$ and $y_{\tilde{r}}$ converge, then calculate $\sigma_{k\tilde{r}} = x_{\tilde{r}}^T A_k y_{\tilde{r}}$ and go to step 5, else go to step 2.

Step 5: If $\tilde{r} = r$ then stop, else increment $\tilde{r}$ by one and update A_k to $A_k - \sigma_{k\tilde{r}} x_{\tilde{r}} y_{\tilde{r}}^T$.

This iteration converges to a locally optimal solution of (2)[3]. We call this algorithm JSVD subsequently in this paper.

III. HIGHER-ORDER SVD

The higher-order SVD (HOSVD)[6] is an extension of the SVD to tensor data. The HOSVD is defined for any order of tensors in which only the third order is relevant in this paper.

If we pile up the number N of $m \times n$ matrices A_k, we get a single third-order tensor $A = [a_{ijk}]$ ($i = 1, ..., m$; $j = 1, ..., n$; $k = 1, ..., N$). Before we define the HOSVD for this A, some terminology and notations are introduced. At first,

[Definition 1] A tensor composed of piled-up matrices can be decomposed into constituent matrices which are then laid end to end to construct a matrix long sideways. This oblong matrix and also this operation is called the unfolding and denoted by $A_{(1)} = [A_1, ..., A_N]$, $A_{(2)} = [A_1^T, ..., A_N^T]$, $A_{(3)} = [B_1^T, ..., B_m^T]$; $B_i = [a_{ijk}]$ ($j = 1, ..., n; k = 1, ..., N$). Conversely the operation restoring A from $A_{(1)}, A_{(2)}, A_{(3)}$ is called folding.

[Definition 2] A tensor obtained by folding of $XS_{(1)}$ which is a product of two matrices X and $S_{(1)}$ which is an unfolding of a tensor S is denoted by $S \times_1 X$ and is called the product of the tensor S and the matrix X. This multiplication is commutative: $S \times_1 X \times_2 Y = S \times_2 Y \times_1 X$

The HOSVD is defined by using this notation as:

[Property 1] An arbitrary third-order tensor with size $m \times n \times N$ can be decomposed as $A = T \times_1 U \times_2 V \times_3 W$ where T is a third-order core tensor with size $m \times n \times N$ and U, V, W are matrices of size $m \times m, n \times n, N \times N$. This decomposition is called the HOSVD[6].

We can reduce the rank of the HOSVD by discarding columns in U except for the first $\tilde{m}$ ones and denoting the remaining $m \times \tilde{m}$ matrix as X, and similarly denoting the first $n \times \tilde{n}$ submatrix of V as Y, and also denoting the first $N \times \tilde{N}$ submatrix of W as Z. Then the tensor is shrunk to $\tilde{A} = S \times_1 X \times_2 Y \times_3 Z$ where S is a third-order core tensor of size $\tilde{m} \times \tilde{n} \times \tilde{N}$. This $\tilde{A}$ is called a truncated HOSVD which is subsequently called the HOSVD shortly in this paper. This (truncated) HOSVD $\tilde{A}$ satisfies the following property:

[Property 2] There holds the inequality $\|A - \tilde{A}\|^2 \leq \sum_{i=\tilde{m}+1}^{m} \sigma_{i(1)}^2 + \sum_{j=\tilde{n}+1}^{n} \sigma_{j(2)}^2 + \sum_{k=\tilde{N}+1}^{N} \sigma_{k(3)}^2$ where $\sigma_{i(l)}$ is a three-mode singular value[6].

This property 2 is formally an extension of the error bound of the SVD. The SVD is, however, the minimum error solution, while the HOSVD is not necessarily so. Nevertheless it is a good approximation of the minimum error solution and coincides exactly with it in many well-posed tensors[6]. Thus the HOSVD is a good approximate solution of

$$\min_{X,Y,Z,S} \|A - S \times_1 X \times_2 Y \times_3 Z\|_F^2$$
$$\text{subj.to} \quad X^T X = I_{\tilde{m}}, \quad Y^T Y = I_{\tilde{n}}, \quad Z^T Z = I_{\tilde{N}} \quad (5)$$

where $I_{\tilde{m}}, I_{\tilde{n}}, I_{\tilde{N}}$ are $\tilde{m} \times \tilde{m}, \tilde{n} \times \tilde{n}, \tilde{N} \times \tilde{N}$ identity matrices. When a set of images are given, we can compress them with the HOSVD by folding them into a single third-order tensor. Superior property of the HOSVD is a direct analytical solution method in contrast to iterative methods such as the JSVD and the GLRA. Hence the HOSVD is expected to be faster than them.

When we practically calculate the HOSVD of a tensor A, we at first unfold A into three matrices $A_{(1)}, A_{(2)}, A_{(3)}$ and decompose them in the SVD form, next construct X, Y, Z from their left singular vectors and finally compute S as $S = A \times_1 X^T \times_2 Y^T \times_3 Z^T$.

IV. DYADIC SVD

In the JSVD in section 2, every data matrix A_k is decomposed as $X\Sigma_k Y^T$ where Σ_k is a diagonal matrix while X and Y are not necessarily orthogonal. Now we relax the diagonality of the central matrix Σ_k while require X and Y to be orthonormal, i.e. we consider the following problem;

$$\min_{X,Y,D_k} \sum_{k=1}^{N} \|A_k - XD_k Y^T\|_F^2$$
$$\text{subj.to} \quad X^T X = I_{\tilde{m}}, Y^T Y = I_{\tilde{n}} \quad (6)$$

where every matrix A_k is of size $m \times n$ and X, Y, D_k are $m \times \tilde{m}, n \times \tilde{n}, \tilde{m} \times \tilde{n}$ matrices.

A. GLRA: Iterative Algorithm

Ye[4] has presented the following iterative method for solving (6), which we call the generalized low rank approximation (GLRA):

Step 0: Initialize X as $X^{(0)} = [I_{\tilde{m}}, \mathbf{0}_{\tilde{m}, m-\tilde{m}}]^T$ where $\mathbf{0}_{\tilde{m}, m-\tilde{m}}$ is a matrix of size $\tilde{m} \times (m - \tilde{m})$ with zero elements. Set $\xi = 0$.

Step 1: Calculate the first $\tilde{n}$ singular vectors of $\sum_{k=1}^{N} A_k^T X^{(\xi)} [X^{(\xi)}]^T A_k$ and array them to construct $Y^{(\xi+1)}$.

Step 2: Calculate the first $\tilde{m}$ singular vectors of $\sum_{k=1}^{N} A_k Y^{(\xi+1)} [Y^{(\xi+1)}]^T A_k^T$ and array them to construct $X^{(\xi+1)}$.

Step 3: If $X^{(\xi+1)}$ or $Y^{(\xi+1)}$ does not converge yet, then increment ξ by one and go to step 1.

Step 4: Set the converged values to $X = X^{(\xi+1)}, Y = Y^{(\xi+1)}$, and calculate $D_k = X^T A_k Y$ ($k = 1, ..., N$).

This iteration converges to a locally optimal solution of (6)[4].

B. DSVD: Analytical Algorithm

We present, in this paper, a direct method for approximately solving (6). The present method is derived from the following relationship:

$$\|A - S \times_1 X \times_2 Y \times_3 Z\|_F^2$$
$$= \|A - D \times_1 X \times_2 Y\|_F^2$$
$$= \sum_{k=1}^N \|A_k - X D_k Y^T\|_F^2 \quad (7)$$

where $D = S \times_3 Z$ and if we denote $D = [d_{\tilde{i}\tilde{j}k}]$ ($\tilde{i} = 1, ..., \tilde{m}$; $\tilde{j} = 1, ..., \tilde{n}$; $k = 1, ..., N$), then $D_k = [d_{\tilde{i}\tilde{j}k}](\tilde{i} = 1, ..., \tilde{m}, \tilde{j} = 1, ..., \tilde{n})$.

The last term in (7) coincides with the objective function in (6). From this observation and the property 2, we get
[Property 3] $XD_k Y^T$ derived from the HOSVD of A is an approximate solution of (6).

We can get an approximate solution of (6) analytically owing to this property by the following procedure:
Step 1: Calculate the first $\tilde{m}$ left singular vectors of the unfolding $A_{(1)}$ of A, and array them to construct X.
Step 2: Calculate the first $\tilde{n}$ left singular vectors of the unfolding $A_{(2)}$ of A, and array them to construct Y.
Step 3: Calculate $D_k = X^T A_k Y$ ($k = 1, ..., N$).

Since this algorithm is a direct solution method instead of iterative ones, it is expected to be faster than the GLRA algorithm. We call this algorithm DSVD subsequently.

V. Experiments of Image Compression

We have experimented the above mentioned JSVD, HOSVD, GLRA and the DSVD algorithms for the compression of various monochromatic square, i.e. $m = n$, images.

A. Computational Time

The computational time of JSVD and that of GLRA with the number $N = 300$ of images are plotted in Fig.1 where the horizontal axis is the image size $mn = m^2$. Convergence of the JSVD is too slow to be used practically. Next the comparison between the GLRA and the DSVD is shown in Fig.2 where note that the vertical axis in Fig.2 is two orders of magnitude smaller than Fig.1. Convergence of the GLSA becomes slow as the size of images increases, while the increasing rate of the DSVD is smaller than the GLRA because the DSVD is a direct solution method. The computational time of the HOSVD is nearly equal to that of the DSVD.

B. Reconstruction Error

Accuracy in image compression can be measured by the error of reconstructed image from its original. It is better that the error is small if the compression rate is the same. We define the compression ratio as $1 - \beta/\alpha$ where α is the size of original image and β is that of compressed one. The compression ratio is $1 - (m + n + 1)r/mnN$ for the JSVD, $1 - (m\tilde{m} + n\tilde{n} + N\tilde{N} + \tilde{m}\tilde{n}\tilde{N})/mnN$ for the HOSVD, and $1 - (m\tilde{m} + n\tilde{N} + \tilde{m}\tilde{n}N)/mnN$ for the GLRA and the DSVD. The reconstruction error per one pixel is expressed as

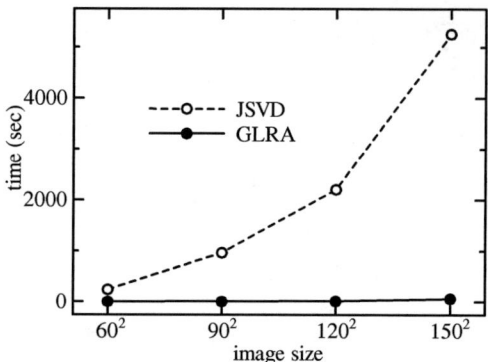

Fig. 1. Computational times of JSVD and GLRA

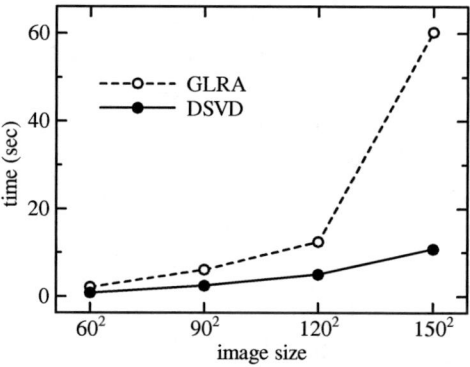

Fig. 2. Computational times of GLRA and DSVD

$\sum_{i=1}^m \sum_{j=1}^n \sum_{k=1}^N |a_{ijk} - \tilde{a}_{ijk}|/mnN$ where a reconstructed image is denoted as $\tilde{A} = [\tilde{a}_{ijk}]$. Figure 3 illustrates the error of the HOSVD and that of the DSVD where the horizontal axis is their compression ratio and the vertical axis is their error. Image size is 60×60, i.e. $m = n = 60$. The error of the HOSVD is larger than that of the DSVD and furthermore increases with the number of images N, while the error of the DSVD is almost invariant with N. Next the errors are plotted for the JSVD, the GLRA and the DSVD in Fig.4. The error of the DSVD is almost the same as that of the GLRA and its difference from the JSVD is small.

C. Overall Valuation

In summary, the computational time is ranked as
DSVD$\simeq$HOSVD$<$GLRA$\ll$JSVD
while the reconstruction error is ranked as
JSVD$\simeq$DSVD$\simeq$GLRA$<$HOSVD
Thus it is concluded that the DSVD outperforms the other three algorithms JSVD, HOSVD and GLRA in total.

VI. Application to Face Recognition

A popular method for face recognition uses the eigenface technique where face images are represented by vectors, i.e. one-dimensional arrays of pixels of which dimension is reduced by using the principal component analysis. We

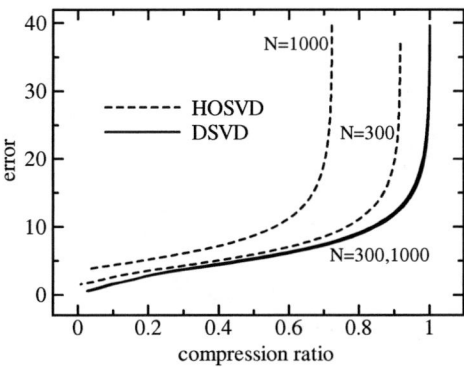

Fig. 3. Reconstruction errors of HOSVD and DSVD

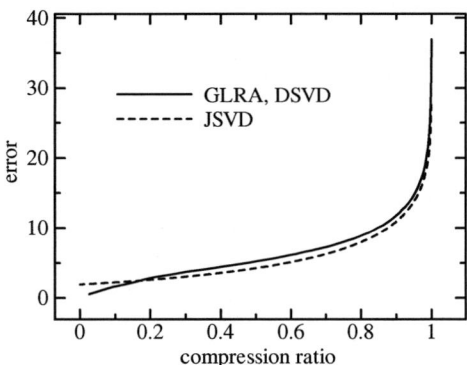

Fig. 4. Reconstruction errors of JSVD, GLRA and DSVD

represent here face images by matrices intactly and reduce their dimension by the DSVD.

Let us be given sample data of N face images $A_k(k = 1,...,N)$ about which persons are known. We apply the DSVD to this dataset and decompose each image as XD_kY^T. A new image of which person is unknown is classified in this projected space in which a new image A is projected to XDY^T with $D = X^TAY$. The distance between this new image A and each sample image A_k is measured by the Frobenius norm $\|A - A_k\|_F^2 \simeq \|XDY^T - XD_kY^T\|_F^2 = \|D - D_k\|_F^2$. The new image A is classified by the nearest neighbor rule in the projected space as follows:
[Learning: Offline]
We apply the DSVD to the sample data $A_k(k = 1,...,N)$ to calculate the projection matrices X and Y, and project each sample datum A_k to $D_k = X^TA_kY$.
[Classification: Online]
When a new datum A is inputted, we project it to $D = X^TAY$ and calculate the distances between D and every D_k. The inputted datum is classified to a class to which its nearest neighbor belong.

We have experimented using a dataset of the UMIST face database (http://images.ee.umist.ac.uk/danny/database.html) of 575 images of 20 persons with various directions and illuminations. Image size is 112×92. The number of 88 images are

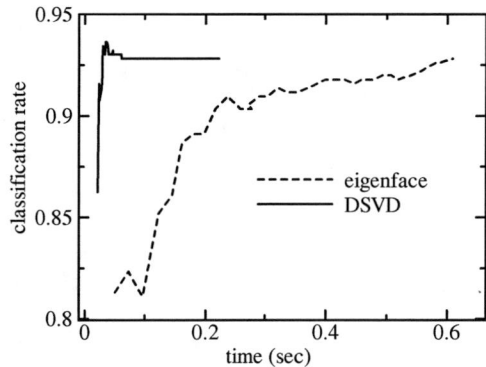

Fig. 5. Relationship between computational time and classification rate

used for learning and the remaining 487 images are used for test, i.e. $m = 112, n = 92, N = 88$ in this example. We fix the size of the matrix Y to 92×92 and vary the number $\tilde{m}$ of columns in X from 2 to 100. The relation between the computational time for classification and the classification rate is plotted by a solid line in Fig.5 where the broken line denotes the result of the eigenface method with the dimension varied from 4 to 37. Note that the computational time of the DSVD is at most 0.22 seconds owing to its high speed. The recognition rate of the DSVD is higher than that of the eigenface method particularly at high compression rates. Additionally the computational time for the learning is 1.0 seconds in the DSVD which is shorter than 1.2 seconds for the eigenface method. Thus the DSVD outperforms the eigenface method in both of the learning and the classification.

VII. CONCLUSION

We have presented a tensor-based dimensionality reduction method DSVD for a set of matrices and have applied it to image compression and face recognition. The DSVD outperforms the JSVD, HOSVD and GLRA in the computational speed and accuracy in image compression. Its face recognition rate is higher than the eigenface method. Its extension to the dimensionality reduction of a set of tensors is under study.

REFERENCES

[1] B. Pesquet-Popescu, J.-C. Pesquet and A. P. Petropulu, "Joint singular value decomposition – A new tool for separable representation of images," Proc. ICIP, vol.2, pp.569-572, Greece, Oct. 2001.
[2] A. Shashua and A. Levin, "Linear image coding for regression and classification using the tensor-rank principle," Proc. CVPR, vol.1, pp.42-49, Hawaii, Dec. 2001.
[3] K. Inoue, T. Hiraoka and K. Urahama, "Compression of multiple images with joint singular value decomposition," J. ITE, 57, 5, pp.624-626(in Japanese), 2003.
[4] J.-P. Ye, "Generalized low rank approximations of matrices," Proc. ICML, Banff, July 2003.
[5] M. Turk and A. Pentland, "Eigenfaces for Recognition," Journal of Cognitive Neuroscience, vol.3, no.1, pp.71-86, 1991.
[6] L. D. Lathauwer, B. D. Moor and J. Vandewalle, "A multilinear singular value decomposition," SIAM J. Matrix Anal. Appl., vol.21, no.4, pp.1253-1278, Apr. 2000.

Approximate Treatment for Calculation of the Rate-Distortion Slope in EBCOT

Yue-xin Zhu, Nan-ning Zheng, Jing Zhang, Zong-ze Wu
School of Electrical Engineering
Institute of Artificial Intelligence and Robotics, Xi'an Jiaotong University
Xi'an, P.R.China
yxzhu@aiar.xjtu.edu.cn

Abstract—Some novel approaches are presented aiming at reducing computation and memory for the EBCOT algorithm in JPEG2000. In current structure of EBCOT, the algorithm for calculating the distortion reduction is still too complex for hardware-oriented design, and the memory space needed for the storage of large amounts of rate-distortion slopes is often unaffordable. In this paper, we proposed a set of simplification schemes based on some approximate treatments by means of quantization median substitution and logarithm expression. These approaches can facilitate calculation of the rate-distortion slopes and reduce the memory by 50% (only 8 bits for each slope). Experimental results show that the quality of decoded images has only slightly decreased, which is negligible for most of the applications.

I. INTRODUCTION

Many of the new merits and characteristics of JPEG2000, especially the low bit-rate lossy compression, should be attributed to the algorithm of *Embedded Block-Coding with Optimized Truncation* (EBCOT)[1][3]. But unfortunately, EBCOT is currently still a software-oriented algorithm that remains troublesome for hardware implementation. For instance, bit-operation must be frequently used in the tier1-part of EBCOT, which makes it difficult to fulfill the bit-plane encoding and to compute the ΔD (distortion reduction) simultaneously. It has been substantiated that EBCOT tier-1 contains a large portion (about 45% to 60%) of computation of the whole system[4][5], within which calculation of the rate-distortion slope (RD-slope) is an obviously strenuous labor and should be simplified for the hardware-oriented design.

What's more, even if the RD-slopes for each coding-pass could be computed smoothly, how to store these slopes would become another spiny problem. In order not to seriously affect the accuracy of the rate-allocation, each RD-slope needs to be expressed by at least 16 bits (floating-point representation must be used, or else the RD-slope is much likely to overflow). This high memory consumption, in addition with the interconversion between fix-point and floating-point, is burdensome for many hardware designs. In order to shoot above troubles, we propose two approximate schemes in this paper:

1. Substitute the quantization median, i.e. 0.5, for the 6 fractional bits of the normalized difference when calculating the distortion reduction. In this way, there is no need for the encoder to load lower bit-planes' data in the course of entropy coding.

2. Use logarithm format to store the RD-slopes. 8 bits, although somewhat sketchy, are enough to represent any slope. This method has two key advantages: Firstly, it guarantees a low consumption of memory elements; secondly, logarithm expression could avoid the multiply-divide operation, which is burdensome for many of the IC designs. The extra cost is only a 5-bit lookup table for logarithm calculation.

We organized this paper as follows: in Section 2, the algorithm of rate control for JPEG2000 is introduced. Section 3 makes a description of the proposed approximate treatment for calculating the ΔD in the tier-1 part of EBCOT; Section 4 discusses how to use logarithm format to express/operate the RD-slopes, and how it can save the memory space. Experimental results are shown in section 5 and the conclusion is given in the final section.

II. RATE-CONTROL ALGORITHM OVERVIEW

The JPEG2000 system mainly subsumes two sections: Discrete Wavelet Transform (DWT) and the Embedded Block Coding with Optimized Truncation (EBCOT). EBCOT is further divided into two parts: tier-1 and tier-2: The tier-1 part chiefly serves to fulfill the entropy coding, as well as the calculation and storage of rate-distortion slopes of each code-block. The Rate-control algorithm works in the tier-2, aiming to meet a particular target bit-rate or transmission time. Rate-control assures that the desired number of bytes is used by the codestream while assuring the highest image quality possible.

The main role of rate-control is to find in the bit-stream of each code-block the optimized truncation point. This

---

Supported by National Science Foundation of China under grant of 60021302, 60205001, 60405004 and High Tech. Program 863, No.2002AA103011.

problem is equivalent to use Lagrange Operators and to find an appropriate λ to minimize the following equation:

$$\sum_i (R_i^{n_i}(\lambda) - \lambda D_i^{n_i}(\lambda)) \quad (1)$$

Where $R_i^{n_i}$ and $D_i^{n_i}$ mean the number of code bytes and the distortion (denoted by MSE) of the *i-th* code-block beyond the truncation point n_i, respectively. According to the JPEG2000 standard, we can calculate the RD-Slope values of all the passes for each code-block by the equation:

$$S_i^k = \Delta D_i^k / \Delta R_i^k \quad (2)$$

Where ΔD_i^k and ΔR_i^k represent respectively the distortion reduction and the increase of the number of code bytes between the *k-th* and *(k-1)-th* truncation point for the *i-th* code-block. Following equations are used to compute the ΔD_i^k:

$$\Delta D_i^k = \omega_i \Delta_i^2 \sum_{m,n} 2^{2p} f(\tilde{v}_i^p[m,n]) \quad (3)$$

$$\tilde{v}_i^p[m,n] = 2^{-p} v_i^p[m,n] - 2 \left\lfloor \frac{2^{-p} v_i^p[m,n]}{2} \right\rfloor \quad (4)$$

Where $\omega_i \Delta_i^2$ is the contribution to distortion in the reconstructed image which would result from an error of exactly one step size in a single sample from the i-th code-block, p denotes the level of the bit-plane currently being encoded, $\tilde{v}_i^p[m,n]$ holds the normalized difference between the magnitude of sample and the largest quantization threshold in the previous bit-plane which was not larger than the magnitude. According to the JPEG2000 standard, six fractional bits must be reserved for $\tilde{v}_i^p[m,n]$, which are obtained from the same position (m,n) at the next six consecutive bit-planes. The function $f(\tilde{v}_i^p[m,n])$ defined in terms of 9/7 wavelet filters is shown as follows:

$$f(\tilde{v}_i^p[m,n]) = \begin{cases} (\tilde{v}_i^p[m,n])^2 - (\tilde{v}_i^p[m,n] - \frac{3}{2})^2 \\ \quad when [m,n] \, has \, not \, yet \, been \, refined \\ (\tilde{v}_i^p[m,n] - 1)^2 - (\tilde{v}_i^p[m,n] - \lfloor \tilde{v}_i^p[m,n] \rfloor - \frac{1}{2})^2 \\ \quad when [m,n] \, has \, already \, been \, refined \end{cases} \quad (5)$$

The calculation of $f(\tilde{v}_i^p[m,n])$ can be accomplished by using some lookup tables. And for each coding-pass, the slope S_i^k acquired from (2) must be stored before we make a global search within the whole image tile to find the optimized λ.

III. SIMPLIFICATION FOR CALCULATING THE DISTORTION REDUCTION

Considerable price has to be paid in order to obtain $\tilde{v}_i^p[m,n]$ and $f(\tilde{v}_i^p[m,n])$ in that $\tilde{v}_i^p[m,n]$'s fractional bits must be acquired from lower bit-planes, which are temporarily free from entropy encoding. In fact, the fractional bits of $\tilde{v}_i^p[m,n]$ only serve to make a tiny contribution to the accuracy of rate-control. Most of the contribution comes from the weight coefficients of $\omega_i \Delta_i^2$, the bit-plane's level, as well as the number of pixels which are involved in the ΔD computation.

Taking these factors into account, we recommend to use the quantization median to simplify the ΔD computation, i.e. to use the constant value 0.5 to replace the $\tilde{v}_i^p[m,n]$'s fractional bits. In this way, $\tilde{v}_i^p[m,n]$ has only two possible values: 0.5 and 1.5, and the calculation of $f(\tilde{v}_i^p[m,n])$ can also be simplified by the following representation:

$$f(\tilde{v}_i^p[m,n]) = \begin{cases} 2.25 & when [m,n] \, has \, not \, yet \, been \, refined \\ 0.25 & when [m,n] \, has \, already \, been \, refined \end{cases} \quad (6)$$

By using this approximate treatment, there is no need to access the lower bit-planes' values in the course of entropy encoding, which is very advantageous for the hardware-oriented design. Slight decrease of PSNR of decoded images does exist, but negligible (see Table I). From experiments it can be observed that the representation precision of ΔD is simply not a prominent factor that affects the quality of reconstructed images. Sometimes higher computation price does not equal to a better performance.

IV. REPRESENTATION OF RD-SLOPE

The rate-distortion slope calculated according to the JPEG2000 standard is often a very large value (e.g. the largest RD-slope of image "couple" is 17,929,645), which means the storage of all these slopes pertaining to each coding-pass may consume a large quantity of memory. As for the hardware design, some special expression of RD-slopes should be considered, so as to make these notoriously large slopes less burdensome to store while more convenient to manipulate.

TABLE I. COMPARISON OF THE SIMPLIFIED AND THE ORIGINAL APPROACH TO CALCULATE THE DISTORTION REDUCTION BY PSNR *(gray-scale standard test images, use 4 levels of 9/7 wavelet decomposition).*

image tested	100:1 (PSNR)		50:1 (PSNR)	
	original	simplified	original	simplified
lena	27.263602	27.259574	29.903445	29.902598
couple	26.691597	26.566064	29.595984	29.620132
camera	21.988388	21.985036	24.863053	24.649592
boats	27.572418	27.557984	30.383226	30.413710
airplane	26.606016	26.590807	29.677997	29.557940

A. use logarithm format to express RD-Slope

Because of the large span of the RD-Slopes in value, fix-point format can hardly meet the requirements. An apparent alternative is to use floating-point representation, but this may bring additional complexity and added cost, such as the indispensable floating-point arithmetic unit. Considering the exponential variation of RD-Slopes, here we recommend to use the logarithm representation, by which 8 bits are enough to express any slope. This expression, although not very accurate, can greatly reduce the memory and simplify the multiply-divide computation. The only extra requirement is a small 5-bit lookup table for the use of logarithm calculation.

In the sense of logarithm expression (based on two), (2) and (3) can be modified as follows:

$$S_i^k = \Delta D_i^k - \Delta R_i^k \quad (7)$$

$$\Delta D_i^k = 2^4 \log_2(\omega_i \Delta_i^2) + 2^4 \cdot 2p + F(i, p, k) \quad (8)$$

Note that 2^4 is multiplied on the former two terms of (8) only because we demand 4 fractional bits be reserved in this fix-point expression. As for each code-block, $\omega_i \Delta_i^2$ is a constant, hence the term $2^4 \log_2(\omega_i \Delta_i^2)$ can be computed in advance. The third term $F(i, p, k)$ is defined as below:

$$F(i, p, k) = 2^4 \log_2(\sum_{m,n} f(\tilde{v}_i^p[m,n])) \quad (9)$$

Simplified equation (6) can also be used here to calculate the function $f(\tilde{v}_i^p[m,n])$. Since $f(\tilde{v}_i^p[m,n])$ has only 2 possible values, we can normalize this function and thus obtain the following representation of $F(i, p, k)$:

$$F(i, p, k) = \begin{cases} \lfloor 2^4 \log_2(A) \rfloor + 51 & \text{when in pass1 or pass3} \\ \lfloor 2^4 \log_2(A) \rfloor & \text{when in pass2} \end{cases} \quad (10)$$

TABLE II. LOOKUP TABLE FOR FUNCTION $\tilde{y}(x)$

x	$\tilde{y}(x) = \lfloor 2^4 \log_2(x) \rfloor$ $(0 < x < 32)$							
00~07	00	00	10	19	20	25	29	2c
08~0f	2f	32	35	37	39	3b	3c	3e
10~17	40	41	42	43	45	46	47	48
18~1f	49	4a	4b	4c	4c	4d	4e	4f

Within a specific coding-pass, A denotes the number of pixels which are involved in the ΔD calculation, i.e. in pass1 or pass3 they are the pixels that firstly become significant while in pass2 any pixel is involved.

The constant number *51* equals to the expression $\lfloor 2^4 \log_2(9) \rfloor$, where the value *9* comes from the fact that the upper value of $f(\tilde{v}_i^p[m,n])$, *2.25*, is nine times as its lower value, *0.25*. 2^4 is multiplied because 4 bits are intended to be reserved for the fractional segment of $F(i, p, k)$. In the hardware-oriented design, the largest size of the code-block is often set as 32×32, which signifies that 10-bit is enough for the accumulator A. Now the question falls into that how we can compute the expression $\lfloor 2^4 \log_2(A) \rfloor$. Since the accumulator A is 10-bit, it seems that we should also use a 10-bit lookup table to perform the function. Indeed, a 5-bit lookup table is adequate without appreciable calculation errors if it is devised in this way: separate the accumulator A into two parts: five bits each, which we name as MSB (more significant bits) and LSB (less significant bits) respectively. So the logarithm function can be expressed in an approximate way as follows:

$$y(x) \approx \begin{cases} \tilde{y}[LSB(x)] & (0 < x < 32) \\ \tilde{y}[MSB(x)] + 80 & (32 \leq x < 1024) \end{cases} \quad (11)$$

$$\tilde{y}(x) = \lfloor 2^4 \log_2(x) \rfloor \quad (0 < x < 32) \quad (12)$$

The constant term *80* in (11) equals to $2^4 \times \log_2(32)$, the unit value of MSB with 4 fractional bits. In this way, the 10-bit logarithm computation could be achieved by a 5-bit lookup table, and each entry of the table holds a 8-bit fixed point representation of $\lfloor 2^4 \log_2(x) \rfloor$. The key idea of this scheme comes from the characteristic of the rate-control algorithm, in which the precision requirement becomes looser when the RD-slope is exponentially increasing. The lookup table for $\tilde{y}(x)$ is shown as Table II.

B. clamping of RD-slope

Obviously a 10-bit register is enough to temporarily save the RD-slope S_i^k, which is calculated by (7) and (8), without any risk of overflow. If the memory is sufficient, we will also use 10 bits to save each slope in the SDRAM, otherwise, in order to further reduce the memory, we can allocate only 8 bits for each slope, i.e. the RD-slopes are clamped. The clamping program is as follows:

```
if ( slope > 255 )
    slope = 255;
else if ( slope < 0)
    slope = 0;
```

In most cases, this clamping procedure will not bring pronounced side-effects. However, if the compression rate is very low (beneath 0.005), it won't work well since many slopes may exceed the value of 255.

C. elimination of abnormal truncation points

In EBCOT, the truncation points are available only under the prerequisite that the corresponding RD-slopes are monotonically decreasing with R. Abnormal truncation points must be eliminated. In the sense of logarithm expression, there is no optimal way to finish this elimination, but an approximate treatment could still be provided as follows:

1) set $N_i = \{n\}$, i.e. the set of all truncation points;
2) set $p = 0$, $S_i^p = 0$;
3) For k = 1,2,3,4,...
   if k belongs to N_i
   set $S_i^k = \Delta D_i^k - \Delta R_i^k$
   if $p \neq 0$ and $S_i^k > S_i^p$
   set $S_i^k = 0.5 \times (S_i^k + S_i^p)$
   then remove p from N_i and go to step (2)
   otherwise, set p = k.

This procedure can be employed when all the passes' entropy encoding has finished. The only related thing that must be done in tier-1 is to calculate and save each pass-ending-point's S_i^k and ΔR_i^k, free from concerning whether or not this point is abnormal. Attention should be paid that we use logarithm expression only for the sake of simplifying the calculation of S_i^k, so the term ΔR_i^k in (7) holds the meaning of $\lfloor 2^4 \log_2(\Delta R_i^k) \rfloor$, whereas, the ΔR_i^k that is used to form the rate-distortion curve has the original sense, without logarithm treatment.

V. Experimental Result

To verify our schemes, some modification has been made on the "Jasper" software[2], a verification model for JPEG2000, which was programmed with C language. Experiments have been carried out so as to make clear how much the quality of decoded images is affected by these approximate treatments. The result is shown in Fig. 1, in comparison with the original method in terms of average PSNR of ten random selected images. The tested compression rate are 0.01, 0.02, 0.05, 0.10, 0.12, 0.15, 0.20, 0.25, and 0.30 respectively.

From Fig. 1 we can see that the decrease of PSNR caused by our approximate treatments is tiny. The most serious decrease of PSNR in our experiment is 0.61dB, which is under the compression rate of 0.35. But in most cases, compression rate is set much lower in order to achieve a small file size. When the rate is 0.01, say, the corresponding PSNR decrease is only 0.06dB. To many applications, this precision loss is expendable considering the fact that the simplifications could greatly facilitate the implementation, retrench the memory space, as well as avoid the complex multiply-divide circuits.

VI. Conclusion

An efficient methodology is used aiming at simplifying the calculation of distortion reduction, as well as saving the memory space needed for the storage of RD-slopes. The main idea is that the fractional bits of normalized difference $\tilde{v}_i^p[m,n]$ only gives a small contribution to the quality of decoded images, and that logarithm expression is more suitable to represent RD-slopes considering the slopes' exponential variation and the complexity of multiply-divide circuits. Even though some approximate treatments have been used, the compression performance remains comparable to the JPEG2000 original method.

References

[1] ISO/IEC JTC 1/SC 29/WG 1 N1646, "JPEG2000 Part I Final Committee Draft Version 1.0," March 2000, pp. 180-184.
[2] Michael D.Adams, "Jasper Software," 2001.
[3] D.Taubman, "High performance scalable image compression with EBCOT," IEEE Trans. Image Proc July 2000.
[4] T.H.Chang, L.L.Chen, C.J.Lian, H.H.Chen, L.G.Chen, "Computation reduction technique for lossy JPEG2000 encoding through EBCOT Tier-2 feedback processing," in Proc of IEEE Int. Conf. on Image Processing, vol.3. pp. 85-88, 2002.
[5] T.H.Chang, C.J.Lian, H.H.Chen, J.Y.Chang, L.G.Chen, "Effective hardware-oriented technique for the rate control of JPEG2000 encoding," Circuits and Systems, 2003. ISCAS'03, vol 2, pp 684-687, 2003.

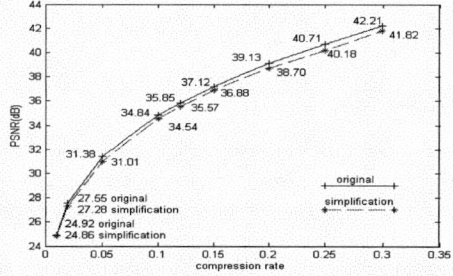

Figure 1. The average PSNR performance of the proposed simplification approach versus EBCOT original method. The PSNR average is obtained from ten random selected images: *airplane, baboon, boats, bridge, camera, couple, goldhill, lena, peppers,* and *goldenears*. The wavelet filter is 9-7 irreversible, with 4 decomposition levels. The code-block size is 32×32, and the clamping procedure is involved.

A Simplified Algorithm of JPEG2000 Rate Control for VLSI Implementation

Qin Xing, Yan Xiaolang, Ge Haitong, Yang Ye
Institute of VLSI Design
Zhejiang University
Hangzhou, P.R.China
{qinx, yan, geht,yangy}@vlsi.zju.edu.cn

Abstract—It is an effective speedup technology in JPEG2000 VLSI implementation to integrate the rate control with Tier-1 encoding. For VLSI implementation, a simplified rate-distortion optimization rate control algorithm is proposed. The algorithm includes component bit allocation based on weights, a simplified delta-distortion estimation model and two-table optimal slope threshold generation. Compared with the existing one, the algorithm has some merits: code-block memory access free, less memory requirement, lower computation complexity, with about 0.3dB decrease of image quality.

I. Introduction

The JPEG 2000[1], the last image compression standard, uses rate-distortion optimization rate control to finish post-compression and can obtain high image quality at any target bit rate. Because the Tier-1 encoding (include bit-plane context modeling and binary arithmetic coder) is a speed bottleneck for JPEG2000 application, it is an effective speedup technology in VLSI implementation to integrate the rate control with Tier-1 encoding [see figure 1]. However, the primitive algorithm [2] is not suitable for VLSI implementation. First, the computational power and the processing time are wasted, since the rate control must be done after all code-blocks finish Tier-1 coding regardless of the target bit rate. Second, a large temporary memory is required to buffer redundant bit-stream and side information for rate control. So it is main topic on how to reduce computational redundancy and memory cost.

Some works about the topic have been done. In the paper [3], a model-based prediction method was used to estimate the number of the need-to-be-coded passes and eliminate the unnecessary computation and memory. Compared with the VM (Verification Model)[4], however, concerning the image quality decrease 1-2dB. The paper [5] applied interpolation of rate-distortion ratio to predicate the possible truncate point and its results, similar to the one in [3], causing about 1dB image quality loss. The paper [6] proposed two algorithms: slope-table method and minimal slope discarding methods, which are both able to reduce the memory cost and computation complexity almost without image quality loss. However, minimal slope discarding method requires that the slope space be searched one time once the total rate is larger than target one, which causes more iteration. The slope-table method must update the whole slope-table to make slope arrayed degressively, which causes high computation complexity.

In this paper, component bit allocation based on weights, a simplified delta-distortion estimation model and two-table optimal slope threshold generation algorithm are introduced in section 3,section 4 and section 5, respectively. Finally, experiment results and conclusions are given.

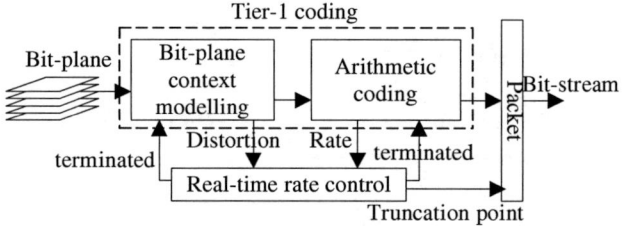

Figure 1. Integrate rate control with Tier-1 encoding

II. Rate Control Algorithm in JPEG2000

After DWT, the quantized wavelet coefficients are written into code-blocks. Each code-block is partitioned into several bit-planes and each bit-plane is orderly scanned by significance propagation pass (SP), magnitude refinement pass (MRP), and cleanup pass (CP). According to [2], final embedded bit-stream of code-block i can be truncated at some candidate pass, n_i, to yield rate, $R_i^{n_i}$ and corresponding distortion, $D_i^{n_i}$. The total rate and distortion of the final bit-stream are $R = \sum_i R_i^{n_i}$ and $D = \sum_i D_i^{n_i}$, respectively. For rate control, the goal is to minimize the distortion while keeping the rate of code-stream smaller than a target bit-rate, R_T. The problem can be converted to Lagrange optimization problem as the minimization of

$$D + \lambda R = \sum_i (D_i^{n_i} + \lambda R_i^{n_i}) \qquad (1)$$

In order to solve (1), the R-D slope, $S_i^{n_i}$, for each code-block at candidate truncation point must be calculated:

$$S_i^{n_i} = \frac{\Delta D_i^{n_i}}{\Delta R_i^{n_i}} = \frac{D_i^{n_i-1} - D_i^{n_i}}{R_i^{n_i} - R_i^{n_i-1}} \quad (2).$$

An optimal threshold λ^* is found among those slopes from most bit-plane to least bit-plane. If the slope is larger than λ^*, the corresponding bit-stream will be written to the final bit-stream.

III. A Simplified Algorithm for JPEG2000 Rate Control

Above rate control algorithm can be divided into three sub-processes: target bit-rate allocation among tiles, slope calculation and optimal slope threshold generation. From a point of VLSI implementation, we improved the algorithms of those sub-processes.

A. Component Bit Allocation

In JPEG2000 coding system, a source image is partitioned into several tiles (typical size, 128×128) and all coding parts take tile as basic unit. So the target file size must be distributed to a tile of a component. The target rate of one tile can be calculated by average method. For a color image, however, it is not easy to obtain the target rate of each component. In the rate control of VM [4], three components (luminance component, Y, and two chrominance components, U and V) in a tile are processed together by component visual weights and global slope threshold. In VLSI implementation, this method will be ineffective since source constraint makes rate control be done as a tile of a component unit. But this problem has not been concerned in other literatures. We have been inspired by sub-sampling of JPEG or MPEG and proposed the flowing method:

$$L_i = \frac{w_i L}{(w_Y + w_U + w_V)N_{tile}} \quad i = Y, U, V \quad (3).$$

Here, L and L_i are target size of the final file and a tile of the ith component, respectively and N_{tile} is total number of the tile in a component. w_Y, w_U and w_V are defined as weights of three components. Let w_{yu} and w_{vu} be weights of Y component and V component renormalized for U component weight as unity. The pair of optimal weights can be gained by the following optimization:

$$f(w_{yu}, w_{vu}) = \frac{1}{3}\sum_{i=1}^{3}|Q_i(w_{yu}, w_{vu}) - \tilde{Q}_i| \quad (4),$$
$$\min f(w_{yu}, w_{vu}),$$
$$s.t \ w_{yu} \in [3.0 \ 5], w_{vu} \in [1.0 \ 2.0].$$

where $f(w_{yu}, w_{vu})$ is an objective function, and at a certain compression ratio, $Q_i(w_{yu}, w_{vu})$ is PSNR of the ith reconstructed component with component bit allocation based on weights (w_U and w_V), while $\tilde{Q}_i$ is PSNR with the method in VM. When $f(w_{yu}, w_{vu})$ is minimum, its corresponding optimal w_{yu} and w_{vu} at a certain compression ratio are found. It is shown in figure 2 that optimal weights (w_{yu} and w_{vu}) at compression range from 0.3bpp(byte per pixel) to 2bpp tend to cluster.

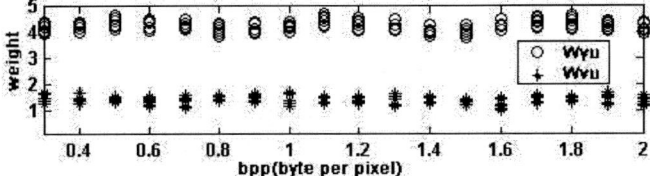

Figure 2. Optimal w_{yu} and w_{vu} at 0.3bpp to 2bpp for different images.

B. A Simplified Delta-distortion Estimation Model

Let $u_j^p = 2^{-p}v_j^p - 2\left\lfloor \frac{2^{-p}v_j^p}{2} \right\rfloor$, where v_j^p is the quantized magnitudes. It is easy to verify that $u_j^p \in [0,2)$. In VM, the delta-distortion of the jth significant sample scanned by SP or CP of the pth bit-plane in the kth code-block is calculated by:

$$\Delta D_{s/c}^{jp} = w^k[(u_j^p)^2 - (u_j^p - 1.5)^2] \quad (5),$$

and the one by MRP is obtained by

$$\Delta D_r^{jp} = w^k[(u_j^p - 1)^2 - (u_j^p - 1.5)^2]|_{u_j^p \geq 1}$$
$$+ w^k[(u_j^p - 1)^2 - (u_j^p - 0.5)^2]|_{u_j^p < 1} \quad (6).$$

Here, $w^k = (W_w^k \Delta^k)^2 2^{2p_{max}}$, W_w^k, Δ^k and p_{max} are the wavelet filter weight, quantized step-size and the maximal bit-plane number of the kth code-block, respectively. Equation (5) and (6) indicate that the delta-distortion of a sample is the function of u_j^p. So a loop-up table is used to estimate the delta-distortion in VM. The size of look-up table is 2×128×8 bits and the loop-up table is indexed by the coefficients of code-block. However, some side information and regularity in bit plane coding could be used to estimate the total delta-distortion.

If $u_j^p \geq 1$, set $u_j^p = 1 + \Delta_j^p$, or else $u_j^p = \Delta_j^p$. Let N denote the total number of significant samples in the SP and CP. N_1 is the total number of scanned refinement samples whose bit is "1", while N_2 is the total number of refinement samples whose bit is "0".

Set $p_s(x)$ and $\hat{\lambda}_s$ are the probability and expected value of Δ_{ij}^{kp} in [0, 1]. The total delta-distortion of SP or CP in the pth bit-plane of the kth code-block is calculated by

$$\Delta D_{s/c}^p = \sum_{j=1}^{N} \Delta D_{s/c}^{jp} w^k = \sum_{j=1}^{N}(0.75 + 3\Delta_j^p)$$
$$= 0.75w^k N + 3w^k N \int_0^1 p_s(x)dx \quad (7),$$
$$= w^k N(0.75 + 3\hat{\lambda}_s)$$
$$= w^k w^{s/c} N$$

where $w^{s/c}$ is expected value.

Similarly, set $p_{r1}(x)$ and $p_{r2}(x)$ are the probability of Δ_{ij}^{kp} in [0, 1], respectively. The total delta-distortion of MRP is calculated by

$$\Delta D_r^p = \sum_{j=1}^{N_1} \Delta D_r^{jp} + \sum_{j=1}^{N_2} \Delta D_r^{jp}$$
$$= w^k [N_1(\int_0^1 p_{r1}(x)dx - 0.25) + N_2(0.75 - \int_0^1 p_{r2}(x)dx)] \quad (8)$$
$$\overset{\Delta}{=} w^k w^r (N_1 + N_2)$$

where w^r are expected value.

From (7),(8), we would find that total delta-distortion is a function of total number of significant or refinement samples scanned and the expected value. The expected value denotes the contribution to the total delta-distortion of a significant or refinement sample. We called it delta-distortion weight. In JPEG2000, the bit-plane coding is based on the optimal principle [2][7] that coefficient bits that are likely to reduce distortion most should be coded first. Namely, each pass has different contribution to the total distortion, so those weights are different between different passes within the same bit-plane. In addition, from most bit-plane to least bit-plane, the total delta-distortion is digressive. Therefore, those weights are different between different passes within the same bit-plane.

From the above analysis, we can see that the delta-distortion of each pass in code-blocks will be estimated if those delta-distortion weights are known. In following model, we uses 5 different weights for different contribution to the total delta-distortion, namely, W_{sr}^H, W_{cr}^H, W_s^V, W_r^V and W_c^V. W_{sr}^H and W_{cr}^H are weights of delta-distortion for significant pass and cleanup pass renormalized for refinement pass weight as unity. While W_s^V, W_r^V and W_c^V are weights for 3 passes in upper bit-plane renormalized for corresponding passes in immediate lower bit-plane as unity.

After we tested more than 20 test images, we found that those weights W_{sr}^H, W_{cr}^H, W_s^V, W_r^V and W_c^V all show a tendency to normal probability distribution. Based on the above analysis, hypothesis and statistic properties, we proposed a delta-distortion model as follows:

For the p th bit-plane of the kth code-block, the total delta-distortion of SP, ΔD_s^p, of MRP, ΔD_r^p, and, of CP, ΔD_c^p, are estimated by:

$$\Delta D_s^p = w^k W_{sr}^H (W_s^V)^{p-p_{\max}} n_s^p \quad (9)$$
$$\Delta D_r^p = w^k (W_r^V)^{p-p_{\max}} n_r^p \quad (10)$$
$$\Delta D_c^p = w^k W_{cr}^H (W_c^V)^{p-p_{\max}} n_c^p \quad (11)$$

respectively. Here, p is the index of bit-planes from most bit-plane, p_{max} to 0 and n_s^p, n_r^p and n_c^p denote the total the number of significant or refinement coefficients of SP, MRP and CP of the pth bit-plane in the kth code-block, respectively.

TABLE I. DELTA-DISTORTION WEIGHTS

W_{sr}^H	W_{cr}^H	W_s^V	W_r^V	W_c^V
4.02	3.18	4.76	3.72	4.31

We obtained the serial weights in Table1. As a result, according to our proposed model, total delta-distortion of each pass is calculated in the following tow steps:

(1) Sum up the number of significant or refinement coefficients which are coded in current pass.

(2) According to (9)-(11), calculate the total delta-distortion of this pass after it is completely scanned.

Compared with the delta-distortion model, ours doesn't require wavelet coefficients so that it is possible to avoid access to code-block memory. Original look-up table is also saved.

C. Two-table Slope Threshold Generation

In VM, an optimal slope threshold is found by repetitious searching all passes of code-blocks. So this algorithm is much too iterative and not real-time one. We used two tables (Table 1 and Table 2) to keep inside information for rate control, as illustrated in figure 3. One is used to store slope, delta-rate and pass index of each candidate truncation point, the other is used to record the address for the last candidate truncation point of each code-block. At initial stage, the threshold and total rate are set to zero. After Tier-1 finishs coding a pass, the corresponding slope is calculated and dealt with convex hull analysis and the total rate is summed up. If the threshold is still zero and total rate exceeds the target rate, the minimum slope will be found according to information from Table 2 and Table 1 and corresponding rate will be discard from total rate. The minimum slope discarding procedure is done until the total rate is not more than the target rate. The threshold is updated to the largest one in discarded slope. If threshold is larger than zero and the slope is smaller than the threshold, the Tier-1coding of current code-block is terminated and the same minimum slope discarding procedure will be executed. When all code-blocks finish Tier-1coding, the final threshold and the information about truncation points is listed in Table2 and Table1. Because the search range of the minimum slope discarding

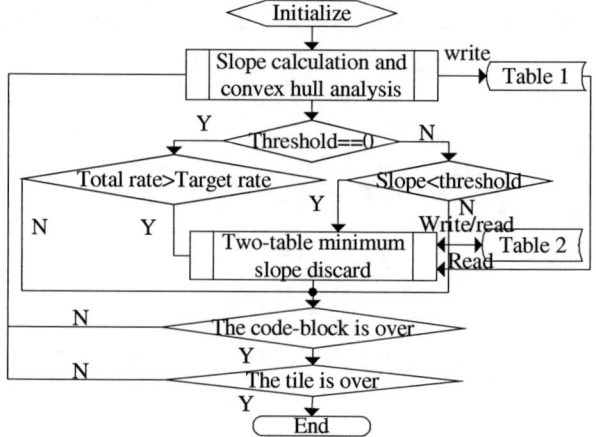

Figure 3. The flow diagram of two-table slope threshold generation.

is determined by the number of encoded code-blocks, the complexity of the algorithm is lower than the one in [6].

IV. EXPERIMENTS AND COMPARASIONS

We chose 20 test images. Through our coder, each image is encoded at 5bpp, 4bpp, 2bpp, 1bpp, 0.8bpp, 0.5bpp 0.3bpp and 0.1bpp with (5x3) and (9x7) wavelet filter respectively. We use "Jasper" as the reference software. Figure 4 shows rate and PSNR (in dB) comparison of our coder to the reference software by taking "lenna" for example. From the results, we can see the curves of ours and VM almost overlap. Table2 gives the differences of rate and PSNR between the results of ours and VM over five images ("lenna", "peppers", "canyon","barara" and "goldhill"), namely, Δb (percent) and $\Delta PSNR$. Our rates are slightly lower than those of VM on average, as well as PSNR. However, the degradation is very slight. So this model should be a low-cost alternative to estimate the delta-distortion for JPEG2000 rate control.

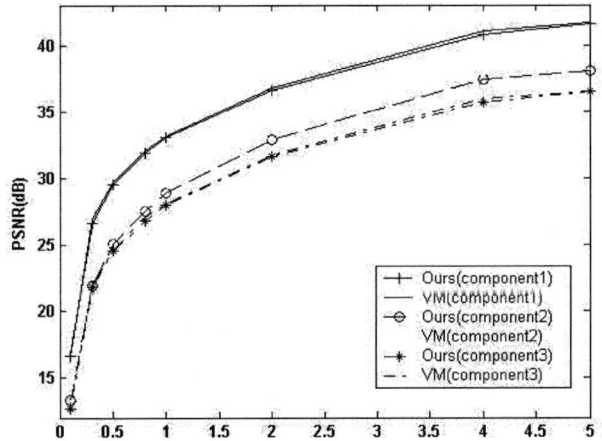

Figure 4. PSNR comparison of our coder to the reference software for 5/3 wavelet filter using "lenna"

TABLE II. COMPARISON RATE AND PSNR WITH THE RESULTS OF VM'S FOR FIVE IMAGES

Ratio	Lenna(5/3)		Peppers(5/3)		Canyon(5/3)		Barbara(9/7)		Goldhill (9/7)	
	Δb	$\Delta PSNR$	Δb	$\Delta PSNR$	Δb	$\Delta PSNR$	Δb	$\Delta PSNR$	Δb	$\Delta PSNR$
4.8	2.83	0.07	-0.96	-0.07	1.23	0.25	1.92	0.12	-1.96	0.01
6	-0.57	-0.02	-1.07	-0.07	1.31	-0.05	-0.60	-0.16	-2.63	-0.29
12	0.16	-0.06	-0.68	-0.13	2.25	0.01	-0.81	-0.13	-1.15	-0.01
24	0.42	-0.07	-0.16	-0.04	-0.20	-0.11	-1.47	-0.27	-0.24	-0.18
30	0.06	-0.07	0.01	-0.11	1.22	-0.07	-0.21	-0.13	0.02	-0.19
48	-0.07	-0.05	-0.09	-0.09	-0.11	-0.14	-0.75	-0.24	-0.48	-0.12
80	0.05	-0.06	-0.27	-0.07	0.33	-0.13	-0.29	-0.20	-0.42	-0.10

Taking "lenna" for example, simulation results (see Table 3) indicate that using our rate control algorithm, the number of processed passes is 33% to 84% of the total passes under different compression ratios, while the number of passes that need to be stored in the table 1 is 18% to 50%, which is fewer than the number of passes that need to be preserved in the final bit-stream. Because the rate control is done by a great deal of comparison operations, computation times were regarded as computation complexity criterion. In Table 3, the comparison times of ours are much fewer than the one in paper [6], while image quality drops 0.3dB compared with results of VM at the same compression ratio (see Table 4).

TABLE III. COMPUTATION COMPLEXITY AND MEMORY EXPENDITURE FOR "LENNA" ON AVERAGE OF A TILE AT DIFFERENT BIT-RATE.

Ratio	Processed	Stored	Finally preserved	Our Compression Times	Compression Times [6]
80	100	54	65	91	245
48	128	70	91	132	425
24	178	98	134	244	825
16	210	118	164	343	1126
12	230	132	185	403	1358
8	254	149	221	408	1673

TABLE IV. COMPARISON IMAGE QUALITY(PSNR) WITH THE RESULTS OF VM FOR "LENNA".

Ratio	80	48	24	16	12	8
VM's	26.44	28.74	32.11	34.46	36.47	40.50
Ours	26.25	28.62	31.83	34.17	36.18	40.24

V. CONCLUSIONS

For VLSI implementation of JPEG2000 rate control, we proposed a simplified algorithm, including component bit allocation based on weights, a simple delta-distortion estimation model and tow-able slope threshold generation. Experiments show that the algorithm can reduce memory cost and computation complexity and is suitable for integration with Tier-1 encoding modules with slight image quality loss. As to the future work, we will apply the algorithm to JPEG2000 codec IC.

ACKNOWLEDGMENT

This work is partly supported by National 863 Plan Foundation of China 2003AA1Z10060 and 2003AA141050.

REFERENCES

[1] JPEG 2000 Part 1 020719 (Final Publication Draft), ISO/IEC JTC1/SC29/WG1 N2678, July, 2002.

[2] D. Taubman, "High performance scalable Image Compression with EBCOT," *IEEE Trans. on Image Processing*, vol. 9, no.7, July 2000, pp.1158-1170.

[3] T.Masuzaki, H.Tsutsui; T.Izumi, T.Onoye and Y.Nakamura, "JPEG2000 adaptive rate control for embedded systems", *IEEE International Symposium on Circuits and Systems*, vol.4, Scottsdale,Arizona,USA May 2002, pp.333 -336.

[4] JPEG2000 verification model 7.2, ISO/IEC JTC1/SC29/WG1, May 2000.

[5] Yeung Y.M.,An O.C, Chang A, "Successive bit-plane rate allocation technique for JPEG2000 image coding",*IEEE International Conference on Acoustics, Speech, and Signal Processing*, vol3, April 2003, pp. 261-264.

[6] C.Te-Hao, L.Chung-Jr, C.Hong-Hui, C.Jing-Ying and C.Liang-Gee, "Effective hardware-oriented technique for the rate control of JPEG2000 encoding", *IEEE International Symposium on Circuits and Systems* , vol. 2 , Bangkok,Thailand , May 2003,pp.684 -687.

[7] E. Ordentlich, M. Weinberger, and G. Seroussi, "A low-complexity modeling approach for embedded coding of wavelet coefficients", *Data Compression Conference*, Snowbird, Utah, USA, Mar. 1998, pp 408-417..

# Quality Improvement Technique for JPEG Images with Fractal Image Coding

Megumi Takezawa,
Hirofumi Sanada
and Kazuhisa Watanabe
Department of Information Network Engineering
Hokkaido Institute of Technology
7-15 Maeda Teine-ku, Sapporo 006-8585, Japan

Miki Haseyama
School of Information Science and Technology
Hokkaido University
N-14 W-9 Kita-ku, Sapporo 060-0814, Japan

Abstract—This paper proposes a quality improvement technique for JPEG images by using fractal image coding. JPEG coding is a commonly used standard method of compressing images. However, in its decoded images, quantization noise is sometimes visible in high frequency regions, such as the edges of objects. Hence, in order for the JPEG coding to become a more powerful image-coding method, the JPEG image quality must be improved. Therefore, our method solves this problem by adding the obtained codes by the fractal image coding to improve the image quality. Some simulation results verify that the proposed method achieved higher coding-performance than the traditional JPEG coding.

I. Introduction

Recently, new-generation coding techniques, such as Joint Photographic Experts Group (JPEG) 2000 and Set Partitioning in Hierarchical Trees (SPIHT), are proposed for digital images. They can encode the images at very low bit-rate with acceptable quality and provide the decoded images with higher quality than the traditional techniques, such as JPEG. However, practically, many digital picture and graphics tools do not support their new image formats as yet and the traditional JPEG format is still used commonly. Therefore, we directed our attention to the JPEG coding. And in this paper, a quality improvement technique is proposed for the JPEG image by using fractal image coding.

The JPEG images are sometimes corrupted because quantization noise is visible in high frequency regions. Especially, in the color images, the quantization noise of the luminance data is more visible than that of the chrominance data. Therefore, in this paper, a quality improvement method for the color JPEG images is proposed by reducing the quantization noise of the luminance data in the high frequency regions.

The proposed method achieves the quality improvement of the color JPEG images as follows: Firstly, the high frequency regions in the image obtained by the traditional JPEG codes are detected. Secondly, for the high-frequency regions, the difference intensity-values between the luminance of the original image and the one of the JPEG image are calculated and then encoded. Finally, the traditional JPEG codes and the new codes representing the difference are transmitted to the decoder side. And in the decoder side, we can obtain the higher quality images by adding the difference intensity-values, which are provided by the new codes, to the intensity values of the JPEG image, which are provided by the traditional JPEG codes. However, even if the image quality is improved by the above process, the proposed method cannot achieve the high coding-performance if the number of bits for the added new codes is large. Therefore, the proposed method encodes the difference intensity-values by fractal image coding[1], [2], [3], which has an advantage of the high image-compression capability. By using the fractal image coding, the bit number required for the new codes are minimized and thus the higher coding-performance than the traditional JPEG is realized. In this paper, we present some simulation results to verify the high coding-performance of the proposed method.

II. Traditional JPEG images

In this section, we illustrate the quantization noise in the JPEG images with some experiments. Here, we compress the image "Lena" (Fig. 1 (a)), whose size is 256×256 pixels and the maximum RGB value is 255, and show its JPEG images in Fig. 1 (b), (c) and (d), where the JPEG quality factor [4] Q is set at 15, 40 and 70.

For each image shown in Fig. 1, we pick up one block around the lena's shoulder and show it in Fig. 2. In this figure, (a) is obtained from the original image (Fig. 1 (a)); and (b), (c) and (d) are obtained from the JPEG images when $Q = 15$ (Fig. 1 (b)), $Q = 40$ (Fig. 1 (c)) and $Q = 70$ (Fig. 1 (d)), respectively. Comparing Fig. 2 (b), (c) and (d), the JPEG block noise is reduced at the lower compression ratios. However, the quantization noise in the high frequency region of the shoulder line is still in the JPEG images even if their compression ratios are lower.

These results verify that the quantization noise is visible in the high frequency regions in the JPEG images. Especially, in the color images, it is well known that the human eye is much more sensitive to the changes in the luminance than those in the chrominance. Therefore, in this paper, we propose the quality improvement method for the high frequency regions in the JPEG images by reducing the quantization noise of the luminance data.

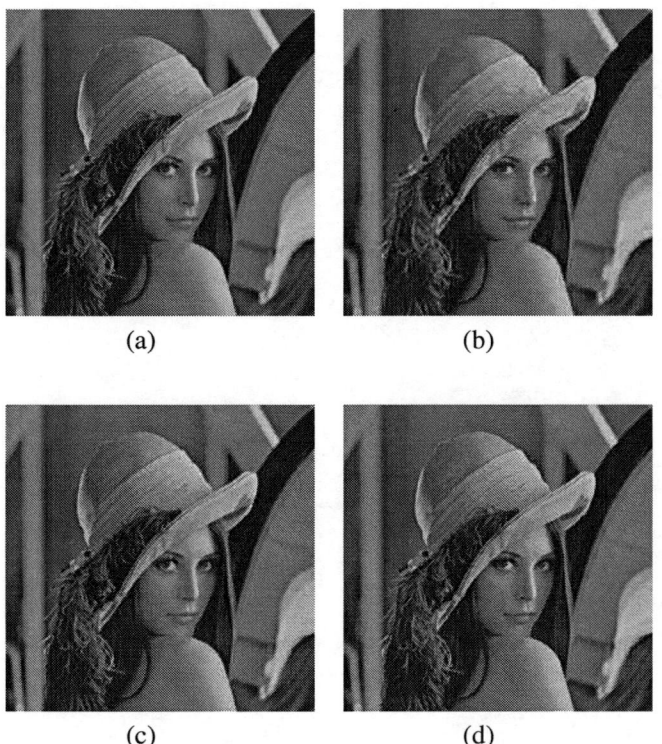

Fig. 1. (a) Original image. (b) JPEG image (Quality factor=15), (c) JPEG image (Quality factor=40), (d) JPEG image (Quality factor=70).

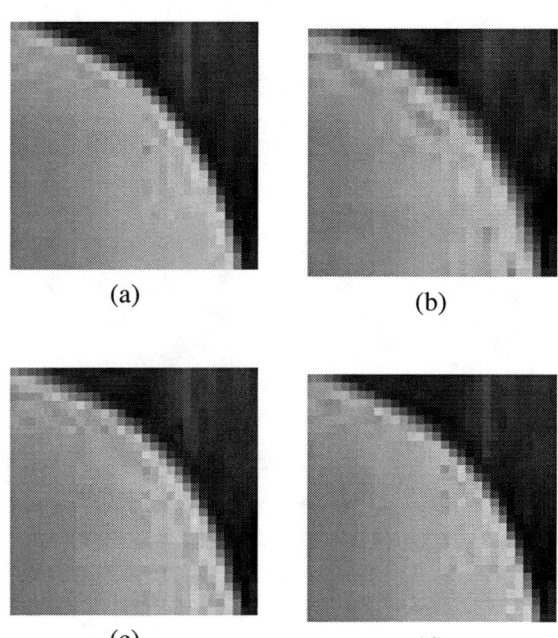

Fig. 2. Enlargements of the images ((a) Original image, (b) JPEG image (Quality factor=15), (c) JPEG image (Quality factor=40), (d) JPEG image (Quality factor=70)).

III. PROPOSED METHOD

As stated in Sect. II, the quantization noise of the luminance data is visible in the high frequency regions in the JPEG images. Therefore, in order to reduce the quantization noise, the proposed method calculates the codes representing the difference intensity-values between the original image and the JPEG image in the high frequency regions and adds them to the traditional JPEG codes. However, even if the image quality is improved by adding the new codes, the proposed method cannot achieve the high coding-performance if the number of bits for the added new codes is large. Therefore, the proposed method makes an image which consists of the difference intensity-values and thus encodes it. Here, we show one example of the difference image in Fig. 3. From this image, the difference intensity-values are random and their correlation is low. Hence, it is difficult to encode the image effectively by the coding techniques, such as the transform coding and the predictive coding. Therefore, the proposed method encodes the image by the fractal image coding[1], [2], [3].

A. Encoding process

(i) An original image is encoded by the traditional JPEG.
(ii) We obtain the JPEG image by the codes provided in (i) and detect the edges in the luminance component of the JPEG image. In the following simulations, we detect the edges by using a laplacian filter.

Fig. 3. An image composing the difference intensity-values between the original image and the JPEG image.

(iii) We scan the luminance component of the JPEG image in the raster order and, for each edge pixel, calculate the difference intensity-values between the luminance of the original image and the one of the JPEG image.
(iv) An image which composes the difference intensity-values calculated in (iii) is encoded by using the fractal image coding. And the obtained codes, which are called the IFS (Iterated Function System) codes, are just the codes for improving the image quality.

The JPEG codes and the IFS codes obtained by the above procedures are transmitted to the decoder side.

B. Decoding process

(i) The JPEG image is obtained by the JPEG codes.
(ii) The difference intensity-values between the luminance of the original image and the one of the JPEG image are obtained by the IFS codes.

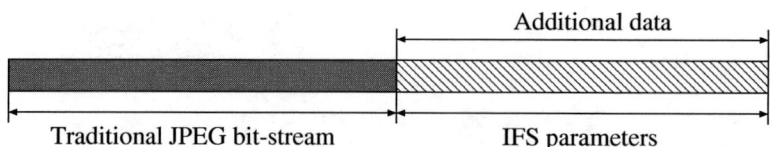

Fig. 4. Structure of the bit stream by the proposed method.

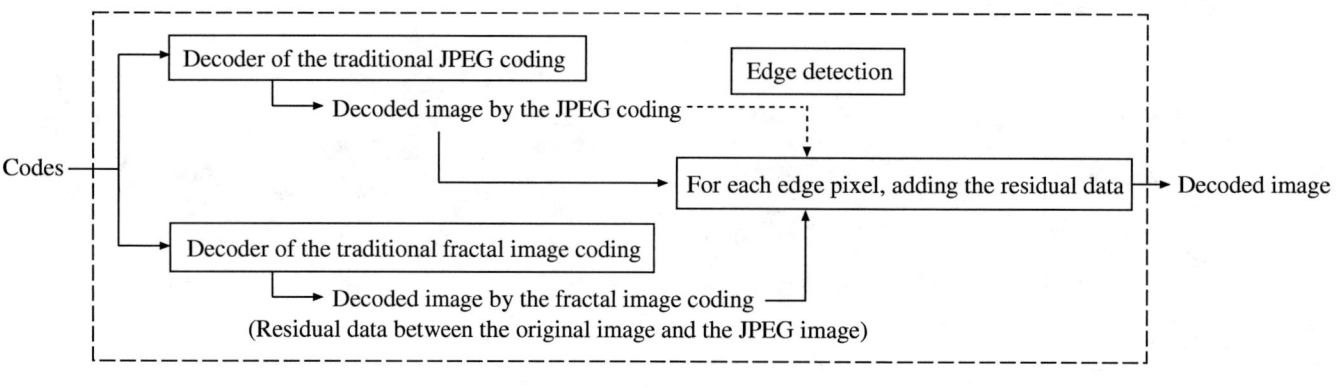

Fig. 5. Decoder of the proposed method.

(iii) We detect the edges in the luminance component of the JPEG image.

(iv) We scan the luminance component in the raster order and, for each edge pixel, add the difference intensity-value to its intensity value.

By the above decoding process, we can obtain the high quality images in which the quantization noise is invisible.

C. Decoder

The bit stream obtained by the encoding process in the Sect. III-A is shown in Fig. 4. In this figure, the bit stream consists of the traditional JPEG bit stream and the bit stream which represents the IFS parameters. Here, we also show the decoder for them in Fig. 5. From the figure, we can see that the decoder is easily designed by utilizing the traditional decoders of the JPEG coding and the fractal image coding, which are embedded in the existing graphics tools.

IV. SIMULATION RESULTS

We verify the effectiveness of the proposed method by encoding a real image: "Lena" (Fig. 6 (a)) of size 256×256 pixels and 24bits/pixel. The JPEG quality factor Q is set at 30. In the fractal image coding, the range block size is 16×16. Also, scaling coefficients and offsets are quantized with 2bits and 3bits, respectively. In the affine transformations of the IFS, the rotation and the reflection are not utilized.

A. Decoded image

Figure 6 shows the decoded images. In this figure, (a) shows the original image; (b) shows the decoded image by the proposed method; (c) shows the decoded image by the traditional JPEG when $Q = 30$; and (d) shows the decoded image, which is compressed at the same ratio as (b), by the traditional JPEG when $Q = 40$. Comparing Fig. 6 (b) and (c), the proposed method improves the peak signal-to-noise ratio (PSNR) [1] of the decoded image by 1.5dB. Further, comparing (b) and (d), at the same compression ratio, the proposed method provides the decoded image with 0.4dB higher quality than the traditional JPEG.

Moreover, in order to visually verify the effectiveness of the proposed method in the high frequency regions, we pick up one block in the high frequency regions in each decoded image shown in Fig. 6 and show it in Fig. 7. In this figure, (a) shows the original image; (b) shows the decoded image by the proposed method; (c) shows the decoded image by the traditional JPEG when $Q = 30$; and (d) shows the decoded image, which is compressed at the same ratio as (b), by the traditional JPEG when $Q = 40$. Comparing (b) and (c), it is clear that the proposed method reduces the quantization noise in the high frequency regions and preserves the edges

[1]The PSNR for the decoded image of size $M \times N$ is defined as follows:

$$PSNR = 10 \log_{10} \frac{255^2}{MSE} \quad (1)$$

and the mean-square error (MSE) is defined as

$$MSE = \frac{1}{MN} \sum_{y=1}^{M} \sum_{x=1}^{N} (I(x,y) - I'(x,y))^2 \quad (2)$$

where $I(x,y)$ and $I'(x,y)$ denote the pixel values at the coordinates (x,y) of the original image and the decoded image, respectively.

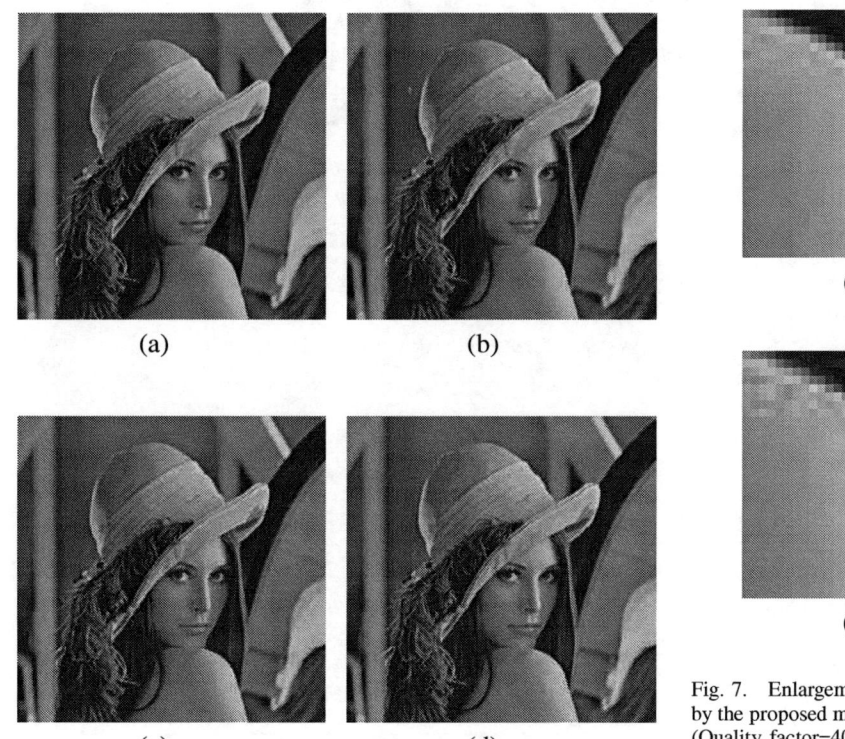

Fig. 6. Decoded images. ((a) Original image, (b) Decoded image by the proposed method, PSNR = 33.21 (0.94 bpp), (c) JPEG image (Quality=30), PSNR = 31.73 (0.80 bpp), (d) JPEG image (Quality=40), PSNR = 32.76 (0.93 bpp))

TABLE I
CALCULATION TIME ((A) PROPOSED METHOD, (B) JPEG CODING).

Methods	Calculation time (s)
(a) Proposed method	0.41
(b) JPEG coding	0.01

successfully. Further, comparing (b) and (d) reveal that, at the same compression ratio, the quantization noise in the decoded images by the traditional JPEG is more visible than that by the proposed method. Based on these results, we can see that the proposed method can effectively reduce the quantization noise and provide the higher quality images than the traditional JPEG.

B. Calculation time

In this simulation, the calculation time required to encode an image is shown in Table I: (a) is the total time required by the proposed method; (b) is the time by the traditional JPEG coding. From these results, we can see that the proposed method encodes the image in the practical time. However, the calculation time by the proposed method is 40 times longer compared to the traditional JPEG coding. Therefore, further study is needed to reduce the calculation time.

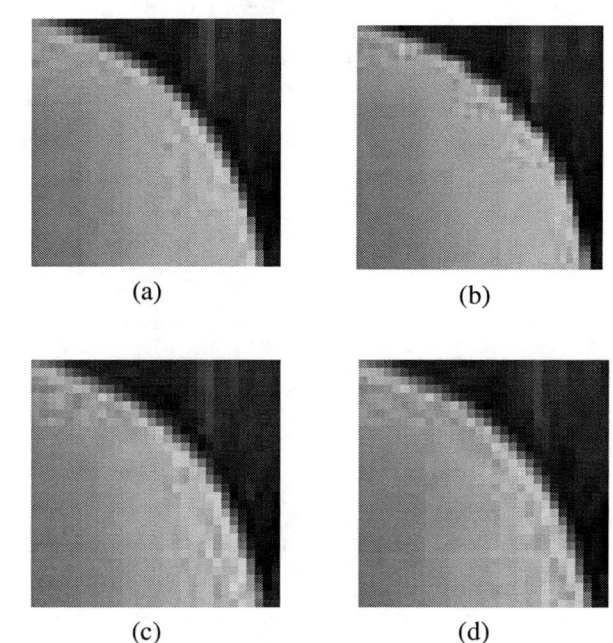

Fig. 7. Enlargements of the images ((a) Original image, (b) Decoded image by the proposed method, (c) JPEG image (Quality factor=30), (d) JPEG image (Quality factor=40)).

V. CONCLUSION

In this paper, we have presented a new JPEG including the fractal image coding for the color images. As stated in Sect. II, in the JPEG images, the quantization noise is sometimes visible in the regions of the high frequencies even if they are compressed at the lower ratios. Therefore, the proposed method improves the image quality by adding the obtained codes by the fractal image coding to reduce the quantization noise in the high frequency regions. The simulation results reveal that the proposed method effectively reduces the quantization noise in the high frequency regions and thus provides the decoded image with 0.4dB higher quality than the traditional JPEG.

REFERENCES

[1] M. F. Barnsley, *Fractals Everywhere*. Boston: Academic Press, 1988.
[2] A. E. Jacquin, "Image coding based on fractal theory of iterated contractive image transformations," *IEEE Trans. on Image Processing*, vol. 1, no. 1, pp. 18–30, Jan. 1992.
[3] Y. Fisher, Ed., *Fractal Image Compression : Theory and Application*. New York: Springer-Verlag, 1995.
[4] W. B. Pennebaker and J. L. Mitchell, *JPEG Still Image Data Compression Standard*. New York: Van Nostrand Reinhold, 1993.

A Scalable Encryption Method allowing Backward Compatibility with JPEG2000 Images

Osamu WATANABE†, Akiko NAKAZAKI‡, and Hitoshi KIYA‡

†Department of Electronics and Computer Systems, Takushoku University
815-1, Tatemachi, Hachioji, Tokyo, 193-0985, JAPAN
‡Department of Electrical Engineering, Tokyo Metropolitan University
1-1, Minami-osawa, Hachioji, Tokyo, 192-0397, JAPAN
Email: owatanab@es.takushoku-u.ac.jp, nakazaki@isys.eei.metro-u.ac.jp, kiya@eei.metro-u.ac.jp

Abstract—
A new method for encryption of JPEG 2000 images, which is referred to as 'scalable encryption', is proposed in this paper. Scalable encryption method makes the encrypted images to have multi-level-encryption and to reduce the computational complexity of encryption, since different encryption algorithms can be simultaneously used in its procedure. Moreover, the encrypted images produced by the proposed method have the complete compliance with JPEG 2000, so that a standard JPEG 2000 decoder can decode the encrypted images and the useful functionalities of JPEG 2000 codestream are preserved after the encryption. For example, the proposed method enables that contents holders have no need of preparing two or more encrypted images for various users who are provided the different access rights. In addition to this, the time for the encryption can be controlled by selection of adequate encryption algorithms for faster processing.

I. INTRODUCTION

With the growth in network technology, exchanging digital images commercially or non-commercially has become very common. Because digital images can be easily duplicated and re-distributed, protecting copyrights an the privacy of digital images, whether they are encoded or not, is an important issue. For example, security aspects of JPEG 2000 [1], which is the new image coding standard, has been considered in JPEG 2000 part 8 (known as JPSEC [2]). Besides, system requirements and specifications for digital cinema, which adopts JPEG 2000 as the compression scheme, stipulate that all contents should be encrypted [3]. Of course, this kind of content protection should be secure against attacks. However, it is considered that decryption of encrypted contents are performed at projectors, which have poor computing environment, as well as the other powerful platforms. For faster processing at such poor environment, the development of a fast method for content protection is desired.

This paper discusses a encryption method for making JPEG 2000 images secure which has scalability on the security of encrypted images and that on the complexity of the encryption. The purposes of the method are (A) to achieve the control of the degree of encryption and (B) to reduce the computational complexity of encryption. Some related works are found in [4]–[10]. In these works, the methods referred to as partial scrambling have been proposed. Partial scrambling(or Partial encryption) also performs encryption. However, it encrypts some parts of the images. For compressed images, this means that the header information of the image is not encrypted so that the encrypted output keeps an image-data format. Although only authorized users can obtain the encryption-key and decrypt the encrypted images, partial scrambling makes it possible for all users to preview the encrypted images at a degraded quality without an encryption key. The degree of quality degradation is controlled by adjusting the amount of the parts that are to be encrypted. Methods described in [7]–[9] can control the degree and the computational complexity of scramble by using some control parameters. However,

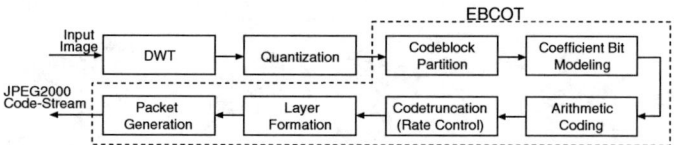

Fig. 1. JPEG 2000 encoder

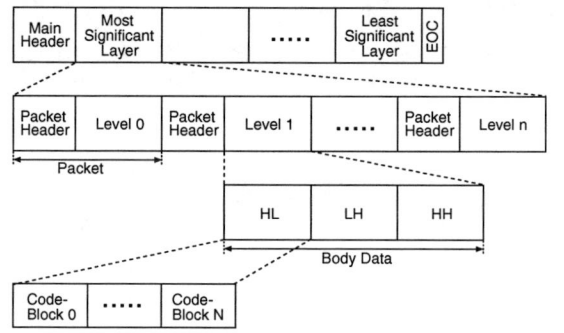

Fig. 2. Structure of JPEG 2000 codestream

those conventional methods encrypt images under a single condition (e.g. encryption algorithms, encryption key, and parameters, etc). This means that the complete decryption of encrypted data is possible once the condition is discovered. Therefore, two or more degrees of encryption cannot be set up with conventional methods.

In this paper, we propose an encryption method for JPEG2000 images which takes multi stage of encryption with different conditions (e.g. secret-key and/or public-key encryption, different keys, etc) so that the method makes it possible to provide more than one level of decryption and to solve the trade-off between the encryption speed and the security level. To examine the effectiveness of the proposed method, we provides some simulations.

II. JPEG 2000 CODING

Before the discussion of the proposed method, it is worth to describe the overview of JPEG2000.

A. JPEG 2000 Coding procedure and codestream structure

Figure 1 shows a block diagram of the JPEG 2000 encoder. The coding procedure is briefly summarized as follows. An input image is decomposed into subbands by using a discrete wavelet transform (DWT). The number of DWT is called the DWT level. Each subband is divided into code blocks, which are the coding unit of the EBCOT algorithm [11]. The EBCOT algorithm encodes each code block individually and produces a compressed codestream. The compressed

TABLE I
MARKER CODES

Value in HEX	Mnemonic	Name
$FF90_h$	SOT	Start of tile
$FF91_h$	SOP	Start of packet
$FF92_h$	EPH	End of packet header
$FF93_h$	SOD	Start of data
$FFD9_h$	EOC	End of codestream

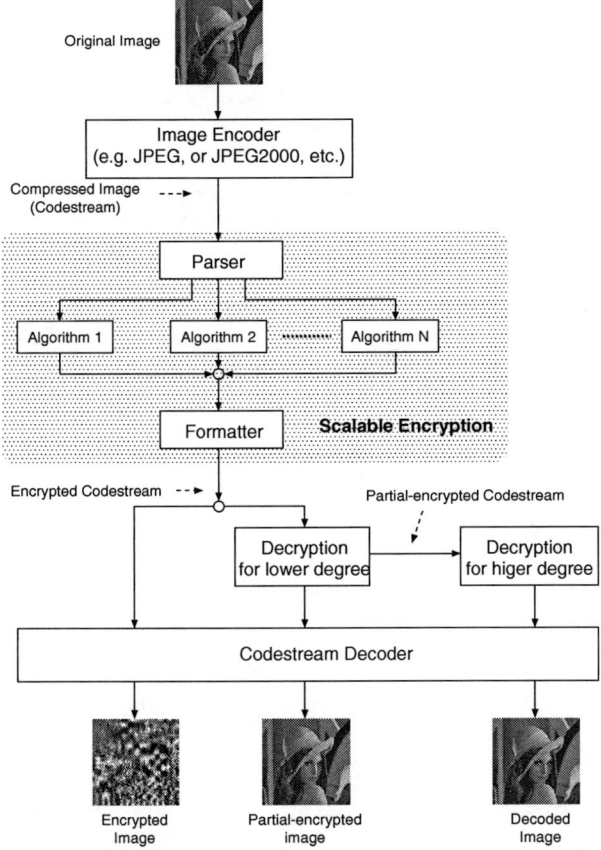

Fig. 3. Procedure of proposed scalable encryption

codestream may have more than one layer. A layer is defined as the set of compressed data of a certain image quality. The structure of the JPEG 2000 codestream is shown in Fig. 2. The codestream has a main header, packet headers, etc. In this paper, these headers are referred to as header information, whereas the rest of the codestream is called body data.

B. Marker codes and encryption

Marker codes in JPEG 2000 codestreams are special codes with values ranging from $FF90_h$ to $FFFF_h$, where "$_h$" indicates hexadecimal notation. All marker codes are represented with two bytes. The upper byte is FF_h, and the lower byte is xx_h. Therefore, the marker codes are represented as $FFxx_h$. An encryption methods for JPEG 2000 codestreams should avoid generating the marker codes shown in Table I because these markers are used to distinguish body data from the rest of the codestream. If not, the decoding process of the encrypted codestream may fail.

III. SCALABLE ENCRYPTION

A. Overview

Because the encryption of images degrades those visual quality, the degree of encryption is reflected in the image quality. Conventional image encryption methods can mainly use only a single condition of encryption. This fact leads to a problem that they cannot be suitable for the use under the situation that various encryption conditions are required. For example, a image contents holder may wants to encrypt images in two levels: one is for free registered users and the other is for authorized users who payed for the images, and he also wants to forbid non-authorized users to view the images. In this case, more than two levels of visual degradation are required. That is, the contents is required to be encrypted in multi levels. The purpose of scalable encryption is to achieve this kind of multi-level-encryption.

Figure 3 shows the concept of the scalable encryption. The procedure is briefly described as follows.

An image is inputted to a certain encoder such as JPEG2000 or JPEG. The resulting compressed image is analyzed by a parser, which reads the format of compressed images. The outputs from a parser are the header information and the target body-data. Then the target part of body-data is encrypted. Note that there are some freedom in the use of encryption algorithms.

The encryption process can be done with the different algorithms; a block cipher and a public-key algorithm (e.g. RSA) can be combined. For example, the important data may be encrypted by a public-key algorithm, however, the less prior data may be encrypted by a block cipher for fast processing.

On the other hand, the decryption process will be done in hierarchical. A block decipher may be used to decrypt the portion of data encrypted by a block cipher. For certain users, it is enough to confirm the contents of images but they can only view the images at degraded quality. After the block decipher, a public-key decryption may be applied to the much prior data that is encrypted by a public-key algorithm, for users who have payed for the contents.

B. Procedure

1) Conditions to avoid generating marker codes: To apply the concept of scalable encryption to JPEG 2000 images, marker codes should not be generated through the encryption processing. Let us assume that the original JPEG 2000 codestream has no marker codes in its body data. In other words, marker codes in the range $FF90_h$ to $FFFF_h$ are not normally generated in the body data itself. Thus, scrambling the body data must not generate any marker codes in this range. The following conditions for avoiding marker codes have been proposed by [7], [8] to avoid this problem.

> **(Condition 1)** If a byte is below $F0_h$, no restriction is set on the value of the scrambled lower half byte.
> **(Condition 2)** If a byte is $F0_h$ or above, the lower half byte must be skipped.

2) Encryption: As shown in Fig. 3, a JPEG 2000 codestream is sent to the parser and decomposed into the header and body data. The layers, DWT levels, subbands, and code blocks are then selected, and the body data (presupposed L bytes) are distilled. A parameter M for controlling the computational complexity is introduced, and the body data is divided into blocks B_k ($k = 1, 2, ..., (L-m)/M, m = mod(L, M)$). Each block has M bytes. The following operation is done to each block B_k byte by byte (Fig. 4).

Step 1. Using an initial value p and an algorithm for generating

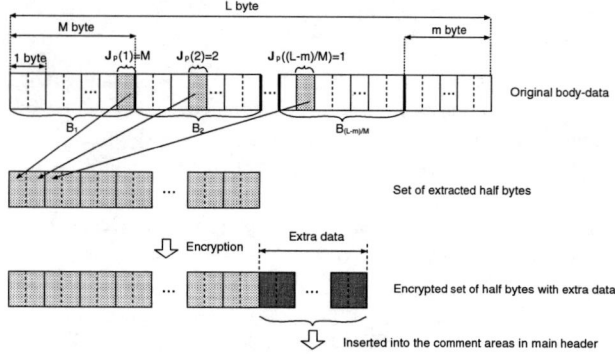

(a) Encryption

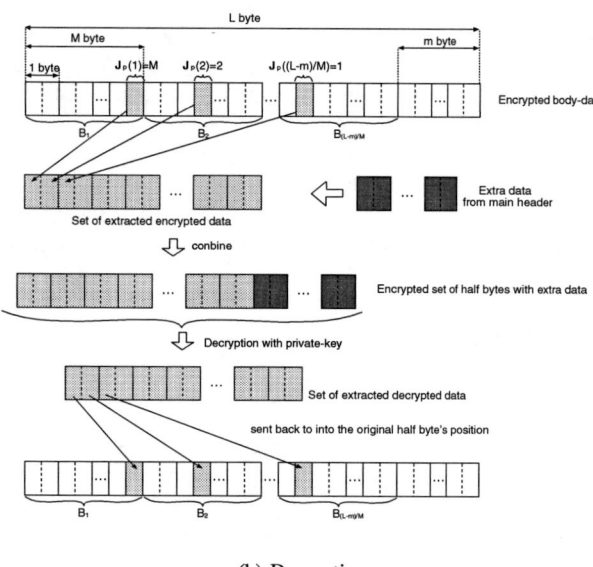

(b) Decryption

Fig. 4. Detailed procedure of encryption and decryption

random integers $J_p(k)$, generate random integers in the range 1 to M:

$$J_{p_1}(k) \in \{1, 2, ..., M\}, k = 1, 2, ... \quad (1)$$

Step 2. Select the $J_p(k)$-th byte from B_k.

Step 3. Divide the selected $J_p(k)$-th byte into upper and lower half bytes on the basis of

(3.1) If the selected byte is below $F0_h$, the lower half byte is selected.

(3.2) If the selected byte is $F0_h$ or above, the lower half byte is skipped (block B_k is excluded from encryption).

Finally, half bytes obtained in Step 3 are gathered in the encryption buffer. Then, the encryption of codestream is performed with a certain algorithm, such as blowfish [12], or RSA [13], etc. When the public-key algorithm is used here, the extra data produced by the encryption is to be inserted into the comment area of the main header [9].

3) Decryption: The decryption operation uses an algorithm for generating random integers ($J_p(k)$), the initial value p, an area of L bytes for the body data (layer, DWT level, subband, or code block), and a parameter M. The encoded data is sent to the parser and decomposed into the header and the body. The decryption is done as follows.

Step 1. Given p, random integers $J_p(k)$ are generated in the range 1 to M.

Step 2. In each B_k, the $J_p(k)$-th byte selected on the scrambled side is located.

Step 3. A half byte of the $J_p(k)$-th byte is selected as the decryption target.

(3.1) If the selected byte is below $F0_h$, the lower half byte is selected.

(3.2) If the selected byte is $F0_h$ or above, the byte is skipped (block B_k is excluded from decryption).

The bytes selected in Step 3 are sent to the decryption buffer. The extra data written in the comment areas of the main header is concatenated with the data in the decryption buffer to obtain the encrypted data. Finally, decryption using the private key of the user is performed on the encrypted data.

C. Features

The features of the proposed method are summarized below.

1) Scalability of encryption: In the proposed method, the hierarchical security levels can be set up, as a result of the encryption based on more than one condition. The different encryption algorithms (or different length of encryption-key etc.) can be used to encrypt a certain images. For example, a block cipher and a public-key encryption can be used simultaneously. Various encryption algorithms can be used and combined.

It is worth notice that public-key algorithm can be used for not only encryption but also digital-signature. Thus, the tamper-resistant encryption of JPEG 2000 images can be possible.

Moreover, by using parameter M, the computational complexity of the encryption can be controlled in the proposed method. For example, it is possible to encrypt visually important data by a public-key encryption and to encrypt the rest by a block cipher.

2) Backward compatibility with JPEG 2000: The scrambled JPEG 2000 codestreams can be decoded by a standard JPEG 2000 decoder because the generation of marker codes is avoided. Therefore, the functionality of JPEG 2000 codestreams is maintained through the encryption/decryption procedure. That is, only specific JPEG 2000 coding units such as layers, DWT levels, subbands, or code blocks are selected as encryption targets. However, at the same time that M reduces the computational complexity, this parameter controls the degree of scrambling.

IV. EXPERIMENTAL RESULTS

The experiment evaluated the effectiveness of the proposed method. "Lena" (512×512, gray scale), which is shown in Fig. 5 was used as the test image. This experiment used JPEG 2000 VM8.6 Software [14] as the JPEG 2000 codec. The coding-rate was 1.0 [bits/pixel], and the number of layers was ten. The layers were numbered zero to ten (layer 0 means the most visually important layer). Each layer had roughly equal amounts of data (= 0.1[bits/pixel]). An RSA [13] based public-key encryption and an Blowfish [12] based block cipher were used for the encryption/decryption. The key length of each algorithm was set to 1024-bit. The platform for this experiment was a Linux (Kernel Ver. 2.4.22) PC that had a 1.2-GHz Celeron processor with 512-KB L2 cache and 512-MB RAM.

Fig. 5. Test image

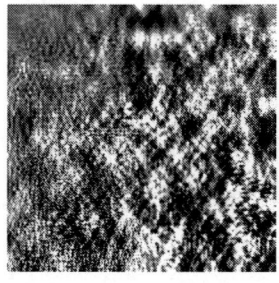

(a) No decryption (Blowfish, $M = 1$)   (b) Layer 0 decrypted

Fig. 6. Hierarchical decryption

(a) RSA only (Partially encrypted, $M = 256$)   (b) (a) and Blowfish ($M = 1$)

Fig. 7. Combination of encryption algorithm

A. Hierarchical decryption

First, the result of hierarchical decryption is shown in Fig.6. We used Blowfish as the encryption algorithm and independently encrypted each layer. Figure 6(a) shows the result of this encryption. Then, only the most significant layer, layer 0, was decrypted and the partially encrypted image shown in Fig. 6(b) was generated. Hence, we confirmed that hierarchical decryption was realized by the fact that each encryption target could be decrypted independently.

B. Combination of encryption algorithm

Second, Layer 0 of the compressed test image was selected as the target of encryption. Then this target layer was encrypted by the complexity controlled RSA encryption with the value of parameter $M = 256$. The value of parameter $M = 256$ makes the computational complexity of the encryption $1/256$ compared with that of $M = 1$. The resulting image of above operations is shown in Fig. 7(a). This result is the example of partially encrypted (visually recognizable). Next, this encrypted layer was encrypted again by the block cipher with $M = 1$. As a result, the completely encrypted image shown in Fig. 7(b) was produced. Finally, we have confirmed that the above encryption based on RSA and Blowfish was approximately 60 times faster than the case that all of the target was encrypted by only RSA with the value of parameter $M = 1$. The encrypted images have the advantages of public-key encryption, however, the completely encrypted images can be produced in much less time than the encryption which only uses the public-key based algorithm.

V. CONCLUSIONS

A scalable encryption method for JPEG 2000 images has been proposed in this paper. The proposed method allows the use of multiple encryption algorithms so that the encrypted codestreams have multiple security levels. This feature realizes the lower computational complexity of the encryption by selecting appropriate algorithms for the particular encryption targets than the methods which are allowed to use only a single algorithm. Moreover, the encrypted images have the backward compatibility with JPEG 2000 so that many functionalities of JPEG 2000 can be utilized.

Finally, we have confirmed the effectiveness of the proposed method from the experimental results.

REFERENCES

[1] "Information technology — JPEG 2000 image coding system – Part 1: Core coding system," Int. Std. ISO/IEC IS-15444-1, 2000 Dec.

[2] "JPSEC Committee Draft – Version 2.0," ISO/IEC JTC 1/SC 29/WG 1 N3397, July 2004.

[3] "Digital Cinema System Specification version 4.2," Digital Cinema Initiatives, LLC Technology Committee, Aug 2004

[4] Howard Cheng and Xiaobo Li, "Partial encryption of compressed images and videos," *IEEE Trans. Signal Processing*, vol. 48, no. 8, pp. 2439–2451, Aug. 2000.

[5] M. Van Droogenbroeck, "Partial encryption of images for real-time applications," in *Fourth IEEE Benelux Signal Processing*, Hilvarenbeek, The Netherlands, April 2004, pp. 11–15, Invited presentation.

[6] Raphaël Grosbois and Pierre Gerbelot and Touradj Ebrahimi, "Authentication and access control in the JPEG2000 compressed domain," in *Proc. of SPIE 46th Annual Meeting, Applications of Digital Image Processing XXIV*, July 2001, vol. 4472, pp. 95–104.

[7] Hitoshi Kiya, Shoko Imaizumi, and Osamu Watanabe, "Partial-scrambling of JPEG2000 Images without Generating Marker Codes(in Japanese)," *IEICE Transaction on Fundamentals*, vol. J-86-D-II, no. 11, pp. 1628–1636, Nov. 2003.

[8] Hitoshi Kiya, Shoko Imaizumi, and Osamu Watanabe, "Partial-scrambling of JPEG2000 Images without Generating Marker Codes," in *Proc. of IEEE International Conference on Image Processing(ICIP)*, Sept. 2003.

[9] Osamu Watanabe, Akiko Nakazaki, and Hitoshi Kiya, "A Fast Image-Scramble Method using Public-Key Encryption allowing Backward Compatibility with JPEG2000," in *Proc. of International Conference on Image Processing (ICIP)*, Singapore, Oct. 2004, (Accepted).

[10] Frédéric Dufaux, Diego Santa Cruz, and Touradj Ebrahimi, "EPFL's Proposal for JPSEC Core Experiment," ISO/IEC JTC 1/SC29/WG1 N3082, Oct. 2003.

[11] David Taubman, "High performance scalable image compression with EBCOT," *IEEE Trans. Image Processing*, vol. 9, no. 7, pp. 1158–1170, July 2000.

[12] Bruce Schneier, *Applied Cryptography Second Edition : protocols, algorithms, and source code in C*, John Wiley & Sons, Inc., 1996.

[13] Ronald L. Rivest, Adi Shamir, and Leonard M. Adleman, "A method for obtaining digital signatures and public key cryptosystems," *Communications of the ACM*, vol. 21, no. 2, pp. 120–126, Feb. 1978.

[14] "JPEG 2000 verification model 8.6 software," ISO/IEC JTC 1/SC 29/WG 1 N1894, 2000.

Lossless Implementation of Motion JPEG2000 Integrated with Invertible Deinterlacing

Takuma Ishida, Shogo Muramatsu, and Hisakazu Kikuchi
Department of Electrical and Electronic Engineering, Niigata University
8050, Ikarashi 2-no-cho, Niigata 950-2181, Japan
Email: takumaro@telecom0.eng.niigata-u.ac.jp, {shogo,kikuchi}@eng.niigata-u.ac.jp

Abstract— A lossless implementation technique of Motion JPEG2000 (MJP2) integrated with invertible deinterlacing is presented. In previous works, we developed an invertible deinterlacing technique that suppresses the comb-tooth artifacts, which are caused by field interleaving of interlaced videos and affect the quality of scalable intraframe-based coding such as MJP2. We suggested a scenario of applying the technique as a prefilter to intraframe-based coding systems. This system suppresses the comb-tooth artifacts, while guaranteeing recovery of the original quality through the inverse process. In this work, we further propose an exact lossless implementation technique by embedding the deinterlacer into MJP2 and modifying header information for appropriate decoding. We demonstrate that the comb-tooth suppression capability can be kept at low bitrates with the standard MJP2 decoder. The experiments also show that the overhead of our proposed method is less than 0.9% compared with the normal field interleaving.

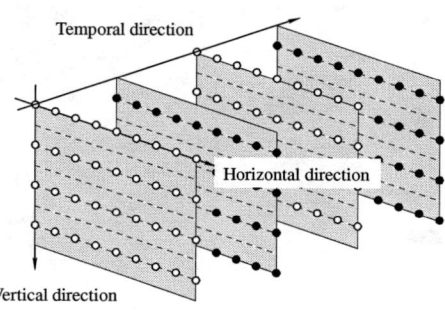

Fig. 1. Interlaced scanning, where the white and black circles are sample points on top and bottom fields, respectively.

I. INTRODUCTION

Interlaced scanning and progressive scanning are known as record and display formats of motion pictures [1]–[3]. For interlaced pictures, such as NTSC signals, intraframe-based coding requires field-interleaving so that any still picture coding is directly applicable. This process, however, causes horizontal comb-tooth artifacts at edges of the moving objects [4]. In the case of wavelet transform-based scalable coding such as Motion-JPEG2000 (MJP2) [5], [6], errors yielded by quantizing vertical high frequency components of the comb-tooth artifacts become recognizable for low bit-rate decoding. Such a behavior for interlaced video sequences is not desirable with respect to the SNR scalability of MJP2.

As previous works, to solve the problem, we developed invertible deinterlacing with sampling density preservation as a pre-process of scalable intraframe-based coding [7]–[10]. With this technique, we can suppress the comb-tooth artifacts while guaranteeing the quality recovery through the inverse process. So far, we mainly investigated a lossy coding system integrated with the invertible deinterlacer. On the other hand, frame pictures are frequently edited and re-compressed both in professional and consumer applications. The reversible, i.e. lossless, conversion therefore is important to exactly reconstruct the interlaced pictures from a code-stream. Lossless compression is appropriate in applications where the images are to be extensively edited and re-compressed so that the accumulation of errors from multiple lossy operations might become unacceptable. This paper is concerned with the SNR scalability of MJP2 with lossless mode for interlaced video sequences.

In this work, we investigate such a deinterlacing technique by taking the lossless mode of MJP2 into account. To achieve this purpose, we consider integrating an MJP2 system with lossless invertible deinterlacing subject to guarantee of the standard code-stream format and proper low bit-rate standard decoding. Since our deinterlacing technique is implemented as lifting structures, it is easily embedded into DWTs of MJP2 [11]. We show that our lossless compression is achieved by controlling the subband gain factors [6] according to the DWT integrated with the invertible deinterlacing. In addition, our proposed method offers a function of decoding a picture of which comb-tooth artifacts are suppressed beforehand. This function is realized within the standard code-stream by modifying a component of header marker in JPEG2000 (JP2). We evaluate performances of the proposed method compared with the normal field interleaving.

II. REVIEW OF INVERTIBLE DEINTERLACING

As previous works, we have proposed a deinterlacing technique that preserves sampling density and has invertibility [7], [8]. In this section, let us briefly review this technique as a preliminary, assuming that the input array $X(\mathbf{z})$ is given as shown in Fig. 1, where $\mathbf{z}$ is a 3×1 vector consisting of variables in a 3-D Z-transform-domain.

A. Invertible Deinterlacer

Figure 2 (a) shows a basic structure of the deinterlacer with sampling density preservation and (b) shows the inverse process, i.e. *reinterlacing*. The invertible deinterlacer constructs

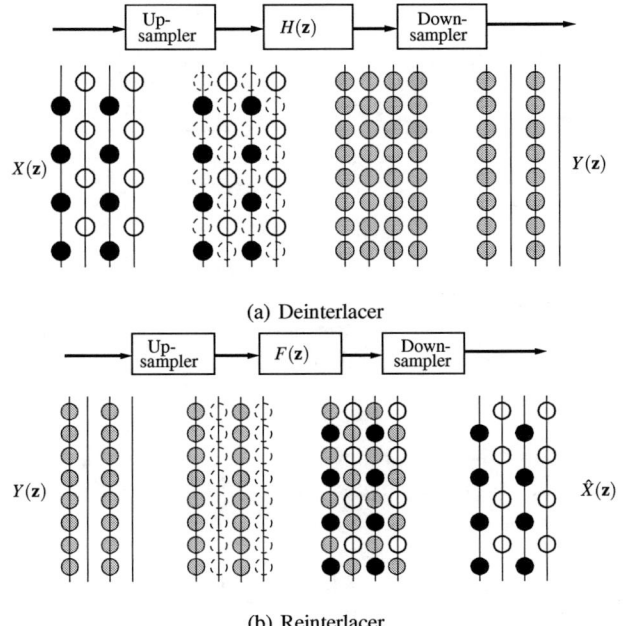

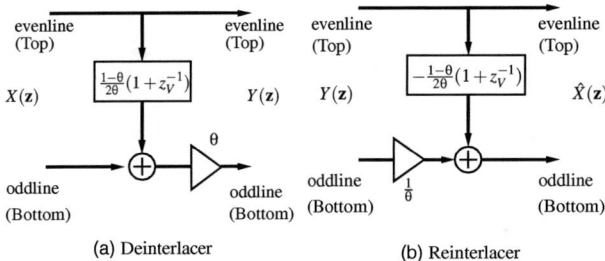

Fig. 4. Lifting implementation.

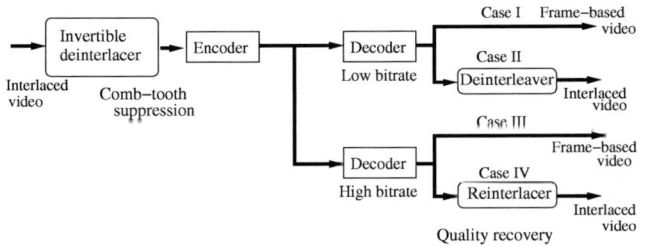

Fig. 2. Basic structure of invertible deinterlacing, where only vertical-temporal plane is shown.

Fig. 3. Intraframe-based coding system with deinterlacer.

a frame picture from two successive top and bottom field pictures as shown in Fig. 2 (a).

This process suppresses the comb-tooth artifacts, while maintaining the perfect reconstruction property. In the articles [7] and [8], we have shown that there exist general solutions and that this process can be regarded as a generalization of the simple field interleaving.

B. Application scenario

An application scenario developed on our suggested codec system is outlined in Fig. 3 [9], [10]. The system offers the compatibility and coexistence of interlaced and progressive scanning video formats, and also provides an extension of decoding options at front-end receivers. An encoded single bit-stream is delivered to decoders, and users select their favorable decoding option among various choices depending on their displaying terminals and available transmission bitrates. Single-source encoding and multiple-destination decoding can be achieved.

The system shown in Fig. 3 uses an invertible deinterlacer as a pre-filter to support both of frame and field-based display.

The comb-tooth artifacts are suppressed for both of field and frame-based pictures at low bitrates, whereas the quality of field-based pictures are maintained with the reinterlacer at high bitrates. For low bit-rate decoding, a simple deinterleaver splits frames into fields in the usual way. Both of interlaced and frame-based videos can be obtained from decoded sequences on demand. Note that producing a progressive scanning video via another interpolative deinterlacing for the reinterlaced result is also possible.

C. (3+1)-tap Deinterlacing Filter

In a special case of deinterlacing and reinterlacing filters [7], [8], the implementation is available in a lifting structure. Figure 4 shows the corresponding lifting implementations. Each structure is composed of one prediction with the simple Haar type 2-tap filter and the scaling factor on the odd line [11]. These lifting structures are equivalently applied to the vertical direction of interleaved frame pictures. We can select any value of the design parameter in the range $0 < \theta \leq 1$ for the pair shown in Fig. 4. We exemplify four frequency characteristics of the deinterlacing filter $H(\mathbf{z})$ in Fig. 5. The characteristics are changed among temporal, vertical-temporal (V-T), and vertical lowpass filters by selecting θ.

In JP2 part I [6], two types of wavelet transforms are standardized. One is the reversible 5/3 transform of lossless compression and the other is the irreversible 9/7 transform of lossy compression. These transforms are implemented as lifting structures of two-tap filtering. Additionally, our deinterlacing can be cascaded to the following wavelet transform. Figure 6 illustrates the lifting structures where the deinterlacing and reinterlacing are embedded into the 5/3 transform, respectively.

III. LOSSLESS IMPLEMENTATION

In this section, we present a lossless compression of MJP2 integrated with the invertible deinterlacer. There are the following two issues in the realization of the lossless compression:

- How to prevent the high frequency subband's fractions due to the output scaling with factor θ
- How to maintain the standard decoding and the capability of comb-tooth suppression

A. Preservation of fractions

As mentioned, the 5/3 transform is employed for lossless compression in JP2. In order to realize the reversible wavelet

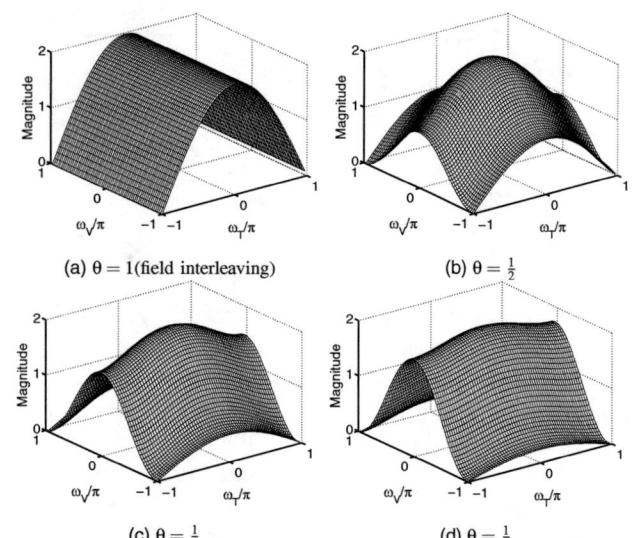

Fig. 5. Characteristics of $H(\mathbf{z})$, where only the vertical-temporal frequency plain is shown.

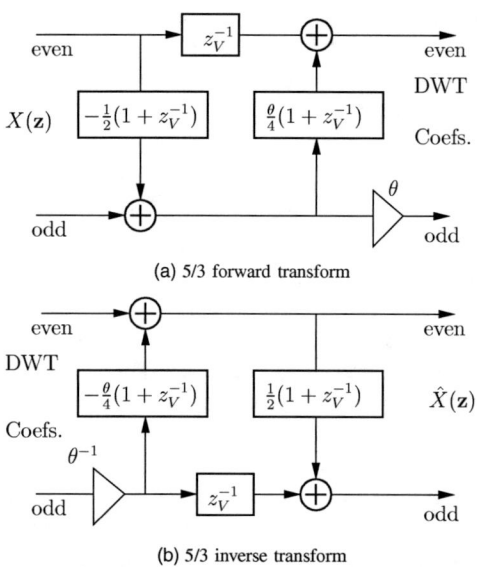

Fig. 6. Lifting implementation of the 5/3 transform integrated with the deinterlacer and its inverse.

TABLE I
ORIGINAL GAIN

Subband b	G_b
LL_b	1
HL_b	2
LH_b	2
HH_b	4

TABLE II
PROPOSED GAIN

Subband 1	G_1
LL_1	1
HL_1	2
LH_1	$2\theta^{-1}$
HH_1	$4\theta^{-1}$

transform, the rounding operations are inserted after lifting steps of "predict" and "update" to make transform coefficients integers [6]. The bit-depth of wavelet coefficients of each subband is represented by the nominal range R_I of the relevant subband and subband gain factor G_b as follows:

$$R_b = R_I + \log_2(G_b) + GB, \quad (1)$$

where G_b is given by the lowpass and highpass filters of the wavelet transforms as shown in Table I, and GB denotes a guard bit. For the lifting structure integrated with the deinterlacing shown in Fig. 7 (a), the scaling process on the odd line generates the fraction parts. As a result, the lossless mode becomes unavailable. In this work, we calculate the gain of highpass filter by taking the scaling factor θ into account in order to preserve the fractions. Table 2 shows the derived subband gain factors. According to the result, by modifying the bit-depth of wavelet coefficients, we can guard the fractions that would have been eliminated in the original gains. Note that parameter θ can be given by 2^{-n} with a positive integer n and that the proposed gains are applied for only the first vertical wavelet decomposition ($b = 1$).

B. Maintenance of the standard decoding

The important issue of our lossless implementation is how to guarantee the appropriate decoding with standard code-stream format and maintain the capability of comb-tooth suppression at low bitrates. Modification of the subband gains makes the high frequency components larger than the original bit-depth range defined in JP2. As a result, distortion occurs in standard decoding. The dynamic ranges of LH_1 and HH_1 subband's wavelet coefficients are increased $\theta^{-1}(>1)$ times as large as the original. The direct application of our proposed gains makes the SNR scalability of MJP2 meaningless.

In JP2, these subband gains are written with a marker QCD (Quantization Default) [6] to the header information. In this work, we forcibly write the original gains in segments of marker QCD in the code-stream despite of applying the proposed gains in Table II. As the dequantization step of JP2 decoders, an appropriate bit-depth of the transform coefficients are obtained by using the original gains in the segments of marker QCD. Such a stream deceives standard decoders as if itself has been normally encoded. What happens in standard decoders is that the LH_1 and HH_1 subband's LSBs (least significant bit), i.e. the fraction parts are just discarded. As a result, we can guarantee standard decoding with the standard code-stream format. It only differs in the way that comb-tooth artifacts are suppressed beforehand.

For the lossless decompression, it is sufficient to dequantize the vertical high frequency components in the proposed subband which is derived by taking the scaling factors of the deinterlacing into account, and the inverse transform integrated with reinterlacing as shown in Fig. 7 (b) except for scaling with factor θ^{-1}.

IV. PERFORMANCE EVALUATION

In this section, we evaluate the comb-tooth suppression capability and the coding efficiency among several values of θ and compare them with the simple field interleaving.

A. Suppression capability at low bit-rate

To demonstrate the effect of deinterlacing filters, a frame in *Football* sequence was encoded with our proposed method with the lossless mode, and then decompressed with the standard JP2 decoder at 0.1 bpp. Figure 7 shows the results at $\theta \in \{1, \frac{1}{2}, \frac{1}{4}, \frac{1}{8}\}$. From the subjective point of view, it is

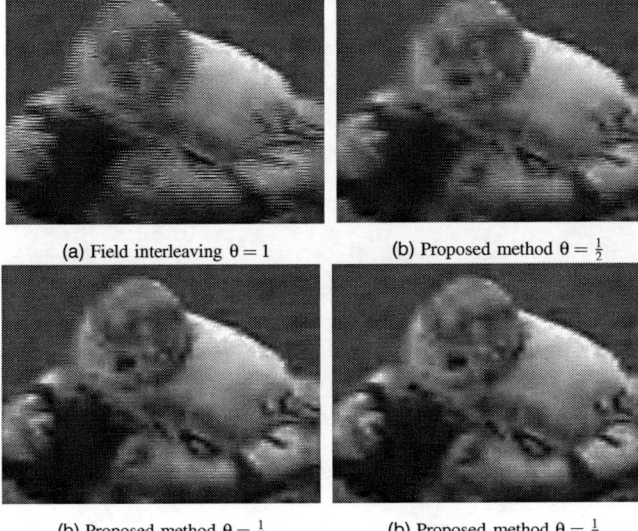

Fig. 7. Simulation results decoded at 0.1 bpp

verified that the comb-tooth suppression capability is improved by using our proposed method from the result of simple field interleaving. The temporal lowpass filter shown in Fig. 5 (a), has less capability to remove the vertical high frequencies than the other filters. As a result, comb-tooth influence is strongly perceived at low bitrates. We verify that the filter shown in Fig. 5 (d) of $\theta = \frac{1}{8}$ significantly removes the comb-tooth artifacts.

B. Coding efficiency in the lossless mode

TABLE III
CODING EFFICIENCY IN LOSSLESS MODE

	Football	Mobile & Calendar
Field interleaving ($\theta = 1$)	4.1288 bpp	5.7779 bpp
Proposed method ($\theta = 2^{-1}$)	4.1658 bpp	5.7867 bpp
Proposed method ($\theta = 2^{-2}$)	4.1988 bpp	5.8006 bpp
Proposed method ($\theta = 2^{-3}$)	4.2143 bpp	5.8067 bpp

In order to show the significance of our proposed method, we evaluate the coding efficiency of our lossless codec system. In this experiment, we used sequences 'Football' (720×486@60Hz, 240 frames, luminance) and 'Mobile & Calendar' (720×576@50Hz, 220 frames, luminance). Every frame picture was encoded with the lossless mode. Table III indicates the results of the average bit-rate of each sequence. For the sequence 'Football', including fast moving objects with fast camera action, one can verify a little increase in bit-rate with any θ compared with simple field interleaving, although the deinterlacing effect suppresses the artifacts at low bit rates. The increasing rates are less than 2.0 % compared with the field interleaving.

In the case of the sequence 'Mobile & Calendar,' including slow moving objects with slow camera action, one can verify that the result of our proposed method are almost the same as that of field interleaving. The increasing rates are less than 0.5 % compared with the field interleaving. Note that our proposed method gives moderate results at both of low bitrates and the lossless mode.

Comparing among the different deinterlacing filters, we can see from Table III that the more fraction bits in the LH_1 and HH_1 subbands, the less the coding efficiency results in. In summary, there is a trade-off between the comb-tooth capability of suppression and coding efficiency.

V. CONCLUSION

We presented a lossless implementation of Motion JPEG2000 integrated with invertible deinterlacing. It was shown that recalculating the subband gain factors and modifying the QCD marker segments realize the lossless compression with deinterlacing. With some experimental results, it was shown that the coding efficiency was almost the same as that obtained by the normal field interleaving, while maintaining the suppression capability of comb-tooth artifacts through standard decoding.

We further would like to investigate applying the lossless implementation technique to adaptive invertible deinterlacing with variable coefficients [10]. We have to decide a region of comb-tooth parts in given pictures and consider the boundary processes between motion and still areas.

ACKNOWLEDGMENT

The authors would like to thank Mr. Tetsuro Kuge for his helpful comments. This work was in part supported by the Grand-in-Aid for Scientific Research No.16-5404 from Japan Society for the Promotion of Science.

REFERENCES

[1] A. Murat Tekalp, *Digital Video Processing*, Prentice Hall, Inc., 1995.
[2] G. de Haan and E. B. Bellers, "Deinterlacing-an overview," in *Proc. IEEE*, Sept. 1998, vol. 86, pp. 1837–1857.
[3] L. Vandendorpe and L. Cuvelier, "Statistical Properties of Coded Interlaced and Progressive Image Sequences," *IEEE Trans. Image Processing*, Vol.8, no.6, pp.749–761, June 1999.
[4] T. Kuge, "Wavelet picture coding and its several problems of the application to the interlace HDTV and the ultra-high definition images," in *IEEE Proc. of ICIP*, WA-P2.1, 2002.
[5] ISO/IEC JTC1/SC 29/WG1 N2117, "Motion JPEG2000 Final Committee Draft 1.0," March 2001.
[6] D. S. Taubman and M. W. Marcellin, *JPEG2000 Image Compression Fundamentals, Standards and Practice*, Kluwer Academic Publishers, 2002.
[7] S. Muramatsu, T. Ishida, and H. Kikuchi, "A design method of invertible de-interlacer with sampling density preservation," in *IEEE Proc. of ICASSP*, May 2002, vol. 4, pp. 3277–3280.
[8] S. Muramatsu T. Ishida and H. Kikuchi, "Invertible deinterlacer with sampling-density preservation: Theory and design," *IEEE Trans. Signal Processing*, vol. 51, no. 9, pp. 2343–2356, Sept, 2003
[9] H. Kikuchi, S. Muramatsu, T. Ishida, and T. Kuge, "Reversible conversion between interlaced and progressive scan formats and its efficient implementation," *Proc. of EUSIPCO*, no. 448, 2002.
[10] T. Ishida, S. Muramatsu H. Kikuchi and T. Kuge, "Invertible deinterlacing with variable coefficients and its lifting implementation," in *IEEE Proc. of ICASSP*, IMSP-L5.3(III-105), April 2003. (also appears in *IEEE Proc. of ICME, ICASSP-8.7(III-177), July 2003*)
[11] T. Soyama, T. Ishida S. Muramatsu H. Kikuchi and T. Kuge, "Lifting architecture of invertible deinterlacing," *IEICE Trans. Fundamentals*, vol. E86-A, no. 4, pp. 779–786, Apr. 2003.
[12] Cannon, EPFL and Ericsson "http://jj2000.epfl.ch," URL.

Improved Fast Encoding Method for Vector Quantization Based on Subvector Technique

Zhibin Pan[1], Koji Kotani[2], and Tadahiro Ohmi[1]

1) New Industry Creation Hatchery Center, Tohoku University, Japan
2) Department of Electronic Engineering, Graduate School of Engineering, Tohoku University, Japan
Aza-aoba 10, Aramaki, Aoba-ku, Sendai, 980-8579, Japan
E-mail: pzb@fff.niche.tohoku.ac.jp

ABSTRACT

The encoding speed of vector quantization (VQ) is a time bottleneck to its practical applications due to it performing a lot of k-dimensional (k-D) Euclidean distance computations. By using famous statistical features of the sum and the variance of a k-D vector to estimate Euclidean distance first, IEENNS method has been proposed to reject most of unlikely codewords for a certain input vector. By dividing a k-D vector in half to generate its two corresponding (k/2)-D subvectors and then apply IEENNS method again to each of subvectors, SIEENNS method has been proposed as well. SIEENNS method is the so far most search-efficient subvector-based encoding method for VQ but it still has a large memory and computational redundancy.

This paper aims at improving state-of-the-art SIEENNS method by introducing a new 3-level data structure to reduce memory redundancy and by avoiding using the variances of two (k/2)-D subvectors to reduce computational redundancy. Experimental results confirmed that the proposed method can reduce memory requirement for each k-D vector from (k+6) to (k+1) and meanwhile improve total search efficiency by 20%~30% compared to SIEENNS method.

1. INTRODUCTION

Vector quantization (VQ) [1] is a popular method for image compression. In a conventional VQ method, an N×N input image is firstly divided into a series of non-overlapping smaller n×n (n<<N) image blocks. Then VQ encoding is implemented block by block in a raster order.

For an image block encoded by VQ, the real distortion is a difference vector as defined below

$$D_i = x - y_i = [D_{i,1}, D_{i,1}, \cdots, D_{i,k}]^T \quad i=1,2,\cdots,N_c \quad (1)$$

where $x=[x_1, x_2, \ldots, x_k]^T$ is the input image block, $y_i=[y_{i,1}, y_{i,2}, \ldots, y_{i,k}]^T$ is the ith codeword in the codebook $Y=\{y_i| i=1,2, \ldots, N_c\}$, the superscript "T" means a transpose operation, k (k=n×n) is the vector dimension and N_c is the codebook size.

For simplicity, the real distortion D_i is usually measured by squared Euclidean distance as

$$d^2(x,y_i) = (\|D_i\|)^2 = \sum_{j=1}^{k}(x_j - y_{i,j})^2 \quad i=1,2,\cdots,N_c \quad (2)$$

where d(.,.) is a function for computing Euclidean distance, $\|.\|$ is L_2 norm of a vector and j is a dimension of the vector.

Therefore, the best-matched codeword y_w (i.e. winner) in the codebook can be determined directly by

$$d^2(x,y_w) = \min_{y_i \in Y}[d^2(x,y_i)] \quad i=1,2,\cdots,N_c \quad (3)$$

Then the winner index "w" instead of y_w is transmitted to the receiver for data compression. This is a full search (FS) over $Y=\{y_i| i=1,2, \ldots, N_c\}$. FS method can achieve the best PSNR for a fixed codebook but it is computationally very expensive due to N_c times Euclidean distance computations.

There exist many fast search algorithms developed for VQ encoding. By using the sum information, ENNS method [2] (i.e. equal-average nearest neighbor search) has been proposed. By using the variance information, EENNS method [3] (i.e. equal-average equal-variance nearest neighbor search) has been proposed. By using both of the sum and the variance simultaneously, IEENNS method [4] (i.e. improved EENNS) has been proposed. All of these three fast search methods are based on the original k-D vectors.

Recently, by first dividing a k-D vector in half to construct its two (k/2)-D subvectors and then applying IEENNS method to each subvector again, SIEENNS (i.e. subvector-based IEENNS) method has been proposed in [5], which is so far most search-efficient but it still has a large memory and computational redundancy. By only using the partial sum information of the two subvectors, SENNS (i.e. subvector-based ENNS) method has been proposed in [6]-[8], which features no extra memory requirement at all.

This paper proposes to combine IEENNS method and SENNS method together to construct a new subvector-based search method to further improve SIEENNS method.

2. RELATED PREVIOUS WORKS

Suppose the so far minimum Euclidean distance found in a winner search process for x is d_{min}. Because VQ encoding only needs to find out the final minimum Euclidean distance so as to determine a sole winner y_w, it allows to introduce an estimation for Euclidean distance. Thus, if this estimation is larger than d_{min}, it can guarantee that the corresponding real Euclidean distance will surely be larger than d_{min} as well. Then, it is safe to reject current y_i as a winner so as to avoid computing Euclidean distance for y_i.

Because a k-D vector can also be viewed as a k-sample set, a vector can be described approximately by its statistical features of the sum and the variance. The previous work [2] proposed an estimation for Euclidean distance as a codeword rejection rule by using the sum information as

$$kd^2(x, y_i) \geq (S_x - S_{y_i})^2 \tag{4}$$

where $S_x = \sum_{j=1}^{k} x_j$ is the sum of an input image block x and $S_{y_i} = \sum_{j=1}^{k} y_{i,j}$ means the same for y_i. If $(S_x - S_{y_i})^2 \geq kd_{min}^2$ holds, then reject y_i safely. This is ENNS method. To use Eq.4, it needs one extra memory to store S_{y_i} for each y_i and one "±", one "×" and one "cmp" (comparison) operation for a rejection test.

To improve [2], the previous work [3] proposed another estimation for Euclidean distance as an auxiliary codeword rejection rule by using the variance information as

$$d^2(x, y_i) \geq (V_x - V_{y_i})^2 \tag{5}$$

where $V_x \stackrel{Def}{=} \sqrt{\sum_{j=1}^{k}(x_j - M_x)^2}$, $V_{y_i} \stackrel{Def}{=} \sqrt{\sum_{j=1}^{k}(y_{i,j} - M_{y_i})^2}$ is the variance of input x and y_i, respectively. $M_x = S_x/k$, $M_{y_i} = S_{y_i}/k$ are the mean of x and y_i. If $(V_x - V_{y_i})^2 \geq d_{min}^2$ holds, then reject y_i safely. This is EENNS method. To use Eq.5, it needs one extra memory to store V_{y_i} for each y_i and one "±", one "×" and one "cmp" operation for a rejection test.

To improve [3], the previous work [4] proposed to combine Eq.4 and Eq.5 to achieve a new estimation for Euclidean distance as a codeword rejection rule by using both the sum and the variance information simultaneously

$$kd^2(x, y_i) \geq (S_x - S_{y_i})^2 + k(V_x - V_{y_i})^2 \tag{6}$$

If $(S_x - S_{y_i})^2 + k(V_x - V_{y_i})^2 \geq kd_{min}^2$ holds, then reject y_i safely. Obviously, Eq.6 is more powerful than either Eq.4 or Eq.5. In practice, Eq.6 uses a 2-step test flow as: (1) $(S_x - S_{y_i})^2 \geq kd_{min}^2$ by Eq.4 and (2) $(S_x - S_{y_i})^2 + k(V_x - V_{y_i})^2 \geq kd_{min}^2$ by Eq.6 in order to avoid a possible overhead of computation, which means to avoid accumulating $k(V_x - V_{y_i})^2$ because a lot of codewords can be simply rejected by the first test. This is IEENNS method. To use Eq.6, it needs two extra memories to store S_{y_i} and V_{y_i} for each y_i and three "±", two "×" and one "cmp" operations for a complete rejection test. Clearly, all of Eq.4, Eq.5 and Eq.6 are derived based on the original k-D vectors.

In order to reduce search space more efficiently, the previous work [5] recently proposed to divide a k-D vector in half and then to apply IEENNS method to each subvector separately again for constructing new rejection tests. Then, the first and the second subvector of x and y_i can be expressed as $x_f = [x_1, x_2, \ldots, x_{k/2}]^T$, $x_s = [x_{k/2+1}, x_{k/2+2}, \ldots, x_k]^T$, $y_{i,f} = [y_{i,1}, y_{i,2}, \ldots, y_{i,k/2}]^T$ and $y_{i,s} = [y_{i,k/2+1}, y_{i,k/2+2}, \ldots, y_{i,k}]^T$, respectively. Similarly, partial sums and partial variances of each subvector can be defined as

$$\begin{aligned}
S_{x,f} &= \sum_{j=1}^{k/2} x_j, & S_{x,s} &= \sum_{j=k/2+1}^{k} x_j \\
M_{x,f} &= S_{x,f}/(k/2), & M_{x,s} &= S_{x,s}/(k/2) \\
S_{y_i,f} &= \sum_{j=1}^{k/2} y_{i,j}, & S_{y_i,s} &= \sum_{j=k/2+1}^{k} y_{i,j} \\
M_{y_i,f} &= S_{y_i,f}/(k/2), & M_{y_i,s} &= S_{y_i,s}/(k/2) \\
V_{x,f} &= \sqrt{\sum_{j=1}^{k/2}(x_j - M_{x,f})^2}, & V_{x,s} &= \sqrt{\sum_{j=k/2+1}^{k}(x_j - M_{x,s})^2} \\
V_{y_i,f} &= \sqrt{\sum_{j=1}^{k/2}(y_{i,j} - M_{y_i,f})^2}, & V_{y_i,s} &= \sqrt{\sum_{j=k/2+1}^{k}(y_{i,j} - M_{y_i,s})^2}
\end{aligned} \tag{7}$$

Partial sums and variances in Eq.7 can be interpreted by Fig.1.

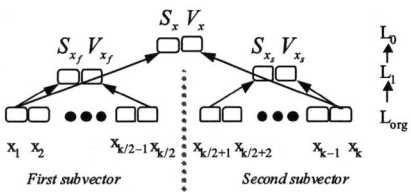

Fig. 1. A 3-level data structure for x. For y_i, it is similar. It requires (k+6) memories in total for each vector.

In addition to using IEENNS method at L_0 level like Eq.6, by directly applying IEENNS method to each subvector at L_1 level, [5] proposed two more estimations for Euclidean distance as

$$\frac{k}{2} \times d^2(x, y_i) \geq (S_{x_f} - S_{y_{i,f}})^2 + \frac{k}{2} \times (V_{x_f} - V_{y_{i,f}})^2 \tag{8-1}$$

$$\frac{k}{2} \times d^2(x, y_i) \geq (S_{x_s} - S_{y_{i,s}})^2 + \frac{k}{2} \times (V_{x_s} - V_{y_{i,s}})^2 \tag{8-2}$$

Then, [5] uses a 6-step test flow as: (1) $(S_x - S_{y_i})^2 \geq kd_{min}^2$ (2) $(S_{x_f} - S_{y_{i,f}})^2 \geq (k/2) \times d_{min}^2$ (3) $(S_{x_s} - S_{y_{i,s}})^2 \geq (k/2) \times d_{min}^2$ (4) $(S_x - S_{y_i})^2 + k(V_x - V_{y_i})^2 \geq kd_{min}^2$ (5) $(S_{x_f} - S_{y_{i,f}})^2 + (k/2) \times (V_{x_f} - V_{y_{i,f}})^2 \geq (k/2) \times d_{min}^2$ (6) $(S_{x_s} - S_{y_{i,s}})^2 + (k/2) \times (V_{x_s} - V_{y_{i,s}})^2 \geq (k/2) \times d_{min}^2$.

If any test among Step1~Step 6 above becomes true, then reject y_i safely. This is SIEENNS method. It needs six extra memories for each y_i and nine "±", six "×" and six "cmp" operations for a complete rejection test. For 4×4 block size, this is a large memory and computational overhead.

On the other hand, if only partial sum information is used, a 3-level memory-efficient data structure [8] is proposed as

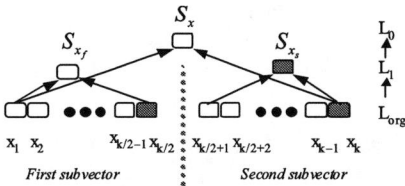

Fig. 2. A 3-level data structure for x. For y_i, it is similar. Because the values of (S_{xs}, $x_{k/2}$, x_k) at the shadowed positions are *not* stored but on-line computed when necessary, it only requires (k) memories in total for each vector.

Then, [8] proposed an estimation for Euclidean distance at L_1 level across both the first and second subvectors as

$$\begin{aligned}
\frac{k}{2} \times d^2(x, y_i) &\geq (S_{x_f} - S_{y_{i,f}})^2 + (S_{x_s} - S_{y_{i,s}})^2 \\
&= (S_{x_f} - S_{y_{i,f}})^2 + [(S_x - S_{y_i}) - (S_{x_f} - S_{y_{i,f}})]^2
\end{aligned} \tag{9}$$

Then, [8] uses a 2-step test flow as: (1) $(S_x - S_{y_i})^2 \geq kd_{min}^2$ and (2) $(S_{x_f} - S_{y_{i,f}})^2 + [(S_x - S_{y_i}) - (S_{x_f} - S_{y_{i,f}})]^2 \geq (k/2) \times d_{min}^2$.

This is SENNS method. It does not need any extra memory for a k-D vector. And it needs four "±", three "×" and two "cmp" operations for a complete rejection test.

3. SOME OBERSERVATIONS

Observation 1: From the conclusion of [5], SIEENNS method can achieve a large 20.5%~26.8% reduction for Euclidean distance computation but it can only achieve a small 2.4~6% reduction for encoding time compared to IEENNS method. This implies that the 6-step SIEENNS method has a rather large computational overhead than the 2-step IEENNS method. The first reason is that SIEENNS method needs additional 2(k–1) "±", (k+2) "×" and two "sqrt" operations to on-line compute two partial variances of V_{xf}, V_{xs} for the input x, which is unavoidable and becomes a large overhead. The second reason is that too many costs for computing test conditions in the 6-step search flow of SIEENNS method introduce a large overhead.

Observation 2: Concerning memory requirement, IEENNS method uses two extra memories, SIEENNS method uses six extra memories but SENNS method does not use any extra memory as demonstrated in Fig.2. Therefore, how to integrate this memory-efficient data structure into SIEENNS method is also rather important.

Observation 3: In the case of codebook size is 256 and block size is 4×4, an experimental comparison among IEENNS method, SIEENNS and SENNS method is made to see how effective each rejection step could be in order to find out at which step the computational redundancy occurs. Search efficiency is evaluated by the number of remaining candidate codewords after a rejection test. A smaller value is better.

Table 1 Comparison of the remaining distance computations per input vector after completing each rejection test (N_c=256)

Method	Test	Lena	F-16	Pepper	Baboon
IEENNS	Step 1	16.27	14.17	18.58	49.60
	Step 2	6.44	6.24	6.94	30.09
SIEENNS	Step 1	16.27	14.17	18.58	49.60
	Step 2	11.69	10.60	13.56	42.38
	Step 3	9.75	8.43	11.14	36.71
	Step 4	5.26	4.70	5.61	24.85
	Step 5	5.02	4.38	5.34	23.88
	Step 6	4.81	4.11	5.08	22.92
SENNS	Step 1	16.27	14.17	18.58	49.60
	Step 2	9.04	7.65	10.29	34.18

From Table 1, it is clear that (1) Step 2 in IEENNS method is most powerful because it uses the sum and the variance simultaneously, which are actually orthogonal to each other [9] in R^k space. (2) Step 5 and Step 6 in SIEENNS method has only a very little contribution to reducing the search space. The reason could be that the remaining search space is already rather small after finishing Step 1~Step 4 rejection tests. (3) SENNS method is not very powerful but it does not need to use any extra memory. For codebook size of 512 or 1024, it has a similar trend.

4. IMPROVED METHOD

Based on the observations above, it concludes that (1) IEENNS method should be used firstly as a core technique. (2) Step 5 and Step 6 in SIEENNS method should be deleted by considering of the memory and computational overhead. (3) As a result, for two subvectors, only their partial sums of (S_{xf}, S_{xs}), ($S_{yi,f}$, $S_{yi,s}$) are necessary. Then it is beneficial to use these two partial sums simultaneously as in SENNS method instead of individually as in SIEENNS method. A new 3-level data structure can be.

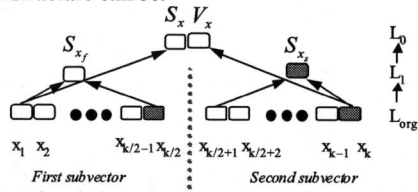

Fig. 3. A 3-level data structure for x. For y_i, it is similar. Based on Fig. 2, it requires (k+1) memories for each vector.

Because $(S_x - S_{y_i})^2 \geq kd_{min}^2$ is always tested first, by using $a^2+b^2=(a+b)^2-2ab$, Eq.9 of SENNS method can be realized more efficiently as

$$\frac{k}{2} \times d^2(x, y_i) \geq (S_x - S_{y_i})^2 - 2(S_{x_f} - S_{y_{i,f}}) \times [(S_x - S_{y_i}) - (S_{x_f} - S_{y_{i,f}})] \quad (10)$$

Eq.10 can save one more "×" operation compared to Eq.9.

Then, an improved subvector-based search method can be summarized as a 3-step rejection test flow shown below:
(1) $(S_x - S_{y_i})^2 \geq kd_{min}^2$ (2) $(S_x - S_{y_i})^2 + k(V_x - V_{y_i})^2 \geq kd_{min}^2$ (3) $(S_x - S_{y_i})^2 - 2(S_{x_f} - S_{y_{i,f}}) \times [(S_x - S_{y_i}) - (S_{x_f} - S_{y_{i,f}})] \geq (k/2) \times d_{min}^2$.

Obviously, this improved search method becomes very concise and compact compared to the 6-step SIEENNS method. It only has a minimum extra memory requirement of one to store V_{y_i} for each y_i. Concerning computational overhead, because V_{xf}, V_{xs} are not used anymore, the on-line cost for constructing them can be cut, which is 2(k–1) "±", (k+2) "×" and two "sqrt" operations. In addition, because the test steps are simplified from six to three, it is also beneficial to reduce computational overhead in a search process.

If all of these three rejection tests fail, a further test by real Euclidean distance must be executed at L_{org} level but in two steps as (1) $\bar{d}^2(x, y_i) = \sum_{j=1}^{k/2-1}(x_j - y_{i,j})^2 + \sum_{j=k/2+1}^{k-1}(x_j - y_{i,j})^2 \geq d_{min}^2$ and (2) $\tilde{d}^2(x, y_i) = \bar{d}^2(x, y_i) + (x_{k/2} - y_{i,k/2})^2 + (x_k - y_{i,k})^2 \geq d_{min}^2$, where $y_{i,k/2} = S_{y_{i,f}} - \sum_{j=1}^{k/2-1} y_{i,j}$, $y_{i,k} = (S_{y_i} - S_{y_{i,f}}) - \sum_{j=k/2+1}^{k-1} y_{i,j}$ will be computed on-line. Since almost all candidate codewords can be rejected by the first test over a total of (k–2) dimensions, it is seldom need to on-line compute $y_{i,k/2}$ and $y_{i,k}$. If $\tilde{d}^2(x, y_i) < d_{min}^2$ becomes true, it means y_i is a better-matched codeword. Then let $d_{min}^2 = \tilde{d}^2(x, y_i)$ to update d_{min}.

5. EXPERIMENTAL RESULTS

Simulation experiments are conducted for three codebook sizes of 256, 512 and 1024, which are generated by using Lena image as a training set. All images are of size 512×512. Block size is 4×4. According to the results of [5], it has been confirmed that SIEENNS method is more powerful than IEENNS method. According to the results of "Observation 3", it is obvious that SIEENNS method is more powerful than SENNS method as well. Thus, only a comparison between the proposed method and SIEENNS method is made. Final results are summarized in Table 2 and Table 3.

Table 2 Comparison of the remaining distance computations per input vector after completing each rejection test

Size	Method	Test	Lena	F-16	Pepper	Baboon
256	FS		256	256	256	256
	SIEENNS method	Step 1	16.27	14.17	18.58	49.60
		Step 2	11.69	10.60	13.56	42.38
		Step 3	9.75	8.43	11.14	36.71
		Step 4	5.26	4.70	5.61	24.85
		Step 5	5.02	4.38	5.34	23.88
		Step 6	4.81	4.11	5.08	22.92
	Proposed method	Step 1	16.27	14.17	18.58	49.60
		Step 2	6.44	6.24	6.94	30.09
		Step 3	5.01	4.35	5.30	23.54
512	FS		512	512	512	512
	SIEENNS method	Step 1	29.81	27.40	35.83	98.29
		Step 2	20.48	20.29	25.81	84.18
		Step 3	16.66	15.95	20.85	73.07
		Step 4	8.64	7.98	10.10	46.26
		Step 5	8.16	7.43	9.54	44.43
		Step 6	7.72	6.91	9.00	42.53
	Proposed method	Step 1	29.81	27.40	35.83	98.29
		Step 2	10.91	10.74	12.83	55.56
		Step 3	8.14	7.32	9.44	43.80
1024	FS		1024	1024	1024	1024
	SIEENNS method	Step 1	52.06	52.55	68.86	189.97
		Step 2	32.26	36.75	46.52	158.63
		Step 3	25.74	28.27	36.83	135.04
		Step 4	12.32	13.71	16.62	86.91
		Step 5	11.48	12.52	15.48	82.89
		Step 6	10.73	11.42	14.36	78.83
	Proposed method	Step 1	52.06	52.55	68.86	189.97
		Step 2	16.21	19.49	22.07	107.57
		Step 3	11.49	12.42	15.39	81.65

From Table 2, it is obvious that the proposed 3-step search method can achieve almost the same reduced search space compared to the 6-step SIEENNS method.

Table 3 Comparison of total computational cost per input vector

Size	Method	Operation	Lena	F-16	Pepper	Baboon
256	SIEENNS method	±	323.4	291.5	342.4	1113.1
		×	176.2	157.3	189.1	628.8
		cmp	69.1	60.6	77.9	249.9
		sqrt	3	3	3	3
	Proposed method	±	256.7	229.6	272.9	954.5
		×	135.2	120.2	144.9	521.8
		cmp	44.0	38.9	49.4	152.8
		sqrt	1	1	1	1
512	SIEENNS method	±	476.5	441.9	547.5	2016.8
		×	272.2	251.9	317.0	1160.0
		cmp	121.3	113.4	147.0	487.1
		sqrt	3	3	3	3
	Proposed method	±	398.8	365.9	458.9	1754.4
		×	216.8	198.7	251.5	969.0
		cmp	78.7	72.9	93.9	295.9
		sqrt	1	1	1	1
1024	SIEENNS method	±	658.0	694.8	857.3	3682.0
		×	392.6	414.0	517.9	2139.7
		cmp	196.6	207.8	267.5	922.2
		sqrt	3	3	3	3
	Proposed method	±	574.6	608.2	751.7	3254.7
		×	320.2	339.4	422.0	1810.0
		cmp	131.8	137.0	175.2	569.2
		sqrt	1	1	1	1

On the other hand, because the encoding time strongly depends on programming skills, the total computational cost that includes all computational overhead in "±", "×", "cmp" and "sqrt" operations is used to evaluate the overall encoding performance. From Table 3, it is obvious that the proposed 3-step search method can reduce the total computational cost further by 20%~30% compared to SIEENNS method. This is a rather large reduction of computational cost.

6. CONCLUSION

In this paper, three issues are made clear through some experimental observations. First, SIEENNS method is most efficient so far based on the subvector technique. But it still has a large computational overhead because it redundantly used the two partial variances of subvectors to construct too many test steps, which just have a very little contribution to reducing search space. Second, IEENNS method is rapidest for reducing search space. Third, SENNS method has the most memory-efficient data structure. Based on these three facts, a new subvector-based search method is proposed by (1) avoiding using the two partial variances of subvectors in SIEENNS method completely; (2) putting the core IEENNS method forward as the second rejection test; and (3) adopting the memory-efficient data structure of SENNS method.

Compared to state-of-the-art SIEENNS method, the proposed search method can reduce memory requirement from (k+6) to (k+1) for each k-D vector. Meanwhile, the proposed search method can reduce the total computational cost by 20%~30% further. It is a more efficient and more practical subvector-based method for fast VQ encoding.

7. REFERENCES

[1] N.M.Nasarabadi and R.A.King, "Image coding using vector quantization: A review," *IEEE Trans. Commun.*, vol. 36, pp.957-971, Aug. 1988.

[2] L.Guan and M.Kamel, "Equal-average hyperplane partitioning method for vector quantization of image data," *Pattern Recognition Letters*, vol.13, pp.693-699, Oct. 1992.

[3] C.H.Lee and L.H.Chen, "Fast closest codeword search algorithm for vector quantization," *IEE Proc.-Vision Image Signal Processing*, vol.141, no.3, pp.143-148, 1994.

[4] S.Baek, B.Jeon and K.Sung, "A fast encoding algorithm for vector quantization," *IEEE Signal Processing Letters*, vol.4, pp.325-327, Dec. 1997.

[5] J.S.Pan, Z.M.Lu and S.H.Sun, "An efficient encoding algorithm for vector quantization based on subvector technique," *IEEE Trans. Image Processing*, vol.12. pp.265-270, March 2003.

[6] Z.Pan, K.Kotani and T.Ohmi, "A hierarchical fast encoding algorithm for vector quantization with PSNR equivalent to full search," *2002 IEEE International Symposium on Circuits and Systems*, vol. I, pp.797-800, May 2002.

[7] Z.Pan, K.Kotani and T.Ohmi, "A fast full search equivalent encoding method for vector quantization by using appropriate features", *2003 IEEE International Conference on Multimedia and Expo*, vol. II, pp.261-264, July 2003.

[8] Z.Pan, K.Kotani and T.Ohmi, "An improved fast encoding method for vector quantization based on memory-efficient data structure," *2004 IEEE International Conference on Multimedia and Expo*, June 2004.

[9] Z.Pan, K.Kotani and T.Ohmi, "A unified projection method for fast search of vector quantization," *IEEE Signal Processing Letters*, vol.11, pp.637-640, July 2004.

Author Index

A

A. S., Satheesh Kumar 5545
Aaltonen, Lasse 5377, 5381
Abadal, Gabriel 4209
Abbasian, Ali 6218
Abbott, Justin 1936
Abdalla, Hisham 4201
Abe, Masahide 264, 528, 968, 972, 3027
Abeysekera, Saman 33, 1734, 5718
Abid, Z. 4649
Aboulhamid, El Mostapha 4767
Aboushady, Hassan 3672
Abrahamsen, Jens Petter 2751
Abramson, David N. 468
Abshire, Pamela 3491, 3495, 5314, 6146
Achar, Ramachandra 3777, 5770
Adachi, Hidekazu 3215
Adalev, Alexei S. 980
Addabbo, Tommaso 892
Adrian, Victor 5405
Afridi, M. Y. 416
Afzal, Behrouz 5858
Afzali-Kusha, Ali 1666, 1674, 5234, 5278, 5858, 6218
Agarwal, A. 584, 5906
Agarwal, Kanak 604
Agarwal, Patrika 3579
Agathoklis, P. 1441, 1718
Agazzi, Oscar E. 1390
Aguado, José 4196
Aguiar, Rui L. 984, 1166
Aguilar-Torrentera, Jorge 3958
Ahmad, M. Omair 836, 868, 1146, 1461, 2405, 3143, 3147, 3753, 4935
Ahmad, T. 133
Ahmadi, Hossein 2381, 4610
Ahmadi, Majid 3559
Ahmadi, Mohammad Mahdi 4261
Ahmed, Syed 4485
Ahmed, Waleed Abd El Maguid 832
Ahn, Hyung Keun 1016, 1614
Ahn, K.Y. 3371
Aho, Eero 1134
Aho, Mikko 4622
Aigner, Manfred 1066
Aihara, Kazuyuki 4791, 6058
Akagi, Miho 192
Akatimagool, Somsak 5970
Akazawa, Takashi 5926
Åkesson, Henrik 852
Akhtar, Muhammad Tahir 264
Akino, Toshiro 452
Akiyama, Satoru 5250
Aksin, Devrim 3007, 3627, 5898
Alarcón, E. 1302, 2473
Alarcon, Eduard 4114, 3926, 4453, 4457
Al-Ardi, Ebrahim 3805
Albonesi, David 2514
Albu, Felix 2283, 4337
Al-Hashimi, Bashir 616, 2815, 4163, 5389
Ali, Mustafa 6226
Alioto, Massimo 892, 2437, 4685
Alippi, Cesare 652, 5142, 5509
Alisafaee, Mohammad 424, 5278
Alizadeh, B. 424
Allam, Ahmed 2120, 2635
Allen, Phillip 3869, 4819, 4823, 5577, 5942, 5946
Allstot, David 2100, 2104, 5457, 6034
Almasganj, Farshad 5730
Al-Mualla, Mohammed 3805
Aloi, Daniel 6198
Alonso, Corinne 724
Alquie, G. 2337
Altun, Oguz 3668, 5577
Alvandpour, Atila 5465
Alves, Luis Nero 984
Aly, Ramy E. 1650
Aly-Mekawi, Walid 1932

Amara, Amara 4094
Amaral, José Nelson 3339
Amelifard, Behnam 1666, 6218
Amemiya, Takashi 2839
Amemiya, Yoshihito 1923, 2535
Amindavar, Hamidreza 5730
Amira, Abbes 3789, 5826, 6252
Amirabadi, A. 1674
Analui, Behnam 2494
Anantrasitichai, N. 2092
Anazawa, Yoshihisa 2044
Anderson, David V. 860, 996
Anderson, John B. 344
Andreani, Pietro 4614
Andreasson, Daniel 2353
Andrejevic, Miona 5182
Andreou, Andreas. 137, 644, 2763, 4142, 4205
Angulo, F. 1485
Aniruddhan, Sankaran 6034
Antoniou, A. 1441, 2615
Antoniou, George 2299
Ao Ieong, Ka-Hou 392
Aoki, Masakazu 5934
Aoyama, Kenichi 3507
Arai, Hiroki 528
Arai, Shin 4010
Arakawa, Kaoru 4010
Araki, Shoko 5882
Aramvith, Supavadee 3785
Aranda, J. G. 400
Aras, Sualp 3627
Arbat, A. 141
Ardö, Håkan 1142
Arekapudi, Srikanth 5055
Arena, Paolo 4102, 5214, 5818
Arian, Peyman 2008
Arik, Sabri 4665
Arimoto, Kazutami 5202
Arlenghi, Leonardo 2763
Armer, John 1194
Arnold, Mark 3954, 5154
Arona, Riccardo 1903
Arora, Deepali 1718
Arrowsmith, David 3773
Arslan, Tughrul 41, 1048, 5274, 5441, 5894
Arteaga, Rubén 3821
Arthansiri, Teerawat 1016
Arungsrisangchai, Itthichai 5345
Arzel, Matthieu 332
Asadi, Ghazanfar 2991
Asai, Hideki 4903, 5774
Asai, Testuya 1923, 2535
Ascia, Giuseppe 4090
Assi, Ali 1952
Astola, Jaakko 2088
Atarodi, Mojtaba 2255, 5031
Ates, Özgur 4345
Athalye, Akshay 1646
Athlaye, Akshay 3511
Atienza, D. 2365
Atoofian, Ehsan 5278
Atti, Venkatraman 848
Au, Oscar C. 2659, 4923, 4991, 5477, 6260
Avedillo, María J. 2647
Averty, Charles 268
Avner, Evgeny 580
Axelrad, V. 1154
Axelrod, B. 1310
Ayatollahi, A. 1406
Ayazi, Farrokh 5461
Azadmehr, Mehdi 2751
Azin, Meysam 6150
Azou, Stephane 3813
Azuara, Luis 2935
Azzolini, Cristiano 1386

B

Baba, Takaaki 2891
Babb, Christopher R. 668
Babic, Hrvoje 3283
Baccelli, Emmanuel 2971
Badaoui, Raoul 5978
Badawy, Wael 1346
Baderna, D. 1266, 1891
Bae, Jaewan 4570
Baek, Jaehyun 4477
Baek, Kwang-Hyun 5621
Baek, Woongki 2715
Bagga, Sumit 5357
Baglin, Thomas 3857
Baglio, Salvatore 5910
Bai, Huihui 4038
Bakhtiar, Mehrdad Sharif 200, 1417, 6150
Bakkaloglu, Bertan 4401
Bakos, Jason D. 1662
Balasinski, A. 1154
Balestra, Michele 1489
Balkir, Sina 1294, 4110, 5322, 5549
Ballapuram, Chinnakrishnan 1867
Balsi, Marco 3295
Bálya, David 5222
Bandi, Sripriya 5103
Bandler, John 5605
Bandyopadhyay, Abhishek 2148, 5318
Banerjee, Soumitro 3367
Banerjee, Suman K. 612
Banh, Xuan-Quang 5485
Bansal, Aditya 1
Bara, G. Iulia 3195
Barajas, Enrique 204, 4273
Barboza, Luciano 5294
Barniol, Núria 5361, 4209
Barrettino, Diego 4795
Barúqui, Fernando A. P. 1012
Baschirotto, Andrea 388, 776, 4614, 5385
Basile, Adriano 5818
Batalheiro, Paulo B. 1114
Bates, Stephen 336, 4513
Baumhof, Christoph 400
Baykal, Buyurman 5746
Baykal, Ertan 5003
Bayoumi, Magdy ... 29, 552, 1650, 4697, 5198, 5437
Becanovic, Vlatko 2767
Beckett, Paul 2329, 2345
Bedeschi, F. 1270
Behjat, Laleh 4196, 6238
Beilleau, Nicolas 3672
Bélanger, Normand 3515
Belot, Didier 2675
Belski, Alexander 5445
Benediktsson, Jon Atli 4241
Benini, L. 2365
Bensaoula, Abdelhak 4767
Beric, Aleksandar 2943
Berisha, Visar 4050
Berkovi, Laurence 220
Berkovich, Y. 1310
Bermak, Amine 1754, 1911, 5306
Bernard, Thierry M. 2227, 5242
Bertuccio, Giuseppe 5569
Berube, Paul 3339
Bharath, Anil 3918
Bhattacharya, Mrinmoy 2599
Bhattacharyya, Asok 3299
Bhatti, Ghulam 3539
Bhuiyan, M. I. H. 4935
Bi, Yun-long 2887
Bian, Jinian 5641, 6230
Biel, Domingo 1318, 3067
Birkenes, Øyvind 1593
Bizzarri, Federico 3403, 4915
Blaauw, David 3563
Blakiewicz, Grzegorz 3015
Blalock, Travis 1843
Blizzard, John 4653
Block, Gary 336
Boahen, Kwabena 5230
Bobda, Christophe 2184
Boffino, C. 121
Bogdanov, Momcilo 2068
Bomel, Pierre 680
Bommalingaiahnapallya, Shubha ... 4393, 4630
Boni, Andrea 1386
Bonizzoni, Edoardo 1270, 1903

Bonnin, Michele 3765, 4669
Booksh, Karl.. 3483
Boonchu, Boonchai................................... 1004
Borghetti, Fausto.. 5569
Borkowski, Maciej 5601
Bose, Nirmal K. 6296
Bose, Tamal........................... 556, 2591, 4325
Bossche, Andre.. 2919
Bottiglieri, Gerardo 2679
Bouchard, Martin.. 4951
Bouganis, Christos-Savvas 532
Bouguezel, Saad.............................. 836, 2405
Boukadoum, Mounir................................... 4767
Bouridane, Ahmed 2313
Bourrier, David 724
Boutayeb, Mohamed...................... 3195, 3415
Bowron, Peter 3291
Boyce, Jill................................... 1246, 4365
Boyer, François-Raymond 3335
Brady, Philomena....................................... 6154
Brahim-Belhouari, Sofiane 1911
Brandon, Tyler 4513
Brea, Victor 3922, 5790
Bregovic, Robert 1098
Brennan, P. V.. 2243
Brooke, Martin A. 1402, 1738
Brown, Charles Grant 5011
Brown, Richard B. 604, 624
Bruno, Marcelo... 4329
Bruti Liberati, Nicola 1702
Bruton, L. T. 2397, 2032
Bruton, Len 1457
Bucci, Marco 3547
Bucolo, M. 296
Bull, David............................. 960, 2092, 2719
Bulsara, Adi.. 5910
Burel, Gilles.. 3813
Burleson, Jeff... 5103
Byun, Hyun-Geun 440, 1847

C

Cabello, D. 3688, 3922, 5798
Cabrini, A. 121, 1266
Cabrini, Alessandro.............................. 1891
Cackov, Nikola 2068
Cai, Hua 5473
Cai, Jianfei 6078
Cai, Li 2325
Cai, Rui 964
Cai, Yici 93, 97, 101, 1871, 6230
Cai, Yuan 616
Calazans, Ney 5866
Callicó, Gustavo 6130
Calvo, Belén 208
Campanella, Humberto 5361
Canagarajah, C. Nishan 2092
Canagarajah, Nishan 960, 2719
Canavero, Flavio 5754
Cannas, Barbara 992
Cantin, Marc-André 3439, 4159
Cao, Bin 656, 664, 1630
Cao, Hong 4405
Cao, Zhiheng 3091
Carastro, Larry 3869
Cardarilli, G.C. 1102
Cardinal, Christian 4159
Cardoso, R. S. 428
Carezzano, Linda 3403
Carley, L. Richard 1162
Caron, Mario 1710
Carrara, Francesco 2679, 3239, 3247
Carrillo, Juan M. 1008
Carro, Luigi 428, 5597, 5866
Caruso, Giuseppe 5637
Carvajal, Ramón G. 1008
Casagrande, G. 1270
Casanova, R. 141
Casanueva, E. 1742
Catania, Vincenzo 4090
Catthoor, F. 2365
Catudal, Serge 3439
Cauwenberghs, Gert.. 1919, 3934, 4106, 4205
Ce, Zhu 4301
Celma, Santiago 208
Cen, Ling 520
Centurelli, F. 448
Cerneková, Zuzana 3849
Cerqueira, Augusto 5417
Cetin, Ediz 4481
Cha, Sungwoo 812
Chakrabartty, Shantanu 3934
Chakraborty, Sudipto 4819
Chakradhar, Srimat 1778
Chalidabhongse, Thanarat 3785
Cham, Wai-Kuen 316
Chan Carusone, Anthony 936, 1521, 5194
Chan, Cheong-Fat 2152, 3279, 4999, 5842
Chan, Pak-Kee 4999
Chan, S. C. 1094, 2607, 2631, 4297, 4321
Chan, Shing Chow 4313, 4333, 1722, 3435, 4317
Chan, Tai-Wai 2659, 4923, 6260
Chan, Yui-Lam 2671
Chana, Jatinder 5071
Chandra, Gaurav 820, 5039
Chandrakasan, Anantha P. 184
Chandramouli, Soumya 4847
Chang, Andy 5477
Chang, Cheng-Po 1242
Chang, Cheng-Ru 4525, 4529
Chang, Ching-Chi 4895, 5794
Chang, Chip Hong 1630, 69, 85, 656, 664, 864, 1823, 3833, 4082
Chang, Chun-Ming 5950
Chang, Ee-Chien 2707
Chang, Feng-Cheng 4550, 4983
Chang, Hsiu-Cheng 6110
Chang, Hung-An 4871
Chang, Hung-An 6284
Chang, J. Morris 2445
Chang, Joseph S. 224, 1326, 4078
Chang, Joseph Sylvester 5405
Chang, Ka-Fai 3817
Chang, Kuo-Hsuan 5573
Chang, Lung-Hao 1074
Chang, Nelson 1806
Chang, Ruey-Feng 6300
Chang, Shih-Chieh 5653
Chang, Shih-Hung 1074

Chang, Shoou-Jinn 3003
Chang, Shu-Ming 1174
Chang, Szu-Wei 684
Chang, Teng-Hung 2555
Chang, Tian-Sheuan 1509, 1806
Chang, Wang-Rong 1533
Chang, Wen-Chieh 3271
Chang, Yao-Wen 4134
Chang, Yuan-Cheng 3781
Chang, Yu-Hao 1730
Chang, Yu-Lin 6082
Chang, Yun-Nan 5019
Chao, Chie-Min 3503
Chao, Her-Chang 3994
Chao, Wen-Hung 156
Chao, Yi-Chih 1802
Charles, Cameron 5457
Charoensak, Charayaphan 33, 5822
Chau, Lap-Pui ... 896, 2144, 3447, 3451, 4006,
    4022, 4357, 4582, 6070
Chau, Wing-San 2659, 4923, 6260
Chauhan, Ojas 2591
Chawla, Ravi 1980, 6106
Cheetham, Andrew 2699
Chen, Bo-Wei 736
Chen, Chang Wen 3527
Chen, Chang-Hung 3676
Chen, Che-Hong 1802
Chen, Chen 1497
Chen, Chien-Fa 2136
Chen, Chien-Han 2603
Chen, Chien-In 6142
Chen, Chih-Kai 444
Chen, Chih-Ming 2084, 3255
Chen, Chi-Kuang 4521
Chen, Ching-Yeh 1798, 5469
Chen, Chun 6022
Chen, Chun-Chi 560
Chen, Degang 784, 792, 1378, 3809, 4285,
    4289, 4831, 4835, 6190
Chen, De-Sheng 6214
Chen, Emery 5115
Chen, Fang-Jiong 3607
Chen, Guanrong .. 284, 880, 1473, 3391, 6038
Chen, Guoqing 596, 2514, 4126
Chen, H. H. 4321
Chen, Ho-Chun 4542
Chen, Homer 1497, 1505, 3801, 4871, 6284
Chen, Hong-Sheng 3419
Chen, Hsiao-Chen 736
Chen, Hsin-Jung 4554
Chen, Hui 2514
Chen, Huifang 1126
Chen, Hun-Chen 6126
Chen, Hung-Yi 780
Chen, Jeng-Huei 2939
Chen, Jen-Wen 5091
Chen, Jian-Chou 4578
Chen, Jianfeng 1682, 1750
Chen, Jiann-Jong 3275, 6170
Chen, Jian-Wen 4525
Chen, Jing 1274
Chen, Jun 3873
Chen, Jun-Cheng 4594
Chen, Jung-Sheng 3861
Chen, Jun-Hong 5254
Chen, Jun-Ning 2481
Chen, Kuan-Hung 4517, 5051
Chen, Kun Lung 3315
Chen, Liang 1421, 3635, 5011
Chen, Liang-Gee 1790, 1798, 2931, 3155,
    5190, 5469, 6082
Chen, Lien-Fei 4578
Chen, Li-Hsun 544, 6118
Chen, Luonan 4791
Chen, Mei 556
Chen, Mei-Juan 2895, 5481
Chen, Ming-Chih 2433
Chen, Ming-Jen 3825
Chen, Oscal T.-C. 544, 6118
Chen, Pao-Lung 4875
Chen, Pinhong 2465
Chen, Poki 560
Chen, Rong-Jian 1690, 3059
Chen, Ru 688
Chen, Sau-Gee 3315
Chen, Shih-Lun 1859
Chen, Shin-Kai 3503
Chen, Shi-Wei 6122
Chen, Siyue 3151

Chen, Sizhong 4959
Chen, Song 1863
Chen, Song 6222
Chen, To-Wei 2931
Chen, Tung-Chien 1790, 2931
Chen, Wei-Ta 2643
Chen, Wensong 1190
Chen, Wen-Tsuen 640
Chen, Weoi-Luen 5119
Chen, Xiaowei 4225, 4229
Chen, Xu 4887
Chen, Yang-Chaun 5099
Chen, Yen-Hsiang 1350, 1366, 2219
Chen, Yi-Hau 1798
Chen, Yi-Jan Emery 212
Chen, Yi-Jen 1258
Chen, Ying-Chi 3655
Chen, Yi-Rung 3307
Chen, Yi-Sheng 5702
Chen, You-Gang 1074
Chen, Yuan 3680, 5589
Chen, Yuan-Peir 3994
Chen, Yu-Chieh 152
Chen, Yuchuan 5666
Chen, Yueh-Hong 4979
Chen, Yu-Han 2931
Chen, Yu-Luen 5119
Chen, Yung-Chang 2084, 3255, 4590
Chen, Yung-Ching 6234
Chen, Yuxin 3749
Cheng, Binjie 2510
Cheng, Chao 1835
Cheng, Chao-Chung 1509
Cheng, Chih-Chi 5190
Cheng, Chung-Han 2811
Cheng, Chung-Kuan 1863, 6222
Cheng, Chun-Hua 5633
Cheng, H. 2631
Cheng, Hui 1722
Cheng, Jierong 149
Cheng, Jun 4405
Cheng, Kuo-Hsing 1070
Cheng, Kuo-Hsing 1174
Cheng, Shyi-Chyi 3455
Cheng, Wen-Chang 3023

Cheng, Wen-Huang 3219, 3829
Cheng, Xu 660, 696
Cheong, Pedro 404
Cherrier, Estelle 3415
Cheung, Hoi-Kok 2903
Cheung, Kwok-Wai 1541
Cheung, Peter Y. K. 532, 692
Chi, Hsiang-Feng 6006
Chi, Ming-Chieh 2895
Chiang, Andy 6238
Chiang, Hsiao-Dong 4907
Chiang, Jen-Shiun 4554
Chiang, Nam-Po 2216
Chiang, Tihao 3267
Chiang, Tsung-Hsi 5674
Chiarulli, Donald M. 1662
Chiba, Toru 5333
Chidanandan, Archana 5437
Chien, Chiang-Ju 512
Chien, Chih-Da 4542
Chien, Chih-Lung 5846
Chien, Hung-En 2393
Chien, Shao-Yi 6082
Chiewchanwattana, Sirapat 5162
Chih, Jen-Chuan 3315
Chio, Kin-Sang 3099
Chiou, Chyi-Tian 5122
Chiu, Cheng-Nan 2132
Chiu, Chun-Wei 4578
Chiu, Julian 3079
Chiu, Shih-Chieh 81
Chiueh, Tzi-Dar 5051
Cho, Chuan-Yu 3419
Cho, Grace Yoona 548
Cho, Namjun 740, 4602, 4763
Cho, Seong-Hwan 6178
Cho, Uk-Rae 440, 1847
Cho, Young-Jae 4054
Choe, Myung-Jun 5621
Choi, Gwan 5782
Choi, Jongsoo 4951
Choi, Sang 3962, 5111
Choi, Seungjin 3595, 5886
Choi, Sujeong 3259
Choi, Yongsoo 5537

Chong, Euhan ... 1537
Chong, Kwen-Siong ... 4078
Chong, Tak-Song ... 2659, 4923, 6260
Chou, Chih-Chung ... 432
Chou, Jyh-Horng ... 2603
Chou, Mei-Fen ... 5063
Chow, Hwang-Cherng ... 736
Choy, Chiu-Sing ... 2152, 3279, 4999, 5842
Chrzanowska-Jeske, Malgorzata ... 3015, 4721
Chu, Chia-Chi ... 1278
Chu, Chia-Yi ... 2983
Chu, Ching-Yun ... 3861
Chu, Chris Chong-Nuen ... 2445, 2461
Chu, Hong-Ming ... 5649
Chu, Shu-Chuan ... 3990
Chu, Wei-Ta ... 3219, 3829
Chu, Xiaodong ... 3906
Chu, Yuan-Sun ... 3825, 5043
Chuang, Chen-Chia ... 2498
Chuang, Cheng-Long ... 3885
Chuang, Hsiao-Chiang ... 3267
Chun, Carl ... 4847
Chun, Luo ... 5662
Chung, Ching-Che ... 4875
Chung, Henry ... 2469, 3079
Chung, Jen-Feng ... 2859
Chung, Kah-Seng ... 3611
Chung, Kyusik ... 4546
Chung, Pei-Ju ... 3163
Chung, Woo Cheol ... 3355
Chung, Yu-Chieh ... 432
Cichocki, Andrzej ... 3599
Cidon, Israel ... 600
Cijvat, Ellie ... 2683
Cilio, Andrea ... 1122
Civalleri, Pier Paolo ... 3765, 4661
Claesson, Ingvar ... 852
Clapp, Matthew ... 4775
Clarke, C. T. ... 3319
Clarke, Chris ... 145, 748
Clausen, Thomas ... 2971
Claveau, David ... 4759
Cohen, Marc H. ... 5314
Cohen, Marc ... 6146
Colodro, Francisco ... 5581

Comeau, Benita M. ... 1738
Cong, Jason ... 696
Conover, David L. ... 576
Constantinides, George A. ... 532, 692
Conti, F. ... 296
Corbishley, Philip ... 1398, 5966
Cordeiro, Viviane ... 5834
Coria-Mendoza, Lino ... 320
Corinto, Fernando ... 4661, 4669
Corporaal, Henk ... 1122
Corre, Gwenole ... 680
Cosp, Jordi ... 3926, 4114
Coulibaly, L. M. ... 1254
Cousseau, Juan ... 4329
Coussy, Philippe ... 680
Cramer, Jeffrey ... 3483
Crocce, Pablo ... 2763
Cruz-Roldan, Fernando ... 1106
Culurciello, Eugenio ... 137, 644, 1919, 4142
Cunha, A. I. A. ... 1020
Cyrusian, Sasan ... 5453
Czarkowski, Dariusz ... 1306

D

da Silva, Eduardo A. B. 2309
Dabrowski, A. 1154
Dabrowski, Jerzy 4843
Dahl, Mattias 852
Dai, Dong 1469
Dai, Foster 380, 1178, 2208, 4397, 5750
Dai, Xin 4831, 4835
Dalcastagnê, André 1944
Dalla Betta, G.-F. 572
Dam, Hai Huyen 856
Dam, Hai Quang 856
D'Amico, Arnaldo 228, 1887
D'Amico, Stefano 388, 776
Daneshrad, Babak 4465
Dang, Hung 6158
Darabiha, Ahmad 5194
Darwazeh, Izzat 3958
Das, Tejasvi 4385
Dasika, Ganesh S. 624
Dasygenis, Minas 3347
Dave, Vibhuti B. 668
Davis, Zachary 5361
de Armas, Valentín 412, 3821
de Barcellos, Luiz 4309
De Bernardinis, F. 1286
de Haan, Gerard 2943
de la Rosa, José M. ... 3103, 4626, 5585, 5609
De Lima, Jader 1988
De Micheli, G. 2365
de Micheli, Giovanni 2361
De Nisi, F. 572
de Oliveira, M. C. G. 1020
de Pinho, Manoel Gomes 2309
De Queiroz, Antonio Carlos M. 236
De Ranter, Frederic 1194
De Sandre, G. 1266
De Santis, F. 1266
De Smet, Herbert 1855
De Vita, Giuseppe 5075
De, Vivek 9, 464, 592
Deen, Jamal 4353
Degnan, Brian 2172, 2441
Del Re, A. 1102
Del Rio, Rocio 4626
Delgado-Frias, José G. 1851
Delgado-Restituto, Manuel 800, 4626, 5609
Demirsoy, Süleyman 536, 540
Demosthenous, Andreas 744, 4566, 4827
Dempster, Andrew 536, 540, 1814
Deng, Guang 2267
Deng, Ke 5738
Deng, Tian-Bo 500, 504, 2004
Dennis, Karla S. 1738
Desai, Uday 900, 3207
Dessard, Vincent 4681
Dhir, Chandra Shekhar 5874
Di Benedetto, Maria-Gabriella 49
di Bernardo, Mario 1485, 3199, 3773
Di Marco, Mauro 4677
Diaz-Carmona, Javier 3733
Dieguez, A. 141
Dietrich, Heidelinde 2919
Díez-Villar, Leila 4626
Dimitriadis, C. A. 2196
Dimitroulakos, Gregory 472, 3519
Dimitrov, Vassil 3914
Dinc, Huseyin 4819, 4823, 5942
Ding, Heping 2279
Ding, J.J. 2036
Ding, Le 5706, 5710
Ding, Li-Fu 6082
Diniz, Paulo S. R. 2309
Diniz, Paulo 4309
Divakaran, Dinesh 4473
Djahanshahi, Hormoz 5071
Dlugosz, R. 1154
Do, M. A. 5027
Do, Manh Anh 376, 4815
Doboli, Alex 1044, 1282, 5629
Doboli, Simona 1044
Dolecek, Gordana Jovanovic 3733
Donaldson, Nick 145, 748
Doñate, Pedro 4329
Donde, Vaibhav 4179
Dong, Jie 316
Dong, Sheqin ... 1210, 1863, 1883, 2999, 6222
Doris, Konstantinos 4062
Dorsch, Rainer 2935

Dosaka, Katsumi 5202
Douglas, Rodney 5150
Doutreloigne, Jan 1855
Drakakis, Emmanuel 3918
Dranga, O. 2477, 4445
Draper, Jeffrey 2951, 3331
Draxelmayr, Dieter 1394
Drazdziulis, Mindaugas 1654
Drechsler, Rolf 4167
Dreyfert, Andreas 5625
Duan, Jiang 6276
Dubrova, Elena 2212
Duch, Marta 4209
Ducharme, Mathieu 4257
Dudek, Piotr 5806
Dugger, Jeff 4441
Duhamel, Pierre 5565
Dung, Lan-Rong 1642, 2555, 5674
Duo, Xinzhong 3982
Duque, Carlos 5417
Duque-Carrillo, J. Francisco 1008
Duster, Jon 3243, 5067
Dyer, Michael 2076

E

Easwaran, Prakash ... 820, 5039
Echeto, Francesc Moll ... 460
Ecker, Allan ... 2100
Edman, Fredrik ... 4489
Eichrodt, Christian ... 1593
Einwich, Karsten ... 5166
Ekpanyapong, Mongkol ... 1867
El Aroudi, Abdelali ... 2835
El Nokali, M. ... 5858
El-Araby, Elsaid ... 5290
Elassal, Mahmoud ... 29
Eleftheriadis, Alexandros ... 4377
El-Gamal, Mourad ... 4389, 4879
Elgamel, Mohamed ... 1650, 5198
Elgharbawy, Walid ... 4697
El-Masry, Ezz ... 1932
Elvira, Luis ... 4273
Elwakil, Ahmed ... 4345
Endo, Tetsuro ... 3399, 6054
Enescu, Mihai ... 6094
Enomoto, Tadayoshi ... 1678
Epassa, Habib Gabriel ... 3335
Erdogan, Ahmet T. ... 41, 5274, 5441, 5894
Eriksson, Henrik ... 1654
Eriksson, Jan ... 5890
Eshima, Nobuoki ... 2056
Esper-Chaín, Roberto ... 3821
Esteve, Jaume ... 4209, 5361
Estibals, Bruno ... 724
Etienne-Cummings, Ralph .. 1919, 4106, 4771, 4775
Evans, G. ... 796
Eversmann, Bjoern ... 3479

F

Facen, Alessio 1386
Fahmy, Hossum A. 69
Falconi, Christian 228, 1887
Falkowski, Bogdan J. 476, 480, 484, 488
Falkowski, Bogdan 4705, 4709, 6202
Fam, Adly.. 2008
Fan, Chen-Wei... 4550
Fan, Huijin... 2044
Fan, Jen-Lin... 1374
Fan, Yi... 2072
Fan, Yulong.. 4533
Fan, Zhengping... 284
Fang, Bo.. 1126
Fang, Jin-Qing.. 288
Fang, Tao................... 3447, 3451, 4022, 6070
Fang, Wai-Chi... 420
Fang, Wen-Hsien....................................... 1726
Fann, Ju-Lung.. 4521
Fanni, Alessandra....................................... 992
Farag, Fathi A... 1413
Fard, Ali... 2691
Farooqui, Aamir... 2377
Farquhar, Ethan... 2172
Farrell, Peter.. 6252
Fattah, Shaikh.. 3147
Fauchet, Philippe....................................... 2514
Fayomi, Christian Jesus B. 2200
Feng, Dagan.. 2903
Feng, Haigang... 4807
Feng, Ja-Shong... 4717
Feng, Jin Wei... 276
Feng, Lei.. 61
Feng, Min... 2703
Feng, Wu-Shiung..................... 736, 1278
Feng, Xiao-Yu... 240
Feng, Xue... 4409
Feng, Yongfeng.. 5186
Fernandes, Jorge....................................... 2635
Fernandes, Jorge....................................... 5353
Fernández, Daniel................... 3926, 4114
Fernández, Francisco V. 3103, 5585
Fernandez, Virgilio..................................... 612
Fernández-Bootello, Juan F. 4626
Fernández-Bootello, Juan Francisco.......... 800
Fernandez-Vazquez, Alfonso 840
Ferragina, Vincenzo 5569
Ferrão, Daniel 3567
Ferrari, Flavio 5569
Feruglio, S. 2337
Fiedler, Alan 2639
Filanovsky, Igor 1561, 2120, 2635
Fink, Wolfgang 2927
Fish, Alexander 580, 5310
Flak, Jacek 3938
Flandre, Denis 4681
Fleshel, Leonid 588
Flynn, Michael 2651
Fok, Chong-Yin 392
Fok, Si-Weng 404
Földesy, Péter 5802
Folkesson, Kalle 5625
Foo, Say Wei 149
Foo, Say-Wei 268
Fornasari, Andrea 5982
Fort, A. 2437
Fort, Ada 892
Forti, Mauro 4677
Fortier, Paul... 6018
Fortuna, L. 296
Fortuna, Luigi 4102, 5214, 5818
Fossas, Enric 3199
Fouad Hanna, V. 2337
Fournaris, Apostolos 4645
Franchi, Eleonora 2108
Frasca, M. 296
Frasca, Mattia 4102, 5214, 5818
Frattale Mascioli, Fabio Massimo 5393
Frey, Alexander 2915
Friedman, Eby G. 464, 596, 612, 620, 2514, 4126
Friggstad, Zac 4637
Fu, Bingmei 1746
Fu, Chang-Hong.. 2671
Fu, Cheng 476, 488, 4709
Fu, Hsin-Chia 4979
Fu, Shujun 2779
FU, Shujun 2779
Fu, Wei-Gang 5525

Fuh, Chiou-Shann	3801
Fujii, Kensaku	256, 272
Fujii, Nobuo	1577, 5561
Fujii, Tetsuro	5918
Fujimoto, Makoto	1362
Fujioka, Yoshichika	1786
Fujisaki, Hiroshi	872
Fujita, Hideki	4737
Fujiwara, Hideo	5686
Fujiwara, Ryosuke	4497
Fujiyoshi, Masaaki	4987
Fukasawa, Atsushi	360
Fukui, Yutaka	1589, 1698, 5413
Fukunaga, Tomiyuki	168
Fukuoka, Takanori	752
Fukuyama, Yoshikazu	4907
Fuller, Arthur	4437
Funaki, Tsuyoshi	3894
Fung, Kai-Tat	908
Furukawa, Toshihiro	2271, 5722
Furuta, Atsuhiro	4751
Furuya, Kiyohi	5722

G

Gabbouj, Moncef 2899, 6074
Gagnon, Yves 196, 4257
Galanis, M. D. 472, 1206
Galanis, Michalis 3519, 4641
Galias, Zbigniew 3407
Galkowski, Krzysztof 2040
Galup-Montoro, Carlos 1413
Gan, Woon S. 276, 2275
Gan, Woon-Seng 3139, 5994
Gan, Zhi-Feng 3435
Gang, Yao 5441
Gansen, Michael 5210
Gao, Guo-Zhang 240, 764
Gao, Wen 304, 3427, 3459, 3467, 3471
Gao, Zhiqiang 384
Gao, Zhi-Wei 4046
García Ortiz, Alberto 1078
Gardini, Laura 3403
Garg, Manish 3551
Gargour, C. S. 2321
Gastaldi, R. 1270
Gattiker, Giorgio 2755
Gaudet, Vincent 2204, 3339, 5790
Ge, Qi-Wei 168
Ge, Tong 224
Gebara, Edward 4847
Geiger, Randall 784, 1618, 1378, 1968,
    3809, 4285, 4289, 4831, 4835, 5373, 6190
Geiger, Randy 792
Genko, N. 2365
Genov, Roman 4433
Gerald, José 2739
Gerfers, Friedel 2539, 2543
Gerna, D. 400
Gerosa, Andrea 2551
Ghadiri, Aliakbar 2421
Ghallab, Yehya H. 1346
Ghidini, C. 400
Ghittori, Nicola 388, 776
Ghoda, Yuji 6194
Gholipour, Morteza 5234
Ghoneima, Maged 592, 4146
Ghouti, Lahouari 2313

Giancola, Guerino 49
Gielen, Georges 1298, 2247, 5170
Gies, Valentin 2227, 5242
Gilli, Marco 3765, 4661, 4669
Ginés Arteaga, Antonio 1976
Ginosar, Ran 600, 5238
Ginsburg, Brian P. 184
Girod, Bernd 3531
Glesner, Manfred 37, 1078, 3970
Gnaedig, David 332
Gnudi, Antonio 2108
Goes, J. 796
Goes, Joao 3123, 4074
Goh, W. L. 2429
Goh, Wang Ling 1525
Gohokar, Vijay 3898
Gohokar, Vinaya 3898
Golconda, Pradeep 4697
Goldberg, H. 4205
Goldgeisser, Leonid 5178
Goldstein, Seth 2329
Golikeri, Adarsh 320
Gonzalez Bayon, Javier 4843
Gonzalez, J. L. 568
Gonzalez, José Luis 4273, 460, 204
Gonzo, L. 572
Gopalan, Anand 4385
Gopalan, Kaliappan 4413
Gopinathan, Venugopal 1390
Gordon, Christal 2172
Goto, Satoshi 2891
Gottardi, Massimo 5902
Gou, Bei 3881, 3902, 5282
Goutis, C. E. 472, 1206
Goutis, Costas 3519, 4086
Graham, David 1980, 2172
Grant, Alex 5790
Grassi, Marco 5385
Grasso, A.D. 1569
Gray, Jordan D. 468
Grazzini, Massimo 4677
Grecu, Cristian 1774
Gregori, Stefano 436, 5898
Gregory, Dennis 4461
Grimm, Christoph 5166

Grivet-Talocia, Stefano 5754
Grosse, Daniel 4167
Gu, Fan 4417
Gu, Jiangmin 69, 4082
Gu, Ming 6248
Gu, Yongru 408
Guan, Xiaokang 4807
Guarino, Giuliano 228
Gueorguiev, Svetoslav 5505
Guerrieri, Roberto 2911
Guerrini, Nicola 3295
Guevorkian, David 2088
Gugang, Gao 5662
Guglielmo, Michele 3547
Guimarães, Leticia 5834
Guinjoan, F. 2473
Guinjoan, Francesc 1318, 3067, 4453, 4457
Günel, Serkan 4899
Gunther, Jacob 4325
Güntzel, José Luís 3567
Guo, Jiun-In 4517, 4542, 6110
Guo, Jun 3631
Guo, Jwin-Yen 2555
Guo, Li 5998
Guo, Xun 3471
Guo, Zhan 4947
Gupta, Amit 4361, 4373
Gupta, Nikhil 4943
Gupta, Pawan 3063
Gupta, Subhanshu 2104
Gupta, Vishal 4245
Gürkan, Hakan 1334
Gustafsson, Oscar 1449, 1453
Guthaus, Matthew R. 624
Güz, Ümit 1334
Gwee, Bah-Hwee 4078, 5405

H

Ha, Dong Sam 1051, 3355, 3962
Ha, Dong 5111
Habuchi, Hiromasa 5425
Haccoun, David 4159
Haddad, M. I. 1686
Haddad, Sandro 3287, 5357
Haeiwa, Kazuhisa 1517
Häfliger, Philipp 2751
Haga, Yasutaka 220
Haigh, David 244, 248, 5966
Haitong, Ge 492, 6316
Hakkarainen, Väinö 4622
Haley, David 5790
Halonen, Kari 2687, 3938, 4122, 4622,
   5059, 5377, 5381, 6114
Halvorsrød, Thomas 1593
Hamada, Nozomu 2847
Hamaguchi, Kiyoharu 3555
Hämäläinen, Saku 6114
Hämäläinen, Timo 1134, 3351, 3463
Hamami, Shy 5310
Hamamoto, Takayuki 564
Hamid, Nor H. 2510
Hamilton, Alister 1028
Hamour, Marwa 5071
Han, Jun-Hua 1557
Han, Wei 5274, 5842
Han, Yinhi 5666, 5670
Hang, Hsueh-Ming 4550, 4983
Hang, Wei 1746
Hanke, Hans-Christian 2915
Hanna, Magdy Tawfik 832
Hannuksela, Miska 2899, 6074
Hansen, Lisa E. 1762
Hanumolu, Pavan Kumar 2807, 3986
Hao, Jinxin 516
Hara, Jun-Ichi 5922
Harada, Youmei 3215
Haratcherev, Ivaylo 3523
Harjani, Ramesh 4381, 4393, 4630
Harmanci, Mehmet 1782
Harpe, Pieter 4062, 4839, 5541
Harrison, Jeffrey 232
Hartin, Olin 612
Hartley, Lee 5369
Harvey, Jackson 4381
Hasan, Masood 5958
Hasan, Mohammed 2259, 2317, 5138, 5274
Hasan, Tawfique 1899
Hase, Hiroyuki 720
Hasegawa, Minoru 3950
Haseyama, Miki 2239, 4030, 4931, 6320
Hashemi, Majid 5107, 5617
Hashiesh, Mohammed A. 1622
Hashimoto, Hideo 2080
Hashimoto, Yasuyuki 5561
Hashizume, Masaki 2995
Hasler, Paul 125, 468, 996, 1980, 2148,
   2168, 2172, 2441, 4106, 4341, 4441, 5318,
   5553, 6106, 6154
Haslett, James 5369
Hata, Toshihiko 952
Hatakawa, Yasuyuki 25
Hatami, Safar 2381, 4610, 5278
Hatayama, Yoshinori 5341
Hattori, Toshihiro 5349
Hatzopoulos, A. A. 2196
Hauer, Johan 3688
Haurylau, Mikhail 2514
Hayakawa, Masashi 980
Hayasaka, Naoto 4281
Hayashi, Yoshiteru 2096
He, Chengming 4835
He, Fei 6248
He, Lei 105
He, Wei-feng 2887
He, Yajuan 69, 4082
He, Yuwen 304
He, Zhenya 5409
He, Zhiyong 6018
Hedberg, Hugo 3431
Hefner, A. R. 416
Hegedûs, Ákos 3375
Hegt, Hans 4062, 4839
Hegt, J. A. 788
Heikkinen, Jari 1122
Heiskanen, Antti 704
Hekmat, Mohammad 5617

Helfenstein, Markus 5003
Hella, Mona Mostafa 1170
Henderson, Clifford L. 1738
Henkel, Joerg 1778
Heo, Deukhyoun 212, 5115
Hermida, R. 2365
Hernandez, Luis 4070
Hernández-Martínez, Luis 3203
Heydari, Payam 2695
Hezar, Rahmi 3668
Higami, Yoshinobu 2987
Higashimura, Masami 1589
Hihara, Hiroki 812
Hikawa, Hiroomi 3623
Hikihara, Takashi 3894
Hinamoto, Takao 256, 508, 2028, 2287, 2401, 6014
Hioka, Yusuke 2847
Hirai, Noriyuki 5722
Hirano, Satoshi 816
Hiratsuka, Seiichiro 2891
Hironaka, Tetsuo 3507
Hirose, Keikichi 5734
Hirose, Tetsuya 1923
Hisakado, Takashi 2803, 2831
Hisamura, Tomiji 760
Hisayasu, Osamu 256
Hiskens, Ian 4179, 5298
Ho, Chien-Peng 3271
Ho, Fat 4154, 5657
Ho, Gavin 5461
Ho, Hsin-Hsien 6214
Ho, K. C. 2295, 3745
Ho, K. L. 4321
Ho, Ka Yau 6086
Ho, Sheng-Feng 3363
Ho, Wen Tsern 4879
Hodge-Miller, Angela 2787
Hoffman, Michael W. 5322, 5549
Hofmann, Franz 2915, 3479
Hogan, S. J. 1485
Holmberg, Johnny 4469
Holzapfl, Birgit 2915, 3479
Homma, Yuya 1786
Honda, Hirohiko 3757, 4509
Hong, Juhyung 4477
Hong, Sangjin 1646, 3511
Hong, Sung Je 4693, 5854
Hong, Xianlong 93, 97, 101, 1210, 1863, 1871, 1883, 2449, 2999, 5641, 6222, 6230
Hong, Yoing-Fu 2979
Hong, Yuru 4533
Hong, Yu-Ting 2983
Hoon, Siew Kuok 3873
Hooper, Mark 125
Hooshmand, A. 424
Hoque, M. R. 133
Horikawa, Yoshiyasu 180
Horiuchi, Timothy 4201, 4217, 4755
Hosokawa, Shinji 6014
Hosoya, Tetsuo 6268
Hosseini, Reshad 4610
Hotti, Mikko 5059
Hou, Hao-Sheng 3003
Hou, Kuen-suey 328
Hou, Ling 3187
Hou, Tom 3535
Howard, G. Michael 3559
Howard, Steve 5103
Hsia, Chih-Hsien 4554
Hsiao, Chin-Chih 5119
Hsiao, Frank 4606
Hsiao, Hsu-Feng 2663
Hsiao, Pi-Chen 3503
Hsiao, Shen-Fu 2433
Hsiao, Ying-Tung 3885
Hsieh, Hao-Yueh 1879
Hsieh, Hsiang-Ji 5007
Hsieh, Ming-Ta 4883
Hsieh, Tsai-Ming 6234, 6244
Hsieh, Wen-Tsan 3591
Hsin, Yi-Chih 3990
Hsu, Chao-Hsin 3043
Hsu, Chao-Yong 5517
Hsu, Chao-Yuan 2619
Hsu, Ching-Ting 2895, 5481
Hsu, Han-Jen 6288
Hsu, Heng-Ming 129
Hsu, Jen-Chien 2643
Hsu, Shaomin 3235

Hsu, Terng-Yin ... 6030
Hsu, Wei-Hung ... 2731
Hsung, Tai-Chiu ... 4867
Hsung, Terng-Yin ... 4606
Hu, Bo ... 2823, 5621
Hu, Chang-Fen ... 1258
Hu, Eric ... 2791
Hu, Jingyu ... 2563, 2567, 2571, 2575
Hu, Qingsheng ... 340
Hu, Sanqing ... 5690
Hu, Shiyan ... 1306
Hu, Weiwu ... 676
Hu, Yi ... 2263
Hu, Zongqi ... 3083
Hua, Chung-Hsien ... 444
Hua, Qiang ... 1421
Huang, Chao-Hui ... 3942
Huang, Chao-Tsung ... 1790, 5190
Huang, Chao-Wei ... 3503
Huang, Chen-Hsiu ... 2711
Huang, Chi-Fu ... 640
Huang, ChingYao ... 3267, 4533
Huang, Chung-Lin ... 1553
Huang, Chun-Hao ... 3643, 3647
Huang, Chun-Ming ... 2883
Huang, De-Shuang ... 1557
Huang, Dong-Yan ... 3135
Huang, Gaoming ... 5409
Huang, Garng ... 4183
Huang, Han ... 1306
Huang, Hao-Jung ... 5122
Huang, Hong-Yi ... 3363
Huang, Hsiang-Cheh ... 3990, 4979, 4983
Huang, Hsin-Hsiung ... 6234
Huang, Ing-Jer ... 6122
Huang, Jiwu ... 4014
Huang, Kuang-Chih ... 3990
Huang, Lin-Chieh ... 4542
Huang, Li-Ren ... 5501
Huang, Li-Wei ... 3363
Huang, Ming-Feng ... 2116, 2811
Huang, Qingming ... 304, 3471
Huang, Shih-Hsu ... 3307, 5633, 6244
Huang, Shih-Way ... 3155
Huang, Shih-Yu ... 2883

Huang, Teng-Yi ... 152, 156
Huang, Wei-Kan ... 3455
Huang, Wei-Sheng ... 1218
Huang, Wenjie ... 1158
Huang, Wen-Kai ... 6122
Huang, William ... 3475
Huang, Xiaoling ... 5186
Huang, Xinping ... 1710
Huang, Xu ... 2699
Huang, Yen-Chuan ... 3692
Huang, Yih-Fang ... 2623, 5998
Huang, Yu-Jen ... 77
Huang, Yu-Wen ... 1790, 2931, 5469, 6082
Huang, Zhang-cai ... 2795
Huang, Zhenan ... 5158
Huertas, José L. ... 2647
Huggett, Clare ... 5513
Hui, John ... 1190
Huijsing, Johan ... 6162
Humayun, Mark ... 2743
Hung, Chung-Chih ... 3131
Hung, Kevin ... 2723
Hung, William N. N. ... 6248
Hur, Youngsik ... 4847
Hurst, Paul ... 4058
Hurtig, Patrik ... 180
Hwang, Jenq-Neng ... 2663, 2667, 3419
Hwang, Jin-Hong ... 4803
Hwang, TingTing ... 3311
Hwang, Tsung-Ming ... 824, 828
Hwang, Wei ... 444
Hwang, Wen-Shyang ... 2124
Hwang, Yin-Tsung ... 6098
Hwang, Yuh-Shyan ... 1040, 3275, 6170
Hyvonen, Sami ... 1202

I

Iannaccone, Giuseppe 5075
Ibrahim, Mohammad 2313
Ibrahim, Youssef 3914
Ibrahimy, M. I. 2385
Ichikawa, Yoshinori 2271
Ichimiya, Masahiro 2995
Ida, Tsukasa 812
Ienne, Paolo 1782, 2361
Ignjatovic, Zeljko 3664
Ihalainen, Tero 3175
Ikebe, Hayato 5341
Ikeda, Masahiro 13
Ikehara, Masaaki 4293
Ikenaga, Takeshi 2891
Ikuta, Akira 260
Imae, Joe 2823, 3183
Imai, Kiyotaka 13
Imai, Yu 4911
Imamura, Kousuke 2080
Inagaki, Jun 2239
Inan, Aziz S. 976
Indiveri, Giacomo 5150
Iniewski, K. 1154
Iniewski, Krzysztof 5071
Inoue, Hirotaka 2502
Inoue, Kohei 6308
Inoue, Michiko 5686
Inoue, Yasuaki 1130, 2795, 3761, 4911
Inouye, Yujiro 3031, 5870
Ioinovici, A. 1310
Iordanov, Ventzeslav 2919
Ip, Henry 3918
Ip, Tak-Piu 2671
Irita, Takahiro 5349
Iroaga, Echere 5055, 5557
Ishida, Koichi 3119
Ishida, Takuma 6328
Ishida, Tsutomu 4903
Ishihara, Hiroaki 812
Ishihara, Manabu 2735
Ishihata, Kyohei 4851
Ishii, Yasuhisa 3055
Ismail, Mohammed 3982
Ismail, Yehea 592
Ismail, Yehia 4146
Isoaho, Jouni 456
Italia, Alessandro 3239
Ito, Kazuhito 6210
Ito, Kiyoto 2389
Ito, Takehiro 160
Itoh, Niichi 73
Itoh, Yoshio 272, 1698, 5413
Iu, Herbert 2481, 4445
Ivanov, André 1774, 6134, 6138
Ivanov, Vadim 1561, 3853
Iwanaga, Nobuyuki 5345
Iwata, Atsushi 192
Iwata, Ken-ichi 2401

J

Jabedar-Maralani, P. 3107, 3231
Jackson, Larry L. 576
Jaeger, Richard 1178, 5750
Jafaripanah, Mehdi 5389
Jahed Motlagh, M. R. 1406
Jain, Lakhmi 4018, 4975
Jain, Vijay 4473
Jalil, Amir Minayi 5730
Jang, Hyung-Wook 2192
Jang, Inseon 3595
Jansson, Jussi 4269
Jantsch, Axel 1770
Jaskulek, S. 137
Jazayeri, Pouyan 4196
Jen, Chein-Wei 1218, 3503
Jeng, Jin-Tsong 2498
Jenkins, William 2291
Jenkner, Martin 3479
Jeong, Chan-Young 4803
Jézéquel, Michel 332
Ji, Cang 3651
Ji, Honghao 3491
Jia, Lin 376
Jia, Qingwei 5786
Jia, Weimin 3159
Jiang, D. 2243
Jiang, Guo-Ping 6038
Jiang, Hanjun 784, 4285
Jiang, Hongtu 1142
Jiang, Joe-Air 3885
Jiang, Minqiang 1501, 4369
Jiang, Shu-Yu 1174
Jiang, Tai-Ying 5682
Jiang, Yingtao 688, 1746
Jiang, Yueming 3877
Jiang, Yung-Chuan 5649
Jiang, Zhong-Ping 2072
Jie, Jin 3359
Jin, Chunfeng 2827
Jin, Le 1378, 3809
Jin, Xin 3251
Jing, Tong 2449
Jing, Zhujun 4791
Jirapong, Peerapol 4175
Jitapunkul, Somchai 5401
Jitsumatsu, Yutaka 884, 2056
Joe, Inwhee 2955
Johansson, Kenny 1449
Johnson, Louis G. 548
Johnson, Tord 4469
Johnston, Matthew 1762, 3483
Jokerst, Nan M. 1738
Jones, Michael 1774
Jong, Ching-Chuen 1823, 5645
Jose, Sajay 5111
Jou, Jing-Yang 4134, 5682
Jou, Shyh-Jye 1055
Jovanovic-Dolecek, Gordana 840
Jozwiak, Phil 1194
Juang, Chia-Feng 3043, 5122
Julian, Pedro 2763, 4205
Jullien, Graham 588, 2522, 3914, 4261, 5326
Jun, Yang 5662
Jun, Zou 2915
Jung, Dyunghoo 4381, 4393
Jung, Moon-Suk 3704
Juntunen, Jari K. 3463
Jurišic, Dražen 808, 3303

K

Kaabi, H. 1406
Kadim, H. J. 1254
Kaewdang, Khanittha 1577
Kagawa, Keiichiro 2923, 3487
Kajitani, Yoji 1210, 1883
Kakani, Vasanth 1178
Kakarountas, Athanasios 4086
Kakiuchi, Yosuke 3555
Kako, Shinya 2080
Kakuda, Yoshiaki 2967
Kale, Izzet 220, 536, 540, 1409, 2547, 4481, 5746
Kaler, K. V. I. S. 568
Kaler, Karan 2755
Kallakuri, Sankalp 1044
Kam, Alvin Harvey 1750
Kamada, Masaru 5425
Kamae, Shoichi 5349
Kamarei, Mahmoud 2381, 4610
Kamath, Anant 5039
Kameda, Seiji 2771
Kameyama, Michitaka 1786
Kameyama, Shingo 2958
Kamuf, Matthias 344
Kanda, Kouichi 3119
Kaneko, Mineo 700
Kang, In Koo 5529
Kang, Jin-Ku 2192
Kang, Kyeongok 3595
Kang, Li-Wei 3998
Kang, Moonseok 1465
Kang, Sang 3259
Kang, Se-Hyeon 5778
Kang, Sung-Mo 5493
Kankanhalli, Mohan 5525
Kao, Chang-Jung 4871
Kao, Hsueh-Wu 5007
Kao, Meng-Ping 1086, 2036
Kao, Wen-Chung 5015
Kaoliang, Chia-Kai 4871
Karahaliloglu, Koray 4110
Karakiewicz, Rafal 4433
Karandikar, Shrirang 3575
Karel, Joël 3287
Karl, Eric 3563
Karthik, Preethi 5629
Karvonen, Sami 4425
Kasemsuwan, Varakorn 1016, 1614
Kashiwabara, Toshinobu 3555
Kashiyama, Hideki 564
Kasnavi, Soraya 3339
Kastha, Debaprasad 2485
Katagiri, Takashi 1706
Kataoka, Yoshihiko 5286
Kato, Aya 1477
Katoch, Atul 4138
Katti, Rajendra 348, 4501
Katyal, Vipul 1618, 1968
Kaul, Alexander 3479
Kaul, Himanshu 604
Kawaguchi, Hiroshi 3119, 4701
Kawahara, Takayuki 17, 632
Kawakami, Hiroshi 6058
Kawakita, Hiroyuki 1358
Kawamai, Hiroshi 6050
Kawamata, Masayuki 264, 528, 968, 972, 3027, 3729, 6264
Kawamoto, Mitsuru 3031, 5870
Kawamoto, Takuji 4509
Kawamura, Takao 6256
Kawazu, Hideki 3499
Kazimierczuk, Marian 708, 712
Ke, Kai-Wei 1533
Keane, John 4058
Kehrer, Daniel 3227
Keller, Gerhard 872
Kelliher, K. 400
Keppens, Bart 1194
Ker, Ming-Dou 1182, 1859, 3861
Keshavarzi, Ali 9
Keshi, Ikuo 5337
Khabiri, Shahnam 1634, 2425
Khademzadeh, Ahmad 1666, 5234
Khaehintung, Noppadol 5162
Khan, Faisal 5154
Khan, Ishtiaq 3721
Khan, Tahir Abbas 2056
Khasawneh, M. 1686

Khatri, Sunil	4130
Khawam, Sami	1048
Khazaka, Roni	1290
Khellah, Muhammad	592
Khoo, Kei-Yong	672
Khorasani, K.	260
Khouas, Abdelhakim	4859, 5246
Khouri, O.	121
Khumsat, Phanumas	5954
Ki, Wing-Hung	1895, 1907, 3071
Kiaei, Sayfe	4401
Kiatisevi, Pattara	2935
Kida, Takuro	6280
Kida, Yuichi	6280
Kikuchi, Hisakazu	6328
Kikuchi, Takafumi	5926
Killat, Dirk	3651
Kim, Bara	3259
Kim, Chang-Su	944, 956
Kim, Changsung	308
Kim, Chang-Wan	3704
Kim, D.H.	3371
Kim, Daeik D.	1402, 1738
Kim, Daewook	1138
Kim, Donghyun	2369, 4546, 4570, 4574
Kim, Gounyoung	4377
Kim, Hyejung	4602
Kim, Hyoung	5537
Kim, Hyoungsoo	4847
Kim, Hyung-Ock	4150, 4713
Kim, Hyung-Seuk	4389
Kim, Hyungsoo	5766
Kim, Jae-Whui	1063
Kim, Jae-Yung	3704
Kim, Jihong	2715
Kim, Jina	1051
Kim, Jong	4693, 5854
Kim, Jong-Tae	5529
Kim, Joungho	5766
Kim, Ju Yeob	5854
Kim, Kwanho	2357, 4602
Kim, Lee-Sup	4546, 4570, 4574
Kim, Manho	1138
Kim, Nam-Seog	440
Kim, Sangki	3595
Kim, Sang-Min	65
Kim, Se-Won	4054
Kim, Shiho	4763
Kim, Sookjeong	5886
Kim, Sung D.	3323
Kim, Sunyoung	740, 4763
Kim, Tae-Hyoung	1847
Kim, Y. B.	584
Kim, Yong-Bin	2527
Kim, YongSin	5493
Kim, Youngbok	5906
Kim, Youngjae	4237
Kimuzuka, Naohiko	13
Kinget, Peter	4221
Kiranon, Wiwat	5162
Kiriyama, Osamu	3215
Kisaka, Tomoo	6206
Kishida, Kuniharu	5694
Kitagawa, Shinji	4907
Kitaguchi, Susumu	5333
Kitajima, Akira	3555
Kitajima, Hideo	4030, 4931
Kitamura, Iwai	5126
Kitamura, Tadashi	816
Kitsos, Paris	4641
Kiya, Hitoshi	4987, 6324
Kiyoyama, Koji	1940
Klar, Heinrich	3087
Kleinberg, Michael	4743
Klumperink, Eric A.M.	6090
Knopik, Vincent	2675
Knuthammar, Björn	5625
Ko, Hung-Yang	6026
Ko, Myeong-Lyong	1063
Ko, Young-Bae	2975
Kobayashi, Akira	2783
Kobayashi, Fuminori	1214
Kobayashi, Haruo	4281
Kobayashi, Hiroyuki	4987
Kobayashi, Nobuaki	1678
Kobayashi, Tetsuya	6268
Kobayashi, Tomoaki	2823, 3183
Kobayashi, Wataru	5345
Kobayashi, Yuhiro	4293
Kocarev, Ljupco	2349, 3375

Kodama, Mei 912
Koeda, Kazuhiko 4729
Koh, Jinseok 5577
Kohda, Tohru 884, 1477, 2056
Kohno, Kiyotaka 3031, 5870
Koide, Tetsushi 3215, 3507, 5202
Koivunen, Visa 2611, 5890, 6094
Koizumi, Hirotaka 1322
Kok, Chi-Wah 3607, 3737
Kokubo, Masaru 3966, 4497
Kollig, Peter 4163
Kolluru, Ramchander 4187
Kolnik, Jan 5103
Kolodny, Avinoam 600
Kompella, Sastry 3535
Komuro, Takanori 4281
Kong, Hao-Song 952
Kon'no, Yoshio 1927
Kontani, Ken 5126
Koo, Kyoung-Hoi 1063
Koolivand, Yarallah 3107
Koolivand, Yarallah 3231
Koon, Suet-Chui 3071
Kornegay, Kevin 3243, 4891, 5067
Korovkin, Nikolay V. 980
Koshita, Shunsuke 968, 972
Koskinen, Lauri 4122, 6114
Kostamovaara, Juha 216, 4269, 4425, 5601
Kot, Alex C. 4002, 4405
Kotani, Koji 6332
Kotropoulos, Constantine 2283, 2871, 3849
Kotturi, Deen 4653
Kou, Y. J. 2615
Koufopavlou, Odysseas 4086, 4641, 4645
Kousaka, Takuji 6050
Kovavisaruch, L. 2295
Kowalczyk, P. 1485
Kozic, Slobodan 888
Koziel, Slawomir 5605
Kratyuk, Volodymyr 2807, 5986
Krenz, René 2223
Krenzke, Rainer 3651
Kreutz, M. E. 428
Kreutz, Marcio 5866
Krishnan, Shankar M. 149
Krishnan, Shoba 2639
Kristensen, Fredrik 3431
Kroupis, Nikolaos 3347
Krupar, Jörg 3383
Krusienski, Dean 2291
Kschischang, Frank R. 5194
Kuan, Kan-Sheng 3660
Kubin, Gernot 1394, 2655
Kubina, Stefan 2767
Kubota, Hidemasa 4903, 5762
Kuchcinski, Krzysztof 2867
Kucic, Matt 125
Kuenzle, B. 2397
Kulmala, Ari 3351
Kumagai, Sadatoshi 756
Kumaki, Takeshi 5202
Kumar, Aatish 3551
Kumar, Ashok 29, 552, 4697
Kumar, Shashi 2353
Kummert, Anton 2040
Kunarak, Sunisa 5146
Kung, David 21
Kung, Wei-ying 3263
Kuo, C.-C. Jay 308, 948, 1545, 3263, 4562, 5533
Kuo, Chien-Hung 2799, 3676
Kuo, Chien-Nan 5099
Kuo, Chung J. 2116, 3781
Kuo, James 5258
Kuo, Jin-Hau 2711, 3219, 3829, 4594
Kuo, Sen M. 276, 5994
Kuo, Shyi-Shiun 3998
Kuo, Sy-Yen 81
Kuo, Te-Son 5119
Kuo, Wu-An 3311
Kuroda, Yasuto 5202
Kurokawa, Atsushi 2795
Kurokawa, Kosuke 1322
Kuroki, Wataru 3761
Kurosawa, Minoru Kuribayashi 1706
Kursun, Volkan 464
Kutila, Matti 1758
Kuusilinna, Kimmo 1134
Kuwahara, Naoki 952
Kwan, Chiman 3745

Kwan, H. K. 844, 1421, 2595, 2759, 3725, 4337
Kwan, Man-Wai 3607, 3737
Kwasniewski, Tadeusz 1158, 1274, 4485
Kwok, Chee Yee 1899
Kwon, Taek-Jun 3331
Kwong, Sam 3607

L

Lacort, J. 141
Lacourse, Alain 4257
Ladino, Pedro 932
Lagendijk, Reginald 3523
Laguna, M. 5581
Lahdenoja, Olli 4118
Lahoz, Tomás 204
Lahti, Juho 3463
Lahuec, Cyril 332
Lai, E.M-K. 352, 496, 5429
Lai, Feipei 684
Lai, Hsin-Ya 1875
Lai, Jin-Shin 5119
Lai, Jui-Lin 1690, 3059
Lai, Li-Chun 3039
Lai, Ming-Hong 1278
Lai, Yen-Tai 1875, 5649
Lai, Yeong-Kang 432, 4578
Lai, Yi-Te 3059
Lai, Zhoa-Hong 6006
Laiho, M. 3922, 5798
Laiho, Mika 3938, 5810
Lam, Chi-Wai 1513
Lam, Kin-Man 3841, 4586
Lam, Stanley 5230
Lam, Yat-Hei 3071
Lambacher, Armin 3479
Lambie, Johan 460
Lameres, Brock 4130
Lampe, Alexander 5003
Lampinen, Harri 188
Landry, Alexandre 3343
Lang, Stephan 4465
Langendoen, Koen 3523
Langlois, Peter 748, 2727
Lao, Chon-In 3095
Larsen, Torben 5083, 5505
Larsson-Edefors, Per 1654
Laskar, Joy 3243, 4847
Latiri, Anis 3672
Lau, Francis 4493, 4955, 4967
Lau, W. H. 2469
Lau, Wing Yi 4333

Laurenson, David 2510
Lauter, Robert 1770
Lawrance, Anthony J. 876, 2052
Lawson, Stuart 6002
Lazar, Aurel 4221
Lazzarini, Marco 1489
Le Cam, Laurent 3551
Le, Charles 420
Lebel, Eric 1952
Leblebici, Yusuf 1059, 1782, 2535
Lee, Byung Geun 1960
Lee, Chen-Yi 940, 1810, 2140, 4558, 4875, 6030
Lee, Chia-Lin 5469
Lee, Chih-Hung 6234, 6244
Lee, Ching-Li 356
Lee, Choong-Hoon 5529
Lee, Chun Yi 57
Lee, Dong-Soo 3167
Lee, Edward 3877
Lee, Hae-Yeoun 5529
Lee, Hanho 1036
Lee, Hee-Sub 4054
Lee, Herng-Jer 1278
Lee, Heung-Kyu 5529
Lee, Hsien-Hsin S. 1867
Lee, Hua-Chin 368
Lee, Hyung-Jin 3355, 3962, 5111
Lee, J. H. 1198
Lee, Jae-Youl 740, 4763
Lee, Jae-Yup 4054
Lee, Jeong Hoo 3323
Lee, Jin 3978
Lee, Joon-Suk 6178
Lee, Jun Wei 4305
Lee, Jungwon 5318
Lee, Kangmin 2357, 2369
Lee, Kong A. 2275
Lee, Kuen-Jong 2983
Lee, Kuo-Cheng 3271
Lee, Kyung-Hoon 4054
Lee, M. J. 728
Lee, Ming-Sui 4562
Lee, Min-Wuk 4602
Lee, Mi-Young 4803

Lee, Peter	4863
Lee, Po-Ming	780
Lee, Sang-Gug	3704
Lee, Sang-Hoon	6178
Lee, Sang-Uk	944, 956
Lee, Se-Joong	2357, 2369
Lee, Seung-Hoon	4054
Lee, Shuenn-Yuh	2116, 2811
Lee, Shyh-Chyang	2811
Lee, Simon	5357
Lee, Soo-Young	5874
Lee, Sung-Sop	2192
Lee, Tai-Cheng	1382, 3692
Lee, Tai-Hsing	129
Lee, Trong-Yen	1040, 6170
Lee, Tsong-Li	3039
Lee, Tsung-Sum	3111
Lee, Wen-Ping	2140
Lee, Wen-Ta	1040, 3275, 6170
Lee, Yinman	3179
Lee, Yu	1150
Lee, Yu-Cheng	1370
Lehmann, Torsten	732, 1899
Lehto, Raija	2012
Lehtoranta, Olli	3463
Lei, Shawmin	1230
Lei, Zhang	4301
Leiwo, Jussipekka	1226
Leon, Walter D.	5549
Leon-Salas, Walter D.	5322
Leou, Jin-Jang	1493
Leou, Jin-Jang	3998
Lerm, André	5294
Leung, Alfred Tze-Mun	1290
Leung, Carina	5914
Leung, Henry	300, 1473, 3151, 3391
Leung, Kelvin	3079
Leung, Michael	4493
Leung, Shu-Hung	6086
Leung, Victor	4042
Leung, Yee Hong	3745
Levacq, David	4681
Levitan, Steven P.	1662
Levy, Bernard	4058
Lewis, M. Anthony	4106
Lewis, Stephen	4058
Li, Chengqing	880
Li, Chi-Fang	5043
Li, Chunguang	288
Li, Day-Uei	5501
Li, DongMei	4995
Li, Feng	4851
Li, Gang	516, 2587
Li, Guolin	3700, 4995
Li, Gwo-Long	5481
Li, Hua	4634, 4637
Li, Hung-Ju	5481
Li, Ji	5670
Li, Jianbing	4449
Li, Jiang	4927, 5473
Li, Jin-Fu	77
Li, Jun	2457
Li, Miao	1158, 1274
Li, Ning	1746
Li, Ruiming	113, 2263
Li, Shengyuan	4823, 5942
Li, Shipeng	304, 1126, 2703, 4927, 6078, 6304
Li, Shujun	880
Li, Sing-Rong	2531
Li, Wei-Chang	3974
Li, Xiang	280
Li, Xiaowei	5666, 5670
Li, XiaoWen	4995
Li, Xinxiao	880
Li, Xuequn	436
Li, Yijun	5198
Li, Yu	2779
Li, Yunhong	3745
Li, Yunlei	5095
Li, Yushan	1314
Li, Zhao	5621
Li, Zheng	928
Li, Zhenyan	3845
Li, Zhuoyuan	6230
Lian, Yong	520, 2004, 2016, 4811
Liang, Chao-Jui	4253
Liang, Chih-Hao	3829
Liang, Hau-Jie	1118
Liang, Li-Han	3163

Liang, Liuhong	1549
Liang, Sheng-Fu	152, 156
Liang, Xiaoyao	1646, 3511
Liao, Guang-Wan	4717
Liao, Lu-Po	6170
Lie, Wen-Nung	2136, 4046
Lillie, Jeffrey J.	1738
Lim, Keng Pang	4233, 4939
Lim, Sung Kyu	1867
Lim, Wei Meng	4815
Lim, Yong Ching	1819, 4305
Lin, Bill	2349
Lin, Bor-Ren	3643, 3647
Lin, Chang-Hua	3655
Lin, Chang-Tzu	6214
Lin, Chen-Fu	1493
Lin, Cheng-Chen	6098
Lin, Cheng-Hung	1032
Lin, Chia-Wen	1242, 3255, 3781
Lin, Chien-Chang	6110
Lin, Chien-Chih	368
Lin, Chih-Hsien	1055
Lin, Chih-Yuan	6244
Lin, Ching-An	5702
Lin, Ching-Yung	1250, 4598
Lin, Chin-Teng	152, 156, 2859, 3023, 3942
Lin, Chung-Yuan	3223
Lin, Fujiang	5087
Lin, Hai	2506
Lin, Heng-Yao	1802
Lin, Hongchin	4253
Lin, Huaizhong	6022
Lin, Huang-Shih	3676
Lin, Hui-Tang	1533
Lin, I-Ting	6122
Lin, Jia-Huang	3825
Lin, Jian-Hung	2373
Lin, Jian-Liang	2667
Lin, Junlong	4014
Lin, Kai-Ping	2219
Lin, Kun-Hsien	1182
Lin, Li-Chun	3503
Lin, Li-Ping	356
Lin, Po-Feng	3825
Lin, Rung-Bin	4717, 4725, 5862
Lin, San-Ho	1040
Lin, Sheng-Yuan	5015
Lin, Shiang-Jiun	1670
Lin, Shu-Fa	1505
Lin, Ta-Wei	2731
Lin, Tay-Jyi	1218, 3503
Lin, Ting-An	940, 1810, 2140, 4558
Lin, Tom	2136
Lin, Wei-Pin	2084
Lin, Weisi	2048, 3793
Lin, Wen-Sheng	5573
Lin, Xiao	3793, 4233, 4939
Lin, Yen-Nan	1638
Lin, Youn-Long	4525, 4529
Lin, Yu	1618, 1968
Lin, Yuan-Pei	3163
Lin, Yung-Yu	5007
Lin, Zhiping	312, 1338, 2024, 2044, 3741, 3769
Lin, Zhiqiang	5322
Lindfors, Saska	5083, 5505
Lindsay, Iain	1048
Ling, Nam	1501, 3443, 4369, 5489
Ling, Zhi-Hong	3781
Liou, J. J.	416
Liou, Yuan-Chang	3043
Litovski, Vanco	5182
Liu, Bin	1871
Liu, Bin-Da	1802, 1972
Liu, Chia-Hsien	3503
Liu, Chien-Nan	3591
Liu, Chien-Nan	5682
Liu, Chih-Wei	1218, 3503
Liu, Chih-Yun	212
Liu, Christianto	2939
Liu, Chun-Nan	2851
Liu, Debing	304
Liu, Der-Jenq	2859
Liu, Derong	3183, 3387, 5690
Liu, Guan-You	5258
Liu, Hai Qi	1525
Liu, Hongmei	4014
Liu, Jia-Hwa	2895
Liu, Jin	113, 4887, 5095
Liu, Jun	4002, 4405

Liu, Kuang-Yu 4225
Liu, Lu 3700, 5023
Liu, Mingjian 2060
Liu, Minxuan 5103
Liu, Patty Z.Q. 1190
Liu, Qiang 660
Liu, Qingyan 5302
Liu, Rong 1883
Liu, Shen-Iuan 368, 3676
Liu, Shih-Chii 4213
Liu, Tsu-Ming 1810, 2140, 4558
Liu, Tung-Kuan 2603
Liu, Wei 3599
Liu, Wenbo 3809
Liu, Wentai 1082, 2743, 2927
Liu, Xiande 3251
Liu, Yilong 4034
Liu, Yongzhi 2024, 3741
Liu, Yuan'An 4026
Liu, Yutian 3889, 3906
Liu, Zhongzhi 5838
Lo, Kam-Fai 4586
Lo, Li-Chu 840
Lo, Te-Jung 4554
Lo, Wing-Hang 5653
Lo, Yu-Lung 1070
Loikkanen, Mikko 216
Long, Di 2999
Long, John R. 5357
Longxing, Shi 5662
López, José Fco. 412
López, José 6130
López, Paula 3688
López, Sebastián 412, 6130
Lotfi, Reza 4618
Lotfi-Kamran, P. 424
Lou, Feiyin 5666
Lou, Jian 316
Lovisolo, Lisandro 2309
Low, Siow Yong 856, 2875
Lozano, Cicilia 480, 4705
Lu, Chi-Chang 3111
Lu, Chun-Shien 5517
Lu, Hoang-Yang 1726
Lu, Hong 1238, 1549, 3837
Lu, Jianming 720, 2783, 3051, 3211, 4851, 4919, 5130, 5726, 6292
Lü, Jinhu 300, 1473, 3391
Lu, Junan 2477
Lu, Kai-Sheng 240, 764
Lu, Meng-Ting 1505
Lu, W.-S. 2615
Lu, Wen-Fu 560
Lu, Wen-Kai 1690
Lu, Wu-Sheng 1831, 2028, 2401
Lu, Xiaoan 1246
Lu, Yan 3471, 6078
Lu, Yan-Chen 4521
Lu, Zhe-Ming 4417
Lu, Zhongkang 3793
Luca, Mihai Bogdan 3813
Lucic, Želimir 2068
Lulli, G. 448
Lun, Daniel Pak-Kong 4867
Lune, Hong-Wen 2643
Luo, Fa-Long 2184, 2559
Luo, Ming 3615
Lutz, Barry 4795
Luzzi, Raimondo 3547

M

Ma, Chun-Yat 4867
Ma, Dongsheng 3083
Ma, J. G. 5027
Ma, Jian Guo 4815
Ma, Jian 1746
Ma, Jian-Guo 376
Ma, Jianguo 384, 5035
Ma, Jia-Shing 5846
Ma, Kaixue 5035
Ma, Lin 4171
Ma, Liying 260
Ma, Min 2453
Ma, Ming 6102
Ma, Siwei 3467
Ma, Yuan 5397
Ma, Yuchun 1863, 6222
Ma, Yue 1469, 6050
Ma, Yung-Cheng 6118
Mabey, Glen W. 556
Macleod, Malcolm 1814
Madanayake, Arjuna 1457
Madanayake, H.L.P.A. 2032
Madden, Patrick 3579
Madhukumar, A.S. 5421
Madoc, Allan 2699
Madrenas, Jordi 3926, 4114, 4453
Madsen, Kaj 5605
Madureira, Miguel A. M. 1166
Maeda, Moto 3507
Maejima, Yuji 508
Maekawa, Shuji 5330
Maeki, Akira 4497
Maeng, Moonkyun 4847
Maggio, Gian Mario 53, 2349, 3375
Maghari, Nima 4265
Magierowski, S. 1154
Mahapatra, Rabi 5782
Maharatna, Koushik 5513
Mahfuz, Ejaz 3753
Mahlke, Scott A. 624
Mahmoodi, Hamid 2421
Mahmoud, Soliman A. 1000
Mahmoud, Soliman 1622
Mahmoudi, Farsheed 2112
Maio, Ivano 5754
Mak, Pui-In 392
Makino, Hiroshi 73
Makino, Shoji 5882
Maksimovic, D. 1302
Maksimovic, Dragan 1314
Makur, Anamitra 4301
Malcovati, Piero 388, 776, 4614, 5385, 5569, 5982
Mallik, Udayan 1919
Maloberti, Franco 1903, 3007, 3873, 5898, 5982
Manaresi, Nicolò 2911
Mandal, Debashis 4277
Mandal, Gunjan 2180
Mandal, Pradip 2180, 3865, 4277
Mandic, Danilo 3599
Mandolesi, Pablo 2763, 4205
Mandrekar, Rohan 5758
Manfredi, Sabato 2235
Mangard, Stefan 1066
Manohar, Rajit 2939
Manoli, Yiannos 2539, 2543, 4066
Mansour, Hassan 320, 4042
Mantooth, A. 133
Mantooth, Alan 5186
Mäntyniemi, Antti 4269
Mao, Shiwen 3535
Mao, Zhi-gang 2887
Marche, David 196
Marcon, Cesar 5866
Marietti, P. 448
Marietti, Piero 3295
Markou, Kleoniki 3347
Marquardt, Brian 2747
Marsili, Stefano 5497
Marsman, Eric D. 624
Martens, Ewout 2247
Martin, Eric 680
Martin, Joel R. 1662
Martin, Maria Elena 4795
Martineau, Baudouin 2675
Martinez, Herminio 4457
Martínez, Juan 1318

Martínez-Salamero, Luis 2835
Martini, Filippo 1702
Martin-Martin, Pilar 1106
Martins, Miguel A. 5353
Martins, R. P. 392, 1581, 1585
Martins, Rui Paulo 3095, 3099
Martins, Rui 404
Martorell, Ferran 4273
Mase, Kenichi 2958
Mashima, Toshiya 752, 2231
Maskell, Douglas 4799
Masoud, Elzinati 6102
Massari, Nicola 5902
Masson, Favio 2763
Masuda, Hiroo 5934
Masuda, Masayuki 6210
Masugata, Kasumi 5126
Masuoka, Sadaaki 13
Masuzaki, Takahiko 2096
Mateo, Diego 4273
Mathis, Wolfgang 5445
Matsubara, Shigeki 3623
Matsubayashi, Akira 1354
Matsumoto, Hiroki 5722
Matsumoto, Naoki 3399
Matsumura, Tomoya 5345
Matsuo, Mitsuhiro 5425
Matsuo, Takashi 5330
Matsuoka, Toshimasa 812
Matsuoka, Yusuke 3411
Matsuura, Kei 452
Matsuura, Tomoyuki 4855
Mattausch, Hans Jüergen 3507, 3215, 5202
May, Tony 6062
Mayaram, Kartikeya 2807, 3986, 5986
Maymandi-Nejad, Mohammad 3684
Mayyas, K. C. 1686
Mazumder, Pinaki 1262, 1948, 2531
Mbaye, Maria 3515
McConaghy, Trent 1298, 5170
McCorquodale, Michael S. 624
McNutt, T. 133
McQuaide, Sarah 4795
Mediratta, Sumit 2951
Medoro, Gianni 2911

Mehta, Dipan 900
Meijer, Maurice 5, 1839
Meldrum, Deirdre 4795
Melgarejo, Miguel 932
Memik, Seda Ogrenci 1222
Méndez, C. 1742
Méndez, Miguel A. 4273
Mendias, J. M. 2365
Meng, Yinkuo 3019, 5710
Mengyuan, Li 4409
Menicocci, Renato 1066
Menon, Shibu 1630
Mergens, Markus 1194
Mérigot, Alain 2227, 5242
Merkwirth, Christian 4657
Mesgarzadeh, Behzad 5465
Meunier, Michel 3696, 4257
Meyer-Ortmanns, Hildegard 280
Meza, Carlos 1318, 3067
Mezher, Kahtan 3291
Michael, Marinos 2299
Michail, Harris 4086
Midkiff, Scott 3535
Mijat, Neven 808, 3303
Milidonis, A. 472, 1206
Milirud, Vadim 588
Miller, William C. 3559, 3914
Minasyan, Susanna 2088
Minch, Bradley 996, 4341
Mine, Takashi 4903
Minematsu, Nobuaki 5734
Mintchev, M. P. 568
Mintchev, Martin 2755
Mirabbasi, Shahriar ... 5071, 5107, 5617, 6134, 6138
Miribel-Català, P. 117
Mishchenko, Alan 4721
Mishima, Yuji 4733
Misischi, B. 1742
Mita, R. 2579
Mitra, Joydeep 4187
Miu, Karen 4743
Miura, Seiji 5250
Miyamoto, Ryusuke 2096
Miyamoto, Toshiyuki 756

Miyanaga, Yoshikazu ... 25
Miyaoka, Yuichiro ... 3499
Miyata, Kouji ... 4538
Miyazaki, Masayuki ... 17, 632, 4497
Mizugaki, Kenichi ... 4497
Mks, Sastry ... 772
Mobini, Nastaran ... 3571
Moe, Andrew ... 2747
Moezzi Madani, N. ... 1674
Moghe, Yashodhan ... 732
Mohd-Yasin, F. ... 2385
Mojarradi, M. M. ... 133
Mokrian, Pedram ... 3559
Molavi, Reza ... 5107
Molla, Md. Khademul ... 5734
Molnar, Goran ... 3283
Mondal, Somsubhra ... 1222
Monsurrò, P. ... 448
Monteiro, Paulo M. P. ... 1166
Monteiro, Rui ... 3123, 4074
Montgomery, David ... 3789
Montisci, Augusto ... 992
Mony, Madeleine ... 372
Moon, Jong H. ... 2947
Moon, Un-Ku ... 2807, 3986, 5986, 6182, 6186
Morgenshtein, Arkadiy ... 600
Mori, Hiroyuki ... 4747, 4751
Mori, Shinsaku ... 1322
Morie, Takashi ... 4237
Morimoto, Takashi ... 3215
Morling, Richard ... 4481
Mortazavi, Y. ... 1674
Moschytz, George S. ... 808, 3303
Mottarella, Alan ... 652
Moule, Eric ... 3664
Movahedian, Hamid ... 200, 6150
Mow, Wai-Ho ... 3359
Muchherla, Kishore ... 2465
Müeller, Kurt ... 5003
Mühlhäuser, Max ... 37
Muir, Dylan ... 5150
Mukai, Ryo ... 5882
Mukaidani, Hiroaki ... 3047, 3055
Mukherjee, Rupam ... 3367
Mukund, P.R. ... 4385, 5103

Mun, Ju -Hyoung ... 1529
Muneyasu, Mitsuji ... 256
Murai, Tadakuni ... 5126
Murakami, Naoyoshi ... 2967
Muramatsu, Naokazu ... 2341
Muramatsu, Shogo ... 6328
Murasaki, Izuru ... 1517
Murata, Hiroshi ... 5678
Muresan, Radu ... 436
Murgan, Tudor ... 1078, 3970
Murji, Rizwan ... 4353
Murmann, Boris ... 5055, 5557
Murray, Alan ... 2510
Murtagh, Fionn ... 3789
Mvuma, Aloys ... 2287
Myung, Hyung G. ... 5830

N

Naderi, M.	424
Naegle, Kristen	2104
Nagai, Akira	3635
Nagaraja, H.N.	2485
Nagasaka, Hiroyuki	3950
Nagase, Yukinori	6256
Nagata, Takeshi	4737
Nagazumi, Yasuo	4421
Nair, Nirmal-Kumar	4183
Naka, Masa-aki	2064
Nakada, Kazuki	1923
Nakamoto, Masayoshi	508
Nakamura, Shinji	4421
Nakamura, Yuichi	628
Nakamura, Yukihiro	2096
Nakanishi, Isao	1698, 5413
Nakano, Hideo	3757
Nakano, Keisuke	2962
Nakashiba, Yasutaka	13
Nakata, Mitsuru	168
Nakayama, Tomoyuki	5365
Nakazaki, Akiko	6324
Nakazawa, Chikashi	4907
Nakhla, Michel	3777, 5770
Nakhla, Natalie	5770
Nam, Byeong-Gyu	4602
Nam, Myung-Ryong	1529
Nam, S.W.	3371
Namgoong, Won	61, 3235
Nandi, Asoke	3031
Nandi, Shuvabrata	3367
Nannarelli, A.	1102
Nara, Koichi	4733
Narendra, Siva G.	464
Narendra, Siva	9
Narihisa, Hiroyuki	2502
Naroska, Edwin	109
Nasiopoulos, Panos	320, 4042
Nassif, Sani	604
Nathawad, Lalitkumar	5557
Nauta, Bram	6090
Navabi, Zainalabedin	424, 5278
Nayak, Deepak	900
Ndjountche, Tertulien	2184, 2559
Nebel, Gerhard	3857
Neeb, Christian	1766
Negishi, Toshihiro	1517
Negroni, Juan	3067
Nekili, Mohamed	3343
Nelson, Graham	5326
Nelson, Nicholas	2514
Neto, João	4074
Netto, Sergio	4309
Neubauer, Harald	3688
Neurohr, N.	924
Neves Rodrigues, Joachim	1330
Neviani, Andrea	2551
Newcomb, Robert	2787, 5397
Ng, King-To	3435
Ng, Michael K.	6296
Ng, See-Kiong	4779
Ng, Wai-Yin	2267
Ngan, King Ngi	6078
Nguyen, Truong T.	1090, 1110
Nguyen, Viet-Anh	904, 1238
Nie, Tingyuan	6206
Niederhöfer, Christian	5218
Nielsen, Jannik	4614
Nielsen, Troels	5083
Nierlich, M.	141
Nii, Kouji	73
Nikolaidis, Nikos	3849
Nilsson, Peter	3431, 4947
Ninomiya, T.	2835
Ninomiya, Tamotsu	2827
Nishi, Tetsuo	768, 3631, 4673
Nishiguchi, Nobuyuki	5938
Nishihara, Akinori	4855
Nishimura, Shotaro	2287
Nishio, Yoshifumi	1481
Nishiyama, Tadashi	1358
Nishizeki, Takao	160
Nitanda, Naoki	4030
Niu, Zhongxia	4449
Niwamoto, Hiroaki	5337
Noceti Filho, Sidnei	1944
Noda, Hideyuki	5202
Noh, Hyun-Chul	4054

Noll, Tobias G. 5210
Nomura, Mitsuru 5918
Nookala, Vidyasagar 608
Nooshabadi, Saeid 2076, 4361, 4373, 5262
Nordholm, Sven 856, 2875
Nossek, Josef A. 524
Nouet, Pascal 5910
Nourani, M. 1674
Nourani, Mehrdad 5234
Nowrouzian, Behrouz 4437
Núñez, Antonio 6130
Nunoshita, Masahiro 2923, 3487
Nuzzo, P. 1286
Nwankpa, Chika 5302
Nyathi, Jabulani 1851

O

Ober, Raimund 1338, 3769
Oberst, Mathias 3688
Obridko, Ilya 5238
Och, Sung K. 3323
Ochoa-Montiel, Marco 4163
Ock, Sungmin 6178
Ogata, Masato 768
Ogawa, Kazuya 5341
Ogawa, Takahiro 4931
Ogorzalek, Maciej 4657
Ogunfunmi, Tokunbo 2843, 3475
Ogura, Munenori 6292
Oh, Do-Kwan 5874
Ohashi, Koji 700
Ohba, Ryoji 3721
Ohlsson, Henrik 1453
Ohmi, Tadahiro 6332
Ohno, Shuichi 6014
Ohta, Atsushi 760
Ohta, Jun 2923, 3487
Ohta, Tomoyuki 2967
Ohta, Yuzo 3191
Ohtani, Yoshihiro 5333
Ohtsuki, Tatsuo 3499
Okada, Hiroyuki 5561
Okada, Minoru 5337
Okawa, Shinichi 5934
Okazaki, Chiho 4509
Okazaki, Hideaki 3757, 4509
Okazaki, Hiroshi 2341
Oki, Nobuo 3708
Oksman, Jacques 5565
Okuda, Masahiro 3721
Okumura, Kohshi 2803, 2831
Olarte, Fredy 932
Olivar, G. 1485
Oliveira, Luís 2635
Oliveira, Vlademir 3708
Oliver, Timothy 4799
Olivieri, Mauro 1066, 5206, 5266
Olleta, Beatriz 784
Olsson, Thomas 1330
O'Mahony, Frank 920
Ong, Eeping 3793
Ong, T. C. 1198
Ongsakul, Weerakorn 4175
Onizawa, Masatoshi 1517
Ono, Goichi 17, 632
Onoda, M. 1940
Onoye, Takao 2096, 5345
Oppermann, Ian 364
Orabi, Mohamed 2835
Oraintara, Soontorn 1090, 1110, 4034, 5990
Orino, Yuichiro 1706
Orlik, Philip 3539
Ortmanns, Maurits 2539, 2543
Oshima, Takashi 3966
Osterberg, Peter M. 976
Otani, Norihisa 3950
Otobe, Eiichiro 3950
Ou, Hsin-Hung 1972
Ou, Shih-Hao 1218
Oulmane, Mourad 2453
Öwall, Viktor 344, 1142, 1330, 2867, 3431, 4489
Oweiss, Karim G. 1342
Oya, Takahide 2535
Özalevli, Erhan 2168, 5553
Özcelik, Izzet 5746
Özoguz, Serdar 4345

P

Paasio, A. 3922, 5798
Paasio, Ari 2819, 3115, 3930, 3938, 4118, 4122, 5810, 6272
Pak, Junsp 5766
Pal, Dipankar 748, 2727
Pal, Siddharth 2291
Palaniswami, Marimuthu 2303
Palesi, Maurizio 4090
Paliouras, Vassilis 3954
Palipana, Rajitha 3611
Palmisano, Giuseppe 2679, 3239, 3247
Palomera-Garcia, Rogelio 252
Palumbo, G. 2579
Palumbo, Gaetano 3583, 4685
Pan, Chia-Ho 1505, 5190
Pan, Feng 312, 4939
Pan, Jeng-Shyang 3990, 4018, 4417, 4975
Pan, Min 2445, 2461
Pan, Zhibin 6332
Pancheri, L. 572, 2437
Pancioni, Luca 4677
Pande, Partha 1774
Pandey, Neeta 3299
Pandey, Sujan 37
Panella, Massimo 5393
Pänkäälä, Mikko 3115, 6272
Panock, Richard 1170
Panovic, Mladen 4566, 4827
Panta, Rajesh 2962
Panwar, Shivendra 2072
Pao, Hsiao-Tien 4979
Papadopoulos, N. 2196
Papantonopoulos, Yiannis 792
Pappalardo, Francesco 5206
Paradis, Ken 5103
Parenti, Matteo 1386
Pareschi, Fabio 4349
Parhi, Keshab 65, 408, 1835, 2373
Park, Hyungju 2417
Park, Hyung-Min 5874
Park, Hyun-min 3243
Park, In-Cheol 3167, 5778
Park, Jaejin 920, 1162
Park, Jinho 2100
Park, Jongbae 5766
Park, Jung-Wook 5298
Park, JunYoung 2651, 4693
Park, Sin-Chong 3978
Park, Sung Min 1529
Parodi, Mauro 4915
Paschero, Maurizio 5393
Pasotti, M. 1266
Pasotti, Marco 1891
Patanavijit, Vorapoj 5401
Patané, Luca 4102, 5214
Patangia, Hirak 4461
Paton, Susana 4070
Patra, Amit 2485, 3063
Patti, Davide 4090
Paul, Kolin 5513
Paul, Sajal 3299
Paulino, N. 796
Paulino, Nuno 3123, 4074
Paulus, Christian 2915, 3479
Pavan, Shanthi 5962
Pavone, Marco 5214
Payton, Michael 4154, 5657
Pazos Escudero, Nuria 1782
Peeters, Ralf 3287
Pei, Soo-Chang 1086, 2036, 2667
Pekau, Holly 5369
Pekhteryev, Georgiy 3539
Peltola, Johannes 3423
Peng, Jingliang 948
Peng, Sheng-Yu 4341
Peng, Yu-Chun 4871
Pennisi, S. 1569, 2579, 2583
Pennisi, Salvatore 1573
Perälä, Pauli 188
Peralias, Eduardo 1976
Pereira, F. A. 1020
Pérez-Murano, Francesc 4209
Pessolano, Francesco 5, 1839
Petchjatuporn, Panom 5162
Petraglia, Antonio 1012, 3587
Petraglia, Mariane R. 1114
Petrescu, Tudor 5565
Petrov, Mihail 1078, 3970

Phan, Anh-Tuan 3704
Phang, Khoman 1537, 1984, 2791
Philipp, Ralf 4771
Philippe, Jean-Marc 4689
Phoong, See-May 3163
Phua, Koksoon 1682
Phuah, Jiunshian 3051, 5130
Phyu, M. W. 2429
Pialis, Tony 2791
Piazza, Francesco 3603, 5134, 5714, 5742
Piccardi, Massimo 1702
Piedade, Moisés 2739
Pierre, Samuel 3515
Pillement, Sébastien 4689
Piloto, Rui 4074
Pineda de Gyvez, Jose 5, 1839
Ping Xu, Yong 396
Pinto, Hugo 3123
Piper, Johan 6166
Pirinen, Tuomo W. 1429
Pitas, Ioannis 3849
Pitchumani, Vijay 6230
Plant, David V. 372
Platen, Eckhard 1702
Plett, Calvin 1936
Plotkin, Eugene 4943
Po, Lai-Man 1513, 1541
Poikonen, Jonne 2819, 3930, 4118
Pomeranz, Irith 616
Pongpalit, Wacharapol 1614
Popescu, Gabriel 5178
Popovich, Mikhail 620
Porta, Sonia 4457
Pottenger, William 5154
Pouladi, Alireza 5262
Pouliquen, P. 137
Pourkamali, Siavash 5461
Poveda, A. 2473, 4453, 4457
Powers, John R. 576
Prakash, Somashekar Bangalore 3495
Prasad, Srinivasa 3865
Prefasi, Enrique 4070
Premkumar, Benjamin 5421
Prete, E. 924
Principi, Emanuele 5714
Prochaska, Marcus 5445
Provost, Ghislain 4159
Pu, Kuo-Hua 5043
Puig, M. 141
Puig-Vidal, M. 117
Pun, Kong-Pang 2152, 2723, 3279, 4999, 5842
Puri, Ruchir 21
Puschini, Diego 2763
Pwint, Moe 2863
Pylarinos, Louie 1984

Q

Qi, Zichu 676
Qiang, Qiang 3327
Qiu, Guoping 6276
Qu, Dongdong 1425
Quan, Shaolei 3327
Quelhas, Mauricio 3587
Quinn, P. J. 788
Quinn, Patrick 1956, 1964, 4062
Quintana, José 2647
Quinton, Bradley 45
Qureshi, Muhammad 5946

R

Radenkovic, Miloje 2591
Radfar, Mohammad Hossein 3571
Radhakrishnan, Shailesh 6142
Radmore, Paul M. 244
Radulov, G. I. 788
Radulov, Georgi 4062
Ragonese, Egidio 3239
Ragot, José 3415
Rahardja, Susanto 480, 3793, 4233, 4705, 4939
Rahkonen, Timo 704
Rahman, Shahidur 2855
Rajaee, Omid 1417
Rajan, P.K. 2409
Rajashekharaiah, Mallesh 5115
Rakhmatov, Daler 4098
Ralph, Stephen E. 1738
Ramachandran, Ravi 2321
Ramachandran, V. 2321
Raman, Sangeeta 2639
Ramos, Rafael 3067
Ran, Lixin 6042
Rana, Ram Singh 4811
Ranade, Satish 4187
Rankov, A. 728
Rao, K.R. 5990
Rao, Rahul M. 604
Rapoport, Eduardo 1012
Ratti, Nicoletta 5569
Rau, Jiann-Chyi 2979, 5846
Raut, Rabin 1601
Ravindran, Rajiv A. 624
Ravindran, Sourabh 860
Re, M. 1102
Reaz, M. B. I. 2385
Reddy, Hari C. 2409
Reddy, Sudhakar M. 616
Redondo, Xavier 2000
Reed, Jeffrey H. 1051
Reggiani, Luca 53
Reis, Ricardo 3567
Reja, Mahbub 2120
Rekeczky, Csaba 5802
Ren, Aifeng 3019
Ren, Haipeng 2827
Renfors, Markku 2611, 3175
Rengachari, Thirumalai 6186
Repo, Heikki 364
Resta, C. 1270
Ribeiro, Moises 5417
Ribeiro, Ricardo 2739
Riddle, Larry 4205
Rieger, Robert 145, 748, 2727
Riikonen, Jaana 4622
Rincon-Mora, Gabriel 4245
Rioux, Simon 4257
Roberts, Gordon W. 2200
Robucci, Ryan 5318
Rocchi, S. 2437
Rocchi, Santina 892
Rodríguez-Navarro, José J. 5210
Rodríguez-Vazquez, Ángel ... 800, 3103, 4626, 5585, 5609
Rodriguez-Villegas, Esther 728, 1398, 2156, 2160
Rogers, John 1936
Romani, Aldo 2911
Rombouts, Pieter 4070
Roopkom, Ittipat 1606
Rosehart, William 4196
Rosenbaum, Elyse 1202
Roska, Botond 5222
Roska, Tamas 5802
Ross, J. Neil 804
Rossini, A. 1266
Rossini, Andrea 5569
Rovatti, Riccardo 1489, 2349, 4349
Roy, Kaushik 1, 2020, 4963
Roy, Scott 2510
Roy, Sébastien 6018
Roy, Sourav 1626
Ruan, Qiuqi 2779
Ruan, Shanq-Jang 109, 684
Ruan, Xiaoyu 348, 4501
Rubio, Antonio 460
Rueda, Adoracion 1976
Ruegg, Michael 5453
Ruiz-Amaya, Jesús 4626, 5609

Rumin, Nicholas 2453
Russ, Christian 1194
Rydberg, Ray Robert III 1851
Ryynänen, Jussi 5059

S

Saas, Christoph 524
Saavedra, Carlos 3639, 5433
Sacco, Vincenzo 5910
Sachdev, Manoj 3684
Sadowski, D. 568
Safarian, Amin 2695
Sahinoglu, Zafer 3539
Sait, Sadiq 2377, 6226
Saito, Kazunori 5202
Saito, Tadashi 3507
Saito, Toshimichi 1927, 2064, 3411
Saiz-Vela, A. 117
Sakagami, Iwata 5047
Sakaguchi, Seishiro 3047, 3055
Sakamoto, Hiroyuki 1698, 5413
Sakamoto, Masaki 756
Sakamoto, Noriaki 5926
Sakata, Kohji 5561
Sakayori, Hiroshi 4281
Sakurai, Takayasu 3119, 4701
Salama, Aly 3011
Salama, C. Andre T. 2112
Salazar, Carlos 916
Salcedo, J. A. 416
Saleh, Res 1774
Salgado, Roberto 5294
Sällberg, Benny 852
Salles, Alain 724
Salminen, Erno 3351
Salo, Teemu 5377
Salomon, Max-Elie 4859
Saluja, Kewal 5686
Samaan, Nader 4191
Samadi, Saed 1146
Sameni, Pedram 5071
Samid, Lourans 4066
Samitier, J. 117, 141
San Sebastian, Iker 3857
San, Chi-Leung 4999
Sanada, Hirofumi 6320
Sánchez-Sinencio, Edgar 2567, 2571, 2575
Sanders, A. 924
Sangiovanni Vincentelli, A. 1286

Sangwan, Abhijeet 868
Santi, Stefano 2349
Santos, Cristiano 3567
Sanz, M. Teresa 208
Sapatnekar, Sachin 608, 3575
Saramäki, Tapio 1098, 1106, 1827, 2008, 2012, 2599
Saraswat, Dharmendra 3777
Sarmiento, Roberto 412, 3821, 6130
Sarmiento-Reyes, Arturo 3203
Sarpeshkar, Rahul 2164
Sarwate, Dilip 5838
Sasaki, Hiroshi 4729
Sasaki, Mamoru 192
Sasaoka, Naoto 272
Sato, Aya 816
Sato, Takahide 5561
Sato, Takahiro 3950
Satoh, Taiji 4733
Sattar, Farook 2863, 5822
Satzoda, Ravi Kumar 85
Saukoski, Mikko 5377, 5381
Savaci, Ferit Acar 4899
Savaria, Yvon 196, 3335, 3343, 3439, 3515, 3696, 4159, 4257, 4859, 6158
Savone, Giancarlo 1887
Sawada, Hiroshi 5882
Sawan, Mohamad 1952, 2200, 4159, 6158
Sawangsri, Teerayoot 5401
Saxena, Prashant 6230
Sayood, Khalid 5549
Scandiuzzo, M. 572
Scarana, Mirko 5266
Schellenberg, Antony 4196
Schemm, Nathan 4110, 5322
Schienle, Meinrad 2915
Schimming, Thomas 888
Schindler-Bauer, Petra 2915
Schlarmann, Mark 1968
Schlegel, Christian 2204, 5790
Schmid, Alexandre 2535
Schmidt, Bertil 4799
Schmitt-Landsiedel, Doris 3479
Schneider, Marcio C. 1413
Schoebinger, M. 924

Scholz, O. 141
Schrader, Jan H. Rutger 6090
Schrom, Gerhard 464
Schulte, Michael J. 89
Schwarz, Wolfgang 3383
Schwiegelshohn, Uwe 109
Scotti, Fabio 5142
Scotti, G. 448, 2583
Scotti, Giuseppe 1066, 3295
Scuderi, Angelo 2679
Scuderi, Antonino 2679
Seavarsson, Birgir Bjorn 4241
Secareanu, Radu M. 612
Sedighi, Behnam 2494
Sedlak, Holger 3857
Seedher, Ankit 5545
Seevinck, Evert 4138
Seguin, Fabrice 332
Seif, Nabila Philip Attalla 832
Seiyama, Tetsuya 2987
Sekar, Swaminathan 85
Seki, Yusuke 4987
Sekiya, Hiroo 720, 6292
Selim, Mohamed 3011
Senan, Sibel 4665
Senevirathna, H.M.S.B. 2506
Senger, Robert M. 624
Sengoku, Masakazu 176, 2962
Sengupta, S. 2251
Sengupta, Susanta 3869, 4823, 5942
Senior, John 6102
Senn, Eric 680
Senthilkumar, T N 268
Sentieys, Olivier 4689
Seo, Jin-Ho 1063
Seo, Jong-Wan 3910
Serbanescu, Alexandru 3813
Serdijn, Wouter 1988, 3287, 5357
Serra-Graells, Francesc 1742, 2000
Serrano, Guillermo 2148
Seth, Bhartendu 3207
Sethuraman, Ramanathan 2943
Setti, Gianluca 1489, 2349, 4349
Setton, Eric 3531
Sewell, John 1992, 1996

Sewter, Jonathan 1521
Sezaki, Kaoru 648
Shahnaz, Celia 3143
Shalabi, R. M. 1686
Shamma, Shihab 4205
Shams, Maitham 1634, 2425, 5613
Shamshiri, S. 424
Shamsi, H. 3231
Shanableh, Tamer 6062
Shanbhag, Naresh R. 636
Shao, Yu 864, 3833
Sharawi, Mohammad 6198
Sharif-Bakhtiar, Mehrdad 2494
Sharifi, Shervin 2381
Sharma, Vivek 6182, 6186
Shawker, Ali 3603
Sheikhaei, Samad 6134, 6138
Shen, Bo 928
Shen, Cheng 6042
Shen, Chi-Yuan 1642
Shen, Chun-Fu 4521
Shen, Ding-Lan 1382
Shen, Guobin 1126
Shen, Haihua 4171
Shen, Meigen 456
Shen, Meiyin 4562
Shen, Wei-Chih 3998, 6300
Shepherd, Leila 5226
Sheu, Meng-Lieh 2731
Sheu, Ming-Hwa 2393, 2879
Shi, Bertram 5230
Shi, C-J Richard 5621
Shi, Minghua 1911
Shi, Rock 4217
Shi, Yin 380, 4397
Shi, Yujie 4449
Shi, Yun Q. 1437, 4971
Shi, Zhiguo 6042
Shibagaki, Takeshi 73
Shibata, Tadashi 2389, 5365
Shibkov, A. 1154
Shida, Masaaki 4497
Shieh, Ce-Kuen 2124
Shieh, Jenn-Jong 3660
Shieh, Ming-Der 5254

Shih, Chi-Huang... 2124
Shih, Huang-Chia... 1553
Shih, Sheng-Yu... 4529
Shih, Yi-Shun... 2799
Shim, Byonghyo... 5830
Shim, Yun-A... 3704
Shimamori, Takao... 3950
Shimamura, Tetsuya... 2855, 6268
Shimizu, Kuniyasu... 3399, 6054
Shimizu, Shinsaku... 812
Shimonomura, Kazuhiro... 1915, 2771
Shin, Myong-Chul... 3910
Shin, Youngsoo... 4150, 4713
Shinoda, Shoji... 176, 2962
Shiosaka, Sadao... 3487
Shiraishi, Yumi... 5337
Shirakawa, Isao... 5345
Shirakawa, Kazuhiro... 5918
Shirataki, Jun... 2735
Shiue, Chih-Chieh... 3591
Shivappa, Shankar... 5962
Shoaei, Omid... 1409, 1565, 3107, 3231, 4265, 4618, 5593
Shojaee, Kambiz... 5234
Shoukry, Ehab... 372
Shoushun, Chen... 5306
Shu, Haiyan... 896, 2144, 4357, 4582
Shubair, Raed... 3805
Shue, Louis... 1682, 1750
Shum, Heung-Yeung... 3435
Sidahao, Nalin... 692
Siddiqi, Umair... 2377
Siek, Liter... 1525, 4249
Signell, Svante... 988, 4429
Sim, Jae-Young... 956
Sin, Sai-Weng... 1581, 1585
Singerl, Peter... 2655
Singh, Chanan... 4191
Singh, Randeep... 3207
Singh, Virendra... 5686
Singhal, Rohit... 5782
Singye, Jigme... 5126
Sips, Henk... 3523
Siskos, S... 2196
Sit, Ji-jon... 2164

Siu, Chris... 5071
Siu, Wan-Chi... 908, 2671, 2903
Sivaprakasam, Mohanasankar... 1082, 2743, 2927
Själander, Magnus... 1654
Sjöland, Henrik... 2683
Smela, Elisabeth... 3491, 3495
Smith, Jl. R... 1198
Smith, Paul... 1980
Smorfa, Simone... 5266
Snoeij, Martijn... 6162
So, Pui-Tak... 2152
Soares, André... 5834
Sobelman, Gerald... 1138, 4883
Soderstrand, Michael A... 548
Sohn, Ju-Ho... 4602
Soliman, Ahmed M... 1622
Sondeen, Jeff... 2951, 3331
Song, Jianshe... 3159
Song, Min-An... 81
Song, Seong-Jun... 740, 4763
Song, Tongyu... 2563, 2567, 2571
Song, Xiaodan... 4598
Song, Xiaoyu... 6248
Sonkusale, S... 584, 5906
Sörnmo, Leif... 1330
Sorrentino, Francesco... 3773
Soudris, D... 1206, 3347
Souza Jr., Adão... 5597
Spady, David... 3853
Spanias, Andreas... 848, 4050
Squartini, Stefano... 3603, 5134, 5714, 5742
Srikanthan, T... 3319
Srikanthan, Thambipillai... 656, 664, 1226, 1630
Srinivasan, Venkatesh... 4441
Stadius, Kari... 2687
Stan, Mircea... 1843
Stanacevic, Milutin... 4205
Starikov, David... 4767
Steele, Craig... 2951
Steiger-Garção, Adolfo... 3123
Steinmetz, O... 141
Stievano, Igor... 5754
Stine, James E... 89, 668

Stitz, Tobias Hidalgo 3175
Stiurca, Dan 2176
Stochino, Giovanni 3295
Stocker, Alan 2767
Stoica, Lucian 364
Stok, Leon 21
Stoppa, D. 572
Storace, Marco 3403, 4915
Stoyanov, Georgi 3729
Strohbehn, K. 137
Stroud, Charles 2208
Su, Borching 3035
Su, Chauchin 2643
Su, Chia-Wei 1070
Su, Chin-Hung 6244
Su, Feng 1895, 1907
Su, Hui-Kai 3825
Su, Jianing 928
Su, Jun-Min 4979
Su, Wei 1437
Su, Xinrong 3135
Su, Yan-Kuin 3003
Su, Yeping 1234
Subramanian, Shyam 996
Suchitra, S. 3319
Sudhakar babu, Chakkirala 3075
Suetsugu, Tadashi 708, 712
Sueyoshi, Tetsuya 3507
Sugahara, Kazunori 6256
Sugawara, Kensaku 4733
Sugimori, Akashi 812
Sugimoto, Yasuhiro 6194
Sugino, Nobuhiko 4855
Sugita, Hiroaki 2096
Sugita, Norihiko 5926
Sugiura, Tsuyoshi 3950
Suh, Changsu 2975
Suhail, Yasir 1342
Sukhwani, Bharat 2518
Sukthankar, S. 3319
Suleesathira, Raungrong 5146
Sumanen, Lauri 4622
Sumi, Keisuke 272
Sumi, Yasuaki 1589
Sun, Hanwu 1682

Sun, Huifang 1234, 3527
Sun, Jiaguang 6248
Sun, Ming-Fu 6030
Sun, Ming-Ting 1230, 4598
Sun, Qibin 2707
Sun, Ronghai 5706
Sun, Ruifeng 920, 1162
Sun, Shijun 1230
Sun, Tai-Hua 5015
Sun, Yan 2783
Sun, Yichuang 5958, 6102
Sun, Yi-Ran 4429
Sun, Yuanyuan 3889
Sunat, Khamron 5162
Sundaresan, Krishnakumar 5461
Sung, Guo-Ming 5573
Sung, Wonyong 1465
Sunwoo, Myung Hoon 4477
Sunwoo, Myung 2947, 3323
Surakampontorn, Wanlop 1004, 1577
Susin, A. A. 428
Susin, Altamiro 5834, 5866
Susuki, Yoshihiko 3894
Suykens, Johan A. K. 5814, 5818
Suzuki, Shunya 912
Sveinsson, Johannes 4241
Svelto, Francesco 1565
Svensson, Christer 5625
Svensson, Henrik 2867
Swaminathan, Madhavan 5758
Swamy, M.N.S. 836, 1146, 2405, 3749, 4437, 4935, 4943
Swamy, Ramkrishna 4513
Sylvester, Dennis 604, 3563
Szolgay, Péter 5802

T

Taal, Jacco 3523
Tabari, Karima 4767
Tadeparthy, Preetam 820, 5545
Taft, Stephanie 420
Tagawa, Kiyoharu 3191
Taguchi, Akira 4538, 6264
Tahara, Toshimitsu 4737
Taherzadeh-Sani, Mohammad 4618, 6218
Tahoori, Mehdi B. 2991
Tai, Cheng-Lun 3994
Tajalli, Armin 2255, 5031
Takafuji, Daisuke 172, 1362
Takagi, Shigetaka 5561
Takahashi, Hiroshi 2987
Takahashi, Norikazu 3631, 4673
Takahashi, Yasuhiro 1445
Takahashi, Yoshitake 6268
Takala, Jarmo 1122
Takama, Yasuhiro 3894
Takamatsu, Yuzo 2987
Takami, Ryotaro 2771
Takata, Hidehiro 73
Takemura, Hiroshi 5341
Takezawa, Megumi 6320
Takizawa, Yumi 360
Taleie, Shahin Mehdizad 4401
Tam, Kam C. 2490
Tam, Kam-Weng 404, 3817
Tamaki, Saneaki 5349
Tamesada, Takeomi 2995
Tamtrakarn, Atit 3119
Tamura, Hiroshi 176
Tan, H. Y. 2385
Tan, Meng Tong 224, 1326
Tan, S. L. 2385
Tan, Yanzhuo 5666
Tan, Yap-Peng 904, 1238, 1549, 3797, 3837, 3845, 5485
Tanaka, Aya 2064
Tanaka, Hideki 5926
Tanaka, Hiroto 3379
Tanaka, Shigeto 6194
Tanaka, Tomoyuki 812
Tanaka, Y. 1940
Tang, Hua 1282, 5629
Tang, Jun 65
Tang, Rui 2527
Tang, Wai-Chung 5653
Tang, Wallace K. S. 6038
Tanigawa, Kazuya 3507
Taniguchi, Kenji 812
Tanji, Kouki 3950
Tanji, Yuichi 5762
Tao, Ji 3797
Tao, Liang 844
Taoka, Satoshi 752, 2231
Tarbell, Mark 2927
Tarczynski, Andrzej 1425
Tarim, Tuna 2775, 5974
Tartagni, Marco 2911
Tasaki, Futoshi 176
Tatschl-Unterberger, Eva 5453
Taubman, David 2076, 4361, 4373
Tavares, Gonçalo 2739
Tawfik, Shadi 5083
Tay, David 2303
Taylor, John 145, 748, 2727
Tayu, Satoshi 180
Tecpanecatl-Xihuitl, J.Luis 552
Teixeira, Thiago 644
Tekalp, Murat 3543
Temes, Gabor 6182, 6186
Tenhunen, Hannu 456, 3982
Tenore, Francesco 4106
Teo, Kok Ann Donny 6014
Teplechuk, Mykhaylo 1992, 1996
Terés, L. 1742
Terés, Lluis 5361
Terreni, P. 1286
Tetzlaff, Ronald 5218
Teva, Jordi 4209
Thanailakis, Antonios 3347
Theis, Fabian 5878
Theodoridis, G. 1206
Thepayasuwan, Nattawut 1044
Thepvilojanapong, Niwat 648
Therrien, Charles 2307, 6010
Theuwissen, Albert 6162

Thewes, Roland 2915, 3479
Thiran, Patrick 2361
Thoka, Sreenath 5373
Thomas, Mikkel A. 1738
Thomas, Olivier 4094
Thomson, Kyle E. 1342
Thongkamwitoon, Thirapiroon 3785
Thul, Michael 1766
Thulasiramn, Krishnaiyan 164
Tian, Dong 2899, 6074
Tian, Hongbo 5738
Tian, Junhua 928
Tiew, Kei Tee 5589, 3680
Tiow, Tay Teng 5270
Titti, Alessio 5134
Tiuraniemi, Sakari 364
Tiwari, Sandip 2939
Tjitrosoewarno, Cindy Bernadeth 360
Tjukanoff, Esa 456
Toba, Yoshikazu 1517
Tobajas, Félix 412, 3821
Tobe, Yoshito 648
Togawa, Nozomu 3499
Togo, Mitsuhiro 13
Tökés, Szabolcs 5802
Tokuda, Naoyuki 3635
Tokuda, Takashi 2923, 3487
Tomabechi, Nobuhiro 1786
Tomassoni, Massimo 5742
Tommasino, Pasquale 3583
Tong, Dong 660
Tong, Paul C. F. 1190
Tong, Shiqiong 4743
Tongprasit, Benjamas 2389
Toprak, Zeynep 1059
Torelli, G. 121, 1266, 1270
Torelli, Guido 1891, 1903
Torikai, Hiroyuki 1927, 3411
Torkzadeh, Pooya 2255, 5031
Torralba, Antonio 1008, 5581
Tortosa, Ramón 3103, 5585
Totev, Emil 1597
Tóth, László 4221
Toumazou, Christofer 57, 5226
Tourapis, Alexis 1246, 4365
Tousaad, Jawad 5597
Toyonaga, Masahiko 6206
Trajkovic, Ljiljana 2060, 2068
Tran, Canh 4701
Tran, Trac D. 916
Triantis, Iasonas F. 744
Trifiletti, A. 448, 2583
Trifiletti, Alessandro 1066, 3295, 3547, 3583
Trindade, Rogério M. 5417
Trisanto, Agus 3051
Troedsson, Niklas 2683
Tsai, Chen-Yu 6098
Tsai, Chia-Chun 1040, 6170
Tsai, Chia-Ming 5501
Tsai, Chia-Sheng 1859
Tsai, Chin-Chung 560
Tsai, Chuan-Yung 1790, 2931
Tsai, Chun-Jen 2132, 2907, 3271, 6066
Tsai, Hwa-Long 2333
Tsai, Jinn-Tsong 2603
Tsai, Kun-Lin 684
Tsai, Meng-guang 328
Tsai, Meng-Ting 5091
Tsai, Ming-Yu 2433
Tsai, Tsung-Han 1032, 2851, 3155, 3223, 4590
Tsai, Yan-Chr 2333
Tsao, Hen-Wai 328, 368
Tschanz, James 9, 592
Tse, Chi K. 716, 1469, 2477, 2481, 2490, 4445, 6050
Tse, K. W. 4317
Tseng, Belle L. 1250
Tseng, Chao- Hsuing 2128
Tseng, Chien-Cheng 824, 828, 3713, 3717
Tseng, Chien-Hsun 6002
Tseng, Chien-Tang 2132
Tseng, Der-Feng 4505
Tseng, Po-Chih 5190
Tseng, Wei-Hsiang 5794
Tseng, Yih-Long 356
Tseng, Yi-Hung 4381
Tseng, Yu-Chee 640
Tsui, Chi-Ying 1895, 1907
Tsui, Chi-ying 3359

Tsui, K. M. 2607, 4297
Tsuji, Kohkichi 760
Tsuji, Shigeki 6058
Tsuji, Toshio 3047, 3055
Tsukamoto, Yasumasa 73
Tsukimori, Akifumi 5349
Tsukutani, Takao 1589
Tsunokawa, Shingo 4747
Tsutsui, Hiroshi 2096
Tu, Shang-Wei 4134
Tu, Shu-Hui 804
Tubío, Óscar 3821
Tufail, Muhammad 3027
Tulunay, Gülin 1294
Twigg, Christopher 468, 5553, 6106

U

U, Seng-Pan 392, 1581, 1585, 3095, 3099
Uchida, Jumpei 3499
Ueda, Toru 5333
Ueno, Shuichi 180, 1358
Ueta, Tetsushi 6058
Ueyama, Teppei 4729
Unbehauen, Rolf 2559
Upadhyaya, Parag 5115
Urahama, Kiichi 6308
Uranga, Arantxa 4209, 5361
Urdaneta, Mario 3491, 3495
Ushio, Toshimitsu 3379
Uwate, Yoko 1481
Uzun, Isa 5826
Uzunov, Ivan 3729

V

V. T, Anuroop 5545
Vachoux, Alain 5166
Vagliasindi, Guido 4102, 5818
Vahdat, Bijan 3571
Vahedi, Haleh 436
Vaidyanathan, Palghat P. 3035
Vainio, Olli 188, 2012
Valkama, Mikko 2611
Van Calster, André 1855
Van Camp, Benjamin 1194
van der Schaar, Mihaela 3543
van Hartingsveldt, Koen 1610, 5353, 5357
van Meerbergen, Jef 2943
van Roermund, A.H.M. 788
van Roermund, Arthur 1956, 1964, 4062, 4839, 5541
van Schaik, André 4213
van Wingerden, Johannes 3551
Van, Lan-Da 81
Vandewalle, Joos 5814
Vanini, Giovanni 652, 5509
Vanne, Jarno 1134
Varela, Elísio 2739
Vasilescu, G. 2337
Vaz, Bruno 3123, 4074
Vázquez-Leal, Héctor 3203
Vecchi, Davide 1386
Veendrick, Harry 4138
Veerachary, Mummadi 3075
Vehkaperä, Janne 3423
Vemuri, Ranga 5978
Verd, Jaume 4209
Verhaege, Koen 1194
Verhoeven, Chris 1597, 1610, 5353
Vermandel, Miguel 1855
Ververidis, Dimitrios 2871
Vesalainen, Laura 3930
Vetro, Anthony 952, 1234, 3527
Viarani, Nicola 5902
Vidal, Eva 3926, 4114, 4457
Vidyarthi, Arvind 3579
Vigna, Andrea 388, 776
Vignoli, V. 2437
Vignoli, Valerio 892
Viitanen, Jouko 1758
Vilariño, D.L. 3922, 5798
Villar, Arturo 412
Villar, G. 2473
Villar, Gerard 3926, 4114, 4453, 4457
Vinod, A.P. 352, 496, 4799, 5429
Violas, Manuel 1166
Virtanen, Kati 3115, 6272
Visalli, Giuseppe 5206
Visschers, Jan L. 6090
Vitali, Stefano 2108
Vitkowski, Arseni 1770
Vogel, Christian 1394
Vogelstein, R. Jacob 1919, 4106
Vouzis, Panayiotis 3954
Vucic, Mladen 3283
Vytyaz, Igor 5986

W

Waho, Takao... 2341
Wakabayashi, Kazutoshi... 5930
Waldhauser, Dirk S. ... 524
Walus, Konrad ... 2522
Wan, Yuanzhong ... 5613
Wang, Albert ... 4807, 6174
Wang, Baohua ... 1948
Wang, Bao-Yun... 5698
Wang, Bei ... 3837
Wang, Chao-Shiun... 3974
Wang, Chien-Shun... 6288
Wang, Chin-Hui... 6234
Wang, Chorng-Kuang .. 368, 3974, 4895, 5794
Wang, Chua-Chin ... 356
Wang, Chunyan ... 1461, 3127, 3753, 4759
Wang, D. ... 960
Wang, Dong ... 2719, 4058
Wang, Feng-Hsing ... 4018, 4975
Wang, Guoxing ... 2743
Wang, Hung-Ming ... 2128
Wang, J. Y. ... 2124
Wang, Janet... 2457, 2465, 2518
Wang, Jhing-Fa... 6288
Wang, Jia-Shung ... 3419
Wang, Jinn-Shyan... 1258, 1670, 4517
Wang, Lei... 1658
Wang, Ling... 688
Wang, Min... 3619, 3946
Wang, Mingzhen... 6142
Wang, N. Y. ... 1441
Wang, Nai-Chung ... 5258
Wang, Peng ... 964, 2188
Wang, Ping-Ying ... 5007
Wang, Qiang... 3467
Wang, Ruiqi ... 4791
Wang, Sen ... 2325
Wang, Sheng-Zen... 1810, 4558
Wang, Shoujun ... 1158
Wang, Shuenn-Shyang... 1118
Wang, Sying-Jyan... 1638
Wang, Teng-Yi ... 6118
Wang, Ting-Chi... 1879
Wang, Tzu-Ya... 1370
Wang, Wei... 2522, 4649, 5449
Wang, Wenbo ... 1714, 4026, 5158
Wang, Wenqia... 2779
Wang, Xiao Fan ... 3395
Wang, Xiaofan... 280
Wang, Xiaomeng... 2048
Wang, Yanjie... 1601
Wang, Yanxin... 3243, 5067
Wang, Yen-I... 1055
Wang, Yi... 1226
Wang, Yibo... 93
Wang, Yi-Chiuan ... 6026
Wang, Yi-Ming... 1258
Wang, Yi-Wen ... 6214
Wang, Yongtao ... 2020, 4963
Wang, Yueh-Yi ... 2907
Wang, Yunfeng ... 5641
Wang, Zhi... 1734
Wang, Zhigong... 340
Wang, Zhihua... 3700, 4995, 5023
Wang, Zhiliang ... 3387
Wang, Zhongfeng... 5786
Wanhammar, Lars... 1449
Washburn, Clyde... 4385, 5103
Watanabe, Kazuhisa ... 6320
Watanabe, Kazuki ... 632
Watanabe, Minoru... 1214
Watanabe, Osamu ... 6324
Watanabe, Takayuki... 4903, 5774
Watanabe, Toshimasa ... 172, 752, 1362, 2231
Weder, Uwe ... 3857
Wedi, Thomas ... 324
Wee, Keng Hoong ... 2164
Wehn, Norbert... 1766
Wei, Cao ... 1794
Wei, Foo Say... 4409
Wei, Gao ... 5035
Wei, Shengfang... 5750
Wei, Xinjie ... 101
Weiland, James... 2743
Welch, Brian... 4891
Wen, Chia-Sheng ... 2433
Wen, Kuei-Ann ... 5063
Wen, Min-Chen ... 1638
Wen, Yuan ... 3139

Weng, Ching-Chih ... 3801
Weng, Ping-Kuo ... 2393
Wennekers, Peter ... 3087
Westra, Ronald ... 3287
Wey, Chin-Long ... 3327
Wey, I-Chyn ... 1074, 5449
White, Neil ... 5389
Wichard, J. ... 4657
Wicheanchote, Phinyo ... 5162
Wijaya, Tandi ... 4461
Wilcock, Reuben ... 2815, 5174
Wild, Guillaume ... 3696
Willms, John ... 1610
Willson, Jr., Alan N. ... 672
Wilson, Denise M. ... 1762
Wilson, Denise ... 2747, 3483, 5914
Wilson, Peter ... 2815, 5174
Wilton, Steven ... 45
Winstead, Chris ... 2204, 5790
Wittmann, Steffen ... 324
Wohlmuth, Hans-Dieter ... 3227
Wolf, Wayne ... 1778
Wong, Alex ... 2723
Wong, Hoi-Ming ... 5477
Wong, Ka-Man ... 1541
Wong, Siu Chung ... 716, 2490
Wong, Stephen ... 4225, 4229, 4787
Wong, Yanyi Liu ... 5314, 6146
Wongsawat, Yodchanan ... 5990
Worapishet, Apisak ... 1606, 5954
Worm, Frédéric ... 2361
Wu, Allen C.-H. ... 3311
Wu, An-Yeu ... 1032, 1074, 5449, 6026
Wu, Chai Wah ... 292, 5521, 6046
Wu, Chang-Ching ... 5063
Wu, Chen-Lung ... 1070
Wu, Chia-Hsin ... 368
Wu, Chia-Ju ... 3039
Wu, Chia-Tsun ... 5449
Wu, Chia-Wei ... 1350
Wu, Chia-Wei ... 1366
Wu, Chien-Hsing ... 2333
Wu, Chien-Ming ... 5254
Wu, Chi-Hao ... 2711
Wu, Chung-Yu ... 5079
Wu, Feng ... 6078
Wu, Ho-Ting ... 1533
Wu, Hsien-Huang ... 2393, 2879
Wu, Ja-Ling ... 2711, 3219, 3829, 4594
Wu, Jieh-Tsorng ... 1374
Wu, Jun ... 516
Wu, K. F. ... 1198
Wu, Meng ... 2627
Wu, Meng-Chiou ... 4725
Wu, Min ... 3527
Wu, Ping-Hao ... 1497
Wu, Po-Han ... 2979
Wu, Qiong ... 4807, 6174
Wu, Ruei-Cheng ... 152, 156
Wu, Shu-Meng ... 6126
Wu, Shunjun ... 3946
Wu, Tian-Luu ... 3455
Wu, Ting ... 3986
Wu, Wen-Rong ... 2619, 3179
Wu, Xiaolong ... 1746
Wu, Xiaoqun ... 2477
Wu, Y. H. ... 1198
Wu, Ying Yih ... 2393
Wu, Yu-liang ... 1210
Wu, Yu-Liang ... 5653
Wu, Zong-Ze ... 6312
Wuen, Wen-Shen ... 5063
Wunderlich, Hans-Joachim ... 2935
Wunderlich, Richard ... 2441
Wybo, Geert ... 1194

X

Xiao, Chi ... 3889
Xiao, Jie ... 340
Xiao, Shiyuan ... 3251
Xiao, Yang ... 3171
Xiao, Yegui ... 260, 2044
Xiao, Ying ... 164
Xiao, You-Neng ... 1238
Xiaolang, Yan ... 492, 6316
Xiaoping, Zhu ... 5270
Xie, Dahua ... 5533
Xie, Xiang ... 4995
Xie, Xudong ... 3841
Xin, Jun ... 1234
Xin, Yan ... 4959
Xing, Hanqing ... 3809, 4289, 4835
Xing, Qin ... 492, 6316
Xing, Xianwu ... 5645
Xotta, Andrea ... 2551
Xu, Fei ... 1823
Xu, Jian ... 3395
Xu, Jiang ... 1778
Xu, Jingyu ... 2449
Xu, Junjuan ... 696
Xu, Li ... 260, 2044
Xu, Ping-Ping ... 1190
Xu, Qing ... 3817
Xu, Rongtao ... 4955, 4967
Xu, Xiaoyin ... 4225
Xu, Yong Ping ... 5087
Xue, Guoliang ... 164
Xue, Ping ... 2048
Xue, Robert ... 3817
Xue, Xiangyang ... 1238, 1549, 3837

Y

Yadid-Pecht, Orly... 568, 580, 588, 5310, 5326
Yagi, Masakazu 2831
Yagi, Tetsuya 1915, 2771
Yahagi, Takashi 720, 2783, 3051, 3211, 4851, 4919, 5130, 5726, 6292
Yalcin, Mustak E. 5818, 5814
Yamada, Isao... 2413
Yamada, Toshinori................................... 1358
Yamagata, Yasushi 13
Yamaguchi, Takahiro 5918
Yamaguchi, Tomohiko 2839
Yamamoto, Heiichi................................... 5337
Yamamoto, Takahiko 6256
Yamamoto, Tetsuya.................................. 2839
Yamamura, Kiyotaka..................... 3761, 4911
Yamasaki, Toshihiko................................ 5365
Yamashita, Noritaka................................. 6292
Yamashita, Takayuki................................. 1517
Yamashitha, Katsumi 2506
Yamazaki, Koji 2987
Yan, Jiangnan 396
Yan, Jin-Tai................ 1350, 1366, 1370, 2219
Yan, Shixing 484, 6202
Yan, Shouli.......1960, 2563, 2567, 2571, 2575, 3091
Yan, Tan 5678
Yan, Wei-Qi.. 5525
Yan, Zhiyuan... 5838
Yanagisawa, Masao.................................. 3499
Yanamanamanda, Satish......................... 2457
Yang, Chenyang 4649
Yang, Chih-Chyau... 81
Yang, Ching-Yuan............................ 1150, 5091
Yang, Chun-Yueh 3131
Yang, Dayu 2208
Yang, Ge... 5493
Yang, Hannal 6230
Yang, Huijuan 4002
Yang, Jar-Ferr............................... 1802, 2128
Yang, Jeong-Hyu 944
Yang, Jingbo.. 1326
Yang, Jun... 3139
Yang, Jung Mo... 3323
Yang, Kyounghoon................................... 2531
Yang, Linlin 5158
Yang, Li-Wu 4807
Yang, Luxi 5409
Yang, Meng-Ta 5007
Yang, Sheng 948
Yang, Shi-Qiang... 964
Yang, Shiyuan.. 2188
Yang, Shuyuan............................... 3619, 3946
Yang, Shyue-Wen 2393
Yang, Tung-Yu... 2879
Yang, Wei-Bin 1174
Yang, Wenxian... 6078
Yang, Xiaokang 3793, 4939
Yang, Y. 4649
Yang, Yang 6304
Yang, Yeong-Yil 4377
Yang, Yun 1130
Yang, Zheng-Zhang 3643, 3647
Yang, Zhi.. 5473
Yao, Chia-Yu.. 512
Yao, Ji... 876, 2052
Yao, Minli 3159
Yao, Ning 3737
Yao, Susu........................... 3793, 4233, 4939
Yao, Yuan 380
Yarman, B. Siddik.................................... 1334
Yasser, Muhammad 5130
Yasuda, Takeo 1024
Yasuda, Yuri... 13
Yasunaga, Akiyoshi................................... 452
Yat-Fong, Yung 1754
Yau, Andy.. 6296
Yaung, Ming-Feng................................... 5674
Yavari, Mohammad 1565, 5593
Ye, Shuiming... 2707
Ye, Yang 492, 6316
Ye, Yibin 592
Ye, Yizheng... 384
Ye, Yi-Zheng 5850
Ye, Yun.. 5718
Yeap, Tet... 4951
Yedidia, Jonathan 3527
Yeh, Chia-Nan... 1875
Yeh, Chingwei 1670

Yeh, Han-Ching	2136
Yeh, Jen-Hao	4594
Yen, Jieh-Hwang	1642
Yen, Jui-Cheng	6126
Yeo, K. S.	2429, 5027
Yeo, Kiat Seng	376, 4815
Yeung, Nang-Ching	3279
Yi, Xiaoquan	3443, 5489
Yin, Peng	1246, 4365
Yin, Qinye	3019, 3615, 5706, 5710, 5738
Yin, S. S.	1094
Yin, Shishu	4313
Yin, Yi	3087
Yip, Shu-Kei	4991
Yiu, Mimi	2204
Yli-Kaakinen, Juha	1827
Yokota, Arimitsu	564
Yokoyama, Michio	1445
Yoneyama, Akio	4562
Yoo, Changsik	4803
Yoo, Hoi-Jun	740, 2357, 2369, 4602, 4763
Yoo, Seong-Moo	4653
Yoon, Suk H.	2947
Yorino, Naoto	4729, 5290
Yoshida, Takashi	564, 192
Yoshimura, Takeshi	628
Yoshimura, Takuya	6054
Yoshioka, Shinichi	5349
Yoshizawa, Shingo	25
Yotsuyanagi, Hiroyuki	2995
Young, Ian	2919
Yousefzadeh, V.	1302
Yu, Chang-Hyo	4574
Yu, Chi-Yao	5079
Yu, Hao	105
Yu, Hongtao	312
Yu, Jianghong	2016
Yu, Jiunn-Der	77
Yu, Jui-Yuan	6030
Yu, Keman	4927
Yu, Lu	316
Yu, Mingyan	384
Yu, Shen-Hau	3111
Yu, Simin	1473, 3391
Yu, X. P.	5027
Yu, Xiaoli	1730
Yu, Xuefeng	4397
Yu, Ya Jun	1819
Yu, Ya-Hui	6066
Yu, Zaihe	1437
Yu, Zhongjun	792, 6190
Yuan, Jiren	6166
Yuan, Shuai	6264
Yuan, Xue	3211
Yue, C. Patrick	920, 1162

Z

Zahabi, Ali ... 3107, 3231
Zaidi, Ali ... 6226
Zanikopoulos, Athon ... 4062, 4839, 5541
Zarándy, Ákos ... 5802
Zare-Hoseini, Hashem ... 220, 1409, 2547
Zarei, Hossein ... 2100
Zella, D. ... 1270
Zemouche, Ali ... 3195
Zeng, Shengke ... 576
Zeng, Wei ... 3459
Zeng, Xuan ... 113
Zeng, Yanxing ... 5710
Zhai, Guisheng ... 2823, 3183
Zhai, Jiefu ... 4927
Zhan, Cheng ... 1048
Zhan, Jian-Ting ... 3111
Zhan, Jing-Hong ... 4891
Zhan, Rouying ... 4807
Zhang, Chengjun ... 1461
Zhang, Chun ... 1995
Zhang, Ci-Xun ... 316
Zhang, Fengming ... 2527
Zhang, Ge ... 676
Zhang, Hao ... 2072
Zhang, Haobin ... 5035
Zhang, Heng ... 4171
Zhang, Hua ... 2263
Zhang, Huaguang ... 3387, 5690
Zhang, Hui ... 1262, 2060, 5629
Zhang, J. ... 2243
Zhang, Jianguo ... 3615
Zhang, Jianmin ... 1750
Zhang, Jianzhong ... 2263
Zhang, Jing ... 6312
Zhang, Jinsuo ... 1746
Zhang, Jun ... 340
Zhang, Ming ... 636
Zhang, Nuo ... 5726
Zhang, Qianling ... 928
Zhang, Qingxiang ... 4249
Zhang, Rumi ... 2522
Zhang, Tong ... 4959
Zhang, Wenjing ... 588
Zhang, Xiaolin ... 4649, 4783
Zhang, Xiaowei ... 5726
Zhang, Xing ... 4026
Zhang, Yan ... 1843
Zhang, Yaxiong ... 1028
Zhang, Ying ... 5710
Zhang, Yiqian ... 97
Zhang, Yiwen ... 3615, 5706
Zhang, Yu ... 688
Zhang, Yuan-Ting ... 2723
Zhang, Zhiguo ... 1722
Zhang, Zhiguo ... 4317
Zhang, Zhongkai ... 4325
Zhang, Zhurun ... 2591
Zhao, Debin ... 3467
Zhao, Hui ... 1714
Zhao, Jichuan ... 5894
Zhao, Qing ... 3902
Zhao, Wenqing ... 1130
Zhao, Xueqin ... 4919
Zhao, Yan ... 2899
Zhao, Yao ... 4038
Zhao, Ying ... 3171
Zhao, Yinqing ... 1545
Zhao, Yutian ... 41
Zhao, Zixue ... 2587
Zheng, Hui ... 3902
Zheng, Jinghong ... 4006
Zheng, Kan ... 1714
Zheng, Li-Rong ... 456, 3982
Zheng, Nan-Ning ... 6312
Zheng, Wei Xing ... 1433, 1694, 5698, 6038
Zheng, Wei ... 5186
Zheng, You ... 3639, 5433
Zheng, Yuanjin ... 396
Zheng, Zengwei ... 6022
Zhi Gang, Mao ... 1794
Zhiwei, Lin ... 5421
Zhong-hai, Wang ... 5850
Zhou, B. ... 2469
Zhou, Bo ... 5246, 6022
Zhou, Dian ... 113, 2263
Zhou, Dongfang ... 4449
Zhou, Hai ... 2461
Zhou, Hao ... 2623

Zhou, Jiong	2587
Zhou, Lei	5087
Zhou, Lili	5621
Zhou, Mingcui	1082
Zhou, Qiang	1871, 6230
Zhou, Tracey	2775
Zhou, Xiangdong	4811
Zhou, Xiao	160
Zhou, Xiaobo	4225, 4787
Zhou, Xiaosong	3263
Zhou, Y.	2631
Zhou, Yi	4333
Zhou, Yufei	2481
Zhou, Zhe	1210
Zhou, Zhi	1230
Zhu, Bin	2703, 6304
Zhu, Ce	4038
Zhu, Feng	2575
Zhu, Jinmin	4225
Zhu, Ronghua	4397
Zhu, W.P.	868
Zhu, Wei-Ping	2627, 3143, 3147, 3749
Zhu, Xiaoqing	3531
Zhu, Xiqun	5397
Zhu, Yue-Xin	6312
Zimmer, Heiko	37
Zipf, Peter	3970
Zoka, Yoshifumi	4729, 5290
Zou, Dekun	4971
Zou, Qiyue	1338, 3741, 3769
Zou, YuanZhi	3427
Zwolinski, Mark	5182

Session Chair Index

A

Abe, Masahide	35
Alarcon, Eduard	6, 55, 79
Alippi, Cesare	222
Ampadu, Paul	196
Arakawa, Kaoru	7
Arena, Paolo	187
Arik, Sabri	157
Arslan, Tughrul	188
Asai, Hideki	120, 197, 231
Au, Oscar	39, 60, 114, 251

B

Badawy, Wael	209, 232
Baglio, Salvatore	168
Barretino, Diego	214
Barry, Mark	134
Bates, Stephan	15
Bayoumi, Magdy	38
Bian, Jinian	228
Biel, Domingo	123

C

Caruson, Anthony Chan	61
Chakrabarti, Chaitali	26
Chakrabartty, Shantanu	99, 122, 147, 248
Chan Carusone, Tony	125, 239
Chan, S. C.	195
Chang, Joseph	56, 77
Chang, Robert C.	188
Chang, Tian-Sheuan	221
Chantraporn, Chantana	163
Chau, Lap-Pui	244
Chen, Chang Wen	198
Chen, Guanrong	13
Chen, John R.	160, 207
Chen, Liang-Gee	71
Chen, Sau-Gee	27, 186
Chen, Yung-Chang	83
Chiang, Tihao	137, 169
Chiueh, Tzi-Dar	219
Chrzanowska-Jeske, Malgorzata	50, 96
Chung, Pau-Choo	62
Cichocki, Andrzej	121, 236
Constantinides, George	28
Coussy, Philippe	249
Czarkowski, Dariusz	52

D

de la Rosa, Jose	185
De Micheli, Giovanni	70
Delgado-Restituto, Manuel	162, 185
Demosthenous, Andreas	238
Desai, U. B.	45
Di Bernardo, Mario	128
Ding, Heping	35
Diniz, Paulo	150

E

El-Gamal, Mourad 40, 176, 216
Endo, Tetsuro 243
Etienne-Cummings, Ralph . 110, 116, 139, 217

F

Fang, Wai-Chi 85, 177
Filanovsky, Igor 11, 63, 196
Foo, Say Wei 12
Friedman, Eby 5, 24, 50, 165
Fujii, Kensaku 12
Fujisaki, Hiroshi 36
Fukui, Masahiro 74
Funaki, Tsuyoshi 156

G

Galias, Zbigniew 197
Genov, Roman 99, 145
Gilli, Marco 151, 164
Goldgeisser, Leonid 208
Gustafsson, Oscar 72
Gwee, Bah Hwee 44

H

Ha, Dong S. 133
Hajj, Ibrafim 54
Hamada, Nozomu 103
Hamamoto, Takayuki 191
Harjani, Ramesh 124, 193
He, Yuwen 14
Helfenstein, Markus 32, 87, 194
Heo, Deuk 84
Hernandez, Luis 101, 147
Hisakado, Takashi 113
Ho, Dominic K. C. 173
Hong, Xianlong 250
Huang, Garng 167
Huang, Hsiang-Cheh 177
Huang, Jiwu 200
Huang, Xinping 104
Huang, Yih-Fang 57
Hyogo, Akira 33, 178

I

Ibaragi, Eitake	41
Ibrahim, Mohammad	94
Ikehara, Masaaki	149, 172
Ikenaga, Takeshi	184
Ikuta, Akira	195
Imamura, Kousuke	39
Indiveri, Giacomo	76
Inouye, Yujiro	229, 236
Ioinovici, Adrian	98
Ismail, Mohammed	32
Ismail, Yehea	25
Itoh, Yoshio	173
Ivanov, Andre	70
Ivanov, Vadim	102, 155

J

Jamali, Mohsin	119
Jitsumatsu, Yutaka	112
Jou, Jing-Yang	142
Jou, Shyh-Jye	73, 211
Jullien, Graham A.	27
Junji, Jshikawa	20, 48

K

Kajitani, Yoji 54
Kamada, Masaru 219
Kambe, Takashi 201
Kaneko, Mineo 28
Kang, Steve 107
Kawahito, Shoji 226
Kawamata, Masayuki 80
Kazimierczuk, Marian 29, 146
Ker, Ming-Dou 47
Ki, Wing Hung 75
Kiaei, Sayfe 176
Kim, Yong 189
Kiya, Hitoshi 192
Kocarev, Ljupco 82, 93
Kohda, Tohru 59
Kok, C. W. 22
Kolodny, Avinoam 24
Kondo, Katsuya 53, 154
Kornegay, Kevin 107
Kot, Alex Chichung 198, 223
Kousaka, Takuji 89
Krishnan, Shoba 204
Kumar, Ashok 46, 180
Kuo, C.-C. Jay 83
Kuroda, Ichiro 108, 246
Kwan, H. K. 103

L

Lach, John	119
Lai, Jui-Lin	108
Lande, Tor Sverre	30
Lau, Francis	220
Lawrance, Tony	36
Lee, Chang-Ho	202
Lee, Chen-Yi	2, 71
Lee, Gwo Giun	221, 253
Lee, Hsien-Hsin	227
Leou, Jin-Jang	129
Leung, Shu-Hung	245
Li, Shipeng	106, 252
Lian, Yong	21, 53, 72
Licks, Vinicius	245
Lim, Y. C.	80
Lin, C. T.	100, 206
Lin, Chia-Wen	14, 49
Lin, Yuan-Pei	67
Lin, Zhiping	91, 95
Linan Cembrano, Gustavo	233
Ling, Nam	60, 153
Liu, Bin-Da	234
Liu, Derong	38, 122, 144
Lu, Jianming	57
Lustenberger, Felix	9, 78

M

Ma, Dongsheng 120
Ma, Siwei 138, 175
Madden, Patrick 74
Maggio, Gian Mario 3, 93, 105
Maharatna, Koushik 222
Makino, Shoji 144
Maloberti, Franco 23
Margala, Martin 4
Mase, Kenichi 118
Masuda, Hiroo 5
Matsunaga, Yusuke 143
Meiling, Janet 51
Moon, Un-Ku 55, 78
Mori, Hiroyuki 190
Mori, Shinsaku 75, 98
Mukund, P.R. 130, 171, 193
Muneyasu, Mitsuji 34, 237
Muto, Cosy 125, 239

N

Nagata, Makoto 51
Nakamura, Yuichi 182
Nakanishi, Isao 168, 218
Nakasako, Noboru 121, 230
Nakashizuka, Makoto................................ 200
Natarajan, Sreedhar.......................... 142, 235
Ndjountche, Tertulien........................ 124, 225
Newcomb, Robert 109, 237
Ng, T.S... 199
Nguyent, Truong 172
Niamat, Mohammad................................... 140
Nishimura, Shotaro 150
Nishio, Yoshifumi 105
Nishitani, Takao 131, 183
Nowrouzian, Behrouz................................. 218

O

Ochi, Hiroshi 199
Ochi, Hiroyuki 189
Ogunfunmi, Tokunbo 16
Ohta, Jun 30
Ohta, Yuzo 128
Okazaki, Hideaki 242
Onoye, Takao 238
Oraintara, Soontorn 149

P

Paasio, Ari	187
Palumbo, Gaetano	10, 25, 79, 212
Pan, Jeng-Shyang	62
Parhi, Keshab K.	117
Park, Jung-Wook	213
Pessolano, Francesco	1
Phang, Khoman	33
Phoong, See-May	127
Pineda de Gyvez, Jose	1
Ping, Xue	154
Prasad, Vinod A.	16

R

Rahardja, Susanto 175
Ramachandran, Ravi Prakas 126
Ramirez-Angulo, Jaime 56
Rana, Ram Singh 215
Renfors, Markku 68
Rincón-Mora, Gabriel 170
Rizzo, Allesandro 145
Roberts, Gordon 240
Rodríguez-Vázquez, Angel 162

S

Sahula, Vineet	97
Saito, Toshimichi	82
Salama, Khaled	116, 139
Saluja, Kewal	143
Sato, Takahide	11
Sawan, Mohamad	7, 19, 235
Secareanu, Radu M.	42
Sekine, Ketaro	224
Sekiya, Hiroo	52, 104, 127, 146, 179
Sengoku, Masakazu	31
Serdijn, Wouter	64, 216
Setti, Gianluca	174
Sheu, Bing	117
Shi, Yun Q.	223
Shibata, Tadashi	86, 148, 191, 217
Shimamura, Tetsuya	126
Singh, Jugdutt Jack	211
Siu, W. C.	34
Sobelman, Gerald E.	73, 96, 158
Sonkusale, Sameer	9, 101, 247
Soudan, Bassel	43, 212
Soudris, Dimitrios	4
Spanias, Andreas	26
Sridhar, Ramalingam	165
Stan, Mircea R.	66, 163
Stocker, Alan	76, 110
Stouraitis, Thanos	140
Suetsugu, Tadashi	6, 29, 123
Sugino, Nobuhiko	58
Sumi, Yasuaki	214
Sun, Yichuang	132
Sunwoo, Myung H.	48, 84, 94
Swamy, M. N. S.	31, 81

T

Taguchi, Akira	241
Takahashi, Atsushi	97
Takai, Nobukazu	77, 102
Takeuchi, Yoshinori	215
Tan, Yap-Peng	49, 152
Tanaka, Hiroto	135
Tanaka, Mamoru	164
Tao, Liang	241
Tarim, Tuna	171, 240
Tasic, Aleksandar	148
Tetzlaff, Ronald	210
Torikai, Hiroyuki	136
Tsai, Chun-Jen	115, 244
Tsai, Tsung-Han	58, 129
Tse, Michael	59
Tsukada, Toshiro	10, 170

U

Uchiyama, Kunio 92, 100
Ueno, Shuichi 88
Ueta, Tetsushi 174

V

Velev, Miroslav 153, 166
Vetro, Anthony 37, 141

W

Wada, Naoya 61
Waheed, Khurram 2, 186
Wang, Albert 194
Wang, Guoxing 111
Wang, Jhing-Fa 152
Wang, Xiaofan 13
Wang, Yuke 65
Wanhammar, Lars 18, 234
Warapishet, Apisak 17
Watanabe, Toshimasa 8
Wey, Chin-Long 85
Wilson, Denise 69
Wohlmuth, Hans-Dieter 130
Wong, Stephen 169, 192
Wu, An-Yeu 209, 232, 242
Wu, Dapeng Oliver 37

X

Xu, Li 81, 95

Y

Yamada, Isao 15
Yamamura, Kiyotaka 151
Yang, Jar-Ferr 246
Yasuaki, Inoue 220
Yorino, Naoto 213
Yoshimoto, Masahiko 166

Z

Zhang, Qian 106, 141
Zheng, Wei Xing 90
Zhu, Ce 131, 161